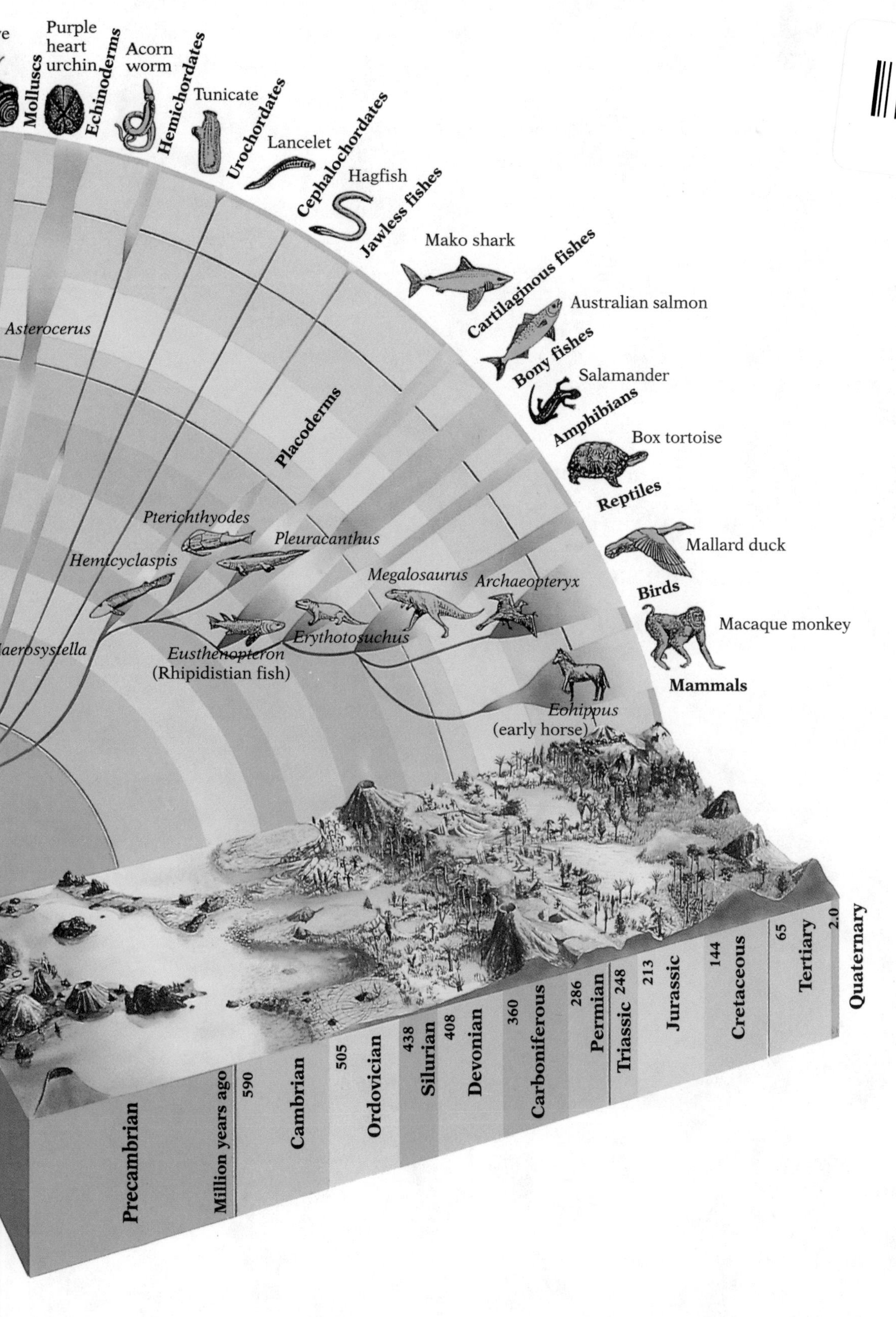

Purple heart urchin
Acorn worm
Molluscs
Echinoderms
Hemichordates
Tunicate
Urochordates
Lancelet
Cephalochordates
Hagfish
Jawless fishes
Mako shark
Cartilaginous fishes
Australian salmon
Bony fishes
Salamander
Amphibians
Box tortoise
Reptiles
Mallard duck
Birds
Macaque monkey
Mammals
Asterocerus
Placoderms
Pterichthyodes
Pleuracanthus
Hemicyclaspis
Megalosaurus
Archaeopteryx
Erythotosuchus
Eusthenopteron
(Rhipidistian fish)
Eohippus
(early horse)
Precambrian
Million years ago
590
Cambrian
505
Ordovician
438
Silurian
408
Devonian
360
Carboniferous
286
Permian
Triassic
248
213
Jurassic
144
Cretaceous
65
Tertiary
2.0
Quaternary

BIOLOGY

AN EXPLORATION OF LIFE

BIOLOGY

AN EXPLORATION OF LIFE

CAROL H. MCFADDEN

WILLIAM T. KEETON

CORNELL UNIVERSITY

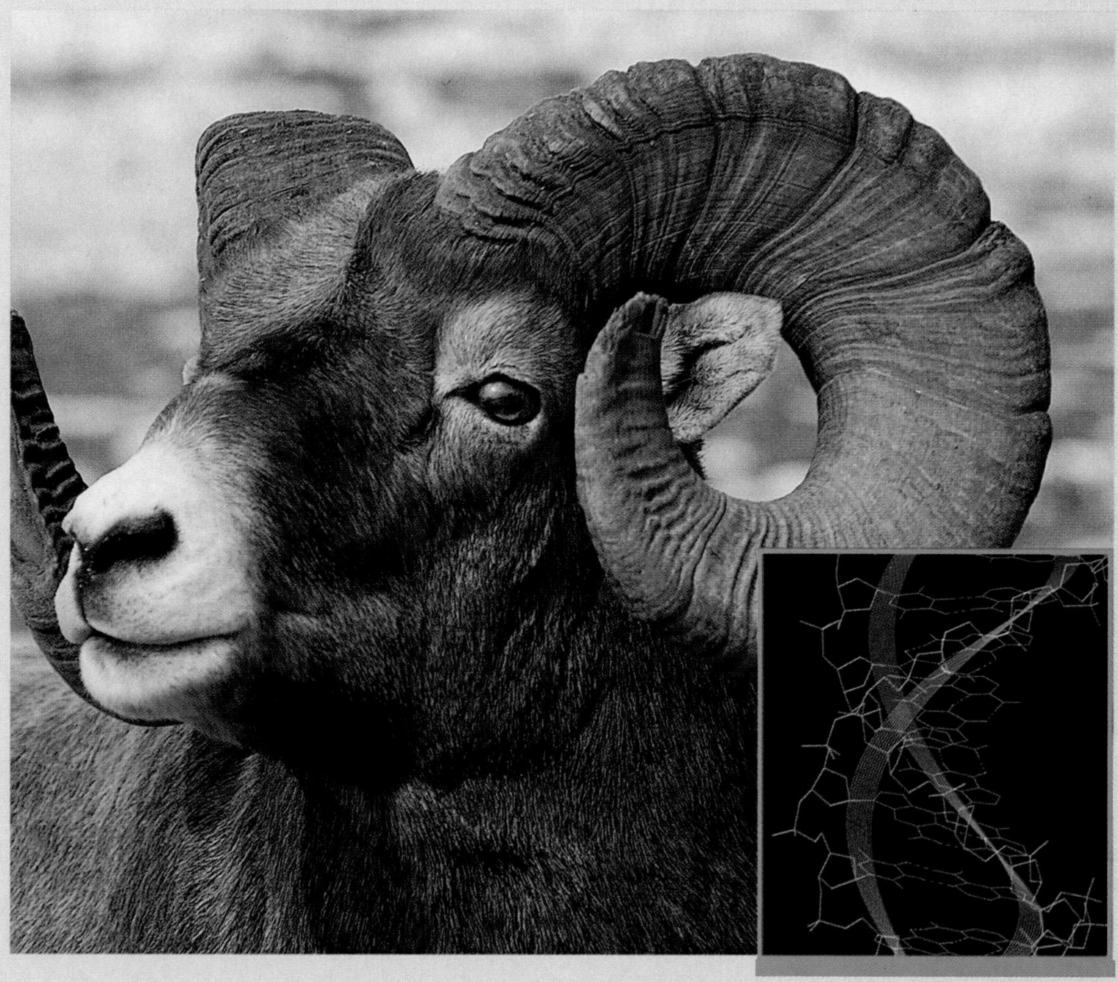

W. W. NORTON & COMPANY

NEW YORK • LONDON

Printed in the United States of America
First Edition
The text of this book is composed in Times Roman with the display type set in Zurich Extra Condensed. Composition by New England Typographic Service, Inc. Manufacturing by R. R. Donnelley & Sons Company.
Book design by M space.

Library of Congress Cataloging-in-Publication Data

McFadden, Carol Hardy.
Biology: an exploration of life / Carol McFadden, William T. Keeton.
p. cm.
ISBN 0-393-95716-0
1. Biology. I. Keeton, William T. II. Title.
QH308.2.M42 1994
574—dc20 93-1530
CIP

W. W. Norton & Company, Inc., 500 Fifth Avenue, New York, N.Y. 10110
W. W. Norton & Company Ltd., 10 Coptic Street, London WC1A 1PU
1 2 3 4 5 6 7 8 9 0

ABOUT THE AUTHORS

Carol Hardy McFadden is a Senior Lecturer in the Department of Physiology at Cornell University. She began her instructional career teaching basic science courses to student nurses at the Evanston Hospital School of Nursing, in Evanston, Illinois. Later, she taught for one year at the University of Dubuque in Dubuque, Iowa and three years at Ithaca College, in Ithaca, New York. Dr. McFadden received her B.S. and Ph.D. from Cornell University, and has been teaching at Cornell since 1968. Over time, she has taught all four of Cornell's introductory biology courses for majors and non-majors, and she is the recipient of Cornell University's Clark Lecturer Award for Distinguished Teaching. For the past ten years she has been alternating between teaching the large non-majors introductory biology course and a self-paced introductory course for majors. She has extensively reorganized the latter course, developing a mastery-based curriculum in which testing is primarily by oral examination.

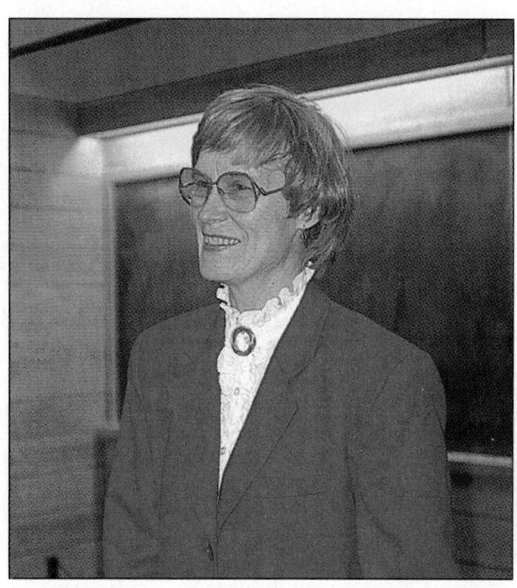

Dr. McFadden has a strong commitment to improving undergraduate education, and has participated in several innovative teaching workshops for faculty, and served on numerous faculty committees, including the Admissions Committee and Faculty Committee on Academics and Athletics. Her avocation is college basketball. She has been the academic advisor to the Men's Basketball Team since 1981, and often travels with the team.

In addition to *Biology: An Exploration of Life,* Dr. McFadden has authored *Elements of Biological Science* and the *Study Guide* accompanying *Biological Science* 5E (Keeton and Gould).

William T. Keeton was the Liberty Hyde Bailey Professor of Biology and Chairman of the Section of Neurobiology and Behavior in the Division of Biological Sciences at Cornell University. He was world-renowned for research on the behavior of homing pigeons. His primary interest was in the mechanisms of pigeon orientation and navigation, especially the ontogeny and sensory basis of these mechanisms. In connection with this work, he served as visiting professor at the Max Planck Institute in Seewiesen, Germany, and at the University of Konstanz, Germany. Professor Keeton earned his A.B. and S.B. degrees at the University of Chicago and his Ph.D. degree at Cornell University.

Shortly after receiving his Ph.D. in 1958, Professor Keeton took charge of Cornell's introductory biology course. Despite an extensive research program, he maintained his interest, as lecturer and administrator, in the course from which his textbooks evolved. Professor Keeton's reputation as a teacher and researcher resulted in invitations to address biology departments throughout North America and Europe. Students at Cornell honored him with the Professor of Merit Award for outstanding teaching. Professor Keeton died suddenly in 1980.

CONTENTS IN BRIEF

Part I The Chemical and Cellular Basis of Life

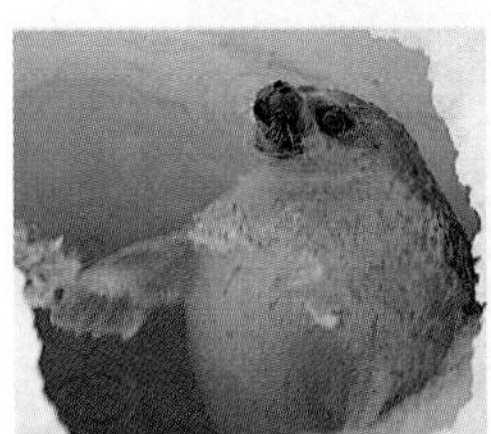

Part II The Perpetuation of Life

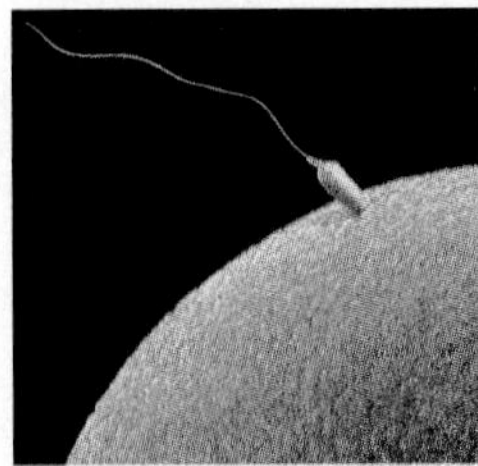

Part III The Biology of Organisms

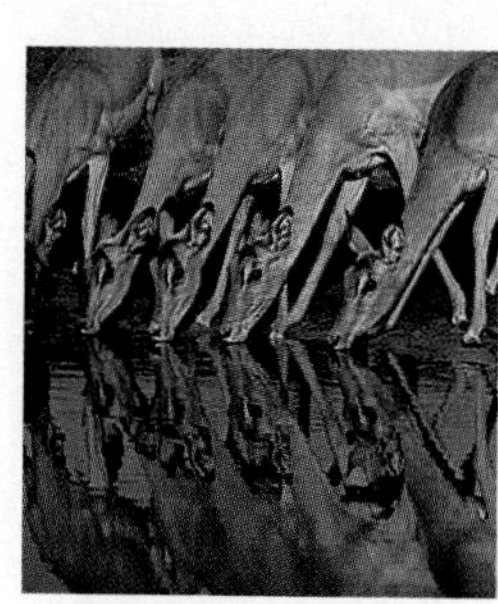

CONTENTS

CHAPTER 5 AT THE BOUNDARY OF THE CELL 89

CHAPTER 6 INSIDE THE CELL 113

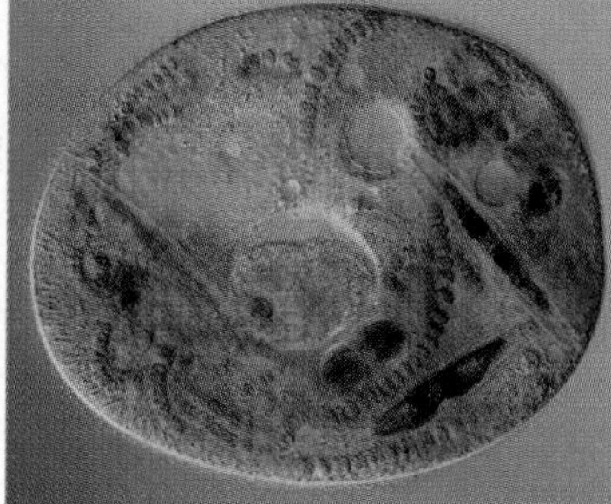

CHAPTER 7 ENERGY TRANSFORMATIONS: PHOTOSYNTHESIS 139

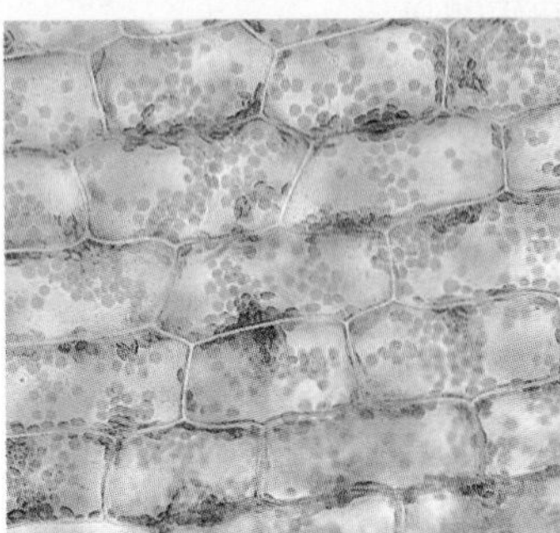

CHAPTER 8 ENERGY TRANSFORMATIONS: RESPIRATION 161

Part II The Perpetuation of Life

CHAPTER 9 HOW CELLS REPRODUCE 179

CHAPTER 13 RECOMBINANT DNA TECHNOLOGY 257

CHAPTER 14 INHERITANCE 277

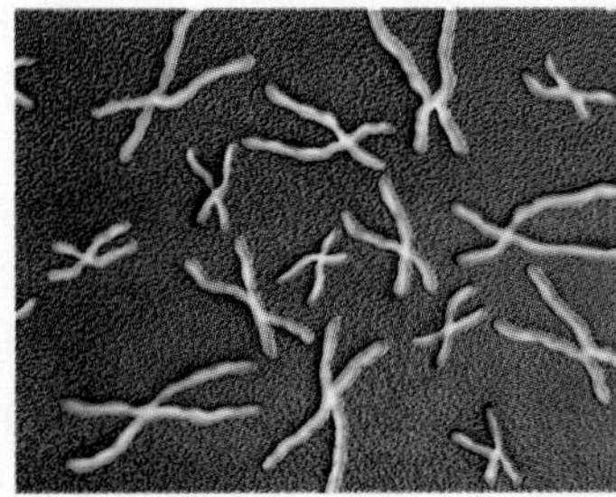

Part III The Biology of Organisms

CHAPTER 15 MULTICELLULAR ORGANIZATION 307

CHAPTER 16 NUTRIENT PROCUREMENT AND GAS EXCHANGE IN PLANTS AND OTHER AUTOTROPHS 323

CHAPTER 19 INTERNAL TRANSPORT IN PLANTS 389

CHAPTER 20 INTERNAL TRANSPORT IN ANIMALS 407

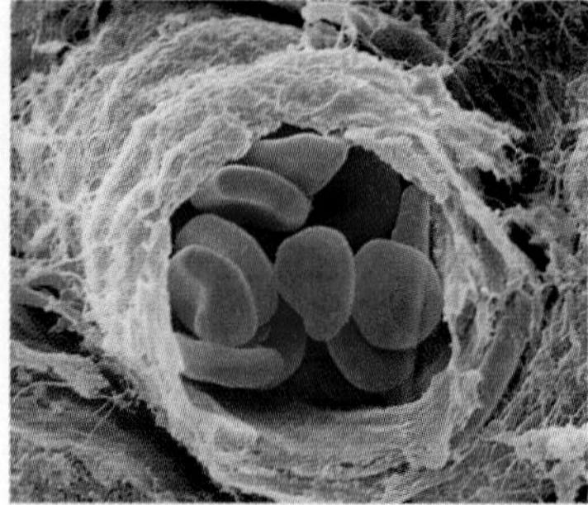

CHAPTER 26 HORMONAL CONTROL OF VERTEBRATE REPRODUCTION 531

CHAPTER 29 SENSORY RECEPTION AND PROCESSING 617

CHAPTER 30 EFFECTORS AND ANIMAL LOCOMOTION 649

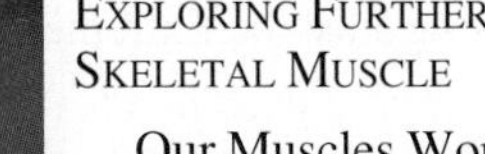

Part IV The Biology of Populations and Communities

CHAPTER 31 EVOLUTION: ADAPTATION 669

CHAPTER 32 EVOLUTION: SPECIATION AND PHYLOGENY 703

CHAPTER 33 ECOLOGY 737

CHAPTER 34 ECOSYSTEMS AND BIOGEOGRAPHY 763

Part V The Genesis and Diversity of Organisms

CHAPTER 36 THE ORIGIN AND EVOLUTION OF EARLY LIFE 839

CHAPTER 37 VIRUSES AND BACTERIA 861

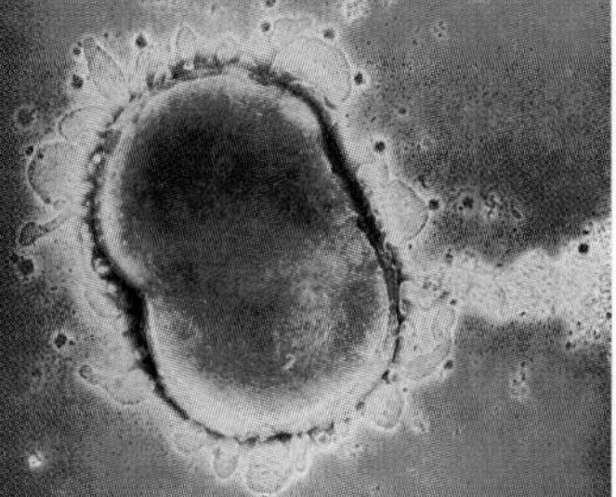

CHAPTER 38 THE PROTISTAN KINGDOM 881

CHAPTER 39 THE PLANT KINGDOM 895

CHAPTER 40 THE FUNGAL KINGDOM 917

CHAPTER 41 THE ANIMAL KINGDOM: THE RADIATES AND THE PROTOSTOMES 927

CHAPTER 42 THE ANIMAL KINGDOM: THE DEUTEROSTOMES 957

CHAPTER 43 THE EVOLUTION OF PRIMATES 979

Appendices

Exploring Further Boxes

PREFACE

Biological science can and should be one of the most stimulating subjects a college student encounters. Nothing else, after all, has such immediate personal relevance as the phenomenon of life; and biological science, as the study of life, sheds light on what every individual experiences in himself and observes around him. Given the inherent excitement of the subject, there is no excuse for an introductory biology course to be dull.

The above paragraph, written by Professor William T. Keeton, the co-author of this text, in the first edition of his textbook *Biological Science* is equally appropriate today. He was a gifted, dedicated and inspiring teacher who taught introductory biology at Cornell University for over 20 years. When he took over the introductory biology course it had about 150 students, and in less than 10 years it grew in popularity until it reached the unwieldy size of over 1,100 students and had to be subdivided into three different courses. I was a graduate student and teaching assistant in "the Keeton course" when the first edition of *Biological Science* was published in 1967. I shall never forget the excitement of the publication of that book, and how it revolutionized the teaching of biology. The students (myself included) brought their books to class and stood in line to have their textbook autographed by him. His enthusiasm for biology and skill in lecturing were legendary. Later, he became a member of my committee for both my masters and doctoral degrees. His untimely death in 1980 was a loss to many, but his influence on biology lives on. I am proud to be able to co-author this text with him, because his view of biological science is the guiding principle upon which this book is based.

This textbook has been a long time in the making. In fact, I started out writing one textbook, the fourth edition of *Elements of Biological Science,* and ended up with another! For three editions, *Elements of Biological Science* was primarily a shortened version of the longer Keeton book, *Biological Science,* but as the revision progressed I found that I was making more and more changes and additions to make the book more accessible to the nonbiology major as well as to those majoring in biology. Some of the more specialized topics from *Biological Science* had to be omitted or simplified. When I finished, the title *Elements of Biological Science* no longer seemed appropriate.

The major reason for the long delay between the publication of the last edition of *Elements* and this book was the difficulty juggling a full-time teaching position in introductory biology at Cornell University and the demands of textbook writing. To make matters worse, the longer I took, the more revisions had to be made due to the phenomenal accumulation of new knowledge. I have written at least two complete manuscripts in the course of this "revision," and have gone through three editors at W. W. Norton & Co.! Never again will I delay so long; it is truly counterproductive.

I am fortunate that my position at Cornell has given me so much classroom experience and student contact, both of which have been invaluable in writing the text. I have been teaching biology at Cornell since 1968, and for the past 15 years have been alternating between two different courses, a large (500 student) non-major introductory biology course, and a self-paced course for 180 biology majors. The latter has been particularly instructive because most of the testing is by oral examination. Nothing compares to giving hundreds of oral tests for finding out where students have problems and need help, and precisely what their misconceptions are! Our introductory biology courses are not team taught; a single professor has the responsibility for teaching the entire two semester sequence. This is useful from the standpoint of integration and continuity (and for writing a textbook), but it means that you must frequently teach in areas far from your expertise—hence the importance of a reliable and accurate textbook. Many of the examples I use in teaching the non-majors course have been incorporated into the text, and many of the handouts I designed for my lectures have been turned into figures. I was able to pretest many sections of the textbook by using them in these courses.

This book, then, is an outgrowth of my teaching experience and my interaction with students at Cornell, and the help provided by numerous teachers and reviewers. Work on the book began with an extensive questionnaire sent to users of the third edition of *Elements.* Analysis of the results inevitably yielded conflicting responses, but several themes recurred. Virtually every teacher thought that introductory biology textbooks were too long (while mentioning that their own specialty did not get enough coverage), and, often, too difficult. Teaching introductory biology today is a real challenge. New information is accumulating at an astonishing rate. In some areas, genetics and photosynthesis for example, the depth of our knowledge has reached awesome proportions. No textbook can ever hope to be completely up to date, but it should provide an understanding of the current state of knowledge (and ignorance) that reflect new discoveries in major areas. Students today must be introduced to the newer areas of biological science, but not at the expense of the older areas that provide the foundation upon which the new information rests. The explosion in molecular biology, particularly in recombinant DNA technology, has meant that we have to include far more molecular information in our courses than ever before, yet the students are not arriving from high school better prepared than they were in previous years. High school chemistry typically focuses on inorganic chemistry, and few students come in with much background in organic chemistry, which compounds the problem. Consequently, most of the material in Chapter 3, "The Molecules of Life," is completely new to students. Yet it is essential that they have a strong knowledge of this material if they are to have sufficient background to comprehend the molecular biology that must be taught to today's students.

As the teachers indicated, one of the major concerns facing any author writing a textbook is length. There is no denying it: modern biology textbooks are long. Thirty years ago an introductory textbook was typically between 700 to 900 pages, including the appendices, glossary, index, etc. Today's texts typically run between 900 and 1400 pages and exclude the end materials in the page count. While it is true that today's textbooks are more lavishly illustrated, the amount of text and the quantity of conceptually difficult information has certainly increased, even though the length of the semester remains the same and student preparation is not noticeably different. I certainly was aware of this as I wrote the text, but there are trade-offs involved. I've found that writing a short text can actually make it more difficult for students to learn. If you do not provide students with enough information so they can *understand* the material, they are forced to memorize isolated facts—which are then forgotten within 24 hours after taking a test. Adding illustrations and study aids, such as "Concepts in Brief," "Study Questions," and "Making Connections" boxes also increase the length of the text, yet these greatly aid student comprehension and concept learning. The length of the text is, therefore, a compromise.

One of the first issues involved in developing the text was to decide on the sequence of topics. The sequence of chapters originally laid out by Professor Keeton alternated cellular and organismal topics, so each semester had some of each. This is the sequence of topics we still use at Cornell in all our introductory biology courses, but our survey of teachers indicated that a majority of courses focus on cellular and molecular topics during the first semester, and organismal biology, evolution, ecology, and diversity during the second semester. In response to this finding, I changed the sequence of topics to conform with the majority, but have taken special care that the new sequence still works for those who prefer the Keeton approach. I have tried to maintain the integrity of the Keeton approach, viewing living organisms in terms of the similarities and differences in the evolutionary adaptations they have undergone. Plants and animals are treated together, comparing and contrasting the problems in living they face and the solutions they have evolved.

Students often view biology textbooks as a vast store of facts to be memorized. There certainly is a great deal of information packed into a textbook, and much of it does need to be learned in order to succeed. To accomplish this, the students need to see the relationships among the facts and concepts they are learning so that it all fits into a framework and "makes sense." Therefore, the text emphasizes various unifying themes which serve as the foundations for new facts and information. The themes are first outlined in Chapter 1, and recur frequently in various contexts throughout the text. These are the basic principles that will still apply when much of the specific information in the text is outdated. I am grateful to Lindsay Goodloe of Cornell University for his help in identifying and clarifying these themes.

It was brought to my attention that students learn more new vocabulary words in an introductory biology course than they learn in the first year of a foreign language. Whether or not this is true, they certainly are exposed to a large vocabulary load. We all know that technical terms play an essential part in the language of science; correctly used they contribute to an economy of language and precision. Nevertheless, biology is overloaded with terms new to the beginning student. With this in mind, I tried to minimize the use of new technical terms, retaining those that are commonly used and part of a biologist's basic vocabulary. The glossary was expanded to provide additional assistance.

Students will be pleased to find that we have retained the brief summaries, "Concepts in Brief," at the ends of the chapters, and have added "Study Questions" that require students to review the basic concepts presented in the chapter. Page references for each question are provided for ease in checking their answers. An important innovation in this edition, which we have found greatly facilitates concept learning, is the use of summary statements at the beginning of each section. Another innovation is the addition of "Making Connections," a series of questions contained in boxes throughout each chapter that encourage the student to relate the chapter material to material discussed in previous chapters. Page references and brief answers have been provided in an appendix to assist the student in making these sorts of connections. In addition, an expanded glossary, a variety of study materials, and an exceptionally thorough index are included at the end of the book. An updated list of suggested readings is now provided in an appendix.

A new, readable page layout allows a synchrony of text and illustrations lacking in many other books. Research suggests that the old Chinese proverb, "One picture is worth a thousand words" is indeed correct. Many investigators have shown that students can pick up and retain more information from pictures than from a verbal description, and that memory from pictures is retained longer than that from text. Thus, the number and quality of illustrations in a textbook plays an important role in its effectiveness as a learning device. There is a significant number of new illustrations, many of which are derived from handouts used in teaching my classes. Many of the more complex figures show step-by-step processes with attached captions (as in the techniques in the Recombinant DNA chapter); others are more schematic, showing negative feedback cycles (as in the hormone chapter). All of these have been class-tested.

ACKNOWLEDGEMENTS

The development of a book of this magnitude would not have been possible without the help of many people. In particular, I wish to thank the many reviewers of the text for their thoughtful comments: Natalie McGucken Barratt, Hiram College; Penelope H. Bauer, Colorado State University; Steven B. Carroll, Indiana University; Suzanne H. Costanza, University of Illinois; Orlando Cuellar, University of Utah; Bruce S. Cushing, Indiana University; Jean DeSaix, University of North Carolina; Max P. Dunford, New Mexico State University; Merrill L. Gassman, University of Illinois at Chicago; James T. Giesel, University of Florida; Sally J. Holbrook, University of California at Santa Barbara; Robert Hurst, Purdue University; Ann S. Lumsden, Florida State University; Mary C. McKitrick, University of Michigan; Scott Meissner, McKendree College; Claude Nations, Southern Methodist University; Stephen Nowicki, Duke University; F. Scott Orcutt, University of Akron; Gregory S. Paulson, Washington State University; Ann E. Reynolds, University of Washington; Wayne Silver, Wake Forest University; Jerry A. Waldvogel, Clemson University; and Jeffrey Wells, Cornell University. And I am also deeply indebted to those who filled out extensive questionnaires on introductory biology instruction: Marjay D. Anderson, Howard University; J. Sherman Boates, Sir Wilfred Grenfell College; Dennis Bogyo, Valdosta State College; Allyn Bregman, The State University of New York at New Paltz; Penni Croot, Clarkson University; Alfred G. Diboll, Macon Junior College; Steven Everhardt, Campbell University; Douglas M. Fambrough, The Johns Hopkins University; Joseph Faryniarz, Mattatuck Community College; Alan Gubanich, University of Nevada Reno; Arnold G. Hyndman, Rutgers University; David T. Jenkins, University of Alabama at Birmingham; Charlene M. Lutes, Radford University; Randy Moore, Baylor University; Joel Ostroff, Brevard Community College; Walter Quevedo, Brown University; Bertha O. Richard, Norfolk State University; Robert Wolff, Boston College; and Thomas E. Wynn, North Carolina State University.

Special appreciation is due to all the people at W. W. Norton & Co. who provided so much assistance and support. Foremost among these was my editor, Stephen Mosberg, without whose encouragement, pushing, and prodding the book would never have been completed. His careful reading and numerous suggestions improved the book in countless ways. I also owe thanks to the copy editor, David Sutter, whose eagle eye and careful attention to detail caught many errors and inconsistencies. Ruth Mandel, our photo editor, has spent many hours ferreting out new photos, and I am grateful to her. Catherine von Novak and Claire Acher also worked hard to find suitable photographs for chapter openers and table of contents. Michael Goodman and Michael Reingold were the artists who produced the numerous new drawings. They are amazing—the art they developed from my crude sketches is truly spectacular, and I thank them for their artistry and patience. Roy Tedoff, production director, and Mary Walsh Kelly, project editor, are responsible for taking an extraordinarily complex manuscript and turning it into a book.

I am especially grateful to my friends and family for their support during the years it has taken to complete the project. John Kramer, Lindsay Goodloe, and Anatol and Carolyn Eberhard have been my mainstays, and my administrative aide at Cornell, Anne Plescia, has efficiently managed the administration of my course and kept me on track. Finally, I thank my son Daniel, and daughter Jean, who have had to put up with a very busy and overburdened mother for many years. Work on "The Book" has consumed a disproportionate share of our lives for far too long, and it will be a relief for all of us to have it over, at least for the short time before I start the new edition.

When all is said and done, the important thing about a textbook is how well it works for you. I hope that, through this textbook, you will become aware of the scope and excitement of modern biological science. Your criticisms, comments, and suggestions are most welcome.

Carol Hardy McFadden
209 Stimson Hall
Cornell University
Ithaca, New York 14853
E-mail: chm7@cornell.edu

ANCILLARIES

Instructor's Manual, by Carol McFadden, Cornell University, and William Kopachik, Michigan State University, provides practical teaching suggestions on how to approach chapter contents, including lecture outlines, key objectives, suggested readings and for further thought questions.

Test Bank, by Carol McFadden and Lindsay Goodloe, Cornell University, contains over 2,700 true-false, multiple choice, and essay questions in printed form or on diskettes for IBM compatible machines and Macintosh.

Full-color transparencies of more than 250 figures in the text.

The Norton Biology Videodisc contains full motion computer animations and film sequences that cover major concepts from the text and hundreds of still images of living organisms.

The Norton Biology Videocassette contains the full motion computer and film sequences available on the Videodisc.

Study Guide, by Carol McFadden and Lindsay Goodloe, Cornell University, contains key concepts, chapter objectives, expanded summaries, suggested readings, study questions to test student comprehension, and concept maps that help students master fundamental principles.

The Norton CD-ROM Classroom Assistant contains much of the line art from the text, the *Instructor's Manual* and computerized *Test Bank on Norton TestMaker,* the *Biology Tutor* software program, hundreds of still images of living organisms, and the computer simulations of fundamental biological principles found on *The Norton Biology Videodisc.*

The Biology Tutor, by Graham Kent, Smith College, is a computer program for Macintosh or Windows that serves as a study companion to the text. Each chapter of the text is covered by brief notecards explaining key concepts, large screen animations and diagrams from the text that help students visualize key concepts, and an interactive quizzing and testing program.

INTRODUCTION

Every day, the sun radiates vast amounts of energy to the earth and our moon, and to all the other planets in our solar system (Fig. 1.1). And every day, the planets reradiate that energy into space. Only on earth is there a unique delay in the reemission of a tiny fraction of that energy: for a brief moment, a minute portion of it is trapped and stored. This minor delay in the flow of cosmic energy powers life. Plants capture sunlight and use its energy to build and maintain stems and leaves and seeds, while animals secure the energy of sunlight by eating plants, or by eating other animals that have eaten plants. At each stage in the many processes of living and dying, waste heat is produced, which joins the pool of energy being relayed back into space.

Biology is the study of this very special category of energy utilization—it is the science of living things. The world is teeming with life: millions of species of organisms of every description inhabit the earth, feeding directly or indirectly off lifeless sources of energy like the sun. What is there in this maze of diversity that unites all biology into one field? What, for instance, do amoebae, redwoods, and people have in common that lends unity to biological science? What differentiates the living from the nonliving? What, in short, *is* life?

Most dictionaries define life as the property that distinguishes the living from the dead, and define dead as deprived of life. These singularly circular and unsatisfactory definitions give us no clue to what we have in common with protozoans and plants. The difficulty for the scientist as well as for the writer of dictionaries is that life is not a separable, definable entity or property; it cannot be isolated on a microscope slide or distilled into a test tube. To early "mechanistic" philosophers like Aristotle and Descartes, life was wholly explicable in terms of the natural laws of chemistry and physics. The "vitalists," philosophers of the opposing school, were convinced that there was a special property, a "vital force," absent in inanimate objects, that was unique to life. Though some scientists continue to think in terms of an unnamed and intangible special property, vitalism has been essentially dead in biology for at least half a century. The more we learn about living things, the clearer it becomes that life's processes are based on the same chemical and physical laws we see at work in a stone or a glass of water.

If life is not a special property, what is it? One answer may be found by comparing living and nonliving things. Organisms from bacteria to humans seem to have several attributes in common. For instance, all are chemically complex and highly organized. All use energy (metabolize), organize themselves (develop), and reproduce. All change (evolve) over generations. So far as we know, no nonliving thing possesses all these

Chapter opening photo: A variety of plant life. Flowering plants seen in this meadow include grasses, bell flowers, bird's foot trefoil, and pink thistle.

1.1 The earth as seen from space

attributes. In addition, and perhaps most important, only the living organism has a set of instructions resident in its genes that directs its metabolism, organization, and reproduction, and is the raw material upon which evolution acts (Fig. 1.2).

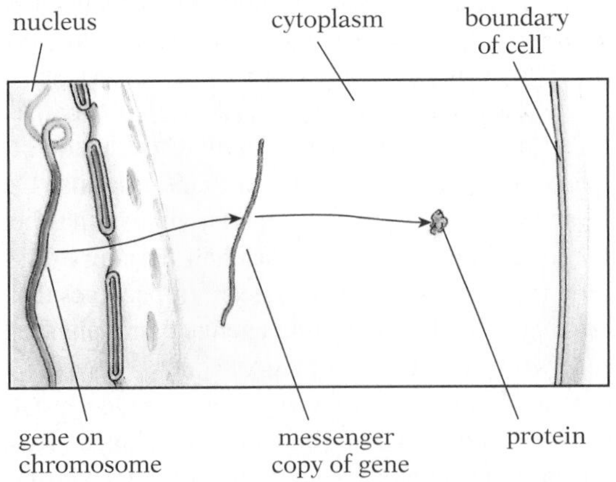

1.2 A simple model of information flow in a cell The internal chemistry of every cell is controlled ultimately by the cell's genes. The genes, located on chromosomes in the cell nucleus, consist of long sequences of four different chemical compounds. A copy of a gene is transcribed into a messenger molecule when needed, and is then transported into the cytoplasm (the portion of the cell that lies outside of the nucleus). There, the messenger is used to direct the construction of a specific protein, which in turn can act as a structural component of the cell or can facilitate and regulate chemical reactions.

THE SCIENTIFIC METHOD

We live in a science-conscious age. Almost every day the news media report some fresh development in agriculture, medicine, or space technology. Commercials use "scientists" in starched, white lab coats to woo us with the latest products of "scientific" research. Yet even among well-educated people there is a surprising lack of understanding of what science really is. Some view it as a sort of magic, as indeed it was in the days when many scientists were at least part-time alchemists or astrologers. The sinister scientist seeking forbidden knowledge whatever the cost to society pervades literature in such characters as Faustus, Dr. Jekyll, and, of course, Frankenstein. This image of science persists in some circles, manifesting itself as an uninformed fear of, for instance, research on recombinant DNA or on nuclear fusion. Equally simplistic is the image of scientists as preoccupied, absentminded recluses shut contentedly in academic ivory towers, divorced from—and not greatly interested in—reality.

For the most part, though, scientists are neither sinister nor reclusive; rather, they are people whose native problem-solving mentality and curiosity about the natural world have simply deepened as they have grown older. It is this active curiosity about the living world, springing in most cases from a profound respect for nature, that binds together those who do research in the biological sciences. The unknown and perhaps unknowable instills in us humility and awe. This sense of mystery, accompanied by a belief in the existence of an underlying order, motivates virtually all scientists.

But motivation is only the first requirement for a scientist. The second is the rigorous and creative application of the ***scientific method.*** Like so many truly great ideas, the scientific method is a basically simple concept and is used to some extent by almost everyone every day. As the English biologist T. H. Huxley put it, the scientific method is "nothing but trained and organized common sense." Its power in the hands of a scientist stems from the rigor and ingenuity of its application.

The Scientific Method Involves Making Observations, Asking Questions, and Formulating Hypotheses

Science is concerned with the material universe, seeking to discover facts about it and to fit those facts into conceptual schemes, called theories or laws, that will clarify the relations between them. Science must therefore begin with observations of objects or events in the physical universe. The objects or events may occur naturally, or they may be the products of planned experiments; the important point is that they must be *observed,* either directly or indirectly. Science cannot deal with anything that cannot be observed.

Science rests on the philosophical assumption (well justified by its past successes) that virtually all events of the universe can be described by physical theories and laws, and that we get the data with which to formulate those theories and laws through our senses. Needless to say, natural laws are descriptive rather than prescriptive; they do not say how things *should* be, but instead how things are and probably will be. Scientists readily acknowledge the imperfection of human sensory perception: the major alterations our neural processing imposes upon our picture of the world around us are themselves a subject of scientific study. In addition, experience has shown that there is often an interaction between phenomenon and observer; however careful we may be, our preconceived notions and even our physical presence may affect our observations and experiments. But to recognize the imperfection of sensory perception and observation is not to suggest that we may get scientific information from any other source (scriptural revelation, for example). No other means are open to us as scientists.

The first step in the scientific method is to formulate the question to be asked. This is not as simple as it sounds: scientists must decide which of the endless series of questions our escalating knowledge inspires are important and worth answering, and which are trivial. The next step is to make careful observations in an attempt to answer the question. Here too there are difficulties: the researcher must decide what to observe and, since measuring everything is impossible, what to ignore. The scientist must also decide how to make the measurements and how to record the data. This is no trivial matter: an oversight or a mistake can render years of work useless. Next, something must be done with the observations. Simply to amass data is not enough; the data must be analyzed and fitted into some sort of coherent pattern or generalization. A formal generalization, or ***hypothesis,*** is a tentative causal explanation for a group of observations. The step from isolated bits of data to generalization can be taken with confidence only if enough observations have been made to give a firm basis for the generalization, and then only if the individual observations have been reliably made. But even when data have been carefully collected, a hypothesis does not automatically follow. Often data can be interpreted in several ways, or may appear to make no sense whatsoever.

Hypotheses Must Be Tested Using Controlled Experiments

The making of a general statement, or hypothesis, is not the end of the process. Scientists must devise ways of testing their hypotheses by formulating predictions based on them and checking to see if the predictions are accurate. Again, this is often not easy: a hypothesis may supply the basis for many predictions, but probably only one can be tested at a time. A scientist must decide which predictions can be most readily tested, and of those, which one provides the toughest challenge for the hypothesis (Fig. 1.3).

Perhaps most difficult of all, when researchers find a discrepancy between a hypothesis and the results of their tests, they must be ready to change their generalization. If, however, all evidence continues to support the new idea, it may become widely accepted as probably true and be dignified by the appellation "theory." It is important to realize that scientists do not use the term "theory" as does the general public. To many people a theory is a highly tentative statement, a poor makeshift for fact. But when scientists dignify a statement by the name of theory, they imply that it has a very high degree of probability and that they have great confidence in it.

A *theory* is a hypothesis that has been repeatedly and extensively tested. It is supported by all the data that have been gathered, and helps order and explain those data. Many scientific theories, like the cell theory, are so well supported by essentially all the known facts that they themselves are "facts" in the nonscientific application of that term. But the testing of a theory never stops. No theory in science is ever absolutely and finally proven. Good scientists must be ready to alter or even abandon their most cherished generalizations when new evidence contradicts them. They must remember that all their theories, including the physical laws, are dependent on observable phenomena, and not vice versa. Even incorrect theories, however, can be enormously valuable in science. We usually think of mistaken hypotheses as just so much intellectual rubbish to be cleared away before science can progress, but tightly drawn, explicitly testable hypotheses, whether right or wrong, catalyze progress by focusing thought and experimentation.

In its simplest form the scientific method begins with careful observations, which must be shaped into a hypothesis. The hypothesis suggests predictions, which must be tested. The

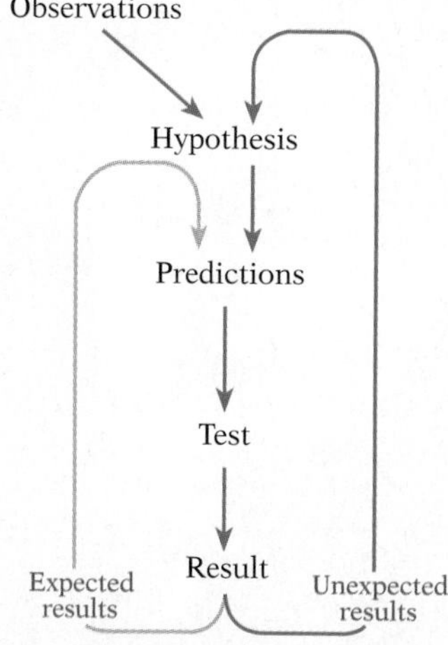

1.3 The scientific method As this diagram suggests, experimental results are used, when necessary, to modify a hypothesis and to test predictions, but the cycle of testing never really ends.

tests, furthermore, must be *controlled,* a matter of vital importance, since only a controlled test has any hope of illuminating anything. Controlling a test, or experiment, means making sure that the effects observed result from the phenomenon being tested, not from some other source. The most familiar way of controlling an experiment is to perform the same process again and again, varying only one minute part of it each time, so that if a difference appears in the result, we can easily trace the cause. When Louis Pasteur, for instance, took his memorable stand against the prevailing theory of spontaneous generation (the idea that life can arise spontaneously from nutrients), he designed his experiment so that there could be no doubt about the outcome, whatever it might be. He took identical flasks, filled them with identical nutrient solutions, subjected them to identical processes, but left some of them open to the air, while the others were effectively sealed. In time bacteria and mold developed in the open flasks, but the liquid in the sealed flasks remained clear. Since exposure to the outside air was the only variable in the procedure, it was then obvious that the bacteria and the mold-producing organisms, rather than being spontaneously generated from the broth itself, must have come from the air.

We can see, then, that in actual practice science is neither easy nor mechanical. Most scientists will readily admit that every stage in the scientific process requires not just careful thought but a large measure of intuition and good fortune. At the end of the last century, for instance, physicists saw many small problems, or "anomalies," in Newtonian physics. Sometimes anomalies in a theory may signal a conceptual error. Albert Einstein (Fig. 1.4) sensed that two of the many irritating anomalies in Newtonian physics were crucial difficulties, and rearranged the pieces of the puzzle to create a revolutionary new theory that accounted for those two anomalies. Like most new hypotheses, Einstein's theory of relativity did not win over the world of science at once.The potential appeal of Einstein's theory, however, was great; it made some very unlikely but testable predictions—that the light from stars, for instance, should bend as it passed near the sun. When these were subsequently proven to be correct, the theory of relativity was accepted rather quickly, loose ends and all, where a far more plausible but less dramatic hypothesis might well have been ignored.

1.4 Albert Einstein

Scientific investigation, then, depends on a combination of subjective judgments and objective tests, a delicate mixture of intuition and logic. Done well, scientific research is truly an art: the ability to make insightful guesses and imagine clever and critical ways in which to test them is usually the distinguishing characteristic of great scientists. But in the final analysis, the basic rules are the same for all: observations must be accurate and hypotheses testable. And as testing proceeds, hypotheses must be altered when necessary to conform to the evidence.

The Scientific Method Has Limitations

The insistence on testability in science severely limits the range of its applications. For example, the idea—widely held by scientists and nonscientists alike—that there is a God working through the natural laws of the universe is simply not testable, and hence cannot be evaluated by science. Science seeks neither to confirm nor to refute it.

Another limitation of science is that it cannot make value judgments: it cannot say, for example, that a painting or a sunset is beautiful. And science cannot make moral judgments: it cannot say that war is immoral. It cannot even say that a river should not be polluted. Science can, however, analyze responses to a painting; it can analyze the biological, social, and cultural implications of war; and it can demonstrate the consequences of pollution. It can, in short, try to predict what people will consider beautiful or moral, and it can provide them with information that may help them make value judgments or moral judgments about war or pollution. But the act of making the judgments is not itself science.

THE RISE OF MODERN BIOLOGICAL SCIENCE

Early Science Began With Ancient Civilizations

Science as we have seen, is the endeavor to understand the natural world. Science must have originated with early humans as they realized that their subjective observations could generate rules of practical utility—when to plant crops, for example, or how to recognize the approach of rain (Fig 1.5). Though a far cry from modern science and the scientific method, this intellectual observation of cause and effect marked the beginning of scientific thought.

The most important advance in early science came with the Greeks: instead of seeing the universe as ruled arbitrarily by a collection of gods who intervened frequently and capriciously in human events, they began to view the world as operating in a consistent, rule-governed fashion with a minimum of supernatural intervention. Philosophers could therefore set about attempting to discover the natural laws—philosophical principles, as they called them—that the universe obeyed.

The Greeks, particularly Aristotle (384–322 B.C.; Fig. 1.6), made systematic observations and from them formed generalizations, or hypotheses, largely divorced from utilitarian goals. They developed and elaborated formal logic as a powerful intellectual tool, and employed it in their pioneering practice of making deductions from their hypotheses—what we would call making predictions, except that no effort was spent on experimental verification. The emphasis of Greek science was philosophical, its goal the creation of a unifying world picture rather than the working out of details. Not all Greek science was quite so metaphysical, however, and one particularly fortuitous combination of observation and deduction that is still with us today is the method of medical diagnosis originated by Hippocrates.

Roman culture, though marked by excellence in literature, history, and the arts, added little to the scientific knowledge acquired by the Greeks, and progress of every sort greatly declined after barbarians from northern Europe sacked Rome in the fifth century and ushered in the Dark Ages. During this period, almost no important scientific advances were made in the Western world, and much of what the ancients had known was forgotten. Greater reliance was placed on religious dogma and on the writings of a few venerated ancient scholars than on observation of the universe itself.

1.6 The Greek philosopher Aristotle The central portion of *The School of Athens,* a mural by the Renaissance painter Raphael depicting philosophers and scholars of all ages. Aristotle is at right; the Greek philosopher Plato is at left.

1.5 Stonehenge, England Stonehenge is a circular setting of large stones that is thought to have been utilized as an astronomical observatory. The information collected was used to predict seasons and appropriate times for planting crops.

Modern Science Began With Discoveries in Astronomy and Physics

Then came a period of intellectual reawakening in Europe. No longer was it beneath the dignity of an educated person to look at the material objects of nature. Thus Nicolaus Copernicus (1473–1543), a Polish astronomer, analyzed the movements of the heavenly bodies and announced that the earth moves around the sun rather than the sun around the earth. But though the intellectual climate may have been changing and becoming more friendly to science, it was not yet ready for such a proposition as this. After all, if the earth moved around the sun, then the earth was not the center of the universe and—outrageous suggestion!—humanity did not stand at the center of creation. When the great Galileo Galilei (1564–1642; Fig. 1.7) embraced Copernicus' theory, he was forced to recant publicly under threat of excommunication.

However, the ideas of Copernicus, Galileo, and the great scientists who followed them could not be suppressed. The conviction grew that the physical universe could be understood in terms of orderly relationships, of universal laws, that physical events have comprehensible impersonal causes, that the

1.7 Galileo Galilei

whims of gods and magicians and evil spirits need no longer be invoked to explain a physical event. No one else stands out so prominently during this period as Isaac Newton (1642–1727; Fig. 1.8), who was born in the year of Galileo's death. His discovery of the Law of Gravitation and his explanation in 1685 of the movements of the planets caused a revolution in thought and carried physical science into a new era. In a very real sense, the work of Newton marks the birth of modern physics.

The science of the new era was restricted largely to physics, astronomy, and chemistry—to the physical sciences. If many were now ready to give up their place at the center of the universe and admit that the earth circles the sun, they were nevertheless not yet ready to admit that they and other living creatures could be understood in terms of impersonal forces. Life seemed too full of purpose, of design, to be studied in the same way as chemicals and moving particles. And if the ideas of Copernicus had been a threat to human dignity and pride, how much more of a threat was an explanation of life processes in mechanistic terms!

The forerunners of modern biological investigation appeared at about the same time as Copernicus and Galileo. Three individuals set the basic course the life sciences were to follow. The earliest was Andreas Vesalius (1514–1564), who made the first serious studies of human anatomy by dissecting corpses. He discovered that the body is composed of numerous complex subsystems, each with its own function, and he pioneered the comparative approach, using other animals to work out the purpose and organization of these anatomical units.

This powerful style of comparative and experimental study was carried forward by the English physician William Harvey (1578–1657), who showed conclusively that the heart pumps the blood, and the blood circulates. The heart, in short, is not in some metaphysical sense the seat of emotions, but a mechanical device with a clear function. As a result of these studies and the anatomical work that followed, an increasingly mechanistic point of view toward life began to develop.

The third of the pioneers was Antony van Leeuwenhoek (1632–1723; Fig. 1.9). Just as Galileo had the brilliant idea of pointing the newly invented telescope at the heavens, so Leeuwenhoek had the idea of using the microscope—with which he inspected cloth, as a draper's assistant—to look at

1.8 Isaac Newton

1.9 Antony van Leeuwenhoek

living things. The most important of his many discoveries were microorganisms (including bacteria), sperm and the eggs they fertilized, and the cells of which all living things seemed to him to be composed.

For biology, unlike physics, centuries of painstaking observation were required to establish the science's fundamental generalizations. The cell theory, for instance, was not given its essentially modern form until 1858, and it wasn't until 1862 that Louis Pasteur (1822–1895; Fig. 1.10) disproved the theory of spontaneous generation. With the realization that Leeuwenhoek's microorganisms might be responsible for disease, the English surgeon Joseph Lister (1827–1912) proved the effectiveness of antiseptics (from *anti-,* "against," and *sepsis,* "decay") in 1865, and Pasteur greatly expanded the use of vaccination.

1.10 Louis Pasteur

1.11 Charles Darwin

By far the most important figure in the history of biology, however, is Charles Darwin (1809–1882; Fig. 1.11). The publication in 1859 of his *The Origin of Species,* presenting the theory of evolution by natural selection, suddenly provided a coherent, organizing framework for the whole of biology. His work sparked the explosive growth of biological knowledge that continues today. As the most important unifying principle in biology, the theory of evolution underlies the logic of every chapter of this book.

DARWIN'S THEORY

The theory of evolution by natural selection, as modified since Darwin, will be treated in detail in Chapter 31. But since we shall be referring to it in earlier chapters, we must examine the essential concepts of the theory at the outset. It consists of two major parts: the concept of evolutionary change, for which Darwin presented a great deal of evidence, and the quite independent concept of ***natural selection*** as the agent of that change.

The Concept of Evolutionary Change

Until only 200 years ago, it seemed self-evident that the world and the animals that fill it do not change: robins look like robins and mice like mice year after year, generation after generation, at least within the short period of written history. This commonsense view is very like our untutored impression that the earth stands still and is circled by the sun, moon, planets, and stars: it accords well with day-to-day experience, and until evidence to the contrary appeared, it provided a satisfying picture of the living world. The idea of an unchanging world also corresponded to a literal reading of the powerfully poetic opening of the Book of Genesis, in which God is said to have created each species independently, simultaneously, and relatively recently—a little over 6,000 years ago by traditional scriptural reckonings.

But problems with the commonly held scriptural theory of creation arose from many sources; scientists attempted first, quite naturally, to discount the evidence as ambiguous, and then, when that proved impossible, to construct a new explanation. Let's look at the evidence for evolution that confronted Darwin and his contemporaries.

The most dramatic findings came from geology. In the 18th

century a picture of a changing earth had begun to emerge. Extinct volcanoes and their lava flows had been discovered; most geological strata were found to represent sedimentary deposits, laid down layer upon layer a millimeter at a time in columns 3,000 meters or more deep; the gradual erosive action of wind and water were seen to have leveled entire mountains and carved out valleys; unknown forces had caused mountains to rise where ocean floors had once been. This latter fact in particular was impressed upon Darwin when he discovered fossilized seashells high in the Andes. Each of these phenomena implied continuous change during vast periods of time.

Another problem for the static view of life was presented by the New World fossils themselves. Many represented plants and animals wholly unknown in Europe, and though theologians had argued that the organisms these fossils represented were alive in the New World, increasingly intensive exploration of the Americas indicated that the hundreds of species of dinosaurs, for instance, were really extinct. In addition, many previously unknown and often bizarre animals inhabited the Americas (Fig. 1.12). As the realization grew that the number of animal species for which evidence was accumulating ran at least into the hundreds of thousands, Noah's ark began to seem very small indeed. In fact, it appeared that the extinct species greatly outnumbered the living, and that new constellations of species had come and gone several times in the past. Moreover, the lowest, oldest rocks contained only the most primitive fossils—seashells, for instance—and these were followed in order by the more modern forms: fish appeared later, for example; reptiles still later; then birds and mammals. The hypothesis of a young earth populated almost overnight by a single bout of creation began to seem very unlikely.

1.13 Jean Baptiste de Lamarck

Lamarck Recognized That Organisms Change Over Time

Jean Baptiste de Lamarck (1744–1829; Fig. 1.13) was the first to offer the major alternative explanation of the fossil record: evolution. Lamarck had arranged fossils of various marine molluscs in order of increasing age; he saw clearly that certain species had slowly changed into others, and concluded that this process of slow change had continued right to the present day. As Lamarck put it in 1809, "it is no longer possible to doubt that nature has done everything little by little and successively," over a nearly infinite period of time. In Lamarck's view, the living world had begun with simple organisms in the sea, which eventually moved onto the land, and evolution had culminated in the appearance of our species, the inevitable result of the gradual trend toward change and "increasing perfection."

A

B

1.12 Challenges to traditional ideas of the origin of species The discovery of fossils of now-extinct species brought into question the static view of life. Shown here are the remains of a baby mammoth (A) that had been preserved in the permafrost in Siberia. The discovery in the New World of organisms unfamiliar to Europeans, such as the anteater (B), also required a reinterpretation of traditional ideas of the origin of species.

Lamarck was basically on the right track, though as we will see his mechanism for evolutionary change was incorrect. But he was ignored for the very understandable reason that he could not offer sufficient evidence for the *fact* of evolution. Darwin, only 50 years later, was in a far better position: there was much more evidence of the sort Lamarck had pointed to, and Darwin, a respected geologist, was well acquainted with it. Furthermore, he had the ability to spot important data in the midst of apparent chaos. He could find powerful support for the idea of evolution where Lamarck and others—if they looked at all—saw only irrelevancies.

Darwin Found Many Lines of Evidence That Supported Evolutionary Change

One of the most important lines of evidence put forward by Darwin was the existence of morphological resemblances among living species (the findings of what we today call comparative anatomy). If, for example, we observe the forelimbs of a variety of different mammals, we see essentially the same bones arranged in the same order (Fig. 1.14). The basic bone structure of a human arm, a cat's front leg, and a seal's flipper

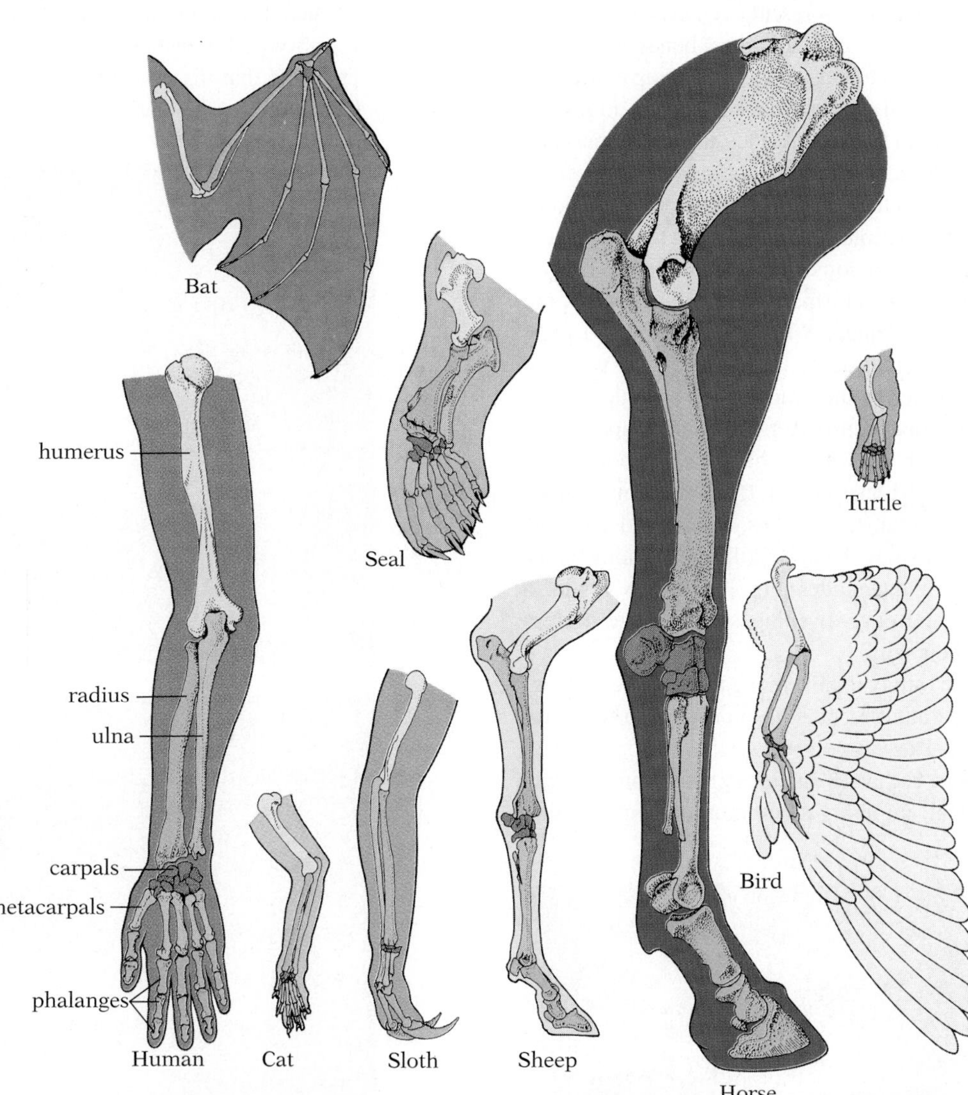

1.14 The bones in some vertebrate forelimbs compared The labeled and color-coded bones of the human arm permit identification of the same bones in the other forelimbs depicted. The cat walks on its phalanges (toes), the metacarpals (hand bones in the human) forming part of the leg. The sloth normally hangs upside down from tree limbs—hence its recurved claws. The horse walks on the tip of one toe, which is covered by a hoof (specialized claw), and the sheep walks on the hoofed tips of two toes (only one hoof can be seen in this side view); the carpals of both the horse and the sheep (wristbones) are elevated far off the ground because the much-elongated metacarpals form a section of the leg. In the seal the bones are shortened and thickened in the flipper. In the bat the metacarpals and phalanges are elongated as supports for the wing. All the animals mentioned so far—human, cat, sloth, horse, sheep, seal, bat—are mammals, but the same bones can be seen in the leg of a turtle and the wing of a bird.

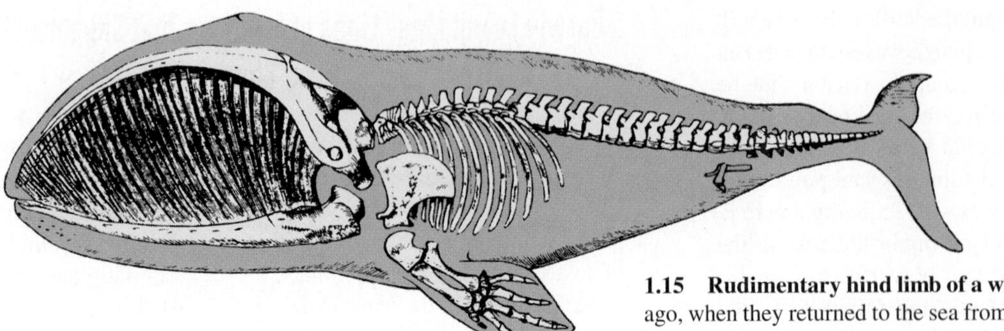

1.15 Rudimentary hind limb of a whale Whales lost their hind limbs long ago, when they returned to the sea from the land, but they retain rudimentary bones that correspond to the pelvic girdle and the thighbone (red).

is the same; the same bones are present even in a bird's wing. True, the size and shape of the individual bones vary from species to species, and some bones may be missing entirely in one species or another, but the basic construction is unmistakably the same. To Darwin the resemblance suggested that each of these species had descended from a common ancestor from which each had inherited the basic plan of its forelimb, modified to suit its present function. The observation that structures with important functions in some species appear in vestigial nonfunctional form in others further convinced Darwin of the reality of evolutionary change. Why otherwise would pigs, which walk on only two toes per foot, have two other toes that dangle uselessly well above the ground? Why would certain snakes, such as the boa constrictor, and many species of aquatic mammals, such as whales, have pelvic bones and small, internal hind-limb bones (Fig. 1.15)? Why would flightless birds such as penguins, ostriches, kiwis, and the cormorants of the Galápagos Islands still have rudimentary wings—or feathers, for that matter (Fig. 1.16)? Why would so many subterranean and cave-dwelling species have useless eyes buried under their skin?

Embryology—the study of how living things develop from eggs or seeds to their adult forms—also provided powerfully suggestive evidence. Darwin pointed out that in marine crustaceans as different as barnacles and lobsters the young larvae are virtually identical, implying a common descent. Telltale traces of their genealogy are obvious in vertebrates as well. Human embryos, for instance, have gill pouches and well-developed tails that disappear before the time of birth (Fig. 1.17). It seemed clear to Darwin that such inappropriate struc-

1.16 Flightless male ostriches from the Namib desert in Africa Ostriches have rudimentary wings and feathers, but are unable to fly.

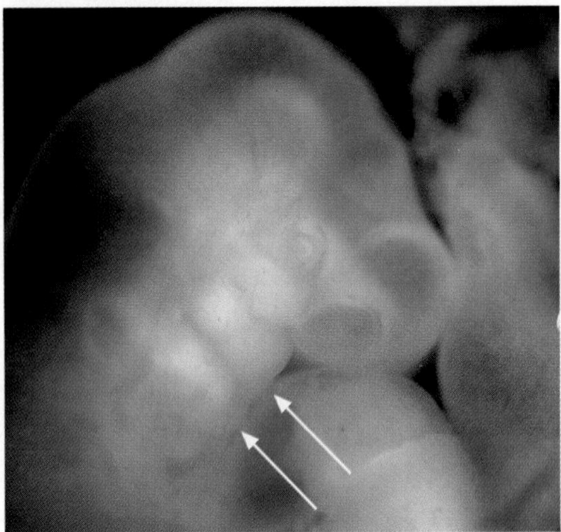

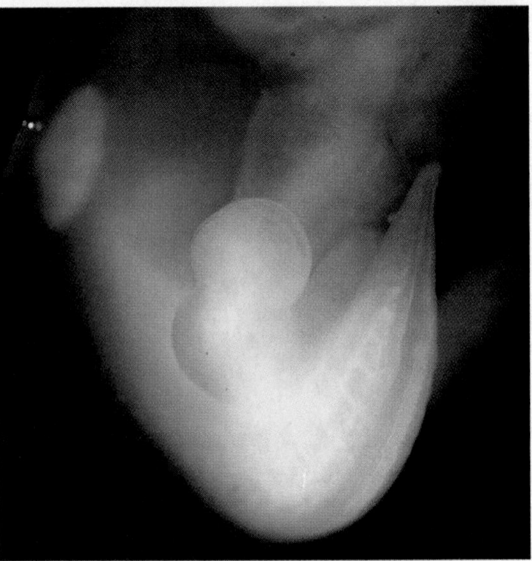

1.17 Embryological evidence of evolution Top: Pharyngeal ("gill") pouches (arrows) in a four-week human embryo. Bottom: Tail in a five-week human embryo.

tures are inherited vestiges of structures that functioned in ancestral forms, and that may still function in other species descended from the same ancestors.

Another particularly convincing line of evidence offered by Darwin was the well-known ability of breeders to produce dramatic changes in both plants and animals. How could anyone contemplating the alterations produced in domesticated species by artificial selection doubt that vast changes are possible, given sufficient time? Great Danes, sheep dogs, Irish setters, Yorkshire terriers, poodles, bulldogs, and dachshunds, for instance, are all members of the same species, bred from tamed wolflike ancestors to look like almost anything breeders have fancied (Fig. 1.18). Similarly, cabbage, brussels sprouts, cauliflower, broccoli, kohlrabi, rutabaga, collard greens, and savoy have all been bred from the same species, the wild form of which looks nothing like its domesticated progeny. The many varieties of chickens, cattle, horses, flowers, grains, and so on have likewise been bred over the years. Who could compare the colors and shapes of the wild rose or jonquil to the many colors and shapes of the far larger domesticated roses or daffodils, with new varieties bred each year, and doubt that a species has the capacity to change enormously even in a hundred years? In everything Darwin looked at—fossils, anatomy, embryology, and breeding—he saw the same message: species can and do change.

Darwin Recognized That Evolutionary Change Is Directed by Natural Selection

But what mechanism accounts for the changes? Lamarck's now-discredited hypothesis was one of the first attempts at a plausible explanation. Lamarck was impressed by how well suited each animal was to its particular position in the web of life, even though the environment had changed enormously again and again over countless millions of years. To account for this ability to adapt, he imagined that God had given each

1.18 Selective breeding of dogs The domestic dog, *Canis lupus familiaris,* is thought to have evolved from a lineage of gray wolves *(Canis lupus)* about 12,000 years ago. Through intense selective breeding, humans have achieved an astonishing variety of modern breeds of dogs that have diverged greatly from their wild, wolflike ancestors. Shown here are a wolf, a sheepdog, a dalmation, a schnauser, and a dachshund.

species a tendency toward perfection that allowed for small alterations in morphology, physiology, and behavior to accommodate changes in the environment, and that these alterations could be inherited by the offspring.

Belief in Lamarck's idea of a natural tendency toward perfection and the inheritance of acquired characteristics required no more faith in his day than did belief in other invisible everyday forces, such as gravity and magnetism. His hypothesis was a perfectly logical extension of the prevalent Western view that God had set things going by creating nature and nature's laws, and had then left things for the most part to run themselves. But where in the vestigial legs of whales or the dangling toes of pigs was there evidence of perfection? Instead the clear mark of compromise was everywhere. Plants and animals were well adapted to their places in the environment, but they were by no means perfect.

Darwin proposed a different mechanism—natural selection—requiring no internal tendency other than the one toward variation so obvious in nature. Darwin had conceived the idea of natural selection two years after his return from his voyage to the Americas on the *Beagle,* but was only goaded into publishing, 20 years later, by the receipt in 1858 of A. R. Wallace's manuscript proposing essentially the same theory. (We normally associate Darwin's name with the theory because of the impressive evidence he presented—he had been collecting it for two decades—and because of his thorough exploration of the theory's many ramifications.)

1.19 Thomas Robert Malthus

In essence, Darwin put together two ideas. The first was that numerous variations exist within species, and that variations are largely heritable. Immersed as he was in the Victorian preoccupation with plant and animal breeding, Darwin knew that while cuttings produce plants identical to the parent, sexual reproduction produces individual offspring that differ both from their parents and from each other. Variation is a fact of life: breeders, as we know, are able to select for desirable traits and create new, morphologically distinct lines of plants and animals.

Darwin's second inspiration came on 28 September 1838, when he reread the *Essay on the Principle of Population* by the economist Thomas Robert Malthus (Fig. 1.19). Malthus pointed out that humans produce far more offspring than can possibly survive; population growth always outruns any increase in the food supply and is held in check largely by war, disease, and famine. Vast numbers of people thus live perpetually on the edge of starvation. Both Darwin and Wallace were struck at once by the consequence of applying the gloomy Malthusian logic to plants and animals: like humans, the creatures of each overpopulated generation must compete for the limited resources of their environment, and some—indeed most—must die. Each female frog, for example, produces thousands of eggs per year, and a fern produces tens of millions of spores, yet neither population is growing noticeably. Any organism with naturally occurring heritable variations that increase its chances in this life-or-death contest will be more likely than others to survive long enough to have offspring, some of which will inherit these variations. They in turn will have an above-average chance to survive the struggle, and so will form an increasingly larger part of the population. As a result of this "selection," the population as a whole will become better adapted—that is, it will evolve—and the never-ending struggle for existence will then turn on the possession of still better adaptations. To distinguish this process from the sort of directed, artificial selection practiced by agriculturalists, Darwin called it *natural* selection.

The contrast between artificial and natural selection that served Darwin so well provides an instructive summary of the evolutionary process. In both, far more offspring are born than will reproduce; in both, differential reproduction, or selection, occurs, causing some inherited characteristics to become more frequent and prominent in the population and others to become less so as the generations pass. But in the breeding of domesticated plants and animals, selection results from the deliberate choice by the breeder of which individuals to propagate. In nature, it takes place simply because individuals with different sets of inherited characteristics have unequal chances of surviving and reproducing. Notice, by the way, that selection does not change individuals. An individual cannot evolve. The changes are in the makeup of populations.

Artificial and natural selection also differ significantly in the *degree* of selection, and its effect on the rate of change. Breeders can practice rigorous selection, eliminating all unwanted individuals in every generation and allowing only a few of the most desirable to reproduce. They can thus bring about very rapid change (Fig. 1.18). Natural selection, which involves a large measure of chance, is usually much less rigorous: some poorly adapted individuals in each generation will be lucky enough to survive and reproduce, while some well-adapted members of the population will not. Hence evolutionary change is usually rather slow; major changes may take thousands or even millions of years, depending on the degree of selection pressure imposed by the environment and by other species.

Darwin's evidence for evolutionary change and the common descent of at least the major groups of organisms was widely accepted in his time, but the idea of natural selection by small steps remained controversial until the 1930s. Some biologists had difficulty seeing how an elaborate and specialized structure like an eye, for instance, could evolve, since the first rudimentary but necessary steps might lack obvious survival value. As we will see in later chapters, an expanded understanding of the nature and organization of genes and their role in development has now made it clear that natural selection does explain most evolutionary change. Darwin extolled the beauty and simplicity of such a system in the final sentence of later editions of *The Origin of Species:* "There is grandeur in this view of life, with its several powers, having been originally breathed by the Creator into a few forms or into one; and that, whilst this planet has gone cycling on according to the fixed law of gravity, from so simple a beginning endless forms most beautiful and most wonderful have been, and are being evolved."

In summary, then, Darwin's explanation of evolution in terms of natural selection depends upon five basic assumptions:

1. Many more individuals are born in each generation than will survive and reproduce.
2. There is variation among individuals; they are not identical in all their characteristics.
3. Individuals with certain characteristics have a better chance of surviving and reproducing than individuals with other characteristics.
4. Some of the characteristics resulting in differential survival and reproduction are heritable.
5. Vast spans of time have been available for change.

All the known evidence supports the validity of these five assumptions.

THE DIVERSITY OF LIFE

Living Things Are Classified on the Basis of Their Evolutionary Relationships and Level of Organization

According to the fossil record, the most primitive organisms known—the bacteria—date back over three billion years, the first land plants and insects over 400 million years, and the first birds and mammals over 180 million years. Since the appearance of the simplest forms of life, innumerable kinds of organisms, increasingly complex and adapted to widely varying environments, have evolved.

Ideally, living organisms are classified on the basis of the evolutionary relationships thought to exist among them, i.e., on their commonality of ancestry. The evolutionary relationships are ofter far from clear, however, and other criteria, such as overall similarity and unique characters, are often used to create logical groupings. The techniques of molecular biology provide additional data. Depending on the criteria chosen, a number of different systems of classification are possible. The system used in this book recognizes six broad categories, or kingdoms: Eubacteria and Archaebacteria (formerly classified together in the Monera), Protista, Plantae, Fungi, and Animalia. A general guide to the classification of organisms is shown in Figure 1.20.

One characteristic used to separate organisms into king-

MAKING CONNECTIONS

1. Contrast the "mechanistic" and "vitalistic" approaches to the study of living things. Which approach has been more rewarding to biologists and is now generally accepted by them?

2. Why are controlled experiments important?

3. What are the limitations of the scientific method?

4. What geological discoveries affected the development of evolutionary thought?

5. Why is Darwin's theory of evolution by natural selection considered the most important unifying principle in biology?

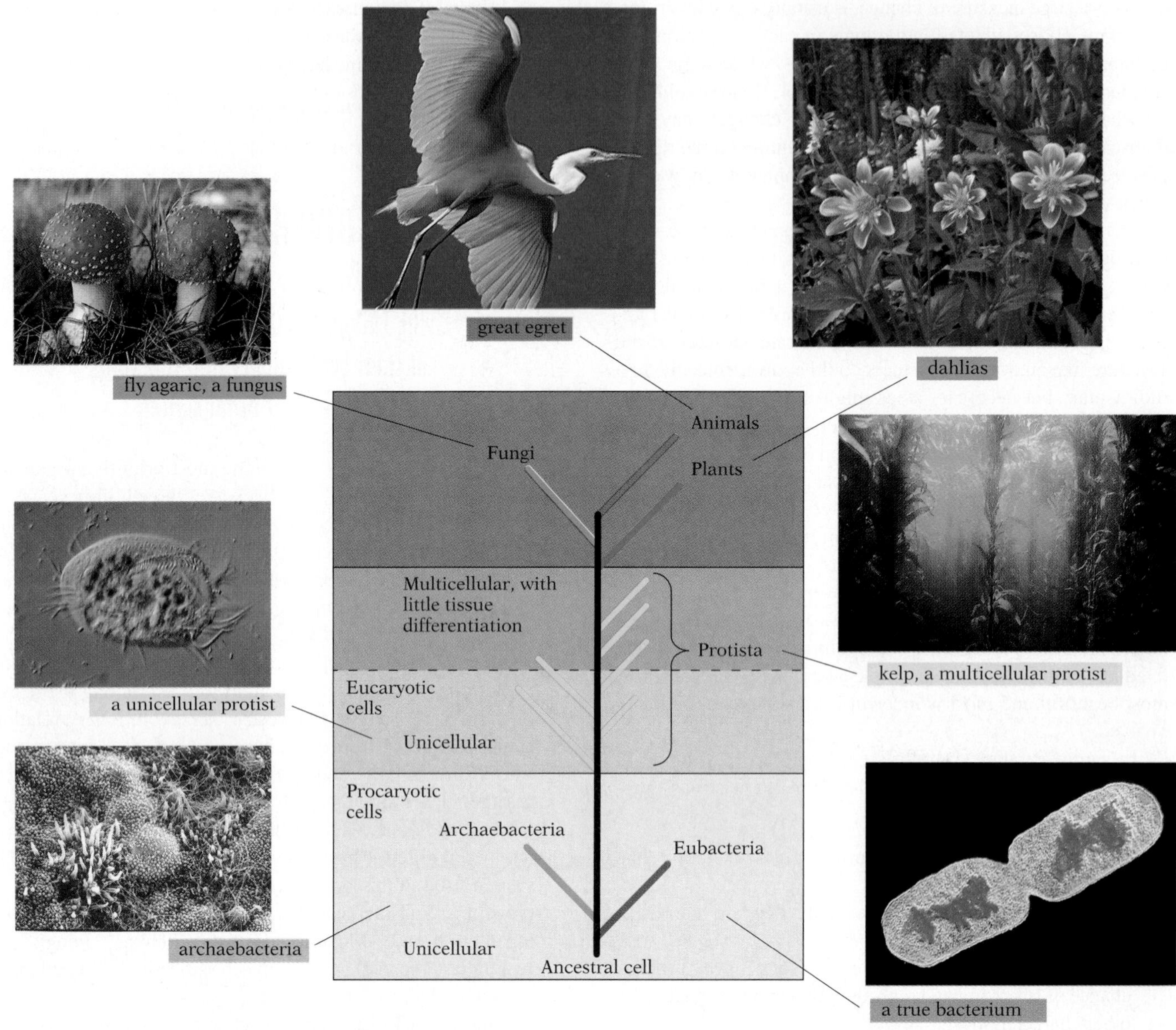

1.20 Relationship among the six kingdoms of life In this representation, the point at which a group's line departs from the central "trunk" of this evolutionary tree indicates the chronological order in which the kingdoms are thought to have arisen. The horizontal lines indicate the grade level of the organisms in that group (e.g., procaryotic or eucaryotic cells, unicellular or multicellular). Notice that the kingdom Protista consists of a number of different lineages. Common names are used for most groups; technically, animals are in kingdom Animalia and plants are in kingdom Plantae.

doms is the type of cell making up the organism. The cells of organisms belonging to the kingdoms Archaebacteria and Eubacteria are ***procaryotic,*** meaning that they do not have a membrane-surrounded nucleus and that they lack other membranous structures present in the cells of all other types of organisms. The cells of the members of the other four kingdoms are ***eucaryotic,*** which means they have a true nucleus and other intracellular membranous structures. Eucaryotic organisms that are primarily unicellular (one-celled) are placed in the kingdom Protista; those that are multicellular are put in the Plantae, Animalia, or Fungi, depending on their primary mode of nutrition. The following survey briefly introduces you to the six kingdoms and their most important subdivisions. We shall refer to these major groups again and again as we discuss the basic problems faced by all living things and the different adaptations these have evolved in response to them. The groups mentioned here, and many others, will be described in far greater detail in Part V.

Members of the Kingdoms Eubacteria and Archaebacteria Are Unicellular and Procaryotic

Procaryotic organisms are the most abundant organisms in the world. They are all single-celled and can be seen only under magnification (Fig. 1.21). Not until the advent of the electron microscope could their internal structure be examined. Most procaryotes have strong cell walls, which protect them from damage. The terms "procaryote" (referring to the type of cell) and "bacteria" are often used interchangeably. Two evolutionary distinct groups of procaryotes exist, the ***archaebacteria*** (*archae-*, "ancient"), and the more common ***eubacteria*** (*eu-*, "true"). The archaebacteria and the eubacteria are thought to have arisen from a common ancestral cell and to have diverged very early in the evolution of life.

The organisms of the kingdom Eubacteria are very diverse. Many are *photosynthetic*—they use the energy of sunlight to manufacture their own food (Fig. 1.22)—while others lack photosynthetic ability and must obtain their nutrients, already synthesized, from other organisms. Some obtain these nutrients from dead organisms; others, from living organisms. Because bacteria can use so many sources of energy, they can survive in many diverse habitats. They have been found living in hot springs, where temperatures range between 55 and 90° C, as well as in glaciers in Antarctica.

Most bacteria are beneficial, though some are disease-pro-

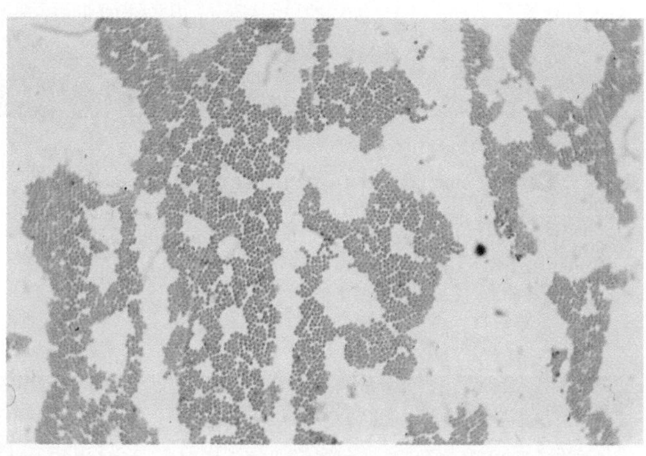

A

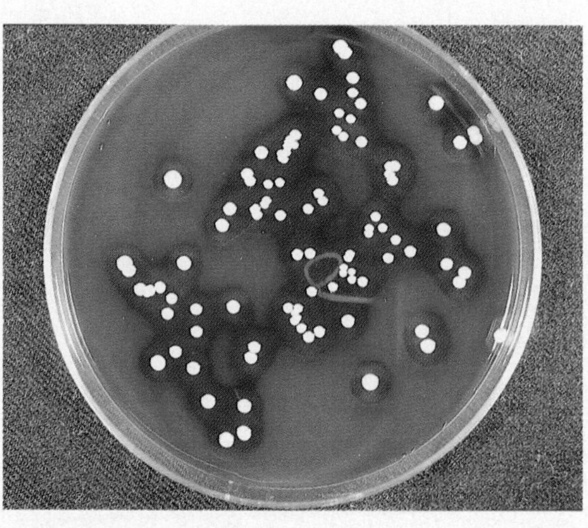

B

1.21 The bacterium *Staphylococcus aureus* (A) High magnification (here ×930) is needed—and the staining is helpful—in making this bacterium detectable by the eye. It is an agent for many infections, including boils and abscesses. (B) The visible colonies that *S. aureus* forms on a blood agar medium each contain many thousands of cells.

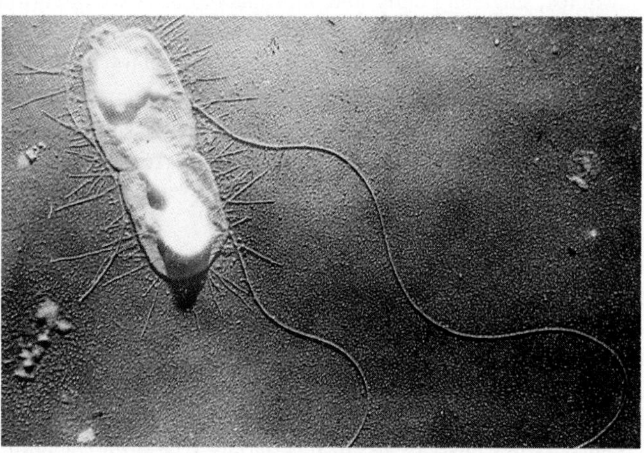

A

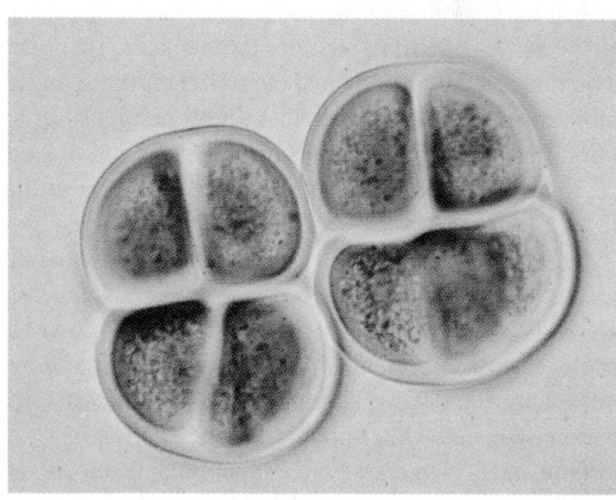

B

1.22 Representative eubacteria (A) The intestinal bacterium *Escherichia coli.* Clearly visible in this electron micrograph are the stiff, helical flagella, which rotate like propellers to push the bacterium. (B) Cyanobacteria *(Chroococcus)* from the Pine Barrens of New Jersey. Small groups of cells are enclosed within common gelatinous sheaths. These cells are photosynthetic.

ducing agents that attack human beings, other animals, or plants. Many act as decomposers, destroying the dead bodies and bodily wastes of other organisms. In the process important compounds from these organisms are recycled into the environment, where they become available to other living things. Bacteria are invaluable to several industrial and manufacturing processes as well as to biological research. By studying bacteria, biologists have learned many important principles about life that have proved applicable to other organisms, including ourselves.

The procaryotes of the kingdom Archaebacteria (Fig. 1.23) were formerly classified as eubacteria, but recent studies of their cells, cell membranes, gene structure, and biochemistry make it clear that the two groups of organisms are very different. The archaebacteria live only in very challenging environments. One group lives only in anaerobic environments such as swamps, stagnant ponds, and intestinal tracts. A second group lives only in extremely salty environments, while a third exists only in extremely hot environments (e.g., above 70° C). Recently archaebacteria have been isolated from submarine hydrothermal vents along the East Pacific Rift, where the vent temperatures reach 250° C and pressures of 265 atmospheres are common.

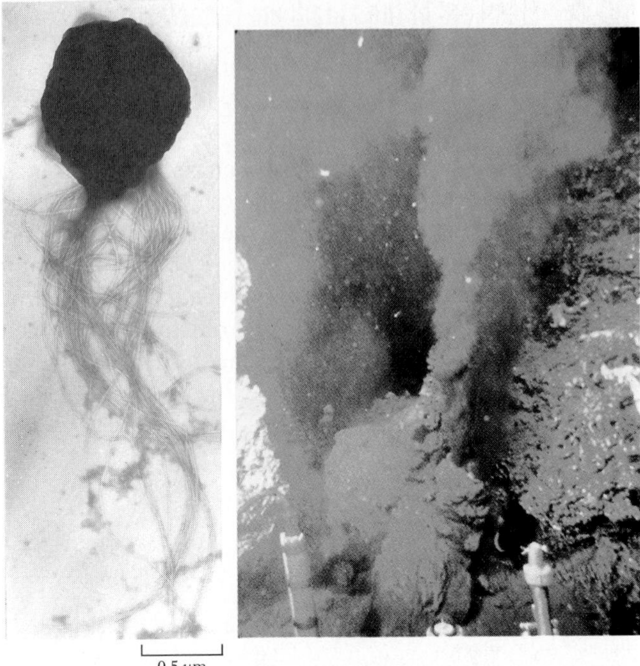

1.23 Scanning electron micrograph of an archaebacterium Archaebacteria typically live in harsh environments. This bacterium, which moves about by means of its flagella, lives in the hot water of a submarine hydrothermal vent in the eastern Pacific (right).

The Kingdom Protista Contains a Diverse Grouping of Organisms

The kingdom Protista includes a variety of groups, or phyla, many of which are probably not closely related evolutionarily. This is a grouping of convenience, placing together all the organisms at a similar level of organization. The kingdom Protista, as we refer to it, includes those eucaryotic organisms that are unicellular during most or all of their lives, and all the plantlike organisms whose bodies show relatively little distinction between tissues. The Protista can be subdivided into two groups: those that are photosynthetic, and those that are not.

Many protistan groups have chlorophyll, the green pigment essential for photosynthesis. Most are unicellular, such as the euglenoids, diatoms (Fig. 1.24), and golden-brown algae. These are very important organisms in the marine environment, where they serve as a chief source of food for other aquatic organisms. Other photosynthetic protists, such as the ***red algae*** and ***brown algae,*** are multicellular. Both the red and brown algae are primarily marine and are commonly known as seaweeds. They are especially prevalent in the intertidal zone along rocky coasts, where they can easily be observed at low tide. Their color is due to brown and red pigments, which often mask the chlorophyll these algae also contain (Figs. 1.25 and 1.26). They are always multicellular, but all of their cells are

1.24 Representative diatoms The diatoms are single-celled photosynthetic organisms common both in freshwater habitats and in the oceans. They characteristically have elaborately ornamented glasslike cell walls.

1.25 A brown alga, *Saccorhiza polyschides* Brownish pigments mask the green pigment chlorophyll also present in this seaweed. The flattened leaflike blades are supported by the water in which the plant is growing; when uncovered at low tide, the blades lie flat on the substrate. Notice the holdfasts, which anchor this alga to the substrate.

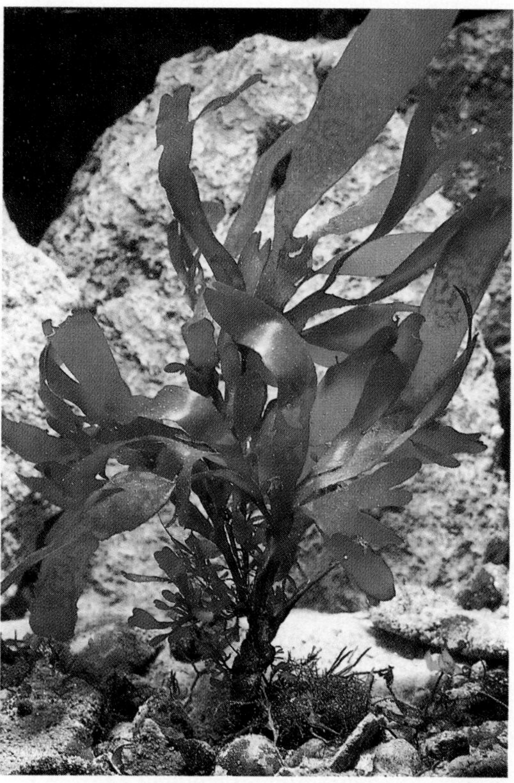

1.26 A red alga, *Rhodymenia palmata* Like the brown algae, the red algae are mostly seaweeds, but they generally grow at greater depths than the brown algae.

similar in structure and function, and form a single continuous tissue, which is the plant body. This is unlike the cells of land plants, which are usually different from one another. Though some brown algae do have many characteristics similar to the land plants, it is believed that they and the land plants evolved independently of each other and are not closely related. Even the brown and red algae are thought not to be closely related, despite many superficial similarities.

Many protistan groups lack chlorophyll and are not photosynthetic. Some of these resemble fungi, while others, known as the ***protozoa,*** have often been viewed as unicellular animals. Protozoa are often highly specialized, their single cell exhibiting a complexity and a separation of functions similar to those observable in multicellular animals. Protozoa are generally much larger than bacteria. Most are very mobile, swimming rapidly in the water in which they live or crawling along the bottom or on submerged objects. Some propel themselves by the whiplike motion of long hairlike structures known as flagella. Others, like *Paramecium,* bear many shorter hairlike structures called cilia, which often function in both locomotion and feeding (Fig. 1.27). Still others, like *Amoeba* and *Pelomyxa,* have neither flagella nor cilia, but move by a complex flowing motion in which the cell, constantly changing shape, sends out extensions into which the rest of the cell contents flows (Fig. 1.28). We shall refer to representative protozoans many times as we examine their fascinating evolutionary adaptations.

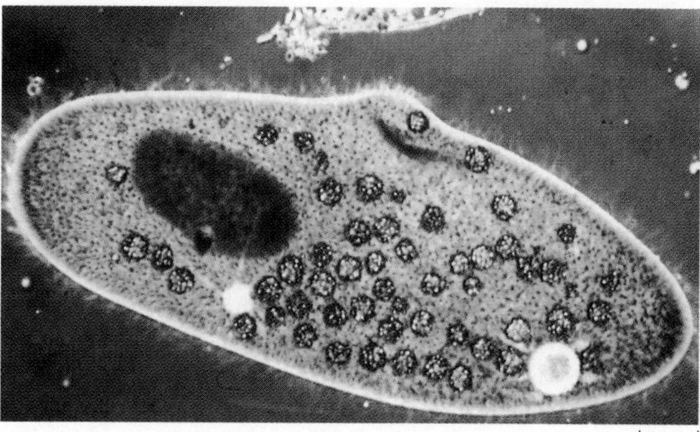

20 μm

1.27 A ciliate protozoan *Paramecium caudatum* The colors seen here, which were produced by optical staining with the interference microscope, help differentiate the component parts of this living specimen; they are not the organism's natural colors. Ciliate protozoans are unusual in having two kinds of nuclei—as seen here, a large brown structure with a small darker one overlapping it. The yellow cilia, which seem to edge the cell, actually cover all of it. The numerous small green circles are food vacuoles, chambers here containing yeast on which the protozoan has been feeding. The two white circles with radiating canals are contractile vacuoles, whose main function is eliminating water from the cell.

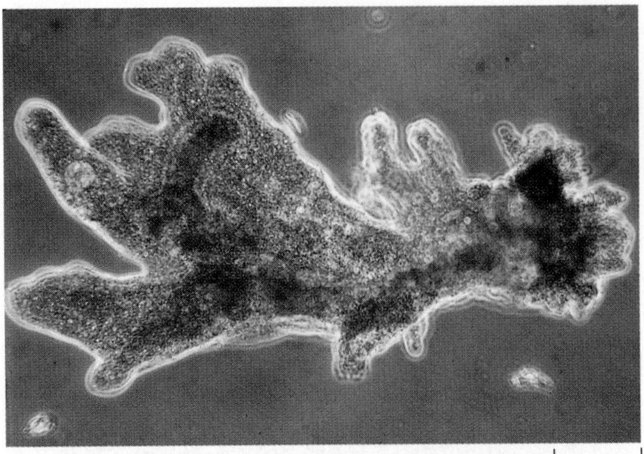

1.28 An amoeboid protozoan, *Pelomyxa carolinensis* "Amoeboid"—from the Greek word for "change"—describes a cell that can alter its shape as it thrusts out or withdraws many armlike extensions.

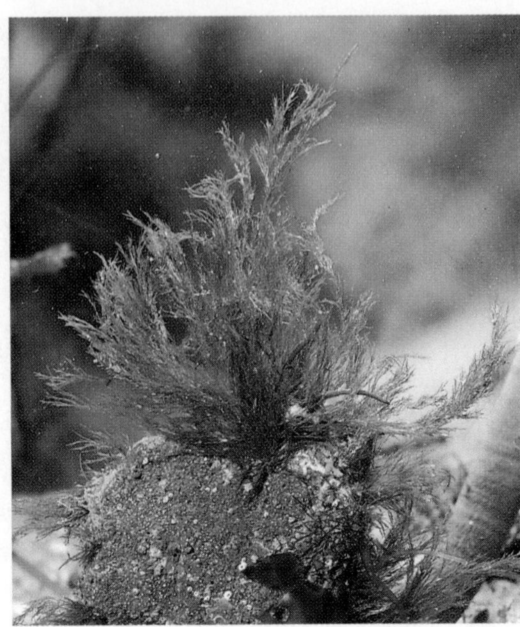

1.29 A marine green alga, *Cladophora rupestris* The green algae include many unicellular and colonial forms. This most common species of multicellular green alga exhibits a branched and filamentous structure. Some less common multicellular algae are bladelike in form.

Members of the Kingdom Plantae Are Photosynthetic and Adapted for Life on Land

The kingdom Plantae includes all those organisms commonly recognized as plants, plus the green algae, the group from which the land plants arose. Common to all members of the plant kingdom are cells with rigid cell walls and chloroplasts (subcellular membrane-surrounded structures containing chlorophyll). Plants, unlike fungi and animals, can synthesize all the food they need for maintenance and growth. Most of the members of the plant kingdom are adapted for life on land; most have waterproof coverings, including cuticle, over the surfaces exposed to air, and special adaptations that allow them to reproduce on land. There are three main groups of plants: (1) the green algae, relatively simple plants that live only in the water or in very moist environments on land, (2) mosses and liverworts, which lack any vascular (internal transport) tissues, and (3) the vascular plants, which, as the name suggests, have a vascular system to transport materials from one part of the plant to another.

Green algae are relatively simple plants. Some are unicellular, others multicellular. In the multicellular forms, most of the cells are similar to one another and are not highly specialized; they form what might be considered a single continuous tissue, which is the plant body (Fig. 1.29). The green algae are similar to the land plants in their biochemical characteristics, and most botanists agree that it was probably from ancestral flagellated green algae that the land-plant groups arose.

Although the ***mosses, liverworts,*** and their relatives live on land, they are not entirely independent of the ancestral aquatic environment. They occur only in very moist habitats (Fig. 1.30). One reason is that parts of their reproductive cycle are dependent on abundant moisture. Another is that, unlike the vascular plants, they have not evolved an efficient internal transport system through which water from the soil can be carried to all parts of the plant.

Of all members of the plant kingdom, the ***vascular plants*** show the greatest internal specialization into tissues and or-

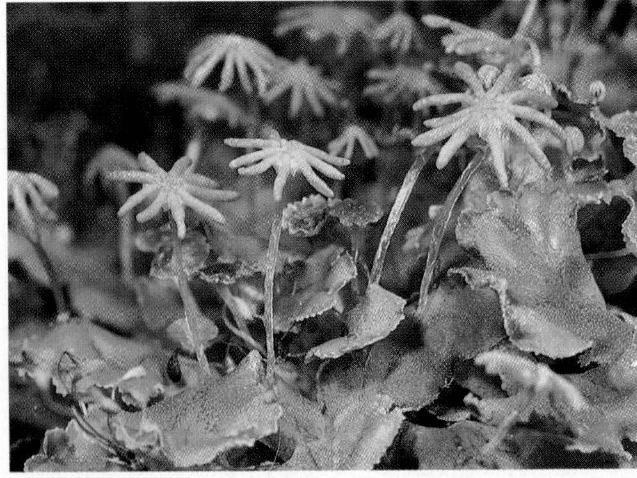

1.30 Liverworts, *Marchantia polymorpha* Growing on a damp forest floor in England, these liverworts bear reproductive structures known as archegonia in receptacles at the ends of their stalks.

gans—roots, stems, leaves, and reproductive organs (cones, flowers). Because they possess vascular (transport) tissue, pipelines through which water and dissolved substances move from one part of the plant body to another, they are less dependent than the other plant groups on water in the surrounding environment. They are the dominant plant group on land today.

The vascular-plant group is usually subdivided into several sections, three of which are the ferns, the conifers, and the flowering plants (or angiosperms). The ferns are the oldest of these three groups. They dominated the land for a long time, but eventually gave way to the other groups and are now largely overshadowed by them (Fig. 1.31).

The conifers and flowering plants are known collectively as the ***seed plants;*** they are more highly specialized for a terrestrial existence than the ferns. Some common conifers are pine, cedar, spruce, fir, and hemlock; all of these bear cones and have needlelike leaves (Fig. 1.32).

The majority of the most familiar land plants are angiosperms, or flowering plants, which come in every shape and size from grasses to cacti, and from tiny herbs and wildflowers to large oak and maple trees. This huge and very diverse group is customarily divided into two subgroups: the *dicotyledons* (dicots, for short), which include beans, buttercups, privets, dandelions, oak trees, maple trees, roses, potatoes, and a great variety of other plants; and the *monocotyledons* (monocots, for short), which include the grasses and grasslike plants such as corn, lilies, irises, and palm trees. The two subgroups differ in many characteristics, including two that are readily observable: (1) the leaves of dicots usually show a network of veins, those of monocots usually show parallel veins; and (2) the petals of dicot flowers occur in fours or fives or multiples of these; the petals of monocots occur in threes or multiples of three (Fig. 1.33).

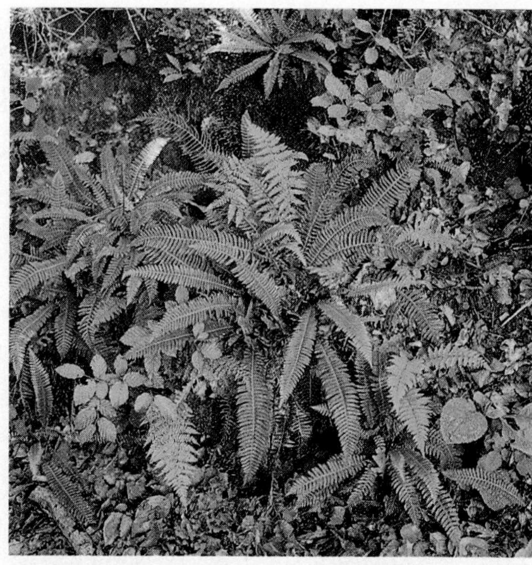

1.31 Ferns Ferns are vascular plants that have large, well developed leaves and underground, horizontal stems. The large leaves are the only parts of the plant normally seen. Most ferns live in moist, humid environments.

A

B

1.32 A conifer The conifers, or evergreens, are seed plants with needlelike or scalelike leaves (A). The reproductive structures of the conifers are cones (B); seeds develop on the inner surface of the scales of female cones.

The Kingdom Fungi Includes Multicellular Organisms That Obtain Their Nutrients by Absorption

Fungi at one time were placed in the plant kingdom, because they are predominantly sedentary and because their cells, like plant cells, have walls. But their cell walls are quite different from those of plants, and fungi lack chlorophyll and cannot manufacture their own food. Like animals, they must obtain the complex high-energy nutrients they need in an already synthesized form. Unlike animals, they cannot ingest large particles of food, and depend entirely on *absorption* of nutrient molecules; hence they must live in or on their nutrient sources,

1.33 Monocot and dicot flowers compared Left: A trillium flower, a monocot, has three petals and parallel veins in its leaves. Right: A nasturtium flower, a dicot, has five petals and a network of veins.

which are either other living organisms or the dead remains of other organisms. Since fungi differ from both plants and animals in so many ways, most recent classifications assign them to a kingdom of their own. Some fungi, such as yeast, are unicellular, but most are multicellular; among the latter are bread molds, fruit molds, mushrooms, toadstools, and bracket fungi (Fig. 1.34).

The Kingdom Animalia Includes Multicellular Organisms That Ingest Their Food

Of the many characteristics that distinguish the animals from the two other main categories of multicellular organisms—the plants and the fungi—we shall here mention but two. First, an-

1.34 Two representative fungi Left: The black bread mold *Rhizopus* is made up of numerous threadlike filaments hung with what look like black and white balls. These are spore-producing reproductive structures; the black ones are ripe and ready to release their spores. Right: The velvet-stemmed *Collybias* growing among lichens on a tree trunk are composed of filaments so densely packed that the unaided eye cannot make them out individually.

imal cells lack rigid cell walls. Second, the principal mode of nutrition in animals is *ingestion* of food particles; most plants, by contrast, depend on photosynthesis, and most fungi on absorption (Fig. 1.34). Below is a brief description of the important groups or phyla (singular: phylum) of animals that are frequently referred to throughout the textbook.

The ***cnidarians*** (formerly called coelenterates) are a large group of primitive aquatic animals whose body plan is ***radially symmetrical;*** that is, the body parts are arranged regularly around a central axis, rather like spokes on a wheel. Their saclike bodies are composed of two distinct tissue layers with a much less distinct third layer between them. There is a digestive cavity, but it has only one opening, which must serve both for the intake of food and the elimination of wastes. Tentacles are often present around the opening and are used in capturing prey. Special stinging cells are found on the tentacles. The nerves and muscles of these simple animals are of an exceedingly primitive type. The group includes the jellyfish, sea anemones (Fig. 1.35), and corals. A representative freshwater form to which we shall refer is the hydra (Fig. 1.36).

The ***flatworms*** are more complex than the cnidarians in some ways, but they too have a digestive tract with only one opening. The body is composed of three layers of tissue, and

1.35 A sea anemone *Epiactis* A veritable thicket of tentacles surrounds the animal's mouth. Many newly budded young anemones are attached to the stalk portion of its body.

1.36 Hydra, feeding The green hydra *Chlorhydra* owes its color to the cells of a green alga that it incorporates into its own cells, thus benefiting from the alga's photosynthetic activity. Note the buds with tentacles on the lower portion of the stalks; hydras reproduce asexually by budding. Left: The hydra is manipulating a small crustacean with its tentacles. Right: The food has been taken into the digestive cavity, where it makes a prominent bulge.

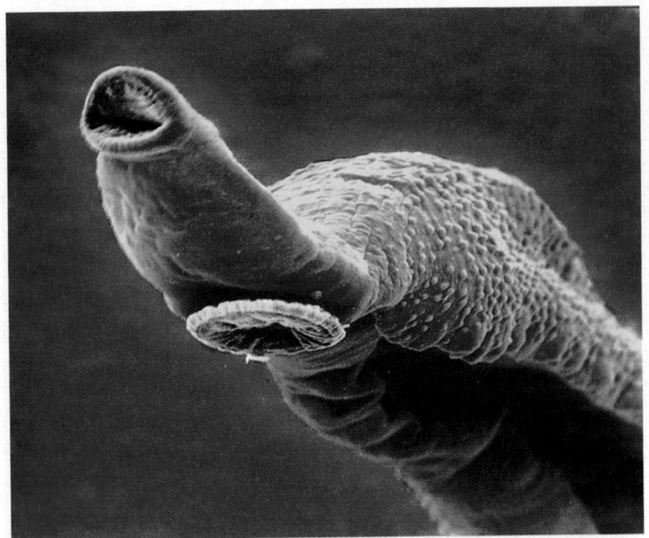

1.37 Parasitic flatworm The blood fluke *Schistosoma mansoni* is responsible for schistosmiasis, a debilitating human disease characterized by dysentery and anemia. The sucker at the tip encloses the mouth.

1.38 Planarians, members of a group of free-living flatworms These animals are free-living (nonparasitic) freshwater scavengers. Note in each the eyespots, which allow a primitive sort of vision, the much-branched digestive cavity (brown), and the centrally located tubular pharynx, which is extruded during feeding.

the symmetry is bilateral. ***Bilaterally symmetrical*** animals have an elongated body with two similar sides, distinct dorsal (upper) and ventral (lower) surfaces, and distinct anterior (front) and posterior (rear) ends. Many flatworms, such as flukes and tapeworms, are parasites and show many interesting specializations for this mode of existence (Fig. 1.37). Others, such as planarians (Fig. 1.38), small animals to which we shall make frequent reference, are free-living aquatic organisms.

Molluscs are fairly complex animals, most of which have shells. They include snails (Fig. 1.39), clams, oysters, and

1.39 Two types of molluscs Left: The majority of molluscs live in the water, but many snails, such as the Florida tree snail, are adapted for life on land. The snail moves by means of a muscular "foot" extruded from the shell. Note the eyes located on the ends of long retractable stalks. Below: Pacific Giant Octopus, *Octopus dofleini*. Note the large suckers on the tentacles.

scallops, as well as octopuses and squids, which do not have obvious shells. These animals are particularly abundant in the oceans, as anyone who has collected their shells along the seashore knows. They are also common in freshwater. Over time, some snails have evolved lungs and have thus become fully terrestrial.

The ***annelids*** are often called the segmented worms. As this term implies, the bodies of these highly evolved organisms are divided into a series of units, or segments, which are often clearly visible externally. Though most annelids are aquatic, some, such as the earthworm, occur on land, always in moist places. We shall generally take the earthworm as representative of this group (Fig. 1.40); occasionally, however, we shall mention *Nereis,* a marine worm with large lobes protruding from each side of the body segments (Fig. 1.41).

The ***arthropods*** constitute an immense group of complex animals that includes more different species than all other animal groups combined. All arthropods have jointed legs and a hard outer skeleton. Spiders, scorpions, crabs, lobsters, crayfish, centipeds, millipeds, and insects all belong to this major group (Fig. 1.42). Of these, the insects are by far the largest

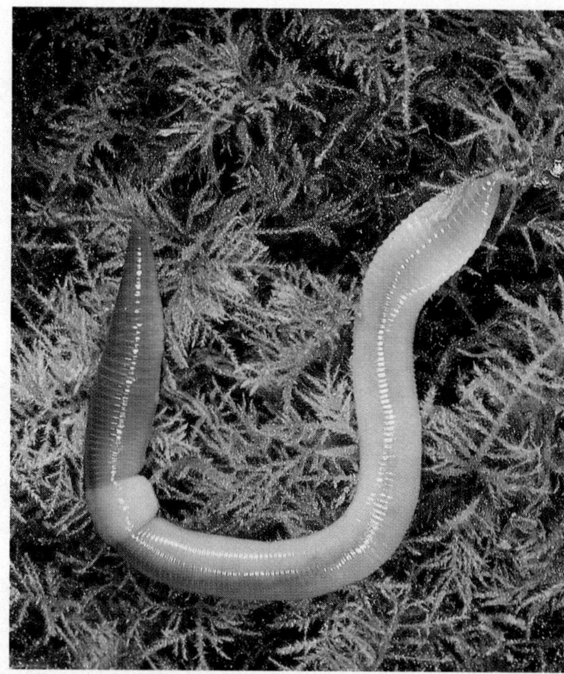

1.40 An earthworm, *Lumbricus terrestris* The animal's body is divided into numerous ringlike segments. Its front end, with the mouth, is at upper right. The girdle-like structure is the clitellum—a group of glandular segments that secrete mucus during mating and produce a cocoon for the eggs.

1.41 A marine annelid worm, *Nereis* This worm has a pair of large flaplike appendages on each of its body segments, which function in gas exchange and in locomotion.

A

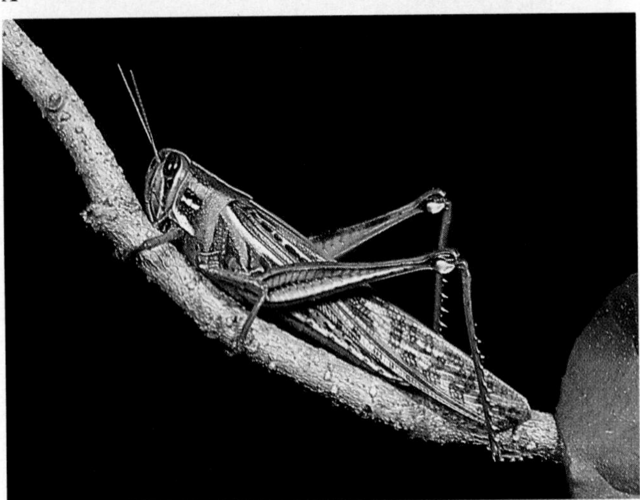
B

C

1.42 Three representative arthropods (A) Centipedes have a segmented body with a pair of jointed legs on each segment. This Australian species is found under tree bark. (B) Like all insects, the lubber grasshopper *(Romalea microptera)* has only three pairs of legs. The abdominal segments are legless. (C) The crab *Neolithodes grimaldii.*

1.43 Three representative echinoderms Horse stars *(Hippasteria phrygiana),* right foreground and left background, a northern sea star (*Asterias vulgaris*), left foreground, and a sea urchin (*Strongylocentrotus droebachiensis*) in a New England tide pool. Sea stars anchor themselves against the ebb and flow of the tide by means of hundreds of tube feet; sea urchins often use their spines to anchor themselves against the tide in crevices.

subgroup. They are among the most successful of all land animals, rivaled only by the mammals, particularly humans.

All ***echinoderms*** are strictly marine; they have apparently never been able to invade either the freshwater or the terrestrial habitats. Though their symmetry is radial, they arose from bilaterally symmetrical ancestors. The group includes sea stars (starfish) (Fig. 1.43), sand dollars, sea urchins, sea cucumbers, and a variety of other forms. The enchinoderms are closely related to the ***chordates,*** the group to which humans belong.

A major subgroup in the chordates is the ***vertebrates,*** which includes all animals possessing an internal bony skeleton, particularly a backbone. Fish, amphibians (e.g., frogs and salamanders), reptiles (e.g., snakes, lizards, turtles, and alligators), birds, and mammals (including humans) belong to this group (Fig. 1.44). We shall pay special attention to the vertebrates throughout this book.

A

B

C

D

1.44 Four vertebrates (A) A white shark cruising for food. (B) A frilled lizard raising its frill and opening its mouth in a threat display. (C) An Osprey, fish hawk. (D) Japanese macaques *(snow monkeys)* are a member of a family of Old World monkeys that includes mandrills and baboons, the Japanese macaque is one of the few primates able to survive in a cold climate.

MAKING CONNECTIONS

1. **What biological mechanism has generated the diversity of life?**
2. **What criteria are used to classify living organisms?**

CENTRAL CONCEPTS IN THE STUDY OF LIFE

As you begin your study of biology, keep in mind that science is much more than a collection of facts. It is the goal of science to discover relationships among facts, or, alternatively, to search for *patterns* in nature. Try to see the connections among the facts and concepts you are learning, just as scientists search for such relationships. By developing the skill of comprehending the "facts" as part of an interconnecting network of information and ideas, you will be able to apply your growing body of knowledge to the analysis of new situations—an activity essentially similar to that of a scientist making predictions on the basis of a well-understood theory.

A list of concepts that have broad applicability to life science is provided to help you in the tasks of making connections among facts and of identifying the central organizing ideas of biology. Although every scientific concept is theoretically subject to modification or refutation, these seem certain to remain central to biology. They are the unifying themes that serve as the solid foundation for modern biology. And you will discover that, like some themes in a musical composition, they will recur frequently in various contexts.

1. *Living organisms are composed of the same chemical and physical components as nonliving things, and all life processes obey the laws of chemistry and physics.*

2. *The cell represents the lowest level of structure capable of performing all the functions of life.* Biological systems are characterized by numerous levels of organization, each more complex than the one below it. Thus organs make up the body of a complex multicellular organism, but are themselves made of tissues, which in turn consist of cells. Within this hierarchy, the cell is the basic structural unit of life.

3. *At all levels from the molecular to the macroscopic, biological structure is closely related to function.* Whether we consider a molecule such as hemoglobin or an organ such as the kidney, we find that an understanding of its structure is essential to a full understanding of how it works. The converse is also true.

4. *All living things must take in energy and materials and expel wastes to maintain their internal organization.*

5. *Living organisms are extremely diverse, but possess an underlying unity.* For example, many biochemical pathways found in bacteria are also present in human beings. Many other aspects of cellular structure and function are universal among living organisms.

6. *Biological systems are maintained by feedback control mechanisms that maintain the constancy and control the activity of the system.* The classic physical example of a feedback control mechanism is the thermostat: It keeps the room temperature constant by detecting when the temperature goes below or above its setting and turning the furnace on or off accordingly. A large array of analogous biological mechanisms enable cells and organisms to maintain constant internal conditions, an ability known as *homeostasis.*

7. *All organisms are capable of reproduction based on a set of instructions known as the genome.* The genome in all cellular organisms is composed of DNA—a striking example of the unity of diverse organisms mentioned above.

8. *Life as it exists today is the product of evolution: the change in the genomic composition of populations of organisms over time. The course of evolutionary change is directed by natural selection.* Evolution by means of natural selection is the most important unifying theme in biology. It accounts for the unity and diversity exhibited by life.

CONCEPTS IN BRIEF

1. Biology is the science of living things. All living organisms are chemically complex, are highly organized, utilize energy, undergo development, and reproduce. In addition, every living organism has a set of instructions resident in its genes that directs its metabolism, organization, and reproduction, and is the raw material upon which natural selection acts.

2. All science is concerned with the material universe, seeking to discover facts about it and to fit these facts into theories or laws that will clarify their relationships.

3. The *scientific method* involves formulating a question and making repeated, careful observations in an attempt to answer it, and then using these observations to generate a *hypothesis*. Hypotheses are then tested, using *controlled experiments*. If the test results do not confirm a hypothesis, the hypothesis must be altered to confirm with the evidence, or rejected. If all evidence continues to support the new idea, it may eventually become widely accepted and called a *theory*. The insistence on testability in science imposes limitations on what it can do. Science cannot make value and moral judgments.

4. The physical sciences underwent explosive growth during the 16th and 17th centuries, but biological science did not enter its modern era until, in 1859, Charles Darwin proposed his theory of evolution by natural selection. To this day the theory remains one of the most important unifying principles in all biology.

5. Darwin postulated that contemporary organisms descended by a slow, gradual process from ancient ancestors that were morphologically quite different. The course of this evolutionary change is determined by natural selection. Darwin's theory depends upon five basic assumptions:

a) Many more individuals are born in each generation than will survive and reproduce.

b) There is variation among individuals.

c) Individuals with certain characteristics have a better chance of surviving and reproducing than individuals with other characteristics.

d) Some of the characteristics resulting in differential survival and reproduction are heritable.

e) Long spans of time are available for slow, gradual change.

6. There are two kingdoms of unicellular procaryotic organisms: the *Eubacteria* and the *Archaebacteria*. *Procaryotic cells* lack the membrane-enclosed nucleus and other membranous structures present in the cells of other organisms, which are called *eucaryotic cells*.

7. The kingdom *Protista* includes a wide variety of organisms, largely unicellular but many multicellular groups as well. Two major groups of Protista are the photosynthetic protists (algae), which synthesize their own food, and the protozoa, which cannot manufacture their own food and must ingest their food already made. The red and brown algae are photosynthetic, multicellular protists whose bodies show little tissue differentiation.

8. The organisms belonging to the kingdom *Plantae* live on land; all have cells with rigid cellulose cell walls and chloroplasts. Thus they are photosynthetic and can synthesize their own food. The members of this kingdom, the mosses, liverworts, and vascular plants, probably evolved from ancestral green algae.

9. Organisms of the kingdom *Fungi* also have cell walls, but of different composition, and they lack chlorophyll and cannot manufacture their own food. They must obtain their nutrients already synthesized. Fungi depend entirely on absorption of nutrient molecules.

10. Members of the kingdom *Animalia* differ from plants and fungi in the lack of rigid cell walls, and in their mode of nutrition; animals ingest their food. Seven animal groups (cnidarians, flatworms, molluscs, annelids, arthropods, echinoderms, and chordates) will be frequently referred to throughout the book.

STUDY QUESTIONS

1. What are the most important attributes that living organisms have in common? (pp. 1–2)

2. Describe the scientific method. What is the distinction between a hypothesis and a theory? (pp. 2–3)

3. **How did Vesalius, Harvey, van Leeuwenhoek, and Pasteur contribute to the growth of biological thought? (pp. 6–7)**
4. **What lines of evidence did Darwin present to support the concept of evolutionary change in living organisms? (pp. 9–11)**
5. **Describe the five assumptions of Darwin's theory of natural selection. Contrast this concept with Lamarck's ideas about the mechanisms of evolutionary change. (p. 13)**
6. **How do eucaryotes differ from procaryotes? Which group is more ancient? Which group is more abundant today (in terms of numbers of individuals)? (p. 14)**
7. **Name and briefly describe the two kingdoms of procaryotes. (p. 15)**
8. **What are the major characteristics of the Protista? Why is this kingdom called a "grouping of convenience"? (pp. 16–17)**
9. **What attributes do members of the kingdom Plantae share? List the distinguishing characteristics of the three main groups of plants. (pp. 18–19)**
10. **In what ways do Fungi resemble plants and animals? What unique features of Fungi have led biologists to place them in a kingdom of their own? (pp. 19–20)**
11. **What are two characteristics of the kingdom Animalia? List the distinguishing features of seven phyla in this kingdom. (pp. 20–24)**

PART ONE

The Chemical and Cellular Basis of Life

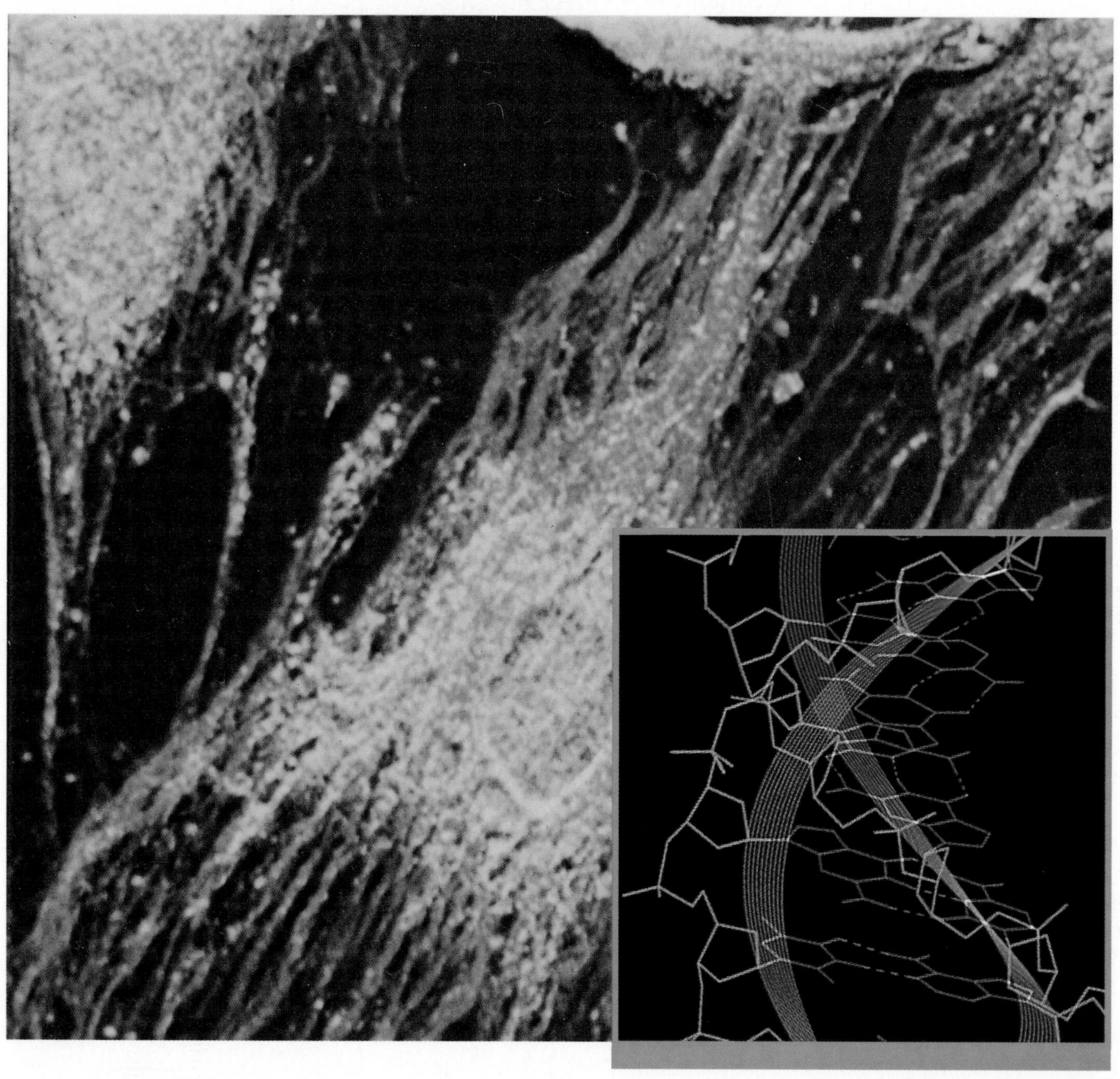

CHAPTER 2

BUILDING BLOCKS OF MATTER: ATOMS, BONDS, AND SIMPLE MOLECULES

In recent decades, biologists have gained a growing appreciation of the contributions the physical sciences can make to biology. Before the 1950s, many crucial processes of life, such as the trapping and storing of solar energy by green plants, the extraction of usable energy from organic nutrients, the growth and development of organisms, and the mechanisms of genetic inheritance, were understood only at a comparatively superficial level. Since then, thanks to a succession of major advances in biochemistry, it has become possible to explore these and many other important biological phenomena at the molecular, atomic, and even subatomic levels. Indeed, much current biological and medical research—from the effort to genetically engineer frostproof strawberry plants to the development of potential AIDS treatments—hinges on the detailed investigation of chemical reactions both within the living cell and between the cell and its external environment.

Clearly, no understanding of modern biology is possible without some knowledge of chemistry. This chapter and the next one are designed to introduce a few basic chemical concepts you will need as background for later chapters. You'll find them easier to assimilate if you focus first on the key ideas, which are summarized throughout this book by the headings above each section.

Certain Elements Are Essential for Life

All matter is composed of a limited number of ***elements,*** basic substances that cannot be decomposed into simpler substances by chemical reactions. There are 92 naturally occurring elements. Each is designated by one or two letters that stand for its English or Latin name. Thus H is the symbol for hydrogen, O for oxygen, N for nitrogen, C for carbon, Cl for chlorine, Mg

Chapter Opening Photo: Water is essential to life. Here rain forms droplets on grass stems due to the high surface tension of water.

for magnesium, K for potassium (Latin, *kalium*), Na for sodium (Latin, *natrium*), and so on.

Not all of the 92 elements found in nature are important in life processes. Table 2.1 lists the important ones, beginning with hydrogen, carbon, nitrogen, and oxygen, which together account for more than 96 percent of the human body by weight. Some of the elements listed are present only in minute (trace) amounts in the cells of certain organisms. Nevertheless they may be necessary for the maintenance of life. Iodine, for instance, is an essential component of the thyroid hormone that controls your metabolic rate, even though the recommended daily allowance is only 0.00015 grams, or only a few millionths of your body weight.

TABLE 2.1 *Elements important to life**

Element	*Atomic number*	*Approximate atomic mass*	*Approximate % of human body by weight*
Hydrogen (H)	1	1	9.5
Carbon (C)	6	12	18.5
Nitrogen (N)	7	14	3.3
Oxygen (O)	8	16	65.0
Sodium (Na)	11	23	0.2
Phosphorus (P)	15	31	1.0
Sulfur (S)	16	32	0.3
Chlorine (Cl)	17	35	0.2
Potassium (K)	19	39	0.4
Calcium (Ca)	20	40	1.5
Iron (Fe)	26	56	Trace
Iodine (I)	53	127	Trace

*All the elements listed have been shown to be essential for one or more species, but not all are essential for every species.

An Element's Atomic Structure Determines Its Properties

The smallest particle of an element that is indivisible by ordinary chemical means and has the same chemical properties as the element in bulk is called an ***atom.*** Although atoms can be considered the basic chemical units of matter, they themselves are composed of still smaller particles. Many of these particles belong to the world of subatomic physics and are of little immediate concern to biologists, but three of them—the proton, the neutron, and the electron—play such a central role in determining the properties of an element that we cannot hope to understand its chemical behavior without first becoming familiar with them.

Protons and Neutrons Occupy the Atomic Nucleus

All the positive charge and almost all the mass of an atom are concentrated in its ***nucleus,*** or core, which consists primarily of ***protons*** and ***neutrons.*** Each proton carries a charge of +1. The neutron, as its name implies, has no charge. The proton and the neutron have roughly the same mass (Table 2.2).

The number of protons in the nucleus is unique for each element. This number, called the ***atomic number,*** is sometimes written as a subscript immediately before the chemical symbol. Thus, ${}_1H$ indicates that the atomic number of hydrogen is 1; its nucleus contains only one proton. Similarly, ${}_8O$ indicates that all oxygen nuclei contain eight protons.

The total number of protons and neutrons in a nucleus approximates the total mass of the nucleus and so is called the ***mass number.*** The mass number is commonly written as a superscript immediately preceding the chemical symbol. For example, most atoms of oxygen contain eight protons and eight neutrons; the mass number is therefore 16, and can be symbolized as ${}^{16}O$ or, if we wish to show both the atomic number and the mass number, as ${}^{16}_{8}O$.

Although the number of protons is the same for all atoms of the same element, the number of neutrons is not always the same, and neither, consequently, is the mass number. For example, most oxygen atoms, as noted, contain eight protons and eight neutrons and have a mass number of 16. Some, however, contain nine neutrons and therefore have a mass number of 17 (symbolized as ${}^{17}O$) and still others have 10 neutrons and a mass number of 18 (symbolized as ${}^{18}O$). Atoms of the same element that differ in mass, because their nuclei contain different numbers of neutrons, are called ***isotopes.*** For example, ${}^{16}O$, ${}^{17}O$, and ${}^{18}O$ are three isotopes of oxygen. Some elements have as many as 20 naturally occurring isotopes; others have as few as two.

The number of neutrons in the nucleus does not affect the chemical properties of an atom; all isotopes of the same element behave in essentially the same way in their chemical reactions with other substances. The number of neutrons, however, does affect an atom's physical properties, such as its mass and its nuclear stability. The nuclei of some isotopes are unstable and tend to break down spontaneously to more stable

TABLE 2.2 *Subatomic fundamental particles*

Particle	*Mass (in units of 10^{-24} grams)*	*Charge (in electronic charge units)*
Electron	0.001	−1
Proton	1.672	+1
Neutron	1.674	0

forms, emitting high-energy radiation in the process. For example, the simplest element, hydrogen, has three different isotopes: ^1_1H, which is the usual form of the element; ^2_1H, a stable, nonradioactive isotope known as deuterium; and ^3_1H, an unstable, radioactive isotope called tritium (Fig. 2.1)

Radioactive isotopes are valuable research tools in many areas of biology and medicine (see Exploring Further: Biological Uses of Radioactive Elements, p. 32), because the radiation they emit can be detected either photographically or with an electronic sensing device. Tritium, for instance, is often used in this way to trace the movements of hydrogen in biochemical reactions. Because they are taken up by tissue as readily as their more stable counterparts, radioactive isotopes can also be used as remote "tracers" to help physicians locate circulatory blockages or pinpoint tumors (Fig. 2.2).

In addition, the known decay rate of certain naturally occurring radioactive isotopes, notably carbon 14 (^{14}C), is the basis of a widely applied method for the dating of rocks and fossils.

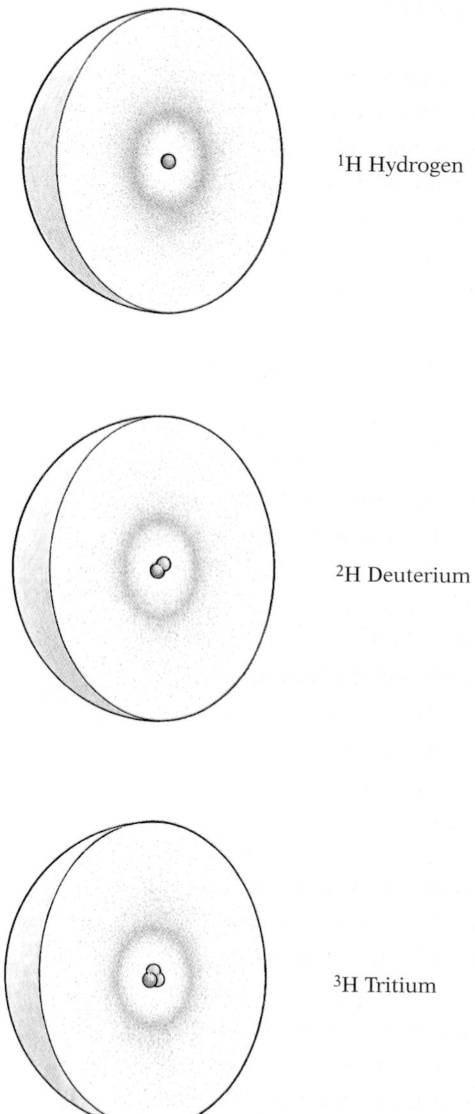

2.1 The three principal isotopes of hydrogen Each of the three isotopes has one proton (blue) in its nucleus and one electron orbiting the nucleus. The isotopes differ in that ordinary hydrogen (^1H) has no neutrons in its nucleus, deuterium (^2H) has one, and tritium (^3H), which is unstable, has two. The volume within which the single electron can be found 90 percent of the time (a sphere) is indicated by stippling; the denser the stippling in this cross-section view, the greater the likelihood that at any given moment the electron will be found within that portion of the sphere.

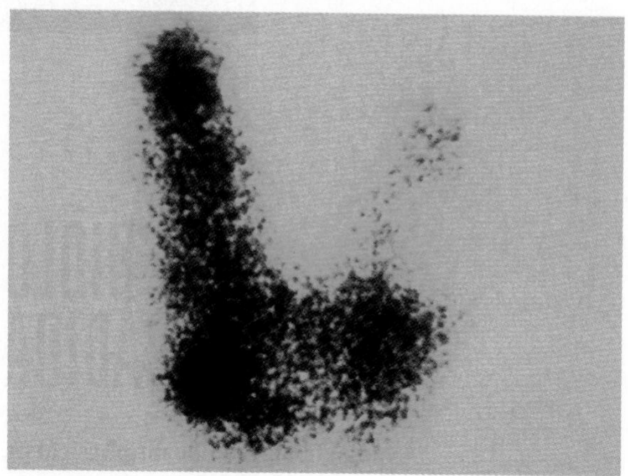

2.2 A thyroid scan using radioactive iodine In this routine clinical test, the patient is given a glass of water containing the radioactive ^{131}I tracer, and the uptake of the tracer by the thyroid is determined by placing a gamma ray counter over the thyroid gland in the neck. Inactive glands will not pick up normal amounts of iodine whereas overactive glands will take up excessive amounts. By moving the counter back and forth over the neck, a radioactive "picture" of the thyroid gland can be made, showing "hot" spots where the tissue is very active, and "cold" spots where the tissue is inactive. In this photograph, the left side of the gland is normal but the top portion of the right side is inactive, failing to concentrate the ^{131}I tracer. Malignant tumors of the thyroid gland frequently appear as cold spots.

The Electrons Occupy an Atom's Extranuclear Volume

The part of the atom outside the nucleus, which makes up virtually all of the atom's volume, contains the ***electrons.*** Compared with protons and neutrons, electrons have little mass (Table 2.2). Each electron carries a charge of −1, exactly the opposite of a proton's charge.

In an electrically neutral atom, the number of electrons around the nucleus is equal to the number of protons in the nucleus. The positive charges of the protons and the negative charges of the electrons cancel each other, making the atom

BIOLOGICAL USES OF RADIOACTIVE ELEMENTS

A biologically significant physical property of some isotopes is their tendency to decay into more stable forms, giving off various particles on their way to physical stability. These isotopes are termed radioactive because in time they spontaneously emit radiation, primarily alpha particles (each particle comprising two protons and two neutrons), beta particles (electrons), or photons, often called gamma radiation. The stability of an isotope is measured by its half-life, i.e., the time it takes half the atoms in a sample to decay. Tritium (^3H), for instance, has a half-life of about 12 years; ^{32}P, roughly 14 days; ^{40}K, 1.2×10^9 years, and so on.

Radioactive isotopes are extraordinarily useful in biology, since an isotope added to a sample emits radiation that scientists can track. With a "labeled" isotope of carbon dioxide (CO_2), for instance, we can trace how plants use carbon to build sugars. Because they are taken up by tissue as readily as their more stable counterparts, radioactive isotopes can also be used as "tracers" to help doctors locate circulatory blockages or pinpoint tumors. For example, an isotope of iodine, ^{131}I, is often used to assess thyroid function and to find tumors in the thyroid gland. In a normal individual, iodine (including ^{131}I) is rapidly taken up by the thyroid gland. In routine clinical tests, the patient is given a glass of water containing the ^{131}I tracer, and the uptake of the tracer by the thyroid is determined by placing a gamma ray counter over the thyroid gland in the neck. Inactive glands will not pick up normal amounts of iodine whereas overactive glands will take up excessive amounts. By moving the counter back and forth over the neck, a radioactive "picture" of the thyroid gland can be made, showing "hot" spots where the tissue is very active and "cold" spots where the tissue is inactive (Fig. 2.2). Tumors of the thyroid gland frequently appear as cold spots.

However, isotopes can also have potentially harmful effects. A radioactive atom, like uranium, in a living cell can decay into another element by losing protons, thereby changing the chemistry of its molecule completely, or, more commonly, radioactive isotopes can produce highly reactive atoms that have too many or too few electrons to balance the electrical charge of their protons, and thus have a net charge. Since the behavior of electrons, as we shall see, determines the chemistry of life, such unpredictable and uncontrolled movement of electrons can disrupt the precisely ordered and carefully regulated workings of the cell. For instance, the redistribution of electrons in a critical part of a cell's DNA can trigger the complicated chain of events leading to cancer. Occurring in the cells of the reproductive system, such a change in the DNA can cause defects in later offspring. Changes of this sort arise in all of us every day from exposure to the sun's radiation and the natural decay of radioactive elements in the earth's atmosphere and crust; however, each cell has a battery of defense mechanisms to counteract these changes and repair them.

neutral. In such an atom, the atomic number represents both the number of protons inside the nucleus and the number of electrons outside. If, then, we see the symbol $^{35}_{17}Cl$, we can tell that a neutral atom of this isotope of chlorine has 17 protons, 18 neutrons, and 17 electrons. Similarly, the symbol $^{39}_{19}K$ means that this isotope of potassium contains 19 protons, 20 neutrons, and 19 electrons.

The electrons are not in fixed positions outside the nucleus. Each is in constant motion, and it is impossible to know exactly where a given electron is at any particular moment. For this reason some illustrations of atoms, such as Figure 2.1, do not show the electron itself but instead show the region where the electron is likely to be most of the time. Physicists define the volume of space within which the electron would be found 90 percent of the time as the electron's ***orbital*** (Fig. 2.3). By a fundamental rule of physics, every orbital can contain a maximum of two electrons.

The distance of any given electron from the nucleus is a function of its energy: the higher its energy, the greater its probable distance from the nucleus. But in any particular atom, only discrete energy levels—like steps of a staircase—are possible. To occupy a specific energy level, an electron must have a certain amount of energy. To achieve a higher energy level, an electron must absorb, from some outside source, additional energy equal to the difference in energy between the two levels. Conversely, when an electron falls into the next lower level, it emits an amount of energy equal to the difference between the two energy levels (Fig. 2.4).[1]

An electron occupying the lowest step available to it in the atom is said to be in the ***ground state.*** Once it has absorbed enough energy to move up to the next energy level, it is said to be in an ***excited state.*** An electron in the excited state has a strong tendency to return to its ground state by emitting, in some form, the additional energy just acquired. Most often the energy is released as light. The decay of excited electrons in the lining of fluorescent tubes, for instance, helps us light our world.

The potential energy of excited electrons during that brief moment before they drop down the energy staircase plays a key role in biology: life on earth is based on the ability of certain organisms to capture and make use of the energy of excited electrons. Green plants have special pigments, called chlorophyll, that absorb light energy from the sun. The absorbed energy stimulates electrons in atoms within the chlorophyll molecules to move to a higher energy level. The energy released by these excited electrons when they return to the ground state is then used by the plant to manufacture food molecules. We shall be discussing this fundamental energy-transformation process, known as photosynthesis, in greater detail in Chapter 7.

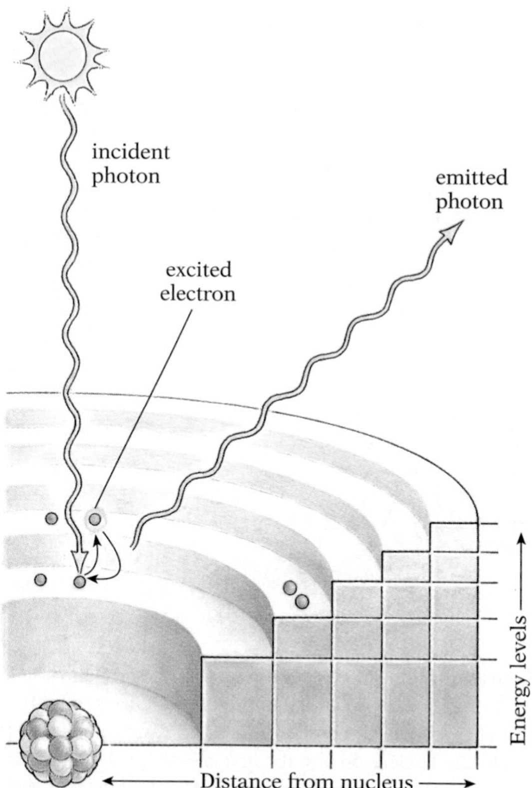

2.4 Energy levels of electrons The electrons in an atom occupy discrete energy levels. If an electron absorbs the right amount of energy (shown as a photon, a discrete particle of light) and there is a vacancy in a higher energy level, it can move up the energy staircase to this level. Normally this "excited" electron quickly reemits the absorbed energy (here again shown as a photon) and returns to its original energy level.

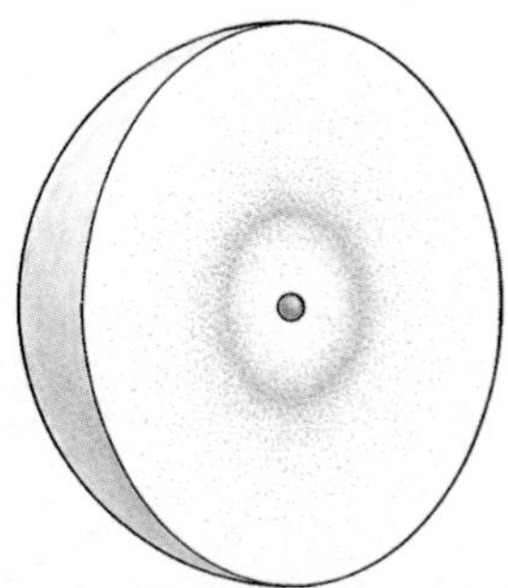

2.3 One way of representing the hydrogen atom Since no one has ever seen the particles that make up an atom, all our knowledge of what atoms look like is indirect, and we can only picture them as models that fit the data. The nucleus is shown here as a central blue area, with the "cloud" around it representing the region where the electron is likely to be. The circle encloses the orbital of the electron—the volume, a sphere, within which the electron will be found 90 percent of the time (see also Fig. 2.1).

[1] In actuality, because of the Second Law of Thermodynamics (p. 78), slightly less energy is emitted than what was absorbed when the electron achieved the higher level, since some energy is lost in the conversion.

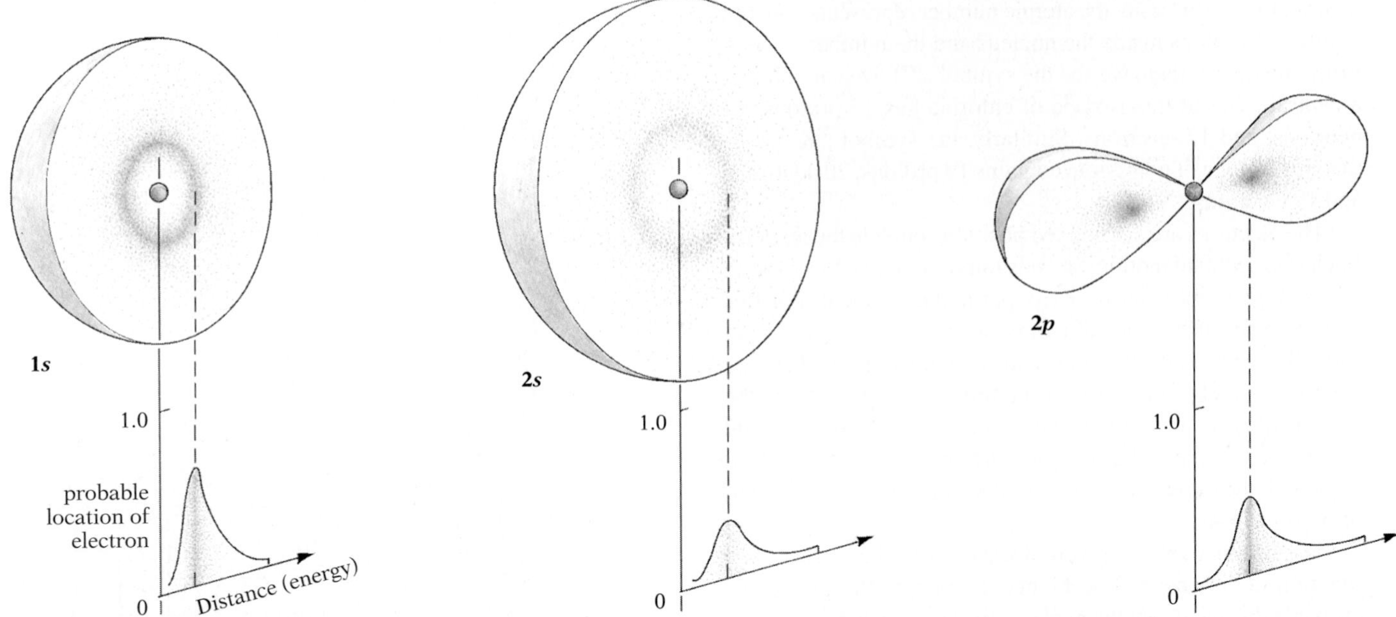

2.5 Representations of electron orbitals The orbitals of *s* electrons are approximately spherical, those of *p* electrons roughly dumbbell-shaped. The numerals before *s* and *p* indicate the energy levels. Thus the 1*s* electron is at the first energy level, that nearest the nucleus; the 2*s* and 2*p* electrons are at a higher energy level, and hence are at a greater average distance from the nucleus than the 1*s* electron. Note that, despite the very different shapes of their orbitals, the 2*s* and 2*p* electrons are at the same energy level—their most probable distances from the nucleus are the same.

Electron Orbitals Come in Various Sizes and Shapes

The energy an electron possesses determines not only its energy level but also the size and shape of its orbital. Every orbital, no matter what its size or shape, can contain a maximum of two electrons. At the lowest energy level, electrons occupy a spherical orbital designated the 1*s* orbital (Fig. 2.5). Accordingly, electrons at this lowest energy level are designated 1*s* electrons. Chemists represent hydrogen, the simplest element, by the following simple shorthand: $1s^1$. The number preceding the orbital designation is the energy level and the superscript following it indicates the number of electrons found in that orbital.

At the second energy level, electrons occupy four possible orbitals. The first orbital of the second energy level, the 2*s* orbital, is spherical around the nucleus like the 1*s* orbital, but because 2*s* electrons are at a higher energy level than 1*s* electrons, they can move farther from the nucleus (Fig. 2.5). The other three orbitals of the second energy level, the *p* orbitals, are dumbbell-shaped (Fig. 2.5). Within a given energy level, *p* orbital electrons are at a slightly higher energy level than *s* electrons, and the planes of their orbitals are at right angles to each other like the *x, y,* and *z* axes of geometry (Fig. 2.6). Since each orbital can hold a maximum of two electrons, the second energy level, with four orbitals, can accommodate a maximum of

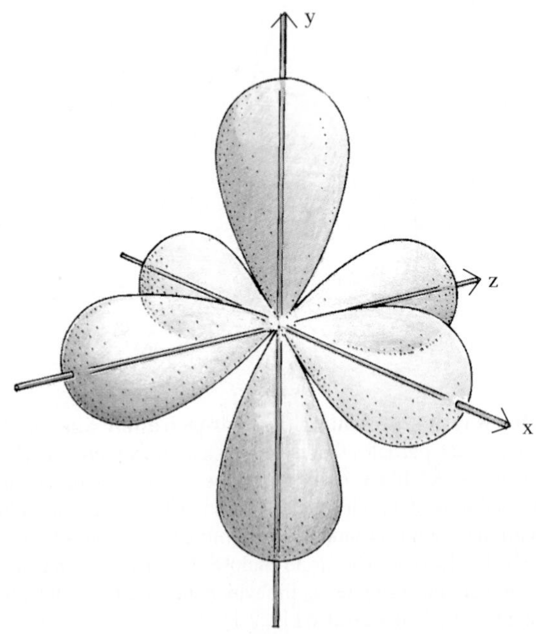

2.6 The three 2*p* electron orbitals Each of the dumbbell-shaped orbitals is oriented in a different dimension of space, at right angles to the other two. The *x, y,* and *z* axes have been inserted to aid visualization. The intersection of these axes is the location of the nucleus.

eight electrons: one *s* orbital containing two electrons and three *p* orbitals containing a total of six electrons.

Oxygen ($_8O$) is a good example of an element essential to life processes that contains both *s* and *p* orbitals. Its shorthand formula of electron configuration, $1s^22s^22p^4$, indicates a total of eight electrons: two *s* electrons at the first level, two *s* electrons at the second, and four (of six possible) *p* electrons at the second level. As we shall see in the following discussion of chemical bonding, the fact that the oxygen atom can accommodate two more electrons in its $2p$ orbitals has special significance for this element's ability to join with other elements in chemical reactions.

When elements have more than 10 electrons (two in the first level and eight in the second), the additional electrons are accommodated at higher energy levels beyond the second. Although the third and successive levels can hold more than eight electrons each, the outermost energy level can never hold more than eight electrons (two *s* and six *p* electrons), no matter how many total electrons an atom may have. This constraint plays an important part in determining the chemical reactivity of such elements, as we shall see in the next section. Because the elements important to life processes have, for the most part, 20 or fewer electrons (Fig. 2.7), we will not be concerned with energy levels beyond the third. The electron configurations of some biologically important elements are shown in Table 2.3.

An Atom's Outermost Electrons Determine Its Chemical Properties

In the 1860s, when Dmitri I. Mendeleev in Russia and Julius L. Meyer in Germany began to organize the known elements in a sequence of increasing atomic numbers beginning with hydrogen, they noticed that elements with very similar chemical properties, such as the tendency to react with other specific elements, appear at regular intervals in the list rather than adjacent to each other. In Figure 2.7, fluorine (atomic number 9) reacts more like the element below it, chlorine (17), for instance, than like an adjacent element with a closer atomic

Number of electrons in outer shell

1	2	3	4	5	6	7	8
$1p^+$ $1e^-$ H							$2p^+$ $2e^-$ He
$3p^+$ $2e^-$ $1e^-$ Li	$4p^+$ $2e^-$ $2e^-$ Be	$5p^+$ $2e^-$ $3e^-$ B	$6p^+$ $2e^-$ $4e^-$ C	$7p^+$ $2e^-$ $5e^-$ N	$8p^+$ $2e^-$ $6e^-$ O	$9p^+$ $2e^-$ $7e^-$ F	$10p^+$ $2e^-$ $8e^-$ Ne
$11p^+$ $2e^-$ $8e^-$ $1e^-$ Na	$12p^+$ $2e^-$ $8e^-$ $2e^-$ Mg	$13p^+$ $2e^-$ $8e^-$ $3e^-$ Al	$14p^+$ $2e^-$ $8e^-$ $4e^-$ Si	$15p^+$ $2e^-$ $8e^-$ $5e^-$ P	$16p^+$ $2e^-$ $8e^-$ $6e^-$ S	$17p^+$ $2e^-$ $8e^-$ $7e^-$ Cl	$18p^+$ $2e^-$ $8e^-$ $8e^-$ Ar
$19p^+$ $2e^-$ $8e^-$ $8e^-$ $1e^-$ K	$20p^+$ $2e^-$ $8e^-$ $8e^-$ $2e^-$ Ca						

2.7 Partial periodic table with electron distributions The first 20 elements are shown arranged according to their positions in the periodic table. Elements in the same column share many chemical properties because they have the same number of electrons in their outer shell. (Helium is placed in column 8 even though it has only two outer electrons because, like neon, argon, and the other so-called noble gases, its outer shell is full, and its chemical properties are therefore those of a noble gas.)

TABLE 2.3 *Electron configurations of some biologically important elements*

Element	Atomic number	Orbitals 1s	2s	2p	3s	3p	4s	Electrons in valence level	Electron configuration
Hydrogen (H)	1	1						1	$1s^1$
Carbon (C)	6	2	2	2				4	$1s^2\,2s^2\,2p^2$
Nitrogen (N)	7	2	2	3				5	$1s^2\,2s^2\,2p^3$
Oxygen (O)	8	2	2	4				6	$1s^2\,2s^2\,2p^4$
Sodium (Na)	11	2	2	6	1			1	$1s^2\,2s^2\,2p^6\,3s^1$
Phosphorus (P)	15	2	2	6	2	3		5	$1s^2\,2s^2\,2p^6\,3s^2\,3p^3$
Sulfur (S)	16	2	2	6	2	4		6	$1s^2\,2s^2\,2p^6\,3s^2\,3p^4$
Chlorine (Cl)	17	2	2	6	2	5		7	$1s^2\,2s^2\,2p^6\,3s^2\,3p^5$
Potassium (K)	19	2	2	6	2	6	1	1	$1s^2\,2s^2\,2p^6\,3s^2\,3p^6\,4s^1$
Calcium (Ca)	20	2	2	6	2	6	2	2	$1s^2\,2s^2\,2p^6\,3s^2\,3p^6\,4s^2$

number, oxygen (8). The tendency for chemical properties to recur periodically throughout the sequence of elements is known as the ***Periodic Law.***

We now know that the chemical properties of elements are largely determined by the electronic configuration of their outermost energy level, and this is what accounts for the recurring pattern of the periodic table of the elements. As a rule, elements with the same electron configuration in their outermost energy level—also known as the ***valence energy level***—have similar chemical properties and are aligned vertically in the periodic table. For example, if the orbitals of the valence energy level are filled, as in the case of the so-called ***noble gas elements***—helium (atomic number 2), which has two *s* electrons, and neon (10) and argon (18), both of which have two *s* and six *p* electrons (a total of eight electrons) in their valence energy levels—the element will have very little tendency to react chemically with other atoms and hence will be extremely stable. In general, when atoms have their valence energy levels complete, with eight electrons (two *s* and six *p*), they are relatively inert.

The electrons in the outermost energy level, often called ***valence electrons,*** can be denoted by an array of dots around the chemical symbol for that element. Thus hydrogen, carbon, nitrogen, and oxygen, which have one, four, five, and six electrons respectively in their valence energy levels, would be represented as follows:

H· ·Ċ· :Ṅ· ·Ö·

The placement of the dots is just a bookkeeping convention and in no way indicates the actual positions of the electrons themselves.

As noted, atoms are in a particularly stable configuration when the *s* and *p* orbitals of the valence energy level are filled, that is, when each atom has eight electrons in its valence level. (Hydrogen and helium, the two smallest atoms, are an exception to the requirement for eight electrons. Their valence levels are complete with just two electrons.) Atoms that lack the requisite eight will tend toward satisfying this requirement. Depending on their electronic configuration, atoms may react with other atoms and give up, take, or share their valence electrons, forming bonds between the atoms. When two or more atoms are bound together in this fashion, the force of attraction that holds them together is called a ***chemical bond.*** The strength of a chemical bond between two atoms is measured by the amount of energy required to break the bond and separate the atoms. The discussion that follows will focus on two types of chemcial bonds particularly important to biology: covalent bonds and ionic bonds.

Strong Covalent Bonds Form When Valence Electrons Are Shared

One way atoms can fill their valence energy levels, and thereby achieve greater stability, is by sharing electrons with other atoms. The strong bond that results from such a sharing of electrons is called a ***covalent bond.*** For example, an isolated hydrogen atom, which has only one electron in its 1*s* orbital, tends to fill its orbital by sharing a second electron with another atom. One possible reaction is for two atoms of hydrogen to bond with each other covalently, thereby sharing their electrons and forming what is called molecular hydrogen (written H_2):

$$H\cdot + H\cdot \longrightarrow H{:}H$$

Indeed, covalent bonding is essential to the very concept of a ***molecule,*** which by definition is the smallest particle of an element or compound that can have a stable, independent existence. Here, each hydrogen atom shares its electron with the other atom, so that each hydrogen has, in a sense, two electrons (Fig. 2.8A). The stability of the resulting molecule is much greater than that of the isolated atoms, since the valence energy levels of both atoms are now filled.

Energy is involved in the formation and breaking of chemical bonds. In the foregoing example, the combination of two hydrogen atoms to form an H_2 molecule releases energy; conversely, the separation of the H_2 molecule into its two component atoms requires energy. Remember that *breaking a chemical bond always requires energy, whereas forming one always releases energy.*

Covalent bonds are not limited to the sharing of one electron pair between two atoms. Sometimes two atoms share two or three electron pairs and form even stronger double or triple bonds. For example, when two atoms of oxygen bond together, they form a double bond (because an oxygen atom needs two electrons to complete its valence energy level), and when two atoms of nitrogen (atomic number 7) bond together, they form a triple bond (because each nitrogen atom needs three more electrons to fill its valence energy level):

$$:\dot{\underset{\cdot}{O}}: + :\dot{\underset{\cdot}{O}}: \longrightarrow :\dot{\underset{\cdot}{O}}::\dot{\underset{\cdot}{O}}:$$

$$:\dot{N}: + :\dot{N}: \longrightarrow :N:::N:$$

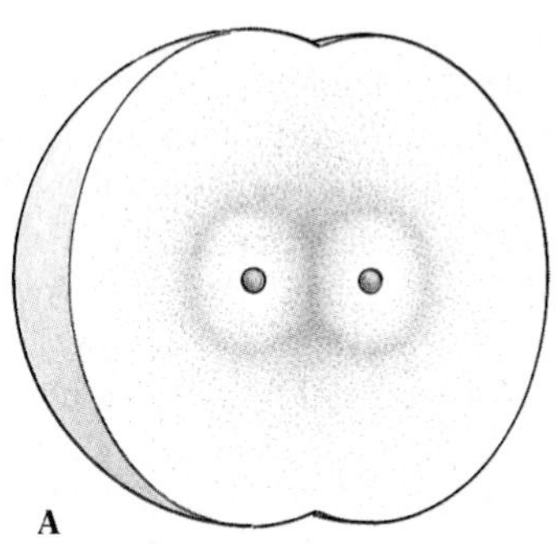

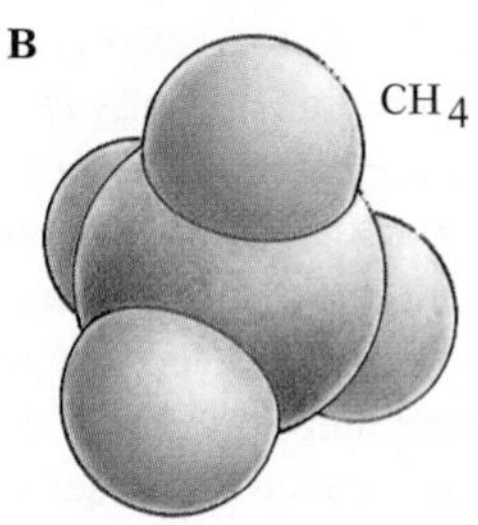

2.8 Covalent bonding in molecular hydrogen and methane (A) The sharing of electrons between two hydrogen molecules is shown by overlapping electron clouds. (B) In the methane (CH_4) molecule, the four hydrogen atoms (blue) share their single *s* electron with a carbon atom (gray) so the electron clouds of the carbon atom overlap with the electron clouds of the four hydrogen atoms.

A covalent bond may be represented simply by a line between two atoms rather than a pair of dots. According to this convention, the molecules H_2, O_2, and N_2 appear as follows:

$$H\text{–}H \qquad O{=}O \qquad N{\equiv}N$$

Double or triple bonds between two atoms are usually stronger than single bonds. The greater stability of such elements as molecules is what accounts for the fact that they normally assume this form in the gaseous state, for instance, in the earth's atmosphere.

The tendency of atoms to form a characteristic number of bonds is referred to as their ***covalent bonding capacity.*** Thus, hydrogen is said to have a covalent bonding capacity of 1; oxygen, a capacity of 2; and nitrogen, a bonding capacity of 3. For life on earth the most important example of an atom with a covalent bonding capacity of 4 is carbon. Indeed, as we shall see in the next chapter, the ability of carbon to form four bonds is what accounts for the vast diversity of ***organic chemistry,*** which by definition is the chemistry of carbon- and hydrogen-containing molecules. The covalent bonding capacities of some biologically important elements are listed in Table 2.4.

TABLE 2.4 *Bonding capacities of certain biologically important elements**

Element	*Bonding capacity*	*Electrons in valence level*
Hydrogen (H)	1	1
Oxygen (O)	2	6
Nitrogen (N)	3	5
Carbon (C)	4	4

*Many atoms fill their valence energy levels, which can hold eight electrons, by sharing electrons with other atoms. The tendency of atoms to form a characteristic number of bonds to fill their valence levels is referred to as their bonding capacity. Thus carbon, with four valence electrons, needs four electrons and has a bonding capacity of 4.

A Covalent Bond's Polarity Depends on Whether Electrons Are Shared Equally or Unequally

When two atoms of the same element bond together, as in the examples cited above, neither will have more attraction for the shared electrons than the other; hence, the electrons will be shared equally. A bond in which the electrons are likely to be no closer to one atom than to another is said to be a ***nonpolar covalent bond.*** Bonds between atoms of the same element are always nonpolar; so are bonds between atoms that have relatively equal attractions for the shared electrons. Carbon and hydrogen atoms, for instance, share electrons relatively equally, making the carbon-to-hydrogen bond nonpolar (Fig. 2.8B).

Suppose, however, that instead of being bonded to each other, two hydrogen atoms are covalently bonded to an oxygen atom, forming water (H_2O):

$$\mathrm{H}\cdot + \mathrm{H}\cdot + :\dot{\underset{\cdot}{\mathrm{O}}}: \longrightarrow {}^{(\delta^+)}\mathrm{H}:\overset{(\delta^-)}{\ddot{\underset{\underset{(\delta^+)}{\mathrm{H}}}{\mathrm{O}}}}:{}^{(\delta^-)}$$

With six electrons in its valence energy level, oxygen can get a full valence energy level by sharing electrons with two hydrogen atoms. At the same time, each hydrogen fills its 1*s* energy level (Fig. 2.9). Molecules such as this, which contain two or more different elements combined in a fixed proportion, are called ***compounds.***

A covalent bond that joins two or more different elements in a compound is somewhat different from one between two hydrogen atoms or between two oxygen atoms. The difference arises from the fact that each element has its own characteristic affinity for electrons, a property known as its ***electronegativity.*** Oxygen and nitrogen atoms, for instance, are highly electronegative and therefore tend to strongly attract electrons. In a bond between a hydrogen atom and an oxygen atom, for example, the oxygen atom is the more electronegative participant—that is, it has a greater tendency to attract the shared electrons than the hydrogen atom does. As a result, the shared electrons that form such a bond spend more time closer to the oxygen atom than to the hydrogen atom, giving the oxygen atom a slightly negative charge and the hydrogen atom a slightly positive charge.

Note that this bond, formed between two atoms that differ in their affinity for electrons, results in a separation of charge: one end positive and the other negative, rather like the north and south *poles* on a magnet. Such a bond, in which the electrons are shared *unequally* between its atoms, is called a ***polar covalent bond.*** Bonds to highly electronegative atoms such as nitrogen and oxygen are likely to be polar, because these atoms tend to attract the shared electrons strongly. Bonds between certain pairs of atoms, such as carbon and hydrogen, are essentially nonpolar, since the two atoms have almost the same attraction for the shared electrons.

The phenomenon of polarity helps explain many properties of molecules in living systems. Whole molecules can be polar because of the polarity of bonds within them. One example is the water molecule, in which the three atoms are arranged in a V-shaped structure, with the oxygen at the apex of the V and the two hydrogen atoms forming the arms. Since the two shared electrons are drawn closer to the oxygen atom, we think of the oxygen atom as having two partial negative (δ^-) charges (where "partial" means less than the full negative charge of an electron) and the two hydrogen atoms, which have the electrons less of the time, as having partial positive (δ^+) charges. Such a molecule, with both negative and positive ends, is said to be a ***polar molecule.***

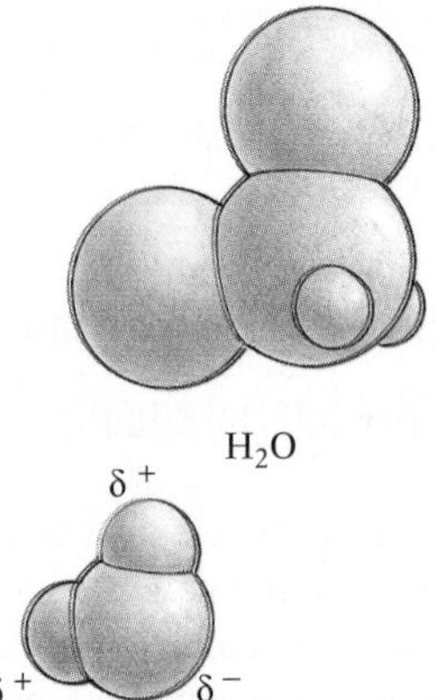

2.9 Polarity of the water molecule In water, two hydrogen atoms are bonded covalently to one oxygen atom, but the shared electrons are pulled closer to the oxygen because of its higher electronegativity. Because the atoms in water are arranged at an angle of 104.5 degrees, the charge distribution is asymmetrical, with negative charge concentrated at the oxygen end and the positive charges at the hydrogen ends; as a result the molecule as a whole is polar.

Ionic Bonds Form When Electrons Are Transferred From One Atom to Another

Some atoms have such a strong attraction for electrons that the electrons are pulled completely from one atom to another. For example, consider the interaction of sodium (atomic number 11) and chlorine (17). An atom of sodium has one electron in its valence energy level and an atom of chlorine has seven. Each chlorine atom, then, needs just one more electron to complete its valence level. Furthermore, chlorine is very electronegative, i.e., it very strongly attracts electrons. When metallic sodium and chlorine gas are brought together, each chlorine atom pulls the single valence electron away from each sodium atom. Each chlorine atom has thus acted as an ***electron acceptor,*** taking one electron from sodium, while each sodium atom has acted as an ***electron donor,*** giving up the one electron in its valence energy level. As a result, each atom ends up with eight electrons in its outermost energy level, thereby achieving a stable electronic configuration (Fig. 2.10).

Once it has given up an electron, the sodium is left with one more proton than it has electrons; it therefore has a net charge of $+1$. Similarly, the chlorine atom that gained an electron has one more electron than it has protons and so it has a net charge of -1. Such charged atoms (or charged aggregates of atoms) are called ***ions.*** They are symbolized by the appropriate chemical symbol followed by a superscript showing the charge. Sodium and chlorine ions, for instance, are written Na^+ and Cl^-. The process by which one or more electrons is transferred from one atom to another to form ions is called ***ionization.***

A

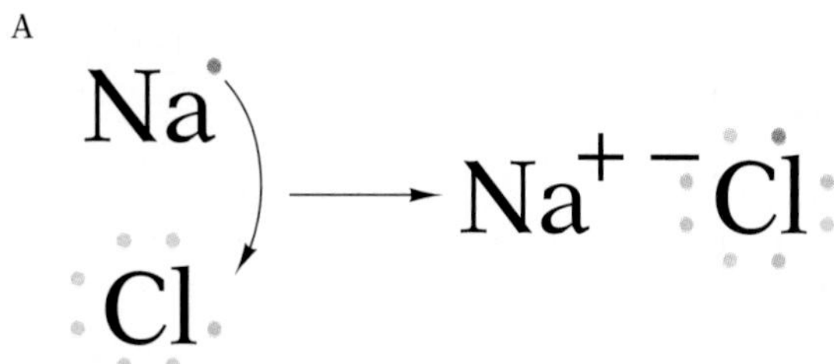

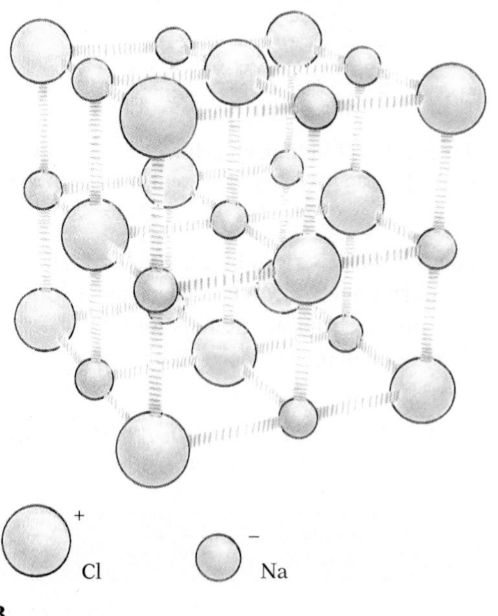

B

2.10 Ionic bonding of sodium and chlorine (A) A convenient way to represent the electronic configuration of the valence level is to symbolize each electron by a dot placed near the chemical symbol for the element. Sodium ($1s^22s^22p^6\mathbf{3s^1}$) has only one electron in its valence energy level (represented in dark brown) while chlorine ($1s^22s^22p^6\mathbf{3s^23p^5}$) has seven. Chlorine, needing one more electron to complete its valence level, is very electronegative and pulls the one valence electron away from sodium. After sodium donates an electron to chlorine, it has one more proton than electrons and a net positive charge. Conversely, the chlorine, having gained an electron, has one more electron than it has protons, and has a negative charge. The two charged atoms, or ions, are attracted to each other by their unlike charges. The result is the ordinary table salt, sodium chloride or NaCl. (B) Ionic compounds such as sodium chloride in the solid state often do not actually form discrete molecules. Instead, many sodium and chloride atoms are bound together by electrostatic attractions to form a large crystal, as depicted by the lattice on the left (stippling represents the electrostatic attraction). Salt crystals—potassium chloride (KCl) and sodium chloride (NaCl)—are shown on the right.

A sodium ion with its positive charge and a chlorine ion (usually called a chloride ion) with its negative charge tend to attract each other. Held together by this ***electrostatic attraction,*** the two ions form sodium chloride (NaCl), the compound we know as table salt (Fig. 2.10). A chemical bond of this type, which involves the complete *transfer* of an electron from one atom to another and the mutual attraction of the two ions thus formed, is termed an ***ionic bond.*** Unlike two atoms that are covalently bonded in a polar molecule, two ions that are bonded in this way retain their distinctive charges even if they are separated.

Ionic bonding may involve the transfer of more than one electron, as in calcium chloride. Calcium (atomic number 20) has two electrons in its valence energy level, and it gives up both to form the calcium ion, Ca^{++} (Fig. 2.11). Chlorine, however, needs only one electron to complete its valence level. Hence, it takes two chlorine atoms to act as acceptors for the two electrons from a single calcium atom, and a total of three ions bond together to form calcium chloride, symbolized as $CaCl_2$. (The subscript 2 shows there are two chlorine atoms for each calcium atom in this compound.) In other words, calcium has a bonding capacity, or valence, of +2, while chlorine has a valence of −1.

Ionic bonds occur only between strong electron donors (atoms on the left side of the periodic table with only a few electrons in their valence energy levels) and strong electron acceptors (atoms near the right side of the periodic table with close to eight electrons in their valence energy levels). Ionic

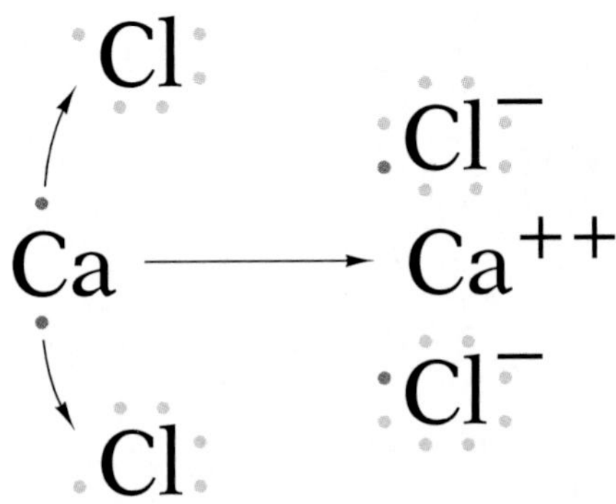

2.11 Ionic bonding of calcium and chlorine Calcium ($1s^2 2s^2 2p^6 3s^2 3p^6 \mathbf{4s^2}$) has two electrons in its outer energy level (represented in dark brown). It donates one to each of two chlorine ($1s^2 2s^2 2p^6 \mathbf{3s^2 3p^5}$) atoms and the two negatively charged chloride ions thus formed are attracted to the positively charged calcium ion to form calcium chloride ($CaCl_2$).

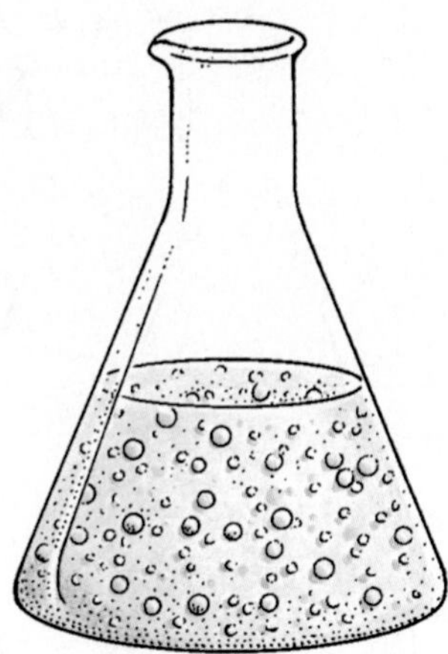

2.12 Dissociation of sodium chloride When in solution, the NaCl dissociates into separate Na^+ (yellow) and Cl^- (green) ions.

bonds generally do not form between elements with an intermediate number of electrons in their valence energy levels, or between two strong electron donors or two strong electron acceptors. Most chemical bonding between such elements is therefore covalent, with a sharing of electrons between the atoms participating in the bond.

Ionic Bonds Split Apart in Solution

Ionic bonds in the solid state, e.g., in dry crystals of salt, are strong, but in aqueous solutions they are relatively weak and easily broken. Accordingly, ionic substances such as sodium chloride and calcium chloride have a pronounced tendency to ***dissociate,*** or split apart, into separate ions in solutions, in which case they no longer exist as molecules. In solution, sodium chloride dissociates into two separate entities, a Na^+ ion and a Cl^- ion (Fig. 2.12). Similarly, calcium chloride in solution forms one Ca^{++} ion for every two Cl^- ions.

Ionic substances, either wholly or partly dissociated in body fluids, play many important roles in the functioning of biological systems. The coordinated movement of sodium ions and potassium ions across nerve-cell membrances, for example, is the key to the propagation of nerve impulses. Such ions behave differently from neutral atoms or molecules. In later chapters we will be concerned with the effects of charge on the movements of materials through the membranes of living cells.

In actuality, there is no sharp distinction between ionic bonds and covalent bonds. Ionic bonds, in which electrons are pulled completely from one atom to the other, are merely one extreme in a range of possible interactions between atoms. Nonpolar covalent bonds, in which the electrons are pulled with equal force by two atoms, are the other extreme. Polar covalent bonds represent a middle ground between these two extremes; in polar covalent bonds, such as those formed in water molecules, the electrons are pulled closer to one atom than to the other, but not all the way (see Table 2.5).

TABLE 2.5 *A summary of important characteristics of chemical bonds*

	Nonpolar covalent	*Polar covalent*	*Ionic bonds*
Example	H – H	H – O – H	Na^+Cl^-
Characteristics	Electrons shared **equally** between atoms H:H	Electrons shared **unequally** between atoms H:Ö:H	Electrons completely **transferred** from one atom to another Na^+:C̈l:$^-$

Weak Bonds Acting in Concert Can Have Significant Biological Effects

Ionic bonds in aqueous solutions are between five and 10 times weaker than covalent bonds, and other kinds of noncovalent bonds are weaker still (Table 2.6). Nevertheless, by acting in concert, such bonds are capable of playing a number of important roles in biological systems. For example, weak bonds play a crucial role in stabilizing the shape of many large molecules found in living matter—proteins in particular—and they often hold together groups of such molecules in orderly arrays. Like the minute hooks and eyes of a Velcro fastening, the bonds are individually quite weak, but the attachment can be strong and

TABLE 2.6 *A comparison of bond energies of different bonds and interactions*

Type of bond	*Average bond energy*[a]
Covalent bonds	50–250 kcal/mole[b]
Ionic bonds (in water)	7–10 kcal/mole
Hydrogen bonds	4–5 kcal/mole
Hydrophobic interactions	1–3 kcal/mole

[a]Bond energy is the amount of energy in kilocalories (kcal) required to *break* the chemical bond in a gaseous substance. A kilocalorie is the amount of heat required to raise the temperature of water one degree Celsius (for example, from 14.5°C to 15.5°C).
[b]A mole is the amount of a substance, in grams, that equals the combined atomic mass of all the constituent atoms in a molecule of that substance. Thus one mole of water is 18 grams since there are two hydrogen atoms (atomic mass = 1.0) and one oxygen atom (atomic mass = 16).

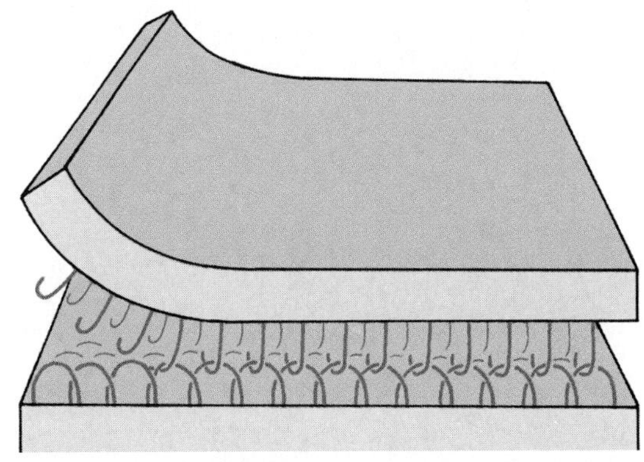

2.13 Stability from weak bonds An array of many individually weak bonds—here represented as the many hooks and eyes of a Velcro fastener—can be surprisingly strong as a unit.

stable when many bonds act together (Fig. 2.13). Just as a Velcro patch is easy to unfasten when you begin at a corner, attacking a few "bonds" at a time, so an array of molecules held together by weak bonds (as opposed to the more powerful covalent bonds) can be disassembled and rearranged with relative ease by forces that dissolve the bonds one by one.

Among the other types of weak bonds with biologically significant cumulative effects are ***hydrogen bonds.*** In such a bond, a hydrogen atom is shared by two electronegative atoms (usually oxygen or nitrogen). The hydrogen atom is covalently bonded to one electronegative atom and simultaneously attracted to another electronegative atom. This attraction is the hydrogen bond (Fig. 2.14A). Such bonds are extremely weak: they are only about half as strong as ionic bonds in aqueous solutions and hence break very easily (Table 2.6).

Water is a good example of a hydrogen-bonded substance. The two hydrogen atoms in each water molecule are covalently bonded to the oxygen atom, but because of oxygen's strong attraction for the shared electrons, each hydrogen has a partial positive charge (Fig. 2.14A). Owing to this charge, a hydrogen atom in one water molecule is weakly attracted to a negatively charged oxygen atom in another nearby water molecule, forming a hydrogen bond (Fig. 2.14B).

Since each of the hydrogens, while remaining covalently bonded to the oxygen atom of its own molecule, can form a weak attachment with the oxygen of another water molecule, and the oxygen can form a weak attachment with two external hydrogens, each water molecule has the potential for being simultaneously linked by hydrogen bonds to four other water molecules (Fig. 2.14C). In a sense, then, a volume of water is a continuous chemical entity, because of the hydrogen bonding between the individual water molecules. As we shall see shortly, it is the polarity of water molecules and hydrogen bonds between them that gives water its special properties.

Much weaker than ionic or hydrogen bonds are the associations known as ***hydrophobic interactions.*** As implied by the name (from the Greek roots *hydro-*, "water," and *phobos*, "fear"), such interactions occur between groups of molecules that are insoluble in water. Molecules of this type, which are nonpolar, tend to clump together in the presence of water, thus minimizing their direct exposure to the water. Hydrophobic interactions are responsible for a wide range of phenomena, from

MAKING CONNECTIONS

1. Why are radioactive isotopes valuable research tools for biologists?

2. How is chlorophyll's ability to capture the energy of excited electrons critical to life?

3. How does carbon's bonding capacity relate to the diversity of organic compounds it can make?

4. Name and briefly describe the atomic structure of two ions that play important roles in biological systems.

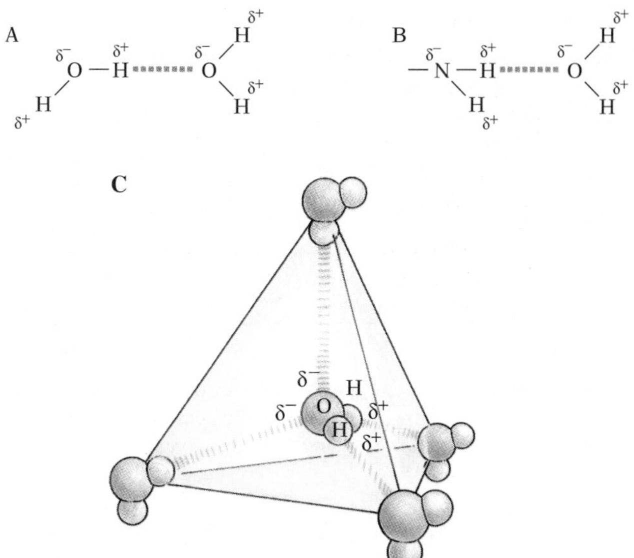

2.14 Hydrogen bonding Hydrogen bonds form when hydrogen is covalently bonded to one electronegative atom and is simultaneously attracted to another electronegative atom. (A) Hydrogen bonding between two water molecules. In water, each hydrogen atom is covalently bonded to oxygen, an electronegative atom. Because of oxygen's strong attraction for electrons, the shared electrons in each bond spend more time around the oxygen atom. The oxygen atom has two partial negative charges and each hydrogen has a partial positive charge. Often, the symbols δ^+ and δ^- are used to represent partial charges, where the δ^+ shows that the hydrogen atoms have partial positive charges, and the δ^- shows that the oxygen has two partial negative charges. The (partially) positively charged hydrogen atom is attracted to the (partially) negatively charged oxygen in another water molecule. This weak attraction (pink band) constitutes the hydrogen bond. (B) Hydrogen bonds often form when a hydrogen is shared between an oxygen and a nitrogen atom. Like oxygen, nitrogen is a strongly electronegative atom. (C) Like the central H_2O molecule shown here, each water molecule can form hydrogen bonds (pink bands) with four other water molecules, because there are four partial charges on each water molecule. The array then assumes the shape of a tetrahedron.

the structure of cell membranes, to the drops that form when a little cooking oil is added to a glass of water, to the formation of a massive oil slick (Fig. 2.15).

Some Inorganic Molecules, Notably H_2O, Are Essential for Life

Chemists refer to molecules that contain the element carbon and have at least one carbon-hydrogen bond as ***organic*** compounds. All other compounds are called ***inorganic,*** a designation that should not mislead you into assuming these compounds play no role in life processes. Many inorganic sub-

2.15 Oil slick covering the surface of Prince William Sound, Alaska after the grounding of the *Exxon Valdez* ship in 1989. Oil, which consists of hydrocarbons, floats on the surface of water because the nonpolar hydrocarbons are insoluble in water and tend to clump together to minimize their exposure to the water.

stances, in fact, play a crucial role in the chemistry of life, the most notable among them being water.

Life on earth is totally dependent on water. Between 70 and 90 percent of living tissue is water, and the chemical reactions that characterize life all take place in an aqueous medium. Life based on some substance other than water could conceivably exist elsewhere in the universe, but such life would be vastly different from anything in our experience—so different, in fact, that we might not recognize it as life even if we should stumble on it.

The Biological Importance of Water Arises in Part From the Polarity of Its Molecules

One reason water is so well adapted as the medium for life is that it is a superb solvent. More different substances dissolve in water, and in greater quantities, than in any other liquid. Water's superiority as a solvent arises from the marked polarity of the water molecule and its resulting readiness to interact with ions and polar molecules. Thanks to this property, both ionic substances and nonionic polar substances are soluble in water (Fig. 2.16).

Consider, for example, what happens when a dry crystal of a salt, such as sodium chloride, is dissolved in an aqueous medium. Within the dry crystal, the ionic bonds between the positive sodium ions and negative chloride ions are very strong, and much energy would be required to pull these ions away from each other. When the crystal is put into water, however, the attraction of the electronegative oxygen end of the water molecules for the positively charged sodium ions and the similar attraction of the electropositive hydrogen ends of the water molecules for the negatively charged chloride ions are

2.16 Dead Sea Because water is polar and sodium chloride is ionic in solution, relatively large amounts of salt can be dissolved in water. In the Dead Sea, nearly 30 percent of the fluid is dissolved salt, and the density is so great that humans float high in the water.

greater than the mutual attraction between the sodium and chloride ions. In water, then, the ionic bonds are broken with extreme ease, because of the competitive attraction of the water molecules for the ions. Consequently, the sodium and chloride ions dissociate, and each becomes surrounded by a sphere of regularly arranged water molecules that are attracted to it (Fig. 2.17). Such an ion is said to be ***hydrated.***

Water is also an excellent solvent for nonionic, polar molecules. Indeed, such molecules are said to be ***hydrophilic*** ("water-loving"). The solubility of such molecules arises from an electrostatic attraction between the charged portions of the solute molecules and the oppositely charged parts of the water molecules.

In short, substances dissolve in water if their molecules can interact with the polar water molecules. The old adage "like dissolves like" is useful in determining a substance's solubility in a particular solvent. Water, because it is polar, can interact with other polar or charged substances, and such substances will readily dissolve in water (Fig. 2.18). Substances that are electrically neutral and nonpolar, however, dissolve poorly in water. They show no tendency to interact electrostatically with water and, indeed, are repulsed by it. We have already noted that such hydrophobic substances are involved in the formation of cell membranes and oil slicks. When a hydrophobic substance such as oil is stirred into water, it will soon begin to separate out, because the water molecules tend slowly to reestablish the hydrogen bonds broken by the physical intrusion of the insoluble material. In a very real sense, the water "pushes" the nonpolar molecules together (Fig. 2.19).

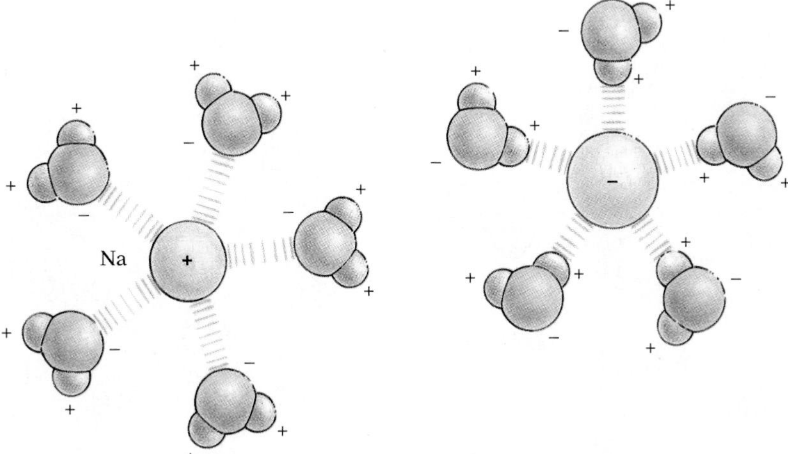

2.17 Hydration spheres of Na^+ and $C1^-$ When dissolved in water, each of the Na^+ and $C1^-$ ions is hydrated—that is, surrounded by water molecules electrostatically attracted to it. Note that the oxygen of the water molecules is attracted to the positively charged Na^+, while the hydrogen of the water molecules is attracted to the negatively charged $C1^-$. Water molecules in a hydration sphere are called bound water. This bonding between ions and polar molecules (pink bands) makes evident the common electrostatic basis of ionic bonds and polar (hydrogen) bonds.

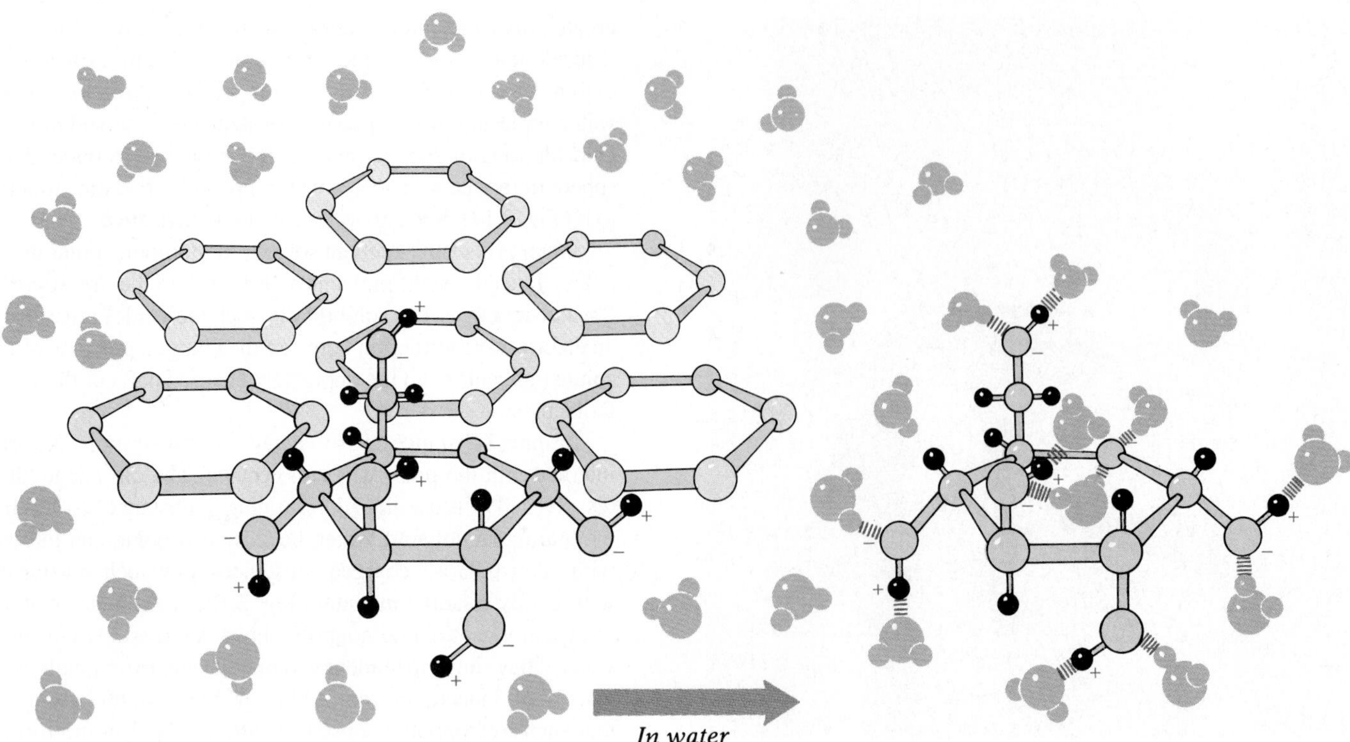

2.18 Polar basis of solubility When a polar substance such as glucose, an energy-rich sugar (left), is placed in contact with water, the water molecules are attracted to the polar atoms of the sugar. (For clarity the polar hydroxyl groups [green and black] are shown for only two of the sugar molecules.) The water forms hydrogen bonds with the substance, surrounding it with water molecules, and so dissolves it (right).

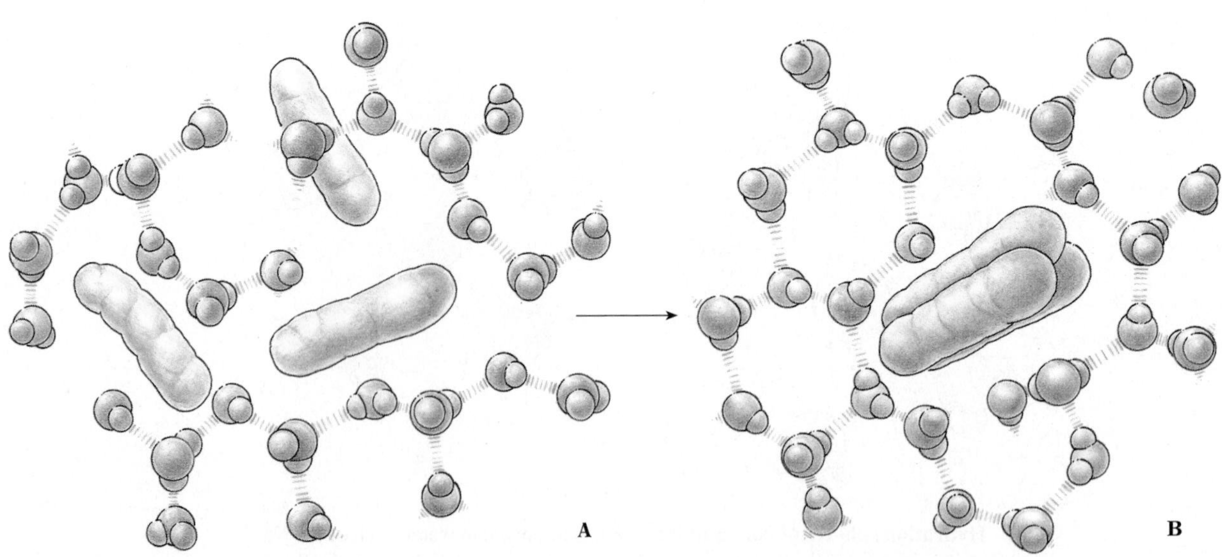

2.19 Water-induced clumping of hydrophobic molecules Dispersed hydrophobic molecules disrupt the polar bonding pattern of pure water, so that few hydrogen bonds can form in the solution (A). As hydrophobic molecules (schematically represented as brown ovals) encounter one another randomly in a solution of water, they tend to become trapped in clumps by polar bonding of water molecules to one another (B). Because there is more polar bonding when hydrophobic molecules are clumped, the solution becomes stabilized in this form.

The Concentration of Hydrogen Ions in an Aqueous Solution Determines Its Acidity

Although the bonds in the water molecule are polar covalent bonds, at any particular time a tiny proportion of water molecules ionize to form a hydrogen (H^+) ion and a hydroxide (OH^-) ion:

$$H_2O \rightleftharpoons H^+ + OH^-$$

In pure water, the concentration of hydrogen ions always equals the concentration of hydroxide ions. The addition of certain substances to it can change that balance. An ***acid*** can be defined simply as any substance that increases the concentration of hydrogen ions in water, and a ***base*** as a substance that decreases the concentration of hydrogen ions. In other words, an acid is a hydrogen-ion donor, and a base is a hydrogen-ion acceptor. In water, the concentration of hydrogen ions times the concentration of hydroxide ions always equals a constant number[2]:

$$[H^+] \times [OH^-] = 1 \times 10^{-14} M$$

(The symbol [ ] means "concentration of.") Therefore, increasing the hydrogen ion concentration always results in a corresponding decrease in hydroxide ion concentration, and vice-versa.

The degree of acidity or alkalinity (i.e., basicity) of a solution is commonly measured by a value known as ***pH***.[3] The pH scale ranges from 0 at the acidic end to 14 at the alkaline, or basic, end. A solution with a pH of 7 is neutral; it has equal concentrations of hydrogen ions and hydroxide ions. A solution with a pH of less than 7 is acidic; it has a higher concentration of hydrogen ions than of hydroxide ions. The lower the pH, the more acidic the solution. Conversely, a solution with a pH of more than 7 is alkaline; it has a higher concentration of hydroxide ions than of hydrogen ions. Because the pH scale is logarithmic, a change of one pH unit means a 10-fold change in the concentration of hydrogen ions. Thus a solution at a pH of 4 has 10^3 or a thousand times more hydrogen ions than a solution at a pH of 7. Figure 2.20 illustrates the range of pHs encountered in everyday life.

Living cells are extraordinarily sensitive to pH (Fig. 2.21) and most cells function best when their interior is at a near neutral pH of 6.8. The blood plasma and other fluids that bathe the cells in our own bodies have a pH of 7.2–7.3. The pH of these fluids must stay within these rather narrow limits; if the pH of

	pH	Concentrations of ions (moles/liter) H^+	OH^-	
Caustic soda (NaOH)	14	10^{-14}	10^0	ALKALINE
	13			
	12	10^{-12}	10^{-2}	
Detergent				
	11			
	10	10^{-10}	10^{-4}	
Baking soda	9			
Seawater Human blood	8	10^{-8}	10^{-6}	
Pure water	7	10^{-7}	10^{-7}	NEUTRAL
Saliva Cell interior Unpolluted rainwater	6	10^{-6}	10^{-8}	ACIDIC
Coffee Typical acid rain	5			
Beer	4	10^{-4}	10^{-10}	
Orange juice	3			
Carbonated soft drink	2	10^{-2}	10^{-12}	
Stomach acid	1			
Hydrochloric acid (HCl)	0	10^0	10^{-14}	

Increasing (OH^-) ⟶

⟵ Increasing [H^+]

2.20 The pH scale The concentration of hydrogen ions in a solution is measured by pH. Since the pH scale is a log scale, a change of one pH unit means a 10-fold change in the concentration of hydrogen ions. At pH 7, the concentration of hydrogen ions (H^+) exactly balances the concentration of hydroxide ions (OH^-), and so the solution is neutral. At lower pHs (corresponding to higher H^+ concentrations) solutions are acidic; at higher pHs (corresponding to lower H^+ concentrations) solutions are alkaline, or basic. Notice that the pH number matches the concentration of H^+ in moles/liter—for example, pH 8 corresponds to H^+ concentration of 10^{-8}.

the blood goes much above 7.35 or below 7.15, death can result. Many special mechanisms aid in stabilizing these fluids so that cells will not be subject to appreciable fluctuations in pH. Foremost among these mechanisms are certain chemical substances known as ***buffers,*** which have the capacity to bond to

[2] Molarity (M) is a common measure of solute concentration and stands for moles/liter. A mole is the amount of a substance, in grams, that equals the combined atomic mass of all the atoms in a molecule of that substance.

[3] Short for "potential of hydrogen," pH defines the negative logarithm of the concentration of hydrogen ions: $pH = -\log [H^+]$.

2.21 Forest damage attributed to acid rain These trees are showing the effect of acid rain on forest soil in the Adirondack Mountains of northeastern New York State.

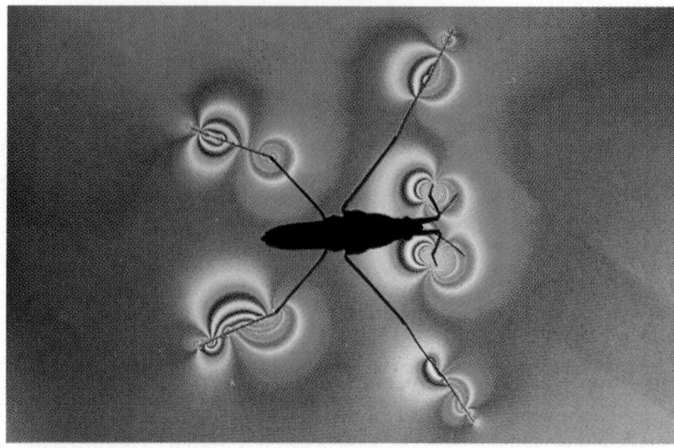

2.22 A water strider on the surface of the water Due to the high surface tension of water, water striders can move rapidly across the surface of still water, where they often congregate in large numbers. Note the dimples in the water surface where each foot rests.

hydrogen ions, thereby removing them from solution whenever their concentration begins to rise, and conversely, to release hydrogen into solution whenever their concentration begins to fall. Buffers thus help minimize fluctuations in pH. For example, many of the blood proteins such as hemoglobin act as buffers. When the concentration of hydrogen ions in the blood is too high, hemoglobin combines with the excess hydrogen ions and removes them from solution. And, when the concentration of hydrogen ions is low, hemoglobin releases the hydrogen ions. In this way the pH can be stabilized.

Water's Hydrogen Bonds Give It Special Physical Properties

The strong tendency of water molecules to form hydrogen bonds with one another and the consequent ordering of the molecules to which this tendency gives rise have important implications for life processes. For example, water has a high ***surface tension;*** in other words, the surface of a volume of water is not easily broken. Surface tension is what enables a water strider or other insect to walk on the surface of a pond without breaking it (Fig. 2.22), and you to fill a glass with water to a point slightly above the top without it spilling.

The reason water has a high surface tension is that hydrogen bonds link the molecules at the surface to one another and to the molecules below them. Before the legs of the water strider (or any other object, for that matter) can penetrate the water's surface, they must break some of these hydrogen bonds and deform the orderly array of water molecules. Similarly, the cohesion that prevents the extra water in an overfilled glass from spilling is due to the hydrogen bonds that bind the extra water molecules to the molecules below them.

Just as water molecules are attracted electrostatically to areas of charge on dissolved molecules, so they are attracted to charged groups on a hydrophilic surface such as glass. Consequently, such surfaces are ***wettable,*** in the sense that water spreads over them and binds loosely to them. By contrast, hydrophobic surfaces, such as those of most plastics and waxes, lack surface charge and hence are not wettable; water on them will form isolated droplets, but will not spread out over the surface.

The readiness of water to bind to hydrophilic surfaces explains the phenomenon of ***capillarity***—the tendency of aqueous liquids to rise in narrow tubes. If the end of a narrow glass

MAKING CONNECTIONS

1. **Describe two weak bonds important in biological systems.**
2. **What are the properties of water that make it biologically important?**
3. **How do buffers stabilize the body fluids' pH?**

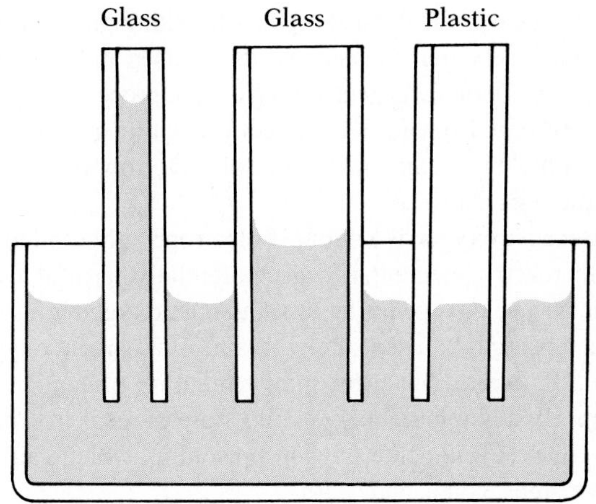

2.23 Capillarity Water rises higher in a glass tube of small bore (left) than in one of large bore (center) because in the smaller tube a higher percentage of the water molecules are in direct contact with the glass and can form hydrogen bonds with charged groups on the glass. By contrast, water cannot "stick" to the surface of a plastic tube (right) because plastic is uncharged.

tube is inserted below the surface of a volume of water, water will rise in the tube to a level well above the water level outside (Fig. 2.23). The reason is that glass is very hydrophilic, having many charged groups on its surface. The water molecules, electrostatically attracted to the glass, tend to creep along the surface of the tube and to pull other water molecules (attached to them by hydrogen bonds) along with them.

Another important physical property of water is that in the liquid state the potential of each water molecule to form hydrogen bonds with four other water molecules is not fully realized, since hydrogen bonds between water molecules are constantly breaking and reforming. As water is cooled, however, the water molecules move more slowly and the extent of hydrogen bonding increases. Hydrogen bonding reaches its full potential when the water has frozen into ice. When all four possible bonds have formed, each is oriented in space with maximal divergence from the other three. The resulting three-dimensional lattice of water molecules in ice is rather open (Fig. 2.24); the packing of the molecules is not as tight as would be possible if they were less rigidly arranged, as they are in liquid water.

When ice is warmed to the melting point, a few of the hydrogen bonds break, and the water molecules become less rigidly oriented. The resulting deformation of the lattice and tighter packing of the molecules make the water denser than ice. This means that ice floats, and that ponds and streams freeze from the top down instead of from the bottom up. The crust of ice that forms on the surface insulates the water below it from the cold air above and thereby often prevents the pond or stream from freezing solid, even in very cold winter weather. The importance of this special property of water for organisms living in ponds or streams is obvious (Fig. 2.25).

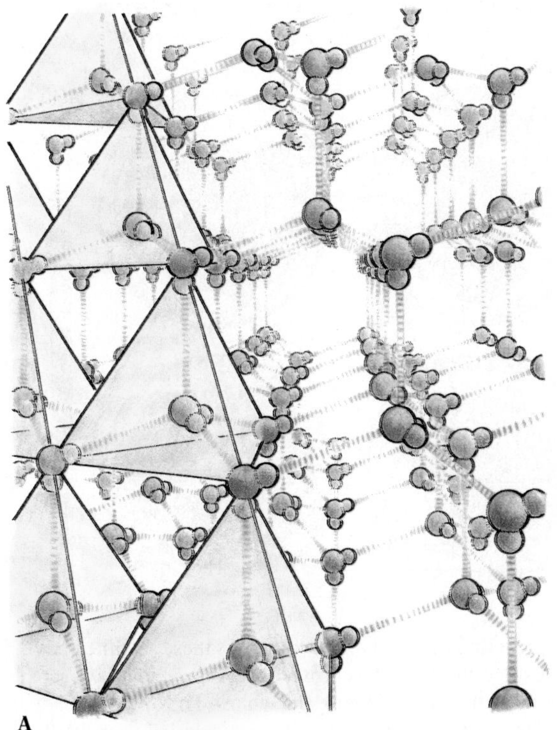

2.24 The molecular structure of ice Because of the tetrahedral arrangement around each water molecule, the lattice is an open one, with considerable space between molecules (A). In liquid water the arrangement is not quite so rigid, and the packing of molecules is therefore slightly denser, but the general lattice arrangement is nonetheless largely preserved. (Planes have been added to help show the three-dimensional disposition of the molecules.) Note the hexagons created by the hydrogen bonding in this bit of ice. This conformation is the basis of the hexagonal shape of most snow crystals (B). Each "snowflake" contains about 10^{16} water molecules.

2.25 Weddell seal swimming under Antarctic ice Ice, because it is less dense than water, floats on the surface and thus water freezes from the top down. The ice crust insulates the water below it from the cold air above and often prevents the ocean from freezing solid.

Water Helps Stabilize Environmental Temperatures

Because it takes energy to break hydrogen bonds, the hydrogen bonds in water enable it to absorb much heat energy without a very large increase in temperature (and, conversely, to release much heat energy without a great drop in temperature). Consequently, water acts as an effective buffer against extreme temperature fluctuations in the environment. In this way water helps stabilize the earth's temperatures within the range favorable to life.

Besides damping fluctuations in temperature, water plays an important role in determining what the temperature is at the earth's surface, because the water vapor (and carbon dioxide) in the atmosphere exerts what has been called a ***greenhouse effect.*** The vapor absorbs much of the solar radiation striking it from above and also much of the radiation reemitted by the earth. The absorbed radiation warms the atmosphere which, in turn, warms the earth's surface.

Molecular Oxygen Is Necessary for Life in Most Organisms

Molecular oxygen (O_2) constitutes approximately 21 percent of the atmosphere. It is a necessary inorganic material for maintenance of life in most organisms, although a few can live without it. Oxygen molecules can be used directly, without change, by both plants and animals in the process of extracting usable energy from nutrient molecules. Although oxygen is not very soluble in water, enough dissolves to supply the needs of aquatic organisms.

Because oxygen is a strongly electronegative and therefore highly reactive element, it is also potentially harmful to life. To some bacteria oxygen is lethal. And pure oxygen is harmful to human beings. You can safely breathe 100 percent oxygen for about 12 hours, but after that symptoms of lung irritation begin. Breathing pure oxygen under pressure, as in scuba diving, can result in acute oxygen poisoning, which can lead to convulsions, coma, and death.

Oxygen also has an important indirect impact on life on earth. Ultraviolet light, present in sunlight, striking oxygen molecules in the atmosphere triggers the formation of ***ozone*** (O_3). Continued over billions of years of earth history, this process has led to the buildup of a comparatively thin layer of atmospheric ozone that shields living organisms from the harmful effects of high-energy ultraviolet radiation from the sun. From about 1974, evidence for the depletion of the ozone layer has mounted (Fig. 2.26). The search for the agent responsible for this ominous change has focused on the release into the atmosphere of man-made chlorofluorocarbons, organic molecules that can promote the conversion of O_3 to O_2. As a result, some countries, including the United States, have banned such compounds in aerosol spray cans. A faster phase-out of the use of such compounds on a worldwide basis may be

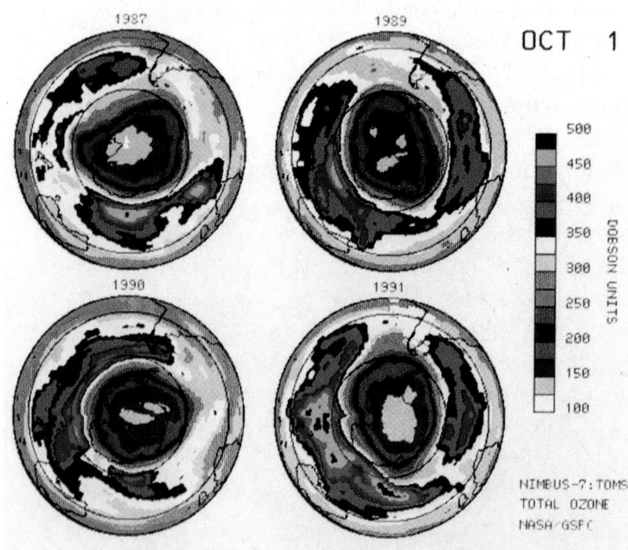

2.26 Satellite photo of ozone layer As these satellite photos from NASA show, the ozone levels over Antarctica (center in each photo) are particularly low (light pink and white). The ozone level over Antarctica varies each year, but was particularly low in 1987 and 1991.

needed. The United Nations Environmental Program estimates that there will be 300,000 additional skin cancers and 1.6 million additional cataracts a year if the ozone level drops by 10 percent.

Atmospheric Carbon Dioxide Is a Major Source of Carbon in Living Things

Carbon dioxide (CO_2) accounts for only a very small fraction of the atmosphere (roughly 0.33 percent). Nevertheless, atmospheric carbon dioxide is a major source of carbon, and carbon is the principal structural element of living matter. Carbon dioxide and water are the raw materials that green plants use in the process of photosynthesis to manufacture complex organic compounds essential to life, as we shall see in detail in Chapter 7.

When these complex compounds are broken down again to carbon dioxide and water, the carbon dioxide is eventually released into the atmosphere. The simple compound carbon dioxide, then, is the beginning and the end of the immensely complex ***carbon cycle*** in nature.

Like water vapor, carbon dioxide absorbs solar radiation and contributes to the greenhouse effect. Many scientists believe the increasing concentration of carbon dioxide in the atmosphere, which is primarily due to the burning of fossil fuels, is enhancing the greenhouse effect, thereby influencing global weather patterns. We shall be considering this topic in more detail in Chapter 34.

CONCEPTS IN BRIEF

1. Of the 92 naturally occurring *elements,* only a few are important for life. Prominent among them are hydrogen, carbon, oxygen, nitrogen, phosphorus, and sulfur.

2. An *atom,* the basic chemical unit of matter, is made up of a *nucleus* that contains positively charged *protons* and uncharged *neutrons* and is surrounded by a cloud of negatively charged *electrons.* The number of protons is equal to the number of electrons. The *atomic number* of an atom represents the number of protons; the *mass number* the total number of protons and neutrons.

3. The electrons are in constant motion outside the nucleus and occupy discrete *energy levels.* The higher the energy of an electron, the greater its probable distance from the nucleus. The volume of space within which the electron would be found 90 percent of the time is its *orbital.* Electrons occupy discrete energy levels in an atom; the lowest energy level has one orbital (holding a maximum of two electrons) and the second energy level has a total of four orbitals (holding a maximum of eight electrons).

4. The chemical properties of an element are determined largely by the number of *valence electrons* in its outermost energy level. Most atoms are particularly stable when their outermost (valence) energy levels contain eight electrons. Atoms tend to complete their outer energy levels by reacting with other atoms, forming *chemical bonds.*

5. There are several types of chemical bonds. *Covalent bonds,* in which electrons are shared between atoms, are strong, hence covalently bonded molecules are stable. When the electrons in a covalent bond are shared equally, the bond is *nonpolar covalent;* when the electrons are pulled closer to one element than to the other, the charge is distributed unequally, and the bond is *polar covalent.*

6. An *ionic bond* is formed when one or more electrons is transferred from one atom to another. The atom that gives up an electron *(electron donor)* becomes a positively charged ion, while the atom that takes the electron *(electron acceptor)* becomes a negatively charged ion. The electrostatic attraction between the oppositely charged ions creates an ionic bond. Ionic bonds are strong in the solid state, but weak in aqueous solutions.

7. Two other types of weak chemical bonds, *hydrogen bonds* and *hydrophobic interactions,* are important in stabilizing the shape of many of the large, complex molecules found in living matter.

8. The water molecule is polar because of the polarity of bonds within the molecule and the V-shaped arrangement of its atoms. Each water molecule can form hydrogen bonds with four other water molecules, permitting an orderly arrangement of molecules and giving water

many special properties. Because of its polarity, water is an excellent solvent for ionic and polar substances. Electrically neutral and nonpolar substances are not soluble in water because they cannot interact with the polar water molecules.

9. *Acidity* and *alkalinity* are expressed in terms of *pH,* a measure of the concentration of hydrogen ions. Acids act as hydrogen-ion donors, bases as hydrogen-ion acceptors. Solutions with a pH less than 7 are acidic, while those with a pH greater than 7 are basic. Living matter is extremely sensitive to changes in pH.

10. Like water, molecular oxygen and carbon dioxide play important roles in life processes; they are basic to the chemistry of life.

STUDY QUESTIONS

1. What are the four most abundant elements in living organisms? (p. 30)
2. Describe the basic structure of an atom. Define nucleus, proton, neutron, electron, atomic number, and mass number. (p. 30)
3. Distinguish between the *energy level* and the *orbital* of an electron. How many electrons can occupy the outermost energy level of an atom? How many can occupy an orbital? (pp. 33–34)
4. What determines the chemical properties of an atom? How can one predict whether or not two elements will have similar chemical properties? What makes some elements, such as neon, chemically unreactive (inert)? (p. 36)
5. Define and distinguish between nonpolar and polar covalent bonds. Describe in general the circumstances in which each type of bond will form. Relate your answer to the concept of electronegativity. (pp. 37–38)
6. How is an ionic bond formed? How many electrons are typically found in the outer energy level of electron donors? Of electron acceptors? Compare the strength of the ionic bonds of dry, crystalline salt with the strength of those bonds when the salt is placed in aqueous solution. (pp. 38–41)
7. Define a hydrogen bond, and draw a diagram to show how it occurs. What two elements (other than hydrogen) are most likely to take part in hydrogen bonding? (p. 38)
8. Which kind of substance is the *least* soluble in water: ionic, polar, or nonpolar? How is the structure of the water molecule related to its properties as a solvent? What happens to nonpolar substances in water? (pp. 43–44)
9. How can each of the following bonds or interactions be explained using the concept of electronegativity: nonpolar covalent bonds, polar covalent bonds, ionic bonds, hydrogen bonds, and hydrophobic interactions? (pp. 38–41)
10. Define an acid and a base. What is the neutral point on the pH scale? Is a pH below the neutral point basic or acidic? What is the difference in hydrogen ion concentration between a solution of pH 5 and a solution of pH 1? (p. 45)
11. Explain why water exhibits capillarity and surface tension. Also explain why ice is less dense than liquid water. What is the significance of this fact for aquatic life? (pp. 46–47)
12. What properties of water result from the fact that it is a polar molecule? (pp. 46–47)
13. In what ways are oxygen and carbon dioxide important for life on earth? (pp. 48–49)

CHAPTER 3

THE MOLECULES OF LIFE

The diversity of form, function, and behavior evident at the macroscopic level in living organisms is matched at the molecular level by an equally diverse cast of characters: the complex organic molecules that give cells their structure and their capacity to control the flows of energy and information necessary for life. The source of the vast diversity of organic molecules found in living things is, as we have already mentioned, the bonding capacity of just one of the 92 naturally occurring elements—carbon (Fig. 3.1).

$^{12}_{6}C \quad \dot{\cdot}\underset{\cdot}{C}\cdot$

3.1 Electronic configuration of carbon The element carbon has the electronic configuration $1s^22s^22p^2$, indicating that it has two electrons in its first energy level and a total of four in its valence level. Carbon therefore has a binding capacity of 4 and can form four covalent bonds.

The key to carbon's versatility lies in its atomic structure: the four unpaired electrons in the outermost energy level of the carbon atom enable it to form strong covalent bonds with as many as four other atoms. Very often the carbon atom is linked to other carbon atoms, creating a comparatively stable "backbone" to which other atoms may be attached, typically in a repetitive pattern. In addition to itself, carbon most commonly bonds to hydrogen, oxygen, nitrogen, phosphorus, and sulfur, leading to an almost endless variety of organic (i.e., carbon-based) molecules.

Hydrocarbons Are the Basis of All Organic Molecules

Of central importance in organic chemistry are the compounds containing only hydrogen and carbon. The number of different compounds of this type, called ***hydrocarbons,*** is immense, ranging from simple natural gases such as methane (CH_4) to plastics such as polyethylene, whose molecules may extend for hundreds or even thousands of carbons and hydrogens. The primary importance of the hydrocarbons to the study of biology is indirect: they are the structural basis for the more complex organic molecules found in living organisms, in which other atoms (or groups of atoms) can substitute for either the carbon or the hydrogen atoms.

One reason for the great variety of hydrocarbons—and, by extension, of all organic molecules—is the ease with which

Chapter Opening Photo: A crayfish (Procambus clarki). Arthropods, such as this crayfish, have a hard outer skeleton composed of chitin, a complex organic molecule.

carbon-to-carbon bonds can form, producing chains of varying lengths and shapes. The chains may be straight or branched, or they may form rings with three, four, five, six, or even more carbons. The capacity of adjacent carbon atoms to form single, double, or triple bonds adds to the variety of both the hydrocarbons and the more complex organic molecules derived from them (Fig. 3.2).

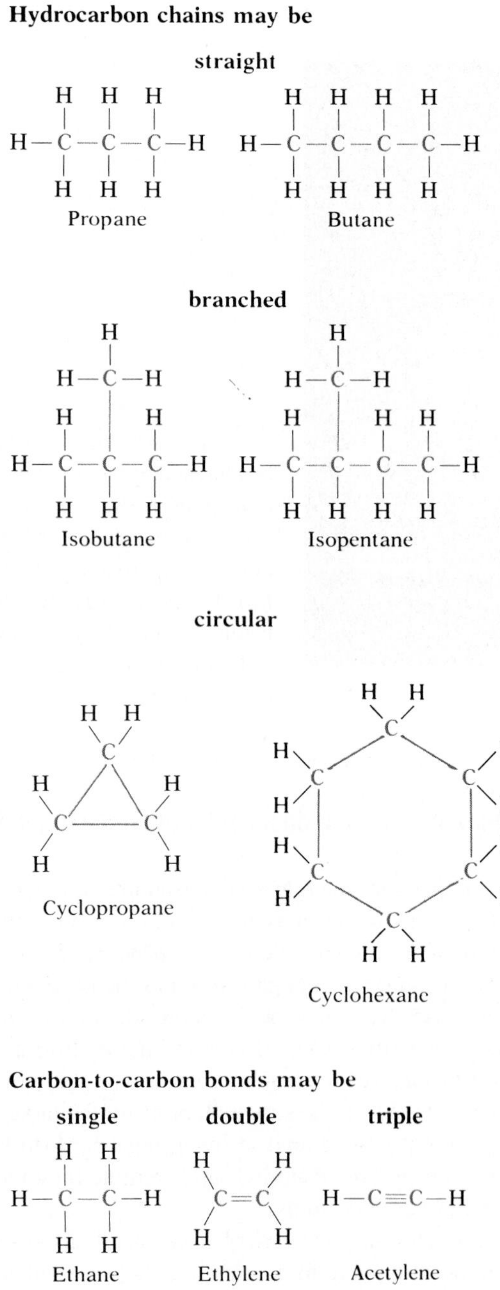

Hydrocarbon chains may be

straight

Propane

Butane

branched

Isobutane

Isopentane

circular

Cyclopropane

Cyclohexane

Carbon-to-carbon bonds may be

single **double** **triple**

Ethane

Ethylene

Acetylene

3.2 Examples of hydrocarbons The molecules appear flat in these conventionally drawn structural diagrams, though they are, in fact, three-dimensional. Adding to the variety of hydrocarbons is carbon's ability to form single, double, or triple bonds.

The more atoms a molecule contains, the more three-dimensional arrangements of those atoms are possible. Compounds with the same atomic content and molecular formula but with different three-dimensional arrangements of the atoms are called *isomers* (Fig. 3.3). Large organic molecules may have hundreds of isomers, each with its own distinctive physical properties. As we shall see, the biological function of some key organic molecules depends as much on their three-dimensional shape as on their chemical composition.

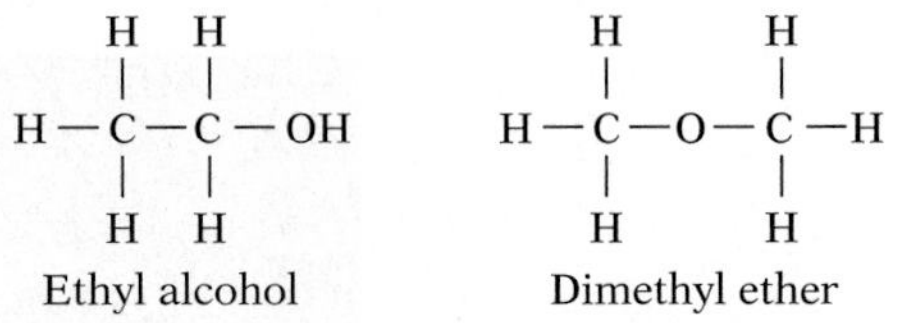

Ethyl alcohol

Dimethyl ether

3.3 Two structural isomers The two isomers differ in the basic grouping of their constituent atoms, one being an alcohol (characterized by a hydroxyl group) and the other an ether (characterized by an oxygen bonded between two carbons). Though the properties of ethyl alcohol and ether are very different, the compounds have the same molecular formula.

The four major classes of complex organic molecules found in living organisms are carbohydrates, lipids, proteins, and nucleic acids. Molecules in each of these classes are often identified on the basis of the subunits, or clusters of atoms, that are attached to their carbon-atom backbones. Each subunit, referred to as a ***functional group,*** has its own characteristic properties, which help determine such chemical traits as the molecule's solubility and reactivity. For instance, the ***carboxyl*** group, written as $—C(=O)—O—H$ (or $—C(=O)—OH$; sometimes abbreviated —COOH), acts as an acid, ionizing to release a hydrogen ion. Organic molecules with the carboxyl group, then, have the properties of acids. Some frequently encountered functional groups are listed in Table 3.1, along with their important properties.

CARBOHYDRATES

Complex Carbohydrates Consist of Simple Sugars Bonded Together

Carbohydrates are compounds composed of carbon, hydrogen, and oxygen. The ratio of hydrogen to oxygen atoms in these molecules is always 2:1, the same as in water. (Indeed, the

TABLE 3.1 *Important functional groups*

Group	*Name*	*Properties, solubility in water*	*Ions (if present) or ionized form*
—OH	**Hydroxyl**	Polar (soluble, because it is able to form hydrogen bonds)	
$-\mathrm{C}(\mathrm{H})(\mathrm{H})-\mathrm{OH}$	**Alcohol**	Polar (soluble)	
$-\mathrm{C}(=\mathrm{O})-\mathrm{O}-\mathrm{H}$	**Carboxyl**	Polar (soluble); often loses its hydrogen, becoming negatively charged (an acid)	$-\mathrm{C}(=\mathrm{O})-\mathrm{O}^-$
$-\mathrm{N}(\mathrm{H})-\mathrm{H}$	**Amino**	Polar (soluble); often gains a hydrogen, becoming positively charged (a base)	$-\mathrm{N}(\mathrm{H})(\mathrm{H})-\mathrm{H}^+$
$-\mathrm{C}(=\mathrm{O})-\mathrm{H}$	**Aldehyde**	Polar (soluble)	
$>\mathrm{C}=\mathrm{O}$	**Ketone**	Polar (soluble)	
$-\mathrm{C}(\mathrm{H})(\mathrm{H})-\mathrm{H}$	**Methyl**	Hydrophobic (insoluble); least reactive of the side groups	
$-\mathrm{P}(=\mathrm{O})(\mathrm{OH})-\mathrm{OH}$	**Phosphate**	Polar (soluble); usually loses its hydrogens, becoming negatively charged (an acid)	$-\mathrm{P}(=\mathrm{O})(\mathrm{O}^-)-\mathrm{O}^-$

name "carbohydrate" means "hydrates," i.e., water, of carbon.) Consequently, the group $-CH_2O$ recurs frequently in carbohydrate molecules; it is diagrammed:

$$\mathrm{H}-\overset{|}{\underset{|}{\mathrm{C}}}-\mathrm{O}-\mathrm{H}$$

Some carbohydrates, such as starch and cellulose, are very large, complex molecules. Like most large molecules, however, they are composed of many simpler "building-block" compounds, arranged in repetitive patterns (Fig. 3.4). Understanding these constituent compounds, or ***monomers,*** is the first step toward understanding more complex substances.

The basic building blocks of carbohydrates are the simple sugars known as ***monosaccharides.*** The carbon chain that forms the backbone of a sugar can be of different lengths, ranging from three carbons to six or more. Although both three-carbon and five-carbon sugars play important biological roles, the six-carbon sugars, notably *glucose* and *fructose,* are the principal structural units for most of the complex carbohydrates. Since all six-carbon sugars have the proportions of oxygen and hydrogen typical of carbohydrates, they all have the same molecular formula: $C_6H_{12}O_6$. In other words, they are isomers of each other (Fig. 3.5).

Glucose, the basic energy source for most cells and the sugar most often used as a quick energy source by athletes, and fructose, the sugar present in many fruits, differ from each other only slightly, but the difference is highly significant. All sugars, when in straight-chain form, contain a $C=O$ group. If the double-bonded oxygen ($=O$) is attached to the terminal carbon of a chain, the combination is called an ***aldehyde*** group; if it is attached to a nonterminal carbon, the combination is called a ***ketone*** group (Table 3.1). This difference in the position of a single atom is all that distinguishes the glucose molecule, an aldehyde sugar, from the fructose molecule, a ketone sugar (Fig. 3.5). Because of this small structural difference, these two otherwise identical six-carbon sugars have quite different properties; they behave differently in certain biochemical reactions, and indeed they even taste different.

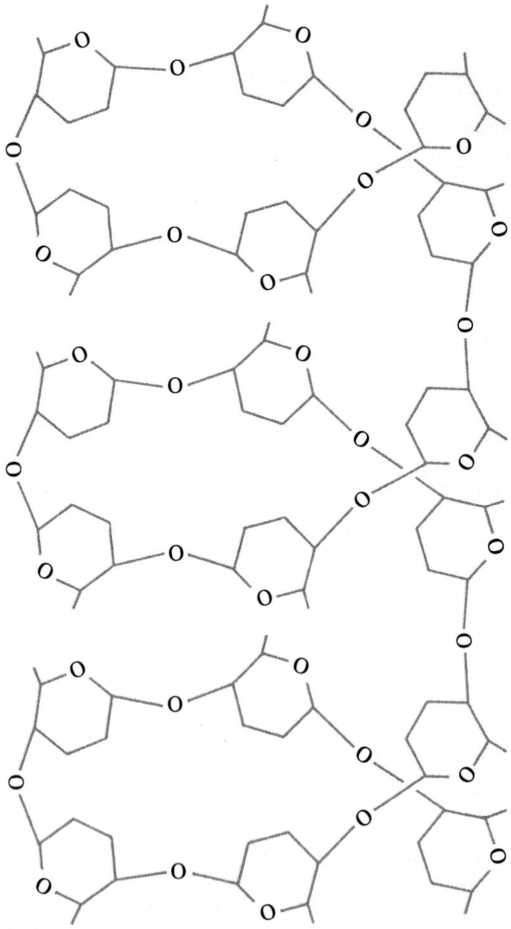

3.4 A moderately complex organic compound One of the largest of the moderately complex organic compounds is starch, the main energy-storage molecule in plants. Despite its apparent complexity, the starch molecule is actually composed of a repetitive string of glucose units, each represented here as a hexagon. Only a small part of one starch chain is shown. A branched molecule composed of units of several different compounds would be far more complex.

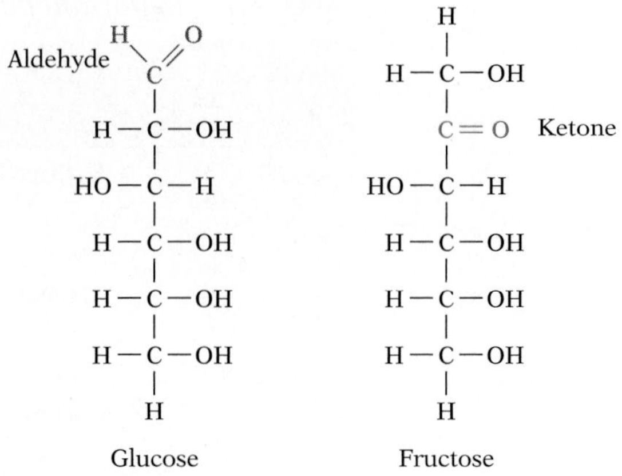

3.5 Two isomeric sugars Glucose and fructose, both six-carbon sugars, have the same molecular formula, $C_6H_{12}O_6$; hence each is an isomer of the other.

Hydroxyl (—O—H; abbreviated —OH) groups are attached to all the carbon atoms in a simple sugar, except those with a double-bonded oxygen. The presence of a hydroxyl group attached to a carbon characterizes a compound as an ***alcohol.*** Accordingly, the hydroxyl group is termed the alcohol functional group (Table 3.1). Note that the hydroxyl group differs from the hydroxide (OH^-) group of a base. The hydroxide group is ionized, whereas the hydroxyl group is not; the latter is instead attached to a carbon atom by a strong covalent bond. Hydroxyl groups are polar, however, meaning that sugars readily form hydrogen bonds with water and hence are soluble, not only in a cup of coffee or tea but also in the aqueous environment of the living cell.

Most sugars in living systems rarely exist in the straight-chain form diagrammed in Figure 3.6. Glucose, for example, most often exists as a ring composed of five carbons and one oxygen, with the sixth carbon sticking out on a branch from one of the other carbons (Fig. 3.6). As noted, glucose plays a unique role in the chemistry of life. Other six-carbon monosaccharides are constantly being converted into glucose or synthesized from glucose. Also, many other classes of organic compounds, such as proteins, can be converted into glucose or synthesized from glucose in living organisms.

In addition to ordinary monosaccharides, there are a variety of derivative monosaccharides containing other elements in addition to carbon, hydrogen, and oxygen. For example, the glucose derivative glucosamine has an ***amino group*** (—NH_2) substituted for a hydroxyl group on one of the carbons of the original glucose framework, whereas glucose-6-phosphate has a ***phosphate group*** (—PO_3^{-2}) substituted for the hydrogen of a hydroxyl group at another position on the glucose molecule (Fig. 3.7).

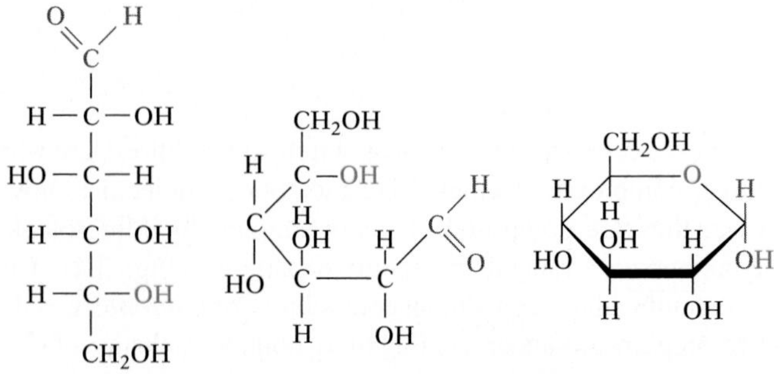

3.6 Two forms of glucose Glucose may exist in the straight chain aldehyde form shown at left, or it may fold up (center) to form a ring structure, as shown at the right. The ring structure is the most common. By convention, the unmarked corners of the hexagon signify carbon atoms.

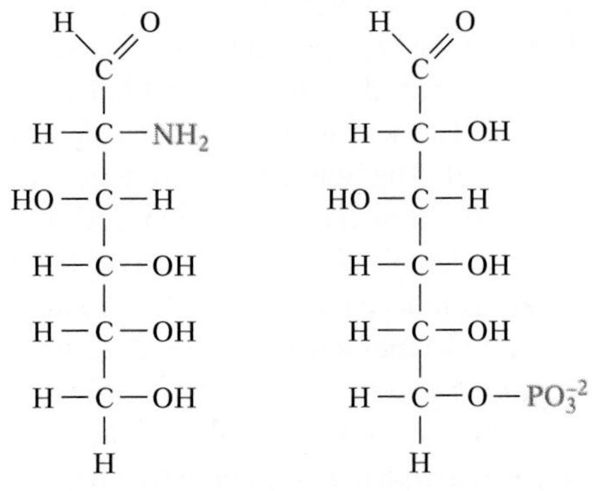

Glucosamine Glucose-6-phosphate

3.7 Two examples of derivative monosaccharides Glucosamine is merely a glucose molecule with an amino group (—NH_2) substituted for a hydroxyl (—OH) group. Similarly glucose-6-phosphorate acid is glucose with a substituted phosphate group.

Such derivative monosaccharides are important in living cells: the amino sugars are combined with certain proteins to form important constituents of cell membranes, and glucose must be converted into glucose-6-phosphate before it can be broken down to release the energy that is stored in the molecule. We shall learn more about the metabolism of glucose in Chapter 8.

Disaccharides Are Formed by the Bonding of Two Simple Sugars

Disaccharides are composed of two simple sugars bonded together through a reaction that involves the removal of a molecule of water. A reaction of this type is called a ***condensation reaction*** or a ***dehydration reaction.*** Such reactions are used by the body to help store energy by creating large complex carbohydrate molecules. The reaction can be described by the following equation, which summarizes several intermediate steps:

$$C_6H_{12}O_6 + C_6H_{12}O_6 \longrightarrow C_{12}H_{22}O_{11} + H_2O$$

sugar + sugar yields disaccharide + water

The equation by itself is not very specific, however, since any simple six-carbon sugar has the formula $C_6H_{12}O_6$ and any double sugar synthesized from such building blocks will have the formula $C_{12}H_{22}O_{11}$. Thus the equation can describe many different reactions involving a variety of isomers. Consider, for example, the double sugar *sucrose,* or table sugar. This compound is synthesized in large amounts in plants such as sugar cane and sugar beets by a condensation reaction between a molecule of glucose and a molecule of fructose.

To understand specifically what is involved in such a condensation reaction, it is helpful to look at a diagram indicating the structures of the molecules. Figure 3.8 shows that the hydrogen atom from a specific hydroxyl group of a molecule of glucose combines with a specific hydroxyl group from a molecule of fructose to form a water molecule. The oxygen from which the hydrogen was removed then forms a bond with the carbon from which the hydroxyl group was removed. As a result, the two monosaccharide units are connected by an oxygen atom shared between them, forming the sucrose molecule.

Maltose, or malt sugar, is a disaccharide composed of two glucose molecules. It is a standard ingredient in the brewing of beer. *Lactose,* the sugar naturally present in milk, is a double sugar composed of glucose and the six-carbon monosaccharide *galactose.* Other complex sugars are also characterized by repeated structural building blocks.

Disaccharides can be broken down into their constituent simple sugars by a reaction that is the reverse of condensation. This reaction, called ***hydrolysis,*** involves the addition of a water molecule:

$$C_{12}H_{22}O_{11} + H_2O \longrightarrow C_6H_{12}O_6 + C_6H_{12}O_6$$

disaccharide + water yields sugar + sugar

Hydrolysis reactions are involved in digestion. They are involved, for example, in the multistep chemical process by which cells break down carbohydrate-rich foods such as bread and pasta, producing the simple sugars such as glucose that are used as energy sources by the cells of the body.

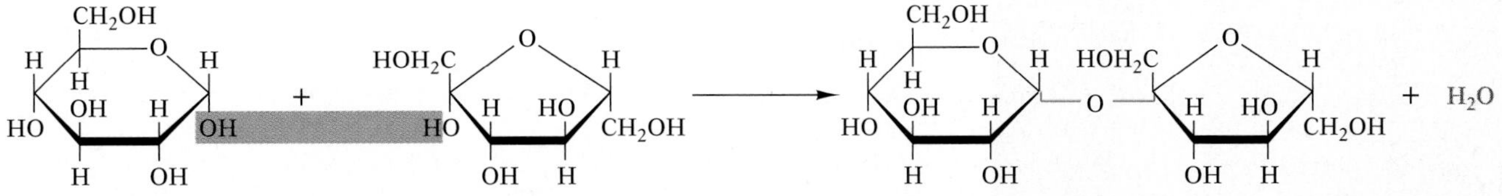

Glucose + Fructose Yields Sucrose + Water

3.8 Synthesis of the disaccharide sucrose Removal of a molecule of water between a molecule of glucose and a molecule of fructose results in the formation of a bond (red) between the two. The double sugar produced is sucrose, or table sugar.

Polysaccharides Are Chains of Many Simple Sugars

Carbohydrates made of many simple sugars bonded into long chains are called ***polysaccharides*** (Fig. 3.9). They are synthesized by exactly the same kind of condensation reaction as the disaccharides and, like them, can be broken down by hydrolysis.

A number of complex polysaccharides are important in living organisms. Starches, for example, are the main carbohydrate energy-storage products of higher plants. They are made of hundreds of glucose molecules bonded together through condensation reactions. *Glycogen,* the principal carbohydrate energy-storage product in animals, is sometimes called animal starch. Its molecules are much like those of starch. They have the same type of bond between adjacent glucose units, but the chains are more branched. *Cellulose* is a highly insoluble, unbranched polysaccharide common in plants, where it is a major supporting material. The bonds between its glucose units are aligned differently from those of starch and glycogen, and most animals are unable to hydrolyze these bonds (see Exploring Further: Cellulose Digestion in Ruminants, p. 57).

Long-chain molecules such as polysaccharides are known generically as ***polymers,*** and the reactions in which small monomers bond together to form molecules of this type are called ***polymerization reactions.*** If a polymerization reaction is to take place, say, between molecules A and B, two requirements must be met. First, one end of molecule A must be capable of interacting with one end of molecule B so as to split out some small group, such as water. Second, each molecule must con-

A

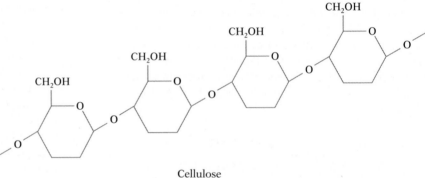

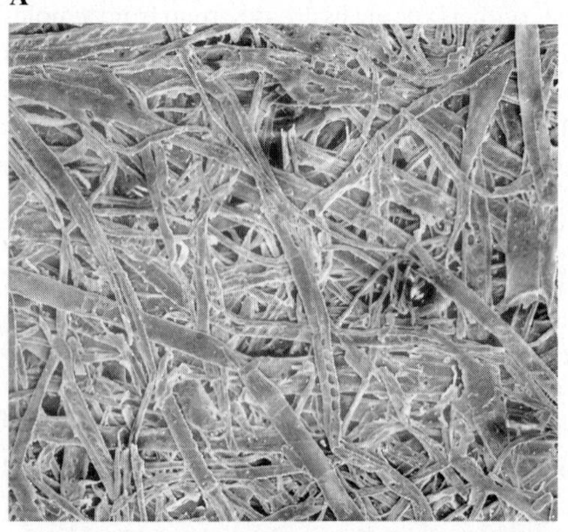

Cellulose

B

3.9 Cellulose and chitin (A) Cellulose, seen here as fibers in uncoated paper, is composed of long chains of glucose molecules that have been joined by condensation reactions. (B) The chitin of insect exoskeletons (visible here in the shed skin of a mantis) is chemically similar and functionally equivalent to the cellulose of plants. Chitin is composed of glucose units containing an amino group.

Chitin

EXPLORING FURTHER

CELLULOSE DIGESTION IN RUMINANTS

Cellulose, the major component of the cell walls in plants, the most abundant organic compound on earth, is also one of the most difficult for animals to digest. Because of its abundance and the store of energy-rich glucose molecules it contains, many animals have evolved special adaptations for digesting it. Like starch and glycogen, cellulose is a polymer of glucose molecules and as such could be an excellent source of glucose, if it were digestible. Cellulose contains linked beta (β)-glucose molecules, while starch and glycogen are composed of alpha (α)-glucose molecules (Fig. A).

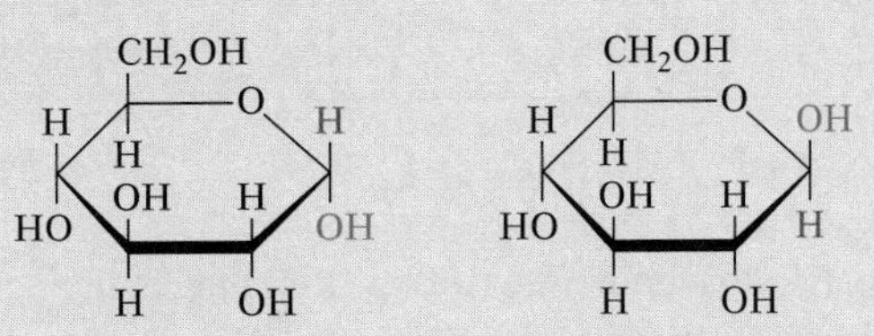

A Two forms of glucose Notice the position in each molecule of the hydroxyl group (red) attached to the first carbon atom.

The difference in the position of the hydroxyl group attached to the first carbon atom accounts for the difference between these two forms of glucose. While the disparity does not seem great, the linkages that result are very different, both structurally and functionally. Most animals manufacture special enzymes that can hydrolyze the bonds between the alpha-glucose units in starch and glycogen, but these enzymes are unable to hydrolyze the beta-glucose bonds in cellulose. As a result, animals are unable to digest cellulose, and it remains in the digestive tract, forming wastes that are eliminated in the feces.

The inedible cell wall surrounding most plant cells is composed largely of cellulose. How do the animals that feed on these plants manage to hydrolyze the bonds in cellulose and thus liberate the glucose molecules for later use as a source of energy? One group, the ruminants (such as cattle, sheep, and deer) solve the problem by harboring in their digestive tract an army of microorganisms (bacteria and protozoa) that are capable of digesting cellulose. These organisms produce the cellulose-digesting enzymes their hosts lack and use them to break the linkages between adjacent molecules of beta-glucose. Ruminants possess four stomachlike chambers (Fig. B), which contain the microbes necessary for the digestion of cellulose. Once the cellulose has been digested to glucose, the glucose molecules can be absorbed and transported to other parts of the body for storage or used for energy-related tasks.

Humans lack both the cellulose-digesting enzymes and the microbes of the ruminants.

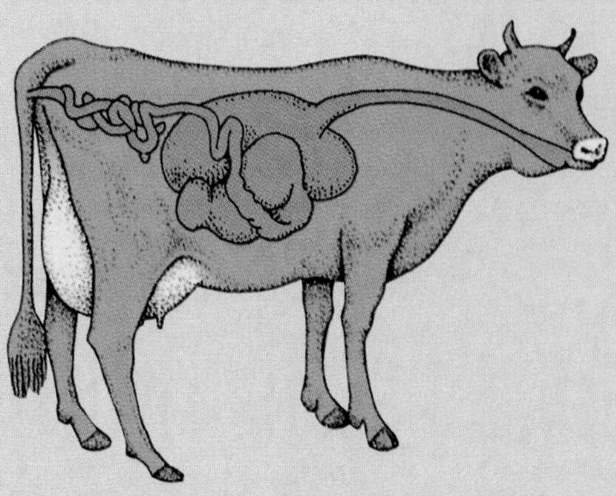

B The digestive system of a ruminant A cow with the various chambers in approximately their correct locations. The large chambers at the anterior end of the digestive tract contain huge numbers of cellulose-digesting microorganisms.

The only way they can utilize the energy in cellulose is to feed it to such ruminants as cows and then consume their milk and meat, a practice that has been of profit to humans for centuries.

tain two functional groups, so that after the first two combine, the free ends can continue to react, extending the polymer:

$$\text{H} - \textbf{A} - \text{O-H} + \text{H} - \textbf{B} - \text{O-H} \longrightarrow \text{H} - \textbf{A} - \textbf{B} - \text{O-H} + \text{H}_2\text{O}$$

molecule A + molecule B yields molecule A-B + water

Starches are polymers of glucose formed by condensation reactions similar to the above reaction. Glucose molecules are like molecules A and B in that they have hydroxyl groups at each end of the molecule that can react with one another through condensation reactions, forming the large, complex starch molecules.

LIPIDS

Lipids Are Generally Nonpolar and Hence Hydrophobic

The fats and fat-like substances known as ***lipids*** are soluble in organic solvents, such as chloroform, but are nonpolar and therefore relatively insoluble in water. Like carbohydrates, lipids are composed principally of carbon, hydrogen, and oxygen, but they may also contain other elements, particularly phosphorus and nitrogen. They differ from carbohydrates in that they contain a higher proportion of hydrogen and a much smaller proportion of oxygen. Unlike the complex carbohydrates, lipids are not composed of repeating subunits and hence are not polymers. We shall consider four classes of lipids here: fatty acids, fats, phospholipids, and steroids.

Fatty Acids Differ in Size and Degree of Saturation

Fatty acids contain the ***carboxyl*** *group* ($-\overset{\displaystyle /\!\!/ \text{O}}{\text{C}} - \text{OH}$), attached to a long hydrocarbon chain:

$$\begin{array}{ccccccccc} & \text{H} & & \text{H} & & \text{H} & & \text{O} & \\ & | & & | & & | & & /\!\!/ & \\ - & \text{C} & - & \text{C} & - & \text{C} & - & \text{C} & - \text{O} - \text{H} \\ & | & & | & & | & & & \\ & \text{H} & & \text{H} & & \text{H} & & & \end{array}$$

There are many different fatty acids. They vary in the number of single or double carbon-to-carbon bonds, and in the length of the hydrocarbon chain (Fig. 3.10). The fatty acids in living organisms usually contain an even number of carbon atoms, and most of them have relatively long chains, generally between 14 and 20 carbons but in some cases even more. The length of the chain affects the fluidity of the molecule; short

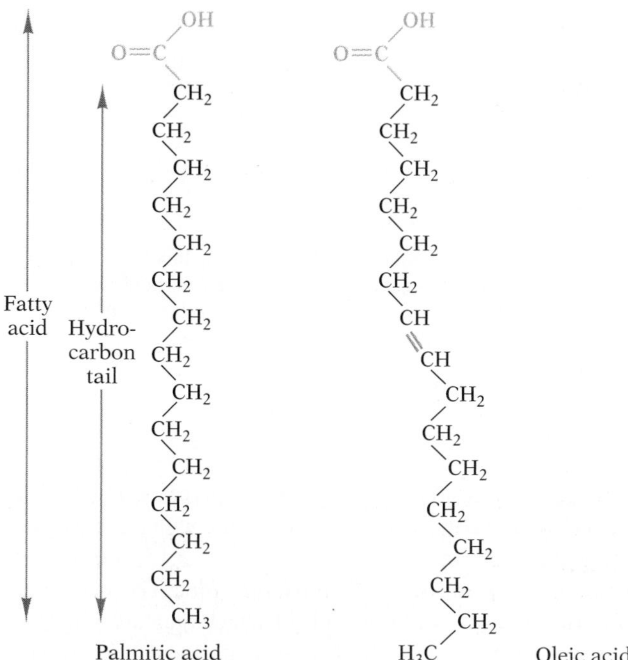

3.10 Examples of saturated and unsaturated fatty acids Palmitic acid is saturated with hydrogen, i.e., it has only single bonds between carbon atoms, and contains the maximum number of hydrogens possible. By contrast, oleic acid, with its one carbon-to-carbon double bond (blue), contains two fewer than the maximum number of hydrogens. The double bond induces a bend in the hydrocarbon chain.

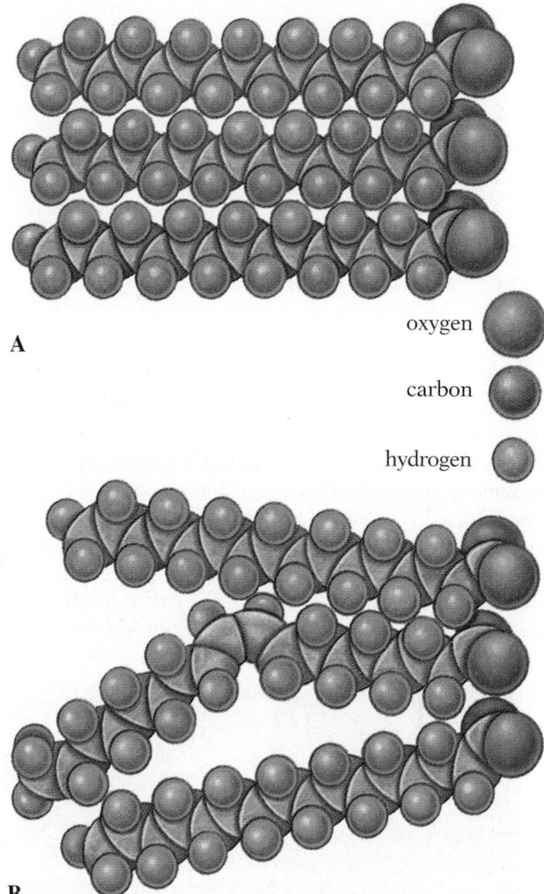

3.11 Space-filling models of three fatty acids In space-filling models each atom is given a symbol based on its size and bonding properties. Carbon atoms are gray, hydrogen white, and oxygen red. Three molecules of palmitic acid (C_{16}), a saturated fatty acid, are shown in A. In B, an unsaturated fatty acid, oleic acid (C_{18}), is shown between two palmitic acids. Notice that the molecules are very tightly packed in A, producing an orderly arrangement, but in B, the double bond produces a kink in the fatty acid, disrupting the orderly arrangement and thereby permitting more movement. Unsaturated fatty acids tend to be more fluid than saturated fatty acids.

chains are fluid at room temperature whereas long-chained fatty acids tend to be solid.

Fatty acids in which there are one or more double bonds between the carbon atoms in the hydrocarbon chain are *unsaturated;* in other words, they are not completely saturated with hydrogen. If there are two or more double bonds in the hydrocarbon chain the fatty acid is said to be *polyunsaturated.* The degree of unsaturation also affects the fluidity of the molecule, another example of how a small difference in molecular structure can drastically affect physical properties: at room temperature, fatty acids with many double bonds are liquid, whereas those without such bonds tend to be solid. The reason for the difference is that the double bond induces a "kink" in the otherwise linear structure of the unsaturated fatty acid (Figs. 3.10 and 3.11), thereby making the molecules less densely packed and able to move about more freely.

Fats Are Used Primarily to Store Energy

Among the best-known lipids are the neutral, or uncharged, **fats** (Fig. 3.12). Important as energy-storage molecules in living organisms, the fats also provide insulation, cushioning, and protection for various parts of the body. Fat molecules have more than twice as much energy per unit weight as carbohydrates, which is why the body's energy requirements are met (and often exceeded) much more readily on a high-fat diet. The body stores excess fat in the form of droplets in the interior of adipose or fat cells (Fig. 3.13), which can lead to unwanted weight gain.

Each molecule of fat is composed of two types of building-block compounds—an alcohol called glycerol and three fatty acids. *Glycerol* (sometimes called glycerin) has a backbone of three carbon atoms, each containing a hydroxyl group (Fig. 3.14). Fatty acids and alcohols have a tendency to combine

```
    H        O              H   H
    |        ||             |   |
H — C — O — C — (CH2)7 — C = C — (CH2)7 — CH3
    |        O
    |        ||
H — C — O — C — (CH2)10 — CH3
    |        O
    |        ||
H — C — O — C — (CH2)12 — CH3
    |
    H
```

3.12 Structure of a fat molecule A fat is a complex molecule that is formed from a condensation reaction between three long-chain fatty acids and a glycerol molecule. The three fatty acids may be the same or different (as here), and they may be unsaturated (having at least one carbon to carbon double bond; top, right) or saturated (no carbon to carbon double bonds; middle, and bottom, right). The chemical shorthand designation $(CH_2)_x$ means that CH_2 is repeated x number of times. Because the fatty acid tails are hydrocarbons, they are nonpolar and fats are strongly hydrophobic.

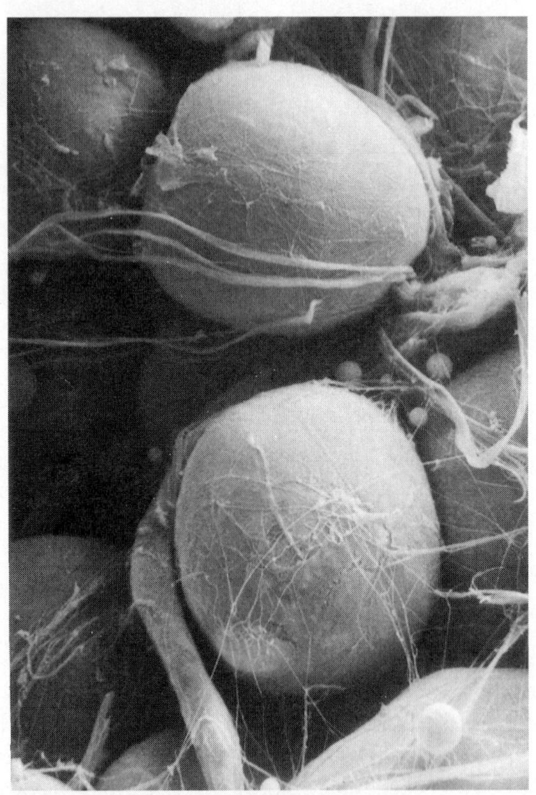

3.13 Scanning electron micrograph of fat cells Fat cells (adipocytes) are fat storage depots; fat droplets occupy up to 95 percent of the cell volume. Fat tissue is well supplied with blood; the vessel going around the lower cell is a blood capillary. Fibers of connective tissue hold the cells in place. Almost all cells of the body can use fatty acids for energy, so the fat stores in these cells are constantly being used and replaced. Dieting causes a reduction in the size of the fat stores and consequently the size of the cells, but does not eliminate the cells.

```
    H
    |
H — C — O — H
    |
H — C — O — H
    |
H — C — O — H
    |
    H
```

3.14 Glycerol Glycerol is one of the building blocks of fats and phospholipids. It is an alcohol; each of the three carbons bears an alcohol (—OH) group. Consequently, it can combine with three molecules of fatty acid.

through condensation reactions. Since glycerol has three hydroxyl, or alcohol, groups, it can combine with three fatty acid molecules to form a molecule of fat (Fig. 3.15). For this reason, fats are sometimes called ***triglycerides.*** The various fats differ in the types of fatty acids of which they are composed. Saturated fats such as shortening, butter, and lard are composed of saturated fatty acids and hence have no carbon-to-carbon double bonds. The fatty acids in unsaturated fats such as corn oil, olive oil, and other vegetable oils are unsaturated, having at least one carbon-to-carbon double bond. At room temperature, fats composed of unsaturated fatty acids are liquid, whereas those with saturated fatty acids tend to be solid. In general, animal fats tend to be saturated and most vegetable fats (oils) unsaturated. Margarines are made from vegetable oils to which just enough hydrogen is added to make them solid at room temperature; they are a mixture of saturated and unsaturated fats (see Exploring Further: Atherosclerosis and the Diet-Heart Hypothesis, p. 63).

Since fats are synthesized by condensation reactions (with the removal of water), they, like complex carbohydrates, can be broken down to their building-block compounds by hydrolysis. Indeed, this is what happens in digestion.

Phospholipids Are the Main Structural Components of Cell Membranes

Some lipids, called ***phospholipids,*** have a phosphate group substituted for one of the three fatty acids. Among the most common phospholipids are those composed of one unit of glycerol, two units of fatty acid, and one phosphate group, which is often linked to a nitrogen-containing group (Fig. 3.16). The phosphate group is bonded to the glycerol at the point where the third fatty acid would be in a fat. Because the phosphate group tends to dissociate and give up a hydrogen ion, one of the oxygens becomes negatively charged. Similarly, the nitrogen tends to attract a hydrogen ion and to become positively charged. In short, the end of the phospho-

$$3\ \text{H–C–OH} + 3\ \text{HO–C(=O)–C–C–C–} \longrightarrow 3\ \text{H–C–O–C(=O)–C–C–C–} + 3H_2O$$

Glycerol + Fatty acids = Fat + Water

3.15 Synthesis of a fat Removal of three molecules of water by condensation reactions results in the bonding of three molecules of fatty acid to a single molecule of glycerol. Only a small portion of the hydrocarbon chains of the fatty acids are shown here. Notice that fats are not polymers; there are no free hydroxyl groups at the ends of the fat molecule that can join additional fatty acids by condensation reactions. Three molecules of water will be added by hydrolysis when this molecule of fat is digested. The carbon chains of the fatty acids are usually longer than shown here.

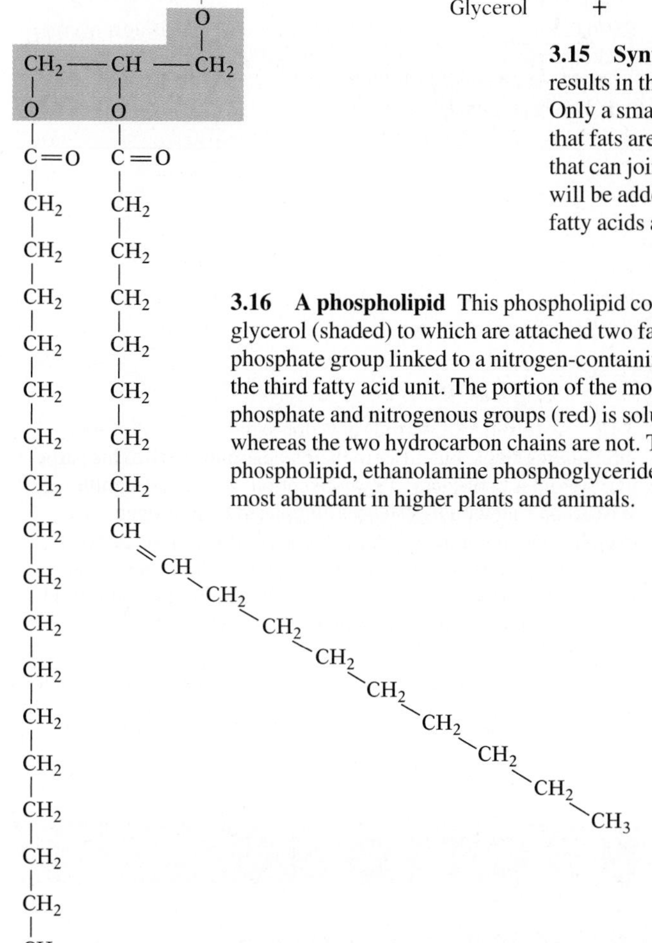

3.16 A phospholipid This phospholipid consists of one unit of glycerol (shaded) to which are attached two fatty-acid units and a phosphate group linked to a nitrogen-containing group in place of the third fatty acid unit. The portion of the molecule with the phosphate and nitrogenous groups (red) is soluble in water, whereas the two hydrocarbon chains are not. This particular phospholipid, ethanolamine phosphoglyceride, is one of the two most abundant in higher plants and animals.

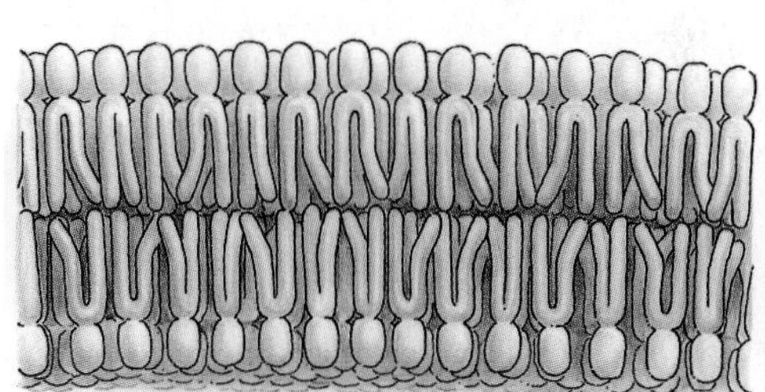

3.17 Phospholipids in the cell membrane Phospholipids are the primary structural elements of membranes. When placed in water, they tend to form bilayers, with the hydrophilic heads (blue) oriented toward the water and the hydrophobic chains (yellow) toward each other, in the interior.

lipid molecule with the phosphate and nitrogen-containing groups is strongly ionic and hence soluble in water. However, the other end, composed of the two long hydrocarbon tails of the fatty acids, is nonpolar and insoluble.

Phospholipids are sometimes called "schizophrenic" molecules because of their split personality—one end is soluble in water, the other end is not. This unusual property makes the phospholipids well suited as major constituents of cell membranes. As will be discussed in detail in Chapter 5, cell membranes are composed principally of two layers of phospholipid molecules, arranged with their ionic (hydrophilic) ends exposed at the two surfaces of the membrane and their nonpolar (hydrophobic) hydrocarbon chains buried in the interior, directed away from the surrounding water (Fig. 3.17).

Steroids Are Four-Ring Molecules With Many Biological Functions

Although commonly classified as lipids, the ***steroids*** differ markedly in structure from fats, oils, and phospholipids (Fig. 3.18). They are not formed from condensation reactions between fatty acids and alcohols. Instead, they are composed of four linked rings of carbon atoms, with various side groups attached to the rings. They are classified as lipids because, like the other lipids, they are soluble in organic solvents and relatively insoluble in water.

Steroids are very important in many biological systems. Some vitamins and hormones are steroids, and steroids often occur as structural elements in living cells, particularly in cellular membranes. The common steroid *cholesterol,* for example, serves to stabilize the fluidity of such membranes. An excess of cholesterol in the diet has been associated in some medical studies with the buildup of cholesterol-rich fatty deposits in blood vessels, possibly leading to an increased risk of heart attack. Table 3.2 lists the ranges of cholesterol levels in human blood.

TABLE 3.2 *Recommended lipid levels in human blood*

Recommended levels	*Men*	*Women*
Total Cholesterol	180–220 mg	180–220 mg
HDL Cholesterol	55+ mg	65+ mg
LDL Cholesterol	less than 130 mg	less than 130 mg
Triglycerides (Fats)	40–200 mg	40–200 mg
Ratio of total	3.43 Below average	3.27 Below average
cholesterol to	4.97 Average risk	4.44 Average
HDL and associated	9.55 2 × risk	7.05 2 × risk
risk of heart attack	23.39 3 × risk	11.04 3 × risk

3.18 A steroid All steroids have the same basic unit of four interlocking rings, but differ in their side groups (red). This particular steroid is cholesterol. (By convention, a hexagon signifies a six-carbon ring with its valences completed by hydrogens; see cyclohexane in Figure 3.2. A pentagon signifies a five-carbon ring, also with hydrogens attached to the carbons.) Cholesterol is an important constituent of mammalian cell membranes and is used for the synthesis of many hormones.

MAKING CONNECTIONS

1. Describe how hydrocarbons form the basis of all organic molecules.

2. Why is glucose considered the most important of the six-carbon monosaccharides?

3. If, instead of synthesizing fat for the storage of energy, you could only synthesize carbohydrate (e.g., glycogen), would you weigh more, less, or the same?

4. What properties of phospholipids make them important constituents of cell membranes?

EXPLORING FURTHER

ATHEROSCLEROSIS AND THE DIET-HEART HYPOTHESIS

In recent years, there has been much debate in medical and nutritional circles about the relationship between atherosclerosis, saturated fats, and cholesterol. Atherosclerosis (hardening of the arteries) is an extremely widespread disease in which excessive cholesterol-rich fatty deposits are formed in the walls of the blood vessels. The smooth muscle and fibrous tissues of the blood-vessel walls then proliferate and add additional layers. Eventually the walls become hardened to form rigid tubes. The hardened blood-vessel walls interfere with normal blood flow and are easily ruptured. In addition, the fatty deposits often protrude into the flowing blood, and the roughness of their surfaces causes blood clots to develop. Over time, the hardened blood vessels may become so congested from fatty deposits that blood flow through the blood vessels is impaired and even a tiny blood clot can block circulation. The result of such a blockage is often a heart attack or stroke.

Though many factors may be involved in the development of atherosclerosis, much attention has been focused recently on high blood cholesterol and diet as major factors. A necessary constituent of cell membranes, cholesterol can be synthesized by most cells in the body, but much of it is manufactured by the liver and carried in the blood as lipoprotein (a combination of protein, cholesterol, fats, and phospholipids). Three types of lipoproteins are particularly important in cholesterol transport: the so-called low-density lipoproteins (LDLs), the very low-density lipoproteins (VLDLs), and the high-density lipoproteins (HDLs). LDLs are the principal carriers of cholesterol in the blood, and it is from the LDLs and VLDLs that cholesterol deposits in the vessel walls are derived. The more LDLs and VLDLs in the blood, the more likely it is that cholesterol deposits will be excessive. Most HDLs, on the other hand, appear to protect against atherosclerosis by helping to remove excess cholesterol from the cells and returning it to the liver for excretion. It has been shown that diet tends to affect cholesterol levels in the blood; both saturated fats and cholesterol in the diet increase the proportion of cholesterol carried by the LDLs, while polyunsaturated fats (those with two or more double bonds) tend to increase the proportion carried by the HDLs, favoring the eventual excretion of excess cholesterol by the liver. Regular rigorous exercise also increases HDL transport.

It must be said that only circumstantial evidence exists for a causal connection between diet, high blood cholesterol, and atherosclerosis, as proposed by the so-called diet-heart hypothesis. According to one study conducted by the National Institutes of Health, the death rate from heart attack and stroke since 1968 has dropped by more than 20 percent. This period of generally improved health seems to correspond with a clear change in the American

diet; the average saturated-fat and cholesterol intake has dropped, while the intake of polyunsaturated fats has greatly increased. Also, the average blood-cholesterol levels for middle-aged males since the 1960s have shown a decline of about 7 percent. None of these data show a rigorous cause-and-effect relationship between diet and atherosclerosis, but the evidence is very strong. That atherosclerosis can develop in the absence of high blood cholesterol and that many individuals with normal blood cholesterol levels nevertheless die of heart attacks support the idea of multiple risk factors, such as age, diabetes, high blood pressure, cigarette smoking, stress, or decreased physical activity.

Most physicians feel that the most important preventative measure against developing atherosclerosis is to eat a low-fat, low-cholesterol diet, and to consume primarily unsaturated fats. Studies have shown reducing the blood LDL levels reduces the risk of dying from atherosclerotic heart disease: for each 1 mg/100 ml decrease in LDL, there is a 2 percent decrease in mortality due to atherosclerosis.

PROTEINS

Proteins Play Crucial Roles in Cell Structure and Function

Far more complex than either carbohydrates or lipids, ***proteins*** play leading roles in both the structure and function of living organisms. Proteins, for instance, are the major components of muscles and are responsible for muscle contraction. They are also important in support, making up the fibers found in skin, bone, tendons, nails, claws, and hair. Still other proteins function in vision, in oxygen transport, in the immune response, and as hormones. As we shall see, proteins are also catalysts of biological reactions. Despite their great diversity, proteins—like carbohydrates and lipids—are composed of basic building-block molecules joined together by condensation reactions.

The Sequence of Amino Acids in a Protein Determines Its Primary Structure

All proteins contain four essential elements: carbon, hydrogen, oxygen, and nitrogen; most proteins also contain some sulfur. These elements are bonded together to form the building-block molecules called ***amino acids,*** so called because each has an amino group and a carboxyl group attached to the same carbon. In addition, each amino acid has a hydrogen atom and a characteristic side chain, designated R, also attached to the same carbon:

```
H     H    O
 \    |   //
  N — C — C — O — H
 /    |
H     R
```

The various amino acids differ only in their side chains, or ***R groups.*** The R group may be simple, as in alanine, where it is only a methyl group (—CH_3), or it may be complex, as in phenylalanine, where it includes a ring structure (Fig. 3.19). Twenty different amino acids are commonly found in proteins; the structural formulas of eight of them are shown in Figure 3.19.

The various R groups give each amino acid distinctive characteristics, which in turn influence the properties of the proteins incorporating them. For example, some amino acids are relatively insoluble in water, because their R groups are nonpolar at a pH in the neutral range, whereas other amino acids are water-soluble, because their R groups are polar or electrically charged.

In the slightly acidic pH range prevailing within cells, a high percentage of the amino acid molecules are ionized. The carboxyl group gives up a hydrogen ion and becomes negatively charged, while the amino group attracts an extra hydrogen ion, and becomes positively charged:

```
    H        H    O
     \ +     |   //
  H — N   —  C — C — O⁻
     /       |
    H        R
```

Proteins are long, unbranched *polymers* of the 20 possible amino acids. The amino-acid building blocks of proteins are joined head to tail by condensation reactions between the car-

NONPOLAR (HYDROPHOBIC)

Alanine (Ala)

Leucine (Leu)

Methionine (Met)

Phenylalanine (Phe)

HYDROPHILIC

POLAR

Serine (Ser)

Cysteine (Cys)

CHARGED

Aspartic acid (Asp)

Histidine (His)

3.19 Structural formulas of eight of the amino acids common in proteins The amino acids are shown in their ionized form. All amino acids have the same basic structure (black): a carboxyl group, an amino group, a hydrogen atom, and an R group are attached to the same carbon. The R group (shown in brown) is the only portion of the amino acid that varies. Rows 1 and 2: the four amino acids have nonpolar R groups that are insoluble in water. Row 3: The two amino acids have polar R groups and are water-soluble. Row 4: The two amino acids, with R groups ionized at intracellular pH levels, are electrically charged and hence water-soluble; the first one, having given up a hydrogen ion, is negatively charged, whereas the second, having accepted a hydrogen ion, is positively charged.

boxyl group of one amino acid and the amino of the next (Fig. 3.20). The covalent bonds between the amino acids are called ***peptide bonds,*** and the chains they produce are called ***polypeptide chains.*** The number of amino acids in a single polypeptide chain within a protein molecule is usually between 40 and 500, although shorter and longer chains sometimes occur. To cite just one example, oxytocin, a hormone that stimulates uterine contractions during birth in humans, is a very small polypeptide—composed of only nine amino acids.

Many proteins contain cysteine, a sulfur-containing amino acid. Two cysteines can join together to form a strong nonpolar covalent bond called a ***disulfide bond*** or ***disulfide bridge*** (Fig. 3.21). Such bonds can link two parts of a single polypeptide chain, maintaining it in a bent or folded shape, or they can link different chains (Fig. 3.22). Because they are strong covalent bonds, disulfide bonds can play an important role in helping to determine the final three-dimensional shape of proteins. Insulin, a hormone secreted by the pancreas of vertebrates to regulate the breakdown of sugar, is an example of a protein whose polypeptide chains are held together by disulfide bond (Fig. 3.22).

The structural characteristics discussed so far—the number, type, and sequence of amino acids in each chain and the location of the disulfide bonds (if present)—constitute what is

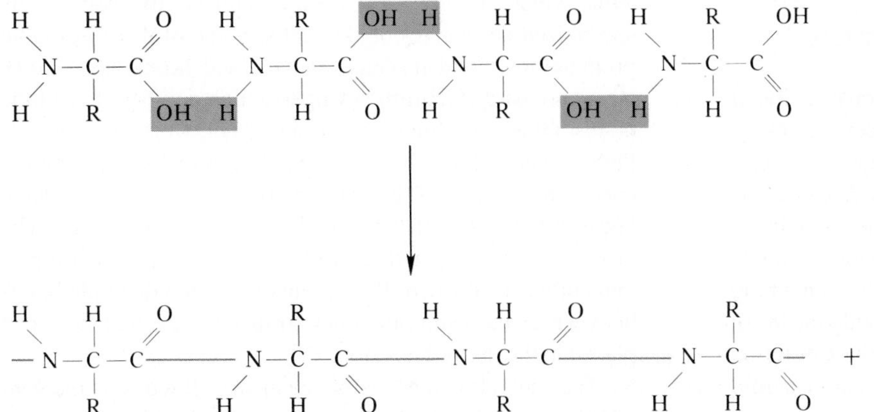

3.20 Synthesis of a polypeptide chain Condensation reactions between the carboxyl and amino groups of adjacent amino acids result in peptide bonds (brown) between the acids, and the release of water as a by-product. The sequence of amino acids in a protein is referred to as its primary structure.

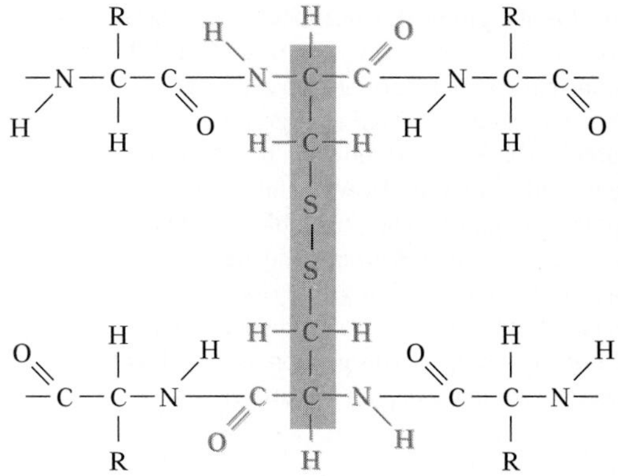

3.21 The structural formula of a disulfide bridge A disulfide bridge (gray) is formed when the R groups of two cysteines (brown) from different parts of a protein are linked by a disulfide bond.

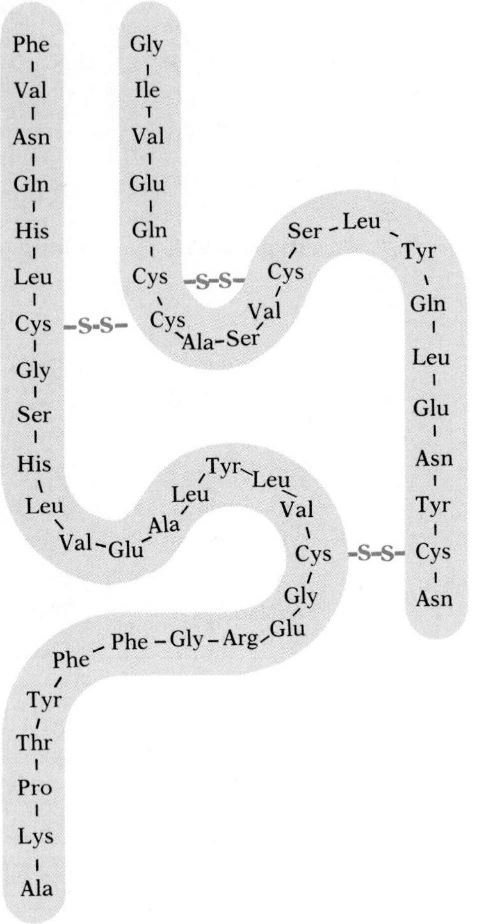

3.22 The structure of beef insulin The molecule consists of two polypeptide chains joined by two disulfide bridges. There is also one disulfide bridge within the shorter chain (right). Because disulfide bonds are strong covalent bonds, the disulfide bridges hold the two chains together and resist breaking. Hydrogen bonds (not shown) between the chains and between segments of the same chain are also present.

called the protein's ***primary structure.*** Since each position in the chain can be occupied by any one of 20 different amino acids, the potential number of different proteins is enormous. That explains how the many millions of different species of organisms can manufacture proteins specific to each of them.

Hydrogen Bonds Between Parts of a Polypeptide Chain Determine a Protein's Secondary Structure

Proteins are not laid out simply as two-dimensional chains of amino acids, as our discussion so far might seem to imply. Instead, they are coiled and folded into elaborate spatial patterns, called ***conformations,*** which play a crucial role in determining the distinctive biological properties of each protein.

First consider the simplest level of conformation, namely the one determined by the spatial relationship of each amino acid to its immediate neighbors in the polypeptide chain. In 1951, Linus Pauling and Robert B. Corey of the California Institute of Technology found that polypeptide chains can fold into regularly repeating structural patterns, constituting the protein's ***secondary structure.*** Two such repeating patterns, the alpha helix and the beta sheet, are particularly common in most proteins.

The ***alpha*** ***(α) helix*** consists of a tightly coiled polypeptide chain (Fig. 3.23). The chain is held in this helical shape by hydrogen bonds formed between the nitrogen of one amino acid and the oxygen of another beyond it in the polypeptide chain. The helical pattern is seen at its simplest in some ***fibrous proteins,*** which provide the structural elements for many of the specialized derivatives of skin cells. Some of these insoluble proteins are found in nails, hooves, and horns (Fig. 3.24), which are hard and brittle owing to a high number of disulfide bonds. Others are found in hair and wool, which are soft and flexible and can easily be stretched (especially when moistened and warmed). The stretching of hair or wool is possible because there are relatively few strong disulfide bonds. The process of breaking and reforming bonds is the basis for permanent waves for hair. Permanents break the disulfide bonds between chains, and new ones form after the hair has been placed in the desired position.

The second major type of secondary structure is the ***beta*** ***(β) sheet,*** also known as the pleated sheet. In this pattern, seg-

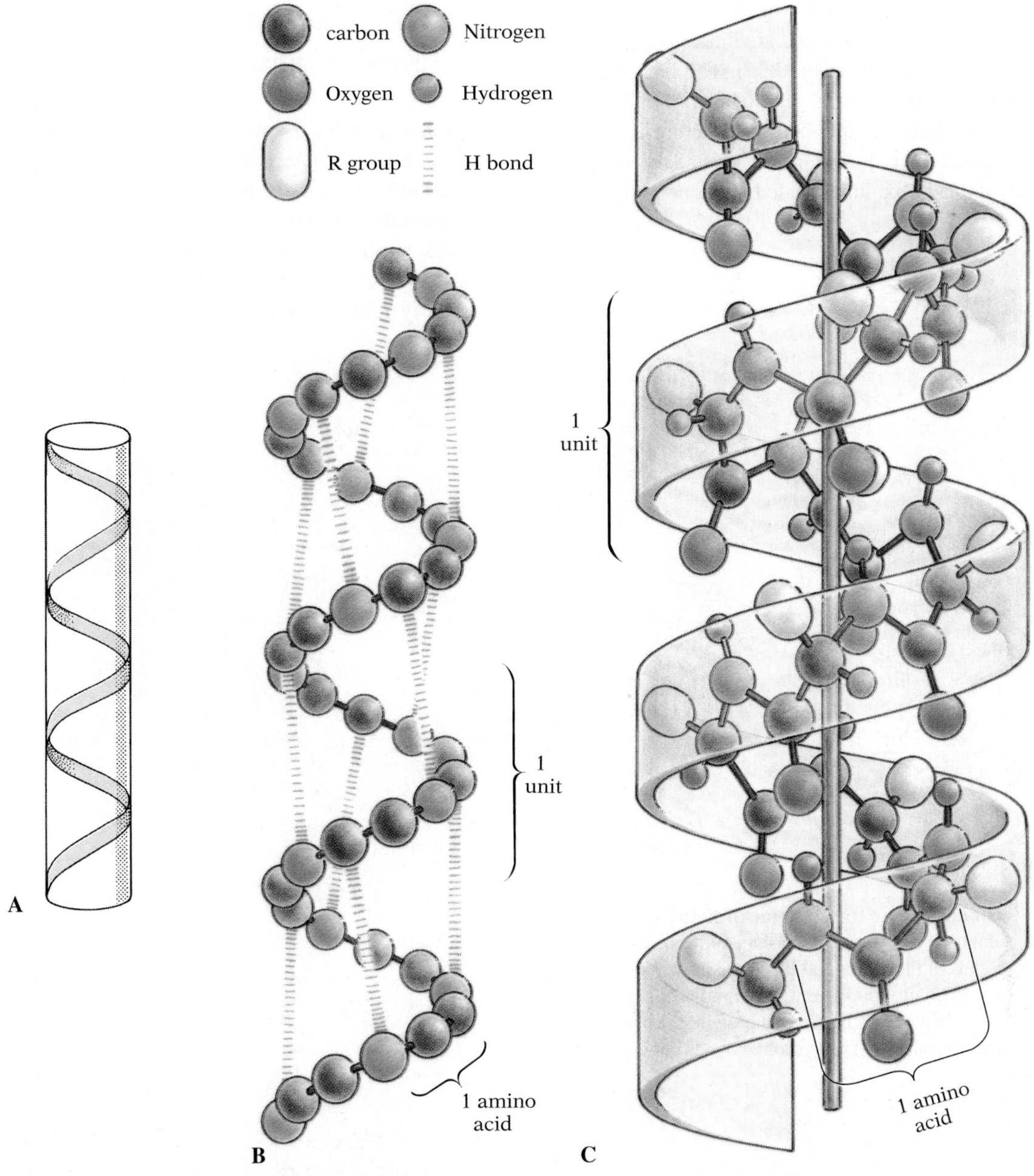

carbon
Nitrogen
Oxygen
Hydrogen
R group
H bond
1 unit
1 amino acid
A
B
C

3.23 Alpha-helical secondary structure of some proteins (A) The helix may be visualized as a ribbon wrapped around a regular cylinder. (B) The backbone of a polypeptide chain (the repeating sequence of N–C–C–N–C–C–N–C–C) is shown coiled in a helix (all other atoms and R groups are omitted). The hydrogen bonds shown extend between the amino group of one amino acid and the oxygen of the third amino acid beyond it in the polypeptide chain. (C) A ball-and-stick model of an alpha-helical section of a protein. The vertical rod and the helical ribbon are meant to aid visualization.

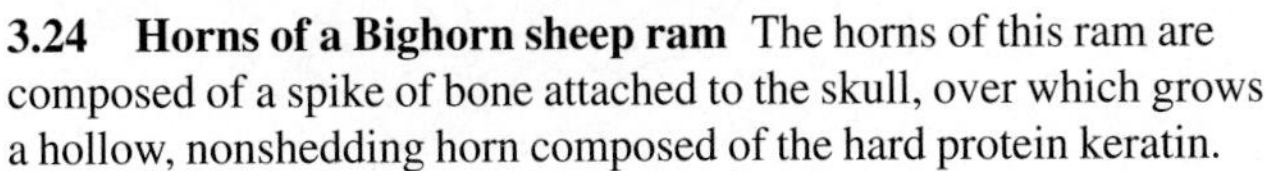

3.24 Horns of a Bighorn sheep ram The horns of this ram are composed of a spike of bone attached to the skull, over which grows a hollow, nonshedding horn composed of the hard protein keratin.

ments of different polypeptide chains (or of a single chain folded back on itself) lie side by side and are linked by hydrogen bonds between the segments (Fig. 3.25). The beta-sheet arrangement is flexible and strong, but resists stretching. Probably the most familiar example of this structure is silk. Many complex proteins have both types of secondary structure—regions of alpha helix interspersed with beta sheets (Fig. 3.26).

Interactions Between Distant Amino Acids Determine a Protein's Tertiary and Quaternary Structures

Far more complex in spatial conformation than the fibrous proteins are the ***globular proteins,*** whose polypeptide chains are folded into elaborate, roughly spherical shapes. Typically, these soluble proteins—which include enzymes, protein hormones, antibodies, and most blood proteins—are made up of sections of alpha helix and/or beta sheet interspersed with irregularly folded regions. The common muscle protein myo-

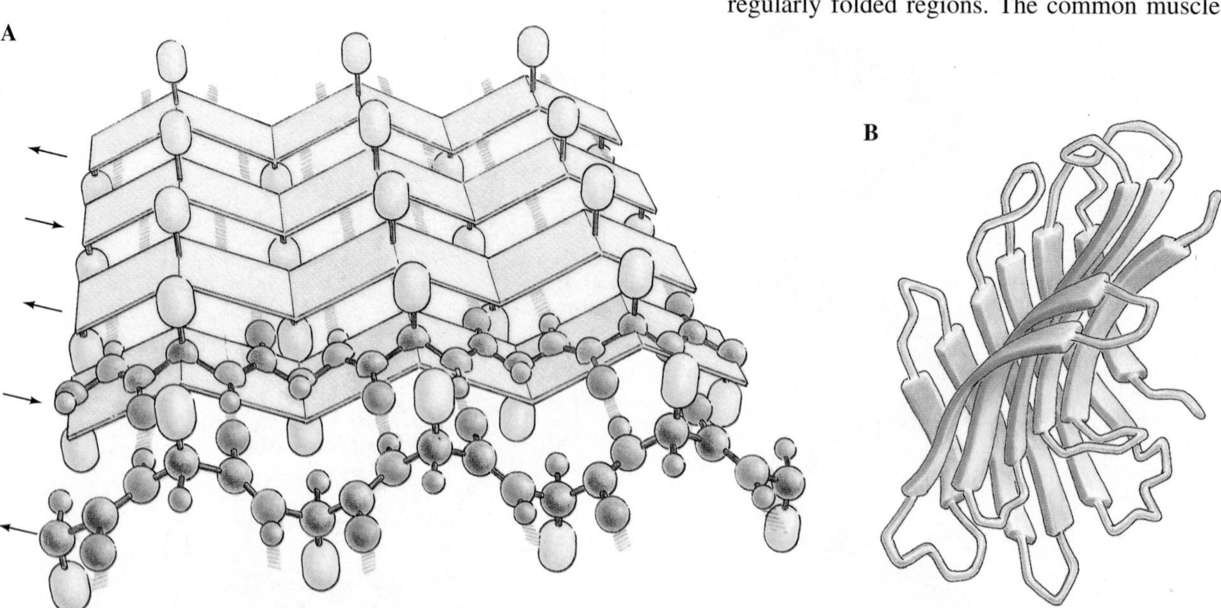

3.25 Beta-sheet secondary structures of some proteins (A) Diagrammatic representation of five parallel polypeptide chains, with ball-and-stick models of the atoms and R groups shown for the first two chains. Hydrogen bonds (pink) between the adjacent chains stabilize the structure. (B) In a ribbon model of a protein, the beta sheets are shown as flattened sheets. The hydrogen bonds between the chain are not shown.

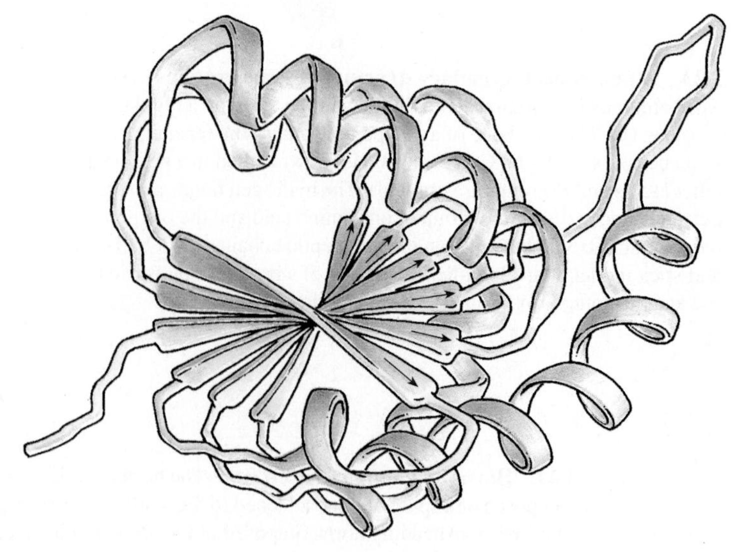

3.26 Ribbon model of a globular protein In the so-called ribbon models of protein conformation, the ribbon represents the polypeptide chain, and the alpha helices are shown as pink coils and the beta sheets as flattened blue ribbons. Notice that globular proteins incorporate elements of secondary structure; here there are regions of alpha helix interspersed with regions of beta sheet. The remainder of the polypeptide segments (yellow) do not have a regular structure.

globin, for example, consists of eight major sections of alpha helix connected by short regions of irregular coiling (Fig. 3.27). At each irregular region, the three-dimensional orientation of the polypeptide changes, giving rise to the protein's characteristic folding pattern. This three-dimensional pattern, which is superimposed on the secondary structure, is called the protein's ***tertiary structure;*** it is largely a consequence of weak interactions between the R groups of the widely separated amino acids making up the chain, and between R groups and the surrounding aqueous environment.

Some globular proteins consist of more than one polypeptide chain, and such chains often come together much like the decorative curly ribbons on a gift package. On a molecular level, the chains are usually held together by numerous weak bonds of the types described previously, namely ionic bonds, hydrogen bonds, and hydrophobic interactions. The manner in which the coiled, folded chains fit together and make up the final three-dimensional structure is called the protein's ***quaternary structure.*** For example, a single molecule of hemoglobin, the red, oxygen-carrying protein in your blood, has a quaternary structure made up of four independently folded polypeptide chains linked by weak bonds between the R groups (Fig. 3.28). Although the bonds between the polypeptide chains are easily broken, their cumulative effect—as in the analogy of the Velcro patch cited in the previous chapter—is to give the hemoglobin molecule a remarkably stable three-dimensional shape. The various levels of protein structure are summarized in Table 3.3.

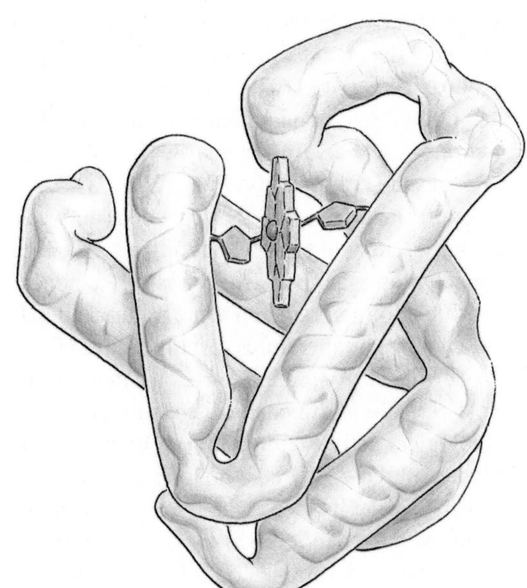

3.27 Spatial conformation of a molecule of myoglobin A protein's tertiary structure—its unique three-dimensional shape—is illustrated here by myoglobin, a globular protein related to hemoglobin. Like hemoglobin, myoglobin is characterized by a strong affinity for molecular oxygen and consists of a single complexly folded polypeptide chain of 151 amino acid units. Attached to the chain is a prosthetic group called heme (represented here by the disk). The polypeptide chain consists of eight sections of helix with nonhelical regions between them. These nonhelical regions are a major factor in determining the way the helical sections fold together, and hence myoglobin's tertiary structure.

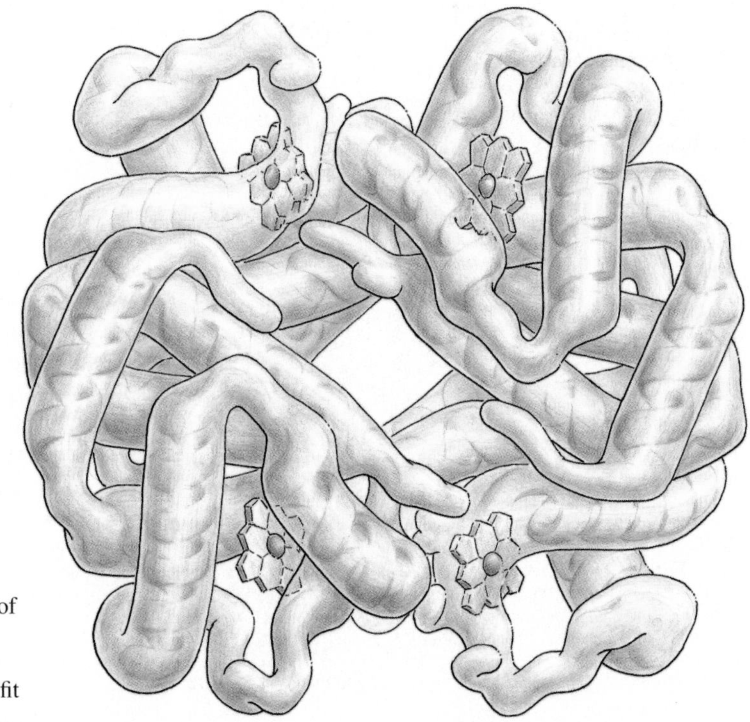

3.28 Quaternary structure of hemoglobin A single molecule of hemoglobin is composed of four independent polypeptide chains, each of which has a globular conformation and its own prosthetic group. The spatial relationship between these four—the way they fit together—is called the quaternary structure of the protein.

TABLE 3.3 *Levels of structure in proteins*

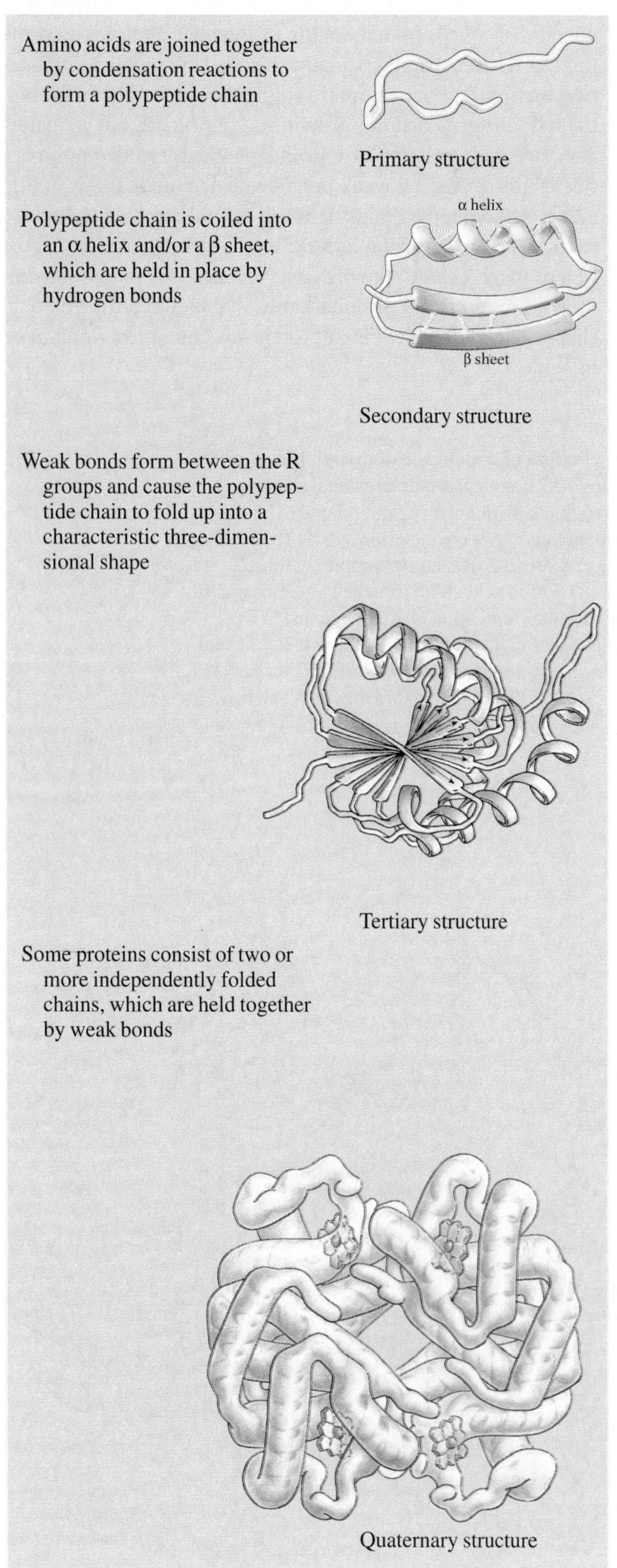

Amino acids are joined together by condensation reactions to form a polypeptide chain
Primary structure
α helix
Polypeptide chain is coiled into an α helix and/or a β sheet, which are held in place by hydrogen bonds
β sheet
Secondary structure
Weak bonds form between the R groups and cause the polypeptide chain to fold up into a characteristic three-dimensional shape
Tertiary structure
Some proteins consist of two or more independently folded chains, which are held together by weak bonds
Quaternary structure

The Final Shape of a Protein Is Inherent in Its Primary Structure

Several aspects of a protein's primary structure contribute to producing its final shape. If, for example, a polypeptide chain contains two cysteine units, the disulfide bond joining them introduces a fold in the chain (Fig. 3.22). The distinctive properties of the various R groups of the amino acids also help determine the three-dimensional shape of the protein. For example, hydrophobic groups tend to be close to each other in the interior of the folded chain—as far away as possible from the surrounding water—whereas hydrophilic groups tend to be on the outside, attracted to the water. Polar R groups tend to assume positions where they can form hydrogen bonds with other such groups; similarly, electrically charged groups tend to be in positions where they can form ionic bonds with oppositely charged groups. Thus the various kinds of weak bonds discussed earlier—hydrogen bonds, ionic bonds, and hydrophobic interactions— all play important roles in stabilizing the three-dimensional shape of proteins (Fig. 3.29).

Since protein conformation is largely dependent on weak bonds (which are very sensitive to temperature and pH), it is easily disrupted by anything that breaks or alters those bonds. It is stable only within a limited range of temperature and pH. Even brief exposure to high temperatures (usually above 60°C) or to extremes of pH will cause a protein to become ***denatured,*** that is, to lose most of its three-dimensional conformation and thus its normal biological activity (Fig. 3.30). Often the protein is irreversibly denatured—that is, it no longer has any biological activity. This is what happens when you cook an egg; the protein is denatured and cannot revert to its original form. Sometimes, however, under favorable conditions in cells, denatured proteins can spontaneously refold into their native three-dimensional conformation and recover their normal biological activity. Since only the primary structure is available to dictate the folding pattern in such cases, it alone must determine all other aspects of the protein's structure.

In short, there are compelling reasons to believe that the primary structure of a protein dictates its three-dimensional structure and hence its biological function. More specifically, it appears that the primary structure determines the most stable possible arrangement of the polypeptide chains. Hence, the question of exactly how three-dimensional form is specified when a protein is being synthesized becomes synonymous with the question of how its amino-acid content and sequence are specified.

Conjugated Proteins Contain Nonprotein Groups

Attached to some proteins are nonprotein groups called ***prosthetic groups;*** an example is the heme group of myoglobin, a

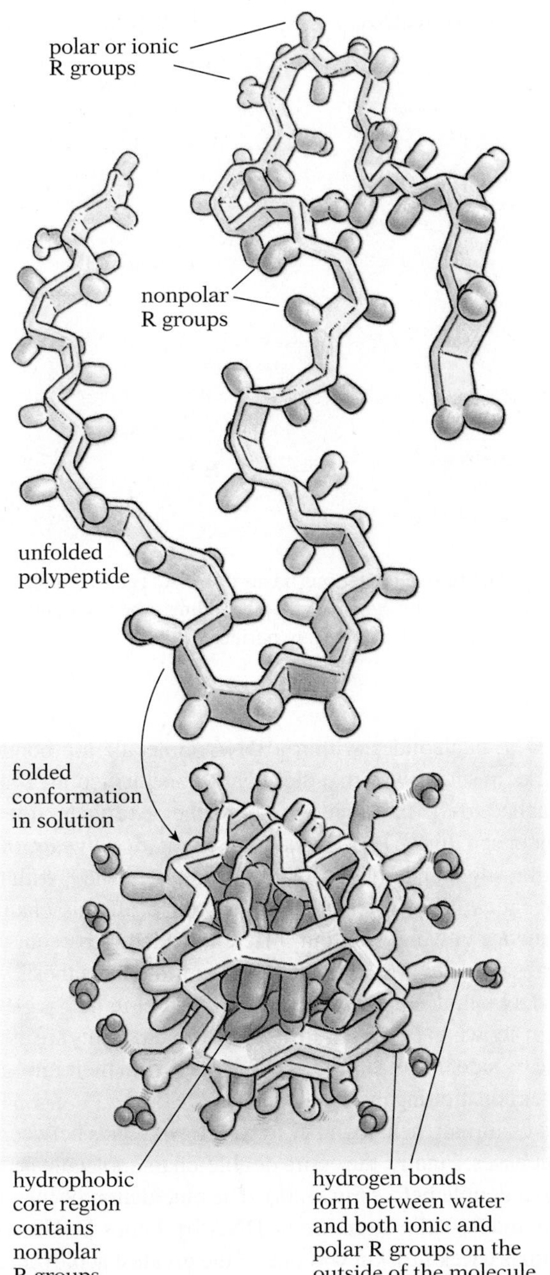

3.29 Folding of a globular protein in solution As water molecules form polar bonds with polar and ionic side groups, polypeptide chains can fold spontaneously so that the nonpolar groups are herded into the middle.

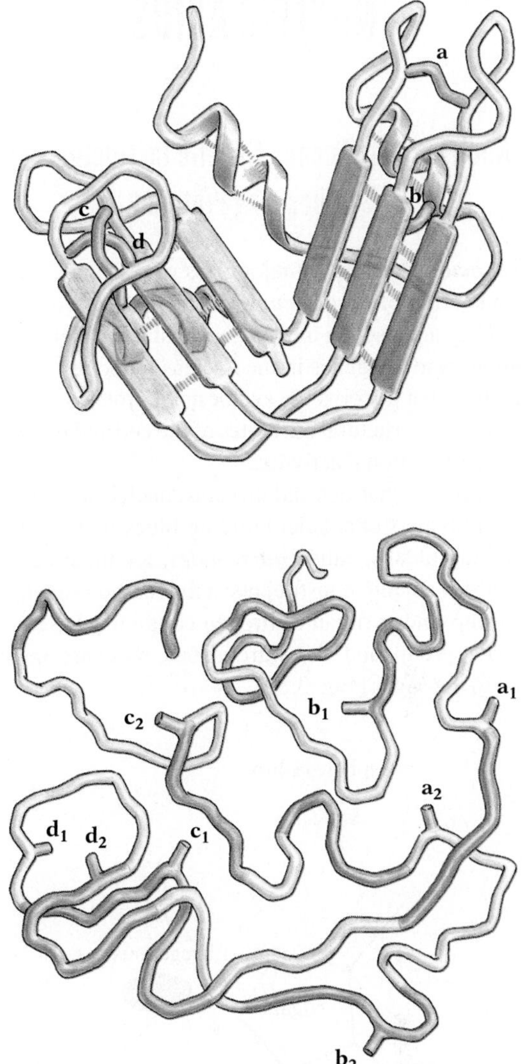

3.30 Denaturation and renaturation of ribonuclease When ribonuclease, a normally globular protein (top), is denatured, with both its weak bonds and its four intrachain disulfide bonds (green) broken, it unfolds into an irregularly coiled state (bottom). In this denatured condition the ribonuclease lacks its usual ability to digest RNA. When the denaturing agents are removed and favorable conditions restored, the protein spontaneously refolds into its native conformation (top) and regains its capacity for biological activity as an enzyme. Even the four disulfide bonds reform correctly (i.e., a_1 joins to a_2, b_1 to b_2, etc.). Since there are 105 possible ways to join the cysteines but the enzyme folds only to bring the correct pairs together, we must conclude that the conformation does not result just from the positions of the cysteines; most disulfide bonds serve to stabilize, rather than to determine structure.

ringlike structure with an iron atom in its center (Fig. 3.27). Prosthetic groups may be as simple as a single metal ion bonded to the polypeptide chain; they may be sugars or other carbohydrate entities; or they may be lipids. Whatever the nature of the prosthetic group, its presence alters the properties of the protein in important ways, and is necessary for its biological activity. Without their heme groups, for example, myoglobin and hemoglobin lose their high affinity for molecular oxygen and cannot bind to it. All proteins that contain nonprotein substances are called ***conjugated proteins.***

NUCLEIC ACIDS

Nucleic Acids Encode Genetic Instructions for Synthesizing Proteins

Nucleic acids, the fourth major class of organic compounds, are the materials of which ***genes,*** the units of heredity, are composed. They are also the messenger substances that convey information from the genes in the nucleus to the rest of the cell, information that governs the synthesis of proteins and thereby determines the structural attributes of the cell and regulates the cell's other functional activities.

Like polysaccharides and proteins, nucleic acid molecules are long polymers of smaller building-block units. In this case the building blocks, called ***nucleotides,*** are themselves composed of still smaller constituents: a five-carbon sugar, a phosphate group, and an organic nitrogen-containing base. Both the phosphate group and the nitrogenous base are covalently bonded to the sugar (Fig. 3.31).

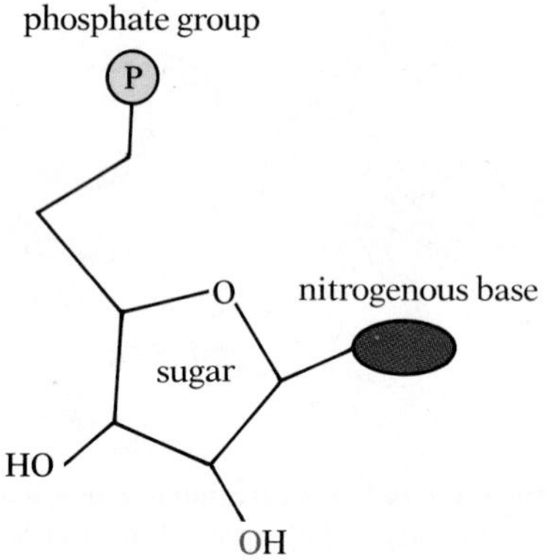

3.31 Diagram of a nucleotide A phosphate group and a nitrogenous base are attached to a five-carbon sugar.

DNA Is the Primary Repository for the Genetic Information

Deoxyribonucleic acid, commonly called DNA for short, is the nucleic acid most genes are made of. Four different kinds of nucleotide building blocks occur in DNA. All have, as the name suggests, ***deoxyribose*** as their sugar component, but they differ in their nitrogenous bases, which may be one of four possible substances. Two of these, ***adenine*** and ***guanine,*** are double-ring structures of a class known as ***purines;*** the other two, ***cytosine*** and ***thymine,*** are single-ring structures known as ***pyrimidines*** (Fig. 3.32).

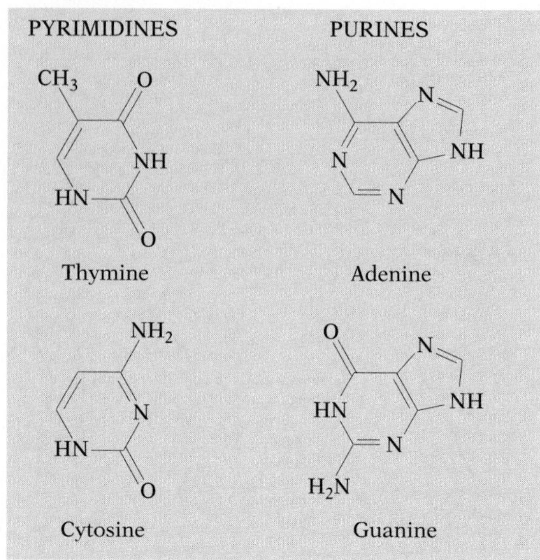

3.32 The four nitrogenous bases in DNA The two single-ring bases, thymine and cytosine, are pyrimidines; the two double-ring bases, adenine and guanine, are purines.

The nucleotides within a DNA molecule are bonded together in such a way that the sugar of one nucleotide is always attached to the phosphate group of the next nucleotide in the sequence (Fig. 3.33). Thus a long chain, or ***polymer,*** of alternating sugar and phosphate groups is established, with the nitrogenous bases oriented as side groups off this chain. The sequence in which the four different nucleotides occurs determines the sequence of amino acids in protein synthesis.

DNA molecules do not ordinarily exist in the single-chain form shown in Fig. 3.33. Instead, two such chains are arranged side by side like the uprights of a ladder, with their nitrogenous bases constituting the cross rungs of the ladder (Fig. 3.34). The two chains are held together by hydrogen bonds between adjacent bases. Finally, the entire double-chain molecule is twisted into a double helix (Fig. 3.35). The elucidation of the distinctive double-helix structure of DNA by James D. Watson and Francis Crick in 1953 was one of the greatest achievements of modern science and marked the beginning of the study of molecular genetics.

The regular helical coiling and the hydrogen bonding between bases impose two extremely important constraints on how the cross rungs of the ladderlike DNA molecule can be constructed. First, each rung must be composed of a purine (double ring) and a pyrimidine (single ring); only in this way will all cross rungs be of the same length, making possible the formation of a regular helix. Second, if the double ring is adenine, the single ring must be thymine, and, similarly, if the double ring is guanine, the single ring must be cytosine; only these two pairs are capable of forming the required hydrogen bonds (Fig. 3.34). Since it does not matter, however, in which order the members of a pair appear (A-T or T-A; G-C or C-

3.33 Portion of a single chain of DNA Nucleotides are hooked together by bonds between their sugar and phosphate groups. The nitrogenous bases (G, guanine; T, thymine; C, cytosine; A, adenine) are side groups.

3.34 Portion of a DNA molecule uncoiled The molecule has a ladderlike structure, with the two uprights composed of alternating sugar and phosphate groups and the cross rungs composed of paired nitrogenous bases. Each cross rung has one purine base (large oval) and one pyrimidine base (small oval). When the purine is guanine (G), the pyrimidine with which it is paired is always cytosine (C); when the purine is adenine (A), the pyrimidine is thymine (T). Adenine and thymine are linked by two hydrogen bonds (striped bands), guanine and cytosine by three. Note that the two chains run in opposite directions—that is, the free phosphate is at the upper end of the left chain and at the lower end of the right chain.

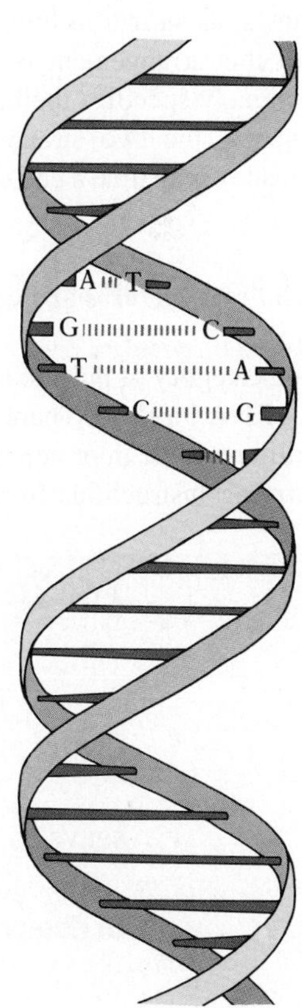

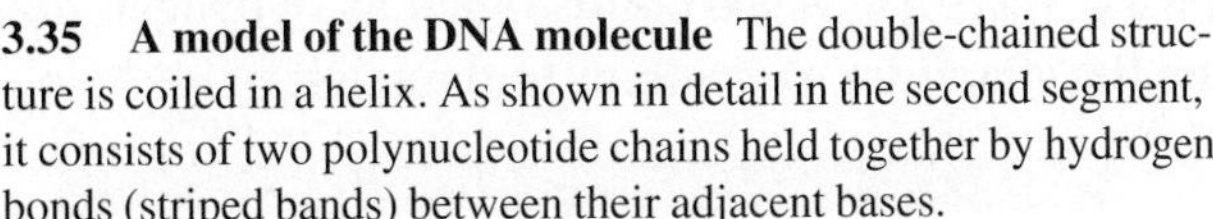

3.35 A model of the DNA molecule The double-chained structure is coiled in a helix. As shown in detail in the second segment, it consists of two polynucleotide chains held together by hydrogen bonds (striped bands) between their adjacent bases.

TABLE 3.4 *The four major classes of organic molecules found in living organisms*

Class	*Properties, functions*	*Building block molecules*	*Types*	*Examples*
Carbohydrates	Polar molecules, hydrophilic; energy source, storage	Polymer of simple sugars	Monosaccharides Disaccharides Polysaccharides	Glucose, fructose Sucrose, lactose maltose Starch, cellulose
Lipids	Nonpolar molecules, hydrophobic; energy storage, membrane components, hormones	Fatty acids, glycerol	Fatty acids Fats Phospholipids Steroids	Palmitic, oleic acid Triglycerides Membrane lipids Cholesterol, steroid hormones
Proteins	Structural (in muscles, connective tissue), enzymes, transport of oxygen, membrane components, etc.	Polymer of amino acids	Fibrous proteins (primary, secondary structure only) Globular proteins, (all levels of structure)	Collagen, muscle, keratin Digestive enzymes (e.g., pepsin, amylase); myoglobin, hemoglobin, enzymes, hormones
Nucleic acids	Carriers of genetic information, controls protein synthesis	Polymer of nucleotides	Deoxyribonucleotides Ribonucleotides	DNA RNA

G), the double-chain molecule can have four different kinds of cross rungs, as shown in Figure 3.34. The biological significance of this arrangement is that the base sequence of one chain uniquely specifies the base sequence of the other. As a consequence, the two strands can be separated and exact copies made every time a cell divides.

RNA Plays Several Key Roles in Protein Synthesis

A second category of nucleic acids comprises the ***ribonucleic acids,*** or RNA for short. There are several types of RNA, each with a different role in protein synthesis. Some act as messengers carrying instructions from the DNA genes to the ribosomes, the structures within the cell that are the sites of protein synthesis. Others are structural components of the ribosomes. Still others transport amino acids to those ribosomes, for assembly into proteins. We shall discuss each of these types of RNA in much more detail in Chapter 11. Suffice it to say here that all types of RNA differ from DNA in three principal ways: (1) the sugar in RNA is *ribose,* whereas that in DNA is *deoxyribose* (it is the sugars that give the two nucleic acids their different first initials); (2) instead of thymine, one of the four nitrogenous bases of DNA, RNA contains a very similar base called ***uracil;*** (3) RNA is usually single-stranded, whereas DNA is usually double-stranded.

A summary of some of the important characteristics of the four major classes of organic molecules is found in Table 3.4.

MAKING CONNECTIONS

1. Why are proteins so much more versatile in function than carbohydrates and lipids?

2. What is it about the structure of nucleic acids that enables them to serve as "information molecules"?

3. In what ways are weak bonds (such as hydrogen bonds) important in determining the structure of proteins and nucleic acids?

CONCEPTS IN BRIEF

1. Organic chemistry is the chemistry of carbon-based molecules, the simplest of which are the *hydrocarbons,* molecules containing only hydrogen and carbon. The four major classes of complex organic compounds found in living organisms are *carbohydrates, lipids, proteins,* and *nucleic acids.* Although the molecules of these classes differ in structure and function, they have the common attribute of being made up of many simpler "building-block" molecules.

2. Building-block subunits are combined to form complex organic molecules by the removal of water molecules in *condensation reactions.* Condensation reactions are reversible; the complex organic molecules can be *hydrolyzed* into the simpler building-block molecules with the addition of water.

3. The basic building-block molecules of *carbohydrates* are the simple sugars known as *monosaccharides.* When two simple sugars are bonded together, a *disaccharide* is formed. When many simple sugars are bonded together in long chains, a *polysaccharide* is formed. Starch, glycogen, and cellulose are examples of polysaccharides. The carbohydrates are a major structural component in plants and an important energy source for all organisms.

4. *Lipids,* the fats and fatlike substances, are soluble in organic solvents and insoluble in water. Fats are composed of glycerol and three fatty acids, which are joined by condensation reactions. *Phospholipids* are derived from the fats; they are important constituents of cell membranes. *Steroids,* the third major class of lipids, have four interlocking hydrocarbon rings, with various side groups attached to the rings.

5. *Proteins* are composed of long chains of *amino acids,* which are bonded together by condensation reactions. The resulting bond is the *peptide* bond and the chains produced are *polypeptide* chains.

6. The *primary structure* of each protein is the sequence of amino acids making up the polypeptide chains and the disulfide bonds. Because hydrogen bonds form between one amino acid and another, the chain often assumes a stable shape known as the *secondary structure,* which may take the form of an *alpha helix* or a *beta sheet.* These regular conformations may in turn be folded into complicated globular shapes by weak attractions between the different R groups within the chain, thus forming the protein's *tertiary structure.* Some globular proteins are made up of two or more polypeptide chains held together by weak bonds; the way these coiled, folded chains fit together determines the protein's *quaternary structure.*

7. Because the three-dimensional shape of a protein depends on weak bonds, it is easily altered, causing a change in its biological function. Changes in temperature and pH can denature proteins.

8. The building blocks of *nucleic acids* are the *nucleotides,* each of which is made up of a five-carbon sugar to which is attached a phosphate group and a nitrogen-containing base.

9. Nucleotide units are joined together through condensation reactions between the sugar of one nucleotide and the phosphate group of the next. There are four different nucleotides in each nucleic acid. It is the different sequences of the nucleotides that encode the hereditary information and dictate the sequence of amino acids in the synthesis of proteins.

10. The two types of nucleic acid, *DNA* and *RNA,* differ in the form of ribose sugar (ribose versus deoxyribose), their nitrogenous bases, the number of strands in the molecule, and their biological roles.

STUDY QUESTIONS

1. What are the four major classes of organic molecules important in living organisms? (p. 52)
2. Name and describe the type of reaction that joins building blocks of complex organic molecules together. Name and describe the reverse reaction. (p. 55)

3. What elements are carbohydrates composed of? In what ratio? What are the building blocks of disaccharides and polysaccharides? Make a diagram showing how the disaccharide sucrose is formed. (pp. 52–55)

4. Describe three functions of carbohydrates. (pp. 53, 55–56)

5. How does the chemical composition of lipids compare with that of carbohydrates? What physical property is shared by all lipids? Describe the three major categories of lipids and the functions of each. (pp. 55–62)

6. Describe how proteins contribute to both the structure and function of living organisms. (p. 64)

7. What are the building blocks of proteins? What is a peptide bond? Explain how it is formed and how it is broken. (pp. 64–65)

8. Describe the four levels of protein structure. (pp. 66–70)

9. Why is the final shape of a protein inherent in its primary structure? What happens when a protein is denatured? (p. 70)

10. What are the building blocks of nucleic acids? Describe and diagram the components of one of these building blocks. (p. 72)

11. Describe how nucleic acids are linked together to form DNA and RNA. (pp. 73–74)

12. What are the three principal ways that RNA differs structurally from DNA? (p. 74)

CHAPTER 4

THE FLOW OF ENERGY IN BIOCHEMICAL REACTIONS

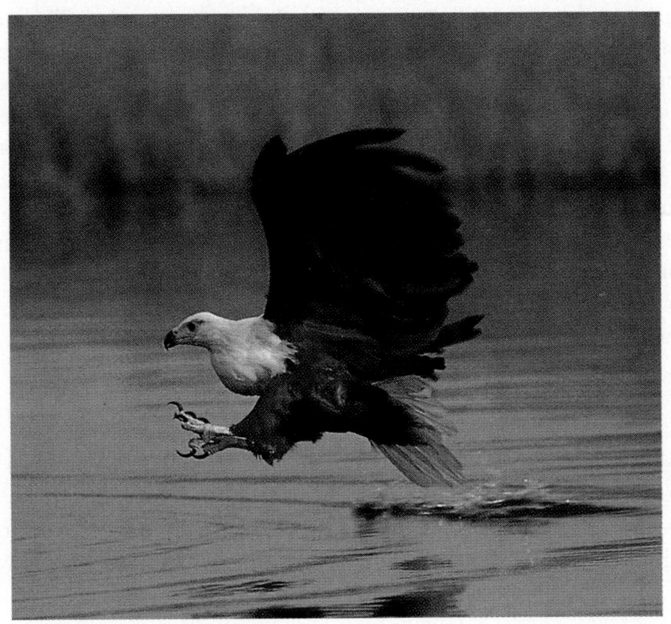

As noted in Chapter 2, virtually all the energy needed for living things comes from the sun. Sunlight is absorbed by chlorophyll molecules in green plants, exciting electrons in those molecules to higher energy levels. The energy released by the electrons as they fall back to lower energy levels initiates a precisely ordered sequence of chemical reactions, in which energy is transferred from molecule to molecule, in effect distributing the original solar energy to the far corners of the biosphere (i.e., the earth and its atmosphere). At various stages along the way the many branching flows of energy are tapped by other chemical reactions to drive all the processes of life.

Examples of such energy-transferring chemical reactions abound in living organisms. In your own muscles, for instance, energy is stored mainly in the molecular structure of glycogen, a highly branched polysaccharide made up of hundreds of glucose subunits. The stored energy in the glucose molecule can be released to power muscular contractions by a series of chemical reactions. Such reactions take place rapidly within living cells, but only very slowly in the test tube—too slowly to support life. Under what conditions within living organisms do these chemical reactions take place? More specifically, what factors influence the rates of chemical reactions? Before we can adequately address such questions, it will be helpful to review briefly some general rules governing the transfer of energy in chemical reactions.

All Energy Conversions Are Subject to the Laws of Thermodynamics

Energy, which is defined as the capacity to do work, comes in many different forms: light, heat, and electricity are all manifestations of energy. Energy can be nuclear, mechanical, or chemical. There are two states in which any form of energy can exist: ***potential*** (stored energy), and ***kinetic*** (energy in motion). For work to be done, the energy involved must be kinetic. For example, a mass of water held behind a dam has potential energy by virtue of its position; when the floodgate is opened, the water flows down a channel, converting its potential energy

Chapter opening photo: Fish eagle swooping down on prey Many birds of prey use gravitational potential energy to power a rapid dive toward the animals they hunt. This eagle is braking from its dive, talons extended to grasp its prey.

into kinetic energy, which is then available to do work—say, by turning the turbine of a hydroelectric generator (Fig. 4.1). Similarly, a lump of coal contains potential energy, which is released mainly in the kinetic form of heat when the coal burns; the heat energy produced by the burning coal can in turn be used to convert water into steam to power a piston engine. As these two examples demonstrate, all forms of energy are at least partly interconvertible.

According to the ***First Law of Thermodynamics*** (also known as the Law of Conservation of Energy), the total amount of energy in the universe is constant. Whenever energy is converted from one form into another, no energy is gained or destroyed. Living organisms draw primarily on chemical energy; they do work by utilizing the energy released by reactions involving complex organic molecules. Viewed on this level, the individual cells of living organisms are miniature ***transducers,*** or energy conversion devices, that turn chemical energy into other forms of energy, or the reverse.

A living organism is a storehouse of potential chemical energy, which can be converted to kinetic energy, as needed, to do work. As this stored energy is used up, however, less and less remains in reserve. A source of usable energy outside the organism must be available to replenish its supply. For many organisms, that outside energy source is other organisms; one living thing obtains new supplies of energy-rich molecules by eating other living things. Since all these energy conversions, in accordance with the First Law of Thermodynamics, are accomplished without reducing the total amount of energy, it might seem at first glance that the same energy could pass continuously from organism to organism and that no source of energy outside the vast system of all living things would be required. This is not the case, however, because energy is constantly being passed from organisms to nonliving matter, as when you throw a rock or move a pencil. All such energy is lost from the life system. Furthermore, the molecules of substances that leave the body retain some energy, and this energy, too, is lost.

But there is another basic reason why energy is constantly lost from the living system. ***The Second Law of Thermodynamics*** states that in the universe as a whole, the amount of energy available to do work is declining with time. This is because with every energy transfer there is a loss of useful energy and an increase in the randomness and disorder of the system. Energy transformations are never 100 percent efficient; some energy is always lost to the surroundings, usually in the form of heat, as when heat from your body warms the air (Fig. 4.2). Heat is the most random and disorganized form of energy; most systems cannot use this form of energy to do work, and it cannot be recycled. The magnitude of this energy loss is considerable, as the need for cooling towers in power plants and radiators in automobiles makes clear. Even in our own bodies waste heat can be a serious problem, and specialized mechanisms (e.g., sweating) have evolved to release excessive amounts of this form of energy to the environment. Because of the Second Law, all things in the universe, living and nonliving, tend to become more and more disordered.

Chemical Reactions Depend on the Flow of Free Energy

With this in mind, let us now consider the following generalized chemical reaction, in which substances A and B, the ***reac-***

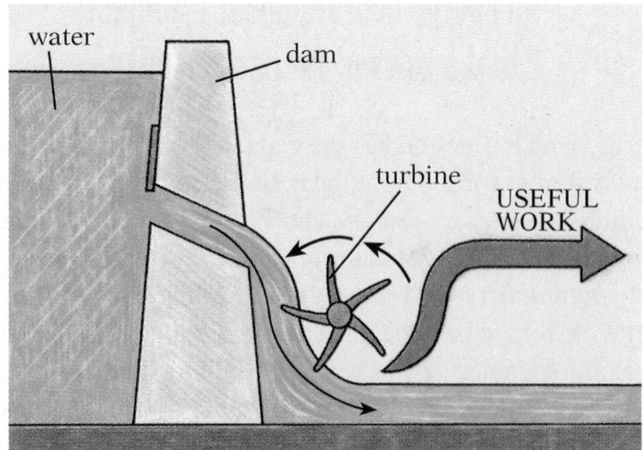

4.1 Kinetic and potential energy Water stored behind a dam has potential energy, but for work to be done this potential energy must be converted into kinetic energy. When the floodgate is opened, the water flows down a channel, converting its potential energy into kinetic energy, which then turns the turbine of a hydroelectric generator.

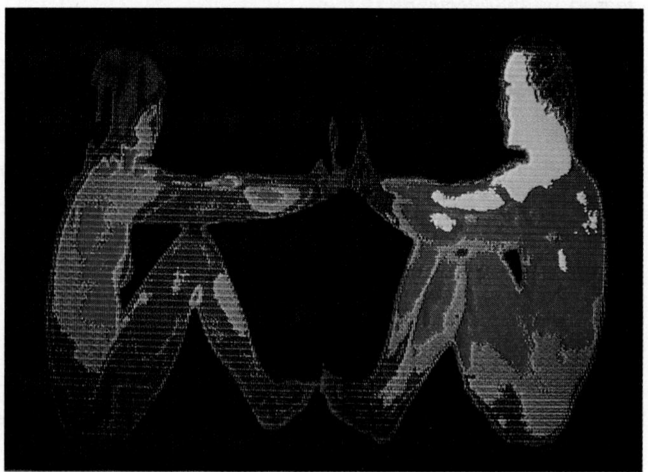

4.2 Thermogram of human male and female As this thermogram shows, not all parts of the human body are at the same temperature. White areas indicate the warmest areas, followed by red, green, purple, blue, and black, with black being the coldest.

tants, react with each other to form two new substances, C and D, the ***products:***

$$A + B \longrightarrow C + D$$

Reactants yield Products

As we said in Chapter 2, breaking a chemical bond always requires energy, whereas the formation of a chemical bond always releases energy. What determines whether this reaction proceeds spontaneously in the direction indicated by the arrow? The answer to this question turns on the concept of ***free energy,*** a term that is taken by chemists and biologists to mean the energy in a system available for doing work under conditions of constant temperature and pressure. Wherever stored energy is available to be tapped, whether it be in the mass of water held behind a dam, in an electron that has been excited into a higher energy level by solar radiation, or in the reactions involving the breakdown of molecules such as glucose, the potential for work is present in the form of free energy.

A universal law of biochemistry states that the course of any reaction depends on whether the breaking and formation of chemical bonds that occur in a chemical reaction result in a net gain or loss of free energy. If the reaction results in a net loss of free energy, that is, more energy is released in the formation of new bonds in the products than is required to break the bonds in the reactants, the reaction runs "downhill" and hence proceeds spontaneously. Since the energy liberated is usually released as heat, reactions of this type are said to be ***exothermic*** (heat-releasing) or, more generally, ***exergonic*** (energy-releasing).

If, on the other hand, the reaction requires a net input of external energy, it cannot proceed spontaneously from reactants to products. An "uphill" reaction of this kind, in which more energy is required to break the bonds of the reactants than is released in the formation of new bonds in the products, is said to be ***endothermic*** (heat-consuming), or ***endergonic*** (energy-consuming). Such reactions require the addition of free energy from an external source. The two alternatives are illustrated schematically in Figure 4.3.

The reaction of A and B to form C and D proceeds spontaneously only if it is exergonic, that is, only if it proceeds energetically downhill, releasing free energy. However, reactions within the cell are not exclusively exergonic. If only exergonic reactions are spontaneous, then, how do essential endergonic reactions occur within organisms? Such reactions as the joining of amino acids to form proteins or the linking of simple sugars into complex polysaccharides, which require a substan-

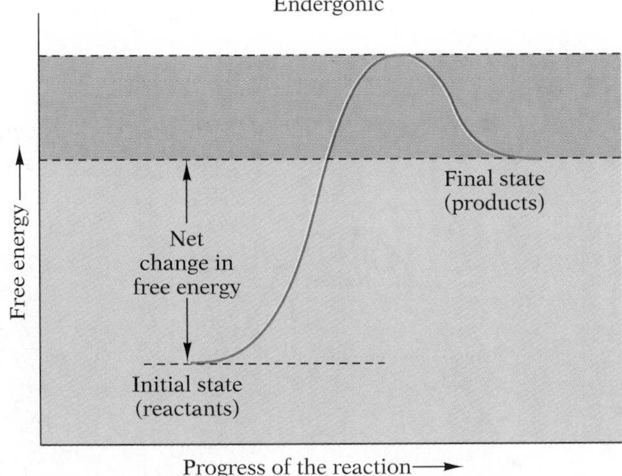

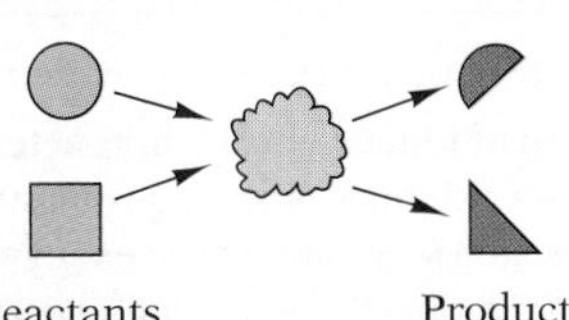

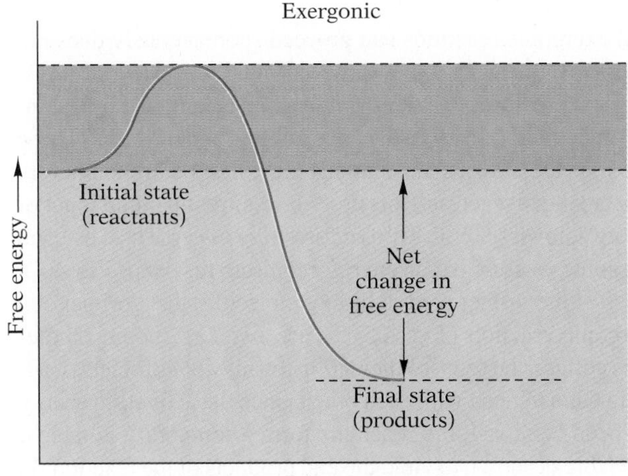

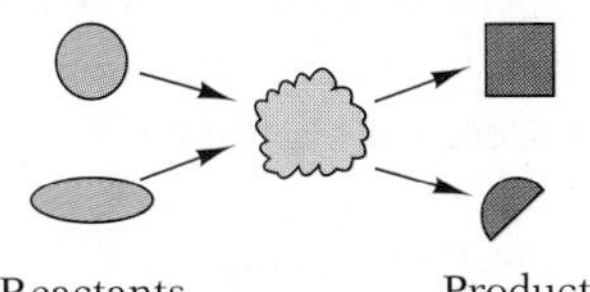

4.3 Endergonic versus exergonic reactions Some reactions require a net input of free energy to proceed; others have a net release of free energy during the course of the reaction. Those reactions whose products have more energy than the reactants require an input of energy to go "uphill" from the initial state to the final state (left). These energy-requiring reactions are said to be endergonic. Those reactions whose products have less energy than their reactants go "downhill" and release energy (right), and are said to be exergonic. Even exergonic reactions require a small amount of energy to get started.

tial input of energy, are possible because they are closely linked with exergonic reactions. In such a ***coupled reaction,*** some of the energy released by the exergonic reaction of the pair is used to drive the endergonic reaction:

(1) A + B ⟶ C + D + energy	Exergonic reaction
Coupled reactions	
(2) D + energy ⟶ E	Endergonic reaction

Taking the two reactions together, the overall reaction is A + B ⟶ C + E, the direction of which is energetically downhill—i.e., exergonic. In coupled reactions, the energy liberated in one reaction is only partially stored in the final products; the remainder, in accordance with the Second Law of Thermodynamics, is dissipated as heat. The net result is that the final products, C and E, have less free energy than the reactants, A and B. Such reactions are very common in living systems, where virtually every endergonic reaction is driven by a coupled exergonic reaction.

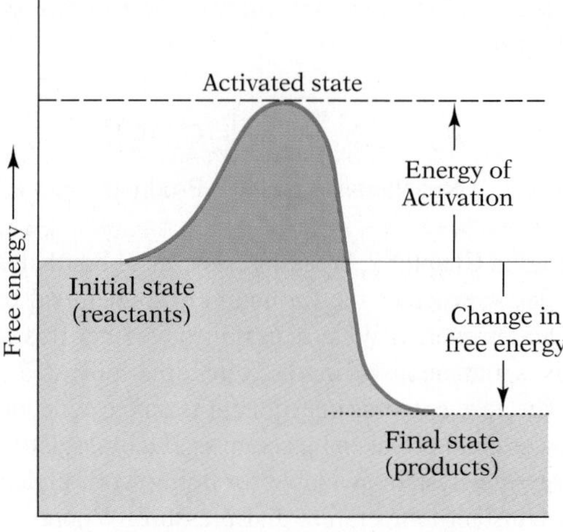

4.4 The energy changes in an exergonic reaction Though the reactants are at a higher energy level than the products, the reaction cannot begin until the reactants have been raised from their initial energy state to an activated state by the addition of activation energy. It is the need for activation energy that ordinarily prevents high energy substances from breaking down, and hence makes them stable; the higher the activation-energy barrier, the slower the reaction and hence the more stable the substance. When activation energy is available, the reactants form a temporary and unstable activated complex, which breaks down to yield the end products of the reaction; in the process both activation energy and free energy are released.

Most Physiological Reactions Require Activation Energy

That exergonic reactions can proceed spontaneously does not necessarily mean that they proceed rapidly. Indeed, at physiological pressures and temperatures most exergonic biochemical reactions proceed at exceedingly slow rates. These reactions are slow because bonds must first be broken before molecules can react, and this process requires an initial input of energy, known specifically as ***activation energy.*** Thus, even an exergonic reaction has an energy-requiring first step (Fig. 4.4). It is the kinetic energy of colliding molecules that "primes" an exergonic reaction (Fig. 4.5). When two fast-moving molecules collide, the force of their collision may be sufficient to activate them to react with each other. Once the activation energy has been achieved, the reactants form a temporary complex, which breaks down to yield the end products of the reaction. In the process, both the activation energy and free energy are released (Fig. 4.4).

Because of the strength of covalent bonds, a substantial amount of energy may be necessary to achieve activation. For example, consider the following exergonic reaction:

$$2\,H_2 + O_2 \longrightarrow 2\,H_2O + \text{energy}$$

Even though this combination of oxygen and hydrogen can release a considerable amount of energy (it happens to provide much of the energy that pushes the space shuttle into orbit!),

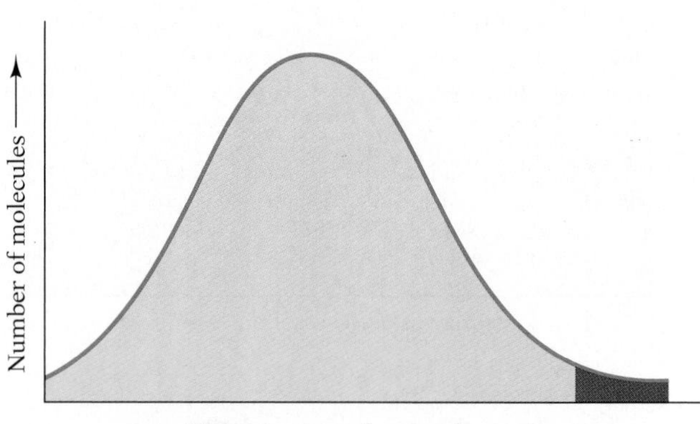

4.5 Distribution of kinetic energy in reactant molecules Reactant molecules exhibit a wide range of kinetic or thermal energy. Only a minute fraction (dark area) have enough energy to react with one another and overcome the activation-energy barrier. When molecules are heated they gain kinetic, or thermal energy, and move faster. The higher the temperature, the more thermal energy the molecules have, causing the molecules to collide more frequently and with enough force to attain activation-energy levels.

the two reactants can coexist as a stable mixture almost indefinitely. It takes only a single spark, however, to provide the necessary activation energy to trigger an explosive reaction.

The same relative stability is characteristic of most reactants in living systems: the energy needed to bring the reactants together and to break their covalent bonds is far greater than the energy of all but the most rapidly moving molecules in a solution. The activation-energy barrier prevents most reactions from taking place at a significant rate (Fig. 4.4). Indeed, without such a barrier, the complex high-energy molecules on which life depends would be unstable and would spontaneously break down into useless forms.

Beneficial as activation-energy barriers may be in conferring stability on living systems, such barriers must be overcome if organisms are to carry out the chemistry of life. How can reactions that are normally very slow take place with the extreme speed often necessary for the organism to respond promptly to change? To return to our earlier example, how can the breakdown of a complex molecule such as glucose be made to proceed rapidly enough to supply the energy needed for vigorous muscular activity? To answer this question, let us now examine several factors that can affect the rates of chemical reactions.

Heating Can Overcome Activation-Energy Barriers

In the laboratory, a standard way of speeding up chemical reactions—of supplying activation energy—is to apply heat. The higher the temperature, the more kinetic energy molecules have and the more likely it is that they will collide more frequently and with enough force to reach the activation-energy level (Fig. 4.5). That is why the application of heat is so effective in speeding up chemical reactions.

Indeed, this procedure is common in everyday life, as, for example, when a match or some other hot object is applied to a piece of paper (or wood or gasoline) to make it burn. Once ignited, the material continues to burn, because some of the exergonically released heat energy of the first molecules to react supplies the activation energy that causes other molecules to also react, releasing additional heat that activates still other molecules, and so on.

Heating is a fine way to accelerate reactions in the chemist's test tube, but it clearly cannot be the principle method of speeding up reactions inside living organisms, since cells literally "cook" at temperatures much above 40° C. The chemistry of life is relatively "cold" chemistry, and we must search further for its way of surmounting the activation-energy barrier.

Increasing the Concentration of the Reactants Can Accelerate the Reaction Rate

Anything that increases the probability of collisions between molecules increases the rate of a reaction. An obvious way to increase the probability of collisions is to increase the concentrations of the reactants: the more molecules per unit volume, the more collisions between them per unit time. In other words, when all other conditions are constant, the rate of a reaction is proportional to the concentrations of the reactants.

Although changes in the concentrations of reactants can play an important role in determining the rates at which reactions occur in living organisms, they cannot explain the extreme rapidity of most biochemical reactions, whose rates far exceed anything that could be produced merely by increasing concentration. Nor, as we have indicated, can heat be used as a way of overcoming the activation-energy barrier, except to a very limited extent. If biochemical reactions are to take place with the rapidity necessary to sustain life, cells must use some other method, one that not only lowers the barrier, but does so selectively, so that some reactions run, while others do not. Cellular chemistry, then, is essentially controlled by the selective lowering of particular activation-energy barriers. How is this crucial task managed?

MAKING CONNECTIONS

1. Trace the flow of energy in living organisms.

2. How do living organisms replenish the energy lost according to the Second Law of Thermodynamics?

3. Explain how coupled reactions provide the energy necessary to synthesize proteins and complex carbohydrates.

4. How does activation energy give stability to living systems?

Catalysts Speed Up Chemical Reactions

As previously stated, a simple mixture of hydrogen and oxygen gases does not react spontaneously, but if one provides the initial activation energy (a spark), the mixture will explode and water will be formed. The same explosion will take place if one adds instead a small quantity of finely divided platinum. At the end of the reaction, the platinum will remain unchanged (Fig. 4.6).

A substance of this type—that is, one that speeds up a reaction but is itself unchanged when the reaction is over—is known as a ***catalyst.*** A catalyst affects only the rate of the reaction: it merely speeds up reactions that are energetically possible to begin with. A catalyst does not alter the direction of a reaction or the amount of energy involved in the reaction, but it does lower the activation-energy barrier that must be overcome for reactions to occur, thereby increasing the proportion of reactants energetic enough to react (Fig. 4.7). Catalysts bind reactants in an intermediate state that properly orients them with respect to each other and puts extra stress on crucial internal bonds. As a consequence, less energy is needed to get the reaction started, making the conditions for the reaction more favorable.

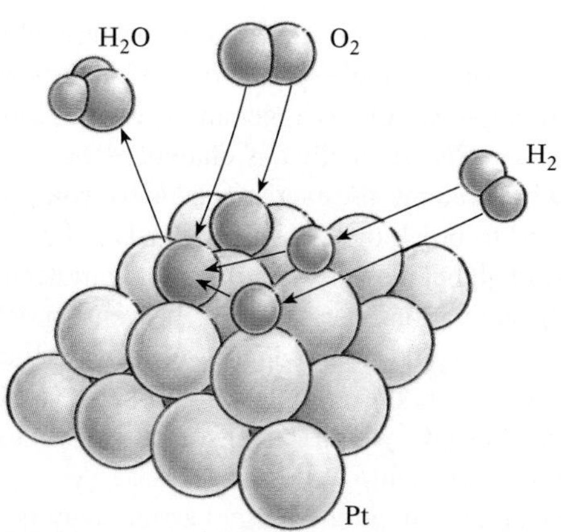

4.6 Action of platinum as a catalyst Because of its loosely bound outer electron, platinum (Pt) is able to form weak temporary bonds with molecules of both hydrogen (right) and oxygen (top). This binding draws the hydrogen and oxygen electrons away from their covalent positions, thus weakening the bonds within their respective molecules. In addition, the spacing of the platinum atoms tends to align the hydrogen and oxygen atoms in such a way that new bonds between hydrogen and oxygen can be more easily formed. Platinum is a catalyst in that it facilitates the reaction without being itself altered.

Enzymes Are the Catalysts of Life

An inorganic catalyst such as platinum is relatively unselective about the reactants it "helps." Living systems, in contrast, have an enormous variety of specialized organic catalysts, called ***enzymes,*** most of them globular proteins of the type described in Chapter 3. Like all catalysts, enzymes reduce the amount of activation energy necessary to start a reaction (Fig 4.7), but cannot induce a reaction that could not take place on its own. However, unlike a typical inorganic catalyst, such as platinum, enzymes are highly specific. Most interact with only one type of reactant or pair of reactants, customarily called ***substrates,*** or occasionally with a few unrelated ones. Because of their specificity, enzymes can steer certain substrates into particular reaction pathways and block them from others, thus guiding the chemistry of life with great precision. For example phosphofructokinase is a key enzyme in a series of reactions in which glucose is broken down to yield energy for running the machinery of the cell. When this enzyme is active, glucose is readily degraded, but when the enzyme is inhibited, this pathway is blocked and instead glucose enters another series of reactions in which glycogen is synthesized. Consequently, the activity of this enzyme determines which reaction pathway glucose will follow—breakdown or synthesis of glycogen.

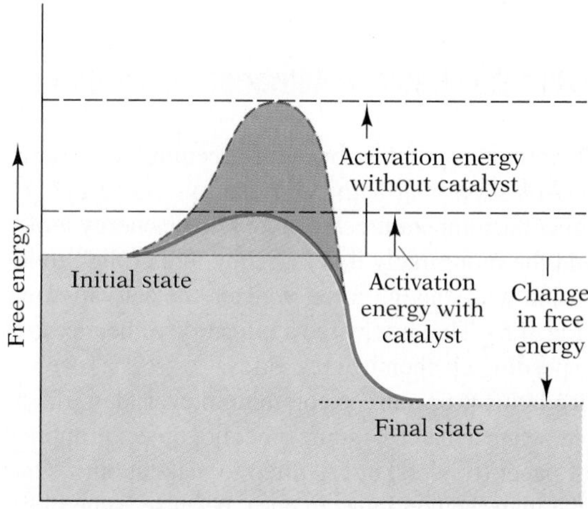

4.7 Reduction of necessary activation energy by catalysts The activation energy necessary to initiate the reaction is much less in the presence of a catalyst than in its absence. It is this lowering of the activation-energy barrier by enzyme catalysts that makes possible most of the chemical reactions of life. Note that the amount of free energy liberated by the reaction is unchanged by the catalyst—it is the same for both the catalyzed and the uncatalyzed reaction—and only the activation energy is changed.

Enzymes are also highly efficient catalysts; usually enzyme-catalyzed reactions occur in milliseconds, and some take place as much as a million times faster than the same uncatalyzed reactions. The enzyme catalase is one of the fastest enzymes known. This enzyme catalyzes the destruction of hydrogen peroxide (H_2O_2), a potentially harmful substance formed in certain reactions in the cell, at the rate of more than a million molecules per second! Most enzymes are not that efficient, generally working in the range of one to 1,000 substrate molecules per second (Table 4.1). And since enzymes are not permanently altered during a reaction, the same enzyme can be used over and over. Some enzymes are incredibly efficient.

TABLE 4.1 *Rates of enzyme catalysis for some enzymes*

Enzyme	*Substrate*	*Rate in molecules/sec*[a]
Carbonic anhydrase	CO_2, HCO_3^-	600,000
Acetylcholinesterase	Acetylcholine	25,000
Penicillinase	Penicillin	2,000
Chymotrypsin	Protein	100

[a]The number of substrate molecules converted into products per second by the enzyme when the substrate is present in excess.

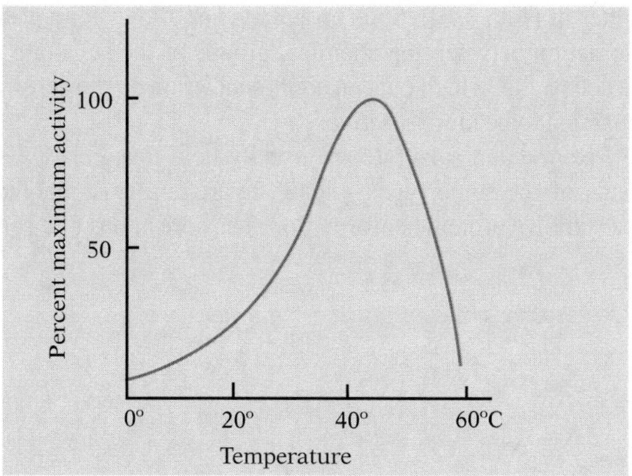

4.8 Enzyme activity as a function of temperature The activity of an enzyme rises steadily with temperature (approximately doubling for each 10°C increase) until about 45°C, when the weak bonds holding the enzyme in its configuration begin to break, and the enzyme starts to lose its activity. The enzyme becomes completely denatured and inactivated at temperatures above 60°C because its three-dimensional conformation is severely disrupted. This enzyme has a temperature optimum (the temperature at which it functions most efficiently) of about 42°C.

The key to an enzyme's function appears to be its surface activity, which is determined by its three-dimensional shape. Enzymes generally have distinctive shapes and interact only with those substrates whose molecular shapes "fit" a specific binding site on that enzyme's surface. This site, known as the ***active site,*** is typically a depression or groove in the surface of the enzyme into which the substrate fits. Thus, the specificity of enzymes can be viewed as a function of their molecular conformation. In short, their form determines their function, or how they work.

The conclusion that the action of enzymes depends on their three-dimensional shape is consistent with the observation that when proteins are heated to high temperatures or subjected to high or low pHs, they become denatured and lose their characteristic biological activity (Figs. 4.8 and 4.9). Such treatment breaks many of the weak bonds that help stabilize the conformation of enzymes, and when their three-dimensional shape is disrupted, their enzymatic properties vanish.

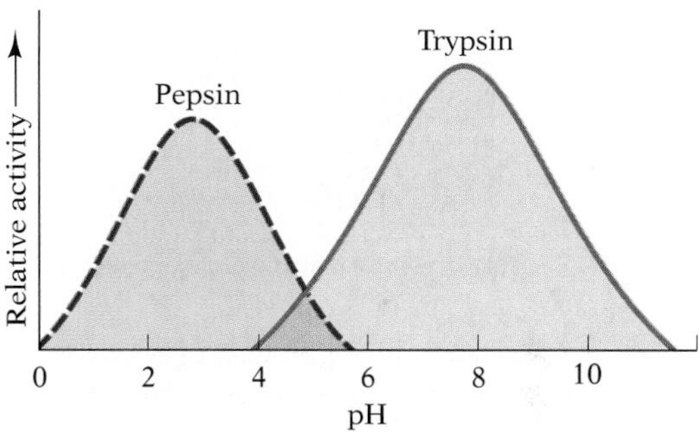

4.9 Enzyme activity as a function of pH Most enzymes are very sensitive to pH and become denatured when subjected to pHs much higher or lower than their pH optima (i.e., the pH at which their activity is the highest). However, these optima differ for different enzymes. For example, pepsin and trypsin are both enzymes that digest protein, but the pH ranges within which they are active overlap only slightly. Pepsin is most active under strongly acidic conditions, trypsin under neutral or slightly alkaline conditions.

An Enzyme's Catalytic Action Requires the Formation of an Enzyme-Substrate Complex

Enzyme and substrate have traditionally been visualized as fitting together as a lock and key, or like the pieces of a puzzle. The two must be roughly complementary if they are to combine temporarily. At the very least, the reactive portion of the substrate molecule and the active site of the enzyme must fit together intimately enough to bond temporarily, forming a transitory entity called the ***enzyme-substrate complex.*** Once the enzyme-substrate (E-S) complex has formed, the reaction can take place and the products leave the enzyme:

$$\text{Enzyme} + \text{Substrate} \longrightarrow \text{E-S Complex} \longrightarrow \text{Enzyme} + \text{Products}$$

Although enzymes, like all catalysts, resume their original conformations when a reaction is over, often they do undergo changes in shape while bonding with a substrate, making for a tighter fit (Fig. 4.10). Such an ***induced fit*** allows for a greater interaction between the chemical groups of the substrate and the active site, which puts an additional strain on the substrate, causing it to be more reactive.

Enzyme and substrate are usually held together by weak bonds of the same types—ionic, hydrogen, hydrophobic—that stabilize protein conformation. They are bonds that can be made and broken rapidly. The type of substrate to which a given enzyme molecule can become bonded depends on the amino acids making up its active site—specifically, on the exposed R groups of these amino acids. The active site of most enzymes is a cleft or groove into which the reactive portion of the substrate must fit. If some of the exposed R groups in this cleft are electrically charged, then the reactive portion of the substrate must be complementarily charged, or polar. A hydrophobic substrate could not react with the active site of such an enzyme, even if it could, by chance, fit into the cleft. Therefore, spatial fit, by itself, does not fully explain enzyme-substrate interactions. The enzyme-substrate complex can form only if the enzyme and the substrate are chemically compatible and capable of bonding with each other. In other words, the concept of enzyme-substrate fit includes not only spatial fit but also chemical fit.

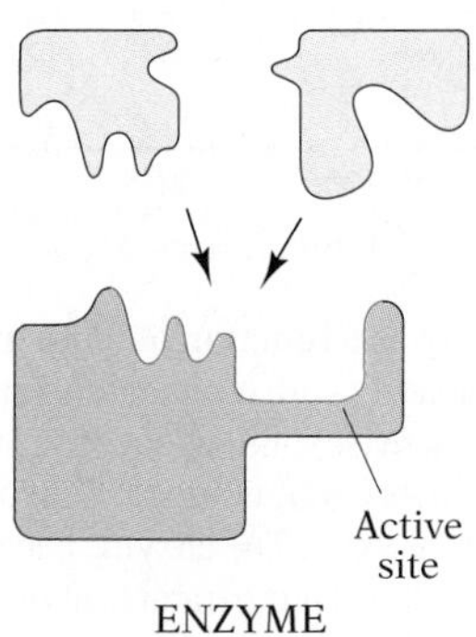

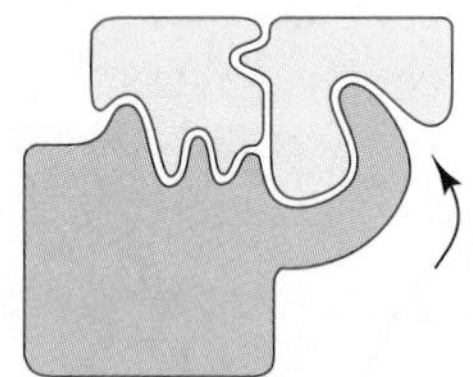

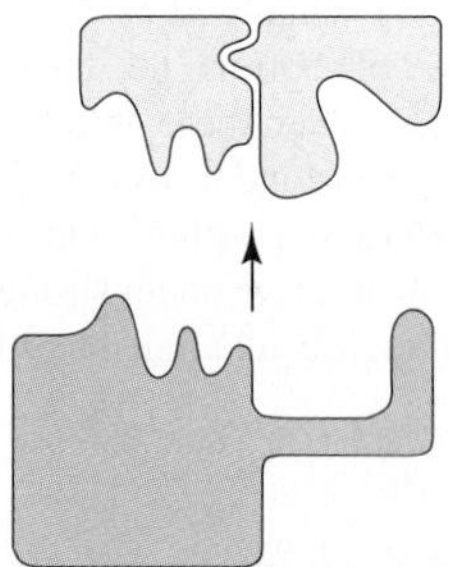

4.10 Induced-fit model of enzyme-substrate interaction The enzyme molecule has an active site onto which the substrate molecules can fit (top), forming an enzyme-substrate complex (middle). The binding of the substrate induces conformational changes in the enzyme that maximize the fit and force the complex into a more reactive state. The enzyme molecule reverts to its original conformation when the product is released (bottom).

Figure 4.11 shows a model of the active site of the digestive enzyme carboxypeptidase. Note that the critical amino acids in the active site are not adjacent to each other in the polypeptide chain. The complex folding of the protein—its tertiary structure—has brought amino acids from several regions of the protein into close proximity to form the active site. The inclusion of nonadjacent amino acids in the active site is the usual pattern—active sites nearly always include some nonadjacent amino acids. This explains why high temperatures or changes in pH, which alter the weak bonds that maintain an enzyme's three-dimensional shape, reduce enzyme activity: any change in the folding pattern of the polypeptide chains is likely to alter the critically important arrangement of amino acids in the active site. Accordingly, most enzymes have a ***temperature optimum*** and a ***pH optimum*** at which they function most efficiently (Figs. 4.8 and 4.9).

Many enzymes require the presence of other "helper" groups for activity. Some groups bind permanently and may be functional parts of the active site itself (Fig. 4.11). Other substances, known as ***cofactors,*** bond only briefly to enzymes during catalysis, momentarily altering the geometry of the active site. Cofactors may be metallic ions, such as magnesium or zinc, or they may be small, nonprotein organic molecules, called ***coenzymes.*** Like enzymes, coenzymes are not permanently altered by the reactions in which they participate, and they can be used repeatedly. Although only very small amounts of such substances are needed, they are critical to life. If the supply of coenzymes falls below the normal level, the health or even the life of the organism may be endangered. That is why the molecules known as B complex vitamins are necessary in the diet—they function as coenzymes.

Enzymes Promote Catalysis by Different Mechanisms

How exactly does the temporary, unstable complex formed by an enzyme and its substrate reduce the amount of activation en-

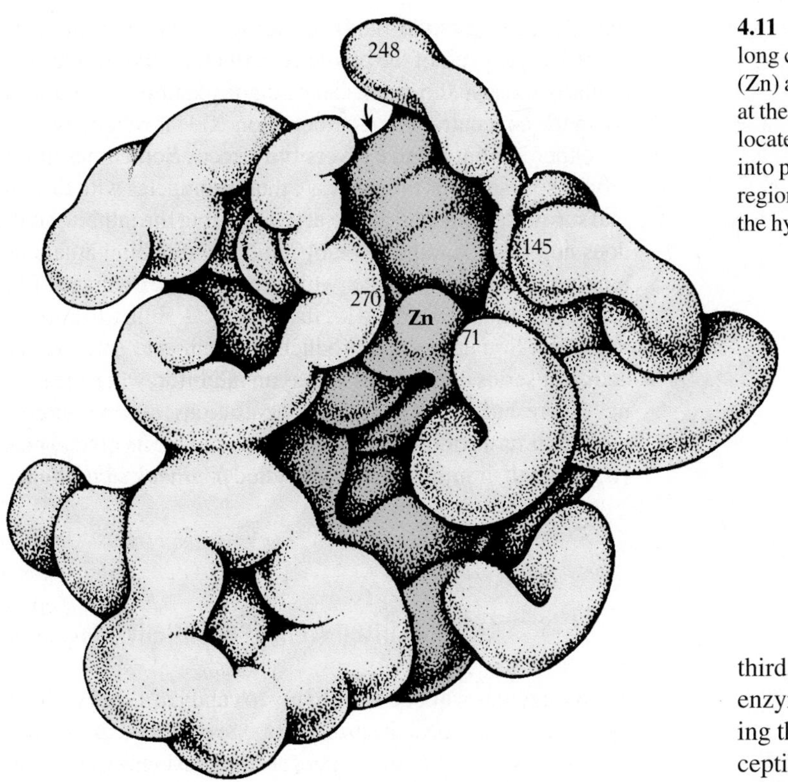

4.11 Location of the active site in an enzyme The folding of the long chain of amino acids of carboxypeptidase brings together a zinc (Zn) atom and three of the four amino acids that bind to the substrate at the active site, even though—as their numbers indicate—they are located at different places in the chain. The fifth part of the site folds into place (arrow) when the substrate binds to the enzyme. The region in dark color is the active site, while the area in light color is the hydrophobic entrance to the cleft.

ergy needed for a reaction? One possible explanation is that a substrate's effective concentration is greatly increased within the active site of the enzyme that binds with it, thereby increasing the probability that the reaction will take place. Second, it is thought that when substrate molecules are bound to the enzyme, the molecules are oriented in such a way that their chemical groups are optimally positioned to react (Fig. 4.12). And third, it is thought that the weak bonds that form between the enzyme and substrate place a strain on the substrate by disturbing the distribution of electrons within its molecules. The susceptible bonds in the substrate molecules may be weakened as a result, becoming more easily broken, meaning less energy is necessary for the reaction to occur. In other words, enzymes perhaps not only orient substrate molecules so that their reactive groups are properly aligned, but may also enhance the reactivity of those groups.

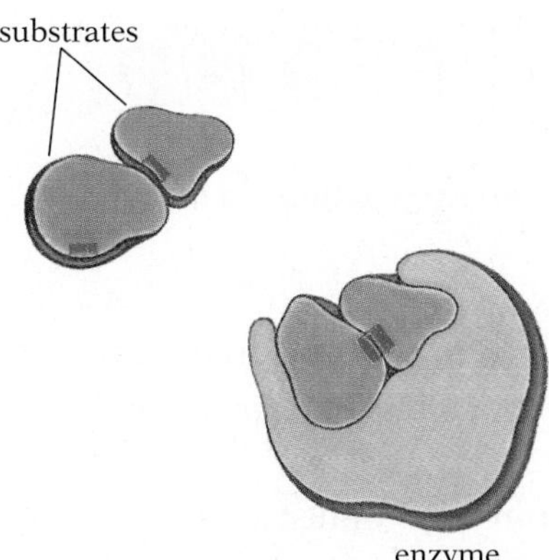

4.12 Orientation of substrate for greatest activity When substrate molecules collide (left), their orientation often fails to bring together their reactive groups (red). When the substrate molecules are bound to an enzyme, they are oriented for maximal interaction of their reactive groups.

Inhibitors Are Substances That Control Enzyme Activity

Since enzymes control all the chemical reactions within living organisms, it is not surprising that a variety of mechanisms have evolved to control the activity of the enzymes themselves. These mechanisms depend not only on physical factors such as temperature, pH, and substrate or enzyme concentration, but also on chemical agents, which may mask, block, or alter the active sites of the enzymes they regulate.

Any substance that blocks or changes the shape of the active site and thereby interferes with the activity or efficiency of the enzyme is called an ***inhibitor*** (Fig. 4.13). The effects of some inhibitors are irreversible; substances of this type bind very tightly to chemical groups in the active sight and prevent the substrate from binding. For example, certain pesticides and nerve gases, classified generically as organophosphates, work this way. They interact with specific amino acids in the active sites of a particular enzyme in the nervous system, effectively blocking those sites. Carbon monoxide poisoning is another example of inhibition. Both oxygen and carbon monoxide bind to the same active sites in hemoglobin, a protein in the blood of vertebrates that normally carries oxygen to the body's cells,

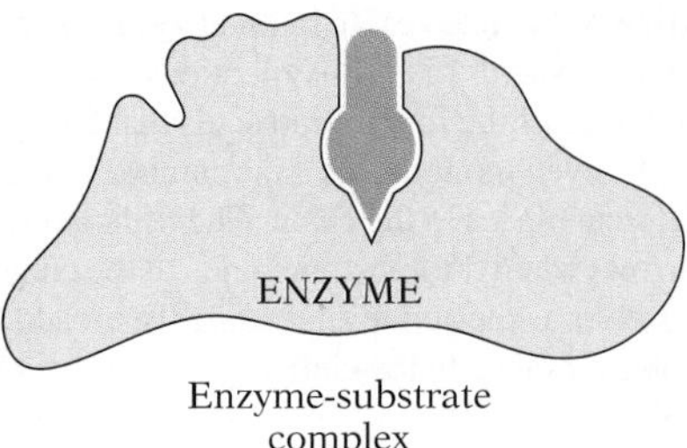

Enzyme-substrate complex

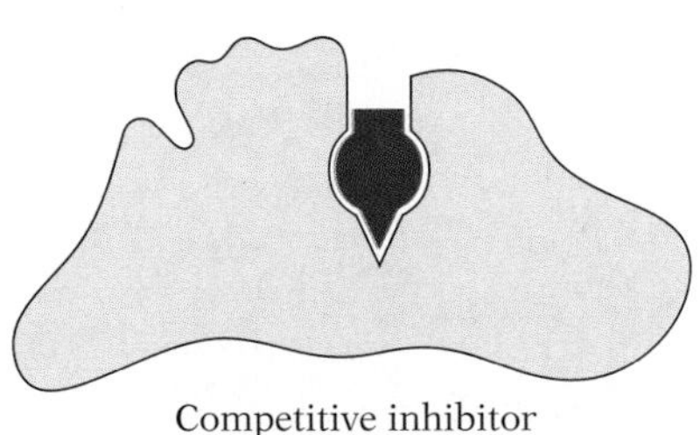
Competitive inhibitor bound to enzyme

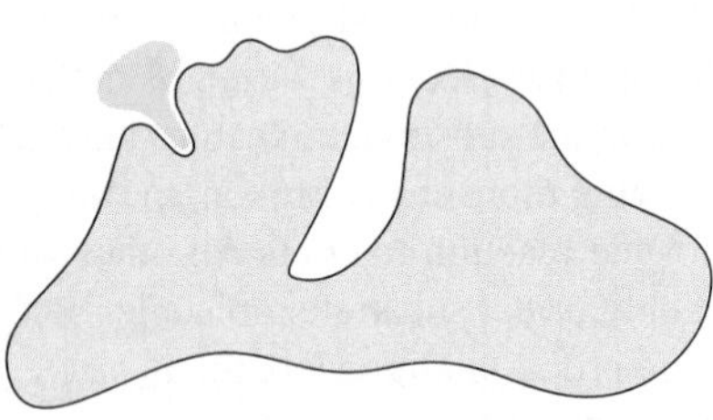
Noncompetitive inhibitor bound to enzyme

4.13 Inhibition of an enzyme Top: The substrate is bound to the active site of the enzyme. Middle: The binding of an inhibitor molecule to the active site prevents the substrate from binding. Bottom: Another type of inhibitor binds to a different site on the enzyme and induces change in the shape of the enzyme that prevents the active site from binding the substrate and catalyzing reactions.

but carbon monoxide binds so strongly that oxygen is excluded. As a result of the oxygen deprivation, living cells, particularly those in the brain, can be damaged or destroyed, even when the concentration of carbon monoxide is relatively low.

Other inhibitors have reversible effects. Sometimes the inhibitor so resembles the substrate that it competes with the normal substrate for access to the active site, but the inhibitor itself does not react. Another type of inhibitor may bind at another binding site on the enzyme, which then alters the shape of the active site and interferes with the enzyme's ability to catalyze reactions. Sometimes the product of an enzyme-catalyzed reaction or series of reactions acts as an inhibitor: when the product is present in such high concentrations that no more is needed, it turns off the process responsible for its own synthesis. This self-limiting strategy is called ***feedback inhibition.***

Enzymes Often Work Sequentially

Many enzymes work consecutively to catalyze a series of reactions. In such a series, the product of one enzyme-catalyzed reaction becomes the substrate for the next enzyme in the sequence, and so on:

$$A \xrightarrow{\text{enzyme 1}} B \xrightarrow{\text{enzyme 2}} C \xrightarrow{\text{enzyme 3}} D$$

Such a chain of reactions, called an ***enzymatic pathway,*** is common in cells. Indeed, it is precisely a pathway of this type that is involved in the breakdown of glucose in animal cells, as we shall see in greater detail in Chapter 8. In this case, as in many others, most of the reactions in the pathway are coupled reactions in which the energy released from an exergonic reaction provides the energy necessary to drive the associated endergonic reaction.

MAKING CONNECTIONS

1. Why would heating not be a good way to speed up reactions in living systems?

2. Explain the role of enzymes as the catalysts of life.

3. That enzymes no longer function when heated to only 60°C demonstrates the importance of three-dimensional shape. What alterations in an enzyme's three-dimensional shape affect its activity?

4. Why are small quantities of various vitamins and minerals necessary in our diet for the proper functioning of enzymes?

5. Describe how feedback inhibition controls the rate of many enzyme-catalyzed reactions?

CONCEPTS IN BRIEF

1. The course of a chemical reaction depends on whether the *energy* required to break the covalent bonds in the *reactants* is greater or less than that released in the formation of covalent bonds in the *products.* Reactions that proceed spontaneously, releasing *free energy,* are said to be *exergonic;* reactions that require a net input of free energy to proceed are *endergonic.* In living systems an exergonic reaction is usually coupled with an endergonic reaction; energy released by the exergonic reaction is used to drive the endergonic reaction.

2. Although exergonic reactions proceed spontaneously, initiating such a reaction still requires *activation energy.*

3. Chemical reactions can be sped up by heat, by increasing the concentrations of the reactants, or by providing an appropriate *catalyst.* In living systems the catalysts are *enzymes;* like all other catalysts, they speed up reactions by lowering the activation-energy barrier.

4. Enzymes are large globular proteins that catalyze the thousands of chemical reactions in living organisms. Because the amino acid sequence of each enzyme is different, every enzyme has a unique three-dimensional structure, which confers great specificity. Most enzymes can interact only with those *substrates* that fit spatially and chemically into the *active site.*

5. When an enzyme reacts with its substrate, a short-lived *enzyme-substrate complex* is formed. This complex facilitates the reaction both by properly aligning the substrate's reactive groups and by enhancing the reactivity of those groups. The substrate is converted into product and leaves the active site.

6. Since the formation of the enzyme-substrate complex requires the enzyme and its substrate to be complementary, anything that alters the shape of the enzyme will alter its activity and efficiency. Heat and changes in pH break the weak bonds stabilizing an enzyme's conformation and alter its shape. If changes in the enzyme are large enough, the enzyme becomes *denatured* and can no longer function.

7. Some *inhibitors* bind irreversibly to the active site of an enzyme and prevent the substrate from binding. Other inhibitors resemble the substrate molecules and compete for the active site. Still others bind to a second site on the enzyme and induce a change in the shape of the active site so the substrate can no longer bind.

8. Enzymes often work consecutively to catalyze a chain of reactions. In such a chain, the product of one enzyme-catalyzed reaction becomes the substrate for the next enzyme in the sequence, and so on.

STUDY QUESTIONS

1. How is energy defined? What are its two fundamental states? What does the First Law of Thermodynamics say about the total quantity of energy in the universe? (pp. 77–78)
2. During a chemical reaction, chemical bonds are broken and formed. Which process requires an input of energy, and which process releases it? (p. 79)
3. Define *free energy.* According to the Second Law of Thermodynamics, what is happening to the amount of free energy in the universe? (p. 78)
4. In terms of free energy, what happens in an exergonic reaction? What happens in an endergonic reaction? Which type occurs spontaneously? Draw graphs of both kinds of reactions. (p. 79)
5. What are *coupled reactions?* What is their significance for the chemistry of cells? (p. 80)
6. Why must activation energy be supplied to initiate even so-called spontaneous reactions? Contrast the roles of heat and catalysts in initiating or speeding up chemical reactions. (pp. 80–82)
7. How does an enzyme catalyze a chemical reaction? Define the terms *substrate* and *active site.* (pp. 82–83)
8. Use the *induced-fit* concept of enzyme function to explain how an enzyme may destabilize a substrate and make it more reactive. (p. 84)
9. Why are enzymes typically most effective as catalysts over a rather narrow range of temperatures and pHs? (p. 84)

10. What is the role of *cofactors* in enzyme function? (p. 84)

11. Describe three specific mechanisms by which an enzyme may reduce the activation energy required for a chemical reaction. (pp. 84–85)

12. What are enzyme inhibitors? Describe two ways in which they may act. (pp. 85–86)

13. How are products and substrates related in an enzymatic pathway? Draw a diagram to illustrate this concept. (p. 86)

CHAPTER 5

AT THE BOUNDARY OF THE CELL

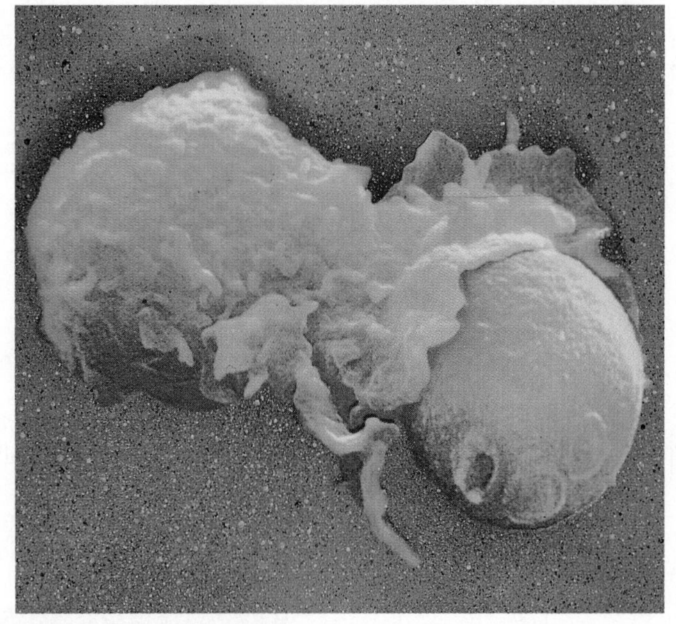

The molecules of life, and the chemical reactions that they undergo, are not randomly dispersed in living organisms. On the contrary, they are elaborately organized and compartmentalized. The fundamental unit of organization is the cell, but even at the subcellular level the processes of life are precisely relegated to an assortment of highly specialized structures. One of the most important of these structures is the cell's ***plasma membrane,*** or flexible outer boundary. This very thin but remarkably versatile envelope separates the delicate internal chemistry of the cell from the vagaries of the external environment, holding some substances in, keeping others out, and passing still others through in one direction or the other. Just as the cell membrane protects a favorable internal climate from the world outside, so the various structures inside the cell keep themselves separate by means of their own membranes. In this way, the cell's specialized chemical processes are effectively partitioned off, so that, for example, food is digested by enzymes in certain compartments whose membranes prevent the enzymes from digesting the rest of the cell.

In this chapter we examine the structure and function of the cell membrane and learn how it works, both actively and passively, to make life possible. In later chapters we shall see how the system of subunits inside the cell uses the same principles to direct cellular chemistry, to extract energy from food, to build new compounds and structures, and, in multicellular organisms, to orchestrate each cell's specialized role in a kind of society of cells.

Compartmentalization and Self-Reproduction Form the Foundation of the Cell Theory

The idea that all living things are composed of cells—the ***cell theory***—is commonly credited to two German investigators, the botanist Matthias Jakob Schleiden and the zoologist Theodor Schwann, who published their conclusions in 1838 and 1839 respectively. Although this idea, one of the most important generalizations of modern biology, was first proposed almost two centuries earlier by Robert Hooke, Schleiden and Schwann stated the principle with particular clarity and thus helped it to gain general acceptance.

Chapter opening photo: Scanning electron micrograph of a phagocytic white blood cell (lymphocyte) surrounding and engulfing a yeast cell (yellow)

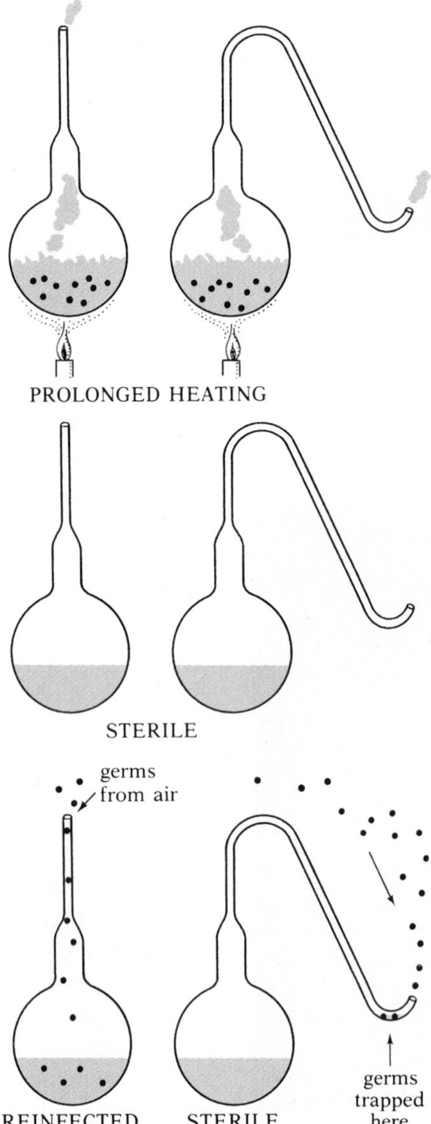

5.1 Pasteur's experiment Nutrient broths in two kinds of flasks, one with a straight neck, the other with a bent neck, were boiled to kill any germs they might contain (top). The sterile broths were then allowed to sit in their open-mouthed containers for several weeks (middle). Microorganisms entering the straight-necked flask contaminated the broth, but those entering the bent neck of the other flask were trapped in films of moisture in the curves of the neck and did not contaminate the broth (bottom). Thus, Pasteur's experiment disproved spontaneous generation by showing that the sugar solution was capable of supporting growth, but could not generate microorganisms; they had to be introduced into the solution. The microorganisms could only gain access to the straight-necked flasks.

An important extension of the cell theory, proposed in 1858 by the German physician Rudolf Virchow, was that all living cells arise from preexisting cells, that there is no spontaneous creation of cells from nonliving matter. The theory of ***biogenesis,*** life from life, contradicted the belief in spontaneous generation, then widely held not only by the general public but also by scientists. In 1862, Louis Pasteur of France lent additional support to Virchow's theory in a series of classic experiments. Pasteur showed that the microorganisms that cause wine or milk in an open container to go bad are spread by the air—they do not arise spontaneously from the nutrient media (Fig 5.1).

The theory of biogenesis has undergone some modification in recent years, as we shall see in a later chapter. Current theory maintains that spontaneous generation of life from nonliving matter, while not occurring under present conditions, probably did occur under the conditions existing on earth when life first arose.

The two chief components of the cell theory—that all living things are composed of cells and that all cells arise from other cells—give us the basis for a working definition of living things: living things are organized chemical systems composed of cells and capable of reproducing themselves. Notice we have said "a working definition." In fact, any attempt to draw a sharp line between the nonliving and the living is essentially arbitrary. Viruses, for example, have some of the essential attributes of living things, even though they are not composed of cells in the usual sense.

A Cell's Size Affects Its Viability

Most cells are very small and can be distinguished only with a microscope (see Exploring Further: Viewing the Cell, p. 91). Some, however, such as bird egg cells, are visible to the naked eye. Others are very small in cross section, but extremely long—a single nerve cell, for example, can be more than a meter in length, but still hard to see (Fig. 5.2). Even among microscopic cells there is a wide range of sizes. The diameter of a

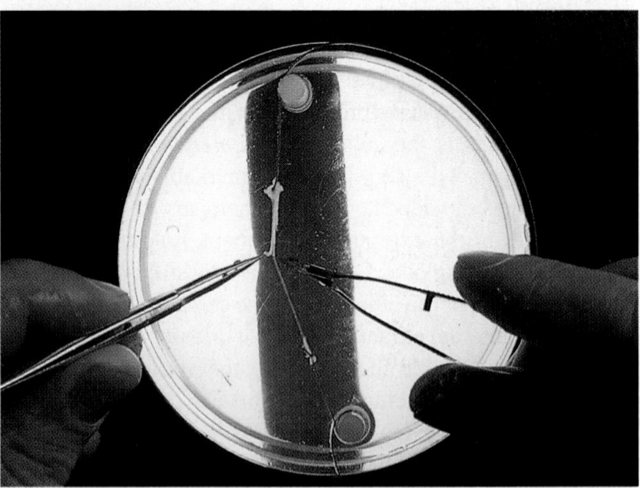

5.2 Giant axon from a squid Although most cells are small, some are quite large, as can be seen in this fiber, which is just a portion of a giant nerve cell. The investigator is removing connective tissue from the fiber, which is called an axon. The squid's axon is 50 to 100 times larger than that of any in the human body.

EXPLORING FURTHER

VIEWING THE CELL

Much of our knowledge of cellular organization has been made possible by the development of better and more powerful microscopes. In the detailed analysis of subcellular structure, three attributes of microscopes are of particular importance: magnification, resolution, and contrast. Magnification is a means of increasing the apparent size of the object being viewed. Resolution is the capacity to separate adjacent forms or objects—to show them as distinct. Contrast is important in distinguishing one part of a cell from another.

Although ordinary light microscopes (Fig. A) can be endowed with very high magnification, their resolving power is limited. The view they provide is about 500 times greater than that of the unaided human eye, but this is still not enough to see some of the smaller subcellular structures (Fig. E). Contrast is often obtained in microscopy by fixing and staining the material being studied. Since different parts of the material often differ in their affinities for various dyes, it is possible to stain these parts different colors to make them stand out from each other.

The electron microscope (EM) has opened up exciting new vistas in the study of the cells. This microscope, as its name implies, uses a beam of electrons instead of light as its source of illumination. The specimen to be examined must be sliced into extremely thin sections to allow a beam of electrons to pass through and fall on a photographic plate, where an image of the specimen is produced (Fig. B). Because the electrons pass through the specimen, the EM is often referred to as the transmission electron microscope or transmission EM. Electron microscopes are capable of resolving objects about 10,000 times smaller than those distinguishable by the unaided human eye (see Fig. F).

Though the transmission EM has very high magnification and resolution, it does not give direct information about the three-dimensional shape of the specimen. Another kind of electron microscope, the scanning EM, gives a surface view of the specimen being studied and a sense of depth lacking in the transmission photograph. In the scanning EM (Fig. C) the electron beam moves back and forth across the specimen in a manner similar to the electron beam in a television tube. The beam does not pass through the specimen but instead causes electrons to be emitted from the surface of the specimen. The scattered electrons are collected to produce a picture with a three-dimensional effect (Fig. G). Another not inconsiderable advantage is that the specimen need not be sectioned and can frequently be studied whole and intact. Though the resolution of a scanning EM is not as great as that of a transmission EM, the three-dimensional image it provides is of great importance for many kinds of studies.

The most recent elaboration of the electron microscope—the scanning tunneling microscope (STM)—uses an ultrafine, electron-emitting tip that is moved back and forth over the sample, systematically scanning it from top to bottom (Fig. D). The tip almost touches

the specimen, and so must be moved up and down to accommodate the contours of the sample. These up/down movements of the tip are used by a computer to reconstruct the topography of the sample, and a picture is created on a video monitor (Fig. H).

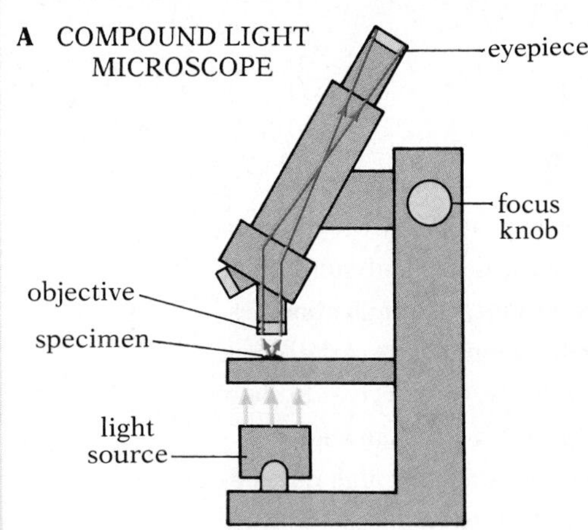

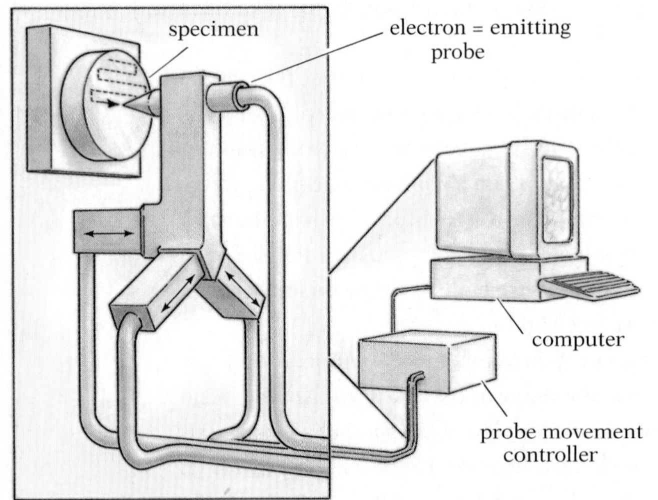

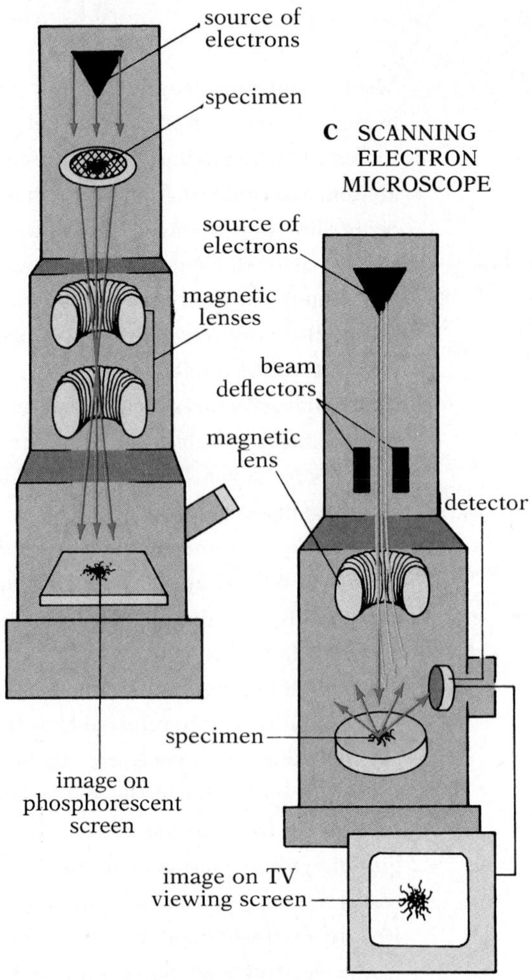

Microscopes (A) In a compound light microscope, light passes through a specimen to the objective lens. Here the light is refracted and directed to the eyepiece, where it is focused for a camera or for the eye. (B) In a transmission EM, electrons provide the illumination, which passes through the specimen and is then focused by magnets to form an image for photographic film or on a phosphorescent screen. (C) In a scanning EM, a focused beam of electrons moves back and forth across the specimen, while a detector monitors the consequent emission of secondary electrons and reconstructs an image. (D) The scanning tunneling microscope moves an electron-emitting tip over the sample, constantly adjusting its height to maintain a fixed distance above the sample; these variations in tip elevation allow reconstruction of the specimen's contours. The minute movements necessary are accomplished by a mounting arm attached to piezoelectric crystals. Since the dimensions of such crystals change as different amounts of voltage are applied, it is possible to control the position of the tip exactly.

E .01 mm

F .01 mm

G .01 mm

H 20 mm

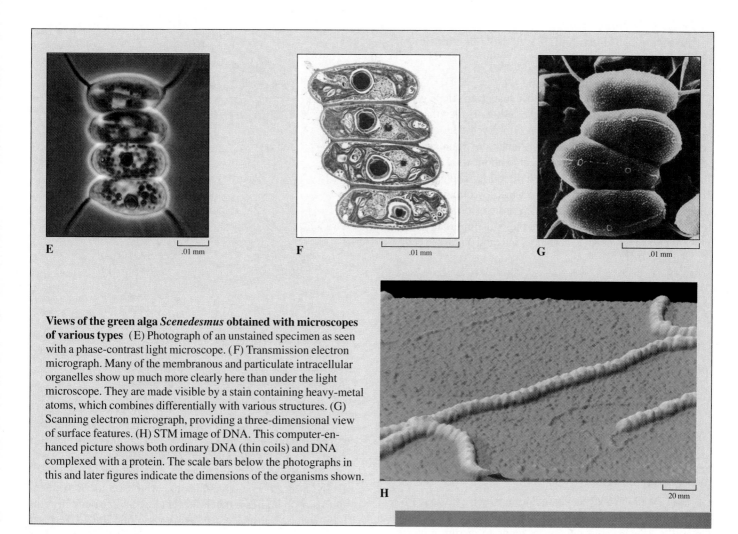

Views of the green alga *Scenedesmus* obtained with microscopes of various types (E) Photograph of an unstained specimen as seen with a phase-contrast light microscope. (F) Transmission electron micrograph. Many of the membranous and particulate intracellular organelles show up much more clearly here than under the light microscope. They are made visible by a stain containing heavy-metal atoms, which combines differentially with various structures. (G) Scanning electron micrograph, providing a three-dimensional view of surface features. (H) STM image of DNA. This computer-enhanced picture shows both ordinary DNA (thin coils) and DNA complexed with a protein. The scale bars below the photographs in this and later figures indicate the dimensions of the organisms shown.

human red blood cell is about 35 times greater than that of some very tiny microorganisms, whereas that of a human egg cell is about 14 times greater than that of a red blood cell. The diameter of an ostrich egg cell, in turn, is about 1,500 times greater than that of a human egg cell. Most cells, however, have diameters that range between 0.5 and 40.0 micrometers.[1]

Throughout the course of evolution, this range of diameters has not varied significantly, one probable reason being that the ratio of a cell's surface area to its volume strongly affects the way it functions. Cells obtain necessary materials such as oxygen and nutrients from the medium surrounding them. These materials must enter across the surface of the cell and leave by the same route. As cell size increases, the volume increases much more rapidly than the surface area. (In a sphere, volume increases by the cube of the radius, surface area by the square of the radius.) Thus, as a cell gets bigger, the requirement for an adequate exchange surface to support its greatly increased volume becomes more difficult to meet (Fig. 5.3).

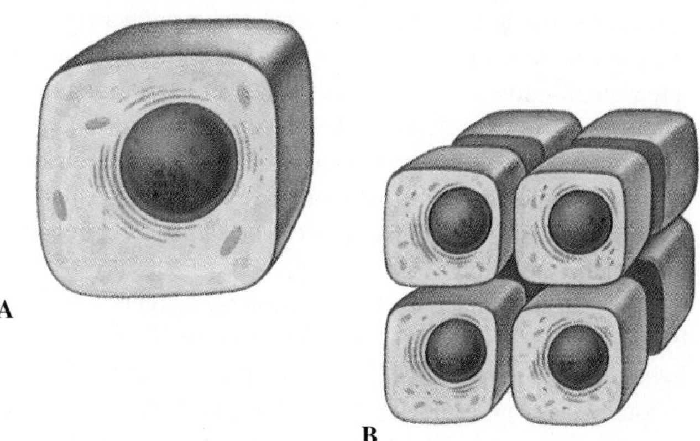

5.3 Relationship between cell surface area and volume As cell size increases, volume increases much more rapidly than the surface area. When the volume present in the cell (A) is partitioned up into a number of smaller cells (B), the total surface area is much larger and the ratio of its surface area to volume in each cell is greater. A high surface-to-volume ratio is important because cells must exchange necessary materials across their surfaces. The larger the cell, the more difficult it is to establish an adequate exchange surface to support its increased volume.

[1] For basic units of measurement, see Table 2, Appendix 2.

The Cell Membrane Regulates the Flow of Materials Into and Out of the Cell

The cell membrane is much more than just an envelope giving mechanical strength and shape to the cell; it also regulates the chemical traffic between the precisely ordered interior of the cell and the potentially disruptive outer environment. The membrane of each cell is quite specific about what is to pass through, at what rate, and in which direction. The cell membrane exercises this control in two ways: by utilizing essentially passive physical processes such as diffusion and osmosis and by taking advantage of special biological mechanisms to actively transport specific substances into and out of the cell.

Diffusion Is the Movement of Particles to Zones of Lower Concentration

Before examining the passive and active movement of particles across the cell membrane in more detail, let us consider movement in general.

Imagine a small rectangular box containing 20 marbles, all placed near each other in one corner (Fig. 5.4A). When the box is shaken, the marbles are scattered almost evenly over the bottom of the box (Fig. 5.4B). Obvious as this result might seem, it is worth a closer look.

First, of all the possible directions in which a given marble might move, more lead away from the center of the cluster than toward it. Hence rapid movement tends to disrupt a cluster rather than maintain it. This is precisely what happens, for instance, when a lump of sugar dissolves in a cup of warm coffee: the sugar molecules move down a concentration gradient, from a region of high concentration (the sugar crystal) to regions of lower concentration. Eventually the sugar molecules disperse throughout the liquid. The warmer the liquid, the more kinetic energy the molecules in solution have on average, and the faster the redistribution takes place. The same effect occurs in the aqueous solution of living cells.

We may now make a generalization based on our example of the marbles in the box and others like it: all other factors being equal, the net movement of the particles of a particular substance is from regions of high concentration to regions of lower concentration. Note that we said the *net* movement. There will always be some particles moving in the opposite direction, but overall, the movement will be away from the centers of concentration. An obvious result is that the particles of a given substance tend to become distributed with relatively uniform density within any available space. When this uniform density is reached, the system is the equilibrium; the particles continue to move, but there is little net change in the system.

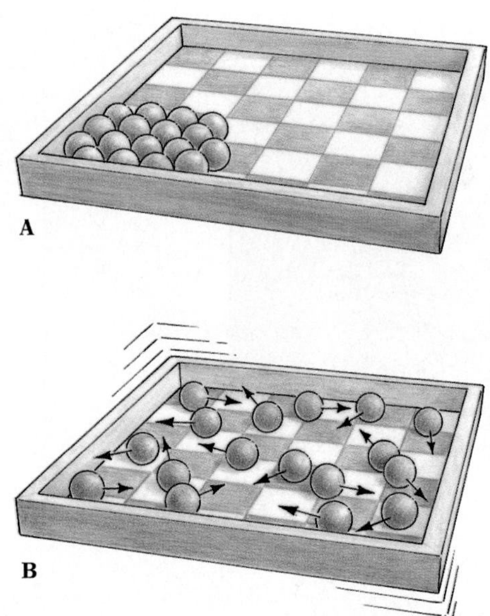

5.4 Mechanical model for diffusion (A) All 17 marbles are placed in a cluster at one end of a rectangular box. (B) When the box is shaken to make the marbles move randomly, they become distributed throughout the box in nearly uniform density.

The net movement of small particles from an area of high concentration to an area of lower concentration of that substance is called ***diffusion.*** Diffusion is fastest by far in gases, where there is much space between particles and hence relatively little chance of collisions, which retard movement. Diffusion in liquids is much slower (Fig. 5.5); in perfectly still, cold water a substance can take a very long time—years, in fact—to move in appreciable quantity only a few feet. Diffusion in solids is, of course, much slower still: there is very little space between molecules of a solid, and collisions occur almost before the molecules get going. In all of these instances, however, regardless of the rate of diffusion, the net effect is movement down the concentration gradient—away from regions of high concentration—as long as all regions are the

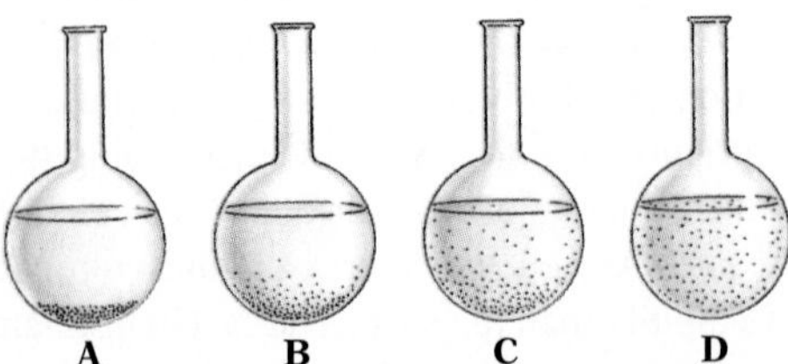

5.5 Diffusion in a liquid Particles of solute are at the bottom of a flask of water (A). The particles slowly diffuse away from the cluster until (D) they are distributed with nearly uniform density through the water. If the water is cold and there are no convection currents to help move the particles, it may take a considerable period of time to reach the uniform distribution shown in D.

same temperature and pressure. In living organisms, where molecules are generally in a warm aqueous solution and the distances involved are measured in fractions of a millimeter, diffusion is a highly significant process.

Diffusion can be related to the Second Law of Thermodynamics, which states that energy is constantly degraded to waste heat that cannot be recycled, and that systems left to themselves tend to become more and more disordered. Just as rocks on a hill tend to move down to the bottom rather than up to the top, all systems tend to lose free energy, and ultimately reach a state in which their free energy is as low as possible. A concentrated substance is a relatively orderly and unlikely arrangement; an orderly arrangement of molecules has more potential for doing work than a disorderly one, and has more free energy. Diffusion is spontaneous because orderly molecules, concentrated together, have greater free energy than dispersed molecules: the movement of particles by diffusion is a downhill reaction, from order to disorder.

Diffusion is an important cellular process because the concentration of organic molecules and ions inside a cell is different from the concentration of substances in the surrounding medium. The various substances therefore have a spontaneous tendency to diffuse into or out of the cell until their concentrations are equal with the surrounding medium. It is the barrier of the cell membrane between the inside and outside of the cell that maintains the proper concentrations of ions and other solutes inside the cell.

Osmosis Is the Diffusion of Water Across a Selectively Permeable Membrane

To visualize how cell membranes control the movement of substances into and out of the cell, consider another model: a chamber divided into halves by a membrane partition. Let us assume, further, that particles of some substances can pass through the membrane, while particles of other substances cannot, much as brewed coffee flows through a cone of filter paper, leaving the grounds behind. Such a membrane is said to be ***selectively permeable.*** How does the membrane affect the diffusion of materials between the two halves of the chamber?

Suppose the chamber is a U-shaped tube divided in half by a selectively permeable membrane (Fig. 5.6). Suppose further that side A contains pure water and side B an equal quantity of sugar solution (sugar dissolved in water), both sides being subject to the same initial temperature and pressure. If the membrane is permeable to water, but not to sugar, water molecules will be able to pass in both directions, from A to B and from B to A.

Since water is already present on both sides of the membrane in the U-tube in Figure 5.6, it might at first seem that the movement of water molecules across the membrane would have no net effect. But consider the differences between the pure water and the sugar solution more carefully. On side A, all the molecules that bump into the membrane during a given

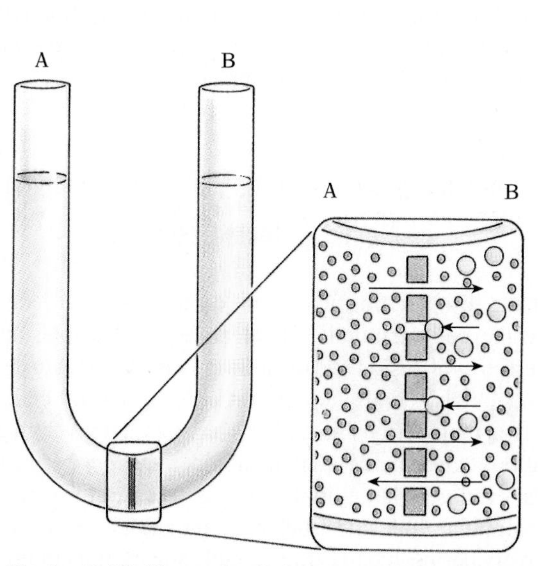

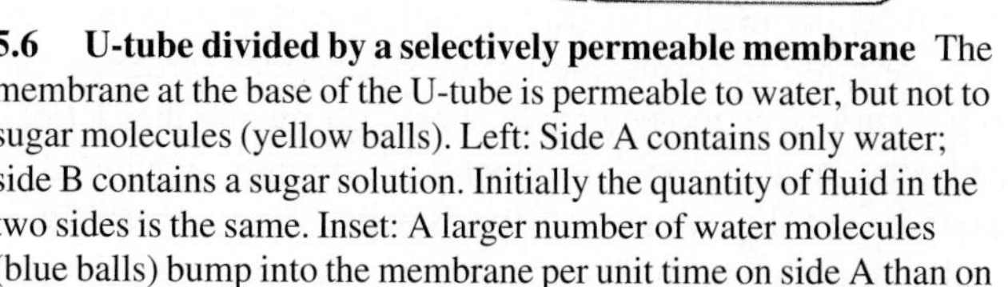
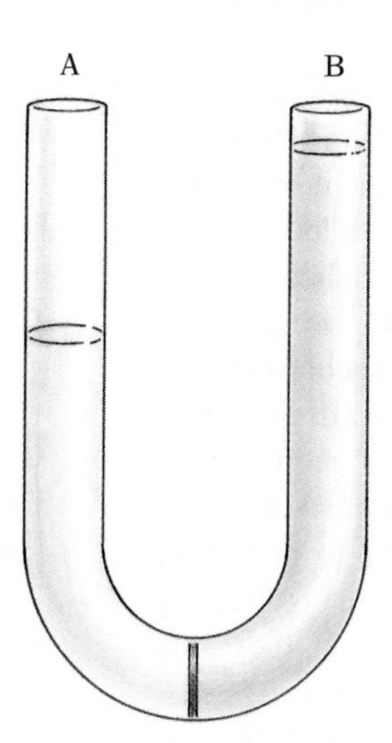

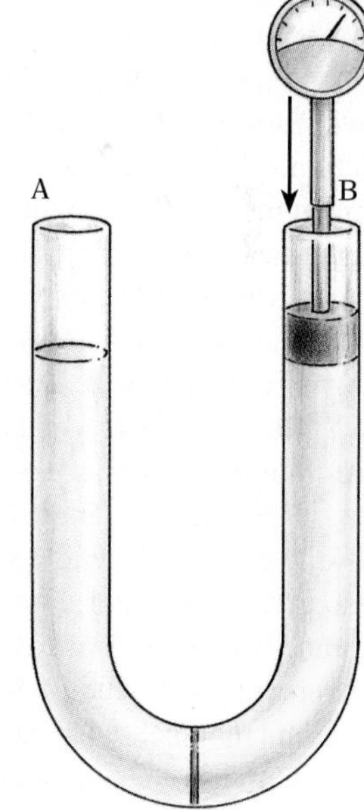

5.6 U-tube divided by a selectively permeable membrane The membrane at the base of the U-tube is permeable to water, but not to sugar molecules (yellow balls). Left: Side A contains only water; side B contains a sugar solution. Initially the quantity of fluid in the two sides is the same. Inset: A larger number of water molecules (blue balls) bump into the membrane per unit time on side A than on side B. Middle: Because more water molecules move from A to B than from B to A, the level of fluid on side A falls while that on side B rises. Right: A gauge can measure the amount of pressure that must be exerted on side B to keep it in balance with side A. This is the osmotic pressure of the solution.

time interval are water molecules, and because the membrane is permeable to water, many of these will pass through the membrane from A to B. By contrast, on side B, some of the molecules bumping into the membrane during the same interval will be water molecules, which may pass through, some will be sugar molecules,which cannot pass through because the membrane is impermeable to them. At any given instant, then, part of the membrane surface on side B is in contact with sugar molecules and part is in contact with water, whereas on side A all of the membrane surface is in contact with water. Hence more water molecules will move across the membrane from side A to side B per unit time than in the opposite direction; the net movement of water molecules will be from A to B.

This movement of water through a selectively permeable membrane from regions of high water concentration to regions of low water concentration is called ***osmosis.*** Biological membranes are selectively permeable, and the movement of water through them can be predicted on the basis of osmosis. As we shall see, some solutes, such as lipid-soluble molecules and small polar molecules, also pass freely through biological membranes by simple diffusion.

You will note that the osmosis of water from A to B is in full accord with our earlier generalization concerning diffusion: the net movement of water molecules is from the region of their greater concentration (side A) to their lesser concentration (side B). But this means that the volume of fluid increases on side B and decreases on side A. How long can this process continue? Will an equilibrium point be reached?

Clearly, the concentrations on the two sides of the membrane will never be equal, no matter how many water molecules move from A to B, because the fluid in B will remain a sugar solution, albeit an increasingly weak one, and the fluid in A will remain pure water, provided that the membrane is completely impermeable to sugar molecules. The net movement of water from A to B might be expected to continue indefinitely. But this is not in fact what happens. Under normal conditions, the fluid level in B will rise to a certain point and then cease to rise. Why? The column of fluid is, of course, being pulled downward by gravity—i.e., it has weight. As the column rises, therefore, it exerts an increasing downward pressure. Eventually the column of sugar water becomes so high, and the pressure so great, that water molecules will be pushed across the membrane from B to A as fast as they move from A to B. When this point is reached—when the water is passing through the membrane in opposite directions at the same rate—the system is said to be in ***dynamic equilibrium.***

The equilibrium point is also influenced by the difference in the number of solute particles per unit volume on the two sides of the membrane; the greater the difference, the higher the column will rise before equilibrium is reached. The total number of osmotically active (i.e., dissolved) solute particles per unit volume is referred to as the ***osmotic concentration.*** If there are several kinds of solutes in the same solution, as is invariably the case in living cells, then the osmotic concentration of that solution is determined by the total (per unit volume) of *all* the particles of all kinds. If a dissolved substance separates into ions, each ion functions as a separate particle; for example, in figuring the osmotic concentration of a saline solution, one dissolved molecule of sodium chloride (NaCl) is counted as two particles—one sodium ion (Na^+) and one chloride ion (Cl^-).

We are now in a position to make an additional generalization: if, under conditions of constant temperature and pressure, two different solutions are separated by a membrane permeable only to water, the net movement of water will be from the solution with fewer osmotically active particles to the solution with more such particles; that is, from the solution with the lower osmotic concentration to the solution with the higher osmotic concentration.

One way to characterize the osmotic properties of a solution is to measure its osmotic concentration, or the number of dissolved particles it contains per unit volume. Another way is to measure the pressure that must be exerted on a solution to keep it in equilibrium with pure water when the two are separated by a selectively permeable membrane. This measured value is called the ***osmotic pressure*** of the solution. In our U-tube example (Fig. 5.6C), the osmotic pressure of the sugar solution in side B is the pressure we must exert on the solution to prevent the fluid column in B from rising. Thus, the osmotic pressure of a solution is a measure of the tendency of water to move by osmosis into it. The more dissolved particles in a solution, the greater the tendency of water to move into it, and the higher the osmotic pressure of the solution. Thus, under constant temperature and pressure, water will move from the solution with the lower osmotic pressure to the solution with the higher osmotic pressure when the two solutions are separated by a selectively permeable membrane.

The Osmotic Concentration of the Surrounding Medium Affects Osmosis

We have discussed diffusion and osmosis at such length because the cell membrane is selectively permeable, and the processes of diffusion and osmosis are fundamental to cell life. Although the membranes of different types of cells vary widely in their permeability characteristics, a few rough generalizations can be made. Cell membranes are very permeable to lipids, even very large lipids, to lipid-soluble substances, and to small polar molecules such as water and glycerol. They are relatively permeable to certain simple sugars and amino acids, but are quite impermeable to polysaccharides, proteins, and other very large molecules. In short, cell membranes let pass only the smaller building blocks of complex organic compounds, not the compounds themselves. The permeability of cell membranes to small inorganic ions varies greatly, depending on the particular ion, but in general, negatively charged

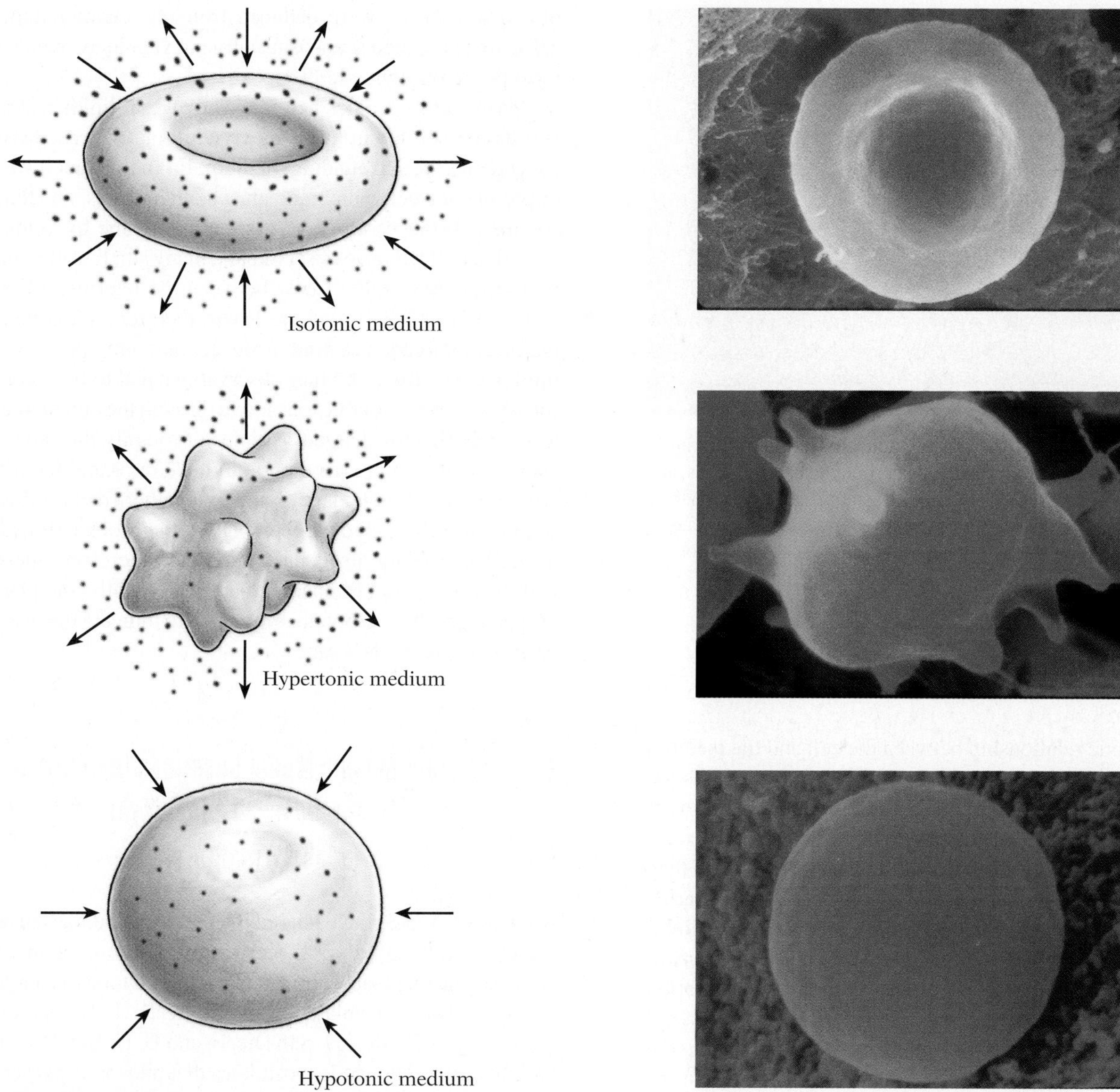

5.7 Osmotic relationships of a cell In an isotonic medium, water gain and water loss are equal, and so the cell neither shrinks nor grows. In a hypertonic medium, there is a net loss of water from the cell, and the cell shrinks. In a hypotonic medium, the cell swells as water moves from the medium inside. The accompanying photographs are of human red blood cells.

ions can cross more rapidly than positively charged ions, although neither can do so as readily as uncharged particles.

What implications do these generalizations hold for life? On the one hand, selective permeability enables cells to retain the large organic molecules they synthesize. On the other, the tendency of water to pass through selectively permeable membranes into regions of higher osmotic concentration can be harmful or even fatal. When a cell is in a medium that contains a higher concentration of solutes (and therefore a lower concentration of water) than the cell does itself, the cell will tend to lose water by osmosis and shrink (Fig. 5.7). If the process goes too far, it may die. That is why watering a plant with seawater, rather than fresh water, can quickly cause the plant to become dehydrated and die.

A medium that has a greater concentration of solutes than the cell—and hence draws water out of the cell—is said to be ***hypertonic*** (Fig. 5.8). Conversely, a medium that contains a lower concentration of solutes (and therefore a higher concentration of water) than the cell is said to be ***hypotonic.*** In a hypotonic medium the cell will gain water and swell, and unless it has special mechanisms for expelling excess water, or special structures that prevent excessive swelling (as most plant cells do), it may burst (Fig. 5.7). A cell in an ***isotonic*** medium has the same solute concentration as the medium and therefore

5.8 Sperm whales living in a hypertonic environment Marine vertebrates, whether they be whales, seals, bony fish, or sea birds, are living in an environment that has a higher osmotic concentration than their body fluids. Because of this they have a tendency to lose water by osmosis and become dehydrated. To survive in such an environment they have evolved adaptations such as impermeable outer coverings to guard against excessive water loss.

neither gains nor loses appreciable quantities of water by osmosis.

The osmotic relationship between the cell and the medium surrounding it is critical to the life of the cell. Some cells are normally bathed by an isotonic fluid and therefore have no serious osmotic problems. Human red blood cells are an example; they are normally bathed by blood plasma,with which they are in relatively close osmotic balance. Most simple marine plants and animals are also in isotonic media; their cellular contents have an osmotic concentration close to that of seawater. All cells, however, have a higher osmotic concentration than fresh water. Freshwater organisms therefore live in a hypotonic medium and face the problem of accumulating too much water by osmosis. Their very existence depends on the evolution of ways to prevent their cells from becoming so turgid—so distended by their fluid content—that they would burst.

Although the cell membrane acts in many respects like a passive osmotic partition, it must also do more. For nutrients to be captured and retained, wastes expelled, and cell volume controlled, the cell membrane must pass many chemicals in only one direction. The key to this critical ability lies beyond mere selective permeability, in the structure of the cell membrane itself.

Lipids and Proteins in the Cell Membrane Determine Permeability

Despite the long-standing certainty that cells are bounded by a plasma membrane, only in the last three decades has a direct proof of its existence been obtained. Most earlier conceptions of the membrane were deduced from the characteristics of cells, since the membrane is usually not visible even under the most powerful light microscopes.

Permeability studies had long shown that lipids and many substances soluble in lipids move with relative ease between the cell and the surrounding medium. From this fact it was deduced that the cell membrane must contain lipids, and that fat-soluble substances move across the membrane by being dissolved in it. It was also observed that many small water-soluble molecules move quite freely between the interior of the cell and its external environment. It was therefore suggested that the cell membrane is a kind of sieve, containing pores or nonlipid patches. But still other observations had to be accounted for. Small, water-soluble ions move through the cell membrane less freely than uncharged particles of roughly the same size. Moreover, different ions do not all have the same facility for crossing the cell boundary. Some pass rather freely; others do so only in very limited numbers. It was therefore assumed that the cell membrane itself has charge and therefore interferes with the movement of charged particles. Finally, the physical properties of the cell boundary seemed to indicate the presence of protein in the membrane.

The Fluid-Mosaic Model Describes the Plasma Membrane as a Lipid Bilayer Studded With Proteins

A series of models have guided the research that has led to our present understanding of the structure of plasma membranes. Of these, the one that is universally accepted today is the ***fluid-mosaic model,*** postulated in 1972 by S. J. Singer of the University of California, San Diego, and G. L. Nicolson of the Salk Institute. The model postulates that phospholipids are the basic structural components of such membranes. Recall from Chapter 3 that phospholipids are unusual molecules; one end of the molecule is charged and thus hydrophilic while the other end has two long, nonpolar hydrocarbon tails (Fig. 5.9). According to the model, the membrane is a ***lipid bilayer*** composed of two continuous layers of phospholipid molecules oriented with their polar (hydrophilic) ends exposed at the inner and outer surfaces of the membrane, and their two nonpolar (hydrophobic) hydrocarbon tails buried in the interior, hidden from the surrounding water (Fig. 5.10). The membrane proteins are distributed in an irregular (mosaic) pattern, both on the surfaces and in the interior of the membrane. The phospholipid bilayer forms the basic structure of the membrane, but the proteins actively carry out a wide range of critical functions, including transporting substances through the membrane.

Of the proteins found on the surfaces, those on the inner surface may differ markedly from those on the outer surface. Indeed, some membranes may have no surface proteins at all. The proteins located within the lipid bilayer may also exhibit a

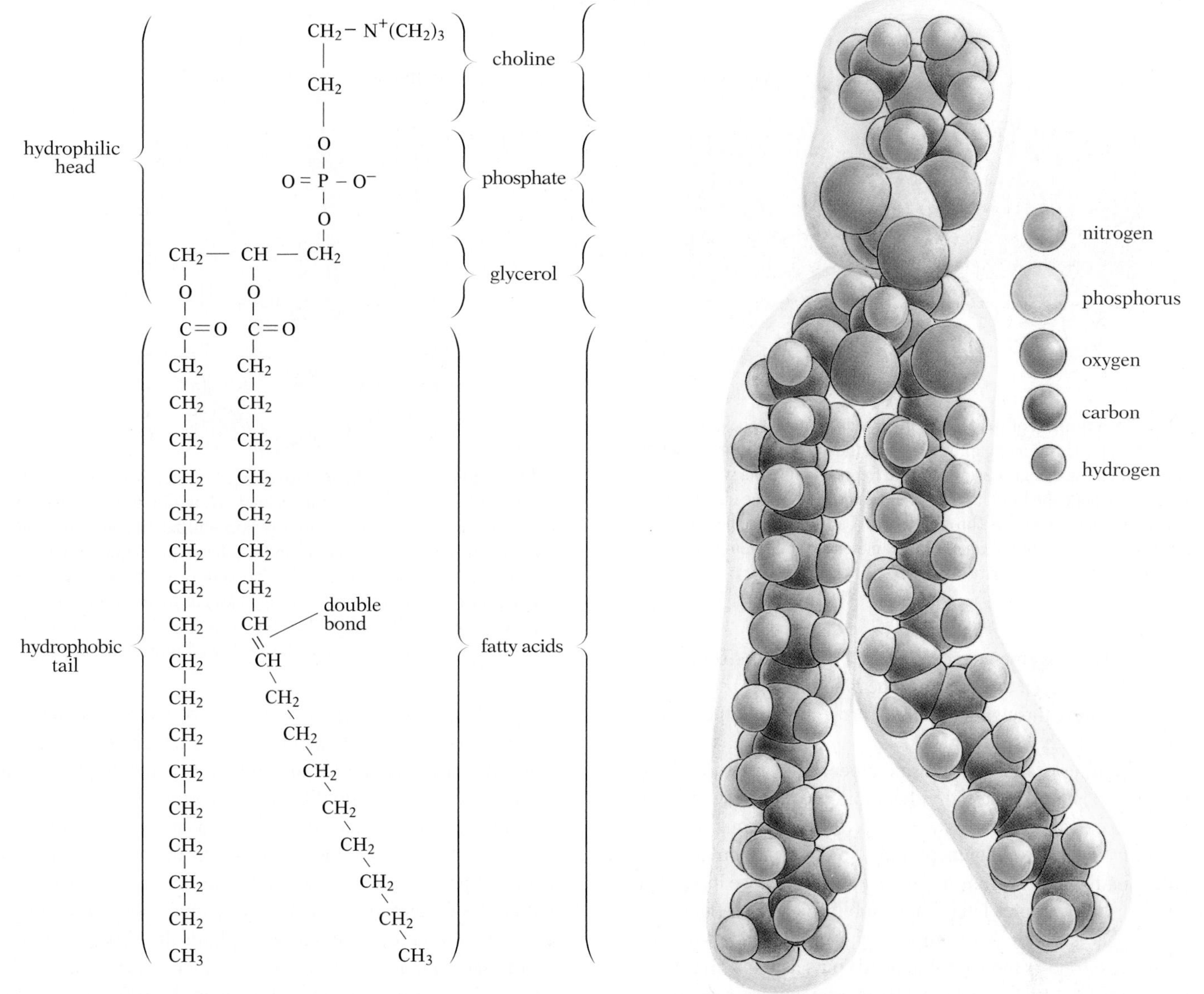

hydrophilic head

$CH_2 - N^+(CH_2)_3$ (choline)
CH_2

$O = P - O^-$ (phosphate)

$CH_2 - CH - CH_2$ (glycerol)

$C=O$ $C=O$

hydrophobic tail (fatty acids)

CH_2 ... CH_3 (saturated chain)

CH_2 ... $CH = CH$ (double bond) ... CH_3 (unsaturated chain)

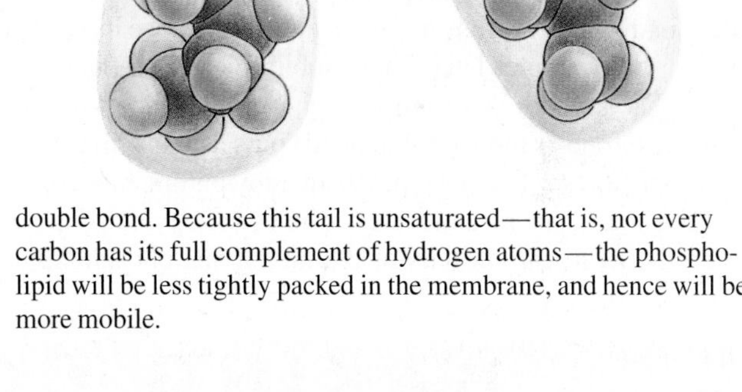

5.9 A phospholipid The cell membrane is made up mostly of phospholipids. Phosphatidylcholine, a common membrane phospholipid, consists of a polar head (a positively charged choline, a negatively charged phosphate, and glycerol) joined to two hydrophobic fatty acid chains. The "kink" in the right tail is created by a carbon double bond. Because this tail is unsaturated—that is, not every carbon has its full complement of hydrogen atoms—the phospholipid will be less tightly packed in the membrane, and hence will be more mobile.

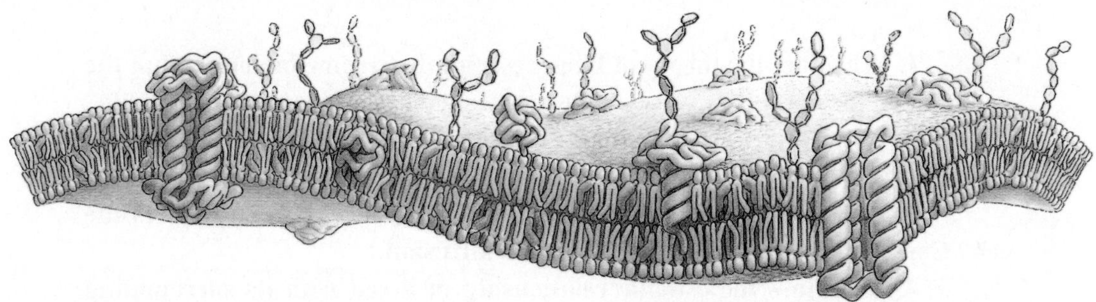

5.10 The fluid-mosaic model of the cell membrane A double layer of lipids forms the main continuous part of the membrane; the lipids are mostly phospholipids, but in plasma membranes of higher organisms cholesterol (brown) is also present. Proteins occur in various arrangements. Some, called peripheral proteins, are entirely on the surface of the membrane, to which they are anchored by a covalent bond with a membrane lipid. Others, called integral proteins, are wholly or partly embedded in the lipid layers; some of these may penetrate all the way through the membrane. Three protein units are joined by covalent bonds (not shown) to form part of a single protein molecule bounding a membrane-spanning pore (right). Proteins make up about half of the weight of membranes. The hexagons represent carbohydrate groups.

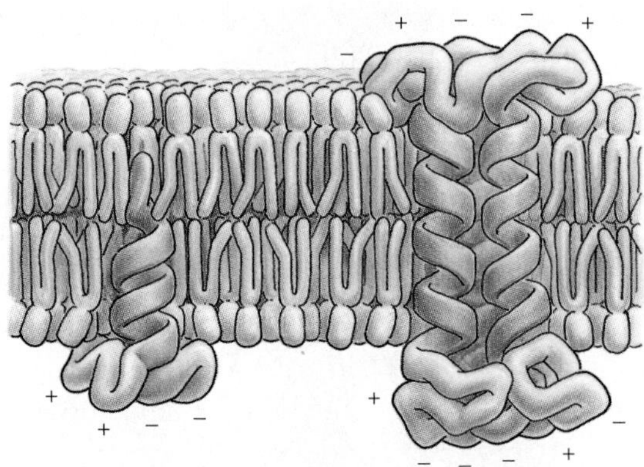

5.11 Orientation of proteins within membranes The parts of the polypeptide chain containing most of the hydrophilic amino acids (polar or charged R groups; blue) tend to project into the watery medium outside the lipid layers, whereas the parts of the chain with hydrophobic amino acids (brown) tend to be folded into the inner, lipid portion of the membrane. The diameter of the protein strands has been reduced for clarity.

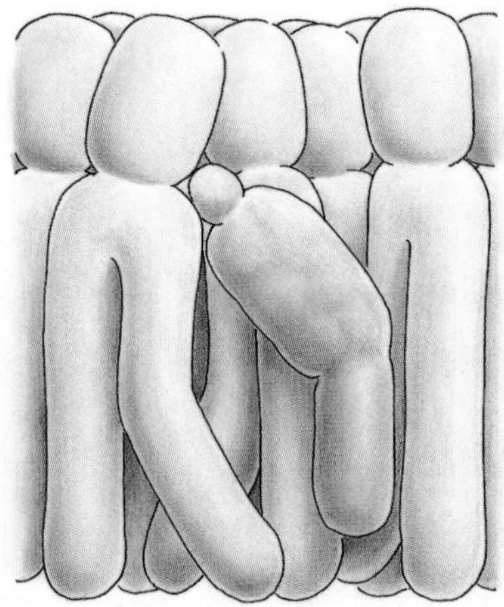

5.12 Cholesterol in the membrane Cholesterol (brown) binds weakly but effectively to two adjacent phospholipids. One possible effect is that the cholesterol partially immobilizes the phospholipids, making the membrane less fluid and mechanically stronger. The amount of cholesterol varies widely according to cell type, with the membranes of some cells possessing nearly as many cholesterol molecules as phospholipids while others lack cholesterol entirely. For the structural formula of cholesterol, see Figure 3.18, p. 62.

variety of arrangements: some are confined to the outer portions of the bilayer, and others to the inner portions; some may extend entirely through the bilayer, projecting into the watery medium on both sides (Fig. 5.11).

According to the fluid-mosaic model, the structure of the membrane is fluid. The individual lipid molecules (which are linked to one another only by weak, hydrophobic interactions) can move laterally in the membrane, so that a particular molecule found in one position at a given moment may be in another position only minutes later. Such mobility is greatest in membranes that are high in unsaturated phospholipids because the presence of one or more double bonds puts a "kink" in the hydrocarbon tail (see Fig. 3.11, p. 59), thereby putting more room between the adjacent molecules. Cholesterol, which fits between the hydrocarbon tails of the membrane phospholipids (Fig. 5.12), also influences the fluidity of the membrane. The proteins, too, can move laterally but to a lesser extent than the lipids. Some of the membrane proteins are anchored in place, thereby limiting the fluidity of the membrane; others form structural and functional complexes too large to move easily.

In the fluid-mosaic model, the pores in the membrane are depicted as passageways through one or a group of protein molecules (Fig. 5.10). The ability of unanchored proteins to drift laterally in the lipid bilayer explains the observed mobility of many membrane pores. The distinctive properties of the various R groups of the amino acids in the proteins give the pores some selectivity; not all ions or molecules small enough to fit in the pores can actually move through them.

MAKING CONNECTIONS

1. Why does the theory of biogenesis imply to some biologists that the evolution of the first cell was the single most crucial event in the history of life?

2. Why has cell size remained restricted throughout the course of evolution?

3. How does the cell membrane regulate what enters and leaves the cell? Consider the processes of osmosis and diffusion.

4. How does the osmotic relationship of a cell with its surrounding medium affect its proper functioning?

5. How is the high permeability of the cell membrane to lipids and its low permeablity to ions related to its structure? Consider the "split personality" of phospholipids.

Channels and Pumps Selectively Transport Substances Across Membranes

The lipid bilayer that constitutes the cell membrane forms a flexible but effective barrier between the inside of the cell and the outside. Moreover, the bilayer provides an anchoring plane for a variety of membrane proteins. The largely lipid composition of the membrane explains why small or lipid-soluble molecules essential to life can diffuse into and out of cells with relative ease (Fig. 5.13). The membrane's permeability to certain life-sustaining substances that do *not* dissolve in lipids, however, must depend on the proteins in the bilayer. The ability of cells to actively transport such substances across the membrane against their concentration gradients (i.e., from low concentration to high concentration) must also be accounted for by the properties of the membrane proteins.

We now know that the transport mechanisms that control molecular traffic into and out of the cell are highly specialized structures called channels and pumps. Each mechanism depends on membrane proteins for its operation; in fact, each is usually fabricated from several "cooperating" membrane proteins. Since they enable specific chemicals to permeate the membrane, channels and pumps are known collectively as ***permeases.***

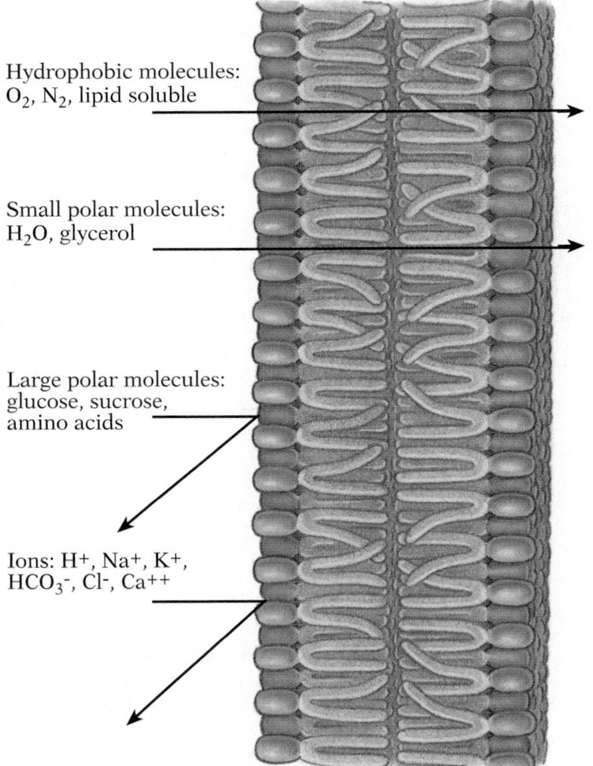

5.13 The relative permeability of the phospholipid portion of the cell membrane Phospholipids form the basic structure of the cell membrane. The phospholipid bilayer is highly permeable to nonpolar and lipid-soluble substances, but small polar molecules such as water and glycerol also pass through readily by squeezing between the phospholipid molecules. The phospholipid bilayer is relatively impermeable to larger polar molecules such as glucose and amino acids, and quite impermeable to charged ions; these substances require permeases for transport through the membrane.

The simplest of the permeases, the ***channels,*** provide openings through which specific substances can diffuse across the membrane. These passageways, although selective, are passive; they simply permit particular chemicals to move down their concentration gradients, that is, from an area of high concentration to an area of low concentration. Because the movement is with a favorable concentration gradient, no input of energy is required. This strategy, known as ***facilitated diffusion,*** is the basis of the membrane's highly specific permeability. The protein channel for potassium ions (K^+) (Fig. 5.14A) illustrates this strategy. For reasons that will be discussed in later chapters, most cells have a much higher concentration of K^+ ions inside the cell than outside. As a charged particle, the K^+ ion cannot diffuse through the lipid portion of the membrane, but a channel specific for K^+ ions allows them to leak out slowly, at a controlled rate. The leakage of the positively charged K^+ ions is important because it helps to maintain the proper balance of electrical charges between the inside and outside of the cell.

Another strategy for controlling movement across the membrane utilizes a transmembrane protein channel that opens and closes like a gate (Fig 5.14B). Such an arrangement, called a ***gated channel,*** is widely used to convert a molecular signal specialized for carrying information between cells into a second signal more suitable for communicating inside the cell. When a molecular signal—a hormone or a transmitter substance that carries messages from one cell to another—binds to an exposed part of the transmembrane protein, known as a ***receptor,*** a change in the conformation of the protein takes place. The change allows the gate to open, and the second signal, usually an ion such as Na^+, K^+, or Ca^{2+} ions, then moves across, carrying the message into the cell. Other channels are voltage-gated—that is, they open and close in response to changes in the electrical gradient across the membrane. Voltage-gated chemicals are important in the conduction of nerve impulses, as we shall see in Chapter 28. The gated-chemical strategy underlies the transmission of many chemical messages in both plants and animals, including, for example, both the nerve impulses that animals use to sense and respond to the outside world, and the movement of muscles.

Another way for molecules to get across the membrane would be for the permease to act as a mobile carrier, taking them through one by one (Fig. 5.14C). Though no definite example of such a permease is yet known, evidence for its existence is mounting.

Protein ***pumps,*** the other major category of permeases, bind to the substance that is being transported and actively pump it across the membrane. Substances are often transported against their concentration gradient, from a region where they are in

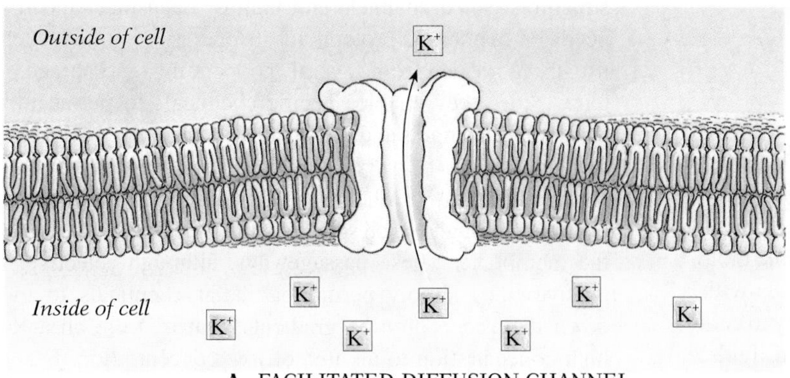

A FACILITATED-DIFFUSION CHANNEL

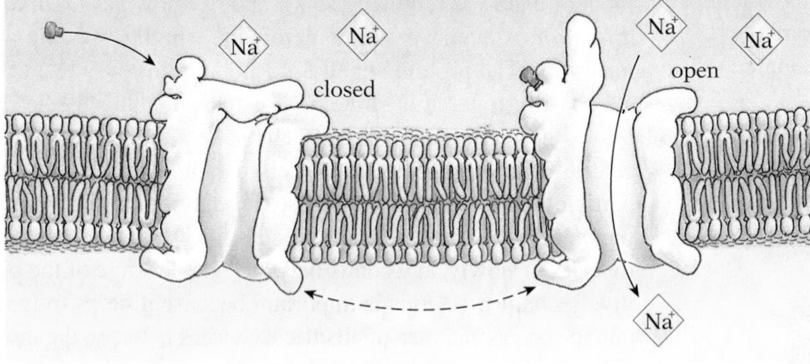

B GATED CHANNEL

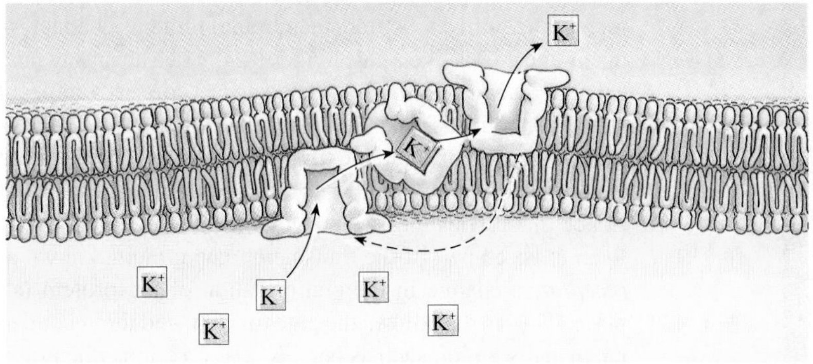

C MOBILE CARRIER

5.14 Models of membrane transport Many strategies for moving substances across the cell membrane are known. (A) In facilitated diffusion of the simplest kind, a protein channel, or pore, embedded in the membrane provides a direct path for the chemical it passes down its osmotic-concentration gradient. The channel's diameter and the chemical environment it creates (hydrophilic or hydrophobic, for example) serve to prevent all but the correct substance from passing. (B) An interaction between a signal molecule (pink) and the gated channel causes the gate to open, so that diffusion can take place down a favorable concentration gradient. Other molecular systems, not shown, then inactivate the signal molecule so that the channel can close again. (C) A mobile carrier would not provide transmembrane channels, but would itself migrate back and forth from one surface to the other. The existence of mobile carriers is controversial.

low concentration to one where they are in higher concentration (Fig. 5.15). Because the movement is against the concentration gradient, the cell must expend energy to perform this work. This process, known as ***active transport,*** is needed not only to move essential ions into and out of the cell but also to transport many building-block materials such as glucose and amino acids. The best known example of a membrane pump is the sodium potassium pump, which we shall discuss in detail in Chapter 28.

The energy source for this pump, and for many other cellular processes, is the energy carrier ATP, or adenosine triphosphate.

Vesicles Convey Substances Into and Out of Cells by Endocytosis and Exocytosis

Substances may enter a cell without actually moving through the cell membrane. By an active process called ***endocytosis,*** the cell encloses the substance in a membrane-bounded vesicle pinched off from the cell membrane. Two types of endocytosis are recognized. When the material engulfed is in the form of large particles or chunks of matter, the process is called ***phagocytosis*** (Fig. 5.16A). Usually, armlike extensions of the cell, called ***pseudopodia*** (Fig. 5.16B), flow around the material, en-

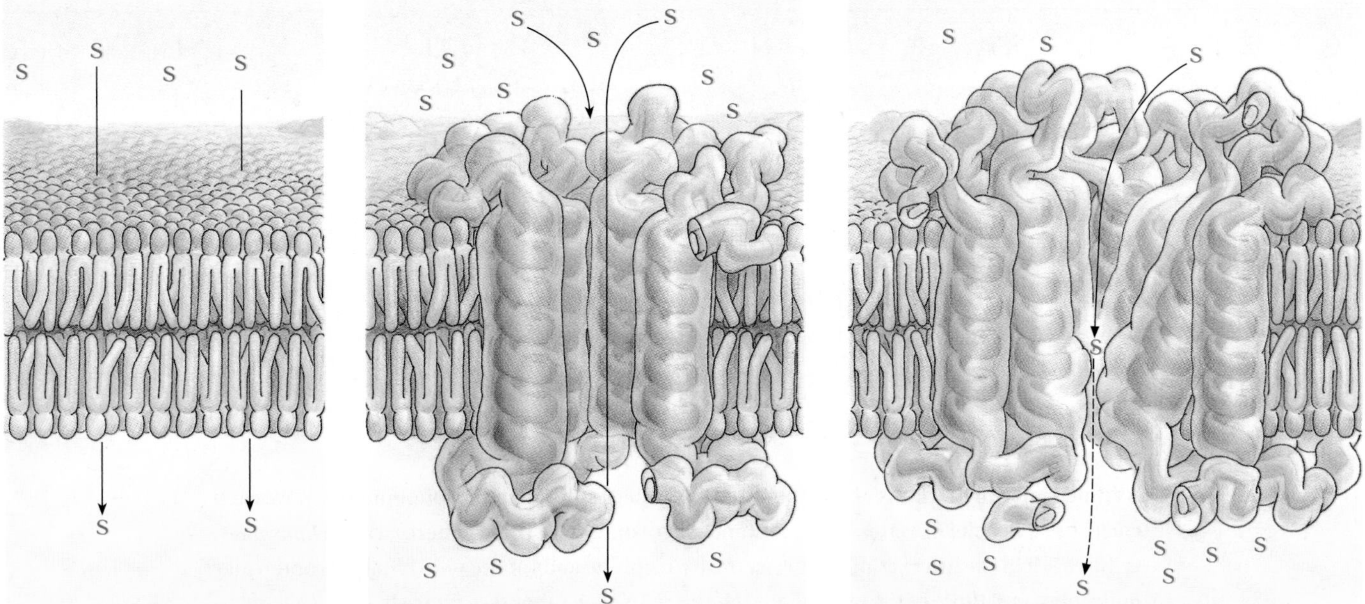

5.15 A comparison of simple diffusion, facilitated diffusion, and active transport (A) In simple diffusion the substrate, S, moves through the membrane without the aid of permease and without expenditure of energy by the cell. (B) In facilitated diffusion a permease transports the substrate through the membrane, but no energy is expended. (C) In active transport a permease is involved, and energy must be expended to change the permease's shape and move the substance against its concentration gradient.

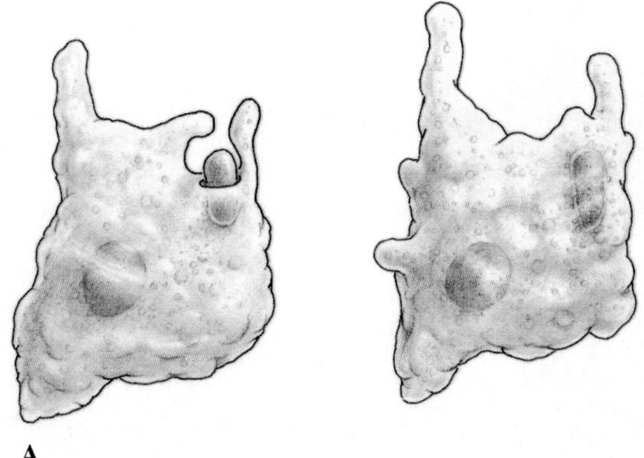

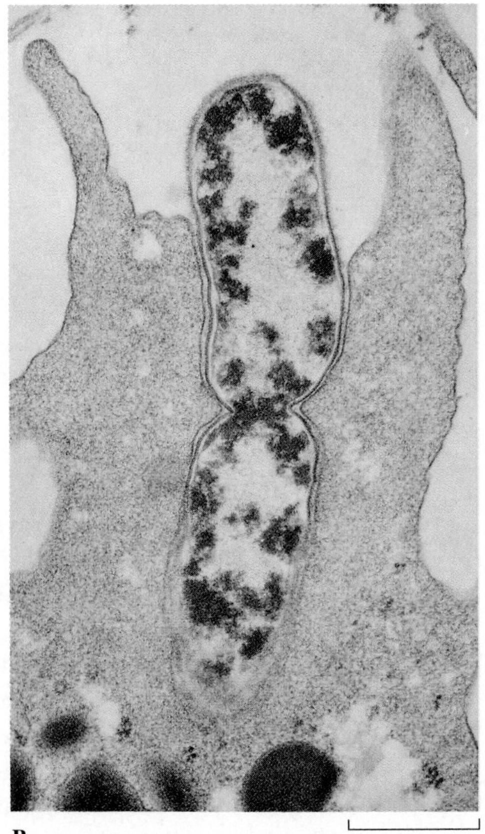

5.16 Phagocytosis (A) In *Amoeba,* pseudopodia flow around the prey until it is entirely enclosed within a vacuole. (B) White blood cells, or leukocytes, in the bloodstream use phagocytosis to capture foreign organisms; here the leukocyte is engulfing a dividing bacterium.

closing it within a vesicle, which then becomes detached from the cell membrane and migrates into the interior of the cell. When the engulfed material is liquid or consists of very small particles, the process is termed ***pinocytosis.*** The material first becomes attached to the cell membrane, probably at selective binding sites. Then the loaded membrane either flows inward to form deep narrow channels, at the end of which vesicles take

EXPLORING FURTHER

THE USE OF LIPOSOMES AS DELIVERY VEHICLES

Adding water to dry films of phospholipids produces spherical phospholipid bilayers, each enclosing a droplet of water. The spontaneous formation of these spheres, called *liposomes,* is the result of the interactions between the hydrophilic ends of the phospholipids and water molecules, and those of the hydophobic chains with each other (see figure below). Cell membranes are structured in exactly this way. Liposomes can be filled with different substances, and can act as delivery vehicles, carrying the substance to the target site. An additional benefit is that the liposomes themselves are nontoxic and biodegradable.

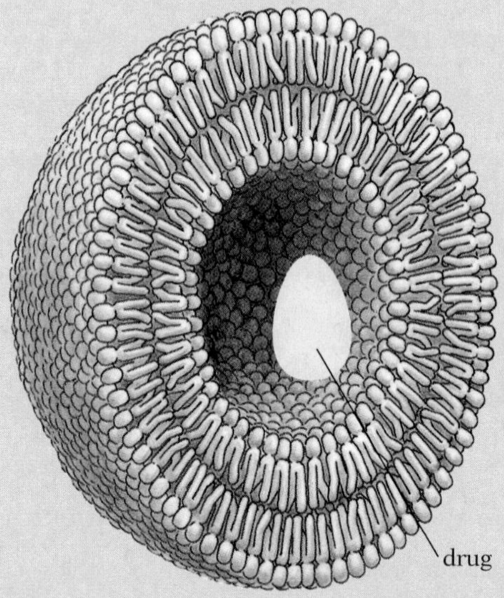

A drug-bearing liposome

Drug-bearing liposomes have been used in the treatment of several diseases, and are particularly useful in cases where the free drug (without a carrier) is toxic. For example, liposomes have been used to deliver anticancer drugs. Such drugs are highly toxic and can cause damage to the heart muscle and suppress the function of the bone marrow. Encapsulating the drugs within the liposomes has decreased these undesirable side effects because the liposomes do not accumulate within these structures. Unfortunately, most of the drug-bearing liposomes injected into the blood are ingested by macrophages, large phagocytic white blood cells. This is useful when the macrophage is the target, as in certain fungal and parasitic diseases, but it is a problem when the target is other cells of the body. Work is proceeding to develop "stealth" liposomes—liposomes coated with substances that

prevent ingestion by macrophages. Clinical tests suggest that these remain in the bloodstream longer, and are much less toxic than unpackaged drugs.

In the future, liposomes may be used routinely as vehicles for delivering drugs to the skin, eyes, or lungs. Liposomes are able to penetrate the superficial layers of the skin, and studies have shown that cutaneous applications of drug-bearing liposomes result in a larger portion of the drug being retained within the skin, and for a longer time than when the drug is given in creams or ointments that often pass through the skin and enter the circulatory system. Work to develop topically applied liposomes for delivering antibiotics and antiinflammatory drugs is currently proceeding. The advantages of liposomal preparations have not gone unnoticed by the cosmetics industry, and liposomes that would moisturize, or deliver sunscreens, vitamins, proteins, and perhaps even perfumes, directly to the deeper layers of the skin are also being developed. It is clear that these little vesicles, which were at one time simply a laboratory curiosity, have many potential applications.

shape, or small vesicles are simply detached directly from the membrane at the cell surface (Fig. 5.17).

In some cases the selective binding site that collects a particular substance before trapping it in a vesicle appears as a "coated pit," with receptor molecules clustering in one spot in the membrane (Fig. 5.18). One example, which illustrates the importance and specificity of receptor-mediated endocytosis, involves cholesterol uptake by cells. As we saw in Chapter 3 (see p. 63), cholesterol is transported in the blood to cells by a carrier complex of low density lipoprotein (LDL). When the cell needs cholesterol—usually to manufacture new membranes—it synthesizes receptors for LDL and inserts them into the cell membrane (Fig. 5.19). The LDL receptors aggregate spontaneously and when LDL molecules bind to the receptors, endocytosis begins. The cholesterol is then transported in endocytotic vesicles for use in the cell's vast complex of membranes. One cause of atherosclerosis—the condition in which the buildup of fatty deposits inside arteries causes them to thicken and lose their elasticity—involves failure of the receptors to bind LDL and remove it from the blood; another

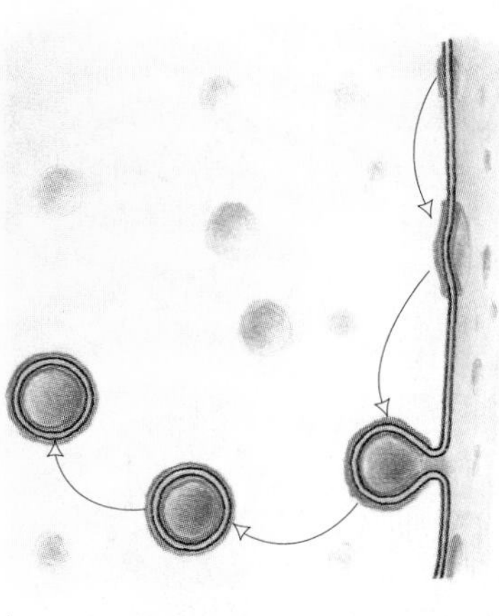

A

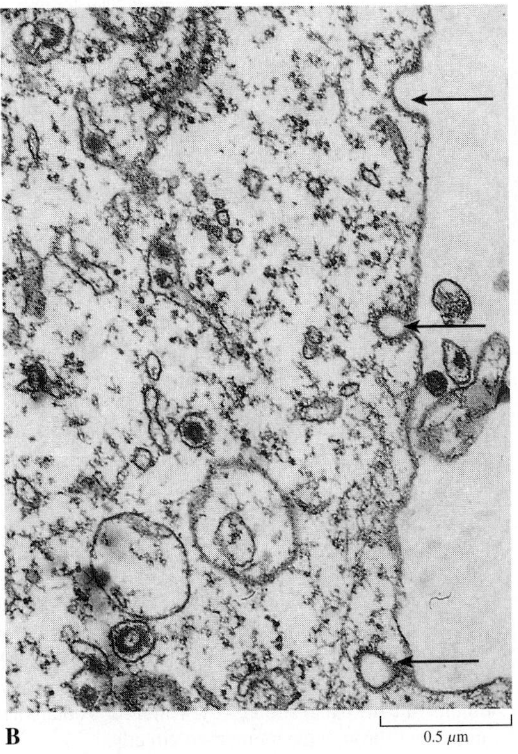

B

5.17 Pinocytosis (A) Particles adsorbed on the membrane are enclosed in vesicles that detach directly from the cell surface and move into the cell. (B) Three stages of pinocytosis by a cultured nerve cell. The material being enclosed is not visible in this photograph.

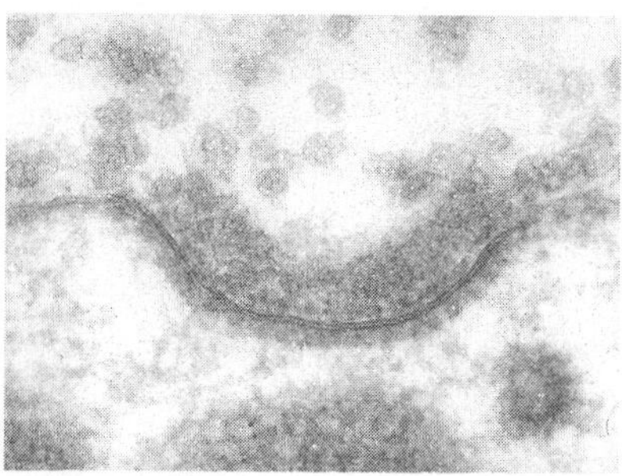

5.18 Endocytosis by means of a coated pit (A) Specialized receptors for lipoproteins have aggregated to form a coated pit in the membrane of an egg cell. (B–D) The pit is subsequently pinched off to form a vesicle. Most endocytotic vesicles are transported to lysosomes, organelles within the cell where the contents are enzymatically altered. The lipoprotein in the vesicle shown here will become part of the yolk.

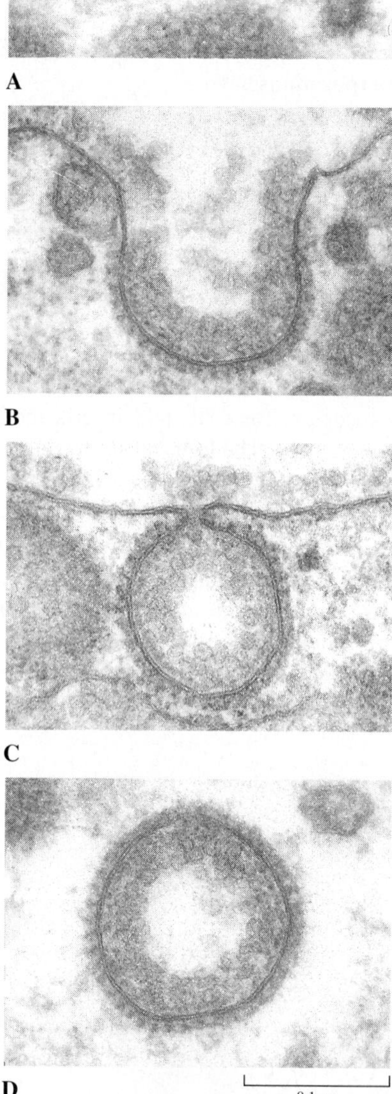

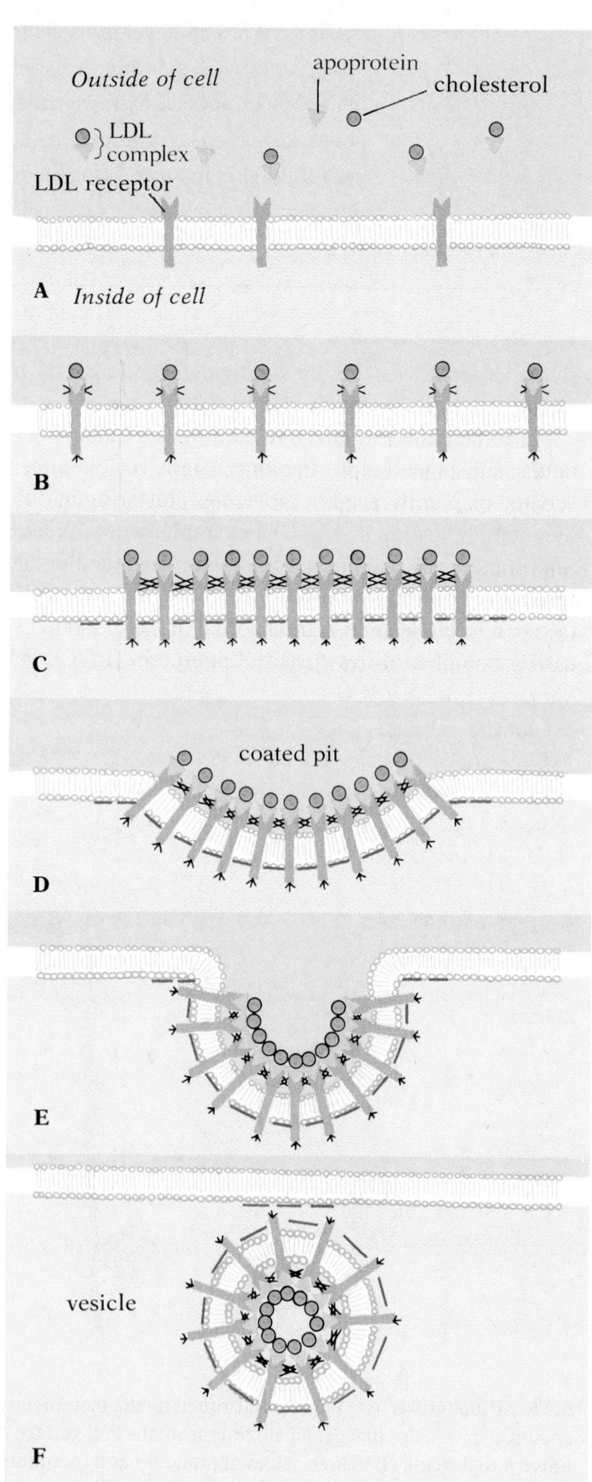

5.19 Endocytosis of cholesterol When a cell needs cholesterol, it synthesizes receptors for low-density lipoprotein (LDL) and incorporates them into the cell membrane, where they are free to migrate (A). The receptors soon bind LDL, a carrier complex that transports cholesterol in the blood (B). The LDL complex includes some 2,000 cholesterol molecules and an associated protein that binds to the LDL receptor. Having bound the LDL, the receptors stop drifting in the membrane and stick to one another (C). Aggregations of LDL receptors trigger endocytosis, the first step of which is the formation of a coated pit (D–E). The resulting vesicle (F) is transported to the site of membrane synthesis. This method of cholesterol uptake may demonstrate the usual strategy by which cells obtain nutrients that cannot pass through the membrane directly.

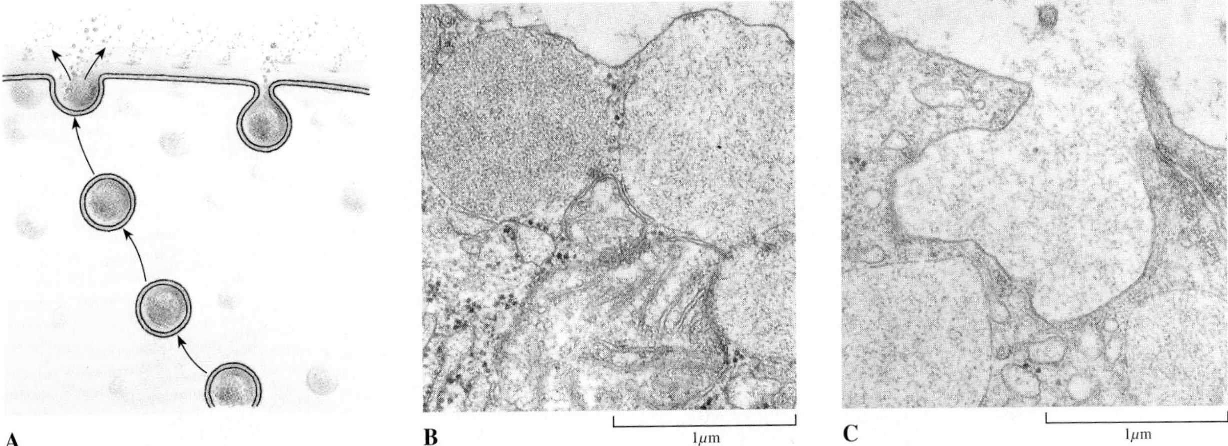

5.20 Exocytosis (A) A membranous vesicle moves to the periphery of the cell, where it bursts, releasing its contents to the exterior. (B) The final steps of exocytosis are seen here, as a vesicle containing fluid fuses with the plasma membrane and bursts (C).

involves failure of the receptors, having bound LDL, to aggregate and initiate endocytosis.

When the material is enclosed within the vesicles, it has not yet fully entered the cell. It is still separated from the interior of the cell by a membrane, and it must eventually cross that membrane (or the membrane must disintegrate) if it is to become incorporated into the cell. Often the material is transported to the lysosomes, organelles within the cell containing digestive enzymes, where the contents are broken down into smaller, simpler substances that can diffuse more easily across the vesicular membranes.

In a process essentially the reverse of endocytosis, called ***exocytosis,*** materials contained in membranous vesicles are conveyed to the periphery of the cell, where the vesicular membrane fuses with the cell membrane and then bursts, releasing its contents to the surrounding medium (Fig. 5.20). Many glandular secretions are released from cells in this way; the hormone insulin, for example, is released by exocytosis from the pancreatic cells that synthesize it. Exocytosis also functions in the secretion of wastes from the cell.

All Cell Walls and Coats Contain Carbohydrates

Biologists have long known that the cells of plants, fungi, and most bacteria have not only cell membranes but also strong, thick outer walls containing many carbohydrates. But only in recent years have they come to realize that most animal cells, too, have a carbohydrate coat on the outer surface of their cell membranes, and that this coat plays an important role in determining certain properties of the cells. Thus, the presence of carbohydrates on their outer surfaces appears to be a universal property of all cells. Nevertheless, the thick, relatively rigid walls of plant, fungal, and bacterial cells, on the one hand, and the thin, nonrigid coats of animal cells, on the other, are among the most striking differences between these groups.

The Cell Walls of Plants, Fungi, and Bacteria Are Stiff

The plant cell wall, which is located outside the cell membrane, has as its principal structural component the complex polysaccharide ***cellulose,*** which is generally present in the form of long threadlike structures called fibrils (Fig. 5.21). The cellulose fibrils are cemented together by a matrix of other carbohydrate derivatives. The spaces between the fibrils are not

5.21 Plant cell wall Electron micrograph of cellulose microfibrils from the cell wall of a green alga. The microfibrils are laid out in parallel lines in two directions; each is about 20 nanometers wide. Water and ions move freely through this meshwork.

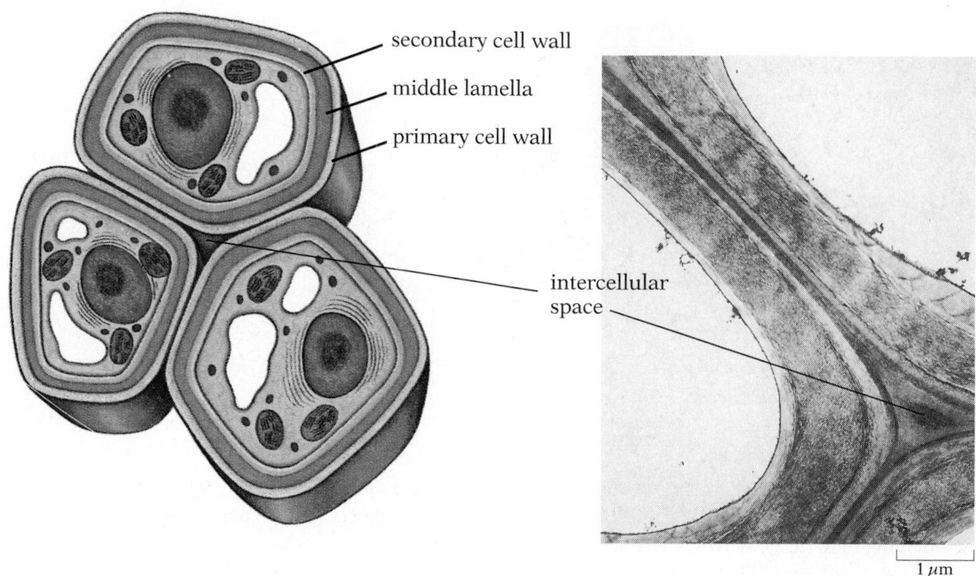

5.22 Cell walls and middle lamella of three adjacent plant cells

entirely filled with matrix, however, and they generally allow water, air, and dissolved materials to pass freely through the cell wall. The wall does not usually determine which materials can enter the cell and which cannot. This function is mostly reserved for the membrane located inside the cell wall.

The first part of the cell wall laid down by a young growing cell is the primary wall. As long as the cell continues to grow, this somewhat elastic wall is the only one formed. Where the walls of two cells abut, a layer between them, known as the ***middle lamella,*** binds them together (Fig. 5.22). ***Pectin,*** a complex polysaccharide generally present in the form of calcium pectate, is one of the principle constituents of the middle lamella. If the pectin is dissolved away, the cells become less tightly bound to one another. That is what happens, for example, when fruits ripen. The calcium pectate is partly converted into other, more soluble forms, the cells become looser, and the fruit becomes softer. Many of the bacteria and fungi that produce soft rots of higher plants do so by first dissolving the pectin, reducing the tissue to a soft pulp that they can absorb.

Cells of the soft tissues of plants have only primary walls and intercelluar middle lamellae. After ceasing to grow, the cells that eventually form the harder, more woody portions of a plant add further layers to the cell wall, forming what is known as the ***secondary wall.*** This cellular structure is located inside the earlier-formed primary wall, lying between it and the membrane (Fig. 5.22). The secondary wall is often much thicker than the primary wall and is composed of a succession of compact layers (composed of cellulose fibrils) oriented at steep angles to one another. This arrangement gives added strength to the cell walls. In addition to cellulose, secondary walls usually contain other materials, such as ***lignin,*** that make them stiffer. Once deposition of the secondary wall is completed, many cells die, leaving the hard tube formed by their walls as a mechanical support and internal transport system for the plant. The cellulose of plant cell walls is commercially important as the main component of paper, cotton, flax, hemp, rayon, celluloid, and, of course, wood itself (Fig. 5.23).

Plant cell walls generally do not form completely uninterrupted boundaries around the cells. There are often tiny holes in the walls through which delicate connections between the internal media of adjacent cells may run. These connections are called ***plasmodesmata*** (Fig.5.24). Thus the interior of an individual cell in a multicellular plant body is not isolated, but is in contact and communication with the interior of other cells by way of plasmodesmata. The complex of cells interconnected by plasmodesmata constitutes a continuous system called the ***symplast.*** A large portion of the intercellular exchange of such materials as sugars and amino acids probably takes place through the plasmodesmata of the symplast.

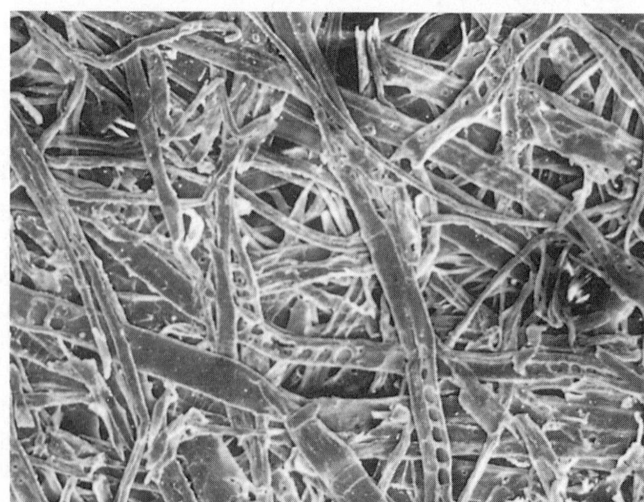

5.23 Scanning electron micrograph of paper fibers. Shown here are the cellulose fibers in uncoated paper.

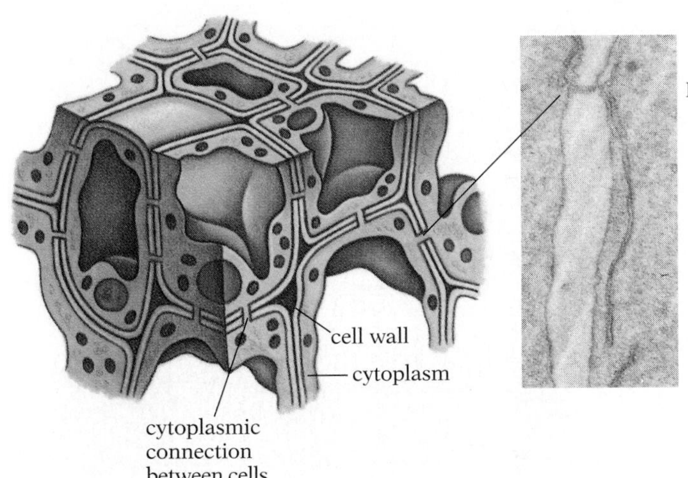

5.24 The symplast The cytoplasm of an individual plant cell in a multicellular plant body is not isolated, but is joined to the cytoplasm of other cells by many tiny cytoplasmic connections called plasmodesmata. The interconnected cells form the symplast, through which water and dissolved materials can move.

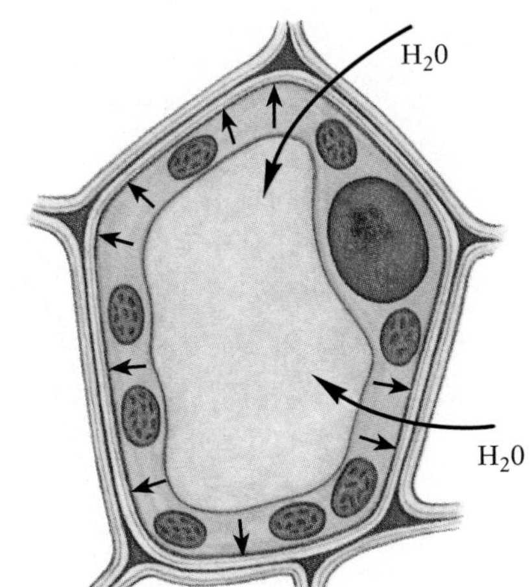

5.25 Plant cell in a hypotonic medium When a plant cell is placed in a hypotonic medium, water moves into the cell, increasing the turgor pressure (arrows) against the cell wall. The walls resist stretching and exert an opposing pressure (wall pressure) against the contents of the cell and prevent bursting.

The cell walls of both fungi and bacteria differ from those of plant cells in that they are not composed of cellulose. In fungi the main structural component of the wall is ***chitin,*** a polymer that is composed of the amino sugar glucosamine (Fig. 3.7, p. 55). In bacteria the cell wall is composed of a compound called ***murein,*** which consists of polysaccharide chains linked together by short chains of amino acids.

Cell walls enable the cells of plants, fungi and bacteria to withstand very dilute (hypotonic) external media without bursting. In such media the cells tend to take up water by osmosis, as a result of the high osmotic concentration of the cell contents. The cell swells, building up ***turgor pressure*** against the cell walls. The walls exert pressure against the swollen cell. The cell wall of a mature cell can usually be stretched by only a minute amount. Equilibrium is reached when the resistance of the wall is so great that no further increase in the size of the cell is possible and, consequently, no more water can enter the cell (Fig. 5.25). Thus, the cells of plants, fungi, and bacteria are not as sensitive as animal cells to the difference in osmotic concentration between the cellular material and the surrounding medium. Because of their walls, such cells can withstand wide fluctuations in the osmotic makeup of the surrounding medium as long as the medium remains hypotonic relative to the cell. If the external medium becomes hypertonic relative to the cell, the cell may lose so much water that it pulls away from the rigid cell wall, a process known as ***plasmolysis*** (Fig. 5.26).

MAKING CONNECTIONS

1. Describe how the different membrane channels convey chemical messages.

2. What is the energy source for active transport?

3. How is the functioning of gated channels related to the action of some of the enzyme inhibitors discussed in Chapter 4?

4. Active transport requires energy to move substances in or out of the cell. How is this requirement of energy related to the Second Law of Thermodynamics and the concept of homeostasis (see Chapter 1, p. •)?

5. Explain why a houseplant wilts if it isn't given enough water, and why watering it enables it to regain its normal healthy appearance (if done in time!).

6. Discuss the role the glycocalyx plays in all recognition and communication.

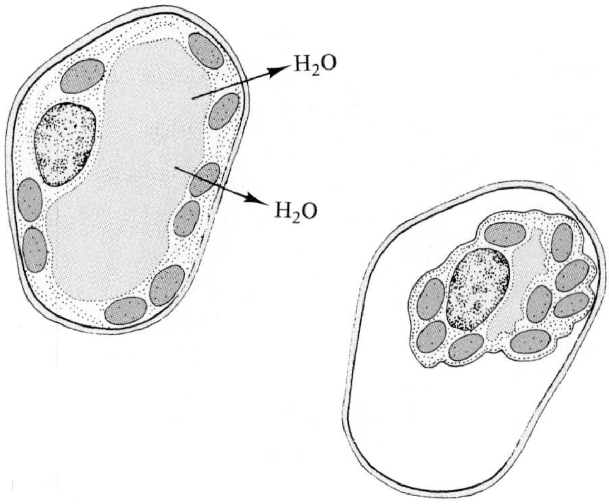

5.26 Plasmolysis A plant cell in a hypertonic medium will lose so much water (left) that, as it shrinks, it will pull away from its more rigid wall (right).

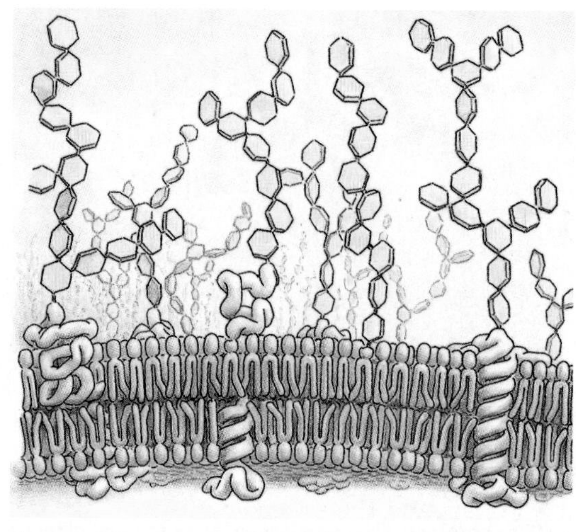

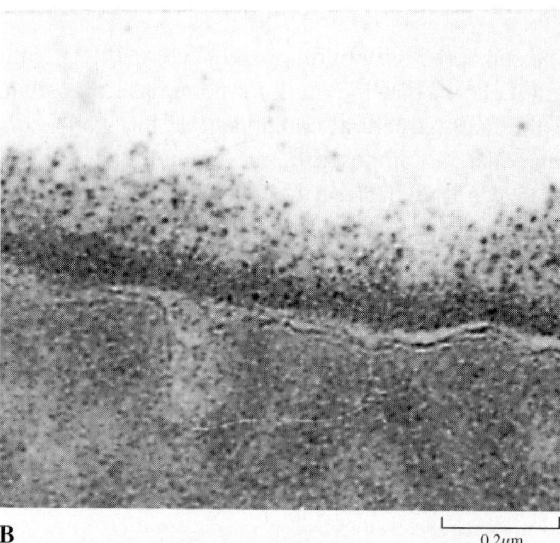

5.27 Plasma membrane with glycocalyx (A) The glycocalyx of an animal cell is composed of oligosaccharides (branching structures) attached to some of the protein and lipid molecules of the outer surface of the membrane. (B) In this electron micrograph, the glycocalyx of a red blood cell gives the outer surface of the membrane a fuzzy appearance.

Animal Cells Have Flexible Coats

In plants, fungi, and bacteria, the cell wall is entirely separate from the cell membrane. By contrast, the cell coat of an animal cell is not an independent entity. The carbohydrates of which it is made are covalently bonded to protein or lipid molecules in the cell membrane (Fig. 5.27). The resulting complex molecules, made up of carbohydrate and protein or carbohydrate and lipid components, are termed glycoproteins and glycolipids, and the cell coat itself is called the ***glycocalyx.*** Glycolipids and glycoproteins are found only on the *outside* of the lipid bilayer.

According to recent research, the glycocalyx provides the recognition sites on the surface of the cell that enable it to interact with other cells. For example, if individual liver and kidney cells are mixed in a culture medium, the liver cells will recognize one another and reassociate; similarly, the kidney cells will seek out their own kind and reassociate. Apparently some property of the glycocalyx enables the cells to distinguish one from the other in such a situation. Cell recognition in the process of embryonic development must also depend, at least in part, on the glycocalyx, and the same is probably true for the control of cell growth. When normal cells grown in tissue culture touch each other, they cease moving and their growth slows down or stops altogether. This phenomenon, known as ***contact inhibition,*** appears to be absent in cancer cells, which continue growing without restraint, probably because an abnormal glycocalyx prevents them from interacting normally. As a means of identification, the glycocalyx is also important in a variety of infectious diseases: malaria parasites, for instance, recognize their host's red blood cells by a distinctive carbohydrate marker on the surface of the cells. Some viruses, including the HIV virus responsible for AIDS, also appear to use carbohydrate markers of the glycocalyx to identify and then invade their target cells.

In all these examples recognition is based on a precise interaction of membrane-based, organic molecules with other molecules. Such molecular recognition mechanisms are common in nature. In animals, for example, they are found in allergic reactions to foreign substances and in the transmission of nerve impulses from one cell to another. Some plants use membrane-based recognition systems in their roots to identify and then nurture beneficial, nitrogen-fixing bacteria.

CONCEPTS IN BRIEF

1. The fundamental organizational unit of life is the cell. According to the *cell theory,* all living things are composed of cells and all cells arise from preexisting cells.

2. Materials required by cells must enter by crossing the surface of the cell, and waste products must leave by the same route. As cell size increases, the volume increases much more rapidly than the surface area, creating the problem of maintaining an adequate exchange surface to support the increased volume.

3. The *cell membrane* is an active part of the cell; it regulates the movement of materials between the ordered interior of the cell and the outer environment. *Diffusion* governs the movement of particles; diffusion is the net movement of particles of a substance from regions of high concentration to regions of lower concentration of that substance.

4. The cell membrane is *selectively permeable;* it allows particles of some substances to pass through while restricting others. The movement of water through a selectively permeable membrane is called *osmosis.*

5. If two different solutions are separated by a selectively permeable membrane, the net movement of water will be from the solution with fewer dissolved particles per unit volume to the solution with the higher osmotic concentration.

6. Because the cell membrane is selectively permeable, the processes of osmosis and diffusion are fundamental to cell life. Cell membranes are relatively permeable to water and to certain simple sugars, amino acids, and lipid-soluble substances; they are relatively impermeable to polysaccharides, proteins, and other very large particles. Their permeability to small particles varies, but in general uncharged particles cross more rapidly than charged ones.

7. A cell in a medium that is *hypertonic* (i.e., a medium with a higher osmotic concentration than the cell) tends to lose water and shrink. Conversely, a cell in a *hypotonic* medium (a medium with a lower osmotic concentration than the cell) tends to gain water and swell, and may even burst. A cell in an *isotonic* medium (a medium whose osmotic concentration is in balance with the cell) neither gains nor loses appreciable amounts of water.

8. According to the *fluid-mosaic model,* the cell membrane is composed of lipids and proteins, with many small pores. The basic structure of the membrane is a *lipid bilayer* composed of two opposed sheets of phospholipid molecules, with their hydrophilic heads oriented toward the surfaces of the membrane and their hydrophobic tails toward the interior. The proteins are distributed both on the surfaces and in the interior of the membrane; some span the entire membrane. The pores are thought to be bounded by protein; the distinctive properties of these proteins make the pores selective as to what can move through them.

9. Transport proteins called *permeases* control molecular traffic into and out of the cell. Some permeases act as *channels,* providing passageways through which specific substances can diffuse passively across the membrane, in a process called *facilitated diffusion.* Other permeases have a *gated channel;* a signal molecule combines with a receptor, which changes shape, opening the channel. Still other permeases, known as *pumps,* carry on *active transport,* using energy to move substances against their concentration gradients.

10. Sometimes substances are taken into the cell by *endocytosis,* a process in which a substance is enclosed in a membrane-bound vesicle pinched off from the cell membrane. The reverse sequence, in which materials within vesicles are conveyed to the surface of the cell and discharged, is called *exocytosis.*

11. The plant cell wall is located outside the membrane and is composed mainly of *cellulose,* a complex polysaccharide. The *primary wall,* the first portion of the wall laid down, consists of a loose network of fibrils. Many plant cells add further layers inside the primary wall, forming a thicker, more compact *secondary wall.* Adjacent cells are bound together by the *middle lamella.*

12. Fungi and bacteria have cell walls made of complex polysaccharides called *chitin* and *murein* respectively. The presence of the cell wall enables the cells of plants, fungi, and bacteria to exist in hypotonic media without bursting.

13. Most animal cells have a cell coat, or *glycocalyx,* composed of carbohydrates covalently bonded to protein or lipid molecules in the cell membrane. The glycocalyx has special *recognition sites* on its surface that enable the cell to interact with other cells and with molecules outside the cell.

STUDY QUESTIONS

1. What are the two components of the cell theory? How did Schleiden, Schwann, Virchow, and Pasteur contribute to its development? (pp. 89–93)
2. Why is diffusion a significant process in living organisms? (pp. 94–95)
3. Define *osmosis, osmotic concentration,* and *osmotic pressure.* Which would have a higher osmotic concentration: distilled water or 1 percent salt solution? Which would have a higher osmotic pressure? (pp. 95–98)
4. Is distilled water hypotonic or hypertonic to 1 percent salt solution?
5. How permeable is the cell membrane to each of the following: lipids and other nonpolar substances, small polar molecules, large polar molecules, and inorganic ions? (p. 98)
6. Describe and make a diagram of the fluid-mosaic model of the cell membrane. Why is "fluid-mosaic" an appropriate name? (pp. 98–100)
7. Compare simple diffusion, facilitated diffusion, and active transport. Which require permeases? Which require the expenditure of energy by the cell? (pp. 101–102)
8. Distinguish between phagocytosis and pinocytosis. Is material taken in by these processes truly in the cell? (pp. 102–107)
9. Describe exocytosis. What kinds of materials may be removed from the cell by this process? (p. 107)
10. Describe the formation and structure of the plant cell wall. Make a drawing to show the positions of the middle lamella, primary cell wall, and secondary cell wall. (pp. 107–109)
11. How are plant cells connected to one another? (p. 108)
12. How do the cell walls of bacteria and fungi differ from those of plants? (p. 109)
13. Describe the structure and function of the glycocalyx in animal cells. (p. 110)

CHAPTER 6

INSIDE THE CELL

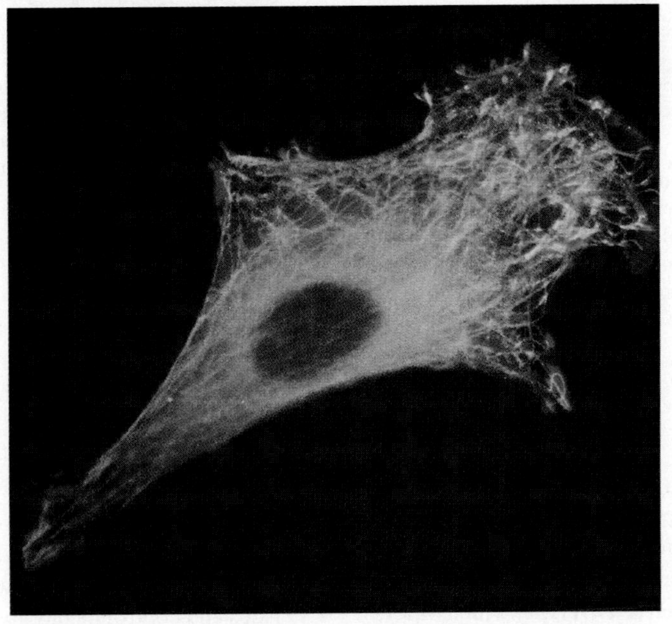

The cell membrane encloses and maintains a chemically fine-tuned, mostly fluid environment called the ***cytoplasm,*** in which a variety of subcellular structures, known as ***organelles,*** are immersed. The organelles are like the divisions of a factory, each with its own specifically assigned task. Some organelles supply energy to other parts of the cell on demand; some use energy to synthesize molecules such as proteins or lipids; still others are responsible for packaging these products for delivery to destinations within the cell or for release beyond the cell's outer boundary. In addition, certain organelles specialize in management or communications.

Because the chemistry of many of the important processes of life is different from that of the cellular cytoplasm as a whole, some organelles serve as miniature containment vessels for reactions requiring unusual conditions, such as high or low pH. Such organelles resemble small cells, complete with their own substructures and bounded by their own protein-studded, lipid-bilayer membrane. Indeed, as we shall see, a current evolutionary theory suggests that some of these organelles originated as independent cells.

The largest and most conspicuous organelle within the cells of most organisms is the ***nucleus.*** The cells of bacteria, however, lack not only well-defined nuclei but most of the other organelles found in the cells of other organisms as well. The structural differences between the cells of bacteria and those of other organisms are so fundamental that bacteria are classified in two kingdoms of their own, and their cells are characterized as ***procaryotic*** ("before a nucleus"), whereas the cells of all other organisms are characterized as ***eucaryotic*** ("having a true nucleus"). In this chapter we first discuss the nucleus and other subcellular organelles of eucaryotic cells, and then take up the special characteristics of procaryotic cells.

The Nucleus Directs a Wide Range of Cellular Activities

The eucaryotic nucleus plays the central role in cellular reproduction, the process by which a single cell divides and forms two new cells. It is from the nucleus of the parent cell that the genetic instructions come forth to guide all the life processes of the succeeding generations of cells, including how they differentiate and what form they will have at maturity. In short, the nucleus serves as the control center not only of its own activities but also determines its progeny's.

The nucleus contains two distinct types of structures: long, threadlike bodies called ***chromosomes*** and dark-staining, generally oval bodies called ***nucleoli.*** With an electron micro-

Chapter opening photo: Cytoskeleton of a cell. A complex array of protein fibers within the cell provides an internal framework and is involved in cell movements. The dark round structure is the nucleus.

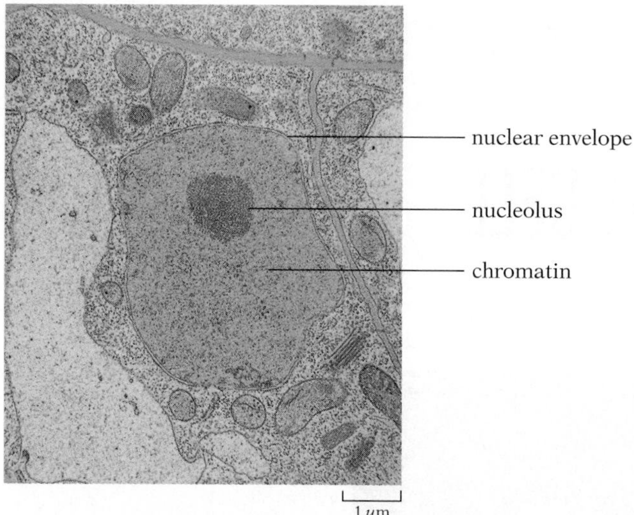

6.1 Electron micrograph of a plant cell nucleus The nucleus of most cells is about one-third the diameter of the whole cell, and so occupies 3 to 4 percent of cellular volume. The nucleus is clearly visible as a distinct membrane-surrounded organelle, and the nucleolus is prominent. However, the chromosomes are not visible in the nondividing cell because they are unwound and too thin and tangled to be recognized as separate entities. They appear only as an irregular granular-looking mass of chromatin.

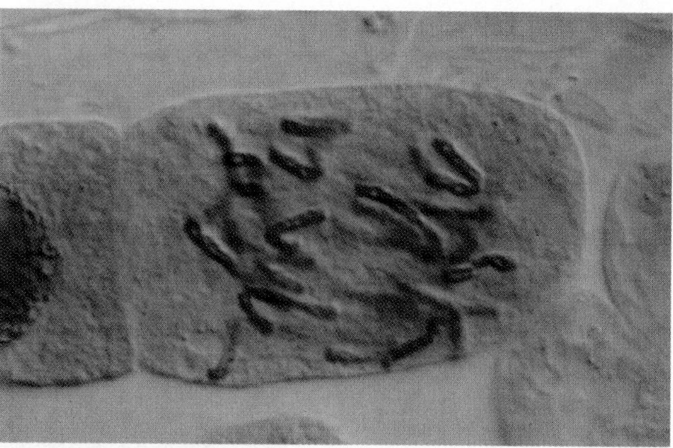

6.2 Chromosomes in a dividing plant cell During cell division the chromosomal material condenses to form elongated, threadlike structures clearly visible under the microscope. In the stage shown here, two groups of chromosomes are moving to opposite ends of the cell.

scope, one can see that both are embedded in a mass of amorphous, granular-appearing material, which is called the ***nucleoplasm*** (Fig. 6.1). The entire nucleus is surrounded by a distinctive double-membrane ***nuclear envelope,*** which differs from other cellular membranes in several important ways (see below).

The chromosomes, which are clearly visible only when the cell is undergoing division (Fig. 6.2), are composed of a complex of DNA and protein called ***chromatin.*** The DNA is the substance of the basic units of heredity, called ***genes,*** while the protein provides spoollike cores on which the DNA is wound to form structures called ***nucleosomes*** (Fig. 6.3). Passed from generation to generation, the genes dictate the cell's characteristics and control its day-to-day activities.

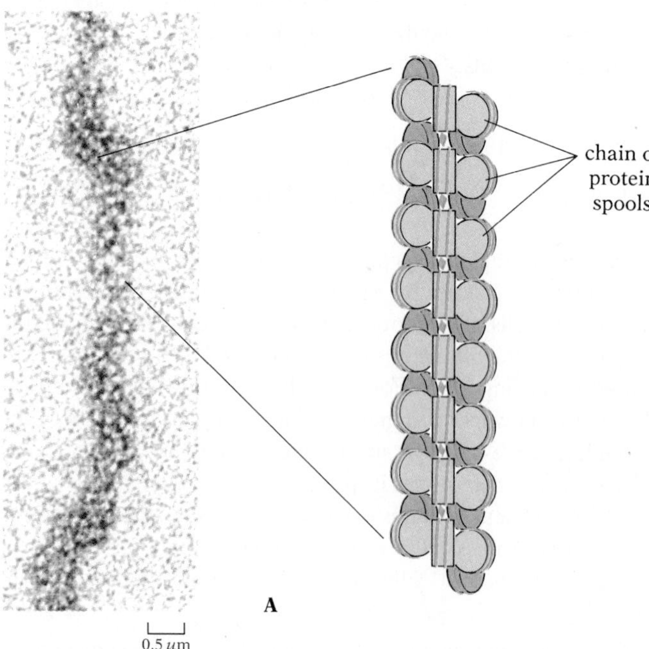

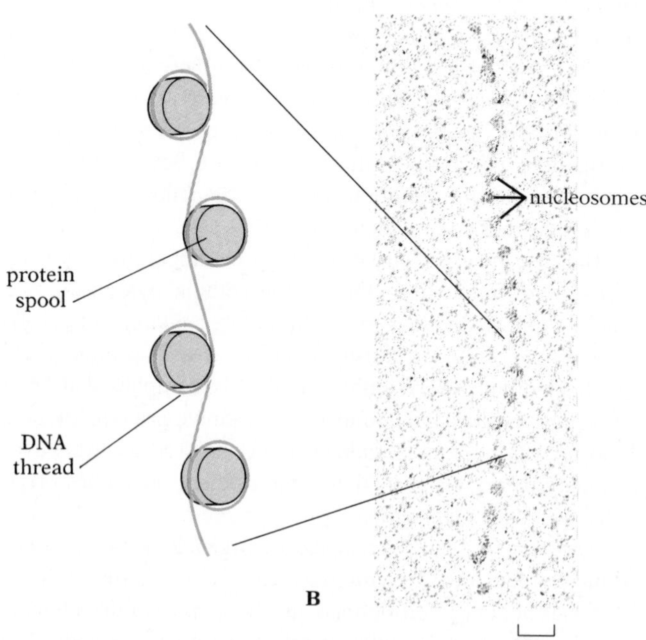

6.3 DNA on nucleosomes The chromosomal DNA of eucaryotes is wound on protein spools, or cores, to form structures called nucleosomes. Normally the spools adhere to one another in a regular way, giving the DNA the appearance of a piece of yarn (A). When the DNA is treated to break the connections between nucleosomes, the individual protein spools and the thread of DNA can be seen (B).

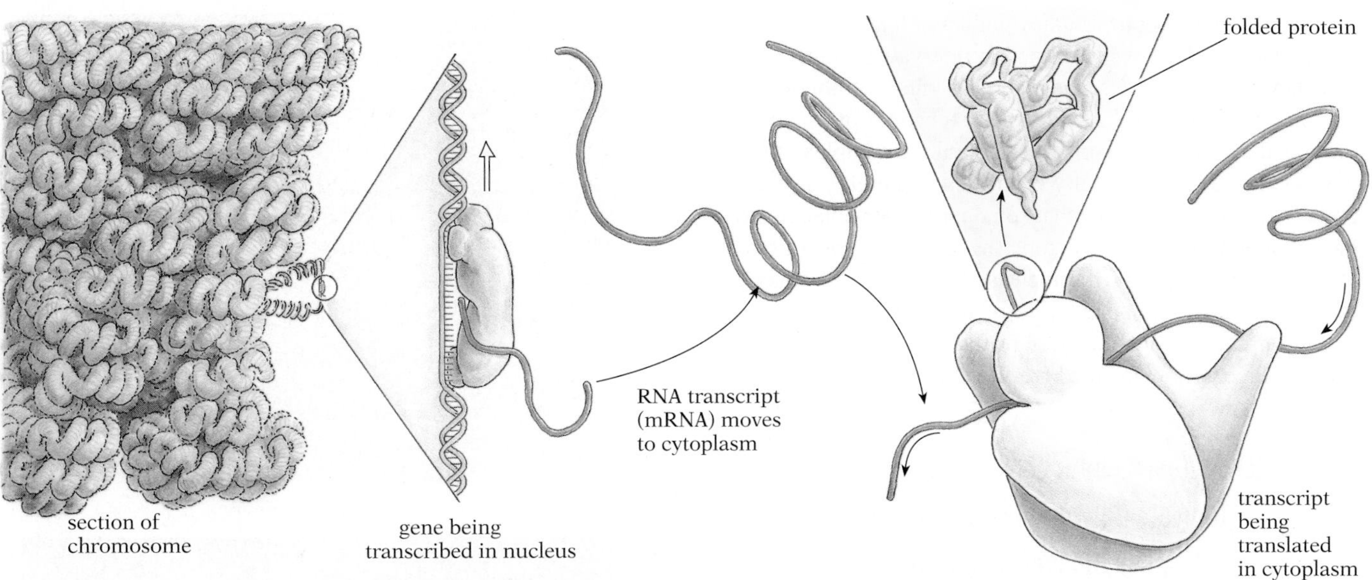

6.4 Flow of information from nucleus to cytoplasm Information encoded in a region of the chromosome (a gene) is transcribed by an enzyme complex to create an RNA copy. This messenger RNA is then exported to the cytoplasm, where it is decoded by a ribosome and the protein specified by the gene is synthesized.

Hereditary information is written in the sequence of the nucleotide building blocks of the DNA molecules. This sequence determines the sequence of amino acids in the proteins (including enzymes) synthesized by the cell. Thus the genes are at the very hub of life. They encode the information needed to synthesize the enzymes, thereby regulating the myriad, interdependent chemical reactions that characterize cells and organisms, and transmit that information from parent to offspring (Fig. 6.4).

The ***nucleoli,*** usually visible within the nuclei of nondividing cells (Fig. 6.1), are, in fact, large loops of DNA coming from a number of chromosomes. The loops contain genes for a particular type of RNA called ribosomal RNA (rRNA). After the rRNA is synthesized, it combines with proteins, leaves the nucleus, and enters the cytoplasm, where it becomes part of the protein-synthesizing organelles called ***ribosomes.*** Accordingly, in cells that carry out little protein synthesis, nucleoli tend to be small or absent.

The nuclear envelope helps maintain a chemical environment within the nucleus different from that in the surrounding cytoplasm. Unlike the cell membrane, the nuclear envelope consists of two distinct lipid-bilayer membranes, an inner one and an outer one, with a fluid-filled space between them (Fig. 6.5).

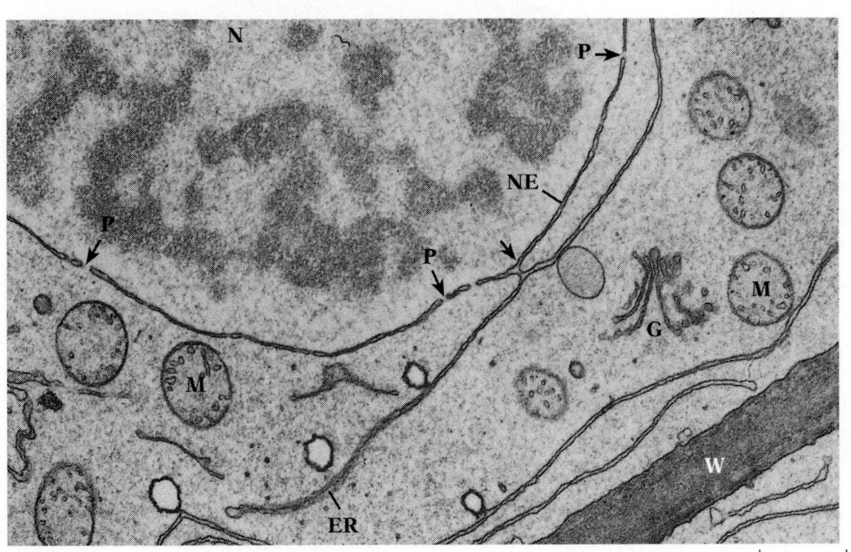

6.5 Electron micrograph showing the nuclear envelope of a cell from corn root The large structure filling the upper left quarter of the picture is the nucleus. The unlabeled arrow indicates a point where the endoplasmic reticulum and the double nuclear membrane interconnect. ER, endoplasmic reticulum; G, Golgi apparatus; M, mitochondrion; N, nucleus; NE, nuclear envelope; P, pore in nuclear envelope; W, cell wall.

Electron-microscope studies indicate that the double-membrane envelope is interrupted at intervals by fairly large and elaborate pores where the outer and inner membranes are continuous (Fig. 6.6). Evidently, exchange of materials through these pores, each of which contains a complex of proteins (Fig. 6.6), is a carefully controlled and highly selective process. It has been shown that some molecules able to cross the cell membrane into the cytoplasm apparently cannot cross the nuclear envelope into the nucleus, even though they may be much smaller than the pores, while other, larger molecules pass readily through in one direction or the other. (See Exploring Further: Freeze-fracture and Freeze-etching, p. 118)

The Endoplasmic Reticulum, the Cell's Internal Delivery Network, Is Also the Site of Much Biochemical Activity

Biologists long wondered if the cytoplasm, the part of the cell contents outside the nucleus, had some sort of invisible structural organization. Faint traces of such a cytoplasmic network were reported at various times. The matter was finally put to rest in 1945 when investigations using a new type of light microscope (a phase-contrast microscope) revealed a complex system of membranes forming a network in the cytoplasm. This system, called the ***endoplasmic reticulum,*** has since been shown to be present to some extent in all eucaryotic cells. Moreover, the electron microscope has revealed a particularly interesting fact about this organelle: it is continuous at some points with the outer membrane of the nuclear envelope (Fig. 6.5).

Although the endoplasmic reticulum varies greatly in appearance in different cells—its components may look like long tubules or round oblong vesicles, or they may form stacks of flattened sacs—it is always a system of membrane-enclosed, fluid-filled spaces. And it seems very likely that in most (and perhaps all) cells, the endoplasmic reticulum forms a single continuous sheet rather than separate segments. In most cells, most of the endoplasmic reticulum's membrane is lined on its outer surface with ***ribosomes,*** small, protein-synthesizing organelles built from rRNA derived from the nucleoli. In this case, the endoplasmic reticulum is spoken of as ***rough ER.*** Where no ribosomes line its membrane, the endoplasmic reticulum is described as ***smooth ER*** (Fig. 6.7). Just as the endoplasmic reticulum may exist without associated ribosomes, so too ribosomes may occur independently of the endoplasmic reticulum and float freely in the cytoplasm.

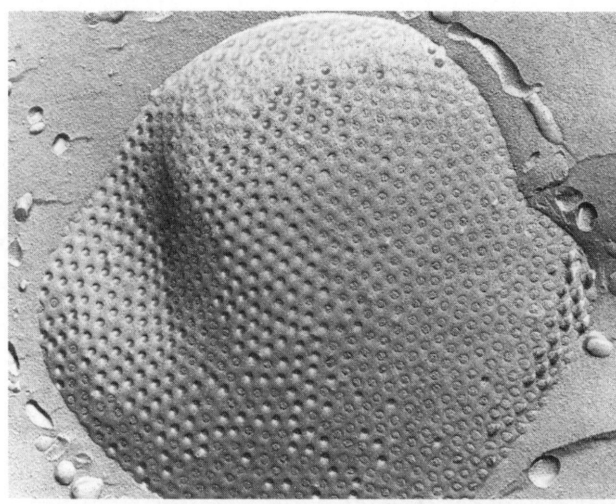
A

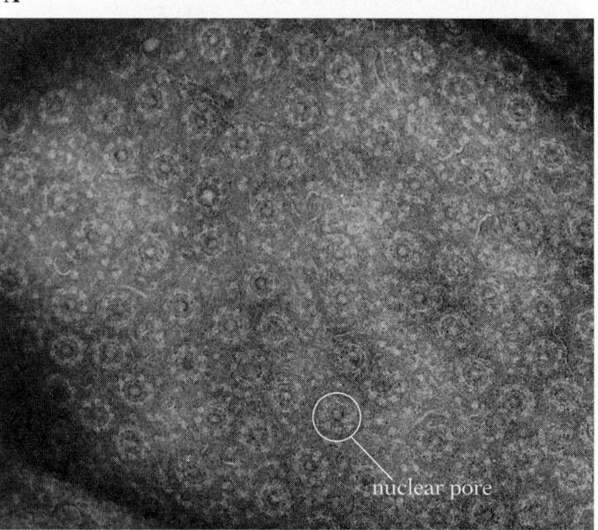

B

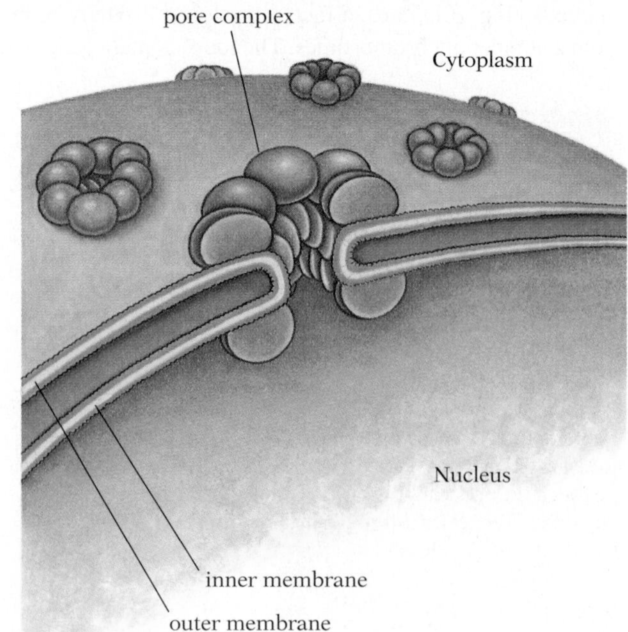

C

6.6 The nuclear envelope (A) The surface of the nuclear envelope with numerous pores. A typical nucleus has a few thousand pores. This electron micrograph was prepared by the freeze fracture technique (see p. 118). (B) An enlargement of the nuclear pores. Each pore complex is composed of eight proteins. (C) Interpretive drawing of the pore complex.

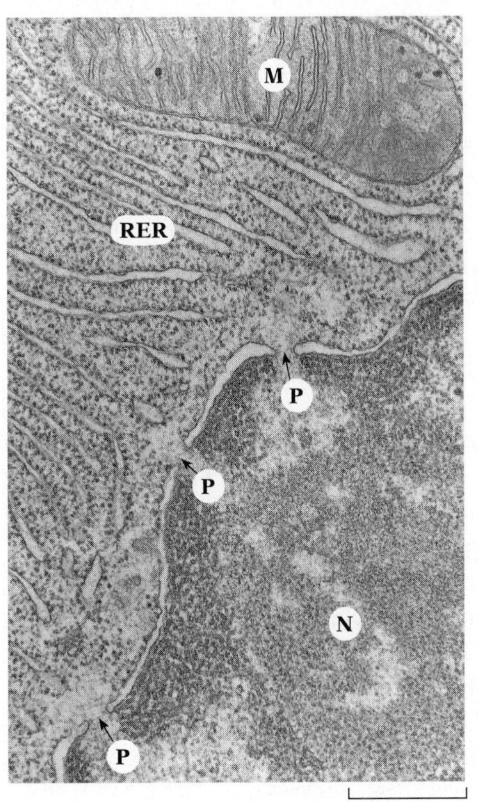

Because the space between the inner and outer membranes of the nuclear envelope is connected to the membrane-enclosed channels and spaces of the endoplasmic reticulum (Figs. 6.7 and 6.8), one possible function of the endoplasmic reticulum is immediately apparent. Its channels can serve as routes for the transport of materials between the nucleus and various parts of the cytoplasm, forming a communication network between the nuclear control center and the rest of the cell. It is thought that most or all of the proteins incorporated into the membrane or transported by the endoplasmic reticulum are synthesized by the ribosomes of the rough ER. Proteins synthesized on free ribosomes in the cytoplasm are not destined for export from the cell or for incorporation into cellular membranes. Instead, they are released to function as enzymes in the ***cytosol,*** the fluid part of the cytoplasm.

The endoplasmic reticulum does not function merely as a passageway for intracellular transport. The membrane of the endoplasmic reticulum has a large protein content, and pro-

6.7 Electron micrograph of endoplasmic reticulum Thin section of a pancreatic cell from a bat, showing many flattened vesicles of rough endoplasmic reticulum (RER); the ribosomes lining the rough ER membranes can be clearly distinguished. In the lower right portion of the micrograph is part of the nucleus (N); note the very prominent pores (P) in the double nuclear envelope. A mitochondrion (M) is at the top.

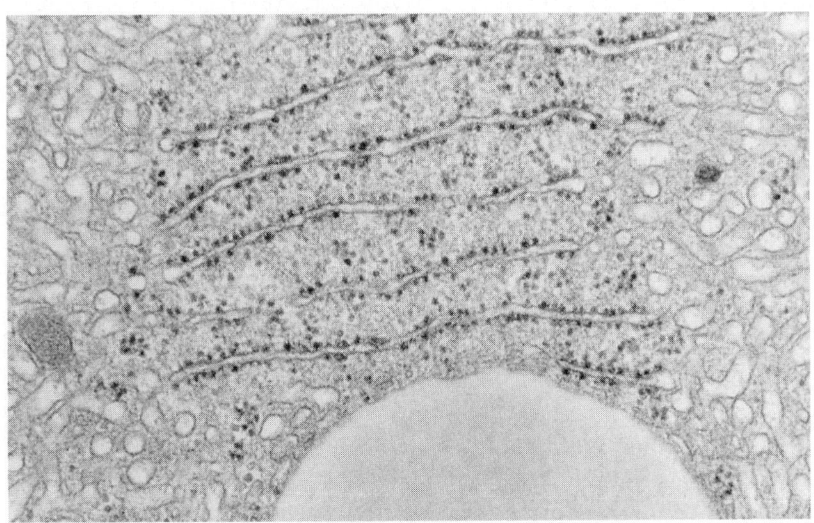

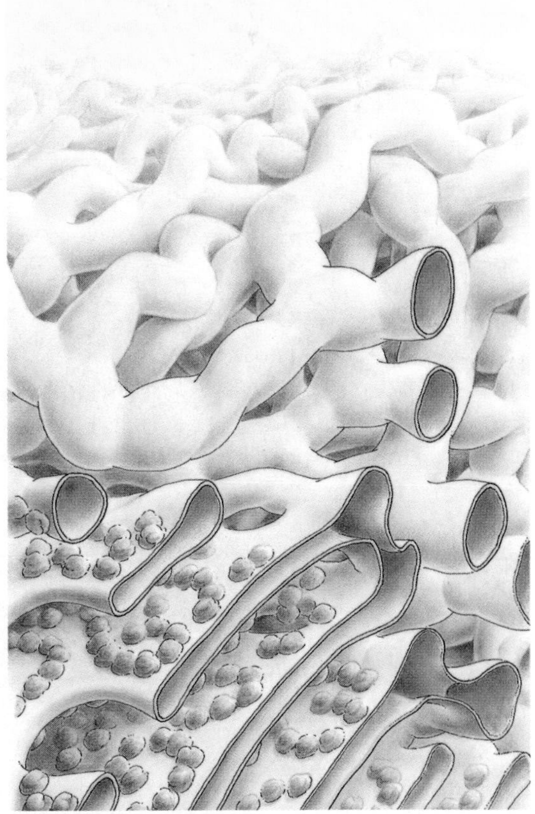

6.8 The endoplasmic reticulum The endoplasmic reticulum from a steroid-producing cell of a guinea pig testis, shown in this electron micrograph, consists of a complex system of membranes including the rough ER (top, center), with associated ribosomes, and smooth ER (sides). The relationship of rough and smooth ER varies from cell to cell; the sketch shows an association more typical of a cell in which macromolecules are synthesized in the rough ER (bottom) and transported to the smooth ER. This representation assumes that the rough and smooth ER are physically continuous; they may actually be separate organelles that communicate via vesicles.

EXPLORING FURTHER

FREEZE-FRACTURE AND FREEZE-ETCHING

Freeze-fracture-etching, a technique for preparing specimens for electron microscopy, has become an indispensable tool for providing detailed confirmation of membrane structure. The specimen is first rapidly frozen and then fractured along the plane of its bilipid membrane (Figs. A and B). Some of the ice is then removed from the specimen by sublimation (conversion directly to vapor), which exposes the inside surface of the membrane and gives the specimen an etched appearance (Fig. C). Carbon and a metal, usually platinum, are then applied at an angle to the specimen (Fig. D) so as to shadow any irregularities in the membrane. Next the original specimen is removed from the platinum cast or surface replica thus formed (Fig. E). The replica can now be examined by microscopy. EMs of freeze-etched cell membranes or other structures in cells usually have a three-dimensional appearance (Fig. F).

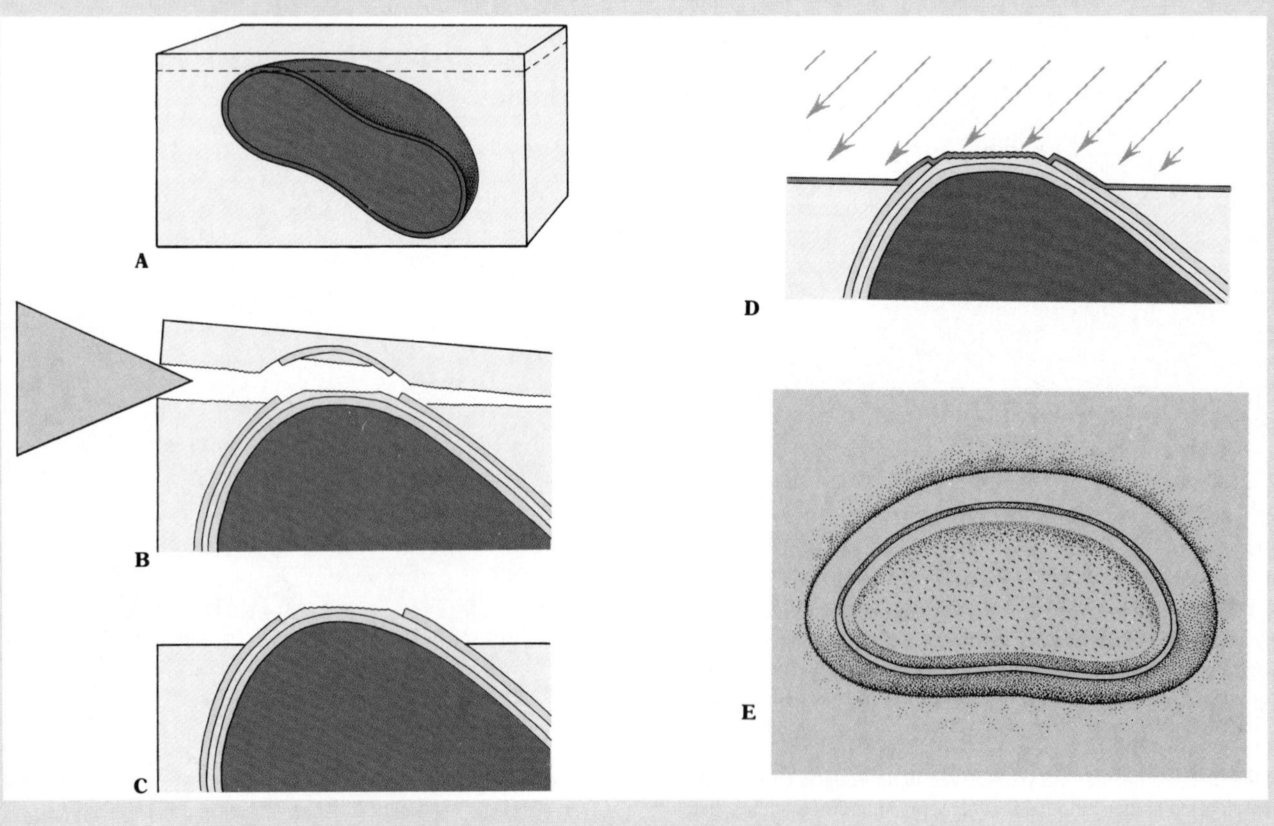

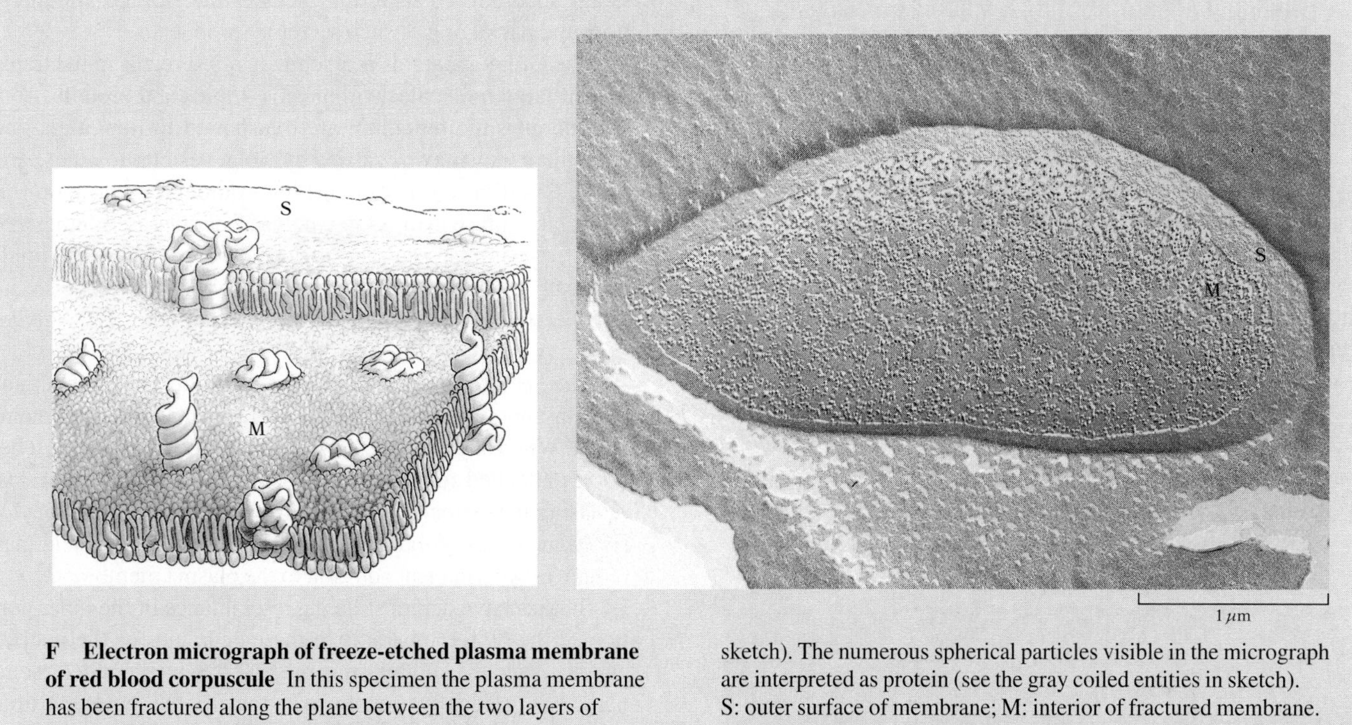

F Electron micrograph of freeze-etched plasma membrane of red blood corpuscule In this specimen the plasma membrane has been fractured along the plane between the two layers of lipids–that is, along the middle of the bilipid core (see sketch). The numerous spherical particles visible in the micrograph are interpreted as protein (see the gray coiled entities in sketch). S: outer surface of membrane; M: interior of fractured membrane.

teins, you will remember, may act as both structural elements in cells and as enzymes for catalyzing chemical reactions. There is now abundant evidence that at least some of the many protein molecules of which these membranes are composed act as enzymes. Some of these enzymes attach small sugar complexes (oligosaccharides) to proteins being transported by the endoplasmic reticulum. These sugar complexes serve as molecular "mailing labels," directing the protein to its destination within the cell (Fig. 6.9). The endoplasmic reticulum also functions as part of a biochemical factory; its proteins catalyze the synthesis of most of the phospholipids and cholesterol used in the manufacturing of new membranes. The membrane proteins synthesized by the ribosomes are inserted into the newly synthesized lipid bilayer. As we shall see shortly, these membrane lipids and proteins are incorporated into the membranes of many other cell organelles, as well as into the plasma membrane.

The most obvious functions of the smooth ER (that is, endoplasmic reticulum lacking ribosomes) are to synthesize membrane phospholipids and to package proteins into membrane-surrounded vesicles for transport to other locations in the cell. Smooth ER (Fig. 6.8) is abundant in cells that specialize in the synthesis of lipids such as steroid hormones. In the vertebrate liver, the membrane proteins of the smooth ER play a crucial role in detoxifying many poisons and drugs, including barbiturates, amphetamines, morphine, and codeine.

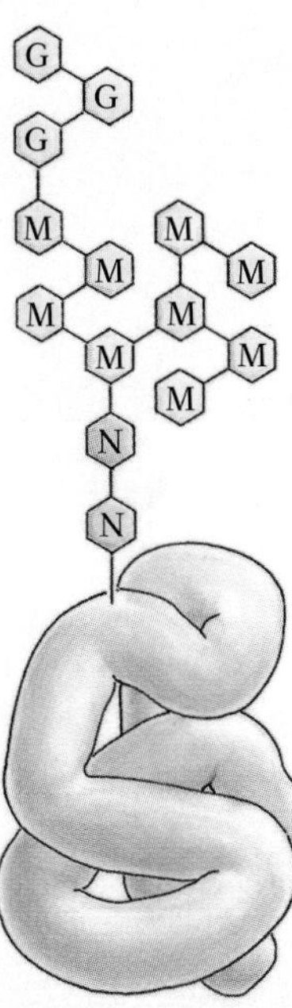

6.9 Carbohydrate mailing label attached to proteins in the endoplasmic reticulum A 14-sugar side chain is attached to nearly all proteins synthesized on the rough ER and serves as a "mailing label." Proteins lacking this label remain in the rough ER. When the four terminal sugars are removed, the protein is exported to the Golgi apparatus in vesicles. The size of the tag in this drawing is exaggerated compared to the protein. G, glucose; M, mannose; N, *N*-acetyl-glucosamine.

The Golgi Apparatus Stores, Modifies, Packages, and Transports Secretory Products

Another cellular network, known as the ***Golgi apparatus*** (after Camillo Golgi, who first described it in 1898), consists of a system of membrane-enclosed sacs arranged approximately parallel to each other (Figs. 6.10 and 6.11). Transport vesicles that "bud" off the endoplasmic reticulum fuse with the sacs of the Golgi apparatus, bringing newly synthesized proteins and lipids along with them. Vesicles that bud off the Golgi apparatus will in turn deliver these membrane components eventually to other organelles and to the plasma membrane.

The Golgi apparatus is particularly prominent in cells that are involved in the secretion of various chemical products, such as the pancreatic cells that secrete insulin or the intestinal cells that secrete mucus. It is now known that the role of the Golgi apparatus in secretion includes the storage, modification, and packaging of such secretory products.

The Golgi apparatus is also the major director in the transport of large molecules within cells. Lipids and proteins from the endoplasmic reticulum are transported to the Golgi, and from there they may be carried by Golgi vesicles to other parts of the Golgi or to other organelles. Although no protein synthesis takes place in the Golgi, polysaccharides are synthesized there from simple sugars and are attached to proteins and lipids to create glycolipids and glycoproteins. Some of these are transported outward as part of vesicle membranes to the glycocalyx.

The secretory vesicles produced by the Golgi apparatus play an important role in adding surface area to the cell membrane. When one of the vesicles moves to the cell surface, it becomes attached to the plasma membrane and then ruptures, releasing its contents to the exterior in the process of exocytosis. Some or all of the membrane of the ruptured vesicle may remain as a permanent addition to the plasma membrane.

Studies have demonstrated a resemblance of the inner portion of the Golgi apparatus to the membranes of the nuclear envelope and the endoplasmic reticulum; however, this resemblance is gradually transformed toward the outer portion, which resembles the plasma membrane on the cell's periphery. It is likely, therefore, that a dynamic relationship exists between the different parts of the cellular membrane system (Fig. 6.11). Thus, a membrane molecule originating in the rough ER could conceivably appear consecutively in the smooth ER, the Golgi apparatus, a vesicle membrane, and finally the cell's plasma membrane. From the plasma membrane it might eventually migrate back to the Golgi apparatus or some other organelle as part of an empty vesicle. Such recycling of the membrane phospholipids is important.

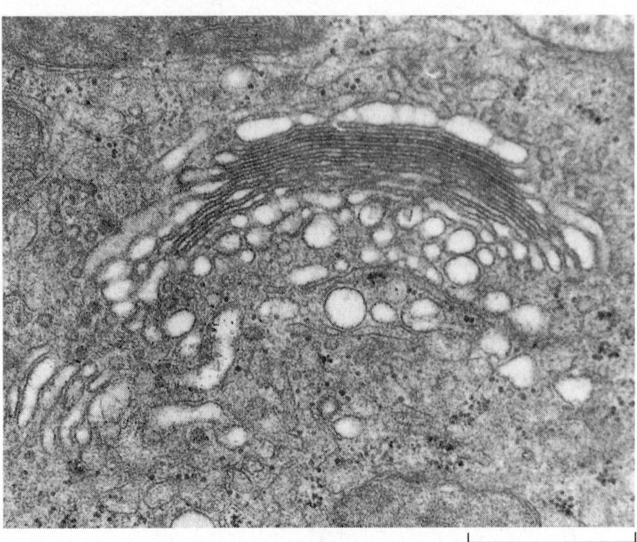

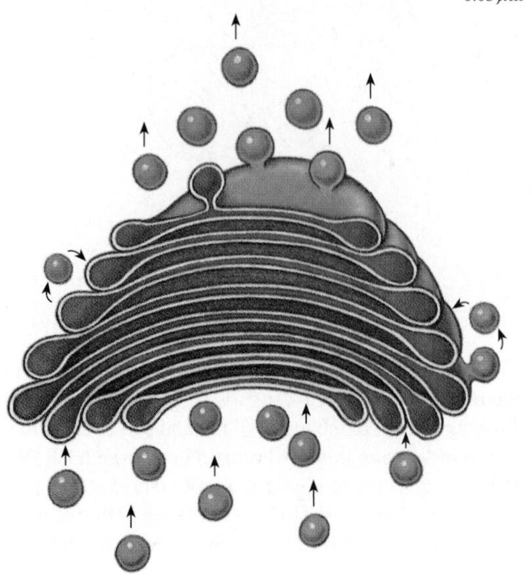

6.10 The Golgi apparatus Electron micrograph of Golgi apparatus from an amoeba and an interpretive drawing. Transport vesicles from the endoplasmic reticulum (bottom) fuse with the sacs of the Golgi, bringing newly synthesized proteins and lipids along with them. The materials move through the various compartments to the outermost sac. The Golgi sorts, modifies, relabels, and packages molecules into vesicles for further transport. Vesicles forming at the ends of some of the sacs can be seen in the electron micrograph.

Lysosomes Store Digestive Enzymes for Delivery to Various Parts of the Cell

Lysosomes are membrane-enclosed bodies that function as storage vesicles for many powerful hydrolytic enzymes used in the digestion of macromolecules within the cell (Fig. 6.12). The lysosome membrane, which comprises a single layer, is impermeable to the digestive enzymes stored within it and ca-

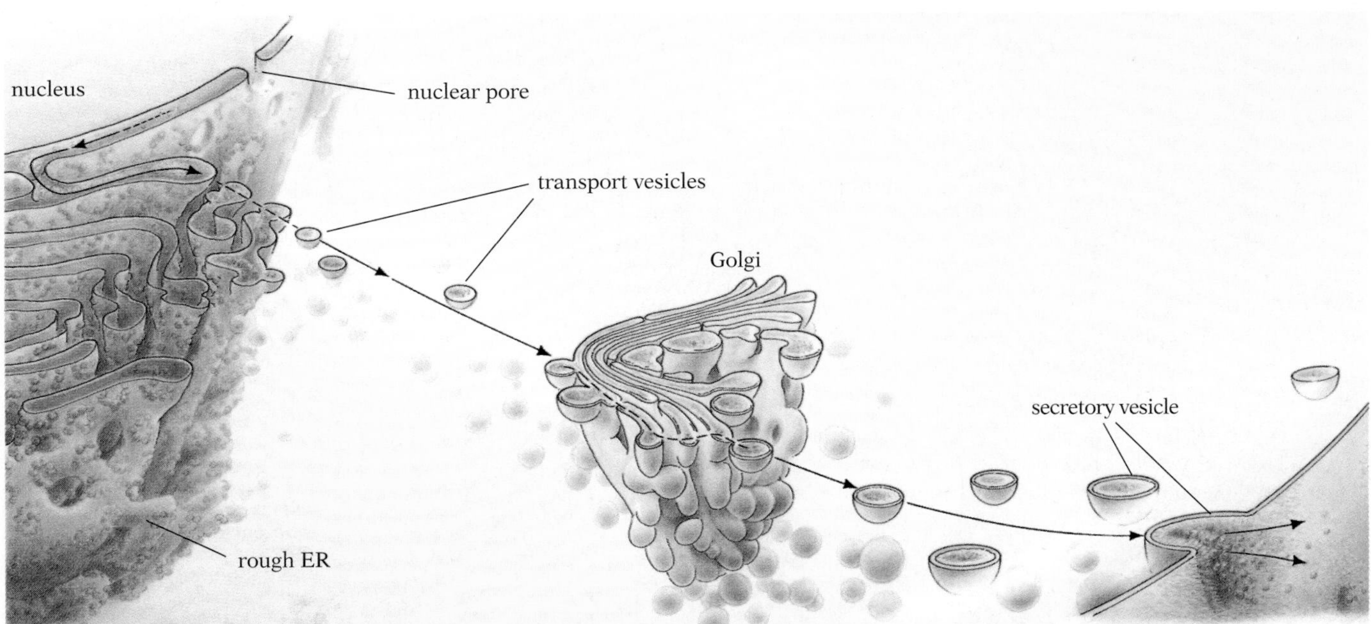

6.11 A probable path of phospholipid movement in the cell Structural molecules of cellular membranes are constantly on the move. The path traced here shows the movement of a phospholipid from the point of synthesis in the nuclear envelope until it is incorporated into the cell membrane. The lipid first moves through the lumen of the rough ER and (either directly or via vesicles) into the smooth ER. There it receives a specific carbohydrate marker, which makes it a glycolipid, and becomes part of a vesicle transporting proteins to the Golgi apparatus; it is then incorporated into the membrane of that organelle. Subsequently the same molecule, perhaps with a new or modified carbohydrate marker, moves to the outer layer of the Golgi, where it becomes part of a secretory vesicle that is carried to the plasma membrane. There, in the process of exocytosis, the vesicle membrane (including the glycolipid molecule) fuses with the plasma membrane of the cell. Other membrane molecules follow different paths.

pable of withstanding their digestive action. If the lysosome membrane is ruptured, the enzymes released into the surrounding cytoplasm begin immediately to break down the macromolecules inside the cell.

Lysosomes act as the digestive system of the cell, enabling it to process the bulk material taken in by endocytosis. Digestive enzymes are synthesized in the rough ER, packaged into transport vesicles in the smooth ER, and carried to the Golgi apparatus. When a region has received an appropriate supply of the enzymes, it buds off from the Golgi membrane as a lysosome, and the enzymes are activated. Proteins mounted on the exterior of the lysosome's membrane serve as recognition sites to assure that the enzymes being carried are delivered to appropriate targets, which include endocytotic vesicles derived from the cell surface; the two bodies often fuse to produce a special class of digestive vesicles. When digestion of the contents of such vesicles is complete, the useful products pass into the cytosol, while the residue is discharged by exocytosis (Fig. 6.13).

Several diseases are caused by lysosome disorders. A particularly devastating one is the nervous disorder known as Tay-Sachs disease, in which lysosomes lack a particular lipid-digesting enzyme. When these enzyme-deficient lysosomes fuse with lipid-containing vesicles, they do not fully digest

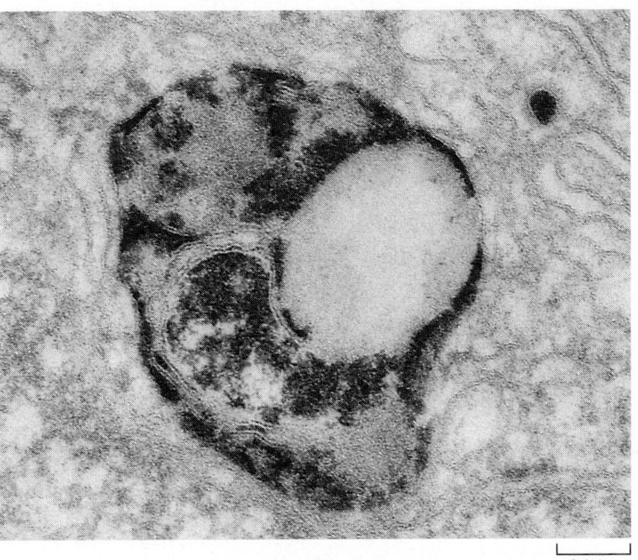

6.12 Lysosomes in a connective-tissue cell from the vas deferens of a rat The small dark body at upper right is a lysosome. The much larger body at left is a digestive vacuole formed by fusion of a lysosome with a phagocytic or pinocytic vesicle. (The dark appearance of the lysosomes results from staining for acid phosphatase, a digestive enzyme whose presence is used as the definitive test for these organelles.)

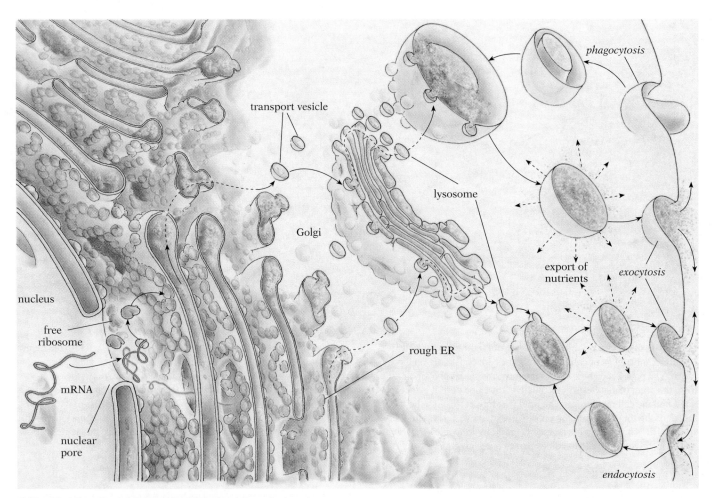

6.13 The digestive enzyme cycle The role of the endoplasmic reticulum and the Golgi apparatus in cellular digestion is representative of their functions in the cell. The DNA in the nucleus that encodes the digestive enzymes is copied into mRNA (left), which is transported to the ribosomes. The ribosomes then bind to the rough ER, where the enzymes are synthesized and passed into the ER. The enzymes then move into the smooth ER where they are packaged into transport vesicles. These vesicles fuse with the Golgi and lysosomes are formed. The lysosomes fuse with endocytotic vesicles (right) and the enzymes digest their contents. Nutrients are exported into the cytoplasm and the residue leaves the cell by exocytosis.

their contents. The resulting defective lysosomes accumulate and block the long, thin parts of nerve cells responsible for transmitting nerve impulses, leading eventually to death.

Peroxisomes Contain Oxidative Enzymes That Transfer Hydrogen

Improved methods for separating cell contents have revealed that certain membrane-surrounded organelles previously confused with lysosomes actually are distinct entities, which have become known as ***microbodies.*** The most thoroughly understood of these microbodies are ***peroxisomes*** (Fig. 6.14).

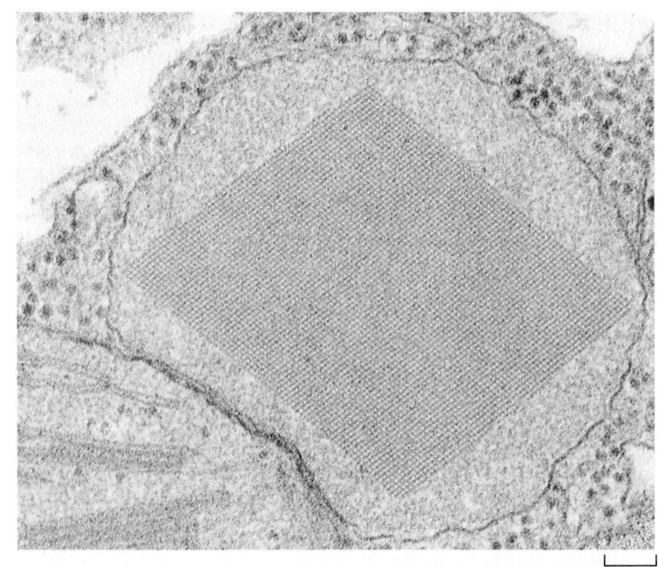

6.14 Electron micrograph of a tobacco-leaf peroxisome The tobacco-leaf cell has been treated with stain, which reveals the crystalline core of this peroxisome and the presence in it of oxidizing enzymes. In mammals, peroxisomal enzymes neutralize hydrogen peroxide, alcohol, and other potentially harmful substances.

Though similar in appearance to lysosomes, peroxisomes are not produced by budding from the Golgi apparatus. Instead, peroxisomes arise by the growth and division of existing peroxisomes. Cells cannot spontaneously generate peroxisomes, and therefore a cell that fails to inherit at least one peroxisome from the cytoplasm of its "parent" will inevitably die.

Like lysosomes, peroxisomes contain powerful enzymes. Unlike the digestive enzymes found in lysosomes, however, peroxisome enzymes are oxidative: they catalyze reactions in which hydrogen atoms are transferred from certain organic substrates (such as formaldehyde and ethyl alcohol) to oxygen, forming hydrogen peroxide (H_2O_2), a compound that is extremely toxic for living cells. Fortunately, peroxisomes have another enzyme, ***catalase,*** that converts the toxic hydrogen peroxide into harmless water and oxygen. Human liver and kidney cells have large peroxisomes; almost half of the ethyl alcohol a person consumes in alcoholic drinks is oxidized by the peroxisomes in these cells.

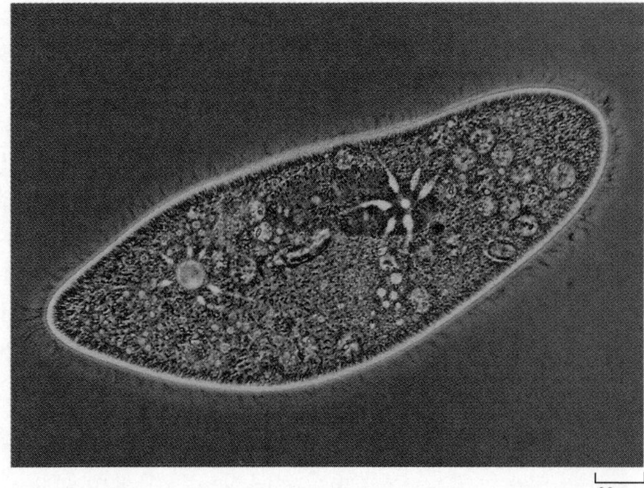

6.15 *Paramecium* with two contractile vacuoles The contractile vacuole (light starlike structure) at the right is in the process of expelling water from the cell. Water is still accumulating in the vacuole on the left.

Vacuoles Store Nutrients and Other Substances and Expel Wastes

Membrane-enclosed, fluid-filled spaces called ***vacuoles*** are found in both animal and plant cells, although they have their greatest development in plant cells. There are various kinds of vacuoles, with a corresponding variety of functions. In some protozoans, specialized vacuoles called ***contractile vacuoles*** play an important role in expelling excess water from the cell (Fig. 6.15). Many protozoans also form special vacuoles for digesting food particles.

In most mature plant cells, a large vacuole generally occupies between 30 and 90 percent of the cell volume. The immature cell usually contains many small vesicles derived from the endoplasmic reticulum and the Golgi apparatus. As the cell matures, these vesicles take in more water and become larger, eventually fusing to form the large, definitive vacuole of the mature cell (Fig. 6.16). This process pushes the cytoplasm to the periphery of the cell, where it forms a thin layer.

The plant vacuole contains a liquid ***cell sap***—primarily water, with a variety of substances dissolved in it. The vacuolar membrane has its own distinctive permeability characteristics and is capable of regulating the direction of movement of substances across it. As the vacuole takes in water by osmosis, it swells and pushes outward against the cytoplasm, which, being essentially fluid, resists compression and transmits the pressure to the cell wall. The wall is strong enough to limit the swelling and prevent the cell from bursting, but the outward push of the vacuolar membrane makes the cell stiff, or ***turgid.*** The turgidity of plant cells is an important source of support for some plant parts; think about what happens to the leaves and stems of a nonwoody plant when there is a loss of turgor (rigidity) when the plant wilts.

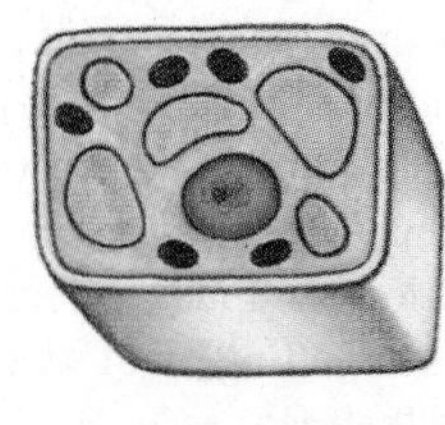

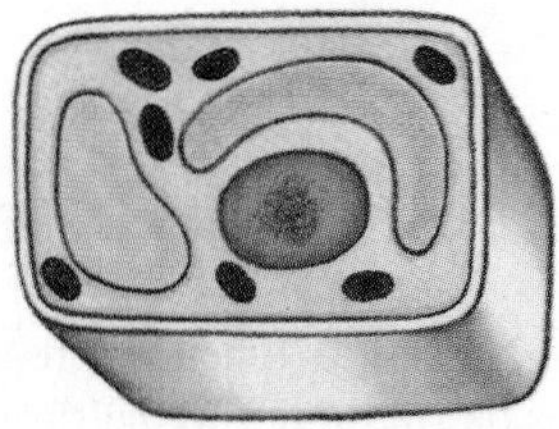

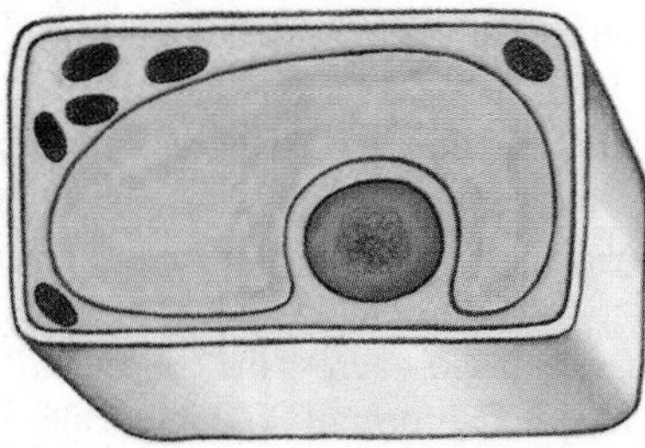

6.16 Development of the plant-cell vacuole The immature cell (top) has many small vacuoles. As the cell grows, these vacuoles fuse and eventually form a single large vacuole, which occupies most of the volume of the mature cell, the cytoplasm having been pushed to the periphery (bottom).

MAKING CONNECTIONS

1. Why are there so many organelles in eucaryotic cells?

2. Why is the nucleus called the control center of the cell?

3. Trace the sequence of events and cellular structures involved in the digestion of food taken into the cell. How do lysosomes carry their digestive enzymes to the appropriate targets?

4. Relate the functions of the contractile vacuole of some protozoans and the vacuole found in most mature plant cells to the osmotic relationships of these cells to their environment.

Many other substances of importance in the life of the plant cell are stored in the vacuoles, among them the red pigments called ***anthocyanins,*** which are responsible for many of the purples, blues, and dark reds commonly seen in flowers, fruits, and autumn leaves. Vacuoles also store high concentrations of ions and soluble organic nitrogen compounds (including amino acids), sugars, various organic acids, and some proteins. The high osmotic concentration of the vacuole enables many plants to absorb water by osmosis from very dry soils. In addition, plant vacuoles often function as dumping sites for metabolic wastes. The latter function is important because plants have no kidneys or other means of getting rid of wastes; they shed their wastes when they shed their leaves.

Mitochondria Supply Energy for Most Cell Functions

For subcellular organelles to function properly—from the building of proteins on ribosomes to the fusion of lysosomes with the plasma membrane in exocytosis—energy is required. Two organelles are responsible for supplying the energy necessary to build and fuel all the others: mitochondria and chloroplasts.

Mitochondria, the main powerhouses of eucaryotic cells, are the sites of the respiratory processes that extract energy from food to synthesize ATP, the cell's energy currency. The energy needed to make muscles contract or to drive the pumps involved in active transport across the cell membrane, for example, is provided in the form of ATP molecules produced in mitochondria. Indeed, the activity level of a cell is directly proportional to the number of mitochondria the cell contains; a heart muscle cell, for example, which pumps 72 times per minute, may contain thousands of mitochondria.

Like the nucleus, each mitochondrion is bounded by a double membrane; the outer membrane is smooth, while the inner membrane has many inwardly directed folds called ***cristae,*** which extend into an amorphous semifluid matrix (Fig. 6.17). We shall discuss the structure and operation of this important organelle more fully in Chapter 8.

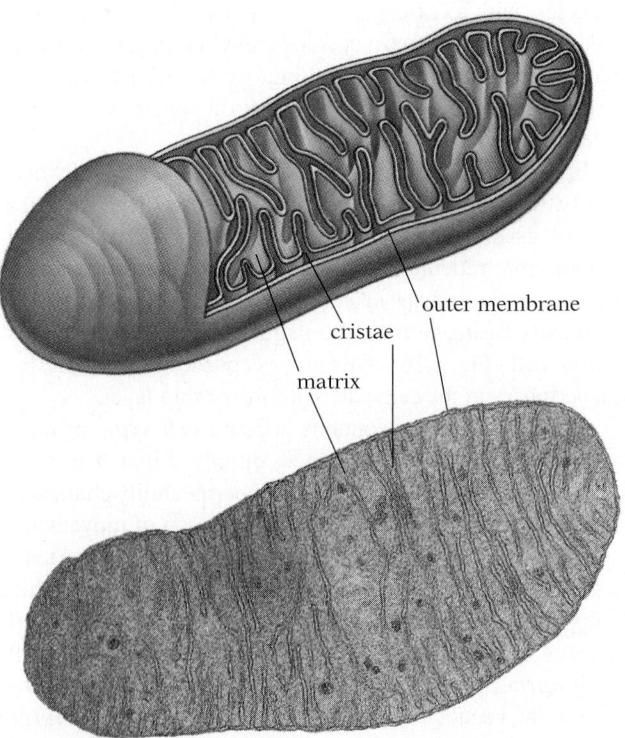

6.17 Diagram and Electron micrograph of a mitochondrion
Note the double membrane: the outer membrane is smooth white and the inner membrane consists of numerous cristae, which apparently arise as folds of the inner membrane.

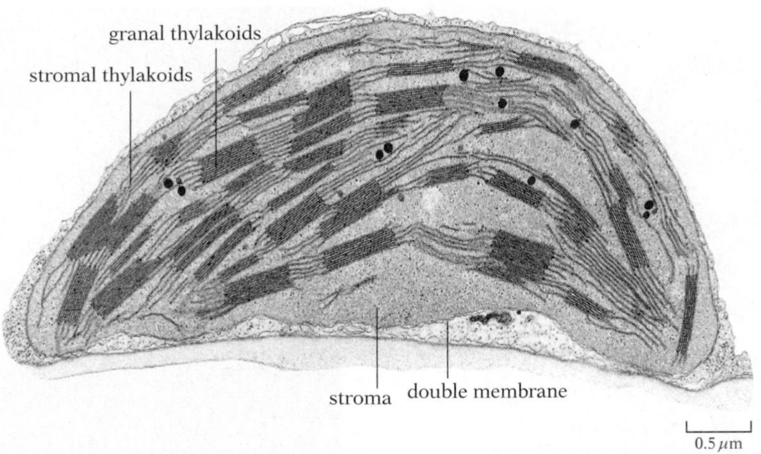

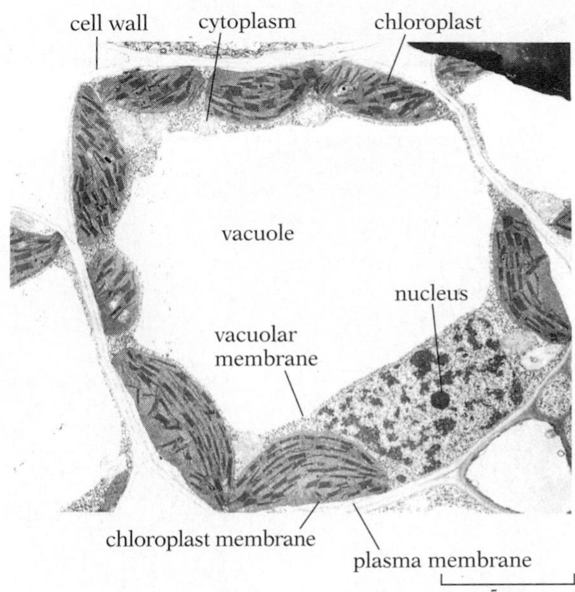

6.18 Electron micrographs of chloroplasts Stacks of disklike thylakoids, forming grana, can be seen in this chloroplast of a corn leaf (top). Numerous chloroplasts lie close to the perimeter of a mature leaf cell of timothy grass (bottom). Notice that the single large vacuole occupies most of the volume of the cell, the cytoplasm with its chloroplasts having been pushed to the periphery.

Plastids Contain Various Molecules, Including Light-Absorbing Pigments, in Plant Cells

Plastids are large cytoplasmic organelles found in the cells of most plants, but not in the cells of fungi or animals. Plastids are clearly visible through an ordinary light microscope. There are two principal categories of plastids: ***chromoplasts*** (colored plastids) and ***leucoplasts*** (white or colorless plastids).

Chloroplasts are chromoplasts that contain the green pigment ***chlorophyll*** along with various yellow or orange pigments called ***carotenoids.*** We shall explore in the next chapter how the radiant energy of sunlight is trapped in chloroplasts by molecules of chlorophyll and then used in the manufacture of complex organic molecules (particularly sugars) from simple inorganic raw materials such as water and carbon dioxide. Oxygen is a by-product of this photosynthetic reaction, upon which nearly all organisms depend.

Like the mitochondrion, the typical chloroplast is bounded by two concentric, lipid-bilayer membranes and has a complex, internal membranous organization (Fig. 6.18). The fairly homogeneous internal matrix, known as the ***stroma,*** has numerous flat membranous sacs, known as ***thylakoids,*** embedded in it. The thylakoids are interconnected; some thylakoids run through the stroma, and other platelike thylakoids form in stacks called ***grana*** (Fig. 6.18). The chlorophyll and carotenoid pigments are embedded in the thylakoid membranes. The membrane structure of chloroplasts will be considered in more detail in the next chapter, when we examine the photosynthetic process.

Chromoplasts lacking chlorophyll are usually yellow or orange (occasionally red) because of the carotenoids they contain. It is these chromoplasts that give many flowers, ripe fruits, and autumn leaves their characteristic yellow or orange color. Some of these chromoplasts have never contained chlorophyll, while others have lost their chlorophyll. The latter are particularly common in structures that were once green, such as ripe fruits and autumn leaves.

The leucoplasts, or colorless plastids, are primarily organelles in which materials such as starch, oils, and protein granules are stored. Plastids filled with starch, called ***amyloplasts,*** are particularly common in seeds, such as rice and corn, and in storage roots and stems, such as carrots and potatoes, although they also occur elsewhere in the plant. The starch, an energy-storage compound, is deposited in the plastid as a grain or group of grains (Fig. 6.19). Plants rich in such starch grains are good high-energy foods.

In addition to playing key roles in photosynthesis and energy storage, plastids are involved in other important biosynthetic processes. For example, it is in the chloroplasts that most amino acids, fatty acids, purines, and pyrimidines are synthesized.

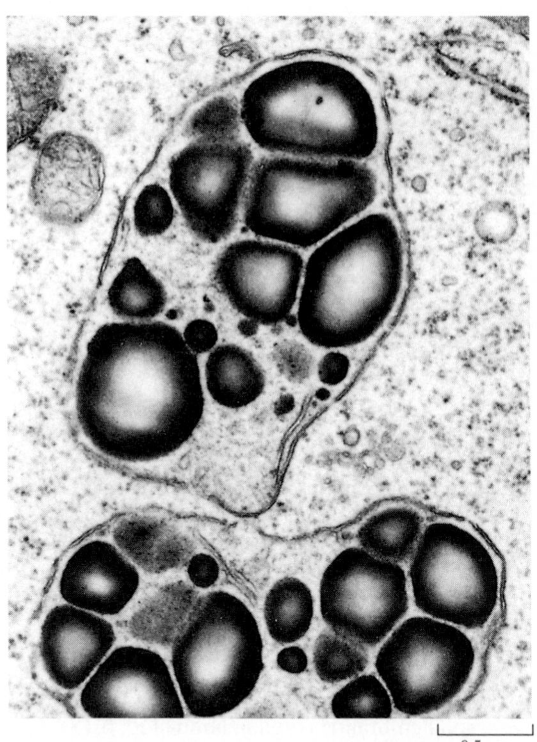

6.19 Electron micrograph of leucoplasts from the root tip of *Arabidopsis* Because they contain numerous prominent starch grains, these leucoplasts from the small desert plant *Arabidopsis* are called amyloplasts. Potatoes also have large numbers of such leukoplasts.

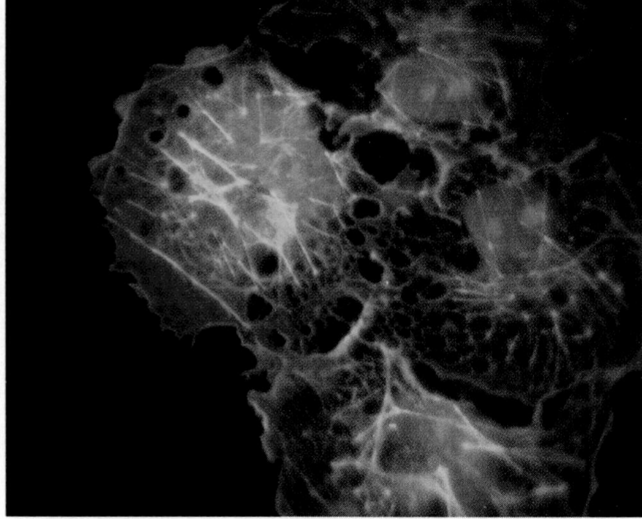

A

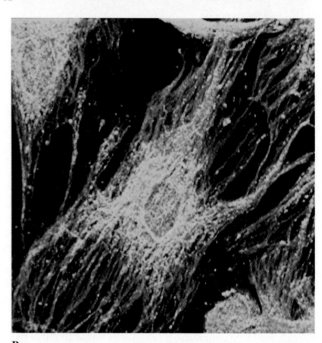

B

6.20 The cytoskeleton This intracellular framework, which is composed of microtubules, microfilaments, and intermediate filaments, helps establish cell shape and is involved in cell movement. (A) The cells have been stained with a special dye that attaches to the actin microfilaments, which stain fluorescent yellow. The nucleus is stained orange. (B) The cytoskeleton in a culture of cells from the brain of a 21-day-old rat fetus. Tubulin, the protein that makes up microtubules, is stained green.

The Cytoskeleton Is an Intracellular Framework Made Up of Protein Filaments

The specialized, membrane-lined organelles we have discussed so far do not all drift about freely in the fluid cytosol of the eucaryotic cell. Instead, they are embedded in a complex internal array of protein fibers, called the ***cytoskeleton*** (Fig. 6.20). This intracellular framework helps to define and control the shape of the cell, is involved in cell movements, and is important in cell division. The three most important components of the cytoskeleton are microfilaments, intermediate filaments, and microtubules. All three components are composed of protein subunits that can assemble or disassemble rapidly to effect changes in cell shape or to produce movement.

Microfilaments are long, thin, helically intertwined polymers of the protein ***actin*** (Fig. 6.21). Microfilaments composed solely of actin play a purely structural role; they interact with intermediate filaments and microtubules to form the cytoskeleton. Other actin filaments are involved in cell movements through a second protein, ***myosin.*** Myosin filaments are long threads much like actin filaments, but with a hinged protein "foot" at one end. The unique feature of the actin-myosin

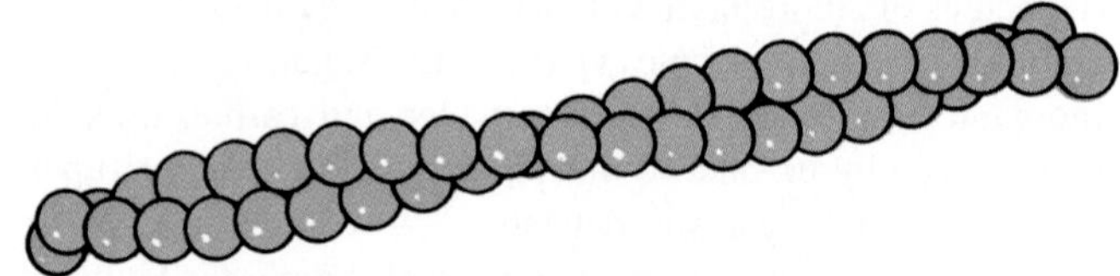

6.21 An actin filament This portion of an actin filament shows the helically intertwined chains of protein subunits.

combination is that the foot of a myosin filament can, when supplied with ATP energy, literally climb along an adjacent actin filament as though it were a ladder (Fig. 6.22). This ability of one filament to move along another accounts for many types of movement, including cell movement (Fig. 6.23), the contraction of muscles, the transport of some or all vesicles inside cells, and the constriction along the midline of cells as they divide.

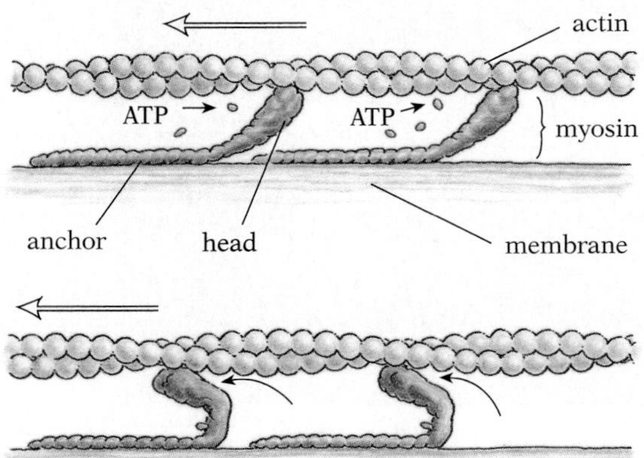

6.22 Action of a myosin cross-bridge during contraction Each foot, or cross-bridge, that forms part of a myosin filament hooks onto an actin microfilament. The cross-bridges bend toward the actin, hook onto it at specialized receptor sites, and then bend in the other direction, pulling the actin with them. They then let go, bend back in the original direction, hook onto the actin at a new active site, and pull again. The energy for this movement is supplied by ATP. The process will be discussed in more detail in Chapter 30.

Many eucaryotic cells also have tough, ropelike bundles of fibrous proteins known as ***intermediate filaments.*** A network of intermediate filaments is found around the nuclear envelope, radiating outward into the cytoplasm; another network is found just inside the nuclear envelope (Fig. 6.24). Because such microfilaments are often observed in abundance in cells that are subjected to mechanical stress (e.g., nerve cells), it is thought that they provide structural support for such cells and their nuclei.

Microtubules are heavy-duty versions of microfilaments. They are long, hollow, cylindrical structures formed by the polymerization of two-part subunits of the globular protein ***tubulin*** into helical stacks (Fig. 6.25 and 6.26). Microtubules, which grow by adding tubulin molecules to only one end of the structure, play a critical role in general cell structure and in cell division. During cell division the microtubules radiate from each end of the cell to form a basketlike arrangement (the spindle), which is instrumental in moving the chromosomes to the new nuclei. Like microfilaments, they may also aid intracellular transport and help provide shape and support for the cell and its organelles. Finally, as we will see presently, microtubules play a key role in the structure and movement of cilia and flagella.

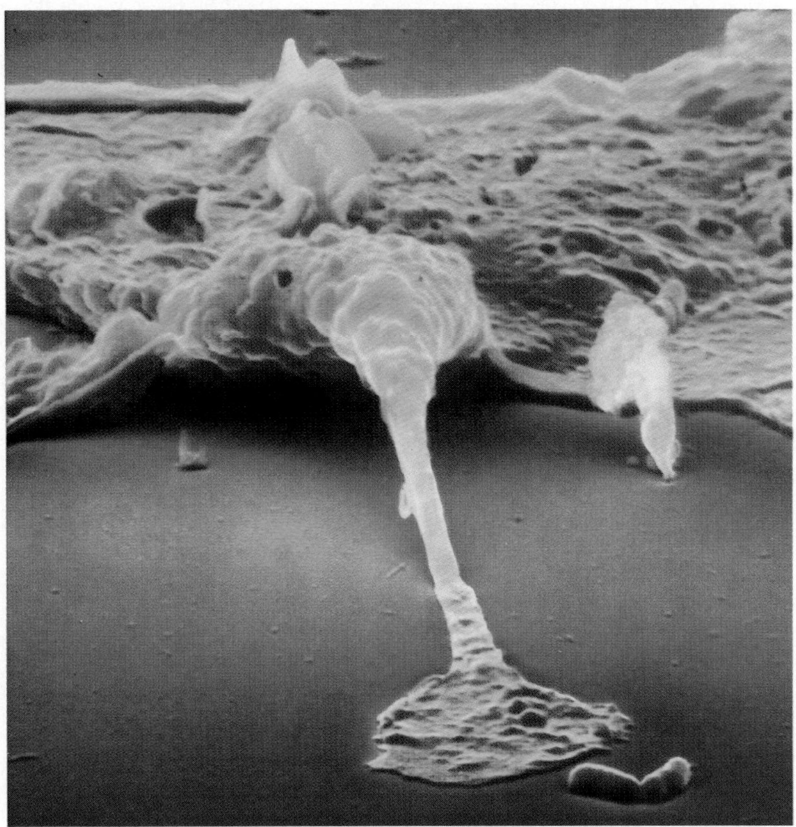

6.23 Amoeboid movement In this dramatic example of actin-based cell movement, a phagocyte (an amoeboid cell in the blood that captures foreign material by endocytosis) is extending a pseudopodium toward a bacterium (lower right), which it has detected on the basis of a waste chemical released by the bacterium.

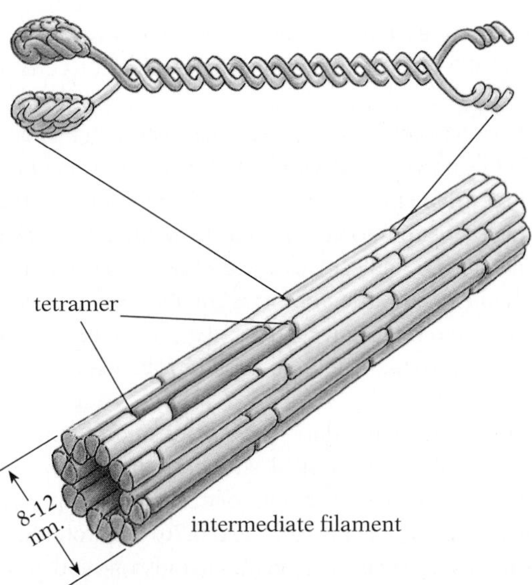

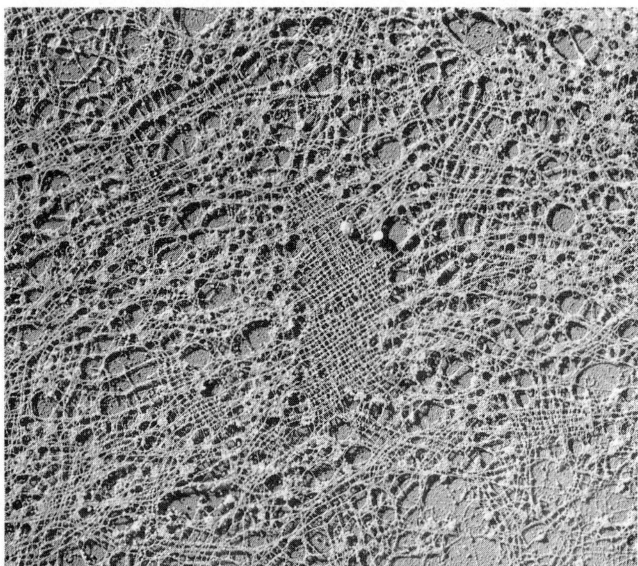

6.24 Intermediate filaments Left: The basic subunit is a two-part protein formed from two identical molecules coiled together. Each two-part protein is paired with another identical, slightly offset protein, to form a four-part protein (a tetramere). These in turn bind head to tail to create ropelike strands. Eight of these strands bind together, forming a hollow tube and creating the structure typical of intermediate filaments. Right: This meshlike array of intermediate filaments lies just inside the inner membrane of the nuclear envelope of a frog egg.

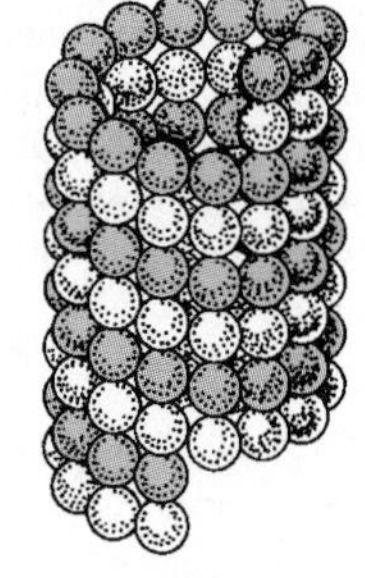

6.25 Structure of portion of a microtubule The subunits of tubulin, each consisting of two proteins (one kind shown blue, the other white), are helically stacked to form the wall of the tubule, which is usually 13 subunits in circumference. The tubules can be assembled and disassembled rapidly.

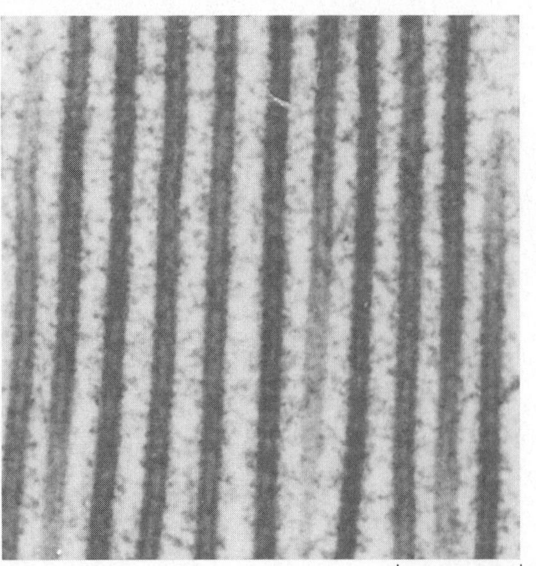

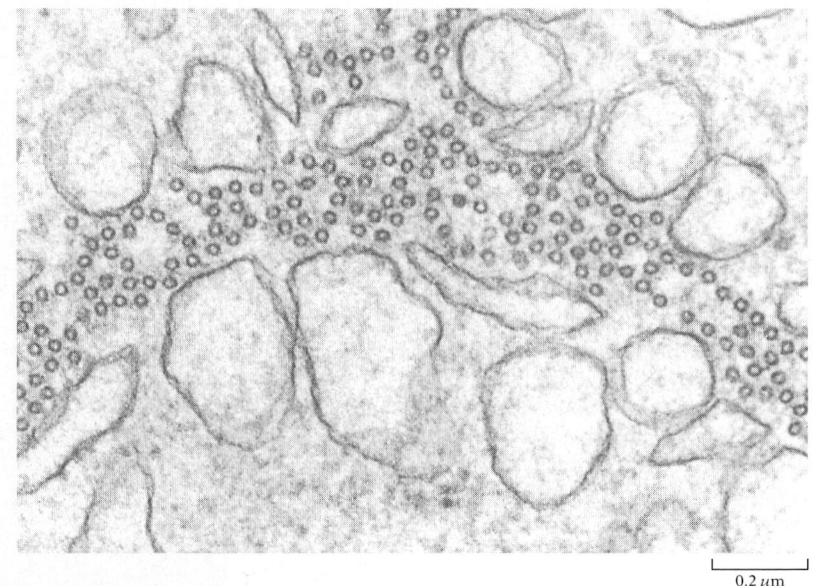

6.26 Electron micrographs of microtubules Left: Longitudinal section, from bovine brain. Right: Cross section, from hamster spermatid.

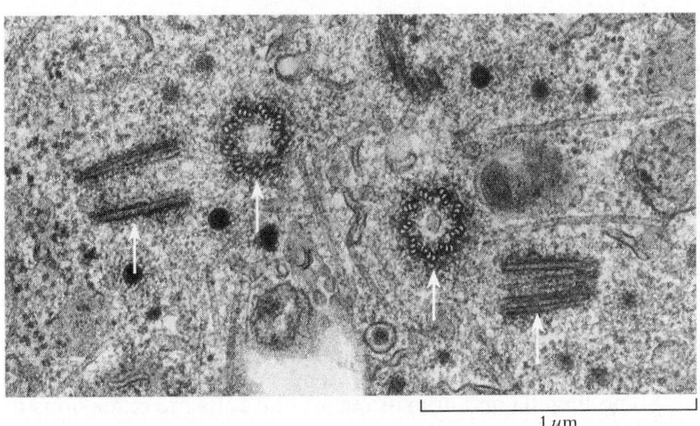

6.27 Centrioles This electron micrograph shows newly replicated centrioles. Since the centrioles of each pair (arrows) lie at right angles to each other, the sectioning of the specimen results in one from each pair being cut longitudinally and one being cut in cross section.

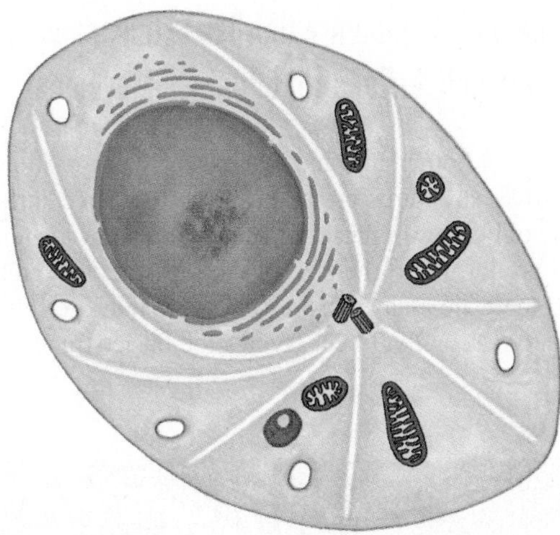

6.28 Microtubules in the cytoskeleton It has been proposed that the microtubules (purple), which often radiate outward from the centrosome, help hold the cell in its characteristic shape, and provide a structural backbone for the other elements of the cytoskeleton.

The Centrosome Serves as an Organizing Center for Microtubules

Just outside the nuclei of animal cells is an area called the ***centrosome,*** which contains the organelles known as ***centrioles.*** The centrioles are always found in pairs, oriented at right angles to each other (Fig. 6.27). The function of the centrosome is to act as the microtubule-organizing center; an array of microtubules project out of the centrosome into the cytoplasm (Fig. 6.28). The microtubules in turn influence the distribution of the microfilaments and intermediate filaments, so the centrosome is considered to be the master architect of the cytoskeleton. Prior to cell division, the centrosome and its newly-replicated centrioles divide, and the two centrosomes serve as the focus of the microtubular spindle during cell division. In cross section, the centrioles display a remarkable uniformity of structure, with nine triplets, composed of three fused microtubules each, arranged in a circle (Fig. 6.29).

Basal bodies, organelles with exactly the same structure as centrioles, are found at the base of hairlike appendages called cilia and flagella. The basal bodies apparently serve as organizing centers for the microtubules in cilia and flagella; the microtubules grow outward from each basal body. In fact, the relationship between centrioles and basal bodies is so close that they are functionally interchangeable. For example, the basal body that anchors the flagellum in many kinds of sperm becomes one of the egg's two centrioles after fertilization. In addition, the centrioles of cells that differentiate to line the oviduct multiply to become the basal bodies of the cilia whose rhythmic beating moves the eggs from the ovary to the site of fertilization.

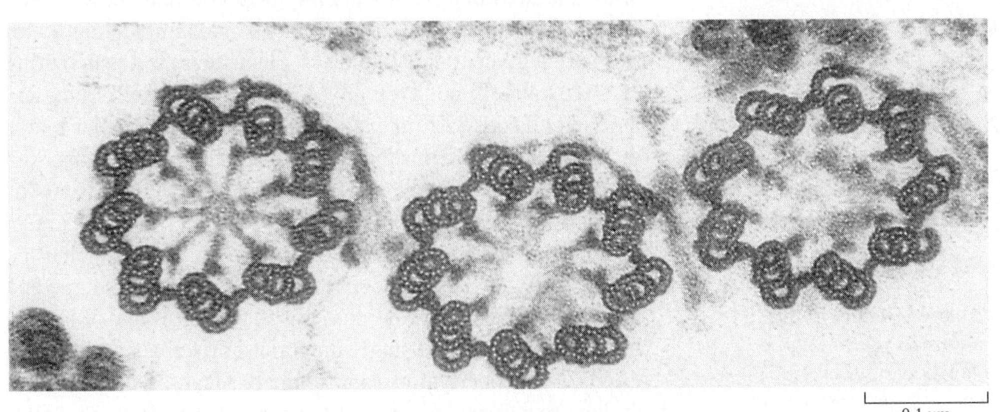

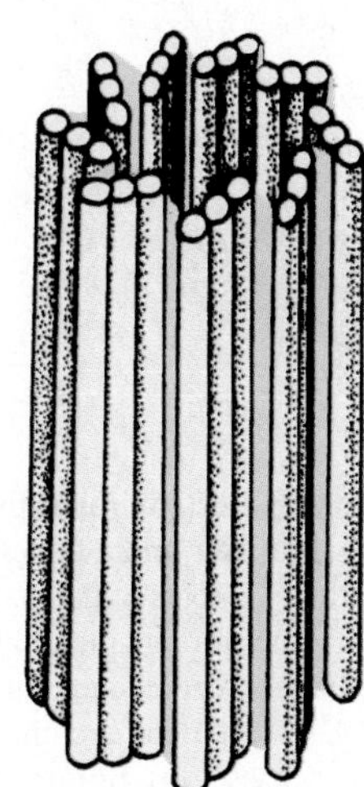

6.29 Basal bodies and centrioles Basal bodies (left), such as the three shown in cross section in this electron micrograph of a protozoan, are essentially identical to centrioles in structure. Centrioles (right) are composed of nine triplet microtubules.

Cilia and Flagella Are Hairlike Extensions From the Cell Surface That Produce Movement

Some cells of both plants and animals have one or more movable hairlike structures projecting from their free surfaces. If there are only a few of these appendages and they are relatively long in proportion to the size of the cell, they are called ***flagella.*** If there are many and they are short, they are called ***cilia*** (Fig. 6.30). Actually, in eucaryotes the basic structure of flagella and cilia is the same, and the terms are often used interchangeably. Both usually function either to move the cell through a liquid or to move liquids (or small particles) across the surface of the cell. They occur commonly on unicellular and small multicellular organisms and on the male reproductive cells of most animals and many plants, in both of which they may be the principal means of locomotion. They are also common on the cells lining many internal passageways and ducts in animals, where their beating aids in moving materials through the passageways. In the human trachea (windpipe), they reach densities of a billion per square centimeter.

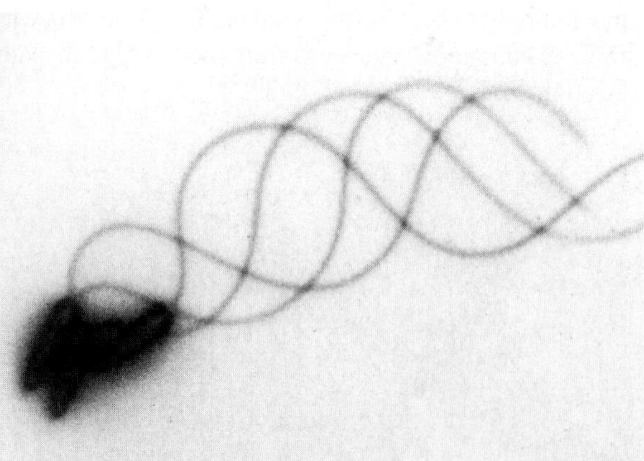

6.30 Scanning electron micrograph of ciliated surface of the trachea of a hamster and a flagellum of a human sperm Top: Coordinated movements of the many cilia function to sweep dust particles and other foreign material from the respiratory surfaces to the mouth. (One effect of smoking cigarettes is that cilia are paralyzed, which results in the accumulation of foreign material within the respiratory passages). Bottom: Successive waves of bending that move along the flagellum, from its base toward its tip, push against the water and propel the sperm forward.

The flagella and cilia of eucaryotic cells are extensions of the cell membrane, and contain a cytoplasmic matrix, with eleven groups of microtubules embedded in the matrix. Invariably, nine of these groups are fused pairs arranged around the periphery of the cylinder, and the other two are isolated microtubules in the center, a pattern often referred to as the 9 + 2 arrangement (Fig. 6.31). Each cilium and flagellum has a basal body (centriole) at its base, and attached to the basal body are rootlike structures that anchor the basal body to the cell's cytoskeleton.

Cilia and flagella move when the microtubules slide past one another (Fig. 6.32). Attached to the nine outer doublets are protein arms that act as movable cross-bridges between the adjacent doublets. The arms from one doublet attach to the adjacent doublet and bend, pulling one doublet past the other (Fig. 6.33). The arm then detaches and binds to a new site, and the process is repeated. The movement is powered by the ubiquitous cellular energy source, ATP.

No Single Eucaryotic Cell Can Be Said to Be Typical

The complexity of cells was not truly appreciated until the advent of the electron microscope and modern biochemical techniques, which together have changed our whole picture of the cell. Four or five decades ago, biology books still regularly included a diagram of a so-called typical cell that showed only five or six simple internal components. No simple diagram of this sort is given today. In the first place, there is no such thing as a typical cell, not even a typical eucaryotic cell. Plant and animal cells differ from one another; cells of particular plants or animals differ from those of other plants or animals; and within the body of any one plant or animal the various cells often differ strikingly in shape, size, and function. This much, of course, has been known for a long time. But now that the number of known cellular components has grown so large and their great variability has been so well demonstrated, it is even more obvious that no single diagram, or even a series of diagrams, can properly portray a "typical" eucaryotic cell. Nevertheless, to help you visualize the arrangement of the organelles discussed in the preceding pages, two diagrams (Figs. 6.34 and 6.35) are given here. As you examine them, keep in mind that not all the components shown are always found together in any one real cell.

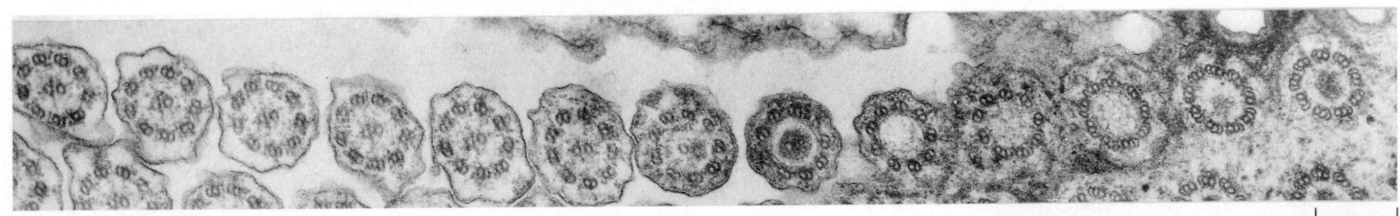

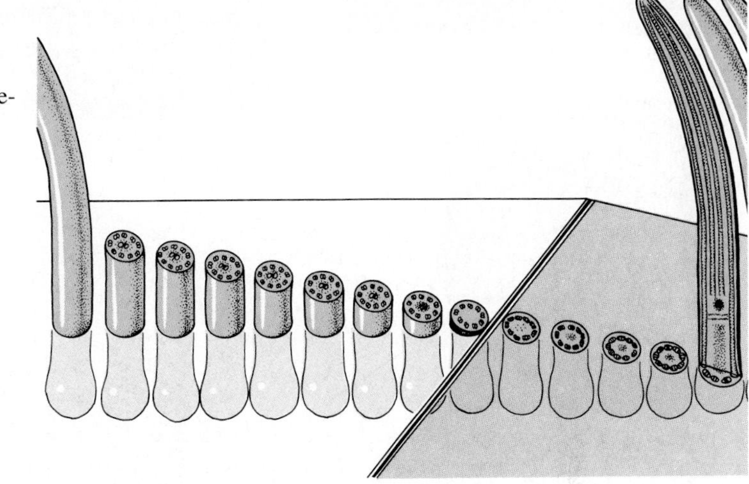

6.31 Cross section of cilia from the protozoan *Tetrahymena* The electron micrograph of an oblique section of surface tissue reveals both the 9 + 2 arrangement of microtubules in cilia and the nine triplet microtubules characteristic of basal bodies. The interpretive drawing at right shows the plane of the section and position of microtubules in a lateral section of an adjacent cilium.

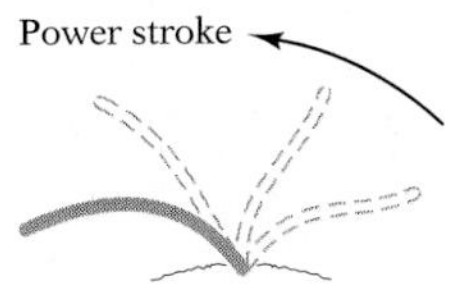

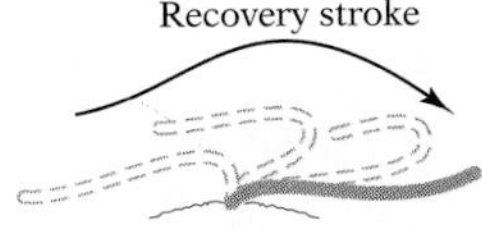

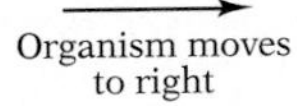

A

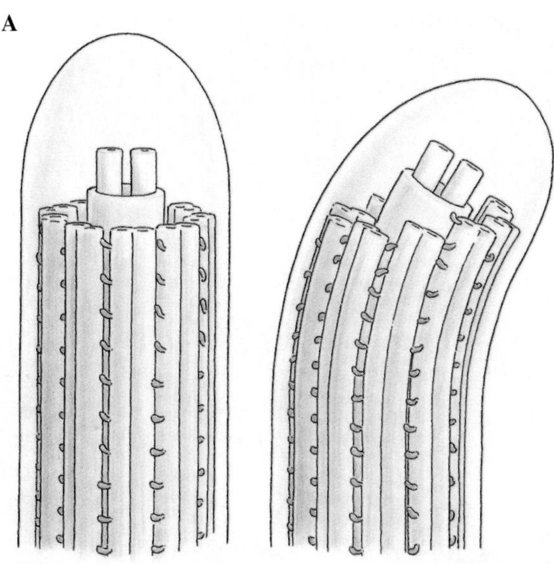

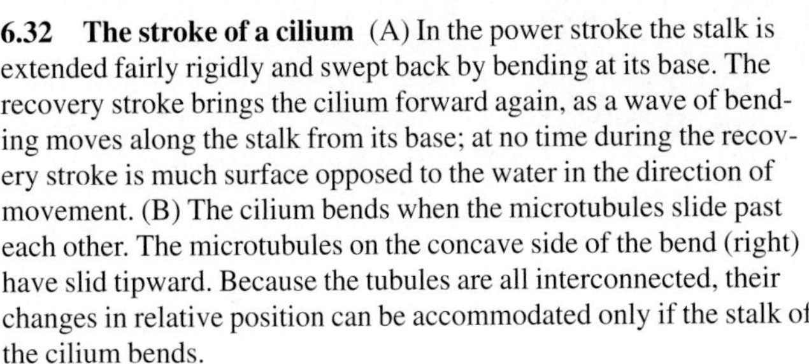

B

6.32 The stroke of a cilium (A) In the power stroke the stalk is extended fairly rigidly and swept back by bending at its base. The recovery stroke brings the cilium forward again, as a wave of bending moves along the stalk from its base; at no time during the recovery stroke is much surface opposed to the water in the direction of movement. (B) The cilium bends when the microtubules slide past each other. The microtubules on the concave side of the bend (right) have slid tipward. Because the tubules are all interconnected, their changes in relative position can be accommodated only if the stalk of the cilium bends.

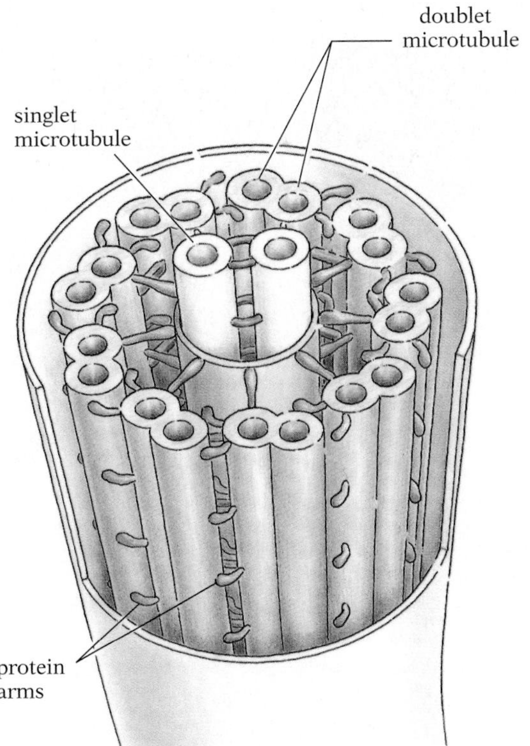

6.33 Internal structure of a cilium The nine microtubule doublets are arranged around two singlet microtubules in the center of the cilium. Movement is generated when the protein arms of one subset of doublets begin to "walk" along the length of other doublets. The doublets moving the most force the other doublets to bend.

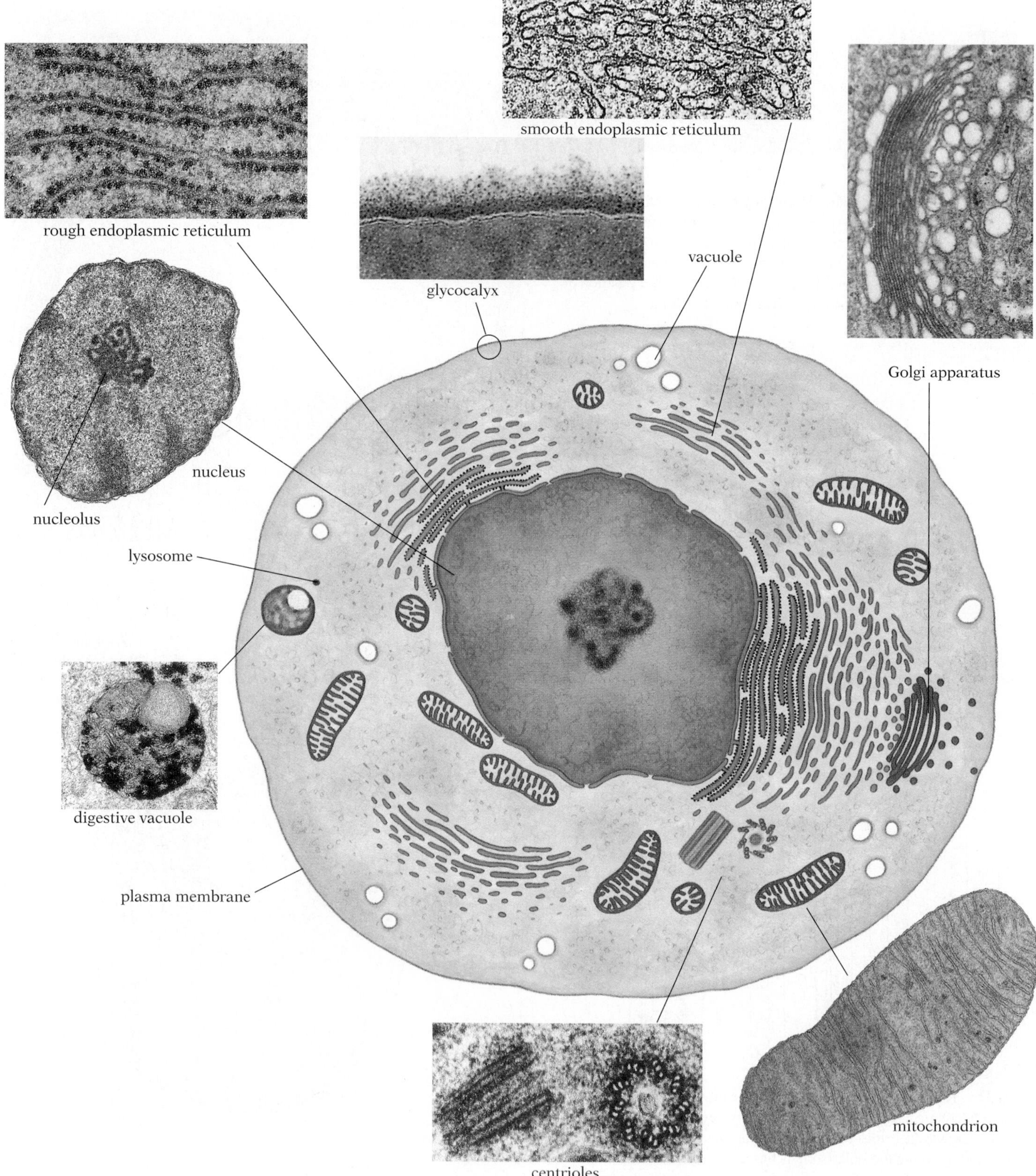

6.34 A "typical" animal cell Not every animal cell includes all the organelles shown here, nor are all the substructures that may occur in an animal cell represented. While the diagrams on this and the facing page indicate relative size, the electron micrographs reproduce some of the organelles much enlarged in comparison to others, in order to show internal detail.

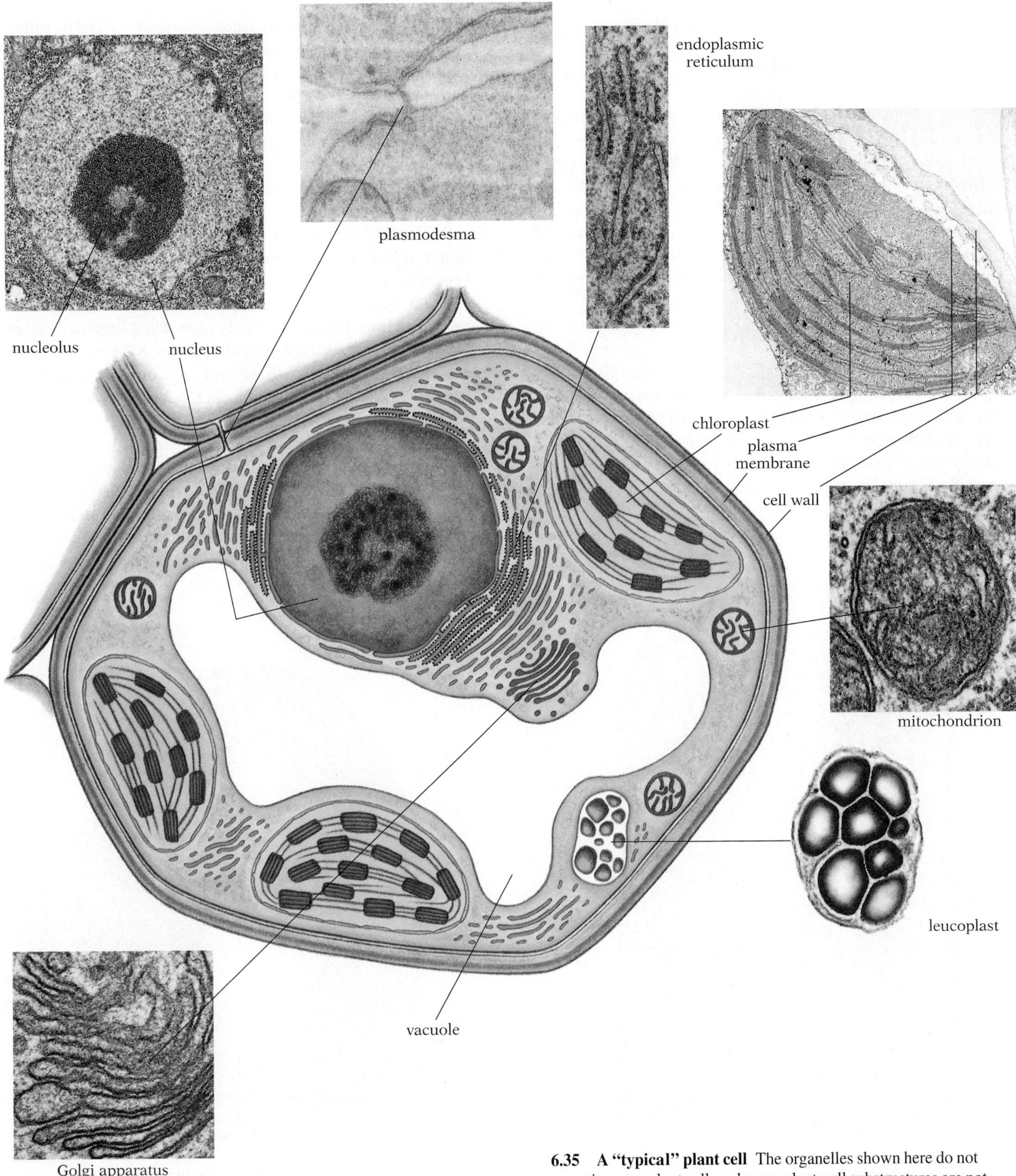

6.35 A "typical" plant cell The organelles shown here do not occur in every plant cell, and some plant-cell substructures are not represented.

Procaryotic Cells Lack Most Internal Membranous Organelles

Most of the organelles present in eucaryotic cells are lacking in procaryotic cells (Fig. 6.36). We have already mentioned that such cells have no nuclear envelope, but they also lack other membranous structures such as an endoplasmic reticulum, a Golgi apparatus, lysosomes, peroxisomes, and mitochondria (although many of the functions of mitochondria are carried out in bacteria by the cell membrane). Most photosynthetic bacteria, however, have membranous structures containing light-absorbing pigments. These membranes are usually in pocketings of the cell membrane rather than being independent membrane-enclosed plastids (Fig. 6.37).

For a long time it was thought that procaryotic cells had no chromosomes. With the advent of the electron microscope, however, it became possible to detect in each procaryotic cell a region containing a single, large DNA molecule (Fig. 6.38). Although this molecule is not closely associated with proteins, as DNA is in eucaryotic cells, it is considered a chromosome. There may also be small independent pieces of DNA, called ***plasmids,*** in the procaryotic cell (Fig. 6.38). Unlike eucaryotic chromosomes, which are linear, the procaryotic chromosomes and plasmids are circular.

Like eucaryotic chromosomes, the procaryotic chromosome bears, in linear array, the genes that control both the hereditary traits of the cell and its ordinary activities. The DNA functions by directing protein synthesis on ribosomes, important cytoplasmic structures in both eucaryotic and procaryotic cells. The ribosomes of procaryotic cells, however, are structurally different from those of eucaryotic cells, and they are somewhat smaller.

Some bacterial cells have hairlike organelles used in swimming, and these have traditionally been called flagella. In most procaryotic species these structures do not have microtubules and their structure and movement are completely different from that of eucaryotic flagella (Fig. 6.39). (The procaryotic flagella rotates like a minipropeller rather than bending, as seen in eucaryotic cilia and flagella.)

Finally, the cell walls of most bacteria are made up of the polysaccharide murein (a polymer of amino sugars), which is found only in procaryotes. Table 6.1 summarizes some of the most important differences between procaryotic and eucaryotic cells.

0.1 μm

6.36 Electron micrograph of part of a bacterial cell The light area in the center of the cell contains the single circular chromosome, but is not bounded by a membrane. Note the prominent cell wall, with the plasma membrane visible just inside it, and the absence of cellular organelles.

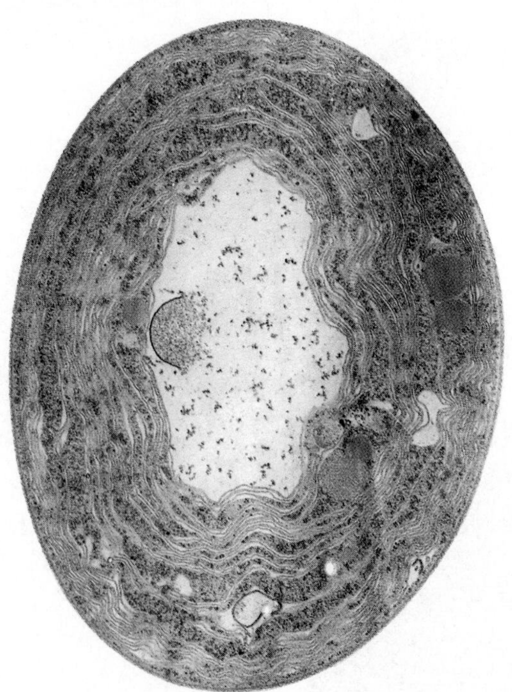

6.37 A photosynthetic bacterium, *Prochloron* This photosynthetic bacterium has its light-absorbing pigments located in membranous structures lying along the periphery of the cell. These membranes are not contained within any membrane-surrounded structure that could be interpreted as a chloroplast.

TABLE 6.1 *A comparison of typical procaryotic and eucaryotic cells, and certain eucaryotic organelles*

Characteristic	*Procaryotic cells*	*Eucaryotic cells*	*Mitochondria and chloroplasts*
Size	1–10 μm	10–100 μm	1–10 μm
Nuclear envelope	Absent	Present	Absent
Chromosomes	Single, circular, with no nucleosomes	Multiple, linear, wound on nucleosomes	Single, circular, with no nucleosomes
Golgi apparatus	Absent	Present	Absent
Endoplasmic reticulum, lysosomes, peroxisomes	Absent	Present	Absent
Mitochondria	Absent	Present	
Chlorophyll	Not in chloroplasts	In chloroplasts	
Ribosomes	Relatively small	Relatively large	Relatively small
Microtubules, intermediate filaments, microfilaments	Absent	Present	Absent
Flagella	Lack microtubules	Contain microtubules	

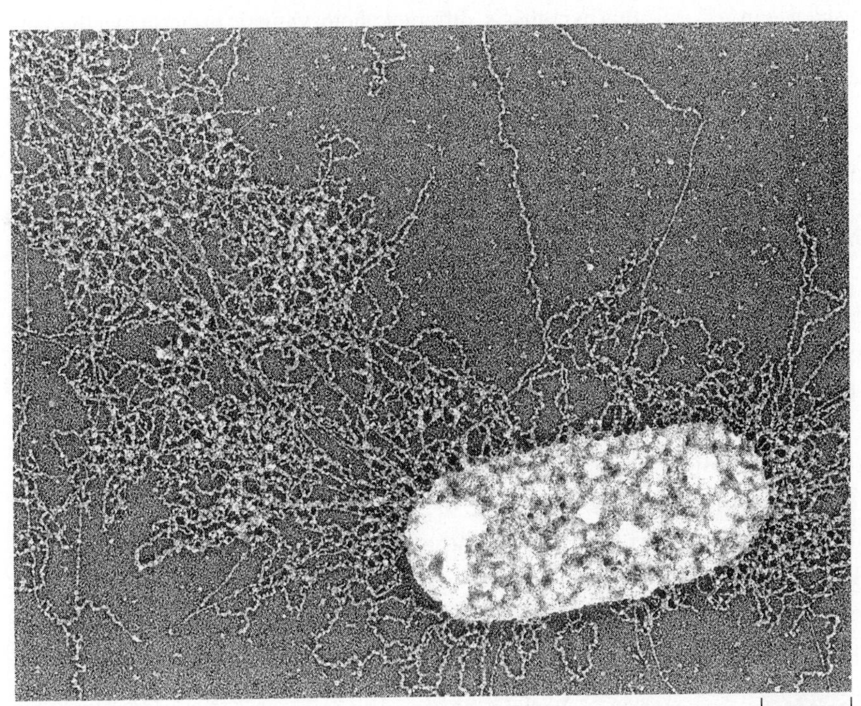

6.38 A disrupted cell of *Escherichia coli* Exposure to detergent releases the DNA of the common intestinal bacterium *E. coli,* most of which can be seen outside the cell in this electron micrograph. Though not readily recognizable here, the main chromosome and the small accessory chromosomes (plasmids) of a bacterium each form a circle rather than individual strands in the cell. The mottled appearance of the surface of the bacterium results from the effects of alcohol, drying, and shadowing with platinum.

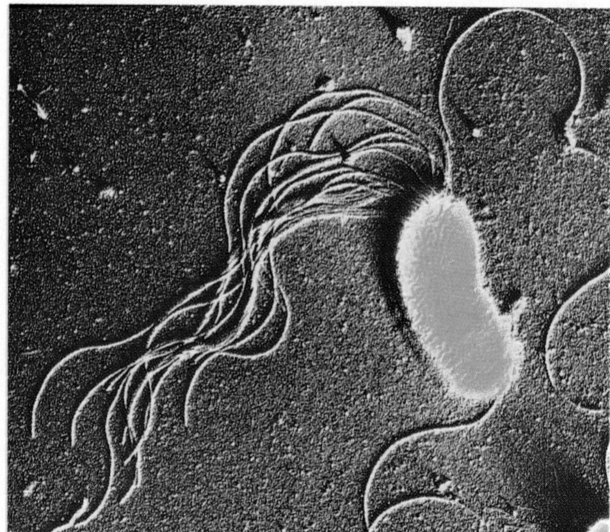

6.39 Scanning electron micrograph showing flagella of the soil bacterium *Pseudomonas fluorescens* This bacteria has hairlike structures, flagella, used in swimming through the water layer surrounding soil particles. These structures do not have microtubules and their structure and movement is completely different from that of eucaryotic flagella.

According to the Endosymbiotic Hypothesis, Some Organelles Originated as Independent Cells

Many biologists now believe that at least two organelles found only in eucaryotes—mitochondria and chloroplasts—are descendants of procaryotic organisms that took up residence inside the host-cell precursors of eucaryotes. This view, termed the ***endosymbiotic hypothesis*** by its chief proponent, Lynn Margulis of Boston University, is supported by several lines of evidence:

- Both mitochondria and chloroplasts contain their own ribosomes and their own chromosomes. Mitochondria and chloroplasts also build their own membranes.
- The chromosomes of these organelles resemble those of procaryotes: both are circular, not wound on proteins, and not enclosed by a nuclear envelope.
- The internal organization of the organelles' genes is similar to that of procaryotes, but very different from that of eucaryotes. We shall examine the details of this organization in later chapters.
- The ribosomes of mitochondria and chloroplasts are more similar to those of procaryotes than to the ribosomes of the eucaryotic cells in which they live.
- Many present-day photosynthetic bacteria live inside eucaryotic hosts, providing them with food in return for shelter. Similarly, certain nonphotosynthetic bacteria live symbiotically within eucaryotic organisms, extracting and sharing energy from foods that their hosts cannot themselves metabolize.

In abbreviated form, the endosymbiotic hypothesis assumes that about 1.5 billion years ago a procaryotic cell captured a bacterium through endocytosis. The bacterium resisted digestion, lived symbiotically inside its host, and divided independently of the host. Later, some of the symbiont's genes were moved to the host nucleus, which then took control from its guest. From this partnership all present-day mitochondrion-containing organisms, including plants, animals, fungi, and protozoans, must have evolved. Subsequently, according to this hypothesis, different photosynthetic bacteria met the same fate, the result being the several lineages of chloroplast-containing organisms we know as algae and plants.

MAKING CONNECTIONS

1. Relate the roles of the mitochondria and chloroplasts in energy conversion to the Second Law of Thermodynamics.

2. What change occurs in the chromoplasts of leaves as they change color in the autumn?

3. Which cells in the human body would you examine if you wanted to study prominent or abundant examples of the following organelles or structures: (1) smooth ER? (2) Golgi apparatus? (3) peroxisomes? (4) mitochondria? (5) actin and myosin? (6) intermediate filaments? (7) cilia?

4. What is the possible evolutionary significance of the fact that both mitochondria and chloroplasts are bounded by a double membrane?

CONCEPTS IN BRIEF

1. The cells of most organisms have a membrane-surrounded *nucleus* and are therefore characterized as *eucaryotes,* to distinguish them from *procaryotes* such as bacteria, which lack a well-defined nucleus.

2. Inside the eucaryotic nucleus are the *chromosomes,* which contain the genes that direct the cell's life processes, and one or more *nucleoli,* small oval bodies from which *ribosomes,* the future sites of protein synthesis, are derived. Separating the nucleus from the cytoplasm is a double-membrane *nuclear envelope* perforated by pores.

3. The *endoplasmic reticulum* forms a system of interconnected membrane-enclosed spaces that are continuous with that of the nuclear envelope. Sometimes the membranes of the endoplasmic reticulum are "rough," with ribosomes on their outer surfaces. When no ribosomes are present, the endoplasmic reticulum is "smooth." The endoplasmic reticulum functions both as a passageway for intracellular transport and as a site for the synthesis of phospholipids and cholesterol for new membranes. Proteins for new membranes and for export are synthesized by ribosomes attached to the endoplasmic reticulum.

4. The *Golgi apparatus* consists of stacks of membrane-enclosed sacs that function in the storage, modification, and packaging of secretory products. Polysaccharides synthesized here are attached to glycoproteins and glycolipids, which are incorporated into the cell's glycocalyx.

5. *Mitochondria* are the powerhouses of the cell. Chemical reactions within the mitochondria provide energy in the form of ATP for the activities of all the other cellular organelles. *Lysosomes* are membranous sacs that function as storage vesicles for powerful digestive enzymes; they may act as the cell's digestive system, hydrolyzing materials taken in by endocytosis. *Peroxisomes* are membranous sacs of oxidative rather than digestive enzymes.

6. Most plant cells have large membranous organelles called *plastids.* The two principal categories of plastids are the colored *chromoplasts* and the colorless *leucoplasts.* Chromoplasts that contain the green pigment chlorophyll are called *chloroplasts;* they capture energy of sunlight and utilize it in the manufacture of organic compounds. The leucoplasts' primary function is storage of starch, oils, or protein granules.

7. Membrane-enclosed, fluid-filled spaces termed *vacuoles* are found in many plant cells and perform a variety of functions. Most mature plant cells have a large central vacuole occupying much of the volume of the cell; it plays an important role in maintaining the turgidity of the cell and in the storage of important inorganic and organic substances, as well as wastes.

8. *Microtubules, intermediate filaments,* and *microfilaments* assist intracellular movement and help support the cell. Microtubules also form the spindle of dividing cells and are essential components of *centrioles* and *basal bodies,* identical cellular structures with different functions. *Cilia* and *flagella* are hairlike projections from the cell's surface that move the cell or move materials across the cell's surface. They contain microtubules in a 9 + 2 arrangement.

9. Although procaryotic cells lack the membrane-surrounded nucleus and all the internal membranous organelles discussed above, they do have ribosomes and a single circular chromosome of DNA. Some bacterial cells also have flagella, but these organelles lack microtubules.

STUDY QUESTIONS

1. **What is an organelle? What organelles are characteristic of both procaryotic and eucaryotic cells? Which are found only in eucaryotes? (p. 113)**
2. **Describe the structure and function of the nucleus, especially the chromosomes and nucleolus. What role does the nuclear envelope play and how does it differ from the cell membrane? (pp. 113–116)**
3. **What are the functions of the endoplasmic reticulum (ER)? How does rough ER differ from smooth ER? (pp. 116–120)**

4. Describe the structure and functions of the Golgi apparatus. (p. 120)

5. Compare lysosomes and peroxisomes with respect to appearance, function, and mode of origin. (pp.120–123)

6. What are the functions of the vacuole found in most mature plant cells? How does it develop? What is the role of the contractile vacuole found in many protozoans? (pp. 123–124)

7. Cyanide, a deadly poison, disrupts the activity of mitochondria. How, specifically, would cells exposed to this substance be affected? (p. 124)

8. You examine two plant cells under the microscope and find that each has one type of plastid. The first cell has many chloroplasts, the other many leucoplasts. What are the most likely functions of the two cells? (p. 125)

9. Compare the structure and function of these components of the cytoskeleton: microfilaments, intermediate fibers, and microtubules. (pp. 126–128)

10. Describe the structure of a eucaryotic cilium. How is it different from a eucaryotic flagellum? (p. 130)

11. Using Figures 6.34 and 6.35, compare the structure of "typical" animal and plant cells. (pp. 132–133)

12. According to the endosymbiotic hypothesis, which eucaryotic organelles originated as independent procaryotic cells? Describe five lines of evidence supporting this viewpoint. (p. 136)

C H A P T E R 7

ENERGY TRANSFORMATIONS: PHOTOSYNTHESIS

According to evidence gleaned from the fossil record, the earth's first organisms arose between 3.5 and 4 billion years ago from organic molecules that had formed in the atmosphere and accumulated in the ancient seas. Because the supply of organic compounds was limited, these primitive life forms apparently gave rise in turn to organisms capable of synthesizing their own food from inorganic substances. To create organic compounds, this second group of early organisms utilized a process known as ***chemosynthesis,*** meaning that they obtained energy from chemical reactions involving inorganic substances such as molecular hydrogen, ammonia, and sulfur. Chemosynthetic organisms are still found today, some of which live in unusual environments such as bogs, hot sulfur springs, and volcanic vents on the ocean floor (Fig. 7.1).

Then, between 3.5 and 3 billion years ago, a momentous biochemical event occurred: certain organisms developed the ability to capture the sun's energy directly and to use it to synthesize complex organic compounds. This was probably the earliest form of ***photosynthesis,*** the biological process by which light energy is transformed into chemical energy. The sun is an ideal energy source. An enormous amount of solar energy reaches the surface of the earth each year—an amount equivalent to the heat released from burning 86 million tons of oil. By tapping into this energy source, organisms were freed from their previous dependence on chemosynthetic organisms and ready-made organic energy sources. From that time on, the continuation of life on earth has depended primarily on the activity of photosynthesizing organisms.

In its original form, photosynthesis did not produce oxygen as a by-product. But more than three billion years ago, some photosynthetic cells developed the ability to use water for photosynthesis, producing molecular oxygen as a by-product. Oxygen was toxic to most of the early organisms; being strongly electronegative it drew essential electrons away from molecules within the cells and disrupted their life processes. However, as it began to accumulate in the atmosphere, some organisms evolved the ability to use oxygen in a series of enzyme-catalyzed reactions to release the energy stored within food molecules. This process, called ***aerobic respiration,*** be-

Chapter opening photo: Mist in a Venezuelan tropical rain forest canopy. Rainforests have enormous photosynthetic productivity due to the large amount of sunlight and rainfall they receive all year long.

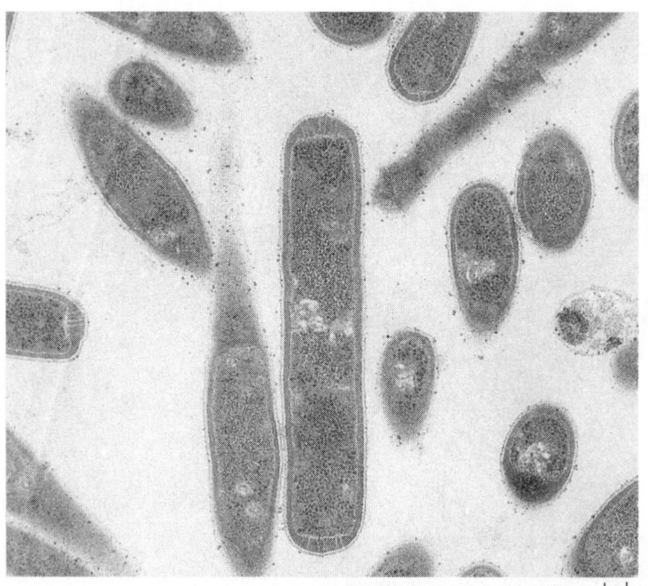

7.1 A chemosynthetic organism This electron photomicrograph shows thin sections of the bacterium *Methanothermus fervidus* which lives at temperatures up to 97°C within Icelandic hot springs. This bacterium combines CO_2 with H_2 to produce methane and water. The reaction is exergonic, and some of the energy released is used to do work in the cell.

came the main mechanism for extracting energy from food. Eventually, photosynthesis and aerobic respiration were linked through an exchange of end products. Plants use the energy of sunlight to synthesize carbohydrates from carbon dioxide and water, releasing molecular oxygen in the process. In most organisms, when the cells need energy, the carbohydrates produced by plants are broken down by aerobic respiration, which releases carbon dioxide and water. *Photosynthesis utilizes the carbon dioxide and water produced by aerobic respiration, and aerobic respiration utilizes the food and oxygen produced by photosynthesis* (Fig. 7.2).

Although life had existed on the earth for almost two billion years in the absence of significant quantities of molecular oxygen, it can hardly be said to have flourished. The earth's atmosphere today is about 21 percent oxygen, virtually all of which has come from photosynthesis. Only after the advent of large-scale aerobic respiration—made possible by the prior development of oxygen-producing photosynthesis—did life become the dominant feature on the face of the earth.

Today, virtually all organisms depend directly or indirectly on photosynthesis to fill their energy needs. ***Autotrophs*** manufacture their own food by synthesizing organic nutrients from inorganic materials almost exclusively by photosynthesis (Fig. 7.3). Plants are predominantly autotrophic. ***Heterotrophs*** must obtain organic nutrients from the environment, so depend indirectly on photosynthesis, since they consume autotrophs, or heterotrophs that have eaten autotrophs, or both. Photosynthesis, then, is a logical starting point for our discussion of the basic energy transformations of life.

Much about photosynthesis is still not known, but in the last 30 years the intricate maze of chemical pathways involved in the process has become much better understood. It is not our purpose here to discuss in detail all the chemistry of photosynthesis, or to cite every reaction and compound now thought to be involved in it, but rather to present only the broad outlines of this process so fundamental to life. Thus, we shall begin our discussion of energy transformations in living systems with a

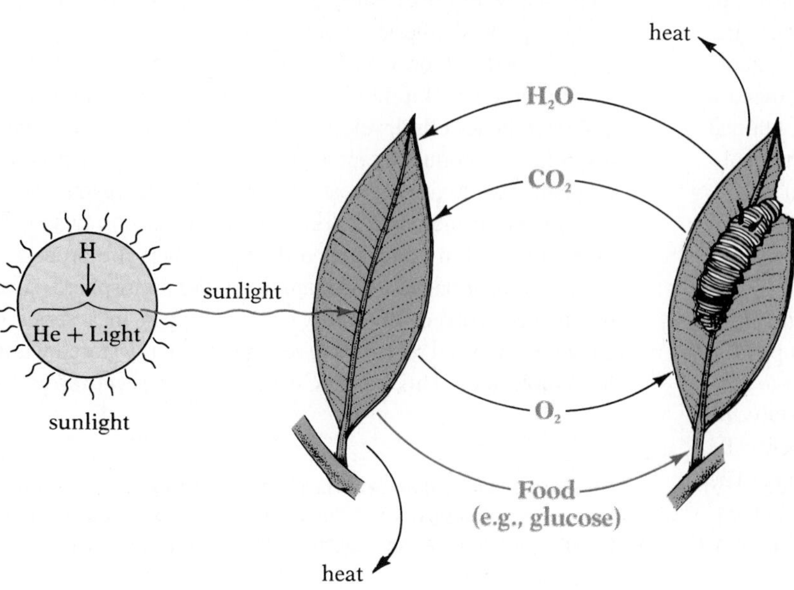

7.2 Energy flow in photosynthesis and respiration Today, nearly all the energy for life originates in the sun. The process of photosynthesis converts the energy of sunlight into chemical energy in the form of sugars. With the evolution of water-based photosynthesis approximately three billion years ago, oxygen began to accumulate in the atmosphere. This led in turn to the spread of organisms capable of aerobic respiration, the process by which the energy stored in sugar is released and made available to fuel the energy needs of the cell. The photosynthetic and respiratory pathways are unified by shared products: organic food, water, carbon dioxide, and molecular oxygen. Respiration utilizes the organic food and oxygen produced by photosynthesis, and photosynthesis utilizes carbon dioxide and water, products of respiration.

7.3 An elephant (heterotroph) consuming a plant (autotroph)

brief survey of the ways in which green plants capture the sun's radiant energy and use it to synthesize complex organic molecules. In the next chapter we shall continue the discussion by looking at the ways in which organisms release the energy stored in these complex organic molecules.

Early Experiments Clarified the Basic Chemical Reactions of Photosynthesis

As early as 1772, the English investigator Joseph Priestley demonstrated that green plants counteract the effects that burning or breathing have on air (Fig. 7.4). His pioneering experiments were the first to show that plants produce oxygen, although he did not realize that this is what was happening, or that light was essential for the phenomenon he observed. Nevertheless, his findings stimulated interest in the process we now know as photosynthesis, and they led to a fairly rapid succession of further discoveries. As a result, by the early 19th century, all of the important components of the photosynthetic process were at least vaguely known; they can be summarized by the following equation:

$$\text{carbon dioxide} + \text{water} + \text{sunlight} \xrightarrow{\text{green plant}} \text{organic material} + \text{oxygen}$$

For a time scientists thought that the oxygen produced during photosynthesis comes from the splitting of carbon dioxide, but we now know that the oxygen comes from the splitting of water molecules:

$$2H_2O + \text{sunlight} \xrightarrow{\text{green plant}} 4H^+ + 4\text{ electrons} + O_2$$

We also know that the energy required to split water molecules comes from the sun and is trapped by molecules of ***chlorophyll,*** the green pigment of plants. The free H^+ ions and electrons produced by splitting water molecules are then used to convert carbon dioxide into carbohydrate and new water is formed:

$$CO_2 + 4H^+ + 4\text{ electrons} \xrightarrow{\text{green plant}} (CH_2O) + H_2O$$

The parentheses around the CH_2O indicate that this combination of atoms does not represent an actual molecule, but rather symbolizes the general formula for carbohydrates. The currently accepted summary equation for the two preceding reactions is:

CO_2	+	$2\ H_2O$	$\xrightarrow{\text{green plant}}$
1 molecule of carbon dioxide		2 molecules of water	yields

(CH_2O)	+	H_2O	+	O_2
1 unit of carbohydrate		1 molecule of new water		1 molecule of oxygen

One of the products of photosynthesis is glucose, a six-carbon

7.4 Priestley's demonstration that plants and animals "restore" the air for each other. A plant alone in a closed jar died, and a mouse alone in another closed jar died, but when plant and mouse were together in the same jar, both lived. The plant produced the molecular oxygen (O_2) required by the mouse, and the mouse produced the carbon dioxide (CO_2) required by the plant.

simple sugar; to show this outcome one simply multiplies the preceding summary equation by 6:

$6CO_2$ +	$12H_2O$	$\xrightarrow{\text{light, chlorophyll}}$
6 molecules of carbon dioxide	12 molecules of water	in the presence of light and chlorophyll yields

$6O_2$ +	$C_6H_{12}O_6$ +	$6H_2O$
6 molecules of oxygen	1 molecule of glucose	6 molecules of new water

It may seem curious that water should appear on both sides of the equation. But water is both a product and a reactant in photosynthesis; water is used as a raw material in one part of the process, and is produced in another.

Although the last equation given above is a convenient summary of the photosynthetic production of carbohydrates, the process is not one grand chemical reaction, as the summary equation might imply. Instead, dozens of reactions are involved, some that require light, known as ***light reactions,*** and others that can take place without it, known as ***dark reactions.***

Photosynthesis Involves Linked Oxidation-Reduction Reactions

Carbon dioxide is an exceedingly energy-poor compound, whereas sugar is energy-rich. Photosynthesis, then, not only converts light energy into chemical energy but also stores it by synthesizing energy-rich sugar from energy-poor carbon dioxide. In chemical terms, the energy is said to be stored by the ***reduction*** process, that is, by the addition of one or more electrons. The converse process, ***oxidation,*** liberates energy from a compound by the removal of one or more electrons. Originally these terms referred solely to the addition and removal of oxygen in a reaction. Today, however, the terms are applied also to reactions in which oxygen is not involved. For example, in biological reactions the transfer of electrons is often accompanied by the exchange of one or more hydrogen atoms; thus oxidation is sometimes accomplished by the removal of a hydrogen atom and reduction by the addition of a hydrogen atom. These relationships are summarized in Table 7.1. The important point to remember is that reduction—the addition of electrons—*stores* energy in the substance being reduced, whereas oxidation—the removal of electrons—*liberates* energy from the substance being oxidized.

TABLE 7.1 *Redox reactions*

Oxidation	*Reduction*
Removes electron(s)	Adds electron(s)
Removes hydrogen	Adds hydrogen
Liberates energy	Stores energy

Since an electron added to one molecule must have been removed from some other molecule, it follows that whenever one substance is reduced, another is oxidized. Because reduction and oxidation reactions must occur together, with an electron added to one reactant and removed from the other, they are known generically as ***redox reactions*** (from *red*uction-*ox*idation).

In the linked redox reactions of photosynthesis, the energy of sunlight causes both the splitting of water molecules and the reduction of carbon dioxide to form energy-rich sugars. In other words, *the H^+ ions and electrons obtained by splitting water molecules are added to the CO_2 to form reduced compounds based on (CH_2O) units, and energy from light is stored in the process.* To understand the process of photosynthesis, then, we shall focus on two key points: the mechanism for trapping and handling the energy of sunlight and the mechanism for transferring hydrogen and electrons from water to carbon dioxide.

MAKING CONNECTIONS

1. Why was the evolution of photosynthesis a critical event in the history of life?

2. How are photosynthesis and aerobic respiration linked? Explain in terms of exchange of end products and the flow of the energy.

3. How are autotrophs and heterotrophs dependent on one another?

4. How do redox reactions release and store energy?

Green Leaves Are the Principal Organs of Photosynthesis

Although photosynthesis can occur in all of the green, chlorophyll-containing parts of the plant, in most plants it is the leaves that expose the greatest area of green tissue to the light and are therefore the principal organs of photosynthesis. Accordingly, we shall examine first the anatomy of leaves (Fig. 7.5).

The leaves of most plants consist of a ***petiole,*** or stalk, and a flattened ***blade.*** (Some leaves lack petioles, in which case the blade attaches directly to the stem.) The blade is usually broad and thin and is traversed by a complex system of veins. Because the blade is flat, the leaf exposes an area that is very large in relation to its volume, making it an efficient structure for capturing light.

Through a microscope it can be seen that the outer surfaces of the leaf are covered by a layer of ***epidermis,*** usually only one cell thick, but sometimes two, three, or more cells thick (Fig. 7.6). A waxy layer, the ***cuticle,*** typically covers the outer surfaces of both the upper and lower epidermis. The chief function of the epidermis is to protect the internal tissues of the leaf from excessive water loss, from invasion by fungi, and from mechanical injury. Most epidermal cells do not contain chloroplasts—the plant's chlorophyll-containing organelles—and hence play no active role in photosynthesis.

The region between the upper and lower epidermis constitutes the ***mesophyll*** portion of the leaf; the cells of the meso-

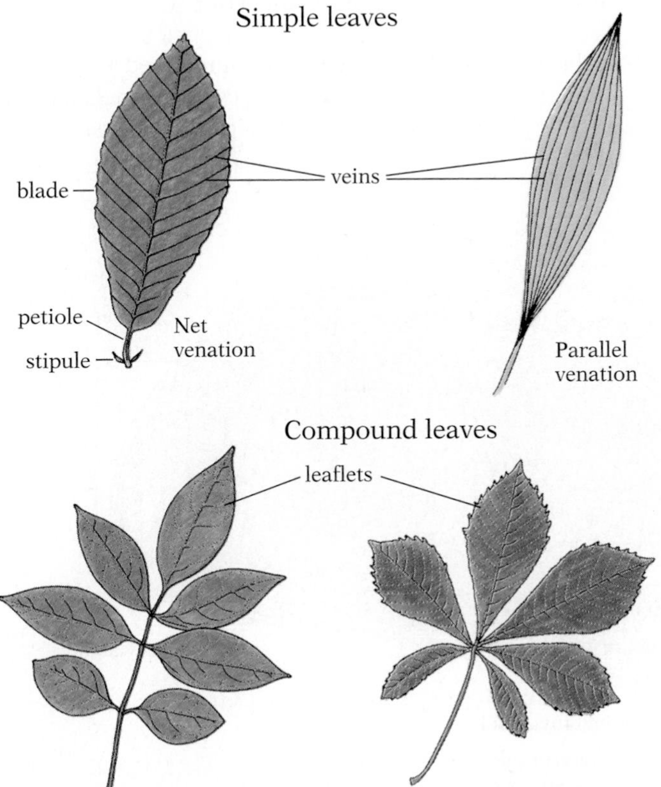

7.5 Leaf types A leaf usually consists of a blade and a petiole. Veins run from the petiole into the blade. The main veins may branch in succession off the midvein (net venation) or they may be parallel. The blade may be simple; or it may be compound—that is, divided into leaflets that may be arranged in two different ways, as shown.

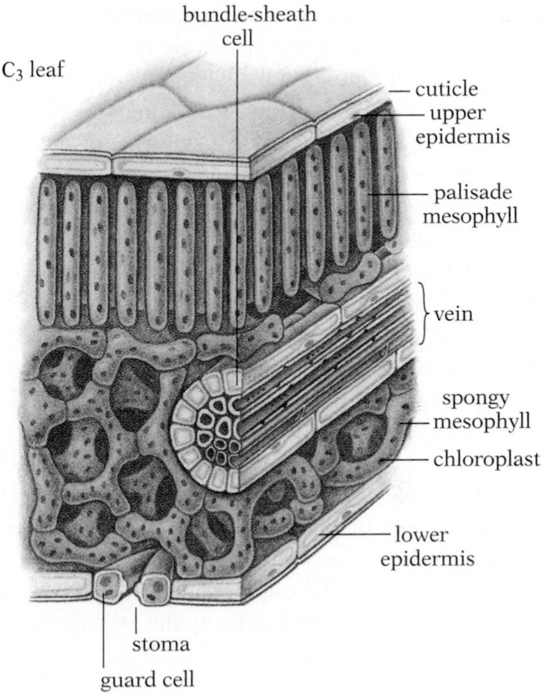

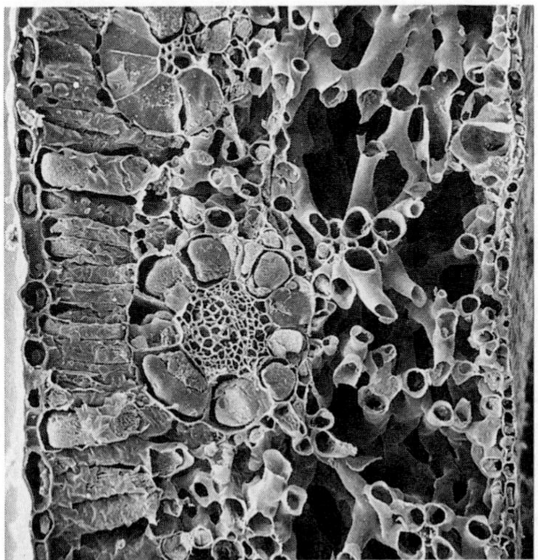

7.6 Leaf anatomy Top: The entire region between the upper and lower epidermis, except for the vein, is the mesophyll; it is the primary photosynthetic tissue of the plant. The mesophyll is divided into two regions, the palisade mesophyll and the spongy mesophyll. Veins conduct materials from one part of the plant to another. Carbon dioxide from the air enters the leaf through the stomata, the size of which is regulated by the guard cells. Bottom: This scanning electron micrograph shows a cross section of a typical dicot leaf. Note that the mesophyll cells are loosely packed and have many spaces between them. The round structure in the center is a vein.

phyll contain many chloroplasts and are the chief sites of photosynthesis in the plant. The mesophyll is often divided into two fairly distinct parts: an upper ***palisade mesophyll,*** consisting of cylindrical cells arranged vertically, and a lower ***spongy mesophyll,*** composed of cells that are both irregularly shaped and oriented more randomly. The cells of both parts of the mesophyll are very loosely packed, having many interconnected spaces between them. These spaces are linked with the atmosphere by way of holes in the lower epidermis called ***stomata*** (sing., ***stoma***). The carbon dioxide required for photosynthesis is obtained directly from the air, which can enter the leaf through the stomata. The size of the stomatal openings is regulated by a pair of modified epidermal cells called ***guard cells.***

The conspicuous system of veins (also called vascular bundles) that branches into the leaf blade from the petiole forms a structural framework for the blade and also acts as a local transport system, connecting the different parts of the leaf with the main transport system of the rest of the plant. Each vein contains cells of the two principal vascular tissues; ***xylem,*** the tissue that transports water and dissolved minerals upward from the roots, and ***phloem,*** the tissue that carries organic materials throughout the plant. Each vein is usually surrounded by a ***bundle sheath*** of tightly packed cells. In most cases the branching of the veins is such that no mesophyll cell is far removed from a vein. The water required for photosynthesis is delivered to the mesophyll cells by the xylem; the carbohydrate end products are transported to the other cells of the plant by the phloem.

Chloroplasts, the Organelles of Photosynthesis, Have Specialized Structures for Different Functions

As noted in the preceding chapter, in eucaryotic cells photosynthesis takes place within the chloroplasts (Fig. 7.7). Recall that the chloroplast is bounded by two concentric membranes and contains a third set of internal membranes that form a series of flattened, interconnected sacs known as thylakoids. The numerous disklike thylakoids are often arranged in stacks called grana, which look like stacks of coins. The thylakoid membrane serves as a barrier between the interior of the thylakoid and the stroma, the internal matrix of the chloroplast. The light reactions of photosynthesis (those in which light energy is trapped and converted into chemical energy) take place in or on the thylakoid membrane. The dark reactions of photosynthesis (those involving the reduction of carbon dioxide to sugars such as glucose) take place in the more fluid stroma that surrounds the thylakoid sacs.

Chlorophyll Absorbs Light Energy and Passes It on to Other Molecules

Light energy must be absorbed if it is to be used for photosynthesis. Certain wavelengths of light, especially red and violet, are absorbed readily by various pigment molecules, while

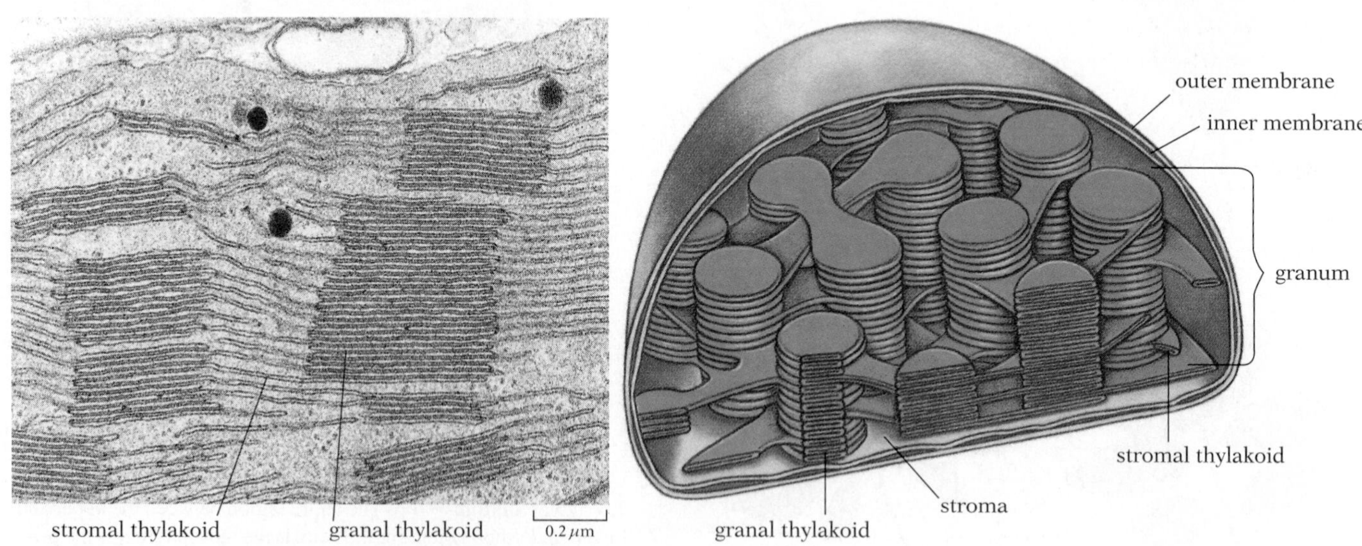

7.7 Structure of a chloroplast Left: Electron micrograph of a section of a chloroplast of timothy grass showing several grana; note the continuity between granal and stromal thylakoids. Right: This cutaway view of a typical chloroplast shows the inner and outer membranes lying close together, enclosing the large compartment known as the stroma. Inside the stroma can be seen a third distinct membrane, which forms the interconnected compartments called thylakoids. The stacks of flat, disklike thylakoids are grana. Chlorophyll molecules and most of the electron-transport-chain molecules are located within the thylakoid membrane.

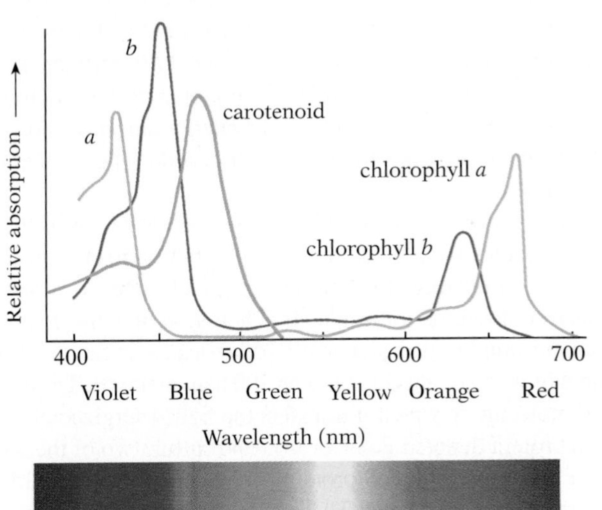

7.8 Absorption and action spectra of photosynthetic pigments Absorption spectra of chlorophyll *a,* chlorophyll *b,* and a carotenoid. Taken together, the absorption spectra of the two chlorophylls and the carotenoid cover more of the range of wavelengths available to the plant than does the spectrum of chlorophyll *a* alone.

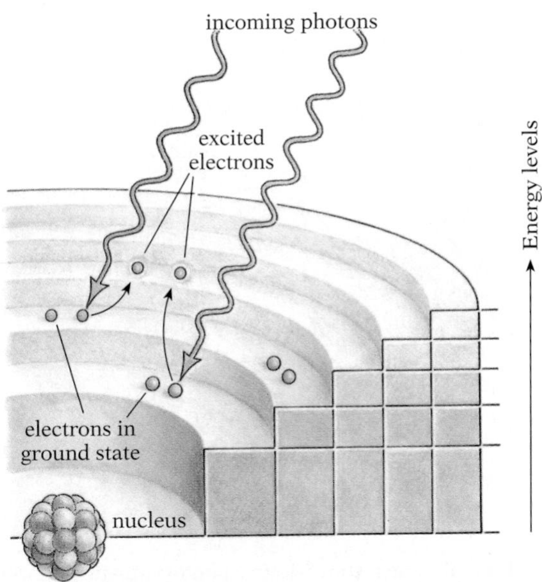

7.9 Effect of light on chlorophyll When a photon is absorbed by a chlorophyll molecule, the photon's energy raises an electron to a higher energy level. In this simplified representation we see a typical distribution of electrons at their lowest available energy levels, and the distribution as it is altered by absorption of a red photon, which raises one electron from the second level to the third.

other wavelengths, notably green, are absorbed only very slightly (Fig. 7.8). What happens when light of a proper wavelength strikes a pigment molecule? The exact answer is not known, but we have some reasonable hypotheses.

The chlorophyll and accessory pigments necessary for photosynthesis are organized into two functional systems of molecules, called ***Photosystem I*** and ***Photosystem II,*** both of which are built into the thylakoid membrane. Each photosystem comprises some 300 pigment molecules, including two slightly different kinds of chlorophyll (chlorophyll *a* and chlorophyll *b*) as well as carotenoids. One group of nine pigment molecules in each photosystem is distinct from all the rest and acts as a ***reaction center.*** Other pigment molecules function like antennas, absorbing light energy of different wavelengths and passing the energy along to the reaction center.

When a ***photon,*** a discrete packet of light energy, strikes a pigment molecule and is absorbed, its energy is transferred to an electron of the pigment molecule, raising this electron from its stable ground state to a higher, relatively unstable energy level (Fig. 7.9). The excited electron subsequently falls back to its ground state, giving up the absorbed energy as heat or light. Isolated chlorophyll in a test tube promptly loses the light energy it captures by reemitting it as visible red light (Fig. 7.10). Evidently, the pigment molecules alone are incapable of converting light energy into chemical energy.

In the functioning chloroplast, once light energy has raised an electron in an "antenna" molecule to a higher energy state, the excited state is passed from pigment molecule to pigment

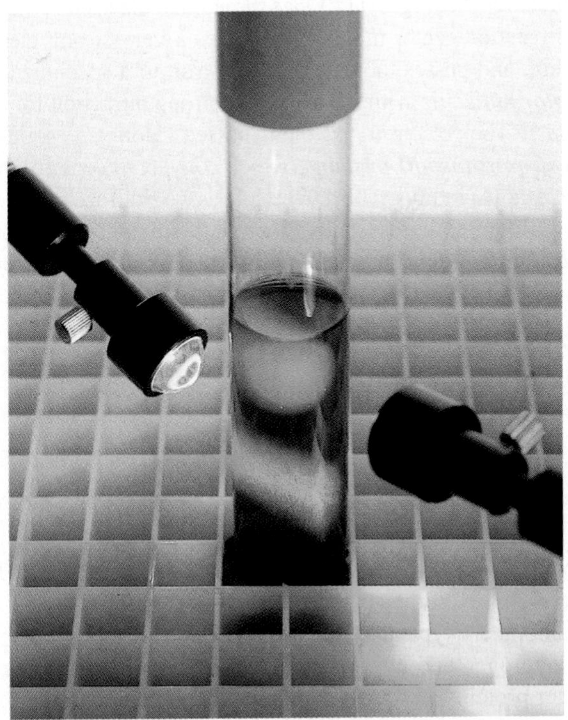

7.10 Fluorescence and transmittance by chlorophyll This extract of chlorophyll from a green leaf fluoreses under reflected light (right light). The light energy it absorbs is reemitted as visible red light. Light that is transmitted through the extract (left light) appears yellow-green.

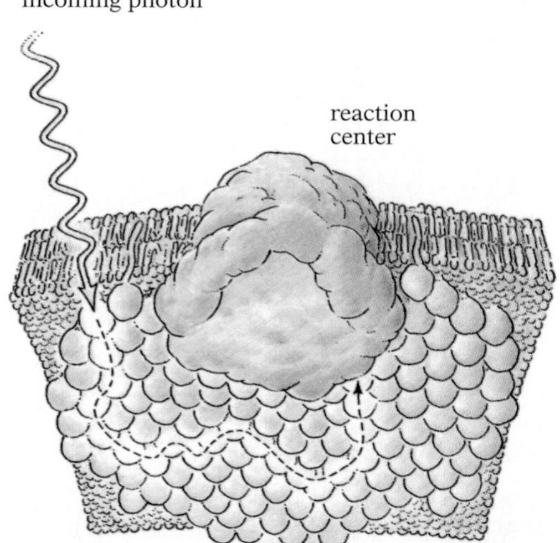

7.11 Flow of excited state within a photosynthetic unit A photon strikes one of the antenna pigments (green circles) and raises an electron in the pigment to a higher energy level. This excited state is then passed from pigment molecule to pigment molecule in a random sequence (dotted pathway) until it eventually reaches the reaction center (large green unit), where it is trapped.

molecule[1] and eventually reaches the reaction-center complex, which traps it (Fig. 7.11). Once the excited electron is trapped, the reaction center molecule becomes a very strong electron donor, and passes the energized electron to a specialized ***acceptor molecule*** with a particularly strong attraction for electrons. The electron is then passed along a series of ***electron-transport proteins,*** which convert the energy into a form more readily usable by the cell, as we shall see below. The two different photosystems found in the thylakoid membrane differ in the number and types of pigment molecules present in each, and in the composition of their reaction-center complexes.

Electrons Energized by Light Are Passed From One Molecule to the Next Along the Noncyclic Pathway

To understand the chemistry of light reactions, we need to look more closely at the anatomy of the thylakoid membrane. The membrane, a typical lipid bilayer, is impermeable to ions such as hydrogen and hydroxide ions. The two photosystems and other proteins and protein complexes are built into the membrane. As Figure 7.12 illustrates, both the light-absorbing pigment molecules in the photosystems and the electron-transport proteins are arranged in complexes within the thylakoid membrane; the complexes are linked by mobile carrier proteins. The electron-transport proteins are arranged in such a way that electrons can be passed from one electron-carrier protein complex to the next along the chain. Also located within the membrane are large protein complexes called ***ATP synthetases.*** As their name indicates, the role of these enzymes is to produce (by a process described below) the energy-currency molecule ATP, which provides the energy for a wide range of cellular processes.

The specialized chlorophyll complex that serves as the reaction center of a photosystem is, as we have already noted, capable of donating the light-energized electron to a special electron-acceptor molecule. In Photosystem I the reaction-center complex is designated ***P700*** (because it cannot absorb light of wavelengths higher than 700 nanometers[2]). The acceptor molecule to which it transfers the light-energized electron is a protein denoted FeS (for iron and sulfur, two of the molecule's atoms). The FeS protein is reduced and P700 oxidized by this electron transfer. The energized electron is then passed from FeS to a second electron-acceptor molecule, which passes it to still another acceptor molecule (Fig. 7.13). As each acceptor in the series receives the electron it becomes reduced, and as it gives up the electron it becomes oxidized. The third electron acceptor in the chain, the FAD complex, passes the electron to a molecule called nicotinamide adenine dinucleotide phosphate, or ***$NADP^+$,*** which is located outside the membrane, in the stroma. Each $NADP^+$ molecule can accept two electrons from FAD and picks up a hydrogen ion from the stroma of the chloroplast, and is thereby reduced to ***NADPH.*** The NADPH, which remains in the stroma, will eventually act as an electron donor in the reduction of carbon dioxide to carbohydrate.

To summarize, the electrons move from the P700 reaction center to the FeS acceptor complex and along a series of electron-transport proteins to $NADP^+$, forming NADPH. We shall refer to the oxidized form of this important electron-transport protein as $NADP^+$ and to the reduced form as NADPH throughout this and subsequent chapters.

Now, if the energized electrons from the chlorophyll of Photosystem I are retained in NADPH and eventually incorporated into carbohydrate, it follows that Photosystem I is left short of electrons; in other words, it is left with positively charged "electron holes" and becomes a very strong electron acceptor. These vacancies are filled by electrons derived from water as a result of a second light event, which we shall examine next.

---

[1] The transfer of energy from one pigment molecule to an adjacent one does not appear to involve the physical transfer of an excited electron. Instead, when the excited electron in one molecule falls back to a lower energy level, an electron in an adjacent pigment molecule is boosted to a higher level, thus taking on the excited state. Some researchers refer to this process as a transfer of excitation energy.

[2] One nanometer (nm) equals 10^{-9} meter.

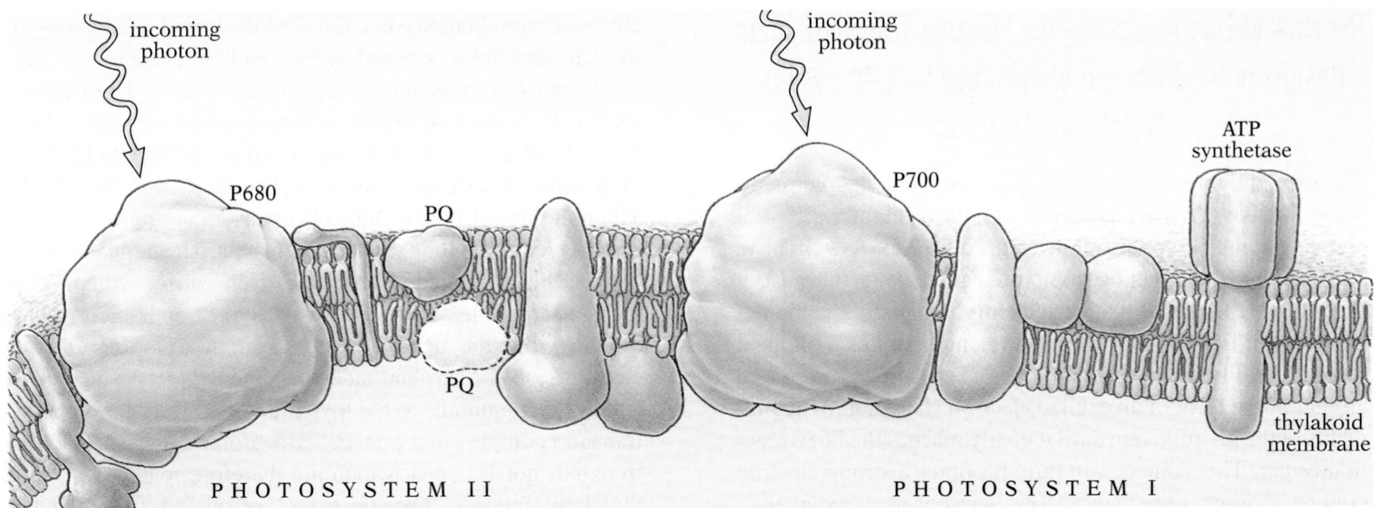

7.12 The thylakoid membrane The light-absorbing pigment molecules of Photosystems I and II (green), the electron-carrier proteins (pink), and the ATP synthetase complexes are built into the thylakoid membrane. The electron-transport complexes are arranged so that electrons are passed from one electron-carrier molecule to the next along the chain. One of the electron-transport molecules (PQ) is mobile and can move within the plane of the membrane. A second mobile electron-transport molecule, PC, moves along the inner surface of the thylakoid membrane and transfers electrons from cytochrome *f* to Photosystem I.

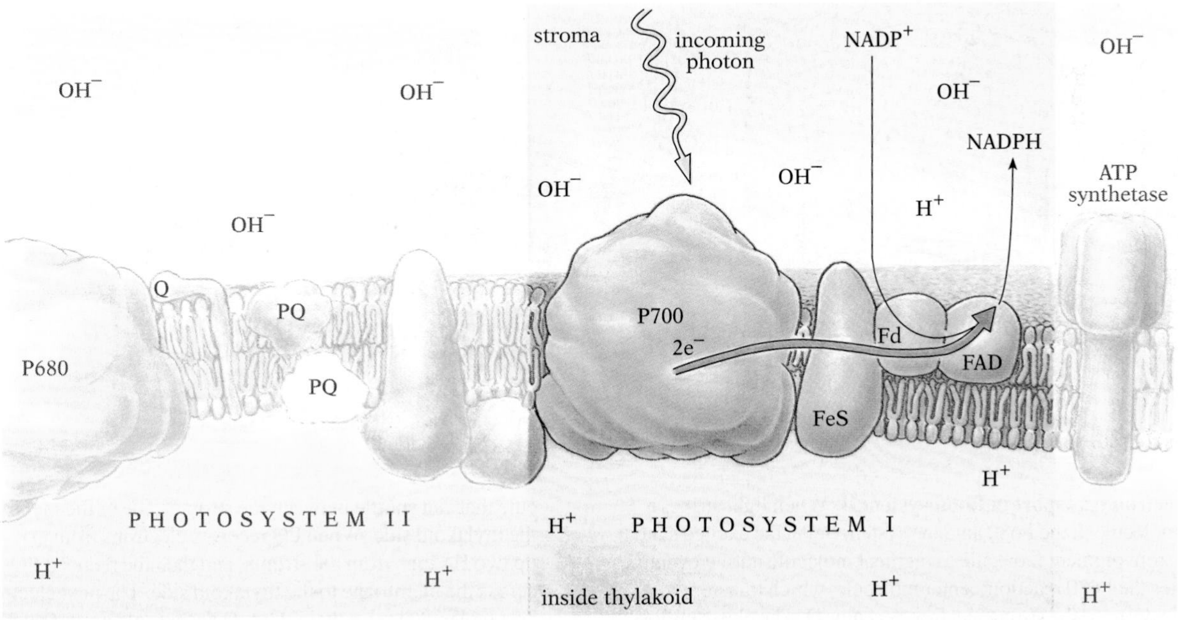

7.13 Electron transport in Photosystem I to form NADPH When light strikes a pigment molecule in the P700 antenna system, the excited state is passed from pigment molecule to pigment molecule until it eventually reaches the special P700 molecule, which is thereby energized. An excited electron is passed through a series of electron transport molecules and finally to NADP+, which also picks up an H^+ ion from the stroma to form NADPH. Notice that the NADPH is formed in the stroma of the chloroplast.

The Flow of Electrons Along the Electron-Transport Chain Results in the Transport of Hydrogen Ions Across the Thylakoid Membrane

The second light event that takes place in the thylakoid membrane involves Photosystem II, which has a slightly different pigment-molecule composition from that of Photosystem I and a different reaction-center complex, called ***P680*** (because it cannot absorb light of wavelengths higher than 680 nanometers). When light of the proper wavelength strikes a pigment molecule of Photosystem II, the energy is absorbed and passed around in the form of an excited electron state within the pigments of the photosystem until it finally reaches the P680 reaction center. This complex, in turn, becomes a strong electron donor and donates the high-energy electrons to a special electron acceptor molecule called Q. Q in turn passes the electrons to a series of electron-transport molecules, which conduct the electrons from Photosystem II to Photosystem I, where they fill the "electron holes" created by the first light event.

As the electrons are passed along the chain from Photosystem II to Photosystem I, a very important process takes place: hydrogen ions are pumped across the thylakoid membrane, from the stroma into the thylakoid interior (Fig. 7.14). The pumping of the H^+ ions occurs as a consequence of the flow of electrons along the transport chain. The molecule PQ is both an electron carrier and an H^+ ion carrier; whenever it picks up electrons from Q, it must also pick up H^+ ions. PQ is mobile within the membrane; it picks up two H^+ ions and two electrons on the stromal side (becoming PQH_2) and moves across the membrane to the thylakoid side. The next electron-transport complex in the chain (cytochrome *f*) can accept electrons but not H^+ ions, which are therefore released into the thylakoid interior. The electrons are passed from the cytochrome *f* complex to the mobile carrier PC (plastocyanin), which transports them to the pigment molecules of

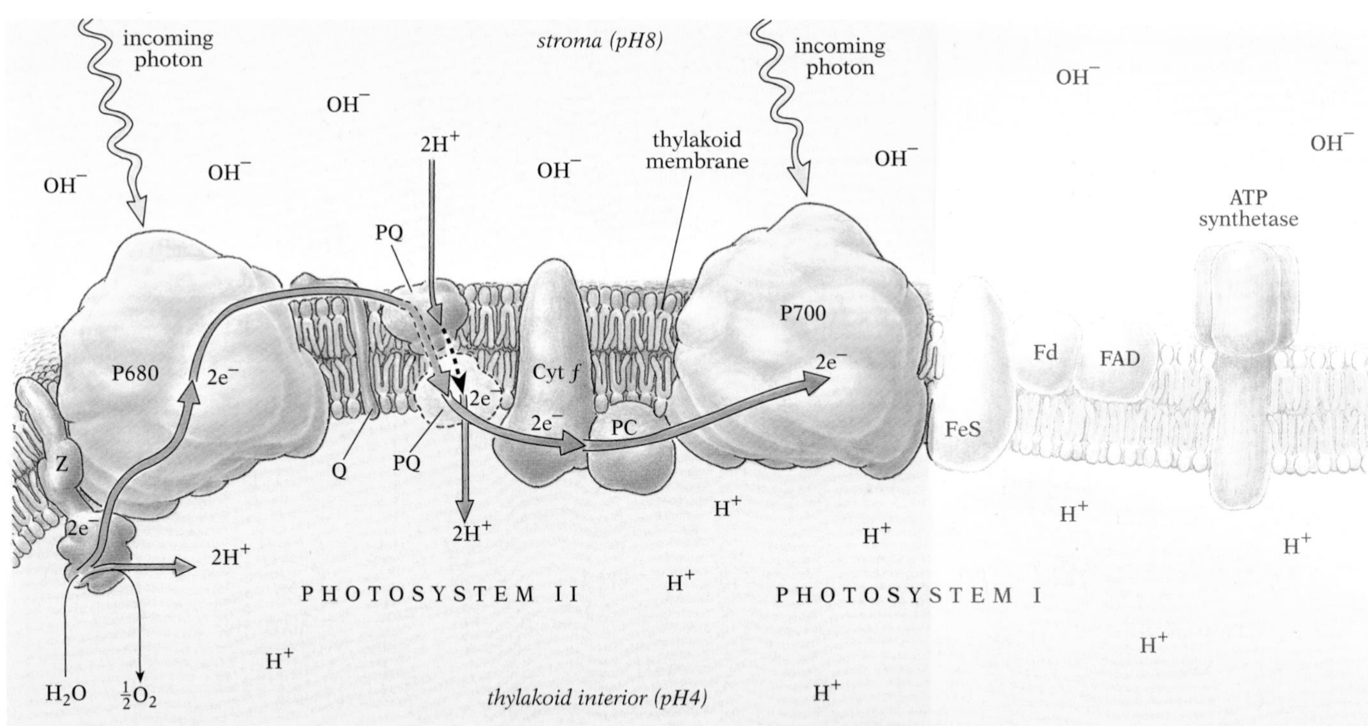

7.14 Electron transport in Photosystem II When light strikes a pigment molecule in the P680 antenna system (left), the excited state is passed from pigment molecule to pigment molecule until it eventually reaches the P680 reaction-center molecule, which transfers excited electrons to the strong electron acceptor Q. The electrons lost from P680 are replaced by electrons pulled away from water; the splitting of water also produces molecular oxygen and releases H^+ ions into the interior of the thylakoid. Q passes the energized electrons to the electron transport molecule PQ, which is a carrier molecule that can shuttle between the stromal side of the membrane and the thylakoid side. When PQ receives electrons from Q it also picks up two H^+ ions from the stroma, and then the reduced PQH_2 migrates across the membrane to the thylakoid side. The next electron transport molecule in the chain (Cyt. *f*) accepts electrons from PQH_2 but not H^+ ions, so the H^+ ions are left behind in the thylakoid interior. The result is that H^+ ions are transported from the stroma into the thylakoid interior. The electrons continue along the chain, eventually reaching the pigment molecules of Photosystem I.

Photosystem I. *The direct result of this flow of electrons through this portion of the electron-transport chain is the active transport of H^+ ions from the stroma into the thylakoid interior.* The H^+ ion concentration increases in the thylakoid space (to a pH of about 4) and decreases in the stroma (to a pH of 8), and a steep pH difference is established across the thylakoid membrane. In addition, an electrical gradient is produced: the stroma, having lost the positively charged H^+ ions, becomes more negatively charged, and the thylakoid space, having gained H^+ ions, becomes more positively charged. This electrical and hydrogen-concentration difference across the membrane, known as an ***electrochemical gradient,*** is a source of energy that can be tapped by a special mechanism (described below) to synthesize ATP.

Thus the "electron holes" created in Photosystem I by the first light event are refilled by electrons moved from Photosystem II by the second light event. This process alone, however, would leave electron holes in Photosystem II; the electron deficit would simply have been shifted from Photosystem I to Photosystem II. It is at this point that the electrons from water, mentioned earlier, play their role. Photosystem II is a large protein complex consisting of antenna pigments, the P680 reaction center, and a water-splitting enzyme complex, which has a cluster of four manganese ions in its catalytic site. The P680, having given up an electron when excited by light, becomes positively charged ($P680^+$) and has a very strong attraction for electrons. As Figure 7.14 indicates, it is thought that $P680^+$ (with the help of the water-splitting enzyme and an enzyme complex referred to as Z) pulls electrons away from water, leaving behind free H^+ ions and molecular oxygen:

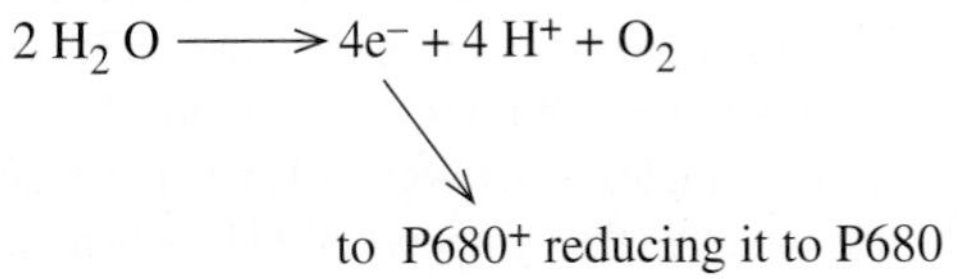

The oxygen is released as a gaseous by-product, which diffuses out of the cell, and ultimately into the atmosphere via the stomata. The H^+ ions remain inside the thylakoid and contribute to the electrochemical gradient across the membrane.

As noted, the electrons involved in the second light event move from water to the P680 molecule of Photosystem II to the electron-transport chain to Photosystem I (Figs. 7.13 and 7.14). If we combine these steps with the electron movement associated with the first light event, as traced above, we obtain the following abbreviated sequence showing the overall movement of electrons:

$H_2O \rightarrow$ Photosystem II $\rightarrow$ electron-transport chain $\rightarrow$
Photosystem I $\rightarrow$ a second electron-transport chain $\rightarrow$
NADPH $\rightarrow$ carbohydrate

This sequence reinforces our earlier statement that the electrons necessary to reduce carbon dioxide to carbohydrate come from water, but it shows that the movement of electrons from water to carbohydrate is an indirect and complex process.

Since some electrons leave the system via NADPH and others enter the system from water as replacements, the electrons follow a one-way, or noncyclic, pathway through the membrane. The process so far has three results: the formation of NADPH, the release of molecular oxygen, and the establishment of an H^+ gradient across the thylakoid membrane.

ATP Is Synthesized Using the Energy Stored in the Electrochemical Gradient

The energy of excited chlorophyll electrons released along the electron-transport chain can be harnessed by the cell to do useful work. How? We saw that the flow of electrons along the electron-transport chain resulted in the splitting of water molecules and the transport of H^+ ions across the thylakoid membrane. Both processes led to the generation of an electrochemical gradient across the thylakoid membrane. The energy stored in this gradient is used in turn to synthesize adenosine triphosphate, or ***ATP.*** This compound, referred to earlier as the universal energy currency of the living cell, plays a key role in most biological energy transformations, including photosynthesis.

The ATP molecule (Fig. 7.15) is composed of a nitrogen-containing compound (adenosine) plus three phosphate groups bonded in sequence:

adenosine —Ⓟ—Ⓟ—Ⓟ

By convention, Ⓟ stands for the entire phosphate group, and the colored lines between the first and second and the second and third phosphate groups represent what are sometimes called "high energy" phosphate bonds.[3] What makes these particular bonds "high-energy" is the ease with which ATP is hydrolyzed to ADP and Ⓟ, and the fact that the hydrolysis releases much usable energy. ATP is constantly being formed and hydrolyzed in the cell. Indeed, it has been estimated that in 24 hours a human being may use as much as 40 kilograms of ATP.

[3]Actually the term "high energy" here is something of a misnomer, since the bond energy of the covalent bonds themselves is not unusually high. The term "high energy" in this context refers to the amount of energy released when a bond is *hydrolyzed,* or broken apart by the addition of water, whereas the bond energy is the amount of energy needed to break a covalent bond.

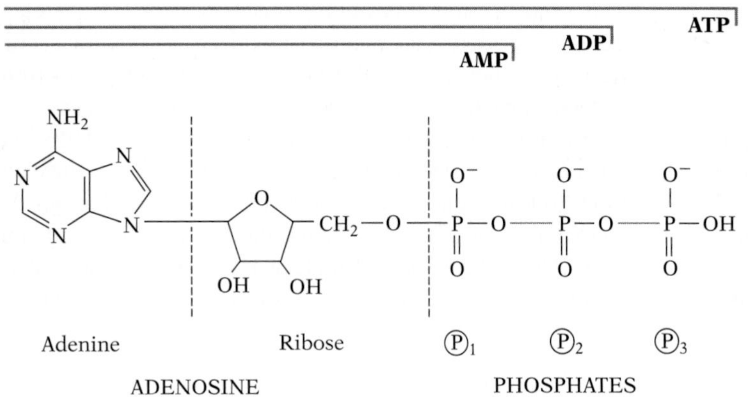

7.15 The ATP molecule As its full name, adenosine triphosphate, implies, this molecule is composed of an adenosine unit (a complex of adenine and ribose sugar) and three phosphate groups arranged in sequence. The cell stores energy by adding a phosphate group to ADP (adenosine diphosphate) to make ATP, and later recovers some of this energy by hydrolyzing ATP into ADP and inorganic phosphate. Occasionally ADP is hydrolyzed to make AMP (adenosine monophosphate), one of the nucleotides of which RNA is composed.

Actually, it is often only the terminal phosphate bond of ATP that is involved in energy conversions. The exergonic reaction by which this bond is hydrolyzed and the terminal phosphate group removed leaves a compound called adenosine diphosphate, or ***ADP*** (adenosine plus two phosphate groups), and inorganic phosphate Ⓟ:

$$ATP + H_2O \xrightarrow{\text{enzyme}} ADP + \text{Ⓟ} + \text{energy}$$

If both the second and third phosphate groups are removed from ATP, the resulting compound is adenosine monophosphate, or ***AMP***.

Conversely, new ATP can be synthesized from ADP and inorganic phosphate if adequate energy is available to force a third phosphate group onto the ADP. The addition of phosphate is termed ***phosphorylation:***

$$ADP + \text{Ⓟ} + \text{energy} \xrightarrow{\text{enzyme}} ATP + H_2O$$

The light-dependent reactions of photosynthesis are often grouped under the general heading of ***photophosphorylation,*** which means *the use of light energy to phosphorylate (add inorganic phosphate to) a molecule,* usually ADP. Like so many other terms, "photophosphorylation" became part of the scientific vocabulary before the process it describes was well understood. We now know that in fact light-energy absorption and phosphorylation are separate reactions.

How then is ATP synthesized within the chloroplast? Of the various hypotheses advanced to explain ATP synthesis, in the mitochondrion as well as in the chloroplast, the most popular one is the ***chemiosmotic hypothesis.*** Proposed in 1961 by Peter Mitchell of Glynn Research Laboratories in England, who was awarded the Nobel Prize in 1978 for his contribution, this hypothesis states that the transfer of electrons along the transport chain results in the pumping of H^+ ions across the membrane, which creates an electrochemical gradient. It is this gradient that provides the energy to drive the synthesis of ATP.

Let us look at the chemiosmotic process in more detail. As noted earlier, H^+ ions accumulate in the interior of the thylakoids as a result of two processes: (1) the pumping of H^+ ions across the membrane (by PQ) in the course of electron transport along the transport chain, and (2) the splitting of water molecules to provide replacement electrons for Photosystem II. The net result is a steep electrochemical gradient across the membrane, with the concentration of H+ ions higher in the thylakoid interior than out in the stroma. The electrochemical gradient functions rather like a battery—the flow of H^+ ions (rather than electrons as in an ordinary battery) back across the membrane drives the synthesis of ATP.

How does this happen? Recall that the thylakoid membrane contains numerous enzyme complexes, the ATP synthetases, which look rather like lollipops inserted into the membrane. The synthetase complexes act both as H^+-ion channels and as enzymes that catalyze the synthesis of ATP. When the H^+ ions move through the channels down their concentration gradient (i.e., from a region of high concentration in the thylakoid to one of low concentration in the stroma), ATP is synthesized from ADP and Ⓟ and is released into the stroma (Figs. 7.16 and 7.17). The precise mechanism involved is not well understood.

Because it is the energy of light that ultimately provides the power to phosphorylate ADP to ATP, the whole process—noncyclic electron flow followed by ATP synthesis—is called ***noncyclic photophosphorylation.*** *The end products of noncyclic photophosphorylation are ATP, NADPH, and molecular oxygen.* The NADPH and ATP are subsequently used to drive the reduction of carbon dioxide in the dark reaction, while the oxygen diffuses out of the cell into the interior of the leaf. The reactions of photophosphorylation have traditionally been known as the light reactions of photosynthesis, even though, as we have seen, only two steps—involving Photosystems I and II—are directly light-dependent.

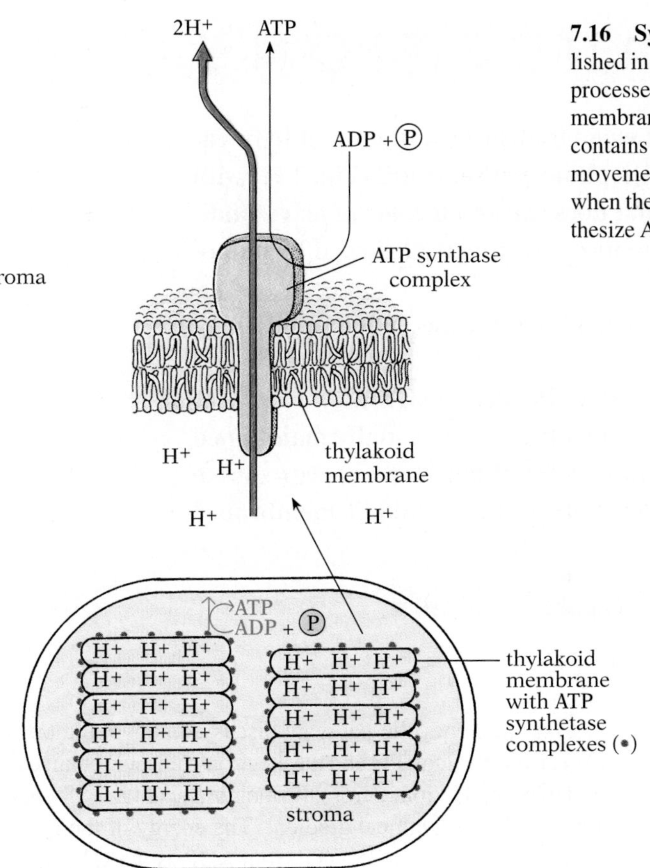

7.16 Synthesis of ATP in the chloroplast A high H^+ ion concentration is established in the interior of the thylakoids (bottom), which occurs as a result of two processes: the splitting of water molecules and the pumping of H^+ ions across the membrane by PQ in the course of electron transport. The thylakoid membrane contains numerous ATP synthetase complexes that act as H^+ ion channels for the movement of H^+ ions back across the membrane down their concentration gradient; when the H^+ ions move through the channels, energy is released and is used to synthesize ATP from ADP and inorganic phosphate.

7.17 Summary of noncyclic photophosphorylation Light energy is used to move electrons from water to P680, to Q, and along an electron transport chain to P700. As the electrons move along the chain, H^+ ions are pumped across the membrane (by PQ). From P700, electrons are excited by light and passed along a second electron transport chain to $NADP^+$, to form NADPH. The energy of the H^+ ion gradient can be used by the ATP synthetase complex (right) to make ATP as the H^+ ions flow through the complex. The two important products of the noncyclic pathway are ATP and NADPH, which make possible the synthesis of carbohydrate from CO_2. Molecular oxygen is released as a by-product.

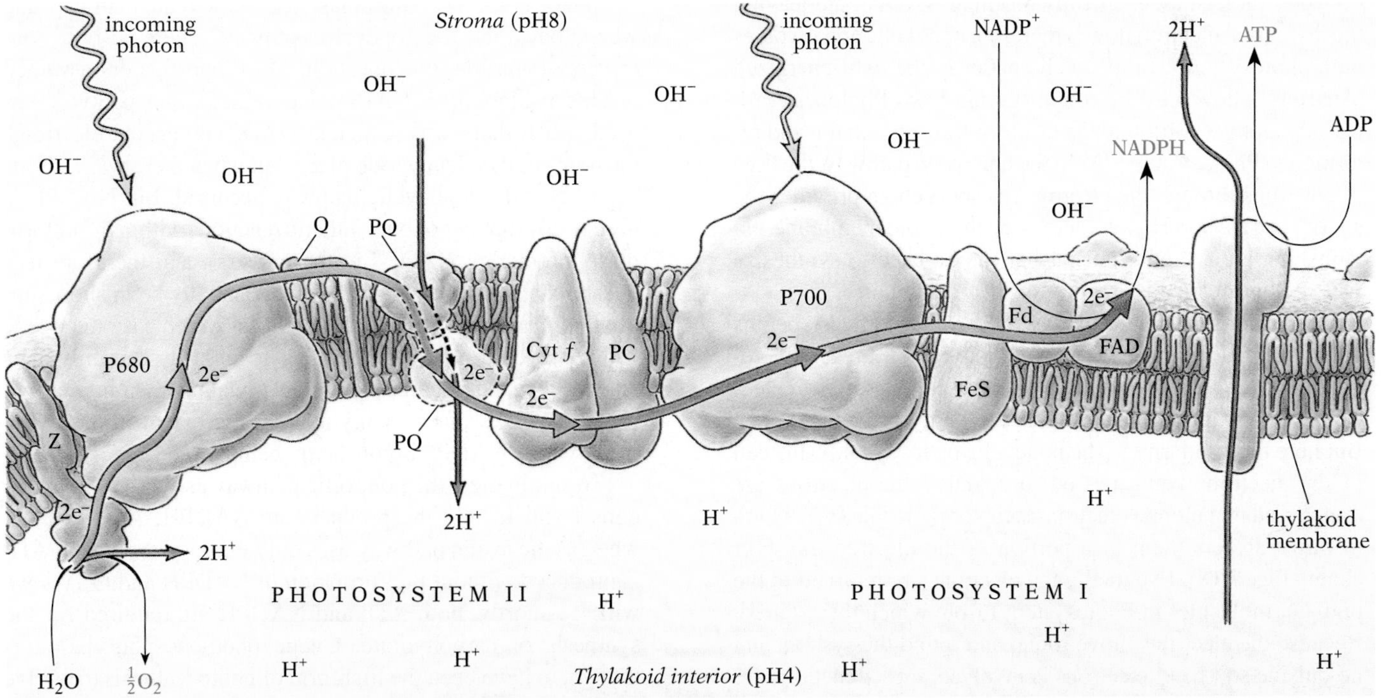

MAKING CONNECTIONS

1. What happens to chlorophyll when it absorbs a photon of light energy? Compare a photon of light striking a photosynthetic unit (Fig. 7.11) with a photon exciting an atom (Fig. 2.4). What does the green color of leaves indicate about which wavelengths of light are most effectively absorbed by photosynthetic pigments?

2. Compare and contrast the thylakoid membrane with the cell membrane pictured in Figure 5.13.

3. What membrane characteristic does PQ demonstrate?

4. Concentration gradients across membranes are usually maintained by the expenditure of energy in the form of ATP. What is the energy source used to establish the gradient of H^+ ions across the thylakoid membrane?

Cyclic Photophosphorylation Produces Only ATP

There are two general pathways by which the energy of light-energized electrons is harvested to produce ATP: the ***noncyclic pathway,*** which we have just discussed, and the ***cyclic pathway,*** which involves only one of the two photosystems found in most plants. In the noncyclic pathway, electrons from water are energized and transported through a series of electron-transport molecules, eventually reaching $NADP^+$ and leaving the thylakoid; electron flow is therefore noncyclic and involves both photosystems. In the cyclic pathway the light-energized electrons follow a cyclic route, starting from Photosystem I, passing along a portion of the electron-transport chain, and returning to Photosystem I. As in the noncyclic pathway, the flow of electrons through the electron-transport chain provides the energy to transport H^+ ions across the thylakoid membrane, establishing an H^+ ion gradient that can be used to synthesize ATP. Let us look at these processes in more detail.

Like the noncyclic pathway, the cyclic pathway begins when light-energized electrons in Photosystem I reach the reaction-center complex P700 and are passed to a series of electron-acceptor molecules, the first one being FeS (Fig. 7.18). But here the similarity to the noncyclic pathway ends. Instead of the electrons being passed to $NADP^+$, the electrons are transferred to a mobile electron-acceptor molecule (b_6), which in turn transfers them to a portion of the electron-transport chain (Fig. 7.18). Eventually the electrons are returned to the pigment molecules in Photosystem I from which they started. Because the electrons move round and round the system, and no outside source of electrons is involved, these reactions are referred to collectively as the cyclic pathway.

Like the noncyclic pathway discussed above, the transfer of electrons through PQ and the electron-transport chain can be used to pump H^+ ions across the membrane (Fig. 7.18), generating an electrochemical gradient. The energy of the gradient can then be used to make ATP from ADP as the H^+ ions flow back through the ATP synthetase complex. This whole process—cyclic electron flow and H^+ ion pumping followed by ATP synthesis—is called ***cyclic photophosphorylation.*** Its only product is ATP.

When does the chloroplast use the cyclic pathway and when does it use the noncyclic pathway? There is some controversy regarding this question. Most investigators assumed that the availability of $NADP^+$ determined which pathway was used; when there was sufficient $NADP^+$ to accept electrons, the noncyclic pathway took place, but when there was a shortage of $NADP^+$, the cyclic pathway occurred. But NADPH is used in any manner of biosynthetic reactions within the chloroplast (producing $NADP^+$ in the process), and therefore it is questionable whether $NADP^+$ would actually be in short supply in the actively metabolizing leaf, in which case the cyclic pathway would not occur. The cyclic pathway certainly does take place in the test tube with intact thylakoid membranes, but it may not take place to any appreciable extent in the living plant, where $NADP^+$ is constantly being produced.

To summarize, the noncyclic pathway uses both Photosystems I and II, and the products are NADPH, ATP, and O_2, whereas the cyclic pathway uses only Photosystem I, and ATP is produced without the formation of NADPH and O_2. As we will see shortly, both ATP and NADPH are required for the synthesis of carbohydrate. Cyclic photophosphorylation is thought to have been the first form of photosynthesis to evolve in primitive organisms, and it is still the only form available to

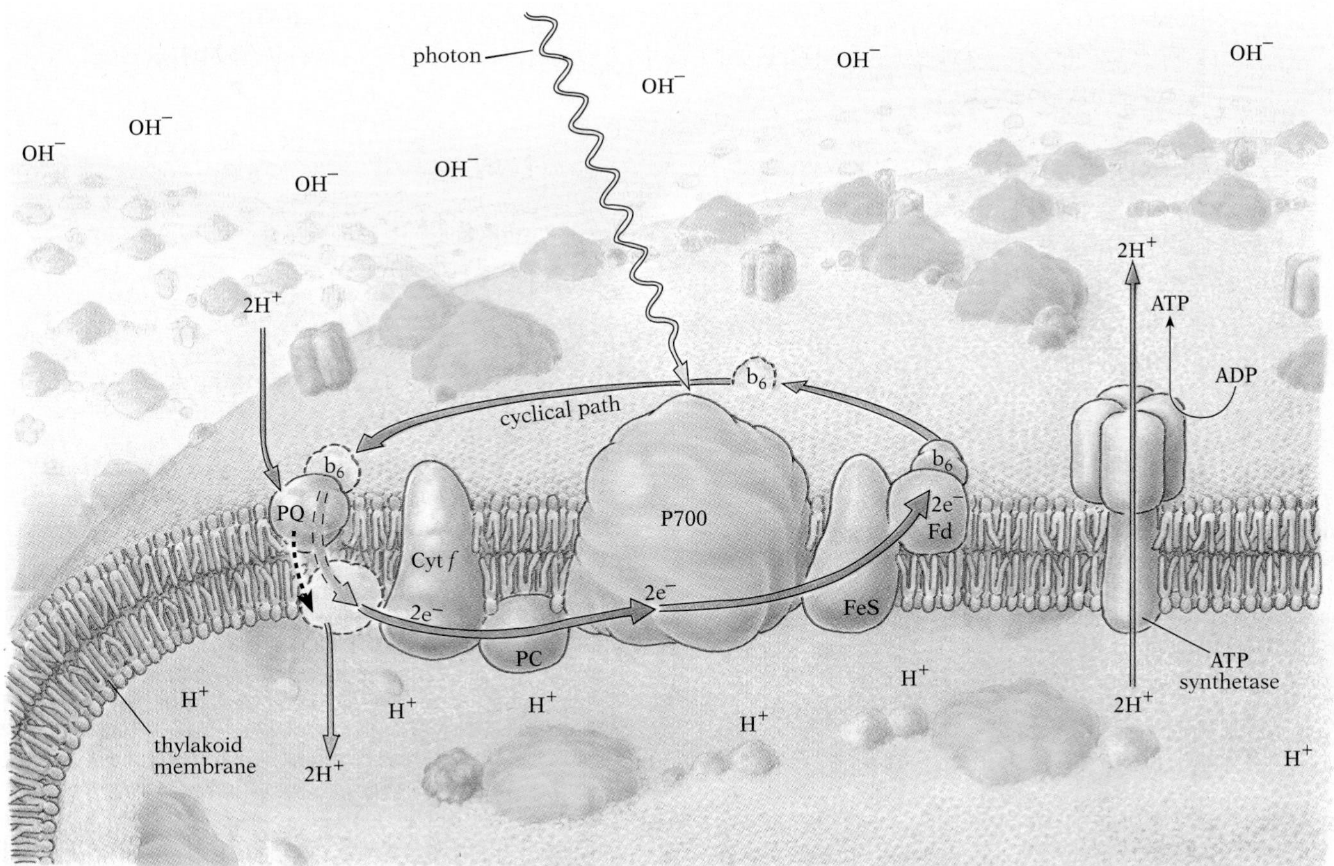

7.18 The cyclic pathway The energy of light is absorbed by a pigment molecule in the antenna system and excites it. The excited state eventually reaches a molecule of P700, which is thereby energized. An energized electron from P700 is then passed from FeS to Fd and then to b_6, a mobile electron acceptor that passes the electron to PQ. The electron-transport proteins of photophosphorylation are generally shown in a linear chain for simplicity, but in the membrane the chain is thought to be folded back on itself, so that b_6 connects Fd and PQ, making possible the cyclic pathway. When PQ receives the electron, it also picks up an H^+ ion from the stroma, which it transports across the membrane. The electron is passed through the two electron transport molecules Cyt.*f* and PC and is returned to the same photosystem from which it originated, completing the cycle. The flow of electrons around the pathway has resulted in a build-up of H^+ ions in the thylakoid interior; the energy of the H^+ ions gradient can be used to make ATP when H^+ ions flow back through the ATP synthetase complexes. The only product of the cyclic pathway is ATP.

most photosynthetic bacteria. In multicellular plants, however, cyclic photophosphorylation is only, at best, a supplement to noncyclic photophosphorylation, and it provides comparatively little additional ATP for the cell.

The Dark Reactions of the Calvin-Benson Cycle Reduce Carbon Dioxide to Carbohydrates

So far, we have discussed the reactions unique to photosynthetic organisms: the synthesis of O_2, ATP, and NADPH by means of light energy. The ATP and NADPH produced in the light reactions are used to power the synthesis of carbohydrate from CO_2. The reactions involved in carbohydrate synthesis are often called the dark reactions because they can be carried out in the dark (as long as there is sufficient NADPH and ATP). These reactions require the products of the light reactions but they do not directly use light. In most plants, however, carbohydrate synthesis occurs only in the daylight, for it is only then that ATP and NADPH are being produced.

The reduction of energy-poor CO_2 to form energy-rich sugars proceeds by many steps, each catalyzed by an enzyme. In effect, CO_2 is pushed up an energy gradient through a series of intermediate compounds, some of them unstable, until the stable, energy-rich carbohydrate end product is formed (Fig. 7.19). An analogy would be a man moving a large and very heavy chest up a flight of stairs from the first floor of his home to the second floor. The man might be able to lift the chest just high enough to get it up one step at a time, balancing it on each step just long enough to marshal his strength before the next heave. If he were to let go (stop applying energy) at any point

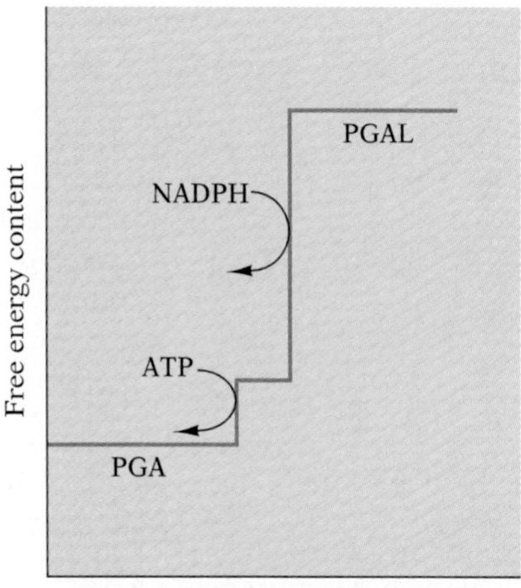

7.19 Changes in free energy content during the synthesis of carbohydrate during the Calvin-Benson cycle During the Calvin-Benson cycle, a three-carbon molecule called PGA is reduced to form a sugar, PGAL. The free-energy content of the molecule is substantially increased during this process. ATP provides the energy for the first step, and NADPH for the second step.

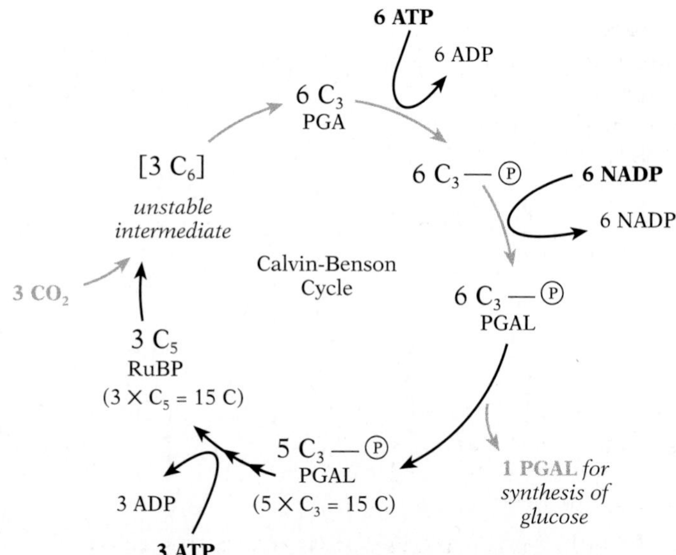

7.21 Synthesis of carbohydrate by the Calvin cycle Each CO_2 molecule combines with a molecule of ribulose bisphosphate (RuBP), a five-carbon sugar, to form a highly unstable six-carbon intermediate, which promptly splits into two molecules of a three-carbon compound called PGA. Each PGA is phosphorylated by ATP and then reduced by NADPH to form PGAL, a three-carbon sugar. Thus each turn of the cycle produces two molecules of PGAL. Five of every six new PGAL molecules formed are used in synthesis of more RuBP by a complicated series of reactions (not shown separately here) driven by ATP. The sixth new PGAL molecule is the end product of the cycle. The path of carbon from CO_2 to glucose is here traced by blue arrows. Since it takes three turns of the cycle to yield one PGAL for glucose synthesis, the diagram begins with three molecules of CO_2; it would require a total of six turns to produce one molecule of glucose, a six-carbon sugar. Note that the cycle is driven by energy from ATP and NADPH, formed by the light reactions of photophosphorylation.

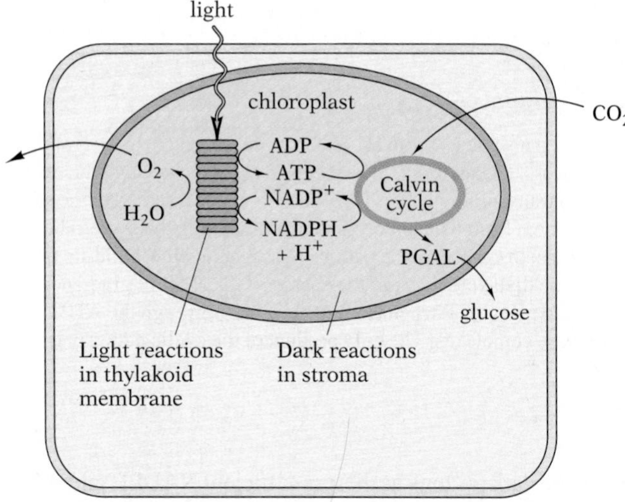

7.20 The light and dark reactions Photosynthesis consists of two physically separate but interlocking sets of reactions. The light reactions—photophosphorylation—use light to generate ATP and NADPH. These reactions take place in the thylakoid membrane. The dark reactions—carbon fixation—use ATP and NADPH to reduce carbon dioxide to form sugars. These reactions take place in the stroma of the chloroplast. Cyclic photophosphorylation may provide additional ATP for the cell.

between the stable level of the first floor and the stable but higher energy level of the second floor, the chest would come crashing to the bottom. The steps, then, make it possible to move the chest up an energy gradient, but they themselves are unstable intermediate levels. In this case, the energy necessary to move the chest through the series of unstable intermediate levels to the stable high energy level at the top is supplied by the man. In the case of synthesis of carbohydrate from CO_2, the energy comes from light via ATP and NADPH (Fig. 7.20). Recall that the ATP and NADPH necessary for these reactions is produced on the stromal side of the thylakoid (Fig. 7.17). The synthesis of carbohydrate also takes place in the stroma.

The stepwise series of reactions in which CO_2 is reduced to carbohydrate is called the ***Calvin-Benson cycle,*** after the researchers who worked out the details of the pathway, Melvin Calvin and Andrew Benson at the University of California at Berkeley. The cycle is shown in abbreviated form in Figure 7.21. The cycle begins when CO_2 from the air combines with a five-carbon sugar called ribulose bisphosphate (abbreviated ***RuBP***) to form a highly unstable six-carbon compound, which is promptly broken into two three-carbon molecules called phosphoglyceric acid, or ***PGA*** (Fig. 7.21). The enzyme that

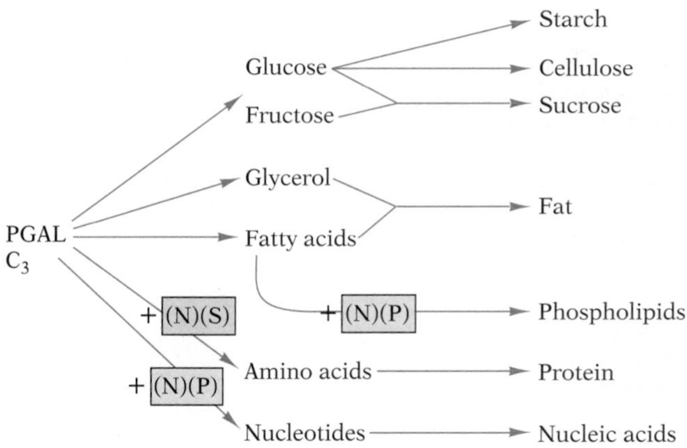

7.22 Central role of PGAL in biosynthesis of organic molecules Some of the PGAL produced by the Calvin-Benson cycle is converted into glycerol, fatty acids, amino acids, and nucleotides. Many of these biosynthetic reactions take place within the stroma of the chloroplast. Glucose is also produced from PGAL but most of the glucose thus produced is converted into disaccharides and polysaccharides. PGAL can also be broken down by aerobic respiration to yield ATP energy for the cell.

catalyzes this reaction, ribulose bisphosphate carboxylase, or ***Rubisco,*** is the most abundant enzyme in the world. *This is the key biosynthetic reaction in photosynthesis.* (See Exploring Further: The C_4 Cycle, p. 156.)

Next, each molecule of PGA has an additional phosphate group added to it by ATP and then is reduced by electrons and hydrogen from NADPH. Here is where the light reactions come into play: the products of the light reactions, ATP and NADPH, are necessary to power these reactions. The resulting energy-rich, three-carbon compound is called phosphoglyceraldehyde, or ***PGAL*** for short. PGAL is a true sugar and is the stable end product of photosynthesis. Most of the molecules of PGAL are used in the formation of new RuBP (by a complicated series of reactions powered by more ATP). But some of the PGAL is synthesized into glucose and then into starch and stored within the chloroplast, and some is exported to the cytoplasm, where it is combined and rearranged in a series of steps to form sucrose for transport to other parts of the plant.

Although glucose has traditionally been regarded as the end product of photosynthesis, very little glucose is actually present in most plant cells. Most of the PGAL produced by plant cells is used to synthesize starch, fatty acids, amino acids, and nucleotides, or is broken down to yield energy for doing work (Fig. 7.22). What glucose is synthesized is normally used almost immediately as a building-block unit for sucrose, starch, cellulose, or other polysaccharides.

While we generally focus our attention on the role of photosynthesis in harnessing light energy to synthesize carbohydrates, it is important not to lose sight of the other important product of photosynthesis—molecular oxygen. The oxygen that we breathe today has accumulated in our atmosphere as a direct result of the process of photosynthesis.

MAKING CONNECTIONS

1. Which evolved first, cyclic or noncyclic photophosphorylation? Which is more significant in higher plants today?

2. Summarize the relationship between the light and dark reactions. What is the source of the energy that drives the Calvin-Benson cycle?

3. What happens to the ADP, Ⓟ, and $NADP^+$ formed during the dark reactions of photosynthesis?

4. Suppose the thylakoid membranes were made freely permeable to H^+ ions. What effect would this have on: splitting of water, electron flow, H^+ transport, synthesis of NADPH, synthesis of ATP, and synthesis of PGAL?

5. If, as some climatologists predict, global warming occurs as a result of increased atmospheric CO_2, would C_3 or C_4 plants be more likely to benefit?

EXPLORING FURTHER

THE C_4 CYCLE

One property of the Calvin-Benson cycle is perplexing in that it has no obvious biological function: apparently *Rubisco,* the enzyme that catalyzes the addition of CO_2 to RuBP at the start of the Calvin-Benson cycle, can also catalyze the addition of O_2 to RuBP. When O_2 is added instead of CO_2, RuBP cannot enter the Calvin-Benson cycle and instead is broken down to CO_2. In other words, O_2 and CO_2 are alternative substrates that compete with each other for the same active site on the Rubisco enzyme. When the concentration of CO_2 is high and that of O_2 is low, the addition of CO_2 is favored and carbohydrate synthesis by the Calvin-Benson cycle proceeds. But when the reverse conditions prevail—when the concentration of CO_2 is low and that of O_2 is high—O_2 is added and RuBP is broken down to CO_2. Under certain conditions, then, oxygen is consumed and RuBP is broken down by a series of reactions initiated by the very same enzyme that under more favorable conditions would facilitate photosynthesis. This oxidative breakdown of RuBP to CO_2 is called *photorespiration.* Unlike other forms of respiration, very little ATP is produced in the process, so photorespiration is a rather wasteful process, short-circuiting the Calvin-Benson cycle while at the same time generating very little energy for use in the cell.

Because photorespiration predominates over photosynthesis at low concentrations of CO_2, plants that depend exclusively on the Calvin-Benson cycle for CO_2 fixation cannot synthesize carbohydrates unless the CO_2 concentration in the atmosphere is above a critical level, and even at normal levels much of the production of photosynthesis is undercut by concurrent photorespiration. The problem is particularly severe in plants growing in hot, dry environments, where excessive loss of water induces the stomates to close. With the stomates closed, the concentration of CO_2 in the air spaces inside the leaf falls as CO_2 is used during photosynthesis, and the concentration of O_2 rises as it is produced by photosynthesis, therefore favoring photorespiration. To avoid this problem certain flowering plants of tropical origin have evolved a different leaf anatomy and a different way of fixing CO_2 initially.

Recall that in the Calvin-Benson cycle, CO_2 is initially "fixed," or incorporated into RuBP, which then splits into two three-carbon molecules (PGA). Accordingly, plants that fix CO_2 in this way are referred to as *C_3 plants.* An alternative method of fixing CO_2 is found in plants with a distinct leaf structure known as *Kranz* anatomy, from the German word *Kranz,* or "wreath," referring to the ringlike arrangement of photosynthetic cells around the leaf veins of these plants (Fig. A). In Kranz plants, unlike other plants, the bundle-sheath cells have numerous chloroplasts, and mesophyll cells are clustered in a ringlike arrangement around the bundle-sheath. Under conditions of high temperature and intense light, the stomates will close and most plants (C_3 plants) will carry on photorespiration, but Kranz plants do not, due to their special way of fixing CO_2 initially. In the latter plants—

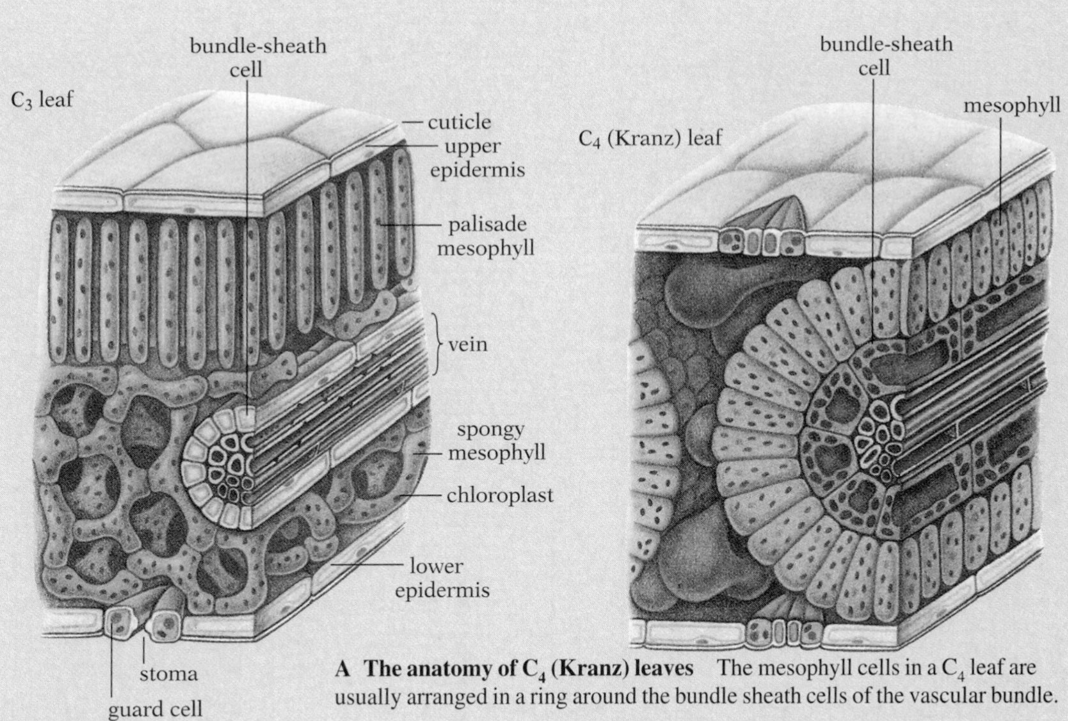

A The anatomy of C_4 (Kranz) leaves The mesophyll cells in a C_4 leaf are usually arranged in a ring around the bundle sheath cells of the vascular bundle.

also known as *C_4 plants*—CO_2 is combined with a three-carbon compound in the mesophyll cells, forming a four-carbon compound (C_4) that passes into the bundle-sheath cells (Fig. B). In the bundle-sheath cells the C_4 compound is then broken down to CO_2 and another C_3 compound. The C_3 compound moves back into the mesophyll cells, where it is converted into another compound and starts the C_4 cycle over again. The CO_2, however, remains in the bundle-sheath cells, where it can enter the Calvin-Benson cycle and be incorporated into carbohydrate. In effect, the mesophyll cells act as CO_2 pumps, transferring enough CO_2 (via the C_4 intermediate) into the bundle-sheath cells to maintain an artificially high CO_2 concentration in which the Calvin-Benson cycle is able to function.

In short, C_4 plants have an advantage over C_3 plants under conditions of high temperature and intense light, when stomatal closure results in low CO_2 and high O_2 in the air spaces inside the leaf. Under such conditions, C_3 plants are unable to use CO_2 effectively because O_2 competes with CO_2 for RuBP. C_4 plants can fix CO_2, however, because the mesophyll cells, acting as CO_2 pumps, can raise the CO_2 concentration in the bundle-sheath cells to a level where the addition of CO_2 to RuBP (leading to the Calvin-Benson cycle) exceeds the addition of O_2 and photorespiration. The Kranz anatomy, with its concentric rings of mesophyll and bundle-sheath cells, facilitates the process of CO_2 pumping. Because some energy is needed to run the cycle, C_3 plants use less energy to fix CO_2 and therefore have a competitive advantage in temperate climates.

The combination of Kranz anatomy and C_4 photosynthesis has evolved independently in a variety of unrelated plants, including a number of major crops such as corn, sugarcane, and sorghum. It is therefore not only an impressive illustration of the intimate relationship between structure and function in living systems, but also an important source of nutrients for human consumption.

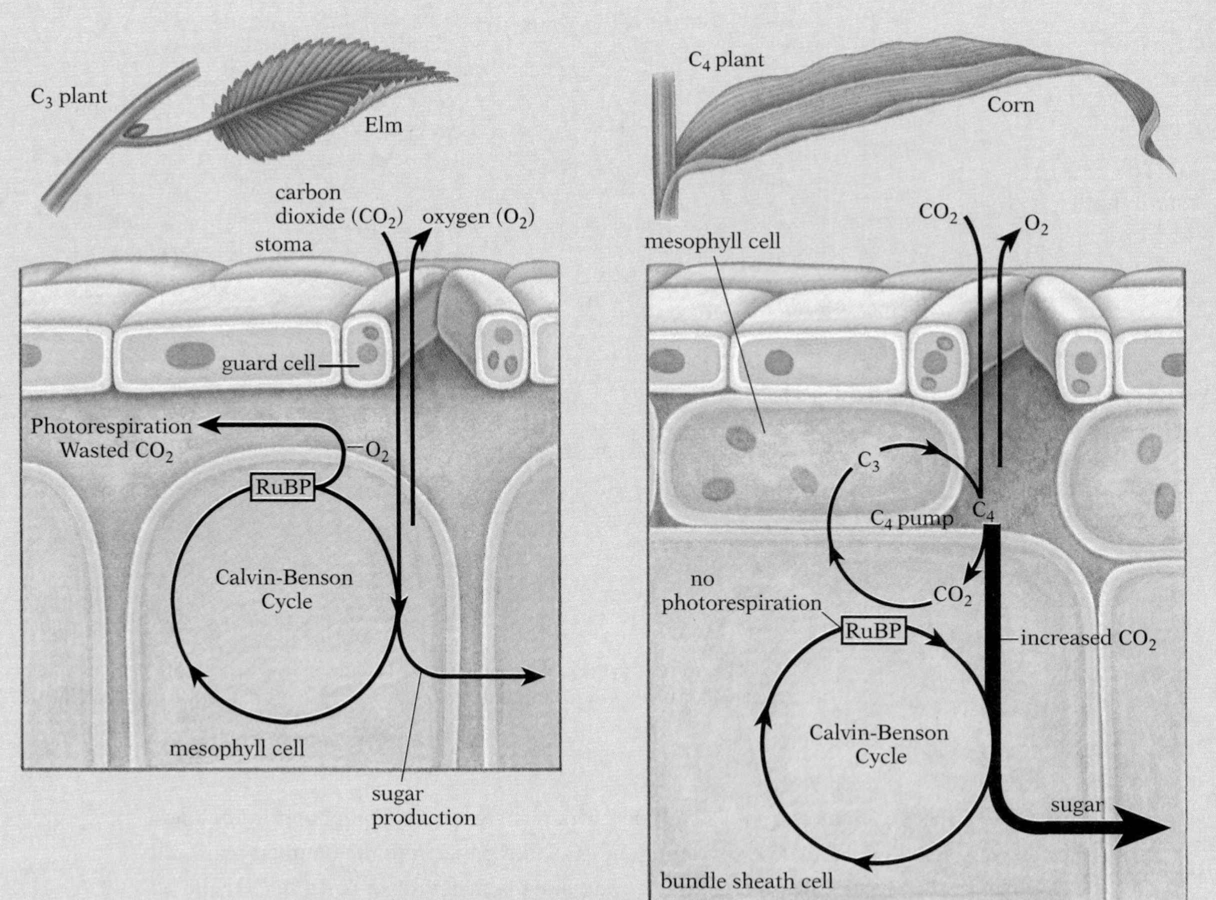

B Carbon dioxide fixation in C_3 and C_4 plants Left: In C_3 plants, when CO_2 is plentiful it combines with RuBP and enters the Calvin-Benson cycle, but when CO_2 is low and O_2 is high, O_2 is added to RuBP instead, and photorespiration occurs. Right: C_4 plants have another way of fixing CO_2, which avoids photorespiration. The CO_2 is combined with a three-carbon compound in the mesophyll cells, forming a four-carbon compound (C_4) that passes into the bundle-sheath cells. Here the C_4 compound is broken down to CO_2 and another C_3 compound, which moves back into the mesophyll cells to start another cycle. The CO_2 enters the Calvin-Benson cycle and is reduced to carbohydrate. The mesophyll cells act as CO_2 pumps, transferring enough CO_2 (via the C_4 intermediate) into the bundle-sheath cells to maintain an artificially high CO_2 concentration in which the Calvin-Benson cycle is able to function.

In recent years we have heard a great deal about global warming and the increasing levels of CO_2 in the atmosphere. Would not the increased CO_2 levels prove to be beneficial to plants? Because CO_2 is often the limiting factor in photosynthesis, with more CO_2 available, wouldn't the plants be able to avoid the problem of photorespiration and synthesize more carbohydrate? Recent evidence suggests that the answer to these questions is no. Some plants do initially appear to show increased growth, but over time the growth slows to normal levels. The reasons for this are complex and still being studied, but evidence suggests that plants are not going to be able to use the increased CO_2 levels productively, and decreasing fossil fuel combustion and deforestation remains a sound policy choice. We will cover much more on global warming in Chapter 34.

CONCEPTS IN BRIEF

1. The ultimate energy source for most living things is sunlight, transformed by green plants into chemical energy in the process called *photosynthesis.* The green plants utilize the energy of light to remove the hydrogen ions and electrons from water and use them to reduce carbon dioxide to sugar. Oxygen is formed as a by-product.

2. The reactions of photosynthesis can be divided into two categories: the *light reactions,* in which light energy is trapped and stored, and the *dark reactions,* in which electrons and hydrogen are transferred from water to carbon dioxide to form carbohydrate.

3. The leaves of higher green plants are the principal organs of photosynthesis. Photosynthesis takes place within the *chloroplasts* of the *mesophyll cells.* The pigments are precisely arranged within the membranes of flattened sacs called *thylakoids.* The light reactions of photosynthesis (cyclic and noncyclic photophosphorylation) take place within the thylakoid membranes, while the dark reactions (Calvin-Benson cycle) take place in the *stroma.*

4. The chlorophyll and accessory pigments necessary for photosynthesis are organized into two functional systems, called *Photosystem I* and *Photosystem II,* both of which are built into the thylakoid membrane, along with the electron-transport chains and *ATP synthetase* complexes.

5. In *noncyclic photophosphorylation,* absorbed light energy causes electrons and H^+ ions to be pulled away from water (forming O_2); the electrons are passed to Photosystem II, where they are energized by light. They are then passed along a chain of transport molecules in the thylakoid membrane to Photosystem I, where there is another light event, and finally, along a second transport chain to $NADP^+$, to form NADPH.

6. As electrons are transferred along the electron transport chain from Photosystem II to Photosystem I, H^+ ions are transported into the interior of the thylakoid. These H^+ ions plus those left behind in the splitting of water establish an electrochemical gradient across the thylakoid membrane. The energy of this gradient can be used by the ATP synthetase complexes to make ATP—the energy currency of the cell—when the H^+ ions move back through the complex, flowing from high to low concentration.

7. In *cyclic photophosphorylation,* light energy is trapped by chlorophyll molecules in Photosystem I, exciting electrons to a higher energy level. The energized electrons are passed along a chain of electron-transport molecules and are returned to the chlorophyll molecules from which they started. Some of the energy released during the transfer of electrons along the chain is used to pump H^+ ions into the thylakoid sac. The energy of the H^+ ion gradient is used by the ATP synthetase complex to make ATP as the H^+ ions flow back through the complex, from high to low concentration.

8. The end products of noncyclic photophosphorylation are ATP, NADPH, and O_2. Cyclic photophosphorylation, however, produces only ATP.

9. The ATP and NADPH produced in the light reactions are used to reduce carbon dioxide to carbohydrate in a series of reactions called the *Calvin-Benson cycle*. First, ribulose bisphosphate (RuBP) is combined with CO_2 to form an unstable molecule that splits into two three-carbon molecules, which are then phosphorylated by ATP and reduced by NADPH to form the sugar PGAL. PGAL may be used to synthesize more RuBP, as a source of energy, or for the synthesis of glucose, fatty acids, and amino acids.

STUDY QUESTIONS

1. Compare the source of energy for autotrophs and heterotrophs. Describe the two processes by which autotrophs can synthesize energy-rich organic molecules. Which evolved first? (p. 140)
2. Write the summary equation for photosynthesis. Why is photosynthesis an oxidation-reduction reaction? (pp. 141–142)
3. Diagram a cross section of a leaf and describe how its anatomy aids its photosynthetic function. (p. 143)

4. Diagram a chloroplast, label its major regions, and explain what processes are associated with each region. (p. 144)
5. How are chlorophyll and carotenoid pigments organized within the thylakoid membrane? What are the respective functions of antenna molecules and reaction centers? (pp. 145–146)
6. Trace the flow of electrons in noncyclic photophosphorylation from their point of origin to their final destination. What is the role of the two light events? (p. 146)
7. Describe chemiosmotic ATP synthesis during noncyclic photophosphorylation. How is the electrochemical gradient of H^+ ions established across the thylakoid membrane? On which side of the membrane is the concentration of H^+ ions highest? How is this electrochemical gradient exploited to generate ATP? (pp. 148–151)
8. Contrast cyclic with noncyclic photophosphorylation. What photosystems does each use and what are their respective products? (p. 152)
9. Are dark reactions generally carried out in the dark? Why or why not? (p. 153)
10. Describe the Calvin-Benson cycle, particularly the roles of CO_2, RuBP, Rubisco, PGA, ATP, NADPH, and PGAL. What happens to the PGAL produced in the cycle? (pp. 153–155)
11. What is photorespiration? Under what conditions does photorespiration predominate over photosynthesis? What is the consequence for the plant? (p. 156)
12. How do C_4 plants overcome the problem of photorespiration? Why do they have Kranz anatomy? (pp. 156–157)

CHAPTER 8

ENERGY TRANSFORMATIONS: RESPIRATION

Within a cell, energy is needed at every stage to drive the reactions that maintain life: to read, copy, and repair the genetic instructions in the chromosomes; to construct, repair, and move organelles; to bring in nutrients, expel wastes, and preserve proper pH and ionic balances; and so on. Without a constant supply of energy, these reactions cannot take place, and life ceases. As noted at the beginning of the preceding chapter, virtually all the energy that fuels life today comes from the sun and is captured in the process of photosynthesis by plants, which use it to build energy-rich compounds like glucose. Most nonphotosynthetic organisms obtain energy by ingesting photosynthetic organisms or other organisms that have themselves ingested photosynthetic organisms.

In cells, the energy stored by the chemical reactions of photosynthesis is usually released through a process known as ***aerobic respiration,*** the multistage reaction sequence in which glucose is broken down to yield a series of intermediate products, leading ultimately to carbon dioxide and water. The free energy made available by this process is then used to fuel a host of other enzyme-mediated reactions in living cells (Fig. 8.1). These reactions, known collectively as ***cellular metabolism,*** are generally divided into two phases: ***anabolism,*** the processes by which complex organic molecules are assembled from simpler building-block molecules, and ***catabolism,*** the processes by which living things extract energy from food by breaking down the complex organic molecules into simpler ones. In this chapter we shall concentrate mainly on the catabolic phase, with particular reference to aerobic respiration.

Respiration, Like Photosynthesis, Involves Redox Reactions

The breakdown of food molecules in living cells, like the build-up of such molecules in photosynthesis, involves a complex sequence of chemical reactions, including some redox reactions. As we pointed out in Chapter 7, all redox reactions combine reduction, the addition of one or more electrons, either alone or in association with one or more H^+ ions (i.e., pro-

Chapter opening photo: Chameleon walking across hot sand. These lizards, whose body temperature varies with environmental temperature, have many behavioral adaptations that enable them to survive in such a hot environment.

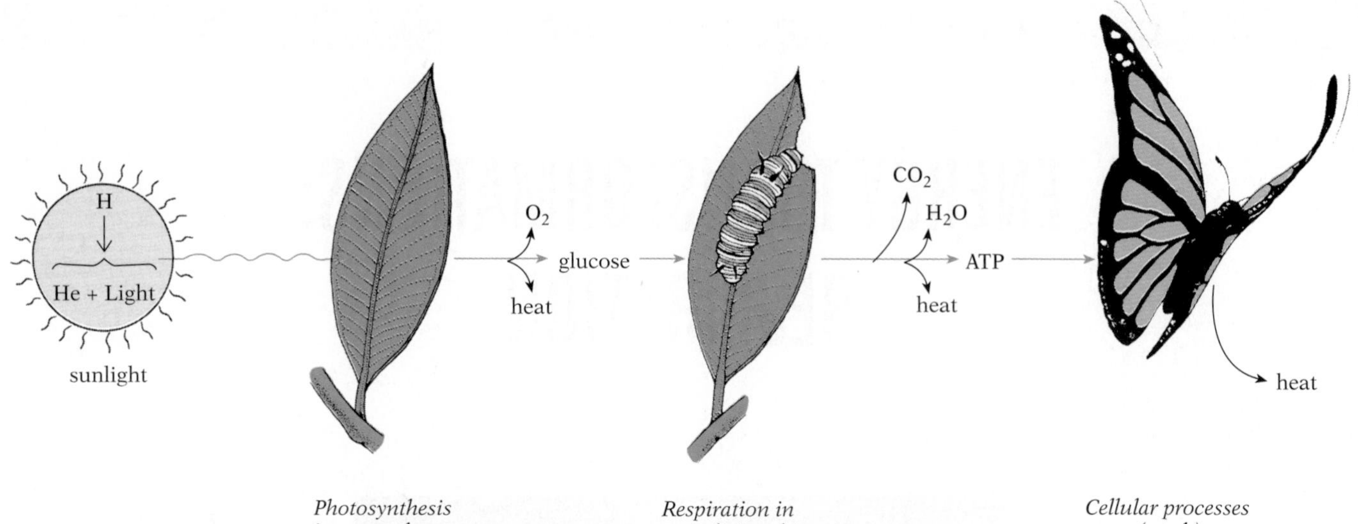

8.1 Summary of biological energy flow Today, nearly all the energy for life originates in the sun, where hydrogen is converted by fusion into helium and light is produced. In the process of photosynthesis, green plants convert the radiant energy of sunlight into chemical energy, which is most often stored initially in glucose. When cells in most organisms need energy, the glucose is broken down and some of its chemical energy recovered by the process of aerobic respiration; the resulting product—ATP—supplies the energy in a more manageable form, making it available for muscular contraction, nerve conduction, active transport, and other work. Aerobic respiration utilizes oxygen, a by-product of photosynthesis, and photosynthesis utilizes carbon dioxide and water, by-products of respiration. With each of the transformations shown, much energy is lost as waste heat.

tons), and oxidation, the removal of one or more electrons, again alone or in association with one or more hydrogens.

The redistribution of energy in redox reactions is probably best illustrated by looking once again at the basic energy equation of photosynthesis. Light energy from the sun is used by photosynthetic organisms to reduce carbon dioxide, thereby creating energy-rich carbohydrates such as glucose:

$$6CO_2 + 12H_2O + \text{light} \longrightarrow 6O_2 + C_6H_{12}O_6 + 6H_2O$$

The details of this reaction are the main subject of the preceding chapter.

The energy stored in substances such as glucose as a result of the reduction of their precursors can in turn be partially recovered when the storage compound is oxidized:

$$C_6H_{12}O_6 + 6O_2 \longrightarrow 6CO_2 + 6H_2O + \text{energy}$$

This reaction liberates free energy and so can proceed spontaneously. The stored energy is only partially recovered for use by the cell because, as in all reactions, some energy is lost as heat. The oxidation of glucose (the main subject of this chapter) provides energy for the creation of ATP and other energy-storage compounds, thus providing energy to power other reactions. Ultimately, it regenerates the raw materials (CO_2 and H_2O) that can then be utilized again in the process of photosynthesis.

Although cells store energy in the form of fats and carbohydrates, the amount of energy in even a single glucose molecule is considerably larger than that required to drive most reactions within the cell. The hydrolysis of ATP to ADP, on the other hand, releases a more useful amount of energy—just enough to power most reactions.[1] In effect, glucose molecules are the hundred-dollar bills of the cellular economy, whereas ATP molecules are the singles. Thus, the characterization of ATP as the universal energy currency of living things is justified.

The cell uses the energy from the breakdown of high-energy molecules such as glucose to synthesize ATP molecules. Once formed, ATP is stored in the cell until needed for other cellular functions. Although some other compounds can supply energy, ATP is the one most often used by cells in the various kinds of work they perform: synthesis of more complex compounds, muscular contraction, nerve conduction, active transport across cell membranes, light production, and so on. The energy price of the work is paid through the energy-releasing hydrolysis of ATP to ADP.

[1]The amount of energy released from the oxidation of a single glucose molecule is 670 kilocalories per mole, whereas the hydrolysis of ATP releases 7.3 kilocalories per mole. (A kilocalorie is the amount of heat required to raise the temperature of 1,000 grams of water by 1°C; a mole is the number of grams of a substance that is equal to the molecular weight of that substance.)

In Most Cells Respiration Is Accomplished in Four Stages

The energy stored in lipids and carbohydrates is not liberated through a single large reaction. Rather, the universal catabolic process by which the molecules are broken down occurs as a series of smaller reactions, each catalyzed by its own specific enzyme. These reactions result in the release of small bursts of energy, some of which is transferred to ATP by the phosphorylation that synthesizes ATP from ADP and inorganic phosphate.

The complete breakdown of an energy-rich compound such as glucose involves a long series of reactions, of which only the more important steps will be examined here. Our discussion will deal, first, with a chain of reactions that can take place whether oxygen is present or not (***anaerobic metabolism***) and, second, with a chain of reactions that are dependent on oxygen (***aerobic metabolism***). In most cells the complete catabolism of glucose involves four stages, divided between anaerobic and aerobic series of reactions.

Glycolysis, an Anaerobic Process, Is the First Stage of Respiration in All Cells

The anaerobic portion of respiration, known as ***glycolysis,*** is the first stage in the breakdown of glucose. Glycolysis, which takes place in the cytosol of all living cells, is thought to be the most ancient series of reactions in the pathway, since it could occur in the anaerobic environment of the early earth.

Glucose is a stable compound—one with little tendency to break down spontaneously into simpler products. If its energy is to be harvested, the glucose must first be made more reactive by the investment of a small amount of energy to activate the molecule. The first steps of glycolysis, therefore, are preparatory, enabling the later steps to extract the stored energy. It is ATP that provides the energy for initiating glycolysis (Fig. 8.2). The initial reactions, like the succeeding ones, are facilitated by specific enzymes. In the preparatory reactions, two molecules of ATP transfer their terminal phosphate groups to

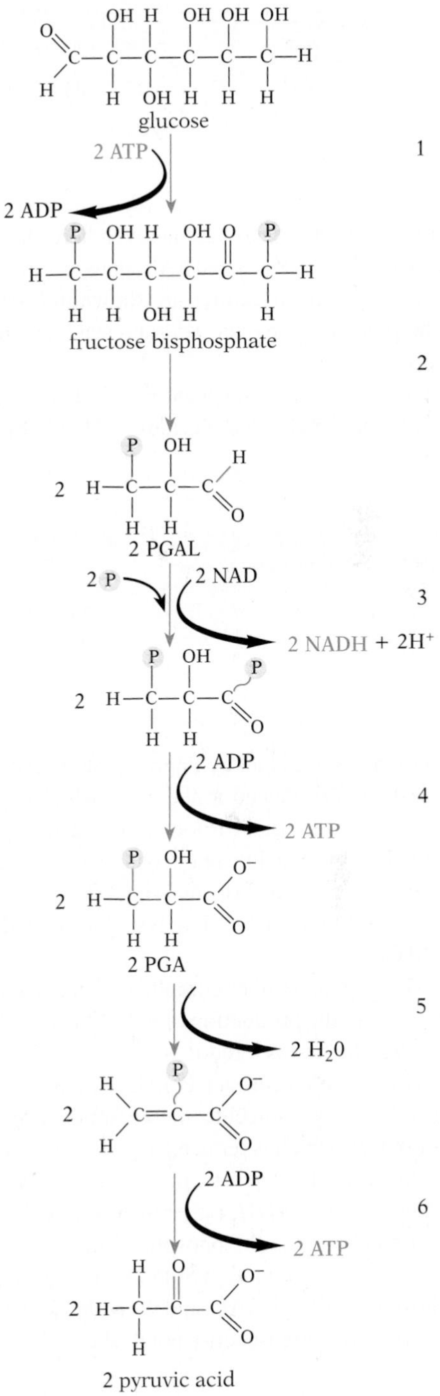

8.2 Main reactions of glycolysis Energy to initiate the breakdown of glucose is supplied by two molecules of ATP (Step 1). The resulting compound is then split into two molecules of PGAL (Step 2). This completes the preparatory reactions. Next, the PGAL is oxidized by removal of hydrogen, which is picked up by NAD^+ to form two molecules of NADH; in the same reaction, inorganic phosphate is added to each of the three-carbon molecules (Step 3). A series of reactions then results in synthesis of four new molecules of ATP, for a net gain of two (Steps 4 to 6).

the glucose, which is then converted into fructose bisphosphate:

$$(1)\quad \underset{\text{glucose}}{\text{C—C—C—C—C—C}} + \underset{+\ 2\ \text{ATP yields}}{2\text{ATP}} \xrightarrow{\text{enzymes}}$$
$$\underset{\text{fructose bisphosphate}}{\textcircled{P}\text{—C—C—C—C—C—C—}\textcircled{P}} + \underset{+\quad 2\ \text{ADP}}{2\text{ADP}}$$

(The simplified summary equations given here show only the carbon skeleton, but oxygens and hydrogens are attached to the carbons. The complete structures are shown in Figure 8.2. As noted in the preceding chapter, Ⓟ represents an entire phosphate group.)

Next, the fructose bisphosphate is split between the third and fourth carbons, and two molecules of PGAL are formed:

$$(2)\quad \underset{1\ \text{fructose bisphosphate}}{\textcircled{P}\text{—C—C—C—C—C—C—}\textcircled{P}} \xrightarrow[\text{yields}]{\text{enzymes}}$$
$$\underset{2\ \text{PGAL}}{2\ \text{C—C—C—}\textcircled{P}}$$

PGAL, you will recall, is a key intermediate in the photosynthetic process; it is produced in the Calvin-Benson cycle and can be used to produce other carbohydrates such as glucose. So far, then, glycolysis looks like photosynthesis in reverse. Up to this point, however, instead of releasing energy from glucose to form new ATP molecules, glycolysis has actually cost the cells two ATPs.

The next reaction, a rather complicated one that begins the changes leading to the production of new ATP, is really two reactions in one. The first is a redox reaction: two electrons and an H^+ ion are removed from each PGAL (which is oxidized) by the electron-acceptor molecule nicotinamide adenine dinucleotide, or ***NAD^+***, which is reduced. NAD is closely related to the NADP found in chloroplasts. In this case, however, the intermediate product is ***NADH,*** rather than NADPH.

The second change is the phosphorylation of PGAL. The energy released by the oxidation of PGAL is used to add an inorganic phosphate Ⓟ to PGAL, and the added phosphate is attached by a so-called high-energy bond (shown here in color):

$$(3)\quad \underset{2\ \text{PGAL}}{2\textcircled{P}\text{—C—C—C}} + 2\ NAD^+ + 2\textcircled{P} \xrightarrow{\text{enzyme}}$$
$$2\textcircled{P}\text{—C—C—C—}\textcircled{P} + 2\ \text{NADH} + 2H^+$$

In the next reaction, the newly added phosphate group is transferred to ADP to form ATP. This process, in which a high-energy phosphate group is transferred from a substrate to ADP to form ATP, is called ***substrate-level phosphorylation.*** The three-carbon product is phosphoglyceric acid, or PGA—another intermediate in the Calvin-Benson cycle, showing again the interrelationships among the two processes:

$$(4)\quad 2\textcircled{P}\text{—C—C—C—}\textcircled{P} + 2\ \text{ADP} \xrightarrow{\text{enzyme}}$$
$$\underset{2\ \text{PGA}}{2\textcircled{P}\text{—C—C—C}} + 2\ \text{ATP}$$

At this point, then, the cell regains the two ATP molecules used in phosphorylating glucose (Step 1) to start glycolysis. The initial energy investment has been repaid.

Next come several reactions that result in the removal of H_2O from the PGA molecule and the subsequent conversion of the remaining phosphate bond into a "high-energy" phosphate bond. Only the net result is shown here:

$$(5)\quad \underset{2\ \text{PGA}}{2\textcircled{P}\text{—C—C—C}} \xrightarrow{\text{enzymes}}$$
$$2\textcircled{P}\text{—C—C—C} + 2\ H_2O$$

In the next step, the remaining phosphate groups are transferred to ADP by substrate-level phosphorylation, which leads to the formation of two more ATP molecules plus two molecules of a three-carbon compound named ***pyruvic acid:***

$$(6)\quad 2\textcircled{P}\text{—C—C—C} + 2\ \text{ADP} \xrightarrow{\text{enzyme}}$$
$$\underset{\text{pyruvic acid}}{2\ \text{C—C—C}} + 2\ \text{ATP}$$

Because the two ATP molecules used up in Step 1 have already been regained (in Step 4), the two additional ATP molecules formed here represent a net gain of two ATP for the cell. The energy stored in the new ATP molecules represents only about 2 percent of the energy initially available in the glucose molecule.

The most important features of glycolysis to bear in mind are as follows:

- Each molecule of glucose ($C_6H_{12}O_6$) is broken down to two molecules of pyruvic acid ($C_3H_4O_3$).
- Two molecules of ATP are used to initiate the process. Later, four new molecules of ATP are synthesized as a re-

sult of substrate-level phosphorylation, for a net gain of two molecules of ATP.

- Two molecules of NADH are formed.
- Because no molecular oxygen is used, glycolysis can occur whether or not O_2 is present. It is a process encountered in *all* living cells, whatever their mode of life.
- The reactions of glycolysis occur within the cytosol of the cell, outside the mitochondria.

Figure 8.3 shows graphically the changes in free-energy content at each successive step in glycolysis, from glucose to pyruvic acid. Notice that the largest drop in free energy is the reaction in which PGAL is oxidized and NAD^+ is reduced. Some of the energy released in the oxidation of PGAL is stored in the NADH molecule.

Fermentation Oxidizes NADH to NAD^+, Which Enables Glycolysis to Continue

We have seen that in glycolysis two molecules of NAD^+ are reduced to NADH. The NAD molecule functions in the cell as an electron-transport compound, shuttling H^+ ions and electrons between one substance and another. Thus NAD^+ is only a temporary acceptor of hydrogen and electrons. The extra hydrogen and electrons are promptly passed to some other compound, and then the NAD^+ goes back for another load. The cell has only a limited supply of NAD^+ molecules, which must be used over and over. If the NADH molecules formed in glycolysis could not quickly unload their hydrogen and electrons (and thus be reconverted to NAD^+), all of the cell's NAD^+ would soon be tied up (i.e., it would all be transformed into NADH). With no NAD^+ available, Step 3 of glycolysis could not take place, and the whole glycolytic process would come to a halt. In short, *the oxidation of NADH to NAD^+ is essential for glycolysis to continue.*

In most cells, if molecular oxygen is available, it becomes the ultimate acceptor of electrons from NADH, by a process we shall describe below. But under anaerobic conditions, with no molecular oxygen to accept the hydrogen and electrons, it is the pyruvic acid formed by glycolysis that accepts the hydrogen and electrons from NADH, in the process known as ***fermentation.*** The metabolic fate of pyruvic acid during fermentation varies in different organisms (Fig. 8.4). In animal

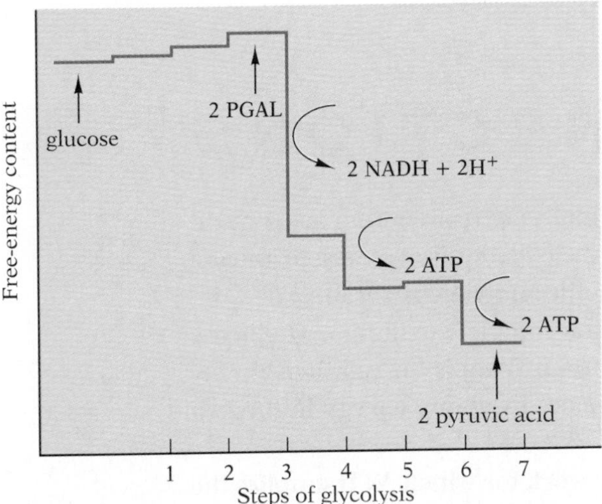

8.3 Changes in free-energy content at successive steps in glycolysis In the preparatory steps, which convert glucose into PGAL, the energy content of the substances is slightly increased owing to the investment of two molecules of ATP. In Step 3 there is a sharp drop in energy associated with the formation of two molecules of NADH. There are also major drops in Steps 4 and 6, each associated with the formation of two molecules of ATP.

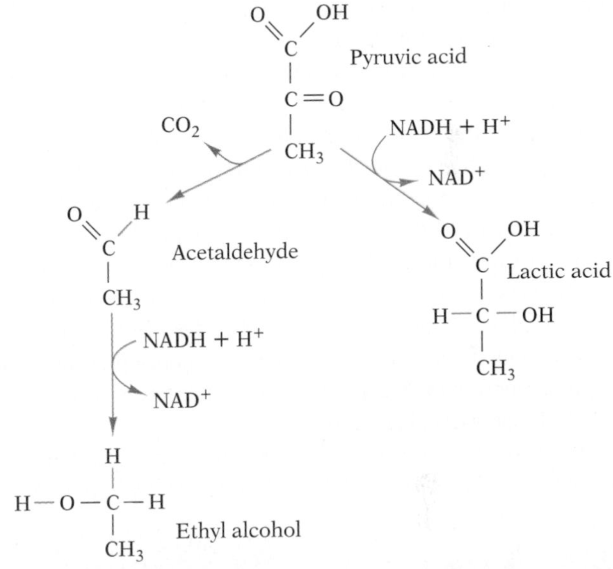

8.4 Fermentation In the absence of sufficient oxygen, the pyruvic acid will undergo fermentation. In some cells pyruvic acid is reduced by NADH to form lactic acid, in others the pyruvic acid is reduced to ethyl alcohol and carbon dioxide. (Left) Yeast cells in the absence of oxygen carry out alcoholic fermentation producing the alcohol in beer. (Right) The muscles of this weight lifter who is strenuously lifting are carrying out lactic acid fermentation because insufficient oxygen is being delivered to the muscle cells. Lactic acid, if it is not removed from the muscles, will damage the muscle fibers and "sore muscles" will result.

cells and many microorganisms the reduction of pyruvic acid results in the formation of lactic acid:

$$(7) \quad 2 \text{ pyruvic acid} + 2\,\text{NADH} + 2\,H^{+} \xrightarrow{\text{enzymes}} 2 \text{ lactic acid} + 2\,\text{NAD}^{+}$$

In most plant cells and in yeasts, a different chemical pathway is utilized, and the end product is ethyl alcohol and carbon dioxide:

$$(7) \quad 2 \text{ pyruvic acid} + 2\,\text{NADH} + 2\,H^{+} \xrightarrow{\text{enzymes}} 2 \text{ ethyl alcohol} + 2\,CO_2 + 2\,\text{NAD}^{+}$$

Thus, under anaerobic conditions, the NAD molecule shuttles back and forth between Steps 3 and 7, picking up hydrogen and electrons (taking the form NADH) is Step 3 and giving up hydrogen and electrons (taking the form NAD^+) in Step 7.

Fermentation is an extension of the glycolytic pathway, whereby glucose is transformed into alcohol or lactic acid under anaerobic conditions. Thus one often refers to alcoholic fermentation or lactic acid fermentation, depending on the end product of the process. Whatever the end product, fermentation enables a cell to continue synthesizing ATP by the breakdown of glucose under anaerobic conditions. But this process extracts only a very small portion of the energy present in the original glucose; the end products of fermentation still contain much of that original energy.

Fermentation by yeast cells and other microorganisms is, of course, the basis for a number of economically important industries, including the manufacture of bread and other bakery products as well as commercial alcohol and alcoholic beverages. Products other than lactic acid and alcohol are produced in certain microorganisms; for example, the distinctive flavor of Swiss cheese is due to the production of proprionic acid, another possible end product of fermentation. Microbial fermentations are essential to the production of most cheeses, yogurt, and a variety of other dairy products.

The Second, Third, and Fourth Stages of Respiration Are Aerobic

Under aerobic as under anaerobic conditions, the breakdown of glucose initially follows the glycolytic pathway to pyruvic acid. But in the presence of molecular oxygen, which can act as the ultimate acceptor of electrons from NADH, pyruvic acid need not act as an electron acceptor and become converted into lactic acid or alcohol. Instead, it can be further broken down and yield energy for synthesis of still more new ATP. In other words, under aerobic conditions, ATP synthesis does not end with the pyruvic-acid step.

Aerobic respiration—the breakdown of nutrients in the presence of O_2, with the accompanying synthesis of ATP—comprises the second, third, and forth stages in the catabolism of glucose, which we shall describe next. Since aerobic respiration begins with the entrance of pyruvic acid into the mitochondrion, where the breakdown of pyruvic acid to CO_2 and H_2O takes place, we shall first review some important aspects of this organelle.

MAKING CONNECTIONS

1. Would the synthesis of a protein molecule from amino acids be an anabolic or catabolic reaction? What about the hydrolysis of starch to glucose molecules? Which reaction would be endergonic and which exergonic?

2. Write the chemical equation for the complete oxidation of glucose and compare it with the photosynthesis equation. What is the relationship between the equations? How do redox reactions redistribute energy in the two processes?

3. Name at least five kinds of cellular work for which ATP supplies the energy. Why is ATP used, rather than glucose or fat?

4. Some microorganisms—such as the bacterium *Clostridium tetani,* which causes tetanus (lockjaw)—are metabolically active only in anaerobic conditions. Why does the inability of such organisms to use oxygen pose a metabolic problem for them? What is their solution?

Mitochondria, the Organelles of Aerobic Respiration, Are Divided Into Two Compartments

The electron microscope shows that each mitochondrion is bounded by an outer membrane and an inner membrane, which divide the organelle into outer and inner compartments (Fig. 8.5). Most of the enzymes necessary for the second and third stages of respiration are found in the inner compartment. The inner membrane, you will recall, has a series of cristae, or folds, which greatly increase its total surface area. Although the outer membrane is permeable to small molecules, the inner membrane is quite impermeable. A variety of protein channels, gates, and pumps are built into the inner membrane to selectively regulate the molecular traffic into and out of the inner compartment. The inner membrane also contains the proteins of the respiratory electron-transport chain, which are analogous to the photosynthetic electron-transport proteins found in the thylakoid membrane.

In Stage II of Aerobic Respiration, Pyruvic Acid Is Oxidized to Acetyl-CoA

The pyruvic acid produced by glycolysis in the cytosol is transported into the inner compartment of the mitochondrion, where the aerobic part of the process begins. In a complicated set of reactions, the three-carbon pyruvic acid is oxidized to

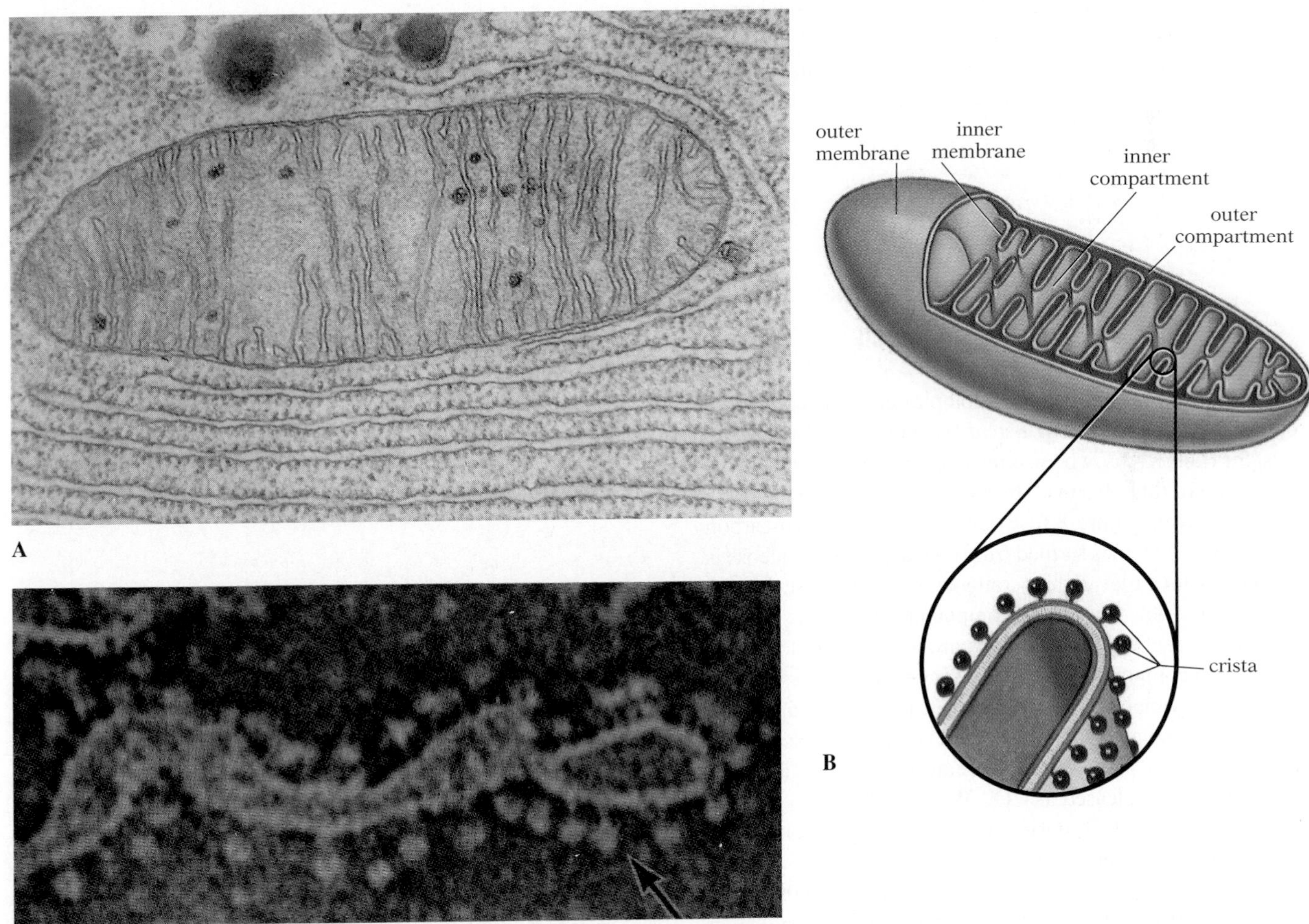

8.5 Structure of a mitochondrion (A) The electron photomicrograph of a typical animal mitochondrion reveals that the mitochondrion is a double-walled vessel; the smooth outer membrane and the folded inner membrane form two separate compartments. (B) In this diagram of a mitochondrion, much of the outer membrane has been cut away, and the interior has been sectioned to show how the inner membrane folds into cristae. The mitochondria of metabolically very active cells have more cristae than those of less active cells. The inner compartment is within the inner membrane, while the outer compartment is the space between the two membranes. As shown in Figure 8.9, much of the activity of cellular respiration, such as electron transport, occurs across the inner membrane between the inner and outer compartments. The knoblike structures are enzyme complexes that synthesize ATP.

CO_2 and a two-carbon acetyl group, which is attached to a coenzyme called coenzyme A, or CoA for short; the complete compound is named ***acetyl-CoA.*** When pyruvic acid is oxidized, electrons and H^+ ions are removed, and once again NAD^+ acts as the acceptor for the electrons and H^+ ions, forming NADH. This complicated series of reactions can be summarized by the following equation:

$$2\ CH_3{-}\overset{O}{\overset{\|}{C}}{-}\overset{O}{\overset{\|}{C}}{-}OH + 2CoA + 2\ NAD^+ \xrightarrow{\text{yields}}$$

2 pyruvic acid + 2 Coenzyme A + $2NAD^+$

$$2\ CH_3{-}\overset{O}{\overset{/\!/}{C}}{-}CoA + 2\ CO_2 + 2\ NADH$$

2 acetyl-CoA + 2 CO_2 + 2 NADH

Note that at the end of Stage II, two of the six carbons present in the original glucose have been released as CO_2. Note also that the newly formed NADH must be oxidized if the breakdown process is to continue; we shall return to this problem shortly.

Stage III Comprises the Reactions of the Krebs Citric Acid Cycle

The acetyl-CoA is next fed into a complex circular series of reactions called the ***Krebs citric acid cycle*** (after the British scientist Sir Hans Krebs, who was awarded the Nobel Prize for his elucidation of this system). The essential features of the cycle are indicated in Figure 8.6. Briefly, each of the two-carbon acetyl-CoA molecules formed from one molecule of glucose is combined with a four-carbon compound already present in the cell to form a new six-carbon compound called citric acid. In subsequent reactions, two carbons are lost as CO_2, leaving a four-carbon compound that is then converted into the original four-carbon compound, and the cycle can begin again. Since each glucose molecule being oxidized yields two molecules of acetyl-CoA, two "turns" of the cycle are required, and a total of four carbons are released as CO_2. With the other two carbons already released as CO_2 during Stage II, all six carbons of the original glucose are accounted for.

In one turn of the cycle (Fig. 8.6), one molecule of ATP is synthesized (by substrate-level phosphorylation) and eight electrons and eight hydrogen ions are removed and picked up by electron-acceptor molecules. Six of these electrons and six hydrogens are used to reduce three molecules of NAD^+ (forming three NADH plus three H^+ ions), and two of the electrons and two hydrogens are accepted by a related compound FAD (forming $FADH_2$). Since the breakdown of one molecule of glucose leads to two turns of the Krebs cycle, a total of two

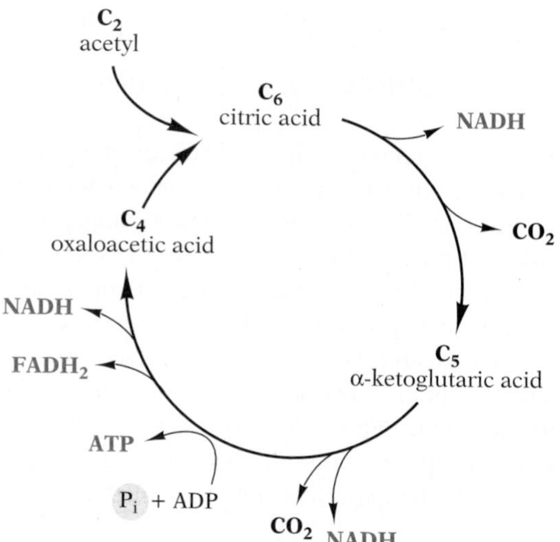

8.6 Simplified version of the Krebs citric acid cycle The two carbons of the acetyl group combine with a four-carbon compound to form citric acid, a six-carbon compound. Removal of one carbon as CO_2 leaves a five-carbon compound. And removal of a second carbon as CO_2 leaves a four-carbon compound, which can combine with another acetyl group and start the cycle over again. In the course of the cycle, one molecule of ATP is synthesized and eight hydrogens are removed; six are used in the reduction of NAD^+ and two in the reduction of FAD. Since one molecule of glucose gives rise to two acetyl units, two turns of the cycle occur for each molecule of glucose oxidized, with the production of four molecules of CO_2, two molecules of ATP, six molecules of NADH, and two molecules of $FADH_2$.

molecules of ATP and eight molecules of reduced carrier (six NADH plus two $FADH_2$) are formed during this stage of the breakdown of glucose (Fig. 8.7).

Stage IV Is Oxidative Phosphorylation

We have so far seen a net gain of only four new ATP molecules (two in glycolysis and two in the Krebs cycle). These represent only about 4 percent of the energy originally available in the glucose. Where has the rest of the energy gone? Much of the remaining energy is stored in NADH and $FADH_2$, both energy-rich substances. Remember, 12 molecules of these substances are synthesized in the process of breakdown of a molecule of glucose: two molecules of NADH in glycolysis, two molecules of NADH in the oxidation of pyruvic acid to acetyl-CoA, and six molecules of NADH plus two molecules of $FADH_2$ in the Krebs cycle.

How is the energy used to synthesize ATP? Under aerobic conditions the regeneration of NAD^+ from NADH is achieved by passage of the electrons from NADH to O_2; in other words, O_2 acts as the ultimate acceptor of electrons and hydrogen, and water is formed:

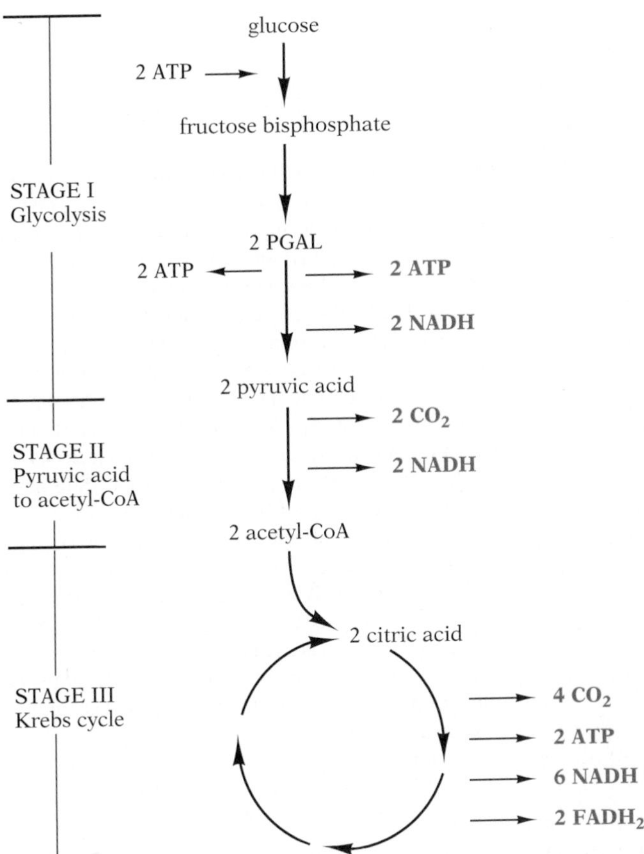

8.7 Summary of the most important products of Stages I, II, and III in the complete breakdown of one molecule of glucose Stage I (glycolysis) begins with expenditure of two molecules of ATP to produce fructose bisphosphate, which is broken down to two molecules of PGAL. After these preparatory steps, the two PGAL molecules are broken down to two molecules of pyruvic acid, a process that first pays back the two ATP molecules originally invested and then yields two molecules each of ATP and NADH (red). Stage II (the breakdown of two molecules of pyruvic acid to two molecules of acetyl-CoA) yields two molecules each of CO_2 and NADH. Stage III, in which the two molecules of acetyl-CoA are fed into the Krebs cycle and further broken down, yields four CO_2 molecules, two ATP molecules, six NADH molecules, and two $FADH_2$ molecules.

$$O_2 + 2\,NADH + 2\,H^+ \longrightarrow 2\,H_2O + 2\,NAD^+$$

The NADH does not, however, pass its electrons directly to the oxygen, as this summary equation might seem to indicate. Instead, the electrons from NADH are transferred through a series of electron-transport compounds; at the end of the chain molecular oxygen accepts the electrons and combines with H^+ ions to form water (Fig. 8.8). The electron-transport proteins, often called the ***respiratory chain,*** are embedded in the inner membrane of the mitochondrion. The transfer of electrons through the respiratory chain in the mitochondrion (which is analogous to the transfer of electrons through the photosynthetic chain in the chloroplast) results in the creation of an electrochemical gradient that is ultimately used in the synthesis of ATP from ADP and inorganic phosphate. This process, which is termed electron-transport phosphorylation or ***oxidative phosphorylation,*** involves a chemiosmotic coupling of electron transport and ATP synthesis. The respiratory electron-transport proteins are essential for the process of oxidative phosphorylation, which generates most of the ATP produced by the oxidation of glucose. *Oxidative phosphorylation, like photophosphorylation, utilizes the flow of electrons along a transport-chain to pump H^+ ions across a membrane; the resultant electrochemical gradient is used to synthesize ATP.*

In respiration, then, there are two methods of producing ATP: (1) substrate-level phosphorylation, in which a phosphate group is transferred from a "high-energy" compound to ADP, and (2) oxidative phosphorylation, in which the energy of the electrochemical gradient is harnessed to produce ATP.

To recapitulate, then, the four stages in the catabolism of glucose are: (1) glycolysis, in which glucose is broken down to pyruvic acid; (2) the oxidation of pyruvic acid to acetyl CoA; (3) the Krebs citric acid cycle, in which acetyl CoA is oxidized; and (4) oxidative phosphorylation, in which ATP is produced through chemiosmosis.

The Chemiosmotic Hypothesis Accounts for Electron Transport and ATP Synthesis in the Mitochondrial Membrane

ATP synthesis in the mitochondrion resembles ATP synthesis in the chloroplast—both involve the process of chemiosmosis, first described by Peter Mitchell in 1961. (Actually, the chemiosmotic hypothesis was initially formulated by Mitchell to account for ATP synthesis in the membrane of the mitochondrion; it was only afterward that it was found to apply as well to the synthesis of ATP in the thylakoid membrane of the chloroplast.) Let us briefly review the hypothesis.

In the last chapter we saw that the functioning of the electron-transport chain in the thylakoid depended on the orderly arrangement of the electron-transport molecules embedded in the inner membrane. We saw too that as electrons flowed through the molecules in the electron-transport chain, energy was released and harnessed to pump H^+ ions from one side of the membrane to the other, resulting in the build-up of an electrochemical gradient across the membrane. The energy of the gradient was then used by the ATP synthetase complex to synthesize ATP from ADP and inorganic phosphate. The mitochondrion is organized in a similar way, with electron-transport molecules establishing and maintaining an electrochemical gradient across the inner membrane.

The electron-transport molecules in the mitochondrion are arranged into four large protein complexes, which are embed-

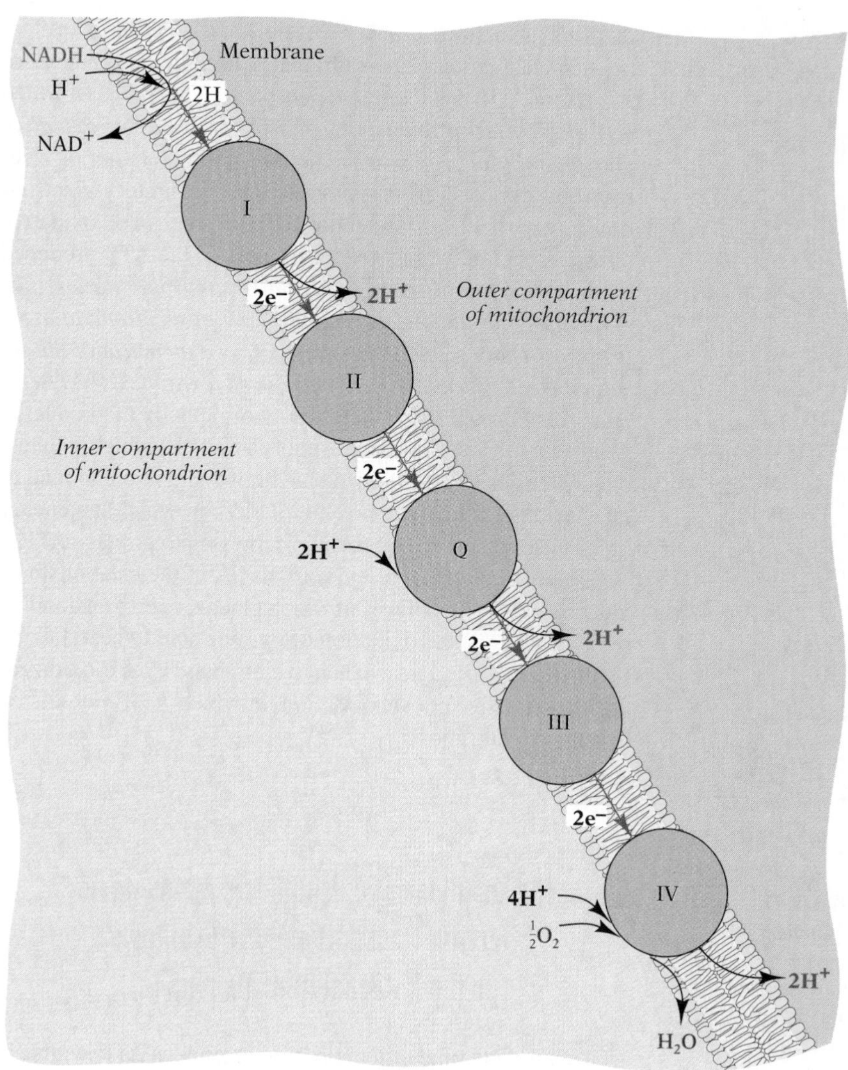

8.8 The respiratory electron-transport chain The reactions summarized in this diagram take place on the inner membrane of the mitochondrion. NADH donates two electrons and an H^+ ion to the electron-transport chain; a second hydrogen ion is drawn from the medium. The H^+ ions are released into the outer compartment and the electrons are passed from one acceptor substance to the next, step by step down an energy gradient from their initial high energy level in NADH to their final low energy level in H_2O. (When available, molecules of $FADH_2$ can also donate their electrons to the electron-transport chain; since these electrons have less energy than those of NADH, they enter lower down on the chain.) Each successive acceptor molecule is cyclically reduced when it receives the electrons, and then oxidized when it passes them on to the next acceptor molecule. At three sites along the chain, some of the free energy released is used to pump H^+ ions into the compartment outside the inner membrane. Later the H^+ gradient generated by the electron-transport chain is used in the synthesis of ATP. For simplicity, steps have been combined wherever possible. A more complete sequence is shown as part of Figure 8.9.

ded in the mitochondrion's inner membrane (Fig. 8.9). Mobile electron carriers (such as the molecules Q and Cyt. *c*) transport the electrons from one complex to the next.[2] The NADH produced in Stages II and III passes its two electrons and an H^+ ion to Complex I, and in the process is oxidized to NAD^+. The electrons are passed from one transport molecule to the next in a precise sequence to oxygen, the final electron acceptor (Fig. 8.9). Here, finally, we see why oxygen is required for aerobic respiration. Without oxygen to accept the electrons, the electron-transport chain would not function, and the NAD^+ necessary for Stage II and III could not be regenerated from NADH. When O_2 accepts electrons, it also picks up H^+ ions from the matrix, forming H_2O, one of the end products of glucose catabolism.

Research has shown that it is only the electrons that are passed down the respiratory electron-transport chain. The H^+ ions do not accompany the electrons. As electrons from NADH are passed along the chain to oxygen, however, three of the large complexes pump H^+ ions from the inner compartment across the inner membrane, into the outer compartment. As a result, the H^+ ion concentration increases in the outer compartment and decreases in the inner compartment, and the outside of the inner membrane becomes positively charged with respect to the inside. Thus a steep electrochemical gradient is created across the inner membrane.

The energy of the electrochemical gradient is used for the formation of ATP from ADP and inorganic phosphate. Special enzyme complexes, called ***ATP synthetases,*** are built into the inner membrane of the mitochondrion, just as they are in the thylakoid membrane of the chloroplast. These complexes act both as H^+ ion channels for the movement of H^+ ions back across the membrane and as enzymes for catalyzing the synthesis of ATP. The mechanism by which this is accomplished is not well understood and is an area of active investigation. *What is known is that the energy of the electrochemical H^+ ion gra-*

[2]The transport molecule Q is a coenzyme that is lipid-soluble and can move through the membrane. Cytochrome *c* (Cyt. *c*) is a water-soluble protein that is found on the outer surface of the inner mitochondrial membrane.

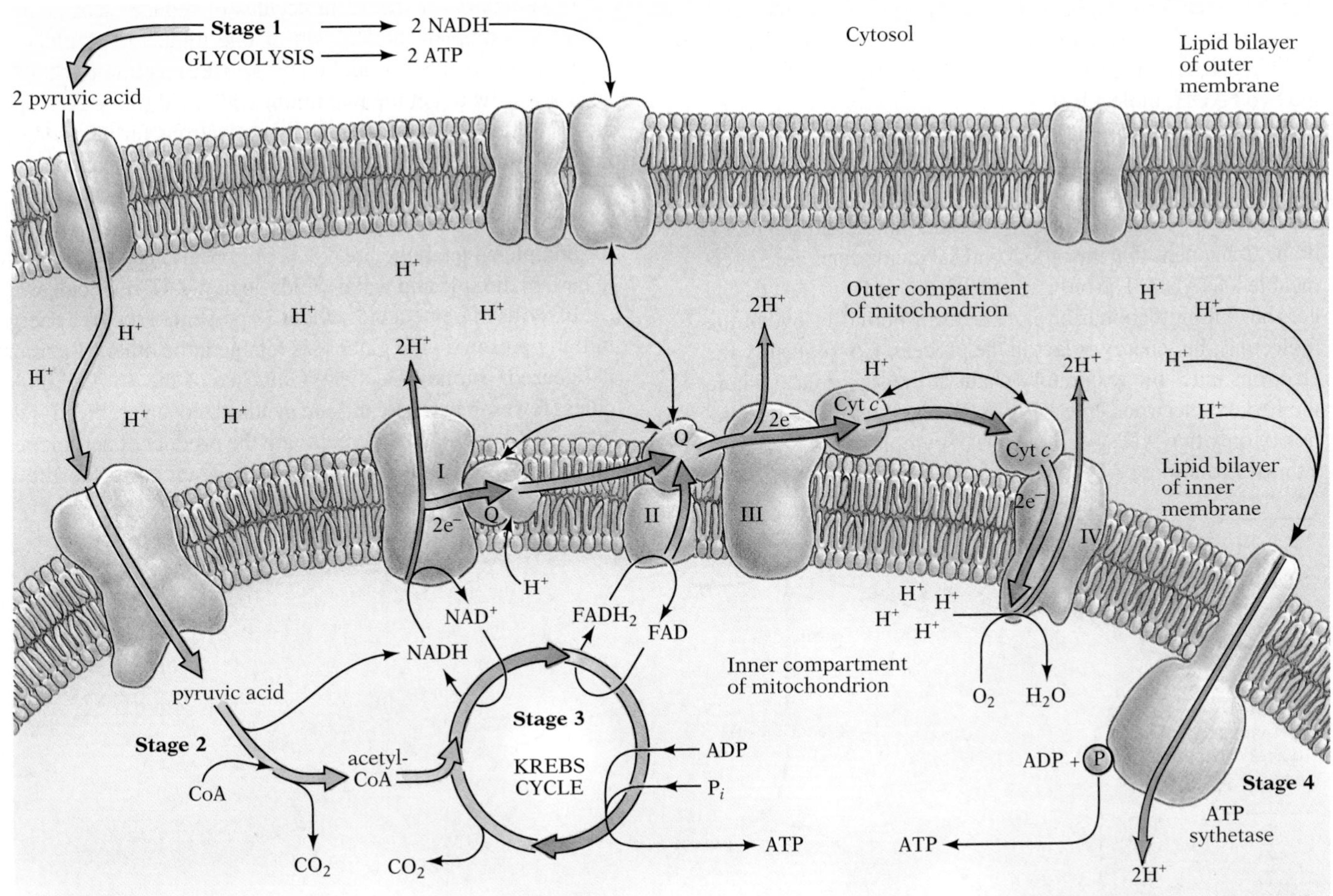

8.9 The anatomy of respiration The pyruvic acid produced in glycolysis is transported into the inner compartment of the mitochondrion. Here it is oxidized to acetyl CoA and CO_2, and the acetyl CoA is fed into the Krebs cycle. The NADH and $FADH_2$ produced in these reactions will transfer their electrons and H^+ ions to the proteins of the electron transport chain, which are mounted in the inner membrane of the mitochondrion. Several electron-transport molecules are grouped together to form four large enzyme complexes (beige), three of which pump H^+ ions across the membrane as electrons flow through them. Two mobile electron carriers (Q and Cyt. *c*) transport the electrons between the complexes. As the electrons from NADH and $FADH_2$ are transported along the chain (red arrow), H^+ ions are pumped across the inner membrane into the outer compartment, establishing a high concentration of H^+ ions in the outer compartment and a steep electrochemical gradient. The H^+ ions can move back across the membrane through the ATP synthetase complexes (right). The movement of H^+ ions down their energy gradient releases energy, which enables the ATP synthetase to catalyze the synthesis of ATP from ADP and inorganic phosphate. Approximately two H^+ ions must pass through the ATP synthetase complex to synthesize one ATP molecule. Because each pair of electrons donated to the chain from the NADH results in the pumping of approximately six H^+ ions into the outer compartment, the oxidation of one NADH gives rise to about three ATP molecules. The electrons from $FADH_2$ enter the chain farther down (at Q), and fewer H^+ ions are pumped outward. Thus the oxidation of the lower-energy compound $FADH_2$ gives rise to just two ATPs. The electrons from the NADHs produced in glycolysis also enter the chain farther down, resulting in the production of two ATPs per molecule of cytoplasmic NADH.

dient is harnessed by the ATP synthetase complex to synthesize ATP. As the H^+ ions move down the gradient through the ATP synthetase, energy is released and ATP is synthesized. The process is in many respects the reverse of active transport across a membrane. The ATP produced in the matrix is transported by carrier proteins through the mitochondrial membranes into the cytosol, where it can be used by the cell.

Studies have shown that at least two H^+ ions must pass through the ATP synthetase complex to create one ATP molecule from ADP. Since each pair of electrons donated to the chain from the NADH results in the pumping of approximately six H^+ ions into the outer compartment (Fig. 8.9), the oxidation of one NADH gives rise to about three ATP molecules. A total of eight NADHs are produced within the mitochondrion (two from the oxidation of pyruvic acid and six from the Krebs cycle), for a total of 24 new ATP molecules. As Figure 8.9 shows, however, the two electrons from $FADH_2$ have less energy than those of NADH; these electrons enter the chain farther down (at Complex II), and fewer H^+ ions are pumped outward. Thus the oxidation of the lower-energy compound $FADH_2$ gives rise to just two ATPs (for a total of four ATP from the two $FADH_2$ molecules).

What happens to the two molecules of NADH produced in glycolysis? Can their energy also be used to make ATP? Remember that glycolysis takes place in the cytosol, whereas electron transport takes place in the inner mitochondrial membrane. It happens that the mitochondrial membranes are impermeable to NADH. The only way the electrons from NADH can get into the mitochondrion is to be transported in by shuttle molecules, but energy is lost in the process. Consequently the electrons enter the respiratory chain at the same place where the $FADH_2$ electrons enter (Fig. 8.9), resulting in the production of only two ATPs per molecule of cytoplasmic NADH (for a total of four more ATP).

In summary, of the 12 molecules of reduced carrier produced by complete breakdown of one molecule of glucose, eight donate their extra electron pairs to the respiratory chain in such a way as to get the maximum ATP yield (three ATPs per NADH); the other four donate their electrons farther along in the chain and get only two molecules of ATP per pair of electrons (Figs. 8.9 and 8.10). The total yield from electron-transport phosphorylation, then, is 32 molecules of ATP, whereas the complete metabolic breakdown of one molecule of glucose to carbon dioxide and water yields 36 new ATP molecules.

In terms of percentages, about 39 percent of the free energy initially present in the glucose is retained; the other 61 percent is released, primarily as heat. Only two of the 36 ATP molecules (fewer than 6 percent) are synthesized anaerobically; the other 34 (more than 94 percent) are the product of aerobic respiration. We now see why oxygen is essential to the life of

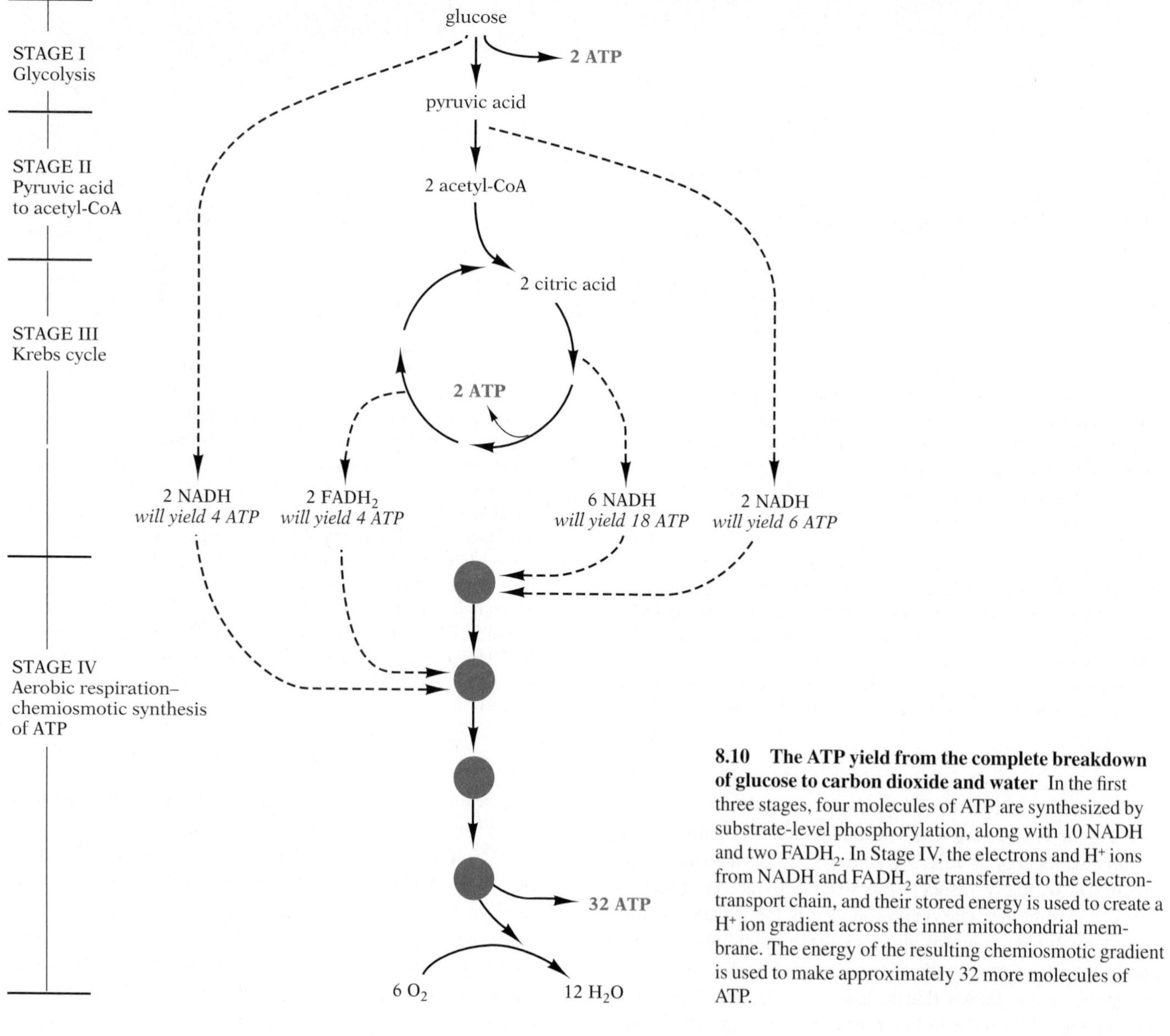

8.10 The ATP yield from the complete breakdown of glucose to carbon dioxide and water In the first three stages, four molecules of ATP are synthesized by substrate-level phosphorylation, along with 10 NADH and two $FADH_2$. In Stage IV, the electrons and H^+ ions from NADH and $FADH_2$ are transferred to the electron-transport chain, and their stored energy is used to create a H^+ ion gradient across the inner mitochondrial membrane. The energy of the resulting chemiosmotic gradient is used to make approximately 32 more molecules of ATP.

MAKING CONNECTIONS

1. What is the function of the folds (cristae) found in the inner membrane of the mitochondrion?

2. All the carbon dioxide in your exhaled breath is produced during respiration. On the assumption that glucose is the only substance you are metabolizing for energy, what fraction of the carbon dioxide is produced in the Krebs citric acid cycle? Where does the remainder come from?

3. Stages II and III of respiration are considered aerobic, and yet neither uses oxygen directly. Why would Stages II and III soon stop functioning in the absence of oxygen?

4. What would happen to ATP synthesis if either the inner or the outer membrane of the mitochondrion were made freely permeable to H^+ ions? How much ATP would be made from one molecule of glucose?

5. Why does intensive exercise such as doing push-ups cause muscle fatigue? What happens during the recovery process?

human beings and most other organisms: *aerobic respiration extracts 18 times as much energy from glucose as does anaerobic metabolism.* Nevertheless, parts of our bodies can operate anaerobically for short periods. During intensive exercise, for instance, muscles often need so much energy that the oxygen supplied from breathing is insufficient. In such cases glycolysis provides the needed energy, although the resulting build-up of lactic acid can cause muscle fatigue. Later, the oxygen debt is paid back by deep breathing or panting, and the lactic acid that accumulated in the muscles as a result of fermentation is removed and transported to the liver, where it is reconverted into glucose.

Of the 34 aerobically synthesized ATP molecules, 32 result from the H^+-ion gradient created by the electron-transport chain, which shows the importance of this process. Cyanide and some other poisons form complexes with certain of the electron-transport proteins and thereby block the transport of electrons along the chain. The lethality of these substances arises from the fact that without a functional electron-transport chain, none of the reactions of aerobic respiration can take place, and the cell derives so little energy from its nutrients that it soon dies.

Chemiosmosis in the Mitochondrion Differs in Some Ways From That in the Chloroplast

We have said repeatedly that the chemiosmotic process in mitochondria and chloroplasts is very similar. In both organelles the flow of electrons through a transport chain results in the pumping of H^+ ions across a membrane, generating an electrochemical gradient. The energy stored in the gradient is then used to synthesize ATP. Although the basic process is similar, there are a number of differences between the two processes.

For one thing, the source of energy that drives the flow of electrons is quite different: in the chloroplast, sunlight provides the energy, whereas in the mitochondrion, it is the energy from food (Fig. 8.11). Also, the direction in which H^+ ions are pumped is different: in the thylakoid, H^+ ions are pumped *into* the thylakoid interior, whereas in the mitochondrion the flow is opposite—out of the inner compartment. Consequently, the H^+ ion concentration is greater in the *interior* of the thylakoid, whereas it is greater in the *outer* compartment of the mitochondrion (Fig. 8.12). Finally, in the chloroplast, the electrons come from water and flow through the chain to the electron carrier NADP and oxygen is produced, whereas in the mitochondrion electrons come from the electron carriers NADH and $FADH_2$ and flow through the chain to oxygen, forming water.

Fats and Proteins Also Supply Energy in the Form of ATP

Cells can extract energy in the form of ATP not only from carbohydrates, on which we have focused so far, but also from fats

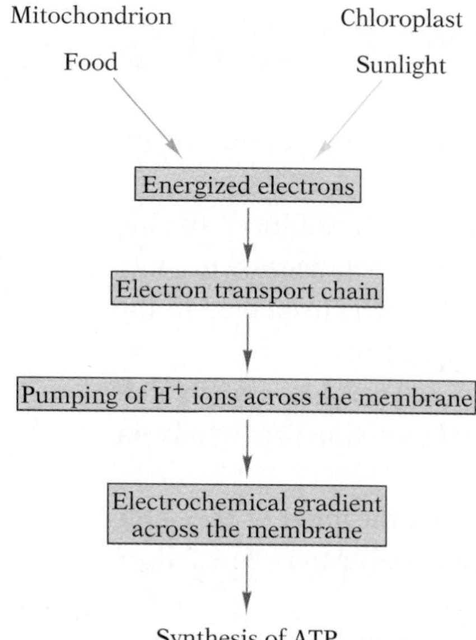

8.11 A comparison of ATP synthesis in mitochondrion and chloroplast The energy for synthesizing ATP in the chloroplast comes from sunlight; in the mitochondrion it comes from chemical energy stored in food.

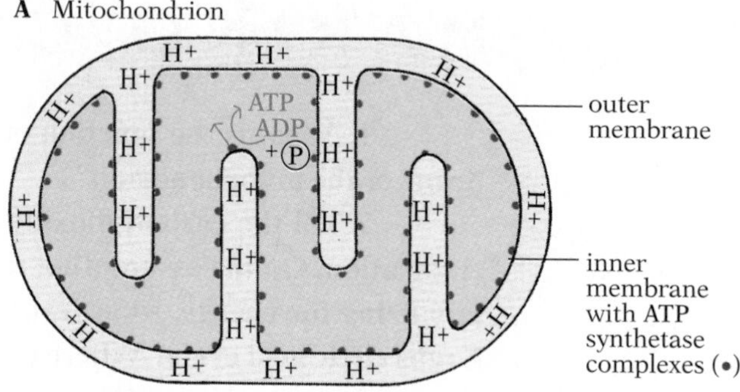

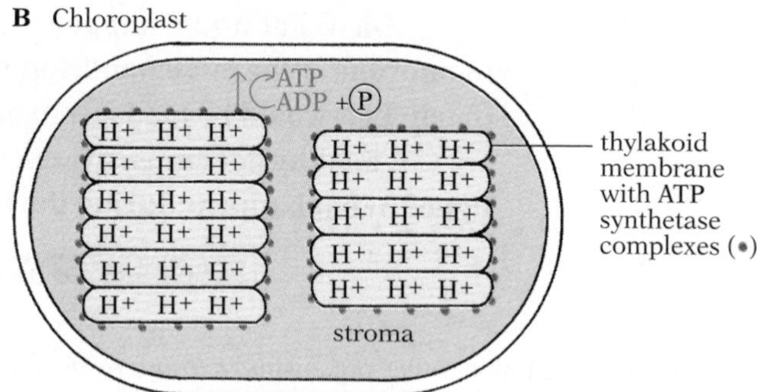

8.12 A comparison of chemiosmosis in the mitochondrion and chloroplast In both organelles the flow of electrons through a transport chain results in the pumping of H^+ ions across a membrane, generating an electrochemical gradient. The energy stored in the gradient is then used to synthesize ATP. The direction in which H^+ ions are pumped is different: in the mitochondrion, H^+ ions are pumped from the inner compartment into the outer compartment, whereas in the chloroplast the flow is opposite—out of the stroma into the thylakoid interior. Consequently, the H^+ ion concentration (yellow) is higher in the *outer* compartment of the mitochondrion, and the *interior* of the thylakoid. Notice that the ATP synthetase complexes (black "lollypops") are oriented in opposite directions in the two organelles, which reflects the different directions of movement of the H^+ ions through them.

and proteins, as Figure 8.13 shows. The metabolism of fats begins with their hydrolysis to glycerol and fatty acids. The glycerol (a three-carbon compound) is then converted into PGAL and fed into the glycolytic pathway at the point where PGAL normally appears. The fatty acids are transported into the inner compartment of the mitochondrion, where they are broken down into acetyl-CoA and fed into the Krebs cycle. Since fats have a higher proportion of hydrogen than carbohydrates, their complete oxidation yields more energy per unit weight; one gram of fat yields slightly more than twice as much energy as one gram of carbohydrate.

The amino acids produced by the hydrolysis of proteins are metabolized in a variety of ways. After removal of the amino group in the form of ammonia (NH_3), certain amino acids are converted into pyruvic acid, others into acetyl-CoA, and still others into one or another compound of the Krebs cycle. The complete oxidation of one gram of protein yields roughly the same amount of energy as one gram of carbohydrate.

Compounds such as pyruvic acid, acetyl-CoA, and the components of the Krebs cycle, which are common to the catabolism of several different types of substances, not only play a crucial role in the oxidation of energy-rich compounds to carbon dioxide and water but also function in the anabolism, or synthesis, of amino acids, sugars, and fats. They serve as biochemical crossroads, at which several metabolic pathways intersect. By investing energy, the cell can reverse the direction in which substances move along some of these pathways. For example, the PGAL and acetyl-CoA produced in the breakdown of carbohydrate can be used to synthesize fats. Similarly, many amino acids can be converted into carbohydrate via the common intermediates in their metabolic pathways. Not all pathways are two-way, however. In animal cells sugars can be converted into fat but fatty acids cannot be converted into sugar, because the reaction from pyruvic acid to acetyl-CoA is irreversible. This has implications for diets in humans. Because glucose is the only fuel for brain cells, it follows that diets must always include carbohydrates to supply sufficient amounts of glucose. If the glucose levels in the blood drops too low, protein, such as muscle protein, must be broken down and some of the resulting amino acids will be converted into glucose to supply the brain cells. The popular high fat-low carbohydrate diets can be dangerous because they do not supply adequate amounts of glucose.

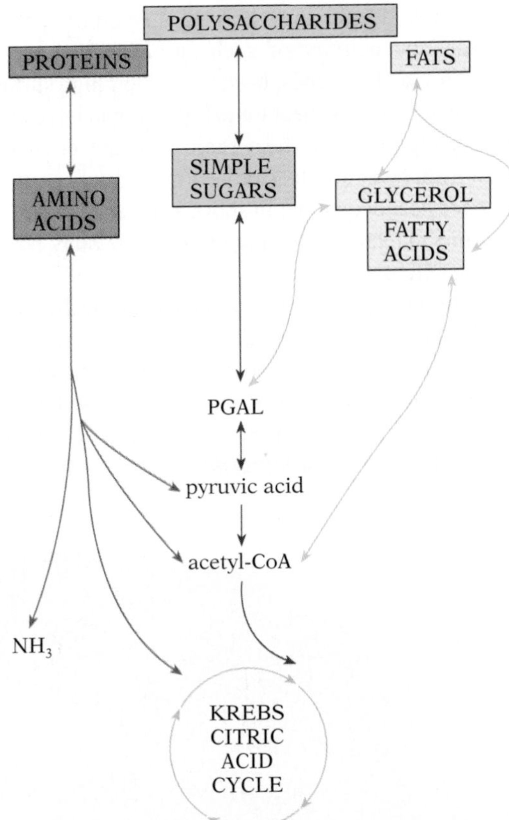

8.13 The relationships of the metabolism of proteins and fats to the metabolism of carbohydrates Compounds such as pyruvic acid, acetyl-CoA, and the components of the Krebs cycle are common to the catabolism of several different types of substances but also function in the anabolism, or synthesis, of amino acids, sugars, and fats. By investing energy, the cell can reverse the direction in which substances move along some of these pathways. Not all pathways are reversible; the reaction from pyruvic acid to acetyl-CoA is irreversible. Consequently, fatty acids cannot be used to synthesize glucose.

Body Temperature Affects Metabolic Rate

We have seen that cellular metabolism captures some of the energy released by the oxidation of carbohydrates, fats, and proteins and converts it into the energy of phosphate groups in ATP. This process, however, fails to capture more than 60 percent of the energy. Most of this energy is released in the form of heat. The vast majority of animals, as well as all plants, promptly lose most of this thermal energy to their environment. Such organisms are said to be ***ectothermic*** ("externally heated"). Because ectothermic animals' heat comes largely from the external environment, their body temperature fluctuates to some extent with the environmental temperature; when they are at rest, it is nearly the same as that of the surrounding medium, particularly if the medium is water.

The metabolism of an organism is closely tied to temperature. Within the narrow range of temperatures to which the active organism is tolerant, the metabolic rate increases with increasing temperature and decreases with decreasing temperature in a very regular fashion (Fig. 8.14). As would be expected, therefore, the activity of ectothermic animals is radically affected by temperature changes in their environment. As the temperature rises (within narrow limits), they become more active; as the temperature falls, they become sluggish and lethargic. Such animals, then, are restricted as to the habitats they can effectively occupy, because they are at the mercy of the temperatures in those habitats. There are, for example, no insects or reptiles in polar regions (though there are fish), and relatively few amphibians and insects in subpolar arctic environments.

Ectothermic animals include amphibians, reptiles, and most fish and invertebrates (Fig. 8.15). Although these animals are dependent on the environment for their heat, most ectotherms do control their body temperature to some degree by

MAKING CONNECTIONS

1. Compare how chemiosmotic synthesis of ATP occurs in mitochondria and chloroplasts. Compare the structure of the two organelles. How does their structure relate to their function?

2. Why do people get fat? Why is it possible to gain fat tissue even if your diet contains no fats?

3. Why would a "crash diet" that is very low in carbohydrates and total calories result in the loss of body protein as well as fat?

4. If a climatic change caused environmental temperatures to fluctuate more widely, would ectotherms or endotherms be more vulnerable to extinction?

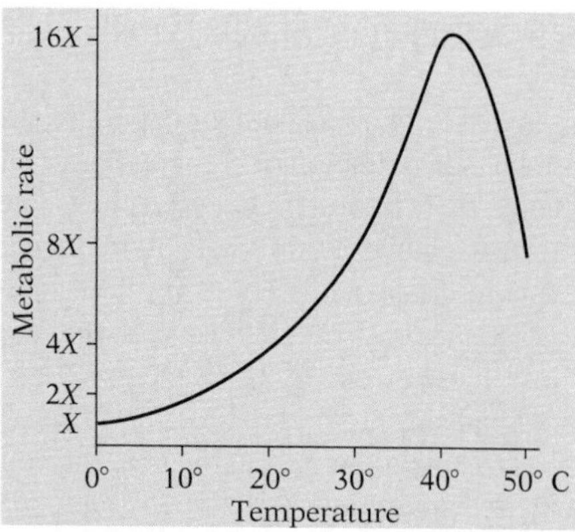

8.14 Graph of changes in metabolic rate with changes in temperature The metabolic rate increases more and more rapidly as the temperature increases; this type of exponential increase produces a curve that becomes steeper and steeper as the temperature rises. This rise is the result of the greater thermal energy of the reactants in the cell and the increasing effectiveness of the cellular enzymes. The abrupt decline above 40 degrees represents the point at which the weak bonds that hold enzymes in their specific active conformations begin to break. As a result the enzymes become denatured and metabolic activity is severely disrupted.

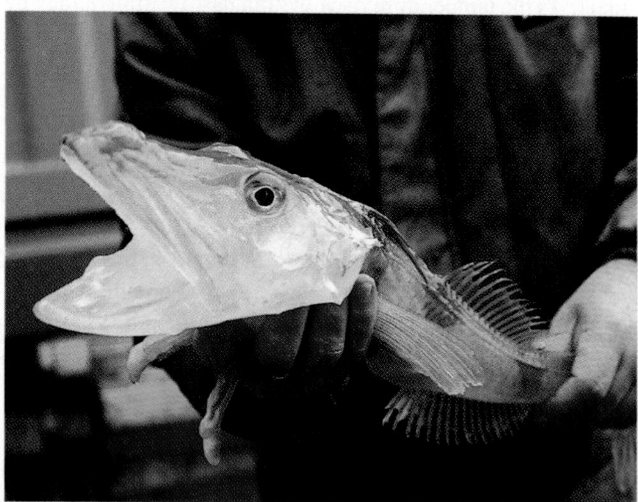

8.15 An antarctic ice fish Ice fish are ectotherms whose body temperature is closely tied to environmental temperature, which in the case of these fish can reach as low as -2°C. The bodies of most marine fishes will freeze at temperatures below 0°C, but ice fish have a special adaptation—organic antifreezes in their blood and tissue fluids—that keeps them from freezing. The fish are pale because their blood lacks the red pigment hemoglobin for transporting oxygen; instead, the oxygen is carried to the cells dissolved in the blood plasma. Notice the very large toothed mouth that is used for capturing small fish and scooping up invertebrates from the ocean floor.

regulating their behavior (e.g., snakes and lizards commonly sun themselves on cold days). Some are highly successful at thermoregulation. For example, desert lizards maintain a very constant body temperature during the day (maintaining their body temperature to within 0.1°C) by moving into and out of the sunlight and shade. At night they control the amount of cooling by how deeply they burrow. In addition, the size, shape, color, and activity of such animals influences thermoregulation. Small animals, because of their high surface-to-volume ratio, tend to lose heat more rapidly than large animals. Although ectotherms lose heat rapidly to the environment, they also absorb heat rapidly. Body color, body shape, and posture influence heat exchange. In short, almost all animals seem to have some adaptations for regulating their body termperature to some extent.

A few animals, notably all mammals and birds, can make use of the heat produced during the exergonic reactions of their metabolism, because they have evolved mechanisms—often including insulation by fat, hair, feathers, etc.—whereby heat loss to the environment is retarded (Fig. 8.16). Such animals are commonly called ***endothermic*** ("internally heated").

8.16 Endotherms living in cold environments Mammals and birds, such as these harp seals (top) and penguins, (bottom) are able to survive in frigid environments because of the insulation provided by fat and hair or feathers. Such insulation retards heat loss to the environment and makes use of the heat produced during their metabolism. Their body temperatures are considerably higher than the environmental temperature, and remain relatively constant even when the environmental temperature fluctuates widely.

Surprisingly, some organisms normally thought of as ectotherms regularly warm their bodies using metabolically produced heat. Mako sharks, tuna, honeybees, moths, and even certain plants, such as skunk cabbage and crocuses, generate heat through increased metabolic activity.

The body temperature of animals that have internally produced heat is fairly high—usually higher than the environmental temperature—and relatively constant even when the environmental temperature fluctuates widely. Endotherms can maintain a uniformly high metabolic rate and remain very active. They are thus less dependent on environmental temperatures than ectotherms, and are freed for successful exploitation of habitats that are too cold for most ectotherms.

Another way in which animals differ with respect to their thermoregulation has to do with whether or not they maintain a relatively constant body temperature at all times. Of course, no animal is able to keep its body temperature perfectly constant. (Your own body temperature, for example, is slightly lower in the early morning than it is in the late afternoon.) Some animals substantially lower their body temperature on a regular basis. Bears and woodchucks, for example, do so during hibernation, whereas hummingbirds and bats have a much reduced body temperature during certain parts of the day. Animals who maintain a more or less constant body temperature are referred to as ***homeotherms*** ("same temperature"). Organisms whose body temperature varies on a regular basis are known as ***heterotherms*** ("different temperature").

In short, organisms can fit into different categories depending on the two methods of thermoregulation. Most birds and mammals (including humans) are both endothermic and homeothermic, but some are endothermic and heterothermic. Conversely, most invertebrates, fish, amphibians, and reptiles are ectothermic and heterothermic, but some reptiles living in tropical environments are ectothermic and homeothermic. As you can see, the old terms "warm-blooded" and "cold-blooded" are outmoded and no longer useful in this context.

Although endothermy has its advantages, it also has considerable costs. The metabolic rate of birds and mammals is about five times that of an ectotherm of equivalent size. This means that endotherms must obtain five times as much food, since roughly 80 percent of their metabolism is devoted just to maintaining their high body temperature. Animals that do not have to spend so much energy on maintaining a high body temperature to sustain their high activity level, such as amphibians and reptiles in the tropics, actually have an advantage over endotherms, because of their greater energy economy.

CONCEPTS IN BRIEF

1. *Metabolism,* a general term embracing the myriad enzyme-mediated reactions of a living cell, can be divided into two phases: *anabolism,* the building-up phase, and *catabolism,* the breaking-down phase.

2. Before the potential energy stored in complex organic compounds can be used by the cell to do work, the compounds must be broken down in a series of chemical reactions and the energy transferred to ATP.

3. The first series of reactions in the degradation of glucose, termed *glycolysis* (Stage I), consists of the breakdown of glucose to two molecules of *pyruvic acid,* with the production of two molecules of *NADH* and a net gain of two ATP molecules. This process, common to all living cells, can take place in the absence of oxygen and hence is called *anaerobic.*

4. The fate of the pyruvic acid depends on the oxygen supply. In the absence of O_2, the pyruvic acid may be reduced by NADH to form CO_2 and ethyl alcohol or lactic acid, in a process called *fermentation.* NAD^+ molecules formed in this way are available to be reused in glycolysis.

5. Under aerobic conditions the pyruvic acid can be further oxidized, with the accompanying synthesis of ATP. This process, called *aerobic respiration,* begins with the breakdown of the two pyruvic acid molecules to form two molecules each of acetyl-CoA, CO_2, and NADH (Stage II).

6. The acetyl-CoA thus formed is fed into the *Krebs citric acid cycle* (Stage III). In the course of the cycle, two carbons are lost as CO_2, a molecule of ATP is synthesized, and eight electrons and eight hydrogens are picked up by carrier compounds, forming three molecules of NADH and one of $FADH_2$. Since one molecule of glucose gives rise to two molecules of acetyl-CoA, two turns of the Krebs cycle occur for each molecule of glucose oxidized.

7. The final stage of respiration involves the passage of the electrons from the carrier molecules down a *respiratory chain* of electron-transport molecules to oxygen, with which the electrons and H^+ ions from the medium combine to form water (Stage IV). As the electrons are passed along the chain, energy is released and is used to

pump H^+ ions from the inner compartment of the mitochondrion into the outer compartment, establishing an *electrochemical gradient* across the inner membrane.

8. Special enzyme complexes, called ATP synthetases, act as H^+-ion channels in the inner mitochondrial membrane. As the H^+ ions move down the electrochemical gradient through the complex, energy is released and can be used to synthesize ATP.

9. The total number of new ATP molecules produced by the complete metabolic breakdown of glucose is usually 36: two from glycolysis, two from the Krebs cycle, and 32 from electron-transport phosphorylation.

10. Cellular respiration captures about 39 percent of the energy of glucose and converts it into ATP; the rest of the energy is released, mostly as heat.

11. Cells can also extract energy in the form of ATP from fats and proteins, which are hydrolyzed to form products that can be fed into the catabolic pathways.

12. *Ectothermic* ("externally heated") animals promptly lose most of their heat to the environment. The body temperature and metabolic rate of such organisms fluctuate with the temperature of the environment.

13. *Endothermic* ("internally heated") animals retain the heat generated by metabolism and hence maintain a constant high body temperature. Their metabolic rate and activity can accordingly be maintained at a uniformly high level.

14. Animals who maintain a more or less constant body temperature are referred to as *homeotherms* ("same temperature"). Organisms whose body temperature varies on a regular basis are known as *heterotherms* ("different temperature").

STUDY QUESTIONS

1. Why is ATP called the "energy currency" of the cell? How does the amount of energy released by the hydrolysis of ATP compare with the amount of energy released by the complete oxidation of a molecule of glucose? (p. 162)
2. What are the four stages of respiration? Where does each occur? Which require oxygen? (pp. 163–172)
3. Which products of glycolysis contain the carbon atoms that were originally in glucose? Which contain energy released from glucose during the course of this series of reactions? (p. 164)
4. What is the principal function of fermentation? Describe the two major fermentation pathways and the organisms in which each occurs. What is oxidized and what is reduced in each case? (pp. 165–166)
5. Draw a mitochondrion and label its parts. Which part or parts of the mitochondrion are associated with each of the stages of aerobic respiration? (p. 167)
6. What happens to the pyruvic acid produced by glycolysis in Stage II of respiration? Which molecule is oxidized and which is reduced? What is the role of coenzyme A? (pp. 167–168)
7. What are the products of the Krebs citric acid cycle, and how much of each is produced per glucose molecule? (p. 168)
8. What molecules donate electrons to the respiratory chain? Where were they produced? What is the final electron acceptor in the chain? (p. 170)
9. What are the two methods by which ATP is produced during cellular respiration? (p. 172)
10. Describe how chemiosmotic synthesis of ATP occurs. Where and how is a high H^+ ion concentration built up? Where can H^+ ions diffuse across the inner membrane of the mitochondrion? With what result? (pp. 170–172)
11. Construct a table accounting for the 36 ATPs generated by glucose oxidation. It should show the amount of ATP, NADH, and $FADH_2$ generated by the first three stages of respiration, and the number of ATPs produced by the oxidation of each NADH and $FADH_2$ molecule. (p. 172)
12. Explain how lipids and proteins can be oxidized to yield energy. (p. 174)
13. Every organism can be categorized as ectothermic or endothermic, and as heterothermic or homeothermic. To which categories does a human belong? A woodchuck? A maple tree? A fish living in a relatively constant environment such as a cave? (pp. 175–177)

P A R T T W O

The Perpetuation of Life

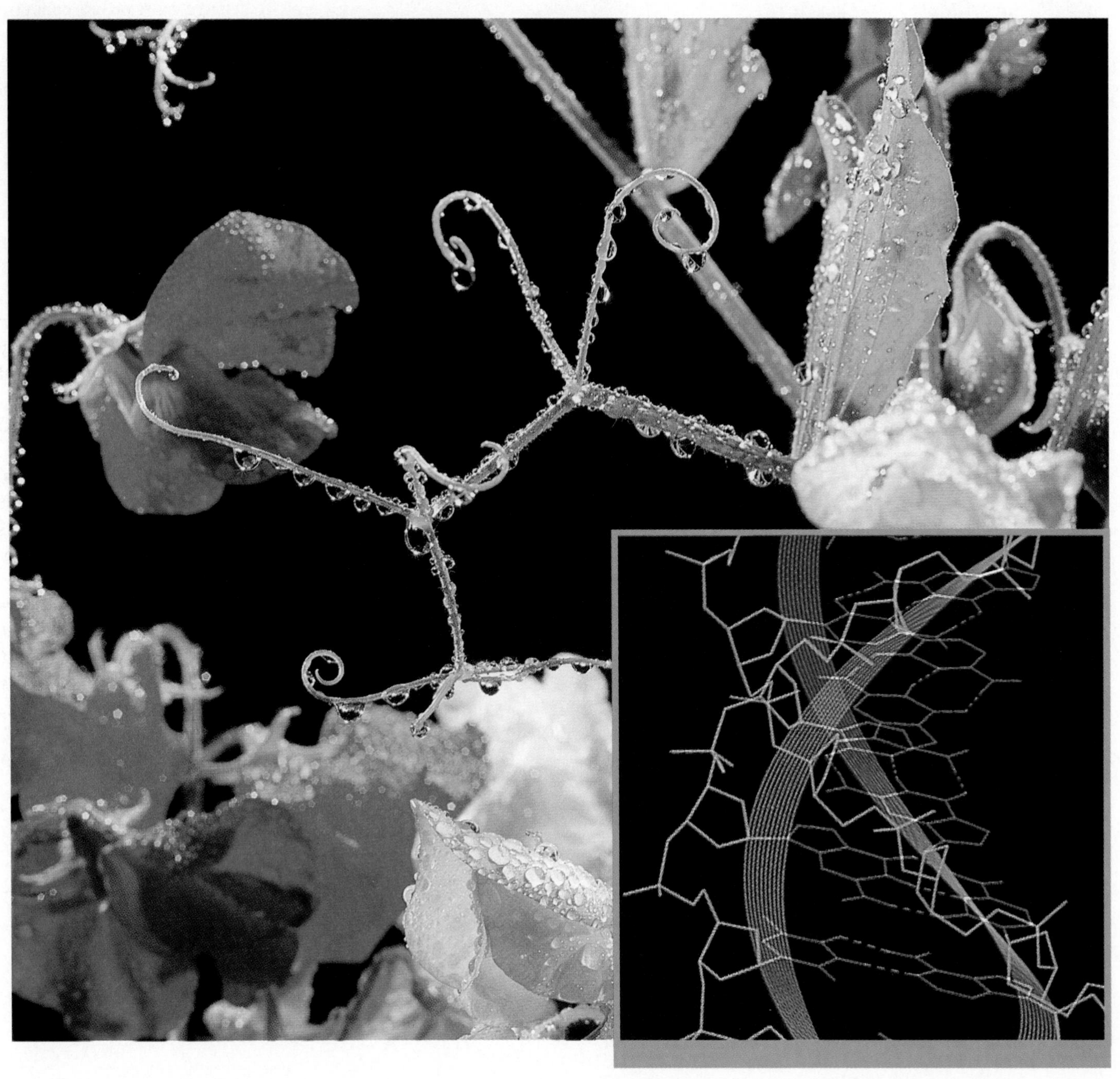

CHAPTER 9

HOW CELLS REPRODUCE

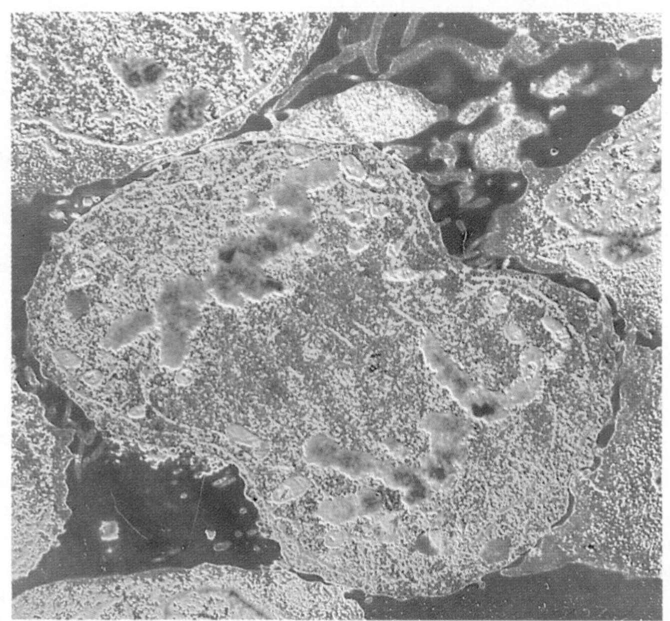

In Chapters 7 and 8 we saw how life is dependent upon photosynthetic organisms, which use the energy of sunlight to build energy-rich macromolecules, and, then, how all organisms break down these macromolecules to release the energy they need to sustain all the activities of life. In this chapter we explore another essential cellular activity: how cells reproduce and pass along hereditary information from one generation to the next. As we have already pointed out, an organism's genes—the units of hereditary information—are incorporated into its chromosomes, the long, often elaborately convoluted strands of DNA and protein present in both procaryotic and eucaryotic cells. Passed from cell to cell through generations of cells and organisms, the genes determine the characteristics of each new organism and direct its many activities. Thus, the replication of chromosomes during cellular reproduction is fundamental to both the continuity and the diversity of life.

Because the chromosomes carry the genetic plan, or blueprint, for the development of the new individual, it follows that when a cell divides to form two new cells, this information must be transmitted in an orderly fashion to both of the new cells. The information cannot be simply split into two halves, because that would give neither of the new cells a satisfactory blueprint. Just as two buildings cannot be erected by cutting one blueprint in two and giving half to each of two contractors, so two new cells cannot develop if each receives only half the information that is contained within the parental cell. If two contractors are to erect two identical buildings simultaneously, each must have a complete blueprint. The same applies to the cell. If a parental cell is to divide and produce two viable new cells, it must first duplicate the genetic information in its chromosomes and then, as it divides, give one complete copy to each daughter cell (Fig. 9.1).

Cell Division Is Simpler in Procaryotic Cells Than in Eucaryotic Cells

Cell division appears to be much less complex in procaryotes, such as bacteria, than in the eucaryotic cells of plants and animals. After the procaryotic cell has elongated sufficiently to provide the basis for two independent daughter cells, the single, circular chromosome replicates, and the resulting second chromosome attaches to a point away from the first one on the expanding plasma membrane (Fig. 9.2). Next, new plasma membrane and wall material begin to form near the midpoint

Chapter opening photo: False-color transmission electron micrograph of a dividing human lymphocyte. This lymphocyte is dividing by mitosis and the two daughter cells that result will be genetically identical to each other and to the parent lymphocyte.

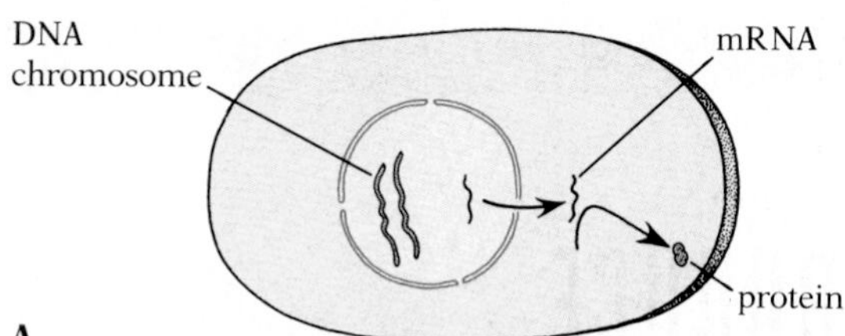

Instructions in the DNA are used to build cellular components and direct biochemical events. In most eucaryotes, there are two copies of each chromosome.

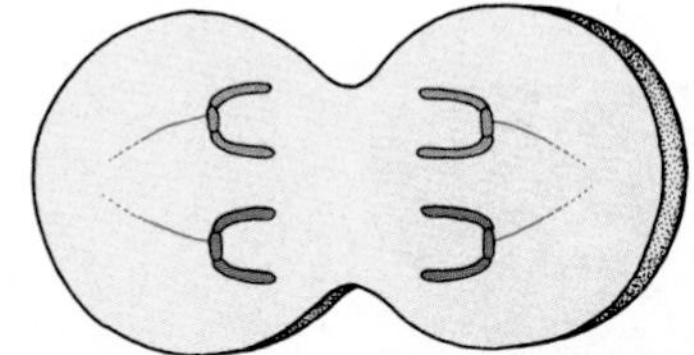

Chromosomes are duplicated prior to cell division and a complete copy of each pair is passed to each new cell.

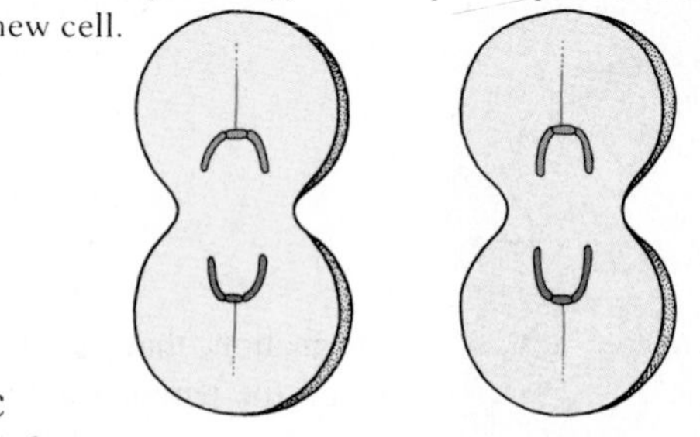

When gametes (sex cells) are formed, they receive only one chromosome from each pair. The normal number of chromosomes is restored by fusion of a male and female gamete.

9.1 Transfer of genetic information The set of instructions in the DNA is transcribed and used to direct events within the cell (A), is duplicated and passed on to daughter cells during normal cell division (B), and is halved and passed to gametes in preparation for sexual reproduction (C).

of the parental cell between the two chromosomal attachment points. The newly formed boundary structure grows inward, cutting through both the cytoplasm and the ill-defined nuclear region (Fig. 9.3). Ultimately, the parental cell splits into two completely separate daughter cells, each with its own chromosome. The comparatively simple process by which procaryotic cells divide is called ***binary fission.***

In eucaryotes, the chromosomes are contained in the nucleus, the control center of the cell, and the mechanism of organizing the chromosomes into new daughter nuclei differs markedly from the corresponding process in procaryotic cells. An elaborate sequence of nuclear changes, observable under the ordinary light microscope, occurs in eucaryotic cell division. The remainder of this chapter is largely an account of these changes as they are currently understood.

In eucaryotic cells, each chromosome consists of a single, long DNA molecule wound around spool-like structures called ***nucleosomes.*** (The DNA in procaryotic cells, in contrast, is

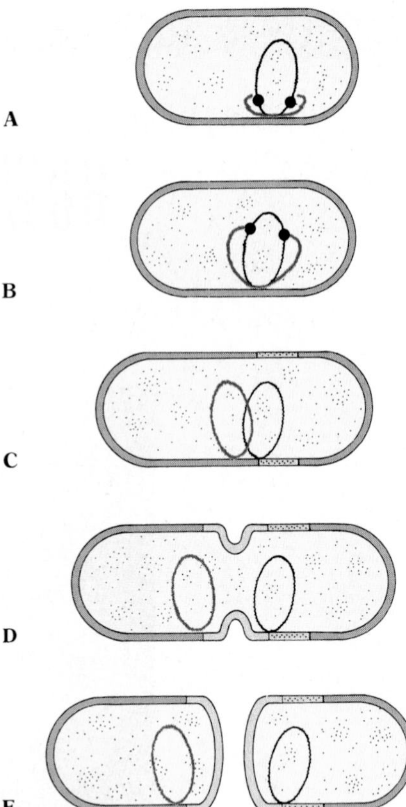

9.2 Binary fission of a procaryotic cell (A) The circular chromosome of a cell is attached to the plasma membrane near one end; it has already begun replication (the partially formed second chromosome is shown in red). (B) Replication is about 80 percent complete. (C) Chromosomal replication is finished, and the second chromosome now has an independent point of attachment to the membrane. During replication additional membrane and wall (stippled) has formed. (D) More new membrane and wall (tan) has formed between the points of attachment of the two chromosomes. Part of this growth forms invaginations that will give rise to a septum cutting the cell in two. (E) Fission is complete, and two daughter cells have formed. The chromosomes, which are actually so long that they must be looped and tangled to fit into a cell, are depicted here as small open circles. At the scale of this drawing, the actual circumference of each chromosome would be about 30 meters.

not only circular but also "naked," meaning it is not wrapped on nucleosomes.) The nucleosome cores, which are composed of histone proteins, are packed into thick strands (Fig. 9.4). During cell division, the chromosomes are further compacted as the strands are organized into loops, creating the characteristic structures visible microscopically only during cell division (Fig. 9.5A).

Unlike procaryotes, most eucaryotes receive chromosomes from two parental cells—a male sperm cell and a female egg cell, for instance—rather than from just one. The chromosomes of these eucaryotes are said to occur in ***homologous pairs,*** one from each parent cell. The chromosomes of a homologous pair have similar shapes, and the genes they contain are in the same order along the DNA molecule. The ho-

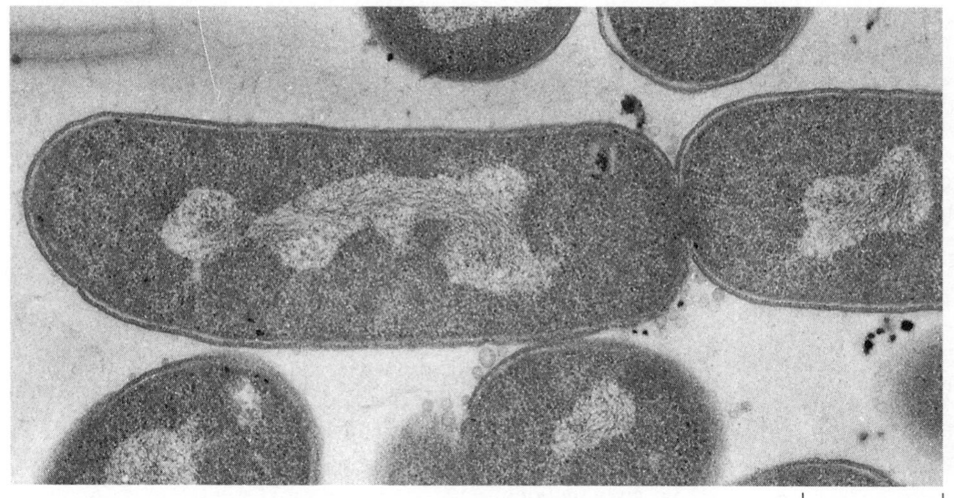

9.3 Electron micrograph of a dividing bacterial cell New wall is growing inward, cutting this *E. coli* in two. The replicated chromosomes, found in the white areas inside cells, have already separated.

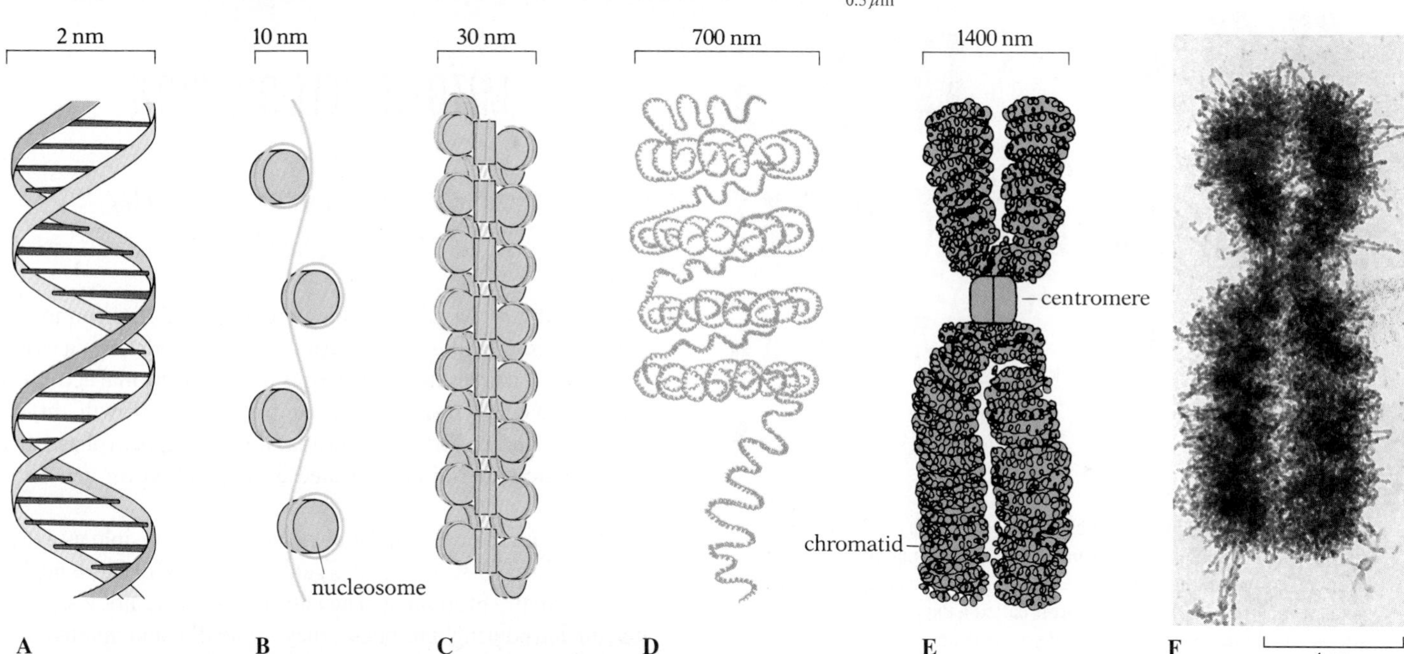

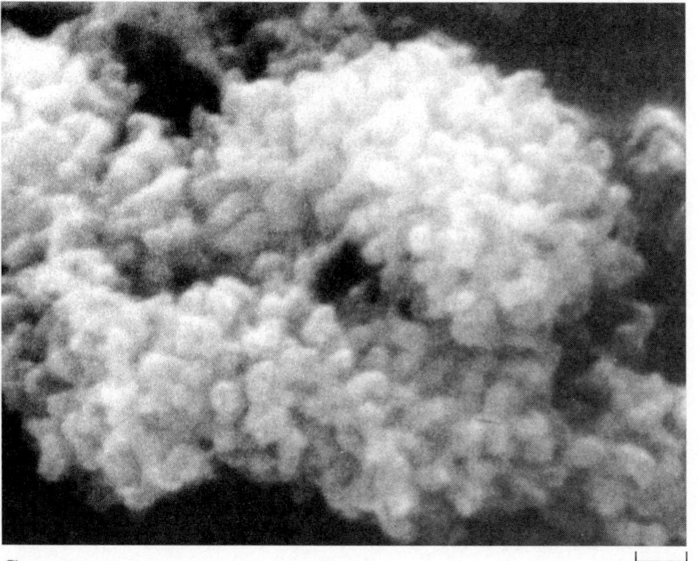

9.4 Packaging of DNA in a eucaryotic nuclear chromosome The double helical strand of DNA (A) is wound around nucleosome cores about 10 nanometers in diameter (B). This chain of spools is itself coiled in some way to produce a thick strand about 30 nanometers in diameter (C). The arrangement shown here is hypothetical. Early in cell division the thick strands are collected into a long series of loops, which are wound into a helix (D). The result is the ragged appearance of the chromosome, as seen in the drawing (E), in the electron micrograph of a human chromosome (F), and in a scanning electron micrograph of a condensed chromosome showing highly condensed loops (G).

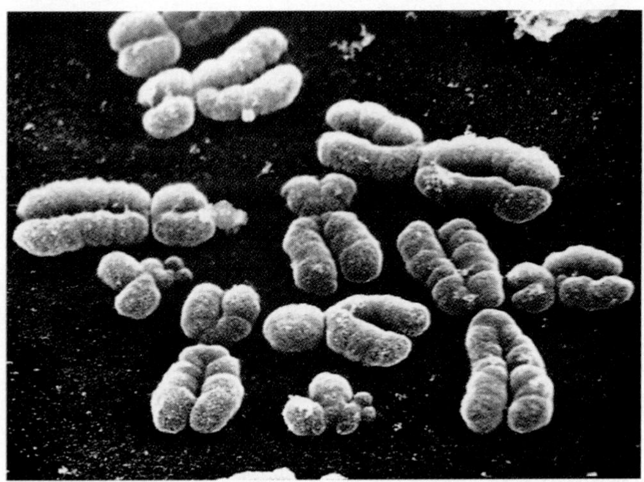

A

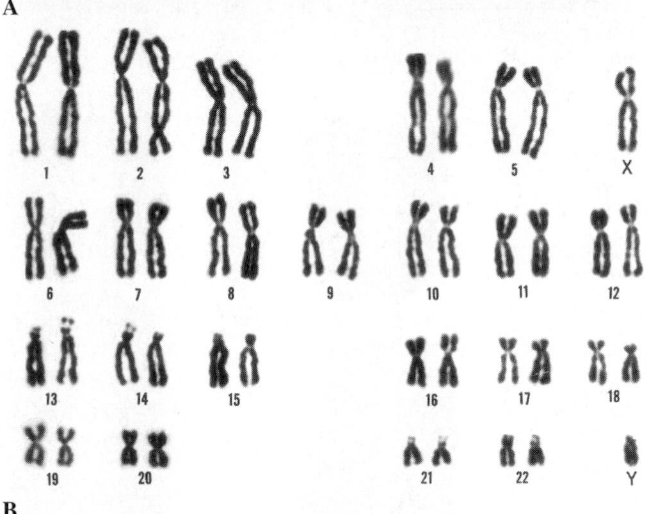

B

9.5 Photographs of human chromosomes (A) Scanning EM of human chromosomes. (B) Chromosomes of a human male are cut out, placed in homologous pairs, and arranged in a chart according to size and shape. The resulting chart is called a karyotype. A human somatic cell (any cell except the egg or sperm cells) contains 23 pairs of chromosomes, including a pair of sex chromosomes (for a male, X and Y).

mologous chromosomes of humans have been subjected to particular scrutiny, and their distinctive features—and abnormalities—are becoming familiar to researchers exploring the causes of human genetic diseases. Because eucaryotes have many more chromosomes than procaryotes, and because eucaryotic chromosomes tend to occur in pairs, eucaryotic cells have much more genetic material than procaryotic cells. For example, it is estimated that a typical human cell has about 1,000 times more DNA than a typical bacterial cell. If you were to take all the DNA present in the chromosomes of one of your cells and place it end to end, the DNA would be about two meters long.

The number of chromosomes is usually constant in individuals of the same species. For example, humans have 23 chromosomal pairs, for a total number of 46 chromosomes. (Fig. 9.5B). The fruit fly *Drosophila melanogaster* has four pairs, or eight chromosomes, and the onion eight pairs, or 16 chromosomes. Cells of this type, with two of each type of chromosome, are said to be ***diploid.*** The somatic[1] cells of most higher plants and animals, including those of the human body, are diploid. If each daughter nucleus is to get a complete set of genetic instructions, it must receive a full set of chromosomes exactly like the set initially present in the parental nucleus. Only in this way can a basic constancy of chromosome number and gene content be maintained.

There are two basic types of cell division in eucaryotes: division as part of the growth of an organism, also known as somatic or mitotic cell division, and division to produce the gametes that give rise to new individuals, also known as meiotic cell division. We will describe each of these in turn, in separate sections.

MITOTIC CELL DIVISION

Eucaryotic Cells Go Through a Complex, Multistage Cell Cycle

Most of the preparation for the process of cell division in eucaryotes goes on during the period of cell growth known as ***interphase.*** An interphase cell carries out all the normal activities of a living cell—respiration, protein synthesis, growth, differentiation, and so forth. In addition, it is during interphase that the genetic material is replicated in preparation for cell division.

In an interphase cell, the nucleus is clearly visible as a distinct membrane-surrounded organelle, and one or more nucleoli are usually prominent. The chromosomes are not visible in the nucleus during interphase; they are so thin and tangled that they cannot be recognized as separate entities. The chromosomes appear only as an irregular granular-looking mass of chromatin material (Fig. 9.6).

In all interphase animal cells (and in a few interphase plant cells), there is a special region of cytoplasm just outside the nucleus, called the ***centrosome,*** which contains a pair of small cylindrical bodies oriented at right angles to each other. These are the ***centrioles,*** mentioned in Chapter 6 (see Figs. 6.27 and 6.29). Right before cell division, the pair of centrioles within the centrosome replicates, and the two pairs of centrioles eventually move apart each taking some of the centrosomal material with them. The two centrosomes with their centrioles then move to the opposite ends of the dividing cell. Once there, the centrosomes organize the cytoskeleton to form the spindle, which is necessary for cell division.

---

[1] "Somatic," as used here, refers to all cells in the body except reproductive cells—eggs and sperm.

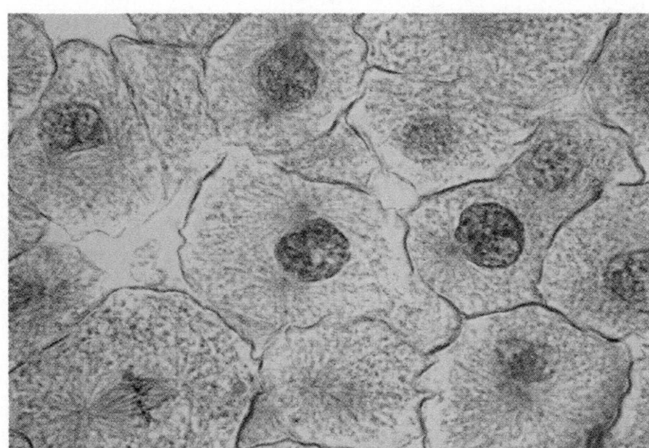

9.6 Cells of a whitefish embryo The cells in the center of this photograph are in interphase of a cell cycle. The nucleus is clearly visible as a membrane surrounded organelle, and nucleoli (dark spots) are visible within the nucleus. Notice that the chromosomes are not visible at this stage. The cell at the bottom left is in the process of dividing.

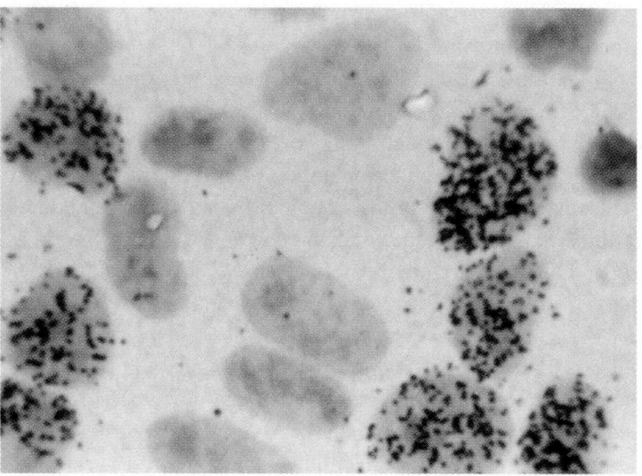

9.7 Nuclei where chromosomal replication is occurring Cells briefly exposed to a radioactive DNA precursor (^{3}H-thymidine) will take up the precursor during S stage and use it to make new DNA, which will then be radioactively labelled. A thin film of photographic emulsion placed over the preparation is used to detect the locations of the radioactive materials. The blackened areas in the photograph are cell nuclei that have incorporated the radioactive materials, indicating that these cells were in S stage.

During interphase, a newly formed cell goes through a precise sequence of events in which all the key components of the cell are doubled, including the genetic material. The cell then divides equally to produce two identical daughter cells. This cycle of growth, followed by division, is known as the ***cell cycle.*** For convenience, it is customary to partition each cycle, from one cell division to the next, into a series of stages, each designated by a special name.

Three Distinct Interphase Stages Precede the Actual Division of a Eucaryotic Cell

The replication of the genetic material during interphase does not begin immediately after completion of the previous cell division. There is a gap in time, designated the ***G*₁** (for first gap) stage, before genetic replication. This is a period of intense biosynthetic activity, during which new ribosomes, organelles, and other cellular structures are synthesized. Next comes the ***S*** (for synthesis) stage, during which new chromosomal DNA is synthesized, along with further duplication of organelles. During this stage the DNA molecule of each chromosome is precisely duplicated. The replication process, which cannot be observed with the light microscope, has been demonstrated through the use of radioactive DNA precursors (Fig. 9.7). Once duplicated, the two identical molecules remain together, connected by a length of DNA known as the ***centromere*** (Fig. 9.8). Each member of the bound pair is called a ***chromatid,*** and

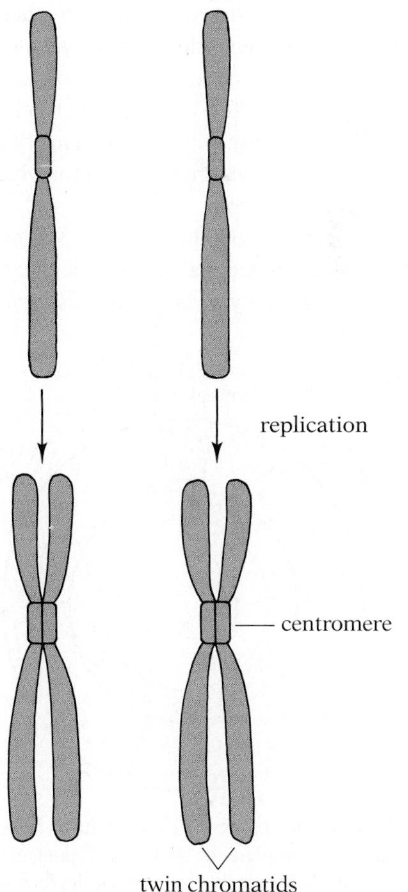

9.8 Production of chromatids through replication During S stage prior to cell division the genetic material of each chromosome duplicates itself. As a result, twin chromatids take form upon condensation, joined to each other by a centromere. For clarity, the unduplicated chromosomes are shown in condensed form, even though condensation normally takes place only after duplication.

the bound pair is referred to as a double-stranded, or twin-chromatid chromosome.

Another time gap, designated G_2, separates the end of DNA replication from the onset of actual division. During G_2 the cell continues to grow and to synthesize the materials necessary for cell division. The actual separation of the cell into two daughter cells occurs during the ***M*** stage (for mitosis) immediately following the G_2 stage. The three subdivisions of interphase (G_1, S, and G_2), together with the subsequent stage of cell division, the M stage, constitute a complete cell cycle (Fig. 9.9).

The duration of a specific cell cycle varies greatly. It usually lasts between 10 and 30 hours in plants and between 18 and 24 hours in animals, but it may be as short as 20 minutes in early embryonic development or as long as a year in an adult human liver cell. Within a cell cycle, the greatest variation by far occurs in the G_1 stage. At one extreme is the slime mold *Physarum,* in which there is no G_1 stage at all, while at the other extreme are some cell types that become "arrested" in the G_1 stage and may never divide again. Differentiated skeletal muscle cells and nerve cells, for example, are arrested in the G_1 stage and normally do not divide further. (There is evidence that the muscle cells of some weightlifters on high-resistance weightlifting programs may sometimes split. For many years it was thought that the increased size of muscles was due solely to an increase in the *diameter* of muscle cells rather than an increase in the *number* of cells. Now it appears that in some cases the fibers may split, but this splitting is thought to be due to injury to the muscle fibers rather than the normal processes of cell division.) There are a few cases in which cells are arrested in G_2 stage; the heart muscle cells of human adults are an example of cells in G_2 arrest, a phenomenon for which there is, at present, no clear explanation.

The arrest of cells in both the G_1 and G_2 stages results from a failure to produce an essential control chemical. The chemicals that control the cell cycle are known as ***cyclins.*** S-cyclin, for instance, is involved in stimulating replication (DNA synthesis), while M-cyclin helps trigger mitosis. The potential link between these control chemicals and the control failures that underlie cancer, where rapid cell growth proceeds unchecked, is being intensively studied. It is widely hoped that studies of the control of replication and mitosis will lead to new techniques for restoring quiescence to tumor cells.

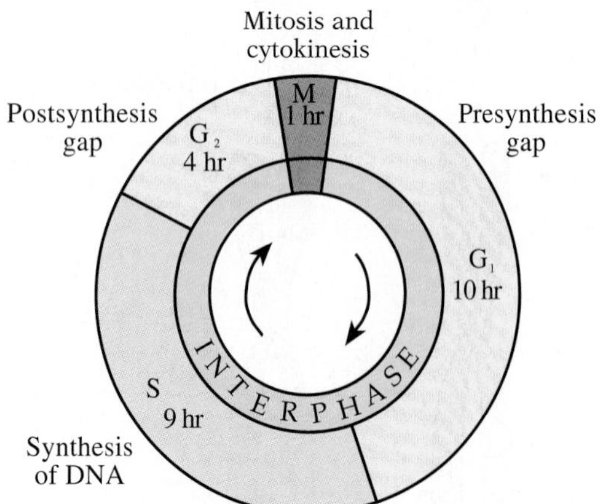

9.9 The cell cycle This particular cycle assumes a period of 24 hours, but some cells complete the cycle in less than an hour and others take many days. Similarly, the ratios of the four stages of the cycle vary, with G_1 exhibiting the most variation.

Mitosis, the Division of the Nucleus, Takes Place in Four Recognizable Phases

Cell division in eucaryotic cells involves two fairly distinct processes that often, but not always, occur together: division of the nucleus and division of the cytoplasm. The process by which the nucleus divides to produce two new nuclei, each with the same number of chromosomes as the parental nucleus, is called ***mitosis.*** The division of the cytoplasm is called ***cytokinesis.*** Mitosis is a complex process that is customarily subdivided into four phases: ***prophase, metaphase, anaphase,*** and ***telophase.*** Although each phase will be discussed separately here, it should be kept in mind that the entire process is a continuum, not a series of discrete events.

During Prophase, Chromosomes Condense, Centrosomes Separate, and Microtubules Appear

Prophase is, in a sense, a preparatory stage that readies the nucleus for the crucial separation of two complete sets of chromosomes into two daughter nuclei. As the two centrosomes with their pairs of centrioles in an animal cell move toward opposite sides of the nucleus, the initially indistinct chromosomes begin to condense into visible threads, which become progressively shorter and thicker and more easily stainable with dyes (hence the name *chromosome,* or "colored body").

Chromosomes first become visible during early prophase as long, thin, intertwined filaments, but by late prophase the individual chromosomes are more compacted and can be clearly discerned as much shorter, rodlike structures. As the chromosomes become more distinct, the nucleoli become less distinct, often disappearing altogether by the end of prophase (Fig. 9.10).

Examined under very high magnification, an individual chromosome from a late-prophase nucleus can be seen to consist of separate twin chromatids connected by their centromeres (Figs. 9.8 and 9.10). The replication that occurs during the S stage results in two identical copies of the DNA molecules of the original chromosome. Hence, the two chromatids of a prophase chromosome are genetically identical.

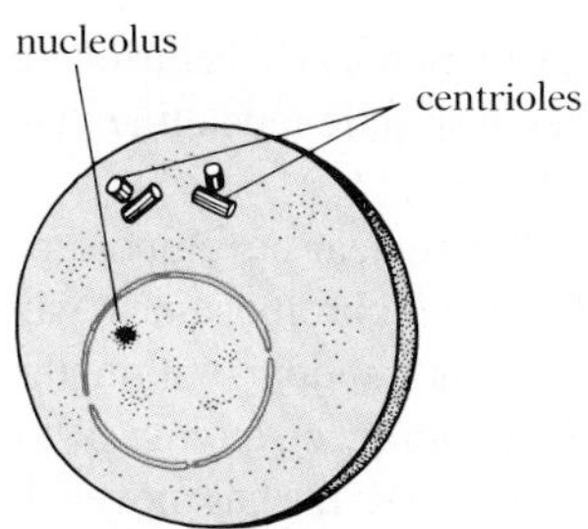

1. INTERPHASE
Chromosomes not seen as distinct structures. Nucleolus visible.

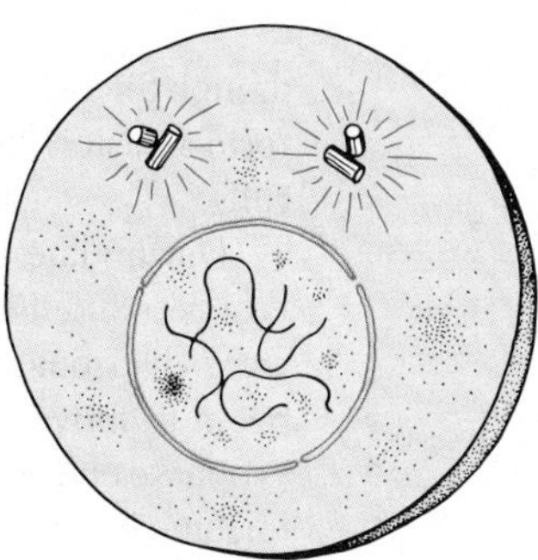

2. EARLY PROPHASE
Centrioles begin to move apart.
Chromosomes appear as long thin threads.
Nucleolus becomes less distinct.

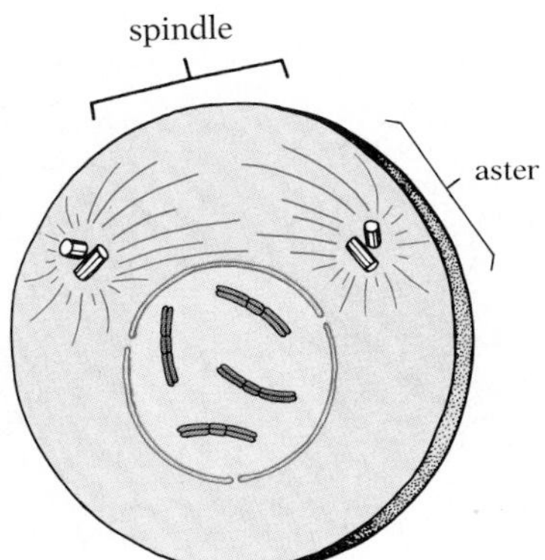

3. MIDDLE PROPHASE
Centrioles move farther apart.
Asters begin to form.
Twin chromatids become visible.

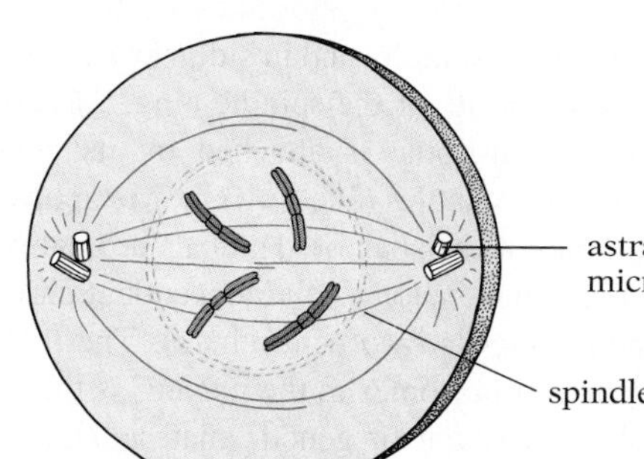

4. LATE PROPHASE
Centrioles nearly reach opposite sides of nucleus.
Spindle begins to form and kinetochore microtubules project from centromeres toward spindle poles.
Nuclear membrane is disappearing.
Nucleolus is no longer visible.

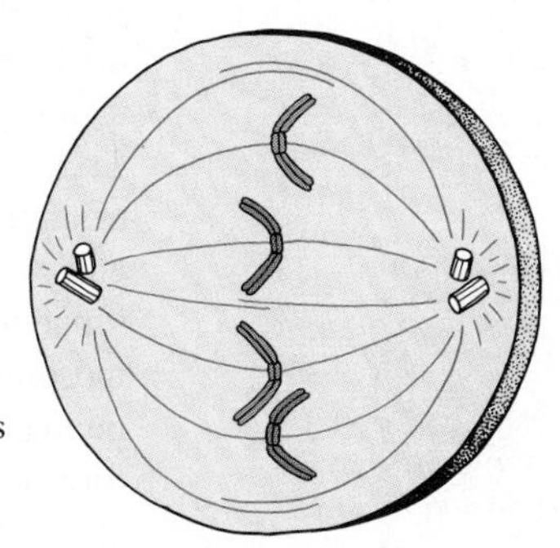

5. METAPHASE
Nuclear membrane has disappeared.
Kinetochore microtubules move each twin-chromatid chromosome to the midline; other spindle microtubules interact with spindle tubules from opposite pole.

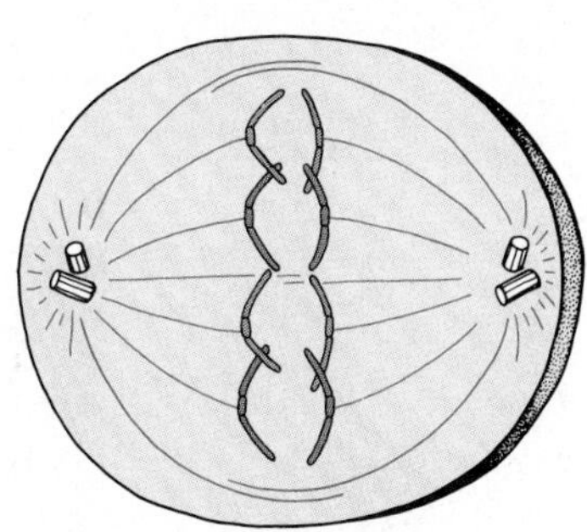

6. EARLY ANAPHASE
Centromeres have split and begun moving toward opposite poles of spindle.
The two sets of single-chromatid chromosomes are being pulled toward their respective poles by spindle microtubules.

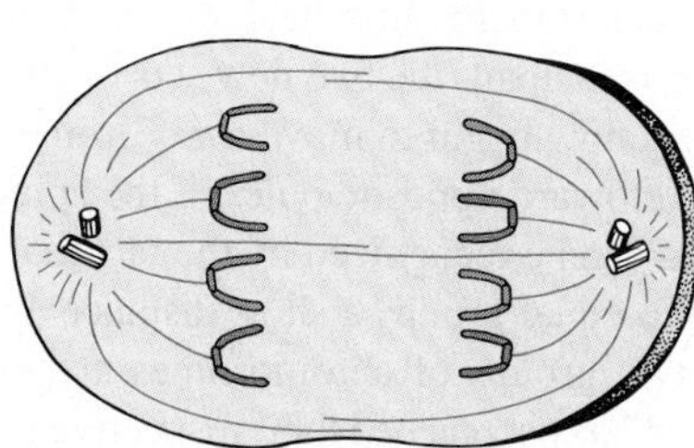

7. LATE ANAPHASE
The two sets of new single-chromatid chromosomes are nearing their respective poles.
Cytokinesis begins.

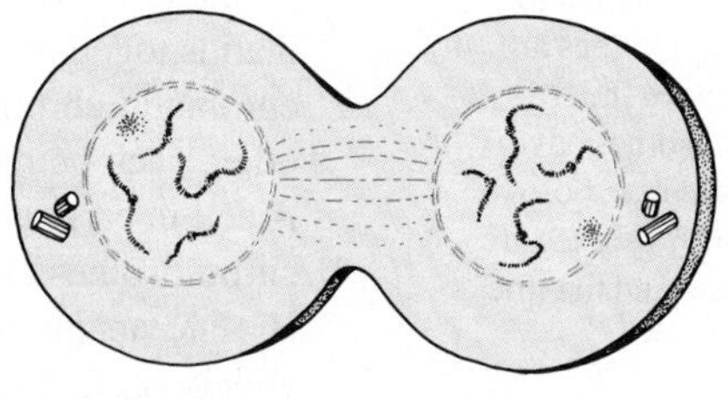

8. TELOPHASE
New nuclear membranes begin to form.
Chromosomes become longer, thinner, and less distinct.
Nucleolus reappears.
Cytokinesis is nearly complete, forming two genetically identical daughter cells.

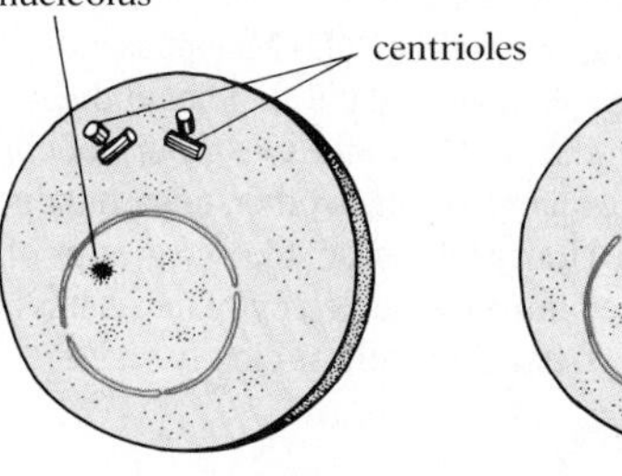

9. INTERPHASE
Cytokinesis is complete.
Nuclear membranes are complete.
Nucleolus is visible in each cell.
Chromosomes are not seen as distinct structures.
Replication of genetic material occurs during the S stage of interphase.

9.10 Mitosis and cytokinesis in an animal cell In these drawings, purple and green are used to distinguish the two pairs of homologous chromosomes.

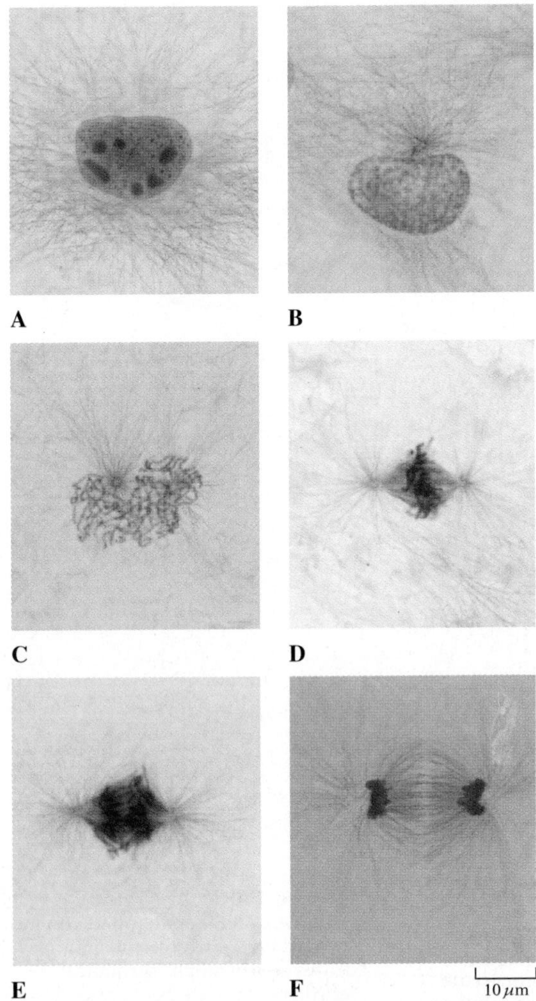

9.11 Mitosis in a cell from the endothelium of a frog's heart (A) Interphase: a diffuse network of microtubules (stained red) is visible. (B) Early prophase: the chromosomes begin to condense; the centrioles have not yet separated. (C) Late prophase: prophase ends as the nuclear envelope disappears; the centrioles have moved apart, and the two asters are prominent. (D) Metaphase: the chromosomes are lined up along the equator of the spindle, midway between the two asters. (E) Middle anaphase: the two groups of chromosomes have begun to move apart from each other, toward their respective poles of the spindle. (F) Early telophase: the two groups of chromosomes are being organized into new nuclei; cytokinesis has begun, and the line dividing the two daughter cells is clearly visible.

As the two centrosomes of an animal cell move apart, the microtubular cytoskeleton present in the cell during interphase breaks down, and a completely different system of ***microtubules,*** the mitotic ***spindle,*** begins to form between each pair of centrioles (Figs. 9.10:4 and 9.11). These microtubules play an important role in the movement of chromosomes during mitosis. The microtubules are divided into two groups, the ***spindle microtubules,*** which form a basketlike structure between the two centrioles, and the outwardly directed ***astral microtubules*** (collectively referred to as an ***aster***), which radiate in all other directions from each centriole pair (Fig. 9.10:4). Even though the cells of most seed plants lack centrioles and do not form an aster, they develop a spindle much like that in animal cells.

During late prophase, the nuclear envelope separates into vesicles, and the twin-chromatid chromosomes, which at first were distributed essentially at random within the nucleus, begin to move toward the ***equator,*** or middle of the spindle. Prophase ends when each chromosome has moved to the equator, attached by their centromeres to a system of spindle microtubules.

During Metaphase, the Chromosomes Are Aligned Along the Spindle's Equator

In the stage termed metaphase, the chromosomes are arranged on the equatorial plane of the spindle, and in side view appear to form a line across the middle of the spindle (Fig. 9.11D). Each twin-chromatid chromosome is attached by its centromeres to the spindle microtubules (Fig. 9.10:5). Metaphase ends when the centromeres of each pair of identical chromatids separate. Each chromatid then becomes an independent chromosome (that is, one having its own centromere). The total number of independent chromosomes in the nucleus is therefore doubled, though the amount of genetic material is unchanged.

During Anaphase, the Duplicate Chromosomes Move to Opposite Poles of the Spindle

Now begins the separation of two complete sets of chromosomes, the critical event for which the previous stages were the preparation. Once the centromeres that held the twin chromatids together have broken apart, the two new sets of single-chromatid chromosomes now begin to move away from each other, each being pulled toward opposite poles of the spindle. A cell in early anaphase can be recognized by the fact that the chromosomes are in two equal groups a short distance apart. By late anaphase the two groups of chromosomes are more widely separated, having almost reached their respective poles of the spindle (Figs. 9.10:7 and 9.11E). Cytokinesis (division of the cytoplasm) often begins during late anaphase.

During Telophase, the Separated Chromosomes Are Incorporated Into Two New Nuclei

Telophase is essentially a reversal of prophase. The two sets of chromosomes, having reached their respective poles, become enclosed in a reassembled nuclear membrane as the spindle disappears (Fig. 9.11 F). The chromosomes then begin to uncoil and to resume their interphase form, while the nucleoli slowly reappear (Fig. 9.10:8). Telophase ends when the new nuclei have fully assumed the characteristics of interphase, thus bringing to a close the complete mitotic process.

Note that when the individual chromosomes fade from view at the end of telophase they consist of a single chromatid, but when they reappear during prophase in the next cell cycle they are twin-chromatid chromosomes. The replication, as we have seen, occurs during the S stage of interphase (Fig. 9.9). What was a single nucleus containing one set of twin-chromatid chromosomes in prophase is now two nuclei, each with its own set of single-chromatid chromosomes. In short, *mitosis results in two daughter cells that are genetically identical to each other, and to the parental cell from which they came.*

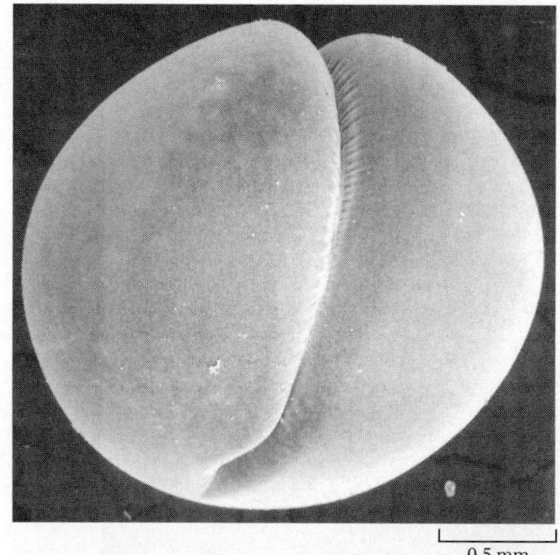

9.13 Scanning electron micrograph of a dividing frog egg Top: The cleavage furrow is not yet complete (see bottom of cell). Bottom: Note the puckered stress lines in the furrow.

Cytokinesis Differs in Plant and Animal Cells

The division of the cytoplasm that often accompanies division of the nucleus usually begins in late anaphase, reaching completion during telophase. Mitosis can occur without cytokinesis, however. For example, unaccompanied mitosis is common in fungi, producing bodies with many nuclei but few, if any, cellular partitions (Fig. 9.12). It is also observed during certain phases of reproduction in seed plants and certain other land plants, and it is common in some invertebrate animals.

In animal cells, cytokinesis normally begins with the formation of a ***cleavage furrow*** running around the cell (Fig. 9.13). When cytokinesis occurs during mitosis, the furrow usually forms in the equatorial region of the spindle (Fig. 9.10:8). The furrow becomes progressively deeper, until it cuts completely through the cell (and its spindle), producing two new cells. Very little is known about how the cleavage furrow forms, but recent evidence suggests that the process involves the contraction of a ring of actin and myosin filaments lying underneath the furrow.

Since plant cells have comparatively rigid cell walls, which cannot develop cleavage furrows, it is not surprising that cytokinesis should differ in plant and animal cells. In many algae, new plasma membrane and wall grow inward at the midline of the wall, until the growing edges meet and completely separate the daughter cells (Fig. 9.14). In higher plants a new cross-

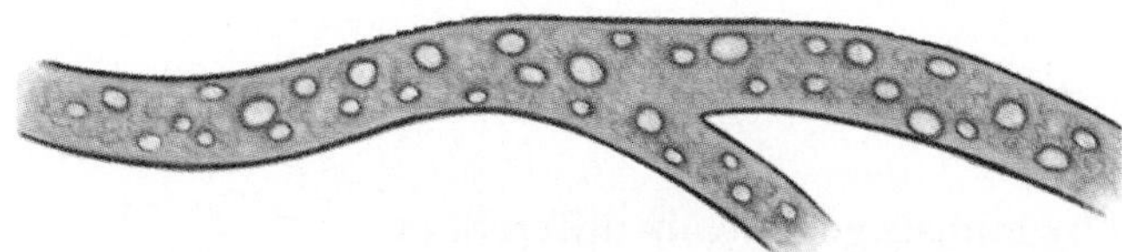

9.12 Mitosis with cytokinesis in a fungus In many fungi nuclei are continuously produced by mitosis but the cytoplasm is not divided by cross walls into separate cells or compartments. Thus, the threadlike fungal body is multinucleate (coenocytic).

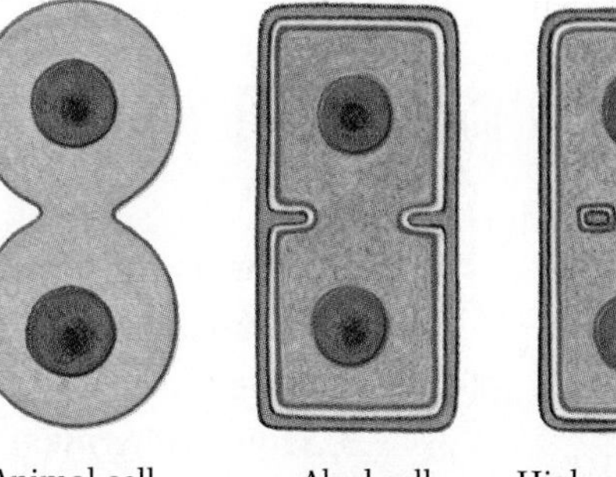
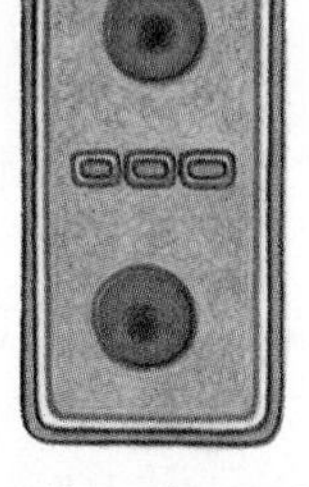

9.14 Three mechanisms of cytokinesis Cytokinesis in animal cells typically occurs by a pinching-in of the plasma membrane. In many algal cells cytokinesis occurs by an inward growth of new wall and membrane. In higher plants cytokinesis typically begins in the middle and proceeds toward the periphery, as membranous vesicles fuse to form the cell plate.

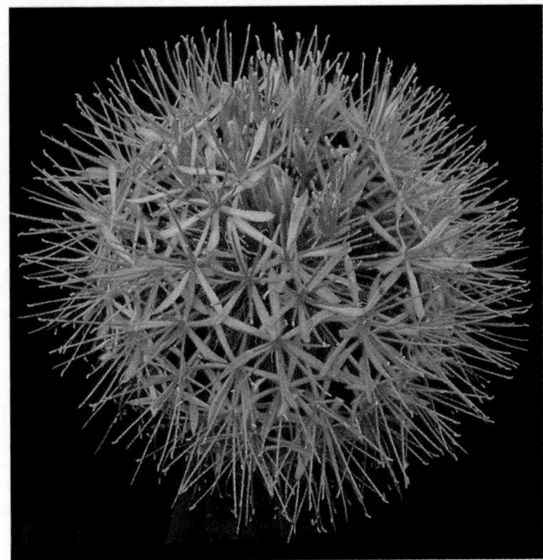

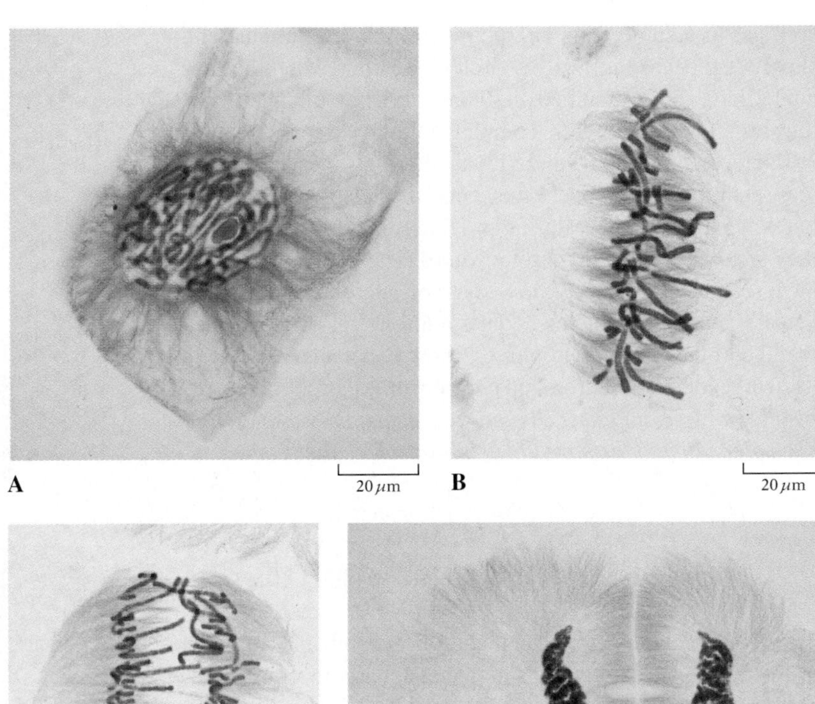

9.15 Cell division in a plant, the African blood lily (above) (A) Prophase: the chromosomes have condensed; microtubules are visible. (B) Metaphase. (C) Anaphase: the two groups of chromosomes are moving to opposite poles of the cell. (D) Telophase: a cell plate has begun to form.

wall, called the ***cell plate,*** forms halfway between the two nuclei—at the equator of the spindle if cytokinesis accompanies mitosis (Fig. 9.15D). The cell plate begins to form in the center of the cytoplasm and slowly becomes larger, until its edges reach the outer surface of the cell and the cell's contents are cut in two. In general, then, cytokinesis in plant cells progresses from the middle to the periphery, whereas in animal cells it progresses from the periphery to the middle.

The cell plate forms from membranous vesicles, derived from the Golgi apparatus and endoplasmic reticulum, that are carried to the site of plate formation by the microtubules that remain after mitosis. There they line up and then unite (Fig. 9.16). When the membranes of the vesicles have fused with one another and, peripherally, with the old plasma membrane, they form the partitioning membranes of the two newly formed daughter cells. The contents of the vesicles give rise to both the

MAKING CONNECTIONS

1. How do the chromosomes of procaryotes and eucaryotes differ?

2. How could you use the DNA content of a cell to determine what stage of the cell cycle it is in?

3. How is chemical control of the events of the cell cycle linked to the medical research effort to find a cure for cancer?

4. Are the daughter cells produced by mitosis genetically different or identical? How does their genetic make-up compare with that of the parent cell?

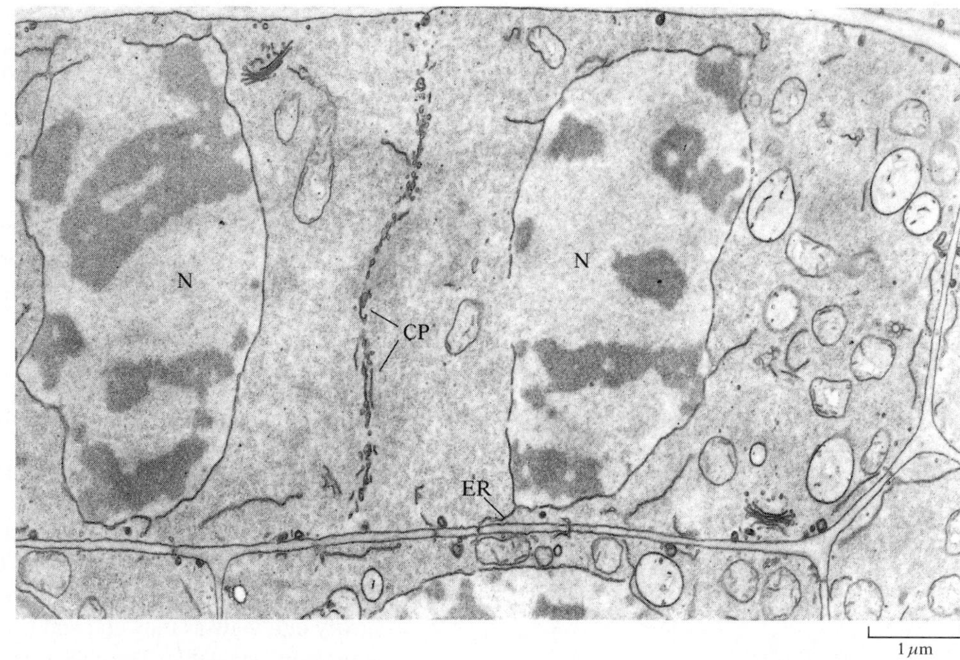

9.16 Electron micrograph of a late-telophase cell in corn root, showing formation of cell plate Mitosis has been completed, and the two new nuclei (N) are being formed; the chromosomes (dark areas in the nuclei) are no longer visible as distinct structures, but the nuclear membranes are not yet complete. A cell plate (CP) is being assembled from numerous small vesicular structures. At the lower end of the nucleus on the right, a length of endoplasmic reticulum (ER) can be seen that appears to run from the nuclear membrane to the cell wall, through the wall, and into the adjacent cell.

middle lamella and the beginnings of primary cell walls. Parts of the endoplasmic reticulum are often caught in the newly forming wall, giving rise to the plasmodesmata.

MEIOTIC CELL DIVISION

In Sexual Reproduction, Cell Division Takes a Specialized Form Known as Meiosis

As we have seen, mitosis serves to maintain a constant number of chromosomes: the two daughter cells produced by mitosis are genetically identical to each other and to the parent cell. What would the effect be if reproductive cells were produced by mitosis? In sexual reproduction, two specialized cells (Fig. 9.17) known as ***gametes*** (egg and sperm) unite to form the ***zygote,*** or first cell, of the new individual. If in human beings or in the hypothetical four-chromosome organisms of Figure 9.10, the two gametes were produced by normal mitosis, the zygote produced by their union would have double the normal number of chromosomes, and at each successive generation the number would again double, until the total chromosome number per cell approached infinity—a clear impossibility.

In actuality, the chromosome number normally remains constant within a species. At some point, therefore, a different kind of cell division must take place, one that reduces the number of chromosomes by half, so that when the egg and sperm cells unite in fertilization, the normal diploid number is re-

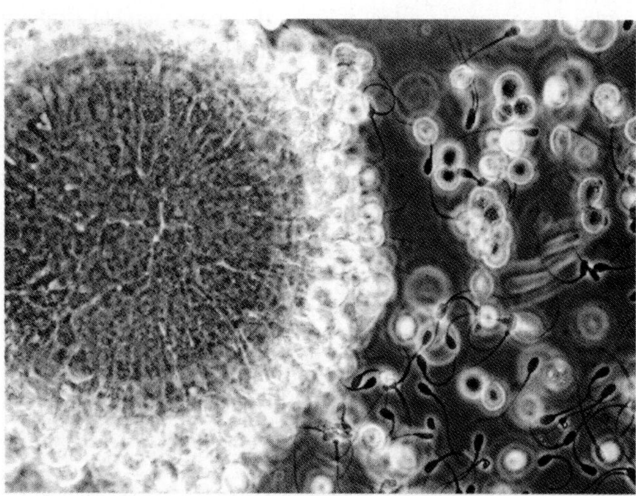

9.17 Human egg and sperm The human egg (left) is enclosed in a protective layer of cells. The tiny, motile sperm (right) must swim to the egg. Sperm contains an enzyme that loosens the cells of the egg's protective covering so the sperm can fertilize the egg.

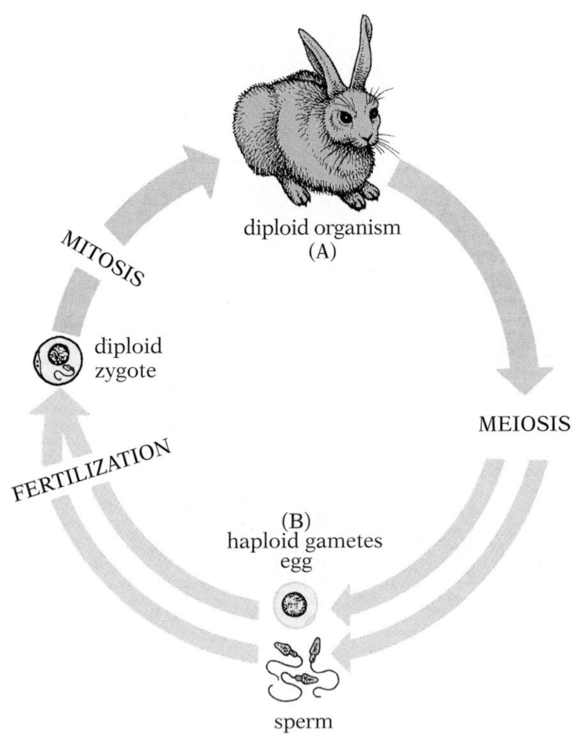

9.18 The sexual cycle Adult animals (A) are diploid and produce haploid gametes (egg and sperm, B) by meiosis. When a sperm fertilizes an egg, the resulting diploid zygote divides by mitosis to produce the large number of cells of the adult organism.

stored. This special form of cell division is called ***meiosis*** (from the Greek word for "diminution").

In all multicellular animals meiosis occurs at the time of gamete production. Consequently, each gamete has only half the number of chromosomes typical of a particular species. It is important to note that in the ***reduction division*** of meiosis the chromosomes of the parental cell are not simply separated into two random halves. Since the diploid nucleus contains two of each type of chromosome (a homologous pair), meiosis divides these chromosome pairs so that each gamete contains one of each type of chromosome. Such a cell, with only one of each type of chromosome, is said to be ***haploid.*** When two haploid gametes unite in fertilization, the resulting zygote is diploid (Fig. 9.18), having received one of each type of chromosome from the sperm of the male parent and one of each type from the egg of the female parent. The symbol commonly used for the diploid number is **2n,** and that used for the haploid number, **n.** In humans, for instance, $2n = 46$ and $n = 23$.

Meiosis Proceeds Through Two Successive, Four-Phase Division Sequences

Complete meiosis involves two successive division sequences, resulting in a total of four new haploid cells. The first division sequence ***(meiosis I)*** accomplishes the reduction in the number of chromosomes. The second ***(meiosis II)*** is similar to mitosis and separates chromatids. The same four phases as in mitosis—namely prophase, metaphase, anaphase, and telophase—are recognized in each division sequence.

During Prophase I, Homologous Chromosomes Synapse

Many of the events in prophase I of meiosis resemble those in the corresponding phase of mitosis. The centrosomes move to opposite poles of the cell and individual chromosomes come slowly into view as they coil and become shorter, thicker, and more easily stainable. The nucleoli slowly fade from view, the nuclear membrane separates into vesicles, and the spindle is organized. Radioactive-tracer studies show that the replication of the genetic material occurs during the S stage of interphase that precedes prophase I, as in mitosis. However, there are important differences between the prophases of meiosis I and mitosis.

The chief difference between the two prophases is that in meiosis the members of each pair of homologous chromosomes move together and come to lie side by side (Fig. 9.19:2). Also, instead of being fastened only by centromeres, as in mitosis, the twin chromatids of each chromosome are held together in meiosis by a pair of long, thin protein axes that run along their entire length (Fig. 9.20). As a result, the two chromatids are held together so tightly as to be indistinguishable. The DNA is gathered into loops as part of the condensation process. The protein axes of the two homologous chromosomes now join by means of protein cross-bridges to form a compound structure, which lines up the four chromatids—often referred to as a ***tetrad***—in perfect alignment (Fig. 9.19:3). This pairing is known as ***synapsis.*** *The process of synapsis is unique to meiosis.* (The homologous chromosomes remain independent in mitosis.)

At this point, an important event known as ***crossing over*** begins. Large protein complexes (recombination nodules) appear along the ladderlike cross-bridges (Fig. 9.20). These molecular structures probably determine the locations at which genetic material will be exchanged between homologous chromosomes. The number of complexes depends on the species and the length of the chromosome; each human chromosome during synapsis has an average of three. Next begins a process by which two of the chromatids, one from each side of the homologous chromosomes, are clipped open at precisely the same place, and the resulting fragments are spliced together

1. EARLY PROPHASE

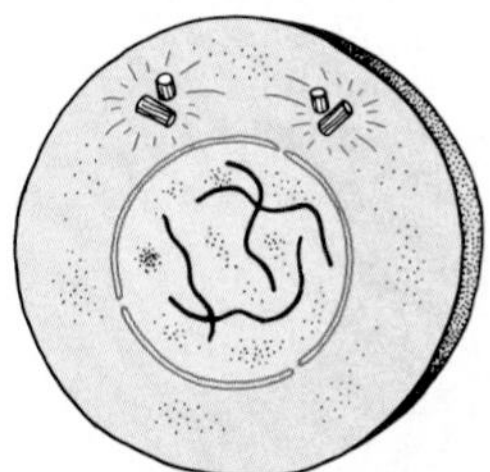

Chromosomes become visible as long, well-separated filaments; replication has already occured.

2. MIDDLE PROPHASE I

Homologus chromosomes become shorter and thicker, and synapse; crossing over takes place.

3. LATE PROPHASE I

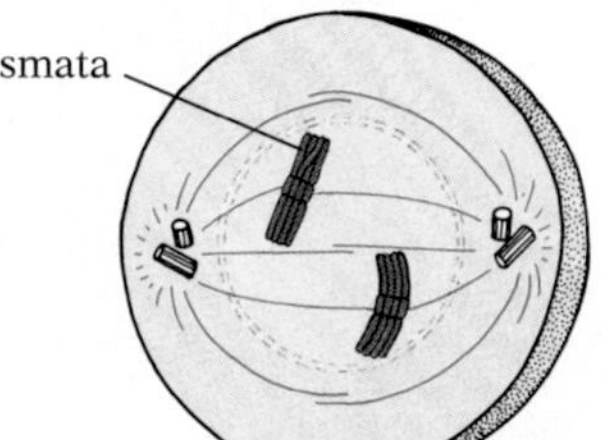

The tetrad structure of the synapsed chromosomes, and the chiasmata created by crossing over, become visible.
Nuclear membrane begins to disappear.
Spindle microtubules arise from the centromeres.

4. METAPHASE I

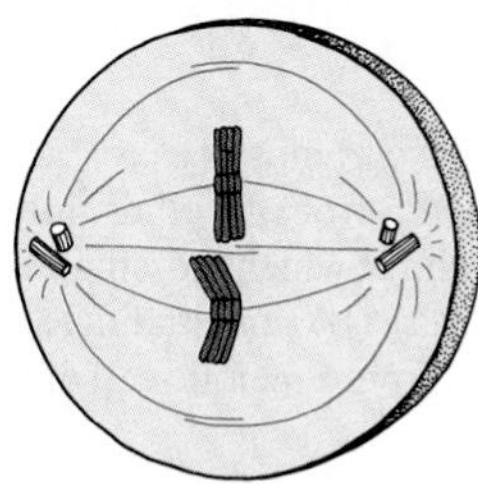

Each chromosomal pair moves to the equator of the spindle as a unit.

5. ANAPHASE I

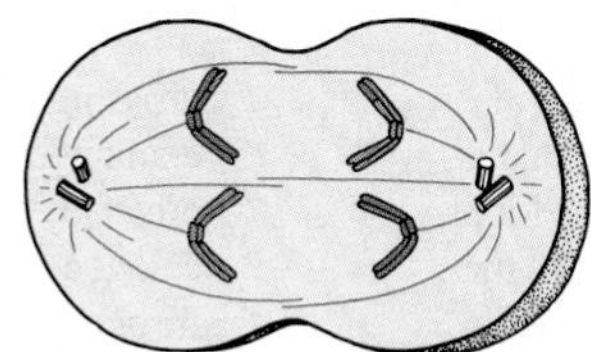

Centromeres do not split.
Twin-chromatid chromosomes move apart to opposite poles.

6. TELOPHASE I

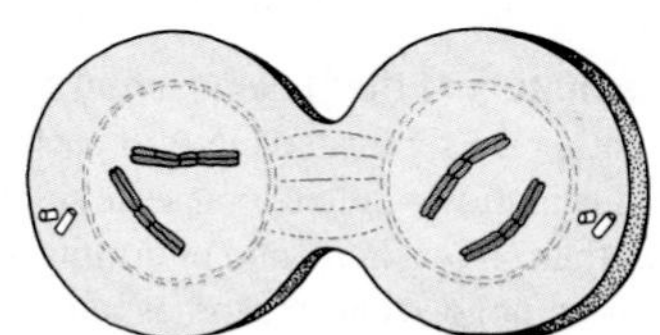

New haploid nuclei form.
Chromosomes fade from view.

7. INTERKINSESIS

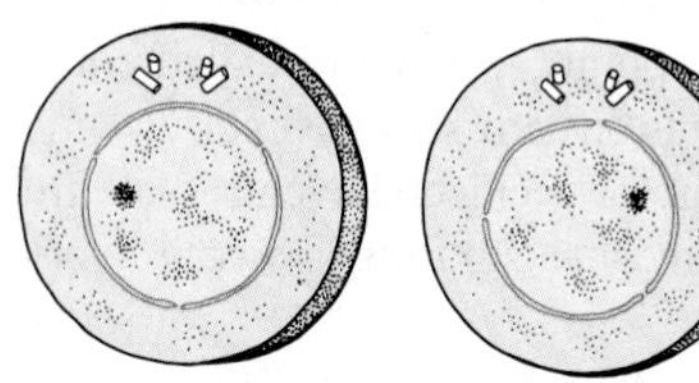

No replication of genetic material occurs.

8. PROPHASE II

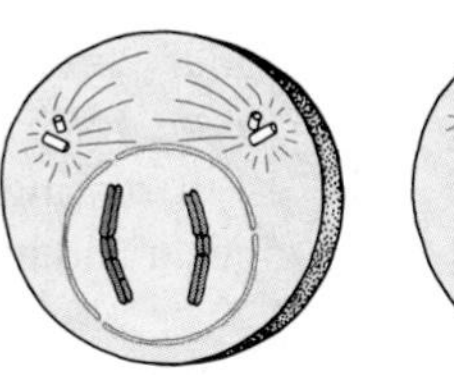

9. METAPHASE II

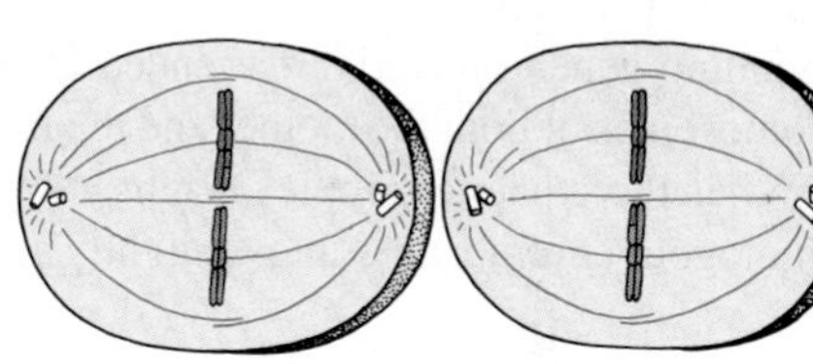

10. ANAPHASE II

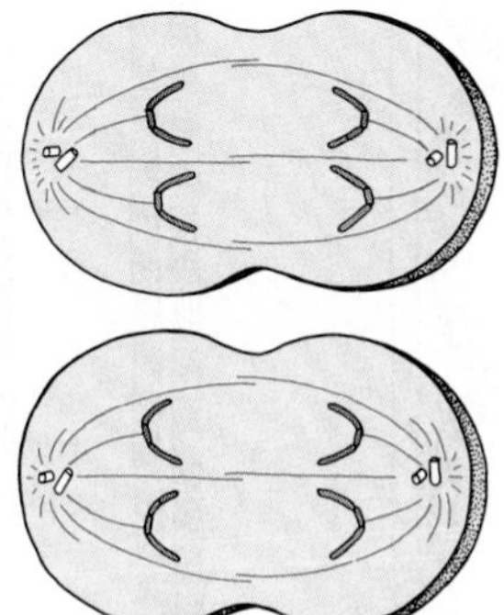

11. TELOPHASE II

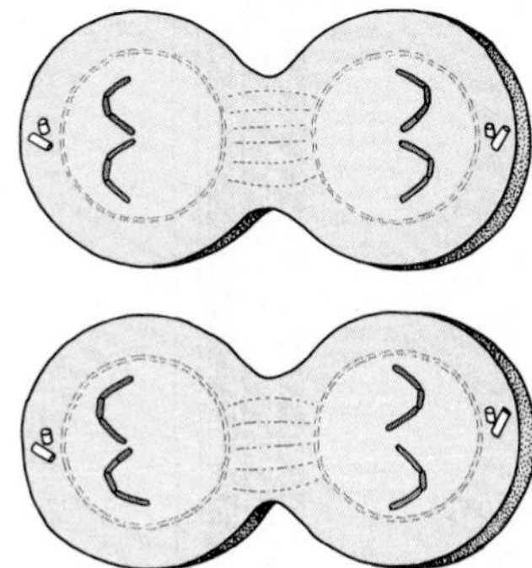

12. INTERPHASE

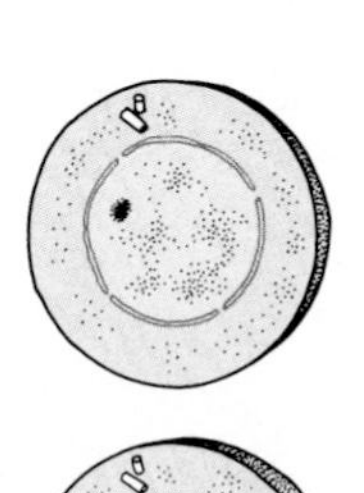
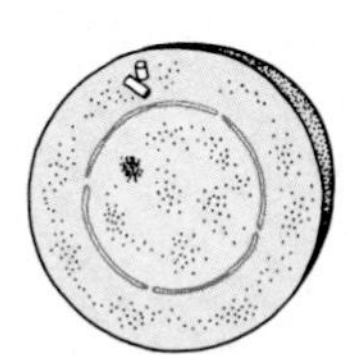
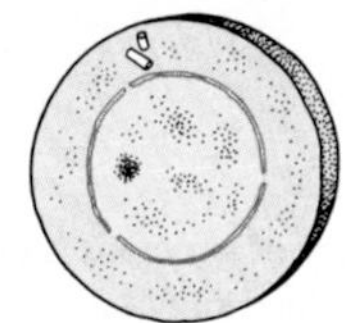
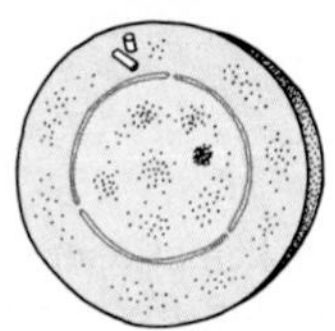

9.19 Meiosis in an animal cell In these diagrams, the members of each pair of homologous chromosomes are shown in different colors to aid in visualizing the results of crossing over.

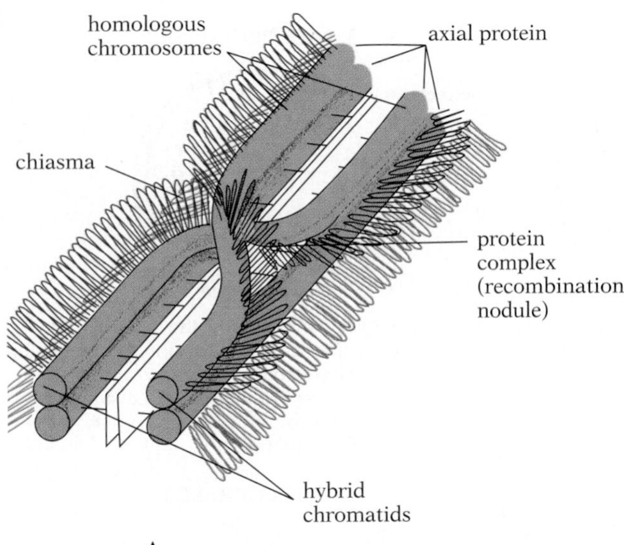

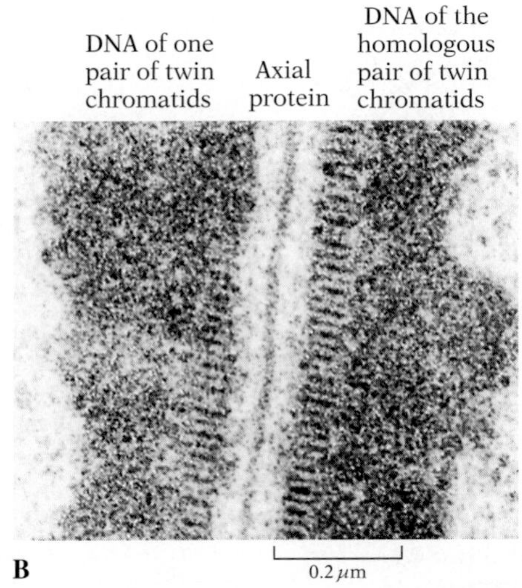

9.20 Formation of the chromosomal synapse (A) The axial proteins gather the replicated DNA of each chromosome—that is, the twin chromatids—into a long series of paired loops. The loops are longer than shown here, and intermingle. The protein axes then bring the homologous chromosomes into alignment. When synapsis is complete, protein complexes (recombination nodules; not shown) appear, which mediate the process known as crossing over. During crossing over chromatids from one chromosome cross over and exchange fragments with chromatids from the other chromosome to create hybrid chromatids. The point at which crossing over occurs is called a chiasma (plural, chiasmata). (B) Longitudinal section through a synaptonemal complex from a meiotic cell of the ascomycete mold *Neottiella.*

(Fig. 9.21). The important result of this cutting and splicing is that the chromatids of the tetrad no longer form two sets of identical twins. The recombined chromatids are now ***hybrids,*** containing genetic material descended from the homologous chromosomes of both the mother and the father.

When the synaptic complex begins to break up and the two homologous chromosomes move slightly apart in a later part of prophase I, the points at which crossing over has taken place begin to become visible (Figs. 9.21 and 9.22). The hybrid chromatids produced by crossing over link the two homologous chromosomes at these points, which are called ***chiasmata.*** Each chiasma represents one crossover event.

Crossing over is not a rare or accidental event; it is a frequent and highly organized mechanism for generating hybrid

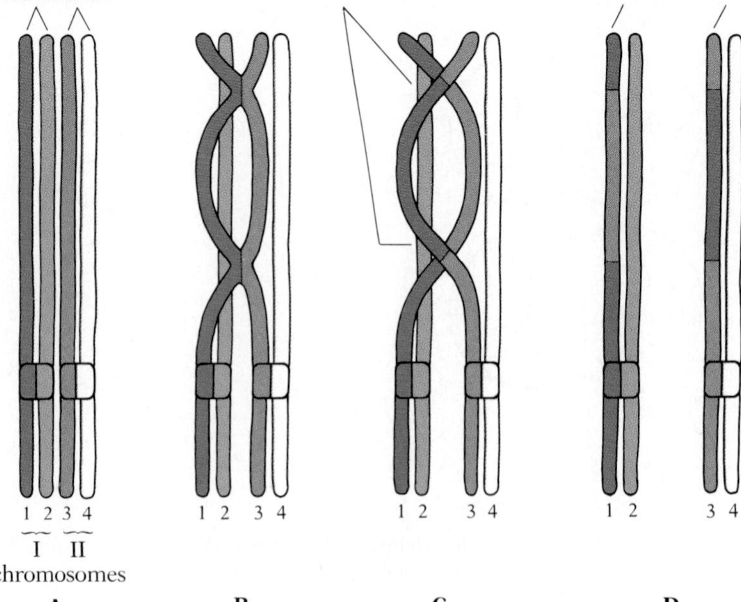

9.21 Schematic summary of synapsis and crossing over (A) Crossing over begins as homologous chromosomes (I and II), each consisting of twin chromatids (1,2 and 3,4), are precisely aligned and synapse. (B) Parts of separate chromatids are spliced together, while the protein axis (not shown) keeps each segment firmly attached to its twin. (C) When the complex breaks up, the homologous chromosomes begin to drift apart, but the chiasmata of spliced hybrid chromatids prevent them from separating. (D) The homologous chromosomes separate fully during anaphase.

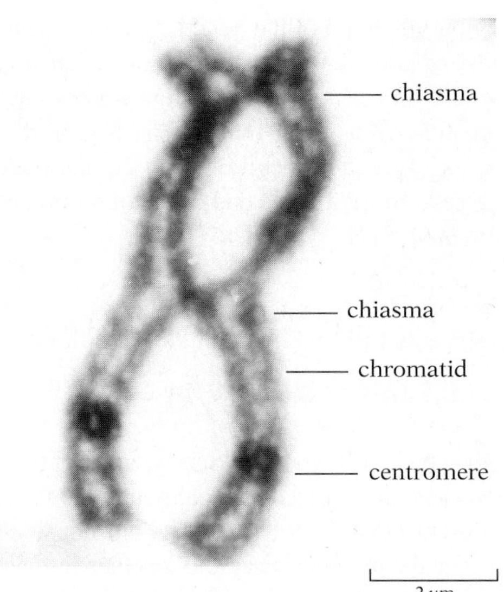

9.22 Photomicrograph showing two chiasmata between homologous chromosomes Each chromosome is clearly recognizable as a pair of chromatids. The centromeres, too, are clearly visible. Crossing over has taken place at two points—the chiasmata. Note that the crossing over involves only one chromatid from each chromosome. These chromosomes, of a salamander from Costa Rica, are from a spermatocyte in late prophase of meiosis I.

chromatids. More to the point, by *increasing the number of different genetic combinations that can be produced, crossing over contributes directly to the diversity of life.*

The gradual disintegration of the protein axis that links the chromatids from the beginning of prophase I allows the individual chromatids to separate. As prophase I ends, a pair of chromosomes, each made up of two chromatids, again becomes visible. Most chromosomes now consist of two hybrid chromatids rather than twin chromatids, since genetic material has been exchanged between the chromatids of homologous chromosomes. Finally, in late prophase I, spindle microtubules appear, radiating from the centrioles at the two poles of the cells. Toward the end of prophase I each pair of homologous chromosomes moves as a unit to the equator of the spindle.

During Metaphase I, Homologous Chromosomes Line up in Pairs on the Spindle

Recall that in mitosis, each chromosome consists of a pair of identical chromatids, and each chromosome moves independently of the other chromosomes to the midline of the cell. The homologous chromosomes do not synapse, and each chromosome moves on its own to a separate set of spindle microtubules. Hence, at metaphase of mitosis in the hypothetical organism of Figure 9.10, each of the four original chromosomes occupies a different set of microtubules. Similarly, each of the 46 double-stranded chromosomes in a mitotically dividing human cell occupies a different set of microtubules.

In meiosis I, in contrast, the homologous chromosomes synapse, as we have seen, and move onto the spindle as a unit. The chiasmata serve to hold the homologous pair together and are necessary for positioning it properly on the equator of the spindle. At metaphase I of meiosis, then, only two sets of spindle microtubules are occupied (Fig. 9.19:4). An entire synaptic pair (two chromosomes, four hybrid chromatids) is attached to each of these two sets of microtubules.

During Anaphase I, the Paired Homologous Chromosomes Separate

In mitosis, you will recall, metaphase ends and anaphase begins when the coupled centromeres of each twin-chromatid chromosome separate, and the two independent single-chromatid chromosomes thus formed move away from each other toward opposite poles of the spindle (Fig. 9.23). In meiosis, however, the separation occurs instead between the two ho-

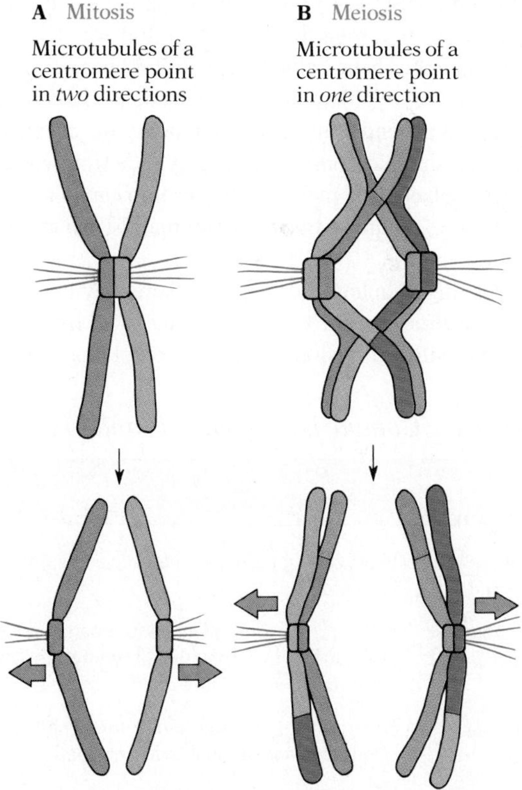

9.23 Mitosis and meiosis (A) In mitosis, each centromere sends microtubules toward both poles. As a consequence, the two chromatids are pulled apart and wind up at opposite ends of the cell during anaphase. (B) In meiosis I, however, a centromere sends microtubules toward only one of the two poles, with the homologous centromere sending microtubules toward the other pole. The result is that in anaphase I the chromatids remain joined, and homologous two-chromatid chromosomes wind up in separate cells.

mologous chromosomes that have been joined since the middle of prophase I (Fig. 9.23). Since each chromosome has its own centromere, there is no separation of centromeres. The twin-chromatid chromosomes of each homologous pair move away from each other toward opposite poles during anaphase I. It follows that in our hypothetical organism, two twin-chromatid chromosomes, each with two hybrid chromatids, move to each pole, in contrast to the four single-chromatid chromosomes that move to each pole in mitosis (Fig. 9.23). Notice, however, that because the synaptic pairing was not random, but involved two homologous chromosomes of each type, the two daughter nuclei get not just any two chromosomes, but rather one of each of the two types (Fig. 9.19:6).

By the End of Telophase I, the Chromosome Number Has Been Halved

Telophase I of meiosis is essentially the same as telophase of mitosis, except that each of the two new nuclei formed in mitosis has the same number of chromosomes as the parental nucleus, whereas each of the new nuclei formed in meiosis I has *half* the chromosomes present in the parental nucleus (Fig. 9.19:6). Meiosis I, then, is a reduction division; the chromosome number has been halved. Each diploid parent cell has divided to form two haploid daughter cells, which are not identical. At the end of telophase of mitosis, the chromosomes consist of a single chromatid when they fade from view; at the end of telophase I of meiosis, the chromosomes are double-stranded, i.e., composed of twin chromatids, when they fade from view.

Following telophase I of meiosis, there may be a short period called ***interkinesis,*** which is similar to an interphase between two mitotic division sequences except that there is no S stage and no replication of the genetic material; hence no new chromatids are formed. Replication is unnecessary in this stage of the meiotic process, since the chromosomes already have two chromatids each when interkinesis begins. Interkinesis may be so brief as to be negligible; in human males for instance, the cells formed at the end of meiosis I proceed immediately into meiosis II.

In the Second Division Sequence of Meiosis, the Twin Chromatids Are Separated

Meiosis II, which follows interkinesis, is essentially mitotic from the standpoint of mechanics, although the functional result is different (Fig. 9.19:8–12). The chromosomes do not synapse, since the nucleus is haploid and there are no homologous chromosomes. Each twin-chromatid chromosome moves to the midline and onto the spindle independently. At the end of metaphase II, the centromeres separate, and the new single-chromatid chromosomes thus formed move away from each other toward opposite poles of the spindle during anaphase II. The new nuclei formed during telophase II are haploid, like the nuclei formed during telophase I, but their chromosomes consist of a single cromatid rather than twin-chromatids (Fig. 9.24).

In summary, then, the first meiotic division produces two haploid cells containing twin-chromatid chromosomes. Each of these cells divides in the second meiotic division. Thus, a total of four new haploid cells containing single-chromatid chromosomes are produced. *Because of crossing over and the reduction division in meiosis I, the four cells are not genetically identical to each other or to the parent cell from which they arose.* Mitosis and meiosis are compared in Figure 9.25 and Table 9.1.

TABLE 9.1 *Comparison of mitosis and first division sequence of meiosis*

Phase	*Mitosis*	*Meiosis I*
Prophase	No synapsis; chromosomes move to spindle individually	Synapsis; homologous chromosomes move to spindle in pairs (tetrads)
Metaphase	Each chromosome attached to separate microtubules; centromeres in chromatids of twin-chromatid chromosomes uncouple	The two chromosomes of each homologous pair attached to the same microtubules; centromeres in chromatids of twin-chromatid chromosomes do not uncouple
Anaphase	Separation of new single-chromatid chromosomes derived from one original twin-chromatid chromosome	Separation of old twin-chromatid chromosomes of each synaptic pair
Telophase	Formation of two new nuclei, each with same number of chromosomes as parental nucleus; chromosomes are single-chromatids when they fade from view	Formation of two new nuclei, each with half the chromosomes present in parental nucleus; chromosomes still twin-chromatids when they fade from view
Interphase, Interkinesis	Replication of genetic material, with formation of new chromatids	No replication of genetic material and hence no new chromatids

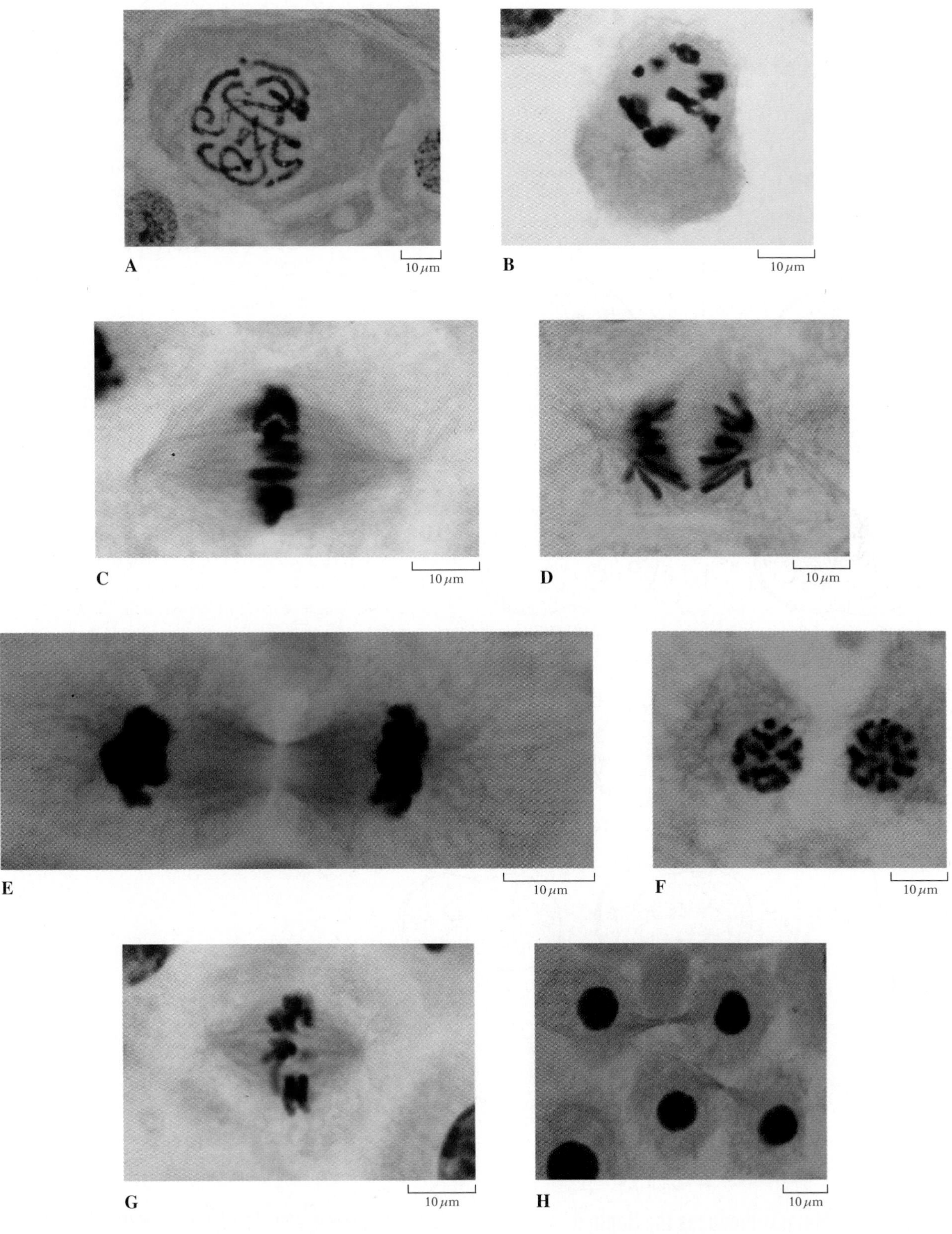

9.24 Meiosis in a cell from the grasshopper *Mongolotetix japonicus* (A) Early prophase I: the chromosomes are seen as long filaments. (B) Middle prophase I: synapsing of homologous chromosomes takes place. (C) Metaphase I: chromosomal pairs line up at the equator. (D) Anaphase I: homologous chromosomes have separated and are moving to opposite poles. (E) Telophase I: division into two haploid nuclei has begun. (F) Prophase II: early in this stage, separate chromosomes are barely distinguishable. (G) Metaphase II: chromosomes in each cell are at the equator, and the spindles are evident. (H) Telophase II: four new haploid nuclei can be seen.

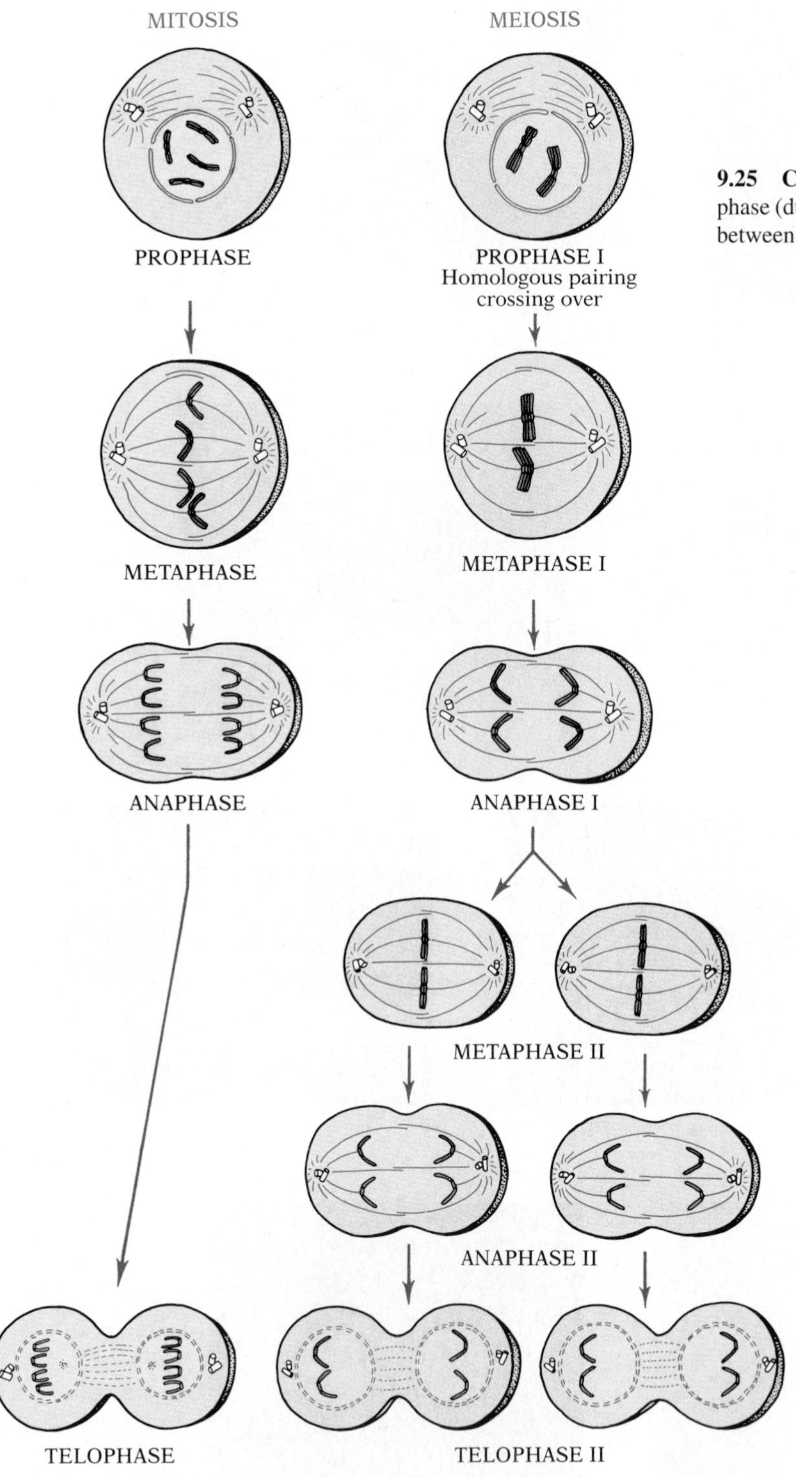

9.25 Comparison of mitosis and meiosis The first step, interphase (during which replication occurs), is omitted, as are the stages between anaphase I and metaphase II of meiosis.

In Animals, Meiosis Produces the Haploid Cells Called Gametes

With rare exceptions, animals exist as diploid multicellular organisms through most of their life cycle. In vertebrates, early in embryonic development certain cells are set aside that will later give rise to gametes. These ***primordial germ cells*** migrate to the gonads (ovaries in females, testes in males) and divide to form either ***oogonia*** or ***spermatogonia,*** depending on whether the gonad is an ovary or a testis. Later, these cells will divide and differentiate to produce special cells, called ***primary oocytes*** and ***primary spermatocytes,*** which are the only cells in the body that undergo meiosis. At the time of reproduction,

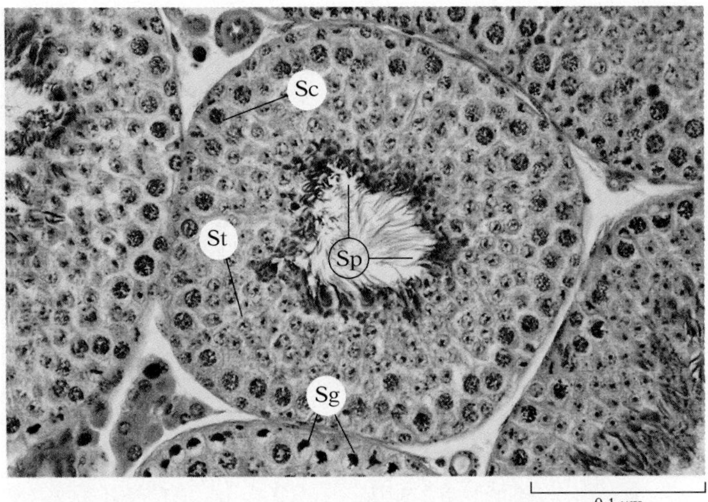

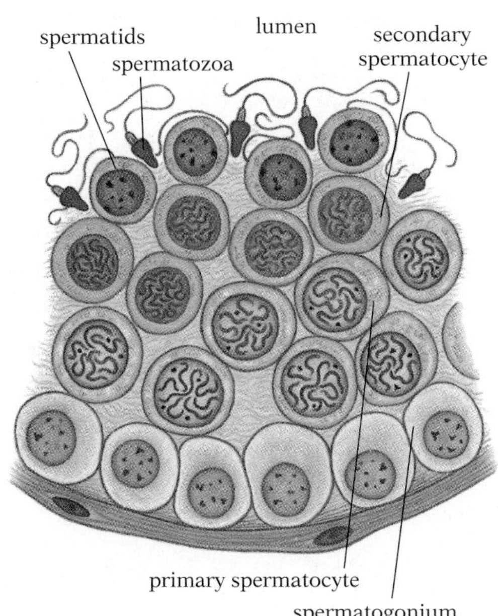

9.26 Photograph and drawing of cross section of rat seminiferous tubule, showing spermatogenesis The dark-stained outermost cells in the wall of the tubule are spermatogonia (Sg), which divide mitotically, producing new cells that move inward. These cells enlarge and differentiate into primary spermatocytes (Sc), which divide meiotically to produce secondary spermatocytes and then spermatids (St). The spermatids differentiate into mature sperm cells, or spermatozoa (Sp), whose long flagella can be seen in the lumen of the tubule in this photograph.

meiosis in higher animals produces mature sperm and egg cells, haploid gametes, which unite in fertilization to form a diploid zygote. The zygote then divides mitotically to produce the new diploid multicellular individual. The gametes are thus the only haploid stage in the animal life cycle (Figs. 9.17 and 9.18).

In male animals, sperm cells (spermatozoa) are produced by the spermatogonia lining the seminiferous tubules of the testes (Fig. 9.26). When the male is sexually mature, the spermatogonia divide continuously (by mitosis) to form new primary spermatocytes, each of which undergoes meiosis I to produce two secondary spermatocytes that undergo meiosis II and produce four haploid cells that are all quite small, but approximately equal in size (Fig. 9.28). All four soon differentiate into functional sperm cells, with long flagella but with very little cytoplasm in the head, which consists primarily of the nucleus. This process of sperm production is called ***spermatogenesis.*** The testes of a young adult male produce about 120 million sperm a day. The sperm can survive about a month in the genital ducts of the male, but ordinarily survive only 24 to 48 hours in the female body.

In female animals, the egg cells are produced within the follicles of the ovaries by a process called ***oogenesis.*** Unlike sperm, which in the human male are produced continuously from puberty on, eggs develop in stages, starting in fetal life. During embryonic development the oogonia produced from primordial germ cells begin to divide and differentiate to form ***primary oocytes,*** which begin the process of meiosis (Fig. 9.27). In human females, the primary oocytes become arrested in prophase I and remain in this stage for many years. At birth a human female has approximately two million primary oocytes, most of which will degenerate by puberty. Under the influence of hormones at sexual maturity, during each menstrual cycle one primary oocyte will be stimulated to complete meiosis I. The meiotic division of the primary oocyte produces two haploid cells that are very unequal in size: one large cell

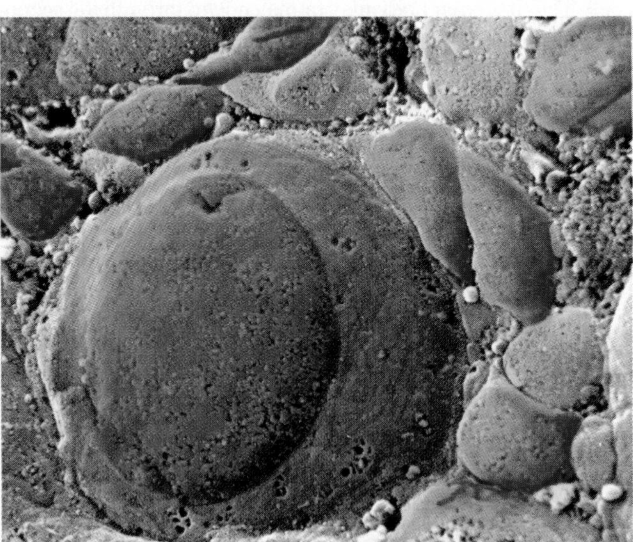

9.27 Scanning electron micrograph of a primary oocyte This primary oocyte (blue) is from a section of an ovary of a developing human embryo. The oocyte consists of the nucleus (deep blue) and cytoplasm (lighter blue). Surrounding the oocyte is a protective layer of follicular cells (red, green). The primary oocyte begins meiosis but becomes arrested in late prophase I. At birth a human female has approximately two million primary oocytes.

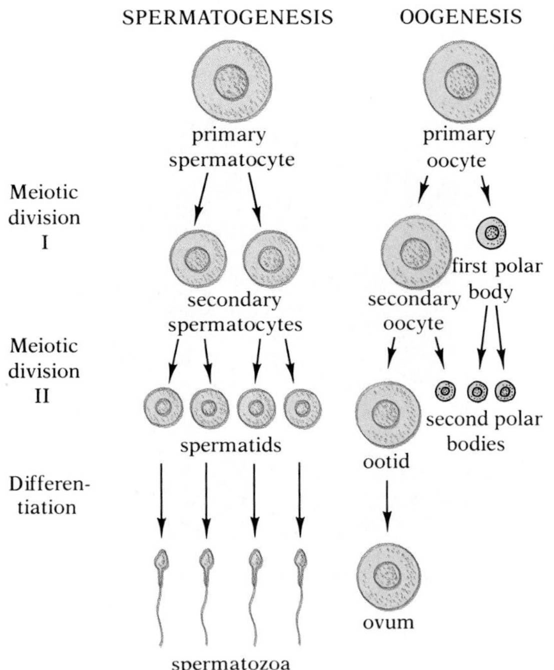

9.28 Schematic illustration of spermatogenesis and oogenesis in an animal In some animals the first polar body does not divide.

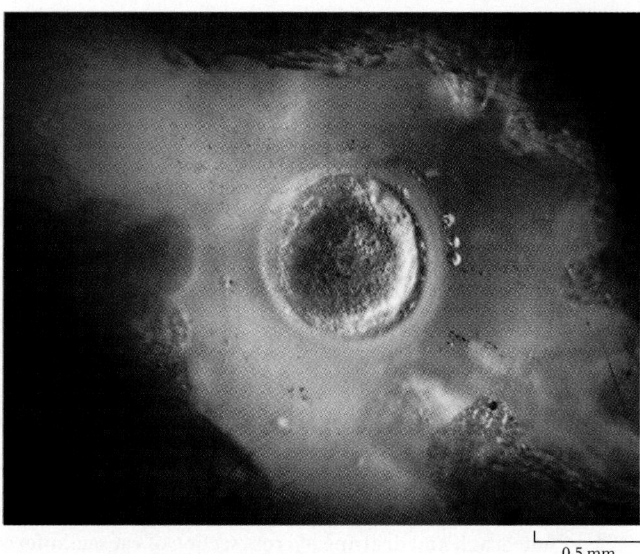

9.29 Human egg cell with polar bodies The polar bodies are the three small circular structures at the right.

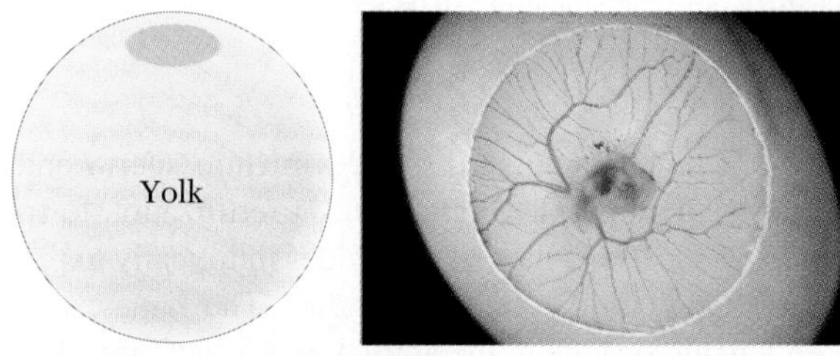

9.30 Egg diagram and photo of early embryo of a chick Embryonic development has begun and the tiny embryo lies on the surface of the yolk. The egg provides almost all the cytoplasm and all the food for the growing embryo.

(the ***secondary oocyte***) and a very small one called a ***polar body*** (Fig. 9.28). In humans, the egg is released from the ovary as a secondary oocyte. Meiosis II is not completed until a sperm penetrates the secondary oocyte. The second meiotic division of the secondary oocyte produces a second polar body and a large cell that soon differentiates into the ***ovum,*** or egg cell. The first polar body may or may not go through the second meiotic division. Thus, when a primary oocyte in the ovary undergoes complete meiosis, only one mature ovum is produced (Fig. 9.29). The polar bodies are essentially nonfunctional.

Unequal cytokinesis in the production of one mature ovum has an important advantage—it yields an unusually large supply of cytoplasm and stored food for use by the embryo. In fact, the ovum provides almost all the cytoplasm and initial food supply for the embryo (Fig. 9.30). The tiny, highly motile sperm cell contributes only its genetic material.

In Plants, Meiosis Produces Haploid Cells Called Spores

That meiosis produces gametes in animals does not mean that it must do so in all organisms. There is no inherent reason why the cells resulting from meiosis must be specialized for sexual reproduction and unite in fertilization to give rise to a diploid zygote, the first cell of a new organism. Indeed, meiotically produced cells are not specialized in this way in most plants. Meiosis in plants usually produces haploid cells called ***spores,*** which often divide mitotically to develop into haploid multicellular plant bodies.

Let us briefly examine the life cycle of a typical plant (a fern in this instance) in which meiosis produces spores rather than gametes (Fig. 9.31). While the fern is in the diploid part of its life cycle, certain cells in its reproductive organs first divide by meiosis to produce haploid ***spores*** (stage 1). These spores in turn divide mitotically and develop into haploid multicellular plants (stage 2). The haploid multicellular plant eventually produces cells specialized as gametes, i.e., eggs and sperm (stage 3). Notice that the gametes are produced by mitosis, not meiosis, because the cells that divide to produce the gametes are already haploid. Two of these gametes unite in fertilization to form the diploid zygote (stage 4), which divides mitotically and develops into a diploid multicellular plant (stage 5). In time, the plant produces spores, and the cycle starts over again. Most plants have a life cycle of this type, with two alternating multicellular stages. We shall examine such life cycles in more

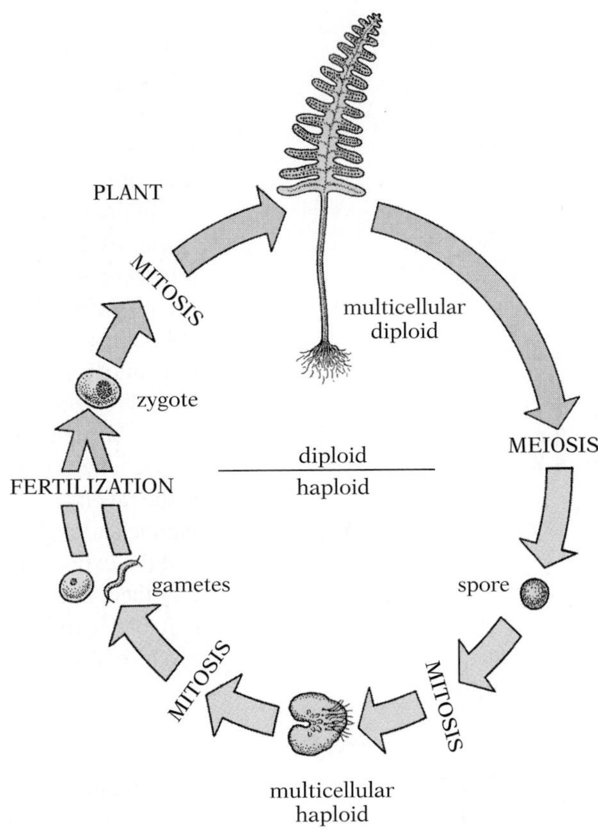

9.31 A typical plant life cycle In most multicellular plants there are two multicellular stages, one haploid and one diploid; the relative importance of these two stages varies greatly from one plant group to another. In ferns and flowering plants the diploid multicellular stage is the major one, and the haploid multicellular stage is much reduced, being represented by a tiny organism with very few cells.

detail in Chapter 39. A few plants have a life cycle like animals in which meiosis produces gametes directly—the spore stage and the haploid multicellular stage is absent (see Fig. 9.18).

Sexual Reproduction, With Its Potential for Genetic Change, Is the Basic Mode of Reproduction in Most Organisms

Sexual reproduction, which is characteristic of the vast majority of both plants and animals, is so widespread that we tend to regard it as a universal attribute of life, although it is not. Many plants and animals reproduce asexually—that is, they give rise to new individuals by mitotic cell division. This type of reproduction is undeniably simpler and more efficient than sexual reproduction, in which the complicated processes of meiosis and fertilization must alternate. Asexual organisms do not need to find mates for reproduction; they simply reproduce themselves. Why, then, has evolution so often favored sexual reproduction over asexual reproduction?

We have seen that mitosis produces new cells with a genetic endowment identical to that of the parental cell. Each new cell gets a set of chromosomes copied from the parental set. Hence, the characteristics of offspring produced by mitosis will be essentially the same as those of the parent. In other words, asexual reproduction does not give rise to genetic variation. Instead, reproduction without meiosis and fertilization gives rise to individuals genetically identical to the parent and to one another. Such a group of genetically identical individuals is called a ***clone.***

What about the chromosomes of a diploid individual produced by sexual reproduction? The first cell (zygote) of this individual was formed by the union of two haploid gametes, an egg cell from one parent and a sperm cell from a second parent. To understand why variation arises in sexual reproduction, consider a hypothetical six-chromosome organism. Three chromosomes came originally from the gamete of the male parent and three from the gamete of the female parent, to form the three pairs of homologous chromosomes. Each time meiosis occurs in an individual producing gametes for reproduction, only one member of each pair goes into one gamete. In our organism with its three pairs of chromosomes, eight different chromosomal combinations are possible in the gametes

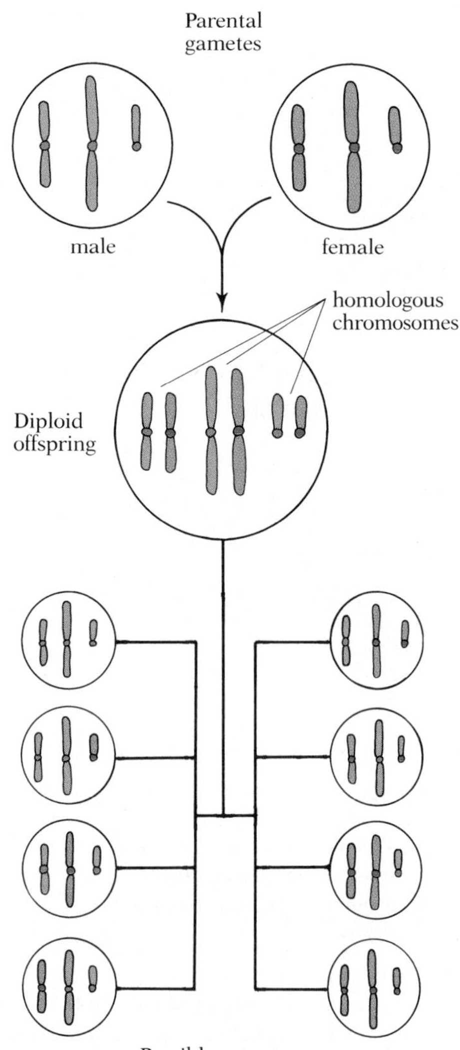

(Fig. 9.32). The formula for determining the number of different chromosomal combinations in a gamete is 2^n where n is the haploid number of chromosomes. In a zygote formed by the union of two gametes, $(2^n) \times (2^n)$ or $(2^n)^2$, different chromosomal combinations are possible. For humans, then, the number of different chromosome combinations possible in the offspring of the same two parents is $(2^{23})^2$, or about 7×10^{13}. Sexual reproduction, then, leads to enormous variation among the offspring. Crossing over during meiosis makes possible still more genetic combinations, and thus increases the variation among individuals. *Unlike mitotic asexual reproduction, sexual reproduction augments variation in the population by producing new chromosomal combinations.*

Genetic variation is clearly advantageous to the species; it is the raw material that natural selection acts upon in promoting the survival and reproduction of well-adapted individuals and the extinction of ill-adapted ones. Without genetic variation there could be no evolution and hence no response to inevitable environmental changes. If a certain type of organism produces genetically varied descendants, the chances that some of these will adapt well to new environmental conditions, and can survive to reproduce, will be much greater than if they are genetically uniform.

William D. Hamilton of Oxford University presents a new argument for the adaptiveness of sexual reproduction: sexual

9.32 Variation in gametes Even without crossing over, a diploid organism produces 2^n different kinds of gametes, where n is the number of pairs of homologous chromosomes. In this hypothetical organism with an n of 3, there are eight (2^3) possible kinds of gametes.

MAKING CONNECTIONS

1. Why is it necessary for meiosis to occur in a sexually reproducing species?

2. Which meiotic division (I or II) is very different from mitosis, and how does it differ? Which is similar to mitosis?

3. Why is it advantageous for oogenesis to yield only one gamete for each cell beginning meiosis instead of the four produced by spermatogenesis?

4. The prevalence of sexually transmitted diseases such as AIDS illustrates one of the many hazards of sexual reproduction. Why, then, is sexual reproduction more widely used than asexual reproduction despite its disadvantages?

reproduction evolved to help host organisms withstand the ravages of parasites. His argument is that sexual reproduction, because it produces genetically varied offspring, increases the chances that some of these will be resistant to parasites infecting the parents, and can survive to reproduce. Asexually produced offspring, on the other hand, would be genetically identical to their parents and would be just as susceptible to parasites. Sexual reproduction, then, with its pool of variability, may have evolved to enable parasite-infected populations to survive.

Asexual reproduction, by contrast, may be advantageous when environmental conditions are constant. Under stable conditions an asexual individual whose genetic endowment suits it particularly well for the exploitation of its environment may rapidly produce a clone of progeny equally well suited to exploiting that environment. Artificial propagation by asexual means of valuable decorative or crop plants has become a standard practice of modern horticulture.

On balance, given the continual fluctuations that characterize most environments both over the short term and the long term, sexual reproduction, with its potential for genetic change and hence for evolutionary adaptation, has evidently been more advantageous than asexual reproduction. It has come to be the basic mode of reproduction in the majority of organisms.

CONCEPTS IN BRIEF

1. Cell division in procaryotic cells proceeds by the comparatively simple process of *binary fission.* In this process, the single, circular chromosome replicates, the two chromosomes attach to different sites on the plasma membrane, and new plasma membrane and wall material grow inward, cutting the cell in two.

2. Cell division in eucaryotic cells involves *mitosis* (division of the nucleus) and *cytokinesis* (division of the cytoplasm). In mitosis, the genetic material is precisely duplicated, and a complete set of the genetic material is distributed to each daughter cell.

3. The interval from one cell division to the next, known as the *cell cycle,* consists of a series of stages. After a cell has completed the division process, there is a time gap, the G_1 stage, before replication of the genetic material begins. Next comes the S stage, during which new DNA is synthesized. Another time gap, the G_2 stage, separates the end of replication from the onset of the M stage (mitosis).

4. During the G_1, S, and G_2 stages, the cell is not dividing; these three stages together constitute *interphase.* After the cell has passed through G_1, S, and G_2, it enters the M stage, during which the actual division takes place.

5. Mitosis is divided into four phases: *prophase, metaphase, anaphase,* and *telophase.* The entire process produces two new cells with exactly the same chromosomal endowment as the parent cell.

6. Cytokinesis frequently accompanies mitosis. In animal cells, cytokinesis begins with the formation of a *cleavage furrow,* which deepens until it cuts the cell in two. In plants, a *cell plate* forms in the center of the cytoplasm and enlarges until it divides the cell.

7. *Meiosis* reduces the number of chromosomes by half so that when the *gametes* (egg and sperm) unite in fertilization the normal number is restored. During meiosis, the chromosomal pairs are divided so that each gamete is *haploid,* that is, it contains one of each type of chromosome. When the two gametes unite in fertilization, the resulting zygote is *diploid,* having received one chromosome of each type from each parent.

8. Meiosis involves two successive division sequences, which produce four new haploid cells. The first division is a *reduction division;* it reduces the chromosome number from diploid to haploid. *Synapsis* and *crossing over* occur during the first division, producing new hybrid chromosomes. The second meiotic division separates the chromatids. As a result of the two successive divisions, four new haploid cells, each containing single-chromatid chromosomes, are produced.

9. Most higher animals are diploid. During reproduction, meiosis produces haploid gametes, which unite to produce the diploid zygote. In male animals, *spermatogenesis* produces four functional sperm cells from each primary spermatocyte. In females, *oogenesis* produces only one mature ovum from each primary oocyte. The *polar bodies* that are produced in the process are nonfunctional.

10. Meiosis in plants produces haploid cells called

spores, which divide mitotically to form haploid multicellular plants. When mature, these haploid plants produce gametes by mitosis. Two of the gametes unite to form a diploid zygote, which then develops into a diploid multicellular plant, completing the cycle. Most multicellular plants have all five stages in their life cycles, but the relative importance of the stages varies greatly.

11. Asexual reproduction produces new cells genetically identical to the parent cell, whereas sexual reproduction increases variation in the population by making possible new genetic recombinations.

STUDY QUESTIONS

1. Describe the process of cell division (binary fission) in procaryotes. (p. 180)
2. What happens in each stage of the cell cycle? Which three stages constitute interphase? Which stage is the most variable in length? Which stage is generally the shortest? (pp. 183–184)
3. Distinguish between *mitosis* and *cytokinesis.* Do they always occur together? How does cytokinesis differ in plant and animal cells? (p. 184)
4. Describe the events of the four stages of mitosis. Make a drawing of each stage of mitosis for a cell with a diploid number of six, for example, to make sure you understand the process. (pp. 184–187)
5. Describe meiosis I and meiosis II and make a drawing of each stage. Which meiotic division halves the number of chromosomes? (pp. 189–194)
6. Define *homologous chromosomes.* Describe how synapsis and crossing over occur. What are recombination nodules? What are chiasmata? What is the function of crossing over and how common is it? (pp. 180, 190–193)
7. Describe how spermatogenesis occurs in male animals, correctly identifying: *spermatogonium, primary spermatocyte, secondary spermatocyte, spermatid,* and *spermatozoa.* (pp. 196–198)
8. Describe oogenesis in female animals, defining the terms *oogonium, primary oocyte, secondary oocyte, ootid,* and *ovum,* and explaining the significant differences between this process and spermatogenesis. What are polar bodies? (pp. 196–198)
9. Diagram the typical life cycles of an animal and a plant and show where meiosis, fertilization, and mitosis occur. How does a spore differ from a gamete? (pp. 198–200)
10. If a species has a diploid chromosome number of 10, how many different chromosomal combinations are possible in its gametes? (Ignore the possibility of crossing over.) How many different chromosomal combinations are possible in the zygote of this species? (p. 200)

CHAPTER 10

THE STRUCTURE AND REPLICATION OF DNA

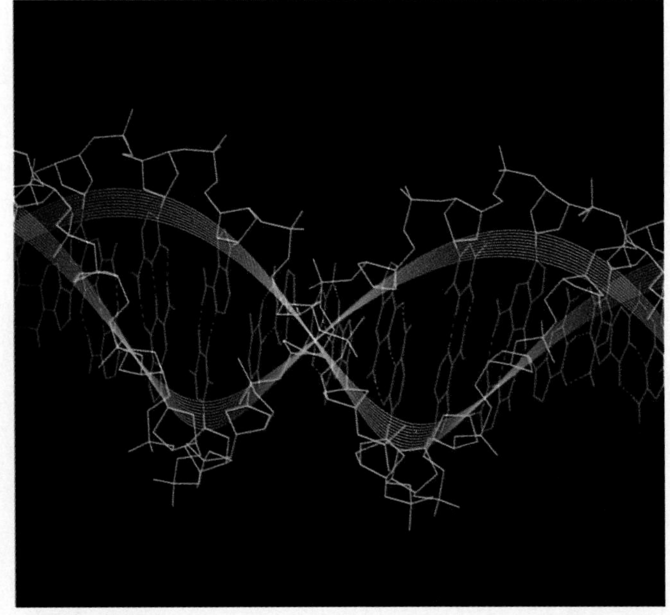

In the last chapter, we learned that when a cell divides, it must first duplicate the genetic information contained in its chromosomes, so that each daughter cell receives its own copy (i.e., a complete set of genes) from the parental cell. We made only passing reference, however, to the chemical nature of the genetic material—DNA—and we did not address the crucial question of how, on a molecular level, the genes are copied. In this chapter, we shall set out to address these issues by first relating how DNA came to be identified as the genetic material and how the ensuing search for its molecular structure also revealed the key to its function as the vehicle of heredity. We shall then take a closer look at the process of chromosome duplication, or as it is more formally known, ***replication.*** In the next chapter, we shall go on to examine how the information encoded in the molecular structure of the genes is "read out" by the machinery of the cell to direct the synthesis of proteins.

A Series of Experiments Led to the Realization That Genes Are Composed of DNA

If the genes are located in the chromosomes, an obvious first step in ascertaining the chemical nature of genes would be to determine the types of compounds present in the chromosomes. That is precisely what early researchers tried to do. Chemical analyses undertaken before 1900 showed that the chromosomes of most organisms contain both protein and DNA. For decades, most biologists assumed that the protein must be the genetic material, since in their view, only protein had the chemical complexity necessary to encode so much information.

Then, in 1928, Frederick Griffith, an English bacteriologist, published some experiments on pneumococci, the bacteria that causes pneumonia. Pneumococci occur in two strains—a disease-causing strain (designated S because it grows in colonies with a smooth surface) and a harmless strain (designated R because the surface of its colonies is rough). Griffith showed that mice injected with live bacteria of strain R survived, mice injected with live S bacteria soon died, and mice injected with heat-killed S bacteria survived. These results were hardly surprising, since one would expect that the disease-causing bacteria must be alive to be pathogenic.

The results of another of his experiments, however, were

Chapter opening photo: Computer generated image of the DNA double helix The nitrogenous bases connecting the two strands are red.

10.1 Griffith's experiments on bacterial transformation When pneumococci bacteria of the rough R strain are injected into a mouse, the mouse survives (A), while mice injected with living bacteria of the smooth S strain die of pneumonia (B). Mice injected with heat killed strain S also live (C). Finally, mice injected with a mixture of live R bacteria and heat-killed S bacteria die (D). Griffith hypothesized that material from the dead S cells transformed live R bacteria into live S bacteria. Later other investigators showed that the transforming material was DNA.

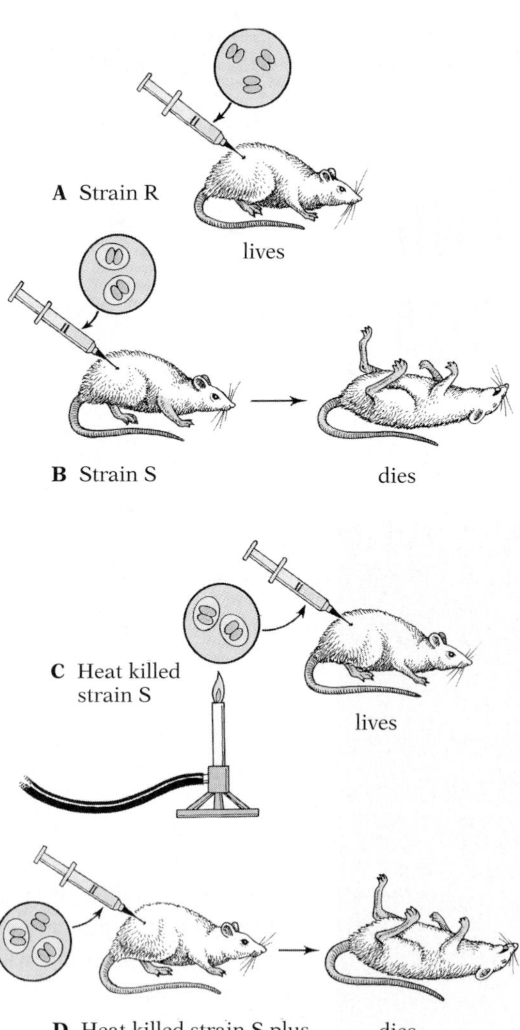

thoroughly perplexing: mice injected with a mixture of live R bacteria and heat-killed S bacteria died (Fig. 10.1). How could a mixture of harmless and dead bacteria have killed the mice? Griffith examined the bodies of the dead mice and found that they were full of live S bacteria! Where had they come from? Many careful experiments convinced Griffith that the live R bacteria had been *transformed* into live S bacteria by material from the dead S cells. Hereditary material from the dead bacteria had entered the live R cells and changed them into S cells. Such a change is now called bacterial ***transformation.***

In the 1940s other researchers showed, by putting DNA extracts from heat-killed S cells into the culture medium, that the transforming agent was DNA. Apparently the live R cells picked up DNA from the surrounding medium and were transformed. No protein was involved. Nevertheless, many scientists remained unconvinced that the genetic material was DNA.

During the next 10 years, evidence for DNA as the genetic material steadily became stronger. At least 30 different examples of bacterial transformation by purified DNA were described. Additional evidence came from studies of another microorganism undertaken by Alfred D. Hershey and Martha Chase of the Carnegie Laboratory of Genetics. They focused their attention on a special type of virus that attacks the bacterium *Escherichia coli,* which is abundant in the human digestive tract. This type of bacteria-destroying virus is called a ***bacteriophage,*** or phage, for short (from the Greek *phagein,* "to eat").

Viruses are tiny infective agents that cannot reproduce themselves. Instead, their genetic material enters a cell, takes over its cellular machinery, and directs it to manufacture new virus particles. It has been established by a variety of methods that all free virus particles, or ***virions,*** are made up of two components: a protein coat and a nucleic acid core. The electron microscope has revealed that the phage virus is structurally more complex than many other types of virus (Fig. 10.2). The protein coat of a phage is divided into a faceted head region,

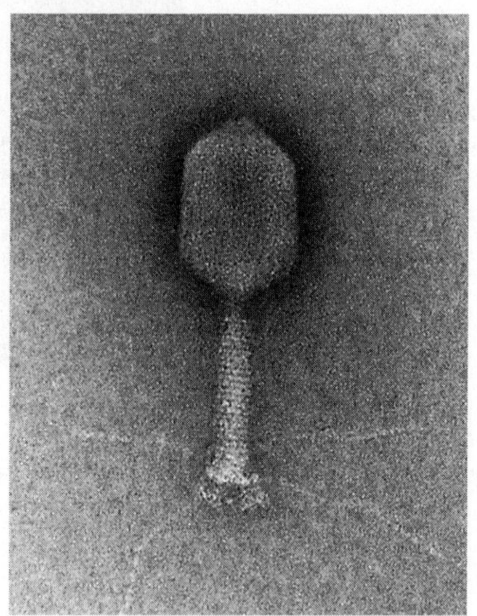

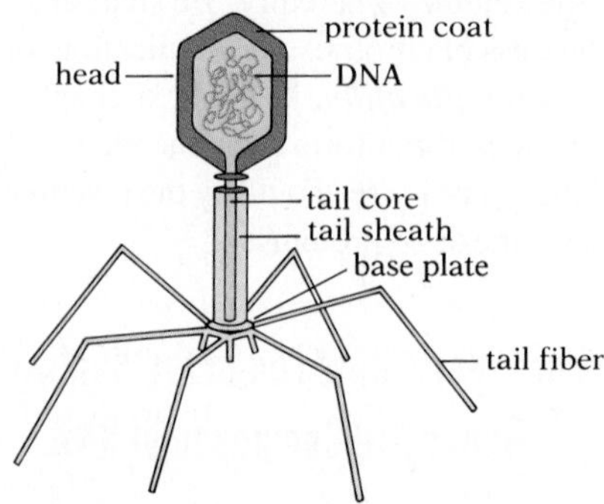

10.2 Electron micrograph and interpretive drawing of a bacteriophage particle Micrograph: × 200,000.

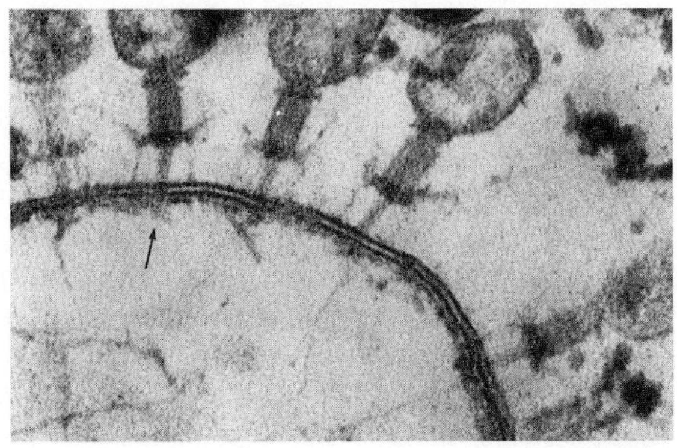

A

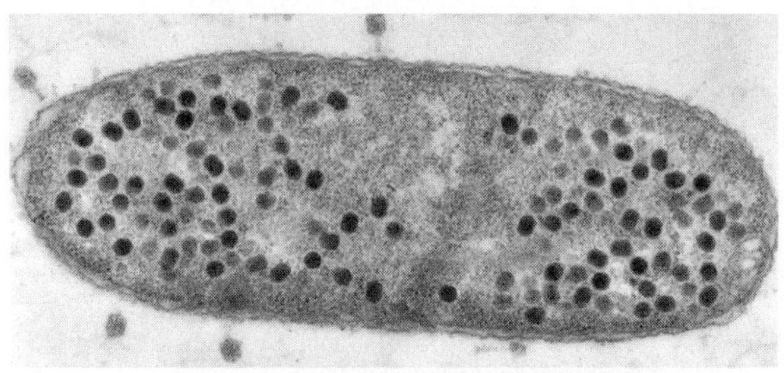

B

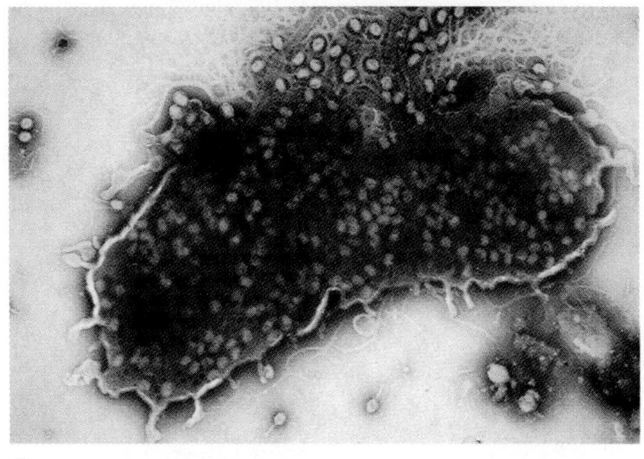

C

10.3 Bacteriophage replication (A) Bacteriophage particles attach to the bacterial cell wall by their tail fibers and inject their genetic material (arrow) into the cell. (B) Once inside, the phage genetic material takes over the metabolic machinery of the cell and puts it to work making new viral DNA and phage proteins. These two components are then assembled into new infective phages. (C) About 20 to 25 minutes after the initial injection of the genetic material, the bacterial cell bursts, releasing the new bacteriophage particles, which may then attack other bacterial cells. The bacterium is destroyed.

which contains all the DNA, and an elongated tail region, made up of a hollow sheath and six leglike fibers.

Electron micrographs show that when a phage particle attacks a bacterial cell it becomes attached by its tail to the wall of the bacterial cell (Fig. 10.3). There is no evidence that the protein coat of the particle ever actually enters the bacterial cell. Yet about 20 to 25 minutes after the phage becomes attached to the cell wall, new phages appear within the bacterium (Fig. 10.3), which soon ruptures, releasing hundreds of new phage particles into the surrounding medium. These new particles are genetically identical with those that initiated the infection, suggesting that hereditary material injected into the bacterial cell by the phage takes over the cellular machinery of the bacterium and puts it to work manufacturing new phages. The reproductive cycle of a bacteriophage is shown in Figure 10.4.

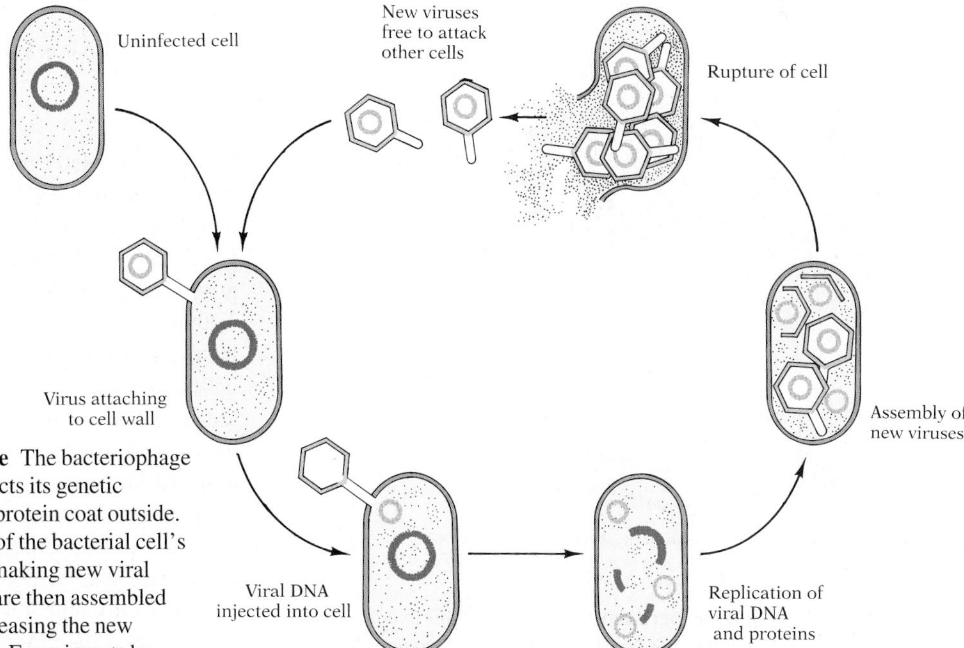

10.4 Bacteriophage reproductive cycle The bacteriophage virus attaches to the bacterial cell and injects its genetic material (DNA) into the cell, leaving the protein coat outside. Once inside, the viral DNA takes control of the bacterial cell's metabolic machinery and puts it to work making new viral DNA and proteins. The two components are then assembled into new viruses. Later, the cell bursts, releasing the new virions, which may then attack other cells. Experiments by Hershey and Chase showed that the genetic material injected is DNA, not protein.

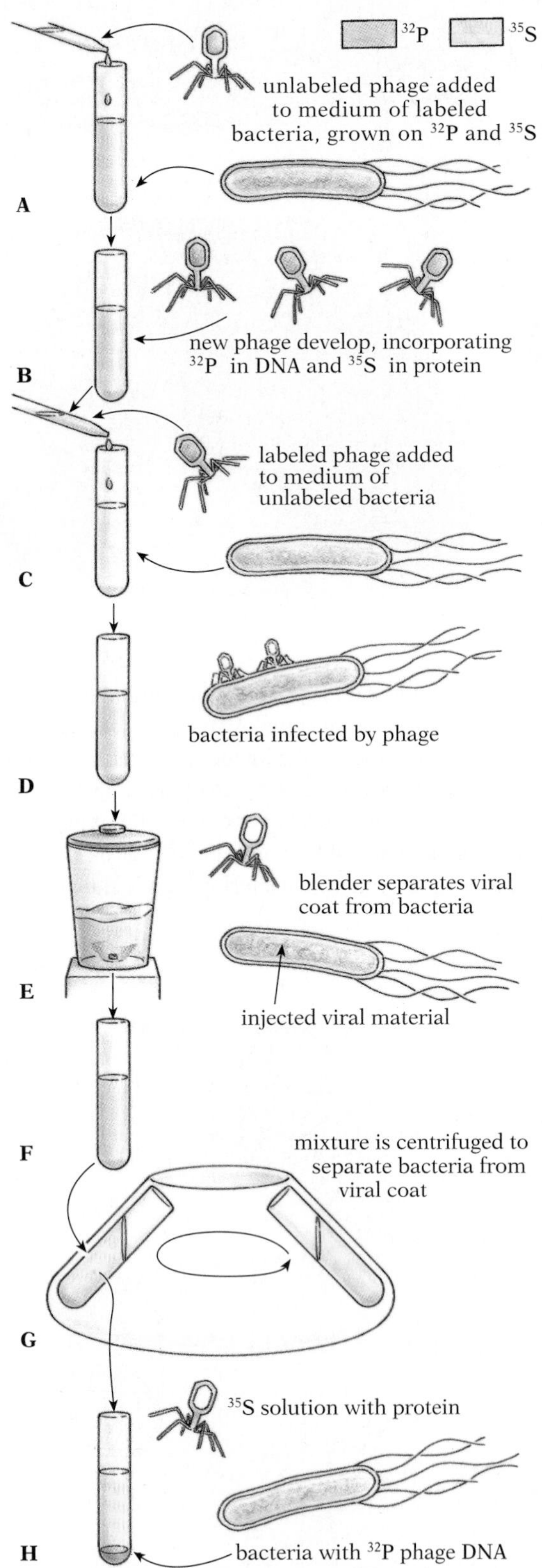

32P
35S
unlabeled phage added to medium of labeled bacteria, grown on 32P and 35S
A
new phage develop, incorporating 32P in DNA and 35S in protein
B
labeled phage added to medium of unlabeled bacteria
C
bacteria infected by phage
D
blender separates viral coat from bacteria
injected viral material
E
F
mixture is centrifuged to separate bacteria from viral coat
G
35S solution with protein
bacteria with 32P phage DNA
H

10.5 The Hershey-Chase experiment For details of this experiment, which demonstrated that DNA rather than protein carries genetic information, see the text.

In 1952, when Hershey and Chase did their work, biologists did not yet know the actual structure of viruses or the details of their reproductive cycle. They knew only that bacteriophages were composed primarily of two components, DNA and protein, and that the phages, by injecting something into bacterial cells, were able to put the cell's machinery to work making new phages. On the basis that DNA contains phosphorus but no sulfur, whereas most proteins contain sulfur but no phosphorus, Hershey and Chase designed an experiment to determine whether the infecting phage injects into the bacterium only DNA, only protein, or some of each (Fig. 10.5). Phages cultured on bacteria grown on a medium containing radioactive isotopes of phosphorus (^{23}P) and sulfur (^{35}S) incorporated the ^{35}S into their protein and the ^{32}P into their DNA. The experimenters then infected nonradioactive bacteria with the radioactive phage, and allowed sufficient time for the phage to become attached to the walls of the bacteria and to inject hereditary material. Then they agitated the bacteria in a blender in order to detach what remained of the phage from their surfaces. Analysis of these remains showed ^{35}S but very little ^{32}P, an indication that the empty protein coat had been left outside the bacterial cell. Analysis of the bacteria showed ^{32}P but no ^{35}S, an indication that only DNA had been injected into them by the phage. Since new phage particles were produced in these bacteria, DNA alone provided the genetic information necessary to make the bacteria produce new phages sufficient to transmit to the bacteria all the genetic information. This experiment supported the earlier conclusion based on transformation experiments that nucleic acids, not proteins, constitute the genetic material.

The DNA Molecule Was Finally Found to Be a Double Helix

By the early 1950s, much information had accumulated on the chemical nature of DNA. The DNA molecule was known to be composed of building blocks called nucleotides, each of which is composed of a five-carbon sugar bonded to a phosphate group and a nitrogenous base (Fig. 10.6). As noted in Chapter 3, there are four different nucleotides in DNA, which differ from one another only in their nitrogenous bases. Two of the bases, adenine and guanine, have double-ring structures and are classified as ***purines;*** the other two, cytosine and thymine, have single-ring structures and are classified as ***pyrimidines*** (Fig. 10.7).

Furthermore, analysis of DNA from many different organisms by Erwin Chargaff and his colleagues at Columbia University had demonstrated that in any DNA sample the amount of thymine nucleotides always equaled the amount of adenine, and the amount of guanine nucleotides equaled the amount of cytosine. Almost nothing was known about the spa-

10.6 Diagram of a nucleotide from DNA A phosphate group and a nitrogenous base are attached to deoxyribose, a five-carbon sugar. By convention a pentagon represents a five-member carbon ring with its valence completed by hydrogen. A hexagon represents a six-member ring; in this case two of the corners are occupied by nitrogen atoms rather than carbon atoms.

10.7 The four nitrogenous bases in DNA The two single-ring bases, thymine and cytosine, are pyrimidines, the two double-ring bases, adenine and guanine, are purines.

tial arrangement of the atoms within the DNA molecule, however, nor was it known how the molecule could incorporate information necessary to replicate itself and to control cellular function.

At about this time, several workers began applying X-ray diffraction analysis to DNA in an effort to determine its three-dimensional structure. In this procedure, X rays are sent through crystalline DNA as it is rotated. The X rays are diffracted, or scattered, by the electrons in the DNA, and the resulting pattern is recorded on a photographic plate. Maurice H. F. Wilkins and Rosalind Franklin of King's College, London, succeeded in obtaining much sharper X-ray diffraction pictures than had previously been obtained (Fig. 10.8). Their diffraction pictures suggested that the DNA molecule is coiled into a helical configuration.

Then began a collaboration that would produce one of the major milestones in the history of science. James D. Watson and Francis H. C. Crick, working at Cambridge University, developed a model of the DNA molecule by combining what was known about its chemical content with the information gained from the X-ray diffraction studies. They were guided in their effort by data on the exact distances between bonded atoms in molecules, the angles between bonds, and the sizes of atoms.

Watson and Crick assumed that the nucleotides were arranged in sequence, held together by covalent bonds between the sugar of one nucleotide and the phosphate group of the next nucleotide, and that the nitrogenous bases were

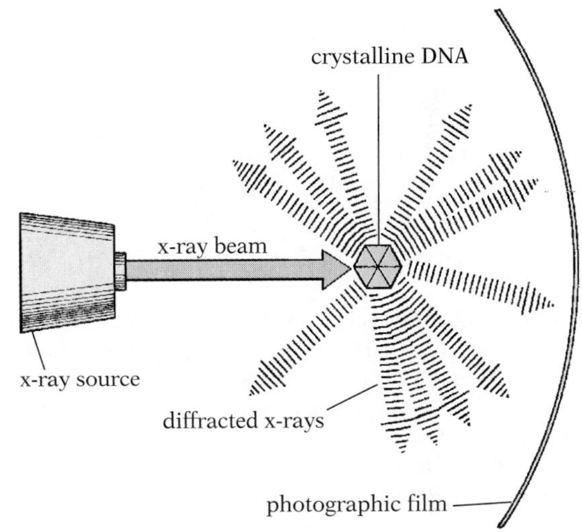

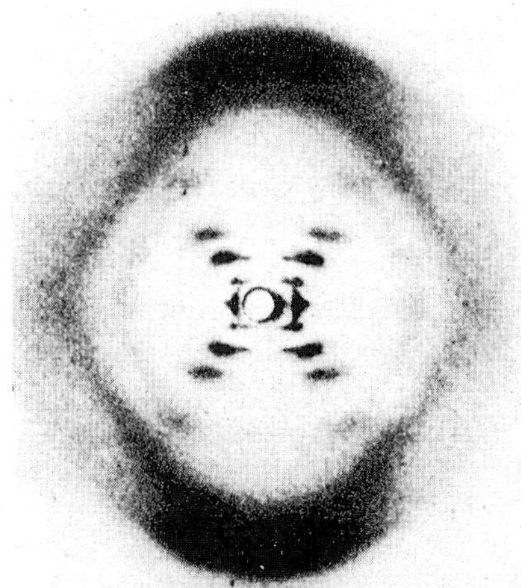

10.8 X-ray diffraction of DNA (A) In X-ray diffraction analysis of DNA, X rays are sent through crystalline DNA as the crystal is rotated. X rays are diffracted by electrons in the DNA. (B) The resulting pattern can be recorded on a photographic plate. This photograph suggested the DNA is coiled into a helix.

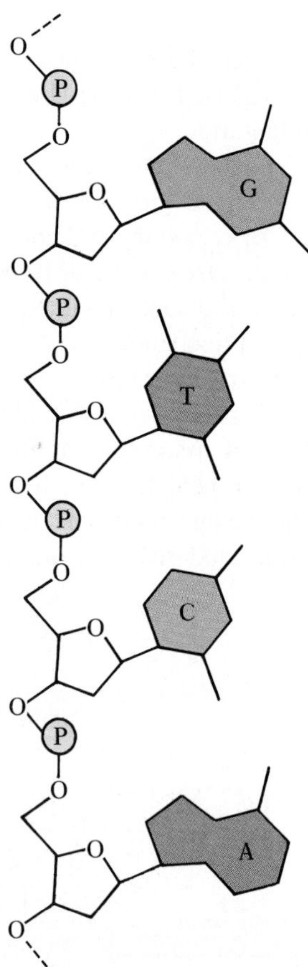

10.9 Portion of a single chain of DNA Nucleotides are linked together by bonds between their sugar and phosphate groups. The nitrogenous bases (G, guanine; T, thymine; C, cytosine; A, adenine) are side groups. In this diagram P represents the main components of each phosphate group—the phosphorus atom with its hydroxyl and the double-bonded oxygen; only the oxygen atoms in the connecting chain are shown separately.

arranged as side groups off the chains (Fig. 10.9). They then attempted to fit the pieces of the molecular puzzle together in a way that would agree with the information from all sources.

The two investigators assumed as well that DNA consisted of a single chain of nucleotides, but they found that a single chain of nucleotides coiled in a helix would have a density only half as great as the known density of DNA. An obvious inference was that the DNA molecule is composed of two nucleotide chains rather than one. Next they had to determine the relationship between the two chains. They tried several arrangements of their jerry-built scale model and found that the one that best fitted all the data had the two nucleotide chains wound in opposite directions around a hypothetical cylinder of appropriate diameter, with the purine and pyrimidine bases oriented toward the inside of the cylinder (Fig. 10.10). With the

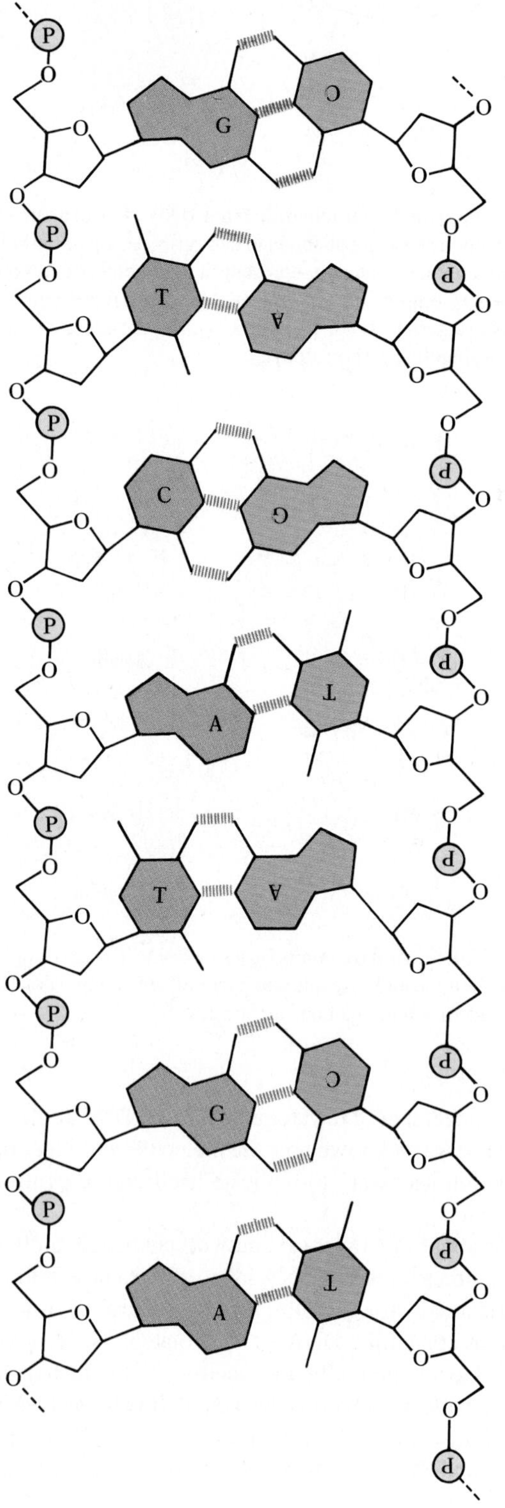

10.10 Portion of a DNA molecule uncoiled The molecule has a ladderlike structure, with the two uprights composed of alternating sugar and phosphate groups and the cross rungs composed of paired nitrogenous bases. Each cross rung has one purine base (a pentagon attached to a hexagon) and one pyrimidine base (a hexagon). When the purine is guanine (G), the pyrimidine with which it is paired is always cytosine (C); when the purine is adenine (A), the pyrimidine is thymine (T). Adenine and thymine are linked by two hydrogen bonds (striped bands), guanine and cytosine by three. Note that the two chains run in opposite directions like the traffic lanes on a highway: they are "upside down" with respect to each other.

bases arrayed in this manner, hydrogen bonds between the bases of opposite chains would supply enough force to hold the two chains together and to maintain the helical configuration. In other words, the DNA molecule, when unwound, would have a ladderlike structure, with the uprights of the ladder formed by the two long chains of alternating sugar and phosphate groups, and with each of the cross rungs formed by two nitrogenous bases loosely bonded to each other by hydrogen bonds (Fig. 10.10).

Watson and Crick soon realized that each cross rung must be composed of one purine base and one pyrimidine base. Their model showed that the available space between the sugar-phosphate uprights was just enough to accommodate three-ring structures. That meant four possible parings: adenine-thymine, adenine-cytosine, guanine-thymine, and guanine-cytosine. Further examination revealed that, although adenine and cytosine were of the proper size to fit together into the available space, they could not be arranged in a way that would permit hydrogen bonding between them. The same was true of guanine and thymine (Fig. 10.11). Therefore, neither adenine-cytosine nor guanine-thymine cross rungs could occur in the DNA molecule. That left only adenine-thymine and guanine-cytosine. Both of these base pairs seemed to fulfill all requirements.

The base pairing discovered by Watson and Crick explained the earlier finding by Chargaff that all samples of DNA contain equal amounts of adenine and thymine nucleotides, and equal amounts of guanine and cytosine nucleotides. The amounts of adenine and thymine are always equal because these two bases are always paired; similarly, the amounts of guanine and cytosine are always equal because these are always paired.

In summary, then, the Watson-Crick model of the DNA molecule shows that each chain of the DNA molecule consists of nucleotides arranged in sequence, held together by covalent bonds between the sugar of one nucleotide and the phosphate group of the next nucleotide, with the nitrogenous bases arranged as side groups off the chains (Fig. 10.12). DNA molecules ordinarily exist as double-chain structures, with the two chains, or strands, held together by hydrogen bonds between their nitrogenous bases. Such bonding can occur only between cytosine and guanine or between thymine and adenine (Fig.

Thymine Adenine

Cytosine Guanine

PYRIMIDINES PURINES

10.11 Bonding of nitrogenous bases in nucleotides The spacing and polarity of these molecules permit thymine to form two hydrogen bonds with adenine, and cytosine to form three hydrogen bonds with guanine. (The asterisk marks the point of attachment of each base to a sugar.)

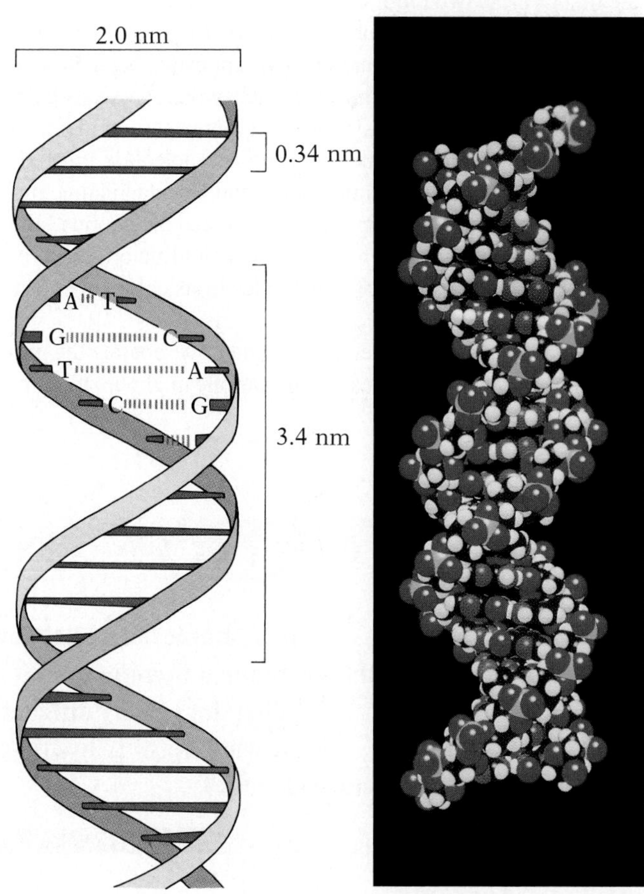

10.12 The Watson-Crick model of DNA Left: The molecule is composed of two polynucleotide chains held together by hydrogen bonds between their adjacent bases. The double-chained structure is coiled in a helix. The width of the molecule is 2.0 nanometers (nm); the distance between adjacent nucleotides is 0.34 nm; and the length of one complete coil is 3.4 nm. Interactions between bases within each chain help stabilize the molecule in the helical shape shown here. Right: Space filling model of DNA. In this model, carbon = dark blue, nitrogen = grey, oxygen = red, phosphorus = orange, and hydrogen = white. Notice that the helix has a solid core.

10.11). Notice that the two strands run in opposite directions; they are "upside down" with respect to each other.

Finally, the ladderlike double-chain molecule is coiled into a ***double helix*** (Fig. 10.12) and stabilized by hydrogen bonds between nearby bases within each chair, much as the alpha helix of a protein is stabilized by hydrogen bonds between the amino acids. This model, in essentially the form in which it was first proposed in 1953, has been consistently supported by later research, and has received general acceptance. Watson, Crick, and Wilkins were awarded the Nobel Prize for their critically important work in 1962. Rosalind Franklin, whose X-ray diffraction studies of DNA were so important in showing DNA's helical nature, did not receive the prize, because the awards are not given posthumously.

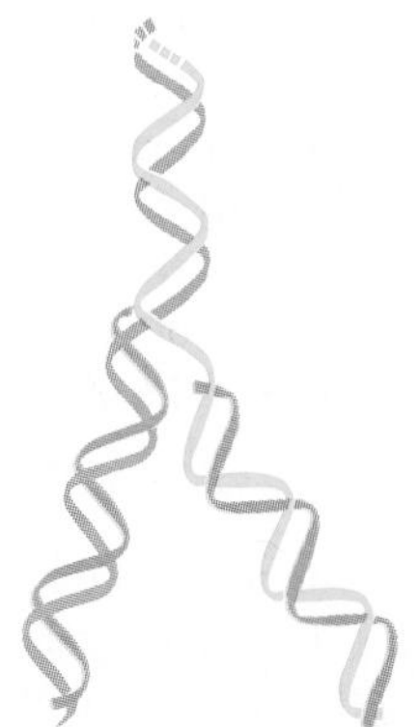

10.13 Replication of DNA As the two polynucleotide chains of the old DNA (yellow and blue) uncoil, new polynucleotide chains (green) are synthesized on their surfaces. The process produces two complete double-stranded molecules, each of which is identical in base sequence to the original double-stranded molecule. Replication actually begins not at the end of the molecule, as shown here for simplicity, but at specific internal points.

The Watson-Crick Model Immediately Suggested a Mechanism for Replication

For DNA to be the genetic substance, it must have built into it the information necessary to copy itself and to control the cell's attributes and functions. One of the most satisfying things about the Watson-Crick model of DNA, as they noted in their landmark 1953 paper in the British scientific research journal *Nature,* is that it immediately suggested a way in which the first of these two requirements might be met.

Watson and Crick pointed out that if the two chains of a DNA molecule were separated, by rupturing the hydrogen bonds between the base pairs, each chain would provide all the information necessary for synthesizing a new partner identical to its previous partner. Since an adenine nucleotide must always pair with a thymine nucleotide, and since a guanine nucleotide must always pair with a cytosine nucleotide, the sequence of nucleotides in one chain would determine precisely what the sequence of nucleotides in its complementary chain must be.

In other words, if the cell could somehow separate the two chains in its DNA molecules—much as one might unzip a zipper—it could line up nucleotides for a new chain next to each of the old chains putting each type of nucleotide opposite its proper partner. Since the nucleotides would be arranged in the proper sequence, they could be bonded together to form the new chain. Thus, separating the two chains of a DNA molecule and using each chain as a template, or mold, against which to synthesize a new partner would result in two complete double-stranded molecules identical to the original molecule (Fig. 10.13).

DNA Replication Is Mediated by an Assortment of Enzymes

The actual process of replication is complex and anything but haphazard. Hydrogen bonds stabilizing the helical shape and linking the two chains of the DNA molecule must first be bro-

MAKING CONNECTIONS

1. In a sample of DNA taken from a particular species, 23% of the nucleotides contained thymine. What percentage would contain cytosine?

2. What does X-ray diffraction tell us about atomic structure?

3. In what ways is hydrogen bonding important in determining the structure of DNA?

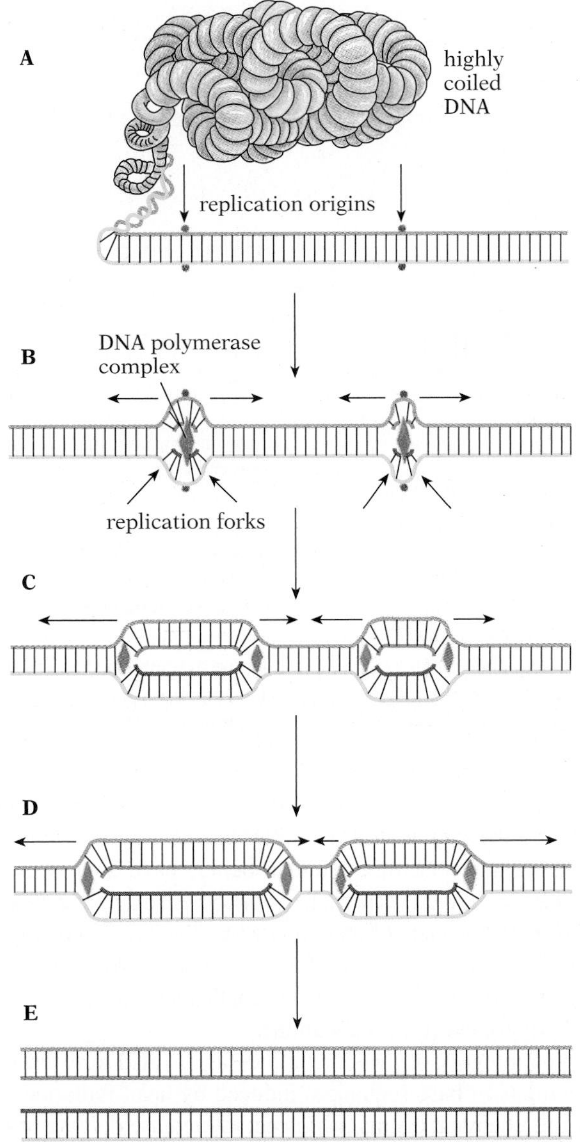

10.14 **Diagrammatic representation of DNA replication**
Replication begins at the replication origins (A) and proceeds in both directions away from the replication origins forming replication forks (B). In the process the two chains of the DNA molecule are unwound and separated. A DNA polymerase complex uses each unwound chain as a template to pair complementary nucleotides to each nucleotide of each existing chain (B–D). The new nucleotides are covalently linked together forming a growing chain complementary to each unwound chain, which eventually results in two complete double-stranded DNA molecules identical to the original (E). This replication occurs for every chromosome in a cell during the S stage of the cell cycle.

ken by enzymes, and the chains unwound and separated. Complementary nucleotides must then be paired with the nucleotides of each existing chain, and then the complementary nucleotides must be covalently linked to one another to form a chain. Each step in this precisely ordered sequence is mediated by specific enzymes, and takes place with amazing speed and accuracy. In *E. coli,* for instance, a complex of several enzymes, known as ***DNA polymerases,*** adds an average of 500 base pairs each second, and there is only one error in every billion pairs copied.

The basic task of DNA replication is the same in procaryotes and eucaryotes. Indeed, there are many similarities between the processes in bacteria and humans. For example, although the chromosomes differ in shape—bacterial chromosomes are circular, whereas human chromosomes are linear—replication begins in both cases at particular sequences in the DNA, called ***replication origins,*** and proceeds in both directions away from the initiation site (Figs. 10.14 and 10.15). Because eucaryotic chromosomes are large (an average human chromosome has 150 million base pairs!), and there are approximately six billion base pairs in human DNA, replication proceeds in many places simultaneously; otherwise it might take weeks or even months for a complex eucaryotic cell to replicate its DNA and divide. In addition, both groups of organisms have special mechanisms that prevent the chains from tangling, and both have special enzyme complexes that locate and correct errors.

Chromosomal replication in eucaryotic cells also involves the assembly of chromosomal proteins. We have already seen that in eucaryotic cells, each chromosome consists of a single long DNA molecule wound around nucleosome cores, which in turn are composed of histone proteins. As in the case of DNA synthesis, the synthesis of new histone proteins takes place only during the S stage of the cell cycle. Although the production mechanism is not well understood, the newly replicated DNA is wound almost immediately onto histone cores (parts of which are newly synthesized) and remains perma-

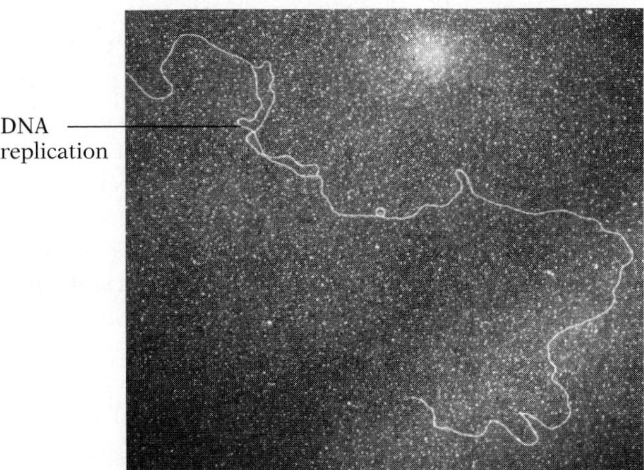

10.15 **Electron micrograph of a DNA molecule in the process of replication** A replication origin region where the two strands of the DNA have uncoupled can be seen at upper left in a so-called eye region. The replication proceeds in both directions; hence the eye grows at both its ends. ×9,800.

nently bound. By the end of the S stage, each chromosome has been completely duplicated, forming two identical strands, which are held together by their centromere regions.

Special Enzymes Help to Repair Replication Errors and Mutations

The precise replication of DNA is essential for normal cell function. Nevertheless, replication errors do occur; they include mistakes in base pairing, additions or deletions of nucleotides, and chemical modifications of some of the nucleotides. Such errors can have serious consequences for the functioning of the cell.

The chromosomes contain the instructions for synthesizing all the proteins necessary to build and run an organism, and random changes in those instructions are far more likely to disrupt a pathway (by destroying the delicate architecture of an essential enzyme, for instance) than to improve it. As an analogy, think about the machinery of a finely tuned watch. If you took out one of the gears and randomly replaced it with a different gear, you would not be surprised if the watch failed to work, or worked less well. So it is with the cell; random mutations are more likely to be harmful than beneficial. In human hemoglobin, for instance, the substitution of an adenine for a thymine in the 17th base of the coding region of the gene results in the production of a hemoglobin with a slightly altered shape, which distorts the contour of the red blood cells and causes the genetic disease known as sickle cell anemia.

Since the instructions for the assembly of the thousands of structural proteins, enzymes, regulatory proteins, and so on are each hundreds or thousands of bases long, an error rate as low as one in 1,000 bases, though it may seem insignificant, is far too great. Not surprisingly, then, special mechanisms have evolved in both procaryotes and eucaryotes that keep uncorrected replication errors at a very low level. As a further safeguard, there are enzymes that locate and repair most of the damage that occurs to the DNA between bouts of replication. Like the other enzymes we have discussed, repair enzymes recognize and bind to specific faulty or damaged areas of the DNA, and proceed to repair the defect. So effective are the DNA repair mechanisms that only rarely is there a ***mutation***—a permanent change in the sequence of nucleotides in the DNA.

The initial error rate of the replication enzymes of both procaryotes and eucaryotes is about three errors in 10^5 base pairs. If the errors produced at this rate were left uncorrected, the result would be a mistake in roughly 3 percent of each cell's proteins, or perhaps 1,000 changed proteins in every human cell after each replication. Fortunately, the DNA polymerase complex includes one or more enzymes that successively "proofread" each base, clipping out mistakes (Fig. 10.16). Other enzymes in the polymerase complex then substitute rematched bases for the excised unit. The complex then moves on without checking a second time. Human beings have approximately 50,000 functional genes, each of which has an average of about 1,500 bases on each of its two strands. The error rate after repair for humans has been calculated to be approximately one error in a gene somewhere in the genetic material of the cells of the body for every 10 replications.

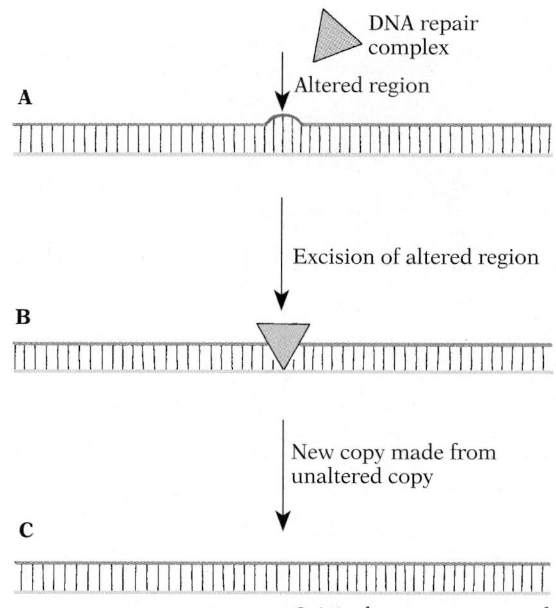

10.16 DNA repair The repair complex (green triangle) binds to a faulty sequence (such as a guanine-adenine base pairing) on DNA (A) and cuts out the faulty sequence (B). A new, correct sequence is generated using the unaltered strand as a template (C).

The integrity of the genetic message is also threatened by alterations in base sequences induced by heat, radiation, and various chemical agents, which act to increase the rate of mutations. The rate at which these mutations occur is astoundingly high. Thermal energy alone, for instance, breaks the bonds between roughly 5,000 purines (adenine and guanine) and their deoxyribose backbones in each human cell every day. Ultraviolet radiation from sunlight fuses together adjacent thymines at a high rate in exposed skin cells, causing mutations and increasing the risk that those cells will become cancerous (Fig. 10.17). (Commercial sunscreens contain compounds that filter out much of the ultraviolet radiation.) And yet, because of the continuous operation of repair enzymes, the rate at which mutations accumulate in cells, on average, is even lower than that of uncorrected errors in replication.

The strategy for repairing chromosomes is basically the same for mutations as for replication errors. Enzymes locate and bind to the faulty sequences and clip out the flaws, and the intact complementary strand guides repair. A remarkable array of specific enzymes is involved. For instance, the chromosomes are scanned for chemically altered bases by fully 20 different enzymes, each specific for a particular class of problem.

10.17 Australian sunbather protecting himself from the sun

Cytosine → Methylated cytosine → Thymine

10.18 An example of a mutation that cannot be correctly repaired Sometimes a methyl group (CH_3) is added by enzymes to cytosine. Later a mutagenic chemical may convert the methylated cytosine into a normal thymine. This change cannot be corrected reliably because the repair system cannot determine whether it is the thymine or the guanine—the base in the complementary strand to which the cytosine was originally paired—that is incorrect.

Another five or so enzymes are specific for faulty covalent bonding between a base and some other chemical, or between adjacent bases on one strand. Other enzymes bind at the sites of missing bases, like the purines so easily lost to thermal energy. In all, some 50 enzymes locate and correct errors.

In view of this elaborate repair system, how do mutations survive to generate the genetic variability needed for species to adapt to changing environmental conditions? Obviously, no enzymatic system is perfect—some errors are missed, and others are repaired incorrectly. Then, too, when a mutation occurs just before or during replication, there may not be time for detection and repair.

Finally, some mutations simply cannot be detected by the repair enzymes. For example, cytosine can easily be chemically modified to produce a thymine (Fig. 10.18). The thymine on the one chain is now mismatched with guanine, the partner of its predecessor. Since thymine is a normal base, however, there is no way to determine whether the incorrect base is the thymine in one chain or the guanine in the other.

Even though some mutations are missed by the DNA repair mechanisms, the overall mutation rate for a gene of average size remains extremely low—approximately one mutation in every million cell generations. Because the repair process is very efficient, the genetic material remains remarkably stable from one generation to the next. As low as the mutation rate is, it is the ultimate source of new genes. We said in the last chapter that genetic variation is advantageous to the species; it is the raw material that natural selection acts upon in promoting the survival and reproduction of well-shaped individuals and the extinction of ill-shaped ones. Without genetic variation there could be no evolution and hence no response to inevitable environmental changes. While the vast majority of the genetic variability driving evolution is due to new combinations formed as a result of sexual reproduction, mutations do increase variability and ultimately provide the raw material for evolution.

MAKING CONNECTIONS

1. What type of chemical reaction covalently links the nucleotides of DNA during replication?

2. Based on the discussion of enzyme function in Chapter 4, what factors would affect the speed and accuracy with which DNA polymerase would add base pairs to a growing strand of DNA?

3. How is chromosome replication in eucaryotes different from the process in procaryotes?

4. In what stage of the cell cycle does DNA replication occur?

5. How do mutations contribute to the evolutionary process?

CONCEPTS IN BRIEF

1. Experiments have demonstrated that DNA constitutes the genetic material. There are four different *nucleotides* in DNA; they differ in their nitrogenous bases, which may be the double-ring purines, *adenine* and *guanine,* or the single-ring pyrimidines, *cytosine* and *thymine.*

2. According to the Watson-Crick model of DNA, the nucleotides are joined by covalent bonds between the sugar of one nucleotide and the phosphate group of the next nucleotide in the sequence; the nitrogenous bases are side groups of the chains. DNA molecules are double-stranded, with the two chains running in opposite directions and held together by hydrogen bonds between adenine and thymine from opposite chains and between guanine and cytosine from opposite chains. The ladderlike double-chained molecule is coiled into a double helix.

3. During *replication* portions of the two chains of the DNA molecule separate, and each chain acts as a template for the synthesis of its new partner. The process produces two complete double-stranded molecules, each identical in base sequence to the original double-stranded molecule.

4. Special mechanisms have evolved in both procaryotes and eucaryotes to locate and correct mutations and replication errors. Enzymes find and bind to faulty or damaged sequences and clip out the flaws. The repair process is not perfect, however; some errors are missed, others cannot be detected, and still others may actually be created by the repair system. Nevertheless, the rate at which mutations accumulate is kept relatively low by the repair system.

STUDY QUESTIONS

1. **What is genetic transformation? How did Griffith discover it? Why was its discovery significant for genetics? (pp. 203–204)**
2. **Describe the life cycle of a bacteriophage. What characteristics of the bacteriophage made it particularly useful for demonstrating that DNA is the genetic material? (pp. 204–206)**
3. **Diagram a nucleotide and label its components. What nitrogenous bases occur in DNA? Which are purines and which are pyrimidines? What is the major difference between these two types of bases? (p. 207)**
4. **Why is DNA said to have a ladderlike structure? From what nucleotide components are the uprights of the "ladder" formed? What forms the "rungs"? (p. 209)**
5. **What base-pairing rules did Watson and Crick discover? Why are only these particular base pairs possible in DNA? (p. 209)**
6. **Describe the basic process by which DNA replication occurs. What are the roles of *DNA polymerases* and *replication origins?* (pp.210–211)**
7. **How is *mutation* defined? What environmental factors can increase the rate of mutation? (p. 212)**
8. **How are DNA replication errors and mutations detected and repaired? (pp. 212–213)**

CHAPTER 11

TRANSCRIPTION AND TRANSLATION

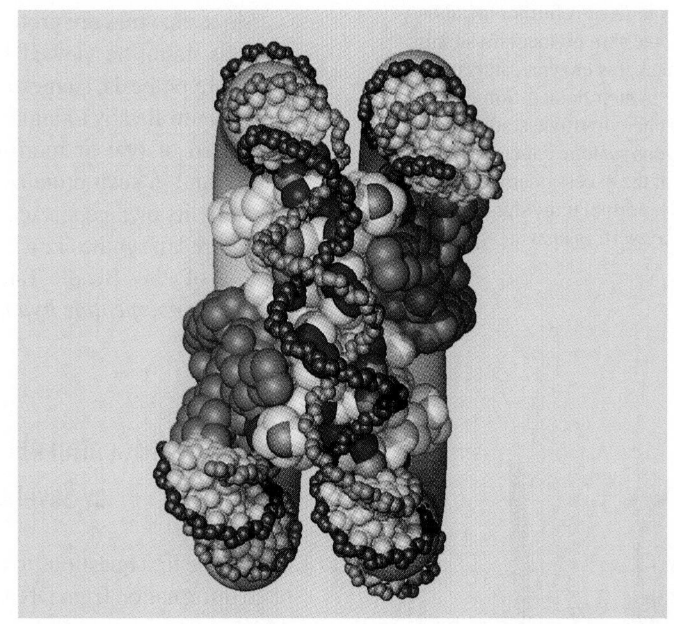

Now that we have seen how a complex set of enzymes makes new copies of the DNA in chromosomes and corrects most of the errors that arise in the process, we turn next to the question of how genes determine the rest of the cell's activities. To put the question another way, how can DNA molecules in the nucleus of a eucaryotic cell control what goes on in the cytoplasm? In this chapter, we shall take a look at the mechanisms by which the information encoded in the nucleotide sequence of DNA is transmitted out of the nucleus to direct the synthesis of other biologically important molecules, specifically the proteins that form the structure of the cell and the enzymes that catalyze the cell's myriad chemical reactions. Again we start with a short historical account to show how we have arrived at our present understanding of how genes work.

Studies of Induced Mutations in Bread Mold Led to the One Gene–One Enzyme Hypothesis

The connection between genes and enzymes is evident in a number of inherited human diseases. In phenylketonuria (PKU), for example, abnormally high concentrations of the amino acid phenylalanine are found in the blood. Normal individuals convert excess phenylalanine into the amino acid tyrosine, which is then further metabolized. Victims of PKU lack the enzyme that catalyzes this reaction, and consequently phenylalanine accumulates in their bodies (Fig. 11.1). Some of the excess phenylalanine is converted into phenylpyruvic acid, which builds up in the cells of the central nervous system, causing mental retardation. Apparently, normal individuals have a gene for producing the phenylalanine-converting enzyme that persons with the disease lack.[1] Although early studies of PKU and other similar diseases suggested a close relationship between genes and enzymes, for many years the true nature of the link was not recognized.

Concepts of gene action remained nebulous until the early 1940s, when George W. Beadle and Edward L. Tatum, then at Stanford University, performed a series of pioneering experiments on the red bread mold *Neurospora crassa* (Fig. 11.2).

Chapter opening photo: Computer generated image of the core of a eucaryotic chromosome. In a chromosome the strands of DNA (grey tubes) are wound around a protein core (blue and white) in a very precise manner. This structure must open like an accordian in order for the DNA to be replicated or expressed.

[1] When PKU is detected within the first four months of life and a diet instituted that contains only a bare minimum of phenylalanine (some is essential for protein synthesis), the neurological effects that lead to mental deficiency can usually be avoided. Some states now require testing for PKU in newborn infants.

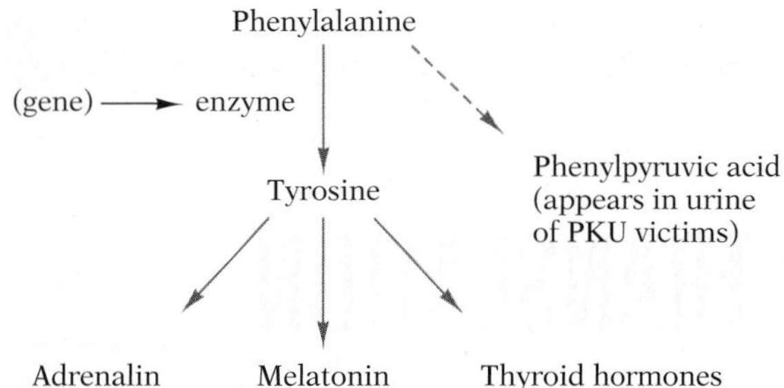

11.1 Relationship between genes and enzymes in phenylketonuria Normal individuals have an enzyme that catalyzes the reaction that converts excess phenylalanine into the amino acid tyrosine, which is used to synthesize proteins or is further metabolized into such substances as adrenalin, the skin pigment melatonin, or thyroid hormones. Victims of PKU lack this enzyme, and consequently they accumulate phenylalanine in their bodies. Some of the excess phenylalanine is converted into phenylpyruvic acid, which builds up in the cells of the central nervous system, damaging them and causing mental retardation. Some of the excess phenylpyruvic acid is excreted in the urine. Apparently, normal individuals have a gene for producing the phenylalanine-converting enzyme that persons with the disease lack.

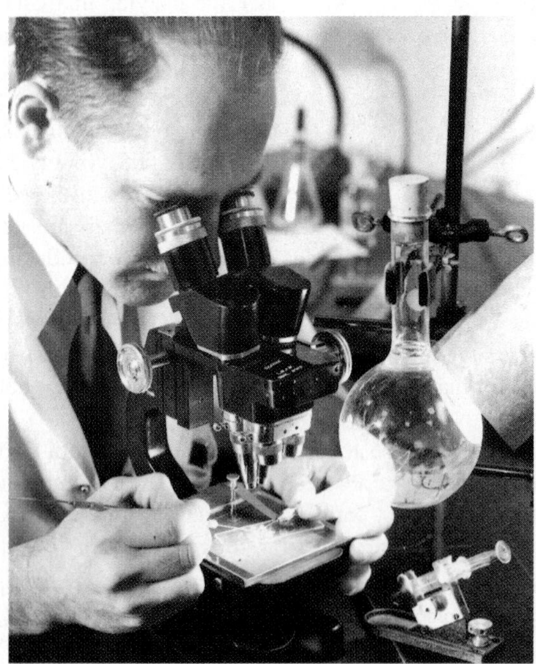

11.2 Nobel Laureate George W. Beadle examining *Neurospora* Beadle and his partner E. L. Tatum examined mutations of the red bread mold *Neurospora crassa* to determine the biochemical differences between mutant and normal strains. They reasoned that a mutated gene could no longer control the production of an enzyme essential for the synthesis of the nutrient in question, and advanced the one gene-one enzyme hypothesis, which states that each enzyme in a cell is determined by a single gene.

They used X rays to induce mutations that interfered with the mold's normal ability to synthesize certain nutrients for itself. In every case where the mutant had lost the ability to synthesize a particular nutrient, they found that the trait was inherited in a pattern that indicated the action of only one gene. Reasoning that the mutated gene could no longer control production of the enzyme essential for the synthesis of the nutrient in question, they advanced the hypothesis that synthesis of each enzyme in a cell is determined by a single gene. In recognition of their contribution, known as the ***one gene–one enzyme hypothesis,*** Beadle and Tatum were awarded a Nobel Prize in 1958.

Since enzymes are proteins, the one gene–one enzyme hypothesis might be viewed as tantamount to a one gene–one protein hypothesis, suggesting that the synthesis of every protein is controlled by a single gene. Some proteins, however, are composed of two or more chemically different polypeptide chains, and in such proteins, each polypeptide chain is determined by its own gene. Accordingly, more than one gene may be required to synthesize a protein. The more accurate generalization of the Beadle-Tatum idea, therefore, is the ***one gene–one polypeptide hypothesis.***

The Genetic Control of Protein Synthesis Is Mediated by Several Types of RNA

One of the first questions to arise when study of the flow of genetic information from DNA to protein began in the 1950s was whether protein synthesis is controlled directly by DNA through the intervention of the appropriate enzymes, or indirectly by means of some intermediary substance. Since nearly all the DNA in eucaryotes is found in the nucleus, whereas protein synthesis was known to take place in the cytoplasm, most researchers suspected that a molecular middleman must exist.

Several investigators had demonstrated a decade or so earlier that cells in tissues where protein synthesis is particularly active, such as the vertebrate pancreas, contain large amounts of ribonucleic acid (RNA). This nucleic acid is found in only limited quantities in cells that do not produce protein secretions. It had long been known that RNA, unlike DNA, is present in the cytoplasm as well as in the nucleus. Radioactive-tracer experiments revealed that RNA is synthesized in the nucleus and moves from the nucleus into the cytoplasm (Fig. 11.3). All these lines of evidence pointed to the possibility that RNA might be the chemical messenger between the DNA of the nucleus and the site of protein synthesis in the cytoplasm.

As we saw in Chapter 3, RNA and DNA are very similar compounds: both are polymers of nucleotides. They differ, however, in three important ways: (1) the sugar in RNA is ribose, whereas that in DNA is deoxyribose (Fig. 11.4); (2)

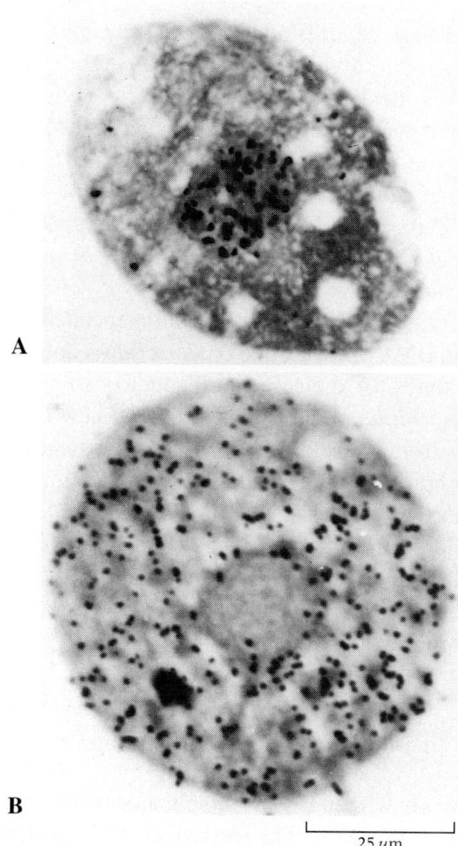

11.3 Radioactive-tracer experiments showing movement of messenger RNA Cells were placed in a medium containing radioactive RNA nucleotides and the radioactively labelled nucleotides were incorporated into messenger RNA. (A) After 15 minutes the cells were removed from the medium and washed. The dark spots in this cell indicate that most of the labelled RNA is in the nucleus. (B) This cell was also removed from the radioactive medium after 15 minutes but was then grown on nonradioactive medium for 90 minutes. The dark spots show that most of the labelled RNA has left the nucleus and is now in the cytoplasm. Studies such as this have shown conclusively that RNA is synthesized in the nucleus and moves from the nucleus into the cytoplasm.

RNA has the ribonucleotide uracil,[2] whereas DNA has thymine (Fig. 11.4); and (3) RNA is normally single-stranded, whereas DNA is usually double-stranded.

Despite these differences, it was obvious that DNA could easily act as a template for the synthesis of RNA, much as it does for the synthesis of new DNA. In this process, we now know, the two strands of the DNA molecule are separated, and RNA is synthesized along *one* of the DNA strands by the enzyme RNA polymerase (Fig. 11.5). For every adenine in the

[2] The use of uracil in place of thymine enables the active site of a wide variety of enzymes to distinguish DNA from RNA, and so be selective in their activity. DNA polymerase, for instance, does not attempt to replicate mRNAs, nor does RNA polymerase try to transcribe messenger RNAs, nor can enzymes in the cytoplasm that digest the DNA of invading viruses chew up the cell's own RNAs.

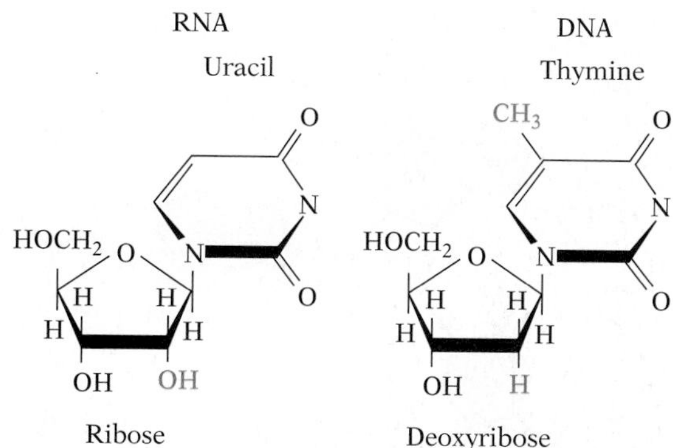

11.4 Uracil in RNA and thymine in DNA compared The five-carbon sugars of RNA and DNA differ only at the site shown in color, where deoxyribose lacks an oxygen atom that is present in ribose. Uracil differs from thymine only in lacking a methyl group (CH_3).

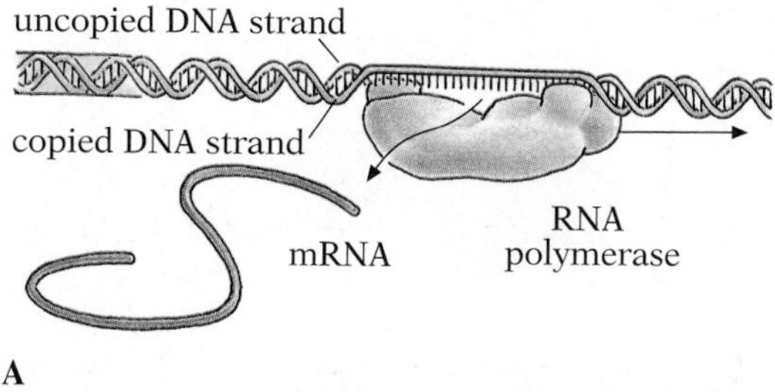

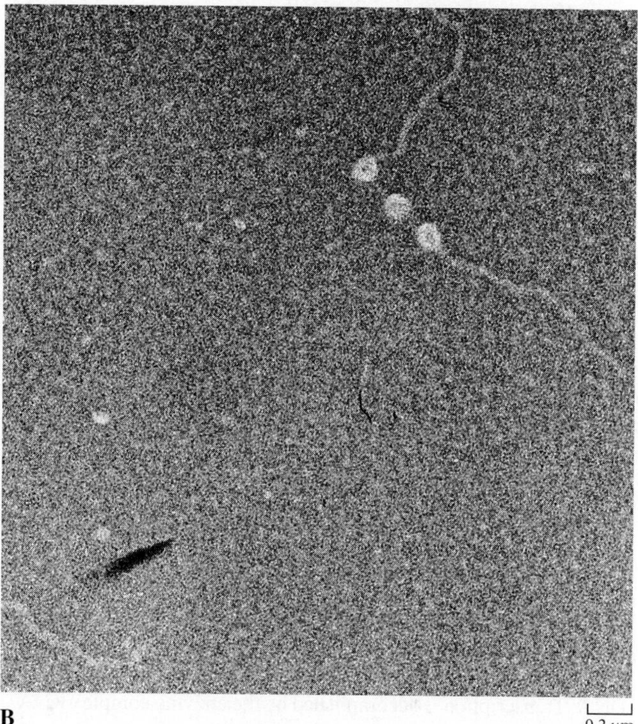

11.5 Transcription of a gene (A) As the RNA polymerase complex moves along the DNA, it catalyzes transcription of only one of the two DNA strands. (B) An electron micrograph showing three polymerases transcribing a gene in *E. coli.*

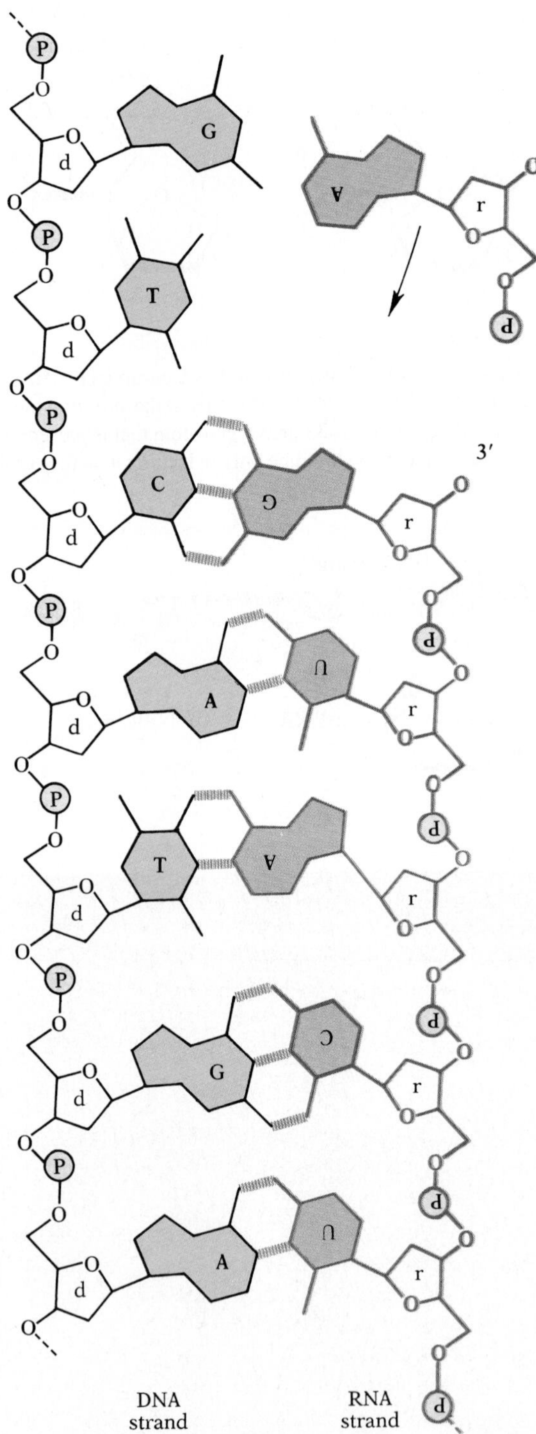

11.6 The synthesis of RNA by transcription of a DNA template The sugar in RNA (ribose, *r*), is slightly different from that in DNA (deoxyribose, *d*), and uracil (U) takes the place of the thymine in DNA. Transcription is accomplished by the enzyme complex RNA polymerase (not shown).

DNA template, a uracil is added to the growing RNA strand (substituting for the thymine in a new DNA strand). Otherwise, the process is identical for the two nucleic acids: for every thymine in the DNA template, an adenine is added to the new strand; for every guanine, a cytosine is added; and for every cytosine, a guanine is added (Fig. 11.6). *In other words, the sequence of deoxyribonucleotides in the single DNA strand determines the sequence of ribonucleotides in the resulting RNA molecule.*

The process by which the information encoded in the language of the DNA gene (the deoxynucleotide sequence) is rendered into the RNA dialect (the ribonucleotide sequence) is called ***transcription.*** Once synthesized, the new RNA molecule travels from its original site on a chromosome in the nucleus to a ribosome in the cytoplasm, where the information it carries is used to synthesize protein. The latter process, which entails a more fundamental change (from the nucleotide sequence of the nucleic acid language to the amino acid sequence of the protein language), is called ***translation*** (Fig. 11.7). Thus the flow of information in most living cells is

$$\text{DNA} \xrightarrow[\text{transcription}]{} \text{mRNA} \xrightarrow[\text{translation}]{} \text{protein}$$

We now know that at least three major types of RNA are synthesized from DNA: (1) ***messenger RNA (mRNA),*** the type just discussed, which carries the information for protein synthesis from the DNA in the nucleus to the ribosomes in the cytoplasm; (2) ***ribosomal RNA (rRNA),*** which functions as part of the ribosomes; and (3) ***transfer RNA (tRNA),*** which brings amino acids to the ribosomes during protein synthesis. We shall examine in greater detail how ribosomes work in a later section. Although the following discussion of transcription focuses on the synthesis of mRNA, the process is similar for all three types of RNA.

In Transcription, the Genetic Message Is Transferred From DNA to RNA

The transcription of DNA, with the resulting synthesis of RNA, is accomplished by a six-protein enzyme complex called ***RNA polymerase,*** which binds to specific sites on the DNA and opens up the helix (Fig. 11.5). Like DNA polymerase, RNA polymerase moves along the DNA, synthesizing a strand of RNA that is complementary to the DNA. How does RNA polymerase know which of the two strands to use as a template, and where to start and stop transcription? The answer lies in the fact that there are certain sequences in the DNA, called ***promoter*** sites, to which RNA polymerase binds, thereby determining where transcription begins. In procaryotic cells such as bacteria, for example, the promoter region is about 40 base pairs long and has two recognition sequences of six bases each

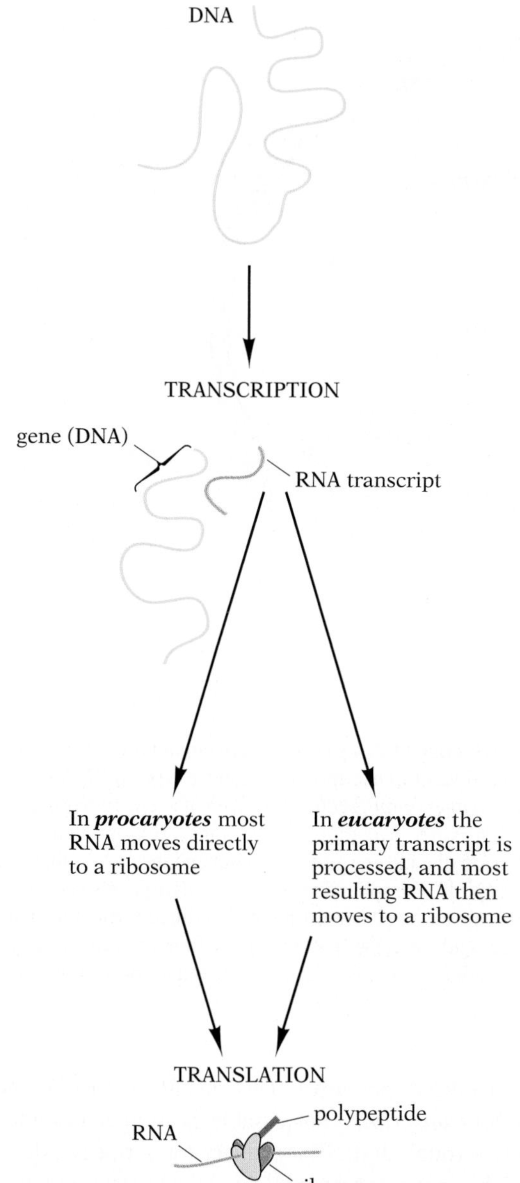

11.7 The flow of genetic information Transcription and translation occur, with some differences, in both procaryotes and eucaryotes: an RNA version of a gene is produced by the process of transcription; most such transcripts bind to a ribosome, which—guided by the information carried by the RNA—then assembles amino acids for the synthesis of a particular protein. In eucaryotes, the transcript is "processed" before moving to the ribosome for translation into polypeptides.

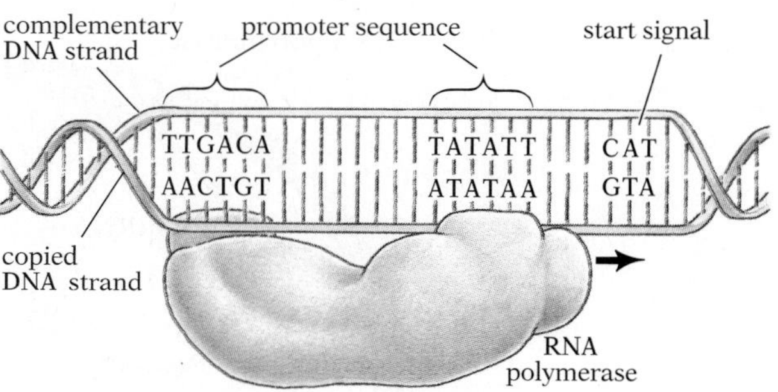

11.8 Transcription signal in *E. coli* DNA RNA polymerase binds to a region of DNA with the sequences shown. Once the polymerase is bound, transcription begins with the GTA signal (corresponding to the CAT sequence on the complementary strand) and proceeds until a termination signal is encountered.

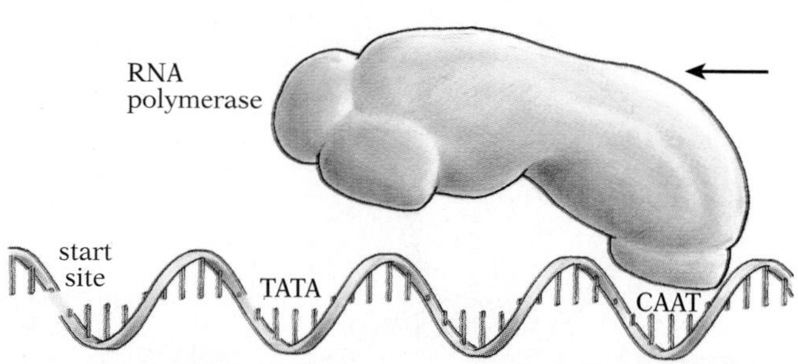

11.9 Eucaryotic promoter sequences In eucaryotic cells RNA polymerase initially binds to a region of DNA with a TATA, or very similar sequence. This sequence appears to be essential but other groupings (such as a CAAT) may also be necessary for binding. Once the polymerase is bound, transcription begins with a special start signal and proceeds until a termination signal is reached.

(Fig. 11.8). In eucaryotic cells the promoter is about 100 bases long and contains several activating sequences including the sequence thymine-adenine-thymine-adenine (TATA), which apparently is important for promoter activity (Fig. 11.9).

Once bound to the promoter, the RNA polymerase unwinds and separates a section of the two DNA strands. Recall that the two nucleotide strands run in opposite "directions" (the "direction" of each strand refers to the orientation of the sugars in that strand, as shown in Figure 10.10). The RNA polymerase can move along the DNA in only one direction, so that once the polymerase binds to the promoter, it is automatically oriented with respect to both the direction it can move and the strand it can copy. Synthesis begins at a specific "start" site; in bacteria it is usually a purine five to eight nucleotides "downstream" of the recognition site (Fig. 11.8). The RNA polymerase complex then moves along the DNA strand, opening up the double helix, complementary base-pairing takes place between the DNA strand and ribonucleotides, and the polymerase catalyzes the joining of one ribonucleotide to the next. The rate of synthesis is quite rapid—20 to 50 RNA nucleotides are added per second. This process continues until a specific "stop" sequence in the DNA is reached. The polymerase and the newly synthesized mRNA then drift free of the DNA.

Transcription in eucaryotic nuclei is more complicated—so much so that its details are just being worked out. One recent finding is that, in addition to promoters, special ***transcription factors*** and ***enhancer elements*** are required to initiate transcription in many eucaryotic cells. Transcription factors are spe-

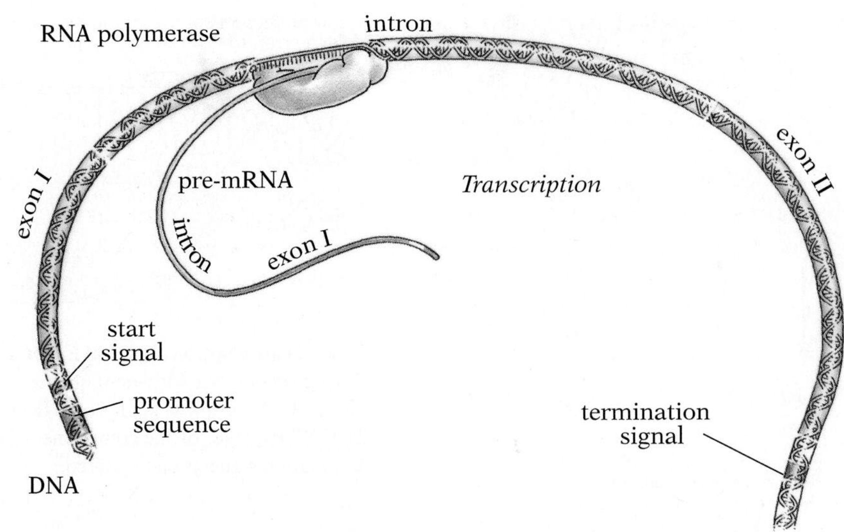

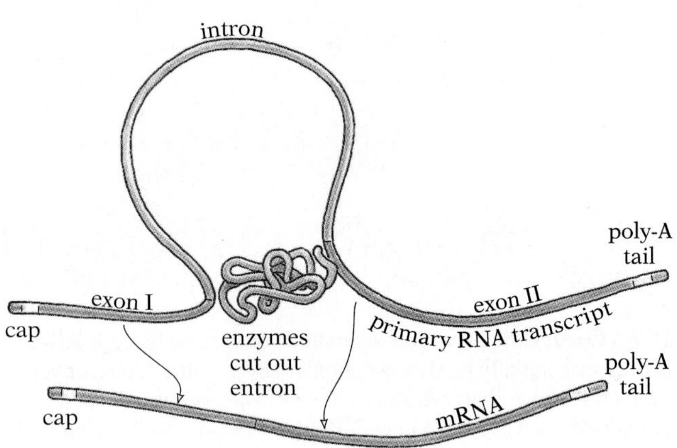

11.10 Messenger RNA processing in eucaryotes Top: The immediate product of transcription of a gene is a pre-mRNA molecule. The introns (intervening sequences) from the gene (gray) are transcribed along with the coding regions (pink). Bottom: In the process of conversion, the intron (gray) is cut out and the two coding regions, the exons (pink), are spliced together. For clarity, only one intron is shown. Finally, a "cap" consisting of a modified form of a guanine nucleotide is added at the front end of the transcript and a string of 100 to 200 adenines is added at the other end to form a poly-A tail.

cific proteins that bind to the promoter regions; without them the RNA polymerase cannot bind to the promoter. Enhancers are sequences of DNA some distance from the promoters that act to regulate transcription of certain genes. We shall have more to say about the role of enhancers in Chapter 12.

In Eucaryotic Cells, the mRNA Message Must Be Modified Before It Can Be Translated

The role of newly produced mRNA in procaryotes differs from that in eucaryotes (Fig. 11.7). In procaryotic cells, such as bacteria, the RNA transcript derived from the DNA template is functional mRNA, that is, it is ready to be used directly by the ribosomes for protein synthesis. In fact, translation of the mRNA by the ribosomes generally begins while transcription is still going on; thereafter the two processes occur simultaneously. In eucaryotic cells, on the other hand, the immediate product of DNA transcription is not functional mRNA but rather an mRNA precursor. This precursor mRNA, the ***primary RNA transcript,*** is not a usable messenger. In some sense it is only a rough draft. The primary transcript is extensively "edited" by enzymes before it leaves the nucleus and moves to the ribosomes. The editing frequently involves the removal of large regions of the primary transcript. The excised noncoding regions, referred to as intervening sequences, or ***introns,*** are cut out by special enzymes in the nucleus, and the rest of the coding sequences, or ***exons,*** are then spliced together to form functional mRNA[3] (Fig. 11.10). In addition, the transcript is modified at both ends: a special nucleotide "cap" composed of a modified form of the guanine is added to the "front" end of the mRNA and a string of 100 to 250 adenine nucleotides is added to the other end to form a ***poly-A tail.*** The purpose of these additions is yet to be discovered. Thus, the mRNA that moves out of the nucleus to the ribosomes is quite different from the original transcript (Fig. 11.10). This "cut and paste"

[3] The length of the average eucaryotic gene is 2,000 nucleotides, of which only 1,200 are in exons.

process is not completely errorfree; mutations do occur occasionally due to improper excision of introns. Intron removal is characteristic only of eucaryotic cells. The DNA of procaryotic cells does not have such sequences that must be removed (Fig. 11.11).

Why eucaryotic cells should have introns while procaryotic cells do not has been a question for molecular biologists ever since the discovery of introns in 1977. Because the first cells were procaryotic and eucaryotic cells evolved from them, it was thought that introns evolved in eucaryotic cells. Evidence now suggests that the opposite is true—that the ancestral procaryotic cells probably had introns, and that the introns were retained in eucaryotic cells but lost from procaryotes. Without introns the DNA can be replicated much faster—an advantage to small unicellular organisms living in environments where rapid reproduction is necessary.

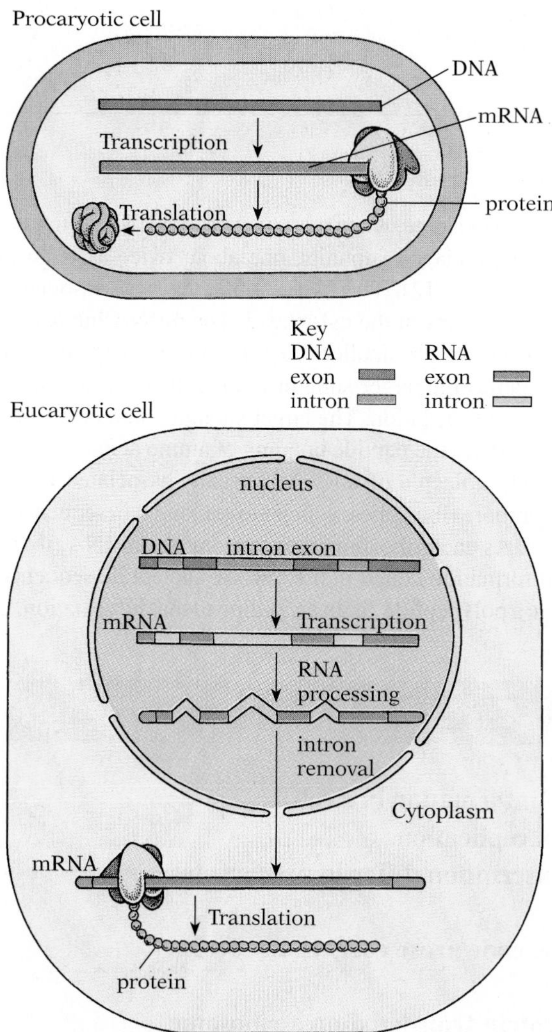

11.11 A comparison of transcription and translation in procaryotic and eucaryotic cells

In Translation, the Genetic Message Is Read Out in Three-Letter Coding Units to Specify Each Amino Acid in a Protein

As we have seen, transcription is initiated when an RNA polymerase binds to a promoter; the polymerase complex begins making an RNA copy at the start sequence, and stops at the termination signal. In procaryotes this primary transcript is immediately usable, but in eucaryotes the message is edited to remove introns. The resulting mRNA, which moves from the DNA to the ribosomes, contains the sequence of bases that specifies the identity and order of the amino acids that will be linked to form the protein the gene encodes. This sequence, which is preceded by a short "leader" and followed by a "trailer" (analogous to the blank margins at the top and bottom of a page), is decoded on a ribosome by the process of translation. During translation, information is translated from one molecular "language" to another (i.e., from the sequence of nucleotides in RNA to the sequence of amino acids in a protein). To understand the process of translation, it is necessary to understand first how information about amino acids is encoded by the nucleic acid sequence in mRNA.

There are only four different ribonucleotides in mRNA (corresponding to four deoxynucleotides in DNA), which must be capable of coding for at least 20 different amino acids. There is now convincing evidence that the ***codon,*** or coding unit, for an amino acid is a group of three nucleotides, a combination sometimes referred to as a triplet.

With a triplet codon, there are 64 different possible combinations (4^3). It is now known that all but three of the possible 64 triplets specify some amino acid. In some cases many triplets have the same meaning. For example, the codons ACU, ACC, ACA, and ACG all specify the amino acid threonine. Thus, there are synonyms in the genetic code. Such a code, in which some words are synonymous, is said to be a degenerate code.

The finding by Francis Crick and Sydney Brenner in 1961 that the genetic code uses three-letter words to designate the various amino acids spurred an intensive effort to construct a dictionary that would give the meaning of each triplet codon. Eventually this effort, which followed many different experimental routes, culminated in the key reproduced in Table 11.1. This code applies to essentially all living organisms.[4] In other words, with a few minor exceptions, the mRNA codons are always translated into the same amino acids, suggesting that the genetic code must have arisen early, as soon as the first cells arose on the primitive earth.

---

[4] Certain minor inconsistencies in the genetic code have been uncovered. For example, in some ciliates the codons UAA and UAG, which are normally termination signals, instead code for glutamine; in some bacteria, UGA, which can also be a termination signal, codes for tryptophan.

TABLE 11.1 *The genetic code (messenger RNA)*

First base in the codon	*Second base in the codon* U	C	A	G	*Third base in the codon*
U	Phenylalanine	Serine	Tyrosine	Cysteine	U
	Phenylalanine	Serine	Tyrosine	Cysteine	C
	Leucine	Serine	*Termination*	*Termination*	A
	Leucine	Serine	*Termination*	Tryptophan	G
C	Leucine	Proline	Histidine	Arginine	U
	Leucine	Proline	Histidine	Arginine	C
	Leucine	Proline	Glutamine	Arginine	A
	Leucine	Proline	Glutamine	Arginine	G
A	Isoleucine	Threonine	Asparagine	Serine	U
	Isoleucine	Threonine	Asparagine	Serine	C
	Isoleucine	Threonine	Lysine	Arginine	A
	Methionine[a]	Threonine	Lysine	Arginine	G
G	Valine	Alanine	Aspartic acid	Glycine	U
	Valine	Alanine	Aspartic acid	Glycine	C
	Valine	Alanine	Glutamic acid	Glycine	A
	Valine	Alanine	Glutamic acid	Glycine	G

[a] Also *initiation* when located at leading end of mRNA.

Ribosomal RNA Plays a Central Role in Protein Synthesis

The ribosomes, which are essential for protein synthesis in living cells, are organelles composed of three different types of ribosomal RNA (rRNA) and 60 to 70 proteins. More than half the mass of the ribosome is rRNA, and the importance of this rRNA in translation is just beginning to be appreciated. Each ribosome appears capable of synthesizing any protein for which it is supplied the necessary information by mRNA. It is thus analogous to a manufacturing machine that receives its instructions from a computer.

Each ribosome, when it is attached to mRNA, is made up of two mitten-shaped subunits, one about twice as large as the other (Fig. 11.12). The two subunits exist as associated but separate entities in the cytoplasm. The mRNA binds first to a specific site on the smaller subunit, causing a structural change that enables the larger subunit to bind, thereby beginning the process of translation. The larger subunit contains the enzyme that catalyzes the peptide bonding of amino acids.

Each molecule of mRNA is usually associated with four, five, or more ribosomes, which move along it in sequence (Fig. 11.13). As each ribosome moves along the mRNA, it "reads" the information coded in the mRNA nucleotide sequence and builds a polypeptide chain according to that information, trans-

MAKING CONNECTIONS

1. Make a diagram showing the flow of information in a cell.

2. Compare RNA transcription to DNA replication.

3. How does the initiation of RNA transcription differ in procaryotes and eucaryotes?

4. What is the evidence that the genetic code arose early in the evolution of life?

5. Trace the possible pathways of a protein translated on a ribosome within a cell.

lating the nucleotide sequence into an amino acid sequence at the rate of about 35 amino acids per second. The cluster of ribosomes associated with a single strand of mRNA is called a ***polyribosome,*** or sometimes simply a polysome (Fig. 11.13).

How does a ribosome read the codons? Evidence shows that the sequence of bases in mRNA is read three bases at a time, sequentially, from a specific starting point. Thus, in the mRNA sequence

...AUGACCAAAUAUUAA...

the first codon is AUG, the second ACC, the third AAA, and so forth. Using the genetic code in Table 11.1, we can see that the first codon, AUG, specifies the amino acid methionine; the

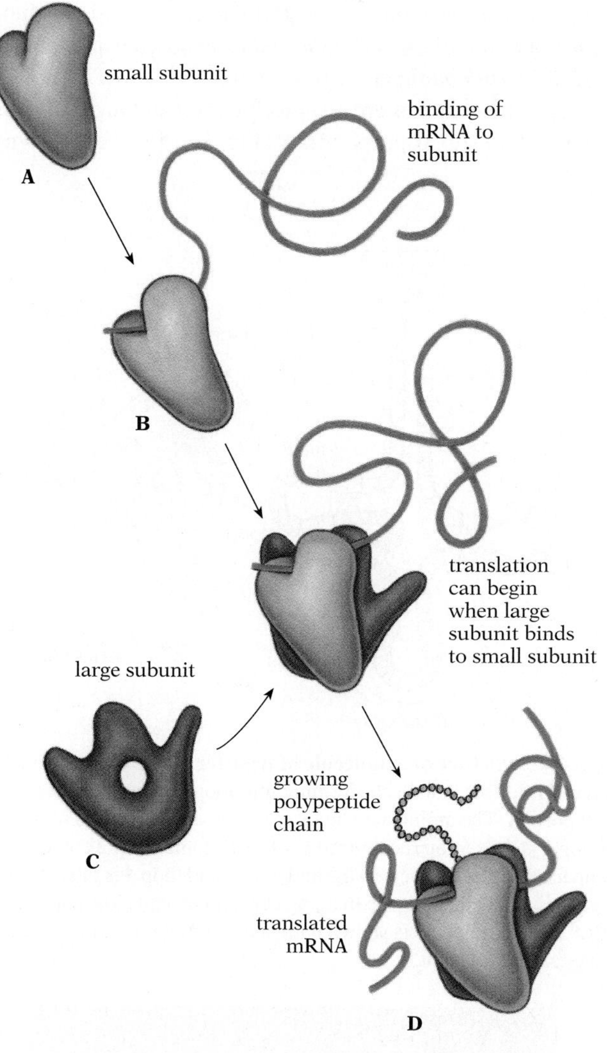

11.12 Synthesis of a polypeptide chain by a ribosome Free ribosomes exist as two separate subunits, one large and one small (A–C). A signal sequence on the mRNA binds to the small subunit, causing a change that permits it to bind the large subunit (B–C). The ribosome then translates the mRNA into a polypeptide chain, producing a protein (D). The growing chain is thought to be threaded out through the tunnel in the large subunit. The mRNA is much longer than shown here, and additional ribosomes can bind and begin translation while the first ribosome is still at work.

11.13 Simultaneous transcription and translation In procaryotes translation can begin as soon as the first part of a message has been transcribed from the chromosome, and several ribosomes can be involved simultaneously (left). The complex of mRNA and two or more ribosomes is often called a polysome. In the micrograph (right), the two thin parallel lines are DNA; the upper strand is being transcribed and, simultaneously, the RNA transcripts are being translated by ribosomes (dark spots). Note that the transcripts become longer from left to right as transcription of the gene proceeds. The growing polypeptide chains are not visible. The boxed area corresponds to the interpretive sketch. In eucaryotes transcription is always complete before translation begins.

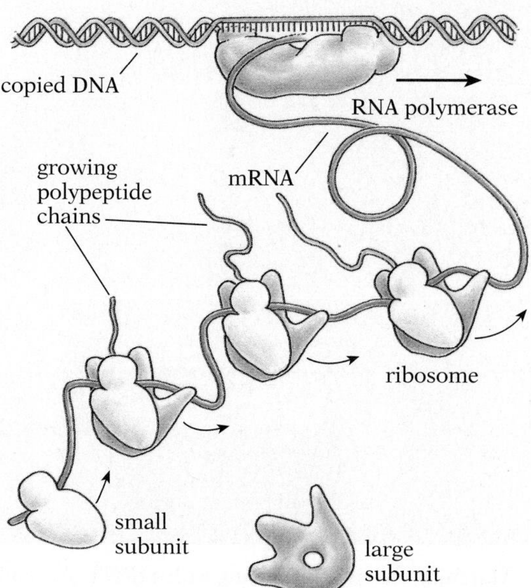

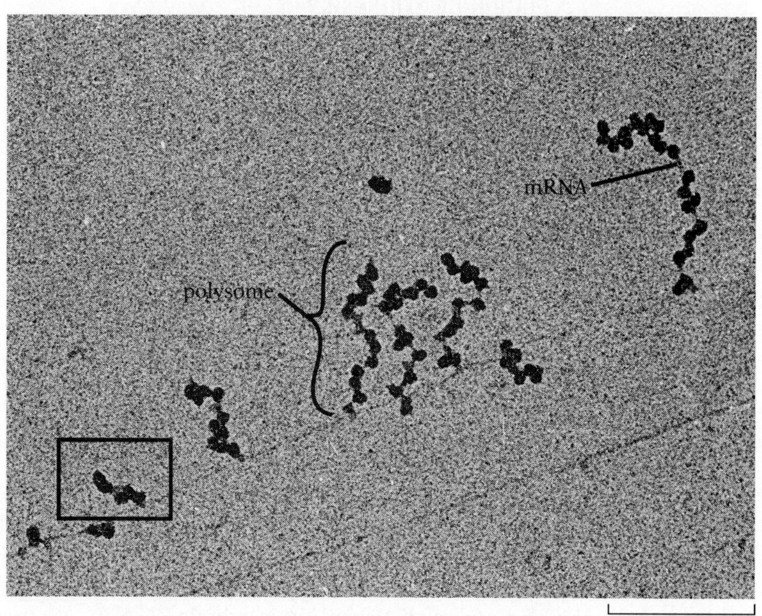

second, ACC, threonine; the third, AAA, lysine; and so on. Notice that the genetic code is given in mRNA codons, not DNA.

How does the ribosome know where to bind along the mRNA? In particular, how does it know where the message begins and where it ends? As you might guess, there are special start and stop signals in the mRNA. In bacteria, for instance, there is a specific sequence of nucleotides in the mRNA that acts as the ribosome binding site. Called the ***Shine-Delgarno sequence*** (after John Shine and Lynn Delgarno, the two investigators who discovered it in 1974), this is a purine-rich (i.e., guanine and adenine) sequence of three to 10 nucleotides near the end of the mRNA in the leader segment. The small part of the ribosome has an rRNA sequence complementary to the Shine-Delgarno sequence, so that the ribosome binds at this site. The actual "start" signal for translation is an ***initiation codon,*** which is the first AUG on the mRNA "downstream" of the Shine-Delgarno sequence. In eucaryotic cells the ribosome simply starts translation at the first AUG it encounters. In both cell types the ribosome begins translating at the AUG initiator and continues along the mRNA, taking each triplet in turn. The end of the mRNA nucleotide sequence coding for a single polypeptide chain is now thought to be indicated by one of three ***termination codons,*** codons that do not code for an amino acid (see Table 11.1) and whose sole function is to signal the end of the portion of the mRNA molecule that contains the instructions for synthesis of a single polypeptide chain. Thus, in the above sequence, UAA is the termination codon that signals the end of the stretch of mRNA specifying a protein. In effect, a termination codon functions as a period in the genetic language.

Transfer RNA Carries Amino Acids to the Ribosomes for Protein Synthesis

We have seen so far that mRNA, synthesized on a DNA template, moves to the ribosomes, where it functions in turn as a template for protein synthesis. The amino acids do not interact with the codons of the mRNA directly, however. Instead, another intermediate agent—tRNA—is involved.

In 1957, Mahlon Hoagland and his associates at Harvard University demonstrated that amino acids become attached to RNA *before* they arrive at the ribosomes. This RNA is neither mRNA nor rRNA, but a third type consisting of relatively small cytoplasmic molecules. It has been shown that each molecule of this RNA binds a single molecule of a specific amino acid, activates it with energy from ATP, and transports it to the ribosome; hence the name *transfer RNA,* or *tRNA* (Fig. 11.14).

At least one form of tRNA is specific for each of the 20 common amino acids. The amino acid arginine, for example, will combine only with tRNA specialized to transport arginine, the amino acid leucine will combine only with tRNA specialized to transport leucine, and so on. Each RNA is a single RNA strand folded back on itself to form an amino acid attachment site at one end of the molecule and an exposed triplet of bases, called an ***anticodon,*** at the other end (Fig. 11.14).

Specific enzymes are responsible for matching each amino acid with the appropriate tRNA (Fig. 11.15). These enzymes,

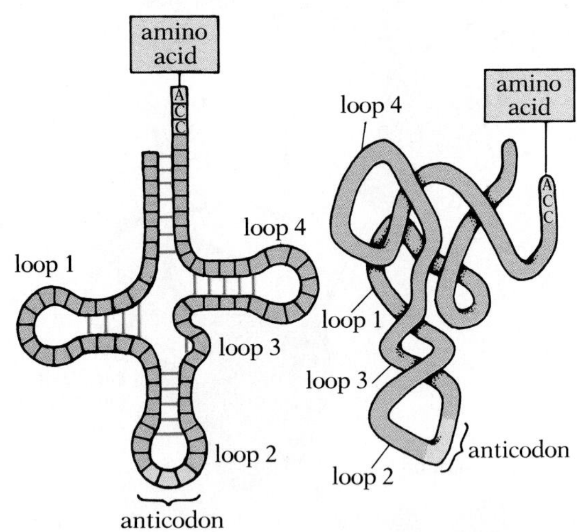

11.14 Structure of a molecule of transfer RNA The single RNA chains folds back on itself, forming four loops and an unpaired terminal portion. The amino acid binds to the unpaired terminal region. Loop 1 is the region recognized by the ribosome, loop 2 contains the unpaired triplet that acts as the anticodon, and loop 4 is the region recognized by the enzyme that joins the amino acid to its appropriate tRNA. The molecule is shown flattened at left; its coiled structure is diagrammed at right.

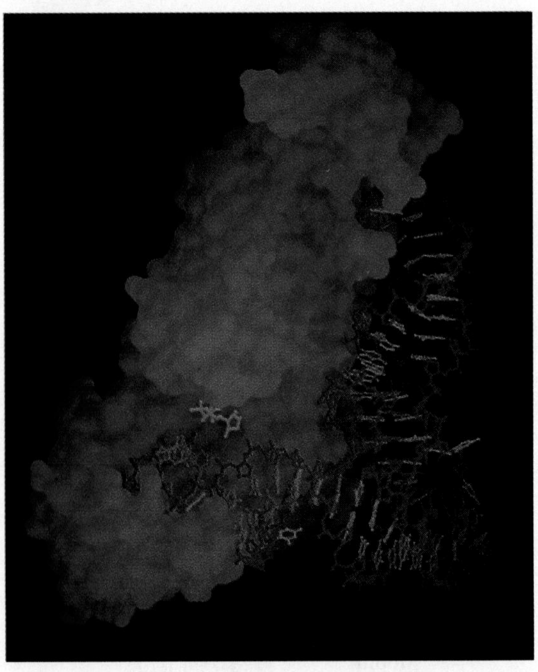

11.15 Matching enzyme interacting with a tRNA A computer-generated image of an enzyme (blue) preparing a tRNA (red and orange) to load a particular amino acid. The green molecule is ATP.

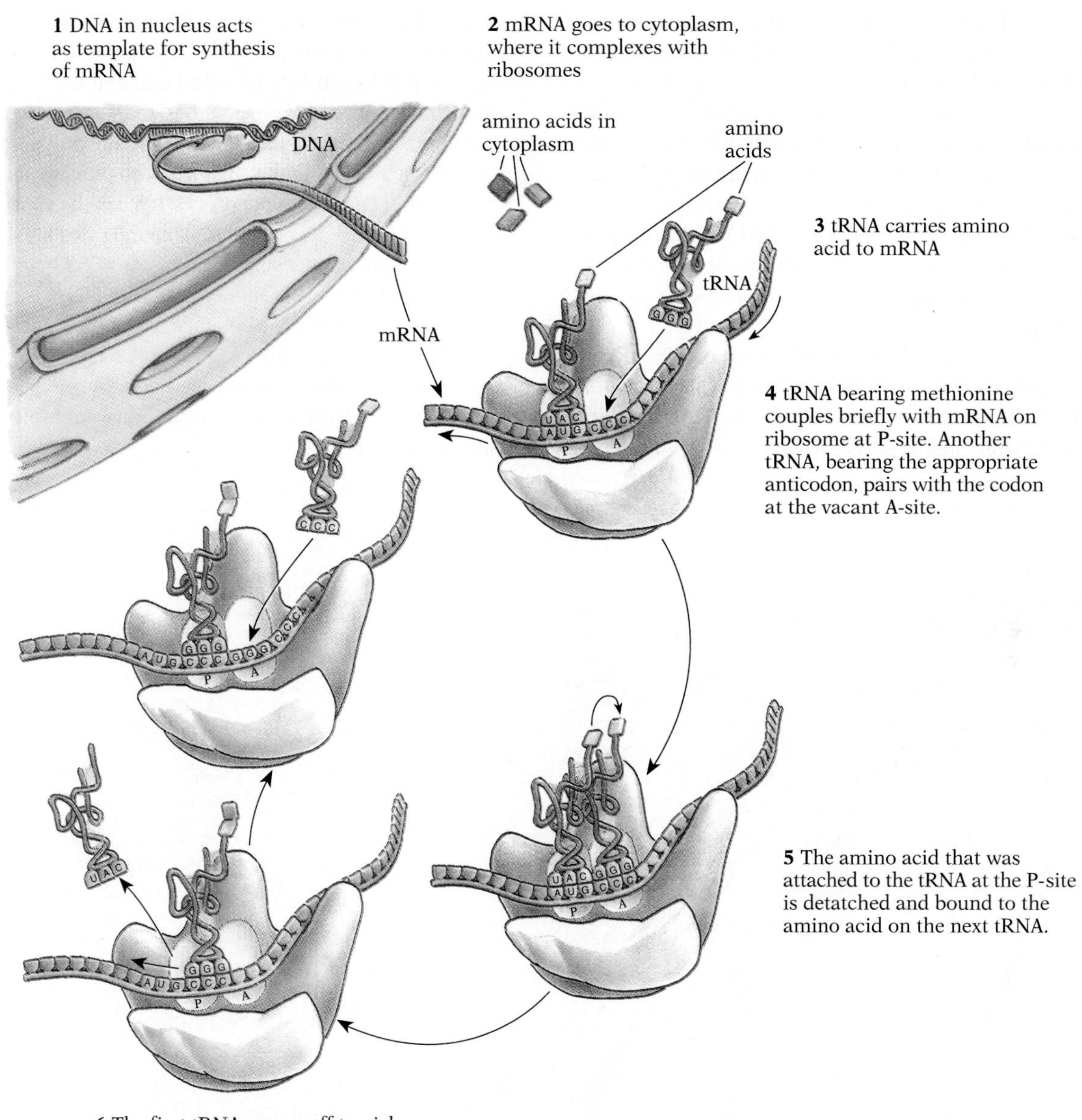
1 DNA in nucleus acts as template for synthesis of mRNA
2 mRNA goes to cytoplasm, where it complexes with ribosomes
DNA
amino acids in cytoplasm
amino acids
3 tRNA carries amino acid to mRNA
tRNA
mRNA
4 tRNA bearing methionine couples briefly with mRNA on ribosome at P-site. Another tRNA, bearing the appropriate anticodon, pairs with the codon at the vacant A-site.
5 The amino acid that was attached to the tRNA at the P-site is detatched and bound to the amino acid on the next tRNA.
6 The first tRNA moves off to pick up more amino acid, and the ribosome moves along the mRNA.

11.16 Steps involved in protein synthesis Messenger RNA is synthesized, one chain of the DNA gene serving as the template (1). This mRNA then goes into the cytoplasm and becomes associated with ribosomes (2). The various types of tRNA in the cytoplasm pick up the amino acids for which they are specific and bring them to a ribosome as it moves along the mRNA (3). Each tRNA bonds to the mRNA at a point where a triplet of bases complementary to an exposed triplet on the tRNA occurs (4). The amino acid attached to the first tRNA is detached and bound by a peptide bond to the amino acid on the second tRNA (5). The ordering of the tRNA molecules automatically orders the amino acids, which are linked by peptide bonds. Synthesis of the polypeptide chain thus proceeds one amino acid at a time in an orderly sequence as the ribosomes move along the mRNA. As each tRNA donates its amino acid to the growing polypeptide chain, it uncouples from the mRNA and moves away into the cytoplasm, where it can be used again (6). The various molecules and organelles shown here are not drawn to scale with respect to one another or to the cell.

which recognize the structure of a particular amino acid and its specific tRNA, attach the amino acid to the tRNA, using ATP as an energy source. The activated tRNA then carries the amino acid to a ribosome that is moving along a strand of mRNA (Fig. 11.16). There the tRNA becomes attached to the mRNA by complementary base pairing.

For example, in the case of the mRNA codon CCG (cytosine-cytosine-guanine), which codes for the amino acid proline, one type of tRNA for proline will have the anticodon sequence GGC, which is complementary to CCG. Whenever a molecule of this type of proline tRNA with an attached molecule of proline approaches a molecule of mRNA, its exposed GGC triplet can bond to the mRNA only at those points where the mRNA has a CCG triplet. Similarly, in the case of GUA (guanine-uracil-adenine), which codes for valine, one type of tRNA for valine will have the anticodon CAU, which is complementary to GUA. The exposed CAU triplet of valine tRNA can bond to the mRNA only at points where a GUA triplet occurs.

In short, the sequence of three-base codons on the mRNA determines the precise order in which the different tRNAs with their attached amino acids will bond to the messenger molecule. This sequential binding of tRNAs to the mRNA automatically brings the amino acids to the ribosomes in the proper sequence. As each amino acid is brought to the ribosome by its tRNA, it is linked by a peptide bond to the length of polypeptide that has already been synthesized. Once a tRNA molecule has donated its amino acid to the growing polypeptide chain, it leaves the mRNA and moves away to pick up another load of amino acid and repeat the process. When the peptide chain has been completed, it is released from the ribosome to assume its functional role in the cell.

To summarize the model then, the DNA of the gene determines the mRNA sequence, which determines protein structure, which controls chemical reactions and produces the characteristics of the organism. The probable pathway for the synthesis of a hydrolytic enzyme for a eucaryotic cell is given in Figure 11.17.

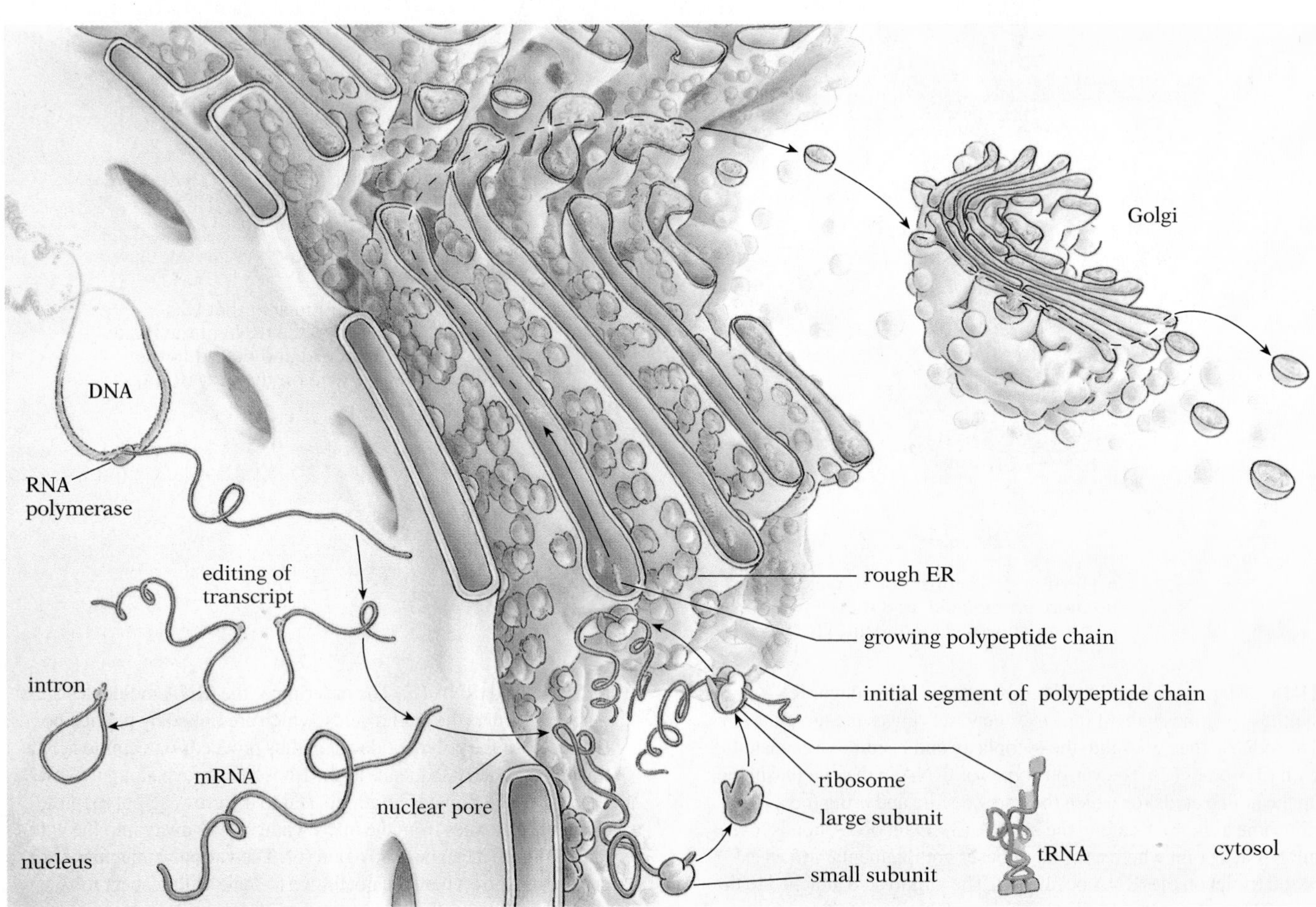

11.17 Synthesis of a hydrolytic enzyme Most mRNA is translated freely in the cytoplasm, but when protein products are destined for the endoplasmic reticulum (ER) lumen, the ER membrane, or secretion, translation of mRNA occurs on the rough ER. A gene is transcribed (left), and the transcript is processed to form mRNA. Next, the mRNA passes through the nuclear pore and binds to a small ribosomal subunit. A large ribosomal subunit then binds, and translation begins as tRNAs deliver amino acids until a special signal sequence of bases is reached. The ribosome complex then binds to the rough ER and translation resumes, with the growing polypeptide chain passing into the lumen of the ER. Finally, the newly synthesized protein, in this case a hydrolytic enzyme, moves to the smooth ER, where it is packaged into a vesicle for transport to the Golgi apparatus.

Many antibiotics function by interfering with protein synthesis on the ribosome. For example, streptomycin inhibits procaryotic ribosomes, causing the ribosome to misread the mRNA message. Tetracycline blocks the binding of the tRNA molecules to the mRNA in procaryotic ribosomes, and puromycin causes early termination of protein synthesis. Such antibiotics are particularly useful because they inhibit only the bacterial (procaryotic) ribosomes—the eucaryotic ribosomes continue to function normally.

Mutations Affect Protein Synthesis by Changing the Genetic Message

It is now apparent why mutations—alterations in the sequence of bases in DNA—can change the information content of the genes and hence interfere with the normal process of transcription and translation. As we know from our discussion of DNA repair at the end of Chapter 10, genes are subject to various mutational events. Here we shall examine briefly some of the consequences of these events.

Two types of mutation are particularly important: ***deletion*** of nucleotides from the DNA sequence and ***addition*** of extra nucleotides to the sequence. Additions or deletions (the latter are the more common) often result in the production of inactive enzymes, since the ribosomes begin to read incorrect triplets from the point at which the addition or deletion occurs and the original meaning of subsequent codons is lost. Suppose, for example, that the message is

...THE BIG RED ANT ATE ONE FAT BUG...

Deletion of the first *E* will make it

...THB IGR EDA NTA TEO NEF ATB UG...

A single deletion or addition changes the reading frame, i.e., it redefines all subsequent codons. Such mutations are referred to as ***frameshift mutations.*** If the frame shift occurs near the beginning of the message, the sequence of the protein is completely disrupted.

The third and most common type of mutation is ***base substitution*** (also called point mutation), in which one nucleotide is replaced by another. A codon that normally has the base composition CGG, say, might be changed to CAG. This new triplet would code for a different amino acid. Base substitution is not as serious as deletion or addition, however, because in most codons a change in the third base does not alter the meaning (notice all the synonymous codons in Table 11.1). Moreover, even when an amino acid is changed, the substitution may have little effect on the activity of the protein. Sometimes, however, a base substitution can have important consequences in the functioning of a protein. For example, the difference between normal hemoglobin and the defective sickle-cell hemoglobin is the substitution of the amino acid valine for glutamic acid at position six in the polypeptide chain. This substitution alters the shape of the hemoglobin molecule so that it forms long fibers and deforms the red blood cells when oxygen is low. The change is produced by a single base substitution in one hemoglobin gene: an adenine is substituted for a thymine in the sixth DNA codon (from CTC to CAC). This substitution of course changes the mRNA codon and the amino acid. We shall have more to say about sickle-cell anemia in Chapter 14.

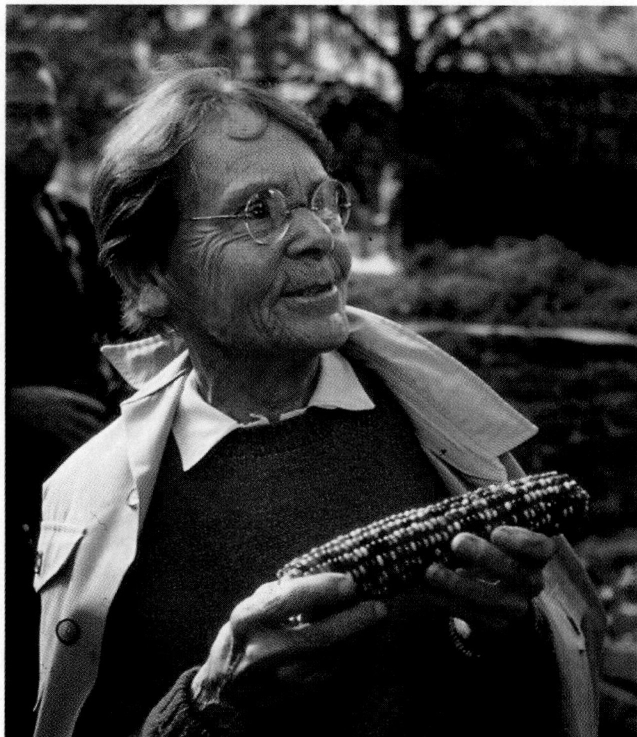

11.18 Barbara McClintock, through the study of maize, discovered that some genes could "jump" within and between chromosomes.

Another very common type of mutation is ***transposition.*** This phenomenon, which we shall discuss further in Chapter 31, results when long stretches of DNA move from one position on a chromosome to another, or to another chromosome. Discovered by Barbara McClintock (Fig. 11.18), who won the Nobel Prize in 1983, and sometimes referred to as "jumping genes," ***transposons*** are mobile genetic elements that can move about and insert themselves into almost any stretch of DNA. When they insert themselves into the middle of a gene, they cause mutations that inactivate or alter the gene's activity. The majority of easily detectable spontaneous mutations in *Drosophila,* for example, are known to result from transposition.

MAKING CONNECTIONS

1. How do antibiotics such as streptomycin and tetracycline attack bacteria? Why don't they harm the cells of the person taking the antibiotic?

2. By referring to Table 11.1, explain how different base-substitution mutations in DNA could cause phenylalanine to be replaced by either serine or leucine in a protein. Which of these substitutions would be more likely to affect the function of the protein? (Hint: consult Figure 3.19 on page 65 and recall the importance of R-group interactions in determining protein conformation.)

Mutagenic Agents Are Common Causes of Cancer

As noted in the previous chapter, high-energy radiation, including both ultraviolet light and ionizing radiations such as X rays, cosmic rays, and emissions from radioactive materials, can cause genetic mutations. In addition, a variety of chemicals have been found to be mutagenic. A normal spontaneous mutation rate for a single gene would be one mutation in every 10^6 to 10^8 replications, but the rate can be greatly increased by unusual exposure to mutagenic agents.

Ionizing radiations sometimes induce simple base-substitution mutations, but they also frequently produce large deletions of genetic material, presumably because their high-energy particles collide with the DNA molecules and cause breaks to occur. This property of radiation can be used to destroy cancerous tissue, but unfortunately also damages normal tissue.

Some mutagenic chemicals produce their effects by directly converting one base into another. For example, nitrous acid (HNO_2), a major component of cigarette smoke, is a very powerful mutagen that converts cytosine into uracil (Fig. 11.19). Other chemical mutagens, called base analogs, are similar enough to one of the normal nucleotide bases that they are incorporated into nucleic acids in place of one of the normal bases. An example is 5-bromouracil, an analog of thymine. When a strand of DNA contains a unit of 5-bromouracil instead of thymine, it is prone to errors of replication. Most of these mutations are detected and repaired before they can exert any effect, but by flooding a cell with 5-bromouracil, it is possible to overwhelm the repair enzymes. This strategy is often used in cancer chemotherapy to kill rapidly dividing tumor cells. An unfortunate side effect is that some highly active normal cells—those producing hair or those lining the digestive tract for instance—are also killed. This is why many chemotherapy patients suffer hair loss and digestive ailments.

There is now a wealth of evidence for a close correlation between the mutagenicity of a chemical and its carcinogenicity, that is, between the potential of a chemical for producing genetic mutations and its cancer-inducing activity. More than 95 percent of the known carcinogens are also mutagens. The tars in cigarette smoke, for instance, are highly mutagenic and carcinogenic. The close correlation between mutagenicity and carcinogenicity strongly suggests that many cancers are caused, at least in part, by mutations in somatic cells. Thus, we see that alterations of the DNA are important not only in germ cells, where they may affect future offspring, but also in body cells, whose metabolism or growth they may disrupt, causing cancer or cell degeneration.

The connection between mutagenicity and carcinogenicity is the basis for the widely used Ames test, in which various chemicals, such as environmental pollutants, reagents used in industrial processes, proposed new drugs, and food additives, are screened for potential carcinogenicity. The compound to be tested is added to special culture of bacteria. When the mixture of bacteria and chemical is grown on a medium deficient in a required amino acid, some bacterial cells may undergo a mutation that enables them to grow on the deficient medium. The increase in the mutation rate of the bacteria over the normal

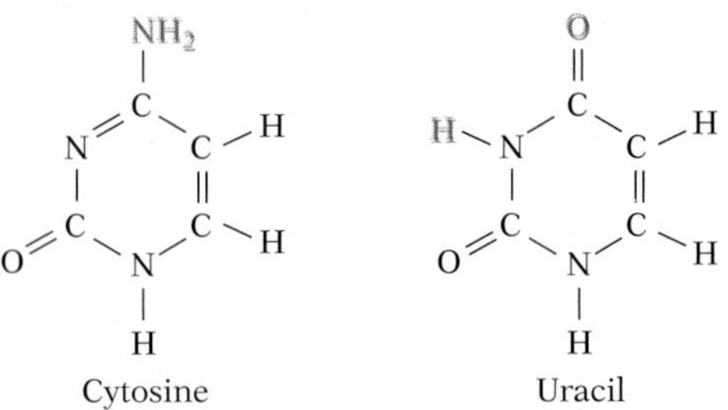

11.19 Deamination of cytosine to form uracil Nitrous acid, a powerful mutagen, causes the removal of an amino group (brown) from cytosine, converting it into uracil.

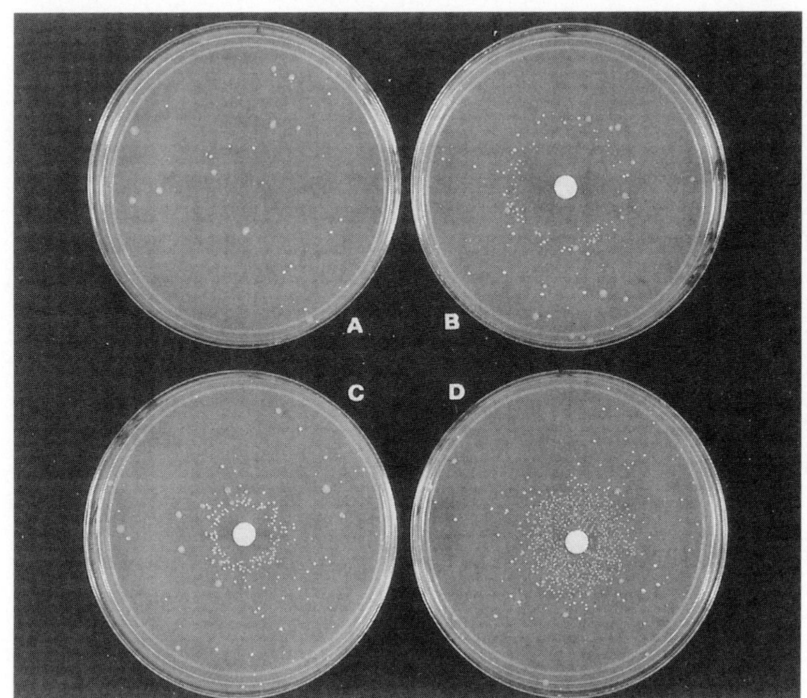

11.20 The Ames test Each of these Petri dishes has vast numbers of histadine-requiring bacteria in a layer of agar. The agar contains everything the bacteria need to grow except histadine. The white spots in (A) indicate colonies descended from a bacterium in which a spontaneous mutation restored the ability to synthesize histadine. The large white disks in the center of the other dishes contain a mutagenic substance that is diffusing out from the disk: furylfuramide (B), aflatoxin (C), and 2-aminofluorene (D). The mutagenic potential of these chemicals is indicated by the large number of colonies (small white spots) able to grow as a result of a genetic change that restored their capacity to synthesize histadine. Note that 2-aminofluorene has the most colonies, and is therefore the most mutagenic.

spontaneous level is then used to predict the likely cancer-inducing potency of the chemical (Fig. 11.20).

What, Then, Is a Gene?

By 1902 it was generally accepted that the factors of inheritance, first called genes by Wilhelm Johannsen, a Danish botanist, in 1911, are associated with the chromosomes. Physically, the genes were thought of as tiny particles arranged in linear sequence on the chromosomes. Throughout this chapter we have used the Beadle-Tatum definition of a gene: the length of DNA coding (i.e., containing the information for) a single polypeptide. Even this definition is not completely adequate, however.

We now know that each chromosome consists of a single, long, unbroken molecule of DNA, and that the genes are arranged sequentially along it. In bacteria the average gene is approximately 1,000 nucleotide bases long. In eucaryotic cells the genes are generally much longer (1,500 to more than 2,000,000 base pairs), since most eucaryotic genes are "split genes" and have one or more introns.

Although our discussion of the gene has focused on the function of determining a polypeptide chain, it must not be overlooked that some functional units of DNA, instead of determining polypeptide chains, code for tRNA, rRNA, and the like. In addition, some parts of the DNA molecule (such as enhancer and promoter regions) serve exclusively in the regulation of genetic transcription. Hence, it would perhaps be more accurate to define the gene as the sequence of nucleotides in DNA codes for an RNA molecule and the nontranscribed regulatory regions that influence its transcription.

Besides the DNA in genes and regulatory regions, there is much DNA of unknown function. In mammalian cells, for example, only about 1 percent of the DNA is actually transcribed into functional RNA. Indeed, there are long stretches of what appear to be nonsense (noncoding) sequences between adjacent genes as well as intervening sequences within a single gene. Given this level of complexity, you can begin to appreciate the difficulty in defining a gene in a eucaryotic organism.

Some Cytoplasmic Organelles Have Their Own Genes

Our attention in this chapter has been focused so far on the chromosomal genes as the hereditary units that determine the characteristics of organisms. But the chromosomal genes are not the only hereditary constituents of cells. It is now known that some cytoplasmic organelles, such as chloroplasts and mitochondria, have their own chromosomes and are self-replicating, and that their characteristics are at least partly determined by genes located within them.

As noted in Chapter 3, mitochondria and chloroplasts share a number of characteristics with procaryotes, including rela-

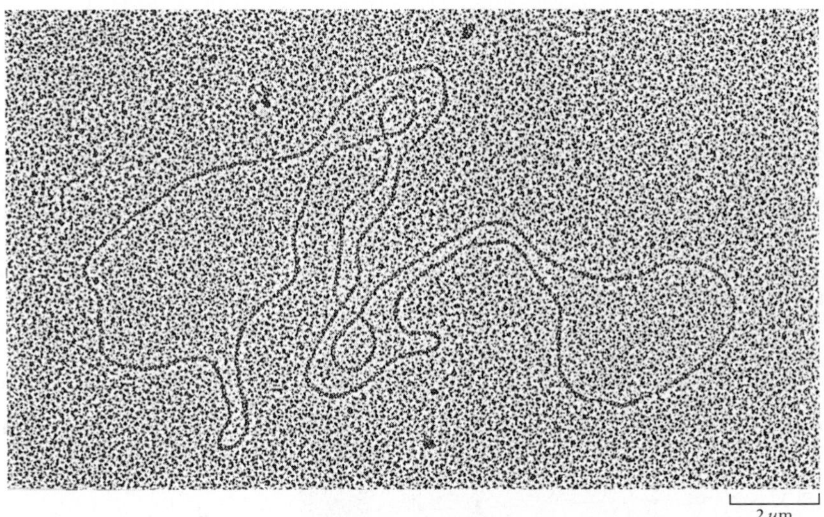

11.21 Organelle DNA Electron micrograph of mitochondrial DNA from a liver cell of a chicken. As in bacteria, the DNA forms a single, circular chromosome.

tively small ribosomes and (usually) circular chromosomes (Fig. 11.21). According to the endosymbiotic hypothesis, these organelles were once free-living procaryotes—probably purple bacteria in the case of mitochondria, and cyanobacteria or prochlorophytes in the case of chloroplasts—that took up symbiotic residence inside primitive eucaryotic cells. Most recent discoveries about organelle heredity are consistent with this view.

The DNA in these organelles is, similar to the DNA in procaryotes, almost always a single, circular molecule; moreover, it is "naked" DNA, in that it is not wound around nucleosome cores, as are eucaryotic chromosomes, and is not enclosed in a nuclear membrane (Fig. 11.21). However, the chromosome is considerably smaller than those of procaryotes, and most of the genes necessary for organelle function are found in nuclear DNA. For instance, the entire base sequence (16,569 nucleotides) of human mitochondrial DNA is now known. The DNA codes for about 40 products, including some of the rRNAs for the ribosomes, all the tRNAs, and some of the organelle proteins. The other mitochondrial RNAs, enzymes, and proteins are encoded by nuclear genes and are transported to the organelles.

The organelle ribosomes are also smaller and much closer in size and composition to those of procaryotes than to eucaryotes. The transcription and translation processes in the mitochondria and chloroplasts are remarkably similar to the corresponding processes in procaryotes, and different from those in eucaryotes.

MAKING CONNECTIONS

1. Why is the thinning of the ozone layer in the atmosphere a matter for concern?

2. Why are hair loss and malfunctions of the digestive system common side effects of cancer chemotherapy?

3. Explain how the Ames test is used to screen chemicals such as drugs and food additives for their cancer-causing potential.

4. Why is the definition of a gene as a length of DNA coding for a single polypeptide not entirely correct?

5. What characteristics of the DNA and protein-synthesis machinery of mitochondria and chloroplasts are consistent with the endosymbiotic hypothesis for the origin of these organelles?

CONCEPTS IN BRIEF

1. According to the *one gene–one enzyme hypothesis,* each gene directs the synthesis of an enzyme that controls a chemical reaction of the cell; the chemical reactions, in turn, determine the phenotypic characteristics. Because some proteins are composed of two or more chemically different polypeptide chains, each determined by its own gene, the hypothesis has been restated as the *one gene–one polypeptide hypothesis.*

2. The sequence of bases in DNA determines the sequence in which amino acids must be linked in protein synthesis. During *transcription,* RNA polymerase binds to the promoter and partially uncoils the two nucleotide chains, and one of the chains acts as the template for the synthesis of single-stranded *messenger RNA (mRNA).* The mRNA then leaves the chromosome and moves into the cytoplasm, where it becomes associated with the ribosomes.

3. In eucaryotic cells the mRNA must be processed before it leaves the nucleus; the noncoding introns must be spliced out and a "cap" of modified guanine and a tail of adenines added to the transcript. Only the exons, the coding sequences, survive the editing process.

4. The mRNA carries information for protein synthesis, but the nucleic acid message must be converted into an amino acid sequence, a process known as *translation.* The coding unit, or *codon,* in the nucleic acids is three nucleotides long; each codon specifies a particular amino acid. The genetic code is essentially universal.

5. During translation, *transfer RNAs* (*tRNAs*) in the cytoplasm—each specific for one of the 20 amino acids—bring amino acids to a ribosome as it moves along the mRNA. Each tRNA attaches to the mRNA at the point where a triplet of mRNA bases is complementary to the *anticodon* on the tRNA. This ordering of the tRNAs along the mRNA automatically orders the amino acids, which are then linked by peptide bonds. When a ribosome reaches a termination codon, it releases the completed polypeptide chain.

6. Mutations are alterations in the sequences of bases in DNA that change its information content. Several types of mutations are possible: *additions, deletions,* and *base substitutions* (point mutations). Most mutations are detected and repaired before they exert any effect. Additions and deletions, because they lead to a shift in the reading frame *(frameshift mutation),* are the most serious.

7. High-energy radiation and a variety of chemicals can cause genetic mutations. There is a strong relationship between the mutagenicity of a chemical and its carcinogenicity, or cancer-inducing activity.

8. Cytoplasmic organelles such as chloroplasts and mitochondria have their own DNA and replicate themselves independently of the nucleus. The DNA in these organelles is usually circular and is not wound on nucleosomes. Transcription and translation are similar to the corresponding processes in procaryotes.

STUDY QUESTIONS

1. **According to the evidence provided by certain inherited human diseases (e.g., PKU) and the experiments of Beadle and Tatum on *Neurospora,* how do genes determine phenotypic characteristics? (pp. 215–216)**
2. **How would you convince someone that "one gene–one polypeptide" is a more accurate hypothesis than "one gene–one enzyme"? (p. 216)**
3. **Suppose you are given a sample of nucleic acid. What three characteristics would you test to determine if the sample was DNA or RNA? What are the three major types of RNA? (pp. 216–218)**
4. **Describe how RNA is transcribed from DNA, including the roles of RNA polymerase and promoter sites. (pp. 218–219)**
5. **What are the roles of transcription factors and enhancer elements? (pp. 219–220)**
6. **In eucaryotes, how is the primary RNA transcript "edited" to form a functional mRNA molecule?**

What is the advantage of the procaryotic system of RNA synthesis, which requires no editing? (pp. 220–221)

7. **Describe the general properties of the genetic code. What is a *codon?* Why couldn't the codon be two nucleotides long? (p. 221)**
8. **Describe the structure of a ribosome and explain how it "reads" a molecule of mRNA during the process of translation. How is translation started and stopped? (pp. 222–224)**
9. **Explain the role of tRNA in translation. What is the minimum number of different tRNA molecules necessary for protein synthesis? What is an *anticodon?* (pp. 224–226)**
10. **Why are mutations involving base substitution generally less harmful than additions or deletions? (p. 227)**
11. **Which is more disruptive to the genetic message, a frameshift mutation at the beginning or end of a nucleotide sequence to be read? Why? (p. 227)**
12. **How can transposons disrupt the structure and function of genes? (p. 227)**
13. **What percentage of known carcinogens are also mutagens? What is the likely explanation for the close correlation between mutagenicity and carcinogenicity? (pp. 228)**
14. **Which organelles contain DNA? Do they contain all the genetic information necessary to function independently of the cell in which they are found? (pp. 229–230)**

CHAPTER 12

THE CONTROL OF GENE EXPRESSION

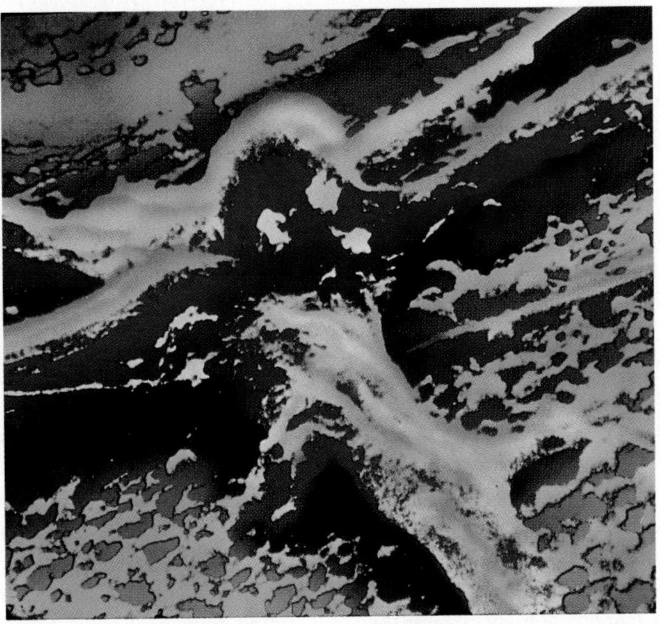

In the process of mitosis, *every cell of a multicellular organism inherits an identical set of chromosomes* from the original, fertilized egg cell. The nucleotide sequences of the DNA in these chromosomes carry in full the genetic information that is every cell's evolutionary endowment. Even though all cells of an organism receive identical DNA, however, individual cells may look entirely different from other cells, and they may behave in entirely different ways. A knowledge of how gene expression is controlled is central to understanding the process of development in multicellular organisms.

In particular, cells differ in the kinds of proteins they manufacture. In the human body, for example, cells of the retina produce light-sensitive pigments, cells that line the stomach make digestive enzymes, and those of the pancreas secrete insulin. All three types of cells contain the complete genetic blueprints for manufacturing all three types of proteins, but the genes for specific proteins are expressed, or turned on, in certain instances and not in others.

In fact, only a particular subset of all the DNA a cell contains will ever be used to generate proteins. Furthermore, only about 1 percent of the genetic material is active at any one time. What determines when and how each cell will act on the genetic instructions it has inherited? This question pertains not only to normal cellular function but also to the abnormal growth and development characteristic of cancer cells. In this chapter we shall examine some of the mechanisms controlling gene expression, a process involving chemicals that interact directly or indirectly with DNA or mRNA.

Gene Regulation Was First Explored in Bacteria

Bacterial cells are genetically far simpler than eucaryotic cells. They have just one chromosome, which is circular and has little protein associated with it, and only about 3,000 genes (compared to between 50,000 and 100,000 in humans). For this reason, and because bacteria can be grown rapidly in huge numbers under controlled conditions in the laboratory, they have been especially useful for studying the regulation of gene expression. Much of our current knowledge of this subject

Chapter opening photo: False-color scanning electron micrograph of a cancer cell invading normal tissue. Cells become cancerous when they develop the ability to invade healthy tissue and to spread from the original site of growth to other parts of the body.

comes from research on the common intestinal bacterium *Escherichia coli.*

Early investigators of gene regulation in bacteria proceeded on the basis of several assumptions. First, they assumed that proper control of gene expression would require that only those genes whose products were needed at any given moment be expressed. Second, since most genes code for an enzyme that controls only a single step in a biochemical pathway, the genes coding for several enzymes in the same pathway might be expected to be regulated as a group. Third, since the function of a pathway is to change a reactant into a product, the availability of the reactant in the cell might be expected to turn on transcription, whereas the availability of the final product might turn it off. These assumptions have proven correct in many cases of transcription in bacteria.

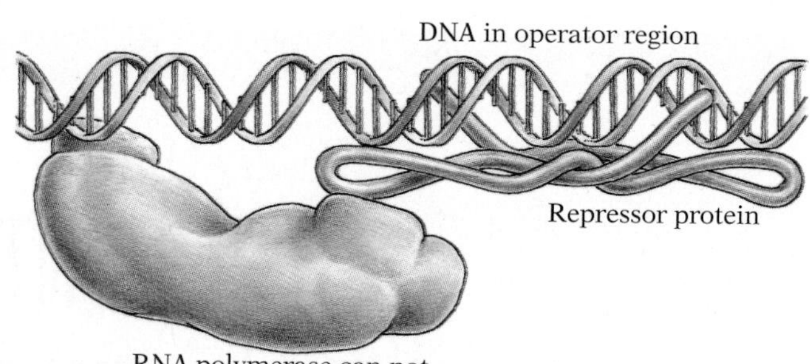

12.2 Binding of a repressor protein to an operator Each repressor protein bears active sites that match one unique sequence of bases on the DNA in a specific operator region. The repressor probably has a shape corresponding to the varying widths of the DNA molecule and polar R groups that form hydrogen bonds with polar groups in DNA. When the repressor protein is bound to the DNA of the operator, RNA polymerase cannot bind to the promoter and transcription is blocked.

In the Jacob-Monod Model, a Group of Genes Is Controlled as a Unit

In the early 1960s, the French biochemists François Jacob and Jacques Monod (Fig. 12.1) formulated a powerful model of the control of gene expression in bacterial cells, based on their investigation of enzyme synthesis in *E. coli.* They worked mainly with the enzyme β-galactosidase, which, along with two other enzymes, catalyzes the breakdown of the double sugar lactose to glucose and galactose. They found that these enzymes are usually produced only when lactose is present in the medium. In other words, lactose acts as an ***inducer,*** turning on the production of the enzymes that degrade it.

Jacob and Monod were eventually able to demonstrate the participation of four genes in the production of β-galactosidase and the two other enzymes involved in lactose metabolism. Three of the genes are ***structural genes,*** each specifying the amino acid sequence of one of the three enzymes. The fourth is a ***regulator gene,*** which controls indirectly the activity of the structural genes. They proposed that the regulator gene, which is located near the structural genes, encodes the information for the synthesis of a ***repressor protein,*** which inhibits transcription of the structural genes.

Jacob and Monod later discovered a special region of DNA next to the structural genes for these enzymes, that functions as an on–off switch, determining whether or not the structural genes will be transcribed. The repressor protein (Fig. 12.2) binds tightly to this region, which they called the ***operator.*** Subsequently it was found that the operator (which is not transcribed and does not code for a specific product) overlaps a portion of the ***promoter,*** the region of DNA to which RNA polymerase binds to start transcription (see Fig. 11.8, p. 219). This overlap explains why, when the repressor protein is bound to the operator, RNA polymerase cannot bind to the promoter, and transcription is blocked (Fig. 12.3). Together, the promoter, the operator, and the structural genes make up what Jacob and Monod referred to as an ***operon.***

A

B

12.1 François Jacob (A) and Jacques Monod (B)

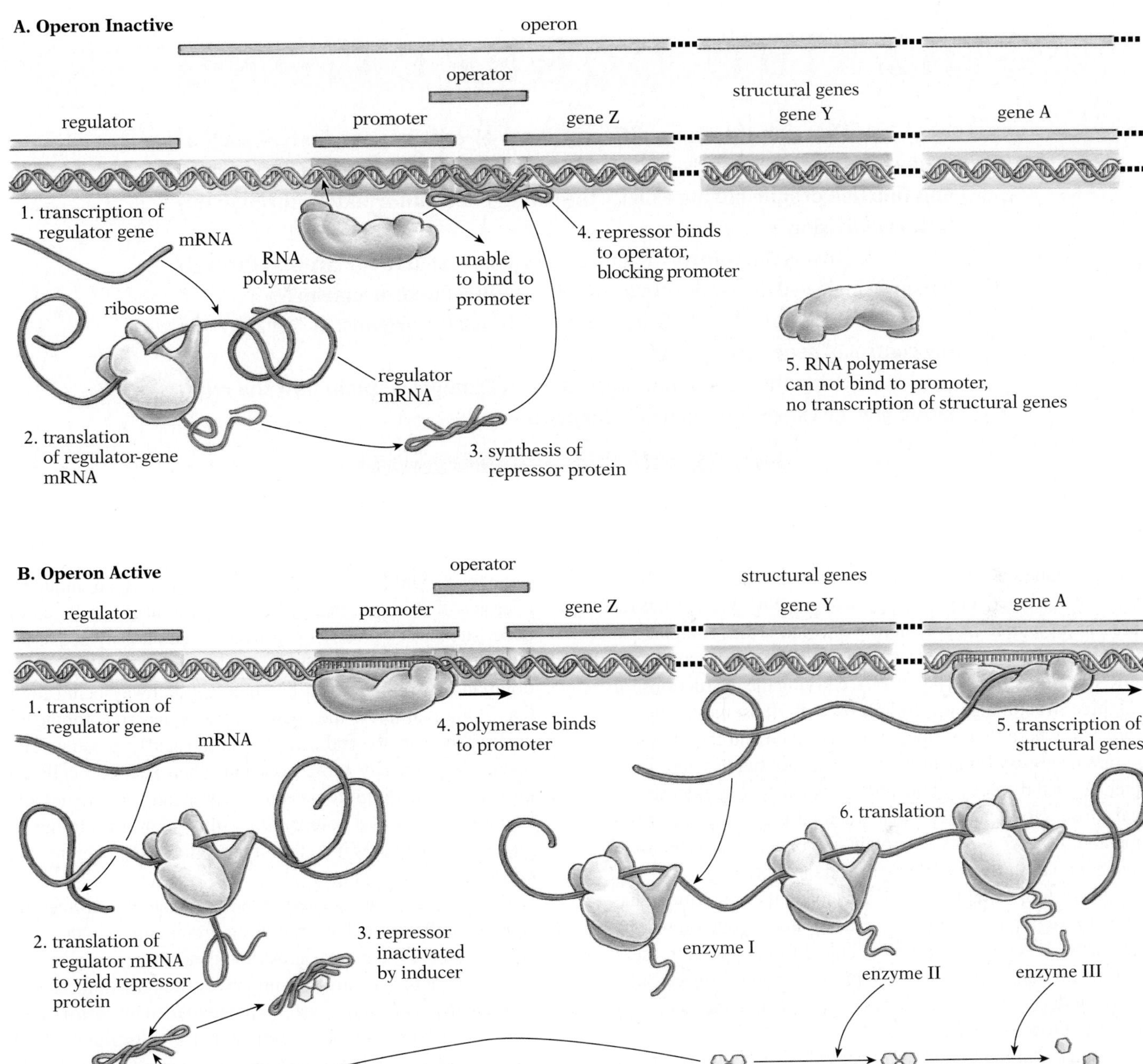

12.3 The *lac* operon in *E. coli:* an inducible operon (A) The operon consists of an operator region, a promoter region, and three structural genes (*Z, Y,* and *A*). The regulator gene codes for an mRNA (1), which is translated on the ribosomes (2) and determines the synthesis of a repressor protein (3). When the repressor protein binds to the operator (4), it blocks the promoter's binding site of RNA polymerase (5) and thus prevents transcription of the structural genes. (B) Binding of the inducer (lactose) to the repressor inactivates the repressor protein (3) so that it cannot bind to the operator. RNA polymerase is now free to bind to the promoter (4) and transcription of the structural genes can proceed (5). The resulting mRNA complexes with the ribosomes and is translated into three enzymes (6), which catalyze the reactions in the breakdown of lactose (7).

For example, in the current version of the Jacob-Monod model, the *lac* operon, which is responsible for the synthesis of enzymes that catalyze the metabolism of lactose, works as follows: before transcription of the three structural genes can begin, RNA polymerase must bind to the promoter region of the operon. This binding is blocked if the repressor protein (coded by the regulator gene) has already bound to the operator region (Fig. 12.3A). As long as active repressor protein is present, no binding of RNA polymerase—and hence no transcription—can occur.

If an inducer substance (lactose in this case) is present, however, it will bind to the repressor protein, causing it to

MAKING CONNECTIONS

1. How can the many different types of cells in your body—such as retinal and pancreatic cells—be so different from one another in appearance and function despite having exactly the same genetic information through mitotic cell division?

2. How is the control of gene expression related to the process through which a single cell develops into a complex multicellular organism?

3. Why have bacteria been especially useful organisms for studying the control of gene expression?

4. Using the tryptophan operon as an example, explain how the regulation of gene expression relates to homeostasis.

change shape so that it no longer binds to the operator. In short, the inducer inactivates the repressor protein. Without the repressor protein, RNA polymerase can bind to the promoter and start transcription of the structural genes and the production of mRNA (Fig. 12.3B). The mRNA, carrying the instructions for all three structural genes, complexes with ribosomes in the cytoplasm, where its information is translated and the three enzymes necessary for lactose metabolism are synthesized. They then begin the process of breaking down lactose to glucose and galactose. Note, however, that when all the lactose has been metabolized, the repressor protein, no longer inactivated, will again bind to the operator, blocking transcription.

According to the Jacob-Monod model, the condition of the operator region is the key to whether or not the operon will be activated. If repressor protein is bound to the operator, there will be no transcription. If the repressor protein is inactivated by the inducer, it no longer binds to the operator and transcription can proceed freely.

Notice that the three jointly controlled structural genes of the *lac* operon specify enzymes with closely related functions. It is characteristic for the structural genes of an operon to code for the enzymes of a single biochemical pathway, enabling the whole pathway to be regulated as a unit. The advantage of such coordinated control is that all the enzymes required for a pathway are there when they are needed, and only when they are needed.

Gene Expression in Procaryotes Is Controlled by Operons in at Least Two Different Ways

In the years since the Jacob-Monod model was first proposed, it has become apparent that not all operons are regulated in the same way. Unlike the *lac* operon, which is an inducible operon—that is, one that is inactive until turned on by an inducer substance—many operons are continuously active unless turned off by a ***corepressor*** substance in the cytoplasm, often an end product of a biochemical pathway being regulated. One example is the operon whose genes code for the enzymes necessary to synthesize the amino acid tryptophan. This operon is ordinarily turned on, but when bacterial cells are grown in a medium containing tryptophan, it switches off. Tryptophan acts as a corepressor—that is, it binds to the repressor and activates it so that it can bind to the operator and block transcription (Fig. 12.4).

Both of the systems cited above are examples of ***negative control,*** in the sense that a repressor protein binds to the operator and blocks transcription. Although negative control of one sort or another is the most common way of regulating gene expression in *E. coli,* some systems are regulated by ***positive control,*** in which proteins act as activators for transcription. Such control proteins, called ***transcription factors (TF),*** bind to the DNA in an activator region a little "upstream" of the promoter and then help the RNA polymerase bind and start transcription (Fig. 12.5).

Some operons are controlled by both negative and positive control mechanisms. In the *lac* operon, for instance, the presence of lactose in the medium inactivates the repressor protein so the RNA polymerase is free to bind to the promoter. Whether or not it binds strongly or weakly depends on the presence of transcription factors to assist binding. When TF are absent, the polymerase binds only weakly to the promoter and transcription is infrequent, but when TF are present, the polymerase binds more readily and transcription is greatly facilitated. As a result, the *lac* operon is most active when lactose is present *and* the transcription factors are bound.

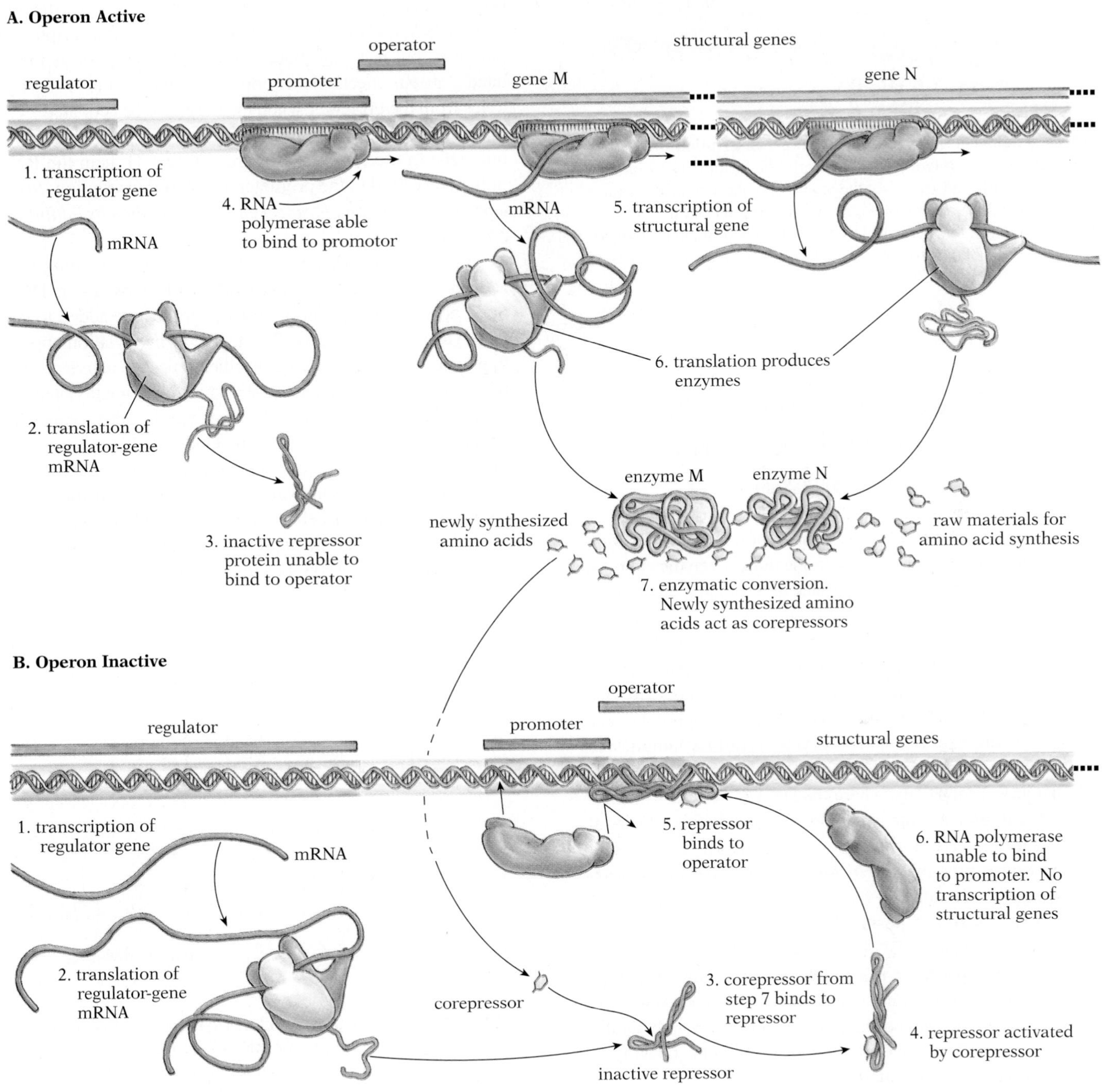

12.4 A repressible operon in *E. coli* (A) The regulator gene codes for an mRNA (1), which is translated on the ribosomes (2) and determines the synthesis of an inactive repressor protein (3) and so RNA polymerase is free to bind to the promoter (4) and transcribe the structural genes (5). Translation of the mRNA takes place (6) and the resulting enzymes catalyze the reactions in a biochemical pathway, and an end product is synthesized (7). The end product (color) may act as a corepressor. (B) If a corepressor molecule (color) is present (3), it binds to the repressor (4) and activates the repressor so it can bind to the operator (5). The binding of the active repressor to the operator blocks the binding of RNA polymerase so there is no transcription of the structural genes (6). When the corepressor becomes scarce, the repressor becomes inactive and no longer binds to the operator. Transcription of the structural genes then resumes.

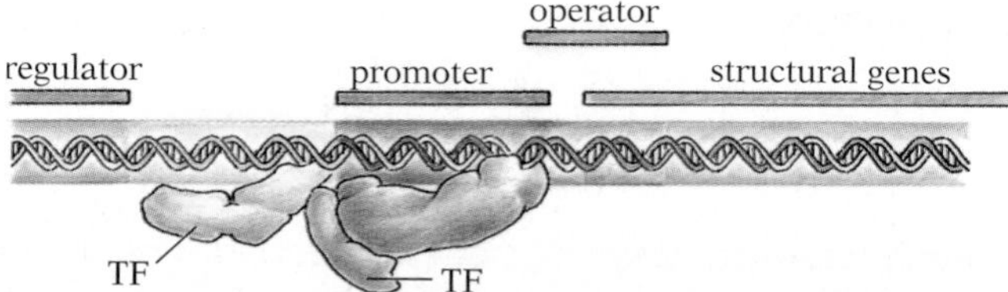

12.5 A model of the regulation of gene transcription by positive control Control proteins, called transcription factors (TF), bind to the DNA a little "upstream" of the promoter in the activator region and help the RNA polymerase bind and start transcription. Transcription factors bend the DNA, causing the promoter region to twist the DNA of the promoter open. RNA polymerase can then bind to the promoter more effectively.

The Elaborate Chromosomal Organization of Eucaryotes Complicates the Control of Gene Expression

The control of gene expression in eucaryotic cells is more complex than in procaryotic cells. For one thing, even the simplest eucaryote has many more functional genes than a procaryote, and most genes need to be turned on or off in an orderly sequence. Moreover, as we have seen, before translation can take place in a eucaryotic cell, the introns in the primary mRNA transcript synthesized from an active gene in the nucleus must be removed by an elaborate enzyme-mediated process. The control of gene expression in eucaryotes is further complicated by the characteristic organization of eucaryotic chromosomes. Unlike procaryotic cells, where the DNA is "naked" (uncomplexed with protein), the single, long double helix of DNA of the eucaryotic chromosome is associated with two types of chromosomal proteins, the ***histone proteins*** and the ***nonhistone proteins.*** The histones, the most abundant chromosomal protein, are small and positively charged, and bind tightly to the negatively charged DNA. In eucaryotic chromosomes, the long DNA double helix is wrapped around cores of histone proteins, forming structural units called nucleosomes (see Fig. 9.4, p. 181). The nucleosomes are then wound into tightly packed coils and organized into loops, forming the nucleoprotein material called ***chromatin.***

If the DNA is tightly bound to the histones in nucleosomes, how then can the RNA polymerase gain access to the DNA for transcription? Evidence indicates that before transcription can take place, a eucaryotic gene is "unpacked," that is, the nucleosome arrangement is loosened in some way. It is not necessary for the DNA to be unwound from the core; the RNA polymerases, once started, apparently can proceed around the cores. However, recent information suggests that transcription cannot take place if the histone proteins are bound to the promoter region. There is evidence that the histones and transcription factors may compete for the same binding sites on the promoter DNA, and that if the histones are bound, transcription is inhibited, but if TF are bound instead, transcription is facilitated. Various mechanisms come into play to move the histones from the promoter when the gene is to be activated.

Transcription factors play a vital role in controlling eucaryotic gene transcription. As in procaryotes, TF help the RNA polymerase bind to the promoter and start transcription. *In eucaryotes, activation of transcription, rather than its inhibition, appears to be the rule.*

Another factor thought to be involved in regulating gene activity is the degree of ***methylation*** of cytosines in DNA. Soon after DNA replication, methyl groups are added to certain cytosines (Fig. 12.6). The degree of methylation appears to play a role in determining whether genes are active or inactive—active mammalian genes have fewer methylated cytosines than inactive genes. In other words, active genes are "undermethylated." There is good evidence that at least some of the problems associated with aging arise from the progressive loss of methylation over time, and hence activation of inappropriate genes in cells. Some cancers also appear to be related to changes in the methylation pattern that may lead to certain genes being wrongly activated or inactivated.

A

m
|
—A—T—C—G—T—C—A—
—T—A—G—C—A—G—T—
|
m

B

m
|
—A—T—C—G—T—C—A— Parental Strand
—T—A—G—C—A—G—T— New Strand
*
enzyme methylating C–G sequence

12.6 Preserving the patterns of methylation of cytosines after replication (A) The degree of methylation appears to play a role in determining whether genes are active or inactive—active mammalian genes have fewer methylated cytosines than inactive genes have. Methyl groups (m) are added to cytosines in certain C-G sequences in *inactive* genes after replication. Because only C-G sequences can be methylated by the enzyme involved, and because C-G on one strand is always paired with C-G on the other strand, the same pattern of methylation exists on both strands of the DNA. (B) After replication, a special enzyme scans the DNA for methylated cytosines in C-G sequences, methylating the corresponding cytosines on the new strand. An asterisk (*) marks the site where the new methyl group will be added. Thus the pattern of methylation—gene inactivation—is passed on intact.

Eucaryotic DNA Is Often Repetitive, Enabling a Cell to Mass-Produce rRNA and Protein

The organization of eucaryotic chromosomes has been the subject of much study in recent years. Research has shown that about 10 percent of most eucaryotic DNA contains base sequences that are found not once but thousands of times in the genome, constituting what is called ***highly repetitive DNA.*** One class of highly repetitive DNA is located near the centromere and is thought to be important in aligning the chromosomes during cell division (Fig. 12.7). Another class of highly repetitive DNA consists of units repeated over and over, one after another, that contain genes coding for one of the RNAs that make up the ribosome. There is a clear reason for having a large number of these genes: growing cells and cells active in protein synthesis need enormous numbers of ribosomes to handle all the protein synthesis. Repeated transcription of a single copy of this gene would be inadequate to meet the needs of the cell for rRNA. Consequently, most eucaryotes have approximately 25,000 copies of this sequence. The function of the rest of the highly repetitive DNA is not known.

About 20 percent of the typical eucaryotic genome consists of ***moderately repetitive DNA.*** Each moderately repetitive DNA sequence is found hundreds rather than thousands of times and comes in one of two varieties. The first is another tandem repeat of three more genes (Figs. 12.7 and 12.10) that code for other rRNAs. The portion of the chromosome with the repeats for the rRNA genes is then itself replicated repeatedly, independent of the rest of the chromosome, producing what is known as a ***polytene*** region, consisting of between 25 and 250 parallel copies of this portion of the chromosome that have remained stuck together. This process results in a total of up to 25,000 copies of each gene in active cells. This region of the chromosome with the many replicated, repeating segments forms the ***nucleolus.*** Other types of moderately repetitive DNA are found scattered throughout the chromosome; their function is not well understood.

Untranscribed Single-Copy Sequences With No Apparent Function Abound in Eucaryotes

About 70 percent of the genome in a typical eucaryotic cell consists of single-copy sequences (Fig. 12.7F). Most of this DNA is never transcribed and its function remains unknown. In mammals, much of it exists as ***pseudogenes***—nearly identical but untranscribed copies of functional genes.

When researchers tally up the amount of functional DNA for different species, taking into account the intervening sequences (introns) that are removed after transcription, they typically find that only about 1 percent of eucaryotic DNA ever codes for mRNA that is subsequently translated. This is dra-

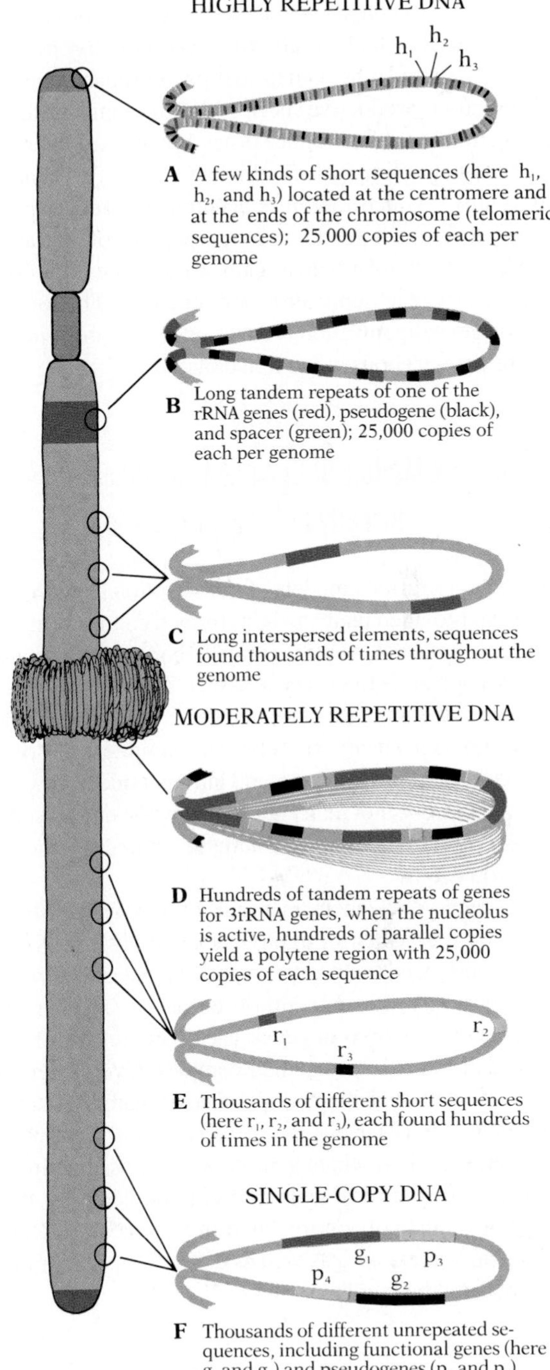

HIGHLY REPETITIVE DNA
h_1 h_2 h_3
A A few kinds of short sequences (here h_1, h_2, and h_3) located at the centromere and at the ends of the chromosome (telomeric sequences); 25,000 copies of each per genome
B Long tandem repeats of one of the rRNA genes (red), pseudogene (black), and spacer (green); 25,000 copies of each per genome
C Long interspersed elements, sequences found thousands of times throughout the genome
MODERATELY REPETITIVE DNA
D Hundreds of tandem repeats of genes for 3rRNA genes, when the nucleolus is active, hundreds of parallel copies yield a polytene region with 25,000 copies of each sequence
r_1 r_2 r_3
E Thousands of different short sequences (here r_1, r_2, and r_3), each found hundreds of times in the genome
SINGLE-COPY DNA
g_1 p_3 p_4 g_2
F Thousands of different unrepeated sequences, including functional genes (here g_1 and g_2) and pseudogenes (p_3 and p_4)

12.7 Organization of the eucaryotic genetic material The short highly repetitive DNA (A), long tandem repeats (B), and polytene tandem repeats (D) are confined to different specialized regions of the chromosome. The long interspersed elements (C), moderately repetitive sequences (E), and single-copy DNA (F) are actually intermixed; for clarity, however, they are shown here in isolation on separate loops.

matically different from procaryotes, in which more than 90 percent of the genetic material is translated at one time or another. In the last chapter we suggested that the procaryotic genetic material, which is almost devoid of introns, spacers, pseudogenes, and other genetic debris, is probably the result of selection for reproductive speed—those organisms that lacked such material could reproduce faster. Some researchers believe that eucaryotes simply lack efficient ways of expunging genetic "junk," which therefore accumulates endlessly; others suspect that much of this vast proportion of untranscribed DNA is actively tolerated as a laboratory for evolutionary experiments in gene duplication and mutation. The logic of eucaryotic gene organization remains one of the most interesting and elusive questions in modern biology.

Procaryotic Models of Gene Control Cannot Be Applied Directly to Eucaryotic Cells

Although the various models of transcriptional control in bacteria have provided many insights to such control in eucaryotic cells, they have not, as was originally hoped, been found to be directly applicable to eucaryotic cells. There is no convincing evidence that a small set of related genes form an operon of the bacterial type in eucaryotic cells. The complications of repetitive sequences, pseudogenes, and introns aside, a lack of similarity of control is not really surprising when one considers the complex structure of the chromatin that makes up the eucaryotic chromosome.

The eucaryotic genetic material, with its need to stretch out and uncoil (decondense) tightly packed loops of DNA before transcribing them, must have an added level of control not required by procaryotes. In addition, eucaryotes must orchestrate multiple constellations of genes, each specific for different tissues and different stages in their complex development from a single cell. Although the "rules" for coordinated activation and repression of genes in eucaryotic cells are still unknown, there are several levels at which gene control is possible. In eucaryotic cells the control of gene transcription is clearly the most important method of control, but different rates of mRNA processing and breakdown as well as different rates of translation also appear to be involved.

Eucaryotic Chromosomes Exhibit Varying Patterns of Activity

Microscopic studies of chromatin suggest that the internal structure of a chromosome is not uniform, and that this variation reflects gene activity. For example, some regions of chromosomes stain only faintly when treated with basic dyes, whereas other regions stain intensely. The nonstaining regions are called ***euchromatin*** and the staining ones ***heterochromatin*** (Fig. 12.8). Chromosomal mapping over the past few years has shown that euchromatic regions contain active genes, whereas heterochromatic regions are inactive.

At first, some investigators thought the heterochromatic regions were simply devoid of genes, functioning merely as structural elements of the chromosome. This is probably true of some of the regions, but most heterochromatic regions do not lack genes. Instead, the genes (often long series of genes) are highly condensed and therefore inactive. Many regions completely heterochromatic in mature organisms were euchromatic at earlier stages in development. The genes for human hemoglobin provide a good example of this. Early in fetal life, two hemoglobin genes, the zeta and epsilon genes, are active, but during the second and third months of fetal life these two genes become permanently inactive (heterochromatic) and the alpha and gamma genes are turned on and remain active until birth. Shortly after birth, however, the gamma gene is permanently inactivated and the beta gene is turned on. From that time on, the two alpha and beta genes remain active, and code for the two alpha and beta chains of adult hemoglobin. Note that there are five genes involved, and that each is activated or inactivated at a particular stage in development in a carefully orchestrated pattern.

In some chromosomes there is direct, visible evidence of differential gene activity. The developing egg cells (oocytes) of many vertebrates, for instance, show such patterns of activity. During maturation, egg cells synthesize large amounts of mRNA for use during the early stages of development, after fertilization, when the chromosomes are so busy with DNA

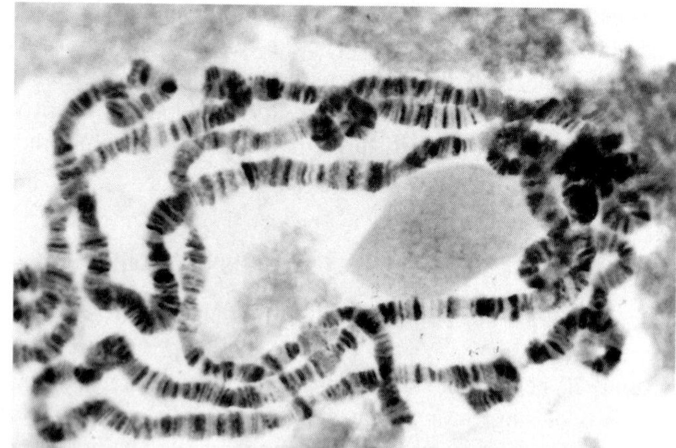

12.8 Photograph of chromosomes from salivary gland of *Drosophila melanogaster* These chromosomes have been stained with a basic dye to produce the various bands. The unstained light bands (euchromatin) correspond to regions where genes are being actively transcribed whereas the dark staining bands (heterochromatin) represent inactive regions. The pattern of banding can be used to identify different parts of the chromosomes. A similar process can be used to stain and identify human chromosomes and parts of chromosomes.

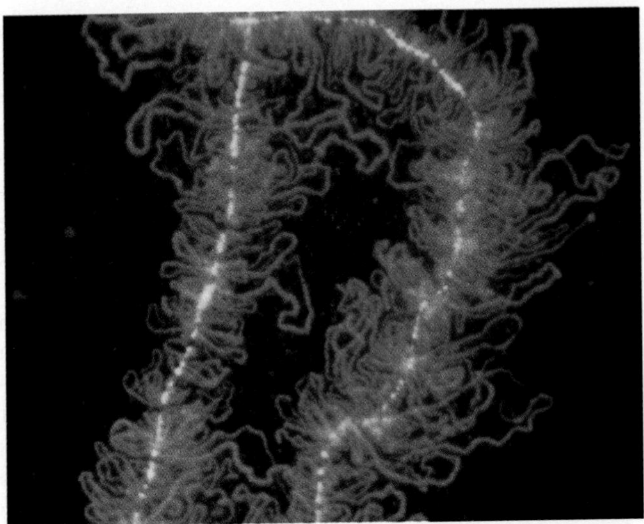

12.9 Lampbrush chromosome from a developing egg cell of the spotted newt *(Triturus viridescens)* The many feathery projections from the chromosome are regions bearing genes that are being repeatedly transcribed.

replication in support of rapid cell division that they are largely unavailable for transcription into mRNA. In vertebrate egg cells, this changing activity can be observed with a phase-contrast microscope. The parts of the chromosome bearing genes that are being repeatedly transcribed—the euchromatic regions—are looped out laterally from the main chromosomal axis, while the parts bearing repressed genes are tightly compacted; chromosomes with many looped-out regions are called ***lampbrush chromosomes*** (Fig. 12.9). The loops, arising in the single-copy DNA, represent those parts of the genetic material that code for the particular proteins the early embryo will have to produce in quantity. Since some of these proteins will not be needed later in development, some of the chromosomal regions so evident as lampbrush loops in the egg cell presumably become heterochromatic for most of the life of the organism.

The giant chromosomes of the salivary glands of larval flies (houseflies, mosquitoes, fruit flies, and the like) provide another example of variable gene activity within the cell. These chromosomes are believed to consist of hundreds of copies of a single chromosome that have replicated and remained stuck together to form a polytene chromosome approximately 200 times the size of a normal chromosome. When certain regions of a giant chromosome are especially active, all the parallel DNA molecules form brushlike loops in those regions (Fig. 12.10). The resulting clusters of lateral loops, called ***chromosomal puffs,*** are clearly visible under the optical microscope. The locations of puffs are different on chromosomes in different tissues, and they are different in the same tissue at different stages of development (Fig. 12.10). At any given time, however, all the cells of any one type in any given tissue show the same pattern of puffing.

If the puffs indicate the location of active genes—as the correlation of puffing pattern with developmental stages suggests—one would expect them to be the primary sites of active synthesis of mRNA. In fact, the unpuffed bands of the chromosome contain mainly DNA and protein, while the puffs contain much more RNA. Furthermore, it can be shown that the RNA made in one puff differs chemically from the RNA made in a puff at a different position on the chromosome, as would be expected if each gene codes for a different mRNA.

Chromosome puffs provide a way of determining visually whether or not changes in the cytoplasm can alter the pattern of gene activity. For example, if ecdysone, the hormone that

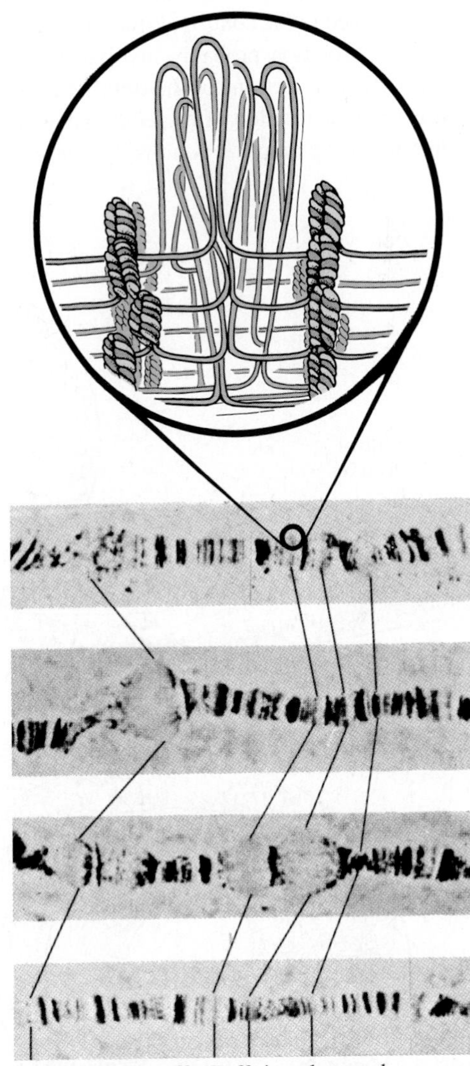

12.10 Chromosomal puffs Puffs in polytene chromosomes consist of hundreds of parallel copies of the chromosomal DNA looped out to expose the maximum surface for synthesis of RNA. The interpretive drawing shows only a few of the loops. The series of photographs indicates how the location of puffs in a *Drosophila* chromosome changes dramatically but predictably over time; we can clearly see the varying activation of four different genes or sets of genes on chromosome 3 (each identified by a series of connecting lines) over a period of 22 hours in larval development.

causes molting in insects, is injected into a fly larva, the chromosomes rapidly undergo a shift in their puffing pattern, taking on the pattern characteristically found at the time of molting in normal, untreated individuals. If treatment is stopped, the puffs characteristic of molting disappear. If treatment is begun again, they reappear. Puffing can be prevented entirely by treatment with actinomycin, an inhibitor of nucleic acid synthesis.

Transcription in Eucaryotes Involves Activation of Genes

From the above account, here is what we can say about the mechanisms involved in the control of transcription in eucaryotic cells. Gene activation probably involves two steps: the loosening of particular sections of the tightly packed nucleosomes, followed by activation of a specific gene or group of genes on the unwound loops. It is unlikely that transcription can take place until the tightly packed chromatin is loosened, allowing the RNA polymerase to bind to the DNA.

As we have already said, eucaryotes rely mainly on positive control of transcription: transcription does not occur without active aid from transcription factor proteins, which help the polymerase bind to the promoter and start transcription. In addition to the promoter, it turns out there are other regions on the chromosome that are involved in controlling the binding of transcription factors. In the last chapter we mentioned the ***enhancer regions,*** which are located a few thousand nucleotides away from the gene being activated. These regions are sites on the DNA to which specific regulatory proteins bind; they apparently exert their effects when the loop folds back on itself to bring the entire region with its regulatory proteins into contact with the promoter region (Fig. 12.11). Some enhancers help load the transcription factors to the promoter, others help the binding of the polymerase, and still others act to inhibit binding.

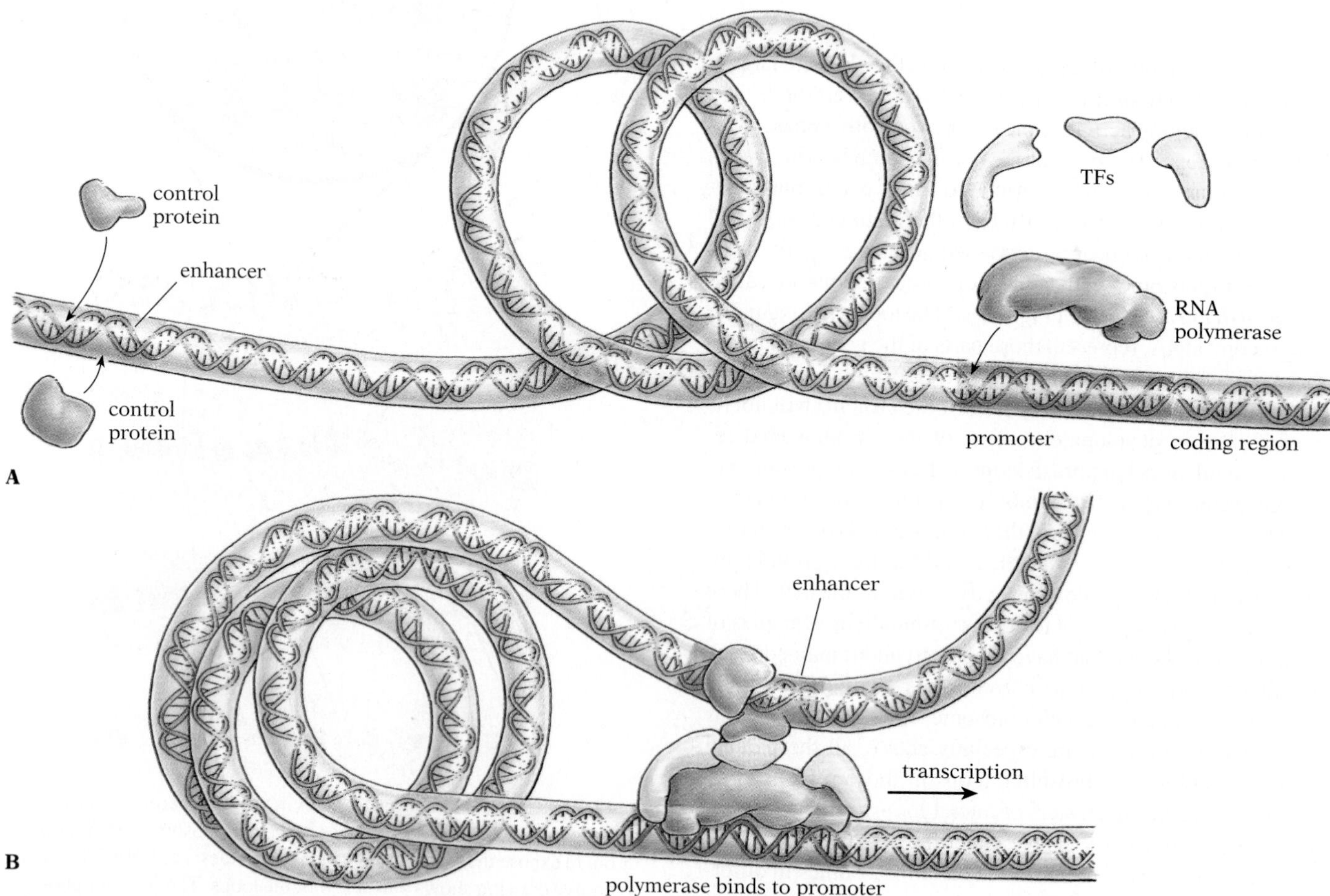

12.11 Control of transcription in eucaryotic cells In eucaryotic cells transcription does not occur without active aid from transcription factor proteins (TF), which help the polymerase bind to the promoter region. (A) The transcription factors are not bound to the promoter so the RNA polymerase (P) cannot bind effectively to initiate transcription. (B) Control proteins that bind to the distant enhancer regions cause the loop to fold back on itself to bring the entire region with its regulatory proteins into contact with the promoter region. The control proteins influence transcription by helping to load the transcription factors on the promoter, thereby helping the polymerase to bind so transcription can occur.

TABLE 12.1 *Comparison of procaryotic and eucaryotic gene control*

	Procaryotes	*Eucaryotes*
Control of DNA polymerase binding	Usually negative, blocked by a repressor	Usually positive, possible only with aid of transcription factors (TF)
Indirect control: enhancer sites	Very rare; exerts negative effect	Very common; can exert positive or negative effect
Levels of control	Usually one; rarely two	Usually at least three; very often four or more

The sex hormones provide an example of how transcription can be regulated through enhancers and regulatory proteins. The genes that control growth and development of the sexual organs, such as the penis, the vagina, the uterus, and the oviducts, and that control distribution of body hair, size of breasts, pitch of voice, and other secondary sexual characteristics are present in individuals of both sexes but are normally expressed only in one sex. Such characteristics are said to be "sex-influenced." The sex hormones apparently influence gene activity by binding to special receptor proteins, which in turn bind to enhancer regions, which in turn stimulate transcription of the appropriate target genes.

This action-at-a-distance system is nearly ubiquitous in eucaryotes, and a few bacterial genes have been found to be controlled in this way as well. Why do most eucaryotic genes have so many alternative ways to regulate transcription? The answer seems to be that eucaryotes are usually multicellular, with distinct tissues and organs, and so must grow in a controlled and directed way to fulfill their specialized roles. In short, gene activity depends not only on immediate metabolic needs, as in bacteria, but on a cell's location and developmental stage as well. The main features of procaryotic and eucaryotic gene regulation are summarized in Table 12.1.

Gene Expression Can Also Be Controlled After Transcription

Our discussion of cellular control mechanisms has so far focused on regulation of the amount of mRNA synthesis by control of DNA transcription. There are, however, numerous other points in the flow of information within the cell where control can be exerted. We have already mentioned that in eucaryotic cells the immediate product of DNA transcription is not functional mRNA but a primary transcript that must be extensively edited in the nucleus. The enzymatic conversion of the primary mRNA transcript to functional mRNA would seem to be a likely point where control might be exerted. For example, editing of the primary transcript might play an important role in determining which proteins are made by the cell. Some mRNA might be discarded during the conversion process. Alternatively, some primary transcripts can be processed in more than one way, to yield slightly different proteins, depending on the cell's needs. Even after splicing, it is estimated that almost half of the mature mRNA produced in the nuclei of most cells never reaches the cytoplasm for translation. There must be a system that can selectively block the export of specific mRNA

MAKING CONNECTIONS

1. What characteristics of the genes and chromosomes of eucaryotes make the control of gene expression more complex in these organisms than in procaryotes?

2. Compare procaryotes and eucaryotes with regard to the percentage of their genomes that is ever translated. Discuss possible explanations for this dramatic difference. Which do you find the most plausible?

3. Trace the steps by which a sex hormone activates genes concerned with sexual development. Why do most eucaryotic genes have so many alternative ways to regulate transcription?

molecules, probably in response to chemical signals that reflect the cell's needs. And even when mRNA reaches the cytoplasm, it may not be translated because inhibitor substances may block either ribosomal binding or complete translation. Finally, control might be exerted by regulation of the rate at which the messenger RNA is broken down: some types of mRNA have long life-spans, while others survive for only minutes or hours.

In addition, the activity of the enzymes created by translation is frequently regulated by activators or inhibitors. Even the assembly of many structural proteins (collagen, for example) is regulated by chemical signals. Hence, a gene's expression can in some cases be controlled at every step from before transcription until after translation (Fig. 12.12). Failure of these control systems can have disastrous results, producing a variety of diseases, chief among which is cancer.

Cancer Results From a Failure of Normal Cellular Controls

Within an animal, cells of different tissues divide at very different rates. Skin and intestinal cells, for example, divide almost continuously. (The lining of the human intestine is normally replaced every four days.) Differentiated nerve cells and skeletal muscle cells, on the other hand, normally do not divide. The processes of cell division are controlled so that cells divide only when new cells are needed to replace others that have been lost. Occasionally, however, the control mechanisms fail and a cell begins to grow and divide without restraint, producing a mass of cells called a ***tumor.*** Alternatively, many cells are programmed to die after a set number of divisions. Some tumors may result from a failure of programmed cell death. In this case the cells also continue to divide without restraint. The tumor may remain localized, in which case it is said to be ***benign,*** or it may continue to grow and invade the surrounding tissues, eventually spreading to other parts of the body, in which case the tumor is said to be ***malignant.*** Malignant, or cancerous, tumors are classified as ***carcinomas*** if they arise from epithelial tissues, and ***sarcomas*** if they arise from connective tissues. ***Leukemias*** and ***lymphomas,*** malignancies of the blood-forming tissues, are two common types of sarcoma.

The most distinctive feature of cancer cells is their unregulated growth. The cells grow to produce a tumor and become malignant when they develop the ability to invade healthy tissue and to spread from the original site of growth to other parts of the body, a process known as ***metastasis.*** Clearly, these are

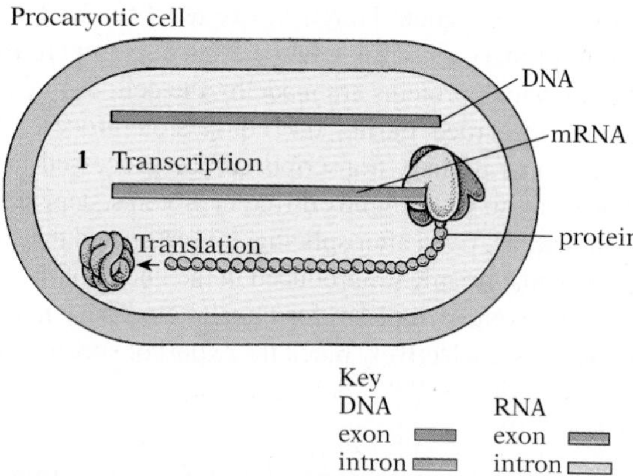

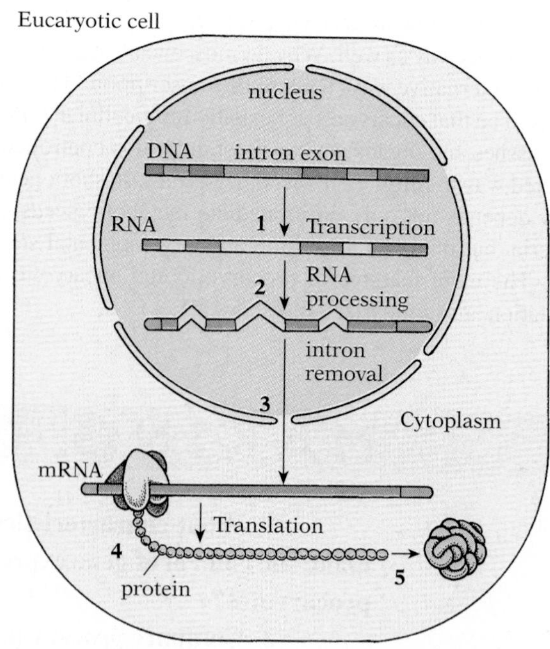

12.12 Control of gene expression In procaryotic cells (left), control is exerted at the level of transcription (1): positive and negative control mechanisms determine which genes are transcribed. In addition, the activity of the proteins and enzymes created by translation may be regulated by activators or inhibitors. The control mechanisms in eucaryotic cells (right) are more complex; there are numerous points in the flow of information within the cell where control can be exerted. Control can be exerted at the level of transcription (1) through use of transcription factors, control proteins, and enhancers. Control can also be exerted at the level of mRNA processing; in eucaryotic cells the primary transcript must be extensively edited by enzymes in the nucleus (2) before the functional mRNA leaves the nucleus and goes to the ribosomes for translation. There may also be a system that can selectively block the export of specific mRNA molecules from the nucleus (3). Even when mRNA reaches the cytoplasm, it may not be translated because inhibitor substances can block either ribosomal binding or complete translation (4). Control might also be exerted by regulation of the rate at which the messenger RNA is broken down. In addition, the activity of the proteins and enzymes created by translation is frequently regulated by activators or inhibitors (5). Hence, a gene's expression can in some cases be controlled at every step from before transcription until after translation.

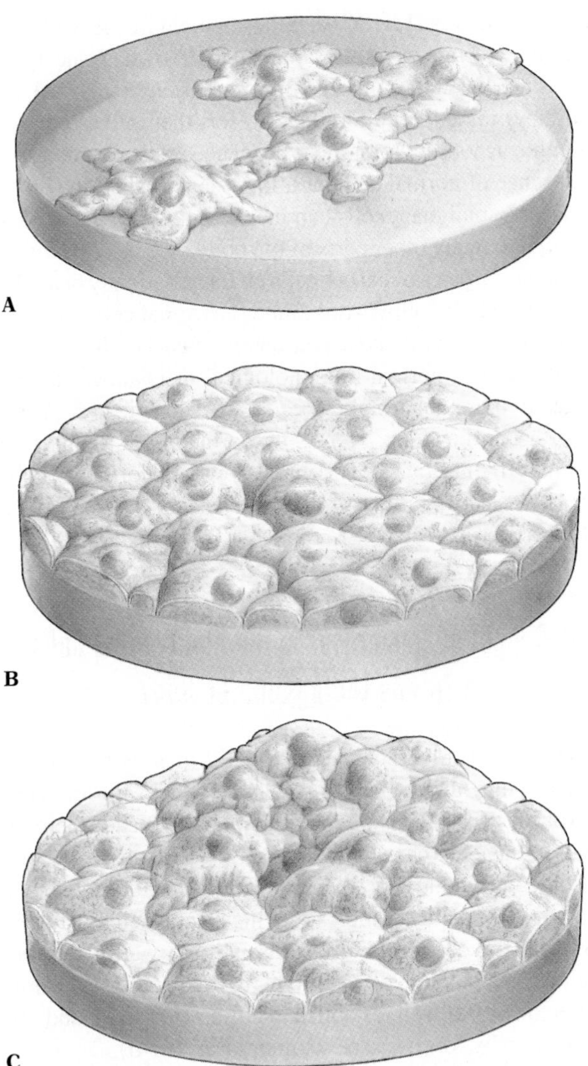

12.13 Inhibition of cell growth When normal cells are placed in appropriate culture medium (A), they will continue to divide and grow until the cells come in contact with one another and with the sides of dish, forming a single layer of cells over the surface of the dish (B). This inhibition of growth due to the cells coming into contact with one another is called contact inhibition. Cancer cells (green), however, do not show contact inhibition. They will continue to divide and grow without restraint and will frequently pile up on one another (C).

cells in which the normal controls are no longer working properly. One of the most profitable ways of studying specific kinds of cells is to remove them from the organism and grow them in tissue culture in the laboratory. The usual procedure is to seed a large number of cells in a sterile culture medium to which a variety of nutrients and growth-stimulating factors have been added. Although a cell culture, no matter how elaborately set up and controlled, is far from the normal environment of the cells, some of the cells may nevertheless survive and engage in many of their usual activities, including cell division (Fig. 12.13). Most lineages of animal cells grown in culture stop dividing when the neighboring cells come into contact with one another (Fig. 12.13B) or, if crowding is prevented, after a set number of cellular generations. Cancer cell lines do not stop dividing—they will grow indefinitely regardless of crowding to create an ever-larger mass of tissue.

Cancer Cells Differ From Normal Cells

The failure of normal control systems can produce a variety of changes in a cell's anatomy, chemistry, and general behavior. Studies of these changes tell us quite a bit about how cancer cells differ from normal ones, and sometimes suggest potential avenues of treatment. We shall mention a few of the characteristics of cancer cells that these studies have revealed.

One property of cancer cells grown in culture is that they almost always have an abnormal number of chromosomes (usually extra chromosomes). For example, the cells of the human cancer line called HeLa,[1] by far the most widely studied line of cultured human cells, typically have between 70 and 80 chromosomes instead of the normal 46 (Fig. 12.14). The presence of extra chromosomes in such cells appears to be associated with the lack of some of the normal constraints on proliferation. In cancerous tissue in organisms, however, such an increase in chromosome number is not typical. Instead, frequently there are numerous self-perpetuating chromosomal segments called "minute chromosomes," or else specific rearrangements of the normal chromosomes. One common rearrangement is a ***translocation,*** where a portion of one chromosome may be broken off and fused to another. For example, in 90 percent of patients with Burkitt's lymphoma (a cancer of the immune system) one tip of chromosome 8 is transferred to the end of chromosome 14. The other 10 percent of patients with Burkitt's lymphoma have other transfers from chromosome 8. In some cancers there are certain patterns of deletions on particular chromosomes. Other nuclear abnormalities are common, and the high rates of mutation in cancer cells means that a given tumor may contain a number of different cell populations, seriously complicating treatment and making drug resistance common.

Besides differences within the nucleus, cancer cells and normal cells display a number of significant differences in cell shape and surface. Cultured cancer cells, for instance, tend to have a rather spherical shape, making these cells more mobile than normal cells at most stages of development. This characteristic is related to another striking feature of cancer cells grown in culture known as anchorage independence: most cells must "cling" to a solid surface in order to grow. Cancer

[1] The HeLa cell line is derived from a carcinoma of the cervix of Henrietta Lacks, who died of her cancer in 1951. This was the first stable, vigorously growing line of cultured human cells used in cancer research.

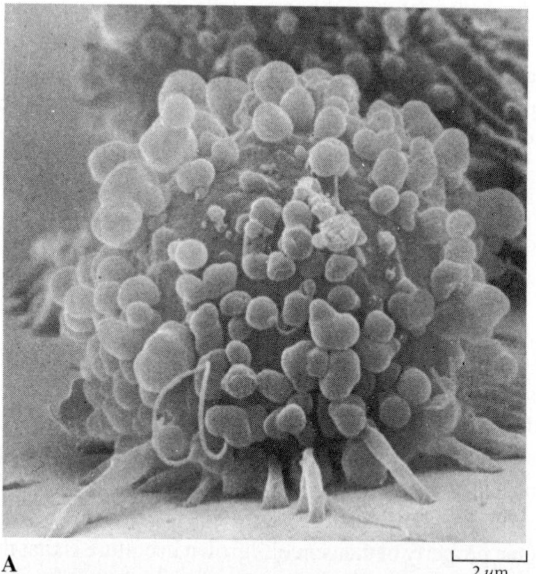

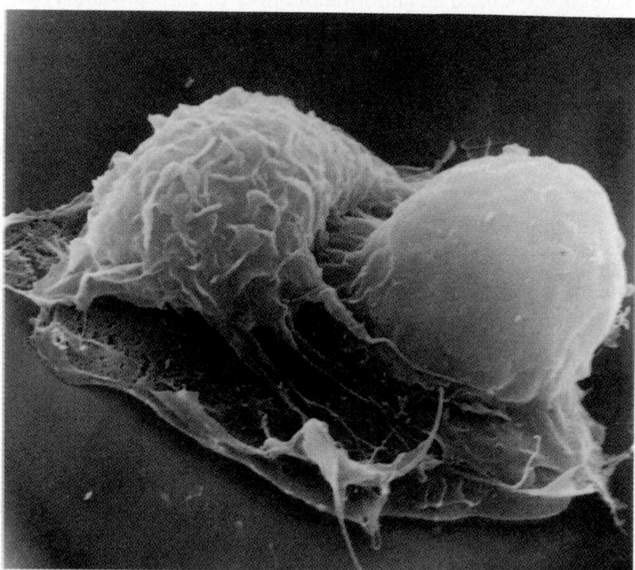

12.14 Typical cancer cells (A) This cell, from the HeLa line, is growing in tissue culture. The cell is rather spherical, and is covered with many small "blisters," called blebs, whose significance is not understood (B) Here an activated macrophage (left) has partially surrounded a cancer cell and is in the process of killing it. The macrophages search out and destroy cancer cells but leave normal tissue unharmed.

cells are not constrained in this way, and so they can grow in liquids or on soft surfaces.

Because a cell's environment must first act upon the cell surface, cancer cells, with their abnormal surfaces, sometimes fail to react or react abnormally to environmental changes. Cancer cells typically have fewer glycolipids and glycoproteins in their cell coats than normal cells, and those they do have are quite different. The anomalous nature of the glycolipids and glycoproteins is probably correlated with both the absence of normal inhibition of cell division when cells become crowded (Fig. 12.13C) and the apparent inability of cancer cells to recognize other cells of their own tissue type. Whereas normal cells from two different tissues (such as liver and kidney) mixed in culture tend to sort themselves out and reassemble according to tissue type, cancer cells do not do so. This absence of normal cellular affinity is probably one of the reasons why malignant cells can metastasize from their tissue of origin into many other parts of the body.

That the surfaces of cancer cells characteristically bear glycoproteins and glycolipids not found on normal cells suggests another abnormality in the development of cancer. In most humans, the body's normal immunological mechanisms probably react to these abnormal molecules, destroying new cancer cells as fast as they arise. Hence, the formation of a cancerous tumor may require not only a breakdown of normal cellular controls but also a failure of immune system to properly respond to abnormal cells.

The Danger of Cancer Lies in the Ability of Its Cells to Spread Throughout the Body

It is the ability of certain cancer cells to ***metastasize***—to spread throughout the body—that makes them life-threatening. Not all cancer cells have the ability to spread; there are myriad changes a cancer cell must go through to become metastatic, and relatively few cells succeed in doing so. The process is remarkably complex: the metastatic cell must break away from tumor mass, burrow through the wall of a blood vessel or lymphatic vessel, travel to a distant site, move through the vessel wall, establish itself in a new area, and proliferate, thereby forming a new tumor (Fig. 12.15).

At several points during the journey the cell must penetrate a basement membrane. The basement membrane is an extracellular layer of material that separates epithelial (surface) tissue from the underlying connective tissue (Fig. 12.16). Such membranes are reinforced by a complex of proteins, which serve as a mechanical barrier to contain the growth of tumors. To penetrate the barrier, the cancerous cell produces special membrane receptors that allow it to bind to portions of the basement membrane. The cell then secretes enzymes that digest the basement membrane, which allows the cell to cross the barrier (Fig. 12.17).

In particular, cancer cells that break through the basement membrane surrounding blood capillaries may spread throughout the body. However, they face additional obstacles. Once the cells have reached the lymphatic vessels or bloodstream, they become the target of the host defenses, the cells of the immune system. Most cancerous cells probably die as a result of the immune attack, but a few may penetrate a vessel wall and invade a new area. Many of the metastatic cells land in a nearby lymph node or capillary bed, but often they enter the general circulation and pass through the heart, frequently ending up in the capillaries of the lungs. Many of the cells that ar-

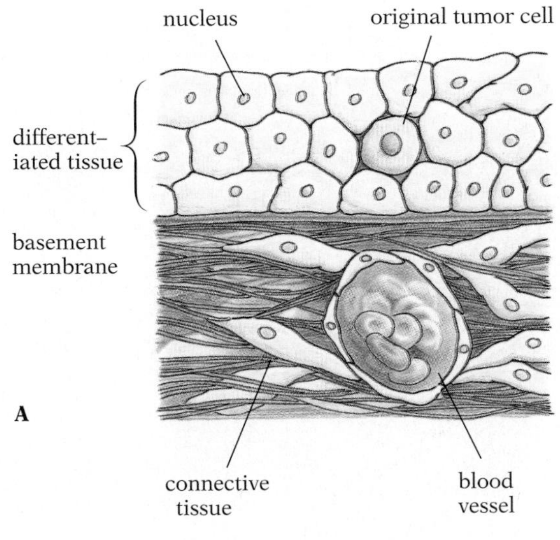
nucleus
original tumor cell
different-
iated tissue
basement
membrane
A
connective
tissue
blood
vessel

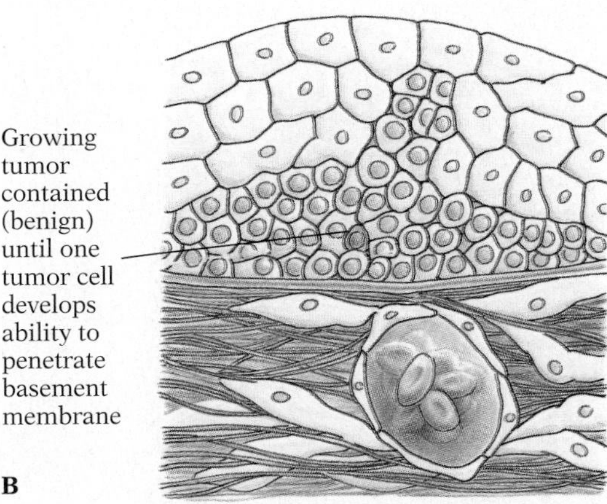
Growing
tumor
contained
(benign)
until one
tumor cell
develops
ability to
penetrate
basement
membrane
B

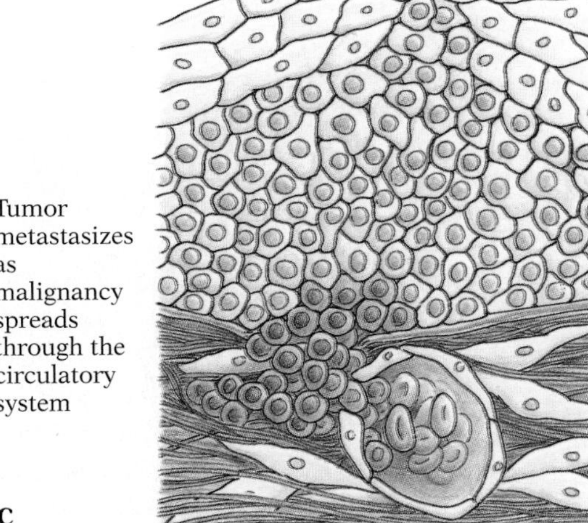
Tumor
metastasizes
as
malignancy
spreads
through the
circulatory
system
C

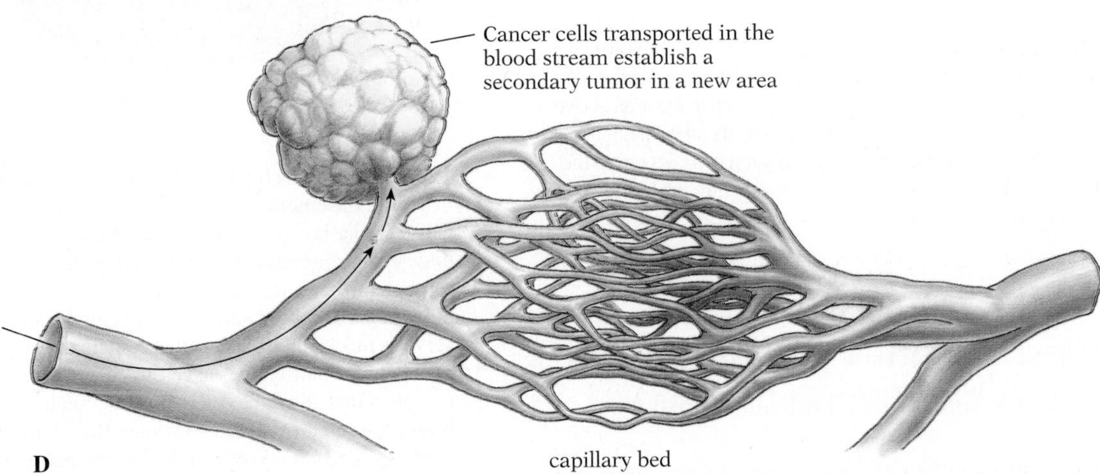
Cancer cells transported in the
blood stream establish a
secondary tumor in a new area
D
capillary bed

12.15 Development of a malignant tumor Normal tissue has differentiated cells with small, quiescent nuclei, separated from the circulatory system and loosely packed connective tissue by a tough basement membrane (A). When a cell incurs enough changes to begin proliferating, its nucleus enlarges, its shape changes, and it begins to replicate its DNA. As the cell reproduces, the clone it forms begins to push other cells back. If it lacks contact inhibition, it will grow regardless of crowding, and if it is not limited to a fixed number of divisions, the clone will grow as long as nutrients (blood supply) permit, but may still be contained by the basement membrane (B). If a cell in the clone develops the ability to penetrate the basement membrane, it will create a breach and its descendants will move across and proliferate. The tumor will metastasize widely when it penetrates the basement membrane surrounding blood vessels (C). A new tumor can form at a secondary site as the metastatic cell penetrates the blood vessel in a new area (D).

rive at the new site will not be able to grow there, however. Cancer cells have sometimes been likened to plant seeds—just as seeds can grow only when they are planted in appropriate soil, so cancer cells can grow only if they reach hospitable "soil"—that is, if the proper growth factors or hormones are present. The process of metastasis is so difficult that it is estimated that fewer than one cell in 10,000 has the ability to leave the primary tumor and successfully establish a secondary tumor in a new area. (See Exploring Further: New Cancer Treatments, p. 249)

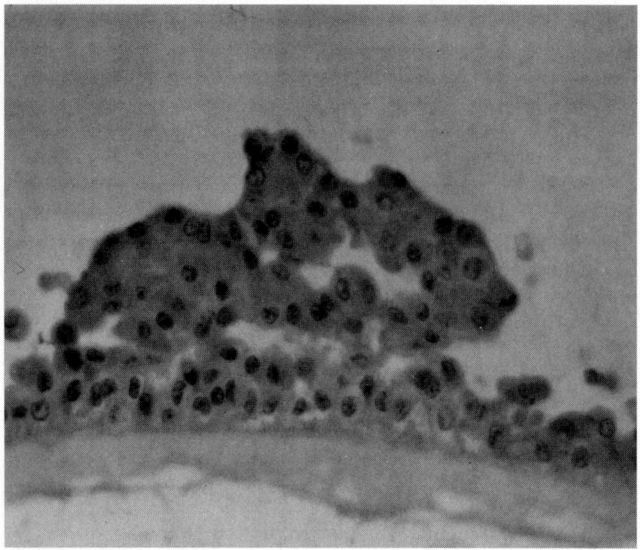
A

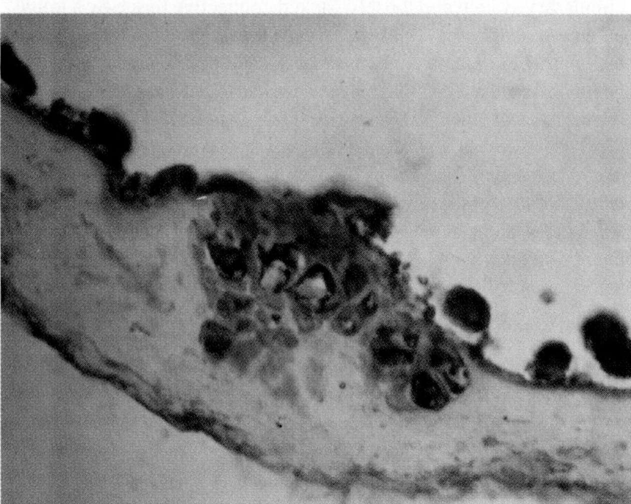
B

12.16 Metastasis of cancer cells Some cancer cells are noninvasive; they multiply and grow but lack the ability to penetrate the basement membrane (A). Such cells simply pile up in the area above the basement membrane. Some cancer cells, on the other hand, have the ability to penetrate the membrane (B). If such cells penetrate the basement membrane in a blood vessel, they can enter the bloodstream and travel to distant sites.

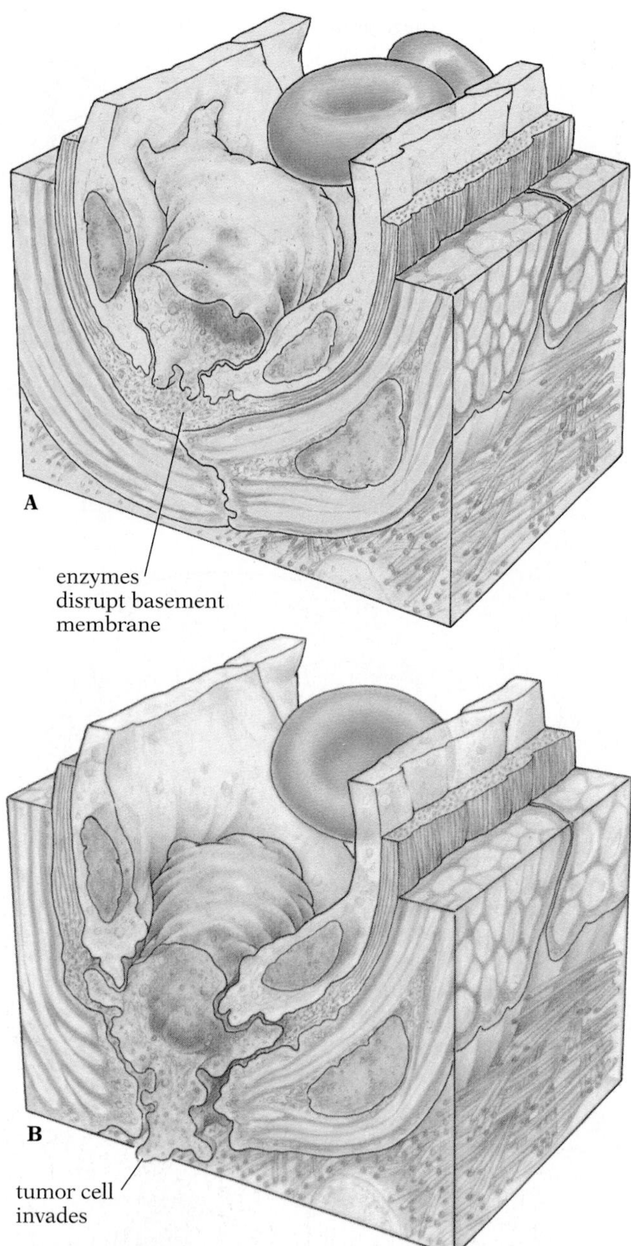

12.17 A crucial step in metastasis is the ability of a tumor cell to cross the basement membrane The basement membrane is an extracellular layer of material that separates surface tissue from the underlying connective tissue. In particular, the capacity to break through the basement membrane surrounding blood capillaries allows cancer cells to spread throughout the body. Such membranes serve as mechanical barriers that contain the growth of tumors. To penetrate the barrier, the cancerous cell produces special membrane receptors that allow the cell to bind to proteins of the basement membrane; the cell then secretes enzymes that digest the basement membrane (A) and allows the cell to cross the barrier (B).

Cancer Results From Several Independent Mutational Events, Whose Effects Are Cumulative

Our current understanding of cancer suggests that most cancers arise from a single abnormal cell. The transformation of a normal cell into a cancer cell may involve several changes: loss of contact inhibition, loss of fixed-number-of-division control, increased mobility, ability to invade the basement membrane, etc. But in many tissues even these changes result only in benign tumors that grow slowly and then stop, apparently because the cells fail to produce the chemical signal or signals that stimulate the establishment of a capillary system to supply oxygen and nutrients to growing tissue. Even when it can re-

EXPLORING FURTHER

NEW CANCER TREATMENTS

Metastases are what kill the cancer patient. All too frequently, many small, undetectable metastases are already present by the time the patient's cancer is first diagnosed. Presently, cancer treatment focuses on surgical removal and treatment with radiation and toxic chemicals (chemotherapy). Unfortunately, the treatments that kill cancer cells also affect normal cells, and there are often unpleasant side effects. Chemotherapy, for instance, has the greatest effect on rapidly dividing cells. This includes cancer cells, but also the cells lining the digestive tract and hair cells, which also divide frequently. Consequently, tumors treated with anticancer drugs may shrink, but most patients also have problems with nausea and diarrhea, and often lose their hair. And, unfortunately, cancer cells typically develop resistance to the drugs over time. If even just a few cells survive, they can proliferate and spread. Many of the new treatment strategies are being developed with the aim of preventing or treating metastases.

For metastasis to occur, some of the tumor cells must become invasive and cross the basement membrane to get to the lymphatic vessels or blood vessels. Studies of human cancers have shown that invasive cells produce higher than normal amounts of protein-digesting enzymes called metalloproteinases. Such enzymes may aid cell movement through the basement membrane by dissolving the proteins, particularly collagen, that make up this extracellular layer. The metalloproteinases are secreted in inactive form and must be converted to the active form. Some investigators think it may be possible to develop a drug that would block the activation of metalloproteinases, thereby suppressing invasion. That is not all there is to the metalloproteinases story, however. Normal tissues and tumor cells themselves also secrete a metalloproteinase inhibitor. Called TIMPs (for tissue inhibitor of metalloproteinases), these substances bind specifically to metalloproteinases and block their activity. The TIMPs, then, act as metastasis-suppressor proteins. Whether or not a cancer cell invades or metastasizes depends on the balance between the activator and inhibitory proteins. Someday, perhaps TIMPs, or drugs that mimic their activity, could be used to alter the balance and prevent invasion or metastasis.

Other potential metastasis inhibiting proteins have been discovered. One such protein, called nm23 (for nonmetastatic 23), is encoded by a gene that is missing or inactive in metastatic tumor cells. Studies of breast cancer patients have shown that patients whose tumor cells have high levels of nm23 have few metastases and a good prognosis, while those with low levels have many metastases and tend to die sooner. If the gene for nm23 could be turned on or replaced in tumor cells, it might be possible to suppress metastasis. Another approach is to use the levels of nm23 in tumors as an indicator for determining the most appropriate cancer treatment.

All of these possible treatments act by interfering with the metastatic process. But what about the patients who have many small secondary tumors at the time of diagnosis?

Chemotherapy and radiation are useful in some cases, but the side effects are unpleasant at best, and life threatening at worst, and moreover the treatment may not help. What is needed are drugs that specifically arrest the growth of already established metastatic tumors without the severe toxic side effects of present chemotherapeutic agents. Clinical trials are beginning with a new group of synthetic compounds, called CAIs, which block the growth of certain types of secondary tumors. These agents are thought to interfere with the signaling pathways within the cell that causes the uncontrolled growth. The results of animal testing are promising, and investigators at the National Cancer Institute are eagerly awaiting the results of the first clinical trials in humans.

Developing new therapies is a long and difficult procedure, but researchers hope that by understanding the metastatic process new and more effective cancer treatments can be developed. Since one out of five people living today can expect to die of cancer, this work has important implications for all of us.

cruit capillaries, the tumor cannot metastasize without being able to bind to and destroy the basement membrane. Do all these changes in the cell take place at once due to a single change in the DNA of the affected cell, or is each change the result of one or more separate events?

Health statistics suggested the answer to this question before the actual mechanisms began to be understood. If just a single genetic event were required, one would expect the probability of an individual's contracting a particular type of cancer to increase proportionally with age. For example, if the chance of contracting skin cancer were 1 percent per year, then the chance of having contracted it would be 2 percent for a two-year-old, 3 percent for a three-year-old, and so on. In fact, however, the incidence of cancers in the population as a whole does not rise in this linear fashion (Fig. 12.18). Instead, cancer is normally a disease of old age.

In order to account for the observed exponential rise of the incidence of cancer, several independent genetic mutations occurring *in the same cell* are evidently required (Fig. 12.19). For example, cancer of the large intestine appears to require five mutational events. If each cancer, then, is a result of several mutations that may have occurred at any time during a patient's life, it can be seen that the development of a cancer is a long-term process, dating from the time the first mutation took place. Thus, it may take decades for a cancer to develop. One (or more) of these mutations may be inherited, which explains

12.18 Incidence of representative cancers If each type of cancer were caused by a single event, the probability of contracting a particular cancer would increase linearly with age; the range of possible curves is indicated. Actually, the incidence of most cancers increases exponentially, indicating that several contributing events are usually involved. The incidence curve for skin cancer closely approximates a three-event curve, while the curve for prostate cancer approximates a five-event curve.

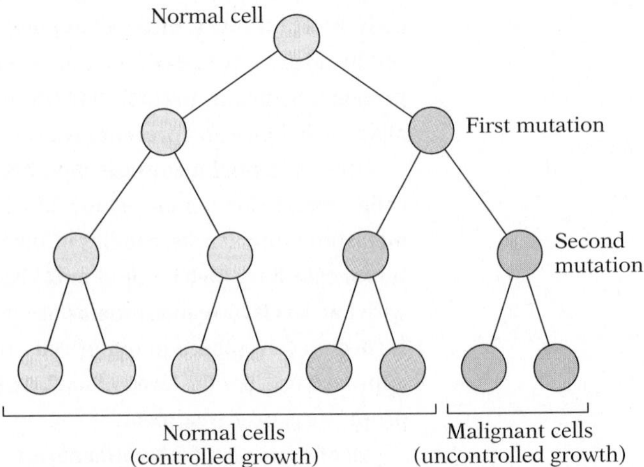

12.19 The multistep hypothesis Several independent genetic mutations occurring in the same cell are evidently required to convert a normal cell into a malignant cell. This diagram illustrates two mutational events occurring in a lineage of cells.

why some families appear to show a susceptibility to a particular form of cancer.

Although statistical studies are suggestive, however, they cannot tell us what constitutes a genetic event leading to cancer. (Among the possibilities are single-base changes, deletions, insertions, or chromosomal rearrangements such as translocations.) Moreover, such studies cannot tell us the precise effects of the events, or what genes are involved. Recent evidence, however, strongly suggests that most steps are directly linked to the loss of or escape from a particular level of control. The best data on this come from studies on oncogenes.

Cancer-Causing Genes Were First Observed in Viruses

Genes that cause cancer—that is, genes that can transform a normal cell into a malignant cell—are called ***oncogenes.*** Such genes were first discovered in certain tumor-causing viruses in animals. It is thought that oncogenes are produced when certain normal genes, called ***proto-oncogenes,*** which are involved in stimulating cell proliferation, undergo a mutation that converts them into cancer-causing genes. Since 1970 nearly two dozen rapidly acting, cancer-inducing viruses infecting a variety of birds and mammals (but not humans) have been discovered, each carrying an oncogene almost identical to the normal host proto-oncogene. Oncogenes, then, are genes that inevitably cause cancer; proto-oncogenes, on the other hand, have the potential to cause cancer, but some change is required to convert them into oncogenes.

The viral cancers are convenient to study, but as yet few forms of cancer in humans have been found to be definitely viral in origin. However, not all oncogenes come from viruses—some are proto-oncogenes that have been converted by mutations or translocations. Many of these proto-oncogenes closely resemble those found in tumor-causing viruses. Some 60 proto-oncogenes have been discovered so far in human cells, each of which could potentially be converted into an oncogene. Although the role of most proto-oncogenes in normal cells is not yet understood, it is thought that they are involved in regulating normal growth and development. Evidence indicates that in humans more than one active oncogene is necessary to transform a normal cell into a malignant cell.

What, then, do oncogenes do? Current research indicates that most oncogenes can exert their effect at any point in the flow of information from the extracellular environment to the chromosomes (Fig. 12.20). Some act to create high levels of

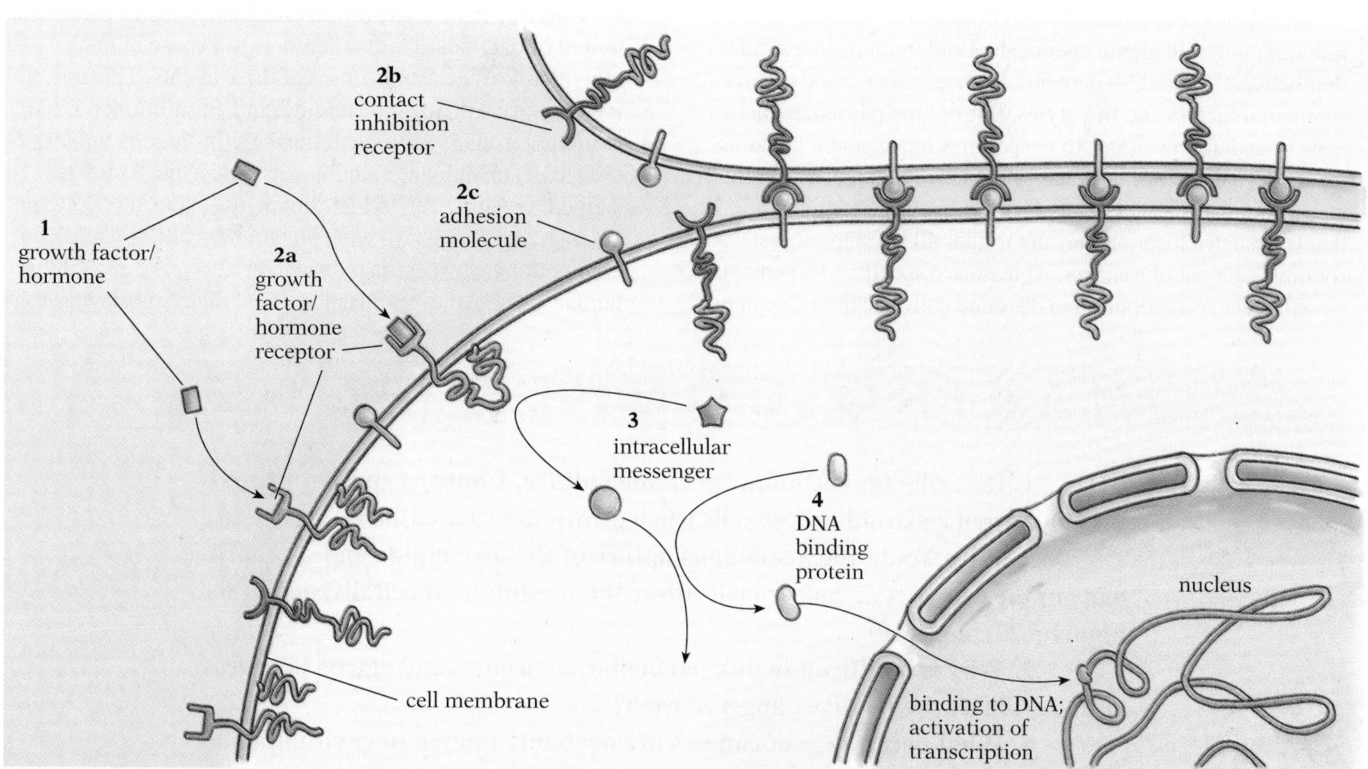

12.20 General sites of oncogene action Oncogenes can exert their effect at any point in the flow of information from the extracellular environment to the chromosomes. Some act to create high levels of growth factor or GF with unusual properties (1). Others alter the sensitivity of membrane receptors for GF or other relevant signals, like those involved in adhesion or contact inhibition (2). Still others alter the communication between the receptors and the intracellular messenger chemicals (3). Finally, some oncogenes alter the DNA-binding proteins that are activated or repressed by the intracellular messengers (4).

growth factors—molecules that stimulate cell proliferation. Others alter the sensitivity of membrane receptors for growth factors, other signals, contact inhibition, or surface adhesion. Still others interfere with the signaling pathways within the cell that communicate information from receptors to enzymes or binding proteins. Finally, some oncogenes alter the DNA binding molecules that regulate transcription of certain genes or replication.

Mutations in proto-oncogenes are not the only way in which a cell may be made cancerous; the loss or inactivation of ***tumor-suppressor genes*** may also be involved. As the name suggests, tumor-suppressor genes inhibit excessive cell growth and division. Mutations in such a gene may alter its activity so that it no longer functions to restrain growth. The role of tumor-suppressor genes in human cancer is not yet well understood.

Environmental Agents Are Related to the Development of Cancer

As we have seen, several factors—including inheritance, viruses, mutations, and translocations—can each play a role in triggering cancer. We also know that many mutations and translocations are caused by environmental agents, such as radiation and mutagenic chemicals. Epidemiologists estimate that between 70 and 80 percent of cancers are related to environmental factors and life-styles. But linking particular chemicals or radiation sources to cancer in humans and evaluating the relative danger of each has proved difficult. Not every mutation occurs in a cancer-causing location, and even a mutation that is in such a location may not trigger all the steps necessary for development of a cancer. At least two specific independent genetic events are required in the same cell. The final step may occur years after the first significant exposure to a mutagenic chemical or radiation.

The absence of strict cause-and-effect relationships between particular mutagenic agents and particular cancers has made the identification and evaluation of causative agents difficult, even in the most obvious cases. For example, although health statistics have shown that cigarette smoking is probably responsible for about 150,000 fatal cancers annually in the United States (and for 25 percent of fatal heart disease), representatives of the tobacco industry deny the link. While granting that smokers develop lung cancer and heart disease far more often than nonsmokers, they point out that many smokers never develop lung cancer, and that a few nonsmokers do.

An informed understanding of how oncogenes work and of the probabilistic nature of their development, however, reveals the weaknesses in these arguments. The genetic events necessary for inducing lung cancer must be greatly facilitated by the highly mutagenic "tar" in cigarette smoke (a deadly combination of chemicals that kills two-pack-per-day smokers eight years earlier than nonsmokers on average), but each of the genetic changes must occur in the correct locations in the same cell, so even with regular exposure to mutagenic smoke, some smokers will escape unscathed. At the same time, other sources of mutation, such as radiation, are also present, and although less potent, they will sometimes cause the changes necessary to trigger cancer even in nonsmokers.

One way to evaluate the carcinogenic (cancer-causing) potential of chemicals and radiation is to expose animals—usually mice—to measured doses of these agents. This procedure is expensive and time-consuming, and, in addition, it is only an assumption that what is carcinogenic in mice is equally carcinogenic in humans. Another assay is the Ames test (described on p. 229), which involves exposing bacteria to potential carcinogens to see if mutations result, and relies on the assumption that what is a mutagen for *E. coli* is a carcinogen for humans. The Ames test has identified many substances as po-

MAKING CONNECTIONS

1. Describe the technique of tissue culture. Contrast the growth pattern of normal cells and cancer cells when grown in tissue culture.

2. Why would the anomalous nature of the glycolipids and glycoproteins in the cell coat of cancer cells affect the inhibition of cell division? (See Figure 5.27, page 110)

3. Why is it difficult to link particular environmental agents to cancer and to evaluate the relative danger of each?

4. What percentage of cancers are probably related to environmental factors and life styles? What can you do to minimize your risk of developing cancer?

EXPLORING FURTHER

RISK FACTORS FOR CANCER

The use of tobacco and tobacco products is by far the most important risk factor for cancer in the world today, accounting for 30 to 40 percent of all cancer deaths. Indeed, if one adds to that total the tobacco-related deaths from cardiovascular disease and pulmonary disease, tobacco use must be considered the world's most serious (and preventable) health problem. One of the difficulties with tobacco use is the long "incubation period" for the development of cancer. Trying to convince a 20-year-old of the danger of smoking is difficult when the effects may not be apparent for decades. If death followed within a few months of use, it would not be hard to perceive the real danger of tobacco use.

Excessive consumption of alcohol has been shown to be related to cancer of the mouth, larynx, and esophagus, especially if coupled with tobacco use. The effect of these two factors on the cancer death rate has been demonstrated by statistical studies of Mormons and Seventh Day Adventists, who have significantly lower cancer death rates than the general U.S. population. Members of both these groups abstain from alcohol and tobacco use. There is no doubt that altering one's life-style to avoid the use of these products would drastically lower the death rate from cancer.

Approximately 4 percent of cancer deaths are thought to be related to occupational hazards, such as exposure to carcinogenic chemicals or radiation. Unfortunately, it may be many years before the first cancers arise to awaken one to the danger, and then it is too late. Often the cancer may not appear until years after the person has retired or changed occupations. Identifying carcinogens in the workplace and minimizing exposure offer the best hope of lowering occupational cancer deaths.

The ultraviolet rays in sunlight are known to cause many skin cancers, and thus exposure to sunlight should be minimized. A comparatively thin layer of atmospheric ozone shields living organisms from most of the sun's high-energy ultraviolet radiation, but, in 1974, scientists noted that the ozone layer has been diminished, particularly over Antarctica. The United Nations Environmental Program estimates that if the ozone layer decreases by 10 percent, there will be a 26 percent increase in the incidence of skin cancers. Over-the-counter sunscreens, which help reduce the levels of damaging radiation reaching the skin, are useful in preventing skin cancers. Although skin cancer is highly curable, many people die of it every year.

Recently there has been a great deal of interest in the effect of diet and obesity on the cancer rate. Those individuals who are overweight, especially more than 40 percent overweight, show higher rates of cancer than those individuals whose weight is within the normal range. Recent nutritional research suggests that 35 percent of all cancer is related to the way we eat. Although hard data are still lacking on the precise effects of diet on cancer, the American Cancer Society has issued a set of nutritional guidelines that may help reduce the chances of getting cancer. Some of these guidelines are summarized below.

- *Reduce total fat intake.* **High fat intake appears to be related to cancer of the breast, colon, and prostate; hence, lowering dietary fat reduces the chances of getting these cancers.**
- *Eat more high-fiber foods.* **Vegetables, fruits, and whole-grain cereals provide high fiber in the diet that may protect against certain types of intestinal cancer. Such foods also provide vitamins A and C, which may also have a protective function.**
- *Eat more cruciferous vegetables.* **Members of the mustard family—cabbage, brussel sprouts, broccoli, cauliflower, turnips, kale, and kohlrabi—appear to be effective in preventing certain cancers, and inclusion of such foods in the diet is advised.**
- *Eat more green and yellow fruits and vegetables.* **Such foods are rich in vitamin A and beta-carotene—chemicals that function as antioxidants and may inhibit the formation of cancer-causing chemicals in the body. Vitamin A and beta-carotene may help protect against cancers of the larynx, lung, and esophagus.**
- *Decrease consumption of salt-cured, smoked, and nitrite-cured meat.* **Nitrite has been shown to be carcinogenic and thus it is wise to limit its consumption. Accordingly, the American meat industry has substantially decreased the amount of nitrite in prepared meats. Smoked meats provide a different threat, however, since these meats absorb cancer-causing tars similar to those in tobacco smoke.**

Obviously it is not possible to prevent exposure to all factors known to cause cancer in humans (for example, sunlight), but establishing a life-style that avoids or minimizes exposure to environmental carcinogens can go a long way toward lowering the chances of one's dying from cancer.

tential carcinogens. The coal-tar components of many hair dyes and cosmetics, as well as hexachlorophene soaps, flame-retardant chemicals in children's sleepwear, the seared protein of grilled meat, the smoke from wood fires, and several chemicals in certain vegetables and spices are just a few of the substances identified by the Ames test as possible carcinogens. When tested on animals, each of these has in fact proven carcinogenic. Although it seems prudent to assume that all mutagens are carcinogenic, avoiding them all is impossible. The only sensible course, then, is to eliminate our exposure to the most potent—cigarette smoke and coal tars—and, weighing the costs and benefits (just as we do each time we decide to take the risk of driving a car), to minimize contact with the rest. (See Exploring Further: Risk Factors for Cancer, p. 253)

CONCEPTS IN BRIEF

1. Although every cell in the body of a multicellular organism has identical genetic information, individual cells have different structural and functional characteristics. Only about 1 percent of this genetic material is expressed at any one time. Various control mechanisms determine when and how each cell will act on its genetic instructions.

2. In bacteria, the Jacob-Monod model of gene induction proposes that three parts of the chromosome are involved in controlling transcription of the *structural genes:* the *regulator* gene, the *operator,* and the *promoter.* The regulator gene codes for the *repressor protein.* When the repressor protein is bound to the operator, it blocks the promoter's binding sites for RNA polymerase and thus prevents transcription of the structural genes. In this inducible system, the genes specifying particular enzymes are inactive until turned on by an inducer substance (usually a substrate).

3. In a *negative control* system such as the *lac* operon, a repressor protein binds to the operator and turns off transcription. When the repressor protein is inactive, the operator is turned on and transcription and translation automatically occur. In a *positive control* system, proteins bind to the promoter and act as activators for transcription.

4. The control of gene expression is more complex in eucaryotes than in procaryotes. Eucaryotes have many more functional genes than procaryotes, and most eucaryotic genes have introns that must be removed before translation. Also, the DNA of the eucaryotic chromosome is tightly wrapped around histone-protein cores to form *nucleosomes,* the basic packing unit of the chromosome.

5. Approximately 10 percent of the genetic material in a eucaryotic chromosome consists of base sequences repeated thousands of times, forming *highly repetitive DNA.* Another 20 percent consists of *moderately repetitive DNA,* in which sequences are repeated hundreds of times. The remaining 70 percent consists of single-copy sequences. Only about 1 percent of the eucaryotic DNA is both transcribed and translated, compared to 90 percent for procaryotes.

6. In eucaryotic cells the control of gene transcription is the most important method of gene control, but different rates of mRNA processing and breakdown and different rates of translation are also involved.

7. *Euchromatic* and *heterochromatic* regions of chromosomes, *lampbrush chromosomes,* and *chromosomal puffs* are visible evidence of differential patterns of gene activity.

8. In eucaryotic cells, gene activation probably involves the unwinding of sections of the tightly packed nucleosomes, followed by activation of a specific gene or group of genes on the decondensed loops.

9. The degree of methylation of cytosines appears to play a role in determining whether genes are active or inactive: active mammalian genes have fewer methylated cytosines than inactive genes.

10. In eucaryotic cells transcription does not occur without active aid from *transcription factor proteins,* which help the polymerase bind to the *promoter region.* Control proteins that bind to distant *enhancer regions* also influence transcription by helping to load the transcription factors on the promoter, helping the polymerase to bind, or, in some cases, by inhibiting polymerase binding.

11. Posttranscriptional controls such as mRNA processing, differential rates of mRNA breakdown, and differential rates of translation may also act as important genetic control mechanisms.

12. The most distinctive feature of cancer cells is their unregulated growth, which often results in invasion of surrounding tissue and the *metastasis,* or spread, of cells from the original site of growth to other sites. To metastasize, cells must break away from tumor mass, burrow through the wall of a blood vessel or lymphatic vessel, travel to a distant site, move through the vessel wall, establish itself in a new area, stimulate blood vessel formation, and proliferate to form a new tumor.

13. Cancer cells have a high rate of mutation and frequently show specific chromosomal rearrangements such as *translocations,* deletions, and minute chromosomes. The cells are more spherical and more mobile than normal cells, and do not show normal contact inhibition of cell division or fixed-number-of-division control.

14. Cancer cells have an abnormal cell surface and do not recognize other cells of their own tissue type as normal cells do. The body's immune system probably destroys most new cancer cells as fast as they arise.

15. Several independent genetic events are required to induce a cell to become cancerous. Thus, the development of a cancer takes many years, dating from the time the first mutational event took place.

16. *Oncogenes* are genes that cause cancer. Sometimes the oncogene is brought into the cell by a virus. Other oncogenes are certain normal cellular genes *(proto-oncogenes)* that have undergone a genetic change to become an oncogene. More than one active oncogene is probably necessary to transform a normal cell into a malignant cell.

STUDY QUESTIONS

1. Suppose you are culturing *E. coli* in a medium high in lactose. Explain how the lac operon of *E. coli* is regulated according to the Jacob-Monod model. Describe the roles of the *inducer, structural genes, regulator gene, repressor, operator,* and *promoter.* (pp. 234–236)

2. Compare negative and positive control systems by describing how they collaborate to regulate the *lac* operon in *E. coli.* Which type of control is more common in eucaryotes? (p. 236)

3. **Explain the role of cytosine methylation in the regulation of eucaryotic genes. (p. 238)**
4. **Contrast *highly repetitive DNA* and *moderately repetitive DNA*. What are the functions of repetitive DNA? (p. 239)**
5. **In addition to the regulation of gene transcription, what other methods are used by eucaryotes to control gene expression? (pp. 243–244)**
6. **Using light microscopy, you examine the chromosomes of various cells to see if you can find evidence of gene activity. (1) You observe cells whose chromosomes were exposed to basic dyes. Some regions of the chromosomes were intensely stained and others unstained. How are these regions related to patterns of gene activity? (2) You examine maturing egg cells and discover loops of chromosomal material extending laterally from the main chromosomal axis. What are these chromosomes called, and what is the significance of the loops? (3) You look at cells taken from the salivary glands of flies in different stages of larval development. The giant chromosomes in these cells show varying patterns of chromosomal puffs. What do the puffs indicate, and why do the locations of the puffs change? (pp. 240–242)**
7. **Explain how *euchromatin, heterochromatin, lampbrush chromosomes,* and *chromosomal puffs* demonstrate differential patterns of gene activity. (pp. 240–241)**
8. **What are the two basic steps necessary to activate eucaryotic genes? What is the function of *transcription factor proteins?* Of *enhancer regions?* (pp. 242–243)**
9. **Distinguish between *benign* and *malignant* tumors and between *carcinomas* and *sarcomas.* (p. 244)**
10. **You are examining a cell that might be cancerous. What characteristics would you look for? Why are they significant? (p. 245)**
11. **What is *metastasis?* What steps must a cancerous cell go through to metastasize? (pp. 244–246)**
12. **Cancer is normally a disease of old age. Why is this fact evidence that several mutational events are necessary to transform a normal cell into a cancer cell? (pp. 248–251)**
13. **How do *oncogenes* arise? How are they related to *proto-oncogenes?* What do they do? (p. 251)**
14. **What assumptions underlie the use of the Ames test and animal tests to evaluate the carcinogenic potential of chemicals? What are the advantages and disadvantages of each type of testing? (pp. 252–254)**

CHAPTER 13

RECOMBINANT DNA TECHNOLOGY

Since the dawn of agriculture humans have practiced "genetic engineering" through the selective breeding of domesticated plants and animals. The ancient farmers saved their hardiest grains for planting, and their best animals for breeding, in the hope that the desired traits would reappear in the next generation. Such selection has been responsible for producing most of our present-day crop plants and domestic animals.

Even under the best of circumstances, however, selective breeding is a slow process. In any successful breeding program, the original plants or animals must be cross-bred, the offspring screened, and those individuals most nearly approximating the desired type selected as breeders for the next generation. Only over many successive generations of repeated selection are organisms having the desired traits developed (Fig. 13.1). For example, fruit breeders at Cornell University recently released to growers a new variety of apples (called Northern Lights) that took 52 years to perfect! Time spans of this length are not unusual in fruit breeding; a typical breeder will probably see only a few of his fruits released over the course of a career.

In addition to the long time spans involved, selective breeding presents other difficulties. Selection for certain characteristics may mean that other advantageous characteristics, such as resistance to insect pests or the ability to withstand drought, are lost. Often the selected crop plants have very high yields but require intensive agricultural practices such as irrigation, the application of fertilizers, and the use of chemical sprays to protect them against various pathogens (Fig. 13.2).

Within the past two decades the ability to alter the genetic characteristics of organisms has been enhanced most dramatically by the advent of modern ***recombinant DNA technology.*** The aim of this technology is to impart some new characteristic or function to an organism by transferring into its cells the DNA of a different organism. This approach offers several advantages over selective breeding: only the desired genes need be transferred, genes from one species can be put into another to create novel combinations, and the whole process can be accomplished relatively quickly, at least compared to standard breeding procedures.

The new technology has far-reaching potential. For exam-

Chapter opening photo: Strawberry plants sprayed with genetically-engineered bacteria. These bacteria inhibit the formation of frost and reduce the incidence of frost-kill in associated plants. Many potato and strawberry fields in California are being sprayed with such bacteria.

13.1 Selective breeding in dairy cows Shown here are Guernsey (brown) and Holstein-Friesian cows. Both have been bred for high milk production. Guernseys originated on the Isle of Guernsey (Great Britain). Holstein-Friesians were developed in the Netherlands.

13.2 Insecticide spraying in a peach orchard Many of our crop plants have very high yields but require intensive use of chemical sprays to protect them against various insect pests.

ple, specific genes coding for the synthesis of the comparatively rare human protein insulin were isolated and inserted into bacteria or yeast, which, reproduce and then manufacture large quantities of insulin for therapeutic use. Genes that code for disease resistance in certain plants or for growth hormones in certain domestic animals can be introduced directly into other species of plants or animals that lack these genes and would benefit from them. In medicine, recombinant DNA technology may enable us someday to replace defective genes that cause various hereditary disorders, such as cystic fibrosis or sickle-cell anemia. In addition, the new techniques are proving invaluable in the chemical industry, for producing rare drugs and hormones to treat diabetes and dwarfism, as well as in agriculture, for producing higher-yielding crops and livestock.

This revolution in genetics is the result of new methods that make it possible to cut up DNA and move the fragments into the cells of another organism. Four basic steps are involved in recombinant DNA technology: (1) DNA from the donor organisms is cut into small, manageable pieces; (2) the DNA fragments are joined to some sort of ***vector,*** or transporting agent, that will carry the donor DNA into the host cell; (3) the host cells and their vectors are allowed to multiply to produce large numbers of descendants; (4) the descendants of these cells are screened for the gene of interest. We shall now examine this process in more detail.

Restriction Enzymes Cut DNA Into Pieces That Have Sticky Ends

Each chromosome consists of a single, long, double-helical molecule of DNA, which encodes an enormous amount of genetic information. Modern recombinant DNA techniques enable molecular biologists to isolate certain genes of interest from the rest of the genetic material. The complexity of this task, given the total length of the genetic material, can be highlighted by a simple analogy. Imagine the DNA molecule as a piece of yarn that is 100,000 times the size of an actual DNA molecule. The length of an average gene on this scale would be five centimeters, and the total DNA content of the nucleus would be approximately 100 miles long. How can the geneticist isolate a single gene? The first step, as noted above, is to cut the long DNA molecules into smaller, more manageable pieces. To do this special enzymes are required.

The enzymes that cut DNA are called ***restriction enzymes.*** There are several hundred of these specialized protein molecules, each of which has the capacity to recognize a specific nucleotide base sequence and to cut each DNA strand when the sequence occurs (Fig. 13.3). For instance, the restriction enzyme named *Eco*RI recognizes the following sequence of bases in the DNA molecule

—G-A-A-T-T-C—
—C-T-T-A-A-G—

wherever it occurs and cuts each strand between the G and the A, leaving single-stranded tails of unpaired bases at the cut ends. These ends are usually referred to as ***"sticky ends,"*** since they can form hydrogen bonds with the complementary bases of other such strands:

—T-A-T-A-C-**G-** and **A-A-T-T-C**-C-G-A—
—A-T-A-T-G-**C-T-T-A-A-** **-G**-G-C-T—

Restriction enzymes generally recognize nucleotide sequences that are palindromic, i.e., sequences that read the same from right to left on one strand and left to right on the other. (An example of an English palindrome is the phrase "Madam I'm Adam.") Thus, a particular restriction enzyme moves along the

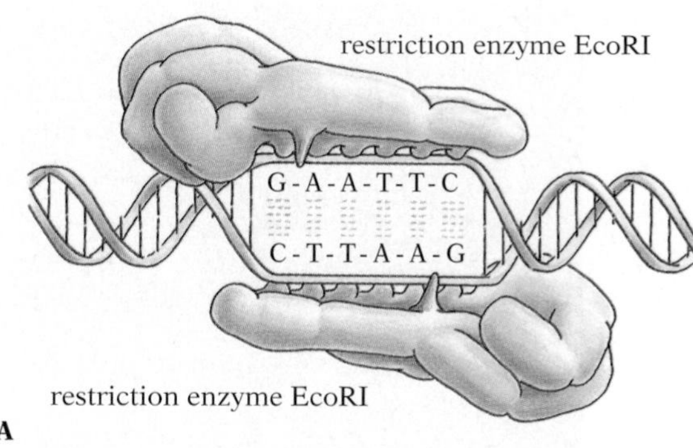

A

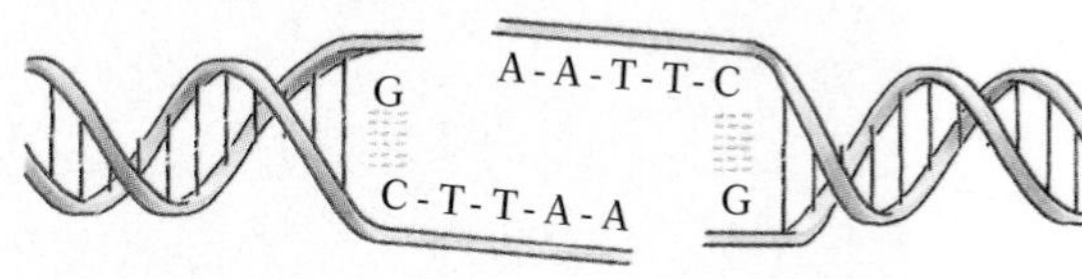

B

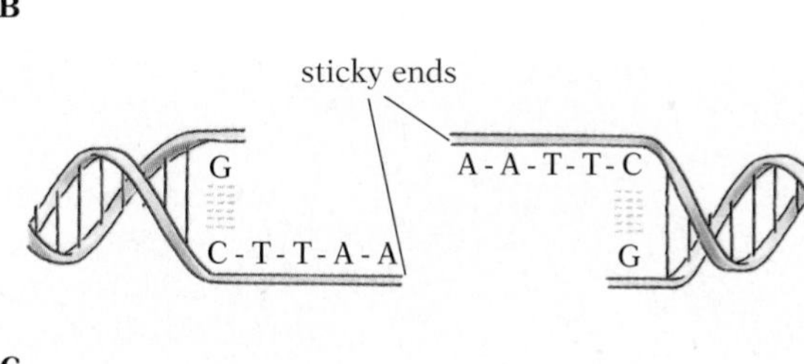

C

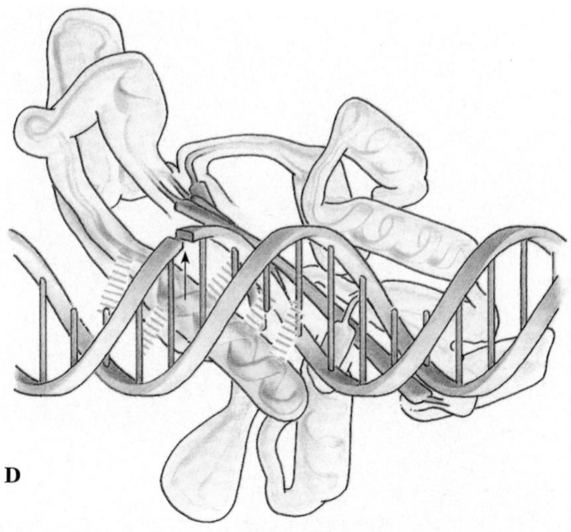
D

13.3 An example of restriction enzyme action (A) Restriction enzymes bind to a pair of palindromic target sequences—GAATTC in the case of the enzyme *Eco*RI. (B) They break a particular phosphate bond in the backbone of each strand of the DNA. (C) The resulting cut ends with unpaired bases are called sticky ends because under favorable conditions they will pair with other cut ends having the complementary sequence of bases. (D) One *Eco*RI attaches to a target sequence; the other endonuclease is omitted for clarity. The blue portion of the enzyme is important for recognizing the target sequence; the binding is indicated by dashed lines. The red portion is involved in weakening the bonds in the DNA; the arrow indicates where the helix is cut. More than 100 restriction enzymes are known, each with its own specific target sequence. The normal function of restriction enzymes is to protect bacteria against invading viruses; such enzymes digest viral DNA and thereby prevent their replication.

DNA molecule until it recognizes a specific palindromic sequence of bases, and it cuts the two strands of the molecule wherever it finds that sequence. The result is a mixture of DNA segments, some short, others quite long, depending on the enzyme used and how often that particular sequence happens to be present. If one assumes, for example, that the sequence G-A-A-T-T-C occurs randomly in the donor DNA, then *Eco*RI should cut the DNA approximately once every 4,096 base pairs.[1] If the geneticist is cutting human DNA, which has three billion base pairs, a great many pieces of DNA will obviously result. Of all these fragments, only one may be of interest. Indeed, the enzyme may well have found its target sequence within a particular gene, and so cut the gene itself, rendering it inactive. This whole procedure is sometimes referred to as the "shotgun approach."

[1] In DNA there is one chance in four that a given base in a sequence is a G, one chance in four that it is an A, one chance in four that it is a T, and so forth. Therefore, the probability that the sequence G-A-A-T-T-C will occur in that order is $1/4 \times 1/4 \times 1/4 \times 1/4 \times 1/4 \times 1/4$, or 1/4,096. It follows that if this sequence is found randomly throughout the chromosome, the enzyme *Eco*RI would be expected to cut the DNA approximately once every 4,096 bases.

The Donor DNA Is Joined to a Vector Before Being Introduced Into a Host Cell

The second step in this technology involves joining the pieces of DNA to a ***vector,*** a length of DNA that acts as a carrier vehicle to transport the gene into a new host cell. In bacteria two types of vectors are commonly used: plasmids and bacteriophages. ***Plasmids*** are small, circular DNA molecules that are found in the cytoplasm of bacteria and are independent of the bacterial chromosome (Fig. 13.4). Some plasmids carry only one or two genes, while others may be as much as one-fifth the size of the main chromosome and carry many genes. As we shall see, plasmids with genes for resistance to various antibiotics, including streptomycin, tetracycline, and ampicillin, are especially important for this technique.

Not just any DNA fragment can be used as a vector; all vectors must have a ***replication origin*** (*ori*) region in their DNA that allows the DNA to be replicated once it gets into the new host cell. Without an *ori* region, neither the vector DNA nor the donor DNA could be replicated. (To produce enough gene product to be of use, there must be many copies of the vector

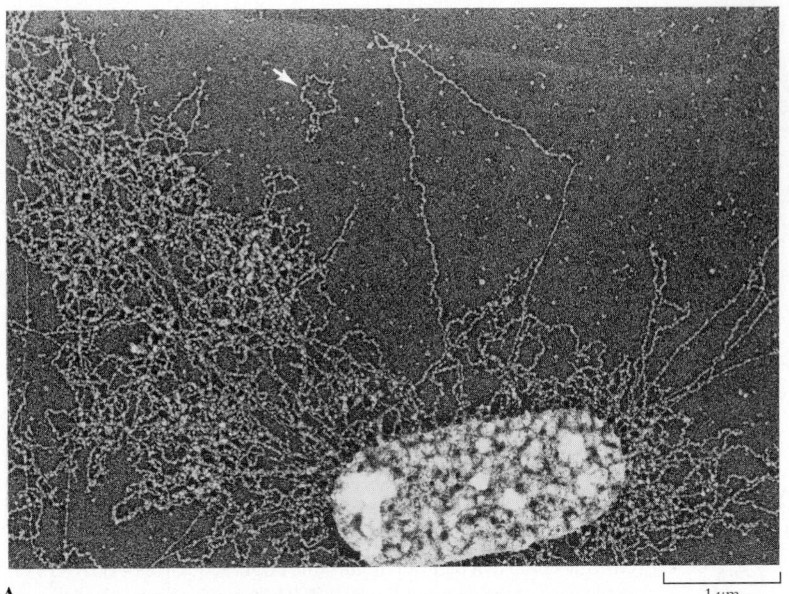

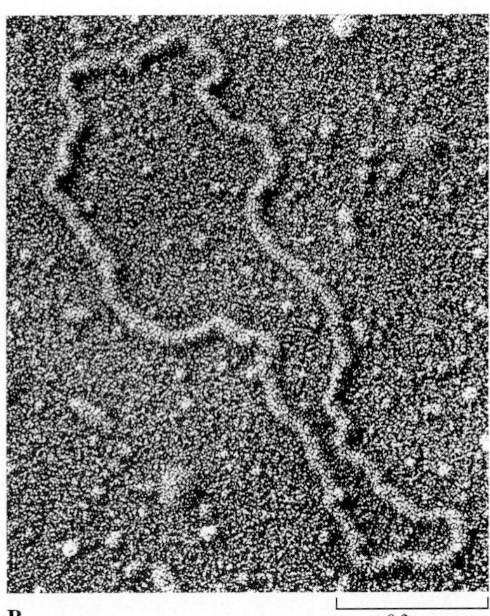

13.4 Electron micrographs of plasmids from *E. coli* (A) A plasmid (arrow) is visible in this *E. coli* cell that has been lysed to expose the DNA of its main chromosome. (B) The isolated plasmid seen here carries a gene encoding a product that confers resistance to the antibiotic tetracycline. (C) Plasmids are replicated independently of the main chromosome. The bacterial enzyme complex DNA polymerase can be seen copying each of these plasmids.

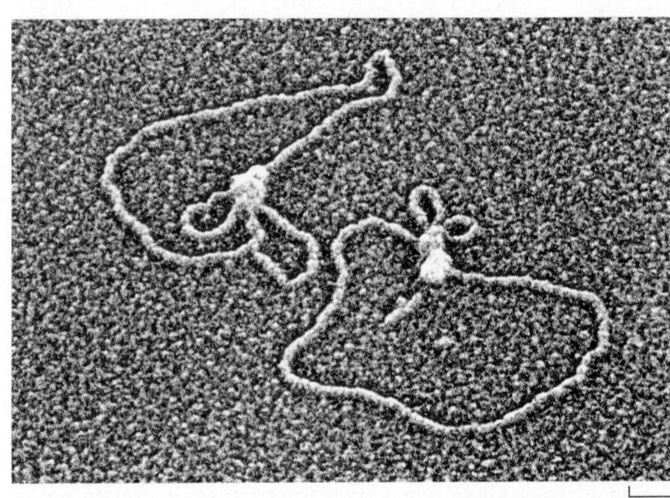

with its inserted DNA.) Plasmid replication occurs independently of chromosomal replication (Fig. 13.4C), and a cell that originally had only two or three plasmids may soon have a thousand or more.

Plasmids isolated from bacterial cells are exposed to a restriction enzyme that cuts the circular DNA molecule, making it a linear molecule with sticky ends (Fig. 13.5B). The same restriction enzyme is used on DNA from another organism, producing a variety of DNA fragments, each with sticky ends complementary to those of the plasmid DNA. The plasmid DNA and the donor DNA are then mixed (Fig. 13.5C). Because all the sticky ends are complementary, the ends will join with one another and a variety of different combinations may result. Plasmids may simply reform, the donor DNA fragments may join together, several plasmids may join, and so on. Some of these combinations may be the desired ones; that is, donor DNA may join with plasmids by complementary base repairing, reforming plasmids containing donor DNA in the process (Fig. 13.5C). At this point the DNA segments are held together only by weak hydrogen bonds between complementary base pairs. A special joining enzyme called a ***ligase*** must be used to seal the gaps in the DNA backbone (Fig. 13.5C).

The next step involves getting the modified plasmid into a bacterial cell. Plasmids have the potential to ***transform*** bacterial cells (Fig. 13.5D). In other words, if bacterial cells are placed in a medium containing plasmids, with special treatment a few (usually about 1/1,000) of the cells may pick up plasmids, just as some of the harmless pneumococci studied by Griffith picked up DNA that had been released into the medium by dead, disease-causing pneumococci (see p. 203). Recombinant DNA technology makes use of the transforming potential of plasmids. Plasmids are modified by the addition of DNA from another organism, and when bacterial cells pick up the modified plasmids, they also acquire the foreign genes.

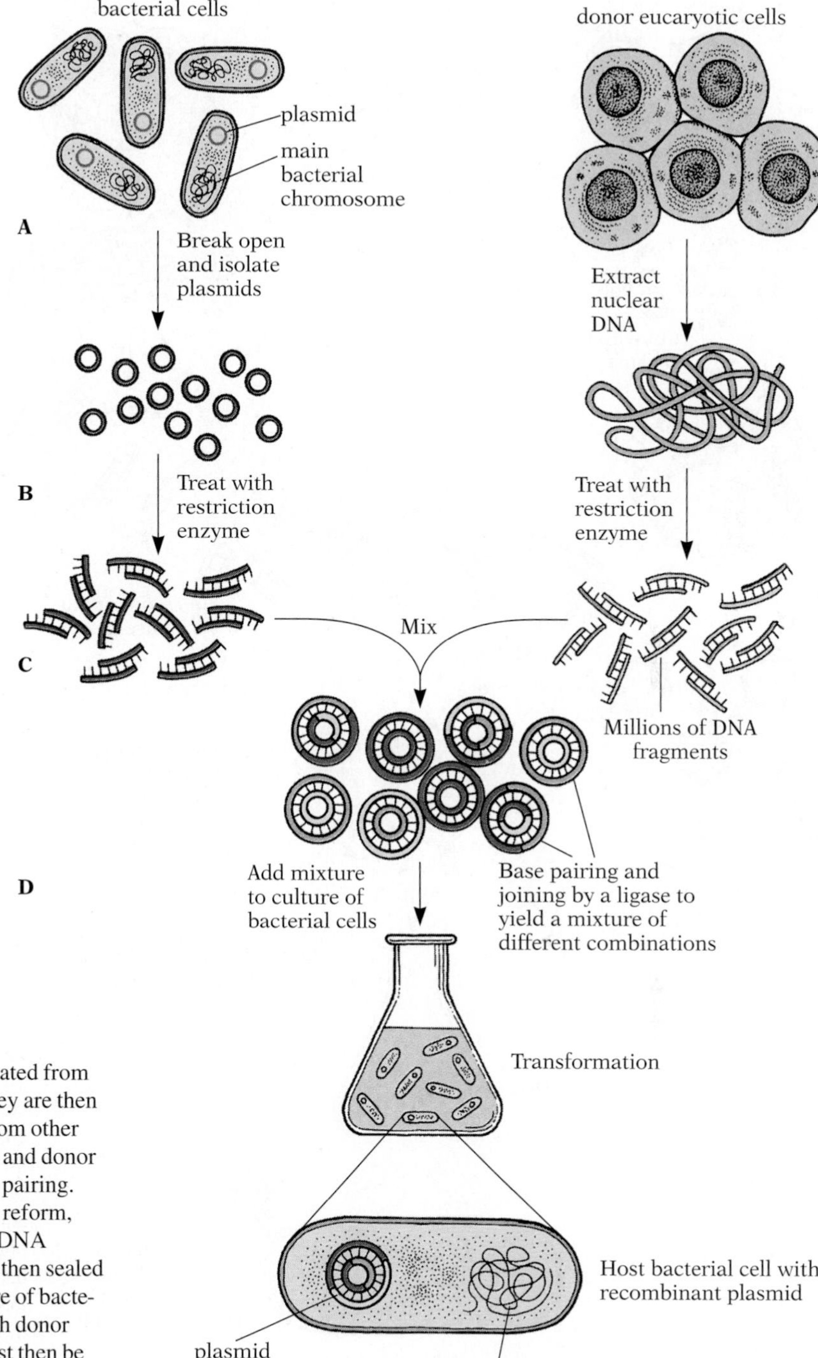

13.5 The recombinant DNA technique Plasmids, isolated from bacterial cells (A), are cut by a restriction enzyme (B). They are then mixed with fragments of donor DNA (blue and yellow) from other cells, produced with the same enzyme. The plasmid DNA and donor DNA can join at their sticky ends by complementary base pairing. Many different combinations are formed; the plasmid can reform, plasmid DNA can join with the donor DNA, or the donor DNA fragments can join to form a circle (C). The DNA ends are then sealed by treatment with a ligase. The mixture is added to a culture of bacterial cells. Some of the cells pick up the plasmids (some with donor DNA) and are thus transformed (D). Other procedures must then be used to find the cells that have the plasmids bearing foreign genes.

Bacteriophages (which, we know from Chapter 10, are viruses that infect bacterial cells) can also be used as vectors to take donor DNA into bacterial cells. The initial procedure is similar to that used for plasmids: the donor DNA and phage DNA are cut with the same restriction enzyme to create the same sticky ends, the two types of DNA are mixed, complementary base pairing occurs between the sticky ends, and the ends are joined by ligase. Bacteriophage proteins are added next, and they combine with the DNA to form bacteriophages (Fig. 13.6). The bacteriophages then infect other bacterial cells by attaching to the cells and injecting their DNA. This process, in which a virus is used to transport genes into a new host, is called ***transduction.*** The transduction procedure is much more efficient than transformation; a higher proportion of the cells obtain donor DNA. Unfortunately, there is a size limit to the DNA fragment that can be incorporated within the phage, and, probably more important, only one or a few closely related species of bacteria (e.g., *E. coli*) can be used for this procedure.

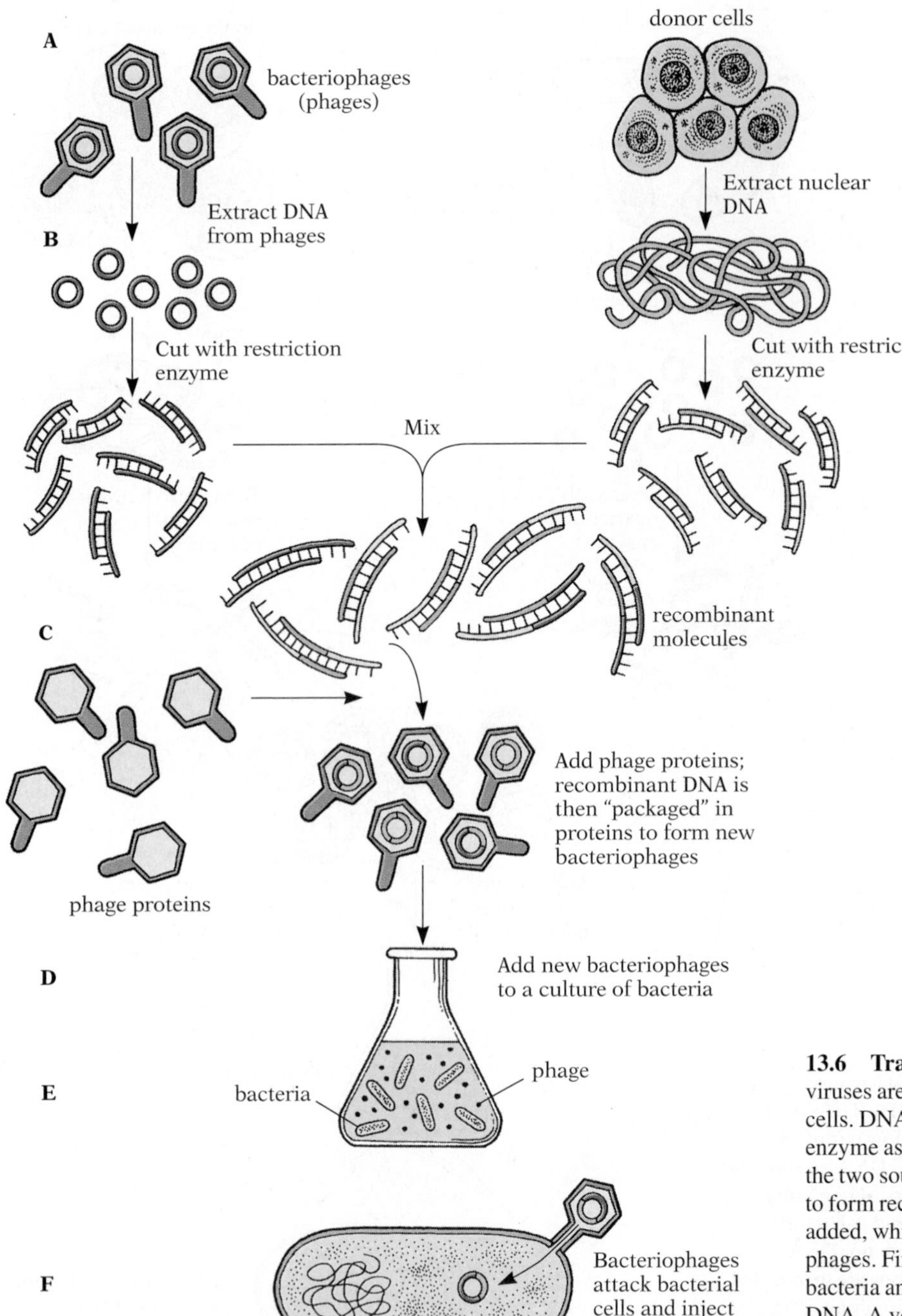

13.6 Transduction In this procedure, bacteriophage viruses are used as vectors to transport DNA into bacterial cells. DNA from host cells is cut with the same restriction enzyme as DNA from bacteriophages (B). The DNA from the two sources is then mixed together and joined by a ligase to form recombinant molecules (C). Phage proteins are then added, which package the DNA, forming new bacteriophages. Finally, the bacteriophages are added to a culture of bacteria and attack the cells, injecting their recombinant DNA. A very high proportion of bacterial cells will receive the recombinant molecules.

The Cells That Have Acquired the Desired Gene Must Be Identified

Once inside the bacteria, the plasmid or bacteriophage replicates autonomously to produce many copies of itself and the foreign DNA it carries. Because thousands of identical copies (called ***clones***) can be produced, the entire process by which foreign genes are introduced into bacteria and replicated is often referred to as ***cloning*** the gene. Each vector will be carrying different fragments of DNA, so there may be thousands of different clones produced. These clones are sometimes referred to as a ***gene library,*** or ***clone bank.*** The entire library must be searched to find the one clone that has the desired gene.

Generally, plasmids containing genes that confer antibiotic resistance are used as vectors. When the host bacterial cells are grown on media containing antibiotics, only the bacteria that have plasmids will survive; bacteria lacking plasmids will be killed or their growth severely retarded. Thus, the researcher will be able to find all the cells that have plasmids. A much more difficult problem, however, is to identify the specific bacterial cells that have the desired foreign gene incorporated within the plasmid. If something is known about the gene of interest, such as a short sequence of nucleotide bases or of the amino acids in the gene's protein, or if the researcher has isolated mRNA from cells expressing that gene, it is possible to synthesize a short single strand of radioactive DNA or RNA to use as an indicator or ***probe*** to find it (Fig. 13.7). These probes are single-stranded nucleic acids that are complementary to sequences found in the desired gene: bacterial cells containing the gene are located when these specially designed probes bind to them. An alternative procedure, if information about the gene is lacking, is to use special proteins called ***antibodies,*** which specifically bind to the protein produced by the desired gene, thereby identifying the clone (Fig. 13.8). A major obsta-

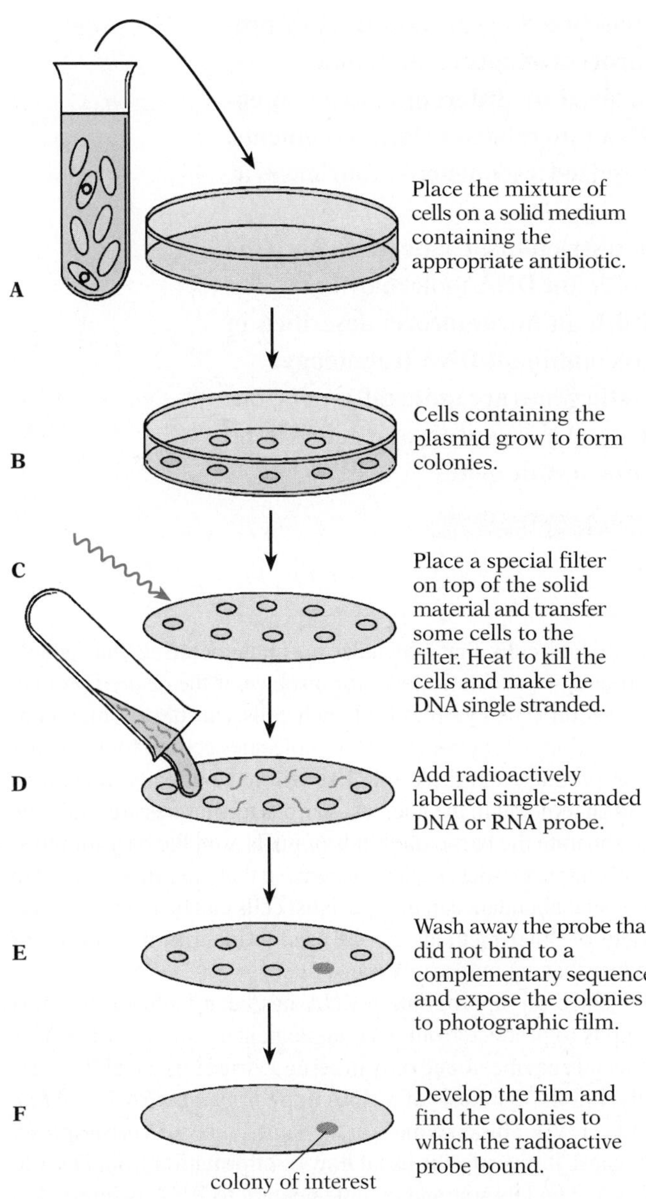
A
Place the mixture of cells on a solid medium containing the appropriate antibiotic.
B
Cells containing the plasmid grow to form colonies.
C
Place a special filter on top of the solid material and transfer some cells to the filter. Heat to kill the cells and make the DNA single stranded.
D
Add radioactively labelled single-stranded DNA or RNA probe.
E
Wash away the probe that did not bind to a complementary sequence and expose the colonies to photographic film.
F
Develop the film and find the colonies to which the radioactive probe bound.
colony of interest

13.7 Using radioactive probes to locate the clone with the desired gene The bacterial cells are plated on a solid medium (A) containing an antibiotic that only the cells containing the plasmid can inactivate. The cells containing the plasmid grow; each cell gives rise to a colony of identical cells (B). A special filter is placed on top of the solid medium that picks up cells from each colony (C), and the cells are heated to separate the two strands of DNA. Radioactively labeled single-stranded DNA or RNA specific for the gene of interest is added to the filter (D). The single-stranded probe will bind to the gene of interest by complementary base pairing, but will not be able to bind to other genes (E). The filter is then rinsed to remove all nonbound probe and photographic film is placed next to the filter (F). The radioactive probe will expose the area on the film where the probe is found, thereby locating the colony with the desired gene.

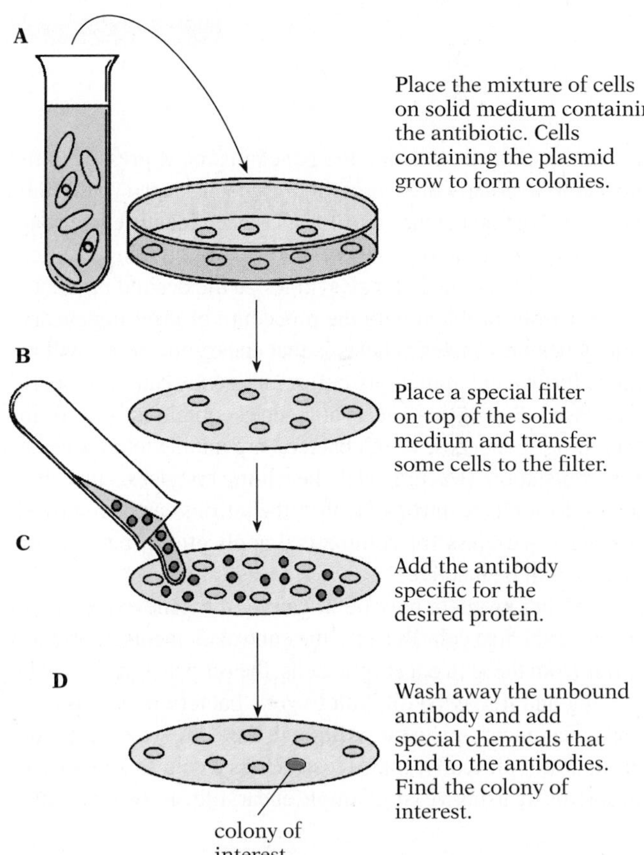
A
Place the mixture of cells on solid medium containing the antibiotic. Cells containing the plasmid grow to form colonies.
B
Place a special filter on top of the solid medium and transfer some cells to the filter.
C
Add the antibody specific for the desired protein.
D
Wash away the unbound antibody and add special chemicals that bind to the antibodies. Find the colony of interest.
colony of interest

13.8 Using an antibody to locate a clone with the desired gene The cells are grown on solid medium to form colonies (A). A copy of the colonies is made by placing a special filter on top of the medium and removing some cells to the filter (B). Next, antibodies specific to the protein coded for by the desired gene are added to the filter and bind to the protein wherever it is found on the filter (C). The filter is rinsed thoroughly to remove all of the unbound antibodies. Special chemicals are added that react with the bound antibodies (D), indicating the colony producing the desired protein.

MAKING CONNECTIONS

1. Compare and contrast the selective breeding process used to produce domestic plants and animals with the process of natural selection.

2. The palindromic sequences recognized by different restriction enzymes vary in length. If you wanted to cut DNA into relatively large fragments, would you use a restriction enzyme that recognized a sequence of four bases or one of eight bases? Explain your answer.

3. Why is it essential to use the same restriction enzyme to cut both the donor and the vector DNA in creating recombinant DNA molecules?

4. How did the experiments of Griffith on pneumococci described in Chapter 10 help to lay the groundwork for recombinant DNA technology?

5. In Chapter 12 we saw that eucaryotic genes are quite different from procaryotic genes. How could these differences cause eucaryotic genes to be less likely expressed in bacterial cells than procaryotic genes?

cle in this procedure is that the gene must be expressed in the host cell; in other words, protein must be produced or the antibody cannot find it. Once the clone has been identified, the bacteria with the desired gene can be cultured, creating limitless numbers of bacteria that may synthesize the desired product.

A serious problem with the procedure of inserting eucaryotic genes into bacterial hosts is that eucaryotic genes will require further manipulations before the genes can be properly expressed. Recall that eucaryotic genes contain introns, or intervening sequences, which bacteria are unable to remove before translation. (Bacteria lack the editing enzymes, since their genes do not have introns.) Unless the introns can be removed, the gene is useless for commercial applications, because the protein will be inactive.

Another problem may lie in getting the gene expressed in its bacterial host cell. Because the control elements in bacteria differ from those in eucaryotic cells, the proper signals may be lacking and it may be difficult to get a bacterium to express a particular eucaryotic gene. Although there are ways of getting around this difficulty, many researchers avoid this and other problems by using yeast, a simple eucaryote, as the host cell.

Complementary DNA Can Also Be Used to Clone a Gene

An alternative to the shotgun approach involves synthesizing a DNA copy of the desired gene and inserting that gene into bacteria. In this procedure, the functional mRNA coding for the desired product is isolated, and this mRNA is used to synthesize a complementary gene. The crucial step in this procedure is to find cells that specialize in manufacturing that gene's product—pancreatic cells, for instance, if the desired product is insulin. The cytoplasm of such cells will have a high concentration of functional mRNA molecules coding for their special product, and these mRNAs will have already undergone intron removal in the nucleus. Various techniques are available to separate the particular kinds of mRNA on the basis of physical characteristics, such as weight, so that the mRNA found in unusual abundance in the specialist cells can be isolated. A variety of other methods can be used to identify the mRNA of genes that are only rarely transcribed.

Once the appropriate mRNA has been isolated, the next step is to produce from it a complementary strand of DNA. A special enzyme, found only in certain viruses, is capable of catalyzing the synthesis of a DNA copy *from an mRNA template* (Fig. 13.9). This enzyme is aptly named ***reverse transcriptase,*** because it reverses the usual flow of information from DNA to RNA. The DNA strand complementary to RNA is referred to as ***complementary DNA,*** or ***cDNA.*** When formed, it is single-stranded. Then the enzyme replicates the DNA strand to make double-stranded DNA, which can then be inserted into a plasmid. The procedures used are similar to those for inserting foreign DNA into plasmids (Fig. 13.4B–D): the plasmid is treated with a restriction enzyme to cut open a plasmid and create a pair of sticky ends. The cDNA, with a complementary set of sticky ends, is mixed with the plasmid DNA, and complementary base pairing takes place. A ligase restores the bonds in the DNA backbone, and the plasmid with its cDNA gene is taken up by a bacterial host (Fig. 13.9).

This procedure has two immediate advantages over the shotgun procedure: the expensive process of sorting the whole

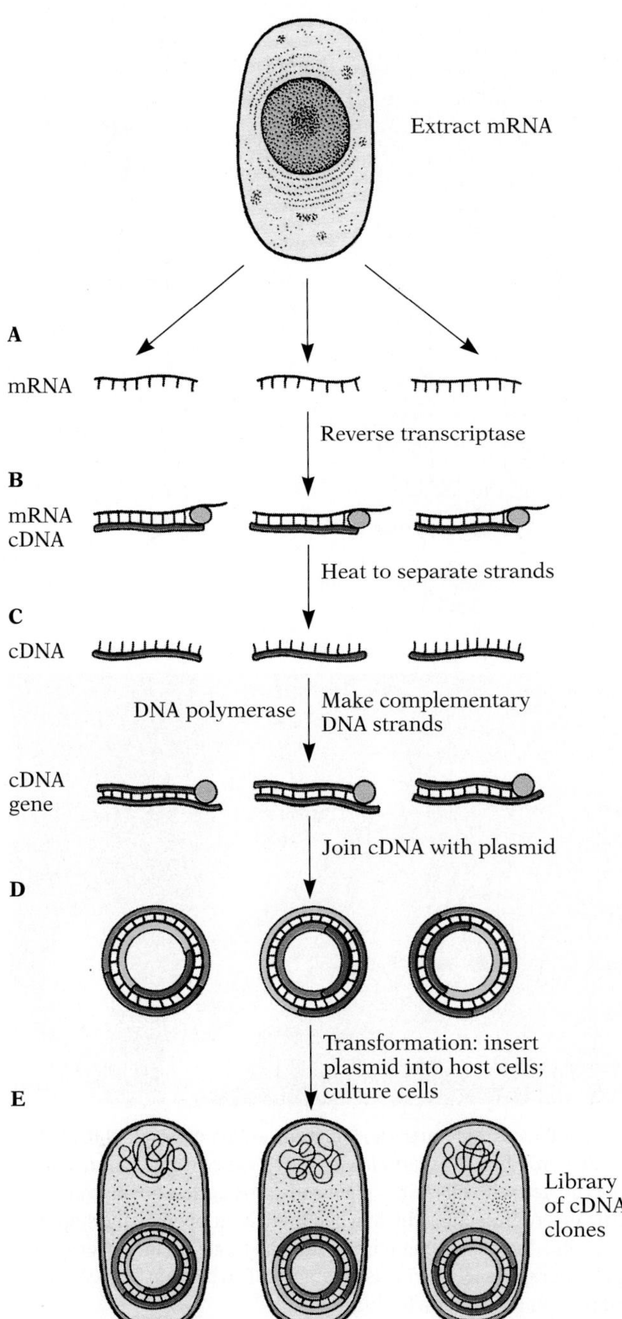

13.9 Using complementary DNA (cDNA) to clone a gene In this procedure, mRNAs are isolated from cells specialized to produce the desired protein (A), and reverse transcriptase is used to make a cDNA transcript of the mRNAs (B). The DNA strand is then replicated to form double-stranded DNA (C), which can be inserted into a plasmid (D) and taken up by a bacterial cell. The procedure creates a small gene library (E) because only the appropriate mRNAs are used. And because the introns have already been removed from mRNA, the inserted gene can be accurately transcribed and translated by bacterial cells.

gene library for bacteria bearing the desired gene is eliminated, since only the appropriate cDNA gene is used, and introns, which cannot be removed by procaryotes, have already been eliminated in the production of the functional mRNA. The transformed bacterial cells grow and divide rapidly, creating vast numbers of bacteria that can synthesize the desired product, particularly if the host plasmid has signals that enhance transcription of the cDNA. Besides insulin, this technique is used to produce large quantities of other hormones—growth hormone is particularly important—that are difficult to synthesize. Recombinant bovine growth hormone is now used to boost milk production by 10 to 40 percent, and synthetic human growth hormone is employed to prevent stunted growth in hormone-deficient children. A naturally produced but poorly understood agent called ***interferon,*** which holds promise in the treatment of cancer and various viral diseases, is also now widely available for research and medical applications as a result of recombinant DNA technology.

Special Procedures Must Be Used to Get Donor DNA Into Eucaryotic Cells

The procedures discussed so far are very useful for getting donor DNA into bacterial cells, and for getting the host cells to produce a particular protein. But what if the goal is to transfer genes into eucaryotic cells to repair a defective gene or to add new genes? Bacteriophages and plasmids are useful vectors to transport DNA into bacterial cells, but such vectors cannot normally enter eucaryotic cells. How can one get the desired gene or genes into a eucaryotic cell?

Several methods involve physically transferring DNA into eucaryotic cells. In one procedure donor DNA is deposited in a coat on the surface of tiny tungsten pellets, which are literally shot (using a shotgun mechanism) into the host cells, and the new DNA is incorporated into the host chromosomes (Fig. 13.10). Crude as this mechanism is, it has been successful in introducing DNA into both plant and animal cells, although the success rate is low. Recently this technique has been used to insert foreign genes into the DNA of chloroplasts—creating possibilities for adding genes to improve photosynthetic efficiency or disease resistance.

Microinjection techniques have also been used to insert genes; here the DNA is injected directly into the nucleus using a very fine needle (Fig. 13.11). The procedure has a very low success rate—probably only about one cell in 10,000 will acquire the new gene. Another new procedure, called electroporation, offers some promise. It involves exposing the host cells to brief pulses of high-voltage electricity. The electricity temporarily disrupts the membrane, creating pores, through which small molecules, such as DNA, can be taken up by the cell. Unfortunately, this procedure kills many of the cells.

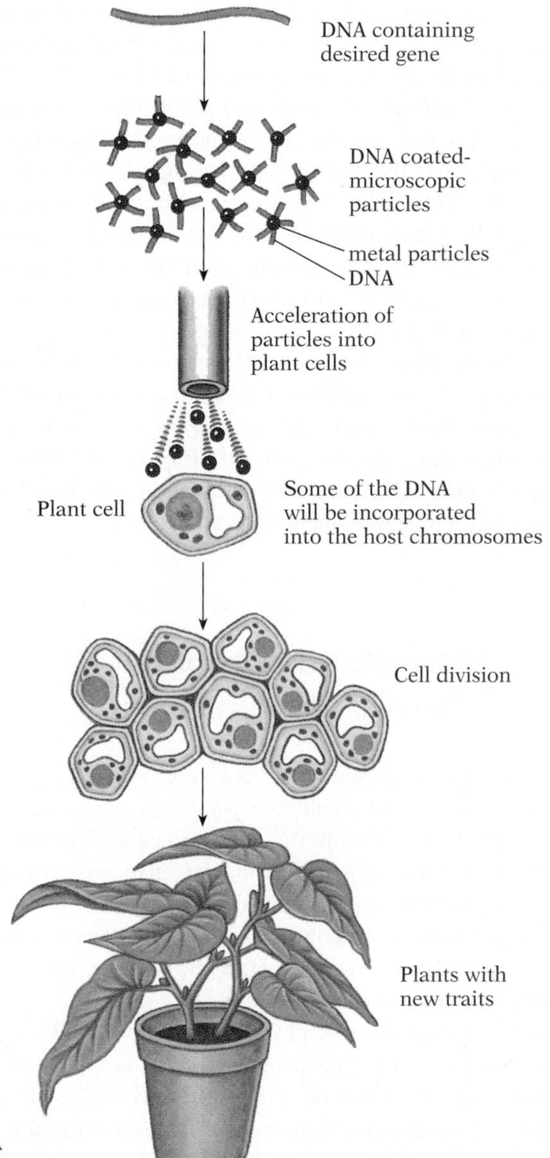

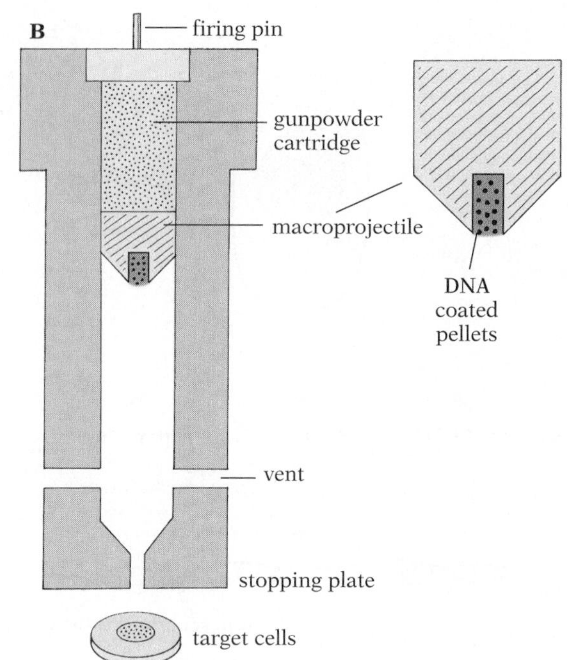

13.10 Use of the DNA particle gun to insert genes in plant cells (A) In this procedure tiny metal particles are coated with DNA containing the genes for the desired traits. The particles are "shot" into the plant cell using a shotgun mechanism and gunpowder charge. Some of the plant cells will incorporate the DNA into one of its chromosomes. The plant cells will divide producing a clone of cells, each of which is capable of giving rise to a new plant that has the genes for the desired traits. (B) Diagram and photograph of the DNA particle gun. The DNA particle gun uses a gunpowder charge to fire pellets coated with DNA into plant cells. The pellets are placed in a plastic macroprojectile which, when fired, is stopped by the stopping plate, while the pellets continue on into the target. In the accompanying photograph, a technician is placing cells to be transformed into the device in preparation for firing.

Gene transfers in plants often make use of the bacterium responsible for crown gall disease (Fig. 13.12A). This bacterium has a 10-gene plasmid that actually induces the crown gall tumor; bacteria lacking this plasmid do not cause tumors. When the bacterium infects a plant cell, it inserts its plasmids into the host's chromosomes (Fig. 13.12B). Any foreign DNA added to the plasmid is inserted right along with the plasmid DNA. Researchers have already transferred genes resistant to herbicides, insects, and diseases into tobacco and petunias using this plasmid. Unfortunately, not all plants are susceptible to the crown gall bacterium, so the search is on for other vectors. Plants are particularly good targets for gene transfer techniques because it is possible in many plant species to generate a whole new plant from a single transformed plant cell (Figs.

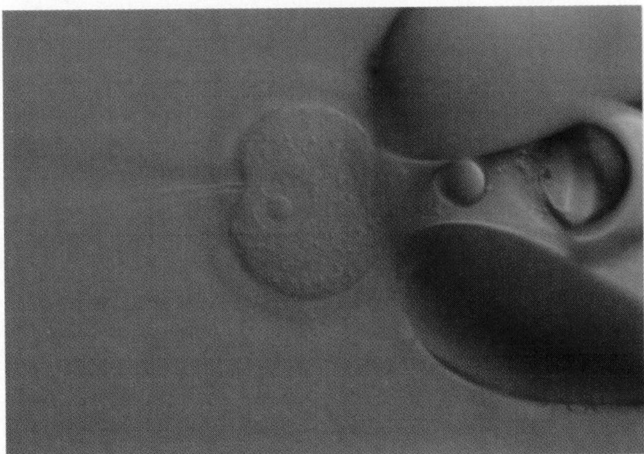

13.11 Microinjection of DNA into mouse zygote A very fine needle (left) is used to inject DNA directly into the male pronucleus before it fuses with the egg nucleus.

13.10 and 13.12). It is therefore not necessary to transform all the cells of the plant body, as might be true in an animal, but simply a single cell, which can then grow into a new plant, every cell of which will carry the transferred gene.

Certain viruses can also be used as vectors to transport genes into a eucaryotic host cell. The viruses most often used are retroviruses, viruses whose genetic material is RNA rather than DNA. This means that the donor gene must be transcribed into RNA before being joined to the viral RNA. Thereafter, when the virus infects a cell and releases its RNA, it also releases the donor RNA gene(s). Each retrovirus contains the enzyme reverse transcriptase, which catalyzes the formation of a cDNA copy of the viral RNA. The cDNA copy is made double stranded by the same enzyme, and is then inserted into a chromosome of the host cell.[2] Retroviruses are a very useful vector for delivering the foreign gene, because they usually do not kill their host cell and because their genetic material is inserted into the host chromosomes with high efficiency. The system is not perfect, however, because the gene at the site of the insertion is usually destroyed by the insertional event.

This technique is already bearing fruit. Richard Mulligan at

[2]Retroviruses have an enzyme that recognizes a particular sequence of DNA in the host chromosomes and catalyzes the insertion of the viral cDNA at those sites, wherever they occur.

A

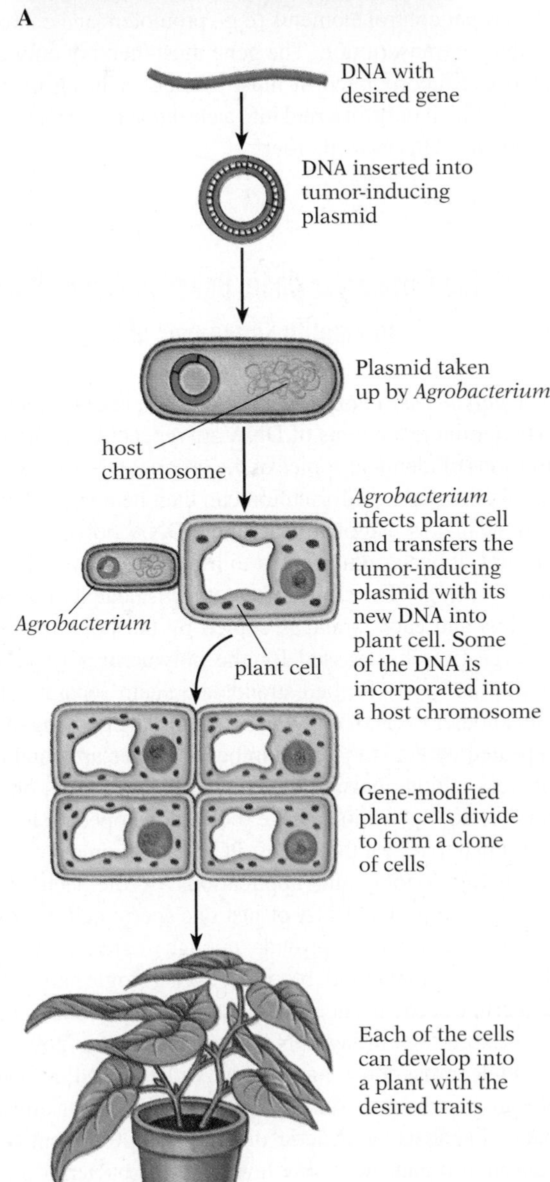
DNA with desired gene
DNA inserted into tumor-inducing plasmid
Plasmid taken up by *Agrobacterium*
host chromosome
Agrobacterium infects plant cell and transfers the tumor-inducing plasmid with its new DNA into plant cell. Some of the DNA is incorporated into a host chromosome
Agrobacterium
plant cell
Gene-modified plant cells divide to form a clone of cells
Each of the cells can develop into a plant with the desired traits
B

13.12 Using the bacterium *Agrobacterium* to insert genes into plant cells (A) The bacterium *Agrobacterium tumefaciens* inserts a plasmid into its host's cells, leading the plant to produce the crown gall tumor. (B) In the *Agrobacterium* procedure, the desired genes are inserted into the tumor-inducing plasmid, which is taken up by the *Agrobacterium* cells. When *Agrobacterium* infects plant cells, the plasmid is transferred from the bacterium to the plant cell, where the inserted DNA may become integrated into the host cell chromosomes. The plant cells will divide, producing a clone of cells, each of which is capable of giving rise to a new plant with the genes for the desired traits.

the Whitehead Institute for Biomedical Research has used retroviruses to transfer a human hemoglobin gene into developing blood cells in a mouse. In addition, James M. Wilson of the Howard Hughes Medical Institute at the University of Michigan in Ann Arbor used retroviruses to transfer a normal gene that codes for a normal protein into pancreatic cells bearing a mutant gene for cystic fibrosis that encoded a defective version of the protein. The normal gene repaired the genetic defect, opening the door to a dramatic new way to treat an otherwise incurable disease.

Of course, it must be remembered that getting the DNA into the cell is only the beginning. If the foreign DNA is to function properly in its new host cell, it must first be inserted into a chromosome, and then it must be transcribed and translated. In the last chapter we saw that eucaryotic cells must have the proper control elements (e.g., promoters and enhancers) to stimulate transcription. The gene must then not only be delivered and incorporated, it must also either bring its promoter along with it or be inserted into a chromosome next to an active promoter. This is a tall order!

The Polymerase Chain Reaction Can Be Used to Amplify Sequences of DNA

The ***polymerase chain reaction,*** or ***PCR,*** is a new technique in which minute amounts of DNA are repeatedly copied to form millions of identical copies of the original DNA sequence. The amplified genetic information can then be analyzed. The procedure makes use of a heat-stable DNA polymerase from a species of bacterium adapted to life in the near-boiling waters of hot springs. The DNA to be copied is heated to separate the strands and each strand is copied by the polymerase. When enough time has passed for the polymerases to finish their work, the newly formed strands are again separated by heat, and another cycle of copying is begun. The copying cycles are repeated until a sufficient number of copies are obtained. The process is done outside of living cells, and takes only a few hours. PCR machines have been developed that are programmed to repeat the cycles. PCR has become a very powerful research tool, and applications of the technology are growing daily. The DNA of just one sperm cell, for example, can be amplified to provide enough material for analysis. Traces of seminal fluid, blood, or even a single hair follicle left at a crime scene is enough to identify the person it came from (see Exploring Further: DNA Fingerprinting, p. 269).

PCR procedures have been widely used in medicine. Because such small samples are required, it is useful for prenatal diagnosis of genetic diseases and for identifying the causative organism in low-level viral or bacterial infections. Recently it has been used for matching organ transplant donors and recipients. It can even be used to help clarify evolutionary relationships among organisms: DNA samples from bones, mummies, woolly mammoths, and other extinct animals have been analyzed in this way.

Recombinant DNA Techniques Can Be Used for Mapping Genes on Chromosomes

The new recombinant DNA techniques make it possible not only to use host cells as chemical factories to produce substances of medical importance, but also to study the sequencing and activity of genes from eucaryotic cells. Indeed, one of many spin-offs from recombinant DNA technology is a method for mapping genes with great precision. Messenger RNAs isolated from different cells are used to make cDNA copies in a medium containing the radioactive isotope ^{32}P. The radioactive DNA is then mixed with a set of chromosomes treated to separate the double helix into single strands. When base pairing is again made possible in the resulting mixture, the radioactive DNA acts as a probe to find the corresponding gene on the chromosome, locating with great precision the position of the gene (Fig. 13.13).

More than 2,000 human genes (including the genes for cystic fibrosis, muscular dystrophy, Huntington's disease, sickle-cell anemia, and some types of cancer) and marker regions have already been mapped on the 46 human chromosomes.

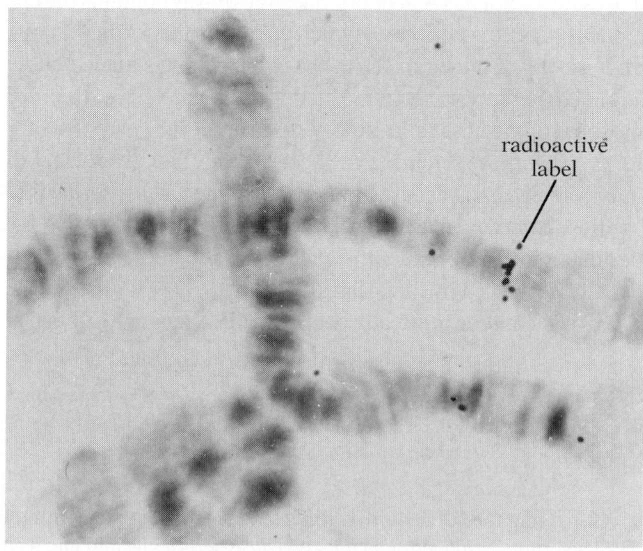

13.13 Locating a gene In this version of the DNA probe technique, salivary-gland chromosomes from the much-studied fruit fly *Drosophila* have been treated to weaken base pairing; a radioactively labeled probe of previously identified DNA has been added and has bound to the complementary region of the desired gene. The dark dots reveal where radioactive decay is taking place, and therefore where this particular gene is located.

EXPLORING FURTHER

DNA FINGERPRINTING

Recently recombinant DNA techniques have been used in new ways—to identify criminals and to determine paternity. For decades, fingerprints, blood samples, hair, seminal fluids, etc., have been analyzed and used to place criminals at the scene of the crime. Unfortunately, the best these tests could do was to exonerate a falsely accused person or suggest a suspect's guilt. Because every person's DNA is unique (with the exception of identical twins), evidence from DNA analysis is different—this information can positively identify a suspect, or completely exonerate them. Although the amount of DNA in a person's cells is enormous, and about 99.9 percent of everyone's DNA is identical, there are certain regions that vary a great deal from person to person, and it is these "hypervariable" areas that can be used to distinguish the DNA of one person from that of another. The most variable regions are certain sequences of two to 300 base pairs that are repeated over and over a variable number of times. For instance, the sequence CAT may appear in one allele once, but another allele might have three repeats (CATCATCAT), another might have six repeats, etc. Population studies are done to determine the frequency of these variable regions in a specific population. Using this information geneticists can calculate the probability of DNA samples from two different individuals matching by chance; estimates often vary from 1:500,000 to 1:738 trillion. The procedure was first developed by British geneticist Alec Jeffreys of the University of Leicester.

Evidence from DNA analysis was first used in the United States in a court in Orlando, Florida, in 1987. The defendant was suspected of 23 incidents of prowling, assault, and rape. Because the rapist was very careful not to let his victims see his face, and wore gloves, there was very little physical evidence for conviction. The prosecutors had heard of the process of "DNA fingerprinting," which was used to solve a rape-murder case in Britain, and decided to try the procedure in the Orlando cases.

Since every person's DNA is unique, analysis of DNA from biological sources found at the crime scene can definitively identify—or exonerate—an individual. The procedures have been well worked out. DNA is extracted from a sample of the suspect's blood and from a tissue sample obtained at the scene of the crime (e.g., a blood or semen sample left on the victim). If the sample is sufficiently large, the analysis can proceed directly, but if the sample is small, the DNA must be amplified using the new PCR technique (see p. 268). The amplified genetic information can then be analyzed. The DNA of a single sperm cell, for instance, can be amplified to provide enough material for analysis. Traces of seminal fluid, blood, or even a hair with cells attached left at a crime scene are enough to identify an individual.

Once the sample is sufficiently large, the DNA in each sample is cut with the same restriction enzyme, producing DNA fragments of varying lengths. In the areas where the DNA is constant from individual to individual, the resulting fragments are identical, but in

the "hypervariable" regions the fragment lengths vary greatly, depending on the number of repeats. It is these varying fragment lengths that must be compared.

The process of *gel electrophoresis* is used to separate the DNA fragments on the basis of charge and size: the DNA samples are placed at one end of a sheet of a special gel, electrodes are placed at each end, and an electric current is applied across the gel (see Fig. A). DNA, because of its many phosphate groups, is strongly negative and will migrate toward the positive electrode. The gel, which is a semisolid material with minute pores, acts like a molecular sieve. The distance the fragments move depends on the charge on the fragments and their size: smaller fragments are slowed down less by the gel, and therefore move farther. At the end of the procedure the DNA fragments are nicely separated by size. Because the investigators are only interested in fragments of the variable regions, they must use specific techniques to locate these regions: the DNA bands are treated in such a way that the two strands of DNA separate, leaving exposed nucleotide bases. (e.g., A-C-T-G-etc.). The DNA pattern is then transferred to a nylon sheet and radioactive DNA probes are added (Fig. B). These probes are radioactive sequences of single-stranded DNA known to be complementary to sequences found in the variable regions of DNA. The radioactive probes will bind tightly to the bands of DNA containing the complementary sequences. Because the DNA fragments for each individual will be of different sizes, and thus will migrate different distances, the probes will attach at different positions on the sheet. X-ray film is placed against the sheet, and the film is developed. Black bands show the location of the probes. This pattern of bands constitutes a "DNA print" and is unique for each individual. Because there are many different hypervariable regions in the human genome, a number of different probes are generally used, and the information is used to produce a "DNA profile" or "DNA fingerprint" of the individual.

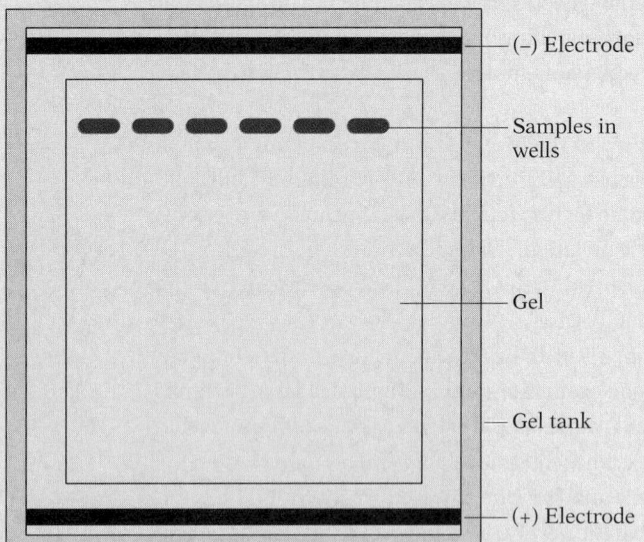

A Gel electrophoresis is used to separate DNA pieces by size DNA, being negatively charged, moves to the positive electrode. Large fragments cannot move through the gel as easily as small fragments, so the DNA is sorted by size, with the small fragments moving farther toward the positive pole than the large fragments.

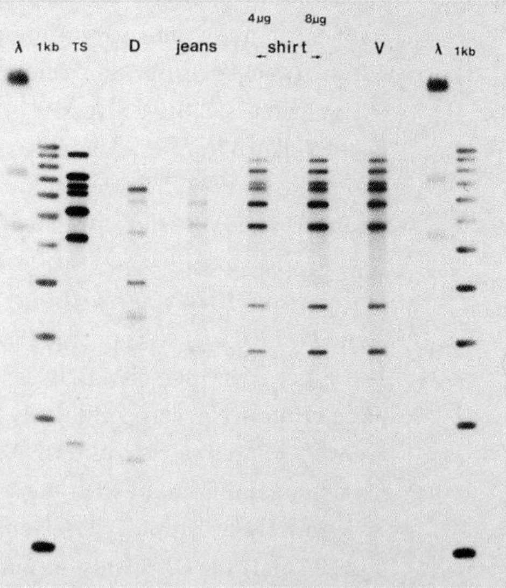

B DNA fragments from gel are transferred to a nylon filter and exposed to radioactive probes that are complementary to the hypervariable regions A photographic film is placed over the filter and the film is developed. In this film, a stabbing victim's DNA profile (V) matches the DNA bands found on the defendant's shirt and jeans. The other bands contain size markers.

The band pattern obtained from a suspect's blood can then be compared with seminal fluid obtained from the victim, and if it matches, the suspect can be positively identified. In the Orlando cases the jury accepted the DNA evidence, and the suspect was convicted. The "print" can be made from almost any biological source—hair and skin cells, seminal fluid, blood—and because each individual has a unique pattern, the potential uses of these techniques in criminal investigations are enormous, promising to revolutionize the way such procedures are conducted. To date more than 2,000 U.S. court cases in 49 states have used DNA testing for forensic purposes. Already the Federal Bureau of Investigation is designing its first computer database of DNA fingerprints of convicted criminals. And California already has a law mandating that blood samples be obtained for DNA profiling from all men convicted of sex felonies. The information is computerized, and whenever a sex crime occurs, a DNA profile will be run on the evidence and checked to see if there is a match with a previous offender.

Despite the wide acceptance of DNA evidence, its use is still controversial. Questions of how well the samples match and the reliability of the procedures in different laboratories continue to be an issue. Technical standards have yet to be developed and implemented. Most investigators insist that "blind-test methodology" must be implemented within the laboratory, so that neither the laboratory nor the examiner knows whether any given sample is actually a test. Congress is, at this writing, considering a DNA Proficiency Testing Act, which will address the issues of quality control.

This technique, together with others, is now being used, in the ***Human Genome Project,*** to map all of the human chromosomes. This ambitious project will first attempt to map the location of each gene on each chromosome, and then determine the nucleotide sequences for all the known map positions. Eventually it is hoped that the complete nucleotide sequences for the entire human genome will be worked out.

Since present estimates suggest that humans have between 50,000 and 100,000 genes and some three billion base pairs in their DNA, this project is immense. The expense is also formidable; at a cost of approximately $1.5 million a year, the total estimated price tag is about $3 billion for the project. There is considerable debate about whether wholesale sequencing is worth the tremendous cost, especially since other projects will necessarily go unfunded to allow this initiative, and it will mean a huge redirection of talent from other genetic engineering projects—all for the task of sequencing a set of chromosomes, 90 to 99 percent of whose DNA appears never to be transcribed. Why map the human genome? Is the expense justified? What will this tell us? Certainly the mapping and sequencing of the genes responsible for genetic diseases will aid in the diagnosis of these diseases, and perhaps enable us to correct the defect. And sequencing will certainly give us more information on how our genes work and interact, and the control mechanisms involved. In addition, comparisons of DNA sequences between different species should aid our understanding of the evolutionary process. Similarities in the DNA sequences of humans and apes, for instance, suggest that we diverged from the apes between four and six million years ago.

Recombinant DNA Technology Has Various Agricultural Applications

Some of the results of recombinant DNA technology are already available for agricultural use. Genetically engineered bacteria that reduce the incidence of frost-kill in plants are being utilized commercially on a limited basis. Researchers have already transferred herbicide-, insect-, and disease-resistance genes into a variety of plants (Fig. 13.14). And now the techniques are being used in novel ways: researchers have transferred genes for such proteins as human albumin and interferon into crop plants like potato and tobacco, enabling plants to synthesize these proteins. As the technology improves, crop plants could be used to manufacture a wide variety of new products—perhaps even become production factories for oil! Looking to the future, some researchers have suggested using this technology to transfer genes for the fixation of atmospheric nitrogen into crop plants, thus eliminating the need for the application of nitrogenous fertilizers. This would, however, be a formidable undertaking, and will not be accomplished in the near future.

Gene transfers to domesticated animals are also possible.

A

B

13.14 Normal and genetically engineered plants (A) The tobacco plant on the right has had a gene inserted that codes for a protein that is toxic to certain insects; the control plant on the left lacks this gene and shows extensive insert damage. (B) The cotton plant on the right has had a gene inserted that codes for a protein that confers resistance to a particular herbicide (bromoxynil), while the plant on the left lacks resistance. When the herbicide is sprayed on the plants, the genetically engineered plant is unaffected, while the control is damaged. Such herbicides could be sprayed on cotton fields to control weeds.

A

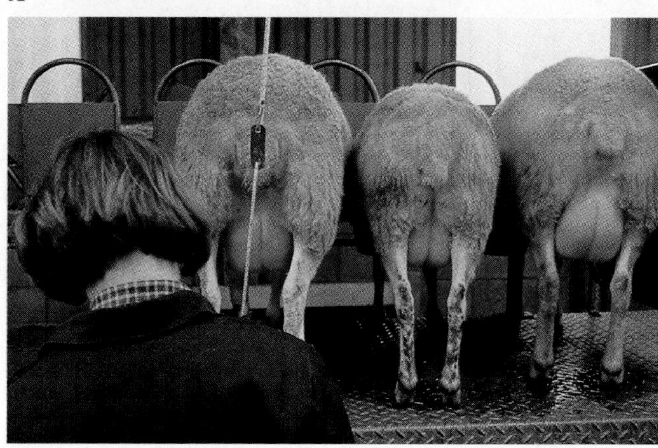

B

13.15 Genetically engineered animals (A) Pig with bovine growth hormone inserted into its DNA. The gene for growth hormone was injected into the zygote of a pig embryo. Scientists hope to produce leaner, faster-growing pigs using less feed. (B) Transgenic sheep awaiting milking. These sheep have had a human gene that codes for the enzyme α-1-antitrypsin (AAT) inserted into their DNA. AAT is produced in mammary cells, and is excreted in the sheep's milk. The AAT is then isolated from milk and is used to treat hereditary AAT deficiency in humans. About 100,000 people in the western world suffer from this deficiency, which leads to the lung disease emphysema.

The germ line is altered by effecting the change in the egg, either with a virus or by microinjecting DNA, which can then recombine with a similar sequence in the chromosomes. Extra genes for growth hormone have been added in this way to pigs and cattle, leading to strains that grow larger and store less fat (Fig. 13.15A). This shift in metabolic priorities, however, overburdens some of the internal organs of these creatures, and much remains to be done to realize the full promise of this technique.

Recombinant DNA Technology Also Has Many Potential Medical Uses

One of the major medical applications of these procedures lies in the diagnosis of genetic disease in fetuses. There are approximately 500 genetic disorders resulting from the mutation of a single gene, of which only a very few can be successfully treated. Early detection of affected fetuses, followed by abortion, has allowed parents to avoid the birth of an afflicted child. For many years biochemical tests for defective proteins or enzymes were performed on fetal cells, but it is now possible to diagnose many of these diseases directly, by DNA analysis. The procedure involves cloning the normal gene and sequencing portions of it, and then devising screening procedures for the defective genes. A number of inherited disorders such as thalassemia, sickle-cell anemia, citrullinemia, and hemophilia can now be diagnosed by DNA analysis. The opportunities to use this technology will surely become more widespread, and it is suggested that within 10 or 15 years it will be possible to screen all of a person's genetic material for abnormal genes and chromosomes—a total gene screen. Similar procedures are being used in ways never before dreamed possible.

Recombinant DNA technology has many other potential applications to medicine. Researchers are developing vaccines for immunizations to prevent diseases caused by certain viruses; because such vaccines use only specific viral proteins and not whole (live, weakened, or dead) viruses, they are purer and therefore safer than those currently available. Unfortunately, the results so far have shown that the vaccines produced by this technology are not as effective as the whole virus in stimulating the immune response. As more is learned about the process, however, it is hoped that this difficulty can be overcome. Vaccines for immunizing individuals against the common cold, serum hepatitis, poliomyelitis, influenza, rabies, and malaria are now being developed.

Gene therapy—the addition of good copies of a gene to organisms with defective genes—may also become possible, and has succeeded in tests in fruit flies and other organisms. In spite of this success, the problems involved in gene replacement therapy, particularly in a postembryonic organism, are formidable. One of the difficulties has been to devise ways to successfully insert the new genes into eucaryotic cells, and to regulate their function once they are in place. Attention has focused on using retroviruses or liposomes (which can fuse with the cell membrane and so deliver their contents; see p. 104) to carry therapeutic genes to target cells.

Preliminary tests of gene therapy in humans are just beginning. Proposals to do such tests must be approved by the Recombinant DNA Advisory Committee and the Human Gene Therapy Subcommittee of the National Institutes of Health, which is composed of scientists, lay persons, and specialists in medical ethics, and then win approval from the Food and Drug Administration. The first approved proposal involves the treatment of advanced melanoma tumors in humans by injecting genetically engineered white blood cells. These cells will have a gene inserted that causes shrinkage of tumors in mice, and it is hoped that they will search out and destroy the melanoma tumors. Another proposal involves the treatment of children with a rare inherited immune disorder. These children lack the normal gene that codes for a crucial enzyme necessary for proper functioning of certain white blood cells. Researchers propose to remove white blood cells from the victim, insert a normal gene for the missing enzyme into their cells, and return the gene-corrected cells to the body. Many other proposals for human gene therapy are under consideration. So far all the therapies utilize similar protocols: cells are removed from the patient, a new gene is inserted, and the modified cells are returned to the body (Fig. 13.16). When the recombinant cells are white blood cells, the procedure must be repeated at intervals as the short-lived cells die. One of the proposed new therapies involves genetically altering comparatively long-lived liver cells. This therapy is for familial hypercholesterolemia, a fatal genetic disease in which the victims lack a protein that removes cholesterol from the blood. Without this protein the cholesterol level grows increasingly higher and early death results. Researchers plan to remove a portion of the patient's liver, insert the gene into the liver cells using a virus, and reintroduce the modified liver cells through a blood vessel leading to the liver. Hopefully the modified liver cells will establish themselves along the blood capillaries and begin producing the vital protein.

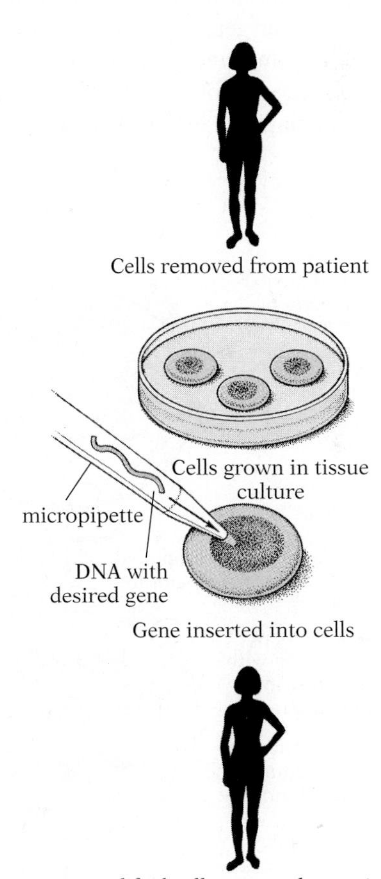

13.16 Human gene therapy In the procedures now undergoing testing, researchers propose to remove cells from the patient, insert a normal gene for the missing or mutated gene into them and return the gene-corrected cells to the body. The problems with this procedure are the same encountered in inserting genes in all eucaryotic cells—i.e., getting the gene into the cells, getting it inserted into the host cell DNA at an appropriate site, and having the gene regulated properly by normal physiological signals.

MAKING CONNECTIONS

1. As you learned in Chapter 12, positive control of gene expression is the rule in eucaryotes. How does this fact complicate the task of inducing a foreign gene to function after it has been introduced into a eucaryotic cell?

2. Describe the possible benefits and costs of the Human Genome Initiative.

3. What are some of the specific ways in which recombinant DNA technology can be applied to agriculture and medicine?

4. What concerns regarding safety and ethics have been raised by the actual or potential applications of recombinant DNA techniques?

New spin-offs of the recombinant DNA techniques keep coming. One new approach is to design drugs that block the expression of a particular gene, such as cancer-causing oncogenes. One class of compounds, called ***antisense*** compounds, are generating a great deal of interest. These compounds are short sequences of single-stranded nucleotides that recognize and bind to specific mRNAs, thereby interfering with translation (Fig. 13.17). Already the first human trials using antisense compounds to treat a form of leukemia have begun.

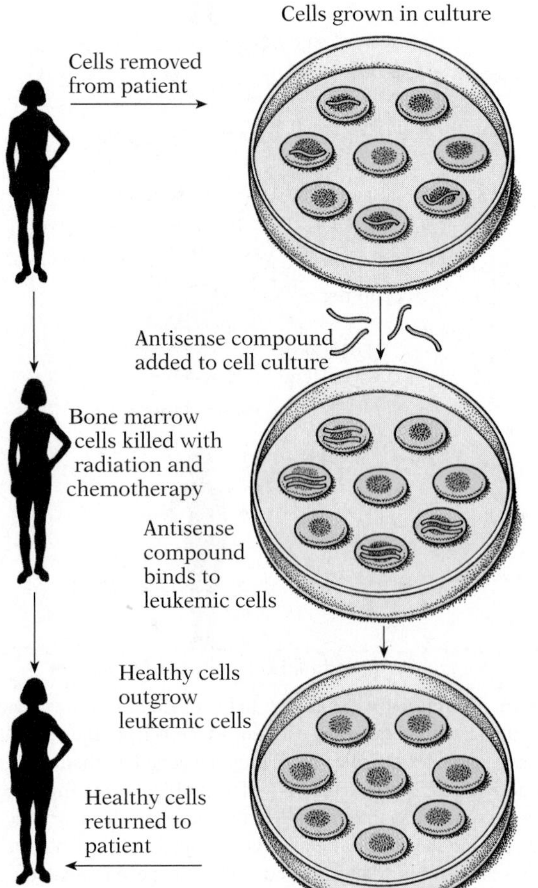

13.17 Using an antisense compound to treat leukemia In this procedure, bone marrow cells, which produce white blood cells, are removed from the patient. Many of these cells are normal and produce normal white blood cells, but others have a mutant gene and produce malignant cells. This technique attempts to separate the diseased from the normal cells. The removed cells are grown in culture and treated with an antisense drug, which will bind to the mRNA of the leukemic cells and interfere with cell function. The normal cells are unaffected and continue to grow. After the patient's bone marrow is completely destroyed by radiation and chemotherapy, the healthy normal cells are placed in the patient.

Safety Considerations Govern the Use of Recombinant DNA Technology

Like any modern industry, recombinant DNA technology carries with it a fair amount of controversy over perceived risks and hazards. Some observers have questioned the safety of certain procedures, and many citizens, including some scientists, are concerned that genetically engineered organisms may pose a threat to the environment. For example, special bacteria have been engineered to destroy pollutants in soil and water. Could such bacteria alter the environment in some unforeseen way? Could plants that have been engineered to grow faster and larger, or to resist diseases and pesticides, grow out of control and alter the ecological balance in a region? Or, an added concern, could these techniques be used to create new microorganisms for biological warfare?

Recognition of the possibility of the undesirable side effects of recombinant DNA technology has led to strict regulation of how the technology is to be used. Laboratories must be equipped with facilities for sterile handling and for physical containment similar to those long used in medical laboratories where dangerous pathogens are handled. In addition, there is a consensus that the microorganisms used in recombinant DNA research should be from strains carrying crippling mutations that make them incapable of surviving outside the laboratory, or they should be organisms that are not potential pathogens.

Recombinant DNA Technology Also Presents Some Ethical Dilemmas

The possibility of human gene therapy raises serious ethical questions. While few would question the benefits of using recombinant DNA technology to produce human insulin or to repair the genetic defect causing inherited disorders, such as cystic fibrosis or sickle-cell anemia, gene therapy could conceivably be used in other ways as well, to alter or enhance various human traits. What about using gene therapy to select a child's height, or weight, or to ensure a high IQ? Should we add genes to the sperm and egg cells, or just to body cells? Transferring genes to a person's blood cells, bone marrow cells, or pancreatic cells affects only that individual, not his progeny. But adding genes to the sex cells will affect all future generations of such an individual. Should such transfers be permitted? Where do we draw the line?

Opponents of using this technology for human genetic engineering question whether we have the wisdom to safely manipulate our genetic material, and indeed if we even have the right to do so. Who would have the authority to decide which genes are the "good" genes that should be added or maintained and which are the "bad" genes that should be eliminated? These are the kinds of questions the public will have to answer in the not too distant future.

CONCEPTS IN BRIEF

1. Recombinant DNA technology is the process by which DNA from one organism is isolated and transferred into the cells of another in order to impart some new characteristic or function to the host organism.

2. First, DNA from the donor organisms is cut into small, manageable pieces using *restriction enzymes.* Next, the DNA pieces are joined to a *vector,* which are most often *plasmids*—small, circular DNA molecules, found in the cytoplasm of bacteria, that replicate autonomously. The foreign DNA pieces join to the plasmid DNA by complementary base-pairing between their "sticky ends." After sealing with a *ligase,* the modified plasmids can be picked up by bacterial cells, which thereby acquire the foreign genes. The last step involves selecting the cells that have acquired the desired gene, using a radioactive probe or specific antibody. The selected clone is stimulated to reproduce repeatedly, producing many copies of the donor gene. The aim is to get the donor gene to function in its new host cell.

3. Bacteriophages can also be used as vectors to transport genes into a bacterial cell. The donor gene is first added to the viral DNA, using restriction enzymes and ligases, and phage proteins are then added to complete the bacterial viruses. The bacteriophages attack bacteria, injecting the viral DNA and the donor DNA into the cell.

4. An alternative procedure is to isolate functional mRNA coding for the desired product and, using *reverse transcriptase,* to synthesize a cDNA copy of the mRNA. The cDNA gene can then be inserted into a plasmid, and the plasmid can be taken up by bacterial cells.

5. Many different techniques are used to get DNA into eucaryotic cells. Some common methods involve microinjection, shooting tiny DNA-coated pellets into the cells, disrupting the cellular membrane, and the use of viruses as delivery systems.

6. The *polymerase chain reaction* can be used to make copies of a DNA sequence for further analysis.

7. The substantial technology that has grown up around recombinant DNA makes it possible to use bacterial cells as chemical factories to produce substances of commercial, agricultural, and medical importance: besides being used to introduce new genes into plants and animals, these techniques are also being used to map human chromosomes, and may soon be used for human gene therapy.

8. Recombinant DNA technology carries with it some risks and hazards. The public has questioned the safety of the procedures, and many individuals, scientists included, are concerned that some of these genetically engineered organisms may pose a threat to the environment.

STUDY QUESTIONS

1. **You are trying to produce a juicier peach. How would you do so through selective breeding? Would recombinant DNA technology offer any advantages? (p. 257–258)**
2. **What are the four basic steps involved in recombinant DNA technology? (pp. 258)**
3. **Explain how restriction enzymes are used to cut DNA into fragments of manageable size. What is a palindromic base sequence? Why do some restriction enzymes produce "sticky ends"? (p. 258)**
4. **What are plasmids? How are they used in recombinant DNA technology? (p. 259)**
5. **What is the function of DNA *ligase?* (p. 260)**
6. **Compare plasmids and bacteriophage as vectors to take donor DNA into bacterial cells. Distinguish between *transformation* and *transduction.* (pp. 260–261)**
7. **You are given a gene library and told to isolate the cells that contain a gene to produce insulin. Describe three methods you could use to identify these bacterial cells. What are some of the difficulties you may encounter in producing insulin from these bacterial cells? (pp. 262–264)**
8. **Explain how complementary DNA (cDNA) is produced and how it is used in recombinant DNA technology. What advantages does it have compared to the shotgun approach? (pp. 264–265)**
9. **Describe four techniques that have been developed to transfer genes into eucaryotic cells. (pp. 265–268)**
10. **How is the polymerase chain reaction (PCR) used to make millions of copies of a particular DNA sequence? Describe some of the applications of this technique. (pp. 268)**
11. **Describe how you would use cDNA and the radioactive isotope ^{32}P to locate the position of a gene on a chromosome. How is this information potentially useful? (pp. 268–271)**

C H A P T E R 1 4

INHERITANCE

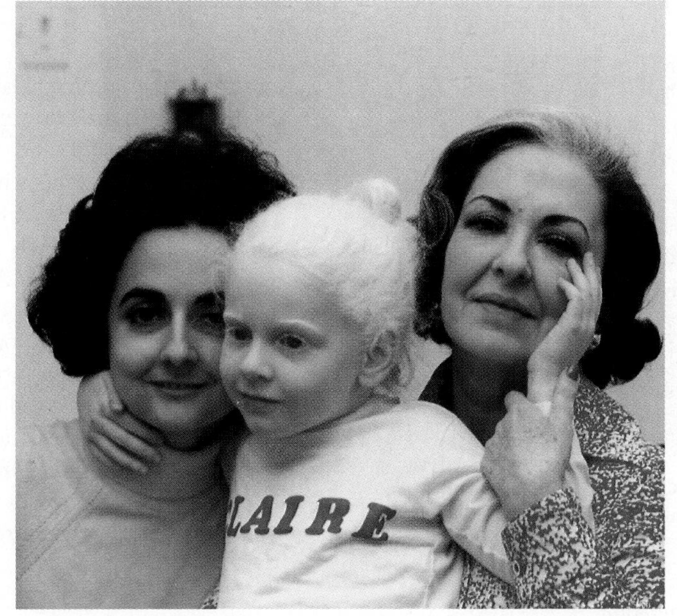

Until now we have considered the genetic material—DNA—at the molecular level, learning how DNA is replicated, how genes influence the characteristics of the organism, how genes are controlled, and how they regulate the many activities of the living cell. In this chapter we are going to discuss the organismal level—how the genes are passed on from generation to generation, and how they interact with each other and the environment to determine the characteristics of organisms. Evidence from breeding experiments led to the modern concepts of chromosomal inheritance, and it is this evidence that we shall examine here.

The Chromosomal Theory of Inheritance Began With Mendel's Work

We begin our examination of the chromosomal theory of inheritance by considering a series of experiments performed on ordinary garden peas by an Austrian monk named Gregor Mendel. Mendel published his results in 1866 (Fig. 14.1).

Mendel began with several dozen strains of peas, mostly purchased from commercial sources. He raised each variety for several years to discover which strains had recognizable variations that bred *true* (i.e., matings between two such strains produce offspring like their parents). Of the numerous characteristics of garden peas that Mendel studied, seven were particularly interesting to him. He noticed that each of these seven characteristics, or traits, occurred in two contrasting forms: the seeds were either round or wrinkled, the flowers were either red or white, the pods were either green or yellow, and so on (Table 14.1). When Mendel cross-pollinated[1] plants with contrasting forms of one of these characteristics, all the offspring (usually referred to as the F_1, or first filial, generation) were alike and resembled one of the two parents (the ***P***, or parental, generation). When these offspring were crossed among themselves, however, some of their offspring (the F_2, or second filial, generation) showed one of the original contrasting traits and some showed the other. In other words, a trait that had been present in the grandparental generation but not in the parental generation reappeared. For instance, when true-breeding plants with red flowers were

[1] Peas normally self-pollinate; that is, pollen from the male part of a flower is transferred to the female part of a flower on the same plant. Mendel cross-pollinated his plants; he removed the male parts of the flower from some plants and transferred pollen from the male part of other plants to the female parts of the flower from which the male parts were removed.

Chapter opening photo: Oculocutaneous albinism in a human being. Albinism is a recessive genetic trait present at birth resulting in a lack of pigment in the hair, skin, and eyes. The lack of pigment in the eyes results in nearsightedness, involuntary movements of the eye, and sensitivity to bright light. The skin of people with albinism burns easily in strong sunlight.

TABLE 14.1 *Mendel's results from crosses involving single character differences*

P characters	F_1	F_2	F_2 *ratio*
1. Round × wrinkled seeds	All round	5,474 round:1,850 wrinkled	2.96:1
2. Yellow × green seeds	All yellow	6,022 yellow:2,001 green	3.01:1
3. Red × white flowers	All red	705 red:224 white	3:15:1
4. Inflated × constricted pods	All inflated	882 inflated:299 constricted	2.95:1
5. Green × yellow pods	All green	428 green:152 yellow	2.82:1
6. Axial × terminal flowers	All axial	651 axial:207 terminal	3.14:1
7. Long × short stems	All long	787 long:277 short	2.84:1

14.1 Gregor Mendel investigated the laws of chromosomal inheritance by performing breeding experiments with pea plants

P generation
Red flower × White flower

F_1 generation 100% red

F_2 generation 75% red, 25% white

14.2 Results of Mendel's cross of red-flowered and white-flowered peas

crossed with true-breeding plants with white flowers, all of the F_1 offspring had red flowers (Fig. 14.2). In the F_2 generation, however, both red and white flowered plants occurred. Similarly, when plants characterized by round seeds were crossed with plants characterized by wrinkled seeds, all the F_1 offspring had round seeds, and in the F_2 generation round and wrinkled seeds appeared. Apparently in the F_1 generation one form of each characteristic was visually expressed while the other was not; i.e., red color was expressed rather than white in the flowers, and round form was expressed over wrinkled in the seeds. Mendel termed the traits that appear in the F_1 offspring of such crosses (in these examples, red flowers and round seeds) ***dominant characters,*** and the traits that were not expressed in such crosses (in these examples, white flowers and wrinkled seeds) ***recessive characters.***

Mendel Concluded That During Gamete Formation, Each of the Two Hereditary "Factors" for Each Trait Separate Into Different Gametes

When Mendel allowed the F_1 peas from the cross involving flower color, all of which were red, to breed freely among themselves, their offspring (the F_2 generation) were of two types; there were 705 plants with red flowers and 224 with white flowers. The recessive character had reappeared in approximately one-fourth of the F_2 plants. Similarly, when F_1 peas from the cross involving seed form, all of which had round seeds, were allowed to breed freely among themselves, their F_2 offspring were of two types: 5,474 had round seeds and 1,850 had wrinkled seeds. Again, the recessive character had reappeared in approximately one-fourth of the F_2 plants. The same was true of crosses involving the other five characters that Mendel studied (Table 14.1); in each case the recessive character disappeared in the F_1 generation, but reappeared in approximately one-fourth of the plants in the F_2 generation. We can summarize the results of the experiment involving flower color as follows:

P	red	×	white
F_1		all red	
F_2	$^3/_4$ red		$^1/_4$ white

From experiments of this type, Mendel drew the important conclusion that each pea plant possesses two hereditary ***factors*** for each character, and that when sperm or eggs are formed the two factors segregate and pass into separate gametes, so that each gamete possesses only one factor for each character. Each new plant thus receives one factor for each character from its male parent and one for each character from its female parent. The fact that two contrasting parental traits, such as red flower color and white flower color, can both appear unchanged in the F_2 offspring indicates that the hereditary factors must exist as separate particulate entities in the cell; they do not blend or fuse with each other. Thus, the cells of an F_1 pea plant from the cross involving flower color contain, according to Mendel, one factor for red color and one factor for white color, the factor for red being dominant and the factor for white being recessive. But their existence together in the same nucleus does not change the factors; the red and white factors do not alter each other. They remain distinct, and segregate unchanged when germ cells (the cells that will produce the gametes) are formed. This principle is referred to as Mendel's first law, the ***Law of Segregation***—each organism possesses two hereditary factors for each character, and during gamete formation the two factors separate (segregate) into different gametes so that each gamete has only one of each type of factor. When the gametes unite during fertilization, the pairs are restored.

Mendel's conclusions are consistent with what we know about the chromosomes and their behavior in meiosis. The diploid nucleus contains two of each type of chromosome. Each of the two chromosomes of any homologous pair bears genes for the same characters; hence the diploid cell contains two copies, which may or may not be identical, of each type of gene; these are the two hereditary factors for each character that Mendel described. Since the members of each pair of chromosomes segregate during meiosis, gametes contain only one chromosome of each type and, therefore, only one copy of each gene, just as Mendel deduced. Mendel's theories seem rather obvious in the light of the events of cell division, but it should be remembered that Mendel did his work before the details of meiosis had been learned—before, in fact, the significance of chromosomes for heredity had been discovered. He arrived at his conclusions purely by reasoning from the patterns of inheritance he detected in his experiments, without any knowledge of chromosomes or meiosis. The chromosomal theory of inheritance, therefore, rests today on two independent lines of evidence, one from breeding experiments, the other from microscopic examination of the nucleus. That these two lines of evidence should agree is testimony to the strength of the theory.

Mendel's Experiments Can Be Interpreted in Modern Terminology

Let us now reexamine Mendel's cross of garden peas of different colors, interpreting his results in modern terminology. Mendel was working with two forms of the gene for flower color, one that produced red flowers and another that produced white flowers. When a gene exists in more than one form, the different forms are called ***alleles.*** Each of these two alleles is located at the same position, or ***locus,*** on each of the two homologous chromosomes (Fig. 14.3). In the gene for flower color in peas, the allele for red is dominant while the allele for white is recessive. It is customary to designate genes by letters, using capital letters for dominant alleles and small letters for recessive alleles. We may thus designate the allele for red flower color in peas as *C* and the allele for white flower color as *c.* Since a diploid cell contains two copies of each gene, one from each parent, it may have two copies of the same allele or it may have one copy of one allele and one copy of the other (Fig. 14.3). Thus, cells of a pea plant may contain two copies of the allele for red flowers (*C/C*), or two copies of the allele for white flowers (*c/c*), or one copy of the allele for red and one copy of the allele for white (*C/c*). Cells with two copies of the same allele (*C/C* or *c/c*) are said to be ***homozygous.*** Those with one each of two different alleles (*C/c*) are said to be ***heterozygous.*** (The slash between letters indicates that the two alleles are on separate chromosomes.)

Note that one cannot tell by visual inspection whether a given pea plant is homozygous for the dominant allele (*C/C*) or heterozygous (*C/c*), because the two types of plants will

14.3 Anatomy of segregation Diploid cells have pairs of homologous chromosomes, a pair consisting of one chromosome from each parent. Each gene is found in two copies, one on each chromosome of the homologous pair, at corresponding loci. When the genes at the corresponding loci are different, they are called alleles. In meiosis, each gamete receives one chromosome from each homologous pair, and hence only one of the two alleles.

look alike; both will have red flowers. In other words, where one allele is dominant over another, the dominant allele takes full precedence over the recessive allele, and a heterozygous organism exhibits the trait determined by that dominant allele; one copy of the dominant allele is as effective as two copies in determining the trait. This means that there is often no one-to-one correspondence between the different possible genetic combinations (***genotypes***) and the possible appearances (***phenotypes***) of the organisms. In the example of flower color in peas, there are three possible genotypes, *C/C, C/c,* and *c/c,* and two of the three genotypes yield the red phenotype.

We can now apply this understanding of genes to Mendel's pea cross and rewrite the summary on p. 279 as follows:

P	*C/C*	×	*c/c*	
F_1	*C/c*	×	*C/c*	
F_2	*C/C*	*C/c*	*c/C*	*c/c*
	red	red	red	white

Here we have shown both the genotypes and the phenotypes of the plants in the three generations. Mendel began with a cross in the parental generation between a plant with a homozygous dominant genotype (red phenotype) and a plant with a homozygous recessive genotype (white phenotype). All of the F_1 progeny had red phenotypes, because all of them were heterozygous, having received a dominant allele for red (*C*) from the homozygous dominant parent and a recessive allele for white (*c*) from the homozygous recessive parent. But when the F_1 individuals were allowed to cross freely among themselves, the F_2 progeny they produced were of three genotypes and two phenotypes; one-fourth were homozygous dominant and showed red phenotypes, two-fourths were heterozygous and showed red phenotypes, and one-fourth were homozygous recessive and showed white phenotypes. Thus the ratio of genotypes in the F_2 was 1:2:1, and the ratio of the phenotypes was 3:1.

How do we figure out the possible genotypic combinations in the F_2? This is an easy matter in a ***monohybrid cross*** (a cross involving only one character) such as this. All individuals in the F_1 generation are heterozygous (*C/c*); they have one of each of the two types of alleles. Each of these two alleles is located on a different one of the two chromosomes of a homologous pair. According to Mendel's Principle of Segregation, the two alleles will separate during gamete formation; one allele will go into each gamete. This means that half the gametes produced by such a heterozygous individual will contain the *C* allele and half the *c* allele. When two such individuals are crossed (Fig. 14.4), there are four possible combinations of their gametes:

C from male parent, *C* from female parent = *C/C* = red
C from male parent, *c* from female parent = *C/c* = red
c from male parent, *C* from female parent = *c/C* = red
c from male parent, *c* from female parent = *c/c* = white

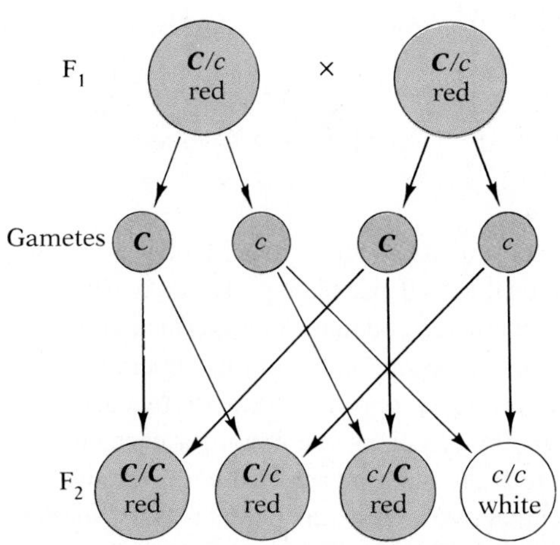

14.4 Gametes formed by F_1 individuals in Mendel's cross for flower color, and their possible combinations in the F_2

Because each of these four combinations is equally probable, we would expect, if a large number of F_2 progeny are produced, a genotypic ratio close to 1:2:1 and a phenotypic ratio close to 3:1, just as Mendel found.

An easy way to figure out the possible genotypes produced in the F_2 is to construct a Punnett square (named for the geneticist R. C. Punnett). Along a horizontal line, write all the possible kinds of gametes the male parent can produce; in a vertical column to the left, write all the possible kinds of gametes the female parent can produce; then draw squares for each possible combination of these, as follows:

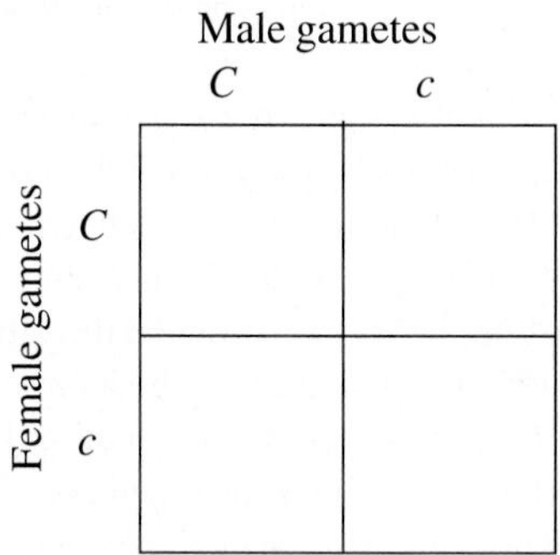

Next, write in each box, first, the symbol for the female gamete and, second, the symbol for the male gamete. Each box will then contain the symbols for the genotype of one possible zygote combination from the cross in question. A glance at the completed Punnett square in Figure 14.5 shows that the cross yields the expected 1:2:1 genotypic ratio and, since dominance is present, the expected 3:1 phenotypic ratio.

Extensive investigation of a vast array of plant and animal species by thousands of scientists has demonstrated conclusively that the results Mendel obtained from his monohybrid crosses, and the interpretations he placed on them, are not limited to garden peas but are of general validity. Whenever a monohybrid cross is made between two contrasting homozy-

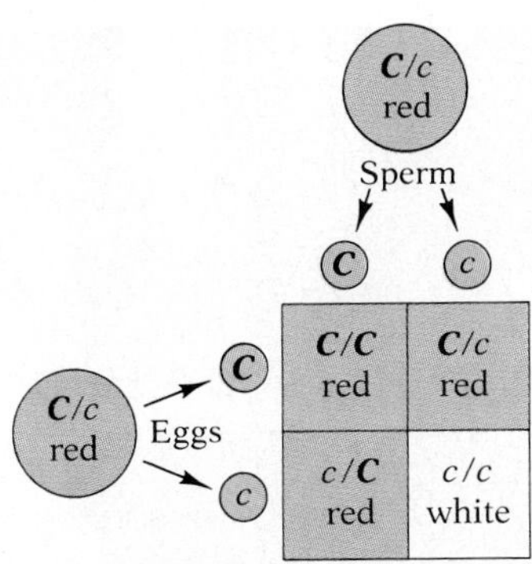

14.5 Punnett-square representation of the information shown in Figure 14.4

gous individuals (*C/C* and *c/c* in our example above), regardless of the character involved, the expected genotypic ratio in the second generation, the F_2, is 1:2:1. And whenever there is dominance, the expected phenotypic ratio is 3:1. If these ratios are not obtained in large samples, it can be assumed that some complicating condition is present, which must be identified.

A Test Cross Is Performed to Determine Whether an Individual Is Homozygous Dominant or Heterozygous

We have seen that in a monohybrid cross involving dominance the homozygous dominant progeny and the heterozygous progeny have the same phenotype and cannot be distinguished by inspection. But it is often of practical importance to distinguish between individuals with these two genotypes. If Mendel, for example, wanted to produce pea plants with only red flowers, he had to know if the plants he used for breeding were homozygous or heterozygous for red flower color. Homozygous individuals breed true, i.e., matings between two such individuals produce offspring all of a single genotype and phenotype, like their parents. But heterozygous individuals do not breed true; as we have seen, a cross between two such individuals may produce offspring of three genotypes and two phenotypes. Consequently, the identification of homozygous individuals is of the utmost importance to an animal or plant breeder who is trying to establish true-breeding strains of animals or plants. This is no problem when the breeder is interested in a recessive character, because homozygous recessive organisms can readily be recognized by their phenotypes. But if the breeder is interested in establishing a strain that is true-breeding for a dominant character, he needs a test that will enable him to tell whether a given individual is homozygous dominant or heterozygous.

The test used for this purpose is called a ***test cross:*** a cross between the individual of unknown genotype and a homozygous recessive individual. Suppose, for example, we want to know whether a particular red-flowered pea plant has a genotype of *C/C* or *C/c*. The most we can say from simple inspection is that it is red and hence must have at least one *C* allele; in other words, half of its genotype for flower color is known, and half is unknown and can be designated by a dash: *C/–*. To find out the unknown genotype we cross the red flower of unknown genotype with a homozygous recessive, i.e., a white-flowering plant, which can only have the genotype *c/c*. The results should give us the information we seek. If the plant in question has *C/C* as its genotype, then the cross will turn out as follows:

C/C	×		*c/c*
red			white
C/c	*C/c*	*C/c*	*C/c*
all red			

If, however, the plant has the other possible genotype, *C/c*, then the cross will turn out as follows:

C/c	×	*c/c*	
red		white	
C/c	*C/c*	*c/c*	*c/c*
red	red	white	white

If we obtain a large number of progeny from this test cross and all of them show the dominant phenotype (in this case red flowers), the chances are great that the test plant's genotype is *C/C* (Fig. 14.6A). If, however, we obtain progeny of which some

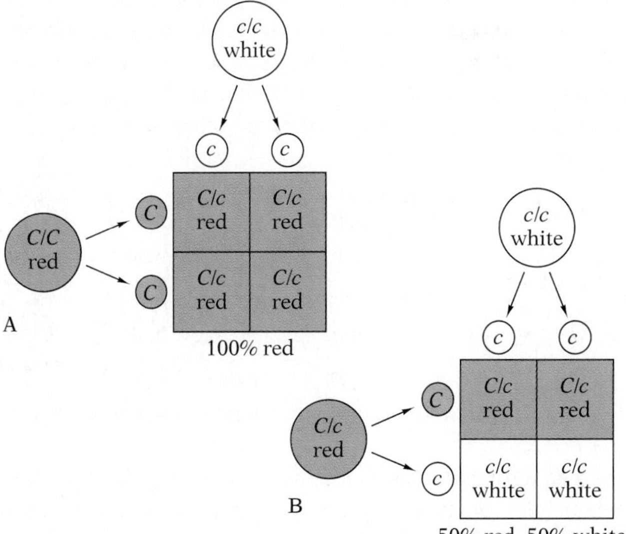

14.6 A test cross Homozygous dominant (*C/C*) and heterozygous (*C/c*) red-flowered individuals are to be distinguished by crossing them with a homozygous recessive (*c/c*) white-flowered individual. (A) If the red-flowered plant is homozygous, all the progeny of the test cross will be heterozygous (*C/c*) and will have red flowers. (B) If the red-flowered plant is heterozygous, 50 percent of the progeny will be heterozygous (*C/c*) and have red flowers, and 50 percent will be homozygous recessive (*c/c*) and have white flowers.

show the dominant phenotype (red) and some the recessive phenotype (white), we know that the test plant's genotype is *C/c* (Fig. 14.6B). If any of the progeny show the recessive phenotype, their genotype must be *c/c,* which means that they received a *c* allele from each parent; hence the test plant must have a *c* allele in its genotype. Since we already know that it has at least one *C* allele, we combine these two pieces of information to write its genotype as *C/c.*

Many Alleles Show Incomplete Dominance or Codominance

The seven characters of peas that Mendel used in his experiments were all of the kind in which one allele shows complete dominance over the other. Many characters in a variety of organisms show this mode of inheritance, but many others do not. When a heterozygous individual clearly shows effects of both alleles, the alleles are referred to as partial, or ***incomplete dominance.***

In cases of incomplete dominance, heterozygous individuals have a phenotype that is actually *intermediate* between the phenotype of individuals homozygous for one allele and the phenotype of individuals homozygous for the other allele. For example, crosses between homozygous red snapdragons or sweet peas and homozygous white varieties yield plants producing pink blossoms (Fig. 14.7). In other words, neither the red nor the white allele is completely dominant; neither one completely masks the other. When these pink plants are crossed among themselves, they yield red, pink, and white offspring in a ratio of 1:2:1, as follows (when there is incomplete dominance both alleles are often designated by a capital letter, the one being distinguished from the other by a superscript, e.g., I^r for red, I^w for white):

P		I^r/I^r red	×	I^w/I^w white
F_1		I^r/I^w pink	×	I^r/I^w pink
F_2	I^r/I^r red	I^r/I^w pink	I^r/I^w pink	I^w/I^w white

The term ***codominance*** is used to describe a situation in which both alleles are expressed independently of one another in the heterozygote. The heterozygotes have a phenotype that is different from either of the homozygotes, but not necessarily intermediate between the two different homozygotes. An example of codominance is found in human A-B-O blood groups. Individuals with type A blood have the A allele and express only the A glycoprotein on their red blood cells, while those with type B have the B allele and express only the B glycoprotein. However, those individuals with type AB blood have both the A and B alleles and express both glycoproteins, so their phenotype is different from either of the other two. Neither the A nor B allele is dominant; both are expressed equally. We shall discuss the human blood groups in more detail later in the chapter.

14.7 Incomplete dominance in sweet peas When plants homozygous for red flowers are crossed with plants homozygous for white blossoms, the heterozygous F_1 sweet-pea plants have pink flowers.

The inheritance pattern in incomplete dominance and codominance differs in the following ways from that in complete dominance: (1) the heterozygotes have a phenotype different from either of the homozygotes, and (2) the phenotypic ratio in a cross between two heterozygotes is 1:2:1 (just like the genotypic ratio) rather than 3:1.

For many years it was assumed that instances of incomplete dominance and codominance were rare, but it is now thought that *at the molecular level,* many alleles show incomplete dominance or codominance. In other words, incomplete dominance or codominance is the rule, not the exception. How can this be? In many genes one allele (the "dominant") codes for an active protein while the other allele (the "recessive") codes for no protein at all, or an inactive protein. An individual homozygous for the inactive allele will lack a working copy of the normal gene, and the resulting phenotype is recessive. Individuals homozygous for the normal allele have two copies of the allele and presumably could direct the production of twice as much protein as a heterozygote, which has just one copy. But, as long as enough normal protein is produced, the phenotype will appear normal. This seems to be the situation with most genes; one copy of the dominant allele is sufficient to ensure a normal phenotype. In situations where one copy is not sufficient, the heterozygote may have a phenotype noticeably different from the homozygotes, and incomplete dominance or codominance results.

Mendel Found That the Alleles for One Gene Are Inherited Independently of the Alleles for Another Gene

We have limited our discussion so far to crosses involving a single phenotypic character; or rather we have so far chosen to ignore all but one character. The latter is a more accurate statement, because all crosses involve many more than one character. Organisms contain thousands of genes, and it would be impossible to set up a cross in which only one character was allowed to vary. Let us now turn to crosses involving the genes for two characters (dihybrid crosses).

Mendel's experiments on garden peas were not limited to single characters, but sometimes involved two or more of the characters listed in Table 14.1. For example, he crossed plants characterized by round yellow seeds with plants characterized by wrinkled green seeds. The F_1 plants all had round yellow seeds. When these plants were crossed among themselves, the resulting F_2 progeny showed four different phenotypes:

315 had round yellow seeds
101 had wrinkled yellow seeds
108 had round green seeds
32 had wrinkled green seeds

These numbers represent a ratio of about 9:3:3:1 for the four phenotypes.

This experiment demonstrated that a dihybrid cross can produce some new plants phenotypically unlike either of the original parental plants (Fig. 14.8). Here the new phenotypes were wrinkled yellow and round green. It demonstrated, in other words, that the genes for seed color and the genes for seed form do not necessarily stay together in the combinations in which they occurred in the parental generation.

Let us examine in somewhat more detail the genetics of this cross, using *R* for the allele for round seed, *r* for the allele for wrinkled seed, *G* for the allele for yellow seed, and *g* for the allele for green seed. In the summary of Mendel's cross, below, the dash means that it does not matter phenotypically whether the dominant or the recessive allele occurs in the spot indicated.

P	*R/R G/G* round yellow	×	*r/r g/g* wrinkled green	
gametes	*RG*		*rg*	
F_1	*R/r G/g* round yellow	×	*R/r G/g* round yellow	
gametes	*RG Rg, rG rg*		*RG Rg, rG rg*	
F_2	9 *R/– G/–* round green	3 *r/r G/–* wrinkled green	3*R/–g/g* round green	*r/r g/g* wrinkled green

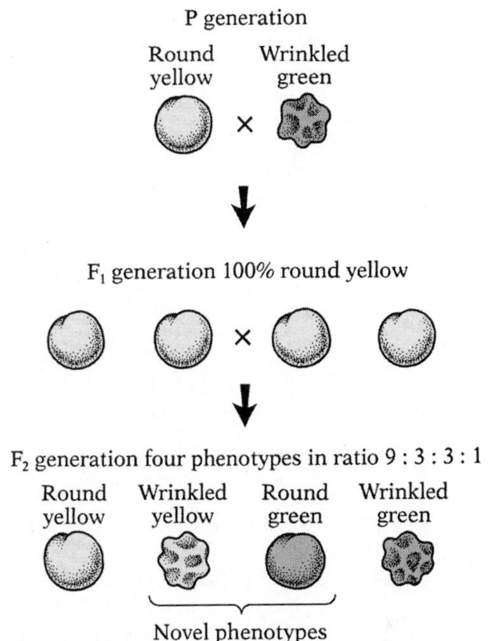

14.8 Expression of novel phenotypes Crosses between plants homozygous for round yellow seeds and plants homozygous for wrinkled green seeds produce two novel phenotypes in the F_2 generation: wrinkled yellow seeds and round green seeds.

The round yellow parent could produce gametes of only one genotype, *RG*. The wrinkled green parent could produce only *rg* gametes. When *RG* gametes from the one parent united with *rg* gametes from the other parent in the process of fertilization, all the resulting F_1 offspring were heterozygous for both characters (*R/r G/g*) and showed the phenotype of the dominant parent (round yellow). Each of these F_1 individuals could produce four different types of gametes, *RG, Rg, rG,* and *rg*. When two such individuals were crossed, there were 16 possible combinations of gametes (4 × 4). One way to determine all the different combinations of gametes in the F_2 is to set up a Punnett square with 16 squares for the different combinations (Fig. 14.9). List the gametes for the male parent along the top, and the female parent along the side. Next fill in each box with the appropriate gametic combinations and count the number of boxes representing each genotype and phenotype (Fig. 14.9).

The 9:3:3:1 phenotypic ratio is characteristic of the F_2 generation of a dihybrid cross (with dominance) in which the genes for the two characters are ***independent*** (located on nonhomologous chromosomes). Each independent gene behaves in a dihybrid cross exactly as in a monohybrid cross. If we view Mendel's F_2 results as the product of a monohybrid cross for seed color (ignoring seed form), we find that there were 416 yellow seeds (315 + 101) and 140 green seeds (108 + 32), which closely approximates the 3:1 ratio expected in a monohybrid cross. Similarly, if we treat the experiment as a monohybrid cross for seed form and ignore seed color, the F_2 results also show a phenotypic ratio of approximately 3:1. The dihy-

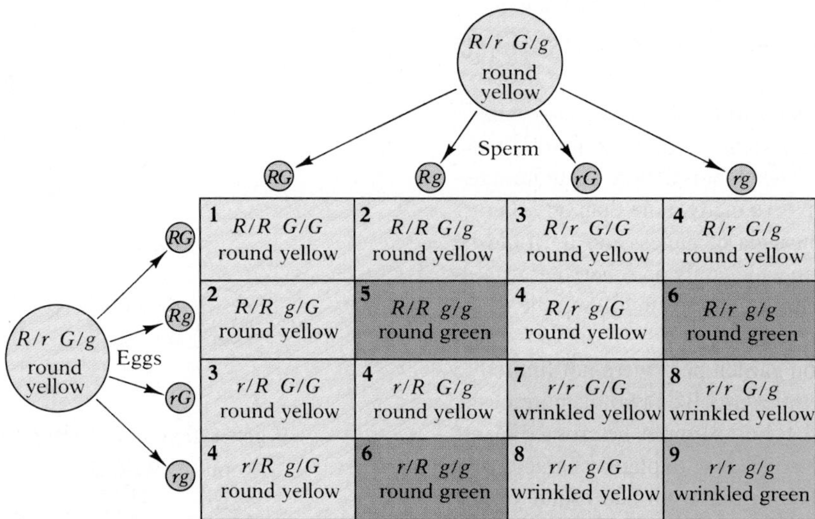

14.9 Punnett-square representation of an F_1 dihybrid cross When two individuals heterozygous for both seed color and shape were crossed, Mendel obtained roughly nine plants that produced round yellow seeds, three that produced round green seeds, and three that produced wrinkled yellow seeds for every one that produced wrinkled green seeds. Underlying these results were nine different genotypes, as indicated by the numbers in the boxes. Combinations of alleles that represent identical genotypes are shown with the same number; combinations of alleles that give rise to identical phenotypes have the same shading.

brid F_2 ratio of 9:3:3:1 is thus simply the product of two separate and independent 3:1 ratios. Upon observing these results Mendel formulated his second law, the ***Principle of Independent Assortment,*** which states that during gamete formation, when two or more genes are involved in a cross, the alleles of one gene are inherited independently of the alleles for another gene.

Probability Can Be Used to Calculate the Results of a Genetic Cross

Constructing a Punnett square to determine the results of a cross is a simple way to do a monohybrid cross and is only moderately laborious in a dihybrid cross, but it becomes prohibitively tedious in a cross involving more than three characters. There is an alternative procedure that is much easier, based on the principle that *the chance that a number of independent events will occur together is equal to the product of the chances that each event will occur separately.*

Suppose we want to know how many of the 16 combinations in Mendel's cross will produce the wrinkled yellow phenotype. We know that wrinkled is recessive; hence it is expected in one-fourth of the F_2 individuals in a monohybrid cross. We know that yellow is dominant; hence it is expected in three-fourths of the F_2 individuals. Multiplying these two separate values ($^1/_4 \times {}^3/_4$) gives us $^3/_{16}$; three of the 16 possible combinations will produce a wrinkled yellow phenotype. Similarly, if we want to know how many of the combinations will produce a round yellow phenotype, we multiply the separate expectancies for two dominant characters ($^3/_4 \times {}^3/_4$) to get $^9/_{16}$.

Not All Genes Show a Mendelian Pattern of Inheritance

Mendel's laws provided the basis for understanding patterns of inheritance, but there are many more complicated situations where these principles do not hold. We have already examined one such situation where one allele is not completely dominant over another, resulting in partial dominance. We shall now examine some fundamentally different, non-Mendelian patterns of inheritance.

Characteristics Are Often Determined by Many Genes Acting Together in Additive Fashion

So far we have discussed characteristics that have only a limited number of relatively distinct phenotypes. Pea flowers are either red or white, pea seeds either yellow or green. True, the boundaries of the classes may be blurred, as in human eye

14.10 Frequency distribution of human height These students at Connecticut Agricultural College are grouped according to differences in height. The shortest student (left) was 4'10" while the tallest (right side) was 6'2". The relative numbers in each height class produces a jagged approximation of a so-called normal or bell-shaped curve centered around 5'8". Human height is thought to be an example of polygenic inheritance.

color, but in general the phenotypes fall into a few clearly recognizable classes. Many traits, however, show much greater phenotypic variation, with indistinct boundaries. Human height, skin pigmentation, and IQ are just three of the many possible examples. What can be said about the genetic basis of characteristics such as these?

One explanation, first put forward by the Swedish geneticist Herman Nilsson-Ehle in 1909, is that two or more separate genes can affect the same character in the same way, in an additive fashion. Because of the additive nature of the interaction, this type of inheritance is sometimes referred to as ***quantitative*** or ***polygenic inheritance.*** An example is provided by human height, which varies over a wide range (Fig. 14.10). Human height is thought to be determined by a number of similarly acting genes, each with active and inactive alleles. The effects of the active alleles are approximately additive; the more active alleles present, the taller the individual. Some individuals have only a few of the active alleles in their genotype and are very short, others have many active alleles and are very tall. The result of quantitative inheritance is continuous variation within the population with respect to that trait (Fig. 14.10).

The Expression of a Gene Depends Both on the Other Genes Present and on the Physical Environment

Just as different alleles of the same gene can interact when both are partially dominant, separate genes can also interact. Indeed, most phenotypic characteristics, whether they are visible, like flower color, or more subtle chemical characteristics, like enzyme pathways, are controlled by several genes interacting together. Probably no inherited characteristic is controlled exclusively by one gene. Even when only one principal gene is involved, its expression is influenced to some extent by countless other genes with individual effects often so slight that they are very difficult to locate and analyze. An example is eye color in human beings.

Human eye color can be regarded as controlled by one gene with two alleles—a dominant allele, *B,* for brown eyes, and a recessive allele, *b,* for blue eyes. Brown-eyed people (*B/B* or *B/b*) have branching pigment cells containing melanin in the front layer of the iris. Blue-eyed people (*b/b*) lack melanin in the front layer; the blue is an effect of the black pigment on the back of the iris as faintly seen through the semiopaque front layer.

This description of the inheritance of eye color on the basis of a single-gene system assumes only two phenotypes, brown and blue, and it is, in fact, possible to assign most people to one or the other of these two phenotypic classes. But we all know that eyes exhibit many variations in hue. It is obvious, then, that an explanation of eye color in terms of a single-gene system is an oversimplification. Many ***modifier genes,*** genes that influence the expression of other genes, are also involved. Some affect the amount of pigment in the iris, some the tone of the pigment (which may be light yellow, dark brown, etc.), some its distribution (even over the whole iris, or in scattered spots, or in a ring around the outer edge of the iris, etc.). Two blue-eyed people may occasionally have a brown-eyed child, because one of them, in whom the lack of pigmentation is due

14.11 Effect of temperature on expression of a gene for coat color in the Himalayan rabbit (A) Normally, only the feet, tail, ears, and nose are black. (B) Fur is plucked from a patch on the back, and an ice pack is applied to the area. (C) The new fur grown under the artifically low temperature is black. Himalayan rabbits are normally homozygous for the gene that controls synthesis of the black pigment, but the gene is active only at low temperatures (below about 33°C).

to the action of modifier genes, actually has the genotype *B/b* (instead of *b/b*). This explanation is based on the following biological principle: *the action of any gene can be fully understood only in terms of the overall genetic makeup of the individual organism in which it occurs.*

The expression of genes may also be affected by environmental influences. For example, Himalayan rabbits are normally white with black ears, nose, feet, and tail (Fig. 14.11), but if the fur on a patch on the back is plucked and an ice pack is kept on the patch, the new fur that grows there will be black; the gene for black color can express itself only if the temperature is low, which it normally is only at the body extremities. Similarly, a human being with genes for great height will not grow tall if raised on a starvation diet.

We see, then, that the expression of a gene depends both on the other genes present (the genetic environment) and on the physical environment (temperature, sunlight, humidity, diet, etc.). We do not inherit characters. We inherit only genes, only potentialities; other factors govern whether or not the potentialities are realized. *All organisms are products of both their inheritance and their environment.* (See Exploring Further: The Dog Genome Project, p. 287).

A Gene May Have Many Different Alleles in the Population, but Each Individual Can Only Possess Two Alleles

From our discussion so far, the impression may result that a gene has only two allelic forms, each of which exhibits either a dominant-recessive or an intermediate relationship to the other. This is, however, an oversimplification. In fact, there may be any number of allelic forms of the same gene present in a population but the maximum number of alleles for each gene that any individual can possess is two, because he has only two copies of each gene.

A well-known example of multiple alleles in human beings—and a relatively simple one, since only a few alleles are involved—is that of the A-B-O blood series, in which four blood types are generally recognized: A, B, AB, and O. The inheritance of the A-B-O groups involves three alleles, here designated I^A, I^B, and *i*. Both I^A and I^B are dominant over *i*, but neither I^A nor I^B is dominant over the other; instead they are codominant. The I^A allele codes for an enzyme (A) that catalyzes a reaction in which a particular sugar is added to a short

MAKING CONNECTIONS

1. New discoveries in biology have often hinged on the ability of a researcher to choose an organism that is well-suited for experimentation on the particular question being investigated. Why was the pea plant a good choice for genetics research?

2. Explain why the results of Mendel's breeding experiments offer strong support for the chromosome theory of heredity.

3. Make diagrams of a cell with a diploid chromosome number of four during metaphase of meiosis I to show how two genes carried on nonhomologous chromosomes will assort independently.

4. What accounts for the variability in the phenotype of individuals that have the same genotype for a particular trait (e.g., eye color)?

EXPLORING FURTHER

THE DOG GENOME PROJECT

The relationship between inheritance and learning and behavior, particularly in humans, has long been controversial. Geneticists studying inheritance in humans typically study the inheritance of a single gene, like the genes for cystic fibrosis or phenylketonuria. Little work has been done on characteristics determined by groups of genes acting together, such as skin color, height, or behavior. But ongoing studies of the behavior of dogs may soon provide new insights.

Purebred dogs, because they exhibit clearly defined behavioral traits that are inherited, provide an excellent model for the study of the inheritance of behavior. Through the years, dogs have been bred for specific behavioral traits: retrievers, for instance, have a natural tendency to find and retrieve land and water birds, border collies have a strong herding instinct, and Newfoundlands love being in water (Fig. A). Through selective breeding experiments between such breeds, geneticists hope to determine which features are passed on, and whether the traits are dominant or recessive. Such investigations should provide a better understanding of the inheritance of complex behavioral patterns. Many of the principles will probably carry over to humans and other mammals. This study, known as the dog genome project, is multifaceted. In addition to determining how different physical and behavioral traits are inherited, researchers are using recombinant DNA techniques to map the dog's 78 chromosomes.

An understanding of the dog genome may help breeders eliminate some single gene disorders from breeding lines. Genetic disease is a particular problem in dogs because champion purebred male dogs usually act as studs, and may father thousands of puppies. When offspring of this father interbreed, the chances are increased that the puppies may receive the same recessive gene from both parents. Many genetic diseases common in dogs (e.g., hip degeneration, blood clotting abnormalities, and epilepsy) are also found in humans. Insights gained from dogs may provide clues about human conditions.

A

B

C

A Chesapeake retriever bringing in a pheasant. **B** German Shorthaired pointer on point. **C** Sheepdogs sorting sheep.

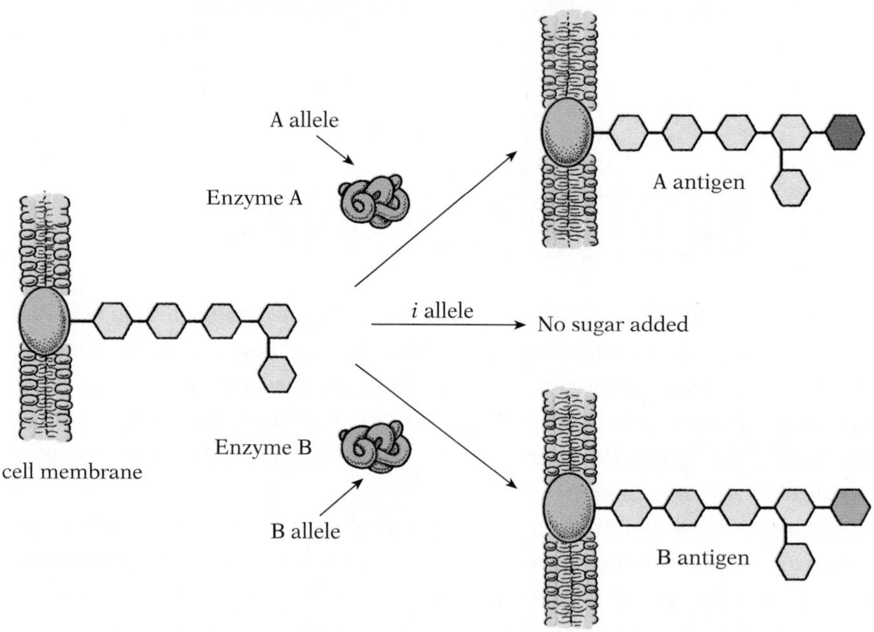

14.12 Synthesis of the A and B blood group antigens The A-B-O blood group antigens consist of a short carbohydrate chain of six sugars attached to a lipid or protein (color) that is embedded in the plasma membrane of the red blood cells. The I^A allele codes for an enzyme that adds a particular sugar to the end of the carbohydrate chain; the I^B allele codes for a slightly different enzyme that adds a different sugar. The *i* allele apparently codes for no enzyme at all or an inactive enzyme because no sugar is added to the carbohydrate chain.

chain of sugars attached to proteins in the cell membrane of red blood cells (Fig. 14.12). The allele I^B codes for a slightly different enzyme (B), which attaches a different sugar. The allele *i* codes for an inactive enzyme so no sugar is added. Individuals having both I^A and I^B alleles have both A and B carbohydrate chains (called A and B antigens) on their red blood cells, while homozygous recessive *(i/i)* individuals have neither A nor B antigens, meaning they have no antigens at all (Table 14.2). Because the red blood cells of different individuals may have different carbohydrate chains on their surfaces, it is very important to match blood types when giving blood transfusions.

Although there are three alleles in the human population, each individual can have only two, because he has only two copies of this gene, one from each parent. Thus, four blood types are possible; Table 14.2 shows the genotypes associated with each of the blood types.

TABLE 14.2 *Antigens and genotypes of the A-B-O blood types*

Blood type	*Blood contains cellular antigens*	*Genotype*
O	None	i/i
A	A	I^A/I^A or I^A/i
B	B	I^B/I^B or I^B/i
AB	A and B	I^A/I^B

Mutations in Genes Occur at Random; Most Mutations That Have a Phenotypic Effect Are Deleterious

A variety of influences can cause slight changes—mutations—in the DNA of a gene. We have already learned that the rate at which any particular gene undergoes mutation is ordinarily extremely low. But mutations are constantly occurring, and since every individual organism has a very large number of different genes, over time every individual will accumulate mutations in his genetic material. Most of these mutations will prove to be harmful—a random change in any mechanism as delicate and intricate as the cell is far more likely to damage it than to improve it. Only very rarely is a new mutation beneficial.

There are about 500 genetic disorders in humans caused by mutations in a single gene. Phenylketonuria, Tay-Sachs disease, cystic fibrosis, and sickle-cell anemia are a few of the hundreds of single-gene disorders. These disorders result from mutations in individual genes such that the recessive alleles specify inactive enzymes or other essential proteins. Individuals homozygous for the recessive allele cannot produce the normal enzyme or protein. If the product is an enzyme, the immediate result is a blockage of the metabolic pathway in which the enzyme acts, and a buildup of the substrate such as we saw in phenylketonuria (see Fig. 11.1, p. 216). Normal cell function is thus impaired, resulting in a host of serious mental and physical problems.

Another example is Tay-Sachs disease, in which the mutation results in a deficiency of the enzyme hexosaminidase A. Without this enzyme lipids are not metabolized properly and a fatty substance accumulates in the cells of the nervous system, leading to progressive blindness, mental retardation, paralysis, and death, generally before the age of 5 years. Although the incidence of this allele is relatively low in the general U.S. population, it is considerably higher in Jews of eastern European origin (perhaps as many as one person in 30 in this population may be a carrier).

Single-gene disorders show the typical Mendelian pattern of inheritance. These diseases run in families and are often diagnosed through pedigree analysis (Fig. 14.13A). For example, among Caucasian Americans, about one person in 20 carries the recessive allele for cystic fibrosis, and one of every 2,000 white babies born has this defect, making it the most common lethal allele in the United States. If two carriers (individuals heterozygous for the cystic fibrosis allele) mate, one-fourth of their children would be expected to have the disease (Fig. 14.13B). In cystic fibrosis the defect involves a faulty version of a protein that regulates the transport of chloride ions into and out of the cell. Because chloride ion transport is impaired, the osmotic concentration in the cells and surrounding fluids is disrupted, resulting in unusually thick and sticky secretions in the lungs, pancreas, and other mucus-secreting organs. These secretions clog the ducts of the organs producing them and provide an ideal environment for microorganisms to live. Repeated bacterial infections destroy healthy tissue, usually leading to death in early adulthood.

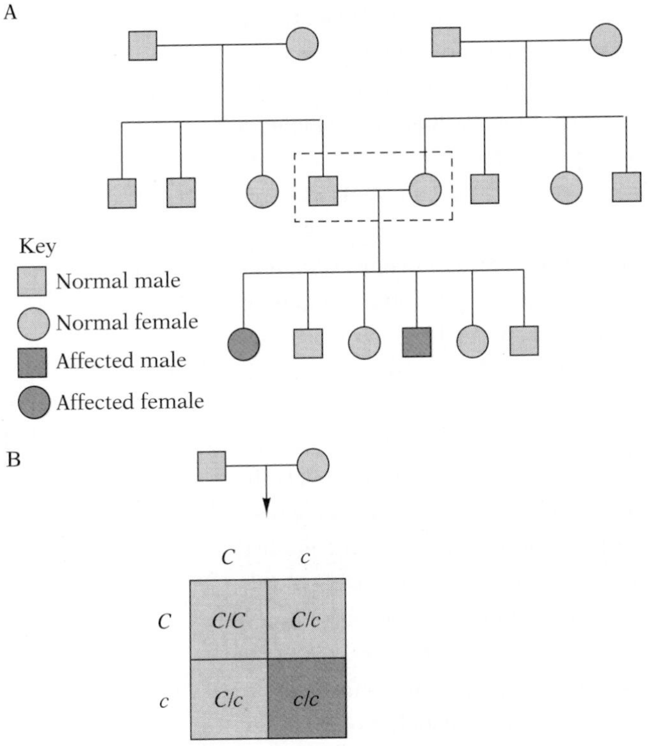

14.13 Pedigree for cystic fibrosis (A) The allele for this condition is inherited as a simple recessive. In a pedigree the red symbols represent individuals with cystic fibrosis; gray symbols show the normal phenotype. In this family pedigree, no one in the first two generations expressed the trait. Neither parent (boxed area) has the trait, but each must be a carrier for two of their children to have the condition. (B) Punnett-square representation of a cross between two carriers for cystic fibrosis. In this cross, each parent has the normal phenotype but is heterozygous, having one copy of the normal allele and one copy of the defective allele (*C/c*). On the average, one-fourth of the children from such parents would be expected to receive both recessive alleles (*c/c*) and have the disease.

The examples we have discussed, phenylketonuria, Tay-Sachs, and cystic fibrosis, are all examples of recessive disorders. The patient inherited a defective gene from both parents. Some disorders, however, are caused by dominant alleles. A single copy of the defective gene is enough to cause the disease. Why is it that some genetic diseases require two copies of defective genes while others need just one? Research suggests that recessive diseases result from defects in genes that code for enzymes, whereas dominant disorders result from defects in genes that code for structural proteins such as collagen. Because enzymes can be used over and over, a single working copy of the normal gene may result in a sufficient amount of enzyme production to ensure a normal phenotype. However, a single copy of a gene that codes for a structural protein may not be enough, and disease symptoms result.

At present there are no cures for these single-gene disorders. Indeed, most result in severe debilitating diseases that are incurable and cause much human suffering. Treatment, when available at all, is generally directed at alleviating the symptoms. PKU, for example, can be controlled through use of a diet low in phenylalanine in childhood. Special physical treatments to clear the airways and antibiotics to treat infections have enabled about 50 percent of children with cystic fibrosis to survive to adulthood; a child born with cystic fibrosis today has a life expectancy of about 40 years. At present the only other alternatives are to diagnose the condition in the embryo and terminate the pregnancy to avoid the birth of an afflicted child, or to counsel the parents about the problems they will face after their child is born (see Exploring Further: Testing the Unborn, p. 290).

Gene therapy may, in the future, provide the ability to replace the defective gene. In the last chapter we discussed some of the various techniques that are being used to introduce normal genes into defective cells. Rapid advances are being made in this area, especially with respect to the cystic fibrosis gene. Both the allele that codes for normal proteins and the one that codes for defective proteins have been identified using recombinant DNA techniques. Two separate laboratory teams have successfully used viruses to insert the normal allele into cultured cells from cystic fibrosis patients. The inserted allele has repaired the abnormality and normal protein is produced within the cells. Inserting the gene into cells grown in tissue

EXPLORING FURTHER

TESTING THE UNBORN

Approximately 4 percent of babies born in the United States have a recognizable defect due to a chromosomal abnormality, single-gene defect, or other heritable factor. Families, often devastated by the birth of a child with a genetic disorder, require genetic information and counseling. Parents are concerned not only about the child, but also about the health of future children. Close relatives may wonder if they too carry the defective gene. Genetic counseling generally includes a careful medical diagnosis, and pedigree analysis. Using a variety of techniques, including the new techniques of recombinant DNA technology, it is now possible to detect many kinds of genetic or chromosomal abnormalities. When the diagnosis is unfavorable, a couple has few options. They can either forego childbearing, or risk the chance of having an afflicted child. In the past many parents facing such a choice opted for adoption, or did not have children. Today, there are more choices available for such parents. If abortion is acceptable, it is often possible to perform a *prenatal diagnosis,* that is, to detect the presence of a defective gene in the fetus during embryonic development.

Two techniques are widely used for obtaining fetal cells for prenatal diagnosis: amniocentesis and chorionic villi sampling. During *amniocentesis,* amniotic fluid containing sloughed-off cells from the fetus is withdrawn from the uterus with a long needle inserted through the woman's abdominal wall (Fig. A). The embryonic cells are examined directly or grown in tissue culture to produce dividing cells that can be examined for abnormalities. If severe ones are found, the parents may have the option of having the pregnancy terminated. Unfortunately, amniocentesis cannot be performed until the 12th to 16th week of the pregnancy, and the test results may not be available for an additional two to four weeks. Consequently, termination of the pregnancy presents a serious health risk to the mother, and often is not possible. In this situation the parents undergo intensive counseling to prepare them for the problems they will face in the care of their child after it is born. A new technique, called *chorionic villi sampling (CVS),* involves taking a bit of embryonic tissue from the chorionic villi (a portion of the placenta) by inserting an instrument through the cervix of the woman's uterus (Fig. B). Though the safety of CVS is still undergoing review, the procedure has many advantages over amniocentesis. It is done earlier (between the ninth and 12th weeks), it is more accurate because it samples only fetal tissue (there is a possiblity of sample contamination by maternal cells in amniocentesis), and the results are generally known within 24 hours (because dividing cells are found in the sample). Weeks of anxiety are avoided, and the parents have the option of terminating the pregnancy to avoid the birth of an afflicted child.

Fortunately, only about 3 percent of the prenatal diagnoses pinpoint the presence of the genetic disorder in question, so most parents can look forward to the birth of a child with no more risk of an abnormality than a random birth. (This procedure screens only for a specific gene; it is not a total gene screen.) The availability of prenatal diagnosis has actually resulted in the birth of more children than would otherwise be the case, because many

parents at substantial risk of having an afflicted child would refrain from having children of their own. Having the option of prenatal diagnosis has enabled such couples to have biological children of their own without the risk of giving birth to an afflicted child.

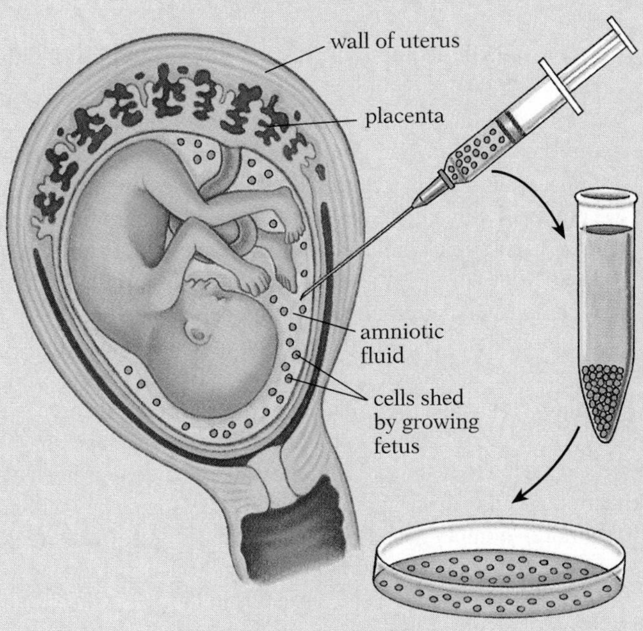
wall of uterus
placenta
amniotic fluid
cells shed by growing fetus
1. Needle inserted through abdominal wall withdraws amniotic fluid containing fetal cells.
2. Sample is centrifuged to separate cells from fluid.
3. Cells are grown in culture and tested for chromosomal abnormalities or enzyme deficiencies. Radioactive probes can be used to find defective genes.

A The amniocentesis procedure During amniocentesis, amniotic fluid containing fetal cells is removed and tested for chromosomal abnormalities, the deficiency of certain enzymes, or the presence of defective genes. The procedure is usually performed after the 12th to 16th week of the pregnancy.

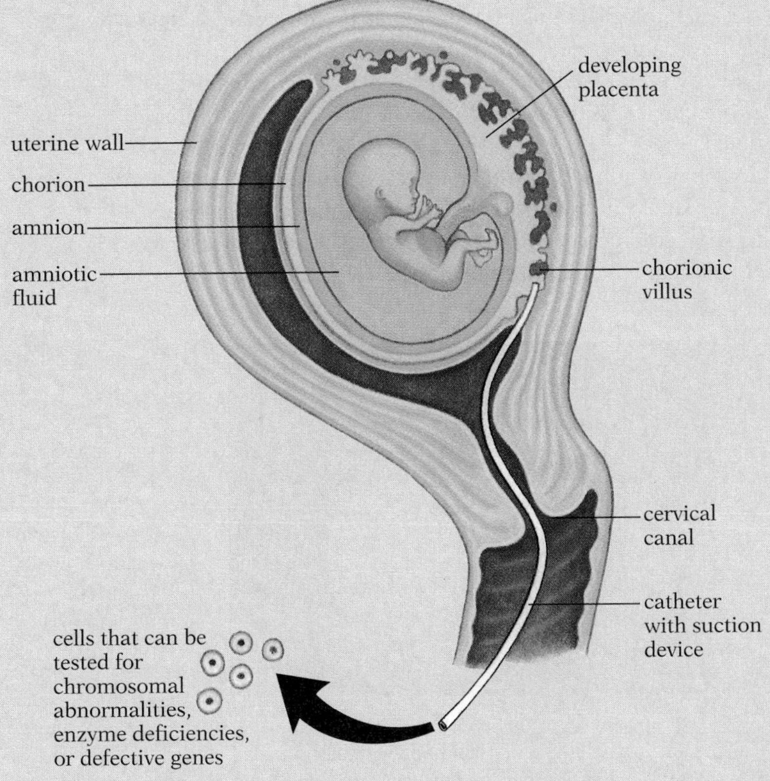
developing placenta
uterine wall
chorion
amnion
amniotic fluid
chorionic villus
cervical canal
catheter with suction device
cells that can be tested for chromosomal abnormalities, enzyme deficiencies, or defective genes

B Chorionic villi sampling An alternative to amniocentesis is choronic villi sampling, in which a special catheter is inserted into the uterus to obtain a sample of fetal cells from the placenta. This procedure has an advantage over amniocentesis in that it can be done earlier, between the ninth and 12th weeks of pregnancy.

culture is one thing, but inserting the gene into cells within the individual is more difficult and awaits further testing. Another procedure currently undergoing testing in mice involves using inhalant sprays to deliver normal genes packaged in liposomes to the passages to the lungs; such tests may be used with cystic fibrosis patients soon.

Natural Selection Can Act Against a Deleterious Allele Only if It Causes Some Change in the Organism's Phenotype

When a new deleterious allele arises by mutation, natural selection can act against it only if it causes some change in the organism's characteristics. Selection acts directly on phenotypes and only indirectly on genotypes. Because dominant deleterious mutations will be expressed phenotypically, they can be eliminated from the population rapidly by natural selection. But many new mutations are recessive to the normal alleles. And since the probability that the same mutation will occur twice in the same individual is extremely slight, most new alleles occur in combination with the normal allele in diploid cells. The diploid cell is heterozygous, containing one normal allele and one new allele produced by mutation. If a new allele is recessive, and if it occurs in heterozygous condition, it can have little immediate phenotypic effect on the organisms that possess it; its deleterious effects cannot be fully expressed. Therefore, natural selection cannot eliminate it from the population very rapidly. Deleterious alleles that are not dominant may be retained in the population in heterozygous condition for a long time.

When two individuals who both carry the same deleterious recessive allele in heterozygous condition mate, about one-fourth of their progeny will be homozygous for the deleterious allele and have the harmful phenotype. The phenotype may even kill the organism (Fig. 14.14). An allele whose phenotype, when expressed, results in the death of the organism is called a ***lethal.*** All of us carry some lethal alleles in our genetic material; present estimates indicate that everyone has between six and 10 such lethal alleles.

14.14 A creeper hen In chickens one allele of a certain gene, when it occurs in a heterozygous condition with the normal allele, causes the chicken to be a "creeper"; the legs are very short and the chicken cannot walk properly. When two creeper chickens are crossed, one-fourth of the progeny are born dead. Apparently the creeper allele is lethal in the homozygous condition.

Deleterious Phenotypes Are a Danger of Inbreeding

The conditions under which genes cause deleterious phenotypes explain the danger of matings between closely related human beings. Everyone probably carries in heterozygous combination many alleles that would cause harmful effects if present in homozygous combination, including some lethals. But because most of these deleterious alleles originated as rare mutations, and are limited to a tiny percentage of the population, the chances are slight that two unrelated individuals will be carrying the same deleterious recessive alleles and produce homozygous offspring that show the harmful phenotype. The chances are much greater that two closely related persons will be carrying the same harmful recessives, having received them from common ancestors, and that, if they mate, they will have children homozygous for the deleterious traits. In short, close inbreeding increases the percentage of homozygosity, as Figure 14.15 shows. You can see from the graph that brother-sister matings and matings between double first cousins cause rapid increases in homozygosity, and that matings between

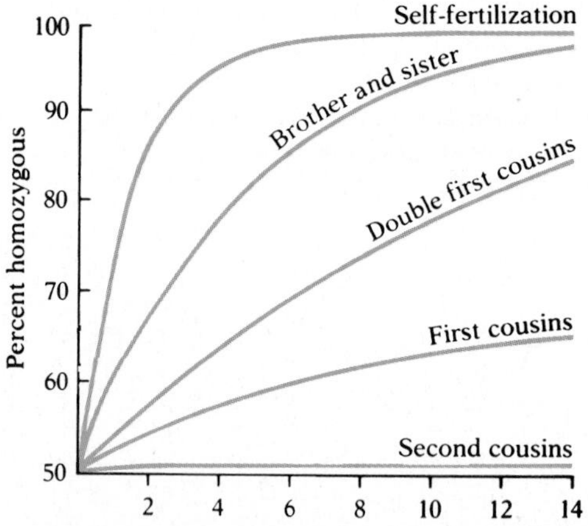

14.15 Graph showing percentage of homozygotes in successive generations under different degrees of inbreeding It is assumed in this graph that the initial condition is two alleles of equal frequency (hence the 50 percent homozygosity at the start). If a similar graph were drawn for very rare recessive alleles, the initial rises in homozygosity would be much steeper, particularly in the three curves for cousin-to-cousin matings. Thus, matings between first or second cousins, though of little effect in the situation graphed here, may greatly increase the percentage of homozygosity of rare, perhaps deleterious recessive alleles. (Double first cousins result when siblings of one family marry siblings of another family—when two brothers marry two sisters, for example. The offspring of such marriages are first cousins through *both* of their parents rather than through just one of them.)

MAKING CONNECTIONS

1. How are the structure and function of the animal cell coat (glycocalyx) described in Chapter 5 related to the alleles of the A-B-O blood system?

2. How are the genetic phenomena of dominance and recessiveness linked to the functions of proteins?

3. How are the recombinant DNA techniques discussed in Chapter 13 being used to develop a treatment for cystic fibrosis?

4. Why does natural selection eliminate harmful dominant alleles from the population more effectively than recessive alleles?

5. Which would have the greater risk of inheriting a genetic disease caused by a homozygous recessive genotype: the offspring of a mating of siblings, or the child of a marriage of third cousins? Why?

first cousins cause slight increases. As we will see, many species have evolved specific behavioral or (particularly in plants) physiological mechanisms that prevent close relatives from mating.

Deleterious Genes Can Be Beneficial When Heterozygous

In numerous instances alleles harmful or even lethal when homozygous are actually beneficial when heterozygous. An example in human beings is the allele for ***sickle-cell anemia*** in Africa. The normal allele codes for two of the four polypeptide chains in hemoglobin, the protein that carries oxygen in the red blood cells. The sickle allele results from a single base substitution in the sixth codon in the DNA of the gene, causing a substitution of the amino acid valine for glutamic acid in the polypeptide chains. This substitution can have profound consequences in the functioning of hemoglobin. When homozygous, the sickle hemoglobin molecules within the red cells crystallize and distort the cells into a sickle shape (Fig. 14.16). These abnormal cells clog the smaller blood vessels. The resulting impairment of the circulation leads to severe pain in the abdomen, back, head, and extremities, and to enlargement of the heart and atrophy of brain cells. In addition, the tendency of the deformed red blood cells to rupture easily brings about severe anemia. Because the number of red cells is decreased, the oxygen-carrying capacity of the blood is reduced and the

A

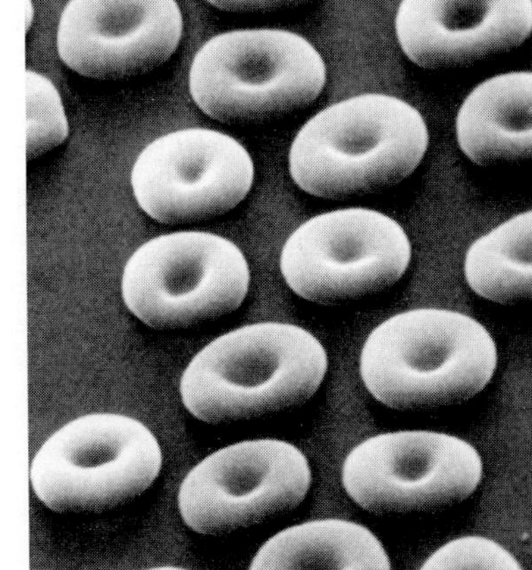

B

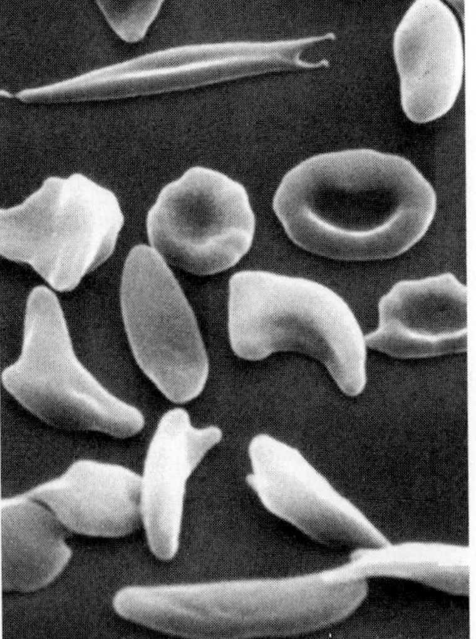

14.16 Scanning electron micrographs of normal and sickled red blood cells Normal red blood cells (A), which are biconcave discs, look dramatically different from sickled cells (B). Some of the sickled cells seen here bear the filamentous processes that may cause clogging of the body's smaller blood vessels.

patient shows symptoms characteristic of anemia—pallor, weakness, headaches, and susceptibility to fatigue. As might be expected, victims of sickle-cell anemia usually suffer an early death, generally by the age of 20.

Individuals heterozygous for the sickle-cell gene are phenotypically normal, because the sickle and normal alleles are codominants and both forms of hemoglobin are produced. Apparently there is enough of the normal hemoglobin produced to enable the individual to function normally. Only under very severe conditions of low oxygen (e.g., at very high altitudes) or perhaps after extremely strenuous exercise do some of their blood cells sickle and cause problems. Indeed, most of the individuals who carry the sickle allele may not know they have it.

It might be supposed that natural selection would operate against the propagation of any allele so obviously harmful and that such an allele would be held at very low frequency in the population. This seems to be true among blacks in the United States. But the allele is surprisingly common in northern Mediterranean regions and many parts of Africa, being carried by as much as 20 percent of the black population in Africa. What is the explanation? A. C. Allison of Oxford, England, has found that individuals heterozygous for this gene have a much higher than normal resistance to malaria. Since malaria is very common in many parts of Africa, the allele must be regarded as beneficial when heterozygous. Thus, in Africa, there is selection for the allele because of its heterozygous effect on malarial resistance and selection against it because of its homozygous production of sickle-cell anemia (Table 14.3). The balance between these two opposing selection pressures determines the frequency of the allele in the population. Another example of this phenomenon may be the Tay-Sachs allele. There is some evidence that individuals heterozygous for this allele have increased resistance to tuberculosis, so this allele may also be regarded as beneficial in areas where the incidence of tuberculosis is high.

The allele for sickle-cell anemia is a dramatic example of a gene that has more than one effect. Such a gene is said to be ***pleiotropic.*** Pleiotropy is, in fact, the rule rather than the exception. All genes probably have many effects on the organism. Even when a gene produces only one perceptible phenotypic effect, it doubtless has numerous physiological effects more difficult to detect.

TABLE 14.3 *Effects of the sickle allele*

Genotype	*Type of hemoglobin*	*Anemia present*	*Resistance to malaria*
HbA/HbA*	Normal	no	no
HbA/HbS	Normal and sickle	no	yes
HbS/HbS	Sickle	yes	yes

* Symbols: HbA is the normal allele that codes for the A chain in hemoglobin; HbS is the allele that codes for the defective (sickle) hemoglobin.

The Sex Chromosomes Differ in Size and Shape and Determine the Sex of the Individual

We have said repeatedly that a diploid individual has two of each type of chromosome, identical in size and shape, and hence two copies of each gene. But we must now qualify that statement somewhat. In most higher organisms where the sexes are separate (i.e., where males and females are separate individuals), the chromosomal endowments of males and females are different, and one or the other of the two sexes has one chromosomal pair consisting of two chromosomes that differ markedly from each other in size and shape. These are the ***sex chromosomes,*** which play a fundamental role in determining the sex of the individual. All other chromosomes are called ***autosomes.***

Let us look first at the chromosomes of *Drosophila* and of human beings. In each case the sex chromosomes are of two sorts: one bearing many genes, conventionally designated the ***X chromosome,*** and one of a different shape and bearing only a few genes, designated the ***Y chromosome.*** Females characteristically have two X chromosomes, and males have one X and one Y. The diploid number in *Drosophila* is eight (four pairs); a female therefore has three pairs of autosomes and one pair of X chromosomes, and a male has three pairs of autosomes and a pair of sex chromosomes consisting of one X and one Y (Fig. 14.17). The diploid number in human beings is 46 (23 pairs); a female therefore has 22 pairs of autosomes and one pair of X chromosomes, and a male has 22 pairs of autosomes plus one X and one Y chromosome (see Fig. 9.5, p. 182).

When a female produces egg cells by meiosis, all the eggs will receive an X chromosome. When a male produces sperm cells by meiosis, however, half the sperm cells carry an X chromosome and half a Y chromosome. In short, all the egg cells are alike in chromosomal content, but the sperm cells are of two types occurring in equal numbers (Fig. 14.17). When fertilization takes place, the chances are approximately equal that the egg will be fertilized by a sperm carrying an X chromosome as by a sperm carrying a Y chromosome. If fertilization is by an X-bearing sperm, the resulting zygote will be XX and will develop into a female. If fertilization is by a Y-bearing sperm, the resulting zygote will be XY and will develop into a male.[2] We see, therefore, that the sex of an individual is normally determined at the moment of fertilization and depends on which of the two types of sperm fertilizes the egg.

[2] The XY system, where XX is female and XY is male, is characteristic of many animals, including all mammals. It is also found in many plants with separate sexes. Birds, butterflies and moths, and a few other animals have just the opposite system, where XX is male and XY is female (to distinguish this system from the usual XY system, the symbols Z and W are often substituted, ZZ being male and ZW being female). A completely different mechanism of sex determination exists in the Hymenoptera (bees, wasps, ants, etc.), where the males hatch from unfertilized eggs and are haploid while the females hatch from fertilized eggs and are diploid.

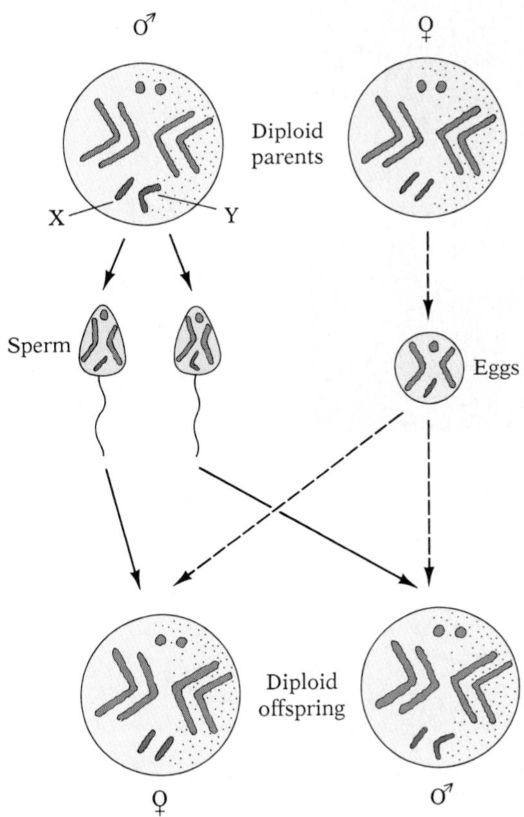

14.17 Chromosomes of male and female *Drosophila melanogaster* There are three pairs of autosomes and one pair of sex chromosomes. Males (symbolized by ♂) have one X chromosome and one Y chromosome; since these separate at meiosis, half the sperm carry an X and half carry a Y. Since females (♀) have two X chromosomes, all eggs have one X. The sex of the offspring depends on which type of sperm fertilizes the egg.

Genes on the X Chromosome Are Sex-Linked and Show Different Inheritance Patterns Than Those on Autosomal Chromosomes

Many genes occur on the X chromosome and not on the Y chromosome. Such genes are said to be sex-linked or "X-linked." The inheritance patterns for the characteristics controlled by sex-linked genes are completely different from those for characteristics controlled by autosomal genes, for obvious reasons. Females have two copies of each sex-linked gene, one from each parent, but males have only one copy of each sex-linked gene, and that one copy always comes from the mother, since the father contributes a Y chromosome instead of an X to his sons. Hence, in the male, all sex-linked characteristics are inherited only from the mother. And since the male has only one copy of each sex-linked gene, recessive genes cannot be masked, and recessive sex-linked phenotypes occur much more often in males than in females.

Sex linkage was discovered in 1910 by the great American geneticist Thomas Hunt Morgan of Columbia University. It was Morgan who began the systematic use of *Drosophila* in genetic studies. The first sex-linked trait observed in *Drosophila* by Morgan was white eye color. This trait is controlled by a recessive allele *r;* the normal red eye color is controlled by a dominant allele *R.* (Because the trait is X-linked, we use the symbols X^R and X^r to represent the two alleles.) If a homozygous red-eyed female is crossed with a white-eyed male, all the F_1 offspring, regardless of sex, have red eyes, because they all receive an X chromosome bearing an allele for red from their mother. The F_1 females also receive an X chromosome bearing an allele for white eyes from their father, but the allele for red, being dominant, masks its presence. The F_1 males, like the females, receive an X chromosome bearing an allele for red eyes from their mother. But unlike the females, they receive no gene for eye color from their father, who contributes a Y chromosome instead of an X. We can summarize this cross in a Punnett square (Fig. 14.18A).

Now let us examine the ***reciprocal cross***, where the parental generation consists of homozygous white-eyed females and red-eyed males. We can summarize this cross in a Punnett square (Fig. 14.18B). Notice that the phenotypic makeup of the F_1 differs both from a normal autosomal cross and from the reciprocal cross for this same sex-linked trait. In the F_1 of this cross (Fig. 14.18B), all females show the dominant phenotype, all males the recessive phenotype. Comparison of the two reciprocal crosses (Figs. 14.18A and B) makes it clear that when a sex-linked trait is involved in a cross the results depend on which parent shows the trait (or carries the allele for the trait). By contrast, in crosses involving autosomal genes it does not matter which parent possesses the allele in question; the results of autosomal reciprocal crosses are identical.

Two well-known examples of recessive sex-linked traits in human beings are red-green color blindness and hemophilia ("bleeder's disease"). Color blindness occurs in about 8 percent of white men in the United States and in about 4 percent of black men. It occurs in only about 1 percent of white women and about 0.8 percent of black women. It makes sense, of course, that more men than women will show such a trait, because a man needs only one copy of the gene to show the phenotype, which he can inherit from a heterozygous mother who is not herself color-blind. But for a woman to be color-blind, she must have two copies of the gene (i.e., be homozygous); not only must her father be color-blind, but her mother must be either color-blind or a heterozygous carrier of the gene. Since the allele is not very common in the population, it is not likely that two such people will mate—hence the low number of color-blind women.

The statement that females have two copies of each sex-linked gene whereas males have only one, though technically correct, requires qualification. In most interphase somatic cells of females, one of the X chromosomes condenses into a tiny dark object called a ***Barr body*** (Fig. 14.19). The genes on

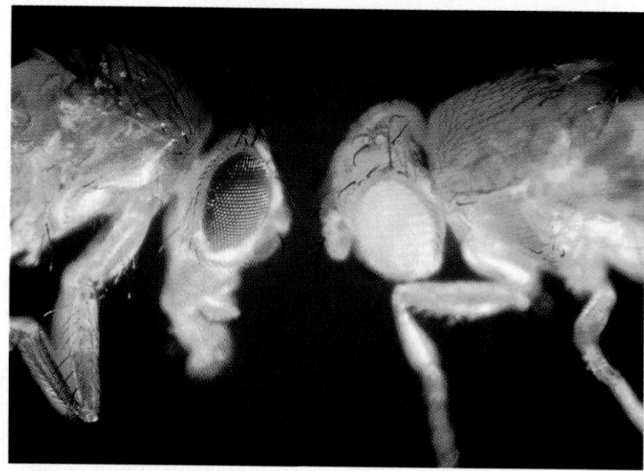

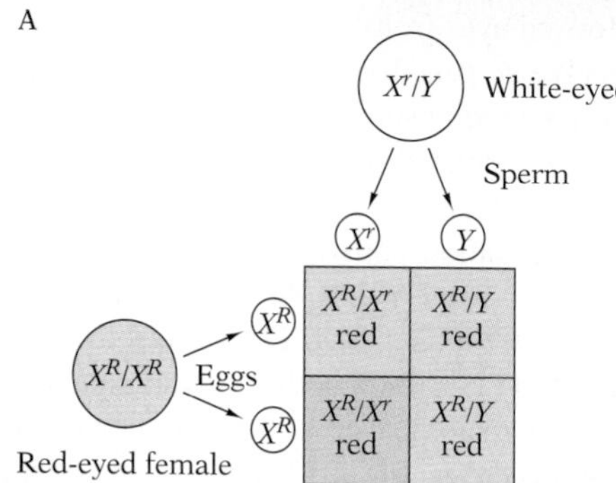

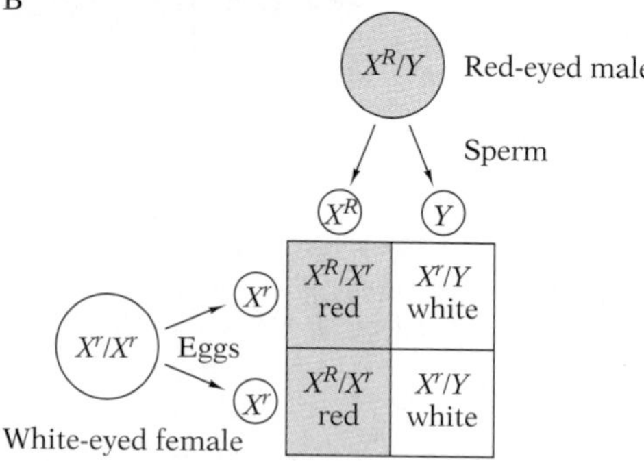

14.18 Sex linkage in *Drosophila* The first sex-linked trait observed in *Drosophila* by Morgan was white eye color. This trait is controlled by a recessive allele X^r; the normal red eye color is controlled by a dominant allele X^R. (A) When a homozygous red-eyed female is crossed with a white-eyed male, all the F_1 offspring, regardless of sex, have red eyes, because they all receive an X chromosome bearing an allele for red from their mother. The F_1 females also receive an X chromosome bearing an allele for white eyes from their father, but the allele for red, being dominant, masks its presence. The F_1 males receive an X chromosome bearing the red-eye allele from their mother and a Y chromosome from their father. (B) In the reciprocal cross, homozygous white-eyed females are crossed with red-eyed males. In the F_1 of this cross, there is a difference in eye color correlated with sex. All females show the dominant phenotype, all males the recessive phenotype. In a cross involving a sex-linked trait, the results depend on which parent shows the trait (or carries the allele for the trait). By contrast, in crosses involving autosomal genes it does not matter which parent possesses the allele in question; the results of autosomal reciprocal crosses are identical. Geneticists studying inheritance of particular traits routinely do reciprocal crosses to determine whether or not a trait is sex-linked.

this tightly condensed X chromosome are inactive. Thus, in a normally functioning female cell, there is normally only one active copy of each sex-linked gene. (Even in cells with abnormal numbers of sex chromosomes—XXX or XXXX—only one X chromosome is functional, the others all being condensed into Barr bodies.) Why, then, are sex-linked recessive traits expressed only in homozygous females? The explanation is that it is not always the same X chromosome that condenses into a Barr body in the different somatic cells of a given individual. Let us call the two X chromosomes X_1 and X_2. Characteristically, about half the cells in any given female show active X_1 chromosomes, the other half active X_2 chromosomes, with no discernible pattern as to which chromosome, X_1 or X_2, is active in which cell (Fig. 14.20). Apparently female cells differ in their effective genetic makeup as far as sex-linked traits are concerned; women are, in a sense, genetic mosaics for sex-linked

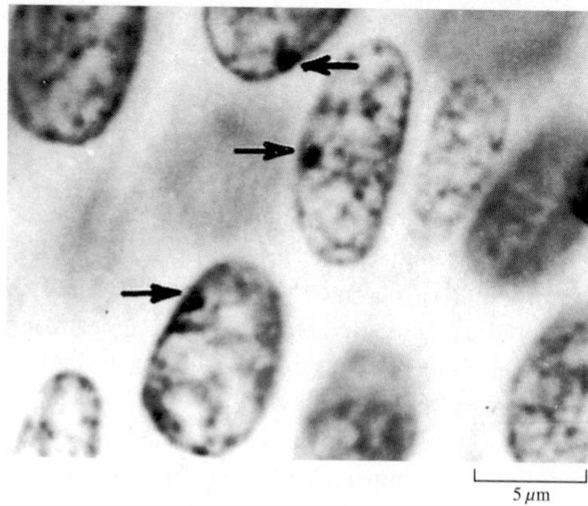

14.19 Nuclei from epidermal cells of a human female The arrows indicate the Barr bodies. Since Barr bodies are present in the cells of female fetuses, the sex of an unborn child can be ascertained by examination of the nuclei of cells sloughed off into the amniotic fluid.

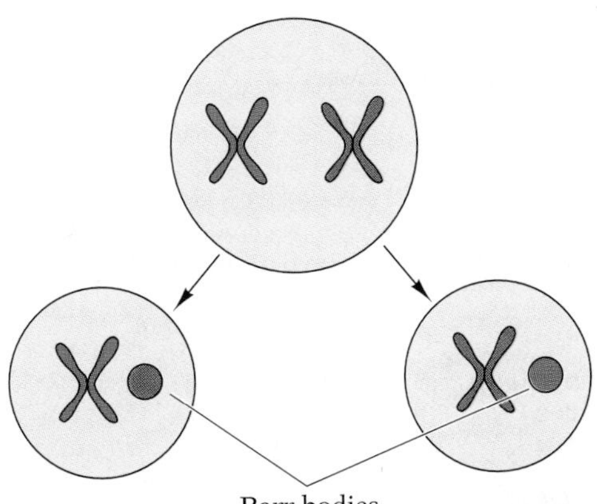

14.20 X chromosome inactivation in mammalian females Early in embryonic life, one of the two X chromosomes in each cell becomes inactive and condenses to form a Barr body. Apparently, it is purely a matter of chance which X chromosome becomes inactive.

traits. But, it seems, as long as half the cells are normal in a woman heterozygous for traits such as red-green color blindness or hemophilia, she will not be color-blind or hemophilic. Careful studies of heterozygotes do show, however, that such women have diminished color discrimination in the case of color-blindness and slightly delayed clotting time in the case of hemophilia, as you might expect. Sometimes the mosaic pattern is clearly expressed. An example is the coat color of tortoise-shell and calico cats (Fig. 14.21).

14.21 X-chromosome inactivation in cats One of the most common natural demonstrations of X-chromosome inactivation is the coat color of tortoise shell (calico) cats. A gene for color found on the cat's X chromosome has two common alleles—black and yellow (or orange). Males can have only one allele, so (in the absence of modifier genes) they are either yellow or black, usually with various white markings. Females, however, can have both alleles, and after inactivation some cells will express the black allele and others the yellow allele. Large patches of cells with one allele or the other develop, and the result is a mosaic of yellow, black, and (usually) white patches. Except for rare XXY individuals, all cats with both yellow and black fur are females.

It has long been known that, because of hormonal imbalances during embryonic development, some human beings exhibit an anatomical and physiological sex different from their genetic sex. For example, an XY individual might appear to be a female. With the discovery of Barr bodies, a simple technique for determining the genetic sex of individuals became available, one of great help in studying such developmental anomalies. If Barr bodies are visible in the nuclei of cells examined microscopically, the individual from whom the cells came can be identified as genetically female; if there are no Barr bodies, the individual can be identified as genetically male. This technique has recently been used to determine the sex of fetuses before birth, where such information would be medically helpful; the cells examined are cells sloughed off by the fetus into the amniotic fluid.

There Are Very Few Genes on the Y Chromosome

Genes on the Y chromosome but not on the X are termed ***holandric*** (Fig. 14.22). The phenotypic traits they control appear, of course, only in males. Apparently, the human Y chromosome bears very few genes but one portion of it does seem to bear genes that determine maleness; if that portion of the chromosome is present the child will be male, if absent, female. One gene in this region (called TDF for testicular determining factor) has been found to be located on the short arm of the Y chromosome.

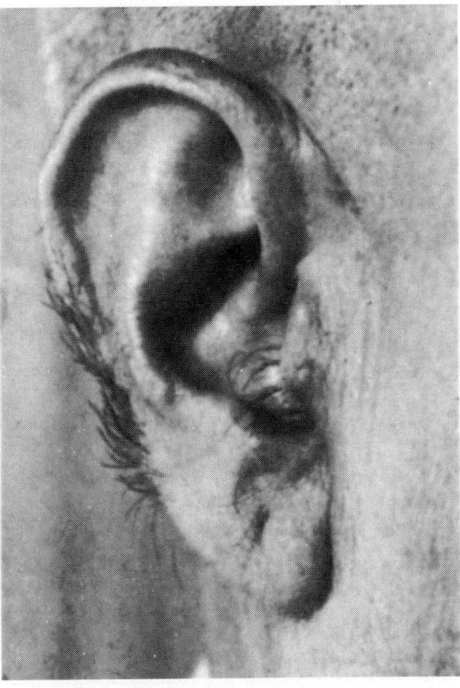

14.22 Hairy pinna This trait is thought to be determined by a gene on the Y chromosome.

P Purple,  long pollen flowers × red,  round pollen flowers

F_1 All purple, long pollen flowers

Test cross: F_1 × red, round pollen flowers

7 purple, long pollen 1 purple, round pollen 1 red, long pollen 7 red, round

How do we explain these unusual ratios?

A. If the genes are on different chromosomes (i.e. unlinked) the test cross and expected results would be:

	B/b L/l	×	*b/b l/l*	
	purple, long pollen		red, round	
gametes:	*BL Bl bL bl*		*all bl*	
Expect:	1 *B/b L/l*	1 *B/b l/l*	1 *b/b L/l*	1 *b/b l/l*
	purple, long	purple, round	red, long	red, round

B. If the genes are linked and no crossing over occurs the test cross and expected results would be:

	BL/ bl	×	*bl/ bl*
	purple, long		red, round
gametes:	*BL bl*		all *bl*
Expect:	1 *BL/L/l* purple, long		1 *bl/bl* red, round

C. If the genes are linked and *crossing over occurs* the test cross and expected results would be:

	BL/bl	×	*bl/bl*	
	purple, long pollen		red, round	
gametes:	*BL bl BL bl*		all *bl*	
Expect:	*BL/bl*	*bl/bl*	*Bl/bl*	*bL/bl*

14.23 Bateson and Punnett's crosses In a test cross, if the genes for flower color and pollen shape are unlinked, i.e., on different chromosomes (A), the two genes would be expected to assort independently and four equal phenotypic classes would likely occur. If the two genes are linked, i.e., on the same chromosome, and no crossing over occurs (B) the results of the test cross should yield two equal classes of offspring that resemble the parents (one-half purple, long pollen and one-half red, round pollen). If the genes are linked and crossing over occurs (C), the results of the test cross should yield four classes of offspring, most of which should resemble the parents (purple, long pollen and red, round pollen) but some would be recombinants: purple, round pollen and red, long pollen flowers. The results of the test cross (7 purple, long pollen:1 purple, round pollen: 1 red, long pollen: 7 red, round pollen) most closely resemble *C,* in which there are two large classes resembling the parents, and two small resembling the recombinants. The conclusion is that the two genes for flower color and pollen shape are on the same chromosome, and that crossing over occurred.

Genes on the Same Chromosome Are Linked; They Remain Together During Meiosis Unless Crossing Over Occurs

All of the crosses that Mendel performed on garden peas supported his second law, the Law of Independent Assortment, which states that when two or more genes are involved in a cross, the alleles of one gene are inherited independently of the alleles for another gene. There are other crosses, however, that fail to obey Mendel's second law, as William Bateson and R. C. Punnett of Cambridge University discovered in 1906. They crossed sweet peas that had purple flowers and long pollen grains with ones that had red flowers and round pollen grains. All the F_1 plants had purple flowers and long pollen, as expected (it was already known that purple was dominant over red and that long was dominant over round). The F_2 plants from this cross did not show the expected 9:3:3:1 ratio, however, but a highly anomalous one. Next, Bateson and Punnett tried a test cross, crossing the F_1 plants to homozygous recessive plants (with red flowers and round pollen). The results of their test cross are shown in Figure 14.23.

In a test cross, the phenotypic ratio of the offspring depends on the genotype of the parent showing the dominant phenotype, since the recessive parent produces only one kind of gamete. Using the symbols *B* for purple, *b* for red, *L* for long, and *l* for round, the homozygous recessive red round parent (*b/b l/l*) could produce only *bl* gametes. Hence it was the gametes of the heterozygous purple long parent that must have determined the phenotype of the offspring. According to Mendel's second law, this parent should have produced four kinds of gametes (*BL, Bl, bL,* and *bl*) in equal numbers. When united with the *bl* gametes from the homozygous recessive parents, *BL* gametes should have given rise to purple long offspring, *Bl* gametes to purple round, *bL* to red long, and *bl* to red round, and these four phenotypes should have occurred in equal numbers, i.e., in a 1:1:1:1 ratio. But the result Bateson and Punnett actually obtained—a ratio of 7:1:1:7—makes it appear that the heterozygous parent produced far more *BL* and *bl* gametes than *Bl* and *bL* gametes.

It was not until 1910 that Morgan, who had obtained similar results from *Drosophila* crosses, provided the explanation accepted today. He postulated that the anomalous ratios were caused by linkage, that the genes for flower color and seed shape were located on the same chromosome, i.e., they were ***linked.*** Hence, we should write the genotypes of the parents in Bateson and Punnett's test cross *BL/bl* and *bl/bl,* to show, by the positions of the slashes, that *B* and *L* are on one of a pair of chromosomes and *b* and *l* are on the other (we would write these genotypes *B/b L/l* and *b/b l/l* if the genes were not linked).

Now, if in Bateson and Punnett's cross the genes for purple and long and the genes for red and round were linked, we might expect the *BL/bl* parent in the test cross to have produced only two kinds of gametes, *BL* and *bl,* and the test cross to have yielded offspring of only two phenotypes, purple long and red round, in equal numbers (Fig. 14.23B). Most of the offspring do conform to this expectation and have the same phenotype as the parents—we refer to these as the ***parentals***—yet the cross also yielded some purple round and red long offspring. How could the *BL/bl* parent have produced *Bl* and *bL* gametes if the genes are linked? Morgan suggested that some mechanism occasionally breaks the original linkages between purple and long and between red and round and establishes in a few individuals new linkage combinations ***(recombinants)*** between purple and round and between red and long, thus making possible the production of *Bl* and *bL* gametes. The mechanism of this recombination is crossing over (Fig. 14.24), a process we discussed in Chapter 9, p. 191 in which homologous chromosomes come to lie side by side, chromatids are clipped, and parts of separate chromatids are spliced together to form new combinations of genes. Crossing over increases the number of

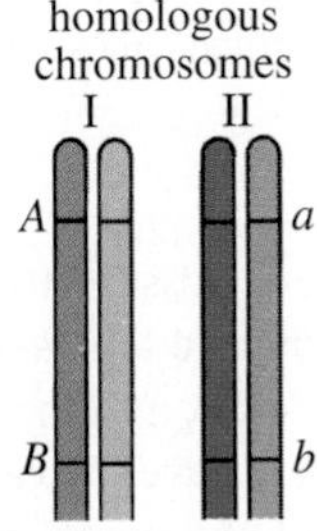

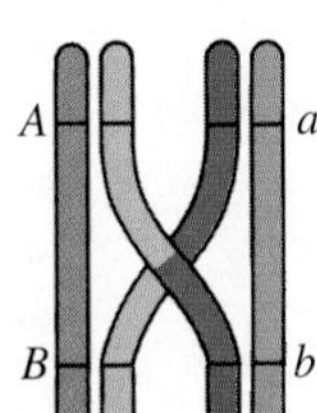

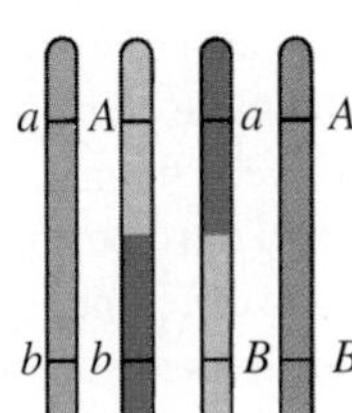

14.24 Linkage and recombination due to crossing over During meiosis, the homologous chromosomes, each consisting of two twin-chromatids, come together during synapsis (left). Crossing over then takes place; chromatids are clipped and parts of separate chromatids are spliced together (middle). When the chromatids separate at the end of meiosis (right), the two genes originally linked as *AB* and *ab* are combined in four ways: *ab, Ab, aB,* and *AB.*

genetic combinations any given cross can produce, and greatly increases the amount of genetic variation within a population.

Crossing Over Can Be Used to Map Genes on the Chromosome

To construct a genetic map we assume that the probability of breakage is approximately equal at any point along the length of a chromosome. Consequently, the greater the distance between two linked genes, the greater the frequency with which they will cross over, because there are more points between them at which a break may occur. Alternatively, the frequency of crossing over between any two linked genes will be proportional to the distance between them. The percentage of crossing over can therefore serve as a tool for mapping the locations of genes on chromosomes.

Though this percentage gives us no information about the absolute distances between genes, it does give us relative distances. By convention, one unit of map distance on a chromosome is the distance within which crossing over occurs 1 percent of the time. The percentage of crossing over can be determined using the following formula:

$$\% \text{ crossing over} = \frac{\text{number of recombinants}}{\text{total number of offspring}} \times 100$$

In Bateson and Punnett's test cross, two of 16 of the offspring were recombinant products of crossing over. Two is 12.5 percent of 16; hence the genes controlling flower color and pollen shape in the sweet peas of this cross are located 12.5 map units apart. By determining the frequency of crossing over between different genes, it is possible to build up a map showing the arrangement of many different genes on a chromosome. Figure 14.25 shows a chromosomal map for *Drosophila.*

That the frequencies of crossing over permit the mapping of genes agrees with a model of the chromosome in which the genes are sequentially arranged in linear fashion along the chromosome, like the beads on a string. If two characters are always linked, and recombination by crossing over never occurs, then we assume that they are controlled by the same locus on the chromosome—by the same gene. If, on the other hand, crossing over does occur between them, even if extremely seldom, then we can say that they must be controlled by different loci on the chromosome—by different genes. Crossing over, then, tests whether two characters are controlled by one chromosomal locus (gene) or by two separate chromosomal loci (genes).

Another approach to chromosomal mapping comes from the study of the giant chromosomes in the salivary glands of the larvae of many flies, including *Drosophila.* When stained appropriately, giant chromosomes have a banded appearance (Fig. 14.26). The bands differ in width and in the spacings between them. It has been possible by detailed comparative studies of chromosome abnormalities to determine the location of individual genes in relation to the bands. Cytological studies of giant chromosomes have provided a second way of mapping the sequence of the genes on the chromosomes. Such mapping has fully corroborated the gene sequences derived from work based on crossing over frequencies, but it has not corroborated the conclusions about the distance between genes (Fig. 14.27). Apparently breaks do not occur with equal facility at all points along the chromosomes, some parts being more susceptible to breakage than others.

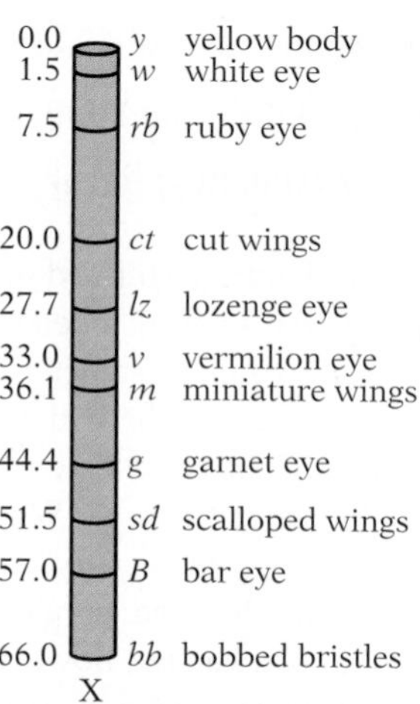

14.25 Map of X chromosome of *Drosophila melanogaster* Only a few of the many known genes are shown. The figures indicate their relative distance in cross-over map units from the zero end of the chromosome. This map tells us, for example, that crossing over between yellow body and vermilion eye can be expected to occur 33 percent of the time.

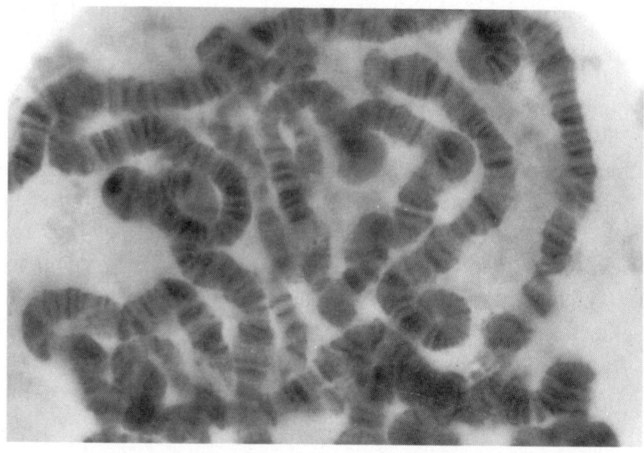

14.26 Photograph of giant chromosomes from salivary gland of *Drosophila melanogaster* Note the pattern of banding by which different parts of the chromosomes can be identified.

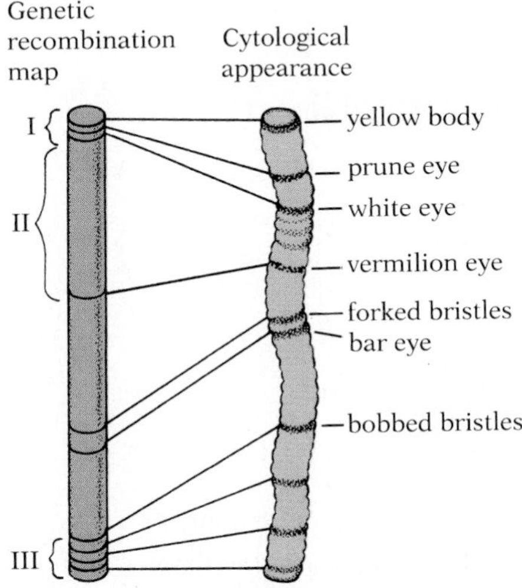

14.27 Comparison of a genetic-recombination map and cytological appearance of a portion of the X chromosome in *Drosophila melanogaster* Staining and photographic techniques now exist to localize the genes on chromosomes directly. When these cytological results are compared with cross-over frequencies (linkage maps), the effects of variations in cross-over rates in different parts of the chromosome become apparent.

Structural Alterations to Chromosomes Sometimes Occur, Resulting in Rearrangements of the Genetic Material

Besides crossing over, in which corresponding segments are interchanged between homologous chromosomes, there are other kinds of chromosomal rearrangements that occur much less frequently. Some involve ***translocations,*** in which a segment is transferred from one chromosome to another nonhomologous chromosome; others involve alterations within a single chromosome, such as ***deletions*** or ***duplications*** of chromosomal material. This happens because occasionally chromosomes break, and some broken-off fragments, which may contain important genes, are lost (deletion) or are added to other chromosomes. If the segment is added to the homologous chromosome, the segment appears twice on the same chromosome, duplicating the genetic material on that chromosome. One harmful effect of intense radioactivity on somatic cells is the large amount of chromosomal breakage it produces, with the consequent loss of important genes or deleted fragments.

Changes in Chromosome Number Occur When the Separation of Chromosomes in Cell Division Does Not Proceed Normally

Occasionally, the separation of chromosomes in cell division does not proceed normally, and the homologous chromosomes that should have moved away from each other to opposite poles of the spindle move instead to the same pole and become incorporated into the same daughter nucleus. The result of such ***nondisjunction,*** as it is called, may be that one of the daughter cells receives one too many chromosomes instead of the normal diploid number. The presence of three chromosomes of one type in an otherwise diploid individual is called ***trisomy*** (see Exploring Further: Trisomy in Humans, p. 302).

Once in a while cell division may be so aberrant that all the chromosomes move to the same pole, giving rise to a daughter cell with twice the normal number of chromosomes. If this happens during meiosis, the gamete produced is diploid ($2n$) instead of haploid. If such a gamete unites at fertilization with a normal haploid gamete, a triploid zygote results; if it unites with another diploid gamete, also produced by aberrant meio-

MAKING CONNECTIONS

1. How does sickle-cell anemia demonstrate that the environment determines the adaptiveness of any characteristic?

2. What are Barr bodies? How are they useful in determining the genetic sex of an individual?

3. Compare how mutations and crossing-over increase the amount of genetic variation in populations. What is the evolutionary significance of this variation?

4. How many alleles for a particular gene could a tetraploid plant have?

5. What is the evolutionary significance of polyploidy?

EXPLORING FURTHER

TRISOMY IN HUMANS

In humans most trisomies are lethal. Trisomy-18* (Edwards' syndrome) and trisomy-13 (Patau's syndrome), for example, produce physical malformations and mental and developmental retardation so severe that most afflicted infants die within a few weeks after birth. Because trisomies of most other autosomes result in spontaneous abortion, they are not found in live births. Two kinds of trisomy—trisomy-21 (Down's syndrome) and trisomies of the sex chromosomes—are exceptional in that their victims may survive.

Down's syndrome, in which three chromosome 21s occur in the individual's cells, was the first clinical condition ever linked to a chromosomal abnormality. It is associated with a variety of characteristic physical features (e.g., broad head, rounded face, perceptible eye folds, a flattened bridge of the nose, protruding tongue, small irregular teeth, short stature) and also mental retardation (the average IQ is about 42). The incidence of Down's syndrome is related, in part, to the age of the mother. It occurs in less than one out of 1,000 births to women under 20; it is more than seven times more common in births to women 35 to 39 years old, more than twenty times more common when the mother is 40 to 44, and more than fifty times more common when she is 45 or older. A similar association with the age of the mother is seen in Edwards' and Patau's syndromes.

Trisomy of the sex chromosomes can take several forms. In one, called Klinefelter's syndrome, the chromosomal result of nondisjunction is XXY, and the individuals are males. Though the symptoms of the condition are variable, and some of those affected are nearly normal, most show a variety of physical abnormalities such as sterility, sparse body hair, and breast enlargement. They are generally very tall, and many are mentally retarded.

Males with a second type of sex-chromosome trisomy, the XYY syndrome, generally show fewer and less severe abnormalities, though they often have poorly developed genitalia, subnormal intelligence, and are excessively tall. Many show some degree of aggressive or antisocial behavior. Because the incidence of XYY individuals in penal institutions is often higher than in the general population, some investigators suggested that men with the XYY condition are predisposed to violence, but the evidence for this conclusion is very weak. Recent studies in Sweden and elsewhere have concluded that XYY individuals are not any more prone to aggressive behavior and violence than normal XY males, but rather that such individuals' subnormal intelligence and excessive height make them more likely to be caught should they commit a crime. This would account for the disproportionate number of such individuals in jail.

Why is it that most trisomies of autosomes are lethal whereas individuals with abnormal numbers of sex chromosomes—XXX, XXY, XYY, XXXX, etc.—not only survive but may lead normal lives? The answer probably lies in the phenomenon of X chromosome inacti-

vation. Recall that in any given cell, only one X chromosome is functional, all others being condensed into Barr bodies. Thus, although extra sex chromosomes may be present, only one X chromosome is active in any one cell (though there is evidence that at least some genes on the tip of the short arm of the inactivated X chromosome may remain active).

*Many trisomies are designated by the number label of the specific chromosome pair involved.

sis, a tetraploid zygote results. Cells or organisms that have more than two complete sets of chromosomes (that are triploid, tetraploid, hexaploid, etc.) are said to be ***polyploid.*** When polyploidy occurs in humans, death results very early in embryonic life. Most of the triploids result when two sperms fertilize a single egg, but some are produced by a diploid sperm fertilizing an egg. Tetraploidy in humans most often results when the zygote undergoes its first cell division. Mitosis occurs properly but cytokinesis fails to occur.

Although instances of viable polyploids are relatively rare in animals, polyploidy has apparently occurred rather often in plants, in which it has sometimes given rise to new species adaptively superior to the original diploid species under certain environmental conditions. Polyploidy can be stimulated in the laboratory by treating plants with certain chemicals that cause nondisjunction during cell division. Polyploidy and its importance in forming new species will be discussed further in Chapter 32.

CONCEPTS IN BRIEF

1. From his breeding experiments on the garden pea, the Austrian monk Gregor Mendel formulated the *Law of Segregation,* which states that each pea plant possesses two hereditary factors (genes) for each character, and that when gametes are formed the two factors separate into separate gametes. Each new plant thus receives one factor for each character from each parent. The hereditary factors exist as distinct entities within the cell; they do not blend or alter each other, and they segregate unchanged when gametes are formed.

2. Mendel was working with two different forms (*alleles*) of the genes for flower color. When both alleles were present, the allele for red flowers was expressed (*dominant*) while the allele for white flowers was masked (*recessive*). A diploid cell may be *homozygous* or *heterozygous.* When a monohybrid cross is made between two contrasting homozygous individuals, the expected genotypic ratio in the F_2 is 1:2:1. When dominance is involved, the expected phenotypic ratio is 3:1. *Genotype* refers to genetic combinations, and *phenotype* to appearances.

3. In a *test cross* the unknown is crossed with a homozygous recessive individual. If all the progeny show the dominant phenotype, the unknown genotype is probably homozygous dominant; if any of the progeny show the recessive genotype, the unknown is heterozygous.

4. One allele is not always completely dominant over the other; in some cases heterozygous individuals show the effects of both alleles and are different from both homozygotes. If the heterozygote's phenotype is intermediate between the parents it is *incomplete dominance;* if both alleles are expressed independently in the heterozygote it is *codominance.* At the molecular level, most alleles probably show incomplete or codominance.

5. Mendel also made crosses involving two characters (a *dihybrid* cross). Mendel found that the F_2 offspring consistently conformed to a 9:3:3:1 phenotypic ratio. From these results Mendel formulated the *Law of Independent Assortment,* which states that during gamete formation, when two or more genes are involved in a cross, the alleles of one gene are inherited independently of the alleles for another gene.

6. Probably no inherited characteristic is controlled by only one gene pair. In polygenic inheritance, many genes interact in additive fashion to affect a particu-

lar characteristic. Even when only one principal gene is involved, other genes may act as *modifiers* to influence its expression. The action of any gene can be fully understood only in terms of the overall genetic makeup of the individual organism. The expression of a gene depends on the other genes present and on the physical environment.

7. Genes may exist in a number of allelic forms (*multiple alleles*) within a population, but each individual can have only two alleles for a given trait. An example of multiple alleles in humans is the A-B-O blood groups, in which there are three alleles: I^A, I^B, and *i*.

8. A variety of influences can cause changes, or *mutations,* in the DNA of genes. Mutations occur constantly; most are deleterious. Natural selection can act against a deleterious gene only if it is expressed phenotypically.

9. In most higher organisms the chromosomal endowments of males and females are different. One chromosomal pair, the *sex chromosomes,* differs in size and shape and determines the sex of the individual. All other chromosomes are called *autosomes.*

10. The genes on the X chromosome are said to be *sex-linked.* Because females have two X chromosomes, they always have two alleles for a sex-linked character, whereas males have only one (one on the X, none on the Y). Consequently, recessive sex-linked genes are always expressed phenotypically in the male but may be masked by the dominant allele in the female.

11. Genes located on the same chromosome are said to be *linked;* they ordinarily remain together during meiosis. However, crossing over can occur between homologous chromosomes during synapsis; this results in new linkages. The frequency of crossing over between any two linked genes will be proportional to the distance between them. The percentage of crossing over can be used to map gene locations.

12. Separation of the chromosomes during meiosis does not always occur normally; sometimes both members of one homologous pair move to the same pole. The result of this *nondisjunction* may be the production of a cell with an extra chromosome (*trisomy*). Cells with more than two sets of chromosomes are *polyploid.*

STUDY QUESTIONS

The best way to gain an understanding of genetics is to work with it. The fundamental principles discussed above will become clearer to you, and you will grasp them more surely, if you carefully do the following problems, which illustrate the various patterns of inheritance treated in this chapter. Additional problems, a discussion of how to do genetics problems, and detailed answers to the problems given here will be found in the Study Guide accompanying this book.

1. In squash a gene for white color *(W)* is dominant over its allele for yellow color *w.* Give the genotypic and phenotypic ratios for the results of each of the following crosses:

   W/W × *w/w*
   W/w × *w/w*
   W/w × *W/w*

2. In human beings, brown eyes are usually dominant over blue eyes. Suppose a blue-eyed man mates with a brown-eyed woman whose father was blue-eyed. What proportion of their children would you predict will have blue eyes?
3. If a brown-eyed man and a blue-eyed woman have 10 children, all brown-eyed, can you be certain that the man is homozygous? If their 11th child has blue eyes, what will that show about the father's genotype?
4. A brown-eyed man whose father was brown-eyed and whose mother was blue-eyed mates with a blue-eyed woman whose father and mother were both brown-eyed. The couple has a blue-eyed son. For which of the individuals mentioned can you be sure of the genotypes? What are their genotypes? What genotypes are possible for the others?
5. If the litter resulting from the mating of two short-tailed cats contains three kittens without tails, two with long tails, and six with short tails, what would be the simplest way of explaining the inheritance of tail length in these cats? Show genotypes.
6. In peas a gene for tall plants *(T)* is dominant over its allele for short plants *(t).* The gene for smooth peas *(S)* is dominant over its allele for wrinkled peas *(s).* Calculate both phenotypic ratios and genotypic ratios for the results of each of the following crosses:

   T/t S/s × *T/t S/s*
   T/t s/s × *t/t s/s*
   T/t s/s × *t/t S/s*

7. In hogs a gene that produces a white belt around the animal's body is dominant over its allele for a uniformly colored body. Another independent gene produces fusion of the two hoofs on each foot (an instance of syndactyly); it is dominant over its allele, which produces normal hoofs. Suppose a uniformly colored hog homozygous for syndactyly is mated with a normal-footed hog homozygous for the belted character. What would be the phenotype of the F_1? If the F_1 individuals are allowed to breed freely among themselves, what genotypic and phenotypic ratios would you predict for the F_2?
8. In the fruit fly *Drosophila melanogaster,* vestigial wings and hairy body are produced by two recessive genes located on different chromosomes. The normal alleles, long wings and hairless body, are dominant. Suppose a vestigial-winged hairy male is crossed with a homozygous normal female. What types of progeny would be expected? If the F_1 individuals from this cross are permitted to mate randomly among themselves, what progeny would be expected in the F_2? Show complete genotypes, phenotypes, and ratios for each generation.
9. A dominant gene, *A,* causes yellow coat color in rats. The dominant allele of another independent gene, *R,* produces black coat color. When the two dominants occur together (*A*/– *R*/–), they interact to produce gray. Rats of the genotype *a/a r/r* are cream-colored. If a gray male and a yellow female, when mated, produce offspring approximately $^3/_8$ gray, $^3/_8$ yellow, $^1/_8$ cream, and $^1/_8$ black coats, what are the genotypes of the two parents?
10. In Leghorn chickens colored feathers are due to a dominant gene, *C*; white feathers are due to its recessive allele, *c*. Another dominant gene, *I,* inhibits expression of color in birds with genotypes *C/C* or *C/c*. Consequently both C/– I/– and c/c –/– are white. A colored cock is mated with a white hen and produces many offspring, all colored. Give the genotypes of both parents and offspring.
11. If the dominant gene *K* is necessary for hearing, and the dominant gene *M* results in deafness no matter what other genes are present, what percentage of the offspring produced by the cross *k/k M/m* × *K/k m/m* will be deaf?
12. Suppose two *D/d E/e F/f G/g H/h* individuals are mated. What would be the predicted frequency of *d/d E/E F/f g/g H/h* offspring?
13. If a man with blood type B, one of whose parents had blood type O, mates with a woman with blood type AB, what will be the theoretical percentage of their children with blood type B?
14. Both Mrs. Smith and Mrs. Jones had babies the same day in the same hospital. Mrs. Smith took home a baby girl, whom she named Shirley. Mrs. Jones took home a baby girl, whom she named Jane. Mrs. Jones began to suspect, however, that her child had been accidentally switched with the Smith baby in the nursery. Blood tests were made: Mr. Smith was type A, Mrs. Smith type B, Mr. Jones type A, Mrs. Jones type A, Shirley type O, and Jane type B. Had a mix-up occurred?
15. Coat color in Labrador retrievers is controlled by three alleles that code for yellow, black, or chocolate. The black allele is dominant over chocolate and both are dominant over yellow. Suppose a black male, whose mother was yellow, was mated with a chocolate female who had a yellow father. What genotypic and phenotypic ratios would you predict in the offspring?
16. Suppose that gene *b* is sex-linked, recessive, and lethal. A man mates with a woman who is heterozygous for this gene. If this couple had many normal children, what would be the predicted sex ratio of these children?
17. Red-green color blindness is inherited as a sex-linked recessive. If a color-blind woman mates with a man who has normal vision, what would be the expected phenotypes of their children with reference to this character?
18. The diagram shows three generations of the pedigree of deafness in a family. Black indicates deaf persons. A square indicates a male, a circle a female. State whether the condition of deafness in this family is inherited as

    a. a dominant autosomal characteristic
    b. a recessive autosomal characteristic
    c. a sex-linked dominant characteristic
    d. a sex-linked recessive characteristic
    e. a holandric characteristic

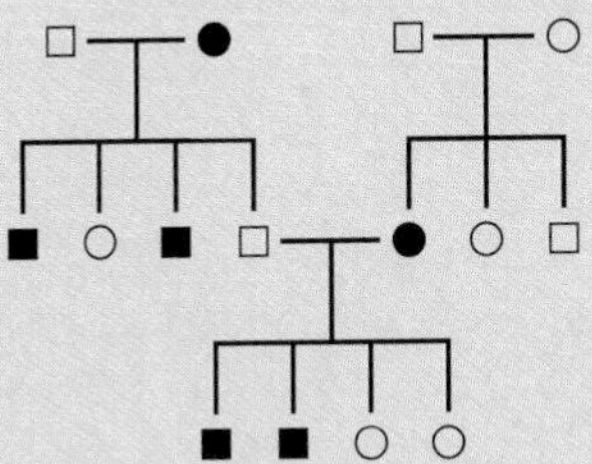

19. **In rabbits a dominant gene produces spotted body color, and its recessive allele solid body color. Another dominant gene produces short hair, and its recessive allele long hair. Rabbits heterozygous for both characteristics were mated with homozygous recessive rabbits. The results of this cross were as follows:**

Spotted, short hair	**96**
Solid, short hair	**14**
Spotted, long hair	**10**
Solid, long hair	**80**

What evidence for linkage is shown in this cross? Give the percentage of crossing over and the map distance between the genes.

20. **In *Drosophila melanogaster,* the genes for normal bristles and normal eye color are known to be about 20 units apart on the same chromosome. Individuals homozygous dominant for these genes were mated with homozygous recessive individuals. The F_1 progeny were then test-crossed. If there were 1,000 offspring from the test cross, how many of the offspring would you predict would show the cross-over phenotypes?**

21. **The cross-over frequency between linked genes *A* and *B* is 40; between *B* and *C*, 20; between *C* and *D*, 10; between *C* and *A*, 20; between *D* and *B*, 10. What is the sequence of the genes on the chromosome?**

22. **Suppose that nondisjunction resulted in the production of new individuals with the following chromosomal abnormalities: XXX, XYY, XXXX, XXXY, XXXXY. Indicate the expected phenotypic sex corresponding to each of these chromosomal combinations if it occurred in a human. How many Barr bodies would there be in human cells showing each of these combinations?**

P A R T   T H R E E

The Biology of Organisms

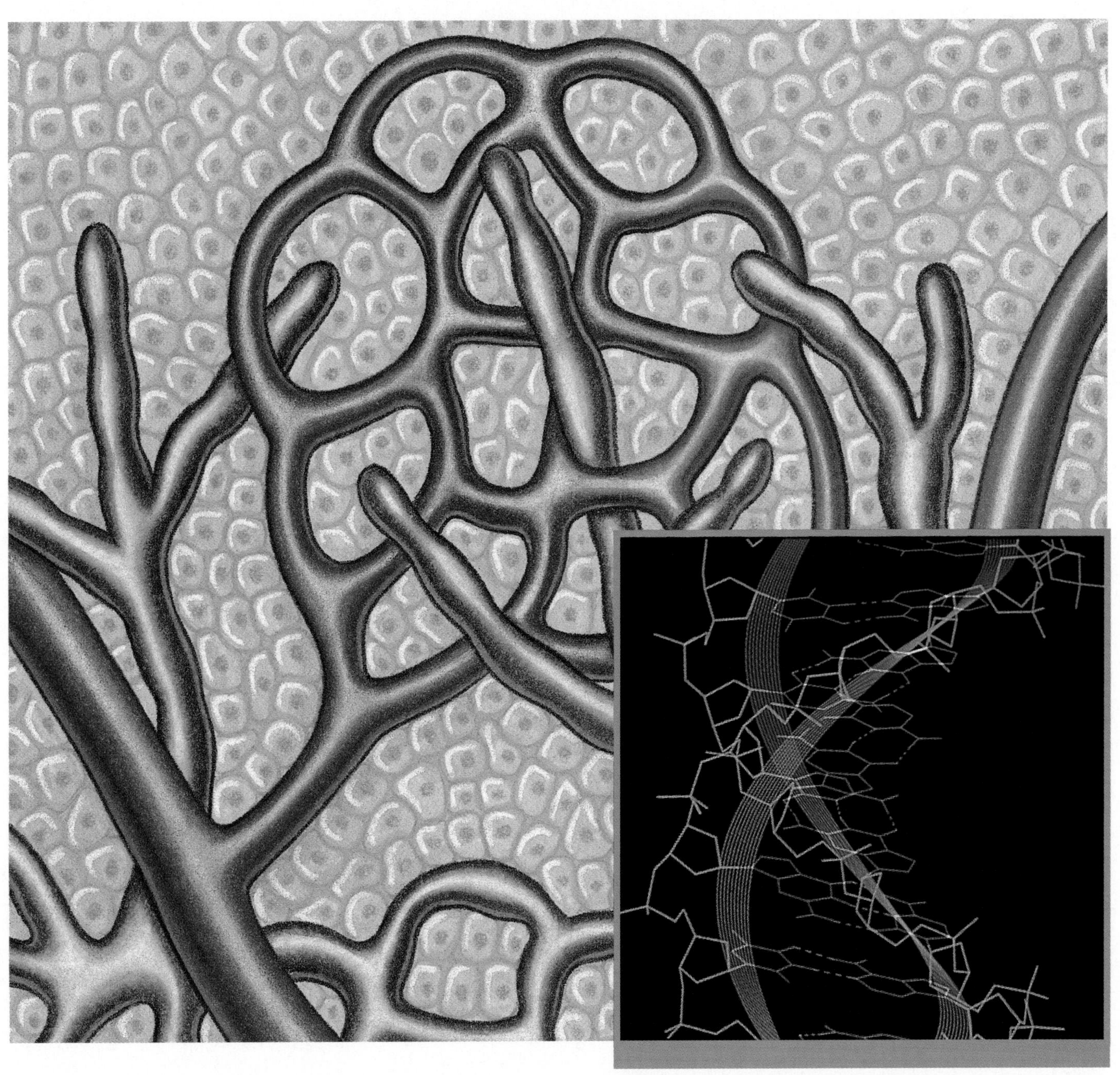

CHAPTER 15

MULTICELLULAR ORGANIZATION

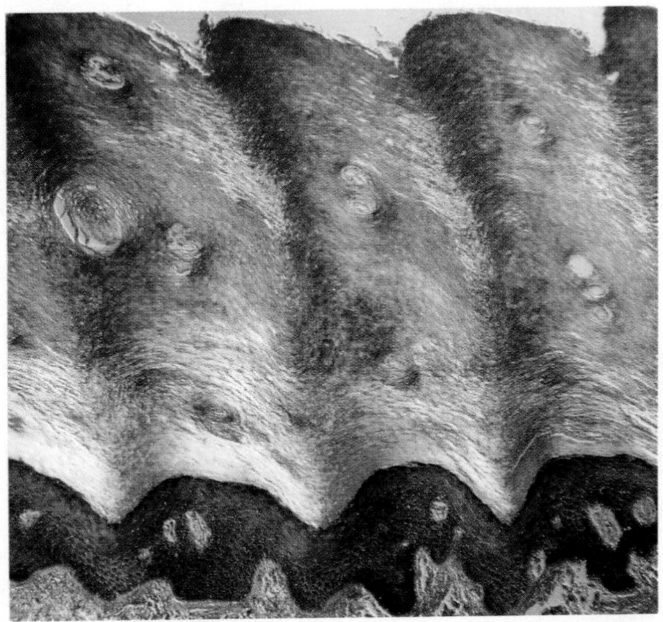

Parts I and II of this book focused primarily on the molecular and cellular levels of life. Our attention now will turn to the structure and function of the whole organism, concentrating on complex multicellular plants and animals. In this chapter we shall look at how the cells of multicellular organisms form the various tissues. In the remaining chapters of Part III we shall focus on how plants and animals obtain necessary nutrients and oxygen, transport materials from one part of the body to another, eliminate waste products, carry on reproduction, and coordinate all the activities of life.

Surface-to-Volume Ratio Limits Cell Size

Though single-celled organisms contribute roughly half of the total weight of all living things on earth, there are enormous benefits to being multicellular. Single cells cannot exploit the great advantages larger size gives an organism: the ability to capture or harvest smaller organisms efficiently, to move farther and faster, and so on. But large size cannot be achieved by simply increasing the size of a single-celled organism indefinitely. A cell must take in its nutrients and oxygen across the membrane. As a cell triples in volume, so does its need for nutrients and oxygen. Yet its membrane surface area does not even double. Since metabolic needs increase faster than the surface area of the membrane, a point arrives at which the membrane can no longer support its contents. The need for efficient diffusion, therefore, puts a strict limit on the surface-to-volume ratio of a cell, and consequently limits cell size.

Many single-celled organisms are extraordinarily complex, because one cell must do everything needed for survival. But in an assembly of cells, specialization is possible, and though simple aggregations of identical cells do exist in nature, the course of evolution has demonstrated that arrangements in which certain cells concentrate on particular functions (movement, feeding, reproduction, and so on) can be far more effective than those in which each cell pursues a "jack-of-all-trades" strategy.

Biological Organization Is Hierarchical: Each Level Builds on the Levels Below It

The bodies of most multicellular organisms are organized on the basis of tissues, organs, and systems. A ***tissue*** is composed of many cells that are usually similar in both structure and function and are bound together by intercellular material. An

Chapter opening photo: Section of skin from the palm of a human. The skin from the palm is exceptionally thick consisting of layers upon layers of epithelial tissue (pink, white regions) that serve a protective function.

organ, in turn, is composed of various tissues (not necessarily similar) grouped together to form a structural and functional unit. Similarly, a ***system*** is a group of interacting organs that "cooperate" as a functional complex in the life of the organism.

The following sections will introduce you to some of the basic plant and animal tissues, organs, and systems. We shall refer to them repeatedly and examine them in more detail in later chapters.

PLANT ORGANS AND ORGAN SYSTEMS

The body of the vascular land plants is customarily divided into two major parts: the ***root*** and the ***shoot*** (Fig. 15.1). These two fundamental organ systems are distinguished on the basis of numerous structural characteristics. Especially important is the arrangement of the vascular (transport) tissues, the formation of lateral roots versus lateral branches in stems, and the presence of leaves on the shoot but not on the root. Yet many of the tissues are essentially continuous throughout the entire plant. In particular, the vascular tissue of root and shoot, despite a somewhat different arrangement in each, forms an uninterrupted transport system.

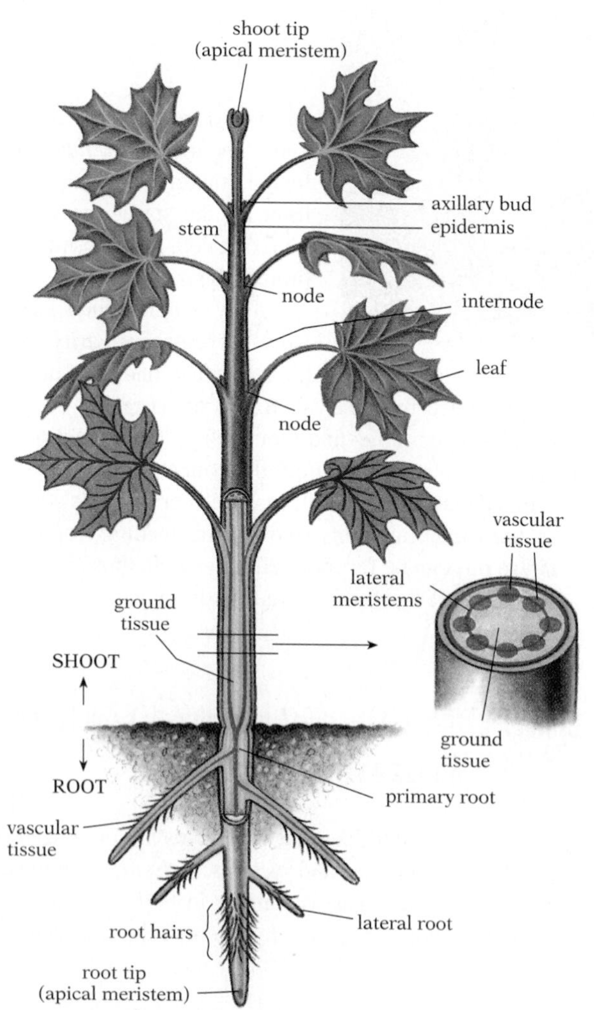

15.1 Diagram of a vascular plant body The vascular tissue is continuous through all parts of the plant. The shoot tip and root tip contain the apical meristems. The vascular tissue and lateral meristems are shown in the inset. Most of the plant body is composed of ground tissue (grey).

Plant roots take up inorganic nutrients such as minerals and water. They function to transport and store nutrients and to anchor the plant to the ground.

The shoot is structurally more complex. It consists of the stem and its appendages, particularly the leaves and the reproductive organs. The stem provides support for the leaves and connects the leaves with the roots. The leaves are organs in which the critical process of photosynthesis takes place.

You will notice that the number of distinct organs mentioned here—root, stem, leaf, and reproductive organs—is much smaller than it would be for most animals. In general, the plant body is not as clearly subdivided into readily distinguishable functional components, or organs, as the animal body. Plant parts grade into each other. We now turn to the various tissues that make up the plant body.

PLANT TISSUES

Tissues of land plants have been classified in a variety of ways by different botanists. The system used here is not necessarily better than alternate systems; it is simply one of several acceptable possibilities. The lack of full agreement on classification is due to the characteristics of the plant cells. There are many intermediate forms between the different cell types, and a given cell may even change from one type to another during the course of its life. Consequently, the various tissues formed from such cells may share structural and functional characteristics. Furthermore, plant tissues may be ***simple,*** that is, contain cells of only one type, or they may be ***complex,*** containing more than one type of cell. In short, plant tissues cannot be fully characterized or distinguished on the basis of any single criterion such as structure, function, location, or mode of origin.

Plant Tissues Can Be Categorized as Meristematic or Permanent

Meristematic tissues are composed of immature cells that actively divide to produce new cells. There are meristematic tissues wherever extensive growth occurs, e.g., at the growing tips of roots and stems, in the bark of trees, and between the wood and bark of trees. The new cells produced by the meristems enlarge and become specialized (differentiated) into mature ***permanent tissues,*** which normally retain their structural and functional characteristics for the remainder of their lives, and do not divide again. However, the distinction between meristematic and permanent tissue is not absolute. Some permanent tissues do revert to meristematic activity under certain conditions.

The permanent tissues fall into three subcategories: surface tissues, ground or fundamental tissues, and vascular tissues. Each of these, in turn, contains several different tissue types. The system of classification used here can be summarized as follows:

I. Meristematic tissue
   A. Apical meristems
   B. Lateral meristems
II. Permanent tissue
   A. Surface tissue
      1. Epidermis
      2. Periderm
   B. Ground (fundamental) tissue
      1. Parenchyma
      2. Collenchyma
      3. Sclerenchyma
   C. Vascular tissue
      1. Xylem
      2. Phloem

It must be emphasized that this classification is based on the more complex land plants—those containing xylem and phloem, the so-called vascular plants. It has little relevance for other plants, where multiple tissue types seldom occur. The characteristics of the various types of plant tissues are summarized in Table 15.1.

Meristematic Tissue Is Composed of Cells Capable of Dividing and Is Restricted to Certain Regions of the Plant Body

Meristematic tissues are composed of embryonic cells capable of dividing. Cell division occurs throughout the very young embryo, but as the plant develops, many regions become specialized for other functions and cease producing new cells. Consequently, cell division becomes largely restricted to certain regions; these are called the meristems.

As long as the plant is living, there are always regions of meristematic tissue at the growing tips of roots and shoots. These ***apical meristems*** produce cells responsible for the increase in length of the plant body. The tissues produced by the apical meristems are referred to as ***primary tissues.*** In many plants, there are also meristematic areas around the periphery of the roots and stems, located between the wood and the bark, and in the bark itself. These ***lateral meristems*** are responsible for growth in diameter, and give rise to ***secondary tissues*** (Fig. 15.1).

Surface Tissue Protects the Plant Body

As the term implies, surface tissues form the protective outer covering of the plant body. In young plants and herbaceous adult plants, the principal surface tissue of the roots, stems, and leaves is the ***epidermis*** (Figs. 15.1 and 15.2); in older woody stems another surface tissue, the ***periderm,*** forms the outer covering.

Most epidermal cells are relatively flat. Often their outer walls are thicker than the other walls. Epidermal cells on the aerial parts of the plant often secrete a waxy, water-resistant ***cuticle*** on their outer surface; this, combined with the thick outer wall, aids in protection against loss of water, mechanical injury, and invasion by parasitic fungi (Fig. 15.3). What makes

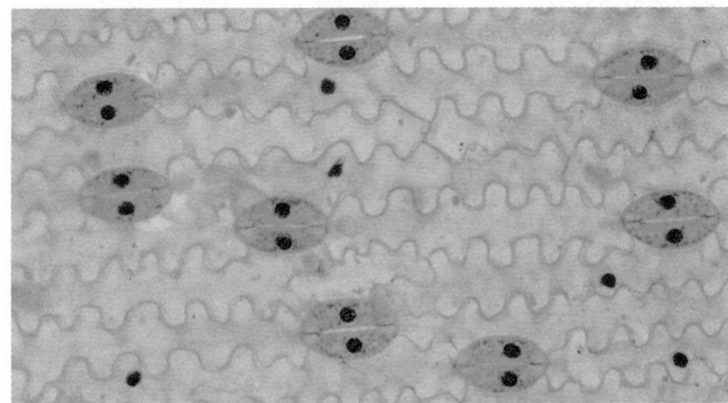

15.2 Surface view of the epidermis of a lily leaf Most epidermal cells are relatively flat. The cells here are irregularly shaped and fit together like pieces of a puzzle leaving no spaces between the cells. Such tight connections help protect against loss of water, mechanical injury, and invasion by parasitic fungi. Some epidermal cells are modified to form guard cells (dark green). The guard cells, which contain chloroplasts, regulate the size of the small openings, stomata, that allow gases to move in and out of the leaf.

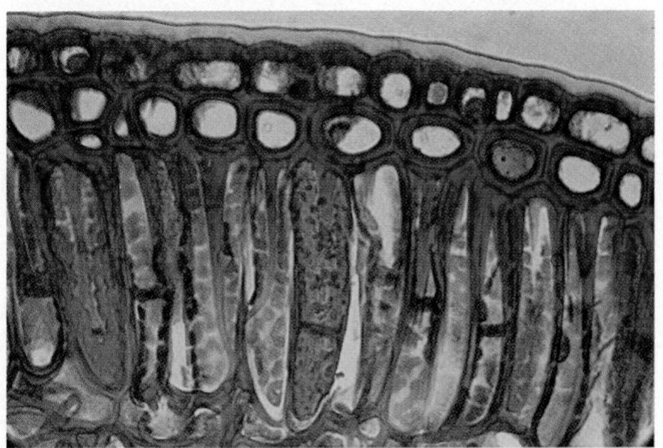

15.3 Epidermis and cuticle of a leaf The epidermis on this leaf is two cell layers thick and serves to protect the interior cells from excessive water loss and mechanical injury. The epidermal cells secrete a waxy, water-resistant cuticle (dark pink) over the surface of the leaves that minimizes water loss. The cuticle is unusually thick on this plant.

TABLE 15.1 *Plant tissues*

Type of tissue	*Distinguishing characteristics*	*Function*
Meristematic tissue	Embryonic cells capable of cell division; small, thin walled cells	Produce new cells
Apical 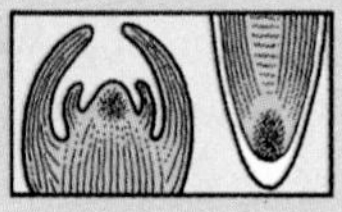	Located at tips of shoots and roots	Increase in length of plant body; produce primary tissues
Lateral 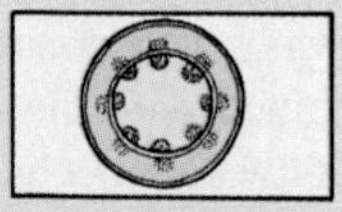	Located near periphery of roots and stems. Examples: vascular and cork cambia	Increase in girth of plant body; produce secondary tissues
Permanent tissue	Composed of more mature, differentiated cells	Protection of underlying tissues
Surface tissue	Cover the outer surfaces of plant body	Protection in young plants
Epidermis 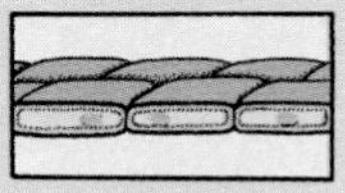	Flat cells, often with thicker outer walls; aerial parts often covered with waxy cuticle	
Periderm 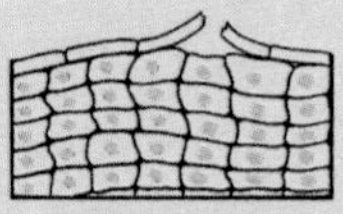	Waterproofed cells with thick cell walls that are dead at maturity	Forms outer bark in trees
Ground tissue	Simple tissues composed of a single type of cell	
Parenchyma 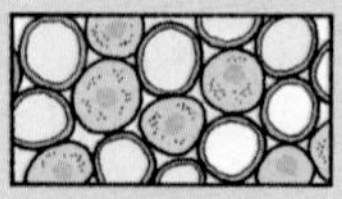	Unspecialized cells with thin primary walls	Photosynthesis, secretion, storage
Collenchyma 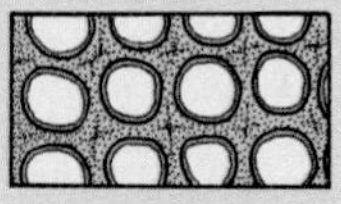	Elongated cells with unevenly thickened primary walls. Living at maturity	Support in young leaves and stems
Sclerenchyma 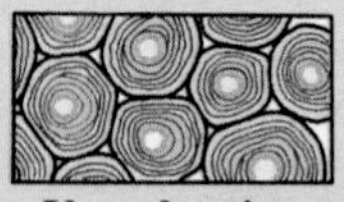	Elongated cells with very thick secondary walls. Dead at maturity	Support
Vascular tissue	Elongated cells, specialized for conduction	Transport of material throughout the plant body
Xylem 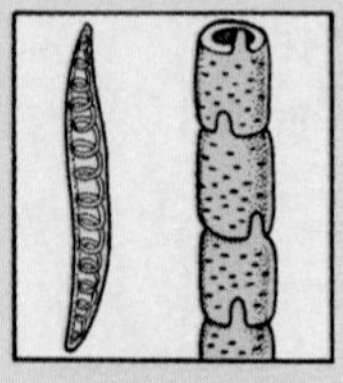	Conductive cells dead at maturity; thick cell walls arranged end to end to form empty passages	Transport of water and minerals upward from roots to leaves, support
Phloem 	Conductive cells living at maturity; sieve cells elongated, arranged end to end for conduction	Transport of organic materials up and down the plant body

the barrier even more complete is the fact that the irregularly shaped cells usually interlock tightly, like pieces of a puzzle, leaving no intercellular spaces between the cells. Often the epidermis is only one cell thick, though it may be thicker, as in some plants living in very dry habitats, where protection against water loss is critical. Some epidermal cells are modified to form hairlike structures that may protect the plant against insects.

Some epidermal cells, particularly of the leaves, are specialized as ***guard cells*** and regulate the size of the stomata, the small holes in the epidermis through which gases move into or out of the leaf (Fig. 15.4). Epidermal cells of the roots, which have no cuticle and function in water absorption, commonly bear long hairlike processes called ***root hairs*** that greatly increase the total absorptive surface area (see Fig. 16.11, p. 333).

As the stems and roots of plants with active lateral meristems increase in diameter, the epidermis is slowly replaced by the periderm (Fig. 15.5). This tissue makes up the corky outer bark so characteristic of old trees. Functional cork cells are dead; it is their waterproofed cell walls that function as the protective outer covering of the plant.

Ground Tissue Is Widespread Throughout the Plant Body

Most of the ground (fundamental) tissues are simple tissues, i.e., each is usually composed of a single type of cell. Often these same types of cells are also found in complex tissues such as xylem or phloem. Ground tissues consist of three basic tissue types: parenchyma, collenchyma, and sclerenchyma.

Parenchyma tissue occurs in all parts of the plant: flowers, fruits, roots, stems, and leaves. Parenchyma cells are produced by both the apical and lateral meristems, so the tissue may be primary or secondary tissue, depending on its source. These cells are relatively unspecialized cells, like those that make up almost the whole body of the nonvascular plants and algae. They have not lost the capacity for cell division, and in some circumstances they take on meristematic activity and begin to divide. They also sometimes undergo further specialization, forming other cell types. Parenchyma cells are living at maturity and usually have thin primary walls and no secondary walls. The cells are ordinarily loosely packed with abundant intercellular spaces (Fig. 15.6). Most of the chloroplasts of

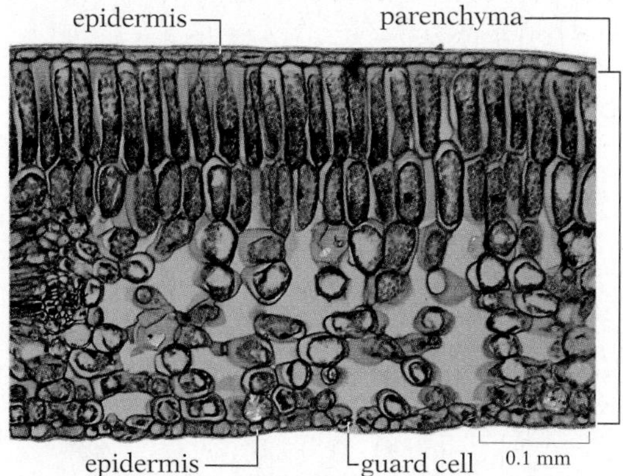

15.4 Photograph of cross section of ivy leaf Guard cells regulate the flow of gases into and out of the leaf.

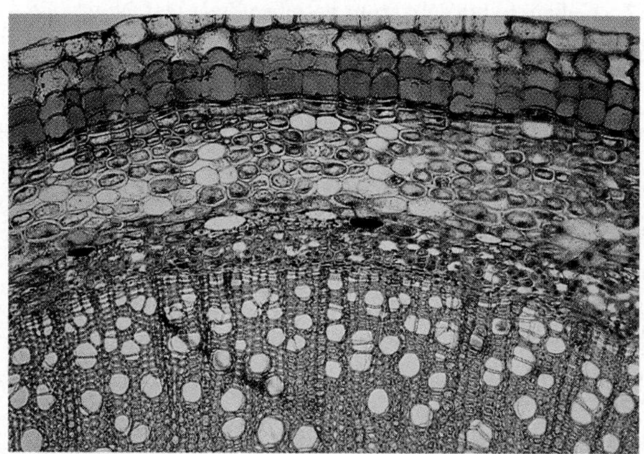

15.5 Cork formation on an elderberry stem As the stem grows in diameter water-proofed cork cells (top, brown) are produced under the epidermis (outer surface). Eventually the epidermis will slough off leaving the corky periderm as the protective covering on the outer surface.

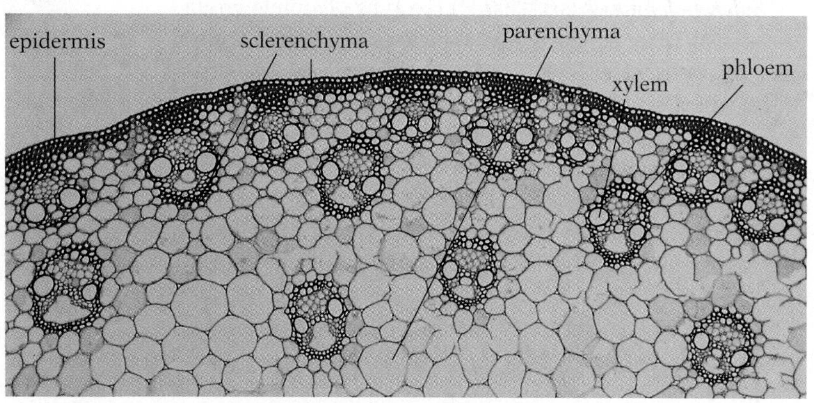

15.6 Photograph of portion of cross section of corn stem showing various plant tissues

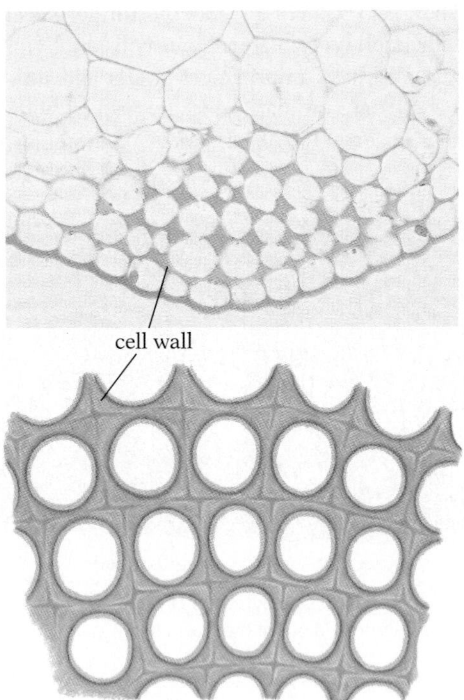

15.7 **Collenchyma cells from petiole of beet leaf** Notice the particularly thick walls at the corners of the cells.

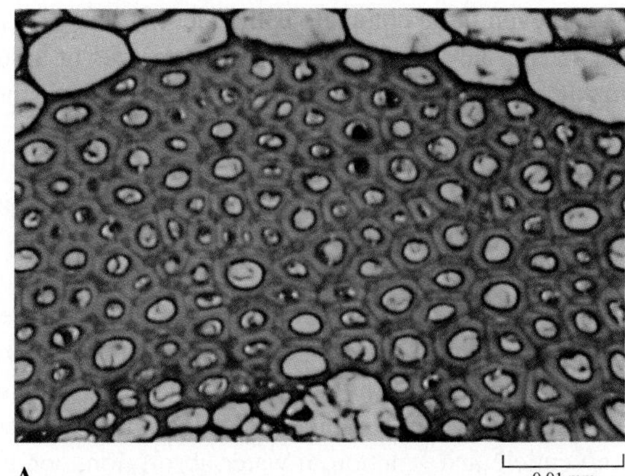

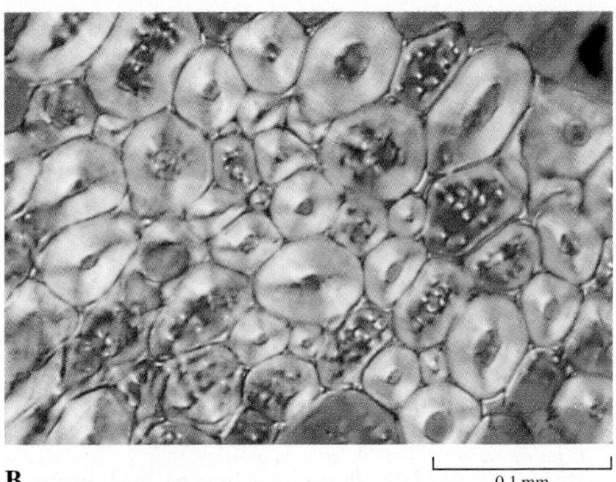

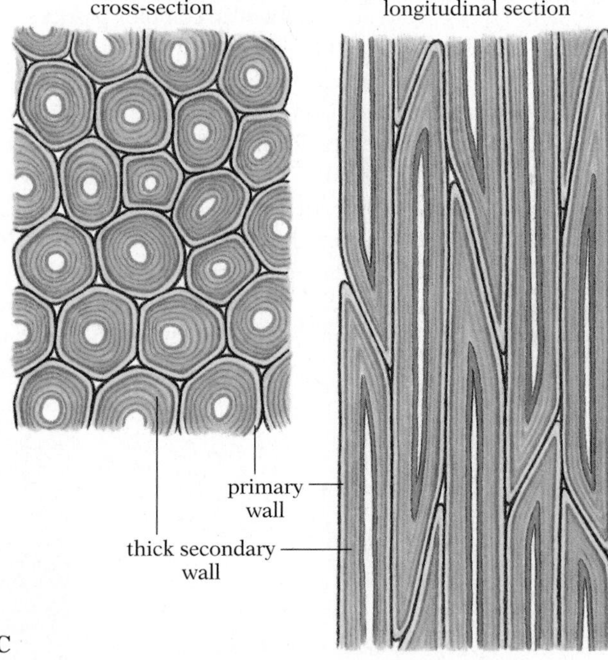

15.8 Sclerenchyma (A) Cross section of fibers (red) from corn stem. (B) Stone cells (sclereids) of pear fruit. (C) Cross section and longitudinal section of sclerenchyma tissue. Note the very thick cell walls. Sclerenchyma cells lose their cytoplasm and nuclei at maturity.

leaves are in the cells of parenchyma tissue, and it is largely here that photosynthesis occurs. Parenchyma cells of stems and roots function in storage of nutrients and water.

Collenchyma (*coll-,* "glue") is a simple primary tissue whose cells function as an important supporting tissue in young plants and leaves. The "strings" on the outer surface of celery, for instance, are composed of collenchyma. Like parenchyma cells, collenchyma cells remain alive during most of their functional existence. They are structurally similar to parenchyma cells, except that they are more elongate, and their primary walls are unevenly thickened. The thickened areas are usually most prominent at the "corners" where the cells meet (Fig. 15.7); it is this feature that makes this tissue function in support.

Sclerenchyma (*scler-,* "hard") is a type of simple ground tissue that, like collenchyma, functions in support. Sclerenchyma cells are characterized by their very thick secondary walls, which give strength to the plant body. Often these walls are so thick that the lumen (internal space) of the cell is nearly obliterated. Unlike collenchyma and parenchyma, the cells are usually dead at functional maturity. Sclerenchyma cells are produced by both the apical or lateral meristems, so may be either primary or secondary tissues, though the latter predominate.

Sclerenchyma cells are customarily divided into two categories: fibers and sclereids (Fig. 15.8). ***Fibers*** are long, thick-walled cells with tapered ends. They are tough and strong, but flexible; commercial flax and hemp are derived from strands of sclerenchyma fibers. ***Sclereids*** are short cells, irregular in shape. The simpler sclereids are frequently called stone cells; they are common in nutshells and the hard parts of seeds, and are scattered in the flesh of hard fruits (the gritty texture of pears is due to small clusters of stone cells; Fig. 15.8B).

Vascular Tissue Is Specialized for the Transport of Materials

Vascular, or conductive, tissue is a distinctive feature of the vascular plants (Figs. 15.6 and 15.9), making possible their extensive exploitation of the terrestrial environment. Vascular tissue has cells that function as tubes or ducts through which water and numerous substances in solution move from one part of the plant to another. There are two principal types of vascular tissue: xylem and phloem. Both of these are produced by the apical and lateral meristems and therefore may be primary or secondary tissues, depending on their origin. These are complex tissues, i.e., they consist of more than one kind of cell.

Xylem is a vascular tissue that functions in the transport of water and minerals *upward* in the plant body. It forms a continuous pathway running through the roots, the stem, and leaves. In flowering plants there are two types of conductive cells unique to xylem, ***tracheids*** and ***vessel elements*** (Fig. 15.9A). Tracheids and vessel elements are laid end to end to form long tubes in which upward movement of materials takes place. Both the cytoplasm and the nuclei of these cells disintegrate at maturity. Only the thick, empty cell walls remain as functional structures. Xylem tissue also includes numerous parenchyma and sclerenchyma cells. The parenchyma cells are the only living cells in mature functioning xylem.

Xylem is also very important in support, particularly of the aerial parts of the plant. The numerous xylem fibers function almost exclusively in this way. The thick-walled conductive cells also provide support as well as function in transport. Xylem is characteristically very strong. Keep in mind that its common name is wood.

The second vascular tissue, ***phloem,*** is unlike xylem in that materials can move both *up and down* in it. Phloem functions particularly in the transport of organic materials such as the carbohydrates synthesized during photosynthesis and amino acids. Like xylem, phloem is a complex tissue and contains both parenchyma and sclerenchyma cells; in addition it contains sieve elements and companion cells, found only in phloem (see Fig. 15.9B). The ***sieve elements*** are the conductive cells of the phloem, and, unlike tracheids and vessel elements, they are living cells at functional maturity, conducting materials up and down the stem.

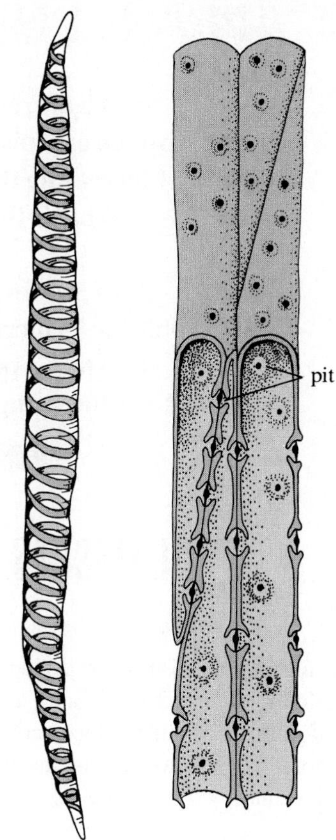

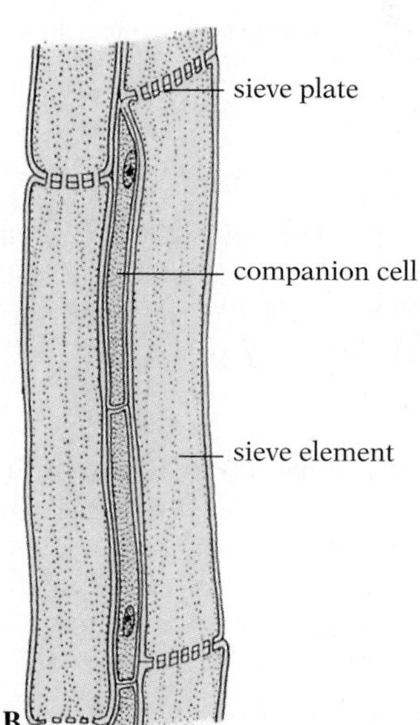

15.9 Vascular tissue (A) The conductive cells in xylem are tracheids and vessels that are laid end to end to form a continuous pathway from the roots to the leaves. Xylem conducts materials upward, from the roots to the leaves. (B) The conductive cells in phloem are sieve elements. Phloem conducts materials up and down the stem.

MAKING CONNECTIONS

1. What are the advantages of large size to an organism? Why must growth of an organism be achieved by increasing the number of its cells rather than by increasing their size?

2. What is the advantage of cell specialization to a multicellular organism?

3. In a growing pine tree, which type of meristem produces cells responsible for its increase in length? For the increase in diameter of its trunk?

4. What type of plant tissue is responsible for the gritty texture of pears? For the stringiness of celery stalks?

ANIMAL TISSUES

Traditionally, all animal tissues are divided into four categories: epithelium, connective, muscle, and nerve (Table 15.2). Each of these, particularly the second, has numerous subtypes. It should be emphasized that the subtype classification is based primarily on vertebrate animals, especially human beings.

The classification of animal tissues used here can be summarized as follows:

I. Epithelium
   A. Simple epithelium
      1. Squamous
      2. Cuboidal
      3. Columnar

TABLE 15.2 *Animal tissues*

Type of tissue	*Distinguishing characteristics*	*Function*
Epithelium 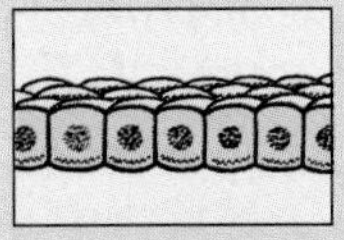	Tightly joined cells at all body surfaces, few spaces between cells	Forms a tight barrier at all body surfaces, protecting underlying cells; often has secretory or absorptive function
Connective 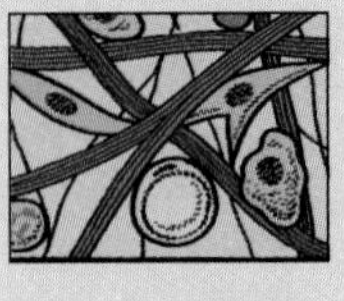	Varying amounts of intercellular matrix in which cells and fibers are embedded. Matrix can be liquid, semisolid, or solid. The matrix determines the physical characteristics of the tissue	Connects and supports the other tissues of the body
Muscle 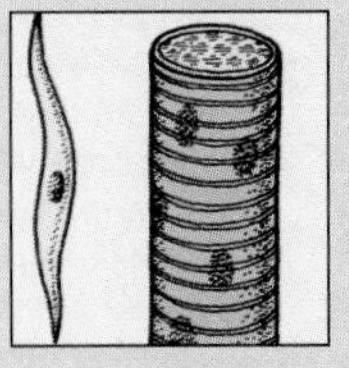	Elongated cells, specialized for contraction. Three types: striated, smooth, and cardiac	Striated muscle is responsible for voluntary movement; smooth muscle produces movements of internal organs; cardiac muscle forms the heart
Nerve 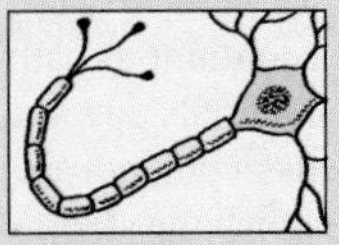	Cells with many elongate processes, specialized to respond to stimuli and conduct messages	Cells receive nerve impulses and transmit impulses to other cells; conduct messages over long distances

- B. Stratified epithelium
  1. Stratified squamous
  2. Stratified cuboidal
  3. Stratified columnar
- II. Connective tissue
  - A. Vascular tissue
    1. Blood
    2. Lymph
  - B. Connective tissue proper
    1. Loose connective tissue
    2. Dense connective tissue
  - C. Cartilage
  - D. Bone
  - E. Adipose
- III. Muscle
  - A. Skeletal muscle
  - B. Smooth muscle
  - C. Cardiac muscle
- IV. Nerve

Epithelium Covers or Lines All Body Surfaces

Epithelial tissue covers or lines all body surfaces, both external and internal, and serves a protective function. The epithelial cells are packed tightly together and therefore provide a continuous barrier protecting the underlying cells from the external world. Anything entering or leaving the body must cross at least one layer of epithelium. Thus, the permeability characteristics of the various epithelial cells are exceedingly important in regulating the exchange of materials between different parts of the body and between the body and the external environment. Epithelial tissues are found in the outer portion of the skin, and in the linings of the digestive tract, the lungs, the blood vessels, the various ducts, the body cavity, and so on.

In order for the epithelial cells to form a tight barrier at body surfaces, the cells must tightly join one another. Three types of specialized cellular junctions occur: spot desmosomes, tight junctions, and gap junctions (Fig. 15.10). ***Spot desmosomes*** are buttonlike junctions between cells that act something like rivets or spot welds. In ***tight junctions,*** specific membrane proteins attach directly to their counterparts in the adjacent cell. The cells are so tightly drawn together that there is no intercellular space. A ***gap junction*** forms when protein channels in the membranes of two cells line up and bind to one another to create a pathway between cells. Gap junctions permit direct movement of ions and some small molecules from cell to cell and thus facilitate rapid communication between cells.

Epithelial cells show significant differences between the surface exposed to air or fluid and the opposite surface resting upon other cell layers. Because the epithelium plays a crucial part in the directional passage of materials, the existence of differences between the "free ends" exposed to air or fluid and the attached ends that touch other cells is understandable. Often highly specialized, the free ends commonly bear cilia, hairs, or short fingerlike processes (Fig. 15.11).

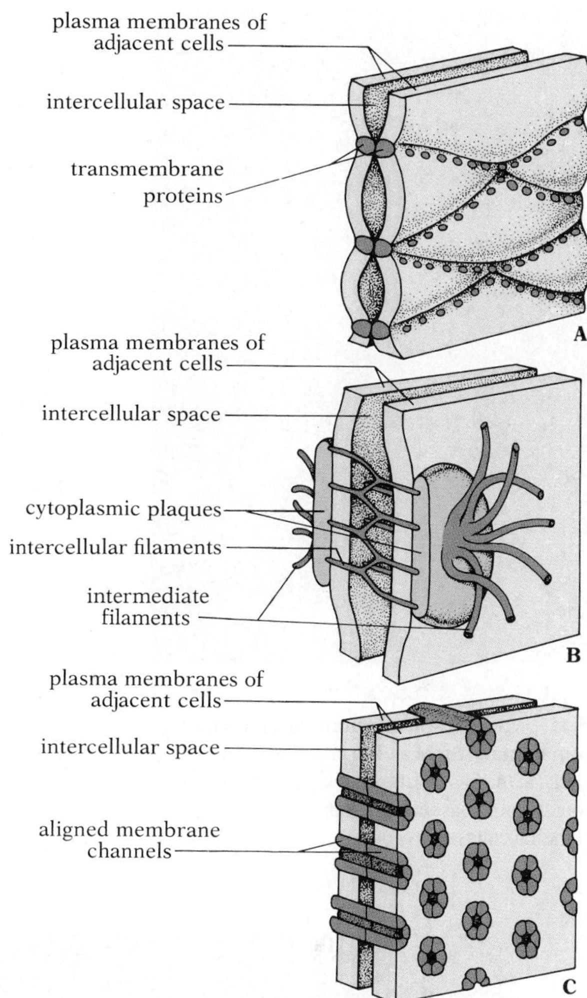

15.10 Varieties of cellular junctions The cells lining the mammalian small intestine are attached by several types of specialized junctions. An example of each is shown in enlarged detail. (A) A tight junction is composed of rows of transmembrane proteins in adjacent cells that bind to each other. (B) A spot desmosome consists of a pair of cytoplasmic plaques, each just inside the cell membrane and connected to the other across the intermembrane space by specialized intercellular filaments. Each plaque is also attached to fibers of the cytoskeleton within its cell. (C) A gap junction is formed by a pair of membrane channels aligned and bound together to create a specialized pathway between cells.

Epithelial cells are usually placed into three categories on the basis of cell types: squamous, cuboidal, and columnar (Figs. 15.12 and 15.13). ***Squamous cells*** are much broader than they are thick and have the appearance of thin, flat plates. ***Cuboidal cells*** are roughly as thick as they are wide and, as their name implies, have a rather square or cuboidal shape. ***Columnar cells*** are much taller than they are wide and in vertical section look like rectangles set on end.

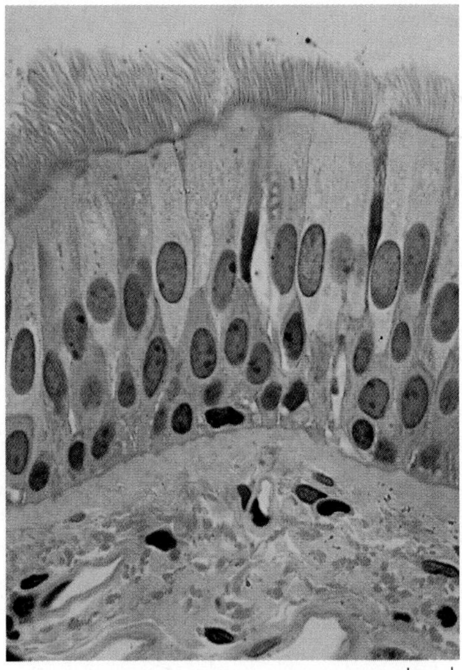

15.11 Micrograph of columnar epithelium from human trachea The cilia and the dark basal bodies at the base of the cilia are plainly visible. Note the pink basement membrane on which the epithelial cells rest.

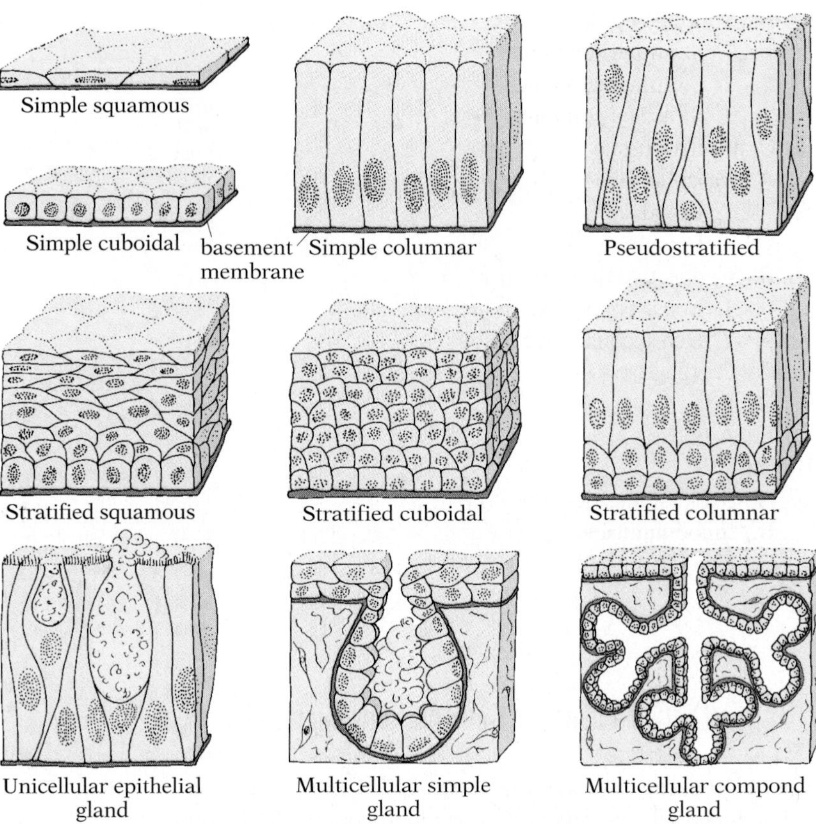

15.12 Epithelial tissues The extracellular basement membrane is shown in brown.

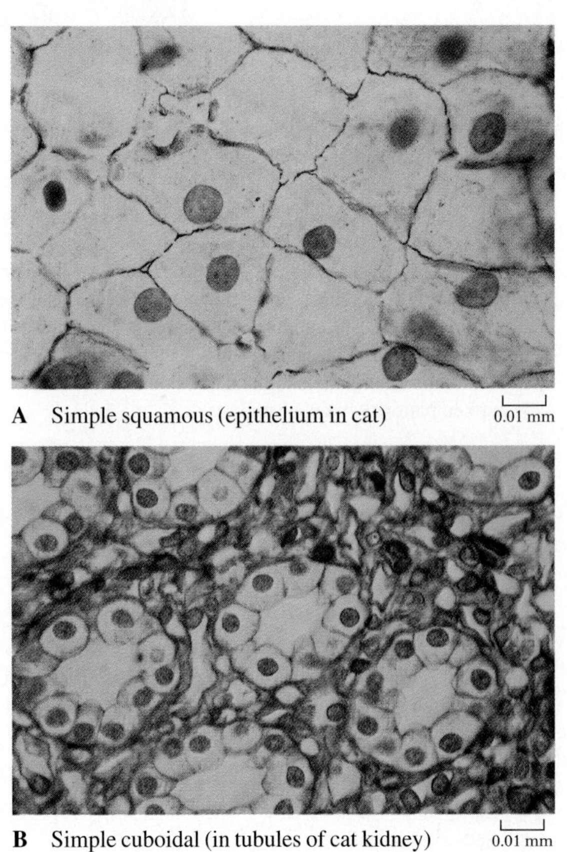

A Simple squamous (epithelium in cat)

B Simple cuboidal (in tubules of cat kidney)

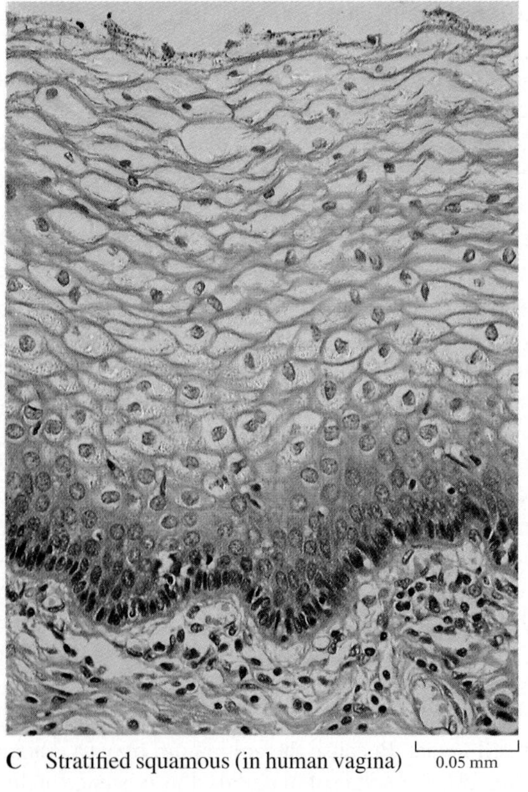

C Stratified squamous (in human vagina)

15.13 Photographs of epithelial tissues

Epithelial tissue may be only one cell thick, in which case it is called ***simple epithelium,*** or it may be two or more cells thick and is then known as ***stratified epithelium*** (Figs. 15.12 and 15.13). Different epithelia are named on the basis of cell type and number of cell layers. The cell type of the outermost layer of stratified epithelium determines whether it will be called squamous, cuboidal, or columnar. Regardless of type, epithelium is usually separated from the underlying tissue by an extracellular ***basement membrane*** containing the fibrous protein collagen (Figs. 15.12 and 15.14).

Epithelial cells often become specialized as gland cells, secreting substances at the epithelial surface (Fig. 15.15). Sometimes a portion of the epithelial tissue folds inward, and a multicellular gland is formed.

Connective Tissue Has Cells and Fibers Embedded in a Matrix

Connective tissue is the most widespread and abundant tissue in the body. Its primary function is to provide structural and metabolic support for the body. Bone and cartilage, for instance, provide rigid skeletal support. And, as the name suggests, some connective tissue binds together groups of cells. However, connective tissue carries on other functions as well. Certain types of connective tissue play an important metabolic role in the storage of fat in the body. Blood, another connective tissue, provides metabolic support by delivering nutrients and oxygen to cells and removing waste products. The blood and

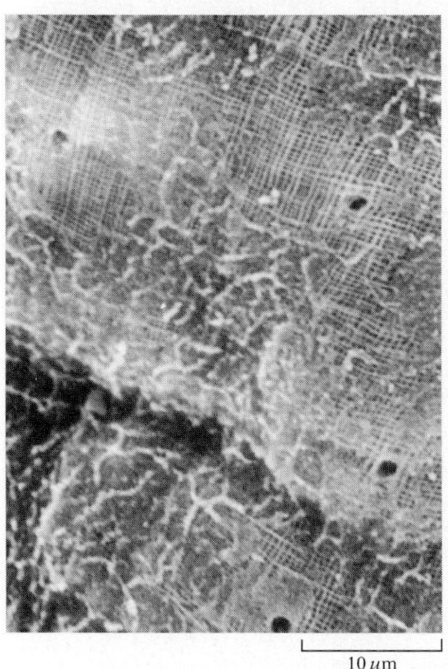

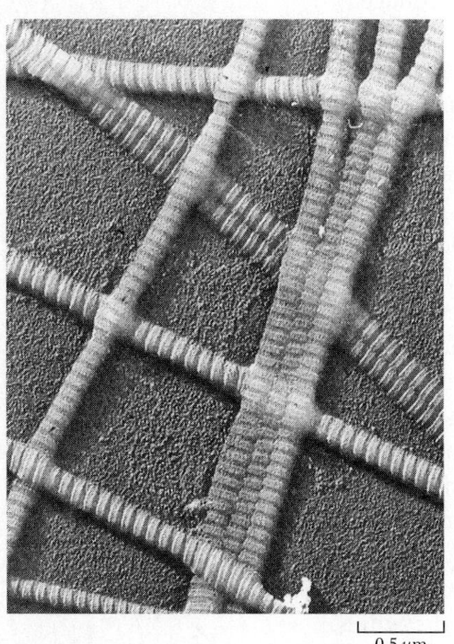

15.14 Electron micrographs of collagen Left: Scanning EM of the network of collagen fibers in the skin of an earthworm. Right: Higher magnification transmission EM of collagen fibrils from calf skin.

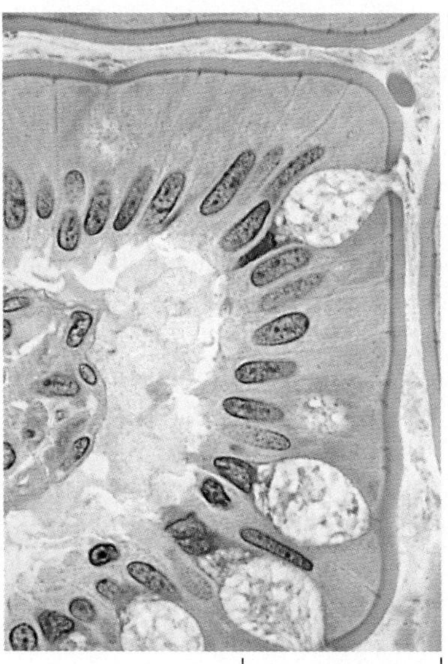

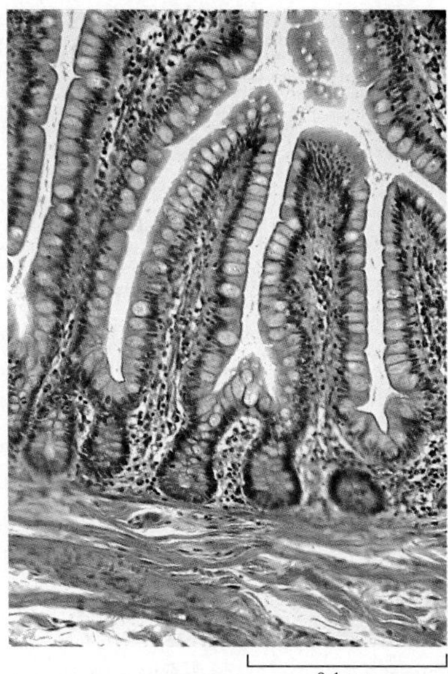

15.15 Photographs of gland tissues Left: Simple columnar, with unicellular glands (in human small intestine). Right: Multicellular compound glands (in villi of small intestine).

lymph also act to defend the body against disease-causing organisms. Finally, some connective tissue serves as a "filler tissue," occupying spaces in the body where there are no specialized cells. It has been said that if all other tissues were destroyed, the connective tissue alone would show the exact contours of the body and the detailed shape of most internal organs.

Microscopically, connective tissue is characterized by cells and protein fibers embedded in large amounts of intercellular (*inter-*, "between") material called ***matrix.*** Much of the total volume of connective tissue is matrix, the cells themselves often being widely separated. The matrix can be liquid, semisolid, or solid. Note that the fibers in connective tissue are quite different from the fibers in plant tissues. Connective tissue fibers are proteinaceous (usually composed of collagen), whereas fibers in plants are cellular.

Connective tissue is often divided into five main types: (1) connective tissue proper; (2) cartilage; (3) bone; and (4) adipose or fat tissue; and (5) blood and lymph, or vascular tissue. The first three are sometimes collectively described as supporting tissues.

Connective tissue proper can appear in many forms, but its intercellular matrix always contains numerous cells and protein fibers (Fig. 15.16), as well as a mixture of water, proteins, carbohydrates, and lipids. There are generally considered to be two kinds of connective tissue proper—loose connective tissue and dense connective tissue—though the distinction is often vague, and intermediate types sometimes occur.

Loose connective tissue (Fig. 15.16) is practically everywhere in the animal body. It supports, surrounds, and connects the elements of all other tissues. For example, it binds muscle fibers together, attaches the skin to underlying tissues, forms the membranes that line the chest and abdominal cavities, functions as packing material in the spaces between organs, and forms a thin sheath around blood vessels. Because of its flexibility, loose connective tissue allows movement between the units it binds or connects.

Dense connective tissue is characterized by the compact arrangement of its many collagen fibers, the decreased amount of matrix, and the relatively small number of cells (Fig. 15.17). The fibers may be irregularly arranged into an interlacing network, as in the lower portion of the skin, or they may occur in a definite pattern—usually parallel bundles oriented to withstand tension from one direction, as in tendons connecting muscle to bone or ligaments connecting bone to bone (Fig. 15.17A).

Cartilage (gristle) is a specialized form of connective tissue, whose intercellular matrix is composed of a protein-carbohydrate complex that has a rubbery consistency. There are relatively few cells, which are located in cavities in the matrix (Fig. 15.17B). Cartilage can support great weight; yet it is often flexible and somewhat elastic.

Cartilage is found in the human body in such places as the nose and ears (where it forms pliable supports), the larynx and trachea (you can feel the rings of cartilage in the front of your neck), the disks between the vertebrae, the surfaces of skeletal joints, and the ends of ribs. Most of the skeleton of the young vertebrate embryo is composed of cartilage; the developing bones follow this model and slowly replace it. Some vertebrate groups, the sharks for example, retain a cartilaginous skeleton even in the adult.

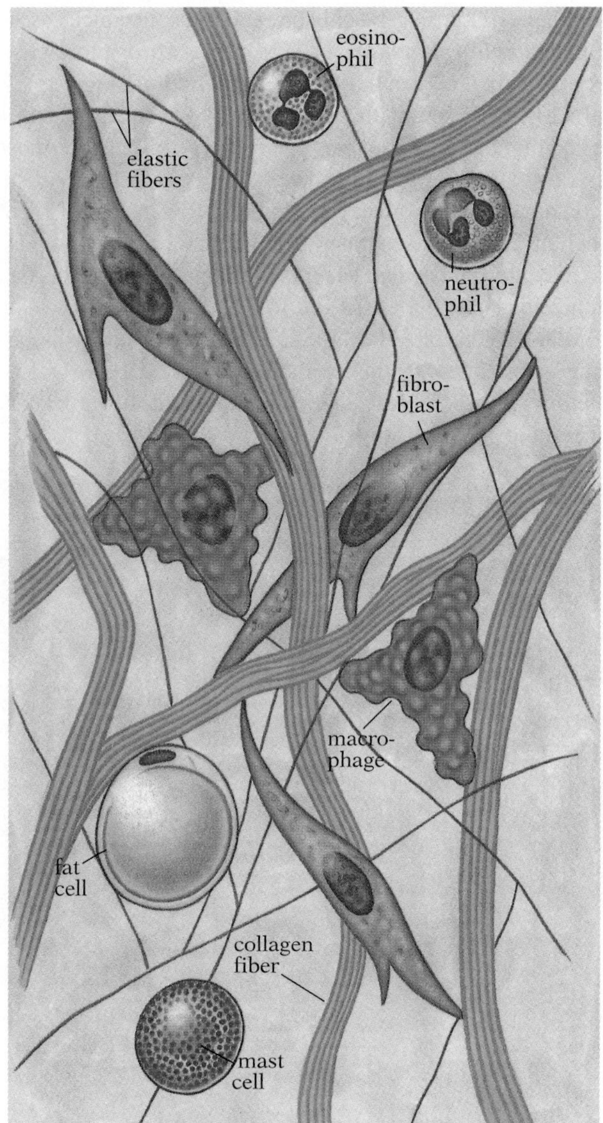

15.16 Loose connective tissue Loose connective tissue is characterized by the loose, irregular arrangement of bundles of collagen fibers and elastic fibers, the large amount of semifluid matrix, and the presence of numerous cells of a variety of types. The matrix in which the cells and fibers lies is composed largely of tissue fluid, a liquid derived from the blood.

Bone is a connective tissue with a hard, relatively rigid matrix. This matrix, which has numerous collagen fibers and a surprising amount of water, is hard because it is impregnated with inorganic salts such as calcium carbonate and calcium phosphate. Bone will be discussed in more detail in Chapter 30.

Adipose, or fat tissue, is a modified type of connective tis-

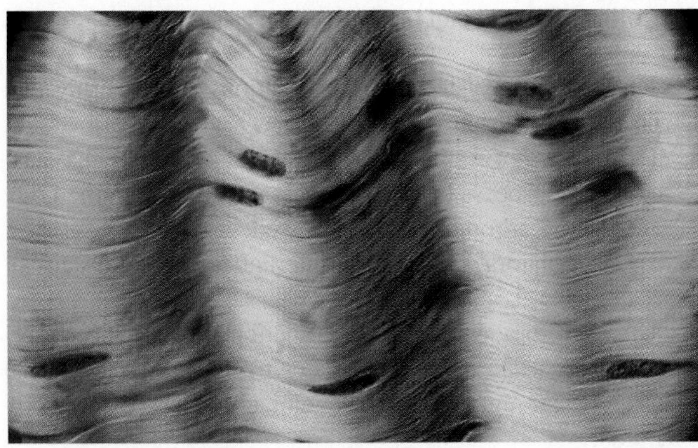

A

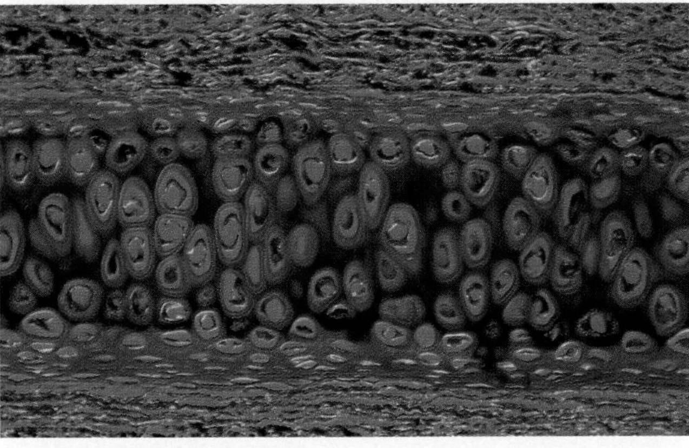

B

15.17 Dense connective tissue and cartilage (A) Scanning electron micrograph of a tendon. Tendons, which connect muscles to bones, and ligaments, which connect bones to other bones, are composed of dense fibrous connective tissue. The fibers are tightly packed together in bundles and are oriented in the same direction. The cells (dark spots) are arranged in rows between the bundles. (B) Hyaline cartilage. Cartilage (in the center) has a firm rubbery matrix (black) containing many tightly packed fibers. Cartilage cells are located in small spaces (blue) scattered throughout the matrix.

sue. Its cells are adapted for the storage of fat (see Fig. 3.13, p. 60). The stored fat droplet occupies a large part of the contents of these cells. Fat tissue is important not only in nutrient storage, but also in protection of other tissue, insulation, and padding.

Blood and ***lymph*** are rather atypical connective tissues with liquid matrixes. They will be discussed in some detail in Chapter 20.

Muscle Tissue Is Specialized for Contraction and Causes Movement

Muscle tissue is responsible for most movement in complex animals. The cells of muscle have greater capacity for contraction than most other cells, although all cytoplasm probably possesses this capacity to some extent. The individual muscle cells are usually elongate and are bound together into sheets or bundles by connective tissue. Three principal types of muscle tissue (see Fig. 30.7, p. 656) are recognized in vertebrates: (1) ***skeletal*** or striated muscle, which is responsible for most voluntary movement; (2) ***smooth*** muscle, which is involved in most involuntary movements of internal organs; and (3) ***cardiac*** muscle, the tissue of which the heart is composed. This classification does not hold for many invertebrates.

Nervous Tissue Is Specialized for the Reception of Stimuli and Conduction of Signals Throughout the Organism

Nervous tissue has the capability to respond to stimuli (changes in the environment). It allows sensing of the outside world and inside the body, and conducts signals to the parts of the body that respond to stimuli. To some extent, all living tissue possesses the property of irritability—the ability to respond to stimuli—but nerve tissue is highly specialized for such response. Nerve cells are easily stimulated and can transmit impulses very rapidly. Each cell comprises a cell body, containing the nucleus, and one or more long, thin extensions called fibers (Fig. 15.18). Nerve cells are thus admirably suited

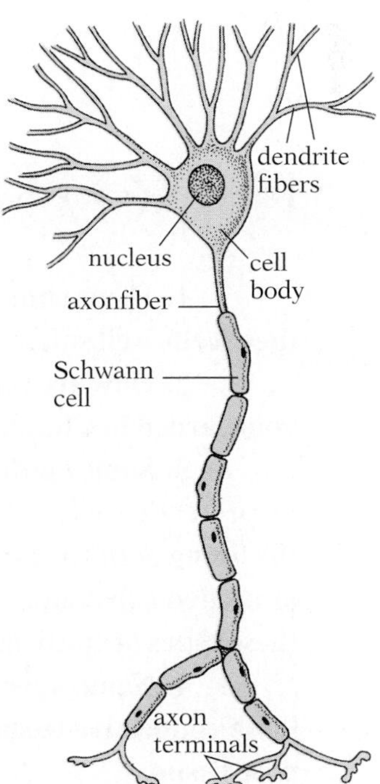

15.18 Nerve cell The dendrites carry impulses toward the cell body; the axon carries impulses away from the cell body. The axons of vertebrates are often sheathed along much of their length by Schwann cells.

to serve as conductors of messages over long distances. An individual nerve cell may be a meter long, or even longer; no other kind of cell even approaches such length. Many nerve fibers bound together by connective tissue make up a nerve.

The functional combination of nerve and muscle tissue is fundamental to all multicellular animals except sponges. These tissues enable animals to move rapidly in response to stimuli.

ANIMAL ORGANS AND ORGAN SYSTEMS

Most large, complex animals have numerous organs that are arranged into various organ systems. These organs and organ systems are more specialized and structurally sophisticated than are those of simpler multicellular animals, which have few distinct organs. Vascular plants, as we have seen, have relatively fewer organs than most animals. This fundamental difference reflects the different modes of life pursued by these two major groups of organisms, particularly their different feeding methods and their different adaptations for active locomotion on the one hand (animals) and a sedentary existence on the other (plants).

The numerous organs that together constitute the body of a complex animal are commonly grouped into organ systems. The following will be of particular concern to us:

1. The digestive system, which functions in procuring and processing nutrients;
2. The respiratory system, which is essential to the gas-exchange process by which oxygen is taken into the body and waste carbon dioxide is released;
3. The circulatory system, the internal transport system of animals;
4. The excretory system, which functions not only in the release of certain metabolic wastes from the body, but also as a critical regulator of the chemical makeup of the body fluids;
5. The endocrine system, whose glands, and the hormones they produce, play an important role in internal control;
6. The nervous system, which coordinates the myriad functions of a complex multicellular animal;
7. The skeletal system, which provides support and determines shape in some animals;
8. The muscular system, which is of fundamental importance in the movement of animals;
9. The reproductive system, which functions in the production of new individuals.

An example of the complex integration of different types of cells and tissues to form an animal organ is human skin (Fig. 15.19). It is clear from the illustration that skin is far more complex than a first impression might convey. Individual organs and organ systems will be discussed in later chapters.

MAKING CONNECTIONS

1. What characteristics of the cell membranes of epithelial cells make these cells well suited to their function?

2. How are the structure and function of gap junctions related to what you learned in Chapter 5 about the permeability of the cell membrane to ions?

3. Some epithelial tissues, such as the lining of the vagina, are primarily protective in function and hence are virtually impermeable. Others, such as the lining of the intestinal tract, must be protective but also permit the passage of needed substances. What difference in structure would you expect between these types of epithelia?

4. Name specific structures in your body where you would find (a) loose connective tissue, (b) dense connective tissue, (c) cartilage, and (d) adipose tissue.

5. Why are most animals structurally more complex than plants?

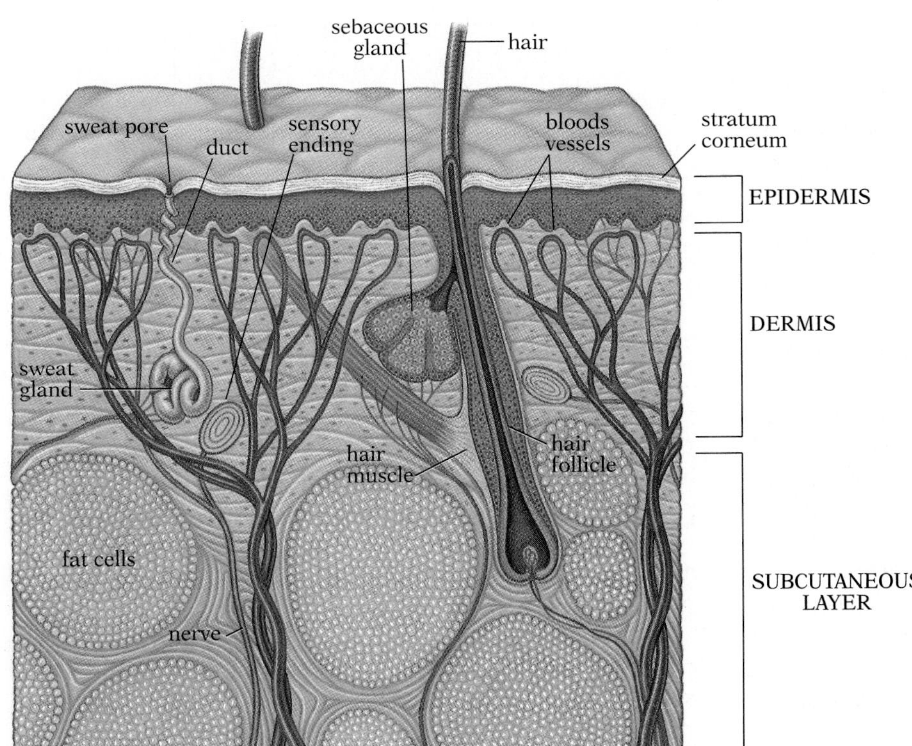

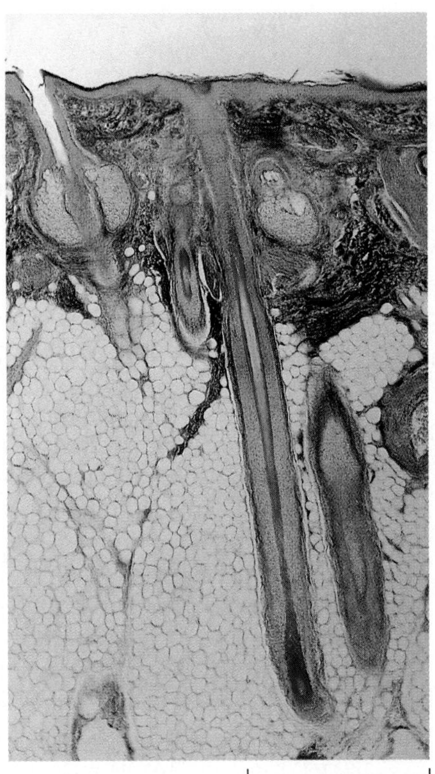

15.19 Human skin in cross section The skin shown in the photograph is from the human scalp. The outer portion of the skin, the epidermis, is composed of stratified squamous epithelial tissue. (Individual epithelial cells are not visible.) Beneath the epidermis is a layer, called the dermis, composed chiefly of connective tissue (stained blue in the photograph). Blood vessels penetrate into the dermis but not into the epidermis. Sweat glands are embedded in the deeper layers of the dermis, and their ducts push outward through both dermis and epidermis to open onto the surface of the skin through sweat pores.

Hairs, and the inner layers of the hair follicles, are also embedded in the deeper layers of the dermis.

Numerous nerves penetrate into the dermis, and a few even penetrate into the epidermis. Among them are nerves to the hair muscles, sweat glands, and blood vessels, and also nerves terminating in the sensory structures for detecting touch, temperature, and pain.

Beneath the dermis, and not sharply delineated from it, is the subcutaneous layer, which is not considered a part of the skin itself. This is a layer of very loose connective tissue, usually with abundant fat cells. It is this layer that binds the skin to the body. The extent and form of its development determine the amount of possible skin movement.

CONCEPTS IN BRIEF

1. The bodies of multicellular organisms are organized on the basis of *tissues* (a group of cells similar in structure and function), *organs* (various tissues grouped together to form a structural and functional unit), and *systems* (a group of interacting organs).

2. All plant tissues can be divided into meristematic tissue (growth tissue—undifferentiated cells capable of dividing) and permanent tissue (mature differentiated cells). Regions of *meristematic tissue* are found at the growing tips of roots and stems *(apical meristems)* and, in many plants, in areas around the periphery of the roots and stems *(lateral meristems)*. Apical meristems produce primary tissues; lateral meristems produce secondary tissues.

3. The permanent tissues fall into three subcategories: surface tissues, ground (fundamental) tissues, and vascular tissues. The surface tissues *(epidermis, periderm)* form the protective outer covering of the plant body.

4. There are three types of ground tissues:

parenchyma (thin-walled, loosely packed cells found throughout the plant body), *collenchyma* (supportive tissue whose cell walls are irregularly thickened), and *sclerenchyma* (supportive tissue with very thick cell walls).

5. The vascular, or conductive, tissue is characteristic of vascular plants and consists of two principal types of complex tissue: xylem and phloem. *Xylem* supports the plant and transports water and dissolved minerals upward. *Phloem* conducts organic materials up and down the plant body.

6. The body of vascular land plants is divided into two major parts: the root and the shoot. The root functions in absorbing water and nutrients while the shoot supports the leaves, which carry out photosynthesis and synthesize their food.

7. Animal tissues are divided into epithelium, connective tissue, muscle, and nerve. *Epithelial tissue* covers or lines the internal and external surfaces of all free body surfaces. Epithelial cells are packed tightly together, with almost no intercellular spaces. Specialized intercellular junctions join the cells to one another. Epithelial tissues provide a continuous barrier protecting the underlying cells from the external medium.

8. *Connective tissue,* composed of cells and fibers embedded in an extensive intercellular matrix, connects, supports, or surrounds other tissues or organs; examples of connective tissue are blood and lymph, connective tissue proper, cartilage, and bone.

9. The three types of *muscle tissue* (skeletal or striated, smooth, and cardiac) consists of cells specialized for contraction and are responsible for most movement in complex animals.

10. *Nervous tissue* is highly specialized for the ability to respond to changes in the environment (stimuli). It carries information from the outside environment and from different body parts to allow the animal to respond to stimuli.

11. Organs are grouped into functional complexes called organ systems.

STUDY QUESTIONS

1. Describe the hierarchical organization of most multicellular organisms and define *tissue, organ,* and *organ system.* (pp. 307–308)
2. How is a vascular plant organized? Name and describe the functions of the four organs of a plant. (p. 308)
3. How is *permanent tissue* different from *meristematic tissue?* Is the distinction absolute? How is *primary tissue* different from *secondary tissue?* (pp. 308–309)
4. Compare the structure of the epidermis of leaves and roots. What would happen to a plant if it lost its cuticle? Guard cells? Root hairs? How does periderm differ from epidermis? (pp. 309–311)
5. Compare and contrast *parenchyma, collenchyma,* and *sclerenchyma* with respect to structure, function, and location in the plant body. (pp. 311–313)
6. Why are xylem and phloem called *vascular tissues?* Which tissue contains living cells at maturity? What materials are transported in each tissue? Distinguish between tracheids and sieve elements. How does the direction of transport differ in them? (p. 313)
7. What is the function of epithelial tissue? How would the stratified squamous epithelium of the skin differ in structure from the simple cuboidal epithelium found in kidney tubules? (pp. 315–317)
8. Describe the four major types of connective tissue and their functions. What shared structural feature allows both blood and bone to be classified as connective tissue? (pp. 318–319)
9. What special property is common to all types of muscle tissue? What are the three types of muscle tissue and where are they found? (p. 319)
10. What functional specialization is exhibited by nervous tissue? How does the shape of nerve cells make them suited for this function? (pp. 319–320)

CHAPTER 16

NUTRIENT PROCUREMENT AND GAS EXCHANGE IN PLANTS AND OTHER AUTOTROPHS

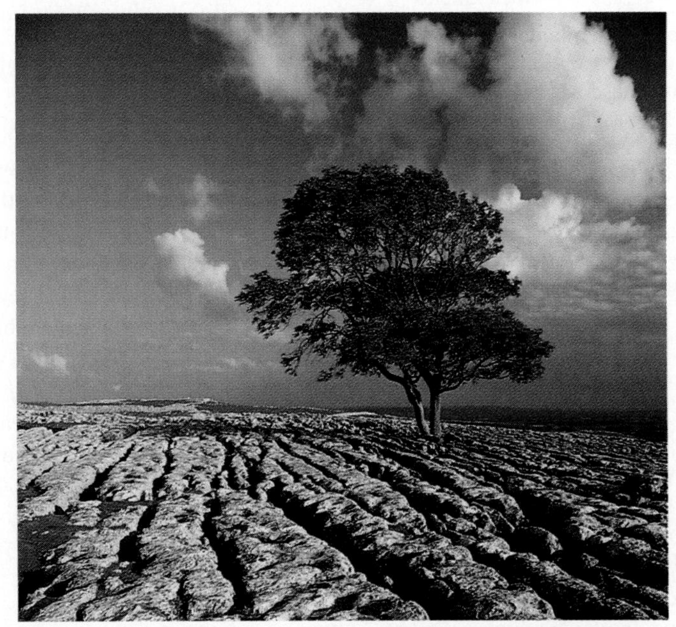

As we have seen, the cells of all living organisms, whether plant, animal, fungus, protist, or bacterium, contain carbohydrates, proteins, lipids, and nucleic acids, which are necessary for their structure and function. During photosynthesis, plants use the energy of sunlight to convert carbon dioxide and water into carbohydrates. But how does the plant obtain proteins, lipids, and nucleic acids?

Plants Synthesize the Organic Nutrients Used by All Animals

Unlike animals, plants synthesize organic molecules themselves from inorganic materials. They possess enzyme systems that enable them to combine inorganic ions, such as nitrate (NO_3^-), sulfate (SO_4^{2-}), and phosphate (PO_4^{3-}, $H_2PO_4^-$, or HPO_4^{2-}), with various carbohydrates to synthesize amino acids and nucleic acids (Fig. 16.1). Other enzymes change carbohydrates into lipids. Because animals, fungi, and many protists and bacteria lack the enzymes required to make many of these conversions, they depend on green plants. It is the plants that synthesize the complex organic nutrients that all animals and other consumers require. *All forms of life, with the exception of the chemosynthetic organisms, obtain their complex organic nutrients either directly or indirectly from photosynthesis.*

The abilities to synthesize new cellular materials and to break down energy-rich compounds by respiration are vital to the well-being of almost all organisms. Both processes require three substances from the environment. These are (1) nutrients—complex organic molecules or the raw materials from

Chapter opening photo: Ash tree growing in crevice in limestone bedrock. Enough soil has accumulated in the cracks in the limestone for the tree to obtain the nutrients it needs for its growth.

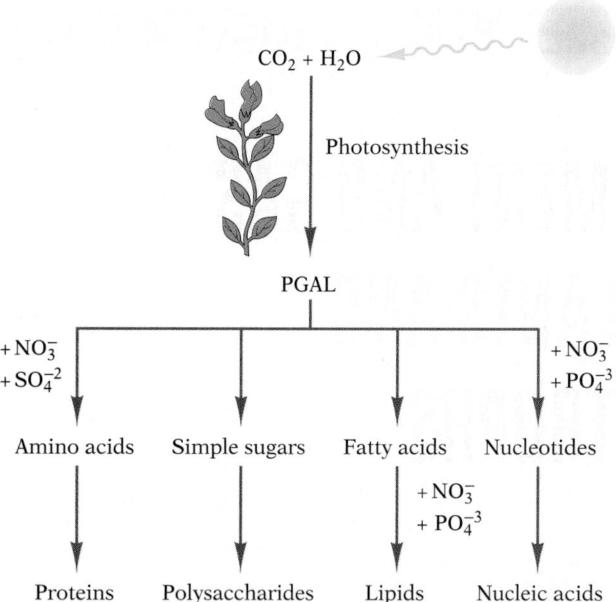

16.1 Biosynthesis of organic compounds in plants The PGAL synthesized during photosynthesis is used to manufacture complex organic compounds. Nitrogen and sulfur (absorbed as ions) are required to synthesize amino acids, which can be joined to make proteins. PGAL can also be used by the plant to synthesize simple sugars and then polysaccharides, and fatty acids and lipids. Finally, nucleotides and nucleic acids can be synthesized. Nitrogen and phosphorus (absorbed as ions) are required to synthesize these complex compounds. Plants have the enzymes necessary to catalyze the synthesis of complex organic nutrients from inorganic materials.

which these molecules and new cellular materials can be synthesized, (2) the oxygen used in cellular respiration, and (3) water, the liquid in which most life processes occur. This chapter will focus on how plants procure these required molecules.

A ***nutrient*** is any substance that an organism obtains from its environment and uses for some phase of its metabolism. Water and minerals are considered nutrients. Nutrients that are required in relatively large amounts in plants are classified as ***macronutrients,*** while those needed in only minute amounts are called ***micronutrients.***

Organisms Get Organic Nutrients by Either Making Them or by Eating Other Organisms That Have Made Them

Organisms can be divided into two classes, ***autotrophs*** and ***heterotrophs,*** on the basis of how they obtain nutrients. Fully autotrophic organisms can live in an exclusively inorganic environment because they create their own organic compounds from inorganic raw materials. Autotrophic organisms do not digest their nutrients before taking them into their cells. The molecules of these raw materials are small enough and soluble enough to pass through cell membranes. As you would guess, most autotrophs are photosynthetic, although a few are chemosynthetic. Photosynthetic plants are the most important of the earth's autotrophic organisms.

Heterotrophic organisms (most bacteria, protozoa, fungi, and animals) cannot manufacture their own complex organic compounds from simple inorganic nutrients; they must obtain already synthesized organic molecules by eating plants or other organisms. Many organic molecules are too large to be absorbed through cell membranes, and must first be broken down into smaller, more easily absorbable molecular units, that is, digested to their building-block constituents, before they can be absorbed. Heterotrophs are the subject of the next chapter.

The Way an Organism Gets Its Nutrients Is Reflected in Body Structure and Function

Autotrophs and heterotrophs differ in both their nutrient requirements and their means to procure nutrients. They have adapted differently to the different selection pressures acting upon them. Plants are sedentary and have evolved an elaborate root system to anchor them and to absorb necessary inorganic nutrients from the soil. They also have an aerial stem with flattened leaves to absorb light for photosynthesis. Animals, on the other hand, are highly mobile and are adapted to ingest their food.

Although we will be treating autotrophic and heterotrophic nutrition in separate chapters, remember that our focus is on the basic problems faced by *all* forms of life, whether plant, animal, fungus, or bacterium.

Plants Require Carbon Dioxide and Water for Photosynthesis

The basic raw materials needed by photosynthetic organisms are water and carbon dioxide. These two compounds supply carbon, oxygen, and hydrogen, the predominant elements in organic molecules. Water is obtained by terrestrial plants from the soil in which they grow. Most plants absorb water through their roots. Carbon dioxide, one of the main gases of the earth's atmosphere, is obtained by terrestrial plants directly from the air through their leaves; submerged aquatic algae absorb dissolved carbon dioxide from the surrounding water.

A very high percentage of the total dry body weight of a plant is carbohydrate, and much of the rest is synthesized from carbohydrate. This fact has some rather startling implications.

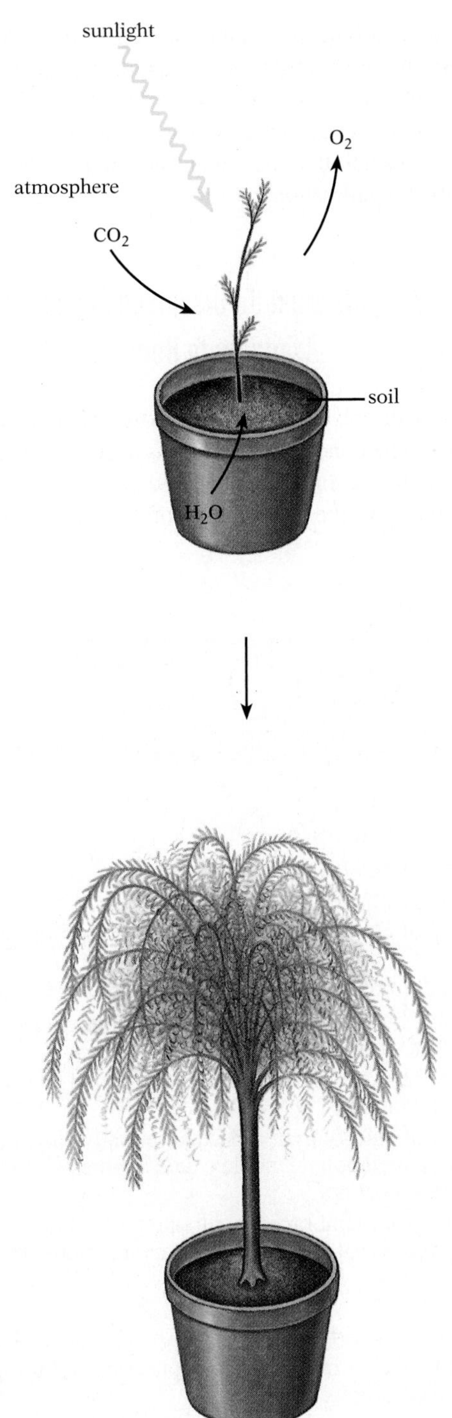

16.2 Major sources of plant mass Most of the dry mass of plants is in the form of carbohydrates (or their derivatives) generated originally by photosynthesis. A growing seedling uses the energy of light to combine the carbon and oxygen of carbon dioxide with hydrogen split off from water to synthesize carbohydrates; the oxygen from water in an actively growing plant is mostly discharged into the air. Thus, approximately 93 percent of the dry mass of a plant is obtained from the atmosphere; the remainder (including, in addition to hydrogen, essential minerals like nitrogen and phosphorus) comes from the soil or from water.

The formula for glucose, which may be taken as the formula for most of any plant's carbohydrate, is $C_6H_{12}O_6$. One molecule of glucose contains six atoms of carbon, twelve of hydrogen, and six of oxygen. Now, the combined weight of the six atoms of carbon and six atoms of oxygen is about 93 percent of the total weight of a glucose molecule. The carbon and the oxygen in glucose come from carbon dioxide, which in turn comes from the air (Fig. 16.2). Thus, about 93 percent of the weight of a large, immensely heavy tree comes initially from the air. The hydrogen in glucose makes up roughly 7 percent of its weight. The hydrogen comes from water. Hence, about 7 percent of the dry weight of the tree comes initially from water. Amazingly, most of the mass of any plant comes from the air, not from the earth in which it grows. The remaining mass comes mostly from water.

Plants Require Minerals From the Soil for Synthesis of Many Complex Organic Molecules

Carbon dioxide and water are not the only raw materials needed by a green plant. These two compounds provide only the three main elements: carbon, oxygen, and hydrogen. Yet we know that other elements, too, occur in plants. Nitrogen, for example, is present in amino acids, the building-block units of proteins, which are essential components of cytoplasm. Sulfur also is present in two very important amino acids. There is phosphorus in ATP, nucleic acids, and many other important compounds. Chlorophyll, the essential light trap for photosynthesis, contains magnesium. These elements, along with others listed in Table 16.1, are macronutrients, which are needed in relatively large amounts. Where does the plant obtain the nitrogen, sulfur, phosphorus, magnesium, and other elements it needs? This is where the soil plays its role as a source of raw materials. Soil minerals provide plants with the inorganic nutrients they need to live. The mineral nutrients are absorbed by the plants primarily as inorganic ions dissolved in water: nitrogen as nitrate (NO_3^-), nitrite (NO_2^-), or ammonium (NH_4^+) ions; phosphorus as $H_2PO_4^-$ or HPO_4^{2-} ions; sulfur as sulfate (SO_4^{2-}) ions; and calcium, potassium, magnesium, and iron as their simple ions (Ca^{2+}, K^+, Mg^{2+}, and Fe^{2+} or Fe^{3+}).

The macronutrients nitrogen, potassium, and phosphorus are required in substantial amounts because the compounds in which they occur make up a significant percentage of the plant body. Consequently, if crops are to flourish, they must be supplied with these nutrients. Fertilizers, either organic (from compost or manure) or inorganic, must be applied each year to replace the lost nutrients. Commercial fertilizers are often designated by their N-P-K weight percentages. The widely used garden fertilizer designated "5-10-5" contains, by weight, 5 percent nitrogen, 10 percent phosphoric acid, and 5 percent soluble potash, a potassium compound. Many modern fertilizers are also fortified by small amounts of the other essential minerals. Today, we hear a great deal about the merits of "organic" fertilizers versus commercial fertilizers. From the standpoint of nutrient procurement, it does not matter to the plant whether the nutrient ions are derived from a bag of commercial fertilizer or from organic wastes—the nitrate, phosphate, and potassium ions are the same to the plant, no matter what their origin. A disadvantage of organic fertilizers is that their nutrient content is lower than that of commercial fertilizers, and it can vary widely. On the other hand, organic fertilizers not only contribute to soil fertility, they also improve the condition of the soil, promoting drainage and aeration. Also, the organic material decomposes slowly, releasing the nutrients gradually and providing a more continuous supply of nutrients.

The micronutrients are essential to plants but are needed only in trace amounts. They are manganese, boron, chlorine, zinc, copper, molybdenum, and iron (Table 16.1). Most of the micronutrients are components of enzymes or act as cofactors. Since enzymes and cofactors can be used over and over, only a very small quantity of each is necessary. Often the amounts required are so small that the seed has enough of the nutrient to supply several generations.

Soil Minerals Must Be Dissolved in Water to Be Available to Roots

Ions must be dissolved in soil water to be available to plants for absorption. The concentration of ions varies according to the fertility and the acidity of the soil and other factors. Chemical analyses that give the total amount of the various ions present

TABLE 16.1 *Essential elements for plants**

Element	*Form in which element is absorbed*	*Concentration in healthy plants (% dry weight)*	*Function*
MACRONUTRIENTS			
Carbon (C)	CO_2	45	Structural component of organic compounds
Oxygen (O)	H_2O, O_2	45	Structural component of organic compounds
Hydrogen (H)	H_2O	6	Structural component of organic compounds
Nitrogen (N)	NO_3^-, NH_4^+	1.5	Structural component of amino acids, nucleotides, many hormones, and coenzymes
Phosphorus (P)	$H_2PO_4^-$, HPO_4^{2-}	0.2	Structural component of nucleotides, phospholipids, ATP, coenzymes
Potassium (K)	K^+	1.0	Plays a role in the ionic balance of cells; cofactor for enzymes involved in protein synthesis and carbohydrate metabolism
Sulfur (S)	SO_2^{2-}	0.1	Structural component of two amino acids—cysteine and methionine—and of several vitamins
Magnesium (Mg)	Mg^{2+}	0.2	Structural component of chlorophyll; cofactor for many enzymes involved in carbohydrate metabolism, nucleic acid synthesis, and the coupling of ATP with reactants
Calcium (Ca)	Ca^{2+}	0.5	Influences permeability of membranes; component of pectic salts in middle lamellae and necessary for cell wall formation; activator for several enzymes
MICRONUTRIENTS			
Iron (Fe)	Fe^{2+}, Fe^{3+}	0.01	Structural component of the iron-containing heme groups, which are important components of electron-transport molecules and some enzymes; plays a role in chlorophyll synthesis
Chlorine (Cl)	Cl^-	0.01	Important in maintenance of osmotic balance
Boron (B)	BO_3^-, $B_4O_7^{2-}$	0.002	Affects Ca^{2+} use. Unknown functions
Manganese (Mn)	Mn^{2+}	0.005	Activates certain enzymes
Zinc (Zn)	Zn^{2+}	0.002	Activates certain enzymes
Copper (Cu)	Cu^{2+}, Cu^+	0.0006	Activates certain enzymes
Molybdenum (Mo)	MoO_4^{2-}	0.00001	Necessary for nitrogen fixation in bacteria in root nodules

*An element must meet three criteria to be regarded as essential: (1) when not supplied the element is needed for normal growth and reproduction of the plant in several different plant species; (2) it is not replaceable by other elements; (3) its function is a direct one, in that it is not needed simply to correct a toxic condition induced by other substances. Often it is difficult to determine whether or not an element is essential because the amounts needed may be so minute that the seed may carry enough of the element to supply several generations.

MAKING CONNECTIONS

1. The water we drink and many of the foods we eat may contain inorganic ions such as nitrate, sulfate, and phosphate. Should these ions be considered human nutrients? Would they be nutrients for a plant?

2. What adaptations have plants and animals evolved to take advantage of their particular modes of feeding?

3. Where does more than 90 percent of the mass of a plant come from? Where does most of the remainder come from?

4. Based on your knowledge of the chemical structure of proteins, lipids, and nucleic acids, why would you expect nitrogen and phosphorus to be major components of commercial fertilizers?

5. What are the benefits and disadvantages of using organic fertilizers instead of inorganic commercial fertilizers?

6. How is the need of plants for micronutrients related to what you learned about enzyme structure and function in Chapter 4 (see pp. 84–86)?

in soils can be misleading because certain proportions of these ions may not be free in solution but rather bound to the soil particles. When soil minerals are bound, they cannot be absorbed by the roots. Many factors, especially acidity, can influence the proportion of ions bound to the particles and those free in the soil solution. Thus, the concentration of various ions available in the soil water is often determined by the pH of the soil itself.

Agricultural soil management often involves changing the soil pH to free more bound minerals for absorption by roots. The addition of lime (CaO), for instance, to very acidic soil raises the pH and may increase the availability of phosphorus, potassium, and molybdenum. An excess of lime, however, may decrease the availability of iron, copper, manganese, and zinc (Fig. 16.3). A proper pH balance, appropriate to the particular crop to be grown, must exist for maximum yield.

Most Ions Are Absorbed From the Soil Solution by Active Transport

The absorption rate of each mineral by roots is essentially independent of the absorption rates of water and other minerals. The rate at which each nutrient moves into the root depends on (1) its concentration both inside and outside the root; (2) the ease with which it can passively penetrate cell membranes; and (3) the extent to which carrier molecules are involved. Some of the inward movement of minerals, like that of water, is from passive diffusion along a concentration gradient. The rate of absorption is, however, generally greater than would be possible by passive diffusion alone. This means that some sort of carrier-mediated transport—facilitated diffusion or active

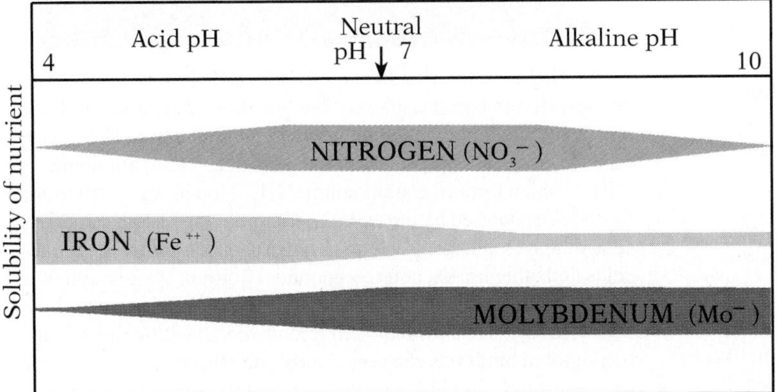

16.3 Solubility of three mineral nutrients as a function of pH The changing width of each band indicates the relative solubility of the mineral between pH 4 and pH 10. Nitrogen is most soluble, and hence most available to plants, in the neutral pH range; iron is most soluble under acid conditions, and molybdenum under alkaline conditions. Other minerals have their own distinctive solubility curves. Since it is obviously impossible for soil to have a pH at which all minerals will be maximally available to plants, farmers and gardeners must adjust the soil pH according to the nutrient requirements of the particular plants they wish to grow.

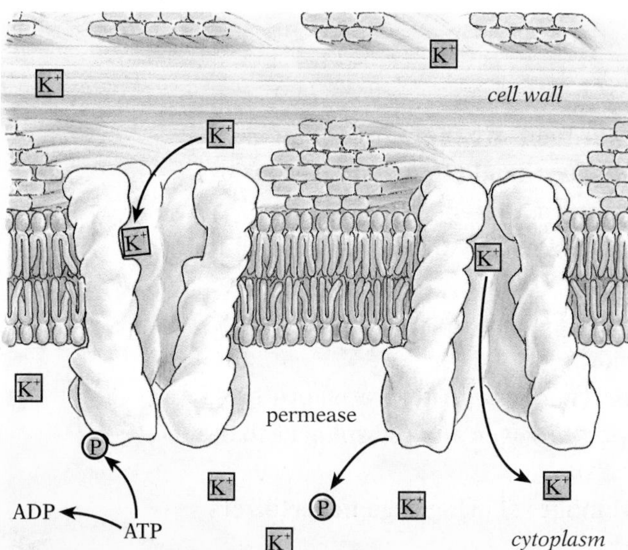

16.4 A model of active transport across a membrane As the permease (which is probably composed of at least two subunits) picks up the substrate at its binding site, it undergoes a conformational change that channels the substrate to the other side of the membrane, where it is released. ATP energy is expended by the cell to cause the conformational change and drive this process.

transport—takes place. Because the soil solution is ordinarily dilute, plants often take in a mineral that is more highly concentrated inside the root cells than in the soil solution, absorbing it against the concentration gradient by active transport (Fig. 16.4). This requires the expenditure of energy by the root cells. Evidence indicates that most ion absorption is by active transport.

Certain Bacteria Convert Nitrogen Gas Into Ammonia That Can Be Used by Plants

Nitrogen is one of the macronutrients needed in substantial amounts. It is usually the major limiting factor in plant growth. Although nitrogen gas makes up roughly 78 percent of the atmosphere, plants cannot use it directly from the air, as they do carbon dioxide. The two nitrogen atoms in one gas molecule have a strong triple bond that plants cannot break; therefore, most plants must absorb nitrogen as ions (usually as nitrate, NO_3^-) from the soil. Where do many of these ions come from? Certain soil bacteria convert ("fix") nitrogen gas to a form (the ammonium ion, NH_4^+) that can be used by plants in a process known as ***nitrogen fixation.***

Some nitrogen-fixing bacteria live in a close relationship inside the roots of certain higher plants, where they occur in prominent ***nodules*** (Fig. 16.5). The legumes (angiosperm dicot plants belonging to the pea or bean family, e.g., bean, clover, alfalfa, lupine, etc.) are particularly well known for their numerous root nodules, but other plant families also have them. Such a plant and its nitrogen-fixing bacteria have a mutualistic relationship. The bacteria provide the plant with fixed nitrogen while the plant provides the bacteria with carbohydrates.

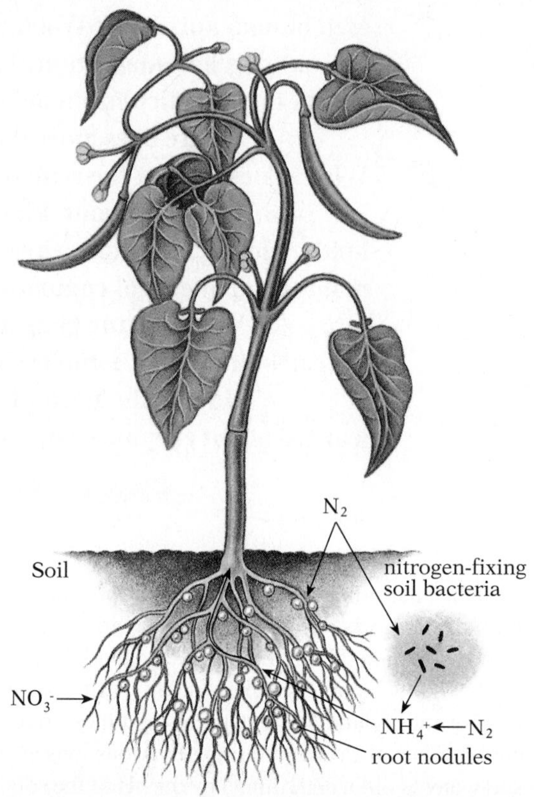

16.5 Nitrogen fixation in root nodules of legumes Top: Certain nitrogen-fixing bacteria interact with the roots of legumes and certain other plants to form nodules in which the bacteria live. These bacteria have enzymes capable of reducing nitrogen gas to ammonia (NH_3), which forms the ammonium (NH_4^+) ion in water. Ammonium ions are transported by the vascular tissues to the photosynthesizing cells of leaves and stems where they are used to synthesize amino acids and other nitrogenous compounds. Bottom: Photograph of roots of a legume (pea) showing nodules. The nodules are pink because they have a form of hemoglobin stored within them. This hemoglobin binds oxygen very tightly and effectively removes free oxygen from the nodule, thereby allowing the anaerobic reactions of nitrogen fixation to occur.

How do the bacteria in the nodules fix nitrogen for the plant? These bacteria have powerful enzymes capable of breaking the triple bonds of nitrogen gas and adding hydrogen ions and electrons to it. This reduces nitrogen gas to ammonia (NH_3). Ammonia in water solution and as a salt often takes the form of ammonium (NH_4^+) ions. Plants use ammonium ions to synthesize amino acids and other nitrogenous compounds. Most synthesis of these compounds actually takes place in the stroma of chloroplasts. Thus, the nitrate and ammonium ions absorbed by the root cells must be transported by the vascular tissues to the photosynthesizing cells of leaves and stems.

One of the most challenging goals for modern applied biotechnology is to develop crop plants that can fix atmospheric nitrogen and thereby be self-sufficient in nitrogen. Unfortunately, there are about 20 enzymes involved in the nitrogen fixation process, which means that every single one of the genes that code for these enzymes must be transferred into the host plant. Moreover, one of the key enzymes operates only under anaerobic conditions, so it will not function in the aerobic environment found in most plant cells. Because of these difficulties some researchers are trying another approach. They are trying to get the nitrogen-fixing bacteria to form root nodules on nonlegume plants. Preliminary experiments on rice and wheat seedlings are promising. Specially treated roots of these plants have been induced to form nodules in the laboratory, but whether or not these nodules actually function in nitrogen fixation is still unknown. Furthermore, inducing nodule formation in plant roots in a laboratory is a far cry from doing it under farming conditions.

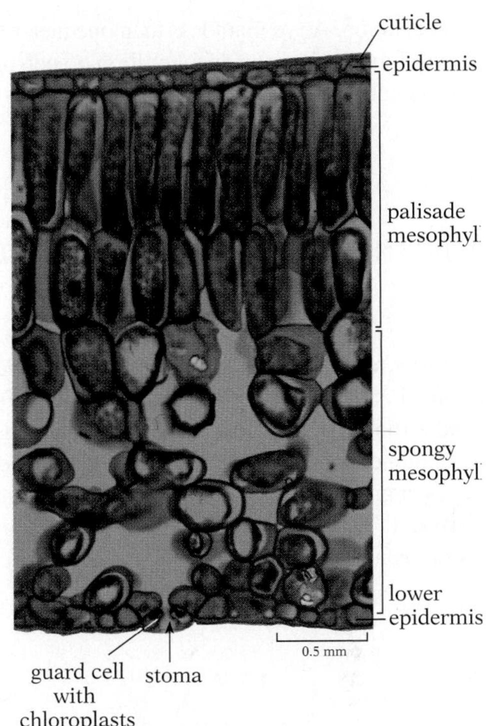

16.6 Cross section of part of a leaf of privet CO_2 enters through the stomata and circulates through the intercellular spaces. Notice that many of the cells of the mesophyll are loosely packed and, therefore, the total surface area of cells exposed to the intercellular spaces is enormous.

Leaves Function as Organs for Nutrient Procurement

Leaves are the principal organs of photosynthesis. Their structure is adapted for this function. Leaf blades are generally broad and flat, to present a large surface area for light absorption. A complex system of veins delivers water and inorganic nutrients to the various leaf cells. The carbon dioxide needed for photosynthesis, together with the other gases in air, moves into internal spaces of the leaf through epidermal openings called ***stomata.*** The mesophyll tissue in a typical leaf contains large intercellular spaces (Fig. 16.6). A high percentage of the total surface area of each mesophyll cell is exposed to the air in these spaces, which are interconnected. Gases can thus move readily between the surrounding atmosphere and the internal spaces of the leaf whenever the stomata are open. Carbon dioxide circulates throughout the intercellular spaces (see Fig. 16.6). It dissolves in a film of water on the surfaces of the mesophyll cells, and then diffuses into the cells, where it is used as a raw material for photosynthesis. Oxygen produced during photosynthesis can be used by the cells for respiration, or it can move into the intercellular spaces and diffuse outward through the open stomata.

Root Systems Present a Large Surface Area for Absorption of Inorganic Nutrients

Plant roots have several functions. They take up inorganic nutrients such as water and minerals, they store and transport these nutrients, and they anchor the plant to the soil.

The root system of a plant normally extends very far, more so than most people think. When we pull up a plant from the soil, we seldom get the entire root system. Most of the smaller roots are so firmly embedded in the soil that they break off and are lost. A large root system anchors the plant and provides a sufficiently large absorptive surface.

In Chapter 5 we mentioned the problem of surface-to-volume ratio—the fact that as a cell or an organism gets bigger, its volume increases much faster than its surface area. A large multicellular organism faces a serious problem. It must have an absorptive surface large enough to obtain the nutrients it needs to support its large volume. This problem of a sufficiently large absorptive surface is more acute if most of the absorption is restricted to a limited region of the body, such as the roots. Many organisms have solved this problem by evolving highly subdivided absorptive surfaces, with far greater total area than those surfaces of an undivided system of the same volume. The great degree of branching in a typical root system exemplifies this kind of adaptation. A rye plant less than one meter tall, for instance, was found to have some 14 million branch roots with a combined length of over 600 kilometers!

The First Formed Primary Root Branches to Produce a Secondary Root System

The first root formed by the young seedling is called the ***primary root.*** Later, ***secondary roots*** branch from the primary root, and a root system is formed. If the branching creates a system of numerous slender roots, with no single root predominating, as in grass or clover, the plant is said to have a ***fibrous root system*** (Fig. 16.7A). If, however, the primary root remains dominant, with smaller secondary roots branching from it, the arrangement is called a ***taproot system*** (Fig. 16.7B). Dandelions, beets, and carrots, among others, are plants with taproots. As these examples suggest, taproots are frequently specialized as storage organs. Storage is a function of all roots, but particularly of taproots.

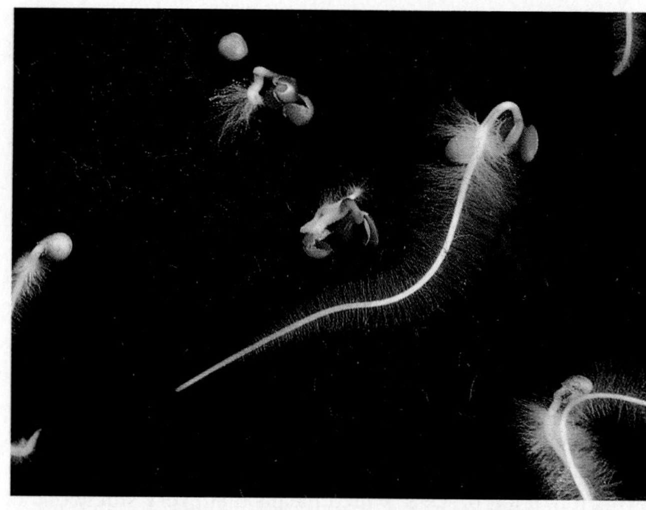

16.8 Root of radish seedling with many prominent root hairs

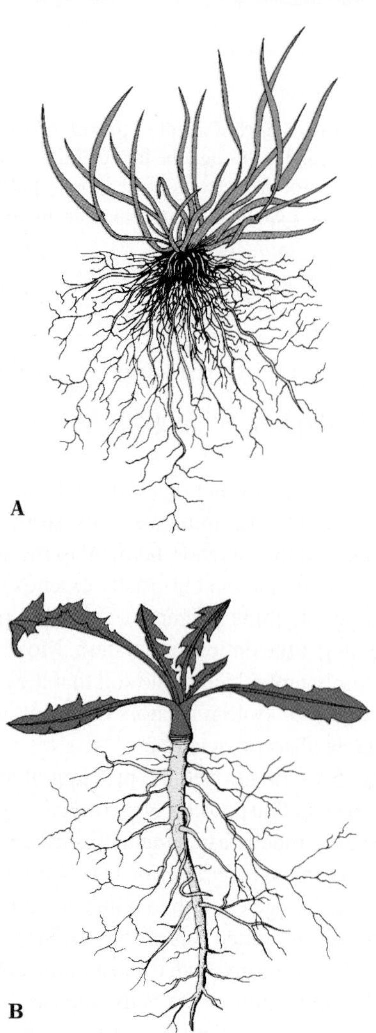

16.7 Two types of root systems (A) Fibrous root system of grass. (B) Taproot system of dandelion.

Roots have yet another means to increase their absorptive capacity. Just behind the growing tip of each rootlet, there is usually an area bearing a dense cluster of tiny hairlike extensions of the epidermal cells (Fig. 16.8). The zone of these ***root hairs*** on each rootlet is not very long, but the number of root hairs on all the many rootlets is so vast that the total absorptive surface they provide is enormous. The rye plant cited above may have had as many as 14 billion root hairs with a total surface area of over 400 square meters. It is in the region of the root hairs that most absorption of water and minerals takes place. Root hairs may also serve to anchor the root so the growing tip can push through the soil.

Mycorrhizae of Fungi Living With Roots Increase Absorption of Water and Minerals

In nature, the absorptive potential of most plant roots is further increased by ***mycorrhizae.*** A mycorrhiza (*myco-*, "fungus," and *rhiza,* "root") is an intimate association between plant roots and certain fungi (Fig. 16.9). The association is often highly specific; the roots of a particular plant species can only form an association with one species of fungus. The threadlike fungi actually penetrate the root cells, and greatly increase the surface area for the absorption of water and other nutrients, especially phosphates, from the soil. In turn, the fungi obtain organic nutrients from the root cells. The relationship is mutually

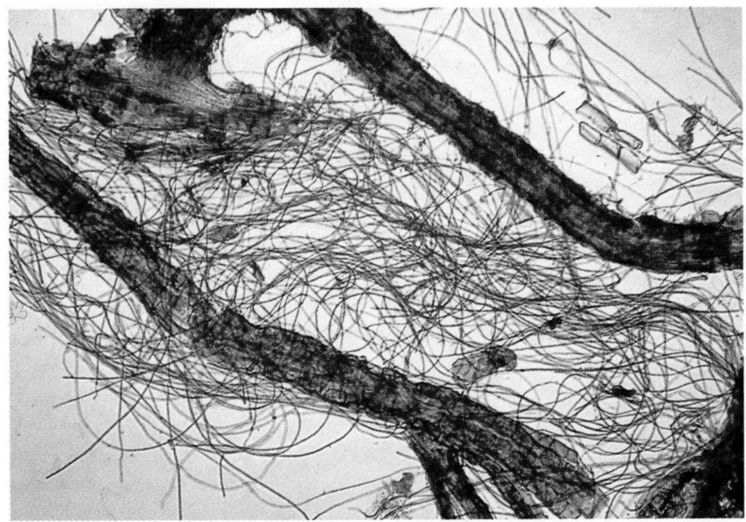

16.9 Mycorrhizae on roots A mycorrhiza (fungus-root) is a mutualistic association between plant roots and certain fungi. The threadlike fungi actually penetrate the root cells, and greatly increase the surface area for the absorption of water and other nutrients from the soil. The fungi benefit by obtaining organic nutrients from the root cells.

advantageous, i.e., a ***mutualistic symbiosis.*** Although mycorrhizal associations were once thought to be rare, research has now established that mycorrhizae are found in almost all vascular plant species, and that many species would be unable to survive without their mycorrhizae. Douglas fir, for instance, forms a mycorrhizal association with a particular species of edible mushroom. If the fungus species were eliminated, which in the northwest United States is a possibility due to overpicking by humans, the Douglas fir in that area could not survive. Mycorrhizal associations are particularly beneficial in inhospitable soils, such as tropical soils where phosphorus is tightly bound, in acidic soils where nitrates are scarce, or in the difficult conditions found in mine slag heaps and landfills. Studies of plant fossils show that mycorrhizal associations were common in primitive plants. Their presence may have been critical in enabling ancient plants to colonize land, because conditions on land at that time were very harsh.

Root Structure Is Similar in Dicots and Monocots

The cross section of a young dicot root has a series of different tissue layers (Fig. 16.10). On its outer surface is a layer of ***epidermis*** that is one cell thick. Unlike the epidermis of the aerial parts of the plant, that of the root usually has no waxy cuticle on its surface. (The root's epidermis absorbs water, while epidermis of aerial parts is a barrier against water loss through

MAKING CONNECTIONS

1. Why is the management of soil pH important to growing crops? Based on your knowledge of plant nutrients, what would happen to plants grown in soil with too much lime? What would you do to counteract the effect of too much lime?

2. A farmer alternates planting a field with corn one year and clover (a legume) the next. The clover is plowed under at the end of the growing season. What is the value of this practice (called crop rotation)?

3. How does the goal of developing crop plants that can fix nitrogen demonstrate the potential and the practical problems of the genetic engineering techniques described in Chapter 13?

4. Describe how mycorrhizal fungi may have helped plants colonize land. What kind of relationship do they have with plants?

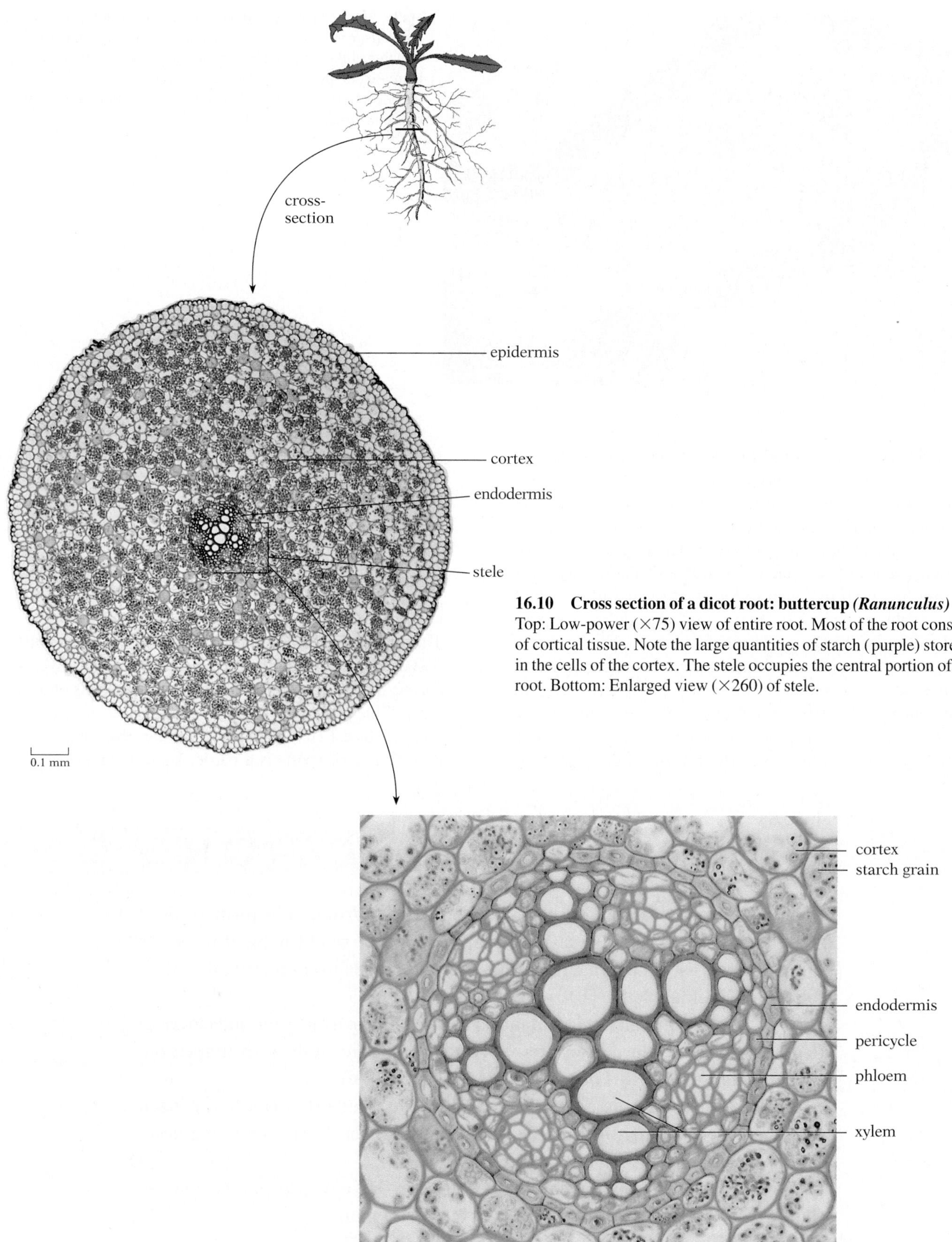

16.10 Cross section of a dicot root: buttercup *(Ranunculus)*
Top: Low-power (×75) view of entire root. Most of the root consists of cortical tissue. Note the large quantities of starch (purple) stored in the cells of the cortex. The stele occupies the central portion of the root. Bottom: Enlarged view (×260) of stele.

outward diffusion.) Some root epidermal cells in back of the growing tip of a rootlet have root hairs (Fig. 16.11).

Beneath the epidermis is the ***cortex,*** a wide area composed primarily of parenchyma tissue, with numerous intercellular spaces. Large quantities of starch are often stored in cortex cells. Cortex is often prominent and important in young roots, but it is frequently much reduced or even lost in older roots. In such roots, both cortex and epidermis may be replaced by a corky ***periderm.***

Inside the cortex is the ***endodermis,*** a layer one cell thick (Fig. 16.10). The endodermal cells are characterized by a waterproof band, the ***Casparian strip,*** which runs through their radial (side) and transverse (top and bottom) walls (Fig. 16.12). The walls of mature endodermal cells are often very thick and strengthened by deposits of lignin. A well-differentiated endodermis is always present in roots.

The endodermis forms the outer boundary of a central core of the root that contains the vascular cylinder. This core is called the ***stele.*** Just inside the endodermis is a layer, often only one cell thick, of thin-walled undifferentiated cells, the ***pericycle.*** These cells are meristematic and can produce new cells that grow outward from the stele to give rise to lateral roots.

In a dicot root, the central portion of the stele is filled with the two vascular tissues, ***xylem*** and ***phloem.*** The thick-walled

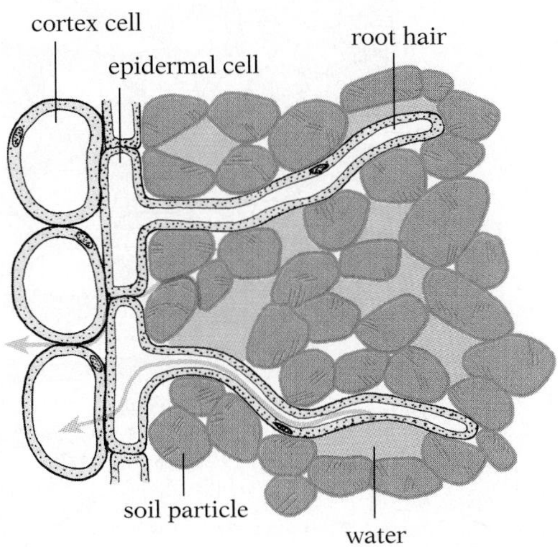

16.11 Root hairs penetrating soil Each root hair, which is an extension of a single epidermal cell, is in contact with many soil particles (brown) and with soil spaces, some of which contain air, some water (blue).

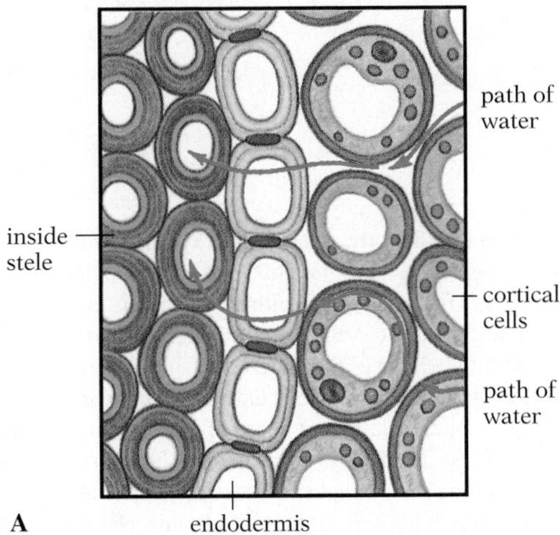

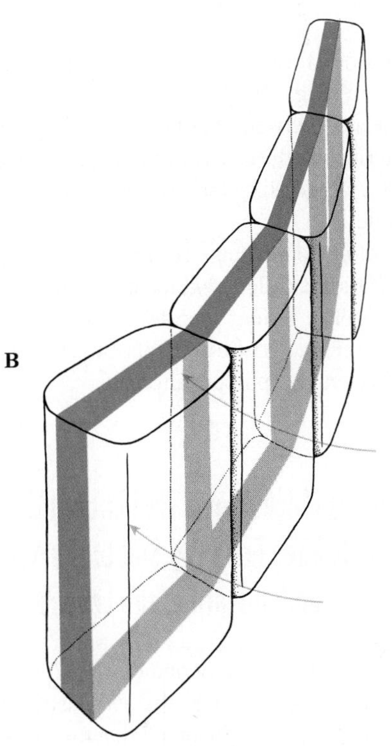

16.12 Endodermis (A) Cross section of root showing endodermis. The path of water flow is shown by the blue arrows. (B) Endodermal cells with Casparian strip. The endodermis marks the outer boundary of the stele. A waterproof band, the Casparian strip (color), runs through the radial (side) and transverse (top and bottom) walls of each endodermal cell, forming a watertight barrier. All water moving through the cortical cells must pass through an endodermal cell to enter the stele. Blue arrows show the direction of water movement.

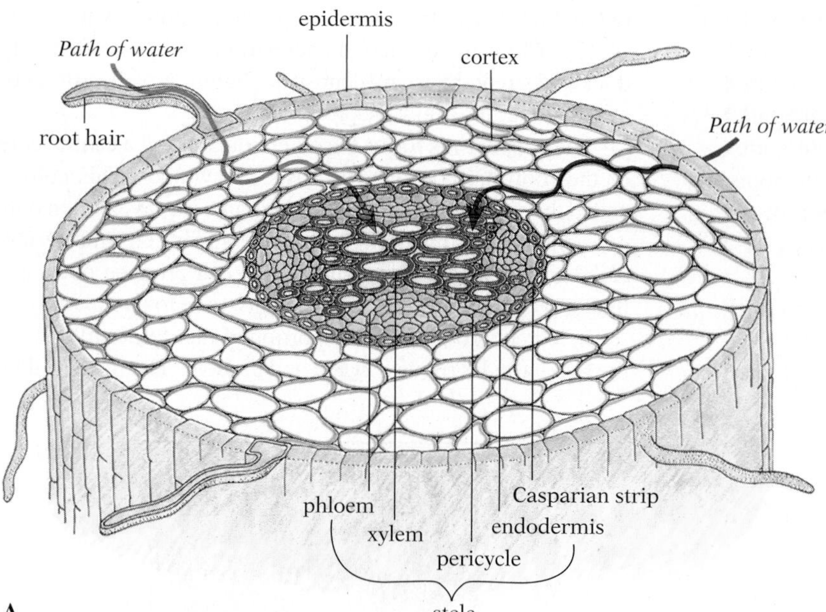

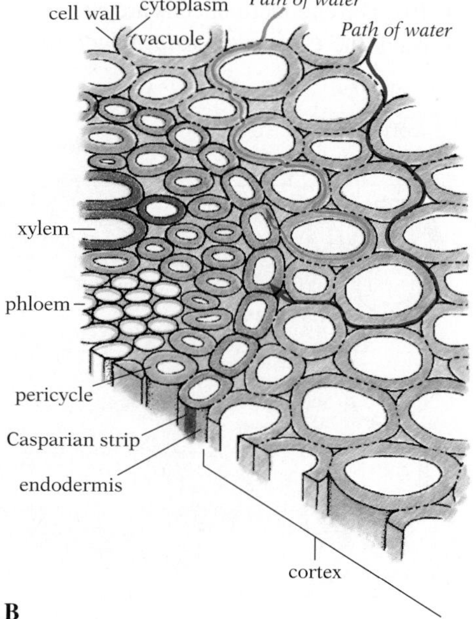

16.13 Movement of water from soil to xylem in root (A) Cross section of a root. Some water (light-blue arrow) is absorbed by the epidermal cells, particularly the root hairs, and moves from cell to cell either by osmosis or by diffusion through plasmodesmata. The interconnected cells form the symplast, through which water and dissolved minerals can move. Most of the water (dark-blue arrow) flows along cell walls and does not cross membranes of living cells until it reaches the endodermis. This interconnected system throughout the plant, composed of porous cell walls and intercellular spaces, is called the apoplast. The Casparian strip (dark brown) of the endodermal cells prevents flow of water along their radial and end walls; hence all water entering the stele must move through the living cells of the endodermis. (B) A region near the endodermis of a root. The pathway through the cortex weaves along cell walls and through intercellular spaces.

xylem cells often form a cross- or star-shaped figure in the center of the stem (Figs. 16.10 and 16.13) and bundles of phloem cells are located between the arms of the xylem. Xylem and phloem alternate in the center of the stele, instead of forming a continuous cylinder like the epidermis, cortex, endodermis, and pericycle.

Large monocots roots commonly have a region of parenchyma tissue, called *pith,* located at the center of the stele (Fig. 16.14). The xylem therefore does not form the star-shaped figure characteristic of dicots, but even in such roots the bundles of xylem and phloem alternate.

Water Always Moves Down Its Concentration Gradient as It Goes From the Soil to the Xylem and Upward to the Leaves

Two pathways exist for water and dissolved minerals to reach the stele—the symplastic and the apoplastic pathways (Fig. 16.13). We mentioned in Chapter 5 that the cytoplasm of adjacent plant cells is interconnected by plasmodesmata to form a continuous system called the ***symplast.*** Water can thus move from an epidermal cell to a cell in the cortex, and from one cortex cell to the next, through the root symplast (Figs. 16.13A and 16.15). The concentration of dissolved substances in an epidermal cell, for instance, is normally much higher than the concentration of dissolved substances in the soil water, which is usually very dilute. With the osmotic concentration higher inside the cell than outside, water will move across the membrane into the cell by osmosis. Once water or dissolved substances have entered an epidermal cell, they can then move to other cells through the plasmodesmata, without having to cross any additional cell membranes. Water will move continuously across the cells of the root cortex and into the stele through the symplast, because the concentration of dissolved substances is greater inside the stele than in the cortex, which, in turn, has a higher osmotic concentration than the epidermis. This concentration gradient is maintained by the active transport of ions from the soil water into the cells of the epidermis and cortex. The ions then move from cell to cell in the cortex through plasmodesmata, until they have crossed the endodermis of the root and entered the stele. (The Casparian strip prevents the ions

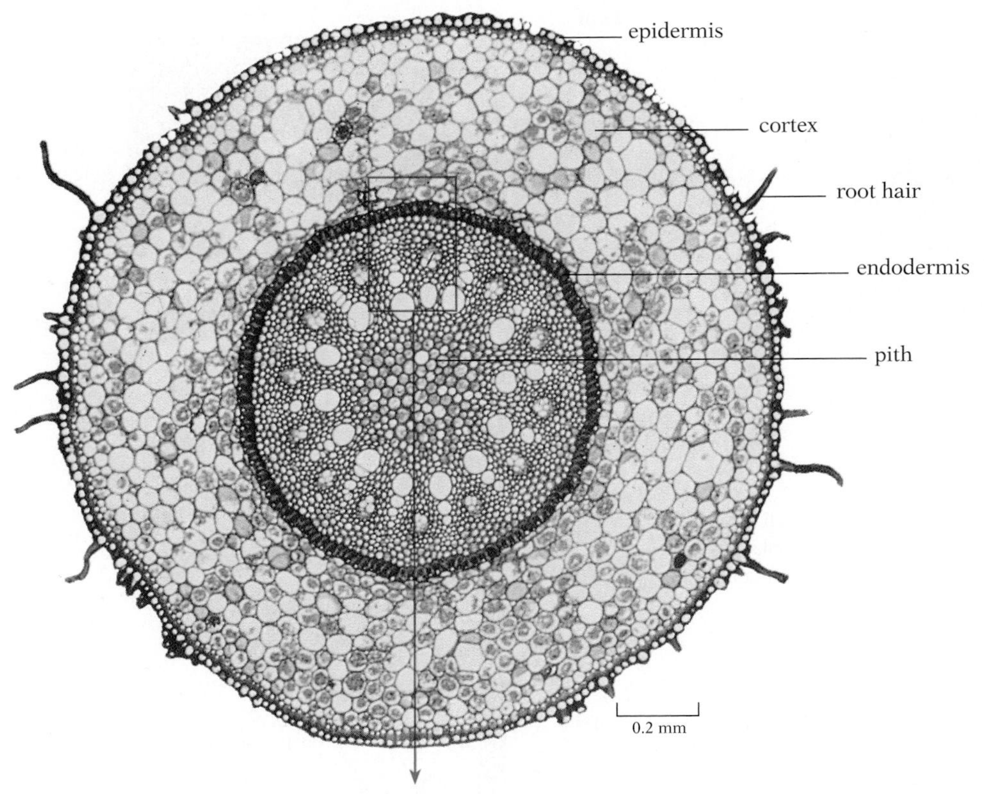

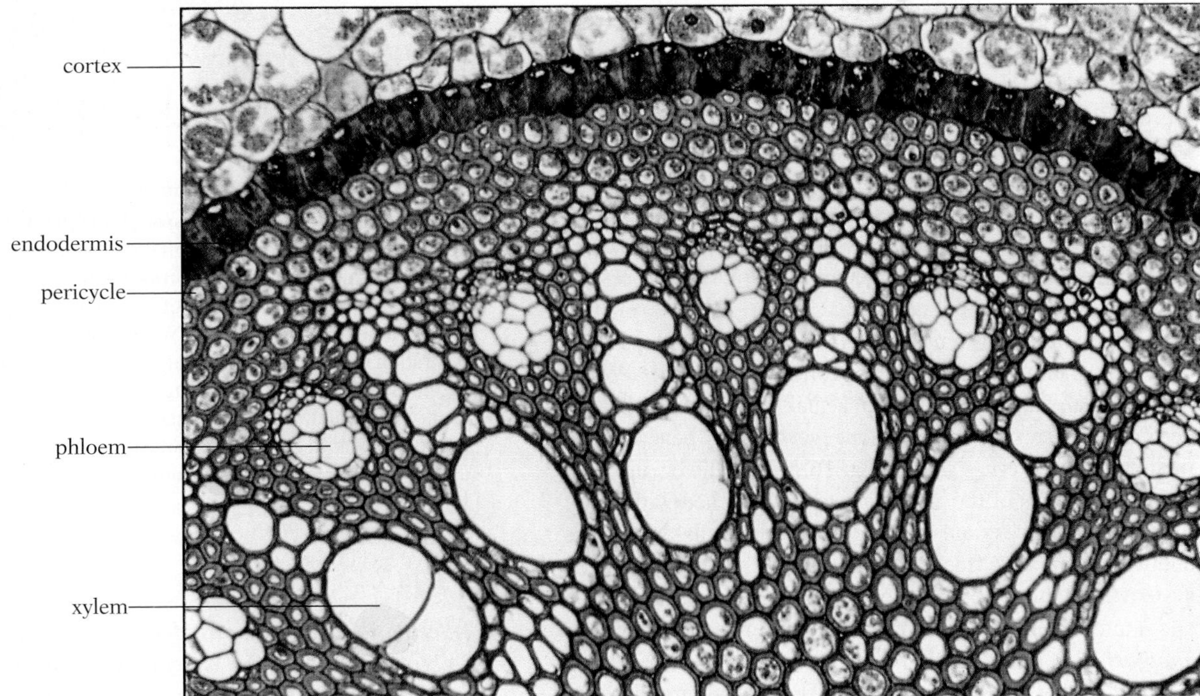

16.14 Cross section of monocot root: *Smilax* Top: Entire root. Bottom: Enlarged portion of stele. The pericycle of this root, unlike that of the buttercup, is many layers thick. Note the very thick walls of the cells in the endodermal layer.

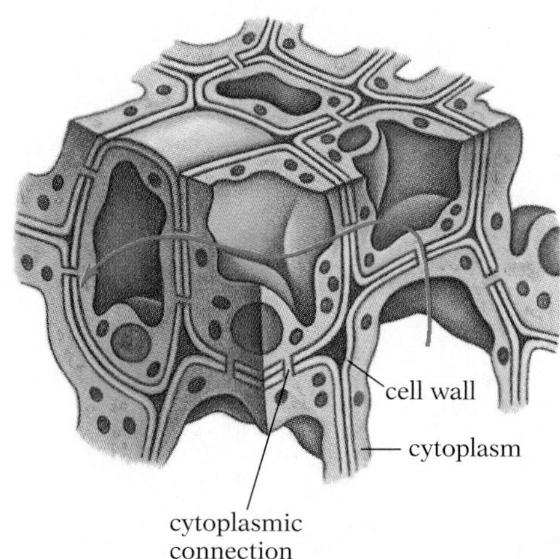

16.15 The symplast The cytoplasm of an individual plant cell in a multicellular plant body is not isolated, but is joined to the cytoplasm of other cells by tiny cytoplasmic connections called plasmodesmata. Plasmodesmata are very numerous—more than one million per square centimeter. The interconnected cells form the symplast, through which water and dissolved minerals can move. Blue arrow shows the path of water and nutrient movement.

from diffusing back into the cortex.) Water follows along passively by osmosis, moving down its concentration gradient from points of high concentration to points of low concentration. Once in the stele, the water and dissolved materials move into the conducting cells of the xylem and are transported to other parts of the plant. Xylem cells form a continuous pathway conducting the water and dissolved minerals from the root through the stem to the leaves. Water is still moving down its concentration gradient as it passes out through the intercellular space of the leaf mesophyll and finally through the stomates to the outside air (Fig. 16.16). Removal of water from the center of the root via the xylem maintains the concentration gradient from the root epidermis to the xylem and allows the process of water absorption to continue.

The apoplastic system provides the other and perhaps more important pathway that nutrients follow to the stele. Water molecules and dissolved ions can actually flow across the epidermis and cortex of a root without ever actually penetrating a membrane or entering a cell (Fig. 16.13B). This occurs both because the cells of the cortex are loosely packed with many intercellular spaces, and because plant cell walls are porous and water can move freely through them. This interconnected system throughout the plant, composed of the porous cell walls and intercellular spaces, forms a functional network called the ***apoplast.*** But the apoplast is interrupted at the endodermis. Water cannot flow between the cells of the endodermis because the Casparian strip acts as a barrier; it prevents water and dissolved materials from flowing between the cells and through the cell walls. *Consequently, all water and dissolved materials entering the vascular cylinder must cross through the living cells of the endodermis.* Water moves freely through the cortex and endodermis to the xylem, while mineral entry is controlled by the endodermis. The plasma membranes of the endodermal and nearby cortical cells control the movement into the stele of all substances dissolved in the water.

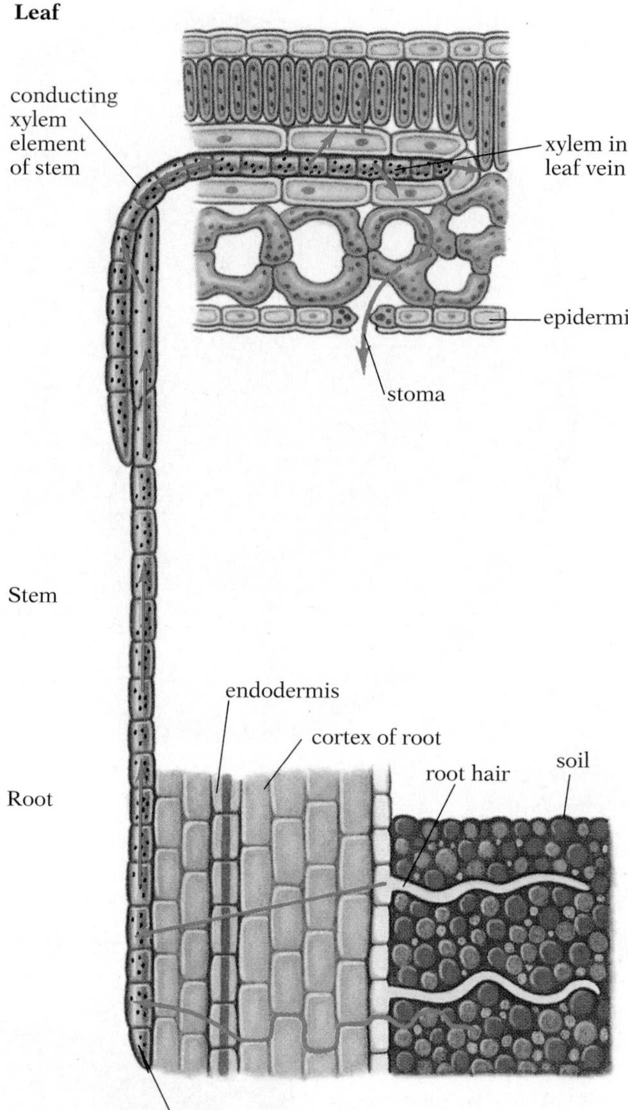

16.16 Movement of water through the plant Water moves continuously by osmosis through the cells of epidermis, cortex, and endodermis into the stele of the root. Once in the stele, the water can move into the conducting cells of the xylem to be transported up the stem to other parts of the plant. Water is still moving down its concentration gradient as it evaporates into the intercellular spaces of the leaf mesophyll and finally through the stomate to the outside air. Removal of water from the center of the root via the xylem maintains the concentration gradient from the root epidermis to the xylem and allows the process of water absorption to continue.

Some Green Plants Get More Nitrogen by Being Carnivorous

A few photosynthetic plants supplement their inorganic diet with organic compounds obtained by trapping and digesting insects and other tiny animals. Such plants can survive without capturing any prey, but when they do capture prey the nutrients thus obtained stimulate more rapid growth. Apparently it is the nitrogenous compounds of the animal's body that are of most benefit to insectivorous plants. This is because these plants often grow in nitrogen-poor soils, particularly acid bogs and heavy volcanic clays, and they have small root systems. Their highly specialized leaves show interesting adaptations for capturing prey (Figs. 16.17–16.19).

16.17 Pitcher plant *(Sarracenia purpurea)* from upstate New York The plants have leaves modified into a tube, or sac, which is partly filled with water. Insects that fall into the sac are prevented from climbing out by numerous stiff downward-pointing hairs. The proteins of the trapped insects are digested by enzymes secreted into the water, and the products of this digestion are absorbed by the inner surface of the leaf.

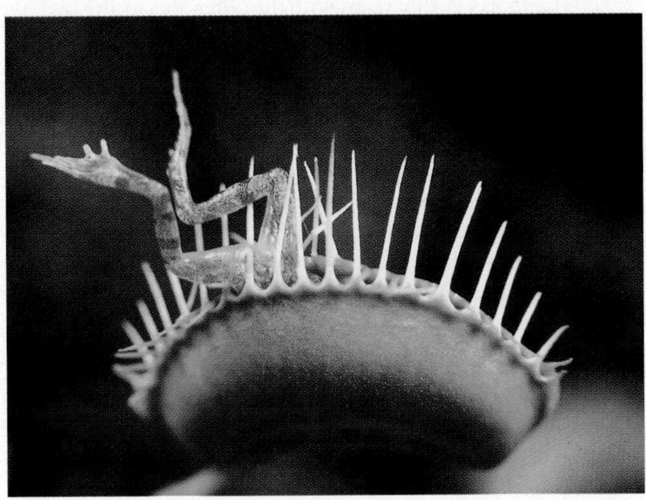

16.19 Leaf of sundew *(Drosera rotundifolia)* The leaf bears numerous hairlike tentacles, each with a gland at its tip. The gland secretes a sticky fluid in which small insects become trapped. The stimulus from a struggling trapped insect causes nearby tentacles to bend over the animal, further entangling it.

Plants Must Obtain Oxygen as Well as Carbon Dioxide

We have seen that in green plants the intake of carbon dioxide for photosynthesis and the release of oxygen occurs across the membranes of the cells of the mesophyll. It is a common misconception that gas exchange in green plants consists exclusively in this exchange of gases. Carbon dioxide is not, however, the only gas needed by plant cells—oxygen is necessary as well. Recall that when nutrient compounds are broken down during respiration, over 90 percent of the energy yield depends on the presence of oxygen, which makes possible the complete oxidation of the compounds to carbon dioxide and water and the production of large amounts of ATP. Thus, plants are constantly taking in oxygen and releasing carbon dioxide as cellular respiration occurs. When a green plant is exposed to bright light, both photosynthetic and respiratory gas exchange are usually taking place in the leaves. However, since the rate of photosynthesis exceeds the rate of respiration during the

16.18 Leaf of Venus's-flytrap *(Dionaea muscipula)* The leaf is composed of two lobes with a midrib between them. Note the row of long stiff teeth along the margin of each lobe. When the frog seen at left touched a trigger hair, the two halves of the leaf quickly moved together, interlocking their marginal teeth and trapping the frog. It will then be slowly digested by enzymes secreted from glands on the leaf surface, and the resulting amino acids will be absorbed. The rapid movement exhibited by the leaves of the Venus's-flytrap involves changes in the relative turgor pressure of certain epidermal cells.

day, there is a net uptake of carbon dioxide and release of oxygen. The reverse is true, of course, in the nonphotosynthetic parts of the plant or when the green plant is in the dark. *Respiratory gas exchange is a requirement for plants as much as it is for animals.*

Gas Exchange Occurs by Diffusion Across a Moist Cell Membrane

Gas exchange between a living cell and its environment always takes place by diffusion across a cell membrane. No active transport or facilitated diffusion is involved. Furthermore, the gases must first dissolve in the film of water that coats the cells if they are to diffuse through the membrane. In photosynthetic protists and many small multicellular autotrophs, particularly those that are aquatic, this requirement poses no serious problem. Each cell in these organisms is either in direct contact with the surrounding water or only a few cells away. These organisms have usually not evolved special respiratory surfaces but simply use their body surface for gas exchange. Many of these organisms are very small, so they have an adequate surface area to support their small volume. Others are large only in one or two dimensions, such as many of the algae. Some of the brown algae, the kelps, may grow to lengths of 60 to 70 meters, but the blades of even the longest kelps remain very thin. As a result, no cell is far from the surface, and the total gas-exchange area is fairly large in relation to the volume of the alga. Because these organisms live in water, their surface is kept moist.

In Terrestrial Plants the Requirements of Water Conservation Conflict With Those of Gas Exchange

The terrestrial environment is in many ways hostile to life. When fluids do not bathe the plant's entire surface, one of its most serious problems is obtaining enough water. The plant must maintain an extensive moist surface for gas exchange, but, at the same time, it must prevent excessive loss of water by evaporation. The large terrestrial plants have a real challenge when it comes to gas exchange. They must balance protection from water loss caused by sun and wind with their need to take in carbon dioxide and oxygen. Simultaneously, they must support the fragile, thin tissues that do the gas exchange, and protect them from tearing. To solve these problems, most terrestrial plants have evolved "compromise" mechanisms that address three basic needs: (1) a respiratory surface area of adequate dimensions; (2) a means of protecting the fragile gas-exchange surface from mechanical injury; and (3) a means of keeping the tissues moist, that is, preventing desiccation.

When large body size is mainly two dimensional (i.e., the organism is very thin), gas exchange is chiefly by direct diffusion between individual cells and the surrounding medium, because no cell is far from the surface. However, when increase in body size involves three dimensions, as it generally does, the first basic requirement, a respiratory surface area of adequate dimensions relative to the body volume, becomes a problem. This is because area (a square function) increases much more slowly than volume (a cube function). This problem of maintaining adequate surface area as volume increases also applies to roots and their absorption of water and minerals. While the absorbing surface of the root is increased by branching, root hairs, and mycorrhizae, the leaf gas-exchange surface is increased by the many intercellular spaces in the mesophyll.

In response to the second basic need, many plants have evolved outer body coverings that are relatively impermeable to water and other substances. Coverings such as the waxy cuticle on leaf epidermis or the bark of tree stems are protective barriers between the fragile internal tissues and the often hostile outer environment. The presence of these coverings, however, makes the maintenance of an adequate exchange area even more difficult.

The third basic need is for direct contact between the moist cell membranes across which gas exchange occurs and the environmental medium. The cells must be directly exposed to the environment, but they must be exposed in such a way as to minimize the tendency to dry out. Because a thin, moist surface is fragile and can easily suffer mechanical damage, the tendency has been toward the evolution of protective devices.

Gas exchange associated with both photosynthesis and cellular respiration takes place at a particularly high rate in green leaves, organs well adapted for this process. We have already seen that most of the outer surface of a leaf, covered as it is by a waxy cuticle, is rather impermeable to water and other substances, and, therefore, ill suited for the exchange of gases. Gas exchange must take place *inside* the leaf, across the membranes of the mesophyll cells. Let us consider how the structures of the leaf help it to meet the three previously stated requirements for respiratory systems.

The surface area available for gas exchange inside the leaf is very large. By comparison with the outer area of the leaf, the total area of cell surface exposed to the intercellular spaces is enormous (Fig. 16.20). The principle involved is a very elementary one: a chamber irregularly shaped and greatly subdivided by partial partitions will have far more wall space than a round or square one of equal volume. Furthermore, the exchange surfaces inside the leaf remain moist because they are exposed to air only in intercellular spaces. With the humidity within those spaces at nearly 100 percent, the walls of the mesophyll cells always retain a thin film of water on their surfaces. Gases dissolve in this water before moving into the cells. The protective epidermal tissues and the layers of waxy cuticle on their outer surfaces act as barriers between the dry outside air and the moist inside air, and also function as a protective covering for the entire leaf.

But these barriers are not complete; if they were, movement of gases between the outside and the inside could not take

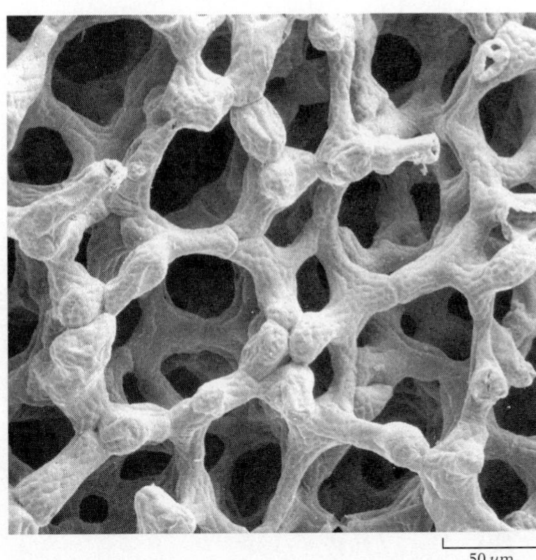

16.20 Scanning electron micrograph of spongy mesophyll in a bean leaf The micrograph gives an especially clear view of the extensive system of interconnecting air spaces.

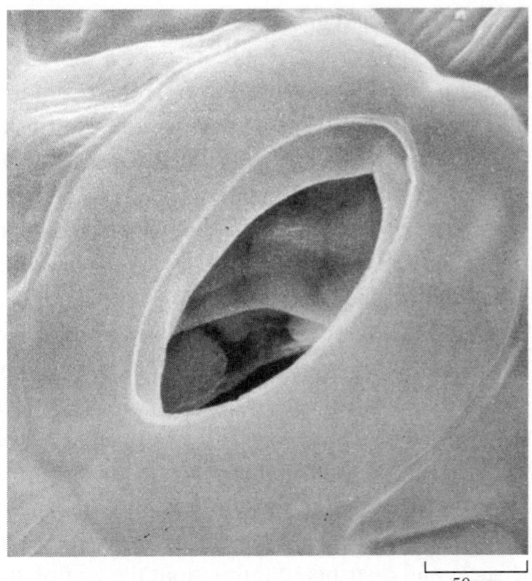

16.21 Scanning electron micrograph of a stoma in a cucumber leaf The stoma is open because the two large crescent-shaped guard cells are turgid and have pulled away from each other. Mesophyll cells in the interior of the leaf can be glimpsed through the opening.

place. So, although openings are essential, the stomata in a sense constitute weak links in the protective armor of the leaf. When the stomata are open, they allow the exchange of gases, but at the same time, they permit the loss of water from the interior of the leaf. Here we see the sort of compromise that has been a constant feature of evolutionary adaptation. Few characteristics, however beneficial, are without possible harmful effects. What determines the evolutionary fate of a characteristic is not whether it is exclusively beneficial or harmful, but whether or not the beneficial effects outweigh the harmful ones. The stomata, then, present advantages for gas exchange that outweigh the danger of desiccation.

Guard Cells Regulate the Size of the Stomata, Permitting Gas Exchange and Regulating Water Loss

The function of the stomata is to balance the leaf's need to perform photosynthesis against the danger of drying out. During the day, when the raw materials—light, water, and carbon dioxide—are available, the stomata are open to allow gas exchange. At night, when there is no light for photosynthesis, the stomata are closed to prevent excessive loss of water by evaporation from the mesophyll cells. But, even during the day, these openings must be closed when water is deficient. Each opening, or stoma, in the epidermis is bounded by the two highly specialized epidermal cells called ***guard cells*** (Fig. 16.21), which, unlike most other epidermal cells, contain chloroplasts. These bean-shaped cells have walls of unequal thickness; the walls next to the stoma are considerably thicker than those on the side away from the stoma (Fig. 16.22). In addition, bands of inelastic fibers run around each cell. When the guard cells take up water and are *turgid* (usually in the light), the thin outer wall of each cell buckles outward, pulling the rest of the cell with it and opening the stoma (Figs. 16.21 and 16.22). Gas exchange then takes place, and the mesophyll cells within the leaf obtains carbon dioxide for photosynthesis. When water is not available, or is being lost too quickly, the leaf cells, including the guard cells, may become *flaccid,* causing the leaf to wilt and the guard cells to close.

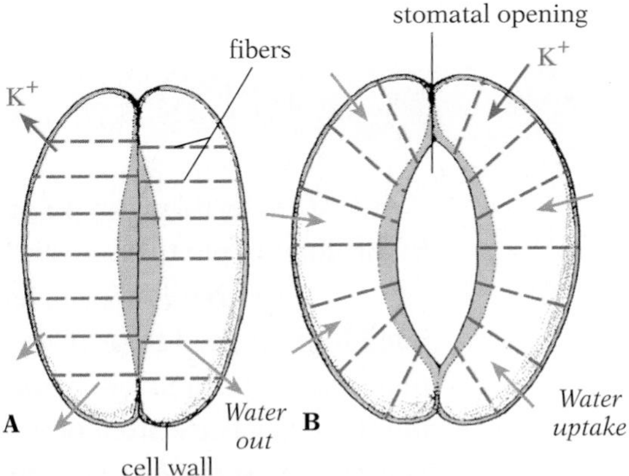

16.22 Guard cells (A) Two unusual features of guard cells account for the way in which they regulate the stomatal opening: the cell wall is thicker on the side close to the stoma than on the side away from the opening; and bands of inelastic fibers (green) run around each cell. (B) When uptake of water causes the cell to become turgid, expansion is limited to the side away from the opening. As a result the guard cells buckle, pulling apart, and the stoma opens. The degree of swelling is largely controlled by the availability of water and sunlight. Water (blue arrows) moves into and out of the guard cells in response to the flow of potassium ions (red arrows).

The rapid changes in guard-cell turgidity are tied primarily to potassium (K^+) ion concentrations within the guard cells. During the day, an ATP-driven pump actively transports potassium ions into the guard cells, increasing their osmotic concentration. Water then moves into the guard cells by osmosis, causing them to swell and open the stomata. At night the process reverses: potassium ions diffuse out of the guard cells, the osmotic concentration drops, water diffuses out, and the guard cells become flaccid, closing the stomata.

When the stomata are open, gas exchange can take place, but at the same time the plant loses water by evaporation through the stomata, in a process called ***transpiration.*** But though most plants lose large quantities of water by transpiration every day, the humidity in the intercellular spaces of the leaf probably does not often drop appreciably, because the lost water is steadily replaced by water drawn up through the stem from the roots and distributed throughout the leaf by the many small veins.

The Living Cells of Stems and Roots Also Exchange Gases

Gas exchange is not confined to the leaves, though these organs are beautifully adapted for this process. Young stems also have stomata through which gases are exchanged, but older stems have impermeable bark on their surface, so gas exchange generally takes place through numerous ***lenticels.*** Lenticels are groups of loosely arranged cells with many spaces between them through which gases move to reach the interior tissues (Fig. 16.23). Since most of the cells in the inner layers of large

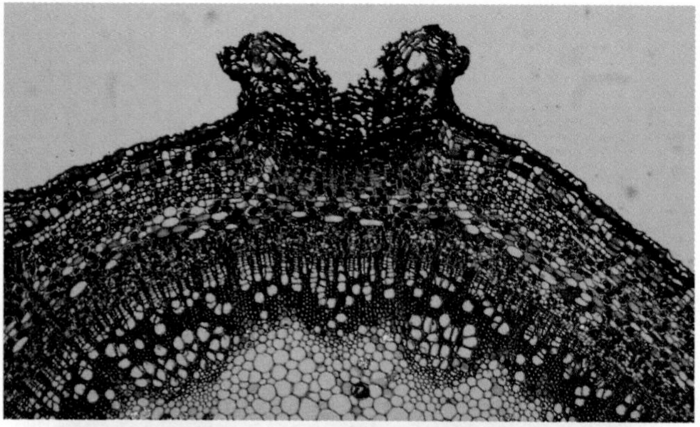

16.23 Lenticels Top: The waterproof outer bark (layer of dark cells on the surface) on this section of elderberry *(Sambucus)* stem is interrupted at the center of the lenticel. Thus the more loosely arranged cell layers beneath, with their numerous intercellular air spaces, are exposed to the atmosphere. Bottom: The individual lenticels can be seen as white areas on the surface of a young sycamore maple *(Acer)* stem.

MAKING CONNECTIONS

1. Why do green plants require oxygen even though it is produced as a by-product of photosynthesis?

2. In Chapter 5, you learned that the upper limit of the size of cells is related to their need to provide a large enough surface area for the exchange of materials with their environment. How are multicellular organisms such as plants—which are often billions of times larger than a single cell—able to provide enough surface area to obtain water and nutrients and to exchange gases?

3. How does the structure of the leaf epidermis demonstrate the compromises inherent in evolutionary adaptation?

4. Describe how the processes of active transport and osmosis regulate the turgidity of guard cells.

5. Why are special gas-transportation mechanisms found in animals but not in plants?

stems are dead, there is little need for oxygen in the intercellular spaces to penetrate deep into the stem.

Root cells also carry out gas exchange. In young roots, gases can diffuse readily across the membranes of root hairs and other epidermal cells, which present an enormous surface area for gas exchange as well as for nutrient procurement. In older roots, the epidermis and cortex is often replaced by periderm; lenticels are present in such roots to allow the exchange of gases in the interior tissues. For all roots, however, the soil in which they grow must be well aerated to provide sufficient oxygen for the cells of the root. One of the benefits of hoeing, raking, plowing, or otherwise cultivating the soil, and of earthworm burrowing, is the increased air circulation these activities make possible.

Plants, unlike animals, do not seem to need any special gas-transporting mechanisms. Most of the intercellular spaces in the tissues of land plants are filled with air, in contrast with those in animal tissues, which are filled with fluid. These air-filled spaces are interconnected to form the intercellular apoplast system that opens to the outside through stomata and lenticels and penetrates to the innermost cells of the plant body. Thus, incoming gases can move through the apoplast directly to the internal parts of the plant without having to cross membranous barriers. They do not have to diffuse long distances through water or cellular fluids, because they do not go into solution until they reach the film of water on the surfaces of the individual cells. Since oxygen can diffuse some 10,000 times faster through air than through liquids, the intercellular air-space system ensures that all cells, even the more internal ones, are adequately supplied.

CONCEPTS IN BRIEF

1. All organisms require proteins, carbohydrates, lipids, and nucleic acids to carry out their life functions. Organisms capable of manufacturing their own organic nutrients from inorganic raw materials are *autotrophic;* those that cannot and require prefabricated organic nutrients from the environment are *heterotrophic.*

2. Most autotrophs are *photosynthetic;* they use light energy to drive the synthesis of organic compounds. Green plants, the largest group of photosynthetic organisms, require carbon dioxide, water, oxygen, and certain mineral nutrients. The minerals are required for the synthesis of the organic molecules needed by the plant.

3. Minerals are absorbed in ionic form from soil water. The availability of inorganic ions depends on the fertility and pH of the soil. Simple and facilitated diffusion and active transport are all involved in the absorption process. Most ions are absorbed by active transport.

4. Leguminous plants form a mutualistic association with certain bacteria, which can convert nitrogen gas into ammonia that can then be used by the plant for synthesizing amino acids and other organic constituents. The plant in turn provides the bacteria with organic compounds necessary for its nutrition and growth.

5. The carbon dioxide required by the plant for photosynthesis is absorbed by diffusion by cells in the interior of the leaf. Tiny openings, or *stomata,* in the *epidermis* allow the carbon dioxide to penetrate to the interior of the leaf, where it is available to individual cells.

6. Most higher land plants take in water and mineral nutrients from the soil through their roots. A typical root is extensively branched and has many *root hairs,* which vastly increase the total absorptive surface. Mycorrhizal associations with vascular plant roots also help plants to absorb water and ions, and may enable the plant to survive in harsh environments. Roots also store organic compounds and anchor the plant to the soil.

7. Water from the soil may be absorbed directly into the epidermal cells and move from cell to cell through the *cortex* to the *stele* by flowing through the *symplast.* Alternatively, it may move across the epidermis and cortex without entering a cell by flowing through the cell walls and intercellular spaces that make up the *apoplast.* However, the *endodermis,* which separates the cortex from the vascular cylinder, presents a barrier to its movement. The side and end walls of the endodermal cells contain a *Casparian strip* that forms a waterproof seal between the endodermal cells. Consequently, all materials must enter the endodermal cells to reach the stele, and this allows the plant to exercise control over the movement of substances into the vascular tissue.

8. A few photosynthetic plants supplement their inorganic diet with organic compounds obtained by trapping and digesting insects and other small animals.

9. Since aerobic respiration is the chief method of respiration in both plants and animals, most organisms, including plants, must have some method of obtaining oxygen and getting rid of waste carbon dioxide. Plants also require carbon dioxide for photosynthesis.

10. Gas exchange between a living cell and its environment always takes place by *diffusion* across a cell membrane. Therefore, there must be a surface area of adequate dimensions, which must be kept moist and protected from mechanical damage.

11. The cells of most small aquatic autotrophs are adjacent or close to the surrounding medium and carry out gas exchange across the whole body surface; no special gas-exchange devices are necessary.

12. In land plants, gas exchange in the leaves takes place across the membranes of the loosely packed mesophyll cells in the interior of the leaf. The epidermis with its waxy cuticle protects the mesophyll from mechanical damage and excessive water loss. Tiny openings—*stomata*—in the epidermis allow the gases to penetrate to the interior, where they circulate through the numerous intercellular spaces.

13. The size of each stoma is regulated by two *guard cells.* When the guard cells are turgid, the stoma is open; when the guard cells lose water, the stoma is closed. During the day, the stomata are open, and gases can move into the interior of the leaf, but at the same time the plant loses water by *transpiration.* The stomata close at night, preventing excessive water loss.

14. Root cells obtain oxygen by simple diffusion across the membranes of the individual cells; air reaches the interior cells via the intercellular spaces of the apoplast system. Gas exchange in older stems and roots takes place through lenticels connected with the extensive apoplast system.

STUDY QUESTIONS

1. Define *autotroph* and *heterotroph.* What is the ecological relationship between these two categories of organisms? (p. 324)
2. Explain briefly why plants need each of the macronutrient elements listed in Table 16.1. (p. 326)
3. In general, what are the roles of micronutrients in plants? How are their functions related to the fact that plants require them in such small quantities? (p. 326)
4. In what form and by what processes do plants absorb mineral nutrients? How are they able to absorb many nutrients even against unfavorable concentration gradients? (pp. 326–328)
5. What is *nitrogen fixation* and why is it important for plant nutrition? What kinds of organisms can fix nitrogen? What is the significance of the nodules found on the roots of legumes? (pp. 328–329)
6. How is the structure of a leaf adapted to the function of obtaining carbon dioxide for photosynthesis? (p. 329)
7. How are roots able to provide a sufficiently large surface area for the absorption of water and nutrients? Distinguish between a *taproot system* and a *fibrous root system.* (pp. 329–330)
8. How does a plant use concentration gradients to move water from the soil to the stomata of a leaf? Be sure to name and describe the function of all regions through which the water would pass. (pp. 334–336)
9. Contrast the *apoplast* and *symplast* as routes for water movement in the root. Describe the structure and function of the endodermis. (pp. 334–336)
10. What is the main nutritional benefit for plants of being carnivorous? Why are such plants principally found growing in acid bogs and volcanic clays? (p. 337)
11. How are gases exchanged between cells and their environment? Is active transport ever involved? Can gases be exchanged without the presence of water? (pp. 337–338)
12. What are the requirements of a gas-exchange surface? Why are these requirements more challenging for a large terrestrial plant than for a single-celled aquatic organism? (pp. 338–339)
13. Describe how a plant regulates the opening and closing of its stomata so as to minimize transpiration while permitting gas exchange. Under what conditions does the plant normally close its stomata? (pp. 339–340)
14. How do stems and roots exchange gases with the environment? How does air reach interior cells in these structures? (pp. 340–341)

CHAPTER 17

NUTRIENT PROCUREMENT IN HETEROTROPHIC ORGANISMS

Heterotrophic organisms cannot manufacture their own high-energy compounds from low-energy inorganic raw materials. Yet they, like all living things, depend on such compounds for both maintenance and growth. They must, therefore, obtain prefabricated high-energy organic nutrients.

There are four main groups of heterotrophic organisms: the nonphotosynthetic bacteria, the protozoa, the fungi, and the animals. The bacteria and the fungi lack internal digestive systems and hence depend mainly on absorption as their mode of feeding; they are considered ***absorptive heterotrophs.*** Such organisms are usually either ***saprophytic*** (living and feeding on dead organic matter) or ***parasitic*** (living on or in other living organisms and feeding on them) (Fig. 17.1). By contrast, the principal mode of feeding for animals and protozoans is ingestion—the taking in of particulate or bulk food, which must then be digested. These organisms are said to be ***ingestive heterotrophs.*** Animals and protozoans may be ***herbivores,*** eating green plants and thereby obtaining high-energy compounds directly from the organisms that first made them, they may be ***carnivores,*** eating the animals that ate the plants, or they may be ***omnivores,*** eating both plant and animal material (Fig. 17.2). Whether a heterotrophic organism is saprophytic or parasitic or herbivorous, carnivorous, or omnivorous, it is clear that its energy-yielding nutrients come originally from green plants, which used radiant energy from the sun to make them.

In many ways, the absorptive and the ingestive organisms are as different from each other as each is from the green plants, with their photosynthetic mode of nutrition. Indeed, much of the diversity among living things can be viewed as reflecting alternative adaptations based on the three major nutritive modes—photosynthetic, absorptive, and ingestive.

Chapter opening photo: Raccoon at garbage can. Raccoons eat both plant and animal material, they have adapted very well to living with humans, often inhabiting the storm sewers in cities and towns and coming out at night to raid garbage cans.

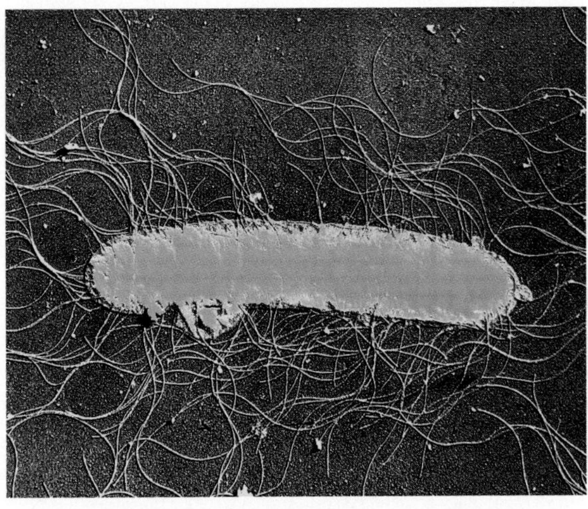

A

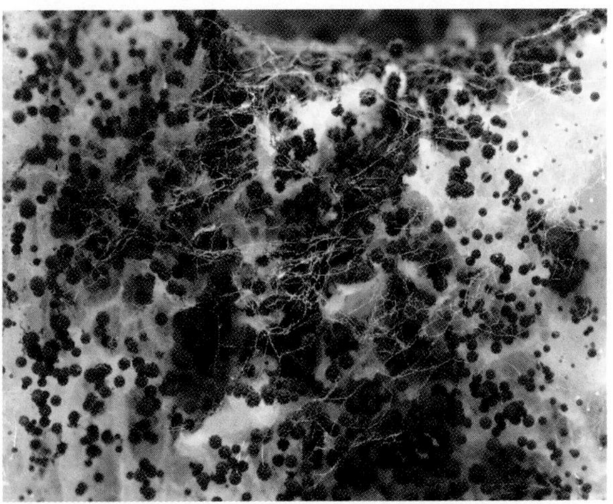

B

17.1 Two absorptive heterotrophs (A) *Proteus mirabilis,* a heterotrophic bacterium. Bacteria of this species are normal inhabitants of the human intestine, where they absorb nutrients from the host's intestines. *Proteus* can cause infections when it gets into the urinary tract or in debilitated patients. It can move about using its flagella, enabling *Proteus* strains to spread rapidly. (B) Black bread mold, *Rhizopus nigrans.* The fungal body usually consists of microscopic, tubular threads (white) that branch and spread over the surface of the food supply. Digestion is extracellular; digestive enzymes are secreted directly into the food supply and the products of digestion are then absorbed. The black spheres are reproductive structures.

A

B

17.2 Two ingestive heterotrophs (A) A hippopotamus, an herbivore. (B) A leopard, a carnivore, with its prey (a Thomson gazelle).

All Heterotrophs Require Certain Nutrients in Bulk

Carbohydrates, fats, proteins, and water are the main macronutrients for heterotrophic organisms. The carbohydrates (Fig. 17.3) serve primarily as energy sources—glucose and other simple sugars provide the fuel for glycolysis and aerobic respiration. But carbohydrates perform a second very important function in addition to being a source of energy; they provide the carbon skeletons for the synthesis of new organic compounds. Some heterotrophs—many bacteria and fungi, as well as a few protozoans—can flourish on a diet consisting solely of carbohydrates and minerals. They need no protein in their diets, because they, like plants, can combine inorganic nitrogen with carbon skeletons from carbohydrates to make amino acids. Similarly, they can synthesize for themselves all the other compounds (e.g., proteins, fats, and nucleic acids) necessary for life. For most heterotrophs, however, a diet limited to carbohydrates and minerals would be inadequate. Animals, especially, are unable to synthesize many of the complex organic molecules necessary for life. Organic nitrogen (i.e., amino acids and proteins) in particular is required by most animals, and for them a diet restricted to carbohydrates would soon be fatal. Animals, then, must consume the macronutrients that are synthesized by plants. Animals have evolved many specific adaptations—body plans, modifications of the digestive tracts, alternative feeding styles, etc.—to maximize the efficiency with which they obtain nutrients from the environment.

In the last few decades the science of nutrition has improved the health of individuals by determining the precise amounts of lipids, proteins, carbohydrates, and vitamins that must be included in the diet to promote health and well-being. The human diet must include an adequate number of calories

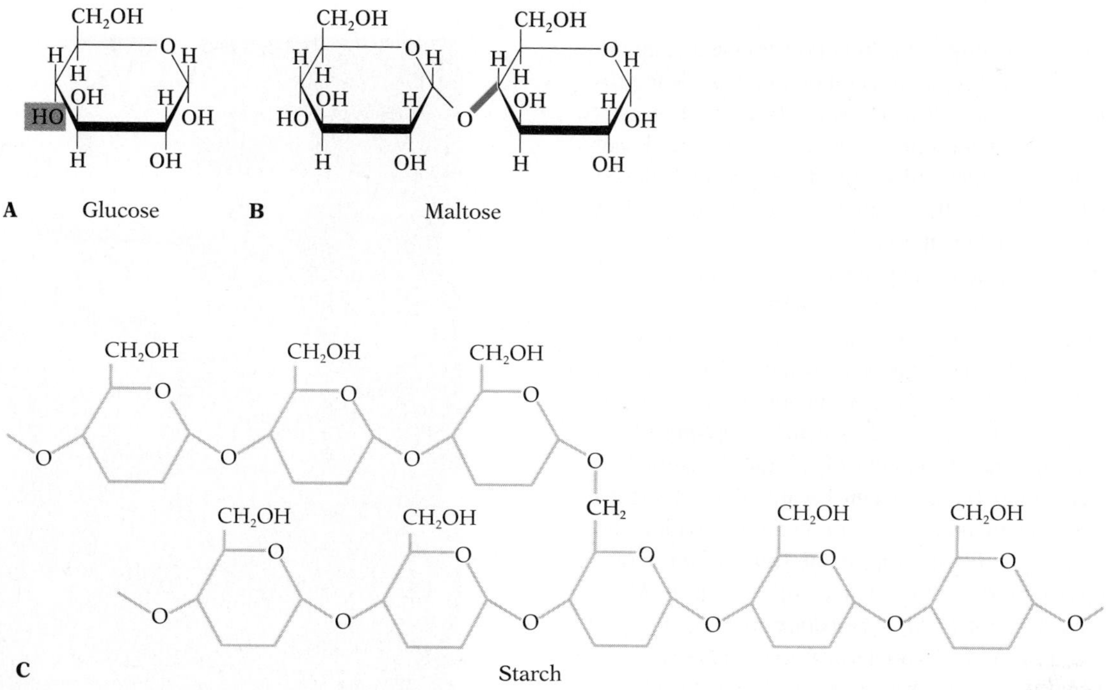

17.3 Some representative carbohydrates The basic building-block molecules of the carbohydrates are the simple sugars, or monosaccharides (A). When two simple sugars are bonded together a double sugar or disacccharide is formed (B). When many simple sugars are bonded together in long chains, a polysaccharide such as starch (C) is formed. The carbohydrates are an important energy source for all organisms.

and sufficient water, protein, fat, vitamins, and minerals. When caloric intake is too low, the body stores of protein and fat are broken down; when it is too high, obesity results. The average middle-class American diet now derives 40 to 50 percent of the calories from carbohydrates, 35 to 40 percent from fat, and 15 percent from protein. The American Heart Association has recently recommended that, for health reasons, we limit our fat intake to no more than 30 percent of calories in the diet.

Animals Require Certain Essential Amino Acids in Their Diet

Among the extensive dietary requirements of animals is a continuous supply of proteins or the amino acids (Fig. 17.4) of which they are composed. Most animals have lost the ability to synthesize certain amino acids and must get them in their diets. These are called the ***essential amino acids,*** a somewhat misleading term, since it seems to imply that the other amino acids commonly occurring in proteins are not essential. What is meant is that the designated amino acids are *essential in the diet,* whereas the others, which are also necessary for life, can be synthesized by the organisms itself from other amino acids or organic nitrogen compounds.[1] A continuous supply of protein is necessary in the diet because there is limited storage of amino acids in the body.

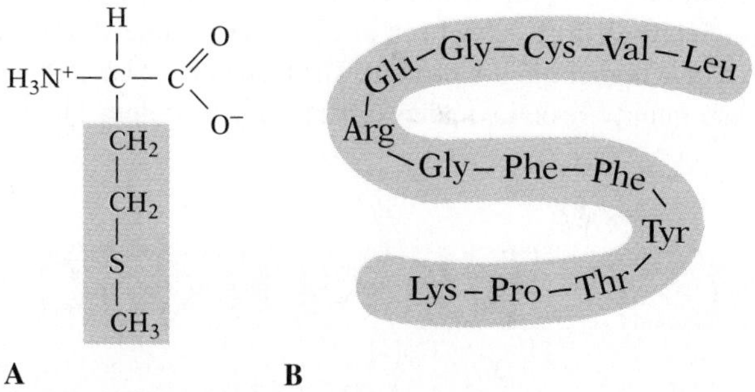

17.4 An amino acid and a portion of a protein Amino acids (A) are the basic building-block molecules of the proteins. All the amino acids have the same basic structure but differ in their side chains, or R groups (beige). Amino acids are joined together by condensation reactions to form a protein (B).

The source of protein is important, since a single protein may not include all of the essential amino acids. Animal proteins, such as meat, eggs, and fish, are considered high-grade proteins since they contain all the essential amino acids and in the proper proportions. Plant proteins, however, supply different proportions of amino acids, and most lack one or more of the essential ones. For this reason, plant proteins are less reliable than animal proteins as a source of essential amino acids for human beings. For example, zein, the main protein in corn, lacks sufficient tryptophan and lysine. Someone who depended exclusively on zein for protein would suffer from a deficiency

[1]The eight essential amino acids for humans are lycine, leucine, phenylalanine, isoleucine, tryptophan, valine, threonine, and methionine. Meat, cheese, eggs, and milk are good sources of these amino acids.

not only of these two amino acids but of other essential amino acids as well since effective utilization of amino acids in protein synthesis requires that *all the essential amino acids be present simultaneously in the correct relative amounts.* If one amino acid is deficient, then utilization of the others is reduced proportionately, and since they cannot be stored they are lost through metabolism and excretion.

One way to avoid a deficiency of an essential amino acid is to include a variety of different proteins in the diet, since it is unlikely that they all will be deficient in the same amino acids. Unfortunately, this is not possible in many regions of the world where food is limited. A large proportion of the world's population suffers from some form of ***protein-energy-malnutrition,*** or ***PEM.*** Sometimes the malnutrition is due to insufficient caloric intake, but more often it occurs because the diet is inadequate with respect to protein and/or vitamins. Within the United States many children living in the slums of our larger cities and in poverty-stricken rural areas suffer from PEM. A particular severe form of PEM is ***kwashiorkor*** (Fig. 17.5), a protein-deficiency disease characterized by degeneration of the liver, sever anemia, and inflammation of the skin. Kwashiorkor is particularly common among children in countries where the diet consists primarily of a single plant material, as in Indonesia, where rice is the main food, and in parts of Africa and Latin America, where corn (maize) is the principal staple. Worldwide, corn is the second or third largest crop and it is a staple for nearly half the world's malnourished people. For this reason intense efforts have been directed at breeding strains of corn with a higher quality protein. The results have been spectacular; quality protein maize (QPM) has a protein quality

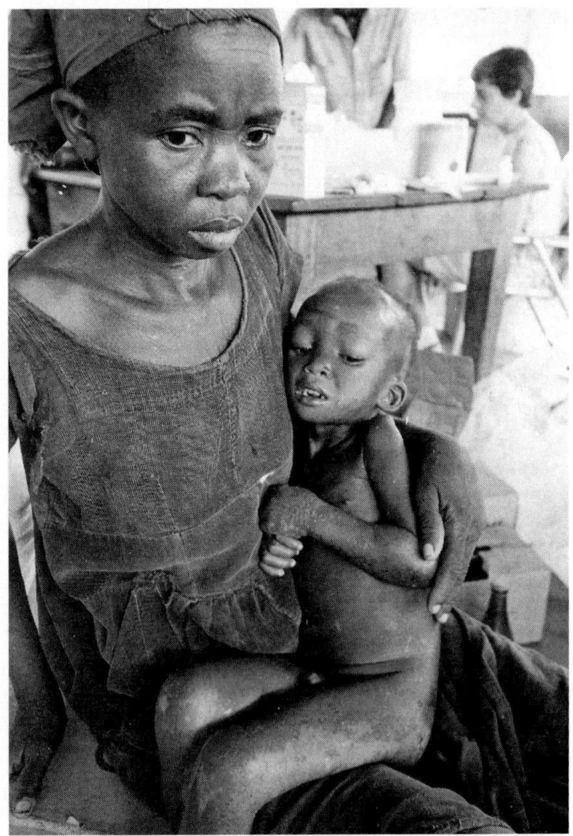

17.5 A child suffering from extreme kwashiorkor Notice the inflammation of the skin, especially on the arms and legs. Kwashiorkor is especially serious among children because it leads to stunted growth.

MAKING CONNECTIONS

1. How are the three major nutritive modes—photosynthetic, absorptive, and ingestive—related to the classification of organisms into various kingdoms, as described in Chapter 1?

2. Why do you require protein in your diet, whereas many bacteria and fungi can flourish on a diet consisting solely of carbohydrates and minerals?

3. Why are only certain amino acids considered "essential," if all 20 are needed to make protein?

4. Why is a strict vegetarian who is not knowledgeable about nutrition more likely to suffer from protein deficiency than a person who eats meat? How do knowledgeable vegetarians avoid this problem?

5. Among the symptoms of kwashiorkor are severe anemia and sores on the skin that fail to heal. Why would protein deficiency produce these symptoms?

twice that of normal corn and almost equal to milk. QPM is now being field tested in different regions for adaptability to different climates, rainfall, temperature, soils, pests, etc. If the field trials are successful, it will be good news for the impoverished populations in Mexico and Central America where maize constitutes 85 percent of the grain consumed.

A human on a vegetarian diet must take care to select a combination of plant proteins that will complement one another, making up for one another's deficiencies. For example, the proteins in beans are deficient in methionine, whereas those in wheat are deficient in lysine; if both beans and wheat (e.g., in bread or rice) are eaten at the same meal, they will complement each other and there will be no deficiency of either methionine or lysine. Also, the protein intake must be larger than if the person were eating meat, because of the wastage of amino acids.

We have stressed the necessity of having enough protein and amino acids in the diet for good reason: proteins play important structural roles in the body, as parts of the cell membrane, as a major component in muscle, skin, bone, tendons, hair, and they are necessary for growth, repair, and maintenance of the body. You may recall also that enzymes are proteins, and that enzymes catalyze all the chemical reactions within the cell. In addition, the organic nitrogen in amino acids can be used to synthesize the necessary nucleic acids, phospholipids, and other nitrogen-containing compounds.

Fats Are a Concentrated Source of Energy

Fats are highly reduced compounds (Fig. 17.6) very rich in energy, having more than twice the caloric content of carbohydrates or proteins (Table 17.1). Because they are such a concentrated source of energy, they are often used for energy storage. In addition, they provide insulation, cushioning, and protection for certain organs within the body.

Most animals can convert carbohydrate into fat and deposit it as a storage product in their bodies. This enables them to survive and grow with little or no fat in their diets. Some animals (including rats and humans) cannot synthesize enough linoleic acid (a common fatty acid), which is necessary for the synthesis of certain fats (Fig. 17.6). For such animals linoleic acid is an ***essential fatty acid,*** which must be included in the diet.

TABLE 17.1 *Caloric value of foods*

Nutrient	*Energy value (kilocalories/gram)*
Carbohydrates	4.1
Proteins	4.3
Fats	9.3

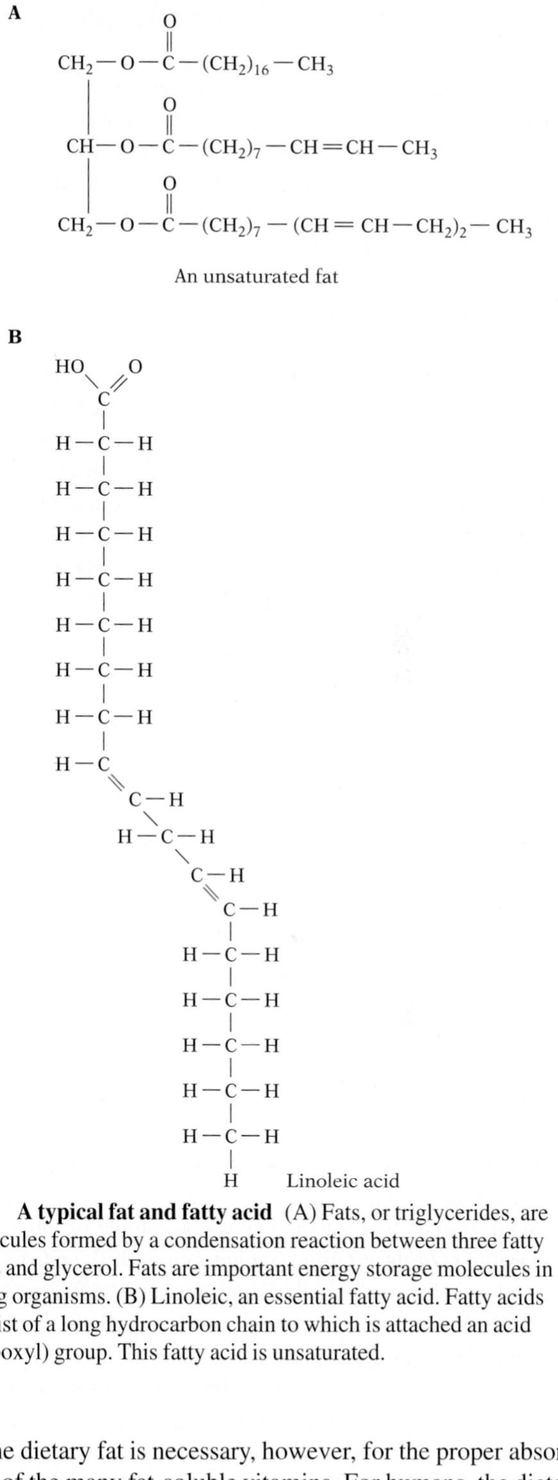

A

$$CH_2-O-C(=O)-(CH_2)_{16}-CH_3$$
$$CH-O-C(=O)-(CH_2)_7-CH=CH-CH_3$$
$$CH_2-O-C(=O)-(CH_2)_7-(CH=CH-CH_2)_2-CH_3$$

An unsaturated fat

B

$$HO-C(=O)-(CH_2)_7-CH=CH-CH_2-CH=CH-(CH_2)_4-H$$

Linoleic acid

17.6 A typical fat and fatty acid (A) Fats, or triglycerides, are molecules formed by a condensation reaction between three fatty acids and glycerol. Fats are important energy storage molecules in living organisms. (B) Linoleic, an essential fatty acid. Fatty acids consist of a long hydrocarbon chain to which is attached an acid (carboxyl) group. This fatty acid is unsaturated.

Some dietary fat is necessary, however, for the proper absorption of the many fat-soluble vitamins. For humans, the dietary requirement for fat is small, since excess carbohydrate, fat, or protein is converted into fat. A diet with only 5 percent of its calories as fat would be sufficient to meet the dietary requirement, yet the average American diet is about 40 percent fat. The type of fat ingested is also important. The evidence suggests that a high ratio of unsaturated to saturated fats may be important in preventing atherosclerosis.

Early Humans Were Probably Omnivorous

The nutrient requirements for our species reinforce most current theories of human evolution, which indicate that early humans evolved as omnivorous hunter-gatherers. They were undoubtedly subjected to regular shortages of food, and those who were able to efficiently store excess energy as fat had a selective advantage. Most of their diet was probably plant material, with periodic animal protein and fat. These humans had no need to synthesize certain of the complex organic molecules necessary for life because they were provided ready-made in the diet. Eventually these pathways have been lost; thus a strict vegetarian diet is now inherently artificial in our species, and care must be taken to balance the amino acids. One can certainly obtain all the required amino acids in a strict vegetarian diet, as long as the person knows the amino acid content of the various vegetables and eats the right proportion of different vegetables at each meal. To be safe, most vegetarian diets include some animal products (usually dairy products).

Vitamins and Certain Minerals Must Be Supplied in the Diet

Vitamins are organic compounds necessary in small quantities for given organisms that cannot synthesize them and must therefore obtain them prefabricated in the diet. A compound may be a vitamin for species A and not for species B, because B can synthesize it. Guinea pigs, for instance, cannot synthesize vitamin C and must obtain it in their diet. Rabbits, on the other hand, can synthesize all the vitamin C they require and do not need to ingest it. Vitamins are necessary only in very small quantities, because they usually function as coenzymes or as parts of coenzymes, which can be reused many times and hence are not needed in large amounts. So, vitamins are important in cellular metabolism.

Vitamins can be classified as water-soluble or fat-soluble. The ***water-soluble*** vitamins include vitamin C and the vitamins of the B complex (riboflavin, niacin, B_6, B_{12}, and folic acid). They function as coenzymes in metabolic reactions that take place in almost all animal cells. Some animals can synthesize one or more of these coenzymes, and so for them, such coenzymes are not vitamins and are not required in the diet. The compounds collectively known as the ***fat-soluble*** vitamins (vitamins A, D, E, and K) are vitamins only for vertebrate animals (vitamin E, which is also required by a few invertebrates, is an exception). The fat-soluble vitamins serve a variety of different functions (Table 17.2) but generally do *not* function as coenzymes.

It has been difficult to establish reliable minimum daily requirements for the vitamins for humans and animals. Those supposedly established are still very much open to question. Little is known, for example, about how requirements alter with age or with changing health. Although much research remains to be done, one thing can be asserted with reasonable confidence: healthy persons who eat a varied diet including meats, fruits, and vegetables will probably get all the vitamins they need, numerous advertisements to the contrary notwithstanding. (See Exploring Further: Vitamin Deficiency and Its Treatment, p. 350.)

Like green plants, heterotrophs require certain minerals, which are usually absorbed as ions. Some minerals, such as sodium, chlorine, potassium, phosphorus, magnesium, and calcium, are macronutrients and are needed in relatively large quantities; others, such as iron, manganese, and iodine, are micronutrients, needed in much smaller quantities. And still others, such as copper, zinc, molybdenum, selenium, and cobalt, though essential to life, are needed only in trace amounts.

Calcium is a major constituent of bones and teeth in vertebrates and plays a variety of other roles in most organisms (e.g., in nerve impulse transmission and muscle contraction). Phosphorus is a component of phospholipids and nucleic acids. Iron is a constituent of the electron-transport molecules and of hemoglobin. Sodium, chlorine, and potassium ions are important components of the body fluids, playing vital roles in maintaining proper osmotic concentration and in such processes as nerve and muscle action. Iodine is a component of the hormones produced by the thyroid gland. But the functions of some of the minerals, particularly those needed only in trace amounts, are less obvious. Apparently most of them act as components of enzymes, or perhaps as cofactors that help catalyze reactions without being actually incorporated into enzymes or coenzymes. Like vitamins, the minerals needed by each species come from the normal diet.

All Fungi Are Absorptive Heterotrophs

The fungi constitute a large and diverse group of sedentary organisms; they are absorptive heterotrophs that live on or in their food supply and digest and absorb nutrients from these immediate sources. The fungal body usually consists of microscopic, tubular threads that branch and spread over the surface of the food supply. Bread mold is a familiar example (Fig. 17.1B). The bread on which the fungus grows is composed mostly of starch, a rich source of energy. But starch is a polysaccharide, whose very large and insoluble molecules cannot move across the cell membranes of the mold. Before absorption can take place, the starch must be broken down to its constituent building-block compounds, the simple sugars; in short, the starch must be digested. ***Digestion*** is simply another name for enzymatic hydrolysis, which involves the addition of water to break the bonds between the simple sugars. In bread mold the hydrolysis takes place outside the cells, and the process is called ***extracellular digestion.*** Digestive enzymes synthesized inside the cells of the mold are released from the cells into the

TABLE 17.2 *Some vitamins needed by human beings*

Vitamin	*Functions*	*Some deficiency symptoms*	*Important sources*
FAT-SOLUBLE			
Vitamin A (retinol)	Necessary to form visual pigments in eye, for normal growth of epithelial cells; "antiinfection" vitamin	Dry, brittle epithelia of skin, respiratory system, and urogenital tract; xerophthalmia and night blindness	Green and yellow vegetables and fruits, dairy products, egg yolk, fish-liver oil, liver, kidney, animal fat
Vitamin D (calciferol)	To increase Ca^{2+} ion absorption from digestive tract	Rickets or osteomalacia (very low blood-calcium level, soft bones, distorted skeleton, poor muscular development)	Sunlight; egg yolk, milk, fish oils
Vitamin E (tocopherol) Need in humans not definitely established	Antioxidant of unsaturated fats	In rats, malfunction of muscular and nervous systems; anemia (from rupture of red blood corpuscles); male sterility	Widely distributed in both plant and animal food, such as meat, egg yolk, green vegetables, seed oils, grains; intestinal bacteria
Vitamin K (phylloquinone, etc.)	Necessary for formation of blood-clotting factors	Slow blood clotting and hemorrhage	Green vegetables; intestinal bacteria
WATER-SOLUBLE			
Thiamine (B_1)	As coenzyme in stage 2 of respiration	Beriberi (muscle atrophy, paralysis, mental confusion, congestive heart failure)	Whole grains, yeast, nuts, liver, meat
Riboflavin (B_2)	As coenzyme (in FAD)	Vascularization of the cornea, conjunctivitis, and disturbances of vision; sores on the lips and tongue; disorders of the liver and nerves in experimental animals	Milk, cheese, eggs, yeast, liver, wheat germ, leafy vegetables, grains
Pyridoxine (B_6)	As coenzyme in reactions involving amino acid and protein metabolism	Convulsions, dermatitis, impairment of antibody synthesis	Whole grains, fresh meat, eggs, liver, fresh vegetables
Pantothenic acid	As coenzyme (acetyl CoA) in stage 2 of respiration	Impairment of adrenal cortex function, numbness and pain in toes and feet, impairment of antibody synthesis	Present in almost all foods, especially fresh vegetables and meat, whole grains, eggs; intestinal bacteria
Biotin	A part of several metabolic enzymes	Clinical symptoms (dermatitis, conjunctivitis) extremely rare in humans, but can be produced by great excess of raw egg white in diet	Present in many foods, including liver, yeast, fresh vegetables
Nicotinamide	As coenzyme (in NAD and NADP)	Pellagra (dermatitis, diarrhea, irritability, abdominal pain, numbness, mental disturbance)	Meat, yeast, grains; intestinal bacteria
Folic acid	Active in synthesis of purines and thymine in DNA; promotes growth	Anemia, impairment of antibody synthesis, stunted growth in young animals	Leafy vegetables, liver, meat; intestinal bacteria
Cobalamin (B_{12})	As coenzyme for reducing ribonucleotides to deoxyribonucleotides; promotes growth; promotes formation and maturation of red blood cells	Pernicious anemia	Liver, meat; intestinal bacteria
Ascorbic acid (C)	Activates enzymes in collagen synthesis; necessary for growth of connective tissue, cartilage, bone, teeth	Scurvy (bleeding gums, loose teeth, anemia, painful and swollen joints, delayed healing of wounds, emaciation)	Fresh vegetables and fruits

bread and hydrolyze the starch into simple sugars, which are then absorbed (Fig. 17.7A).

Mold living on bread exemplifies a saprophytic way of life, but many fungi are parasitic (Fig. 17.7B). Indeed, bread mold itself is not restricted to saprophytic nutrition; it is one of the most common destructive fungi on fresh fruits and vegetables. The parasitic fungi employ basically the same mode of nutrition as saprophytes: enzymes are secreted onto the food supply

EXPLORING FURTHER

VITAMIN DEFICIENCY AND ITS TREATMENT

Successful treatment of vitamin deficiency can be much more complicated than might be imagined. In fact, recognition of the deficiency is only the first and perhaps the simplest step toward treatment, as the case of vitamin A deficiency in the Philippines will show.

In humans, a mild deficiency in vitamin A, the chief component of the light-sensitive pigment in the visual cells of the eye, can lead to *night blindness,* a marked impairment of vision in dim light (Fig. A). But the most serious effect of vitamin A deficiency is *xerophthalmia* ("dry-eye"), a condition in which the secretions of the eye glands dry up and the epithelial tissues of the eye are progressively destroyed, resulting in partial or complete blindness. Indeed, in many developing countries vitamin A deficiency is the most common cause of childhood blindness. It occurs particularly in areas where white corn, rice, or cassava forms the basic staple in the diet, because these foods contain no vitamin A or carotenoids (plant pigments that the body can convert into vitamin A). The World Health Organization considers vitamin A deficiency one of the four major nutritional problems in low-income countries and has given high priority to its elimination.

Because carotenoids are found in green leafy vegetables and some fruits, and because vitamin A is inexpensive to produce and administer, it should be easy, theoretically, at least, to eliminate vitamin A deficiency. In recent years control programs have been established in a number of countries, using either periodic mass dosing with vitamin A or fortification of suitable foods. The generally recommended method for mass dosing has been to give large doses of vitamin A every six months to children under six years of age, since they tend to be most seriously affected by the deficiency. The doses can be spaced, because 30 to 50 percent of the ingested vitamin is stored in the liver, available for later use. However, this approach is not as effective as one might expect. Just reaching the high-risk children to administer the regular doses is difficult; they usually come from the poorest families, which tend to move often and live in inaccessible places. Because of these problems, the alternative method, fortification of foods with vitamin A, is being tested in certain areas.

Fortification has the advantage of reaching all who eat the food, and does not require the active participation of the population or an expensive delivery system. Because vitamin A can be produced in stable liquid or dry forms, it can be added to a wide variety of foods, such as margarine, dairy products, cereals, sugar, and salt. The problem is to select a food that can easily be fortified and is consumed regularly by all members of the population. A pilot study in Cebu, the Philippines, was conducted to evaluate the effectiveness of a proposed fortification program. Unfortunately, in some areas of the Philippines, very few store-bought foods are widely consumed, since the people are accustomed to raising most of their own food and buying little. A careful study of local diets indicated that salt, monosodium glutamate (MSG), and flour products are the most frequently purchased

foods. Salt and flour were eliminated from consideration because each is marketed by dozens of small, independent manufacturers. In contrast, MSG seemed an ideal product for fortification; it is consumed regularly by over 95 percent of Philippine families, variation in the amount consumed is not large, and nearly all the MSG used in the Philippines is produced by one manufacturer. In the pilot project, fortified MSG was distributed to families in the study area. The results were encouraging; most signs of night blindness and early xerophthalmia were eliminated, and levels of vitamin A in the blood were significantly improved. The product is now being distributed in three provinces in the Philippines, and if the results are satisfactory, a national fortification program will be established. Such a program will go a long way toward eliminating vitamin A deficiency and the blindness that results.

Experiences with vitamin A mass dosing and fortification of foods have highlighted the difficulties associated with resolving nutritional problems. As is so often the case, the presence or absence of one nutrient influences the utilization of another. For example, the storage of vitamin A in the liver requires vitamin E; without sufficient vitamin E, vitamin A is not stored properly. For this reason, when mass dosing is used, vitamin E has to be given along with vitamin A. Another complication arises if the vitamin A deficiency is accompanied by a protein deficiency. When protein intake is inadequate, the two proteins necessary for the transport of vitamin A out of the liver may not be synthesized by the body in sufficient amounts. The vitamin will then remain stored in the liver, unavailable to other body tissues, and signs of vitamin A deficiency may appear. These are two illustrations of the point that in treating any vitamin deficiency, control measures must be aimed at improving the whole diet; treating one aspect may not solve the problem. And of course nutrition education and food-production activities must play a prominent role in every nutritional program. (For an indication of the major vitamins needed by humans, and their sources, see Table 17.2.)

A Night blindness Left: Road as seen by a normal individual. Right: Road approximately as seen under the same lighting conditions by an individual with a vitamin A deficiency. That person cannot see the road sign at all.

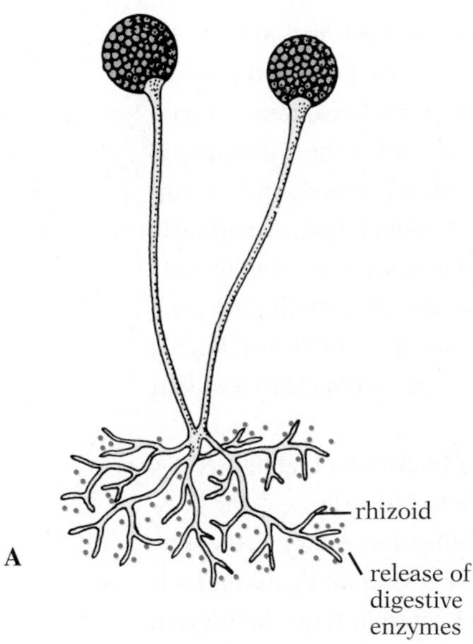

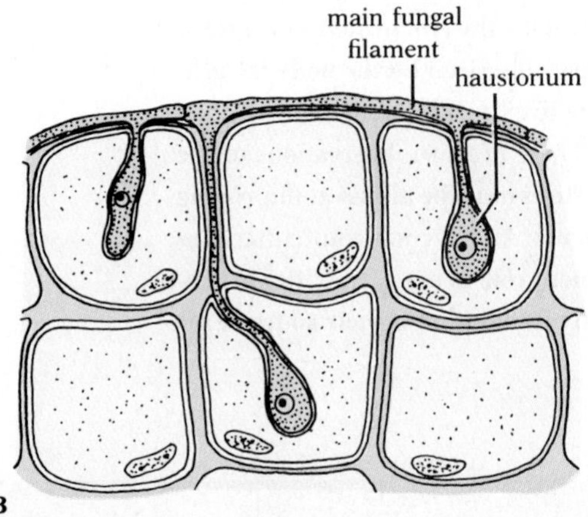

17.7 Nutrient-procurement structures of fungi Some of the threads that make up the body of a fungus are specialized for absorption. (A) The rhizoids of a saprophytic fungus. Digestive enzymes (red dots) are released from rhizoid cells and starch is hydrolyzed to simple sugars, which can then be absorbed. (B) The haustoria of a fungus parasitic on a multicellular plant. The body of the fungus (gray) is filamentous and can grow between the cells of the plant host. The haustoria penetrate the cell walls and make deep invaginations in the membranes of host cells, through which they absorb nutrients. Note that the haustoria are not in direct contact with the cytoplasm of the host cells, because the haustorial invaginations are lined with host-cell membrane.

on (or in) which the fungus lives; digestion takes place extracellularly; and the products of digestion are then absorbed by the fungus. Note that fungi, unlike most animals, have no internal cavity where bulk food can be digested; they simply release digestive enzymes onto their surroundings and absorb organic nutrients across the body surface, much as plant roots absorb inorganic nutrients.

Protozoans and Animals Are Ingestive Heterotrophs

Nutrient procurement by protozoans and animals usually involves much more activity than it does in plants and fungi. Animals must often resort to elaborate methods of locating and trapping their food. Like the fungi already examined, the ingestive heterotrophs must digest their food before it can cross the membranes of their cells. Usually their food material is in the form of polysaccharides, fats, proteins, and other large molecules, which must be digested into their building-block molecules before absorption. But animals rarely secrete digestive enzymes directly onto their food in the matter of fungi. The vast majority ingest particles or lumps of food into some sort of digestive structure, in which enzymatic digestion takes place. In some other organisms—the protozoans, in particular—food is ingested directly into a cell by phagocytosis or a similar process, and then digested in a food vacuole. Though this process is classified as ***intracellular digestion,*** the food material is actually separated from the rest of the cellular material by a membrane that it cannot cross until after digestion has occurred. Thus, intracellular digestion and extracellular digestion are alike in that the *digestion always precedes the actual absorption of complex foods across a membrane.*

Though both the nutritional requirements and the basic processes of digestion are essentially alike in protozoans and all types of animals from worms to human beings, the body plans of these organisms vary so greatly that the structures involved in food processing and the details of that processing are often very different. The larger the animal, and the more active it is, the greater its need for a larger or more efficient digestive system to meet its increased metabolic needs. In the following sections we shall briefly examine the digestive mechanisms of protozoans and a variety of animals to see how their digestive systems are adapted to their body plan and way of life.

Protozoans Digest Food Intracellularly

Since protozoa, as single-celled organisms, have a body plan obviously very different from that of multicellular animals, we would expect their adaptations for food procurement to be likewise markedly different. And the differences are, in fact, considerable. But a more interesting point, one with important biological implications, is that the similarities are often more striking than the differences. Even at the unicellular level, food must be digested in a specialized structure before it can be absorbed across a membrane into the cytoplasm.

For instance, an amoeboid protozoan, which constantly changes shape as its protoplasm flows along, pushes out new armlike pseudopodia and withdraws others (Fig. 17.8). When an amoeba is stimulated by nearby food, some of the pseudopodia may flow around the food until they have completely surrounded it. This is the process known as phagocyto-

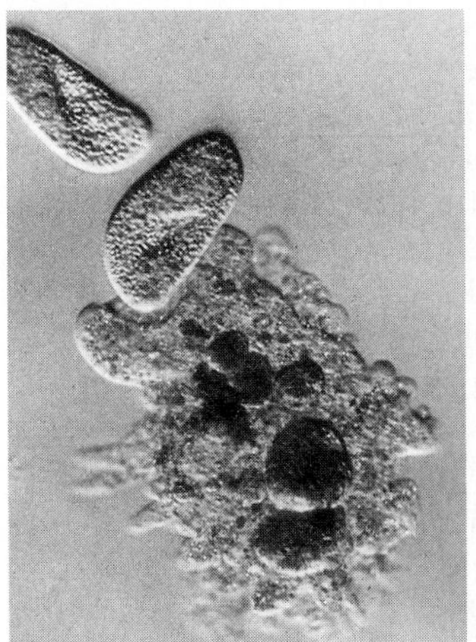
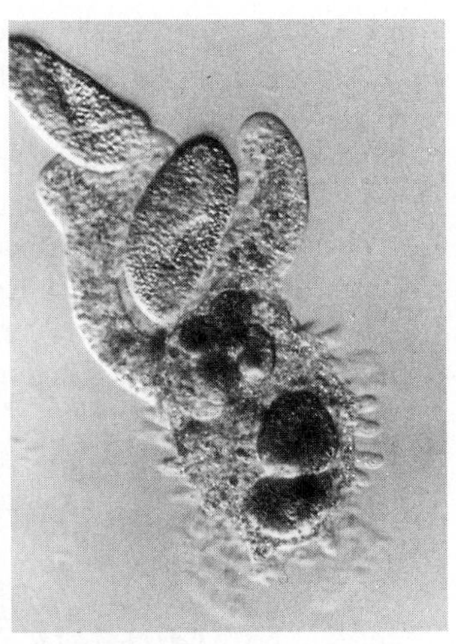
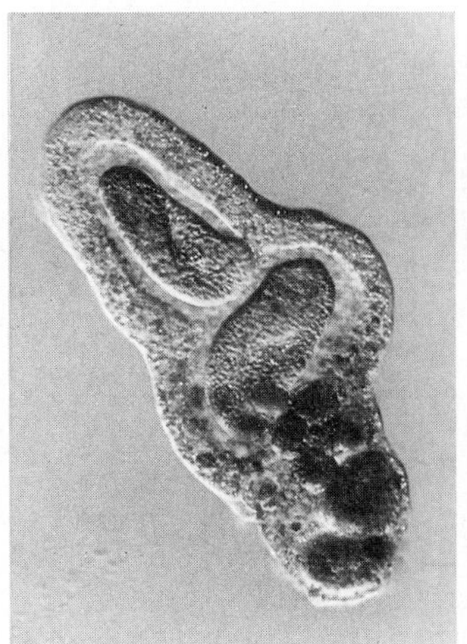
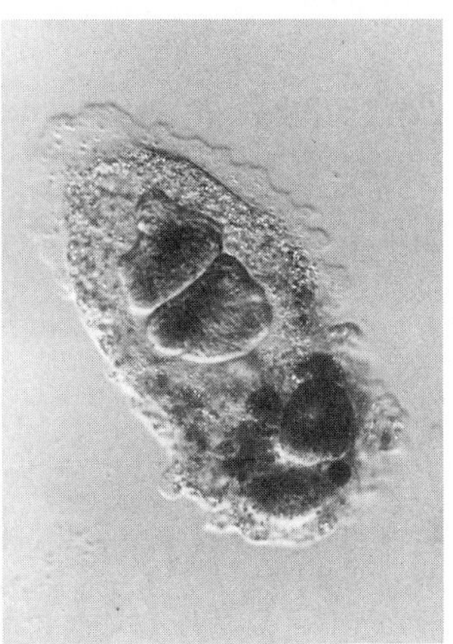

17.8 Phagocytosis of food by *Amoeba* Pseudopodia flow around the prey (two cells of *Paramecium*) until it is entirely enclosed within a vacuole.

sis (see Fig. 5.16, p. 103). The food is completely engulfed by the cytoplasm and is enclosed in a ***food vacuole*** (Fig. 17.9). Once a food vacuole is formed, a lysosome containing digestive enzymes soon fuses with it. Food materials and the digestive enzymes are mixed in the resulting digestive vacuole, and digestion begins. As digestion proceeds, the vacuole circulates in the cytoplasm, and the products of digestion (simple sugars, amino acids, etc.) diffuse across the membrane of the vacuole into the cytoplasm. The vacuole eventually fuses with the cell membrane and then ruptures, expelling the indigestible materials to the outside. Amoebas exemplify protozoans that lack specialized permanent digestive structures, though their food vacuoles correspond in function to the digestive systems of higher animals.

The ciliates are characterized by innumerable cilia covering the surface of their bodies (see Fig. 6.30, p. 130). Like all protozoans, they are commonly regarded as unicellular. Though they lack actual subdivision into recognizable cellular units, the more complex ciliates show much of the internal specialization usually associated with multicellularity. Unlike the amoeba, the *Paramecium* has a permanent structure, an organelle, that functions in feeding. Food particles are swept into an ***oral groove,*** a ciliated channel located on one side of the cell (Fig. 17.10), by water currents produced by the beating of the cilia, and are carried down the groove into a ***cytopharynx.*** As food accumulates at the lower end of the cytopharynx, a food vacuole forms around it. Eventually the vacuole breaks off and begins to move toward the anterior end of the cell. A lysosome fuses with it, digestive enzymes are secreted into the vacuole, and digestion begins. As digestion proceeds, the products are absorbed and the vacuole begins to move back toward the posterior end of the cell. When the vacuole reaches a tiny specialized region of the cell surface called the anal pore, it attaches

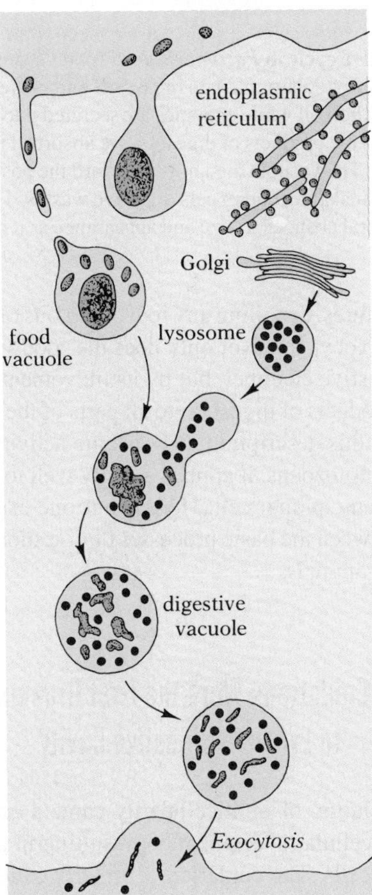

17.9 The role of lysosomes in intracellular digestion The digestive enzymes are synthesized on the ribosomes, move through the endoplasmic reticulum to the Golgi apparatus, and there become surrounded by a membrane to form the lysosome. Food material (brown) that the cell takes in by phagocytosis is enclosed in a food vacuole, which fuses with a lysosome containing the digestive enzymes. Digestion takes place within the composite structure thus formed (digestive vacuole), and the products of digestion are absorbed across the vacuolar membrane. The vacuole eventually fuses with the cell membrane and then ruptures, expelling digestive wastes to the outside.

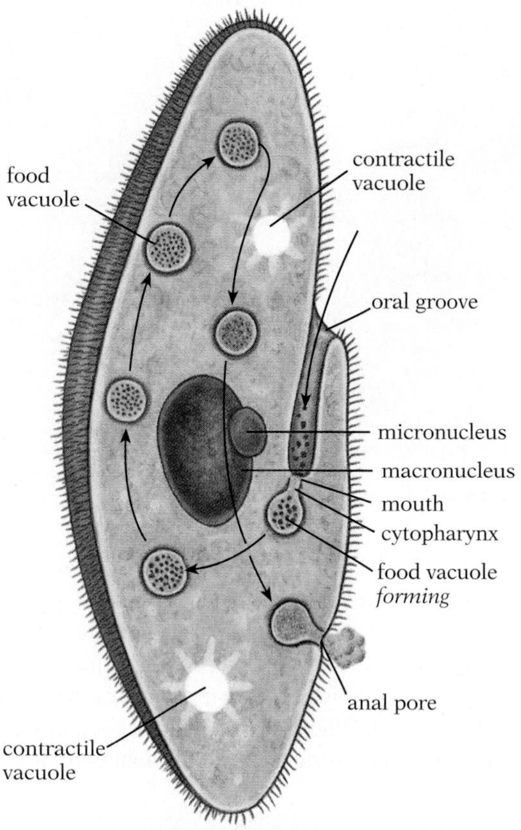

17.10 Digestive cycle in *Paramecium* A food vacuole forms at lower end of the cytopharynx, then breaks off and moves toward the anterior end of the cell while enzymes are secreted into it; digestion takes place, and the products of digestion are absorbed into the general cytoplasm. The vacuole then moves toward the posterior end, attaches to the anal pore, and expels digestive wastes. The vacuole undergoes several changes in size and appearance as it moves.

there and ruptures, expelling any remaining bits of indigestible material by exocytosis. Not only does the food vacuole function as a digestive chamber, but by its movement it helps distribute the products of digestion to all parts of the cell.

Although this description of lysosome activity pertains to digestion in protozoans, it applies equally well to intracellular digestion in any animal cell. Here is but one example of the similarity between the basic processes of digestion in protozoa and in higher animals.

The Cnidarians Were the First Organisms to Evolve a Digestive Cavity

With the evolution of multicellularity came a corresponding evolution of cellular specialization, resulting in a division of labor among cells. The cnidarians (formerly called coelenterates) provide a comparatively simple example of this phenomenon. These radically symmetrical animals have a saclike body composed of two principal layers of cells, with a jellylike layer, called mesolamella ("middle layer"), between them (Fig. 17.11). Within the central cavity of this saclike body, extracellular digestion takes place. This cavity has only one opening to the outside, which is surrounded by mobile tentacles. A digestive cavity of this sort, with a single opening that functions as both mouth and anus, is called a ***gastrovascular cavity.***

Cnidarians are strictly carnivorous. Embedded in their tentacles are numerous stinging structures called ***nematocysts.*** Each nematocyst consists of a slender thread coiled within a capsule, with a tiny hairlike trigger penetrating to the outside (Fig. 17.12). When appropriate prey comes in contact with the

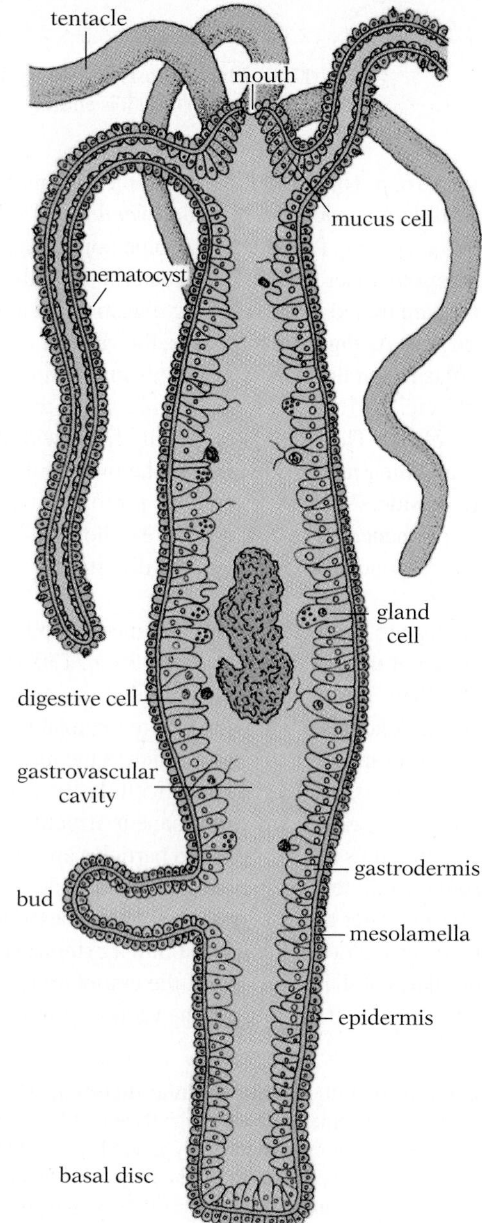

17.11 Hydra, showing gastrovascular cavity The cavity contains food material (red). The mesolamella layer in the body wall is much more extensively developed in some other cnidarians.

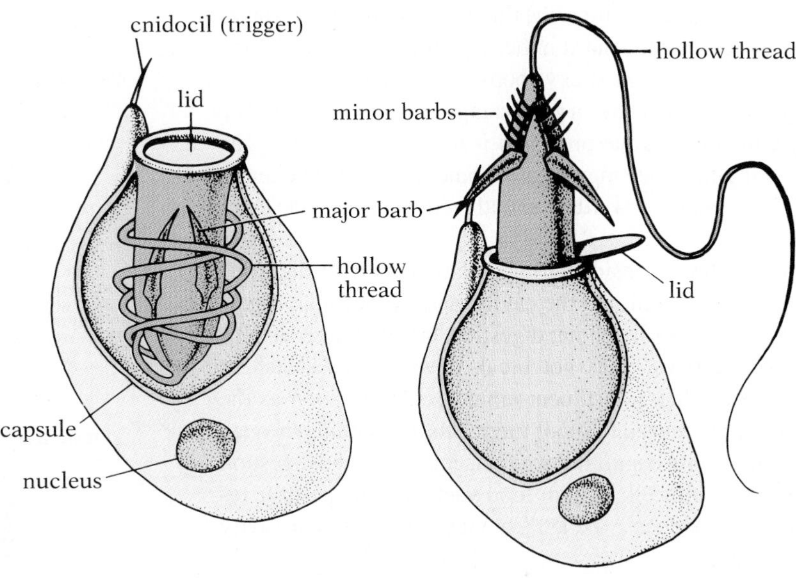

A B C

17.12 Nematocyst A Portuguese man-of-war has captured a fish (A). Numerous nematocysts along the tentacles function to sting and immobilize prey; human swimmers occasionally suffer such painful stings. Nematocysts are organelles contained in specialized cells known as cnidoblasts. This example of a nematocyst is shown before discharge (B) and after discharge (C). The minor barbs are part of a hollow thread that is folded back inside itself before discharge. Most nematocysts fire only when the trigger has been stimulated simultaneously in two ways—for example, by touch and by exposure to chemicals characteristic of the cnidarian's prey. The exploding nematocyst propels the major barb at the target, and after it has struck, the hollow thread begins to be everted, and the minor barbs are exposed and dig themselves into the target. The rest of the hollow thread is similarly ejected, and the poison flows through it into the victim. This is the source of the jellyfish's painful sting. Nematocysts are found all over the cnidarian body but are found in greatest numbers on the tentacles.

MAKING CONNECTIONS

1. Many nutritionists believe that the American diet would be healthier if it included less fat. What would be the consequences for your health if you carried their recommendation to extremes and completely eliminated fat from your diet?

2. What do the nutrient requirements of our species indicate about human evolutionary history? Why have humans lost many synthetic pathways that might occasionally be useful (e.g., for vegetarians) if they had been retained?

3. Which of the vitamins listed in Table 17.2 (p. 349) are essential participants in cellular respiration, discussed in Chapter 8? What is their general role?

4. Compare the general role that most mineral micronutrients play in animals with their role in plants. (See Table 16.1, p. 326)

5. How are the size and activity level of an animal related to the complexity and efficiency of its digestive tract?

6. Relate the process of intracellular digestion in a protozoan such as *Paramecium* to the process of phagocytosis and the action of lysosomes described in Chapters 5 and 6.

trigger, the nematocyst fires, the thread turns inside out, spines on its surface unfold, and it either penetrates the body of the prey or entangles it in sticky loops. Many nematocysts also eject poisons, which have a paralyzing action on the prey. The tentacle then grasps the prey, and if it continues to struggle, neighboring tentacles may also become involved. The tentacles draw they prey toward the mouth, which opens wide to receive it.

Once the food is inside the gastrovascular cavity, digestive enzymes are secreted into the cavity, and extracellular digestion begins. This extracellular digestion, largely limited to proteins in cnidarians, does not break down these substances completely to their constituent amino acids. As soon as the food has been reduced to small fragments, cells lining the gastrovascular cavity engulf them by phagocytosis, and digestion is completed intracellularly in food vacuoles. Indigestible remains of the food are expelled from the gastrovascular cavity via the mouth.

The evolution of the additional process of extracellular digestion provides cnidarians with an adaptive advantage: intracellular digestion severely limits the size of the food the organism can handle. Extracellular digestion enables it to utilize much larger pieces of food; even whole multicellular animals become possible prey. *Extracellular digestion is the rule rather than the exception in multicellular animals.*

Most Flatworms Also Ingest Food Into a Gastrovascular Cavity for Digestion

Unlike the radially symmetrical cnidarians, the flatworms are bilaterally symmetrical; they have distinct anterior (front) and posterior (rear) ends, and also distinct dorsal (upper) and ventral (lower) surfaces. Their bodies are composed of three well-formed tissue layers. Many flatworms are parasitic, but some are free-living, such as the planaria (see Fig. 17.13).

The mouth of planaria is located on the ventral surface, near the middle of the animal. It opens into a muscular tubular ***pharynx,*** which can be protruded through the mouth directly onto prey. The pharynx leads into an ***intestine,*** where both extracellular and intracellular digestion and absorption take place. The intestine is a gastrovascular cavity and has only one opening to the outside. As with the cnidarians, both incoming food and outgoing wastes pass through the mouth. The intestine is much more highly branched than that of the cnidarians. The extensive branching greatly increases the total absorptive surface of the cavity and serves to transport food to all parts of the body. We saw with plants that as organisms increase in size, and particularly as their volume increases, the problem of sufficient absorptive surface becomes more acute. Many organisms have evolved greatly subdivided absorptive surfaces, thereby compacting much total surface area into relatively little space. The root hairs of plants were one example, and the branched gastrovascular cavity of planaria is another; we shall encounter many more examples in this and later chapters.

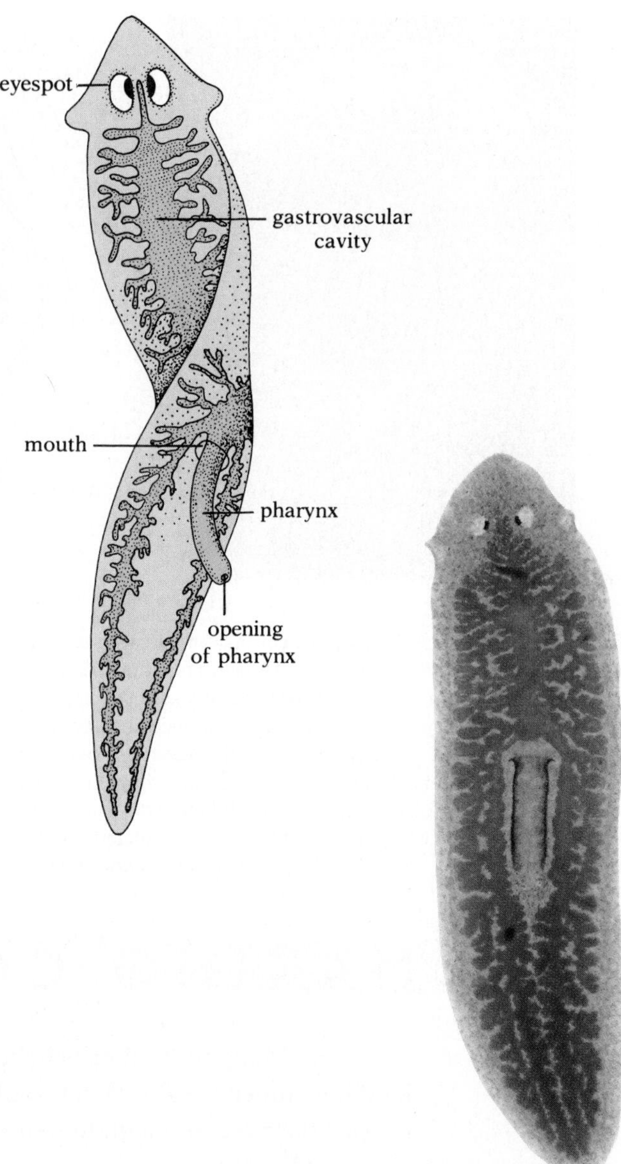

17.13 Planaria, showing much-branched gastrovascular cavity and extruded pharynx Notice that the mouth, through which the pharynx is extruded, is not at the anterior end of the animal, but rather down the animal's body. The pharynx has very powerful muscles and acts somewhat like a muscular straw, sucking up food.

The tapeworms, one class of flatworms, have become so highly specialized as parasites living in the digestive tracts of other animals that in the course of their evolution they have lost their own digestive systems (Fig. 17.14). They are constantly bathed by the products of the host's digestion and can absorb them without having to carry out any digestion themselves. Evolutionary adaptation can involve the loss of structures as well as their acquisition.

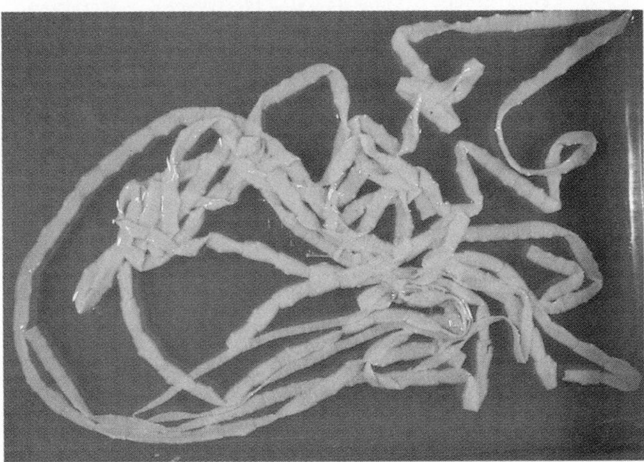

17.14 An adult beef tapeworm, *Taenia saginata* The beef tapeworms are highly specialized parasites; the adult form lives in the digestive tracts of humans. They completely lack digestive systems but are constantly bathed by the products of the host's digestion and can absorb them without having to carry out any digestion themselves. This specimen is about 10 meters long.

A Complete Digestive Tract Is an Important Evolutionary Innovation

Animals more complex than cnidarians and flatworms have a ***complete digestive tract***—one with two openings, a ***mouth*** and an ***anus.*** In these organisms incoming food can move in one direction from mouth to anus through a tubular system, which can be divided into a series of distinct sections or chambers, each specialized for a different function. As the food passes along this assembly line, it is acted upon in a different way in each section. The sections may be variously specialized for mechanical breakup of bulk food, temporary storage, enzymatic digestion, absorption of the products of digestion, reabsorption of water, storage of wastes, and so on. The overall result is a much more efficient digestive system, as well as a potential for special evolutionary modifications fitting different animals for different methods of obtaining nutrients.

The digestive system of an earthworm is a good example of subdivision into specialized compartments (Fig. 17.15). Food, in the form of decaying organic matter mixed with soil, is drawn into the mouth by the sucking actions of a muscular chamber called the ***pharynx.*** It passes from the mouth through a short passageway into the pharynx and then through a connecting passage called the ***esophagus,*** after which it enters a relatively thin-walled ***crop*** that functions as a storage chamber. Next, it enters a compartment with thick muscular walls, the ***gizzard,*** where it is ground up by a churning action; the grinding is often facilitated by small stones in the gizzard. The pulverized food, suspended in water, now passes into the long ***intestine,*** where enzymatic digestion and absorption take place. Finally, in the rear of the intestine, some of the water involved in the digestive process is reabsorbed, and the indigestible residue is eliminated from the body through the anus.

Notice that earthworms use only extracellular digestion. Glandular cells in the epithelial lining of the intestine secrete hydrolytic enzymes into the intestinal cavity, and the end products of digestion—the simple building-block compounds—are absorbed into the blood and are distributed to all parts of the body.

We have already seen that extracellular digestion is an adaptation for eating sizable pieces of food; the gizzard in

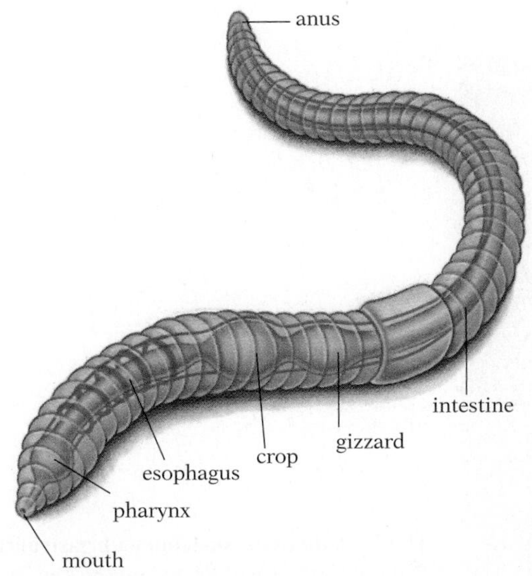

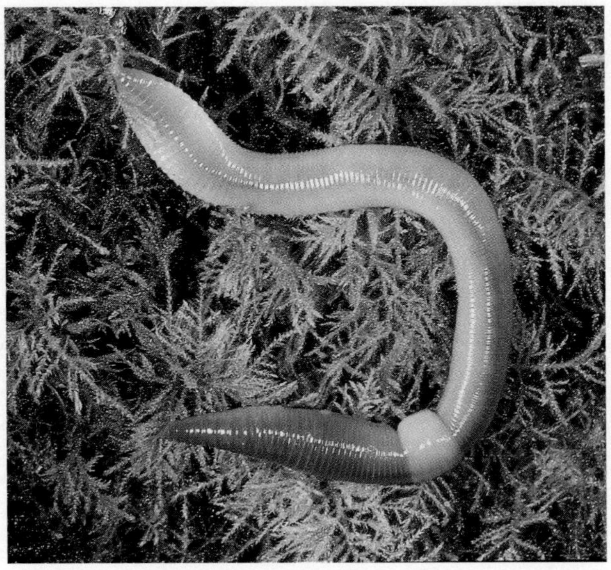

17.15 Digestive (pink) and circulatory (red) systems of an earthworm

earthworms is obviously another such adaptation. Mechanical breakup of bulk food is common among animals, and numerous structures have evolved to serve this function. In our own case, food is torn and ground by the teeth, which creates smaller pieces and more surface area for the digestive enzymes to work on. Note that the grinding or chewing device need not be in the first section of the digestive tract, as with our teeth. In earthworms, the grinding chamber comes after the crop, which, like our stomach, functions as a storage organ, and mechanical breakup thus follows temporary storage instead of preceding it. The same arrangement exists even in some vertebrates. Birds, for instance, have a muscular gizzard, posterior to the less-specialized stomach (Fig. 17.16), in which hard food is ground with pebbles (often called grit).

It is thought that dinosaurs, like birds, had grinding mills in their guts to help them digest their food, which in the case of plant-eating dinosaurs consisted of ferns and conifers. However, rather than eating grit to line their gizzards as birds do, dinosaurs ate rocks, called gastroliths (Fig. 17.17). These rocks were used to crush food in what may have been two separate grinding chambers—one at the base of the esophagus and the other between the stomach and small intestine. Anatomical evidence linking dinosaurs and birds is not surprising; many scientists argue that birds are the living descendants of dinosaurs, though this remains a controversial and ongoing research topic.

The functional significance of the earthworm's crop, which stores food, is that it enables the animal to take in large amounts of food in a short time, when it is available, and then to utilize this food over a considerable period of time. Such discontinuous feeding makes it possible for the animal to devote much of its time to activities other than feeding, such as searching for a mate, mating, egg laying, and, in some animals, caring for young. Our own stomach functions as a storage organ analogous to the earthworm's crop; it enables us to live well on only three or four meals a day and to devote the rest of our time to other pursuits. A human can survive if his stomach is removed surgically, but he is unable to eat more than a few bites at a time and must therefore eat very frequently. It is not sur-

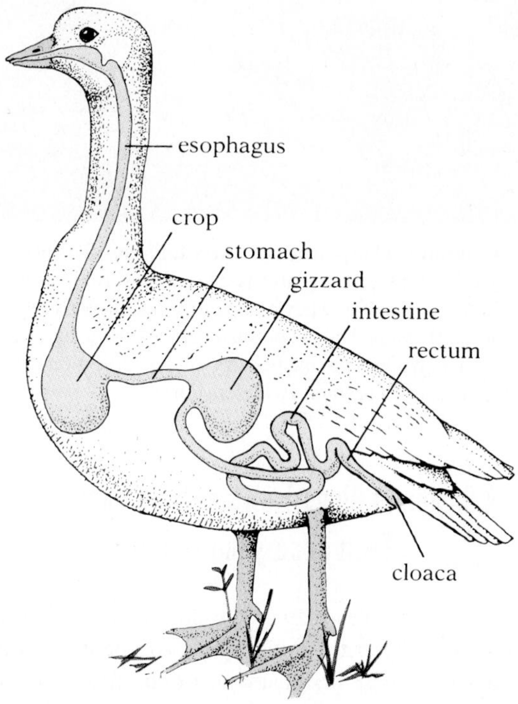

17.16 Digestive system of a bird The chamber for mechanical breakup (gizzard) is located posterior to the stomach. The crop, which is a specialized portion of the esophagus, is the principal chamber for temporary storage of food, the cloaca is a common passageway for the digestive, excretory, and reproductive systems.

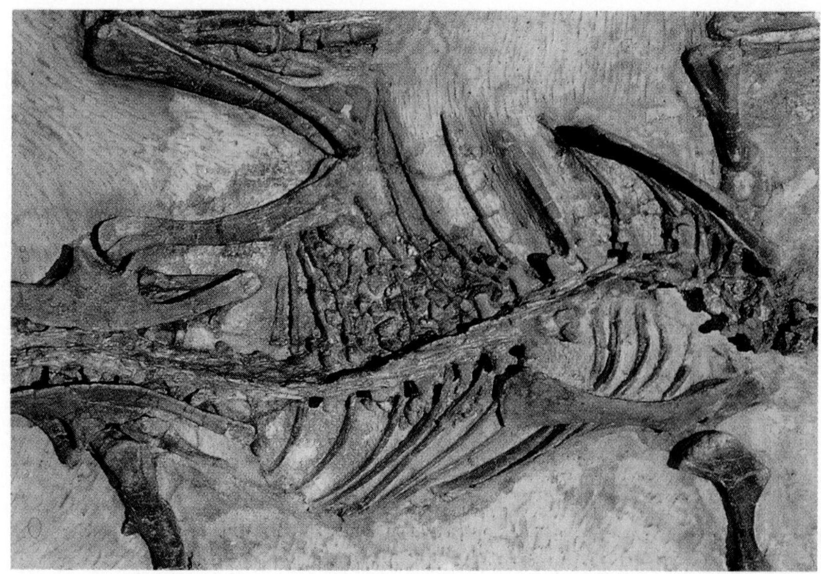

17.17 A dinosaur skeleton with gastroliths Inside this rib cage of *Psittacosaurus* can be seen small rocks called gastroliths, which dinosaurs used to crush their food.

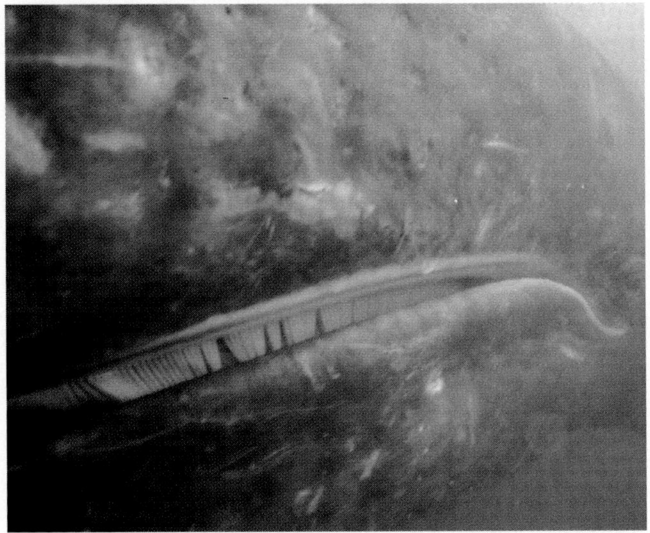

A

B

17.18 Structures used for filter feeding (A) Strips of baleen hanging down from the top jaw of an Atlantic right whale, one of the filter-feeding whales. The baleen allows the whale to take a mouthful of water and strain out the organisms—primarily shrimp and small fish—that form its diet. (B) On a much smaller scale, the isolated food brush of a mosquito larva. Its motion causes water to flow into the mouth.

prising that the vast majority of higher animals have evolved adaptations for discontinuous feeding, thereby gaining time for a behaviorally more varied existence.

Our emphasis on special masticating devices is not meant to imply that all except the tiniest animals eat large pieces of food, which is not true. Some animals, such as bloodsucking and sapsucking insects, have liquid diets. Other animals are ***filter feeders,*** straining small particles of organic matter from water (Fig. 17.18). Some of the largest present-day vertebrates—certain species of whales, for example—are filter feeders, straining small plants and animals from the vast quantities of water they take into their mouths. Clams and many other molluscs filter water through tiny pores in their gills; microscopic food particles, trapped in streams of mucus that flow along the gills, are moved along by beating cilia until they enter the mouth.

The Vertebrate Digestive Tract Is Highly Compartmentalized

Like the earthworm, the vertebrate has a digestive tract divided into compartments that are specialized to mechanically and chemically process the ingested foods. Humans, because they are omnivores and consume both plant and animal foods, have a digestive tract that has adapted to a wide variety of foods.

The digestive tract of an adult human is a tube approximately nine meters long. Throughout its length the wall of the tube us composed of four layers of tissue: an inner mucous lining; a submucous coat of connective tissue containing blood vessels; a muscular layer; and a thin layer of connective tissue, the serosa (Fig. 17.19). Although these four layers are present throughout, their structure is modified in different organs. For example, the mucosal lining of the stomach is arranged in folds to allow for distention, and the muscle layer of the stomach is much thicker than it is in the rest of the tract.

Mechanical Breakup and Digestion Begin in the Oral Cavity

The first chamber of the digestive tract is the mouth, or ***oral cavity.*** Located here are the teeth, which function in the mechanical breakup of food by both biting and chewing. Human teeth are of several different types, each adapted to a different function (Fig. 17.20).

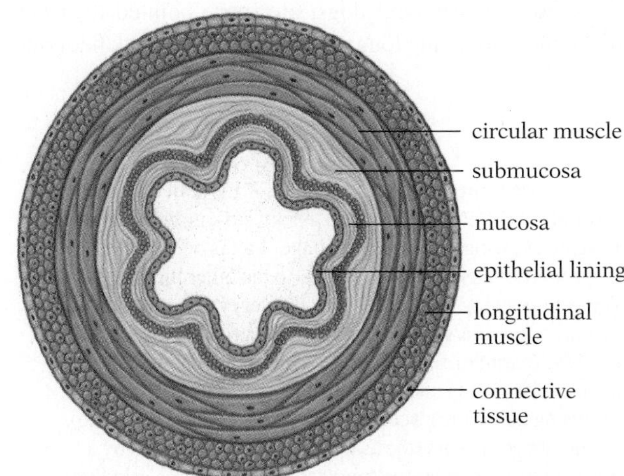

17.19 The four layers of tissue of the vertebrate digestive tract

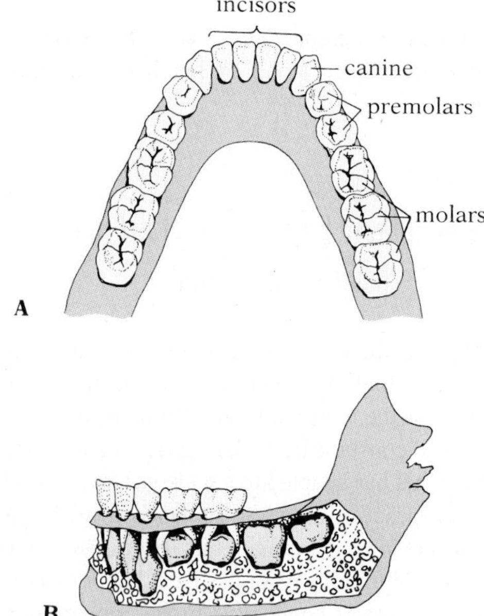

17.20 Human teeth (A) Lower jaw of an adult. In front are the four chisel-shaped *incisors,* which are used for biting. Then come the more pointed *canine* teeth, one on each side of each jaw, which are specialized for tearing food. Behind each canine are two *premolars* and three *molars;* these have flattened, ridged surfaces, and function in grinding, pounding, and crushing food. (B) Lower jaw of child. A child's first set of teeth does not include all of those mentioned above; the first (or milk) teeth are lost as the child gets older, and are replaced by the permanent teeth that have been growing in the gums. Humans, being omnivorous, have incisors and canines capable of tearing meat, and broad, flat molars capable of grinding plant cell walls.

The teeth of different species of vertebrates are specialized in a variety of ways and may be quite unlike human teeth, in number, structure, arrangement, and function. For example, the teeth of snakes are very thin and sharp and usually curve backward (Fig. 17.21A); they function in capturing prey, but not in mechanical breakup, because snakes do not chew their food, but swallow it whole. The teeth of carnivorous mammals, such as cats and dogs, are more pointed than human teeth; the canines are long, and the premolars lack flat grinding surfaces, being more adapted to cutting and shearing (Fig. 17.21C). On the other hand, such herbivores as cows, horses, and deer have very large flat premolars and molars with complex ridges and cusps; the canine teeth are often absent in such animals (Fig. 17.21D).

Notice that sharp pointed teeth, poorly adapted for chewing, seem to characterize meateaters like snakes, dogs, and cats, whereas broad, flat teeth, well adapted for chewing, seem to characterize vegetarians. How can this difference be ex-

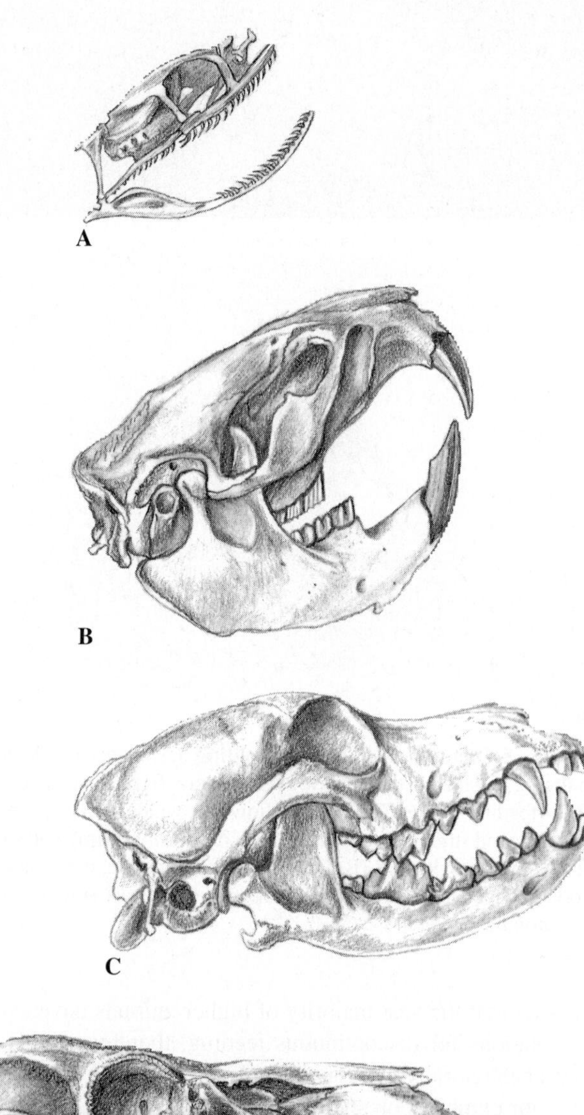

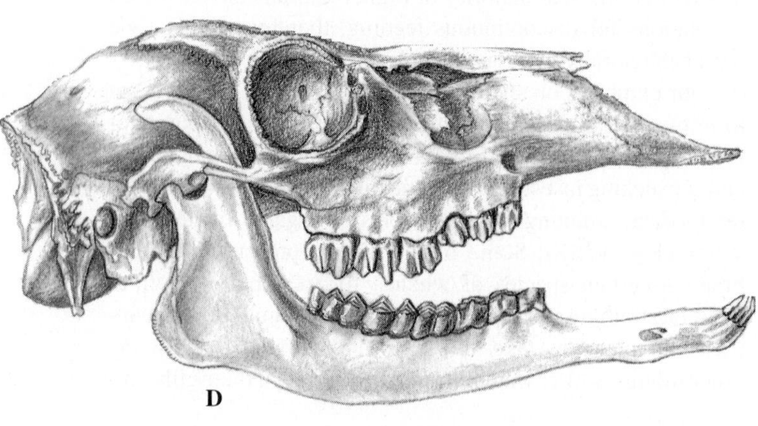

17.21 Structure and arrangement of teeth in different animals (A) Snake: thin, sharp, backward-curved teeth that have no chewing function (the snake skull is here shown disproportionately large in relation to the other three). (B) Beaver (gnawing herbivore): few but very large incisors; no canines; premolars and molars with flat grinding surfaces. (C) Dog (carnivore): large canines; premolars and molars adapted for cutting and shearing. (D) Deer (grazing and browsing herbivore): six lower incisors (three on each side), but no upper incisors (these are functionally replaced by a horny gum); premolars and molars with very large grinding surfaces. Notice the large gap between the incisors and premolars.

plained? Remember that plant cells are enclosed in a cellulose cell wall. Very few animals can digest cellulose; most of them must break up the cell walls of the plant they eat in order to release the cell contents, which can be digested. Animals cells like those in meat do not have any such nondigestible armor and can be acted upon directly by digestive enzymes. Therefore chewing is not as necessary for carnivores as for herbivores. You have probably seen how dogs gulp down their food, while cows and horses spend much time chewing. But carnivores have problems of their own. They must capture and kill their prey, and for this, sharp teeth capable of piercing, cutting, and tearing are well adapted. Humans, being omnivores, have teeth that belong, functionally and structurally, somewhere between the extremes of specialization attained by the teeth of carnivores and herbivores.

The oral cavity has other functions besides those associated with the teeth. Here food is tasted and smelled, activities of great importance in food selection, and the food is mixed with saliva secreted by several sets of salivary glands. The saliva dissolves some of the food and acts as a lubricant, facilitating passage through the next portions of the digestive tract. Human saliva also contains a starch-digesting enzyme, ***amylase,*** which initiates the process of enzymatic hydrolysis.

The muscular tongue manipulates the food during chewing and forms it into a mass, called a bolus, in preparation for swallowing. It then pushes the bolus backward through a cavity called the ***pharynx*** (throat) and into the ***esophagus*** (Fig. 17.22). The pharynx functions also as part of the respiratory passageway; the air and food passages cross here, in fact (Fig. 17.22). Swallowing, therefore, involves a complex set of reflexes that close off the opening into the nasal passages and trachea (windpipe), thereby forcing the food to move into the esophagus. As you know, these reflexes occasionally fail to occur in proper sequence, and the food, entering the wrong passageway, causes choking.

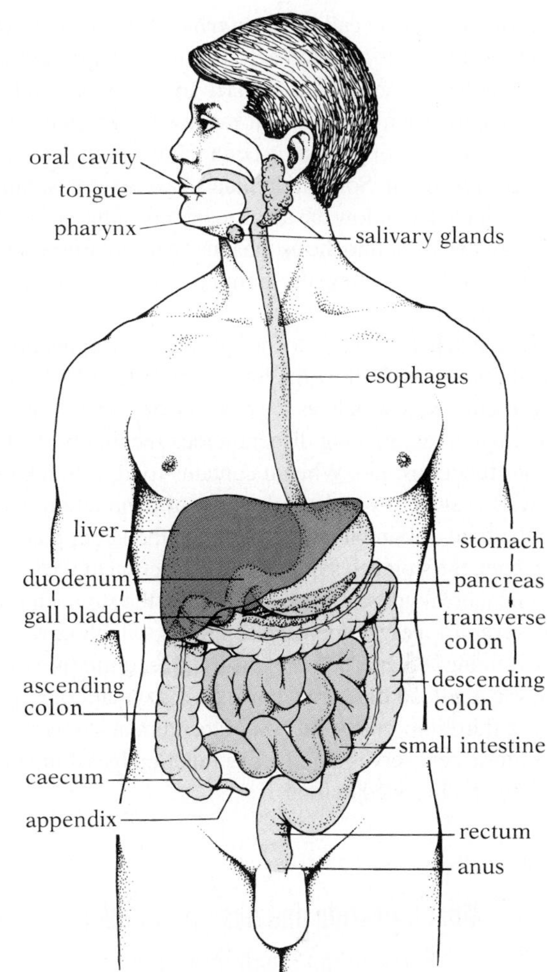

17.22 Human digestive system The small intestine has been shortened for clarity.

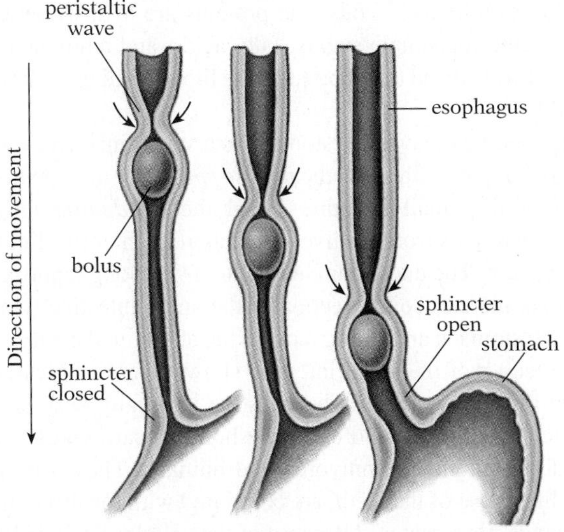

17.23 Peristalsis The wave of muscular contraction pushes the bolus of food ahead of it. The sphincter muscle at the junction between the stomach and esophagus normally remains closed, and opens when a wave of peristaltic contraction coming down the esophagus reaches it.

The Stomach Functions in Storage, Mechanical Breakup, and Digestion

The esophagus is a long tube running downward through the throat and thorax and connecting to the stomach in the upper portion of the abdominal cavity (Fig. 17.22). Food moves quickly through the esophagus, pushed along by waves of muscular contraction in a process called ***peristalsis.*** Circular muscles in the wall of the esophagus contract just behind the food bolus, squeezing the food forward (Fig. 17.23). As the food moves, the muscles it passes also contract, so that a region of contraction follows the bolus and constantly pushes it forward, much as though you were to keep a ball moving through a soft rubber tube by giving the tube a series of squeezes, with your hand always just behind the ball.

At the junction between the esophagus and the stomach is a

special ring of muscle called a ***sphincter.*** When it contracts it closes the entrance to the stomach. It is normally closed, thus preventing the contents of the stomach from moving back into the esophagus when the stomach moves during digestion. It opens when a wave of peristaltic contraction coming down the esophagus reaches it. Some individuals have a weak sphincter muscle, which does allow the acidic contents of the stomach to be regurgitated back into the esophagus (particularly when the stomach is full), creating the burning sensation commonly called heartburn.

The stomach lies slightly to the left side in the upper portion of the abdomen, just below the lower ribs (Fig. 17.24). It is a large muscular sac, which, as we have already seen, functions as a storage organ, making discontinuous feeding possible. It has other functions, too. When it contains food, its thick muscular wall is swept by powerful waves of contraction, which churn the food, mixing it and breaking up the larger pieces. In this manner, the stomach supplements the action of the teeth in mechanical breakup of food. The glands of the stomach lining are of several types. Some secrete mucus, which covers the stomach lining; others secrete gastric[2] juice, a mixture of hydrochloric acid and digestive enzymes. Enzymatic digestion, then, is a third important function of the human stomach. The hydrochloric acid serves a protective function by killing many of the bacteria ingested in food.

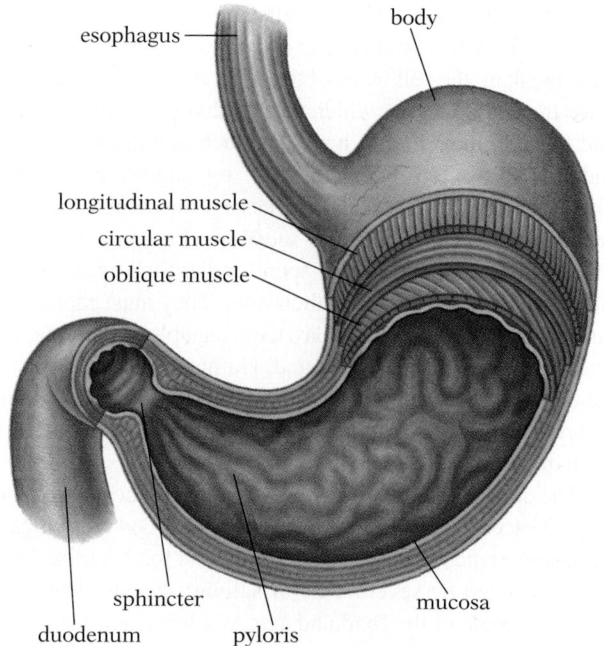

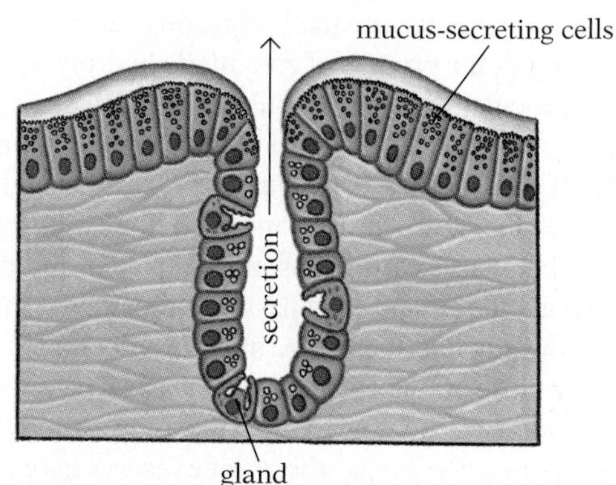

17.24 The stomach When food is present numerous muscular bands contract, aiding mechanical breakup. Glands lining the stomach secrete mucus (green), which protects the lining, and gastric juice, which contains digestive enzymes and hydrochloric acid.

Most Digestion and Absorption Take Place in the Small Intestine

The small intestine is the portion of the digestive tract where most of the digestion and absorption of foodstuffs takes place. Here the carbohydrates, fats, and proteins are hydrolyzed by enzymes into the simple sugars, fatty acids, and amino acids that are absorbed and transported by the blood to all the cells of the body.

The food that leaves the stomach is a soupy mixture, very low in pH. It passes through the ***pyloric sphincter*** into the first section of the small intestine called the ***duodenum*** (Fig. 17.22). Secretions from the liver and pancreas enter the digestive tract here. The duodenum leads into a very long (approximately six meters) coiled section of the small intestine lying below the stomach and filling most of the abdominal cavity.

The length of the small intestine shows interesting variations in different animals. The intestine is usually very long and much coiled in herbivores, much shorter in carnivores, and of medium length in omnivores like humans. These differences, like those of the teeth, are correlated with the difficulty of digesting plant material because of the cellulose cell walls. Even if the cellulose has been well broken up, it remains mixed with the digestible portions of the cells and tends to mask them from the digestive enzymes. This interference makes digestion and absorption of plant material much less efficient than the processing of animal material, with the result that a longer intestine is an adaptive advantage in extracting a maximum amount of nutrients from an herbivorous diet.

Since the small intestine is where absorption of the products of digestion occurs, we would expect it to have special structural adaptations that increase its absorptive surface area. Clearly, its great length plays a role here. But examination of the internal surface of the vertebrate small intestine reveals other modifications that vastly increase its surface area over that of a smooth-walled tube of equal length and girth. First,

[2]The adjective *gastric* and the prefix *gastro-* always refer to the stomach.

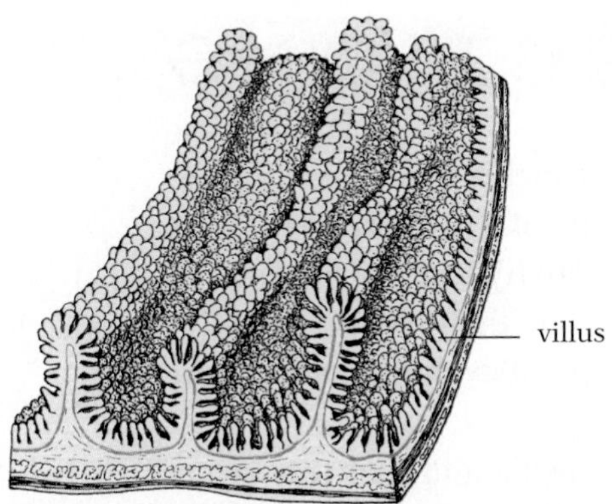

17.25 Longitudinal section through the wall of the human intestine Three folds of the lining of the intestine are shown, each bearing numerous villi, which greatly increase the absorptive surface area. The villi are not stationary; they have smooth muscle fibers that enable them to move back and forth. Such movement increases after a meal.

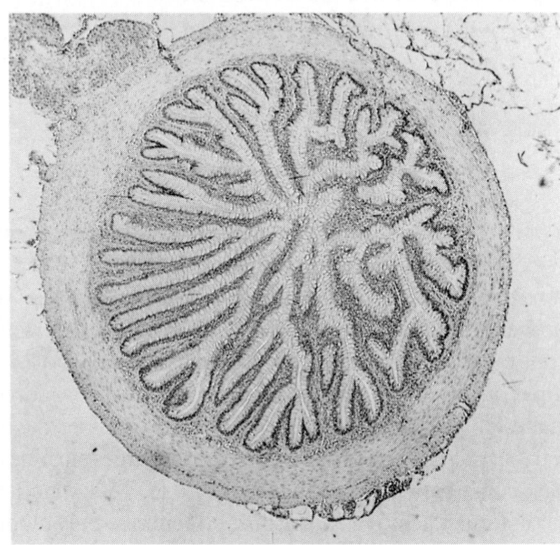

17.26 Cross section of intestine of calico bass, showing extensive folding

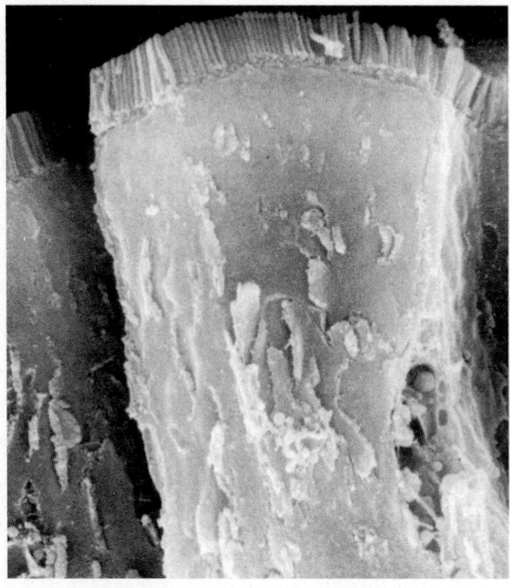

A

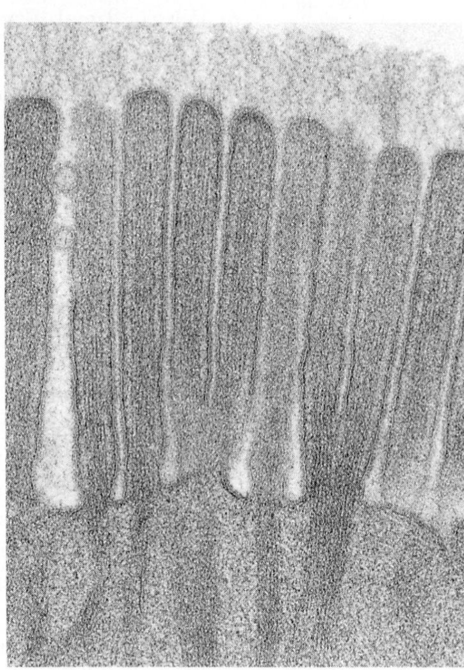

B

17.27 Electron micrographs of microvilli (A) Scanning electron micrograph of an intestinal cell with a brush border. (B) Transmission electron micrograph of part of the brush border on an epithelial cell of the intestinal lining of a cat. Notice the prominent glycocalyx covering the ends of the microvilli. A bundle of microfilaments runs into each microvillus.

the epithelial tissue lining the intestine has numerous folds and ridges (Figs. 17.25 and 17.26). Second, small fingerlike outgrowths, called ***villi,*** cover the internal surface. And third, as the electron microscope reveals, the individual epithelial cells covering the folds and villi have what is called, for obvious reasons, a brush border, consisting of countless, closely packed, cylindrical processes, the ***microvilli*** (Fig. 17.27). Thus the total internal surface of the small intestine, including folds, villi, and microvilli, is incredibly large.

In addition to providing a large surface area, the membranes of the cells of the microvilli contain enzymes important in the digestive process. These enzymes are embedded in the plasma membrane of the epithelial cells, with their active sites oriented toward the lumen, or opening. The epithelial cells are subject to wear and tear and the entire lining is replaced approximately every five days. Because these cells divide so rapidly, the lining of the intestine is particularly sensitive to damage by radiation and anticancer drugs. Consequently, cancer patients undergoing chemotherapy or radiation to the abdominal area often suffer nausea and severe diarrhea.

MAKING CONNECTIONS

1. What is the advantage of the extensive branching of the gastrovascular cavity of planaria compared to the saclike cavity of hydra? Does its more branched digestive cavity make planaria "superior" to hydra?

2. The tapeworms, which live as parasites in their host's digestive tract, have lost their own digestive system in the course of evolution. What does this specialization demonstrate about evolution?

3. Suppose a person contracted stomach cancer and had this organ surgically removed. How would the person's life be affected?

4. Why is a chamber specialized for mechanical breakup of food important for digestion for most animals?

5. How are the digestive systems of a bobcat and a buffalo adapted to their respective diets?

6. Why is the ability to smell and taste food adaptive?

7. Compare the structure and function of root hairs, described in Chapter 16, and intestinal microvilli. Why are microvilli so much smaller than root hairs?

Water Reabsorption Takes Place in the Large Intestine

In humans, the junction between the small intestine and the large intestine that follows it is usually in the lower right portion of the abdominal cavity. A blind sac, the ***caecum,*** projects from the large intestine near the point of juncture (Fig. 17.22). In humans there is a small fingerlike process, the ***appendix,*** at the tip of the caecum, which often becomes infected and must be surgically removed.

In humans the caecum is small and functionally unimportant, but in some mammals, particularly herbivores, it is large and contains many microorganisms (bacteria and protozoa) capable of digesting cellulose. Since the mammal cannot itself digest cellulose, it benefits from the action of these microbes.

From the caecum, the large intestine, or ***colon,*** of humans ascends on the right side to the midregion of the abdominal cavity, then crosses to the left side, and descends again (Fig. 17.22). The surface area of the large intestine is considerably smaller than that of the small intestine because its lining does not have folds and villi. The large intestine is approximately half as long as the small intestine, and larger in diameter.

One of the chief functions of the large intestine is reabsorption of much of the water used in the digestive process. If all the water in which enzymes are secreted into the digestive tract as well as the water ingested in food were lost with the feces, dessication would be a severe problem in terrestrial animals. Most of the water is absorbed in the first half of the large intestine. Occasionally, the intestine becomes irritated, and peristalsis moves material through it too fast for enough water to be reabsorbed; this condition is known as diarrhea. Conversely, if material moves too slowly, too much water is reabsorbed and constipation results. A proper amount of roughage (indigestible material, primarily cellulose) in the diet provides the bulk needed to stimulate enough peristalsis in the large intestine and prevent constipation. Many laxatives (e.g., Metamucil and other "natural" laxatives) use this principle—they contain fiber that absorbs water and creates the bulk necessary to stimulate peristalsis. Other laxatives (e.g., Milk of Magnesia, $MgSO_4$) increase the osmotic concentration within the large intestine, thereby causing water to be retained and creating softer feces.

A second function of the large intestine involves the excretion of certain salts, such as those of calcium and iron, when their concentration in the blood is too high. The salts are excreted from the blood into the large intestine and are eliminated from the body in the feces. Great numbers of bacteria normally inhabit the large intestine; their presence is required for normal function. Occasionally, treatment with antibiotics destroys some of this bacterial flora and the resulting disturbance results in diarrhea.

A third function of the large intestine is storage of fecal material until defecation (elimination of the feces from the body). Material is moved slowly (taking perhaps eight to 15 hours) through the first portion of the large intestine. A modified type of peristalsis, called a mass movement, occurs in the remainder of the large intestine. Here successive portions of the large intestine constrict as a unit and force material *en masse* down the lower portion of the large intestine into the ***rectum.*** The need to defecate is felt when the feces are forced into the rectum. The feces are eliminated from the rectum through the ***anus.***

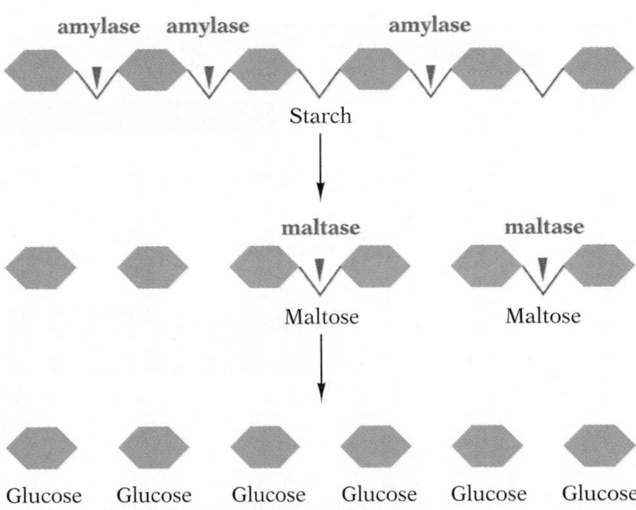

17.28 Digestion of starch Amylase in the saliva and in the pancreatic juice hydrolyzes some of the bonds between glucose units, producing small amounts of free glucose, but much larger quantities of the double sugar maltose. The maltose is then digested to glucose by maltase secreted by intestinal glands.

ENZYMATIC DIGESTION IN HUMANS

Starch Is First Hydrolyzed Into Double Sugars

Having traced the human digestive tract from mouth to anus, let us next consider the chemical changes that occur in a carbohydrate meal as it passes through this complex tubular system. There are three main sources of carbohydrates in the human diet: sucrose or cane sugar, a disaccharide; lactose, the sugar in milk; and starches. Enzymatic digestion starts in the mouth. Saliva contains the enzyme amylase,[3] which begins but does not complete the hydrolysis of starch to glucose. Although amylase produces some glucose, it yields primarily the double sugar maltose (Fig. 17.28), which must be further digested in the intestine. The saliva of many mammals contains no amylase; dogs are an example. Doubtless there was little selection pressure for the evolution of such a starch-digesting enzyme in animals that, at least ancestrally, were almost entirely carnivorous.

Once in the stomach, food is exposed to the action of gastric juice secreted by numerous gastric glands of the stomach wall, and hydrolysis of starch stops. Probably only about 3 to 5 percent of the ingested starches are hydrolyzed by salivary amylase. Gastric juice contains much hydrochloric acid, which makes the contents of the stomach very acidic (pH between 1.5 and 2.5). Note that, advertisement for many patent medicines to the contrary, an acid stomach is both normal and necessary for proper function. The low pH rapidly inactivates the amylase so little or no carbohydrate digestion takes place in the stomach.

The Double Sugars Are Then Hydrolyzed Into Simple Sugars

It is in the small intestine that carbohydrate digestion is completed. When partially digested food passes from the stomach into the duodenum, its acidity stimulates, through special hormones, the release of secretions into the duodenum from two organs, the pancreas and liver. The ***pancreas*** is a large glandular organ, lying just below the stomach (Fig. 17.22), that originates in the embryo as an outgrowth of the digestive tract and retains a connection to the duodenum called the ***pancreatic duct.*** When food enters the duodenum, the pancreas secretes a mixture of enzymes that flows through the pancreatic duct into the duodenum. Included in this mixture are enzymes that digest all three principal classes of foods—carbohydrates, fats, and proteins—as well as some that digest nucleic acids. One of the pancreatic enzymes is ***pancreatic amylase,*** which, as its name implies, acts like salivary amylase, splitting starch into the double sugar maltose. It is far more important than salivary amylase, for it carries out most of the starch digestion.

In summary, then, the action of the amylases in the mouth and small intestine results in a splitting of starch into the double sugar maltose, but it does not produce very many free glucose molecules. Other enzymes must complete the digestive process. These enzymes are actually extrinsic proteins built into the membranes of the intestinal cells lining the tract, oriented with their active sites facing the lumen. The double sugars are digested into simple sugars as they contact the enzymes. The enzyme ***maltase*** catalyzes the splitting of maltose (Fig. 17.28), ***sucrase*** the splitting of sucrose, and ***lactase*** the splitting of lactose. The final products of carbohydrate digestion are all simple sugars. The absorption of the simple sugars across the membranes of the intestinal cells is by active transport. Each sugar has a specific carrier molecule. From there the simple sugars are transported into the blood capillaries and carried to different parts of the body. (See Exploring Further: The Digestion of Lactose in Human Adults, p. 366.)

[3] The names of most enzymes end with the suffix *-ase,* which designates enzymes by international agreement. The first part of the name usually indicates the substrate upon which the enzyme acts; thus *amyl-* (from *amylum,* Latin for "starch") indicates that amylase acts upon starch.

EXPLORING FURTHER

THE DIGESTION OF LACTOSE IN HUMAN ADULTS

Digestive capabilities vary not only among species (we have already mentioned the absence of salivary amylase in dogs), but also within species. The digestion of lactose,* a sugar found only in milk, provides a striking example.

Secretion of milk by the mammary glands of female mammals evolved as a way of feeding the young. The only food provided very young mammals, milk is a nearly complete food, containing, in most species, carbohydrate (in the form of lactose), fat, and protein, as well as important minerals. But except for humans, adult mammals do not use milk as a food. It is not surprising then, that production of lactase, the lactose-digesting enzyme, usually greatly diminishes or even ceases altogether once an animal is past the age of weaning.

It has only recently been realized, however, that this pattern applies also to most human beings; in most parts of the world, humans older than four years produce little or no lactase. Indeed, of the various peoples studied to date, only those of European ancestry and those belonging to a few pastoral tribes in Africa have been found to produce enough of the enzyme to be able to digest the lactose in large quantities of milk (see Graph). When people of other ancestries drink more than modest quantities they will often become ill, getting a bloated feeling, cramps, and diarrhea. One reason is that the undigested sugar in the intestine upsets the normal osmotic balance to the point where an excessive amount of water moves into the intestinal lumen from the cells; another is that fermentation of the lactose by bacteria in the large intestine produces large quantities of acids and carbon dioxide. The lactose tolerance (i.e., continued production of lactase in adults) of Europeans and pastoral Africans must have evolved during the roughly 10,000 years since the milking of domestic animals began.

How widely peoples living near one another may differ is shown by the major tribes of Nigeria. The Ibo and Yoruba live in the southern part of the country, where conditions are unfavorable for cattle; milk has not traditionally been a part of their diet after weaning, and they cannot tolerate lactose. By contrast, the nomadic Fulani in northern Nigeria have been raising milk cattle for thousands of years, and they are lactose-tolerant (see Graph).

Most American blacks are descended from nonpastoral tribes of western Africa, and they are relatively intolerant of lactose, though not so much as native Africans. Their somewhat greater tolerance may be due in part to evolutionary change that occurred during the generations they have lived in dairy regions and in part to admixture of European genes. In view of the widespread lactose intolerance in most underdeveloped countries, it has become

*There is no lactose in the milk of seals and their close relatives.

clear that the former large-scale distribution of powdered milk in these countries was ill-advised. If milk is sent, it should be as powder from which the lactose has been removed or as products such as cheese, in which the lactose has been broken down by microbial action, or yogurt, which contains lactase produced by bacteria.

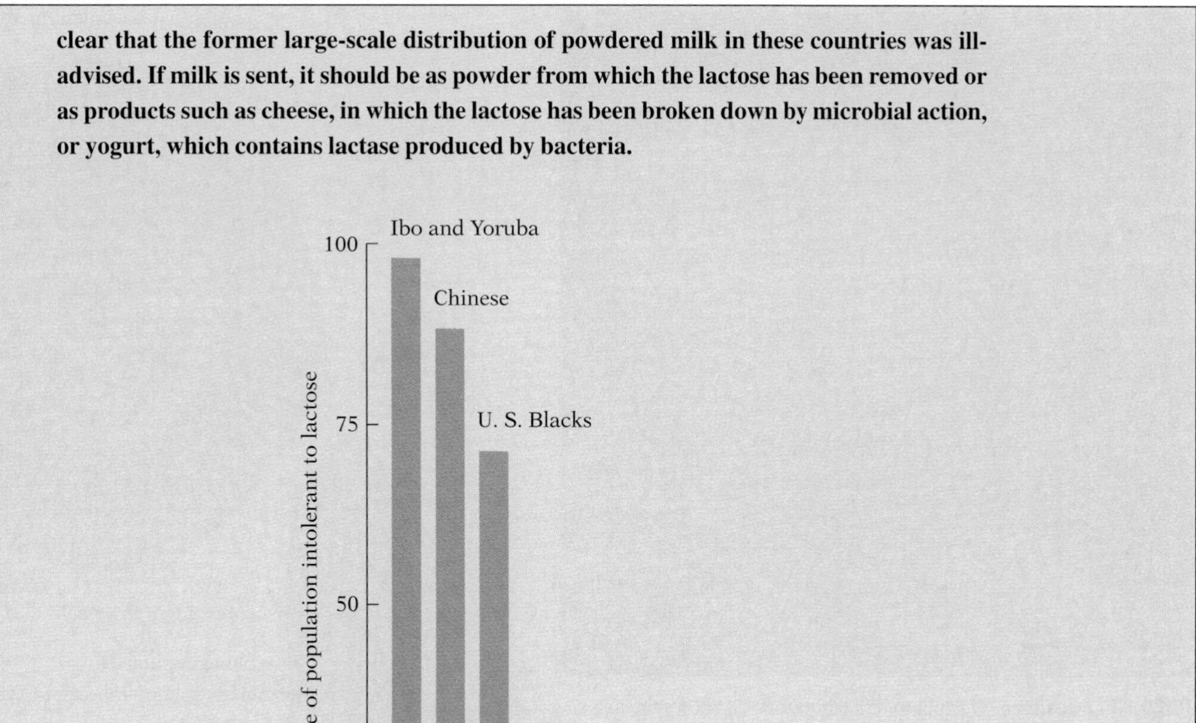

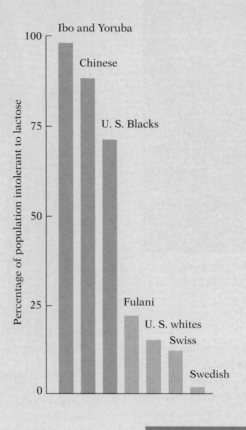

Proteins Are Digested by Endopeptidases and Exopeptidases

Protein digestion begins in the stomach and is completed in the small intestine. The principal enzyme of the gastric juice is ***pepsin,*** which digests protein. Pepsin, which functions only under acid conditions, does not hydrolyze protein all the way to its amino acid components. Instead, it splits the peptide bonds adjacent to only a few specific amino acids, with the result that the long polypeptide chains are cut into shorter chains.

Why, then, isn't the wall of the digestive tract, which is composed primarily of protein, digested by the protein-digesting enzymes? The first line of defense appears to be mucus that covers the wall of the digestive tract and apparently shields it from enzymes. When this defense breaks down, the enzymes do begin to eat away a portion of the lining; the resulting sore is known as an ulcer (Fig. 17.29). The major symptom of an ulcer is a burning, gnawing pain in the upper part of the abdomen. Most ulcers appear to develop from infections of a particular bacteria, and treatment with an appropriate antibiotic may effect a cure.

When the acid food reaches the small intestine, the pancreas releases other enzymes that act in protein digestion. Like pepsin, ***trypsin*** and ***chymotrypsin,*** two of the protein-digesting enzymes of the pancreas, break only the peptide linkages adjacent to certain amino acids. In summary, then, the action of pepsin in the stomach and of trypsin and chymotrypsin from the pancreas results in a splitting of proteins into fragments of varying lengths, but does not produce many free amino acids. These enzymes are known as ***endopeptidases***—enzymes that catalyze the hydrolysis of specific peptide bonds between amino acids located *within* the protein, not bonds linking terminal amino acids to the chain. Other enzymes, called ***exopep-***

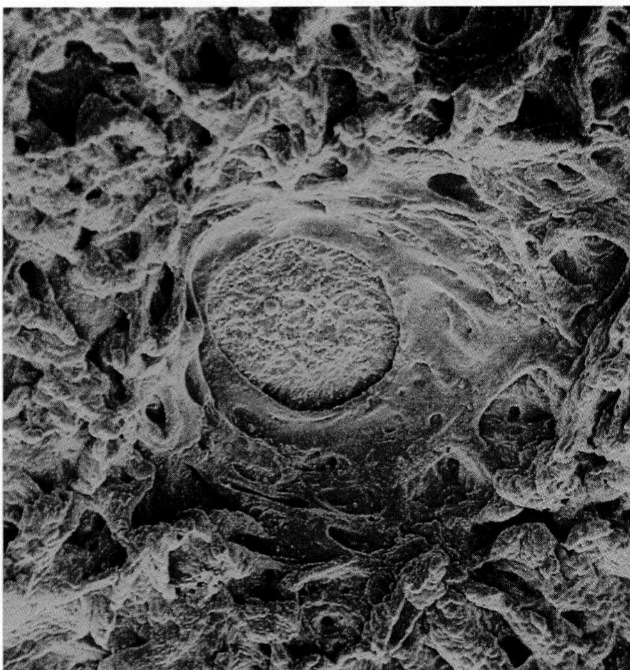

17.29 Peptic ulcer This photo of a laboratory rat shows the loss of folds lining the stomach that is characteristic of an ulcer (center).

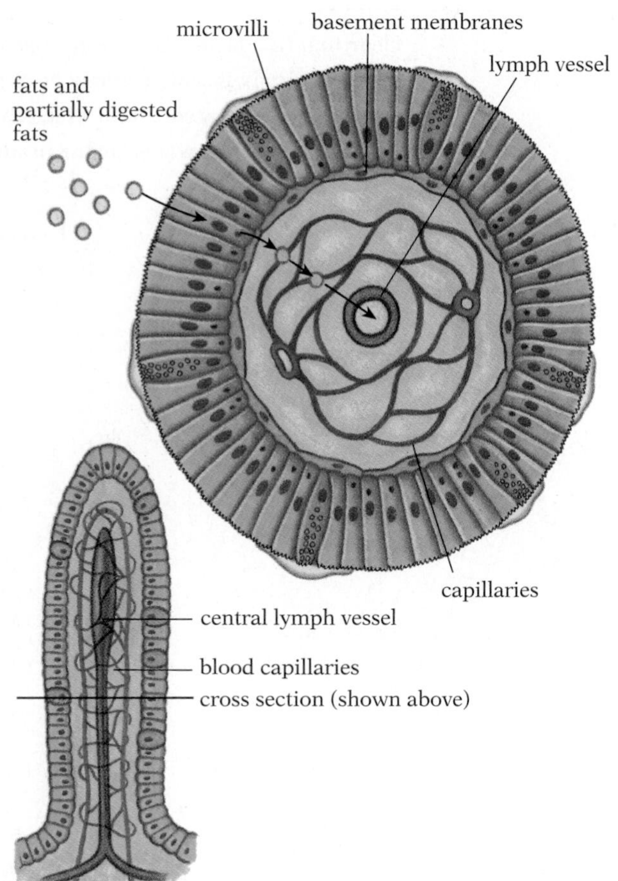

17.30 Absorption of fat in the small intestine Shown here are longitudinal (bottom) and cross sections (top) of a villus in the small intestine. Lipase digests fat into monoglycerides and fatty acids. Some fat remains undigested. Because all these are lipid-soluble, they can diffuse through the membrane of the microvilli of the intestinal cell and enter the cytoplasm. There the partially digested fats are taken up by the smooth endoplasmic reticulum and new fats are synthesized. These combine with cholesterol and phospholipids, are coated with proteins, and are released from the cell by exocytosis. From there the lipoproteins move into the lymphatic vessels (green).

tidases, catalyze the removal of amino acids from the ends of the chains, thereby completing the digestive process. There are a great variety of exopeptidases, each highly specific in its action. One is produced in the pancreas but the others, like the enzymes that hydrolyze the double sugars, are extrinsic proteins that protrude from the surfaces of the epithelial cell membranes. The proteins are digested when they contact the enzymes. The products of protein digestion are free amino acids, dipeptides (two joined amino acids), and tripeptides (three joined amino acids).

Like the simple sugars, the dipeptides, tripeptides, and amino acids are absorbed across the membranes of the intestinal cells by active transport. Several different transport molecules are available for this transport process, which requires ATP energy. The dipeptides and tripeptides are hydrolyzed to free amino acids within the intestinal cells. Only free amino acids are transported into the blood capillaries.

Not All Fat Is Completely Digested Before Absorption

The digestion and absorption of fat is quite different from that of carbohydrate and protein. Only a small percentage of the fat is completely digested to glycerol and three fatty acids. Most of the fat is partly digested (e.g., usually by removal of two of the three fatty acids), and some is not digested at all. But because fats, and the products of the partial digestion of fats, are lipid-soluble, they can be absorbed across cell membranes without complete digestion, something not possible for proteins and carbohydrates.

One secretion that greatly aids the process of fat digestion and absorption is a fluid called ***bile,*** which is produced by the liver. The ***liver,*** a critically important organ about which much will be said in later chapters, is a very large organ occupying much of the space in the upper part of the abdomen. On its surface is a small storage organ, the ***gallbladder*** (Fig. 17.22). Bile, produced throughout the liver, is collected by a series of branching ducts and emptied into the gallbladder. When food enters the duodenum, a hormone is produced that causes the muscular wall of the gallbladder to contract, and the bile is forced down the bile duct into the duodenum.

Bile is *not* a digestive enzyme; it is not even a protein. It is a complex solution of bile salts, bile pigments, and cholesterol. The bile salts act as emulsifying agents, causing large fat droplets to be broken up into many tiny droplets suspended in water. This action is much like that of a good detergent. The many small fat droplets expose much more surface area to the digestive enzymes than a few large droplets would.

The bile pigments and cholesterol play no direct role in digestion. The pigments, produced through the destruction of red blood cells in the liver, give feces their characteristic brown color. The cholesterol, a relatively insoluble compound, sometimes becomes concentrated into hard gallstones, which may block the bile duct and interfere with the flow of bile.

Lipase, also secreted by the pancreas, is the body's principal fat-digesting enzyme. Large quantities of this enzyme are present in pancreatic juice. Most of the fat is partially digested, and tiny fat droplets are absorbed directly across cell membranes of the intestinal cells. Once inside the intestinal cell, the various lipids are recombined to form fats, and coated with proteins, to form lipoproteins. The lipoprotein globules are then secreted by exocytosis into the tissue fluid and enter the lymphatic vessels of the villi (Fig. 17.30). Notice that fat and the products of fat digestion are first absorbed into the lymphatic system, whereas the products of protein and carbohydrate digestion are absorbed into the blood. The lymphatic vessels will eventually deliver the lipids into the blood supply, as we shall see in Chapter 22.

Figure 17.31 summarizes the digestive process.

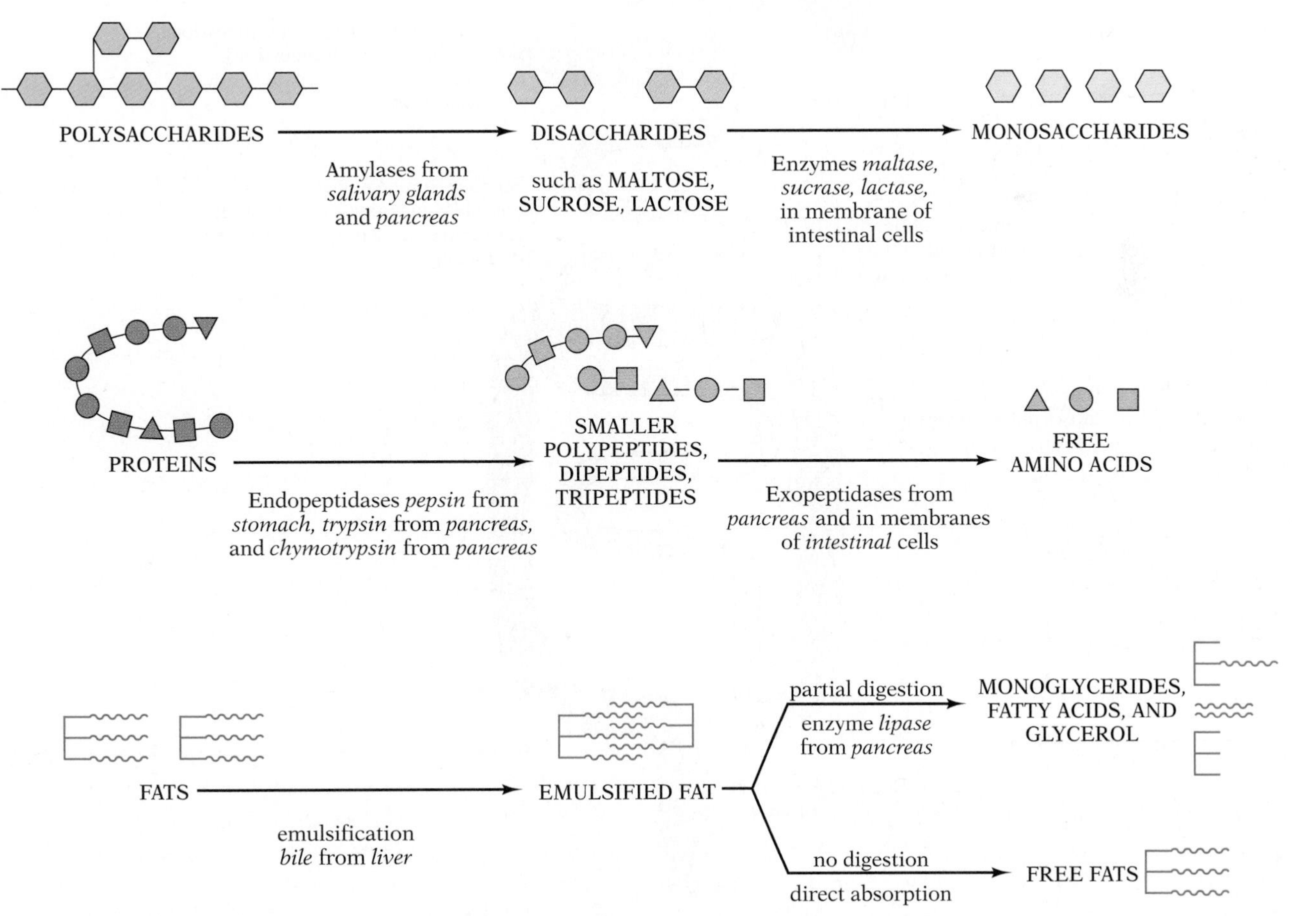

17.31 Action of digestive enzymes.

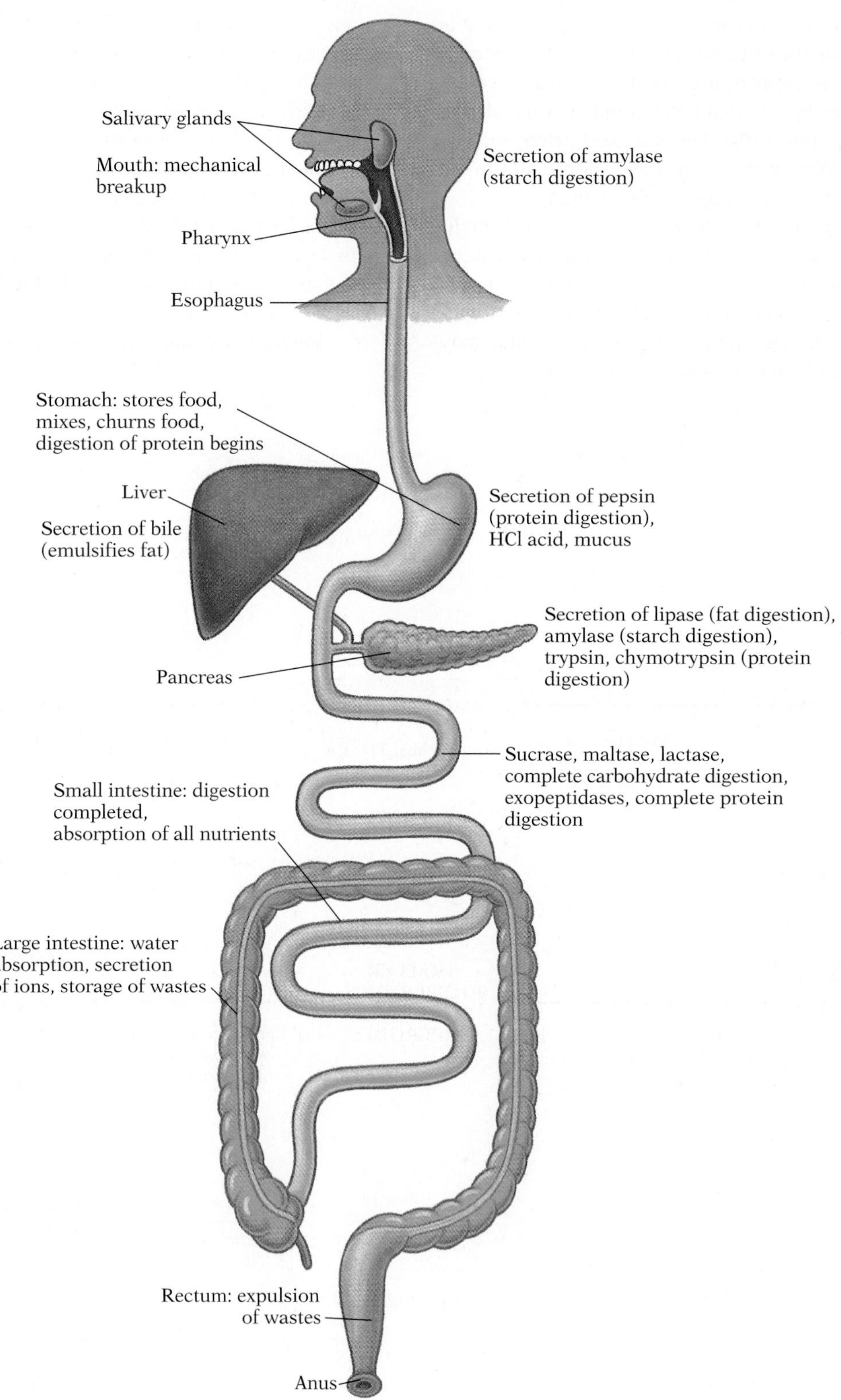
Salivary glands
Mouth: mechanical breakup
Pharynx
Esophagus
Secretion of amylase (starch digestion)
Stomach: stores food, mixes, churns food, digestion of protein begins
Liver
Secretion of bile (emulsifies fat)
Secretion of pepsin (protein digestion), HCl acid, mucus
Secretion of lipase (fat digestion), amylase (starch digestion), trypsin, chymotrypsin (protein digestion)
Pancreas
Sucrase, maltase, lactase, complete carbohydrate digestion, exopeptidases, complete protein digestion
Small intestine: digestion completed, absorption of all nutrients
Large intestine: water absorption, secretion of ions, storage of wastes
Rectum: expulsion of wastes
Anus

17.32 Schematic diagram of the human digestive system

MAKING CONNECTIONS

1. Why are nausea and severe diarrhea frequent side effects of cancer therapy?

2. Why do people who contract diseases—such as cholera—that cause severe diarrhea frequently die from dehydration?

3. Describe two examples of how natural selection has affected the evolution of digestive enzymes.

4. Why is a greater variety of enzymes required to digest proteins than carbohydrates or lipids?

5. Why are people who have had their gallbladders surgically removed likely to have severe indigestion if they consume a lot of fat at a single meal?

CONCEPTS IN BRIEF

1. Since many of the organic molecules found in nature are too large to be absorbed unaltered through cell membranes, they must first be hydrolyzed (i.e., digested) by enzymes into their constituent building-block molecules, which can then be absorbed.

2. There are four main groups of heterotrophic organisms: bacteria, fungi, protozoans, and animals. Bacteria and fungi are *absorptive heterotrophs;* they lack internal digestive systems and depend mainly on absorption as their mode of feeding. In contrast, protozoans and animals are *ingestive heterotrophs;* they take in particulate or bulk food and digest it within chambers inside their body.

3. Many bacteria and fungi thrive on a diet containing only carbohydrates because they can synthesize other necessary organic compounds from carbohydrates. However, most heterotrophic organisms require water, carbohydrates, fats, and proteins in bulk. In addition, they require certain vitamins and minerals in small quantities.

4. Digestion in fungi is always *extracellular;* digestive enzymes are secreted directly into the food supply and the products of digestion are then absorbed.

5. Like fungi, protozoans and animals must digest their food before it can cross their cell membranes. In protozoans and some animals, digestion is *intracellular;* the food is ingested directly into the cell by endocytosis, then hydrolyzed in a food vacuole, and the products of digestion are absorbed. In others, digestion is *extracellular,* either in the environment or in a specialized digestive structure.

6. The radially symmetrical cnidarians have a saclike body consisting of two principal cell layers surrounding a *gastrovascular cavity.* The specialized cells lining the cavity carry on both extracellular and intracellular digestion. Extracellular digestion allows the organism to eat larger pieces of food than it could handle using only intracellular digestion; it is the general rule in multicellular animals.

7. The free-living flatworms are bilaterally symmetrical, elongate animals with a gastrovascular cavity. They carry out some extracellular digestion, but most of the food is digested intracellularly.

8. Animals more complex than cnidarians and flatworms have a *complete digestive tract,* each section of which can be specialized for a different function—mechanical breakup of bulk food, temporary storage, enzymatic digestion, absorption of products, reabsorption of water, storage of wastes, etc.

9. The digestive tract of humans and other vertebrates consists of a series of chambers specialized for different functions. The first chamber is the *oral cavity,* where the teeth break up the food mechanically and the tongue

manipulates the food and mixes it with *saliva.* The food is pushed backward through the *pharynx,* into the *esophagus,* and down to the *stomach.* Pepsin secreted in the *gastric juice* begins protein digestion. The food moves from the stomach into the *duodenum,* where ducts carrying secretions from the liver and pancreas enter. The rest of the *small intestine* is long and coiled. Most of the digestion and the absorption of the products of digestion occurs in the small intestine, where several structural adaptations increase the absorptive surface area.

10. Indigestible material, water, and unabsorbed substances move on into the *large intestine,* or *colon.* A *caecum* projects from the junction of the small and large intestines. Two important functions of the large intestine are the reabsorption of water and the excretion of certain salts when their concentration in the blood is too high. The last portion of the large intestine functions as a storage chamber for the feces until they are released.

11. Enzymatic digestion, the complete hydrolysis of organic nutrients into their building-block molecules, occurs mainly in the mouth, stomach, and small intestine. It is generally a two-step process. First the complex nutrients are hydrolyzed enzymatically into smaller fragments; then other enzymes complete the hydrolysis into the building-block compounds. Absorption of the simple sugars and amino acids into the blood involves active transport.

12. In the digestion and absorption of fat, fat and partially digested fat are absorbed directly since they are lipid-soluble and can cross the cell membrane. The liver produces bile that acts to emulsify the fats so they can be hydrolyzed more effectively by lipase.

STUDY QUESTIONS

1. Classify each of the following heterotrophic organisms using the terms *absorptive, ingestive, saprophytic, parasitic, carnivorous, herbivorous,* and *omnivorous:* (a) athlete's foot fungus; (b) a rabbit; (c) a human being. (p. 343)
2. What are the dietary requirements of an animal? Which of these are required in bulk and which in only small quantities? (pp. 344–345)
3. What characteristics do all vitamins have in common? Why may a compound be a vitamin for one species but not for another? Is vitamin C a vitamin for all vertebrates? (p. 348)
4. Explain the role(s) of each of the following mineral micronutrients in the body: calcium, phosphorus, iron, sodium, chloride, and potassium. (p. 348)
5. What type of chemical reaction is catalyzed by digestive enzymes? (pp. 348–349)
6. Compare intracellular and extracellular digestion. Which is characteristic of fungi? Of protozoans? What do the processes have in common? (pp. 348–354)
7. How do cnidarians capture prey? Where do they digest their food? What is the advantage of extracellular digestion for these (and other) animals? (pp. 354–356)
8. Diagram the digestive system of an earthworm. What is the advantage of its complete digestive tract compared to the gastrovascular cavity of cnidarians and flatworms? (p. 357)
9. Diagram the pathway of a sandwich through the human digestive tract, describing the structure and function of each region. (p. 370)
10. Describe the process of enzymatic digestion of a roast beef sandwich with lettuce and mayonnaise. What enzymes are involved, where are they produced, where do they act, and what are the products of the reactions they catalyze? (p. 370)
11. Describe the composition of bile, its sites of production and storage, and its digestive function. Why is it not considered an enzyme? (pp. 368–369)
12. How does the absorption of fat differ from that of proteins and carbohydrates? (pp. 368–369)

CHAPTER 18

GAS EXCHANGE IN ANIMALS

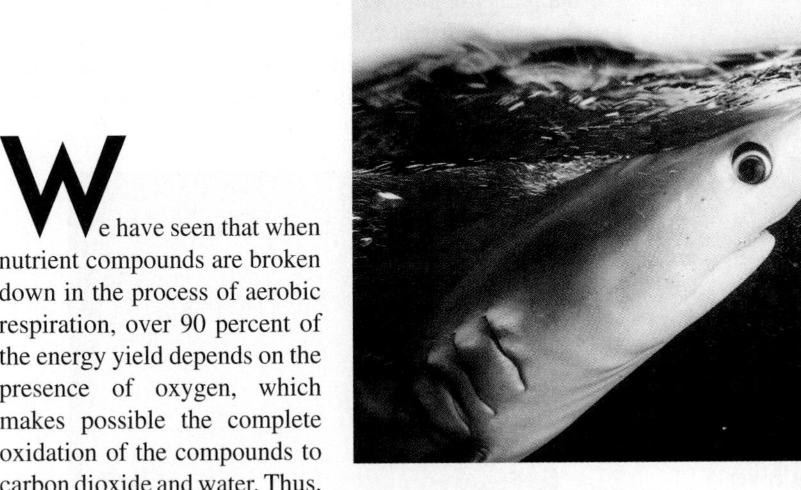

We have seen that when nutrient compounds are broken down in the process of aerobic respiration, over 90 percent of the energy yield depends on the presence of oxygen, which makes possible the complete oxidation of the compounds to carbon dioxide and water. Thus, a basic problem for the great majority of living organisms, including both plants and animals, is obtaining oxygen for cellular respiration and eliminating the waste carbon dioxide (Fig. 18.1).

Animals differ greatly with respect to their oxygen requirements. Very active animals with high metabolic rates consume energy rapidly and therefore require more oxygen per gram of body weight than sedentary organisms with low metabolic rates. Recall that an organism's metabolic rate is closely tied to temperature: within the narrow range of temperatures to which an organism is tolerant, its metabolic rate increases with increasing temperature and decreases with decreasing temperature. Thus, an endothermic organism such as a mammal, which maintains a high body temperature, needs much more oxygen per gram of body weight than an ectothermic organism such as a fish or amphibian living at a colder temperature. Different organisms therefore require different strategies to meet their unique oxygen requirements. Diffusion of gases across the body surface provides a large enough gas-exchange surface for small, thin organisms with low metabolic rates. But, larger, more complex organisms require a very large and specialized gas-exchange surface to obtain sufficient oxygen. Without the evolution of efficient means of gas exchange to fuel the process of cellular respiration, many organisms could not exist. Mammals, for instance, cannot produce enough energy to keep their body cells alive for more than a few minutes without a continuous supply of oxygen.

All Gases Are Exchanged by Diffusion Across a Moist Membrane

We learned in Chapter 16 the three basic requirements for a gas-exchange surface.

1. *A gas-exchange surface of adequate dimensions relative to the volume of the organism.* Because the exchange of oxygen and carbon dioxide between a living cell and its environment *always* takes place by diffusion, from an area of high concentration to an area of low concentration of either gas, there must be a sufficient surface area for the respiratory gases to diffuse across the membrane. A small or

Chapter opening photo: A blue shark *(Prionace glauca)* Like all fishes, sharks carry out gas exchange across the surface of their gills. Many sharks have no way of ventilating their gills and must swim all the time.

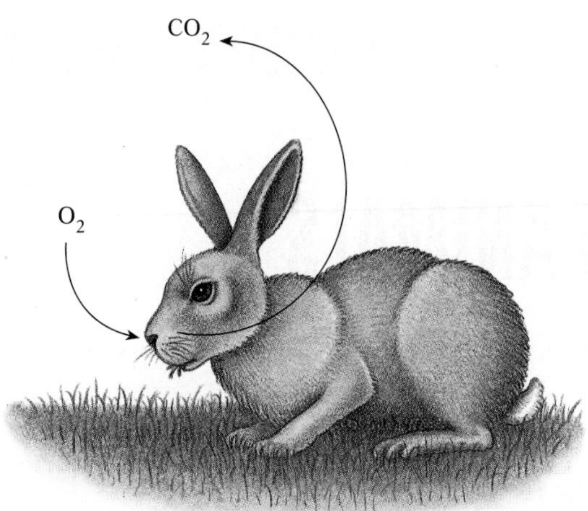

18.1 Basic pattern of gas exchange in plants and animals Most heterotrophs require oxygen for respiration and produce waste carbon dioxide (and water), whereas most autotrophs require carbon dioxide (and water) for photosynthesis, generating waste oxygen.

very thin organism has a large enough surface area compared to its volume that no special exchange surface is necessary (Fig. 18.2). However, as an organism increases in size, the maintenance of a respiratory surface of adequate dimensions relative to the volume becomes a problem, because area (a square function) increases much more slowly than volume (a cube function). The problem is most acute for the more active animals, whose rapid utilization of energy demands a large amount of oxygen per unit of body volume per unit of time. An additional complicating factor is that like terrestrial plants, many animals have evolved relatively impermeable outer body coverings. Coverings such as animal skin with its derivative scales, feathers, and hair function as protective barriers between the fragile internal tissues and organs and the often hostile outer environment, but their presence, which demands that the gas-exchange surface be confined to a restricted region of the body, makes the problem of adequate exchange area even more critical.

2. *A means of protecting the fragile gas-exchange surface from mechanical injury and desiccation.* The need for direct contact between the moist membranes across which gas exchange occurs and the environmental medium (e.g., water or the atmosphere) also poses serious difficulties, especially for terrestrial organisms. The moist membranes must be exposed to the environment to exchange gases, but they must be exposed in such a way as to minimize their chances of drying out. Also, a large, thin, moist surface is often fragile and easily suffers mechanical damage, so the tendency has been toward the evolution of protective devices.
3. *A means of keeping the surface moist.* If gases are to move across the cell membrane, they must be in solution, and therefore the membrane must be kept moist. In aquatic organisms, such as sponges, hydra, and flatworms, this is not a serious problem, because each cell is either in direct contact with the surrounding water or only a few cells away. For organisms living on land, the need to keep the exchange surface moist is a problem since water is constantly lost by evaporation from the surface. The lost water must be replaced or the animal will dry out.

To these three basic requirements, a fourth is added for most animals:

4. *A method of transporting gases between the area of exchange with the environment and the more internal cells.* Another complication brought on by large size in animals is that many cells are deep within the body of the organism, far from the gas-exchange surface. Diffusion alone is too slow to move gases in adequate concentrations across the immense number of cells that may intervene between these more distant cells and the exchange surface. In general, simple diffusion suffices for movement of substances through aqueous media only when the distances are less

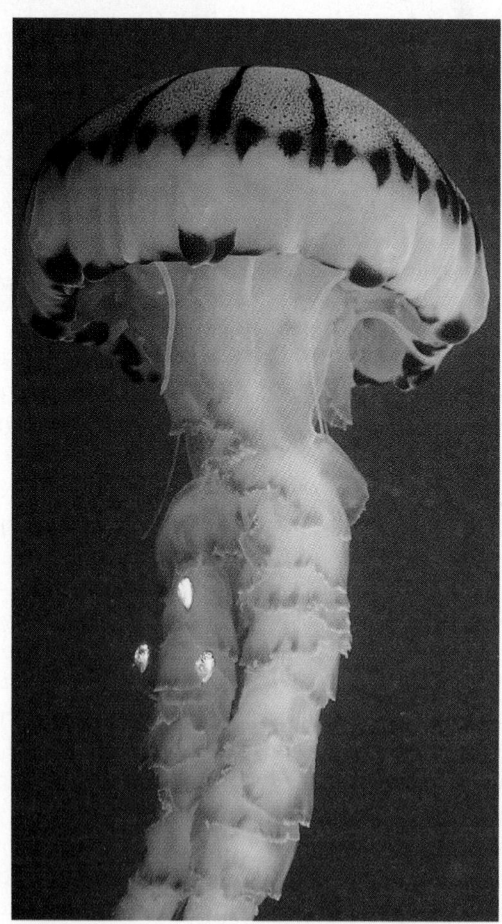

18.2 A jellyfish *(Pelagia)* In a simple solution to the problem of gas exchange, the jellyfish's thin body provides a relatively large surface for the diffusion of gases, with every cell in the organism close to the inner or outer surface.

than one millimeter. Some other mechanism for conveying gases to and from the individual cells of the organism therefore becomes essential. Very often this mechanism is a blood circulatory system, which transports the respiratory gases between the exchange surface and the cells. Typically, the exchange surface has a rich supply of blood vessels that are very close to the surface. Oxygen moves by diffusion from the water or air across the surface cells into the bloodstream, which then transports the oxygen to the individual cells of the body. Carbon dioxide produced by respiration moves in the opposite direction, from the cells, into the bloodstream, and back to the exchange surface. Both gases must be dissolved in water before they can be absorbed by the blood.

In general, specialized gas-exchange surfaces in animals may be grouped in two categories: inward-oriented, or ***invaginated,*** and outward-oriented, or ***evaginated,*** extensions of the body surface (see Fig. 18.3). Each category embraces a diversity of form and detail, but the diversities become less bewildering if one bears in mind that each type of gas-exchange system represents merely one way of meeting the four basic requirements listed above. As we examine a variety of gas-exchange systems, try to see how each meets these four requirements, and how each has adapted to meet that organism's oxygen requirement.

The Availability of Oxygen in Aquatic Environments Is Low

While organisms living in an aquatic environment have no problem keeping their gas-exchange surfaces moist, they do have a problem obtaining enough oxygen, for three reasons. First, oxygen has a low solubility in water—a liter of seawater contains only five milliliters of dissolved oxygen and a liter of freshwater only seven milliliters (versus a liter of air, which contains 210 milliliters). Second, the solubility of gases such as oxygen in water is inversely proportional to temperature: the higher the temperature, the lower the solubility of gases. (The same priniciple applies to a bottle of cola. When the bottle is warm, the gases come out of solution, and the soda bubbles over when opened. At lower temperatures, gas is more soluble in the liquid and it remains in solution.) Organisms living in warm water, then, have very little oxygen available in the surrounding water. Third, the diffusion of oxygen is 10,000 times slower in water than in air, so the oxygen available for cellular respiration is quickly depleted. To continuously renew the supply of oxygen, many organisms have evolved mechanisms to actively move water over (or ventilate) their gas-exchange surfaces.

Aquatic organisms utilize two basic solutions to the problem of gas exchange: the body surface and gills.

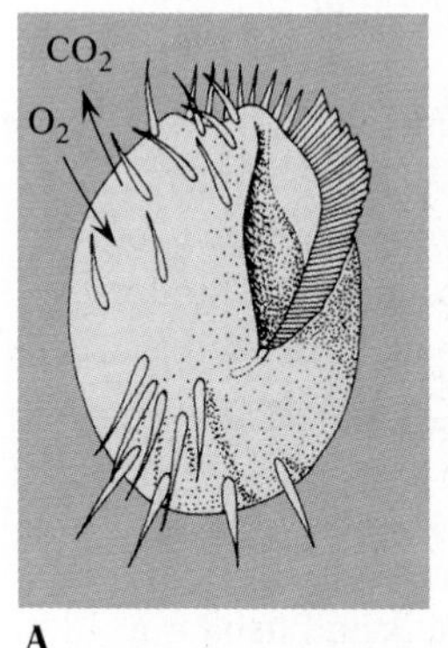

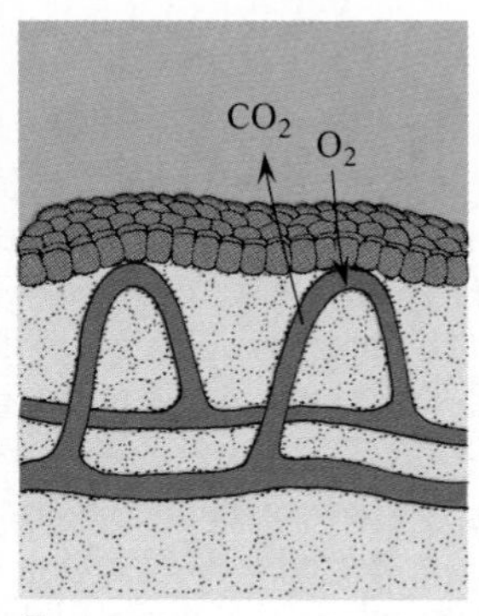

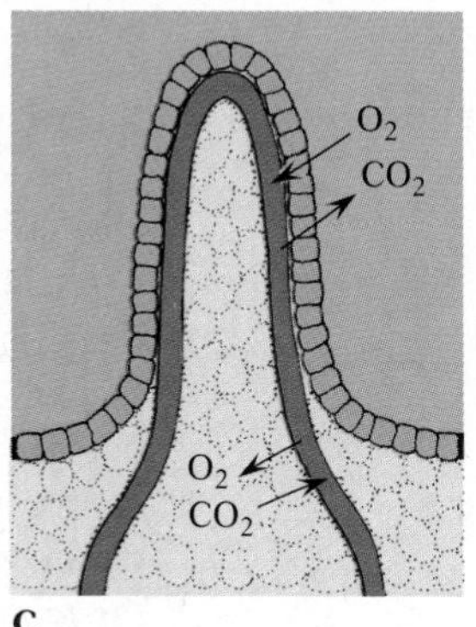

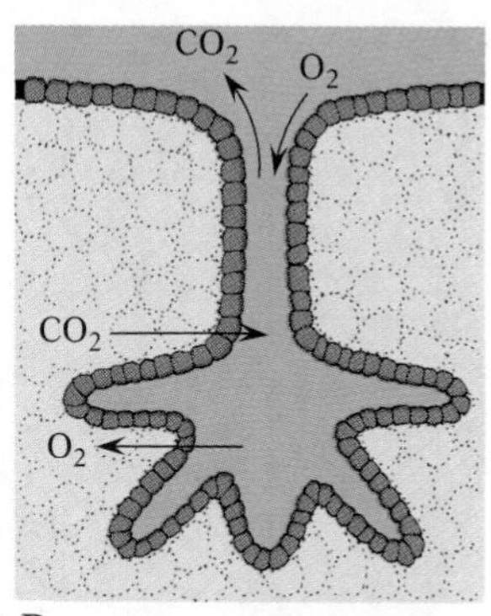

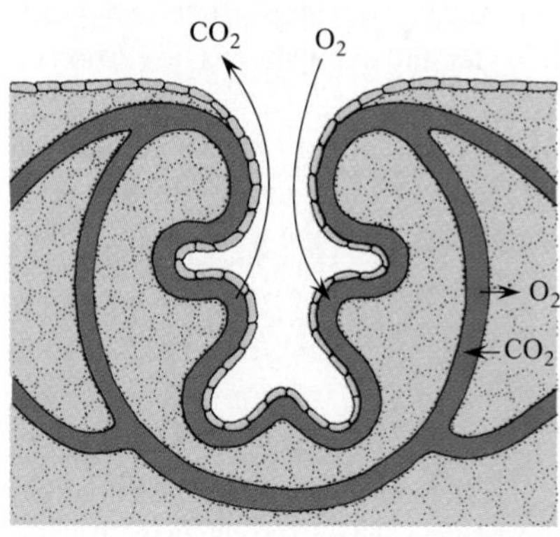

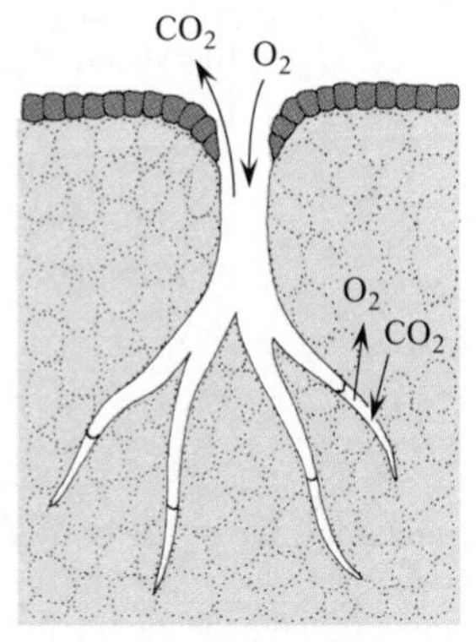

18.3 Types of gas-exchange systems in animals (A) Unicellular organisms exchange gases with the surrounding water directly across the general cell membrane. (B) Some multicellular aquatic animals use the general body surface as an exchange surface; the blood (red) transports gases to and from the surface. (C) Many multicellular aquatic animals have specialized evaginated gas-exchange structures (gills). (D) A few aquatic animals, such as the sea cucumber, use invaginated exchange areas. (E) Most true air breathers have lungs, specialized invaginated areas that depend on a blood transport system. (F) Land arthropods have tracheal systems, invaginated tubes that carry air directly to the tissues without the intervention of a blood transport system.

Many Organisms Use Their Body Surface for Gas Exchange

For many organisms, the body surface provides a large enough surface area for adequate gas exchange and no special gas-exchange surface is necessary. For example, unicellular organisms have no special gas-exchange devices; simple diffusion across their cell membranes is sufficient (Fig. 18.3A). Some of the smaller and simpler multicellular animals, like jellyfish, hydra, and planaria, also rely on their body surface for gas exchange, although their multipurpose gastrovascular cavities do facilitate the exposure of the more internal cells to the environmental water (containing dissolved oxygen) that they draw in through the mouth; no cell in these animals is far from the water medium (Fig. 18.4). A few larger aquatic animals, particularly some of the marine segmented worms, lack special gas-exchange systems and use the skin of the general body surface, which is usually richly supplied with blood vessels (Fig. 18.3B). The oxygen diffuses through the skin and into the blood, which then delivers the oxygen to all parts of the body.

The use of the body surface for gas exchange is not limited to small aquatic animals, however; some animals living in moist environments on land, such as earthworms, use only the moist skin of their body surface for gas exchange. Even among vertebrates, skin-breathing can be an important mechanism for gas exchange. In adult amphibians, approximately 30 percent of the oxygen is taken up across the surface of the skin, and the proportion is even higher in the larvae. Many have evolved thinner skin and special circulatory adaptations to increase oxygen uptake across it. Certain fishes and a number of reptiles also obtain a significant amount of their total oxygen across their skin. Even human beings carry out some gas exchange across their skin. Skin cells, for instance, exchange respiratory gases directly with the air. For most larger multicellular animals, however, skin breathing acts only as a supplement to other mechanisms of gas exchange, and most have evolved specialized gas-exchange systems.

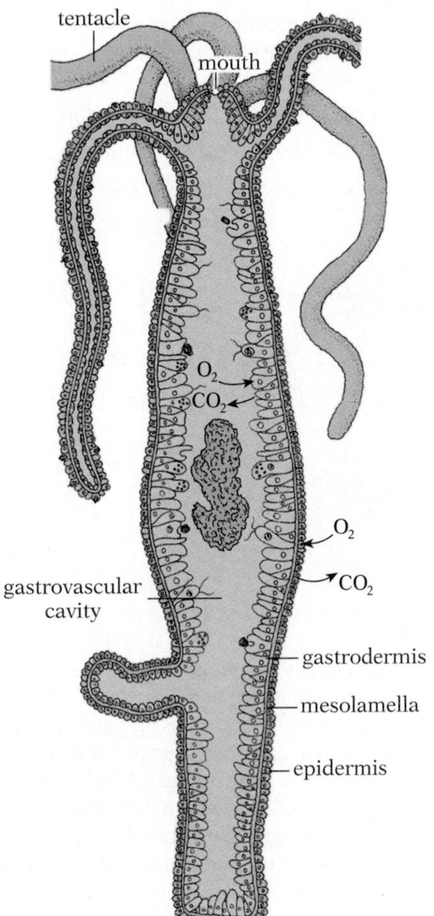

18.4 Gas exchange across the body surface in hydra Gas exchange in hydra occurs across the body surface by simple diffusion across the cell membranes. Hydra also draw environmental water, through the mouth, into the gastrovascular cavity, so no cell in these animals is far from the water medium.

Most Multicellular Aquatic Organisms Use Gills for Gas Exchange

Most multicellular aquatic animals with an active way of life and hence high metabolic rates require a more efficient gas-exchange system than their exterior surfaces provide. Consequently, they have evolved specialized respiratory systems that involve evaginated exchange surfaces, usually known as gills. Gills vary in complexity from the simple bumplike skin gills of some sea stars (Fig. 18.5), the flaplike ***parapodia*** of some segmented marine worms (Fig. 18.6), and the mantle-protected gills of clams and squids (Fig. 18.7) to the minutely subdivided gills of lobsters and fish (Fig. 18.8). Thus, these diverse aquatic animals have independently evolved similar specialized gas-exchange structures—gills—to extract oxygen from water and expel the waste carbon dioxide.

Most gills, particularly those of very active animals, have such finely divided surfaces that a few small gills may expose an immense total exchange surface to the water. Hence, though

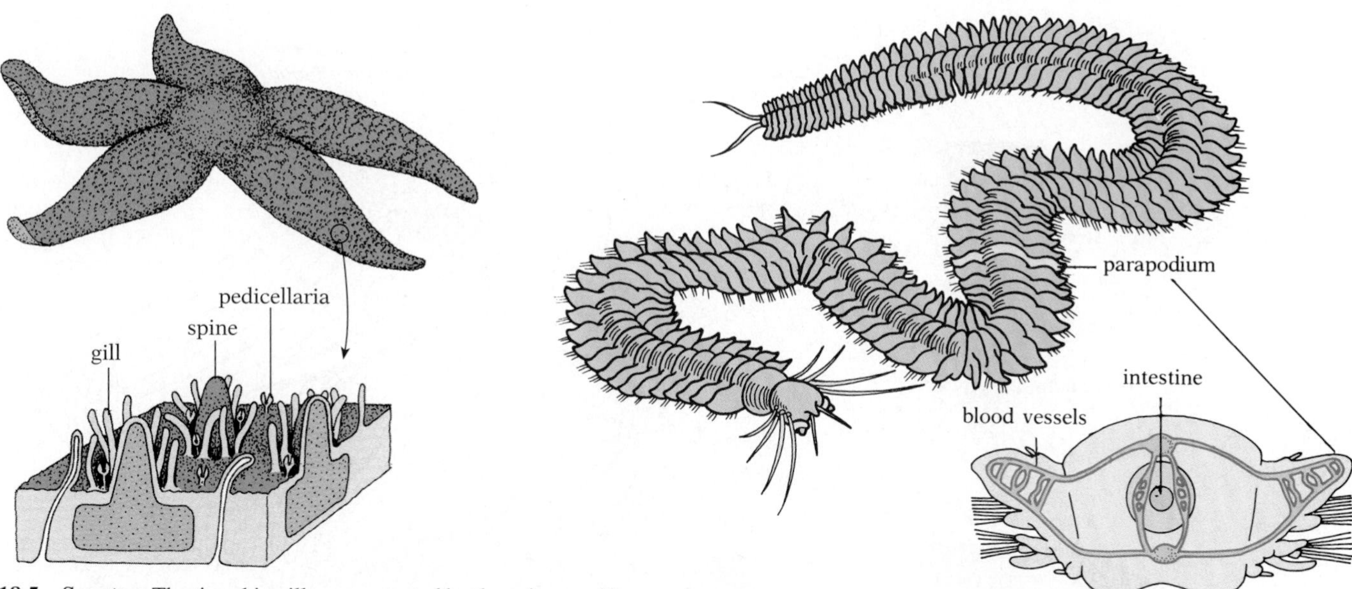

18.5 Sea star The tiny skin gills are protected by the spines and by the pincerlike pedicellariae, which repel (or capture) small animals that might otherwise settle on the surface of the sea star.

18.6 Gills of marine segmented worm *(Nereis)* As the cross section shows, the flaplike extensions of the body surface on each segment, the parapodia, are richly supplied with blood vessels. The oxygen diffuses through the skin and into the blood, which then delivers the oxygen to all parts of the body.

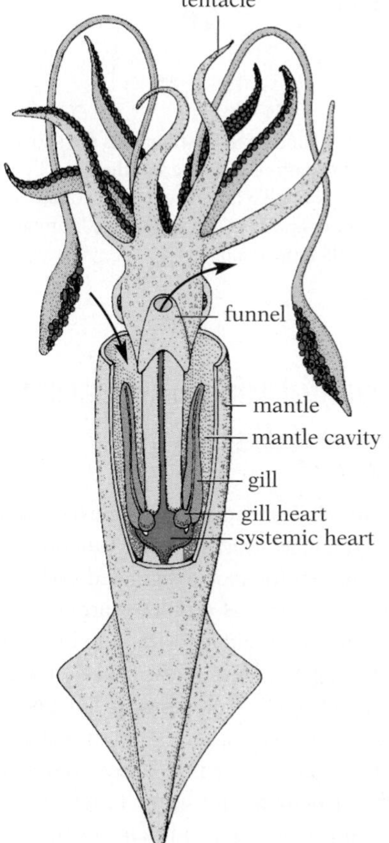

18.7 The gills of a squid Part of the protective mantle has been cut away to expose the gills within the mantle cavity. Water containing oxygen (left-hand arrow) flows into the mantle cavity when the mantle is relaxed, providing O_2 to the capillaries of the gills. When the mantle contracts, its collar seals the opening and the CO_2-containing water is forced out of the funnel.

the gas-exchange surface takes up a very limited part of the animal body, most of which can thus be protected by relatively impermeable coverings, the surface-to-volume ratio of the exchange surface remains high. The gills themselves often have some type of protective device. The spines and pincerlike structures on the sea star or the flaplike operculum (Fig. 18.8) of bony fishes serve to protect the fragile gills from physical damage.

Another characteristic of most gills is that they contain a rich supply of blood vessels. Often the blood in these vessels is separated from the external water by only two cells: the single cell of the wall of the vessel and a cell of the gill surface. Oxygen moves by diffusion from its area of high concentration, in the water, across the intervening cells, and into the blood, where the concentration of oxygen is low. The oxygen is ordinarily picked up by a carrier pigment in the blood (transport by the blood will be discussed in Chapter 20). The blood then distributes the oxygen throughout the body to the individual cells. Carbon dioxide produced by cellular metabolism moves in the opposite direction, along its concentration gradient, and is transported by the blood from the cells to the gills where it diffuses into the surrounding water. The movement of oxygen and carbon dioxide is always down their concentration gradients, according to the rules of diffusion.

One intriguing feature of the exchange of oxygen and carbon dioxide between water and blood in fish gills deserves special mention. Most fish actively pump water into the mouth, across the gill filaments, and out behind the operculum, thereby ventilating the gills. The water flows over the surface

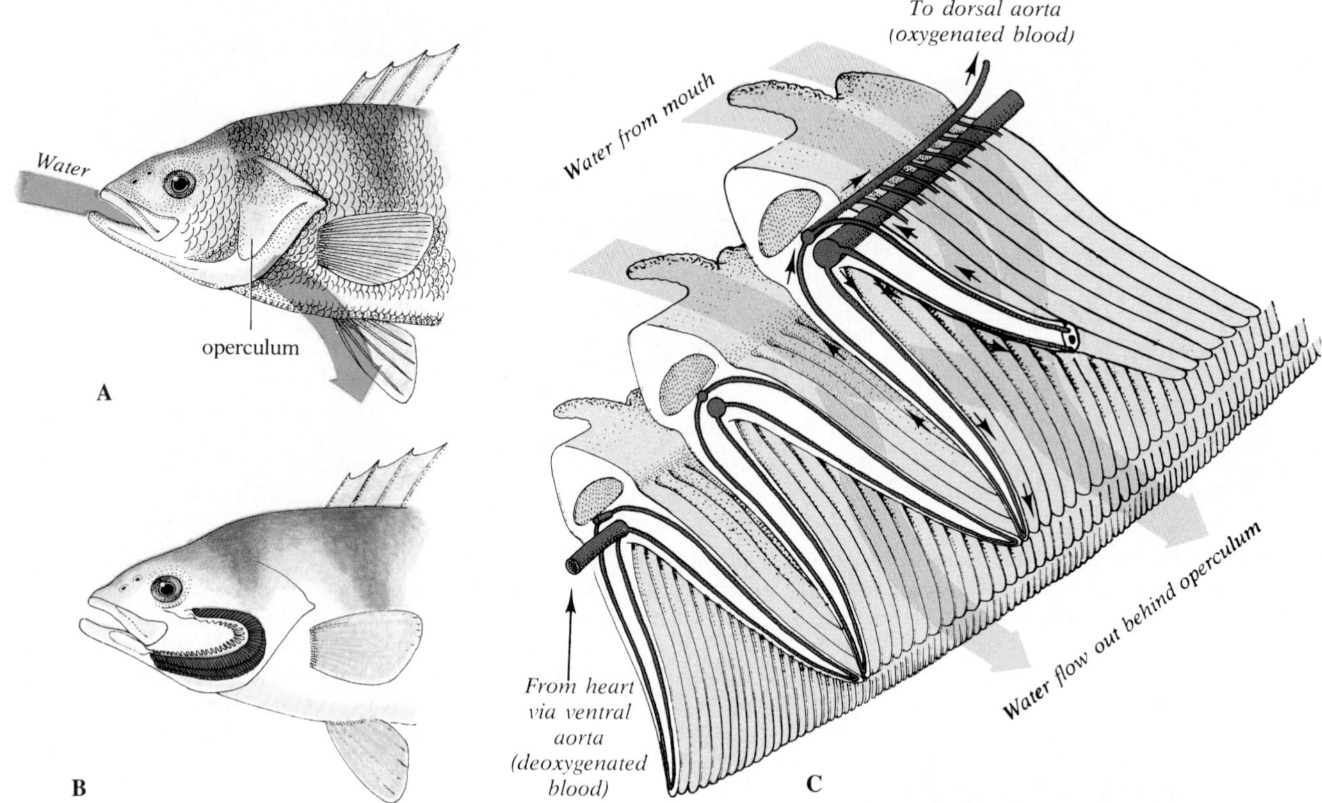

18.8 Gills of fish (A) Head with operculum covering gills. Water carrying oxygen is drawn in through the mouth, flows across the gills, and exits behind the operculum. (B) Head with the operculum cut away and the gills exposed. (C) Portions of three adjacent gill arches. Each arch bears two rows of primary filaments. The main paths of blood flow to and from the filaments are shown in red; the blue arrows trace the path of water across the gills. Each primary filament bears many disklike lamellae, which contain capillaries that run from one artery to another. The end of one filament has been cut off to show a lamella more clearly.

of a gill lamella in a direction *opposite* (counter) to the flow of blood in the vessels of the lamella (Figs. 18.8 and 18.9). This ***countercurrent exchange*** system creates favorable concentration gradients for gas exchange and thereby maximizes the amount of oxygen the blood can pick up from the water. This would not be the case if the two fluids had the same direction of flow (Fig. 18.9B).

So, all of these features, a large surface area, active ventilation, and a blood transport system, maximize an aquatic animal's ability to obtain oxygen—that is, gills make the most of the low solubility of oxygen in water and slow diffusion rates. Also, ventilation of the gills constantly brings in new oxygen-containing water and removes the waste carbon dioxide.

Not all aquatic animals with special respiratory systems use evaginated gills, and not all animals that live in water are fully aquatic. Many insects and some mammals that live in water, for instance, must periodically come to the surface to breathe air. Of these, aquatic spiders that construct an underwater web in which they store a large bubble of air show a particularly interesting adaptation (Fig. 18.10).

Most Terrestrial Animals Have Invaginated Gas-Exchange Surfaces

A few land animals (like spiders) have evolved highly modified gill-like exchange structures that function in air. But the hazards of drying out for such evaginated surfaces are considerable, and major structural problems are associated with an array of filaments or a branched structure both strong enough to maintain its shape against surface tension and gravity and thin-walled enough to allow easy passage of gases. It is not surprising, therefore, that most terrestrial animals have evolved invaginated respiratory systems. These systems are of two principal types, ***tracheae*** and ***lungs*** (Figs. 18.3E and F). In both, the air is moistened as it comes inside the system, and the cells of the exchange surface are covered by a film of water in which gases can dissolve. Thus, the process of gas exchange has remained essentially aquatic in land animals.

The problems associated with gas exchange in aquatic and terrestrial organisms are quite different. In water, where the

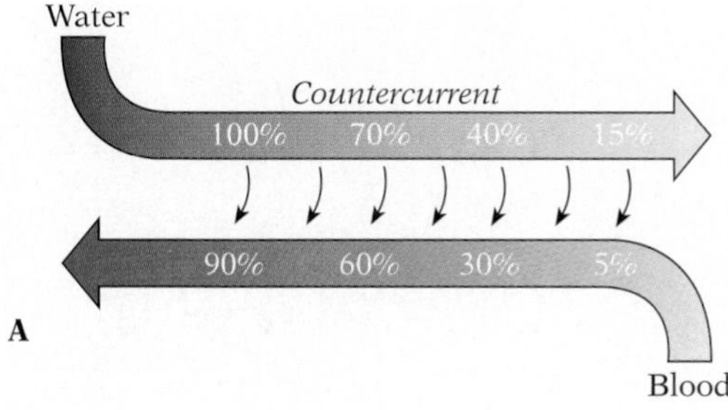

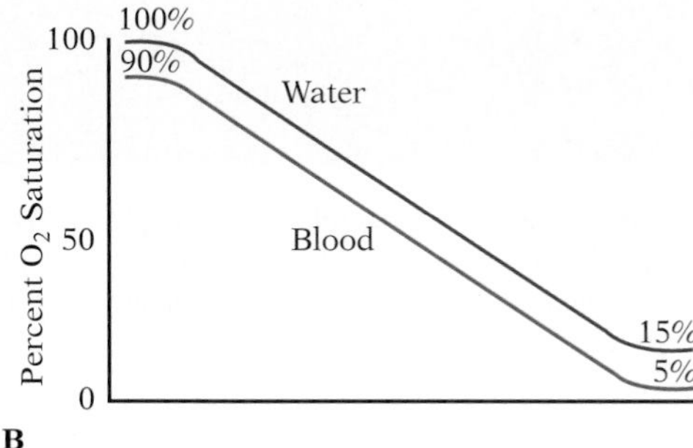

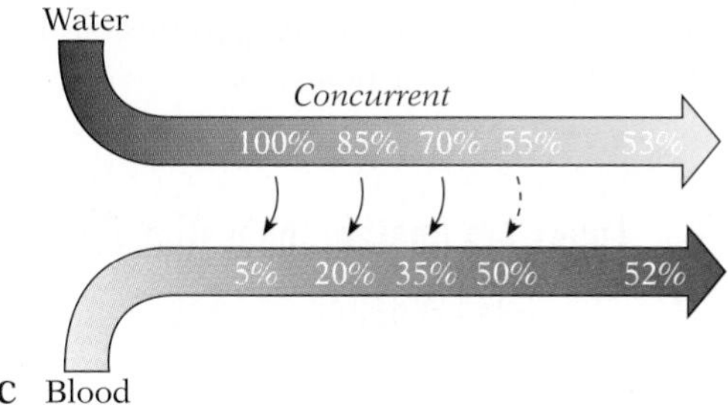

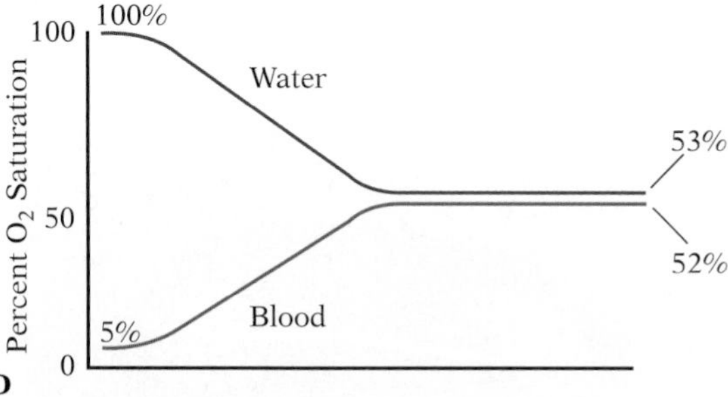

18.9 Countercurrent exchange system in lamella of fish gill Figures indicate the degree of oxygen saturation for both water and blood. Because of the counterflow (A), the oxygen gradient between water and blood always favors diffusion of oxygen (vertical arrows) from water to blood, and the blood can extract a high percentage of the oxygen from the water (B). If the flow were parallel (concurrent; C), initially there would be a large concentration gradient favoring the diffusion of oxygen from the water into the blood, but the gradient is not maintained and overall the blood could extract much less oxygen and would leave the blood far from fully loaded (D).

18.10 The water spider *Argyroneta aquatica* The male spider is beneath his underwater web, which holds a large bubble of air and to which he has returned to breathe. He transports the air there by trapping it among hairs on his abdomen.

oxygen supply is low but there is no problem in keeping the exchange surface moist, gills are a very efficient mechanism for exchange. But on land, where the oxygen concentration is high and water loss is a problem, lungs or a tracheal system are more efficient. The ease of obtaining oxygen from air instead of water is reflected in the energy expenditure involved in carrying out gas exchange: fish devote nearly 20 percent of their total energy expenditure to ventilate their gills, whereas air breathers devote only 1 to 2 percent of their energy to ventilate their lungs.

Most Land Arthropods Have Evolved a Tracheal System for Gas Exchange

The first principal type of invaginated respiratory system evolved for air breathing is the tracheal system. It is one solution to the problem of bringing oxygen close to the cells, and carrying away the waste carbon dioxide. Tracheal systems are typical of most terrestrial arthropods (e.g., insects). Here we find no localized gas-exchange organ and little or no significant transport of gases by the blood. Instead, the system is composed of many small tubes, called tracheae, that open to the outside and branch throughout the body (Fig. 18.11). The tracheae and the smaller tracheoles into which they branch carry air from the outside *directly* to the individual cells. But,

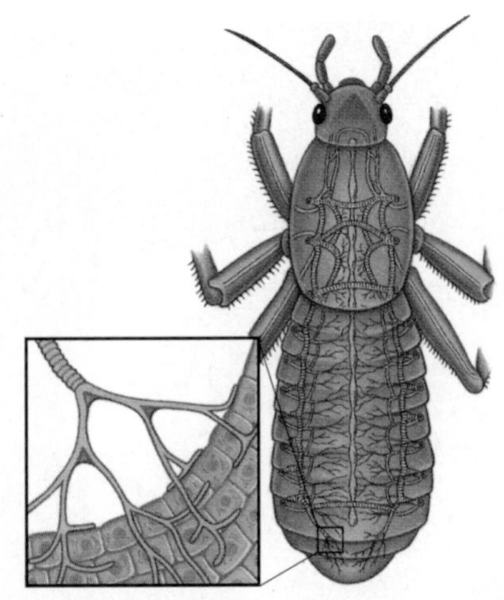

18.11 The tracheal system of insects (A) The tracheal system consists of many small internal tubes called tracheae that branch throughout the insect body. The tracheae open to the outside through holes called spiracles. The tracheae and the smaller tracheoles (insert) into which they branch carry air directly to the individual cells, where diffusion across the cell membranes takes place. (B) At top center is a spiracle (brown), from which numerous branching tubes—the tracheae—can be seen running to many parts of the insect's body.

in order to diffuse across cell membranes, the oxygen must first dissolve in the fluid at the end of the tracheoles. This is similar to other gas-exchange systems.

Air enters the tracheae by way of the ***spiracles,*** pores in the body wall that can usually be opened and closed by valves (Fig. 18.12). Some of the larger insects actively ventilate their tracheal systems by muscular contraction, but most small insects and some fairly large ones do not, although normal muscle movements do aid the movement of air. Calculations have shown that the rate of diffusion of oxygen in air through the tracheae is rapid enough to maintain at the tracheal endings an oxygen concentration only slightly below that of the external atmosphere. But diffusion, remember, is effective only over short distances (i.e., less than one millimeter), so this type of respiratory system, which depends on diffusion through the tracheal tubes, limits the size of insects.

Lungs Are Another Innovation for Gas Exchange on Land

Lungs, which are invaginated gas-exchange organs limited to a particular region of the animal and dependent on a blood trans-

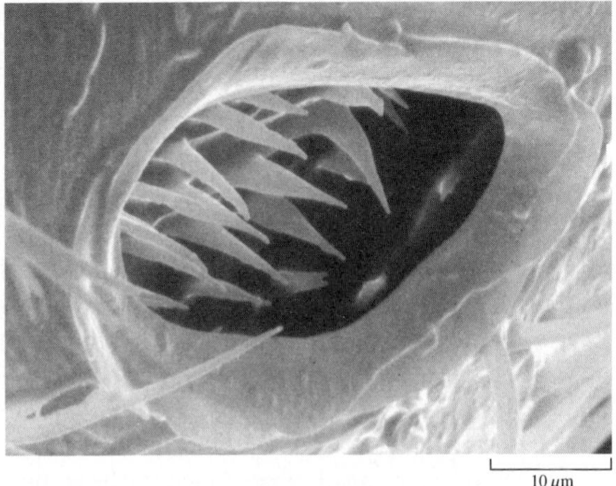

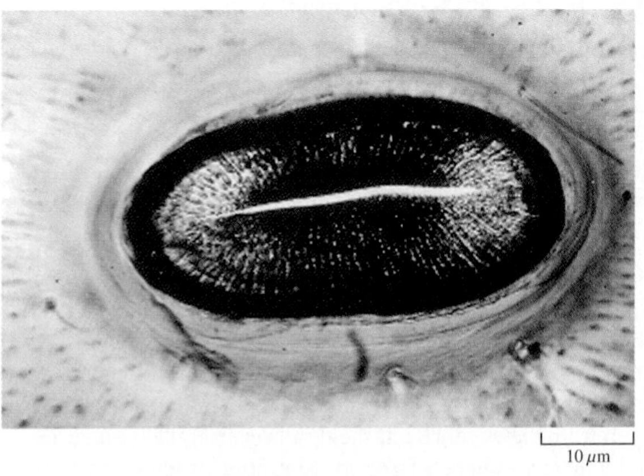

18.12 Spiracles of two insects Left: Scanning electron micrograph of a fully open ant spiracle. The pointed projections are sensory hairs that monitor external conditions and can trigger spiracle closing when necessary. Right: A nearly closed grasshopper spiracle; the black areas are the valves. Note the resemblance to the stoma of a leaf (Fig. 16.21, p. 339).

port system, are most typical of two unrelated animal groups, the land snails and the vertebrates, including some fish, most amphibians, and all reptiles, birds, and mammals. In their simplest form, as in some primitive salamanders, lungs are little more than sacs with a slightly increased blood supply and with some sort of passageway leading to the outside. From such a rudimentary beginning, the evolution of the lung has tended toward a greatly increased surface area, by subdivision of its inner surface into many small pockets or folds, and toward an increased blood supply to its exchange surface (Fig. 18.13).

A lungfish

B toad

C reptile

D mammal

18.13 Evolution of vertebrate lungs The lungs of vertebrates show a progressive increase in surface area from lungfish to amphibians, reptiles, and mammals. (A) The ancestral vertebrates had lungs that were little more than sacs supplied with blood. Some fish and amphibians have lungs of this type. (B) Other amphibians, such as the frog, have increased partitioning within the lung, which gives them a greater gas-exchange surface. (C) Reptiles, which evolved from amphibians, have lungs with still more partitions. (D) Mammals, which evolved from reptiles, have lungs that are very finely divided into alveoli, which provide an enormous surface area for gas exchange.

The Human Gas-Exchange Surface Consists of Very Finely Subdivided Lungs

Let us look at the human gas-exchange system (Fig. 18.14) in some detail, as an example of the mammalian type. The system is composed of the passageways connecting the lungs to the outside, the actual gas-exchange surface (the lungs), and a pump (the rib cage) that ventilates the lungs. While at rest, a normal human breathes (ventilates the lungs) about 12 to 15 times per minute. About six to eight liters of air enter and leave the body per minute, so over a 24-hour period approximately 10,500 liters of air are exchanged! The tremendous exchange of gases reflects the fact that humans are homeothermic endotherms, and therefore have a high metabolic rate and a correspondingly high oxygen requirement. The procurement of oxygen and its delivery to the individual cells is one of the crucial functions of the human body.

During breathing, air is drawn in through the ***external nares,*** or nostrils, and enters the ***nasal cavities,*** which function in warming and moistening the air, filtering out dust particles, and smelling. It then passes into the ***pharynx*** (throat). You will recall that the pharynx also functions as a part of the digestive

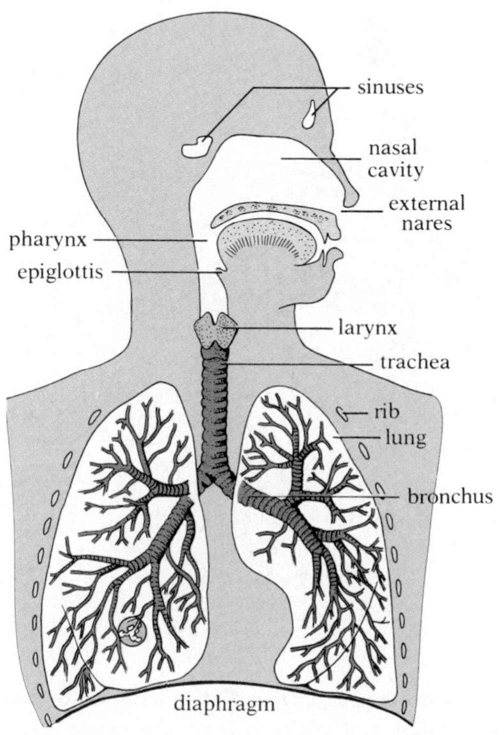

18.14 Schematic drawing of the human gas-exchange system

MAKING CONNECTIONS

1. Why is oxygen required by the vast majority of organisms? What factors determine how much oxygen an animal needs?

2. Why is an invaginated respiratory system better suited to a terrestrial animal than an evaginated system?

3. Compare the problems and methods of gas exchange of terrestrial animals with those of terrestrial plants, described in Chapter 16.

4. Compare the spiracles of insects with the stomata of plants, described in Chapter 16. How might each play a homeostatic role in the organism?

5. How has the vertebrate lung evolved to be more efficient in terrestrial habitats?

system; the air and food passages cross here (Fig. 18.15). During inhalation, air leaves the pharynx via a ventral opening, the ***glottis,*** and enters the larynx. (In humans, the term "ventral" refers to the front of the body.) A flap of tissue, the ***epiglottis,*** covers the glottis during swallowing, thus preventing food from entering the larynx. It is the functioning of the epiglottis that is affected when you choke. Choking occurs when food, usually improperly chewed meat, becomes stuck in the lower pharynx and prevents the epiglottis from opening (see Exploring Further: Heimlich Maneuver, p. 384).

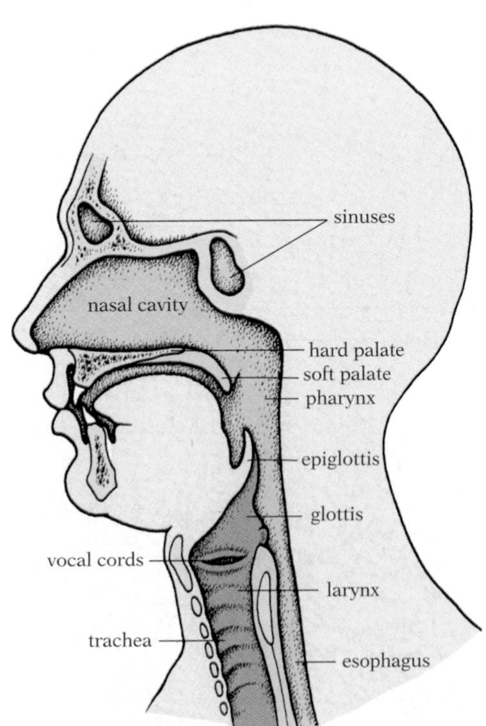

18.15 Detail of upper portion of human respiratory and digestive system

The ***larynx*** is a chamber surrounded by a complex of cartilages (commonly called the Adam's apple). In many animals, including humans, the larynx functions as a voice box. It contains a pair of vocal cords—elastic ridges stretched across the laryngeal cavity that vibrate when air currents pass between them; changes in the tension of the cords result in changes in the pitch of the sounds emitted.

The ***trachea*** is an air duct leading from the larynx into the thoracic cavity. A series of C-shaped rings of cartilage are embedded in the walls of the trachea and prevent it from collapsing upon inhalation. At its lower end, it divides into two ***bronchi,*** tubes that lead toward the two lungs. Each bronchus branches and rebranches, and the ***bronchioles*** thus formed branch repeatedly in their turn, forming smaller and smaller ducts that ultimately terminate in tiny air pockets, each of which has a series of small chamberlike bulges in its walls, termed ***alveoli,*** where the actual gas exchange takes place. (No gas exchange takes place across the walls of the conducting passageways—the trachea, bronchi, etc.) The conducting tubes that deliver the air to the alveoli serve two other important functions: (1) warming and moistening the incoming air, and (2) filtering the air and removing particulate matter. Incoming air must be warmed and moistened in order to prevent the drying out of the fragile alveolar surfaces. By the time inspired air reaches the alveoli its temperature has reached 37°C (body temperature) and is at almost 100 percent humidity. The second function is served by the combined action of cilia and mucus secreted by the epithelial cells lining the con-

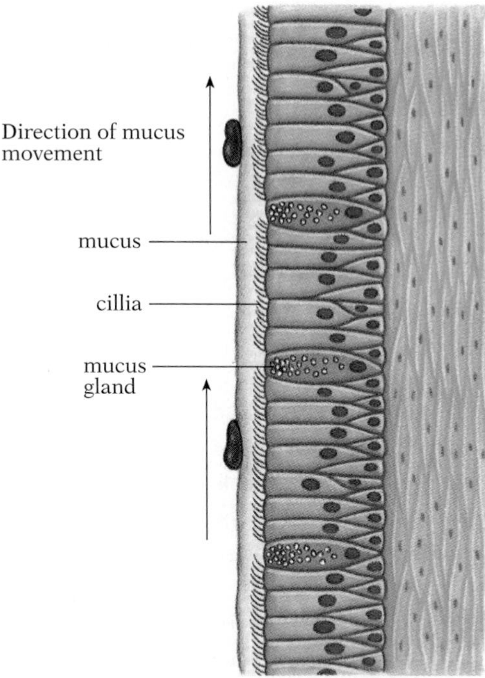

18.16 Epithelial lining of the conducting passageways in the human respiratory tract The conducting passages of the human gas-exchange system are lined with ciliated epithelial cells and mucus-secreting glands. Foreign particles are trapped in a flow of mucus that is being constantly moved upward to the pharynx by the coordinated beating of the cilia lining these passageways. In the pharynx the mucus is automatically swallowed or coughed up. Cilia are very sensitive to chemicals; a single cigarette can paralyze the cilia for hours, leaving the mucus stagnant.

ducting tubes (Fig. 18.16). Foreign particles are trapped in a flow of mucus that is constantly moved upward to the pharynx by the coordinated beating of the cilia lining these passageways. In the pharynx the mucus is automatically swallowed or coughed up. Cilia are very sensitive to chemicals; a single cigarette can paralyze the cilia for hours, leaving the mucus stagnant.

There are 300 million alveoli in the lungs and the total alveolar surface is enormous: about 100 square meters—many times greater than the total area of the skin. The walls of the alveoli are exceedingly thin, being usually only one cell thick, and each alveolus is surrounded by a dense bed of blood capillaries (Fig. 18.17). The alveoli are the site of the actual gas exchange and may therefore be regarded as the primary functional units of the lungs. The air in the alveoli has a higher concentration of oxygen than the blood, so the oxygen dissolves in the film of water on the alveolar wall and then diffuses across the intervening cells to the blood, moving down its concentration gradient. The dissolved oxygen is picked up by the carrier pigment hemoglobin in the blood (transport by the blood will be discussed in Chapter 20), which then distributes

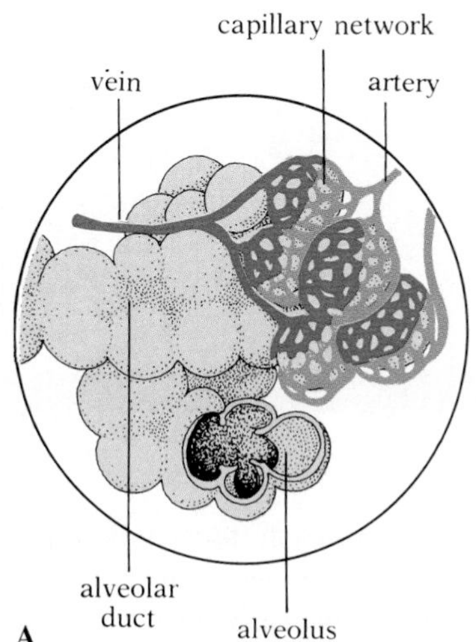

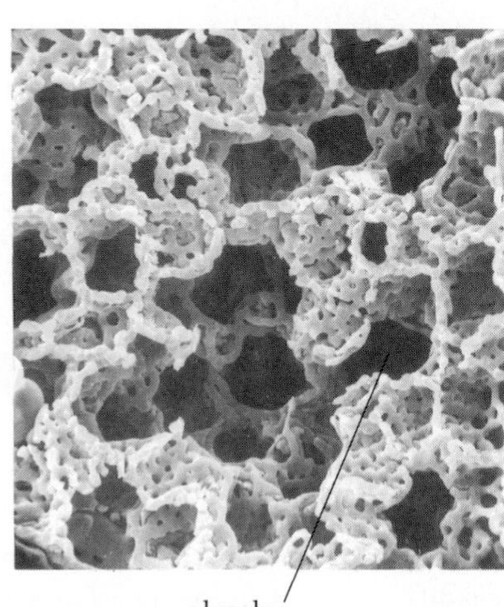

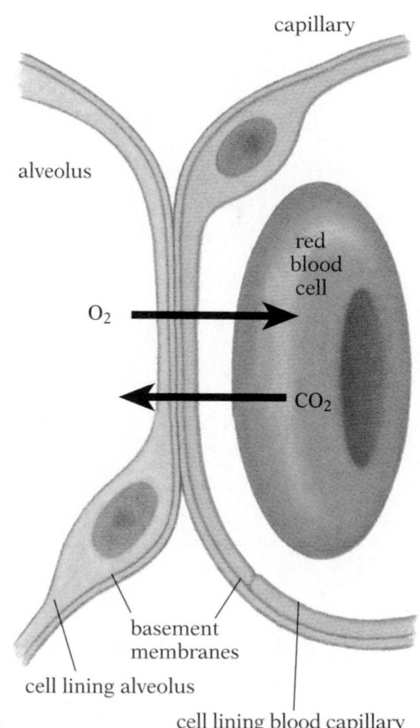

18.17 Gas exchange across the walls of the alveoli (A) The drawing shows a few sectioned alveoli and other alveoli surrounded by blood vessels. Actually, all alveoli, including those lying along alveolar ducts, are surrounded by networks of capillaries. (B) The walls of the alveoli are very thin, usually only about one cell thick. Each alveolus is surrounded by a capillary bed. To reach the blood, oxygen must diffuse across the alveolar cells, the basement membrane, and the single layer of cells that lines the blood capillary. Carbon dioxide must diffuse in the opposite direction. The oxygen is transported by hemoglobin in the red blood cells to the various parts of the body.

EXPLORING FURTHER

HEIMLICH MANEUVER

About 4,000 persons die each year because they choke on food, usually improperly chewed meat. Choking occurs when food becomes stuck in the lower pharynx and prevents the epiglottis from opening. When the airway is completely blocked, the person will be unable to cough, speak, or breathe, and suffocation soon follows. A simple treatment, called the *Heimlich maneuver,* is usually effective in dislodging the food. It should be used only when a person cannot speak or has choked and becomes unconscious.

To perform this maneuver, stand behind the victim and place the thumb side of the fist on the abdomen, above the waist and below the rib cage (Figure A). Grasp the fist with your other hand and give four hard inward and upward thrusts. These thrusts should dislodge the blockage. If it does not, repeat three more times.

If you are alone and choke, you can give yourself thrusts by pressing your fist between the rib cage and waist and giving quick inward and upward thrusts. Alternatively, you can lean forward and quickly press your abdomen over a firm object such as the back of a chair.

the oxygen throughout the body to the individual cells. The carbon dioxide produced by the cells during respiration diffuses into the blood, which conveys it to the lungs, where it diffuses from the blood into the alveoli (Fig. 18.17).

A number of debilitating human diseases interfere with proper lung function. Sometimes the air flow through the bronchioles is obstructed as a result of excessive or abnormally thick mucus secretion (as in infections or cystic fibrosis), inflammation (due to infection or toxic chemicals), or constriction of the smooth muscles in the walls of the bronchioles (as in asthma). All of these conditions narrow the bronchioles and therefore interfere with air flow and increase resistance, making breathing difficult. In emphysema, the alveolar tissue itself is destroyed, resulting in fewer, but larger alveoli. This condition greatly decreases the surface area for gas exchange and also leads to a collapse of the bronchioles during exhalation. The most common cause of emphysema is heavy cigarette smoking. Research indicates that a component in cigarette smoke stimulates certain white blood cells to secrete enzymes that destroy alveolar tissue. Severe emphysema can lead to high blood pressure in the blood vessels supplying the lungs, and to eventual heart failure.

The Lungs Must Be Ventilated to Bring Oxygen In and Take Carbon Dioxide Away

During the mechanical process of breathing, the lungs are ventilated—air is drawn into and then expelled from the lungs. In mammals this process generally involves muscular contractions of two regions, the ***rib cage*** and the ***diaphragm.*** The latter is a muscular partition separating the thoracic (chest) and abdominal cavities (Fig. 18.18). Inhalation, or inspiration, occurs whenever the volume of the thoracic cavity, which contains the lungs, is increased; such an increase reduces the air pressure within the chest to below that of the atmosphere, and air is drawn into the lungs. The increase in thoracic volume is accomplished both by contraction of the rib muscles, which pulls the rib cage up and out, and by contraction, or a downward pull, of the normally upward-arched diaphragm (Fig. 18.18). The first mechanism is often called chest breathing, while the second is called abdominal breathing. Normal exhalation, or expiration, is a passive process; the muscles relax, allowing the rib cage to fall back to its resting position and the diaphragm to arch upward. This reduction of thoracic volume, combined with the elastic recoil of the lungs themselves, causes a rise in the pressure inside the lungs to a level above that of the outside atmosphere, which drives the air out.

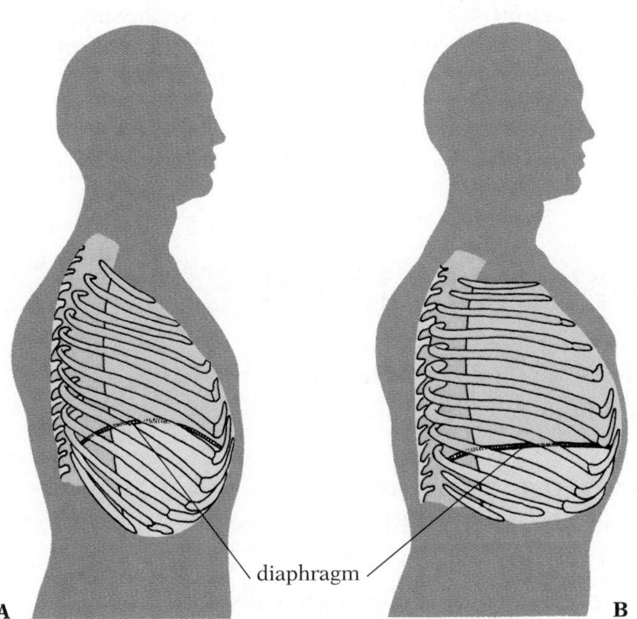

18.18 The mechanisms of human breathing (A) Resting position. (B) Inhalation. The rib cage is moved up and out by the intercostal muscles between the ribs, and the diaphragm is pulled downward by contraction of diaphragmatic muscles. Both these motions increase the volume of the thoracic cavity. The consequently reduced air pressure in the cavity causes more air to be drawn into the lungs.

18.19 Respiratory system of a bird Attached to the lungs are many air sacs (light brown), some of which even penetrate into the marrow cavities of the wing bones. As in a mammal, the respiratory system has bilateral symmetry (not obvious in this side view).

The pattern of air flow in the respiratory system of birds differs fundamentally from that in mammals. In addition to paired lungs, birds possess several (most commonly eight or nine) thin-walled air sacs that occupy much of the body cavity and even penetrate some bones (Fig. 18.19). The air sacs are poorly supplied with blood vessels and do not themselves absorb oxygen or release carbon dioxide. The arrangement of air sacs makes possible a continuous one-way flow of air through the lungs: during inhalation most of the air goes into the posterior air sacs, then into the lungs, followed by the anterior air sacs, and finally out the trachea.

Birds are far more efficient than mammals in extracting oxygen from air, both because of the continuous one-way flow of air through their lungs, and because blood in the capillaries associated with the gas-exchange surface in the lungs flows in a direction opposite to the flow of air. This provides some of the same benefits as the countercurrent exchange system of fish gills. Because of the countercurrent flow, favorable oxygen concentration gradients between the blood and lungs are established (Fig. 18.9), maximizing oxygen absorption by the blood and thereby supplying sufficient amounts of oxygen to the cells

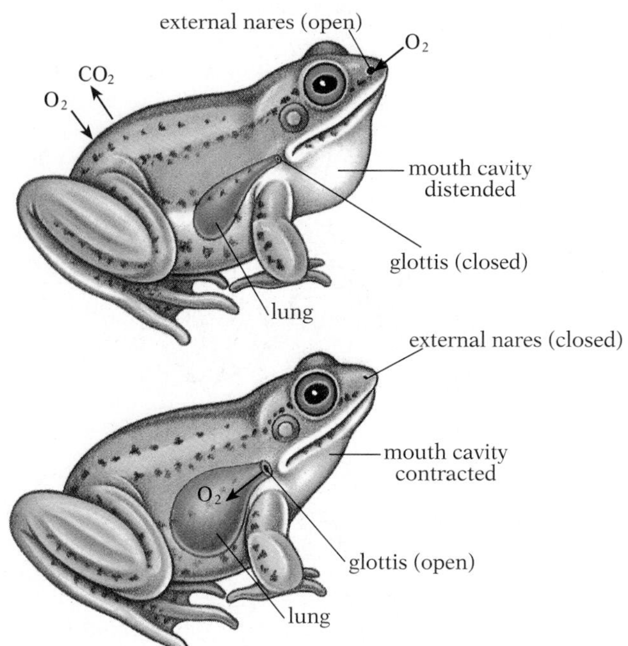

18.20 Gas exchange in the frog (A) With the mouth closed, nostrils open, and the glottis closed, the frog lowers the floor of its mouth, thereby drawing air into the mouth cavity. (B) Then it closes the nostrils, raises the mouth floor, and opens the glottis. This reduction in the volume of the mouth cavity exerts pressure on the imprisoned air and forces it into the lungs (positive-pressure breathing). Gas exchange can occur across the surface of the lungs, which are only occasionally filled, and across the exchange surface provided by the soft moist skin.

even when the oxygen concentration in the air is low. (The absorption of oxygen in the lungs of a mammal is, by contrast, directly proportional to the oxygen concentration in the air.) This superior efficiency enables birds to fly at high altitudes, where there is less oxygen in the atmosphere.

The mammalian, reptilian, and avian method of breathing, in which air is *drawn* into the lungs by a pressure within the lungs that is less than atmospheric pressure (i.e., negative pressure), is known as ***negative-pressure breathing,*** by contrast with ***positive-pressure breathing,*** where air is *forced* into the lungs. Both methods are used by adult frogs (Fig. 18.20). With the mouth closed and nostrils open, the frog lowers the floor of the mouth, thereby drawing air into the mouth cavity (negative-pressure method). Then it closes the nostrils and raises the mouth floor; this reduction in the volume of the mouth cavity exerts pressure on the imprisoned air and forces it into the lungs (positive-pressure method). The frog, then, is an excellent example of an animal that uses a variety of gas-exchange mechanisms. Gas exchange can occur across the surface of the lungs, which are only occasionally filled, and also across the exchange surface provided by the soft moist skin.

MAKING CONNECTIONS

1. How do the ciliated cells lining the respiratory tract cleanse and protect it? What is the effect of cigarette smoke on these cells? How might this affect gas exchange?

2. How is cigarette smoking believed to cause emphysema?

3. Both cystic fibrosis and asthma cause breathing difficulty. Compare their effects on the respiratory tract.

4. In cystic fibrosis, what is the underlying genetic defect that leads to the secretion of abnormally thick mucus in the respiratory tract? (See Chapter 14, p. 289)

5. Suppose you are examining the lungs of a frog, a lizard, and a mouse, all approximately the same size. Predict which would have the greatest amount of surface area in its lungs. Which would have the least? Why?

CONCEPTS IN BRIEF

1. Because aerobic respiration is the chief method of respiration in both plants and animals, the vast majority of organisms must have some method of obtaining oxygen and getting rid of waste carbon dioxide.

2. Gas exchange between a living cell and its environment always takes place by *diffusion* across a moist cell membrane. The gases must be dissolved in water before they can cross the cell membrane.

3. With the evolution of large three-dimensional organisms and more active organisms came the necessity for a corresponding evolution of specialized gas-exchange surfaces to meet the increased oxygen requirements.

4. Each gas-exchange system in animals must provide a gas-exchange surface of adequate dimensions compared to its volume, a moist exchange surface, protection for the fragile exchange surface, and, in most animals, some method of transporting the gases to and from the cells.

5. Most large multicellular animals have evolved some type of specialized gas-exchange surface. These may be outward-oriented *(evaginated)* or inward-oriented *(invaginated)* extensions of the body surface.

6. The cells of most small aquatic organisms are in close contact with the surrounding medium and carry out gas exchange across their whole body surface; no special gas-exchange structures are necessary.

7. Most multicellular aquatic animals utilize evaginated gas-exchange surfaces called *gills,* which have finely subdivided surfaces that expose an immense exchange surface to the water. Most gills contain a rich supply of blood vessels, which transport oxygen to the individual cells. Because of oxygen's low solubility and slow diffusion rate in water, most aquatic organisms must constantly ventilate—move water across—the exchange surface.

8. Most terrestrial animals have evolved invaginated respiratory systems, either *lungs* or *tracheae.*

9. The tracheal system typical of land arthropods consists of many small air ducts, called *tracheae,* that run from the *spiracles* in the body wall and carry air directly to the individual cells. Because oxygen is delivered directly to each cell by the system, there is no significant transport of oxygen by the blood.

10. Lungs are invaginated gas-exchange organs well supplied with blood vessels. The evolution of the lung in higher vertebrates has tended toward increased surface area and an increased blood supply to the exchange surface.

11. In human beings, air is drawn through the *external nares* into the *nasal cavities,* then moves into the *pharynx,* a common passageway for food and air. After leaving the pharynx through the *glottis,* the air enters the *larynx.* The air then moves into the *trachea,* which divides at its lower end into the two *bronchi,* which lead into the two lungs. Each bronchus branches repeatedly, eventually forming small *bronchioles,* which branch into smaller ducts that terminate in the *alveoli.* Each alveolus is surrounded by a dense bed of blood capillaries; the diffusion of oxygen and carbon dioxide takes place here.

12. In birds, the special arrangement of the lungs and their associated air sacs permits a continuous one-way flow of air through the lungs during both inhalation and exhalation. Because of the one-way flow and the countercurrent arrangement of blood vessels, birds extract oxygen from air more efficiently than mammals.

13. In mammals and birds, air is drawn into the lungs in the process of *negative-pressure breathing,* by contrast with *positive-pressure breathing* in frogs, where air is forced into the lungs.

STUDY QUESTIONS

1. By what process are gases exchanged between cells and their environment? Is active transport ever involved? (pp. 373–375)
2. What are the requirements of a gas-exchange surface? Why are these requirements more challenging for large animals than for single-celled organisms? What problem is posed by the terrestrial environment? (pp. 373–375, 378–379)

3. **Why must most animals have a method of transporting gases to and from the respiratory surfaces? How is this requirement usually met? (pp. 374–375)**
4. **Distinguish between invaginated and evaginated gas-exchange surfaces. Which type is generally found in aquatic organisms and which in terrestrial organisms? (p. 375)**
5. **Describe three disadvantages of water as a source of oxygen as compared to air. (p. 375)**
6. **How would you design an animal to provide enough surface area for gas exchange without having any specialized exchange surfaces? (p. 376)**
7. **What common feature is shared by all gills? Describe other features of gills that adapt them for their function. (pp. 376–378)**
8. **Make diagrams of both concurrent and countercurrent flow systems and use them to explain how a countercurrent exchange system maximizes gas exchange in the gill of a fish (pp. 378–379)**
9. **Compare the problems associated with gas exchange in aquatic and terrestrial organisms. (pp. 378–379)**
10. **Describe the tracheal system of insects and other land arthropods. In what respects do tracheae differ from lungs? Describe one advantage and one disadvantage of tracheae compared to lungs. (pp. 379–380)**
11. **Trace the path of an oxygen molecule from its point of entry in the nostrils to its point of absorption into the blood stream of a human being. Describe the functions of all structures in the pathway. (pp. 381–384)**
12. **Explain the difference between positive-pressure and negative-pressure breathing, using the frog and a human being to illustrate the two methods. (p. 386)**
13. **Describe the pattern of air flow and blood flow in the respiratory system of a bird. How does the efficiency of the bird's system compare with that of mammals? (p. 385)**

CHAPTER 19

INTERNAL TRANSPORT IN PLANTS

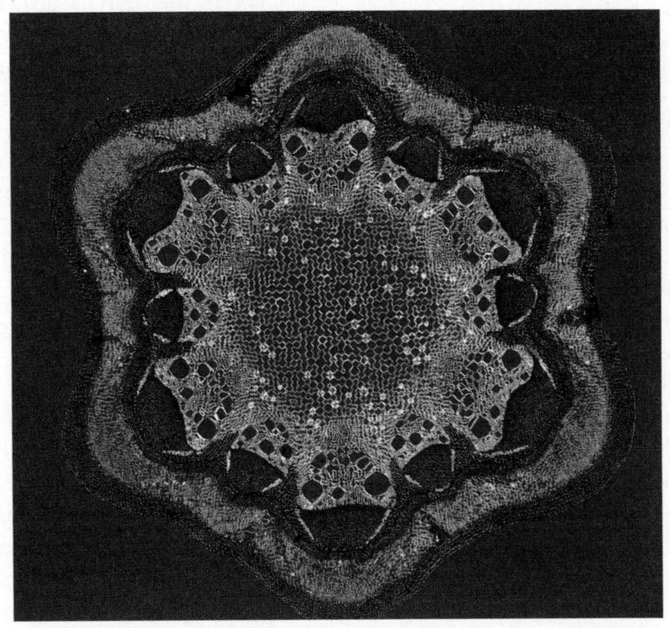

Every living cell, whether it makes up an entire single-celled organism or part of a multicellular one, must perform its own metabolic activities. To do this it requires a continuous supply of nutrients and, if it uses aerobic respiration, oxygen. In both unicellular and multicellular algae, each cell is adjacent or close to the environmental medium and extracts from it the raw materials it needs to survive. But in larger and structurally more complex multicellular plants, many centrally located cells are relatively far from the environmental medium. Furthermore, the processes of nutrient procurement, photosynthesis, and gas exchange are often restricted to regions of the body specialized for those functions, such as the leaves and roots. Some mechanism is needed to transport nutrients from the specialized regions to individual living cells throughout the organism.

Organisms Without Special Transport Systems Rely on Diffusion to Transport Materials

In unicellular algae and within individual cells in multicellular organisms, diffusion is important in both the movement of materials across cell membranes and the transport of materials within any one cell. It is also important in the movement of materials from cell to cell within the body of a multicellular organism. For example, we have examined diffusion's role in the uptake of water by plant roots. Intercellular diffusion is facilitated in plants by plasmodesmata, which interconnect the cytoplasmic contents of adjacent cells to form the symplast. Diffusion of water along cell walls and of gases through the intercellular spaces of the apoplast is also important.

But diffusion is effective only over short distances. Even in unicellular and small multicellular organisms, diffusion is often supplemented by other transport mechanisms. We have seen that the endoplasmic reticulum provides an intracellular pathway for some substances, such as lipids and proteins. In addition, the cytoplasm itself is seldom motionless; it often flows rapidly in definite currents along the surface of the cell vacuole (Fig. 19.1). Such flow causes a characteristic directional movement in many plant cells called ***cytoplasmic streaming,*** or cyclosis. Sometimes the streaming is restricted to local regions of the cell. At other times most of the cytoplasm becomes involved, and a general circulation results. This flow can transport substances from one part of a cell to another many times faster than simple diffusion.

Among multicellular autotrophic organisms, it is not only the very small ones that lack a specialized internal-transport

Chapter opening photo: Cross section of *Clementis* stem. The vascular bundles are found in a ring around the central pith (purple).

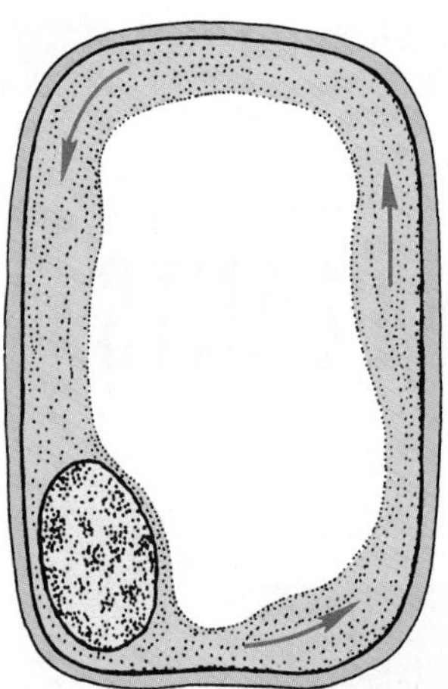

19.1 Cytoplasmic streaming in a plant cell The cytoplasm flows around the large central vacuole.

system. Many algae, particularly the brown and red ones, have large multicellular bodies, yet usually lack specialized vascular tissue.[1] Cells of such plants are seldom far from the surrounding water or from water in intercellular spaces, which is continuous with the external medium. Nutrient and gas procurement is not limited to specialized regions of the body, and neither is photosynthesis. Consequently, each cell gets the nutrients it needs from the surrounding water, and carries out its own photosynthesis, making long-distance transport rarely necessary.

TRANSPORT IN PLANTS

The Successful Adaptation of Plants to Land Depended on the Evolution of a Transport System

Two divisions[2] of plants are adapted for life on land: the nonvascular and the vascular plants. The ***nonvascular*** plants, the mosses, hornworts, and liverworts, lack specialized vascular tissues. They do not have an internal system to transport water to all parts of the plant. Nonvascular plants can live only in moist environments where no cell is far from water. The ***vascular*** plants are the plants most familiar to you—the ferns, the conifers (pines, spruce, and the like), and the flowering plants. These plants have the two principal types of vascular tissue, ***xylem*** (from the Greek *xylem,* "wood") and ***phloem*** (from the Greek *phloios,* "bark"); both were described briefly in Chapter 15.

The vascular tissues serve two vital functions: support and transport. Aquatic organisms are supported by the buoyancy of water, whereas on land, strong rigid structures are needed to counteract the force of gravity. The thickened, strengthened cell walls of the xylem are important because they provide the necessary support against gravity and enable the plant body to enlarge in all dimensions. The evolution of specialized internal-transport tissue has also permitted far greater specialization of parts. Water and mineral uptake can be restricted primarily to the roots, while photosynthesis can be restricted largely to the leaves. In some very tall forest trees, the distance between the roots and the leaves may be enormous; yet the xylem and phloem form continuous pathways between them, and they can exchange materials with relative ease. The successful exploitation of the land environment by plants was dependent on the evolution of such a transport system.

[1] A very few brown algae do possess limited conducting tissue.

[2] Botanists have traditionally used the term "division" for the major groups that zoologists call phyla.

Herbaceous and Woody Plant Stems Differ in Structure

Though plant stems serve many functions, we shall discuss them here as organs of transport and support, and examine the structural adaptations associated with these two functions. Keep in mind that the vascular tissue of the stem is continuous with that of the roots and the leaves, and that internal transport is as important in those organs as it is in stems.

We shall examine two basic types of stems, ***herbaceous*** and ***woody*** stems (Fig. 19.2). Herbaceous stems remain soft and succulent; woody stems and roots are hard and rigid.

There are two basic types of herbaceous stems: monocots and dicots (Fig. 19.2). Monocots are mostly ***annuals*** (that is, they live only a single season and then die). All the grasses (which include important food crops such as wheat, corn, oats, barley, and rice) are monocots, as are tulips, lilies, and palms. As we saw in Chapter 1, monocots and dicots have distinct appearances (see Fig. 1.33, p. 20). The major veins in the leaves of monocots are roughly parallel, whereas those of most dicots have a netlike arrangement. Monocot flowers have petals in threes or multiples of three, rather than the fours or fives of dicot flowers. Herbaceous dicots are mostly annuals, too; the entire plant dies after one season of growth. The bean plant is an example of an economically important herbaceous dicot. Some herbaceous dicots are ***perennials.*** The aboveground part of the plant dies each year but the roots survive for many seasons. Woody dicots are all perennials; the whole plant lives for many years. The deciduous trees and many flowering plants are woody dicots.

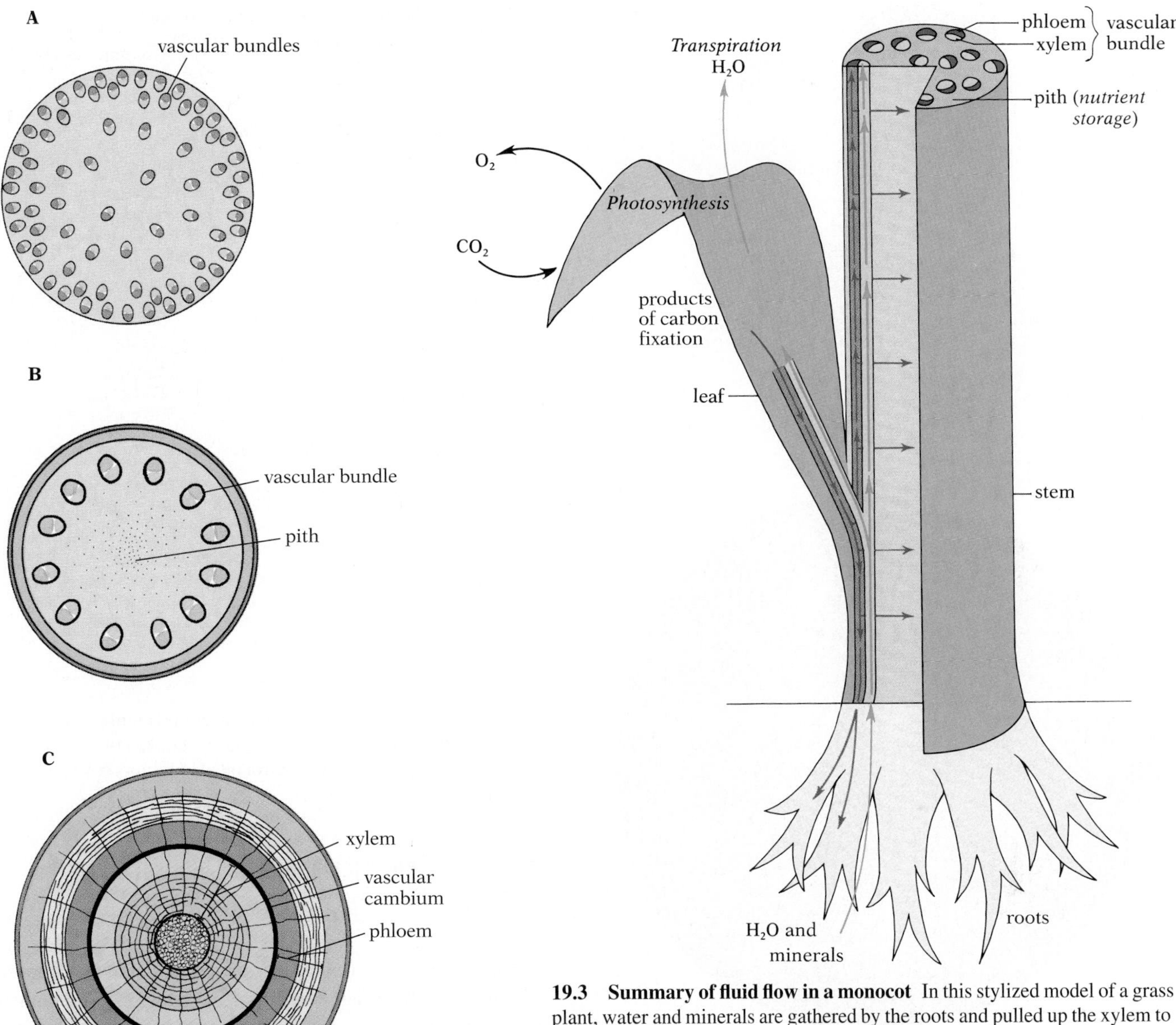

19.2 Basic types of herbaceous and woody stems (A) Cross section of herbaceous monocot stem showing scattered vascular bundles. The vascular tissues of both herbaceous (B) and woody dicot (C) stems are arranged in a circle.

19.3 Summary of fluid flow in a monocot In this stylized model of a grass plant, water and minerals are gathered by the roots and pulled up the xylem to the leaves by a process to be described later. Some of the water and the minerals are then used in photosynthesis and for making organic compounds. The nutrients and other materials thus generated are transported through the phloem to build or nourish cells, or to be stored in the pith or roots. The phloem and xylem of monocots are organized into discrete bundles (only one of which is shown here in vertical section) surrounded by pith. Nutrients to be stored simply diffuse out of the phloem into the pith, as indicated by the red arrows. There is no distinct outer cortex in monocots.

Herbaceous Monocot Stems Have Scattered Vascular Bundles

The outer tissue layer of a monocot stem is epidermis. The stem's vascular tissue forms in discrete vertical bundles scattered throughout the parenchyma tissue of the stem (Figs. 19.3 and 19.4). Each bundle contains both xylem and phloem, and is surrounded by a supportive ***bundle sheath*** (Fig. 19.5). As monocot stems grow in diameter, new bundles are formed toward the periphery, so that no tissue is far away from a source of nutrients. All this tissue originally derives from meristematic tissue, located in the ***apical meristem.*** We saw in Chapter

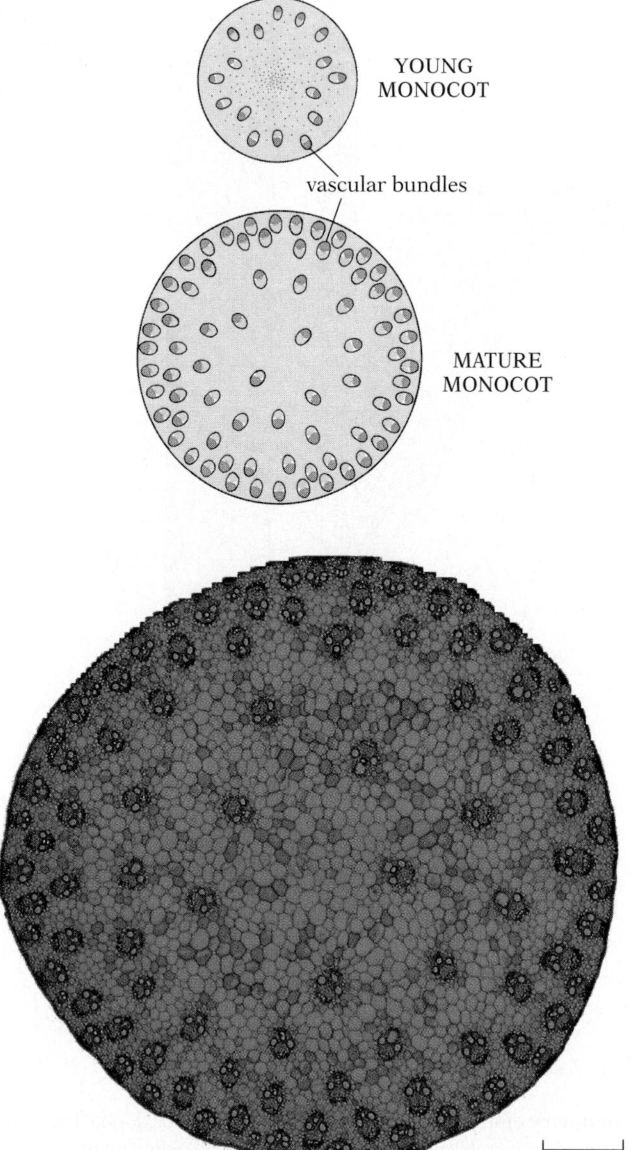

19.4 Cross sections of monocot stems A young monocot has a ring of discrete vascular bundles, each with phloem and xylem vessels (top). As it grows, new vascular bundles develop, until they are distributed throughout the stem (middle). The rest of the stem is pith, the tissue used for nutrient storage. At bottom is a photograph showing a cross section of a monocot stem (corn).

15 that in the bud the apical meristem is at the tip of the shoot, while in the root it is near the root tip. *All tissue produced by the apical meristem is called* **primary tissue;** *it is primarily responsible for the plant's growth in length.* The new cells produced in the apical meristem soon begin to differentiate, some forming epidermis, some forming parenchyma tissue, and some forming the primary phloem, primary xylem, and bundle sheath. The tissues of most monocots are composed of primary tissue.

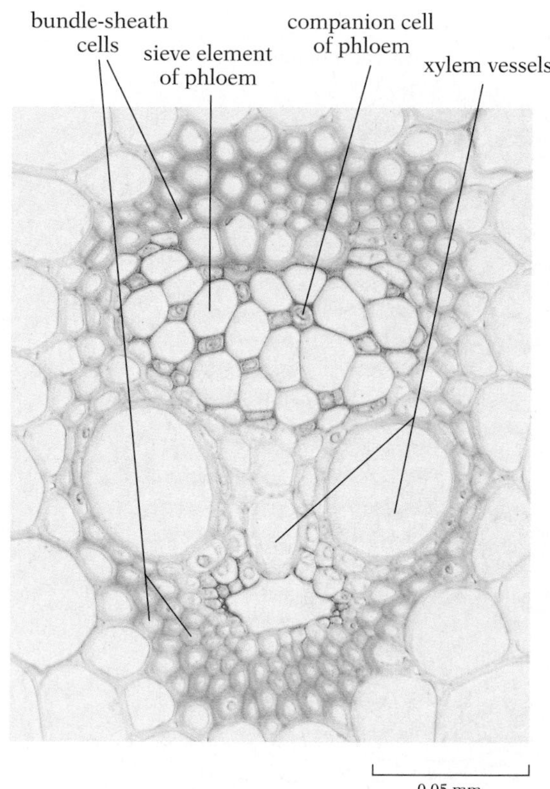

19.5 Cross section of a monocot vascular bundle (corn) A single bundle is shown here. Each one contains a few xylem vessels and a mass of phloem, and is surrounded by a supportive bundle sheath.

The Vascular Bundles of Dicot Stems Are Arranged in a Ring

Figure 19.6 shows a cross section of the stem of a sunflower, an herbaceous dicot. The outer tissue layer of an herbaceous stem is epidermis. Next comes the cortex, comprising an area of thick-walled collenchyma cells just beneath the epidermis followed by an area of parenchyma cells more internally. Inside the cortex lies the vascular tissue. As in young monocots, the xylem and phloem of herbaceous dicots are organized in discrete bundles (Fig. 19.6). However, the differences between the two groups of plants are striking. The vascular bundles in dicots are always arranged in a ring separating the central pith from the outer cortex, rather than scattered, as in monocots. In a dicot vascular bundle, the phloem always lies toward the outside of the stem, while the xylem lies toward the center. Between them is a layer of meristematic tissues, called ***vascular cambium*** (Fig. 19.6). New vascular tissue, if any, is added by the vascular cambium rather than by the formation of new bundles, as it is in monocots. The center of the dicot stem is filled with pith, which is parenchyma tissue that functions as a storage area.

The cambium of many herbaceous dicots never becomes

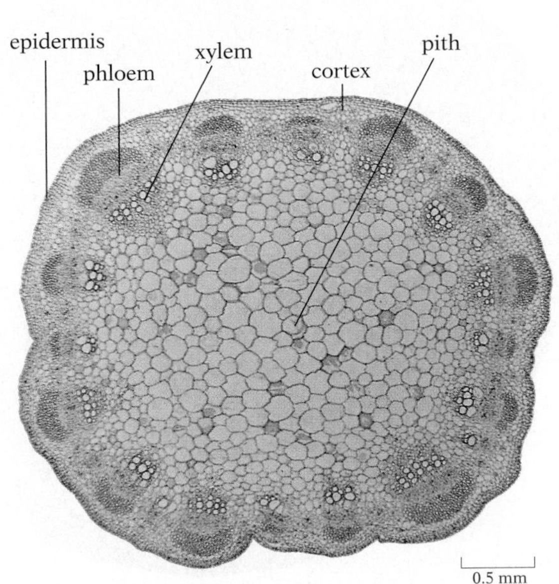

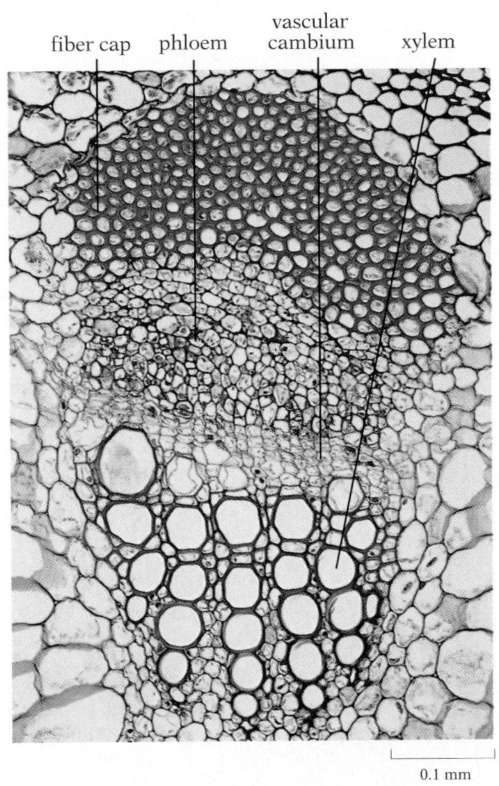

19.6 Cross sections of herbaceous dicot stems Left: Whole stem of sunflower *(Helianthus)*. Right: Vascular bundle of sunflower. The vascular bundles in dicot stems are arranged in a circle.

active and never produces additional phloem or xylem cells. In such plants all the plant tissue is primary tissue. In some species of herbaceous dicots, the cambium does become active, however. As the cells of the cambium divide, they give rise to new cells both to the inside and to the outside of the cambium. The new cells formed on the outer side differentiate to form ***secondary phloem;*** those formed on the inner side differentiate to form ***secondary xylem*** (Fig. 19.7). ***Secondary vascular tissue,*** then, is tissue derived from this lateral meristem, the cambium, that increases the stem's *diameter* (rather than its length, which is the apical meristem's responsibility). As secondary phloem is produced by the cambium, it pushes the older, primary phloem away, toward the outside of the stem. Similarly, as secondary xylem is produced, the cambium is pushed farther from the primary xylem, which is left in the inner portion of the vascular cylinder, toward the center of the stem. In an herbaceous stem that has undergone secondary growth, therefore, the sequence of tissues (moving from the outside toward the center) is: epidermis, cortex, primary phloem, secondary phloem, cambium, secondary xylem, primary xylem, and pith (Fig. 19.7).

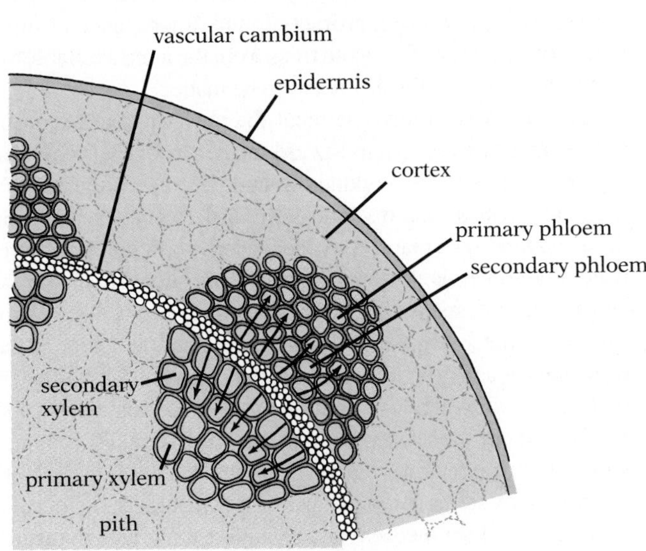

19.7 Cross section of part of a herbaceous dicot stem after secondary growth The vascular tissue of herbaceous dicots is arranged in discrete bundles, with the xylem separated from the phloem by vascular cambium. As the stem grows, the cambium produces new (secondary) xylem and phloem cells, which push away the original (primary) xylem and phloem. The arrows indicate the directions of growth.

Woody Stems Consist Mostly of Secondary Xylem

A cross section of a woody stem early in its first year of growth does not appear very different from that of an herbaceous stem;

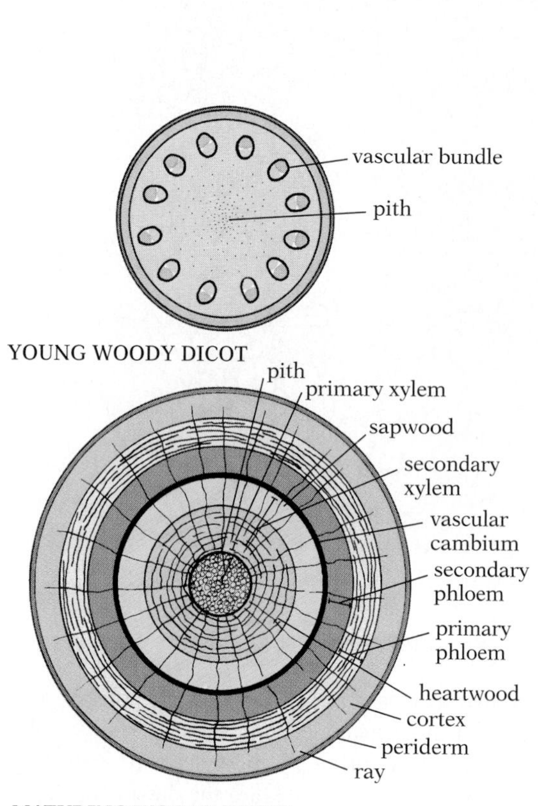

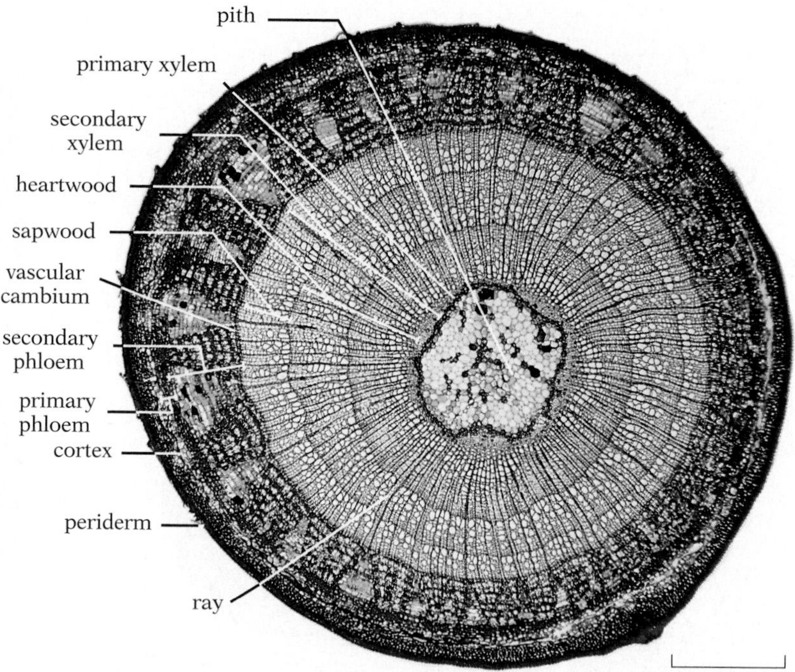

19.8 Cross sections of woody dicot stems In a young woody dicot the phloem and xylem are located in discrete vascular bundles (left, at top). After several seasons the primary xylem and phloem of the first year have been supplanted by new cells—secondary xylem and phloem—produced by the cambium, and the vascular bundles have fused to form a cylinder. Rays carry nutrients laterally from the phloem to other parts of the stem (left, at bottom). Above is a photograph showing a cross section of a woody dicot stem (basswood) at the end of three years of growth. Only the most recently produced xylem—the sapwood—is still active in transport, while older xylem—the heartwood—functions in support.

in both, the primary vascular tissue is arranged in a continuous ring in some species and in discrete bundles in others. Secondary growth, however, soon makes the bundles continuous in all woody stems. This occurs because cells in the ground tissue between vascular bundles take on meristematic activity, dividing to make new xylem cells toward the center and new phloem cells toward the periphery of the stem. These newly dividing cells join up with the vascular cambium inside the bundles to make a continuous cambial ring all around the stem. As secondary growth continues, the stem looks less herbaceous (Fig. 19.8). The secondary xylem inside the vascular cambium becomes thicker until it makes up almost the entire stem. This secondary xylem is commonly called wood.

New xylem cells produced early in the growing season have the best growing conditions and abundant water. They grow larger than cells produced later in the season, when water is limited. This difference in cell size, large in the spring and small in late summer, makes up one annual ring (Fig. 19.9). The repeat of this large-to-small pattern through the year produces a series of concentric annual rings, clearly visible in cross sections of the stem (Fig. 19.8). A fairly accurate estimate of the age of a tree can be made by counting the annual rings. Tree-ring dating has been used in archaeological studies. By matching the rings in wooden artifacts found among ancient ruins with cross sections of very old trees from the area, a reliable estimate of the age of the artifacts can be made.

In older trees various chemical and physical changes occur in the older rings of xylem toward the center of the stem. The parenchyma cells die, conducting cells become plugged, and pigments, resins, tannins, and gums fill the cell spaces. As these changes take place, the older xylem no longer transports water and minerals. It remains as a strong supportive component of the tree, and is known as ***heartwood,*** or inactive xylem. The newer outer rings, which still function in transport, are the ***sapwood,*** or active xylem. Heartwood often is darker in color than sapwood. A tree can continue to live after its heartwood has burned or rotted away, but it is much weakened and cannot withstand strong winds.

As woody stems (or roots) grow in diameter, a layer of cells near the outside of the stem takes on meristematic activity and

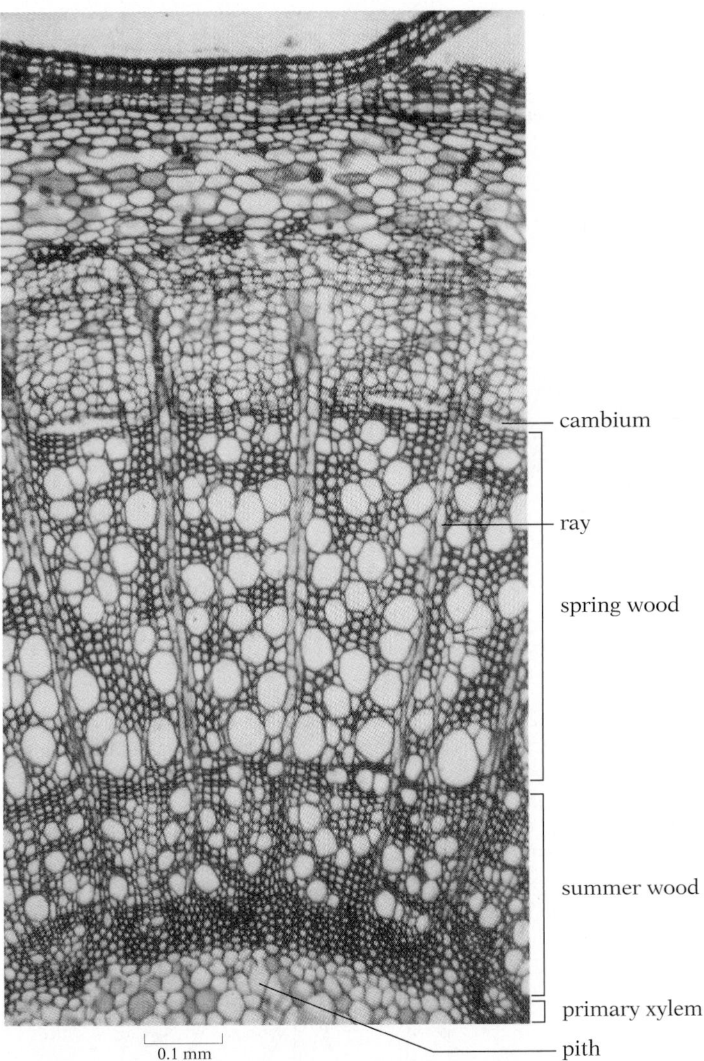

19.9 Cross section of ivy *(Hedera)* stem In this photograph (taken during the plant's spring growth), the larger, thinner-walled xylem cells of spring wood are readily distinguished from the smaller, thick-walled cells of summer wood from the previous year. The primary xylem was produced in the plant's first spring. Several rays can also be seen.

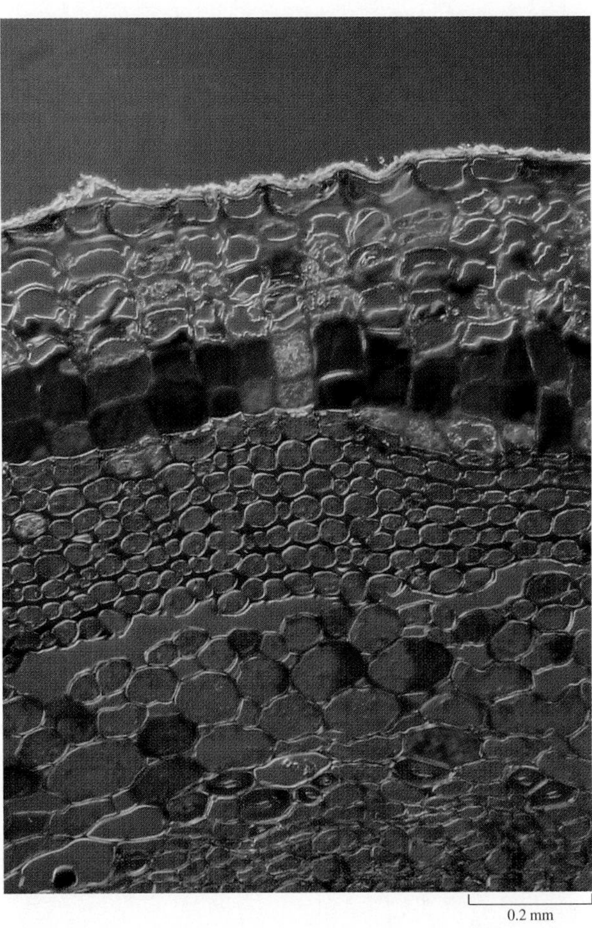

19.10 Periderm of young stem of elderberry *(Sambucus)* The periderm layer, here consisting of the cork cambium and several rows of flattened cork cells (rod cells on top) overlying collenchyma tissue, provides a protective layer on roots and stems that undergo secondary growth.

forms the ***cork cambium.*** The cells of the cork cambium divide to produce ***cork cells*** on its outer surface and parenchyma cells on its inner surface. The cork cambium and the cell layers derived from it are collectively termed the ***periderm*** (Fig. 19.10). As growth continues, the original epidermis and portions of the cortex outside the cork cells flake off. The cork cells are dead; it is their waterproof cell walls that function as the protective ***outer bark*** of the older stem or root. In some plants the cork cambium forms in concentric rings around the stem and the resulting bark is relatively smooth. In other stems the cork cambium is discontinous and/or overlapping, making the bark rough or furrowed.

The ***inner bark*** of a stem is the phloem tissue. The phloem layer never becomes thick like the xylem, because many fewer cells are produced as compared to the secondary xylem, and because the thin-walled phloem cells are crushed as the stem enlarges. Annual rings are difficult, if not impossible, to find. Also, unlike the xylem, the phloem of an older woody plant does not does not support the plant, but it is very active and important in transport.

In summary, then, the old woody stem of a tree has no epidermis or cortex. Its surface is covered by an outer bark of cork tissue. Beneath the cork cambium is the thin layer of phloem, or inner bark, and beneath this is the vascular cambium, which is usually only one cell thick. The rest of the stem is mostly secondary xylem, or wood, of which only the outer annual rings, or sapwood, still function in transport (Fig. 19.11).

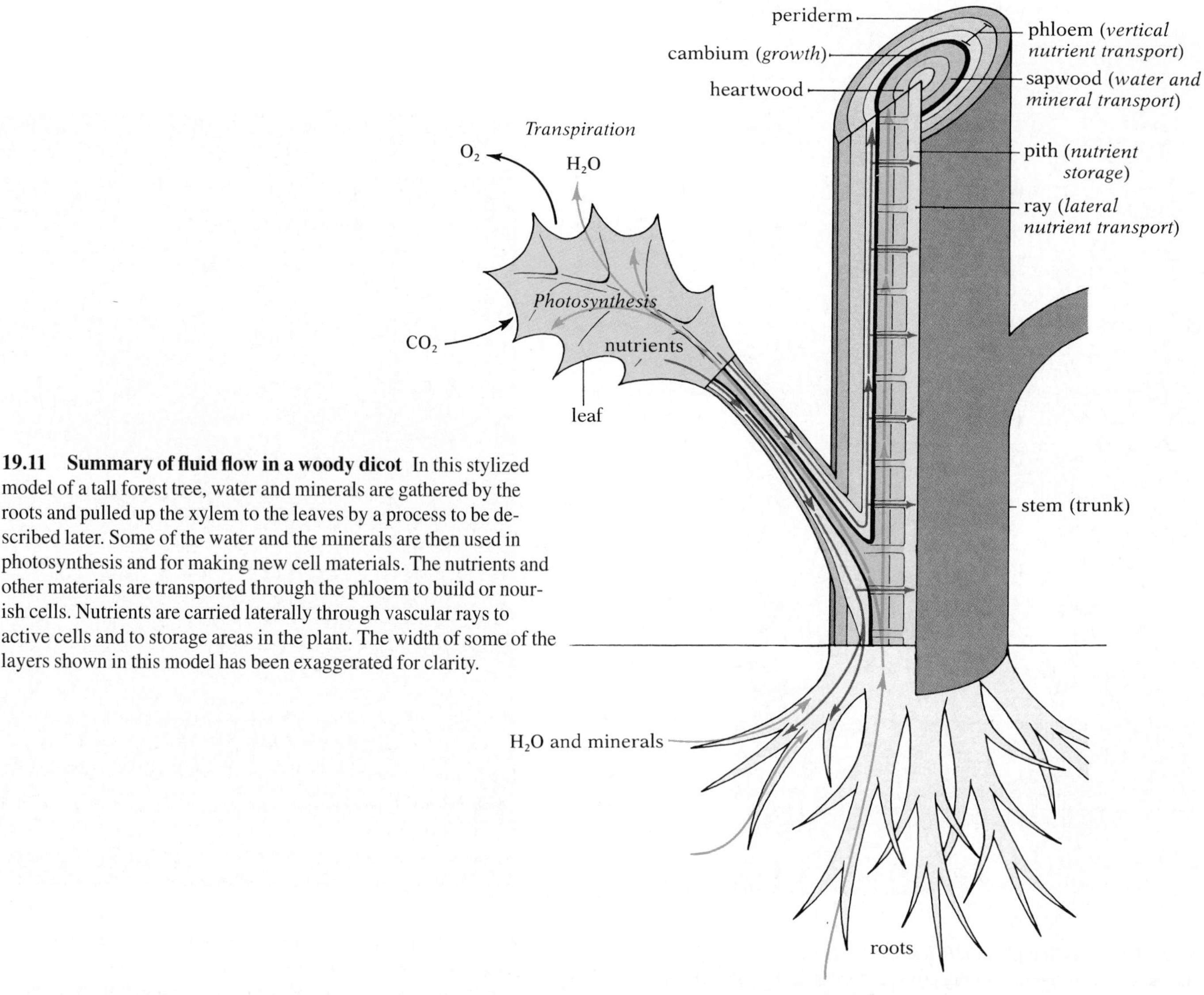

19.11 Summary of fluid flow in a woody dicot In this stylized model of a tall forest tree, water and minerals are gathered by the roots and pulled up the xylem to the leaves by a process to be described later. Some of the water and the minerals are then used in photosynthesis and for making new cell materials. The nutrients and other materials are transported through the phloem to build or nourish cells. Nutrients are carried laterally through vascular rays to active cells and to storage areas in the plant. The width of some of the layers shown in this model has been exaggerated for clarity.

MAKING CONNECTIONS

1. How do size and structural complexity influence the need for a specialized transport system in plants? What influence does the plant's environment (aquatic or terrestrial) have?

2. What two vital roles do xylem and phloem serve that enable vascular plants to grow so much larger than nonvascular plants such as mosses?

3. How is the differentiation of cells produced in the apical meristem and the cambium related to the genetic regulatory mechanisms described in Chapter 12?

4. You see the stump of a recently felled tree and decide to determine its age. What causes the annual growth rings in the wood? What tissue constitutes the wood?

TRANSPORT IN THE XYLEM

The Conducting Cells of the Xylem Are Tracheids and Vessels

Xylem is a complex tissue containing several types of cells: tracheids, vessel elements, fibers, and parenchyma cells. Two of these, tracheids and vessel elements, become important in conduction after they mature. Actually, the cell walls are all that remain of a tracheid or vessel element functioning in transport. The cellular contents, both cytoplasm and nucleus, have disintegrated. The main transport in the xylem occurs, then, in tubular remnants of cells rather than in actually living cells.

Tracheids are elongate, tapering cells whose thick secondary cell walls are heavily reinforced with a compound called lignin (Fig. 19.12). These walls are important in support. In secondary xylem, the tracheid walls are interrupted by numerous ***pits,*** which are particularly abundant along the tapering ends of the cells (Fig. 19.12C). Pits are not holes in the walls as the name implies; rather they are thin areas where there has been no secondary-wall formation. The primary walls and middle lamella remain, forming a pit membrane. Water and dissolved substances move easily from tracheid to tracheid through the pits but air bubbles are unable to pass. This is important because if the tracheids were damaged and air could get in (e.g., if a woodpecker pecked a hole or a branch or twig broke), the pit membrane would prevent the air from moving into undamaged tracheids, preserving their function.

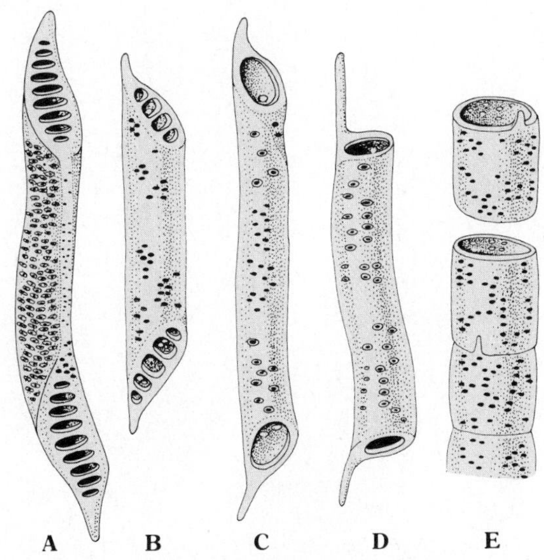

19.13 Vessel elements Five different types of xylem vessel elements are shown—those thought to be the more primitive on the left, those thought to be the more derived (advanced) on the right. The last example (E) shows a single vessel element on top, three elements linked in sequence to form a vessel below. The evolutionary trend seems to have been toward shorter and wider elements, larger perforations in the end walls retained, and less oblique, more nearly horizontal ends.

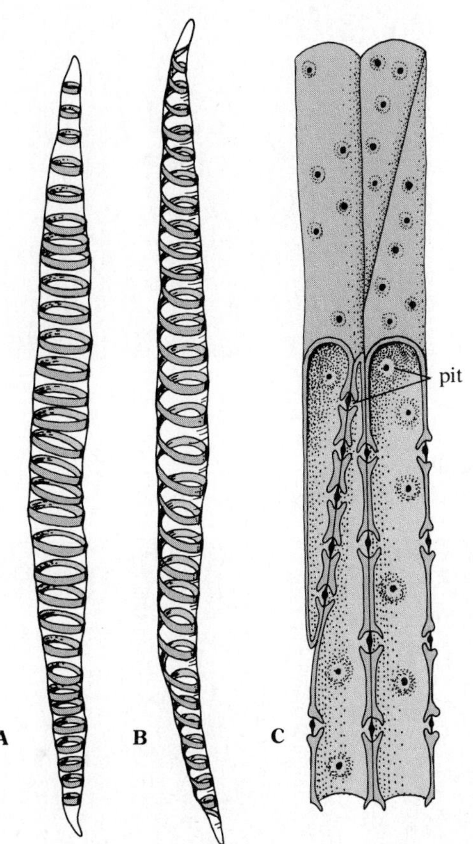

19.12 Tracheids (A) Primary tracheid with secondary walls in the form of rings. (B) Primary tracheid with spiral secondary wall. (C) Secondary tracheids. Parts of four cells are shown, with one wall cut away from portions of three of the cells to expose their lumina and give a clearer view of the junction between cells. Notice the pits, which are particularly abundant along the tapering ends of the cells.

Vessel elements are more highly specialized conductive elements than tracheids, from which they probably evolved. They are characteristic of the flowering plants and do not occur in most conifers (evergreens). In general, vessel elements are shorter and wider than tracheids. They are arranged vertically end-to-end, forming a pipeline called a vessel. Vessel elements have pits along their sides, through which some lateral movement of substances occurs. The ends of vessel elements have extensive perforations and holes that lack membranes. The ends may even be entirely open (Figs. 19.13 and 19.14). Most transport occurs through these ends. Because the perforations, unlike the pits, are completely unobstructed, material moving from one vessel element to the next in a vertical sequence forms a continuous column and transport is more efficient.

In addition to tracheids and vessel elements, xylem contains fiber cells and parenchyma cells. The fibers are elongate, very thick-walled cells that function in support. Some of the parenchyma cells of the xylem are scattered among the other cells, but many of them are grouped together to form ***vascular***

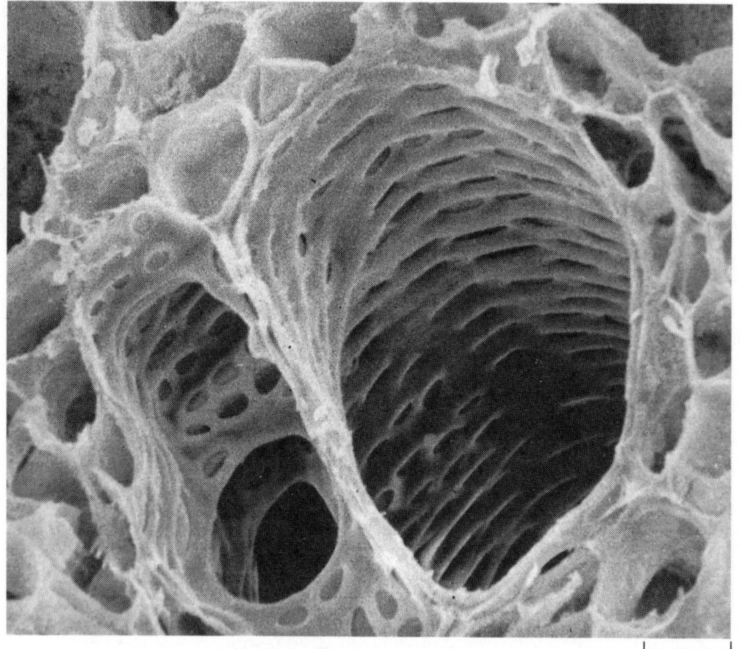

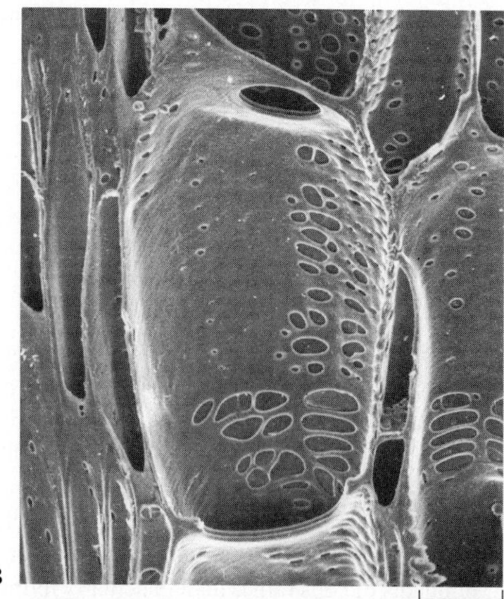

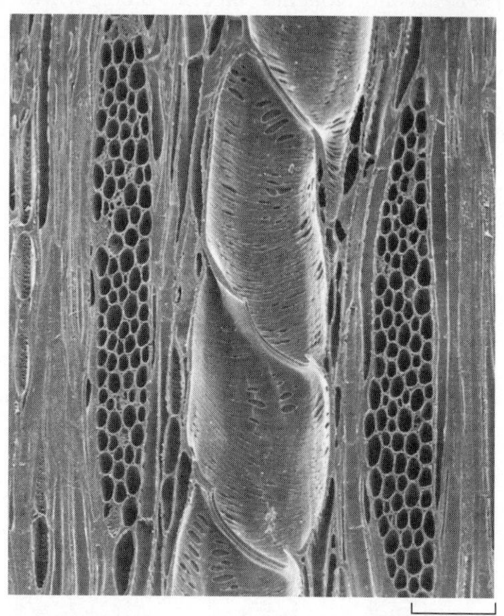

19.14 Scanning electron micrographs of xylem vessels (A) Portion of cross section of corn root showing two large xylem vessels. × 600. (B) Inner surface of a single vessel element. Numerous pits can be seen in the side walls. The large opening at the lower end is the perforation through which water moves from one element to the next in vertical sequence. × 680. (C) Lateral view of four vessel elements stacked one above the other. × 360.

rays that run through the xylem and phloem in a radial direction (Figs. 19.8 and 19.9). They function as pathways for lateral movement of materials, and as storage areas. Water and materials brought upward in the xylem move outward through the vascular rays to living cells in the phloem and cortex.

The number, form, and distribution of tracheids, vessels, fibers, and parenchyma cells vary from species to species. This variation creates differences in appearance and properties in woods of different species (Fig. 19.15)

19.15 Cross sections of woods Left: Red oak. Middle: Tulip tree. Right: Sugar maple. Note the differences in the number and size of vessels.

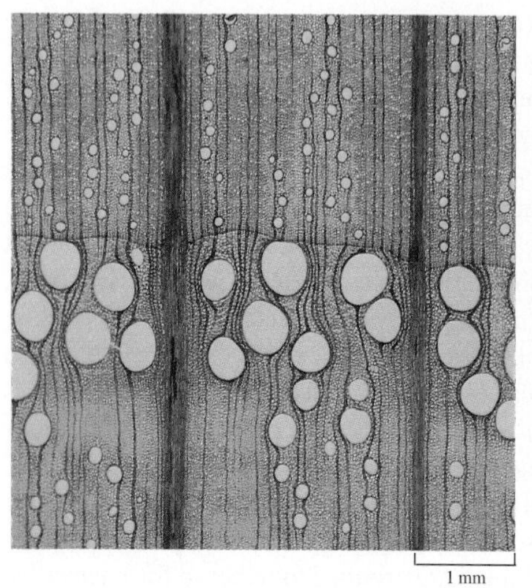

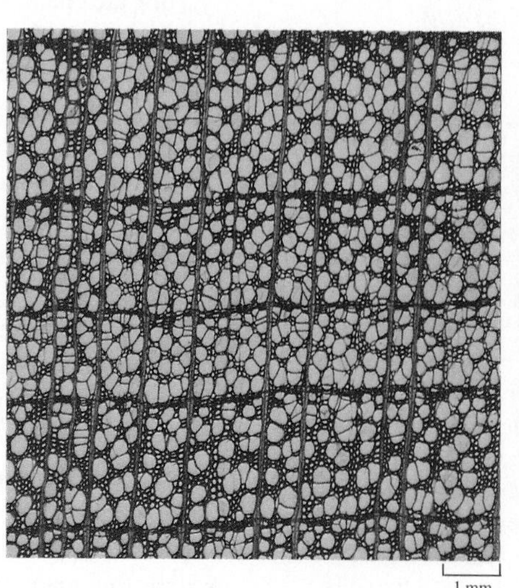

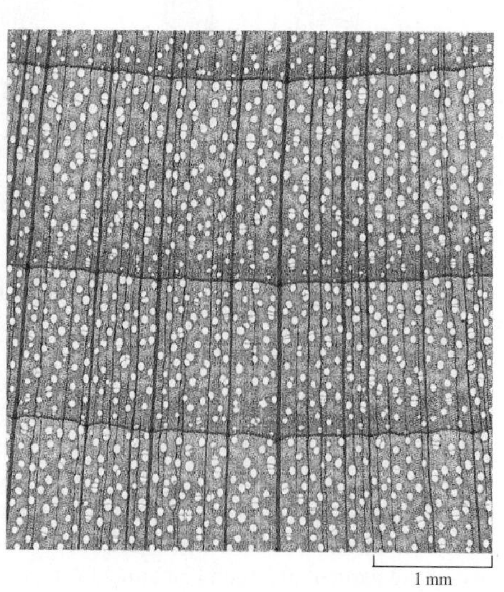

Large Forces Are Necessary to Move Sap Upward in the Plant Body

Sap—the water and dissolved inorganic nutrients absorbed by the roots—moves upward through the plant body in the mature tracheids and vessels of the xylem. That the upward movement is primarily in the xylem can be easily demonstrated. If the cork, phloem, and cambium are removed from a ring around the trunk of a tree, the leaves remain turgid, even though they are connected to the roots only by xylem. This shows that water and minerals reach the leaves via the xylem.

Any general explanation for the ascent of sap in xylem must identify the forces capable of raising water to the tops of the tallest trees, which may be 90 to 120 meters high. An air pressure of one atmosphere (atm) can support a column of water 10.4 meters high at sea level (less than 10.4 meters at higher altitudes) (Fig. 19.16). It follows that an air pressure of about 12 atm is needed to support a column 120 meters high. But the column must be more than supported; the water must be moved upward at a rate that may sometimes be as fast as one meter or more per minute, and this movement must take place in a system that is frictionally resistant to it. It has been calculated that an additional pressure of at least 18 atm is necessary to achieve this, making a total force of at least 30 atm necessary in the tallest trees. Any general theory of water movement in the xylem must account for forces of this magnitude. The driving force might be at the base of the plant and push the water upward; or it might be at the top of the plant and pull the water up; or a combination of the two might be involved.

Root Pressure Pushes Sap Upward in Certain Plants

First we will consider the possibility of a force applied as a push from below (positive pressure). When the stems of certain species of plants are cut, sap flows from the surface of the stump for some time, and if a tube is attached to the stump, a column of liquid a meter high or more may rise in it. Similarly, when conditions are optimal for water absorption by the roots, but the humidity is so high that little water is lost by transpiration, water under pressure may be forced out at the ends of the leaf veins, forming droplets along the edges of the leaves (Fig. 19.17). This process of water secretion is called ***guttation.***

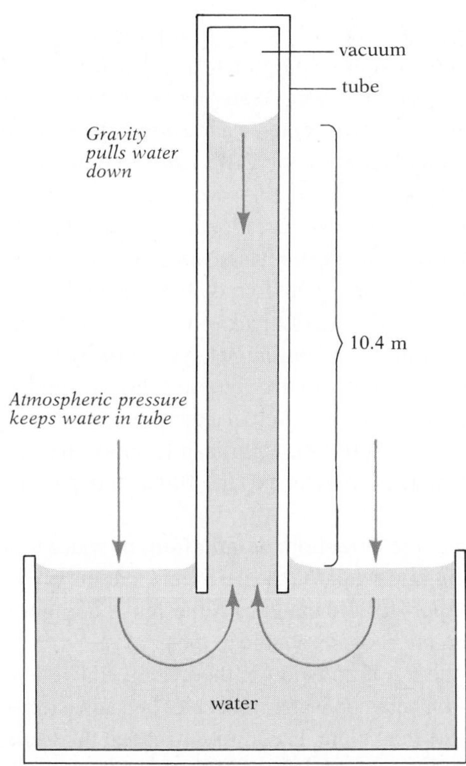

19.16 Role of atmospheric pressure in supporting a column of water Atmospheric pressure—the weight of the air above us pressing down—is sufficient to support a column of water 10.4 meters high at sea level. In the same way, the weight of the atmosphere supports the column of mercury in the familiar mercury barometer, but because mercury is so much denser than water, the column reaches only about 76 centimeters.

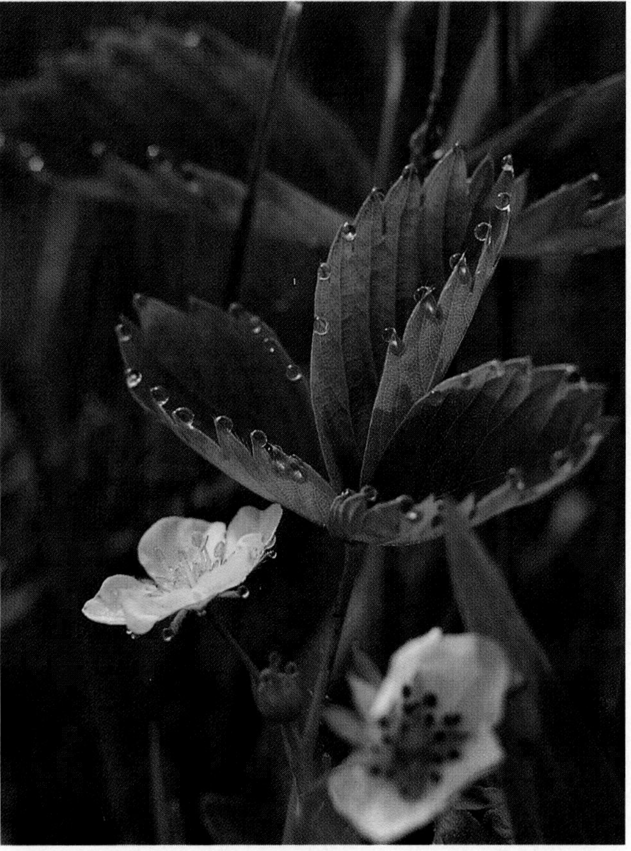

19.17 Guttation by strawberry leaves During the night, root pressure forces tiny droplets of water out special openings at the tips and margins of the leaves.

When the water in the xylem is under pressure, as in these instances of bleeding and guttation, the pushing force involved comes from the roots, and is called ***root pressure.***

How is root pressure built up? We have already seen in Chapter 16 that water moves from the soil, through the epidermis, cortex, endodermis, and pericycle of the root into the xylem, in which it then flows upward to the rest of the plant. But the weight of a tall column of water in the xylem exerts a strong downward pressure. This force might be expected to drive water out of the root xylem, yet water continues to move into the stele in sufficient quantity to force the column upward. If the roots are killed, all root pressure disappears. Likewise, if the roots are simply deprived of oxygen, the root pressure ceases. These observations indicate that development of root pressure needs energy from respiratory production of ATP.

ATP drives the active transport of ions from the soil water into the cortex cells. The ions then move from cell to cell in the cortex through plasmodesmata, until they have crossed the endodermis of the root and entered the stele. (The presence of the Casparian strip in the endodermis prevents the ions from diffusing back into the cortex.) According to this explanation, ATP energy is expended in the active pumping of ions into the stele, and water follows along passively by osmosis in such quantity as to build up root pressure in the xylem.

But does root pressure become strong enough to raise water to the tops of the tallest trees? The answer appears to be no. Some plants, particularly the conifers and their relatives, do not develop any root pressure. Root pressure in species in which it does occur rarely exceeds 1 or 2 atm. This may be enough to raise water to a height of two to three meters—about the height of a corn plant. There is another reason for doubting that root pressure is the principal motive force for the ascent of sap. When a puncture is made in a xylem vessel during the summer, it is uncommon to find water under pressure, i.e., water is seldom forced out of the wound. Instead, one can often hear a short hissing sound as air is drawn into the vessel, indicating that the water in the vessel is under negative pressure instead of positive pressure.

In short, root pressure is not the main upward force of water in plants. It may be involved in the ascent of sap in some, particularly very young, plants, some of the time, especially in early spring when water is plentiful.

Transpiration Pulls the Sap Upward in the Xylem

What about the alternative possiblity that water is pulled from above? This mechanism works as follows. As water in the walls of parenchyma cells in the leaves (or other shoot parts) is lost by transpiration, it is replaced by water from the cell contents. Because water is removed, the osmotic concentration of the leaf cells rises, and water is taken up from adjoining cells—which, in turn, withdraw water from cells adjacent to them. In this way a gradient extends to the xylem in the veins of the leaf and eventually all the way to the root xylem (see Figs. 19.11 and 16.6, p. 336). The parenchyma cells next to the xylem withdraw water from the xylem column, and pull the column upward.

Notice that the column is not pulled up by air pressure or vacuum; the mechanism is not analogous to sucking up a liquid through a straw. Air pressure of 1 atm could support water only 10.4 meters, which is clearly insufficient in a tree that may be over 100 meters high. What is assumed is a continuity between the water on the evaporating surfaces of the leaf mesophyll cell wall and the water in the xylem, and a continuous column of water in the xylem from the top of the column down to the roots. If this continuity of water from leaf cell to root were broken by the entrance of air anywhere into the system, that particular xylem pathway would stop working.

The TATC Theory Relies on Cohesion Between Water Molecules and Adhesion of Water to the Xylem Cell Walls

The idea of pull from above by transpiration depends on several factors. First, transpiration relies on ***cohesion*** between individual water molecules. Water molecules moving out of the xylem into the leaves produce a ***tension*** or ***pull*** on other water molecules up behind them; there can be no break, no separation between the molecules. As we have already seen, the polar water molecules exhibit great cohesion, because of hydrogen bonding. Recall that each water molecule has the potential to form hydrogen bonds with four other water molecules (see Fig. 2.14C, p. 42). The theoretical cohesive strength of an entire column of water—a continuous system of hydrogen-bonded molecules from roots to transpiration surfaces in the leaves—is as high as 15,000 atm. Actual experimental values are lower, but may reach 300 atm, significantly more than the 30 atm needed to make sap ascend to the 100 meters previously mentioned.

Second, the attraction, or ***adhesion,*** of water molecules to the hydrophilic walls of tracheids and vessels provides additional support for the column. Adhesion is thought to play an important, but secondary, role in the process.

This theory of transport in the xylem, which depends both on cohesion between water molecules and adhesion of water to the walls of the xylem, is commonly called the ***transpiration-adhesion-tension-cohesion (TATC) theory*** or the ***transpiration theory.*** Although the theory has not been tested under conditions duplicating those in very tall trees, some interesting experiments have been made on a smaller scale. For example, plant physiologists have boiled water to remove dissolved air, put it in a thin tube to which water molecules adhere, and

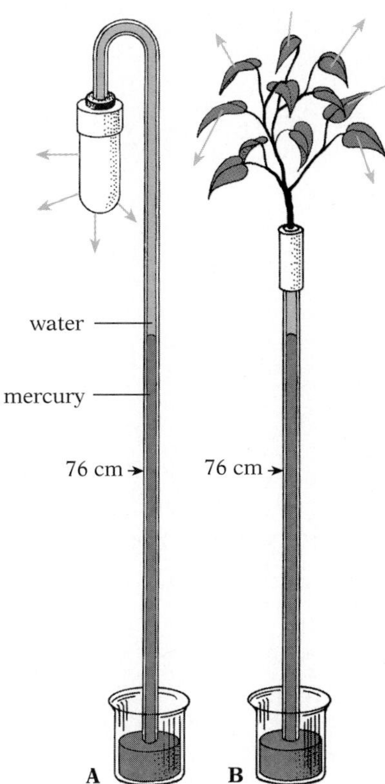

19.18 Demonstrations of rise of water by pull from above (A) Water is evaporated from a clay pot attached to the top of a thin tube whose lower end is in a beaker of mercury. The water in the tube rises and pulls a column of mercury to a point well above the 76 centimeters that atmospheric pressure can support. (B) The same results are obtained when transpiration from the leaves of a shoot is substituted for evaporation from a clay pot. The highly porous hydrophilic clay (A) substitutes for the mesophyll cells of leaves (B).

placed the bottom of the tube in mercury. As water evaporates from the top, it pulls up a column of mercury far higher than a vacuum could (Fig. 19.18). If the base of a cut branch is inserted tightly in the upper end of the tube and the leaves of the branch become the site of evaporation, similar results are obtained. Water molecules have adhered to each other and to the walls of the tube tightly enough to lift a heavy column of mercury to a height above what the barometric pressure would yield.

More evidence comes from measurements of the girth of tree trunks at different times of day. During daylight hours, when xylem should be under maximum tension because of high rates of transpiration, the trunk actually diminishes in girth, just as rubber tubing does when its contents are under subatmospheric pressure.

If the water in the xylem is under tension, the question of how additional water moves into the xylem in the roots presents less difficulty. We said earlier that a water column under positive pressure (as it would be when there is root pressure) would exert a downward force opposing the entrance of more water in the stele. But if the water in the xylem is under tension, it will exert no such downward force, but will, instead, exert a pull tending to draw water from the soil, across the root tissues, and into the xylem (Fig. 19.11).

Sunlight is the ultimate source of energy for the movement of sap in the xylem. It causes the evaporation of water from the leaves, which in turn produces the tension that pulls replacement water up the xylem. The plant itself does not expend any metabolic energy in the ascent of sap (although it does use energy to actively transport individual ions into the root cells).

TRANSPORT IN THE PHLOEM

All Vertical Movement in the Phloem Is Through the Sieve Tubes

Like xylem, phloem is a complex tissue containing supportive fibers as well as parenchyma; the parenchyma of the phloem rays is continuous with that of the xylem rays. The principal vertical conductive elements in phloem are the ***sieve elements,***

MAKING CONNECTIONS

1. You notice that the center of a tree trunk consists of much darker wood than the periphery. What is the significance of the color difference?

2. What is the role of active transport in generating root pressure? Under what circumstances might root pressure contribute to the ascent of sap in the xylem?

3. If water were a nonpolar molecule, would sap be able to rise to the top of a tree?

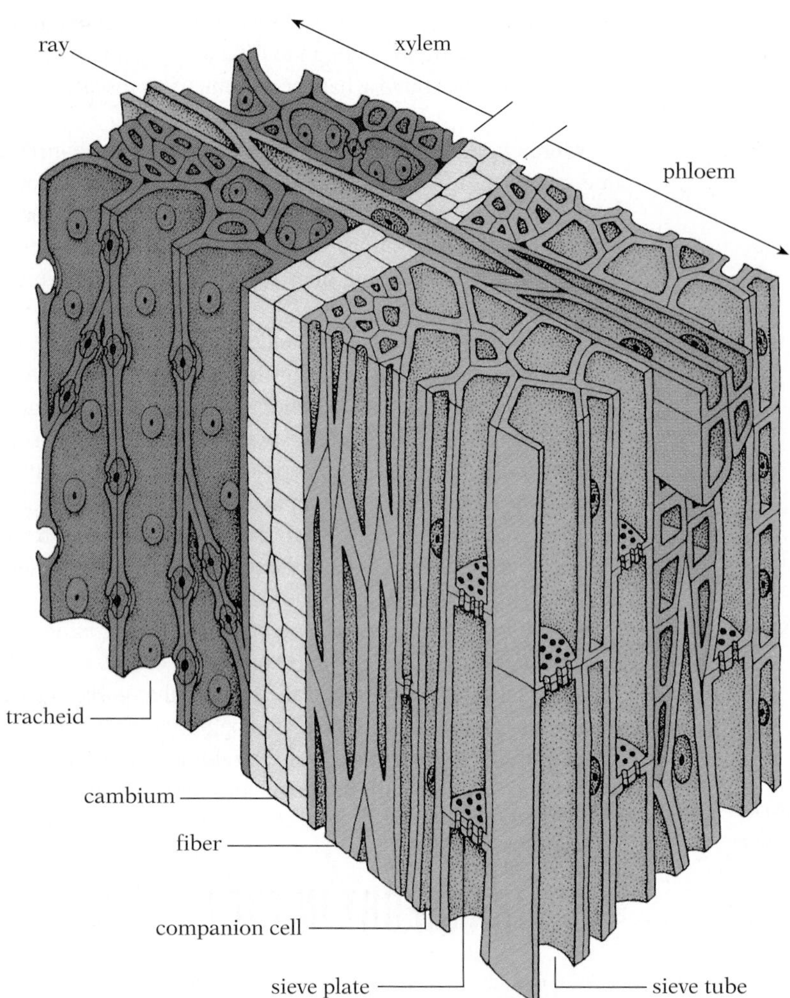

19.19 Phloem Some nearby xylem tissue is also shown.

which lie end-to-end in a series to form a ***sieve tube.*** In their most advanced form, sieve elements are elongate cells with specialized areas on their end and side walls, called ***sieve plates*** (Figs. 19.19 and 19.20). As the name implies, a sieve plate is an area with numerous pores; the cytoplasm is continuous through the pores, so that the contents of one cell are connected with those of the next.

Unlike the tracheids and vessels of xylem, the sieve elements retain their cytoplasm at maturity, so transport through the sieve tubes involves movement through living cells. The cytoplasm of these cells at maturity, however, is quite different from that of typical plant cells. As the sieve elements mature, the cytoplasm becomes rather sparse. The ribosomes, vacuolar membrane, and plastids disappear, the mitochondria degenerate, and, most importantly, the nucleus itself disappears.

Closely associated with the sieve elements of most flowering plants are usually one or more specialized, elongate ***companion cells*** (Fig. 19.19). Mature companion cells have both cytoplasm and a nucleus, and are connected to adjacent sieve elements by numerous plasmodesmata. Some biologists have suggested that the nucleus of the companion cell controls both its own cytoplasm and the cytoplasm of the adjoining sieve element after the nucleus of the latter has disintegrated.

Both Inorganic and Organic Solutes Are Translocated in the Phloem

The transport of solutes within the phloem is called ***translocation.*** Both organic and inorganic solutes move through the phloem. Let us consider the organic solutes first, which can be conveniently divided into two principal types: carbohydrates (sucrose in many species, but other sugars as well) and organic nitrogen compounds.

The classical picture of solute translocation was of upward movement through the xylem and downward movement through the phloem. About 1920, however, this view was revised when experiments that involved cutting a ring of bark, including phloem, from stems and branches showed both upward and downward transport of carbohydrates. When a ring of bark was removed from a tree trunk, the supply of carbohydrates to all parts of the plant below the ring was cut off. Those parts eventually died when their stored reserves were depleted. Downward movement of carbohydrates was clearly through the phloem, and not through the xylem, which had been left intact. But when a branch was ringed a short distance behind a growing bud, the supply of carbohydrates moving to the bud was similarly cut off. Again, the movement must have been through the phloem, but in this case the movement was upward. From numerous similar experiments, the majority of botanists have come to believe that *most carbohydrate movement, whether upward or downward, is through the phloem, and that this movement is fairly rapid.*

Phloem Transports Nitrogen Compounds Up and Down but Xylem Carries Only Materials Upward

Whether xylem or phloem transports organic nitrogen compounds is less clear. Again, the classical idea was that nitrogen, absorbed by the roots primarily as nitrate, was carried upward through the xylem to the leaves. There it was incorporated into organic compounds, which were then transported down through the phloem to other parts of the plant. The up-xylem down-phloem sequence probably holds true for some plants. But there is now good evidence that the roots of many species promptly incorporate incoming nitrogen into organic compounds, such as amino acids. In some species, especially trees, these organic nitrogen compounds do move upward in the xylem, but in others, they move upward in the phloem. Thus, not all *upward* transport of organic compounds is in the phloem. Most of it does appear to be, however, and all *downward* transport of organic compounds is in the phloem.

Xylem Transports Inorganic Ions Upward, and Phloem Transports Some Ions Downward

Inorganic ions such as calcium, sulfate, and phosphate ions are transported upward from the roots to the leaves primarily through the xylem. When radioactive forms of these minerals are followed throughout the plant, some are found to be quite mobile. They travel rapidly back through the phloem down the plant, or move out of older leaves into the newer, more actively growing ones. Phosphate, for instance, easily moves upward in the xylem and downward in the phloem, often circulating rapidly throughout the plant in this manner. Calcium, on the other hand, is not mobile, and hence cannot move from old leaves to newer ones. Consequently plants must constantly take up new calcium from the soil. Well-designed fertilization programs take into account such differences in the properties of the different mineral nutrients.

Solutes Are Transported Through the Sieve Tubes Along a Turgor-Pressure Gradient, From Source to Sink

We have seen that most transport of organic solutes, both up and down, is through the phloem; most downward transport of minerals is also through the phloem. Movement of materials in the phloem appears to follow a ***source-sink*** relationship, where the source is any place where more food is being produced than is being used. Photosynthesizing leaves during the day, or storage cells of the root in early spring, for example, both contain surplus carbohydrate. The sink is any place where more food is being used than produced. Examples are a storage cell in a root or stem in the autumn, when they accumulate winter reserves, or rapidly growing plant parts such as young leaves, roots, stems, and flowers. Notice that at one time a root cell may be a source, later it may be a sink, depending on what each cell is doing at a particular time.

There are several facts that any hypothesis about the transport mechanism in phloem must take into account. (1) The movement is often rapid, much more so than simple diffusion alone could make it. (2) The movement takes place through sieve-tube elements that, unlike xylem, retain their cytoplasm, though the cytoplasm is more sparse than that of most other cells. (3) The ends of the individual sieve-tube elements are penetrated only by the pores of the sieve plates. (4) Energy in the form of ATP is necessary for this transport mechanism (to actively transport sugars into and out of the sieve elements). Clearly, we are dealing with transport through active, living cells, not merely with movement through dead tubes by purely mechanical processes, as is the case with mature xylem.

The hypothesis most widely supported by botanists today involves ***pressure flow,*** or ***mass flow***. According to this hypothesis there is a mass flow of water and solutes through the sieve tubes along a turgor-pressure gradient: during the day, cells like those of the mesophyll of the leaf accumulate high concentrations of osmotically active substances such as sugar. The sugar molecules, usually in the form of sucrose, diffuse from cell to cell through the plasmodesmata until they reach cells in the vicinity of the sieve tubes in the small veins of the leaf. They then diffuse out of the cells into the apoplast (extracellular spaces) from which they are then *actively transported* into the sieve tubes. The addition of sugars to the sieve elements increases their osmotic concentration, and water from the mesophyll cells and xylem follows the sugar by osmosis, causing the pressure within the cell (the turgor pressure) to rise. The turgor pressure within the sieve element often reaches enormously high levels; pressures of 300 pounds per square inch (psi) are not uncommon. By contrast, the pressure within an automobile tire is about 30 psi. The pressure forces substances from these cells into adjacent cells. (Remember, the cytoplasm is sparse and continuous through openings in the sieve plate.) Thus, substances under pressure are forced en masse from cell to cell through the sieve tubes. At the same time, the

MAKING CONNECTIONS

1. Rabbits eat the inner bark of tree saplings, often killing them. Why would the inner bark of the tree be a nutritious food source, and why would the tree die if a ring of inner bark is destroyed?

2. What have radioactive tracers revealed about the relative mobility and directionality of movement of plant nutrients?

3. Contrast the sources of energy used to transport materials in xylem and phloem.

4. Why would a sieve element in phloem tissue be able to withstand 300 pounds per square inch of pressure while an automobile tire would burst?

osmotic concentration falls in the storage organs (sugar is converted into starch) or the actively growing tissues (sugar is respired or used in synthesis), where sugars are being removed from the sieve tubes and used up. Those sections of the tubes, therefore, tend to lose water, and their turgor pressure drops.

In other words, the contents of the sieve tubes in one portion of the plant (the "source") are under considerable turgor pressure while the contents in another portion of the plant (the "sink") are under lower turgor pressure, so a turgor-pressure gradient is established (Fig. 19.20). The result is a mass flow of the contents of the sieve tubes from the region (source) under high pressure (usually the leaves, but sometimes storage organs when reserves are being mobilized for use, as in early spring) to the region (sink) under lower pressure (usually actively growing regions or storage depots). The whole process depends both on the massive uptake of water by cells at the source end, because of their high osmotic concentrations, and on the massive loss of water by cells at the sink end, because their osmotic concentrations are lowered by their loss of sugar (Fig. 19.21).

The movement of water into and out of phloem is made easier by the physical proximity of the xylem. Whether in vascular bundles in herbaceous plants or in continuous sapwood in woody plants, the functional xylem is always located close to the functional phloem.

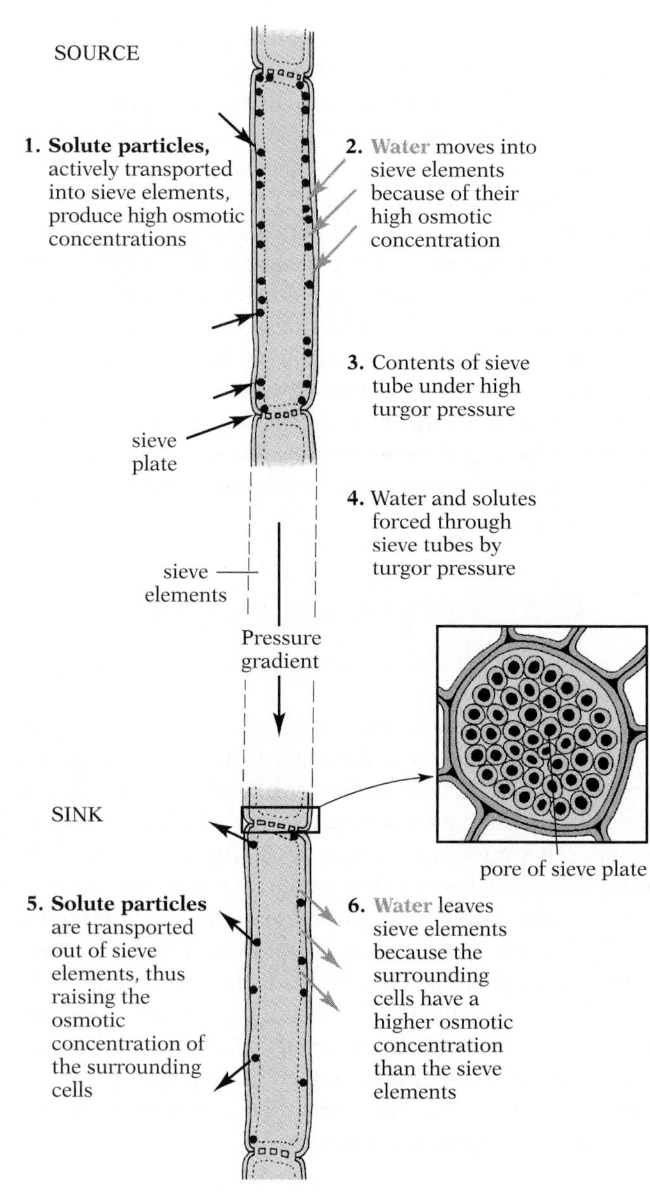
SOURCE
1. Solute particles, actively transported into sieve elements, produce high osmotic concentrations
2. Water moves into sieve elements because of their high osmotic concentration
3. Contents of sieve tube under high turgor pressure
sieve plate
4. Water and solutes forced through sieve tubes by turgor pressure
sieve elements
Pressure gradient
SINK
pore of sieve plate
5. Solute particles are transported out of sieve elements, thus raising the osmotic concentration of the surrounding cells
6. Water leaves sieve elements because the surrounding cells have a higher osmotic concentration than the sieve elements

19.20 Pressure-flow model of phloem transport According to this model, there is a mass flow of water and solutes from an area of high turgor pressure (the "source") to an area of low turgor pressure (the "sink"). At the source, solute particles (mainly sucrose) are actively transported into sieve elements, increasing their osmotic concentration. Water from the surrounding mesophyll cells and nearby xylem then moves into the cells by osmosis, creating a high turgor pressure. At the sink, solutes diffuse out of the cells, lowering the osmotic concentration, and water leaves, thus decreasing the turgor pressure. The higher turgor pressure at the source forces materials along the tubes to the sink.

The Xylem and Phloem Together Constitute an Effective Transport System

We have seen that the xylem conducts most of the water and inorganic nutrients upward from the roots, through the stem, to the leaves and buds. From the xylem the photosynthesizing cells receive the water they require for photosynthesis. The sugars synthesized during photosynthesis are transported by the phloem to all other parts of the plant, including the roots where they are stored as starch. The xylem conducts primarily water and inorganic nutrients only in one direction, up, whereas phloem conducts both organic and inorganic materials in either direction, up or down, depending on the needs of the different parts of the plant body.

Sunlight
xylem
phloem
(3) Movement out of xylem by osmosis
(4) Transpiration: evaporation from mesophyll; diffusion out of stomata and some movement of water into phloem
(2) Ascent in xylem, powered by transpiration
(5) Movement in phloem by pressure flow from source to sink
(1) Absorption of water into root by osmosis

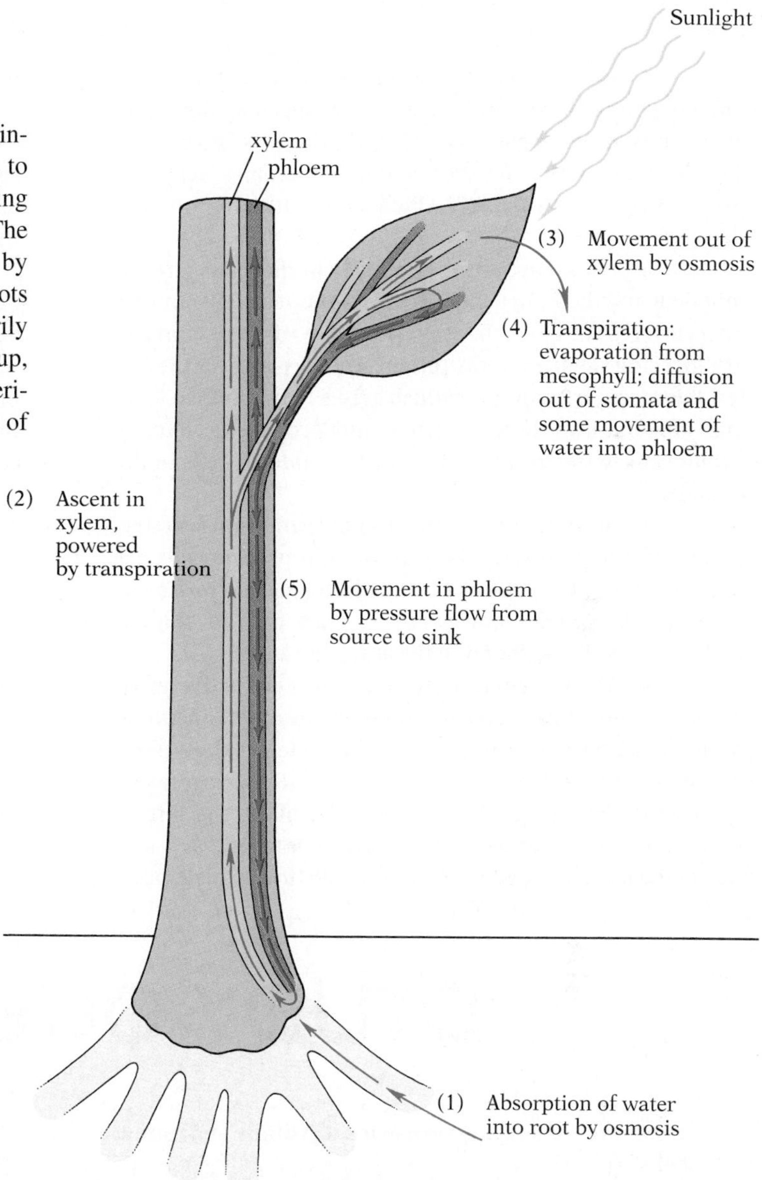

19.21 Summary of water movement in plants Water enters roots by osmosis (1) and is pulled up the xylem (2) by transpiration. Water leaves the xylem in leaves and moves from cell to cell by osmosis (3), where it either evaporates from the gas-exchange surfaces (4) or is drawn into the phloem. Material in the phloem is pushed from the source (producer regions) of the solutes (mainly sucrose) by pressure flow (5). The solutes are transported out at a sink (consumer region), and water follows by osmosis. Some of the water moves back into the xylem. The energy for moving water is generated by active transport of ions in the roots (1), and transpiration in the leaves (4).

CONCEPTS IN BRIEF

1. In general, only very small organisms can rely exclusively on the processes of diffusion and intracellular transport for the movement of substances; most complex organisms require a specialized transport system to deliver nutrients and oxygen to the cells.

2. The large multicellular land plants, the *vascular plants,* have evolved two types of specialized conducting tissue: *xylem* and *phloem.* These tissues form continuous pathways running through the roots, stems, and leaves, and transport materials from one part of the plant body to another.

3. Stems function as organs of transport and support. In young stems new tissues are produced by the *apical meristem* and the stem grows in length; these tissues are *primary tissues.* If the *vascular cambium* becomes active, it produces *secondary tissues,* which contribute to growth in diameter.

4. There are two types of herbaceous stems: monocot and dicot. In a monocot stem the vascular bundles are scattered; in an herbaceous dicot the vascular bundles are in a ring separating the pith from the cortex. In the latter, the vascular cambium, which arises between the xylem and phloem, may become active and produce secondary xylem cells to the inside and secondary phloem cells to the outside.

5. In woody stems the vascular cambium is active, and each year new secondary xylem and phloem cells are produced. As woody stems grow in diameter, the *cork cambium* becomes active and produces cork cells on the periphery of the stem, forming the outer bark.

6. Xylem contains two types of conductive cells: *tracheids* and *vessel elements.* Both are dead at functional maturity. Upward transport in the xylem takes place within empty cell walls that form tubes with perforated or open ends. Numerous *pits* in the walls allow the lateral movement of materials. Xylem also contains *fiber* and *parenchyma cells;* the latter often form *rays,* which function in lateral movement of materials across the stem.

7. Phloem, too, contains both fiber and parenchyma cells, in addition to *sieve elements* and *companion cells.* The elongate sieve elements, with their *sieve plates,* are arranged end to end, forming a *sieve tube.* Transport in the phloem occurs through living cells of the sieve tube.

8. Sap moves upward through the plant body in the tracheids and vessels of the xylem. According to the *transpiration-adhesion-tension-cohesion (TATC) theory,* water lost by transpiration from the aerial parts of the plant is replaced by withdrawal from the water column in the xylem, thereby pulling the whole column of water molecules upward.

9. In some small plants, especially in early spring, *root pressure* may contribute to the movement of sap in the xylem.

10. The most widely supported hypothesis for explaining phloem function is the *pressure-flow hypothesis,* according to which there is a mass flow of water and solutes under pressure through the sieve tubes from a region of high turgor pressure (the "source") to a region of low turgor pressure (the "sink"). The source is any place where more food is being produced than is used (e.g., photosynthesizing leaves during the day). The sink is any place where more food is being used up than produced (e.g., a storage cell in a root or stem).

STUDY QUESTIONS

1. How are materials transported within individual cells? (p. 389)
2. Define these contrasting terms: *xylem* versus *phloem; herbaceous* versus *woody; annual* versus *perennial.* (p. 390)
3. You find a plant that is about two feet tall and has a succulent stem. How could you use a cross section of its stem to determine if it is a monocot or dicot? What are the other major distinguishing features of these groups? (pp. 391–393)
4. Starting from the outside, name and describe the regions found in an herbaceous dicot stem that shows no secondary growth. (p. 393)
5. Compare the structure of an old woody stem with an herbaceous stem that has undergone secondary growth. Which new layers would be present in the woody stem, and which original layers would be missing? (pp. 393–396)
6. What are the two major functions served by xylem tissue? What cell types are found in xylem? How are they adapted to serve their functions? (pp. 397–398)
7. How are materials transported laterally in stems? Why is lateral transport necessary? (p. 398)
8. Describe the *transpiration-adhesion-tension-cohesion (TATC)* theory of water movement in the xylem. What evidence supports this theory? What supplies the energy for the movement of water against the force of gravity? (pp. 400–401)
9. Describe the structure of phloem tissue. What unique cell types are characteristic of phloem? What are their functions? (pp. 401–402)
10. According to the *pressure flow* hypothesis, how are organic materials translocated through phloem tissue? What are sources and sinks? How is a turgor-pressure gradient created in sieve tubes? (pp. 403–404)

CHAPTER 20

INTERNAL TRANSPORT IN ANIMALS

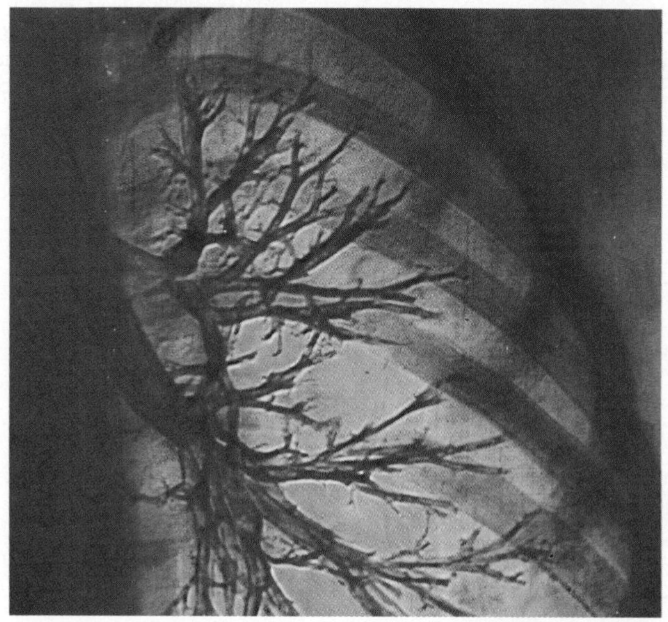

Every living cell in an animal's body requires a supply of nutrients and oxygen, and must eliminate certain waste products such as carbon dioxide and nitrogenous waste compounds. In some animals, each cell is adjacent or very close to the environmental medium from which it extracts needed materials and into which it dumps wastes. But the larger and structurally more complex animals may have trillions of cells that are some distance from the body surface and environmental medium. These cells depend on one another for their very existence, because they have divided the labor of growth and maintenance. We have seen that diffusion is too slow to transport materials efficiently more than 1 millimeter, even when supplemented by intracellular transport mechanisms. In short, most animals require some type of specialized transport (circulatory) system to deliver the materials to the individual cells and remove wastes. The problem is especially acute in very active animals, which have high metabolic rates, rapidly consume energy, and therefore require an almost continuous supply of nutrients and oxygen.

We have also seen that animal cells are extremely sensitive to changes in the osmotic concentration of their environment. In a multicellular animal, the cellular, or internal, environment is a thin film of fluid around each cell. This fluid bathes the cells and acts as a "middleman" between the blood and the cells. The blood transports the materials to the tissue fluid, and the tissue fluid delivers them to the cells. This fluid must be relatively constant, despite changes in the external environment, for the cells to function properly. This stability, this tendency to maintain equilibrium—the technical term is ***homeostasis***—is essential to the normal function of the organism. One of the important functions of a transport system is to help maintain the homeostasis of the extracellular fluids by delivering nutrients of all kinds to the tissues and removing the accumulated waste products (Fig. 20.1).

A Variety of Mechanisms Have Evolved to Transport Materials

Even small animals that lack a circulatory system exhibit some adaptations for transport. The body wall of hydra (see Fig. 17.11, p. 354), for example, is basically two cells thick, but even the cells of the inner layer are exposed directly to water containing dissolved oxygen, which is drawn into the gastrovascular cavity. And the gastrovascular cavity reaches into the tentacles, thereby delivering food to all parts of the body. In

Chapter opening photo: Arteriograph showing arteries supplying the right lung of a human. In this procedure contrast material is injected into the main pulmonary artery and X rays are taken 3 to 6 times per second to observe the blood flow.

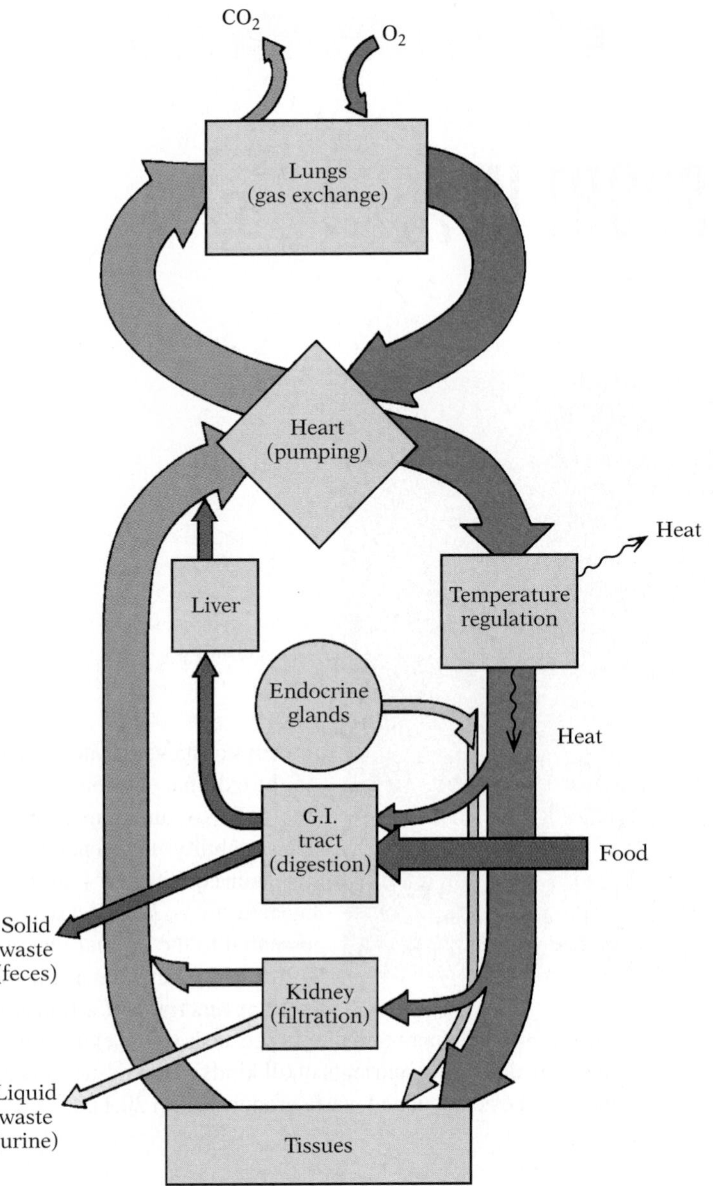

20.1 Major functions of the circulatory system This highly schematic diagram traces the movement of gases, hormones, nutrients, wastes, and heat through the circulatory system. Oxygen enters through the lungs, is carried by the blood, and is exchanged for carbon dioxide in the tissues; the carbon dioxide is carried to the lungs where it is exchanged for more oxygen. Nutrients enter from the digestive system and are carried to the liver, where most remain, and then to the tissues, where those left over are delivered. Wastes from the tissues enter the blood for transport to the kidneys. The circulatory system also transports hormones and is used for heat exchange. One of the important functions of the circulatory system is to help maintain homeostasis of the body fluids.

planaria, the profusely branching gastrovascular cavity penetrates all parts of the body, functioning as both a digestive cavity and a primitive transport system (see Fig. 17.13, p. 356). The success of hydra, planaria, and similar organisms is due to their high surface-to-volume ratios. A relatively large proportion of the internal cells in these organisms are close to the body surface. In short, though animals like hydra and planaria lack a specialized transport system, they do have compensatory adaptations that free them from complete dependence upon diffusion and intracellular transport.

Most animals, like hydra and planaria, are adapted for active movement. Many, however, cannot rely on such a slow process as diffusion to sustain the high metabolic rate required for their active life. Most of the larger, more complex animals that are the subjects of this chapter have some sort of circulatory system for transporting substances between the specialized systems of procurement, synthesis, and elimination and the individual cells throughout the body.

In complex animals this system is most often truly circulatory; blood is moved round and round through the body along a fairly definite path. By contrast, upward transport in the xylem of plants, as we have seen, is unidirectional; it depends on loss of water from the upper end of the system. Similarly, transport in the phloem depends on loss of large quantities of materials from one end. Neither the xylem nor the phloem is therefore a circulatory system in the strict sense, though substances may move from the xylem to the phloem, or the reverse, and circulate in this manner.

Animal circulatory systems usually include some sort of pumping device, called a ***heart***, and a series of vessels to dis-

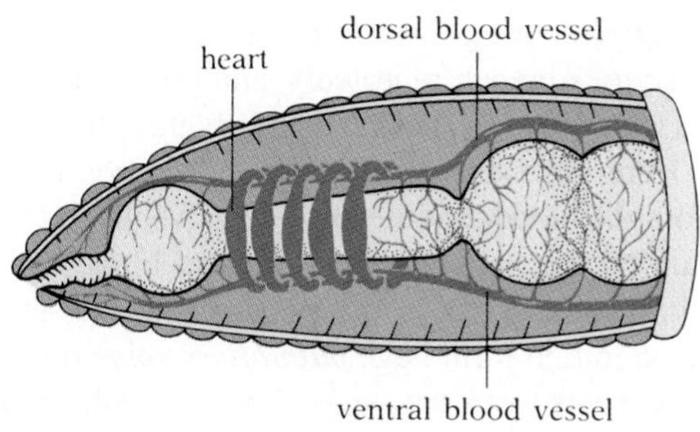

20.2 The circulatory system of an earthworm Ten hearts, five on each side, pump blood through the longitudinal vessels, which also pulsate and help move the blood.

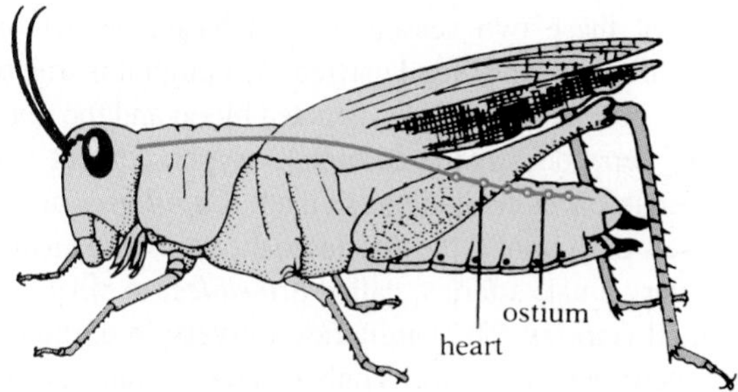

20.3 The dorsal heart of a grasshopper Blood enters the heart through the ostia and is pumped forward and out at its open anterior end.

tribute the blood. There may be only one heart, as in our own case, or a number of separate hearts, as in earthworms, where five pairs of "hearts"—enlarged blood vessels on each side of the animal—pulsate, pumping blood from the main dorsal longitudinal vessel into the main ventral longitudinal vessel (Fig. 20.2). Many insects have both a large general heart and a series of smaller accessory hearts at the bases of their legs and wings. These accessory hearts increase the blood supply to the most active appendages.

The one-way pumping action of the heart, usually combined with a system of one-way valves, moves the blood in a regular fashion through the circuit. This circuit may be a continuous network of well-defined vessels, in which case it is called a ***closed circulatory system***, or the circuit may have sections without vessels, where the blood flows through large, open spaces known as sinuses to bathe the cells directly; such a system is called an ***open circulatory system***. Blood flow through the sinuses is rather slow and sluggish. Closed circulatory systems are characteristic of a great variety of animals, including earthworms and all vertebrates. Open circulatory systems are characteristic of most molluscs (snails, oysters, clams, etc.) and all arthropods (insects, spiders, crabs, crayfish, etc.).

THE INSECT CIRCULATORY SYSTEM

Insects Have an Open Circulatory System

Since movement of the blood through an open system is not as fast, orderly, or efficient as movement through a closed system, it may be surprising that such active animals as insects, which have high metabolic rates, should have open circulatory systems. However, insects do not rely on the blood to carry oxygen to their tissues; that function is reserved for the heavily branched tracheal system. Consequently, it is not vital for insects that their blood flow very fast and in a precise pathway. This is a good example of the way the interrelated systems of a living creature adapt to its needs.

The circulatory systems of insects are less elaborate than those of most other arthropods. Ordinarily, the only definite blood vessel in an insect is a contractile longitudinal vessel, often designated as the heart, which runs through the dorsal portion of the animal's thorax and abdomen (Fig. 20.3). The posterior portion of the vessel or heart is pierced by a series of openings, or ostia, each regulated by a valve that allows only blood into the vessel. When the "heart" contracts it forces blood out of its open anterior end into the head region. When it relaxes again, blood is drawn in through the ostia. Once outside the heart, the blood is no longer in vessels; there are no veins, capillaries, or arteries other than the heart itself with its valve segments and the short, so-called artery that forms its anterior end. The blood simply moves through the spaces between the internal organs of the insect, thus bathing the cells directly.

The action of the heart causes the blood to move sluggishly through the body spaces from the anterior end, where it was released, to the posterior end, where it will again enter the heart. The movement of the blood is accelerated by the stirring and mixing action of the muscles of the body wall and gut during activity. This accelerated blood flow during an animal's more active states, such as running or flying, allows for a more efficient and rapid delivery of nutrients and removal of wastes.

THE HUMAN CIRCULATORY SYSTEM

Humans Have a Closed Circulatory System

Humans, like all vertebrates, have a closed circulatory system, which consists basically of a heart and numerous arteries, capillaries, and veins. The network of blood vessels is very extensive; 60 miles of vessels are estimated to be in the human body. ***Arteries*** are blood vessels that carry blood away from the heart, while ***veins*** carry blood back to the heart. Note that the definitions of these two vessels are not based on the oxygenated condition of the blood carried. Although it is true that the majority of arteries carry oxygenated blood and the majority of veins carry deoxygenated blood, oxygen content is not always a reliable way to distinguish them. ***Capillaries*** are tiny blood vessels that connect the arteries to the veins; they usually run from very small arteries, called ***arterioles***, to very small veins, called ***venules***. The capillaries are very important because it is across their thin walls only that the exchange of materials between the blood and the other tissues takes place.

The Right Ventricle Pumps Blood Through the Pulmonary Circuit; the Left Ventricle Through the Systemic Circuit

We shall trace the movement of blood through the human circulatory system by starting with the return of blood to the heart from the legs or arms. Such blood enters the upper right chamber of the heart, called the ***right atrium*** or right auricle (Fig. 20.4), by way of one of two large veins, the anterior or posterior vena cava. Next the blood moves through a valve (the tricuspid valve) into the ***right ventricle***, the lower right chamber of the heart. Now, this blood, having just returned to the heart from its circulation through tissues, contains little oxygen and much carbon dioxide. Simply to pump this deoxygenated blood back out to the tissues would be of little value to the body. Instead, contraction of the right ventricle sends the blood through a valve (the pulmonary semilunar valve) into the ***pulmonary trunk***, which soon divides into the ***right*** and ***left pulmonary arteries***, one going to each lung. In the lungs the pulmonary arteries branch repeatedly, and each terminal branch connects with dense beds of capillaries lying in the walls of the alveoli. Here gas exchange takes place, carbon dioxide diffuses from the blood into the air in the alveoli, and oxygen is picked up by the hemoglobin in the red blood cells of the blood. From the capillaries, the blood passes into small veins, which soon join to form large ***pulmonary veins*** running back toward the heart from the lungs. The four pulmonary veins (two from each lung) empty into the upper left chamber of the heart, called the ***left atrium*** or left auricle. From the left atrium, blood moves through a valve (the bicuspid or mitral valve) into the ***left ventricle***, which is the lower left chamber of the heart. The left ventricle, then, is the pump for recently oxygenated blood. When it contracts, it pushes the blood through a valve (the aortic semilunar valve) into a very large artery called the ***aorta***, which distributes blood to arteries supplying all parts of the body.

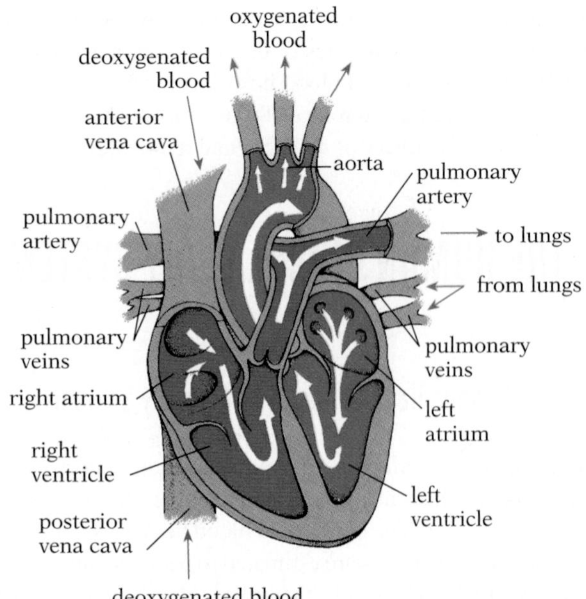

20.4 The human heart The arrows show the direction of blood flow (oxygenated blood, red; deoxygenated blood, blue).

After the aorta emerges from the anterior portion of the heart (the upper portion, in humans standing erect), it forms a prominent arch and runs posteriorly along the middorsal wall of the thorax and abdomen (Fig. 20.5). Numerous branch arteries arise from the aorta along its length. Each of these arteries, in turn, branches into smaller arteries until eventually the smallest arterioles connect with the tiny capillaries embedded in the tissues. Capillaries are so numerous that no cell is more than a few cells away from one. It is only through the capillaries that oxygen, nutrients, hormones, and other substances move out of the blood and into the tissues. Such waste products as carbon dioxide and nitrogenous wastes are picked up by the blood in the capillaries, and substances to be transported, such as hormones secreted by the tissues and nutrients from the intestine and liver, are also picked up. The blood then runs from the capillary bed into tiny venules, which fuse to form larger and larger veins, until eventually one or more large veins exit from the organ in question. These veins, in turn, empty into one of two very large veins that empty into the right atrium of the heart: the ***anterior vena cava*** (sometimes called the superior vena cava), which drains the head, neck, and arms, and the ***posterior vena cava*** (inferior vena cava), which drains the rest of the body. Very little, if any, exchange of materials between the blood and the other tissues occurs across the walls of the arteries or veins themselves. The thick walls of these vessels are largely impermeable to the substances in the blood and tissue fluid. In contrast, capillaries have walls composed of only a thin layer of endothelial cells; it is across these walls that the exchange of materials takes place (Fig. 20.6).

The complete circuit traveled by blood can be summarized as follows. Blood from the body enters the right side of the heart and is pumped to the lungs, where it picks up oxygen and gives up carbon dioxide; then it returns to the left side of the heart. This portion of the circulatory system is called the ***pulmonary circulation*** (Fig. 20.7). Note that in the pulmonary circuit the pulmonary arteries carry *deoxygenated* blood from the heart to the lungs, and the pulmonary veins carry *oxy-*

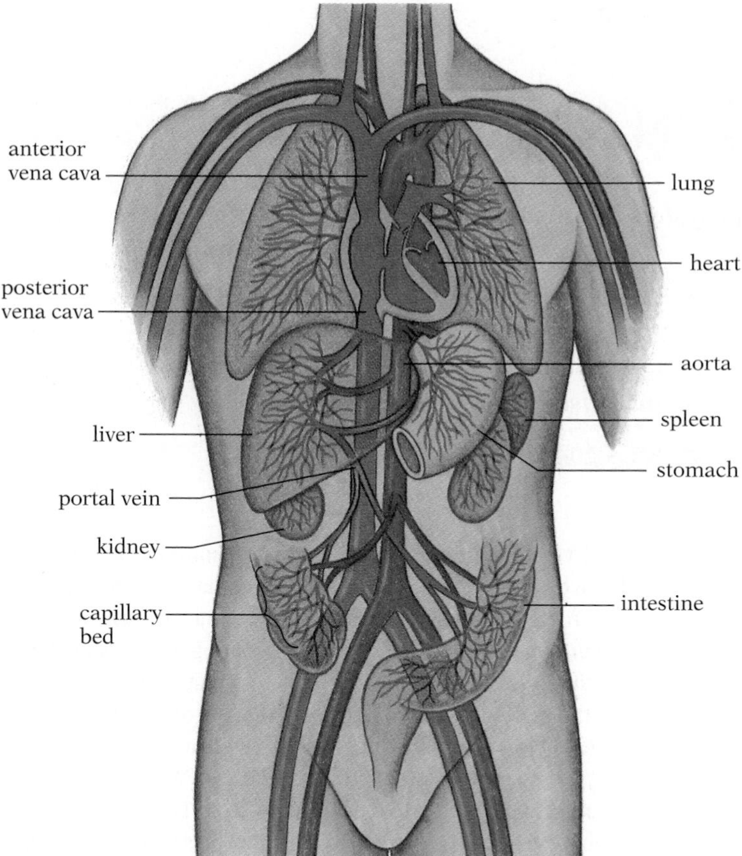

20.5 Diagram of the human circulatory system Red vessels contain oxygenated blood; blue vessels, deoxygenated blood. The transfer of oxygen and nutrients from the blood to tissues, as well as the transfer of carbon dioxide and other wastes from cells to the blood, takes place in each of the capillary beds. The capillaries, which are of necessity represented here as being relatively thick, are actually microscopic. Only a very few of the vast number of arteries and the capillary beds they supply are shown.

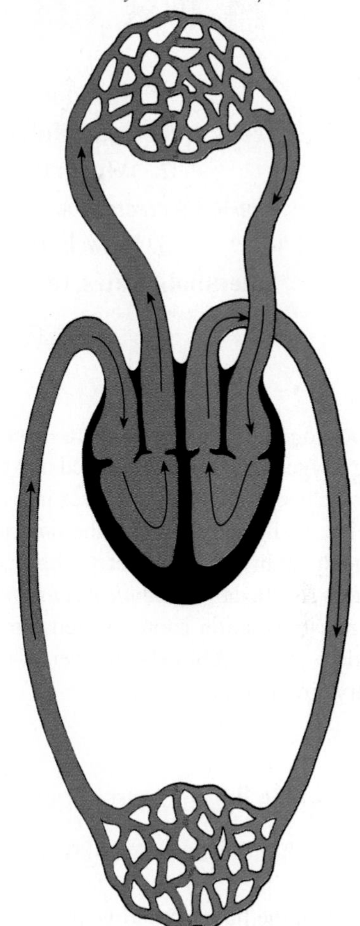

20.7 Diagram of mammalian circulation Red vessels contain oxygenated blood; blue vessels, deoxygenated blood. In an adult, blood passes from the right ventricle to the lungs, where it picks up oxygen. It then returns to the left side of the heart, from which it is pumped into the systemic circulation. Finally, having largely given up its oxygen to cells lining the capillary beds, the blood returns to the right side of the heart.

20.6 Arteries and veins Walls of arteries and veins are composed of three layers: (1) an outer layer of connective tissue; (2) a middle layer of smooth muscle, which can change the size of the vessels; and (3) an inner layer of simple squamous endothelium. Capillaries have thin walls composed only of thin endothelium, through which materials are exchanged. Note that arteries and veins have the same three layers in their walls, but the walls of veins are thinner and less rigid. The walls of veins readily change shape when muscles press against them. The scanning electron micrograph shows an artery (left, bottom) and a vein (right).

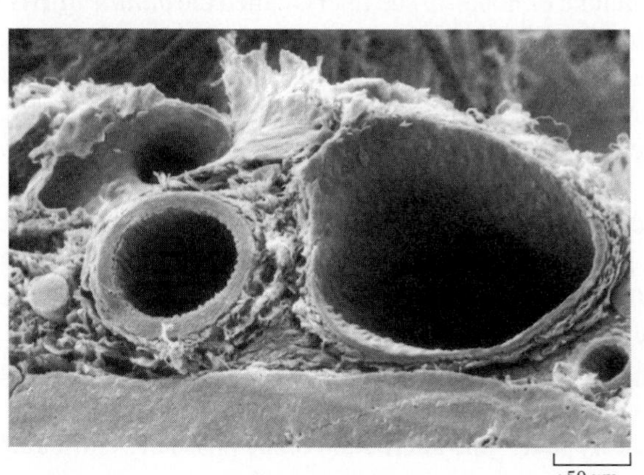

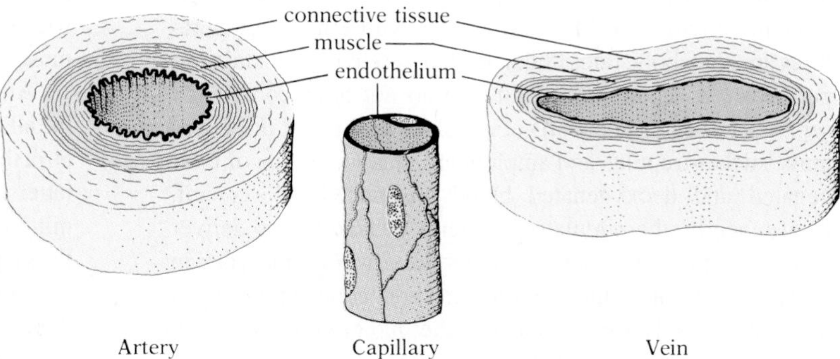

MAKING CONNECTIONS

1. How does the circulatory system in animals maintain homeostasis of cells and extracellular fluids?

2. Why is the transport system of most animals, unlike that of plants, called a *circulatory* system?

3. How is it possible for insects, which are active animals with high metabolic rates, to have an open circulatory system?

genated blood from the lungs to the heart. From the left side of the heart, the oxygenated blood is pumped into the aorta and its numerous branches, from which it moves into capillaries, and then into veins, and finally back to the anterior or posterior vena cava and the right side of the heart. This portion of the circulatory system is called the ***systemic circulation***. The arteries of the systemic circulation carry oxygenated blood and the veins carry deoxygenated blood—a reversal of their roles in the pulmonary circulation.

The Parts of the Human Heart Normally Contract in an Orderly Sequence

We have seen that the human heart is, in effect, two hearts in one, since blood in the left side of a normal heart is completely separated from blood in the right side. This type of heart—four-chambered, with complete separation of sides—is characteristic of mammals and birds, the two groups of endothermic vertebrates. These animals are also homeothermic, i.e., they maintain high, relatively constant body temperatures regardless of fluctuations in the environmental temperature. Such organisms have high metabolic rates and very precise internal control mechanisms to help regulate their body temperature. The constant supply of oxygen-rich blood to their tissues is essential if they are to maintain a high rate of metabolism. It would be highly disadvantageous to such animals if the oxygen-rich blood from the pulmonary circulation were mixed with the oxygen-poor blood returning from the systemic circulation. Most vertebrates whose body temperatures and metabolic rates fluctuate with the environmental temperature (i.e., ectothermic animals) do not have completely separate left and right hearts (Fig. 20.8). In some of these animals, such as reptiles and amphibians, small amounts of oxygenated and deoxygenated blood are mixed as a result. Consequently, the circulatory systems of ectotherms deliver less oxygen to the tissues. Such a system would not work in birds or mammals, which have a much greater need for oxygen per gram of body weight than ectothermic organisms, which live at a colder temperature.

Even though the human heart has a complete separation of its two sides, the two halves beat essentially in unison, and in an orderly sequence. Certain cardiac muscle cells are ***autorhythmic***, i.e., they are capable of beating spontaneously in a rhythmic fashion on their own, without stimulation from the central nervous system. However, the rate of beat is partly regulated by stimulation from two sets of nerves, one that accelerates the heartbeat and one that decelerates it. During physical exercise, when the body's demand for oxygen and nutrients is high, the nervous system causes the heart rate to increase; when the body is at rest, it causes the rate to slow. We shall discuss the control of heartbeat further in Chapter 28. If all nerve connections to the heart are cut, the heart will continue to beat in a normal manner, although the rate of beat may change. One extreme example of this is seen in the heart of a frog or turtle, which, if it is placed in a solution with the proper osmotic concentration, continues to beat even after being removed from the animal's body. If the individual cells are separated, the cells will continue to beat spontaneously, but they will beat at different rates.

The initiation of the heartbeat normally comes from the sino-atrial node, or ***S-A node***, often called the pacemaker of the heart; it is a small mass of ***nodal tissue*** on the wall of the right atrium near the point where the anterior vena cava empties into it (Fig. 20.9). Nodal tissue, which is unique to the heart, has the contractile properties of muscle but can transmit impulses like nerve. A second mass of nodal tissue, the atrio-ventricular node, or ***A-V node***, is located in the partition between the two atria. A bundle of nodal-tissue fibers (called the ***bundle of His***) runs from the A-V node into the walls of the two ventricles, with branches (Purkinje fibers) to all parts of the ventricular musculature (Fig. 20.9).

At regular intervals, the S-A node produces an excitatory impulse that spreads across the walls of both atria, causing the atria to contract. When this impulse reaches the A-V node, the latter is stimulated and excitatory impulses are rapidly transmitted from it down the fibers of the bundle of His to the Purkinje fibers within the ventricular walls. These fibers conduct impulses rapidly to all parts of the ventricles, inducing them to contract simultaneously.

The heart beat—the alternation of the phase of ventricular

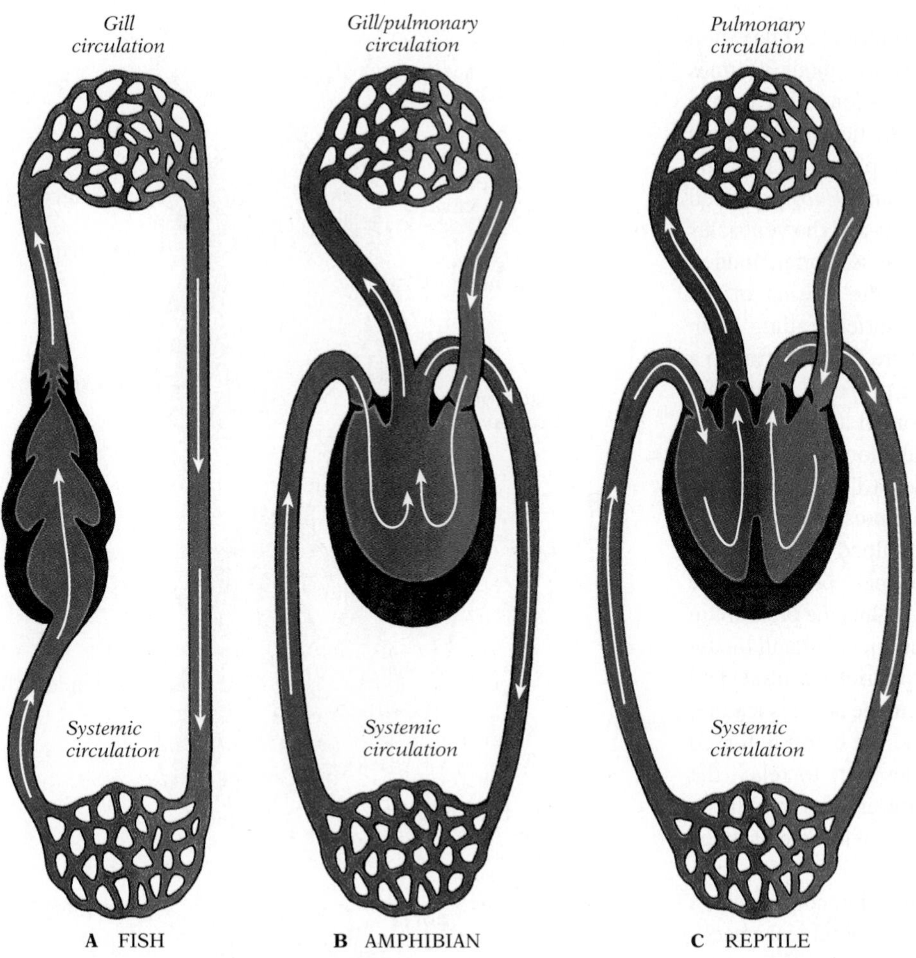

20.8 A schematic comparison of vertebrate hearts and circulatory plans (A) In modern fish the blood is pumped through a linear, multichambered heart to the gills, where it picks up oxygen. The oxygenated blood (red) then passes without further pumping to the systemic circulation, where it gives up its oxygen before returning to the heart. This one-pump heart has a drawback: because of the high resistance of the gill capillaries, blood leaving the gills is under low pressure and moves sluggishly through the systemic circuit. This problem is solved in the amphibians, reptiles, and mammals by the returning of blood that has picked up oxygen to the heart, from which it is pumped into the systemic circulation. (B) The amphibian heart has two atria and a single ventricle, but the spongy nature of the ventricle and arrangement of the vessels exiting it minimize the mixing of pulmonary and systemic blood. (C) In most reptiles the ventricles are partially divided, so less mixing takes place. By contrast, the four-chambered mammalian and bird hearts keep the deoxygenated and oxygenated blood completely separate in the systemic and pulmonary circuits (Fig. 20.7).

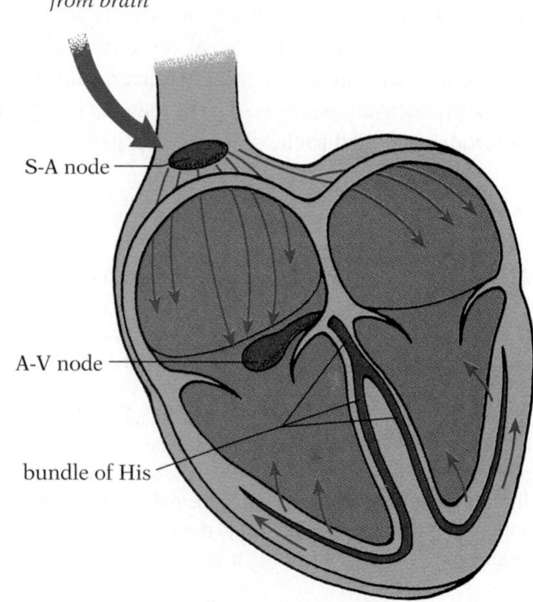

20.9 Electrical control of the heart The pacemaker of the heart is the S-A node. Its spontaneous firing rate can be accelerated or decelerated by nerve impulses from the brain. Each time the S-A node fires, a signal spreads across the two atria, generating a beat and stimulating the A-V node. From there, after a brief delay, the signal is relayed rapidly down the fibers of the bundle of His to the Purkinje fibers, generating a beat in the ventricles.

contraction (***systole***) with the phase of ventricular relaxation (***diastole***)—in a normal human being at rest is about 70 times per minute, but there is much individual variation. In the course of the beat, the heart emits several characteristic sounds, which can be heard easily through a stethoscope placed against the chest. First, there is a long, low-pitched sound (*lub*) produced by the closing of the valves between the atria and the ventricles when the ventricles contract. Then there is a shorter, louder, higher pitched sound (*dup*) produced by the closing of the valves between the ventricles and the arteries leading from them. The normal heart sounds (*lub-dup*) are an indication to a physician that all valves are functioning properly. A normal heart valve opens when the pressure in front of it is greater than the pressure behind it. During late diastole, blood flows into the heart and fills both the atria and ventricles. (The bicuspid and tricuspid valves are open because the blood pressure in the atria is higher than that of the ventricles due to blood entering from the veins.) Contraction of the atria forces more blood into the ventricles. As soon as the atria begin to relax and the pressure in them falls below that in the ventricles, the tricuspid and bicuspid valves snap shut. Similarly, when the ventricles contract and the pressure in them exceeds the pressure in the arteries leading from them, the semilunar valves open and the blood is forced into the arteries; as soon as the ventricles start to relax, the valves snap shut and prevent the blood in the arteries from flowing backward into the ventricles. A summary of the events in the cardiac cycle is shown in Figure 20.10.

If a valve has been damaged and cannot shut completely, a hissing or murmuring sound can be heard as blood leaks backward through the damaged valve; this condition is called a diastolic heart murmur. Sometimes a damaged valve partly obstructs blood flow during systole, and the resulting sound (due to the turbulence in the blood flow) is called a systolic murmur. Heart murmurs are a common result of rheumatic fever and some other diseases. The more extensive the damage to the valve, the less efficient the heart action and the greater the strain placed on the heart. Sometimes a valve is so damaged that it must be replaced with an artificial valve (Fig. 20.11).

In the course of contraction, the heart muscle undergoes a series of electrical changes. These changes can be detected by electrodes attached to the skin and graphed by a device called

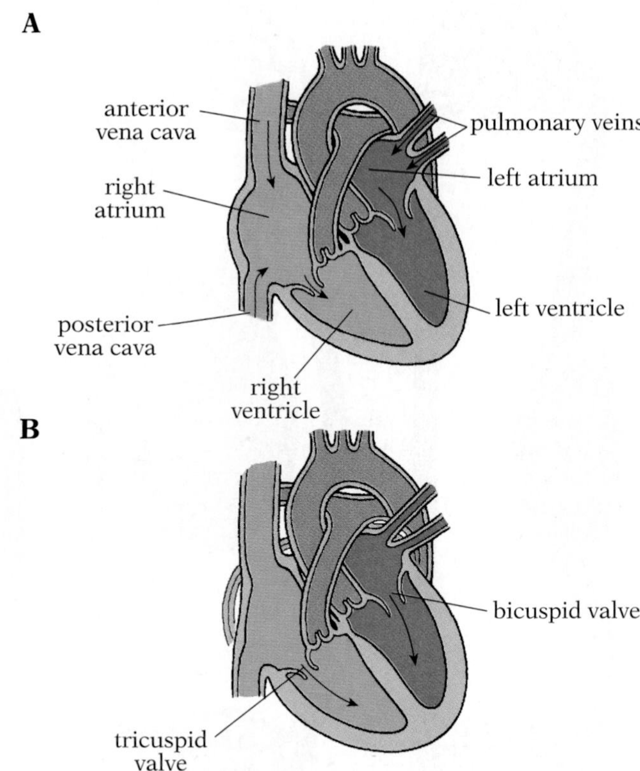

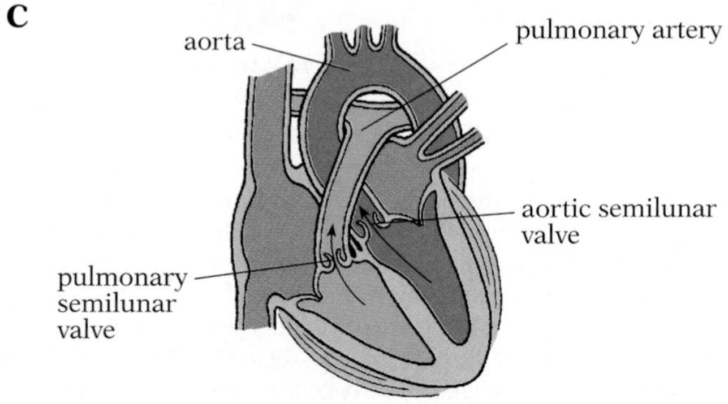

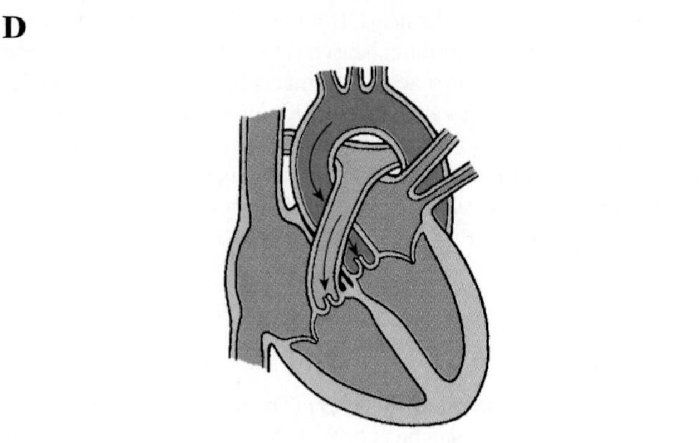

20.10 A summary of the events of the cardiac cycle (A) Following diastole, the atria and ventricles are relaxed. Blood enters the right side of the heart from the two venae cavae and the left side from the pulmonary veins. Both the tricuspid and bicuspid valves are open, so blood flows from the atria into the ventricles. (B) Additional blood is forced into the ventricle when the atria contract. (C) The bicuspid and tricuspid valves close and the valves in the base of the pulmonary arteries and aorta open as the two ventricles contract simultaneously, forcing blood into the pulmonary arteries and the aorta. (D) The ventricles relax and the valves in the base of the pulmonary arteries and aorta close to prevent backflow.

20.11 An artificial heart valve Severely damaged heart valves can be replaced with artificial ones such as this ball-and-cage valve. When the heart contracts, the blood can pass upward by forcing the ball against the top of the cage. When the heart relaxes, the blood starts to flow downward and forces the ball against the ring, closing the valve.

an electrocardiograph. Abnormalities in heart action alter the pattern of the graph, which is called an electrocardiogram (ECG) (Fig. 20.12). These alterations aid a trained physician in diagnosing the abnormality. For example, heart tissue may be damaged by disease, impairing electrical conduction through a portion of the heart and producing a partial or complete heart block. The effect on the heart of such a block depends on the location of the defect in the conducting system.

Another group of abnormalities of the conducting system involves continuous and irregular contractions, or fibrillation, of the atria or ventricles. Ventricular fibrillation (often a consequence of a heart attack) can be fatal because the blood is no longer being effectively pumped. Closed-chest cardiac massage can be used to maintain circulation in the patient until the fibrillation can be stopped by electrical shocks administered to the chest. The shocks send a brief flow of electrical current through the ventricles, which stops all impulse conduction. The heart remains quiet for three to five seconds, and then (if

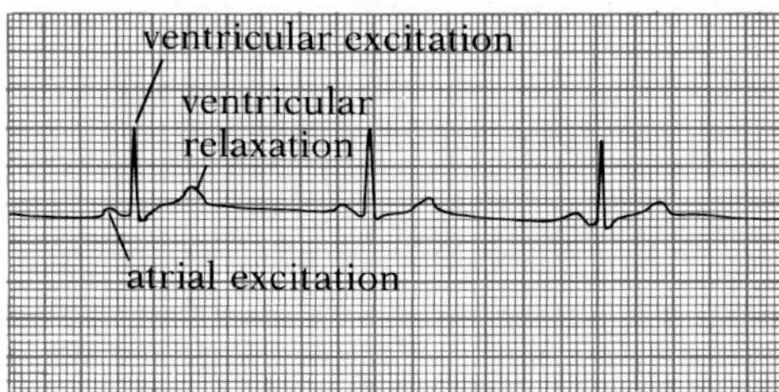

20.12 An electrocardiogram By recording the electrical activity of the heart under various conditions of rest and exercise, electrocardiograms can sometimes provide important information about potential heart problems. Notice the regular order, strength, and spacing of events in this normal ECG. Since atrial relaxation normally takes place during ventricular excitation, it is not visible in this recording.

the procedure has been successful) starts to beat again in a coordinated fashion, with the S-A node functioning as the pacemaker.

The heart of a resting human adult pumps about five liters of blood every minute, which is approximately equal to the total amount of blood in the body. During exercise both the rate of contraction and the amount of blood pumped per beat (the stroke volume) increase greatly. The combination of elevated heart rate and increased stroke volume may raise cardiac output (total amount of blood pumped per minute) to a level four to seven times the resting level.

The Pumping of the Heart Creates the Blood Pressure

When the left ventricle contracts, it forces blood under high pressure into the aorta, and blood surges forward in each of the arteries. The walls of the arteries are elastic, and the surge of blood stretches them, causing a pressure wave that travels along the arteries at about five to eight meters per second. This is the arterial pulse, which can be detected in the large arteries close to the body surface. During diastole, the relaxation phase of the heart cycle, the heart is not exerting pressure on the blood in the arteries and the pressure in them falls, but elastic recoil of the previously stretched artery walls maintains some pressure on the blood. There is thus a regular cycle of pressure in the larger arteries, the pressure reaching its high point during systole and its low point during diastole.

In humans arterial blood pressure in the systemic circuit is usually measured in the upper arm, where systolic values of about 120 mm Hg (see Exploring Further: Measuring Blood Pressure, p. 416) and diastolic values of about 80 mm Hg are normal in young adult males at rest. Note that these pressures apply to the upper arm only; the values would not be the same for the lower arm, the leg, or any other part of the body. In general, the blood pressure decreases continuously as the blood moves farther away from the heart. An exception is found in the arteries of human feet, where the average pressure is considerably higher (by 80 mm Hg) than in the heart. This is due to the weight of the column of blood in the vessels. The average blood pressure in the head is considerably lower (by 40 mm Hg) than the pressure in the heart, due to the effect of gravity. Sufficient blood pressure is required to overcome the effect of gravity to get blood to the brain. Think about the pressure necessary to supply blood to the brain of a giraffe, whose head may be two meters above the heart. Indeed, the average blood pressure of the giraffe measures about 260 mm Hg, more than twice as high as humans. As you might expect, the giraffe's arteries are much thicker than ours, an adaptation to withstand high blood pressure.

As a rule, the blood pressure is greatest in the part of the aorta close to the heart, and it falls off steadily in the more distant parts of the aorta and its branches. It falls even more

EXPLORING FURTHER

MEASURING BLOOD PRESSURE

Both the systolic pressure and the diastolic pressure are important diagnostic indicators to the physician. Ordinarily these pressures are measured in the artery of the arm with a sphygmomanometer (Fig. A). This instrument has a rubber cuff that wraps around the upper arm and into which air is pumped until the pressure exerted blocks all blood flow through the artery. A graduated column of mercury, known as a manometer, attached to the cuff indicates the pressure being exerted on the arm, which is expressed as a rise in millimeters of mercury (mm Hg). Next, the bell of a stethoscope is placed against the artery just below the cuff, and the air in the cuff is gradually released, with consequent reduction of pressure on the arm. When the pressure of the cuff has fallen to a value slightly lower than the maximum systolic pressure in the artery, a small stream of blood squirts through the artery for an instant during each pulse, producing vibrations that can be heard through the stethoscope. The value shown on the column of mercury the instant before this sound is first detected is taken as the systolic pressure. As the pressure in the cuff is gradually lowered beneath this value, the sound produced as more blood surges through the artery at each systole becomes louder and louder. Eventually the pressure in the cuff drops to only slightly above that in the artery at diastole, at which point the sound becomes muffled; the flow of blood through the artery is now continuous. The level in millimeters of the column of mercury at this point is taken to express the diastolic pressure. The systolic and diastolic pressures are frequently written together as a fraction, 120/80, for example.

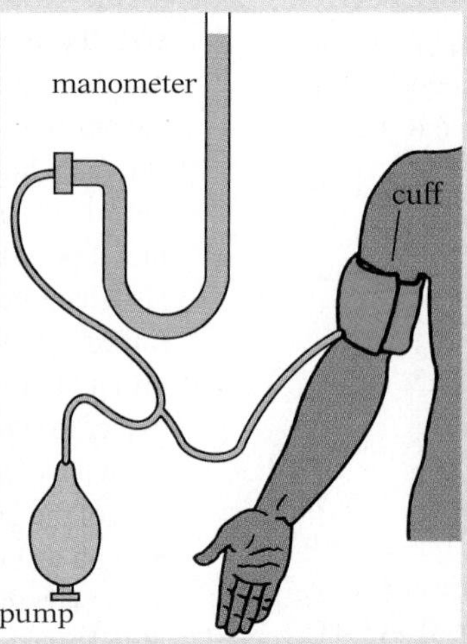

A A sphygmomanometer

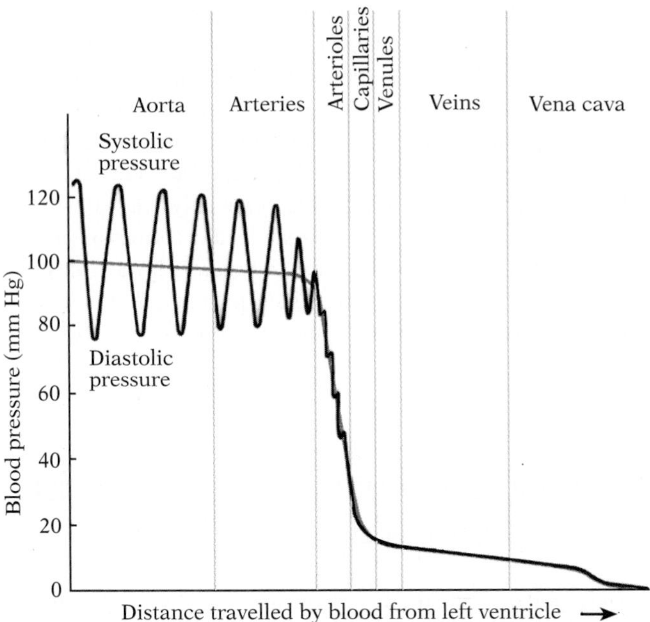

20.13 Graph of blood pressure in different parts of the human circulatory system The red curve traces the mean pressure values. As shown by the black curve, there is considerable fluctuation between the systolic and the diastolic pressure in the arteries. This fluctuation diminishes in the arterioles and no longer occurs in the capillaries and veins, because the elasticity of the vessel walls has effectively damped out the oscillations. The most rapid fall in pressure is in the arterioles and capillaries.

rapidly in the arterioles and capillaries, due to the decreased diameter of these vessels. The blood pressure continues to decline in the veins, and reaches its lowest point in the veins nearest the heart (Fig. 20.13). The decline of the blood pressure in successive parts of the circuit is the result of friction between the flowing blood and the walls of the vessels. Such a gradient of pressure is essential if the blood is to continue to flow; the fluid can only move from a region of higher pressure toward a region of lower pressure.

Breathing and Muscle Action Aid Venous Return

Because the average blood pressure in the veins is low—around 2 mm Hg in the vena cavae near the heart—there must be mechanisms other than blood pressure to return blood to the heart, especially the blood from the lower extremities, which must move against the force of gravity. The movements of the chest wall and diaphragm during breathing greatly aid the blood's return to the heart. During inhalation, the contraction of the diaphragm simultaneously enlarges the thoracic cavity and compresses the abdominal cavity. The negative pressure created within the thoracic cavity during inhalation causes the veins of the thorax to expand, drawing blood into these vessels and hastening its return to the heart. At the same time, the compression of the abdominal cavity increases the pressure there, which forces the blood upward along the posterior vena cava. Deeper and more frequent respirations thus increase venous return and improve circulation—an important principle. Skeletal muscle movements and the presence of valves within the veins also play a significant role in returning blood to the

MAKING CONNECTIONS

1. Why is a four-chambered heart with complete separation of the left and right sides characteristic of birds and mammals but not of other vertebrates?

2. What heart defect causes a heart murmur? What determines how dangerous a heart murmur is?

3. Describe two major types of abnormalities that can occur in the electrical conducting system of the heart.

4. What factors determine the amount of blood that your heart pumps per minute (cardiac output)? How is your cardiac output increased during exercise? How much can it increase during strenuous exercise?

5. Compare and contrast the mechanisms by which pressure differences are used to move fluids through the human circulatory system, the phloem of plants (see Chapter 19), and the human respiratory tract (see Chapter 18). (Recall that "fluids" include gases.)

heart. The walls of veins are relatively thin (Fig. 20.6) and easily collapsible. When nearby muscles contract as the body moves, they put pressure on the veins, compressing their walls and forcing the fluid in them forward. The fluid can move only toward the heart, because the numerous one-way valves with which veins are equipped prevent it from flowing backward into the section from which it just came (Fig. 20.14). When standing very still for a long period, you may have noticed your feet beginning to swell and a sudden onset of fatigue. That happens because there is not enough muscle action in your legs to push the fluids upward against the pull of gravity. If you can manage, while standing, to keep moving your feet and legs, or to regularly contract and relax the leg muscles, the unpleasant symptoms will not be so pronounced. Soldiers who must stand at attention for long periods of time are masters of this muscular manipulation. Similarly, to prevent pooling of blood in the lower extremities, hospitalized patients are encouraged to begin walking as soon as possible, often only a day or two after surgery; such pooling of blood may lead to formation of a clot in a leg vein (thrombophlebitis), a potentially fatal development all too common in bedridden patients. (See Exploring Further: Thromboembolic and Hypertensive Disease in Humans, p. 419.)

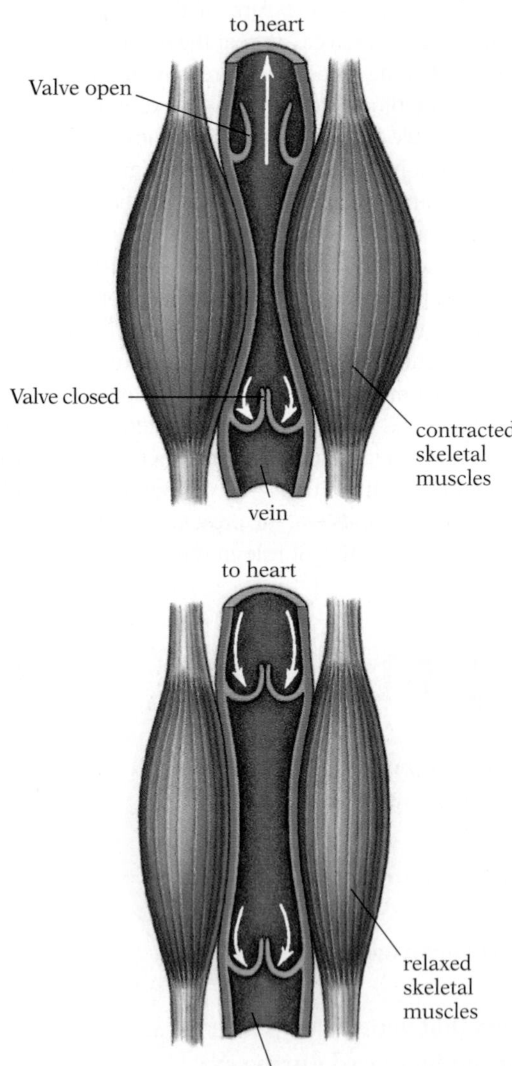

20.14 Movement of blood in the veins (A) Contraction of skeletal muscles helps pump blood toward the heart. When the skeletal muscles push against the thin-walled veins, the blood is propelled toward the heart because the blood forces the flaps of the valves open and moves through. The blood cannot move in the opposite direction, away from the heart, because the lower valves will close. (B) When the skeletal muscles relax, the pressure declines and the blood is pulled down by gravity, automatically closing the flaps of the valves to prevent backward flow.

The Balance Between Blood Pressure and Osmotic Pressure Governs Capillary Exchange

The capillaries penetrate all parts of every tissue; no cell is far removed from at least one capillary, which makes for efficient diffusion of materials between the tissues and blood. It is estimated that in muscle tissue there are as many as 60,000 capillaries per square centimeter of cross section. The diameters of the capillaries are very small, being seldom much larger than those of the blood cells that must pass through (Fig. 20.15). The extensive branching and small diameters of individual capillaries not only allow them to reach all portions of all tissues, but ensure as well that an adequate capillary surface area is available for the exchange process. The branching also increases the total cross-sectional area of the system, which makes blood flow more slowly in the capillaries than in the arteries or veins. This slower flow allows more time for the exchange process. The very small bore of the capillaries also contributes to the time available for exchange, by increasing the frictional resistance to blood flow and decreasing the blood pressure within the capillary bed.

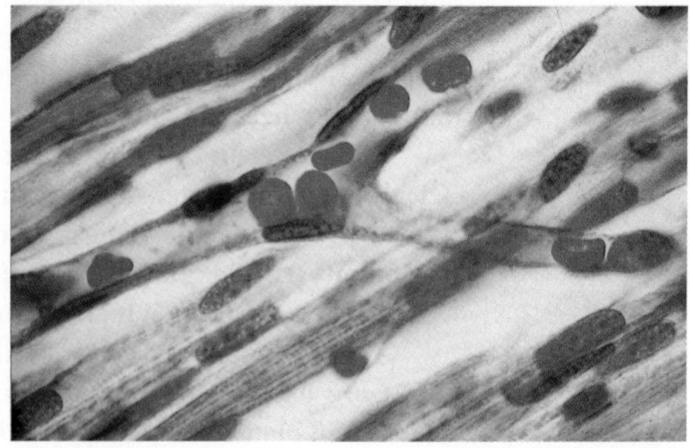

20.15 Blood vessels and capillaries In this longitudinal section of a human capillary, individual red blood cells are seen moving single file.

EXPLORING FURTHER

THROMBOEMBOLIC AND HYPERTENSIVE DISEASE IN HUMANS

Thrombosis is the formation of a solid mass or plug of blood constituents in a blood vessel. The blood clot, or *thrombus,* may block (wholly or only in part) the vessel in which it forms (see Fig. A), or it may become dislodged and be carried to some other location in the circulatory system, in which case it is called an embolus. Thromboembolisms are among the leading causes of serious illness and death in Western societies.

Many factors can predispose a person to formation of a thrombus. These include irritation or infection of the lining of a blood vessel and a reduced rate of blood flow through a vessel, which may result from disease or merely from long periods of inactivity. For example, formation of a thrombus in a vein, especially of the leg, a condition known as thrombophlebitis, is particularly common in postoperative patients who remain immobilized in bed; it is also common in the elderly whose leg muscles may have lost much of their pumping action. The thrombus often leads to an inflammatory reaction and pain.

A thrombus that becomes detached from its site of formation and moves in the bloodstream as an embolus is extremely dangerous, because it may become lodged in a vessel of some essential organ, such as a lung, the heart, or the brain, and cut off its blood supply. Such emboli often lodge in the lung (pulmonary embolism) and cause death (necrosis) of a portion of the lung tissue. Except for pneumonia, pulmonary embolism is the most common acute disease of the lungs in hospitalized patients in the United States.

When an embolus (or a locally formed thrombus) blocks a blood vessel in the brain and causes necrosis of the surrounding neural tissue (owing to lack of oxygen), the condition is

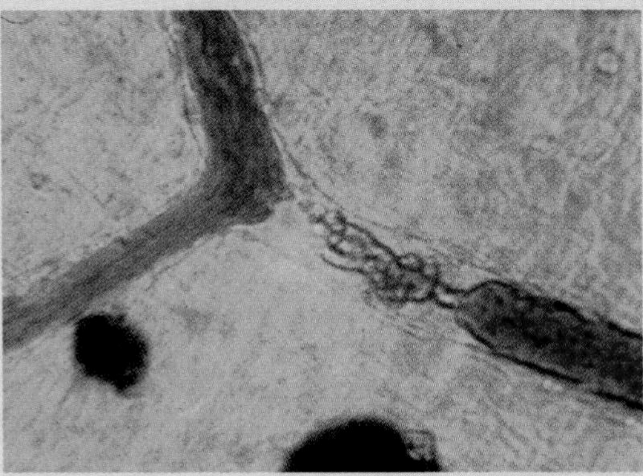

A A thrombus in a small blood vessel The thrombus (tangled red mass) has blocked blood flow near a point where the vessel branches. The blood has pulled away from the left end of the thrombus and is beginning to pull away from the right end also.

known as a stroke, or *cerebral infarction*. The symptoms of stroke vary, depending on which part of the brain has been damaged. When the infarction is in the cerebral hemispheres, there is usually some loss of muscular control in part of the body, sometimes sensory impairment, and often some loss of language ability.

The structure of the circulatory system helps guard against frequent damage from small emboli. Most parts of the body are reached by capillaries from two or more arterioles, a strategy known as collateral circulation. Hence, to cause damage an embolus must be of sufficient size to block the circulation upstream, in the relatively large blood vessel that has not yet branched into alternative pathways, or separate small emboli must block each of the branches independently. However, some parts of the body lack effective collateral circulation; these include the retina and, unfortunately, the heart.

Blockage of a blood vessel in the heart by an embolus (or by a locally formed thrombus) causes necrosis of a portion of the heart muscle, a condition familiarly known as a heart attack (more technically, as a *myocardial infarction*). A high percentage of the deaths that occur in the first few hours after a heart attack result from disruptions of the control system of the heart, with accompanying rhythmic irregularities, especially ventricular fibrillation (rapid, unsynchronized contractions that prevent effective pumping of the blood).

In the vast majority of cases of cerebral or myocardial infarction, the patient already suffers from atherosclerosis, a condition in which fatty deposits in the arteries and thickening of their walls diminish the size of the lumen (Fig. B); the reduced blood flow makes it easier for an embolus to become lodged in the vessel. Indeed, conditions caused wholly or in part by atherosclerosis are the leading cause of death in the United States; they are responsible for more deaths than the next two leading causes, cancer and accidents, combined. Medical authorities are virtually unanimous in recommending a minimum of cholesterol and saturated fat in the diet as the best way of avoiding atherosclerosis. The most reliable indicator of risk is the circulating concentration of low-density lipoprotein (LDL), the cholesterol-transport complex discussed in Chapter 4. Curiously enough, a less common cholesterol complex, high-density lipoprotein (HDL), seems actually to aid in removing cholesterol deposits.

Atherosclerosis is not the only major condition predisposing a person to a heart attack. Another common precondition is damage to the lining of the arteries due to prolonged high blood pressure (hypertension). Hypertension can also lead eventually to weakening of the heart muscle (which becomes thickened owing to the continuing strain imposed on it) and to declining efficiency of its pumping action. Blood may then back up in the heart and lungs, an often fatal condition called congestive heart failure.

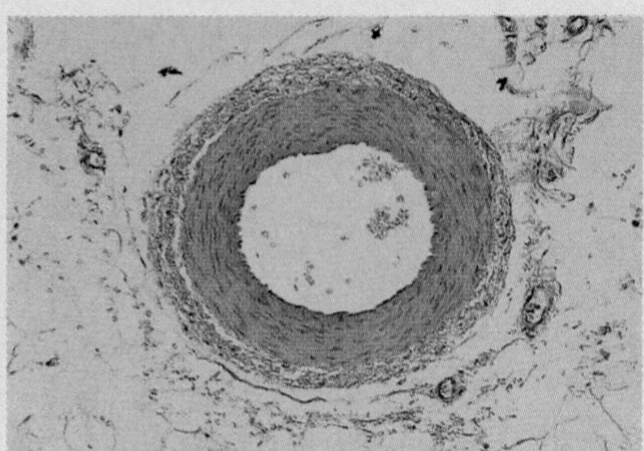

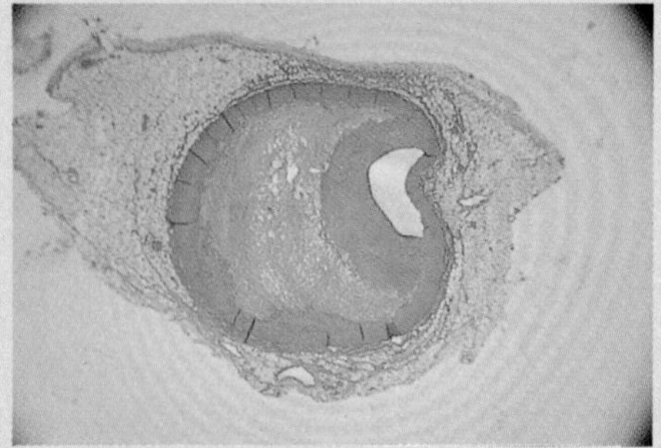

B Atherosclerosis Left: Normal artery. Right: An artery clogged with fatty deposits in the wall.

Exchange of materials between the blood in the capillaries and the tissue fluids outside the capillaries occurs in one of at least three ways: (1) the materials may move entirely by diffusion through the membrane of an endothelial cell in the wall of the capillary, across the cytoplasm of the cell, and out through the cell membrane on the other side; (2) movement may occur through large numbers of vesicles in the endothelial cells that apparently pick up materials by endocytosis on one side of the cell, move across the cell, and then expel the materials by exocytosis on the other side (Fig. 20.16); (3) spaces (clefts) between adjacent endothelial cells in the capillaries of most parts of the body (the central nervous system is an exception) permit filtration of water and most dissolved molecules, but not proteins (Fig. 20.16). Note that the second and third mechanisms—transport in vesicles and filtration between cells—do not require movement of materials through cell membranes.

Let us examine the exchange process in more detail. At the arteriole end of a representative capillary bed, the blood pressure is, on the average, about 36 mm Hg higher than the pressure of the tissue fluid outside the capillaries (Fig. 20.17). The pressure exerted by fluids, such as blood and tissue fluid, against the walls of the capillaries is called hydrostatic pressure. The source of the hydrostatic blood pressure is the pumping action of the heart. Due to the frictional resistance of the capillaries and the increased cross-sectional area, the blood pressure falls to about 15 mm Hg by the time the blood reaches the venule end of the capillary bed. The hydrostatic blood pressure tends to force materials *out* of the capillaries and into the surrounding tissue fluid. If this were the only force involved, there would be a steady loss from the blood by filtration of both water and those dissolved substances that could readily be carried by the water through the clefts in the capillary walls. However, there is relatively little net loss of water from the blood in the capillaries. Clearly, some other force acts in opposition to the hydrostatic force.

This other force derives from the difference between the

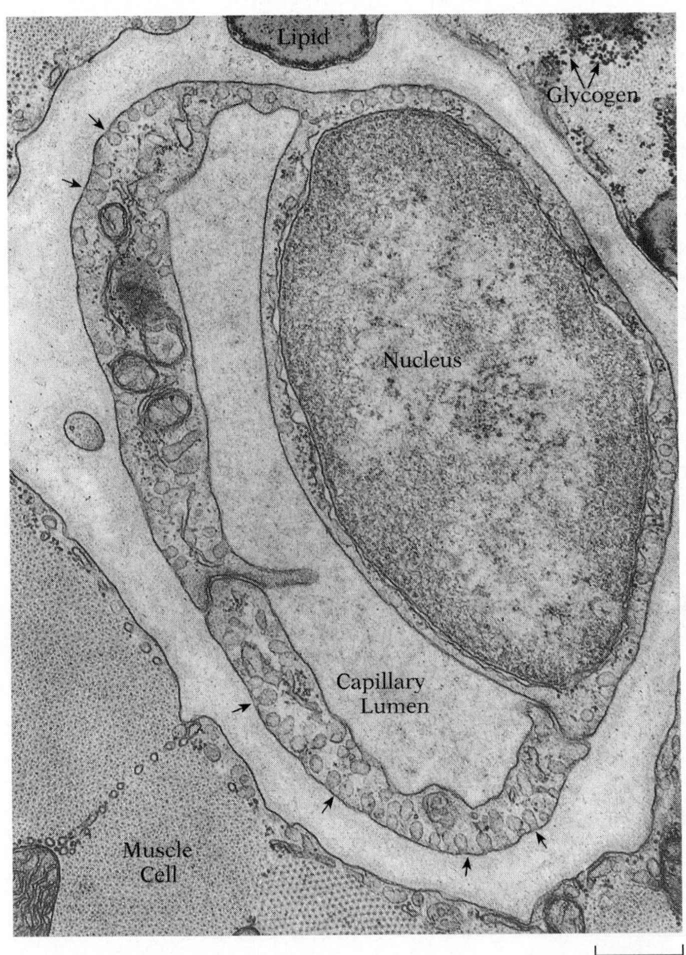

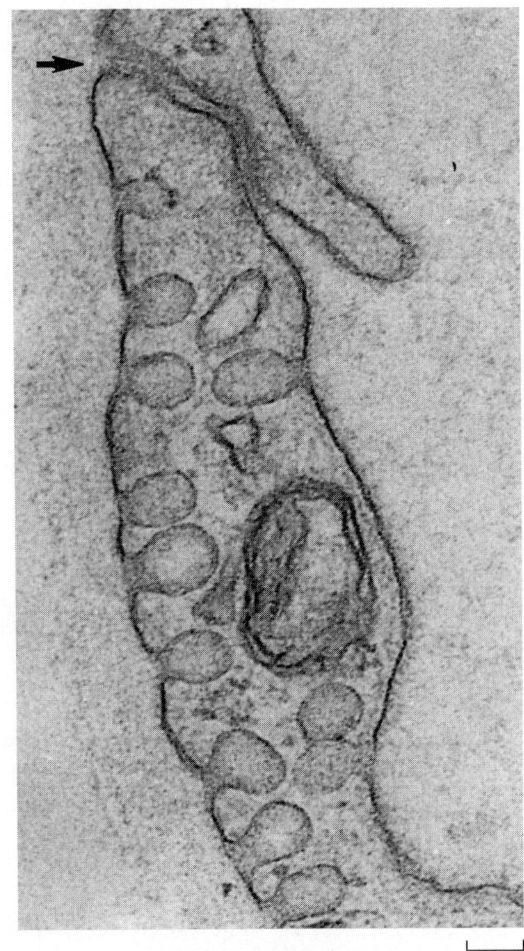

20.16 Electron micrographs of cross section of capillary Left: Two endothelial cells (the section shows the large nucleus of one of them) make up the capillary wall. Note the clefts at each of the two junctions between the cells and the numerous pinocytic vesicles (arrows) in the cytoplasm. These vesicles may transport materials from outside the capillary, across the endothelial cell, into the lumen of the capillary, or they may take the reverse route. Right: Enlarged view of wall of a capillary, showing the cleft where two endothelial cells join (arrow), as well as numerous pinocytic vesicles opening on the outer face of the lower cell.

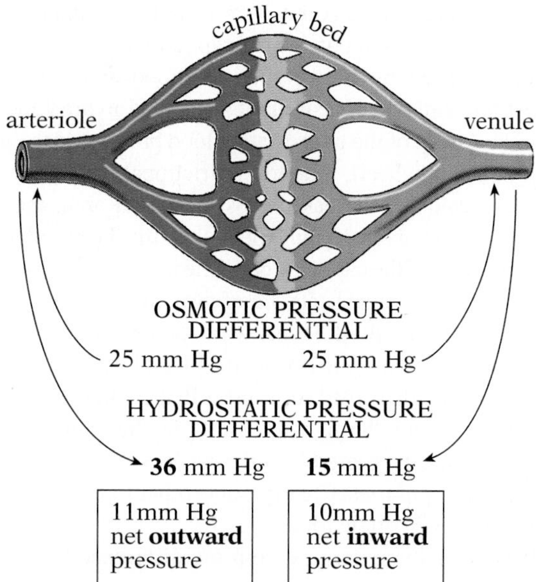

20.17 Diagram of forces involved in the filtration of materials across capillary walls The blood in the capillaries has both a greater hydrostatic and a greater osmotic pressure than the surrounding tissue fluid. The hydrostatic pressure differential tends to force water and dissolved materials out of the capillaries into the tissue fluid; the osmotic pressure differential has the opposite effect, causing the capillaries to take up water and dissolved materials from the tissue fluid. At the arteriole end of a characteristic capillary bed, the hydrostatic pressure differential of 36 mm Hg is greater than the osmotic pressure differential of 25 mm Hg; the difference of 11 mm Hg favors the outflow of materials. At the venule end, the osmotic pressure differential, which remains 25 mm Hg, is greater than the hydrostatic pressure differential, which has dropped to 15 mm Hg; here the difference of 10 mm Hg favors the movement of materials from the tissue fluid back into the blood.

osmotic concentrations of the blood and the tissue fluid. The blood of mammals contains a relatively high concentration of proteins (7 to 9 g/100 ml), and these large molecules cannot easily pass through the capillary walls. The same kinds of proteins occur in the tissue fluids, but in much lower concentration (2 g/100 ml). Because of the different protein concentrations on the two sides of the capillary wall, the blood and tissue fluids have different osmotic pressures. Normally, the osmotic pressure of the blood is about 25 mm Hg higher than that of the tissue fluid. Thus, the blood is hypertonic to the tissue fluid, with the result that water tends to move *into* the capillaries from the tissue fluid by osmosis.

We have, then, a system in which hydrostatic blood pressure, exerted by the heart, tends to force water *out* of the capillaries and osmotic pressure, reflecting a difference in protein concentrations, tends to draw water *into* the capillaries. The net movement of water will be determined by the relative magnitudes of these two opposing forces. Notice that at the arteriole end of our representative capillary bed the hydrostatic pressure differential is 36 mm Hg and the osmotic pressure differential is 25 mm Hg. Subtracting one from the other, we find that there is a net pressure of 11 mm Hg tending to force water out of the capillaries. At the venule end of the capillary bed, the hydrostatic pressure differential has fallen to 15 mm Hg, while the osmotic pressure differential has not changed greatly.[1] There is, therefore, now a net pressure of at least 10 mm Hg tending to draw water into the capillaries. In summary, the balance between hydrostatic blood pressure and osmotic pressure is such that water is forced out of the capillaries at the arteriole end and into the capillaries at the venule end. Since the water carries

[1] Loss of water from the blood has, of course, slightly increased the concentration of protein in the blood and raised the osmotic pressure accordingly, but this change is relatively slight and can be ignored for our purposes here.

MAKING CONNECTIONS

1. Individuals who are required to stand on their feet for long periods of time may find that their feet and ankles swell. Pregnant women, in particular, often experience this problem. What is the basis of this, and why should it be a common complaint of pregnancy?

2. Why are surgical patients in a hospital encouraged to begin walking as soon as possible after their operation?

3. What role is played by diffusion in the exchange of materials across capillary walls? By exocytosis and endocytosis?

4. One of the symptoms of kwashiorkor, the protein-deficiency disease discussed in Chapter 17, is swelling of the abdomen and legs due to edema. How are these symptoms related to the process of capillary exchange?

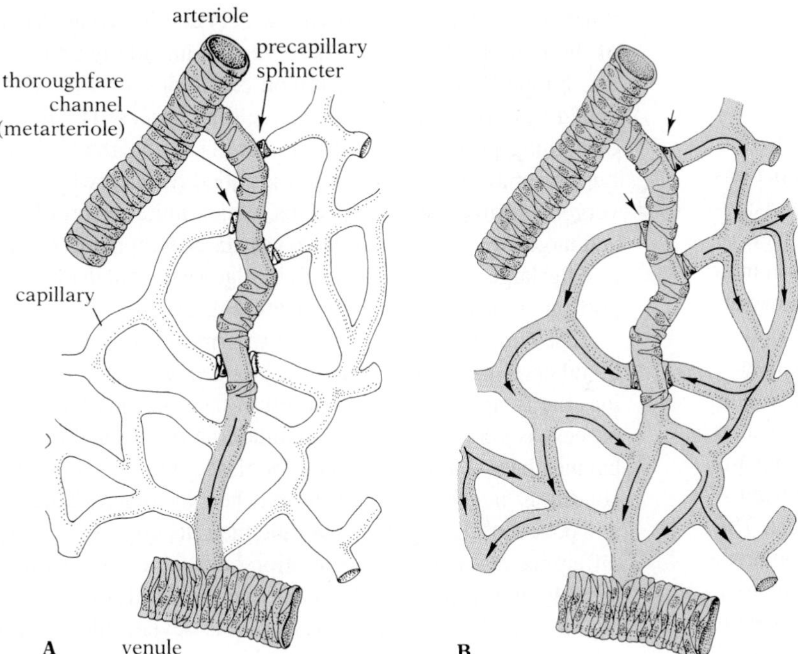

20.18 The vessels of a capillary bed When the precapillary sphincters are closed (A), blood flows only through the capillaries known as thoroughfare channels. When the sphincters are open (B), blood flows through all the capillaries.

with it molecules of many dissolved substances, we can say that the blood in the capillaries first unloads materials for the tissues at the arteriole end and then picks up materials for transport at the venule end. In the process, there is normally only a slight net loss of water from the blood.

The balance of hydrostatic and osmotic pressures in the capillaries is very delicate. Since this balance plays such an important role in the exchange of materials between the blood and tissue fluid, disturbing it may have profound effects on the organism. For example, an increase in hydrostatic blood pressure in a capillary would tend to increase loss of fluid from the blood, while a decrease would have the opposite effect. Such changes in blood pressure could be produced by a variety of factors, such as changes in the rate or strength of heart action, increase or decrease in total blood volume, changes in the elasticity of the walls of the arteries, or increased dilation or constriction of arterioles and capillary sphincters. The capillaries of the body are never fully open all at the same time; many capillaries are usually closed by constriction of rings of sphincter muscle at their bases (Fig. 20.18). The opening and closing of sphincters in different tissues allow blood to be redistributed to areas in the body that require a greater flow of blood at that time, e.g., skeletal muscles during exercise or the digestive tract during a meal. Increased flow through skin capillaries often gives the skin a reddish hue, seen in blushing, while constriction of the sphincters controlling these same capillaries gives the skin a bleached, whitish look. A major form of heat loss from the body is by radiation from the blood in the superficial capillaries of the skin (Fig. 20.19); as we will see, changes in the amount of blood flow in these capillaries are an important factor in helping to regulate heat loss. In general, the more active the tissue, the more capillary sphincters are open, and the greater the flow of blood through that tissue.

Changes in the relative concentrations of proteins in the blood and in the tissue fluids can also severely alter the balance of forces operating in the capillaries. Numerous experiments have been performed in which the protein concentration in the blood supply to a limb of a frog, cat, or dog was artificially regulated. As predicted, increasing the protein concentration in the blood decreases the loss of fluids from the blood and increases absorption from the tissue fluid. Conversely, decreasing the protein concentration in the blood increases the loss of fluids from the blood and decreases reabsorption of fluid from the tissues, the result being an abnormal accumulation of fluid in the tissues that causes swelling, a condition known as edema.

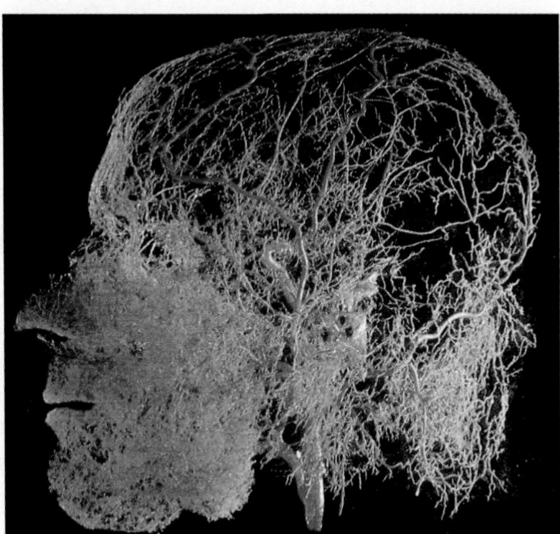

20.19 Blood vascular network in the human head The human face, particularly the area around the lips, contains a dense array of capillaries.

The Circulatory System Plays a Critical Role in Heating and Cooling the Body

Temperature control is especially important for endotherms, which must maintain an optimum temperature for proper enzyme function. Temperature regulation systems are often concentrated in localized regions of the body, and warmed or cooled blood is then pumped through the circulatory system to change the body temperature. In other cases the regulatory efforts are aimed at specific structures.

Endotherms are "internally heated," that is, they are heated by the exergonic reactions of their metabolism. The body temperature of such animals is fairly high—usually higher than the environmental temperature. When environmental temperature drops much below body temperature, shivering begins. Shivering is a way of generating "waste" metabolic heat. The result is a heating of the blood passing through the capillaries serving the muscles; this warmth is carried by the circulatory system to the rest of the body. But when the environmental temperature is low, some heat loss is inevitable. Insulation—fur, feathers, and fat—can minimize but not altogether stop heat loss, and it can create problems on warm days or during heavy exercise when excess heat must be disposed of.

Most animals cool themselves by evaporating water. For humans this means sweating. Blood is shunted to the skin where it can radiate heat directly into the air and simultaneously benefit from evaporative cooling. In species covered with insulating fur or feathers, however, evaporation must be more localized. The most frequently used surface in these cases is the tongue. The familiar panting of dogs provides an excellent illustration. During panting, the outgoing air is blown across the wet tongue, carrying with it air from the sides of the mouth and evaporating water. As a result, the blood in the capillaries of the mouth is cooled.

THE LYMPHATIC SYSTEM

The Lymphatic System Returns Materials to the Blood

We have seen that most of the fluid that leaves the capillaries at the arteriole end is normally reabsorbed at the venule end. But what about the remaining fluid? Can it return to the blood by any means other than direct reabsorption into the blood capillaries? The answer is yes. Vertebrates have a special system, the lymphatic system, which functions to return water and dissolved materials from the tissues to the blood. Approximately two to four liters of lymph are returned during a 24-hour period. The lymphatic system consists of an extensive network of thin vessels (lymphatics) that are widely distributed to all parts of the body (Fig. 20.20). These vessels include lymph veins and lymph capillaries. The lymph capillaries are tiny, blind-ended vessels located in the intercellular spaces. Tissue fluid containing proteins and other materials is absorbed into the lymph capillaries (whereupon it is called ***lymph***), which converge to form small lymph veins, which unite to form larger and larger veins until finally two very large lymph ducts empty into the large veins of the blood circulatory system in the upper portion of the thorax, near the heart (Fig. 20.20).

The lymph capillaries are highly permeable to proteins; any blood proteins that leak out of the blood capillaries can diffuse into the lymph vessels, which return them to the blood. This process is very important in maintaining the normal osmotic balance between the blood and the tissue fluid. Under certain conditions major lymph vessels may become blocked; the protein concentration in the tissue fluid then steadily rises, and the difference in osmotic concentration between it and the blood steadily diminishes, which means that less fluid is reabsorbed by the blood capillaries. The result is severe edema (Fig. 20.21).

The lymphatic system performs many other functions besides returning excess tissue fluid and proteins to the blood. It plays a major role in the immune response, as we shall see in Chapter 21. And in Chapter 17 we learned that the fat absorbed from the intestine is picked up by lymph vessels in the villi rather than by blood capillaries. Absorption of fats thus differs from the absorption of sugars and amino acids, which are picked up by blood capillaries.

Lymph nodes are present in the lymphatic systems of mammals and some birds, but are absent in most other vertebrates. Located along major lymph vessels and composed of a meshwork of connective tissue harboring many phagocytic white blood cells, they act as filters and are sites for certain types of white blood cells to monitor the passing fluid for signs of infection. As the lymph trickles through the nodes, it is filtered, and such particles as dead cells, cell fragments, cancer cells, and invading bacteria are trapped and many destroyed by the phagocytic cells. Cancer cells often travel through the lymphatics and may become lodged in the nodes, forming new tumors if not destroyed. For this reason the lymph nodes near a malignant tumor are often surgically removed when the cancer is removed. Nondigestible particles such as dust and soot, which the phagocytic cells cannot destroy, are stored in the nodes. Since the nodes are particularly active during an infection, they often become swollen and sore, as the lymph nodes at the base of the jaw are apt to during a throat infection.

Since the lymphatic system is not connected to the arterial portion of the blood circulatory system, the lymph does not move by pressure developed by the heart. Its movement, like that of blood in the veins, results from changes in pressure induced by breathing movements, and from contractions of skeletal muscles that press on the thin-walled lymph vessels and push the lymph forward past one-way valves (Fig. 20.22).

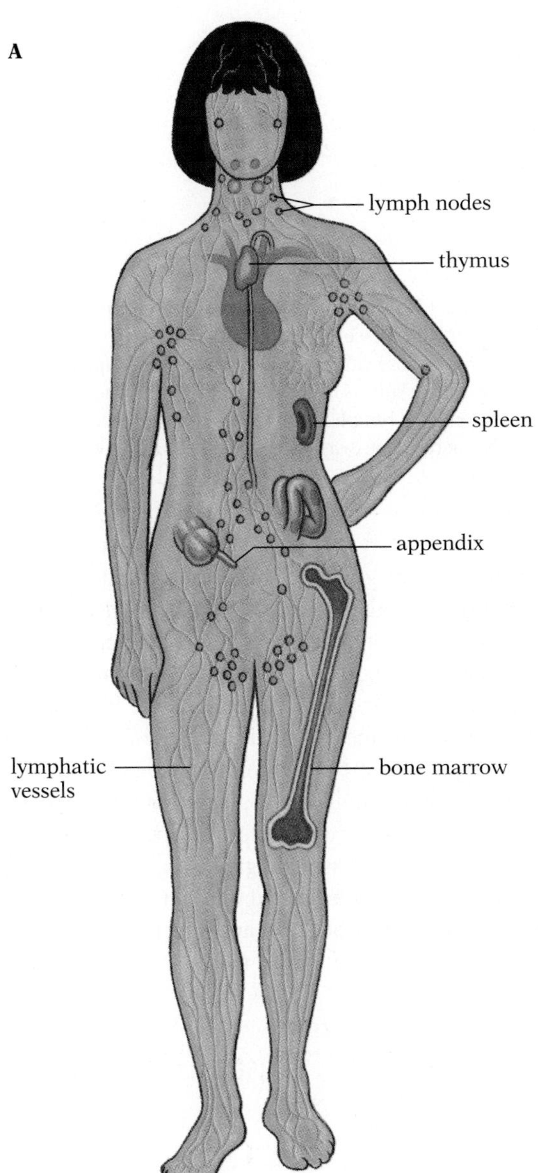

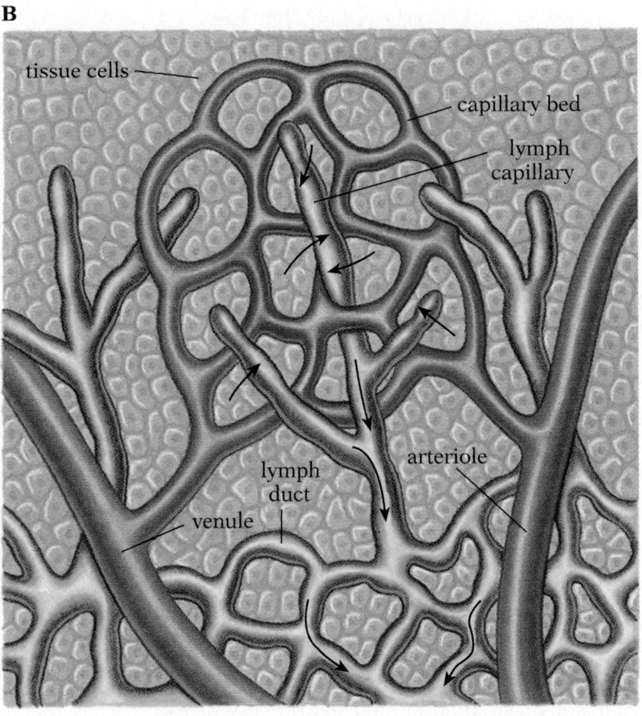

20.20 The human lymphatic system (A) The lymphatic system consists of a series of vessels that drain excess tissue fluid from the tissues and return it to the blood circulatory system. The primary organs of the lymphatic system, the thymus and bone marrow (only one bone is shown), produce lymphocytes that circulate from extremely fine vessels, called lymph capillaries, into the lymphatic system. The lymph capillaries are blind-ended tubes that form networks within the intercellular spaces of most organs (B). They are close to the blood capillaries and absorb the excess tissue fluid that has been filtered out of the blood capillaries. The lymph capillaries merge to form larger lymphatic vessels, which eventually unite to form two large ducts that empty into the blood circulatory system near the heart. Located along the lymphatic vessels are lymph nodes, which contain phagocytic cells.

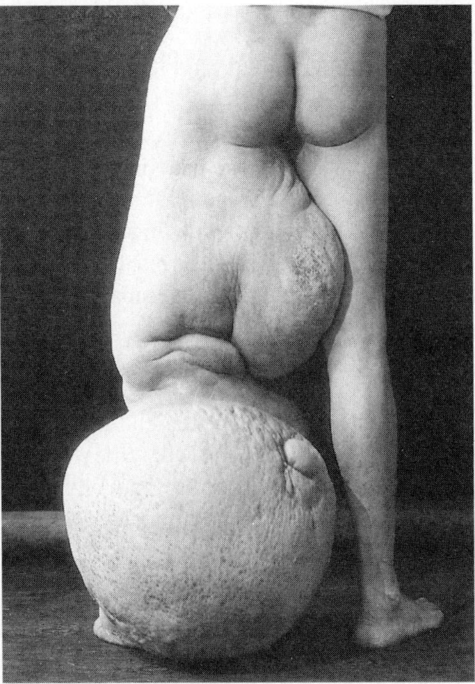

20.21 Elephantiasis Elephantiasis, which occurs in tropical and subtropical regions throughout the world, is a condition of extreme edema that occurs when lymph vessels become blocked by microfilaria worms. Here the left leg is swollen with the fluids accumulated in the tissues as a result of the blockage.

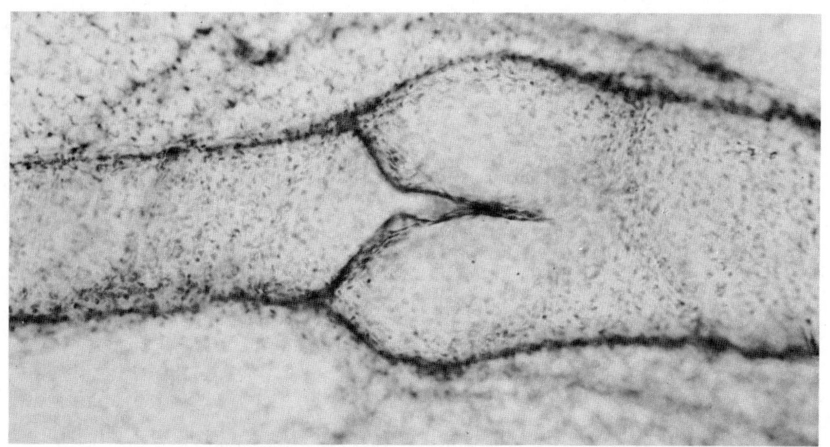

20.22 Photograph of a valve in a lymph vessel Lymphatic vessels are thinner walled than veins and contain more valves. Movement in the lymphatic vessels occurs as nearby skeletal muscles contract, forcing the lymph forward through the vessels. Numerous one-way valves prevent backflow.

BLOOD

We have so far discussed the routes followed by the circulating blood, the mechanism of circulation, and the process of exchange of certain materials with the tissue fluid bathing cells. We now examine the blood itself in more detail. It is one of the most important and unusual tissues in the animal body.

Human Blood Consists of Plasma and Cells

In Chapter 15 we classified blood as a type of connective tissue with a liquid matrix. The extracellular liquid matrix of blood is called ***plasma***. Suspended in the plasma are the cellular elements, which are of three major types in vertebrates: (1) the red blood cells, or ***erythrocytes;*** (2) the white blood cells, or ***leukocytes;*** and (3) the ***platelets,***[2] which are small disk-shaped fragments of cells. All three types of cells are derived from special connective tissue cells called ***stem cells,*** which are located in the red bone marrow in adults (Fig. 20.23).

If whole blood, treated to prevent clotting, is left standing in a test tube, the cellular elements will settle slowly to the bottom, leaving the fluid plasma above. Normally, the cells constitute about 40 to 50 percent of the volume of whole blood, while the plasma constitutes the other 50 to 60 percent.

The basic solvent of the plasma is water, which constitutes roughly 90 percent of the plasma. A great variety of substances are dissolved in the water; the relative concentrations of these vary with the activity of the organism and from one portion of the system to another (e.g., plasma in an artery after a meal will contain higher levels of glucose, amino acids, and fats than plasma in a vein). For convenience, let us divide these solutes into six categories: (1) inorganic ions and salts, which make up about 0.9 percent by weight of the plasma of mammals; (2) plasma proteins (e.g., albumin, globulins, and fibrinogen), which are synthesized by the liver and constitute 7 to 9 percent by weight of the plasma; (3) organic nutrients, including glucose, fats, phospholipids, amino acids, lactic acid, and cholesterol; (4) nitrogenous waste products; (5) special products, such as hormones; and (6) dissolved gases, including oxygen, carbon dioxide from metabolism, and nitrogen, which diffuse into the blood in the lungs but appear to be physiologically inert.

[2]True platelets are found only in mammals. The blood of most other vertebrates contains cells called thrombocytes, which function in blood clotting in a manner similar to platelets.

Blood Clots Form When Fibrinogen Is Converted Into Fibrin

Blood clotting is an evolutionary adaptation for emergency repair of the circulatory system, and for preventing excessive loss of body fluids when blood vessels are damaged. The immediate response of a blood vessel to an injury is to constrict, slowing the flow of blood. Platelets in the area also adhere to the damaged tissue and to each other, forming a loose plug of aggregated platelets. The platelet plug may slow or even stop the bleeding from the damaged vessel, but it is easily dislodged. The plug is stabilized by the formation of a blood clot—a meshwork of threads that forms around the platelets on the damaged tissue (Fig. 20.24). Other blood cells may become enmeshed in the threads, strengthening the clot. The threads are composed of the protein ***fibrin***. The fibrin threads are formed during the clotting process, when the soluble plasma protein ***fibrinogen*** is converted into fibrin, which is insoluble. Though the process is complex and consists of a whole series of reactions, for simplicity it can be condensed into two summary reactions:

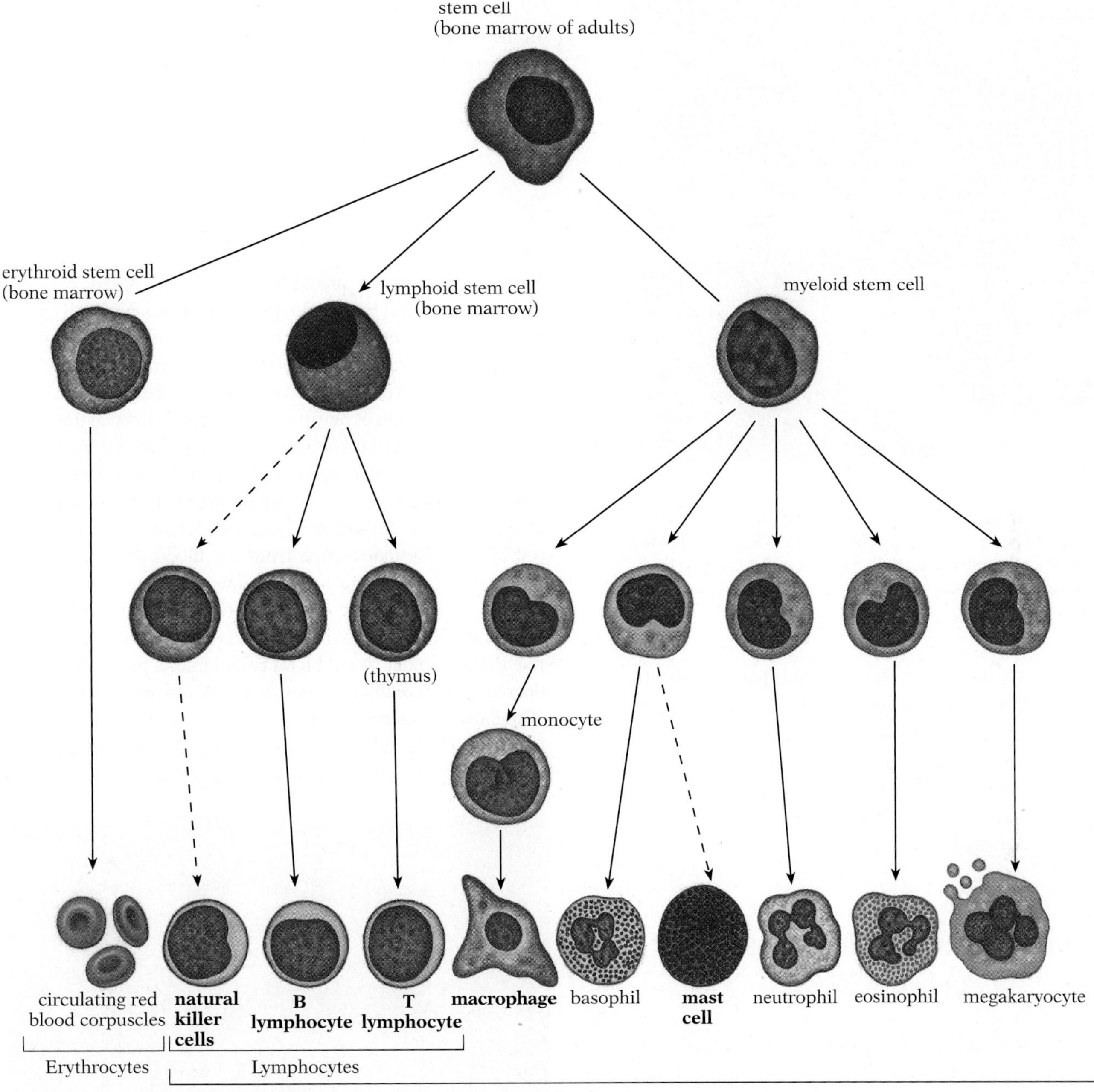

20.23 The origin of blood cells Uncommitted stem cells give rise to at least three kinds of more highly determined stem cells: erythroid stem cells, lymphoid stem cells, and myeloid cells. All four of these stem cells have the ability to regenerate themselves. Each of the three more highly determined stem cells in turn gives rise to more specialized cells: erythroid stem cells produce red blood corpuscles; lymphoid stem cells give rise to three yet more highly determined cell types, which in turn produce the natural killer cells, B lymphocytes, and T lymphocytes (which we will discuss further in Chapter 21) and myeloid stem cells produce at least five classes of more highly determined cells.

$$\text{(1)} \quad \text{prothrombin} \xrightarrow[\text{Ca}^{2+}\text{ ions}]{\text{thromboplastin}} \text{thrombin}$$

$$\text{(2)} \quad \text{fibrinogen} \xrightarrow{\text{thrombin}} \text{fibrin}$$

The process begins when the surfaces of the damaged vessel release a substance known as ***thromboplastin,*** which combines with other blood proteins to form an activated complex. This complex converts the plasma protein ***prothrombin*** into ***thrombin.*** Calcium ions and a specific phospholipid on the surface of platelets are required for this reaction to occur. In the last step in the sequence, thrombin converts the fibrinogen into fibrin.

Unfortunately, a lack of any one of the required compo-

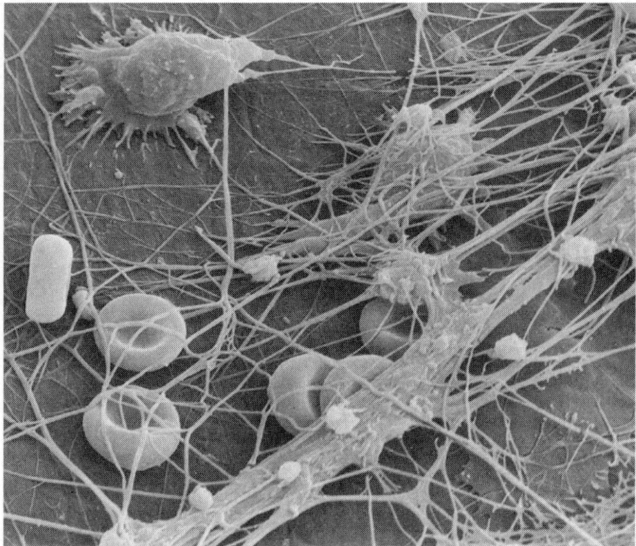

20.24 Blood clotting Blood clots as the various components of the blood become trapped in a network of fibrin fibers. Several red blood cells, which look like concave disks, are visible in this electron micrograph. A large mobile phagocytic white blood cell (macrophage) can be seen at the upper left.

nents can block the entire series of reactions. For example, individuals with the clotting disorder hemophilia (see p. 295) produce defective factor VIII, which is one of the blood proteins necessary to form the activated complex that converts prothrombin into thrombin. Without this protein blood does not clot properly. As a consequence, even small cuts or bruises may be life threatening for hemophiliacs.

The Leukocytes Defend the Body Against Toxins and Foreign Cells

Human white blood cells, or ***leukocytes***, have large, often irregularly shaped nuclei (Fig. 20.23). They are produced from specific stem cells in the bone marrow and are released into circulation. Leukocytes are present in the blood, but are even more abundant in the lymphatic system. They are also capable of traveling freely in loose connective tissues and occasionally in other tissues, such as the wall of the digestive tract. Some are capable of amoeboid movement, and can escape from the blood and lymph vessels by squeezing through the vessel walls at the junctions between the endothelial cells. They act as the body's first line of defense against invading bacteria. In essence, the white blood cells travel in one continuous system comprising the blood, lymph, and other connective tissues. The various white blood cells play very important roles in the body's defense against disease-causing microorganisms, as we shall see in the next chapter.

Malignancies of the various leukocytes do occur, resulting in the different leukemias. Such patients may have enormous numbers of neutrophils, monocytes, or lymphocytes, but are prone to infection because the cells are immature and do not function normally.

The Erythrocytes Carry Oxygen to the Tissues

Human erythrocytes, or red blood cells, are small, concave, disk-shaped cells (Fig. 20.25) that lack nuclei. Normally, there are roughly five million of them per cubic millimeter of blood. Though the number of red blood cells remains amazingly constant from day to day, there is continual destruction of some cells and formation of new ones; the normal survival time of an erythrocyte is 120 days. More than two million red blood cells are destroyed every second, chiefly in the liver and the spleen, where they are engulfed by large phagocytic cells. There are also phagocytic cells in the lymph nodes, which destroy any red blood cells that escape from the blood and get into the lymph. Obviously some type of homeostatic mechanism is operating to maintain the balance between production and destruction. The mechanism involves feedback control, i.e., a rise in the number of red blood cells inhibits production and a decrease in the number stimulates it. A low oxygen level also stimulates production; individuals living at high altitudes have more red blood cells than those living at low altitudes. Red blood cell production is regulated by a hormone called ***ery-***

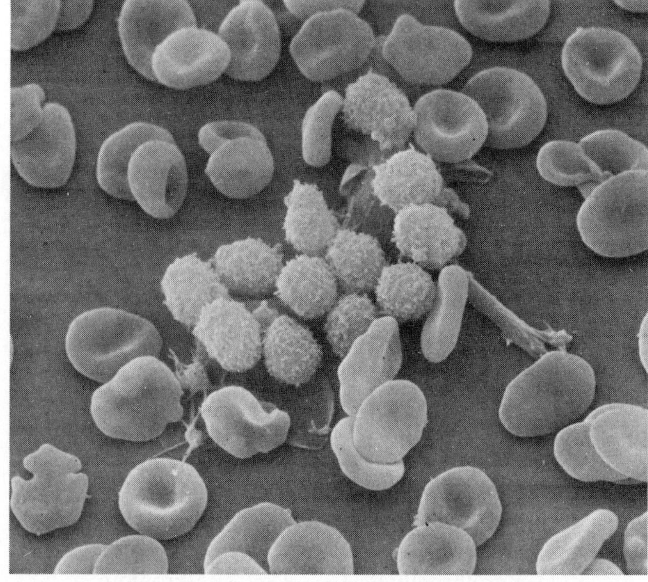

20.25 Scanning electron micrograph of blood cells and corpuscles The red blood cells look like biconcave disks. In the center is a cluster of lymphocytes, a type of white blood cell important in immunologic defense against diseases. A large phagocytic macrophage can be seen under the lymphocytes.

thropoietin, which is secreted primarily by the kidneys in response to low oxygen levels. The hormone stimulates the bone marrow to increase red cell production.

In the adult, the erythrocytes arise from special, rapidly dividing stem cells in the bone marrow (Fig. 20.23). The immature cells have nuclei, mitochondria, Golgi apparatus, etc., but toward the end of their development they lose their nuclei and other organelles, and accumulate the red oxygen-carrying pigment ***hemoglobin,*** a protein with iron-containing prosthetic groups. They then enter the circulating blood. In the red blood cells of vertebrates other than mammals the nuclei are retained. It has been suggested that the evolutionary loss of the nuclei in mammals has the adaptive advantage of leaving room for more hemoglobin in each cell. Extra hemoglobin might, in turn, be correlated with the high metabolic rates, and therefore high oxygen demands, of the tissues of endothermic animals.

Some invertebrates have hemoglobin as well. In those that do, it is sometimes in special cells, as in vertebrates, but in many it is simply dissolved in the plasma. Though hemoglobin molecules can function just as well in the plasma as in red blood cells, their location in cells has a decided adaptive advantage in animals with high metabolic rates; more pigment molecules can then be carried per unit volume of blood, and the oxygen-transporting capacity is correspondingly increased. A single human erythrocyte usually contains about 280 million molecules of hemoglobin. If all this hemoglobin were in the plasma, the concentration of plasma protein would be about three times higher, with profound effects on the osmotic balance between the blood and the tissue fluid. Red blood cells, then, are a convenient method of packaging large amounts of hemoglobin with relatively little disturbance of the osmotic concentration of the blood, and of providing a chemical environment especially favorable for hemoglobin function.

Many of the invertebrates that lack hemoglobin have other oxygen-transporting pigments, which, like hemoglobin, combine a metal with protein. For example, many molluscs and arthropods have a pigment called hemocyanin, which contains copper instead of iron; when oxygenated, it is blue instead of red. Hemocyanin never occurs in cells, but is dissolved in the plasma. Such invertebrates, being ectotherms, do not sustain the high metabolic rates of endotherms such as birds and mammals. Consequently, their requirement for oxygen is lower, and they do not require large amounts of hemocyanin packaged into cells.

Each Hemoglobin Molecule Can Transport Four Oxygen Molecules

The hemoglobin molecule is a globulin protein composed of four independent polypeptide chains (see Fig. 3.28 p. 69). Each of the four chains enfolds a complex prosthetic group called ***heme,*** which has an iron atom at its center (Fig. 20.26). Each of the four iron atoms in a hemoglobin molecule can combine loosely with one molecule of oxygen. In the lungs the compound formed by the union of one molecule of hemoglobin (Hb for short) with four molecules of oxygen is called ***oxyhemoglobin*** ($Hb{:}4O_2$). Hemoglobin picks up oxygen (forming oxyhemoglobin) when there is a relatively high concentration of oxygen in the surrounding medium, as in the lungs. It releases its oxygen in the capillaries of the systemic circulation, where the concentration of oxygen is relatively low. The formation of oxyhemoglobin makes the blood redder; it is on account of the oxyhemoglobin that arterial blood in the systemic circulation is more crimson than venous blood.

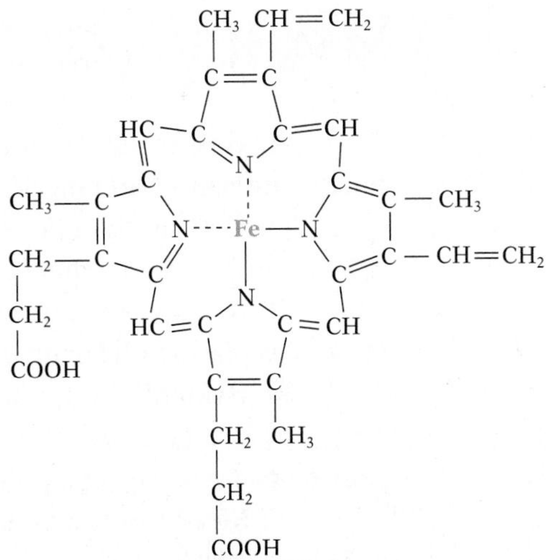

20.26 Structure of the heme group A single molecule of hemoglobin has four of these iron-containing prosthetic groups and can thus carry four 0_2 molecules.

There are other gases besides oxygen that will bind loosely to hemoglobin. One that binds even more readily than oxygen is carbon monoxide (CO). This gas, common in coal gas used for heating and cooking, in the exhaust from automobiles, and in tobacco smoke, is a dangerous poison, because even when its concentration in the air is relatively low, such a high percentage of the hemoglobin may bind with it that not enough is left to carry sufficient oxygen to the tissues. Severe symptoms of asphyxiation (impairment of vision, hearing, and thought) or even death may result from exposure to carbon monoxide.

The ability of the blood to transport oxygen depends on the number of erythrocytes in a given volume, on their size, and on the quantity of hemoglobin in each. The condition known as ***anemia*** results when the total blood hemoglobin is low. It can be due to decreased number of red blood cells (each with the normal amount of hemoglobin), or to low levels of hemoglobin within each cell. Thus, anemia can result from a number of causes, such as blood loss, dietary deficiencies of iron and certain vitamins, the formation of abnormal cells and hemoglobin (as in sickle-cell anemia), and damage to the red bone marrow due to disease, radiation, or toxic chemicals.

MAKING CONNECTIONS

1. How is the lymphatic system involved in the absorption and transport of nutrients from the digestive tract, as discussed in Chapter 17?

2. Why are lymph nodes located near a malignant tumor often surgically removed when a cancer is removed?

3. In Chapter 2, water's superb qualities as a solvent were described. How does plasma illustrate this?

4. Hemophilia, a sex-linked disease discussed in Chapter 14 (see p. 295), causes a failure of the blood to clot properly. Which specific step in the clotting process are hemophiliacs unable to carry out?

5. Describe the homeostatic mechanism that controls the production of red blood cells.

6. How would the osmotic balance of the blood and tissue fluids be affected if hemoglobin were free in the plasma instead of enclosed in red blood cells? Why is it not necessary for hemocyanin, the oxygen-transporting pigment found in many invertebrates, to be contained in cells?

7. How is the reaction of carbon monoxide with hemoglobin related to the action of enzyme inhibitors discussed in Chapter 4 (see p. 85)?

Most of the Carbon Dioxide Is Transported as Bicarbonate Ions

The blood not only transports oxygen from the lungs to the tissues but also has the very important function of transporting carbon dioxide in the reverse direction, from the tissues to the lungs. Some of this carbon dioxide is carried as gas dissolved in the plasma and some in loose combination with hemoglobin in the red blood cells, but most of it is carried as bicarbonate ions in the red blood cells and plasma.

The carbon dioxide released from the tissue cells combines readily with water in the presence of an enzyme found in red blood cells to form carbonic acid (H_2CO_3), most of which ionizes into hydrogen (H^+) and bicarbonate (HCO_3^-) ions:

$$CO_2 + H_2O \xrightarrow{\text{enzyme}} H_2CO_3 \longrightarrow H^+ + HCO_3^-$$

Most of the carbon dioxide is thus transported as bicarbonate ions, and the hydrogen ions are combined with hemoglobin and other plasma proteins. This is an example of the buffering action of proteins, without which the pH of the blood would drop to life-threatening levels. In the lungs, where carbon dioxide pressure is low, the reactions shown above are reversed, and carbon dioxide moves from the blood into the alveoli.

CONCEPTS IN BRIEF

1. Most animals, with their active way of life and high metabolic rates, rapidly consume energy and therefore require an almost continuous supply of nutrients and oxygen. One of the important functions of a transport system is to help maintain the *homeostasis* of the cells and extracellular fluids of the body by delivering materials to the tissues, and removing the accumulated waste products.

2. Most complex animals have a true circulatory system in which blood is moved around the body along a fairly definite path. *Closed circulatory systems,* in which the

blood is always within vessels, are characteristic of earthworms and vertebrates. *Open circulatory systems,* in which the blood moves freely through tissue spaces, are characteristic of most molluscs and all arthropods.

3. The open circulatory system of insects consists of a dorsal longitudinal vessel, or heart, which contracts, forcing the blood out at the anterior end. The blood then flows backward through large sinuses and finally reenters the dorsal vessel.

4. The closed circulatory system of vertebrates consists of a *heart* and *arteries, veins,* and *capillaries*. The actual exchange of materials between blood and other tissues takes place in the capillaries.

5. The human heart is a double pump; each side is divided into two chambers: an upper *atrium,* which receives the blood and pumps it into the lower chamber, and a lower *ventricle,* which then pumps the blood away from the heart to the various parts of the body.

6. The right heart receives deoxygenated blood from all over the body and pumps it via the *pulmonary arteries* to the lungs, where it picks up oxygen and gives up carbon dioxide. The oxygenated blood then returns to the left atrium by the *pulmonary veins.* This portion of the circulatory system is called the *pulmonary circulation.*

7. The left ventricle pumps the blood into the *aorta* and its numerous branches, from which it moves into capillaries, where the exchange of materials takes place, then into venules and into veins, and finally back via the *anterior* or *posterior vena cava* to the right side of the heart. This portion of the circulatory system is called the *systemic circulation.*

8. The heartbeat is initiated when a wave of contraction spreads out from the *S-A node* in the right atrium to the *A-V node,* which sends excitatory impulses down the *bundle of His* to the *Purkinje fibers,* stimulating both ventricles to contract simultaneously. During *systole* (ventricular contraction) the blood is forced out of the heart and into the arteries under high pressure. During *diastole* (ventricular relaxation) the blood pressure in the arteries falls. Breathing movements, one-way valves and skeletal-muscle action aid in moving blood in the veins back to the heart.

9. The movement of materials into and out of the capillaries is governed by the balance between hydrostatic blood pressure and osmotic pressure; water and dissolved materials are forced out of the capillaries at the arteriole end and into the capillaries at the venule end.

10. The *lymphatic system* helps maintain the osmotic balance of the body fluids by returning excess tissue fluid and proteins to the blood. Lymph nodes act to filter out particles and also as sites of formation of some white blood cells.

11. Blood consists of *plasma* (the liquid portion) and *cellular elements* (the *red blood cells, white blood cells,* and *platelets*). Blood clotting is initiated when damaged tissue and disintegrating platelets release *thromboplastin*, which helps convert the protein *prothrombin* into *thrombin*. The thrombin then converts *fibrinogen* into *fibrin*, which forms the clot.

12. The *leukocytes* defend the body against disease and infection. Some leukocytes carry on phagocytosis, others produce proteins that detoxify dangerous substances, and still others produce *antibodies* that destroy or inactivate certain kinds of foreign substances called *antigens.*

13. The *red blood cells* contain the oxygen-carrying pigment *hemoglobin,* which transports oxygen from the lungs to the tissues. Most carbon dioxide is carried from the tissues to the lungs in the form of the bicarbonate ion.

STUDY QUESTIONS

1. Compare open and closed types of circulatory systems. In general, which is better adapted to the needs of an active animal with a high metabolic rate? Describe the open system of insects. (p. 409)
2. Are the vessels that carry deoxygenated blood from the heart to the lungs considered to be arteries or veins? What about the vessels that carry oxygenated blood from the lungs to the heart? (p. 409)
3. A red blood cell enters the right atrium. Trace its route through the pulmonary and systemic circulatory systems until it returns to the same point, mentioning all the chambers of the heart and all major arteries and veins. (pp. 409–412)
4. Describe the sequence of events occurring during the contraction of the human heart. Where does the stimulus for contraction originate? What causes the sounds associated with the heart beat? How is the heart rate controlled? (pp. 412–415)
5. You have your blood pressure taken and are told that it is 118/75 (mm Hg). What do these numbers mean? Why would the pressure be lower in a small artery or arteriole than in a major artery in your arm? Where

in the circulatory system would the pressure be the lowest? (pp. 415–417)

6. **What mechanisms other than blood pressure aid the venous return of blood to the heart? Why are the walls of veins relatively thin (compared to arterial walls)? (pp. 417–418)**
7. **Explain how the balance of hydrostatic (blood) and osmotic pressures governs the movement of materials in and out of capillaries. (pp. 418, 421–423)**
8. **Describe the role of the circulatory system in heating and cooling the body. Why is a dog's panting an efficient cooling mechanism? (p. 424)**
9. **What are the components of the lymphatic system? What are the major functions of this system? (p. 424)**
10. **Describe the composition of the blood. What are its three cellular elements and the basic function of each? What are the six categories of plasma solutes? (p. 426)**
11. **Suppose you just got a paper cut in turning to this page? What steps will your body go through to prevent you from bleeding excessively? (pp. 426–428)**
12. **Compare the mechanisms by which oxygen and carbon dioxide are carried in the blood. (pp. 429–430)**

CHAPTER 21

DEFENSE OF THE HUMAN BODY: THE IMMUNE SYSTEM

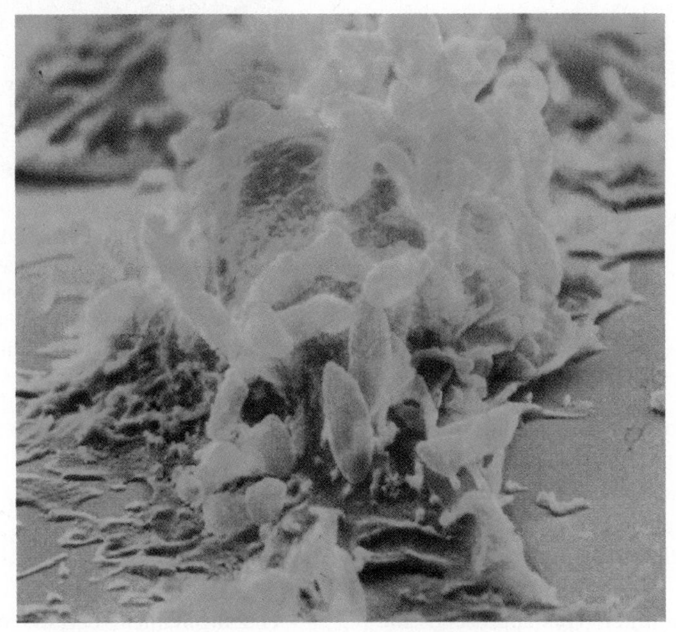

The vertebrate body is surrounded by potential pathogens. Bacteria, viruses, fungi, and parasites are everywhere—in the food we eat, the water we drink, and the air we breathe. They are in the soil, covering the surfaces of everything we touch; they even live on and in our bodies. Most microorganisms are capable of causing harm when introduced into the vertebrate body, but fortunately an elaborate defense system, the immune system, acts to destroy such foreign microorganisms. In humans, the immune system is decentralized and includes several organs, such as the thymus, tonsils, adenoids, lymph nodes, bone marrow, spleen, and appendix (Fig. 21.1). In most vertebrates, immune cells arise from stem cells that form in the embryo and migrate to specific tissues and organs and give rise to the red and white blood cells (see Fig. 20.23, p. 427). Certain of the white blood cells—the natural killer (NK) cells, B and T lymphocytes, macrophages, and mast cells—play a crucial role in the immune response.

The human immune response involves two interacting defense systems called nonspecific and specific immunity. As the names suggest, the ***specific*** immune response involves the recognition and destruction of a specific foreign substance. It also "remembers" that specific foreign substance and can prevent it from causing disease later. The ***nonspecific*** immune response nonselectively defends the body against foreign invaders: it acts as the first line of defense against infectious agents and most potential pathogens. This type of immunity is innate; humans are born with the ability to recognize and destroy certain pathogens on the first encounter.

The cells of the specific immune response work with the cells of the nonspecific response to protect the host organism. This coordination is extremely important in regulating the immune processes and in efficiently attacking foreign substances. The circulatory system plays a vital role in both responses; it is this system that produces and delivers the various white blood cells so important to the body's defense.

THE NONSPECIFIC IMMUNE RESPONSE

Epithelial Tissues Present a Barrier to Pathogens

The first lines of defense are the barriers presented by the skin and the epithelial linings of the respiratory, digestive, and uro-

Chapter opening photo: Scanning electron micrograph of macrophage ingesting bacteria. Macrophages move through the tissue fluid and attack and destroy invading viruses and bacteria.

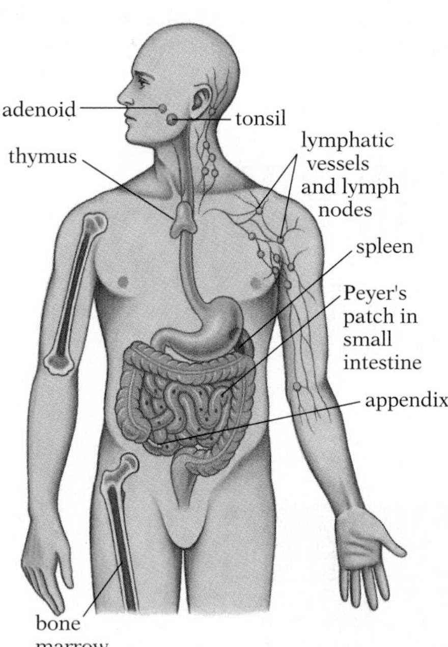

21.1 The human immune system The immune system is dispersed throughout the body. It includes the lymph nodes and vessels, as well as lymphatic tissue located in the tonsils, adenoids, appendix, and intestine. The red bone marrow, spleen, and thymus are also integral parts of this system.

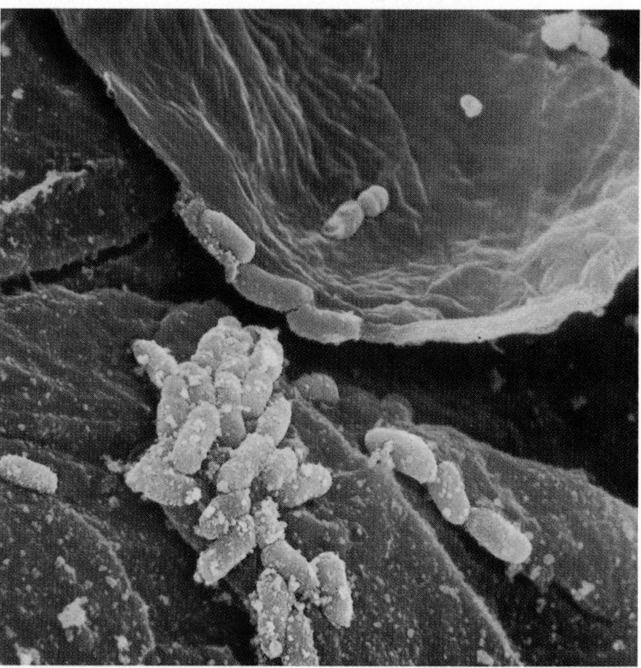

21.2 Bacteria on skin Bacteria are found in abundance on the skin and the parts of the human body open to the exterior. These bacteria make up the normal flora—the bacteria that normally inhabit the human body—and play a protective role because a potential pathogen must compete with and displace the normal flora if it is to become established and cause disease.

genital tracts (Fig. 21.2). When intact, these are quite effective in preventing the entry of harmful microorganisms. The normal flora—the bacteria that normally inhabit the human body—play a role here. The skin and the parts of the human body open to the exterior (mouth, nasal cavity, rectum, ear) abound with bacteria. In saliva, for instance, there are approximately 1×10^8 bacteria per milliliter. This approaches the density of bacteria living in broth culture media. And the skin of humans contains about 1×10^6 bacteria per square centimeter. We are populated by huge numbers of microorganisms! These bacteria do not harm the body, but rather protect it by occupying all the available niches in which bacteria can live. A potential pathogen must compete with and displace the normal flora if it is to become established and cause disease.

Antimicrobial chemicals in body secretions supplement the mechanical barriers presented by the skin and epithelial tissues lining the body cavities. For instance, sweat, tear, and salivary glands secrete chemicals that destroy certain bacteria. Also, the acid secreted by the stomach destroys bacteria ingested in food and water, while mucus in the respiratory passages traps pathogens breathed in the air.

The protective cells that form an important part of the nonspecific immune system as the phagocytic white blood cells, including neutrophils, macrophages, and monocytes (see Fig. 20.23, p. 427). These cells mediate a complex set of reactions known as the ***inflammatory response***.

The Inflammatory Response Destroys Invading Microbes

The mechanical barriers of the body are constantly being breached (Fig. 21.3). Every time you shave, brush your teeth, scratch yourself, etc., tiny breaks are produced that allow the entry of microorganisms such as bacteria and viruses. Once inside, the inflammatory response is initiated, the "classic" signs of which are *calor* (heat), *rubor* (redness), and *dolor* (pain). These symptoms are the result of chemicals released from damaged tissue and from invading bacteria. The chemicals cause the blood vessels in the area to dilate (increasing the blood supply), make the capillary walls more permeable to white blood cells and tissue fluid, and attract white blood cells to the area. Dilation of the blood vessels in the infected area produces the local rise in temperature and reddening characteristic of inflammation, and the increased blood flow in the region results in the presence of more white blood cells there. Many of these, particularly neutrophils and macrophages, are phagocytic and engulf and destroy bacteria and remnants of damaged tissue cells (Fig. 21.4). Others produce powerful proteins that help destroy bacteria and detoxify foreign proteins and other potentially dangerous substances. In a severe infection the

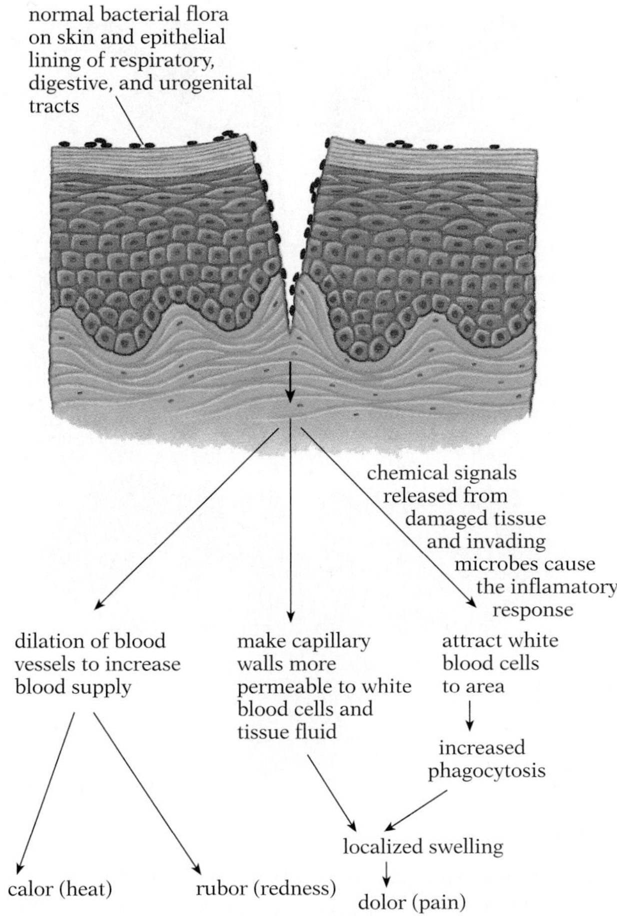

21.3 The nonspecific immune response Breaks in the protective barriers provided by the skin and epithelial linings of the various systems allow bacteria to penetrate, initiating a set of reactions known as the inflammatory response.

number of white blood cells in the blood and lymph increases enormously, and vast numbers may invade the infected area. The resulting accumulation of dead tissue, bacterial cells, and living and dead white blood cells at the site of the infection is commonly known as pus.

THE SPECIFIC IMMUNE RESPONSE

The Specific Immune Response Acts Against Specific Foreign Substances

All animals have phagocytic cells that ingest bacteria and dead cells. The basic lines of defense have evolved in vertebrates into a highly specific immune system that is probably an adaptation to larger body size and longer life. The specific immune responses involve two coordinated systems: humoral, or antibody-mediated immunity, and cell-mediated immunity. Both systems work together to detect and destroy ***antigens***, which are substances that are ordinarily foreign to the organism's own body. Antigens are almost always large molecules, usually either proteins or polysaccharides. They may be free in solution, like the toxins secreted by some microorganisms (e.g., tetanus or diphtheria toxin), or built into the outer surfaces of viruses or foreign cells such as bacteria and pollen. The human immune system recognizes as foreign not only nonhuman cells but also nearly all cells from other individuals, which explains why it is so difficult to successfully transplant organs.

The ***humoral immune system*** (antibody-mediated immunity) defends the body against invading antigens through the secretion of special proteins called ***antibodies***, which inactivate or destroy antigens on the surfaces of bacteria, fungi, parasitic protozoans, and viruses that are *outside* living cells, as

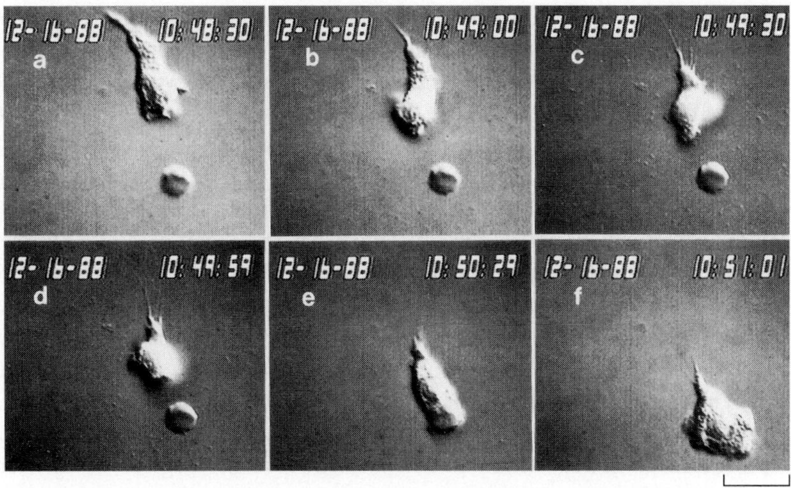

21.4 Neutrophil Neutrophils home in on invading organisms (usually bacteria) by detecting some of their waste products. The neutrophil seen here crawls toward its target, releases toxic chemicals and enzymes onto it, and then envelopes it by phagocytosis.

well as toxins free in the blood and plasma. The ***cell-mediated system***, on the other hand, destroys invading antigens directly, through the production of specialized *cells* that attack the invaders; it primarily defends the body against pathogens such as viruses that are living *inside* cells. Cell-mediated immunity is also involved in regulating the activity of the humoral system.

Lymphocytes Mediate the Specific Immune Response

The cells that respond to the presence of an antigen are specialized white blood cells called lymphocytes, which occur in the blood, in the lymph and lymphoid tissues of the lymph nodes, spleen, liver, tonsils, thymus, and bone marrow, and in the more limited lymphoid areas associated with the lungs and the intestinal tract. There are two different types of lymphocytes: ***B lymphocytes***, which are produced in the bone marrow, and ***T lymphocytes***, which develop in the thymus (Figs. 20.23 and 21.1). Lymphocytes are constantly circulating from the blood stream to the tissue fluid to the lymph, and then back to the blood, but the specific immune response takes place in the nodes, which are packed with B and T lymphocytes.

B Lymphocytes Mediate the Humoral Immune Response

Each of the millions of different B lymphocytes, or B cells, generated during embryonic development has thousands of identical antibody molecules mounted in its cell membrane. Each antibody molecule is Y-shaped and consists of four polypeptide chains: two identical "heavy" chains and two identical shorter "light" chains linked by disulfide bonds (Fig. 21.5A). The amino acid sequences of about three-fourths of each heavy chain and half of each light chain are similar in all antibodies and make up what is known as the constant region. The remaining portions of the chains—at the ends of the two arms of the Y—are highly variable. The binding sites for antigens are in these two variable regions. Antigens can bind only to the arms of antibodies that are specific for it (Fig. 21.5). The interaction between antigens and antibodies is similar to the binding of an enzyme and its substrate; the two fit together precisely, and the interaction is highly specific. Each of the millions of kinds of antibodies has a different binding site in the variable region, so each binds to a different antigen or antigen region.

When an appropriate antigen binds to a B-cell antibody, the bound antigen is engulfed by endocytosis and the antigen is "processed" (digested) within the cytoplasm into smaller fragments. Some of the fragments are then attached to special proteins (which will be described later) and transported to the B-cell surface, where they are "displayed" (Fig. 21.6) for other cells to recognize. This *activates* the lymphocyte, causing it to grow and begin dividing. Over a period of several days the stimulated B cell gives rise to numerous ***plasma cells***, which serve as veritable antibody factories, producing and secreting thousands of antibody molecules, which are identical to the antibodies on their surface (Fig. 21.7). The antibodies circulate freely in the blood and lymph and attack the stimulating antigens wherever they are found. A stimulated B cell also gives rise to other lymphocytes like itself, which serve as ***memory cells*** (Fig. 21.7). They remain in circulation for months or years after the exposure to the antigen. The memory cells "remember" the antigen, and when it is encountered again, they

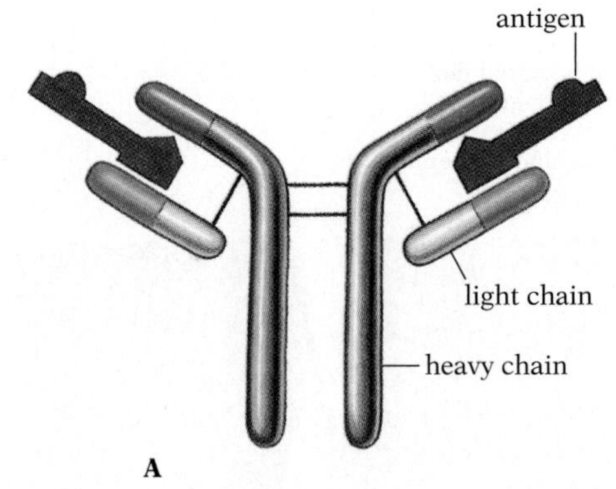

21.5 The B-cell antibody molecule (A) B-cell antibodies consist of two identical pairs of polypeptides; each pair has a heavy chain and a light chain, which is readily seen in this schematic representation. The sections shown in gray have relatively constant sequences, while the colored portions vary greatly from one B cell to another. Antigens bind in the cleft between the heavy and light chain of each pair. B-cell antibodies can exist as free circulating molecules (as shown here) or mounted in the membrane of B lymphocytes or mast cells. (B) A space-filling model showing the three-dimensional shape of antibodies. Each sphere represents one amino acid.

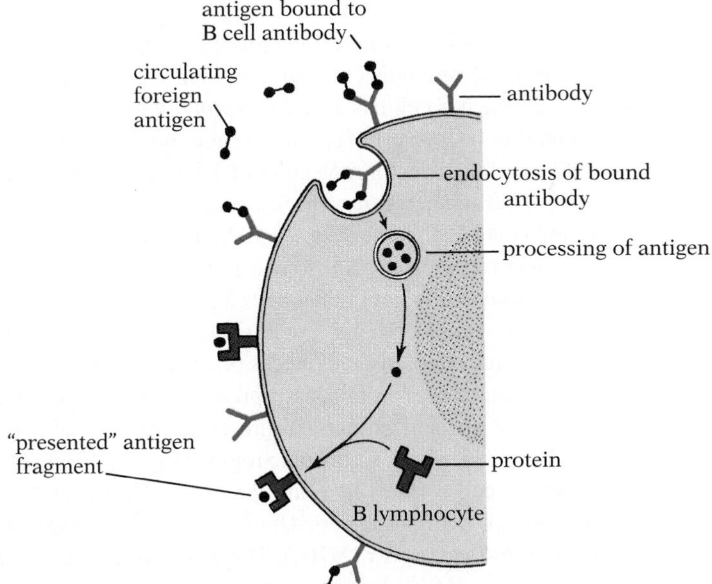

21.6 Presentation of antigen on the surface of a B cell When the membrane-mounted antibodies of a B cell bind their particular antigen, the antibodies are taken in through endocytosis. Antigens are removed in special protein-digesting vesicles and, if necessary, broken into convenient lengths, and these fragments are then bound by special molecules in the cytoplasm and displayed on the cell surface for other cells to recognize.

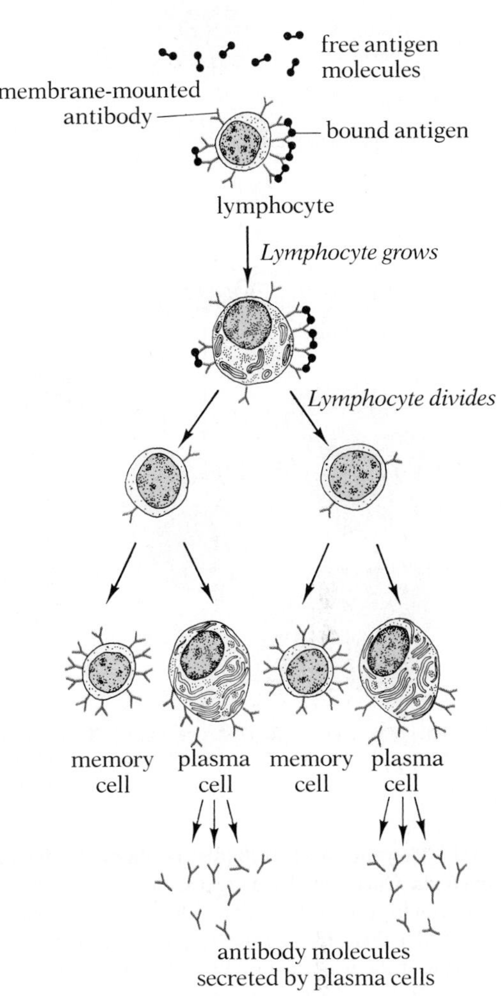

21.7 Stimulation of B lymphocyte by antigen When the membrane-mounted antibodies of a B cell are able to bind a particular antigen, the lymphocyte first grows larger and then begins a series of cell divisions (only two are shown here). Some of the cells produced by this proliferation are memory cells that resemble the original lymphocyte; others become specialized as plasma cells, which secrete antibodies that attack the invader. The antigen in this example is a toxin.

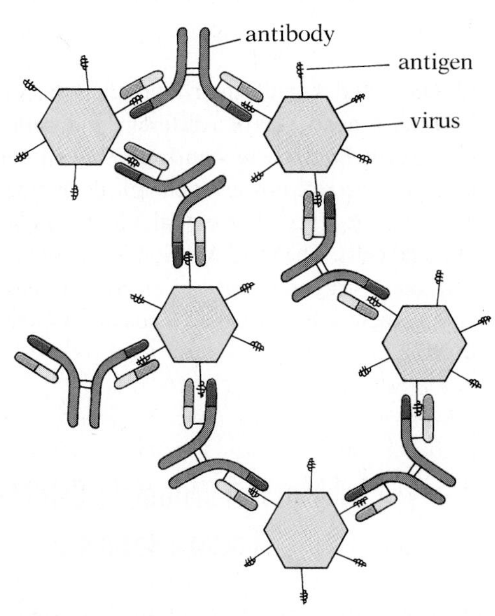

21.8 Agglutination by antibodies bound to antigens Each antibody molecule can bind to two antigen molecules; hence the microorganisms or viruses bearing the antigens can be held together in large clumps. Here the antigens are surface proteins or carbohydrates on invading viruses. (For clarity, the antibodies and antigens have been enlarged. They are in fact much smaller than viruses.) This agglutination aids in the destruction of antibody-bearing bacteria or viruses by macrophages, NK killer cells, or lysis.

immediately form plasma cells and make possible a greater and more rapid response than that in the first exposure.

Because each antibody molecule can bind to two identical antigen molecules, the antibodies tend to agglutinate, or clump together the free antigens or the bacteria or viruses bearing them (Fig. 21.8). Agglutination can trigger three reactions. First, agglutination makes it easier for large phagocytic ***macrophages*** in the lymph to recognize the antigen-bound an-

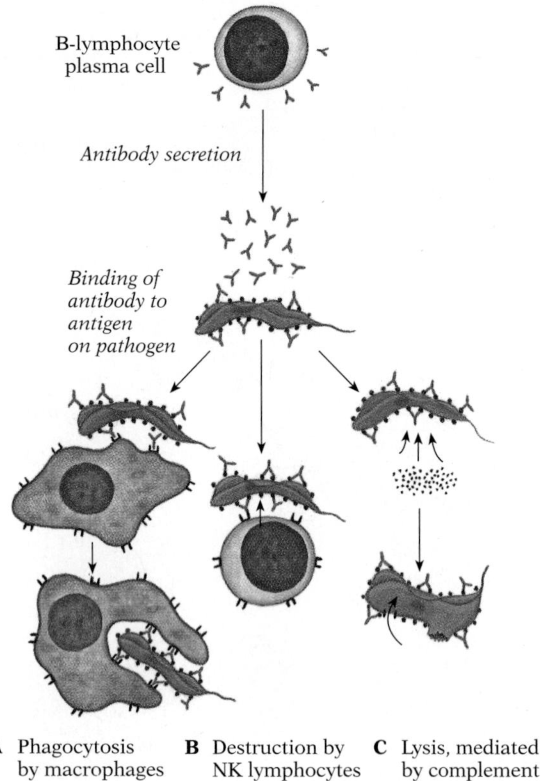

21.9 How humoral antibodies facilitate the destruction of bacteria Once bound to antigens, the antibodies secreted by plasma cells can trigger three reactions. (A) Bound antibodies, and the cell attached to them, are ingested by phagocytic macrophages. This is also the primary mechanism by which toxins and viruses are eliminated. (B) Bound antibodies are recognized by NK cells, which destroy the abnormal cell. (C) Bound antibodies trigger a complex series of reactions (the complement cascade) in which channels are made in the cell membrane, and the cell lyses.

tibodies, and engulf them and their targets (Fig. 21.9A). Second, lymphocytes of a different kind, called ***NK*** (for natural killer) ***lymphocytes***, recognize the bound antibodies, bind to them, and secrete powerful chemicals that destroy the foreign cell (21.9B). NK cells are particularly effective against virus-infected cells. The third reaction is complex in that the bound antibodies trigger a series of enzymatic reactions (called the complement cascade) that leads to lysis of the cell (Fig. 21.9C).

T Lymphocytes Control the Cell-Mediated Immune Response

Another set of immune reactions is mediated by T lymphocytes, which are produced in the thymus. The T lymphocytes, or T cells, look almost identical to the B lymphocytes, but they do not have antibodies mounted on their membrane surfaces. Instead, each T cell has ***T-cell receptors***, which are similar to antibodies but consist of two different polypeptide chains. The antibody-like receptor molecules of T lymphocytes are not secreted, but remain instead firmly attached by their tails to the lymphocyte membrane (Fig. 21.10A). Like B cells, each T cell produces receptors specific to one antigen, but T-cell receptors differ from antibodies in that they can bind only one antigen at a time. And unlike B cells, T cells cannot recognize the antigen in its natural state. Instead, pieces of antigen must be attached to a special marker protein, and "presented" to the T cell on the surface of specialized antigen-presenting cells. Each arm of the T-cell receptor has a specific region (Fig. 21.10) that binds to these cell-surface marking proteins. These membrane-mounted proteins are encoded by the genes of the ***major histocompatibility complex (MHC)***. There are two classes of MHC molecules: MHC II, which are found on the membranes of B cells, cytotoxic T cells, and macrophages, and MHC I, which are found on the membranes of all other cells of the body. This dichotomy reflects the dual role of the T cells as modulators of B-cell activity and assassins of disease-infected cells. Cells bearing MHC-II proteins (e.g., B and T cells and macrophages) participate in regulation of the immune system, while those with MHC-I molecules can be killed by cytotoxic T cells.

The MHC molecules, like the immune-system antibodies and receptors we have already discussed, are similar in structure to antibodies and T-cell receptors (Fig. 21.10B). When a cell is infected with a virus, antigens, usually fragments of proteins from the virus, are attached to MHC-I proteins in the cytoplasm. (If the antigen is too large to fit in the MCH pocket, it is digested into smaller pieces.) The MHC-I/antigen complexes are then transported to the membrane surface of the infected cell for presentation to other cells (Fig. 21.10). T cells with matching receptors will eventually bind to the MHC-I/antigen complexes on the infected cell (Fig. 21.10). The binding stimulates the T cell to grow and divide many times, generating a large number that will differentiate into memory T cells and at least three different varieties of T lymphocytes: helper T cells, suppressor T cells, and cytotoxic T cells (Fig. 21.11). ***Helper T cells*** play a central role in the immune response, as we shall see. When stimulated, they produce a family of chemicals called ***cytokines*** that activate other cells of the immune system—the macrophages and other T and B cells, amplifying the response. They also stimulate antigen-bound B cells to divide and produce antibodies. The ***suppressor T cells*** have the opposite role: they inhibit the activity of macrophages and other lymphocytes. They are important because they stop the immune response once the invader has been destroyed, ensuring that the body does not continue to produce new lymphocytes when they are no longer needed. Ordinarily the helper and suppressor T cells operate simultaneously, together controlling the precise level of each immune response. They are sometimes referred to as modulating T cells because of

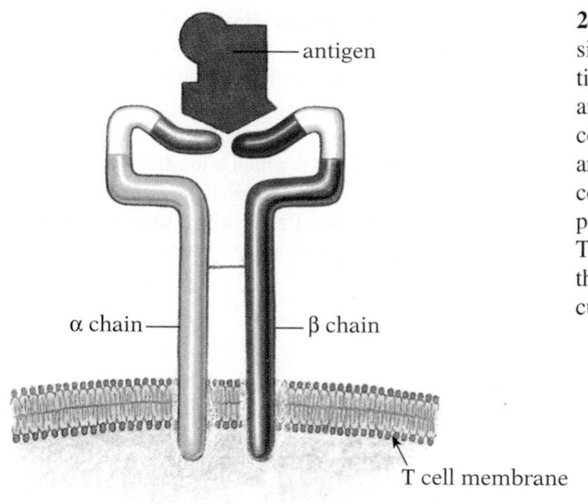

21.10 The T-cell receptor and MHC molecules (A) T-cell receptors consist of two peptide chains, each with its own sequence. Each chain has a relatively constant section (gray) and a highly variable region (purple) to which an antigen can bind. Another section in each chain (white) binds to a portion of complementary MHC molecules. (B) One class of MHC molecules (MHC II) are similar to T-cell receptors, having a pair of distinct chains, each with a constant region (gray), a variable region that binds an antigen (blue), and a portion (white) that can bind to a corresponding T-cell receptor. Notice that the T-cell receptor and MHC molecule bind to different portions of the antigen they have in common. Notice also the great similarity between these two molecules. It is likely that T-cell receptors evolved from MHC proteins.

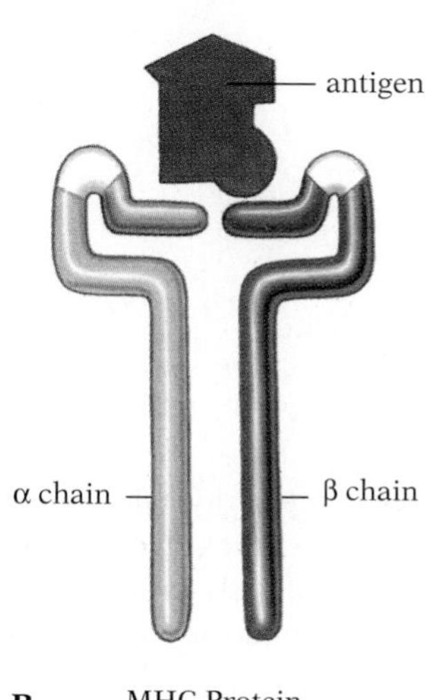

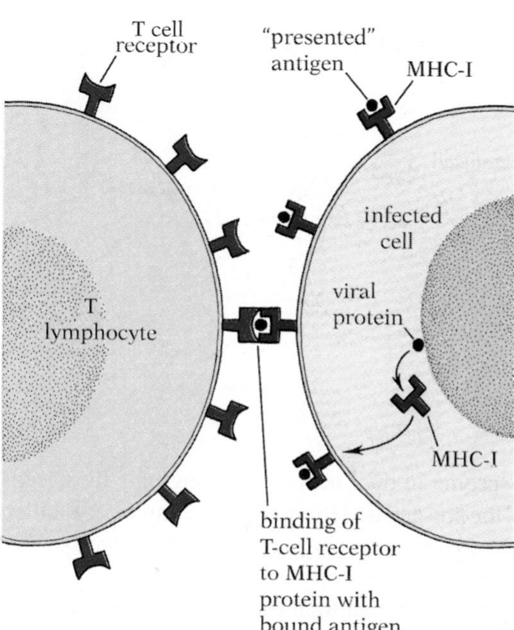

21.11 Antigen display by infected cell and binding of T-cell receptor In this model, proteins from the infecting virus are attached to MHC-I proteins in the cytoplasm and transported to the surface of the cell for display (right). A T cell with a specific affinity for the antigen being presented (left) will simultaneously bind to the MHC-I/antigen complex and become activated. T-cell receptors cannot recognize free antigens; to be recognized antigens must be attached to MHC proteins and "presented" to the T cell as we see here.

their regulatory role. The ***cytotoxic T cells***, as their name suggests, search out and destroy abnormal cells such as cancer cells or virus-infected cells that have the displayed MHC-I/antigen marker. (Fig. 21.12). Like the NK cells, the cytotoxic T cells bind to and release powerful chemicals that kill the target cell. Cytotoxic T cells differ from NK cells in that they are

MAKING CONNECTIONS

1. How does the normal microbial flora of the human body protect us from pathogens?

2. How is the interaction of an antibody with its antigen similar to the interaction of an enzyme and its substrate?

3. How is the immune response regulated?

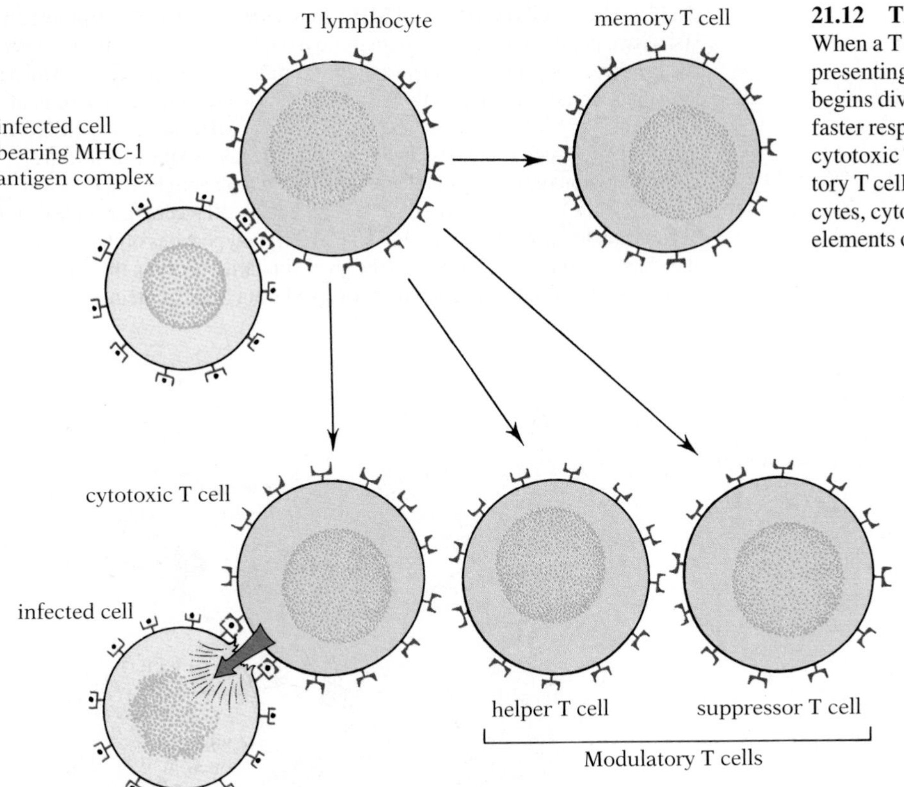

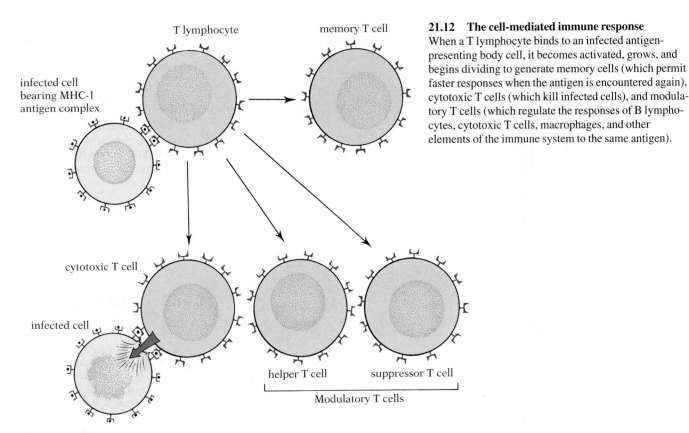

21.12 The cell-mediated immune response When a T lymphocyte binds to an infected antigen-presenting body cell, it becomes activated, grows, and begins dividing to generate memory cells (which permit faster responses when the antigen is encountered again), cytotoxic T cells (which kill infected cells), and modulatory T cells (which regulate the responses of B lymphocytes, cytotoxic T cells, macrophages, and other elements of the immune system to the same antigen).

highly specific in their action, attacking only those cells that possess the presented MHC-I/antigen complex. NK killer cells are nonspecific, destroying any cells that have been coated with bound antibodies.

Note that in the humoral response, it is the antibodies secreted by plasma cells that mediate the destruction of the invading antigen, whereas in cell-mediated immunity the T cells, especially cytotoxic T cells, mediate the destruction.

In most cases of infection, both the humoral and the cell-mediated systems are triggered, leading us to believe that interaction between B cells and T cells is considerable. However, the mechanisms of that interaction are not well understood.

Clonal Selection Produces Large Numbers of Lymphocytes Specific for a Particular Antigen

The human body is capable of making antibodies against virtually any foreign chemical group. When such a group—an antigen—is introduced, antibodies and T cells specific for the antigen are produced, and the antigen is eliminated. How does the presence of a particular antigen stimulate production of the appropriate antibody or T cell?

The most widely accepted hypothesis, the hypothesis of ***clonal selection***, postulates the existence in the human body of an enormous number of slightly different B and T lymphocytes, each of which has antibodies or receptors for a specific antigen on its surface. Each antigen binds only to those lymphocytes whose receptor molecules have binding sites for it. Thus the antigen "selects" the appropriate lymphocyte on the basis of its receptors and binds to the receptors, thereby activating the cell (Fig. 21.13). The activated lymphocyte, whether of the B or T type, begins to grow and divide, producing a ***clone*** of cells—a group of genetically identical cells descended from a common cell.

When an antigen stimulates a given B lymphocyte, it is calculated that each of the plasma cells in the clone to which the lymphocyte gives rise may secrete 2,000 identical antibody molecules per second. Thus, within a week or two there will be billions of antibody molecules synthesized and released into circulation, each one capable of inactivating the antigens.

Macrophages Initiate the Specific Immune Response

Let us follow, in some detail, the human immune response to the influenza virus. Once the virus invades the body it attaches to and enters a target cell, where it releases its genetic material

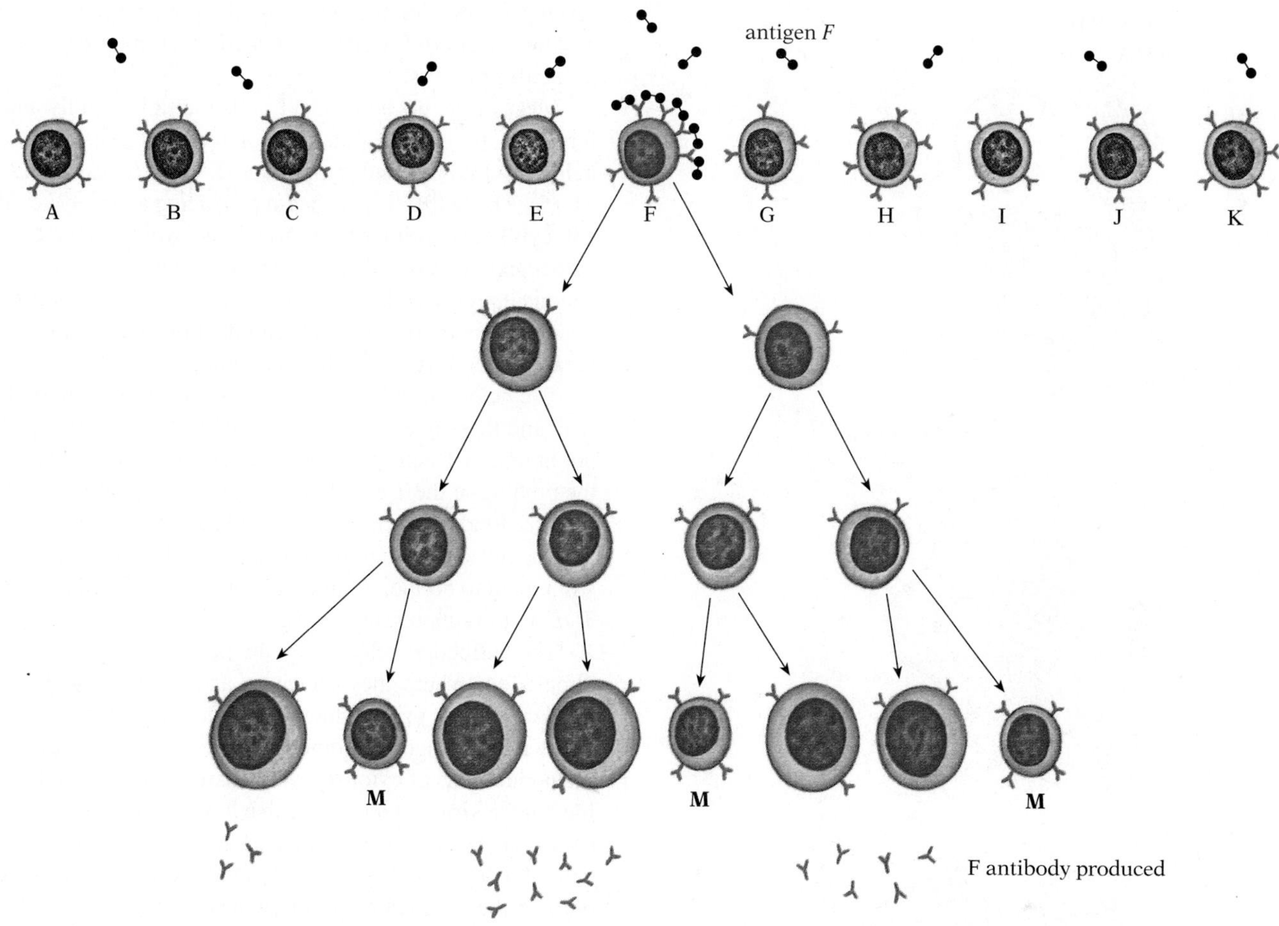

21.13 Clonal selection Antigens are able to react with the membrane-mounted antibodies of only one or a few very specific lymphocytes from among the billions of kinds of lymphocytes in the organism's body. In this example antigen *F* can be bound only by the B lymphocyte of type F; it does not affect the other lymphocytes (top row). Lymphocyte F, stimulated by the binding of the antigen, proliferates to form a clone of genetically identical cells. Some cells of the clone are memory cells (M); others are plasma cells that actively secrete F antibody. The cross-linking of the antibodies on lymphocyte F greatly facilitates the proliferation of the F clone. The diagram shows clonal selection in B cells. T cells are thought to be selected in on analogous manner.

(see Fig. 37.3, p. 864). The viral genetic material takes control of the cellular machinery and puts it to work making new viruses. Eventually the new viruses are released from the infected cell and go on to attack other cells. The immune response begins when circulating phagocytic white blood cells, especially neutrophils and macrophages, encounter the viruses and damaged cells and engulf and destroy them. In addition, some of the viruses encounter B cells in the area and bind to their surface antibodies, causing the B cells to process them (Fig. 21.6).

When macrophages[1] engulf viruses, they, like the B and T cells already described, process antigens from the virus and display the MHC-II/antigen complex on their surfaces (Fig. 21.14A). Thus activated, the macrophage will begin to mobilize the rest of the immune system, using several mechanisms. First it binds to compatible helper T cells—those that have specific receptors on their surfaces that exactly fit the MHC-II/antigen complex displayed on the macrophage (Fig. 21.14B). Once coupling has taken place, the activated macrophage begins secretion of a protein called ***interleukin-1*** or ***IL-1***. This protein has a variety of effects—it stimulates the inflammatory response, it stimulates other T cells to grow and divide, and it acts on the control centers in the brain to produce the symptoms associated with a fever—high body temperature, lack of appetite, aches, pains, feelings of malaise and fatigue, etc.—which is a protective response. Many viruses and bacteria do not reproduce well at high temperatures, and the feelings of fatigue and malaise force the patient to rest, con-

[1] A number of cells of the immune system act as antigen-presenting cells (APCs). They take in material from the surrounding tissue fluid by endocytosis, process it, and display the fragments on their surface MHC molecules. Macrophages are one type of antigen-presenting cell.

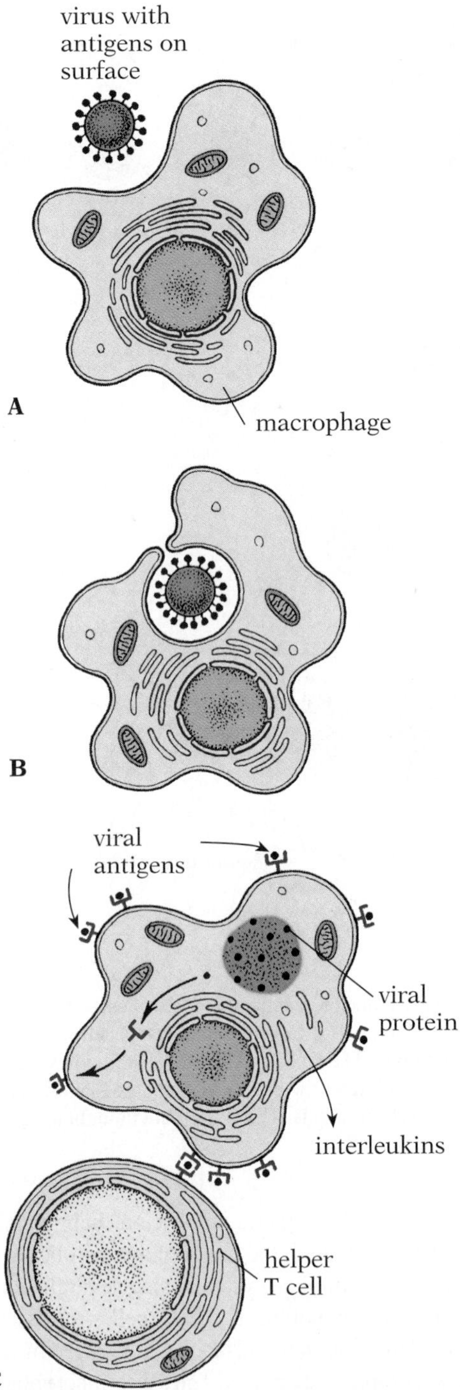

21.14 The role of macrophages in initiating the immune response A bacterial or viral infection initiates the inflammatory response at the site of the infection. (A) Macrophages migrating through the body ingest the foreign invader, in this case, a virus. (B) The macrophage processes the virus, stripping off the antigens, combining them with MHC-II proteins, and placing the complex on its surface. (C) Macrophages displaying markers couple with helper T cells that have receptors on their surface for that particular MHC/antigen complex. The binding activates the helper T cell, which in turn activates other cells in the immune system to join the fight.

serving energy for the business of fighting off the invader. Taking aspirin to lower fever may, in fact, interfere with this protective process.

Meanwhile, the activated T cells multiply rapidly and differentiate to form populations of memory cells and active helper, cytotoxic, and suppressor T cells (Figs. 21.12 and 21.15). The helper T cells begin producing a variety of different cytokines, primarily interleukins, which activate other helper and cytotoxic T cells and nearby macrophages. Another chemical produced by the helper T cells is gamma ***interferon***, which acts both to slow viral reproduction and to stimulate increased phagocytosis by the macrophages.

The helper T cells also bind to B cells of the humoral system, and thereby activate them (Fig. 21.15). B cells that have not bound to T cells produce very few antibodies. Apparently the binding of the T and B cells is necessary to stimulate the B cells to divide to form memory cells and plasma cells, which then begin secreting antibodies. The T cell, for its part, is also stimulated to divide, producing more helper T cells, cytotoxic T cells, and suppressor T cells.

The antibodies released by the plasma cells enter into the circulation, where they attach to and bind extracellular influenza viruses, making them more susceptible to phagocytosis by macrophages and other lymphocytes. At the same time, the populations of cytotoxic T cells are increasing and searching out the virus-infected cells that have MHC-I/antigen markers on their surfaces as a result of the infection. Cytotoxic T cells attach to infected cells and produce chemicals that induce pores in the cell membrane, allowing the cells' contents to leak out and destroying the cell.

Note the central role played by the helper T cells in stimulating the specific immune response—they activate B cells, cytotoxic T cells, macrophages, and other helper T cells. Unfortunately, it is the helper T cells that are attacked by the HIV (AIDS) virus. It is precisely because these cells are infected that the virus is so deadly (see Exploring Further: Acquired Immune Deficiency Syndrome, p. 444).

Once the virus has been destroyed, the immune response must be shut down so that the body no longer continues to produce the various T and B cells that are no longer required, thereby conserving resources so the body can more efficiently respond to a new threat. The suppressor T cells inhibit the activity of macrophages and other lymphocytes, ending the immune response.

ACTIVE AND PASSIVE IMMUNITY

A knowledge of the immune response helps us to understand how we recover from infections and why we develop a long-

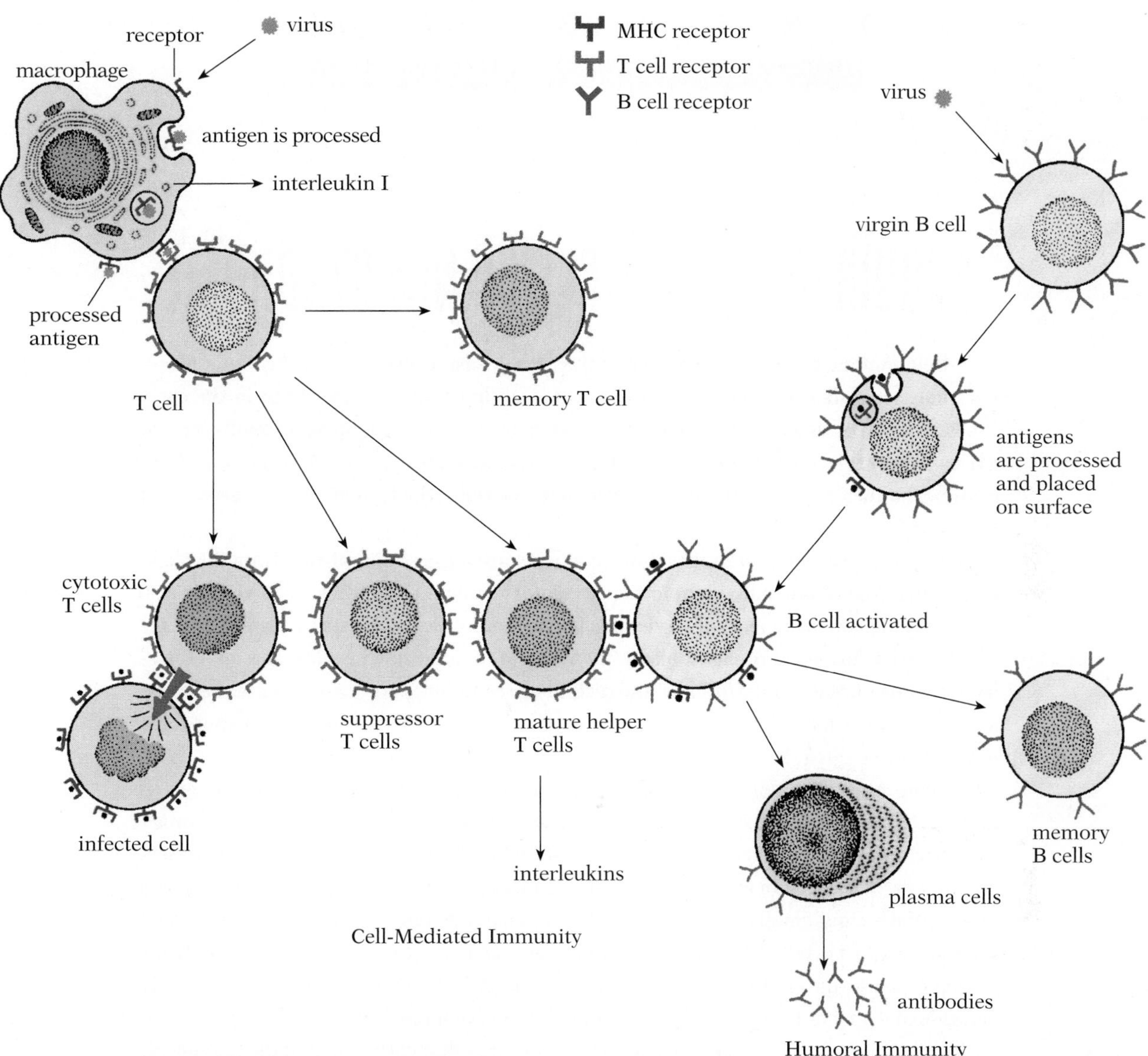

21.15 A summary diagram of the specific immune response The invading virus or bacteria is engulfed by a macrophage (left), and also bound to the appropriate B-cell receptor antibody (right). The antigens are stripped off, combined with MHC II, and placed on the cell's surface. The activated macrophage couples with an matching T cell, thereby stimulating it to divide and give rise to other helper T cells, cytotoxic T cells, suppressor T cells, and memory cells. The helper T cells also secrete interleukins and other chemicals that activate the cytotoxic T cells, suppressor T cells, and other helper T cells. The helper T cells also activate the appropriate B cells, which divide and mature into plasma cells, which produce antibodies. The antibodies and the cytotoxic T cells attack the foreign antigen. Once the foreign antigens have been eliminated, the suppressor T cells turn off the cytotoxic T cells and B cells, ending the immune response.

lasting immunity to certain microorganisms. Production of antibodies the first time a host individual is exposed to a particular antigen is rather slow; it takes at least five days or even weeks to develop a sufficient antibody response. During this time the body must rely on the nonspecific immune response to control the infection while the stimulated lymphocytes proliferate and produce clones of activated T cells, antibody-producing plasma cells, and memory cells of the B- and T-cell types. The cytotoxic T cells and antibodies act to eliminate the antigens, and the memory cells remain in the bloodstream, ready to mount a rapid defense the next time the same antigen is introduced. The memory cells confer on the individual a relatively long-lasting immunity to that particular antigen. For example, a person who has once had whooping cough or chicken pox usually remains immune for life to further infection by the pathogens of that disease.

EXPLORING FURTHER

ACQUIRED IMMUNE DEFICIENCY SYNDROME

Acquired immune deficiency syndrome (AIDS) is a devastating disease of the immune system that, according to the best available evidence, first appeared in Africa in the early 1960s. The AIDS virus, HIV (human immunodeficiency virus), has spread rapidly, first in Africa (where, given the many prevalent diseases, it was not recognized as novel until recently), and then to other parts of the world, notably Haiti, the United States, Europe, and East Asia.

Careful epidemiological studies show that transmission is almost always from the blood or seminal fluid of one individual to the circulatory system of another. (The virus is also present in the saliva and tears of its victims, but at very low levels.) Hence it readily spreads through transfusions of diseased blood, the reuse of hypodermic needles, and anal intercourse (where breaks in the lining of the rectum are common). The virus can also be transmitted through vaginal intercourse, though the rate of this form of transmission appears to be much lower.

Among American homosexuals and intravenous drug users, the infection rates are as high as 70 percent in urban areas like San Francisco and New York City, where contact rates are very high. On average, each infected individual communicates the disease to several others. In less urban regions, where the contact rate is lower and the susceptible pool of potential victims smaller, the spread is slower. Among heterosexuals, both transmission rates and contact rates are lower. At present it is not clear whether there will be an epidemic among heterosexuals in Europe or the Americas. In Africa, HIV transmission is primarily heterosexual, and risk correlates with the number of sexual partners.

The AIDS virus, HIV, is a retrovirus—an RNA virus that carries with it the enzyme reverse transcriptase, which makes a cDNA copy of the viral chromosome and inserts the cDNA into the host genetic material. The virus attacks those cells that have specific receptors (called CD4 receptors) to which the virus can bind. Such receptors are found in abundance on T4 lymphocytes (one subclass of helper T cells), but are also located on certain cells in the digestive tract, brain, and skin. Once the virus enters the cell its genetic material is integrated into the host DNA and remains hidden, evading the body's immune defense. The viral DNA is replicated along with the host DNA whenever the cells divide. This latent period can last for weeks, months, or years until its virulent phase is triggered. From careful studies most researchers now conclude that the average time from infection until symptoms occur is eight years. During this period the individual is infected with the virus and is capable of transmitting it to others, but does not have any symptoms of the disease and, unless the person is tested, may not even know that he or she is infected. Present estimates suggest that all carriers of the virus will eventually develop AIDS sooner or later.

Once a person has the disease symptoms, it is inevitably fatal, but there is hope that new treatments may at least extend the average survival time.

The disease is so devastating because it severely compromises the ability of the immune system to defend the body against invading pathogens (Fig. A). The preferred host cells for the virus, the T4 helper cells, play a central role in coordinating the immune response. They activate the cytotoxic T cells, macrophages, and B cells, recruiting them to enter the battle against the invading pathogens. They also produce a number of chemicals that stimulate the proliferation of other T cells.

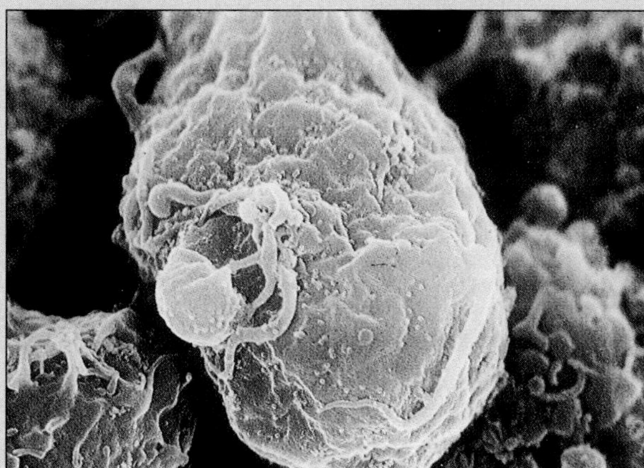

A HIV viruses budding from a cultered lymphocyte

Initially, the immune system responds normally to HIV, with B cells releasing suitable antibodies, helper T cells amplifying the response, and macrophages consuming free viruses. But the virus lives on in infected cells. When infected, the T4 helper cells simply cannot carry out their role—some of them die, and those that remain do not work very well. The population of helper T cells remains normal for about a year, but then begins to drop, reaching, on average, about 25 percent of normal three years after infection. To make matters worse, the suppressor T cells inhibit the activity of what T4 cells there are. That is not the end of the misery for an AIDS patient, however. The macrophages are massively stimulated and produce large amounts of interleukin-1. IL-1 leads to symptoms of infection: high fever, loss of appetite, chills, sweating, fatigue, malaise; in short, AIDS victims feel wretched most of the time, and are incapable of doing much of anything. But the death of helper T4 cells or the disruption of their normal function is not the direct cause of AIDS fatalities; instead, victims succumb to subsequent, normally minor infections that rage unchecked, the immune system being unable to successfully mount a proper defense because the helper T cells responsible for activating the appropriate immune defenses are either missing or no longer effective.

At this writing the World Health Organization has 44,652 documented cases of AIDS in 99 countries, but estimates that there have been at least 100,000 cases in 127 countries since 1975. The death rate is expected to rise beyond 10 million by the end of the decade. In the United States alone there have been 31,982 documented cases. Present estimates indicate that 1.5 to two million Americans may be infected with the virus. Recent surveys on college campuses in the United States have disclosed a disturbingly high incidence of HIV-infected college students. Nearly one in 300 college students is thought to be infected—an incidence

that is nearly as high as that of prisoners (whose infection rate is one in 250). It is fortunate that the virus is fragile and not easily contracted. AIDS is far less contagious than measles, polio, malaria, cholera, tuberculosis—all diseases that kill more people per year than AIDS. (Two million children die each year of measles alone.) There is no evidence that the virus can be spread by ordinary casual contact. How then to protect yourself from this terrible disease? The best protection is abstinence or to maintain a monogamous relationship with an uninfected individual. Condoms used in conjunction with spermicides offer some defense but they are about as effective in preventing HIV transfer as they are in preventing pregnancy, which means about 90 percent effective. Intravenous drug use with shared needles and unprotected (i.e., without a condom) anal intercourse should be avoided in all cases.

Vaccines Stimulate Long-Lasting, Active Immunity

Modern medicine often takes advantage of the body's antigen-antibody reaction to induce prophylactic immunity—immunity that prevents a first case of disease. The patient is inoculated with a ***vaccine***, which contains antigen from the pathogen. Sometimes the antigen consists of dead or weakened microorganisms, and sometimes it consists of weakened or inactivated bacterial toxin.[2] A sufficient amount of antigen is introduced to induce formation of antibodies, but not enough to produce disease. Vaccines stimulate the body to produce its own antibodies and memory cells. Immunity in which the individual actively forms antibodies and memory cells in response to contact with an antigen is called ***active immunity***. It can be induced in two ways: by having the disease, or by immunization with a vaccine. Childhood vaccinations for polio, measles, mumps, and German measles are used to produce an active immunity against the pathogens that cause these diseases.

[2] Some bacterial cells produce poisons called toxins. Often it is the toxin produced by the bacteria that causes the symptoms of disease, not the bacteria themselves.

Passive Immunity Is Borrowed Immunity, and Is Short-Lasting

Another type of immunity is ***passive immunity***. Passive immunity is acquired when the patient is inoculated with an ***antiserum*** containing antibodies previously synthesized by another immune individual. An antiserum is made by injecting antigen into some other animal, usually a horse, waiting until the animal has produced antibodies specific for that antigen, and then removing them from the animal blood serum. Passive immunity takes effect almost immediately after inoculation, but lasts only a short time because the injected antibodies are soon used up and destroyed.

The immunity that a newborn baby acquires from its mother is an example of a natural passive immunity. The milk

MAKING CONNECTIONS

1. What causes the typical symptoms of the flu—fever, achiness, fatigue, etc.? Why may taking a medication to reduce the fever be unwise?

2. How does the HIV (AIDS) virus damage the immune system?

3. The venom produced by poisonous snakes contains powerful protein toxins that the victim's immune system recognizes as antigenic. If you were bitten by a poisonous snake, which would be the more effective treatment, a vaccination or an inoculation with an antiserum?

produced in the first few days after birth consists of ***colostrum***, a thin, yellowish fluid that is rich in protein and contains numerous antibodies. These maternal antibodies provide the baby with immediate short-term protection against certain microorganisms; babies fed exclusively on bottle formulas lack this protection.

RECOGNITION OF "SELF"

The Immune System Distinguishes Between "Self" and "Nonself"

The immune system plays an important role not only in resistance to disease but also in the body's ability to tell which substances are part of itself and which are not. Tissue-transplantation experiments demonstrate that this self-tolerance is acquired early in the embryonic development of the immune system. Grafts and transplanted organs introduced into the adult organism are almost always recognized as "nonself" and are rejected by the immune system. Rejection occurs because the foreign tissue bears many different MHC proteins. Every person has a unique array of these proteins, so when tissue is transplanted from one person to another, most of the donor tissue proteins will be novel to the host's immune system, causing an immune response to be mounted. Hence, when a transplant is performed, X rays or potent chemicals are used to suppress the immune reaction. But even with the techniques now available, it is very difficult to suppress the immune reaction sufficiently to prevent rejection of the transplanted organ without at the same time making the patient vulnerable to infection. In any such attempt, the patient must be given continual massive doses of antibiotic drugs, because his own defenses against infection have been suppressed. Transplant rejection can be minimized if the tissue of the donor is closely matched with that of the recipient. Tissue typing, or histocompatibility testing, is done to match the MHC proteins in the donor and recipient as closely as possible. The donor is often a close relative of the patient, because they may share some of the same MHC proteins.

IMMUNOLOGIC DISEASE

Though we normally think of the immunologic mechanisms of the body as a defense against disease, there is increasing evidence that these very mechanisms are at times responsible for disease symptoms. In some cases the invading microorganisms or other foreign materials do not themselves cause much, if any, harm to the host. Strangely enough, it is the immune response of the host that actually produces the damage. Allergies are examples of this type of disease (see Exploring Further: Allergy: The Immune System Gone Awry, p. 448). The skin eruption caused by exposure to poison ivy provides a good example. The toxin found in poison ivy does not itself harm the skin, but the first exposure activates the cellular immune system, and produces activated T cells. Upon subsequent exposure to the toxin, the activated T cells, cytotoxic T cells, and macrophages invade the tissues and produce the itching, oozing rash characteristic of poison ivy.

Occasionally, an animal's immunologic mechanisms may become so out of control that the animal becomes sensitized to some part of its own body, causing interference with or even destruction of that part. In such cases the ability to distinguish between "self" and "nonself" is impaired, and an ***autoimmune disease*** results, with the immune system attacking the tissues of the body instead of protecting it. During autoimmune reactions, parts of the animal's body may be partially or completely destroyed. A well-known example is myasthenia gravis, a neurological disorder involving loss of tolerance of the body's billions of receptors for acetylcholine, a chemical that is used for communicating between nerve cells and muscles. In this disease the patient makes antibodies against his own acetylcholine receptors. The antibodies prevent the normal functioning of the receptors so the muscle cells no longer receive the normal signals to contract, resulting in a progres-

MAKING CONNECTIONS

1. Why is breast-feeding thought to be better than baby formulas in protecting newborn babies against disease?

2. Why is it important for the MHC proteins of an organ transplant recipient and the donor to match as closely as possible?

3. How is the process of capillary exchange described in Chapter 20 related to allergy symptoms such as hives?

EXPLORING FURTHER

ALLERGY: THE IMMUNE SYSTEM GONE AWRY

The immune system undoubtedly evolved to protect the individual from foreign invaders, but in some cases it overreacts, producing symptoms that actually harm the body. Allergy, or *hypersensitivity*, is an abnormal immune response to antigens, which in this case are referred to as *allergens*. Allergens such as grass and ragweed pollen, dust, and molds are not normally harmful to the body, but the body may react to them and mount a strong defense against them. In many cases the allergic reaction is simply a matter of overkill. If a foreign object enters the eye, it waters temporarily, a reaction that protects the eye and washes away the foreign object. Breathing in irritating dust leads to sneezing and a runny nose, which stops when the substance has been flushed away. These are normal responses to irritants, but for an allergy sufferer, these symptoms may go on for hours, days, weeks, or months.

We now know that hypersensitivity is due to the production of a special type of antibody, called immunoglobulin E or IgE in response to allergens. The IgE type of antibody is usually formed in response to parasites, such as certain flatworms, that penetrate through the epithelial tissues of the body. When challenged with allergens, normal individuals generally produce only a trace of IgE, but hypersensitive individuals produce high levels of IgE. Like all antibodies IgE is Y-shaped, but in this molecule the base of the Y (the constant region) has a special affinity for receptors found on the surface of basophils and special connective tissue cells called mast cells. These cells are found in abundance in the dermis of the skin and in the tissues surrounding the respiratory and digestive tracts. Thus the IgE molecules do not normally circulate freely in the blood; instead they are attached to the receptors on the surfaces of mast cells and basophils. These two types of cells are large cells that are packed with storage vesicles containing histamine and other substances. When the hypersensitive individual encounters the allergen for which the specific IgE antibodies are produced, the IgE antibodies on the surface of the cell combine with the allergen and this antigen-antibody reaction triggers the mast cells or basophils to suddenly discharge their vesicles of histamine (Figs. A and B). Histamine in turn dilates the blood vessels in the region and causes an outpouring of plasma into the area. It also stimulates the mucus-producing cells to secrete large amounts of mucus to wash away the invaders. This leads to the classic symptoms of hay fever: violent sneezing and runny, itchy nose and eyes. A similar scenario is involved in producing the systemic reaction provoked in individuals sensitive to bee stings. In this case the allergen in insect venom is absorbed into the bloodstream and transported to mast cells throughout the body. The combination of the allergen with the IgE antibody on the surface triggers an explosive release of histamine from mast cells, causing a variety of symptoms throughout the body. The most common symptom is hives: itchy, reddened areas caused by the outpouring of plasma from the blood into localized areas of

the skin. Other symptoms can be life-threatening. Dilation of the large blood vessels, for instance, can cause a severe drop in blood pressure and shock can result. In some cases the vocal cords swell and suffocation can occur. Therapy usually involves the immediate injection of adrenalin, which dilates the airways and raises blood pressure.

Treatment of allergies is still only partially successful, as any hay fever sufferer can tell you. Antihistamines are widely used to counteract the effects of histamine, decreasing some of the annoying symptoms. Individuals with serious hypersensitivities often undergo months of desensitization, which may provide relief. During this process, patients are given a lengthy series of injections of gradually increasing doses of the allergen to which they are sensitive. The mechanism by which this therapy induces resistance to the allergen is not well understood, but it is often effective in ameliorating symptoms resulting from allergens produced by dust mites and grass and ragweed. Unfortunately, it does not work in all pa-

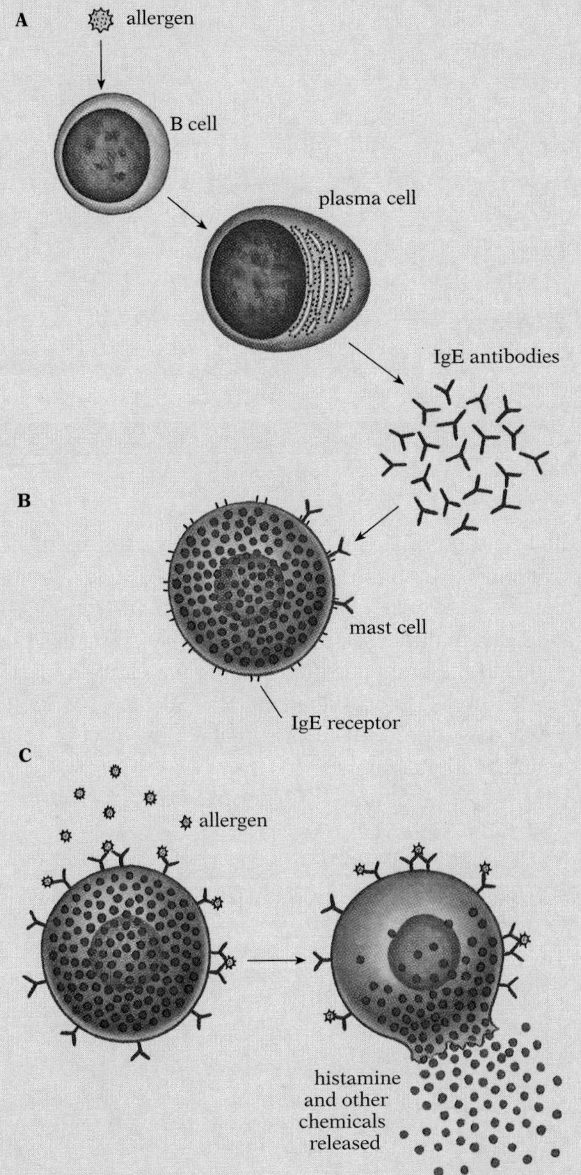

A The allergic response (A) Allergens (antigens that provoke an allergic response) such as pollen enter the body and stimulate B cells that have the appropriate receptors. The stimulated B cells mature into plasma cells and begin secreting a special class of antibodies, IgE antibodies. (B) The IgE antibodies attach to receptors on the surfaces of mast cells. (C) When the allergen is again introduced, the binding of the allergens to the mounted IgE antibodies causes the mast cell to release histamine and other chemicals that produce the symptoms of allergy.

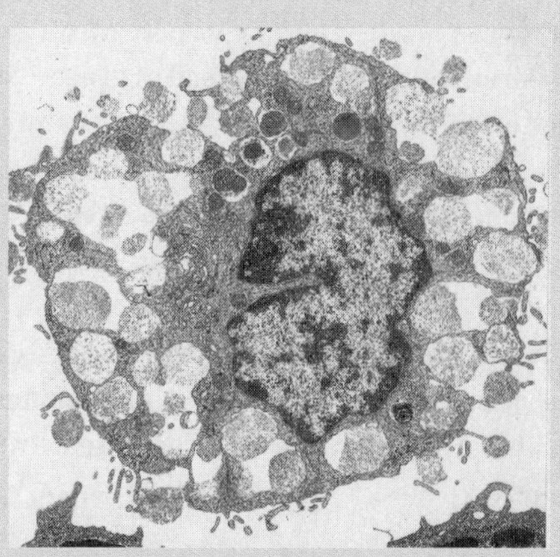

B Mast cell Massive exocytosis of histamine by a mast cell from a rat.

tients and it is lengthy, expensive, and unpleasant. In recent years the desensitization process had been improved, but other approaches are also being explored. The ultimate treatment would be to somehow curtail IgE production, but this is a complex problem. Researchers are also trying to stabilize the mast cells and prevent the release of histamine. Still another approach is to try to modify the IgE receptors so IgE can no longer bind to mast cells. The solutions to these problems are many years off, so hypersensitive individuals are still being told the same thing they have been told for years: "Avoid the substances to which you are allergic." Sound advice, but difficult when the allergens are ubiquitous in the air you breathe.

sive deterioration of muscle control. Symptoms of muscular weakness appear first, especially in the facial muscles: talking, chewing, and swallowing may be difficult. Later, the arms and legs may be affected, and death can result when the muscles involved in breathing become weak. How self-tolerance is overcome in such ailments is not yet understood. Other human diseases thought to be due to an autoimmune response are systemic lupus erythematosus, in which T cells attack the skin, kidneys, and other organs; rheumatoid arthritis, in which T cells attack the lining of joints; insulin-dependent (juvenile) diabetes, in which T cells attack the insulin-secreting pancreatic cells; multiple sclerosis, in which T cells attack the white matter of the brain and spinal cord; and Grave's disease (the thyroid disease that afflicted both President and Mrs. Bush), in which the thyroid gland is stimulated to produce excessive amounts of its hormones.

CONCEPTS IN BRIEF

1. The immune response involves two interacting defense systems called nonspecific and specific immunity. The *specific* immune response involves the recognition and destruction of specific foreign substances, whereas the *nonspecific* immune response nonselectively defends the body against foreign invaders.

2. The nonspecific immune response acts as the first line of defense against infectious agents and most po-

tential pathogens. The epithelial tissue barrier with its normal flora, bodily secretions, phagocytic cells, and the inflammatory response are all part of the nonspecific response.

3. In the specific immune response, vertebrate animals manufacture highly specific cells and *antibodies* to inactivate or destroy invading *antigens*, large molecules ordinarily foreign to the organism's body. Two interacting systems of specific immunity protect the vertebrate body from foreign antigens: humoral and cell-mediated immunity. Humoral immunity defends the body against pathogens living outside human cells whereas cell-mediated immunity defends against pathogens found inside human cells.

4. When activated, B cells divide to form plasma cells that produce circulating antibodies, resulting in *humoral immunity*. In *cell-mediated immunity* three types of T lymphocytes (helper, suppressor, and cytotoxic T cells) respond to the antigen. Cytotoxic T cells destroy infected cells while helper and suppressor cells regulate the immune response.

5. Antibodies are Y-shaped proteins. Each molecule consists of four polypeptide chains: two identical "heavy" and two "light" chains. There are two binding sites for the antigen on each antibody molecule. The two binding sites are specific in their recognition of a particular antigen.

6. Each lymphocyte, whether of the B or the T type, has just one kind of receptor on its surface. B cells have membrane-mounted antibodies that can bind two antigens. T cells cannot respond to the antigen directly; the antigen must be first joined to an MHC protein and displayed on the surface of a cell. An antigen can bind only to those lymphocytes whose surface receptors are specific for it. According to the *clonal selection hypothesis*, the antigen can bind to only those B or T cells that are capable of responding to it. Antigen binding stimulates cell division, producing a clone of cells that can respond to the antigen.

7. When a macrophage engulfs an invader, it processes the antigen, attaches pieces of it to an MHC protein, places the complex on its surface, and presents it to a compatible T cell. In doing so the macrophage is activated and produces IL-1, which produces symptoms of a fever and turns on T-cell proliferation.

8. The binding of the presented MHC/antigen complex to the appropriate T cell induces division of that cell to form a clone. The activated T cells differentiate to form memory cells and active helper, cytotoxic, and suppressor T cells. The helper T cells secrete substances that activate other helper and cytotoxic T cells and macrophages. They also activate B cells of the humoral system and antibodies are produced. The antibodies and cytotoxic T cells destroy the pathogen.

9. Once the antigen has been eliminated, the suppressor T cells inhibit the activity of macrophages and lymphocytes, ending the immune response. The memory cells remain in circulation, ready to mount an immediate defense should the antigen be reintroduced.

10. Some diseases can be prevented by immunization with a vaccine or an antiserum. A *vaccine* consists of antigenic substances from the pathogen; these produce a long-lasting *active immunity*. An *antiserum* contains antibodies against a specific antigen that have been synthesized by another animal; these produce an immediate, short-term, *passive immunity*.

11. Early on in the development of the immune system, the body learns to distinguish between "self" and "nonself." Grafts and transplanted organs are recognized as "nonself" and are rejected. Occasionally the immunologic mechanisms get out of control, and the ability to distinguish between "self" and "nonself" is impaired. In such cases the body begins to destroy itself (autoimmune diseases), or mount an inappropriate response to an antigen (allergy).

STUDY QUESTIONS

1. Which organs constitute the human immune system? Which types of white blood cells play important roles in the immune response? (p. 433)
2. Distinguish between the specific and nonspecific immune responses. How does the nonspecific immune response protect the body against infection? (pp. 433–434)
3. Describe the two components of the specific immune response, and define the terms *antigen* and *antibody*. (p. 435)

4. **Describe the structure and function of the antibody molecule. Which parts are identical to one another? Which parts bind to antigens? What accounts for the specificity of antibodies? (pp. 435–436)**
5. **Compare the structure and function of antibodies and T-cell receptors. (p. 438)**
6. **Describe the sequence of events leading to the production of antibodies by plasma cells. What is the role of memory cells? (p. 436)**
7. **Describe the cell-mediated immune response. What is the role of MHC proteins? What are the functions of the three major classes of T cells? (pp. 438–440)**
8. **Explain the *clonal selection* theory of the immune response. (p. 440)**
9. **Trace the sequence of events by which the body responds to infection by a virus (such as influenza). What roles do macrophages, MHC-II protein, interleukin-1, T cells, and B cells play? (pp. 440–442)**
10. **Contrast active and passive immunity. Which is faster-acting? Which longer-lasting? (pp. 442–443, 446–447)**
11. **What happens if an animal's immune system fails to distinguish "self" from "nonself"? (p. 447)**

CHAPTER 22

REGULATION OF BODY FLUIDS

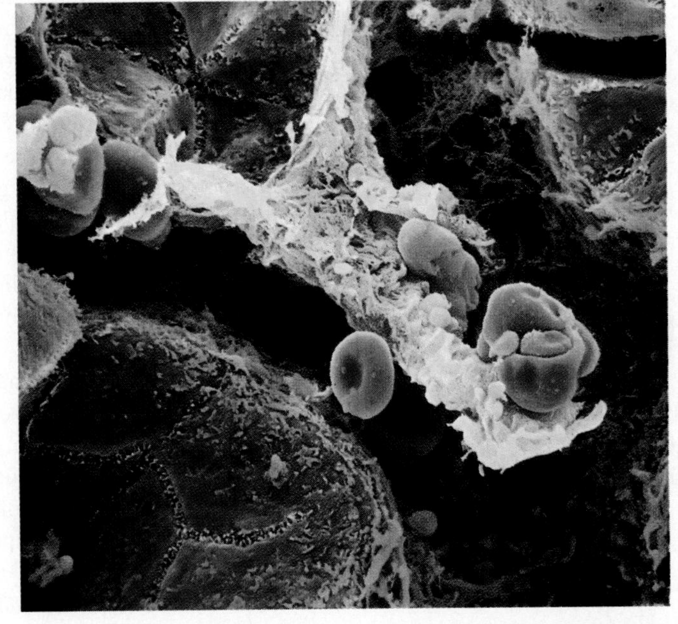

We have already seen that a living cell interacts constantly with its environment. Such critical functions as nutrient procurement, gas exchange, metabolism—indeed, life itself—are closely dependent on the properties of the surrounding medium. Biologists believe that life had its origin in the ancient seas, because the seawater environment is extremely constant—far more stable than freshwater or the land environment. In terms of temperature, acidity, and salt concentration, the seas fluctuate remarkably little over immense spans of time, their vast bulk making any change very gradual and slow. Under such stable conditions, early life was able to flourish.

It is not surprising that the cytoplasm of early cells had many characteristics in common with the seawater that bathed them, and that the sea's inherent stability was crucial to the evolution of life. Nor is it surprising that the evolution of complex multicellular marine animals involved the development of body fluids—tissue fluid, blood, and the like—able to provide even the innermost body cells with a relatively nonfluctuating aquatic environment, and that the internal body fluids of those primitive marine animals resembled the seawater that had been the cradle of life.

As the ages passed and evolution continued, the body fluids of different organisms evolved in different ways, just as their other characteristics did. Present-day marine animals, for example, differ noticeably from one another in the chemical makeup of their body fluids, though these fluids are more similar to one another and to seawater than they are to the body fluids of freshwater or terrestrial animals (Table 22.1), not to mention those of plants. Despite these differences, all these fluids have much in common, and, as Ernest Baldwin of Cambridge University has said, "The conditions under which cell life is possible are very restricted indeed and have not changed substantially since life first began."

We learned in Chapter 20 that multicellular organisms must maintain within themselves a constant fluid environment favorable to the continued life of their cells, despite changes in the external environment. This tendency to maintain equilibrium, or ***homeostasis,*** is essential to the normal function of the organisms. The evolution of immense diversity among living organisms has necessarily involved the concomitant evolution of diverse mechanisms for maintaining homeostasis in their body fluids. We have already mentioned the role of the circulatory system in maintaining homeostasis of the extracellular fluids by delivering required materials to the tissues and removing waste products, and the role of hemoglobin and the other blood proteins in buffering changes in pH in the blood.

Chapter opening photo: Scanning electron micrograph showing the removal of red blood cells (red) by special phagocytic cells (yellow) in the liver. One of the functions of the liver is to remove the old, worn out red blood cells.

TABLE 22.1 *Concentrations of ions in sea water and in body fluids (millimoles/liter)*

	Na^+	K^+	Ca^{2+}	Mg^{2+}	Cl^-
Sea water	470	9.9	10.2	53.6	548
Marine invertebrates					
Jellyfish *(Aurelia)*	454	10.2	9.7	51.0	554
Sea urchin *(Echinus)*	444	9.6	9.9	50.2	522
Lobster *(Homarus)*	472	10.0	15.6	6.8	470
Crab *(Carcinus)*	468	12.1	17.5	23.6	524
Freshwater invertebrates					
Mussel *(Anodonta)*	14	0.3	11.0	0.3	12
Crayfish *(Cambarus)*	146	3.9	8.1	4.3	139
Terrestrial animals					
Cockroach *(Periplaneta)*	161	7.9	4.0	5.6	144
Honey bee *(Apis)*	11	31.0	18.0	21.0	?
Japanese beetle *(Popillia)*	20	10.0	16.0	39.0	19
Chicken	154	6.0	5.6	2.3	122
Human	140	4.5	2.4	0.9	100

THE EXTRACELLULAR FLUIDS OF PLANTS AND OTHER PHOTOSYNTHETIC AUTOTROPHS

Plants Regulate Only the Composition of Their Intracellular Fluids

Multicellular marine algae differ markedly from multicellular marine animals in the sort of fluid environment to which their cells are exposed. The cells within the body of a complex animal are bathed by extracellular fluids—tissue fluid, lymph, or blood plasma—but these fluids are kept completely separate from the environmental water by cellular barriers, and differ in composition from both the cytoplasm within the cells they bathe and the water that surrounds them. By contrast, the fluid between the cells of a multicellular alga is essentially continuous with the environmental water and cannot be regarded as separate or distinct from it. The alga thus has no fluid fully analogous to the tissue fluids and blood of an animal. Hence, unlike the animal, which must regulate the composition of both intracellular and extracellular fluids, the alga must regulate only the composition of its intracellular fluids to maintain homeostasis.

A similar contrast appears between an animal and a large vascular land plant. Such a plant contains much extracellular fluid in the form of xylem sap and the water in the apoplast. But this fluid is not as fully distinct from the environmental water as is the tissue fluid and blood of an animal. You will recall that water can penetrate far into the cortex of a root by flowing along cell walls without having to cross any membranous barrier and that the film of fluid around the cells of the leaf or stem is continuous with the air through stomates or lenticels. Thus, much of the fluid that directly bathes the plant cells, even those far inside the plant body, is essentially continuous with the environment and is therefore not fully analogous to animal tissue fluid, which is separated from the environmental medium by cellular barriers.

For marine algal cells, the inability to regulate the composition of the fluid that bathes them poses no serious problem for the life of their cells. That fluid, after all, is essentially the same osmotic concentration as seawater, which we have described as the cradle of life. (It is true that modern seawater has a composition rather different from that of the ancient seas, but the change was so very gradual that the organisms living in the seas had ample time to evolve with their evolving environment.) Because the marine algal cells are isotonic with their environment, they do not gain or lose water by osmosis, although they still must regulate the concentration of individual ions in their cells to maintain homeostasis. Carrier proteins in the membranes of such cells transport specific ions into or out of the cells, thereby maintaining the proper osmotic concentrations.

What about a plant living in freshwater or on land? Because the extracellular fluid surrounding the internal cells of these plants is continuous with an environment that is, unlike the sea, inherently unstable, it is much more difficult for such a plant to regulate the composition of its extracellular fluid than for an animal to regulate the composition of its tissue fluids. The fluids to which plant cells are exposed fluctuate much more than the tissue fluids of animals—far greater fluctuations than could be tolerated by most animal cells.

Cell Walls Enable Plant Cells to Live in a Hypotonic Environment

Animal cells are seriously affected by changes in the osmotic concentration of the extracellular body fluids, which under normal conditions are kept approximately isotonic with the cells; osmotic shifts often severely alter the physiology of animal cells or even kill them (Fig. 22.1). Cells from your body, for example, would burst if they were placed in a hypotonic environment. Unlike animal cells, the cells of a land or freshwater plant almost always exist in a watery medium that is much more dilute than the cells' contents. In other words, plant cells are decidedly hypertonic relative to the fluid that bathes them. In such a situation animal cells would take in so much water by osmosis that they would burst, unless they had some special mechanism, such as a contractile vacuole, for expelling the excess water. But each plant cell is surrounded by its cell wall, and as the cell takes in more water and becomes more turgid, its wall resists further expansion. Eventually the resistance of the

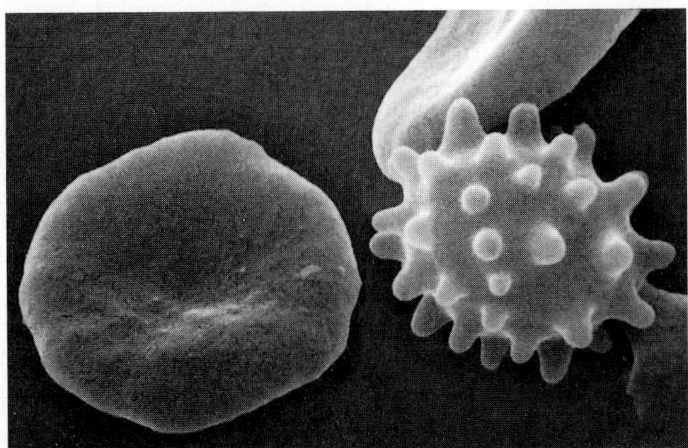

22.1 Red blood cells in isotonic and hypertonic solutions The red blood cell on the left is in an isotonic medium so there is no net gain or loss of water from the cell, and it maintains its normal shape. The cell on the right was placed in a hypertonic medium, and the water moves out of the cell, causing it to shrink and assume this abnormal appearance. Animal cells are very sensitive to changes in osmotic concentration.

wall is as great as the tendency of water to enter the cell by osmosis, and no further net gain of water is possible.

Therefore, plant cells can withstand rather pronounced changes in the osmotic concentration of the surrounding fluids as long as those fluids remain more dilute than the cells' contents—as long as the fluids remain appreciably hypotonic relative to the cell. If the external fluids become decidedly hypertonic relative to the cell, the cell may lose so much water and shrink so grievously that it pulls away from its more rigid wall; such a cell is said to be plasmolyzed, and the phenomenon is called ***plasmolysis*** (Fig. 22.2). The presence of the cell

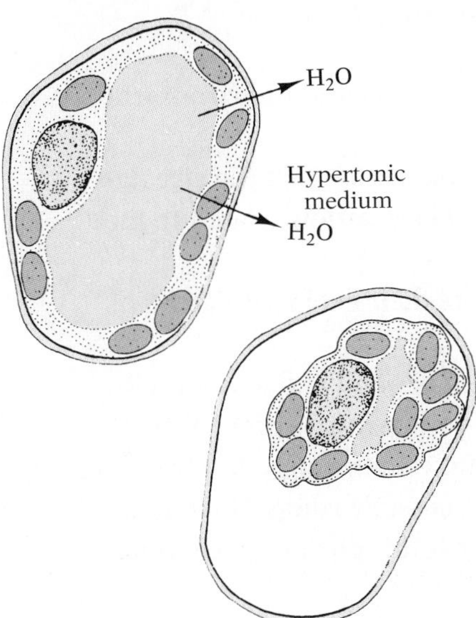

22.2 Plasmolysis A plant cell in a hypertonic medium (left) loses so much water that, as it shrinks, it pulls away from its more rigid wall (right).

wall in plants and its absence in animals thus make the problem of regulating salt and water balance quite different in the two.

That plants can tolerate great changes in the osmotic concentration of the fluid that bathes their cells does not mean they are unaffected by changes in the concentration of individual ions in the surrounding medium. Such changes sometimes have pronounced effects on their health and growth. The effects, however, are usually attributable to an alteration in the chemical makeup of the plant, rather than to a disruption of its osmotic balance.

THE VERTEBRATE LIVER

Animals must regulate the osmotic concentrations of their body fluids and keep them constant despite fluctuations in the environment. As a first example of the problems involved in maintaining homeostasis in a complex vertebrate animal, such as a human, consider the blood leaving the intestinal capillaries shortly after a meal. Digestion takes place in the small intestine, and the products of digestion move in large quantities into the capillaries of the intestinal villi. This means that the blood leaving the intestines contains high concentrations of such compounds as simple sugars and amino acids—concentrations considerably greater than those normally found in the blood in most parts of the circulatory system. But wholesale addition of these materials to the blood, if not controlled, would make the maintenance of a relatively nonfluctuating fluid environment for the cells impossible.

Vertebrates meet this difficulty with a very important homeostatic organ, the liver. Blood coming from the intestine, spleen, pancreas, and stomach is collected in the ***hepatic portal vein,*** which does not empty into the vena cava, as might be expected, but goes to the liver, where it breaks up into a second network of capillaries in the liver tissue (Fig. 22.3). The liver is one of only three places in the human body possessing a special type of circulation, called a ***portal system,*** in which blood flows from capillaries → arteriole or vein → capillaries → vein.[1] In a portal system, then, the blood flows through *two* sets of capillaries before returning to the heart; all other blood circuits involve only a single capillary bed. This structural arrangement has a very important function. It means that all substances absorbed into the blood from the digestive tract must first go to the liver before entering the general circulation. The liver then is in an ideal position to remove and to monitor the levels of biologically important substances.

The liver also plays a role in protecting the body against infection. The capillary beds of the liver are lined with phagocytic macrophages that remove from the blood entering the liver in the hepatic portal vein the bacteria that were picked up

[1] The others, to be discussed later, are the kidney and the pituitary gland.

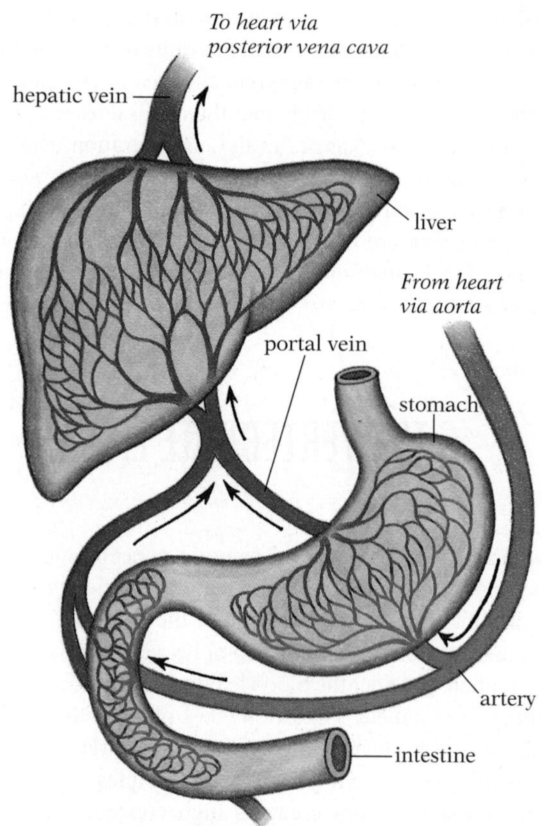

22.3 The hepatic portal circulation Blood from the capillaries of the digestive tract and stomach is carried by the hepatic portal vein to a second bed of capillaries in the liver. It then flows via the hepatic vein into the posterior vena cava, which takes it to the heart.

from the intestinal capillaries. The macrophages efficiently ingest these bacteria so that the blood leaving the liver is essentially cleansed of bacteria.

The Liver Helps Regulate the Blood-Sugar Level

The liver is the largest gland in the human body and the most versatile. We have already discussed the role of bile, which is produced by the liver, in the digestion of fats. The liver carries on many other vital functions; we shall focus our attention on how the liver metabolizes carbohydrates, fats, and amino acids, and how it functions to detoxify foreign chemicals.

The liver helps to regulate blood-sugar levels, either by removing glucose when the level is too high, or by releasing it when the level is too low. After a meal the blood coming to the liver via the hepatic portal vein has a higher than normal concentration of glucose. Under these conditions the liver removes most of the excess, converting it into glycogen, which is the principal storage form of carbohydrate in animal cells (Fig. 22.4).

Once the liver has stored its full capacity of glycogen, the liver begins converting glucose into fat, which is then released to be stored in the various regions of adipose tissue throughout the body. Thus, in spite of the great quantities of sugar absorbed by the intestine, the blood-sugar level in most of the circulatory system is not greatly raised and homeostasis is maintained.

The whole process is reversed if the blood-sugar level in the hepatic vein drops below a certain level, as during exercise,

MAKING CONNECTIONS

1. How is the need of present-day cells for a stable fluid environment related to the conditions in which the first cells evolved?

2. Explain how homeostasis links cellular respiration and the circulatory system. How are both cellular metabolism and circulation related to the subject of this chapter?

3. Why is it impossible for plants to regulate their extracellular fluids in the same way that animals do?

4. Contrast the normal osmotic environment of plant and animal cells. Which can withstand greater fluctuations in its osmotic relationships? Why?

5. Trace a red blood cell in the hepatic portal vein through the circulatory system back to the same point by the shortest possible route. How many sets of capillaries would the cell pass through? Why is this portal system functionally important?

6. How does the liver participate in the nonspecific immune response discussed in Chapter 21?

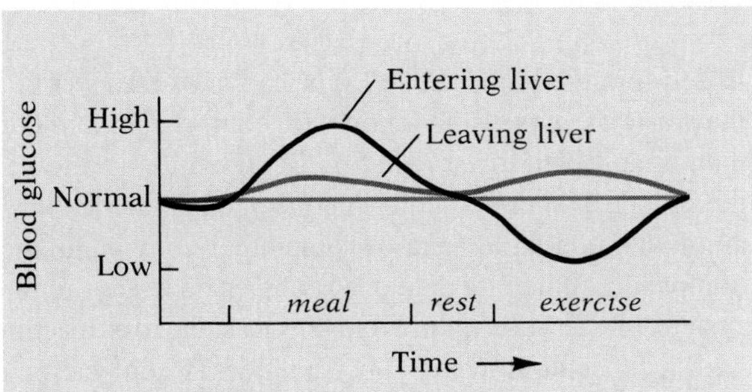

22.4 Regulation of blood sugar by the vertebrate liver The liver acts to maintain constant levels of many substances in the blood. In this example, the liver removes and stores excess glucose carried by the blood from the intestine after a meal (left), but later liberates glucose into the blood during exercise, when muscles begin using up this source of energy at an unusually high rate (right). The slight excess of glucose in the blood leaving the liver is reduced to the normal level when it mixes in the heart with other returning blood.

when muscle cells actively take up the blood's supply of glucose. Under these conditions the liver converts some of its stored glycogen into glucose and adds it to the blood (Fig. 22.4). Before the blood reenters the heart, its blood-sugar concentration has been returned to its normal level (Fig. 22.5B).

The human liver is capable of storing enough glycogen to supply glucose to the blood for a period of about four hours. What happens if at the end of that period no new glucose has come to the liver from the intestine? A drop in the blood-sugar concentration to a level much below normal would soon be fatal; the brain cells (and cells of the retina of the eye and the gamete-producing cells of the gonads) are particularly sensitive to such a drop, because they cannot store adequate amounts of glucose themselves or use fats or amino acids as energy sources. They are wholly dependent on a regular supply of glucose from the blood. Under such conditions the liver begins converting other substances, such as amino acids, into glucose, and in this way maintains the normal blood-sugar level (Fig. 22.5B).

The liver's activity in carbohydrate metabolism is regulated in a complex fashion by several hormones, as will be described in Chapter 25. An abnormal balance of these hormones may result in an unusually high blood-sugar level or an unusually low one. Either condition can be dangerous. If the blood sugar is too low, the brain cells cannot function; if it is too high, the osmotic concentration of the body fluids will be affected.

The Liver Metabolizes Lipids and Amino Acids

The liver also synthesizes, processes, and modifies fatty acids and other lipid materials. For example, it manufactures some plasma lipids, including cholesterol, and plays a major role in

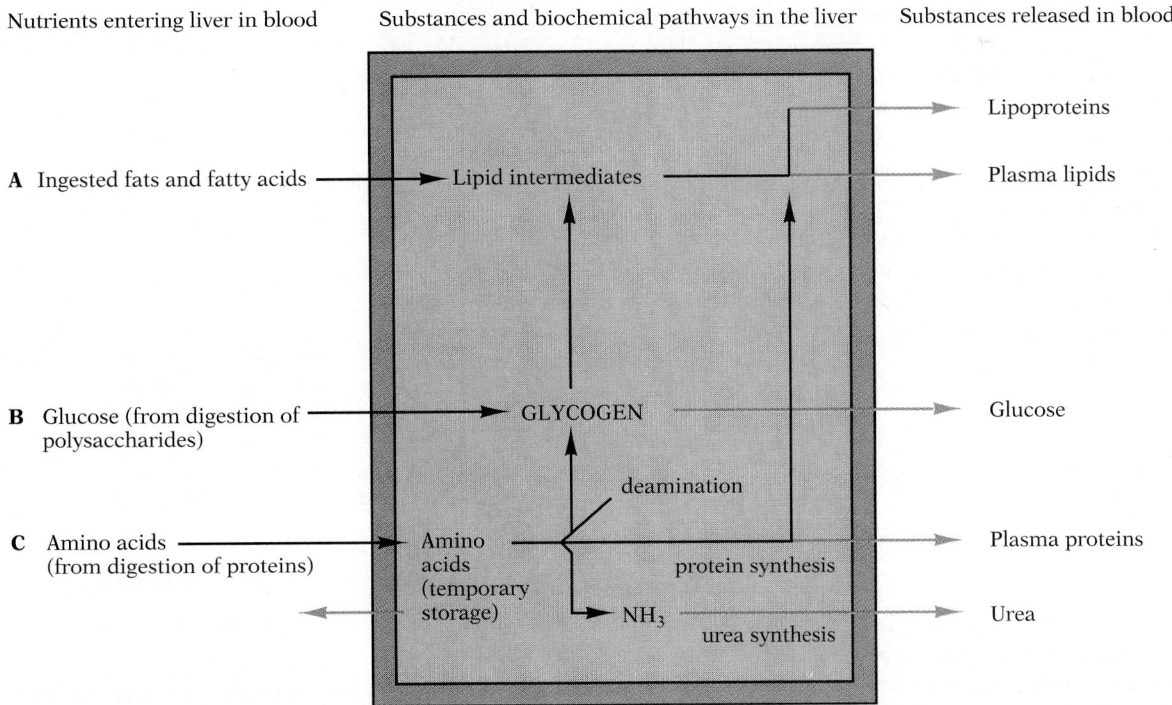

22.5 Chemical pathways in the human liver The liver is the center of metabolism for carbohydrates, lipids, and proteins. (A) Lipids and fats in the blood are converted into intermediates and bound to proteins for transport in the blood. (B) Excess glucose is converted into glycogen for storage; when blood sugar is low, the glycogen is hydrolyzed into glucose and released into the blood. Some of the excess glucose is converted into fat. (C) Some of the amino acids are used for protein synthesis; others are released into the blood for use by other cells; and still others are deaminated to form ammonia (NH_3) and an organic acid that can be used for synthesizing other compounds. The ammonia is combined with carbon dioxide to form urea, which is released into the blood.

the conversion of carbohydrates into fats and the conversion of ingested fats into fats typical of the organism's own body. (Fig. 22.5A) The liver acts as a "biochemical crossroads" for carbohydrate and fat metabolism—it removes excess sugar and fat from the blood when their concentrations are too high and releases them when their concentrations are too low, and it converts sugars into fats and ingested fats into other fats and lipids. The liver's metabolic function plays an integral role in maintaining homeostasis of sugars and lipids.

The liver also processes ingested amino acids. Like glucose, the amino acids absorbed by the villi of the intestine pass into the portal vein and then to the liver. The liver removes many of these amino acids from the blood, and uses them to synthesize plasma proteins such as fibrinogen, prothrombin, albumin, and some globulins. Small quantities of amino acids are also stored temporarily and later returned to the blood, which carries them to other tissues for use in the synthesis of enzymes, hormones, or new cytoplasm. But the usual diet contains far more amino acids than can be used in such syntheses. Unlike plants, the animal body is capable of minimal long-term storage of amino acids, proteins, or other nitrogenous compounds. Therefore, the excess amino acids must be converted into other substances, such as glucose, glycogen, or fat, which takes place in the liver (Fig. 22.5C). Here again, the liver is functioning as a biochemical crossroads to maintain homeostasis of the body fluids.

Because amino acids differ from most carbohydrates and fats in containing nitrogen, in the form of an amino group ($—NH_2$), the first step in converting amino acids into these other substances is ***deamination,*** or the removal of the amino group. In the deamination reaction the amino group is converted into ***ammonia*** (NH_3), a poisonous compound that must be excreted from the body. In some animals the liver simply releases the ammonia into the blood, which is quickly removed by the organism's excretory mechanisms. In many other animals, including humans, the liver first combines the ammonia with carbon dioxide to form a more complex but less toxic nitrogenous compound called ***urea*** (Fig. 22.6), which is then released into the blood. In still other animals the liver converts the waste ammonia into a nontoxic compound more complex than urea, called ***uric acid*** (Fig. 22.6), which is released into the blood. In short, whether the nitrogenous waste product is ammonia, urea, uric acid, or some other compound, the liver dumps it into the blood, and it becomes necessary for another system of the body to prevent the wastes from reaching too high a concentration in the body fluids. A comparison of the characteristics of the three nitrogenous waste products is shown in Table 22.2.

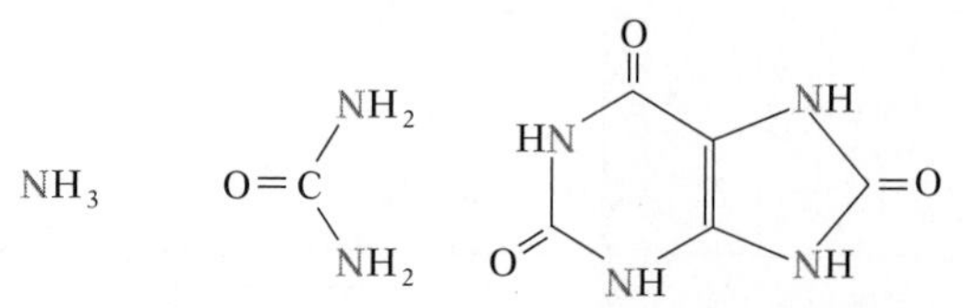

NH_3 — Ammonia

$O=C(NH_2)_2$ — Urea

Uric acid

22.6 Three important nitrogenous waste compounds Because the ammonia produced by deamination of amino acids can be toxic in relatively low concentrations and requires considerable water for its excretion, some animals combine it with carbon dioxide to produce urea, which is less toxic. Other animals go a step further by converting ammonia into uric acid, which is insoluble and nontoxic. Though the conversion of ammonia to uric acid requires a considerably greater expenditure of ATP than conversion to urea, the greater energy investment may be worthwhile for certain land animals, since less water is required for uric acid excretion.

Animals differ markedly from green plants in being unable to reuse much of the nitrogen from the amino acids they metabolize. For animals, this nitrogen is poisonous, and must be removed from the body. Green plants, however, can use this nitrogen, because they can freely shift nitrogen-containing groups from one organic compound to another. And, as we saw in Chapter 16, they can also use inorganic nitrogen to synthesize their own amino acids. Excretion of nitrogenous wastes is therefore essentially an animal activity.

TABLE 22.2 *A comparison of nitrogenous waste products*

	Nitrogenous waste products		
	Ammonia	*Urea*	*Uric acid*
Toxicity of product	High	Moderate	Very low
Product's solubility in water	High	Moderate	Insoluble
Amount of water lost during excretion	Large	Moderate	Very little
Amount of energy required to synthesize product	Small	Moderate	Large
Typical animals using product	Aquatic: both vertebrates and invertebrates	Terrestrial: amphibians, mammals	Terrestrial: insects, birds, reptiles (egg laying animals)

The Liver Detoxifies and Biotransforms Foreign Chemicals

The liver detoxifies a great variety of injurious chemical compounds and is thus one of the body's most important defenses against poisons such as alcohol, environmental pollutants, and drugs. A host of enzymes and chemical pathways are involved, but most of them transform nonpolar (hydrophobic) substances into more polar (hydrophilic) substances that can be excreted by the kidney. (Hydrophobic substances are not read-

$$CH_3CH_2OH + NAD^+ \xrightarrow[\text{enzyme (1)}]{\text{ADH}} CH_3C(=O){-}H + NADH$$

Ethyl alcohol → Acetaldehyde

ALDH enzyme (2)

$$CH_3C(=O){-}OH + \text{Coenzyme A} \dashrightarrow \text{Acetyl CoA}$$

Acetic acid

22.7 Metabolism of ethyl alcohol Ethyl alcohol is metabolized in the liver in two enzymatically controlled reactions. First it is oxidized by NAD^+ to acetaldehyde, using the enzyme alcohol dehydrogenase, or ADH. Acetaldehyde is then further oxidized to acetic acid using a second enzyme, acetaldehyde dehydrogenase (ALDH). The enzymes vary in their efficiency; in some individuals the first reaction takes place rapidly and the second slowly, in others the reactions may be more balanced. The acetic acid formed in the reactions combines with coenzyme A to form acetyl-coenzyme A, which is metabolized via the Krebs cycle. Acetaldehyde, the intermediate compound formed, is toxic to liver cells. Years of alcohol abuse can lead to liver damage.

ily excreted by the kidney.) Generally, the process of biotransformation makes the compound less toxic, but sometimes the process works in reverse: a nontoxic substance is made toxic by enzyme biotransformation. For example, many cancer-causing chemicals become carcinogenic only after biotransformation in the liver.

Ethyl alcohol is an example of how biotransformation produces toxicity. Damage is done to liver cells as a result of alcohol metabolism, not by the alcohol itself. Special enzymes, alcohol dehydrogenases, located in liver cells catalyze the oxidization of alcohol to acetaldehyde and hydrogen (Fig. 22.7). Acetaldehyde is toxic to liver cells and excess amounts of it damage the mitochondria. The hydrogen and electrons produced in this oxidation reaction are accepted by NAD^+, forming NADH. When alcohol intake is high for long periods, so much NADH is produced that there is not enough NAD^+ available for the normal metabolism of fat, and unmetabolized fat is deposited in the liver, where it hinders normal function (Fig. 22.8). Acetaldehyde is further oxidized to acetic acid in a second enzymatically controlled reaction (Fig. 22.7). These reactions are relatively slow; the average person metabolizes one-half ounce of alcohol per hour. If alcohol is ingested at a faster rate than this, the excess alcohol remains circulating in the blood until the liver can metabolize it. (A small amount is removed by the lungs and kidneys.) Generally, the first reaction in the sequence (Fig. 22.7) takes place more rapidly than the second, with the result that acetaldehyde accumulates in the liver cells and enters the blood. It is the acetaldehyde that causes the nausea, dizziness, flushed face, and increased heart rate associated with drinking. (See Exploring Further: Alcoholism: A Serious Health Problem, p. 460.)

The major functions of the liver are summarized in Table 22.3.

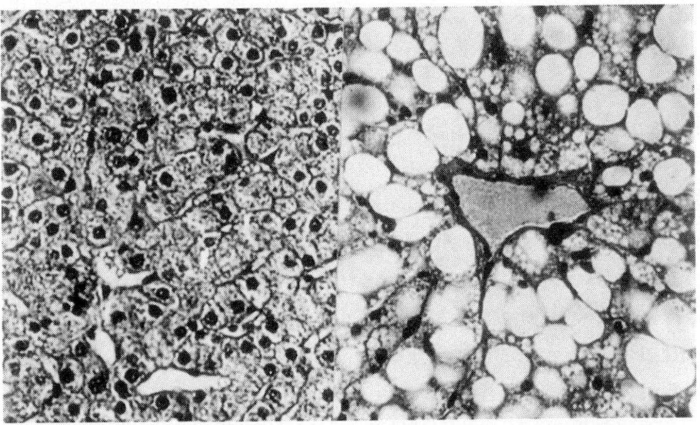

22.8 Normal (left) and fatty liver cells Prolonged high intake of alcohol interferes with normal fat metabolism by the liver cells with the result that tiny fat droplets are deposited within the cells. The droplets coalesce forming large globules that occupy almost all of the cell's volume. The cells in the left photograph are normal liver cells while those on the right are enlarged and have large empty spaces that formerly contained fat globules. (The fat was removed during the process of fixing and staining.)

TABLE 22.3 *A summary of the major functions of the human liver*

Role	*Function*
Production of bile	Synthesis of bile and pigments
Cleansing of blood	Bacterial cells from the intestine are removed by macrophages in liver
Carbohydrate metabolism	Removal of excess glucose from the blood and its conversion into glycogen and fat; conversion of glycogen and amino acids into glucose; secretion of glucose into blood when blood glucose is low
Protein metabolism	Removes excess amino acids from the blood; temporary storage of amino acids; synthesis of plasma proteins; deamination of amino acids and synthesis of urea; synthesis of lipoproteins
Lipid metabolism	Conversion of ingested fats into fats typical of the organism's body; synthesis of plasma lipids, including cholesterol, lipoproteins
Detoxification	Conversion of alcohol, environmental pollutants, drugs, etc., into substances that can be excreted by kidneys; synthesis of urea from ammonia

EXPLORING FURTHER

ALCOHOLISM: A SERIOUS HEALTH PROBLEM

The National Council on Alcoholism defines alcoholism as a "chronic, progressive, and potentially fatal disease" that is marked by "repeated drinking that causes trouble in the drinker's personal, professional or family life." Present estimates indicate that there are about 10 million severe alcoholics and seven million alcohol abusers in the United States (4.4 million of whom are teenagers). The line between heavy drinking and alcoholism is fine, and our concept of alcoholism has changed in recent years. The traditional view of an alcoholic—someone who drinks large amounts of alcohol daily and is never quite sober—has given way to the understanding that someone who drinks heavily only on the weekends, or someone who is sober for long periods but has heavy drinking binges lasting for hours, days, or weeks, may also be an alcoholic.

The role of inheritance in alcoholism is controversial. Children of alcoholics have a much greater tendency to become alcoholics than children of nonalcoholics. For example, sons of alcoholic fathers have about one chance in four of becoming alcoholic themselves. Whether the increased risk is due to inheritance or environmental influences is not yet clear. Further evidence of a genetic link comes from adoption studies by psychiatrist Donald W. Goodwin of the University of Kansas. His studies showed that children of alcoholic parents who were brought up by nonalcoholic parents have an increased risk of becoming alcoholic themselves. At the molecular level, researchers have found some potential "biological markers"—genes, enzymes, or other substance unique to the affected individuals—in families where there is alcoholism. Lines of evidence such as these suggest a genetic predisposition toward alcoholism, at least in families where there is an extensive family history of alcoholism. Most researchers urge caution in interpreting these findings, suggesting that, while there may be a genetic role in some cases, environmental pressures probably play a dominant role in determining alcohol abuse. Just what the environmental pressures are remains a subject of intense study.

On the other hand, some individuals inherit a resistance to alcohol abuse. Many Asians have different, less-active forms of the enzymes that catalyze alcohol breakdown (Fig. 22.7). Because their enzymes, particularly their acetaldehyde dehydrogenase (ALDH) enzymes, are not as efficient, acetaldehyde accumulates in the blood, resulting in painful facial flushing and nausea whenever the affected individuals drink more than a small amount. The symptoms are so unpleasant that such individuals tend to avoid drinking alcohol.

There is also a difference between the way men and women metabolize alcohol. If similarly sized men and women are given the same amounts of alcohol, the average blood-alcohol levels in women exceed those of men. Moreover, women tend to develop liver disease and other alcohol-related complications more quickly than men. Why this gender difference? The answer is not yet known, but recent research targets differences in the location of the alcohol dehydrogenase enzymes. The main site of alcohol metabolism is the cells of the liver, but recently it was discovered that, in many individuals, there are also alcohol dehydrogenase enzymes in the wall of the stomach, and that perhaps 20 to 30 percent of the alcohol degradation begins while the alcohol is still in the stomach. However, this finding held true only for men; women break down a far smaller percentage of the alcohol in this manner. And alcoholics, both male and female, show little or no alcohol degradation in the stomach. The significance of this finding is yet to be explored, but it does indicate that what may be moderate drinking for a man is not moderate for a woman.

Alcohol, consumed in excess, has a variety of long-term physical effects on the body. The primary effect is on the liver, because this is where alcohol is metabolized, but every system is affected. Liver disease begins with the formation of a "fatty liver," resulting from all the unmetabolized fat that is deposited within the cells (see Fig. 22.8). This may progress to alcoholic hepatitis, in which there is inflammation and death of liver cells. Both of these conditions are reversible. About one alcoholic in five will go on to develop alcoholic cirrhosis of the liver. In cirrhosis, the structure of the liver changes and scar tissue forms, particularly around the blood vessels (Fig. A). The flow of blood through the hepatic portal system is impeded, which places a greater burden on the circulatory system. Heart disease and other circulatory problems often result. Gastritis and pancreatitis (inflammation of the stomach and pancreas) are common. Alcohol also destroys brain cells and is toxic to bone marrow cells, causing anemia and damage to the immune system. The reproductive system of alcoholics is affected as well. Both alcoholic men and women experience infertility, a loss of sex drive, and a loss of secondary sex characteristics. Finally, alcoholism is related to certain forms of cancer. Cancers of the mouth, pharynx, esophagus, and stomach occur more often in alcoholics than in the general populace. Autopsies of patients who died from alcoholic cirrhosis showed that almost 30 percent of these individuals had undiagnosed liver cancer.

Not only does alcohol harm the individual; it may also directly affect the next generation. Pregnant women who drink heavily predispose their unborn babies to a lifetime of physical, mental, and behavioral disabilities. The alcohol in the bloodstream of the mother passes through the placenta into the fetus. Because the fetus lacks the enzymes for degrading alcohol, it is very sensitive to it. The association between maternal drinking and developmental abnormalities in the child is called fetal alcohol syndrome, or FAS (Fig. B). It is the leading cause of mental retardation in the United States. The prenatal brain damage may permanently alter the ability to think abstractly and to concentrate. Such children show behavioral problems and poor judgment, and are unable to function independently.

Alcoholism and alcohol abuse ranks with heart disease and cancer as one of our nation's major health problems. The dollar cost of alcoholism is also high; it has been estimated at about 120 billion dollars annually as a result of lost wages, health and welfare benefits, medical expenses, insurance, property damage, accidents, etc. The personal cost to the individual, family, and community is even higher.

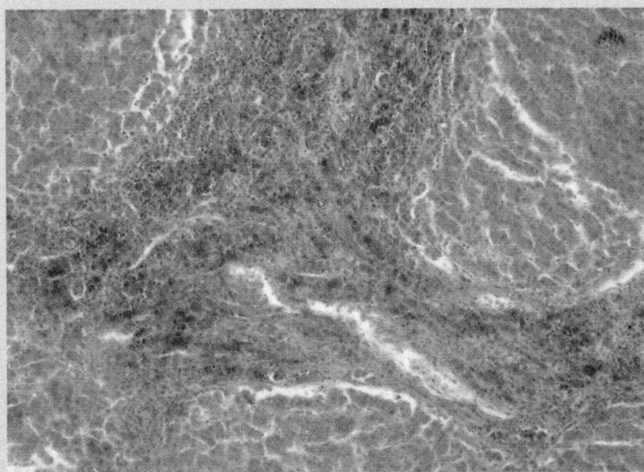

A Human liver with cirrhosis (orange areas)

B Three year old child with fetal alcohol syndrome

THE PROBLEM OF EXCRETION AND SALT AND WATER BALANCE IN ANIMALS

If animals are to maintain a relatively nonfluctuating fluid environment for their cells, they must have mechanisms for ridding their bodies of metabolic wastes—particularly nitrogenous ones, but many others as well. The process of releasing such useless, even poisonous, metabolic wastes is called ***excretion.*** It should not be confused with ***elimination*** (defecation). Whereas excretion is the release of wastes that have been produced inside the cells, tissue fluids, or blood of the organism, elimination is release of unabsorbed wastes from the digestive tract.

In general, excretory mechanisms also serve a second very important function: they help with ***osmoregulation***—that is, controlling the water and salt balance of the organism. Our examination of excretion will focus on both these aspects, which are in most instances inextricably intertwined.

Aquatic Animals Generally Use Ammonia as Their Nitrogenous Waste Product

As we saw, the first nitrogenous waste formed by deamination of amino acids is ammonia. Ammonia is an exceedingly poisonous compound; no organism can survive if the ammonia concentration in its body fluids gets very high. But the small, highly soluble molecules of ammonia readily diffuse across cell membranes, and there is no great difficulty in getting rid of them if an adequate supply of water is available. The water keeps the solution dilute while the ammonia is in the body, acts as a vehicle for its expulsion from the body, and flushes it rapidly away from the vicinity of the animal. In view of the plentiful supply of water available to aquatic animals, it is not surprising that for many of these the characteristic nitrogenous excretory product is ammonia.

Many marine invertebrates lack special excretory systems, relying instead on release of wastes across the general surface membranes. Such organisms seldom have any problem with water balance, because they are essentially isotonic with the surrounding seawater, and hence neither take in nor lose much excess water.

Maintenance of the proper nonfluctuating internal fluid environment is relatively simple for marine invertebrates as long as they remain in the sea; it is quite a different matter when they move into hypotonic media such as the brackish water of estuaries or the fresh water of rivers and lakes. Many marine animals are incapable of moving into such habitats, because their body fluids always lose salts until they have about the same salinity and osmotic concentration as the environmental fluids. Since their cells generally cannot tolerate much change in the makeup of the fluids bathing them, these animals soon die when they are put into brackish or fresh water. For this reason transporting and maintaining marine organisms in marine aquaria presents a challenge, because it is difficult to maintain the constant conditions found in seawater where waves and currents keep the water circulating and constantly renewed (Fig. 22.9).

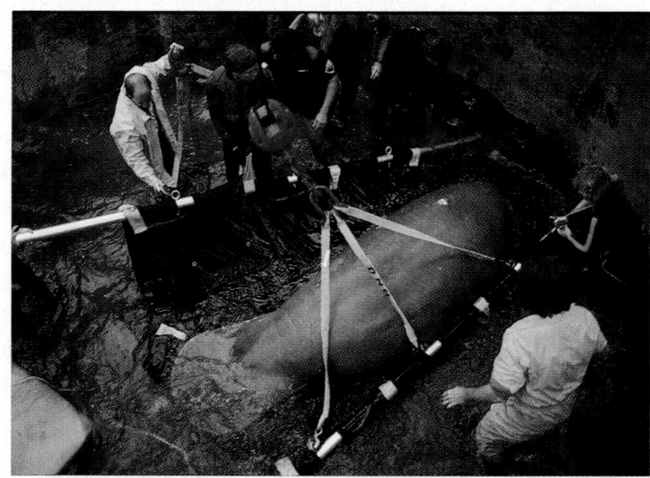

22.9 Moving Rosie Transporting marine organisms like Rosie, a manatee, can be difficult, though the problems are minimized by the fact that manatees have lungs and are air breathers. Transportation is more challenging in the case of marine fishes and invertebrates which may lack the ability to osmoregulate and must obtain their oxygen from the surrounding water. Maintaining the constant conditions found in seawater requires elaborate tanks and circulating devices.

Some marine animals, however, have evolved adaptations that enable them to move into hypotonic media. The adaptations may be of an evasive character, as in oysters and clams, which simply close their shells and thereby exclude the external water during those parts of the tidal cycle when the water in the estuaries is very dilute. But by far the most important adaptations for survival in dilute media—and the ones that have played the principal role in the evolutionary movement of animals into fresh water—are those that enable animals to regulate the osmotic concentrations of their body fluids and keep them constant despite fluctuations in the external medium. Organisms with this ability are said to have the power of osmoregulation. Their adaptations are generally in the form of mechanisms for taking up salt from the surrounding medium and bailing out the excess water that constantly pours in.

Freshwater Animals Gain Water From Their Environment

Once the ancestors of modern freshwater animals had made the transition to the freshwater environment, there was no great advantage to their descendants in continuing to maintain body fluids as concentrated as seawater, as long as they remained in their new environment. Such excessively hypertonic internal conditions simply aggravated the problems of obtaining enough salt and bailing out excess water. Natural selection, therefore, favored a reduction of the osmotic concentration of the body fluids within the bounds possible for the continuance of the life of the tissues, with the result that modern freshwater animals, both invertebrate and vertebrate, have osmotic concentrations decidedly lower than seawater (Table 22.1). The body fluids are not as dilute as fresh water, for no organism is actually isotonic with its freshwater medium.

Now, if freshwater animals are hypertonic relative to the surrounding medium, there will be a strong tendency for water to move into the organisms and for salts to be lost from the organisms to the surrounding water. To compensate, freshwater animals usually have excretory organs that can pump out water as fast as it floods in—preferably through the production of urine more dilute than the body fluids—and/or special secretory cells somewhere on the body that can absorb salts from the environment and release them into the blood. Both corrective measures—production of dilute urine and absorption of salts—entail movement of materials against concentration gradients and therefore necessitate the expenditure of energy.

An examination of the water and salt regulation typical of modern freshwater bony fishes illustrates these processes. The blood and tissue fluids of the fish are more concentrated than the environmental water. Although the outer covering of the fish is relatively impermeable and the fish almost never drink, there is a constant osmotic intake of water across the membranes of the gills and of the mouth, and a constant loss of salts across the same membranes. Correction of the resulting imbalance occurs in two ways: excess water is eliminated in the form of large amounts of very dilute urine produced by the kidneys, and salts are actively absorbed by specialized cells in the gills (Fig. 22.10). Energetically speaking, this process is very expensive; almost half of the ATP produced by cellular respiration is used to drive the active transport pumps for osmoregulation.

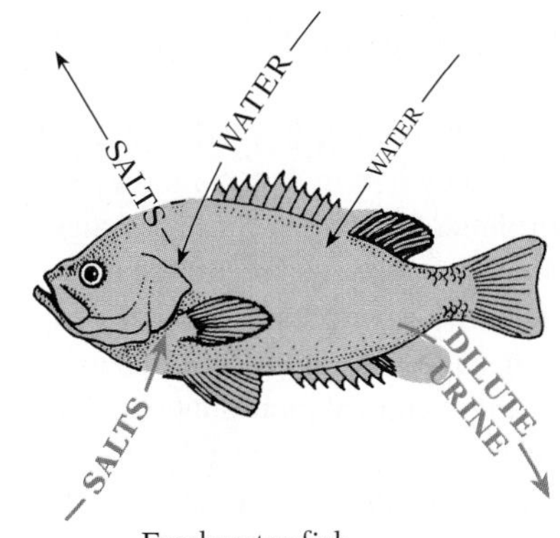

Freshwater fish
(hypertonic relative to medium)

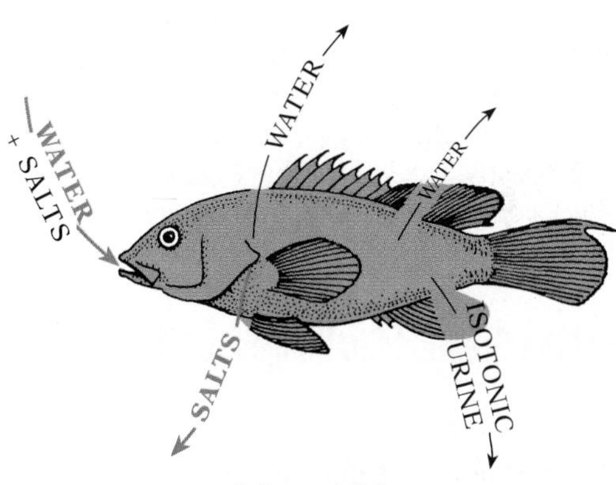

Saltwater fish
(hypotonic relative to medium)

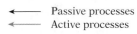

Passive processes
Active processes

22.10 Osmoregulation in bony fish Freshwater fish tend to take in excessive amounts of water and to lose too much salt. They compensate by seldom drinking, by actively absorbing salts through specialized cells in their gills, and by excreting copious dilute urine. Saltwater fish tend to lose too much water and to take in too much salt. They compensate by drinking constantly and by actively excreting salts across their gills. They cannot produce hypertonic urine; hence the kidneys are of little aid to them in osmoregulation.

MAKING CONNECTIONS

1. Why is blood-sugar homeostasis vitally important?

2. Why is the disposal of nitrogenous wastes a problem for animals but not for plants?

3. Why can a high consumption of alcohol lead to deposition of fat in the liver?

4. What adaptations have played a role in the movement of animals into fresh water?

Marine Vertebrates Lose Water to Their Environment

Curiously enough, bony fishes living in the sea have the reverse problem: they live in water, yet they steadily lose water to their environment and are in constant danger of dehydration, rather like living in a desert. The explanation is that the ancestors of the modern bony fishes evolved in fresh water, not in the sea, and when some of their descendants moved to the marine environment they retained their dilute body fluids. Hence marine bony fishes are hypotonic relative to the surrounding water, and they have the problem of excessive water loss and excessive salt intake. They use two corrective measures: they drink almost continuously to replace the water they are constantly losing, and by means of specialized cells in the gills, they actively excrete the salts they unavoidably take in with the water (Fig. 22.10). Most of the nitrogenous wastes are excreted as ammonia through the gills; only a small quantity of urine is produced by the kidneys, and little water is lost in this manner. Apparently, fish kidneys have not evolved the capacity to produce concentrated urine, and they are consequently of no help in salt elimination.

One of the Greatest Challenges to Life on Land Is Desiccation

One of the conditions of animal life in fresh water is a relatively impermeable covering for all but certain portions of the body surface, as an aid in preventing excessive absorption of water. For this reason freshwater animals had an important preadaptive advantage over the marine invertebrates in colonizing the terrestrial environment. Evidence suggests that the movement to land was by way of fresh water, not directly from the sea.

On land life is threatened by desiccation. Water is lost by evaporation from gas exchange surfaces (lungs, tracheae, etc.), by evaporation from the general body surface, by elimination in the feces, and by excretion in the urine. The lost water must be replaced by drinking, by eating foods containing water, and by the oxidation of nutrients (remember that water is one of the products of cellular respiration).

We saw that ammonia is a satisfactory nitrogenous excretory product for aquatic animals, because there is sufficient water to wash it out. It is far from satisfactory for terrestrial ones, however, because of the difficulty of getting rid of this highly toxic substance, where an unlimited water supply is not available. Amphibians and mammals rapidly convert ammonia into urea (Fig. 22.6), which, though quite soluble, is relatively nontoxic. Urea can remain in the body for some time before being excreted, and we can regard its production as an adaptation to the conditions of water shortage characteristic of terrestrial existence (Table 22.2).

Though urea is a far more satisfactory excretory product than ammonia for land animals, it has the disadvantage of draining away some of the critically needed water; being highly soluble, it must be released in an aqueous solution. If, however, uric acid (Fig. 22.6), a very insoluble compound, is excreted instead of urea, almost no water need be lost. Many terrestrial animals—most reptiles, birds, insects, and land snails—excrete uric acid or its salts. The excretion of this substance not only allows them to conserve water, but has another advantage, which may have been even more important in the evolution of uric acid metabolism. Most of these animals lay eggs enclosed within a relatively impermeable shell or membrane. If the embryos excreted ammonia, they would rapidly be poisoned, and if they produced urea, the concentration in the egg by the latter part of development would become decidedly harmful. Uric acid, on the other hand, is so insoluble that it precipitates out in almost solid form and can be stored in the egg without exerting harmful toxic or osmotic effects. In the nitrogen metabolism of fully terrestrial animals, uric acid excretion is correlated with egg laying, while urea excretion is correlated with giving birth to living young (Table 22.2).

EXCRETORY MECHANISMS IN ANIMALS

Contractile Vacuoles Expel Excess Water and Some Wastes

Special excretory structures are absent in many unicellular and simple multicellular animals. Nitrogenous wastes are simply excreted across the general cell membranes into the surrounding water. Some protozoa do, however, have a special excretory organelle, the contractile vacuole (Fig. 22.11). Each vacuole goes through a regular cycle consisting of a stage in which it fills with liquid and become larger and larger, followed by a contraction stage in which the contents of the vacuole are ejected from the cell. Though contractile vacuoles excrete some nitrogenous wastes, most of this waste diffuses across the cell surface, and it seems clear that the primary function of contractile vacuoles is elimination of excess water.

Flame-Cell Systems Are Characteristic of Flatworms

The beginnings of a tubular excretory system can be seen in the flatworms (planaria, flukes, tapeworms, etc.). These animals are relatively small and lack a functional body cavity—their bodies are solid and they have no major break in the tissue mass between the outer epithelium of the body and the gastrovascular cavity exists, nor do they have a circulatory system.

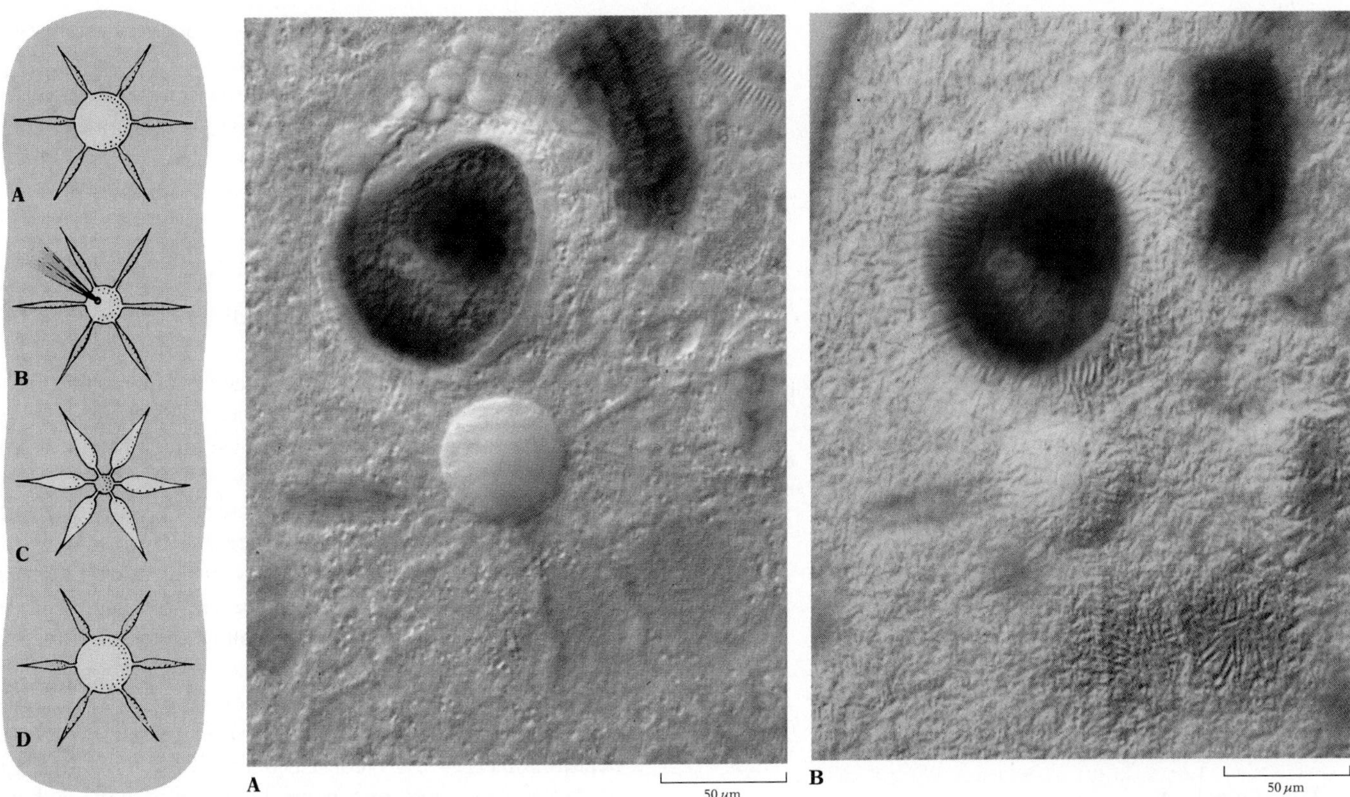

22.11 Contractile vacuole of *Paramecium caudatum* The sequence shown diagrammatically above can also be observed in the three photographs of a live organism taken in the course of the expansion-contraction cycle of its contractile vacuoles (only one vacuole is seen here). The large green object and the brown ones are remains of other microorganisms ingested by the *Paramecium.* (A) The vacuole is full. As shown in the diagram, a system of radiating canals brings fluid from the cytoplasm to the vacuole. (B) The vacuole is in the process of expelling its contents; in the photograph the opening to the outside can be seen as a small circle on the surface of the vacuole. (C) The vacuole is nearly empty, but the radiating canals are collecting more fluid from the cytoplasm and fill the reservoir again (D). Photographs taken with Nomarski optics.

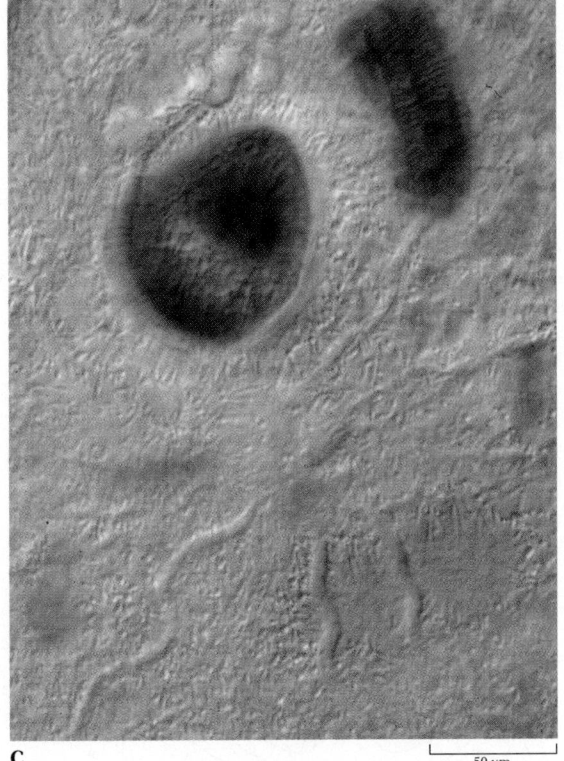

Flatworm excretory systems usually consist of two or more longitudinal branching tubules running the length of the body (Fig. 22.12). In planaria and its relatives, the tubules open to the body surface through a number of tiny pores. The critical portions of these systems are many small bulblike structures located at the ends of side branches of the tubules. Each bulb has a hollow center into which a tuft of long cilia projects. The hollow centers of the bulbs are continuous with the cavities of the tubules, allowing water and some waste materials to move from the tissue fluids into the bulbs. The constant waving of the cilia creates a current that moves the collected liquid through the tubules to the excretory pores, where it leaves the body. Because the motion of the tuft of cilia resembles the flickering of a flame, this type of excretory system is often called a ***flame-cell system.***

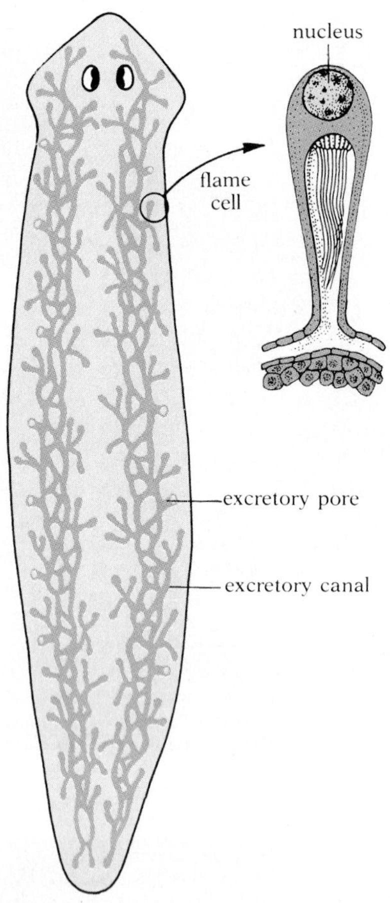

22.12 Flame-cell system of a planarian Each of the two excretory canals consists of a longitudinal network of tubules, some ending in flame cells (one is shown enlarged at right) and others in excretory pores. The cilia in the flame cells create currents that move water and waste materials through the canals and out through the pores.

Like the contractile vacuoles discussed earlier, flame-cell systems seem to function primarily in the regulation of water balance; most metabolic wastes of flatworms are excreted from the tissues into the gastrovascular cavity and eliminated from the body through the mouth.

Nephridia of Earthworms Function in Both Excretion and Osmoregulation

Flame-cell systems, because they function in animals without circulatory systems, pick up substances only from the tissue fluids. In animals that have evolved closed circulatory systems, the blood vessels have become intimately associated with the excretory organs, making possible direct exchange of materials between the blood and the excretory system.

The critical role of the circulatory system in excretion can be observed in the earthworm. The earthworm's body is composed of a series of segments internally partitioned from each other by membranes. In general, each of the compartments thus formed has its own pair of excretory organs, called ***nephridia,*** which open independently to the outside. A typical nephridium (Fig. 22.13) consists of an open ciliated funnel, or nephrostome (which corresponds functionally to the bulb of a flame-cell system), a coiled tubule running from the nephrostome, a bladder into which the tubule empties, and a nephridiopore through which materials are expelled from the bladder to the outside. Blood capillaries form a network around the coiled tubule. Materials move from the body fluids into the nephridium through the open nephrostome, but materials are also picked up by the coiled tubule directly from the blood in the capillaries. There is probably also some reabsorption of materials from the tubule into the capillaries. The principal ad-

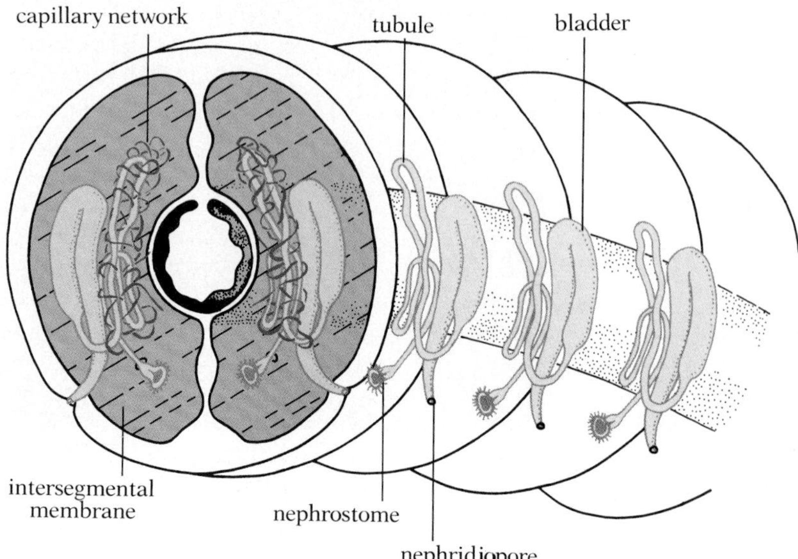

22.13 Nephridia of an earthworm Each segment of the worm's body contains a pair of nephridia, one on each side. The open nephrostome of each nephridium is located in the segment ahead of the one containing the rest of the nephridium. The tubule from the nephrostome penetrates through the membranous partition between the segments and is then thrown into a series of coils, with which a network of blood capillaries is closely associated (the capillaries are shown here only on the nephridia of one segment). The coiled tubule empties into a storage bladder that opens to the outside through a nephridiopore.

vance of this type of excretory system over the flame-cell system, then, is the association of blood vessels with the coiled tubule, which allows waste substances to be transferred directly from the blood to the tubule, and useful substances to be reabsorbed from the tubule into the blood.

Malpighian Tubules Are Excretory Organs in Insects

Insects probably evolved from an ancestral form similar to the ancestor of the modern segmented worms, and this ancestor, like the earthworm, probably had nephridia. Yet insects do not have nephridia, nor have their excretory organs evolved from nephridia. The evolution of an open circulatory system in insects, with their consequent lack of blood capillaries, probably accounts for the evolutionary loss of nephridia, which are dependent on capillaries. Instead, insects and many of their relatives have evolved an entirely new excretory system, one that functions well in association with an open circulatory system.

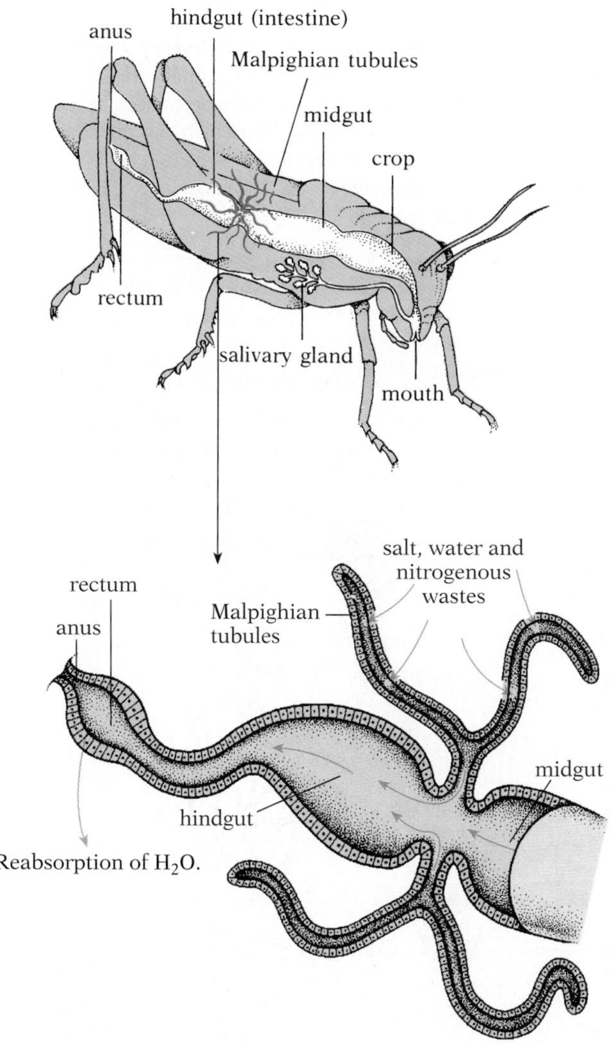

22.14 Malpighian tubules of an insect These excretory organs arise as outgrowths of the digestive system at the junction between the midgut and hindgut. The blind-ended tubules are bathed by the blood; fluid from the blood is secreted into the tubules and passes into the hindgut.

The excretory organs of insects are called ***Malpighian tubules.*** They are outgrowths of the digestive tract located at the junction between the midgut and the hindgut (Fig. 22.14). These thin, blind-ended tubules, variable in number, are bathed directly by the blood in the open sinuses of the animal's body. Fluid moves from the blood into the distal end of the Malpighian tubules. As the fluid moves through the tubules, the nitrogenous material is precipitated as uric acid and much of the water and various salts are reabsorbed into the blood. The concentrated, but still fluid, urine next passes into the hindgut and then into the rectum. The rectum has very powerful water-reabsorptive capacities, and the material leaving the rectum is dry.

THE VERTEBRATE KIDNEY

Like the nephridial system of earthworms, the excretory systems of vertebrates are closely associated with the closed circulatory system. When an efficient circulatory system can bring wastes to the excretory organs, the functional excretory units no longer have to be scattered throughout the body tissues, as in planaria. And the absence of internal segmentation of the body obviates the need for a series of individual excretory organs, as is found in earthworms. Vertebrates have typically evolved compact discrete organs, the kidneys, in which the functional units are massed. In humans the kidneys are located in the back of the abdominal cavity.

The functional units of the kidneys of higher vertebrates are called ***nephrons.*** Each nephron includes a cuplike bulb, called a ***Bowman's capsule,*** that opens into a long, coiled tubule. The tubules of the various nephrons empty into collecting tubules, which in turn empty into the central cavity of the kidney, the ***pelvis.*** From the pelvis, a large duct leaves each kidney and runs posteriorly. In mammals, the ducts, called ***ureters,*** empty into the ***urinary bladder.*** This storage organ drains to the outside via another duct, the ***urethra*** (Fig. 22.15).

The blood capillaries, capsules, and tubules of the nephrons are intimately associated in the modern vertebrate kidney. No longer are materials picked up from the general body fluids; exchange of substances takes place only between blood capillaries and nephrons. Blood reaches each kidney via a ***renal artery,*** a short vessel leading directly from the aorta to the kidney (Fig. 22.15). The renal artery enters the kidney and breaks up into many tiny branch arterioles, each of which leads into a Bowman's capsule. Within each capsule, the arteriole breaks up into a tuft of capillaries called the ***glomerulus*** (Fig. 22.16). Blood leaves the glomerulus via an arteriole formed by

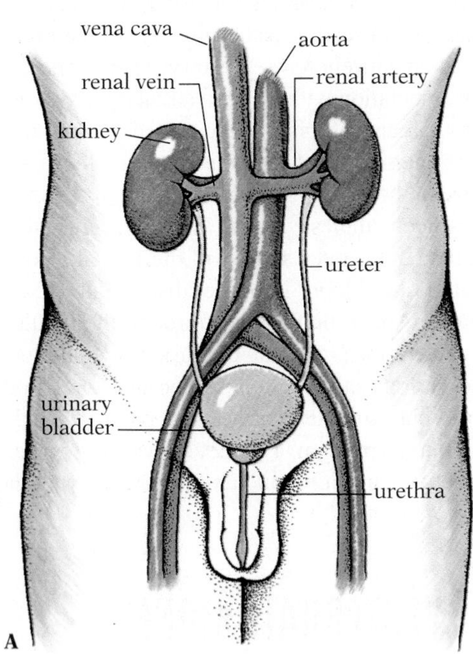

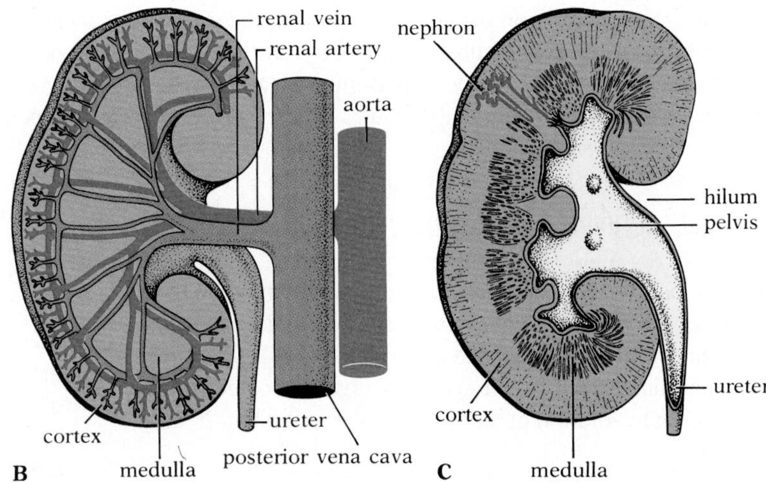

22.15 The human excretory system and sections of the human kidney (A) The organs and vessels are shown larger relative to the body than they actually are. (B) The blood circulation of the kidney. (C) The kidney consists of an outer portion, called the cortex, an inner medulla, and the large renal pelvis (yellow) into which the collecting tubules of the nephrons empty. One nephron is shown (red). The glomerulus, Bowman's capsule, and convoluted tubules are in the cortex, while the loop of Henle and the collecting tubules run down into the medulla.

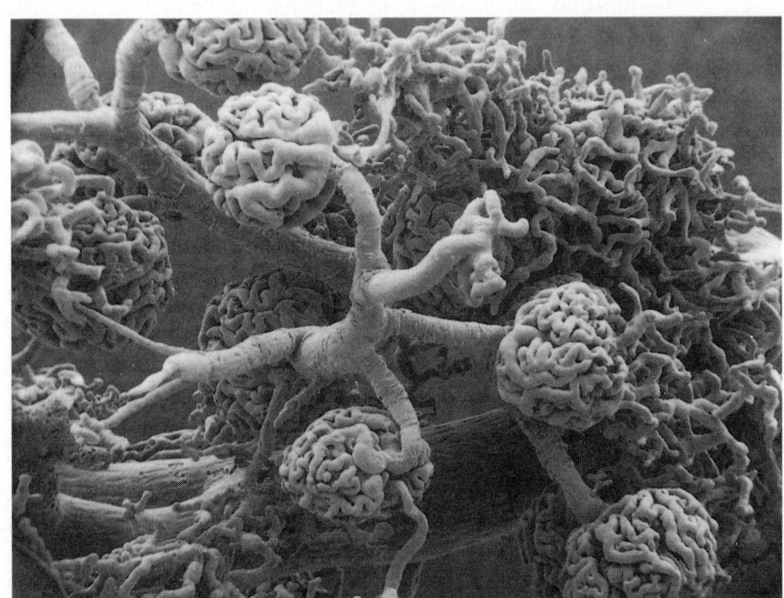

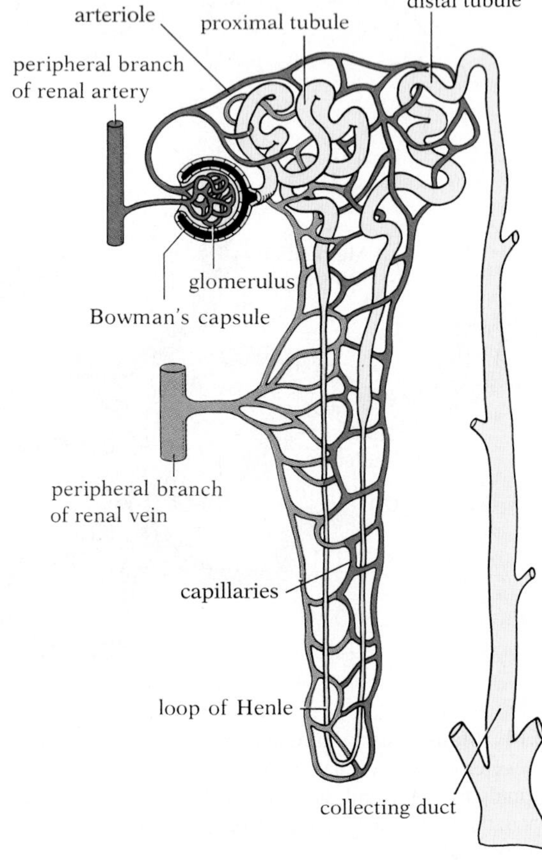

22.16 Structures of the human kidney Left: Blood vessels within the kidney. Blood enters the kidney via the renal artery, which breaks up into progressively smaller branches. Each branch divides (center) to form a series of arterioles that lead into the glomerulus. The Bowman's capsule is not shown in this scanning electron micrograph. ×168. Right: The nephron. High hydrostatic blood pressure within the glomerulus forces small molecules from the blood into the Bowman's capsule. As the filtrate moves through the tubules, water and most of the substances needed by the body are reabsorbed into a second set of capillaries surrounding the tubules.

the rejoining of the glomerular capillaries. After emerging from the capsule, the arteriole promptly divides again into many small capillaries that form a second dense network, around the tubule of the nephron. Finally, these capillaries unite once more to form a small vein. The veins from the many nephrons then fuse to form the ***renal vein,*** which leads to the posterior vena cava. The mammalian kidney, like the liver, is the second place where blood circuits incorporate two sets of capillaries (i.e., a portal system).

The Formation of Urine Involves Filtration, Reabsorption, and Secretion

Knowing the structural relationships of the human kidney, we can now examine the mechanism of urine formation. Three processes are involved in urine formation: filtration, reabsorption, and tubular secretion.

In 1844 the German physiologist Carl Ludwig suggested that a glomerulus acts as a simple mechanical filter—that molecules small enough to pass through the capillary walls, and through the thin membranous walls of the capsule, filter from the blood into the nephron as a result of the high blood pressure in the glomerulus. Modern microscopy has provided evidence as to how such filtration might take place. Pores can be detected in the walls of both the glomerular capillaries (Fig. 22.17) and the Bowman's capsule. The high blood pressure presumably forces ions, water, and other small molecules (such as glucose, amino acids, and urea) through the pores from the capillaries into the capsule, where the pressure is low. Research has shown that the percentage composition of the liquid entering the nephron (its composition expressed by percentages of constituents) is basically the same as that of blood; it lacks only the blood cells and the plasma proteins, both of which are too large to filter through the membranes to any appreciable extent. The cells of the glomerular capillaries and of the Bowman's capsules do not actively transport materials from the glomeruli into the capsules; instead, the materials are forced out by the beating heart as it pumps the blood under high pressure into the glomeruli. Because the filtrate is formed by pressure, the process is referred to as ultrafiltration. The caffeine in coffee, tea, and colas acts to increase glomerular filtration, and therefore increases the volume of the filtrate.

The second process of urine formation, reabsorption, prevents the essential substances contained in the filtrate from being lost. Water alone accounts for 170 liters of filtrate every day, which would need to be replaced if it were removed from the body. (That would mean drinking 50 gallons of water a day.) Selective reabsorption of most of the water and many of the dissolved materials is one of the functions of the tubules of the nephrons. In humans the filtrate passes first through the ***proximal convoluted tubule,*** then through the long ***loop of Henle,*** then through the ***distal convoluted tubule,*** and finally

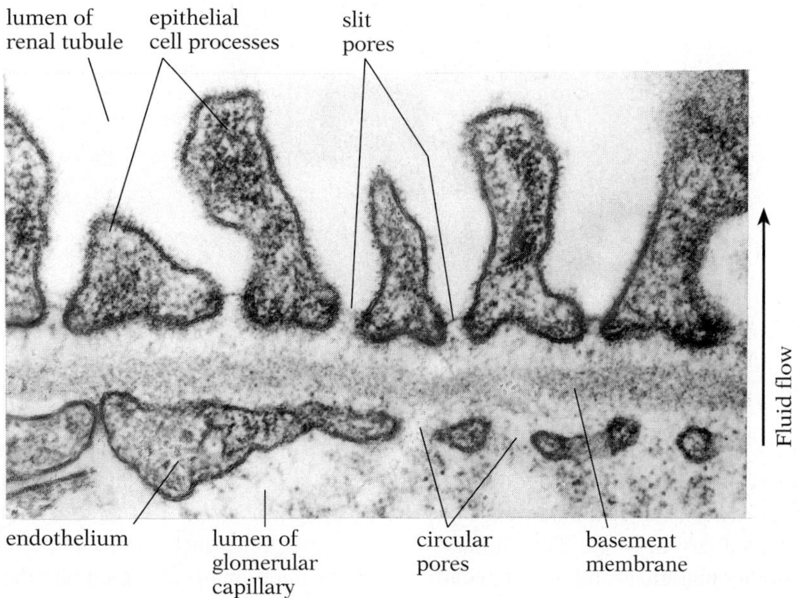

22.17 Electron micrograph of a section through the glomerular filtration barrier of a rat The glomerular filtration barrier has a complex structure consisting of three layers: a highly attenuated endothelium perforated with numerous circular pores, a continuous basement membrane made up of fine filamentous meshwork, and the epithelial cell processes, which are separated by elongated spaces called slit pores. The actual site of filtration is thought to be the basement membrane, which may contain much smaller pores (not visible here) just large enough to allow the passage of water, electrolytes, and other small molecules, and yet small enough to retain albumen and other plasma proteins.

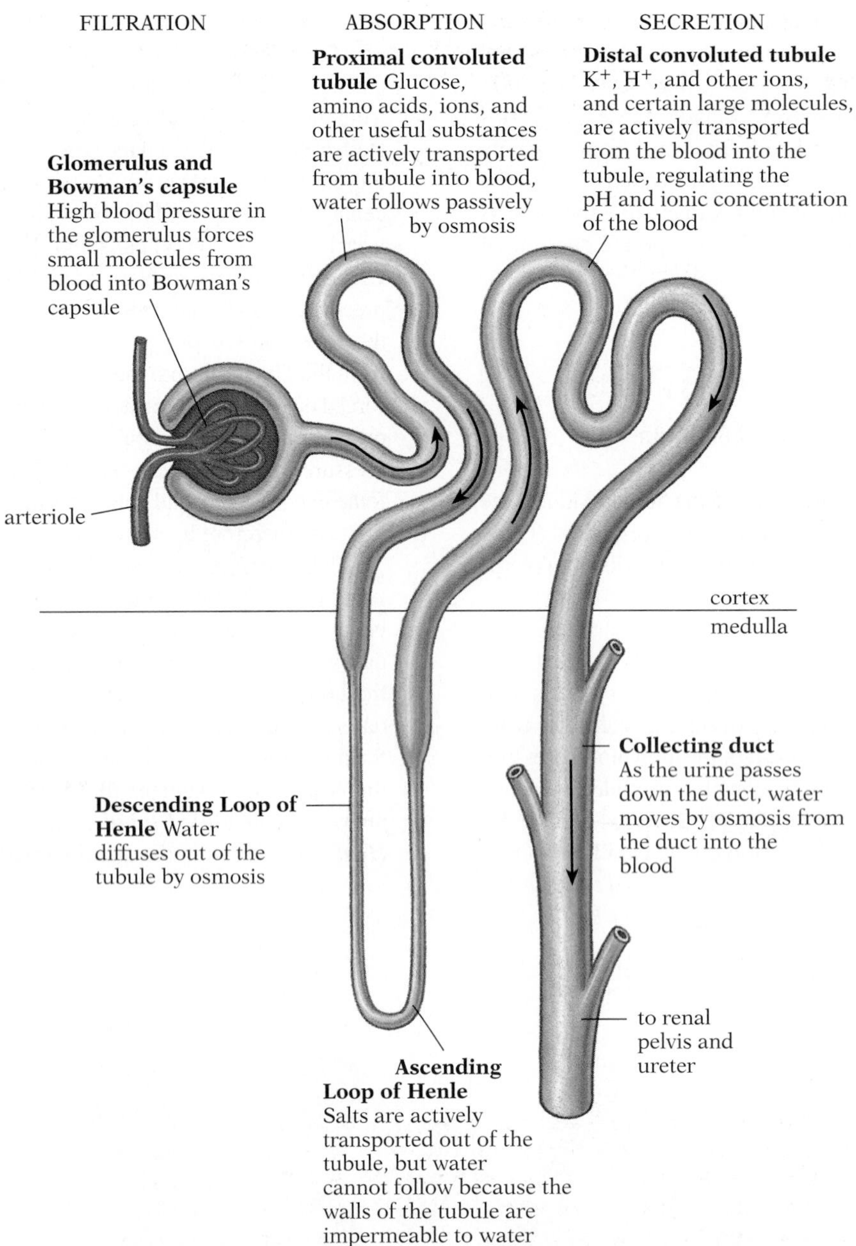
FILTRATION
ABSORPTION
SECRETION
Proximal convoluted tubule Glucose, amino acids, ions, and other useful substances are actively transported from tubule into blood, water follows passively by osmosis
Distal convoluted tubule K^+, H^+, and other ions, and certain large molecules, are actively transported from the blood into the tubule, regulating the pH and ionic concentration of the blood
Glomerulus and Bowman's capsule High blood pressure in the glomerulus forces small molecules from blood into Bowman's capsule
arteriole
cortex
medulla
Collecting duct As the urine passes down the duct, water moves by osmosis from the duct into the blood
Descending Loop of Henle Water diffuses out of the tubule by osmosis
to renal pelvis and ureter
Ascending Loop of Henle Salts are actively transported out of the tubule, but water cannot follow because the walls of the tubule are impermeable to water

22.18 A summary of the structure and function of a human nephron

into the ***collecting tubule*** (Fig. 22.18). As the filtrate moves through the tubules, as much as 99 percent of the water may be reabsorbed by osmosis through the cells of the tubule walls and returned to the blood in the capillary network. Thus human kidneys (and the kidneys of other mammals and of birds) can produce concentrated urine—urine that is hypertonic relative to the blood plasma—even though the initial filtrate was nearly isotonic. Whether the collecting tubule finally releases dilute or concentrated urine depends on whether or not there is a deficiency or an excess of water in the body at the moment.

The pituitary gland responds to changes in the amount of water in the body by releasing a hormone, ***vasopressin,*** that regulates the permeability of the collecting tubule to water. When it is present, the ducts are more permeable and more water leaves the ducts and is absorbed into the blood. As a result, less urine is produced, and it is more concentrated. When vasopressin is absent, the collecting tubule becomes quite impermeable to water, and water remains in the ducts and passes down the ureter to the bladder, producing large volumes of dilute urine. Ethyl alcohol inhibits vasopressin release, so it acts as a diuretic—that is, it tends to increase the volume of urine produced.

Water is not the only substance reabsorbed by the tubules of the nephrons. In a normal healthy person, glucose, almost all the amino acids, and most of the inorganic ions are also reabsorbed and returned to the blood. Most of this reabsorption involves active transport, and thus energy expenditure by the tubule cells. The cells of the proximal convoluted tubule are important in this regard. The membranes of these cells are packed with protein pumps, and the glucose, amino acids, and many inorganic ions are actively absorbed using these pumps. Water reabsorption is passive, by osmosis. Almost 85 percent of the filtered water and ions are reabsorbed in the proximal convoluted tubule and the first part of the loop of Henle. Salts are actively transported out of the filtrate by the cells lining the ascending loop of Henle, but water cannot follow by osmosis because walls of the ascending limb are impermeable to water. Much of the remainder of the water is reabsorbed in the collecting duct (Fig. 22.18). Notice that, in general, the strategy followed by the kidney in urine formation is to filter small molecules out of the blood in the glomerulus, and then to reabsorb into the capillaries surrounding the tubules only what is to be saved. Hence the kidney effectively and automatically removes many toxins and chemicals by simply not transporting them back into the blood.

In spite of the extensive reabsorption that may take place, urine is more than a concentrated solution of urea. Most substances have what is called a kidney threshold level. If the concentration of such a substance in the blood exceeds its kidney threshold level, the excess is not reabsorbed from the filtrate by the tubules but instead appears in the urine. Glucose is an example of a substance with a high threshold value. Ordinarily all glucose in the filtrate is reabsorbed, because the threshold level for glucose is higher than the normal blood-glucose level. If, however, the blood-sugar level is abnormally high, as in diabetes, the kidneys cannot absorb all the excess sugar, and it appears in the urine. This elimination of excess sugar by the kidneys points up once again that excretory organs do far more than remove nitrogenous wastes; they play a critical role in maintaining the relatively nonfluctuating internal fluid environment of the organism. In this case, when the liver and/or the peripheral tissues are not functioning properly and the blood-sugar level rises, the kidneys act as a second line of defense. The nervous system and hormones from various endocrine glands (discussed in Chapter 25) regulate the function of the kidneys. Under their influence, the kidneys control the relative concentrations of such inorganic ions as sodium, potassium, and chloride in the blood plasma so the composition of the blood is maintained at a nearly constant level. The latter function is particularly important. When we think of the kidney we tend to think only of its role in the excretion of waste products, but its role in osmoregulation is even more crucial. A person without functioning kidneys could survive for weeks before the concentration of urea got high enough to cause death, but would die in a matter of days due to an osmotic imbalance.

The movement of materials between the tubules and the capillaries surrounding them is not completely one-way. Some chemicals are actively transported *from* the blood *into* the distal convoluted tubules and deposited in the urine. Certain ions, particularly hydrogen and potassium ions, and large molecules, such as uric acid and foreign chemicals that have undergone biotransformation in the liver, enter the urine by this

MAKING CONNECTIONS

1. In what respect were freshwater animals well prepared for life on land?

2. How are the excretory systems of earthworms and insects related to the type of circulatory system (open or closed) of each animal?

3. Describe the location and function of the portal system of the kidney.

4. Why does a person who gets drunk and wakes up with a hangover almost always feel extremely thirsty?

5. How does the occurrence of glucose in the urine of a diabetic demonstrate the homeostatic role of the kidneys?

6. Suppose you are adrift in a life raft in the middle of the ocean without a supply of fresh water. Why will drinking seawater cause you to become dehydrated even faster than if you drink no water at all?

process. This is ***tubular secretion,*** the third process involved in urine formation, which supplements glomerular excretion and increases the efficiency of the overall excretory regulation of blood composition.

The urine that enters the ureter has a maximum salt concentration of about 2.2 percent compared to the blood, which has a salt concentration of about 0.9 percent. The human kidney then, is capable of producing hypertonic urine, but notice that the maximum concentration is 2.2 percent. This maximum explains why humans cannot drink seawater, which is 3.5 percent salt. In order for this salt to be removed from the body, water from the body tissues will move by osmosis into the nephron to dilute the salt in the urine from 3.5 percent to 2.2 percent. Consequently humans drinking seawater will excrete more water than they drink. Human kidneys are simply not adapted to life at sea or to life in very dry habitats.

Damage to the kidneys can result from nephritis (inflammation of the kidney), high blood pressure (which damages the glomeruli), various drugs and toxic substances, or physical injury. When some nephrons are damaged, other nephrons must work harder to compensate. Depending on the amount of damage, the remaining nephrons often cannot function adequately. Persons with kidney failure are often placed on dialysis machines, which do the work usually performed by the kidneys (Fig. 22.19). Kidney stones are another, rather common problem. They are usually composed of calcium salts, which are insoluble and precipitate out in the pelvis of the kidney. Such stones can be prevented by drinking sufficient amounts of water, which will tend to wash the crystals out before a large stone can form.

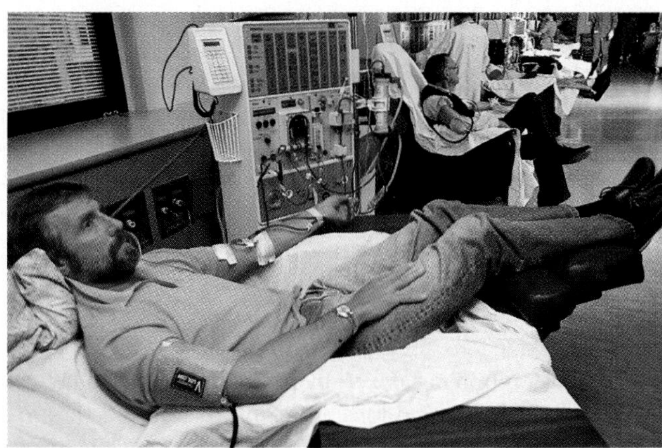

22.19 Patient undergoing kidney dialysis Blood from the patient flows into the artificial kidney where unwanted substances are removed. The blood passes through tiny cellophane channels that are surrounded by dialyzing fluid. Substances move by diffusion through the porous cellophane.

CONCEPTS IN BRIEF

1. Living cells require a relatively constant environment to carry out their life functions. As complex multicellular animals arose, body fluids developed that could provide a stable environment for the internal cells. A variety of mechanisms evolved for maintaining the *homeostasis* of the body fluids despite changes in the external environment.

2. The extracellular fluids of plants are not separate from the environmental medium. Consequently plants only regulate the composition of their intracellular fluids, whereas animals must regulate the composition of both intracellular and extracellular fluids. Plant cells, because of their rigid cell walls, can survive in a hypotonic medium. However, if the external fluids become hypertonic, the cell will lose water and shrink.

3. In vertebrates, the liver is important in helping to maintain a constant environment for the body cells. Because of the hepatic portal system, the products of digestion are brought directly to the liver cells, where the levels of these digestive products in the blood can be regulated.

4. The liver regulates the blood-sugar level and is the center for fat and amino acid metabolism. Excess amino acids are *deaminated,* and the amino group (—NH_2) is converted into *ammonia* (NH_3). The ammonia may be converted into *urea* or *uric acid.* The liver also modifies foreign substances, changing them from nonpolar to polar substances, which can then be more easily excreted from the body.

5. Animals need mechanisms to rid their bodies of foreign chemicals and metabolic wastes *(excretion)* and to regulate their salt and water balance *(osmoregulation).*

6. The nitrogenous excretory product characteristic of most aquatic animals is ammonia. Many marine invertebrates release their nitrogenous wastes across their body surfaces. These organisms are isotonic with their environment and so do not have problems with water balance.

7. Freshwater animals have an osmotic concentration that is less than that of seawater but greater than that of fresh water. A freshwater organism tends to gain water and lose salts to the surrounding water. To survive, the organism must get rid of the excess water and must absorb salts. Freshwater fishes almost never drink water, produce copious amounts of urine, and have special cells on their gills to actively absorb salts.

8. Marine bony fishes are hypotonic relative to seawater. They compensate for a tendency to lose water and take in salt primarily by drinking all the time and actively transporting salts out of the body.

9. Amphibians and mammals produce urea as their nitrogenous waste product; most reptiles, birds, insects, and land snails excrete uric acid. Uric acid excretion is correlated with egg laying, whereas urea excretion is correlated with giving birth to living young.

10. Many aquatic unicellular and simple multicellular animals simply excrete their nitrogenous wastes across membrane surfaces. Some protozoa have a *contractile vacuole,* which collects excess fluid and expels it from the cell. Flatworms have a *flame-cell system* to get rid of excess water.

11. In animals that have evolved closed circulatory systems, the blood vessels have become intimately associated with the excretory organs, making possible direct exchange of materials between the blood and the excretory system. The *nephridial system* of earthworms consists of a series of tubules (a pair for each segment of the body) that are closely associated with the blood vessels.

12. The excretory organs of insects are called *Malpighian tubules.* As outgrowths of the digestive tract, they are bathed directly by the blood. Fluid from the blood moves into tubules, where uric acid is precipitated. The concentrated urine then moves into the hindgut and rectum, where water is reabsorbed.

13. The vertebrate excretory organ is the kidney, which is composed of *nephrons.* High blood pressure within the *glomerulus* forces small molecules from the blood into the *Bowman's capsule.* As the filtrate moves through the tubules most of the water and useful substances are reabsorbed into a second set of capillaries associated with the tubules. In addition, some chemicals are actively transported from the blood into the tubules.

14. The kidneys play a vital role in maintaining homeostasis; they help to regulate the blood-sugar level and the concentration of various inorganic ions in the blood, and they rid the body of waste products.

STUDY QUESTIONS

1. What is a portal system? What does the hepatic portal system connect? What is the function of this system? (p. 455)
2. Suppose you ate two candy bars for breakfast and then ate nothing else until late in the afternoon. Describe how your liver would regulate your blood-sugar level during breakfast and the subsequent fast. (pp. 456–457)
3. Explain why the liver is a "biochemical crossroads" for carbohydrate, fat, and protein metabolism. Why is deamination the first step in converting amino acids into carbohydrate or fat? (p. 458)
4. Describe three functions of the liver in addition to its role in nutrient metabolism. (p. 459)
5. Is urine *eliminated* or *excreted?* What about feces? What is the distinction between the terms? (p. 462)
6. Contrast the osmoregulatory problems of freshwater and saltwater bony fish. How does each solve its problem? (pp. 462–464)
7. Explain how the nitrogenous waste products of animals are adapted to their habitats and reproductive methods. (p. 464)
8. Describe the excretory mechanisms and structures found in protozoa, flatworms, earthworms, and insects. Which of these structures are primarily osmoregulatory? (pp. 464–467)
9. Trace the pathway of a molecule of urea from the point where it leaves the blood stream and enters a kidney nephron to the point where it is excreted from the body. (p. 467)

10. How and where does filtration occur in a nephron? What is the energy source for this process? Compare the composition of the filtrate with that of the blood. How much filtrate is produced by the kidneys each day? How do coffee, tea, and colas affect the filtration rate? (p. 469)

11. How and where does most reabsorption occur in a nephron? What is the primary energy source for the reabsorption of nutrients and salts? How is water reabsorbed? (pp. 469–470)

12. How does the permeability of the collecting tubules of the nephrons determine whether a dilute or concentrated urine is produced? How is their permeability affected by vasopressin? (pp. 470–471)

13. What kinds of substances are subject to tubular secretion by the kidney? In which section of a nephron does most secretion occur, and what is its function? (pp. 471–472)

CHEMICAL CONTROL IN PLANTS

Most organisms respond to changes in their environment: bacteria swim up a food gradient, plants bend their leaves toward the sun, animals generally move toward food and away from predators. Whatever the organism, three steps are involved in the flow of information: reception of the relevant stimuli by the organism, communication of the information from the receptor site to the area where responses are generated, and the response itself. In unicellular organisms, all three functions are carried out in one cell, but most multicellular organisms have tissues specialized for each function. Coordination of these processes depends on chemical control mechanisms, which are found in all organisms, and nervous control mechanisms, which are found only in multicellular animals. This chapter and the next three will be concerned with the first of these mechanisms—chemical control.

Plants respond to external stimuli such as light, temperature, gravity, the length of day, etc., but because they lack a nervous system and cannot move about as animals do, their response is slower, and not as obvious. Because they are sedentary, plants generally respond to stimuli by growth—e.g., stems and leaves grow toward the sun, roots grow downward, toward the pull of gravity, and stems grow upward, away from the pull of gravity. Sometimes the response is quite rapid; for example, minutes after a plant is deprived of water it begins to wilt, the stomata close, and certain genes are activated. Rapid transcription and translation follow, producing proteins that help the plant withstand the effects of drought.

Plants pass information between cells almost exclusively by chemical means. Once the stimulus is received, the ***hormones,*** special control chemicals by which plants transmit signals, are produced. The hormones may act locally, or they may be transported to target cells in another part of the plant, where they influence physiological processes. Hormones act as intercellular messengers and are effective in very low concentrations.

For communication to take place between cells, each hormone must deliver its message to its special target, and must influence a specific physiological process. The various hormones have different ways to elicit particular cellular responses. For instance, hormones can affect specific structures (as we shall see, plant auxins indirectly alter the cell wall), modify biochemical pathways, influence the production and /or activity of enzymes, or interact directly with the chromosomes to turn particular genes on or off (the likely mode of action of plant gibberellins). In the next four chapters, we shall be concerned with the details of how particular hormones function, and with the role of each in integrating the activities of the many specialized tissues and organs in complex multicellular plants and animals.

Chapter opening photo: The persimmons on the left were placed in a paper bag with slices of ripe apple which release the gaseous hormone ethylene, ripening the fruit.

Plant Hormones Act on a Variety of Tissues and Mediate a Great Variety of Functions

Most plants have specialized organs—roots, stems, leaves, and reproductive organs—whose activity must be coordinated if the plant is to grow and reproduce. Chemical control plays a role in orienting and regulating the growth of stems and roots, in timing both reproduction and shedding of leaves, in initiating the germination of seeds, and in regulating many other functions.

Plant hormones, at least those so far known, are produced most abundantly in the actively growing parts of the plant body, such as the apical meristems of the shoot and the root, young growing leaves, and developing seeds and fruits. These tissues not only produce hormones, however; they carry on other functions as well. There are no specialized hormone-producing organs in plants analogous to the endocrine glands of higher animals. Furthermore, plant hormones are predominantly involved in regulating growth, differentiation, and development (they are often called growth regulators), while animal hormones mediate a great variety of functions involving homeostasis as well as growth and development.

There are a number of other differences between plant and animal hormones. Animals have evolved a host of different hormones, each with a specific target (organ or tissue) and function. Plants utilize a different strategy. There are relatively few plant hormones, but each hormone may act on a variety of tissues and mediate a great variety of functions. The physiological effect of a particular plant hormone depends to a great extent both on the concentration of that hormone and on the relative concentrations of other hormones that may be present. For instance, auxin in low concentrations stimulates root growth, but in high concentrations it inhibits growth. Here a single hormone can act as either a growth promoter or a growth inhibitor; it is the concentration of the hormone and the tissue involved that determines its effect. The presence of other hormones can also modify the physiological effect of a particular hormone. In some situations plant hormones act cooperatively to control various aspects of development; in other situations the hormones act in opposition to one another, and it is the balance of hormones that determines the effect.

Five classes of plant hormones are known: ***auxin, gibberellin, cytokinin, abscisic acid,*** and ***ethylene.***

Auxin Plays an Important Role in Controlling Cell Elongation

One of the earliest investigated and best-known groups of plant hormones, or growth regulators, includes those hormones collectively known as auxins. Plants grow both by cell division, which takes place principally in the meristems, and by the elongation and enlargement of cells already present, particularly in the stem and root. The auxins have an important effect on the latter sort of growth—cell elongation.

Since plant cells, unlike animal cells, are enclosed in a box-like cell wall, cell growth is possible only if the walls can be extended. It is in this process—the elongation of the walls—that auxin is so important. The walls are composed primarily of polysaccharides, of which cellulose is the one we have mentioned most often. In primary walls of growing cells, the cellu-

MAKING CONNECTIONS

1. What are the three steps involved in the process by which an organism responds to a stimulus?

2. Contrast the mechanisms and speed with which plants and animals respond to stimuli.

3. Compare the characteristics of plant and animal hormones, including number, sites of synthesis, specificity of target tissues, role in maintaining homeostasis, and role in growth and development.

4. How does the acid growth hypothesis demonstrate the sensitivity of enzymes to changes in pH? Based on what you learned about enzymes in Chapter 4, why would a change in pH cause a change in enzyme activity?

5. Relate the role of auxin in cell elongation to the concept of genetic control of growth and development.

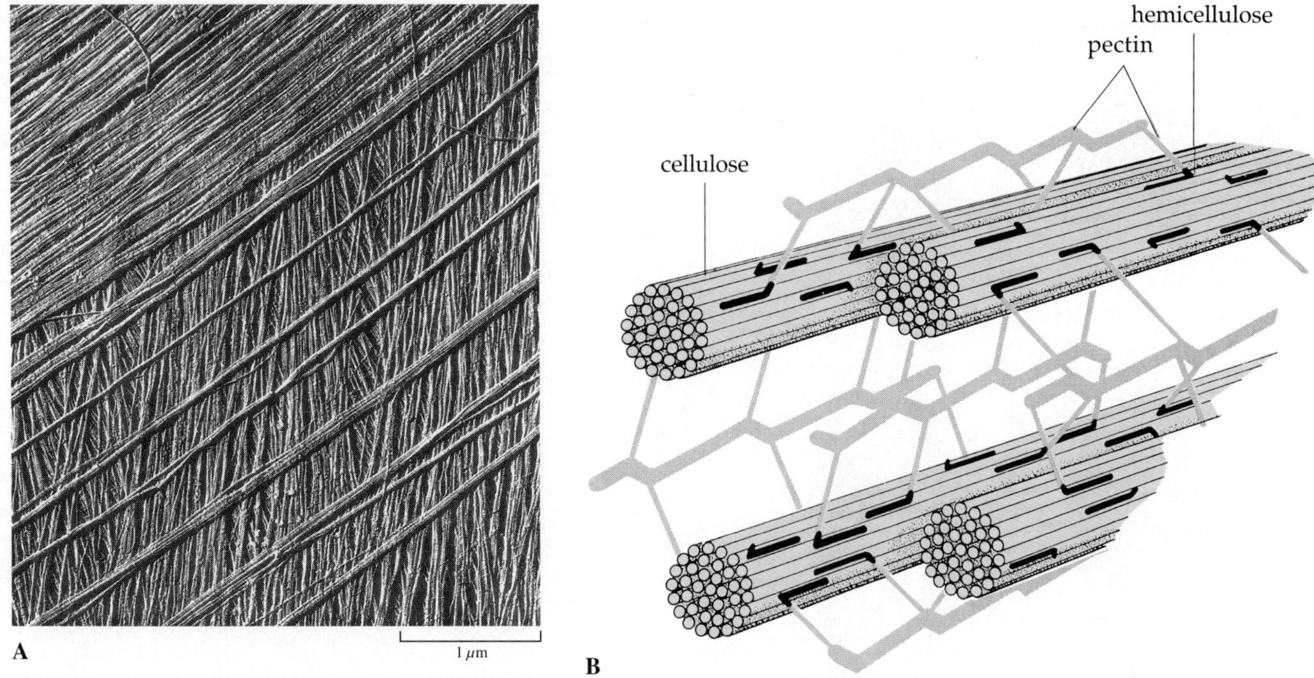

23.1 Molecular arrangement of polysaccharides in a primary cell wall (A) Each cellulose microfibril is a bundle of cellulose chains laid out in parallel lines in two directions. (B) The fibrils, which are cross-linked by the carbohydrates pectin and hemicellulose, form a relatively rigid network. If the cell is to enlarge, some of the cross-linkages must be broken.

lose is present in long fibrils that are linked to other polysaccharides that make up the matrix of the wall (Fig. 23.1). These matrix polysaccharides hold the fibrils in a more or less rigid pattern. For growth to occur, two things must happen: some of the cross-linkages must be temporarily broken, so that the wall will become more plastic (extensible), and new wall material must be inserted. Auxin is believed to play a fundamental role in both aspects of growth.

Auxin's effect on cell wall plasticity is rapid (about 10 minutes). According to the ***acid growth hypothesis***, auxin exerts its effect indirectly, by acidifying the cell wall. Auxin is thought to regulate the pH of the cell wall by inducing the active transport of hydrogen ions to it from the cytoplasm. The acidic pH, in turn, appears to activate enzymes in the wall that break the cross-linkages between the cellulose fibrils, the fibrils loosen, and the walls become more extensible (Figs. 23.1 and 23.2). Plant cells are hypertonic compared to their extracellular fluids, and water moves into the cell by osmosis, creating a high turgor pressure within the cell. When the wall becomes more plastic, it is less resistant to stretching. Water continues to move into the cell by osmosis, and the increasing turgor pressure inside the cell causes more and more stretching of the wall until the volume of the cell has been increased (Fig. 23.3). But note that this increase in cell size has been achieved with little or no synthesis of new cytoplasm.

The effect of auxin on the second aspect of wall enlargement—insertion of new wall material—is much slower. It is thought to depend on the activation of certain genes, causing the synthesis of specific mRNAs, which, in turn, code for the enzymes that catalyze the addition of new polysaccharide units to the wall.

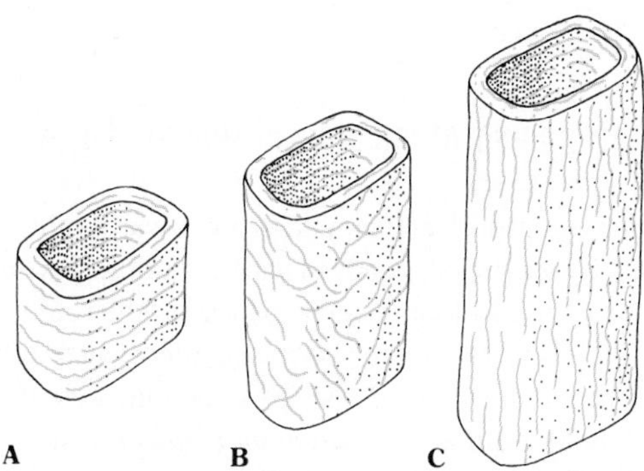

23.2 Change in the orientation of fibrils of the cell wall as a plant cell elongates (A) Before the start of elongation the cellulose fibrils are arranged horizontally. (B) As elongation begins, and the wall is stretched, the fibrils are displaced and tipped toward a more vertical alignment. (C) In the fully elongated cell the fibrils in the outer, older wall layers are oriented vertically. The recently deposited fibrils in the inner newer layers are horizontal.

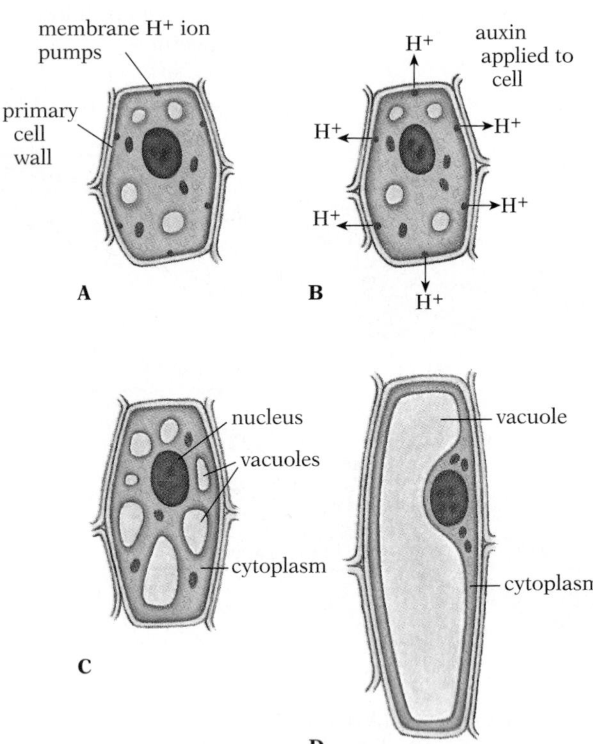

23.3 Elongation growth of a plant cell When auxin is applied to a plant cell, it activates hydrogen ion pumps in the plasma membrane (A). Hydrogen ions are actively transported out of the cytoplasm into the wall. The increase in hydrogen ions activates enzymes that break some of the cross linkages between the polysaccharides making up the wall, and the wall becomes more pliable (B). As more and more water moves into the cell vacuoles, the wall is stretched, but in only one dimension. Almost no synthesis of new cytoplasm occurs during this kind of growth. The increased volume of the cell is taken up by the expanding vacuoles (C), which eventually fuse into a single large vacuole (D); in the mature cell the thin band of peripheral cytoplasm may constitute less than 10 percent of the cell's volume.

Phototropism of the Stem Is Mediated by Auxins

Plants have a strong tendency to grow toward light. A potted plant in the living room bends toward a window; if you turn the plant so that it will look nicer to people in the room, in a disconcertingly short time the shoot is again oriented toward the light of the window (Fig. 23.4). This phenomenon of responding to light by turning is called ***phototropism,*** from the Greek words for "light" and "turning." (Other tropisms involve turning responses to other stimuli. Gravitropism is a turning response to gravity, hydrotropism is a turning response to water.) The turning response results from differential growth; one side of the plant stem (or root) grows faster than the other, causing it to bend. In shoots the phototropism is positive, a turning *toward* the stimulus; roots, on the other hand, exhibit negative phototropism, a turning *away* from a light stimulus. The adaptive significance of the phototropic response of stems and leaves is obvious; by turning toward the light the leaves maximize light absorption for photosynthesis.

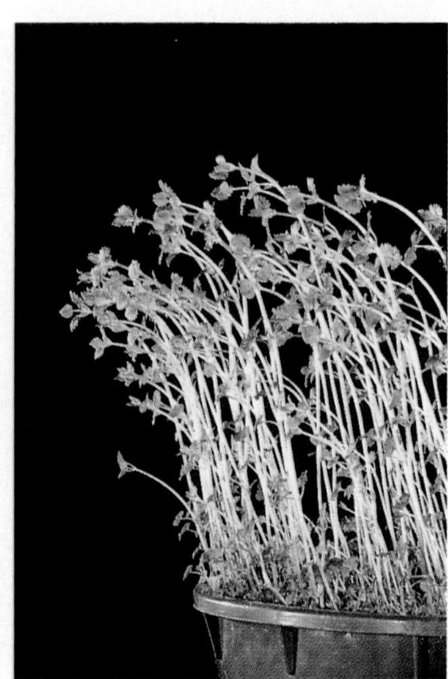

23.4 Phototropism in plants These plants are bending toward the light source.

Among the first to investigate the phototropism of plants was the wide-ranging Charles Darwin, who worked on the problem with his son Francis, about 1880. They, like many who followed them, performed their experiments on the conical sheath that encloses the first leaves of grass seedlings and their relatives (Fig. 23.5). This sheath, the ***coleoptile,*** grows

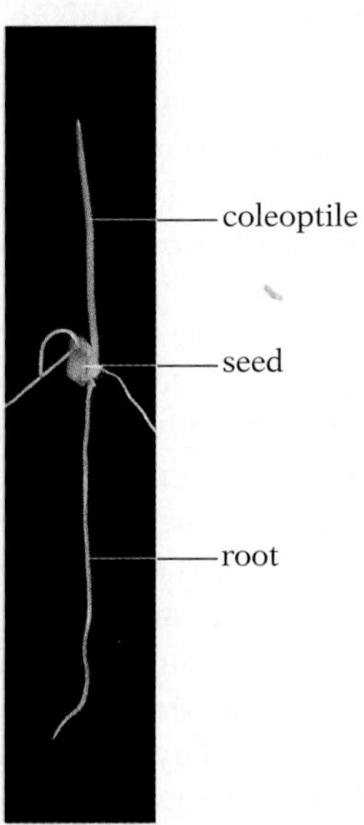

23.5 Photograph of corn coleoptile The shoot of a monocot seedling is covered by a protective sheath called the coleoptile. The first leaves are rolled up inside this sheath. The coleoptile exhibits a very strong phototropic response.

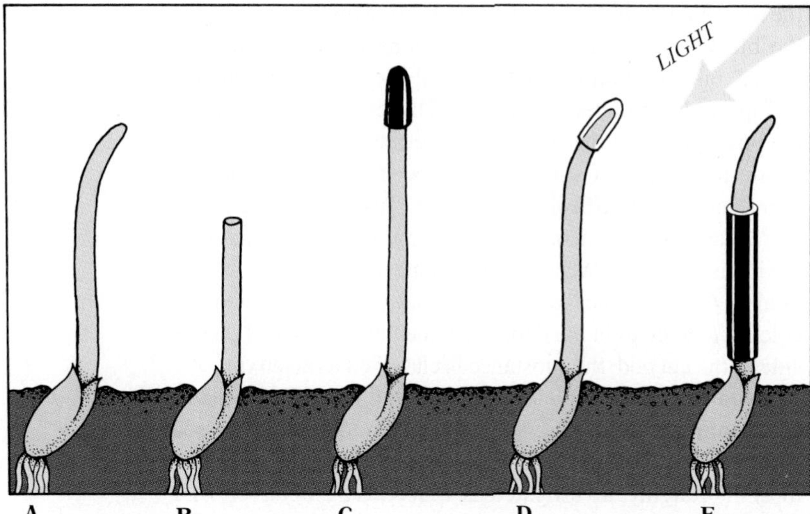

23.6 The Darwins' experiments on phototropism (A) A coleoptile of canary grass grows toward the light. (B, C) The coleoptile does not bend if its tip is removed or is covered by an opaque cap. (D) The coleoptile does bend if its tip is covered by a transparent cap. (E) It also bends if its base is covered by an opaque substance (represented here as a tube; the Darwins used black sand).

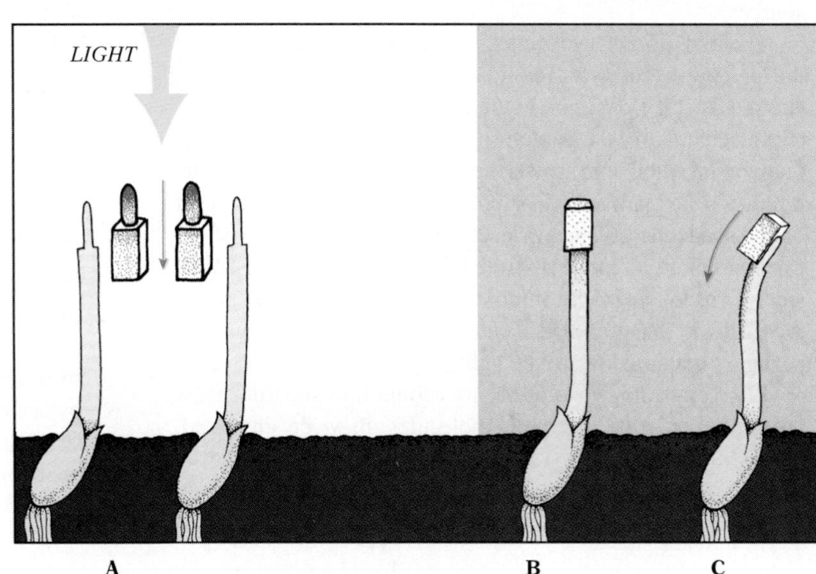

23.7 Went's experiment When the tips of coleoptiles are cut off and placed on blocks of agar for about an hour (A), and one of the blocks alone is then put on a stump (B), the stump will resume growing even in the dark. If a block is placed off-center on a stump in the dark (C), the stump will grow so as to bend away from the side on which the block rests. Apparently a hormone has diffused from the tips into the blocks, and this hormone can then diffuse from the blocks into the stumps.

principally by cell elongation and exhibits a very strong positive phototropic response. The Darwins showed that if the tip of the coleoptile is covered by a tiny black cap, it fails to bend toward light coming to it from one side, while coleoptiles in the control group, with their tips exposed or covered by transparent caps, bend, as expected, toward the light (Fig. 23.6). They observed that a black tube placed over the base of the coleoptile, but not covering the tip, fails to prevent bending. It seemed to be the tip of the coleoptile, therefore, that played the key role in the phototropic response. This was confirmed by experiments in which the Darwins cut off the tip and found that the coleoptile failed to bend, even though control coleoptiles damaged in other ways, but with their tips intact, bent normally. Clearly, it was the absence of the tip and not a reaction to wounding that blocked the phototropic response. The Darwins concluded that light is detected by the tip of the coleoptile and that "some influence is transmitted from the upper to the lower part, causing the latter to bend."

Experiments conclusively demonstrating that phototropism is the result of a growth chemical moving downward from the tip were reported in 1926 by Frits Went, then in Holland. He removed the tips from coleoptiles and placed these isolated tips, base down, on blocks of agar for about an hour (Fig. 23.7). (Agar is a gelatin-like material, made from seaweed, often used as the base for laboratory culture media.) He then put the blocks of agar, minus the tips, on the cut ends of the coleoptile stumps. The stumps behaved as though their tips had been replaced; they resumed growth, responded to lateral light by bending toward it, and, if the agar blocks were put on off center, could be made to bend even in darkness. Plain agar blocks used as controls produced none of these effects. Apparently a growth-stimulating substance had diffused out of the tips and into the blocks of agar while the tips were sitting on the blocks. When the blocks containing the chemical were placed on the stumps, the chemical moved down into the stumps and stimulated elongation. This experiment ruled out the possibility that the stimulus was electrical or nervous, because these types of stimuli cannot be stored in agar blocks. Went named the diffusible substance "auxin" (from a Greek word meaning "to grow"). To this day, the identification of auxins is based on Went's experiment: if an agar block containing the substance in question causes a decapitated oat coleoptile to bend in the dark when the block is placed on one side of the cut end, the substance is classified as an auxin.

Many chemicals, some of them found naturally in plants and some synthesized only in the laboratory, have passed Went's bioassay and are commonly called auxins. The one most thoroughly investigated is ***indoleacetic acid,*** or ***IAA*** (Fig. 23.8), which has been isolated from numerous natural sources.

The experiments we have discussed showed that the tip of the coleoptile releases auxin, which moves downward and stimulates cell elongation in the coleoptile. However, the precise mechanism of the phototropic response is still unclear. It is known that when light strikes the tip of the plant from one side, it reduces the plant's auxin supply on that side and increases it on the shaded side. The evidence indicates that there is active transport of auxin in the tip from the illuminated to the shaded side. Consequently, the illuminated side of the plant grows more slowly than the shaded side, and this asymmetric growth produces bending toward the light (Fig. 23.9).

Plant physiologists are still not certain how the tip detects the light or what receptor molecules may be involved. Apparently light causes a lateral redistribution of auxin at the tip such that auxin accumulates on the side of the shoot away from the light. Unlike most plant hormones, auxin is not transported through the vascular tissue but rather moves downward from cell to cell. The movement is unidirectional and probably by active transport because metabolic energy is required. In a stem auxin moves from the shoot-tip downward to the base of the plant, whereas in a root, the movement is upward from the root-tip to the shoot.

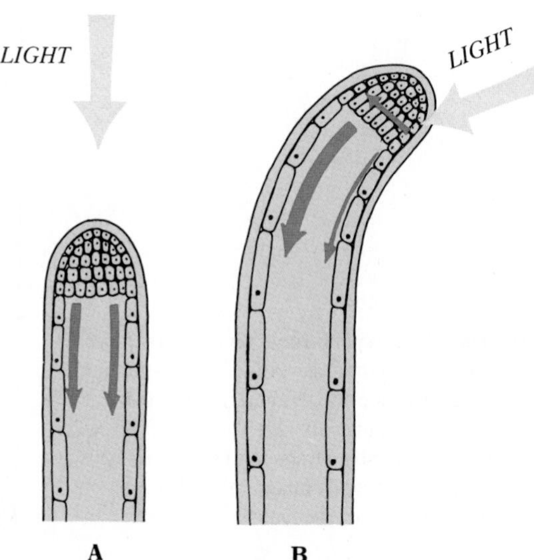

23.9 Auxin-mediated response to a light source in a young shoot Light receptors in the growing tip of the plant are thought to respond to light by altering auxin distribution. In A, with light from directly above, auxin is equally distributed to the growing tissues on all sides. In B, light from the right causes a decrease in the supply of auxin to the right side of the plant, which results in a relative elongation of cells on the left side, and a bending toward the light. Only the cells in the growing tip and the periphery of the shoot are shown individually.

A

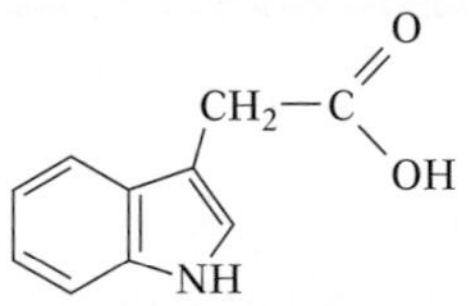

Indoleacetic acid
(IAA)

B

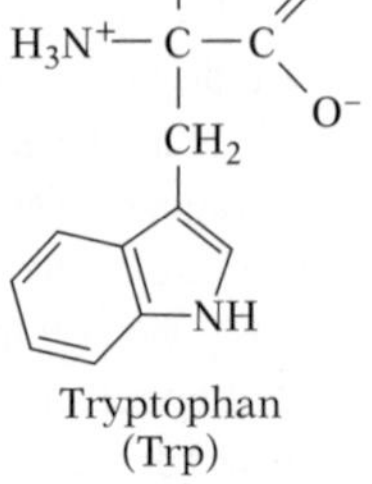

Tryptophan
(Trp)

23.8 Indoleacetic acid The most common auxin is the chemical indoleacetic acid (A). Indoleacetic acid is chemically similar to the amino acid tryptophan (B) and is synthesized from it.

Gravitropism of Shoot and Roots Is Influenced by Auxin

If you lay a potted plant on its side and leave it for a few hours, you will find that the shoot has begun to bend upward (Fig. 23.10). This is a negative gravitropic response; the shoot turns away from the pull of gravity. (How could you prove that the shoot is responding primarily to gravity and not to some other stimulus, such as light?) Why does the shoot grow upward and the roots downward in response to the same stimulus? How does the plant sense the pull of gravity? Plants, like many multicellular organisms, have cells that are sensitive to changes in position and respond to gravity. In the stem the gravity receptors are in the meristematic tissue in the shoot-tip, and in the root, they are in cells in the central portion of the root cap.

Herman Dolk in Holland discovered that the concentration of auxin in the lower side of a horizontally placed shoot in-

23.10 The gravitropism of shoot and root When a growing plant is left lying on its side, the shoot will grow upward and the roots will grow downward in response to auxin gradients.

creases, while the concentration in the upper side decreases. This unequal distribution of auxin stimulates the cells in the lower side to elongate faster than the cells in the upper side, and the shoot therefore turns upward as it grows. Again, the external stimulus, in this case gravity, is apparently detected by the meristematic tissue in the shoot tip. The mechanism by which these cells detect the stimulus is not yet understood. Like the phototropic response we have just discussed, the gravitropic response is probably mediated by auxin.

Roots, unlike shoots, turn toward the pull of gravity (Fig. 23.10). When a plant is turned on its side, the cells on the upper portion of a horizontal root grow more than the cells on the lower side, so the root turns downward (Fig. 23.11). Cells in the central portion of the root cap are thought to detect the gravitational stimulus (Fig. 23.12). Within these cells are high concentrations of specialized starch-storing plastids called ***amyloplasts*** (Fig. 23.13). The amyloplasts, and other organelles, respond to the pull of gravity by settling to the bottom of the cells, and in a few hours, the root starts to grow downward. Many botonists believe that the amyloplasts act as gravity detectors, but others, such as Randy Wayne of Cornell University, are not convinced. Wayne's work suggests that specialized receptors located at the ends of cells sense the weight of the entire cell contents (not just the amyloplasts) which causes a compression at the bottom of the cell and a tension at the top. The receptors sense the compression/tension differences within the cell and respond accordingly. The

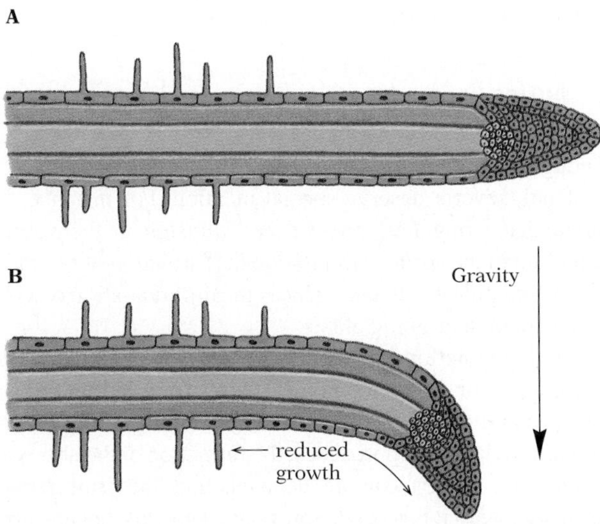

23.11 Gravitropism in roots When a root is placed on its side (A), the cells on the upper portion of the root begin to elongate more than the cells on the lower portion of the root (B), so the root turns downward.

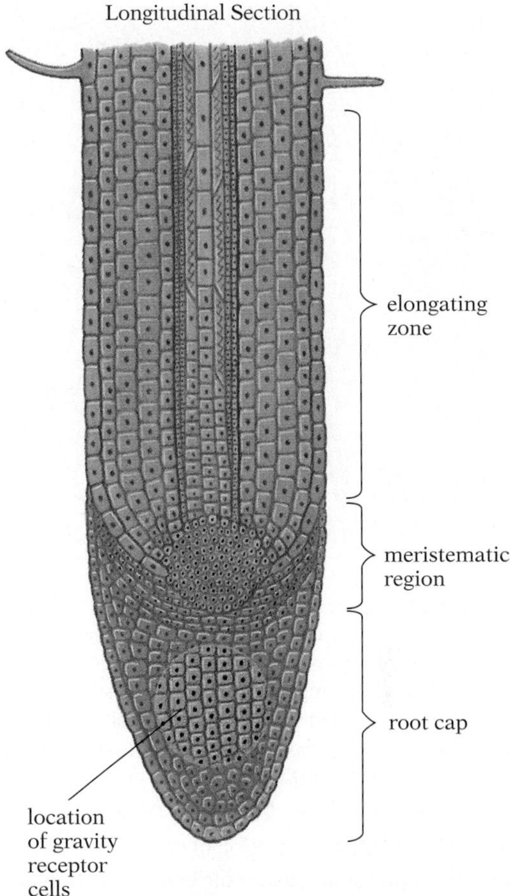

23.12 Location of the gravity-sensitive cells in the root tip The gravity-detecting cells in roots are located in the central portion of the root cap. Just above the root cap is the meristematic region where new cells are formed. Above the meristematic region is the elongating zone; this is the region where the gravitropic response is mediated.

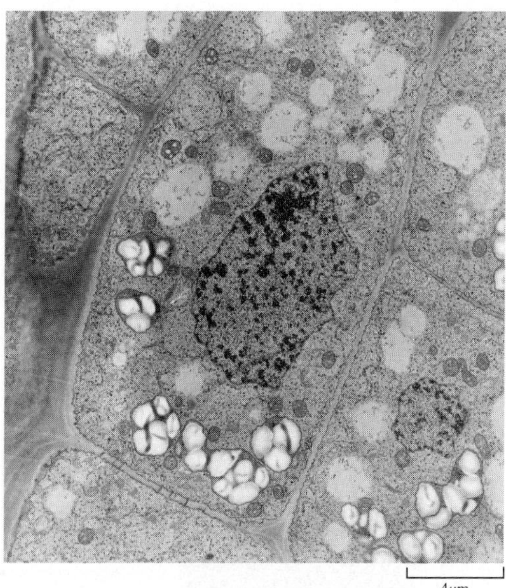

23.13 Amyloplasts in a root-cap cell Since the starch they contain is denser than the surrounding cytoplasm, amyloplasts will move through the cytoplasm in response to gravity and accumulate within minutes at the bottom of a repositioned cell. Organelles of the same density as cytoplasm remain spread throughout the cell. The amyloplasts may act as gravity sensors, initiating a series of responses that culminates in the root growing downward.

mechanism by which the detection of gravity, whether by amyloplasts or compression/tension differences, is translated into changes in auxin gradients is not yet understood.

Auxin Inhibits Growth of Lateral Buds

Auxin also has an inhibiting effect on lateral buds in many plants, slowing or preventing their growth. Auxin produced in the terminal bud moves downward in the shoot and represses development of the lateral buds, while at the same time stimulating elongation of the main stem. The terminal bud thus exerts ***apical dominance*** over the rest of the shoot, ensuring that the plant's energy for growth will be funneled into the main stem and produce a tall plant with relatively short lateral branches. Longer branches usually develop only from buds far enough below the terminal bud to be partly free of the inhibiting effect of auxin. If the terminal bud is removed, however, apical dominance is temporarily destroyed, and several of the upper lateral buds will begin to grow, producing branches whose terminal buds soon exert dominance over any buds below them (Fig. 23.14). Flower and shrub growers frequently pinch off the terminal buds and young leaves of their plants one or more times each season in order to produce bushy well-branched plants with many flowering points instead of tall spindly ones with sparse flowers.

23.14 Inhibition of lateral buds by the terminal bud in chrysanthemum As long as the terminal bud is intact, the lateral buds marked by arrows will grow very little, if at all. But if the terminal bud is removed, those buds are released from inhibition and grow rapidly, forming the new leaders of the plant.

Auxins Have a Variety of Commercial Applications

Though we cannot discuss the other known functions of auxins in detail, several deserve special mention. For instance, it is auxin that stimulates renewed cell division in the vascular cambium in the spring, stimulating the formation of secondary xylem and phloem. It also induces the differentiation of xylem and phloem in growing plants.

Several functions of auxins have important commercial applications. For example, auxins inhibit fruit and leaf abscission (separation from the stem), so by spraying auxins in citrus orchards as harvest approaches, the formation of the abscission layer (Fig. 23.15) can be delayed, and the fruit remains on the tree until it is picked. Sprays of chemicals that are auxin-antagonists are often used for the opposite purpose. Such sprays are commonly applied to the leaves of cotton just before harvest. The anti-auxins induce formation of abscission layers at the base of leaves, causing the leaves to fall prematurely.

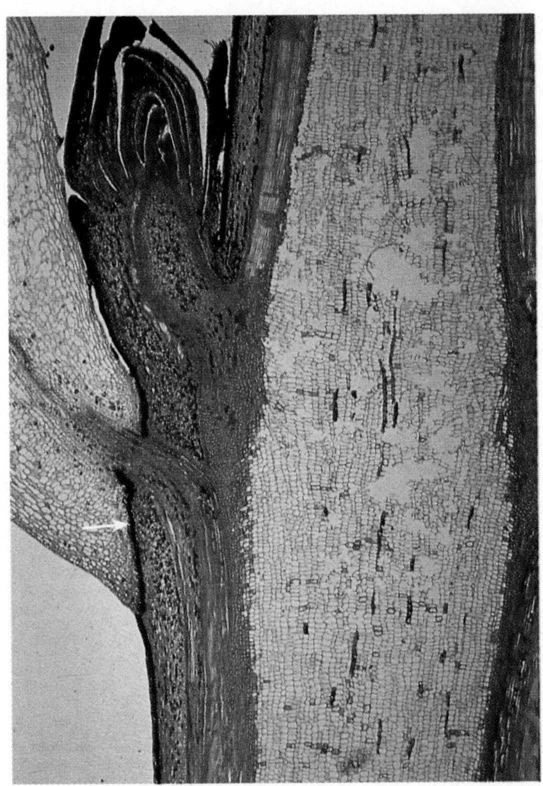

23.15 Photograph showing abscission layer at the base of a petiole of a maple leaf Diminished production of auxin in the leaf leads to the formation of an abscission (separation) layer (arrow). The cell walls in this region have been weakened by an increasing concentration of cellulase, an enzyme that digests cellulose. A protective layer of cells forms underneath the separation layer. The actual break can be initiated by any slight strain like the pressure of a gentle wind. The protective layer forms the leaf scar on the stem.

This makes it easier for mechanical pickers to move through the fields and harvest the bolls (Fig. 23.16).

The role of auxin in initiating the formation of new lateral roots has also proved to be commercially useful. Cuttings of such organs as stems or leaves can sometimes be induced to develop roots, thereby producing whole new plants. While some plants, such as geraniums and willows, will readily root in water or soil without any application of hormone, others will not. Application of auxins (in rooting powders) often stimulates formation of roots in such cases, making it possible to propagate vegetatively many valuable strains of plants that might otherwise be lost.

Auxin-like chemicals such as 2,4-D are widely used for the eradication of broad-leaved weeds in lawns. Dicots are much more sensitive to this compound than are monocots, so applications of low dosages of 2,4-D to lawns will not harm the grass but will kill the dicot weeds. The synthetic auxins apparently achieve their herbicidal effect by stimulating abnormal growth (Fig. 23.17). Agent orange, an herbicide widely

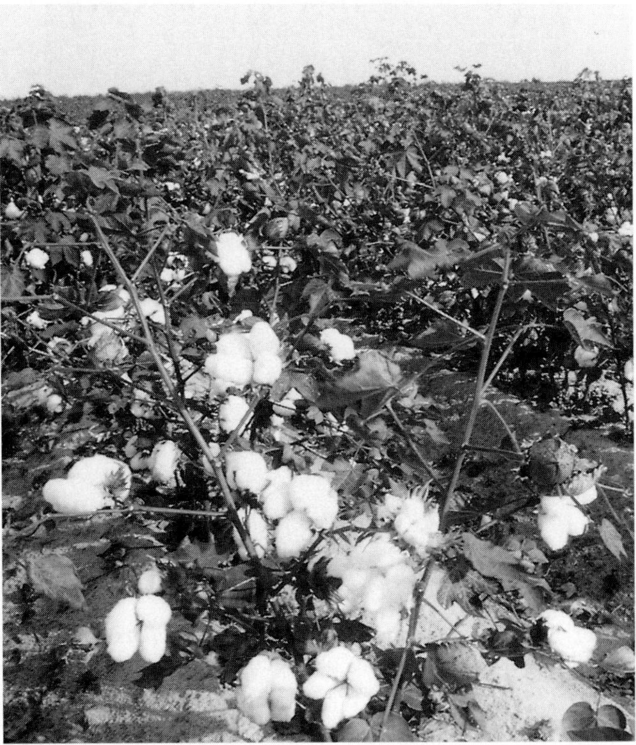

23.16 Effect of auxin antagonists on leaf abscission Top: A cotton field before treatment. Bottom: A typical field after treatment with auxin antagonists. Treated cotton plants drop their leaves, allowing mechanical pickers to remove bolls with ease and at the optimum time.

23.17 Effect of 2,4-D on a dandelion 2,4-D is an auxinlike chemical widely used for the control of broad-leaved weeds. It apparently achieves its effect by stimulating abnormal growth. Grasses and other narrow-leaved plants are not sensitive to this chemical.

sprayed on forests in Vietnam during American military involvement there, is another synthetic auxin. Some veterans of the Vietnam conflict link contact with agent orange with increased incidence of cancer and other health problems, but the connection has not been definitively established.

Gibberellins Are Growth Promoters

The Japanese have long been familiar with a disease of rice that they call foolish-seedling disease. Afflicted plants grow unusually tall, but seldom live to maturity. In 1926 a Japanese botanist, E. Kurosawa, found that such plants are infected with a fungus named *Gibberella fujikuroi*. He showed that when the fungus was moved to healthy seedlings, they too developed the typical disease symptom of rapid stem elongation. He could also produce the symptoms with an extract made from the fungus, and even with an extract made from culture media on which the fungus had grown. Clearly, some chemical was involved.

Several Japanese scientists working on the problem of foolish-seedling disease during the 1930s succeeded in isolating and crystallizing a substance from *Gibberella,* now known as ***gibberellin,*** that produced typical disease symptoms when applied to rice plants. In the last 30 years more than 70 different substances isolated from fungi and from flowering plants have been classified as gibberellins (GAs). More will doubtless be discovered in years to come. The gibberellin most often used in experimental work is called GA_3 or ***gibberellic acid,*** but the active form is probably GA_1.

GA_1 causes rapid stem elongation by stimulating both cell division and cell elongation. The most dramatic effect is on stem elongation in certain dwarf plants and other plants that normally undergo little stem elongation (Fig. 23.18). They have much less effect on most normal-sized plants. The mechanism by which gibberellins induce stem elongation is not well understood, but is clearly different from that of auxin-induced elongation since acid growth is not involved.

Though both gibberellins and auxins stimulate stem elongation, their effects on the plant are quite different. Because

23.18 Effect of gibberellic acid on cabbage The plants at left are normal. Those on the right were treated with gibberellic acid once a week for two months. The loss of normal levels of gibberellic acid from ancestral plants that resembled those on the right, and the consequent development of the desirable compact growth form of cabbage, was the result of intense selective breeding.

gibberellins are transported through both the xylem and the phloem, whereas auxin movement is from cell to cell and unidirectional, they tend to affect the whole plant, and do not include the localized bending movements characteristic of auxin-induced responses.

Gibberellins are produced primarily in young tissues of the shoot and in developing seeds. There are many different gibberellins but all are derived from the same biosynthetic pathway as vertebrate steroid hormones (like estrogen and testosterone) and, like them, are lipid-soluble. As a result, they can easily cross cell membranes, and probably function by entering the cell and turning particular genes on and off.

Gibberellins play a role in a host of other developmental processes besides stimulating cell division and stem elongation. They can (1) induce germination of some seeds; (2) induce the embryos in certain seeds to produce amylase enzymes that hydrolyze starch reserves; (3) induce certain plants to grow very rapidly and flower; and (4) stimulate fruit development in some species.

Gibberellins Have Commercial Applications

There are a number of practical applications for gibberellins in agriculture. One of the important uses is to increase malt production from barley for the manufacture of beer. In this process GA is applied to germinating barley seeds to accelerate the production of enzymes that hydrolyze the stored food in the seed into the amino acids and sugars that make up the malt extract. GAs have also proved useful in increasing the size and quality of seedless grapes, and have also been used to stimulate development in certain fruits (Fig. 23.19). Another use of GA has been to delay fruit ripening in lemons and other citrus species. GA applied in November or December delays the harvest date, increasing the availability of these fruits during the summer months when the demand for them is high, but the supply is normally low. Other important uses for these important hormones will undoubtedly be found in the future.

23.19 Effect of gibberellic acid on grapefruit In addition to promoting growth, gibberellic acid is being used to prevent fruit fly infestations of grapefruit. Caribbean fruit flies (above) respond to the yellow color of a grapefruit to lay their eggs and bypass the gibberellic acid-treated grapefruit in the background. Both grapefruit are fully ripe. Gibberellic acid is also used by citrus growers to improve shipping quality by keeping the peel tough and therefore resistant to molds and mechanical injury.

Cytokinins Promote Cell Division

Though auxins play a role in cell division, that role is an accessory one. The main promoters of cell division in plants are gibberellins and a class of hormones called the ***cytokinins***, which are produced in root tips and developing seeds and are transported via the xylem from the roots to the shoots. The action of cytokinins on cells growing in tissue-culture media depends on the simultaneous presence of auxins. Indeed, the ratio of cytokinin to auxin appears to be of fundamental importance in determining the differentiation of the new cells. Thus, when there is more cytokinin than auxin in the tissue-culture medium, stems and leaves are formed; when there is more auxin, root growth is initiated. In short, the balance between these two hormones determines what sort of tissue will develop, and so is responsible for controlling plant morphology.

In the normal growing plant, cytokinins and auxins act cooperatively in some situations and antagonistically in others. For example, they act cooperatively in promoting cell division, but antagonistically in influencing the growth of lateral buds. Both hormones influence cell growth, but auxin primarily stimulates elongation, whereas cytokinin promotes cell division. This sort of hormonal interaction is a recurrent theme in hormonal control in plants, and in vertebrates as well.

Among their numerous other functions, cytokinins (1) stimulate conversion of immature plastids into functional chloroplasts; (2) release the lateral buds from apical dominance; and (3) retard the onset of senescence (aging), especially in leaves. As this list and the similar one for auxins and gibberellins suggest, all the major plant hormones seem to participate in some fashion in nearly all aspects of plant growth and development.

Abscisic Acid Is a Growth Inhibitor

The hormone abscisic acid was originally thought to control leaf abscission (separation) in some plants, hence the name. It was also believed to induce dormancy in buds and seeds in autumn, preventing their growth until spring. It now appears that abscisic acid plays little role in either leaf abscission or bud dormancy, though it does seem to play a role in seed dormancy, at least in some species. Although abscisic acid applied to a plant will inhibit growth and the hormone has long been thought of as an inhibitor, in many situations it actually induces

MAKING CONNECTIONS

1. Why does removal of the terminal buds of flowering plants such as petunias produce bushier plants?

2. How is the action of gibberellins related to that of steroid hormones in vertebrate animals? Are these hormones polar or nonpolar molecules?

3. Describe how auxins and cytokinins act both cooperatively and antagonistically in the control of plant growth.

4. What is the role of abscisic acid in the homeostatic regulation of the water balance of plants?

growth and development. For instance, it promotes the transport of the products of photosynthesis to the growing embryos in seeds and induces the synthesis of proteins for storage in seeds. It has other important effects as well. For instance, it can cause the stomata to close when there is a shortage of water. Within minutes after a plant begins to wilt, concentrations of abscisic acid increase 10-fold. The hormone apparently triggers a number of responses, including stomatal closure, that help that plant resist the effects of drought.

Abscisic acid is produced in mature leaves and is transported through the phloem.

The Hormone Ethylene Is a Gas

"One rotten apple spoils the lot." That piece of folk wisdom rests on a familiar fact: when one apple in a barrel goes bad, most of the other apples in that barrel soon go bad too. We now know that the bad apple affects other fruit by producing a gas called ethylene (C_2H_4). Ethylene plays a variety of roles in the life cycle of a normal plant, including growth, development, and senescence (aging). Because ethylene is a gas, it is transported through air. The hormone is synthesized by tissues that are undergoing stress, particularly those in the process of aging or ripening.

One of the best-studied effects of ethylene is stimulation of fruit ripening. Once the fruit has attained its maximum size, it becomes sensitive to ethylene, and a host of chemical changes cause it to ripen. The ripening process starts with a sudden burst of metabolic activity triggered by an approximately 100-fold increase in the concentration of ethylene. Inhibition of ethylene production, or removal of the ethylene as fast as the fruit produces it, prevents ripening. Many commercial fruits are now picked and transported while they are still green and firm, and therefore resistant to damage. They can then be ripened with ethylene gas when ready for sale (Fig. 23.20).

Ethylene has also been shown to promote the abscission of leaves, fruits, and flowers and contributes to various other changes characterizing senescence in a plant or parts of a plant. In addition, it can aid in breaking dormancy in the buds and seeds of some species, and it can initiate flowering in some plants (e.g., pineapple).

Most Plant Hormones Regulate Growth and Development

What conclusions can be drawn from our discussion of chemical control of plant growth so far? (1) Cell division, the first phase of growth, is stimulated by gibberellin, cytokinin and

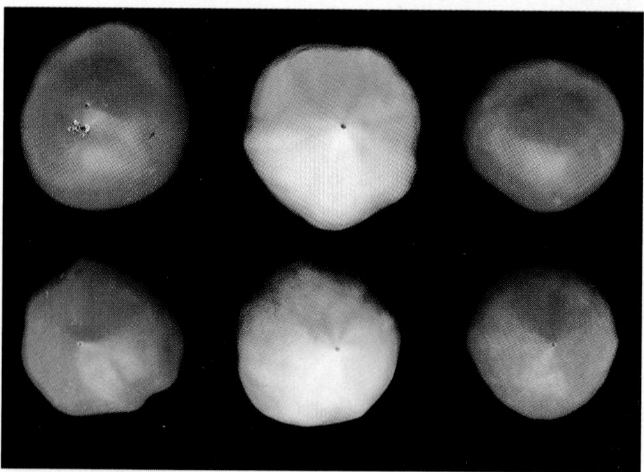

23.20 Effect of ethylene on tomato ripening Tomatoes normally ripen in response to internally produced ethylene (left). Insertion of a genetically engineered gene prevents ethylene production, producing tomatoes that fail to ripen (middle) unless treated with ethylene (right). The ability to manipulate ripening enables growers to produce full-sized green tomatoes that travel well and can be held in this state until needed.

auxin; (2) control of cell enlargement, the second phase of growth, involves substances such as auxins and gibberellins, which promote elongation; (3) the various plant hormones, by their mutual interactions and their differential effects on various parts of the plant body, help integrate and coordinate the development of form and function; and (4) the aging process, leading to death, is brought on by various senescence-inducing substances, of which ethylene is one of the most important. Table 23.1 provides a summary of the important plant hormones.

TABLE 23.1 *Summary of the important plant hormones*

Hormone Structure	*Site of synthesis*	*Mode of transport*	*Major effects*
Auxin 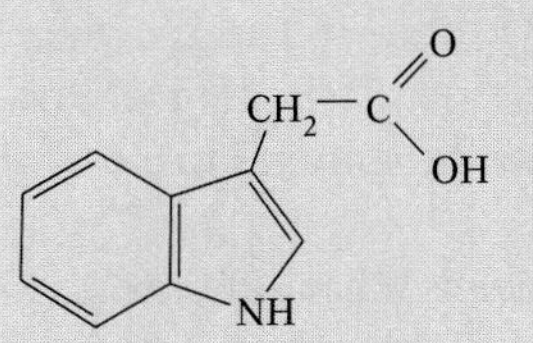	Synthesized from tryptophan in young leaves and rapidly growing parts of plant	Cell to cell	Promotes cell enlargement and cell growth, cell division in cambium, vascular tissue differentiation, root initiation, tropisms, and apical dominance; delays leaf aging and leaf and fruit abscission
Gibberellin 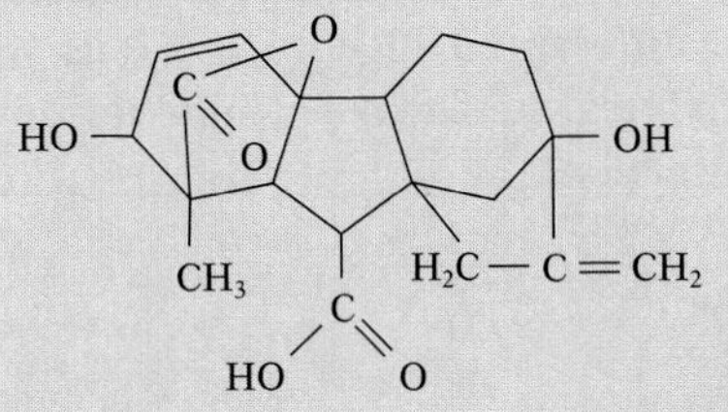	Young tissues of shoot	Xylem and phloem	Promotes cell division at the apical meristem, stem growth, bolting in long day plants, seed germination, enzyme production during germination
Cytokinin 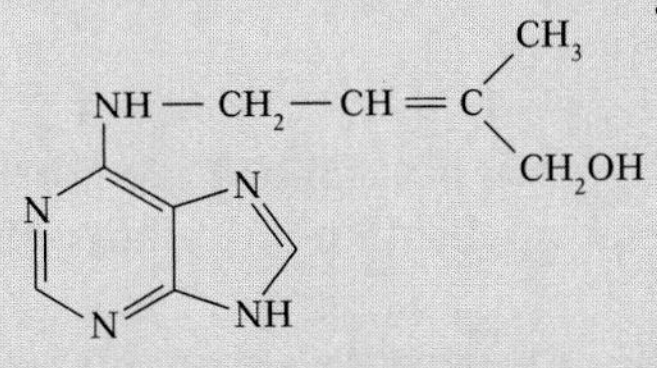	Synthesized from adenine in root tips and developing seeds	Xylem, from roots to shoots	Promotes cell division, growth of lateral buds; delay leaf aging
Abscisic acid 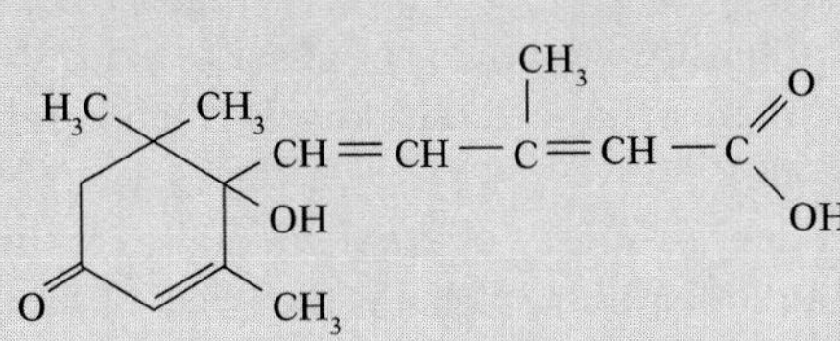	Mature leaves	Phloem	Closure of stomates; transport of sugars, amino acids to seeds, storage of proteins in seeds
Ethylene 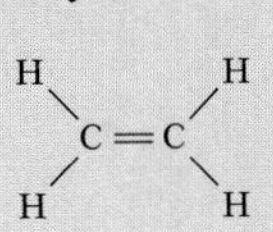	Aging and wounded tissues	Diffusion through air	Fruit ripening, flower and leaf aging, leaf and fruit abscission, wound healing

CONCEPTS IN BRIEF

1. Plant hormones are produced most abundantly in the actively growing parts of the plant body: the apical meristems, young growing leaves, and developing seeds and fruits. The known plant hormones are primarily involved in regulating growth and development.

2. *Auxins*, which are produced by the apical meristems, move unidirectionally; they promote cell elongation. According to the *acid growth hypothesis,* auxins induce the transport of hydrogen ions from the cytoplasm to the cell wall. The acid pH activates enzymes in the wall that break the cross-linkages between the cellulose fibrils, and the walls become more extensible. Auxins produce a wide variety of other effects as well, among them phototropism, gravitropism, apical dominance, and stimulation of vascular tissue differentiation.

3. *Gibberellins* stimulate rapid stem elongation, particularly in dwarf plants, and are important in inducing seed germination.

4. *Cytokinins,* together with auxins, promote cell division. In the normal growing plant, cytokinins and auxins act cooperatively in some situations and antagonistically in others.

5. The hormone *abscisic acid* promotes the transport of the products of photosynthesis to the growing embryos in seeds and induces the synthesis of proteins for storage in seeds. It also can cause the stomata to close when there is a shortage of water.

6. The hormone *ethylene* is a gas that induces the ripening of fruit in mature plants. It also contributes to leaf abscission and various other changes involved in the aging process.

7. The various plant hormones, by their mutual interactions and their differential effects on various parts of the plant body, help integrate and coordinate the development of form and function.

STUDY QUESTIONS

1. Where are plant hormones generally produced? What is their general role in the life of the plant? (p. 476)
2. According to the acid growth hypothesis, how does auxin cause cell elongation? (pp. 476–477)
3. What is a *tropism?* Describe the experiments of the Darwins and Went that demonstrated the role of auxin in phototropism. (pp. 478–480)
4. Explain how the effect of auxin on cell elongation underlies the phenomena of phototropism and gravitropism in shoots. (pp. 478–480)
5. Describe at least three commercial applications of auxins. (pp. 482–483)
6. Compare and contrast the sites of production, methods of transport, and effects on plant growth of gibberellins and auxins. What are some of the other effects and commercial applications of gibberellins? (pp. 478–485)
7. How would a plant be affected if it were unable to produce a normal amount of cytokinins? Where are these hormones produced and how are they transported? (p. 485)
8. Describe at least three effects of abscisic acid. Where is this hormone produced and how is it transported? (pp. 485–486)
9. What physical property of ethylene makes it unique among plant hormones? What is the effect of this hormone on fruit? How does it contribute to the aging process of plants? (p. 486)

C H A P T E R 2 4

HORMONAL CONTROL OF REPRODUCTION AND DEVELOPMENT IN FLOWERING PLANTS

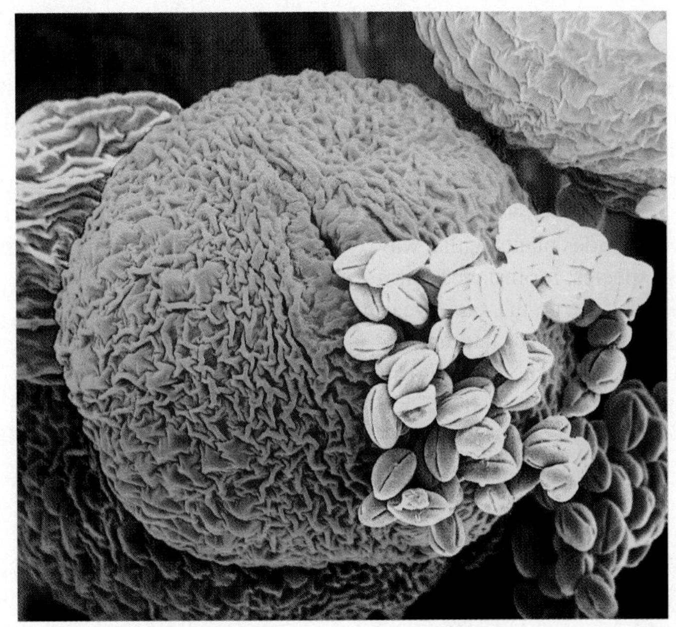

The angiosperms, or flowering plants, make up the most abundant and dominant group in the plant kingdom. The majority of the land plants familiar to you belong to this group, which includes plants of every shape and size from grasses to cacti, from tiny herbs and wildflowers to large oaks and maple trees. We shall look first at the process of reproduction in these plants, then consider the processes of growth and development.

In the angiosperms, the structure in which sexual reproduction takes place is the flower. Flowering is not a random process: some plants flower early in spring, others flower in midsummer, and still others, like chrysanthemums, flower in the fall. These simple facts have been known for centuries, but only since 1920 has anything been known of the control mechanisms involved, and some aspects are still not well understood.

Flowering In Many Plants Is Affected by the Duration and Timing of the Light-Dark Cycle

Modern research in the flowering process dates from the investigations of W. W. Garner and H. A. Allard of the U.S. Department of Agriculture, the results of which were first published in 1920. Working in Beltsville, Maryland, they found that a new mutant variety of tobacco, called Maryland mammoth, grew unusually large (as much as 10 feet tall), but would not flower, thus making it useless for breeding experiments. They propagated the new variety and discovered that it would flower in the greenhouse in winter. Though flowering was not the subject Garner and Allard were originally investigating, they became interested in the question of why Maryland mammoth would flower in the greenhouse in winter, but not in the fields in summer. Accordingly they began a series of experiments that were to open a whole new area of botanical research. Often in science, important leads are uncovered almost accidentally through research originally devoted to a different subject. Good scientists do not overlook such leads, but recognize and pursue them, even though they may have to change the course of their research or discard old hypotheses for new ones that better fit the observed data.

Garner and Allard realized that winter greenhouses and summer fields differ in temperature, moisture, light intensity,

Chapter opening photo: Scanning electron micrograph of anther (orange) and pollen (yellow) from a crown of thorns plant.

day length, and so on. They began experiments that painstakingly eliminated one after another of these environmental factors, until only one was left as the probable controlling factor in flowering—day length. They concluded that the short days of late autumn and early winter induced flowering in Maryland mammoth tobacco. They could get the plants to bloom in summer if they shielded them from the light at the *beginning* or *end* of each day, thereby exposing them to a shorter day. Conversely, they could prevent blooming in the greenhouse in winter by extending the day length with electric lights.

Experiments with other species revealed that most plants can be placed in one of three groups: (1) short-day plants, which flower when the day length is below some critical value, such as in spring or fall (examples are chrysanthemum, poinsettia, dahlia, aster, cocklebur, goldenrod, ragweed, and Maryland mammoth tobacco); (2) long-day plants, which bloom when the day length exceeds some critical value, usually in summer (beet, clover, petunia, larkspur, black-eyed Susan); and (3) day-neutral plants, which are not affected by day length and can bloom whether the days are long or short (sunflower, carnation, pansy, tomato, corn, string bean). This response to the duration of days versus nights, which Garner and Allard called ***photoperiodism,*** provides plants with a way of timing their reproduction. Some species, for instance, form seeds early; seedlings of these plants emerge the same year, and overwinter as young plants. Others form seeds late, and the seeds themselves overwinter. There are distinct costs and benefits to each strategy, including the availability of water and light for germinating seedlings at different times of year (tradeoffs that vary, depending on the climate and habitat involved).

24.1 Maryland mammoth tobacco Tobacco normally flowers in the summertime when the days are long and the nights short, but this mutant variety grows abnormally large but will not flower in the fields in summer. Maryland mammoth will flower, however, when the days are short and the nights long. Investigators discovered that these plants use day and night length as a cue for flowering, and that Maryland mammoth will flower only when the days are shorter than 12 hours.

If the critical element in the photoperiodism of flowering is day length, as the terms "long day" and "short day" imply, then we should be able to prevent a long-day plant from flowering at the proper season by shielding it from light for an hour or so during the *middle* of the day. But if this is done, nothing happens; the plant flowers normally. If, however, a short-day plant is illuminated by a bright light for a few minutes, or even seconds, in the middle of the night during the normal flowering season it will not bloom (Fig. 24.1). It is clear, then, that the critical element of the photoperiod is actually the length of the night, not the length of the day. Instead of speaking of long-day and short-day plants, it would be more accurate to speak of short-night and long-night plants.

The difference between long-day and short-day plants does not hinge on the absolute length of the night at the time of flowering. The difference is that a long-day (short-night) plant will flower only when the night is *shorter* than a critical value, whereas a short-day (long-night) plant will flower only when the night is *longer* than a critical value (Fig. 24.2). The critical night length is thus a maximum value for flowering by long-day plants and a minimum value for flowering by short-day plants.

Florigen, Produced by the Leaves in Response to the Proper Photoperiod, Initiates Flowering

How does the photoperiod affect flowering? The first convincing evidence that a floral hormone might be involved was found in 1936, from the experiments of the Russian scientist M. H. Chailakhian. Chailakhian removed the leaves from the upper half of chrysanthemums (which are short-day plants), but left the leaves on the lower half (Fig. 24.3). He then exposed the lower half to short days while simultaneously exposing the defoliated upper half to long days; the plants flowered. Next, he reversed the procedure, exposing the lower half to long days and the defoliated upper half to short days; the plants did not flower. He concluded that day length does not have an effect directly on the buds, but causes the leaves to manufacture a hormone that moves from the leaves to the buds and induces flowering. This hypothetical hormone has been named ***florigen.***

The explanation of flowering that emerges from numerous experiments is that an inducing photoperiod causes the leaves to produce florigen, which moves to the buds and stimulates development of the flower. A noninducing photoperiod causes the leaves of many, but not all, plants to inhibit florigen in some way. Under natural conditions flower development is triggered when, as the season changes, the photoperiod passes a critical value and production of florigen by the leaves exceeds the inhibiting tendency.

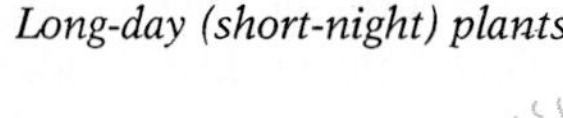

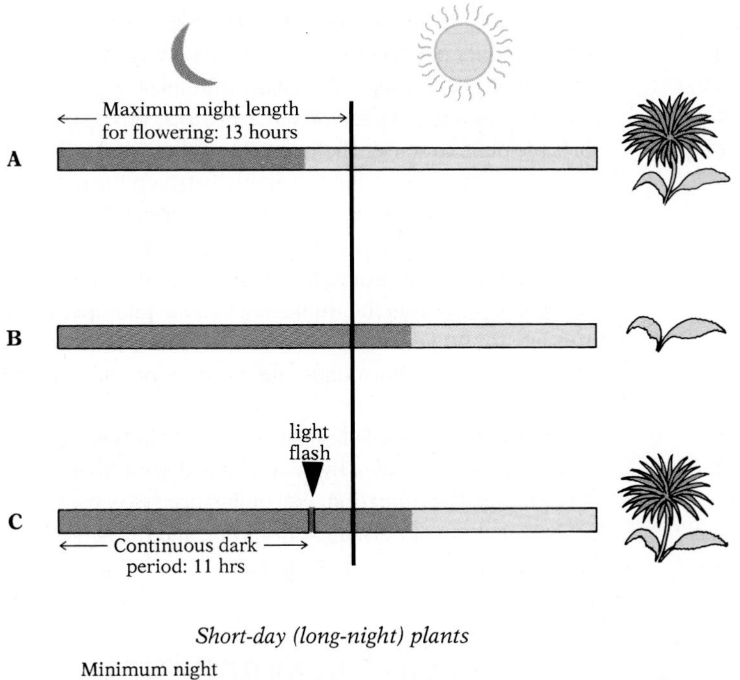

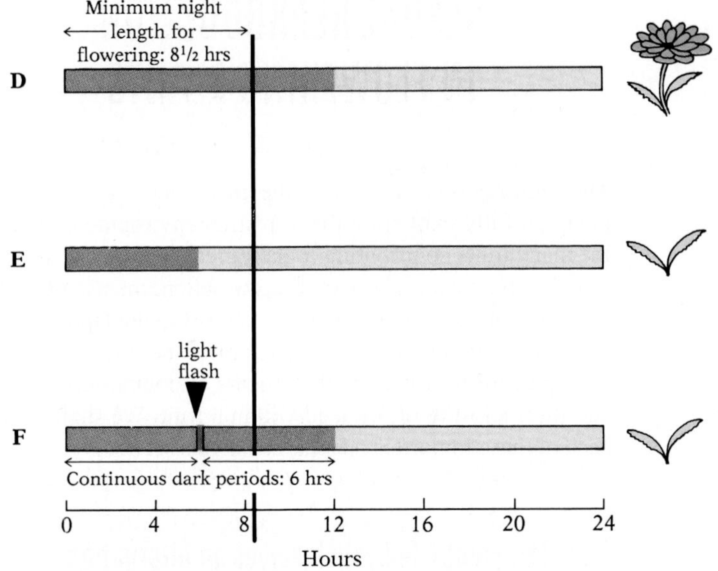

24.2 Comparison of long-day and short-day plants Yellow bars indicate days and blue bars nights. The hypothetical long-day (short-night) plant shown here has a rather long critical night length of 13 hours, and the hypothetical short-day (long-night) plant has a rather short critical night length of $8^1/_2$ hours. In other words, in this example the critical night length for the short-night plant is actually longer than that for the long-night plant. The difference is that the critical night length is a *maximum* value for the short-night plant and a *minimum* value for the long-night plant. The short-night plant will flower when the night length is *below* the critical value (A) but will not flower when it is above the critical value (B); the plant will flower, however, if a long night is interrupted by a flash of light that reduces the period of *continuous* dark below the critical value (C). Conversely, the long-night plant will flower when the night length is *above* the critical value (D), but will not flower when it is below the critical value (E); the plant will not flower if a long night is interrupted by a flash of light that reduces the period of continuous dark below the critical value (F).

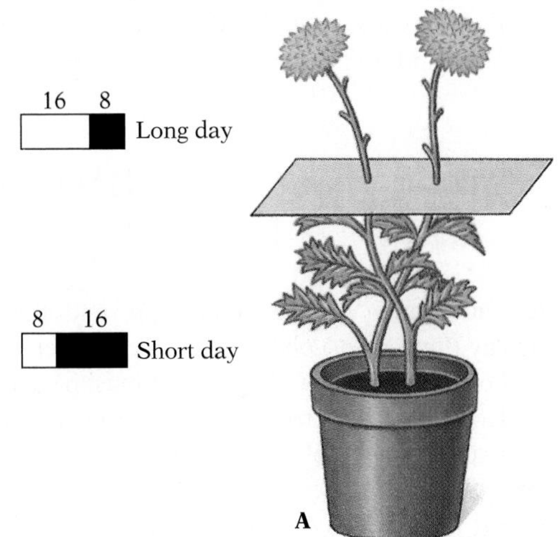

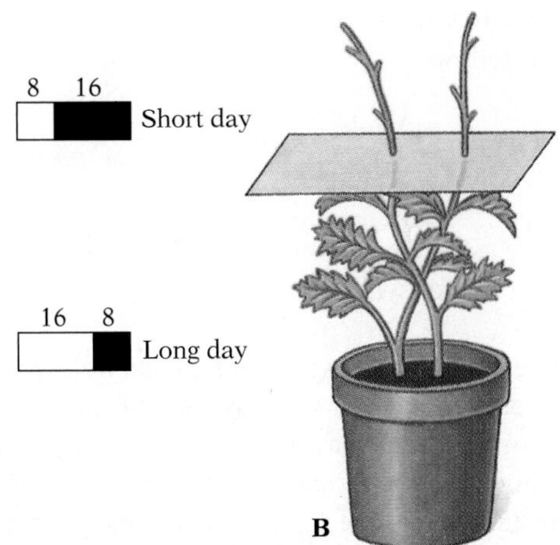

24.3 Chailakhian's experiment (A) Chailakhian removed the leaves from the top half of a chrysanthemum (a short-day plant) and then exposed the top half of the plant to long days and the bottom half to short days. The plant flowered. (B) When he did the reverse experiment, the plant did not flower.

While the florigen hypothesis is a compelling one, it will not rest on firm foundations until a flower-inducing hormone is actually isolated and identified. So far, all attempts to isolate such a hormone have failed. Many botanists working in this area now doubt that a separate flowering hormone exists. The flowering stimulus may involve gibberellin, cytokinin, and certain carbohydrates as factors, so it may be made up of multiple compounds rather than a single "florigen."

Phytochrome Enables the Plant to Sense Light or Darkness

Since interrupting the dark period prevents flowering by a short-day (long-night) plant and induces flowering by a long-day (short-night) plant, the light itself must be detected by the plant. How do plants detect light and measure the photoperiod? Investigators have isolated a light-sensitive pigment, ***phytochrome,*** which is a protein found in the nucleus and cytoplasm of plant cells in minute concentrations. Phytochrome exists in two forms: one that absorbs red light (P_r) and one that absorbs far-red light (P_{fr}). When P_r absorbs red light, it is rapidly converted into P_{fr}. Conversely, absorption of far-red light by P_{fr} rapidly converts it into P_r. The P_r form is the more stable form; in darkness some of the P_{fr} reverts to P_r and some is enzymatically destroyed. We can summarize these conversions as follows:

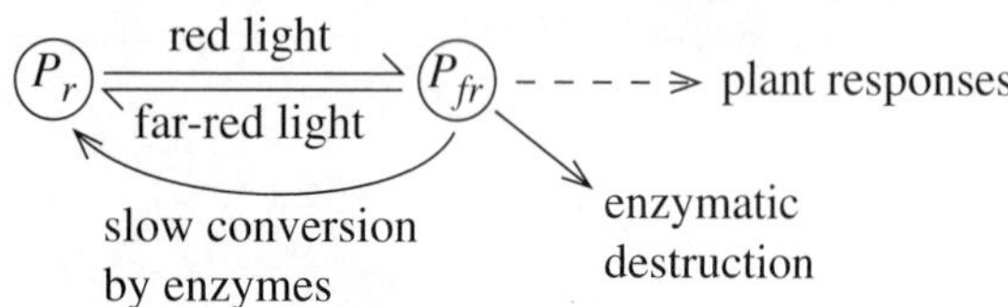

Because sunlight and light from ordinary electric lamps contains more red than far-red wavelengths, most of the P_r phytochrome during the day will be converted to the P_{fr} form. At night, however, the P_{fr} supply dwindles as it reverts to P_r or is destroyed. The ratio of P_{fr} to P_r signals the plant as to whether it is day or night: if most of the pigment is in the P_{fr} form, it is day; if the ratio of the two forms decreases, it is night.

The phytochrome molecules may allow the plant to detect whether it is day or night, but how do plants measure the timing of the photoperiod? It is, after all, the length of the dark period that is fundamental to the control of flowering. For many years it was thought that the time it took to convert P_{fr} back to P_r provided a measure of the length of night. In other words, the system would work like an hourglass. Light would convert all the pigment into P_{fr}. Then during the night the amount of P_{fr} converted back into P_r before the next light would tell the plant how long the night had lasted. Unfortunately, none of the available evidence supports this hypothesis. The mechanism by which the plant measures the length of night is apparently tied to an "internal clock." Plants, like all other organisms, have an internal time-measuring system that produces regular and persistent activity patterns on an approximately 24-hour schedule. We shall discuss biological time measurement at greater length in a later chapter. Apparently the phytochrome pigment enables the plant to sense whether it is in light or darkness, but the actual measurement of the time lapse between the moment the plant senses onset of darkness and the moment it senses the next exposure to light must depend on an internal clock.

Once the phytochrome mechanism and the internal-clock mechanism have together indicated that the photoperiod is appropriate for flowering, the leaves must initiate the next step—synthesizing the hormones necessary for the flowering process.

Phytochrome molecules can powerfully influence a variety of cell activities besides those associated with flowering. For example, seed germination, cell elongation, expansion of new leaves, breaking of dormancy in spring, and formation of plastids in cells also involve this pigment.

SEXUAL REPRODUCTION IN FLOWERING PLANTS

The angiosperms have been the most successful group of plants in fully exploiting the terrestrial environment. Among the many traits contributing to their great success are the evolution of the flower for sexual reproduction, the use of pollen (which contains the male gamete) and special adaptations for transferring pollen from the male part of the flower to the female part for fertilization, the process of double-fertilization, and the enclosure of the seed within a fruit. We shall look at each of these important adaptations in this section.

The Plant Life Cycle Involves an Alternation of Gametophyte and Sporophyte Generations

In the angiosperms, the plant that we see, be it a petunia or a maple tree, is in the diploid stage of the life cycle, and its reproductive structure is the flower. A flower is generally thought to be a short length of stem with modified leaves attached to it. The modified leaves of a typical flower (Fig. 24.4) occur in four sets attached to the enlarged end (receptacle) of the flower stalk. (1) The ***sepals*** enclose and protect all the other floral parts during the bud stage. They are usually small, green, and leaflike, but in some species they are large and brightly colored. (2) Internal to the sepals are the ***petals,*** which together

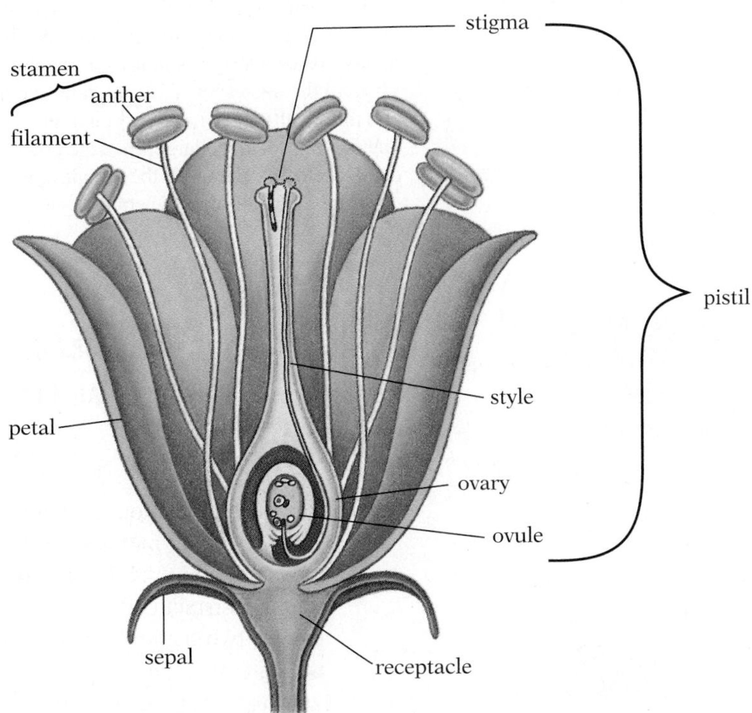

24.4 The parts of a flower

form the ***corolla.*** The petals are often quite showy, to attract insects, birds, and other animals to the flower. (3) Just inside the circle of the corolla are the ***stamens,*** which are the male reproductive organs. Each stamen consists of a stalk called a ***filament*** and a terminal pollen-producing structure called an ***anther.*** (4) In the center of the flower is the female reproductive organ, the ***pistil*** (some species have more than one pistil per flower). Each pistil consists of an ***ovary*** at its base, a slender stalk called a ***style,*** which rises from the ovary, and an enlarged top called a ***stigma.*** The pistil is derived from one or more modified spore-bearing leaves called ***carpels.***[1] Many flowers have all four kinds of floral organs—sepals, petals, stamens, and pistils—but some flowers lack one or more of them.[2]

Recall that the life cycle of plants differs from that of animals in having an alternation of generations (see Fig. 9.31, p. 199) in which there is both a multicellular diploid and a multicellular haploid stage. Meiosis in plants, including the angiosperms, produces haploid reproductive cells called ***spores,*** which divide mitotically to form haploid multicellular plants, known as ***gametophytes.*** The gametophytes, as the name suggests, eventually produce cells specialized as gametes. The haploid male and female gametes unite in fertilization to produce the diploid zygote, which then divides mitotically and develops into a diploid multicellular plant, the ***sporophyte.*** In the flowering plants, the spores and gametophyte plants are formed within the flower.

Meiosis Within the Ovary Produces Megaspores, Which Mature Into the Female Gametophyte Plant

Within the ovary of the flower are one or more ***ovules,*** which are attached by short stalks to the wall of the ovary. Each ovule contains a special spore-producing cell, a sporangium, in which meiosis takes place (Fig. 24.5). Meiosis occurs once in each ovule, with formation of four large haploid cells called ***megaspores,*** three of which usually soon disintegrate. The remaining megaspore then divides mitotically several times, producing, in many species, a structure composed of seven cells, one of which is much larger than the others and contains two nuclei, called polar nuclei (Fig. 24.5E). This haploid, seven-celled, eight-nucleate structure is a haploid multicellular plant, the female gametophyte (often called an embryo sac). The embryo sac remains within the ovary and is completely dependent on it for its nourishment. One of the cells at the lower end of the embryo sac is the egg cell.

[1] A simple pistil is composed of only one sporophyll, or carpel. A compound pistil is composed of several fused carpels.

[2] In some species, such as corn, willow, oak, and walnut, the stamens and pistils are in separate flowers.

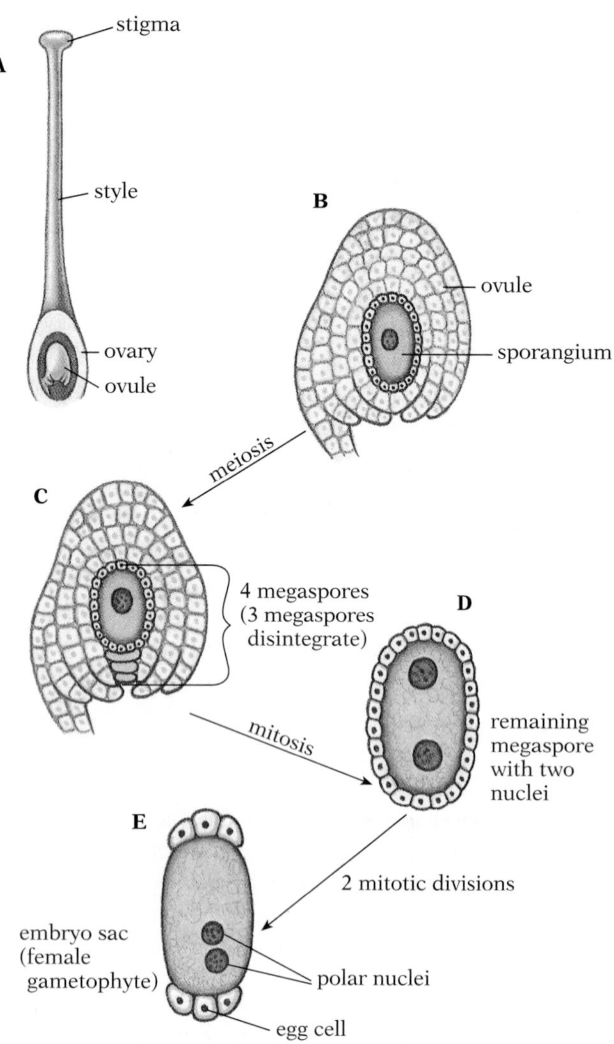

24.5 Formation of the female gametophyte plant Each ovule within the ovary (A, B) contains a special spore-producing cell, a sporangium, in which meiosis takes place, producing four haploid cells called megaspores (C). Three of the megaspores disintegrate, and the remaining megaspore divides (D) mitotically three times, producing a structure composed of seven cells, one of which is much larger than the others (E). This haploid, seven-celled, eight-nucleate structure is the female gametophyte or embryo sac. One of the cells at the lower end of the embryo sac will become the egg cell.

Meiosis Within the Anther Produces Microspores, Which Mature Into the Male Gametophyte Plant, the Pollen Grain

Each anther has four pollen sacs (Fig. 24.6), in which special cells undergo meiosis to produce numerous haploid ***microspores*** (Fig. 24.7), each of which is then surrounded by a tough, resistant cell wall. Each microspore divides once, producing two haploid nuclei, and the structure develops into a pollen grain (Fig. 24.8). The pollen grain is actually the male gametophyte plant. Pollen is released from the anther when the mature pollen sacs (Fig. 24.6) split open.

The next step is ***pollination,*** the transfer of pollen grains from the anther to the stigma of a carpel so that fertilization can occur. Because plants cannot move about, they depend on external agents such as wind, insects, or birds to transfer the pollen grains from the anther to the stigma (Fig. 24.9). The blossom is more than a simple receptacle for pollen. Its coloration and markings, time of opening, and the nature of its scent are adapted to a particular class of pollinators. Often parts of the flower mature at different times so that no pollen is produced during the period in which the egg is capable of being fertilized. This reduces the chance of inbreeding. Most plants

MAKING CONNECTIONS

1. How could a florist induce a long-day flower such as the petunia to flower in the winter?

2. How is phytochrome involved in many of the phenomena of plant growth and development (such as seed germination) controlled by the hormones discussed in Chapter 23?

3. What reproductive adaptations evolved by the flowering plants have contributed to their success in exploiting the terrestrial environment?

4. How do the life cycles of plants and animals differ? What are the products of meiosis in each? (See Chapter 9.)

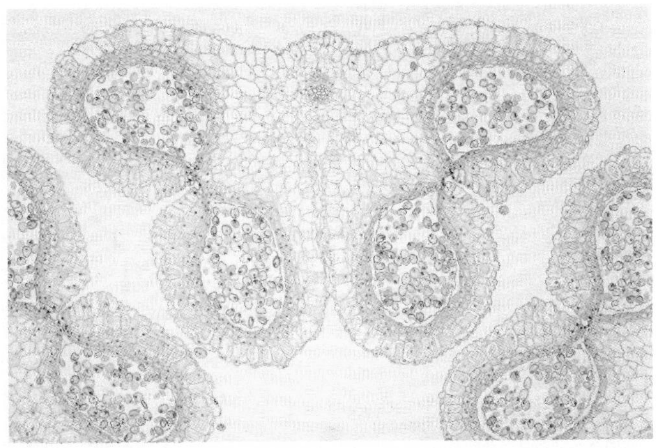

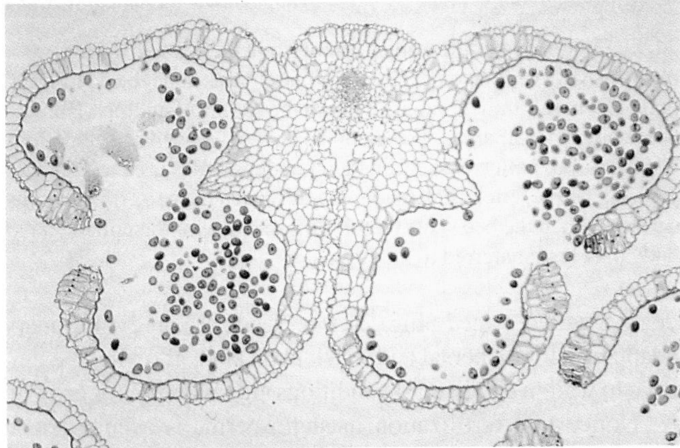

24.6 Photographs of the pollen sacs of an anther When the pollen sacs with their pollen grains are mature, the sacs open (bottom) and the numerous pollen grains within it will be released.

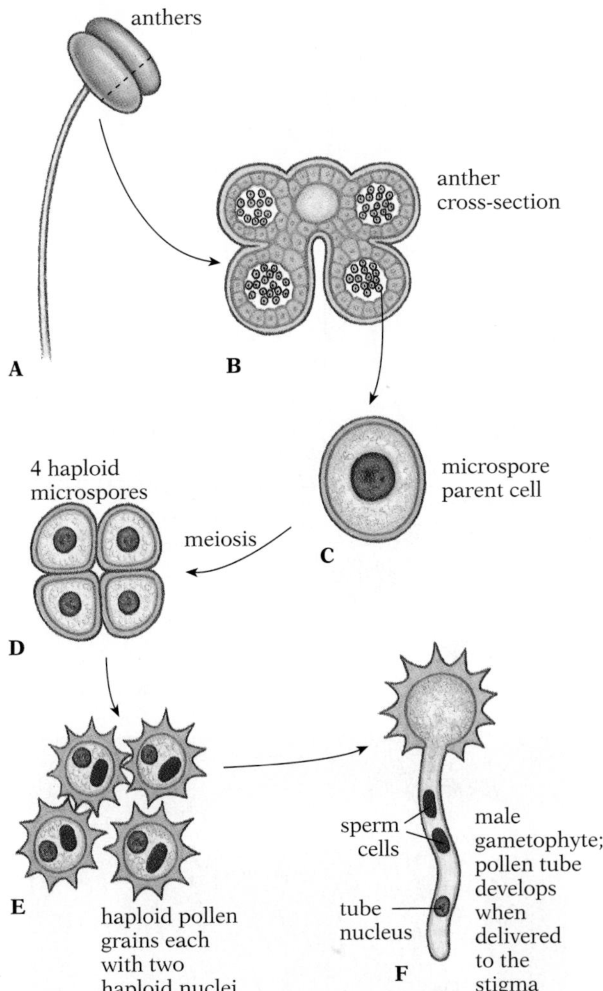

24.7 Formation of the male gametophyte plant Special cells within the anther undergo meiosis (A–D), producing numerous hapoid *microspores,* each of which is then surrounded by a tough, resistant cell wall. Each microspore divides once by mitosis, producing two haploid nuclei, and the structure develops into a pollen grain (E)—a male gametophyte—that is released from the anther when the pollen sacs mature. When the pollen grain is delivered to the stigma, a pollen tube grows out; its growth is governed by one of the two nuclei, the tube nucleus. The second nucleus divides once by mitosis to produce two sperm cells (F).

are cross-pollinated rather than self-pollinated—i.e. the pollen from one plant is deposited on the stigma of another plant of the same species.

Angiosperms Carry Out "Double Fertilization"

The pollen grain germinates when it falls (or is deposited) on the stigma of a pistil, which is usually rough and sticky. A pollen tube begins to grow, and the two nuclei of the pollen grain move into it. One nucleus, the tube nucleus, controls the growth of the pollen tube. The second nucleus divides, giving rise to two sperm cells, which are the male gametes (Fig. 24.7F). The pollen tube grows down through the tissues of the stigma and style and enters the ovary (Fig. 24.8). When the tip of the pollen tube enters an ovule, it discharges the two sperm cells into the female gametophyte (embryo sac). ***Double fertilization*** then occurs: one sperm fertilizes the egg cell, and the zygote thus formed undergoes a series of mitotic divisions and develops into a tiny sporophyte ***embryo.*** The second sperm combines with the two polar nuclei to form a triploid nucleus. This nucleus undergoes a series of mitotic divisions, and a triploid tissue, called ***endosperm,*** is formed. The endosperm functions in the seed as a very rich source of stored food for the embryo. The embryo, together with the endosperm, becomes enclosed in a tough protective ***seed coat.*** The resulting composite structure, made up of embryo, endosperm, and seed

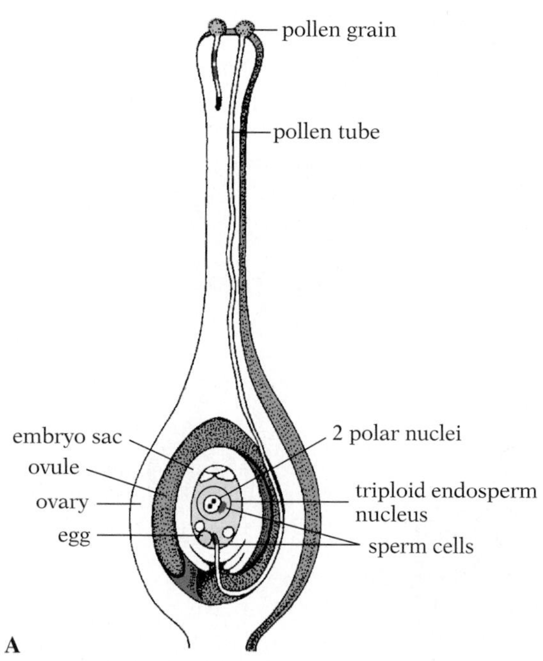

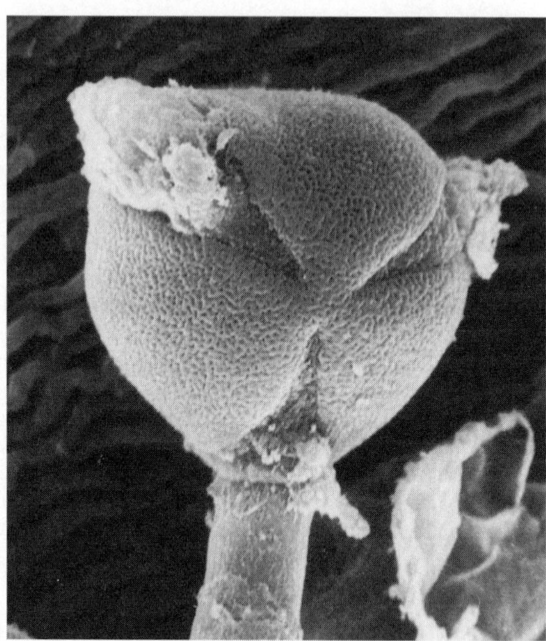

24.8 Fertilization of an angiosperm (A) Pollen grains land on the stigma and germinate forming pollen tubes that grow downward through the style and into the ovary. One of the pollen tubes shown here has reached the ovule within the ovary and discharged its sperm cells into the embryo sac. One sperm will fertilize the egg cell, and the other will unite with the two polar nuclei to form a triploid nucleus, which will give rise to the endosperm. (B) Germinating pollen grain with pollen tube. The tough, resistant wall of the pollen grain protects the male gametophyte on its difficult journey from the anther to the stigma. The pollen tube (below) will grow down through the tissues of the pistil to reach the ovule where it will discharge its sperm.

24.9 Honey bee pollinating a chamomile flower Flowering plants usually depend on external agents, such as honeybees, to carry pollen from the anther of one flower to the stigma of another. Honey bees collect nectar and pollen from the flowers they visit. The pollen is packed into pollen baskets on their rear legs and is used to feed the growing larvae, but it is also liberally dusted on the hairs on the body surface. When the bee visits the next flower some of the pollen on its body may be transferred to the stigma.

coat, is called a ***seed*** (Fig. 24.10). Seeds are an evolutionary adaptation for dispersal and survival on land, allowing the embryo to wait in safety until conditions are favorable for growth.

Notice that fertilization in angiosperms is accomplished without the use of water. The pollen is carried to the stigma by the wind or by an animal pollinator, and the sperm is delivered

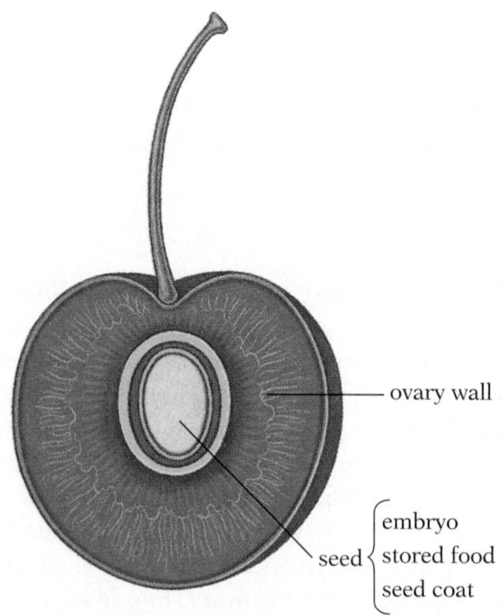

24.10 Diagram of a fruit (a cherry) The skin and fleshy part of the cherry plus the hard covering surrounding the seed is the greatly enlarged ovary wall. Inside the hardened inner wall of the ovary is the seed of the cherry.

directly to the egg by the pollen tube. This method of fertilization does not require flagellated sperm to swim through the water to reach the egg, making this an important adaptation for life on land.

The Ovary Develops Into a Fruit That Protects the Seeds and Facilitates Dispersal

While each ovule is developing into a seed, the surrounding ovary develops into a ***fruit,*** the wall (pericarp) of which usually enlarges greatly in the process (Fig. 24.10). Sometimes other floral structures associated with the ovary are also incorporated into the fruit. The ripe fruit may burst, expelling the seeds, as in peas (where the pod is the fruit). Or the ripe fruit with the seeds still inside may fall from the plant, as in tomatoes, squash, cucumbers, apples, peaches, and acorns. The fruit not only helps protect the seeds from drying out but often also facilitates their dispersal by various means—the wind, for example (Fig. 24.11), or an animal, which, attracted by the fruit, carries it to other locations or eats both fruit and seeds and later releases the unharmed seeds in its feces.

24.11 Fruit of the dandelion Each dandelion "seed" is actually a seed enclosed in a small dry fruit. On one end of the fruit is a parachute, which aids in dispersal by the wind.

The Seed Is an Adaptation to Protect the Fragile Embryo

During the development of the seed, food stores are accumulated and the embryo undergoes an initial period of rapid growth and development, which is followed by an arrest of growth and development and a loss of water. In many plants there is a period of dormancy before active growth of the embryo resumes. The hormones cytokinin, auxin, gibberellin, and abscisic acid are found in relatively high concentration within the seed at different stages of development and are thought to play an important role in controlling the developmental processes.

The first cell division of the zygote may occur a day or so after the egg cell has been fertilized. It invariably gives rise to two cells of unequal size: a smaller terminal cell and a larger basal cell (Fig. 24.12A). The terminal cell will give rise to the embryo, and the basal cell will form an elongate ***suspensor*** structure (Fig. 24.12B), which functions only while the embryo is in the seed. Nutrients move from the endosperm into the embryo through the suspensor. Let us follow the embryonic development of *Capsella,* a much-studied dicot commonly called shepherd's purse.

The terminal cell divides many times to form a globular embryo, in which three types of tissues begin to differentiate: (1) a surface layer of ***protoderm,*** which will form epidermal tissue; (2) an inner core of ***provascular tissue,*** which will form cambium and the vascular tissues; and (3) a middle layer of ***ground tissue,*** which will form the cortex (Fig. 24.12C). Shortly thereafter two mounds arise on the portion of the embryo opposite the suspensor (Fig. 24.12D). These mounds, which give the embryo a heart-shaped appearance, will become the ***cotyledons,*** or embryo leaves. The cotyledons are modified leaves that function primarily in the digestion and absorption of the food stored in the endosperm. They generally do not resemble the mature leaves of the plant. The part of the embryo proper below the point of attachment of the cotyledons is called the ***hypocotyl.*** As the cotyledons and hypocotyl of *Capsella* continue to elongate within the very limited space available to them in the seed, the embryo begins to curve back on itself (Fig. 24.12E). Cytokinins are found in relatively high concentrations in the period of early seed growth and it is thought that this hormone is responsible for stimulating rapid cell division in the embryo. After the early stages, the cytokinin levels drop and auxin and gibberellin levels increase.

As most of the cells take on more and more of the characteristics of the tissues to which they will give rise, small clumps of cells at each end of the embryo are set aside to remain forever embryonic and capable of dividing (meristematic). One clump, located just above the point of attachment of the cotyledons, will become the ***apical meristem of the shoot.*** The other clump, at the opposite pole, will become the ***apical meristem of the root tip*** (Fig. 24.12E).

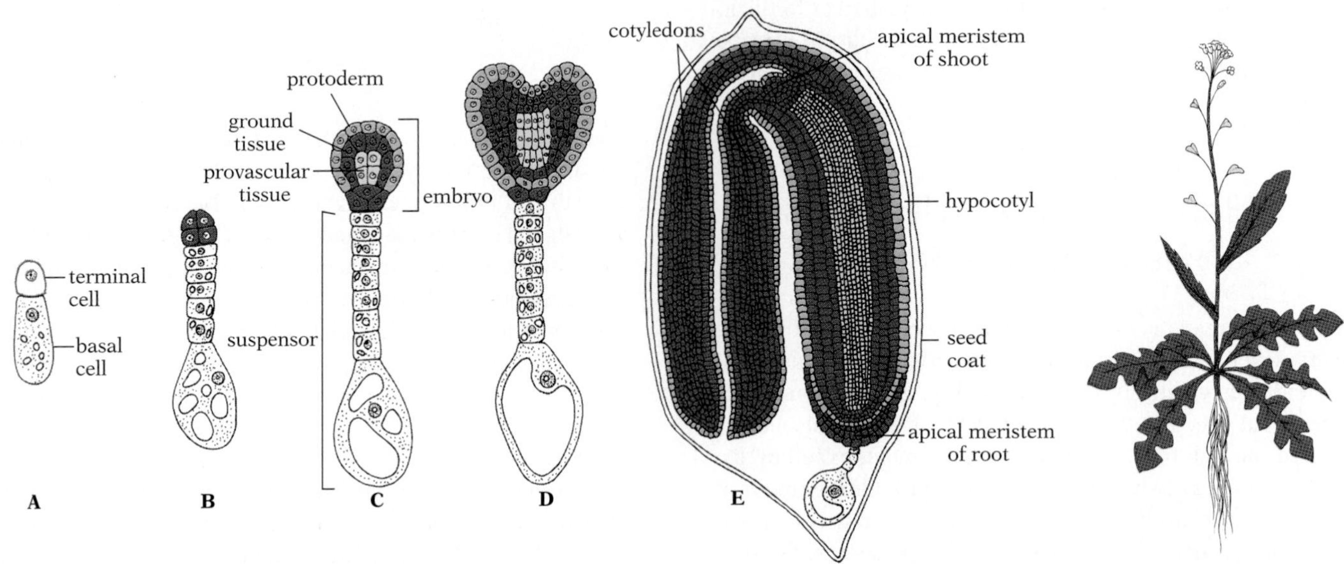

24.12 Embryonic development of shepherd's purse (*Capsella*) (A) The first division of the zygote produces a smaller terminal cell (green) and a larger basal cell. (B, C) Divisions of the terminal cell give rise to a globular embryo, whereas divisions of the basal cell give rise to a stalklike suspensor. Cells of the globular embryo soon differentiate to form three tissue types: protoderm on the surface, provascular tissue in the center, and ground tissue between the other two. (D) The formerly globular embryo becomes heart-shaped as two mounds that will develop into cotyledons begin to form. (E) Elongation of the cotyledons and the hypocotyl within the confines of the seed causes the embryo to fold back on itself. A fully grown plant is also shown.

Thus an embryo consists of either one (monocot) or two (dicot) cotyledons, the hypocotyl, and the apical meristems of the shoot and root. In some seeds cell divisions in these meristems continue during embryonic development and give rise to an ***epicotyl***—a region of shoot above the point of attachment of the cotyledons, frequently bearing the first young leaves—and, at the lower end of the embryo, to a ***radicle,*** which will develop into the primary root (Fig. 24.13).

Notice that the fully developed plant embryo does not possess, even in rudimentary form, all the organs of the adult plant. Unlike animals, plants continue to grow and form new organs throughout their lifetime. New roots, branches, leaves, and reproductive structures are formed during each growing season.

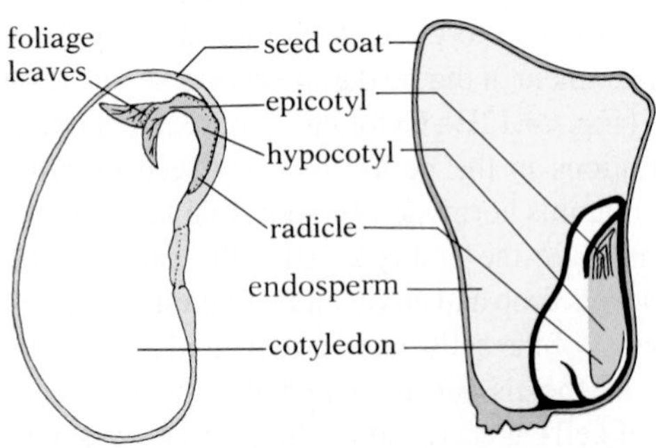

24.13 Diagram of dicot and monocot seeds The embryos in these seeds consist of epicotyl, hypocotyl, radicle, and cotyledon. Left: A dicot seed (bean) in which one cotyledon has been removed to reveal the remaining parts of the embryo. Right: A monocot seed (corn) has only one cotyledon, but a large endosperm.

Finally, embryonic growth and development stops, and the seed begins to lose water and become desiccated. These responses also appear to be under hormonal control. The later stages of seed development are characterized by decreasing levels of auxin and gibberellin and increasing levels of abscisic acid. Abscisic acid is thought to promote the maturation of the embryo and, at the same time, inhibit germination of the seed. The levels of this hormone begin to drop as water is lost during desiccation. The embryo cannot germinate, i.e., grow and emerge from the seed, while in a desiccated condition. The seed may last for months or years in this quiescent state, which is why seeds may be sold in packets at garden shops over long periods of time.

The Uptake of Water by the Seed Induces the Release of Gibberellin, Promoting Germination

Many seeds must go through a period of quiescence before growth can resume. Environmental factors such as water, temperature, and oxygen also affect the germination process. Germination begins with a massive absorption of water by the seed, which greatly increases its volume, sometimes by as much as 200 percent. The resulting hydration causes the embryo to release gibberellin, which in turn induces the synthesis

MAKING CONNECTIONS

1. If you wanted to study meiosis in an angiosperm, where would you look?

2. Which part of a bird's egg would be analogous to the endosperm in a seed?

3. Compare the adaptations of angiosperms and terrestrial vertebrates for accomplishing fertilization without using water to transport sperm.

4. When you eat a cherry, what part of the flower are you eating? What roles do fruits play in the angiosperm life cycle?

5. Which of the plant hormones described in Chapter 23 causes rapid cell division in the early embryonic development of a plant? Which hormones increase in concentration later in development?

6. Contrast how organs develop in animals and plants.

of several hydrolytic enzymes including amylase. These enzymes catalyze the hydrolysis of food stored in the endosperm, providing energy for the growing embryo. The metabolic rate of the embryo increases sharply, and the higher metabolic rate makes possible resumption of active cell division, synthesis of new cytoplasm, and increase in cell size by uptake of water. The growing embryo soon bursts out of the seed coat and rapidly assumes a characteristic plant form, with distinguishable shoot and root.

When a Seed Germinates, the Hypocotyl Emerges First

The hypocotyl (with attached radicle) is the first part of the embryo to emerge from the seed. It promptly turns downward regardless of the orientation of the seed. By the time the epicotyl begins its rapid development, the radicle, at the lower end of the hypocotyl, has already formed a young root system capable of anchoring the plant to the soil and of absorbing water and minerals. In some dicots, the upper portion of the hypocotyl elongates and forms an arch, which pushes upward through the soil and emerges into the air (Fig. 24.14). Once the hypocotyl arch is exposed to light, it straightens, the cotyledons and the epicotyl thus being pulled out of the soil. The epicotyl then begins to elongate.

Other dicots, of which the garden pea is an example, show a slightly different pattern of germination (Fig. 24.15). In these plants no hypocotyl arch forms and the cotyledons are never raised above ground. Instead, the epicotyl begins to elongate soon after the young root system has begun to form; it always grows upward and soon emerges from the soil. A similar pattern is seen in corn, a monocot, which has only one cotyledon but a large endosperm (Figs. 24.15 and 24.16).

Growth and Differentiation of the Plant Body

The seedling plant grows in length rather slowly at first, then enters a longer period of much more rapid growth, and finally slows down again or even stops as it approaches maturity. ***Perennial plants*** (those that live from year to year) continue to grow to some extent throughout their lives while ***annual plants,*** such as sweet peas, marigolds, and petunias, cease growing after they reach maturity and die at the end of the growing season.

As we have seen, growth of the shoot or root of a young plant involves both cell multiplication and cell elongation, and both aspects of growth are influenced by the various growth hormones, particularly auxin, gibberellin, and cytokinin. In plants with only primary tissue, both cell division and cell elongation are ordinarily restricted to the apical meristems found at the tips of the shoot and root. Let us look first at the root.

New Cells Are Constantly Produced by the Apical Meristem of the Root, Pushing the Root Tip Through the Soil

The apical meristem at the tip of the root is protected by a conical root cap consisting of a mass of nondividing cells (Fig.

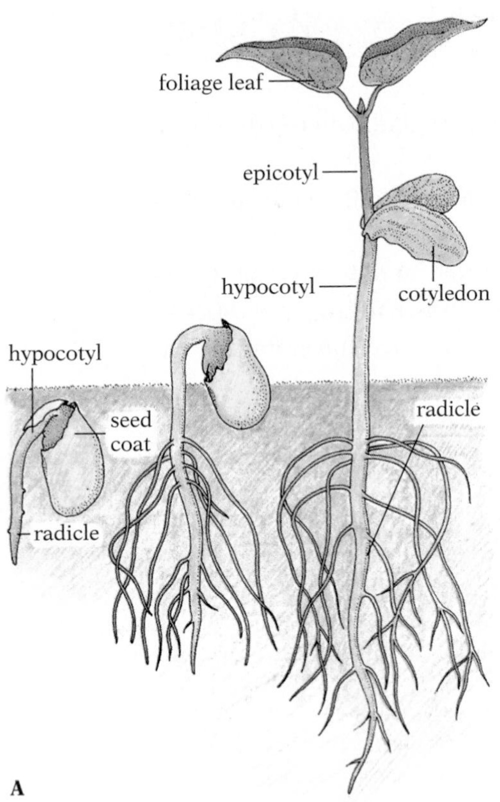

24.14 Germination and early development of a bean plant (A) The hypocotyl and radicle emerge from the seed first (left). As the upper portion of the hypocotyl elongates, it forms an arch that pushes out of the soil into the air; the radicle gives rise to the first root system (middle). The hypocotyl straightens, pulling the cotyledons out of the ground as the epicotyl begins its development (right). (B) The photograph shows a germinating bean plant.

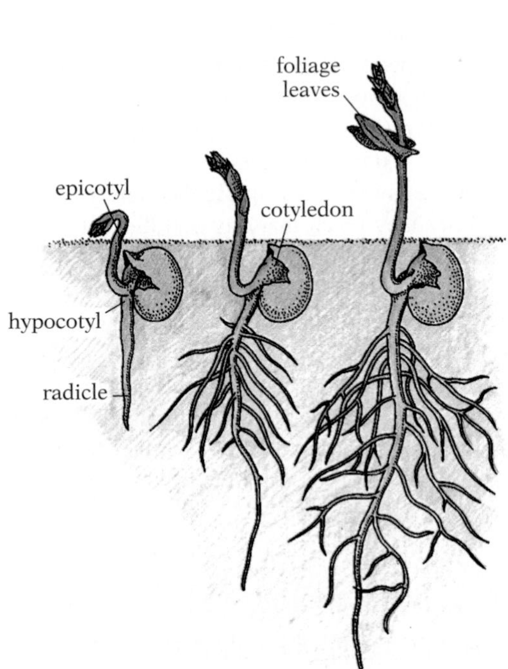

24.15 Germination and early development of a pea plant The development of peas differs from that of beans in that no hypocotyl arch is formed and the cotyledons remain beneath the soil.

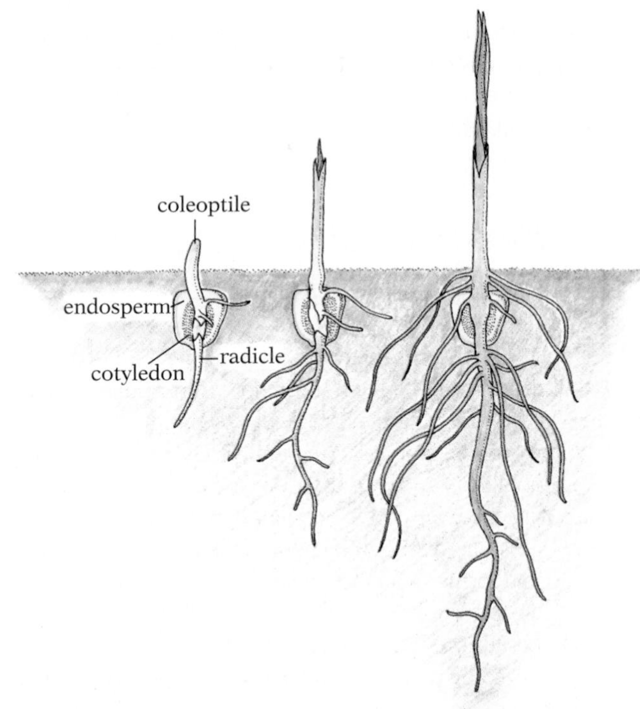

24.16 Germination and early development of a corn plant In corn, the epicotyl begins to elongate as soon as the root system is formed. Notice that the young shoot is initially enclosed with a coleoptile, a tubular protective sheath.

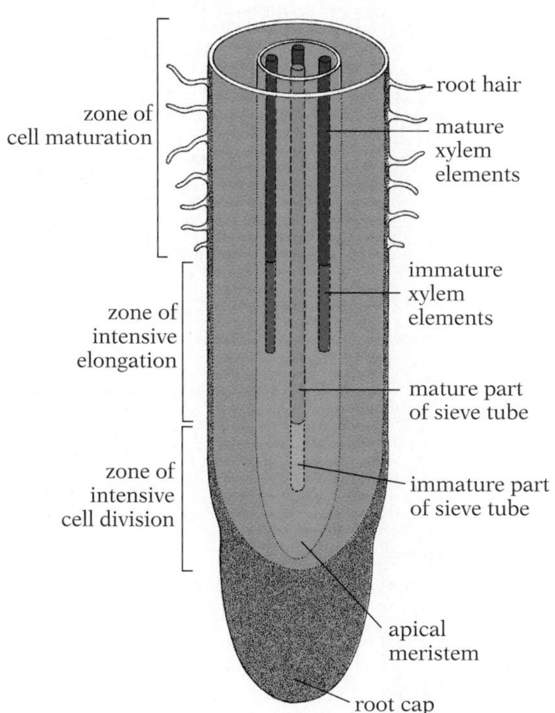

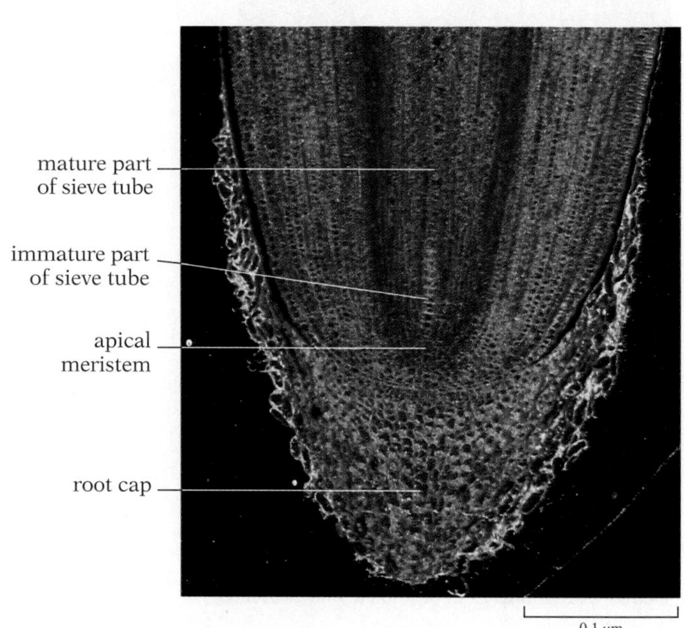

24.17 Longitudinal sections of root tips New cells are produced by mitotic divisions in the meristematic region just behind the root cap, as shown in this drawing of a tobacco root. There is a zone of especially intensive cell elongation not far behind the meristem. Both phloem and xylem are well differentiated in the region of the root hairs, back of the zone of intensive elongation. Many of the same structures are visible in the photograph of the distal part of the root tip of a corn plant.

24.17). As the root elongates and the tip moves through the soil, some of the cells on the surface of the root cap are damaged, and then replaced by new cells added to the cap by the apical meristem. Located just behind the cap, the apical meristem is usually restricted to a tiny region and is composed of relatively small, actively dividing cells. Most of the new cells it produces are laid down on the side away from the root cap. These cells are left behind as the meristem lays down additional new cells in front of them and the tip continues to move through the soil. It is these new cells, derived from the apical meristem, that will form the primary tissues of the root.

As the new cells become further removed from the meristem by the deposition of additional intervening cells, cell division in most of them slows down and cell enlargement becomes the dominant process. Most of this enlargement, which accounts for the greater part of root growth, results from increasing length rather than increasing width. As we saw in Chapter 23, cell elongation is under the control of hormones, particularly auxin and gibberellin. The zone where cell elongation predominates in the root is just behind the meristem, and usually extends only a few millimeters along the root (Fig. 24.17). The zone of cell maturation is next; in this region the cells begin to mature and take their final form. This zone can be recognized by the production of root hairs by the epidermal cells (Fig. 24.17).

Cell Division in the Apical Meristem of the Stem Gives Rise to Primary Tissues and Leaf Primordia

New cells are produced by an apical meristem near the tip of the shoot, and these cells then elongate, pushing the tip upward (Fig. 24.18). Growth of the stem differs from growth of the root in that there is a lateral production of leaves at the tip of the stem. At regular intervals an increase in the rate of cell division under a localized region of the shoot's apical meristem gives rise to a series of swellings called leaf primordia, which will give rise to new leaves (Fig. 24.18). The point at which each leaf primordium arises from the stem is called a ***node,*** and the length of stem between two successive nodes is called an ***internode.*** Most increase in length of the stem results from elongation of the cells in the young internodes.

At the tip of the stem is a series of internodes that have not yet undergone much elongation. The tiny leaf primordia that separate these internodes curve up and over the meristem, with

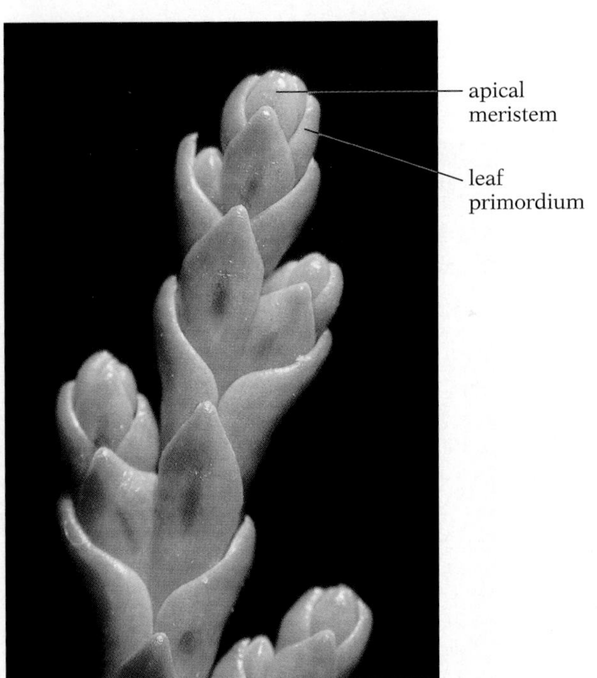

24.18 A leaf bud Leaf buds on a juniper branch.

the older, larger ones enveloping the younger, smaller ones (Fig. 24.19). The resulting compound structure, consisting of the apical meristem and a series of unelongated internodes enclosed within the leaf primordia, is called a ***bud.*** In shoots that have periodic growth, the bud is protected on its outer surfaceby overlapping scales, which are modified leaves that grow from the base of the bud.

When a dormant bud "opens" in the spring, the scales curve away from the bud and then fall off, and the internodes that were contained within the bud begin to elongate rapidly. As the nodes become farther and farther separated, mitotic activity in the leaf primordia gives rise to young leaves. Before a leaf is fully formed, a small mound of meristematic tissue usually arises in the angle between the base of the leaf and the internode above it. Each of these new meristematic regions gives rise to a lateral or ***axillary bud,*** which has the same essential features as the terminal buds (Fig. 24.20). Elongation of the internodes of the lateral buds during the next growing season will produce branch stems.

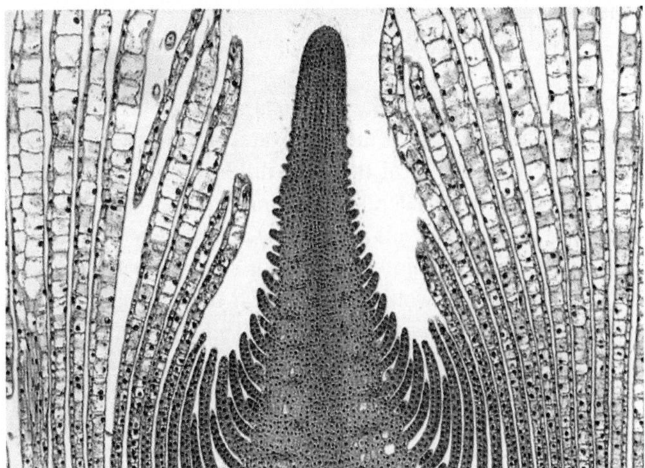

24.19 Photograph of sectioned *Elodea* bud The apical meristem of the shoot is located near the tip of the shoot. The peglike structures located along the sides of the tip are the youngest leaf primordia, which are enclosed by the older, larger leaf primordia.

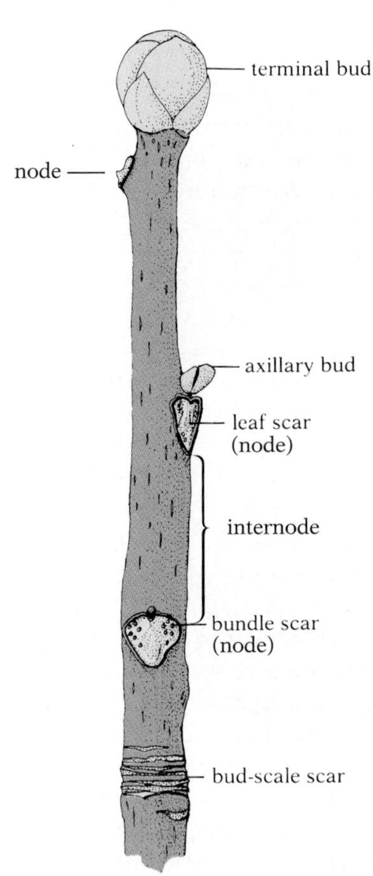

24.20 Portion of a stem The bud-scale scar shows where the dormant terminal bud of the previous winter was; the length of stem between the bud-scale scar and the terminal bud is one year's growth. Leaf scars show where petioles were attached to the stem. The bundle scars within each leaf scar show where the vascular bundles passed into the petiole. The axillary bud will give rise to a branch stem during the next growing season.

As Development Proceeds, Cells Become Committed to One Developmental Pathway

All the new cells produced by the apical meristems are fundamentally alike. Yet some of these cells will become collenchyma, some will become xylem vessels, some will become

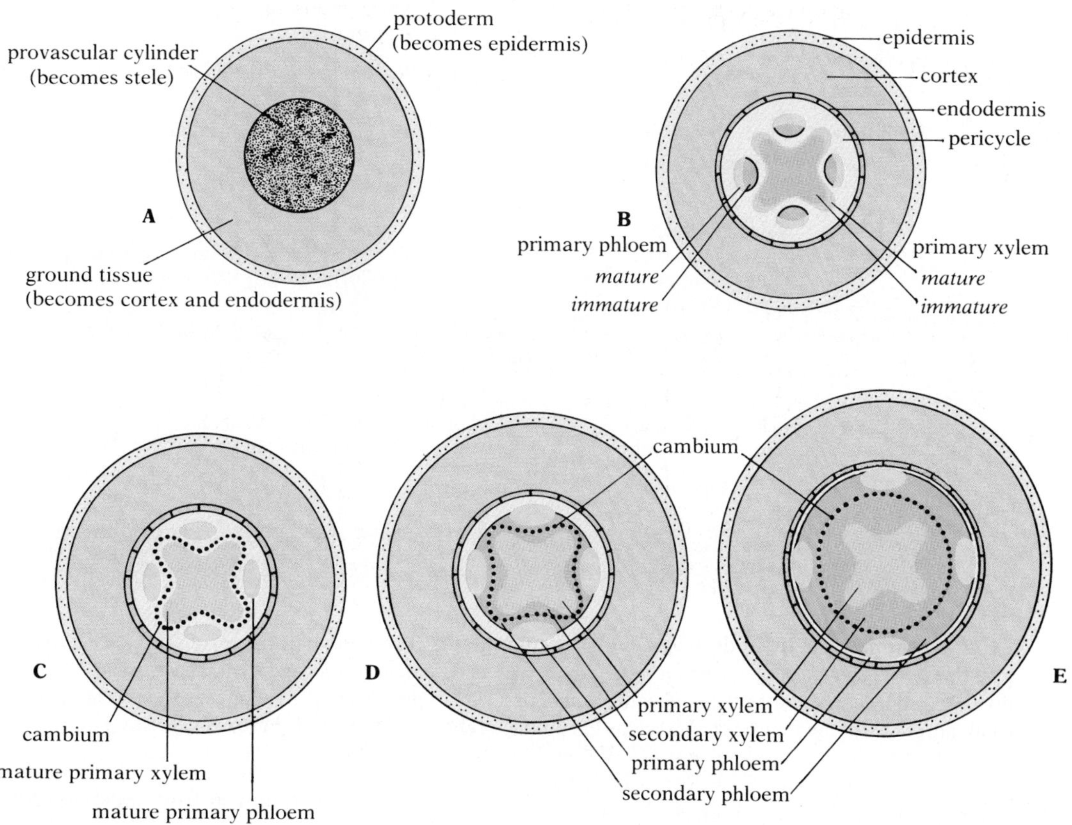

24.21 Differentiation in a young root (A) Cross section just back of the meristem. Three distinct concentric areas can already be detected. (B) At a slightly later stage of development the protoderm has differentiated into epidermis; the ground tissue has differentiated into cortex and endodermis; and the provascular cylinder has begun to differentiate into primary xylem and primary phloem. (C) Differentiation in the provascular cylinder is complete, and the cambium is about to become active. (D) Divisions in the cambium have given rise to secondary xylem and secondary phloem, which are located between the primary xylem and primary phloem. (E) The areas of secondary tissue continue to thicken as more and more new cells are produced by the cambium.

sieve elements, and so forth. The process whereby a cell changes from its immature to mature form is called ***differentiation.***

In the growing root or stem, cells begin to differentiate into the various tissues of the plant while they are still within the meristematic region (Fig. 24.17). After the processes of cell division and cell elongation are completed, the cells begin to mature into their final form. Three concentric areas can be distinguished even in the region just behind the meristem of a root when it is viewed in cross section (Fig. 24.21A). These are (1) an outer protoderm, (2) a wide area of ground tissue (composed of parenchyma cells) located beneath the protoderm, and (3) an inner core of provascular tissue composed of elongated cells. Just as in the embryo, the protoderm rapidly matures into the epidermis, the ground tissue into the cortex and endodermis, and the provascular core into the primary tissues of the stele: primary xylem, primary phloem, pericycle, and vascular cambium (Figs. 24.21B, C and 24.22). Differentiation in the growing stem follows a similar pattern, except that there are usually two areas of ground tissue—one between the protoderm and the provascular cylinder, which gives rise to the cortex and endodermis, and a second inside the provascular cylinder, which becomes the pith (Fig. 24.22).

As we saw in Chapter 19, increase in diameter of the root or stem depends on formation of secondary tissues composed of cells derived from lateral meristems, particularly the vascular cambium. As cells of the vascular cambium undergo mitosis, many new cells are produced on the outer face of the cambium and these, under the influence of auxin, differentiate into secondary phloem, while other new cells are produced on the inner face of the cambium and differentiate into secondary xylem (Fig. 24.21D, E).

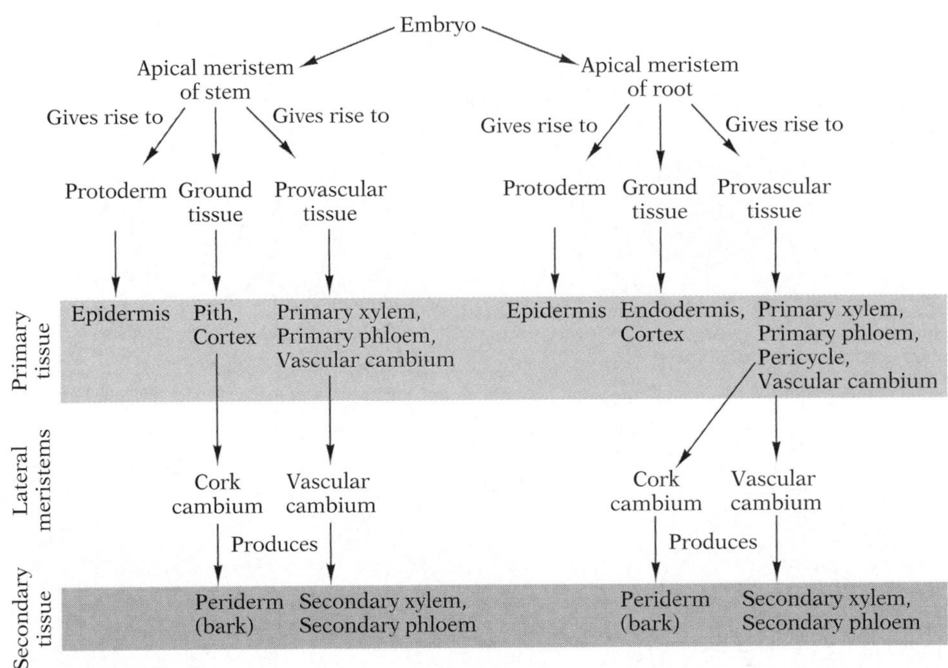

24.22 A summary of tissue development in vascular plants The embryonic tissues of the plant embryo give rise to two clumps of meristematic tissue: the apical meristem of the root and stem. These tissues divide to produce cells that will mature into primary tissues. In some plants, the lateral meristems (vascular cambium and cork cambium) become active and produce cells that differentiate into secondary tissues.

In Many Plant Cells, Differentiation Is Reversible—Some Differentiated Cells Can Resume Totipotency

We have seen that as development proceeds, the individual cells become more and more committed to one particular course of differentiation. Is this differentiation process reversible in plants?

Some years ago F. C. Steward and his colleagues at Cornell University succeeded in growing whole carrot plants from single mature cells removed from a carrot root. Similar experiments on other plants provide evidence that even fully differentiated plant cells (except cells, such as sieve elements, that have lost their nuclei) can be made to revert to the embryonic state and resume ***totipotency*** (unlimited developmental potential).

The technique of growing whole plants from single mature cells is now widely used to create clones; hundreds of genetically identical plants can be grown from the cells of a single plant. Plants such as tobacco, cabbage, potato, tomato, and carrot are grown using this technique.

MAKING CONNECTIONS

1. What property of seeds enables them to remain viable during months or years of dormancy? Which hormone inhibits germination of the seed?

2. If you wanted to study mitosis in a young, growing plant, where would you look?

3. How is the reversibility of differentiation in plant cells related to the processes of gene regulation described in Chapter 12?

4. What are the practical and scientific benefits of the ability to create clones of particular plants, such as tomatoes and carrots?

Plant Hormones Play a Key Role in Nearly All Phases of Development

In the last two chapters we have seen that a plant tissue's response to a hormone depends on the tissue involved, the concentration of the hormone, and the presence and relative concentrations of other hormones. Some plant hormones work together in promoting growth and differentiation; others work antagonistically. Other factors may also influence a plant's response to hormones. For example, pith cells grown in a thin layer on a culture medium containing two parts per million of auxin do not divide, but grow to an unusually large size. When the same pith tissue is molded into a cylindrical shape, and the same concentration of auxin is inserted into the top portion of the cylinder, the cells undergo many divisions and some of them differentiate into xylem. The difference in response to the same inducing stimulus must be attributed to the different locations—superficial or internal—of the cells.

Physical factors from the environment (such as temperature, light, gravity, and humidity) also affect the developmental process through hormones. We have seen that light, through its effects on auxin and phytochrome, plays a critical indirect role in such developmental phenomena as increase in shoot length and initiation of flowering. In plants, however, the ranges of relevant stimuli and possible responses are limited. Plants do not see, hear, smell, or taste in the usual sense, but they do have the ability to respond to particular stimuli. Plant responses usually consist of changes in patterns of growth, and plant hormones integrate and coordinate many phases of this growth and development.

CONCEPTS IN BRIEF

1. In many plants flowering is initiated by *photoperiodism,* the response to the length of light and dark periods. Most plants belong to one of three groups: (1) short-day plants, which flower only when the night is *longer* than a critical value; (2) long-day plants, which flower only when the night is *shorter* than a critical value; and (3) day-neutral plants, which flower independent of day- and nightlength.

2. It is believed that an inducing photoperiod causes the leaves to produce the hormone *florigen,* which moves to the buds and stimulates flowering. A noninducing photoperiod causes the leaves of many plants to inhibit florigen synthesis and/or action.

3. Light is detected by a pigment called *phytochrome,* which exists in two interconvertible forms. One absorbs red light (P_r), and one absorbs far-red light (P_{fr}). During the day most of the phytochrome is in the P_{fr} form; at night, the P_{fr} is converted back to P_r, or is enzymatically destroyed. The ratio of P_{fr} to P_r gives the plant a way to detect whether it is day or night.

4. The mechanism that enables the plant to measure the length of night is apparently tied to an "internal clock." The phytochrome mechanism determines whether it is day or night, while the internal clock measures the length of night. Once these have indicated to the plant that the photoperiod is appropriate, flowering is initiated.

5. In flowering plants meiosis within the ovary produces a megaspore, which divides several times to produce an *embryo sac,* the female gametophyte plant. Meiosis within the anther gives rise to microspores, which mature into pollen grains, the male gametophyte plant.

6. The egg cell of an angiosperm plant is fertilized in the ovary of the maternal plant by a sperm nucleus from a pollen grain; a second sperm fertilizes the two polar nuclei, forming a triploid cell that will give rise to the *endosperm,* the food storage tissue.

7. The zygote divides to form a small embryo. Three types of tissues differentiate in the growing embryo: the outer *protoderm,* which will form epidermal tissue; an inner core of *provascular tissue,* which will form the cambium and vascular tissues; and the middle *ground tissue,* which will form the cortex and pith. Next the *cotyledons* arise. The part of the embryo below the cotyledons is the *hypocotyl,* the part above the cotyledons is the *epicotyl.* Small clumps of tissue at each end of the embryo become the *apical meristems of the shoot and root.* The embryo, together with the endosperm, becomes enclosed in a seed coat, forming the seed.

8. When the seed germinates, the hypocotyl turns downward; the *radicle* at its lower end forms the root. The epicotyl then elongates; it forms most of the shoot. Growth of the shoot and root involves the production of new cells by

the apical meristems, and then the elongation and differentiation of these cells.

9. In stems, certain areas of the apical meristem give rise to *nodes,* swellings where leaf primordia arise. At the tip of the stem is a *bud,* which consists of the apical meristem and unelongated internodes enclosed within leaf primordia.

10. The cells produced by the apical meristems differentiate to form the various primary tissues of the plant. Increase in diameter of a root or stem depends upon the formation of secondary tissues derived from the lateral meristems.

11. Developing cells are influenced by hormones and by the physical environment (e.g., light, temperature, gravity). As development proceeds, the individual cells become more and more committed to one particular course of differentiation. The various growth patterns are coordinated and integrated by plant hormones.

STUDY QUESTIONS

1. Define *photoperiod* and distinguish among *short-day, long-day,* and *day-neutral* plants. How could you demonstrate that the critical aspect of the photoperiod is actually the length of the night and not the length of the day? (pp. 490–491)
2. What is the presumed role of *florigen* in a flowering plant? Describe an experiment that supports its existence. (pp. 489–491)
3. Explain how plants use the combination of phytochrome and their internal clock to measure the photoperiod. (p. 492)
4. Diagram and describe the functions of all the major structures in a flower. (pp. 492–493)
5. Make a diagram of alternation of generations in an angiosperm. Which stages are haploid and which diploid? What cells give rise to the male and female gametophytes? Where are the gametophytes formed and what are they called? (pp. 493–494)
6. Distinguish between *pollination* and *fertilization.* Describe double fertilization in angiosperms. What is the function of endosperm? (pp. 494–496)
7. Diagram the early embryonic development of a dicot plant and describe the developmental fate or function of the following structures: *suspensor, protoderm, provascular tissue, ground tissue, cotyledons,* and *hypocotyl.* (pp. 497–498)
8. Describe seed germination. What are the roles of water, gibberellin, and hydrolytic enzymes? What parts of the plant are derived from the *epicotyl* and *radicle?* (pp. 498–499)
9. Contrast the growth patterns of annual and perennial plants. (p. 500)
10. Compare the growth and cell-differentiation patterns of roots and stems. Distinguish between *nodes* and *internodes.* Describe the structure of a bud. How do *axillary buds* differ from *terminal buds?* (pp. 501–502)
11. Distinguish between primary and secondary growth in plants. Which meristematic regions are responsible for each? (pp. 502–503)
12. Describe an experiment demonstrating that even fully differentiated plant cells can resume totipotency. (p. 504)
13. Describe an example of how environmental factors interact with hormones to affect the growth of plants. (p. 505)

CHEMICAL CONTROL IN ANIMALS

Plant hormones are usually produced by the actively growing parts of plants, and each serves to control growth, development, or both in the tissues it affects. The general pattern in animal hormones is markedly different in several ways (Table 25.1). First, complex multicellular animals have organs specialized for hormone production, whereas plant hormones are produced not by specialized organs, but by the actively growing parts of the plant body. Second, animal hormones perform a wide variety of functions and, being primarily regulatory, are involved in maintaining homeostasis. Plant hormones, as we have seen, are involved in controlling growth and development. Third, animal hormones tend to be very specific in their action, affecting particular tissues or organs. Each plant hormone, on the other hand, may control growth and development in several different ways and act on a variety of tissues. A further difference is that in multicellular animals, coordination of processes also depends on nervous control, a subject we shall address in later chapters. This chapter and the next will be concerned with chemical control mechanisms in multicellular animals.

HORMONES IN MAMMALS

Though hormonal mechanisms have been found in a wide variety of invertebrates, including arthropods, annelids, molluscs, and echinoderms, far more is known about hormones in mammals than about those in invertebrates or those in any other group of vertebrates. Consequently it is the mammalian hormonal system (especially the human system) that we shall emphasize here (Fig. 25.1, Table 25.2). As in our discussion of plant hormones, we shall try to give you some idea of the way animal hormones regulate the diverse functions of a complex organism.

Hormones are specific chemical messengers that produce an effect away from their sites of production. In vertebrates, hormones may diffuse through the tissue fluid from one place to another, but, as would be expected in animals with well-developed circulatory systems, most hormone transport is by the blood. The tissues that produce and release hormones are termed ***endocrine tissues.*** The use of the word "endocrine," which means "secreting internally," is intended to convey that the hormones are secreted directly into the blood via the capillaries supplying the endocrine tissues, and that no special ducts are involved. In fact, endocrine glands are often called the ductless glands. By contrast, glands such as the liver or salivary glands that discharge their secretions directly into ducts are referred to as ***exocrine glands.***

Chapter opening photo: A greater frigate bird from the Galapagos Islands During breeding season the male bird inflates his bright red throat pouch which he uses as a signal to attract mates.

TABLE 25.1 *Comparison of animal and plant hormones*

Characteristic	*Animal hormones*	*Plant hormones*
Site of production	Specific endocrine glands specialized for hormone production	Produced by actively metabolizing tissues that have other functions
Target tissues	Each hormone acts on a specific target tissue or organ	Each hormone acts on a variety of tissues
Number of hormones	Many, each with a specific function	Relatively few, each with a variety of functions
Primary function	Affect homeostasis and are regulatory in action; effects are reversible	Affect growth and development; effects are permanent

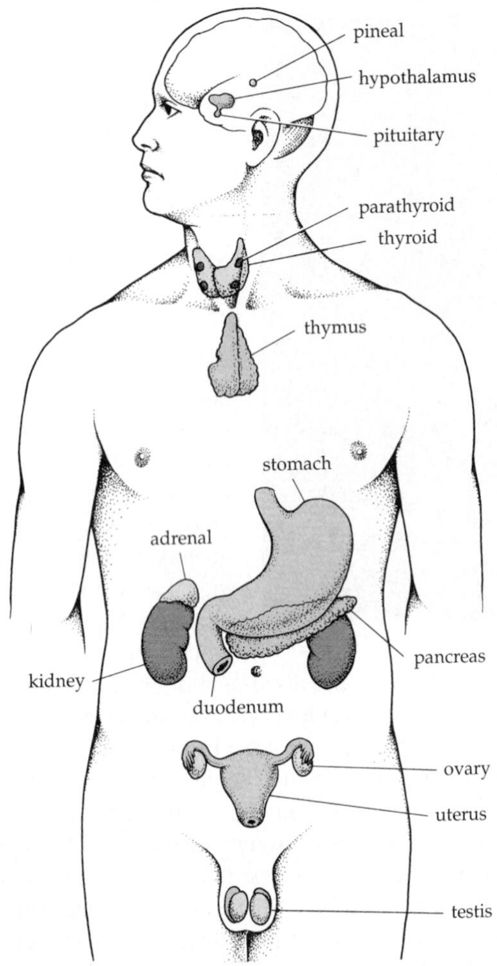

25.1 The major human endocrine organs

Animal Hormones Can Be Classified on the Basis of Their Solubility in Water

The majority of animal hormones are water-soluble, and are derived from amino acids, small peptides, or larger proteins. Because these molecules are hydrophilic, and often large, they cannot easily pass through the hydrophobic lipid bilayer of the cell membrane. Instead they interact with receptor molecules on the membrane surface; the hormones themselves do not enter the cell. Other hormones, such as the steroid and thyroid hormones, are hydrophobic in nature, and as such can readily diffuse through the cell membrane and bind to specific receptors inside the cell. Let us look at the mechanisms of action in more detail.

Most Hydrophilic Hormones Act on Target Cells Indirectly

The 1960s saw some exciting progress in the attempt to learn how hormones act on their target cells. A major contribution was that of E. W. Sutherland and T. W. Rall of Case Western Reserve University, who were investigating the mechanism by which the hormones glucagon and adrenalin stimulate liver cells to release more glucose into the blood. They discovered that these hormones, which are both hydrophilic, stimulate an increase in the concentration within the cell of a compound called cyclic adenosine monophosphate—***cyclic AMP,*** or ***cAMP*** for short. They found that this compound leads, in turn, to activation of an enzyme necessary for breakdown of glycogen to glucose. Subsequent research has demonstrated that a large number of other hormones act on their target cells either to increase or to decrease the concentration of cAMP. This work, far-reaching in its implications, earned Sutherland, now at Vanderbilt University, a Nobel Prize in 1971.

Cyclic AMP is a compound related to ATP (adenosine triphosphate) (see p. 150). Very widely distributed in nature, it has been found in almost all animal tissues studied (vertebrate and invertebrate) and in bacteria. It is synthesized from ATP in living cells by a reaction catalyzed by an enzyme called ***adenylate cyclase,*** which appears to be built into the cell membrane.

As evidence accumulated that many protein or protein-derived hormones, among them the hormones glucagon and adrenalin, do not actually enter their target cells, but rather form weak bonds with receptor sites on the cell membrane, a ***second-messenger model*** of hormonal control was proposed. According to this model, the hormone itself acts as the first

TABLE 25.2 *Important mammalian hormones*

Source	*Hormone*	*Principal effects*
Endocrine glands secreting *hydrophilic* hormones		
Pancreas	Insulin	Stimulates glycogen formation and storage; stimulates glucose oxidation; stimulates cellular uptake of amino acids and fatty acids; synthesis of protein and fat
	Glucagon	Stimulates conversion of glycogen into glucose leading to increased blood glucose
Adrenal medulla	Adrenalin	Stimulates elevation of blood-glucose concentration; stimulates "fight-or-flight" reactions
	Nonadrenalin	Stimulates reactions similar to those produced by adrenalin, but causes more vasoconstriction and is less effective in conversion of glycogen into glucose
Thyroid	Calcitonin	Prevents excessive rise in blood calcium
Parathyroids	Parathyroid hormone (PTH)	Regulates calcium-phosphate balance; acts to increase blood calcium ion level
Hypothalamus*	Releasing hormones	Regulate hormone secretion by anterior pituitary
Posterior pituitary (storage organ for hormones produced by hypothalamus)	Oxytocin	Stimulates contraction of uterine muscles; stimulates release of milk by mammary glands
	Vasopressin	Stimulates increased water resorption by kidneys; stimulates constriction of blood vessels
Anterior pituitary	Growth hormone (STH)	Stimulates growth; stimulates protein synthesis, hydrolysis of fats, and increased blood-glucose concentration
	Prolactin (PRL)	Stimulates milk secretion by mammary glands; participates in control of reproduction, osmoregulation, growth, and metabolism
	Thyrotropic hormone (TSH)	Stimulates the thyroid
	Adrenocorticotropic hormone (ACTH)	Stimulates the adrenal cortex
	Follicle-stimulating hormone (FSH)	Stimulates growth of ovarian follicles and of seminiferous tubules of the testes
	Luteinizing hormone (LH)	Stimulates ovulation and conversion of follicles into corpora lutea; stimulates secretion of sex hormones by ovaries and testes
Pineal	Melatonin	Helps regulate production of gonadotropins by anterior pituitary, perhaps by regulating hypothalamic releasing centers
Endocrine glands secreting *hydrophobic* hormones		
Adrenal cortex	Glucocorticoids (corticosterone, cortisol, cortisone, etc.)	Stimulate formation of carbohydrate from protein, thus elevating glycogen stores and helping maintain normal blood-sugar levels
	Mineralocorticoids (aldosterone, deoxycorticosterone, etc.)	Stimulate kidney tubules to reabsorb more sodium and water and less potassium
	Cortical sex hormones	Stimulate development of secondary sexual characteristics, particularly those of the male
Thyroid	Thyroxine and triiodothyronine (together called TH)	Stimulate oxygen uptake and oxidative metabolism; help regulate growth and development
Testes	Testosterone	Stimulates development and maintenance of male accessory reproductive structures, secondary sexual characteristics, and behavior; stimulates spermatogenesis
Ovaries	Estrogen	Stimulates development and maintenance of female accessory reproductive structures, secondary sexual characteristics, and behavior; stimulates growth of the uterine lining
	Progesterone	Prepares uterus for embryo implantation and helps maintain pregnancy

*As will be seen (p. 522), the hypothalamus consists of nervous tissue, being part of the brain. It is discussed in this chapter because it is the source of important hormones that have a direct influence on the endocrine system.

messenger, going from an endocrine gland to the target cell and binding to a receptor on the membrane. The binding of the hormone to the receptor in turn stimulates the production *inside* the cell of a second messenger, which is often cAMP. Specifically, the binding of the hormone with a highly specific receptor site on the membrane of the target cell activates a second membrane protein referred to as ***G-protein.*** The G-protein in turn influences the activity of the enzyme ***adenylate cyclase,*** which catalyzes production of cAMP from ATP on the inner surface of the membrane (Fig. 25.2). The cAMP then activates specific enzymes inside the cell and thus initiates the cell's characteristic responses to the hormonal stimulation. Therefore, the initial extracellular signal (the hormone, or first messenger) is converted into an intracellular signal (cAMP, or second messenger) that the chemical machinery of the cell can more readily understand.

As we shall see, many different hormones (including glucagon, adrenalin, parathyroid hormone, calcitonin, and the tropic hormones of the anterior pituitary) are believed to act through the adenylate cyclase system. These hormones are all derived from amino acids, small peptides, or proteins, and are hydrophilic. If all of these hormones simply act to regulate the adenylate cyclase system of their target cells, what then is the basis for hormonal specificity? The question is relatively easy to answer. It is the presence or absence of specific hormonal ***receptors*** on a cell membrane that determines whether or not a particular cell type is influenced by a given hormone—that is, whether or not it is a target of that hormone. Different types of cells have different hormone-specific receptors.

Though most protein and protein-derived hormones use cAMP as the second messenger, there are exceptions. Insulin, for example, directly activates intercellular enzymes (Fig. 25.3). Several other control chemicals act to convert GTP (guanine triphosphate) to cGMP, but the details of the cGMP second-messenger system are not yet well understood.

While some hydrophilic hormones exert their effects on

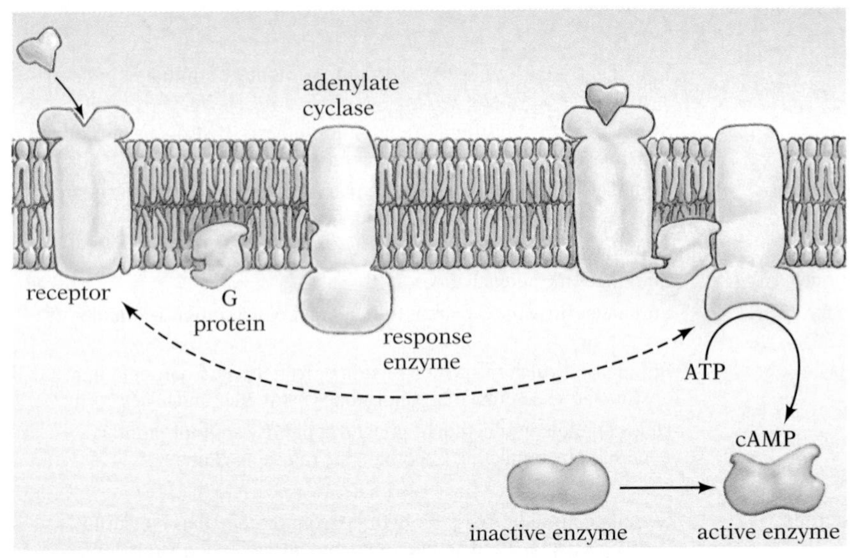

25.2 Model for control of hormonal activity via the adenylate cyclase system The hormone molecule can bond loosely with a specific receptor protein built into the plasma membrane. Formation of the hormone-receptor complex activates the G-protein, which in turn activates the enzyme adenylate cyclase, which is also an integral part of the plasma membrane. The activated adenylate cyclase catalyzes synthesis of cAMP from ATP inside the cell. The cAMP, in turn, triggers activation of an enzyme that catalyzes a particular chemical reaction within the cell.

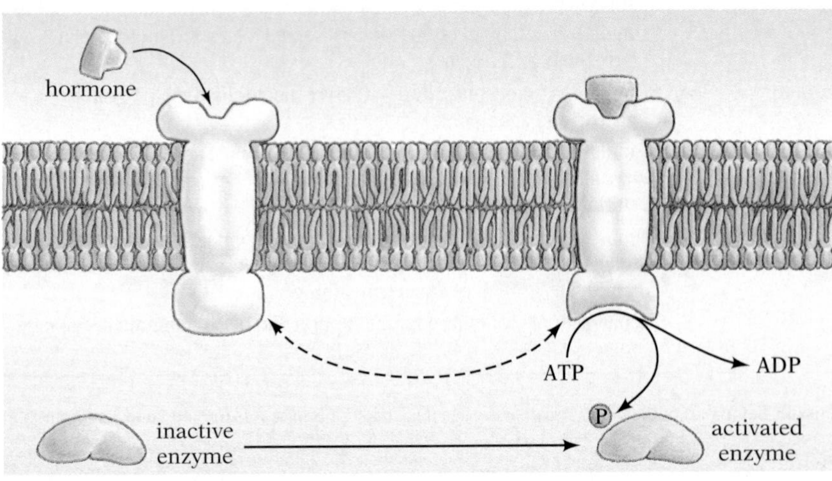

25.3 Model for control of hormonal activity via enzyme phosphorylation An intracellular enzyme may be activated directly by a hormone (usually by phosphorylation), as happens when a hormone such as insulin binds to its receptor protein.

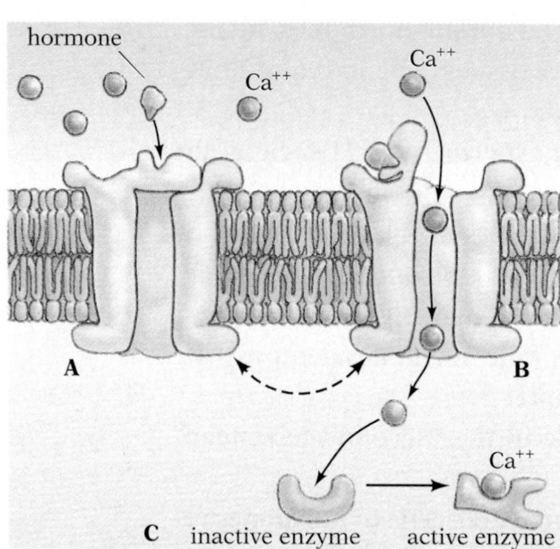

25.4 Ions as second-messengers within the cell The binding of a hormone to its membrane receptor (A) opens a chemically gated channel (B) through which specific ions such as calcium ions can flow. The ion binds to a specific enzyme within the cell (C), thereby activating it.

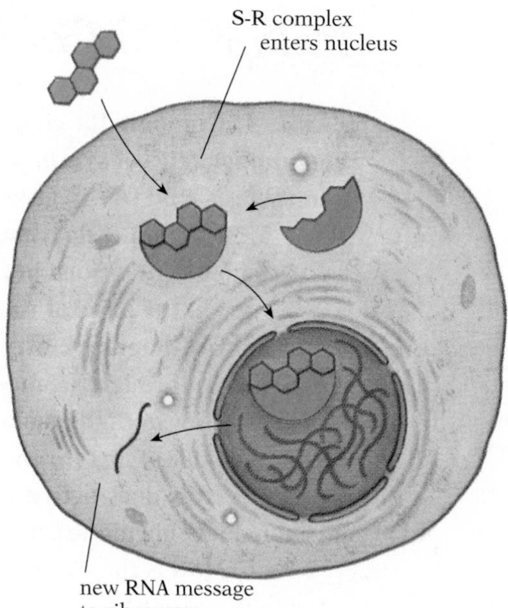

25.5 Model for the mode of action of steroid hormones The hormone, being lipid-soluble, can penetrate the plasma membrane and enter the cytoplasm and nucleus. There it binds with a receptor protein, and the complex binds to certain genes in the nucleus, where it influences synthesis of RNA by the genes. A new message (in the form of new mRNA) is thus sent to the ribosomes, which begin synthesizing the protein coded by the new mRNA. The new protein, perhaps an enzyme, will then influence the chemical activity of the cell.

cell metabolism by activating a second messenger such as cAMP, others open a specific ion channel and the ion acts as a second messenger inside the cell (Fig. 25.4). Calcium ions are by far the most common second-messenger ions. Calcium ions can serve this function because their concentration in the cytoplasm is normally kept low by active transport pumps that move them out of the cell or into the endoplasmic reticulum. As a result, when a hormone binds to its specialized receptor, thereby opening a calcium-ion channel in the membrane, the strongly favorable electrochemical gradient causes calcium ions to rush in. These ions then bind to and activate particular enzymes within the cell. The enzymes regulated vary by cell type.

Hydrophobic Hormones Regulate Gene Expression

Hydrophobic hormones, such as the thyroid and steroid hormones produced in vertebrates by the adrenal cortex and the gonads, have a mode of action that does not involve a second-messenger system. Instead of reacting with a receptor on the outer surface of the target-cell membrane, the steroid hormones, which, because they are hydrophobic, can easily move through membranes, enter the cytoplasm of the target cell. There, or in the nucleus itself, the steroid (S) binds to a specific receptor molecule (R), which is probably a protein. The complex (S-R), having been formed within the nucleus or having moved into the nucleus, then binds to particular sites on the DNA and regulates the activity of specific genes (Fig. 25.5). In other words, by interacting with the genetic material of the target cells, steroid hormones help determine what instructions for protein synthesis (especially enzyme synthesis) are sent from the nucleus to the ribosomes. Whether or not a given cell responds to a particular hormone depends on whether or not it has receptors specific for that hormone. As we shall see, the thyroid hormones also enter the cell, though their mode of action is somewhat different.

The Effects of Hydrophobic Hormones Are Slower but Longer-Lasting Than Those of Hydrophilic Hormones

Most of the protein- and amino acid–derived proteins are in circulation only for a matter of minutes before they are degraded, usually by the liver and kidneys. The steroid hormones, which are synthesized from cholesterol, may remain in the blood longer, perhaps for many hours, before being destroyed. The liver is the most frequent site of steroid hormone catabolism. Because the steroid hormones act inside the cell to

MAKING CONNECTIONS

1. Compare the characteristics of plant and animal hormones, including number, sites of synthesis, specificity of target tissues, role in maintaining homeostasis, and role in growth and development.

2. How do endocrine glands differ from exocrine glands such as the salivary glands described in Chapter 17?

3. On the basis of the structural characteristics of the cell membrane described in Chapter 5, why would you predict that hydrophobic hormones could diffuse through the membrane more easily than hydrophilic hormones?

4. How does the mode of action of hydrophilic hormones demonstrate the importance of membrane proteins? (See Chapter 5.)

5. How does active transport play a role in the "second-messenger" mechanism of hormone action (in some cases)?

6. How is the specificity of response of particular cells to hormones related to the specificity of enzymes?

regulate gene activity and influence protein synthesis, their effects are longer-lasting than those of hydrophilic hormones. On the other hand, the effects of steroid hormones occur much more slowly. It may take several days for enough proteins to be synthesized to initiate the cells response to hormonal stimulation. Once present, however, the proteins can continue to act for a long time.

Having looked at the general classification of hormones and their mode of action, let us now turn our attention to a brief examination of some of the more important endocrine glands. We shall emphasize the mammalian hormonal system, and especially the human system, in this chapter.

THE PANCREAS

The Pancreas Acts as Both an Exocrine and Endocrine Organ

The pancreas is a compound organ, meaning that it is composed of two quite different types of glandular tissue. Most of the cells of the pancreas are exocrine in function; they produce and release (through ducts) digestive enzymes. Other, quite different cells, called ***islet cells,*** or islets of Langerhans (Fig. 25.6), are involved in the gland's endocrine function. The islets are collections of cells scattered throughout the pancreas, composing approximately 1 percent of the pancreatic tissue. Each islet has a large blood supply. Several different types of cells can be differentiated within the islets; two important cell types are the α (alpha) cells, which secrete the hormone ***glucagon,*** and the β (beta) cells, which secrete the hormone ***insulin.*** These two hormones are both important in the regulation of carbohydrate, protein, and fat metabolism, but they have oppo-

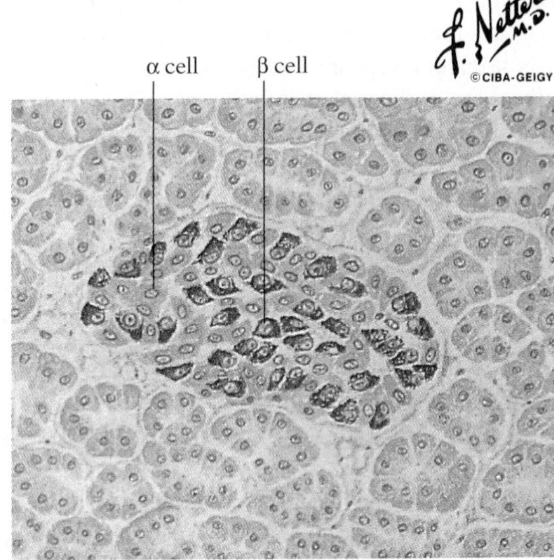

25.6 Drawing of an islet of Langerhans The endocrine cells of the pancreas form an islet, clearly distinct from the surrounding cells whose function is secretion of digestive enzymes. The islet cells make up about 1 percent of the pancreatic mass. The α islet cells secrete the hormone glucagon; the β islet cells, insulin. Two other minor classes of cells in the islets secrete hormones that, respectively, encourage and inhibit release of pancreatic enzymes. The pancreas contains about a million islets, each with approximately 3,000 cells.

site effects; insulin promotes the storage of carbohydrate, protein, and fat, whereas glucagon induces the breakdown of these stores and the transport of glucose, amino acids, and fatty acids into the blood. Both of these hormones are small proteins and work by a second-messenger system.

Insulin Reduces the Blood-Glucose Concentration

Diabetes (or, more precisely, diabetes mellitus), a disease that causes a large amount of sugar to be excreted in the urine, has been recognized for centuries, but its cause did not begin to be understood until the latter half of the 19th century. In 1889 two German physicians, Johann von Mering and Oscar Minkowski, who were interested in the role of the pancreas as a producer of digestive enzymes, surgically removed the pancreas from a dog. A short time later it was noticed that the dog's urine was attracting an unusual number of ants. Analysis showed that the urine contained a high concentration of sugar. Furthermore, the dogs soon developed other symptoms strikingly like those of human diabetes. von Mering and Minkowski removed the pancreas from other dogs, and diabetes invariably developed. To eliminate the possibility that the extensive damage resulting from so severe an operation might be the causal factor, they performed operations in which all the damage usually associated with the pancreas removal operation was produced, but the pancreas was left in place. These dogs did not develop symptoms of diabetes. Clearly, the onset of diabetes was directly correlated with removal of the pancreas. But operations in which the pancreatic duct was destroyed without producing the disease made it clear also that diabetes was not correlated with absence of the pancreatic digestive enzymes. The unavoidable inference was that the pancreas participated in other body functions besides the digestive one.

Mounting evidence pointed to secretion by the pancreas of a substance that prevents diabetes in the normal animal, but all attempts at proof failed. It seemed likely that the hormone so many people were searching for was produced by the islet cells (more specifically, by the β islet cells). In a critical experiment that supported this hypothesis and opened the way for isolation of the hormone, it was shown that tying off the pancreatic duct results in atrophy of most of the pancreas but not in development of diabetes. Examination of the atrophied pancreas revealed that it was the enzyme-producing portion that had atrophied, while the islet cells had remained essentially intact. Therefore, the hormone that prevented diabetes must have come from this portion of the pancreas.

That hormone, ***insulin,*** was finally isolated in 1922 by F. G. Banting and C. H. Best, working in the laboratory of J. J. R. MacLeod at the University of Toronto. They tied off the pancreatic ducts of a number of dogs, waited until the enzyme-producing exocrine tissue had atrophied, removed the degenerated pancreas, and froze and macerated it in an isotonic medium; then they filtered the solution and quickly injected the filtered material into diabetic dogs. The dogs showed marked improvement. Banting and Best also obtained good results with extracts prepared from the pancreases of embryonic animals; since the islet cells develop in the embryo before the enzyme-producing cells, there are no enzymes to destroy the insulin during the extraction procedure. Banting and MacLeod received the Nobel Prize in 1923 for this important work.

High concentration of sugar in the urine is a major symptom of diabetes. How is insulin related to this symptom? Before attempting an answer, we must examine the symptom further. The presence of sugar in the urine of a diabetic means that the blood-sugar concentration is higher than normal and that the kidneys are removing part of the excess to maintain homeostasis. You will recall that the liver plays a critical role in regulating blood-sugar levels. When blood coming to the liver via the portal vein from the intestines contains a higher than normal concentration of sugar, the liver removes much of the excess and stores it as glycogen. Conversely, when blood coming to the liver is low in sugar, the liver converts some of its stored glycogen into glucose, which it adds to the blood. Other parts of the body, particularly the muscles and adipose (fatty) tissue, are also important elements in this homeostatic regulatory system. When the blood-sugar concentration rises after a carbohydrate meal, part of the excess glucose is stored as glycogen in the muscles, and part is converted into fat and stored by adipose tissues; the rate of oxidation of glucose in the liver and muscles may also increase under these conditions.

This brief outline of the interplay between liver, muscles, and adipose tissues—all three of which are target tissues for insulin—helps explain insulin's role in the high concentration of sugar found in diabetics' urine. Insulin's mode of action is not well understood, but it is known that insulin binds to specific receptor proteins on the cell membranes of muscle, adipose tissue, and connective tissue, which initiates physiologic activity inside the cell. This activity includes an increase in the number of glucose-transport pumps in these cells. With more pumps available, more glucose can be transported from the blood into the cell. Insulin also promotes, indirectly, the uptake of glucose into liver cells and stimulates the conversion of glucose into glycogen and fat. Together these actions lower the level of glucose in the blood. In addition, insulin plays a role in protein metabolism by promoting the uptake of amino acids and their synthesis into protein. The net effect of insulin, then, is to decrease the supply of glucose and amino acids circulating in the blood, and to increase the stores of glycogen, fat, and protein.

Notice that the action of insulin is not only on carbohydrate metabolism but fat and protein metabolism as well. But the promotion of fat and protein synthesis forces the cells to rely more heavily on glucose as a source of metabolic energy—the result being a reduction in the supply of free glucose. Thus we see once again that the metabolic pathways for all classes of nutrients form an interlocking system; alteration of one pathway unavoidably influences the others (Fig. 25.7).

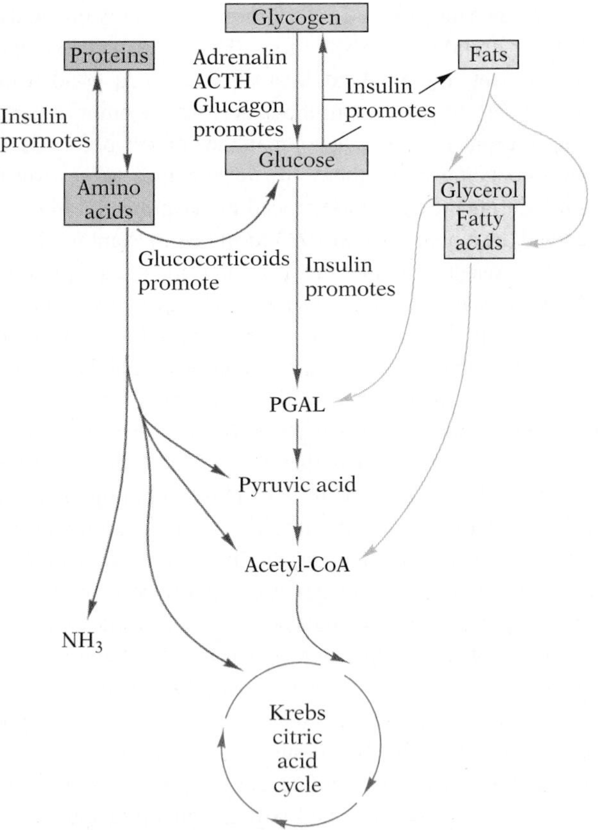

25.7 Role of certain hormones in controlling metabolism Insulin plays a major role in many aspects of metabolism; it stimulates glucose uptake by muscle and fat cells and promotes the oxidation of glucose and the conversion of excess glucose into glycogen or fat for storage. Insulin also favors the uptake of amino acids and protein synthesis in liver and muscle cells. All of these effects lower blood sugar. Certain other hormones act in opposition to insulin; adrenalin, ACTH, and glucagon raise blood sugar by stimulating the conversion of glycogen into glucose. The glucocorticoids from the adrenal cortex also cause an increase in blood sugar by promoting the formation of carbohydrate from protein and fat.

Too much insulin in the system, as from an overactive pancreas, or from administration of too large a dose to a diabetic, can produce a severe reaction called insulin shock. The blood-sugar level falls so low that the brain, which has few stored food reserves of its own, becomes more excitable, convulsions may result, followed by unconsciousness and often death. Far more common is a deficiency of insulin (or an insensitivity of the tissues to insulin), which is the condition referred to as diabetes. The liver and muscles do not convert enough glucose into glycogen, the liver produces too much new glucose, and utilization of carbohydrate is impaired. The blood-sugar level rises above normal, and part of the excess glucose appears in the urine. More water is lost by osmosis because of the high glucose level in the urine, and the diabetic thus tends to become dehydrated. The glycogen reserves become depleted as more and more glucose is poured into the blood and lost in the urine; yet the body still lacks sufficient energy, because of the impairment of carbohydrate metabolism. As the body begins to metabolize its reserves of proteins and fats—particularly the latter—the diabetic becomes emaciated, weak, and easily subject to infections. Further complications arise from the excessive but incomplete metabolism of fats, which produces fatty acids and toxic substances that seriously disturb the delicately balanced pH of the body, and often play a major part in the eventually fatal outcome of the untreated disease. Fortunately, these symptoms can be alleviated by daily injections of insulin. (see Exploring Further: Diabetes, p. 516)

Glucagon and Certain Other Hormones Raise Blood-Glucose Concentration

The polypeptide hormone glucagon is produced by the α islet cells in the pancreas. Glucagon's effect is the opposite of insulin's; it causes an increase in blood-glucose concentration, by causing the breakdown of glycogen to glucose in the liver. As we shall see, several hormones produced by the adrenal glands also cause a rise in the blood-sugar level, and certain hormones from the pituitary may do so as well. Again we see that the normal functioning of an organism depends on a delicate balance between opposing control systems; if one of these systems is disturbed and the proper homeostatic balance destroyed, abnormalities result.

Insulin and Glucagon Secretion Are Regulated by a Negative Feedback Loop

After a meal the glucose level in the blood rises, which stimulates the β cells of the pancreas to secrete insulin and, at the same time, inhibits the secretion of glucagon by the α cells (Fig. 25.8). The increase in insulin secretion results in a lowering of the blood sugar. When the blood sugar drops too low, glucagon secretion by the α cells is stimulated and insulin secretion is inhibited, and the blood-sugar level rises. The interaction between the secretion of these two hormones and the concentration of glucose in the blood is an example of ***negative feedback*** control, a homeostatic strategy very common in living systems. Just as the familiar household thermostat senses when the temperature falls below a desired level—specified by its setting—and takes appropriate action by switching on the furnace, so an organic feedback mechanism senses variation from a desired norm and takes appropriate action in response. And just as the thermostat senses when the required temperature has been attained and switches off the furnace, the organic feedback mechanism senses when the desired blood-sugar level has been attained and switches off the action.

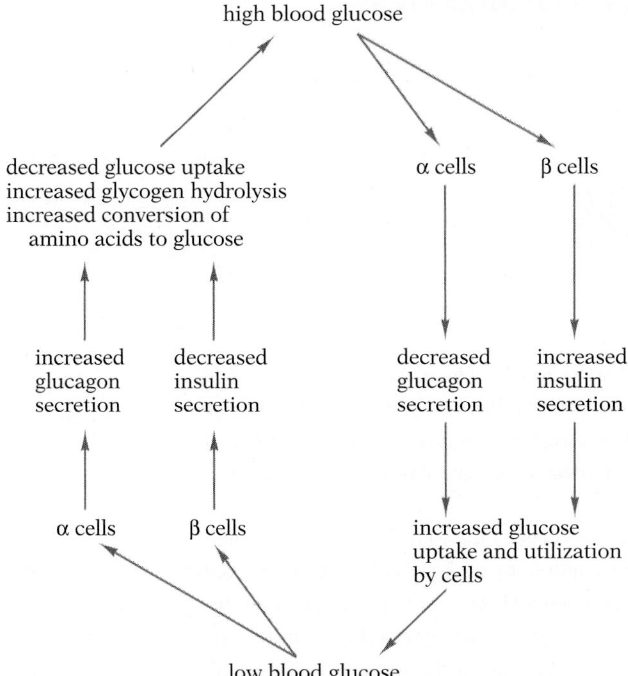

25.8 Control of blood glucose concentration by insulin and glucagon The secretion of insulin and glucagon by the β cells and α cells of the islets is regulated by blood-glucose concentration in a negative feedback loop. When blood glucose concentration is high, insulin secretion is stimulated and glucagon secretion is inhibited. A low blood-glucose concentration stimulates glucagon and inhibits insulin secretion.

Negative feedback acts to compensate for change, restoring a lost homeostasis.

Another pancreatic hormone influences insulin and glucagon secretion. The hormone ***somatostatin*** is produced by the δ (delta) cells of the pancreas in response to high levels of blood glucose, amino acids, or fatty acids. This hormone appears to be primarily inhibitory in nature. It inhibits secretion of insulin and glucagon, slows the movement of the stomach and duodenum, and decreases secretion and absorption of the digestive tract. Its role may be to increase the time that food is absorbed into the blood, making it available for a longer time.

THE ADRENAL GLANDS

The Adrenal Glands Consist of Two Separate Endocrine Glands

The two adrenal glands, as their name implies (*ad-* means "at" and *renal* refers to the kidney), lie very near the kidneys (Fig. 25.1). Each adrenal in mammals is actually a double gland, composed of an inner corelike ***medulla*** and an outer barklike ***cortex*** (Fig. 25.9). The medulla and cortex arise in the embryo from different tissues, and their mature functions are unrelated. We shall therefore discuss the two components of the mammalian adrenals separately.

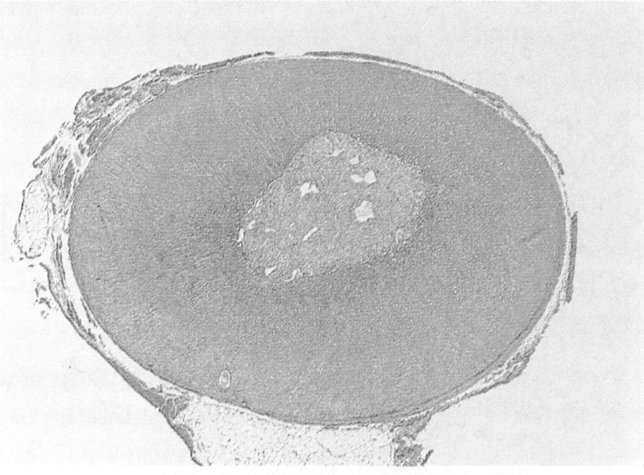

25.9 Cross section of the adrenal gland of a cat The medulla (central region) is entirely surrounded by cortex.

The Adrenal Medulla Secretes Hormones That Help to Prepare the Body to Meet Emergencies

The adrenal medulla secretes two amino acid–derived hormones, ***adrenalin*** (also known as epinephrine) and ***noradrenalin*** (norepinephrine), whose functions are similar but not identical (Fig. 25.10). Adrenalin causes a rise in blood pressure, acceleration of the heartbeat, increased conversion of glycogen into glucose and release of the glucose into blood by the liver, increased oxygen consumption, release of reserve red blood cells into the blood from the spleen, increased blood flow in skeletal and heart muscle, decreased blood flow in the

Adrenalin Tyrosine

25.10 Structural formulas of adrenalin and tyrosine The hormone is derived from the amino acid tyrosine. Adrenalin is a hydrophilic hormone and works by the second-messenger system.

DIABETES

It is estimated that about one person in 20 in the United States has diabetes mellitus. The cause of diabetes is a deficiency in the action of insulin at the cellular level, but there may be several reasons for this deficiency. Two major forms of diabetes occur in humans, *Type I* or *juvenile-onset diabetes* and *Type II,* or *maturity-onset diabetes.* The latter, maturity-onset diabetes, was not recognized as a distinct disease until relatively recently.

As the name suggests, juvenile-onset diabetes usually develops in the young, often during childhood. This form of diabetes is caused by deficient insulin synthesis by the β cells of the pancreas, and the onset is usually sudden. Evidence suggests that juvenile-onset diabetes may be an autoimmune disease, since most individuals with this form of the disease have antibodies against the β cells in their own pancreas. These antibodies destroy the β cells, resulting in insulin deficiency. Just what precipitates the autoimmune response is still in question, but there is probably a genetic predisposition. Most (though not all) of the individuals with this form of diabetes possess two particular genes. Having these genes does not automatically mean that one will become diabetic; in fact, the majority with this genotype will not. Infections with particular viruses appear to be a precipitating factor. This type of diabetes is treated with injections of insulin, so it is sometimes referred to as insulin-dependent diabetes mellitus (IDDM).

The more common form of diabetes is maturity-onset diabetes, which has a slow onset. These individuals generally have normal β cells, and insulin levels in their blood are often normal or even above normal. Here the defect usually has to do with the insulin receptors. Most patients with this form of diabetes are obese, and have low levels of insulin receptors on their adipose cells. These cells are unable to respond to normal levels of insulin, and for this reason maturity-onset diabetes is also known as non-insulin-dependent diabetes mellitus (NIDDM). It has been suggested that this disease occurs when individuals chronically overeat, which stimulates high insulin production and high insulin levels in the blood. Prolonged exposure of adipose cells to insulin appears to lead to a "down regulation" in the number of receptors on the cells. In many such individuals, loss of weight restores the normal concentration of insulin receptors, and the symptoms disappear. In most cases, diet alone can control the disease. There appears to be a genetic component to this form of diabetes; about one-third of such diabetics have a relative with the disease.

Untreated diabetes mellitus can have serious consequences, both short-term and long-term. Juvenile-onset diabetics risk going into diabetic coma or becoming unconscious due to the accumulation of toxic substances and changes in the pH of the blood (see p. 514). Other complications affect all types of diabetics, and occur many years after the onset of the disease. Pathologic changes occur in the blood vessels, which causes diabetics to have a higher susceptibility to kidney failure, scarring of the retina (diabetic retinopathy) with loss of vision, and circulatory problems in the legs. Diabetics also have a higher incidence of atherosclerosis, with its attendant risk of high blood pressure, strokes, and heart attacks. Many of these complications are treatable, so early detection and control of the disease is important.

skin and in the smooth muscle of the digestive tract, inhibition of intestinal peristalsis, erection of hairs, production of "gooseflesh," and dilation of the pupils.

At first glance, this list may seem to include a curious assortment of unrelated effects, but a more careful examination shows that these reactions occur together in response to intense physical exertion, pain, fear, anger, or other heightened emotional states; they have sometimes been called fight-or-flight reactions. It has been suggested, therefore, that adrenalin and, to a lesser extent, noradrenalin help mobilize the resources of the body in emergencies (by stimulating reactions that combine to increase the supply of glucose and oxygen carried by the blood to the skeletal and heart muscles) and help inhibit those functions not immediately important during an emergency (such as digestion, which might otherwise compete with the skeletal muscles for oxygen). But complete removal of the adrenal medullae causes little noticeable change in an animal; the animal still shows normal fight-or-flight reactions to appropriate stimuli. As we shall see in a later chapter, the portion of the nervous system called the sympathetic system stimulates the same fight-or-flight reactions, and most of the available evidence indicates that the sympathetic nervous system is far more important in an animal's response to emergencies.

It is probably as an antagonist to insulin that adrenalin fulfills its most important normal function. It elevates blood sugar by stimulating the liver to produce glucose from its glycogen reserves, and it acts on muscles to transform their glycogen stores into lactic acid, which is converted into glucose after being transported to the liver by the blood. The action of adrenalin is similar to that of glucagon, and both hormones activate the adenylate cyclase system and cause the production of cAMP as a second messenger. The cAMP in turn activates the enzymes necessary for hydrolizing the stored glycogen to glucose.

The Adrenal Cortex Secretes Steroid Hormones That Play Vital Regulatory Roles in the Body

A person can live normally without the adrenal medullae, but not without the cortices. These are essential for life, and their removal is soon fatal. Death is preceded by a severe disruption of ionic balance in the body fluids, lowered blood pressure, impairment of kidney function, impairment of carbohydrate metabolism with a marked decrease in both blood-glucose concentration and stored glycogen, loss of weight, general muscular weakness, and a peculiar browning of the skin. These same symptoms are seen in varying degrees in individuals whose adrenal cortices are insufficiently active, a condition known as Addison's disease.

The numerous symptoms of adrenal cortical insufficiency listed here are not related to a single hormone. The adrenal cortex is, in fact, an amazing endocrine factory, producing so many different hormones that scientists still have no idea of the total number. All the cortical hormones are extremely similar; chemically all are steroids, often differing from each other by only one or two atoms of hydrogen or oxygen (Fig. 25.11). Yet

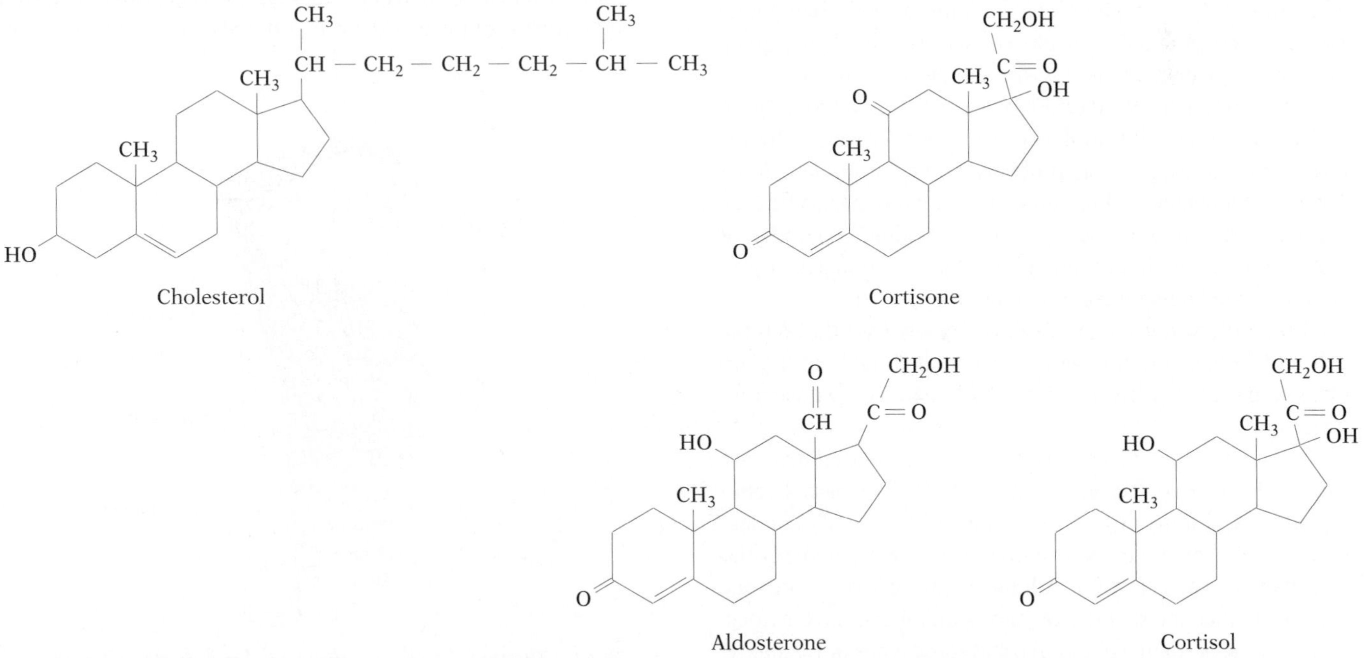

25.11 Some steroids secreted by the adrenal cortex, and the steroid—cholesterol—from which all are synthesized Very slight differences in side chains can result in markedly different properties.

these differences, minor as they may appear, give the various hormones strikingly different properties.

Note that these hormones are chemically unlike the other hormones we have discussed. In mammals, *only the hormones of the adrenal cortex and the gonads are steroids.* The steroid hormones act by entering the cells, combining with receptors, and interacting with the genetic material (Fig. 25.5).

Cortical steroids may be grouped into three categories on the basis of their functions: (1) those that act primarily in regulating carbohydrate and protein metabolism, called glucocorticoids; (2) those that act primarily in regulating salt and water balance, called mineralocorticoids; and (3) those that function primarily as sex hormones.

Hormones in the first category, ***glucocorticoids,*** such as cortisone, cause a rise in blood sugar and an increase in liver glycogen; both effects are probably due to an increased rate of conversion of proteins into carbohydrate. Their effects are opposite to those of insulin.

Hormones in the second category, ***mineralocorticoids*** (especially aldosterone), are part of a complex system for regulating blood volume, pressure, and salt balance. When the sodium ion concentration of the blood drops below a certain level, the adrenal cortex immediately secretes aldosterone into the blood. This hormone binds to cells in the kidney and promotes increased reabsorption of sodium ions by the distal convoluted tubules, which leads also to increased absorption of water. The absorption of these substances, in turn, causes a rise in blood volume and blood pressure. The mineralocorticoids are essential to the proper maintenance of sodium and water homeostasis within the fluids of the body. Animals deprived of these hormones soon begin excreting large quantities of urine containing high concentrations of sodium ions; their blood volume decreases and their blood pressure falls. If not given hormone-replacement therapy, the animals quickly die.

Hormones in the third category are very similar both chemically and functionally to the sex hormones produced by the testes and ovaries. They are probably not very important under normal circumstances, but tumors or other disturbances of the adrenal cortex may cause excessive secretion, especially of male hormone, with masculinizing effects on females or precocious sexual development of males.

The cortical steroids of all three classes pass through the cell membrane, combine with receptors and move to the nucleus, and then act directly on the DNA to influence gene transcription.

The glucocorticoid, cortisone and its chemical relatives are frequently used to facilitate healing or to give partial relief from the symptoms of arthritis and other diseases of connective tissues (where they apparently cause changes in the collagen fibers of such tissues). Administered topically, cortisone can also reduce the severity of some skin rashes. Severe allergic diseases, particularly asthma, and some types of lymphatic diseases are sometimes treated with cortisone. When administered over a long period of time, however, cortisone may cause side effects such as high blood pressure, excessive growth of hair, mental aberrations, lowered resistance to certain infections such as tuberculosis, peptic ulcers, cataracts, and brittle bones that are easily fractured.

The dilemma presented by the cortical hormones serves as an example of a general problem faced by physicians every day. Most drugs—and other treatments, for that matter—have potential harmful side effects. Physicians must therefore always balance possible good against possible harm, and they must remember that even the safest drugs are dangerous when used in excessive quantity or at the wrong time. The body is, after all, a finely tuned machine, with interactions between its parts so intricate that they still largely defy analysis. There is a risk of damage to the machine when it is subject to treatment with chemicals that almost always affect more functions than can be predicted.

THE THYROID GLAND

Iodine Is Required for the Synthesis of the Thyroid Hormones

Most vertebrates have two thyroid glands, located in the neck; in humans the two have fused to form a single gland (Fig. 25.12). Years ago a condition known as ***goiter,*** in which the thyroid may become so enlarged that the whole neck looks swollen and deformed (Fig. 25.13), was very common in some inland areas of the world, such as the Swiss Alps and the Great

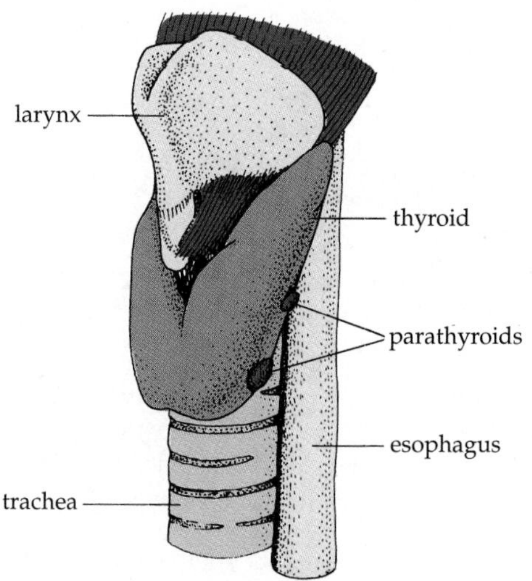

25.12 The thyroid and parathyroid glands in three-quarters view The four parathyroid glands are usually found embedded in the posterior surface of the thyroid gland.

25.13 A Bangladeshi woman with goiter

Lakes region of the United States. Goiter is often associated with a group of other symptoms, including dry and puffy skin, loss of hair, obesity, a slower than normal heartbeat, physical lethargy, and mental dullness. Then in 1883 a Swiss surgeon, who believed that the thyroid had no important function, removed the gland from a number of his patients. Most of these patients developed all the symptoms usually associated with goiter, except the swelling of the neck. The result suggested that the normal thyroid must secrete some chemical that prevents these symptoms. The curious fact that patients with no thyroid and patients with the excessively large thyroid of a goiter showed the same complex of symptoms could be explained if the malformed gland of the goiter was, despite its large size, secreting too little hormone. By the 1890s patients with goiters or the other symptoms of an inactive thyroid gland were being successfully treated with injections of thyroid extract or with bits of sheep thyroid in their diets. But nothing more specific was known as yet about the hypothetical thyroid hormone itself. True, a German chemist, E. Baumann, discovered in 1896 that the thyroid gland contains iodine, an element previously unknown in the body. But little attention was paid to his discovery.

In 1905 David Marine of Case Western Reserve University noticed that many people in Cleveland had goiters. A high percentage of the dogs also had goiters. Even many of the trout in the streams had goiters. Marine suspected that the goiters might be caused by an insufficiency of iodine in the food and water. (Iodine is relatively high in seawater and in soils along the seacoast. But the soil in inland areas is often low in iodine and food grown there is iodine-deficient. Eating seafood, which is high in iodine, once a week would provide sufficient iodine but was often unavailable in these areas.) When Marine administered tiny traces of iodine in water to his experimental animals, their goiters and other symptoms disappeared. In 1916 Marine tried his treatment on approximately 2,500 schoolchildren in Akron, Ohio. He fed these children iodized salt (salt with small amounts of potassium iodide added). Another 2,500 children, used as controls, were fed uniodized salt. At the end of a specified period he found only two cases of goiter among the children who had eaten iodized salt, whereas there were 250 cases among the controls. Though it took years to convince a skeptical public, use of iodized salt finally became widespread, and hypothyroidism caused by insufficient iodine in the soil and water now seldom occurs in the United States or Europe.

The requirement for iodine became more understandable when a thyroid hormone, now known as ***thyroxine*** or ***T_4***, was isolated in 1914 and synthesized in the laboratory in 1927. It is a derivative of the amino acid tyrosine and contains four atoms of iodine (Fig. 25.14). Later, another thyroid compound, identical to thyroxine except that it contained only three atoms of iodine, was found. This substance, called ***triiodothyronine*** or ***T_3***, is three to five times more active than thyroxine, but is secreted in smaller amounts. The T_4 molecule is thought to be the storage form and metabolically inactive; when needed, T_4 can be rapidly converted into the more active T_3 by the removal of one iodine atom. Because of the ease of this conversion, the two hormones are usually considered together, under the designation "thyroid hormones" or ***TH***.

Although the thyroid hormones are derived from the amino acid tyrosine, they resemble the hydrophobic steroid hormones in their ability to easily move through membranes and enter the cytoplasm of the target cell. Once in the cytoplasm they move through the nuclear envelop into the nucleus where they bind

I H I H
NH2
HO — O — CH2 — CH — C(=O)OH
I H I H

Thyroxine, T4

NH2
HC — C(=O)OH
CH2
OH

Tyrosine

25.14 Structural formulas of thyroxine (T_4) and the amino acid tyrosine from which it is derived Triiodothyronine (T_3) has the same formula as thyroxine, except that the iodine atom shown at upper left (red) is replaced by a hydrogen.

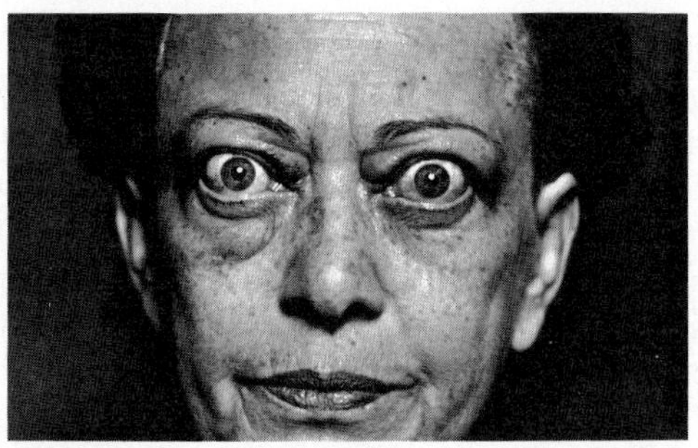

25.15 Exophthalmia About one-third of the patients with hyperthyroidism develop this protrusion of the eyeballs.

to specific proteins in the chromatin and act to stimulate mRNA and rRNA synthesis from certain genes.

The thyroid hormones increase the oxygen consumption and metabolic rate of almost all the cells in the body. They also influence carbohydrate and lipid metabolism and are necessary for normal growth and maturation. Thus hyperthyroidism[1]—excessive secretion of TH—produces a higher than normal metabolic rate, increased body temperature, profuse perspiration, high blood pressure, loss of weight, irritability, and muscular weakness. It also produces one other symptom that you might not predict: exophthalmia, a startling protrusion of the eyeballs (Fig. 25.15). Though hyperthyroidism can sometimes be controlled with antithyroid drugs, it is more often treated by surgical removal of part of the gland or by partial destruction of the gland with radioactive iodine. Hypothyroidism—decreased TH secretion—leads, in general, to the opposite symptoms. It can be caused by malfunction of the thyroid gland itself or by dietary iodine insufficiency. The thyroid gland actively transports iodine from the blood into the gland where it uses it to synthesize the thyroid hormones. When dietary iodine is low, TH cannot be synthesized in adequate amounts and secretion decreases, resulting in the hypothyroid condition. The thyroid gland then begins to enlarge in an attempt to capture what little iodine is available in the blood. The gland may become quite large and is often referred to as an iodine-deficiency goiter. When the untreated hypothyroid condition appears in newborn children, it is called cretinism; its victims show retarded physical, sexual, and mental development. Prevention of cretinism by early administration of thyroid hormone to babies showing deficiency symptoms is surely a triumph of modern medicine.

TH also plays an important role in regulating the synthesis and distribution of protein within the body, and assumes a more general role in the regulation of many aspects of development. Most vertebrates cannot develop normal adult form and function without the hormone. TH is necessary not only for the protein synthesis required for proper growth but for functional maturation of the testes and ovaries, and, together with growth hormone from the pituitary gland, is essential in promoting skeletal development.

[1]The prefix *hypo-* means "less than normal," while the prefix *hyper-* means "more than normal." Hence hypothyroidism means less than normal thyroid activity, and hyperthyroidism means more than normal thyroid activity.

The Thyroid Also Secretes Calcitonin, Which Lowers Blood Calcium Ion Concentration

In 1961 the thyroid hormone called ***calcitonin*** was discovered. It is chemically and functionally unrelated to TH, with its chief effect being the prevention of an excessive rise of calcium concentration in the blood. It thus acts as an antagonist to the parathyroid hormone, discussed below.

THE PARATHYROIDS

The parathyroid glands in humans are small pealike organs, usually four in number, located on the surface of the thyroid (Fig. 25.12). They were long thought to be part of the thyroid or to be functionally associated with it. Now, however, it is known that their close proximity to the thyroid is misleading; both developmentally and functionally, they are totally separate.

The Parathyroid Hormone Regulates the Calcium-Phosphate Balance in the Body

The ***parathyroid hormone,*** usually designated ***PTH,*** is a small protein-derived hormone that acts through a second-messenger system involving the production of cAMP in the target cells. PTH is necessary for life and functions in regulating the calcium-phosphate balance between the blood and the other tissues. Consequently it is a critically important element in maintaining the homeostasis of the internal fluid environment of the body (we have already seen that such hormones as insulin, mineralocorticoids, and calcitonin are also important in this regard). PTH increases the concentration of calcium ions, and decreases the concentration of phosphate ions in the blood by acting on at least three organs: the kidneys, the intestines, and the bones. It inhibits excretion of calcium ions by the kidneys and intestines, and it stimulates release of calcium ions into the blood from the bones. But calcium in bone is bonded

MAKING CONNECTIONS

1. How do the effects of insulin demonstrate the interlocking nature of nutrient metabolic pathways, as described in Chapter 8?

2. Based on the discussion of how the kidney works in Chapter 22, why would a diabetic tend to produce an abnormally large amount of urine and to become dehydrated?

3. How does uncontrolled diabetes demonstrate the importance of maintaining homeostasis of body fluids?

4. Why is the micronutrient iodine needed in the diet? How is it related to goiter?

5. In what respect is the thyroid hormone (TH) similar in its effects to the plant hormones discussed in Chapter 23?

with phosphate, and breakdown of bone releases phosphate as well as calcium ions. PTH compensates for this release of phosphate into the blood by stimulating excretion of this material by the kidneys. Actually, it overcompensates, causing more phosphate to be excreted than is added to the blood from bone; the result is that the concentration of phosphate in the blood drops as the secretion of PTH increases.

PTH secretion is regulated by blood calcium ion levels acting directly on the cells of the gland in a negative feedback loop (Fig. 25.16). When the diet is high in calcium and calcium is absorbed into the blood, PTH secretion decreases, and calcium ions are excreted by the kidneys and intestines and deposited in bone. The blood calcium ion level then falls, and when it gets too low, the parathyroids are stimulated to increase PTH secretion. PTH stimulates the bone to release calcium and phosphate ions, and inhibits excretion of calcium ions by the intestines and kidneys, causing calcium ion levels to increase in the blood. When the level gets too high, PTH secretion will again be inhibited. Although calcitonin from the thyroid lowers blood calcium and phosphate levels when the calcium levels get too high, its action is thought to be relatively minor in controlling blood calcium levels in the adult organism. Individuals whose thyroids are removed show no disturbance in calcium metabolism.

Naturally occurring hypoparathyroidism—diminished secretion of PTH—is very rare, but the parathyroids are sometimes accidentally removed during surgery on the thyroid. The result is a rise in the phosphate concentration and a drop in the calcium concentration in the blood (as more calcium is excreted by the kidneys and intestines and more is incorporated into bone). This change in the fluid environment of the cells produces serious disturbances, particularly of muscles and nerves. These tissues become very irritable, responding even to very minor stimuli with tremors, cramps, and convulsions. Untreated, the patient goes into the condition called tetany, which is characterized by skeletal muscle spasms, including the muscles of the larynx, causing obstruction of the airway.

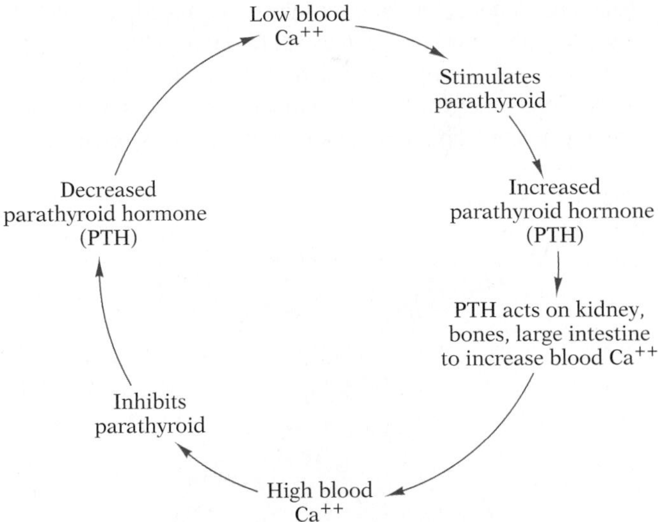

25.16 Control of calcium ion metabolism by the parathyroid hormone Calcium ion levels in the blood act directly on the cells of the parathyroid gland in a negative feedback loop. Low calcium ion levels stimulate the gland to secrete PTH, which inhibits excretion of calcium ions by the intestine and kidneys and also stimulates the bone cells to release calcium ions. Calcium ion levels in the blood then begin to increase; when they get too high the secretion of PTH by the parathyroid is inhibited and calcium ion levels begin to decrease as the kidneys and intestines excrete calcium ions.

Complete absence of PTH is usually soon fatal unless very large quantities of calcium are included in the diet. Injections of PTH are effective in preventing the symptoms.

Hyperparathyroidism—the opposite of hypoparathyroidism—sometimes occurs naturally when the glands become enlarged or develop tumors. PTH is then produced in such quantity that calcitonin is not able to sufficiently lower blood calcium levels to maintain a proper balance. The most obvious symptom of this condition is bones that are weak and easily bent or fractured, because of excessive withdrawal of calcium from the bones.

THE PITUITARY AND THE HYPOTHALAMUS

The pituitary is a small gland lying just below the part of the brain called the ***hypothalamus.*** Like the adrenals, the pituitary is a double gland (Fig. 25.17). It consists of an anterior lobe, which develops in the embryo as an outgrowth from the roof of the mouth, and a posterior lobe, which develops as an outgrowth from the lower part of the brain. The two lobes eventually contact each other as they grow, and the anterior lobe partly wraps itself around the posterior lobe. In time, the anterior lobe loses its original connection with the mouth, but the posterior lobe retains its stalklike connection with the hypothalamus. Although the anterior lobe is not directly connected to the hypothalamus, it is connected through its blood supply, as we shall see shortly. Because of its location and functional connection to the pituitary, the hypothalamus plays an important role in regulating the activities of both of these glands.

Despite their intimate spatial relationship, the two pituitary lobes remain fully distinct functionally, and we shall consider them separately here.

The Posterior Pituitary Is Intimately Associated With the Hypothalamus

Two hormones, ***oxytocin*** and ***vasopressin,*** are released by the posterior pituitary (Fig. 25.18). Oxytocin acts to stimulate the release of milk from the mammary glands and on the muscles of the uterus, causing them to contract. The hormone is necessary for childbirth. Injections of oxytocin are sometimes used to induce labor when pregnancies have gone long past term.

Vasopressin causes constriction of the arterioles, and a consequent marked rise in blood pressure. It also stimulates the kidney tubules to reabsorb more water. A human totally lacking vasopressin would have to excrete more than 20 liters of urine daily! The well-known diuretic effect of ethyl alcohol, found in beer and wine, is due to its tendency to suppress vasopressin release.

We said earlier that the posterior pituitary originates as an outgrowth of the hypothalamus of the brain. Even in the adult it retains a stalklike connection with the hypothalamus, which is composed of nervous tissue (Figs. 25.17 and 25.19). There is now persuasive evidence that oxytocin and vasopressin are produced in the hypothalamus and flow along nerve cells in the stalk to the posterior pituitary, where they are stored (Fig. 25.19). The posterior pituitary releases the hormones on stimulation by nerve impulses from the hypothalamus. In this case, the hypothalamus, not the posterior pituitary, is the true endocrine organ; its action is an example of hormonal secretion by nervous tissue.

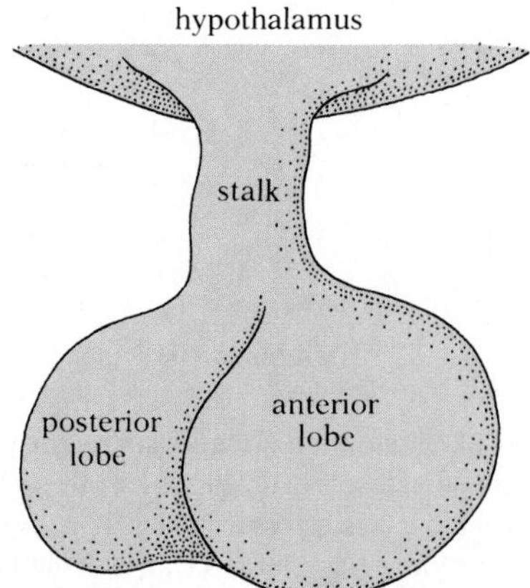

25.17 The pituitary gland The pituitary gland is connected to the region of the brain known as the hypothalamus by a stalk. It is a double gland; the two lobes are anatomically and functionally separate.

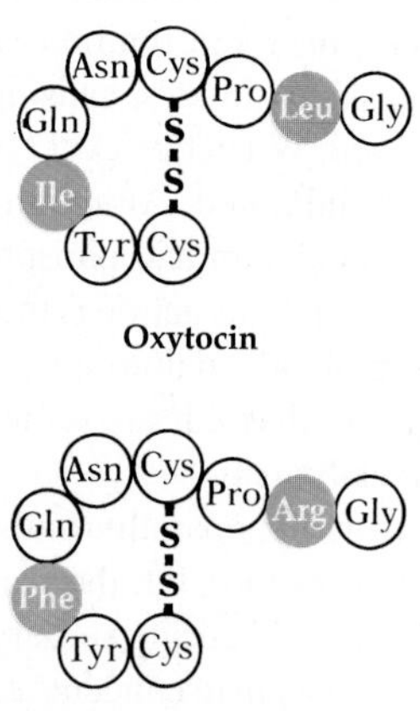

25.18 Oxytocin and vasopressin The two hormones are small proteins that differ by only two amino acids, but this difference accounts for their distinctive activities.

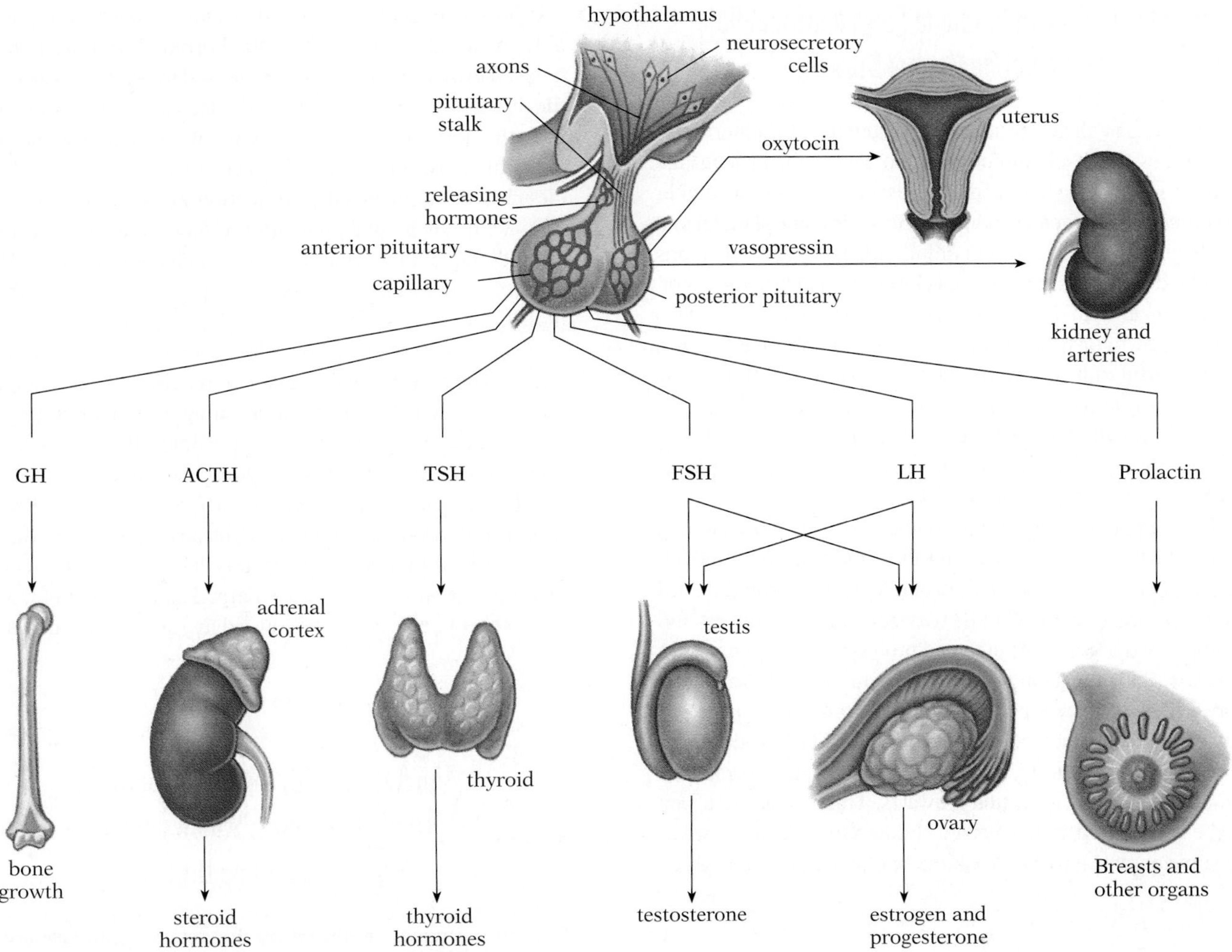

25.19 Hormones of the posterior and anterior pituitary The pituitary and the hypothalamus are intimately associated. Certain nerve cells in the hypothalamus produce hormones that move down their axons to the posterior pituitary; the hormones are released directly into the capillaries there. The cells of the posterior pituitary capture these hormones from the blood and store them until their release into the bloodstream is triggered by electrical signals from another set of nerve cells in the hypothalamus. Other nerve cells in the hypothalamus secrete hormones directly into a capillary bed, which then empties into another capillary bed in the anterior pituitary. The hormones transported in this manner regulate the release of the six important hormones synthesized in the anterior pituitary, four of which control the function of other endocrine glands and are called tropic hormones. The hormone ACTH stimulates the adrenal cortex to secrete its hormones, TSH the thyroid gland, and FSH and LH the ovaries and testes. Two other hormones act directly on target tissues; GH stimulating the growth of the long bones of the body, and prolactin the breasts and various other organs.

The Anterior Pituitary Produces Hormones of Widely Varying and Far-Reaching Effects

At least six hormones are secreted by the anterior pituitary in humans (Fig. 25.19). All of these hormones are hydrophilic proteins or glycoproteins and work by some sort of second-messenger system. ***Prolactin*** is the most versatile of the pituitary hormones. It stimulates milk production by the female mammary glands shortly after birth of a baby; in its absence milk production soon ceases. It also plays a variety of roles in reproduction, osmoregulation, growth, and metabolism of carbohydrates and fats.

The Human Growth Hormone Plays a Critical Role In Promoting Normal Growth

The human growth hormone (hGH) secreted by the anterior pituitary does not itself directly promote growth; instead, the hormone induces the liver and other tissues to secrete a number of protein ***growth factors,*** which in turn stimulate skeletal and muscle growth and the growth of individual tissues and organs (Fig. 25.20). For example, there is an epidermal growth factor, an ovarian growth factor, a nerve growth factor, etc. Also through growth factors, the pituitary growth hormone acts as a very powerful inducer of protein synthesis. In addition, it has effects opposite to those of insulin, acting directly on stored carbohydrates and fats to stimulate hydrolysis, thereby elevating blood-glucose concentration and increasing fat breakdown.

If the supply of growth hormone is seriously deficient in a child, growth will be stunted and the child will be a midget. Precise injections of hGH will promote normal growth. Until recently the only source of hGH was from human cadavers and its supply was severely limited and prohibitively expensive. In the early 1980s the gene for growth hormone was cloned and expressed in the bacteria *E. coli.* In 1985 a biotechnology company called Genentech began marketing the hGH produced by the recombinant bacteria. Although expensive, the hormone is available for treating pituitary dwarfs. Unfortunately, it can also be used to treat normal-sized children to make them taller, and some parents have been asking that their children be given this hormone to make them taller and stronger, and, presumably, better athletes. Because the hormone also promotes increased protein synthesis and increased muscle mass as well as the breakdown of fat, some athletes have also been using growth hormone, with or without medical supervision. Unfortunately, this hormone affects not only growth and muscle mass; it stimulates the production of different growth factors with such wide-reaching effects that the result is unpredictable. Physicians are concerned about the use of such hormones in normal individuals, especially since recombinant hGH hormone is available on the black market.

Oversupply of the growth hormone in a child results in a giant. (The tallest pituitary giant on record reached a height of 8 feet 11.1 inches.) Both pituitary midgets and pituitary giants have relatively normal body proportions. If, however, oversecretion of growth hormone begins during adult life, only certain bones, such as those of the face, fingers, and toes, will resume growth. The result is a condition known as acromegaly, characterized by disproportionately large hands and feet and distorted features—a greatly enlarged and protruding jaw, enlarged cheekbones and eyebrow ridges, and a thickened nose.

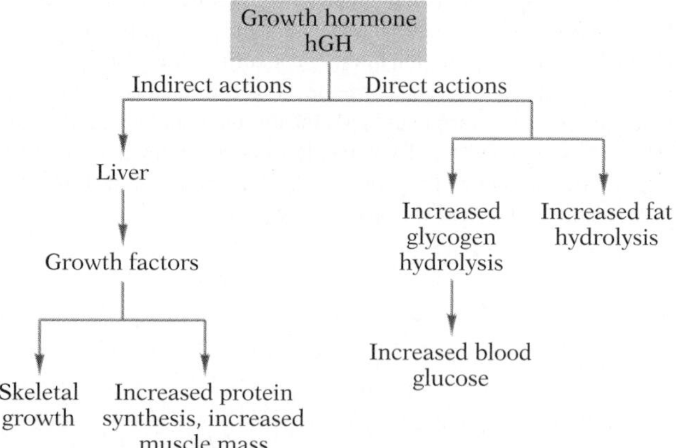

25.20 The human growth hormone The human growth hormone (hGH) has effects opposite to those of insulin, tending to elevate blood-sugar concentration and increase fat breakdown. The hormone also acts indirectly on growth by inducing the liver and other tissues to secrete a number of protein growth factors that in turn stimulate protein synthesis, skeletal and muscle growth, and the growth of individual tissues or organs.

The Anterior Pituitary Also Secretes Tropic Hormones, Which Control Other Endocrine Organs

Certain hormones produced by the anterior pituitary are unusual in that their function is to control the activity of other endocrine glands. These hormones, called ***tropic hormones,*** include ***thyrotropic hormone (TSH),*** which stimulates the thyroid; ***adrenocorticotropic hormone (ACTH),*** which stimulates the adrenal cortex, and at least two ***gonadotropic hormones,*** or ***gonadotropins***—follicle-stimulating hormone ***(FSH)*** and luteinizing hormone ***(LH)***—which act on the gonads (testes and ovaries). Proper growth and development of these endocrine glands depend on adequate secretion of the appropriate tropic (i.e., stimulatory) hormone from the pituitary; if the pituitary is removed or becomes inactive, these organs atrophy and function at very low levels. It is easy to understand why the pituitary is often called the master gland of the endocrine system.

The interaction between the anterior pituitary and the other endocrine glands over which it exerts control is another example of negative feedback. When the concentration of TH in the blood is low, the anterior pituitary begins to secrete TSH, which stimulates the thyroid to increase TH production. The resulting rise in concentration of TH in the blood then inhibits secretion of more TSH by the pituitary. In other words, the pituitary responds to a low TH level in the blood by sending a chemical messenger that stimulates increased activity by the thyroid. Once the thyroid becomes more active, the increasing amount of TH signals the pituitary that TSH production can now be reduced. There is thus a feedback of information from the thyroid to the pituitary. The pituitary exerts control over the thyroid, and the thyroid, in turn, exerts some control over the pituitary. Note that the message from the pituitary to the thyroid is a stimulatory one, while the return message from the thyroid to the pituitary is an inhibitory one; the feedback is negative. The pituitary tends to speed up the system, and the thyroid tends to slow it down. The interaction between the two opposing forces produces a delicately balanced system.

The interaction of the pituitary with the adrenal cortex and with the gonads is similar to its interaction with the thyroid. The pituitary responds to low levels of cortical hormones by secreting more ACTH and to low levels of sex hormones by secreting more gonadotropic hormone. The resulting rise in concentration of cortical hormones or of sex hormones inhibits further secretion by the pituitary.

The anterior pituitary, important as a regulator of other endocrine glands, does not, so far as is known, participate in the control of the pancreas, the adrenal medullae, or the parathyroids. We have seen that the adrenal medullae are controlled in part by the nervous system, and that the parathyroids and the pancreas are regulated by their own negative-feedback mechanisms.

The Hypothalamus Exerts Control Over the Anterior Pituitary Through Secretion of Releasing Hormones

The delicately balanced feedback interaction between the anterior pituitary and other endocrine glands is not the only factor that regulates the anterior pituitary's activity. The hypothalamus also plays an important role in this function.

The hypothalamus is located just above the pituitary. Although there is no direct physical connection between it and the anterior pituitary, there is an unusual connection between their blood supplies (Fig. 25.19). Arteries to the hypothalamus break up into capillaries, and these capillaries eventually join to form several veins leading away from the hypothalamus. But unlike most veins, these do not run directly into a larger branch of the venous system; instead, they pass downward into the anterior pituitary and there break up into a second capillary bed (Fig. 25.19). We have encountered two other places in the body where the circulation depends on two beds of capillaries arranged in sequence: the kidney nephrons, where one bed forms the glomerulus and the other envelops the tubules, and the hepatic portal system, where one bed is in the wall of the intestine and the other is in the liver. In both places the special type of circulation reflects an important functional arrangement. In the hepatic portal system, for example, many substances picked up by the blood in the first capillary bed are removed from the blood in the second capillary bed. The portal system linking the hypothalamus and the anterior pituitary seems to function in a similar fashion. The hypothalamus, when appropriately stimulated, secretes special proteins called ***releasing hormones*** into capillaries in the hypothalamus. These are carried by the portal vessel directly to the anterior pituitary, where they regulate the secretory activity of the pituitary cells. A variety of releasing hormones are produced; each is named according to the anterior pituitary hormone whose secretion it regulates. Thus corticotropic releasing hormone (CRH) from the hypothalamus stimulates the release of ACTH from the pituitary, thyrotropic releasing hormone (TRH) stimulates release of thyrotropic hormone, gonadotropic releasing hormone (GnRH) stimulates release of gonadotropic hormones, and so on. We shall not attempt here to list all the known hypothalamic hormones, but you should realize that there are specific releasing hormones produced for each of the anterior pituitary hormones. A few of the hypothalamic hormones are inhibitory rather than stimulatory; growth hormone–inhibiting hormone (GIH), for instance, inhibits release of growth hormone by the anterior pituitary. Notice that the hypothalamus, which is part of the nervous system, plays an important role in regulating the hormonal secretion of the anterior pituitary. The secretion of ACTH, TSH, FSH, LH, growth hormone, and prolactin are all controlled by releasing hormones from the hypothalamus.

The discovery that the nervous system can influence the endocrine system is an important one. Because the hypothalamus acts as the main control center for gathering and integrating neural information, changing stimuli from the environment results in changes in hormonal secretion. For example, in many birds and mammals the lengthening of days in the spring stimulates, through the hypothalamus, the secretion of gonadotropic releasing hormone, which, in turn, stimulates the gonads and the reproductive cycle begins. And one region of the hypothalamus generates the daily rhythm that helps regulate our activities. Thus there is a close interrelationship between the nervous system and the endocrine system. As we shall see in more detail in Chapter 27, nervous and hormonal control are parts of a single integrated control system.

Let us take the previously discussed negative feedback effect of a rise in thyroid hormones on the anterior pituitary as an example of the interrelationship between the two systems. To some extent, probably, the thyroid hormones exert negative feedback on the pituitary directly, by inhibiting its secretion of the TSH, but to a large extent it does so indirectly, by inhibiting the hypothalamus from secreting TRH (thyrotropic releasing hormone) (Fig. 25.21). Similarly, much of the negative feedback action of other hormones is via the hypothalamus.

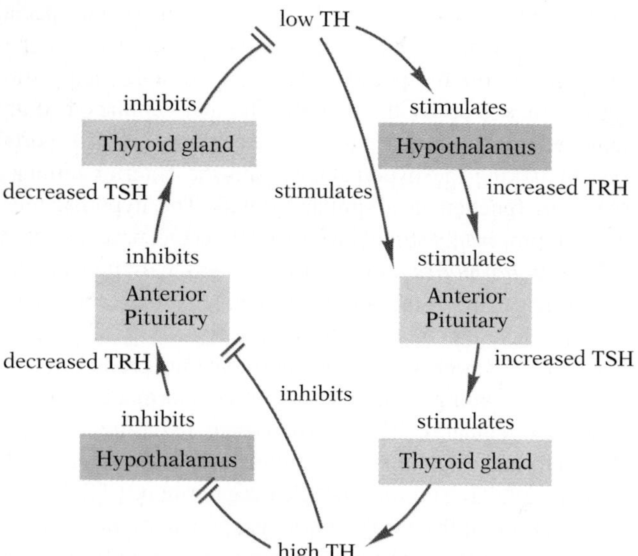

25.21 Feedback control of the thyroid gland When the level of thyroxine (TH) in the blood is low, the hypothalamus secretes thyrotropic releasing hormone (TRH) into the blood; this stimulates the anterior pituitary to release thyrotropic hormone (TSH), which in turn stimulates the thyroid gland to secrete thyroxin (TH). The thyroxine stimulates target cells throughout the body, but it also inhibits both the hypothalamus and the anterior pituitary. (Arrows indicate stimulatory influences; barred lines, inhibitory influences.) Without the stimulatory effect of TSH on the thyroid, the secretion of thyroxine decreases. When the thyroxine level gets too low, the hypothalamus is again stimulated.

THE PINEAL GLAND

The Pineal Secretes Melatonin, Which May Influence Reproductive Activity and Biological Rhythms

The pineal, a lobe at the top of the rear portion of the forebrain, has long intrigued investigators by its glandular appearance. Only recently, however, has it been demonstrated to have an endocrine function.

In some lower vertebrates the pineal is eyelike and responds to light by secreting a hormone called ***melatonin,*** which lightens the skin by concentrating the pigment granules in melanophores (pigment-containing cells). The secretion of melatonin is known to increase at night and decrease during the day. It has been shown that in these animals the pineal is intimately involved (presumably through melatonin) in the control of circadian rhythms—cycles of activity repeated approximately every 24 hours.

The mammalian pineal, too, secretes melatonin, but it has no light-sensitive cells. Though the pineal is a part of the brain, its principal nerve connection originates outside the skull cavity in nervous tissue in the neck. Evidence accumulated in the last few years suggests that the pineal may function in converting neural information about light conditions into hormone output. Animals often behave as though some sort of timer is repeatedly switching their behavior patterns from the set appropriate for one stage of their life cycle to the set appropriate for the next. Often the physiological changes associated with reproduction, migration, or hibernation in many species are tied to factors such as the length of day. As with flowering of plants, daylength is often the critical cue for many species for the timing of reproduction, so the young can be born at a time when conditions are appropriate and adequate food is available. In mammals, it is thought that information about the light-dark cycle received by the eyes goes to the brain, where it is relayed to the nervous tissue in the neck and then to the pineal. The pineal responds by secreting melatonin in inverse proportion to the amount of light. Thus, secretion of melatonin is high at night and drastically decreased during the day. In some mammals there is good evidence that melatonin inhibits the secretion of FSH and LH from the anterior pituitary, with the result that the gonads are "turned off." Thus, in autumn and winter, when nights are long, increased melatonin secretion turns off FSH and LH production, and the gonads become inactive and the animal enters a nonreproductive phase. The sequence of events is different in animals that breed in autumn.

The role of melatonin in regulating human sexual activity is still not well understood, but the photoperiod does appear to affect human reproductive systems. For example, records show that many traditional Eskimo women do not menstruate during the four months of darkness during the long Alaskan winters. There is also some evidence that the pineal may be involved in inhibiting growth and maturation of the gonads until puberty.

Melatonin may exert effects on parts of the brain other than the hypothalamus, particularly those concerned with feeding rhythms and biological rhythms. The hormone is implicated in at least one syndrome. A substantial number of humans suffer from "winter depression," a consequence of overproduction or heightened sensitivity to melatonin. Various treatments with bright lights in the morning appear to help this debilitating condition.

LOCAL CHEMICAL MEDIATORS

According to our working definition, animal hormones are substances produced by organs specialized for hormone synthesis; they are transported through the circulatory system; and they exert highly specific effects on target tissue. There is a group of control chemicals that does not fit this definition, but which are similar to hormones in their action. These substances are secreted by cells that are not part of specialized endocrine organs, and in most cases are so rapidly taken up or destroyed that they affect only cells in their immediate neighborhood. Normally they do not enter the blood in significant amounts. Such control chemicals are called ***local chemical mediators.*** The chemical histamine, which, as we saw in Chapter 21, mediates the allergic response, is an example of a local chemical mediator. It is released by mast cells and signals nearby capillaries to dilate and leak, allowing more blood to reach the site.

Prostaglandins Are Potent Local Chemical Mediators That Have a Wide Array of Actions

Since 1957 much attention has been focused on a group of hormonelike substances called prostaglandins, which are secreted by most animal tissues (Fig. 25.22). Their actions include stimulation or relaxation of smooth muscle, dilation or constriction of blood vessels, stimulation of intestinal movement, control of nerve-impulse transmission, stimulation of inflammation responses, and enhancement of the perception of pain. The effectiveness of aspirin in combating inflammation and pain is due, at least in part, to its inhibition of prostaglandin synthesis. Other prostaglandins stimulate uterine contraction and are thought to be important in childbirth. These prosta-

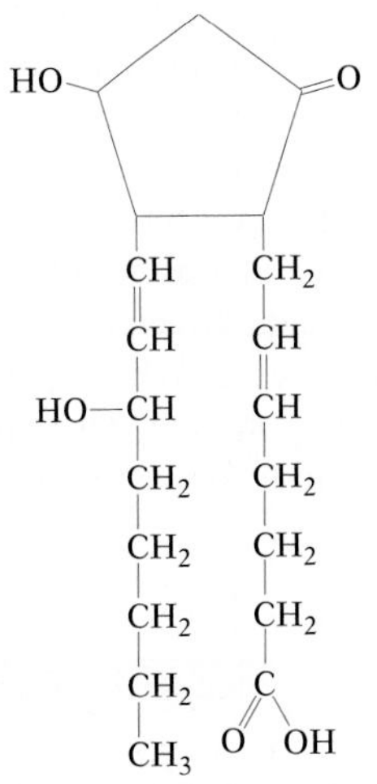

25.22 Prostaglandin E, one of the prostaglandins

MAKING CONNECTIONS

1. How does the production and use of recombinant human growth hormone (hGH) illustrate both the benefits and the ethical issues stemming from the recombinant DNA technology discussed in Chapter 13?

2. How does the function of the portal system between the hypothalamus and the anterior pituitary compare with that of the hepatic portal system described in Chapter 22?

3. How does the development of an enlarged thyroid gland (goiter) in individuals suffering from an iodine deficiency demonstrate the homeostatic control of this gland by the anterior pituitary and the hypothalamus?

4. How do the functions of the hypothalamus demonstrate the close interrelationship of the endocrine and nervous systems?

5. Compare the detection of, and response to, the photoperiod in vertebrates and plants (see Chapter 24).

6. Explain why some people who suffer from depression during the winter find that they can be helped by exposure to bright light during the morning.

7. How are the effectiveness of aspirin and morphine in the relief of pain related to prostaglandins and endorphins, respectively?

glandins are often used to induce abortions as well.

Derived from phospholipids in the cell membrane, prostaglandins are extremely potent. In some cases they may circulate in the blood like normal hormones and exert their effects at distant locations. In other cases they may act at distant target sites but without blood transport; an example is the prostaglandins in semen (secreted by the seminal vesicles), which cause contractions of the uterine muscles that may aid in the transport of sperm to the uterine tubes. In still other cases the prostaglandins may exert their principal effect within the cells where they are produced.

Prostaglandins bind to receptors on target cells. Studies of the effects of prostaglandins on a variety of target cells have revealed that some mimic the effects of certain hormones, apparently by stimulating the adenylate cyclase system or some other second-messenger system. They are continually being synthesized but are rapidly destroyed by enzymes in the body fluids.

Endorphins Bind to Opiate Receptors in the Brain, and Are Involved in Regulating Pain Perception and Mood

The discovery of endorphins resulted from research on opiates, a group of highly addictive pain-killing drugs. It eventually became evident that opiates like morphine bind to specific receptors on nerve cells in the area of the brain concerned with regulating the perception of pain and mood. The body is now known to synthesize more than a half-dozen polypeptides that bind to the opiate receptors, producing decreased pain perception and an elevation of mood. Some of these chemicals are produced by nerve cells throughout the nervous system and are carried by the blood stream to the brain; others are produced in the anterior pituitary. The so-called "runner's high" results from endorphin production. Acupuncture, too, may exert its pain-killing effect by causing release of endorphins.

CONCEPTS IN BRIEF

1. Animal hormones are substances produced by tissues specialized for hormone synthesis (endocrine tissues); they are usually transported through the circulatory system; and they exert highly specific effects on target tissue.

2. Animal hormones can be classified as water-soluble (hydrophilic) or water-insoluble (hydrophobic). Most hormones are water-soluble, and are derived from amino acids or proteins. They interact with receptor molecules on the membrane surface. The hydrophobic hormones (steroid and thyroid hormones) readily diffuse through the cell membrane and bind to specific receptors inside the cell.

3. The hydrophilic hormones cannot pass through the cell membrane. According to the *two-messenger model,* the hormone acts as an extracellular first messenger; it binds to a specific receptor site on the outer membrane surface of the target cell. The binding activates *adenylate cyclase,* which catalyzes the production of a second messenger, often *cAMP,* inside the cell. The increased cAMP then interacts with cytoplasmic enzyme systems to initiate the cell's response to the hormone.

4. The hydrophobic hormones—the thyroid and steroid hormones—can easily pass through the cell membrane. Steroid hormones (S) bind to a receptor molecule (R) and within the nucleus the complex (S-R) interacts directly with specific genes to influence transcription.

5. The pancreas secretes insulin, glucagon, and somatostatin, which regulate the blood-sugar level. *Insulin* acts to reduce the blood-glucose concentration; *glucagon* causes an increase in the blood-glucose concentration. *Somatostatin* inhibits glucagon and insulin secretion and prolongs the time food remains in the digestive tract.

6. The adrenal medulla secretes *adrenalin* and *nonadrenalin.* Both help to prepare the body for emergencies by stimulating reactions that increase the supply of glucose and oxygen to the skeletal and heart muscles ("fight-or-flight" response).

7. The adrenal cortex produces many different steroid hormones, which may be grouped into three functional categories: (1) those regulating carbohydrate and protein metabolism, the *glucocorticoids;* (2) those regulating salt and water balance, the *mineralocorticoids;* and (3) those that function as sex hormones.

8. The *thyroid gland* is located just below the larynx. The *thyroid hormones (TH)* stimulate the oxidative metabolism of most tissues in the body. The thyroid also secretes *calcitonin,* which prevents the excessive rise of calcium ions in the blood.

9. The *parathyroids* are four small, pealike or-

gans located on the surface of the thyroid. The *parathyroid hormone (PTH)* regulates the calcium-phosphate balance between the blood and other tissues; it acts primarily on the kidneys, the intestines, and the bones to lower blood calcium levels.

10. The *posterior pituitary* is connected to the hypothalamus by a stalk. It stores and releases two hormones, which are produced in the *hypothalamus* and flow along nerves in the stalk to the posterior pituitary. The hormones are released upon nervous stimulation from the hypothalamus. *Oxytocin* stimulates the contraction of uterine muscles. *Vasopressin* stimulates the kidney tubules to reabsorb more water.

11. The *anterior pituitary* produces many hormones with far-reaching effects. *Prolactin* stimulates milk production by the mammary glands and also participates in reproduction, osmoregulation, growth, and metabolism of carbohydrates and fats. *Growth hormone (hGH)* promotes normal skeletal and muscle growth.

12. The anterior pituitary also secretes a number of tropic hormones that regulate other endocrine organs. *Thyrotropic hormone (TSH)* stimulates the thyroid, *adrenocorticotropic hormone (ACTH)* stimulates the adrenal cortex, and the two *gonadotropic hormones (FSH* and *LH)* act on the gonads. The secretion of these glands is controlled by a negative feedback loop involving the gland, the anterior pituitary, and the hypothalamus.

13. The activity of the anterior pituitary is regulated by the hypothalamus, which produces special peptide *releasing hormones* that are carried by a portal system to the anterior pituitary, where they stimulate its secretory activity. The hypothalamus is the point at which information from the nervous system influences the endocrine system and is also one of the major sites of feedback from the endocrine system.

14. The pineal, a lobe in the forebrain, secretes a hormone called melatonin, which in lower animals is important in regulating skin color and in circadian rhythms. In mammals melatonin may inhibit the secretion of gonadotropic hormones.

15. Local chemical mediators (such as prostaglandins and endorphins) are control chemicals secreted by cells that are not part of specialized endocrine organs. In most cases they are so rapidly taken up or destroyed that they affect only cells in their immediate neighborhood. Normally they do not enter the blood in significant amounts.

STUDY QUESTIONS

1. Based on their solubility characteristics, which are the two major categories of hormones? How do they differ in their mode of action? In general, which type produces the faster response? Which produces the longer-lasting response? (pp. 508–511)
2. Describe the "second-messenger" model of hormone action, involving adenylate cyclase and cyclic AMP. Explain how calcium can serve as a second messenger. (p. 508)
3. Which endocrine glands produce hydrophobic hormones? Describe how hydrophobic hormones influence their target cells. What do the target cells for both hydrophilic and hydrophobic hormones have in common? (pp. 510–511).
4. Describe the effects of insulin, glucagon, and somatostatin. Where are they produced? Explain how the blood-sugar level is controlled by a negative-feedback mechanism involving insulin and glucagon. (pp. 513–515)
5. How does a deficiency of insulin (or an insensitivity of the tissues to insulin) lead to the various symptoms of diabetes? (p. 513)
6. Describe the "fight-or-flight" response produced by adrenalin. Which part of the adrenal gland produces adrenalin and nonadrenalin? What is the functional relationship of adrenalin and insulin? (pp. 515, 517)
7. What would be the effect on the body if the adrenal cortex stopped producing glucocorticoids? Mineralocorticoids? Sex hormones? (pp. 517–518)
8. How is thyroid hormone (TH) similar to the steroids? What are the basic effects of TH? How are they related to the symptoms of hyper- and hypothyroidism? What is a goiter and what is its usual cause? (pp. 517–520)
9. How are the blood calcium and phosphate levels controlled by parathyroid hormone (PTH) and calcitonin (particularly the former)? (p. 521)
10. How do the anterior and posterior lobes of the pitu-

itary gland differ in embryological origin, function, and means of connection to the hypothalamus? Name and describe the effects of two hormones secreted by the posterior pituitary. (pp. 522–524)

11. **Which two hormones of the anterior pituitary act directly on target tissues? What would be the effects on the body if either could not be produced? (pp. 523–524)**
12. **Which endocrine glands are controlled by tropic hormones secreted by the anterior pituitary? Draw and describe a negative-feedback loop for the control of one of these glands, such as the adrenal cortex. Include in your description the roles of the hormones produced by both the hypothalamus and the anterior pituitary. (pp. 524–525)**
13. **How is the secretion of melatonin by the pineal related to daylength? How is it believed to regulate reproductive behavior in many mammals? (p. 526)**
14. **Describe the effects of prostaglandins and endorphins. How do these local chemical mediators differ from hormones? (pp. 527–528)**

CHAPTER 26

HORMONAL CONTROL OF VERTEBRATE REPRODUCTION

From an evolutionary standpoint, reproduction is the ultimate goal of life. All the other aspects of living discussed in this book—nutrient procurement, gas exchange, internal transport, waste excretion, osmoregulation, growth, hormonal and nervous control, and behavior—can be viewed, in a sense, as processes that enable organisms to survive to reproduce. It has been said that the hen is the egg's way of producing another egg, and the idea is equally applicable to human begins; we are, in a way, elaborate devices for producing eggs and sperm, for bringing them together in the process of fertilization, and for giving birth to young. This chapter explores the physiology of vertebrate (particularly mammalian) reproduction as an illustration of the complex interplay between a variety of different control mechanisms.

SEXUAL REPRODUCTION

Sexual Reproduction in Higher Animals Always Involves the Union of Two Gametes

We have already seen that at the time of reproduction, meiosis occurs and haploid gametes—egg cells (ova) and sperm (spermatozoans)—are produced. Typically the egg cell is large and nonmotile while the sperm are small and motile. The flagellated sperm swims to the egg, and fertilization occurs—the nuclei of the egg and sperm unite to form the diploid zygote (Fig. 26.1), which is the first cell of the new embryo. Occasionally the same individual produces both gametes, which unite in a process known as self-fertilization; it is most common among internal parasites, such as tapeworms, whose chances of locating another individual for cross-fertilization are often poor. However, most animals use cross-fertilization, even when, as in earthworms, each individual is ***hermaphroditic***—that is, each of them possesses both male and female sexual organs (Fig. 26.2). Cross-fertilization produces offspring with novel combinations of alleles and genes. Such genetic variation increases the species' chances of surviving and reproducing in a fluctuating environment. Sexual reproduction among vertebrates, which always involves cross-fertilization, will be the focus of this chapter.

Chapter opening photo: Human egg (oocyte) traveling down the oviduct After release from the ovary, the oocyte with its surrounding cells moves towards the uterus through the oviduct, a tube about 11 cm long. Fertilization usually occurs within the upper portion.

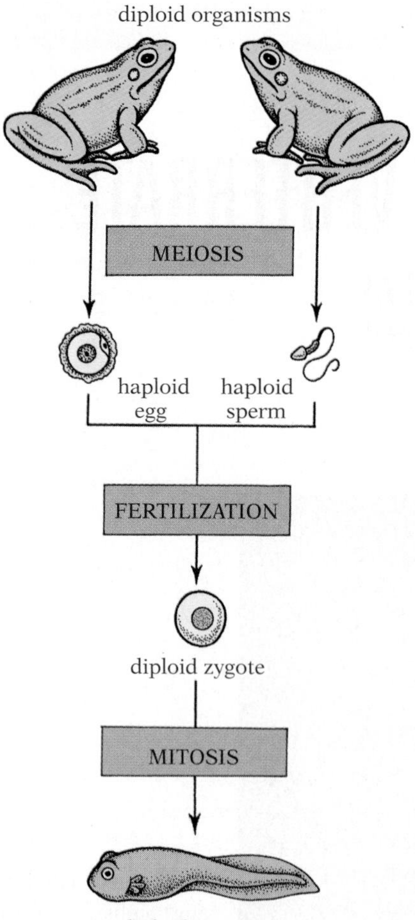

26.1 Typical animal life cycle Animals exist as diploid multicellular organisms through most of their life cycle. At the time of reproduction, special cells undergo meiosis to produce haploid gametes—egg and sperm. The flagellated sperm swim to meet the egg and fertilization takes place, forming a diploid zygote, which is the first cell of the new organism. The zygote divides mitotically many times to produce a new diploid multicellular organism.

Most Aquatic Animals Use External Fertilization, Whereas Most Land Animals Use Internal Fertilization

There are two basic ways in which egg cells and sperm cells are brought together: ***external fertilization,*** where both types of gametes are shed into the surrounding medium and the sperm swim or are carried by water currents to the eggs; and ***internal fertilization,*** where the eggs are retained within the reproductive tract of the female until after they have been fertilized by sperm inserted into the female by the male.

For two reasons, external fertilization is limited essentially to animals living in aquatic environments: (1) the flagellated sperm must have fluid in which to swim; and (2) the eggs, which lack a hard outer shell to permit penetration by sperm, would dry out on land if not continuously bathed in fluid. Almost all aquatic invertebrates, most fishes (but not sharks), and many amphibians use external fertilization. Shedding short-lived eggs and sperm into the water of a lake or stream is an uncertain method of fertilization; many of the sperm never locate an egg, and many eggs are never fertilized, even if both types of gametes are shed at the same time and in the same place, as is usually the case. Consequently animals using external fertilization generally release vast numbers of eggs and sperm at one time (Fig. 26.3). Behavioral sequences and courtship displays (in which hormonal control is very important) have evolved to ensure simultaneous release of eggs and sperm and increase the likelihood that fertilization will occur.

Most land animals, both invertebrate and vertebrate, use internal fertilization, in which the sperm are deposited directly into the reproductive tract of the female (Fig. 26.4). In effect, the sperm cells of land animals are provided with the sort of fluid environment that is no longer available to them outside the animals' bodies. The sperm can therefore remain aquatic, swimming through the film of fluid always present on the walls of the female reproductive tract. Internal fertilization has advantages over external fertilization in that the sperm can be concentrated and protected inside the body of the female, and that the gametes are in very close proximity, making fertilization more likely. And because internal fertilization entails much less wastage of egg cells than external fertilization, fewer egg cells are released during each reproductive season. Once fertilized, the egg is either enclosed in a protective shell and released by the female or held within the female's body until the embryonic stages of development have been completed. Internal fertilization requires very close physiological

26.2 Two earthworms mating The worms are hermaphroditic, each possessing both male and female reproductive organs. They are coupled at two mating points; at one the upper worm is acting as male and the lower worm as female, while at the other they reverse these roles.

26.3 Toads spawning The smaller male clasps the female in an embrace called amplexus and sprays semen over the eggs she releases (round, white structures). Many hundreds of eggs are produced.

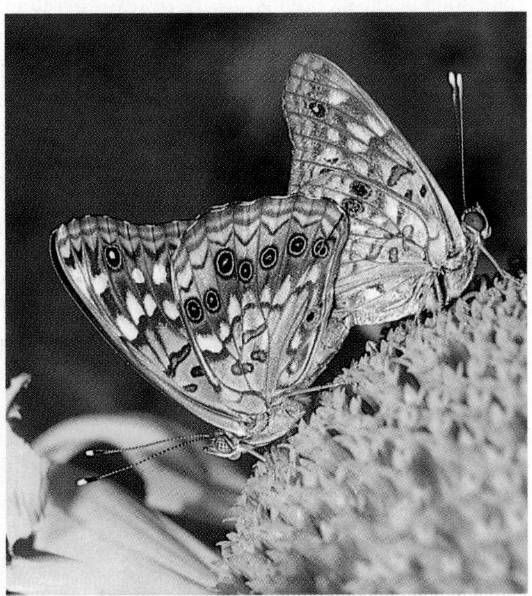

26.4 Internal fertilization by a terrestrial invertebrate The male butterfly inserts sperm into the reproductive tract of the female. Fertilization occurs within the body of the female, and the eggs are invested with a protective shell before being laid.

and behavioral synchronization of the sexes for the mating process to occur, and thus extensive hormonal and nervous control.

Let us summarize briefly the characteristic reproductive methods employed by the major classes of vertebrates. Fish, being aquatic, almost always use external fertilization, in which huge numbers of eggs and sperm are released at each mating; the wastage of gametes is enormous. Amphibians, such as frogs and salamanders, evolved from fish, and they too generally use external fertilization; they must therefore return to the water or to a very moist place on land to lay their eggs. Reptiles, including snakes, lizards, and turtles, evolved from ancestral amphibians. They were the first vertebrates to be fully emancipated from the ancestral dependence on the aquatic environment for reproduction. They use internal fertilization, and they lay eggs enclosed in protective membranes and shells. Birds evolved from one group of ancient reptiles, and they too employ internal fertilization and lay eggs with shells. Mammals evolved from a different group of ancient reptiles, and the internally fertilized egg is retained within the female reproductive tract until embryonic development is completed.

Evolution of the Amniotic Egg Was an Important Evolutionary Innovation for the Conquest of Land

An important "invention" of the reptiles was the ***amniotic egg.*** This egg, with its protective shell and membranes, could be laid on land, thereby freeing the animal from a dependence on environmental water for reproduction. The amniotic egg of land vertebrates such as reptiles and birds has four different membranes inside the outer ***shell.*** These are the amnion, the allantois, the yolk sac, and the chorion (Fig. 26.5A). The ***amnion*** encloses a fluid-filled chamber housing the embryo, which can thus continue to develop in an aquatic medium just as embryos of fishes and amphibians do, even though the egg as a whole may be laid on dry land. In a very real sense, the developing embryo has its own "private pond" in which to grow, and is protected by the membranes and a tough shell. The ***allantois*** functions as a receptacle for the urinary wastes of the developing embryo, and its blood vessels, which lie near the shell, function in gas exchange. The ***yolk sac,*** as its name indicates, encloses the yolk, which is food material used by the developing embryo. The ***chorion*** is an outer membrane surrounding the embryo and the other membranes. Evolution of the amniotic egg—often called the land egg—was a major evolutionary innovation in vertebrate history. No longer did such animals have to return to water for reproduction, but were free instead to exploit the various environments on land. Neither the adult nor the young requires adaptations for living in an aqueous environment. Another advantage is that development in animals with amniotic eggs is direct, from embryo to adult. There is no larval stage—the young hatch or are born fully formed, like miniature adults, and ready to survive on land (Fig. 26.6). The free-living larval stage has been replaced by a long embryonic period within the protective egg shell or uterus.

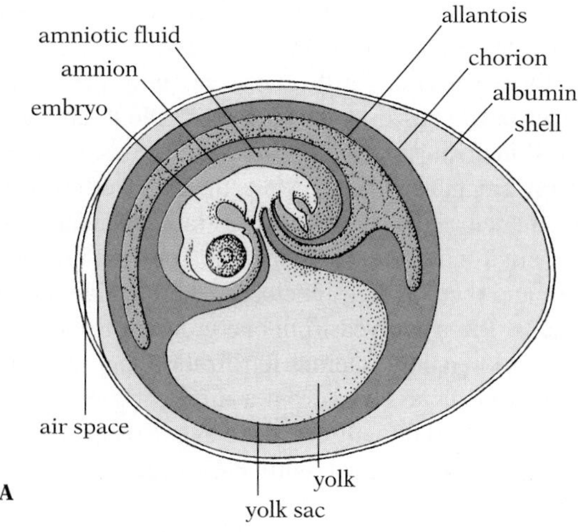

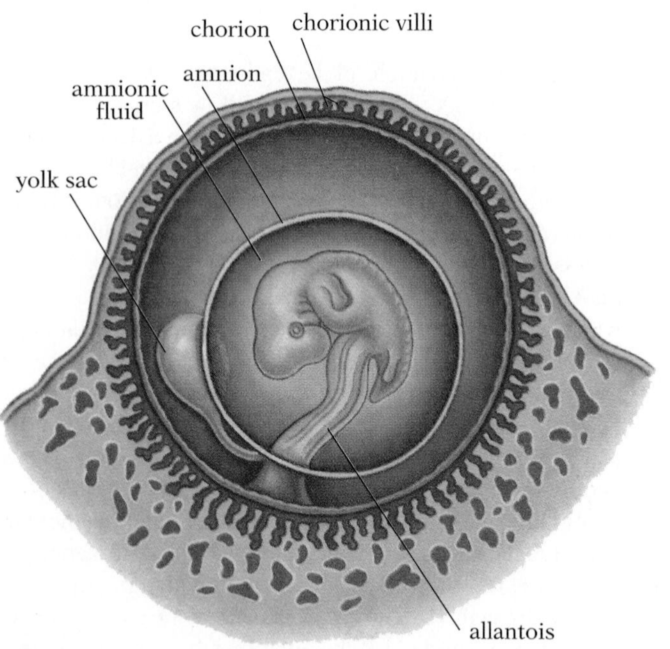

26.5 The embryonic membranes in bird and mammalian eggs (A) Although everything shown here in this bird's egg is commonly referred to as the "egg," the chorion is the outer boundary of the true egg cell. Inside the chorion are the yolk sac, which stores food for the growing embryo; the allantois, which functions in waste storage and gas exchange; and the amnion, which surrounds a fluid-filled chamber housing the embryo. The thick layer of albumin, a protein, is outside the cell. The outer shell is porous so oxygen for aerobic respiration can diffuse through it. (B) The mammalian embryo has the same four membranes as the birds, but the yolk sac and allantois are considerably smaller. There is no shell; instead the embryo with its extraembryonic membranes burrows into the uterine lining. The embryo obtains its nutrients and oxygen from the mother's body through the placenta.

Like reptiles and birds, mammals use internal fertilization. The mammalian embryo possesses the same four membranes as the reptilian and bird embryo (Fig. 26.5B), but (with a few rare exceptions) no shell is deposited around it and it is not laid. Instead, the early mammalian embryo and its membranes are retained within a specialized chamber of the female genital tract, and embryonic development is completed there, in a protected environment. The young animal is then born alive. The remainder of our discussion here will be concerned with mammalian, and in particular human, reproduction.

THE HUMAN AND OTHER MAMMALIAN REPRODUCTIVE SYSTEMS

The Genital System of the Human Male

The male gonads, or sex organs, are the ***testes***—ovoid glandular structures that form in the dorsal portion of the abdominal cavity from the same embryonic tissue that gives rise to the ovaries in females. In the human male the testes descend about the time of birth from their points of origin into the ***scrotal sac*** (scrotum), a pouch that is initially continuous with the abdominal cavity via a passageway called the ***inguinal canal*** (Figs. 26.7 and 26.8). After the testes have descended through the inguinal canal into the scrotum, the canal is slowly plugged by

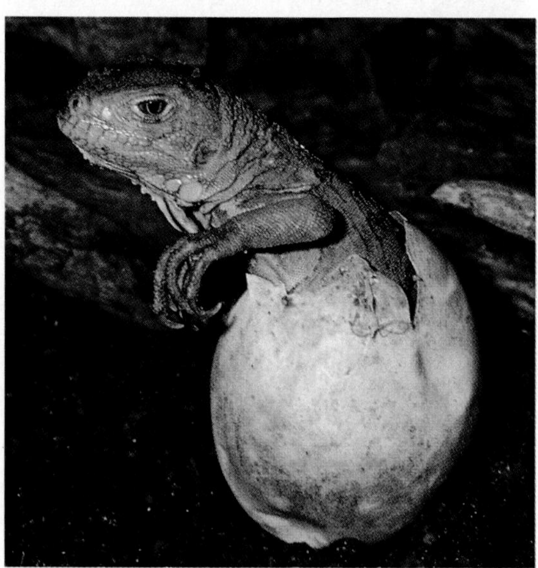

26.6 Green Iguana hatching from its egg The amniotic egg of reptiles, with its protective shell and membranes, allows the embryo to develop within the egg in an aquatic medium even though the egg is laid on dry land. Reptiles no longer have to return to water for reproduction as amphibians do.

growth of connective tissue, so that the scrotal and abdominal cavities are no longer continuous.

Sometimes the inguinal canal fails to close properly; even when it does, it remains a point of weakness and is easily broken open again when subjected to excessive strain, as when a man lifts a heavy object. The opening resulting from insufficient closure or from later rupture is known as an inguinal hernia; it is the most common type of hernia in males and is generally only a human problem, resulting from the strains caused by bipedal (two-legged) posture. If the hernia is large, it must be repaired surgically to prevent a loop of the intestine from slipping through the opening into the scrotal sac, where the intestine may become caught so tightly that its blood supply is cut off and gangrene results.

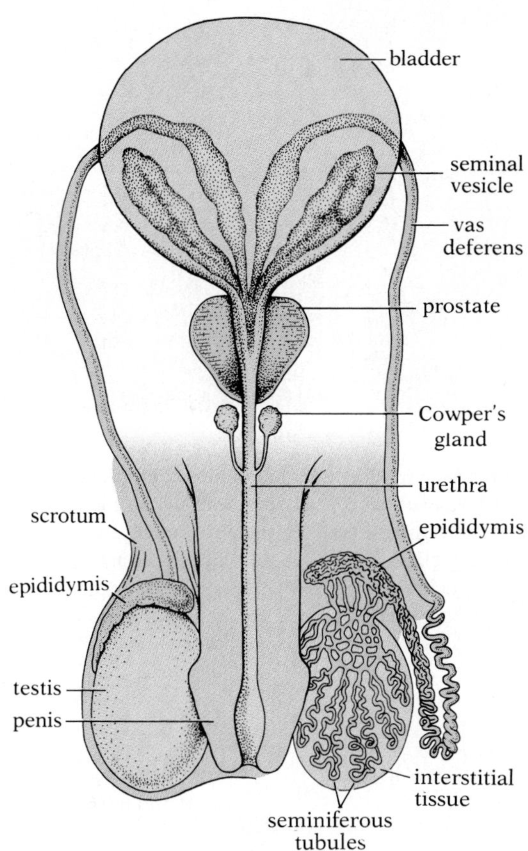

26.7 Reproductive tract of the human male: frontal view

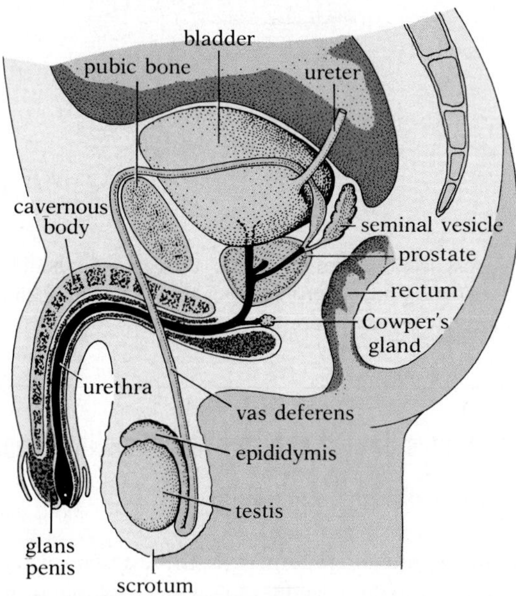

26.8 Reproductive tract of the human male: lateral view

Each testis has two functional components: the ***seminiferous tubules*** (See Fig 9.26, p. 197), in which the sperm cells are produced, and the ***interstitial cells of Leydig,*** which secrete male sex hormones. The seminiferous tubules of the human are not functional at the temperatures characteristic of the abdominal cavity; if the testes fail to descend, the germinal epithelium of the tubules (consisting of the cells that produce the sperm) eventually degenerates. If, however, the testes descend normally into the scrotal sac, where the temperature is approximately 1.5°C cooler, the germinal epithelium becomes functional at the time of puberty. The seminiferous tubules are veritable sperm factories, producing about 120 million sperm cells per day, and eight trillion sperm over a lifetime. Mature sperm cells pass from the seminiferous tubules via many tiny ducts into a much-coiled tube, the ***epididymis,*** which lies on the surface of the testis (Figs. 26.7 and 26.8). The sperm then move up to a region around the prostate gland where they remain until they are activated by secretions produced by it and other glands and released during copulation.

A long sperm duct, the ***vas deferens,*** runs from each epididymis through the inguinal canal and into the abdominal cavity, where it loops over the bladder and joins with the ***urethra*** just beyond the point where the urethra arises from the bladder. The urethra, in turn, passes through the ***penis*** and empties to the outside. The urethra in the mammalian male is a common passageway used by both the excretory and reproductive systems; urine passes through it during excretion and sperm pass through it during sexual activity.

Before ejaculation, however, as sperm pass through the vasa deferentia and the urethra, seminal fluid is added to them to form ***semen.*** The seminal fluid is secreted by three sets of glands: the ***seminal vesicles,*** which empty into the vasa deferentia just before these join with the urethra; the ***prostate,*** which empties into the urethra near its junction with the vasa deferentia; and the ***Cowper's glands,*** which empty into the urethra at the base of the penis (Figs. 26.7 and 26.8). Seminal fluid has a variety of functions: (1) it serves as a vehicle for transport of sperm; (2) it lubricates the passages through which the sperm must travel; (3) as an effectively buffered fluid, it helps protect the sperm from the harmful effects of the acids in the female genital tract; and (4) it contains much sugar (mostly fructose), which the active sperm can use as a source of energy. The tiny sperm cells can store very little food themselves; they depend on an external source of nutrients for the respiratory production of the ATP necessary to keep their flagella active. The

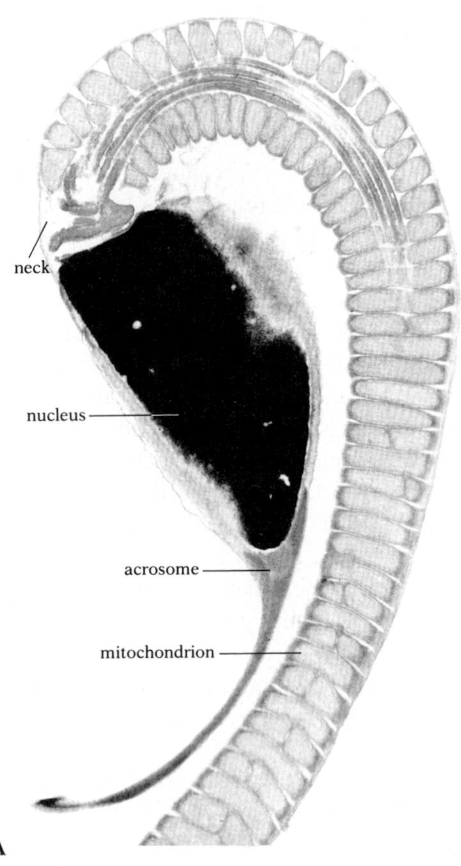

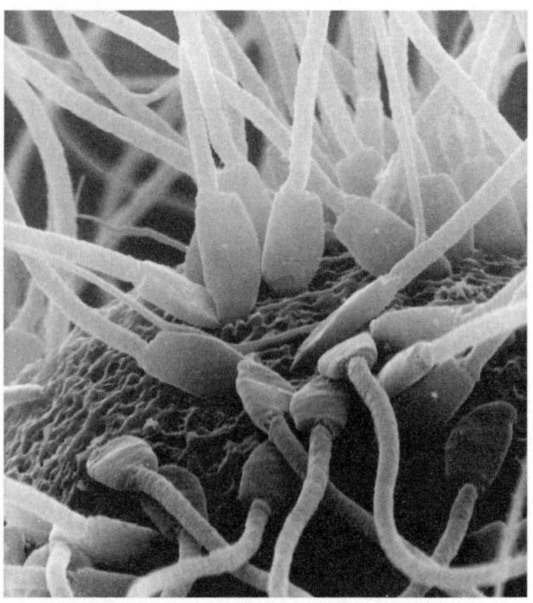

26.9 Mammalian spermatozoa (A) Electron micrograph of longitudinal section of a sperm cell from a kangaroo rat. The sperm head, which contains the acrosome (an enzyme-filled organelle) and the nucleus, is connected by a short neck to a portion of the flagellum tightly packed with spirally arranged mitochondria. There are no mitochondria in the long distal portion of the flagellum, which is not shown here. (B) Scanning electron micrograph of an Israeli sand rat egg surrounded by sperm. Only one sperm will fertilize the egg.

sperm consists of a head (containing the nucleus), and a body that is capable of flagellar movement (Fig. 26.9). The body of a sperm is an amazing power plant, packed with mitochondria, which extract energy from the sugar in the seminal fluid and in the female reproductive tract.

During sexual excitement the arteries leading into the penis dilate, and the veins from the penis constrict, in response to stimulation by nerves. Much blood is pumped under considerable pressure through the arteries into the spaces in the spongy erectile tissue of which the penis is largely composed (Figs. 26.8 and 26.10). The engorgement of the penis by blood under high arterial pressure causes it to increase greatly in size and to become hard and erect, thus preparing it for insertion into the female vagina during copulation (also called coitus).

When the glans penis, the tip of the penis (Fig. 26.8), is sufficiently stimulated by friction, nervous reflexes cause waves of contraction in the smooth muscles of the walls of the epididymides, vasa deferentia, seminal glands, and urethra. These contractions move sperm through the vasa deferentia, combine seminal fluid from the various glands with the sperm, and expel the semen from the urethra. An average of about 400 million sperm cells in about 3.5 milliliters of semen are released during one ejaculation by a human male. As a contraceptive measure, the sperm can be prevented from entering the female reproductive tract by use of a condom, a rubber sheath worn over the penis.

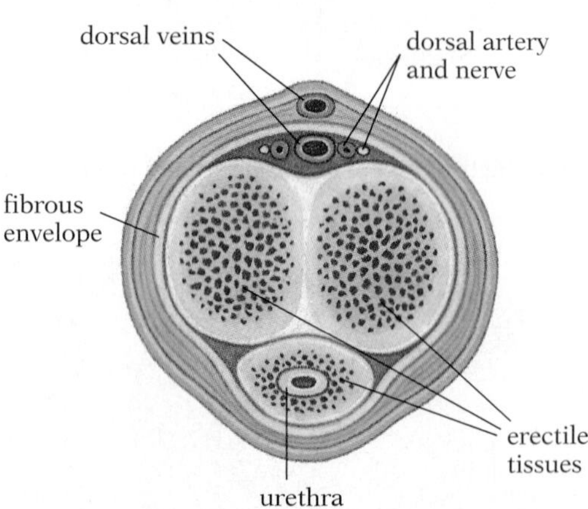

26.10 Cross section of human penis The cavernous bodies of the penis consist of erectile tissue that, when filled with blood, causes the penis to increase greatly in size.

The Testes Begin Secreting Male Sex Hormones During Early Embryonic Development

In addition to producing sperm cells, the testes are also the primary source of male hormones known as androgens, without which the production of sperm cells would not occur.

Testosterone

Progesterone

26.11 The structural formulas of male sex hormone (testosterone) and female sex hormone (progesterone) Differing only in one side group (color), these two steroids carry out very different functions in the male and female. Both hormones are synthesized from cholesterol.

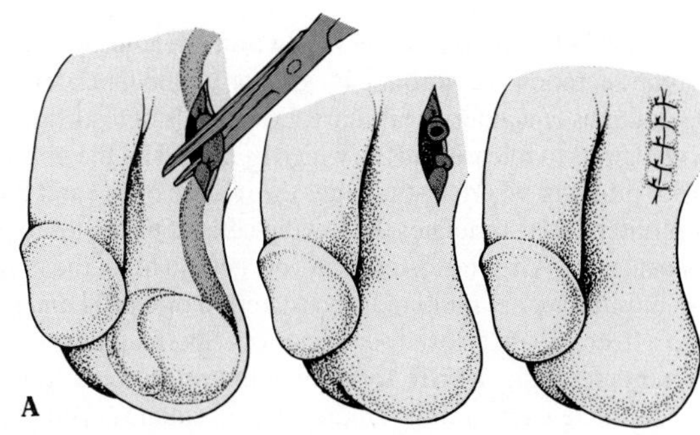

26.12 Vasectomy Vasectomy is a surgical procedure performed to prevent movement of sperm into the urethra and is a method of permanent birth control. A small incision is made in the wall of the scrotum (on each side): the vas deferens (brown) is cut and tied, and the incision is sutured. The procedure causes no retrogression of sexual characteristics because there is no alteration of hormone levels.

Testosterone (Fig. 26.11) is the principal androgen produced and its presence in the embryo is crucial to the development of the male genitalia. Early in development the level of testosterone production remains low and no sperm cells are produced until the time of puberty.

Testosterone Stimulates and Maintains the Secondary Sex Characteristics in Males

Once testosterone appears in appreciable quantity at the time of puberty, it stimulates maturation of the male reproductive structures and the development of the secondary sexual characteristics normally associated with puberty; growth of a beard and pubic hair, deepening of the voice, development of larger and stronger muscles, and the like. If the testes are removed (castration) before puberty, these changes in the secondary sexual characteristics never occur. Until this century castration of young Italian boys with beautiful soprano voices was done to preserve their high voices for the cathedral choirs and grand opera. Many operas have roles written for the "castrati." (Today these are sung by women!) If castration is performed after puberty, there is some retrogression of the adult sexual characteristics, but they do not disappear entirely. Castration after puberty eliminates the sex urge in many animals, but not in man, where psychological factors are of much greater importance than in other animals. Unlike castration, cutting the vasa deferentia (vasectomy)—an operation that prevents movement of sperm into the urethra and is sometimes used as a relatively safe and convenient birth-control measure (Fig. 26.12)—causes no retrogression of sexual characteristics, because there is no alteration of hormone levels.

Testosterone Levels Are Regulated by Negative Feedback Through the Hypothalamus and Anterior Pituitary

Testosterone secretion is controlled by a negative-feedback loop involving the hypothalamus and anterior pituitary (see Fig. 25.19, p. 523). During childhood even minute amounts of testosterone in the body have a strong inhibitory effect on the hypothalamus, and it therefore secretes very little gonadotropic releasing hormone ***(GnRH).*** Without GnRH stimulation, the anterior pituitary does not release the two gonadotropic hormones, follicle-stimulating hormone ***(FSH)*** and lutenizing hormone ***(LH),*** that are necessary for maturation of the testes and sperm production. The factors governing puberty are not fully understood, but it is thought that at the time of puberty the hypothalamus begins to respond to the low levels of circulating testosterone by releasing large amounts of GnRH, which is carried by the portal system to the anterior pituitary, which in turn is stimulated to release LH and FSH. These two hormones will continue to be produced throughout the man's lifetime. The LH from the anterior pituitary is carried by the blood to the testes, where it stimulates the Leydig cells to produce testosterone. The resulting rise of testosterone in the blood, however, inhibits secretion of more GnRH and LH by negative feedback (Fig. 26.13), and testosterone secretion then declines. But when the testosterone level gets too low, the hypothalamus will again be stimulated to produce large amounts of GnRH. This delicately balanced negative-feedback system keeps testosterone within normal levels in the blood.

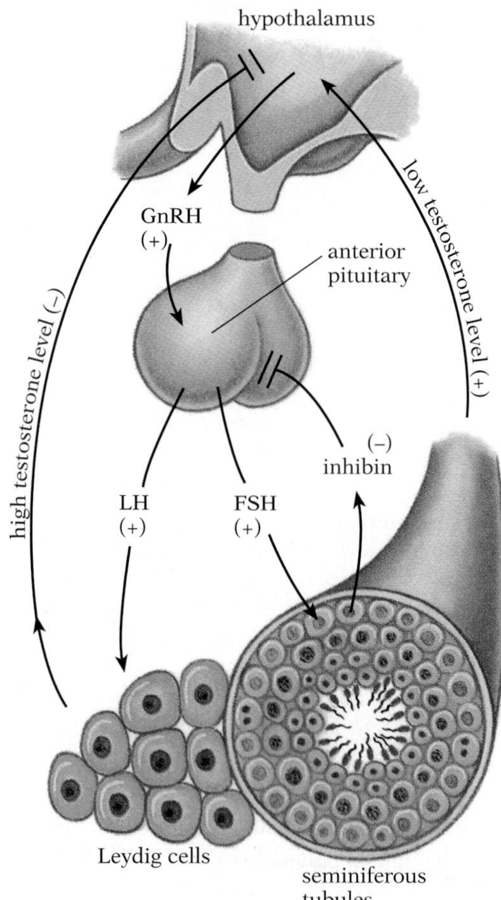

26.13 Control of testosterone secretion and spermatogenesis Testosterone secretion is controlled by a negative-feedback loop involving the hypothalamus and anterior pituitary. Low levels of testosterone stimulate the hypothalamus to release GnRH, which is carried by the portal system to the anterior pituitary where it stimulates the release of LH and FSH by the anterior pituitary. LH stimulates the Leydig cells of the testes to produce more testosterone. When testosterone levels get too high, the hypothalamus and anterior pituitary are inhibited, and secretion of GnRH and LH decreases, leading to a decrease in testosterone secretion. Spermatogenesis is regulated by a different mechanism; FSH, together with testosterone, stimulates spermatogenesis in the seminiferous tubules. The hormone called *inhibin,* which is a protein secreted by the testes, inhibits FSH production by the anterior pituitary, thereby inhibiting spermatogenesis.

The FSH produced by the anterior pituitary, together with testosterone, stimulates sperm production (spermatogenesis). However, FSH secretion is not thought to be inhibited by a negative-feedback metabolism, but rather by a hormone called ***inhibin,*** a protein secreted by the testes (Fig. 26.13). The protein is of great interest to researchers since it may have some use as a male contraceptive.

The anabolic-androgenic steroids (see Exploring Further: The Use of Steroids in Sports, p. 539), which have been taken by many athletes to promote growth, are testosterone analogs. High levels of such drugs act to decrease FSH and LH production (through negative feedback on the anterior pituitary and hypothalamus), with the result that the testes are inhibited, and normal testosterone secretion and sperm production is decreased.

MAKING CONNECTIONS

1. What is the significance of reproduction from an evolutionary standpoint?

2. Why do hermaphroditic organisms usually practice cross-fertilization even when self-fertilization is possible? Under what conditions is self-fertilization favored?

3. Describe the structure of the amniotic egg and explain why its "invention" was a crucial event in vertebrate evolution.

4. How does seminal fluid provide for the energy needs of the sperm cell? How does it enable sperm to survive in the hostile pH of the female genital tract?

5. Distinguish between castration and vasectomy. How do the effects of castration demonstrate that hormones can affect behavior via the nervous system?

6. As steroid hormones, how would the sex hormones produced by the gonads be expected to produce their effects on cells? (See Chapter 25.)

EXPLORING FURTHER

THE USE OF STEROIDS IN SPORTS

Despite being banned by sports-governing bodies worldwide (including the Olympics since 1976), the use of steroids by athletes has remained a serious problem since Soviet weightlifters were accused of using testosterone to enhance performance in 1954. Steroids, the so-called "breakfast of champions," have been used principally by body builders, weightlifters, football players, swimmers, and track and field athletes to increase size, strength, and speed. There are many types of steroids but the ones used in sports are anabolic-androgenic steroids, meaning that they have both anabolic (tissue-building) and androgenic (masculinizing) effects. These steroids are all testosterone analogs, i.e., they are chemically derived from testosterone and have similar chemical and physical properties. Many drug companies have tried to eliminate the androgenic effect of these compounds, and although some of the synthetic steroids are less androgenizing than natural testosterone, their effect remains substantial.

There has been much controversy over the years concerning the effectiveness of anabolic-androgenic steroids in enhancing strength and athletic performance. Many physicians and researchers claimed that there was no consistent scientific evidence to show that steroids improved strength, whereas the athletes who took the drugs knew otherwise. Part of the discrepancy came from the fact that the dosages allowed for research purposes were considerably lower than those used by athletes. (Athletes typically use massive doses, and use a mixture of different ones.) Also, diet, age of subjects, intensity of training, etc., influence results. Recent studies have concluded that anabolic-androgenic steroids do produce gains in strength and performance, but only when taken by athletes already in an intensive weight-training program, who continue that program, and who are also on a high-protein diet. By stimulating protein synthesis, steroids have a growth-promoting (anabolic) effect, but unfortunately they also affect ion and water balance and cause changes in the kidney, liver, and circulatory system.

Intense physical training produces a great deal of wear and tear on the body, and requires a lot of energy. Once the carbohydrate stores are used up, the body begins to break down protein (muscle) and fat to get the energy it needs. Thus an athlete doing severe muscular exercise is constantly prone to protein loss, i.e., he or she is in a state of negative nitrogen balance. The anabolic-androgenic steroids appear to reverse this, improving the utilization of protein and inducing protein synthesis in muscle cells. Anabolic-androgenic steroids, like testosterone, are hydrophobic molecules that enter their target cells by diffusing through the cell membrane. Once inside the cell they bind to specific androgen receptors within the cytoplasm; the hormone-receptor complex then enters the nucleus, where it activates transcription of certain genes and thereby stimulates protein synthesis. Muscle cells have relatively large numbers of testosterone receptors, and so respond strongly to

these substances, which induce synthesis of the two contractile proteins, actin and myosin. Used in conjunction with intensive weight training and a high-protein diet, the drugs result in increased muscle mass and strength, a reduction in fat stores, and shorter recovery time (permitting more frequent and more intensive training sessions). The drugs also have behavioral effects, giving a feeling of euphoria and making the athletes more aggressive and less easily fatigued.

Unfortunately, there are a number of harmful side effects associated with prolonged steroid use, some of which are permanent. In high doses all the anabolic-androgenic steroids cause masculinization in both sexes (the effects are more severe in women). Acne, deepening of the voice, increased facial and body hair, and male pattern baldness are common. The psychological effects are also a universal finding; in both men and women, the drugs cause feelings of euphoria and well-being, as well as increases in self-esteem, energy level, tolerance to pain, mental intensity, aggressive behavior, and, often, hostility and violence. Other possible side effects in females are menstrual irregularities, uterine atrophy, shrinkage of breast size, and a permanent enlargement of the clitoris. In males, other undesirable reactions include temporary shrinkage of the testes (because the testes are no longer stimulated to produce their normal hormones), sterility, and adverse changes in blood cholesterol leading to atherosclerosis and an increased risk of heart disease. And because some tissues in the body convert testosterone into estrogen, there may be enlargement of the breasts and nipples in males. Some of the life-threatening effects, though rare, include liver disorders and cancer, especially liver cancer. Unfortunately, it is not possible to predict which individuals will be most affected or the severity of the reactions, especially since dosage and duration of use vary widely. Most athletes using these drugs over a long period of time employ massive doses and "stack" the drugs (use several different steroids in different combinations) to avoid building up a tolerance. Particularly dangerous is the use of these drugs by young athletes who are still growing, because the drugs can cause stunted growth and because of the emotional alterations (particularly aggressiveness).

Canadian sprinter Ben Johnson who was disqualified from the 1990 Olympic Games for using anabolic steroids.

With such a long list of possible health risks, some even life-threatening, one might ask why an athlete would take such a risk. One problem is that the side effects of steroid use, especially the life-threatening ones, may not show up for years. That, coupled with the natural feelings of invincibility of youth ("It won't happen to me"), makes it difficult to convince many athletes of the dangers. Some would tell you they do not care; if it helps them win a gold medal or make a professional football team, then it is worth the risk. Then too there is the perception that "everyone else is doing it," and that drug use is necessary to be competitive. And, the fact is, the rewards—both financial and emotional—of competitive sports are very high, and many individuals will do anything to become "the best." Because of the ethics of fair play and the risks to health, the sports-governing bodies have banned the use of all anabolic-androgenic steroids by competitors, and have instituted drug-testing programs to control their use. Still, there are ways to beat the system. The problem of drug use in sports is not new, and it is a difficult one to solve. But there have been important gains in the area of sports medicine in recent years, and most experts feel that properly designed, year-round training programs will yield more lasting results. Proper training, effective coaching, balanced nutrition, and old-fashioned hard work are the real answer.

The Genital System of the Human Female

The Ovaries Produce Egg Cells and Secrete Sex Hormones

The female gonads, the ***ovaries,*** are located in the lower part of the abdominal cavity, where they are held in place by large ligaments (Fig. 26.14A). At the time of birth, a girl's ovaries already contain a huge number of primordial egg cells, or ***primary oocytes;*** estimates range from 100,000 to 1,000,000. Over the approximately 35 years of her reproductive life, a woman ovulates about 13 times per year, producing one ovum each time; therefore usually fewer than 450 oocytes are stimulated to mature and leave the ovaries. The rest eventually degenerate, and ordinarily none can be found in the ovaries of women past the age of 50.

Each oocyte is enclosed within a cellular jacket called a ***follicle.*** The oocyte fills most of the space in the small, immature

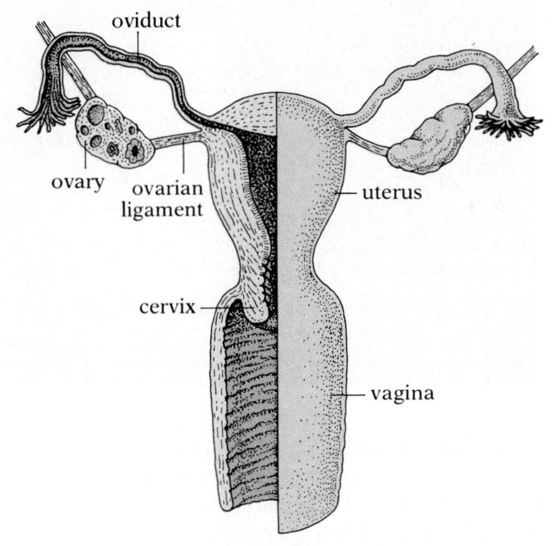

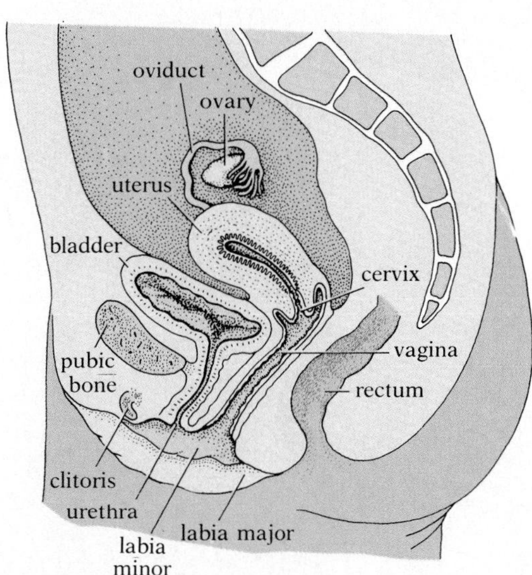

26.14 Reproductive tract of the human female (A) Frontal view. (B) Lateral view. The wall of one side has been dissected away to reveal the internal structure.

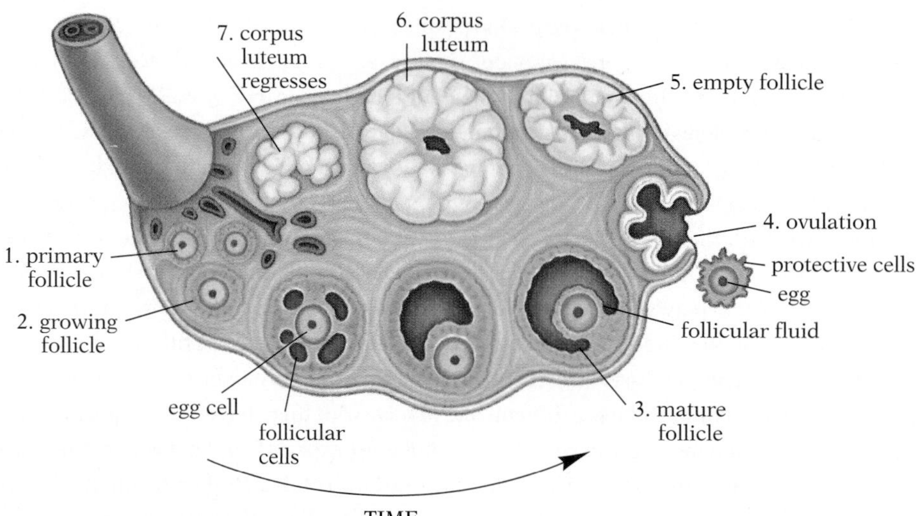

26.15 Schematic diagram of a human ovary showing various stages of egg development Each oocyte is enclosed within a cluster of cells that forms a follicle (1). The oocyte fills most of the space in the small, immature follicle. As the follicle matures, it grows bigger relative to the oocyte and develops a large fluid-filled cavity (2, 3). The oocyte, with its follicular cells, protrudes into the cavity. The outer wall ruptures and both the liquid and the detached oocyte with its surrounding cells are expelled during ovulation (4). The empty follicle is converted into a corpus luteum, which regresses at the end of the cycle (5–7).

follicle (Fig. 26.15). In the process of maturation, the follicle grows bigger relative to the oocyte and develops a large fluid-filled cavity. The oocyte, embedded in a mass of follicular epithelial cells, protrudes into the cavity (Fig. 26.16). A ripe follicle bulges from the surface of the ovary; when ***ovulation*** occurs its outer wall ruptures and both the fluid and the detached oocyte are expelled.

At the time of ovulation the oocyte bursts out of the follicle (Fig. 26.16) and is usually drawn into the large funnel-shaped end of one of the ***oviducts*** (uterine tubes), which partly surround the ovaries but are not continuous with them (Fig. 26.14A). Cilia lining the funnel of the oviduct produce currents that help move the oocyte into it. If sperm are present to meet the oocyte while it is still in the upper third of the oviduct, the

A

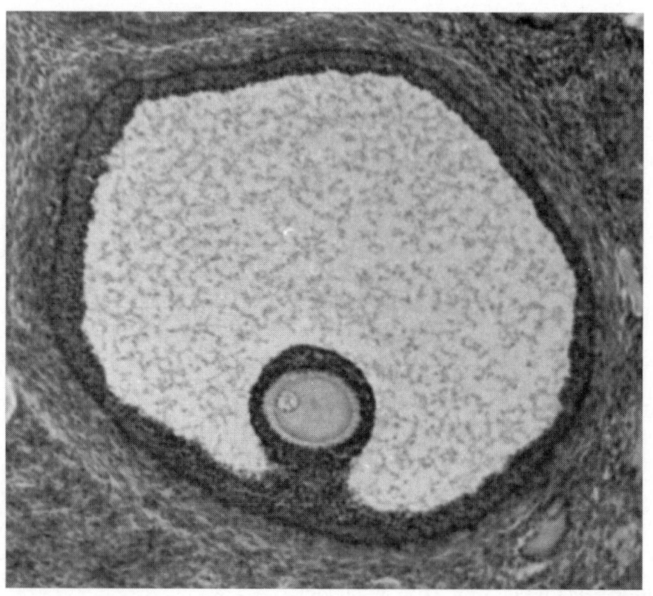

B

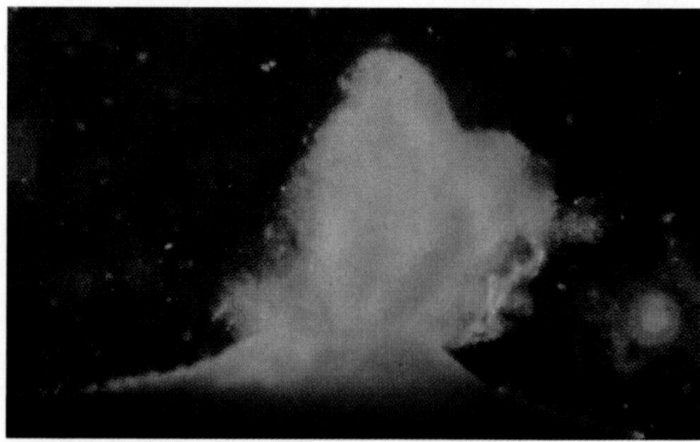

26.16 Photograph of a section of cat ovary (A) The mature follicle shown in this enlarged view has an egg cell embedded in a pedestal of epithelial cells that projects into the cavity. The egg cell itself is surrounded by a jacket of follicular cells. (B) An ovum after being expelled from the ovary.

penetration of a sperm through the membrane of the oocyte stimulates it to complete its maturation into a true egg cell (a fully mature ovum), and almost immediately thereafter the nuclei of the sperm and ovum fuse in the process of fertilization.

Each oviduct empties directly into the upper end of the ***uterus*** (womb). This organ, which is about the size of a fist, lies in the lower portion of the abdominal cavity just behind the bladder (Fig. 26.14B). It has very thick muscular walls and a mucous lining containing many blood vessels. If an egg is fertilized as it moves down the oviduct, it becomes implanted in the wall of the uterus, and there the embryo develops until the time of birth.

A method of permanent sterilization of females, sometimes used as a birth-control measure, is to cut and tie the oviducts (tubal ligation), so the sperm cannot reach the oocyte and no ovum can move down the oviduct into the uterus (Fig. 26.17). Like vasectomy of the male, this operation causes no change in hormone production.

Another method of birth control, which is not permanent, involves insertion of a plastic ring or spiral into the uterus (Fig. 26.18A). Such intrauterine devices (IUDs) seem to be very effective in preventing pregnancy, probably by preventing implantation in the uterus. However, these devices sometimes cause irritation and/or bleeding in the uterus; hence some women cannot tolerate them, and their safety for prolonged use is questionable.

At its lower end the uterus connects with a muscular tube, the ***vagina,*** which leads to the outside. The vagina acts as the receptacle for the male penis during copulation. The great elasticity of its walls makes possible not only the reception of the penis, but the passage of the baby during childbirth. (See Exploring Further: Sexually Transmitted Diseases, p. 544.)

The uterus and vagina do not lie in a straight line, as Figure 26.14A might seem to indicate. Instead, the uterus projects forward nearly at a right angle to the vagina, as shown in Figure 26.14B. The ***cervix,*** a fibrous ring of tissue at the mouth of the uterus, protrudes into the vagina. Devices that block the mouth of the uterus by covering the cervix are widely used in birth control. One such device, the diaphragm, is a shallow rubber cup with a spring around its rim. It is inserted into the vagina and positioned so that it covers the entire cervical region (Fig. 26.18B). It is very effective in preventing sperm from entering the uterus if it is used in conjunction with a spermicidal jelly or cream.

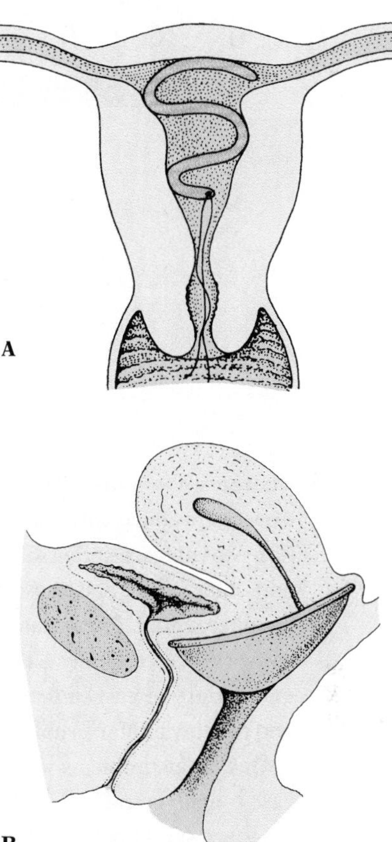

26.18 Two birth-control devices (A) An IUD in place in the uterus. The strings that run through the cervix permit the woman to make sure the IUD has not been expelled. (B) Diaphragm in position in the vagina. The device covers the mouth of the cervix. It is very effective in preventing sperm from entering the uterus when used with spermicidal jelly or cream.

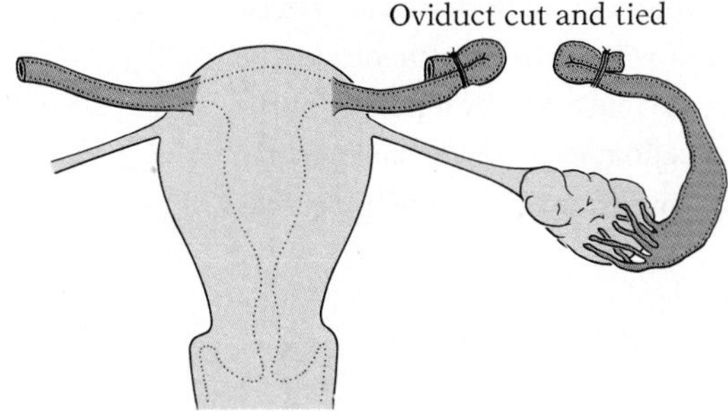

26.17 Tubal ligation This surgical procedure is performed to prevent movement of the egg into the uterus as a permanent method of birth control. Each of the oviducts is cut and tied, so that eggs cannot descend and fertilization cannot occur. The procedure causes no retrogression of sexual characteristics because there is no alteration of hormone levels.

The opening of the vagina in young human females is partly closed by a thin membrane called the ***hymen.*** Traditionally the hymen has been regarded as the symbol of virginity, to be destroyed the first time sexual intercourse takes place. Frequently, however, the membrane is ruptured during childhood, by disease or by a fall or as a result of strenuous physical exercise.

The external female genitalia are collectively termed the ***vulva*** (Fig. 26.19). The vulvar region is bounded by two folds of skin, the labia minor and the labia major, that enclose the space known as the vestibule. The vagina opens into the rear portion of the vestibule, and the urethra opens into the midpor-

EXPLORING FURTHER

SEXUALLY TRANSMITTED DISEASES (STDs)

There are now more than 50 diseases identified as being sexually transmitted. Some of these diseases produce only minor symptoms; others can be fatal. The diseases can be divided into two categories: those that are curable (e.g., syphilis, gonorrhea, trichomonas, vaginitis, and chlamydia), and those that are noncurable (AIDS, genital herpes). It should be remembered that, just because a disease is curable does not mean that it is not serious. The consequences of many STDs can be very serious indeed—infertility, sterility, pelvic inflammatory disease, inflammation of the prostate gland and epididymis, and neurological and cardiovascular problems are just some of the complications of untreated STDs.

One of the problems with STDs is that you can be infected and not know you have a disease—and you can pass it on to others during that period. Unfortunately, many of the diseases are symptomless—but contagious—in the early stages. For example, 80 percent of women with the bacterial infection gonorrhea are symptomless, and, although most men do have symptoms (painful urination or penile discharge), some do not. The first symptom of syphilis is a painless open sore, or chancre, which goes away of its own accord after about six weeks, often fooling the individual into thinking the condition is cured. And, if the chancre occurred on the cervix in a female, she may never even know she had it until other, more serious symptoms occur. The noncurable diseases show the same pattern; we have already said that it can be up to a year after infection before the test for the presence of the AIDS virus turns positive.

Because the consequences of STDs can be so very serious, it is imperative that one seek treatment whenever symptoms such as sores on the genitalia, a rash, discharge, or unusual odor are present. Unfortunately, the presence of any one STD increases the chances of others being present. Most of the diseases are treated with antibiotics, but both partners must be treated or reinfection occurs. In some cases (i.e., genital warts) the treatment may be very prolonged, unpleasant, and expensive. Prevention is obviously the best policy: abstinence, or having only one (healthy) partner is the best protection; each new sexual partner increases the risk of STD. The use of condoms is another important way to help prevent transmission.

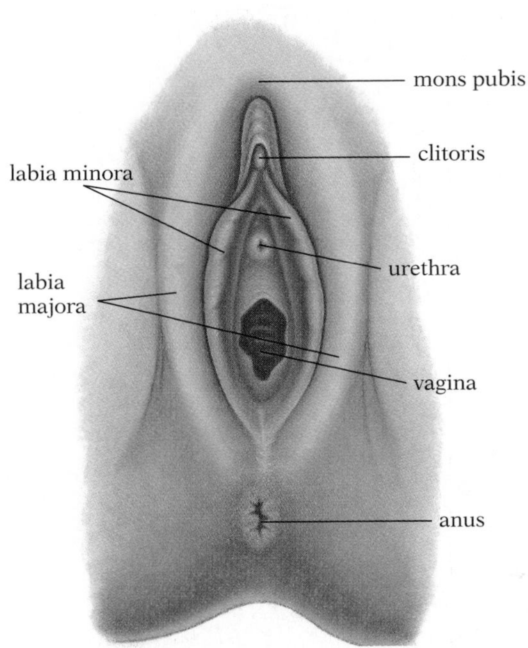

26.19 External female genitalia

tion of the vestibule. Note, then, that in the adult mammalian female there is no interconnection between the excretory and reproductive systems, and that the urethra carries only excretory materials.

In the anterior portion of the vestibule, in front of the opening of the urethra, is a small erectile organ, the ***clitoris,*** which forms from the same embryonic tissue that gives rise to the penis in the male. Like the penis, it becomes engorged with blood during sexual stimulation.

Hormonal Control of the Female Reproductive Cycle

The Female Sex Hormones Stimulate the Maturation of the Reproductive Organs and Female Sexual Characteristics

As in the male, puberty in the female is thought to begin when the hypothalamus becomes sensitive to the low levels of sex hormone in childhood and starts secreting more GnRH, which stimulates the anterior pituitary to release increased amounts of FSH and LH. These gonadotropic hormones cause maturation of the ovaries, which then begin secreting the female sex hormones, ***estrogen*** and ***progesterone.*** Estrogen stimulates maturation of the uterus and vagina and development of the female secondary sexual characteristics: broadening of the pelvis, development of the breasts, change in the distribution of body fat, and some change in voice quality. The changing hormonal balance also triggers the onset of ***menstrual cycles.*** We shall be particularly concerned here with the menstrual cycles as an example of the complex interplay between several hormones and between the endocrine and nervous systems.

The Menstrual Cycle Is Regulated by a Complex Negative-Feedback Mechanism Involving the Ovaries, Anterior Pituitary, and Hypothalamus

Rhythmic variations in the secretion of gonadotropic hormones in the females of most species of mammals lead to what are known as ***estrous cycles***—rhythmic variations in the condition of the reproductive tract and in the sex urge. The females of most species will accept the male in copulation only during those brief periods of the cycle near the time of ovulation when the uterine lining is thickest and the sex urge is at its height. During such periods the female is said to be "in heat," or in estrus. Many mammals have only one or a few estrous periods each year, but some, like rats, mice, and their relatives, may be in estrus as often as every five days. If fertilization does not occur, the thickened lining of the estrus is gradually reabsorbed by the female's body; ordinarily no bleeding is associated with this process.

The reproductive cycle in humans and other higher primates differs from that of other mammals in that there is no distinct heat period, with the female being receptive to the male throughout the cycle. And the thickened lining of the uterus is not completely reabsorbed if no fertilization occurs; instead, part of the lining is sloughed off during a period of bleeding known as ***menstruation.*** Human menstrual cycles average about 28 days; there are consequently about 13 of them each year. This is an extremely rough average; extensive variation occurs from person to person and from period to period in the same person.

During the Follicular Phase, the Follicles Begin to Grow and Secrete Estrogen

We now trace the sequence of events in the menstrual cycle, assuming a period of 28 days. It is customary in medical practice to consider the first day of menstruation as the first day of the cycle, as shown in Figure 26.20 and 26.21. From a biological point of view, however, it is more appropriate to regard the end of the period of bleeding as the beginning of the new cycle. At

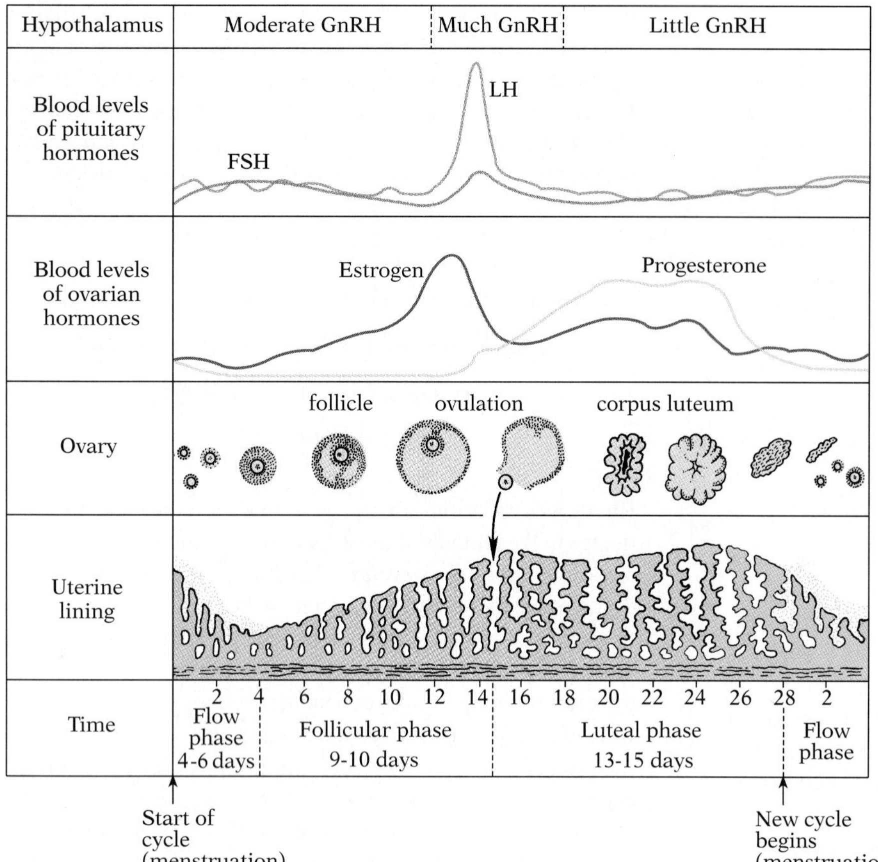

26.20 The sequence of events in the human menstrual cycle (A) The varying levels of FSH and LH during the cycle are shown. Note the surge of LH at the midpoint. (B) Blood levels of estrogen and progesterone vary: Estrogen levels are highest during the follicular phase; progesterone levels during the luteal phase. (C) The follicle develops under the influence of FSH. After ovulation. LH converts the follicle into a corpus luteum. (D) The uterine lining thickens under the influence of estrogen and progesterone. During flow, the thickened lining and associated tissue and blood are sloughed off.

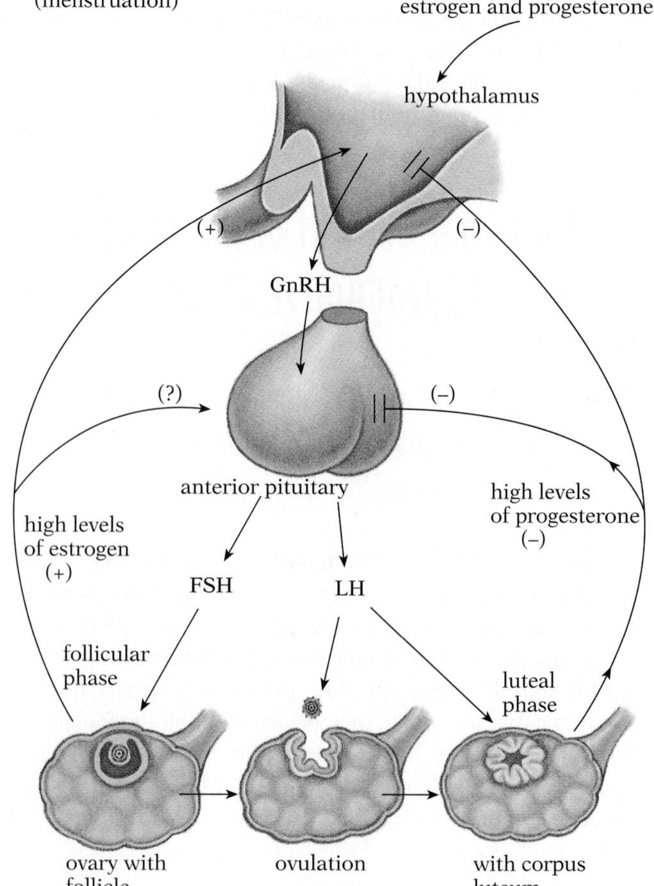

26.21 Hormonal control of the menstrual cycle When the levels of sex hormones are low during the flow phase, the hypothalamus is stimulated to secrete GnRH, which stimulates the anterior pituitary to begin secreting FSH (and some LH). Under the influence of FSH the follicle begins to grow and produce estrogen. The rise in estrogen toward the end of the follicular phase causes a surge of LH secretion, thereby triggering ovulation. Under the influence of LH, the empty follicle is converted into a corpus luteum, which secretes high levels of progesterone and estrogen. These have a negative feedback effect on the hypothalamus and anterior pituitary, and FSH and LH secretion is inhibited, the corpus luteum atrophies, the levels of the sex hormones drop, and menstrual flow occurs. In humans it is not yet certain whether the feedback effects of estrogen and progesterone are directed at the hypothalamus or anterior pituitary.

this point, the uterine lining is thin and there are no ripe follicles in the ovaries (Figs. 26.15 and 26.16). Under the influence of FSH (follicle-stimulating hormone) from the anterior pituitary, several follicles in the ovaries begin to grow and, influenced by the combined action of FSH and LH, they begin to secrete the first of the two female sex hormones, estrogen. One of the follicles soon becomes dominant. It continues to grow and secrete estrogen, while the others atrophy. The estrogen, in turn, stimulates the lining of the uterus to thicken. This ***follicular phase*** (growth phase) of the cycle lasts, on the average, about nine to 10 days after cessation of the previous menstrual flow.

As the follicle grows, it produces more and more estrogen (Figs. 26.20 and 26.21). In the early stages of follicular growth estrogen acts in a negative-feedback manner to inhibit FSH and LH secretion, but toward the end of the follicular phase the level of estrogen gets high enough that it apparently stimulates, by a *positive-feedback* mechanism, an abrupt surge of LH secretion from the pituitary. This ***LH surge*** triggers the release of the egg ***(ovulation)*** from the mature follicle, which by this time is large and bulging, looking much like a blister on the surface of the ovary (Figs. 26.15 and 26.16). Ovulation marks the end of the follicular, or growth, phase of the menstrual cycle.

During the Luteal Phase, the Corpus Luteum Secretes the Hormones Estrogen and Progesterone

Following ovulation, the old follicle is stimulated by LH (luteninzing hormone) to become a new structure, the ***corpus luteum*** (Latin for "yellow body") (Fig. 26.15). From its location on the ovary, the corpus luteum continues secreting some estrogen, but begins to secrete the second female sex hormone, the steroid progesterone (Fig. 26.11).

Progesterone Functions in Preparing the Uterus to Receive the Embryo

Acting on the uterine lining, which has already become much thicker under the stimulation of estrogen during the follicular phase, progesterone causes maturation of the complex system of mucus glands in the lining, and the glands begin secreting a clear fluid. The ***luteal phase*** of the menstrual cycle is, in fact, sometimes called the *secretory* phase. Repeated experiments have shown that implantation of a fertilized ovum in the uterus cannot occur in the absence of the changes in the uterine lining produced by progesterone. Progesterone truly is the hormone of pregnancy.

High Levels of Estrogen and Progesterone Inhibit the Development of New Follicles

The combined high levels of estrogen and progesterone act to inhibit FSH and LH secretion through some type of negative-feedback mechanism, the nature of which is not entirely clear. Estrogen and progesterone may act on the hypothalamus to suppress GnRH release, thereby limiting FSH and LH secretion by the pituitary (Figs. 26.20 and 26.21). But many researchers now feel that the feedback effect may be directly on the anterior pituitary itself. The result, in either case, is that FHS and LH secretion is inhibited; without these hormones, no new follicles can develop. If no fertilization occurs during a normal cycle, the corpus luteum begins to atrophy about 11 days after ovulation, and its secretion of estrogen and progesterone falls. When this happens, the thickened lining of the uterus can no longer be maintained. Part of the lining is reabsorbed but much of it is sloughed off during the ***flow phase*** (menstruation), which lasts about four to six days.

The fall of estrogen and progesterone levels, resulting from atrophy of the corpus luteum, frees the hypothalamus and/or the anterior pituitary from inhibition. FSH secretion by the anterior pituitary increases, and a new cycle begins. The sequence of events in a normal menstrual cycle is depicted diagrammatically in Figures 26.20 and 26.21.

From the above account it is apparent that the critical event in resetting the system is probably the fall in estrogen and progesterone levels as a result of atrophy of the corpus luteum. But what causes this atrophy? In cows a hormone (luteolysin) produced by the uterus causes the regression of the corpus luteum, but no such hormone has yet been isolated in humans. There is some evidence that prostaglandins may play a role in this process in humans. Not knowing what causes the atrophy of the corpus luteum means that a fundamental aspect of the timing of menstrual cycles remains in doubt.

Decreased Levels of Sex Hormones During Flow Can Cause Stress

During the flow phase, particularly in the first few days, secretion of progesterone and estrogen by the old corpus luteum is at a low level, and the follicles of the new cycle have not yet begun producing significant amounts of estrogen. Since a woman's body is accustomed from puberty to functioning in the presence of sex hormones, their withdrawal at the end of the luteal phase of each menstrual cycle is often accompanied by physiological and psychological disturbance, including irritability, depression, and sometimes nausea; abdominal cramps, caused by strong contractions of the uterus, are also common. This condition is commonly referred to as premenstrual syndrome (PMS).

Emotional stress sometimes also accompanies the ***menopause,*** a period lasting a year or two at the end of a woman's reproductive life. The menopause usually comes sometime between the ages of 40 and 50. It is apparently attributable mainly to declining sensitivity of the ovaries to the stimulatory activity of gonadotropins. The ovaries atrophy, the remaining follicles and oocytes disappear, and secretion of estrogen and progesterone falls to low levels. Consequently there is no cyclic thickening of the uterine lining, and hence no menstruation. The changing hormonal balance during menopause may cause physiological and psychological disturbances until a new physiological balance has been established. Often a menopausal woman is given prescribed doses of estrogen and progesterone to prevent some of the unpleasant side effects of this hormonal imbalance. The estrogen also appears to prevent much of the bone loss in the osteoporosis that occurs in many postmenopausal women.

In Humans the Timing of Ovulation Cannot Be Accurately Predicted

We have seen that ovulation in human beings occurs roughly midway in the menstrual cycle. This ovulation is spontaneous; it does not depend on a copulatory stimulus. In some mammals, such as rabbits and cats, however, ovulation is reflex-controlled; the nervous stimulus of copulation triggers the release of LH by the pituitary that will lead to ovulation. In such reflex ovulators it is possible to predict with great precision just when ovulation will occur; in the rabbit, for example, it occurs about 10 1/2 hours after copulation. Such precision is not possible with spontaneous ovulators like humans. Yet predictions of the time of ovulation are important in the practice of the rhythm method of birth control.

The rhythm method of contraception is based on the premise that fertilization can take place only during a very short period in each menstrual cycle. Copulation without risk of pregnancy should be possible during all other parts of the cycle. Present evidence indicates that human egg cells begin to deteriorate about 12 hours (maximum 24 hours) after ovulation and can no longer be fertilized after that time. In other words, fertilization can occur only if fertile sperm are in the upper third of the oviduct during the 12 to 24 hours immediately following ovulation; conception is not possible during the other 27 days of a 28-day cycle.

Immediately, the question of the fertile life of sperm in the female reproductive tract becomes important. We have said that one ejaculation releases about 200 million sperm cells into the vagina. Sperm counts of 20 million per milliliter are necessary for conception to occur, and the sperm must be healthy—that is, about 40 percent must be vigorous swimmers. Conditions in the vagina are very inhospitable to sperm, and vast numbers are killed before they have a chance to pass the cervix. Millions of others die or become infertile in the uterus and oviducts, and millions more go up the wrong oviduct or never find their way into an oviduct at all. Current evidence indicates that, in the female genital tract, human sperm cells remain fertile only about 48 hours or less after their release.

Since the fertile life of the human egg cell lasts at most one day and that of the human sperm at most two days, there is a period of no more than three days during which copulation can result in conception (the day when the egg is fertile and the two preceding days). But which three days? The inability to answer this question precisely makes the rhythm method of birth control unreliable. It is known that the three days come about midway through the menstrual cycle, but how long a given cycle will be cannot be predicted accurately. The cycles of many women are very irregular, varying by as many as eight to 15 days or more. Even women whose cycles are very regular will have some that vary by as many as four or five days during the course of a year. Sickness or emotional upset frequently delays ovulation and prolongs the cycle by altering the hypothalamic control of gonadotropin secretion. The least variable part of the cycle seems to be the luteal phase, which averages between 14 and 15 days. But the fact that the duration of the luteal phase is relatively stable is of little help in predicting the day of ovulation. Since there is no way of knowing beforehand precisely when the next menstrual flow will begin, it is not possible to count back 14 or 15 days to determine the ovulation date.

Recently it has been found that the character of the cervical mucus can be used to determine the timing of ovulation. Normally the cervical mucus is very thick and acts as a barrier to prevent the sperm from moving up into the uterus. But at the time of ovulation, the mucus changes (it becomes much thinner and breaks up into strands), and this feature can be used as an ovulation detector.

Birth-Control Pills Prevent Follicular Growth and Ovulation

Birth-control pills contain synthetic compounds similar to progesterone and estrogen. Taken regularly, they inhibit secretion of FSH and LH and thus prevent follicular growth and ovulation and, consequently, conception. The woman's cycle is like a false luteal phase, with high levels of estrogen and progesterone, and low levels of FSH and LH. The pill is taken for three weeks to prevent conception, and then stopped for one week so that a menstrual period occurs.

The "pill" is an extremely effective means of birth control, but it sometimes has undesirable side effects. These include migraine headaches, increased chance of blood clots, and a slightly increased incidence of cancer of the uterus (but lowered risk of breast cancer). However, the risk of death from use of birth-control pills is far less than the risk resulting from complications of pregnancy. Most of the risk appears to be

MAKING CONNECTIONS

1. Deer typically breed once a year, when the does come into estrus in the autumn. How is the timing of this cycle most likely determined? (See Chapter 25.)

2. As you learned in both Chapter 25 and this chapter, negative feedback is an important mechanism for the regulation of many hormones, including those controlling reproduction. What is meant by *positive feedback,* and what is its role in the menstrual cycle?

3. What brings about menopause in women? How can much of the bone loss (osteoporosis) observed in many post-menopausal women be prevented?

4. How would the immune system respond to a gonorrhea infection? (See Chapter 21.)

5. What are the disadvantages of the rhythm method of birth control?

confined to women who smoke and those over age 35; for this reason such women should not take the pill. Women considering the use of birth-control pills should become familiar with the possible harmful side effects, and should review their own medical histories for potential health risks. (See Exploring Further: RU 486, p. 550).

Hormonal Control of Pregnancy, Birth, and Lactation

Our discussion so far has assumed that conception did not occur and that each cycle was terminated by a menstrual flow. Let us now assume that the egg cell is fertilized sometime during the 12 hours after ovulation. Only one of the millions of sperm cells released into the vagina actually penetrates the egg cell and fertilizes it. As soon as the one cell has fertilized the egg, the outer membrane of the egg changes in consistency and becomes impenetrable to the other sperm cells, which soon die. The fertilized egg, or ***zygote,*** moves down the oviduct, probably carried by fluid whose movement is caused by contractions of the circular muscles in the oviduct walls (Fig. 26.22). During the days of transit, cell divisions begin and an embryo is formed. We shall look at human embryology in the next chapter.

The human embryo becomes implanted in the wall of the uterus eight to 10 days after fertilization. During the interval between fertilization and implantation, the embryo is nour-

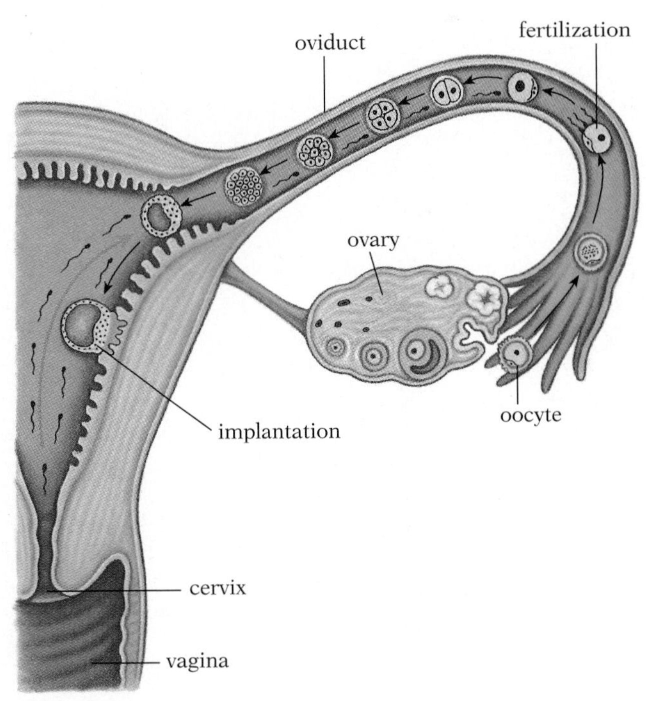

26.22 Fertilization of the egg and development of the human embryo Ovulation occurs about once a month in human females. Cilia lining the oviduct create a current that sweeps the egg into the oviduct, where fertilization occurs. The zygote begins to divide as it moves down the oviduct. It reaches the uterus and becomes implanted in the lining of the uterus about the sixth day after ovulation.

EXPLORING FURTHER

RU 486

RU 486, which has come to be known as the "French abortion pill," is a steroid (Fig. A) that has a high affinity for the progesterone receptor. It is the first available "antiprogesterone" and has been used successfully as a medical alternative for voluntary early pregnancy interruption. It is also effective in treating endometriosis, uterine fibroids, breast and ovarian cancer, and other hormone-related conditions. Recent research has provided some insight into the cellular and molecular mechanisms of action of both RU 486 and progesterone.

Progesterone is essential for maintenance of the uterine lining during implantation and pregnancy. RU 486, like progesterone and the other steroids (Fig. A), is lipid soluble and thus able to penetrate the plasma membrane and enter the cytoplasm of the target cell. Once inside the cell, both progesterone and RU 486 interact with specific progesterone receptors in the nucleus. The complex consists of two proteins, one that has a binding site for progesterone, and one referred to as the heat shock protein *(hsp)* (Fig. B). Both RU 486 and progesterone compete for the same binding site on the progesterone receptor, but RU 486 actually has a higher affinity for the receptor than progesterone. When progesterone is bound to the receptor, the progesterone-receptor complex is able to recognize and bind to a

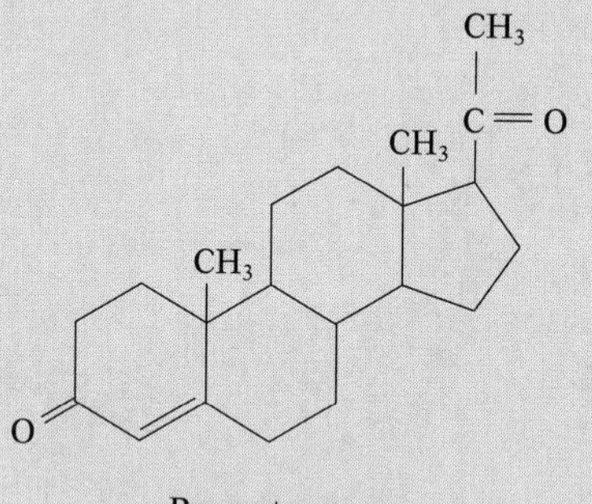

Progesterone

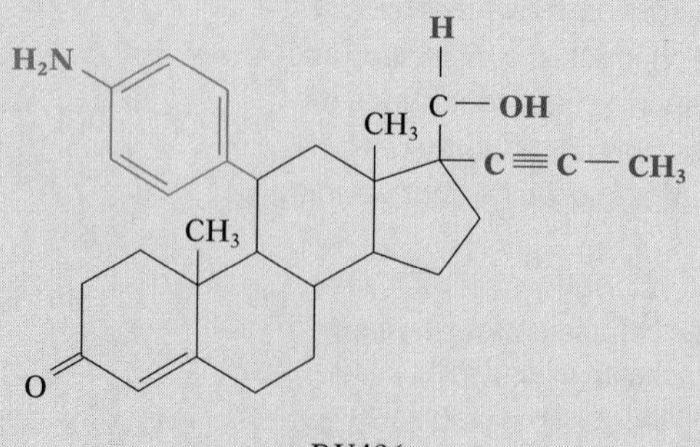

RU486

A A comparison of progesterone and RU 486

specific site on DNA in the nucleus, thereby facilitating transcription of specific genes (Fig. B) that code for proteins necessary for the pregnancy to continue. When RU 486 is bound, the normal response is not triggered. It is thought that hsp may not separate normally from the receptor-binding protein so the RU 486-receptor complex cannot interact properly with the DNA (Fig. B) or, if it does, the complex may be altered in some way as to interfere with transcription (Fig. B). Thus RU 486 has been termed an antiprogesterone, because it binds to and occupies the receptor without triggering a cellular response.

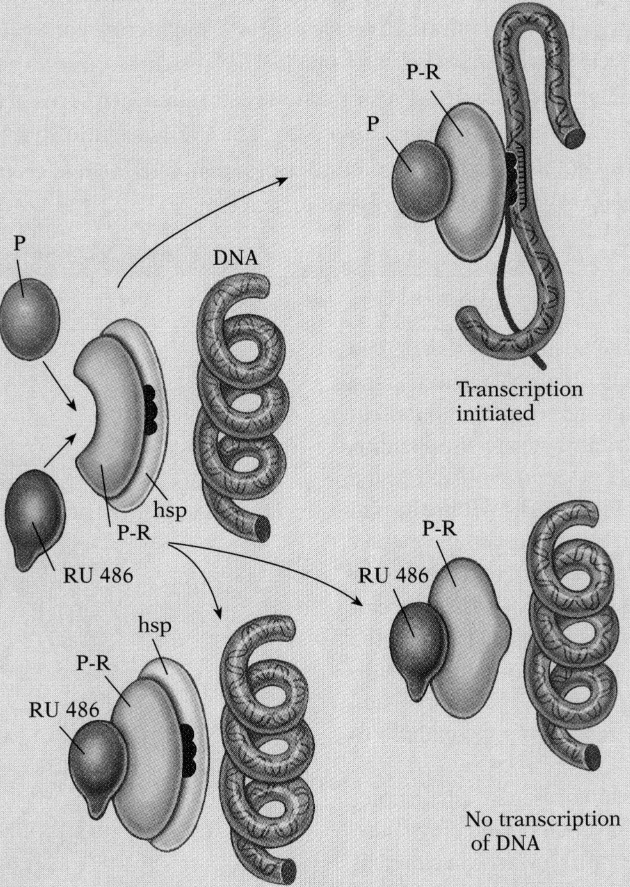

B Both progesterone (P) and RU 486 (RU) can bind to the progesterone-receptor proteins (P-R) in the nucleus of the cell. Attached to P-R is another protein (hsp). When progesterone binds to P-R (top, right), the hsp dissociates and the P-R complex binds to a specific site on the DNA, thereby facilitating transcription of DNA. However, when RU 486 binds, the RU 486-receptor complex either changes shape such that it cannot interact properly with the DNA to facilitate transcription (middle, right) or, it may bind even more strongly to the hsp and there is no interaction with the DNA (bottom, right). Thus the RU 486 binds to and occupies the binding site of the receptor without triggering a hormonal response.

It was hypothesized that a short interruption in the supply of progesterone to the pregnant human uterus would cause irreversible damage to the lining, leading to a sloughing off of the uterine lining and termination of the pregnancy. Clinical trials and experience for several years in different countries have confirmed this hypothesis. In addition, the administration of RU 486 has been found to produce in an increase in the secretion of

prostaglandins, hormonelike substances that are produced by a variety of animal tissues, including the uterine lining. Prostaglandins in the uterus promote a dilation and softening of the cervix and stimulate uterine contractions.

RU 486 has been used successfully for pregnancy termination by more than 100,000 women in France, and is being used in Great Britain and Sweden. The standard treatment, RU 486 pills followed by prostaglandin injection, is 95.5 percent effective. A new "pill only" treatment involving RU 486 pills followed by prostaglandin pills is even more effective. The new RU 486-prostaglandin treatment could be particularly useful in underdeveloped countries where surgical facilities are limited; most women would be able to avoid instrumental intervention, with its risk of infection, cervical injury, and uterine perforation. In addition, this treatment will provide privacy to all women. The Supreme Court has ruled that RU 486 can be brought into the United States for personal use. It has been licensed to the non-profit U.S. Population Council and testing is underway. The Clinton administration is supporting the development of this drug for pregnancy termination, treatment of certain cancers, and many other medical uses related to progesterone action.

ished by its limited supply of yolk and by materials secreted by the glands of the female genital tract. After implantation in the uterine lining, the embryonic membranes form the ***umbilical cord,*** through which blood vessels contributed by the allantois run to a large structure, the ***placenta,*** a separate organ formed from both maternal and fetal cells (Fig. 26.23). Within the placenta the blood vessels of the embryo and those of the mother lie very close together, but they are not joined and there is no mixing of maternal and fetal blood (Fig. 26.24). Exchange of materials takes place in the placenta by diffusion between the blood of the mother and that of the embryo; nutritive substances and oxygen move from the mother to the embryo, and urinary wastes and carbon dioxide move from the embryo to the mother.

We saw earlier that progesterone is essential for maintenance of the uterine lining during implantation and pregnancy. But we saw also that in a normal menstrual cycle, when there is no fertilization, the corpus luteum soon begins to atrophy and cuts off the supply of progesterone, with the result that menstruation occurs. Clearly, this sequence of events cannot be allowed to take place after conception, or the uterine lining with the implanted embryo would be sloughed off and lost. It can be shown that once conception occurs, the corpus luteum does not atrophy, but lasts through most of the term of pregnancy. How is this possible? Apparently the chorionic portion of the placenta soon begins secreting a gonadotropic hormone (human chorionic gonadotropin, or ***hCG***) that is similar to LH. This hormone stimulates the corpus luteum, which continues to secrete progesterone and thus sustains the pregnancy. After about three months, the placenta itself begins secreting increasing amounts of progesterone and a small amount of estrogen.

So much hCG is produced in a pregnant woman that much

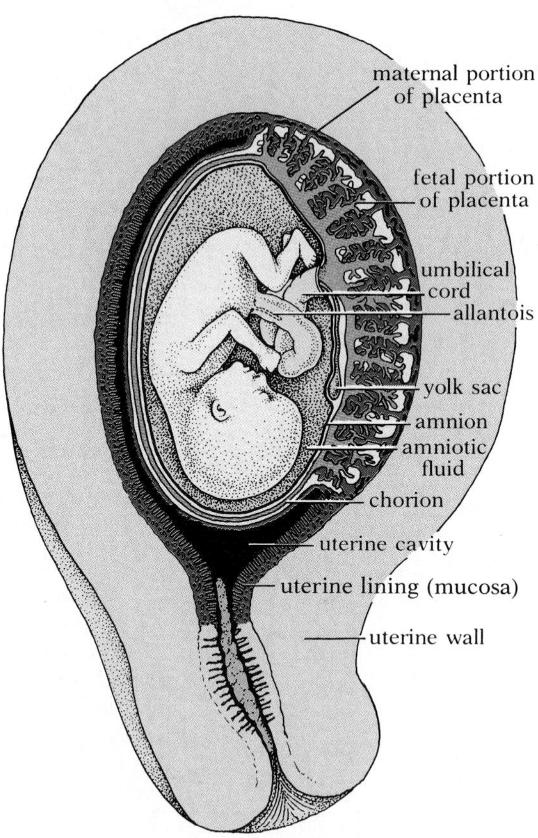

26.23 Uterus of a pregnant woman The placenta consists of both maternal and fetal components, the latter derived mainly from the chorion. Maternal and fetal capillaries lie side by side, allowing the diffusion of oxygen nutrients into the fetal circulation. Nutrients and oxygen absorbed by the fetal capillaries are carried by the umbilical cord to the embryo, which lies in the amniotic sac, bathed by amniotic fluid.

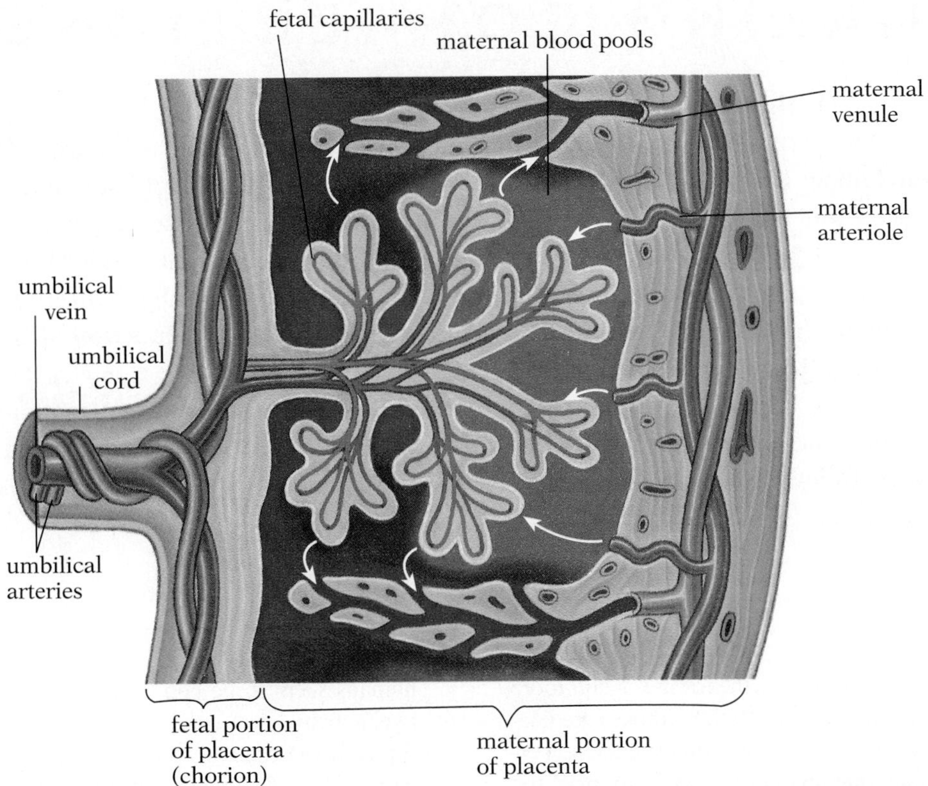

26.24 Diagram of placental circulation Since the fetal lungs are breathing fluid rather than air, gas, nutrient, and waste exchanges between the mother and the fetus take place in the placenta, where fetal blood passes through capillaries alongside those containing maternal blood. The transfer takes place in thousands of blood pools, some of which are shown above. The umbilical artery brings part of the fetal blood supply to the fetal capillaries in the placenta where carbon dioxide and other wastes diffuse into the material capillaries, and oxygen and nutrients move into the fetal blood. The blood then returns to the fetus through the umbilical veins.

of it is excreted in the urine. Many commonly used tests for pregnancy, among them some available over the counter in drugstores, are based on this phenomenon. In these tests a sample of urine is added to a solution containing an hCG antibody (a protein specific to hCG). If the urine contains hCG, a faint color will develop. Correctly used, these tests can yield reliable results in less than an hour.

Both Prostaglandins and Oxytocin Stimulate Uterine Contractions During Birth

Much research is still needed to clarify the complex interactions of hormones that control the birth process. The best evidence to date indicates that the fetus itself determines the time of birth through the production of certain hormones. Experiments with monkeys and sheep have shown conclusively that towards the end of pregnancy, the fetal brain signals the hypothalamus to begin secreting releasing hormones (i.e., CRH) that stimulate the anterior pituitary to produce ACTH. ACTH, in turn, stimulates the fetal adrenal gland to grow and begin producing ever increasing amounts of cortical steroid hormones. The cortical hormones go through the fetal circulatory system to the placenta where they play a crucial role; they cause more estrogen and less progesterone to be produced by the cells of the placenta. This rise in estrogen (which stimulates uterine muscle contractility) and decrease in progesterone (which quiets the uterine muscle) is the initiating event for the birth process in the mother.

In addition to exciting uterine muscle, estrogen also stimulates the cells lining the uterus to increase production of ***prostaglandins,*** which act directly on the uterine muscle to stimulate contractions. The change in levels of circulating hormones causes more and more prostaglandins to be produced, which causes (1) the cervix to soften and dilate, and (2) the uterine muscle to begin the strong, coordinated contractions of labor.

Once the birth process has begun, the mother begins releasing the hormone ***oxytocin*** from the posterior pituitary. Oxytocin has two roles: (1) it stimulates the uterine muscles to

MAKING CONNECTIONS

1. How do birth-control pills prevent pregnancy?

2. Based on what you learned about the immune system in Chapter 21 and about Rh disease in Chapter 14, why is it essential that no mixing of maternal and fetal blood occur through the placenta during fetal development?

3. In Chapter 22 you learned that the kidney helps to maintain homeostasis by allowing substances that are found in abnormally high concentrations in the blood to "spill over" into the urine. How is this function of the kidney utilized in pregnancy tests?

4. Explain why nursing a baby helps to stimulate uterine contractions in the mother and thus reduce her risk of hemorrhage in the period immediately following birth.

contract strongly, and (2) it stimulates the release of still more prostaglandins from the uterine cells. Both hormones are necessary for labor in humans; together they stimulate the powerful contractions of the uterine muscles necessary to expel the baby from the uterus.

Lactation Involves the Complex Interaction of Many Hormones and Neural Mechanisms

Like the hormonal control of the birth process, that of milk secretion is complicated and not completely understood. Growth and development of the mammary glands during pregnancy in humans seem to be controlled by a complex interaction between many hormones, especially estrogen and progesterone. Prolactin, a hormone secreted by the anterior pituitary, stimulates milk production after birth. Lactation is induced when the high levels of sex hormones, which inhibit prolactin secretion, disappear at the time of birth.

The actual release of milk from the mammary glands involves both neural and hormonal mechanisms. Suckling, or, in the case of conditioned cows, seeing the calf or hearing rattling milk pails, causes nervous stimulation of that part of the hypothalamus that instigates release of oxytocin stored in the posterior pituitary. The oxytocin, in turn, induces constriction of the many tiny chambers in which the milk is stored in the mammary glands, forcing the milk into ducts that lead to the nipple.

CONCEPTS IN BRIEF

1. Sexual reproduction in higher animals involves bringing together two *gametes,* an egg and a sperm, which is referred to as fertilization and forms the first cell of the new individual.

2. Most aquatic organisms use external fertilization; the gametes are shed directly into water, and the sperm must swim to the egg.

3. Most land animals use internal fertilization, in which the egg cells are fertilized within the female reproductive tract by sperm inserted by the male. The fertilized egg is then either surrounded by protective membranes and shell and released, or held within the female's body until embryonic development is completed.

4. In the human male each of the *testes* has two functional components: the *seminiferous tubules,* in which the sperm are produced, and the *Leydig cells,* which secrete

male sex hormone. Mature sperm are activated in the *epididymis.* The sperm then move into the *vas deferens,* which conducts them to an area near the prostate gland. During orgasm the sperm move into the *urethra,* which passes through the *penis* and empties to the outside. Seminal fluid from the *seminal vesicles,* and *prostate,* and the *Cowper's glands* is added to form the *semen.*

5. During embryonic development the testes begin secreting small amounts of the male sex hormone, *testosterone,* which is crucial to the differentiation of male genitalia. At puberty, the hypothalamus sends more releasing hormone (GnRH) to the anterior pituitary, stimulating it to release *FSH* and *LH.* LH induces the testes to produce more testosterone; FSH induces the maturation of the seminiferous tubules and causes sperm production to begin.

6. Testosterone stimulates the development of the secondary sexual characteristics. Testosterone secretion is regulated by a negative-feedback loop involving the hypothalamus and anterior pituitary. The hormone inhibin inhibits FSH production and thereby limits spermatogenesis.

7. In the human female the *ovaries* produce the egg cells *(oocytes)* and secrete sex hormones. Each oocyte is enclosed within a small *follicle.* When ovulation occurs, the follicle wall ruptures and the oocyte and fluid are expelled. If sperm are present, fertilization occurs in the oviduct.

8. Each oviduct empties into the muscular *uterus.* If the egg is fertilized it becomes implanted in the uterine wall, where the embryo develops. At its lower end the uterus connects with the tubular *vagina,* which leads to the outside. The vagina is the receptacle for the penis during intercourse.

9. At puberty in the female, the hypothalamus sends GnRH to the anterior pituitary, stimulating it to release FSH and LH. These hormones cause maturation of the ovaries, which begin secreting the female sex hormones, *estrogen* and *progesterone.* Estrogen stimulates maturation of the reproductive structures and development of the secondary sexual characteristics. The changing hormonal balance triggers the onset of the menstrual cycles.

10. At the beginning of the menstrual cycle the uterine lining is thin and the level of sex hormones low. The hypothalamus responds to the low level of sex hormones by releasing GnRH, which stimulates the pituitary to increase FSH secretion. FSH stimulates the maturation of the follicles, which begin to secrete estrogen. The estrogen stimulates the uterine lining to thicken. The high level of estrogen apparently positively stimulates a surge of LH from the pituitary, which triggers ovulation.

11. The LH converts the empty follicle into the *corpus luteum,* which continues to secrete estrogen and begins to secrete progesterone. Progesterone prepares the uterus to receive the embryo.

12. The high level of sex hormones suppresses the growth of new follicles. If no fertilization occurs, the corpus luteum atrophies and progesterone secretion falls. When this happens, the thickened uterine lining can no longer be maintained, and *menstruation* occurs. The hypothalamus again responds to the low level of sex hormones and another cycle begins.

13. If fertilization occurs, the *zygote* moves down the oviduct and becomes implanted in the uterine wall eight to 10 days after fertilization. The embryonic membranes then develop and the *placenta* is formed from both maternal and fetal cells. The placenta soon begins to secrete hCG, which maintains the corpus luteum and its secretion of progesterone, thus sustaining the pregnancy. Later the placenta secretes estrogen and progesterone directly.

14. In late pregnancy, hormones from the fetus cause an increase in estrogen secretion by the placenta. Both oxytocin (secreted by the posterior pituitary) and prostaglandins (secreted by the uterus) are necessary to stimulate uterine contractions in the birth process.

15. The development of the breasts is regulated primarily by estrogen and progesterone. Initiation and maintenance of lactation by mature mammary glands after birth seems to be controlled primarily by prolactin.

STUDY QUESTIONS

1. Distinguish between external and internal fertilization in animals. Compare them with regard to number of gametes and zygotes produced. What is the typical habitat of animals using external fertilization? Internal fertilization? (p. 532)
2. Trace the pathways of human sperm and egg cells from their sites of production to the site of fertilization and explain the function of all structures through which they would pass. (pp. 535–536, 541–543)
3. What are the sites of production and the effects on

male sexual development of GnRH, LH, FSH, and testosterone? Explain how the production of these hormones is regulated by a negative-feedback loop. What is the role of inhibin? (pp. 537–538)

4. **How do GnRH, LH, FSH, estrogen, and progesterone collaborate to bring about sexual development in the female? Which hormone causes the development of female secondary sexual characteristics? (p. 547)**
5. **Compare the human menstrual cycle with the estrous cycle characteristic of most other mammals. (p. 545)**
6. **What are the three phases of the menstrual cycle? How long does each phase last on the average? When does ovulation occur? (p. 546)**
7. **Why does the uterine lining increase in thickness during the follicular phase of the menstrual cycle, and what triggers ovulation? Describe the roles of GnRH, FSH, LH, and estrogen. (p. 547)**
8. **Which hormone converts the empty follicle into the corpus luteum? What is the effect on the uterine lining of progesterone produced by corpus luteum? Why is the uterine lining eventually shed in menstruation? (p. 547)**
9. **If an egg is fertilized, where and when will it implant? What is the function of the placenta? What tells the corpus luteum to continue producing progesterone and maintain the pregnancy? (pp. 549, 552)**
10. **What hormonal changes are believed to trigger uterine contractions at the time of birth? What would be the effect of oxytocin if administered during labor? (pp. 553–554)**
11. **Describe the hormonal control of lactation, including development of the mammary glands during pregnancy, milk production, and milk ejection. (p. 554)**

CHAPTER 27

DEVELOPMENT OF MULTICELLULAR ANIMALS

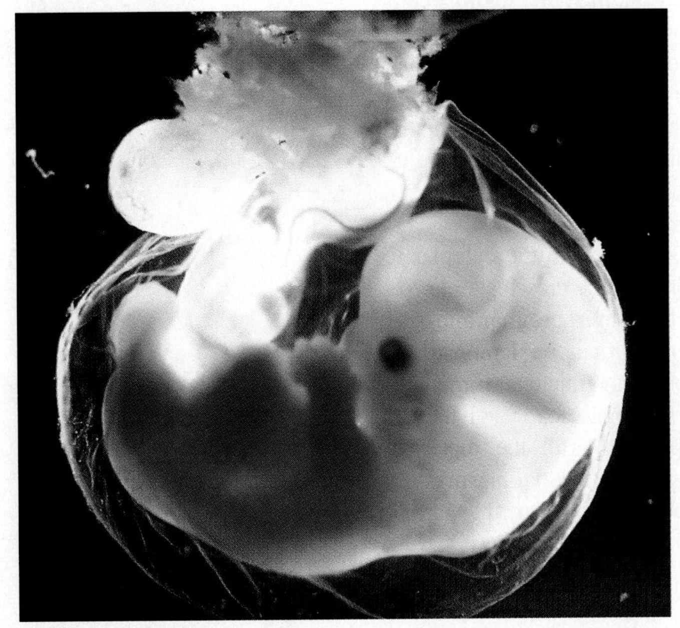

To biologists and nonbiologists alike, probably no aspect of biology is more amazing than the development of a complete multicellular organism from a single cell, a cell created in most species by the fusion of two meiotically produced gametes. The process of development is so precisely controlled that the entire, intricate organization of cells, tissues, organs, and organ systems that will characterize the functioning adult comes into being with rarely a flaw. We have previously examined the genetic information that controls development and programs a mouse zygote to become a mouse, an oak zygote an oak, and an earthworm zygote an earthworm. We have also discussed possible cellular control mechanisms in development—how an individual gene may be turned on or off in its course. And in the last chapter we learned about the process of sexual reproduction in human beings. We now briefly examine a few representative patterns in animal development, relating them to the control mechanisms we considered earlier, but making no attempt to discuss them in great detail or even to mention all the important events they entail. Our purpose is simply to survey the kinds of events that any model of developmental control must seek to explain.

We considered the process of plant development in Chapter 24, and will concentrate in this chapter on patterns and mechanisms of animal development. There are a number of reasons for considering plant and animal development separately, most of which concern the unique living requirements each has. The vast majority of plants are autotrophic, and are adapted to remain rooted to a specific location and to collect sunlight. Since plants are not adapted for locomotion, they can enjoy the structural advantages of more rigid cell walls, a specialization that makes impossible a major defining feature of animal development—the migration of cells during the development of the embryo.

Furthermore, because few plants feed on other organisms, they do not need nervous systems and muscles to guide and power the capture or harvesting of food; nor do they need a mouth, stomach, or digestive system, or the more than 300 specialized cell types that make up the kidneys, bladder, rectum, and other specialized organs and tissues of animals. Plants

Chapter Opening Photo: Five to six week old human embryo within the amnion. The embryo has arm and leg buds, and is roughly 8 to 10 mm long. The hands and feet are mittenlike, with no separation between the individual fingers and toes.

compete with each other by growing taller, or broader, and by producing new organs like leaves and roots whenever they are needed throughout their life cycle. As we shall see, animals usually generate their entire array of tissues and organs early in development, and devote their later energies to the maintenance and growth of these organs and tissues.

The Process of Development Begins With the Maturation of the Egg

We saw in Chapter 9 that certain cells are set aside early as egg primordia in the ovary of a female animal. When these cells undergo meiosis, the divisions are unequal and almost all the cytoplasm is retained in the ripe ovum, the other haploid cells being the tiny polar bodies that soon deteriorate (see Fig. 9.28, p. 198). The sperm cell, on the other hand, is unusually small and has very little cytoplasm. The ovum thus furnishes most of the initial cytoplasm and organelles for the embryo. During the maturation of the egg, large numbers of mitochondria and ribosomes are produced along with stockpiles of proteins, rRNA, tRNA, and special, long-lasting mRNAs. Most eggs also have a supply of stored food or ***yolk*** in the cytoplasm, which provides energy and nutrients for the growing embryo.

The importance of the egg cytoplasm in programming the early stages of embryonic development can be demonstrated using a frog egg: even after removing the nucleus from a mature ovum, it is possible to induce the egg to go through the initial stages of development. Apparently, all the materials needed to begin development are stored in the cytoplasm of the egg. No wonder the egg has sometimes been termed "nature's masterpiece."

The Fusion of a Sperm Cell and an Egg Induces Embryonic Development

It is the fusion of a sperm cell with an egg cell that stimulates the egg to begin development into an embryo. Note that the triggering depends on the *fusion of the cell membranes,* not the fusion of the sperm nucleus with the egg nucleus, even though the latter fusion is the actual event of fertilization.

In mammalian species (including humans), the sperm must undergo a process known as ***capacitation*** to be able to penetrate the egg. Secretions from the female reproductive tract strip a layer of proteins off the heads of the sperm, which prepares the sperm to release enzymes that will enable one of them to fertilize the egg. The egg is initially enclosed in a thin protective layer of follicle cells. This layer, a barrier to the sperm, is loosened by an enzyme secreted by the capacitated sperm. But even after the follicular barrier has been loosened, the sperm cells encounter yet another barrier, a jelly coat that surrounds the egg. At this point the acrosome of the sperm comes into play. The acrosome is a membrane-surrounded vesicle in the head of the sperm cells (see Figs. 26.9 and 27.1) containing enzymes that help the sperm penetrate the jelly coat. When a

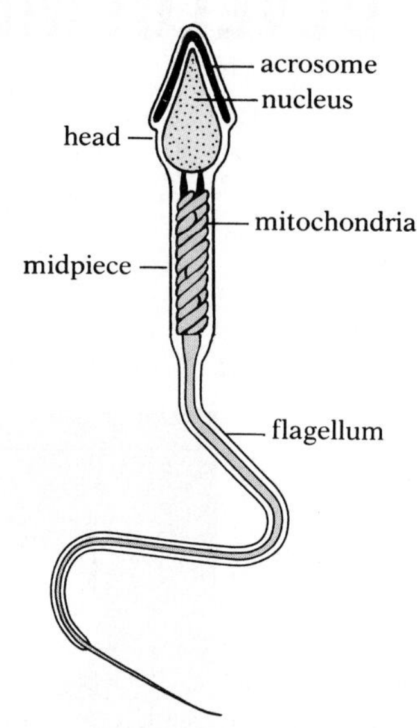

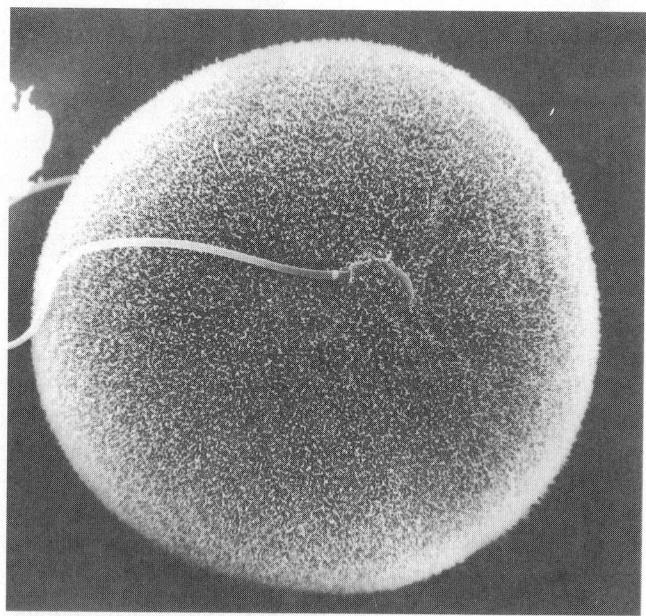

27.1 Human sperm (Top) Three sections are easily distinguishable; the head, containing the nucleus; the midpiece, tightly packed with mitochondria, which produce the ATP that fuels the sperm's swimming; and the long flagellum. In the apex of the head, immediately in front of the nucleus, is the acrosome, a membrane-surrounded vesicle that contains enzymes necessary for fertilization. There is only a tiny amount of cytoplasm in the cell. The photograph (Bottom) shows a human sperm penetrating the egg's membrane.

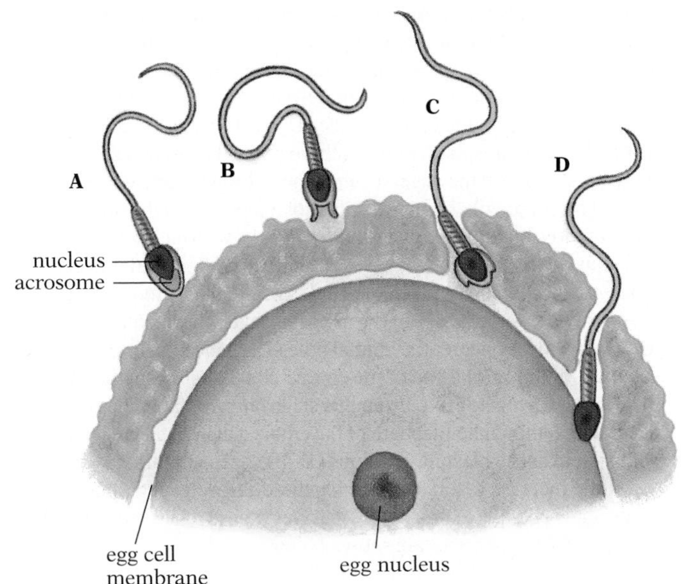

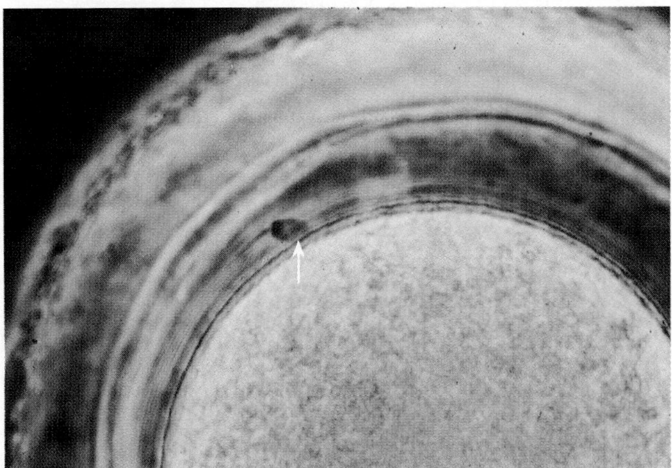

27.2 The fertilization process in mammals Left: (A) A sperm cell comes into contact with the jelly coat (zona pellucida) surrounding an egg cell. (B) The membrane of the acrosome ruptures, releasing the acrosomal contents, which include enzymes that act on the jelly coat and on the membrane of the egg cell. (C, D) The plasma membrane of the sperm fuses with that of the egg. Notice that it is the side of the sperm that makes contact with the egg membrane. In mammalian sperm, fusion typically occurs at the side; in many other organisms, it occurs at the sperm's tip. The sperm pronucleus then moves into the egg cell. Right: Human ovum being fertilized. The sperm can be seen at the egg cell membrane (white arrow).

sperm cell binds to the jelly coat, it releases these acrosomal digestive enzymes, which dissolve a region of the jelly coat and enable the sperm to bore through. The membranes of the sperm and egg fuse and the sperm nucleus, now called a ***pronucleus,*** moves into the cytoplasm of the egg (Fig. 27.2).

After Fusion, a Fertilization Membrane Forms and the Egg Is Activated

When the membranes of a sperm cell and egg fuse and the sperm pronucleus moves into the egg, an influx of calcium ions triggers vesicles in the outer region of the egg cytoplasm to discharge their contents into the region around the cell, forming a ***fertilization membrane.*** The fertilization membrane and the plasma membrane detach other sperm from the surface and act as a barrier against entry of additional sperm cells. Sperm penetration also activates the development of the egg, bringing about intense biochemical activity, such as increased metabolism and altered membrane permeability.

Important as they are, the events discussed so far do not constitute fertilization in the genetic sense; true fertilization is the union of the two gamete nuclei. This union depends on some attraction of the sperm nucleus to the egg nucleus, the nature of which is still unknown.

Medical technology has made it possible to fertilize human eggs ***in vitro.*** In this procedure, a woman is treated with gonadotropic hormones to induce the maturation of several follicles, and the mature eggs are surgically removed from the ovary. The eggs are placed in suitable culture media and fertilized with the father's sperm. The egg or eggs that begin development are then transplanted into the woman's uterus where, hopefully, normal development will take place. Many women whose oviducts are blocked or who are otherwise infertile have been able to have normal children using this procedure.

During Cleavage the Cytoplasm of the Zygote Is Rapidly Partitioned Into Many Smaller Cells

In normal development the zygote begins a rapid series of mitotic divisions immediately after fertilization has taken place. These early cleavages are usually not accompanied by cytoplasmic growth. They produce a grapelike cluster of cells called a ***morula,*** which is little if any larger than the single egg cell from which it derives (Fig. 27.3). The cytoplasm of the one large cell is simply partitioned into many new cells that are much smaller.

In many species, as cleavage continues, the newly formed cells begin to secrete a fluid into the center of the mass of cells. As a result, these cells, referred to as blastomeres, come to be arranged in a sphere surrounding a fluid-filled cavity called a ***blastocoel*** (Fig. 27.3). The embryo at this stage is termed a

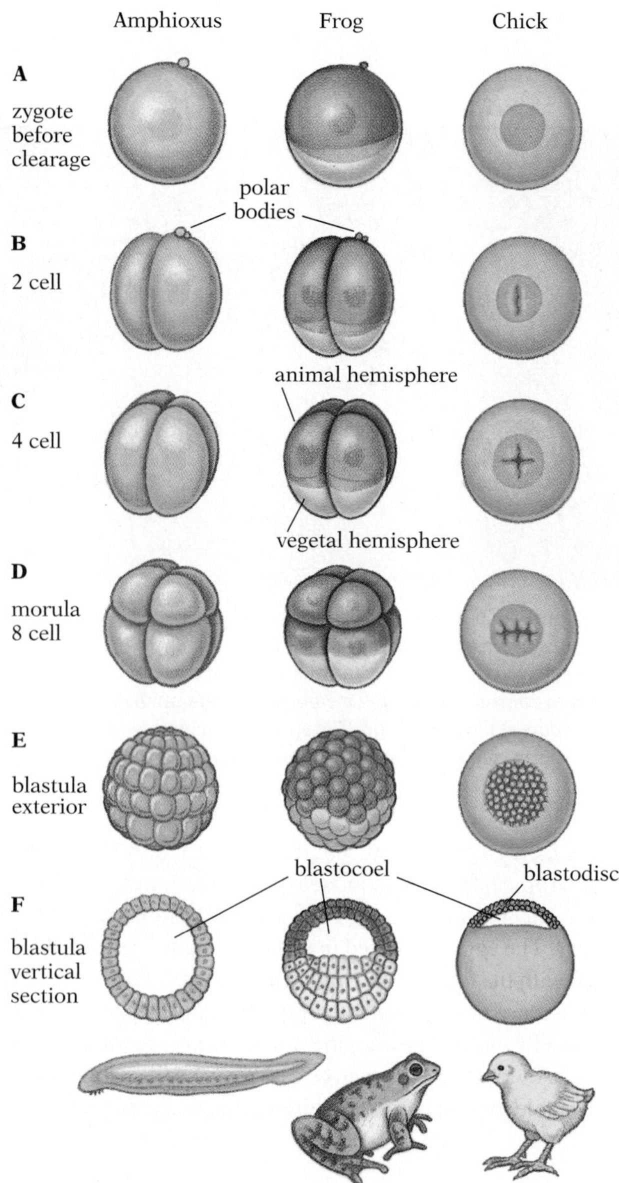

27.3 Cleavage in amphioxus, frog, and bird The pattern of cleavage is greatly influenced by the amount of yolk in the egg. In amphioxus, whose eggs have little yolk, cleavage results in the formation of a blastula in which the cells at one side are only slightly larger than those on the opposite side (D–F). The smaller cells make up the animal hemisphere of the embryo while the larger cells with more yolk make up the vegetal hemisphere. Frogs' eggs, which contain far more yolk than those of amphioxus but much less than those of most birds, serve as examples of eggs with an intermediate yolk mass. The first two cleavages, which are perpendicular to each other, cut through both the animal and vegetal poles, producing cells of roughly the same size. But the next cleavage is horizontal and located decidedly nearer the animal pole (D); hence the four cells produced at the animal end of the egg are considerably smaller than the four at the vegetal end. From this stage onward, more cleavages occur in the animal hemisphere of the embryo than in the vegetal hemisphere as the blastula develops (D–F). As in amphioxus, there is very little increase in total size during these early cleavage stages. Birds' eggs contain so much yolk that cleavage is restricted to the blastodisk. Mature forms are shown below F.

blastula. Note that the blastula is approximately the same size as the zygote, and that there is no opening into the blastocoel. The formation of the blastula marks the end of cleavage.

The Cleavage Pattern Is Influenced by the Amount of Yolk in the Egg

The pattern of cleavage is greatly influenced by the structure of the egg and the amount of yolk (stored food) in the egg. The yolk is composed of lipid and protein and is usually found in the lower region of the egg. When present in large amounts it physically interferes with cell division. The amount of yolk and its distribution result in major differences in developmental patterns in vertebrates.

In amphioxus, a tiny marine chordate, the egg has relatively little yolk and cleavage results in the formation of a blastula in which the cells at one side are only slightly larger than those on the opposite side (Fig. 27.3D–F). The smaller cells make up the ***animal hemisphere*** of the embryo while the larger cells with more yolk make up the ***vegetal hemisphere.*** The differences in cell size are not very great in amphioxus embryos; they are more pronounced in many other animals. Many eggs have far more yolk in their vegetal hemisphere, and this deposit of stored food imposes complications and limitations on the process of cleavage. Generally, the more yolk an egg contains, the more cleavage tends to be restricted to the animal hemisphere.

Frogs' eggs, which contain far more yolk than those of amphioxus but much less than those of most birds, may serve as an example of eggs with an intermediate yolk mass. The first two cleavages, which are perpendicular to each other, cut through both the animal and vegetal poles, producing cells of roughly the same size (Fig. 27.3A–C)—rather like cutting an apple into quarters. But the next cleavage is horizontal and located decidedly nearer the animal pole (Fig. 27.3D); hence the four cells produced at the animal end of the egg are considerably smaller than the four at the vegetal end. From this stage onward, more cleavages occur in the animal hemisphere of the embryo than in the vegetal hemisphere. As in amphioxus, there is very little increase in total size during these early cleavage stages (Fig. 27.4).

Reptile and birds' eggs contain so much yolk that the small disk of cytoplasm on the yolk's surface is dwarfed by comparison. No cleavage of the massive yolk is possible, and all cell division is restricted to the small cytoplasmic disk, the ***blastodisk*** (Fig. 27.3). (Note that the yolk and the small cytoplasmic disk on its surface constitute the true egg cell. The white albumen of the egg is outside the cell.) The large yolk is an adaptation for egg-laying on land; the yolk provides an abun-

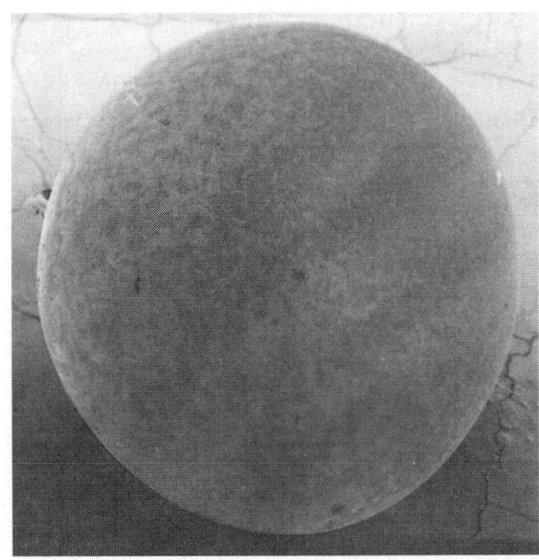

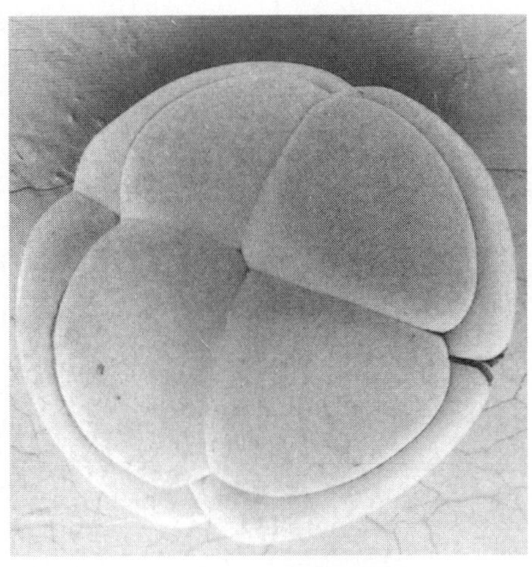

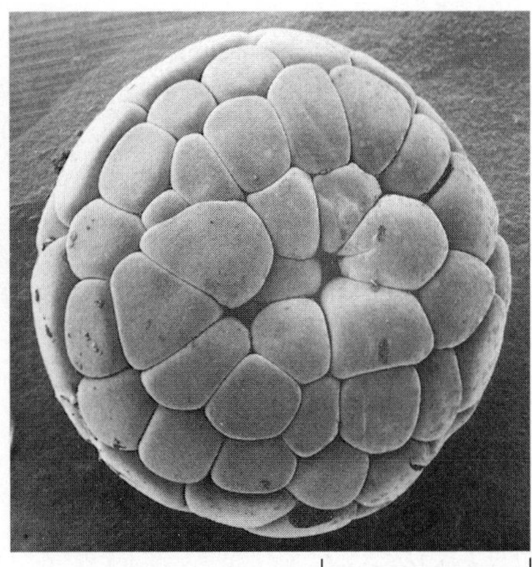

27.4 Scanning electron micrographs of frog egg and some early cleavage stages Left: Unfertilized egg. Middle: Eight-cell stage. Right: 32- to 64-cell stage. All three micrographs are at the same magnification: ×46. Note that there has been no overall growth in size during these cleavage stages—the 32- to 64-cell embryo is no larger than the egg cell.

dant food supply so that the young can complete its embryonic stages within the protected environment of the egg, and hatch out as a miniature adult, ready to survive on land. Animals whose eggs have comparatively small amounts of yolk generally spawn in an aquatic environment. The yolk provides only enough food to get the young to a free-living ***larval stage*** (as in the tadpole of a frog). The larval stage is common in many animal groups. Often it functions as a transitional feeding stage, during which the organism survives on what food stores it has left and what it can find to eat while accumulating enough food energy to build into a more complex adult. However, mammalian eggs, which also have a very small amount of yolk, follow a different developmental pattern: the yolk provides enough nutrients to allow the growing embryo to survive until it is implanted in the uterine wall. There is no larval stage in mammals.

Early cleavage in the human embryo is similar to that in amphioxus, since the human egg has relatively little yolk (Fig. 27.5). The morula is converted into a hollow ball of cells, which in mammals is referred to as a ***blastocyst.*** Unlike amphioxus, however, cleavage produces two different types of cells: those of the ***inner cell mass*** (Fig. 27.5), which will give

MAKING CONNECTIONS

1. How are the developmental differences between plants and animals related to their structural and nutritional differences?

2. How is the fact that the egg is far more important than the sperm in programming the early stages of embryonic development related to the differences in egg and sperm production described in Chapter 9?

3. Describe the process of in vitro fertilization. What is the role in this process of the gonadotropic hormones discussed in Chapter 26?

4. Which stage of the cell cycle (G_1, S, or G_2) would be abbreviated or eliminated in the series of mitotic cell divisions that follow fertilization? (See Chapter 9.)

5. Why is it advantageous for cleavage of the zygote to produce a hollow ball of cells instead of a solid mass?

6. Why is more yolk usually found in the eggs of land animals than in those of aquatic animals? Why are mammals an exception?

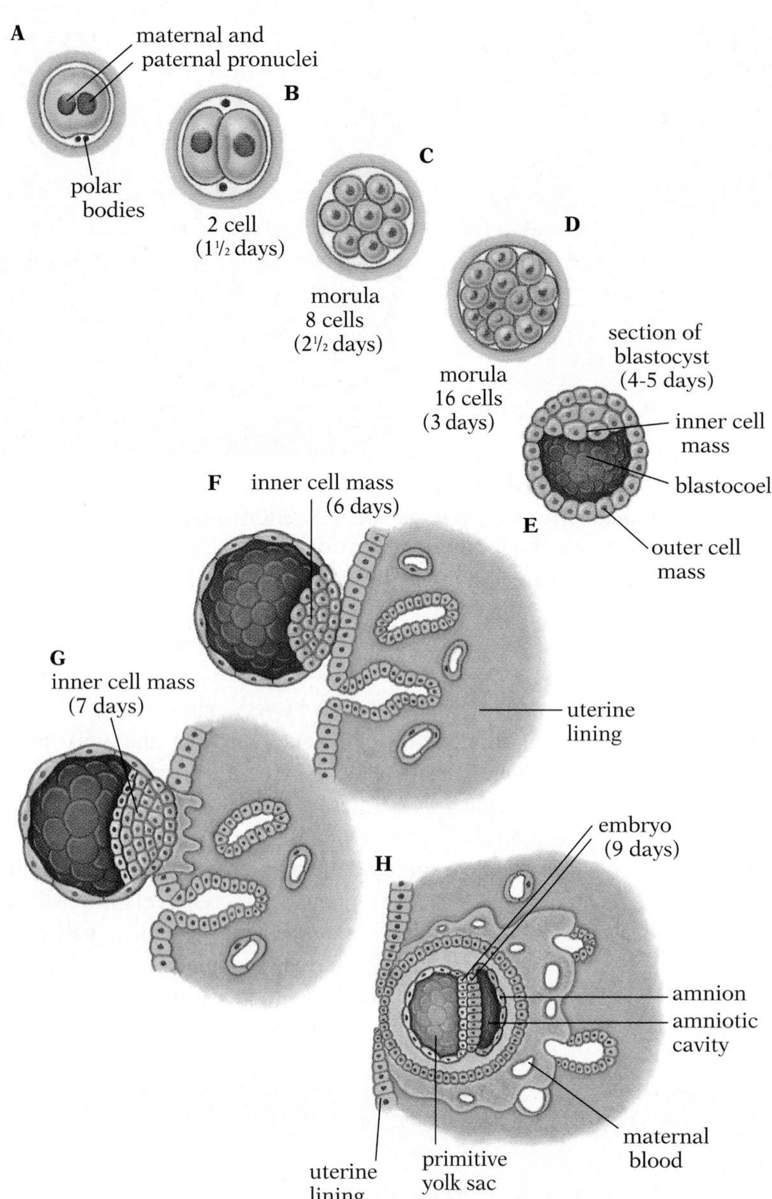

27.5 Early human embryology The process of cleavage in the human embryo, which leads to a hollow ball of cells called a blastocyst, is shown in A–F. Cleavage produces two different types of cells: those of the inner cell mass, which will give rise to the embryo proper, and the surrounding epithelial lining, which will make contact with the maternal tissues of the uterus and form the placenta. On the sixth day, the epithelial cells of the outer cell mass make contact with the endometrial lining of the uterus and begin to burrow (F, G). By the ninth day (H) the embryo is implanted in the uterus. The amniotic cavity is developed and the placenta is beginning to form. The embryo at this point consists of two layers of cells.

rise to the embryo proper, and those of the surrounding epithelial lining, which will make contact with the maternal tissues of the uterus and form the placenta. In humans, fertilization and the early cleavage stages occur while the egg is still in the oviduct (see Fig. 26.22, p. 549).

Gastrulation Creates an Embryo With Three Germ Layers

Next begins a series of complex movements important in establishing the definitive shape and pattern of the developing embryo, which is called ***morphogenesis*** (meaning "the genesis of form"). Morphogenetic movements of cells in large masses always occur during the early developmental stages of animals. The blastula undergoes a complex series of changes to create a ***gastrula.*** This process, called ***gastrulation,*** creates an embryo with three primary cell layers.

At the end of cleavage, the cells of the blastula look very similar. Yet the fate of some of these cells has already been sealed; some will become the eye, some the ear, some the digestive tract, and so on. The fate of other cells has not yet been determined. Following cleavage, the process of morphogenesis begins to transform this tiny hollow ball of cells into a young animal capable of surviving on its own, by positioning masses of cells properly for their later differentiation into the principal tissues of the adult body. In effect, these movements

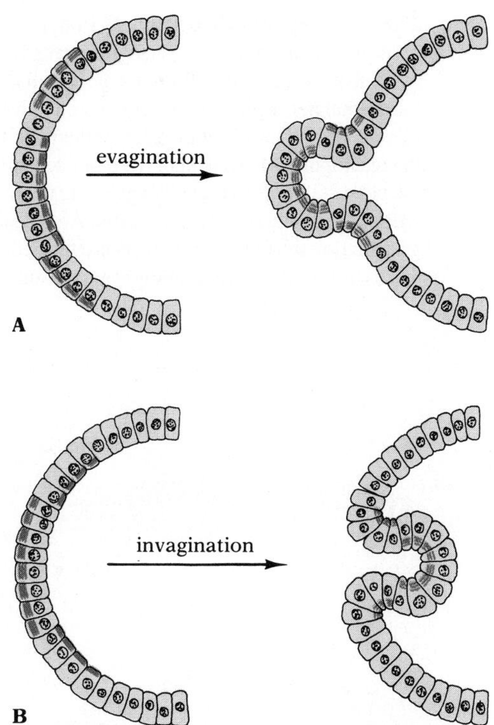

27.6 The mechanism of some morphogenetic movements in cells Contraction of microfilament-microtubular complexes (red), asymmetrically positioned in the cells, may change the shapes of the cells and produce evaginations (A), invaginations (B), or other alterations of the arrangement of cells in a developing organ. The contractile interactions are of the same type as those involved in amoeboid movement and cytokinesis.

mold the embryonic mass into the structural configuration on which differentiation will superimpose the finer detail of the finished organism.

The mechanism of these morphogenetic movements is still poorly understood. There are often changes in the shapes of the cells, probably effected by contractile microfilaments (Fig. 27.6) or by some microtubular apparatus. Possibly important in some of the movements are changes, perhaps only temporary, in the adhesive attachments of the cells for neighboring cells. It may be relatively easy for a group of cells that adhere tightly to each other, but have very little attraction for a layer of cells lying under them, to slide, as a group, over the surface of that underlying layer.

The Pattern of Gastrulation Is Influenced by the Amount of Yolk

Just as the pattern of cleavage is greatly influenced by the amount of yolk in the egg, so too is the pattern of gastrulation, because the heavy, yolky cells physically impede mass movement of cells. We shall examine first the pattern in an animal whose eggs have little yolk, and then the pattern in animals whose eggs have more yolk.

In amphioxus (Fig. 27.7), which has little yolk in its egg, the process of gastrulation begins when a small depression, or invagination, starts to form at the vegetal pole of the blastula. More and more cells move to the point of invagination and then fold inward, and the invagination becomes larger and larger. Eventually the invaginated cell layer comes to lie almost against the outer layer, thus nearly obliterating the old blastocoel (Fig. 27.7E). The resulting gastrula is a two-layered cup, with a new cavity, called the ***archenteron,*** that opens to the outside via the ***blastopore,*** the point where invagination first began. The archenteron will become the cavity of the digestive tract, and the blastopore will become the anus. A very similar process occurs in nearly all animals, though with a few fundamental differences—for instance, in the vast majority of animal phyla (but not the chordates, to which we belong) the blastopore becomes the mouth.

Gastrulation, as it occurs in amphioxus, first produces a

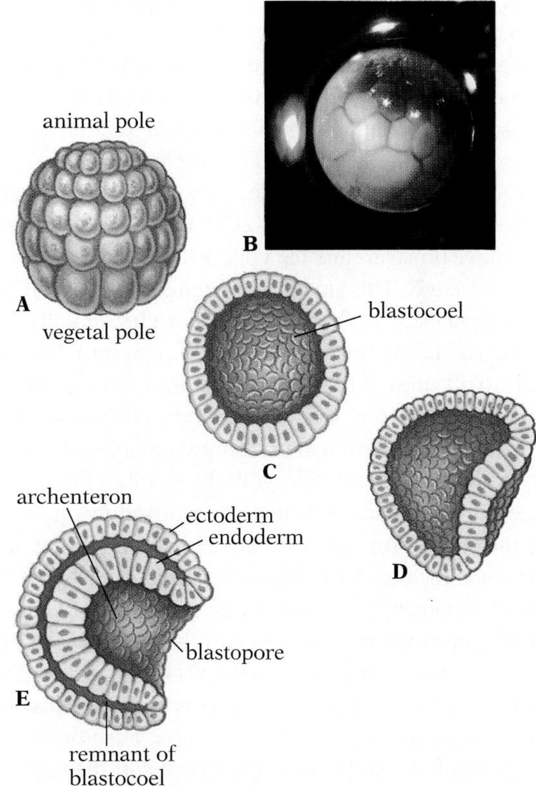

27.7 Gastrulation in amphioxus (A) Blastula. (B) Photograph of an early cleavage showing that the cells in the vegetal hemisphere are enlarged with yolk (yellow). (C) Longitudinal section through a blastula, showing the blastocoel. (D, E) Longitudinal sections through an early and a late gastrula. Notice that the invagination is at the vegetal pole of the embryo, where the cells are largest.

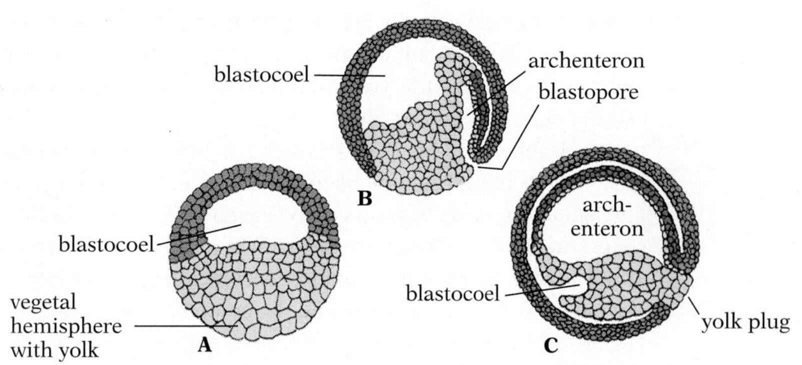

27.8 Gastrulation in a frog (A) Longitudinal section of a blastula. (B, C) Longitudinal section of two late gastrula stages. Portions of the cell layer of the animal hemisphere move down around the yolk-filled cells and turn in at the edge of the yolk, forming the blastopore. Movement of the other cells around the yolk-cell mass eventually enclosing the material within the cavity of the archenteron. At the end of gastrulation the three germ layers are formed: ectoderm (blue), endoderm (yellow), and mesoderm (red).

cuplike embryo with two primary cell layers, an outer ***ectoderm*** and an inner ***endoderm*** (Fig. 27.7E). A third primary layer, the ***mesoderm,*** soon begins to form between the ectoderm and the endoderm.

In the amphioxus egg, where the distinction between animal and vegetal hemispheres is only slight owing to the small amount of yolk in the vegetal hemisphere, gastrulation can occur in a direct and uncomplicated manner. Many eggs have far more yolk in their vegetal hemisphere, which imposes complications on the process of gastrulation. Generally, the more yolk an egg contains, the more gastrulation departs from the pattern in amphioxus.

Frogs' eggs have a moderate amount of yolk and thus the pattern of gastrulation is somewhat different from that in amphioxus. The frog embryo begins gastrulation during its second day of development. Simple invagination at the vegetal pole is not mechanically feasible, because of the large mass of inert yolk-filled cells. Instead, cells from the animal hemisphere move down around the yolk-cell mass and then turn inward at its edge. This involution begins at what will be the dorsal (upper) side of the yolk mass, forming initially a crescent-shaped blastopore there (Fig. 27.8). This infolding slowly spreads to all sides of the yolk, so that the crescent blastopore becomes circular. Movement of the other cells around the yolk eventually encloses this material almost completely within the cavity of the archenteron. The yolk-filled cells will become the endoderm, the surface cells that have moved inward will become the mesoderm, and the cells remaining on the surface will become the ectoderm. The yolk stored in each cell serves as a source of energy for the growing embryo.

Birds' eggs contain so much yolk that the gastrulation process is of necessity greatly modified (Fig. 27.9). Neither invagination of the vegetal hemisphere, as in amphioxus, nor involution along the edge of the yolk mass, as in the frog, can occur. Instead, the cells in the blastodisk sort themselves out into two layers; the lower layer will become the endoderm, and the upper the ectoderm (Fig. 27.9C). Next, certain cells along the midline of the blastodisk begin to flow inward (Fig. 27.9D), forming a primitive streak, which is essentially an elongate blastopore. Many of these cells from the primitive streak will form the mesoderm, but some will form additional endodermal cells.

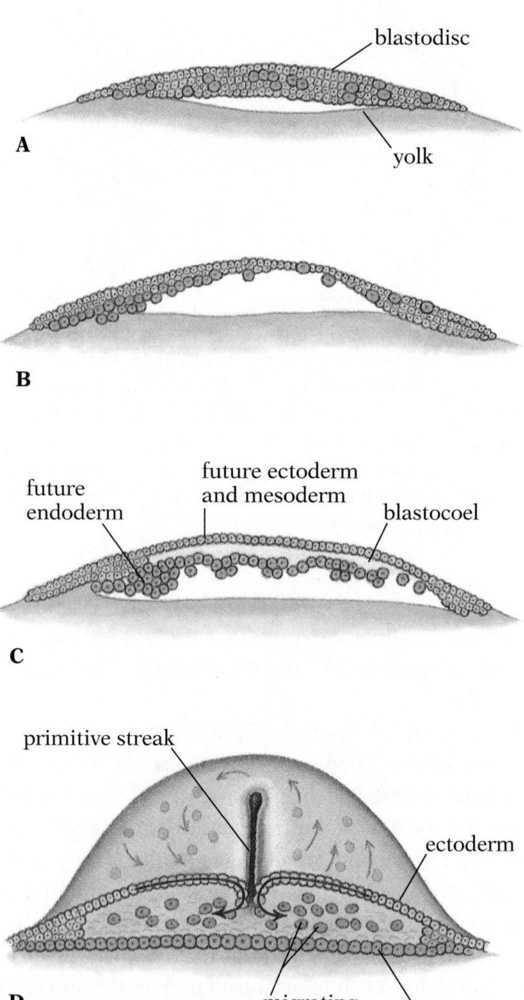

27.9 Gastrulation in the chick embryo (A) Longitudinal section through a blastula. Larger yolk-laden cells are intermixed with smaller cells. (B) The larger cells begin to accumulate on the lower surface of the cell mass. (C) The layer of larger cells separates from the layer of smaller cells to become the future endoderm; the cavity between the two layers is the blastocoel. (D) Surface view of a gastrula. Involution of cells along the midline of the embryo during gastrulation (arrows) produces a clearly visible primitive streak, which is essentially a very elongate blastopore. Some cells of the primitive streak move downward to form the mesoderm; others help form the endoderm.

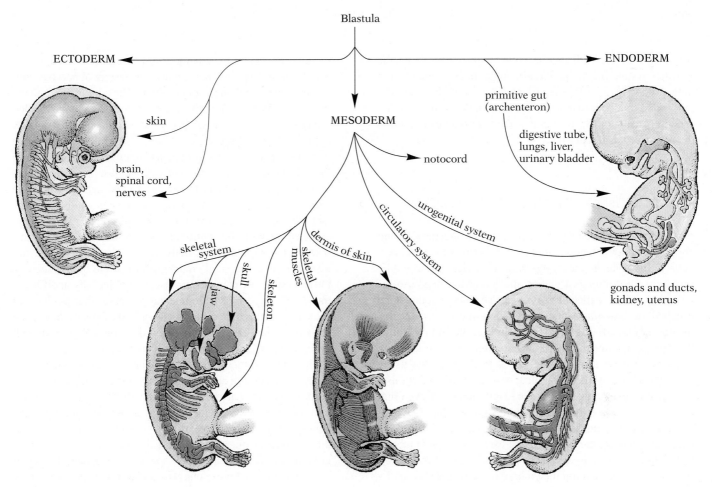

27.10 Fate of the three germ layers

Interactions Among the Three Germ Layers Determine the Formation of Adult Structures

The fates of cells in different parts of the three primary layers of vertebrates have been ascertained by staining them with dyes of different colors and then following their movements. As you might expect, the cells on the outer surface of the embryo, the ectoderm, eventually give rise to the outermost layer of the body—the epidermal portion of the skin—and to structures derived from the epidermis, such as hair, nails, the eye lens, and the nervous system (Fig. 27.10). As you might also expect, the yolky inner layer of cells belonging to the endoderm gives rise to the innermost layer of the body—the epithelial lining of the digestive tract and of other structures derived from the digestive tract, such as the trachea and lungs, the salivary glands, the liver, the pancreas, the thyroid, and the bladder (Fig. 27.11). The mesoderm gives rise to most of the

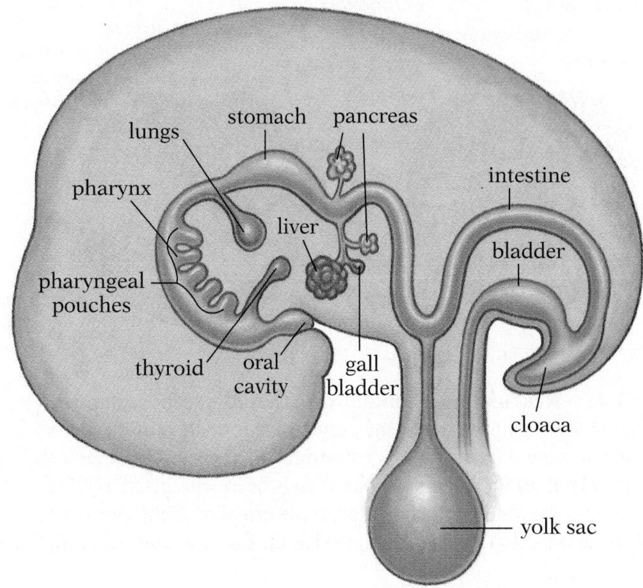

27.11 Structures derived from the digestive tract The embryonic digestive tract or archenteron has an inner lining of endoderm. Parts of this tract will develop into the oral cavity, pharynx, esophagus, stomach, and intestines. Various organs form as outpocketings of the digestive tract. These include the liver, gall bladder, pancreas, urinary tract, and thyroid gland. Also developing from the digestive tract are the lungs and respiratory passages. Thus all of these structures are at least partially derived from endoderm. Notice that a connection is maintained between the digestive tract and yolk sac where the food for the growing embryo is stored.

tissues in between, such as connective tissue, muscle, bone, the vascular system (including blood), urogenital system, and the notochord (Fig. 27.10). The ***notochord*** is a rod-shaped structure found in all chordates at some time in their life cycle. Its function is support. In vertebrates, the vertebrae form around the notochord, and replace it.

The Nervous System Forms During Neurulation

One major tissue located topographically between the skin and the gut does not develop from the mesoderm. This is the nervous tissue, which, curiously enough, is derived from the ectoderm. Soon after gastrulation, the neural tube begins to develop in a process called ***neurulation.*** The embryo at this stage is called the ***neurula.*** A thickened region of ectodermal cells, the ***neural plate,*** forms along the midline of the embryo above the developing notochord. Tissue along the edges of the plate begins to roll up forming the ***neural folds.*** Next the tissue along the center of the neural plate bends inward, and forms a long groove extending most of the length of the embryo (Fig. 27.12). The neural folds that border this groove then move toward each other and fuse, converting the groove into a long tube lying beneath the surface of the back. This neural tube in time differentiates into the spinal cord and brain (Fig. 27.10). At the end of neurulation the embryo has a neural tube and a rudimentary digestive tract; it has a recognizable anterior (front) end and a posterior (rear) end, and a dorsal (upper) surface and a ventral (lower) surface.

There are still many events that must happen to convert a neurula into a fully developed young animal ready for birth: the individual tissues and organs must be formed; an efficient circulatory system must begin to function; in a vertebrate the four limbs must develop; the elaborate system of nervous control must be established; and so forth. The complexity and the precision characterizing these developmental changes (called organogenesis) are staggering to contemplate. For example, approximately 43 muscles, 29 bones, and many hundreds of nervous pathways must form in each human arm and hand. To function properly, all these components must be precisely correlated. Each muscle must have exactly the right attachments; each bone must be jointed to the next bone beyond it in a certain way; each nerve fiber must have all the proper synaptic connections with the central nervous system and must terminate on the right effector cells. Incredibly sensitive mechanisms of developmental control must operate before such an intricate structure can arise from a mass of initially undifferentiated cells. Yet the developmental processes that produce all these later embryonic changes are the same ones we have seen at work in the early embryo—cell division, cell growth, cell differentiation, and morphogenetic movement.

It is beyond the scope of this book to discuss in detail the many events that occur during embryonic development. Yet it is these events that mold morphologically similar gastrulas into a fish in one instance, a rabbit in another, and a human being in still another, depending an the genetic endowment of the gastrula in question; the developmental events are programmed differently for each species. An understanding of how such different programs arise and how they are carried out is one of the important goals of developmental biologists.

One interesting aspect of the developmental programs of different species is that the early embryos of most vertebrates closely resemble one another. For example, the early human embryo, with its well-developed tail and a series of pouches in the pharyngeal region (where the neck will appear), looks very

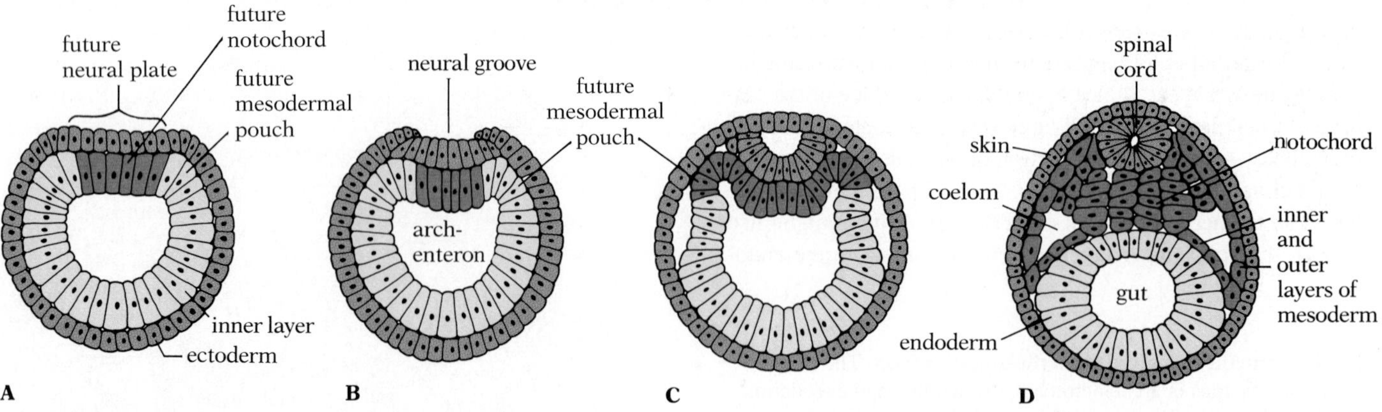

27.12 Neurulation in amphioxus The cross sections through amphioxus embryos show the progressive formation of mesoderm and the neural tube. (A) When gastrulation has been completed, the dorsal part of the inner layer has already been segregated as mesoderm—the future notochord and mesodermal pouches. Similarly, part of the ectoderm has segregated as the future neural plate. (B, C) The notochord and mesodermal pouches form as evaginations from the inner cell layer, and the neural plate invaginates from the ectoderm as it begins to form the spinal cord. (D) In this later embryo, both the spinal cord and the mesoderm are taking their definitive form. Notice that there is a cavity (the coelom) in the mesoderm.

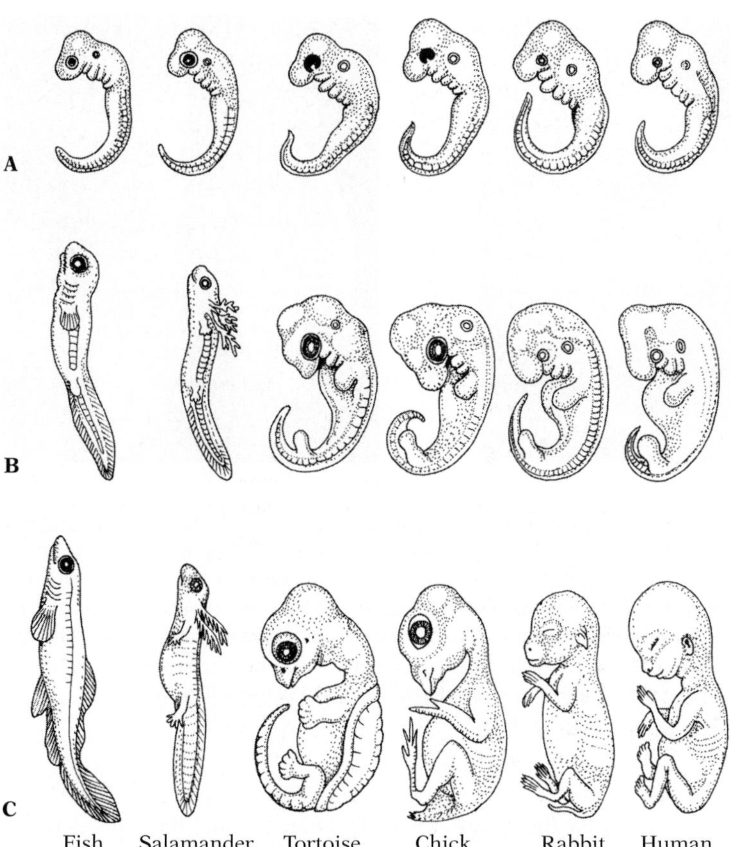

27.13 Vertebrate embryos compared at three stages of development At stage A, all the embryos—whether fish, amphibian, reptile, bird, or mammal—strongly resemble one another. Later, at stage B, the fish and salamander are noticeably different, but the other embryos are still very similar; note the pharyngeal pouches in the neck region and the prominent tail. By stage C, each embryo has taken on many of the features distinctive of its own species.

much like an early fish embryo (Fig. 27.13), and it looks even more like an early rabbit embryo. Only as development proceeds do the distinctive traits of each kind of vertebrate become apparent. About a hundred years ago, the German scientist Ernst Haeckel took this observation beyond its general interpretation as vague evidence of common descent, and proposed that the development of each organism retraces in detail its evolutionary history—that is, that "ontogeny recapitulates phylogeny." According to this hypothesis, early human embryos resemble fish because mammals are the evolutionary descendants of fish. However, we now know that the general developmental pattern of a species can skip the steps of ancestors, or create new structures *de novo.* There is even evidence to suggest that novel groups of organisms can arise through major and relatively rapid changes in developmental programs.

Development of the Human Embryo

In human beings the egg is fertilized soon after ovulation, while it is still in the upper portion of the oviduct. It immediately begins cleavage and has reached the blastocyst stage (Fig. 27.5) by the time it becomes implanted in the uterine wall eight to 10 days after fertilization. During the implantation process the blastocyst burrows into the uterine tissue and the cells of the outer cell mass (developing chorion cells) eventually form rootlike structures (villi) that invade the uterine tissue, much as a plant forms roots in the ground. The cells of the inner cell mass (Fig. 27.5) divide to form the embryo proper, as well as the other extraembryonic membranes (amnion, allantois, and yolk sac).

By the 23rd day the amnion has formed, enclosing the embryo in a fluid-filled chamber, and two other embryonic membranes—the chorion and the allantois—have intertwined with maternal tissues of the uterus to give rise to a functional placenta; the chorion contributes the fingerlike villi of the fetal portion of the placenta, and the allantois contributes the blood vessels (see Fig. 26.24, p. 553). By this time, the neural groove is complete, and the mesoderm is well developed. The mesoderm on either side of the neural tube separates into individual segments called somites, which can easily be distinguished. (The somites will later give rise to the vertebrate, ribs, limbs, and the body muscles.) The tubular embryonic heart has begun to pulsate weakly.

By the end of the first month the embryo, which now exhibits arm and leg buds, is roughly 5 millimeters long. The hands and feet are still mittenlike, with no separation between the individual fingers and toes (Fig. 27.14A).

During the second month the embryo grows to a length of approximately 30 millimeters (a bit more than one inch) and to

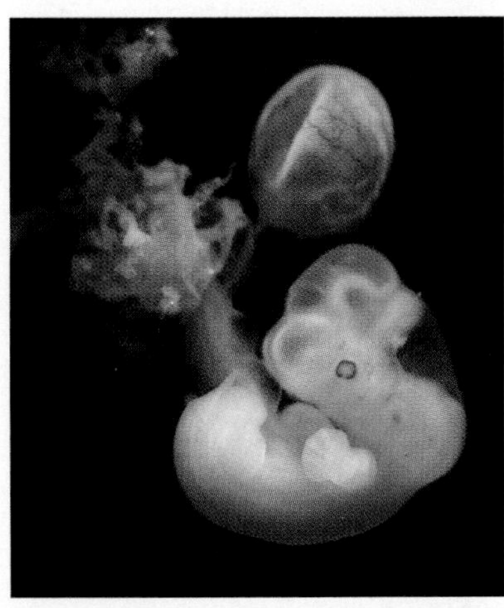

A

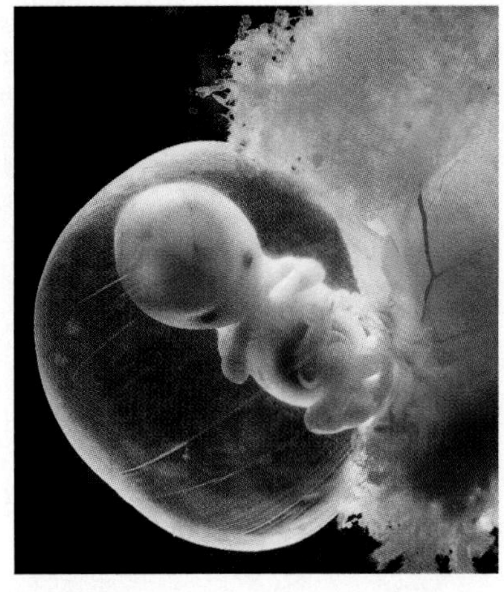

B

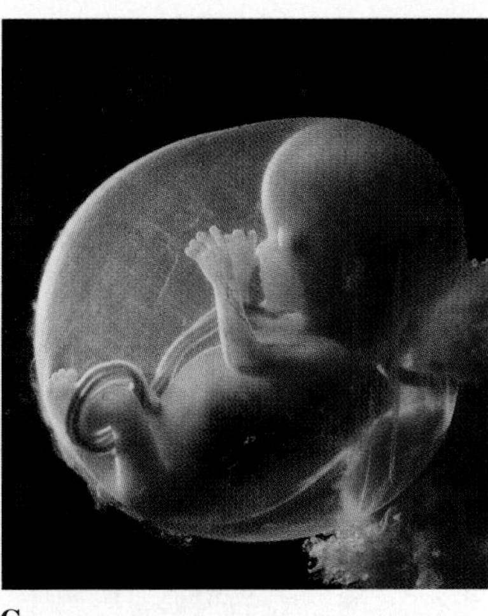

C

27.14 Human embryos at successive stages of development (A) The five-week embryo, 1 centimeter long, shows the beginnings of eyes but no distinctive face; note its mittenlike hands and feet, with no separation between the digits. (B) By contrast, the seven-week embryo, 2 centimeters long, has a distinct face, and there is separation between its fingers. (C) At 13 weeks the fetus is over 7 centimeters long and weighs about 30 grams, 15 times more than at seven weeks. (D) By 17 weeks the fetus is over 15 centimeters long and seven times heavier than it was a month earlier; all the internal organs have formed.

D

a weight of about one gram (0.03 ounce). By the end of the month (i.e., after eight weeks of development), the embryo, now called a ***fetus,*** is recognizable as a human. It has a flat face with widely separated eyes, and the fingers and toes are well separated (Fig. 27.14B). The cerebral cortex has begun to show cellular differentiation, and organization of the sense organs is well advanced. The muscles are sufficiently differentiated and some movement is possible. Deposition of bone in the skeleton has begun. The liver, which is serving as the chief producer of blood cells, is proportionately much larger than it will be later.

It is during the first two months of development that the embryo is especially sensitive to a wide variety of factors that can cause serious malformations. For example, if during that period the mother should contract "German measles" (more properly called rubella), this normally mild viral disease can result in a malformed heart, cataracts of the eyes, or deafness in the fetus. Similarly, the tranquilizer thalidomide taken by mothers early in pregnancy led to the widely publicized deformities of thalidomide infants. We have already mentioned the effects of alcohol on the fetus (resulting in fetal alcohol syndrome).

By the end of 13 weeks (Fig. 27.14C) the fetus is over 7 centimeters long, and the body proportions approach those expected at birth. The sex of the fetus can be determined through the use of ultrasound (Fig. 27.15). The growing cerebral hemispheres have begun to overlap the rest of the brain, and the sense organs are almost complete. The heart, too, has taken on nearly its final form. Spontaneous movements are frequent.

At this point the basic pattern of all the physiological systems has been established. The remainder of fetal development is largely a combination of further refinement and elaboration and growth in overall size (over 90 percent of fetal weight gain takes place in the last four months). In time, the bone marrow assumes the primary responsibility for the production of blood cells; the spinal cord and then the brain become myelinated; the eyes become light-sensitive, and later the ears become sensitive to sound; many new brain cells are produced; and numerous new neural circuits are established and become functional up to and past the time of birth.

By the end of 24 weeks the fetus has developed sufficiently to have some chance of survival outside the uterus if it is given respiratory assistance and kept in an incubator. However, an infant born at this stage is so small—about 0.66 kilograms—and so poorly developed that it will remain subject to many life-threatening medical problems for months. If, on the other hand, birth occurs at full term (on the average, about 280 days after the beginning of the mother's last regular menstrual period), the infant's chances of survival are high.

Several very important developmental changes in the circulatory system occur at the time of birth: the placental circu-

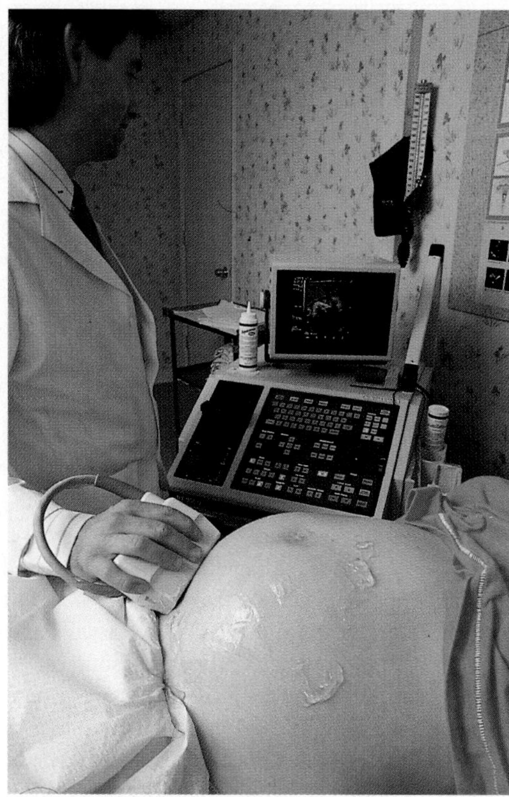

27.15 Ultrasound diagnostic test on pregnant woman In this technique, bursts of ultrasonic energy are beamed into the uterus. When this beam strikes a tissue of different density, such as the fetus and placenta, sonic echoes are reflected back and are converted into electrical impulses that can be displayed on an oscilloscope screen. A pregnancy can be observed from the 4th or 5th week on, and fetal growth can be monitored for abnormalities without the hazards of radiation. It is also possible, with a high degree of accuracy, to determine the sex of the fetus.

lation is cut off; the ductus arteriosus (shunt between the pulmonary artery and the aorta) is closed, the lungs are inflated for the first time; blood is forced into the pulmonary circulation; and production of fetal hemoglobin soon gives way to production of adult hemoglobin.

Growth is a Predominant Mechanism in Postembryonic Development

Though postembryonic development does involve some cell multiplication and differentiation, it rarely includes any major morphogenetic movements. Development from the postembryonic stage in many animals is characterized mainly by growth in size. Usually growth continues slowly, becomes more rapid for a time, and then slows down again or stops. This pattern yields the characteristic S-shaped growth curve shown

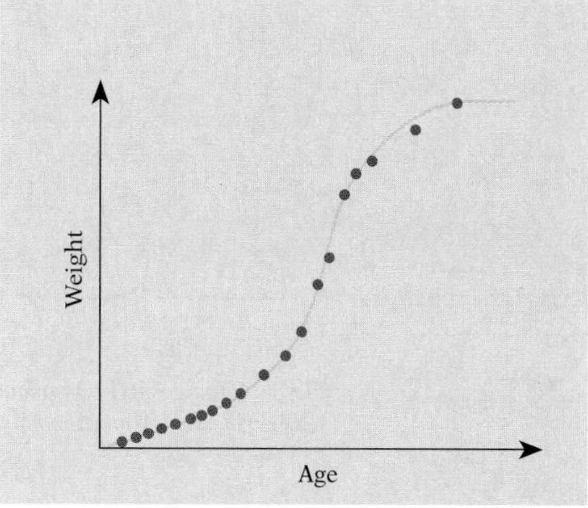

27.16 A typical S-shaped growth curve

in Figure 27.16. The shape of the curve is seldom as smooth as it appears in a generalized growth curve, because so many factors can affect the rate of growth—for example, in most mammals growth slows down for a while immediately after weaning, and it often varies greatly during puberty. An especially marked departure from the smooth generalized curve is seen in the growth of arthropods (Fig. 27.17), which at each molt show a sharp burst of growth during the short period after the old exoskeleton has been shed and before the new one has hardened.

Growth does not occur at the same rate and at the same time in all parts of the body. The differences between a baby chick and an adult hen or rooster, or between a newborn baby and an adult human being, are differences not only of size, but also of proportion. The head of a young child is far larger in relation to the rest of its body than that of the adult. And the child's legs

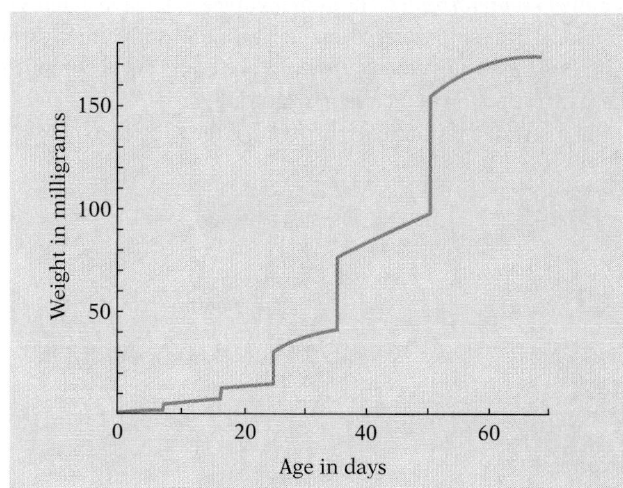

27.17 Growth in weight of an insect The growth spurts of the water boatman, an aquatic insect, occur at the time of molt, when the old exoskeleton has been shed and the new one has not yet fully hardened.

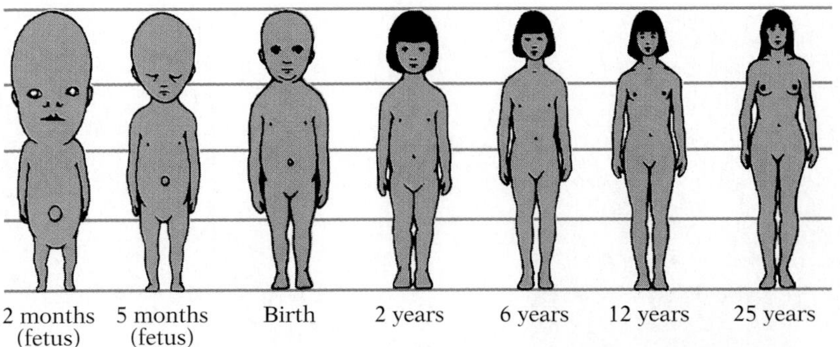

27.18 Changes in body proportions during human fetal and postnatal growth The head grows proportionately much more slowly than the limbs.

are much shorter in relation to its trunk than those of the adult. If the child's body were simply to grow as large as an adult's while maintaining the same proportions, the result would be a most unadultlike individual (Fig. 27.18). Normal adult proportions are achieved because the various parts of the body grow at quite different rates or stop growing at different times.

Metamorphosis Converts a Larval Animal Into an Adult Form

Growth in size is not always the principal mechanism of postembryonic development. Many animals, particularly those leading sessile (nonmotile) lives as adults, go through larval stages in which they bear little resemblance to the adult (Fig. 27.19). The series of sometimes drastic developmental changes that converts an immature animal into the adult form is called ***metamorphosis.*** It often involves extensive cell division and differentiation (specialization), and sometimes even morphogenetic movement; growth alone could not accomplish the transformation of a larva into an adult.

In many aquatic animals dispersal of the species to new environments depends on the larval stage; the tiny larvae either swim or are passively carried by currents to new locations, where they settle down and metamorphose into sedentary adults. In other species, such as frogs, where the adult is not sedentary, the adaptive significance of the larval stage seems more a matter of exploiting alternative food sources.

The most familiar larvae are those of certain groups of terrestrial insects, including flies, beetles, wasps, butterflies, and moths. The young fly, wasp, or beetle is a grub that bears no resemblance to the adult. The young butterfly or moth is a caterpillar. In the course of their larval lives, these insects molt several times and grow much larger, but this growth does not bring them any closer to adult appearance; they simply become larger larvae. Finally, after they have completed this larval development, they enter an inactive stage called the ***pupa,*** during which they are usually enclosed in a case or cocoon. During the pupal stage most of the old larval tissues are destroyed, and new tissues and organs are generated from small disks of cells (imaginal disks) that were present but undeveloped in the larva. The adult that emerges from the pupa is therefore radically different from the larva; it is almost a new organism built from the raw materials of the larval body (Fig. 27.20).

Insects with a pupal stage and the type of development just described are said to undergo ***complete metamorphosis.*** The

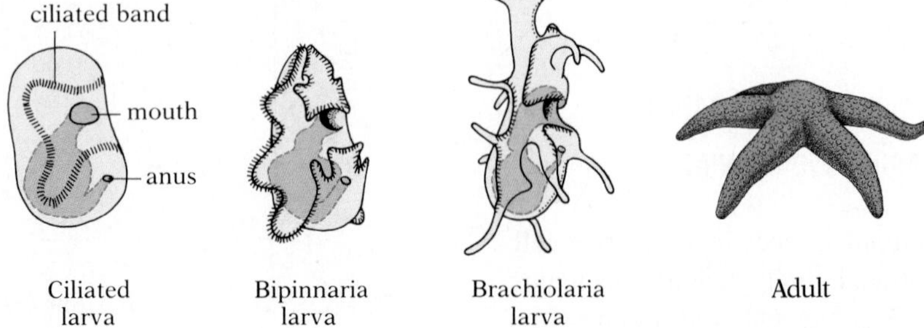

27.19 Three larval stages and adult of the sea star *Asterias vulgaris* The gastrula develops into the ciliated larva, which changes into the bipinnaria larva, which changes into the brachiolaria larva, which metamorphoses into the characteristically shaped sea star, with five arms.

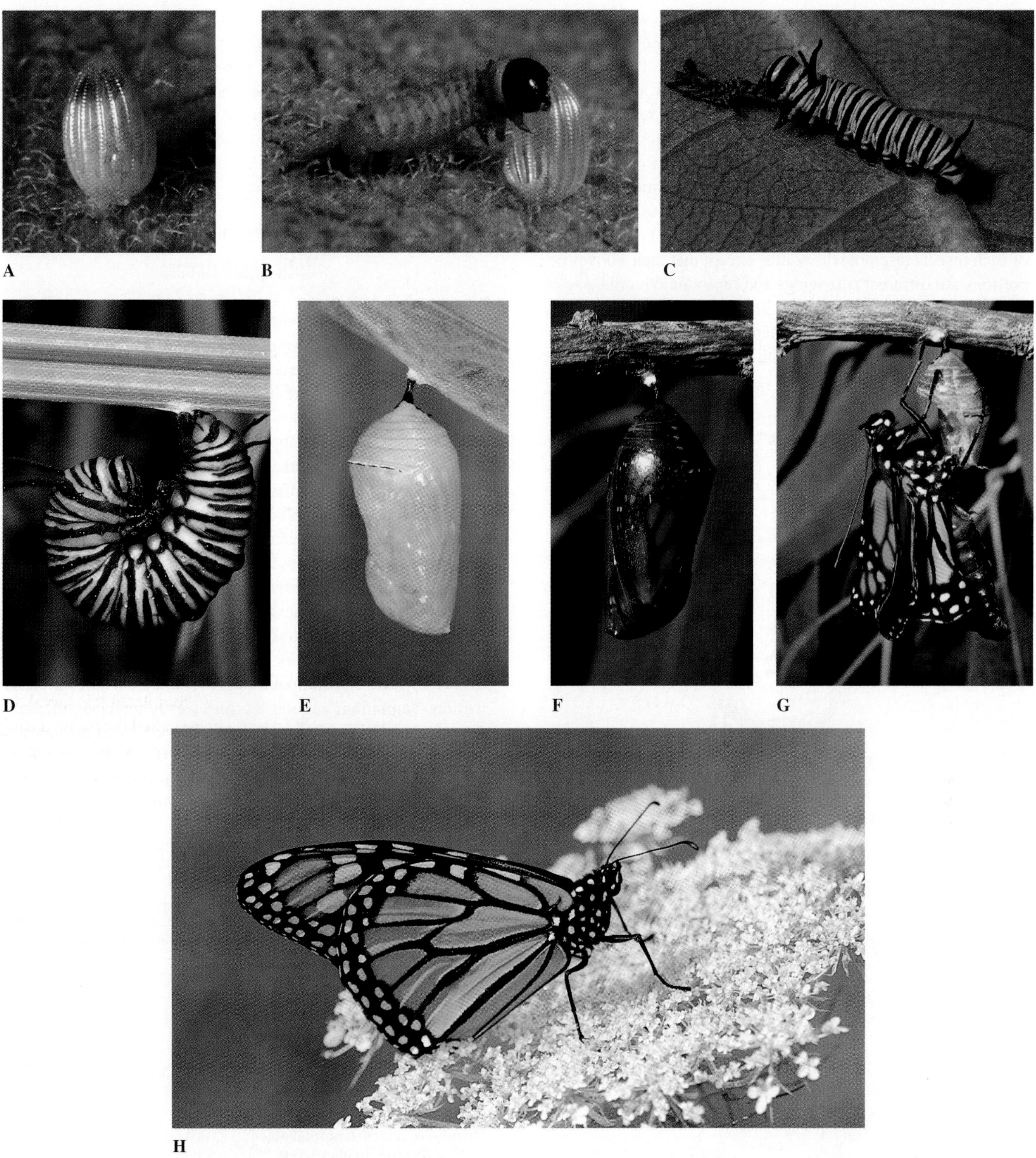

27.20 Developmental stages of a monarch butterfly, an insect with complete metamorphosis From the egg (A) hatches a small larva (B), which goes through a series of molts, growing larger at each (C). Eventually the full-grown larva attaches to a plant (D) and goes into pupation, forming a case around itself (E); in moths and butterflies the encased pupa is called a chrysalis. During pupation most of the larval tissues are broken down and new adult tissues are formed; in the monarch some of the adult structures can be seen through the case of the late chrysalis (F). The newly formed adult emerges from the pupal case and rests while it inflates its wings (G). It then flies off to feed at flowers—here milkweed (H)—before mating and laying eggs (almost always on milkweed) to start the cycle over again.

sharp distinction between the larval and adult stages in such insects has meant evolution in two markedly different directions; in general, the larva is more specialized for feeding and growth, while the adult is more specialized for actively dispersing the species to new environments, and reproduction. The addition of a pupal stage provides more adaptive flexibility in the life cycle; many insects winter as pupae.

Complete metamorphosis is not characteristic of all insects. Many, such as grasshoppers, cockroaches, bugs, and lice, undergo ***gradual metamorphosis*** (Fig. 27.21). The young of such insects resemble the adults, except that their body proportions are different (the wings and reproductive organs, especially, are poorly developed). They go through a series of molts during which their exoskeletons are shed and there are spurts of growth (Fig. 27.17). With each molt their form gradually changes and becomes more and more like that of the adult, largely as a result of differential growth of the various body parts. They have no pupal stage and experience no wholesale destruction of the immature tissues. Because the young of such insects resemble the adults, they are able to move about to find new sources of food.

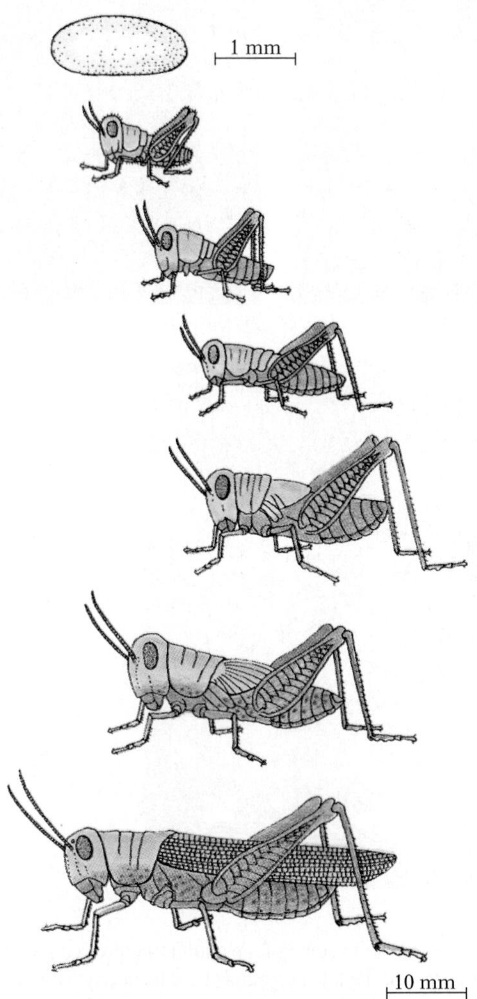

27.21 Gradual metamorphosis of a grasshopper The insect that emerges from the egg (top) goes through several stages that bring it gradually closer to the adult form (bottom).

The Adult Organism Continues to Develop Until Death Occurs

Discussions of development often stop with the completely matured adult. But development in its full biological sense does not cease then. The adult organism is not a static entity; it continues to change; and hence to develop, until death brings the developmental process to an end.

The term "aging" is applied to the complex of developmental changes that lead, with the passage of time, to the deterioration of the mature organism (Table 27.1) and ultimately to its death. It has been established that the human body loses 0.8 percent per year of its functional capacity after age 30. Though aging has become a major field of investigation in recent years, much remains to be learned about the processes involved. Modern scientific progress and improved medical techniques have greatly increased our ability to protect ourselves against certain diseases, starvation, and some of the destructive forces of the physical environment. Improved medical care, good nutrition, sanitation, regular exercise, etc., have significantly lengthened human life expectancy. Evidence indicates that early humans generally lived between 20 and 40 years; today most people can expect to live into their mid-70s. And as the

TABLE 27.1 *Average decline in a human male from ages 30 to 75 years*

Characteristic	*Percent decline*
Weight of brain	44
Number of nerve cells in spinal cord	37
Velocity of nerve impulse	10
Number of taste buds	64
Blood supply to brain	20
Output of heart at rest	30
Speed of return to normal pH of blood after displacement	83
Number of filtering subunits in kidney	44
Filtration rate of kidney	31
Capacity of lungs	44
Maximum oxygen uptake during exercise	60

life expectancy increases and the proportion of the population in the upper age brackets rises, the changes associated with aging become more obvious and more important to all of us.

Aging seems to be related to the ability of a cell to specialize—that is, to serve only one or a few highly specific functions. Cells that remain relatively unspecialized—that remain more versatile—and continue to divide do not age as rapidly (if at all) as cells that have lost the capacity to divide. Cancer cells, of course, divide continually and are essentially immortal. Hence bacteria and some other unicellular organisms cannot be said to age, for any cell that is not destroyed eventually divides to produce two young cells; division is thus a process of rejuvenation. Within the body of a multicellular animal, tissues like muscle and nerve, which do not normally divide, slowly deteriorate, whereas tissues like those of the liver and pancreas, in which active cell division continues, age much more slowly. Furthermore, animals, such as lobsters and many fishes, that grow as long as they live seem to show fewer symptoms of aging than, for example, mammals and birds, which cease growing soon after they reach maturity.

Aging of the whole organism is not simply a matter of the death of its cells. Rather, it involves the deterioration and death of those cells and tissues that cannot be replaced. What makes some irreplaceable tissues age? Scientists cannot as yet give a satisfying answer to this question. Some investigators have suggested that aging of cells results from the cumulative effect of radiation damage and mutations, or from the accumulation of metabolic wastes that cannot be expelled from the cells, either of which could lead to disturbances in cell function. Other investigators put more emphasis on intrinsic factors, suggesting that aging and death is genetically programmed just as other aspects of cell function are programmed. They point out that most normal mammalian cells, when placed in culture, seem to die after a set number of divisions; the number of divisions tends to be specific for the species and tissue of origin. Thus different cell types may have different life spans, and once they have lost the ability to divide, they cannot be replaced.

Whatever the processes of aging, it seems clear that we shall not understand them fully until we know much more about how development processes in general are regulated by the interaction of inherited and environmental influences. Aging may be simply another aspect of the general phenomenon of development.

MECHANISMS OF ANIMAL DEVELOPMENT

Differentiation and Pattern Formation Are Important Processes in Development

The zygote, as we have seen, has the genetic information necessary to orchestrate the precise series of events that give rise to a blastula, then a gastrula, which becomes (as the notochord forms) a neurula, and so on as the growing embryo turns itself into a complex multicellular organism. Throughout this process, cells need to know the organism's developmental age, their own location in the embryo, and what their immediate neighbors are doing. Chemical signals are the major source of all this information, which is used to modify each cell's own pattern of gene expression. The genes and the products they

MAKING CONNECTIONS

1. How do the microfilaments and microtubules described in Chapter six contribute to morphogenesis?

2. What is the evolutionary significance of the embryological similarities of all vertebrates?

3. How may an animal benefit by developing into a larva before metamorphosing into an adult, rather than developing "directly"?

4. What does the fact that cancer cells can be "immortal" indicate about the aging process?

5. Given that all the cells in an embryo arise through mitosis, and so have exactly the same genetic material, how is it possible for the cells to differentiate?

encode guide the repeating cycles of growth, cell movement, pattern formation, and differentiation that characterize development. ***Differentiation*** refers to the process of change that occurs in developing cells and tissues to make them specialized for particular functions, whereas ***pattern formation*** is the spatial arrangement of those tissues to form particular structures. We shall examine a few of the more compelling models for differentiation and pattern formation.

We have already seen that, because mitosis gives each new daughter cell a complete set of chromosomes exactly like that in the parental cell, *all the somatic cells in a multicellular organism ordinarily have identical sets of genes.* The genetic endowment of a nerve or muscle cell is exactly the same as that of a liver or bone cell in the same individual. It is therefore not surprising to find that differentiation and pattern formation depend on the differential activity of various genes, on signals from other cells, and on information derived from various environmental factors.

The Polarity of Cells and Plane of Cleavage Influences Differentiation

If embryonic cells cannot be distinguished on the basis of genetic content, it is logical to look for differences in their cytoplasm. We have already seen that the cytoplasm of the unfertilized egg is often not homogeneous. Most animal eggs contain stored food material, or yolk, which, since it is usually concentrated in one part of the cell, establishes a distinction between animal and vegetal hemispheres; i.e., it brings about a polarization of the cell. There are corresponding differences between the concentrations of various proteins and mRNAs in the two hemispheres. In time, these chemical differences destine the top of the animal hemisphere in most species to become the dorsal part of the organism; in such cases the animal/vegetal distinction defines the dorsal/ventral axis. Another distinction between the two hemispheres is that the animal hemisphere often has much more pigment in the cytoplasm than the vegetal hemisphere. There is abundant evidence that other materials are similarly restricted to certain regions of the cytoplasm—not only in the egg cells of animals but also in those of plants. When an egg cell divides, therefore, the cytoplasmic materials it contains may be distributed unevenly between the two daughter cells, depending on the orientation of the plane of cleavage.

As an example, consider the fertilized egg cell of a leopard frog. Immediately following fertilization, some of the egg contents shift and a crescent-shaped grayish area appears on the egg opposite the point where the sperm entered; these two points define the embryo's anterior/posterior axis. The material in this so-called ***gray-crescent*** will play a very prominent role throughout embryonic development. The first cleavage of the frog zygote passes through the gray crescent, so that each daughter cell receives half (Fig. 27.22A). If these two cells are separated, each will develop into a normal tadpole since each cell would retain the information specifying both axes. But if the plane of the first cleavage is experimentally made to pass to the side of the gray crescent, the result of separating the daughter cells will be very different; the cell that contains the gray crescent will develop into a normal tadpole, but the other cell

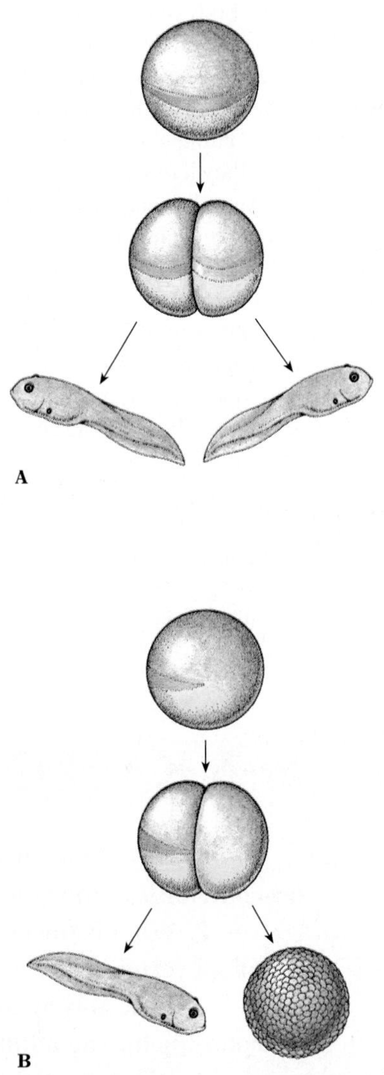

27.22 Importance of the gray crescent in the early development of a frog embryo (A) If the two cells produced by a normal first cleavage, which passes through the gray crescent, are separated, each develops into a normal tadpole. (B) If the first cleavage is experimentally oriented so that it does not pass through the gray crescent, and the two daughter cells are separated, the cell with the crescent develops normally, but the other cell develops into an unorganized cellular mass.

will form only an unorganized mass of cells (Fig. 27.22B). Therefore, the way in which the material of the gray crescent is distributed is of utmost importance to the developmental potential of the cells.

As the above experiment confirms, the normal first cleavage of the frog zygote is ***indeterminate;*** both of the new cells have the full developmental potential of the original zygote and all pathways of differentiation remain open to them. Such cells are said to be ***totipotent.*** Indeterminate cleavage is characteristic of most echinoderms and vertebrates, including humans. The cells of mammalian embryos generally remain totipotent until at least the eight-cell stage. (It is only animals with this kind of development that can give birth to identical twins.[1]) But in some other groups of animals, such as annelids and molluscs, the normal first cleavage divides critical cytoplasmic constituents unequally, and is therefore a ***determinate*** cleavage (Fig. 27.23). When the daughter cells of determinate cleavages are separated, they are not totipotent; the developmental fate of these cells has already been sealed and they will take different developmental paths.

Even in the species where the first few cleavages are indeterminate, a stage must eventually be reached where unequal partitioning of cytoplasmic substances gives rise to cells with different developmental potentialities. For example, in frogs the second cleavage is vertical and oriented at right angles to the first cleavage plane, giving rise to a four-celled embryo with two cells containing material from the gray crescent, and two lacking such material (Fig. 27.3B, C). This second cleavage is therefore determinate.

It is clear, then, that the cytoplasmic substances to which a nucleus is exposed must play a prominent role early in embryological development by activating some genes and repressing others. Because the egg cell contains different substances in different regions of its cytoplasm, cells with different cytoplasmic environments for their nuclei are produced early in embryological development (at the first cleavage in some eggs, at the second cleavage or later in others). By producing different regulatory effects on the genes in the nuclei, the cytoplasmic substances restrict the future course of development of the cells and their progeny. Unequal distribution of cytoplasmic substances in dividing cells, together with the orientation of the planes of cleavage, therefore becomes the key to the initial pattern of development.

But what factors determine the original distribution of substances in the cytoplasm of the egg cell? The detailed mechanisms responsible for patterning the egg cytoplasm are still largely unknown, but physical factors in the environment (such as light, temperature, pH, and gravity) have been shown to be important elements in early differentiation.

[1]Identical twins develop from separated cells derived from the same zygote. Nonidentical (fraternal) twins develop from different zygotes when two egg cells are released from the ovaries at the same time and are fertilized by different sperm cells. Consequently, identical twins are genetically identical, whereas fraternal twins are no more alike genetically than any two siblings. Fraternal twins are much more common than identical twins.

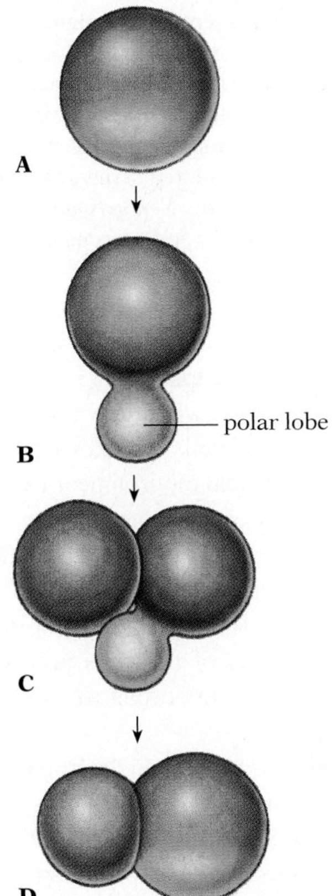

27.23 Cleavage of the fertilized egg of the sea snail *Ilyanassa* (A) At one pole of the zygote is a region of clear cytoplasm (yellow). (B) Just before the first cleavage this polar cytoplasm moves into a large protuberance, the polar lobe. The first cleavage partitions the zygote in such a way that the entire polar lobe goes to one cell. Only the cell that receives the polar-lobe material can give rise to the external structures seen in normal larvae. Therefore, this cleavage was determinate.

Developing Cells Are Influenced by Neighboring Cells

The immediate environment of the nucleus of a developing cell is the cell's cytoplasm, which, as we have seen, exerts a profound influence on differentiation. But developing cells are subject to many influences external to their own cytoplasm as well. These influences include environmental factors acting on the complete organism and also various factors to which only certain of the developing cells may be exposed, depending on their location in the embryo.

Experiments show that developing cells are influenced by

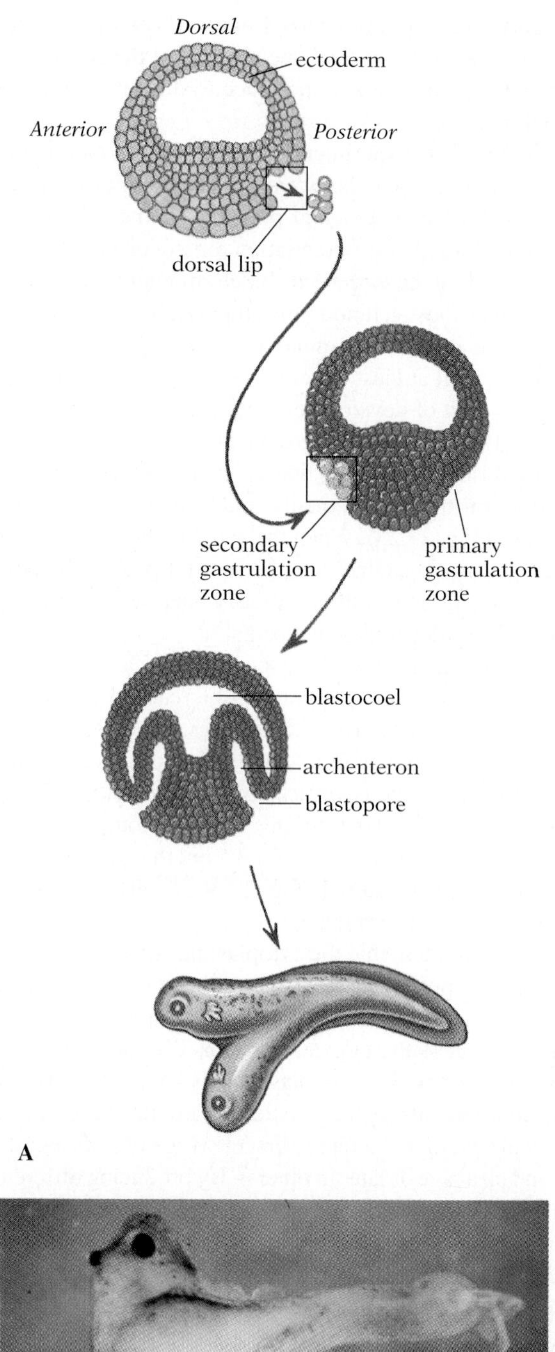

27.24 Spemann's experimental transplantation of the gray crescent (A) When Spemann and Mangold transplanted dorsal-lip material from a light-colored salamander embryo to a dark-colored one, the blastula with two dorsal lips proceeded to form an embryo with two gastrulation zones—one at the original dorsal lip of the blastopore, and one at the transplanted dorsal lip. A double larva of mostly dark-colored tissue was the result. (B) A similar transplantation experiment in the frog *Xenopus laevis* produced an embryo with two body axes, including two heads and a second spinal cord.

neighboring cells. For example, they interact via diffusible chemicals, and both their capacity for movement and their mitotic activity are inhibited when they come into contact with one another. Developing cells are likewise influenced by hormones and by the physical environment (e.g., light, temperature, gravity).

An important experiment showing the effects of tissue interactions in development involves the gray crescent of the frog embryo. Not only is the gray crescent important in determining the potentialities of the cells produced during early cleavage, it also plays an important role in the course of later development.

The gray crescent is located on the frog egg near the boundary between the animal and vegetal hemispheres. Cells move inward along this boundary during gastrulation. Thus in the early gastrula the cells derived from the gray-crescent portion of the egg become the ***dorsal lip*** of the blastopore (Fig. 27.8). These cells soon move inward and form the ***chordamesoderm,*** which will later become the notochord and other mesodermal structures. The chordamesoderm, in turn, influences the ectoderm, which lies over it, to become the neural tube. Neurulation does not occur if chordamesoderm is missing. In fact, mesodermal tissue throughout the embryo seems to play the dominant role in influencing the development. It migrates to new locations and influences adjacent endodermal and ectodermal cells to differentiate.

In a series of classic experiments performed in 1924, Hans Spemann of the University of Freiburg and his colleague Hilde Mangold turned their attention to the dorsal lip of the blastopore in the early gastrula stage of a salamander embryo. They transplanted the dorsal lip from its normal position on a light-colored embryo to a different region on a darker colored embryo. In each such experiment, after the operation, gastrulation occurred in two places on the recipient embryo—at the site of its own blastoporal lip and at the site of the transplanted lip. Eventually two nervous systems were formed, and in some of the experiments, even two complete embryos developed, joined together ventrally (Fig. 27.24). Most of the tissue in both embryos was darkly colored, an indication that the transplanted blastoporal lip had altered the course of development of cells derived from the host. Similar transplants of tissues from other regions of the embryos failed to produce comparable results. The dorsal lip of the blastopore must play a crucial role beyond taking the lead in gastrulation. Signals from these cells induce the formation of the neural tube, which in turn is important in establishing the longitudinal axis of the embryo and in inducing formation of other structures.

Spemann and Mangold called the dorsal lip of the blastopore the ***organizer.*** They envisioned the entire developmental process as one in which a succession of principal organizer regions, each taking over where the previous one left off, induces and controls the differentiation of the major tissues and organs. We now know that induction is not limited to a small number of organizer regions, but is a general phenomenon; inductive tissue interactions are the rule rather than the exception.

Cells of One Tissue May Induce the Pattern of Differentiation of the Cells in Another Tissue

Some of the first definitive studies on embryonic induction in an animal were performed in 1905 by Warren H. Lewis of Johns Hopkins University. Lewis worked on the development of the eye lens in frogs. In normal development the eyes form as lateral outpockets from the brain. When one of these outpockets, or optic vesicles as they are called, comes into contact with the epidermis on the side of the head, the contacted epidermal cells promptly undergo a series of changes and form a thick plate of cells that sinks inward, becomes detached from the epidermis, and eventually differentiates into the eye lens (Figs. 27.25 and 27.26). Lewis cut the connection between one of the optic vesicles and the brain before the vesicle came into contact with the epidermis. He then moved the vesicle posteriorly into the trunk region of the embryo. Despite its lack of connection to the brain, the vesicle continued to develop, and when it came into contact with the epidermis of the trunk, that epidermis differentiated into a lens. The epidermis on the head that would normally have formed a lens failed to do so. Clearly, the differentiation of epidermal tissue into lens tissue depends on some inductive stimulus from the underlying optic vesicle. Later experiments have shown that if a barrier is inserted between the vesicle and the epidermis no lens develops.

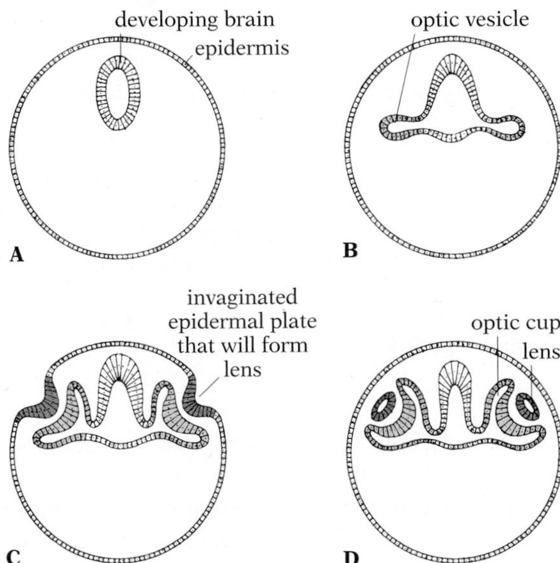

27.25 Development of optic vesicles and their induction of lenses in a frog The proximity of the optic vesicle induces the nearby cells of the epidermis to fold inward and form a lens. The infolding lens tissue, in turn, helps mold the optic vesicle into a two-layered structure called the optic cup. The cup cells then differentiate, the layer adjacent to the lens forming visual receptor cells and nerve cells.

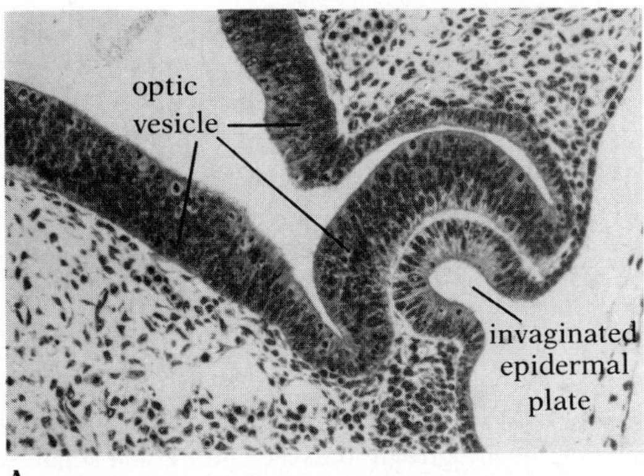

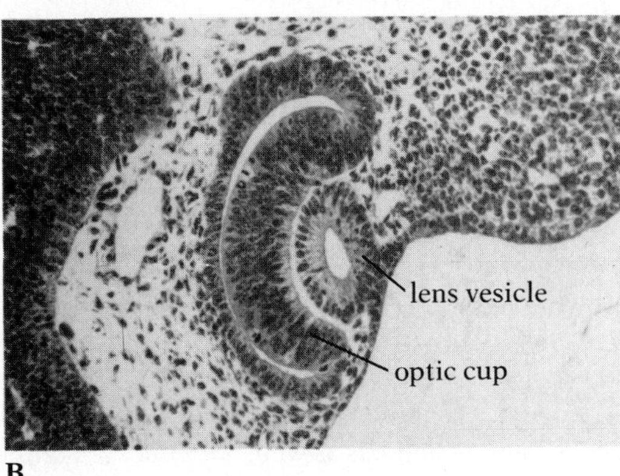

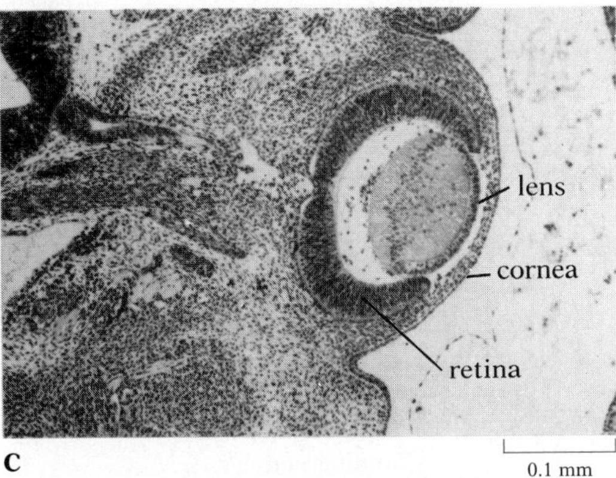

27.26 Development of a mammalian eye Photographs of mammalian eye development over a four-day period: The contact with the tip of the optic vesicle causes the epidermis to invaginate (A). The epidermal region differentiates to form the lens vesicle, while the optic vesicle develops into the optic cup (B), which ultimately becomes the retina—the tissue at the rear of the eye that contains light receptors (C).

Most Developmental Interactions Are Chemically Mediated

Some cell-to-cell interactions may be an effect of cell contact, as in contact inhibition, but most are probably chemically mediated; one cell is influenced by a substance secreted by another. Such chemical mediation can be demonstrated experimentally. For example, if the future epithelial and connective tissues of an embryonic mouse pancreas are separated, the epithelial tissue will not differentiate properly. But if the two groups of cells are grown in tissue culture and separated by a filter that allows large molecules to pass, then the epithelium will differentiate into normal pancreas tissue. Obviously, the necessary induction has occurred, and it must be attributed to a diffusible chemical that is able to pass through a filter.

Hormones Play a Key Role in Inducing Differentiation

In Chapter 24 we saw that plant hormones play an important role in nearly all phases of development: Tissue response to a hormone depends on the tissue involved, the concentration of the hormone, and the presence and relative concentrations of other hormones. In animals as well, hormones are important in development. In vertebrates, the importance of hormones is particularly well illustrated by the development of the gonads (Fig. 27.27) and the genitalia in humans (Fig. 27.28). In every human embryo the gonadal tissue is "indifferent" that is, it has the potential to form either ovaries or testes. The course of development depends on whether the male sex hormone testosterone is present at the crucial stage of embryonic development. It is the Y chromosome (specifically a region near the tip of the short arm) that has the genes for determining maleness. If this region is present the gonads will develop as testes, testosterone will be secreted, and the embryo will develop as a male (Fig. 27.29). In the absence of male hormone, the individual develops as a female, no matter what its genetic sex (XX or XY). Normally, the hormonal condition of the embryo will be appropriate to its genetic sex, but in rare instances this is not the case, and an individual will develop organs that, though ap-

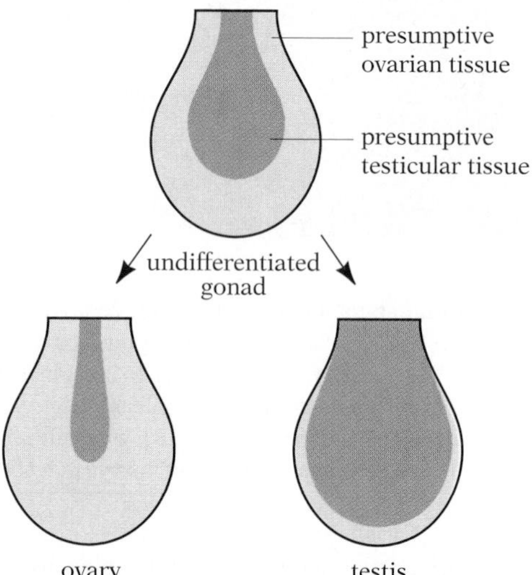

27.27 Development of ovary or testis from the undifferentiated gonad of an amphibian Depending on the hormonal condition of the embryo, one or the other of the two types of tissue in the undifferentiated gonad gains developmental ascendancy and the other type is repressed.

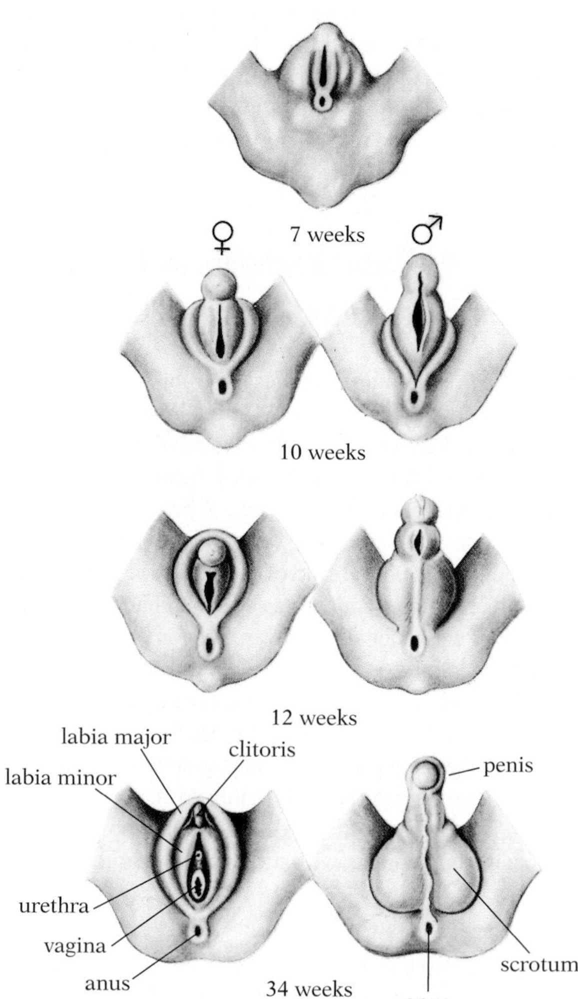

27.28 Development of the external genitalia of human beings At seven weeks the genitalia of male and female fetuses are virtually identical. At 10 weeks the penis of the male is slightly larger than the clitoris and labia minor, which form from the same primordium in the female. At 12 weeks these differences are more pronounced, and the male scrotum has formed from the tissue that becomes the labia major in the female. At 34 weeks the distinctive features of the genitalia of the two sexes are fully apparent. It is largely the concentration of male sex hormones that determines which of these developmental pathways will be followed.

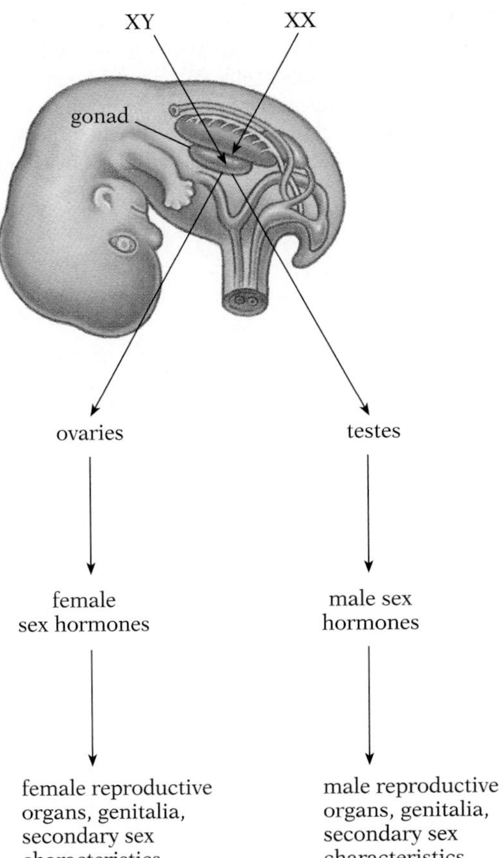

27.29 Sex determination in humans The gonads and genitalia are undifferentiated the first six weeks of embryonic development. Genes on the Y chromosome cause the gonads to develop into testes and begin secreting male sex hormones, and the internal organs, external genitalia, and secondary sex characteristics appropriate for a male develop. In the absence of the male determining genes, female sex hormones will be secreted and the internal organs, external genitalia, and secondary sex characteristics appropriate for a female develop. Femaleness is the "default" state.

propriate to its genetic sex, are poorly formed, or it may develop organs that, though nearly normal, belong to the other sex.

Cell Migration Plays an Important Role in Morphogenesis

One of the most remarkable phenomena of development is the organized movement of cells and the folding of tissues into new shapes (morphogenesis). The dorsal lip cells are able to crawl in the right direction (dragging other parts of the blastula behind them) and stop at the correct spot. The cells of the optic vesicle know when to begin pushing outward toward the ectoderm, when they have reached their goal, and how then to create a suitable cup. The various cells must have a way of telling where they are: some type of spatial information is necessary to give cells a sense of position and to guide them as development proceeds.

If cells from two different developing organs—the liver and spleen, for example—are separated from one another and the two groups mixed together, they slowly but accurately sort themselves out into two clumps. Under the microscope we can see each of the cells extending several pseudopods, touching and adhering to neighboring cells, and then pulling themselves toward some and not others. What then, is the mechanism behind these movements? We now know that each cell membrane is richly supplied with one or (usually) several kinds of cell-adhesion molecules (CAMs). Each kind of CAM attaches itself to other molecules of the same sort of CAM (Fig. 27.30). The membranes of various cells contain several different types of CAMs, but liver cells, for instance, have different proportions than spleen cells. As a result, liver cells stick slightly to spleen cells, but more firmly to their own kind. When a migrating liver cell pulls back on its pseudopods, it drags itself toward the cells it is most strongly bonded to, and loses its tenuous molecular grip on spleen cells. It is this differential affinity that permits the two classes of organ cells to sort themselves out, and to stick together.

A very similar process occurs when the dorsal lip cells begin their wanderings. At the time for gastrulation, a new set

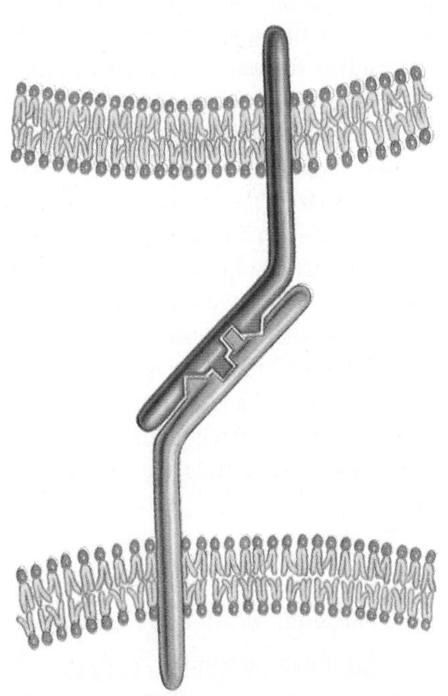

27.30 Binding of like cell-adhesion molecules The complementary structure of the arms of CAMs allows each type to bind to others of the same class. Several dozen kinds of CAM are thought to exist, though fewer than 10 have yet been characterized.

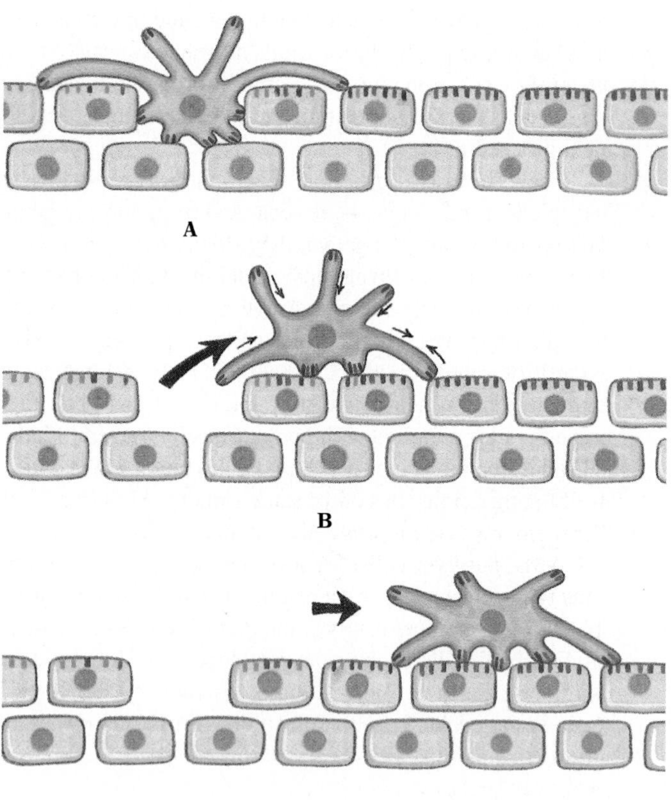

27.31 A model of cell migration (A) A cell induced to begin searching for a better CAM match extends pseudopods and attaches more or less strongly to other cells. (B) When the pseudopods are periodically retracted, the cell is pulled in the direction of the best match. After many such steps (C), the cell reaches a point from which there is no available improvement in CAM matching.

of CAMs is inserted in the membrane. Pseudopods begin to form and seek compatible attachments. Pulled along by the discovery of even better matches further on, the dorsal lip cells drag themselves forward until their CAMs find the best matches available; when no direction offers the prospect of any improvement, migration stops (Fig. 27.31). But when the time comes for further changes—neurulation, for instance—the constellation of CAMs in the membrane is changed, and the cells again begin their search for compatible tissues.

Analyses of the structure of CAMs have yielded a major surprise: they are closely related to the MHC molecules, T-cell receptors, and antibodies discussed in Chapter 21. Since CAMs are found in such organisms as the fruit fly, fishes, and amphibians, all of which lack an immune system, it seems likely that this developmental recognition system provided the basis for the evolution of the molecules of the immune system of mammals.

Positional Information Guides the Developmental Process

Once the embryo has undergone gastrulation and neurulation, different organs begin to appear up and down its length. The strategy of further development, from insects to humans, is one of subdivision of the embryo into a series of domains, followed by largely independent development of each domain. We now turn to how the domains are established in vertebrates.

Once the spinal cord has fully formed in a vertebrate embryo, the most dorsal region of mesoderm begins to form blocks of tissue called ***somites*** (Fig. 27.32). Each somite goes on to produce a vertebra, the ribs (if any) associated with it, the muscles unique to that vertebra (most notably, those serving the limbs), and the dermis (the layer of cells just below the skin). Each somite apparently "knows" which set of bones, nerves, and muscles to construct on the basis of its location, and becomes ***determined***—that is, committed to that fate.

Once a particular somite is instructed to help produce a limb or organ, there comes the major developmental task of orchestrating its construction. It is crucial, for instance, in the development of a vertebrate forelimb, that the bones be arranged in proper order, from humerus at the base to the digits at the distil end, and that the muscles attach to these bones in particular ways. The various cells must have a way of telling where they are: some type of spatial information is necessary to give cells a sense of position and to guide them as development proceeds. Such information permits similar cells to form different patterns.

Often the positional information depends on chemical signals. For example, specific regions of the embryo produce diffusible chemicals called ***morphogens,*** which spread into

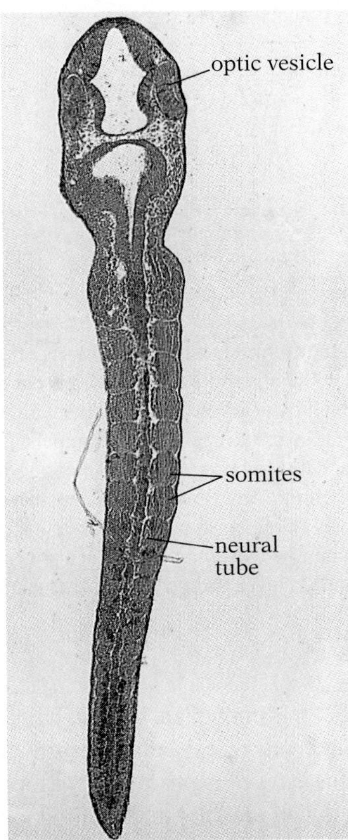

27.32 Somite formation The dorsal portion of embryonic mesoderm differentiates into a series of cell blocks, called somites, which give rise to vertebrae and dermis, as well as (depending on the somite) ribs and muscles of the limbs.

A

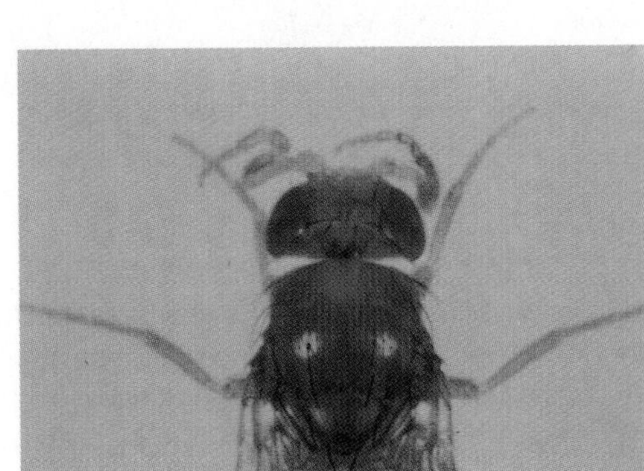

B

27.33 Homeotic mutations of the fruit fly *Drosophila* (A) Extra set of wings and (B) legs in place of antennae.

neighboring regions and then degrade over time. A concentration gradient is established, with the chemical being most concentrated nearest the source, and less so farther away from it. The concentration of the morphogens therefore gives cells information as to where they are relative to the source. Such gradients are believed to be a common strategy for giving cells positional information. The cells apparently "remember" this positional information, and it guides their subsequent behavior.

Various experiments with mutant strains of the fruit fly *Drosophila melanogaster* have allowed researchers to identify certain genes that control the development of the overall body plan. The products of these ***homeotic genes*** are generally control substances that bind to DNA and orchestrate the operation of many other genes. In this way they cause each segment to express its unique character. Mutations of these genes, called homeotic mutations, caused alterations of the body plan such as the production of half a body, legs in place of antennae, or extra sets of wings (Fig. 27.33). The mutations apparently caused whole regions of cells to misinterpret their position and produce an inappropriate structure.

One particularly intriguing discovery is the existence of a 180-nucleotide control sequence called the ***homeobox*** in each of the homeotic genes. Similar sequences have been found in such diverse organisms as earthworms, fungi, frogs, rats, and humans. Though the function of homeoboxes is still largely unknown, the elucidation of the workings of the homeotic genes in *Drosophila* will probably have much to tell us about how our own embryos organize themselves.

Determination Usually Precedes Differentiation

An early embryonic cell may follow any of a large number of different developmental pathways, but as development proceeds, its developmental potential increasingly narrows (Fig. 27.34). Long before any observable change takes place within a cell of a developing embryo, the fate of each cell becomes sealed, or determined. ***Determination*** refers to a restricting of a cell's developmental potential such that it can develop only as a certain cell type. This developmental change appears to be relatively permanent, and is passed down to all descendants of

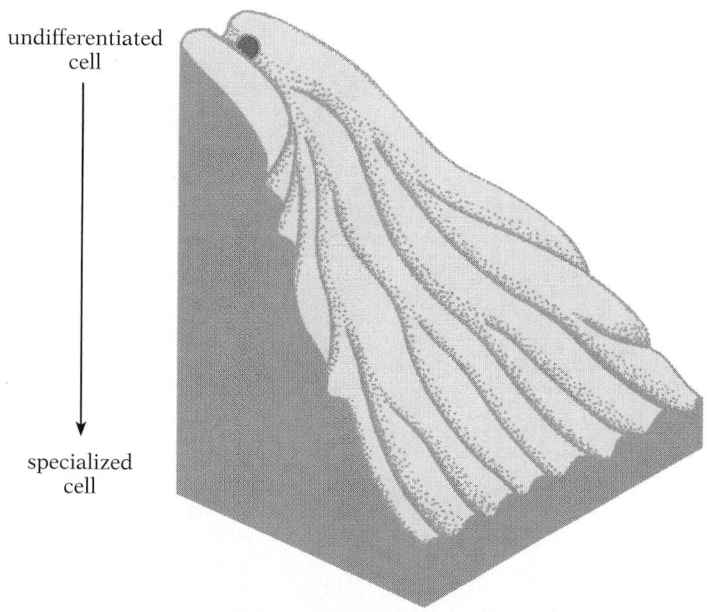

27.34 Developmental landscape model Differentiation can be compared with balls rolling down a slope cut by many valleys. Each ball rolls into one of the valleys. This valley soon branches into separate valleys, and each of these branches in its turn. As the ball passes each intersection, the number of alternative pathways still open to it diminishes. Since the ball cannot roll uphill, it cannot normally retract its course and take a different route. Finally it reaches the bottom of the hill at a point determined by the particular alternative pathways it took at each point of branching.

that cell. Once a determined cell has acquired structures or chemicals by which we can recognize what type of cell it is going to be, the cell is said to be ***differentiated.*** Note that determination and differentiation, though closely related, are not identical. Determination refers to a cell's fate; differentiation refers to a cell's specialized chemistry and morphology. The fertilized egg is totipotent, but as development occurs, cells lose totipotency and gradually become committed to a particular developmental pathway. For example, once a blastula cell has been determined as ectoderm, it cannot ordinarily go back and form a mesodermal or endodermal structure (Fig. 27.10). It has passed the first branching point in the developmental landscape. Now it can form any ectodermal structure. It may sink inward as the neural groove forms and differentiate as nervous tissue, or it may remain on the surface and differentiate as epidermis. As each branching point is passed, the total number of alternatives still ahead diminishes, until finally the cell has become one kind of fully differentiated cell. *Differentiation is thus a matter of progressive determination, a gradual restricting of development to one of the many pathways initially possible.*

What picture of the course of development emerges from the facts and ideas discussed so far? Assuming that all the cells of a single organism usually have the same genetic potential, it follows that other factors determine which of the many potentialities are expressed. The first restrictions on the potential of an embryonic cell are often the result of differences in the early cleavages, which give the nuclei of the different cells different cytoplasmic environments that activate different genes. As development proceeds, the extracellular environment of the cells becomes less uniform. For example, some cells are located more internally and are therefore exposed to different factors than cells on the surface of the embryo. Moreover, the various cells differ from their immediate cellular neighbors, and these different cells exert different influences on each other. Such differences in the environments of the various cells intensify the differences in their developmental directions.

As the cells and tissues become more and more differentiated, their influence on other nearby cells, via chemicals they secrete, becomes more pronounced. Some of the chemicals block pathways the neighboring cells might otherwise have followed; other chemicals tend to induce neighboring cells to follow alternative pathways.

As cells and tissues respond developmentally to the host of influences impinging on them, they in their turn alter the environment of the cells and tissues in their vicinity, causing a snowball effect. Each step in the development of one cell alters the influence that cell has on all other cells. The environment of each cell is constantly changing as development proceeds, and the changes in the environment profoundly affect the activity of the genes.

In Animals, Differentiation Is Sometimes Reversible

We have seen that as development of a multicellular organism proceeds, the individual cells become more and more committed to one particular course of differentiation. But though differentiation usually follows determination, certain cells are designed to become differentiated *before* determination fully fixes their fate. When such cells are transplanted to a new location, they can lose their differentiated appearance and take on a morphology appropriate for their new location. In mammals and birds, dedifferentiation (incomplete determination) of cells is usually restricted to embryos. In amphibians, however,

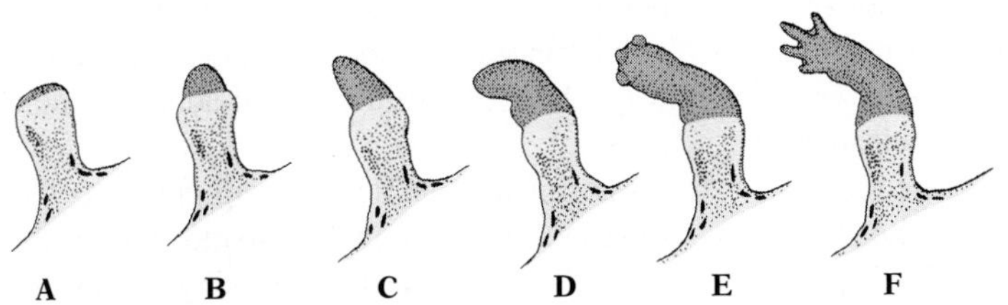

27.35 Regeneration of a salamander arm

this ability to maintain incomplete determination in at least some cells into adulthood leads to remarkable powers of regeneration. For example, if the leg of a salamander is amputated, cells near the wound begin to dedifferentiate under the epidermis at the tip of the stump. Gradually, as mitotic activity and cellular dedifferentiation take place within this area, the stump comes to resemble the normal limb of a salamander (Fig. 27.35).

Dedifferentiation is most common in animals that grow throughout their life cycles: species like many fish, reptiles, and amphibians that have no typical adult size but simply grow larger and larger until they die. With the need to be able to grow indefinitely, it may be equally necessary to keep the developmental options of at least some cells open forever.

Although dedifferentiation is possible for some cells in some species, can determination itself be reversed? In fact, we have already seen that it can: cancerous cells suffer mutations that remove some of the determining constraint, and return a cell to what often appears to be an earlier, less differentiated, rapidly growing stage. However, there are as yet few cases to suggest that a determined state can be altered in animals.

We have already seen that the changes undergone by the nuclei of plant cells during differentiation appear to be reversible. But under what conditions is differentiation and determination reversible in animals? One way to find out is to perform nuclear-transplant experiments, in which a nucleus from a differentiated cell is inserted into an undifferentiated cell from which the nucleus has been removed. If the transplanted nucleus has not undergone any permanent change, it should be able to resume unlimited developmental potential (i.e., be totipotent). Experiments of this kind were performed on frog embryos in 1953 by Robert W. Briggs and Thomas J. King of the Institute for Cancer Research in Philadelphia. Working with nuclei taken from embryos at various stages and transplanted into enucleated egg cells (cells from which nuclei were removed), they found that when the transplanted nucleus was from a cell in the animal pole of a blastula, the egg usually developed normally. This seemed to indicate that nuclei in blastula-stage cells had not been fully determined and therefore retained the capacity to direct all aspects of development. But if the transplanted nucleus was from a late gastrula, the eggs usually did not develop normally, indicating that the nuclei of gastrula cells had undergone some stable change, i.e., determination. Did this mean that the nuclei of differentiated frog cells, unlike those of plant cells, permanently lose some of their former developmental potential?

This is difficult to answer because, even though most nuclei from late gastrula cells in Briggs and King's experiments seemed unable to direct development of a normal embryo, a very few such nuclei could do so. Moreover, in later experi-

MAKING CONNECTIONS

1. Why can human beings give birth to identical twins, whereas snails cannot?

2. Can you suggest a way in which temperature or pH could influence cell differentiation?

3. What does the similarity of cell-adhesion molecules (CAMs) to antibodies, T-cell receptors, and MHC molecules suggest about the evolutionary origin of these immune-system molecules?

4. How are homeotic genes related to the genetic control mechanisms discussed in Chapter 12?

5. How do the nuclear transplantation experiments performed with amphibian embryos bear on the possibility of creating clones of identical human beings?

ments with the African clawed toad (*Xenopus*), J. B. Gurdon of Oxford University found that about 2 percent of the nuclei taken from intestinal cells of swimming tadpoles could direct completely normal development when injected into enucleated egg cells. The fate of these nuclei from fully differentiated cells should have been far more rigidly determined than that of the gastrula nuclei of Briggs and King, yet they seemed able to revert completely to the egg-stage condition and begin development all over again. Possibly these few nuclei had suffered mutations that had dedetermined them, or perhaps there really is a small probability of spontaneous loss of determination.

THE ORGANIZATION OF NEURAL DEVELOPMENT

Many aspects of development are illustrated with particular clarity in the nervous system. Nerve cells are "born," migrate to their proper places, send out fibers called axons to specific target locations, and so come to form a highly integrated functional network that is more complex than any other system in the body.

Many Neurons Must Migrate From Their Place of Origin to Their Final Location

The first step in the life of a newly formed neuron, like many other cells in a developing organ, is usually movement from where it was formed to where it is supposed to be. The cells that give rise to the adrenal medulla, for instance, must migrate from a spot just above the spinal cord down to specific spots below the notochord. At least three mechanisms seem to be involved in guiding the migrating cells along the proper pathway: chemical gradients, cell-adhesion molecules, and tactile cues.

Diffusing chemicals set up gradients that help guide neuronal cells by causing them to move in an amoeboid fashion toward the source of the diffusing chemical, much as a white blood cell moves toward a damaged cell that is releasing histamine. But most pathfinding seems to involve the other two mechanisms—CAMs and tactile cues. Migrating cells apparently follow their respective paths by means of the particular ratio of various CAMs they encounter on the surface of special "guide" cells along the path. Migrating cells move along the guide cells maintaining intimate tactile contact. When a migrating cell encounters the optimum CAM arrangement on the cells it has touched, it stops moving and proceeds to form cell-to-cell attachments that will anchor it in place.

Once a neuronal cell has reached its permanent place in the nervous system, it must send out long thin processes specialized for transmitting information to specific target cells. Here again, both chemical and tactile information seem to play a role.

Cell Death Plays an Important Role in Development

In most animal nervous systems, many more cells are born than are actually put to use. In vertebrates, for instance, identical groups of neural cells develop next to each of the vertebrae. However, it is only the many muscles and sensory receptors of the arms and legs that require as many cells as these groups contain. In other areas, the extra cells die and leave no progeny. Apparently it is easier or more efficient for the developmental program to build all segments alike initially, and then to allow functionless cells to die.

CONCEPTS IN BRIEF

1. The process of development in a sexually reproducing multicellular animal begins with the maturation of the egg. Penetration of the sperm into the ovum activates the developmental program. *Fertilization* occurs when the two gametic nuclei fuse.

2. The zygote then undergoes *cleavage;* the cytoplasm of the one large cell is partitioned into many new smaller cells. Cleavage continues until a hollow ball, the *blastula,* is formed.

3. Next begins a series of complex movements that establish the shape and pattern of the developing organism *(morphogenesis).* The blastula is converted into a *gastrula.* Gastrulation first produces an embryo with two layers, an outer *ectoderm* and an inner *endoderm.* A third layer, the

mesoderm, forms between them. The ectoderm gives rise to the outermost layers of the body, the nervous system, and the sense organs; the endoderm to the lining of the digestive tract and associated structures; and the mesoderm to the supportive tissue—muscles and connective tissues.

4. The morphogenetic movements of gastrulation and neurulation give shape and form to the embryo, and bring masses of cells into proper position for their later differentiation into the principal tissues of the body. The developmental processes of cell division, cell growth, cell differentiation, and morphogenetic movement convert the gastrula into a young animal ready for birth.

5. The predominant factor in postembryonic development in most animals is growth in size. Growth does not occur at the same rate and at the same time in all parts of the body.

6. Many aquatic animals and certain groups of terrestrial insects undergo *metamorphosis.* Some insects show a *complete metamorphosis,* which begins with a wormlike larval stage. The larva then enters the *pupal stage,* during which it is reorganized to form the adult. Other insects undergo *gradual metamorphosis;* the young go through a series of molts that makes them more and more like the adult.

7. Aging is the complex of developmental changes that leads to the deterioration of the mature organism and ultimately to its death. The aging process seems to be correlated with the degree of cellular specialization; cells that remain relatively unspecialized and continue to divide do not age as rapidly as cells that have lost the capacity to divide.

8. Though the genetic content of all embryonic cells is identical, their cytoplasm is not. The daughter cells produced by cleavage of the egg cell may not share equally in all cytoplasmic materials, depending on the orientation of the first plane of cleavage. If the cleavage divides critical cytoplasmic constituents equally, the cleavage is *indeterminate;* the new cells have full developmental potential and are totipotent. If, however, the first cleavage divides critical cytoplasmic constituents unequally, the cleavage is *determinate;* when the daughter cells are separated they will take different developmental paths.

9. Developing cells are influenced by neighboring cells, by diffusible chemicals they secrete, by hormones, and by the physical environment (e.g., light, temperature, pH, gravity). Certain areas of the embryo may induce the pattern of differentiation of cells in another tissue. Most developmental interactions are probably chemically mediated.

10. As development proceeds, the individual cells become determined, committing to one particular course of differentiation. Differentiation is thus a matter of progressive determination; development is gradually restricted to one of the many initially possible pathways.

11. As cells become more differentiated, the activity of the cells' genetic material changes. In plants, the changes undergone by cell nuclei during differentiation appear to be entirely reversible. In animals, the situation is less clear, but at least in some cases differentiation may be reversible.

12. Nerve cells are produced, migrate to their proper places, send axons to specific targets, and form a highly integrated functional network. The migration is oriented by diffusing chemicals, cell-surface chemicals, and tactile cues. In most animal nervous systems many more cells are produced than are used; those not needed die.

STUDY QUESTIONS

1. Describe how fertilization occurs. What is *capacitation?* What is the function of the fertilization membrane? Which event triggers development of the egg: fusion of the cell membranes of the egg and sperm, or fusion of their nuclei? Which is the actual fertilization event? (pp. 558–559)
2. What happens to the zygote during cleavage? What developmental stage marks the end of cleavage? Using the eggs of amphioxus, the frog, and the bird as examples, explain how cleavage is affected by the amount of yolk in the egg. (pp. 559–562)
3. What is meant by *morphogenesis* and how is it brought about in animals? Describe the formation of the gastrula. How is gastrulation influenced by the amount of yolk in the egg? (pp. 562–564)
4. Describe how the nervous system is formed during neurulation. Which other structures are derived from ectoderm? Which tissues and structures are derived from the other two primary germ layers? (pp. 565–567)
5. Diagram the shape of the normal growth curve. What factors can cause deviations from this pattern? In humans, do all parts of the body grow at equal rates? (pp. 569–570)

6. Describe the developmental stages in the life of an insect, such as a butterfly, that undergoes complete metamorphosis. Contrast this pattern of development with that of a grasshopper. (pp. 570–572)

7. How is the aging process related to cell specialization? What hypotheses have been proposed to explain why cells age? (pp. 572–573)

8. Distinguish between *differentiation* and *pattern formation.* Using the leopard frog as an example, explain how the distribution of cytoplasmic constituents in the egg and the pattern of early cleavage can influence the developmental potential of the cells of the early embryo. (pp. 573–574)

9. Describe how researchers have demonstrated experimentally the inductive influence on cells of hormones and neighboring cells. (pp. 575–578)

10. Explain how cell-adhesion molecules (CAMs) guide embryonic cells during morphogenesis. (pp. 579–580)

11. Distinguish between the *determination* and the *differentiation* of cells. Do these processes occur suddenly or gradually? Are they reversible in animals? (pp. 581–582)

12. Explain how chemical gradients, CAMs, tactile cues, and cell death all play a role in the development of the nervous system. (p. 584)

CHAPTER 28

NERVOUS CONTROL

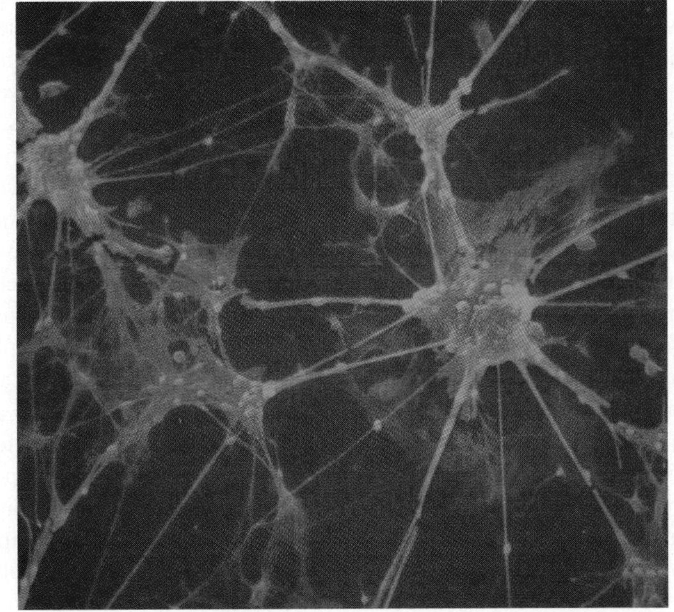

As we have seen, there is an intimate relationship between endocrine and nervous control systems in multicellular animals. Together these systems make possible animals' complex physiological and behavioral functioning. Although plants are complex, dynamic organisms that grow, change, and respond to external stimuli, they are fundamentally different from animals, above all in that they are sedentary while animals are mobile. Much plant behavior depends on variations in growth rates or changes in the turgidity of cells, both rather slow ways of bringing about movement. Animal behavior does not rely on such processes, for animals have evolved tissues specialized for production of rapid movement, notably the muscles. Animals, with their active mobile way of life, depend on quick responses to stimuli and rapid movements to guide and power the capture or harvesting of food, to seek mates, and to escape from predators. Correlated with this difference in speed—a result of the very different ways movement is produced—are basic differences in the control systems involved.

Hormonal control is a relatively slow process—too slow for the many rapid responses required by most animals. While a hormone is being transported in the phloem of plants or the bloodstream of animals, there is an appreciable delay between the release of the hormone and its arrival at the target cells. Response to the stimulus that induced secretion of the hormone is therefore not immediate; there is a lag of seconds, or minutes. For plants, which are sedentary and do not require rapid movement, the delay involved in hormonal control is insignificant. Hormonal control is likewise adequate for animals when instantaneous response is not needed, as in control of digestion, salt and water balance, metabolism, and growth. But when rapid response is needed, as in the movements produced by skeletal muscles, hormonal control is inadequate and nervous control is essential. A nerve impulse can move nearly 100 meters per second, thus reducing the interval between stimulus and response to milliseconds. The evolution of nervous and muscle tissues, then, allowed animals to respond quickly to external stimuli, such as available prey or a stalking predator, and was basic to the evolution of active multicellular animals as we know them today.

EVOLUTION OF NERVOUS SYSTEMS

Irritability Is a Universal Characteristic of Living Cells

The biological definition of irritability is quite different from the common usage of the word. ***Irritability,*** as we shall use the

Chapter opening photo: Surface view of neurons from the spinal cord of a human being. The interneurons and neurons have been stained to show their axons (thicker extensions), dendrites (thinner extensions), and cell bodies.

term, is the capacity to respond to ***stimuli***—an environmental change of some sort. Any manifestation of irritability—any reaction to a stimulus—ordinarily involves four principal components: (1) reception of the stimulus, (2) conduction of a signal, (3) "processing" of the signal, and (4) response. These stages are evident in all organisms, from relatively simple procaryotes to complex multicellular organisms like ourselves. In the *Paramecium,* for instance, specialized membrane proteins detect a stimulus, a change in membrane permeability spreads quickly around the cell, and a coordinated response is initiated—in this case a brief reversal of the cilia that propel the organism.

Any environmental change is potentially a stimulus, but whether it actually becomes one depends, first, on the capacity of the cells to detect that change. All living cells are irritable, but many environmental changes never function directly as stimuli, so far as we know, because no cells can detect them.

Conduction of a signal, the second component in a reaction to stimulus, is a capacity inherent in living cells. If a stimulus can produce a change in the cellular substance at some point (i.e., if it can be received), neighboring regions will be influenced to some extent; the initial change will spread. The spread may be limited and slow, or it may be extensive and rapid. It is this tendency for changes to spread from the point of origin that was probably the raw material for evolution of nervous conduction. The change in membrane permeability that occurs in a *Paramecium* after stimulation resembles that of a nerve impulse in animals. It is from this sort of mechanism of intracellular communication in early eucaryotes that specialized nerve cells evolved. All cells probably have some capacity to respond to stimuli, but animals have cells that have become specialized for this function. But the reception of stimuli and conduction of information would be useless without some information-processing system, the third component, to integrate the incoming information and coordinate the response.

The response, the fourth component, is the action prompted by receipt of a stimulus. The parts of the organism that do things, that carry out the organism's response to stimuli, are the ***effectors.*** The effectors in animals are most often muscles, but glands, such as the salivary or sweat glands, may also be effectors, responding to stimuli by increasing secretion.

Neurons Are Specialized for Reception and Conduction of Signals

The nerve cells, or ***neurons,*** are specialized for conducting information from one cell to another; they form a communications network that integrates and coordinates the activities of the organism. The anatomy of the various neurons often reflects their particular roles in detection, conduction, processing, and response control. Some neurons have many long thin processes (some as long as a meter in length), others are branched profusely rather like a tree, and still others have a spiny appearance (Fig. 28.1). Regardless of size or appearance, however, all neurons have the same functional organization, which enables them to collect information—from the environment directly (as sensory cells), or from other neurons, or from both—and to transmit it to target cells, such as other neurons, muscle cells, and secretory cells. The typical nerve cell consists of an enlarged region, the ***cell body,*** which contains the nucleus and other cytoplasmic organelles, and one or more long processes, or nerve fibers. The fibers on which information is generally received are the ***dendrites;*** those on which it is generally transmitted to other neurons are the ***axons*** (Fig. 28.1). The axons have specialized terminals to transmit the signals to other neurons.

Dendrites are usually rather short and tapering, and neurons characteristically have many of them, receiving input from perhaps several thousand other cells. Most dendrites are profusely branched and have a spiny appearance. As we shall see, it is on the fingerlike dendrites and the cell body that much of neural processing takes place.

In contrast to the multiplicity of dendrites, there is usually only one axon per neuron, and it is usually longer and thicker than the dendrites, and has a constant diameter. An axon may branch extensively at its terminal end (Fig. 28.1B–D), but does not have the spiny appearance of dendrites. These differences reflect the basic functional distinction between dendrites, which receive information from other cells, and axons, which transmit to other cells (though there are some exceptions).

Neurons usually do not touch each other directly; their fibers come very close but (except in a few special cases) a tiny gap called the ***synapse*** remains between them. As we shall see, synapses are of great importance to the processing of information in the nervous system.

Glial Cells Surround Neurons

In addition to neurons, the nervous system of vertebrates contains vast numbers of a second class of cells, the ***neuroglia,*** or simply ***glia*** for short. In the brain, glial cells outnumber neurons by 10 to one, and they occupy about half the cranial volume. Some glia provide the neurons with nutrients, and may absorb substances secreted by the neurons. Much of this absorbed material is then cycled back to the neurons for reuse. In at least some areas of the brain and spinal cord during development, glia provide a framework along which neurons migrate and axons grow to reach their targets. One class of glial cells found in vertebrates wraps around the axons of many neurons in the brain and spinal cord to form a heavily lipid ***myelin***

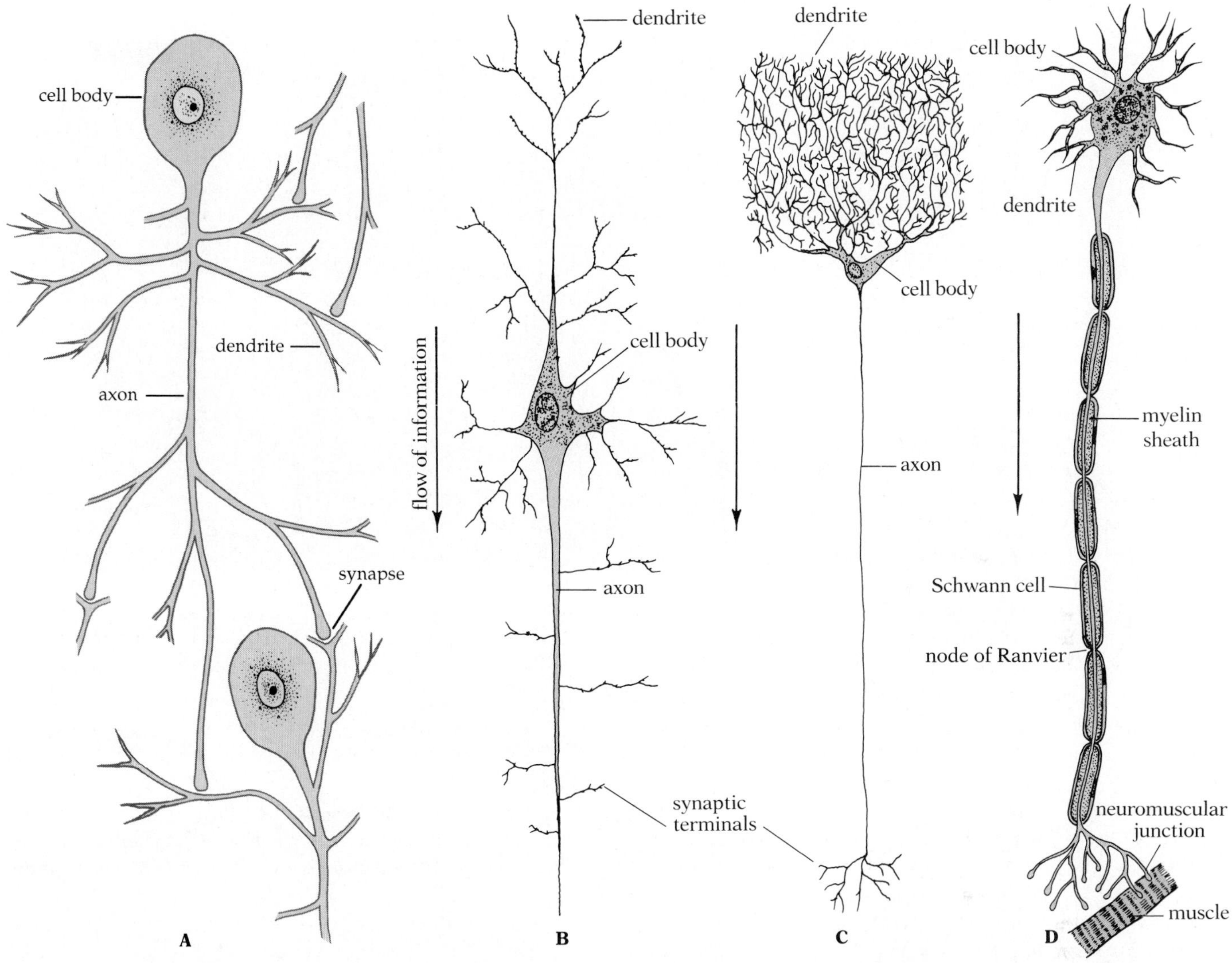

28.1 A variety of neurons Information is normally collected by dendrites, conducted along an axon, and transmitted to other cells by terminals, at information links called synapses. (A) In invertebrates, the cell body of the neuron usually lies out of the information pathway. (B) The dendrites, unlike the axon, often have a spiny look. (C) The dendrites of certain brain cells of vertebrates branch profusely, giving the cell a treelike appearance. (D) Vertebrate motor neurons have long axons that run from the brain and spinal cord to the effector (in this case muscle); these axons are frequently insulated by a series of myelin sheaths.

sheath (Figs. 28.2 and 28.3), which insulates against "cross talk" between adjacent axons and speeds up conduction. Axons with a myelin sheath are known as myelinated axons; those without are said to be unmyelinated.

The processes of vertebrate neurons leading to the periphery of the body are enveloped by special glial cells called ***Schwann cells*** (Fig. 28.2), which often give rise to myelin sheaths in much the same way as the glial cells within the brain do (Fig. 28.3). The sheath is not continuous—there are gaps called ***nodes*** (Fig. 28.1D), points where one glial or Schwann cell ends and another begins. Myelin sheaths here, as in the brain, serve to speed up the conduction of impulses in axons that have them, in addition to preventing "cross talk."

Neurons May Be Classified According to Their Function

Neurons conducting information from sensory receptors *into* the brain or spinal cord are called ***sensory*** or ***afferent*** neurons, while those conducting impulses *from* the brain or spinal cord to effector cells are called ***motor,*** or ***efferent*** neurons (Fig. 28.1D). Those neurons lying between the sensory and motor neurons—the middlemen of the nervous system—are called ***interneurons.*** They are specialized for the crucial role of processing information. An interneuron typically collects and processes input from many neurons (often thousands), and

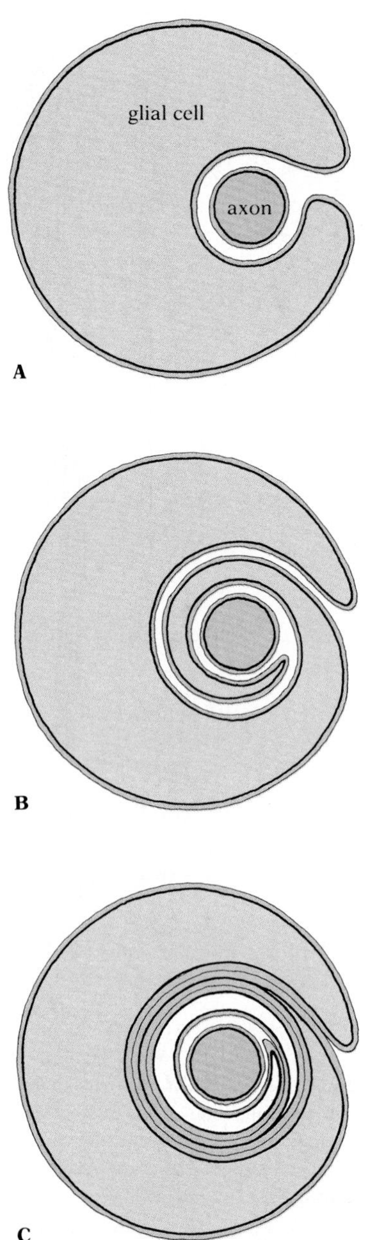

28.2 Development of the myelin sheath Initially the unmyelinated axon lies in a pocket of the glial cell (A). The glial-cell membrane begins to coil around the axon (B). The membrane is wound tightly around the axon, forming what is known as a myelin sheath (C).

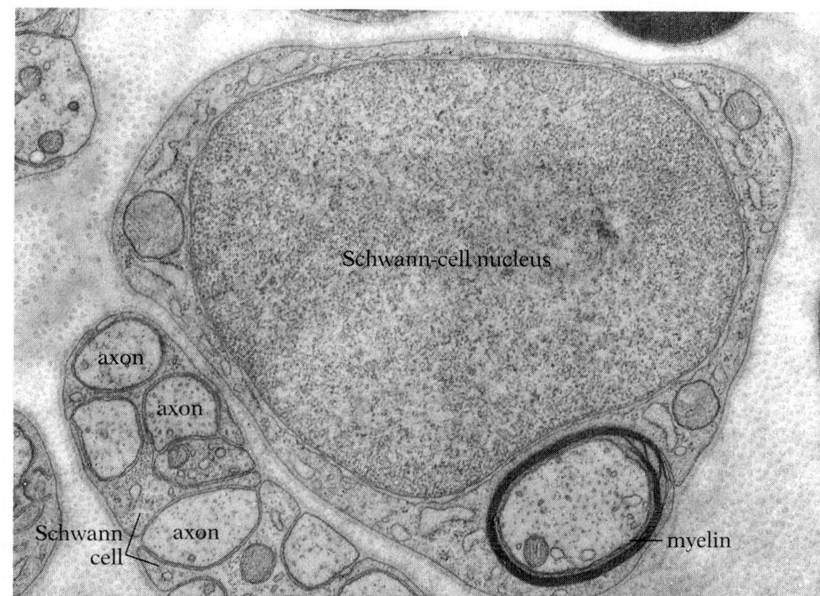

28.3 Electron micrographs of myelin sheath (A) The axons of neurons are enveloped by Schwann cells. At lower left a single Schwann cell envelopes several unmyelinated axons. The axon at lower right has a myeline sheath formed from an invaginated coiled portion of the Schwann-cell membrane that is tightly wound around the axon. Myelinated axons conduct impulses faster than unmyelinated axons. (B) The freeze-fracture electron micrograph shows many transversely fractured myelinated axons in the central nervous system. From *Freeze Fracture Images of Cells and Tissues* by Richard L. Roberts, Richard G. Kessel, Hai Nan Tung, Oxford University Press, NY © 1991.

passes on the resulting commands to its target neurons (Fig. 28.4). The signals an interneuron processes can be either excitatory or inhibitory. Indeed, ***inhibition,*** the ability to counteract excitatory input from other cells, is essential to this information processing, since the nervous system usually employs an antagonist strategy: contradictory signals are sent, and the ratio between them determines the cell's response. The importance of interneurons in this processing of contradictory signals would be hard to overemphasize: the brains of animals consist almost entirely of interneurons arranged in complex, highly specialized networks. The human brain is the most elaborate data processing center available—so much so that some computer scientists are trying to understand brain organization to build better computers.

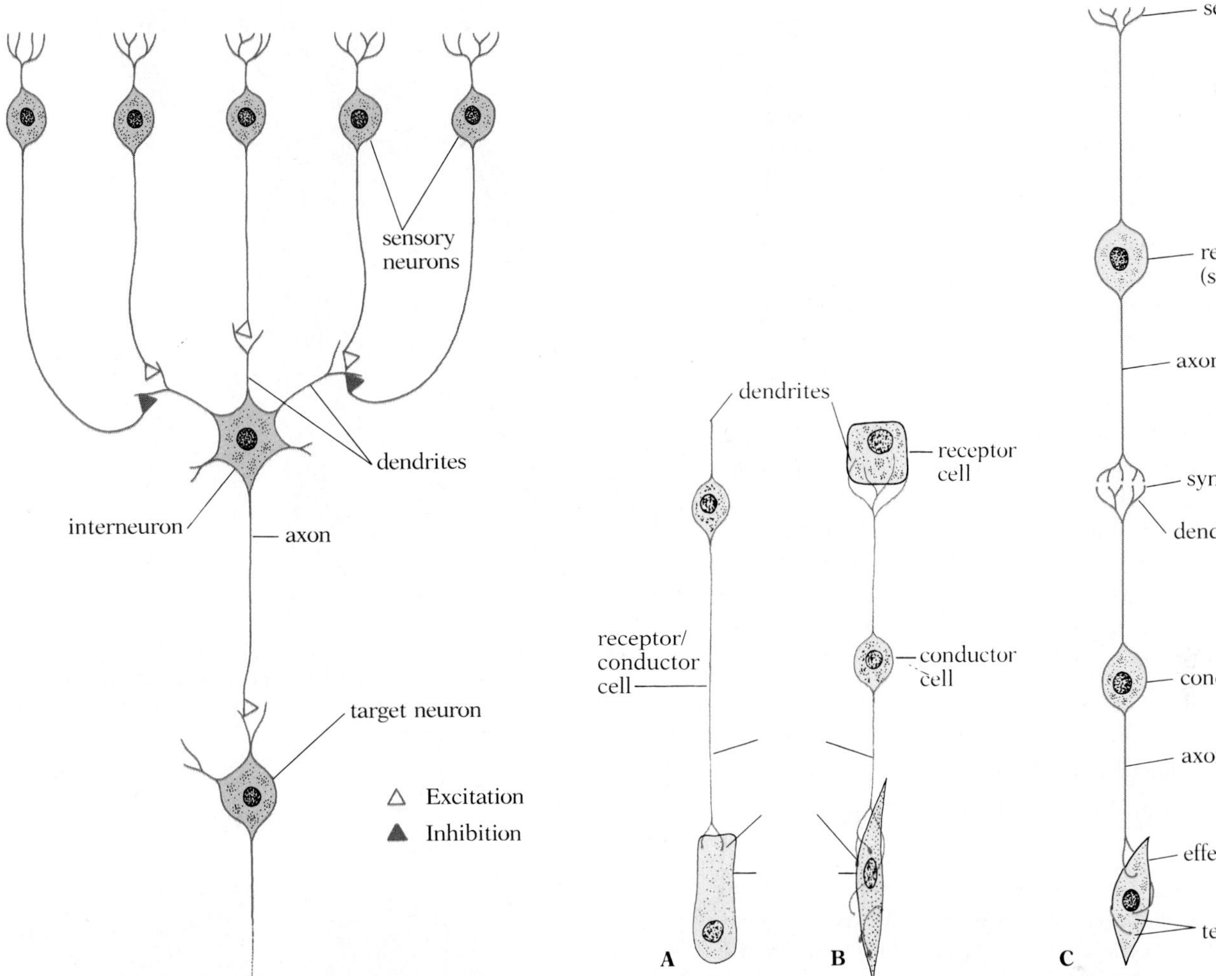

28.4 An interneuron Interneurons combine information from many cells to produce a single output. In this example five sensory neurons communicate with the interneuron, three that excite it and two that inhibit it. The relative strength of the various incoming signals determines whether excitation or inhibition predominates.

28.5 Simple nervous pathways (A) In the simplest pathway—the one generating tentacle withdrawal in cnidarians, for example—only two cells are necessary. A single neuron acts as both sensory receptor and conductor of information; the other cell is the effector, and no processing stage is required. Such pathways are called reflex pathways. (B, C) Slightly more elaborate pathways may involve a specialized sensory receptor cell, itself sometimes a neuron, and a separate conductor cell. Again, however, there is no processing. The conductor cells shown here display typical neuron anatomy: a large cell body containing the nucleus, and two narrow projections—the dendrite (usually branched) for collecting information, and the axon for carrying it to other cells.

Simple Nervous Pathways Allow for Little Information Processing

All animal groups above the level of the sponges have some form of nervous system, though in some groups it is very primitive. In the tentacles of some cnidarians, we find the simplest possible type of true nervous pathway—one composed of only two specialized cells, a receptor-conductor cell and an effector cell (Fig. 28.5A). Such a pathway generates strictly automatic behavior—tentacle withdrawal, for instance—because the processing stage is essentially missing. We refer to such automatic behavior as a pure ***reflex,*** because there is only one input and there are no alternative pathways for the information to take. This type of behavioral response involves no coordination of effectors, and because there is no interconnection between the various circuits of this sort, there is no possibility of central neural control.

And yet, simple as such a circuit looks, it does have one kind of flexibility: as a result of repeated stimulation the sensory neuron can become less sensitive, through a process known as ***adaptation.*** In cnidarians, for instance, adaptation allows the tentacles to adjust to the constant background level of stimulation produced by water currents, so that the defensive reflex is triggered only by an extraordinary stimulus. In humans, adaptation explains why we gradually acclimate to the cold water of a swimming pool or the hot water of a bath. Sensory adaptation is a widespread phenomenon.

Most nervous pathways, however, even in cnidarians, involve at least three separate cells: a ***receptor*** cell specialized for reception of a particular kind of stimulus, a ***conductor*** cell specialized for conducting impulses over long distances, and an ***effector*** cell (often a muscle cell) specialized for giving a response (Fig. 28.5B). More complex pathways may involve any number of additional conductor cells interposed between the receptor cell and the effector cell (Fig. 28.5C). Once the pathways include several conductor cells, more processing of information becomes possible and the response can be more flexible, because more than one route is usually open to the impulse coming from the receptor. Any one of several alternative effectors, or all of them, may be activated. *In general, the more conductor cells in the circuitry, the more flexible the response.*

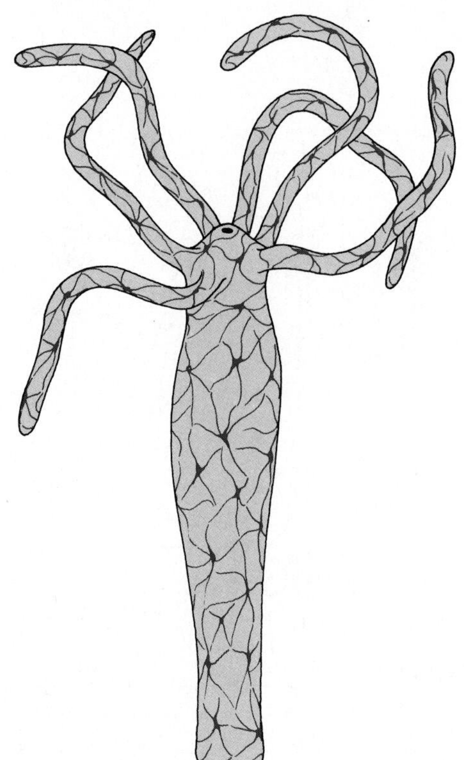

28.6 Nerve-net system of hydra Conductor cells in organisms with nerve nets are not organized into specialized pathways. As a result, there can be no centralized control; only localized responses to stimuli are possible.

Radially Symmetrical Cnidarians Have a Diffuse Nerve Net With Little or No Central Control

The simplest form of organized nervous system, seen in cnidarians of the hydra type, consists of separate receptor, conductor, and effector cells. However, instead of forming definite pathways, the conductor cells interlace, forming a diffuse ***nerve net*** that runs throughout the body (Fig. 28.6). There is no central control: impulses simply spread slowly from the region of initial stimulation to adjacent regions, and in general they can move in either direction along the fibers. The stronger the initial stimulus, the farther the impulses will spread. This sort of organization suits a sedentary organism whose only means of escape involves moving the tentacles and the side of the body that has been touched. Nerve-net reactions generally take the form of localized movements, and in some creatures, the discharge of stinging cells (nematocysts) into potential prey. Such a system, lacking the potential for central coordination of complex reactions, can produce only a limited behavioral repertoire.

Other cnidarians display a degree of centralization. Jellyfish, for instance, have a nerve ring in the "bell" portion of the body (Fig. 28.7). Other neurons funnel into the ring and conduction from one part of the animal to another is more rapid than is possible with a simple nerve net. This centralization is reflected in the rhythmic, coordinated swimming movements of jellyfish.

Radial symmetry, such as that in the cnidarians, is often associated with a sedentary way of life. To an animal that remains

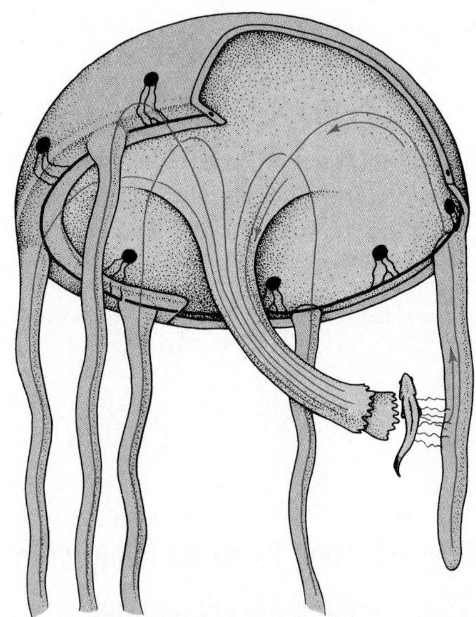

28.7 Nervous system of jellyfish Jellyfish display a degree of centralization among radially symmetrical animals. Here neurons are organized into a primitive ring system that serves to synchronize the contractions of the swimming muscles of the bell. In addition, sensory neurons on each of the peripheral tentacles send axons to muscles on the central stalk. When the firing of a nematocyst stimulates certain of these cells, the resulting signals direct the creature's mouth (which is at the base of the stalk) to the affected part of the tentacle for a possible meal.

attached to some immovable object or moves about slowly, there are obvious advantages to being equally receptive to stimuli on all sides. However, this type of symmetry restricts the development of highly specialized structures like eyes, because it requires multiple repetitions of such structures. Thus, radial symmetry severely restricts the evolutionary potential for extensive centralization of nervous systems.

The Bilaterally Symmetrical Body Form Favors Centralization of the Nervous System

The major trends in the evolution of nervous systems in bilaterally symmetrical animals can be summarized as follows:

1. The nervous system became increasingly centralized by formation of major longitudinal nerve cords (the ***central nervous system***) through which most pathways between receptors and effectors had to pass.
2. Conduction along nervous pathways became restricted to only one direction. A distinction thus developed between sensory fibers leading *toward* the central nervous system (afferent fibers) and motor fibers leading *away* from the central nervous system (efferent fibers).
3. Nervous pathways within the central nervous system became increasingly complex by the addition of large numbers of interneurons—a development that allowed increased information processing and flexibility of response.
4. Cells performing different functions became increasingly segregated within the nervous system, with the result that distinct functional areas and structures developed.
5. Increasing development of the front end of the longitudinal nerve cords led to formation of a ***brain*** at the anterior end of the animal, which became more and more dominant over the rest of the system—a process called ***cephalization.***
6. The number and complexity of sense organs increased.

These trends are not yet very distinct in the most ***primitive*** flatworms (those thought to be most like the ancient ancestral forms). Such flatworms have only a nerve net much like that of hydra. Some slightly more ***advanced*** flatworms (those less like the ancestral forms; derived) show the beginnings of a condensation of major longitudinal cords within their nerve nets (Fig. 28.8A). In somewhat more advanced flatworms the cords are more developed but still numerous; there are often as many as eight, located ventrally, dorsally, and laterally (Fig. 28.8B). Still more advanced flatworms show a reduction in the number of longitudinal cords (Figs. 28.8C, D), and the most advanced representatives, such as a freshwater planarian, have only two, both located ventrally (Fig. 28.8E).

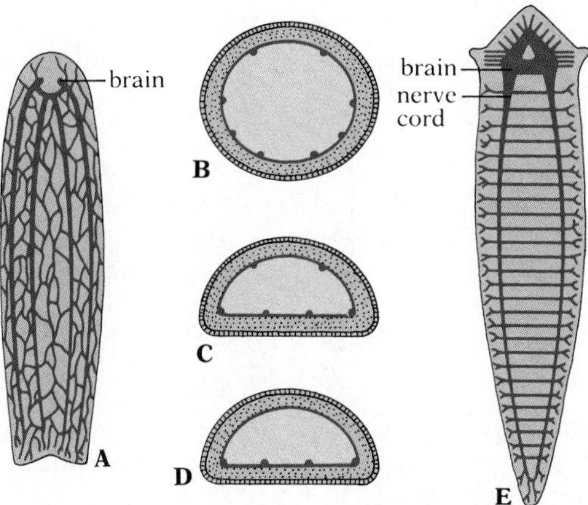

28.8 Nervous systems in flatworms (A) Nervous system of a primitive flatworm. This is essentially a nerve net, but several longitudinal cords are present within it; the so-called brain is only a very tiny thickening at the anterior end of these cords. (B) Cross section of a somewhat more advanced flatworm with eight well-developed cords. (C, D) Cross sections of more advanced flatworms, showing the progressive reduction in the number of cords. (E) Ventral nerve plate of planaria, with two cords and a moderately developed brain.

Those flatworms with the most primitive development of longitudinal cords show very little evidence of any special structure at the anterior end that could be called a brain (Fig. 28.8A); biologists have, however, charitably labeled as a "brain" the tiny swellings present there. Flatworms at more advanced stages show a much better developed brain (Fig. 28.8E), though even this brain exerts only limited dominance over the rest of the central nervous system.

The ancestral version of the brain was probably almost exclusively concerned with funneling impulses from the sense organs into the cords. Then, because of the adaptive advantage of shortening the pathway these impulses have to follow before reaching the main coordination areas of the central nervous system, natural selection favored grouping those areas toward the anterior ends of the cords.[1] Thus the brain came to be more than a sensory funneling area; as coordination became increasingly concentrated in it, the brain became more and more dominant over the rest of the central nervous system. This dominance has its greatest development in mammals, especially human beings.

The evolutionary trends whose beginnings can be seen so clearly in flatworms, where intermediate stages between a nerve-net system and a centralized system can be studied in living animals, culminate in the vertebrates and, among invertebrates, in the annelids and arthropods. In all these animals there is a high degree of centralization. In annelids and arthropods the central nervous system consists of a pair of ventrally

[1] Some dinosaurs, whose immense size made nerve transmission torturously slow, had auxiliary brainlike ganglia near the tail, to facilitate rapid response in the rear parts.

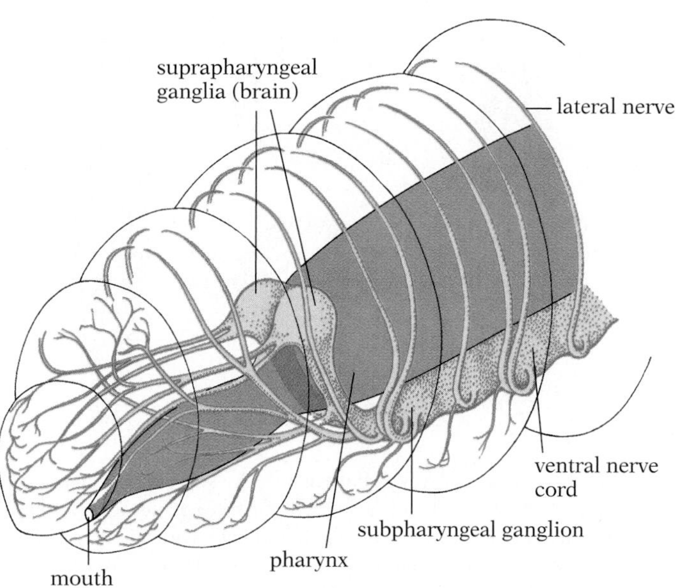

28.9 Earthworm neural organization Many of the sensory receptors in the earthworm are located in the head, and carry information to the brain (the suprapharyngeal ganglia) for processing. From there information is conducted along the ventral cord and ultimately to the muscles. The earthworm brain serves as a device for deciding what action is to be taken and where to direct it; the subpharyngeal ganglion is the center for motor control.

located longitudinal cords, in which the cell bodies of the neurons are grouped together to form masses called ***ganglia,*** and the neuron fibers are gathered into bundles (Fig. 28.9). Thus in the ancestral annelids and arthropods there are prominent ganglionic masses, a pair in each body segment, which are connected by bundles of fibers running between the segments. The brain is simply another ganglion located in the animal's head. It is little, if any, larger than the segmental ganglia; its dominance over the other ganglia is noticeable, but limited in comparison to that of the vertebrate brain. The more complex annelids and the arthropods have increased numbers of interneurons in their nerve cords, ganglia, and brain, with the result that there is more information processing and increased flexibility of response.

The Central Nervous System of Vertebrates Differs From Invertebrates

While the nervous systems of many annelids and arthropods show some centralization and cephalization, none show the extensive development of a brain that is characteristic of vertebrates. The vertebrate brain exerts far more dominance over the entire nervous system than does the brain of any other animal group. In short, the vertebrate brain is the master control center for all bodily functions. Nowhere is this more true than in humans, whose large cerebral cortex is the major coordinating center for sensory and motor functions involving all senses and all parts of the body, as well as having areas devoted to memory formation and storage.

The central nervous system (brain and spinal cord) of vertebrates differs in several other important ways from those of annelids and arthropods:

1. The vertebrate spinal cord is single, and found along the dorsal wall of the body. Recall that the nerve cord develops from the embryonic neural tube, which has a hollow central canal. A remnant of that central canal survives in the adult, so the nerve cord is said to be "hollow" (see Fig. 28.23). In comparison, the nerve cords of annelids and arthropods are double (two cords lying side by side and often partly fused); ventrally located, and solid (e.g., no central cavity or canal).
2. The vertebrate spinal cord is not so obviously organized into a series of alternating ganglia and connecting tracts.
3. Although many coordinating functions in vertebrates are still performed by the spinal cord, the enlarged brain exerts dominance over the entire nervous system.

TRANSMISSION OF NERVE IMPULSES

The transmission of impulses along neural pathways may be as fast as 30 to 90 meters per second. To explain how nerve impulses are transmitted, we shall consider separately conduction along nerve cells and transmission across the synaptic gaps between cells.

Neurons Respond to a Stimulus in an All-or-None Fashion

Neurons will respond to a great variety of stimuli, such as a mild electric shock, a pinch, or an abrupt change of pH. Various types of sensory neurons and sensory cells are specialized to respond to light, odors, movement, and so on. However, mild electrical stimuli are most often used in research because the intensity and duration of such stimuli can be precisely controlled, and because they do little or no damage to the nerve cell. Consider the following experiment.

Suppose we were working with an isolated axon and we had placed two electrodes several centimeters apart on its surface (Fig. 28.10A). The electrodes are connected to recording equipment, so we can detect any electrical changes that occur at the points they touch on the axon. Next we apply an extremely mild electrical stimulus. Nothing happens; our recording equipment shows no change (Fig. 28.10B). We increase the intensity of the stimulus and try again. This time our equipment tells us that an electrical change occurred at the point in

MAKING CONNECTIONS

1. Contrast the mechanisms used by plants and animals to respond to external stimuli.

2. Compare the rapidity of the responses mediated by hormonal and nervous control. Why is hormonal control adequate for plants but not for animals?

3. You go swimming and find that although the water feels very cold initially, it gradually becomes quite comfortable to you, even though the water temperature has not changed. What is the explanation for this phenomenon?

4. Why is radial symmetry advantageous to a sedentary animal? Why is this type of symmetry an obstacle to the evolution of a centralized nervous system?

5. How does the nervous system of vertebrates differ from those of annelids and arthropods? How is the structure of the vertebrate nerve cord related to the process of neurulation described in Chapter 27?

contact with the first electrode and that a fraction of a second later a similar electrical change occurred at the point in contact with the second electrode (Fig. 28.10C). We have succeeded in stimulating the axon, and a wave of electrical change has moved down the axon from the point of stimulation, passing first one electrode and then the other at a rate of 30 to 90 meters per second. Next we apply a still more intense stimulus, and again we record a wave of electrical change moving down the fiber (Fig. 28.10D), but the intensity and speed of this electrical change are the same as those recorded from the previous milder stimulation.

We could draw several important conclusions from this experiment: (1) a nerve impulse can be detected as a wave of electrical change moving along an axon; (2) a potential stimulus must be above a critical intensity to stimulate an axon; this critical intensity is known as the ***threshold value,*** and it differs for different neurons; and (3) increasing the intensity of the stimulus above the threshold value does not alter the intensity or speed of the nerve impulse produced; the axon fires maximally or not at all, a type of reaction commonly called an ***all-or-none response.***

Immediately an important question comes to mind: if an axon exhibits the all-or-none property with respect to the intensity and speed of an impulse, how do animals normally detect the intensity of a stimulus? How does, for instance, a snake

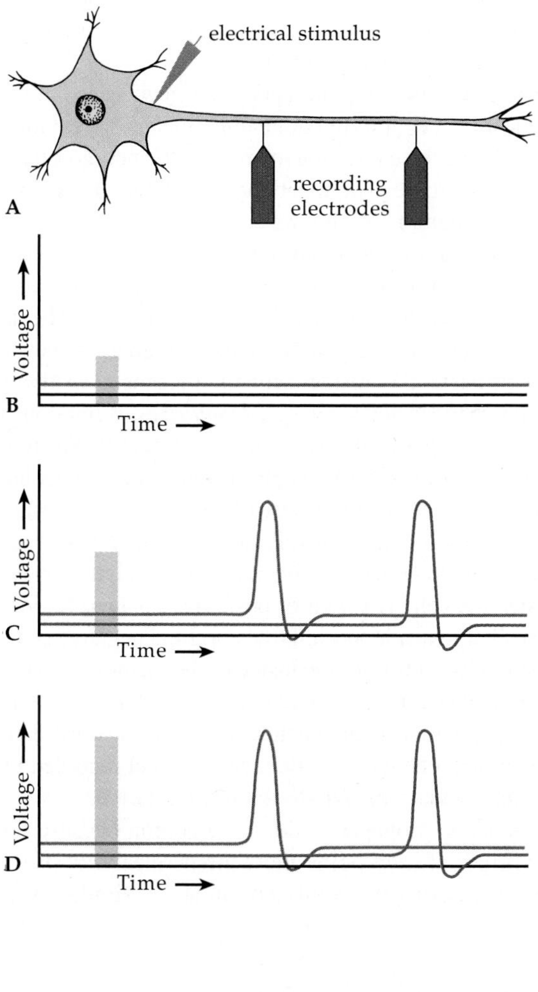

28.10 Initiation and propagation of an impulse (A) By inserting an electrode into the cell body of a neuron to deliver precise amounts of current, and then monitoring the axon with recording electrodes, we can study the initiation and propagation of an impulse. In B–D the magnitude of the stimulus is shown in green, at left. (B) If the stimulus is below the cell's threshold for triggering an impulse, neither recording electrode registers any change. (C) If the stimulus is adequate, however, an impulse is registered as a spike first at electrode 1 (red) and then at electrode 2 (blue) as it moves along the axon. Note that the spike has the same magnitude at both places. (D) Even when a still larger stimulus is used, the speed of movement and the intensity of the impulse are not affected: the impulse is an all-or-none response.

warming itself in the sun sense when its body temperature gets too high, prompting it to move into the shade? There are two ways in which information about stimulus intensity is commonly coded. First, as the intensity of the stimulus increases, the number of impulses produced per unit time—that is, the frequency of impulse generation—goes up. Second, neighboring cells may have different thresholds, so as the intensity of the stimulus increases, the thresholds of more and more cells are exceeded and more neurons fire. The brain can interpret both the higher firing rate of individual neurons and the greater number of active neurons as indicating a more intense stimulus.

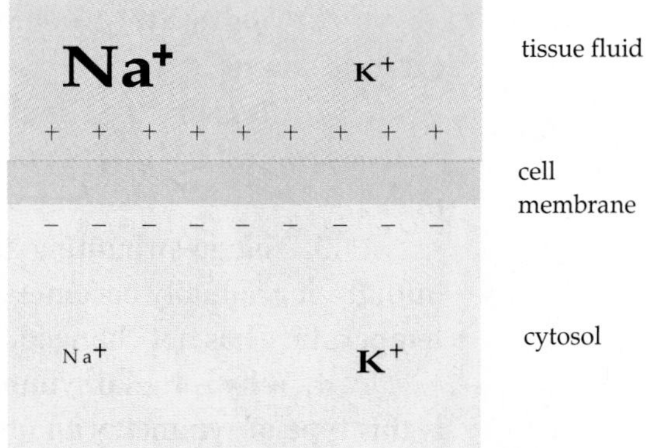

28.11 The polarization of the nerve-cell membrane The concentration of sodium ions (Na^+) is much greater in the tissue fluid outside the cell, and the concentration of potassium ions (K^+) is greater inside the cell. Because the excess of Na^+ ions outside is larger than the excess of K^+ ions inside, the cell has a deficit of positively charged ions. As a result, the inside of the cell is negative (about -70 millivolts) relative to the outside. (The positive charge of the tissue fluid exerts an electrostatic force on the negative ions in the cell, drawing many of them to the membrane, as indicated here by the line of minus signs.)

A Nerve Impulse Is a Wave of Electrical Change Moving Along a Nerve Fiber

When it was discovered over a century ago that a nerve impulse involves electrical changes, scientists assumed that the impulse was a simple electric current flowing through a nerve, just as other currents flow through wires. It was soon shown, however, that the speed of a nerve impulse is far slower than the speed of electricity in a wire and, further, that the cytoplasmic core of an axon offers so much resistance to simple electric currents that the currents die out after moving only a few millimeters. The fact is that any resistance at all, however low, would cause a simple electric current to diminish in strength as it moved. Yet if we measure a nerve impulse at various points along the axon of a neuron, we find that it remains the same; its strength does not decrease with distance. On the other hand, crushing or poisoning an axon may destroy its ability to conduct impulses even though its electrical conductivity has not been altered. In short, impulse conduction depends on activity by the living cell. The impulse is not a simple electric current, but an electrochemical change propagated along the neuron.

The basic outlines of the modern theory of nerve action were proposed in 1902 by the German physiologist Julius Bernstein. Called the Hodgkin-Huxley model of impulse conduction (named after two of the scientists who worked on it), the theory rests upon the fact that the concentrations of certain ions inside a nerve cell and in the surrounding fluid are very different: inside the cell, the concentration of sodium ions (Na^+) is very low and the concentration of potassium ions (K^+) and negatively charged organic ions is very high (Fig. 28.11). The opposite conditions exist outside the cell. This unequal distribution of ions results in a difference in charge across the cell membrane; the inside of the cell has a negative charge of about −70 millivolts (mV) compared to the outside. Bernstein suggested that the permeability of the nerve-cell membrane differs for various ions, and that it is the great selectivity of the membrane that maintains the separation of ions and the resulting electric potential. An important factor in maintaining the electrical potential across the membrane is the presence in the membrane of special transport proteins called ***potassium leak channels.*** These channels allow K^+ ions to diffuse out of the cell, down its concentration gradient, thereby contributing more positive charge to the outside and increasing the potential difference. (Only relatively small numbers of K^+ ions leave, however, because the negative charge inside the cell attracts the K^+ ions, and holds them inside.)

In the resting state then, the nerve-cell membrane is relatively permeable to K^+ ions but quite impermeable to Na^+ ions. Stimulation, however, causes the membrane to undergo a large but short-lived increase in permeability to Na^+ ions. These ions rush across the membrane into the cell, both because of their natural tendency to diffuse from regions of their higher concentration to regions of their lower concentration and because they are attracted by the negative charge inside the cell. The inward flux of positively charged Na^+ ions is so great, however, that for a moment the inside actually becomes positively charged relative to the outside, i.e., it ***depolarizes.*** A fraction of a second later, the permeability of the membrane to Na^+ ions has returned to normal, while its permeability to K^+ ions has increased greatly. The K^+ ions now rush out of the cell because their concentration is higher inside the cell than outside, and because they are repelled by the momentary high positive charge inside the cell. This exit of positively charged K^+ ions restores the charge inside the cell to its original negative state; the cell ***repolarizes.*** In short, the inside of the membrane is initially negative; it becomes positive when the Na^+ ions flood inward and then negative again when K^+ ions rush outward. This cycle of electrical changes can be measured and recorded by inserting a microelectrode into the axon (Fig. 28.12).

The impulse is propagated along the neuron because the cycle of changes at each point alters the permeability of the membrane at the adjacent point and initiates a similar cycle

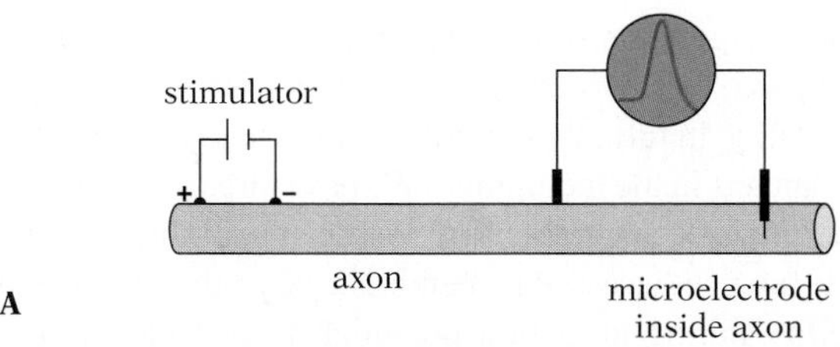

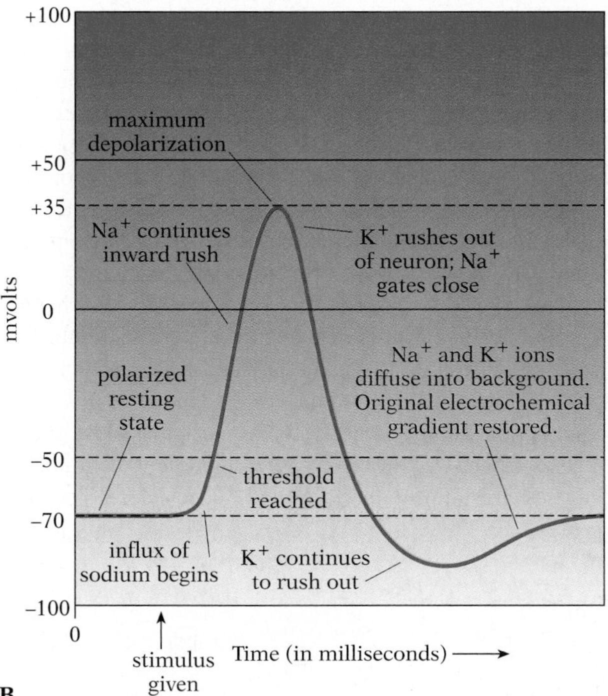

28.12 Recording of an action potential using an oscilloscope (A) The cycle of electrical changes in an action potential is measured using an oscilloscope. (B) The inside of a resting neuron is negative (-70 millivolts) compared to the outside. When a stimulus (arrow) above threshold is given, depolarization occurs: Positively charged Na^+ ions rush in and the inside of the neuron now becomes positive (+35 millivolts) compared to the outside. Next repolarization occurs: Positively charged K^+ ions rush out, restoring the resting potential. The sodium-potassium pump is continuously exchanging the Na^+ and K^+ ions.

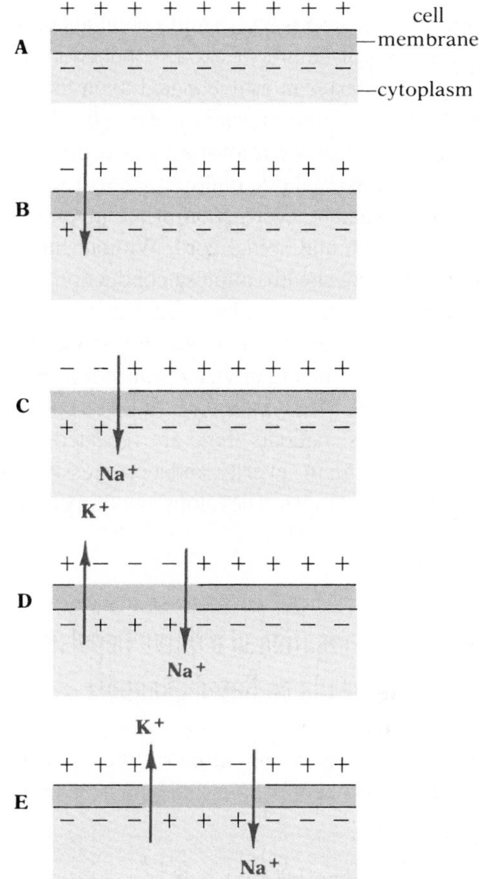

28.13 Model of propagation of a nerve impulse (A) The interior of a resting nerve fiber is negative relative to the exterior because the concentration of positive ions is higher outside the cell. The interior has a high concentration of potassium ions (K^+), the exterior an even higher concentration of sodium ions (Na^+). (B) When a fiber is stimulated, the membrane, previously relatively impermeable to Na^+ ions, becomes highly permeable to them at the point of stimulation, and a large number rush into the cell (arrow). The result is a reversal of polarization at that point, the inside of the fiber becoming positive relative to the outside. (C) Meanwhile, because the change in membrane potential at the point initially stimulated has altered the potential at adjacent points, the same cycle of permeability changes is initiated, and Na^+ ions begin rushing inward at those points. (D) An instant later, the membrane at the initial point of stimulation becomes highly permeable to K^+ ions; a large number of these rush out of the cell, and the inside of the fiber once again becomes negative relative to the outside. (E) The cycle of changes at each point alters the potential (and hence, the permeability) of the membrane at adjacent points and initiates the same cycles of changes there; Na^+ ions rush into the cell, and K^+ ions rush out a moment later. The movement of this cycle of changes along the nerve fiber is called a nerve impulse.

there (Fig. 28.13); this in turn starts the cycle farther along the axon, thus creating a domino effect. Each such cycle of electrical changes, known as an ***action potential,*** occurs at successive points in sequence. It is the action potential that constitutes the nerve impulse as it flows along a neuron.

The Myelin Sheath Insulates the Axon and Speeds Up Conduction

We said earlier that myelinated fibers conduct impulses faster than unmyelinated fibers of the same diameter. This occurs because the myelin sheath insulates the axon and prevents the movement of Na^+ and K^+ ions across the membrane. Consequently, the inward movement of Na^+ ions and the outward movement of K^+ ions that together constitute the action potential can occur only at the nodes where the sheath is absent. Thus the action potentials "jump" from node to node down the axon. Because fewer action potentials need be produced along a given length of axon, conduction may be as much as 50 times faster in a myelinated axon than in an unmyelinated axon. The importance of the myelin sheath to normal conduction is clearly demonstrated in people who suffer from the degenerative disease multiple sclerosis (MS). In this disease there is a patchy destruction of the myelin sheaths in neurons in the brain and spinal cord. Without the insulation provided by an intact sheath, impulse conduction is impaired. Progressive loss of function accompanies degeneration of the myelin. The disease begins with temporary symptoms such as weakness of a limb, blurring of vision, slurred speech, and difficulties in urination. The symptoms usually disappear spontaneously, but in most patients there are repeated attacks and remissions, increasing in severity, and a progressive weakness of the limbs and loss of vision develops, resulting in severe disability.

The Propagation of a Nerve Impulse Depends on Gated Channels

The propagation of a nerve impulse, or action potential, is basically a membrane phenomenon. It depends on an initial difference in charge (an ***electrostatic gradient***) across the membrane, followed by a coordinated series of ion-specific changes in permeability. In turn, these permeability changes depend on special channels in the membrane. We have already mentioned the potassium leak channels and their role in contributing to the charge difference across the membrane. But the changes in permeability during an action potential depend on different types of channels, called ***gated channels,*** that allow Na^+ or K^+ ions to cross the membrane. We know from Chapter 5 that in some channels a signal molecule binds to a channel protein causing it to change shape and open a gate that allows ions to cross through the membrane (see Fig. 5.14, p. 102). The membrane proteins responsible for creating the action potential in axons are a special type of gated channel called ***voltage-gated channels.*** Such channels open and close in response to changes in the charge across the membrane. Once an electrical stimulus has caused the membrane to depolarize by a precise amount and the electrostatic interactions are thereby weakened, the gates swing open, exposing specific ion channels (Fig. 28.14). The Na^+ gates open when the stimulus has reached the threshold for firing the neuron in question. Once enough Na^+ ions have flowed through to depolarize this area of the membrane almost completely, the Na^+ gates close and the K^+ gates open. As the electrostatic gradient continues to change as a result of ionic flow, the K^+ gates close. The nega-

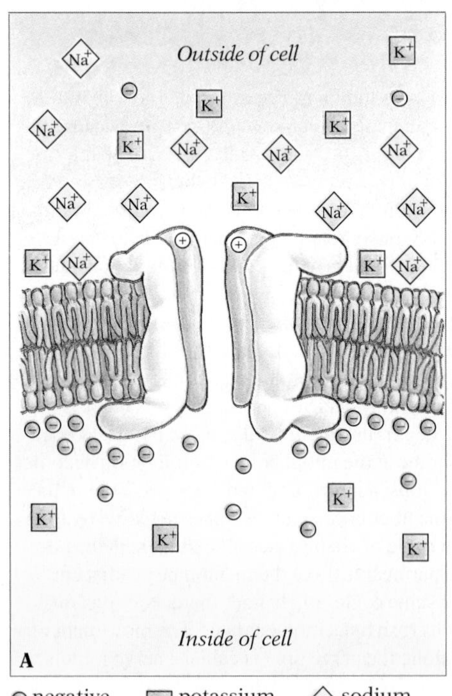

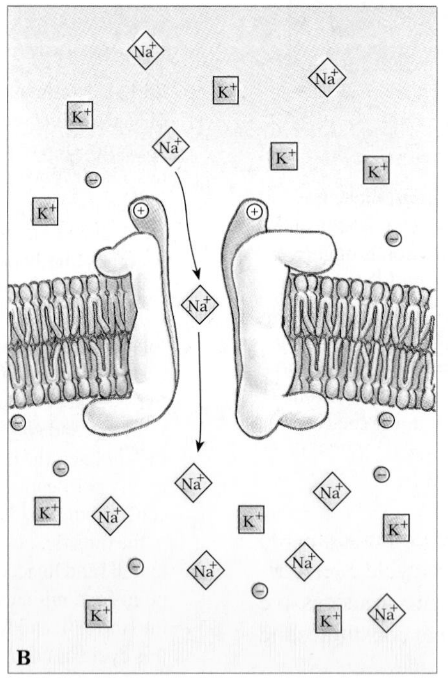

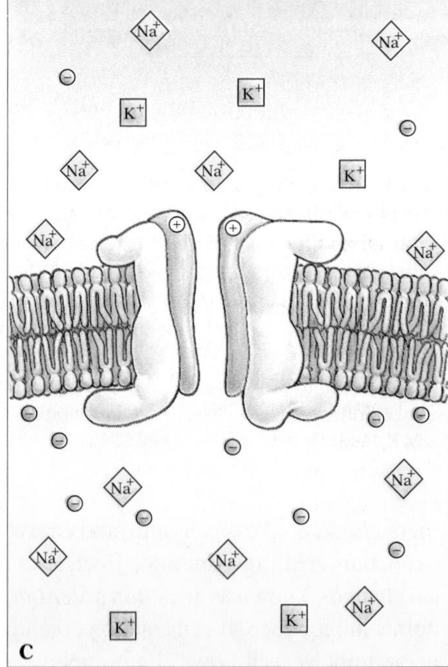

28.14 Voltage-gated channels (A) Axons have voltage-controlled gates held closed by the electrostatic repulsion of positive ions outside the cell and the attraction of negative ions inside. (B) When the membrane depolarizes, these gates swing open to expose specific ion channels. (C) As the electrostatic gradient continues to change during the course of the action potential, the gates close. The ions that cross the neural membrane during an action potential rapidly diffuse away from the membrane into the fluid inside and outside the cell.

tively charged organic ions do not move; they remain inside the cell.

The Ionic Balance of Na^+ and K^+ Ions in the Neuron Must Be Maintained

If impulse conduction involves inward flow of Na^+ ions followed by outward flow of K^+ ions, how does the neuron reestablish its original ionic balance? In other words, how does it get rid of the extra Na^+ ions and regain the lost K^+ ions? If the initial ionic distribution were not restored, the neuron would lose its ability to conduct impulses. But, we know that a normal neuron can continue to conduct impulses indefinitely, with only a very brief ***refractory period*** (on the order of 0.5 to 2 milliseconds) after each impulse, during which time it cannot be stimulated. (See Exploring Further: The Use of Anesthesia, p. 600.)

Three mechanisms work to keep neurons functioning: diffusion, the potassium leak channels that allow K^+ ions to diffuse out, and a sodium-potassium exchange pump. In the short run, as an impulse passes and the membrane is depolarized, diffusion restores the electrochemical balance between Na^+ ions outside and K^+ ions inside the cell almost instantaneously. Despite the dramatic nature of the events at the neural membrane, in reality only minute quantities of Na^+ and K^+ ions enter and leave the cell during an action potential. The net effect of a single action potential on the movement of Na^+ ions inside the cell as a whole is therefore negligible. The tiny alteration in charge that results is rapidly absorbed as the Na^+ ions that crossed into the neuron and the K^+ ions that moved out diffuse into the fluid on either side of the membrane. It is analogous to adding a drop of ink to a pond: for a moment there is a dark patch, but this rapidly dissipates as the ink diffuses into the larger volume of water. With the added Na^+ and K^+ ions thus dissipated, the polarity of the membrane is restored to the resting potential (Fig. 28.14).

The reserve of ions both inside and out is so large that a recently killed neuron will continue to conduct action potentials for some time. Eventually, however, the action potentials begin to deteriorate and conduction comes to a halt. If we were to continue polluting our pond with ink, at some point it would begin to change color. Similarly, at some point the internal fluid of the nerve cell can no longer sufficiently dilute the doses of Na^+ ions that have been repeatedly allowed to enter, and replace K^+ ions.

For a neuron to continue functioning in the long term, the proper ionic concentrations of Na^+ and K^+ ions inside and outside the cells must eventually be restored. Because expelling the Na^+ ions means moving them against their concentration and electrical gradients, and likewise regaining the lost K^+ ions means transporting them against their gradients, the neuron must carry out active transport, which requires energy. It has been demonstrated that the membrane of the neuron incorporates an ATP-driven ***sodium-potassium exchange pump.*** We know now that the membrane of the average neuron contains approximately a million sodium-potassium exchange pumps, and that their power is supplied by ATP. In fact, the sodium-potassium pumps are major energy-users in the cells of the human body; about 33 percent of the energy consumed is used to fuel these pumps. Such pumps enable cells to actively extrude Na^+ ions and take up K^+ ions. *The development of sodium-potassium pumps, combined with the evolution of ion-specific voltage-gated channels, is the basis for the evolution of neural transmission.*

The sodium-potassium pump has been the subject of intensive research in recent years. Though much about it is still unknown, enough facts have been garnered for a model to be proposed (Fig. 28.15). A protein complex in the cell membrane apparently acts both as the permease for the transport of Na^+ and K^+ ions and as the enzyme for breaking down ATP to ADP and inorganic phosphate. The protein has been designated ***Na^+K^+-ATPase*** to indicate that it is an ATPase active only in the presence of Na^+ and K^+ ions, for which it functions as a transmembrane pump.

According to the model, the Na^+K^+-ATPase extends through the entire thickness of the membrane. On the inner surface, facing the cytoplasm, it has a binding site for ATP and three binding sites for Na^+ ions. On its outer surface it has two binding sites for K^+ ions. It is thought that the binding of one ATP molecule to the protein and the subsequent hydrolysis of the ATP molecule causes the protein complex to change its shape, which transports three Na^+ ions outside and the two K^+ ions inside. The whole process depends on conformational changes in the Na^+K^+-ATPase that permit it to act as a channel through the membrane.

A powerful sodium-potassium pump would explain how the cell can maintain its low concentration of Na^+ ions and its high concentration of K^+ ions. *The pump exchanges Na^+ and K^+ ions according to a 3:2 ratio; for each three Na^+ ions pumped out, only two K^+ ions are pumped in.* The net effect is a flow of positive charge out of the cell, a flow that is augmented by the diffusion of K^+ ions through the potassium leak channels. In other words, the K^+ leak channels and the pump build up a separation of charge across the membrane by maintaining an internal charge that is negative with respect to the outside, much as a car battery maintains a voltage difference between the positive and negative terminals.

Transmission Across Synapses Usually Involves Transmitter Chemicals

We have said that a synapse is a gap at the junction between two neurons. Most often, the axon of one neuron synapses with

EXPLORING FURTHER

THE USE OF ANESTHESIA

Whether in a dentist's chair, in a doctor's office, or on an operating table, the patient's first concern is often "how much will it hurt?" Fortunately, there are a large number of chemicals, known as *anesthetics,* that result in a loss of awareness and make us insensitive to pain. Two basic types are available; local anesthetics and general anesthetics.

Local anesthetics such as novocaine and xylocaine are used when the operation is short and only a small part of the body is involved. They allow the patient to remain fully awake, yet the perception of pain is eliminated. How do the local anesthetics act to eliminate pain? Actually, the pain is not eliminated—the sensory receptors are still stimulated—but the *perception* of pain is eliminated. Local anesthetics act by blocking the sodium gates in the neuronal membrane. Because the sodium gates do not open, an action potential cannot be generated in the affected neurons. The pain receptors are stimulated, but nerve impulse transmission is prevented and the information is not conducted to the pain centers in the brain.

Local anesthetics can also be used to eliminate the perception of pain from fairly large areas of the body. In "spinal anesthesia" a local anesthetic is injected into the spinal canal to temporarily block nerve impulse transmission in regions below the injected area. The body parts in that region are temporarily paralyzed and insensitive. Unlike general anesthesia, where the patient is rendered unconscious, the patient with a spinal anesthetic is fully awake, but the perception of pain is eliminated in the affected body parts. And, because it is a local anesthetic, the patient suffers fewer complications and recovers more quickly.

General anesthetics are used for longer, more complicated surgery where it is advantageous for the patient to be unconscious and insensitive to pain. The gaseous general anesthetics such as ether, cyclopropane, and halothane have been widely used for many years, but we still do not know exactly how they work. Because these substances are nonpolar, it is assumed that they dissolve in the nonpolar portions of the membrane and reduce ionic permeability of the membranes, but experimental evidence detailing the precise action is still lacking. The evidence shows that even at low anesthetic concentrations nerve impulse transmission is depressed, but this could be due to reduced ionic permeability of the neuronal membrane, or to changes in synaptic transmission, or both. Anesthetics may, in some way, uncouple synaptic transmission from nerve impulse transmission. Alternatively, general anesthetics may act by interfering with enzyme activity or cell metabolism, thereby reducing the energy available to maintain the ionic gradients or to synthesize transmitter substances.

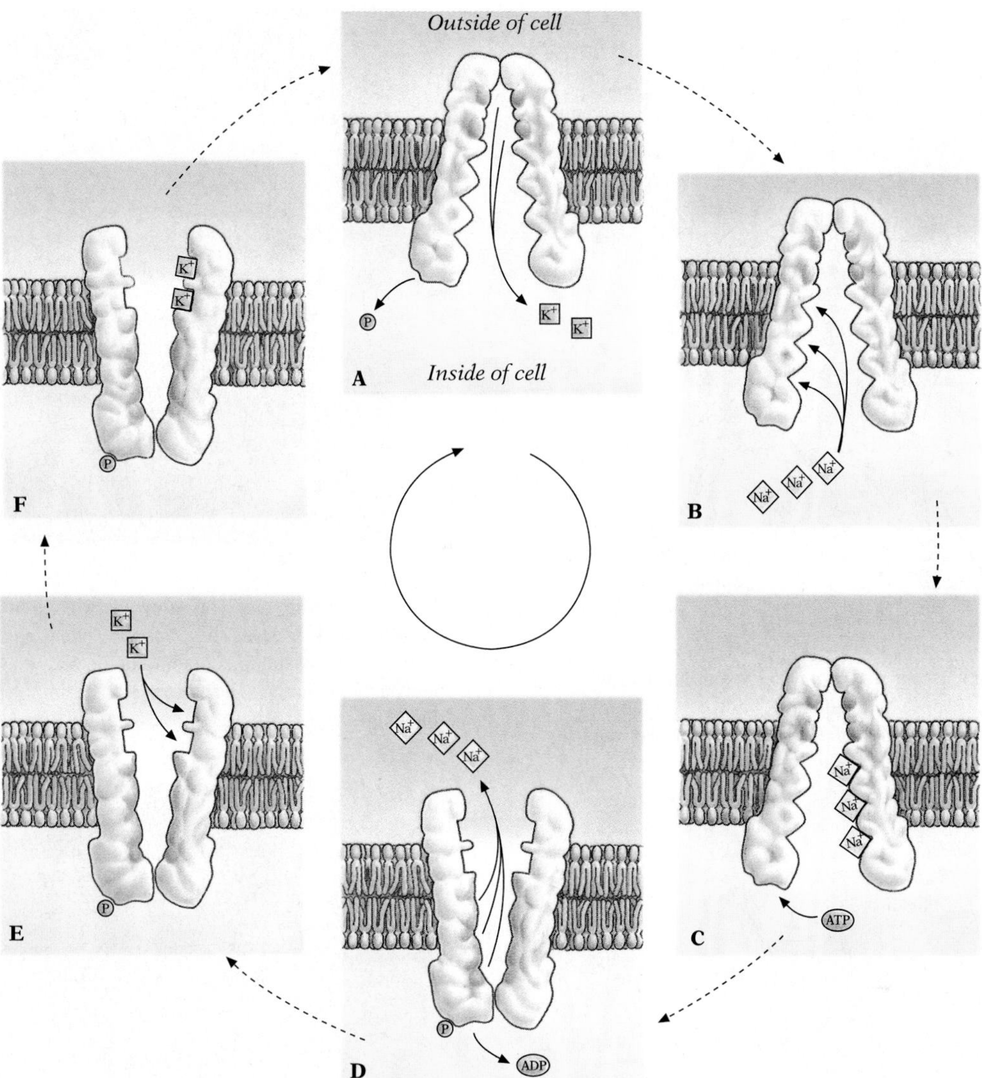

28.15 Model of the sodium-potassium pump The pump consists of proteins embedded in the cell membrane. When the pump complex is open to the inside of the cell (A, B), the K^+ binding sites no longer bind potassium, so those ions drift free, along with the phosphate group that powered the preceding cycle. The Na^+ sites, on the other hand, become active. Once the Na^+ ions have been bound, ATP binds to its site, thus phosphorylating the complex and causing a conformational change that opens the pump to the outside (C, D); this change severely reduces the protein's affinity for Na^+ ions (which in consequence drifts free into the extracellular fluid), while it activates the K^+ sites. The binding of K^+ ions causes the complex to return to its former conformation (A) and to release both the phosphate group obtained earlier from the ATP and the K^+ ions. The extent of movement of the pump subunits has been greatly exaggerated for clarity.

the dendrites or cell body of other neurons. Since the terminal portion of an axon ordinarily branches repeatedly, a single axon may synapse with many other neurons, and it usually synapses at numerous points with each of these neurons (Fig. 28.16). Each tiny branch of an axon usually ends with a small swelling called a ***synaptic terminal,*** or a ***synaptic bouton*** (Fig. 28.17).

In a few cases a special connection between the membrane of the synaptic terminal and the membrane of the adjoining cell permits electrical impulses to pass directly from one neuron to the next. Such ***electrical synapses*** allow action potentials to move rapidly, without delay, from neuron to neuron. Because of the speed of transmission, electrical synapses are often found at sites mediating responses where speed is important, such as the escape response of certain insects.

The vast majority of synapses, however, are not electrical, but ***chemical.*** A space about 20 nanometers wide, called the ***synaptic cleft,*** occurs between the synaptic terminal of the first (presynaptic) neuron and the membrane of the second (postsynaptic) neuron. Transmission across this cleft is by a diffusible ***transmitter chemical*** released from tiny ***synaptic vesicles*** in the terminal of the presynaptic neuron (Fig. 28.18). The trans-

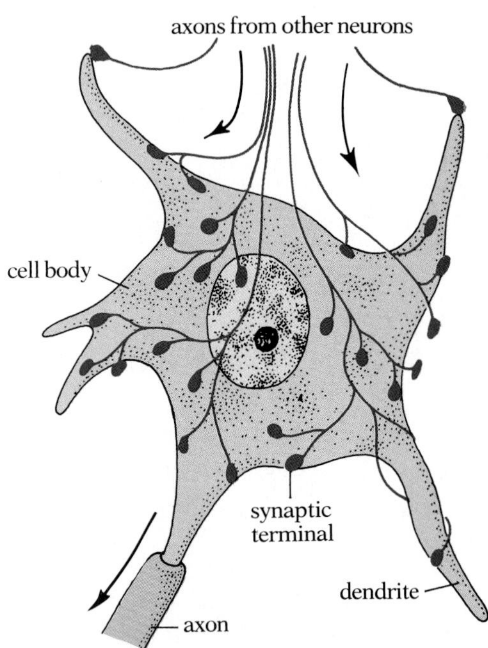

28.16 Synapses on a motor neuron Many axons, each of which branches repeatedly, synapse on the dendrites and cell body of a single motor neuron. Each branch of an axon ends in a swelling called a synaptic terminal.

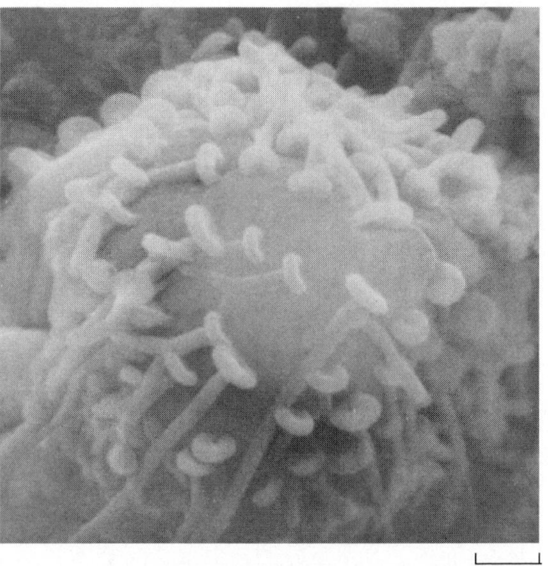

28.17 Scanning electron micrograph of synaptic terminals from the shell-less sea slug *Aplysia* The synaptic terminals of the numerous axons are in contact with the cell body of a postsynaptic neuron. Note that it is the edge of the terminal, not its flattened end, that characteristically forms the synapse.

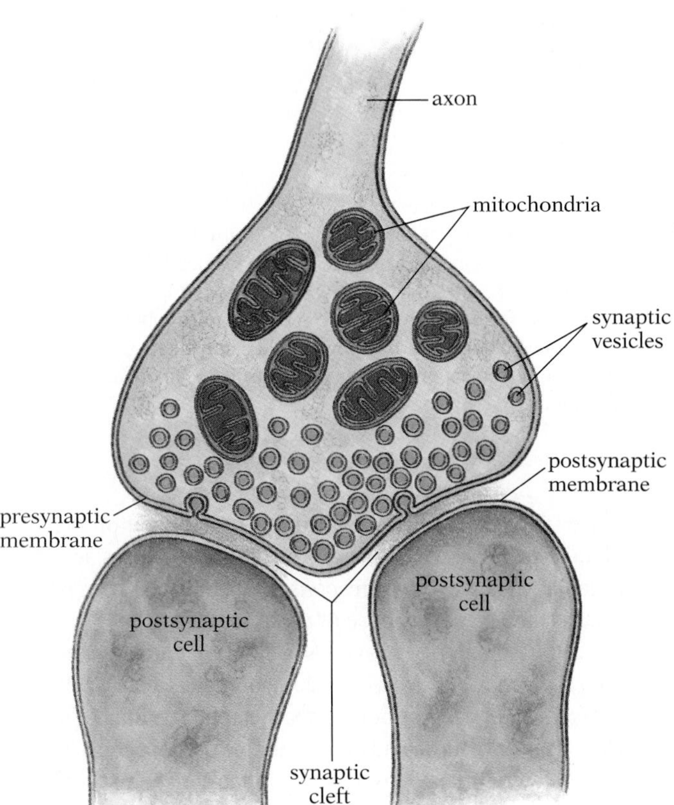

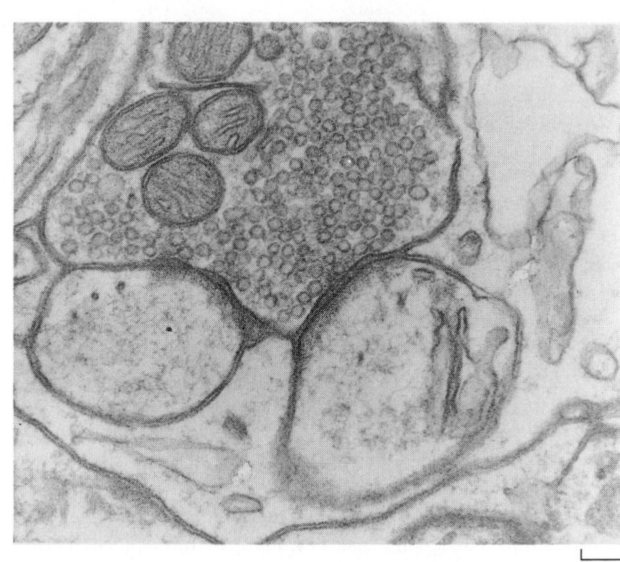

28.18 The synapse (Left) Each synaptic terminal at the end of an axon encloses numerous synaptic vesicles containing transmitter substance. When vesicles release this substance into the synaptic cleft, the substance diffuses across the cleft and alters the polarization of the postsynaptic membrane of the dendrites or cell body of the next cell. (Above) Transmission electron micrograph shows a synaptic terminal packed with vesicles. In this case two postsynaptic cells are involved in the synapse. From Freeze Fracture Image of Cells and Tissues by Richard L. Roberts, Richard G. Kessel, Hai-Nan Tung, Oxford University Press, NY © 1991.

mitter chemical at synapses outside the brain and spinal cord is most often a chemical called ***acetylcholine.*** Because of the time it takes for a chemical to diffuse across the cleft to affect a response in the postsynaptic neuron, chemical synapses are much slower than electrical synapses.

When an impulse traveling down the axon of the presynaptic neuron reaches the synaptic terminal, the impulse temporarily opens voltage-gated calcium ion (Ca^{2+}) channels in the membrane, and Ca^{2+} ions diffuse into the synaptic terminal from the surrounding fluid. The Ca^{2+} ions activate enzymes that cause synaptic vesicles in the synaptic terminal to move to the terminal membrane, fuse with it, and then rupture, releasing transmitter chemical into the synaptic cleft (Fig. 28.19). As we shall see in Chapter 30, Ca^{2+} ions play a similar role in triggering muscle contraction.

The transmitter molecules released into the cleft diffuse across it and bind weakly to receptor proteins on the ***postsynaptic membrane*** of the next neuron. This binding opens the gates of a channel, allowing specific ions to pass through the membrane (Fig. 28.19). The movement of ions into or out of the neuron results in a change in the electrical charge across the membrane. In an ***excitatory synapse,*** the change in ion permeability in the postsynaptic cell brings the membrane closer to threshold for generating an action potential in that neuron. In an ***inhibitory synapse,*** the change in ion permeability stabilizes the neuron at, or slightly below, its resting potential, inhibiting the generation of an action potential in the postsynaptic neuron.

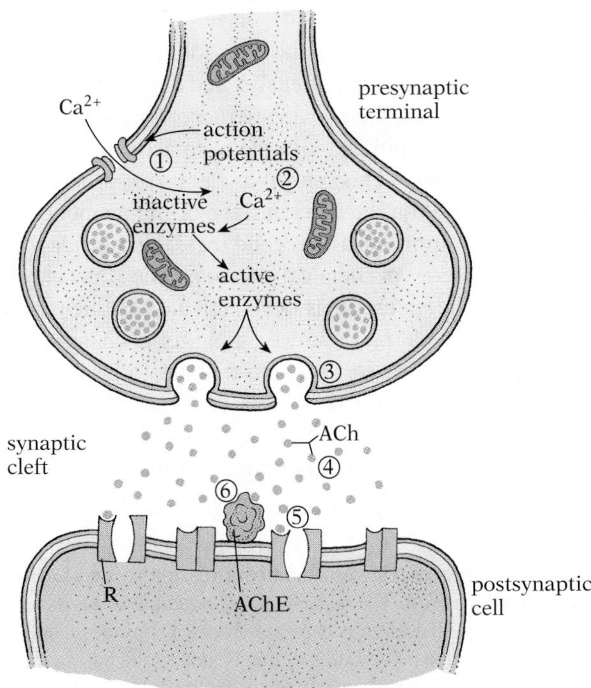

28.19 Events in synaptic transmission An action potential sweeping along the axon causes the voltage-gated Ca^{2+} gates to open (1) and calcium ions move into the synaptic terminal. The Ca^{2+} ions activate certain enzymes (2) that in turn cause the synaptic vesicles to move to and fuse with the presynaptic membrane (3) and release their stored transmitter chemical (4), acetylcholine (ACh). Acetylcholine diffuses across the synaptic cleft and combines with receptor proteins (R) in the postsynaptic membrane (5) and opens or closes a specific gated channel. The acetylcholine is then destroyed by the enzyme acetylcholinesterase, (AChE) (6).

The fact that chemical transmission of impulses involves an influx of calcium ions into the synaptic terminal, followed by movement of the vesicles, and then diffusion of transmitter across the cleft means that synaptic transmission is much slower than impulse conduction along the neurons. In general, the more chemical synapses in a neural pathway, the slower the average speed of transmission per unit distance along the pathway.

Chemical synapses result in one-way transmission of impulses along the neural pathways even though an individual neuron can conduct impulses in both directions. If, for example, we stimulate an axon at a point between its base and its terminus, an impulse will move in *both* directions along the axon from the point of stimulation. But the impulse moving back toward the cell body and dendrites will die when it reaches the end of the cell; it cannot bridge the gap to the next cell because dendrites cannot release transmitter chemicals. Only axons secrete transmitter chemicals, so impulses can only go from synaptic terminals of the axon to the next neuron.

The story does not end when an impulse has been communicated to a postsynaptic cell by the diffusion of transmitter across the cleft. If the transmitter remained, the postsynaptic receptors would be stimulated indefinitely by the arrival of a single action potential; a mechanism to remove the transmitter is therefore required. Some transmitters simply diffuse away from the synaptic cleft, others are taken back up for reuse by the presynaptic neuron, and still others are destroyed by specific enzymes. For example, the transmitter acetylcholine is removed from the synapse by an enzyme called ***acetylcholinesterase*** (Fig. 28.19). By destroying the transmitter, this enzyme makes it possible for the next impulse, with new information, to be transmitted. Many insecticides, such as the organophosphates (also known as nerve gases), are acetylcholinesterase inhibitors. They block the enzymatic destruction of acetylcholine, with the predictable result that an insect exposed to them becomes permanently active. Given in high enough doses, acetylcholinerase inhibitors affect major physiological processes, and the animal dies.

Acetylcholine is only one of perhaps as many as 50 transmitter chemicals found in the brain. Others include ***adrenalin*** and ***noradrenalin*** (substances also produced as hormones by the adrenal medulla), ***serotonin, dopamine, glycine,*** and ***glutamate,*** all of which usually act as excitatory transmitters, and ***gamma-aminobutyric acid (GABA),*** an important inhibitory transmitter. Many transmitters, e.g., glutamate and glycine, are common amino acids in the body. As we shall see, certain disorders such as schizophrenia and severe depression, once blamed on vague emotional disturbances, are now known to be triggered by biochemical malfunctions of transmitters and receptors. These welcome discoveries are beginning to open the way to relatively precise physiological treatments for certain emotional disorders.

Transmitter Chemicals Alter the Permeability of the Postsynaptic Membrane

We now focus on the effects of transmitter substances on postsynaptic membranes. When such a substance has diffused across the synaptic cleft, how does it affect the polarization of the postsynaptic membrane of the next neuron? First, consider a transmitter substance that has an excitatory effect. Apparently the binding of such a transmitter to the receptor opens Na^+ gates in the postsynaptic membrane. The resulting increased inward flow of Na^+ ions slightly decreases the polarization of the neuron; i.e., the inside becomes less negative relative to the outside, a condition known as an ***excitatory postsynaptic potential (EPSP)*** (Fig. 28.20). If the EPSP is sufficient, it may spread to the base of the cell's axon and, if it is above threshold, trigger an action potential, which will move down the axon to the next synapse.

The transmitter substances released by the axons of some neurons have the opposite effect—an inhibitory one. They increase the polarization of the postsynaptic membrane and thus make the neuron harder to fire. Often these substances produce their inhibitory effects by opening chloride ion (Cl^-) gates in the postsynaptic membrane. The Cl^- ion concentration is always higher outside the cell than inside, so Cl^- ions enter the cell by diffusion when the Cl^- channels are open. The influx of the negatively charged Cl^- ions causes the membrane to become more polarized; i.e., the inside of the cell becomes even more negative relative to the outside, a condition known as an ***inhibitory postsynaptic potential (IPSP).*** More than the usual number of excitatory impulses would be needed to reduce the polarization of such an inhibited neuron to the threshold level for triggering an impulse (Fig. 28.20). *The balance between EPSPs and IPSPs underlies all neural processing.*

Notice that the transmitter substance itself is not excitatory or inhibitory; it is the ion specificity of the gated channels in the postsynaptic membrane that determines the effect. If the transmitter binds to receptors that open Na^+ gates, the neuron will be easier to fire, but if the transmitter binds to receptors that open Cl^- gates, the neuron will be more difficult to fire. Thus the receptor makes the synapse excitatory or inhibitory.

Some Transmitters Produce Slow, Long-Lasting Responses in Neurons

Not all transmitters act directly on ion channels to produce short-term effects on membrane permeability. Some transmitters, such as serotonin, noradrenalin, and dopamine, bind to receptors in the postsynaptic membrane, which stimulates a second membrane protein, G-protein, to activate the enzyme ***adenylate cyclase.*** This enzyme catalyzes production of cAMP from ATP on the inner surface of the membrane. The cyclic AMP in turn acts as a second messenger inside the cell and triggers a series of chemical reactions. The effect may be to regulate certain ion channels or to activate specific enzymes. These transmitters cause slower, long-term changes in the cell—changes of the sort that are thought to be involved in learning and memory. This system, involving G-protein and cAMP, is the same system that we discussed in Chapter 25 with respect to the mechanism of hormonal action. As we shall see,

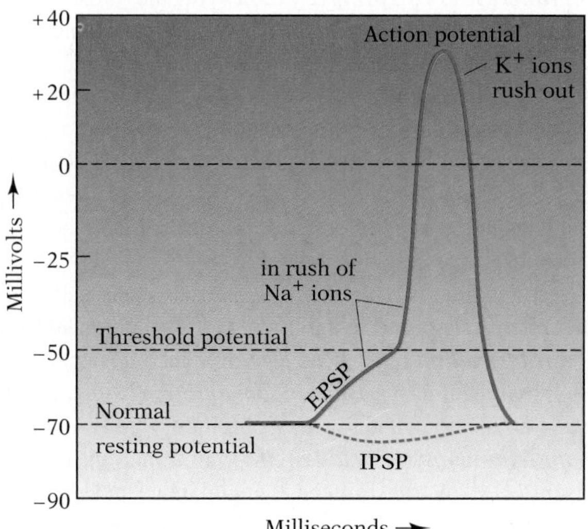

28.20 The effect of transmitter substance on the membrane potential of a neuron The normal resting potential of a typical neuron is about -70 millivolts. An excitatory transmitter substance slightly reduces that polarization—that is, makes the inner surface of the membrane less strongly negative—thereby creating an excitatory postsynaptic potential (EPSP). If the EPSP reaches the threshold level, which is usually about -50 millivolts, an impulse is triggered; a sudden inrush of Na^+ ions causes the inside of the cell to become positive. A fraction of a second later, K^+ ions rush out of the cell, and the inside of the cell again becomes negative. This cycle of electrical changes constitutes the action potential.

If the transmitter substance had been inhibitory, the membrane could have become more polarized (to perhaps -75 millivolts), a condition called an inhibitory postsynaptic potential (IPSP) (dashed curve), and no action potential would have resulted; the neuron would slowly have returned to its resting potential after release of the transmitter had ceased.

MAKING CONNECTIONS

1. How do the symptoms of multiple sclerosis (MS) demonstrate the importance of the myelin sheath for normal conduction of nerve impulses?

2. How is the structure and function of ion channels in the neuron membrane related to the structure and permeability characteristics of the cell membrane, as described in Chapter 5?

3. Which facet of nerve impulse conduction actually requires the neuron to expend energy?

4. How is the effect of some transmitters on postsynaptic cells similar to the "second messenger" mechanism of hormonal action described in Chapter 25? What types of cellular changes do these transmitters cause?

the nervous and endocrine systems share many similarities in their modes of action.

The Cell Integrates All the Signals That Converge on It, and Either Fires an Impulse or Remains Silent

Synapses are points of resistance in nervous pathways. Hence an impulse may travel to the end of the axon of one neuron but die there, because not enough excitatory transmitter is released to imitate an impulse in the next neuron of the pathway. In such a case, the transmitter molecules released as a consequence of the impulse do slightly decrease the polarization of the postsynaptic neuron, producing a slight EPSP, but not enough to cause the neuron to fire. Ordinarily, excitatory transmitters from many different synapses must impinge on the neuron within a short space of time if a sufficient EPSP is to be built up to trigger an impulse. This additive phenomenon is called ***summation.***

If both excitatory and inhibitory transmitters impinge on a single cell at the same time, their effects are added according to sign. Each excitatory molecule causes a slight decrease in polarization (plus), bringing the cell closer to the threshold potential, and each inhibitory molecule causes a slight increase (minus), removing it further from the threshold potential. The net result—EPSPs making the cell more likely to fire, or IPSPs making it less likely to fire—depends, then, on the cell's integrating all the excitatory and inhibitory information it is receiving. The cell responds not to any single unit of incoming information, but to the whole pattern of information impinging on it (Fig. 28.21). This process is called ***integration.*** The cell integrates all the signals that converge on it and either fires or remains silent. In short, all the incoming information is in-

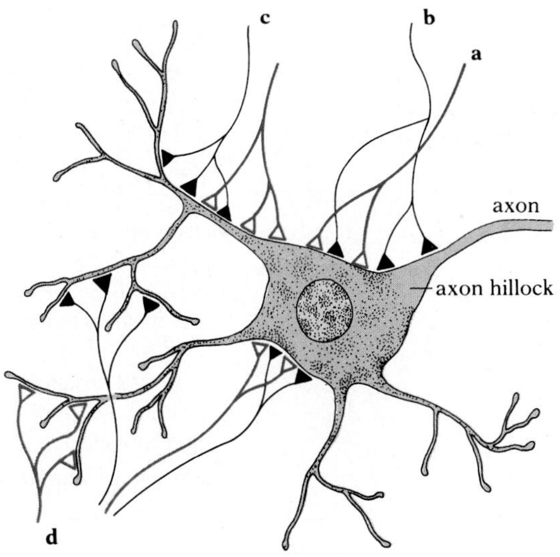

28.21 The integrative function of a neuron The neuron receives both excitatory synapses (open triangles) and inhibitory synapses (solid triangles) from many different sources. The synapses vary in their distance from the base of the axon (the axon hillock), where impulses are generated. Whether or not the neuron will fire an impulse is determined at any given moment by the algebraic sum of all the individual EPSPs and IPSPs arriving at the axon hillock from the various synapses. But unlike an axonic impulse, which shows no decrement with distance, depolarizations (EPSPs) or hyperpolarizations (IPSPs) decrease in magnitude as they spread along a dendrite or the cell body. Hence impinging interneurons such as **a** and **b,** which synapse near the axon hillock, can more easily influence the neuron's firing than interneurons such as **c** and **d,** which synapse on the cell at a distance from the hillock; only if these latter interneurons fire at a very high rate are they likely to have a major effect. In short, the geometry of the synapses on a neuron biases the integration process; inputs from some interneurons are given greater weight than inputs from other interneurons.

tegrated to produce a simple yes-no decision by the cell. The nervous system processes the information it receives by combining many such yes-no decisions from a host of different cells.

Many Neurological Drugs Exert Their Effects at Synapses

Because synapses act as control valves in the nervous system, and because their proper function depends on a very delicate balance between transmitter substance, deactivating enzyme, and membrane sensitivity, it is not surprising that synaptic malfunctions have been implicated in several mental disorders—among them schizophrenia—or that many neurological drugs exert their effects at synapses. Used with caution and under medical supervision, some of these drugs can give relief from anxiety and tension or from specific neurological diseases. But used improperly, the same agents can induce symptoms strikingly reminiscent of certain mental disorders, and in some cases the symptoms may be long-lasting or even permanent.

Neurological drugs can alter synaptic function in a variety of ways. They may turn off certain synapses by (1) interfering with synthesis of the appropriate transmitter substance; (2) blocking uptake of the transmitter into the synaptic vesicles; (3) preventing release of the transmitter from the vesicles; or (4) blocking the receptor sites on the postsynaptic membranes, so that the transmitter has no effect even if released. Other drugs can induce excessive and uncontrolled firing of postsynaptic cells by (1) stimulating massive release of transmitter substance from the vesicles; (2) mimicking the effect of the transmitter; or (3) inhibiting destruction of the transmitter once it has done its job. Cocaine, for instance, binds to and inhibits the protein responsible for removing the transmitter dopamine from dopamine synapses. Dopamine therefore remains active, producing overstimulation of postsynaptic neurons. A summary of the physiological mode of action and effects of some neurological drugs is shown in Table 28.1.

Several previously misunderstood neurological disorders have now been tied to transmitter problems. Severe depression has been firmly linked to a defect in the serotonin transport system: the brains of many suicide victims have about half the normal serotonin level and only two-thirds the usual number of binding sites. Abnormal dopamine levels are now tied to one form of schizophrenia, and perhaps one form of alcoholism. The chemical basis of the manic-depressive syndrome is sufficiently well understood to be treated chemically (with lithium) rather than through psychoanalysis, and recent evidence suggests that severe autism, a disability that affects a person's power to communicate and to relate to others, is a biochemical defect.

TABLE 28.1 *A summary of the effect of certain neurological drugs*

Drug	*Proposed mode of action*	*Physiological effect*
Amphetamine	Stimulates release of noradrenalin and dopamine at noradrenalin and dopamine synapses in brain	Stimulant
Nicotine	Mimics the effect of acetylcholine	Stimulant
Cocaine	In addition to anesthetic effect below, it inhibits uptake of dopamine at synapses, leaving dopamine in the synapses and producing overstimulation of those neural pathways	Stimulant
Local anesthetics, e.g., novocaine, xylocaine, cocaine	Prevent Na^+ gates from opening in injected area	Decreased perception of pain in injected area
Reserpine	Blocks uptake of noradrenalin into synaptic vesicles and thereby prevents its release	Tranquillizer
Chlorpromazine	Binds to both acetylcholine and noradrenalin receptors on postsynaptic membranes	Tranquillizer
Benzodiazepines, e.g., Valium	Interact with inhibitory transmitter GABA to open chloride channels and inhibit synaptic transmission; they increase effectiveness of GABA	Tranquillizer
LSD	Mimics the transmitter serotonin; combines indiscriminately with serotonin receptors	Derangement of mental functions
Curare	Binds to acetylcholine receptors but does not open gates	Paralysis
Strychnine	Blocks glycine receptor in postsynaptic membrane	Spastic paralysis

Transmission From Neuron to Muscle Also Involves Transmitter Chemicals

Just as there is a gap at the synapses between successive neurons in a neural pathway, there is also a gap between the terminals of an axon and the effector it innervates (i.e., makes neural connections to). When the effector is skeletal muscle, the gap is usually located within a specialized structure, the ***neuromuscular junction*** or motor end plate, formed from the end of the axon and the adjacent portion of the muscle surface (Fig. 28.22). Transmission across this gap is via transmitter chemical. The transmitter at neuromuscular junctions of vertebrate skeletal muscle is acetylcholine. A variety of drugs that cause paralysis—e.g., the poison produced by the bacterium responsible for botulism and the famous neuromuscular blocker used on the poison arrows of South American Indians, curare—do so by blocking transmission between the motor neurons and the muscles. Curare is now known to act as an acetylcholine mimic, binding to the receptor proteins in the muscles without opening the gates. To make matters worse, acetylcholinesterase cannot inactivate it. The receptor sites are permanently blocked, and neuromuscular transmission ceases. The action of botulism toxin, on the other hand, is not well understood, but it is thought initially to cause the presynaptic terminals at neuromuscular junctions to be overactive, which results in muscular tremors and paralysis. After a period of time it has a second effect—it silences these same terminals, and death results. The neuromuscular junction is also the site of at least one deadly viral disease: rabies. The extreme virulence of rabies virus and its wide range of host mammals seem to be the consequence of the virus's specific affinity for acetylcholine receptors.

NERVOUS PATHWAYS IN VERTEBRATES

Even though some of the most exciting current neurobiological research is carried out on invertebrates, especially arthropods and molluscs, our models in this discussion of basic nervous

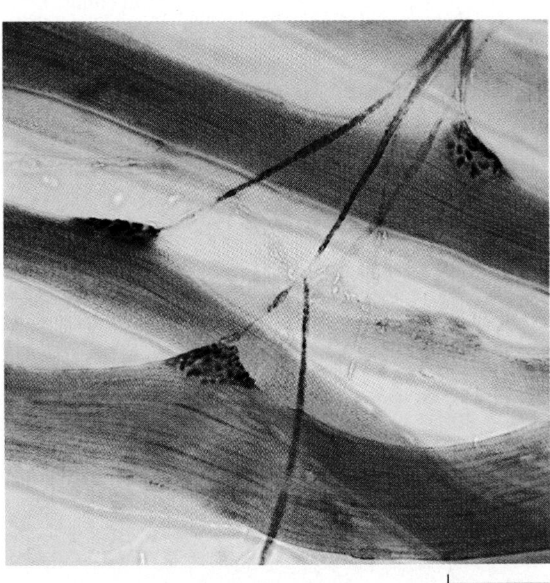

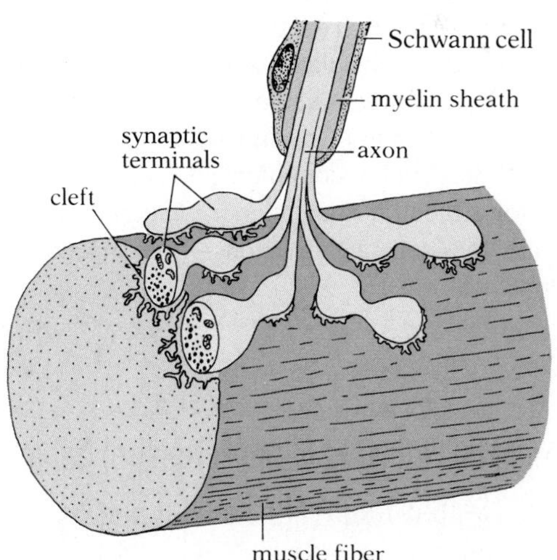

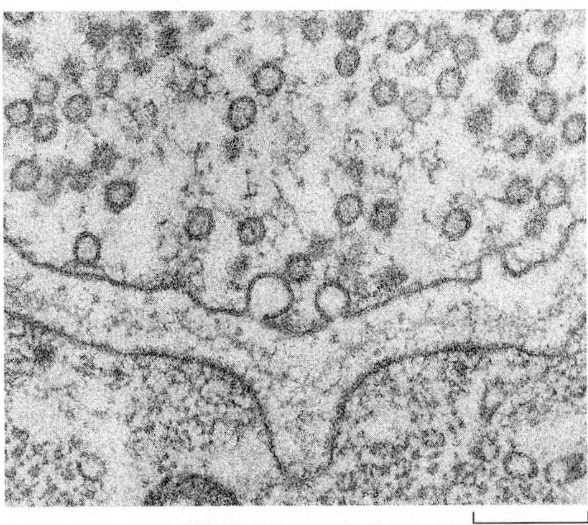

28.22 Neuromuscular junctions Top left: Toward its end, an axon supplying a muscle in a snake branches extensively and forms neuromuscular junctions on individual muscle fibers. Top right: A close-up sketch of one neuromuscular junction. The junction is formed by branches of the axon, with their terminals, and the specialized adjacent portion of the muscle fiber. Bottom right: As in a synapse between two neurons, there is a cleft between muscle and nerve cells. The upper half of the electron micrograph of a neuromuscular junction of a frog shows part of the terminal of an axon containing numerous synaptic vesicles, some of them releasing neurotransmitter. The lower half shows part of a muscle cell. There is a distinct cleft between the two cells.

pathways will be vertebrates—in great part because of the fascination inherent in the system that more than any other makes human beings human.

The human nervous system is organized as follows:

I. Central nervous system (CNS)
   A. Brain
   B. Spinal cord
II. Peripheral nervous system (PNS)
   A. Somatic nervous system
      1. 12 pairs of cranial nerves
      2. 31 pairs of spinal nerves
   B. Autonomic nervous system
      1. Sympathetic division (SANS)
      2. Parasympathetic division (PANS)

The ***central nervous system*** (CNS), which consists of the brain and spinal cord, is the coordinating center, organizing all the incoming and outgoing nervous information. The nerves that connect the brain and spinal cord to the periphery of the body make up the ***peripheral nervous system*** (PNS). A ***nerve*** is a compound structure consisting of a number of neuron fibers bound together. Although there may be thousands of fibers in a single nerve, each is insulated from the others and conducts impulses independently. A nerve, therefore, is much like a telephone cable containing many functionally separate telephone wires, with independent communications pathways packaged together for structural convenience. The peripheral nervous system has two separate divisions—the somatic and autonomic systems.

Somatic Pathways of the Peripheral Nervous System Innervate Skeletal Muscle

Somatic pathways, exemplified by the reflex arcs discussed below, usually innervate skeletal (voluntary) muscle and include both sensory and motor neurons. They involve, at least potentially, some conscious control of the reflex or an awareness that the reflex has occurred.

A ***reflex arc*** is a simple neural circuit linking a sensory receptor to an effector; it is the simplest form of neural control in vertebrates. The reflexes they produce, which are responses to specific stimuli, are usually rapid and automatic. Reflex arcs control behavioral responses that must occur quickly, such as emergency reactions and the automatic maintenance of some kind of equilibrium.

A good example of a familiar emergency reaction is the withdrawal reflex. When we touch something hot, our hand jerks back automatically. The sensory neurons involved in this response run from the hand to the spinal cord; the cell bodies of these neurons are located in a ***dorsal-root ganglion*** that lies just outside the spinal cord near its dorsal (back) side (Fig. 28.23). The axons enter the spinal cord and synapse with interneurons within the gray matter of the spinal cord. The interneurons in turn synapse with motor neurons, the axons of which exit the cord and run to the muscles. In this reflex, a strong signal from the appropriate sensory cells both fires the flexor muscles and inhibits the motor neurons to the extensor muscles. This crucial motor response is well under way before the signals responsible for the conscious sensation of pain (which exit the reflex pathway in the spinal cord) ever reach the brain.

The kind of circuit that automatically maintains equilibrium is exemplified by the well-known knee-jerk reflex, a part of the postural control system. Stretch receptors within the muscle measure the degree to which the muscle is stretched. As the force against which the muscle must act—the amount of weight on one leg for instance—increases, the muscle is stretched, and the receptors signal this fact through sensory

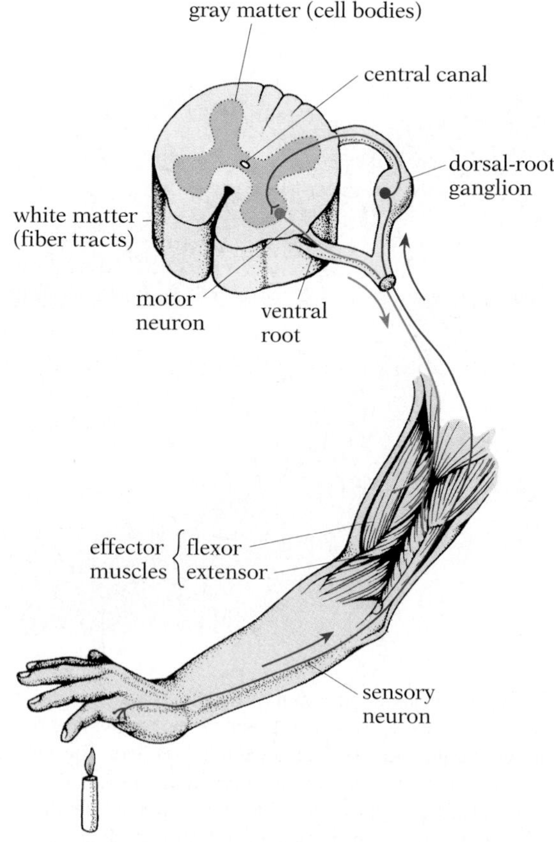

28.23 Hand-withdrawal circuit When the hand touches something hot, signals from pain receptors reach interneurons and motor neurons in the spinal cord, which in turn stimulate muscles that quickly retract the arm. These motor neurons both activate the flexor muscles and inhibit the extensors. (The central canal through the gray matter is a remnant of the hollow nerve cord evident early in development.)

neurons to the spinal cord. As in the previous example, the information is sent both to motor neurons for immediate action and to the brain for analysis. The arrival of signals from the receptor increases the firing rate of the motor neurons, and the muscles tighten to accommodate the added load they perceive. Without the organism's being aware, the knee-jerk reflex automatically tunes posture. Doctors regularly test for this response by tapping a patient's knee with a special rubber hammer (Fig. 28.24). The response ascertains whether certain nerves and a portion of the spinal cord are functioning normally. When the physician taps the knee, a stretch receptor is stimulated and impulses travel up a sensory neuron to the spinal cord and back down a motor neuron to the leg, where they stimulate muscle fibers to contract, causing the leg to jerk. A minimum of three cells are involved in this reflex arc: a receptor-sensory neuron, a motor neuron, and an effector cell (muscle).

Using this very simple reflex as a model, we can make several generalizations about the reflex arcs of the somatic system:

1. For a particular reflex arc there is never more than one sensory neuron, however long it must be, to carry the sensory information from the receptor to the spinal cord (there may be many such neurons running side by side serving the same function).
2. The cell body of the sensory neuron is always outside the spinal cord in a dorsal-root ganglion.
3. The axons of sensory neurons always enter the spinal cord dorsally whereas the axons of motor neurons always leave the spinal cord ventrally.
4. There is a single motor neuron in the pathway carrying information from the spinal cord to the effector.

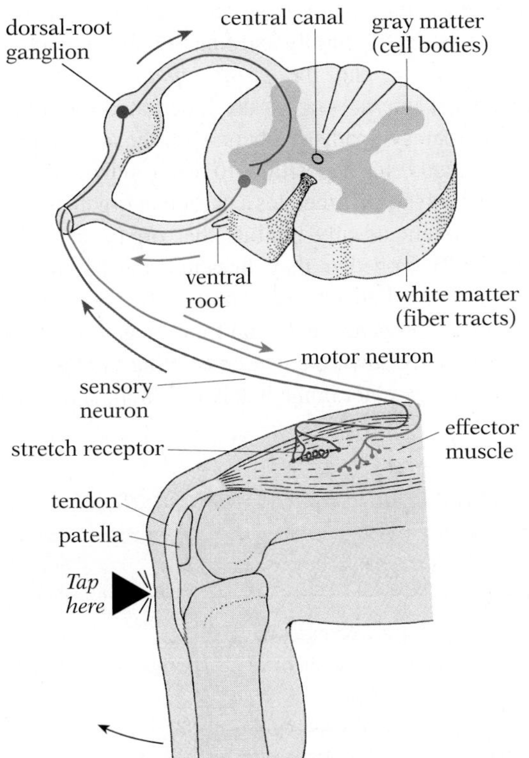

28.24 The knee-jerk reflex This reflex, which is part of the automatic postural control system, measures the load on the muscle (as reported by stretch receptors) and then adjusts the firing rate of the motor neurons to that muscle to compensate for any changes. The knee jerk itself results from the brief overloading of the muscle and its receptor when the tendon is tapped sharply.

The sensory and motor neurons of the knee-jerk reflex run through the same nerve, even though they carry impulses in opposite directions. A nerve containing both sensory and motor fibers is called a ***mixed nerve.*** All the nerves connected to the spinal cord, the spinal nerves, are mixed. There are 31 pairs in humans, all of which branch repeatedly after leaving the spinal cord, giving rise to smaller nerves that innervate most parts of the body below the head. Some nerves, on the other hand, connect directly to the brain rather than to the spinal cord. In humans there are 12 pairs of these cranial nerves, some consisting of only motor neurons, some of only sensory neurons, and some of both types of neurons.

Very few reflex pathways involve only two neurons, one sensory and one motor, in series. At least one interneuron is usually interposed between the sensory neuron and the motor neuron (Fig. 28.25), and it is common for many interneurons to be involved even in relatively simple reflex arcs. It is important to keep in mind that a reflex arc, whether it includes few cells or many, makes two sorts of connections with other neural pathways. First, it almost always sends information to the brain, where instructions to counteract or augment the behavioral reaction can be issued. If you know that the doctor is going to strike your knee, for instance, allowing sufficient time to issue neural commands to modify the reaction, you can consciously either inhibit or exaggerate the response. The second sort of connection is with other reflex arcs which help coordinate the response.

Reflex circuitry is able to control and coordinate a variety of simple responses and automatically fine-tune behavior such as walking, whose details must constantly be adjusted as body weight rhythmically shifts.

Notice that the peripheral nerves transmit information between the body and the spinal cord, and that the spinal cord itself conducts information between the brain and the peripheral nerves. This allows one to detect sensations and control the movements of the body. If however, the spinal cord is injured, the portions of the body below the injury will be affected, since the nerve pathways are disrupted and neurons have limited capabilities of regeneration. Depending on the severity of the injury, there may be numbness, or paralysis of all muscles below the injury, including those that control the bladder and rectum.

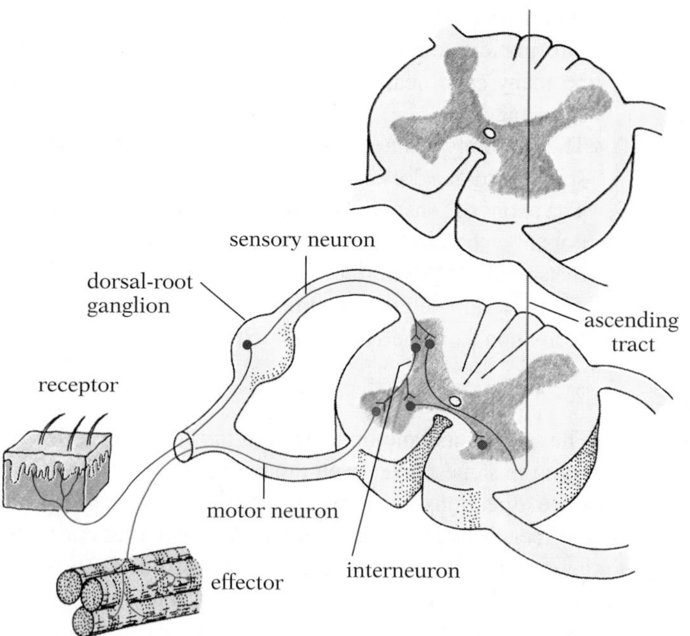

28.25 Diagram of a reflex arc with interneurons The sensory neuron synapses with several interneurons in the gray matter of the cord. Some of these interneurons may synapse directly with motor neurons on the same side, but some cross to the other side of the cord and there synapse with other motor neurons and with additional interneurons that run in ascending tracts through the cord to the brain.

The Autonomic Pathways of the Peripheral Nervous System Innervate Smooth and Cardiac Muscle

In contrast to the somatic system, autonomic pathways are not ordinarily under conscious control and usually function without our being aware of them. They innervate the heart, the smooth muscle in the walls of the digestive tract, the gas-exchange system, the excretory system, the reproductive system, blood vessels, and some glands. Unlike the somatic system, which conducts information along a single motor neuron from the spinal cord to the muscle, the pathways of the autonomic nervous system have two motor neurons in the pathway. The axons of the first motor neurons exit from the brain or spinal cord and run to ganglia lying outside the central nervous system. There they synapse with the second motor neurons, which innervate the target organs.

The autonomic nervous system (ANS) is separated into two parts, both structurally and functionally. These are called the ***sympathetic*** and the ***parasympathetic systems*** (Fig. 28.26). Most internal organs are innervated by both sympathetic and parasympathetic fibers, with the two systems functioning largely in opposition to each other. Thus, if the sympathetic system stimulates a particular organ, the parasympathetic system usually inhibits that organ, and vice versa (Table 28.2). Together they fine-tune the balance between what could be called active and passive behavior. The sympathetic system prepares an animal for emergency action during a crisis: it shuts down oxygen-consuming processes, such as digestion, that are not immediately essential, and prepares the

TABLE 28.2 *A comparison of the effects of the sympathetic and parasympathetic divisions on target organs*

Target organ	*Sympathetic effect*	*Parasympathetic effect*
Heart	Increases rate and strength of contraction	Decreases rate
Blood vessels	Constriction of most vessels; dilation of vessels to heart and skeletal muscle	Dilation of vessels of the digestive tract
Digestive tract	Inhibits movement, inhibits secretion	Stimulates movement, stimulates secretion
Liver	Stimulates glycogen breakdown to glucose and its release into blood	No effect
Lungs and bronchioles	Dilates bronchioles and inhibits secretion of mucus glands	Constricts bronchioles; stimulates secretion of mucus glands
Eye	Dilates pupil	Constricts pupil
Adrenal medulla	Stimulates secretion of hormones	No effect

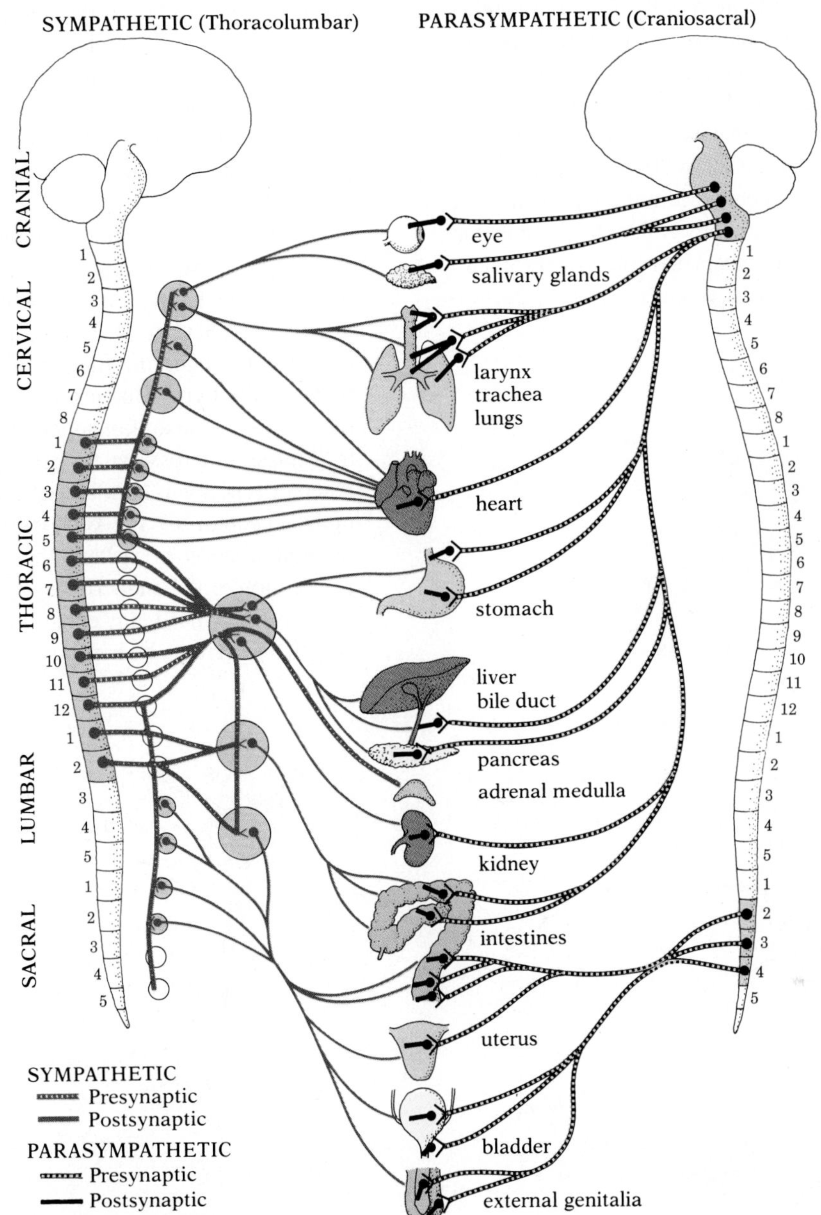

28.26 The autonomic nervous system Of the 12 cranial and 31 spinal nerves, four cranial nerves (see gray area of brain at upper right, where they emerge) and about half the spinal nerves (see brown and gray segments, where they emerge from the cord) contribute neurons to the autonomic nervous system, which innervates internal organs.

The ANS is customarily divided into two parts: the sympathetic and the parasympathetic systems. The pathways of both usually have two motor (efferent) neurons; a first (presynaptic) neuron exits from the central nervous system and synapses with a second (postsynaptic) neuron that innervates the target organ. Most but not all internal organs are innervated by both the sympathetic and the parasympathetic system.

machinery that may be necessary for defense, i.e., the "fight-or-flight" response (increasing heart rate, blood pressure, blood sugar, etc.). At the other extreme, the parasympathetic system restores order or passivity after a crisis has ended, restarting the important but not immediately critical processes that have been shut down.

Both the Nervous System and Endocrine System Depend on the Secretion of Chemicals

In general, the sympathetic system gives rise to the same effects as the hormones of the adrenal medulla—the fight-or-flight reactions—but does so more rapidly. Present evidence suggests that this nervous mechanism is far more important than the endocrine system in preparing an animal for emergency situations, as indeed it should be considering how much faster a pathway it is. (Recall that nerve impulses are conducted along neurons at a rate of 30 to 90 meters per second, whereas a hormone must be secreted into the blood and then transported throughout the circulatory system to reach its target organ.) The reason the sympathetic nervous system and the hormones of the adrenal medulla produce similar effects is clear: the transmitter at the neuromuscular junctions of both somatic and parasympathetic pathways is acetylcholine, but the transmitter at the sympathetic neuromuscular junctions is noradrenalin (or, in a few cases, adrenalin). The effect on the target organs is identical because the transmitters of the sympathetic system and the hormones of the adrenal medulla are identical.

Here, then, is another example of the close interrelationship between the nervous system and the endocrine system that we noted in Chapter 25, and of the similarity in their mechanisms of action (Fig. 28.27). Neurosecretory activity is not uncommon to endocrine function (we saw it in the interaction between the hypothalamus and the posterior pituitary). Neurosecretion is also fundamental to nerve action—impulse transmission across synapses depends on it. And transmitter receptor function is similar to the mechanism of action of hormonal receptors. It is likely, therefore, that natural selection favored a nervous system that functions as a high-speed elaboration of the endocrine system. This would explain why nervous tissue has many endocrine functions, and many of the chemicals secreted by nervous tissue are basically hormones.

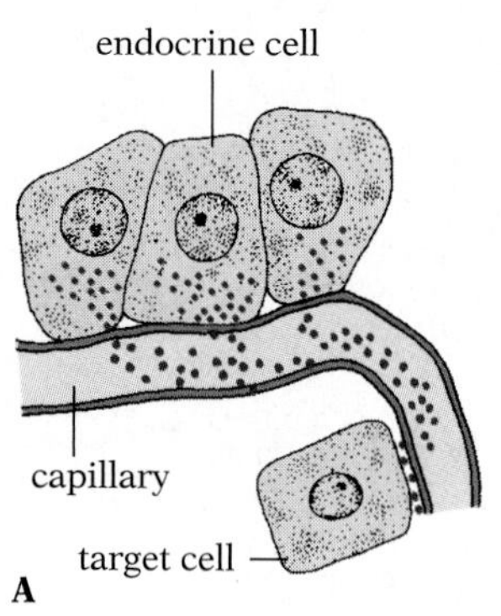

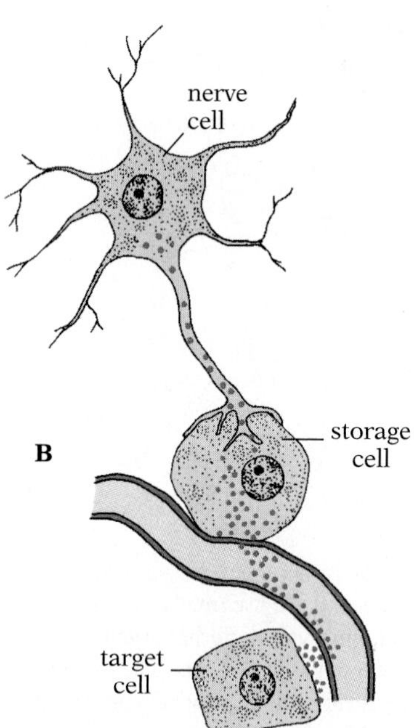

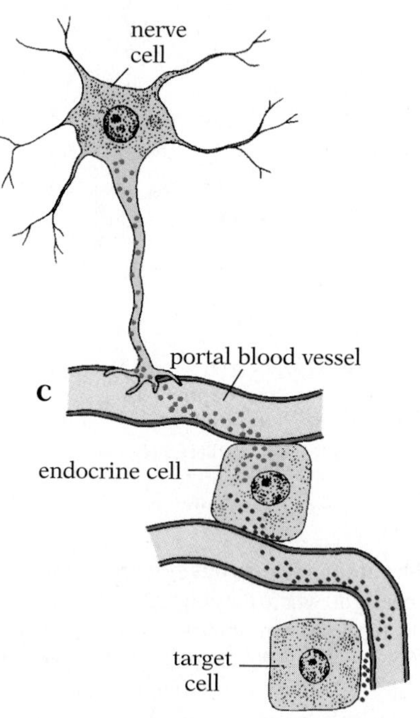

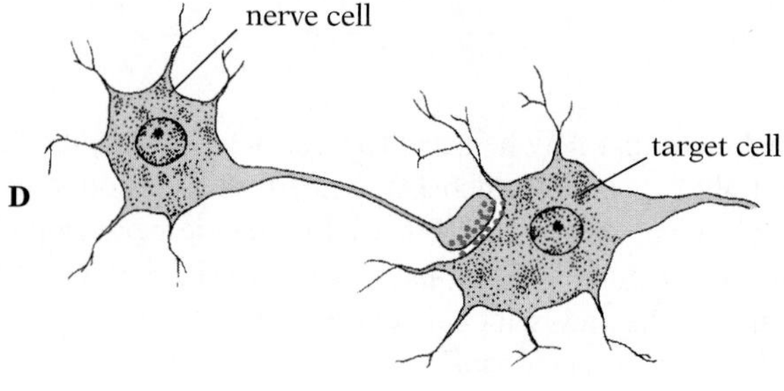

28.27 Four arrangements for chemical control compared
(A) A typical endocrine cell secretes its hormone directly into the blood, which carries the hormone to the target cells. (B) Some hormones secreted by nerve cells in the hypothalamus are stored in cells of the posterior pituitary, and later released into the blood. (C) Releasing hormones secreted by nerve cells in the hypothalamus are carried by the blood in a special portal system to endocrine cells of the anterior pituitary, which respond by secreting their own hormones into the general circulation. (D) A more typical nerve cell secretes its transmitter substance directly onto the target cell; no transport by the blood is involved.

THE NEURAL CONTROL OF MORE COMPLEX BEHAVIOR

Not long ago, most scientists believed that complex behavior—behavior that integrates various sensory inputs and is expressed through coordinated movements of several muscles in sequence—was organized through interacting reflex arcs. Each reflex arc presumably monitored a single sensory receptor and controlled a single muscle (or a pair of antagonistic muscles, as in the hand-withdrawal reflex). Learned behavior—maze-running in rats, for instance—was supposed to come about through the linking of such reflex circuits into chains. This plausible model of the organization of behavior dominated the thinking of many researchers through most of this century.

Many Complex Behaviors Are Controlled by Motor Programs

The idea that complex behavior is built out of chains of reflexes was effectively disproven about 1960 by the research of Donald Wilson of Stanford University, who was studying the circuitry underlying locust flight. By tracing out the circuitry, Wilson discovered a special circuit in the thoracic ganglia that generated all the commands necessary to produce flight (Fig. 28.28). The information from the associated stretch receptors is used simply to fine-tune flight behavior. Since Wilson's dis-

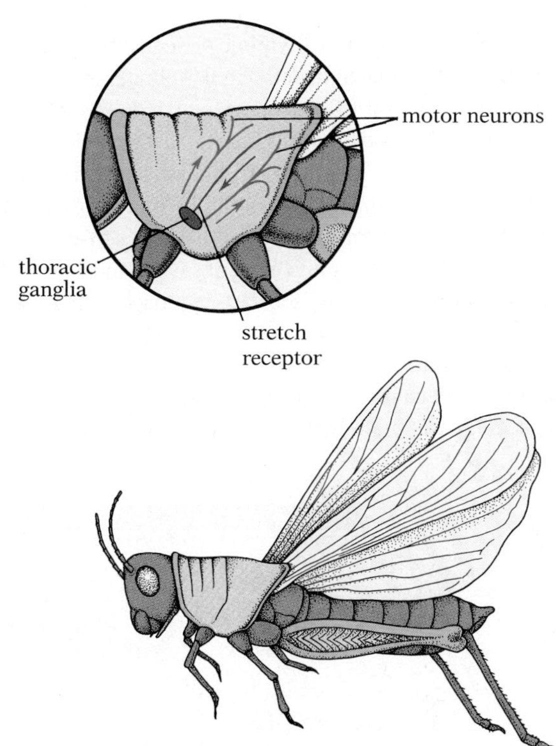

28.28 Flight circuitry in a locust The motor neurons for the flight muscles are located in the thoracic ganglia, and the stretch receptors for the flight muscles send the information they collect about muscle position to the same ganglia. When flight is triggered by outside cues (loss of foot contact with the ground after jumping, for instance), a circuit (red) in the thoracic ganglia is activated that generates instructions for all the flight muscles. The information from the stretch receptors (blue) is used to fine-tune flight behavior.

MAKING CONNECTIONS

1. How are the effects of many neurological drugs (such as cocaine) and some mental disorders (such as depression) related to synaptic function?

2. Why would curare, an arrow poison used by South American Indians, cause paralysis of skeletal muscle? What would be the effect of an acetylcholinesterase inhibitor (such as an organophosphate nerve gas) on skeletal muscle?

3. Why can a nerve—but *not* a neuron—be aptly compared to a telephone cable?

4. Does the antagonistic mode of action of the two branches of the autonomic nervous system have any parallels among the hormones discussed in Chapter 25?

5. As intercellular communication systems, how are the nervous and the endocrine systems similar?

covery, many of the complex rhythmic or sequential behaviors that have been investigated in detail in vertebrates and invertebrates have been found to rely on specialized self-contained neural circuits known as ***motor programs.*** Such circuits orchestrate muscle movement in the proper order and with the correct timing; no complex behavior based on reflex chains is known. Typically a motor program is run by a self-contained circuit that coordinates the movement of several muscles and is automatically fine-tuned by sensory feedback. Motor programs are under direct control of the brain, which can activate or inhibit a program.

The list of behaviors controlled by motor programs includes feeding, walking, swimming, scratching, swallowing, building, attack, courtship and mating, and so on. As we shall see in Chapter 35, the strategy that has evolved for orchestrating even more complex behavior, such as the sequence by which a cat stalks, chases, catches, kills, disembowels, and eats a mouse, has been to construct many separate motor programs and then to devise a higher-level circuit to coordinate the interactions of these individual programs.

The realization that so much of an organism's behavior is accomplished by interacting groups of neural circuits is a major achievement for modern neurophysiology. In the next chapter we shall focus on the structure of the human brain and consider how it functions to coordinate the various neural circuits.

CONCEPTS IN BRIEF

1. *Irritability*—the capacity to respond to stimuli—involves reception of the stimulus, conduction of a signal, processing of the signal, and response by an effector.

2. Most nervous pathways have at least three separate cells: receptor, conductor, and effector cells. More complex pathways have additional conductor cells; these increase the flexibility of response.

3. The typical *neuron* consists of the *cell body,* which contains the nucleus, and one or more long *nerve fibers* that extend from it. The fibers on which information is generally received are the *dendrites;* those on which it is generally transmitted to other cells are the *axons.* Many nerve processes have a *myelin sheath* composed of glial cells wrapped around them. Myelin sheaths speed up conduction.

4. *Sensory neurons* conduct information from the sensory receptors to the central nervous system, and *motor neurons* conduct information from the central nervous system to effector cells. *Interneurons* lie between the sensory and motor neurons. Junctions between neurons are called *synapses.* A *nerve* consists of a number of neuron fibers bound together.

5. All multicellular animals except sponges have evolved some form of nervous system. Radially symmetrical cnidarians have a diffuse nerve net with little central control. The nervous systems of bilaterally symmetrical animals show evolutionary trends toward more centralization and cephalization, increased complexity of pathways, one-way conduction, and better-developed sense organs.

6. Vertebrates have a single dorsal, hollow nerve cord whereas annelids and arthropods typically have two ventral, solid nerve cords. The vertebrate brain is more highly developed and exerts more dominance over the entire nervous system.

7. A nerve impulse is a wave of electrical change moving along a nerve fiber. The potential stimulus must be above a *threshold* to initiate an impulse. If the axon fires, it will fire maximally or not at all—an *all-or-none response.*

8. The inside of a resting nerve fiber is negatively charged compared to the outside. When a fiber is stimulated, Na+ gates open and Na+ ions rush into the cell, making the inside positively charged compared to the outside. An instant later, the Na+ gates close and the K+ gates open, and K+ ions rush out of the cell, restoring the original charge. This cycle of changes is the *action potential.* The action potential at the point of stimulation alters the permeability at adjacent points and initiates the same cycle of changes there. Diffusion, potassium leak channels, and the *sodium-potassium pump* restore the original ion and charge distribution.

9. When an impulse traveling along the axon reaches the *synaptic terminal,* special calcium gates open and calcium ions diffuse into the terminal, causing the *synaptic vesicles* to discharge their stored *transmitter chemical* into the synaptic cleft. The transmitters diffuse across the cleft and combine with receptor proteins in the postsynaptic membrane, opening or closing specific ion gates, thereby altering the membrane potential. Synaptic trans-

mission is slower than impulse conduction along the neuron.

10. Synapses can be excitatory or inhibitory. An *excitatory transmitter* opens Na+ gates, slightly reducing the polarization of the postsynaptic membrane and creating an excitatory postsynaptic potential *(EPSP)*. If the EPSP reaches threshold, it triggers an impulse. An *inhibitory transmitter* makes the neuron harder to fire by increasing the polarization of the postsynaptic membrane, a condition called an inhibitory postsynaptic potential *(IPSP)*. The postsynaptic neuron integrates all the excitatory and inhibitory signals it receives, and then either fires or remains silent.

11. Once the transmitter has been released, it must be promptly removed from the synaptic cleft. Some transmitters are lost by diffusion, others are taken back up for reuse by the presynaptic neuron, and still others are destroyed by specific enzymes.

12. Transmission across the *neuromuscular junction* is also via transmitter chemicals. Acetylcholine is the transmitter at neuromuscular junctions in vertebrate skeletal muscles.

13. The human nervous system can be divided into two main parts: the central nervous system (brain and spinal cord) and the peripheral nervous system (nerves that connect the brain and spinal cord to the periphery of the body). The central nervous system is the coordinating center, organizing all the incoming and outgoing information.

14. A *reflex arc* is a simple neural pathway of the somatic nervous system linking a receptor and an effector. Most reflex arcs begin with a *sensory neuron* that conducts the impulse to *interneurons* in the spinal cord. These in turn synapse with *motor neurons* in the spinal cord, and the impulses are conducted to the effectors (usually skeletal muscles), which respond to the stimulus. Reflex arcs always interconnect with other neural pathways.

15. The *autonomic nervous system* consists of nervous pathways that conduct impulses from the *central nervous system* to various internal organs: these pathways usually involve two motor neurons. The autonomic nervous system regulates the body's involuntary activities.

16. There are two divisions in the autonomic nervous system, the *sympathetic* and *parasympathetic systems.* Most internal organs are innervated by both, with the two systems usually functioning in opposition to each other. The sympathetic system prepares an animal for emergency action during a crisis (fight or flight) whereas the parasympathetic system restores order or passivity after a crisis has ended.

17. There is a close relationship between the nervous and endocrine systems. The sympathetic system and the adrenal medulla have similar effects because they both release adrenalin and noradrenalin. Synaptic transmission also depends on the release of chemicals.

18. Many complex behaviors are based on motor programs that are self-contained neural circuits that coordinate muscle movement. Such programs are fine-tuned by sensory feedback and are under the control of the brain.

STUDY QUESTIONS

1. Someone throws a ball to you and you catch it. How does your action demonstrate the four basic components of *irritability?* (p. 588)
2. Draw and label the regions of a "typical" neuron. What are the functions of each part? (pp. 588–589)
3. Describe the functions of glial cells. What is a myelin sheath and what are its functions? (pp. 588–589)
4. Compare the functions of sensory neurons, motor neurons, and interneurons. How are they linked to one another? (pp. 589–590)
5. In most multicellular animals, which type of cell is associated with each facet of irritability? How can the flexibility of the response be increased? (pp. 591–592)
6. What trends are shown in the evolution of the nervous systems of bilaterally symmetrical animals? What is *cephalization?* (p. 593)
7. Suppose you pinch yourself twice, first gently and then more strongly. Will the nerve impulses generated by these stimuli differ in strength? How do you detect the difference in stimulus intensity? (pp. 594–595)
8. Describe the distribution of ions and of electrical charge on the inside and outside of a resting neuron. What causes the membrane of the axon to depolarize when it is stimulated? What causes it to become repolarized? How does the action potential travel along the axon from its starting point? (pp. 596–597)

9. How is the sodium-potassium exchange pump believed to work? Why is it vital to impulse conduction? Does it actually participate in any of the events of an action potential? (p. 599)

10. Describe the structure of a chemical synapse and the sequence of events occurring during synaptic transmission. (pp. 599, 601, 603)

11. How does a postsynaptic potential (PSP) differ from an action potential? Compare an EPSP and an IPSP. To which ion does the postsynaptic membrane become more permeable in each case? What is the effect of each on the membrane potential of the cell? How are their effects related to the neuron's *threshold potential* and to the processes of *summation* and *integration?* (p. 604)

12. In which branch of the nervous system (autonomic or somatic) do reflex arcs occur? Describe the withdrawal reflex that would occur if you stepped on a tack. Where are the cell bodies of the neurons involved? Would this reflex arc make connections to any other neural pathways? (p. 608)

13. How does the autonomic nervous system (ANS) differ in structure and function from the somatic system? Describe how the two branches of the ANS collaborate to regulate the activity of internal organs, such as the heart. What transmitter does the second motor neuron of each branch release? (pp. 610–612)

14. Contrast reflex chains and motor programs as methods of controlling complex rhythmical behavior such as walking. Which method is believed to be more prevalent among animals? (pp. 613–614)

CHAPTER 29

SENSORY RECEPTION AND PROCESSING

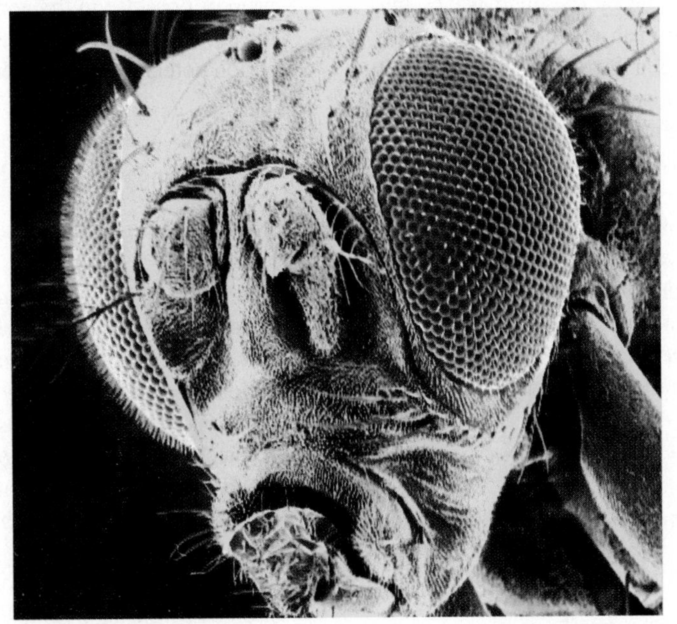

Flying high above a field, a hawk sees a mouse far below and dives to capture it. A male silkworm moth senses molecules of sex attractant in the air and flies upwind to locate the female that emitted the scent, while a bat sends out high-frequency vocalizations to detect echoes from objects in its path. How is it that each of these animals is able to detect and respond to physical changes in their environment? In this chapter we shall explore various mechanisms for detecting such mechanical and chemical changes and translating them into nerve impulses, a process known as ***sensory transduction.*** We shall also examine what is currently known about how nervous systems process the impulses produced by environmental stimuli, and how this processing results in the mental experiences we call ***sensations.*** Finally, we shall turn our attention to the question of how the vertebrate brain is organized.

Specialized receptor cells are the body's principal means of gaining information about conditions within the body and in the surrounding environment. They are our "windows on the world"—all information about the external world must enter our brain through our sensory receptors. They are the first elements in the reflex arcs whose structure and function we discussed in the last chapter. The sensory information collected by the receptor cells is converted into neural signals and conducted along sensory neurons to the spinal cord and brain, which then process the information and initiate an appropriate response.

Each Sensory Receptor Is Maximally Responsive to One Type of Stimulus

There are two basic types of sensory receptors: some are modified terminal branches of sensory neurons, as are many sensory endings in the skin; others are separate, specialized cells that are in close contact with the ends of sensory neurons, as are the taste cells of the tongue. In general, each type of receptor, whether a specialized cell or modified sensory neuron, is maximally responsive to just one kind of stimulus: stretching,

Chapter opening photo: The compound eye of a fruit fly. Each eye of the fly actually consists of about 28,000 individual eyes called ommatidia, each with its own lens and receptor pigment.

pressure, heat or cold, certain kinds of chemicals, vibrations, light, etc. Most receptors are highly selective and will not respond to stimuli other than those for which they are specialized, unless the stimuli are unusually strong. For instance, the visual cells of the eye are far more sensitive to light than to any other type of stimulus, but a strong mechanical stimulus such as a blow to the eye can stimulate these cells and the person will "see" light. The receptors in the ear, on the other hand, are highly sensitive to mechanical disturbances but are insensitive to light. Over the course of evolution each type of receptor has become highly specialized for detecting a particular kind of energy, and are therefore exquisitely sensitive to that stimulus.

Sensory receptors function as transducers, converting the energy that constitutes the particular stimulus to which they are attuned into a nerve impulse. An analogy is a microphone, which converts the mechanical energy of sound waves into electrical signals. And although each receptor is maximally sensitive to respond to a particular form of energy, a given stimulus may stimulate more than one type of receptor. For example, a firm handshake may stimulate both touch and pressure receptors, and an excessively strong handshake may stimulate pain receptors as well.

Receptors can be classified on the basis of the form of energy to which each is especially sensitive, i.e., ***chemoreceptors,*** sensitive to chemicals; ***photoreceptors,*** to light rays; ***mechanoreceptors,*** to mechanical disturbances; ***thermoreceptors,*** to heat or cold; and ***electroreceptors,*** to electrical fields. In this chapter we shall focus our attention on the first three categories of receptors.

Most Stimuli Act by Opening or Closing Gated Channels in Receptor Cells

Sensory reception begins in the membranes of the receptor cells. In some cells there are receptor proteins specific for a particular stimulus embedded within the membrane. When stimulated, these proteins undergo a conformational change that alters membrane permeability. In other cells, the structure of the membrane itself is altered by particular environmental stimuli. In both types of receptor cells, the end result is the same: the stimulus leads to the opening or closing of gated ion channels, which alters the charge across the cell membrane.

The taste receptors on the tongue provide a good example of sensory cells that have specific receptor proteins embedded in their membranes. Each taste receptor has proteins that are sensitive to a particular class of substances—sweet, bitter, salty, or sour. Each chemical binds to its specific receptor protein by weak bonds, and the binding causes gated sodium ion channels to open, altering membrane permeability. For example, when enough sugar molecules bind to receptor proteins on a sugar-sensitive cell in the tongue, enough sodium ion channels open that the sensory cell depolarizes beyond its threshold level to generate a nerve impulse in the sensory neuron. An adjacent cell that has receptor proteins for bitter chemicals like quinine will not respond, because its ion channels are not affected.

Stretch receptors in muscles and tendons function in a different manner. The membranes of these receptors are stimulated by physical distortion or stretch (as in the knee-jerk reflex). When the membrane is mechanically stretched, sodium ion channels open and sodium ions rush in, changing the polarity of the membrane. These ion channels are thought to be mechanically gated—i.e., physical distortion of the membrane directly opens the channels.

The change in polarization across the membrane of the receptor cells—the voltage change—is called the ***generator potential,*** and an increase in the intensity of the stimulus causes a proportional rise in the generator potential. When the generator potential reaches the sensory neuron's threshold level, an action potential is generated.

Sensations Are the Brain's Interpretation of Incoming Stimuli

The sensations of sweet and bitter, or stretching of a muscle, are all conveyed by identical means—the opening of ion channels, followed by depolarization, followed by the generation of action potentials in sensory neurons. They are distinguished as different sensations because the various sensory neurons are connected to different targets in the brain. It does not matter where the impulses originate or what stimulus initiates the impulses. It only matters what part of the brain is stimulated. Specific sensations are the brain's interpretation of incoming stimuli. What an animal senses, then, is shaped and constrained by the nature of its sensory receptors and the wiring of its nerve cells.

The brain is also responsible for the localization of the sensation. Each part of the body has its own sensory area in the brain. Thus fibers from your big toe run to one area in your brain, fibers from your ankle run to another area, and fibers from the thumb run to still another. Again, it is the part of the brain to which the impulses go, not the stimulus or the receptor or the message itself, that determines localization of the sensation. If the fibers from the pain receptors in your big toe were crossed with those from your thumb, and your big toe were then pricked with a needle, you would experience a sensation of pain in your thumb and would promptly examine your thumb for the cause of the trouble. The sensation of pain is a creation of your brain, and it only exists in your brain. The brain refers it back to a specific location, where it then seems to you to exist.

Phantom limb pain provides a striking example of how the quality and localization of sensation are determined by the brain. Amputees sometimes complain of pains, often very severe, in a limb that is no longer there. The phenomenon apparently occurs when the nerve fibers that ran from the amputated limb to the brain become irritated, triggering a response in the brain. The perception of pain, however, is not localized to a particular location in the brain, because neither cutting the peripheral nerves to the limb nor removing appropriate sensory areas of the brain eliminates the pain. The perception of pain is a creation of the brain, apparently involving widely dispersed areas.

Sensory Receptors Gather Information About Both the External and Internal Environment

The sensory receptors in the skin (see Fig. 15.19, p. 321, which shows some receptors in place) are concerned with touch, pressure, heat, cold, and pain. Some skin receptors, especially pain receptors, are simply dendrites of sensory neurons. Others are nets of nerve fibers surrounding, in mammals, the bases of hairs (Fig. 29.1). Other skin receptors are more complex, consisting of nerve endings surrounded by a capsule of specialized connective-tissue cells. The relative abundance of the various receptor types differs greatly in humans. For example, pain receptors are about 30 times as abundant as cold receptors.

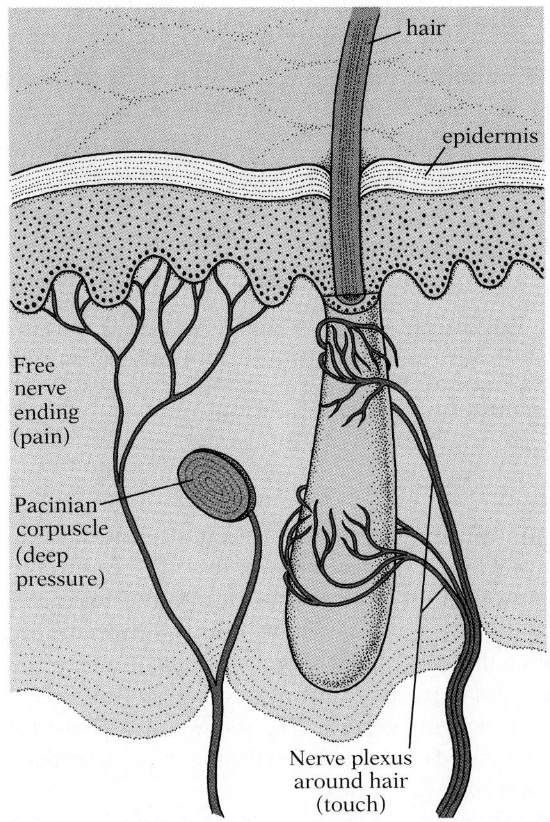

29.1 Three receptors of the skin and the senses they mediate

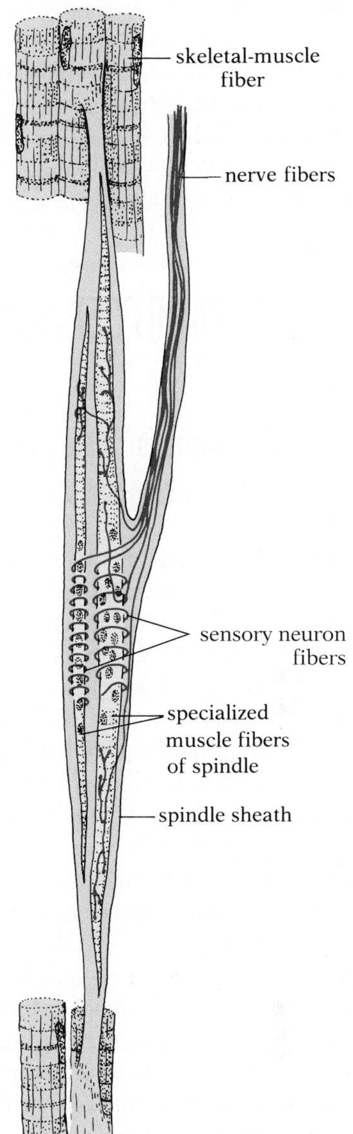

29.2 A stretch receptor in skeletal muscle The terminal branches of sensory nerve fibers are intimately associated with several specialized muscle fibers that form the apparatus called a neuromuscular spindle.

Furthermore, the receptors are not evenly distributed; touch receptors, for instance, are much more densely packed in the lips than on the back, while the fingertips have an intermediate concentration.

Inside the body, receptors are widely dispersed to receive information about internal conditions. Among these are the stretch receptors in the muscles, tendons, and joints, which we mentioned earlier when discussing the knee-jerk reflex (Fig. 29.2). Other dispersed receptors include those of the visceral senses, located in the internal organs. The stretch receptors located in the aortic arch, for instance, are sensitive to changes in blood pressure and help regulate the automatic control of heartbeat. When the blood pressure rises above the normal range, the aorta walls stretch and these receptors are stimulated. When the opposite occurs—when the blood pressure drops below normal—the receptors fire less frequently. The firing of visceral receptors seldom results in sensation (i.e., we are not aware of their action); the responses to their stimulation are usually mediated by the autonomic nervous system.

CHEMORECEPTORS

The Receptors for Taste and Smell Are Chemoreceptors

The receptors for taste and smell are stimulated by certain types of chemicals. The two senses are much alike, and when we speak of a taste sensation we are often referring to a compound sensation produced by stimulation of both taste and smell receptors. One reason why hot foods often have more "taste" than cold foods is that they vaporize more: the vapors pass from the mouth upward into the nasal passages and stimulate smell receptors. And one reason why a person with a cold cannot "taste" foods well is that, with nasal passages inflamed and coated with mucus, the smell receptors are essentially nonfunctional. Conversely, some vapors entering our nostrils pass across the smell receptors and enter the mouth, where they stimulate taste receptors.

In each case—taste and smell—chemicals must go into solution in the film of liquid coating the membranes of receptor cells before they can be detected. The major functional difference between the two kinds of receptors is that taste receptors are specialized cells that detect chemicals present in quantity in the mouth itself, while smell receptors are modified sensory neurons in the nasal passages that detect vapors coming from distant sources. The smell receptors can be as much as 3,400 times more sensitive than the taste receptors.

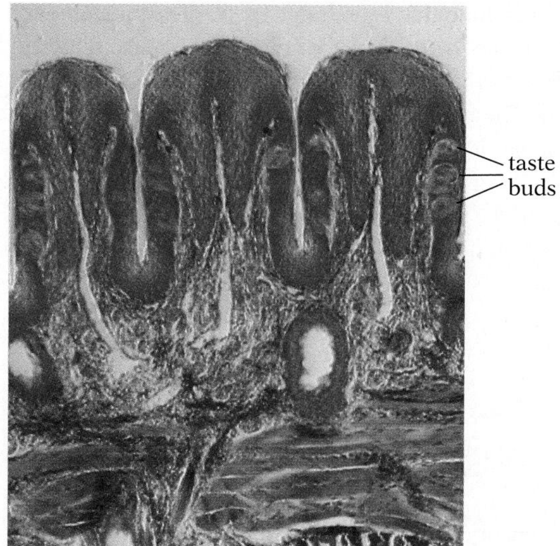

29.3 Photograph of section of mammalian tongue The taste buds are located in the walls of the deep narrow pits.

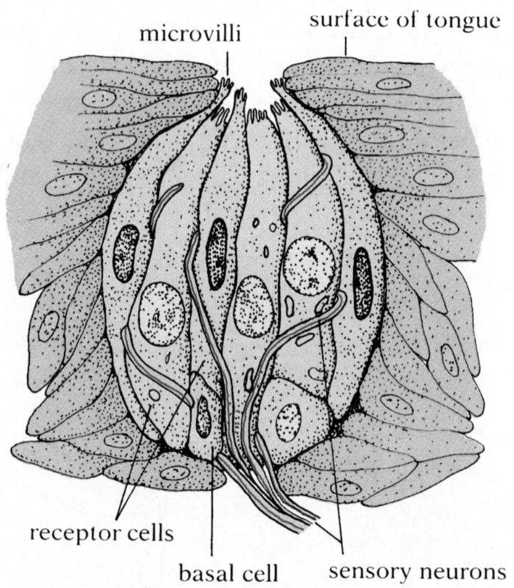

29.4 The structure of a taste bud Each taste bud contains specialized receptor cells bearing sensory microvilli that are exposed in pits on the tongue surface. The ends of sensory neurons (blue) are closely associated with the receptor cells.

Taste Buds Have Specialized Receptor Cells

In terrestrial vertebrates the receptor cells for taste are located in ***taste buds*** (Fig. 29.3). The majority of taste buds are located on the upper surface of the tongue but there are some on the surface of the pharynx and larynx. The receptor cells themselves are not neurons, but specialized cells with microvilli on their outer ends (Fig. 29.4). The ends of sensory neurons lie very close to these receptor cells, and when a receptor cell is stimulated, it generates impulses in the associated neuron. For example, when salt is detected by a salt-receptor cell, an electrochemical change takes place within the receptor cell, triggering an action potential in the associated sensory neuron.

In humans there are four basic taste senses: sweet, sour, salt, and bitter. The receptors for these four basic tastes have their areas of greatest concentration on different parts of the tongue—sweet and salt on the front, bitter on the back, and sour on the sides. A few substances stimulate only one of the four types of receptors, but most stimulate two, three, or four types to varying degrees. The sensations we experience are thus produced by a blending of the four basic sensations in different relative intensities.

Some aquatic amphibians and fishes (e.g., catfish) have taste receptors in the skin on the exterior of the body, particularly on their heads and anterior fins, as well as in the mouth.

They therefore can "taste" the environment as they swim through it—an adaptive advantage for detecting food. The fish called whitings, for instance, will turn and capture a piece of food placed near their tails.

Smell Receptors Are Modified Sensory Neurons

The receptor cells for the sense of smell (olfaction) in terrestrial vertebrates are located in two clefts in the upper part of the nasal passages (Fig. 29.5A). Unlike the receptor cells for taste, which are specialized receptor cells, olfactory receptors are modified sensory neurons. The cell bodies of most of these neurons lie embedded in the epithelial layer of the walls of the olfactory area of the nasal chamber (Fig. 29.5B). Dendrites run from the cell bodies to the surface of the epithelium, where they bear a cluster of modified cilia, which function as the receptor sites.

Since the late 1800s, many attempts have been made to isolate a group of primary odors from which all more complicated odors might be derived. Many groups of fundamental odors were proposed during these years, but all proved unsatisfactory. The olfactory system has proved to be extraordinarily complex, and appears to be structurally quite different from taste reception. While there appear to be four basic types of taste receptors, recent evidence indicates that there may be more than 1,000 different receptor proteins for smell! Each receptor is thought to have a unique shape and to respond only to certain types of odor molecules, just as a key fits its lock. Work on isolating the receptors and determining how they work is just beginning.

From an evolutionary point of view, the olfactory sense is very old—the earliest vertebrates are thought to have had a well-developed sense of smell, and olfactory organs are found in all vertebrates. Odors can be used not only for locating food, but for recognizing family members, marking territories, and finding mates. To attract mates, for example, many different animals, including certain insects, aquatic invertebrates, and mammals, release special scents that contain information about their species, sex, reproductive readiness, and location. Some moths have incredibly sensitive smell receptors (Fig. 29.6). The males are capable of detecting sex-attractant molecules released by females several miles away! Also, some animals use odors as alarm signals. For example, ants, snails, bees, and certain fishes release an alarm substance when injured as a warning to other members of that species.

Most animals depend to a far greater extent on olfaction than humans can fully appreciate. Although humans do have a good sense of smell—we can detect about 10,000 different

29.5 Human olfactory receptors The receptor cells are located in the olfactory epithelium, which is located on the wall of the upper portion of the nasal cavity. The olfactory bulb of the brain receives information from the receptor cells. Inset: The cell bodies of most of the receptor cells are in the olfactory epithelium; their sensory cilia protrude into a layer of mucus on the surface of the epithelium.

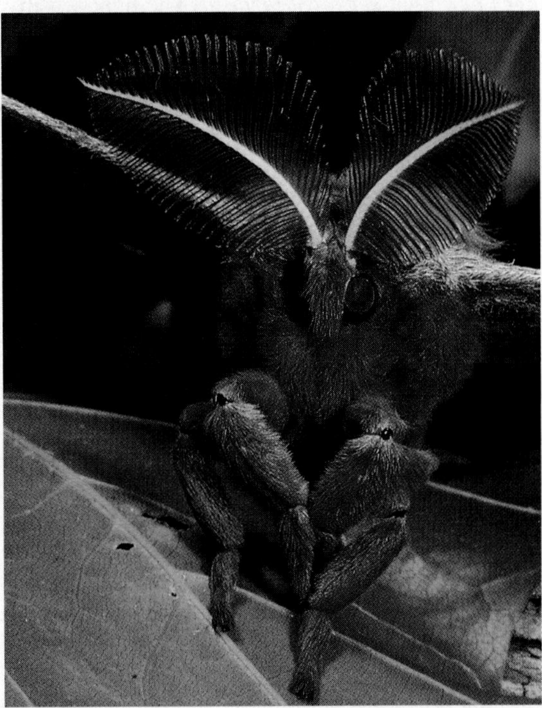

29.6 Antennae bearing olfactory receptors of an adult male polyphemus moth *(Antheraea polyphemus)* The thousands of sensory hairs extending perpendicularly from the combs of the male moth's antennae bear specialized cells that respond to minute amounts of the female's pheromone. While millions of molecules may be required to activate a human olfactory neuron, just a single molecule of pheromone at a receptor cell on the surface of a sensory hair triggers the moth's cell.

odors—our olfactory capability is not as good as those of many vertebrates, especially fish and other mammals. Olfactory stimuli have many advantages over other sensory stimuli. Odors can be transmitted over relatively long distances, are not blocked by obstacles in the environment, and are useful both day and night. Ancestral humans may well have depended on smell more so than modern humans; for instance, smell conveys information in determining whether a food is spoiled, or poisonous.

VISUAL RECEPTORS

Almost all animals respond to light stimuli. Even single-celled protozoans react quickly to changes in light intensity, often moving away from brightly lit areas. In fact, many unicellular organisms, such as *Euglena* (see Fig. 38.11, p. 887), have a special region that serves as a sensitive detector of light. This region contains a pigment that undergoes chemical changes when exposed to light energy, changes that "tell" the protozoan that light is present. Most multicellular animals have evolved specialized light-receptor cells, but the basic mechanism of detection is the same as in protozoans: *light energy produces changes in a light-sensitive pigment.* The light-sensitive pigment common to many animals is a protein to which a portion of a carotenoid molecule is attached. Carotenoids are yellow- and orange-colored pigments found in plants that absorb certain wavelengths of light. Animals, including humans, must obtain carotenoids in their diet as a vitamin (vitamin A), since it can be synthesized only by plants.

The light receptors of many invertebrates do not function as eyes, in the usual sense of that word. They do not form visual images, but simply indicate whether or not light is present. They also indicate changes in light intensity. Some of these receptors give the animal no clue as to where the light comes from, and the animal responds by essentially random movements. However, many light receptors are arranged in such fashion that they do indicate direction. The simple eyespots of planaria (see Fig. 17.13, p. 356) are an example; light receptors are arranged within a cup-shaped organ called an ***eye cup;*** this arrangement conveys information on the direction of the light source (Fig. 29.7). For many animals, being able to detect the presence or absence of light and or its direction is sufficient. Animals that have evolved more complex eyes, with a lens capable of concentrating light on the receptor cells, are able to make better use of this information. Lenses increase the sensitivity of the eye to light of weak intensity (Fig. 29.8), and also greatly increase the eye's ability to detect direction and movement by focusing the light from each source onto only a few receptor cells at a time. Lenses made possible the evolution of image-forming eyes found in some molluscs and most arthropods and vertebrates. With well-developed image-forming eyes, the animal is able to determine not only the structure and texture of the object, but where it is located, and whether it is moving. Furthermore, because the speed of light is 3×10^8 meters per second, the reception is almost instantaneous.

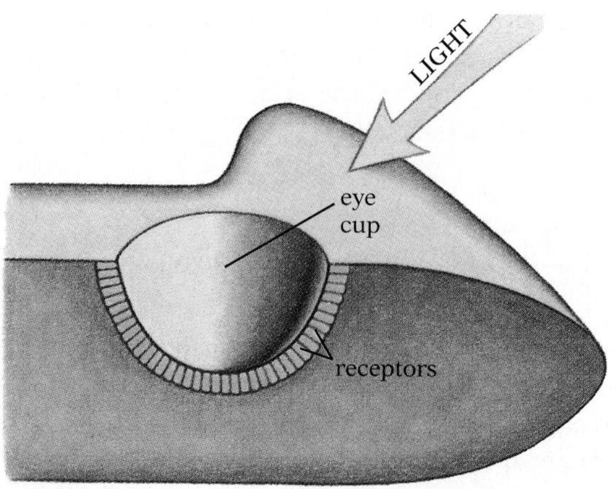

29.7 A planarian eye cup The planarian eye cup provides directional information that the organism can use to maneuver in the environment; it does not form images. The direction of a light source is indicated by the location of the shadow cast by the cup's opaque edge onto receptors within the organ, as shown in this cross section.

The independent evolution of image-forming eyes by different groups of animals has produced two fundamentally different kinds of eyes: ***compound eyes,*** such as those of insects and crustaceans; and the ***camera-type eyes,*** such as those found in some molluscs (e.g., octopus, squid) and vertebrates.

29.8 Two animals with relatively simple lensed eyes Top: The jumping spider has two large eyes directed forward, and smaller ones on each side of its head. Bottom: The scallop has two rows of eyes along the margin of its mantle, just inside the shell. The spot on each eye is the cornea. The eyes of the spider and the scallop, though structurally simpler than those of most vertebrates, have lenses capable of focusing incoming light on the receptor cells of the retina.

MAKING CONNECTIONS

1. Why are microphones, television, and sensory receptors (e.g., the photoreceptors of the eye) all classified as transducers?
2. What does the "phantom limb" pain experienced by amputees, or "referred pain," in which a pinched nerve is perceived as a real pain in another part of the body, demonstrate about the quality and localization of sensation?
3. How is the role of visceral receptors related to the function of the autonomic nervous system, discussed in Chapter 28?
4. How is the detection of tastes and odors similar to the mechanism by which an EPSP is produced in a postganglionic cell?
5. Compared to other sensory modalities, what advantages dose olfaction offer? What role might it have played in human evolution?
6. What steps have probably occurred in the evolution of image-forming eyes from simple ancestral photoreceptors? What is the basic mechanism of light detection found in all photoreceptors?

Compound Eyes Are Composed of Many Closely Packed Ommatidia

A ***compound eye*** consists of an array of individual eyes, each modified into a tube, called an ***ommatidium.*** Light entering through the lens and crystalline cone of each ommatidium is focused onto light-sensitive pigments located in the central portion of the membranes of the receptor cells (Fig. 29.9). Image formation depends on the light pattern falling on the surface of the compound eyes. This light pattern determines which ommatidia will be stimulated and at what intensity. Since each ommatidium points in a slightly different direction, each is stimulated by light coming from different points in the surrounding area (Fig. 29.10).

The picture produced by an insect brain from the information it receives is probably a grainy mosaic of the world rather like needlepoint, with far less precise delineation of objects in the visual field than we experience (Fig. 29.11). However, a compound eye has several advantages that compensate for this low spatial resolution. Besides being very efficient at absorbing light, they are very small and lightweight, which is important for a flying insect. In addition, because each ommatidium

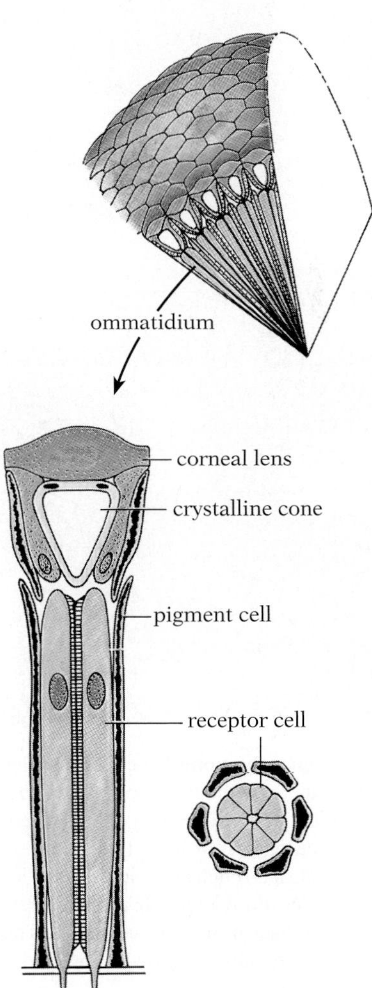

29.9 Ommatidia from the compound eye of a typical daytime insect Top: Section of compound eye. Bottom: Longitudinal and cross sections of one ommatidium. The lens and crystalline cone focus incoming light onto the rhabdom, a translucent cylinder formed by the highly folded membranes in the central portion of the eight receptor cells. The light-sensitive pigments are located in this central portion. The pigment cells surrounding the ommatidium contain a dark pigment that prevents the passage of light from one ommatidium to another.

29.10 The compound eyes of a horsefly Each eye is composed of a huge number of ommatidia. Notice that each ommatidium points in a slightly different direction and therefore receives light coming from slightly different points in the surrounding area.

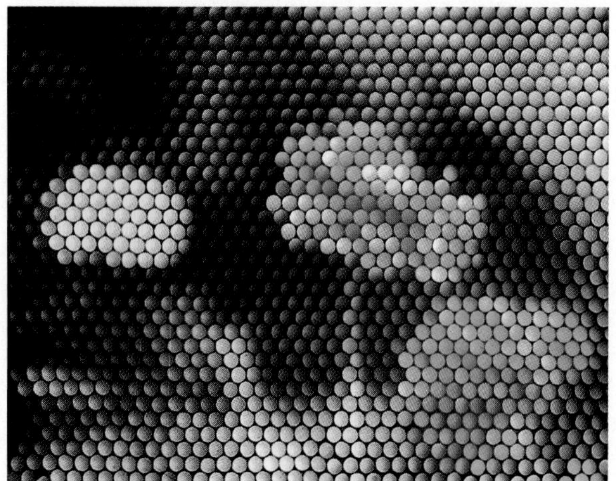

29.11 View through a compound eye The visual world of animals with compound eyes is thought to be broken up into a mosaic of spots. Here dahlias are shown as seen by a human being (top) and as a honey bee might see them from about 10 centimeters away (bottom). This picture, taken through an optical device, is not a perfect representation: bees see ultraviolet light and not red, while this photograph does not reproduce ultraviolet and includes red. Moreover, the circles should be vertically elongated ellipses, to reflect the peculiar anatomy of the bee eye.

recovers rapidly from exposure to light, compound eyes permit arthropods to detect movements that are far too rapid for our eyes. Some flies, for instance, are capable of detecting light flickering at a rate of 250 flashes per second, whereas our eyes cannot detect light flashing faster than 50 times per second. The ability to detect movement is useful in helping an insect escape its predators (and fly swatters!). Flies are even able to see the flickering of fluorescent light.

In a Camera Eye, the Image Is Projected Onto the Retina

A camera-type eye uses a single lens to focus light on a surface containing many photoreceptor cells packed close together (Fig. 29.12). The receptor surface, called the ***retina,*** functions in a manner analogous to a piece of film. The light pattern focused on the retina produces differential stimulation of different receptor cells, just as a light pattern focused on a piece of film produces different amounts of chemical reaction at different points on the film.

The adult human eye, which is a camera-type eye, is globe-shaped and has a diameter of approximately one inch (Fig. 29.13). It is encased in a tough but elastic coat of connective tissue, the ***sclera*** (the "white" of the eye). The anterior portion of the sclera, called the ***cornea,*** is transparent and more strongly curved, and functions as an important part of the light-focusing system of the eye. Inside the sclera is a layer of darkly pigmented tissue, the ***choroid,*** which is richly supplied with blood vessels. The choroid is important both in providing blood to the rest of the eye and as a light-absorbing layer that (like the black inner surface of a camera) helps prevent internally reflected light from blurring the image. The *retina,* which contains the photoreceptor cells, is a thin tissue covering the inner surface of the choroid. It is composed of several layers of cells.

The anterior portion of the choroid is modified to form a ring of pigmented tissue, the ***iris.*** The iris contains smooth muscle fibers arranged in both circular and radial directions. When the circular muscle fibers contract, the opening in the center of the iris, called the ***pupil,*** is reduced; when the radial muscles contract, the pupil is dilated. The iris thus regulates the size of the opening admitting light (the pupil) just as the diaphragm of a camera regulates the lens aperture. Just behind the junction of the white part of the sclera and the cornea, the choroid becomes thicker and has smooth muscles embedded in it; this portion of the choroid is called the ***ciliary body.***

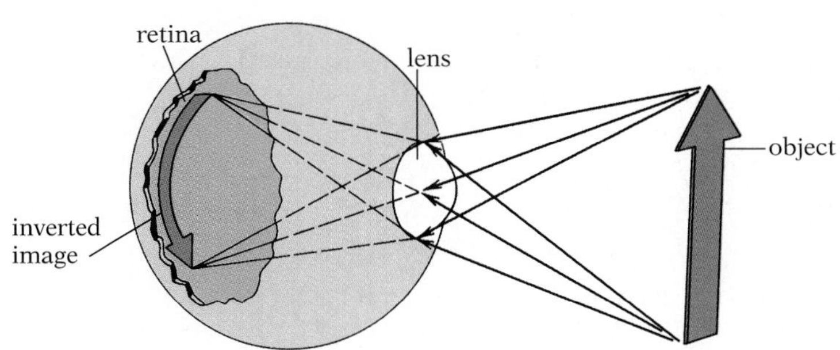

29.12 How a camera eye works Light reflected from objects in the environment is bent by the cornea and lens and is focused on the retina. Note, however, that the image is inverted.

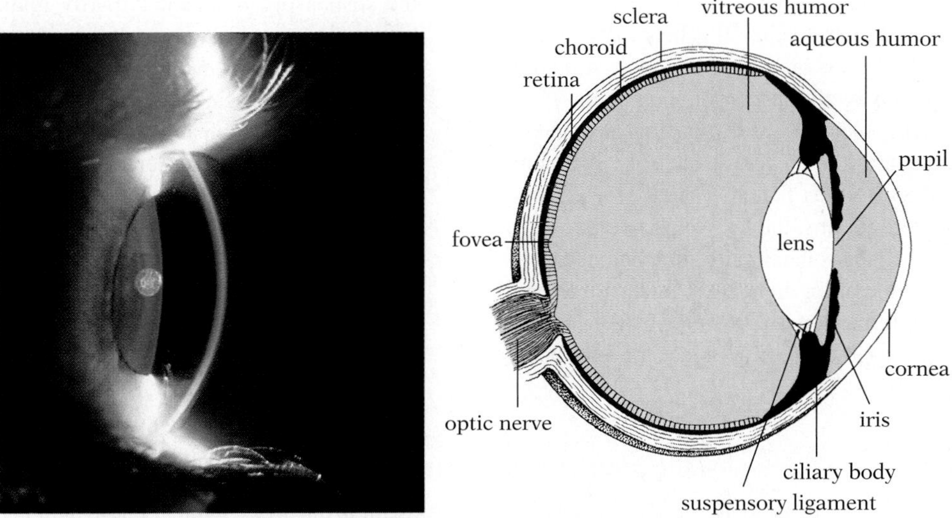

29.13 The human eye The drawing shows a diagrammatic section through the left eye, as seen from above.

The ***lens*** is suspended just behind the pupil by a ***suspensory ligament*** attached to the ciliary body. The lens and its suspensory ligament thus divide the cavity of the eyeball into two chambers; the anterior chamber is filled with a transparent watery material (aqueous humor), and the posterior chamber with a viscous semifluid material (vitreous humor).

Both the Cornea and the Lens Focus the Image on the Retina

Since the pattern of illumination of retinal receptor cells is fundamental to image formation, high-resolution image vision depends on precise focusing of incoming light on the retina. If focusing is not good, the image is blurred. In short, the projection of a true image of the observed object onto your retina requires a lens system capable of bending incoming rays of light and focusing them on the retina (Fig. 29.13), just as the lens system of a movie projector focuses light that has come through the film onto the screen.

The lens system of the human eye has two principal components, the cornea and the lens. The cornea, not the lens, does most of the bending of incoming light rays. The great importance of the lens stems from the fact that it is the flexible part of the system; it makes adjustments in the focus depending on the distance of the object being viewed.

Consider the differences between the focusing of light coming from close objects and from distant objects. Light rays reflected from a given point travel away from that point in all directions (Fig. 29.14). If the cornea of an eye is located near the point source (Position 1), it will be struck by many light rays, some of which will be traveling at strongly divergent angles from each other. If, however, the cornea is located farther away from the point source (Position 2), many of the most divergent light rays, which would have struck it at Position 1, will miss it entirely, and the rays that do strike it will be traveling at only very slight angles to each other. If the cornea is six meters (20 feet) or more from the point source, the rays of light that strike it can be considered as traveling parallel to each other. Therefore, light from a near object must be strongly bent by the lens system, while much less bending is necessary to bring into focus the rays coming from a distant object.

Most ***refraction*** (bending) of light from distant objects is performed by the cornea. But the cornea cannot refract the

29.14 Difference in degree of divergence of incoming light rays from near and far sources Light rays from a point source travel outward in all directions. A cornea near the point source (Position 1) will be struck by many strongly divergent light rays (rays A through I). A cornea farther away (Position 2) will be missed by the most divergent rays (A–C and G–I) and will be struck by rays traveling at much smaller angles to each other (D–F). If the cornea is six meters or more from the point source, the light rays reaching it will be traveling almost parallel to each other.

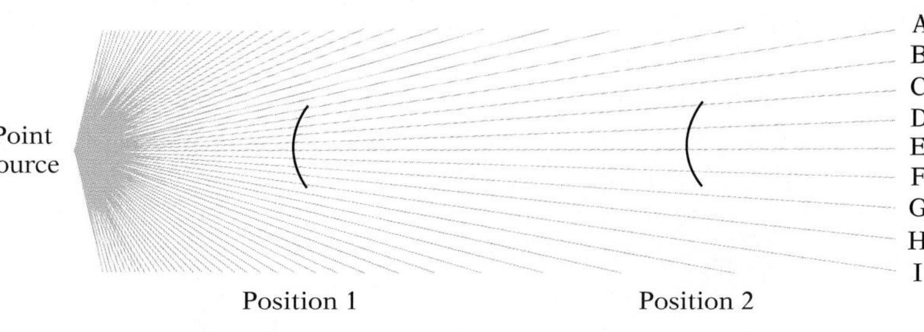

strongly divergent light from near objects sufficiently to bring the image into clear focus on the retina. It is here that the lens becomes important. The lens is an elastic biconvex structure attached to the ciliary body by the suspensory ligament. When the eye is viewing distant objects (objects more than six meters away), a considerable tension on the suspensory ligament stretches the lens, which is thus flattened and becomes less convex. Now, the flatter the lens, the less refractive power it has. Therefore, when the lens is maximally stretched, it exerts very little influence on incoming light rays; under these circumstances most of the refraction is performed by the cornea. When, however, the eye is viewing a near object, the tension on the suspensory ligament is partly relaxed, and the lens becomes more round and convex, because of its natural elasticity. The refractive power of the lens supplements the refractive power of the cornea, and the image is brought into clear focus on the retina. The nearer the viewed object is to the eye, the more the tension on the lens is relaxed, and the more convex the lens becomes. This process of correcting the focus of images of near objects by changes in the shape of the lens is called ***accommodation.*** With age, the lens loses elasticity, with a resulting loss of accommodation, a condition known as ***presbyopia.*** By the age of 45 to 50 years, the lens no longer rounds up and remains flattened as with viewing distant objects. The loss of accommodation is such that reading and close work becomes difficult, and corrective lenses must be used.

Structural defects in the shape of the eye are fairly common in human beings. Eyeballs that are either too short or too long result in the image being focused behind or in front of the retina (Fig. 29.15). Another common defect, astigmatism, stems from uneven curvature of the cornea; it can be corrected with a lens ground unequally to compensate for such irregularities.

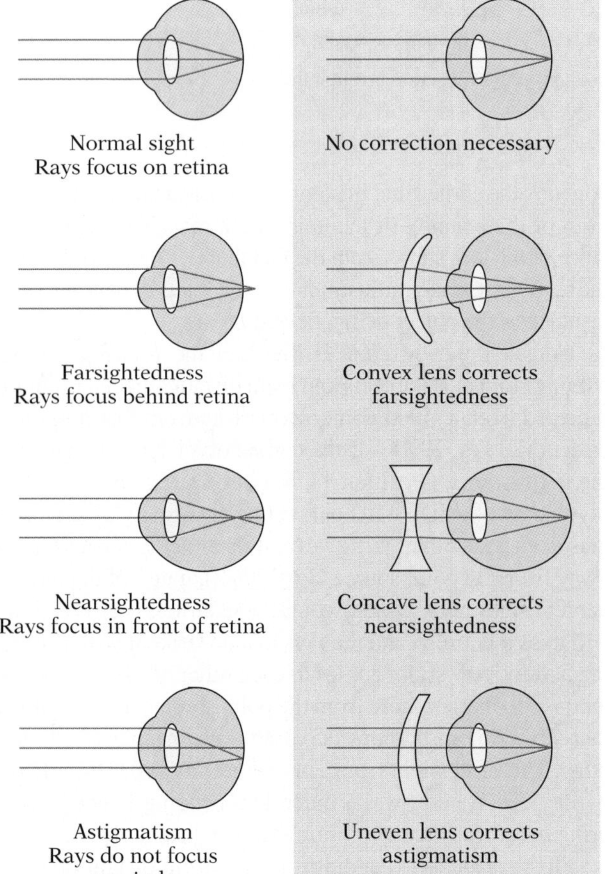

29.15 Farsighted and nearsighted eyes The farsighted eye is so short that the focal point would be behind the retina; the cornea and lens cannot bend the light rays enough to bring them together on the retina. Since rays from a distant source are almost parallel, they need less bending than the divergent rays from a near source; hence the person can see distant objects more clearly. Farsightedness is corrected by a convex lens (one that is thicker in the center than at the edges). Such a lens bends the incoming light rays before they reach the cornea, thus aiding the cornea in bringing them together on the retina.

The nearsighted eye is so long that the focal point is in front of the retina, and the rays have started to diverge again by the time they reach the retina, producing a blurred image. The strongly divergent rays from a near source require more bending; the focal point for them is therefore nearer the retina than the focal point for the almost parallel rays from a distant source. Hence the person can see near objects more clearly. Nearsightedness is corrected by a concave lens (one that is thinner in the center than at the edges). Such a lens spreads the incoming parallel rays from a distant source, thus making them strike the cornea at divergent angles just as the rays from the near source would do. Astigmatism results from uneven curvature of the cornea; it can be corrected by a lens ground unequally to compensate for these irregularities.

Rods and the Cones Contain Light-Sensitive Pigments

The photoreceptor cells of the retina are of two types: rod and cone cells (Fig. 29.16). The ***rod cells*** are more abundant toward the edge of the retina. They are exceedingly sensitive and function only at low light intensities—in light too dim to stimulate cone cells. Rods cannot detect colors, and the images they generate are coarse and poorly defined, much like a very sensitive photographic film that produces a grainy photograph. The ***cone cells,*** which are more abundant in the central portion of the retina, function only when there is good illumination. They generate detailed images (i.e., good visual acuity), and they enable us to detect color. Notice that the rods and cones are oriented with their tips pointed toward the ***back*** of the eyeball, ***away*** from the source of light (Fig. 29.17). This does not seem to be a very logical organization for the retina, but all vertebrates have this type of "inverted retina." Squid, which also have a camera-type eye, have their receptor cells pointed outward, toward the light source. Functionally both eyes are capable of producing good visual resolution.

The rods and cones synapse in the retina with short sensory neurons (bipolar cells), which synapse in the retina with longer neurons (ganglion cells) whose axons, bundled together as the optic nerve, run to the visual centers of the brain (Fig. 29.17). Other specialized neurons connect receptor cells to other receptor cells and interconnect ganglion cells (Fig. 29.17). The presence of several sets of synapses within the retina enables the eye to extensively modify the information transmitted from receptor cells to the brain.

Both rods and cones contain light-sensitive pigments. In rods the pigment, which is built into the membranes of the flattened vesicles in the outer segment (Figs. 29.16 and 29.17), is called ***rhodopsin.*** Rhodopsin is composed of a protein (opsin) bonded to a light-sensitive prosthetic group called retinal, which is synthesized from vitamin A (Fig. 29.18). A deficiency of vitamin A in the diet can result in insufficient production of rhodopsin, which leads to night-blindness. If the deficiency is prolonged, anatomic changes in both the rods and cones occur, which can result in a permanent loss of vision.

Once the rhodopsin in the rod cell has absorbed the light, it dissociates into retinal and opsin, and is inactive; rhodopsin must be regenerated before the rod will respond again to light. Enzymes carry out this conversion, which is much slower than the light-driven hydrolysis of rhodop-sin. Consequently, in good illumination, the rhodopsin is "bleached," i.e., converted into the inactive form, as soon as it is formed. Cone vision, however, does not involve rhodopsin, and is therefore used in bright light, allowing us to see objects in color. At night, when there is less light, rod vision predominates and we see a coarse image in black and white. We have all had the experience of going from the bright sunlight into a darkened room and finding that we are unable to see at first. After a period of time we become "dark adapted"—that is, we can see because rhodopsin has been synthesized in the dark, and the rods can once again respond to light.

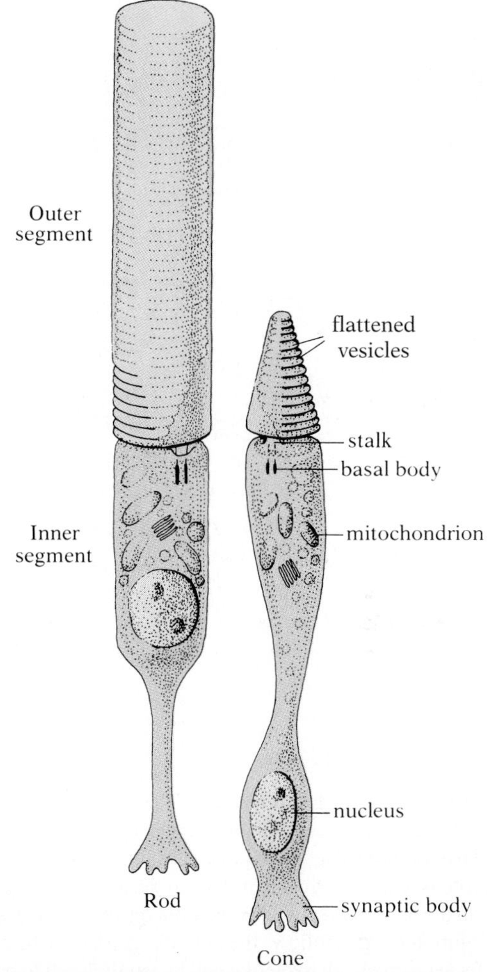

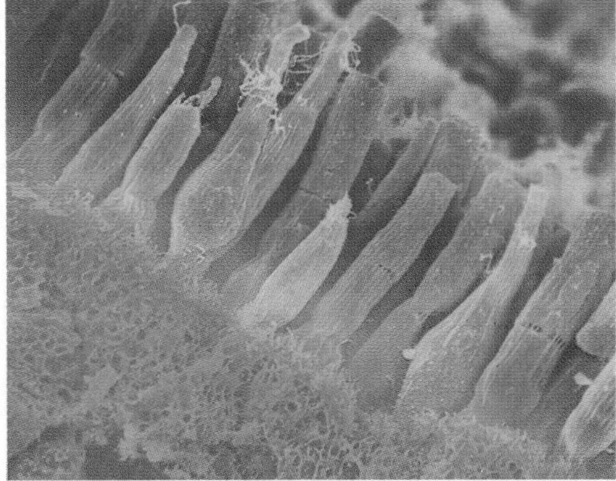

29.16 Rod and cone cells from a human retina Top: Cells from a human retina. The outer segment and the stalk connecting it to the inner segment develop in both rods and cones as a highly specialized cilium. The visual pigment is located in the numerous flattened vesicles of the outer segment. The pigments of rods and cones are different; those of cones are responsible for color vision. Bottom: Scanning electromicrograph of rods (blue) and cones (greenish-blue). At their bases, the rods and cones synapse with various types of neurons (pink).

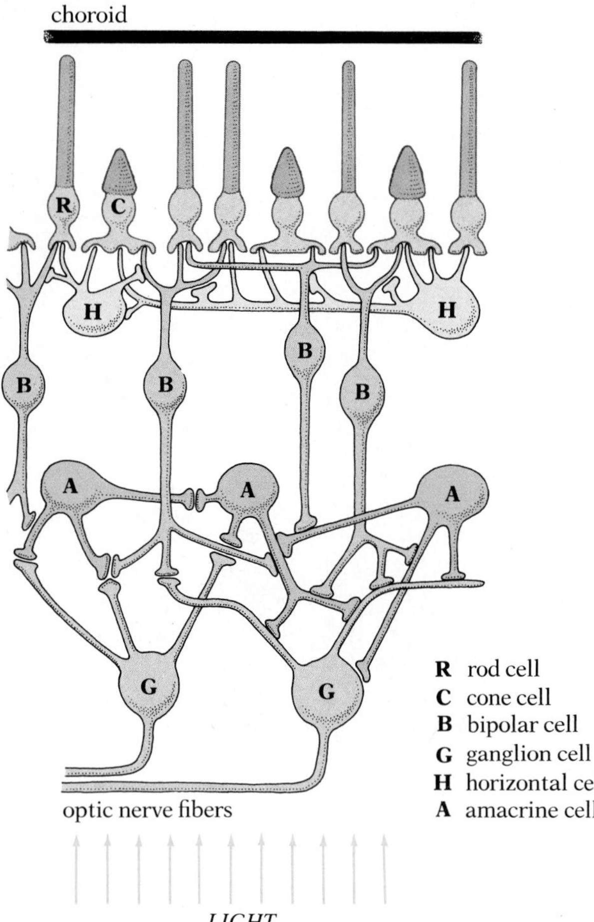

29.17 The cells of the human retina The receptor cells (rods and cones) synapse at their base with bipolar cells, which make up the middle layer of the retina. The bipolar cells, in turn, synapse with ganglion cells, which form the third layer; the axons of the ganglion cells form the optic nerve, which runs from the eye to the brain. Hence information that follows the most direct route to the brain moves from receptor cell to bipolar cell to ganglion cell to brain. Processing of information can occur within the retina because often several receptor cells synapse with a single bipolar cell and several bipolar cells synapse with a single ganglion cell. There is also lateral transfer of information from pathway to pathway via lateral processing cells (**H** and **A**). Note that the three layers of the retina are arranged anatomically in reverse order from what might be expected; the receptor cells are pointed toward the back of the retina, and light must pass through the ganglion-cell and bipolar-cell layers to reach them.

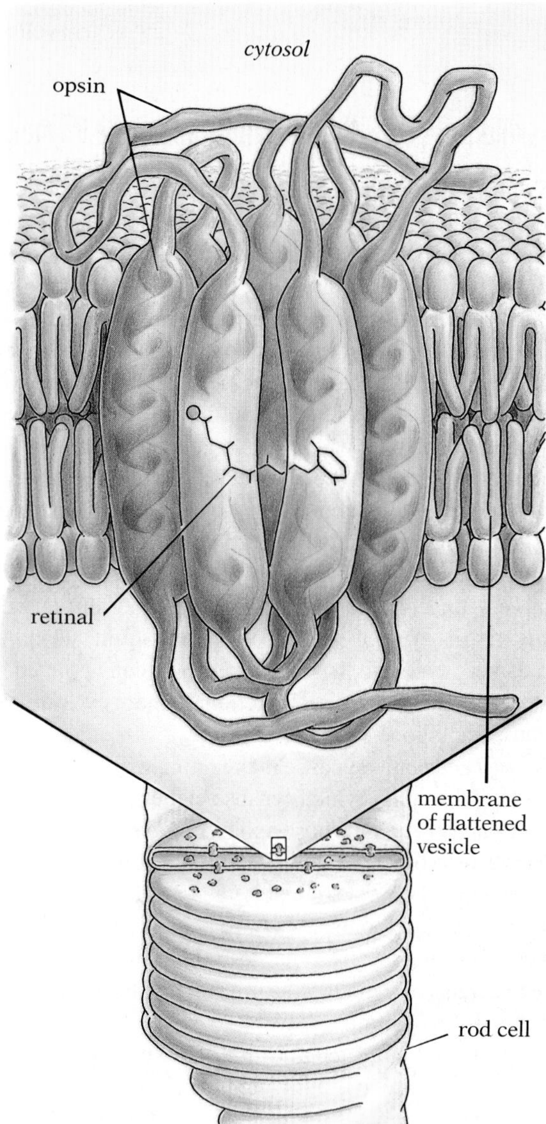

29.18 The light-driven changes of retinal The visual pigment rhodopsin is composed of a protein called opsin (brown) which holds a light-sensitive group called retinal. When a molecule of rhodopsin is struck by light, the retinal absorbs the light and is converted into a slightly different shape. This change in shape induces a change in ion permeability of the membrane, which ultimately results in stimulation of the sensory neurons leading to the brain. Once retinal has been converted by light into this form, bleaching occurs: the retinal dissociates from the opsin. The rods are unable to function again until rhodopsin is regenerated. A series of enzymatic reactions within the rod cell combines the retinal and opsin to form active rhodopsin.

Animals Have Widely Varied and Very Selective Color Perceptions

Our understanding of color reception is still elementary. There are three functional types of cones in humans, each containing a different pigment. Each of the three pigments is sensitive to wavelengths of light covering a broad band of the visible spectrum, but each has its maximum absorption in a different portion of that spectrum. The three pigments can be designated as blue-absorbing, green-absorbing, and red-absorbing. This evidence for a three-color, three-receptor mechanism of cone reception agrees with a theory of color vision long supported by psychological experiments.

Human beings are so accustomed to their own color vision that they tend to assume that other animals see colors in the same way. Apparently many fishes, reptiles, and most birds do have color vision (though it differs from human color vision), and some even detect light in the ultraviolet (UV) range. Humans and other primates are, however, rather unusual among mammals in possessing both good visual acuity and color vision. Most mammals, having evolved as primarily nocturnal animals, have sensitive, pure rod retinas and probably

29.19 Primrose willow *(Ludwigia peruviana)* as seen by humans (top) and with its ultraviolet pattern revealed (bottom) When photographed with ultraviolet-sensitive film, flowers often reveal a pattern of markings that form a bull's-eye in the center, where the reproductive parts (and the nectar) are found. Since insects, unlike humans, see ultraviolet wavelengths, the reproductive parts are underscored for them, so to speak. The device fulfills a critical function for the plant, in that an insect drawn to the reproductive parts may serve the plant as a pollinating agent, carrying any pollen adhering to its body to the next flower it visits. [Courtesy Thomas Eisner, Cornell University.]

see only in shades of gray, with rather low visual acuity. Such eyes are well adapted for vision in dim light, but not for living in bright daylight. Many mammalian groups do have acute vision, but primates are unique among mammals in having well-developed color vision. The cones in primates, which permit both color vision and acute vision, probably evolved from the pure rod retinas found in nocturnal mammals. Indeed, the cones in mammals are different from those in other vertebrates with color vision, and are thought to have evolved independently in response to similar environmental pressures.

Insects, too, often have color vision—a very important attribute for the ones that feed on flowers. Many of them, however, do not have the same range of color vision as humans. Their eyes usually cannot detect red light; hence a room in which there is only pure red light will be in total darkness to many insects. But these insects can see light of wavelengths in the ultraviolet band, which humans cannot see (Fig. 29.19). Work on insect eyes is beginning to suggest that their eyes may also be involved in detecting electromagnetic radiation.

MECHANORECEPTORS

Insect Mechanoreceptors Detect Air Movement

Like vision, hearing is an important sensory ability among animals. For many species, sounds in the environment provide crucial information about predators, prey, and other species' members. A variety of strategies for detecting them have therefore evolved.

Sound is produced when an object vibrates, setting in motion the particles of the medium (usually air) and thus generating alternating bands of high and low pressure (Fig. 29.20).

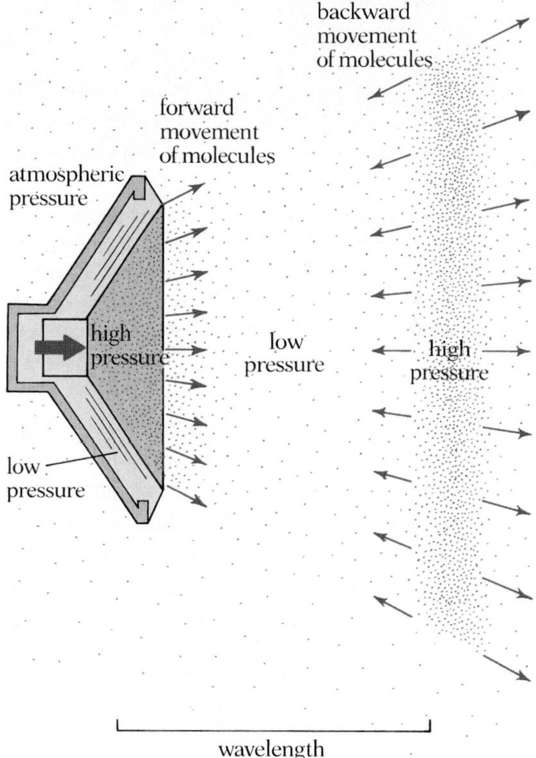

29.20 Sound generation by a loudspeaker By vibrating and setting air molecules into motion, a loudspeaker creates waves of high and low pressure; these are the physical basis of sound. Various kinds of ears have evolved, which respond either to particle movements or to waves of pressure. What we hear is the result of the processing of sensory input in the nervous system. In this diagram, air pressure is indicated by the density of dots, while the directions of particle movement are shown by arrows. The distance between one band of high pressure and the next is one wavelength.

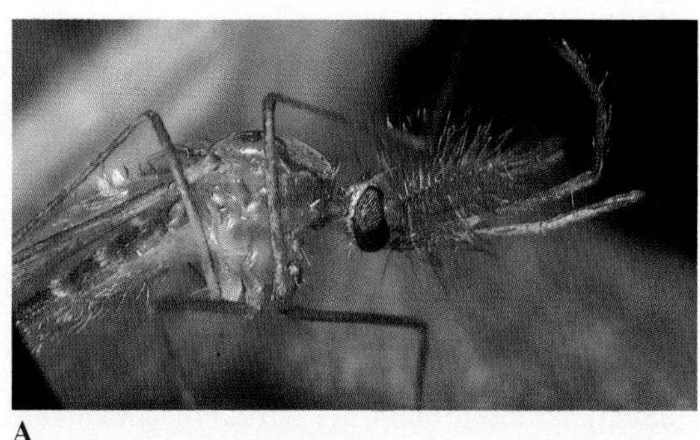

A

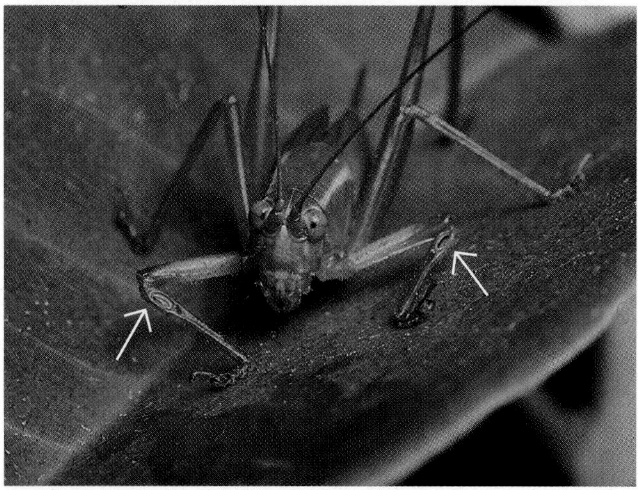

B

29.21 Two sound detectors in insects (A) The antennal hairs on the head of this male mosquito can resonate, and are therefore very sensitive to sounds of the proper frequencies. (B) The membrane-covered organs (arrows) on the forelegs of a katydid, by contrast, detect the pressure changes associated with sound.

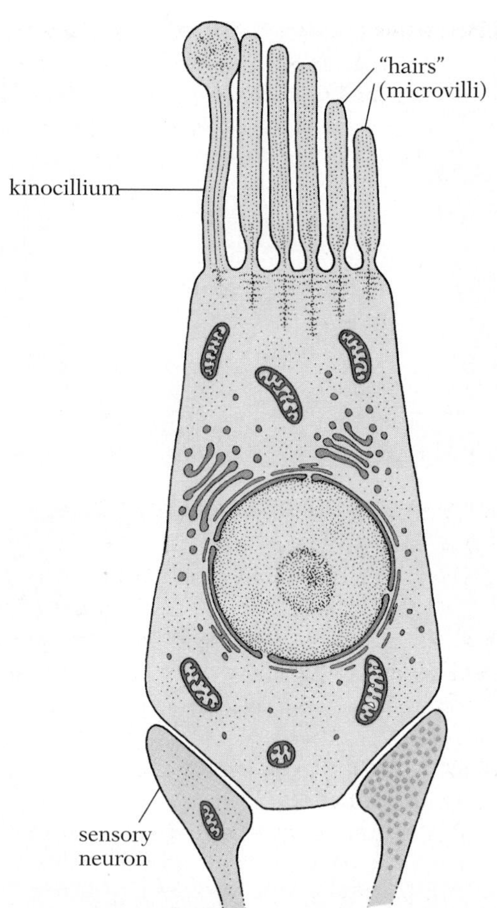

29.22 A typical hair cell. A typical hair cell may contain a large modified cilium, called a kinocilium, and 30 to 150 "hairs," which are actually enlarged microvilli. The base of the hair cell is in contact with a sensory and motor neuron. The "hairs" respond to deflection in one direction or another by opening or closing gated ion channels in the membrane, thereby altering the membrane permeability and producing a generator potential within the hair cell. If the generator potential is above threshold, an action potential is produced in the associated sensory neuron, and the information is transmitted to the brain.

The number of vibrations per second is the sound's frequency, and is given in hertz (Hz), which simply means cycles per second. Many insects have long hairs that are moved back and forth by the air movement (Fig. 29.21A). Like tuning forks, these hairs have strong resonance, so that they respond best to vibrations within a narrow range—typically the frequencies emitted by other members of the species. Most vertebrates, on the other hand, are sensitive to the pressure changes associated with sound waves.

Vertebrate Hair Cells Are Mechanoreceptors That Detect Sound Vibrations and Gravity

The hair cells that are found in the lateral-line system (see below) of fishes and in the inner ear of many vertebrates are highly sensitive mechanoreceptors. Hair cells are so named because each has a hairlike tuft of very large microvilli projecting from one end (Fig. 29.22). Often there is a single large modified cilium, called a ***kinocilium,*** present as well. Bending of the "hairs" opens or closes gated ion channels and alters the permeability of the membrane, producing a generator potential within the hair cell. If the generator potential is above threshold, an action potential is produced in the associated sensory neuron, and the information is transmitted to the brain.

The ***lateral-line system*** of fishes consists of clusters of hair-cell receptors located in fluid-filled canals along the head and sides of the body (Fig. 29.23). Water currents striking the side of the fish cause the fluid in the lateral-line canals to move, deflecting the hairs and stimulating the hair cells. The location and arrangement of the groups of hair cells are important in enabling fish to detect the direction of waves bouncing off nearby

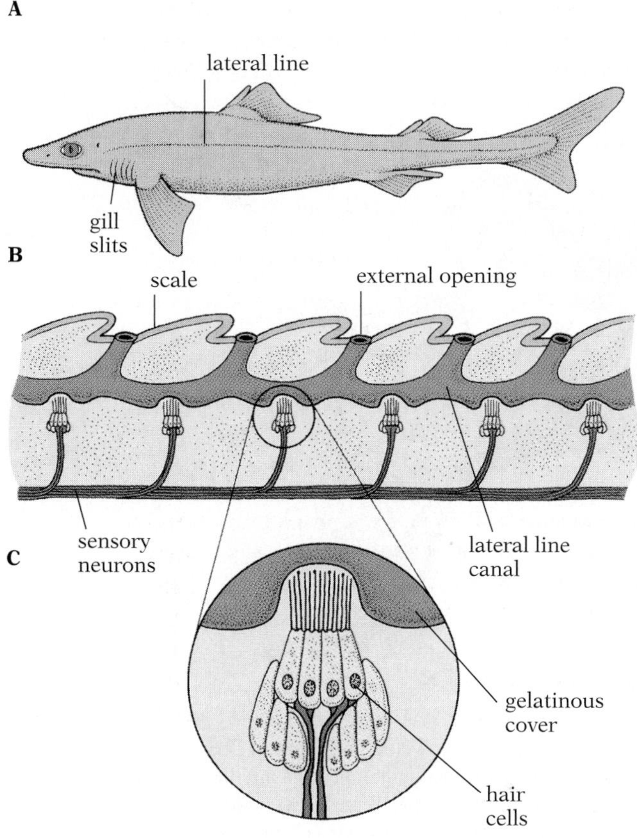

29.23 Lateral-line system of fish (A) The lateral line consists of a series of openings along the side of the fish. (B) Longitudinal section through the lateral-line system of a fish. The openings along the side of the fish lead into a fluid-filled lateral-line canal that contains hair cells. (C) Enlargement of the hair cells. The tips of the hair cells are embedded in a gelatinous mass. Movements of the water in the lateral-line canal moves the gelatinous mass and deflects the hairs, generating an action potential in the associated sensory neurons. With this system the fish can detect water currents, nearby animals, etc.

objects and the direction of water currents. In ancestral fishes, some of the lateral-line organs in the head apparently sunk inward and evolved into the inner ear, whose original function was primarily for equilibrium.

Being human, we naturally think of the ear as the organ of hearing, but in our vertebrate ancestors its primarily function was probably equilibrium, and the inner ear remains an organ of equilibrium in all vertebrates from fish to humans. However, many fishes are able to hear; in such fishes the bones of the skull conduct sound vibrations to the inner ear for sound reception. But it is in the amphibians that the inner ear took on the function of sound reception in addition to equilibrium. One of the gill slits found in the ancestral amphibians was modified to form a middle ear and a covering formed over it that could be used to pick up sound vibrations from the air. Tiny bones conducted sound vibrations to the inner ear. Most terrestrial vertebrates have an ear that functions for both sound reception and equilibrium. Moreover, the sensory receptors for both hearing and equilibrium have remained the same: the all-purpose hair cells.

The Human Ear Is Divided Into the Outer Ear, the Middle Ear, and the Inner Ear

In mammals the outer ear consists of the ear flap, or pinna, and the auditory canal (Fig. 29.24). At the inner end of the auditory canal is the ***tympanic membrane,*** more commonly called the eardrum. On the other side of the tympanic membrane is the chamber of the middle ear. This chamber is connected to the pharynx via the ***Eustachian tube.*** The connection allows the equalization of air pressure between the outer and middle ear. For example, when you climb a high hill, the reduced pressure at the higher altitude results in a lower pressure in the auditory canal than in the middle-ear chamber, and the tympanic membrane is stretched outward. The pressure is equalized when air escapes from the middle ear through the Eustachian tube into the pharynx. When you descend quickly from a high altitude, the reverse process occurs. Swallowing, yawning, or coughing facilitates the passage of air through the Eustachian tube.

Three small bones[1] arranged in a lever-like arrangement extend across the chamber of the middle ear from the tympanic membrane to a membrane called the ***oval window.*** Another membrane, the ***round window,*** lies just below the oval window in the wall of the middle-ear chamber.

Hair Cells of the Inner Ear Are Important in Establishing Equilibrium

The organs of equilibrium are quite similar in all vertebrates, and we shall focus our attention on the structure of the human inner ear. The human inner ear is a complicated labyrinth of interconnected fluid-filled chambers and canals. The upper portion of the labyrinth of the inner ear is composed of three ***semicircular canals*** and a large vestibule that connects them to another portion of the ear, the cochlea (Fig. 29.25). Inside the vestibule are two chambers, the ***utriculus*** and the ***sacculus.*** Each contains a bed of hair cells, upon which rests a gelatinous membrane containing embedded crystals of calcium carbonate, called ***otoliths*** (Fig. 29.26). Because gravity pulls objects downward, any change in the position of the head or change in

[1] These are the malleus, incus, and stapes, also commonly known respectively as the hammer, anvil, and stirrup.

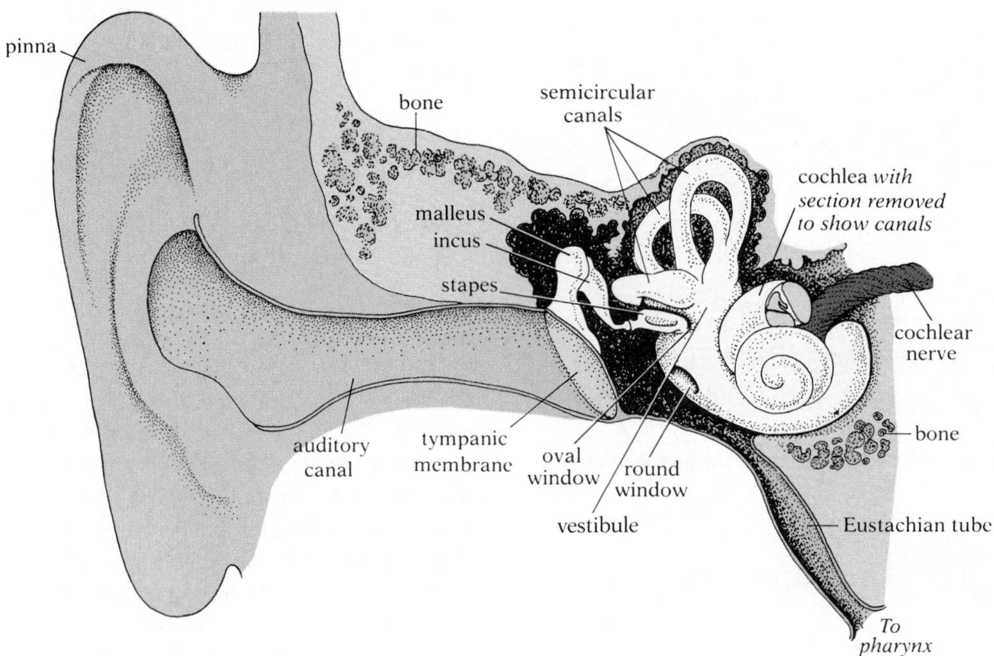

29.24 The human ear The major parts of the outer, middle, and inner ear are shown in this diagram. The outer ear consists of the ear flap, and the auditory canal, which ends at the tympanic membrane. The middle ear chamber, which is connected to the pharynx by the Eustachian tube, contains three small bones that conduct sound waves from the tympanic membrane to the inner ear. The inner ear consists of two groups of interconnected chambers. The upper group, the vestibule and semicircular canals, is concerned with equilibrium, while the lower portion, the cochlea, is concerned with hearing.

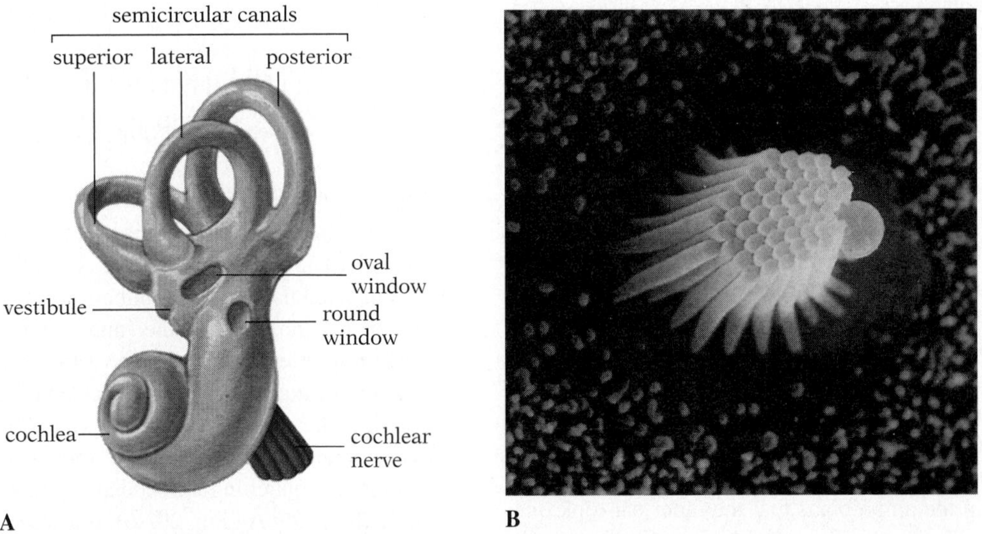

29.25 The inner ear (A) The left inner ear of a human. The inner ear consists of a number of interconnected fluid-filled chambers and canals. The upper portion consists of three semicircular canals and a large vestibule that contains the utricle and saccule. The lower portion of the inner ear, which is coiled like a snail, is the cochlea. (B) Scanning electron photograph of the hairs of a hair cell from the vestibule of a frog ear.

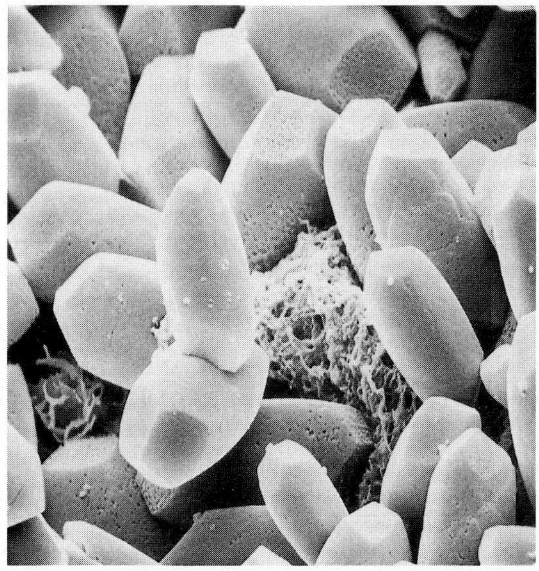

A

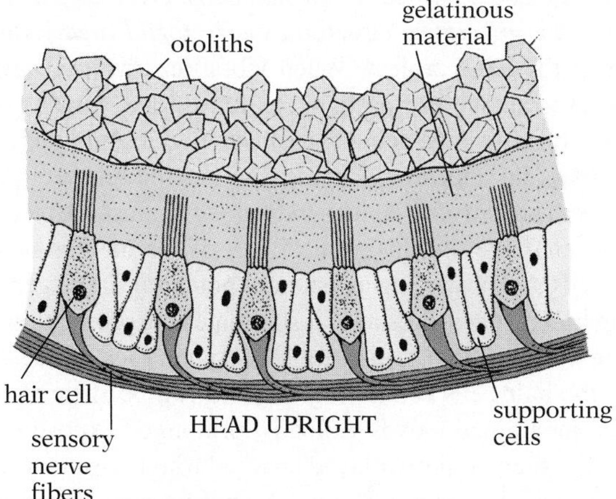

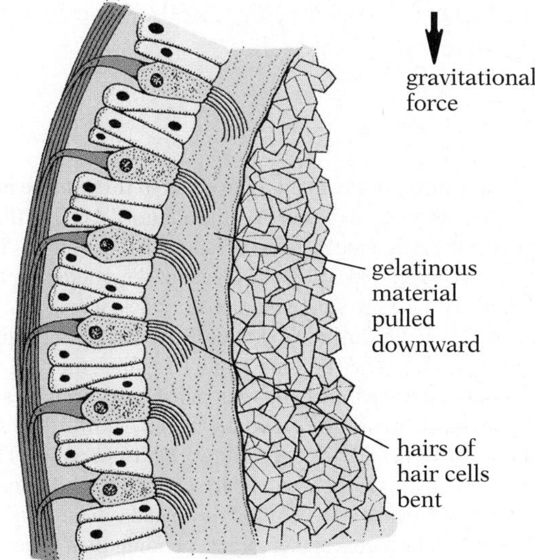

B

29.26 Otoliths in the inner ear The vestibule of the human ear is lined with a bed of hair cells, upon which rests a gelatinous membrane containing embedded crystals of calcium carbonate, called otoliths (A). Since gravity pulls dense objects downward, any change in the position of the head causes the membrane and its otoliths to move and exert more pressure on some hairs than on others. When the head is in an upright position (B, top), the weight of the otoliths presses directly down on the sensitive hair cells, the tips of which are embedded in the membrane. When the head is tilted (B, bottom), the altered pull of the otoliths on the hairs generates signals to the brain.

speed of motion (i.e., linear acceleration) of the head causes the membrane and its otoliths to move and exert more pressure on some hairs than on others (Fig. 29.26). The deflection of the hairs opens potassium ion channels, thereby altering the ion permeability and the polarization of the hair cell membrane. If the change is above threshold, an action potential is generated in the associated sensory neuron, and the information is transmitted to the brain. The input from the various sensory neurons is processed within the cerebellum of the brain to determine the position of the head relative to gravity at any given moment. Similar sensory devices are found in invertebrates. For example, crayfish and lobsters have organs of equilibrium called ***statoliths,*** which consist of sand grains resting on beds of sensory hair cells (Fig. 29.27).

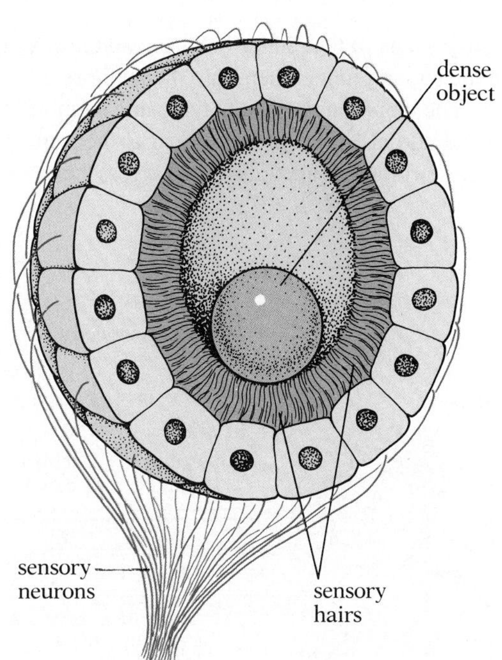

29.27 Diagram of the interior of a statolith Statoliths are the equilibrium organs of invertebrates; each consists of a chamber lined with sensory hairs enclosing one or more dense objects, such as sand grains. Since gravity pulls dense objects downward, the location of the sensory hairs that are stimulated by the weight of the object indicates the direction of gravitational pull.

Hair Cells in the Semicircular Canals Sense Rotational Acceleration

In vertebrates, the design of the semicircular canals makes it possible to sense rotational acceleration (change in the rate or direction of rotation of the head). Like the axes of a three-dimensional grid, each of the three canals is oriented at right angles to the other two (Fig. 29.25). At the base of each canal is a small chamber containing a tuft of hair cells. When the head is moved or rotated in any direction, the fluid in the canals lags behind, because of its inertia, and thus exerts increased pressure on the hair cells. This pressure stimulates the hair cells to initiate impulses to the cerebellum of the brain. The brain, by integrating the different amounts of stimulation it receives from each of the three canals, can then determine very precisely the direction and rate of the acceleration. The system is not perfect. When the head is rotated for an extended period and then stopped abruptly, the fluid will continue to circulate. The result is the kind of dizziness we feel after some amusement park rides.

A Lever-like Arrangement of Bones Conducts Sound Vibrations From the Outer to the Inner Ear

The cuplike shape of the pinnae helps to capture sound vibrations in the air, and funnel them into the auditory canal of the outer ear. The vibrations strike the tympanic membrane, causing it to vibrate at the same frequency as the impinging air waves. These vibrations are transmitted across the cavity of the middle ear to the oval window by the three small middle-ear bones, which are arranged in a lever-like fashion to increase the force of the vibrations on the oval window (Figs. 29.24 and 29.28). This increase in the force of incoming stimuli is very important in enabling us to detect very faint sounds.

Hair Cells in the Cochlea Detect Sound Vibrations

The lower portion of the inner ear consists of a long tube coiled like a snail shell (Figs. 29.24, 29.25, and 29.29). This is the ***cochlea,*** which is the organ of hearing. The cochlea is divided into three fluid-filled canals (Fig. 29.29A, B), the middle one of which, the cochlear canal, contains the hair cells, the auditory receptors. The receptors are located in a structure called the ***organ of Corti,*** which consists of a layer of epithelium on which lie rows upon rows of specialized hair cells that lack kinocilia (Fig. 29.29B, C). Dendrites of sensory neurons terminate on the undersurfaces of the hair cells. Overhanging the hair cells is a gelatinous structure, the ***tectorial membrane,*** into which the hairs project. When vibrations from the oval window are conducted through the fluid-filled canals, the membrane on which the hair cells sit also vibrates, moving the sensory hairs up and down against the less mobile tectorial membrane. The hairs are bent as they move against the tectorial membrane, altering the permeability of the membrane of the hair cells and producing a generator potential. If the generator potential reaches threshold, it stimulates the production of an action potential in the adjacent sensory neurons. Since the hairs or the hair cells cannot regrow, and can be sheared off when the membrane moves violently, prolonged exposure to loud noises (such as music played loudly through earphones or at a rock concert) can result in permanent hearing loss, typically the higher frequencies (Fig. 29.30). Table 29.1 provides several different sound intensities.

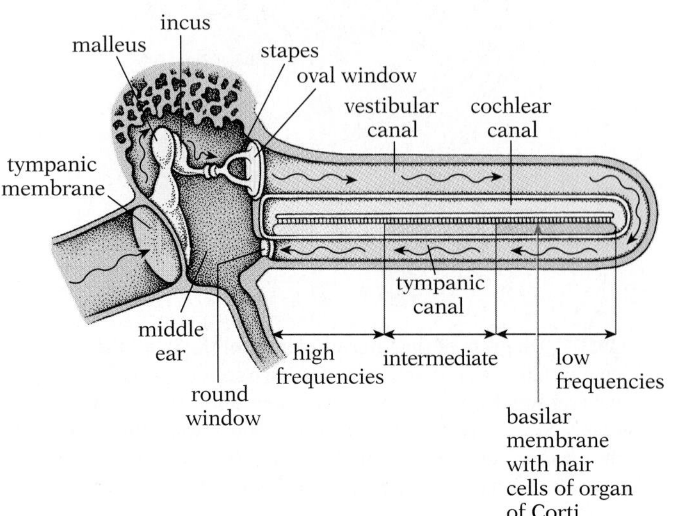

29.28 A summary of the hearing process in the human ear Sound vibrations in the air pass down the auditory canal of the outer ear and strike the tympanic membrane, causing it to vibrate. These vibrations are transmitted across the cavity of the middle ear to the oval window by the three small middle-ear bones, which are arranged in a lever-like fashion to increase the force of the vibrations on the oval window. When the stapes moves against the oval window, the fluid in the vestibular and tympanic canals oscillates (wavy arrows); the oscillation is made possible by the flexible round window, which moves in and out, permitting relief of pressure. The pressure waves in the fluid of the cochlea cause the basilar membrane with its organ of Corti to move up and down and rub the hairs of the hair cells against the tectorial membrane, bending them. Low-frequency vibrations stimulate hair cells near the apex of the cochlea; high-frequency vibrations stimulate hair cells near the base of the cochlea; and intermediate frequencies stimulate hair cells in the intermediate regions of the cochlea.

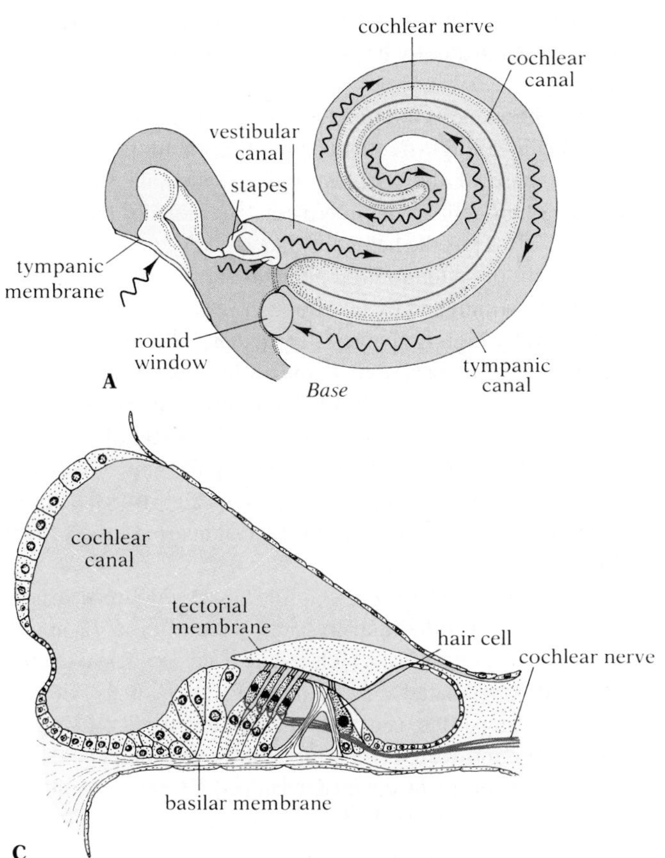

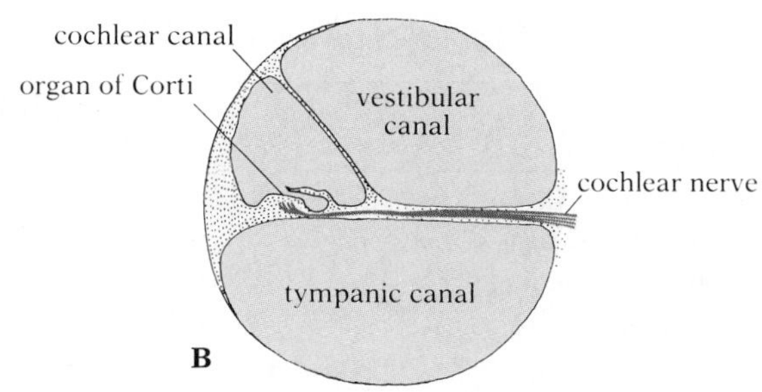

29.29 The cochlea of the human ear (A) Inside the cochlea are three canals: the vestibular canal, which begins at the oval window; the tympanic canal, which connects the vestibular canal to the round window; and the cochlear canal, which lies between the other two. All three canals are filled with fluid. (B) Enlarged cross section of one coil of the cochlea, showing the relationship between the vestibular, tympanic, and cochlear canals and the location of the organ of Corti. (C) Cross section of the cochlear duct showing the organ of Corti. The organ of Corti rests on the membrane that forms the base of the cochlear duct. Sound vibrations result in movements of the fluid of the canals of the cochlea, causing the membrane to move up and down and rubbing the hair cells against the overhanging tectorial membrane. The resulting deformation of the hairs produces a generator potential in the hair cells, which triggers impulses in sensory neurons running from the organ of Corti to the brain.

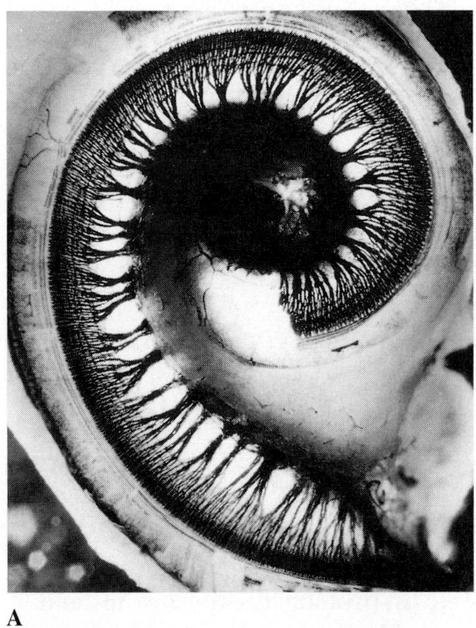

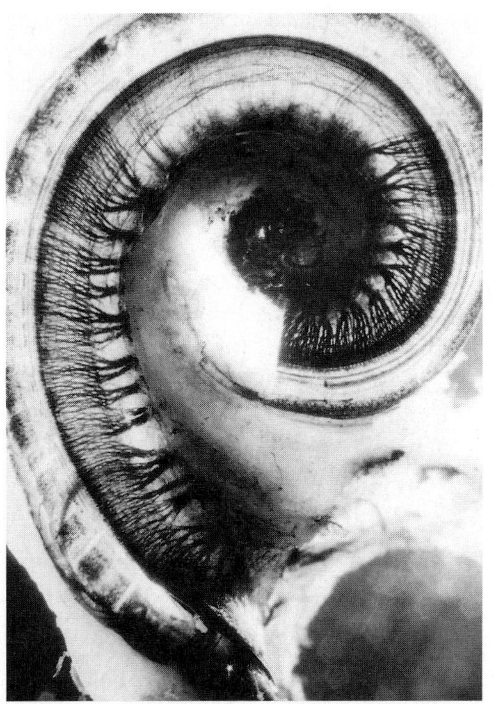

29.30 Normal and noise-damaged cochlea (A) Longitudinal section of the cochlea showing normal distribution of the hair cells and associated sensory neurons. (B) Longitudinal section of noise-damaged cochlea. The hair cells and sensory neurons have been permanently damaged, resulting in a serious hearing loss. Individuals subject to excessive noise should wear ear protectors. Many musicians in rock bands have suffered loss of hearing from their occupation.

TABLE 29.1 *Examples of sound intensities*

Sound	*Intensity*[a]	*Intensity level*[b]
Loudest noise achieved in laboratory	10^{9}	210
Rupture of eardrum	10^{4}	160
Jet engine (at 30 m)	10	130
Threshold of pain	1	120
Thunder (loud)	10^{-1}	110
Subway train (New York City)	10^{-2}	100
Heavy street traffic	10^{-5}	70
Normal conversation	10^{-6}	60
Whisper	10^{-10}	20
Normal breathing	10^{-11}	10
Threshold of hearing	2.5×10^{-12}	4

[a]Intensity is measured in watts per meter squared (W/m^2).
[b]Intensity levels are measured in decibels (dB).

Vibrations in the Cochlear Fluid Stimulate the Hair Cells

Vibrations of the tympanic membrane, then, cause movements of the middle-ear bones, which in turn produce movements of the membrane of the oval window. These in their turn produce movements of the fluid in the canals of the cochlea and in the membrane of the round window (Fig. 29.28). The movements of the fluid in the cochlea are at the same frequencies as those of the air that entered the outer ear. The pressure waves in the fluid of the cochlea cause the organ of Corti to move up and down, which rubs the hair cells against the tectorial membrane, bending them. Thus stimulated, the hair cells activate the sensory neurons that carry impulses to the auditory centers of the brain.

Low-frequency vibrations stimulate hair cells near the apex of the cochlea; high-frequency vibrations stimulate hair cells near the base of the cochlea; and intermediate frequencies stimulate hair cells in the intermediate regions of the cochlea (Fig. 29.28). Thus the hair cells, like the keys of a piano graduated from low pitch to high pitch, are arranged in sequence, from those stimulated by low frequencies at the apex to those stimulated by high frequencies at the base. The neurons from each region along the length of the cochlea lead to slightly different areas in the brain. The pitch sensation we experience depends on which of these areas in the brain is stimulated. With increasing age, the membrane supporting the hair cells begins to stiffen at the base of the cochlea, and so high-frequency sensitivity is progressively lost. This type of hearing loss is found in almost all older people.

Clinical deafness may be classified as conduction deafness, due to impaired sound transmission in the external or middle ear, or as nerve deafness, which results from damage to the hair cells or the associated sensory neurons. Individuals with impaired hearing often use hearing aids to amplify incoming sounds. The individual's hearing at different frequencies and levels is tested, and the deficit determined. The hearing aids are designed to amplify signals to address the particular deficit. Unfortunately, hearing aids amplify everything in that range—background noise as well as the desired signals—with the result that the wearer may still have difficulty distinguishing sounds. Some hearing aids have added filters to block signals below a certain frequency or noise level, but in many cases these have not proved too successful, and the problem remains.

The human ear is sensitive to vibrations over an amazing

MAKING CONNECTIONS

1. What structural defect in the eye causes nearsightedness? Farsightedness? Astigmatism? Presbyopia?

2. How is the ability of human beings (and other primates) to see colors related to our evolutionary history?

3. Trace the evolution of the inner ear from its origin in the lateral line system of fish. What was its original function? How did it evolve into an organ of hearing in terrestrial vertebrates?

4. In what respect are the organs of equilibrium of many invertebrates similar to those of vertebrates?

5. Would the loss of hearing caused by prolonged exposure to loud noise be classified as conduction or nerve deafness?

6. What does the existence of an echolocation system in bats and dolphins suggest about evolution?

range of frequencies—from about 16 to 20,000 hertz (cycles per second) in young people. Some children can even hear frequencies as high as 40,000 hertz (higher than ultrasonic dog whistles) but ability to hear high frequencies declines steadily with age. Many animals are more sensitive to sound than humans. Some animals can hear much higher frequencies; dogs respond to whistles at 30,000 hertz, which few humans can hear, and bats and some moths can hear frequencies of 100,000 hertz or higher. Bats have incredibly sensitive hearing and use echolocation to capture prey (Fig. 29.31). A bat produces a variety of ultrasonic vocalizations that are reflected by objects in its path, and the returning echoes give the bat information about the size and position of the objects. Many bats use echolocation to detect insects, which they capture in flight. By no means restricted to bats, echolocation is also used by many marine animals. Dolphins are an example. When they swim through a tank of turbid water, they emit frequent high-pitched clicking sounds and, while they do, can readily avoid all submerged obstacles, even while blindfolded.

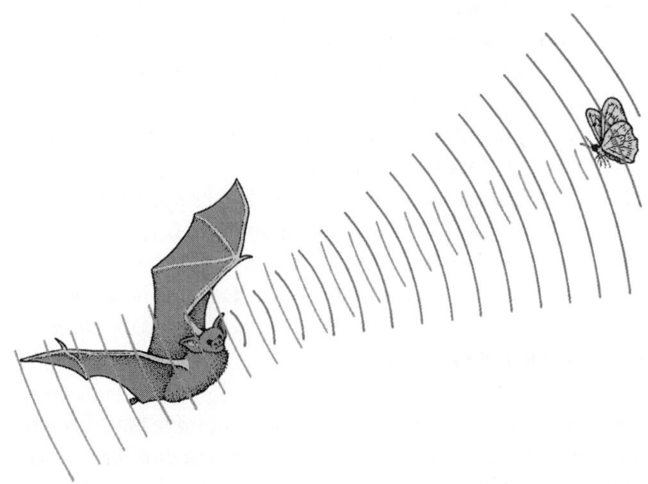

A

B

29.31 Echolocation (A) Bats emit high-frequency chirps that are reflected back by objects in its path. The returning echos give the bat information about the size and position of objects. (B) The bat's well-developed ears can hear frequencies of 100,000 hertz or higher.

THE VERTEBRATE BRAIN

The Vertebrate Brain Is Dominant Over the Rest of the Nervous System

The brain of invertebrate animals is usually much smaller in relation to the body size than the brain of vertebrates, and its dominance over the rest of the central nervous system is usually less pronounced. In many invertebrates, such as earthworms and insects, the brain consists of a ring of nervous tissue encircling the anterior portion of the digestive tract (see Fig. 28.9, p. 594). The two small ganglia lying on the upper side of the digestive tract are the only parts of this ring to which the name "brain" is given. Such a brain is usually not much larger than the other ganglia of the longitudinal ventral nerve cords.

Nevertheless, our understanding of how brain cells interact to produce behavior patterns has been greatly aided by study of invertebrates, especially some of the large sluglike marine molluscs of the genera *Aplysia* and *Tritonia.* Many of the brain cells of these molluscs are large enough, and sufficiently constant in their position, to be mapped visually (Fig. 29.32) and then probed with microelectrodes. In this way it has been pos-

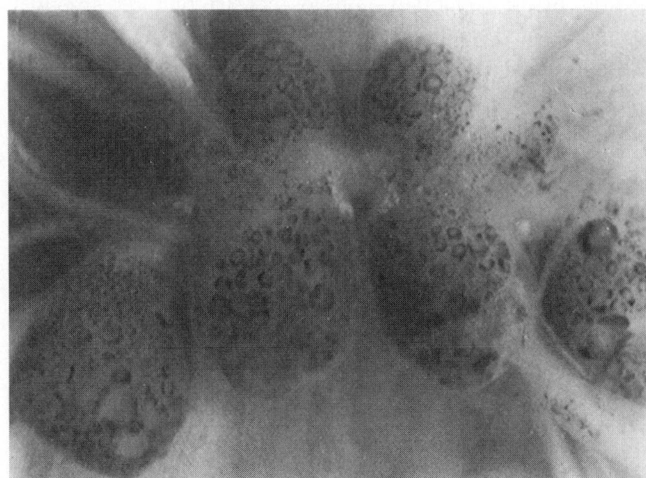

29.32 Brain of *Tritonia diomedia,* a marine mollusc The brain of the mollusc is composed of a group of ganglia, and within each ganglion the individual cells can be distinguished by their markings. So consistent are the positions of the cells from animal to animal that it has been possible to map them with precision and, with the help of microelectrode probes, to study the firing pattern of each cell in detail under different types of stimulation. In this way much can be learned about the circuitry of the central nervous system in these animals, and the interaction of nerve cells in producing their behavior. ×140.

sible to work out, for example, a "wiring diagram" for the feeding response that indicates which neurons control the manipulation and swallowing of food, and how the neurons influence one another during the behavior.

The brains of the ancestral vertebrates probably were not much more dominant than those of invertebrates, but they do show the beginnings of the evolutionary trends that have made extensive brain development one of the most prominent characteristics of vertebrates. The following is a brief introduction to some of the main aspects of the vertebrate brain.

During the Evolution of the Vertebrate Brain, the Midbrain Decreases in Size, While the Forebrain Increases

The brain of the earliest vertebrates was probably just a modest enlargement of the anterior end of the spinal nerve cord, but vertebrate evolution is marked by ***cephalization,*** the tendency for the sense organs and neural control to be concentrated in an anterior head. The brain becomes the coordinating center and gradually assumes control over the rest of the nervous system. We mentioned in the last chapter that the location of the brain at the anterior end of the animal was not an accident. The anterior end is usually the first to encounter new environmental stimuli. Consequently, natural selection favored development of the major sense organs in this region, which, in turn, led to the enlargement of the anterior end of the spinal cord.

The ancestral vertebrate brain and the partly developed brains of all vertebrate embryos consist of three irregular swellings at the anterior end of the spinal cord. These three regions, designated the ***forebrain,*** the ***midbrain,*** and the ***hindbrain*** (Fig. 29.33), underwent much modification in the course of evolutionary development of the more complex vertebrates. Often specially thickened areas form in their walls and distinctive enlargements and outgrowths occur in other places. Despite these changes, however, the original three divisions of the brain can still be recognized even in the mature forms of the most elaborated vertebrates, including humans.

The trend in vertebrate evolution has been toward more

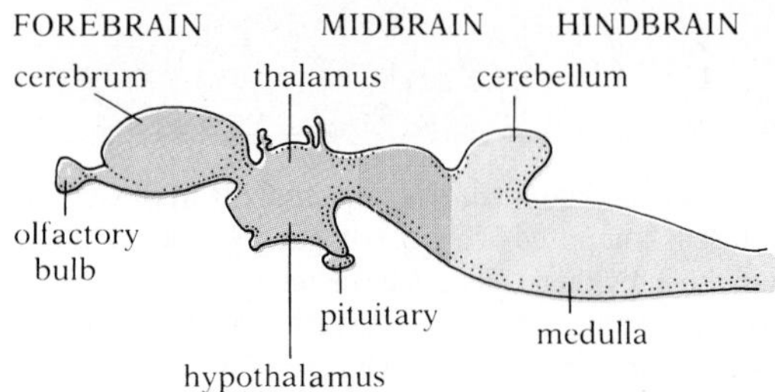

29.33 Diagram of the principal divisions of the vertebrate brain

complex neural pathways within the central nervous system, and increasing dominance by the brain. And because active animals require accurate sensory information about the environment around them, natural selection favored the elaboration of the sense organs and the development of complex neural networks within the brain to process and integrate the sensory information. The brain became an area for analysis and integration. With this enhanced integrative skill came the ability to perform more complex and more flexible behavioral patterns.

Very early in its evolution, the vertebrate brain underwent modifications that set the stage for later evolutionary trends. Briefly, the modifications were these:

1. The hindbrain became divided into a ventral portion, called the ***medulla oblongata,*** and a dorsal portion, the ***cerebellum,*** and the anterior ***pons.*** The medulla became specialized as a control center for some autonomic and somatic pathways concerned with vital functions (such as breathing, blood pressure, and heartbeat) and as a connecting tract between the spinal cord and the more anterior parts of the brain. The pons is above the medulla and also acts as a connecting tract. The cerebellum enlarged and became a structure concerned with balance, equilibrium, and muscular coordination.
2. The midbrain became specialized as the ***optic lobes,*** visual centers associated with the optic nerves.
3. The forebrain became divided into an anterior portion consisting of the ***cerebrum,*** with its prominent olfactory bulbs, and a posterior portion consisting of the ***thalamus*** and ***hypothalamus.***

During the course of vertebrate evolution, there have been few changes in the hindbrain, though the cerebellum has become larger and more complex in many animals. The truly major evolutionary change has been the steady increase in size and importance of the cerebrum, with a corresponding decrease in relative size and importance of the midbrain (Fig. 29.34).

The ancestral cerebrum was only a pair of small smooth swellings concerned chiefly with the sense of smell. The gray matter (cell bodies and synapses) of the brain was mostly internal, as it is in the spinal cord. The synapses functioned predominantly as relays between the olfactory bulbs and more posterior parts of the brain; little, if any, processing of sensory information occurred in the cerebrum. The cerebrums of modern fishes are still little more than relay stations, although the areas of gray matter are more massive. In amphibians, which evolved from ancestral fish, there was an expansion of the gray matter and a multiplication of synapses between neurons. No longer was the cerebrum only a relay station; it now functioned as a processing center for impulses coming to it from various sensory areas of the brain. Slowly, much of the gray matter moved outward from its initially internal position, until it came to lie on the surface of the cerebrum. This layer is known as the ***cerebral cortex.*** Finally, in certain advanced reptiles a new

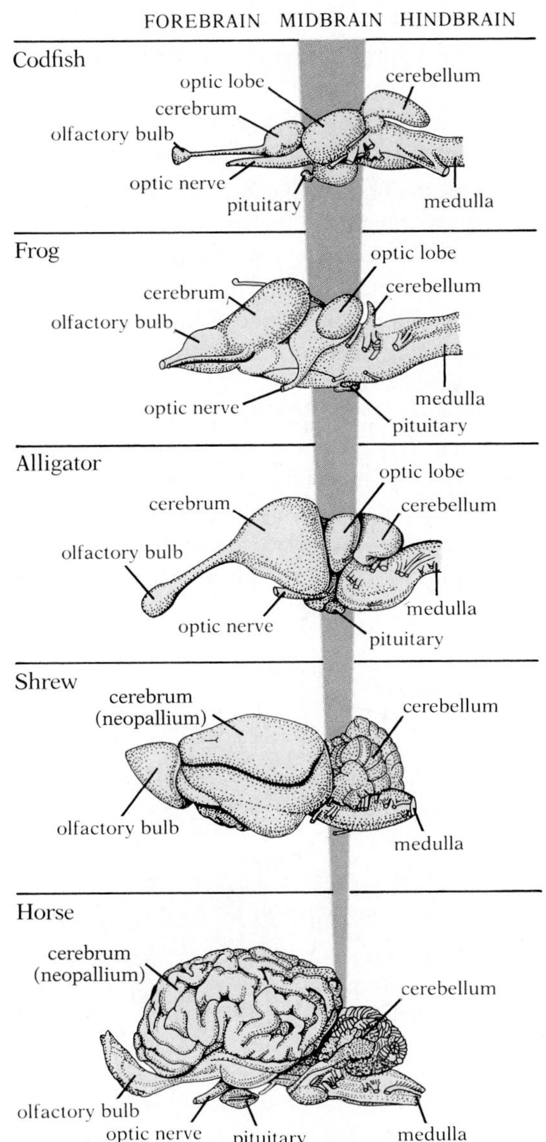

29.34 Evolutionary change in relative size of midbrain and forebrain in vertebrates In this evolutionary sequence the relative size of the midbrain shows a marked decrease, and that of the forebrain a very considerable increase.

component of the cortex, called the ***neopallium*** (or neocortex), arose at a point on the anterior surface of the cerebrum. Mammals, which evolved from reptiles of this type, show the greatest development of the neopallium.

The Neopallium Became the Major Coordinating Center for Sensory and Motor Functions

Even in ancestral mammals the neopallium had expanded to form a surface layer covering most of the forebrain. This does not mean that the old cortex of the ancestral brain has been reduced; it has simply been pushed to an internal position by the growth of the neopallium. Throughout mammalian evolution there has been a steady increase in the relative size of the neopallium. In advanced mammals it dominates the entire cerebrum and becomes the major coordinating center for sensory and motor functions involving all senses and all parts of the body, and is the site of analysis, memory, and integration. In humans and other primates the neopallium has grown to such immense size that it has been folded into convolutions, thereby increasing the total volume of gray matter.

As the neopallium continued to expand in size, it became more and more dominant over the other parts of the brain. The midbrain had been the chief control center in the earliest vertebrates. Then the thalamus portion of the forebrain became a major coordinating center, first sharing this function with the midbrain, then becoming dominant. Finally, with the rise of the neopallium and its preempting of many control functions from both the midbrain and the thalamus, the midbrain was left as a small connecting link between the hindbrain and the forebrain. In humans it still controls a few local reflex mechanisms, some of the simpler visual functions, and is involved in the control of emotions.

This increase in brain size and complexity from fish—the vertebrates with the simplest brains and smallest cerebrums—through amphibians and reptiles to mammals, suggests the likely evolution of the vertebrate brain. But this evolutionary pathway does not mean that the brain of each type of organism has now ceased to evolve. On the contrary, the fish brain has continued to evolve since the rise of amphibians, and the amphibian central nervous system has likewise continued to evolve since reptiles diverged into their own evolutionary line. Though the most primitive vertebrate brains are by and large found in fish, the brains of some species of modern fish are relatively large and complex. The size and complexity of the brains of present-day vertebrate species are determined both by the evolutionary history and the selective pressures that they face in their varying environments.

In mammals, the portion of the total area of the cerebral cortex (including the neopallium) devoted to sensory and motor functions differs greatly from one species to another (Fig. 29.35). In general, the larger and the more convoluted the cerebral cortex, the smaller the proportion devoted exclusively to sensory and motor activities. We represent the extreme example of this trend; in human beings the so-called ***association areas*** constitute by far the largest proportion of the cerebral cortex. The association areas are regions in which information from different sensory systems converges and memory formation and storage occur, making more flexible and complex behavior possible.

Because of the obstacles to human study, much of our information about the function of the various parts of the brain is based on studies performed on rats, cats, monkeys, and chimpanzees. A considerable body of information has accumulated

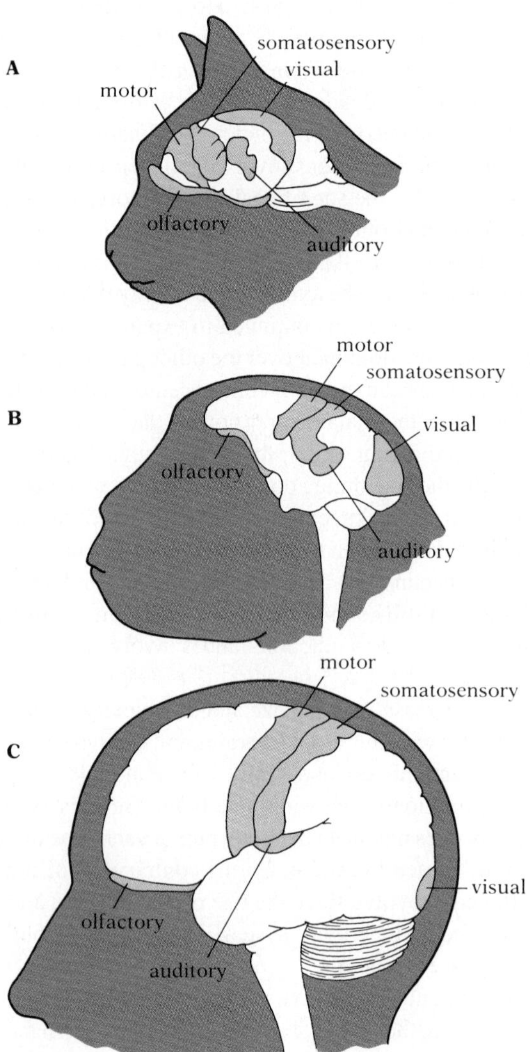

29.35 Proportion of cerebral cortex devoted to sensory and motor functions in three mammals (A) In cats sensory (tan) and motor (pink) areas constitute a major portion of the cortex. (B) In monkeys the proportion of cortex devoted to association areas (white) is much greater than in the cat. (C) In humans the sensory and motor areas occupy a relatively small percentage of the cortex, most of the cortical area being devoted to association. Only the primary visual cortex is shown; as we will see in Figure 29.37, secondary areas where more complex processing and associations occur occupy large areas of cortex.

about the human brain too, most of it derived from electrical stimulation during brain surgery and from observation of the effects of tumors and of accidental damage or destruction of parts of the brain. Though we will not summarize the whole fund of current knowledge of the human brain, we will take a close look at the forebrain, which plays so vital a part in all our lives.

The Reticular System Acts as an Arousal System and a Filter for Incoming Information

The reticular system (Fig. 29.36A) is an extremely important interconnected network of neurons that runs through the medulla, pons, midbrain, thalamus, and hypothalamus. Every sensory pathway running to higher centers of the brain sends side branches to the reticular system, as does every descending motor pathway. In this way, the reticular system is able to "listen in" on whatever is coming into or leaving the brain. It also sends a great many fibers of its own to areas of the cortex, brainstem, and spinal cord.

The ***thalamus*** (Fig. 29.36A) forms an important part of the reticular system. It is a major sensory-integration center in fish and amphibians, but in mammals it has become in large part a sensory relay station on the way to the cerebral cortex, where complex integration takes place. Nevertheless, the thalamus continues to play an integrative role even in humans.

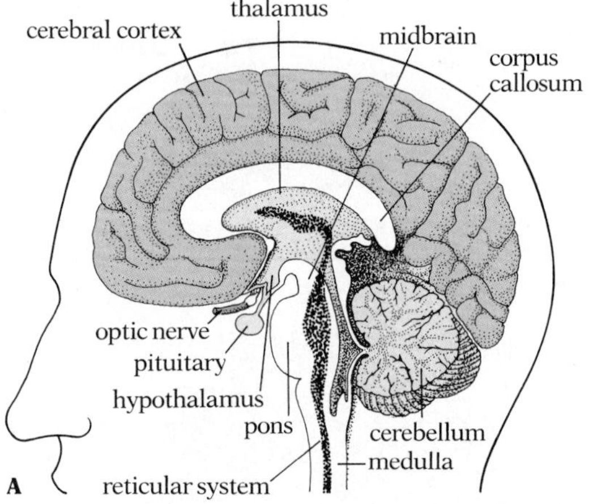

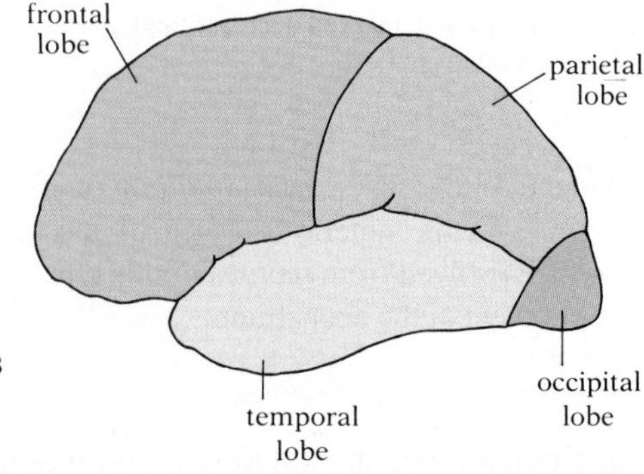

29.36 Human brain (A) A longitudinal section through the midline showing the major parts of the brain. (B) Diagram of the left hemisphere of the cerebral cortex showing its four main lobes. These lobes are duplicated in the right hemisphere of the brain.

The major function of the reticular system in humans seems to be sensitization—the activation of other parts of the brain upon receipt of appropriate stimuli from sensory receptors, of whatever type or location. It acts as the brain's arousal system, much as an alarm clock arouses you in the morning; in fact, the alarm clock awakens you only if the reticular system is sufficiently stimulated. One reason falling asleep is easier in a dark, quiet bedroom is that fewer signals from the sensory receptors reach the reticular system, so it is quieter and does not arouse the brain. Barbiturates, which were once commonly used in sleeping pills, block the reticular activating system and thus facilitate deep sleep. Destruction of the reticular system suspends the brain's ability to respond to sensory stimuli, and results in a permanent comatose state.

The reticular system does not, however, arouse animals indiscriminately. It selectively enhances or suppresses incoming sensory information on the way to the cortex to be processed. In a very real sense, the reticular system "decides" what an animal will be most aware of—when it should pay more attention to sounds, say, than to its touch receptors, and vice versa. Such filtering is essential, since hundreds of millions of sensory receptors continually flood the brain with irrelevant information. The brain simply does not have the capacity to deal with even a tiny fraction of this material at any given time. The system also appears able to modulate motor commands issued by the cortex, amplifying some and attenuating others.

The Hypothalamus Is the Major Integrating Center for Both Visceral and Emotional Responses

The hypothalamus is the part of the brainstem just ventral to the thalamus (Figs. 29.36 and 29.37). Among its major functions, as we saw in Chapter 26, are synthesis of the hormones stored in the posterior pituitary and secretion of the releasing hormones that help regulate the anterior pituitary. Since many endocrine feedback loops run through it, the hypothalamus is a crucial link between the neural and endocrine systems.

The hypothalamus is also the most important control center for the emotional and visceral functions of the body. Stimulation with microelectrodes has enabled us to identify centers in the hypothalamus that control hunger, thirst, sexual desire, body temperature, water balance, blood pressure, reproductive behavior, pleasure, hostility, pain, and so forth.

Located near the hypothalamus is the ***limbic system,*** which is also involved in behavioral responses and in memory. The limbic system is a functionally related set of structures that form a ring around the anterior end of the brainstem (Fig. 29.37). A portion of this system is involved in the sense of smell, but most of it is concerned with rage, fear, aggression, motivation, and sexual behavior—functions that also involve the hypothalamus. Both the limbic system and hypothalamus appear to form the neural basis for these emotional states.

In Humans the Cerebral Cortex Is Involved in Higher Mental Functions

The cerebrum consists of ***right*** and ***left hemispheres,*** which are connected internally by the ***corpus callosum*** (Fig. 29.36A). The inner portion of the cerebrum consists of white matter, and the outer portion, the cerebral cortex, is composed of gray matter, which in humans is about two to four millimeters thick.

Because the cerebral cortex (which includes the neopallium) has been identified with intellectual capacity, we tend to think of it as synonymous with the brain. But, as we have discovered, other parts of the brain play a critical role in many of our activities. The cortex has, in fact, been viewed by some workers as an organ of elaboration and refinement of functions that, in its absence, could be performed to some extent by other

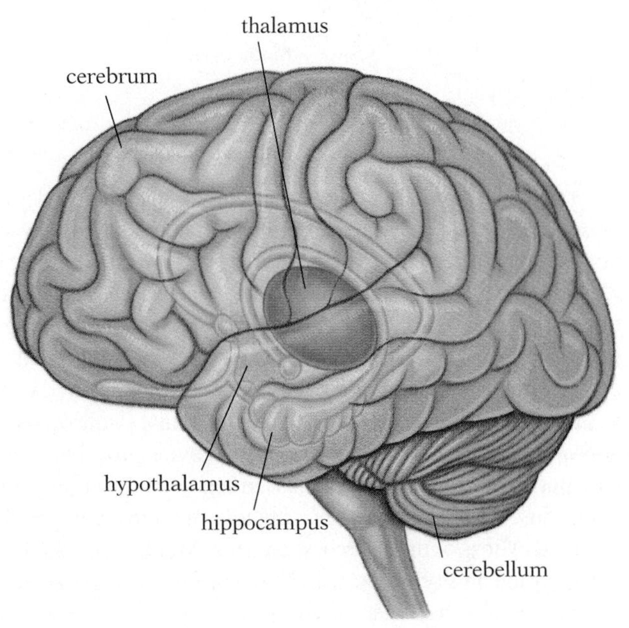

29.37 The limbic system of the human brain The limbic system (blue) is not an anatomically distinct structure, but rather a group of brain areas that are related functionally in giving rise to feelings and emotions. One portion of the limbic system, the hippocampus, is important in memory. The limbic system and the hypothalamus apparently form the neural basis for emotional states like rage, fear, aggression, and sexual arousal.

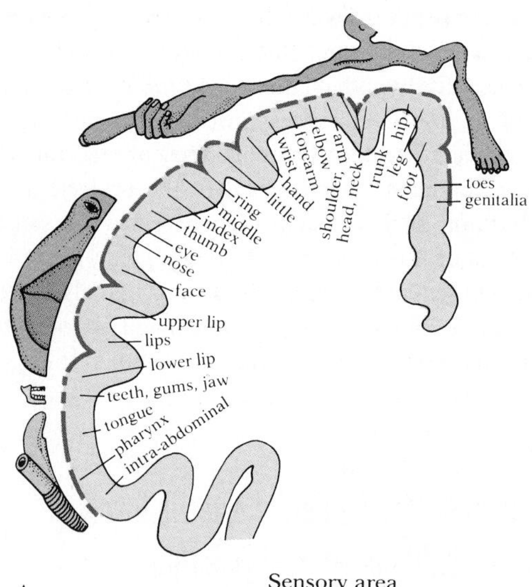

A

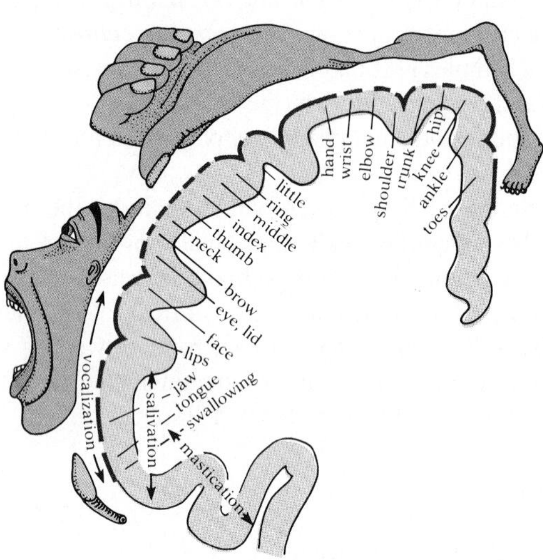

B

29.38 Functional map of the sensory and motor cortex (A) The somatosensory strip (see Fig. 29.35C) of the human cerebral cortex receives sensory input from all parts of the body. On this strip adjacent parts of the body are usually represented next to each other. Parts of the body with large numbers of sensory receptors, such as the lips, tongue, and fingers, are represented by correspondingly large areas on the cortex. (B) A strip from the adjoining motor-control area of the cortex shows similar organization: mapping proceeds from toe to head; adjacent parts of the body are usually represented next to each other; and those with many small muscles for fine motor control occupy especially large amounts of space.

parts of the brain. This certainly seems to be true in most vertebrates. A frog whose entire cerebrum has been removed shows almost no behavioral changes and can see as well as before. It can still catch flies, have sex, swallow, etc. And a rat with its cortex removed shows no obvious motor defects, and though its ability to distinguish complex visual patterns is impaired, it can tell light from dark and can respond to movement. A cat without its cerebral cortex can move sluggishly, swallow, react appropriately to pain stimuli, meow, and even purr, but it has the appearance of an automaton, seeming to be essentially unconscious. Monkeys and human beings are almost completely disabled by loss of their cerebral cortices. Monkeys retain only a very crude ability to detect light and are severely paralyzed. Human beings become totally blind and suffer extensive paralysis. Though they can carry out such vegetative functions as breathing and swallowing, they usually soon die. Apparently, then, no bodily processes vital to life are controlled exclusively by the cortex, but as the cortex evolved, older parts of the brainstem established new connections with it, each part coming to be assisted in its functions by its own piece of the cortex. Gradually, more and more of the function of some of these older parts was transferred to the cortex until, as in human vision, the role of the cortex became predominant.

We said earlier that the proportion of the cortex taken up by purely motor and sensory areas is smaller in humans than in other animals (see Fig. 29.35). However, probing with electrodes shows that within these limited areas each part of the body is represented by its own control area. The area of the cortex devoted to each part of the body is proportional not to the size of the part, but to its sensory or motor capabilities (Fig. 29.38). As you might expect, allocation of brain area is different in different animals and reflects the special characteristics of each. For example, the skin around the nostrils of a horse has nearly as much corresponding cortical area as all the rest of the body put together.

Although the two cerebral hemispheres appear similar, they differ in function. The left hemisphere of the brain controls the right side of the body and the right hemisphere the left side, because the neuron fibers running between the brain and the spinal cord cross to the opposite sides in the corpus callosum. The corpus callosum also transfers information from one side of the brain to the other. In most humans, the left hemisphere is specialized for verbal and analytic ability—understanding the spoken and written word, speech, writing, and analytical processing. The right hemisphere is specialized for visual and spatial relations; it is this hemisphere that is concerned with recognizing faces, spatial relationships, and musical themes. This lateralization of functions is true for almost all humans, whether they are right-handed or left-handed. Virtually all right-handed persons (approximately 91 percent of the population) and 70 percent of left-handed persons show this specialization of functions. Some of the other 30 percent of left-handed individuals have verbal and analytic ability located in the right hemisphere, while others have these abilities divided between the two hemispheres. It is interesting to note that a disproportionately high proportion of artists and musicians are left-handed. The degree of localization is also affected to some degree by sex.

Localized brain specialization is evident from studies of stroke victims who, because a clot has cut off the blood supply to a portion of the brain, have lost the use of specific neural regions. Some patients, for instance, can identify a dog as an animal, but cannot decide if it is a canine or, say, a horse. Others cannot recognize familiar faces, but can recognize the same people by the sound of their voice. These syndromes involve consistent locations in the brain.

The Language Areas of the Brain Are Located in the Left Hemisphere

Much remains to be discovered about language processing, but research indicates that the human brain is already wired at birth with the neural circuitry necessary for language. Language processing is a complex process involving several discrete areas of the left hemisphere (Fig. 29.39). For written language the processing begins in the visual cortex, where what we see is analyzed before being sent to the so-called ***angular gyrus;*** there written words are translated into auditory form, which are then passed to ***Wernicke's area.*** Spoken language is sorted from other sounds in the midbrain, which sends the lowest frequencies of speech to Wernicke's area. Wernicke's area, then, is the destination for both written and spoken language. Higher frequencies go to the right side of the brain, where the emotional overtones of the speech are ascertained. So segregated are the intellectual and the emotional functions of the brain that many people with right-hemisphere damage can understand the *meaning* of a spoken sentence, but cannot say whether the speaker was happy or sad, angry or ironic. Conversely, people with left-hemisphere lesions (any damage from injury or disease that interrupts the flow of information) can often judge the mood and intention of a speaker, and yet have no idea of what has been said.

Wernicke's area is also involved in the individual's own spoken and written expression. It is here that thoughts are formulated into crude linguistic structures before being sent to ***Broca's area*** for grammatical refinement. Neurons from Broca's area convey information to the motor cortex, which coordinates the muscles in the lips, tongue, and larynx involved in speech. Lesions in Broca's area often leave a patient knowing what he wants to say, but unable to express it according to the accepted rules of tense, declension, number, gender, and so on. They may also rob the speaker of such linguistic signposts as pronouns, conjunctions, and prepositions. Severe lesions in Wernicke's area, on the other hand, may leave the patient able to understand the spoken or written word but unable to interpret or express the thought.

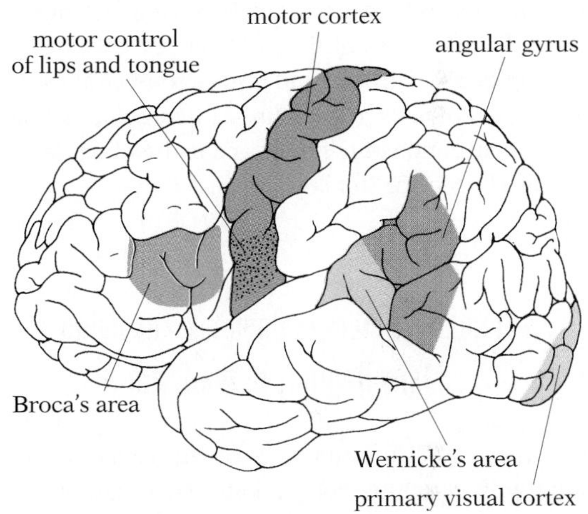

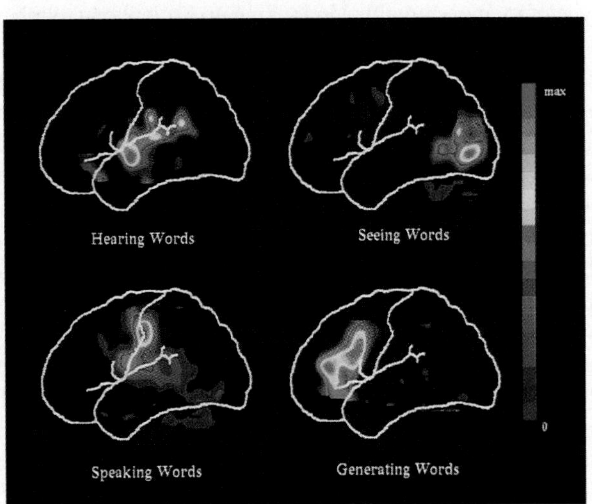

29.39 Anatomy of language (A) The human brain contains many specialized centers concerned with language, particularly in the left cerebral hemisphere (shown here). Sounds are detected by the ears and processed in the midbrain (not shown). Spoken language is then sent to Wernicke's area. Written language is somehow abstracted from the other material processed in the primary visual cortex and sent to the angular gyrus, where it is translated into sound. It too is sent to Wernicke's area, which extracts meaning from incoming language, whatever the source. Language production also begins in Wernicke's area. Thoughts are encoded there into crude linguistic outlines, and these are sent to Broca's area, which refines them into grammatical sentences. Finally, Broca's area transmits directions to the adjacent motor cortex, which controls the organs of speech, to produce utterances. (B) Patterns of blood flow in the brain during language-related activity confirm the anatomical inferences drawn from studies of stroke and accident victims.

Human Male and Female Brains Are Different

In recent years evidence has accumulated that indicates the brains of human males and females are not identical with respect to certain behavioral and anatomical traits. The differences seem to lie not in overall intelligence, but in different patterns of ability. Men tend to be better at spatial tasks, mathematical reasoning, throwing objects at targets, and navigating, whereas women often perform better at identifying matching objects, mathematical calculations, verbal fluency, and fine motor skills. Not only are there behavioral variations, but anatomic dissimilarities as well. Some researchers have found differences in the size of several parts of the hypothalamus that correlate with sex. For instance, Simon LeVay of the Salk Institute in San Diego has found that one particular area of the hypothalamus is significantly larger in human males than females. And, though the data are preliminary, this area is also small in homosexual men, which suggests that sexual orientation may have a biological basis.

The differences between the male and female brains are ascribed to the influences of the different sex hormones on the developing brain. Estrogen and testosterone produced during certain critical periods in the fetus are thought to permanently alter the organization of the brain and its function, thereby establishing a "male" or "female" brain.

Human Memory Is Complex, Involving Many Regions of the Brain

What is memory? Unfortunately we still cannot answer this question with certainty, but we know there are at least two kinds of memory in humans. The so-called ***short-term memory (STM)*** is just that—it lasts for minutes to hours. This is the type of memory involved in looking up a telephone number in a telephone book; we can remember the number long enough to dial it, but promptly forget it. As the name suggests, ***long-term memory (LTM)*** involves permanent storage of information. The ***hippocampus*** (Fig. 29.37), part of the limbic system, is known to be involved with both short-term memory and the conversion of short-term memory into long-term memory as the memories are consolidated and filed away in other parts of the brain. The hippocampus also functions to retrieve old memories from storage. One reason why you should take notes in class is to facilitate the transfer of information from short-term into long-term memory. When you are listening in class the information is going into STM, but most or all of it will be lost within a day or two. If, however, you have taken good notes, and read them over shortly after class, you begin the process of consolidating the information and transferring it into permanent storage. Storage of the initial information is thought to involve structural changes in the neurons and their synapses, probably involving G-proteins and the second-messenger system.

We assume that complex (long-term) memories involve the simultaneous presence of different sensory qualities—that is, sounds, smells, sights, and the like—each stored as memory traces in a particular part of the brain. There are at least two classes of complex memories: the declarative memory, which involves the memory of facts and is associated with the hippocampus, and the procedural memory, which involves the acquisition of skills and memory acquired by classical conditioning and repetitive practice. This type of memory is independent of the hippocampus. The two types of memory involve different parts of the brain, and apparently use different storage mechanisms. The mechanisms by which a permanent memory trace is produced are not well understood, but it is thought that it involves activation of particular genes, synthesis of new proteins, and formation of new connections. Memory traces can be disrupted by certain drugs. Evidence shows that, in teenagers, marijuana abuse damages short-term memory. The effects persist for as long as six weeks after drug use is stopped.

The Brain Undergoes Structural and Functional Changes During Aging

Starting in the 50s and 60s, a number of age-associated structural and chemical changes occur in the brain of humans. For instance, beginning at about age 30, both the overall weight of the brain and the total number of brain neurons decline with age; the average brain has lost 44 percent of its weight by age 75. The loss of neurons is not uniform throughout the brain, however. The loss of neurons is particularly pronounced in the areas of the brain associated with emotions, learning, reasoning, and memory—i.e., the cerebral cortex, thalamus, hippocampus, and limbic system. Not only is there a decrease in the overall number of neurons, but there are also changes in the structure of the neurons that survive. Many neurons shrink or atrophy, and axons or dendrites may degenerate. Another change associated with aging in healthy elderly persons is the appearance of tangles of twisted protein fibers within the cytoplasm of certain neurons. In addition, small plaques of proteins called ***beta amyloid proteins*** are deposited between the cells in the hippocampus and cerebral cortex. Although these age-linked changes would seem to be severe, in actuality they appear to have only a mild effect on intellect and thinking for most elderly adults. The speed of intellectual processing or remembering may decline, but the ability to learn and remember remains nearly the same.

The fatal and debilitating disease known as Alzheimer's stands apart from the normal process of aging. It involves a severe and progressive death of brain cells, resulting in a steady and irreversible dementia and loss of memory and ability to

MAKING CONNECTIONS

1. Compare the brain of an invertebrate such as an insect with that of a vertebrate with respect to size and dominance over the rest of the central nervous system.

2. What evolutionary trends have occurred in the evolution of the vertebrate brain with regard to its size and dominance over the rest of the nervous system? How have the relative sizes of the cerebrum and midbrain changed?

3. Why is it impossible to determine precisely how the vertebrate brain evolved by a comparison of the brains of present-day vertebrate species?

4. Why does the hypothalamus have a central role in the maintenance of homeostasis?

5. How does the difference between the brains of males and females demonstrate the influence of the endocrine system on the nervous system?

6. How may the second-messenger system described in Chapters 25 and 28 be involved in memory?

7. What role may genetic defects play in the development of Alzheimer's disease?

reason. Eventually brain function is destroyed and death results. Estimates indicate that almost half of the population aged 85 and older in the United States are affected, yet the cause of this malady remains unknown. Upon autopsy, the brains of the victims show an excessive accumulation of beta amyloid plaques as well as an extensive formation of bundles of twisted protein fibers (tangles) within certain neurons (Fig. 29.40). The neurons connecting the cerebrum with the hippocampus are particularly affected—the areas of the brain that are involved in memory and reasoning. Neuroscientists are pursuing

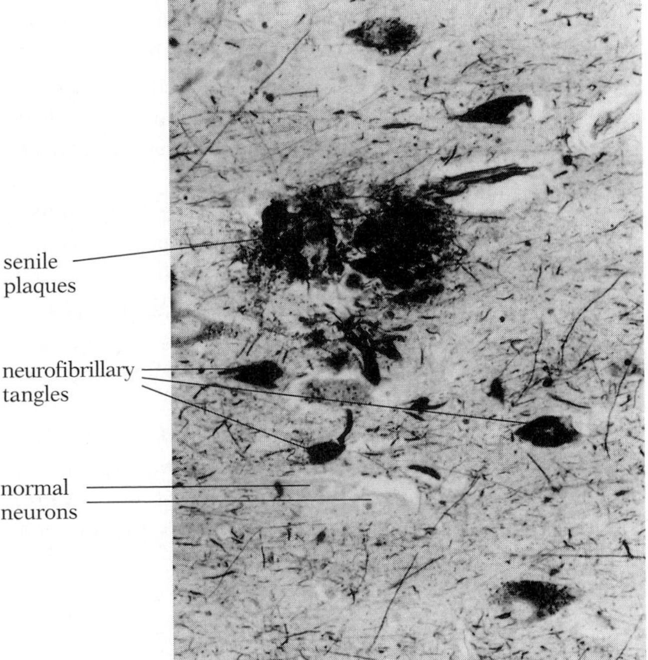

29.40 Brain of Alzheimer's patient Hallmarks of this disease are the presence of neurofibrillary tangles within cells and senile plaques deposited outside cells. The tangles are partially composed of a form of a protein normally assoicated with microtubules, while the senile plaques are composed of a core of β-amyloid protein, which may be produced in abnormal quantities. The cause of these abnormalities remains unknown.

a variety of paths in their attempt to understand this disease. One form of Alzheimer's, which accounts for 15 to 20 percent of the cases, appears to be inherited. Scientists have recently isolated a defective gene on chromosome 21 that may play a role in this form of the disease. Mutations in DNA of the gene that encodes the production of beta amyloid protein appear to be at fault in some cases. Other possibilities being considered as causative agents are genetic mutations in the mitochondrial DNA, abnormalities in membrane lipids and proteins, and malfunctioning of the immune system. So far an immense amount of information has been gathered, and researchers hope that the riddle of Alzheimer's can be solved within this decade.

CONCEPTS IN BRIEF

1. Sensory receptor cells function as *transducers,* converting the energy of a stimulus into the electrochemical energy of a nerve impulse. Each type of receptor is responsive to a particular kind of stimulus.

2. Stimulation of a sensory receptor opens or closes ion gates, producing a local *generator potential* in the receptor cell. When the generator potential reaches threshold level, it triggers an action potential in the sensory neuron.

3. Each receptor sends impulses to a particular part of the brain. It is the part of the brain to which the impulses go, not the stimulus, the receptor, or the message itself, that determines the quality and location of the sensation.

4. The skin contains sensory receptors for touch, pressure, heat, cold, and pain. These receive information from the outside environment. Receptors inside the body receive information about the condition of the body itself.

5. The receptors of taste and smell are chemoreceptors; they are sensitive to solutions of different kinds of chemicals, which can bind to them by weak bonds, thereby opening sodium ion gates. Taste receptors are specialized receptor cells whereas smell receptors are modified sensory neurons.

6. Almost all animals respond to light stimuli. Most multicellular animals have evolved specialized photoreceptor cells containing a pigment that undergoes a chemical change when exposed to light, which in turn opens or closes gated ion channels.

7. There are two types of image-forming eyes: *compound eyes* (in insects and crustaceans) and *camera-type eyes* (in some molluscs and vertebrates). A compound eye is made up of many closely packed *ommatidia,* each of which acts as a separate receptor. A camera-type eye uses a single-lens system to focus light on the many receptor cells that make up the *retina.*

8. The light rays coming into the human eye are focused by the *cornea* and *lens* on the light-sensitive *retina,* which contains the *rods* and *cones.* The sensitive rod cells function in dim light, the cones in bright light. The cones enable us to detect color.

9. Well-defined image vision depends on precise focusing of the incoming light on the retina by the cornea and lens. The shape of the lens is alterable; it makes possible adjustments in the focus *(accommodation)* depending on whether the object being viewed is close or distant.

10. Receptors of the senses of both equilibrium and hearing are mechanoreceptors—specialized hair cells. Hair cells lining the *utriculus* and the *sacculus* send information to the brain about changes of speed or the position of the head relative to gravity, while hair cells within the semicircular canals send information on direction of rotation.

11. In humans, vibrations in the air pass down the *auditory canal* of the outer ear and strike the *tympanic membrane,* causing it to vibrate. The vibrations are amplified as they are transmitted by three small bones across the middle ear to the *oval window.* The movement of the bone against the oval window produces movement of the fluid in the canals of the *cochlea,* causing the membrane on which the hair cells of the *organ of Corti* are located to move up and down, which rubs the hair cells against the overlying *tectorial membrane.* The stimulation of the hair cells open ion gates, which generates an action potential in the associated sensory neurons, which carry impulses to the auditory centers of the brain.

12. The brains of invertebrate animals are much smaller in relation to the size of their bodies than those of vertebrates, and their dominance over the rest of the nervous system is usually less pronounced. The brains of primitive vertebrates consist of three regions, the *forebrain, midbrain,* and *hindbrain.* These have been much modified in the evolution of the vertebrates.

13. Early in its evolution the forebrain was divided into the *cerebrum* and the more posterior *thalamus* and *hypothalamus*. The midbrain became specialized in the *optic lobes*, and the hindbrain was modified to form the *medulla oblongata*, the *pons*, and the *cerebellum*.

14. The hindbrain has changed little over the course of evolution; the medulla continues to be a control center for some autonomic and somatic pathways, and the cerebellum is still concerned with equilibrium and muscular coordination. The major evolutionary change has been the enormous increase in the size and relative importance of the cerebrum, with a corresponding decrease in the midbrain.

15. In certain reptiles, a new area of the cerebral cortex, the *neopallium*, evolved. In mammals the neopallium expanded and dominates the other parts of the brain.

16. The mammalian thalamus is a major relay center for sensory information and forms an important part of the reticular system. The *reticular system*, which also runs through the medulla, pons, and midbrain, activates the brain upon receipt of stimuli and is an indispensable filter that lets only a few of the major sensory inputs reach the brain's higher centers.

17. Besides serving as a crucial link between the nervous and endocrine systems, the *hypothalamus* is also the control center for the visceral functions of the body and a major integrating region for emotional responses. The *limbic system*, a functionally related set of structures that form a ring around the anterior end of the brainstem, is also involved; both the limbic system and hypothalamus appear to form the neural basis for such emotions as rage, fear, aggression, motivation, and sexual behavior.

18. The proportion of the cerebral cortex taken up by purely motor and sensory areas is smaller in humans than in other animals; the association areas occupy the greatest proportion of the cortex. The area of the cortex devoted to each body part is proportional to the importance of that part's sensory or motor activities.

19. The right and left cerebral hemispheres differ in function. In almost all humans, the left hemisphere is specialized for verbal and analytic ability, whereas the right hemisphere is specialized for visual and spatial relations. For example, the language areas of the brain are located in discrete, well-defined areas in the left hemisphere.

20. There are different types of memory in humans—short-term memory and long-term memory (permanent storage). The region of the brain known as the *hippocampus* is involved in both short-term memory and the conversion of short-term memory into long-term memory.

STUDY QUESTIONS

1. Which are the major categories of receptors and to which forms of energy does each respond? Regardless of its source of stimulation, what chain of events occurs in a sensory receptor when it is stimulated? What is a *generator potential?* (p. 618)
2. Compare and contrast taste and smell receptors with respect to structure, function, location, and sensitivity. (pp. 620–621)
3. Compare the basic structure and functional properties of the compound eye of an insect and the camera-type eye of a vertebrate. (pp. 623–625)
4. Describe the structure and function of the human eye, including the *sclera, cornea, choroid, retina, iris, pupil, lens, ciliary body,* and *suspensory ligament.* How is *accommodation* accomplished by the eye? (pp. 624–626)
5. Describe the structure of the retina and the properties of rod and cone cells. What is the function of *rhodopsin?* How do cones produce color vision? (pp. 627–629)
6. Explain how hair cells produce a generator potential. Why are they classed as mechanoreceptors? (pp. 629–631)
7. How are hair cells in the human inner ear involved in the sense of equilibrium (i.e., detection of gravity and of accelerations)? (pp. 631–634)
8. Describe how sound is detected by the human ear. What are the roles of the *auditory canal, tympanic membrane, middle-ear bones, oval window, round window, cochlea, organ of Corti, tectorial membrane, basilar membrane,* and *hair cells?* (pp. 634–637)
9. What are the three divisions of the brain of primitive vertebrates? Identify the division to which each of these brain regions belongs and describe its basic function: *medulla oblongata, cerebellum, pons,* and

optic lobes. (p. 638)

10. Describe the evolution of the cerebrum. How did the _neopallium_ arise and what is its role in advanced mammals? (p. 639)

11. Identify the part of the brain principally involved in each of these situations:

(a) You are so engrossed in studying for an impending biology exam that you are oblivious to the sounds coming from your roommate's radio. (pp. 640–641)

(b) Rushing to class because you are late for the exam, you almost fall down on the icy pavement but manage to keep your balance. (pp. 641–642)

(c) You do fall down on the next patch of ice and rip your pants; you feel angry and embarrassed. (p. 641)

(d) You take the exam and find that because you regularly reviewed the textbook (answering all of the study questions, of course) and your lecture notes, you are able to recall nearly all of the factual information you need to answer the questions. (p. 644)

12. Describe the functions and structural organization of the _sensory cortex_ and _motor cortex_. What is the role of the association areas of the cerebral cortex? How do the cerebral hemispheres differ in function? For example, if a stroke left a person unable to speak, which hemisphere was probably damaged? (pp. 641–643)

EFFECTORS AND ANIMAL LOCOMOTION

A ranging mountain lion sees a herd of grazing deer and silently stalks its prey. When close enough it attacks with a burst of speed, leaping onto a deer's back for the kill. In the previous chapter we saw how visual information enters the eyes and is processed in the brain, but we have not yet considered the mechanisms involved in an animal's response. How do the signals from the brain that are conducted along the motor neurons cause the lion's various muscles to contract, enabling it to capture and kill its prey? How are the muscles and bones of its skeleton arranged to permit movement? In this chapter we will consider the ***effectors,*** those parts of the organism that carry out the organism's response to stimuli. The actions of effectors are as various as glandular secretion, production of light by fireflies, phototropic and gravitropic responses in plants, cytoplasmic streaming in both plant and animal cells, and, most familiarly, muscular movements in animals. Many effectors are controlled by the nervous system, but effector cells are not themselves part of the nervous system, even though they are the last components of reflex arcs. Other effectors, such as nematocysts of cnidarians, and cilia and flagella, are not under nervous control.

Animals rely on muscles for rapid movement, but plants must rely on other mechanisms. As we saw in an earlier chapter, plants are capable of slow movements in response to light and gravity, movements produced by differential growth rates controlled by hormones. Many plants are, however, also capable of some types of rapid movement. For example, some leaves droop or fold at night and expand again in the morning. The flowers of many plants open and close in a regular fashion at different times of day. The leaves of sensitive plants *(Mimosa)* fold and droop within a few seconds of being touched (Fig. 30.1). The leaves of the Venus's-flytrap close rapidly around insects that land on them (see Fig. 16.18, p. 337).

All these movements are far too rapid to depend on differential growth changes. Another mechanism, turgor-pressure change, is involved. Leaves droop when certain of their cells lose so much water that they are no longer turgid enough to give rigidity. Flowers fold when cells arranged in rows along the petals lose their turgidity, and they open again when these

Chapter opening photo: Molting crab emerging from its old shell. Crabs and other arthropods have a hard outer exoskeleton that must be periodically shed if the animal is to increase in size. The new, highly wrinkled exoskeleton is laid down beneath the old and will harden after expansion.

30.1 An example of rapid movement in plants Within a few seconds of being touched, the sensitive plant (*Mimosa pudica*) can respond by folding its leaves. Special touch-sensitive cells respond to touch by producing electrical signals similar to the action potentials produced by animal neurons. These electrical signals move from cell to cell through plasmodesmata, eventually reaching special cells at the base of the leaflets. Rapid decreases in turgor pressure in these cells affect the rigidity of the leaves, causing them to droop or close.

particularly sensitive cells regain their turgidity. Rapid changes in turgidity in special effector cells located along the hinge of the leaf of the Venus's-flytrap are responsible for that plant's curious behavior. Similarly, rapid turgidity changes in specialized effector cells at the bases of the leaflets and petioles of the *Mimosa* are responsible for that plant's response to a touch or other mechanical stimulus.

Active movement is clearly not an exclusive characteristic of animals. But since the most elaborate mechanisms for producing locomotion are found in the animal kingdom, it is on these that we shall concentrate (disregarding the many effector actions, such as glandular secretion, that occur without gross movement of the animal).

The underlying mechanism of the various kinds of effector action—from cytoplasmic streaming to muscular movement—depends on either microfilaments or microtubules. The movement of cilia and flagella is based on a system of microtubules. Muscular movement, on which we shall focus in this chapter, is produced by the action of microfilaments.

THE STRUCTURAL ARRANGEMENTS OF MUSCLES AND SKELETONS

The Most Prominent Effectors in Animals Are the Muscles

The first multicellular animals, excluding the sponges, must have been small, perhaps on the order of one millimeter in length. They probably swam by means of cilia, as do the smallest flatworms and the tiny larvae of many cnidarians. But cilia are practical only in very small organisms, because they cannot effectively move a large mass. As animals became larger, the cilia were supplemented by contractile tissue, which eventually became the main mode of locomotion. The contractile fibers of cnidarians are very primitive, but rhythmic contractions of these fibers enable a jellyfish to swim weakly. Jellyfish also use contractile fibers to move their tentacles. Hydras are able to move by turning somersaults (Fig. 30.2), a rather surprising type of movement in an animal with such primitive nerve and muscle cells.

As multicellular animals grew and became more complex, and as division of labor among their cells and tissues increased, they evolved elongate cells specialized for contraction, which became the muscles. The muscles are the principal effectors of movement. However, muscles alone are not sufficient; there

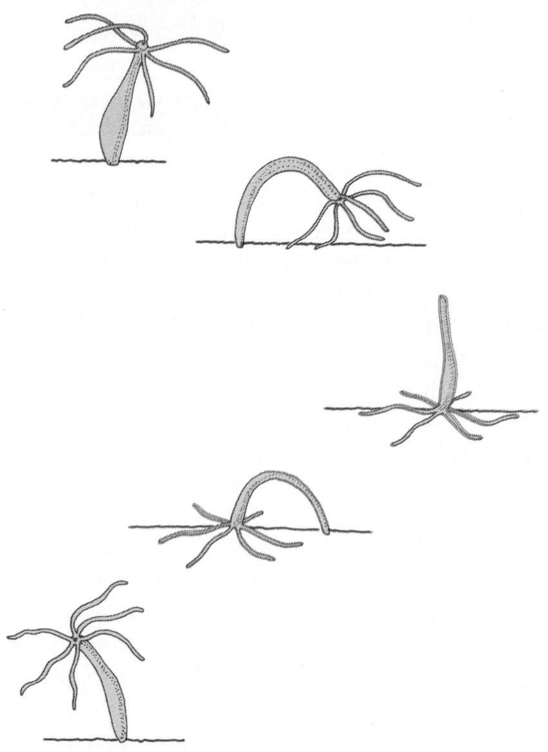

30.2 The somersaulting locomotion of hydra

must also be some kind of framework or skeleton to provide a mechanical support against which muscles can pull. Animals living in an aquatic environment are buoyed up and supported by the surrounding water, so structural support is often not a problem. But when animals moved to land their bodies had to be supported against the pull of gravity, and additional skeletal support was required.

There are three basic types of skeletons in the animal kingdom: a hydrostatic skeleton, an exoskeleton, and an endoskeleton.

The Hydrostatic Skeleton Uses an Incompressible Fluid-filled Cavity for Support

Many animals, such as worms, sea stars, and octopuses, use their body fluids to transmit force, rather like the hydraulic system of an automobile. In such systems, the body fluid is enclosed within a cavity, and when muscles surrounding the enclosed fluid contract, pressure is produced within the fluid. Because the semifluid body contents are incompressible, it functions as a ***hydrostatic skeleton,*** giving the muscles something to work against. In many worms, for instance, the muscle fibers of the body wall are arranged in prominent longitudinal and circular layers. The fibers in these two layers are antagonistic to each other, producing opposite actions. The contraction of the longitudinal muscles produces pressure on the fluid contents, stretching the circular muscles and shortening the worm. When the circular muscles are contracted, the pressure on the fluid contents stretches the longitudinal muscles, lengthening the worm. The shortening that results from longitudinal muscle contraction is accompanied by a compensating increase in body diameter, while the lengthening that results from circular muscle contraction brings a decrease in diameter.

The most complete exploitation of the potentialities of hydrostatic skeletons is seen in certain annelid worms, such as earthworms. Here the body cavity is partitioned into a series of separate, fluid-filled chambers (see Fig. 22.13, p. 466). Correlated with this ***segmentation*** of the body cavity is a similar segmentation of the musculature; each segment of the body has its own circular and longitudinal muscles. It is thus possible for the animal to elongate one part of the body while simultaneously shortening another, an adaptation particularly useful in burrowing. Hard bristles known as ***setae*** are protruded to provide traction (Fig. 30.3) and prevent backsliding.

The boneless appendages of many complex animals are also moved using this principal—our tongues, for example, the elephant's trunk, and the tentacles of a squid. Although the principal is the same—using an incompressible fluid for the

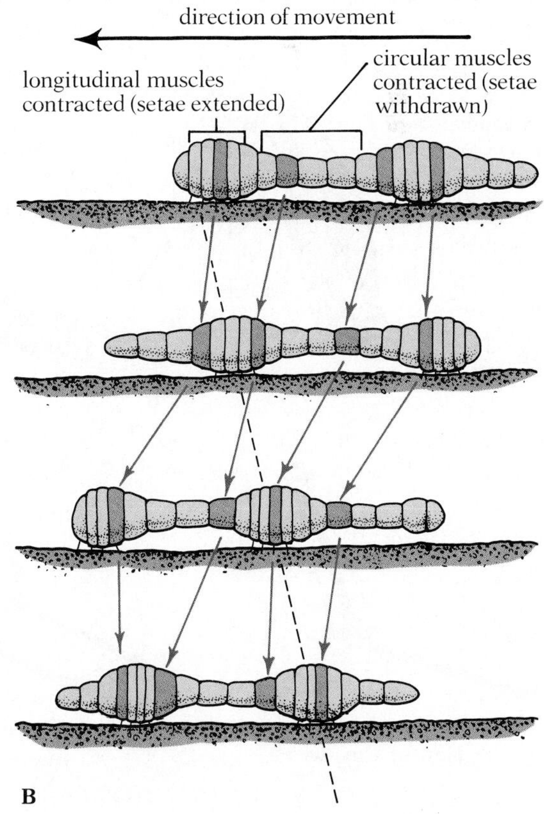

30.3 The use of setae, together with peristalsis, to create movement in earthworms (A) This scanning EM of an earthworm shows the bristlelike setae of each segment. (B) The earthworm, represented here with 20 segments, uses its hydrostatic skeleton to generate movement. Some segments are shown in darker color and connected by arrows for easier identification. As the longitudinal muscles of a segment contract, the segment becomes short and thick, and its setae are extended to anchor the worm to the substrate. As the circular muscles of a segment contract, the opposite occurs; the segment becomes long and thin, extending forward as it loses contact with the substrate. Peristalsis—alternating waves of contraction of circular and longitudinal muscles—thus enables the earthworm to move. The progress of one wave of peristaltic contractions is indicated by the dashed line.

muscle to work against—the actual mechanism is somewhat different. There is no enclosed cavity in these structures to act as a hydrostatic skeleton; instead the muscle itself acts as both the mover and the skeletal support. Muscles contain a great deal of water, and the volume of this fluid remains constant. Thus when the fibers within a muscle contract to shorten the muscle, the muscle grows fatter, and when other fibers contract and decrease the diameter, the muscle lengthens. This type of arrangement, called a muscle hydrostat, may be important in many types of muscle systems. Certainly where they are employed, they are capable of generating very precise manipulative movements.

Both Arthropods and Vertebrates Have Evolved Hard, Jointed Skeletons

The arthropods and the vertebrates are by far the most mobile of the multicellular animals. Both groups possess paired locomotory appendages—legs and sometimes wings—and both have evolved a hard, jointed skeleton. In these animals, skeletal muscles are attached by one end to one section of the skeleton and by the other end to a different skeletal section (Fig. 30.4). When the muscle contracts, it causes the skeletal joint between its two points of attachment to bend or straighten, depending on the placement of the muscle at the joint. In many ways, then, the skeletal and muscular systems of arthropods and vertebrates show striking functional similarities, even though these two groups of animals evolved from entirely different ancestral lineages. Of course, the differences between them are equally as significant.

The most obvious difference between the skeletal systems of arthropods and of vertebrates is that the arthropods have an ***exoskeleton***—a hard body covering with all muscles and organs located inside it—whereas vertebrates have an ***endoskeleton***—an internal framework to which muscles attach. Besides functioning as structures against which muscles can pull, both types of skeleton are important in providing shape and structural support for animals, particularly land animals, which do not rely upon the buoyancy of water for support. Exoskeletons, which are composed of noncellular material secreted by the epidermis, function also as a protective armor for all the softer body parts.

From the standpoint of mechanics, a hollow tube can support more weight than a solid rod of equal weight, so if weight is a consideration, exoskeletons would seem to have an advantage over endoskeletons. The exoskeletal system does, however, limit the possible size of the animal. Since the weight of an animal is a function of its volume (length × width × height), a doubling of an animal's linear dimensions increases its weight by a factor of 8 (that is, 2 × 2 × 2). To support this added weight, the cylindrical exoskeletons of arthropods must become disproportionately thicker and heavier with increasing length. The immense bulk that would be required in an exoskeleton strong enough to support an insect as large as a human being would pose insurmountable mechanical problems. This is certainly one reason why arthropods, as varied and successful a group as they are, have never even approached the size of many vertebrates. Further, rigid exoskele-

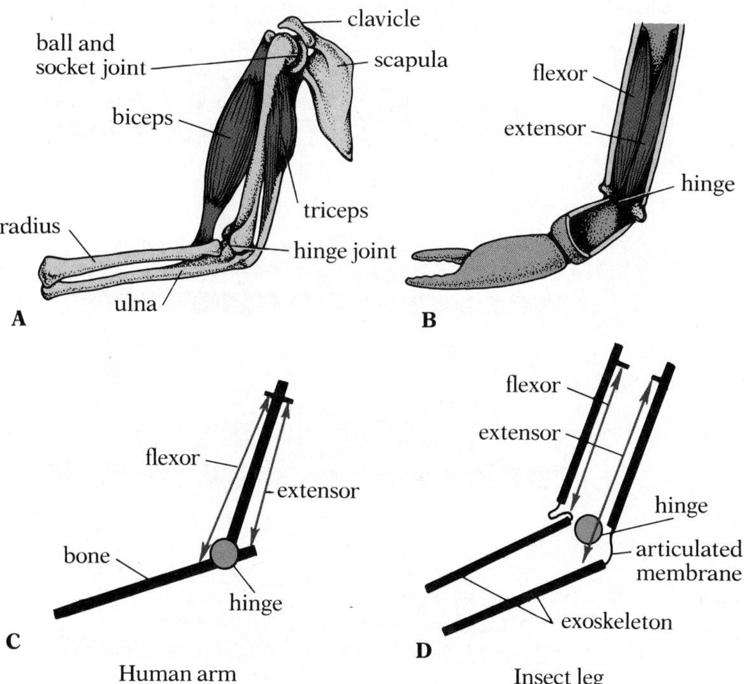

30.4 Mechanical arrangement of muscle and skeleton in a human and an insect leg (A) When the biceps of the human arm contracts, the arm is flexed (bent) at the elbow. The triceps has the opposite action; when it contracts, the lower arm is extended. (B) The comparable flexor and extensor muscles in the insect leg have the same action, even though the muscles are inside the skeleton. The joints in this example are so arranged that each segment, when bent, is perpendicular to its neighbors. Notice that only one of the two "fingers" of the claw can move. (C, D) The similarities between these two simple lever-joint strategies can be seen in these diagrams.

tons impose difficulties in overall growth, such as the periodic molting and deposition of a new covering that are necessary to permit increases in size.

While not as drastic, endoskeletons also impose limits on size. Endoskeletons, which provide support and protection for an animal's internal organs, must be strong enough for the disproportionate increase of weight in larger animals. Horses must have thicker bones than antelope, and elephants must have thicker bones than horses. The structural support and muscular mass required for large animals imposes constraints on size. It should come as no surprise that the largest vertebrates, the whales, are aquatic. The buoyancy of the water in which they live provides much of the support their great weight requires.

Endoskeletons have two advantages over exoskeletons: they are made of stronger materials, and they are located inside the attached muscles instead of enclosing them. Without the confining walls of an exoskeleton, the muscles of vertebrates can be large enough to support a body of relatively large volume. But in small animals, exoskeletons and endoskeletons are about equally effective, and in very small ones exoskeletons are probably superior.

THE VERTEBRATE SKELETAL AND MUSCULAR SYSTEMS

Vertebrate Skeletons Are Composed Primarily of Bone and/or Cartilage

Cartilage is a connective tissue in which the intercellular matrix consists of very thin proteinaceous fibers. It is firm, but not as hard or brittle as bone. In all vertebrate embryos the skeleton is composed of cartilage. A cartilaginous skeleton remains in some adult forms, for example, the sharks and rays. But in most vertebrates bone progressively replaces the cartilage as development proceeds, with cartilage being retained in areas that require a combination of firmness and flexibility, such as at the ends of ribs, on the bony surfaces in skeletal joints, in the walls of the larynx and the trachea, in the external ear, and in the nose.

Bone consists of a proteinaceous matrix that is impregnated with inorganic ions, especially calcium phosphate. Bone is a living tissue; it contains cells and has a well-developed blood supply. There are two basic types of bone, spongy and compact. ***Spongy bone,*** which is found in the center of bones, consists of an interlacing network of hardened bars; the spaces between them are filled with marrow. The outer surfaces of bone consist of the harder ***compact bone,*** their hard parts appearing as an almost continuous mass with only microscopic cavities in them. Compact bone is composed of structural units called ***Haversian systems*** running lengthwise throughout the bone (Fig. 30.5). Each such unit is roughly cylindrical and is composed of concentrically arranged layers of hard, calcified intercellular matrix surrounding a microscopic central ***Haversian canal.*** Blood vessels and nerves pass through this canal. The bone cells lie in small cavities (called lacunae) in the intercellular matrix and are connected by a system of tiny canals, the ***canaliculi,*** which penetrate and cross the layers of the intercellular material. Exchange of materials between the bone cells and the blood vessels in the Haversian canals is by way of these canals.

Bone is constantly being reabsorbed and reformed. During the first two decades of life the rate of bone formation exceeds reabsorption and growth occurs, but in later life bone reabsorption may exceed bone formation. In many older adults so much bone mass is lost that the bones become soft and weak, and prone to fracture. Fractures of the wrist and hip are particularly common. This bone deterioration is referred to as ***osteoporosis.*** Women are more susceptible to developing severe osteo-

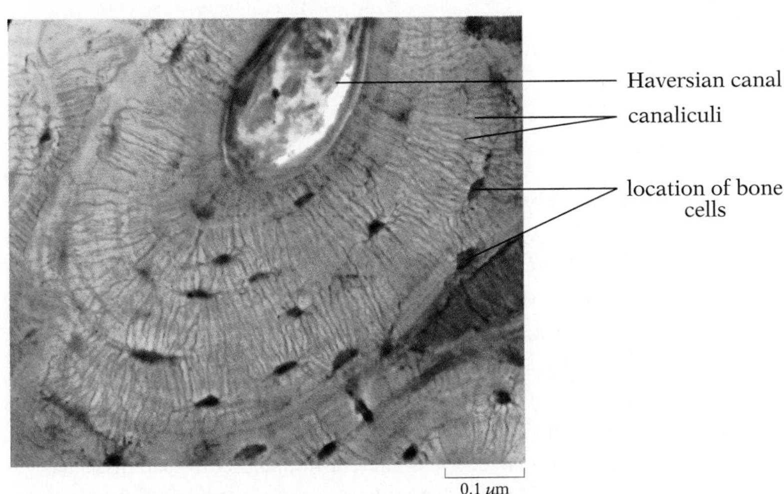

30.5 Photograph of cross section of bone, showing Haversian systems Each Haversian system is seen as a nearly round area. The light circular core of each system is the Haversian canal, through which blood vessels pass. Around the Haversian canal is a series of concentrically arranged layers of hard, inorganic matrix. The elongate dark areas between the layers are cavities called lacunae, in which the bone cells are located. The numerous very thin dark lines running radially from the central canal across the layers of matrix to the lacunae are canaliculi through which tissue fluid can diffuse.

porosis than men, because women have a smaller bone mass than men and because they lose bone more rapidly after menopause, due to the lowered estrogen levels. Osteoporosis can be prevented by small doses of estrogen and progesterone.

Bone is a storehouse for minerals, especially calcium and phosphate. We have already seen in Chapter 25 that the levels of calcium and phosphate ions in the blood are precisely regulated by negative feedback involving the hormone calcitonin and the parathyroid hormone.

Vertebrate skeletons are customarily divided into two components: (1) the ***axial skeleton,*** which is the main longitudinal portion, composed of the skull and the vertebral column with its associated ribcage; and (2) the ***appendicular skeleton,*** which includes the bones that are attached to the axis as appendages (e.g., fins, legs, wings) and their associated pectoral and pelvic girdles (Fig. 30.6). Some bones are joined by immovable joints, or sutures, as are the numerous small bones that together constitute the skull. But many others are held together at movable joints by ***ligaments.*** (Ligaments attach bone to bone.) Skeletal muscles, attached to the bones by means of ***tendons,*** produce their effects by bending the skeleton at these movable joints. The force causing the bending is always exerted as a *pull* by contracting muscles; muscles cannot actively push. Straightening or reversal of the direction in which a joint is bent must be accomplished by contraction of a different set of muscles that function antagonistically.

If a given muscle is attached to two bones with one or more joints between them, contraction of the muscle generally causes one of the bones to move, behaving like a lever system with the fulcrum at the joint (Fig. 30.4). A single muscle sometimes has multiple attachments, which may be on the same or on different bones. The action resulting from contraction of any specific muscle depends primarily on the exact positions of its attachments and on the type of joint between them.

Actually, under normal circumstances, muscles do not contract singly. The nervous system does not send impulses to one muscle without sending impulses to other nearby muscles. Thus the various muscles operate in antagonistic groups; if one group of muscles is strongly contracted, an antagonistic group is exerting a weaker opposing pull, and these stretched muscles are ready to reverse the direction of the movement. In addition, other muscles (synergists) serve to guide and limit the principal movement. Even the simplest action, e.g., taking a step, involves a complicated pattern of activity by a large number of muscles.

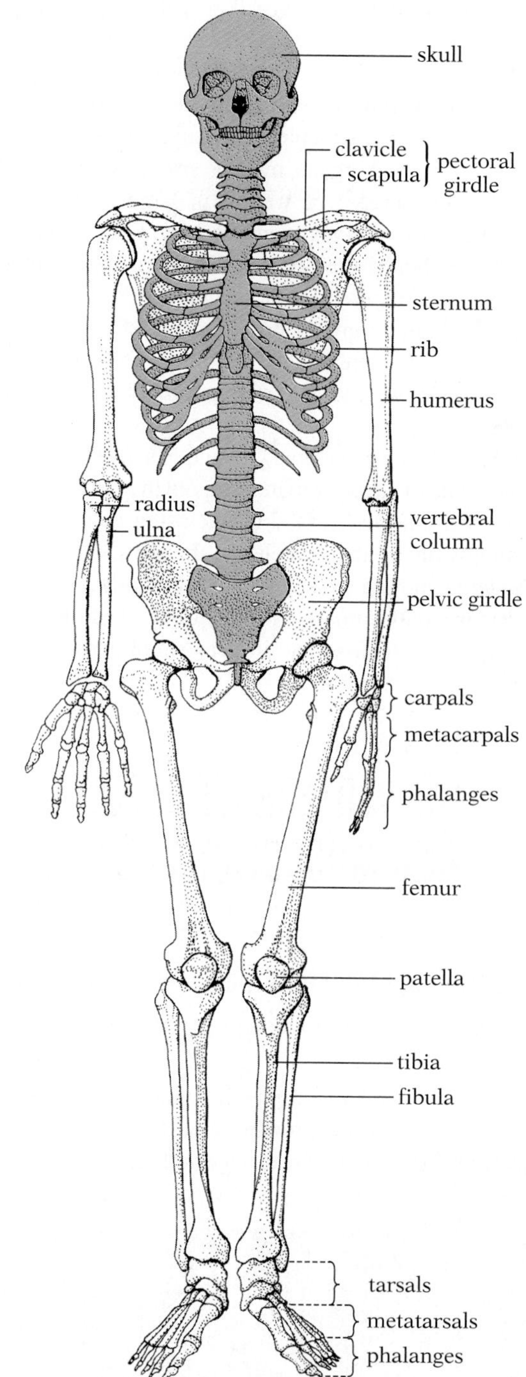

30.6 The human skeleton The axial skeleton (green) consists of the skull, vertebral column, and ribs. All the other bones belong to the appendicular skeleton.

Skeletal Muscle Makes Adjustments to the External Environment; Smooth Muscle to Internal Changes

Three types of muscle tissue are recognized in vertebrates: skeletal muscle, smooth muscle, and heart or cardiac muscle (Table 30.1).

Skeletal muscle is responsible for most voluntary movement; it produces the movements of the limbs, trunk, face, jaws, eyeballs, etc. It is by far the most abundant tissue in the vertebrate body. Most of what we commonly refer to as meat is skeletal muscle. Each skeletal-muscle cell—or fiber, as it is usually called—is roughly cylindrical, contains many nuclei,

TABLE 30.1 *A comparison of skeletal, smooth, and cardiac muscle*

Characteristic	*Skeletal*	*Smooth*	*Cardiac*
Movement	Voluntary	Involuntary	Involuntary
Location	Attached to bones	Internal organs	Heart
Shape of fibers	Cylindrical	Spindle-shaped	Irregular, branching
Nuclei/fiber	Multinucleate	Single nucleus	Single nucleus
Striations	Present	Absent	Present
Fiber arrangement	Fibers in bundles	Fibers interlace	Fibers interlace
Innervation	Somatic nervous system	Autonomic nervous system: SANS and PANS	Autonomic nervous system: SANS and PANS
Contraction	Rapid, short lasting	Slow, long-lasting	Intermediate

SANS: sympathetic autonomic nervous system.
PANS: parasympathetic autonomic nervous system.

and is crossed by alternating light and dark stripes called ***striations*** (Fig. 30.7A). The fibers are usually bound by connective tissue into bundles, which in turn are bound together by more connective tissue to form a muscle. A muscle, then, is a composite structure made up of many bundles of muscle fibers, just as a nerve is composed of many nerve fibers bound together (Fig. 30.8). Skeletal muscle is stimulated by the somatic (voluntary) nervous system and thus is under conscious control. There are at least two types of skeletal muscle: the dark red muscle (or slow-twitch muscle) and the paler white muscle (or fast-twitch muscle). The two types differ both in appearance and properties (see Exploring Further: Red and White Skeletal Muscle, p. 657).

Smooth muscle (also called visceral muscle) forms the muscle layers in the walls of the digestive tract, bladder, various ducts, and other internal organs. It is also the muscle present in the walls of arteries and veins. The individual smooth-muscle cells are spindle-shaped, and lack striations (Fig. 30.7B). Each has a single, centrally located nucleus. Smooth-muscle fibers, which interlace to form sheets of muscle tissue rather than bundles, are stimulated by the autonomic nervous system and are not directly under conscious control.

Many of the differences in function between vertebrate skeletal muscle, which effects adjustments to the organism's external environment, and vertebrate smooth muscle, which brings about responses to internal changes, are reflected in their different nervous-system connections. Cells of skeletal muscle are innervated by only one nerve fiber from the somatic system; they contract when stimulated by nerve impulses and relax when no such impulses are reaching them. Smooth-muscle cells, by contrast, are usually innervated by two nerve fibers, one from the sympathetic system and one from the parasympathetic system; they contract in response to impulses from one of the fibers and are inhibited from contracting by im-

MAKING CONNECTIONS

1. As described in Chapter 5, plant cells normally exhibit turgor pressure. How do some plants use changes in turgidity to produce rapid movement?

2. As discussed in Chapter 6, microtubules and microfilaments help to form the cytoskeleton of eucaryotic cells and are involved in cellular movement. Which type of movement in animals is based on microtubules? Which type is produced by the action of microfilaments?

3. Compared to a gazelle, why does an elephant require disproportionately thick legs for support?

4. Why are women more likely to develop osteoporosis than men? How does this condition demonstrate the influence of sex hormones on metabolism?

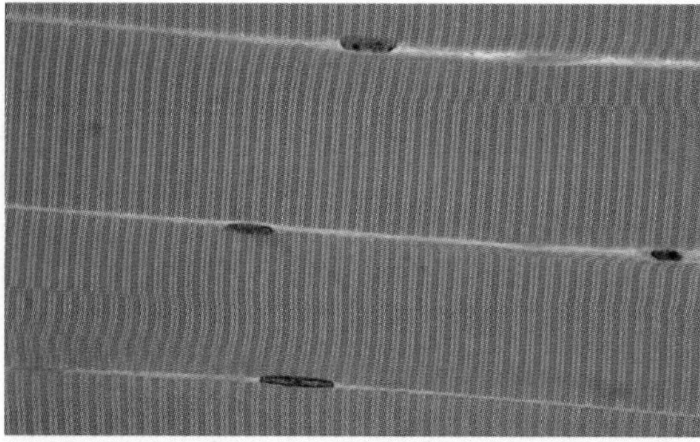
A

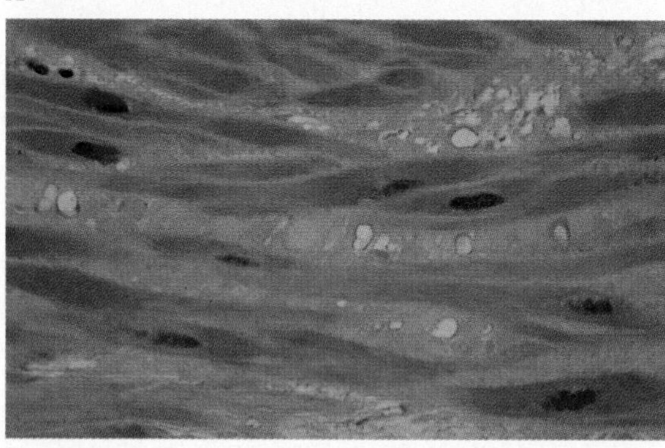
B

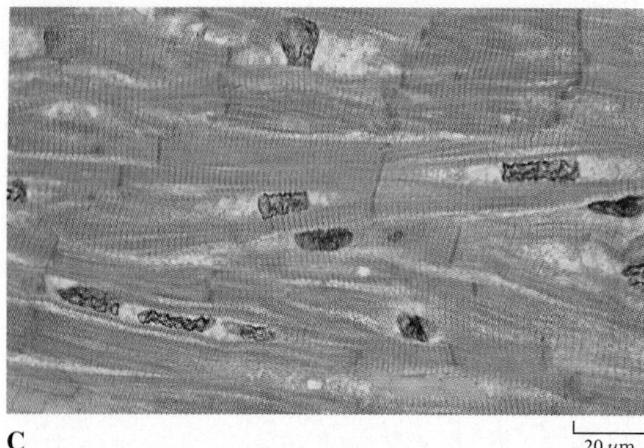

30.7 Photographs of three kinds of vertebrate muscle fiber (A) Portions of four skeletal-muscle fibers from a monkey. Each fiber has several nuclei, located on its outer sheath, and is crossed by alternating light and dark bands, or striations. (B) Spindle-shaped smooth-muscle fibers from a human blood vessel. (C) Human cardiac muscle. The thick, dark lines are intercalated disks, where one cell ends and another begins. Cardiac muscle cells often fork, producing a complex, three-dimensional network.

pulses from the other. Skeletal muscles cannot function normally in the absence of nervous connections and actually degenerate when deprived of their nerve supply, but smooth muscle can often contract without any nervous stimulation, as is commonly the case in peristaltic contractions of the intestine.

Cardiac muscle is classified separately because it shows some characteristics of skeletal muscle and some characteristics of smooth muscle. Its fibers, like those of skeletal muscle, are striated, though the striations are not as obvious. But like smooth muscle, it is stimulated by the autonomic nervous system, and it can contract without any nervous stimulation; also, its activity is more like that of smooth muscle. Where two separate cardiac-muscle fibers (cells) meet, their adjacent membranes are so tightly pressed together and so complexly intertwined that for many years these areas were not recognized as cellular junctions. The sites of these junctions (which are visible under light microscopes as dark-colored disks) are called intercalated disks (Fig. 30.7C).

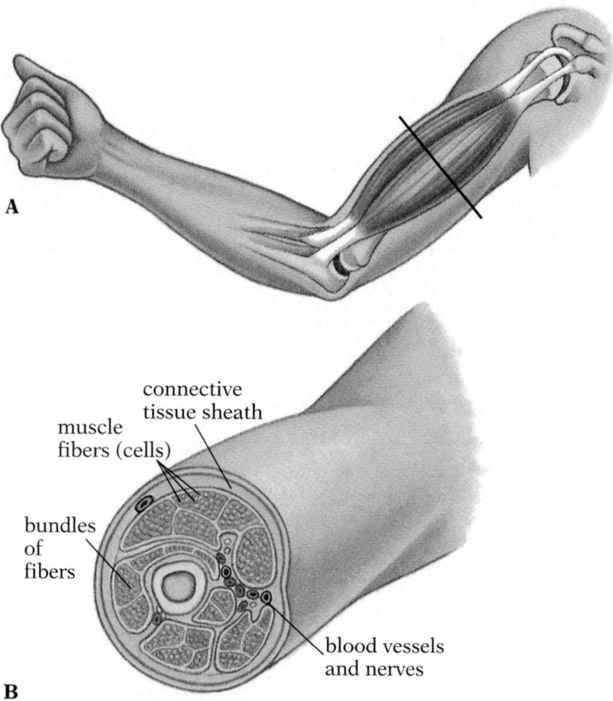

30.8 Skeletal muscle (A) Diagram of a typical skeletal muscle. (B) Cross section of a skeletal muscle. The individual muscle cells or fibers are bound together by connective tissue into bundles, which in turn are bound together by a dense connective tissue sheath.

Since more is known about the processes involved in contraction of skeletal muscle than of smooth muscle, our discussion will concern skeletal muscle.

A Skeletal Muscle Fiber Exhibits an All-or-None Response But a Muscle Gives a Graded Response

Individual muscle fibers and individual nerve cells are alike in that both will fire only if an impinging stimulus is of threshold intensity and duration. Like nerve cells, vertebrate muscle fibers also seem to exhibit the all-or-none property. If an excised vertebrate muscle fiber is administered a stimulus above

EXPLORING FURTHER

RED AND WHITE SKELETAL MUSCLE

Two types of skeletal muscle are often recognized: red muscle (slow-twitch muscle) and white muscle (fast-twitch muscle). *Red muscle* has a rich blood supply, numerous mitochondria, and much myoglobin, a compound similar to hemoglobin that forms a loose combination with oxygen and stores it in the muscle. Red muscle derives its color from the high concentration of myoglobin it contains. It oxidizes fatty acids as its primary source of energy. Though it contracts rather slowly, it is capable of long-term activity without appreciable fatigue. By contrast, *white muscle* has a more limited blood supply, fewer mitochondria, and a low myoglobin content. It depends almost entirely on anaerobic breakdown of glycogen for its energy supply and is capable of very strong, rapid contractions for a short period of time. Because its fibers have fewer mitochondria than those of red muscles, their ability to synthesize ATP through chemiosmosis is limited and they fatigue rapidly. Most skeletal muscles are a combination of the two types, but one type usually predominates. A familiar example is the "dark meat" and "white meat" of a chicken or turkey. An interesting contrast is seen between the breast meat of a chicken or turkey and that of migratory birds such as ducks and geese. The breast meat of a duck, for instance, is red muscle and has much stored fat, while that of a chicken or turkey is white, with little stored fat. The difference correlates with their life cycle. Migratory birds rely on sustained muscle contractions for their long flights; therefore, red muscle predominates in their wings and breasts. Chicken or turkeys use their wings only for quick bursts of flight, and therefore white muscle predominates.

Clearly, the fibers of the two types of muscle are adapted for different purposes and perform different functions. In humans, the leg and back muscles, which support the weight of the body and maintain posture, must be able to contract for long periods of time without fatigue. These muscles tend to have a high proportion of red fibers. Other, more intermittently used muscles, such as those of the arms, must be able to produce strong, rapid contractions for a short period (as in lifting a heavy weight). Here white fibers may predominate. This difference is found in other animals as well. Fish such as the tuna, which swim long distances, have many red fibers.

Though the percentage of each of these muscle types varies with the individual, microscopic examination of the muscle fibers of highly proficient athletes has shown some interesting correlations. Muscles with a high percentage of red fibers, and thus with a capacity for sustained contractions and resistance to fatigue, seem to be associated with activities requiring endurance, such as long-distance running or swimming. For example, about 82 percent of the quadriceps muscles of male marathon runners are red fibers, compared to 45 percent for average nonrunners, and 37 percent for sprinters and high jumpers. By contrast, a higher proportion of white fibers gives a muscle the ability to generate high peak forces; such a muscle is better suited for sports where explosive speed, power, and quick-

ness are required (i.e., sprinting, jumping). Can training convert red fibers into white fibers, and vice versa? The answer appears to be no. The proportion of red and white fibers is determined by inheritance. Some of us are born to be marathoners, with high proportions of red fibers in our leg muscles, while others, with high proportions of white fibers, are destined to become sprinters.

The proportions of each muscle-fiber type is also different for each sex. Women usually have fewer white fibers than men. A female athlete may perhaps lack the explosive strength of the male, but will have more endurance because of the higher proportion of red fibers. Correlated with this difference is a difference in total body fat; women have an average of 20 percent body fat compared with 15 percent for men. During activities requiring endurance women's red fibers can use this fat for fuel. Men, whose muscles contain more white fibers, depend more on stored carbohydrate for fuel, and therefore may have less endurance.

It is clear that the gap between male and female athletes is narrowing in some areas. In marathon running, for example, women are beginning to compete favorably with men. The same is true for distance swimming, where the effect of body-size differences is reduced by the water. Some physicians have suggested that in the future, with improved training, women will be superior to men in running and swimming races of thirty or more miles.

the threshold value, the same degree of contraction is obtained whatever the value of the stimulus, provided that it is not so strong as to damage the cell.

Although a muscle fiber gives an all-or-none response, a muscle, which is composed of many fibers, does not. For instance, you can use the same muscles to perform tasks as different as lifting a pencil and lifting a 10-kilogram weight.

When a whole muscle is stimulated by a single stimulus above threshold, it contracts briefly and then relaxes. Such a muscular response is called a ***simple twitch.*** Increasing the strength of the stimulus increases the strength of the twitch. It is easy to demonstrate in the laboratory that an individual muscle is capable of giving graded responses, depending on the strength of the stimulation (Fig. 30.9): if a muscle is given a stimulus barely above the threshold intensity, the muscle gives a very weak twitch. If a slightly stronger stimulus is applied after a few seconds' delay, the muscle gives a slightly stronger twitch. Increasing the strength of the stimulus elicits an ever stronger contraction from the muscle, until the point is reached where further increases in the stimulus do not increase the strength of the response. The muscle has reached its maximal response (Fig. 30.10).

How can these results be explained if muscle fibers give all-

30.9 Apparatus for studying muscle contraction In this preparation, an isolated frog muscle has been attached to a recording device so that every time the muscle, stimulated by an electric shock, moves a lever, the movement is recorded. The stronger the contraction of the muscle, the higher the lever will be pulled.

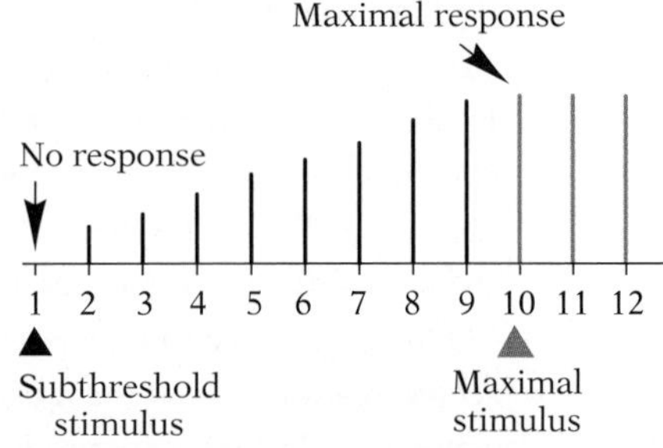

30.10 Response of a muscle to stimuli of various intensities The numbers indicate the intensities at which stimuli are administered, and the height of each bar shows the strength of muscle response. Stimulus 1 is very weak and elicits no response; it is subthreshold. Stimulus 2 is somewhat stronger and proves to be above threshold, for the muscle contracts. Each stimulus from 3 to 10 is slightly stronger than the preceding, and each elicits a correspondingly stronger muscle contraction. Stimuli 11 and 12 are stronger than 10, but the muscle gives no greater response, indicating that 10 elicits a maximal response.

or-none responses? One possible explanation is that the threshold values of the different muscle fibers of which a muscle is composed are not the same. It must be kept in mind, too, that different muscle fibers may be innervated by different nerve fibers, and that these may not all fire at the same time. Consequently, though single fibers give an all-or-none response to stimuli, an increase in the strength of the stimulus above the threshold level may elicit a greater response from the whole muscle by stimulating more muscle fibers. Ultimately, however, all the fibers will be stimulated to respond, and the muscle will thus have reached a maximal response; increasing the intensity of the stimulus further will not increase the response, there being no more fibers to stimulate. It must be stressed that the description given here applies only to vertebrate skeletal muscles. The striated-muscle fibers of invertebrates seldom exhibit the all-or-none property; the strength of their contraction is proportional to the frequency of stimulation.

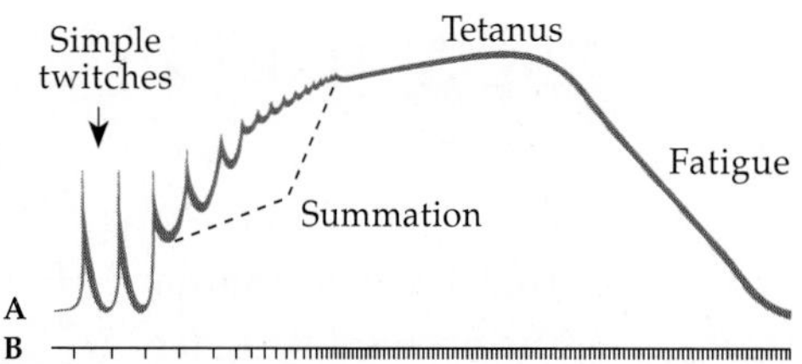

30.11 Summation and tetanus in muscle response When the stimuli (line B) are widely spaced, the muscle has time to relax fully before the next stimulus arrives, and simple twitches result. Each simple twitch is recorded as a sharp strike on the trace. As the frequency of the stimuli increases, the muscle does not have time to relax from one contraction before the next stimulus arrives and causes it to contract again. The result is summation—contractions that are stronger (and hence produce taller spikes in the trace) than any single simple twitch. If the stimuli are very frequent, the muscle may not relax at all between successive stimulations; the resulting strong sustained contraction is called tetanus. If the very frequent stimulation continues, however, the muscle may fatigue and be unable to maintain the contraction.

Our Muscles Work Using Tetanic Contractions

Simple twitches require a time delay between stimuli. Each response is induced by a single brief stimulus, after which the muscle fibers require sufficient time to relax fully before being stimulated again. Suppose that a series of very frequent stimuli is applied to the muscle. In this case the muscle will not have completely relaxed after responding to one stimulus before the next stimulus arrives. When this happens, a contraction is elicited that is greater than either simulus alone would produce (Fig. 30.11). There has been a ***summation*** of contractions, the second adding to the first to produce a more powerful contraction. Each twitch builds on the one before it, adding to the contraction already present. When stimuli are given with extreme rapidity, the individual contractions become indistinguishable and fuse into a single sustained contraction known as ***tetanus*** (do not confuse this normal muscular response with the bacterial infection known also as lockjaw). A tetanic contraction is approximately four times greater than a maximal simple twitch of the same muscle. Normally, our muscle actions involve tetanic contractions, not simple twitches. An example of a simple twitch is the "tic" we sometimes get in our eye. Such twitches are involuntary—we cannot induce them voluntarily, nor can we stop them.

If a tetanic contraction is maintained too long, the muscle will begin to fatigue, and the strength of the contraction will fall, even though the stimuli continue at the same intensity (Fig. 30.11). Muscle fatigue is due to the depletion of stored glycogen reserves, to lactic acid accumulation, and to other chemical changes.

A muscle contraction in which the muscle shortens doing work is called an ***isotonic contraction*** ("same strength") because the force of contraction is relatively constant throughout the contraction period (Fig. 30.12A). Contractions in which the muscle generates force but does not appreciably shorten, as in lifting a very heavy object (Fig. 30.12B), are called ***isomet-***

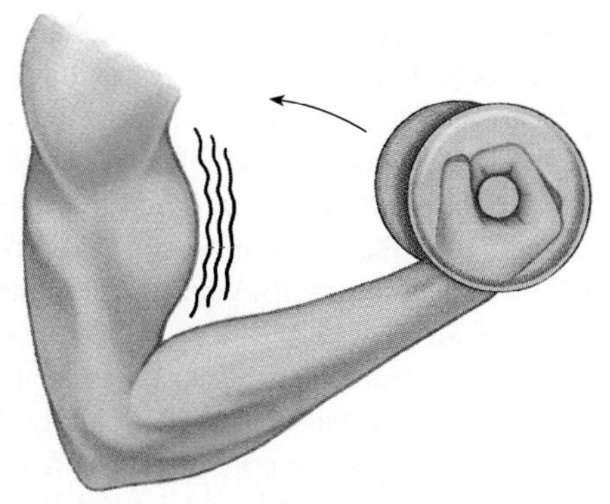

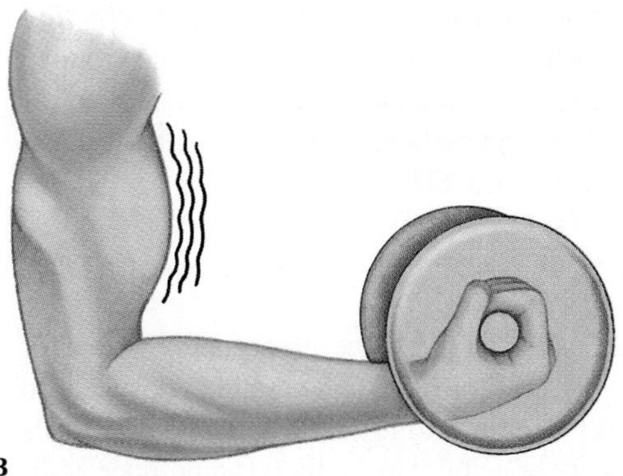

30.12 Isometric versus isotonic contractions of muscle (A) Isotonic contraction. The biceps muscle shortens, lifting the weight. The force of contraction is relatively constant throughout the contraction period. (B) Isometric contraction. The biceps contracts and produces force, but is unable to shorten and raise the weight.

MAKING CONNECTIONS

1. What do the responses of individual skeletal muscle fibers and neurons have in common? How are graded responses produced in muscles and in the nervous system?

2. What metabolic processes discussed in Chapter 8 contribute to muscle fatigue?

3. What happens to muscle tissue when an athlete or body-builder "bulks up" (i.e., adds muscle mass)?

ric contractions ("same length"). Athletes make use of both types of contractions in strength-training programs. The strength of human skeletal muscle depends on its cross-sectional area. At birth humans have all the muscle cells that they will ever have, so the only way to increase muscle mass, and thereby its strength, is to increase the size of the individual cells. The goal of strength-training programs is to increase the diameter of the muscle by increasing the amount of cytoplasm and the contractile microfilaments per cell.

Some of our muscles are never completely relaxed, but are kept in a state of partial contraction called muscle tone or ***tonus.*** Tonus is maintained by alternate contraction of different groups of muscle fibers, so that no single fiber has a chance to fatigue.

Muscle Contraction Requires Large Amounts of ATP Energy

The energy for muscle contraction comes from ATP, which in turn comes from the metabolism of glucose and fatty acids. But so little ATP is actually stored in the muscles that just a few twitches could quickly exhaust the supply. How do muscles overcome this limitation? Although there is little ATP, there is another phosphate compound stored in the muscles, ***creatine phosphate,*** which is formed by linkage of a phosphate group to the substance creatine. Creatine phosphate cannot be used directly to power muscle contraction, but it can transfer its phosphate group to ADP to form ATP:

$$\text{Creatine phosphate} + \text{ADP} + \text{H}^+ \longrightarrow \text{Creatine} + \text{ATP}$$

The newly formed ATP then acts as the direct energy source for contraction. The muscle stores enough creatine phosphate to enable it to contract strongly during the several seconds it takes before the machinery of cellular respiration can produce additional ATP.

If the demands on the muscles are not great, much of the energy used to replenish the supply of creatine phosphate and ATP may come from the complete oxidation of glucose and/or fatty acids to carbon dioxide and water, which requires oxygen. During the unavoidable delay before adjustments of the gas-exchange and circulatory systems increase the oxygen supply to the active muscles, some of the oxygen necessary for aerobic respiration in red muscles may come from oxygenated ***myoglobin.*** Myoglobin, you may recall from Chapter 3, is a special oxygen-storage protein in muscle. Like hemoglobin, it forms a loose combination with oxygen while the oxygen supply is plentiful, and stores it until the demand for oxygen increases. Consequently, muscle has its own built-in oxygen supply.

But during rigorous muscular activity, such as strenuous exercise or the lifting of a very heavy object, the energy demands of the muscles (especially white muscles) are great (Table 30.2) and the oxygen from myoglobin is quickly used up. Because sufficient oxygen cannot be gotten to the tissues fast enough, the muscles obtain the extra energy they need from anaerobic processes. This is accomplished by producing lactic acid through fermentation, and incurring what physiologists call an ***oxygen debt.*** Some of the lactic acid accumulates in the muscles, but much of it diffuses into the muscle capillaries and is transported in the blood to the liver. When the rigorous activity is over, a period of hard breathing or panting helps supply the liver with the large quantities of oxygen it requires for aerobic respiration (Fig. 30.13), thereby paying back the oxygen debt. In the liver, the lactic acid is converted back into pyruvic acid, most of which is oxidized to carbon dioxide and water. The ATP energy thus obtained is used to replace the ATP and creatine phosphate stores, and to synthesize glucose and glycogen from the remaining lactic acid. Note that it is the liver cells, not the muscle cells, where lactic acid is reconverted into pyruvic acid.

Lactic acid, precisely because it is an acid, can damage the muscle fibers if it is not removed promptly. This is why a "cool down" period is so important after strenuous exercise. The continued circulation of blood through the muscle aids in lactic acid removal. "Sore" muscles are the result of damage to the muscle proteins due to lactic acid accumulation.

Endurance-training programs for athletes are designed to

TABLE 30.2 *Energy sources used in various sports*

Source of energy	*Sport*
Stored creatine phosphate	High jump, pole vault, 100-m dash, weightlifting, diving, football (short runs)
Creatine phosphate and anaerobic metabolism (glycolysis and lactic acid fermentation)	Basketball, 200-m dash, football (long runs), hurdles (short distances), 50-m swim
Anaerobic metabolism (glycolysis and lactic acid fermentation)	400-m dash, 100-m swim, tennis, soccer, hockey
Both anaerobic and aerobic metabolism	800-m dash, 200-m swim, 2,000-m rowing, 1,500-m run, 400-m swim, 1,500-m speed skating
Aerobic metabolism	Jogging, marathon, cross-country skiing

increase the oxygen availability in muscles and thereby encourage aerobic metabolism. During such training, the number of mitochondria within the muscle fibers increases, the stores of myoglobin enlarge, and the growth of new blood capillaries within the muscle is stimulated, thereby increasing blood flow through the muscle. As a result, trained athletes are capable of carrying out more strenuous activity without greatly increasing their lactic acid production and accumulation.

Muscle spasms and cramps result from involuntary strong contractions of a muscle. The cramp or spasm is accompanied by sudden pain, which probably results from mechanically stimulating pain receptors within the muscle or from compressing the blood vessels and interfering with the delivery of oxygen to the fibers. Most cramps and spasms will clear up of their own accord within a few minutes. Muscle cramps are sometimes associated with a calcium deficiency, especially in pregnant women, but this does not appear to be true in all cases. The mechanism of cramping is still not understood.

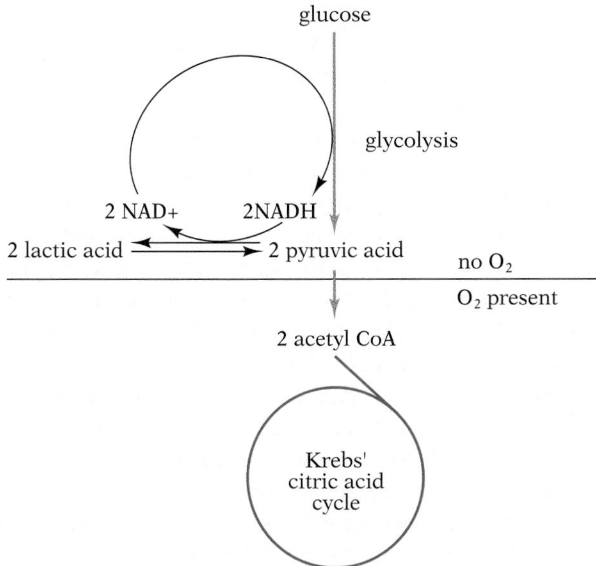

30.13 Lactic acid fermentation During glycolysis, glucose is oxidized to two pyruvic acid molecules. If oxygen is available, the pyruvic acid is oxidized to two acetyl coenzyme A molecules that enter the Krebs cycle and are further metabolized to carbon dioxide and water. If there is an insufficient amount of oxygen, the pyruvic acid instead undergoes fermentation: the pyruvic acid is reduced by NADH to lactic acid. In the process, NAD^+ is produced; it can be reused in the process of glycolysis, thereby enabling glycolysis to continue. When sufficient oxygen is available, the lactic acid is converted back into pyruvic acid, which can then be further oxidized. The conversion of lactic acid to pyruvic acid takes place in the liver.

The Thick and Thin Filaments of Muscle Are Arranged in an Orderly Fashion

We have seen that a skeletal muscle is characterized by striations of light and dark bands and that it is composed of numerous muscle fibers bound together by connective tissue. Examination of these fibers under very high magnification reveals that they, in turn, are composed of numerous long, thin structural units called ***myofibrils,*** each about one to two micrometers in diameter, with mitochondria in the cytoplasm between them. The myofibrils show the same pattern of cross striations as the fibers of which they are a part (Fig. 30.14). There is an alternation of fairly wide light and dark bands, which are identified by letters. The light bands are called

30.14 Electron micrograph of skeletal muscle from a rabbit The myofibrils run diagonally across the micrograph from lower left to upper right; each looks like a ribbon crossed by alternating light and dark bands. The wide, light bands are I bands; there is a narrow Z line in the middle of each. The wide, dark bands are A bands, each with a lighter H zone across the middle.

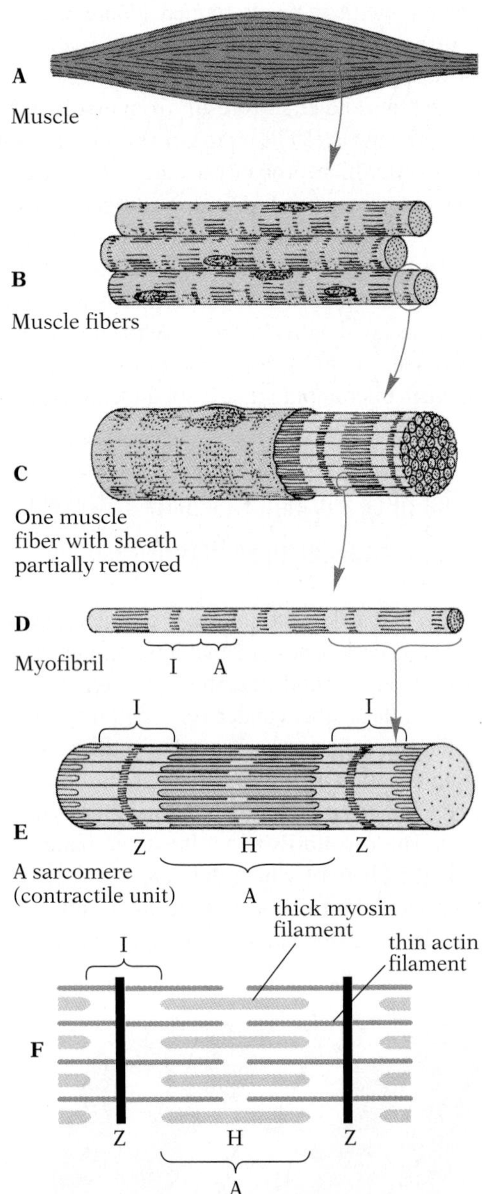

30.15 The component parts of skeletal muscle The pattern of light and dark bands visible in myofibrils under high magnification results from the interdigestion of actin and myosin filaments in each myofibril. As shown in F, the A band corresponds to the length of the thick myosin filaments; the lighter H zone is the region where only the thick filaments occur, while the darker ends of the A band are regions where thick and thin filaments overlap. The I band corresponds to regions where only thin actin filaments occur. The Z line is a structure to which the thin filaments are fastened at their midpoints.

I bands and the dark bands, ***A bands.*** In the middle of each dark A band is a lighter region—the ***H zone.*** In the middle of each light I band is a very dark, thin line called the ***Z line.*** The region from one Z line to the next is called a ***sarcomere.***

It is now believed that the characteristic striations of skeletal muscle are actually a structural reflection of the arrangement of the contractile units. Sarcomeres are the functional units of muscular contraction. As the fiber contracts, the sarcomeres become shorter and the relative widths of the bands change; the I bands and H zones become narrower, but the A bands change very little, with the result that the A bands are moved closer together.

Studies of muscle with the electron microscope reveal that a myofibril contains two types of filaments, ***thick*** ones and ***thin*** ones, arranged in a precise pattern. The thick filaments are found in the A bands, while the thin ones occur primarily in the I bands, but extend into the A bands. This distribution explains the appearances of the A bands, I bands, and H zones. Each dark A band is precisely the length of one region of thick filaments; it is darkest near its borders, where the thick and thin filaments overlap, and lighter in its midregion, or H zone, where only the thick filaments are present (Fig. 30.15). Each light I band corresponds to a region where only the thin filaments are present. The Z line is a disk-shaped structure to which the thin filaments are anchored at their midpoints and against which they exert their pull during contraction; it also functions to hold the filaments in proper position. Analysis shows that the thick filaments are composed of the protein ***myosin*** and the thin filaments primarily of the protein ***actin.***

During Contraction, Actin and Myosin Molecules Slide Past Each Other

These observations led to a hypothesis of muscle contraction—that the filaments telescope together by sliding past each other. If the filaments slide together, the zone of overlap between thick and thin filaments will increase until the thin filaments from the I bands on the two sides of an A band actually meet and even overlap slightly; this sliding together would reduce the width of the H zone, obliterating it entirely if the thin filaments meet (Fig. 30.16). The movement of the filaments would also pull the Z lines closer together and greatly reduce the width of the I bands. But the width of the A bands would only be minimally altered, since these correspond to the full length of the thick filaments, which remains the same (except, perhaps, for a slight crumpling due to contact with the Z lines under conditions of extreme contraction). Thus the ***sliding-filament theory*** accounts for the changes observed in sarcomeres. Still to be answered, however, was the question of how the sliding is brought about.

If the thick filaments are composed of myosin and the thin filaments primarily of actin, some sort of connection must exist between them for sliding to occur. Electron micrographs show what appear to be small cross-bridges between them (Fig. 30.17). These micrographs suggest that a single thick filament is a bundle of myosin molecules, each of which is composed of an elongated tail portion and a pair of globular heads (Fig.

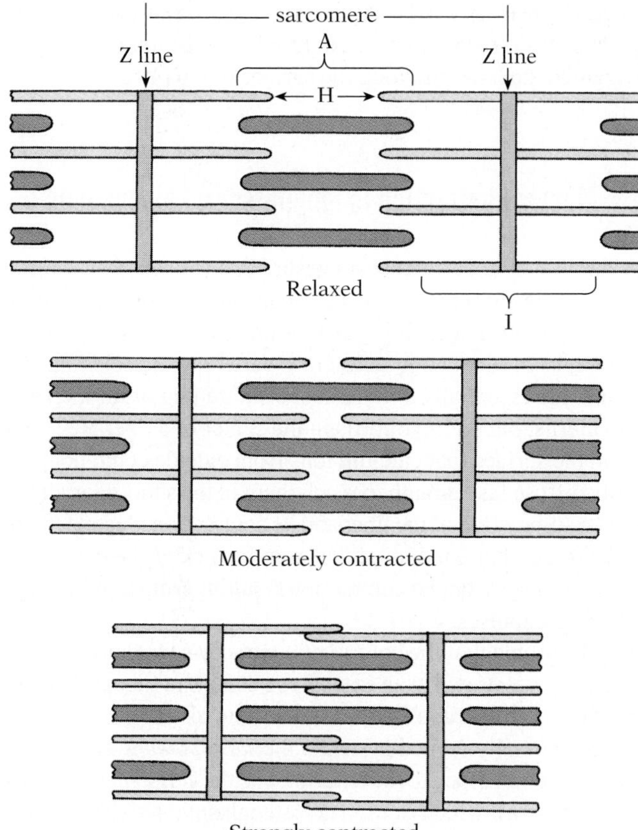

30.16 Arrangement of actin (tan) and myosin (brown) filaments in a sarcomere in relaxed and contracted states The sarcomere is the region between the two Z lines. During contraction, the actin molecules slide along the myosin molecules and the zone of overlap between thick and thin filaments will increase until the thin filaments from the I bands on the two sides of an A band actually meet and overlap slightly. This sliding together will reduce the width of the H zone and pull the Z lines closer together, greatly reducing the width of the I bands.

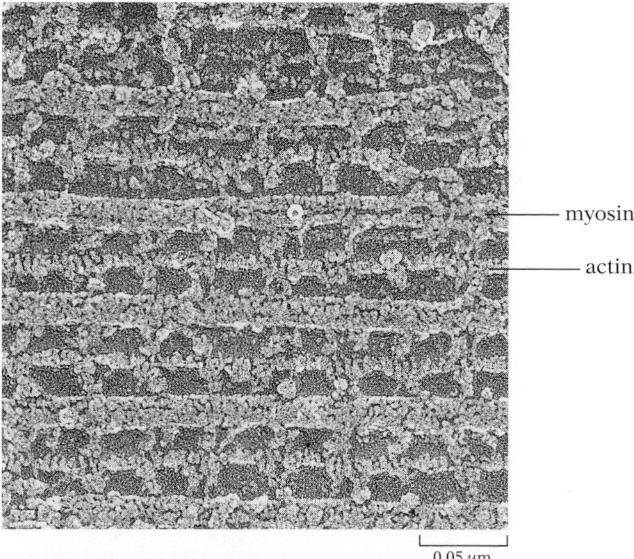

30.17 Electron micrograph of insect flight muscle, showing cross-bridges between filaments The thick (myosin) and thin (actin) filaments are connected by cross-bridges composed of myosin heads. Essentially the same structural pattern is seen in the myofibrils of skeletal muscle.

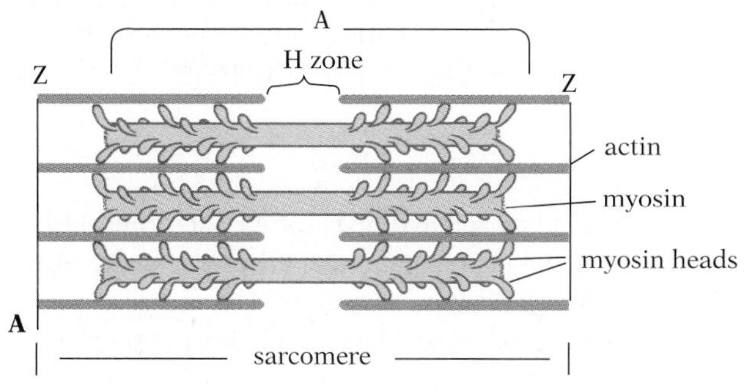

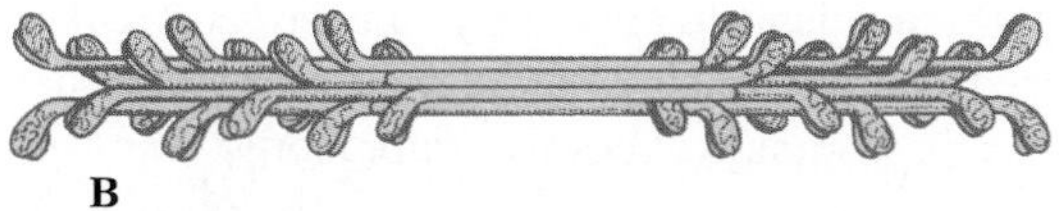

30.18 Molecular structure of myosin filaments (A) Each myosin filament is linked to the adjacent thin filaments by numerous cross-bridges. (B) The myosin filament is composed of a bundle of elongate molecules, each with a double club-shaped head, which acts as the cross-bridge. Each thick filament has about 500 heads.

30.18B). Other research has shown that the globular head regions exhibit both actin-binding activity and ATPase activity—a strong indication that they are the cross-bridges.

According to the sliding-filament theory, the cross-bridges act as hooks or levers that enable the myosin filaments to pull the actin filaments inward by a sort of ratchet mechanism (Fig. 30.19). The myosin heads bend toward the actin, hook onto it at special receptor sites forming a cross-bridge, and then bend in the other direction, pulling the actin with them. The heads then let go, bend back in the original direction, hook onto the actin at a new receptor site, and pull again.

The energy necessary for the movement of the myosin cross-bridges comes from the hydrolysis of ATP. When ATP binds to the myosin heads, the heads detach from actin (Fig. 30.19C), the ADP is hydrolyzed, and the heads rotate about 45 degrees in a movement called the ***recovery flip*** (Fig. 30.19D). The ADP and inorganic phosphate (P_i) remain tightly bound to the myosin. When the muscle is stimulated, the myosin heads *attach* to actin, releasing the ADP and P_i, and the myosin heads again rotate 45 degrees and *pull* the bound actin filament past the myosin in the ***power stroke*** (Fig. 30.19C, D). If ATP is available, it binds to the head, and the head detaches from actin, ready for another cycle of recovery flip, attachment, and power stroke. Notice that the ATP hydrolysis is involved in the recovery flip rather than the power stroke as might be ex-

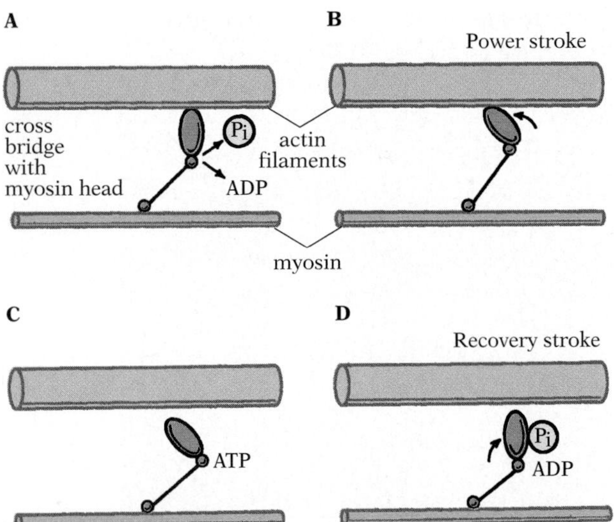

30.19 The action of a myosin cross-bridge in muscle contraction When the muscle is stimulated, the myosin head binds to actin, and ADP and P_i are released from the molecule (A). The myosin head then rotates 45 degrees and pulls the actin filament past the myosin in a power stroke (B). If ATP is available, it binds to the myosin head and the head detaches from actin. The ATP is then hydrolyzed to ADP and P_i, both of which remain tightly bound to the myosin. This hydrolysis causes the myosin heads to rotate about 45 degrees in a movement called the recovery flip and the heads are ready for another cycle of attachment, power stroke, and recovery flip.

pected. The binding and hydrolysis of ATP activates the heads and "cocks" them in preparation for the power stroke. A good analogy is shooting a bow and arrow. The energy required is used to pull back the bow (like cocking the myosin head); releasing the bowstring allows the potential energy stored in the bow to shoot the arrow (myosin power stroke).

If no ATP is available to bind to myosin, the actin remains tightly locked to the myosin in a state of rigor. This is what happens after death, a condition known as rigor mortis.

THE CONTROL OF CONTRACTION

Muscle Cell Membranes Are Polarized and Can Conduct Action Potentials

Like the membrane of a resting neuron, the membrane of a resting muscle fiber is polarized, the outer surface being positively charged in relation to the inner one. The stimulatory transmitter substance acetylcholine is released by a nerve axon at a neuromuscular junction (see Fig. 28.22, p. 607) and causes sodium-ion gates to open; sodium ions then move into the fiber, momentarily reducing this polarization. If the reduction reaches the threshold level, an impulse, or action potential, is triggered that sweeps along the surface of the fiber.

Muscle Contraction Is Regulated by Calcium Ions

A number of experiments suggested that calcium ions caused the cross-bridges to become active and initiate the contraction. When this idea was first put forward, two major objections were raised. First, contraction of a vertebrate muscle fiber requires the essentially simultaneous shortening of all its many myofibrils, but the myofibrils in the center of a fiber are so far from the surface that calcium ions from outside could not possibly diffuse fast enough to reach them in the short interval between stimulation of the fiber and its contraction. Second, there is evidence that not enough calcium ions enter the cell to account for the sustained contraction resulting from rapid volleys of nerve impulses.

The problem of the control of contraction was solved when attention was drawn to an extensive network of tubules in muscle fibers. These tubules were found to make up two separate but functionally related systems: the ***sarcoplasmic reticulum,*** which does not open to the exterior, and the ***T system*** (for transverse system), which does. The sarcoplasmic reticulum is the muscle cell's highly specialized version of the endoplasmic reticulum. Its membranous canals form a cufflike network around each of the sarcomeres of the myofibrils (Fig. 30.20). The sarcoplasmic reticulum at one end of a sarcomere and

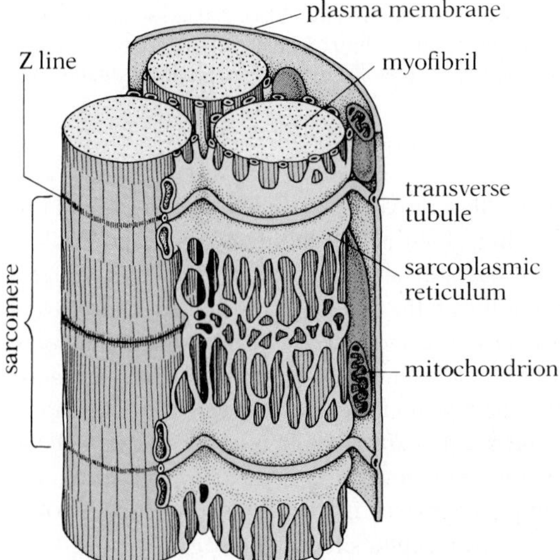

30.20 Sarcomeres with associated sarcoplasmic reticulum Shown here are three myofibrils. The lacelike channels of the sarcoplasmic reticulum surround each myofibril. The T (transverse) tubules are tubular infoldings of the plasma membrane occurring in the region of the Z line. Their membranes are in close contact with those of the sarcoplasmic reticulum.

those at the adjacent end of the sarcomere next to it are very close together, but lying between them, at the level of the Z line, is usually a tubule of the T system. Though the membranes of the sarcoplasmic reticulum and the T tubules are in direct contact, there is no interconnection between their cavities and hence no mixing of their contents.

The tubules of the T system are deep inpocketings of the cell membrane (Fig. 30.20), that allow an action potential sweeping across the cell surface to penetrate into the interior of the muscle fiber. The contraction stimulus reaches all the myofibrils at nearly the same instant so that the myofibrils near the surface and those in the center of the fiber can contract at the same time. This allows for rapid and powerful muscular contractions.

The intimate association of the T tubules and the sarcoplasmic reticulum suggested that an action potential moving along the membrane of a T tubule could somehow alter the properties of the adjacent membranes of the sarcoplasmic reticulum. This is critical because the sarcoplasmic reticulum contains very large amounts of calcium ions, which initiate the contraction of the myofibrils. As it turns out, the action potential in the T tubules induces a sharp increase in the permeability of the reticular membranes to calcium ions, allowing these to escape in large numbers (Fig. 30.21). It is these suddenly released intracellular calcium ions that are the direct stimulant for contraction.

How do calcium ions trigger contraction of the muscle fibers? A close look at the structure of the thin filaments provides an answer. In addition to actin, the main protein in the thin filaments, there are two important ***regulatory proteins.*** In the resting muscle these regulatory proteins prevent the actin from binding with the myosin heads—probably by blocking the binding sites for myosin on the actin molecule. Whenever calcium ions are released from the sarcoplasmic reticulum, they bind to the regulatory proteins, which undergo conformational changes that uncover the myosin-binding sites. The myosin heads can now bind to actin, and the contraction process is initiated (Fig. 30.22).

Now we have all the elements of the current model of cross-bridge action and its stimulation. In a resting muscle the cross-bridges—the globular myosin heads of the thick filaments

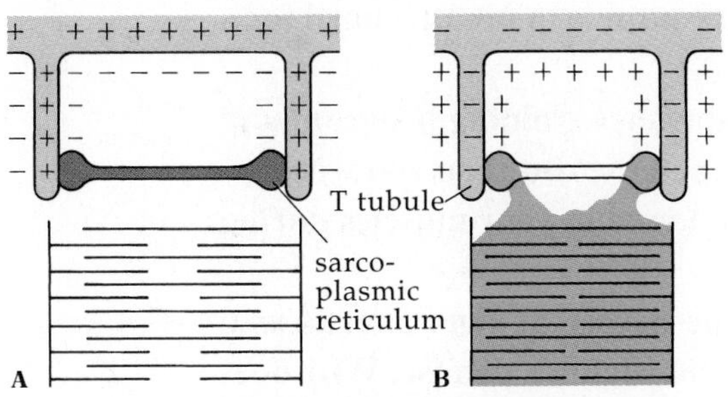

30.21 The role of calcium in stimulation of muscle contraction (A) A sarcomere in the resting (relaxed) condition. Calcium ions (dark brown) are stored in high concentration in the sarcoplasmic reticulum. (B) The polarization of the membranes of the T tubules is momentarily reversed during an action potential (impulse), and this reversal of polarization induces release of the calcium ions, which spread over the sarcomere and stimulate contraction.

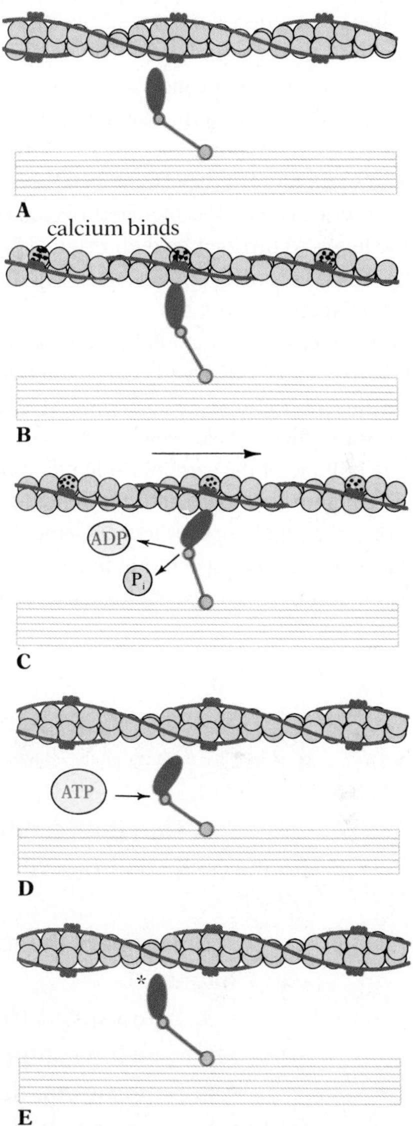

30.22 Model for the stimulation of muscle contraction (A) In a resting muscle the myosin cross-bridges that ATP has already activated cannot bind to the actin in the thin filament, because the binding sites are masked by regulatory proteins. (B) The binding of calcium ions to the regulatory protein causes a conformational change. The active sites of the actin are thus exposed, and the cross-bridges bind to the actin. (C) The binding of each myosin cross-bridge to actin, with the concomitant release of ADP and phosphate, initiates a conformational change in the cross-bridge, whose bending—the power stroke—forces the filaments to slide along each other. (D) ATP binds to the myosin head, which then dissociates from actin. (E) The ATP is hydrolyzed and the myosin head is "cocked" in preparation for a new stroke.

—have picked up and hydrolyzed ATP, but cannot bind to the thin filaments, because the regulatory proteins are blocking the binding sites on the actin molecules. When stimulation from a motor neuron triggers an action potential, the action potential sweeps along the muscle surface and into the T tubules, thereby stimulating the sarcoplasmic reticulum to release calcium ions. Some of the ions bind to the regulatory proteins, which then change position and expose the myosin-binding sites of the actin to the myosin heads. The energized myosin heads can now bind to the actin and, once bound, will bend in the power stroke, pulling the actin filaments with them, causing the filaments to slide along each other. When new ATP molecules bind to the myosin heads, the heads break their attachment to the actin. The ATP is then hydrolyzed to ADP and P_i, causing the heads to flip back to their original position (the recovery flip). The ADP and P_i remain attached to the head and facilitate its attachment to actin.

As long as free calcium is available, the cycle of attachment, power stroke, and recovery flip can occur over and over again, and the muscle continues to contract. But when nervous stimulation ceases, the muscle relaxes, because a calcium pump in the membrane of the sarcoplasmic reticulum actively transports the calcium ions back into the reticulum. Without the calcium ions, the regulatory proteins resume their original position and again block the myosin-binding sites on the actin, inhibiting contraction. The process of muscle contraction is summarized in Table 30.3.

TABLE 30.3 *Steps in skeletal muscle contraction*

1. Nerve impulse causes release of acetylcholine at neuromuscular junction.
2. Acetylcholine binds to receptors in the membrane of the muscle fiber, opening ion gates. Ions rush across the membrane, depolarizing it and generating an action potential.
3. The action potential sweeps along the muscle cell surface and down into the T tubules.
4. The action potential in the T tubules alters the permeability in the membranes of the sarcoplasmic reticulum causing the release of stored calcium ions.
5. Calcium ions bind to regulatory proteins, which change position, uncovering the myosin-binding sites on actin.
6. Myosin heads, which have been activated by the hydrolysis of ATP, *attach* to actin. The ADP and P_i are released and the myosin heads bend in a *power stroke,* pulling the actin filaments past the myosin.
7. New ATP binds to the myosin heads, which then detach from actin. ATP is hydrolyzed causing the heads to "recock" in a *recovery flip* in preparation for a new stroke.
8. The ADP and P_i remain bound to the myosin head, which can then bind once more to the actin.
9. This cycle of attachment, power stroke, and recovery flip can continue as long as there is nervous stimulation (leading to the release of calcium ions) and sufficient ATP (to detach the myosin heads from actin and recock them). There are about five to ten cycles per second in a contracting muscle.

MAKING CONNECTIONS

1. How does creatine phosphate collaborate with the machinery of cellular respiration described in Chapter 8 to supply contracting muscles with ATP?

2. Compare the structure and function of myoglobin and hemoglobin (see Chapter 3).

3. If you sprint 100 meters, why do you continue to breathe hard for a while after you have stopped running?

4. Suppose you undergo a rigorous endurance training program and find that you are much less winded after a 100-meter sprint than you were before you "got in shape." What changes have occurred in your muscles during your training?

5. Police investigating a murder use the degree of rigor mortis as a means of determining the approximate time of the victim's demise. Why does this condition occur after death?

6. In Chapter 25 you learned that accidental removal of the parathyroid glands, which regulate the blood-calcium level, is fatal because of disturbances of nerve and muscle function. What role does calcium play in the activity of neurons and muscles?

CONCEPTS IN BRIEF

1. The *effectors* are the parts of the organisms that carry out the organism's response to a stimulus. Movement in animals depends on either microfilaments or microtubules. The principal effectors of movement in higher animals are the muscle cells. Muscles require some type of skeleton to act as a mechanical support.

2. Movement in many invertebrates is produced by the alternating contraction of the longitudinal and circular muscles against the incompressible fluids in the body cavity. The fluids function as a *hydrostatic skeleton,* providing the mechanical support against which muscles can contract.

3. Both arthropods and vertebrates have evolved a hard, jointed skeleton, paired locomotory appendages, and elaborate musculature. Arthropods have an *exoskeleton,* a hard body covering with all muscles and organs inside it. Vertebrates have an *endoskeleton,* an internal framework composed of bone and/or cartilage, with the muscles outside.

4. Some bones are connected by immovable joints; others are held together at movable joints by *ligaments.* Skeletal muscles, attached to the bones by *tendons,* contract and bend or straighten the skeleton at their joints. The action of any specific muscle depends on the position of its attachments and on the type of joint between them. The muscles operate in antagonistic and synergistic groups.

5. Vertebrates possess three different types of muscles: skeletal, smooth, and cardiac. *Skeletal muscle* is responsible for most voluntary movement. Each muscle fiber is long and cylindrical, contains many nuclei, and is crossed by *striations.* Skeletal muscle is stimulated by the somatic nervous system.

6. *Smooth muscle* is found in the walls of the viscera and the blood vessels. The spindle-shaped cells interlace to form sheets of tissue, stimulated by the autonomic nervous system. Smooth muscle is primarily responsible for movements in response to internal changes, while skeletal muscle is concerned with making adjustments to the external environment.

7. *Cardiac muscle* is found only in the heart. It has striated fibers, but is stimulated by the autonomic nervous system, and it acts like smooth muscle.

8. Individual skeletal-muscle fibers contract only if they receive a stimulus of at least threshold intensity and duration. In vertebrates, muscle fibers seem to exhibit the all-or-none property but muscles give a graded response. When a single threshold stimulus is applied to a muscle, a *simple twitch* occurs. When frequent stimuli are applied to a muscle, the muscle does not have time to relax between contractions and the contractions add together *(summation).* If the stimuli are very frequent, the muscle may not relax at all between successive stimulations; the resulting strong sustained reaction is called *tetanus.* If the frequent stimulation continues, the muscle may fatigue.

9. The energy for muscle contraction comes from ATP. Creatine phosphate, which is stored in muscle, can be used to generate ATP for immediate contraction. For continued contraction, ATP must be produced from the complete oxidation of glucose and/or fatty acids to carbon dioxide and water. During strenuous muscular activity, lactic and fermentation takes place. When oxygen is available, the lactic acid is converted back into pyruvic acid and metabolized.

10. Analysis of muscle shows that its contractile components are the proteins *actin* and *myosin.* The contractile unit is called the *sarcomere.* According to the sliding-filament model, cross-bridges from myosin filaments hook onto actin filaments at specialized receptor sites and bend, pulling the actin along the myosin. The necessary energy comes from the hydrolysis of ATP by the myosin cross-bridges.

11. The membrane of a muscle fiber is polarized, with the outside charged positively in relation to the inside. Acetylcholine released by the neuronal axon reduces this polarization, triggering an action potential that sweeps across the muscle fiber. When the action potential penetrates into the interior of the fiber via a *T tubule,* calcium ions are released by the *sarcoplasmic reticulum.*

12. Calcium ions stimulate contraction by binding to regulatory proteins on the actin and causing a conformational change, which exposes the myosin-binding sites of actin. Myosin heads attach to actin, bend in a power stroke, and the filaments slide past each other. ATP then binds to the head and is hydrolyzed, causing the head to "recock" in a recovery flip.

13. The cycle of attachment, power stroke, and recovery flip can continue as long as there is nervous stimulation (leading to the release of calcium ions) and sufficient ATP (to detach the myosin heads from actin and recock them).

STUDY QUESTIONS

1. How is an *effector* defined? Describe some examples of effectors other than muscles. Do any effectors act intracellularly? (p. 649)
2. How does a skeleton help muscles to work more efficiently? What other function does a skeleton serve in land animals? Describe how an earthworm moves using the combination of its hydrostatic skeleton, antagonistic muscle layers, and segmented body. (pp. 650–651)
3. Compare *endoskeletons* and *exoskeletons.* What are the advantages and disadvantages of each? Which groups of animals have each type? (p. 652)
4. Compare the structure of *cartilage, spongy bone,* and *compact bone.* Diagram and explain the functioning of a Haversian system in compact bone. (pp. 653–654)
5. Why do muscles operate in *antagonistic* groups? What is the role of *synergistic muscles?* What is the role of *tendons?* How do they differ from *ligaments?* (p. 654)
6. Compare the structure, function, and innervation of *smooth* and *skeletal muscle.* What characteristics does *cardiac muscle* share with each? (pp. 654–656)
7. How does a *tetanic contraction* differ from a *simple twitch?* Which occurs during normal muscle action? How do *isotonic* and *isometric* contractions differ? (pp. 658–660)
8. Diagram the arrangement of thick and thin filaments in striated muscle. Of which proteins are they composed? What is a *sarcomere?* What do the A and I bands represent? What happens to filaments during contraction? What is the role of the cross-bridges during contraction? (pp. 661–662)
9. Explain how the contraction of a striated muscle fiber occurs, describing the roles of: the nerve impulses in a motor neuron, acetylocholine, T tubules, the sarcoplasmic reticulum, calcium ions, regulatory proteins, myosin, actin, and ATP. (pp. 662–664)

P A R T F O U R

The Biology of Populations and Communities

C H A P T E R 3 1

EVOLUTION: ADAPTATION

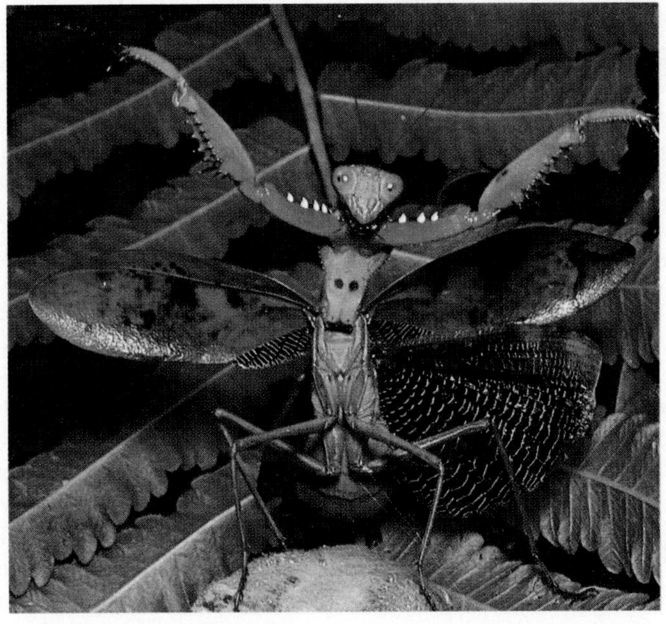

Darwin's theory of evolution, first introduced in Chapter 1, is the unifying principle of modern biology and has informed the discussion of every topic in this book so far. This central concept explains how the vast array of organisms populating the earth (Fig. 31.1) evolved from simple ancestral forms that arose more than three billion years ago, and provides a framework that allows biologists to tackle such fundamental issues as the interactions of organisms with their environments, and their means of surviving, growing, and reproducing. It also explains why some organisms have become extinct, to be found only in the fossil record, while others have flourished for millions of years and are abundant today. In this chapter and the next we examine the mechanisms of evolution. Studies of the fossil record show two evolutionary patterns, phyletic evolution and branching evolution. ***Phyletic evolution,*** or anagenesis, refers to the gradual, steady change of an ancestral population over time. The second process, called ***branching evolution,*** or cladogenesis, describes the branching of an ancestral species into two or more descendant species. This chapter will be primarily concerned with mechanisms involved in the small-scale changes characteristic of phyletic gradualism; branching evolution will be covered in the next chapter.

Natural Selection Is the Driving Force for Evolution

In Chapter 1 we made a distinction between evolution—the gradual change of species over time—and natural selection, which was Darwin's brilliant explanation for that change. As we saw, the mechanism of natural selection rests on five concepts: (1) more offspring are produced than can survive to reproduce (*excess progeny*); (2) the characteristics of living things differ between individuals of the same species (*variability*); (3) many differences are the result of inheritable, genetic differences (*inheritability*); (4) some differences affect how well adapted an organism is to its environment (*differential adaptiveness*); and (5) some differences in adaptedness are reflected in the number of offspring successfully reared (*differential reproduction*).

It is important to keep in mind that evolution does not refer to change in an individual during its lifetime, though change is a universal and important attribute of life; evolution is change in the characteristics of ***populations***—groups of individuals belonging to the same species living in a particular geographic area—over the course of many generations. *An individual cannot evolve, but a population can.* The genetic inheritance of an individual is set from the moment of conception and so is the developmental potential of its cells; they cannot evolve. But populations can evolve over time. *The change in the genetic makeup of a population in successive generations constitutes evolution.* At its simplest, evolution can occur through the emigration or immigration of members of a population. We

Chapter opening photo: A female praying mantis (*Prohierodula congica*) rearing up when disturbed. The two dark spots on the thorax serve as imitation eyes to startle enemies.

31.1 Diversity of life in the hydrothermal-vent communities Thermal springs on the ocean floor from which hot, hydrogen sulfide–laden water flows provide energy for more than 300 animals species living around them. Most of these species were previously unknown, having evolved in isolation from other communities. The harsh conditions under which they live and the unpredictability of the habitat has led to the evolution of a number of exotic organisms that are found nowhere else.

will concentrate here, however, on the roles played by chance and natural selection in evolution.

Genetic Variation Provides the Raw Material for Evolution

A population is composed of many individuals belonging to the same species. With rare exceptions, no two of these are exactly alike. In human beings we are well aware of the uniqueness of the individual, for we are accustomed to recognizing different persons at sight, and we know from experience that each has distinctive anatomical and physiological characteristics, as well as distinctive abilities and behavior traits. We are also fairly well aware of individual variation in such common domesticated animals as dogs, cats, and horses, but tend to believe that less familiar species, such as robins, squirrels, earthworms, sea stars, dandelions, and corn plants (Fig. 31.2), are all the same. In fact, individual variation exists in all species, even though in some it is less obvious to the unpracticed eye.

The members of a population, then, share some important features, but differ from one another in numerous ways, some rather obvious, some very subtle. Natural selection acts on genetic variation that is expressed phenotypically. In other words, a completely recessive allele never occurring in the homozygous condition would be masked from the action of natural selection. Though individuals do not evolve, *individuals, not populations, are the target of natural selection.* If there is selection against individuals with certain variations within a population and selection for individuals with other variations within it, the overall makeup of that population will change (evolve) with time.

Genetic variation arises from three main sources: crossing over in meiosis, the union of two haploid gametes in fertilization (sexual recombination), and mutation (Fig. 31.3). The first two processes, which we examined earlier, do not lead to new alleles but to a recombination of existing ones. Mutation, also discussed earlier, is a potential source of entirely new alleles that give rise to small-scale changes. DNA is a relatively stable molecule that undergoes many small, random mutations, most of which are repaired. The few mutations that remain are usually recessive, which means they are unexpressed (thus hidden from selection) when present in only one copy. Once a variety of alleles is in existence, however, ***recombination*** due to crossing over and sexual reproduction becomes the primary mechanism providing the genotypic variation in the population. But it was hard for biologists to imagine how the few unrepaired mutations, and the reshuffling of the same old genetic deck, could result in the evolution of major new functional proteins in a comparatively short period of time.

The discovery that eucaryotic chromosomes contain introns, exons, seemingly functionless genes, and repetitive DNA provided a clue. As we saw in Chapter 12, approximately 90 percent of eucaryotic DNA is not transcribed and translated and lies apparently functionless and dormant. Most biologists expected that natural selection would have operated to weed out apparently unnecessary DNA, as it seems to have done in procaryotes such as bacteria. Apparently the genetic material is not as stable as had been thought, and the extra DNA provides an additional source of variability. We shall explore how the odd organization of the eucaryotic genetic material may facilitate evolution, looking at some of the more potent mechanisms of gene evolution—notably duplication of genes and gene segments.

Gene Duplication May Facilitate Evolution

Gene duplication may be an essential mechanism for genetic evolution. The basic idea is that an ancestral gene is duplicated, and that small-scale mutations occur in each of the duplicate genes. Since mutations are random, it is unlikely that each gene will experience the same changes, so they slowly acquire differences, and the proteins for which they code are correspondingly different.

The evidence for the role of duplication is particularly clear in the genes for myoglobin and hemoglobin. Myoglobin, the oxygen-storage protein in muscles, consists of a single polypeptide chain, while hemoglobin, which carries oxygen in the blood, has two pairs of chains, for a total of four. The amino-acid sequences of the alpha and beta chains in adult hemoglobin are similar, and they also resemble the single chain

31.2 Selective breeding of pigeons By practicing rigorous selection, breeders can achieve major changes in relatively few generations. The domestic breeds shown here are (A) fantail, (B) reverse wing pouter, (C) Norwich cropper, (D) pigmy pouter, (E) black splash trumpeter, (F) Schmalkalden Moorhead.

31.3 Sources of variation One source of variation is spontaneous mutation. Most wisterias, like the one shown here from the Princeton University campus, have lavender flowers (A). Several decades ago, however, the famous white-flowering Eno wisteria appeared behind the Evolution, Ecology, and Behavior Building on the Princeton campus (B). The white wisteria is almost certainly the result of a spontaneous mutation in a pigment gene of an offspring of one of the many lavender wisterias nearby. Another source of variation is recombination. Offspring of the same parents frequently do not resemble each other or either of their parents.

in myoglobin (Fig. 31.4). The three-dimensional structure of these molecules is dramatically similar. Furthermore, humans make a number of different hemoglobins, i.e., embryonic hemoglobin, fetal hemoglobin, and adult hemoglobin, and the genes for these are lined up on the chromosomes in the chronological order in which they are used during development. Also present are a number of "pseudogenes"—genes that are similar in sequence to hemoglobin genes, but no longer transcribed and translated. The genes for myoglobin, the various chains of hemoglobin, and the pseudogenes are thought to have evolved by duplications from a single ancestral gene, forming multiple copies, and then to have gone their separate evolutionary ways. This conclusion is reinforced by the discovery that introns are located in the same places in each of these genes. The duplicates, as they underwent mutations, would come to code for slightly different products. Most of these new genes would be less adaptive than the original gene, but some would make the animal better adapted to changing conditions.

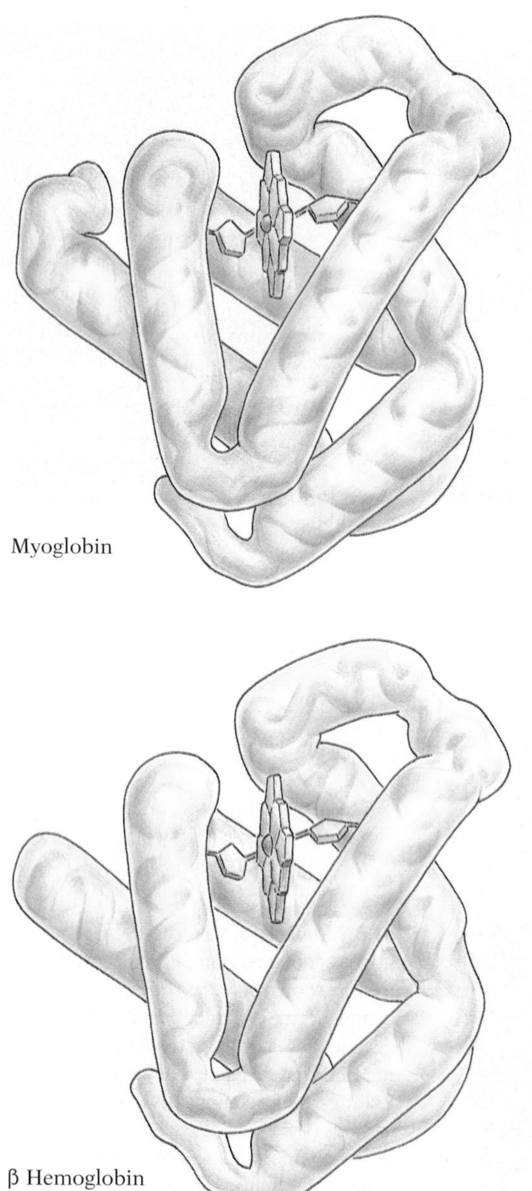

31.4 Myoglobin and the β chain of hemoglobin compared The similarity in conformation between these peptide chains is evident from this representation. The genes for these two polypeptides are thought to have arisen from a duplication event.

Comparisons of the amino-acid sequences of myoglobin and the hemoglobins suggest that the duplication and divergence that produced them may have occurred some 700 million years ago, while the duplication and divergence that produced the alpha- and beta-type hemoglobin chains probably occurred about 500 million years ago. It may be that most of the 100,000 genes in the human genome may have arisen from a relatively few genes by this process.

The same pattern is seen in many groups of enzymes. For example, the digestive enzymes trypsin and chymotrypsin and the blood-clotting enzyme thrombin have different functions, but the genes for them are nearly identical. It is highly unlikely that each evolved independently into a near duplicate of the others. Most researchers believe that the genes for these enzymes began as duplications of the gene for some primordial enzyme, and then evolved separately. We saw in Chapter 17, for instance, that the pancreatic enzymes trypsin and chymotrypsin function as endopeptidases—enzymes that catalyze the hydrolysis of specific peptide bonds between amino acids located within the protein being digested. The differences in function between these two enzymes pertain to their active sites. Trypsin has an amino acid with a negatively charged R group in one position in its active site whereas chymotrypsin has an amino acid with an uncharged R group in the same position. This seemingly small variation accounts for the affinity of the two enzymes for different amino acids in the substrate.[1] The enzyme thrombin, which is found in the blood rather than the digestive tract, also acts as an endopeptidase, hydrolyzing the blood protein fibrinogen into fibrin.

There is now considerable evidence, then, that *eucaryotic gene evolution depends at least in part on gene duplication followed by small-scale changes in base sequence that give rise to functionally different proteins.* But how do such duplications occur? Gene duplication can probably occur in several ways. The simplest is nondisjunction, the complete duplication of chromosomes that results when both members of homologous pairs move to the same pole during meiosis (see Chapter 14, p. 301). Another well-established mechanism, also discussed in Chapter 14, involves chromosomal breakage and fusion during meiosis, in which one egg or sperm cell ultimately gets an extra bit of DNA that the other has lost (see Chapter 14, p. 301). The genes on the segment that breaks off and fuses to the homologous chromosome are duplicates of those already resident there. Still another duplication mechanism is the phenom-

[1] Trypsin hydrolyzes bonds at the carboxyl end of lysine and arginine, and chymotrypsin hydrolyzes bonds at the carboxyl end of tyrosine, tryptophan, phenylalanine, methionine, and leucine.

enon of transposition, which we discussed in Chapter 11. Sometimes referred to as "jumping genes," transposons are mobile genetic sequences that can move themselves to new locations in the genome or duplicate themselves and insert the copies into another location. Transposons are found in all cells, both procaryotic and eucaryotic, and often result in duplications of genetic material.

Reverse transcription can also lead to gene duplications. You may recall from Chapter 13 that certain RNA viruses—the retroviruses (e.g., the HIV virus that causes AIDS)—have an enzyme that catalyzes transcription of their RNA into cDNA, which is then inserted into the host chromosomes. Sometimes this transcriptase does the same to the host's mRNA. The resulting cDNA—complementary to the host mRNA from which it was copied—becomes integrated into the host's chromosomes, and duplicate genes result. If this occurs in a gamete, the resulting duplication can be inherited. A small percentage of the human genome shows evidence of being produced by reverse transcription.

The duplication of genes or segments of genes followed by small-scale evolutionary changes is more likely to lead to functional genes with novel properties than random changes alone would be. Imagine how rarely we could generate a meaningful sentence by randomly arranging letters and spaces, whereas if we *began* with a meaningful sentence (a gene for a functional protein) consisting of words and spaces (exons and introns) and changed a few existing letters, the odds of ending up with an intelligible sentence with a new meaning would be fairly high. Then too, having more than one copy of a gene may be advantageous to an organism. If an individual had but a single copy of a gene that coded for an important protein and a mutation occurred such that the protein was nonfunctional, the individual would be at a severe disadvantage. If, however, there were several copies of that gene, each could evolve independently of the others, and the chances of having at least one of these genes coding for a functional protein would be increased. It is rather like making a back-up copy of a computer diskette to protect against loss of or damage to the original.

Though eucaryotic gene evolution depends in part on gene duplication followed by changes in base sequence that give rise to functionally different proteins, such changes occur slowly, and most random mutations result in genes coding for proteins of reduced function—or no function at all. For every new gene that produces a functional protein, there may be tens or hundreds of incomplete or failed "experiments" involving duplications. Unless some process is at work to edit out useless duplications, the eucaryotic chromosome should be full of nonfunctional base sequences with clear similarities to those of functional genes. And indeed, that is exactly what we find—well over 90 percent of the mammalian genetic material is not transcribed and does not code for functional products. It may be that the enormous number of nonfunctional sequences in eucaryotic chromosomes is evidence of past duplications that never evolved (or have not yet evolved) into functional genes.

We have seen how new genes may arise by duplication of preexisting genes, followed by mutation and separate evolution. We also saw in earlier chapters how new alleles of existing genes arise through mutation. And quite independent of mutational changes we saw that the processes of meiosis, crossing over, and recombination at fertilization provide almost endless variation in the population. Together, these

MAKING CONNECTIONS

1. Why aren't the changes that occur in an organism during development—such as those described in Chapter 27—considered evolution?

2. Why haven't all lethal genetic diseases caused by recessive alleles been eliminated from the population by natural selection?

3. How does crossing over in meiosis affect evolution? (See Chapter 9.)

4. The genes for myoglobin and the three types of hemoglobin probably evolved from a single ancestral gene. How could this have happened, and how was it advantageous?

5. How does the minor structural difference between trypsin and chymotrypsin demonstrate the importance of the properties of the active site in determining enzyme specificity, as explained in Chapter 4?

6. As explained in Chapter 12, most eucaryotic DNA is never transcribed and translated. What is the possible evolutionary significance of the vast number of nonfunctional sequences in eucaryotic chromosomes?

processes generating new genes, new alleles, and new combinations of alleles provide the genetic variability on which natural selection can act to produce evolutionary changes in populations.

Exclusively Phenotypic Variation Does Not Lead to Evolutionary Change

Any phenotypic variation within a population may give rise to differences in reproductive success among individuals, whether or not the variation reflects corresponding genetic differences. Thus variations produced by exposure to different environmental conditions during development, or by disease or accidents, are subject to natural selection. But even though the action of natural selection on all types of variations alters the immediate makeup of a population, it is only its action on variations reflecting genetic differences that has any long-term effect on the population. Variation that is exclusively phenotypic is not raw material for evolutionary change.

An understanding of genetics makes it clear that development of athletic prowess by extensive practice, or development of intellectual powers by education, or maintenance of health by correct diet and prompt medical treatment of all ailments cannot alter the genes in the germ cells. The gametes will carry the same genetic information they would carry in the absence of athletic or mental training or proper health care. In short, selection that acts on variations produced exclusively by practice, education, diet, or medical treatment cannot bring about biological evolution.

Not even all genetic variation provides material for evolutionary change. Somatic mutations are a good example of genetic changes that do not lead to evolution. Suppose, for example, an important mutation occurred in the DNA of an ectodermal cell of an early animal embryo. All the cells descended from the mutant cell would be of the mutant type. The result might be a major change in the animal's nervous system, but the change could not be passed on to the animal's offspring. The ectodermal cells are not the ones that give rise to gametes. *Mutations in somatic cells cannot alter the genes in the germ cells, which are the cells that produce the gametes, and therefore responsible for evolutionary change. Selection that acts on variations produced by somatic mutations cannot result in evolutionary change.*

Lacking genetic data, many prominent biologists of the last century and of the early part of this century assumed that exclusively phenotypic variation could serve as evolutionary raw material; that it cannot is far from obvious to many nonbiologists even today. The theory of ***evolution by natural selection*** proposed by Darwin and Wallace had an influential rival during the 19th century in the concept of ***evolution by the inheritance of acquired characteristics***—an old and widely held idea often identified with Jean Baptiste de Lamarck (1744–1829), who was one of its more prominent supporters in the early 1800s.

The Lamarckian hypothesis was that somatic characteristics acquired by an individual during its lifetime could be transmitted to its offspring. Thus the characteristics of each generation would be determined, in part at least, by all that happened to the members of the preceding generations—by all the modifications that occurred in them, including those caused by experience, use and disuse of body parts, and accidents. Evolutionary change would be the gradual accumulation of such acquired modifications over many generations. The classic example is the evolution of the long necks of giraffes (Fig. 31.5).

According to the Lamarckian view, ancestral giraffes with short necks tended to stretch their necks as much as they could to reach the tree foliage that served as a major part of their food. This frequent neck stretching caused their offspring to have slightly longer necks. Since these also stretched their necks, the next generation had still longer necks. And so, as a result of neck stretching to reach higher and higher foliage, each generation had slightly longer necks than the preceding generation.

The modern theory of natural selection, on the other hand, recognizes that ancestral giraffes probably had short necks, but adds that the precise length of the neck varied from individual to individual because of their different genotypes. If the supply of food was somewhat limited, then individuals with longer necks had a better chance of surviving and leaving progeny than those with shorter necks, because they were better able to reach the foliage at the tops of trees. As a result, the proportion of individuals with genes for longer necks increased slightly with each succeeding generation, causing a long-term genetic change in the population.

Though the hypothesis of evolution by inheritance of ac-

A

B

31.5 An okapi and a giraffe, two related African herbivores The okapi (A) and giraffe (B) are thought to have had a common ancestor with a relatively short neck. The long-necked giraffe of today can reach food unavailable to shorter individuals.

quired characteristics is rejected by modern biologists, it was logical and reasonable when first proposed. It has simply not stood the test of further scientific research. In Lamarck's day (as in Darwin's), nothing was known about the mechanism of inheritance, and the results of Mendel's experiments on garden peas were not published until 37 years after Lamarck's death.

THE GENE POOL AND FACTORS THAT AFFECT ITS EQUILIBRIUM

Evolution Can Be Defined as a Change in the Allelic Frequencies Within a Gene Pool

Synthesis of Mendel's work on inheritance with Darwinian theory leads us to think about evolution in terms of the genetic composition of populations, or ***population genetics.*** To understand evolution as change in the genetic makeup of populations in successive generations, it is necessary to know something about population genetics. Our study of the genetics of individuals in Chapter 14 was based on the concept of the genotype, which is the genetic constitution of an individual. Our study of the genetics of populations will be based in a similar manner on the concept of the gene pool, which is the genetic constitution of a population. The ***gene pool*** is the sum total of all the alleles of all the genes possessed by all the individuals in the population.

We know that in a diploid individual, there are two of each kind of chromosome, and thus two copies (alleles) of each gene. Therefore, under normal circumstances the maximum number of alleles for each gene that any individual can possess is two. But there is no such restriction on the gene pool of a population; it can contain any number of different allelic forms of a gene. For example, in Chapter 14 we learned about human A-B-O blood types, which are determined by three alleles (I^A, I^B, and i). Each individual has only two alleles, but all three are present in the human population.

The gene pool is characterized with regard to any given gene by the frequencies, or ratios, of the alleles of that gene in the population. Suppose, to use a simple example, that gene *A* occurs in only two allelic forms, *A* and *a,* in a particular sexually reproducing population. And suppose that 90 percent of all the alleles of this gene are allele *A* and 10 percent are allele *a.* The frequencies of *A* and *a* in the gene pool of this population are therefore 0.9 and 0.1. If those frequencies were to change with time, the change would be evolution. Such small-scale changes in allelic frequencies are sometimes referred to as ***microevolution.*** Evolution, as change in the genetic makeup of populations, is—more precisely, then—*a change in the allelic frequencies (or genotypic ratios) within gene pools.* It is possible to determine what factors cause evolution by determining what factors can produce a shift in allelic frequencies.

Let us examine more carefully our hypothetical population in which allele *A* has a frequency of 0.9 and allele *a* has a frequency of 0.1. How can we calculate the genotypic ratios that will be present in this population in the next generation? If we assume that all possible genotypes have an equal chance of surviving, this calculation is easy to make. If allele *A* has a frequency of 0.9 in the population and allele *a* a frequency of 0.1, then 0.9 of the sperm cells and 0.9 of the egg cells will carry allele A and 0.1 of the sperm cells and 0.1 of the egg cells will carry allele *a.* Using this information, we can set up a Punnett square much like those we used for crosses between two individuals:

		Sperm	
		0.9 *A*	0.1 *a*
Eggs	0.9 *A*	0.81 *A/A*	0.09 *A/a*
	0.1 *a*	0.09 *A/a*	0.01 *a/a*

Notice that the only difference between this square and one for a cross between individuals is that here the sperm and eggs are not those produced by a single male and a single female, but those produced by all the males and all the females in the population, with the frequency of each type of sperm and egg shown on the horizontal and vertical axes respectively. Filling in the square (by combining the indicated alleles and multiplying their frequencies) tells us that the frequency of the homozygous dominant genotype (*A/A*) in the next generation of this population will be 0.81, that of the heterozygous genotype (*A/a* or *a/A*) 0.18, and that of the homozygous recessive genotype (*a/a*) 0.01. This means that 81 percent of this generation will be homozygous dominant, 18 percent heterozygous, and 1 percent homozygous recessive. Now we want to know whether the frequencies we have found will change in successive generations—in short, whether the population will evolve.

THE HARDY-WEINBERG LAW AND CONDITIONS FOR GENETIC EQUILIBRIUM

The Hardy-Weinberg Law States the Conditions Under Which No Evolution Can Occur

We learned in Chapter 14 that if allele *A* is dominant to *a,* whenever a homozygous dominant (*A/A*) is crossed with a homozygous recessive (*a/a*), all the offspring show the dominant phenotype, and that when two heterozygotes (*A/a*) are crossed, 75 percent of the offspring show the dominant phenotype. Looking at these results it is easily assumed that the more frequent allele (in our hypothetical case, *A*) will automatically increase in frequency while the less frequent allele (*a*) will automatically decrease in frequency and eventually be lost from the population. This perplexed the geneticist R. C. Punnett (of Punnett square fame), who recognized that the allele for blue eyes was recessive to the allele for brown eyes, yet there were many blue-eyed people in England. Why didn't the blue-eyed allele disappear from the population? When he could not arrive at a satisfactory explanation he asked a fellow cricket player, G. H. Hardy, for a solution. Being a mathematician, Hardy recognized that the distribution of alleles and genotypes in a population fit a simple algebraic expression (see Exploring Further: The Hardy-Weinberg Equilibrium, p. 677). Working independently, W. Weinberg, a German physician, arrived at the same conclusion, and in 1908 the Hardy-Weinberg Law was formulated. They both recognized that the rarity of a particular allele in a population does not doom it to automatic disappearance, and that the allelic and genotypic frequencies in the population should remain constant from generation to generation. We can see this for ourselves by using the known genotypic frequencies for one generation to compute the allelic frequencies for the next generation.

We have said that the *genotypic frequencies* in the gene pool of the second generation of our hypothetical population will be 0.81, 0.18, and 0.01. We can use these figures to compute the *allelic frequencies* of the *A* and *a* in this generation. Since the frequency of the *A/A* individuals is 0.81, the frequency of their gametes in the gene pool will be 0.81. All these gametes will contain the *A* allele. Likewise, the frequency of the gametes of the *a/a* individuals will be 0.01, and each gamete will contain an *a* allele. The frequency of the heterozygous (*A/a*) individuals is 0.18 and the frequency of their gametes in the gene pool will be 0.18, but their gametes will be of two types, *A* and *a,* in equal numbers (0.09*A* and 0.09*a*). Hence the frequency of the *A* and *a* alleles in the gametes of the second generation can be calculated as follows:

Frequency of genotypes	*Frequency of A alleles*	*Frequency of a alleles*
0.81 *A/A*	0.81	0
0.01 *a/a*	0	0.01
0.18 *A/a*	0.09	0.09
Totals	0.90	0.10

Since the allelic frequencies (0.9 and 0.1) are the same as those in the first generation, the genotypic ratios in the succeeding generation will again be 0.81, 0.18, and 0.01; and in turn the allelic frequencies will be 0.9 and 0.1. We could perform the same calculation for generation after generation, always with the same result; neither the genotypic ratios nor the allelic frequencies would change. Evolutionary change is not automatic: it occurs only when something disturbs the genetic equilibrium. This was what Hardy and Weinberg first noted in 1908. According to the Hardy-Weinberg Law, *under certain conditions of stability both genotypic ratios and allelic frequencies remain constant from generation to generation in sexually reproducing populations.*

The "certain conditions" that the Hardy-Weinberg Law says must be met if the gene pool of a population is to be in genetic equilibrium are as follows:

1. The population must be large enough to make it highly unlikely that chance alone could significantly alter allelic frequencies.
2. Mutations must not occur, or else there must be mutational equilibrium (i.e., the rate of mutation of one allele into another must be equal to the rate of mutation back).
3. There must be no migration of members into or out of the gene pool.
4. Mating must be totally random.
5. Reproductive success (that is, the number of offspring and the number of their eventual offspring) must be totally random.

The Hardy-Weinberg Law demonstrates that variability and inheritability, the two bases of natural selection, cannot alone cause evolution. Despite variability and inheritability, if all five of the above conditions are met, allelic frequencies will not change, and evolution cannot occur. But, in fact, these conditions are *never* completely met, and so evolution does occur. The value of the Hardy-Weinberg Law is that it provides a baseline against which to judge data from actual populations. By defining the criteria for genetic equilibrium, it also indicates when a population is not in equilibrium, and helps to define the possible causative agent of evolution. We shall look at each of the five factors in turn, and see which ones are most likely to disturb the equilibrium.

Genetic Drift May Disturb the Hardy-Weinberg Equilibrium

With regard to the first condition, allelic frequencies in small isolated populations of, say, fewer than 100 breeding-age members are highly susceptible to chance events, which can easily lead to the loss of an allele from the gene pool even when that allele is adaptively superior. Suppose, for example, in an isolated population of peas there were nine tall pea plants and one short pea plant. If, by chance, the one short pea plant failed to reproduce, the allelic frequencies of the population would change drastically. In such small populations, in fact, there are relatively few alleles with intermediate frequencies; apparently the tendency is for most alleles either to be soon lost or to

EXPLORING FURTHER

THE HARDY-WEINBERG EQUILIBRIUM

On page 675 we used a Punnett square to calculate the ratios of the genotypes produced by alleles *A* and *a,* whose respective frequencies in a hypothetical population were given as 0.9 and 0.1. The same results can be obtained more rapidly by using an algebraic formula.

In this discussion, the symbol p represents the frequency of one allele (in our case, *A*) and q the frequency of the other allele (*a*). We can then use the expression $p + q = 1$ to represent the relationship between the two alleles, since all individuals must have either an *A* or *a* allele, and the two frequencies must add up to 1. If we square both sides of the equation, $(p + q)^2 = (1)^2$, we obtain the formula for the Hardy-Weinberg equilibrium:

$$p^2 + 2pq + q^2 = 1$$

Substituting the allelic frequencies 0.9 and 0.1 for p and q respectively, we obtain

$$\begin{array}{ccccccc} p^2 & + & 2pq & + & q^2 & = & 1 \\ (0.9)(0.9) & + & 2(0.9)(0.1) & + & (0.1)(0.1) & = & 1 \\ 0.81 & + & 0.18 & + & 0.01 & = & 1 \end{array}$$

The three terms of the Hardy-Weinberg formula indicate the frequencies of the three genotypes:

$$\begin{aligned} p^2 &= \text{frequency of } A/A = 0.81 \\ 2pq &= \text{frequency of } A/a = 0.18 \\ q^2 &= \text{frequency of } a/a = 0.01 \end{aligned}$$

These are, of course, the same results we obtained using the Punnett square.

In this example we have assumed that we know the allelic frequencies and want to compute the corresponding genotypic frequencies. But the Hardy-Weinberg formula permits many other sorts of calculations as well. Suppose, for example, that we know a certain disease caused by a recessive allele *d* occurs in 4 percent of a certain population, and we want to find out what percent are heterozygous carriers of the disease. Because the disease occurs only in homozygous recessive individuals, the frequency of the *d/d* genotype is 0.04. Letting q^2 stand for the frequency of *d/d* in the formula, we can write

$$q^2 = 0.04$$

The frequency of allele *d,* then, is the square root of 0.04:

$$q^2 = \sqrt{0.04} = 0.2$$

If the frequency of allele *d* is 0.2, the frequency of allele *D* must be 0.8, because the two frequencies must always add up to 1 (that is, $p + q = 1$). Substituting the frequencies of both alleles in the Hardy-Weinberg formula, we can compute the frequencies of the genotypes:

$$\begin{array}{ccccccc} p^2 & + & 2pq & + & q^2 & = 1 \\ (0.8)(0.8) & + & 2(0.8)(0.2) & + & (0.2)(0.2) & = 1 \\ 0.64 & + & 0.32 & + & 0.04 & = 1 \end{array}$$

Since the term $2pq$ stands for the frequency of the heterozygous genotype, which is what we wanted to know originally, our answer is that 0.32 or 32 percent of the population are heterozygous carriers of the allele *d* that causes the disease we are studying.

become fixed as the only allele present. Thus chance may cause evolutionary change in small populations, but since this change, called ***genetic drift,*** is not much influenced by the relative adaptiveness of the different alleles, it is essentially random evolution, as likely to take one direction as another (Fig. 31.6). A population would have to be infinitely large for chance to be completely ruled out as a causal factor in the changing of allelic frequencies. In reality, of course, no population is infinitely large, but any population with more than 10,000 members of breeding age is probably not significantly affected by random events.

Genetic drift is a real consideration in establishing various programs to save endangered species. Often their numbers have been reduced to such low levels that genetic drift can become an important factor in causing change within the population, thereby complicating the process of reestablishing the species.

31.6 Possible genetic drift in a cichlid fish *Pseudotropheus zebra,* one of hundreds of species of cichlid fish living in the rift lakes of Africa, is divided into numerous isolated populations, many of which have evolved their own distinctive morphology. As there is no known selective force that accounts for this diversity, the varied colors are thought by many researchers to be the result of genetic drift in each small population.

Mutations Are Random and Always Occurring

The second condition for genetic equilibrium—either no mutation or mutational equilibrium—is rarely met in populations. Mutations are always occurring. There is no known way of stopping them. Most genes probably undergo mutation once every million to a hundred million replications, but the rate of mutations for different genes varies greatly. As for mutational equilibrium, very rarely, if ever, are the mutations of alleles for the same character in exact equilibrium, i.e., the number of forward mutations per unit time is rarely exactly the same as the number of backward mutations.[2] If *A*, for instance, is more

[2] By convention, the mutation from the more common allele to the less common one is called the forward mutation, and the reverse is called the backward mutation.

likely to mutate to *a* than *a* is to mutate to *A*, then *a* will increase in frequency. The result of this difference is a ***mutation pressure*** tending to cause a slow shift in the allelic frequencies in the population. The more stable allele will tend to increase in frequency, and the more mutable allele will tend to decrease in frequency.

Even though mutation pressure is almost always present, it is seldom a major factor in producing changes in allelic frequencies in a population. Acting alone, mutation would take an enormous amount of time to produce much change (except in the origin of polyploidy, which we shall discuss later). *Mutations increase variability and are thus the ultimate raw material of evolution, but they seldom determine the direction or nature of evolutionary change.*

Migration Can Influence Evolutionary Change

According to the third condition for genetic equilibrium, a gene pool cannot accept immigrants from other populations, for these might introduce new alleles; and it cannot suffer loss of alleles by emigration. A high percentage of natural populations, however, probably experience at least a small amount of gene migration, generally called ***gene flow,*** and this factor, which enhances variation, tends to upset the Hardy-Weinberg equilibrium. But there are populations that experience no gene flow, and in many instances where flow does occur it is probably sufficiently slight to be essentially negligible as a factor causing shifts in allelic frequencies. We can conclude, therefore, that this third condition for genetic equilibrium is sometimes met in nature.

Mating and Reproductive Success Are Never Totally Random

The final two conditions for genetic equilibrium in a population are that mating and reproductive success be totally random. Among the vast number of factors involved in mating and reproduction are choice of a mate (Fig. 31.7), physical efficiency and frequency of the mating process, fertility, total number of zygotes produced at each mating, percentage of zygotes that lead to successful embryonic development and birth, survival of the young until they are of reproductive age, fertility of the young, and even, in some cases, survival of postreproductive adults when their survival affects either the chances of survival of the young or their reproductive efficiency. For mating and reproductive success to be totally random, all these factors must be random, i.e., they must be independent of genotype, so that natural selection cannot operate. This condition is probably never met in any real population. The factors mentioned here are always correlated in part with an organism's genotype, and therefore *nonrandom reproduction is the universal rule.* And nonrandom reproduction, in the broad sense in which the term has been defined here, is synonymous with natural selection. Natural selection, then, is always operative in all populations; there is always ***selection pressure*** acting to disturb the Hardy-Weinberg equilibrium.

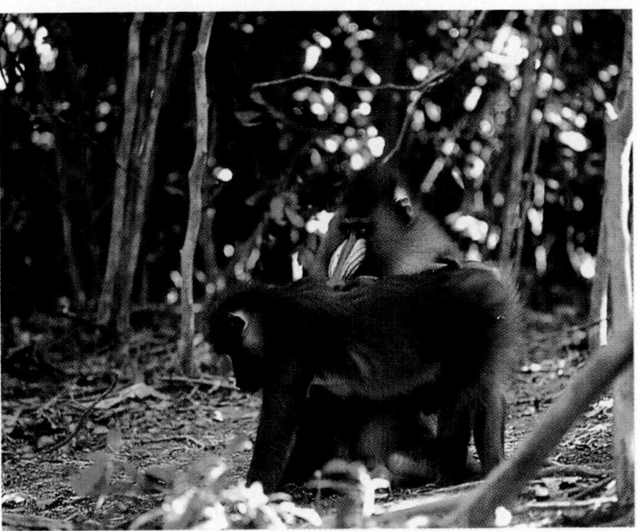

31.7 Female choice in mandrills Animals do not mate at random, but instead choose mates on the basis of certain characteristics. In most species of vertebrates, including mandrills, the female chooses the mate. The male mandrill's face is colored with red and blue markings, while the female's is not. Females choose to mate with males with the most brightly colored faces.

Evolutionary Change Is a Fundamental Characteristic of All Populations

In summary, of the five conditions necessary to achieve the genetic equilibrium described by the Hardy-Weinberg Law, the first (large population size, i.e., no genetic drift) is met reasonably often. The second (no mutation pressure) is never met; mutation is always at work, but in the short run is rarely significant. Like genetic drift, the occurrence of mutation is independent of natural selection. The third (no migration, i.e., no gene flow into or out of the population) is met sometimes. The fourth and fifth (random mating and reproductive success, i.e., no selection pressure) are almost always important in the evolutionary history of a species, and are never met. It follows that complete equilibrium in a gene pool is not expected. *Evolutionary change is a fundamental characteristic of the life of all populations, including human populations.*

Why go to so much trouble to explain the Hardy-Weinberg Law only to show that it describes a situation that does not occur in nature? For a compelling reason: the Hardy-Weinberg Law sets up a null hypothesis—no evolution—which, because it can easily be disproved, provides an indirect demonstration that all populations must constantly be evolving. It should be mentioned, however, that even though the gene pool of a population is not in equilibrium, it is possible for some *alleles* to be at near equilibrium in natural populations.

Natural Selection Is Not the Only Mechanism Involved in Evolution

We shall concentrate on natural selection as by far the most influential evolutionary mechanism, but there are alternative evolutionary scenarios. Let us return again to the evolution of long-necked giraffes. We have already seen that the Lamarckian concept of inheritance of acquired characteristics has been discredited, but the Darwinian explanation is not the only possibility. The most plausible alternative to natural selection is that giraffes have long necks as a result of genetic drift (Fig. 31.8). Suppose that the ancestral population had a wide variety of neck lengths. If we assume that long necks are adaptively neutral—neither advantageous nor disadvantageous—there is no reason according to the Hardy-Weinberg Law for selection pressure to cause neck length in the population as a whole to change. But suppose that the population suddenly decreased because of some severe environmental crisis like disease or bad weather, so that only a few individuals survived the crisis. If, by chance, a large proportion of the survivors were long-necked, the trait could become established without the intervention of natural selection. The new population will have very different allelic frequencies than the parental population; this is a special form of genetic drift called the ***bottleneck effect.*** We shall learn more about the bottleneck effect in the next chapter.

As this hypothetical example indicates, evolution does not depend exclusively on any single particular mechanism, whether natural selection, genetic drift, mutation, or migration. It also underscores the potential error of assuming that all the traits of the living things around us are necessarily the adaptive result of natural selection. However, most of the available evidence indicates that natural selection is by far the most important factor in evolution.

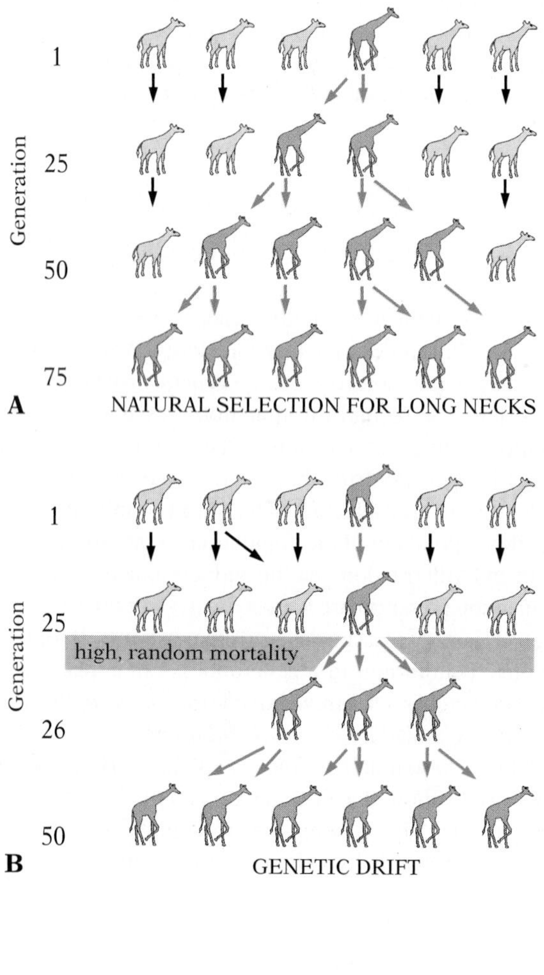

Generation 1, 25, 50, 75
A NATURAL SELECTION FOR LONG NECKS

Generation 1, 25, high, random mortality, 26, 50
B GENETIC DRIFT

31.8 Natural selection versus genetic drift In this hypothetical example, a small group of long-necked individuals arises in an otherwise stable population. (For simplicity, each animal in the drawing represents a fraction of the total population.) In evolution by natural selection (A), long-necked individuals become prevalent because they are able to reach more vegetation, and so survive to have proportionately more offspring in succeeding generations. In evolution by genetic drift (B), long-necked individuals become prevalent by chance: the frequency of long-necked individuals does not change until a chance catastrophe—fire, flood, heavy predation, or disease, for example—kills most members of the population. Because only long-necked individuals happen to survive, their offspring multiply to fill the habitat even though their distinctive trait is of no selective advantage.

THE ROLE OF NATURAL SELECTION

Natural Selection Causes Changes in Allelic Frequencies

Consider again our hypothetical population in which the initial frequencies of the alleles *A* and *a* are 0.9 and 0.1 and the genotypic frequencies are 0.81 for *A/A,* 0.18 for *A/a,* and 0.01 for *a/a.* We have seen that though these frequencies will not change automatically with the passage of time—instead changing only when something disturbs the genetic equilibrium—mutation pressure (to a slight extent) and selection pressure (to a greater extent) are always disturbing this equilibrium; other factors often at work are gene flow and, if the population is small, genetic drift. We shall here restrict discussion to what is by far the most important of these factors in guiding evolution: selection pressure.

Suppose that natural selection acts against the dominant

phenotype in our example such that individuals possessing this phenotype tend to survive in fewer numbers, and that this negative selection pressure is strong enough to reduce the frequency of *A* in the present generation from 0.9 to 0.8 before reproduction occurs. Of course there will be a corresponding increase in the frequency of *a* from 0.1 to 0.2, since the two frequencies must total 1. Now let us set up a Punnett square and calculate the genotypic ratios that will be present in the zygotes of the second generation:

		Sperm	
		0.8 *A*	0.2 *a*
Eggs	0.8 *A*	0.64 *A/A*	0.16 *A/a*
	0.2 *a*	0.16 *A/a*	0.04 *a/a*

We find that the genotypic ratios of the zygotes in the second generation are different from those in the parental generation; instead of 0.81, 0.18, and 0.01, the ratios are 0.64, 0.32, and 0.04 (Fig. 31.9). If selection again acts against the dominant phenotype in this generation, and thereby again reduces the frequency of *A*, the genotypic ratios in the third generation will be different from those of both preceding generations; the frequency of *A/A* will be lower and that of *a/a* higher. If this same selection pressure were to continue for many generations, the frequency of *A/A* would fall very low and the frequency of *a/a* would rise very high. Thus natural selection would have caused a change from a population in which 99 percent of the individuals showed the dominant phenotype and only 1 percent the recessive phenotype to a population in which very few showed the dominant phenotype and most showed the recessive phenotype. This evolutionary change of the phenotype most characteristic of the population would have occurred, without the necessity for any new mutation, simply as a result of natural selection.

Next, we turn to an actual situation in which selection has produced a radical shift in allelic frequencies. Soon after the discovery of the antibiotic activity of penicillin, it was found that *Staphylococcus* bacterial species that can cause numerous infections, including boils and abscesses, quickly developed resistance to the drug. At first the bacteria were killed at low doses of penicillin, but over a number of years, higher and higher doses of penicillin were required to kill these bacteria. "Staph infections" due to resistant bacteria became a serious problem in hospitals. Under the influence of the strong selection pressure exerted by the penicillin, the bacteria had evolved increased resistance to it. How did this change occur? Many studies have shown that the drug does not induce mutations for resistance; it simply selects against (inhibits) the growth of nonresistant bacteria (Fig. 31.10). Within a given staphylococcal population there may be some bacteria present that already have genes that confer resistance to penicillin as a result of spontaneous mutations. When this population is exposed to

WITHOUT SELECTION

First Generation

Genotype	*A/A*	*A/a*	*a/a*
Frequency	0.81	0.18	0.01
Gametes		*A* = 0.9	*a* = 0.1

Second Generation

		Sperm	
		0.9 *A*	0.1 *a*
Eggs	0.9 *A*	*A/A* 0.81	*A/a* 0.09
	0.1 *a*	*a/A* 0.09	*a/a* 0.01

Genotype	*A/A*	*A/a*	*a/a*
Frequency	0.81	0.18	0.01

WITH SELECTION

First Generation

Genotype	*A/A*	*A/a*	*a/a*
Frequency	0.81	0.18	0.01
Gametes		*A* = 0.8	*a* = 0.2

Second Generation

		Sperm	
		0.8 *A*	0.2 *a*
Eggs	0.8 *A*	*A/A* 0.64	*A/a* 0.16
	0.2 *a*	*a/A* 0.16	*a/a* 0.04

Genotype	*A/A*	*A/a*	*a/a*
Frequency	0.64	0.32	0.04

31.9 Comparison of genotypic frequencies under selection and no selection Top: In the absence of selection, the genotypic frequencies in a Hardy-Weinberg population are the same in the second and all later generations as in the first generation. Bottom: When there is selection, the genotypic frequencies change from one generation to the next. In this hypothetical case, discussed in more detail in the text, the frequencies of *A* and *a* in the gene pool of the first generation were initially 0.9 and 0.1, but natural selection altered these to 0.8 and 0.2 before reproduction occurred. Consequently the initial genotypic frequencies in the second generation are different from the initial ones in the first generation.

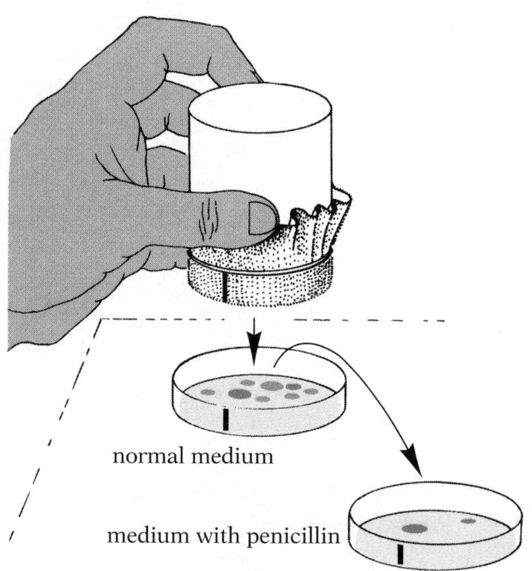

31.10 An experiment showing that mutation for resistance to penicillin is spontaneous Bacterial cells were cultured on a normal agar medium; many colonies developed (upper culture dish). Then a block wrapped with velveteen was pressed against the surface of the culture to pick up cells from each of the colonies. The block was next pressed against the surface of a second culture dish, containing sterile medium to which penicillin had been added; care was taken to align the transfer block and the culture dishes according to markers on the block and dishes (black lines). The cells from most of the colonies on the original dish failed to grow on the penicillin medium, but those of a few colonies (two are shown here) did grow. Had the cells of those two colonies spontaneously become penicillin-resistant before being transferred to the penicillin medium, or did exposure to the penicillin induce a mutation for resistance?

Because the transfer block had been aligned with each dish in the same way, according to the markers, it was possible to tell precisely which original colonies had given rise to the two colonies on the penicillin medium. Cells could therefore be taken from those original colonies, which had never been exposed to penicillin, and tested for resistance. They were found to be resistant. Hence the mutation for resistance must have arisen spontaneously; it was not induced by exposure to the drug.

penicillin, the bacteria possessing the genes for resistance are ***preadapted*** to survive the treatment, while most of the nonresistant bacteria will be killed. Because it is the resistant bacteria that reproduce and perpetuate the population, the next generation shows a marked resistance to penicillin. (If such genes were not already present in a bacterial population exposed to penicillin, no cells would survive and the population would be wiped out.)

Evolution of drug resistance in bacteria is not entirely comparable to evolution in sexually reproducing organisms, be-

MAKING CONNECTIONS

1. Contrast the Lamarckian and Darwinian explanations for the long neck of a giraffe.

2. How are the concepts of the *genotype* (described in Chapter 14) and the *gene pool* related?

3. About 100 years ago, the population of the northern elephant seal was reduced by hunting to about 20 individuals. The population has substantially increased in recent years, but it now contains an abnormally small amount of genetic variation compared to other seal populations. How would you explain this phenomenon?

4. Suppose that a population has been mating at random with respect to a particular gene, which is at Hardy-Weinberg equilibrium. If all members of the population now choose to mate with close relatives instead of selecting mates at random, would the frequency of the homozygotes increase, decrease, or remain the same? Relate your answer to the genetic risks of inbreeding, as explained in Chapter 14. (Hint: consider the consequence of the most extreme type of inbreeding, self-fertilization.)

5. An illness that is caused by a virus, such as the flu, generally cannot be cured by treating the patient with antibiotics. Why should a physician be cautious about prescribing antibiotics "just in case" it might help a patient who has been diagnosed with a viral disease?

cause bacteria are haploid and reproduce asexually, and intense selection can change gene frequencies much more rapidly in haploid asexual organisms than in sexually reproducing ones. (Remember, haploid organisms have only one set of genes so all genes can be expressed and thus selected for or against.) The recombination that occurs at every generation in a sexually reproducing species often reestablishes genotypes eliminated in the previous generation; this does not happen in asexual organisms where daughter cells are identical to the parent cell from which they arose. Nevertheless, even very small selection pressures can produce major shifts in gene frequencies in sexually reproducing populations when the time scale is on the order of thousands of years. J. B. S. Haldane showed that if individuals carrying a given dominant allele consistently benefit by as little as 0.001 in their capacity to survive (that is, if 1,000 *A/A* or *A/a* individuals survive to reproduce for every 999 *a/a* individuals that survive to reproduce) then the frequency of the dominant allele could increase from 0.00001 to 1.0 in fewer than 24,000 generations. Now, 24,000 generations may sound like an incredibly large number, but remember that many plants and animals have at least one generation a year. Fruit flies, for example, have more than 30 generations a year. In very few species is the generation time more than 10 years (humans are among the few exceptions). Hence 24,000 generations often mean less than 2,400 years and rarely more than 240,000 years. Recent evidence suggests that many selection pressures in nature are much larger than 0.001; hence major changes in allelic frequencies sometimes take less than a century, perhaps even less than a decade.

Directional Selection Causes a Population to Evolve Along a Particular Line

So far, we have discussed situations in which we have seen only two clearly distinct phenotypes, determined by two alleles of a single gene. But in reality the vast majority of characters on which natural selection acts are influenced by many different genes (i.e., they are polygenic characters), most of which have multiple alleles in the population. Human height and skin color are thought to be examples of polygenic inheritance—people are not short or tall, or black or white; they come in different heights and have varying skin colors. The expression of many such characters, moreover, is influenced considerably by environmental conditions. Polygenic characters generally have frequency distributions that, when graphed, approximate the so-called normal, or bell-shaped, curve (Fig. 31.11).

If there is a shift in the selection pressure on a population, we would expect the phenotypic variation to shift as a result of changing allelic frequencies. An illustration of this is provided by a long-term selection experiment performed on corn by agronomists at the University of Illinois. These investigators selected for high oil content of the corn kernels. At each generation they selected for breeding only those plants with the highest oil content, and continued this selection process for 50 generations. There was a steady increase in oil content throughout most of the period (Fig. 31.12). The kernels of the original stock of corn plants averaged about 5 percent oil; those of the plants in the 50th generation after selection averaged about 15 percent (higher than any individual in the first generation!), and there was no indication that a maximum had been reached.

The intense selection for oil content is an example of what is called ***directional selection,*** the evolution of a population along a particular functional line (Fig. 31.13A). In this case, the selection was artificial, imposed by the agronomists, but directional selection is common in nature when environmental conditions change in a particular direction (e.g., when the climate in a particular area gets colder, warmer, wetter, drier,

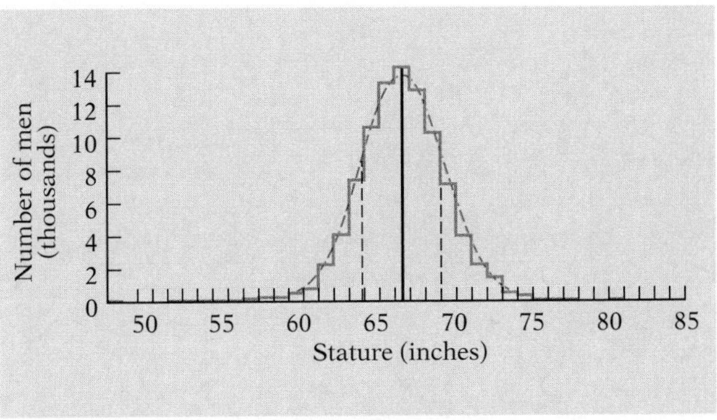

31.11 Frequency distribution of human height This distribution shows the variation in height of young Englishmen called up for military service in 1939. The pattern of variation (shown by vertical bars) approximates, but does not exactly fit, the bell-shaped normal curve.

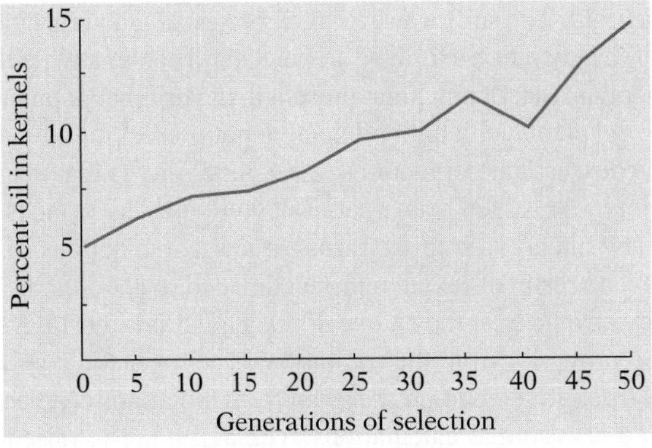

31.12 Evolutionary change of a corn population in response to directional selection for high oil content in corn kernels At each generation the agronomists selected for breeding the plants with the highest oil content. This selection for high oil content exerts directional selection on the plant population, producing a steady increase in oil content throughout most of the period. Notice that the kernels of the original stock of corn plants averaged only about 5 percent oil while those of the plants in the 50th generation averaged about 15 percent.

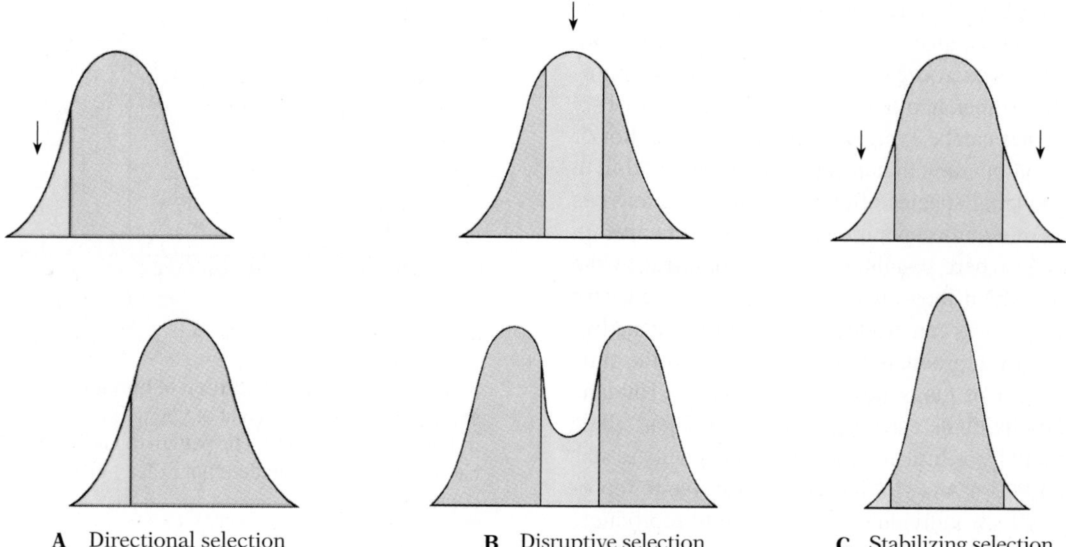

31.13 Directional, stabilizing, and disruptive selection compared Each graph indicates the relative abundance of individuals of various heights in a population. In each instance the original condition of the population is shown above, the later condition, after the specified selection, below. (A) Directional selection acts (arrow) against individuals exhibiting one extreme of a character (here the shortest individuals—represented by the blue area under the upper curve). The eventual result (bottom curve) is that the distribution of heights in the population has shifted to the right, indicating that the population has evolved in the direction of greater height.

(B) By contrast, disruptive selection acts against individuals in the midpart of a distribution, thereby favoring both extremes; in our example both the shortest and the tallest individuals would be favored, but individuals of medium height would be selected against. The result is a tendency for the population to split into two contrasting subpopulations.

(C) Stabilizing selection acts against both extremes, culling individuals that deviate too far from the mean condition and thus decreasing diversity and preventing evolution away from the standard condition.

etc.). The rapid evolution of resistance of *Staphlococcus* populations to penicillin provides another example of directional selection.

In our corn population directional selection caused the extremes as well as the mean (average) of the population to shift to the right. The shift was eventually so great that the plants in the 50th generation averaged 15 percent oil content, which was higher than that of any plant present in the original population. How did plants with high oil content not present initially arise in the descendant populations? One possibility is that the new phenotypes resulted from a series of mutations; a second is that the new phenotypes arose through new combinations of old genes as a result of sexual reproduction and selection. That this steady change over the course of 50 generations must have resulted primarily from the formation of new genetic combinations through selection rather than from mutations can be seen from a few simple calculations. The agronomists raised between 200 and 300 corn plants in each generation. In 50 generations, then, they raised between 10,000 and 15,000 plants. But the usual rate of mutation per gene in corn is never greater than one in 50,000 plants, and it is usually lower. Hence it is unlikely that even one mutation contributing to an increase in oil content occurred in any particular gene affecting this phenotype during the experiment. The gradual increase in oil content during the 50 generations of directional selection must have resulted largely or entirely from the formation of new genetic combinations. Selection, even in the absence of mutations, can produce new phenotypes and play a creative role in evolution.

In sexually reproducing populations, then, selection determines the direction of change by altering the frequencies of genes that arose through gene duplication, transposition, and random mutations many generations before, thereby establishing new gene combinations and gene activities that produce new phenotypes. *Mutations and the other processes that create new alleles and genes are not usually a major directing force in evolution; the principal evolutionary role of genetic changes (mutations) consists of replenishing the variability in the gene pool and thereby providing the potential upon which future selection can act.*

Disruptive Selection Favors Individuals on Both Extremes of a Phenotype Over the Intermediates

Sometimes a polygenic character in a population is subject to two (or more) opposing directional selection pressures. Suppose, for example, that a certain population of birds shows much variation in bill length. Suppose, further, that as conditions change there are increasingly good feeding opportunities for the birds with the shortest bills and also for those with the longest bills, but decreasing opportunities for birds with bills of intermediate length. The effect of such ***disruptive selection*** (also called diversifying selection) will be to divide the population into two distinct types, one with short bills and one with long bills. The combined action of the opposing directional pressures is thus to produce two different variants within the population (Fig. 31.13B).

An actual example of disruptive selection is seen in the picture-wing fruit fly *Rhagoletis pomenella.* This species mates and lays its eggs on hawthorn berries and the larvae feed on the fruit, pupate, and emerge to continue the cycle. The hawthorn was the native host of this fly but about 150 years ago a few *Rhagoletis* began to lay their eggs on apples (probably by mistake). At some stage in their life the flies learn the odor of the plant on which they develop, and use this memory to locate a similar host plant. Those that develop on hawthorn search out mates and deposit eggs on hawthorns, whereas those that grew up on apples are committed to apples. Over time they have evolved adaptations to their different host plants. There is, for instance, a difference in breeding time between the two subgroups. Further, the *Rhagoletis* larvae grow only about half as well on apples, but they suffer far less parasitism, and about six times as many of the eggs laid on apples produce adult flies as compared with hawthorn. The result of this host specificity is that the *Rhagoletis* population through disruptive selection has divided in two, with one subgroup living on apples, the other on hawthorn.

Stabilizing Selection Acts Against the Extremes in a Population

Most often, when a polygenic character is subject to two or more opposing directional selection pressures operating simultaneously, the pressures select against the two extremes of the distribution—plants that are too tall to resist high winds, for example, and those of the same species that are so short that they are shaded by other plants, and so lose essential sunlight. When selection operates against individuals at the two ends of the distribution for a polygenic trait, the process is called ***stabilizing selection*** (Fig. 31.13C).

Stabilizing selection goes on all the time, and plays an extremely important conservative evolutionary role. Each species, in the course of its evolution, comes to have a constellation of genes that interact in very precise ways in governing the developmental, physiological, and biochemical processes on which the continued existence of the species depends. Anything that disrupts the harmonious interaction of its genes is usually harmful to the species. But in a sexually reproducing population, favorable groupings of genes tend to be disrupted and new groupings formed by the recombination that occurs when each generation reproduces. Most of these new groupings will be less adaptive than the original groupings (though some may be more adaptive). And the vast majority of new mutations tend to disrupt rather than enhance the established harmonious relationships among the genes. If unchecked, recombination and random mutation would therefore tend to destroy the favorable gene groupings on which the success of members of the population rests. Stabilizing selection, by constantly acting to eliminate all but the most favorable gene combinations, works against the disrupting, disintegrating tendency of recombination and mutation and is thus the chief factor maintaining stability.

A good example of stabilizing selection at work is provided by the horseshoe crab *Limulus,* the fossils of which suggest that the species has remained essentially unchanged for the last 225 million years. Horseshoe crabs are shore-dwelling, burrowing animals, and although the environment may have changed over geological time, their habitats have persisted relatively unchanged. The horseshoe crab may therefore have experienced few novel selection pressures and the ancient way of life is still possible. Selection has thus acted in a stabilizing fashion, against individuals with new variations, and favoring those with conservative traits. Stabilizing selection is common in stable environments that have persisted for long periods of time.

Effective Selection Pressure Is the Sum of Separate Selection Pressures

Probably many characteristics benefit the organisms that possess them in some ways and harm them in others. The evolutionary fate of such characteristics depends on whether or not the various positive selection pressures produced by their advantageous effects outweigh the negative selection pressures produced by their harmful effects. If the algebraic sum (an addition taking into account plus and minus signs) of all the separate selection pressures is positive, the trait will increase in frequency, but if the algebraic sum is negative, the trait will decrease in frequency.

As an example of the determination of a complex character having both beneficial and harmful effects, consider the selection pressures on the plumage in ducks. Many closely related species of dabbling ducks (mallard, pintail, gadwall, etc.) occur together in most of North America. Hybridization takes place among them because the females sometimes err in their selection of mates.[3] Since the hybrids apparently survive less well than the parental type, there has been strong selection for showy male plumage, distinctive for each species, that helps

[3] In most species of birds, including the ducks, it is the female that chooses the mate. The male displays until some female chooses him. Hence it is more important that the female be able to recognize the males of her own species than that the male be able to recognize the females.

reduce the number of mating errors. No such sexual selection has operated on the females. These brownish, nondescript birds closely resemble one another and are probably much less easily seen by predators than the males, whose bright plumage doubtless makes them more subject to predation—a liability that must cause strong selection pressure against such plumage. But the positive selection pressure resulting from reduced mating errors is apparently greater than the negative selection pressure resulting from predation, and the showy plumage of the males has been maintained (Fig. 31.14). The situation is different, however, on some isolated islands where only one species of dabbling duck exists and where, therefore, no mating errors can occur. Here the negative selection pressure due to predation predominates, and the males have lost their showy plumage and resemble the more protectively colored females.

Just as a trait controlled by several genes, such as plumage in ducks, often has both advantageous and disadvantageous effects, the alleles of a single gene also may have multiple effects (pleiotropy), and it is most unlikely that all of them will be advantageous. Whether an allele increases or decreases in frequency is determined by whether the sum of the various positive selection pressures acting on it is greater or smaller than the sum of the negative selection pressures.

Many instances are known in which the effects of a given allele are more advantageous in the heterozygous than in the homozygous condition. An example in humans is the allele for sickle-cell anemia. In an individual homozygous for the sickle-cell allele, the hemoglobin is abnormal and crystallizes under certain conditions, distorting the red blood cells such that they clog smaller blood vessels. The resulting impaired circulation causes damage to different parts of the body, a severe anemia, and (usually) an early death. Individuals heterozygous for the sickle-cell allele sometimes show mild symptoms of the disease. In some parts of Africa, however, the allele for sickle-cell anemia occurs in human beings much more often than we might expect in view of its highly injurious effect when homozygous (Fig. 31.15). This is because the allele, when heterozygous, confers on the possessor a partial resistance to malaria. The equilibrium frequency of the sickle-cell allele is thus determined by at least four separate selection pressures: (1) the strongly negative selection pressure on the recessive homozygotes, who suffer the full debilitating effects of sickle-cell anemia; (2) the weaker negative selection pressure on the heterozygotes as a result of their mild anemia; (3) the negative selection pressure on the dominant homozygotes, who are more susceptible to malaria; and (4) the fairly strong positive selection pressure on the heterozygotes as a result of their resistance to malaria. In Africa then, the heterozygotes are better adapted than the homozygotes, a condition known as heterozygote superiority, or heterosis.

31.14 A pair of wood ducks The showy coloration of the male contrasts with the drab brown of the female.

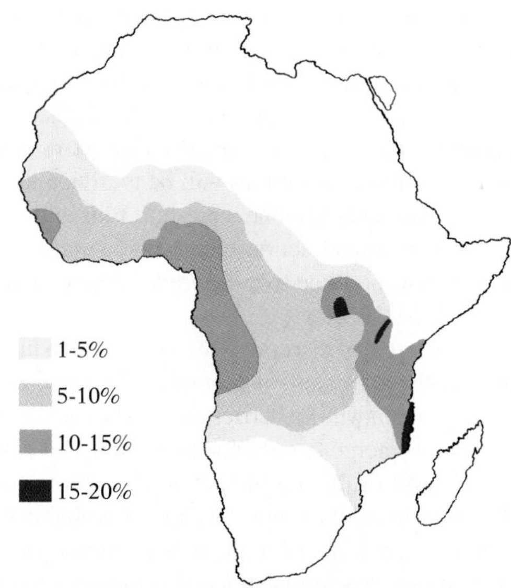

31.15 The distribution of sickle-cell anemia in Africa The various colors indicate the percentage of the population in each area that has sickle-cell anemia. Malaria is also common in regions where the incidence of sickle-cell anemia is high.

Other single-gene disorders may also have the effect of being more advantageous in the heterozygous than in the homozygous condition. Evidence suggests that carriers of the cystic fibrosis allele may be more resistant to the effects of cholera and other diseases that cause severe diarrhea, because they do not suffer the excessive fluid loss that individuals homozygous for the normal allele do. And carriers of the Tay-Sachs allele, which is lethal when homozygous, may possess superior resistance to tuberculosis.

Sexual Selection Operates on Traits Used to Attract and Keep Mates

Darwin distinguished between selection that affects physical survival and selection that operates on traits used exclusively to attract and keep mates. He noted that many male organisms have exaggerated features and behaviors that apparently have nothing to do with increasing the survival of the individuals that have them. Why, for instance, are male cardinals so brightly colored when this makes them conspicuous to predators? Why do male widow birds have such long tails (Fig. 31.16)? The answer, as indicated earlier, is that these features must enhance the individual's chance of mating and producing offspring. In evolutionary terms, the failure to reproduce is just as fatal as early death, but because selection for characters that enhance an individual's chance of mating are often very different for the two sexes, Darwin considered ***sexual selection*** to be nearly independent of natural selection.

Darwin distinguished two basic varieties of sexual selection, male contest and female choice. ***Male contest,*** the more common form, involves contests between members of one sex (usually the males) for access to the other sex. These duels can take the form of dominance fights, which establish a male hierarchy and thus access to reproductively ready females (Fig. 31.17). More often the males fight for the best territories in the habitats females favor; the strongest males obtain the highest quality territories, and therefore the largest number of matings. Male-specific features like large-sized offensive weapons (including the horns of mountain sheep) and defensive structures (like the lion's mane, which protects his neck from bites inflicted by other males) are each the result of sexual selection.

Darwin proposed that other uniquely male features exist only to attract females—the elaborate tail of the peacock, for instance, inhibits the ability to fly and is not used to attack males, but rather is displayed whenever a peahen is nearby. A variety of experiments in which these male features have been altered demonstrate that females of some species do indeed select males on the basis of, for example, the length of their tails (Fig. 31.16). The concept of ***female choice*** has, however, aroused considerable controversy. Skeptics argue that such a system could not evolve if it did not enhance female reproduction, and a male's long feathers are unlikely to help females. But since many male characteristics exaggerate a species-specific recognition sign, the female benefits from the increased certainty of choosing a mate of the right species, one that also carries genes that will enhance the attractiveness (and hence reproductive potential) of her sons. In other cases the female's attention to the male's exaggerated characteristics may serve to enhance her own reproductive success. By mating with "good quality" males, she is more likely to produce strong offspring that themselves will survive and reproduce. The presence of bright colors, long tails, and the like may provide information on the health of the male. For example, in one species of guppy fish, females choose males on the basis of the redness of the male's tail. It turns out that males with bright red tails are free from parasites whereas infected males are dull colored. If resistance to parasites is inheritable, the female should choose the brightest male, so that her offspring will inherit genes for parasite resistance from their father and in turn produce more offspring. The mating systems of many species appear to include the elements of both male-contest and female-choice sexual selection, and it is hard to tell them apart in many cases.

31.17 A dominance ritual in mountain sheep Male mountain sheep work out a dominance hierarchy through a ritualized duel in which a run on hind legs culminates in a loud collision, followed by a head-turning display. Dominant males obtain unchallenged access to females.

31.16 A normal male widow bird in flight Experimental shortening or lengthening of tails demonstrates that females select males with the longest tails.

ADAPTATION

We have seen that organisms become adapted to their local environment through the process of natural selection: those organisms that are best adapted to their environment tend, on average, to leave more offspring than those that are less well adapted. If the adaptations are heritable, then the offspring in turn have a better than average chance to survive and leave offspring. The result of this differential reproduction is that the population will become increasingly better adapted to its particular environment. Throughout the preceding chapters we have discussed a host of different adaptations that enable organisms to survive and reproduce—adaptations that help organisms grow, acquire energy, eliminate wastes, maintain homeostasis, etc., all of which increase the chances of leaving offspring (see Exploring Further: Adaptations of the Desert Ant, p. 690). From an evolutionary perspective, an ***adaptation*** can be defined as any genetically controlled characteristic that increases an organism's fitness. ***Fitness*** refers to an individual's (or allele's or genotype's) probable genetic contribution to succeeding generations. An adaptation, then, is a characteristic that enhances an organism's chances of perpetuating its genes, which usually means the leaving of descendants. Notice that we did not say adaptations increase the organism's chances of surviving, as is sometimes erroneously stated. We should also point out that an adaptation is never 100 percent effective; adaptations increase the probability, not the certainty, of genetic contribution to the next generation.

Adaptations may be structural, physiological, or behavioral. They may be genetically simple or complex. They may involve individual cells or organelles, or whole organs or organ systems. They may be highly specific, of benefit only under very limited circumstances, or they may be general, of benefit under many and varied circumstances.

Populations May Become Adapted Very Rapidly

A population may become adapted to changed environmental conditions with extreme rapidity. A good example is provided by a study begun in the early 1980s by Paul Klerks of the University of Southwestern Louisiana and Jeffrey S. Levinton of the State University of New York at Stony Brook. Klerks and Levinton studied the evolution of cadmium resistance in invertebrates in Foundry Cove on the Hudson River. Industrial plants in the area had dumped more than 100 tons of nickel-cadmium wastes into the river from 1953 until the late 1970s, and the bottom sediments contained high levels of cadmium. In spite of this the bottom-dwelling invertebrates were just as abundant in this area as in nearby unpolluted areas, and Klerks and Levinton were interested in finding out how the animals were able to survive in this polluted environment. They studied the cadmium tolerance of segmented worms belonging to the species *Limnodrilus hoffmeisteri* and found that *Limnodrilus* from Foundry Cove thrived and reproduced in the polluted sediments, while specimens from nearby unpolluted coves did poorly. They then raised specimens from Foundry Cove in unpolluted sediments and studied their progeny for several generations. Tolerance studies showed that the progeny were also able to survive and reproduce when placed in cadmium-polluted sediments, suggesting that the resistance to cadmium was genetically determined. Klerks and Levinton then performed the reverse experiment: they took worms from an unpolluted site and raised them in cadmium-polluted sediments. After several generations they found that the worms had evolved tolerance to cadmium. This evolutionary change was rapid, occurring in two to four generations. This is similar to the rapid evolution of resistance of bacteria to penicillin. The pace of evolution can be very rapid.

We now turn to some other particularly striking examples of adaptation, which will help clarify the processes by which adaptations come into being.

Coevolution Often Occurs Between Flowering Plants and Their Pollinators

Many characteristics of organisms result from the process of ***coevolution,*** a process in which two or more different organisms are evolving together, each in response to the other. There is a reciprocal interaction between the two groups where each organism adapts to the selection pressures imposed by the other. The relationship between the flowering plants and their pollinators presents particularly clear examples of coevolution.

The flowering plants depend on external agents, often insects or birds, to carry pollen from the male parts in the flowers of one plant to the female parts in the flowers of another plant. The flowers of each species are adapted in shape, structure, color, and odor to the particular pollinating agents on which they depend, and they provide an especially clear illustration of the adaptiveness of evolution (Fig. 31.18). Evolving together, the plants and their pollinators become more and more finely tuned to each other's peculiarities in the process termed coevolution.

There is indeed a striking correlation between the pollinators and the species they pollinate (Fig. 31.19). Bee flowers have showy, brightly colored petals that are usually blue or yellow but seldom red (bees can see blue and yellow light well, but they cannot see red at all); they usually have a sweet, aromatic, or minty fragrance to which bees are attracted; they are generally open only during the day (bees are active only then); and they often have a special protruding platform on which the bees can land. Flowers pollinated primarily by hummingbirds are usually red or yellow (hummingbirds can see red well, but blue only poorly), are nearly odorless (the birds have a poor sense of smell), and lack any protruding landing platform, which the birds do not need since they usually hover in front of the flowers while sucking the nectar. Flowers pollinated by moths, in contrast, are usually white, are open during late afternoon and night (times when moths are active), and often have a heavy fragrance that helps attract the moths to them.

The bases of the petals of flowers pollinated by bees, birds, and moths are often fused to form a tube whose length corresponds closely to the length of the tongue or bill of the particular species most important as the pollinator of that plant (Fig. 31.19).

A particularly dramatic example of adaptation for pollination is seen in some species of orchids, where the flowers resemble in both shape and color the females of certain species of flies, wasps, or bees (Fig. 31.20). The male insect is stimulated to attempt to copulate with the flower and becomes covered with pollen in the process. When he later attempts to copulate

A

B

31.18 A variety of plant pollinators (A) A bat (*Leptonycteris*) at a flower, its face liberally dusted with pollen. (B) A butterfly inserting its long proboscis into a flower to obtain nectar. The proboscis is likely to be dusted with pollen, which the butterfly may carry to the next flower it visits.

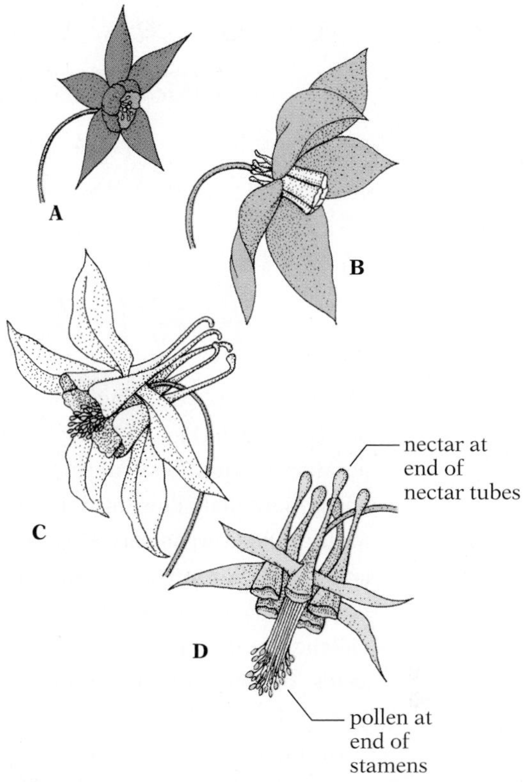

31.19 Characters of columbine flowers correlated with their pollinators (A) *Aquilegia ecalcarata,* pollinated by bees. (B) *A. nivalis,* pollinated by long-tongued bees. (C) *A. vulgaris,* pollinated by long-tongued bumble bees. (D) *A. formosa,* pollinated by hummingbirds. The length and curvature of the nectar tubes of the flowers are correlated with the length and curvature of the bees' tongues and the hummingbirds' bills.

31.20 An orchid flower that resembles a fly The flowers of this species (*Ophrys insectifera*) look enough like female flies to attract some male flies to land on them. The males thus become dusted with pollen, which they may carry to other flowers.

with another flower, some of the pollen from the first flower is deposited on the second. So complete is the deception that sperm have actually been found inside the orchid flowers after a visit by the male insect.

We see, then, that the characteristics of flowers are not simply pleasing curiosities of nature that serve no practical function. Rather, they are important adaptations evolved in response to fundamental selection pressures, namely, to enhance their reproductive success.

EXPLORING FURTHER

ADAPTATIONS OF THE DESERT ANT

The desert ants belonging to the genus *Cataglyphis* inhabit one of the harshest environments on earth (Fig. A). They live on the hot sands of the Sahara Desert, where the temperature reaches 140°F during the middle of the day. The ants have evolved a number of adaptations that allow them to withstand high ground temperatures while foraging about for heat-killed organisms. Their legs, for instance, are a quarter of an inch long, long for an ant, which allows them to hold their body above the hot sands. And they do not merely run, they sprint across the sand, traveling a distance of 100 body lengths per second, a speed unmatched by any other animal. Their feet rarely touch the ground as they dash across the hot sands. To conserve moisture during their sprints, they keep the openings into their tracheal system closed.

During their foraging the desert ants may travel a third of a mile from their burrows across a terrain essentially devoid of landmarks, yet they are able to unerringly find their way home. Superb navigators, they have evolved the ability to detect the polarized light patterns in the sky and use these patterns as a compass to find their way back.

All of these adaptations enable the desert ants to move about and hunt at temperatures that no other organism can withstand—a superb example of adaptation.

Fig. A Saharan silver ant Notice the long legs that allow the ant to hold their body above the hot sands.

31.21 A bombardier beetle (*Brachinus*) spraying its defensive secretion Note how accurately the beetle can aim its spray at the offending object, in this case forceps grasping a rear leg (left) or a front leg (right).

Many Arthropods and Plants Have Evolved Adaptive Noxious Secretions to Repel Predators

Many arthropods possess glands whose secretions act as repellents against predators. In some species the secretions are merely released as a liquid ooze when the animal is disturbed, while in others the secretion is forcibly expelled as a spray that may be aimed very precisely toward the source of the disturbance. The secretions are usually odorous and irritating, particularly if they hit a sensitive part of the predator, such as the mouth, nose, or eyes. For example, the bombardier beetle *Brachinus* sprays a noxious compound from the tip of its abdomen, which it can aim precisely enough to hit a single ant attacking one of its legs (Fig. 31.21). Ants hit by the spray are instantly repelled. The spray is also effective against some vertebrate predators.

Animals are not the only organisms to produce defensive secretions. Plants have evolved a host of toxic chemicals to repel predators. For instance, when the leaves of a tomato plant are eaten by a tomato hornworm, the plant makes chemicals that interfere with the hornworm's digestion (Fig. 31.22).

A Cryptic Appearance Is Often Adaptive

Many animals are cryptically colored—that is, they blend into their surroundings so well as to be nearly undetectable. Frequently their color matches their background almost perfectly (Fig. 31.23). In some cases the animals even have the

31.22 A tomato plant being eaten by a hornworm When a hornworm ingests the leaves, the tomato plant makes chemicals that inhibit digestion. The caterpillar continues eating, but because of impaired digestion, it does not get the nutrients it needs and its growth is retarded.

31.23 Cryptic coloration of aquatic animals The sargassum crab, which lives in dense growths of the brown alga *Sargassum* off Bermuda, is the same color as the alga, and its rounded body resembles the floats of the alga.

31.24 Flounders on two different backgrounds The fish can change color to match the background, whether it is light-colored (top) or dark (bottom).

31.25 Color change by the frog *Hyla versicolor* Individuals of these species are able to change color to match either vegetation or a tree trunk.

ability to alter the condition of their own pigment cells and thus change their appearance to harmonize with their background (Figs. 31.24 and 31.25). Often, rather than match the color of the general background, the animals may resemble inanimate objects commonly found in their habitat, such as leaves (Fig. 31.26), twigs, or sticks (Fig. 31.27). When the shape or color of an animal offers concealment against its background, it is said to have a ***cryptic*** appearance.

Careful studies have shown that cryptic appearance is an adaptive characteristic that helps animals escape predation.

31.26 Leaflike animals The mantis (at center in picture) looks strikingly like the green leaves below it. Even the relationship of its thorax and abdomen reflects the way the leaves often occur in pairs.

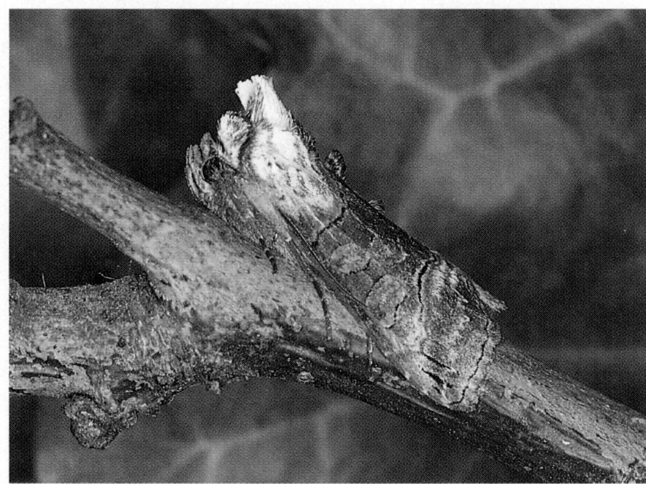

31.27 A moth that looks like a broken twig The perfection of the cryptic appearance of this moth (*Abrostola trigemina*) is a triumph of evolutionary adaptation.

31.28 Cryptic coloration of peppered moths Top: Light and dark forms of *Biston betularia* at rest on a tree trunk in unpolluted countryside. Bottom: Light and dark forms on a soot-covered tree trunk. Here the light form is easier to see.

One of the most extensively studied cases of cryptic coloration is the industrial melanism of moths. This is actually a case of ***polymorphism***—the occurrence in a population of two or more forms of a genetically determined character. Many species of moths exist in two forms, light and dark. Since the mid-1800s the originally predominant light form in these species of moths has given way, in industrial areas, to the dark (melanic) form. In the Manchester area, in England, the first black specimens of the species *Biston betularia* were caught in 1848 and constituted only a small percentage of the population. But by 1895 the dark forms constituted about 98 percent of the total population in the area.

In 1937 E. B. Ford of Oxford University proposed the following explanation for this striking evolutionary change. The various species of moths exhibiting the rapid shift to the dark forms, though unrelated to one another, all habitually rest during the day in an exposed position on the trunks of trees or on rocks, being protected from predation only by their close resemblance to their background. In former years the tree trunks and rocks were rather light-colored and often covered with light-colored lichens. Against this background the light forms of the moths were astonishingly difficult to see, whereas the dark forms were quite conspicuous (Fig. 31.28). Under these conditions it seems likely that predators such as birds captured the dark moths far more easily than the cryptically colored light moths. The light forms were thus more likely to survive and leave offspring, and they would have occurred in much higher frequency than the dark forms. But with the advent of extensive industrialization, tree trunks and rocks were blackened by soot, and the lichens, which are particularly sensitive to such pollution, disappeared. In this altered environment the dark moths resembled the background more closely than the light moths. Thus selection was reversed and favored the dark forms, which consequently appeared in increasing frequency.

With the introduction of pollution controls in recent years, the amount of air pollution in the region has decreased, and consequently the lichens are reappearing. The light form of moth is also now reappearing in larger numbers.

Warning Coloration Often Signals Danger to Predators

Whereas some animals have evolved cryptic coloration, others have evolved colors and patterns that contrast boldly with their background and thus render them clearly visible to potential predators (Fig. 31.29). Many of these animals are in some way disagreeable to predators; they may taste bad, or smell bad, or sting, or secrete poisonous substances. In other words, they are animals that a predator will usually reject after one or two unpleasant encounters. Such animals benefit by being gaudily colored and conspicuous because predators that have experienced their unpleasant features learn to recognize and avoid them more easily in the future. Their flashy appearance is protective because it warns potential predators that they should stay away. The warning is sometimes so effective that after unpleasant experiences with such insects, some vertebrate predators simply avoid all brightly colored insects altogether, whether or not they resemble the ones encountered earlier. G. D. H. Carpenter demonstrated this by offering over 200 different species of insects to an insect-eating monkey. The monkey accepted 83 percent of the cryptically colored insects but only 16 percent of the warning-colored ones, even though many of the insects belonged to species the monkey had probably not previously encountered.

31.29 Warning coloration The bright color of the poison-arrow frog (from which South American Indians obtain poison for their arrow tips) makes it easily recognizable by predators, which carefully avoid it.

Species not naturally protected by some unpleasant character of their own may closely resemble (mimic) in appearance and behavior some warning-colored unpalatable species (Fig. 31.30). Such a resemblance is adaptive; the mimic species suffers less predation because predators cannot distinguish them from their unpleasant models. This phenomenon is called ***Batesian mimicry.***

A second kind of mimicry, called ***Müllerian mimicry,*** involves the evolution of a similar appearance by two or more distasteful species. One striking case of Müllerian mimicry involves the monarch butterfly (*Danaus plexippus*) and the unrelated viceroy (*Limenitis archippus*). These two species look very much alike (Fig. 31.31) and both are distasteful to birds, though they achieve their distastefulness in different ways:

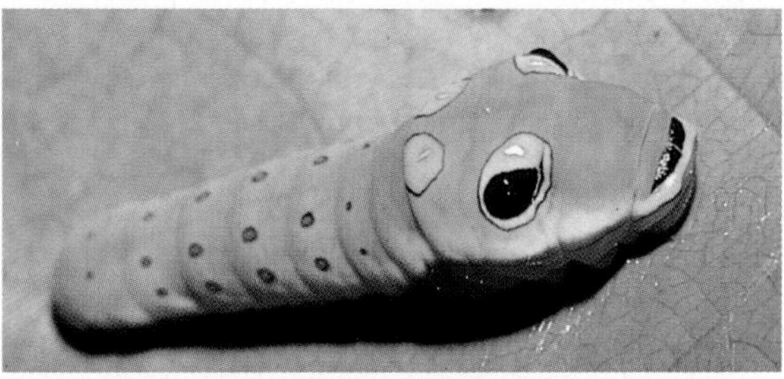

A

B

31.30 Examples of aggressive and Batesian mimicry (A) The prominent imitation eyes and mouth of the spicebush swallowtail larva give it the appearance of a predator, to be avoided. (B) The markings of the harmless syrphid fly resemble those of a stinging bumble bee. (Upon close inspection, flies are readily distinguished from bees because their antennae are very short and they have only a single pair of wings.)

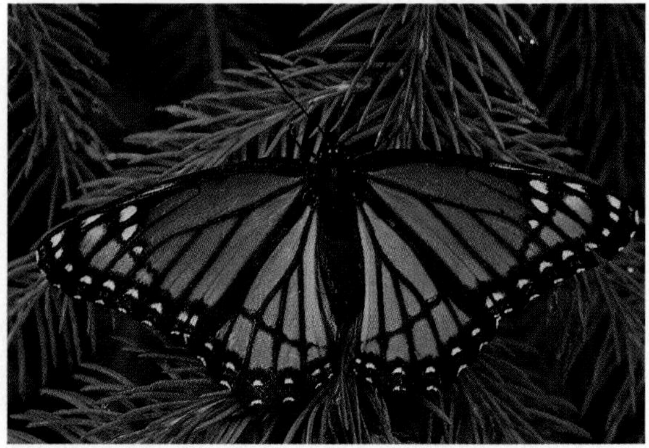

31.31 An example of Mullerian mimicry The monarch butterfly (top) is a distasteful species. The viceroy (bottom) is also distasteful and mimics the monarch; other species in the group to which the viceroy belongs ordinarily have a quite different appearance.

monarchs store poisons from the milkweed plant upon which their caterpillars feed, while viceroys synthesize their own bad-tasting chemical. Birds experiencing one member of either species learn to avoid them both. (Until recently, the viceroy was thought to be a Batesian mimic of the monarch because blue jays, the predators used in early tests, are one of the few species of birds that find viceroys only slightly distasteful.)

The selective advantage of Müllerian mimicry may explain why many unrelated species of bees and wasps have yellow and black stripes on their abdomens, which help them avoid predation (Fig. 31.32). In Mullerian mimicry each species is both model and mimic. Each species has some defensive mechanism, but if each had its own characteristic appearance, the predators would have to learn to avoid each species separately; the learning process would thus be a more demanding one, and some individuals of each prey species would have to be sacrificed to the learning process. If, however, several protected species evolve toward one appearance type, they come to represent a single prey group from the standpoint of the predators, which learn to avoid all the similar appearing groups.

A

B

31.32 Mullerian mimicry in bees and hornets The honeybee (A) and the yellow jacket hornet (B) have evolved similar color patterns—black with yellow stripes. Both species give nasty stings and animals (especially humans) quickly learn to avoid insects of this size that are black with yellow stripes.

Different Species Often Live Together Symbiotically

Symbiosis ("living together") is the intimate association between members of two different species. We shall recognize three categories of symbiosis. The first is ***commensalism,*** a relationship in which one species benefits while the other receives neither benefit nor harm; the second is ***mutualism,*** in which both species benefit; and the third is ***parasitism,*** in which one species benefits and individuals of the other species are harmed (Table 31.1).

In commensalistic symbiotic relationships, one species benefits from the association while the other apparently neither benefits nor is harmed. The advantage derived by the commensal species from its association with the host often involves shelter, support, transport, or food, or several of these. For example, in tropical forests numerous small plants, called epiphytes, usually grow on the branches of the larger trees or in forks of their trunks. These commensals, among which are many species of orchids, are not parasites. They use the host trees only as a base of attachment and do not obtain nourishment from them. They apparently do no harm to the host except when so many of them are on one tree that they stunt its growth or cause limbs to break. A similar type of commensalism is the use of trees as nesting places by birds (Fig. 31.33).

A particularly dramatic example of commensalism is the relationship between certain species of fish and sea anemones.

TABLE 31.1 *Varieties of symbiotic relationships*

Relationship	*Species A*	*Species B*
Commensalism	+	0
Mutualism	+	+
Parasitism	+	−

+ = benefit − = harm 0 = no effect

31.33 A commensalistic relationship between a ruby-throated hummingbird and the tree in which it nests The hummingbird uses the tree only as a base of attachment for its nest and does no harm to the host.

These fish obtain protection and shelter by living in association with anemones, and sometimes steal some of their food (Fig. 31.34). The fish swim freely among the tentacles of the anemones, even though those tentacles quickly paralyze other fish that touch them. The anemones regularly feed on fish; yet the particular species that live as commensals with them sometimes actually enter the gastrovascular cavity of their host, emerging later with no apparent ill effects. The physiological and behavioral adaptations that make such a commensal relationship possible must be quite extensive.

Mutualism, which describes symbiotic relationships beneficial to both species, is common. Figure 31.35 illustrates two instances of the widespread phenomenon called cleaning symbiosis, which is clearly mutualistic. Other examples of mutualism include the relationship between a termite or cow and the cellulose-digesting microorganisms in its digestive tract, or between a human being and the bacteria in the intestine that synthesize vitamin K. The plants we call lichens are actually formed of an alga and a fungus united in such close mutualistic

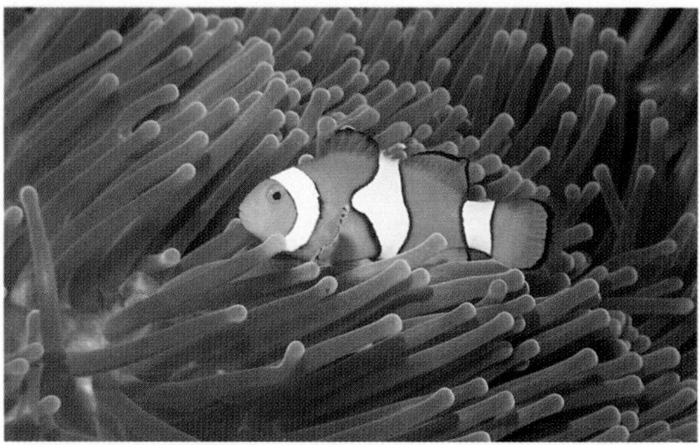

31.34 An anemone fish living among the tentacles of a sea anemone from the Palau Islands

31.35 Two examples of cleaning symbiosis Top: A grouper being cleaned by cleaner fish. The cleaner actually enters the mouth of the predatory grouper and removes parasites. Bottom: Yellow-billed oxpeckers search for parasitic insects on a warthog. Oxpeckers spend almost all their time gleaning blood-sucking flies and ticks from zebras, wart hogs, Cape buffalo, impalas, and giraffes living in eastern Africa. The oxpeckers are completely dependent on the mammals for their food, but the mammals also benefit. It has been estimated that a single tick could remove enough blood from a young mammal to decrease its growth rate by a pound a year. The mammals' behavior in the presence of the oxpeckers suggest that they alter their activity to assist in the cleaning process. Zebras, for instance, stop their feeding and stand still and erect, even raising their tails so the anal region can be cleaned. The cleaning activity of both the cleaner fish and the oxpeckers is mutualistic; the cleaner obtains food, and the host gets rid of parasites that could endanger its health.

symbiosis that they give the appearance of being one plant (see Fig. 40.8 A, B, p. 992). Apparently the fungus benefits from the photosynthetic activity of the alga, and the alga benefits from the water-retaining properties of the fungal walls.

It really is not very important which label—commensalism, mutualism, or parasitism—we apply to most of these cases. The categories are only devices to help us organize what we know about nature, and to form testable hypotheses. What is important to keep in mind is how commensalism, mutualism, and parasitism grade into each other, and to recognize that each case of symbiosis is unique.

In Parasitism, One Species Is Harmed While the Other Benefits

Just as there are no sharp boundaries between the categories of symbiosis, so there is no strict distinction between parasitism and predation. Mosquitoes and lice both suck the blood of mammals, yet we usually call only the latter parasites. Foxes and tapeworms may both attack rabbits, but foxes are called predators and tapeworms are called parasites. The common definitions suggest that a ***predator*** eats its prey quickly and then goes on its way, while a ***parasite*** passes much of its life on or in the body of a living host, from which it derives food in a manner harmful to the host (Fig. 31.36). Parasites generally do not kill their hosts; those that eventually do so are called ***parasitoids.***

Parasites are customarily divided into two types: ***external parasites*** and ***internal parasites.*** The former live on the outer surface of their host, usually either feeding on the hair, feathers, scales, or skin of the host or sucking its blood. Internal parasites usually live in the various tubes and ducts of the host's body, or they may bore into and live embedded in tissues such as muscle, or, in the case of viruses and some bacteria and protozoans, they may actually live inside the individual cells of their host.

Internal parasitism is usually marked by much more extreme specializations than external parasitism. The habitats available inside the body of another living organism are completely unlike those outside, and the unusual problems they pose have resulted in evolutionary adaptations entirely different from those seen in free-living forms. For example, internal parasites have often lost organs or whole organ systems that would be essential in a free-living species. Tapeworms, for instance, have no digestive system. They live in their host's intestine, where they are bathed by the products of the host's digestion, which they can absorb directly across their body wall without having to carry out any digestion themselves.

Because of their frequent evolutionary loss of structures, certain parasites are often said to be ***degenerate.*** "Degenerate" implies no value judgment, but simply refers to the absence, common in internal parasites, of many structures present in their free-living ancestors. From an evolutionary point of view, loss of structures that are useless in a new environment is an instance of adaptation. Such loss is just as much an evolutionary advance, a specialization, as the development of increased complexity in some other environment. ***Specialization,*** then, does not necessarily mean increased structural complexity; it only means the evolution of characteristics particularly suited to some special situation or way of life.

31.36 A caterpillar (tomato hornworm) with numerous pupae of a parasitoid wasp attached to its body The wasp laid her eggs inside the body of the host caterpillar, and the larvae fed on it until they emerged and pupated.

Internal Parasites Often Have Complex Life Histories

Perhaps the most striking of all adaptations of internal parasites involve their life histories and reproduction. Individual hosts do not live forever, and rarely can a parasite move directly from one host to another of the same species. Some mechanism is needed for changing hosts; often this takes the form of an elaborate life cycle. For example, consider the life cycle of the beef tapeworm, an organism that is a problem in areas of poor sanitation.

The eggs of a beef tapeworm living in a human being's intestine are shed in the host's feces (Fig. 31.37); a cow eats plants contaminated with human feces (in areas where human feces are used for fertilizer or fields are irrigated with contaminated water); the tapeworm eggs hatch in the cow's intestine; the larvae burrow through the intestinal wall into a blood vessel and are carried by the blood to a muscle where they encyst (become surrounded by a case and lie inactive). If a person then

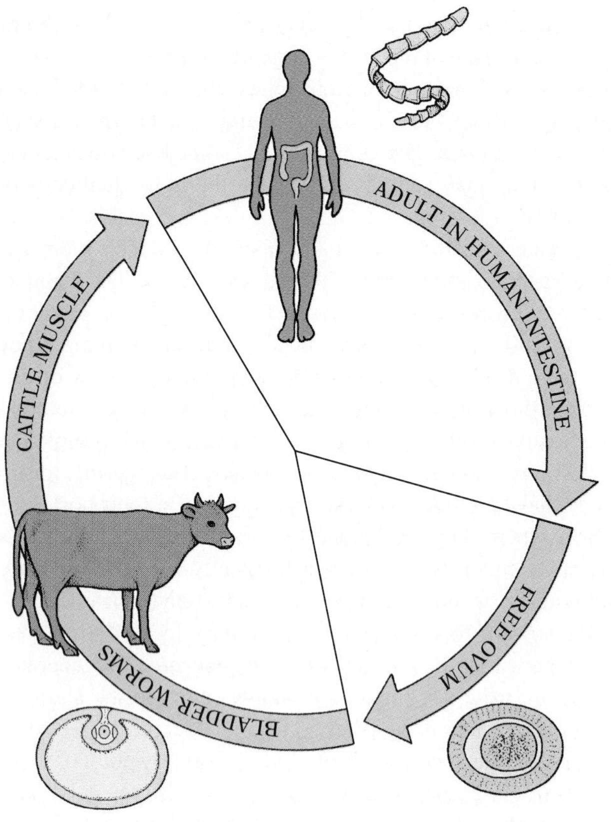

31.37 Life cycle of the beef tapeworm Humans infected with beef tapeworms pass ova in their feces. Cattle, which act as intermediate hosts, swallow the eggs present on contaminated plants, and the ova hatch in the duodenum and burrow through the intestinal wall. They then enter the blood stream or lymphatics to reach the skeletal muscles. There the embryos develop into the larval form, the bladder worms. Humans become infected by eating poorly cooked beef that contains bladder worms. Once the bladder worms reach the intestine, they develop into adult worms in about two months and ova are released, completing the cycle.

eats the raw or insufficiently cooked beef, the cyst is digested away in the person's intestine and the tapeworm head attaches itself to the intestinal wall and a mature worm develops and begins producing eggs. Like most internal parasites, the beef tapeworm produces huge numbers of eggs. The beef tapeworm thus passes through two hosts during its life cycle: an intermediate host (cow), in which it undergoes its early development, and a final host (human being), in which it matures and reproduces. As life cycles of internal parasites go, the one just described is rather simple. It is not unusual for a life cycle to include two or three intermediate hosts and/or a free-living larval stage.

In the course of their evolution, parasites usually develop special features of behavior and physiology that make them better adjusted to the particular characteristics of their host. A particularly nice example of the complexity of host-parasite interactions is shown in Figure 31.38. However, as the parasites evolve, the host species is also evolving, and there is strong selection pressure for its evolving more effective defenses against the parasites. This constant interplay between host and parasite is at the heart of the red-queen hypothesis. The ***red-queen model*** takes its inspiration from the scene in *Through the Looking Glass* in which Alice and the Red Queen are running as quickly as they can and yet make no progress: however fast they move, they cannot outrun their surroundings (Fig. 31.39). The red-queen hypothesis suggests that the need for variation arises from the need to keep up with a rapidly evolving "background" of predators, prey, and parasites, as well as changing conditions. Those individuals of the host species with superior defenses will be better able to survive and reproduce. Correspondingly, those individuals of the parasite species with the best ways of counteracting those defenses will be most likely to prosper. In turn, their counteractions will lead to pressure on the host to evolve still better defenses, against which the parasite may then evolve new means of surviving, and so forth. Though this sort of coevolution will continue as long as the host-parasite relationship exists, a dynamic balance

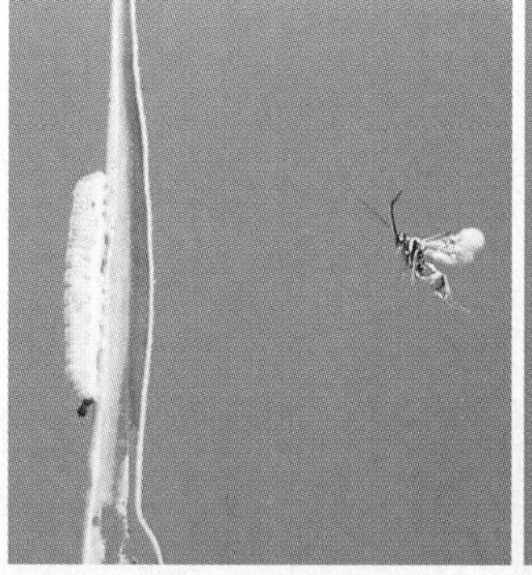
A

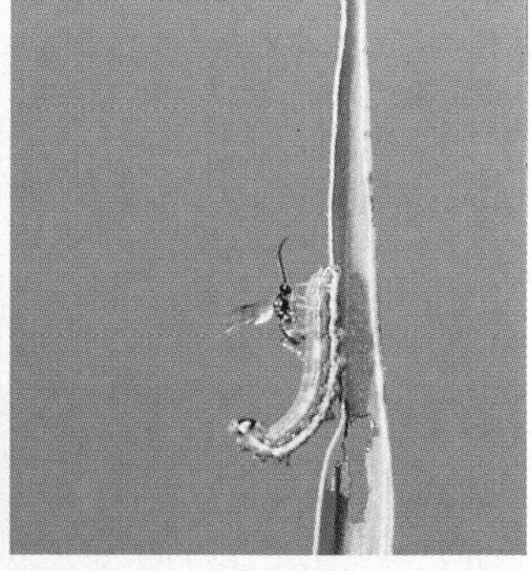
B

31.38 An example of a complex host-parasite interaction (A) Beet armyworm (*Spodoptera exigua*) caterpillar feeding on a corn seedling. While it is feeding the caterpillar releases an oral secretion that interacts with the damaged corn tissue, which in turn induces the plant to emit a volatile chemical into the air. (B) The chemical attracts a parasitoid wasp (*Cotesia marginiventris*) which follows the chemical trail to the seedling and locates the caterpillar. The caterpillar is the natural host for the wasp's eggs, and the female wasp deposits her eggs inside the caterpillar, where the larvae hatch and feed on the caterpillar, eventually killing it. That the release of chemical attractant requires both the caterpillar's oral secretion and a substance from the plant is demonstrated by placing the oral secretion on undamaged corn seedlings, with the result that there is no subsequent release of the volatile chemical attractant. Applying the secretion to damaged tissue on the seedling, however, results in a release of the chemical, and the attraction of the wasps to the seedling.

31.39 Alice and the Red Queen, who were unable to outrun their environment, no matter how hard they tried

is usually reached eventually, the host surviving without being seriously damaged and the parasite prospering moderately well too. Relationships that result in serious disease in the host are usually relatively new, or they are relationships in which a new and more virulent form of the parasite has recently arisen, or in which the host showing the serious disease symptoms is not the main host of the parasite. American Indians, for example, suffered severely when exposed to diseases first brought to North America by European colonists, even though some of those same diseases caused only mild symptoms in the Europeans, who had been exposed to the disease-causing organism for many centuries and in whom the host-parasite relationship had nearly reached a balance. Many examples are known in which humans are only occasional hosts for a particular parasite and suffer severe disease symptoms, though the wild animal that is the major host shows few ill effects from its relationship with the same parasite.

It is worth keeping in mind, however, that relatively benign interactions between the host and a parasite may never evolve. The reason is that the optimal strategy for a parasite depends on the life histories of itself and its host, and for some combinations there is no advantage to the parasite in achieving a balance, and no way for the host to impose one. In some diseases (for example, rabies, which if not treated in time is always fatal to humans and many other mammals), the parasite apparently benefits from a massive attack on the host, which enables it to spread its offspring rapidly. A slower, longer-lasting release of progeny produces lower reproductive success. Much recent work has shown that an evolutionary perspective on host-parasite dynamics can lead to optimal designs for disease treatment and prevention, whereas many intuitively attractive alternatives that fail to consider these interactions turn out to be futile.

MAKING CONNECTIONS

1. Researchers are racing to collect seeds of many endangered genetic strains of our crop plants, and of their wild relatives, found in various regions of the world. These are placed in "seed banks" for use in later breeding experiments. Why is it important to preserve the genetic variability of these plants?

2. How does the allele for sickle-cell anemia demonstrate *heterozygote superiority?*

3. Suppose you have a flower garden and want to attract hummingbirds but discourage visits by bees. What color flowers should you plant?

4. What mutualistic relationships between plants and microorganisms are important for plant nutrition? (See Chapter 16.)

5. What kind of symbiosis is illustrated by the relationship between human beings and domesticated plants and animals, such as wheat and cattle?

6. To control the population of the introduced European rabbit, the Australian government introduced a viral parasitic disease, myxomatosis, in 1950. The disease, to which this species of rabbit had no previous exposure, killed 99.5 percent of the population within one year of its introduction. What would you expect to happen to the mortality rate over time? Why?

Population Genetics Problems

The best way to gain an understanding of population genetics is to do problems. The fundamental principles of population genetics discussed in this chapter will become clearer to you, and you will grasp them more surely, if you work the following problems. Additional problems, and a discussion of how to do problems, will be found in the Study Guide accompanying this book. Answers are provided below.

1. Ten percent of the genes for coat color in a rabbit population are albino (*b*), while 90 percent code for a black coat (*B*). What percentage of the rabbits are heterozygous if the Hardy-Weinberg assumptions hold true?
2. In corn, yellow kernel color is governed by a dominant allele and white by a recessive allele of the same gene. A random sample of 1,000 kernels from a population in Hardy-Weinberg equilibrium reveals that 910 are yellow and 90 are white. What are the frequencies of the yellow and white alleles in the population? What is the percentage of heterozygotes in the population?
3. Suppose 20 percent of a population is homozygous recessive (*b/b*) for a given trait. A sudden catastrophe exerts a selection pressure such that this frequency is reduced to 16 percent in a single generation. After a Hardy-Weinberg equilibrium is established, what will be the genotypic frequencies for all genotypes (*B/B, B/b,* and *b/b*) in the succeeding generations?
4. Suppose *A* is the allele for unattached ear lobes and *a* the allele for attached ear lobes. *A* is dominant to *a*. Suppose 9 percent of a human population had attached ear lobes. What percent of the population would be homozygous for unattached ear lobes assuming a Hardy-Weinberg equilibrium?
5. In the United States, approximately one person in 225 is susceptible to hemochromatosis, a metabolic disorder inherited as a simple autosomal recessive. In this disorder, the body absorbs more iron than it needs and the excess iron is deposited in different tissues, causing a variety of symptoms. What is the frequency of the hemochromatosis allele in the U.S. population, assuming a Hardy-Weinberg equilibrium? How many persons in a town of 10,000 might be expected to show symptoms of the disease? How many would be carriers?
6. Among Caucasians, hair straightness or curliness is thought to be governed by a single pair of alleles showing intermediate inheritance. Individuals with straight hair are homozygous for the I^S allele while those with curly hair are homozygous for the I^C allele. Individuals with wavy hair are heterozygous. In a population of 1,000 individuals, 245 were found to have straight hair, 393 had curly hair, and 362 had wavy hair. Give the allelic frequencies of the I^S and I^C alleles. Is this population in Hardy-Weinberg equilibrium? Explain your answer.

(Answers: 1. 18 percent; 2. Yellow = 0.7, white = 0.3, heterozygotes = 42 percent; 3. *B/B* = 0.36, *B/b* = 0.48, and *b/b* = 0.16; 4. 49 percent; 5. Hemochromatosis allele = 0.067, normal allele = 0.933, 45, 1,250 carriers; 6. I^S = 0.43, I^C = 0.57. Population is not in equilibrium (it does not meet $p^2 + 2pq + q^2 = 1$.)

CONCEPTS IN BRIEF

1. Evolution is the change in the genetic makeup of a population in successive generations. Fundamental to the modern theory of evolution by natural selection are five concepts: (1) more offspring are born than survive to reproduce; (2) there is always variation among members of the same species; (3) many differences are heritable; (4) some differences affect how well adapted an organism is; and (5) some differences in adaptedness are reflected in the number of offspring successfully reared.

2. New genes may arise by duplication of preexisting genes, followed by small-scale mutations in the duplicate genes that give rise to functionally different products. Mutations also produce new alleles of existing genes. The processes of meiosis, crossing over, and recombination at fertilization provide almost endless variation in the population. Together, these processes provide the genetic variability on which natural selection can act to produce evolutionary change.

3. Natural selection can act on genetic variation only when it is expressed phenotypically. Not all genetic variation, however, provides material for evolutionary change; mutations in somatic cells do not lead to evolution.

4. Population genetics is based on the concept of the *gene pool,* the sum total of all the genes possessed by all the individuals in the population. Evolution is a change in the allelic frequencies within gene pools.

5. Evolutionary change is not automatic; it occurs only when something disturbs the genetic equilibrium. The Hardy-Weinberg Law describes the conditions under which there would be no evolution—infinitely large population, no mutations, no migration, and random mating and reproductive success. Since these conditions are not met in nature, it follows that evolution is always occurring. Evolutionary change is a fundamental characteristic of all populations.

6. All populations are subject to *selection pressure,* which disturbs the genetic equilibrium. Even very slight selection pressures can lead to major changes in allelic frequencies over time. Environmental conditions can give rise to *directional, stabilizing,* or *disruptive selection.*

7. Natural selection, even in the absence of new mutation, can produce new phenotypes by combining old genes in new ways. Mutation is not usually a major directing force in evolution; its role is to provide new variations upon which future selection can act.

8. A single allele may influence multiple characteristics. The evolutionary fate of such an allele depends on whether the sum of all the various positive selection pressures acting on it is greater or less than the sum of the negative selection pressures. Sometimes the effects of a given allele are more advantageous in the heterozygous than in the homozygous condition (*heterozygote superiority*).

9. Adaptations are genetically controlled characteristics that enhance an organism's fitness—its chance of perpetuating its genes in succeeding generations. Adaptations can be structural, physiologial, or behavioral; genetically simple or complex; highly specific or general.

10. Flowers are adapted in shape, structure, color, and odor to their particular pollinating agent. The plants and their pollinators have evolved together; such evolutionary interaction is called *coevolution.*

11. Many animals have evolved characters, such as *cryptic appearance, warning coloration,* and *mimicry,* that help them escape predation.

12. Many organisms have evolved adaptations for *symbiosis*—for living together. There are three types of symbiosis: *commensalism* (where one species benefits while the other neither benefits nor is harmed), *mutualism* (where both species benefit), and *parasitism* (where one species benefits while the other is harmed). Over time the host species and the parasite may undergo coevolution, eventually reaching a dynamic balance in which both can survive without serious damage.

STUDY QUESTIONS

1. **What is the distinction between *evolution* and *natural selection?* Describe the five concepts on which natural selection rests. Can an individual evolve? (p. 669)**
2. **What processes contribute to genetic *recombination?* As sources of genetic variability in a population, how do recombination mechanisms differ from *mutation?* Which is the primary mechanism providing genotypic variation in a population? (p. 670)**
3. **What evidence would you cite to persuade a colleague that duplication of preexisting genes has been an important process in eucaryotic gene evolution? How can such duplications occur? Once a duplication has occurred, how do the duplicate genes become different? (pp. 670–672)**
4. **Explain why these categories of variation are not raw material for evolutionary change: (a) exclusively phenotypic variation; (b) somatic mutation. (p. 674)**
5. **How is evolution defined with reference to gene pools? According to the Hardy-Weinberg Law, under what conditions will evolution *not* occur? Are all of these conditions ever met? (pp. 675–676)**
6. **Describe these evolutionary mechanisms: *genetic drift, mutation pressure, gene flow,* and *selection pressure.* Which of these is the most important cause of evolutionary change under most conditions? (pp. 676–679)**
7. **Distinguish the concepts of *directional, stabilizing,* and *disruptive selection* and sketch graphs to show their effects on a population. Give an example of each type of selection. (pp. 683–685)**

8. What are the principal roles of mutation and natural selection in evolution? How does natural selection play a creative role in evolution? (pp. 685–686)

9. What determines whether natural selection works for or against a characteristic? Describe an example of how the environment influences the adaptive value of a trait. (pp. 687–688)

10. What distinction did Darwin make between *sexual selection* and other forms of natural selection? Contrast and give examples of the *male contest* and *female choice* forms of sexual selection. Can you think of possible examples of each form in human society? (pp. 686–687)

11. What is wrong with this definition: *An adaptation is any characteristic that increases an organism's chances of surviving?* (pp. 687–688)

12. Compare the relationship of flowering plants and their pollinators with that of hosts and their parasites as examples of *coevolution.* (pp. 688–689)

13. Why is a cryptic appearance adaptive for some animals whereas conspicuous coloration is advantageous for others? Are Batesian and Mullerian mimics typically cryptic or conspicuous in their appearance? Why? (pp. 691–693)

CHAPTER 32

EVOLUTION: SPECIATION AND PHYLOGENY

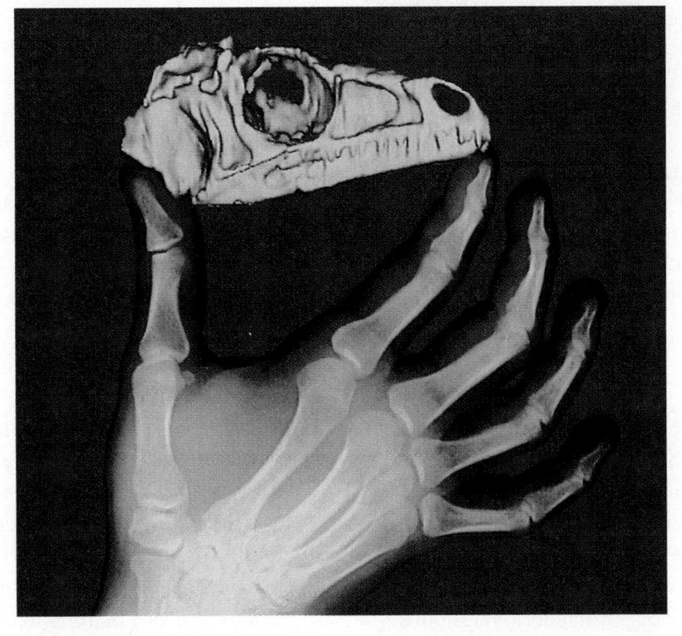

One of the most striking aspects of life is its extreme diversity. A bewildering array of organisms occupies the earth. E. O. Wilson, of Harvard University, estimates that there are 1.4 million species known to currently exist, but that the actual number could be closer to 10 million or perhaps as high as 100 million. According to Wilson, if each of the world's extant species were described on one page of a 1,000-page book, then the "Great Encyclopedia of Life" would comprise enough volumes to fill six kilometers of shelving, equal to the space within a medium-sized public library! Furthermore, the fossil record shows that the species now living represent only a tiny fraction (probably less than 0.1 percent) of those that have ever lived. How did this enormous diversity come about? We have so far discussed only one major aspect of evolution—microevolution—the gradual change of a given population through time. Now we turn to another of its major aspects, the process by which a single population may split, giving rise to two or more different populations. But before we can discuss this topic meaningfully, we must explore more fully what we mean by populations. We can define a ***population*** as a group of individuals of the same species living in a particular geographic area at a particular time that share a common gene pool (i.e., that tend to interbreed).

UNITS OF POPULATION

Members of Demes Resemble Each Other

A ***deme*** is a small local population, such as all the deer mice or all the red oaks in a certain woodlot or all the perch in a given pond. Though no two individuals in a deme are exactly alike, the members of a deme are usually more similar to one another than they are to members of other demes (Fig. 32.1), for at least two reasons: (1) they are more closely related genetically, because matings occur more frequently between members of the same deme than between members of different demes; and (2) they are exposed to more similar environmental influences and therefore to more similar selection pressures.

Note that demes are not clear-cut permanent units of a population. Though deer mice in one woodlot are more likely to mate among themselves than with deer mice in a nearby wood-

Chapter opening photo: Composite image of a hand holding a tiny *Eoraptor* skull. *Eoraptor* was a carnivore and is the earliest known dinosaur. This specimen was found in Argentina and is 228 million years old.

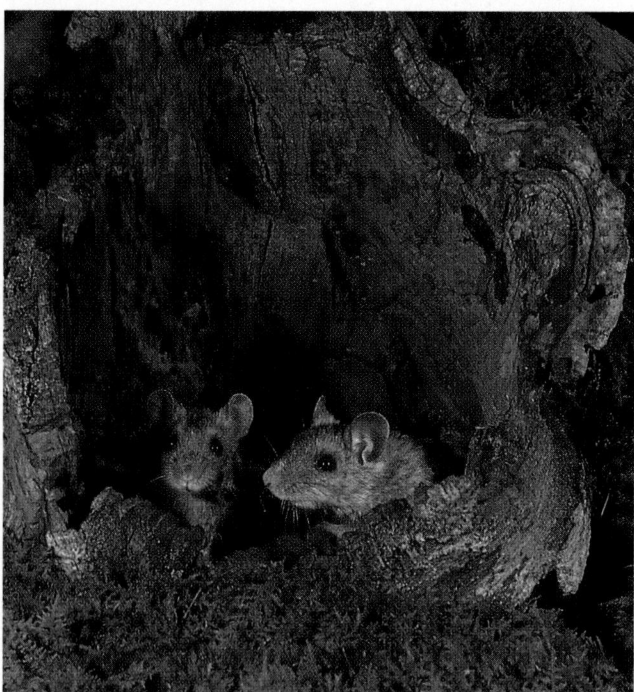

32.1 Deer mice Deer mice in the same deme tend to resemble one another both because they are more closely related genetically and because they are exposed to similar environmental influences and similar selection pressures.

lot, there will almost certainly be occasional matings between mice from different woodlots. Similarly, though the female parts of a particular red oak tree are more likely to receive pollen from another red oak tree in the same woodlot, there is an appreciable chance that they will sometimes receive pollen from a tree in another nearby woodlot. In fact, the woodlots themselves are neither clear cut nor permanent. Neighboring woodlots may fuse after a few years, or a single large woodlot may become divided into two or more separate smaller ones. Such changes in a woodlot will produce corresponding changes in the demes of deer mice and red oak trees that inhabit it. Demes, then, are usually temporary units of population that merge with other similar units.

No matter what the boundary that separates "similar" demes, interbreeding between them is always possible, and we always expect some interbreeding between deer mice from adjacent demes to occur. However, we do not expect interbreeding between deer mice and house mice or between deer mice and black rats or between deer mice and gray squirrels. Nor do we expect to find crosses between red oaks and sugar maples or even between red oaks and pin oaks, even if they occur together in the same woodlot. In short, we recognize the existence of units of population larger than demes and both more distinct from each other and longer-lasting than demes. One such unit of population is that containing all the demes of deer mice. Another is that containing all the demes of red oaks. These larger units are known as ***species.***

Members of One Species Cannot Successfully Interbreed With Members of a Different Species

For centuries it has been recognized that plants and animals seem to divide naturally into many separate and distinct "kinds," or species. This does not mean that all the individuals of any one species are precisely alike—far from it. Any two individuals are probably distinguishable from each other in a variety of ways. But it does mean that all the members of a single species share certain biologically important attributes and that, as a group, they are genetically separated from other such groups.

But though the existence of discrete clusters of living things that can be called species has long been recognized, the concept of what constitutes a species has changed often over time and even today remains controversial. Originally, a species was thought of as *a group of organisms that more closely resemble each other, with respect to their physical appearance (morphology), physiology, behavior, and reproductive patterns, than they resemble any other organisms.* This concept, based on close resemblance, is still the basis used by most systematists for classifying organisms into particular species. But this definition is incomplete; it takes into account shared biological attributes but fails to account for the genetic separation between groups, i.e., that the groups do not interbreed in nature. The so-called ***biological species concept*** takes reproduction into account, defining a species as a *genetically distinctive group of populations whose members are able to interbreed freely under natural conditions and are reproductively isolated from all members of other such groups.* Notice that the biological species concept says nothing about how different from each other two populations must be to qualify as separate species. The final criterion is always *reproduction*—whether or not there is actual or potential gene flow between the populations. If no interbreeding exists between two outwardly almost identical populations (i.e., if there is no gene flow between them), then those populations belong to different species despite their great similarity (Fig. 32.2A). On the other hand, if two populations show striking differences, but there is interbreeding (effective gene flow) between them, then those populations belong to the same species (Fig. 32.2B). Anatomical, physiological, or behavioral characters simply serve as clues toward the identification of reproductively isolated populations; they are not in themselves regarded as determining whether a population constitutes a species.

The biological species concept is not without its problems. Because the definition assumes interbreeding, it obviously does not apply to asexual organisms (e.g., bacteria), yet asexual organisms do seem to form recognizable groups or kinds. This definition is also of little use in determining fossil species, because the criterion of interbreeding cannot be used when comparing a modern organism with its ancestor of a million years ago. Finally, the definition is difficult to apply when two populations are closely related but completely separated geo-

A

B

32.2 Biological species concept (A) Sometimes members of two different biological species look very similar. The alder flycatcher (left) and willow flycatcher (right) are so similar in appearance that even experienced ornithologists cannot always tell them apart. Despite their great similarity in appearance, they are different species and do not interbreed. (B) Sometimes members of the same species look very different. The Clydesdale horse is 17 hands high, yet is the same species as the miniature horse that stands only 32 inches.

A

B

32.3 North-south clines in size of bodies and extremities Many species of mammals and birds that live year round in warmer climates, such as the black-tailed jackrabbit in California (A), have larger extremities such as ears and tails than do species that live in colder climates, such as the showshoe hare (B) that lives in northern North America. Many mammalian species also show north-south clines in average body size, being larger in the colder climates farther north and smaller in the warmer climates farther south. The snowshoe hare, for instance, is a large hare that reaches 18–19 inches in length and weighs between three and four pounds. Larger body size and smaller extremities are adaptive in cooler climates because they produce a lower surface-to-volume ratio which is important in conserving body heat. Although these hares belong to different species, similar variations are seen within members of the same species that live in different climates.

graphically. Because they are separated, they are obviously not exchanging genes. But *could* they interbreed if both were living in the same area? Are they truly "reproductively isolated" or are they simply "geographically isolated"? These questions and those posed by fossil species and asexual organisms make it difficult to apply the biological species concept in all situations. For this reason, other species concepts have been formulated and many biologists think that an alternative should be adopted, though there is at yet no universal agreement as to what that concept should be.

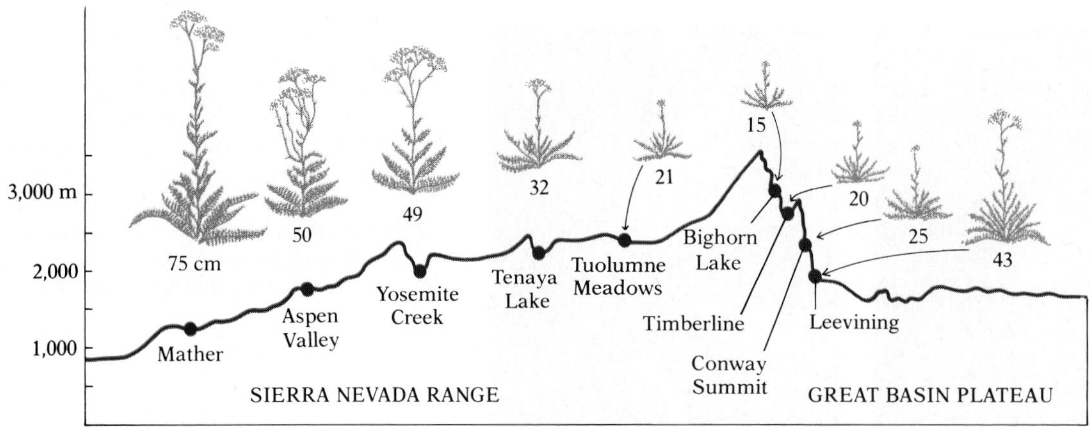

32.4 Altitudinal cline in height of the herb *Achillea lanulosa* The higher the altitude, the shorter the plants. The environments in the different areas vary considerably. Plants grown at the higher altitude are subjected to a shorter growing season and temperatures that drop below freezing, so winter dormancy is necessary and the plants are much smaller. With such different selection pressures, it is not surprising that different traits have been selected for, and that the plants have evolved genetic differences over time. The variation was shown to be genetic (as opposed to merely environmental) by collecting seeds from the locations indicated and planting them in a test garden at Stanford University where all were exposed to the same environmental conditions. The differences in height were still evident in the plants grown from these seeds.

Intraspecific Variation Is Often Correlated With Geography

There is usually so much gene flow between adjacent demes of the same species that differences between them are slight. For example, the ratio of allele *A* to allele *a* may be 0.90 to 0.10 in one deme and 0.89 to 0.11 in the adjacent deme. But the farther apart geographically two demes are, the smaller the chance of direct exchange of genetic material between them, and the greater the probability that genetic differences will be more marked. If, for example, we collect 500 deer mice each from Plymouth County, Massachusetts, Crawford County, Pennsylvania, and Roanoke County, Virginia, we will find numerous differences that enable us to distinguish between the three populations quite readily—much more readily than we could between populations from three adjacent counties in Massachusetts or Pennsylvania. Some of the genetic differences between demes may reflect chance events such as genetic drift or the occurrence in one deme of a mutation that would be favorable in all demes, but has not yet spread to them. But much of the genetic variation probably reflects differences in the selection pressures operating on the populations due to varying environmental conditions in their respective ranges. Geographic variation is therefore adaptive, where each local population, or deme, tends to adapt to the specific environmental conditions in its own small portion of the species range. Such geographic variation is found in the vast majority of animal and plant species.

Environmental conditions generally vary geographically in a regular manner. There are changes in temperature with latitude or with altitude on mountain slopes, or changes in rainfall with longitude, as in many parts of the western United States. Such environmental gradients are usually accompanied by genetic-variation gradients in the species of animals and plants that inhabit the areas involved. Most species show north-south gradients in many characters, and east-west gradients are not uncommon (Fig. 32.3). Altitudinal gradients in various characters are also often found (Fig. 32.4).

When a character of a species shows a gradual variation correlated with geography, that variation is referred to as a ***cline.*** For example, many mammals and birds exhibit north-south clines in average body size,[1] being larger in the colder climates farther north and smaller in the warmer climates farther south (Fig. 32.3). Larger body size is adaptive in cooler climates because larger organisms have lower surface-to-volume ratios (see Chapter 5, p. 93), which is important in conserving body heat. (The larger the surface area, the more rapidly heat is lost by radiation.) Similarly, many mammalian species show north-south clines in the size of such extremities as the tails and ears. These parts are usually smaller in northern clines; smaller surface areas lose less heat to the environment.

Sometimes geographically correlated genetic variation is not as gradual as in the clines discussed above. There may be a rather abrupt shift in some character in a particular part of the species range. When such an abrupt shift in a genetically determined character occurs in a geographically variable species, some biologists designate the populations on the two sides of

[1] Increase in average body size with increasing cold is so common in warm-blooded animals that this tendency has been formally recognized as Bergmann's rule; the tendency toward decrease in the size of the extremities with increasing cold is called Allen's rule. The adaptive significance of these clines reflects the role of surface-to-volume ratios in heat exchange.

32.5 Two subspecies of giraffe These two subspecies of giraffe are found in different areas of Africa and have different breeding ranges. Left: The Masai giraffe *(Giraffa camelopardalis rothschildi),* found in Tanzania, East Africa. Right: The reticulated giraffe *(Giraffa camelopardalis reticulata),* from Kenya. They are partly isolated from each other reproductively because they have different ranges, but have not yet reached the stage of being considered separate species because the two subspecies will mate in zoos and produce fertile hybrids, and because hybrids are occasionally found in nature.

the shift as ***subspecies.*** This term is also sometimes applied to more isolated populations—such as those on different islands or in separate mountain ranges or, as in fish, in separate rivers—when the populations are recognizably different genetically but are believed potentially capable of interbreeding freely. Subspecies may be defined, then, as groups of natural populations within a species that differ genetically and that are partly isolated from each other reproductively because they have different ranges (Fig. 32.5).

Note that two subspecies of the same species cannot, by definition, long occur together geographically, because it is only the limitation of interbreeding imposed by distance that keeps them genetically distinctive. If they occurred together, they would interbreed and any distinction between them would quickly disappear. Many biologists have argued against the formal recognition of subspecies, often referred to as races. One reason is that the distinctions between them are often made arbitrarily on the basis of only one character—the fact that other characters may form entirely different patterns of variation being ignored (Fig. 32.6). Another reason is that most units so recognized probably only exist as separate entities temporarily and the subspecies do not go on to become fully separate species. Still another reason is because subspecies designations are often applied to parts of a cline and there has been little standardization of the definition of subspecies.

32.6 Human diversity Widespread species often tend to become subdivided into geographic subspecies or "races," and *Homo sapiens* is no exception. Regional human populations are often recognizably diferent, as we see in this group of children. Designation of "races," in most species is arbitrary; some authorities recognize only three races in the human population, Caucasoid, Negroid, and Mongoloid, while others recognize as many as 30. For this reason (and others), many researchers think the term "race" is meaningless with regard to humans.

SPECIATION

Evolution involves two main patterns: phyletic evolution and branching evolution. ***Phyletic evolution,*** or anagenesis, involves the gradual change of a lineage over time. Sometimes the population changes so substantially that it is regarded as a new species, although there is still only one living species (Fig. 32.7). ***Branching evolution,*** which is also called ***cladogenesis*** or ***allopatric speciation,*** involves the separation of an ancestral species into two or more descendant species. In considering how species originate—the phenomenon of speciation—we will concentrate on the process of *allopatric speciation,* in which two or more species derived from a common ancestor grow increasingly unlike (diverge) as they evolve (Fig. 32.7).

The fundamental question of speciation is how do two sets of populations that initially share a common gene pool come to have completely separate ones? That is, how does reproductive isolation and the subsequent elimination of effective gene flow between the two sets of populations come about? How do barriers to the exchange of genes arise?

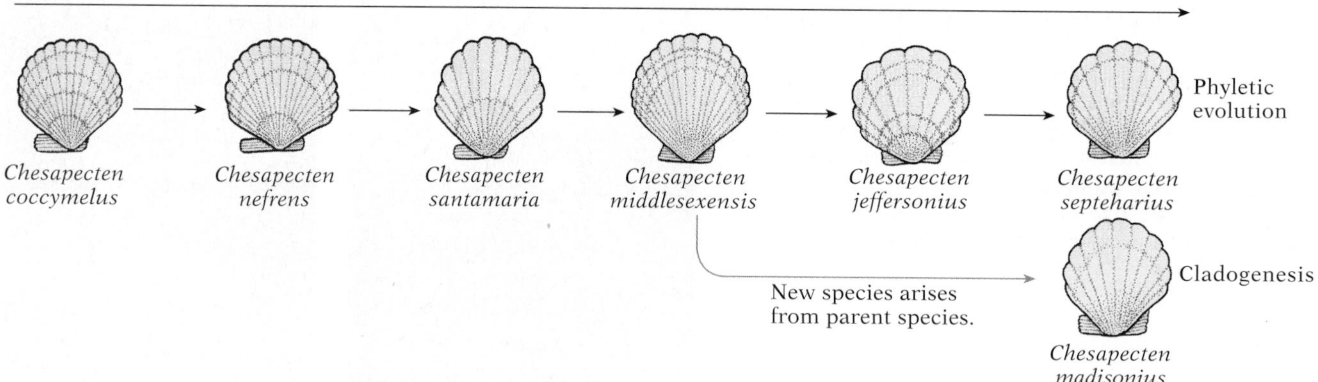

32.7 Phyletic and branching evolution Fossil scallops (genus *Chesapecten*) are found along the shores of the Chesapeake Bay and date back about 14 million years. For much of their evolutionary history they have evolved phyletically, i.e., gradual changes have accumulated in the population over time to the point that it comes to be considered a new species. Branching evolution occurred in *C. middlesexensis* when a new species arose from a population by budding off from the parent species. Note that in phyletic evolution there is only one living species: all ancestral forms have evolved into the new species. In cladogenesis the ancestral species branches into two or more descendant species. Therefore, unlike phyletic evolution, more than one living species is possible.

The First Step in Speciation Is Usually Geographical Separation

In most cases the initiating factor in speciation is geographic separation of populations of the ancestral species. Populations that are geographically separated and occupy different ranges are said to be ***allopatric,*** and thus this model of speciation is sometimes termed allopatric speciation. As long as all the populations of a species are in contact, gene flow will continue throughout the system and speciation will not occur, though there may be much intraspecific variation as populations become adapted to their local environments (i.e., clines and subspecies may develop). But if the initially continuous system of populations is divided by some permanent ***geographical barrier,*** then the separated populations will no longer be able to exchange genes and their further evolution will be independent. At first, the reproductive isolation of each group will be geographic only—they will still be potentially capable of interbreeding—and therefore they would be considered to belong to the same species. Eventually, however, they may become genetically so different that they can no longer successfully interbreed even if they should again come into contact. When this point in their gradual divergence has been reached, the two population systems constitute two separate species.

Once Separated, Populations Will Diverge Over Time

There are at least four factors that will make geographically separated population systems diverge over time.

1. *Chances are that the two populations will start out with different allelic frequencies.* Because most species exhibit geographic variation, it is highly unlikely that a geographic barrier would divide a variable species into groups exactly alike genetically; it would be much more likely to separate populations already genetically different, such as the terminal portions of a cline. Or, as often happens, a small number of individuals manage to cross an already existing barrier and found a new geographically isolated colony (e.g., a few birds being blown by a storm to an isolated island). These founders will carry with them in their own genotypes only a small percentage of the total genetic variation present in the gene pool of the parental populations, and the new colony will thus have allelic frequencies very different from those of the parental population. This is a special form of genetic drift called the ***founder effect.*** Since from the moment of their separation the two populations have different genetic potentials, their future evolution is likely to follow different paths.

2. *Genetic drift can cause random changes in small populations.* Very often the population that becomes geographically isolated from the ancestral population is small. As we have seen, small populations are subject to genetic drift—random changes in allelic frequencies in their gene pool. Thus chance alone can cause evolutionary change in small populations, unlike the ancestral population, which tends to be large and less subject to genetic drift. Consequently, as a result of the founder effect and a greater tendency for genetic drift, the smaller population usually diverges more than the ancestral population.

3. *Separated population systems will probably experience different mutations.* Mutations are random and it is unlikely that the same mutations will occur in both populations. Since there is no gene flow between the populations, a new mutant gene arising in one of them cannot spread to the other.

4. *Isolated populations will almost certainly be exposed to*

different environmental selection pressures, because they occupy different ranges. The chances that two separate ranges will be identical in every significant environmental factor are virtually nonexistent, and each population will evolve to become better adapted to its particular environment.

Barriers Can Be of Many Different Types

A barrier is any physical or ecological feature that prevents the passage of species. What is a barrier for one species may not be a barrier for another. Thus a prairie is a barrier for forest species but not, obviously, for prairie species. A mountain range is a barrier to species that can live only in lowlands, and a desert is a barrier to species that require a moist environment. On a grander scale, oceans, glaciers, and indeed the actual movement of continents, which we will discuss in Chapter 34, have played a role in the speciation of many plants and animals. Let us look at a few actual examples of geographic isolation leading to speciation.

There are innumerable examples, but one of the most frequently cited is that of the Kaibab squirrel, which occurs on the north side of the Grand Canyon, and the Abert squirrel, which occurs on the south side (Fig. 32.8). The two are clearly very closely related and likely evolved from the same ancestor, but they almost never interbreed at present, because they do not cross the Grand Canyon. Biologists are not agreed whether these two squirrels have reached the level of full species or whether they should be considered well-marked geographic variants of a single species, but the fact remains that the Grand Canyon has acted as a barrier separating the two sets of populations, and that those populations have, as a result, evolved divergently until they have at least approached the level of fully distinct species.

32.8 Squirrels of the Grand Canyon Two populations of squirrels that live in different ranges of the Grand Canyon in Arizona are morphologically distinct. The Kaibab squirrel (top), which lives on the Kaibab Plateau on the northern rim of the canyon, is darker than the Abert squirrel (bottom), a related population that inhabits a range on the southern rim.

Intrinsic Reproductive Isolation Completes the Speciation Process

According to the model of allopatric speciation outlined above, the initial factor preventing gene flow between two closely related populations is ordinarily an ***extrinsic*** one—a physical aspect of the environment. Then, as the two populations diverge, they accumulate differences that will lead in time to the development of ***intrinsic*** isolating mechanisms—biological characteristics such as morphology, physiology, chromosomal incompatibility—that prevent the two populations from occurring together or from interbreeding effectively when (or if) they do again occupy the same range. In other words, allopatric speciation is initiated when external barriers cause the two populations to become entirely allopatric, but is not completed until the populations have evolved intrinsic mechanisms that will keep them allopatric or that will keep their gene pools separate even when they are ***sympatric*** (with overlapping ranges) (Fig. 32.9). There are many kinds of intrinsic isolating mechanisms that may arise. One way of classifying them is shown in Table 32.1.

1. Mechanisms that prevent mating

A. Ecogeographic isolation: the two species maintain their geographic separation

Two populations initially separated by some extrinsic barrier may in time become so specialized for different environmental conditions that even if the original extrinsic barrier is removed they may never become sympatric, be-

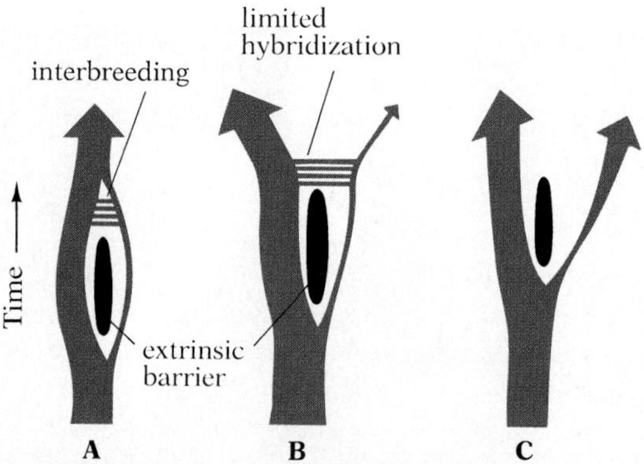

32.9 Model of geographic speciation An ancestral population is split by an extrinsic (geographic) barrier for a time, and diverges with regard to various traits (only one of which is considered in these hypothetical examples). (A) If the barrier breaks down before the two subpopulations have been isolated long enough to have evolved intrinsic reproductive isolating mechanisms, the populations will fuse back together. (B) Two populations are isolated by an extrinsic barrier long enough to have evolved incomplete intrinsic reproductive isolating mechanisms. When the extrinsic barrier breaks down, some hybridization occurs. But the hybrids are not as well adapted as the parental forms; hence there is a strong selection pressure favoring forms of intrinsic isolation that prevent mating, and the two populations diverge more rapidly until mating between them is no longer possible. This rapid divergence is called character displacement. (C) Two populations are isolated by a geographic barrier so long that by the time the barrier breaks down they are too different to interbreed. In some cases, one population is much smaller than the other, as indicated by the width of the "branches" where they separate. The smaller population—often quite small—usually diverges more from the common ancestor than does the larger population. This greater divergence is the result of the founder effect, a greater tendency for genetic drift, and a smaller and perhaps more specialized habitat.

TABLE 32.1 *Intrinsic isolating mechanisms*

Effect	*Mechanism*	*Individuals affected*
Mating prevented	1. Ecogeographic isolation 2. Habitat isolation 3. Seasonal isolation 4. Behavioral isolation 5. Mechanical isolation	**Parents:** fertilization prevented (1–6)
Production of hybrid young prevented	6. Gametic isolation 7. Developmental isolation 8. Hybrid inviability	
Perpetuation of hybrids prevented	9. Hybrid sterility 10. Selective hybrid elimination	**Hybrids:** success prevented (7–10)

cause neither can survive under the conditions where the other occurs.

B. Habitat isolation: the two species occupy different habitats within the same range

When two sympatric populations occupy different habitats within their common range, the individuals of each population will be more likely to encounter and mate with members of their own population than with members of the other population. Their genetically determined preference for different habitats thus helps keep the two gene pools separate. For example, in California the ranges of *Ceanothus thyrsiflorus* and *C. dentatus,* two species of wild lilacs, overlap broadly, but *C. thyrsiflorus* grows on moist hillsides with good soil, while *C. dentatus* grows on drier, more exposed sites with poor or shallow soil.

C. Temporal isolation: the two species breed at different times

If two closely related species are sympatric, but breed during different seasons of the year, or different times of the day, interbreeding between them will be effectively prevented. For example, five species of frogs are sympatric in much of eastern North America, but the period of most active mating is different for each species (Fig. 32.10).

D. Behavioral isolation: behavior is important in species recognition

As we shall see in Chapter 33, behavior is immensely important in courtship and mating, particularly with respect to males and females recognizing members of their own species. For example, ducks often have elaborate courtship displays, usually combined with striking color patterns in the males (see Fig. 31.14, p. 686). These species-specific displays, by minimizing the chances that a female will se-

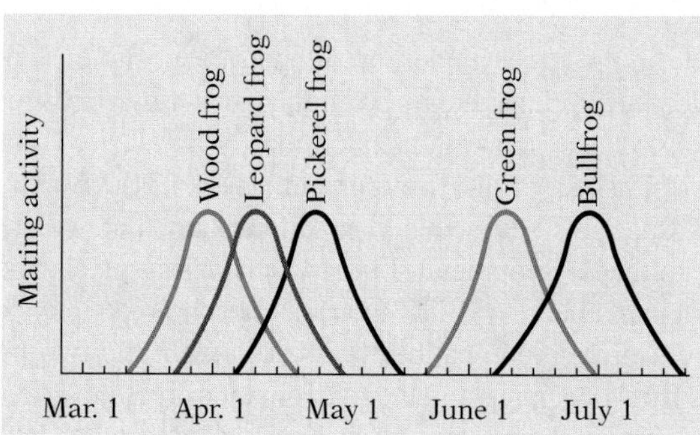

32.10 Mating seasons at Ithaca, New York, for five species of frogs of the genus *Rana* The period of most active mating is different for each species. Where the mating seasons for two or more species overlap, different breeding sites are used.

32.11 A male fiddler crab *(Uca)* The animal waves its large cheliped in the air as a courtship display. Critical details of the display differ among the various species of fiddler crabs.

lect a male of the wrong species as a mate, maintain reproductive isolation (Fig. 32.11).

E. Mechanical isolation: structural differences may prevent mating

Structural differences between two closely related species often make it physically impossible for matings between males of one species and females of the other to occur. In such cases the two populations will obviously not exchange genes. If, for example, one species of animal is much larger than the other, or if the genitalia of the male and female do not fit, matings between them may be very difficult, if not impossible.

2. Mechanisms that prevent production of hybrid young after mating

A. Gametic isolation: fertilization cannot occur

Even if individuals of two animal species mate or the pollen from one plant species gets into the female part of another, actual fertilization may not take place. For example, if cross-insemination occurs between the two species of fruit flies *Drosophila virilis* and *D. americana,* the sperm are rapidly immobilized by the unsuitable environment in the reproductive tract of the female and they never reach the egg cells.

B. Developmental isolation: the embryo may not be viable

Even when cross-fertilization occurs, the development of the embryo is often irregular and may cease before birth. The eggs of fish can often be fertilized by sperm from a great variety of other species, but development usually stops in the early stages. Crosses between sheep and goats produce embryos that die long before birth. This probably occurs because of different developmental systems that do not cooperate.

C. Hybrid inviability: hybrids may die before reproduction

Two species may mate, but the resulting hybrids are often weak and malformed, frequently dying before they reproduce. Consequently there is no actual gene flow through them from the gene pool of one parental species to the gene pool of the other. An example of hybrid inviability is seen in certain tobacco hybrids, which develop tumors in their vegetative parts and die before they flower.

3. Mechanisms that prevent perpetuation of hybrids

A. Hybrid sterility: hybrids are unable to successfully reproduce

Some interspecific crosses produce vigorous but sterile hybrids. The best known is the cross between a female horse (diploid number of 64) and a male donkey (diploid number of 62), which produces the mule (diploid number of 63) (Fig. 32.12). Mules have many characteristics superior to those of both parental species, a condition referred to as ***hybrid vigor,*** but they are normally sterile because their chro-

A

B

C

32.12 A cross between a male donkey (A) and female horse (B) produces a mule (C) Mules have many characteristics superior to those of both parental species, a condition referred to as hybrid vigor, but they are normally sterile because their chromosomes do not occur in homologous pairs and thus cannot correctly pair during meiosis. No matter how many mules are produced, the gene pools of horses and donkeys remain distinct, because there is no gene flow between them.

mosomes do not occur in homologous pairs and thus cannot correctly pair during meiosis. No matter how many mules are produced, the gene pools of horses and donkeys remain distinct, because there is no gene flow between them.

B. Decreased fitness of hybrids: hybrids are less well adapted than parents
The members of two closely related populations may be able to cross and produce fertile offspring. If those offspring and their progeny are as vigorous and well adapted as the parental forms, then the two original populations will not remain distinct for long if they are sympatric, and it will no longer be possible to regard them as distinct species. But if the fertile offspring and their progeny are less well adapted than the parental forms, they will soon be eliminated. There will be some gene flow between the two parental gene pools through the hybrids, but not much. The parental populations are consequently regarded as separate species.

Situations in which only one of the various isolating mechanisms described above is operative are extremely rare. Ordinarily two, three, four, or more all contribute to keeping the species apart.

Sympatric Speciation May Also Occur

The model of allopatric speciation discussed above involves the divergence of *geographically separated* populations, but there are other ways in which new species may arise. Speciation that does not involve geographic isolation is called ***sympatric speciation.*** An important example is speciation by ***polyploidy,*** the condition of having more than two sets of chromosomes, making each cell triploid, tetraploid, etc. This condition can arise so quickly that it is entirely possible for a parent to belong to one species and its offspring to another. In polyploidy, nondisjunction or some other abnormal event in cell division occurs that results in extra sets of chromosomes in the offspring.

One important type of polyploidy, called ***allopolyploidy,*** involves the formation of a hybrid between two species, followed by a sudden multiplication of the number of chromosomes in the hybrid as a result of some abnormality in cell division (Fig. 32.13). A hybrid between two species is normally sterile, because its chromosomes do not occur in homologous pairs and thus cannot correctly pair during meiosis. The polyploids, however, will have paired chromosomes and are fertile. They can breed with each other, but cannot successfully cross with the parental species from which they arose, because

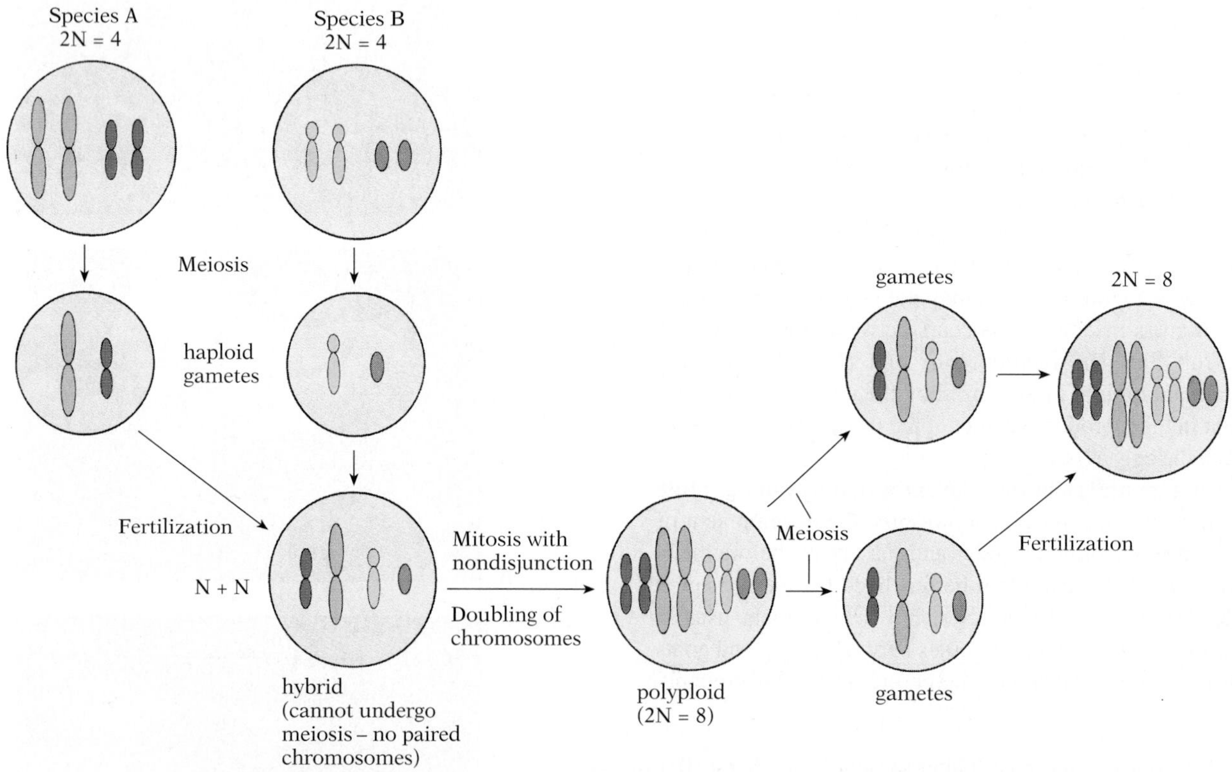

32.13 Allopolyploidy This type of speciation occurs when a hybrid forms between two different species. The hybrid is sterile, because its four chromosomes are all different and cannot form homologous pairs during meiosis. If nondisjunction occurs during mitosis within the hybrid cell, then the chromosome number in that cell would be doubled, and a polyploid results. The polyploids do have paired chromosomes and can carry out meiosis, producing gametes. The polyploids can breed with each other, but cannot cross successfully with either of the parental species from which they arose, because they have different numbers of chromosomes (four versus two).

they have different numbers of chromosomes. These polyploid populations fulfill all the requirements of our definition of species—they are genetically distinctive and reproductively isolated.

Production of polyploid individuals is rare in animals but common in plants—over half of the species of flowering plants are polyploids. Why is it more common in plants? Undoubtedly it is because many plants can reproduce more easily than animals, i.e., they reproduce vegetatively—by runners, suckers, and underground stems—and because they are often self-fertilized. A polyploid animal, on the other hand, normally must find another polyploid individual like itself with which to mate, a much more unlikely scenario.

Allopolyploidy has been important in the development of many valuable crop plants. For instance, bread wheat, *Triticum aestivum,* which has a diploid number of 42, is thought to have arisen about 8,000 years ago as a spontaneous hybrid between a species of wild grass (diploid number of 14) and another wheat species (diploid number of 28). As soon as it was realized that many of our most useful plants, such as oats, wheat, cotton, tobacco, potato, banana, coffee, and sugar cane, are polyploids, plant breeders began to stimulate polyploidy, and have obtained many new varieties.

32.14 Sympatric speciation in tree hoppers Two sympatric populations of the tree hopper *Enchenopa binotata* have evolved adaptations to different host plants. The tree hopper at the top lives on bittersweet, while the one on the bottom lives on butternut. Host specificity may take the place of physical separation (allopatry) in preventing these two populations from interbreeding.

Though most speciation not involving chromosomal changes is allopatric (requiring a period of geographic isolation), there is evidence to suggest forms of sympatric speciation other than polyploidy; however, the subject remains controversial. Reproductive isolation is just as essential to these types of sympatric speciation, but is effected by means other than geographic isolation or polyploidy. For example, relatively small changes in habitat preference may produce reproductive isolation in the absence of true geographic separation. Two populations of the tree hopper *Enchenopa binotata* have adapted to different host plants, one population living on bittersweet, while the other lives on butternut trees (Fig. 32.14). These insects often mate on the plants on which they feed, so individuals that are adapted for a particular species of host plant will tend to inbreed. Selection may favor those individuals that breed strictly on the host species, and over time intrinsic isolating mechanisms may develop that would isolate the tree hoppers that have adapted to different host plants. Host specificity may take the place of physical separation in preventing these two populations from interbreeding.

MACROEVOLUTION

Thus far we have considered microevolution, the gradual change of a population through time and the processes of speciation. Now we turn to ***macroevolution,*** the changes in the kinds of species over time. It involves large-scale evolutionary changes above the level of species, and includes the processes of adaptive radiation, the development of evolutionary trends, the origin of innovative features, and the effect of extinctions.

Bursts of Rapid Evolutionary Divergence May Produce Many Separate Descendant Species

Biologists are in agreement that the enormous diversity of life on earth today evolved from simple organisms that came into existence over three billion years ago. Clearly, divergent evolution—the evolutionary splitting of species into many separate descendant species—has been exceedingly frequent. Some ancestral species undergo slow, gradual change for millions of years, only occasionally giving rise to other species. Other species undergo bursts of evolutionary activity, producing many separate descendant species that display diverse ways of life even though they are in the same geographical area. The latter process is known as an ***adaptive radiation.*** Adaptive radiations are common in periods of environmental change or where new ecological opportunities exist. They often occur when new areas open up for colonization, such as isolated island chains, or when some organisms evolve novel

MAKING CONNECTIONS

1. Gray squirrels inhabit North America from Maine to Florida. Where would you expect the largest squirrels with the shortest tails and ears to be found? Why?

2. Why does the *hybrid vigor* of a mule not constitute *fitness*, as defined in Chapter 31?

3. What stages of meiosis would be disrupted in a mule (in which chromosomes do not occur in homologous pairs)? (See Chapter 9.)

4. How is speciation by polyploidy related to nondisjunction, discussed in Chapter 14?

5. Why is speciation by polyploidy more common in plants than in animals?

6. Suppose two plant species hybridize and produce a new species by allopolyploidy. If the diploid chromosome numbers of the two parent species are 14 and 18, what will be the diploid chromosome number of the new species (at a minimum)?

structural adaptations allowing them to exploit a new way of life (e.g., insect wings or amphibian legs). Adaptive radiation within the insects, for example, has produced more than 750,000 different living species. Is it possible to account for such a degree of evolutionary radiation by the models outlined above? In particular, could opportunities for geographic isolation have been sufficient to lead to all the speciation not caused by polyploidy? How can speciation occur in a very small area if geographic isolation is a necessary factor? To answer such questions, we consider a particularly instructive example—the finches on the Galápagos Islands.

Adaptive Radiation Occurred Among Finches in the Galápagos

The Galápagos Islands lie astride the equator in the Pacific Ocean roughly 950 kilometers west of Ecuador, the country to which they now belong (Fig. 32.15). These islands have never been connected to South America or to any other land mass; nor have they been connected to each other. They apparently rose from the ocean floor as volcanoes more than five million years ago. At first they were completely devoid of life, con-

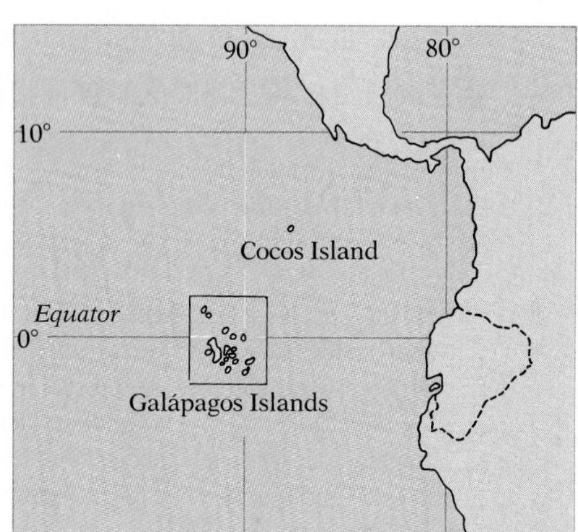

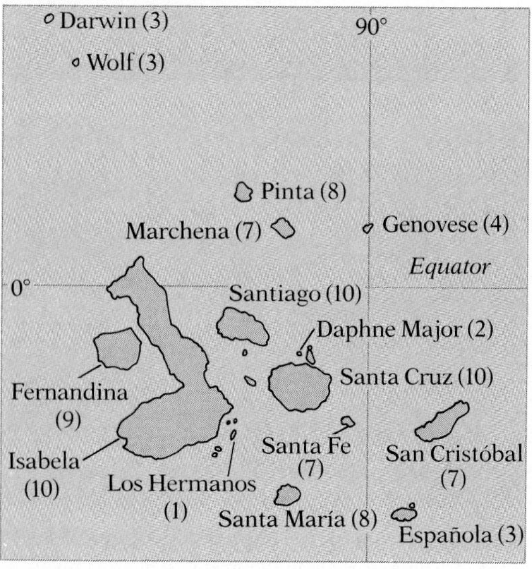

32.15 The Galápagos Islands Left: The islands are located about 950 kilometers off the coast of Ecuador. Cocos Island is about 700 kilometers northeast of the Galápagos. Right: The islands shown in greater detail. The number in parentheses after each name indicates the number of species of Darwin's finches that occur on that island.

taining an environment open to exploitation by whatever species that might reach them. Relatively few species ever did so. The only land vertebrates present on the islands when human beings arrived were six species of reptiles (one snake, a huge tortoise, and four lizards, including two very large iguanas), two species of mammals (a rice rat and a bat), and a limited number of birds (including two species of owls, one hawk, one dove, one cuckoo, one warbler, two flycatchers, one swallow, four mockingbirds, and the famous Darwin's finches).

The 14 species of Darwin's finches (Fig. 32.16) are found

32.16 Darwin's finches Darwin's finches fall into four genera: birds 1, 3, 4, 5, 6, and 10 are the tree finches *(Camarhynchus)*; birds 7, 8, 11, 12, 13, and 14 are the ground finches *(Geospiza)*; bird 2 is the unfinchlike warble finch; bird 9 is the one species inhabiting Cocos Island.

1. Vegetarian tree finch *(C. crassirostris)*
2. Warbler finch *(Certhidea olivacea)*
3. Large insectivorous tree finch *(C. psittacula)*
4. Medium insectivorous tree finch *(C. pauper)*
5. Mangrove finch *(C. heliobates)*
6. Small insectivorous tree finch *(C. parvulus)*
7. Large cactus ground finch *(G. conirostris)*
8. Cactus ground finch *(G. scandens)*
9. Cocos finch *(Pinaroloxias inornata)*
10. Woodpecker finch *(C. pallidus)*
11. Large ground finch *(G. magnirostris)*
12. Sharp-beaked ground finch *(G. difficilis)*
13. Medium ground finch *(G. fortis)*
14. Small ground finch *(G. fuliginosa)*

nowhere in the world other than the Galápagos Islands (except one species that also inhabits Cocos Island, 700 kilometers to the northeast). They are believed to have evolved on the Galápagos Islands from some unknown finch ancestor that colonized the islands from the South American mainland. It is not surprising that the descendants of the geographically isolated colonizers underwent evolutionary change to become, in time, very unlike their mainland ancestors. More surprising is the manner in which the descendants of the original immigrants split into 14 different species.

Much of the finches' great speciation is due to their having colonized more than 15 separate islands. The finches will not readily fly across wide stretches of water, and they show a strong tendency to remain near their home area. Hence a population on any one of the islands is effectively isolated from the populations on the other islands. It is thought that the initial Galápagos colony was established on one of the islands where the colonizers, perhaps blown by high winds, chanced to land. Later, individuals from this colony wandered or were blown to other islands and founded new colonies. The allelic frequencies in the new colonies must have differed from those in the original colonies from the moment they started, resulting in the founder effect. In time, the colonies on the different islands diverged even more, according to geographic speciation (different mutations, different selection pressures, and, in such small populations as some of these must have been, genetic drift). What we might expect, therefore, is a different species, or at least a different subspecies, on each of the islands. But this is not what has actually been found; most of the islands have more than one species of finch, and the larger islands have 10 (Fig. 32.15). What is the explanation?

Suppose that form *A* evolved originally on Santa Cruz and that the closely related form *B* evolved on Santa Maria. If, later, Form A spreads to Santa Maria and *A* and *B* had been separated long enough to have evolved major differences, then *A* and *B* might have become intrinsically isolated from each other. Though coexistence of *A* and *B* on the same island without interbreeding would be possible, it would be highly unlikely if they both used the same food supply or the same nesting sites; the ensuing competition would be very severe, and the less well adapted species would tend to be eliminated by the other unless it evolved differences that minimized the competition. Wherever two or more closely related species occur together, natural selection will often favor ***character displacement,*** the tendency of closely related species to diverge rapidly, which reduces the chance of hybridization and/or minimizes competition between them (Fig. 32.17). This is precisely what we find in Darwin's finches—in this case the evolution of different feeding and nesting habits. There are two main groups, those that live primarily on the ground, and those that live in the trees, and there is considerable specialization with respect to diet within each group. Correlated with the differences in diet among the species are major differences in the size and shape of their beaks: some large and thick for cracking heavily walled seeds, others smaller and adapted for eating buds or fruits, and still others for eating small soft insects (Fig. 32.18). Variations in beak size and shape among birds thought to share a common ancestry can be found in other island habitats as well, as shown, for example, in Figure 32.19. These characteristics of the beak are apparently the principal means by which the birds recognize other members of their own species and vary their diet.

Now, if on Santa Maria selection favored character displacement between species *A* and *B,* the population of species *A* on Santa Maria would become less and less like the population of species *A* on Santa Cruz. Eventually these differences might become so great that the two populations would be intrinsically isolated from each other, forming separate species. We might now designate as species *C* the Santa Maria population derived from species *A*. The geographic separation of the two islands would thus have led to the evolution of three species *(A, B,* and *C)* from a single original species. The process of island-hopping followed by divergence could continue indefinitely and produce many additional species, a process that undoubtedly led to the formation of the 14 species of Darwin's finches.

Darwin's finches provide an excellent example of adaptive radiation—an ancestral species introduced to a new environment giving rise to many new species, each adapted for a particular habitat and way of life. Adaptive radiation, therefore, is central to the immense diversity among living things.

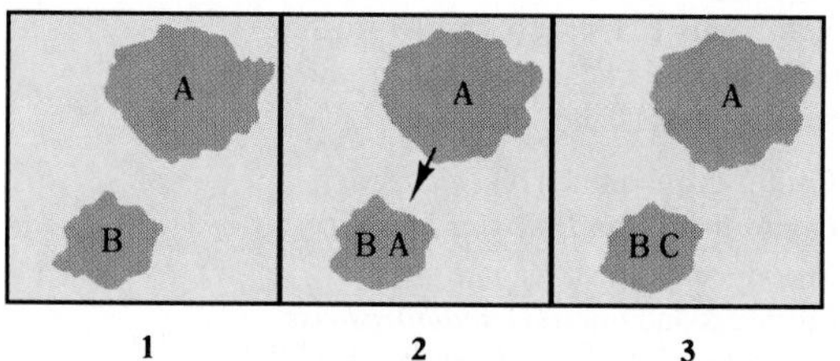

32.17 Model of speciation on the Galápagos Islands An ancestral form colonized the larger of these two hypothetical islands. Later, part of the population dispersed to the smaller island. (1) Eventually the two populations, being isolated from each other, evolved into separate species *A* and *B*. (2) Some individuals of *A* dispersed to *B*'s island. The two species coexisted, but intense competition between them led to rapid character displacement. (3) This rapid evolution of the population of *A* and *B*'s island caused it to become more and more different from the original species *A,* until eventually it was sufficiently distinct to be considered a full species, *C,* in its own right. At the same time, the selection pressure imposed by the small invading population caused the large population of species *B* to evolve to a small degree as well.

32.18 Beak differences in Darwin's finches on the central islands The differences may appear slight at first glance, but they have important functional implications for the birds' diets and for species recognition in mating. Top row (diagonally downward from left to right): *Geospiza magnirostris, G. fortis, G. fuliginosa.* Second row: *G. difficilis, G. scandens, Camarhynchus crassirostris.* Third row: *C. psittacula, C. parvulus, C. pallidus.* Fourth Row: *C. heliobates, Certhidea olivacea, Pinaroloxias inornata.*

32.19 Beak differences in Hawaiian honeycreepers Differences in beak size and shape are apparent in related species of Hawaiian honeycreepers, which, like Darwin's finches, are thought to have evolved from a common finchlike ancestor.

The Rate of Evolutionary Change Often Varies

Darwin pointed out that the rate of evolutionary change is not always constant. When the first colonizing finches reached the Galápagos Islands, they encountered environmental conditions quite unlike those they had left behind in Central or South America. Differences in available resources, for example, likely led to selection for different physical form, physiology, and behavior. When the new habitat became saturated with finches, causing intense competition between species for available resources, selection pressures must have led to very rapid divergence from the ancestral population. Later, as the finches became increasingly well adapted to conditions on the Galápagos, the rate of evolutionary change probably decreased.

In general, when conditions change radically and organisms have new evolutionary opportunities for which they are at least modestly preadapted, they may undergo an evolutionary burst—a period of rapid adaptational change—which may then be followed by a more stable period during which any further evolutionary changes are merely a fine-tuning of their already well-adapted characteristics. Such bursts of rapid evolutionary divergence probably characterized the tremen-

32.20 Niche rule The first mammals were probably small, secretive creatures that fed primarily on insects. They remained a relatively unimportant part of the fauna while the dinosaurs were dominant. The great radiation of mammals dates from the beginning of the Cenozoic era as the mammals rapidly filled the many niches left open by the disappearance of the dinosaurs.

dous radiation of amphibians when they first moved onto land and the explosive radiation of mammals when the demise of the large terrestrial dinosaurs left many adaptive slots unoccupied (Fig. 32.20).

Competition and Chance May Play Important Roles in Evolutionary Change

In the last chapter we contrasted the roles played by genetic drift (chance) and natural selection in the evolution of populations, emphasizing that evolution is a vital, ongoing phenomenon, while genetic drift (potentially important only in small populations) and selection are two contributing—and clearly demonstrated—mechanisms by which it occurs. Speciation follows a similar theme: intrinsic barriers that contribute to reproductive isolation can arise by chance (primarily genetic drift), by natural selection, or by a combination of the two. Competition may play an important role in the formation of species: even reproductively isolated populations may not be able to coexist indefinitely if they compete for precisely the same food, since just a slight but systematic superiority of one will tend to lead to the extinction of the other.

G. F. Gause of the University of Moscow, who first observed this phenomenon in the laboratory in the 1930s, formulated the competitive exclusion principle, an early version of what we now refer to as the ***niche rule:*** *no two species occupying the same niche can long coexist* (Fig. 32.21). ***Niches*** are not preordained "slots" into which species fit; rather, they constitute the functional role and position of an organism in the ecosystem; the way the members of the individual species actually maintain themselves, including for an animal not only what it eats, but when, where, and how it obtains food, where it lives, etc. The ecological attributes of each species define its niche; no two species with identical ecological requirements can coexist indefinitely. The character displacement that natural selection produces, resulting in changes in food preference, habitat choice, and the like, enhances the likelihood that competing species will coexist. This Darwinian interpretation of how separate species form, with its emphasis on reproductive isolation and character displacement in the face of competition, may account for much of what we can observe of speciation. But chance is now seen as playing an important evolutionary role.

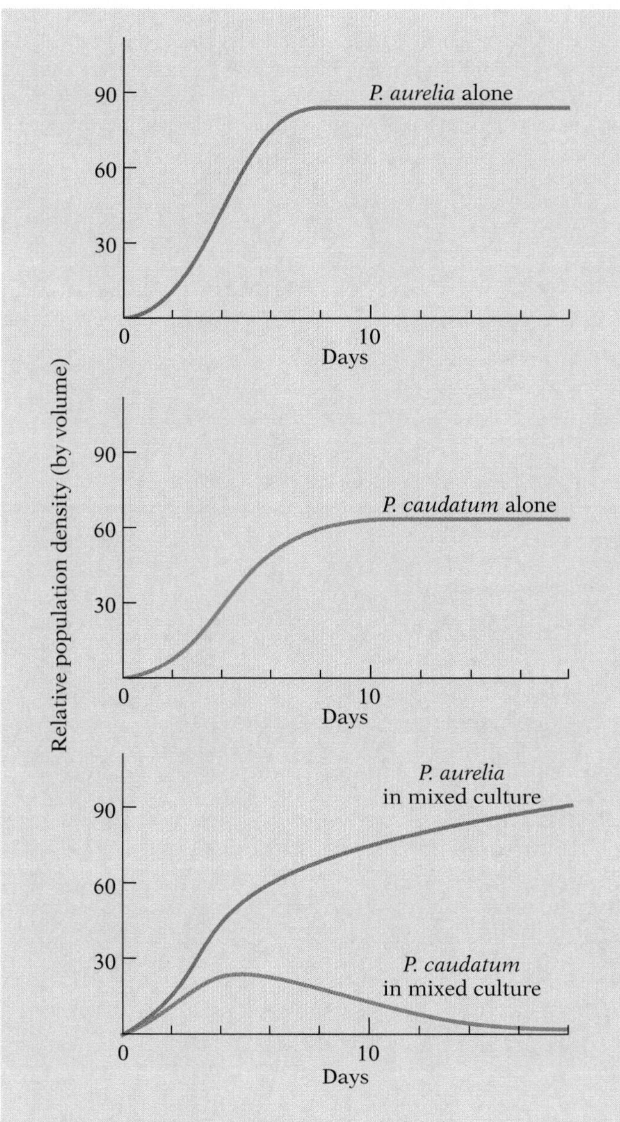

32.21 Competition and extinction When a few individuals of either species of *Paramecium* are introduced into a tank alone (top and middle), they multiply until they reach a limiting density. But when individuals of both species are introduced together (bottom), they multiply independently for the first three days, then begin to compete for resources. *P. aurelia* is in some way more efficient under these conditions, and drives *P. caudatum* to extinction in only three weeks.

Rare Environmental Crises May Cause Evolutionary Change

There are at least two ways in which chance can take precedence over competition in causing species to diverge. The first we have already discussed: in small populations, genetic drift can be very powerful, driving alleles to extinction and producing changes in allelic frequencies. The other process, however, is more fundamental to our understanding of evolution. Most Darwinian analyses tacitly assume that the selection pressures organisms face change relatively slowly, and that large continuous populations therefore have ample time, generation by generation, to evolve specialized adaptations to their environments. But do conditions really remain sufficiently constant for a species to achieve genetic equilibrium—a stable set of allelic frequencies—in the face of all the selection pressures affecting it?

Apparently, many populations are not always at equilibrium. In one study Peter Grant of Princeton University tagged 1,500 ground finches on one of the smaller Galápagos islands, Daphne Major. In 1977 only 20 percent of the normal amount of rain fell, and the plants on the island produced many fewer seeds than usual. Relatively dry or wet seasons, a surprisingly early or late hard frost, or especially warm or cold years are conditions to which organisms can generally adapt, but the effects on plants and animals of extreme once-a-century events can be drastic.

On Daphne Major, the consequences of the reduction in food were dramatic for the medium ground finch: with only 35 percent of the usual amount of food available, 387 of the 388 nestlings of 1976 died, as did the majority of adults. Furthermore, the survivors were not a representative cross section of the former population: roughly 180 of the 500 adult males survived, but only about 30 of the 500 females, and the birds that did survive the year of starvation were significantly larger than those that succumbed. Exceptionally intense natural selection in favor of large body size had occurred.

Thus one of several rare, unpredictable, but inevitable crises finches face sharply altered the selection pressures for a year, and reduced the population to a level at which genetic drift suddenly became a factor. An adaptive equilibrium was upset by a chance disturbance, and a population with altered characteristics resulted. The finch population had been forced through a period of environmental crisis. When such a crisis is so severe as to cause major changes in allelic frequencies in a population, it is called an ***evolutionary bottleneck*** (Fig. 32.22). Another example of a species that may have been forced through an evolutionary bottleneck is the African cheetah, *Acinonyx jubatus* (Fig. 32.23). Cheetahs once ranged throughout Africa, the Near East, and southern India, but about 10,000 years ago some disaster, perhaps overhunting by humans, drastically reduced the population, which now survives only in scattered pockets in sub-Saharan Africa. One of the effects of the population reduction was a 90 percent loss of their genetic variation. This means that the cheetah population is genetically rather uniform, and may therefore lack the diversity to respond to future severe environmental changes and disease epidemics. Despite the lack of diversity, cheetahs appear vigorous, and it is hoped that the limited amount of diversity they do have may be enough for the survival of these magnificent cats.

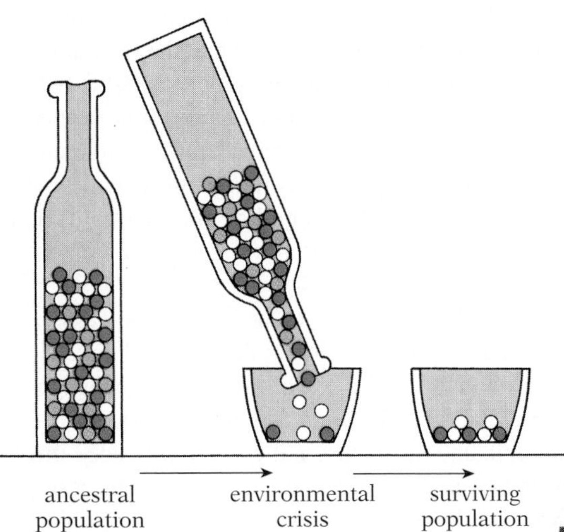

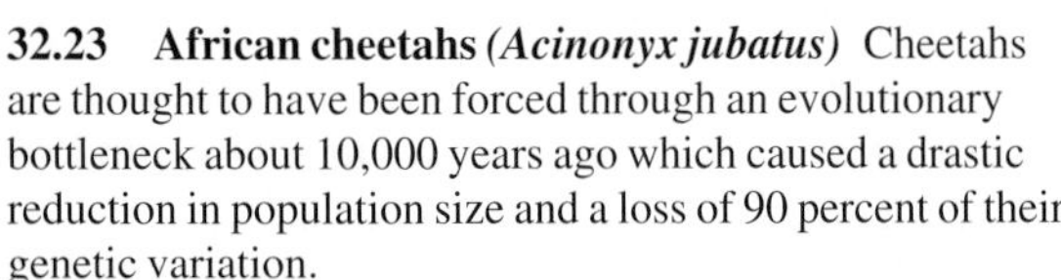

32.22 Model of an evolutionary bottleneck An ancestral population, here represented by colored beads in a bottle, can be drastically and unselectively reduced when some extraordinary environmental crisis (represented by the neck of the bottle) forces the population through a bottleneck. The surviving population is much smaller and has a different composition than the ancestral population.

32.23 African cheetahs *(Acinonyx jubatus)* Cheetahs are thought to have been forced through an evolutionary bottleneck about 10,000 years ago which caused a drastic reduction in population size and a loss of 90 percent of their genetic variation.

MAKING CONNECTIONS

1. Describe the interplay of natural selection, mutations, the founder effect, and genetic drift in the evolution of Darwin's finches.

2. How are the specialized feeding adaptations (beak size, etc.) of Darwin's finches related to the concept of the *niche?*

3. How may the reduction in genetic diversity resulting from an evolutionary bottleneck make a species more vulnerable to extinction?

4. In Chapter 31, you learned that chance events (e.g., mutation, genetic drift) can cause changes in gene frequencies in a population. How can chance events influence speciation and other phenomena of macroevolution?

5. What geological evidence supports the view, held by many biologists, that the evolution of human beings was a matter of chance, rather than an inevitable result of the evolutionary process?

Competition, then, leading to evolutionary divergence and to new species, is not the only major mode of speciation, since chance crises may occur often enough to upset whatever stability exists in the allelic frequencies within a species. Crises can be caused by any environmental factor—disease, or extreme weather, for example—that severely affects an isolated population, or that itself serves to isolate one part of a population from the remaining body. During such a crisis, an isolated population passes through the evolutionary bottleneck, and one character or another may gain ascendancy. Such a character is still at risk if the surviving population is small, because it may decline in frequency or even become extinct through genetic drift. Alternatively, if a character that was not adaptive prior to the crisis proves adaptive thereafter, it may increase in frequency in the population. Though most researchers still believe that competition is the predominant force leading to speciation over time, studies such as Grant's show that the catastrophic effects of rare crises may have played a crucial role in the evolution of certain populations.

Certainly environmental crises have played an important role in the history of life. Evidence shows that earth's climate has undergone periodic transformations throughout its existence, alternating between "greenhouse" and "icehouse" phases. The changes appear to be episodic, with short periods of rapid change alternating with longer periods of relative stability. And the geological time scale is itself divided into different eras on the basis of certain recognizable events (Table 32.2). For example, nearly 600 million years ago during the period called the Cambrian there was a "sudden" appearance of many groups of complex multicellular animals in the fossil record. This rapid proliferation marks the end of the Archeozoic era and the beginning of the Paleozoic era. Likewise, the Paleozoic-Mesozoic boundary is recognized by the so-called Permian crisis—the mass extinction of as many as 95 percent of the marine invertebrate species. Finally, the Cenozoic-Mesozoic boundary—between the Cretaceous (K) and the Tertiary (T) periods—is marked by another crisis that saw the mass extinction of the dinosaurs as well as many invertebrate species, and thereby opened the door for the adaptive radiation of the mammals (Fig. 32.20).

The geologic record suggests there were at least five major periods of mass extinctions that have punctuated the history of life on earth. Three basic hypotheses have been proposed to explain these extinctions: (1) geologic events such as volcanic action or mountain building, (2) changes in climate and large fluctuations in the sea level and patterns of ocean circulation, and (3) impacts by comets or asteroids (see Exploring Further: The Mass Extinctions, p. 722). The massive bouts of extinction cleared away thousands of species at a time, leaving the survivors the opportunity to radiate and exploit novel lifestyles in the absence of competition. According to the increasingly popular view that many large-scale evolutionary developments depend on chance, simple luck may often be more important than natural selection. Had the cards fallen differently, the argument goes, the dominant species of animals today might be something quite different.

TABLE 32.2 *The geologic time scale*

Era	*Period*	*Age (millions of years to present)*
Cenozoic	Tertiary	
		65
	K-T extinction	
Mesozoic	Cretaceous	
		144
	Jurassic	
		208
	T-J extinction	
	Triassic	
		245
Paleozoic	Permian extinction	
	Permian	
		286
	Carboniferous	
		360
	Devonian extinction	
	Devonian	
		408
	Silurian	
		438
	Ordivician extinction	
	Ordivician	
		505
	Cambrian	
		570–600
Archeozoic		

According to Punctuated Equilibrium, Long Periods of Gradual Evolutionary Change Are Punctuated by Short Periods of Rapid Change

Though Darwin did not view the rate of evolutionary change as constant, in general he did believe evolution was characterized by the gradual accumulation of very small changes over long

EXPLORING FURTHER

THE MASS EXTINCTIONS

There has been endless speculation as to what caused the five major episodes of mass extinction. Interest in the mass extinction at the Cretaceous-Tertiary (K-T) boundary in particular has been greatly stimulated by the work of Walter Alvarez and his associates at the University of California at Berkeley. They discovered that the concentration of specific elements, particularly iridium, is 30 times higher than normal in a thin layer of sediments at the K-T boundary, and they argue that the only plausible source for this element is extraterrestrial. Their calculations suggest that 65 million years ago an asteroid or comet 10 kilometers wide struck the earth and exploded, releasing so much energy from the impact that an enormous volume of vapor and dust was thrown into the atmosphere. Enough iridium would have dispersed to account for its sudden worldwide distribution in the sediments. The most likely impact site identified to date for the K-T extinction is a (now buried) crater 180 kilometers in diameter straddling the shoreline of the Yucatán peninsula of Mexico (Fig. A). Another source of evidence for the Yucatán impact is the presence of tiny fragments of glass in nearby sediments. These are thought to be hardened droplets of rocks, melted by the impact and ejected into the atmosphere, that cooled into glass as they rained down.

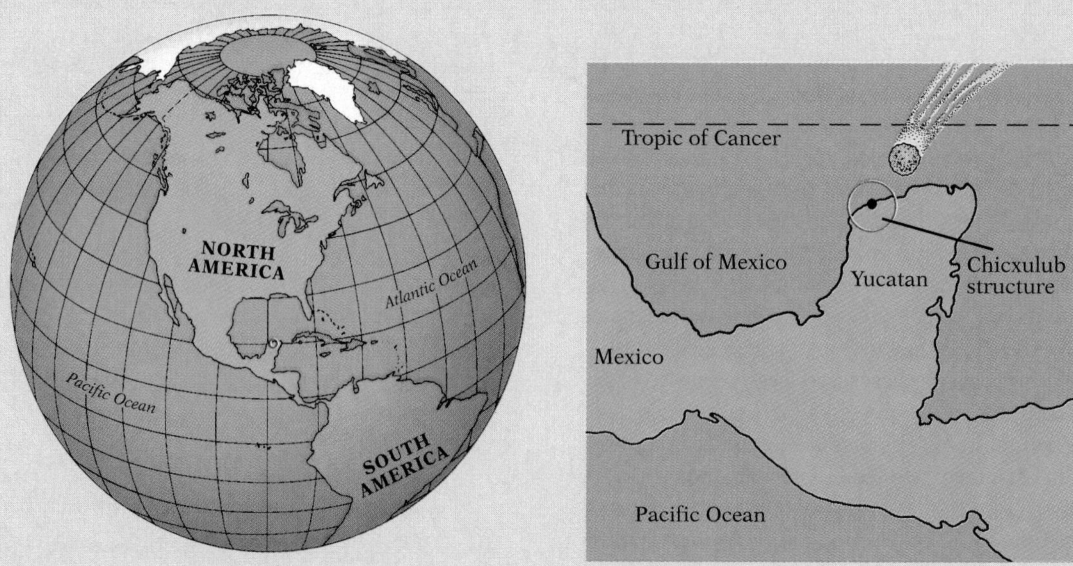

A Possible site of asteroid impact in the Yucatan

How could asteroids or comets cause mass extinctions? If a large extraterrestrial object splashed into the ocean, an enormous tidal wave would have swept around the world and devastated coastal life. There is evidence that at the K-T boundary such an event occurred in the Gulf of Mexico. An impact on land might well have darkened the skies with so much dust as to reduce photosynthesis below the level necessary to support most plants and animals. It is also thought that the movement of the comets or asteroids through the atmosphere would ionize air molecules, which would fall to the ground as acid rain. The rain, which might have exceeded pH 1, could make the ocean's surface acidic enough to kill off many tiny marine animals (and thus the animals that feed on them) by dissolving their calcium-based shells. This would help explain why species with calcium-based shells suffered at the K-T boundary far more than those with silica-based shells. There is even some evidence of a massive fire that could have produced enough soot to have blackened the skies for a very long time. In addition to the effect this would have had on photosynthesis, such a layer of soot could prevent the sun's warming rays from reaching the earth's surface, resulting in a prolonged and extremely severe winter. The combination of darkness, cold, and fires is thought to have wiped out the plants in the western United States at the K-T boundary.

Not all the large-scale extinctions in the fossil record are thought to have had an extraterrestrial cause; indeed, no sharp iridium layer has been found in sediments associated with other extinction events. Many scientists theorize that widespread volcanic activity and upwellings of lava could have been responsible. They point out that volcanic activity could also produce enough dust and smoke to alter world climate dramatically and reduce photosynthesis. Furthermore, there is evidence of extensive volcanic activity in India at the time of the supposed asteroid impact at the K-T boundary. While Alvarez and his supporters believe that the impact alone caused the Cretaceous extinction, others argue that the impact—if it occurred—was simply the final blow in an environment already weakened by other stresses.

Alternating changes in the earth's climate from "greenhouse" to "icehouse" conditions, for example, could lead to wholesale extinctions of organisms unable to live under the new conditions. Greenland once had breadfruit trees and other tropical plants similar to those found today on South Sea islands! Global temperature changes are associated with the melting and freezing of the polar ice caps. Large-scale fluctuations in global sea level would have destroyed habitats and led to extinctions in the marine environment; changes in ocean circulation patterns may have led to a reduction in the amount of oxygen available and the demise of many marine species. Any of these events, the impact of an asteroid or comet, a cycle of intense volcanic activity, changes in climate, alterations in sea level or changes in ocean circulation, would almost certainly have created an evolutionary crisis of some sort, and would have initiated a series of widespread changes in the tempo and direction of evolution. An increasing number of scientists therefore take these hypotheses quite seriously; nevertheless, much about the mass extinctions and their evolutionary impact remains to be explained.

If Alvarez is right and cosmic impacts have wrought such havoc over the eons, do they present a danger to life on earth today? Many astronomers say yes. They point out that at least 130 impact craters have been found to date on the earth's surface (Fig. B). Asteroids one kilometer in diameter and larger are estimated to hit the earth once every 300,000

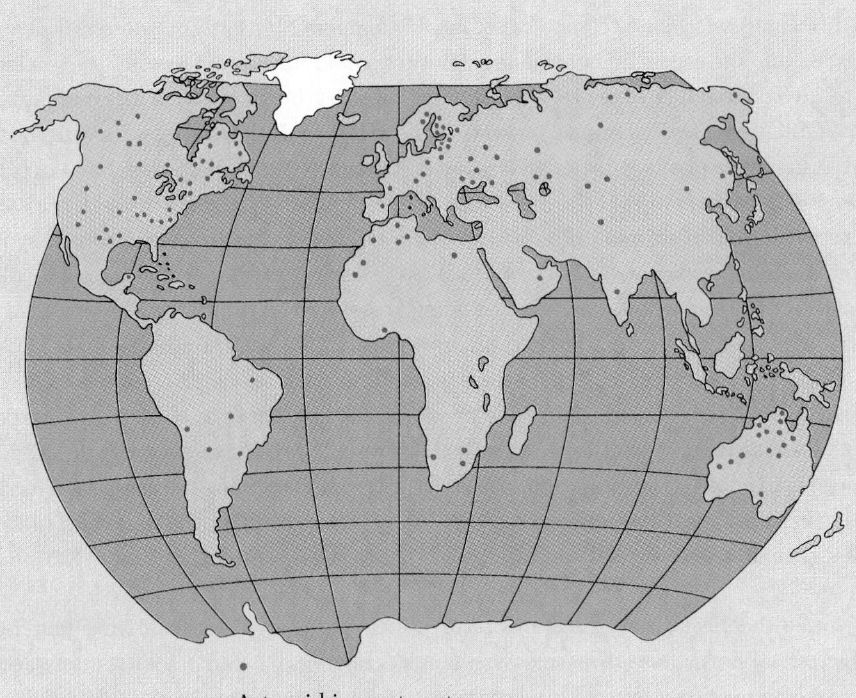

B One hundred thirty identified impact craters

years. Such an asteroid would hit the earth at about 16 miles per second, and the force of impact would be equivalent to the explosion of a million H bombs. So much pulverized rock and dust would be thrown into the atmosphere that sunlight could not reach the earth's surface, crippling agriculture and leading to mass starvation. What is the risk of a collision any time soon? NASA estimates that there are between 1,050 and 4,200 earth-threatening asteroids, and that an individual has about a one in 7,000 chance of seeing an impact during his lifetime. The risk is small but not insignificant, say some scientists. The collisions are rare, but have critical consequences for life when they happen. Accordingly, a government-appointed team has recently proposed a $50-million early warning system. Other scientists call the risk ridiculously small and the warning system a waste of taxpayers' money. The decision to implement the plan will be a political one, and you are likely to see many articles on this subject in future years.

periods of time (Fig. 32.24). This theory, sometimes referred to as ***gradualism,*** has been challenged by Niles Eldredge of the American Museum of Natural History and Stephen Jay Gould of Harvard University. They argue that the fossil record, for the most part, does not support gradualism. Though millions of fossils have been found, rarely is there a complete series of specimens demonstrating the gradual transition from one species to another. More often the fossil record shows one fossil species persisting relatively unchanged for perhaps millions of years, and then another, quite different species "suddenly" appearing in the rocks just above it, with the intermediate fossils often missing. Some scientist believe these gaps will eventually be filled by new finds, but Eldredge and Gould reason that the gaps in the record represent natural phenomena of evolution. In 1972 they proposed a new theory, called ***punctuated equilibrium,*** to account for these gaps. Their theory states that

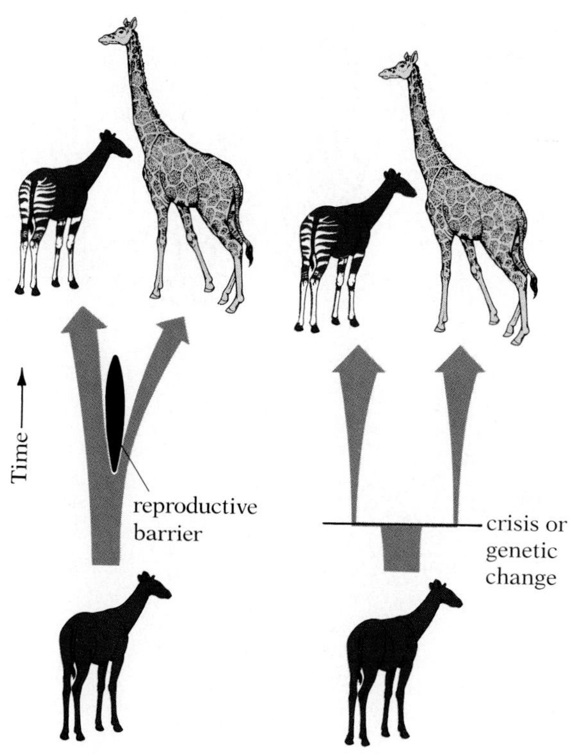

32.24 Gradualism versus punctuated equilibrium These drawings represent two hypotheses about how different species—here, the okapi and the giraffe—might have evolved from a common ancestor, the pre-okapi. Forms intermediate between the pre-okapi and the okapi and giraffe are not shown. (A) In the conventional interpretation of natural selection, a population is imagined to evolve gradually (as suggested by the gentle leftward movement of the lower part of the arrow, indicating a slow change in one trait) until a reproductive barrier (usually geographic isolation) separates two parts of the population. Generally one of the subgroups is very small, as indicated by the narrowness of the right-hand branch; it may diverge slowly from the larger subgroup through selection or genetic drift or both. (B) According to the punctuated-equilibrium model, there is little or no gradual change of a trait over time. Instead, a crisis or other event selects or gives rise by chance to one or more populations with traits very different from those of the ancestral population.

most evolution is marked by long periods of equilibrium, when the organisms change very little, punctuated by short periods of rapid change in small populations. In others words, they reasoned, evolution occurs by fits and starts, not by the slow, gradual process envisioned by Darwin. If speciation does occur very rapidly, and in small isolated populations, as Eldredge and Gould believe, then the chances of finding fossil evidence of intermediate forms is extremely small. Gaps in the fossil record are therefore to be expected, and the new species, once formed, would change very little with time, and would quickly reach equilibrium.

One of the most spectacular cases of rapid evolutionary radiation occurred in the geologic time division called the Cambrian period, about 500 to 570 million years ago. The "sudden" appearance of complex animals is a potentially compelling instance of punctuation. Two sites with well-preserved fossil remains have been discovered, one the Burgess shale deposit in the mountains of western Canada, the other in southern China. Because many of the fossils are similar, the animals probably had a worldwide distribution, and must have evolved over the course of a few million years.

The animals that typify a given place or, in this case, a particular period are known collectively as ***fauna.*** The Burgess fauna consist mainly of arthropods, but also contain a group of segmented worms and the oldest known chordate (Fig. 32.25). None of the Burgess arthropods look like any modern species. The diversity of design among the various animals is staggering, and poses an evolutionary problem: how did they come into being over such a relatively brief period? Most biologists believe that the explosion in diversity was the natural consequence of a biological breakthrough: the invention of a novel developmental strategy. The breakthrough came with the appearance of segmentation: the embryo divides along the anterior-posterior axis into a number of nearly identical units, each of which can differentiate to produce specialized organs and appendages. The ancient body plans that evolved during the Cambrian persist, with modifications, to this day; few new body plans have appeared since that time.

One of the reasons for the remarkable diversity that followed is the absence of competition in an environment with nothing occupying the many niches available to multicellular organisms. As the number of different species began to saturate the environment, however, natural selection would necessarily become more intense, and less efficient designs and less stable developmental programs probably went extinct. This scenario, therefore, postulates a kind of punctuation, followed by more gradual evolution and selective pruning.

Whatever the fate of the punctuation theory, the evolution of organisms may be less uniform and gradual than many biologists have supposed. The usual tempo of speciation probably lies somewhere between the gradual-change and the punctuated-equilibrium models.

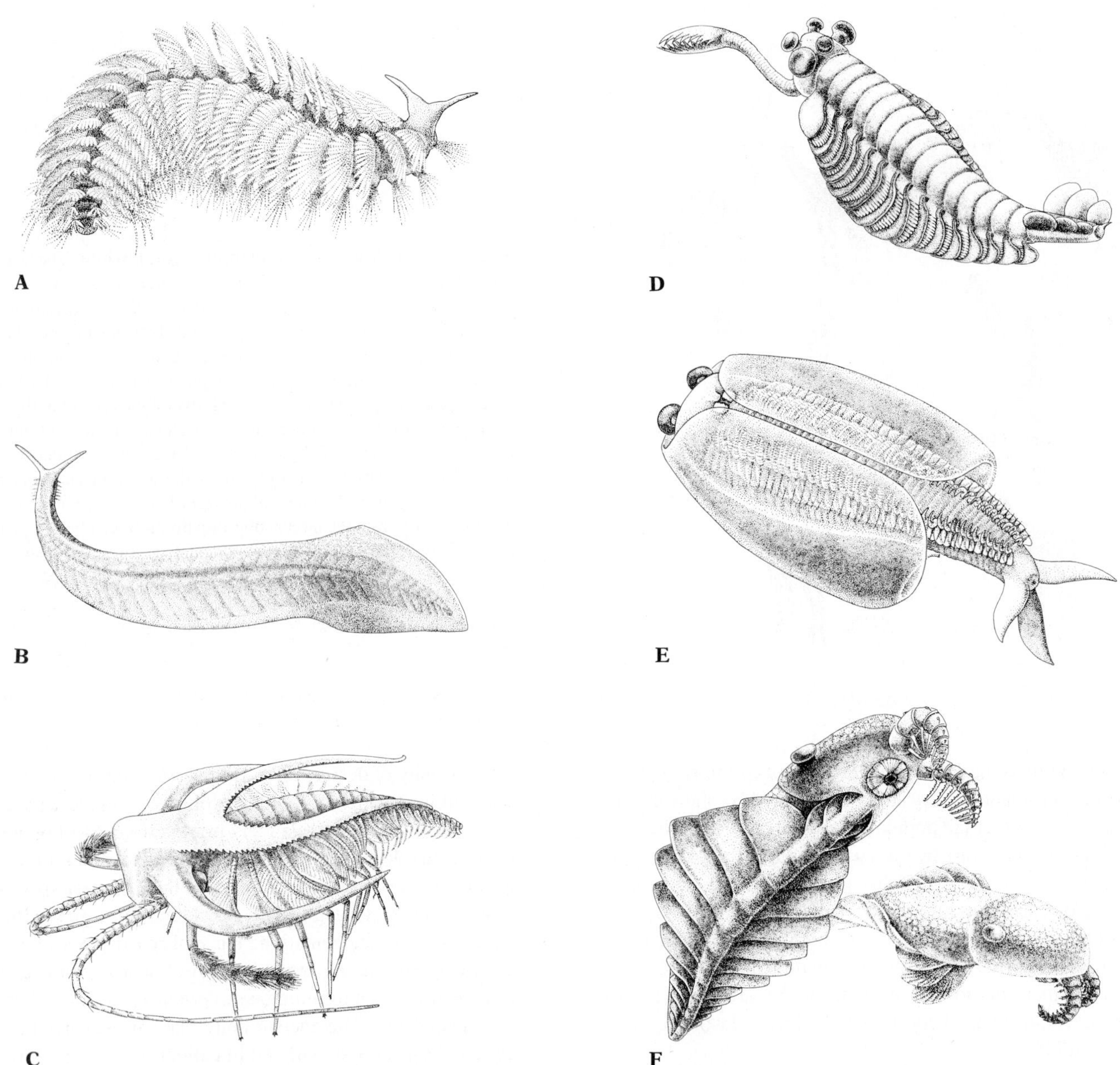

32.25 Some species from the Burgess shale The Burgess fauna include annelids (segmented worms) like the polychaete worm shown here (A) and the oldest preserved chordate (B), whose notochord and somite-produced muscle bands are clearly visible. Among the many kinds of arthropods are curious versions of modern groups (not shown). The remaining species, of which only a small fraction are illustrated here (C–F), represent 20 or more extinct groups of arthropods.

THE CONCEPT OF PHYLOGENY

Evolution implies that many unlike species have a common ancestor and that all forms of life probably stem from the same remote beginnings. One of the challenges the theory of evolution holds for biologists is to discover the relationships among the species alive today and to trace the ancestors from which they descended. Two species are said to be related when they share a common ancestor; the more recent the common ancestor, the more closely related the two organisms are.

Systematics is the science of comparative biology in which the diversity of organisms is studied. It includes taxonomy and

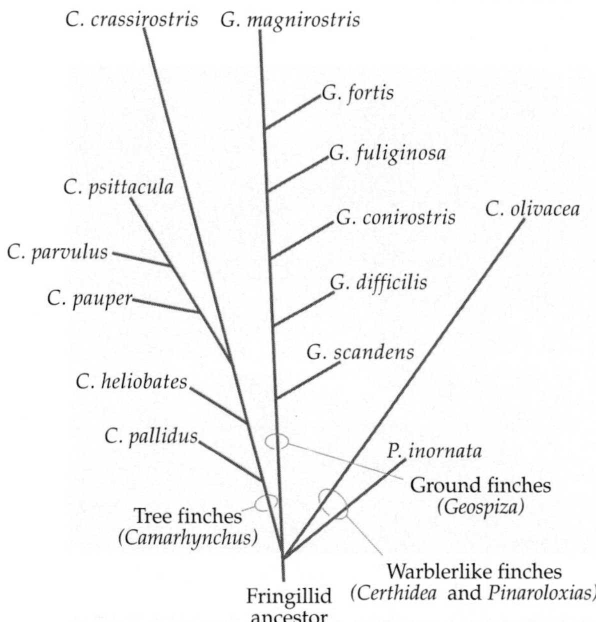

32.26 A phylogenetic tree for Darwin's finches This tree is based on both the degree and the nature of the morphological differences between species. The distances along branches from any one species to another reflect inferred degrees of relatedness.

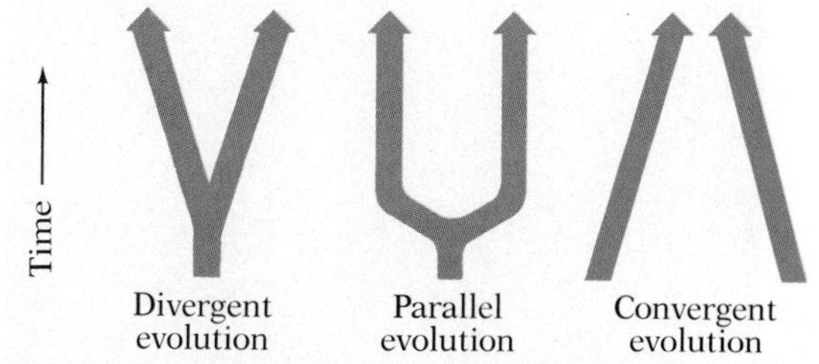

32.27 Patterns of evolution In branching (divergent) evolution one stock splits into two, which becomes less and less like each other as time passes. In parallel evolution two related species evolve in much the same way for a long period of time, probably in response to similar environmental selection pressures. Convergent evolution occurs when two groups that are not closely related come to resemble each other more and more as time passes; this is usually the result of their occupation of similar habitats, and so the presence of similar selection pressures.

classification as well as the study of evolutionary relationships. To understand these relationships, systematists must rely on a comparison of living organisms, examining the distribution of characters in various groups. But because so many different kinds of evidence must be weighed, reconstructing the evolutionary history—the ***phylogeny***—of any group of organisms always entails a strong element of speculation (Fig. 32.26).

Similarities Among Organisms Can Be Due to Relatedness or Convergence

The most widely studied characters used for classification pertain to morphology—including external morphology, internal structure, histology (tissue structure), and the morphology of the chromosomes in cell nuclei. Comparisons of organisms based only on morphology, however, can sometimes be misleading. When different species resemble each other, it can be for one of two reasons: they may share a common ancestor and therefore be related, or they may live in similar environments and, though unrelated, be subject to similar selection pressures. For example, both whales and fish, though not closely related, have fins and a streamlined shape—both adaptations for swimming through water. When organisms that are not closely related become similar in one or more characters because of independent adaptations to similar environmental conditions, they are said to have undergone ***convergent evolution,*** and the phenomenon is called ***convergence*** (Fig. 32.27). Whales, which are mammals descended from terrestrial ancestors, have evolved flippers from the legs of their ancestors; those flippers superficially resemble the fins of fish, but the resemblances are due to convergence, and do not indicate a close relationship between whales and fish. Both arthropods and terrestrial vertebrates have evolved jointed legs and hinged jaws, but these similarities do not indicate that arthropods and vertebrates have evolved from a common ancestor that also had jointed legs and hinged jaws; there is ample evidence that these two groups of animals evolved their legs and jaws independently and that their legless ancestors were not closely related. The "moles" of Australia are not true moles but marsupials (mammals whose young are born at an early stage of embryonic development and complete the development in a pouch on the mother's abdomen); they occupy the same habitat in Australia as do the true moles in other parts of the world and have, as a result, convergently evolved many startling similarities to the true moles (e.g., rudimentary eyes, no external ears, smooth fur, broad digging claws). The marsupial mole is but one of a vast array of Australian marsupials that are strikingly convergent with placental mammals of other continents (Fig. 32.28).

When systematists attempt to determine the evolutionary relationship between two species, they must try to determine whether the similarities are ***homologous*** (inherited from a common ancestor, i.e., similarity by descent) or merely ***analogous*** (similar in function and often in appearance, but of different evolutionary origins). *Homologies are similarities shared by two species because of relatedness whereas analogies are functional similarities due to convergent evolution.* The legs of lizards and those of cats are considered homologous because the evidence indicates that both were derived

32.28 Australian marsupials that are convergent with placental mammals of other continents (A) A marsupial mouse. (B) A marsupial glider, convergent with placental flying squirrels. (C) The tiger cat *(Dasyurus)*, a marsupial carnivore. (D) A cuscus, a marsupial monkey of New Guinea with a prehensile tail.

from the legs of a common ancestor. The legs of lizards and grasshoppers are analogous, because, though they are functionally similar structures, they were not inherited from a common ancestor with legs. Instead, they evolved independently and from very different ancestral structures.

Thus investigators are often faced with the problem of interpreting the similarities and differences they find. They must always ask themselves whether close similarities in a particular character really indicate close phylogenetic relationship or whether they simply reflect similar adaptation to the same environmental situation. Convergence is common in nature and is a frequent source of confusion in phylogenetic studies.

Morphology and the Fossil Record Provide Data for Constructing Phylogenies

When biologists set out to reconstruct the phylogeny of a group of species believed to be related, they usually have nothing more before them than the species living today and (sometimes) the fossil record. To reconstruct phylogenetic history as accurately as possible, they must make inferences based on observational and experimental data. The difficulty is that what can be measured is *similarity,* whereas the goal is to determine *relatedness.* Morphological characters are the most widely

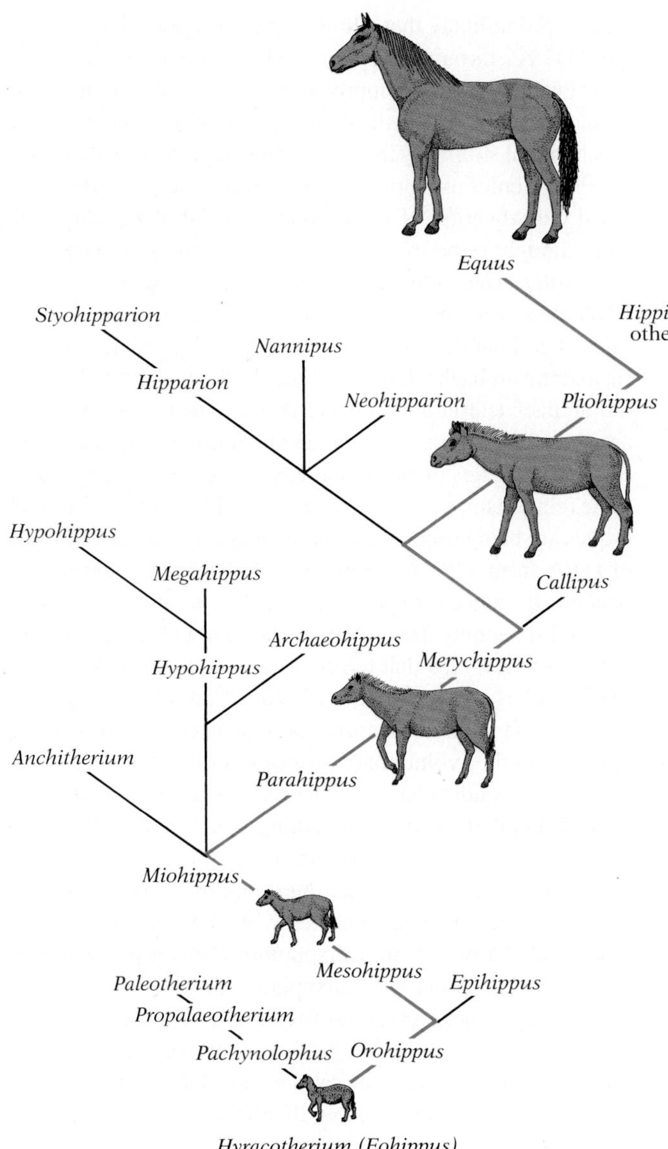

32.29 Presumed evolution of horses The fairly complete fossil record of horses has enabled paleontologists to work out a reasonable picture of the evolutionary history of the group. The emphasis here is on the direct ancestors of the modern horse; many major branches left no modern descendants. *Hyracotherium* lived in the Eocene epoch about 55 million years ago. It was a small animal, weighing only a few kilograms. It had four toes on each front foot and three on each rear foot. It was a browser, feeding on trees and bushes. *Mesohippus,* which lived during the Oligocene epoch about 35 million years ago, was a bit larger, and its front feet, like the rear feet, had only three toes. *Merychippus,* a grazer, lived during the Miocene about 25 million years ago. It had three toes on each foot, the middle one much larger than the others, which were short and thin and did not reach the ground. *Pliohippus,* of the Pliocene, often had only one toe on each foot, though in some individuals tiny remnants of other toes persisted. *Equus,* the modern horse, is much larger than the ancestors shown here. It has only one toe on each foot.

used features or traits used by systematists to understand evolutionary relationships. It is particularly helpful when morphological characters of living species can be compared with those of fossil forms.

Fossils are the remains of organisms, or the direct indications of the former presence of organisms. They may be stems of plants, shells of molluscs, skeletons of vertebrates, imprints of leaves, tracks of animals, etc. The fossil record is the most direct source of evidence about the stages through which ancient forms of life passed. Unfortunately that record is usually incomplete, and for many groups of organisms there is no suitable fossil record at all. Fossilization is a hit-or-miss business, usually requiring the presence of hard parts (e.g., bone, shells, teeth) and rapid burial. Most soft-bodied organisms, therefore, have left no fossil record, and there are few representatives from environments (such as on dry land) where rapid burial is unlikely. At best, fossils may suggest the broad evolutionary outlines of major groups. Fossils also give us a sense of how much time it takes for groups to evolve and diversify. In some groups, notably the horses, the fossil record has provided much phylogenetic information that could have been obtained from no other source (Fig. 32.29). The fossil record of the evolution of humans is more abundant than that of many other species.

Embryology Often Provides Evidence for Determining Phylogeny

Morphological characters are often easier to interpret if the manner in which they develop is known. For example, if a particular structure in organism *A* and a structure of quite different appearance in organism *B* both develop from the same embryonic structure, the resemblances and differences between those structures in *A* and *B* take on a phylogenetic significance that they would not have had if they had developed from entirely different embryonic structures. Embryological evidence often allows biologists to trace the probable evolutionary changes that have occurred in important structures and helps them reconstruct the probable chain of evolutionary events that led to the modern forms. For example, the fact that pharyngeal gill slits appear briefly during the early embryology of mammals, including humans, is thought to indicate that the distant ancestors of land vertebrates were aquatic.

Homology Can Be Determined at the Biochemical and Molecular Levels

The morphology of the adult and embryo, combined when possible with information from the fossil record, has traditionally been the basic source of data on which phylogenetic hypotheses have been founded. Today, differences among organisms are also studied on the molecular level. Comparisons of macromolecules (proteins, DNA, ribosomal RNA) in living organisms can be used to determine evolutionary relationships. For instance, amino acid sequences in proteins of different species can lead to a measure of the evolutionary distance between the species; the more alike the sequences, the more closely related the species. Likewise, the similarity of nucleotide sequences in DNA or ribosomal RNAs from different species can give a measure of their degree of genetic similarity. When the sequence of amino acids in a protein or the nucleotides in DNA or RNA from different organisms are very similar, it is thought to be due to homology, i.e., a common evolutionary origin. It is considered unlikely that a high degree of molecular similarity would occur through convergence because mutations are random events, and it is improbable that two different species would experience identical mutations. Consequently, close biochemical similarity indicates some degree of relatedness.

Phylogenies using molecular characters have resulted in revised classification of microorganisms. Until recently, there were thought to be two cellular lineages, the procaryotes and eucaryotes. The comparative sequencing studies of ribosomal RNA, however, have showed that there are at least three main lines of cellular descent, two of them procaryotic (the eubacteria and the archaebacteria), and the third eucaryotic. Although the archaebacteria and eubacteria are both procaryotes, they are as different from each other as either is from the eucaryotes. Similarly, studies of ribosomal RNA have been used to determine relationships among the various animal groups. The techniques are being used in interesting ways. For instance, studies of DNA from a 25- to 30-million-year-old fossil termite embedded in amber suggest that termites and cockroaches evolved separately from a common ancestor (Fig. 32.30). (The long-held view was that termites evolved from cockroaches.) DNA studies on other amber-entombed insects and preserved bones from tar pits, such as the La Brea tar pits in Los Angeles, may clarify the evolutionary histories of other extinct animals. Sequencing studies have also helped systematists make some very difficult decisions concerning species classification—e.g., settling a heated debate over the ancestry of the Tasmanian wolf. Molecular data support classical taxonomists' placement of this species with the Australian line of marsupial carnivores; other taxonomists had classified it with a group of South American marsupials.

Although many researchers tend to give more credence to molecular than to morphological data, molecular data are not without difficulties and limitations, and DNA/RNA homologies are not any more reliable than morphological evidence.

32.30 Ancient termite embedded in amber The amber that contains this specimen of an extinct species of termite is from the Dominican Republic and is 25 to 30 million years old. Researchers from the American Museum of Natural History in New York City extracted and sequenced DNA from a similar specimen to clarify the evolutionary history of these and related organisms.

MAKING CONNECTIONS

1. How did the evolution of a novel developmental strategy probably contribute to the rapid radiation of multicellular animals during the Cambrian period?

2. Unlike all other mammals, giant pandas and humans enjoy opposable thumbs. How could you use the two species' embryological development to determine whether opposable thumbs are homologous or analogous structures?

3. How might you use recently developed techniques for sequencing proteins and nucleic acids to determine the evolutionary relationship of the giant and lesser pandas?

4. How have biologists attempted to construct a *molecular clock* based on the occurrence of mutations? What difficulties have they encountered?

One cannot therefore consider one type of data "better" than another. All characters should be carefully analyzed and evaluated before they are included in a phylogenetic analysis. A particular hypothesis regarding phylogenetic relationships is most strongly supported (or rejected) if data from a variety of sources, both molecular and morphological, all lead to the same conclusion.

Another early hope for sequence analysis was that the number of differences in the molecules being analyzed could be used as a ***molecular clock*** to precisely date the points at which branching of one group from another occurred. Molecular biologists postulated that if the sequences of nucleotide bases in DNA or RNA and the amino acids in proteins mutate at approximately constant rates, then the differences between the sequences could provide an accurate measure of how long ago different lineages separated from one another. However, it has become clear that different genes within the same organisms can change at very different rates; the same is true for the same gene in different species. Even dates inferred by averaging many different genes must at present be considered approximate.

PHYLOGENY AND CLASSIFICATION

Over a million species of animals and over 248,000 species of plants are known, and large numbers of new species are discovered each year (Fig. 32.31). To deal with this vast array of organic diversity, some system to classify species in a logical and meaningful way is needed. Our system of classification began with the work of the great Swedish naturalist Carolus Linnaeus (1707–1778). Linnaeus grouped organisms solely on the basis of morphological similarities. Today, by contrast, nearly all modern systematists agree that a classification of organisms should reflect their evolutionary relationships.

Whenever new techniques for studying organisms are devised, the data they produce provide another source of information for the systematists. It is the systematists' task to correlate all that is known about the organisms under investigation, and to construct, in the light of modern evolutionary theory, some sort of intelligible picture of these organisms' relationships with one another. Systematists are thus just what the term implies: they are the ones who try to put into order all the information gathered about organisms by anatomists, paleontologists, cytologists, physiologists, geneticists, embryologists, ethologists, ecologists, biochemists, and still other specialists. There are a number of different approaches used in systematics. We shall consider three of these—classical evolutionary taxonomy, phenetics, and cladistics.

Classical evolutionary taxonomy is the traditional approach and uses elements of the other two. It depends more than any other approach on intuition, experience, and subjective judgment. The most widely used classification characters pertain to morphology—including external morphology, internal anatomy, embryology, and histology. The fossil record, when present, provides additional information. The usual procedure in reconstructing phylogenies by the classical method is to examine as many independent characters of the species in question as possible and to determine which characters in these species are most important in showing phylogenetic relationships.

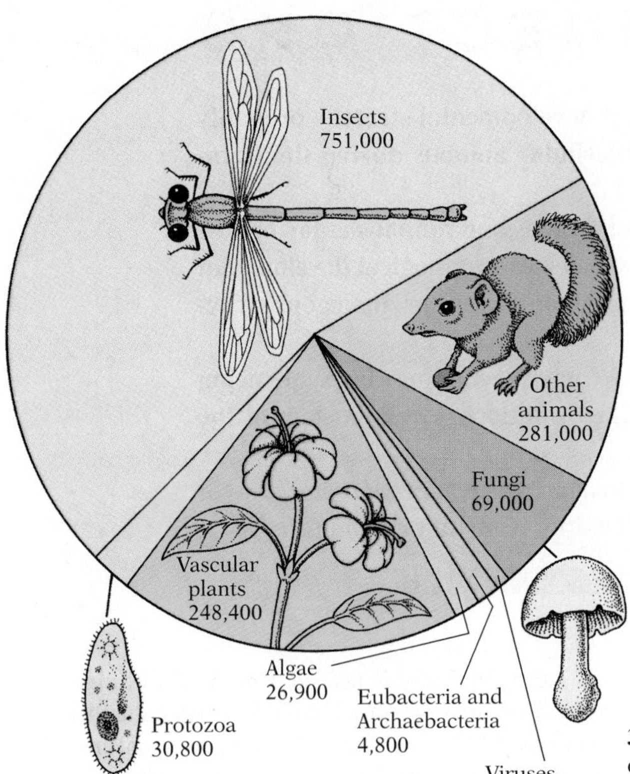

32.31 Diversity of known living species To date, about 1,413,000 species of organisms have been described. Note that in terms of numbers, the vascular plants and insects predominate.

Cladistics Uses Shared Derived Characters for Classification

Recently, two new disciplines have emerged that challenge the traditional method: phenetics and cladistics (phlylogenetic systematics). ***Phenetics,*** often called numerical taxonomy, estimates taxonomic affinities by looking at as many characters as possible and then applying mathematical procedures to determine the overall similarities of various organisms. Soon after phenetics was developed, it was realized that, because of the problem of convergence, "overall similarity" was not the best grouping criterion, and another method—cladistics—was formulated. The foundation of ***cladistics*** is that biological classification should only reflect phylogenetic relationships in contrast to other systems where other factors play a role. Cladists recognize that the similarities between organisms can be of two types: ***ancestral*** (similarities inherited with little or no change from a common ancestor) or ***derived*** (recently evolved traits or features). All species are a mixture of ancestral (sometimes called primitive) and derived (sometimes called advanced) characters. In humans, for instance, our five-digit limbs are an example of an ancestral character since the number of digits has remained unchanged from the ancestral amphibians, but our large cerebral cortex is a derived character. Cladistic methodology requires that groups (clades) be based only on ***shared derived characters***—traits that not only are common to the several species in question, but that are of relatively recent rather than ancestral origin as well (Fig. 32.32). Shared ancestral characteristics are not used to define groups. Hence, in classifying two species of bats, the traits shared by mammals in general would not be considered. Instead, the cladist would attempt to identify unique characters that differentiate one species from a related species. Cladistic methodology dominates systematics today.

The Classification System Uses a Hierarchy of Categories

Suppose you had to classify all the people on earth according to where they live. You would probably begin by dividing the entire world population into groups based on country, thus separating the inhabitants of the United States from those of France and Argentina. Next, you would probably subdivide the population of the United States by state, then by county, then by city or village or township, then by street, and finally by house number. The same would be done for Mexico, England,

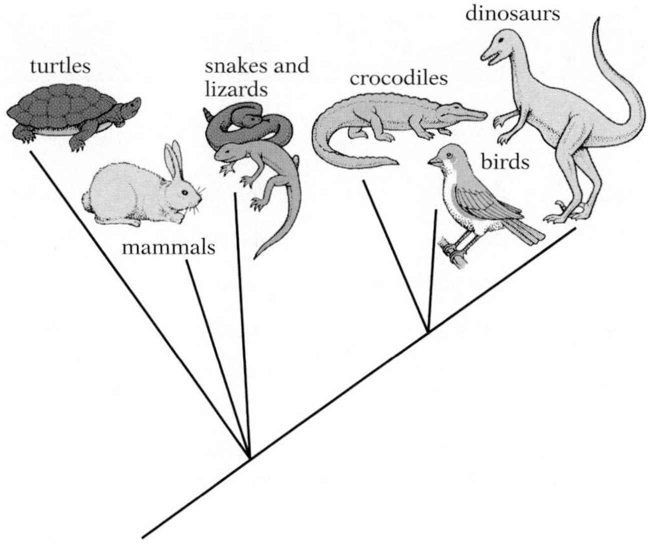

32.32 Cladistic analysis of the phylogeny of higher vertebrates The major feature of this reconstruction is that reptiles (turtles, snakes, lizards, and crocodiles) do not belong to a single class. Depending on how a cladist chooses to group species, mammals, snakes, and turtles could form one class, with birds and crocodiles in the other; alternatively, they could be five different classes of reptiles, with birds and mammals in their own classes. What is important to remember is that cladists classify groups based on shared derived characteristics.

Australia, and all the other countries, using the political subdivisions in those countries that correspond to states, counties, etc., in the United States. This procedure would enable you to place every individual in an orderly system of hierarchically arranged categories, as follows:

Country
  State
    County
      City
        Street
          Number

Note that each level in this hierarchy is contained within and is partly determined by all levels above it. Thus, once the country has been determined as the United States, a Mexican state or a Canadian province is excluded. Similarly, once the state has been determined as Pennsylvania, a county in New York or California is excluded.

The same principles apply to the classification of living things on the basis of phylogenetic relationships. Again, a hierarchy of categories is used, as follows:

Kingdom
  Phylum or Division
    Class
      Order
        Family
          Genus
            Species

Each category (taxon) in this hierarchy is a collective unit containing one or more groups from the next-lower level in the hierarchy. Thus a genus is a group of closely related species; a family is a group of related genera; an order is a group of related families; a class is a group of related orders; and so on. The species in any one genus are believed to be more closely related to each other than to species in any other genus; the genera in any one family are believed to be more closely related to each other than to genera in any other family; the families in any one order are believed to be more closely related to each other than to families in any other order; and so on.

Table 32.3 gives the classification of six species. Notice that the table shows us immediately that the six species are not closely related, but that a human being and a wolf are more closely related to each other, both being Mammalia, than either is to a bird such as the herring gull. And it shows us that the mammals and the bird are more closely related to each other than to the housefly, which is in a different phylum, or to the moss or the red oak, both of which are in a different kingdom. These relationships are correlated with similarities and differences in morphology, physiology, and ecology.

The six species of Table 32.3 are all well known, and the relationships between them are probably intuitively clear to you. But many species are not so familiar, and the relationships between them are not so obvious. Much research may be necessary before such species can be fitted into the classification system with any degree of certainty, and their assignment to genus or family (or even order) may have to be changed as more is learned about them. (An outline of a modern classification of living things is given in the Appendix.)

The modern system of naming species also dates from Linnaeus. Before him, there had been little uniformity in the designation of species. Some species had one-word names,

TABLE 32.3 *Classification of six species*

Category	*Haircap moss*	*Red oak*	*Housefly*	*Herring gull*	*Wolf*	*Human*
Kingdom	Plantae	Plantae	Animalia	Animalia	Animalia	Animalia
Phylum or Division	Bryophyta	Anthophyta	Arthropoda	Chordata	Chordata	Chordata
Class	Musci	Dicotyledones	Insecta	Aves	Mammalia	Mammalia
Order	Bryales	Fagales	Diptera	Charadriiformes	Carnivora	Primata
Family	Polytrichaceae	Fagaceae	Muscidae	Laridae	Canidae	Hominidae
Genus*	*Polytrichum*	*Quercus*	*Musca*	*Larus*	*Canis*	*Homo*

*The name of a particular species consists of the genus name and the species designation, both customarily in italics, as shown in this table.

others had two-word names, and still others had names consisting of long, descriptive phrases. Linnaeus simplified things by giving each species a name consisting of two words: first, the name of the genus to which the species belongs and, second, a designation for that particular species. Thus the honeybee is called *Apis mellifera.* A certain species of carnation is called *Dianthus prolifer;* other species in the genus *Dianthus* have the same first word in their names, but each has its own species name (e.g., *Dianthus caryophyllus, Dianthus barbatus*). No two species can have the same name. Notice that the names are always Latin (or Latinized) and that the genus name is capitalized while the species name is not. Both names are customarily printed in italics (underlined if handwritten or typed). The first time the organism is referred to in a manuscript the full scientific name is used, thereafter it is referred to using the initial of the genus (e.g., *D. prolifer*) or by its genus name (e.g., *Dianthus*).

The same Latin scientific names are used throughout the world. This uniformity of usage ensures that each scientist will know exactly which species another scientist is discussing. There would be no such assurance if common names were used, for not only does a given species have a different common name in each language, but it often has two or three names in a single language.

CONCEPTS IN BRIEF

1. According to the biological species concept, a *species* is a genetically distinctive group of natural populations *(demes)* that share a common gene pool and are *reproductively isolated* from all other such groups.

2. *Speciation* is the process by which one ancestral species gives rise to two or more descendant species, which grow increasingly unlike as they evolve.

3. The initiating factor in allopatric speciation is usually geographic separation; if a population is divided by some barrier, the separated *(allopatric)* populations will no longer be able to exchange genes. In time, the populations will evolve in different directions because of different allelic frequencies, genetic drift, different mutations, and different selective pressures.

4. Eventually, the populations may become genetically so different that they develop *intrinsic isolating mech-*

anisms—biological characteristics that prevent effective interbreeding should they again become *sympatric.*

5. Speciation that does not involve geographic isolation is called *sympatric speciation.* One important example is speciation by allopolyploidy, in which there is a multiplication of the number of chromosomes in a hybrid between two species.

6. Divergent evolution—the evolutionary splitting of species into many separate descendant species—results in an *adaptive radiation.* The rate of evolutionary change is not constant; when conditions change rapidly and organisms have new evolutionary opportunities, they may undergo a rapid evolutionary burst. Adaptive radiation like that of Darwin's finches on the Galápagos Islands helps account for the tremendous diversity among living things on earth today.

7. Although competition is believed to be the predominant force leading to speciation over time, the catastrophic effects of rare environmental crises may be important in the evolution of certain populations.

8. Two opposing views of evolutionary change have been proposed: *gradualism,* which states that evolution occurs by the gradual accumulation of very small changes over long periods of time, and *punctuated equilibrium,* which states that most evolution is marked by long periods of equilibrium followed by short bursts of rapid change.

9. The morphology of the adult and embryo, combined when possible with information from the fossil record, has traditionally been the basic source of data for reconstructing the evolutionary history (phylogeny) of organisms. More recently, techniques have been developed to compare the proteins, DNA, and mRNA of different species, and these provide additional information on relatedness.

10. When different species of animals resemble each other, it may be that they share a common ancestor and therefore are closely related, or they may be unrelated organisms that live in similar environments and over time have come to resemble each other (converge) because they are subject to similar selection pressures. Systematists must try to determine whether similarities between organisms are *homologous* or merely *analogous* before speculating about the degree of relationship between the organisms.

STUDY QUESTIONS

1. How is a *population* defined? What unit of population is represented by all of the largemouth bass in a pond? According to the biological species concept, why do all largemouth bass belong to a single species? (pp. 703–704)
2. Describe two patterns of intraspecific variation. To which is the term *subspecies* applicable? Why is this term controversial among biologists? (pp. 706–707)
3. Distinguish between *phyletic* and *branching* evolution. Which is exemplified by *allopatric speciation?* (p. 707)
4. How does *allopatric speciation* occur? What factors cause geographically separated populations to diverge in time? Distinguish between extrinsic and intrinsic barriers to gene flow. (pp. 708–712)
5. Which category of intrinsic isolating mechanism is represented by each of these examples? (a) Fireflies use species-specific flashing patterns to identify a mate of their own species. (b) When artificially crossed, two tree species produce vigorous and fertile hybrids, but the two species are adapted to different climates and therefore are completely allopatric. (c) Two species of plants cannot interbreed because their flowers differ in size and shape and thus require pollination by different species of bees. (d) Two species of butterflies mate almost randomly and produce fertile offspring when they hybridize, but the hybrids are less viable than the parental forms. (pp. 709–712)
6. How does *sympatric speciation* most often occur, especially in plants? What other mechanisms for sympatric speciation have been proposed? (pp. 712–713)
7. Is the evolution of Darwin's finches an example of *microevolution* or *macroevolution?* (pp. 714–716)
8. Using Darwin's finches of the Galápagos Islands as an

example, explain how *adaptive radiation* can produce a diverse array of descendant species from a single ancestral form. What is the role of *character displacement* in this process? (pp. 714–716)

9. **Compare *gradualism* and *punctuated equilibrium* as theories of evolutionary change. How does each theory interpret the gaps in the fossil record? (pp. 721, 724–725)**
10. **Why is the distinction between *homologous* and *analogous* structures important in studying evolutionary relationships? What evolutionary process produces analogous structures? (pp. 727–728)**
11. **What information and techniques are used to construct the *phylogeny* of a group of organisms? Is any single source of data necessarily better than another? (pp. 726–727)**
12. **Describe the *cladistic* method of systematics. How does cladistics differ from *phenetics* and *classical evolutionary taxonomy?* (pp. 732–734)**
13. **Describe the hierarchical system used to classify organisms. Why was Linnaeus' system of naming species a major advance over previous methods? (pp. 732–734)**

ECOLOGY

In June 1992 delegates and diplomats from some 178 countries attended the United Nations Conference on Environment and Development in Rio de Janeiro. The conference was convened to address environmental problems, such as human population growth, species extinction, global warming, ozone loss, and pollution, that are increasingly recognized to have global implications, and which therefore will require co-operation on an international level to solve. Two important conventions were signed by most countries attending the conference, one on preserving biodiversity, and the second on preventing climate change. But why are these issues so important? Does it matter if species become extinct? Why is global warming a concern? Is ozone loss a serious problem? This chapter and the next will attempt to answer some of these questions as we consider the subject of ecology, and many of the ecological problems we face. ***Ecology*** is the study of the interactions between organisms and their environment (Fig. 33.1). "Environment," in broad terms, embraces everything external to an organism that affects it, including physical (abiotic) factors such as light, temperature, rainfall, humidity, various pollutants, and topography as well as biotic factors such as parasites, predators, mates, and competitors.

Life is characterized by many different levels of organization. Much of this book has dealt with life on the molecular, cellular, tissue, organ, organ-system, and individual levels. Although ecology is often concerned with phenomena on these levels (e.g., the osmotic interactions between an organism and its environmental medium), the present chapter will deal primarily with three higher levels of organization: ***populations,*** which are groups of individuals belonging to the same species living in a particular geographic area at a particular time; ***communities,*** which are units composed of all the populations of all the species living in a given area; and ***ecosystems,*** which are communities and their physical environments considered together (Fig. 33.2). Each of these designations may be applied to a local area or to a widespread one. Thus, the sycamore trees in a given woodlot may be regarded as a population, and so may all the sycamore trees in a county. Similarly, a small pond and its inhabitants, or the pond and the forest in which it is located, may be treated as an ecosystem.

The different ecosystems are linked to one another by biological, chemical and physical processes. Inputs and outputs of energy, gases, inorganic chemicals, and organic compounds can cross ecosystem boundaries by way of meteorological factors such as wind and precipitation, geological factors such as running water and gravity, and biological factors such as the

Chapter opening photo: Landslide hazard to areas below Cantagalo Rock, Rio de Janeiro. As the human population continues to grow and expand, development has taken place in areas with little or no thought to city planning or geologic hazards. The areas below the rock were damaged by a serious landslide in 1966.

33.1 A fishing bat Organisms interact with other organisms and their environment in various ways. Here a fishing bat locates its prey by echolocation, sending out pulses of sound that reflect back from ripples or fish fins sticking out of the water. Once located, the bat sweeps down and grasps the fish with its enlarged feet. Almost immediately the fish is transferred to the mouth and devoured.

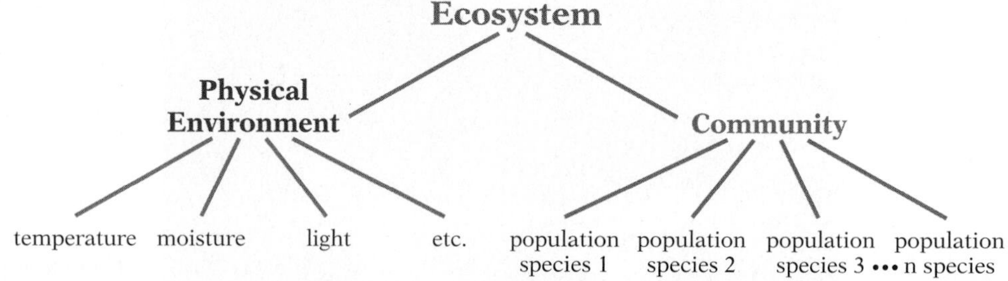

33.2 Ecosystem An ecosystem encompasses a community (itself made up of several populations) within any given area and the biotic (living) and abiotic (physical) factors (which together constitute the environment) that act upon it.

movement of animals. The earth itself is an ecosystem, in that no part is fully isolated from the rest. The global ecosystem is ordinarily called the ***biosphere*** (Fig. 33.3).

The biosphere contains all living organisms and their environment. It forms a relatively thin shell around the earth, extending only a few kilometers above and below sea level. Except for energy from the sun, it is self-sufficient; all other requirements for life, such as water, oxygen, and nutrients, are supplied by the utilization and recycling of materials already contained within the system.

33.3 The Apollo 17 view of earth The portion of the earth containing living organisms, the biosphere, is a relatively thin shell around the earth, extending only a few kilometers above and below sea level.

POPULATIONS AS UNITS OF STRUCTURE AND FUNCTION

We begin our study of ecology at the level of populations. Of special interest are the dynamics of populations and the environmental factors that help regulate them. That is, we want to know whether a population increases or decreases over time, and, if so, how and why such changes occur. Ecologists also strive to explain spatial patterns of distribution.

Population Size and Distribution of Populations Are Important Characteristics in an Ecosystem

Ecologists sometimes need to know the number of individuals in a population. When they consider endangered species, for example, the size of the surviving population is of crucial importance in the design of appropriate management procedures. Conservation authorities must have an accurate count (or at least a good estimate) of just how many whooping cranes or blue whales still exist on the earth.

But more often ecologists are concerned not with the total number of living individuals of a species, but rather with the density of the population in a given region—the number of individuals per unit area or volume (e.g., 50 pine trees per acre, or 5,000 diatoms per liter of water). In some situations, especially when the size of the individuals in a population is extremely variable, ecologists find that measures of biomass (the total mass of all the individuals), or its energy equivalent in calories, per unit area or volume is a more useful index to assess population size.

Individuals Are Distributed Uniformly, Randomly, or in Clumps

Within any given area the individuals of a population can be distributed in different ways (Fig. 33.4). Uniform distributions are not common; they occur only where environmental conditions are fairly uniform throughout the area and where there is

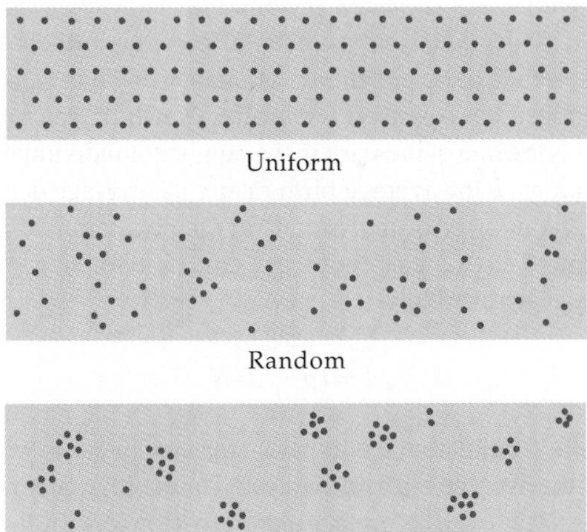

A

B

C

33.4 Uniform, random, and clumped distributions. (A) New Zealand tussock grass plants are spaced uniformly because there is intense competition between individuals and each plant inhibits the germination of seedlings in a circular zone around its base. (B) These *Crocus* flowers on the floor of a New Zealand forest are randomly distributed. (C) Flamingos exhibit behavior patterns that lead to active congregation in flocks, forming clumped distributions.

intense competition or antagonism between individuals. For example, creosote bushes are often spaced almost uniformly over a desert area because the roots of each bush give off toxic substances that prevent germination of seedlings in a circular zone around the base of the bush, and animals of many species frequently create a nearly uniform distribution of individuals by establishing and defending exclusive territories. Random distributions are also relatively rare. They occur where environmental conditions are uniform, where there is no intense competition or antagonism between individuals, and where there is no tendency for the individuals to aggregate. Clumping is by far the most common distribution pattern for both plants and animals in nature, for several reasons:

1. Environmental conditions are seldom uniform throughout even a relatively small area, so organisms tend to occur in those spots where conditions are most favorable for their development.
2. Reproductive patterns in both plants and animals often favor clumping. This is particularly true in plants that reproduce vegetatively (asexually) and in animals whose young remain with the parent.
3. Animals often exhibit behavior patterns that lead to active congregation in loose groups or in more organized colonies, schools, flocks or herds. Such behavior functions in defense, food-getting, and care of young.

A clumped distribution may increase competition for nutrients, space, or light, but this harmful effect is often offset by beneficial ones. For example, trees growing together in a hedgerow on the Great Plains may compete more intensely for nutrients and light than if they were widely separated, but they may be better able to withstand strong winds. And the cluster, which has less surface area in proportion to mass than an isolated tree, may be better able to conserve moisture. Aggregations of animals often reduce the rate of temperature change in their midst—an effect that is particularly important in cold weather. A group of animals may also have an advantage in locating food and in avoiding predation. The dynamics of such groups are complex. For instance, barnacle geese at the edge of a flock spend more time watching for predators, but they get to a new grazing area first, and have better quality food than those in the center of the flock, which are safer from predators (Fig. 33.5).

33.5 Barnacle geese feed in flocks in the winter Flocks of barnacle geese may reach 2,000 individuals in winter. The geese at the edge of a flock must watch for predators, but because the birds land in the center of a field and move outward, those at the edge get to a new grazing area first and have high-quality food. Those at the center of the flock are safer from predators but have poorer quality food available. Despite the risk from predators, research shows that those at the edge have greater breeding success due to their better diet, and birds compete for positions at the edge of the flock.

Populations Living Under Ideal Environmental Conditions Show Exponential Growth

One way to understand the dynamics of real populations is to determine the growth pattern of a population under ideal conditions and then to examine how conditions in nature modify this pattern. Consider a population under conditions that promote maximal growth, where, for instance, food and space are not limiting and there are no predators or parasites. The population growth under such ideal conditions can be enormous. For example, one pair of houseflies starting to breed in April would have 1.91×10^{20} descendants by August if all their eggs hatched and if all the resulting young survived to reproduce.

Calculations such as this led Charles Darwin to formulate his theory of natural selection. Darwin calculated the reproductive potential of the elephant, among the slowest breeders of known animals, and concluded that, under such ideal conditions as those described above, "after a period of 740 to 750 years there would be nearly 19 million live elephants descended from the first pair."

Besides houseflies and elephants, the same considerations apply to all organisms, whether plant or animal, unicellular or multicellular. All have the potential of explosive population growth; under ideal environmental conditions their growth curves would be exponential, as shown in Figure 33.6. Examination of this so-called J-shaped (exponential) curve shows that if I is the rate of increase in the number of individuals in the populations, b the average birth rate, d the average death rate, and N the number of individuals in the population at a given moment, then we can write the equation for a population growth curve as follows:

$$I = (b - d)\,N$$

From this formulation it is apparent that a population will grow only if the average birth rate exceeds the average death rate, so that the term $(b - d)$ is greater than zero. Conversely, the population will decline if the average birth rate is less than the aver-

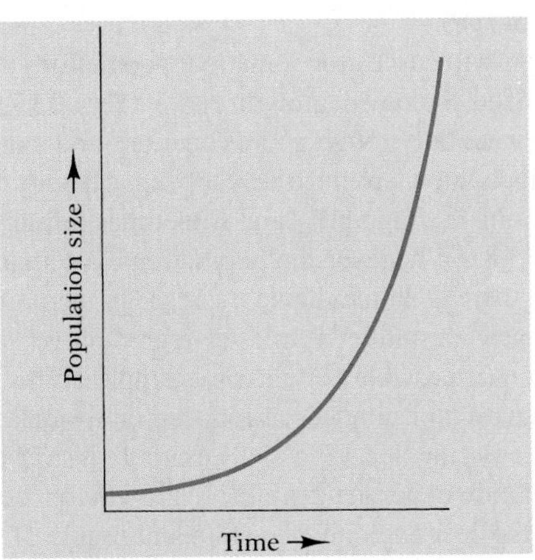

33.6 The J-shaped (exponential) growth curve The rate of increase steadily accelerates until, in theory, the population increases at an infinitely high rate. Clearly, no real population can continue increasing exponentially for long.

age death rate, that is, when the term $(b-d)$ is less than zero. If the birth and death rates are equal, i.e., $b=d$, then $(b-d)=0$, and therefore $I=0$; such a population will be neither growing nor decreasing. In summary, then, *it is the value of* (b − d), *the difference between the birth and death rates averaged over individuals in the population, that determines whether a population will grow, be stable, or decline.* This value, the ***intrinsic rate of increase*** of the population, is designated ***r***. Recognizing that $(b-d)=r$, we can rewrite the equation for the exponential growth curve as follows:

$$I = rN$$

In the hypothetical housefly or elephant population living under ideal environmental conditions and undergoing a population explosion, births exceed deaths and r is therefore greater than zero. But the rate of population growth is a function not just of r but also of N, the population size. Since N becomes larger in each successive generation, the rate of increase, I, will also be larger with each generation. It is this that makes the growth rate exponential.

Exponential population growth is much like an interest-bearing savings account where the interest is compounded. Every time interest is added to the principle, the amount drawing interest increases. As time goes on, the balance grows and the amount of interest earned also grows. So it is with a population—every time an individual is added, the number of reproducing individuals increases, and the population increases at an ever-increasing rate. An easy way to determine how fast a savings account, or a population, is growing is to determine its doubling time using the compound interest equation[1]:

$$\text{Doubling time} = \frac{0.7}{\text{growth (or interest) rate}}$$

In this equation 0.7 is a constant and the growth rate is equal to r, which can be calculated if you know the average birth and death rates. For example, in the United States the birth rate is 16 per 1,000 and the death rate is 8.5 per 1,000, so $r=0.016-0.0085$, or 0.0075. This means that our population is growing at a rate of 0.75 percent per year and will double in about 93 years (doubling time $=0.7/0.0075=93.3$). If the current growth rate for the world's human population is 1.7 percent, what is the doubling time?

Many Populations Show an S-Shaped Growth Curve

However, no real population keeps expanding at an exponential rate; otherwise, we would be buried in houseflies and elephants. The exponential growth of real populations is curtailed. In many instances populations show an ***S-shaped*** (logistic) growth curve, where there is an initial rapid expansion of the population when its density is low, followed by decelerating growth at higher densities, and an eventual leveling off as the density approaches what is called the ***carrying capacity*** of the environment (Fig. 33.7). The carrying capacity, or ***K***, as it is usually designated, is the maximum number of individuals in a population that the environment can support over a sustained period without its ability to support the same species in the future being reduced.

This changing growth pattern of a population as it ap-

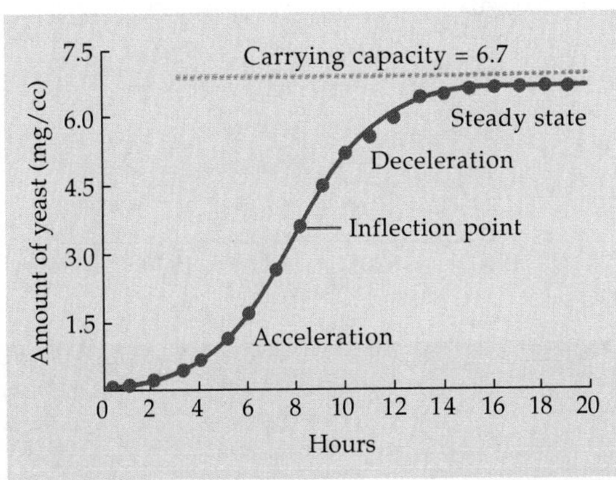

33.7 A representative S-shaped (logistic) growth curve Shown here is the growth curve of a laboratory population of yeast cells. Sometimes called a logistic growth curve, the S-shaped curve exhibits an accelerating rate of growth at low densities, but eventually reaches an inflection point, where the rate of change shifts from acceleration to deceleration. The deceleration continues as the population density approaches the carrying capacity of the environment. When the carrying capacity is reached, there is no further increase in density, and the population continues in steady state. This curve closely approximates the hypothetical logistic curve.

[1]The equation for computing the doubling time is inaccurate at growth rates above 10 percent (i.e., where $r = 0.1$).

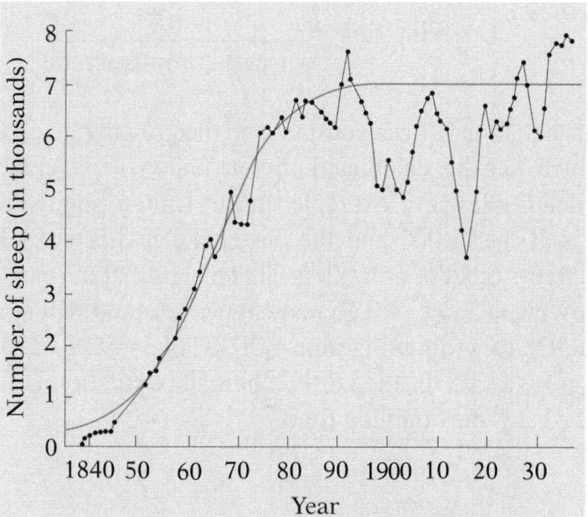

33.8 Growth curve of the sheep population of South Australia The smooth curve (color) is the hypothetical s-shaped curve about which the real curve seems to fluctuate.

proaches maximum density reflects one or several limiting factors that affect ever larger percentages of the individuals in the population as its size increases. Such a limitation is said to be a ***density-dependent*** factor. A population that has reached the steady-state level, at the tail end of an S-shaped curve, where births and deaths are in balance, is said to have ***zero population growth*** (or ***zpg***).

The growth curves for some real populations approximate the idealized S-shaped growth curve (Fig. 33.7), but more often they are only rough approximations and exhibit considerable fluctuation around the carrying capacity, sometimes overshooting it temporarily and sometimes falling well below it (Fig. 33.8). Whenever the population exceeds the carrying capacity, there is an inevitable decrease in the population due to increased mortality, decreased reproductive success, and emigration (if possible). A classic example of this is provided by the situation of reindeer introduced on a small Alaskan island in the Bering Sea. (Fig 33.9). Initially just 29 individuals, the population underwent explosive growth to reach a level of 2,000 as the deer exploited the abundant supply of slow-growing lichens that had been accumulating for centuries. With the deer population so large, the lichens were eaten faster than they could regrow. Without sufficient food, the population declined to fewer than 50 individuals as the deer starved to death. Furthermore, the huge population explosion of the deer disrupted the island's previously stable vegetation and actually lowered its carrying capacity for the deer to a level below what it had been before they were introduced.

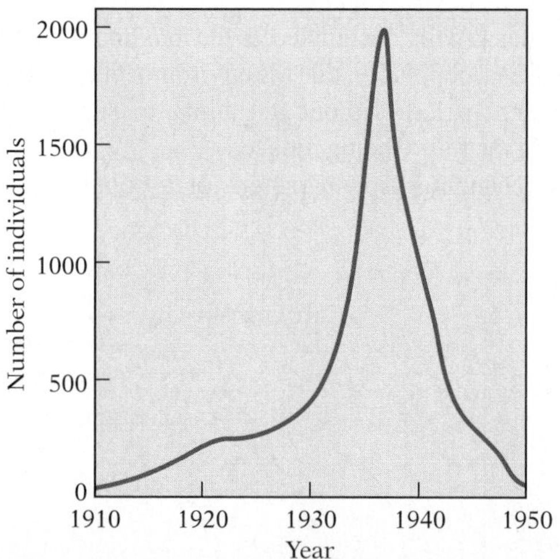

33.9 Reindeer population on an Alaskan island The reindeer population began with 29 individuals in 1910 and underwent explosive growth to reach a population of 2,000 by 1938. With the population at this level the deer severely depleted their food supply (lichens) and without sufficient food, the reindeer population precipitously declined to less than 50 individuals as the deer starved to death. The population explosion disrupted the previously stable vegetation of the island and lowered the carrying capacity of the island for the deer to a level below what it had been before the reindeer were introduced.

Some Populations Show a Boom-and-Bust Growth Curve

Not all growth curves of real populations assume the S-shaped form of Figures 33.7 and 33.8. The population densities of many small, short-lived organisms, and of organisms that live in disturbed or temporary habitats, often do not approach the carrying capacity of the environment before they crash, often abruptly (Fig. 33.10). For example, a population of aphids (small insects that feed on the juices of plants) (Fig. 33.11) in a field of alfalfa in spring may grow at an exponential rate if the weather is cool and moist, but if the weather then becomes hot and dry most of the aphids will die, causing the population density to fall precipitously. This crash will occur if the weather changes even though the number of aphids may be far below what the alfalfa field can sustain. The weather here is exerting a ***density-independent*** limitation on the aphid population; its operation does not in any way depend on the density of the

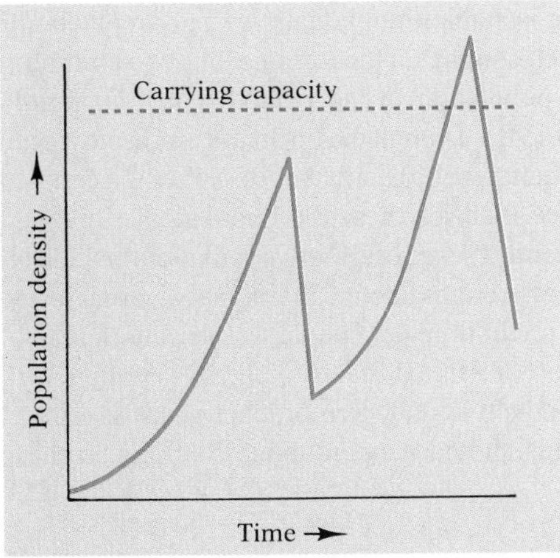

33.10 A growth curve in which an exponentially growing population suddenly declines independent of its density This type of growth curve, sometimes called a boom-and-bust curve, is most characteristic of populations of small organisms limited by density-independent factors such as weather.

33.11 A population of aphids Aphid populations grow at an exponential rate if the weather is cool and moist, but when the weather becomes hot and dry most of the aphids will die, causing the population density to fall precipitously. This crash will occur when the weather changes even if the density of aphids is far below the carrying capacity of the alfalfa field. The weather exerts a density independent limitation on the aphid population; its operation does not in any way depend on the density of the aphids.

aphids. Any physical disruption of habitat, such as flood, fire, or drought, may exert density-independent limitations on population growth. In general, density-independent limitations are associated with physical factors in the environment whereas density-dependent limitations are associated with biological factors such as competition for limited resource, disease, and predation. It is not always possible to isolate these two types of limitations; for instance, a density-independent factor such as inclement weather can alter food supplies (lowering the carrying capacity), thereby precipitating density-dependent factors that cause the actual decline of the population.

Mortality and Survivorship Influence Population Growth

In addition to the birth and death rates, the *potential life span,* the *average life expectancy,* and the *average age of reproductive maturity* are important factors in determining the growth of a population. For example, if the average life span is long and the age of reproductive maturity low, more generations will be living concurrently than if the life span is short and the age of reproductive maturity high.

Population dynamics are affected by patterns of reproduction and mortality for various age groups. Such data show what stages in the life cycle are most susceptible to environmental control, and make it possible to compute the percentage of individuals that will still be alive at the end of each age interval. Ultimately, we can calculate the overall growth of the population.

Patterns of survivorship by age and sex are often graphed as a survivorship curve. The curves in Figure 33.12 illustrate several survivorship patterns. Curve I approaches the pattern that would be expected if all the individuals in the population lived until old age. There would be full survival through all the early age intervals (as shown by the horizontal portions of the curve), and then all the individuals would die more or less at once and the curve would fall suddenly. Curve III approaches the other extreme, where the mortality is exceedingly high among the very young but any individual surviving the earliest life stages has a good chance of surviving for a long time thereafter. Between these two extremes is the condition represented by curve II, where the mortality rate is constant at all ages.

The survivorship curves for most wild-animal populations are probably intermediate between types II and III, and the curves for most plant populations are probably near the ex-

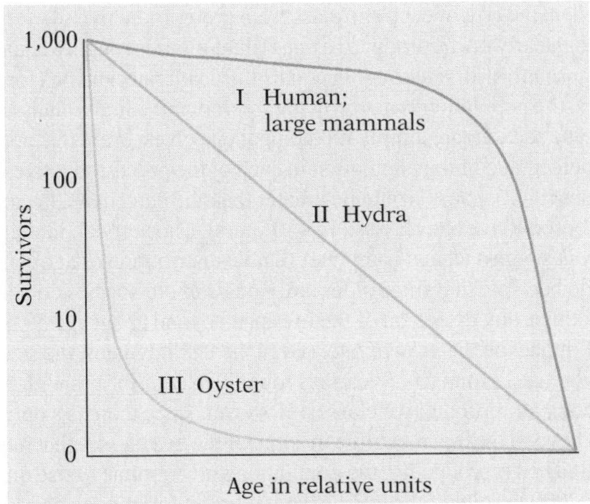

33.12 Three types of survivorship curves For an initial population of 1,000 individuals, the curves show the number of survivors at different ages from birth to the maximum possible age for that species. The curves represent three basic types of survivorship. The human curve is for societies with virtually no infant mortality.

treme of III. High mortality among the young is the general rule in nature.

Changes in environmental conditions may radically alter the shape of the survivorship curve for any given population, and the altered mortality rates, in turn, may have profound effects on the dynamics of the population and its future size. For example, the chief cause of the enormous increase in the human population has been the great reduction in mortality during the early life stages, a result of improvements in sanitation, nutrition, and medical care. These advances have caused the human survivorship curve to approach type I in industrialized societies while it lags between types I and II in underdeveloped countries. There is no evidence that this shift toward a type I curve is due to a rise in the human birth rate per reproductive-age female (in fact, the birth rate has fallen) or, despite all the advances of modern medicine, to an increase in the potential life span.

Knowledge of population dynamics enables us to implement policies of population control if desired. China, for instance, prohibits early marriage and limits family size, thereby slowing population growth.

Age Distribution Also Influences Population Growth

Populations dynamics is also affected by age distribution. Consider the age distribution of three human populations. In Sweden, the percentage of the population in each age class is approximately equal except for the oldest classes (Fig. 33.13A). The birth rate equals the death rate, and the population size is stable; individuals are only replacing themselves since each couple, on the average, has two children.

The population in the United States is slightly different (Fig. 33.13B). Though the birth rate is now down almost to the replacement level (the number of children a couple must have to replace themselves in the population), this has happened only recently. As a result, the population includes a high percentage of prereproductive individuals, born when the average number of children per couple was greater than two. The population of the United States, then, will continue to grow until the reproductive and prereproductive classes reach equilibrium. This should happen about 2030, when the population

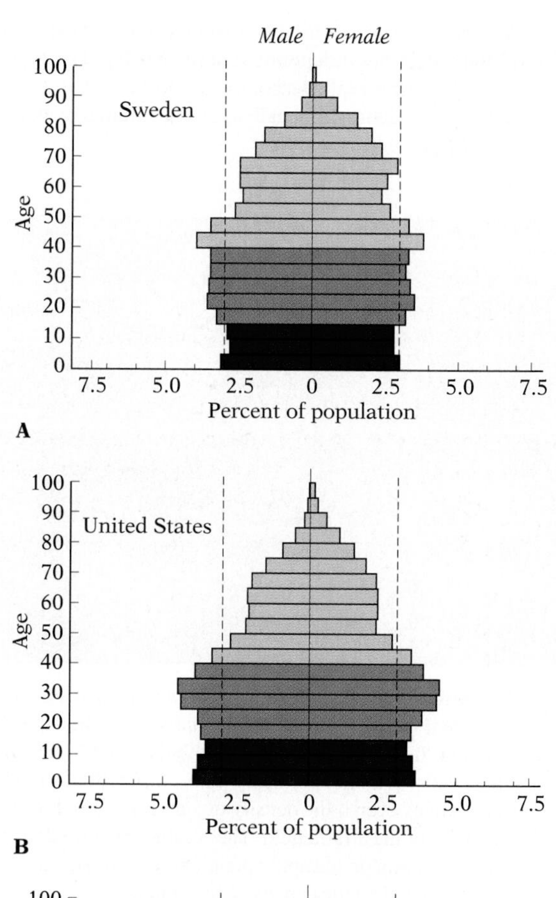

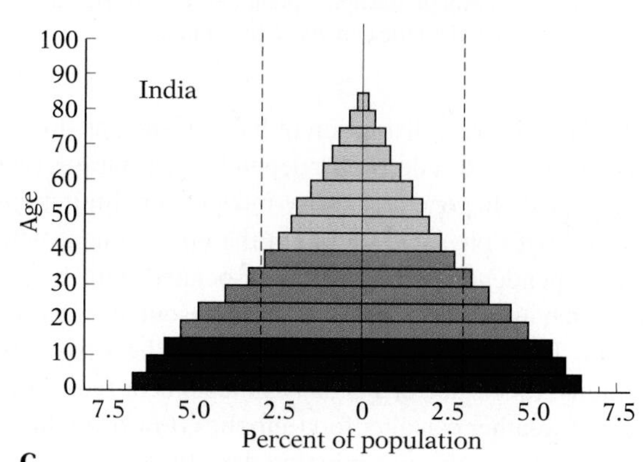

33.13 Age distribution in three human populations (A) In Sweden, the size of each age class from prereproductive (dark blue) through early postreproductive ages (light blue) is roughly equal. Because the birth rate roughly equals the death rate, and has for many years, the population is not growing; reproductive individuals (middle blue section) are simply replacing themselves. Were this not so, the percentage of the population in each of the prereproductive age classes (0–15 years) would be greater than the percentage in each of the reproductive age classes (15–40 years). The vertical lines at 3 percent suggest what a completely stable distribution of age classes would be. Note that since older individuals of our species do not reproduce, any extension of the life span beyond about age 50 has little impact on the growth rate. (B) In the United States, the size of the youngest group (0–5 years) is roughly in equilibrium with that of the younger reproductive classes. However, since there is a bulge well beyond the 3 percent line in some of the age classes that have not finished reproducing, the population will continue to rise until at least 2000 in spite of the low birth rate. Immigration is also swelling the population. (C) In India, the population is heavily weighted toward the younger age classes as a result of a high birth rate. Even if the birth rate were to fall to two children per woman, the population would continue to grow for two generations.

reaches 300 million. This estimate may be overly optimistic, however, due to the high immigration rate. Currently between 750,000 and one million individuals are immigrating annually to the United States. If a majority of these are of reproductive and prereproductive ages we can expect our population to continue to grow well past 2030.

In India, by contrast (Fig. 33.13C), a large percentage of the population is very young due to a high birth rate. This population will continue to grow rapidly even if the birth rate drops to two children per family because there are so many young who will be reproducing in the near future. In technologically advanced countries the average percentage of the population under age 15 is about 21 percent, but in underdeveloped countries (excluding China) it averages 39 percent. In Kenya it is 50 percent! Populations with such high percentages of young have tremendous growth momentum. Demographers estimate that even if India could reduce its average family size to 2.2 children (replacement-level fertility) in the next 33 years, its current population of 870 million would continue to grow until it reaches two billion by 2100. It should be clear, as Stanford ecologist Paul Ehrlich pointed out more than three decades ago, that in the absence of a dramatic reduction in birth rate, advances in reducing infant mortality in developing countries resulting from improved sanitation, nutrition, and medical care will inevitably result in explosive population growth, accompanied by widespread famine and disease.

As Figure 33.14 shows, the human population has increased very slowly for many thousands of years, even though the birth rate was probably high. (Primitive societies put great value on large families, which were essential in view of the very high mortality rates, particularly among infants.) It has been estimated that there were approximately five million people on the entire earth 10,000 years ago, and that the number had risen to only about 250 million by A.D. 1 and to about 500 million by the year 1650. From these figures, one can see that until about 300 years ago the human population doubled approximately every 1,600 years. Now, by contrast, the population of the world is doubling every 41 years. At the present rate, the population of the world, estimated at 5.5 billion in 1992, will reach 8.5 billion in 2025 and, because this level of growth cannot be sustained, will probably level off at about 11.6 billion around 2150.

How much longer the human species can continue with such a rate of increase, with such an imbalance between births and deaths, is one of the most pressing questions of our time. Some have argued, in fact, that it is *the* most important question, since all other aspects of the present day environmental crisis—hunger, poverty, crowding, pollution, accumulation of wastes, destruction of the environment on which all life depends—are the unavoidable consequences of a continuing rise in the number of human beings (see Exploring Further: Achieving Sustainability for the Human Population, p. 747). We will examine the biological bases of some of these problems in the next chapter.

Many Populations Are Limited in Size Primarily by Density-Independent Factors; Others by Density-Dependent Factors

As we have seen, populations of living organisms cannot grow indefinitely; one or more factors always work to limit their size. Ecologists have described two major life scenarios for species depending on their patterns of population growth and regulation. For example, near one extreme are populations that grow very quickly because they have a high intrinsic rate of increase, *r*, and for this reason they are often referred to as **r-*selected species.*** They have evolved the capacity to reproduce rapidly to exploit variable, disturbed, or otherwise unpredictable environments. The rapid multiplication of mosquitoes in springtime typifies the behavior of an *r*-selected species. Fortunately for us, the size of the mosquito population is generally held below saturation by environmental factors like drought and cold.

The lifestyles of *r*-selected species are usually accompanied by a host of behavioral and physiological adaptations (Table 33.1). For example, individuals usually produce large

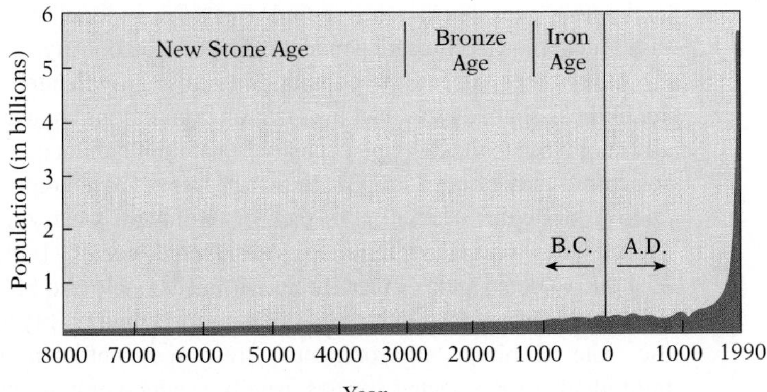

33.14 Growth of the world's human population The growth was slow for many thousands of years, but has become very rapid in the past century. The dip around 1350 corresponds to the outbreak of bubonic plague.

MAKING CONNECTIONS

1. Is any ecosystem isolated from all others? Is the global ecosystem—the biosphere—completely isolated and self-sufficient?

2. Describe how such factors as competition, reproduction, nutrient procurement, and the nature of the environment influence how the individuals of a population are distributed.

3. How does the reproductive potential of elephants shed light on the principle of natural selection discussed in Chapter 31?

4. The population of China in 1992 was estimated to be approximately 1.2 billion. Its birth rate was 22 per thousand and the death rate 7 per thousand. By what percent is it growing per year? At that rate of growth, what is the doubling time of the population?

5. How have improved standards of nutrition, medical care, and sanitation altered the human survivorship curve? What is the implication of this change for efforts to stabilize human population size?

6. The size of the human population reached about 500 million by 1650. By 1850 it had doubled to one billion, by 1930 it had doubled again to two billion, and by 1975 it had reached four billion. Has the growth of the human population since 1650 been exponential?

numbers of small, quickly maturing offspring, since the fitness of such individuals is heavily dependent upon their producing as many young as rapidly as possible before some environmental disruption brings on a sudden crash. The survivorship curves for such species usually resemble type III (Fig. 33.12).

TABLE 33.1 *A comparison of the life histories of* r*-selected and* K*-selected species*

Characteristic	*r-selected species*	*K-selected species*
Body size	Small organisms	Larger organisms
Life span	Short	Long
Age at first reproduction	Early reproduction	Slow growing, mature later
Number of offspring	Many	Few
Amount of parental care	Little	Much
Typical growth curve	Boom or bust	S-shaped (logistic)
Limitations on population growth	Density independent	Density dependent
Typical survivorship curve	Type III	Type I or II
Typical environment	Variable, unpredictable	Stable, persistent

The density of the population in *r*-selected species is generally too low for its growth to be limited by the availability of essential resources like food, water, or nesting sites. Though they may be subject to severe predation, they reproduce so quickly and profusely that this factor is also not greatly limiting. More important are the previously discussed density-independent factors, such as weather, fire, drought, or flooding, that can cause the population to crash periodically (Fig. 33.10). As the name suggests, the effect of density-independent factors is not correlated with population size. These factors are physical factors in the environment that affect the same percentage of individuals in a population whether the population is large or small. Lack of rain and high temperatures, for instance, that cause the pools in which mosquito larvae grow to dry up will decimate groups whether they include just a few individuals or thousands; the effect is independent of population density.

At the other extreme are populations whose growth curves are of the S-shaped type (Fig. 33.7). Such populations are usually at or close to the carrying capacity *(K)* of their habitats, and so grow slowly or not at all. Because they have evolved reproductive strategies in relation to their environment's carrying capacity, they are often referred to as ***K*-selected species** (Table 33.1). *K*-selected species usually live in fairly stable and predictable environments. Their fitness depends on their ability to use the limited environmental resources efficiently. Individuals of *K*-selected species usually produce only a few large, longer-lived, slowly maturing young, with substantial

E X P L O R I N G   F U R T H E R

ACHIEVING SUSTAINABILITY FOR THE HUMAN POPULATION

It seems clear that in the next century, the earth will have to support twice as many humans as it does today. Can our planet do this? The concept of the carrying capacity is just as valid for the human population as it is for other organisms. Is it possible for the earth to support twice as many people as it does now without reducing its ability to sustain the human population in the future? Clearly, the answer is no. We are maintaining our present population of 5.5 billion only by rapid depletion of our resources: groundwater, topsoil, tropical forests, biodiversity, fossil fuels, clean air, etc. Today, approximately 40 percent of the earth's photosynthetic productivity is used or influenced by human activities. If our population doubles, the exploitation would be expected to double, and this would certainly mean a decimation of the very resources upon which we depend. Animal populations that have exceeded their carrying capacity face widespread famine and disease, reduced fertility, increased predation, etc., until deaths exceed births and the population declines. The human population is no different, and we too will have to face the consequences of overpopulation and degradation of our environment.

Some have argued that the solution lies in increasing the carrying capacity of the earth for humans (1) by altering our lifestyles to reduce consumption, and (2) through improved technology. Both remedies appear unlikely to provide the necessary changes. While all humans in the future might become ecologically sensitive vegetarians who would practice conservation, recycle their materials, use only renewable resources (e.g., wind energy, solar energy, trees), and limit family size, it seems unlikely. And though technological innovations may improve efficiency and develop new resources, there is still an upper limit to the carrying capacity. The laws of thermodynamics cannot be repealed: there is a finite amount of sunlight striking the earth, and a limit to photosynthetic efficiency and energy transfer from organism to organism. In addition there is only so much oil and natural gas to burn, mineral-rich ore to mine, and suitable land on which to live and grow food. Ultimately there is a finite amount of each resource, and when the resources are gone, we will either have to find alternatives, or do without.

Can we build a "sustainable society," in which all human activity is maintained within the carrying capacity of the environment so that the needs of future generations are not compromised? To do so will require profound changes on a global level. Humanity will have to develop international policies to regulate critical resources such as fresh water, forests, fossil fuels, and the atmosphere, and take immediate steps to halt or minimize the damage that has already taken place. Once the biodiversity in tropical rain forests is destroyed, for instance, there is no way to restore it. If we fail to achieve sustainability, human activities will continue to eat away at the life-support capabilities of the environment, permanently lowering its carrying capacity for our species. Now is the time to tackle these difficult problems, because to delay will have dire consequences.

33.15 Elephant caring for young Elephants are examples of *K*-selected species. Their fitness depends less on rapid reproduction than on their ability to use the limited environmental resources efficiently. They reproduce comparatively late in life, producing only a few large, long-lived and slowly maturing young that require a great deal of parental care.

parental care. In short, they emphasize quality rather than quantity (Fig. 33.15). The survivorship curve for these species approximates type I.

The two extremes of the population growth spectrum illustrate dramatically the ways in which the various sorts of population controls operate. As we have seen, populations with a high *r* are generally regulated by density-independent factors such as weather or physical disturbance. A population of *K*-selected organisms, on the other hand, is usually kept from getting too much larger than the carrying capacity by such factors as the limited availability of one or more critical resources. Hence these species are subject to both density-independent and density-dependent factors.

We now examine the various factors that can limit populations in a density-dependent fashion.

Predation and Parasitism Act as Density-Dependent Factors Limiting Population Growth

Predation, disease, and parasitism usually influence the prey or host species in a density-dependent manner. As the population of prey increases, individuals become easier to find and attack, and a higher percentage of the prey population is usually victimized (Fig. 33.16). In the same way, the probability that a parasite will find a suitable host is also density-dependent.

When the relative densities of prey shift, predators that take a variety of prey species tend to alter their hunting patterns so as to concentrate on the most common species. If the density of a prey species increases, the density of the predators feeding on it often increases also. This increase in predators, together with their increased concentration of the particular prey species, may be one factor that causes the density of the prey to fall again. But as the density of the prey falls, there is usually a corresponding fall in the density of the predator, which is slightly delayed in time. The result may be a series of fluctuations like those shown in Figure 33.16: the density of the predators closely mimics that of the prey, but with a characteristic time lag. Such linked fluctuations of predator and prey seem to indicate that the major limiting factor for the predator is the availability of its food, and that predation is probably one significant limiting factor for the prey.

One example of population control by predation is a well-known case study involving the moose population of Isle Royale, an island in Lake Superior. In 1908, a small group of moose chanced to walk the 25 kilometers from Canada across the frozen surface of Lake Superior to Isle Royale. With no predators to limit their numbers, the moose population on this 544-square-kilometer island grew steadily to about 3,000 in 1935. Since the vegetation could not support such numbers, the result was mass starvation: about 90 percent of the moose died. As the vegetation recovered, the moose population began slowly to increase again, until in 1948 it again peaked at 3,000 and crashed. In 1949 a group of wolves chanced to cross the ice to Isle Royale. The wolves began to prey on the moose, for the

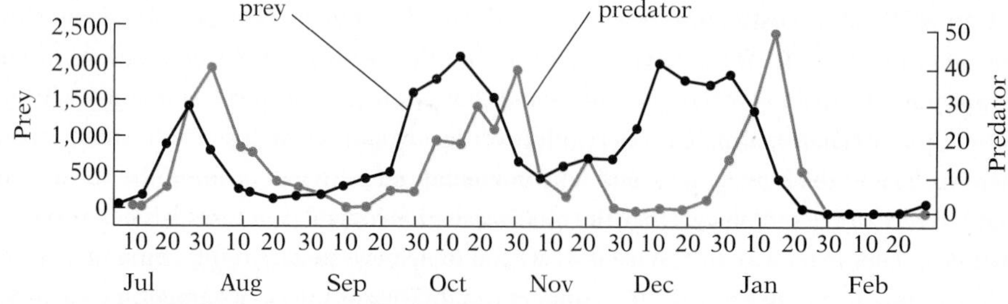

33.16 Linked fluctuations in predator and prey populations The size of the population of a predatory mite tends to reflect (after a delay) fluctuations in the population of its prey (another mite). Predation is only one of many density-dependent limiting factors operating on the prey. The predator population, however, is limited mainly by the availability of prey.

most part attacking the sick, the old, and the young, constantly searching the herd for signs of weakness (Fig. 33.17). A stable balance of two dozen wolves, 800 moose, and a healthy crop of plants was achieved. As this example demonstrates, prey-predator relationships can lead to a dynamic balance, with predation an important factor in keeping the prey population below the carrying capacity of the habitat.

In stable predator-prey systems, predation is often decidedly beneficial to the prey population, even though it is destructive to individuals. This cardinal fact of ecology is frequently overlooked by those to whom parasites are repugnant and predators evil. When people set out to protect the prey from their "enemies" by killing the predators, the results are often very different from those expected. Released from the density-regulating influence of the predators, the prey species may experience such a population explosion that it damages the environmental resources on which its own continued existence depends, with consequent wholesale extermination of the "protected" species through starvation or disease.

Similar difficulties have sometimes arisen when predator-prey stability has inadvertently been destroyed by pesticides. For example, application of certain insecticides to strawberries to destroy cyclamen mites that were damaging the berries killed both the cyclamen mites and the carnivorous mites that preyed on them. While the cyclamen mites quickly reinvaded the strawberry fields, the predatory mites did not. The result was that the cyclamen mites, now free of their natural predators, rapidly increased in density and did more damage to the strawberries than if the insecticides had never been applied.

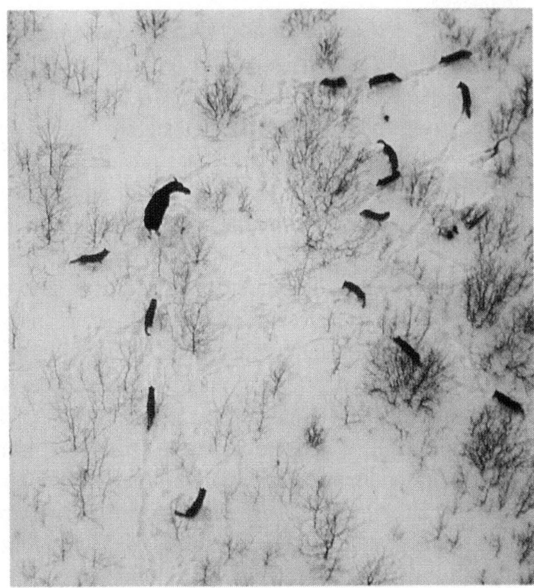

33.17 Population control by predation A pack of wolves on Isle Royale corners an adult moose. Despite the long chase, however, this individual was too strong and healthy to be worth the risk of attacking at close quarters. The wolves eventually left to find a more likely target.

Both Intraspecific and Interspecific Competition Act as Density-Dependent Limitations on Population Growth

Competition among members of the *same* species (i.e., ***intraspecific*** competition) is one of the chief density-dependent factors limiting population size. The continued healthy existence of most organisms depends on utilization of some environmental resources that are in limited supply, such as food, water, space, and light. Each member of a population shares the same basic requirements: each needs the same sort of food, shelter, and mates. Unless some other force, such as predation, holds a population below the carrying capacity of its environment, individuals in such populations must inevitably compete for resources.

To take a familiar example, if flowers are planted too close together in a flower bed the plants will be weak and spindly and will produce few blossoms. Only if they are thinned, either artificially or by the natural death of the weakest individuals, will they grow well. The same sort of competition for space, light, water, and nutrients operates in a forest to reduce the density of trees.

As another example, consider competition among individuals for breeding territories. Though territories tend to be larger when population density is low, and smaller when it is high, there is usually a minimum territory size below which successful reproduction will not occur. The minimum size varies greatly from species to species, being rather large in animals that are only mildly social and sometimes quite small in highly social animals such as nesting gulls. But whatever the minimum size, suitable territories are clearly a limited resource, and competition for them, which will become increasingly intense as population density rises, acts as a density-dependent brake on population growth (Fig. 33.18).

Competition between members of *different* species (i.e., ***interspecific*** competition) also works to limit the size of at least one of the competing populations. Ordinarily, interspecific competition occurs when two or more species use the same limited resources, such as food, water, sunlight, shelter, space, or nesting sites. Clearly, it will become more intense if the number of resources available diminishes or if the number of individuals depending on the same limited resource increases; hence the effect of interspecific competition is density-dependent.

In the last chapter we saw that if two species compete for precisely the same limited resources and have identical ecological characteristics, the superiority of one species may lead to the extinction of the other; this is the competitive exclusion principle developed by Gause.

A species' ***niche*** refers to the ecological role of a species in the community, i.e., the way an organism uses its environment to make a living. The niche includes such factors as what it

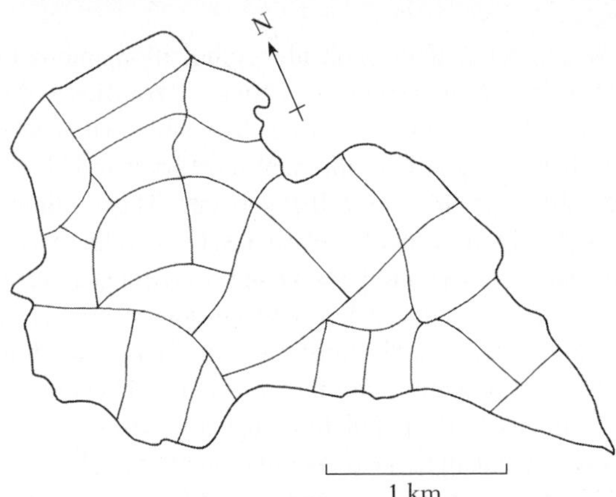

33.18 Owl territories Pairs of tawny owls have divided this patch of habitat into a matrix of well-defined territories whose number and boundaries are relatively stable from year to year. The habitat supports an adult population of 50 individuals.

eats, how and where it finds and captures its food; what extremes of temperature, humidity, sunlight, and other climatic factors it can withstand, and what values of these factors are optimal for it; what its parasites and predators are; where, how, and when it reproduces; and so forth. In addition, the time of year and the time of day when it is most active are important: the niche of night-hunting owls preying on small rodents, for instance, is different from that of small day-hunting hawks or ground-dwelling mammalian carnivores. Under certain conditions, therefore, these species can coexist, even though all prey on small rodents. In short, every aspect of an organism's existence helps define that organism's niche. Ecologists, however, frequently focus on only the more important variables of an organism's niche, such as food, times and sites of activity, and a few key physical variables like temperature.

No Two Species Can Occupy the Same Niche at the Same Time

The more similar two species are, the more likely it is that both species will compete for at least one limited resource (food, shelter, nesting sites, etc.). According to the ***principle of limiting similarity***—a modification of the competitive exclusion principle—there is a limit to the amount of niche overlap compatible with coexistence. Competition for the one most limited resource therefore usually leads to one (or two) of four possible outcomes:

1. The competitive superiority of one of the rival species may be such that the other is driven to extinction.
2. One species may be competitively superior in some regions, and the other may be superior in other regions under different environmental conditions, with the result that one is eliminated in some places and the other is eliminated in other places; i.e., sympatry disappears, but both species survive in allopatric ranges.
3. One species may be superior under normal conditions, but at a strong disadvantage during periodic crises. The crises, then, will reduce the population size of the superior species, and the competition will begin anew when the crisis passes.
4. The two species may rapidly evolve in divergent directions under the strong selection pressure resulting from their intense competition. In other words, the two species evolve greater differences in their niches, thereby reducing competition to the point that continued coexistence becomes possible. This is the phenomenon of character displacement, discussed in the preceding chapter.

It is not always easy to detect the differences between the niches of two or more closely related sympatric species. At first glance, the species may appear to be occupying the same niche in a stable way and thus to disprove the exclusion principle. But closer study usually reveals differences of fundamental importance. When Robert MacArthur, of Princeton University, studied a community where several closely related species of warblers (small insect-eating birds) occurred together, he found that their feeding habits were significantly different. Yellow-rumped warblers fed predominantly among the lower branches of spruce trees; bay-breasted warblers fed in the middle portions of the trees; and Cape May warblers fed toward the tops of the same trees and on the outer tips of the branches (Fig. 33.19). We noted similar differences in feeding habits between sympatric Galápagos finches.

Cape May warbler

Bay-breasted warbler

Yellow-rumped warbler

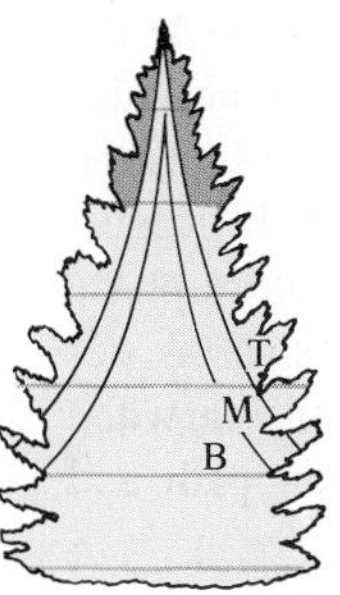

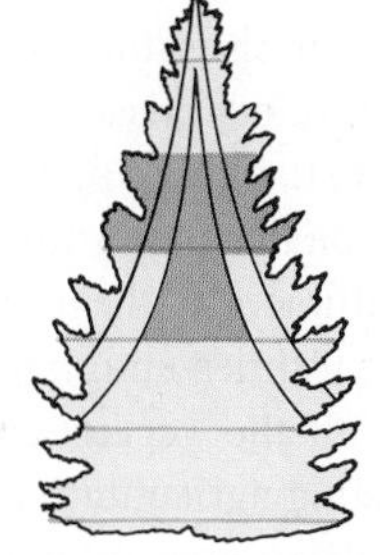

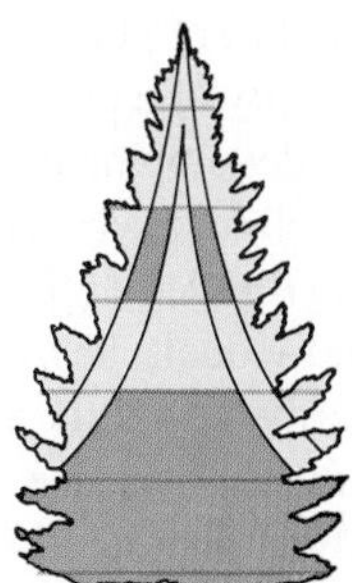

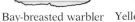

Cape May warbler Bay-breasted warbler Yellow-rumped warbler

33.19 Differences in the feeding niches of three species of warblers living in the same community Above: The Cape May warbler feeds on the tips of the highest branches of the spruce tree; the bay-breasted warbler feeds in the middle part of the tree; and the yellow-rumped warbler feeds predominantly (though not exclusively) among the lower branches. Top right: The dark green areas are the parts of the tree where the birds spend half their total feeding time. T: terminal parts of branches; M: middle; B: base. The height zones (1–6) are measured in 3-meter units.

Behavioral and Physiological Changes Reduce Population Density

In some animals crowding induces physiological and behavioral changes that result in increased emigration from the crowded region. Such changes can be observed in many species of aphids. During the seasons of the year when conditions are favorable, aphids are represented largely by wingless females that can reproduce asexually. But when conditions deteriorate and competition becomes intense, sexually reproducing winged females develop and these may move out of the area in which they were born.

One of the best-known examples of the changes induced by crowding is that of the solitary and migratory phases seen in several species of locusts, particularly *Locusta migratoria* in Eurasia. Individuals of the migratory phase have longer wings, a higher fat content, a lower water content, and a darker color than solitary-phase individuals. They are also much more gregarious and more readily stimulated to marching and flying by the presence of other individuals (Fig. 33.20). The solitary phase is characteristic of low-density populations, while the

33.20 Locusts in the migratory phase, in Ethiopia

migratory phase is more common in high-density ones. As the density of a given population rises, the proportion of individuals developing into the migratory phase also rises; the sight and smell of other locusts seem to play an important role in triggering this development. When the proportion of migratory-phase individuals has risen sufficiently, the infamous swarms described as early as biblical times emigrate from the crowded area, consuming nearly all the vegetation in their path and often completely devastating agricultural crops.

Emigration from crowded areas is not limited to insects. Many animals, including small mammals such as arctic lemmings, tend to emigrate in search of more favorable habitats when the population density or some effect of crowding exceeds a certain threshold.

Some ecologists have suggested that physiological phenomena may be important in limiting populations. It has long been known, for instance, that very dense populations often experience severe disease epidemics. This proneness to epidemics may not be due solely to the greater ease with which the pathogens spread; there is evidence to suggest that crowding induces physiological changes in host resistance. Reproduction may also be affected by severe overcrowding. Numerous experiments with laboratory mice and rats have shown that crowding induces stress disease, the symptoms of which include greatly enlarged adrenal glands and decreased reproductive success. Whether or not overcrowding can induce stress disease in nature remains controversial, and research continues in this area.

THE CONCEPT OF THE COMMUNITY

In nature it is rare to find only one kind of organism living in a region. Usually a number of different plant and animal species occupy the same general area, sharing some resources and competing for others. This association of interacting populations forms a community, which has its own characteristic structure and complex array of interrelationships. In this section we will look at the dynamics of communities and how they respond to disturbances; in the next chapter we will consider other aspects of community structure and organization.

Ecological Succession Is an Orderly Process of Species Replacements

When plants invade a newly formed habitat, such as a lava flow or bare rock, or when a community is disturbed in some way—a forest is cut, a field plowed, an area burned, a sand dune formed—a more or less orderly, predictable sequence of replacements will occur in that community (Fig. 33.21). This orderly series of species replacements is called ***succession.*** If a farmer's field is plowed and left untended, a crop of annual weeds will grow in it during the first year. Many perennial herbs appear in the second year and become even more com-

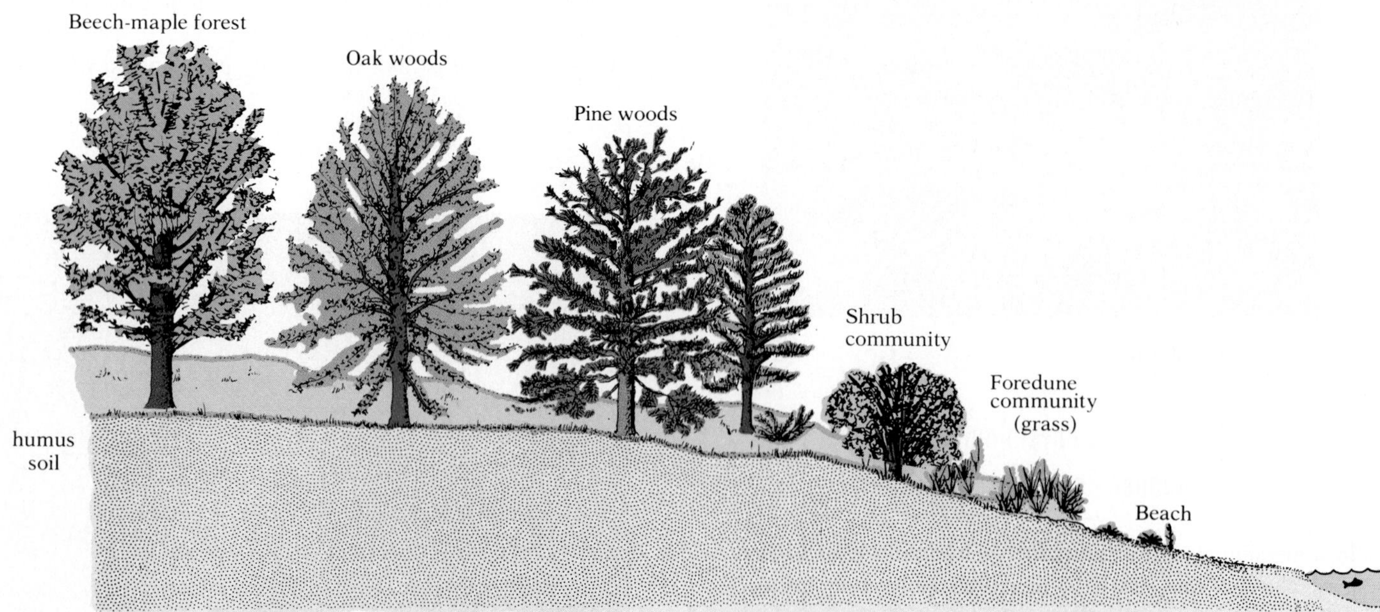

33.21 Ecological succession The orderly series of species replacement during succession can be seen in this sequence of plants from the sandy beach to the beech-maple forest. The first colonizers on the beach are annual weeds, which are replaced by grasses. Later, the grasses are replaced by small shrubs, which in turn are succeeded by pines. Finally, the pines are replaced by deciduous trees and a stable, long-lasting beech-maple forest.

mon in the third year. Soon, however, these are replaced by shrubs, which may in turn be replaced eventually by trees. As another example, if a lake dries up, its sandy bottom may become covered with grass, and later by trees. Ecologists want to understand the processes and patterns of these changes because all communities undergo localized disturbances. Some of it is human-induced, some of it natural.

Why does succession occur? The cause cannot be year-to-year changes in weather, because succession will occur even if the climate remains the same. Climate may be a major factor in determining what sorts of species will follow one another, but the succession itself must result from other causes. Two factors are thought to be particularly important: (1) the modification of the physical environment produced by the community itself over the course of the succession, and (2) the rate at which different species invade a new area. The traditional view has been that the most important cause is the modification of the environment. In this view, successional species tend to alter the area in which they occur in such a way as to make it less favorable for themselves and more favorable for other species. In effect, each set of species in the succession sows the seeds of its own destruction.

Consider the alterations initiated by the ***pioneer community,*** the earliest community to become established on a site. The pioneer species are adapted to be good invaders; they tend to be *r*-selected species, adapted for rapid growth, reproduction, and dispersal. Typically they are rapidly growing annuals that, when they die, produce a layer of litter on the surface of the soil. The accumulation of litter affects the runoff of rainwater and the soil temperature, and results in the formation of ***humus***—organic material that consists mostly of decomposition products from cellulose. The humus, in turn, contributes to soil development, altering the water availability, the pH and aeration of the soil, the availability of nutrients, and the sorts of soil organisms that will occur. But the organisms characteristic of the pioneer communities that produced these changes may not prosper under these new conditions, or they make the area more suitable for competitors than it was before. Pioneer species are often replaced by invading species that are more successful under the new conditions.

In recent years, however, ecologists have come to realize that the rate at which species disperse and the element of chance also play important roles in succession. Some species are more easily dispersed than others, or grow and reproduce more rapidly. Therefore, it is only logical that rapid-growing annual herbs that produce many small, easily dispersed seeds will be far more likely to become established on recently abandoned farmland than the seedlings of slow-growing trees, even though the trees may eventually become the dominant plants. The course of the succession may be influenced too by the species that chanced to arrive first. If ragweed, for instance, is one of the invading species, it may be many years before the succeeding species can displace these plants because ragweed is such a good competitor. Eventually it will be displaced, but the course of succession may take several years longer than it would if other, less competitive annual weeds happened to have arrived first.

Patterns of succession in different places and at different

MAKING CONNECTIONS

1. Suppose a species of Darwin's finch exists on two islands. On one island it has no close competitors, while on the other there is a finch with a very similar niche. How would the evolutionary effect of the intraspecific competition on the niche of Darwin's finch differ from that of the interspecific competition?

2. How are the resources for which plants and animals compete related to their different modes of nutrition?

3. How is interspecific competition related to the evolutionary phenomenon of character displacement, discussed in Chapter 32? What are the other possible evolutionary results of interspecific competition?

4. What are the possible effects of severe overcrowding on the immune and endocrine systems discussed in Chapters 21 and 26?

5. Why do you suppose that humus consists mostly of decomposition products of cellulose? How does humus contribute to soil development?

times are not identical; the species involved are often completely different, and the climatic and soil conditions vary widely. Moreover, the sequence of changes in ***primary succession,*** in which communities are established in newly formed habitats (e.g., on sand dunes or bare rock), is often longer and slower than in ***secondary succession,*** in which communities are reestablished in areas where they or another type of community once existed (e.g., in fields where the original forests were cleared for farming).

Primary Succession Occurs in Newly Formed Habitats

Succession may begin in areas where there have been no previous communities—bare rock, sand dunes, or the wall of an abandoned building. Consider a bare rock surface (Fig. 33.22A). The first pioneer plants may be lichens, which grow during the brief periods when the rock surface is wet and lie dormant during periods when the surface is dry. The lichens release acids and other substances that erode the rock. Dust particles and bits of dead lichen may collect in the tiny crevices thus formed, and pioneer mosses may gain anchorage there. The mosses grow in tufts or clumps that trap more dust and debris and gradually form a thickening mat. A few fern spores or seeds of grasses and annual herbs may land in the mat of soil and moss and germinate. These may be followed by perennial herbs. As more and more plants survive and grow, they catch and hold still more mineral and organic material, and the soil layer thus becomes thicker. Later, shrubs and even trees may start to grow in the soil that now covers what once was a bare rock surface. Succession may also be observed in miniature on a log lying in a forest (Fig. 33.22B).

A much-studied type of primary succession occurs in ponds (Fig. 33.23). Sediments washed from the surrounding land begin to fill the pond, and the dead bodies of pond organisms add organic material. Soon pioneer submerged vascular plants appear in the shallower water near the edges of the pond. Their roots hold the silt, and the pond bottom is built up faster where they grow. In addition, as these plants die, their tissue accumulates faster than decomposers can break them down. Soon the water is shallow enough for broad-leaved floating pondweeds to displace the submerged species, which now become established in a zone farther out in the pond, where conditions are more favorable for them. But as the bottom continues to build up, the floating pondweeds are in their turn displaced by emergent species (plants that have their roots in the mud of the bottom, but their shoots extending into the air above the water), such as cattails, bulrushes, and reeds. These plants grow very close together and hold the sediment tightly, and their great bulk results in rapid accumulation of organic material. Soon conditions are dry enough for a few terrestrial plants to gain a foothold. Now an area that was formerly part of the pond is newly formed dry land. This entire sequence can sometimes be seen as a nearly continuous series of zones girdling a pond or lake. With the passage of the years, the pond becomes smaller and smaller as the zones move nearer and nearer its center. Eventually nothing of the pond remains.

33.22 An early successional stage on a bare rock surface (A) Lichens, shown here bearing fruiting bodies, are often the first multicellular organisms to gain a foothold on rock. Chemicals produced by the lichens corrode the rock surface and help prepare the way for later successional stages. (B) A community of mosses and lichens growing on a fallen log.

A A newly formed pond near the beach has sandy borders bare of vegetation

B After two years such a pond is ringed by low vegetation, including cottonwood saplings.

C A 50-year-old pond is bordered by mature cottonwood trees. So much sediment is produced by organisms growing in the pond that only a small area of water, choked with weeds, remains.

D After 150-250 years an area that was once a pond has become a meadow.

33.23 Succession in ponds in Presque Isle, Pennsylvania, a peninsula in Lake Erie

Secondary Succession Occurs When Communities Are Reestablished in an Area

Secondary succession occurs in habitats that have been disturbed, as in abandoned croplands, unused railway rights-of-way (Fig. 33.24), plowed grasslands, or forests damaged by storms or timber harvesting. This succession often proceeds faster than primary succession, partly because secondary successions are often closer to sources of colonizers, and partly because the effects of the previous communities have not been wholly erased and the physical conditions are not as desolate as those on a beach or a bare rock surface. The long, slow build-up of soil that occurs in a primary succession is no longer necessary. Consider, for instance, succession in an abandoned cornfield in Georgia. The very first year it will be covered with the fast dispersing annual weeds, such as ragweed, horseweed, and crabgrass. In the second year ragweed, goldenrod, and asters will probably be common, and there will be much grass. The grass will usually be dominant for several years, and then more and more shrubs and tree seedlings will appear. The first tree seedlings to grow well in the unshaded field will be pines,

33.24 Secondary succession on an abandoned railway right-of-way The ties are rotting, and vegetation is taking over in the formerly cleared area.

and eventually a pine forest will replace the grass and shrubs. But pine seedlings do not grow well in the shade of older pines. Seedlings of oaks, hickories, and other deciduous trees are more shade-tolerant, and these trees will gradually develop in the lower strata of the forest beneath the old pines, eventually replacing them. The deciduous forest thus formed is more stable and will ordinarily maintain itself for a very long time.

As Figure 33.25 indicates, changes corresponding to the succession of dominant plants also take place in the animal populations of the abandoned cropland community.

Succession Is Characterized by a Number of Trends

Despite numerous differences between various types of succession, especially between primary and secondary succession, some generalizations tend to hold true in most cases where both autotrophs and heterotrophs are involved:

Time in years	1	3	15	20	25	35	60	100	150-200
Dominant plants	Weeds	Grass	Shrubs		Pines				Oak-hickory
Grasshopper sparrow									
Eastern meadowlark									
Yellowthroat									
Field sparrow									
Yellow-breasted chat									
Rufous-sided towhee									
Pine warbler									
Cardinal									
Summer tanager									
Eastern wood pewee									
Blue-gray gnatcatcher									
Crested flycatcher									
Carolina wren									
Ruby-throated hummingbird									
Tufted titmouse									
Hooded warbler									
Red-eyed vireo									
Wood thrush									

33.25 Bird succession on abandoned upland farmland in Georgia The bars indicate when each of the bird species was present in a density of at least one pair per 10 acres. In the early (weed and grass) stages grasshopper sparrows and eastern meadowlarks were the dominant bird species. During the shrub stage yellow-throats and field sparrows became dominant. Pine warblers and rufous-sided towhees dominated the young pine forests, and red-eyed vireos, wood thrushes, and cardinals were the most common birds in the oak-hickory forests.

1. The first colonizers in a disturbed or new habitat are rapidly growing r-selected plant species capable of surviving in a harsh, exposed environment. Later these species are replaced by slower-growing, more competitive K-selected species.
2. The species composition changes continuously during the succession, but the change is usually more rapid in the earlier stages than in the later ones.
3. The total number of species represented increases initially, then becomes more or less stabilized in the older stages.
4. Successional species may alter the environment in such a way as to make it more favorable for competitors or less favorable for themselves.
5. Gross primary productivity (the amount of energy converted into products of photosynthesis) increases until it reaches a stable level (Fig. 33.26).
6. The store of inorganic nutrients held in the organisms and soil of the ecosystem increases, and an increasing proportion of this store is held in the tissues of plants.
7. Both the total biomass (the total weight of all the organisms) in the ecosystem and the amount of nonliving organic matter increase during succession until a more stable stage is reached (Fig. 33.27).
8. The size of the plants in the community increases.
9. The communities become more diverse and complex, and the food webs (to be discussed in the next chapter) become more complicated.

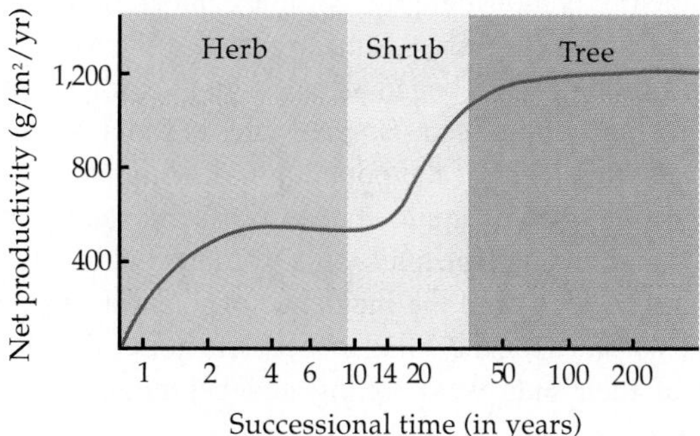

33.26 Change in net primary productivity during plant succession on an area cleared of an oak-pine forest in Brookhaven, New York The first rise represents the invasion of the area by herbs. The later rise (after about 14 years) reflects the entrance of larger woody plants into the community.

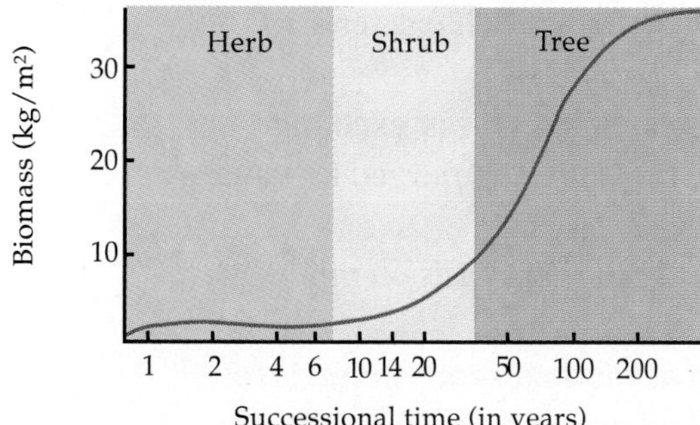

33.27 Change in biomass during succession in the Brookhaven study area The total biomass remained low during the early years, when herbs were the dominant plants in the community, but increased later when shrubs and trees became more prominent.

According to these generalizations, the trend of most successions is toward a more diverse, complex, and longer-lasting community.

Succession May Eventually Reach a Stable Climax Stage

If some disruptive factor does not interfere, most succession eventually reaches a stage in which the species of the community reproduce themselves and the community is far more stable than those that preceded it. A community at this stage of succession is called a ***climax community;*** it is more complex, and is dominated by a few species that came late in the succession. The community becomes self-perpetuating and its appearance remains the same even though there is constant replacement of individuals. The nature of the climax is determined by environmental conditions such as temperature, humidity, soil characteristics, topographic features, and so on. A climax community has much less tendency than earlier successional communities to alter its environment in a manner injurious to itself. In fact, its more complex organization, larger plant structure, and more balanced metabolism enable it to come into equilibrium with its physical environment to such an extent that it can be self-perpetuating. Maple trees replace maple trees, wood thrushes replace wood thrushes, etc., and so the community perpetuates itself. Consequently it may persist for centuries, not being replaced by another stage so long as climate and other major environmental factors remain essentially the same (Fig. 33.28).

Although succession ends with the establishment of a climax community, this does not mean that a climax community is static; it does change slowly even when the climate is constant, and will change rapidly if the community is disturbed in some way. For example, only 60 years ago chestnut trees were among the dominant plants in the climax forests of much of eastern North America, but they have been almost completely eliminated by a fungal blight, and the present-day climax forests of the region are now dominated by beech and maple trees. There can be no absolute distinction between climax and the other stages of succession; the difference between them is relative to current environmental conditions. The characteristics of a community at any specific place are uniquely determined by a combination of local physical conditions, biotic factors, species distributions, and a considerable element of chance. A climax community in many ways resembles a patch work quilt in that scattered throughout there are many areas

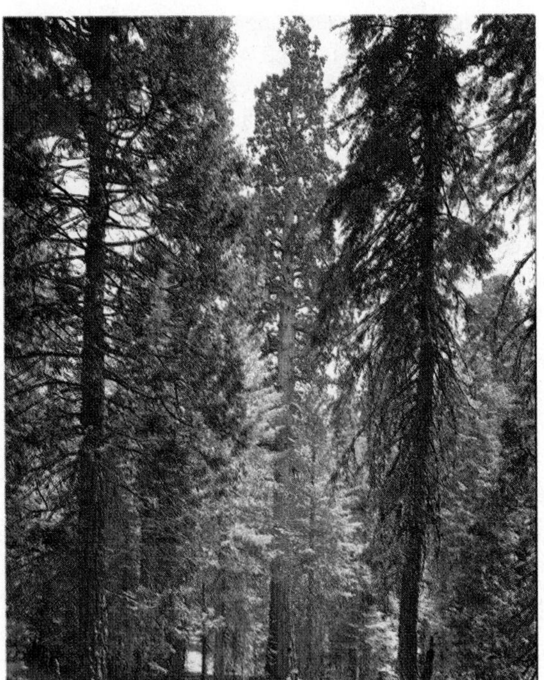

33.28 Sequoia forest The long-term stability of a climax forest is illustrated by this sequoia forest in south-central California. The exact age of the forest is not known, but some of the trees still standing are more than 2,000 years old.

with different climax communities as well as areas at different successional stages that have resulted from local disturbances. By studying the effect of disturbances and the processes of succession, ecologists are better able to understand the structure and regulation of communities.

Human Activities May Cause Community Disturbances

Human intervention in biological communities often has the effect of reducing both the species diversity and the complexity of species interactions. The result is to make the communities far more prone to extreme fluctuations in response to changing conditions. Prime examples are the highly artificial communities created by modern agriculture, which has tended to emphasize ***monoculture***—the planting of a single crop species in enormous fields. These communities can be maintained only at the price of constant vigilance and the investment of much energy in curbing insect infestations and outbreaks of disease, in maintaining soil fertility, and in cleaning out or killing invading weeds. As experiments have demonstrated, multiple-species crops have greater resistance to pests and disease (see Exploring Further: Corn: Gift of the Native Americans, p. 759); but the difficulty and expense of adapting conventional farm machinery to the planting of such crops has made this approach impractical for large-scale farming.

Efforts to eliminate undesirable species from a community often reveal hidden linkages to other organisms, and may dramatically demonstrate the complex interactions on which community stability rests. Some years ago the World Health Organization (WHO) began a campaign to eradicate malaria-carrying mosquitoes in the Borneo states of Malaysia, where as many as 90 percent of the population of some areas suffered from the disease. Mosquito control was achieved by spraying the insides of the village huts with DDT and dieldrin, two powerful insecticides, and the incidence of malaria dropped precipitously. But soon the villagers began to notice that the thatch roofs of their huts were rotting and beginning to collapse.

MAKING CONNECTIONS

1. Why are lichens particularly suited for the pioneer stage of succession on bare rock surfaces? (See Chapter 31.)

2. How is the order of the later stages of bare-rock succession (mosses, annual plants, perennial plants) related to the structural characteristics of these plants, as described in Chapters 19 and 24?

3. From what you learned about the principles of gas exchange in Chapter 18, why would you expect plants whose roots are submerged in water to have special adaptations for gas exchange?

4. What ecological problems are created by the emphasis on monocultures in modern agriculture?

5. Ecologists have a saying that "you can never do just one thing." How is the point of this maxim illustrated by pest-control programs such as the campaign to eradicate mosquitoes in Borneo?

E X P L O R I N G F U R T H E R

CORN: GIFT OF THE NATIVE AMERICANS

When the European explorers reached the New World, they found maize—corn—growing all over the Americas. The wild ancestor of corn probably originated in Mexico or Central America, and by the time Christopher Columbus arrived it was cultivated all the way from Montreal, Canada, down to Santiago, Chile. The Native American farmers practiced plant breeding for several thousand years, breeding and selecting those corn plants that had desirable characteristics. Six basic types of corn were developed: pod, flour corn, popcorn, dent, flint, and sweet corn. The corn varieties most widely planted today are hybrids formed from crossing strains originally developed by farmers of the Iroquois Nation in New York State with strains bred by southeastern Native Americans.

The Iroquois are often thought of as a woodland people who survived by gathering wild plants and hunting, but in actuality, they were serious farmers who had been growing corn and other agricultural crops for more than 800 years. The first Europeans arriving in the Finger Lakes region of New York State found extensive fields of corn under cultivation in large, fertile fields, and ample stores of grain in established villages. Their agricultural system was complex. The Iroquois understood the importance of spacing plants, controlling weeds, managing soil, timely planting, and plant breeding. Unlike the European system of monoculture, the Iroquois practiced "intercropping" (planting more than one crop in the same area), and did not plow their fields. Instead, they used a hill planting system in which corn was planted three or four kernels to a hill, with the hills spaced three feet apart. Two to three weeks after the corn sprouted they planted pole beans at the base of the hill. Recall that beans have root nodules in which nitrogen-fixing bacteria live, so the bean plants with their mutualistic bacteria added essential nitrogen compounds (i.e., fertilizer) to the soil of the hill. The cornstalks in return provided support for the growing bean vines. Between the hills they planted low growing crops such as pumpkin or squash, to help control weeds. They also weeded the hills, throwing the weeds on the planting hill to return the nutrients to the soil. At the end of the season all the stalks were returned directly to the hill, again returning nutrients to the soil. These three crops—corn, beans, and squash—were called the "three sisters." They provided both a dietary and agricultural complement.

The hill system provided not only proper spacing for the plants, but controlled weeds, prevented soil erosion, and maintained soil fertility. The Iroquois also had a system of field abandonment for maintaining soil fertility. When the yield began to decline, the field was abandoned and allowed to regrow into grasses, bush, and trees as succession took place. After about 10 years the larger trees would be killed, the fields burned, and the land once again used for corn.

The Iroquois were superb plant breeders. They commonly grew between 10 and 40 corn

varieties each year. Varieties were planted separately, to prevent cross-pollination, and only the seed from the best plants were used for planting the next year. Periodically they crossed different varieties to increase vigor.

The Iroquois system of agriculture was stable and sustainable. Contrast this system with our modern farming methods, which require extensive machinery and annual applications of pesticides, herbicides, and inorganic fertilizers. In the 1950s and 1960s there was a great deal of interest in a "new" system of no-till planting to reduce soil erosion, a system used by the Iroquois 800 years ago! A more recent development, now used in parts of the Midwest, is ridge-tillage, in which the field is not plowed but instead special implements form the soil into ridges in which corn is planted—a mechanized hill system! Finally, some agronomists at Cornell University are studying the feasibility of intercropping. The Iroquois system of agriculture was based on sound agricultural principles, ones that are still used for growing good corn today. We should be grateful to the Native Americans for their magnificent gift of corn, which is now a worldwide food staple.

Investigation showed that the deterioration, which occurred only in huts sprayed with DDT, was due to the larvae of a moth that normally lives in small numbers in the thatch roofs. Whereas the thatch-eating moth larvae avoided food sprayed with DDT, the moth's natural enemy, a parasitic wasp, was adversely affected by it. The net result was a substantial increase in the population of the thatch-eating larvae because of the near eradication of the wasp.

That would have been an interesting story in itself, but there was yet another potentially serious side effect. Cockroaches and a small house lizard, the gecko, are two normal inhabitants of the village huts. DDT-contaminated cockroaches were eaten by the geckos, which were in turn eaten by house cats (as were some cockroaches). The cats, poisoned by the accumulation of the insecticide, died. What ensued was a population explosion of rats, which are potential carriers of such diseases as typhus and plague. In an attempt to restore the cat population, WHO and the Royal Air Force undertook a remarkable venture, Operation Cat Drop, in which they parachuted cats into the villages. With the cat population restored, the rats and the consequent threat of serious disease subsided.

Fortunately, not every pest-control effort entails such complications. Insecticides rapidly broken down in the environment and more selective in their toxicity have been developed, and these are less disruptive of biological communities. A much more preferred approach, however, is biological control, e.g., the use of the natural predators, parasites, or pathogens of the pest rather than chemical treatments.

CONCEPTS IN BRIEF

1. Ecology is the study of the interactions between organisms and their environment. There are three higher levels of organization: *populations,* groups of individuals belonging to the same species living in the same place at a particular time; *communities,* units composed of all the populations of all the species living in a given area; and *ecosystems,* the sum total of the communities and their physical environments considered together. The various ecosystems are linked by biological, chemical, and physical processes. The entire earth's ecosystem is called the *biosphere.*

2. All organisms have the potential for explosive growth; under ideal conditions their growth curve would be J-shaped or *exponential.* However, the exponential growth of many real populations begins to level off as the density approaches the *carrying capacity* (K) of the envi-

ronment. Such a growth curve, called an *S-shaped or logistic growth curve,* results from a changing ratio of births to deaths. Environmental limitations become increasingly effective in slowing population growth as the population density rises (i.e., the limitations are *density-dependent*).

3. The populations of many short-lived animals, or those living in variable environments, grow exponentially and then suddenly crash (boom-and-bust curve). The crash is usually due to a *density-independent limitation* such as the effects of weather or other environmental factors.

4. In addition to the birth rate and death rate, the potential life span, the average life expectancy, the average age of reproductive maturity, and the age distribution are important determinants of population structure. Determining the mortality rates for the various age groups in the population gives a survivorship curve, of which there are three major types.

5. Organisms with S-shaped growth curves, whose population limitation is primarily density-dependent, are called *K*-selected species. The maximum density for population is determined largely by the environment's carrying capacity. *K*-selected species tend to be large, longer-lived organisms producing only a few slow-maturing young.

6. In organisms with boom-and-bust curves, population growth is limited primarily by density-independent limitations. These organisms are called *r-selected species* because they produce large numbers of small, quickly maturing offspring; i.e., they have evolved high intrinsic rates of increase (high *r*).

7. *Predation* and *parasitism* can influence the prey (or host) species in a density-dependent manner. In general, the density of predators and parasites fluctuates in direct proportion to the changes in the density of the prey or host.

8. *Intraspecific competition* is one of the chief density-dependent limiting factors. As population density increases, competition for limited environmental resources becomes increasingly intense and acts to slow population growth.

9. *Interspecific competition* can also act as a density-dependent limitation. The more the niches of the species overlap, the more intense the interspecific competition. *Niche* denotes an organism's role and position in an ecosystem. Every aspect of an organism's existence helps define its niche. According to the *competitive exclusion principle,* two species cannot simultaneously occupy the same niche in the same place for very long.

10. Intense interspecific competition can lead to extinction, range restriction, character displacement, or a combination of these factors.

11. In some animals, crowding induces physiological and behavioral changes that result in increased emigration from the crowded region. Crowding may also result in decreased resistance to disease and in hormonal changes that may reduce the reproductive rate.

12. The species that make up a community interact with each other and with the physical environment. The biotic community they form can be considered a unit of life, with its own characteristic structure and functional interrelationships.

13. *Ecological succession* is an orderly series of community change involving the replacement, over time, of the dominant species within a given area by other species, until a mature stage is reached. The sequence of changes in *primary succession* (i.e., changes occurring in newly formed habitats) is longer and slower than in *secondary succession* (i.e., changes occurring where previous communities were destroyed).

14. Succession occurs because successional communities modify the physical environment in such a way as to make it less favorable for themselves and more favorable for others, and because of differences in dispersal and growth patterns among species. The species that predominate in the early stages tend to be *r*-selected species that are rapid growers and easily dispersed.

15. The trend of most succession is toward a more diverse, complex and longer-lasting community. Most successional sequences reach a stage that is more stable than those preceding it. This stage is called the *climax community.* It will persist as long as the climate and other environmental factors remain the same.

STUDY QUESTIONS

1. What level of organization is represented by: (a) all the rainbow trout in a lake? (b) all representatives of all the species of organisms living in a lake? (c) all the living organisms and the physical environment (e.g., water, nutrients, light), of a lake? (p. 737)

2. Compare density-dependent and density-independent limitations on population growth. Which type is associated with an S-shaped growth curve? Which is associated with physical factors in the environment, such as weather? (pp. 741–743)

3. **Compare the age distribution of a nearly stable population (such as the United States) with that of a rapidly growing population (such as India). If the birth rate in each country immediately fell to the replacement level, which population would increase by a greater percentage before growth ceased? (pp. 744–745)**
4. **Contrast *r*-selected and *K*-selected species with respect to reproductive potential, mortality rate, maximum life span, and typical environment. Which type of species is most likely to have its population growth limited by density-dependent factors? (pp. 745–748)**
5. **How do predation and parasitism (including disease) act as density-dependent limiting factors on a prey or host population? How is the density-regulating influence of a predator sometimes beneficial to the prey population? (pp. 748–749)**
6. **Compare intraspecific and interspecific competition as density-dependent limiting factors. Rank these relationships according to the severity of the competition expected between each of the following pairs: (a) two zebras; (b) a zebra and a gazelle; (c) a zebra and a lion. (pp. 749–750)**
7. **How does crowding affect the physiology and behavior of migratory locusts? How are these changes adaptive? (pp. 751–752)**
8. **What is *ecological succession*? Are the species characteristic of the earliest stage of succession (the *pioneer community*) more likely to be *r*-selected or *K*-selected? Describe the two most important causes of succession. (pp. 752–754)**
9. **How is primary succession different from secondary succession? Which kind usually proceeds faster? Describe an example of each category of succession. (pp. 754–755)**
10. **Can a *climax community* change? Why or why not? How would the climax and pioneer stages of succession probably differ with respect to complexity, efficiency of energy utilization, and other community attributes? (pp. 753, 757)**

ECOSYSTEMS AND BIOGEOGRAPHY

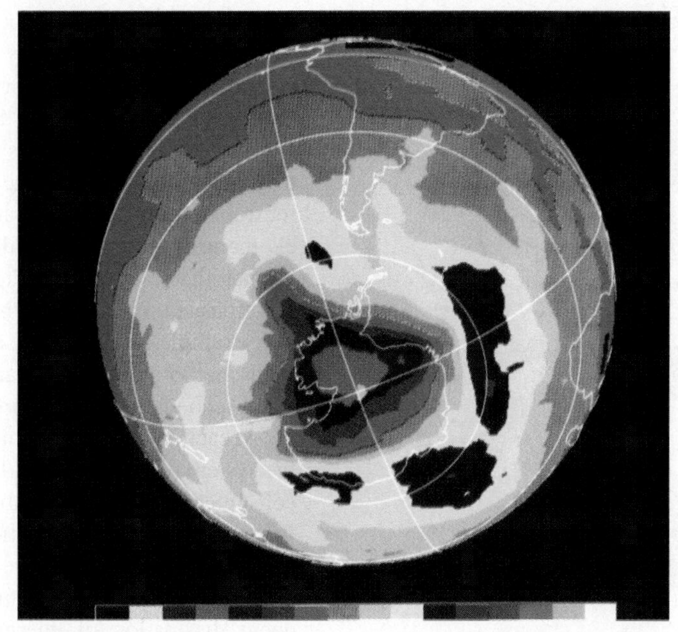

The discussion of ecology in the last chapter was concerned primarily with interactions between individuals within a population, between species within a community, and between the community and its environment. But how do communities come to be located where they are, how do they spread, and how do energy and nutrients essential to the continued survival of communities cycle between their members and the environment? Further, how does the physical environment and the cycling of nutrients influence the organization of communities and the distribution of organisms around the globe? This chapter explores how communities and their physical environments, linked by the movement of energy and matter, function together as a system; it is concerned with the energy and nutrient "economics" of ecosystems and the biogeography of communities.

THE ECONOMY OF ECOSYSTEMS
The Flow of Energy

Life Depends, Ultimately, on Energy From the Sun

Making use of the photosynthetic and catabolic pathways described in Chapters 7 and 8, all forms of life, with the exception of chemosynthetic organisms, obtain their organic nutrients either directly or indirectly from photosynthesis. The total amount of energy that photosynthesis converts into organic matter is called ***gross primary productivity.*** Plants use from 15 to 70 percent of their gross productivity in their own metabolism. What remains after metabolism is known as ***net primary productivity.*** The total net primary productivity of the biosphere provides the energy base for heterotrophic life on earth. Heterotrophic organisms—most bacteria and protists, and all fungi and animals—obtain the energy they need by consuming green plants, feeding on other heterotrophic organisms that fed on green plants, or feeding on the ***detritus*** (dead bodies or waste products) of other organisms.

All Food Webs Begin With Autotrophs and End With Decomposers

The sequence of organisms through which energy moves in a community is called a ***food chain*** (Fig. 34.1). In most communities, many complexly intertwined food chains together form

Chapter opening photo: False color image of Antarctic ozone hole. The ozone hole typically occurs over Antarctica from late August through early October.

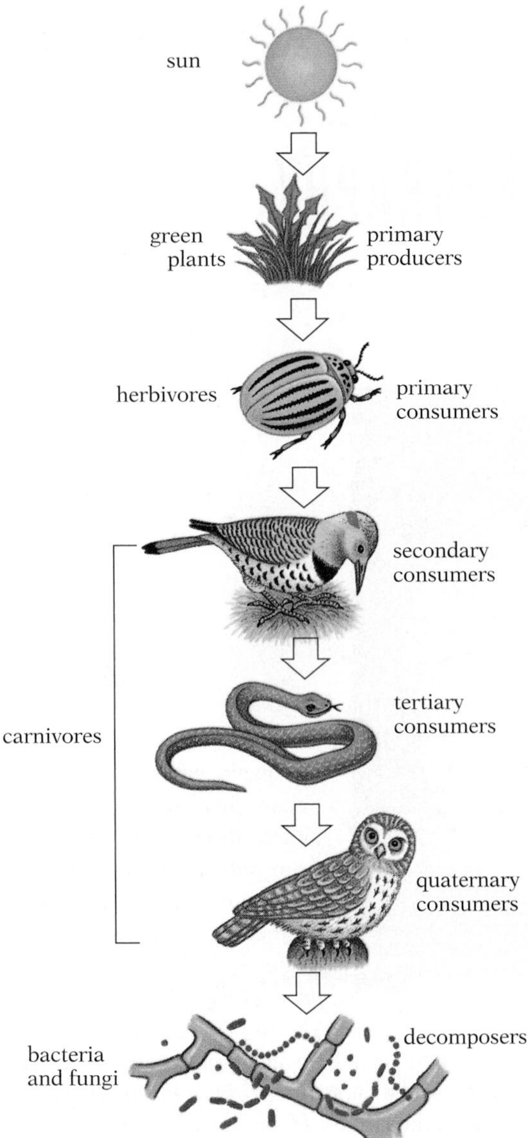

34.1 A food chain The food chain begins with autotrophic organisms (green plants primarily) and ends with decomposers (fungi and bacteria primarily). The energy that primary producers convert by photosynthesis into organic matter is passed along the food chain, from primary consumers to subsequent consumers.

a community *food web* (Fig. 34.2). No matter how long a food chain or how complex a food web may be, certain basic characteristics are always present: 1) every food chain or web begins with the autotrophic organisms (green plants primarily) that are the ***producers*** for the community, and 2) every food chain or web ends with ***decomposers,*** the organisms of decay, which are usually bacteria and fungi that release simple substances—carbon dioxide and ammonia for example—reusable by the producers. The links between the producers and the decomposers are more variable. The producers may die and be acted upon directly by the decomposers, in which event there are no intermediate links. Or the producers may be eaten by ***primary consumers,*** the herbivores. These, in turn, may be either acted upon directly by decomposers or fed upon by ***secondary consumers,*** such as carnivores, parasites, or scavengers (Fig. 34.3).

The Number of Trophic Levels in an Ecosystem Is Limited

Ecologists speak of the successive levels of nourishment in the food chains of a community as ***trophic levels.*** All the producers together constitute the first trophic level. The primary consumers (herbivores) make up the second trophic level, the herbivore-eating carnivores (secondary consumers) the third trophic level, and so on. The species on each trophic level differ from one community to another. Moreover, the trophic levels themselves are not hard-and-fast categories, since many species that eat a varied diet—especially omnivores—may function at two or more trophic levels within a single food web. For example, small birds such as chickadees—which eat seeds, herbivorous insects, and carnivorous insects—function at the second, third, and fourth trophic levels.

At each successive trophic level there is loss of energy from the system (Fig. 34.4). The loss is due to three factors: (1) the consumer population's inability to harvest any more than a fraction of the food population; (2) the failure of assimilation (most animals cannot utilize the cellulose walls of plant cells, for example); and (3) movement and respiration and the consequent dissipation of energy as heat (the Second Law of Thermodynamics states that every energy transformation involves loss of some usable energy; see p. 78). As a result, only about 10 percent of the energy at one trophic level is usually passed on to the next. Therefore, the productivity (energy bound into new organic matter per unit area per unit time) at each trophic level is only about 10 percent of that at the preceding level. There is less productivity from the herbivores of a community than from the plants of that community, there is less productivity from the carnivores than from the herbivores, and so on. This results in a distribution of productivity within a community that can be represented by a pyramid, with the first trophic level (producers) at the base and the last consumer trophic level at the apex (Fig. 34.5). Because of the rapid fall in productivity from one trophic level to the next, there are seldom more than four or five levels in a food chain; the fifth level has no more than about 0.0001 the productivity of the first, making the productivity possible for any subsequent levels too low to be effective at sustaining a viable population of higher level consumers. This is why ***top predators*** (predators at the top of their food chains, i.e., at the top of the pyramid in our example), such as lions or wolves, are not themselves preyed on: there are too few of them, they are too widely scattered, and,

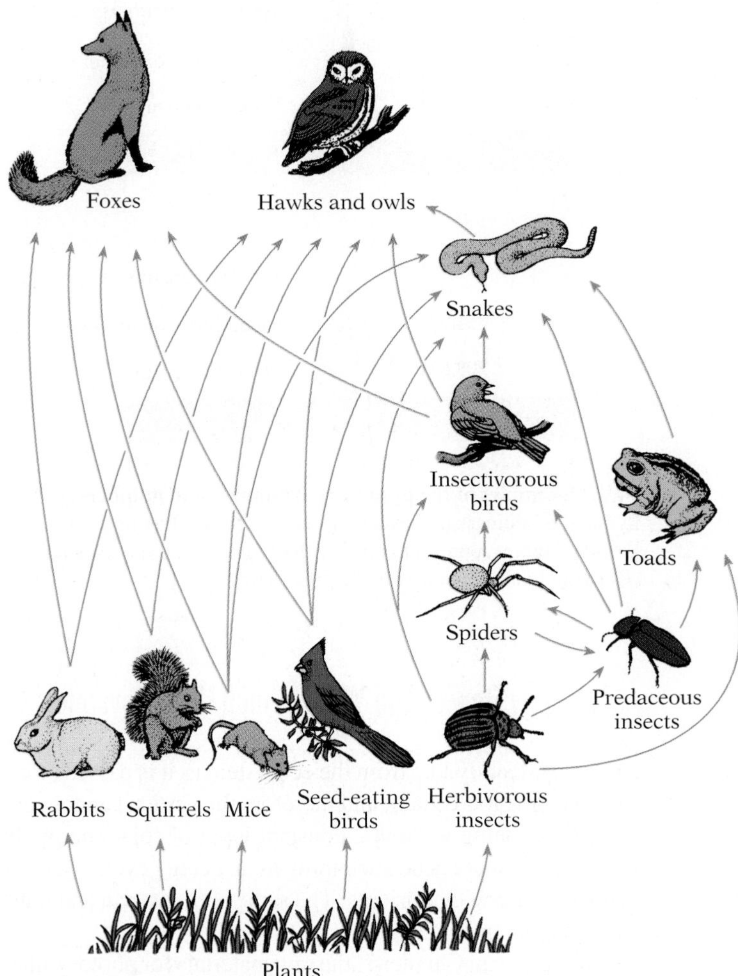

34.2 A hypothetical food web This diagram illustrates the interdependence of living organisms, though no real food web would be as simple as this one. Parasites, disease-causing organisms, and decomposers are omitted.

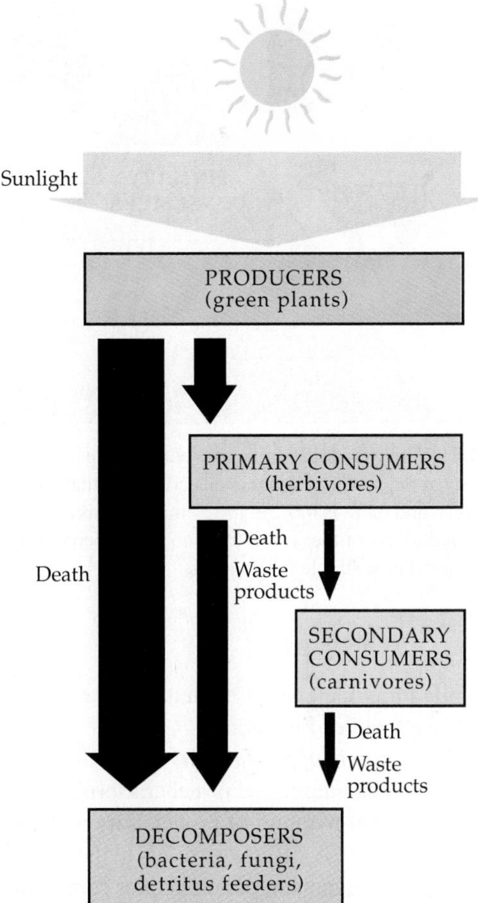

34.3 Flow of energy through the principal trophic levels in an ecosystem The green plants are the producers, which are eaten by the herbivores, the primary consumers. The primary consumers may in turn be eaten by parasites or carnivores, the secondary consumers. Producers may die and become food for decomposer organisms; consumers generate waste products utilized by decomposers, and eventually "contribute" their own bodies when they die.

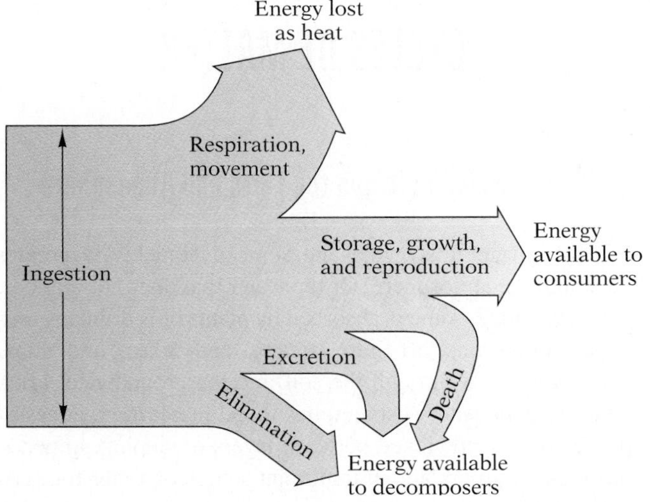

34.4 Energy use by consumers Energy ingested in food is either digested and assimilated or passed through and eliminated in feces. The assimilated energy is used for the functioning of the body (e.g., movement, respiration), excreted, or stored and used for the growth of new tissues and reproduction. When the organism dies the energy stored in tissues is used by the decomposers. Only the stored materials are available to organisms at the next trophic level.

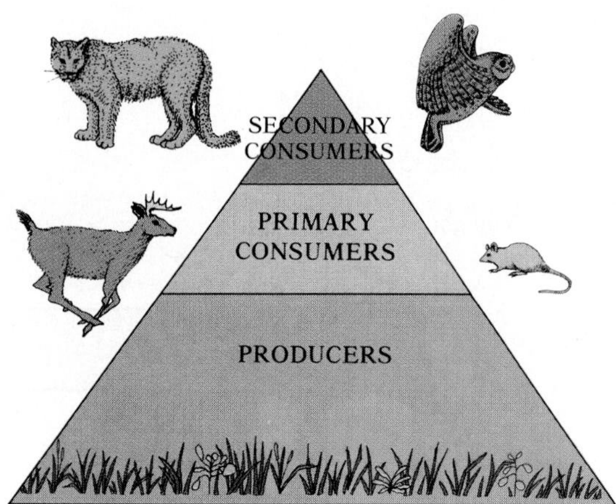

34.5 The pyramid of productivity There is much more productivity at the producer level in an ecosystem than at the consumer levels, and there is more at the primary consumer level than at the secondary consumer level. Only about 10 percent of the energy at one level is available for the next.

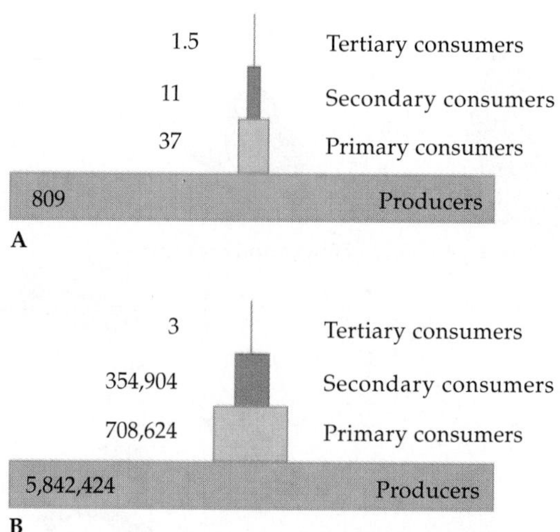

34.7 Examples of the pyramids of biomass and numbers (A) Pyramid of biomass in the aquatic ecosystem of Silver Springs, Florida. Figures represent grams of dry biomass per square meter. (B) Pyramid of numbers in a bluegrass field.

population-wide, they contain too little energy to make the effort worthwhile (Fig. 34.6).

The ***pyramid of productivity*** (also called the pyramid of energy flow) just described is a characteristic of *all* ecosystems. Related to the pyramid of productivity is the ***pyramid of biomass*** (Fig. 34.7A). Biomass refers to the weight of living materials in an organism, population, or community; it is usually expressed as mass per unit area. Because the decrease of energy at each successive trophic level means that less biomass can be supported at each level, the total mass of carnivores in a given community is almost always less than the total mass of herbivores. Some communities show a ***pyramid of numbers,*** there being fewer individual herbivores than plants, and fewer individual carnivores than herbivores (Fig. 34.7B).

34.6 Lion with prey Lions are the top predators in the food chain in this ecosystem.

Energy Flow is One-Way Through the Ecosystem

Energy is steadily lost from the ecosystem as it is passed along the links of a food chain, and therefore the system cannot continue functioning without a constant input of solar energy. In other words, there is no such thing as an energy cycle. But this is not the case with nutrients. The same nutrients can and must be used over and over again.

As we saw in Chapter 7, the raw materials for photosynthesis are light, water (which is also essential for transpiration), and carbon dioxide. In addition, growing plants need substantial quantities of nitrogen, phosphorus, potassium, sulfur, magnesium, and calcium (see Table 16.1, p. 326). We will trace a few of these cycles here, and see how human activity affects the availability and distribution of these materials.

CYCLES OF MATTER

Water Cycles Between the Earth and Atmosphere

When rainwater falls on the land, some of it quickly evaporates back into the atmosphere. Of the water that does not immediately evaporate, some is absorbed by plants or is drunk by animals, some runs off into streams and lakes, and some percolates down through the soil into the groundwater (Fig. 34.8). Groundwater is sometimes found in ***aquifers,*** zones of porous rock sandwiched between layers of sloping, impervious rock. Aquifers are an important source of water for agricultural irrigation and domestic and industrial use.

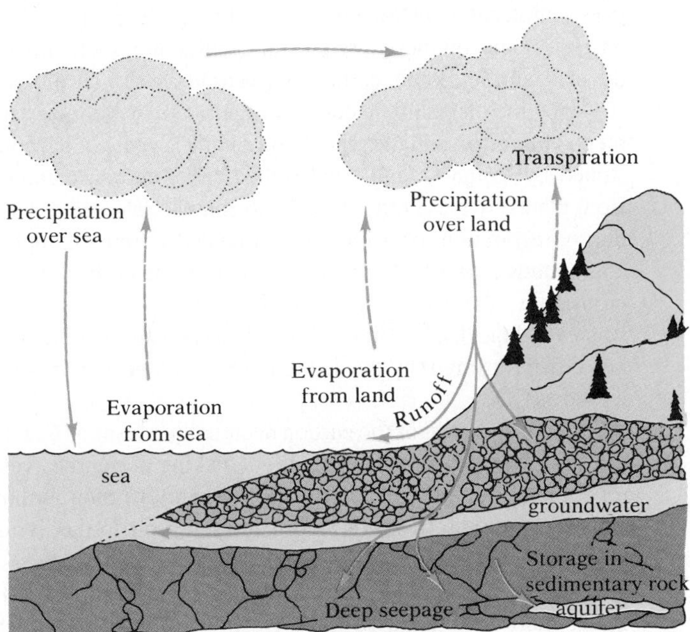

34.8 The water cycle This diagram shows most of the major pathways of water movement through the ecosystem—but not the more recent ones created by human beings.

Unfortunately, many aquifers are being depleted faster than they can be replenished by natural processes, resulting in a lowering of the water table in the area. Water levels in wells in these areas keep dropping and the wells must be periodically drilled deeper to keep the water flowing. Eventually the water will be used up; the rate of replacement is so slow, in fact, that it could take centuries to restore the water levels. And in many areas human activities have so altered the surface runoff patterns that the water may not even be replaced slowly.

The water in streams, lakes, and the ground eventually finds its way to the sea. Most of the water in the atmosphere has come from evaporation off the surface of the oceans, but there is also constant evaporation from lakes, streams, soil, and from plants and animals. Tremendous amounts of water are lost by plants in the process of transpiration. For example, if all the water lost by a forest by transpiration during a year were spread over the forest floor, it would be over a meter deep! The energy for most of this evaporation comes either directly or indirectly from solar radiation.

The endless cycling of water to earth as rain, back to the atmosphere through evaporation, and back again to earth as rain, maintains the various freshwater environments and supplies the vast quantities of water necessary for life on land. There is an estimated 1.36 billion cubic kilometers of water on the earth, over 97 percent of which is found in the oceans, another 2 percent in glaciers and snowfields, and the remainder (less than 1 percent) in lakes, streams, the ground, and the atmosphere. Although the percentages of water in the latter sources are low compared to the oceans, the amount of water involved is enormous: each year about 380,000 cubic kilometers of water fall to the earth as rain and snow. The tremendous importance of rainfall for terrestrial and freshwater life is apparent by contrasting a desert with a lush tropical forest. Glaciers and snowfields store much water on land; for instance if all the glaciers were to melt, the sea level would rise enough to submerge many coastal areas.

The water cycle is also a major factor in modifying temperatures, and it provides for the transport of many chemical nutrients through ecosystems. Unfortunately, the downward percolation of water into the ground through dumps, landfills, industrial wastes, septic tank fields, and the like has caused it to be contaminated (Fig. 34.9). Toxic chemicals and bacteria

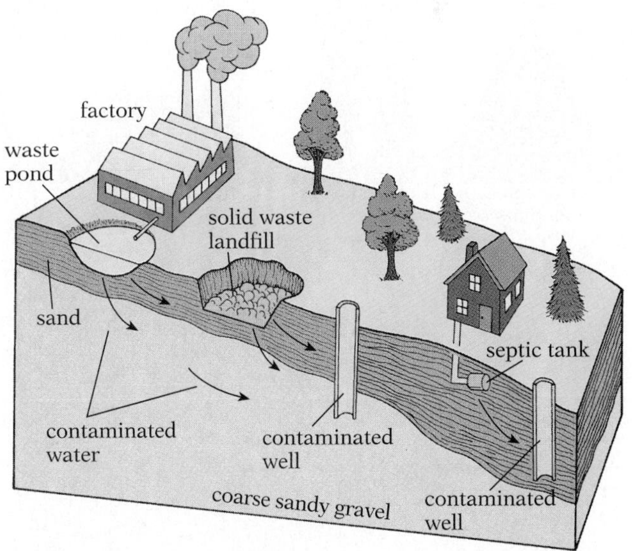

34.9 Contamination of groundwater Groundwater pollution is not visible and often goes undetected. Pollutants from industrial wastes, solid waste landfills, untreated municipal sewage, pesticides, septic tank fields, etc., percolate through the soil and rock and eventually contaminate the groundwater and affect a wide area.

from these stores can make the groundwater unfit for human use. The pollution problems associated with landfill wastes in the United States have become increasingly important. In 1980 the Environmental Protection Agency estimated that each year 150 million tons of municipal wastes and 240 million tons of industrial wastes were emptied into landfills and dumps. Further, an estimated 10 trillion gallons of liquid industrial wastes are stored in basins, lagoons, and pits. An adequate supply of clean water is essential to life on earth, and in the coming years we must take steps to protect our water supplies to ensure a sufficient supply of safe, good-quality water.

Carbon Cycles From Its Inorganic Reservoir Through the Living System and Back

The carbon dioxide contained in the atmosphere and dissolved in water constitutes the reservoirs of inorganic carbon from which almost all organic carbon is derived. It is photosynthesis, largely by green plants, that extracts the carbon from this inorganic reservoir and incorporates it into the complex organic molecules characteristic of living substance (Fig. 34.10). Some of these organic molecules are soon broken down again, and their carbon is released as carbon dioxide (CO_2) by the plants in the course of their own respiration. But much of it remains in the plant bodies until they die or are consumed by animals, fungi, and protists. The carbon obtained from plants by these other organisms may be released as CO_2 during respiration, or it may be eliminated in more complex compounds as wastes, or it may remain in the bodies of heterotrophic organisms until they die. Usually these organic wastes and the dead bodies of both plants and animals are broken down (respired) by the decomposers, and the carbon is released as CO_2.

Notice that most of the carbon eventually returns as CO_2 to the air or water from where it started, making this a true cycle (or rather a complex of interlocking cycles). Carbon is thus constantly moving from the inorganic reservoir to the living system and back again.

Carbon moves rather rapidly through pathways just outlined. There are alternative pathways, however, that take much longer. Leaves, twigs, seeds, and other vegetation deposited in swampy environments occasionally fail to decompose

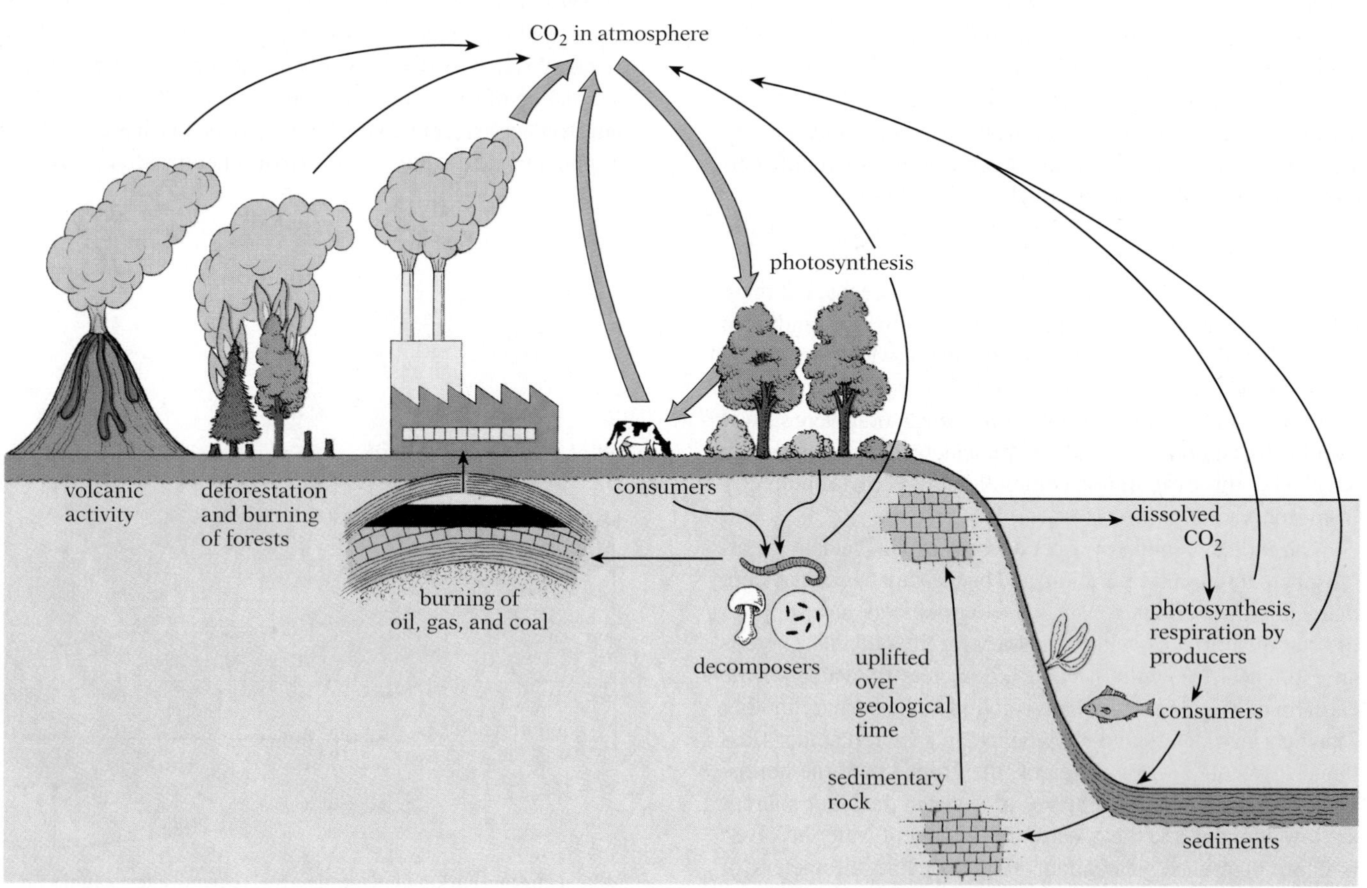

34.10 The carbon cycle The atmosphere contains about 740×10^{12} kilograms (kg) of carbon, while the oceans hold approximately $43{,}000 \times 10^{12}$ kg. Deforestation and burning of fossil fuels contribute about 1×10^{12} and 5×10^{12} kg annually, respectively, of which about 3×10^{12} accumulates in the atmosphere. Some of the remaining 3×10^{12} kg is dissolved into the oceans, but the fate of much of this carbon dioxide is yet to be traced.

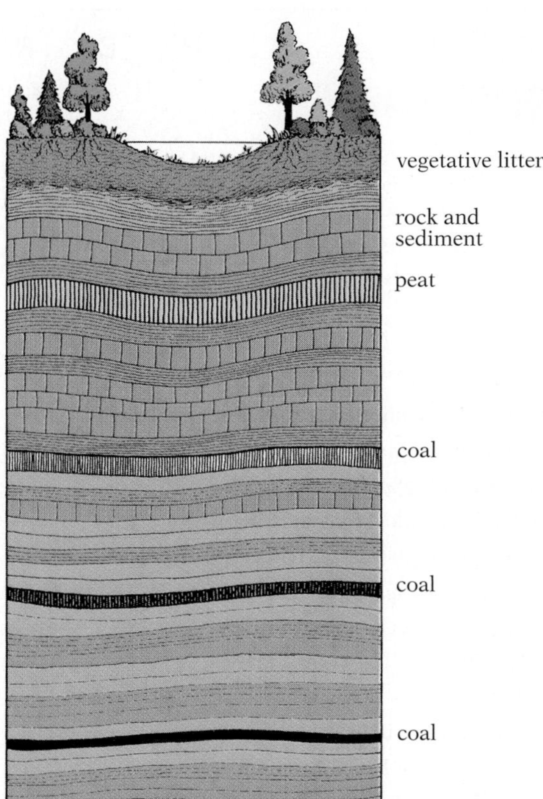

34.11 Coal formation Leaves, twigs, branches, seeds, and other vegetation deposited in swampy environments are protected from complete decay by the watery environment and rapid burial. The partly decayed buried deposits are compressed to form peat. Continued burial of this material under hundreds or thousands of meters of sediment and chemical transformations convert it into coal, oil, gas, or diamonds.

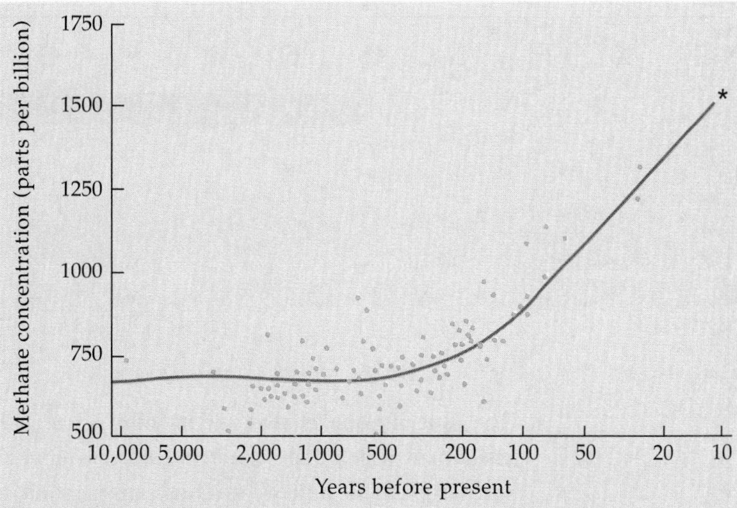

34.12 Increase in atmospheric methane over the past 10,000 years Samples of air trapped in glaciers can be used to reconstruct the composition of the earth's atmosphere in the past. The methane level (which is strongly correlated with carbon dioxide concentration) has risen dramatically over the past 200 years. Most methane is released by anaerobic organisms in bogs, but ruminants, including domestic cattle, are also major sources.

promptly and are eventually compressed to form peat. Continued burial of this material under hundreds or thousands of meters of sediment converts it into coal, oil, gas, or diamonds (Fig. 34.11). Carbon in these forms may be removed from circulation for very long periods, perhaps permanently; but some of it may eventually return to the inorganic reservoir if the coal, oil, or gas (i.e., fossil fuel) is burned.

Human beings have of course greatly accelerated the return of such carbon to the active cycle through the use of fossil fuels and other activities. Of the CO_2 released by the burning of fossil fuel, about half remains in the atmosphere and the rest is absorbed by the ocean waters. The CO_2 reservoir in the atmosphere is also being increased through the cutting down and clearing of forests, particularly those in the tropics, which are an important "sink" for CO_2. These human activities have increased atmospheric CO_2 and methane (CH_4) levels by 15 percent in the last hundred years (Fig. 34.12); it is entirely possible that the CO_2 and CH_4 concentrations will double in the next hundred years (see Exploring Further: Global Change, p. 770).

Carbon dioxide and methane play an integral role in the regulation of temperature on the surface of the earth. Heat radiated from the earth is absorbed by CO_2 and CH_4 in the atmosphere and, when radiated back towards the surface, tends to warm the atmosphere in a so-called ***greenhouse effect*** (Fig.

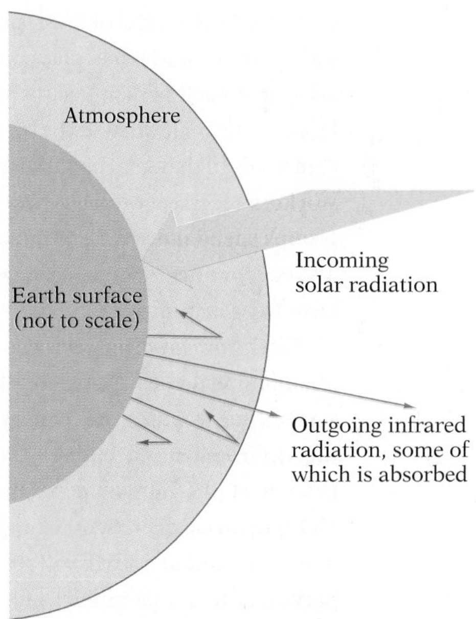

34.13 The greenhouse effect The sun's radiation consists of relatively short wavelengths of visible light, which is absorbed by the earth and radiated back as heat—long-wavelength infrared rays. Carbon dioxide in the atmosphere absorbs many of these rays, and this trapped radiation heats the atmosphere. Burning of fossil fuels and other processes add carbon dioxide and other gases (e.g., methane, nitrogen oxides, sulfur dioxide, and ozone-destroying chlorofluorocarbons) to the atmosphere, trapping even more infrared radiation, and warming the earth even more.

EXPLORING FURTHER

GLOBAL CHANGE

The issue of global change and its influence on the atmosphere and biosphere is a complex one, about which there is much debate. Global change is frequently equated with global warming and the link to fossil fuel consumption, but this is a misleading and narrow view, for global warming is only one part of the picture. The release of chlorofluorocarbons (CFCs) into the atmosphere, for instance, and their effect in ozone depletion clearly contributes to atmospheric change. And the burning of fossil fuels adds nitrous oxides and sulfur dioxide to the atmosphere, which contributes to acid rain (p. 778) and air pollution. The cutting of forests, erosion of topsoil, extinction of species, and pollution of our rivers, lakes, and oceans all affect the biosphere and contribute to global change.

Just how much and how fast the atmosphere, biosphere, and climate will be altered, and what the long-term effects will be, is as yet unclear. The predictions are based on models and statistics, some of which are founded on limited data and uncertain assumptions. Unfortunately we cannot wait for more complete data and models, for if we delay it may be too late to halt or prevent irreversible damage to our planet. We already have some clear indications of the effect of one type of pollution on human health. Studies of the health effects due to ozone, and particulate matter in the air of the South Coast Air Basin of California, show that each person faces a one in 10,000 increased risk of death each year due to high levels of air pollutants in the form of particulate matter, and that meeting the air pollution standards in that area could save 1,600 lives a year. Children, college students, and outdoor workers receive a disproportionately large share of the exposure, probably because these groups spend more time outdoors. As our population continues to grow, the health effects and economic costs of pollution will become more obvious, and more stringent controls will have to be instituted.

Environmentalists point out that the solution to the problem of global change involves abandoning the self-destructive practices occurring today in favor of building a sustainable society, one in which the economic growth is managed so as to do no irreparable damage to the environment. A major obstacle to sustainable development is the alarming rate of growth of the human population, increasing the demands for resources and services. Holding the environmental impact constant or (hopefully) decreasing it in the face of enormous population growth will require reform and new social and economic policies. Poverty plays a crucial role in this; someone who is worried about where his next meal is coming from is unlikely to be concerned about the environmental impact of his activities. The solutions will be costly. There is no doubt but that there will be severe economic and social consequences, and that the quality of life will change, governments will fall, human health will be affected, laws will become more restrictive, and a more global outlook will have to prevail. We must act to achieve sustainability in the interest of *all* organisms on the planet, not just humans.

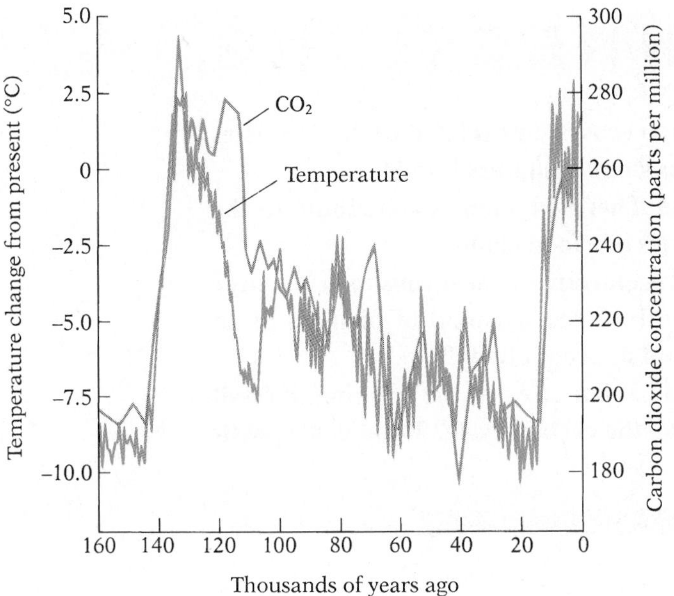

34.14 Correlation between atmospheric CO_2 concentration and average temperature Global temperature (measured in terms of the amount of ice deposited in Antarctic glaciers) correlates well with atmospheric levels of carbon dioxide over the last 160,000 years. The interpretation of this fact is controversial: in many cases the carbon dioxide levels seem to *follow* temperature changes; if there were a strict cause-and-effect relationship, carbon dioxide changes would precede temperature fluctuations.

34.13). Thus, a rise in the CO_2 and CH_4 levels in the atmosphere could cause the temperature of the earth to increase. Other factors—among them cloud patterns, the amount of water vapor, and the number of solid particles in the atmosphere—also influence the temperature balance of the earth. Because total cloud cover and distribution, as well as the amount of water vapor and atmospheric particulate matter, are also changing as a result of human activities, and can act to cool the atmosphere by reflecting solar radiation away from the earth, it is difficult to predict what human activity is going to do to the climate. Historical evidence, however, indicates a strong link between CO_2 and CH_4 (which is about 50 times as active as CO_2) concentration and temperature (Fig. 34.14). The warming trend is already apparent—records indicate that global temperatures have increased by 0.6–0.7°C since the turn of the century, and the five warmest years have occurred in the 1980s. Some current estimates are that the average temperature of the earth will rise by 1.5 to 4.5°C by the year 2050 due to greenhouse warming. This small increase in global temperature would be enough to create significant climate changes and melt a substantial amount of polar and other ice, thereby raising the sea level. The consequences for coastal areas, particularly major port cities like New York, could be serious (Fig. 34.15).

Oxygen Is Maintained at a Steady State in the Atmosphere

The oxygen in our present atmosphere came from two processes: photosynthesis and the dissociation of water vapor in the upper atmosphere. As we learned in Chapter 7, plants use the energy of sunlight to split water, and the oxygen thus formed is released into the atmosphere. Most of the free oxygen in our present atmosphere came from photosynthesis by green plants, but water dissociation is also thought to have contributed a small amount. This process occurs in the upper atmosphere, above the ozone shield, where ionizing radiation from the sun provides energy to split water into molecular hy-

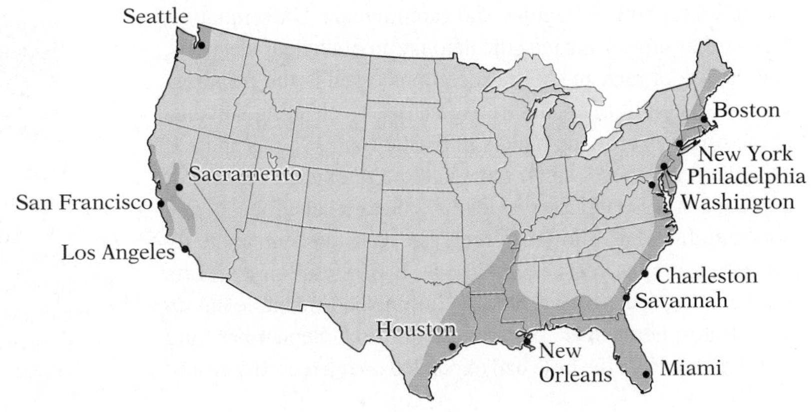

34.15 Effect of melting the earth's polar ice sheets and glaciers If the climate should warm significantly, the earth's ice sheets would begin to melt, adding their water to the oceans and causing the sea level to rise. If all the ice melts, the sea level could rise as much as 75 meters, flooding coastal areas and many major cities.

MAKING CONNECTIONS

1. How is the energy flow through ecosystems related to the processes of photosynthesis and respiration, discussed in Chapters 7 and 8?

2. How does the Second Law of Thermodynamics contribute to the limitation on the number of possible levels in a food chain?

3. Can the earth support more vegetarians or meat consumers? Why?

4. Why is it possible to have an inverted pyramid of biomass in an ecosystem, but never an inverted pyramid of productivity?

5. How were fossil fuels formed? What effect is the burning of fossil fuels and the clearing of forests having on the carbon cycle? What is the possible consequence to the earth's climate?

drogen and molecular oxygen. Hydrogen gas, being light, escapes into outer space, while the oxygen remains behind (Fig. 34.16).

The oxygen level in the earth's atmosphere is maintained at a near steady state of 21 percent. How is this accomplished? To a certain extent, the level of oxygen is linked to that of carbon (Figs. 34.10 and 34.16); the oxygen produced during photosynthesis is balanced by that used in respiration. Forests are particularly important in maintaining a balance of the two gases in the atmosphere.

Oxygen in the outer atmosphere is found as ozone gas (O_3). A layer of ozone gas encircles the earth in the outer two-thirds of the stratosphere (12 to 30 miles above the earth's surface). Ultraviolet (UV) light from the sun splits the ozone molecule into O and O_2, but the ozone molecule quickly reforms, giving off heat in the process:

$$\text{UV light} + O_3 \dashrightarrow O + O_2 \dashrightarrow O_3 + \text{heat}$$

In this matter the ozone layer converts much of the harmful ultraviolet light into heat. Without the ozone shield, the surface of the earth would be virtually uninhabitable because the ultraviolet light is both mutagenic and carcinogenic. Unfortunately, high-flying supersonic jets, the detonation of nuclear weapons, and the use of man-made fluorocarbons (used in the compressors of air conditioners and refrigerators, as cleaning solvents in factories, as blowing agents in plastic foam, and as an inert propellant in aerosol cans) have had the unexpected effect of depleting atmospheric ozone levels. The average global ozone concentration has shown a decrease of 1 percent annually for the past eight years (Fig. 34.17). In 1985 the first reports were made of an "ozone hole"—diminished ozone levels in the stratosphere over Antarctica—during September and October. By 1993 the hole had expanded over a record area and

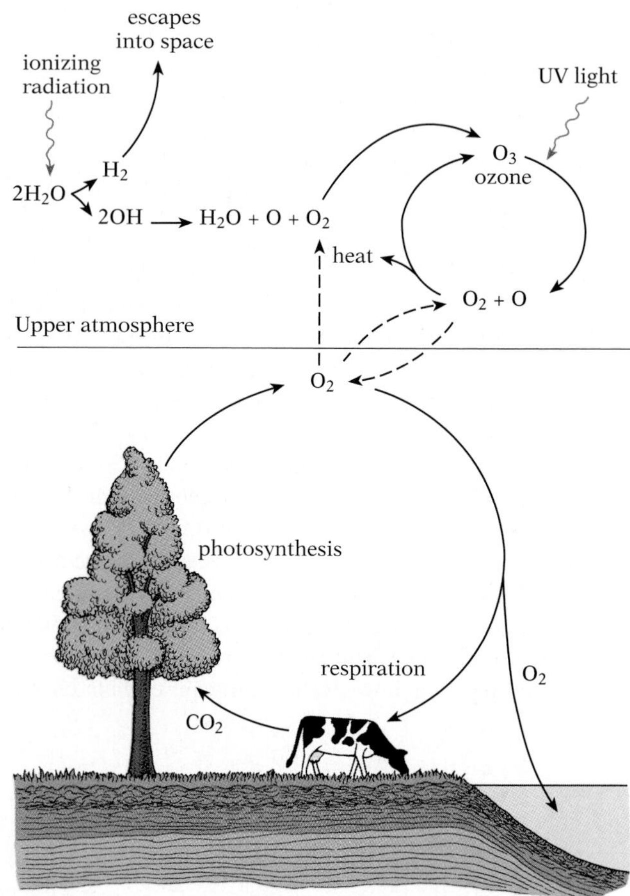

34.16 The oxygen cycle Most of the oxygen (O_2) in the atmosphere today comes from photosynthesis by green plants. In the outer atmosphere, O_2 is converted into ozone that protects the earth from harmful ultraviolet radiation. Ozone molecules absorb ultraviolet light from the sun, forming O and O_2 in the process. The ozone molecule quickly reforms, giving off heat, and can then reabsorb more ultraviolet light. In this way harmful ultraviolet light energy is converted into heat energy.

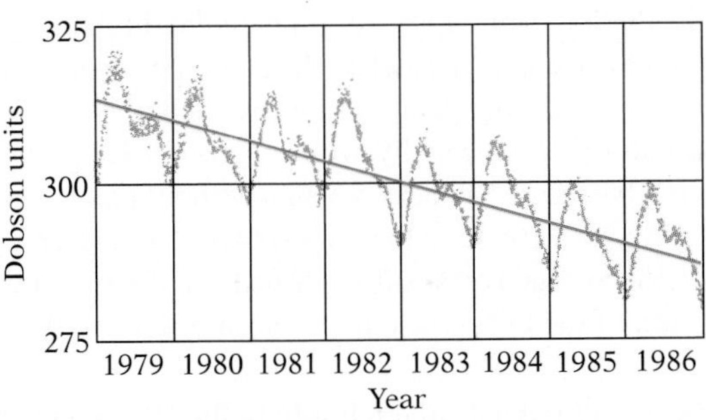

34.17 Decline in atmospheric ozone level The global mean ozone concentration is shown for each day over the course of eight years. There is both a seasonal rhythm and a long-term downward trend of about 1 percent per year.

new reports came in announcing diminished ozone levels over much of the northern United States, Canada, Europe, and Russia (see Fig. 2.26, p. 48). Strong international efforts to eliminate fluorocarbon use and to recycle the gas in existing compressors (particularly automobile air conditioners) may slow the rate of decrease dramatically, but fluorocarbons are very stable, and so will go on catalyzing the breakdown of ozone for decades. A rise in ultraviolet exposure (particularly outside the tropics) seems inevitable. Medical researchers suggest that for each 1 percent depletion in the ozone shield, a 2 percent increase in skin cancer is expected. Other organisms are also affected by high levels of ultraviolet light. Plants, for instance, may show leaf damage, inhibition of photosynthesis, mutations, stunted growth, or even death when exposed to high levels of ultraviolet light. Manufacturers of the fluorocarbons are attempting to produce less stable fluorocarbons, develop good substitutes, and promote recycling to lower the amounts of fluorocarbons in use. Already, considerable progress has been made in these areas.

Nitrogen Cycles Between the Atmosphere and the Living System

Another critical element in nutrient movement through communities is nitrogen (N_2), which is vital for the synthesis of proteins and nucleic acids. Gaseous N_2 makes up roughly 78 percent of the atmosphere; however, this enormous reservoir of inorganic nitrogen cannot be used directly by most organisms. It must first be *fixed* into ammonia (NH_3), nitrate (NO_3-), or amino acids before it can be assimilated. Only certain bacteria (particularly cyanobacteria) are capable of carrying on ***nitrogen fixation;*** they convert N_2 into other nitrogen compounds that can be used by other organisms. Though some nitrogen fixation may also occur as a result of electrical discharges, such as lightning, and though nitrogen is now supplied in commercial, synthetic fertilizers, it is biological nitrogen fixation by microorganisms that provides most of the usable nitrogen for the earth's ecosystems (Fig. 34.18).

Some nitrogen-fixing bacteria live in a close mutualistic relationship with the roots of certain vascular plants, where they

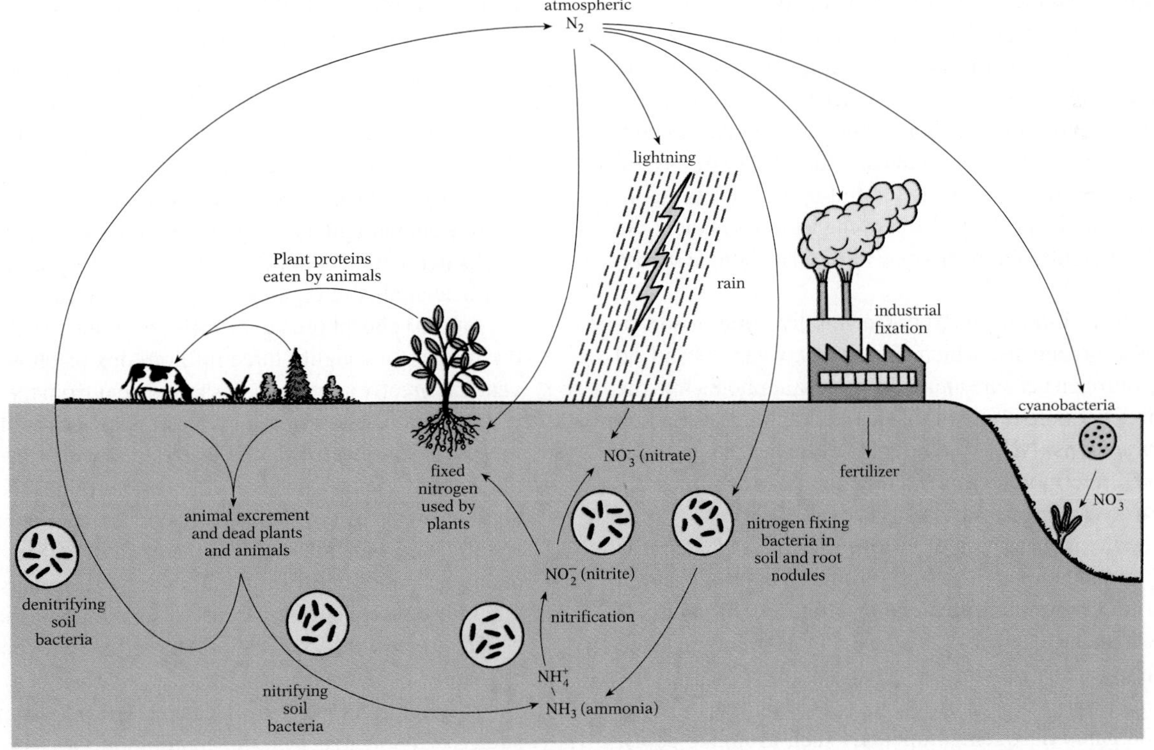

34.18 The nitrogen cycle

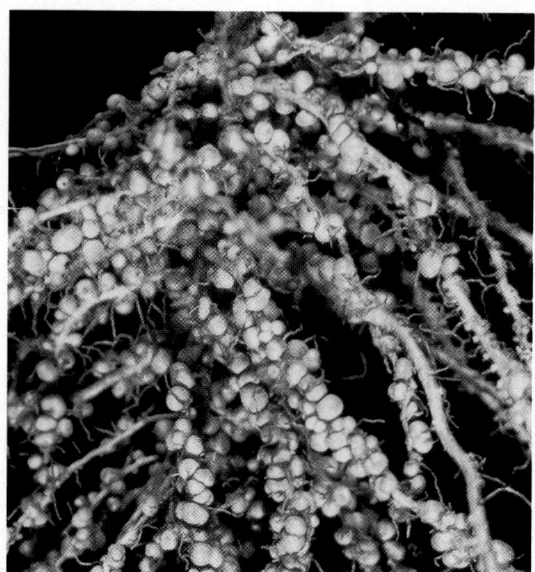

34.19 Photograph of roots of a peanut plant, showing nodules containing nitrogen-fixing bacteria

occur in prominent structures called ***nodules*** (Fig. 34.19). The legumes (plants belonging to the pea family—bean, clover, alfalfa, lupine, etc.) are particularly well known for their numerous root nodules, but plants of some other families also have them. The bacteria housed in root nodules release as much as 90 percent of the fixed nitrogen they produce into the host plant's cytoplasm, primarily in the form of amino acids. Consequently, legumes can grow well in soils very poor in available (fixed) nitrogen. The bacteria in their nodules not only supply the plants with all the fixed nitrogen they need, but actually produce a surplus, some of which is excreted from the roots of the legumes into the soil. Thus plants with nitrogen-fixing microorganisms tend to increase the fertility of the soil in which they grow. Farmers often build up the nitrogen content of their fields by periodically planting legumes in them. As much as 400 pounds of nitrogen may be fixed in a single season by an acre of alfalfa.

Other nitrogen-fixing microorganisms live free in soil or water. These organisms, which may fix between 20 and 50 pounds of nitrogen per acre annually, release ammonia into the surrounding medium. Some of this free ammonia is picked up as ammonium ions (NH_4^+) by the roots of higher plants and is incorporated into more complex compounds. Most flowering plants, however, absorb nitrate instead of ammonia. Nitrate seems to be the main source of nitrogen for these plants. It is produced from ammonia in the soil by other groups of bacteria through a process called ***nitrification*** (Fig. 34.18). Animals are unable to fix nitrogen at all; they obtain their nitrogen in the form of amino acids from the foods they eat.

Most of the nitrate taken up by the roots is quickly incorporated into organic nitrogen compounds such as amino acids, proteins, and nucleic acids, and is then either stored or transported to other parts of the plant body. The nitrogen compounds in the plant body may again be broken down to ammonia by decomposers when the plant dies or when an animal that has eaten the plant dies or excretes it. Notice that nitrogen can cycle repeatedly from plants to decomposers to nitrifying bacteria and back to plants without having to return to the gaseous N_2 state in the atmosphere (Fig. 34.18). In this respect, the nitrogen cycle differs from the carbon cycle, where every turn of the cycle includes a return of CO_2 to the atmosphere.

Though nitrogen need not return to the atmosphere during every turn of the cycle, there is nevertheless a steady drain of some nitrogen away from the soil or water and back to the atmosphere. This is because some bacteria carry out a process of ***denitrification,*** converting nitrite (NO_2-) or nitrate into N_2 and releasing it. In short, the denitrifying bacteria remove nitrogen from the soil-organism part of the nitrogen cycle and return it to the atmosphere. Nitrogen-fixing microorganisms do the reverse, taking nitrogen from the atmosphere and adding it to the soil-organism part of the cycle. Bacteria are crucial in this cycle; without the activities of the nitrogen-fixing bacteria and decomposers, life as we know it could not exist.

Phosphorus Cycles in the Terrestrial Ecosystem

Phosphorus, another mineral essential to life, is required for the synthesis of organic compounds such as phospholipids, nucleic acids, and ATP. It is one of the chief ingredients in commercial fertilizers. Unlike carbon and nitrogen, phosphorus has its reservoir in rocks. When rock erodes, small amounts of inorganic phosphate ions are released into the water, which can then be absorbed by plants and made available to the organisms that eat the plants (Fig. 34.20). When the plant or animal dies, decomposers release phosphate ions, which can then be reused by the plants. Thus phosphate can cycle around the living system without having to return to rock.

Under natural conditions much less phosphorus than nitrogen is available to organisms; in natural waters, for example, the ratio of phosphorus to nitrogen is about 1 to 23. However, the mining of roughly three million tons of phosphorus each year has greatly accelerated its movement from the rocks to the water-organism part of the cycle. Phosphorus is often a limiting factor for the growth of algae in freshwater lakes—that is, the amount of phosphorus available determines the amount of algal growth. Today, phosphorus is being poured into the aquatic environment in enormous quantities through sewage, detergents, and runoff from inorganic fertilizers used in farming. One consequence is extensive algal blooms that cover the water surface with scum and foul the shores with masses of rotting organic matter (Fig. 34.21).

You might expect that the increased photosynthetic productivity associated with the algal blooms would make more food available for higher trophic levels in food webs, and thus be of benefit to the biotic community. But excessive growth of

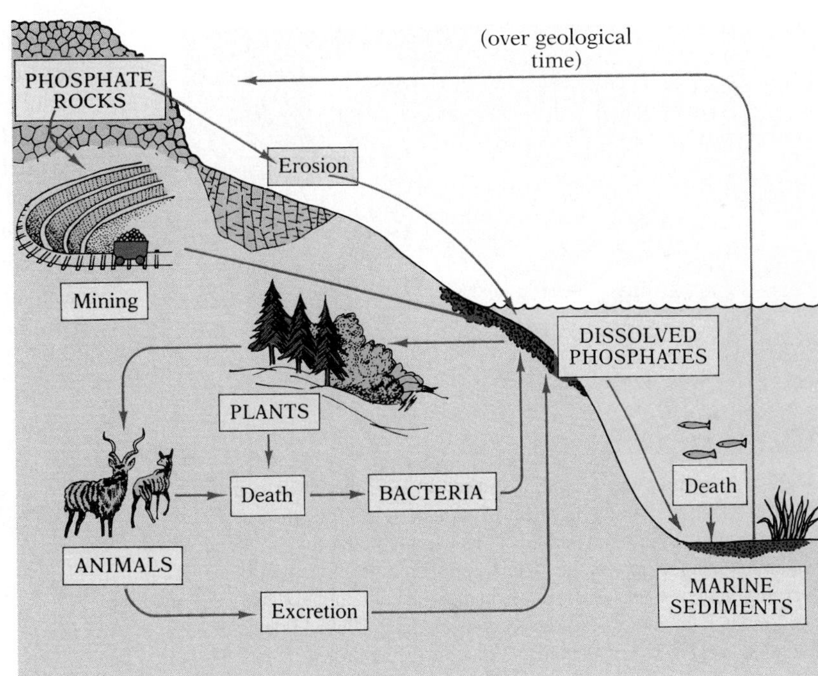

34.20 A simplified version of the phosphorus cycle Phosphate from rock dissolves very slowly (unless the process is speeded up by human intervention). The dissolved phosphate ions are absorbed by plants and synthesized into organic compounds, which are available to the animals that eat them. Some of the phosphate is excreted by animals and goes immediately into the dissolved pool. When plants or animals die, phosphate ions are released by bacteria from organic compounds like nucleic acids that are present in the bodies.

Each year huge quantities of dissolved phosphate ions are carried into the sea in runoff water. Though the formation of new rocks from marine sediments, where the phosphorus eventually comes to rest, is a very slow process, it is unlikely that we will soon run out of phosphate rock, because the known reserves are large. Nevertheless, supplies in soils are readily depleted, and so phosphorus is a limiting nutrient more often even than nitrogen.

34.21 A field experiment demonstrating the limiting nature of phosphorus in the eutrophication of a lake The two basins of a lake were separated by a plastic curtain. The far basin was fertilized with phosphorus, carbon, and nitrogen. The near basin, used as a control, received only carbon and nitrogen. Within two months the far basin had a heavy bloom of algae, whereas the control basin showed no change in organic production.

algae actually causes destruction of many of the higher levels in the food web. At the end of the growing season, many of the algae die and sink to the bottom, where they stimulate massive growth of bacteria the following year. This produces so much bacterial decomposition that the oxygen in the deeper, colder layers of the lake becomes depleted, with the result that cold-water fish such as trout, whitefish, pike, and sturgeon die and are replaced by less valuable species such as carp and catfish. These changes accelerate ***eutrophication,*** or nutrient enrichment and associated "aging," of the lake.[1]

[1] The term "eutrophication" was originally applied to the accumulation of nutrients and increase in organic matter that are a natural part of the aging of lakes. Recently it has been applied not so much to the natural process as to the greatly accelerated one resulting from human interference.

Commercial Chemicals Affect the Ecosystem

Modern industry and agriculture have been releasing vast quantities of new or previously rare chemicals into the environment. The U.S. Food and Drug Administration has estimated the number of these chemicals in the environment at about half a million, with 400 to 500 new chemicals created each year. The pathways taken by these chemicals through ecosystems are known for only a few. Some will be incorporated into the natural biogeochemical cycles and degraded into harmless, simpler substances. But many are so different from any naturally occurring substances that we as yet have no idea what their eventual fate may be, or what effect they may have on the biosphere. Many of the by-products of industrial

34.22 A victim of mercury poisoning In 1953 there was an industrial discharge of mercury near Minimata, Japan. The mercury was incorporated into the bodies of algae living in the contaminated water and fish feeding on the algae developed high concentrations of mercury in their muscles. Japanese eating the contaminated fish accumulated the mercury and were poisoned. By 1960, 43 persons had died and 116 were affected. Mercury affects the central nervous system and the damage is irreparable.

processes have so far not even been fully characterized chemically, and it is likely that some will prove harmful to life. But exactly which ones will be harmful, and how harmful, and to what organisms? Current ecological research is attempting to answer these questions.

The matter is complicated by the fact that a substance may not be harmful in the form in which it is released, but may subsequently be changed by microorganisms or natural physical processes into some other substance with vastly different properties. Mercury pollution of water located near some plastics factories provides a good example. Mercury was originally released from these facilities in an insoluble and nontoxic form that was thought to be chemically stable. When it settled in the bottom mud, however, microorganisms converted it into methyl mercury, a water-soluble compound that accumulates in organisms. The mercury poisoning that resulted was most severe in human beings and other predators at the top of food chains (Fig. 34.22).

Persistent Chemicals May Become Concentrated in the Food Chain

The harmful effect on top predators of the mercury released from plastics factories and from certain fungicides is an example of a common phenomenon called ***biological magnification.*** If a persistent chemical[2] is retained in the body when ingested rather than excreted, then that chemical will tend to become more and more concentrated as it is passed up the food chain. Figure 34.23 shows why this is so.

DDT is a persistent insecticide that has become so pervasive in the biosphere that it can now be found in the fatty tissues of nearly every living organism. This chemical has had more severe effects on predatory birds such as the bald eagle and peregrine falcon than on seed-eating birds because of biological magnification. Some investigators have reported that the reproductive rates of eagles and falcons have been drastically reduced because DDT—and its metabolite DDE—interfere

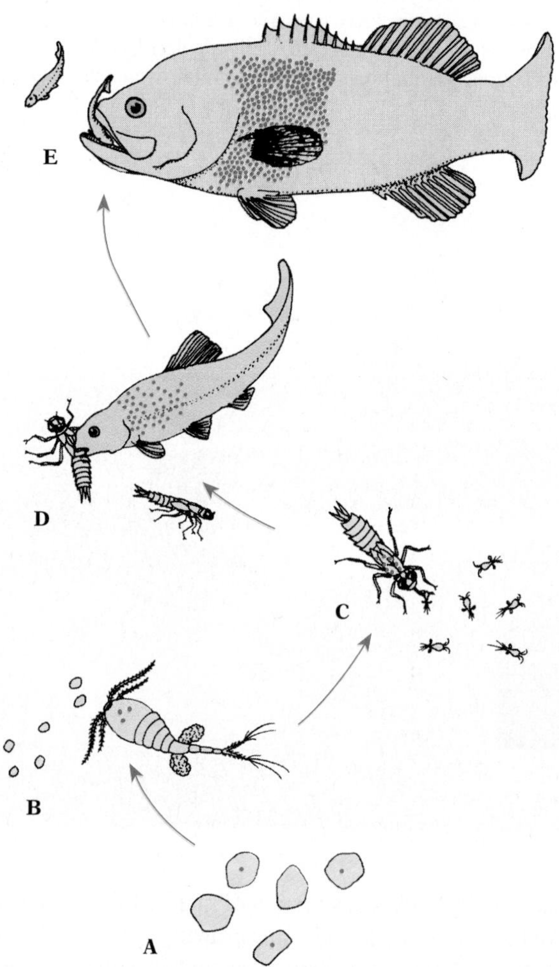

34.23 An example of biological magnification (A) Some individuals of a single-celled plant species at the bottom of a food chain have picked up a small amount of a stable nonexcretable chemical (red). (B) *Cyclops,* a small crustacean, incorporates the chemical from the plants it eats into its own tissues. Like the other organisms in the chain, it lacks the biochemical pathways necessary to metabolize or excrete the novel substance. (C) A dragonfly nymph stores all the chemical acquired from the numerous *Cyclops* it eats. (D) Further magnification occurs when a minnow eats many of the dragonfly nymphs that have stored the chemical. (E) When a bass, the top predator in this food chain, eats many such minnows, the result is a very high concentration in its tissues of a chemical that was much less concentrated in the organisms lower in the chain.

[2] A persistent chemical is one comparatively stable under natural biological and environmental conditions.

with deposition of calcium in the eggshells, with the result that the thin-shelled eggs are easily broken by the incubating parents and few young birds hatch (Fig. 34.24).

PCBs (polychlorinated biphenyls) are other persistent chemicals that are toxic to humans and other animals. PCBs were used in electrical transformers and were for many years discharged as wastes into rivers. They are nearly indestructible in the natural environment and undergo biological magnification in the food chain. For example, PCB contamination in the Hudson River of New York State has resulted in a banning of nearly all commercial fishing in the lower Hudson and the closing of a million dollar shellfish industry. Estimates for the cost of cleanup range as high as $30 million. PCB production was banned in 1977, but large quantities of the PCBs released for 20 years prior to the ban remain in the environment.

34.24 Effect of DDT on osprey In 1950 there were over 200 mating pairs of osprey nesting at the mouth of the Connecticut River, where these pictures were taken. By 1970 only six mating pairs were observed, the decline in the local population being attributed to the detrimental effects of DDT and related hydrocarbons on the calcification of eggshells produced by these birds. The hydrocarbons had been introduced into the runoff of local streams and rivers in insecticides and consumed by the fish that were the osprey's prey. A high percentage of the weakened eggs broke during incubation; approximately 10 eggs, or two to three nestings, were needed to produce a single offspring. A nest with two eggs and a broken shell is seen at top; at bottom is a female osprey with some of the few young successfully hatched during this period. Since 1970 the local osprey population has grown considerably, largely because of a ban on the use of the chemicals.

This brief summary of some of the biogeochemical cycles dramatizes the complex nature of the movement of matter through an ecosystem, the interdependence of the different species within that system, and, in particular, the fundamental and essential role played by microorganisms in these processes. Because the microorganisms are not observed as easily as the large plants and animals, we tend to forget about them, or to think only of the harmful ones, especially those that cause diseases. We thus tend to overlook the others, many of which are indispensable for our continued existence.

Soil Properties Determine Availability of Nutrients to Plants

Soil is essential to plants, not only as a substrate, but also as a reservoir for water and essential minerals—including nitrogen and phosphorus, as well as calcium, sulfur, potassium, and other ions whose cycles we have not discussed. Each of these minerals comes to plants dissolved in soil water. It follows that the properties of soils, including particle size, amount of organic material, and pH, play a very important role in helping to determine the availability of water and minerals to plants, as well as the rapidity with which these materials move through the soil.

Most soils are a complex mixture of mineral particles, organic material, water, soluble chemical compounds, and air. By far the dominant components are the mineral particles, which are composed largely of silicon and aluminum compounds. They vary in size from tiny clay and silt particles to coarse sand grains. The proportions of clay, silt, and sand particles in any given soil determine many of its other characteristics. For example, very sandy soils, which contain less than 20 percent of silt and clay particles, have many air-filled spaces. Unfortunately, such soils are so porous and their particles have so little affinity for water that it rapidly drains through them and they are unsuitable for growth of many kinds of plants. As the percentage of clay particles increases, the water retention of the soil also increases until, in excessively clayey soils, the drainage is so poor and the water is held so tightly to the particles that the air spaces become filled with water; few plants can grow in such waterlogged soil. Though different species of plants are adapted to different soil types, most do best in soils of the type known as ***loam,*** which contains fairly high percentages of particles from each size class (Fig. 34.25). In loamy soils, drainage is good but not excessive and there is good aeration; the soil particles may be surrounded by a shell of water, but there are numerous air-filled spaces between them.

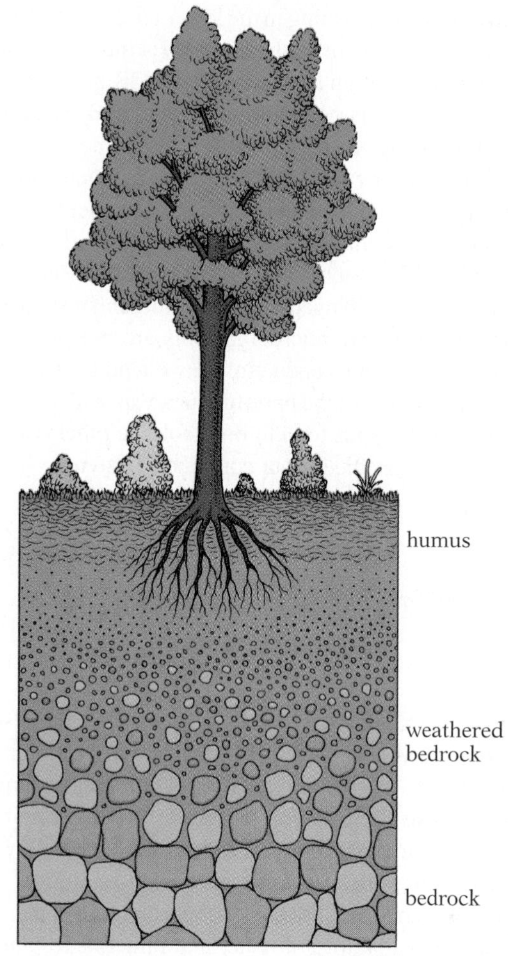

A

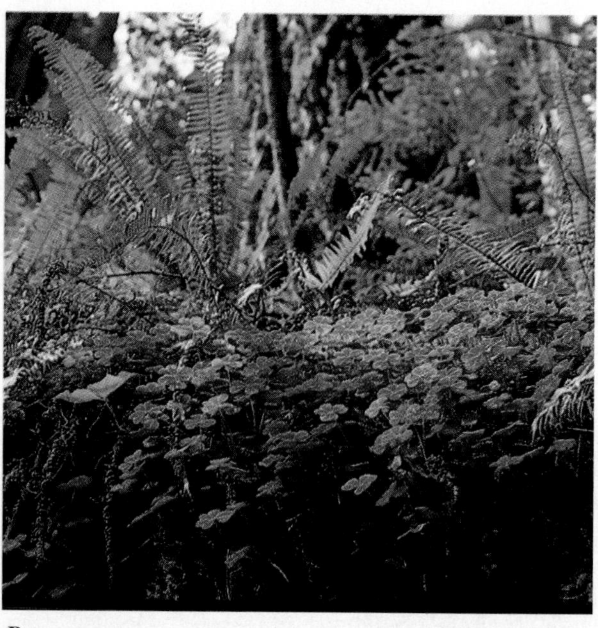

B

34.25 Plants growing in soil The cutaway drawing in (A) shows a soil profile. The plant roots extend throughout the soil but not down into the weathered rock or bedrock areas. Plants grow best in loamy soil that is rich in humus. Humus consists of organic decomposition products that contain high quantities of cellulose and lignin. It improves the drainage and aeration of the soil and provides inorganic substances for the living plants the soil supports. The photo in (B) shows plants growing in a dense layer of humus formed from decaying woody plants on a forest floor in Olympic Park Rain Forest in the state of Washington.

Loams usually also contain considerable amounts of organic material, often in the form of ***humus*** (Fig. 34.25), which consists mostly of products from the decomposition of cellulose and lignin. Humus contributes to soil fertility. Since it usually has a rather porous spongy texture, it helps loosen clayey soils and increases the number of pore spaces, thus promoting drainage and aeration.

The proportion of sand, silt, and clay particles affects not only the physical structure of the soil and its aeration and water-holding capacity, but also the amounts and the availability of certain minerals used by plants. If, for example, water percolates downward through the soil very rapidly and in large quantities, it will tend to remove many important ions from the soil, carrying them deep into the underlying rock layers, where roots cannot reach them. Nitrate ions are especially susceptible to removal by water, as are sulfate, calcium, and potassium ions.

Chemical analyses that give the total amount of various ions present in soils can be misleading, since a certain proportion of these ions may not be readily available to the plants. A complex equilibrium generally exists between free ions in the soil water and ions adsorbed on the surface of colloidal clay or organic particles. The acidity of the soil can shift this equilibrium; acidity influences the solubility, and consequently the availability, of calcium, aluminum, iron, magnesium, phosphate, and some other ions, as well as the activity of soil organisms, many of which are inhibited by high acidity.

Acid Rain Induces Nutrient Imbalances in the Soil

In 1970 the United States Congress passed the Clean Air Act, which established air quality standards to protect the public health. While some industries installed scrubbers on their smokestacks to remove pollutants such as sulfur dioxide, most complied by simply building higher stacks to disperse the pollutants (using "dilution as the solution to pollution!"). Locally, air quality improved, but unfortunately the prevailing winds simply moved the problem farther away, and rainfall hundreds of miles away became more acidic. (Normal rainfall has a pH of approximately 5.6; rainfall in the Adirondack Mountains in New York State now averages 4.2.) Acid rain occurs because the burning of fossil fuels, particularly the less expensive high-sulfur coal, releases sulfur and nitrogen oxides and other pollutants into the atmosphere. These pollutants then react with water to form acids. The acid precipitation is doing billions of dollars of damage to buildings each year (e.g., the Washington

34.26 One effect of acid rain Statues and buildings composed of limestone are severely affected by atmospheric pollution. The acid rain that forms can dissolve limestone, as can be seen here in this statue at a church in Surrey, England.

34.27 Forest damage attributed to acid rain These trees are showing the effect of acid rain on forest soil in the Adirondack Mountains of northeastern New York State.

Monument and the Capitol Building are being eaten away by acid; see Fig. 34.26) and has acidified over 3,000 lakes and 23,000 streams across the country, killing fish and other aquatic life. The effect of such precipitation on forests, particularly coniferous forests, is becoming increasingly clear. Over time, the soil in some affected areas[3] becomes acidified, which in turn influences the solubility of nutrient ions, such as calcium, iron, potassium, and magnesium ions, with the result that water removes these nutrients more rapidly from the soil into streams and lakes. Other ions, such as aluminum, become more available and are preferentially taken up by plant roots instead of calcium ions. Calcium and magnesium ions in particular are necessary for proper plant growth. Compounding the problem is an increase in the level of nitrogen compounds in the soil, again the result of air pollution, in this case mostly from automobile exhausts. The result is a nutrient imbalance in the forests, with increased nitrogen compounds acting as fertilizers to stimulate growth, and deficiencies of calcium and magnesium ions suppressing growth. In this stressed state, the trees are weakened to the point where they are highly susceptible to damage from insects, fungal infections, and drought (Fig. 34.27). There is increasing evidence that the loss of soil fertility as a consequence of acid rain is distressingly widespread and represents a serious blow to our environment.

[3] Not all soils are affected by acid rain; those areas where the bedrock is limestone are not affected because limestone $[(CaCO)_3]$ dissolves and buffers the acid. Thus the soil of the Finger Lakes region of upstate New York (limestone bedrock) remains unaffected while the Adirondack region soil (granite bedrock) is severely affected.

Soil Plants and Animals Affect Soil Properties

Soils cannot be fully understood without considering the effects plants and animals exert on them. Plant roots break up the soil in which they grow, remove substances from the soil, and add other substances to it. The plant shoots shield the soil beneath them, thereby altering the patterns of rainfall, humidity, light, and wind to which the soil is subject. And when the plants die, their substance adds organic material to the soil, changing both its physical and chemical makeup. Microorganisms in the soil alter its composition profoundly. Soil animals, such as earthworms and millipeds, also have a marked effect; they constantly work the soil, breaking down its organic components and moving materials between soil layers. Organisms, then, are not only influenced by their physical environment, they in turn modify that environment.

Some of the effects of vegetation on the soil were dramatically demonstrated in an extensive study of an experimental forest at Hubbard Brook, New Hampshire. The investigators, after first obtaining accurate measurements of the nutrient input and output of a particular area over a period of several years, cut down all vegetation in that area (Fig. 34.28), and again monitored input and output. They found not only a marked rise in the volume of runoff water (during one period it was actually 418 percent greater) but also an extraordinary loss of soil fertility. The runoff output of nitrate was as much as 45

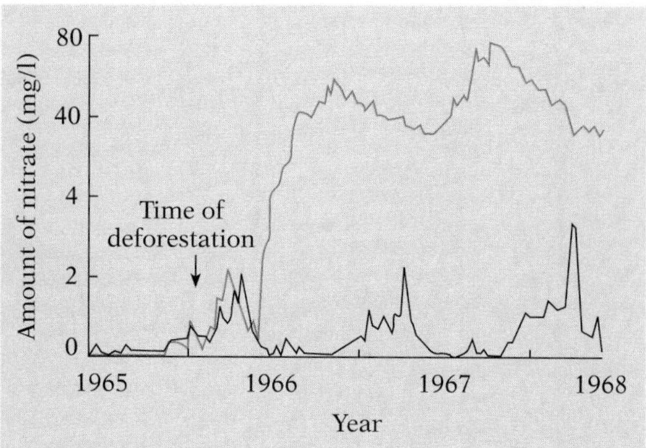

34.28 Change in the runoff of nitrate as a result of deforestation in the Hubbard Brook experimental forest The brown curve indicates the output of nitrate in stream water in the deforested area, and the black curve the output in an undisturbed area. Shortly after deforestation the output in the experimental area rose to about 40 times its previous level, whereas the output in the undisturbed control area remained at about the same level. (Note that the vertical axis is not linear.)

times higher than in undisturbed watersheds, and net losses of potassium and calcium were 21 and 10 times greater than normal, respectively. Apparently, removal of the vegetation had so altered the chemistry of the soil that nutrients were bound less tightly to soil particles and were rapidly lost.

Human destruction of vegetation results not only in increased erosion by wind and water but also in severe loss of fertility in the soil that remains. The stability of the physical part of an ecosystem clearly depends on the production and decomposition of organic matter, and on an orderly flow of nutrients between the living and the nonliving components of the system.

Poor Farming Practices Damage Soil

Cutting down forests is not the only way of ruining the soil; overgrazing and other poor farming practices have caused permanent damage in areas where forests never stood. The valleys of the Tigris and the Euphrates, referred to as the Fertile Crescent, once supported the Sumerian civilization; later they were the granary of the great Babylonian Empire. Poor farming practices led to such extensive erosion and buildup of salt in the soil that the amount of cultivable land today, 6,000 years later, has declined by over 80 percent and the ancient irrigation works are filled with silt. Similar conditions prevail in Syria, where human activity has reduced more than a million acres to desert. Closer to home, overgrazing and the plowing under of native grasses led to the dustbowl conditions of the 1930s in parts of the central United States (Fig. 34.29), when clouds of topsoil were blown hundreds of kilometers from their sources, leaving vast areas barren and useless for cultivation. Such examples from the past show us that our current ecological crisis is not entirely the result of modern technology, but it is disconcerting that, despite increased knowledge of soil dynamics, wholesale destruction of the earth's soils should continue unabated.

34.29 Dust Bowl in the 1930s Drought, overgrazing and poor agricultural practices led to the topsoil being blown away. Vast areas were left barren and useless for cultivation; other areas were buried in dust.

Irrigation Can Lead to Long-term Problems With the Soil

Irrigation has been viewed as a way of greatly increasing the productivity of dry areas (Fig. 34.30). However, it is all too often a short-term remedy, which is likely to be extremely de-

34.30 Irrigation of cotton field The six-week-old cotton plants in this field in Fresno, California, will require a continuous supply of irrigation water to reach maturity. Each day about 83 billion gallons of water are used in the United States for irrigation. Most of this is lost to evaporation, transpiration from plants, or leakage from pipes, and does not return to the aquifers or streams from which it came. Consequences of heavy irrigation include lowering of the water tables, saltwater intrusion, and salinization.

structive in the long run. In many cases it leads simply to accelerated erosion; in the United States, for example, an estimated 2,000 irrigation dams are now useless because of accumulations of silt, sand, and gravel. In other cases irrigation leads to rapid deposition of salts (including sodium, magnesium, and calcium chlorides) in the soil—a process called ***salinization***—until eventually plants cannot grow. The salinization may occur because adding water to land overlying a salty water table causes the groundwater to rise, the salts thus being carried into the topsoil, or because salts originally present in low concentrations in the irrigation water accumulate in the soil as the water evaporates (Fig. 34.31). In the Indus Valley of Pakistan, the largest irrigated region in the world, over 23 million acres watered by canals has fallen victim to salinization. The irrigation system there seemed very promising; the soil was good, and the addition of adequate water was expected to make it produce abundant crops. But as one observer has said, "The result was tragically spectacular. In flying over large tracts of this area one would imagine that it was an arctic landscape because the white crust of salt glistens like snow." Closer to home, it has been estimated that a substantial amount of California's farmland may be lost by the year 2000 as a result of salinization.

One promising development is the use of "drip irrigation," a method of reducing salinization by dripping water onto the soil from overhead pipes at a rate calculated to ensure that virtually all of it penetrates into the soil instead of being lost by evaporation (Fig. 34.32). Of particular value in drip irrigation is the use, where possible, of water from treated sewage. Mineral nutrients are thus recycled to the land, where they are needed, rather than released into streams and lakes, where they are ecologically damaging.

34.31 Signs of salinization in Iran A river on the island of Hormuz in the Persian Gulf shows obvious signs of salinization. Irrigation waters raised the water table of salty groundwater and brought in additional salt, much of which was eventually deposited in the topsoil. The salinization is made obvious here by evaporation, which has left an encrustation of salt on the banks.

34.32 Drip irrigation in Kuwait Water is dripped from the pipe at a calculated rate so that all the water is absorbed.

Soils illustrate how complex and fragile the earth's ecosystems are. In the past, people generally did not worry about the ecological damage they were causing by manipulating them. Such indifference must become an attitude of the past. Gaining knowledge of how ecosystems function and properly using such knowledge have virtually become ethical imperatives if human beings are to continue to inhabit the earth and create a sustainable environment.

THE BIOSPHERE

The myriad array of communities and their environments, bound together by interrelated biological, chemical, and physical forces, make up the ecosystem of planet earth, referred to as the biosphere. In this section we shall consider how physical forces create regional variations in climate, and profoundly influence the dispersal of species and distribution of communities.

Climate and the Sun

Solar Radiation Is Distributed Unequally Across the Earth

The sun is the ultimate energy source for life, and the distribution of its energy in large part determines the distribution of living things. The sun's radiation warms the earth, but because the earth's surface curves away from the path of incident light, areas at different latitudes receive different amounts of sunlight. Consequently, there is latitudinal variation in ranges of temperature. The tropics, for instance, receive almost five times as much energy per unit area as the midpolar latitudes. Moreover, because the earth's axis of rotation is tilted with respect to the solar plane in which it orbits, midtemperate latitudes receive more than twice as much solar energy at the beginning of summer than at the beginning of winter (Fig. 34.33).

In addition to influencing the latitudinal gradient of temperature, the uneven distribution of sunlight affects the distribution of rain. The warm humid air of the tropics tends to rise, drawing behind it cooler air from the temperate zones along the surface. The capacity of air to hold moisture decreases as the temperature falls, and the temperature of the atmosphere falls

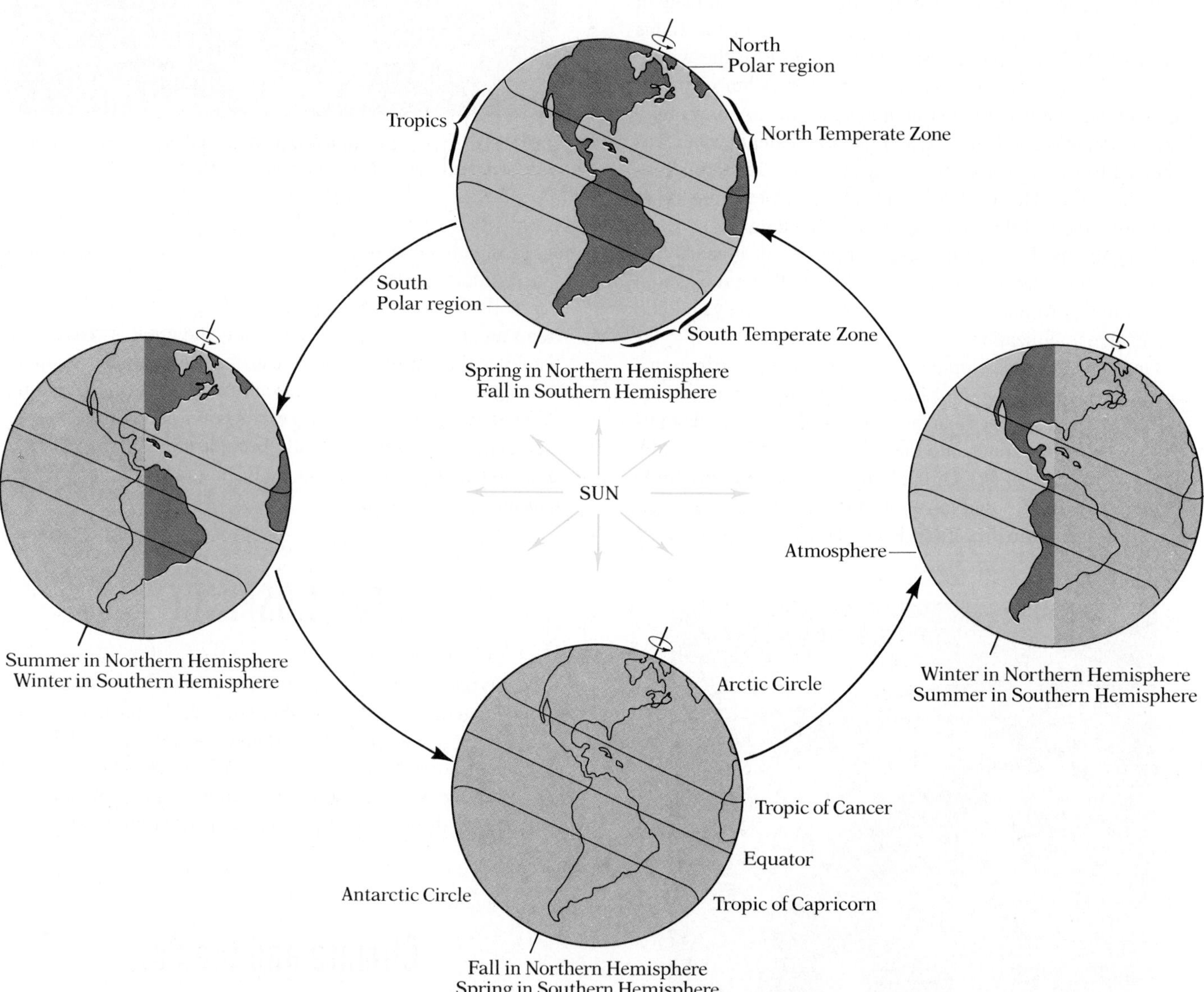

34.33 The source of the seasons Because the earth's axis of rotation is inclined 23.5 degrees with respect to its orbital plane, the northern hemisphere is tipped toward the sun during the summer (left) but away during the winter (right). The part of the earth tipped toward the sun is illuminated more vertically, so the energy received per unit area is higher in summer than in winter. The number of hours of sunlight is also greater in summer. The tropics, by contrast, receive strong, relatively vertical illumination all year, while the polar regions, on average, receive their sunlight more obliquely, and so spread out over more surface area. The absorption of sunlight by the atmosphere magnifies these seasonal and latitudinal differences in illumination: sunlight falling on the polar regions must travel through more air than sunlight in the tropics; hence more of its energy is dissipated before reaching the surface.

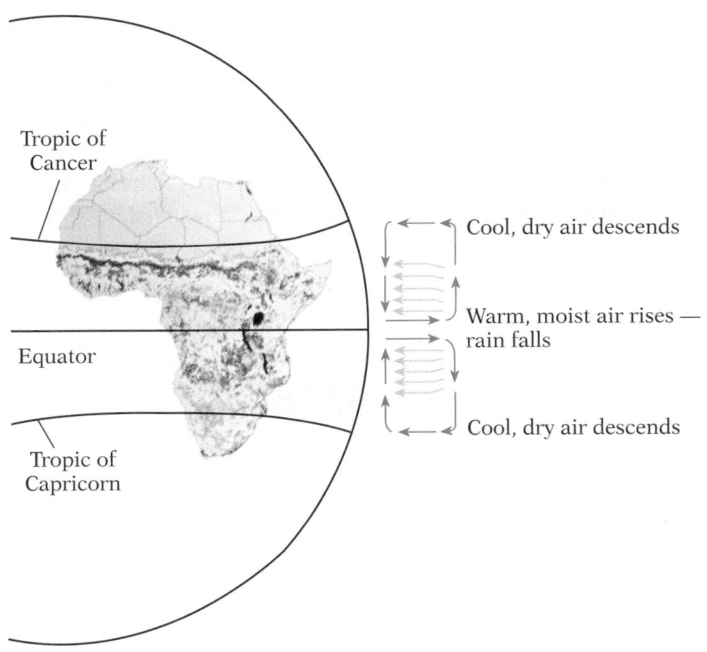

34.34 Latitudinal patterns of rainfall Warm, moist air rises near the equator, cools at higher altitudes, and—still over the tropics—releases its moisture as rain. As a result, when this air descends (roughly at the Tropic of Cancer and the Tropic of Capricorn), it is unusually dry, and contributes to the formation of deserts. This circulation of air is shown in cross section at right. Its effect on the continent of Africa is evident in the ultraviolet NASA photograph. Vegetation is heaviest in the white areas, still abundant in the red areas, and somewhat less so in the green ones. The brown areas are very dry and often barren.

with increasing altitude. As a result, the moisture in the rising tropical air condenses, bathing the tropics in rain. As this air cools and loses its moisture, it moves up and away from the equator, finally descending near the tropics of Cancer or Capricorn (Fig. 34.34).

The presence of mountains can give rise to similarly radical variations in climate. Just as the moisture in tropical air condenses when it is carried up to cooler altitudes, so the moisture in winds blowing up and across a mountain tends to condense at higher altitudes on the windward side. As a result, the side of the mountain facing the prevailing winds is usually much more lush than the side swept by the dry descending air (Fig. 34.35).

Ocean Currents Influence Climate

The horizontal movements of surface water in the ocean, called currents, often have a profound influence on climate. Like the prevailing winds, ocean currents follow a distinct pattern. Currents usually flow parallel to the coast along the edge of continents, with warm currents generally moving from the equator to higher latitudes along the eastern margins of continents, while cold water from higher latitudes flows toward the equator along the western continental margins (Fig. 34.36). The warm and cold currents have marked effects on tempera-

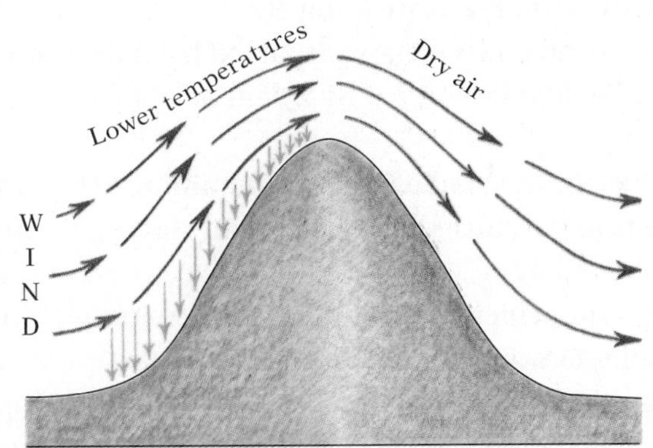

34.35 Effects of mountains on local climate When moist air is forced up to cooler altitudes by mountains, the moisture frequently condenses as rain, and the result may be lush vegetation on the windward side, as found, for example, along the western border of the Sierra Nevada in California (left). The dry air descending on the other side of the mountain creates a more arid environment on the leeward side; the Mojave Desert, on the eastern border of the Sierra Nevada (right), is typical of the deserts that may occur downwind of a mountain range.

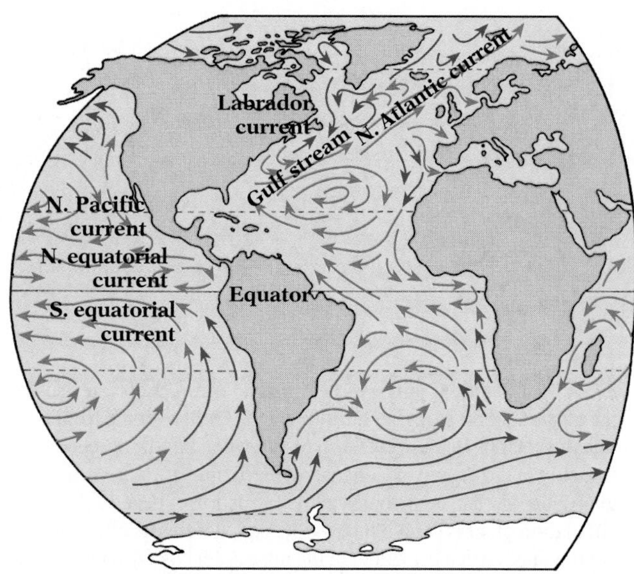

34.36 Surface ocean currents The ocean currents—horizontal movements of surface water in the ocean—often have a profound influence on the climate of the nearby land masses. Currents usually flow parallel to the coast along the edge of continents, with warm currents (red) generally moving from the equator to higher latitudes along the eastern margins of continents, while cold water (blue) from higher latitudes flows toward the equator along the western continental margins.

tures of nearby land masses. For instance, the North Atlantic Drift, a northern continuation of the warm Gulf Stream, keeps Great Britain and northeastern Europe much warmer than their northern location would dictate: the average January temperature in London, England, which is at 51 degrees north latitude, is 4.5°C higher than that of New York City, which is 11 degrees farther south. And cold ocean currents flowing along the west coasts of Chile and Peru convert the tropical deserts in this region to relatively cool, damp, foggy areas.

Biomes

The many interactions of temperature, rainfall, soil, and topography result in large climatic regions with distinctive vegetation called biomes. Each type of ***biome*** has its own distinctive combination of plants and animals, and is characterized by the predominant form of vegetation, such as grass or deciduous forest. Temperature and precipitation appear to be the most important factors in determining the distribution of biomes (Fig. 34.37). As we shall see, the regions differ greatly in their species diversity and productivity (Table 34.1). Let's briefly survey some of the world's major biomes.

MAKING CONNECTIONS

1. Why may a decrease in the ozone layer in the atmosphere lead to an increase in the rate of skin cancer?

2. Explain why decomposers and the bacteria of the nitrogen cycle are critical to the existence of life as we know it.

3. In what form is nitrogen transported from the roots to other parts of the plant body? Would it be carried by xylem or phloem tissue? (See Chapter 19.)

4. How has the use of phosphorus-containing products such as fertilizers and detergents led to eutrophication of many lakes? What are the harmful effects of eutrophication?

5. What is the principal cause of acid precipitation? Describe its harmful effects on bodies of water and forest soils.

6. Explain how such practices as overgrazing and improperly managed irrigation have caused—and continue to cause—the destruction of agricultural soils.

7. What would be some consequences for life if the earth's axis of rotation were not tilted?

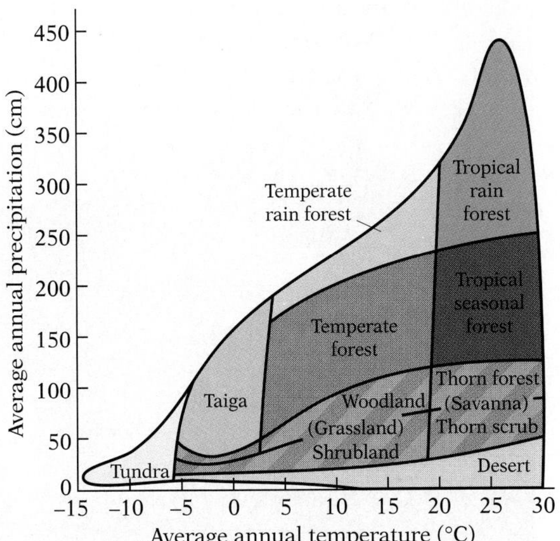

34.37 Temperature and annual precipitation determine the distribution of the various biomes In this graph of annual precipitation versus average temperature, we see that in areas where the temperature is high, vegetation varies from tropical rain forest to tropical seasonal forest to desert. Intermediate climates support deciduous forests and grasslands, and colder regions the taiga and tundra. Which biome predominates depends on how the rain is distributed over the year as well as other local conditions.

TABLE 34.1 *Average net primary production of the earth's major biomes*

Habitat	*Net primary production (g/m²/year)*
Terrestrial	
Tropical rain forest	1,800
Deciduous forest	1,250
Taiga	800
Grassland (tropical)	700
Grassland (temperate)	500
Tundra	140
Desert	—
Marine	
Algae beds and reefs	2,000
Estuaries	1,800
Continental shelf	360
Open ocean	125

The Tundra Has Permanently Frozen Subsoil

In the higher latitude portions of North America, Europe, and Asia is the ***tundra.*** It is the most continuous of the earth's biomes, forming a band around the North Pole that is interrupted only briefly by the North Atlantic and Bering Sea. Most of the tundra is located north of the Arctic Circle, but there are small regions in the southern hemisphere and tundralike regions on the peaks of the tallest mountains at all latitudes. The tundra corresponds roughly to the region where the subsoil is permanently frozen and the average temperature is in the range of −10°C. The land has the appearance of a gently rolling plain, with many lakes, ponds, and bogs in the depressions.

Even though the Russian word tundra means "north of the timberline," and trees and other tall perennial plants are generally absent, a few trees may gain a foothold in protected regions. Much of the ground is covered by mosses (particularly sphagnum), lichens (particularly reindeer moss), and grasses. There are numerous small perennial herbs, which are able to withstand frequent freezing and which grow rapidly during the brief, cool summers, often carpeting the tundra with brightly colored flowers.

Reindeer, caribou, arctic wolves, foxes, hares, and lemmings are among the principal mammals (Fig. 34.38); polar bears are common on parts of the tundra near the coast. Vast numbers of birds, particularly shorebirds (sandpipers, plovers, etc.) and waterfowl (ducks, geese, etc.), nest on the tundra in summer, but they are not permanent residents and migrate south for the winter. Insects, particularly flies (including mosquitoes), are abundant during the summer. In short, far from being a barren lifeless land, the tundra teems with life at certain times of the year. Despite the huge number of individual organisms on the tundra, the number of species is quite limited.

The Taiga Is Dominated by Coniferous Forests

South of the tundra, in both North America and Eurasia, is a wide zone dominated by coniferous (evergreen) forests. This is the ***taiga*** (Fig. 34.39), also called the boreal forest. Like the tundra, it is dotted by lakes, ponds, and bogs. And like the tundra, it has very cold winters. But unlike the tundra, the taiga has more precipitation and longer and somewhat warmer summers, during which the subsoil thaws and vegetation grows abundantly.

The number of species living in the taiga is larger than on the tundra, but smaller than in biomes at lower latitudes. Though drought resistant needle-leaf conifers (including spruce, fir, and tamarack) are the most characteristic of the larger plants, some deciduous trees (such as paper birch) are also common. The principal herbivores of the forests are insects. Moose, black bear, wolves, lynx, wolverines, martens, porcupines, and many smaller rodents are the principal mammals in taiga communities, with many types of birds abundant in summer.

The biomes at lower latitudes are more variable; there is more variation in the amount of rainfall in these latitudes, and consequently more east-west variation in the types of communities that predominate.

34.38 The tundra Left: A tundra in Alaska, seen in spring. Right: A caribou bull, distinctive animal of the tundra.

34.39 The taiga Left: Coniferous forests, such as this one in Norway, cover extensive areas in the northern part of North America, Europe, and Asia. Right: Wolves, one of the important mammals of the taiga.

Deciduous Forests Exist in Temperate Areas With Abundant Rainfall

In those parts of the temperate zone[4] where rainfall is abundant, the summers are relatively long and warm, and the winters cold, as in most of the eastern United States, most of central Europe, and part of eastern Asia, communities are frequently dominated by broad-leafed trees. Such areas of seasonal change, in which the foliage changes color in autumn and drops, constitute the deciduous-forest biome (Fig. 34.40). This biome is less homogeneous than that of the taiga or tundra, because the local climatic conditions, particularly patterns of rainfall, vary considerably. The deciduous forest biome characteristically includes more plant species than does the taiga. Flying insects and birds are abundant, and many amphibians are present (these are absent from the taiga). Among the common mammals present are squirrels, deer, foxes, and bear.

[4] The temperate zone includes those regions where the climate is relatively mild and the temperature is neither excessively hot nor cold.

34.40 A temperate deciduous forest along the Housatonic River, Connecticut, in early autumn

Tropical Rain Forests Require Warm Temperature and Plentiful Rainfall

Tropical areas with abundant rainfall usually produce rain forests, which include some of the most complex communities on earth. The diversity of species is enormous; a temperate forest is composed of two or three, or occasionally as many as 10, dominant tree species, but a tropical rain forest may be composed of a hundred or more. One may actually have difficulty finding any two trees of the same species within an area of many acres.

The dominant trees in a rain forest are usually very tall, and their interlacing tops form a dense canopy that intercepts much of the sunlight, leaving the forest floor only dimly lit even at midday (Fig. 34.41). The canopy likewise breaks the direct fall of rain, but water drips from it to the forest floor much of the time, even when no rain is actually falling. The canopy also shields the lower levels from wind so that the rate of evaporation is greatly reduced and the lower levels of the forest are very humid. Temperatures near the forest floor are nearly constant. The pronounced differences in the environmental conditions at different levels within such a forest result in a striking degree of vertical stratification; many species of animals and epiphytic plants (plants growing on the large trees) occur only in the canopy, others only in the middle strata, and still others only on the forest floor. Some vertical stratification is found in any community, particularly any forest community, but nowhere is it so extensively developed as in a tropical rain forest.

In spite of the lush vegetation, the soil of the tropical rain forest is surprisingly low in fertility; the dead organisms that fall to the forest floor are rapidly decomposed and the nutrients quickly reabsorbed by living plants. Thus, the nutrients cycle tightly through the living system, leaving nutrient levels in the soil low and the soil acidic. When the forests are cut and burned, enough nutrients flow into the soil to support plant growth for two or three years, but cannot continue to be used for crops without substantial inputs of fertilizer. Some tropical soils, when exposed to rain and heat, turn as hard as cement and lose the ability to support most vegetation, including agricultural crops.

The massive destruction of the rainforest, as is currently occurring in Southeast Asia and the Amazon basin of South America, has a number of ecological consequences. The cutting and burning results in increased carbon dioxide production. Further, these forests act as important carbon dioxide

34.41 Tropical rain forest Left: A forest in Queensland, Australia. Right: Epiphytic bromeliads (distinctive plants of the tropics) growing on a branch of a tree in Costa Rica.

"sinks," consuming large amounts of carbon dioxide during photosynthesis. Deforestation thus leads to increased carbon dioxide levels in the atmosphere—at a time when we are already concerned about their high levels and the greenhouse effect. Another serious problem is the loss of species diversity (see Exploring Further: How Important Is Biological Diversity, p. 790). More than half of all species of plants and animals on earth live in the tropical rainforest, and most of these have never been studied. With the destruction of the rainforest occurring at a rate of about 1 to 2 percent per year, thousands of species have or soon will become extinct. Some of these undoubtedly have potential uses in pest control, as new crops, fibers, medicinal drugs, and the like.

Grasslands Exist Where Temperatures Are Moderate but Rainfall Low

Huge areas in both the temperate and tropical regions of the world are covered by grassland. Typical of this biome are areas where either relatively low total annual rainfall (25 to 30 centimeters) or uneven seasonal occurrence of rainfall makes conditions inhospitable for forests, but suitable for the often luxuriant growth of grasses. The grasslands of temperate regions characteristically undergo an annual warm-cold cycle, whereas the grasslands of the tropics (often called savannas) undergo a wet-dry cycle instead.

Temperate and tropical grasslands are remarkably similar in appearance, though the particular species inhabiting them may be very different. Both usually contain vast numbers of insects and large and conspicuous herbivores, often including ungulates (e.g., bison and pronghorn antelope in the United States) (Fig. 34.42). Burrowing rodents or rodentlike animals are often common (e.g., prairie dogs in the western United States).

Deserts Show Extreme Temperature Variations and Low Rainfall

In places where rainfall is often less than 25 centimeters (10 inches) per year, not even grasses can survive as the dominant vegetation. It is here that deserts occur. Deserts are subject to the most extreme temperature fluctuations of any biome; during the day they are exposed to intense sunlight, and the temperature of both air and soil may become very high (to 40°C or higher for the air and to 70°C or higher for the soil surface). But heat is rapidly lost at night, and within a short while after sunset the searing heat has usually given way to bitter cold.

Some deserts, such as parts of the Sahara in Africa, are nearly barren of vegetation, but more commonly deserts contain scattered drought-resistant shrubs (e.g., sagebrush, greasewood, creosote bush, and mesquite) and succulent plants that can store much water in their tissues (e.g., cacti in North American deserts and euphorbias in deserts of the Middle East) (Fig. 34.43). In addition, there are often many small, rapidly growing annual herbs with seeds that will germinate only when there is a hard rain; once the seeds germinate, the young plants shoot up, flower, set seed, and die, all within a few days or weeks.

Most desert animals are active primarily at night or during the brief periods in early morning and late afternoon when the heat is not so intense. Among the animals often found in deserts are rodents (e.g., the kangaroo rat), snakes, lizards, a few birds, spiders, and insects. Most show numerous remarkable physiological and behavioral adaptations for life in this hostile environment.

Biotic Communities Change With Altitude

We have seen that moving north or south from the equator on the earth's surface, one may pass through a series of different

34.42 A grassland in Kenya, with zebras, characteristic animals of this biome

A

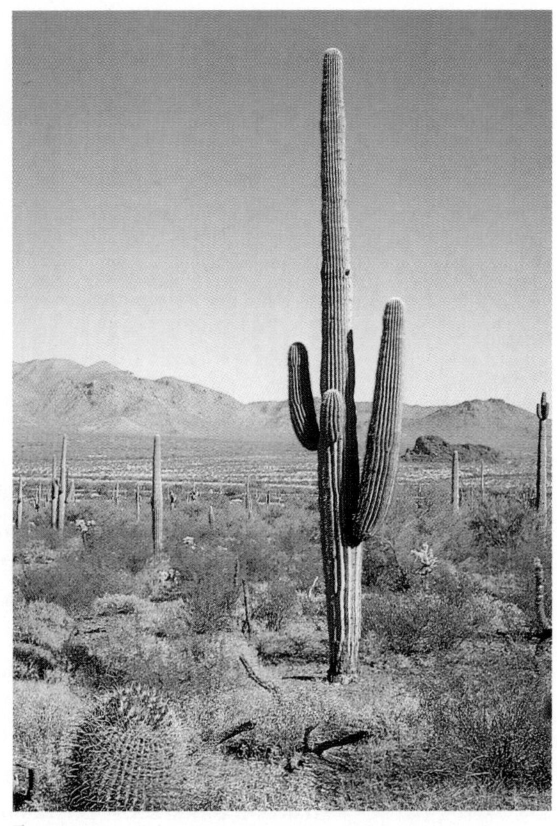

B

34.43 Deserts (A) Death Valley, California; vegetation is extremely sparse. (B) Cacti and other thorny, drought-resistant plants, abundant in many deserts, growing in Picacho Park, Arizona.

biomes. The same is true if one moves vertically on the slopes of tall mountains. Climatic conditions change with altitude, and the biotic communities also change correspondingly (Fig. 34.44). Thus isolated pockets of the taiga extend far south into the United States along the slopes of the Appalachian Mountains in the east, and on the slopes of the Rockies and

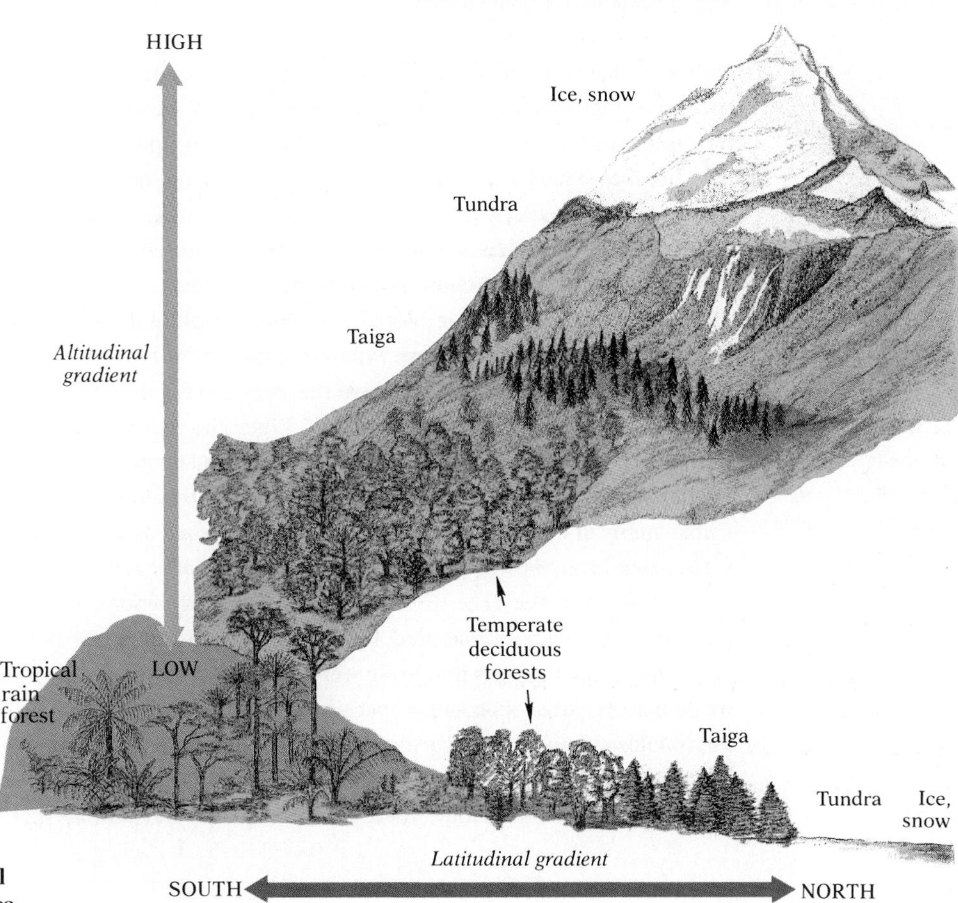

34.44 The similarity between latitudinal and altitudinal life zones in North America

EXPLORING FURTHER

HOW IMPORTANT IS BIOLOGICAL DIVERSITY?

From the time when life first evolved, some 3.8 billion years ago, to the present, biological diversity has slowly increased, despite occasional setbacks due to the five major periods of mass extinctions. Today we do not know even to the nearest order of magnitude exactly how many species there are on the earth; the total number of species could be close to 10 million or as high as 100 million. More than half of these species live in tropical forests. By studying the extinction rates of species there biologists can estimate the loss of biological diversity on a global basis.

The diversity of species within a tropical rain forest is enormous—the exact numbers are still only educated guesses. These forests are currently being destroyed at a rate of between 1 and 2 percent per year. Unlike the forests of the temperate regions, the rain forests, when cut down, regenerate very slowly (taking centuries), if at all. If the present rate of destruction continues, it is estimated that the rain forests will be gone in less than 50 years. This destruction has an enormous impact on biological diversity, driving the extinction of countless unique plants and animals. Using field observations, E. O. Wilson of Harvard University has calculated that when a given habitat is reduced by 90 percent, the number of species in that area eventually drops 50 percent. If we continue to destroy the tropical forests at current rates, Wilson estimates that about 0.5 percent of the species in the forests may become extinct each year. This may not seem like much, but if there are 10 million species in the tropics, then 50,000 species could become extinct in a year! It is clear that we are in the midst of a sixth mass extinction, this one caused by human activities rather than by volcanoes, meteorites, climatic changes, or the like.

Most tropical forests are located in poorer, rapidly developing nations, whose fast-growing, human populations are expanding into the surrounding forests. There they kill the wild animals for food, cut and burn the forests to use the land, and plant crops and pasture animals ill-suited to the environment. When the cleared area is no longer able to support human populations, new areas are exploited and another round of destruction occurs, further diminishing biological diversity. What does this loss of biological diversity mean? Unlike many of the other environmental problems such as toxic pollution and destruction of the ozone layer, the loss of biological diversity cannot be reversed. When species are gone, they are gone forever. True, after each of the mass extinctions, new species arose from the survivors and diversity increased, but that took between 10 and 100 million years to happen! And the new species that arose were different from those that died out. And because we do not even know how many species there are on the earth, where they are located, or their biology, we have no accurate idea of what this loss might mean. Some of these species might have had potential uses in pest control, as food, as new crop plants, fibers, medicinal drugs, and the like. If we know more about the plants and animals that are present, we can

make better decisions about whether to protect them, extract products from them, or perhaps allow the habitat to be destroyed. Plants are superb biochemists, synthesizing a variety of chemicals used in their defense against predators and in their own metabolism. Many of these chemicals may have medicinal value. Aspirin, for instance, is synthesized from salicylic acid from the meadowsweet plant; taxol, useful in the treatment of ovarian cancer, is extracted from the bark of the Pacific yew tree; and an extract from the guggal tree of India shows promise in lowering blood cholesterol levels. About 40 percent of the drugs we use today are of plant, animal, or microbial origin. It is crucial to identify the organisms harboring useful drugs and chemicals before the forests are destroyed and the diversity is gone.

The loss of genetic diversity is also a concern. Traditional crop strains, which are genetically diverse, are being abandoned in favor of newer high-yield strains. Unfortunately the highly productive strains are genetically homogenous and are usually more vulnerable to insect pests and various diseases than the wild species. The predominant strain of rice grown in Asia, for instance, was devastated by a virus, but one wild strain of rice from India had genes for resistance. Through intensive plant breeding a resistant hybrid was created, which is now widely grown. Maintaining genetic diversity is important if we are to have sufficient resources for crop improvement.

The use of wild plants and animals for food is another underutilized biological resource. It is estimated that 30,000 species of plants have edible parts, but 90 percent of the world's food presently comes from just 20 species. One example of a wild species that can be used for food is the winged bean, a tropical legume. Every part of this plant is edible, from the leaves (which taste like spinach), to the young pods (like green beans), to the young seeds (like peas), to the mature seeds (like soy beans), and, finally, to the underground stems, which have more protein than potatoes. In addition, the plant grows very fast and has root nodules with nitrogen-fixing bacteria so it requires little fertilizer. Another crop native to the Americas with great potential is amaranth, which has spinachlike leaves and yields a nutritious grain. These, and other plants like them, could help feed millions of people in the tropical countries. A variety of fruits, nuts, and seeds could be harvested without destroying the forests. Wild animals have also been underutilized. Amazon river turtles raised on the river floodplains yield 400 times the meat that cattle raised on nearby pastures do. Likewise, the green iguanas, the "chicken of the trees," can be cultivated in the forest. Both conservation and economic growth could potentially be served by cultivating native species in their natural ecosystems.

One of the important ways to save biological diversity is to promote sustainable development of the tropical forests. If it is economically feasible to save the forests, biological diversity will be maintained. Right now it is still more profitable to harvest all the trees in an area and move on to a new area. But if the rain forest could be used for the sustained harvesting of fruits, nuts, seeds, oils, latex, medicines, etc., it could be made more profitable in the long-term than clear-cutting the forests. In addition to the potential economic value of diversity, we should not forget ethical concerns. We have an obligation to preserve species, both for future generations and for their ecological roles. It is possible to have both conservation and economic development, but population control, social change, and education must be an important part of the process.

34.45 An alpine meadow in Wyoming Trees can be seen in the lower, more protected areas. Most of the meadow is above the timberline. The upper limit of tree growth in mountainous regions (usually between 3,000 and 3,500 meters) is determined largely by temperature and exposure to winter winds; it is also influenced by soil conditions and rainfall. More barren areas can be seen on the higher peaks in the distance.

Coast Ranges in the west. There are even tundralike spots on the highest peaks in these areas (Fig. 34.45). A rough rule of thumb is that a 100-meter increase in elevation is equivalent to a 50-kilometer increase in latitude.

The Aquatic Environment Contains a Variety of Diverse Habitats

So far, we have concentrated on terrestrial ecosystems, but many of the earth's biotic communities are found in aquatic environments, of which there are two main types, freshwater and marine ecosystems. These too vary in type with varying physical conditions. Freshwater habitats are extremely diverse; variation in oxygen level, temperature, water movement, size, and depth of the body of water or stream create very different environments. Thus the communities in freshwater lakes differ from those in the flowing waters of rivers and streams, and even those in a single stream differ from one another, depending on whether they are in rapids where fast-flowing water is made turbulent by rocks and sudden falls, or in water flowing slowly and calmly over a smooth bottom.

The largest bodies of water, the oceans, occupy three-fourths of the earth's surface. The ocean floor is divided into the ***continental shelf,*** which extends seaward from the coast and joins the steeper ***continental slope,*** and the deep ocean basins called the ***abyssal plains*** (Fig. 34.46). Continuous underwater mountain chains called ***ocean ridges*** arise from the flat abyssal plains and encircle the earth. Sometimes the peaks of these underwater mountains extend above sea level; Iceland is an example of such a mountain top.

Organisms living along the ocean bottoms constitute the so-called ***benthic division,*** the organisms living in the water above the bottom are called the ***pelagic division.*** Many organisms in the pelagic division are free-swimming (the nekton), while others float about and drift with the water currents (the ***plankton***). The photosynthetic plankton are referred to as phytoplankton; they provide food for the herbivorous zooplankton, which in turn are eaten by carnivorous zooplankton, and these in turn are preyed on by small fishes.

Only the upper region of the ocean is lighted. Called the ***photic zone,*** the lighted region gives way to a deeper, lightless ***aphotic zone*** (usually below 200 meters). The photic zone is home to a variety of phytoplankton, zooplankton, and fish species, but except for the isolated colonies of chemosynthetic autotrophs found near underwater volcanic vents (see Fig. 7.1, p. 140), benthic communities in the lightless aphotic zone are made up only of heterotrophs (including some animals, fungi, and heterotrophic protists and bacteria). They depend on the photosynthetic organisms of the photic portion of the pelagic

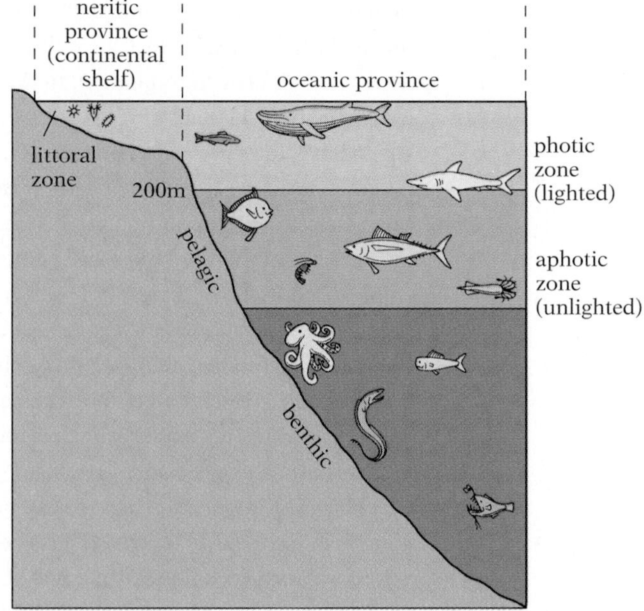

34.46 Ocean ecosystems and some of the organisms dwelling within them There are several ways of classifying the ecosystems of the ocean. The ecosystem comprising the ocean bottom together with all bottom-dwelling organisms is referred to as the benthic division, while the water above the bottom and all the organisms within it constitute the pelagic division. It is also possible to distinguish between the upper, lighted, photic zone where photosynthesis can take place, and the deeper, lightless, aphotic zone (stretching from 200 meters down to the deep ocean bottom). The neritic province lies above the continental shelf and is very rich in living organisms whereas organisms are much more scarce in the oceanic province, the main ocean basin. (Organisms are not drawn to scale.)

division above them for food; they receive their nutrients as a rain of dead organisms from the water above.

The open ocean, or oceanic province, is in many ways similar in productivity to a desert, because the relative scarcity of mineral nutrients (particularly phosphorus) limits the growth of phytoplankton, the primary producers of this ecosystem.

The most complex communities of the marine environment occur in the shallow waters above the continental shelf (often called the neritic province), especially in the regions close to shore. High primary productivity by both free-floating and bottom-anchored algae makes possible a multitude of niches for various herbivores. Because this zone is subject to far more variation in temperature, water turbulence, salinity, and lighting than any other portion of the ocean, the communities vary greatly. Other regions of high productivity are ***estuaries,*** places where large freshwater rivers flow into the ocean. The river provides a continuous supply of nutrients eroded from land to the estuary, as well as fresh water, which dilutes the salinity of the sea water. Salt marshes and mudflats often adjoin estuaries, and they too contribute nutrients. Consequently, photosynthesis occurs at a very high rate in the producers of the estuary, the phytoplankton and algae. They form the base of a large and productive food web. Approximately two-thirds of all shellfish and fish spend some part of their life cycle in this region. Unfortunately, rivers also deliver pathogenic organisms, sediments, and pollutants from land, sewage treatment plants, and industries to the estuaries. Over 40 percent of estuarine communities in the United States have been destroyed, and the destruction continues despite new laws designed to protect these sensitive and productive areas.

The shallow regions of the tropical zone support the ***coral reef*** community, one of the most diverse and productive communities on earth (Fig. 34.47). The cornerstones of these communities are the corals, cnidarians that have formed a mutualistic association with photosynthetic algae that live inside their cells. Stony corals lay down calcareous skeletons, which constitute the basic structure of a reef and provide a multitude of microhabitats for the organisms it supports. Like the rain forests to which they are often compared, coral reefs are now inhabited by animals of virtually every phylum, as well as representatives of most plant divisions. And like the rain forests, they are threatened by local human activities and by human interference with the biosphere. If the burning of fossil fuels results in a heating of the earth's surface and the melting of the polar ice caps, the consequent changes in water temperature and level may irreversibly damage these fragile ecosystems. We are already seeing widespread "bleaching" of the coral, which results when the mutualistic algae leave the coral, thought to be related to pollution and global warming.

BIOGEOGRAPHY

Biogeography is the study of the geographical distribution of plants and animals on the earth's surface. In a given biome the ranges of each species are determined by a complex set of environmental factors. Variations in temperature, rainfall, and soil, for instance, explain some patterns of geographical distribution, while other factors, such as a species' ability to cross barriers, are also important. Since *r*, the intrinsic rate of in-

A

B

34.47 Corals Hard corals growing in abundance in tropical waters provide diverse habitats for marine creatures. Among the corals found in waters surrounding Fiji are leather and alapora corals (A). All corals are colonial assemblies of individual polyps as shown in a close-up view (B).

crease, invariably leads to the growth of a population, a population is under pressure to expand its niche or to spread to new territories when the individuals become too abundant. The geological or ecological zones that separate any two regions will block some species more effectively than others. While a wide expanse of ocean prohibits the movement of horses or elephants, coconut palms may cross it in fair numbers, because their large, water-resistant seeds can float in seawater for many weeks without harm. A grassland separating two forested areas may be an almost insurmountable barrier for some forest animals, preventing their movement from one forest to another. Other forest animals that have difficulty crossing the grassland may nevertheless do so occasionally, and still others may be able to cross the grassland freely. Therefore, what is a barrier to dispersal for one species may be a possible but difficult route for another, and an easily negotiated path for a third. A knowledge of the sorts of dispersal routes and barriers that are effective for different species is fundamental to understanding the distribution patterns of organisms on the earth's surface.

BIOGEOGRAPHIC REGIONS OF THE WORLD

Even a thorough understanding of the ecology of living species and a detailed knowledge of the present geography of the earth do not fully explain plant and animal distributions; the historical elements must also be considered. The earth and its organisms are constantly changing, and present distributions are largely the result of past conditions, which were often very different from those now prevailing. An account of existing conditions would by itself be insufficient to explain why certain animals occur in South America, Africa, and southern Asia but nowhere else. Only by combining knowledge of current conditions with evidence from the fossil record and with geological evidence of the past shapes, connections, and climates of the earth's land masses and oceans can ecologists hope to gain insight into the present geography of life.

The Earth's Continents Are Moving

For many years both geologists and biologists assumed that the distribution of earth's major land masses existed as it does now throughout the history of life. That assumption has been proven false by geologists as evidence began to accumulate during the 1960s that these land masses (or plates, as geologists call them) have changed positions nearly continuously throughout the earth's history. According to the theory of ***continental drift,*** the earth's surface is divided up into a dozen or so massive, rigid plates that rest on the earth's more fluid mantle. Convection currents running along the top of the mantle, created by the upwelling of new crust in some places and the sinking of old crust in others, move the plates at roughly 2.5 centimeters per year (Fig. 34.48). Collisions between plates result in the deformation of the earth's crust and the uplift of mountains, and cause severe earthquake activity as well. Lateral slip between the plates also gives rise to earthquakes, as we see along the western coasts of North and South America (Fig. 34.49).

Evidence suggests that about 300 million years ago all the continents had drifted together to form a single massive supercontinent, called ***Pangaea*** (Fig. 34.50A). Approximately 200 million years ago Pangaea began breaking apart due to continental drift. The first major break was an east-west one, separating a northern supercontinent called ***Laurasia*** (composed of the land masses that would later become North America, Greenland, and Eurasia minus India) from a southern supercontinent called ***Gondwanaland*** (composed of the future South America, Africa, Madagascar, India, Antarctica, and Australia). Soon thereafter Gondwanaland began to break up, and the land masses began to move slowly apart; India drifted off to the north, and an African-South American mass separated from an Antarctic-Australian mass. Roughly 135 million years ago, the distribution of the continents was probably like that shown in Figure 34.50B.

Later, by about 65 million years ago, South America had split from Africa and was drifting westward; India had moved northward, but had not yet collided with the rest of Asia. By about 40 million years ago Australia had split from Antarctica and had begun drifting northeastward into the warmer regions

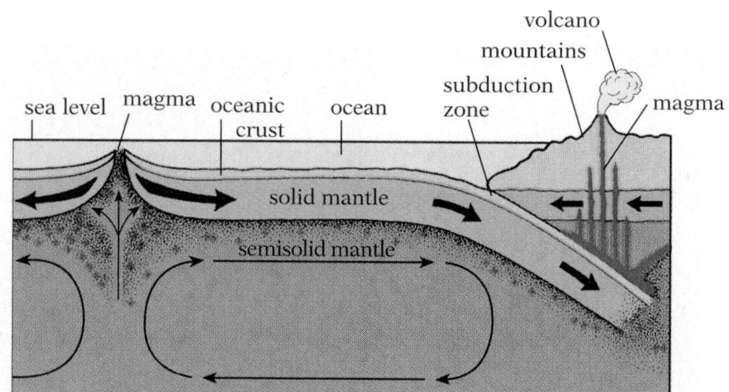

34.48 Movement and deformation of the earth's outer layers The crust and upper portion of the mantle are solid and rest upon a partly molten, semisolid layer of mantle. The solid portion is broken up into smaller, rigid plates that can move about over the semisolid and plastic region. When the plates move apart (left) in the process known as sea floor spreading, the magma wells up into the gap from below and solidifies, creating new sea floor. The sea floor plate moves as if it were on a conveyer belt, and when it meets a continental plate moving toward it (right), the sea floor plate is pushed under, forming a subduction zone. The old sea floor is remelted in the subduction zone. Areas above the subduction zone may be pushed upwards to form mountains and are prone to earthquakes and volcanic activities.

34.49 San Andreas fault in California The fault marks the junction between two different plates that are moving side by side in opposite directions. Earthquakes occur when the plates scrape past one another.

of the southwest Pacific (Fig. 34.50C). About the same time India collided with Asia, giving rise to severe earthquakes in the surrounding area and pushing up the Himalayas and the Tibetan plateau. The division of Laurasia into North America and Eurasia was one of the last major continental changes to take place (Fig. 34.50D), occurring about 25 million years ago (though the continents remain connected at times by a land bridge).

As continents moved over the course of millions of years, and their distances from the earth's poles and the equator changed, their climates must have undergone major shifts. India, for example, moved from a position next to Antarctica all the way across the equator to its present location in the tropics of the northern hemisphere. Australia, too, moved steadily northward. But climatic changes due to causes other than con-

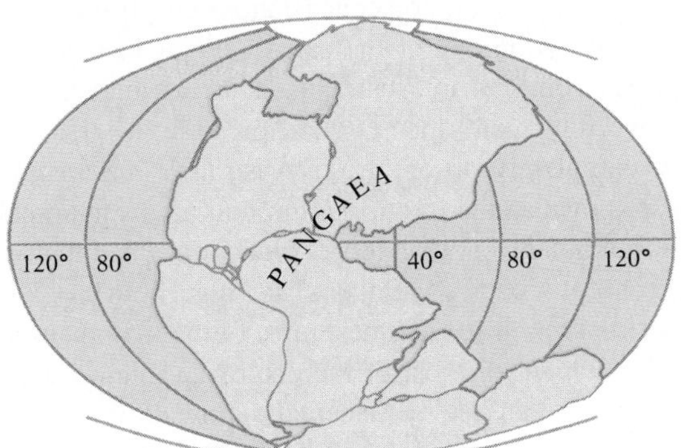

A 225 million years ago

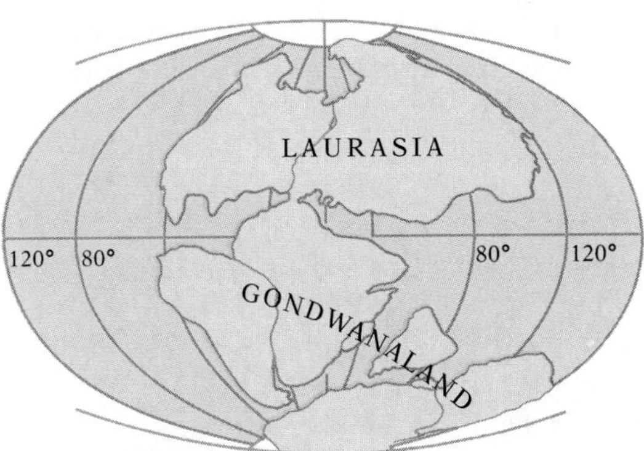

B 135 million years ago

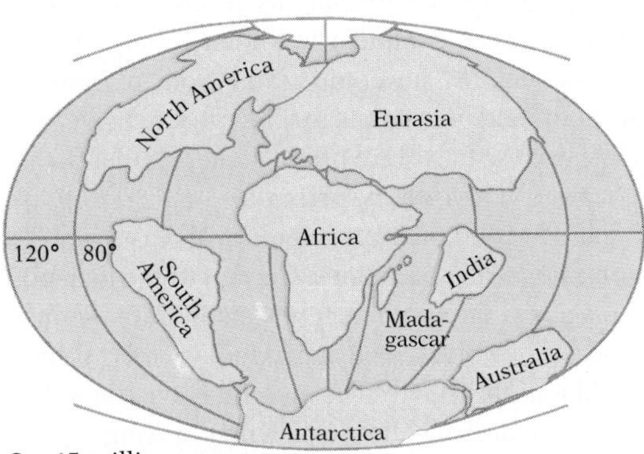

C 65 million years ago

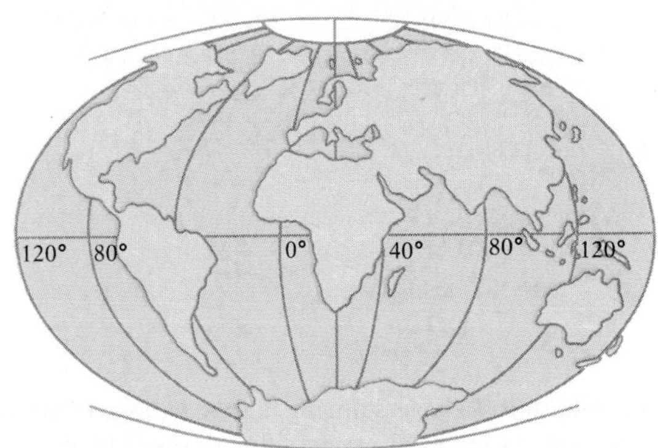

D Present

34.50 The origin of the modern distribution of continents through continental drift (A) Early in the Mesozoic era, about 225 million years ago, all the earth's major land masses were united in a simple massive supercontinent called Pangaea. (B) About 135 million years ago, at the start of the Cretaceous period, Pangaea had broken into a northern supercontinent, Laurasia, and a southern supercontinent, Gondwanaland; Gondwanaland itself had also begun to break up. (C) By 65 million years ago the breakup of Gondwanaland was complete, and the future South America, Africa, Madagascar, India, Antarctica, and Australia were drifting apart. (D) The present continental arrangement.

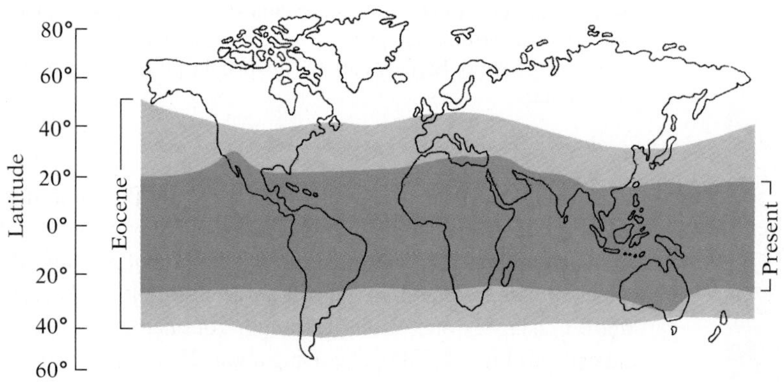

34.51 Distribution of tropical and subtropical forests during the Eocene (about 50 million years ago) and now, shown on a modern map Warm conditions extended much farther north and south during the Eocene. The possible causes of global warming and cooling cycles are hotly debated.

tinental drift have also occurred during the history of earth. Thus Antarctica, though probably always near the South Pole, has not always been the bleak, ice-covered land it is today. Indeed, fossils of temperate and tropical amphibians and reptiles have been found there. During at least part of the Mesozoic, Antarctica must have been reasonably warm, and it was probably warm again about 50 million years ago, when tropical and subtropical climates were far more widespread on the earth than they are today (Fig. 34.51). By contrast, the earth was much colder only a few thousand years ago, during periods of extensive glaciation (Fig. 34.52). Thus, both the past configurations of the land masses and their past climates are important in explaining the distributions of organisms on the earth.

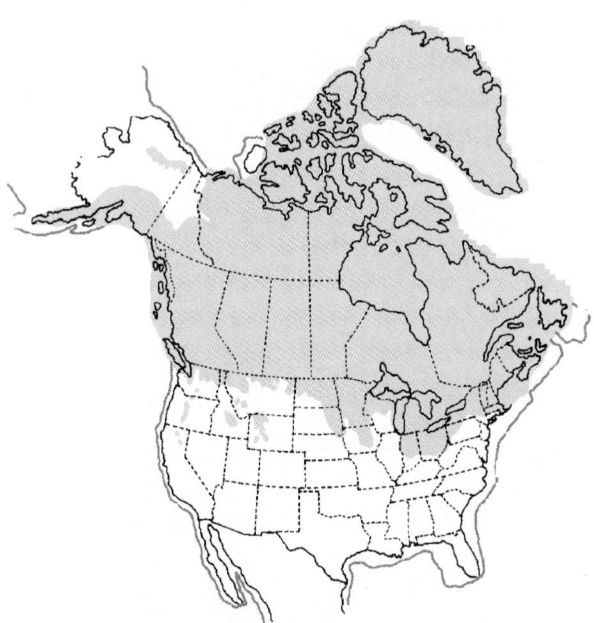

34.52 Distribution of glaciers during the last ice age, 15,000 to 20,000 years ago An extensive ice sheet formed over eastern Canada and moved south and west toward the United States. Other ice sheets formed over Scandinavia and moved southward over northwestern Europe, and over the arctic regions of Eurasia and North America. Meanwhile the ice sheets in Greenland and Antarctica grew larger. So much water was locked up in ice that the world sea level was lowered about 100 meters.

Australia Was Isolated As an Island Continent

The plants and animals of the ***Australian region*** (Australia, New Zealand, and adjacent islands) constitute the most unusual found on any of the earth's major land masses. Many species common in Australia occur nowhere else on earth. Conversely, many species widespread in the rest of the world are completely absent from Australia. Perhaps Australia's best-known and most curious animals are its mammals, which are completely unlike those found on any other continent. Except for wild dogs (dingoes) and pigs, both probably human imports of prehistoric times, most of the mammals are marsupials,[5] which fill most of the ecological niches that on other continents are filled by placental mammals.

The marsupials of Australia are thought to have migrated there some 70 million years ago by way of land bridges from South America and across Antarctica to Australia (arrow, Fig. 34.53). Because the marsupials reached Australia long ago, and because the continent was then isolated from all others, they encountered no competition from placental mammals. This permitted marsupials to undergo extensive adaptive radiation. Because they were filling niches similar to those filled in the rest of the world by placentals, and were thus subject to similar selection pressures, they evolved striking convergent similarities to the placentals. Certain of the marsupials resemble placental shrews, others placental jumping mice, weasels, wolverines, wolves, anteaters, moles, rats, flying squirrels, groundhogs, bears, etc. The uninitiated visitor to an Australian zoo finds it hard to believe when first looking at an assemblage

[5]Whereas the young of placental mammals undergo their entire embryonic development in the mother's uterus, the young of marsupial mammals develop only a short while in the uterus; after birth, they move to a pouch on the mother's abdomen, attach to a nipple, and there complete their development.

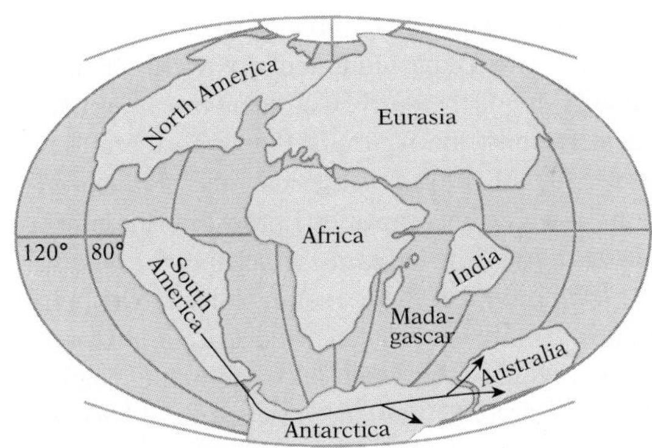

34.53 The marsupial trail The marsupials of Australia are thought to have migrated from South America and across Antarctica to Australia (arrow) by way of land bridges, some 70 million years ago. Because the marsupials reached Australia long ago, and because the continent then became isolated from all others, the marsupials encountered no competition from placental mammals and underwent extensive adaptive radiation.

of these animals that he is not seeing close relatives of the mammals familiar to him from other parts of the world (see Fig. 32.28, p. 728). Not all marsupials look like ecologically equivalent placentals, however; the kangaroos are markedly different (Fig. 34.54), though some of them play an ecological role very similar to that of horses and other large placental grazers.

Once Australia had drifted northward, species from Asia and the islands of the southwestern Pacific could reach it by crossing a water barrier. Because the water barrier was dotted by islands (the East Indies, now the nation of Indonesia), organisms could have spread from one island to the next over millions of years through a process of "island hopping." Furthermore, most of the western islands of the East Indies were probably connected by land bridges to the Asian land mass several times in the not too distant past when the sea level was far lower; hence the distance from Asia to Australia has not always been as great as it is today.

South America Has Been Isolated Through Much of Its History

The South American continent, which forms the major portion of the area known to biologists as the ***Neotropical region*** (meaning "new tropics"), has a history similar to that of Australia. It, too, has been an island continent, unconnected to the other major land masses of the world throughout much of its history. And it, too, had a great variety of marsupials, as the fossil record shows. The prevalence of marsupials in both Australia and South America probably means that these organisms were present throughout the region comprising Australia, Antarctica, and South America at a time when there were only small water breaks (if any) in this land mass (Fig. 34.53) and when the climate was warmer than it is now.

South America also had a variety of placentals, which probably reached it during a short period of connection to North America via a Central American land bridge present during the early part of the Age of Mammals (about 60 million years ago). After this land bridge disappeared, both the marsupials and the placentals evolved, in isolation, many characteristics convergent to those evolved by placentals on other continents (see Fig. 32.28, p. 728). There were times during this period of isolation (which lasted some 55 million years) when the water barrier between South America and what is now northern Central America was not so wide. A few additional placental mammals chanced to migrate into South America at such times; among these were the ancestors of the modern New World monkeys and a number of rodents.

Although there were 23 families of mammals in South America by five million years ago, not one of these was repre-

34.54 Kangaroos Though their appearance and behavior do not immediately suggest it, kangaroos are Australia's ecological equivalents of the ungulate grazers of other continents.

sented in North America. But then a land connection to North America (the isthmus of Panama) was reestablished, and many additional immigrants arrived in South America from North America. Some species also moved in the opposite direction, from South America to North America; the opossum, the porcupine, and the armadillo are examples. The migration was predominantly southward, though, apparently because the groups that had originated in the north were competitively superior and could displace the established South American animals, and many of the South American species suffered extinction.

Central America has never been more than a narrow bridge. Because its climate and rugged terrain have not been hospitable to all species, many groups of organisms have never been able to move between the North American and South American continents. The Central American land bridge has been a selective or filter route, along which some species but not others could pass. For this reason, South America, like Australia, has been an island continent throughout much of its history. However, its nearness to North America (Mexico) and its recent direct connection to North America via the Central American land bridge have given it a diverse group of organisms.

The Plants and Animals of the World Continent Are Similar

Europe, Asia, Africa, and North America formed a relatively continuous land mass, the so-called ***World Continent,*** throughout most of geological time. Consequently their plants and animals are more alike in many aspects than they are like those of the two island continents. Nevertheless, biologists customarily divide the World Continent into four main biogeographical regions: the ***Nearctic*** ("new northern"), which includes most of North America; the ***Palaearctic*** ("old northern"), which encompasses Europe, northernmost Africa, and northern Asia; the ***Oriental,*** which includes southern Asia; and the ***Ethiopian,*** i.e., Africa south of the Sahara (Fig. 34.55).

Figure 34.55 shows the Palaearctic, Oriental, and Ethiopian regions as an essentially continuous land mass, but you may wonder why the land masses of the Nearctic (North America) are also considered continuous. The answer is that North America and Eurasia have been connected through much of their geological history (Fig. 34.50B, C). Even after they broke apart as the northern part of the Atlantic Ocean formed, a new connection, the Siberian land bridge between what are now Alaska and Siberia, provided a link; part of this bridge is beneath water at the present time, and the continents are separated by about 90 kilometers of water. Because the climate of Alaska and Siberia is so forbidding, you might not expect many organisms to use a bridge in that region. But once again present conditions are misleading. Fossils of many temperate and even subtropical species of plants and animals are abundant in Alaska. Indeed, all the fossil evidence points to much movement, on many occasions, between Asia and North America via the Siberian land bridge. Humans crossed this bridge from Siberia into Alaska about 50,000 years ago.

Other land bridges, such as the North Sea bridge to Britain, the New Guinea-Australia bridge, and the Sunda bridge between Asia and Indonesia, have greatly influenced the present distribution of plants and animals on earth. But what caused these bridges to form and then sink? Geologic evidence shows

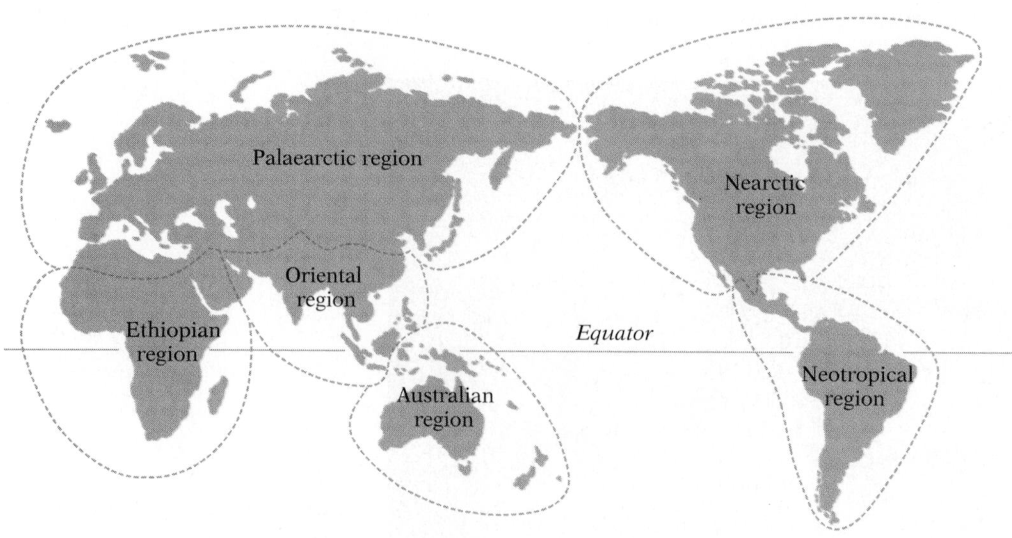

34.55 Biogeographic regions of the world

that during the last few million years the earth has undergone periodic transformations, alternating between "greenhouse" and "icehouse" climates. About 30,000 years ago, during the last glacial age, ice sheets covered about 29 percent of the earth's present land area (Fig. 34.52); by contrast, only about 10 percent is covered today. The moisture needed to form and sustain these ice sheets comes ultimately from the oceans, and when the sheets form there is a substantial lowering of sea level on a worldwide basis. During the last glacial age, for instance, it is estimated that the world sea level was lowered about 100 meters, which was sufficient to join France to Britain, establish the Siberian land bridge, and connect Indonesia to Asia. Plants and animals were able to migrate freely between these areas. As the climate warmed, the glaciers melted and the sea level rose, once again covering the bridges. Geologists suggest that at the present time we are approaching the period of maximum warmth in the glacial-interglacial cycle, after which the climate will begin to cool and the earth will enter another glacial age.

The history of the major land masses and of their changing climates helps explain the present distribution patterns of organisms. Consider the distribution referred to earlier, in which a species occurs in South America, Africa, and southern Asia. If the species belongs to a group of animals of very ancient origin (e.g., cockroaches), this distribution may indicate that the species occurred throughout the old Gondwanaland supercontinent and continued to survive in South America, Africa, and India after these drifted apart. But if the species belongs to a more recently evolved group (e.g., the majority of modern mammalian and bird families), which arose after Gondwanaland had broken up, it may be assumed that the species moved between the New World and the Old World via either the North Atlantic or the Siberian land bridge and between North and South America via Central America, and that it then became extinct in the north, either because of climatic changes or because of intense competition. The fossil record shows that this pattern of dispersal between southern regions by way of the northern continents has occurred again and again. For example, members of the camel family occur today in South America (llamas, alpacas, vicunas, etc.), northern Africa, and central Asia, but fossils indicate that the family originated in North America, spread to South America via Central America and to the Old World via Siberia, and later became extinct in North America. Charles Darwin used the fossil record as one of his lines of evidence for the theory of evolution, suggesting that the extinct creatures whose remains are

MAKING CONNECTIONS

1. Using your knowledge of soil properties, explain why tropical rain forest soils are poorly suited for agriculture. How does the destruction of these forests contribute to the greenhouse effect? Why is the loss of species diversity that accompanies this destruction a matter for practical concern?

2. In Chapter 24, you learned that many plants use photoperiodism to time reproductive processes such as flowering and seed germination. How is this system modified in many desert plants?

3. Why are estuaries such as the Chesapeake Bay particularly vulnerable to pollution? Why is damage to estuarine communities by pollution a matter of great economic as well as ecological concern?

4. Marine fossils are found in the Himalayas in Nepal at an altitude of 4 kilometers above sea level and 800 kilometers from the ocean. What does this tell you about earth's geological history?

5. Why are Australia's mammals unlike almost all of those found on any other continent? What roles have geological history, adaptive radiation, and convergence played in their evolution?

6. Explain the migration patterns of families of mammals between North and South America by the competitive exclusion principle covered in Chapters 32 and 33.

preserved in the rocks as fossils represent the ancestors from which the organisms living today are descended.

It is impossible to end our discussion of ecology without reflecting on the future. Once our species was but a minor member of the biological cast on our planet. Two technological breakthroughs—domestication first of plants and then of animals—erased the major limiting factors that controlled our population size and distribution. The ability to manage crops and herds, and to store grain, allowed the development of cities and wealth and a self-perpetuating series of cultural and agricultural advances. As should be ominously clear from these last two chapters, human population size is continuing to increase almost exponentially, and the effort to feed and otherwise support these billions of individuals is leading inexorably to the rapid extinction of other species, and the destruction of forests, croplands, estuaries, water supplies, and even the air we breathe. Further scientific advances can ameliorate these problems and slow the destruction of the biosphere, but we must build a sustainable society—one that can be maintained over time without compromising the needs of other species and future generations. To do this we need to conserve our resources, recycle materials, and use renewable resources such as wood, solar radiation, or wind energy. But in the end only a dramatic end to human population growth offers any long-term hope for survival. Whether that end comes from a demoralizing and debilitating series of wars, diseases, and famines or from conscious and informed decisions will be the choice of the 21st century.

CONCEPTS IN BRIEF

1. The flow of energy and materials knits a community together and binds it with the physical environment as a functioning system. Almost all forms of life obtain their high-energy organic nutrients directly or indirectly from photosynthesis.

2. The *food chain* is the sequence of organisms, including the *producers* (autotrophic organisms), *primary consumers* (herbivores), *secondary consumers* (herbivore-eating carnivores), and *decomposers,* through which energy may move in a community. In most communities the food chains are complexly intertwined to form a *food web.* The successive levels of nourishment in the food chains are called *trophic levels.*

3. The flow of energy through the ecosystem is one-way. At each successive trophic level there is loss of energy from the system; about 10 percent of the energy at one trophic level is available for the next, so a *pyramid of productivity* results. In general, the decrease of energy at each successive trophic level means there is a limit to the number of trophic levels in an ecosystem and that less biomass can be supported at each level; thus many communities show a *pyramid of biomass.*

4. The endless cycling of *water* to earth as rain, eventually back to the atmosphere through evaporation, and back to earth again as rain maintains the freshwater environments, supplies water for life on land, and transports chemical nutrients through ecosystems.

5. *Carbon* cycles from the inorganic reservoir to living organisms and back again. Carbon dioxide is converted into organic compounds by photosynthesis; the resulting organic compounds may be consumed by animals or decomposers, or respired for energy. Eventually the carbon will be released as CO_2 and the cycle will continue.

6. Increasing levels of CO_2 resulting from the burning of fossil fuels may cause the earth's temperature to rise due to the greenhouse effect.

7. Biological *nitrogen fixation* by microorganisms in the soil or root nodules of plants provides usable nitrogen for the earth's ecosystems. The microorganisms reduce atmospheric N_2 to compounds that can be absorbed by plant roots. The plants convert the inorganic nitrogen into amino acids. When the plant dies, nitrogen is returned to the soil as NH_3, which can be recycled.

8. *Phosphorus* moves from the inorganic reservoir in rocks to living organisms. Phosphate from rock dissolves and is absorbed by plants, which pass it to animals and other heterotrophs. Some is excreted by animals; the rest is released when the organism dies. Sewage, detergents, and fertilizer runoff have greatly increased phosphate levels in the environment, which accelerates the *eutrophication* (aging) process of lakes.

9. Modern industry and agriculture have been releasing new chemicals into the environment. Many (e.g., DDT and PCBs) are persistent chemicals and show *biological magnification;* they become increasingly concentrated as they are passed up the food chain to the top predator.

10. The properties of soils—particle size, amount of organic material, and pH, among others—determine how rapidly water and minerals move through the soil and how available they will be to plants. Acid rain affects soil fertility by influencing the solubility of ions.

11. Most biologists recognize a number of major regional climax formations, called *biomes.* Each biome has a distinctive combination of plants and animals, and is characterized by the dominant form of vegetation. The distribution of the biomes is influenced primarily by temperature and the amount of annual precipitation.

12. The *tundra is* the northernmost biome of North America, Europe, and Asia. The subsoil is permanently frozen. There are many organisms but relatively few species. South of the tundra lies the *taiga,* dominated by coniferous forests. More different species live in the taiga than on the tundra.

13. The biomes south of the taiga show much variation in rainfall and hence more variation in climax communities. The *deciduous forests* predominate in temperate zones with abundant rainfall and long, warm summers. Tropical areas with abundant rainfall are usually covered by *tropical rain forests;* here the diversity of species is enormous. In both the temperate and tropical regions, huge areas where rainfall is low or uneven are covered by *grassland.* Places where the rainfall is very low form the *deserts.*

14. To understand the present geography of life, we must combine knowledge of present conditions with evidence from the fossil record and with geological evidence of past configurations of the earth's land masses and their climates. Geological evidence indicates that 225 million years ago all the earth's land masses were combined in a single supercontinent called *Pangaea.* Pangaea broke up and the land masses began to move slowly apart. As the continents moved, their climates changed, altering the distribution of organisms.

15. The plants and animals of the *Australian region* are most unusual; this can be explained by the long isolation of Australia from the other continents.

16. The South American continent, forming most of the *Neotropical region,* has also been an island continent through much of its history, but its nearness to North America and its recent connection via the Central American land bridge led to increased species diversity; many of its plants and animals resemble those of the World Continent (Europe, Asia, Africa, and North America).

17. The World Continent can be divided mainly into the *Nearctic* (most of North America), *Palaearctic* (Europe, northernmost Africa, northern Asia), *Oriental* (southern Asia), and *Ethiopian* (Africa south of the Sahara). North America and Eurasia have been connected through much of their history by the Siberian land bridge.

STUDY QUESTIONS

1. How does energy flow through a community? Describe the roles of *producers, consumers,* and *decomposers* in this process. (p. 764)
2. If the net primary productivity of a forest is 2,000 kcal/m^2/yr, what would be the expected productivity of the *secondary* consumers? (p. 764)
3. Using the laws of thermodynamics, explain why there is no such thing as an energy cycle. (p. 766)
4. Make a diagram of the water cycle, indicating the percentage of the earth's water found in each location. What supplies energy for the cycle? (pp. 766–768)
5. What is carbon's inorganic reservoir? What process removes it from its reservoir? How is it passed through the community and then returned to its reservoir? (pp. 768–771)
6. Compare the nitrogen and phosphorus cycles, answering the following questions: (a) What is the element's main inorganic reservoir? (b) How is it removed from its reservoir by organisms? (c) How is it used in organisms? (d) How is it returned to its reservoir? (pp. 773–775)
7. Which would be most likely to suffer adverse effects from the runoff of DDT into a lake—microscopic

plankton, fish that feed on the plankton, or ospreys that feed on the fish? Explain your answer. (pp. 776–777)

8. **What properties of *loam* make it a suitable soil for growth of most kinds of plants? How do organic matter and clay affect its characteristics? (pp. 777–778)**
9. **What is a *biome?* Describe the location, climate, soil conditions, vegetation, and animal life of these biomes: tundra, taiga, deciduous forest, tropical rain forest, grassland, and desert. (pp. 784–788)**
10. **Describe the provinces and zones into which the oceanic environment is divided. Which areas of the ocean are particularly productive? Why? (pp. 792–793)**
11. **Explain the theory of *continental drift.* Beginning with the breakup of Pangaea, describe the movements of land masses that have resulted in the present positions of the continents. (pp. 794–796)**
12. **Describe the locations and major characteristics of six biogeographical regions of the world. Which four make up the World Continent? Why is it so named? (pp. 798–799)**
13. **How have climatic change and land bridges influenced the present distribution patterns of organisms? (pp. 798–800)**

CHAPTER 35

ANIMAL BEHAVIOR

Every fall monarch butterflies from all over North America migrate south to spend the winter in more hospitable climates. Enormous numbers of them congregate on specific trees in central Mexico and coastal California, even though they may never have been there before. In the spring, the survivors migrate north to breed. The behavior of these butterflies is highly adaptive: they are leaving the northern regions with its severe winter conditions where they undoubtedly would perish. How do they come by the knowledge of when to leave and where to go? That branch of biology that seeks to understand behavior such as this and the mechanisms and evolution of animal behavior is known as ***ethology.*** The behavior of an animal—what it does and how it does it—constitutes one of its distinctive attributes. It is, as we saw in the preceding chapters, the product of the functions and interactions of the animal's various control and effector mechanisms.

Behavior Should Be Interpreted in Terms of the Simplest Neural Mechanisms

For centuries humans have observed animal behavior in terms of their own experience. Observing an earthworm squirm when pierced by a fishhook, they have explained that the hook hurts the worm, causing it to writhe in pain. Observing adult birds feeding their young, they have declared that the birds love their babies and want to feed and protect them. But such descriptions are unacceptable; we have no evidence that being afraid, feeling pain, loving, wanting to do something, as those expressions apply to human beings, are meaningful when applied to insects, earthworms, or birds. Such descriptions are an unwarranted ***anthropomorphism***—the projection onto animals of the sensations human beings experience in similar circumstances. Insects and earthworms have a brain so different from the human brain that extrapolations about awareness, or even sensation, from one to the other demand caution.

What, then, can we say about earthworms that squirm on the fishhooks and birds that feed their young? Restricting ourselves to what is observable and testable, we may say that the hook stimulates receptors that initiate impulses along nerve circuits, and that these impulses result in squirming. As a reflex response to stimulation of sensory receptors by fishhooks or other objects, squirming has an obvious adaptive advantage; it may help the worm avoid further physical damage. No con-

Chapter opening photo: Monarch butterfly tree in Mexico. Enormous numbers of them cover the branches of this fir tree in their wintering grounds in Sierra Madre, Mexico.

scious awareness need be associated with such an adaptive response. Even the more complex behavior of a mother bird feeding her young can be explained in terms of responses to stimuli within the context of the physiological condition of the bird at the time; no "love" need be assumed.

Interpreting the behavior of other animals in terms of the simplest neural mechanisms that can explain the observed action rules out conscious thought, deliberate decision, purposive determination, and foresight as explanations of animal behavior; it prohibits anecdotal descriptions and anthropomorphic interpretations.

Some Behavior Is Inherited

The idea that something as intangible as behavior could be inherited has always been controversial. But by the 18th and 19th centuries, naturalists had cataloged thousands of observations, and Darwin could point to many adaptive behaviors that were clearly inherited. Perhaps the example most compelling to Darwin was the life history of the European cuckoo, which lays its eggs in the nests of other birds. The cuckoo chick hatches in the nest of one of several possible host species, and even before its eyes are open, it ejects the hosts' own eggs and chicks (Fig. 35.1A). The foster parents feed and care for the chick until it is fledged, by which time it is considerably larger than they are (Fig. 35.1B). Even though it has probably never seen or heard another cuckoo of either sex, the young bird is able to find and recognize a suitable mate the following spring, to court, and to copulate, all in a manner typical of its species.

How, Darwin wondered, is a cuckoo able to do precisely the right thing at the right time having had little or no opportunity to learn; and when something must be learned, how does the bird "know" to ignore a world full of distracting information and focus on exactly what must be memorized? Amazing as it seems, the baby cuckoo must inherit the essential "instructions" in its genes—genes that direct the wiring of its nervous system during development. The underlying instructions that direct learning and behavior are known popularly as ***instinct;*** behavior like that of the cuckoo, which depends largely on inherited mechanisms, is usually referred to as ***innate*** or instinctive. A ***learned behavior,*** on the other hand, is a change in behavior as a result of experience.

As we shall see, it is extremely unlikely that any behavior can be classified as strictly innate or strictly learned: even the most rigidly automatic behavior depends on the influence of the environmental conditions for which it evolved, while most learning, flexible as it seems, appears to be guided by innate mechanisms. ***Instinct,*** then, can be defined as the heritable, genetically specified neural circuitry that organizes and guides behavior. The behavior that is thereby produced can reasonably be called at least partially innate.

A

B

35.1 Egg ejection by a fledgling cuckoo (A) A newly hatched cuckoo rolls the eggs of its host out of the nest. As a result, the young bird does not have to share any of the food its foster parents collect. (B) The unwitting foster parent continues to feed the cuckoo even when it has grown to several times the parents' size.

FUNDAMENTAL COMPONENTS OF BEHAVIOR

Taxes, Kineses, and Reflexes Are Simple Responses to Stimuli

Some animals orient themselves so that a critical stimulus registers equally on their left and right sides. Planarians will move toward a light by orienting so that both eyespots are equally stimulated. If two equally bright lights a short distance apart are placed near a planarian, the animal will orient toward a point midway between them, thus attaining equal stimulation of the two eyespots. A simple, continuously oriented movement in animals is called a ***taxis;*** the behavior of the planarian with respect to lights is called phototaxis.

Responses to simple stimuli are not always oriented relative to the stimulus. For example, the response may be a change in rate of motion, as we see in the tendency of sow bugs (small crustaceans that normally live in moist places under stones or logs) exposed to very dry conditions. Under such circumstances the animals become very active and move about randomly. If they enter moister regions their movement slows; in very moist places they may cease moving altogether. Such behavior, where specific directional movement is not caused by the eliciting stimulus (as it is in a taxis), is called a ***kinesis.*** The result of this kinesis is that the animals tend to congregate in moist spots, where they are in little danger of drying out (Fig. 35.2).

In Chapter 28 we learned about ***reflexes,*** which are automatic acts consisting of a simple response to a stimulus, as when a tap on the knee elicits a knee jerk. Most taxes and kineses involve several reflexes, because the whole body is usually turned in response to the stimulus. However, more complicated behavior patterns, which cannot profitably be interpreted simply as taxes, kineses, or reflexes, predominate in higher animals. In the higher vertebrates, especially the mammals and birds, behavior also becomes increasingly modifiable by learning. Taxes and kineses, as they are usually understood, are almost nonexistent in the higher mammals, and simple reflexes, though still important, constitute a very small portion of the total behavioral repertoire of higher animals.

35.2 Sow bugs congregating in moist areas

Sign Stimuli and Releasers Trigger Specific Behavioral Patterns

In the early 1930s a Bavarian naturalist, Konrad Lorenz, began to realize that the brains of different animals interpreted incoming sensory information differently. Lorenz noticed that he was attacked by his pet jackdaws whenever he carried something black hanging from his hand. The birds, which are themselves black, seemed to interpret any dangling black object as a fellow jackdaw in distress. It seemed to Lorenz that the birds, though evidently capable of recognizing each other as individuals in other situations, were, in this situation, ignoring most of what they could see and focusing instead on a small (and in this case misleading) subset of cues.

Lorenz and the ethologist Niko Tinbergen examined this phenomenon further, and found that many animals are highly responsive to specific stimuli. For example, both males and females of the common stickleback (a minnowlike fish) recognize breeding territorial males by the red coloration on their ventral surfaces. Sticklebacks are so thoroughly attuned to the red color that they are oblivious to additional cues that might otherwise be useful, such as the size and shape of the object displaying the color (Fig. 35.3). In fact, Tinbergen reported that his territorial males would attempt to attack passing British mail trucks (which are red) that were visible through the sides of their aquariums. In behavioral terms a ***sign stimulus*** is any simple signal, such as the red belly, that elicits a spe-

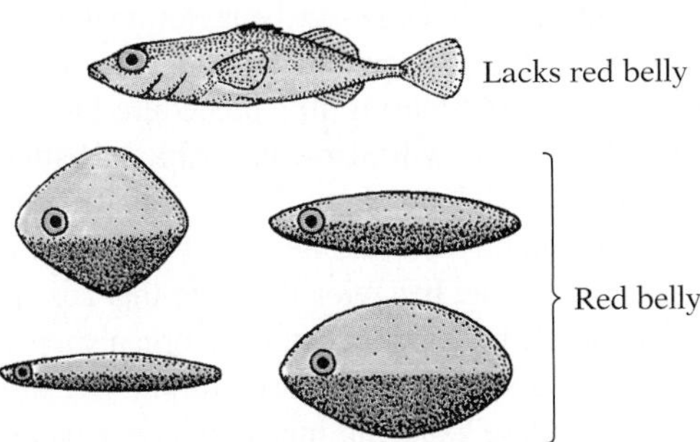

35.3 Models of male stickleback The realistically shaped model lacking the red belly was attacked by male sticklebacks much less often than the oddly shaped models with red bellies.

35.4 Difference in response by a hen to her chick's visual and vocal distress signals (A) The hen ignores the chick if she cannot hear its calls, even though its actions are clearly visible. (B) Distress calls elicit a vigorous reaction from the hen even when she cannot see the chick.

35.5 Hawaiian Laysan albatross attempting to incubate a buoy from a fishing net Albatross instinctively rotate their eggs and carefully roll any eggs that escaped during this procedure back into their nests. But these birds will also roll a variety of other rounded objects, such as this fish buoy, into their nests, apparently mistaking them for eggs.

cific behavioral response. The specificity of a sign stimulus for a particular behavioral reaction is illustrated by Lorenz', observation that a mother hen can recognize a chick in trouble only by its special distress call, even when the chick is clearly visible but unable to have its distress call heard (Fig. 35.4).

Sign stimuli seem to organize behavior throughout the animal world. Ground-nesting birds such as gulls and geese instinctively rotate their eggs—a behavior that prevents the embryos from sticking to their shells—and then carefully roll any eggs that escaped during this procedure back into their nests. But these birds will also roll flashlight batteries, golf balls, beer bottles, and a variety of other rounded objects into their nests, apparently mistaking them for eggs (Fig. 35.5). Such examples of rote behavior illustrate that certain stimuli trigger a particular behavior because the organism possesses a heightened neural sensitivity to these stimuli. Because certain sign stimuli "release" specific innate behavior patterns, they are frequently called ***releasers.***

Depending on the sensory world of the organism in question, behavior may be released by different sensory cues. Odors may serve as releasers. A number of animals secrete chemicals called ***pheromones,*** which influence the behavior of other members of the same species. When a honeybee worker stings an intruder, for instance, the bee releases an alarm pheromone that attracts other workers to the source. Male moths ready to mate can detect species-specific pheromones released by a female at an enormous distance. Sounds, too, may act as releasers. The high-frequency sounds produced by the wingbeat of a female mosquito attracts males of the same species for mating (Fig. 35.6). An example of a visual releaser

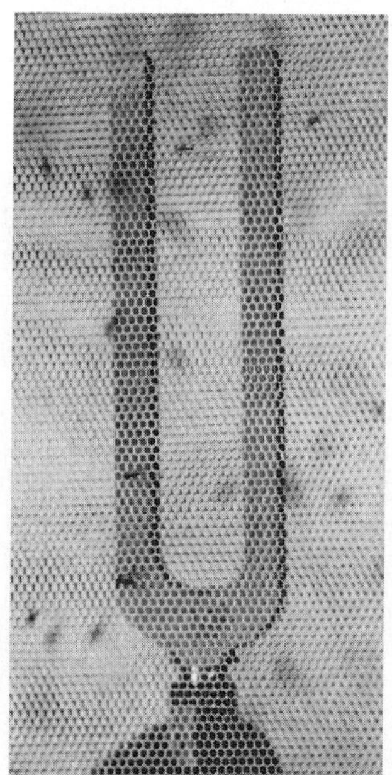

35.6 Photographs showing response of male mosquitoes to a tuning fork Left: The fork is silent. Right: The fork is vibrating at about the same frequency as the female mosquito's wingbeat, and its sound is attracting males.

is the ultraviolet light bull's-eye pattern present on many flowers, which attracts bees (Fig. 35.7).

A sensitivity to species-specific releasers is present in most animals at birth, and generates a variety of stereotypical behaviors. The young of most species are also born with the releasers necessary to direct and trigger parental care. The success of the cuckoo chick, for instance, is assured because it possesses at birth an orange throat patch and specialized peeping, the same releasers that cause its surrogate parents to recognize and feed their own offspring. Releasers have a great advantage: they can initiate certain critical behavioral responses automatically, thus bypassing the time-consuming and error-prone process of learning. But releasers have at least one major disadvantage as well, in that they may be triggered by crude and inappropriate stimuli. As we shall see, behaviors that exhibit a combination of learning and releasers frequently provide animals with the best features of both.

A

B

35.7 Flowers in visible and ultraviolet light To us, the primrose willow appears to be a nearly uniform yellow (A), but to bees, which can see ultraviolet light, the center is marked by a dark bull's-eye (B).

35.8 Oystercatcher reacting to giant egg She chooses to incubate it instead of her own egg (foreground) or a herring gull's egg (left).

Once the sign stimulus for a particular behavior pattern has been carefully analyzed, it is often possible to design ***supernormal stimuli*** that are even more effective than the natural one. For example, Tinbergen showed that the size of oystercatcher eggs is important in determining the releasing properties of the eggs for the adult birds. An adult oystercatcher provided with normal oystercatcher eggs and with larger eggs of other species will usually react preferentially to the largest egg, even if she cannot possibly hatch it (Fig. 35.8); for her the large egg provides a supernormal stimulus.

Though the behaviors released by sign stimuli may be rigid and stereotyped, they nonetheless depend on proper motivation. Motivation, in turn, depends on health, maturation, hormone levels, previous learning experience, recent sensory stimulation, and so on. This is an important point, since a potent sign stimulus will not invariably elicit a response. For instance, a well-fed animal involved in nest building is most unlikely to respond to an otherwise powerful sign stimulus of food.

MOTOR PROGRAMS

Fixed-Action Patterns Are Innate, Stereotyped Responses

Lorenz and Tinbergen noticed in the 1930s that some behaviors appear to be all-or-none stereotyped responses to releasers: they tend, once begun, to run to completion regardless of the situation. A dramatic example is the egg-rolling response of geese. If a goose notices an egg outside her nest, she rises, extends her neck until the egg is touching the underside of her bill, rolls it gently back into the nest, and then settles down to continue brooding (Fig. 35.9). To the casual observer

35.9 Egg rolling by a goose When a goose sees an egg outside her nest she rises, touches the egg with her beak, and then rolls it back in. She completes the same recovery behavior when the object she sees is a beer bottle, or when the egg is removed after she has begun to reach for it.

this behavior appears to show thought on the goose's part: she has recognized a problem and solved it. But if the egg is removed while she is reaching for it, the goose will go on as if nothing had happened, rolling the nonexistent egg carefully into the nest. In short, egg rolling is an independent behavioral unit which, once initiated, proceeds to completion with little or no need for further feedback. Lorenz and Tinbergen called such units ***fixed-action patterns,*** or what we now call motor programs (see p. 807).

Maturation Often Affects Motor Learning

It is difficult to know whether a motor behavior has been learned unless it is not actually exhibited at birth. In some cases, though, particularly among the invertebrates, opportunities for learning are so slight that even many behaviors seen only in adults must be regarded as innate. A wasp that specializes in capturing honeybees, for instance, must be equipped from birth to spin a cocoon, emerge and dig out of its particular kind of burrow or chamber, groom itself, fly, court and mate, pounce on bees, sting them in an unarmored patch under the neck without being itself stung, squeeze the abdomen to obtain nectar the victim has collected, carry the paralyzed prey in flight, dig a burrow, lay an egg, seal the burrow, and so on. The wasp simply has no opportunity for learning between hatching and its first task. But many other invertebrate behaviors are varied and complex, and change with time. And vertebrates, which lead longer and perhaps more leisurely lives, are frequently able to take advantage of experience.

Nevertheless, even in vertebrates many behaviors that appear to be learned are actually innate. A classic example was provided by Eckhard Hess of the University of Chicago in an elegant experiment that demonstrated the maturation of a motor program. He fitted newborn chicks with tiny goggles that deflected their vision seven degrees to the right, and recorded the accuracy of their pecks by providing a target (a nailhead that, like seed, acted as a sign stimulus for pecking) set in soft clay. The pecks of both the normal chicks and those with the goggles were scattered, but the marks of the chicks with goggles were well to the side of the target (Fig. 35.10A,D). A few days later, chicks of both groups were able to produce a tight cluster of pecks, but those with the goggles were still missing the target by as much as before (Fig. 35.10B,E). Apparently the chick's circuitry for aiming and pecking is already hard wired at birth, and the normal improvement in accuracy is a simple consequence of increased nerve and muscle coordination rather than of learning.

By contrast, many behaviors we know to be learned take on the appearance of fully innate motor programs. Walking, for instance, which is innate in most species, must be learned initially in ours. Though the alternation of the legs and the interacting reflex arcs responsible for walking come prewired at birth—a properly supported human infant will perform walking motions on the delivery table—humans must still learn to balance, once they have matured enough to support their own weight. Yet after the difficult process of learning has been completed, simple walking becomes automatic. Swimming and bicycle riding present the same story, and as everyone knows, once painstakingly learned, neither is ever completely forgotten. The same pattern may be seen in other animals: learned behavior can become stereotyped and largely automatic—that is, take on the characteristics of an innate motor program. This apparently occurs as new motor-program circuits are wired up in the brain. In fact, studies of the effects of localized brain damage in humans reveal that special areas of the cortex and cere-

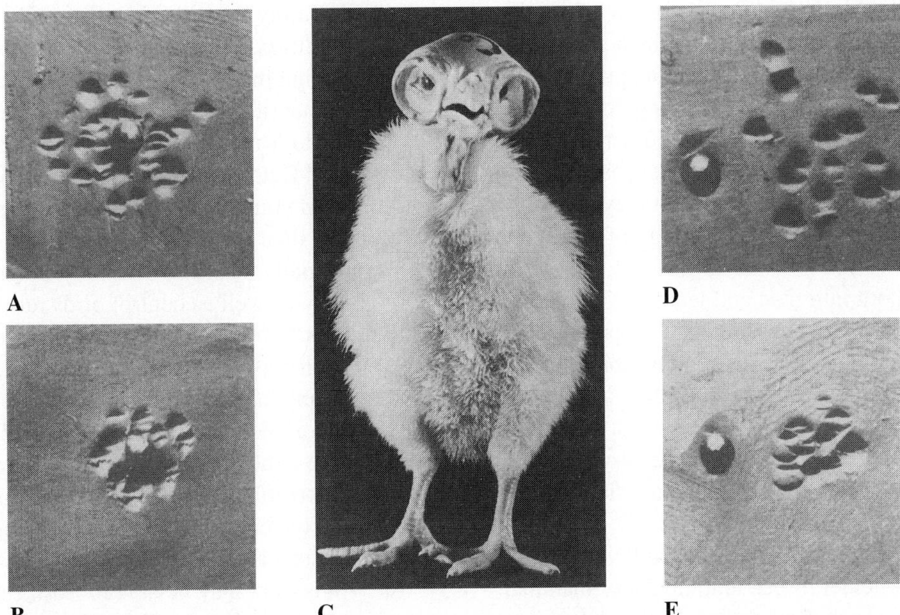

35.10 Maturation of pecking behavior in chicks Newborn chicks peck at a target with fair accuracy (A), but their aim improves with age until at four days the pecks are tightly clustered (B). This improvement could be the result of some sort of maturation—better vision, perhaps, or strengthened neck muscles—or of learning, by which the chick recognizes and corrects its errors. Eckhard Hess pitted these alternatives against one another by raising chicks with goggles that deflected their vision to the right (C). As newborns, such birds produce the usual set of scattered pecks, but the pecking is well to the right of the target (D). By the fourth day, the pecks are tightly clustered but still misdirected (E), indicating that chicks are unable to learn to adjust their aim. The coordination of beak and eye involved in pecking must therefore be a wholly innate behavior, which matures without benefit of learning.

bellum are reserved for learned motor programs. These areas are highly localized in the brain, with one region of the cortex, for example, constructing and storing motor programs requiring fine finger control—typing, writing, weaving, playing the piano, tying shoes, and the like—while another is devoted to programs involving the kind of limb movements necessary for swimming and kicking a ball. This freeing of learned behavior from detailed conscious control allows conscious attention to focus on new problems—an advantage for humans and for other animals too. Thus, for instance, a bird that has learned to shell seeds automatically can devote its attention to watching for predators while it eats.

LEARNING AND BEHAVIOR

Inheritance and Learning Interact in Determining Behavior

Psychologists and biologists increasingly recognize that inheritance and learning are both fundamental in determining the behavior of higher animals and that the contributions of these two elements are inextricably intertwined in most behavior patterns. Particular nervous pathways and effectors are inherited, and an animal can exhibit only these behavior patterns for which it has the appropriate neural and effector mechanisms. Furthermore, the ability to learn depends on inherited neural pathways, and if the necessary connections are lacking, no amount of experience can establish a given behavior pattern.

Inheritance Provides Limits for Learning

One way of viewing the interaction of inheritance and learning in animal behavior is to regard inheritance as determining the limits within which a particular type of behavior can be modified, and to regard learning as determining, within those limits, the precise character of the behavior. In some cases, as in the simplest reflexes, the limits imposed by inheritance leave little room for modification by learning; the available neural pathways and effectors rigidly determine the response to a given stimulus. In other cases the inherited limits may be so wide that learning plays the major role in determining the behavior elicited by a given stimulus.

A dramatic example of the interaction between inheritance and learning is the song of the European chaffinch. W. H. Thorpe of Cambridge University raised chaffinches in isolation and found that such birds were unable to sing a normal chaffinch song. One might immediately conclude that the song is wholly learned. The matter is more complicated, however.

Thorpe demonstrated that young chaffinches raised in isolation and permitted to hear a recording of a chaffinch song when about six months old would quickly learn to sing properly. But young chaffinches permitted to hear recordings of songs of other species that sing similar songs did not ordinarily learn to sing those other songs. Apparently chaffinches must learn to sing by hearing other chaffinches, and to do this they inherit the ability to recognize and respond to the songs of their own species.

Chaffinch song is both inherited and learned. Chaffinches inherit the neural and muscular mechanisms responsible for chaffinch song, and they inherit, apparently, the ability to recognize a chaffinch song when they hear it; they also inherit severe limits on the type of song they can learn. But the experience of hearing another chaffinch sing is necessary to trigger their inherited singing abilities, and in this sense their song is learned.

The limits for different behavior patterns will vary among species; each animal, whether insect, amphibian, or mammal, has some behavioral traits that are rather rigidly determined by inheritance, with very little possibility of modification by learning, and other behavioral traits capable of much modification. Such a difference can be observed in herring gulls. The adult birds nesting in colonies learn to recognize their own young about five days after they hatch; thereafter, if young of the same age from another nest are substituted for their own, the adults will not accept them and will neglect or even kill them. Yet these same gulls show amazingly little aptitude for learning to recognize their own eggs. They can be given substitute eggs quite different in color, pattern, shape, or size and will accept them without hesitation. These gulls, then, exhibit great aptitude for learning to recognize individual young so alike in appearance that human beings can tell them apart only with great difficulty, if at all, but they show very little aptitude for learning to recognize eggs so different that human beings can distinguish them at a glance. From an evolutionary point of view, the difference is readily understandable. There must have been far more selection pressure for the evolution of offspring recognition, since the young might stray from their own nest under normal circumstances, than for recognition of individual eggs, which in nature seldom wander from nest to nest.

Many Factors Complicate the Studying of Learning

It is probably impossible to separate completely what is inherited and what is learned in any behavior pattern. Behavior is not a simple combination of these two elements, but rather the outcome of their fusion. Before we examine some commonly recognized categories of learning, we should mention several factors that complicate its study.

1. It is often difficult to determine whether improvement in the performance of a behavior pattern is due to experience or simply to greater maturity and a different physiological condition. For example, observations that young birds just leaving the nest cannot fly well, but improve rapidly over the next few days, have led to the widespread belief that the birds must learn to fly and that they improve with practice. But, as repeated experiments have demonstrated, when young birds are restrained and prevented from flapping their wings, and are then released at an age when they normally would have already "learned" to fly, they are able to fly as well as control birds raised under normal conditions. In other words, it is not practice that causes the flight of a newly fledged bird to improve, it is greater maturity.
2. An animal may readily learn something in one context and be completely incapable of learning the same thing in another context. For example, foraging honeybees can learn the position of the hive entrance on the first flight of the day, but cannot learn the same thing in the context of later flights that same day.
3. An animal may be able to learn certain behavior patterns only during a rather limited ***sensitive phase*** (also known as the ***critical period***) in its life. If it does not encounter the necessary learning situation during this period, it may never learn the behavior. Exposure to the learning situation before or after the sensitive phase may not produce learning. For example, as Thorpe demonstrated in his studies of the development of chaffinch song, unless young chaffinches hear a chaffinch song during a certain period in their development, they never learn to sing properly. Sensitive phases are seldom so rigid in human beings, but there is abundant evidence that various learning abilities are greatest at certain ages. For example, children between the ages of two and ten can learn languages far more easily than can adults.
4. It is not always possible to tell immediately whether or not learning has taken place. There may be considerable delay between the ***latent learning*** that occurs on exposure to the learning situation and the performance of a behavior pattern that shows the effects of learning. For example, if young chaffinches only a few weeks old are allowed to hear a tape recording of a singing adult for a few days and are then raised in isolation, they will sing a nearly normal chaffinch song the following spring. Exposure to the song during their first summer, long before they are old enough to sing, results in latent learning, but confirmation does not come until months later.
5. Comparisons between the learning capabilities of different species are often misleading. Many papers have been published purporting to measure the relative intelligence of animals as different as fish, pigeons, rats, and monkeys by exposing them to the same problem-solving situations. Results from such experiments are highly suspect, because these animals have different lifestyles and are likely to have evolved radically different levels of response to any given type of stimulus, depending on its importance in their lives under natural conditions.

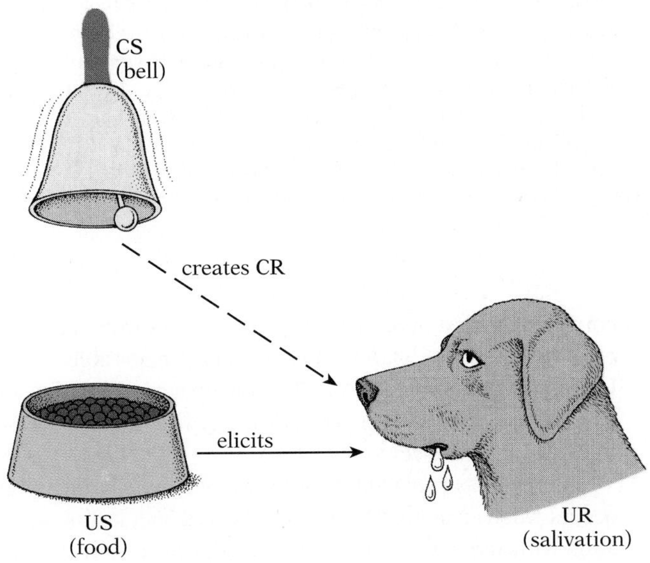

35.11 A conditioned response Classical conditioning begins with an unconditioned response (UR) that follows an unconditioned stimulus (US). In this case the unconditioned stimulus is food and the unconditioned response is salivation at the sight of food. A novel conditioned stimulus (CS), i.e., a ringing bell, is then paired with the US (food), eliciting the salivation response. After this process has been repeated sufficiently, the CS (the bell) alone will elicit the response. In short, the animal comes wired with the sequence US → UR; the researcher provides the relationship CS + US → UR; and the animal generalizes from a series of individual experiences so that CS → UR.

There Are Many Different Types of Learning

So far in our discussion we have focused primarily on innate behavior. Such behavior frequently lacks adaptive flexibility to meet changing environmental and social conditions. Behavioral flexibility is provided by learning, which can be classified according to the following categories.

One of the simplest types of learning is ***habituation,*** a gradual decline in response to "insignificant" stimuli that occurs with repeated exposure lacking any positive ***reinforcement*** (reward) or punishment. A young squirrel in a park, for instance, might initially run away from people, but soon learns to go about its daily life in spite of the presence of humans. In effect, habituation is learning to ignore stimuli that are unimportant in the life of the animal. The durability of the lowered response distinguishes it from mere sensory adaptation or fatigue.

Much of what an animal learns comes as a result of conditioning, in which the organism learns from its experiences in the environment what to do in order to survive. Psychologists recognize two different types of conditioning. The first, discovered by the Russian physiologist Ivan Pavlov, is called ***classical conditioning.*** It is the association, as a result of reinforcement, of a response with a stimulus with which it was not previously associated. Pavlov encountered this form of learning during his studies on digestion. He noticed that if he rang a bell each time he fed meat to a group of dogs, eventually the salivary reflex of the dogs became conditioned to the auditory stimulus of the ringing bell (Fig. 35.11). Pavlov could then ring the bell, and the dogs would salivate even if they could not see, smell, or taste meat. A new reflex had been established; a new stimulus elicited a response that it had never elicited before the training. The new stimulus (the sound of the ringing bell) had apparently been associated in the dogs' nervous system with the original stimulus (sight, smell, or taste of meat), and the same response was now given to both.

Conditioning is not restricted to behavior patterns as simple as reflexes. Animals may be conditioned to perform such complex activities as running, pushing levers, opening doors, and performing complicated tricks (Fig. 35.12). Much of the

A

B

35.12 Sheepherding (A) A sheepherder using a whistle to signal his dogs. (B) This well-trained New Zealand sheep dog is herding sheep following the hand signals given by the man at left. The herding instinct is inherent in these dogs, but they must be conditioned to respond correctly to their master's signals.

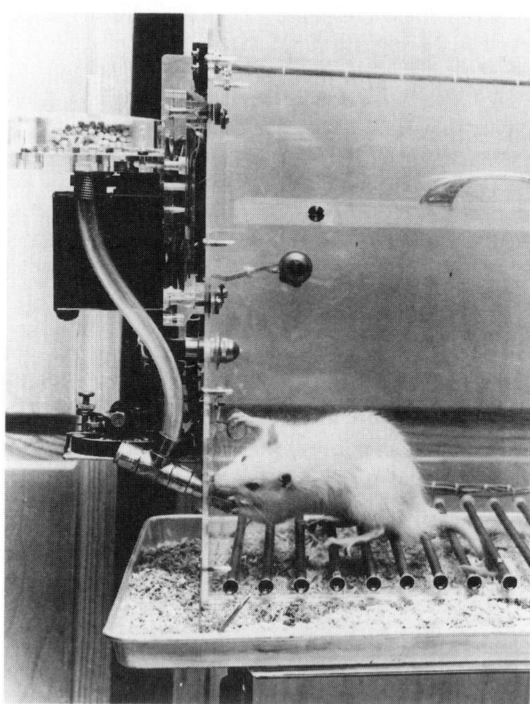

35.13 Trial-and-error learning in the laboratory A pellet of food is dispensed from the apparatus when the rat presses the bar in response to the correct stimulus. This behavior is shaped by rewarding ever-closer approximations of the desired performance. The rat may be fed at first for being in the correct end of the box, and then only for accidentally touching the bar. Then the reward threshold may be raised to require actually pressing it. Finally the task of pressing in response to a particular stimulus is added.

training of domesticated animals such as cats, dogs, and horses is based on conditioning; the animals learn to associate stimuli such as whistles or spoken commands with responses not normally elicited by such stimuli.

The second type of conditioning is called ***operant conditioning,*** or ***trial-and-error learning.*** If an animal does something by chance and the result is rewarding, it may do the same thing again (Fig. 35.13). If the result is not rewarding, or is disagreeable, it may, after several trials, learn not to do this thing anymore. The capacity for operant conditioning confers a considerable advantage on animals, since it allows them to acquire motor behaviors that are not innate. A seed-eating bird, for instance, does not have innate motor programs for dealing with various kinds of seeds. Instead the bird has an innate ability to recognize seedlike objects, and an instinct to experiment with anything that looks like a seed. Opening up a sunflower seed may take a finch several minutes of manipulation with its beak and tongue, but the kernel the bird eventually harvests is the reward that motivates it to try another. By trial and error, the finch discovers the most effective way to get the seed. Experiences in opening many seeds shape the harvesting behavior into an efficient series of movements that becomes automatic—a learned motor program.

Another type of learning, called ***insight learning,*** is most prevalent in primates,[1] particularly human beings; some workers prefer to call it reasoning, and to distinguish it from learning as such. Essentially, insight is the ability to respond correctly the first time to a situation different from any previously encountered. Through insight, an animal is able to apply learning gained in particular situations to a new situation and, in effect, solve the new problem mentally without the necessity of overt trial and error (Fig. 35.14).

A highly specialized type of learning is ***imprinting,*** which occurs when an animal learns to make a strong, lasting association with another organism or object during a short, sensitive phase early in life. The concept of imprinting was first formulated in 1935 by Konrad Lorenz. He was studying birds such as geese, chickens, and partridges whose young are able to move around and feed themselves soon after hatching. He found that the young of such species will follow the first moving object they see, and form a strong and lasting attachment to it, particularly if the object emits a sound. In effect, they adopt this object as their parent. Ordinarily the first moving, vocal object such a young bird sees is its mother, and imprinting on her has obvious survival value. But under experimental conditions the young bird may be imprinted on a toy train or a box pulled around by a string (especially if the box contains a loudly ticking clock or some other sound-producing device), or even on a dog, cat, or human being (Fig. 35.15). Once the sensitive phase, usually only about 36 hours, has passed and the young birds have been imprinted on such surrogate mothers, they cannot be imprinted on any other object, including their true mother. Such ***parental imprinting*** is important in establishing a bond between the young and their mother under natural conditions.

Imprinting plays an important role in establishing proper species recognition and interaction, especially with regard to mate choices. The lasting effects of ***sexual imprinting*** were demonstrated in experiments by Klaus Immelmann, now of the University of Bielefeld in Germany, in which young zebra finch males were raised by foster parents of another species (Bengalese finch). Upon reaching sexual maturity, the male zebra finches invariably courted female Bengalese finches when given a choice between them and females of their own species (Fig. 35.16). Such imprinting lasted throughout the animal's life.

There are many other examples of innately guided learning. Female ducks, for instance, imprint on nest height on their second day of life, and subsequently build their own nests at the same level. Salmon imprint on the odor of their home stream at the time they begin their journey to the sea, and use

[1] Insight learning has been documented in some cetaceans and birds as well.

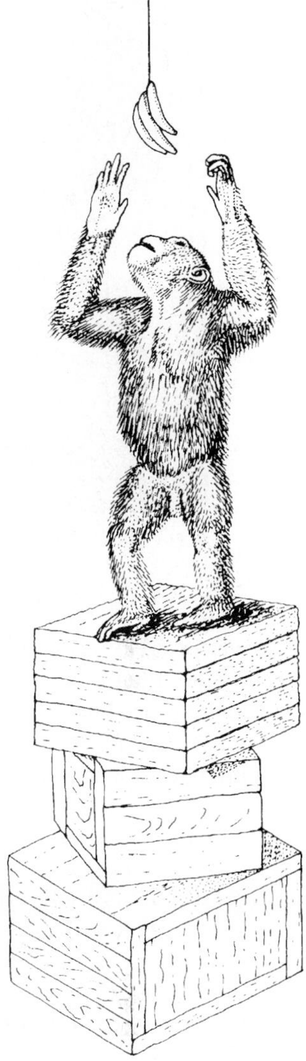

35.14 Insight learning Not all animals are capable of insight learning. At top, the raccoon is tied to a stake and cannot quite reach the food dish as long as its leash is looped around the stake. In this situation a human or a chimpanzee would immediately turn, walk around the stake, and go to the food, without any previous experience of such a situation. The raccoon will not perform the task correctly at first; it must find the solution by trial and error, though once it has done so it will learn very quickly. Bottom: Insight learning is more prevalent in primates. Here a hungry chimpanzee, released in a room with various boxes scattered around the floor and a bunch of bananas hanging from the ceiling above his reach, will often survey the situation for a short time and then begin gathering the boxes and piling them on top of each other under the bunch of bananas. He can then climb on top of the boxes and reach the bananas.

A

B

35.15 Imprinted goslings (A) Young goslings stay close to their own parent, whom they readily distinguish from the other geese. (B) Having been imprinted on Lorenz during their first day of life, these goslings follow him as if he were their parent.

35.16 A male zebra finch (left) courting a female Bengalese finch (right) The male was raised by Bengalese finch foster parents, and was imprinted on them.

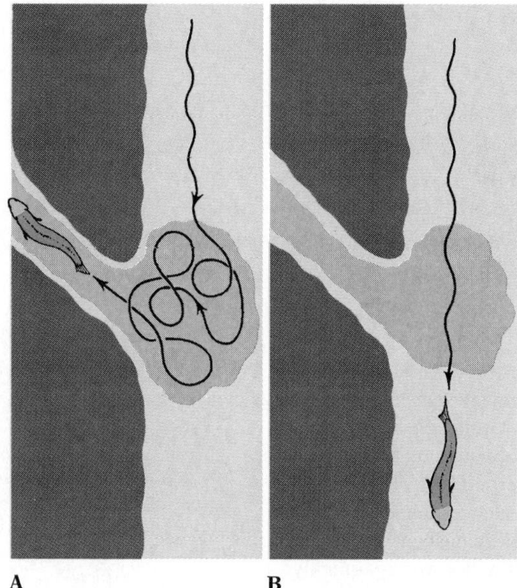

35.17 Tracks of salmon when they encounter morpholine In one series of experiments (not discussed in the text), Arthur D. Hasler and his associates at the University of Wisconsin imprinted young salmon in the laboratory on the odor of a chemical called morpholine. Later, they used ultrasonic tracking to follow the adult salmon as they swam southward along the shore of Lake Michigan in search of a spawning stream. Morpholine was released in the area indicated in gray. Fish previously imprinted on morpholine stopped their southward migration there, began to circle, and swam up the morpholine-scented stream (A). Those not previously imprinted on the chemical typically swam through the morpholine-scented area without pausing (B).

this memory years later on their way upstream from the ocean to find the tributory in which they were born (Fig. 35.17). Homing pigeons imprint on the location of their home loft as fledglings and will return to it even after years of life in a cage hundreds of kilometers away.

MOTIVATION

Different species behave differently—for instance, weaverbirds build nests quite different from those of robins or geese. In addition, the same animal may behave differently at different times—birds are not always building nests, flying south, or courting potential mates. The animal may be motivated to do one thing now and another later. The general health of the animal, its maturational state, hormonal level, activity of the central nervous system, sensory stimuli, and previous experience that led to learning are all involved in determining an animal's current behavior.

MAKING CONNECTIONS

1. Compare the importance of *kineses, taxes,* and *reflexes* (described in Chapter 28) in the total behavioral repertoire of invertebrates and vertebrates.

2. Herring gull chicks beg for food by pecking their parent's bill. The releaser for the pecking behavior is the long shape of the bill with a contrasting red spot near the tip. A bill-like model that is longer and thinner than the gull's actual bill acts as a supernormal stimulus. Why haven't herring gulls evolved bills that are as long and thin as a supernormal stimulus?

3. How are fixed-action patterns related to the motor programs discussed in Chapter 28? As defined in Chapter 28, are all motor programs innate?

4. Considering their longevity, their size, and the organization of their nervous system (discussed in Chapter 28), why would you predict that insects would exhibit less learning capacity than most vertebrates?

5. You are sitting in the back of a lecture hall, trying to get some sleep. At first you are disturbed by your instructor's voice, but eventually you are able to ignore it and sleep almost as comfortably as in your own bed. What kind of learning have you experienced in this lecture?

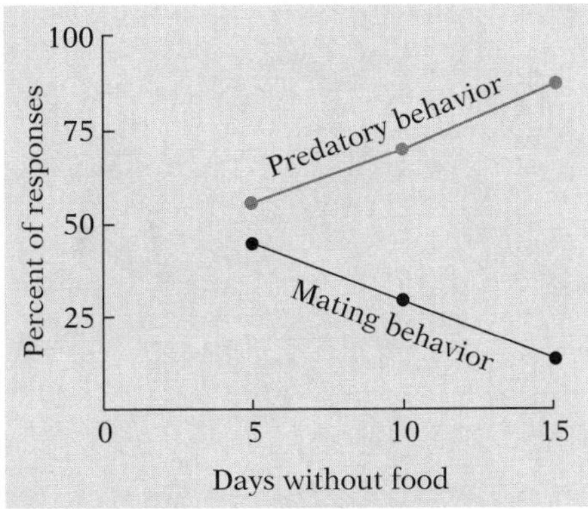

35.18 Change in response of a male jumping spider to a model as a function of the number of days he has gone without food Among the prey of male spiders are some insects that closely resemble female spiders. It is therefore possible to construct models that can release either prey capture or mating behavior, depending on the motivational state of the spider. The longer he has gone without food, the more likely he is to exhibit predatory behavior instead of mating behavior.

Behavioral Priorities Can Be Modulated

An animal must often choose among several behaviors, such as searching for food, searching for water, attempting to attract a mate, grooming itself, patrolling its territorial boundaries to guard against intruders, repairing its nest or burrow, or even playing. At any given moment, some of these behaviors will be more important than others. Thus, when an animal is being stalked by a predator, escape behavior takes precedence over all others. Current priorities can help the animal select among the many behavioral possibilities that may be competing for its attention, a process often referred to as "time-sharing." These priorities determine, for instance, whether a spider will feed or mate (Fig. 35.18). What behavior an animal chooses depends not only on its motivation, but on opportunity as well: an animal only mildly hungry but very thirsty will prefer water to food if both are present, but will probably eat if no water is available.

We now turn to a brief examination of some major classes of animal behavior, in particular communication, orientation, and social behavior.

ANIMAL COMMUNICATION

The ability to communicate is not restricted to members of species that live in societies, such as bees, ants, termites, and human beings; animals that live in less complex social groupings also communicate, and even the least social of animals must communicate with other individuals at certain critical times, such as the time of mating. In fact, communication probably evolved first for mating and reproduction, and from the mechanisms of sexual communication evolved the wide variety of signals that are now used to communicate other useful information (Table 35.1). The highly varied methods of communication that animals have evolved utilize particularly the senses of hearing, sight, and smell.

TABLE 35.1 *Advantages and disadvantages of the major modes of communication*

	Mode of communication		
	Chemical	*Sound*	*Visual*
Useful range	Large	Large	Medium
Useful at night, or in murky places	Yes	Yes	No
Rate of transmission	Slow	Fast	Fast
Interrupted by barrier?	No; moves around	No; moves around	Yes
Locatability of sender	Difficult	Easy	Easy
Energetic cost of transmission	Low	High	Low or moderate

CHEMICAL COMMUNICATION IS THE MOST PRIMITIVE FORM

Since most communication between human beings involves auditory or visual signals, we tend to think of these as inherently the most appropriate methods of communication in animals. But we should not let the limitations of our own senses prevent us from recognizing that the olfactory sense is immensely important in the lives of many animals, constituting a basis for effective communication.

We have already mentioned that many animals secrete substances called ***pheromones,*** which influence the behavior or physiology of other members of the same species. Most pheromones can be classified into one of two groups: those that act as ***releasers,*** triggering a more or less immediate and reversible behavioral change in the recipient, and those that act as ***primers,*** initiating profound physiological changes in the recipient but not necessarily triggering any immediate behavioral reaction.

Releaser Pheromones Trigger an Immediate Behavioral Response

Releaser pheromones, which must have arisen long before auditory or visual communication, probably represent one of the most primitive forms of biological communication. Many species of unicellular organisms depend on chemoreceptors to recognize when they have bumped into another individual of their species, while others, like certain slime molds that aggregate periodically to reproduce, locate each other through pheromone (odor) trails. Only slightly more elaborate is the mating system of most moths: females release a species-specific pheromone, and males follow the odor upwind to its source. Pheromones also play an important role in groups as diverse as beetles, aquatic invertebrates, and mammals. In each case the role is basically the same: the odor informs potential mates of the species, sex, reproductive readiness, and location of an appropriate mate.

The trail substances of ants constitute another class of releaser pheromones (Fig. 35.19). A foraging ant returning to the nest from a food source intermittently touches the tip of its abdomen to the ground and secretes a tiny amount of trail substance. Other worker ants can follow the trail to the food source; these ants, too, will lay trail substance as they return with food to the nest. The better the source of the food, the more ants are attracted to the trail, and the trail substance grows stronger. Workers that do not find food do not lay trail; hence when the food has been consumed no more trail is laid, and since the trail pheromone evaporates quickly, the trail disappears within a few minutes.

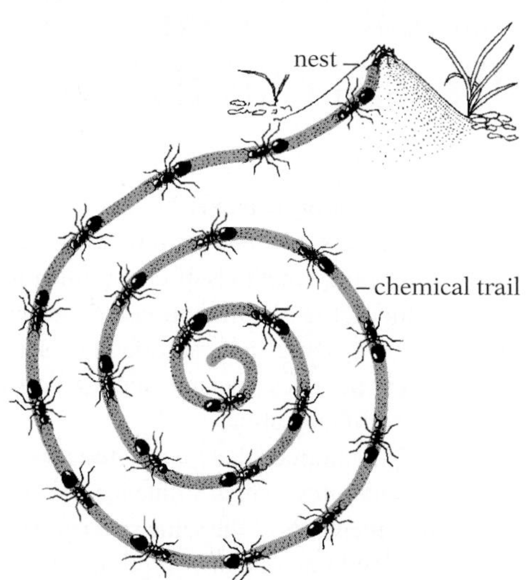

35.19 Ants following a spiral path of trail substance laid down by an experimenter The substance evaporates quickly so the trail will last only a few minutes. Trail substances are species-specific; no two species secrete the same substance. This specificity is adaptive, because it ensures that workers will not mistakenly follow trails of other ant species that may cross their own.

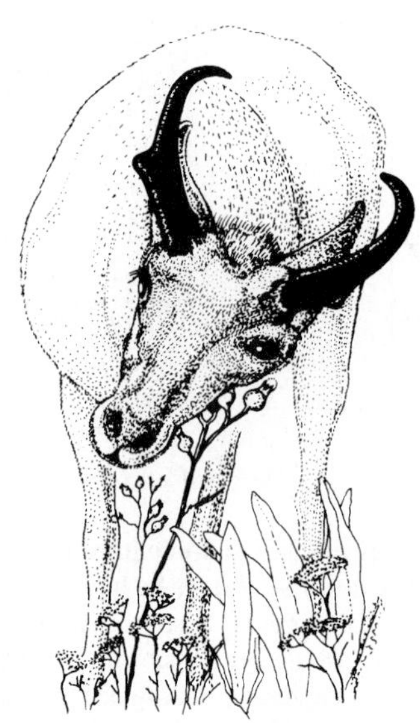

35.20 A pronghorn buck scent-marking a plant with his facial glands.

Other releaser pheromones of ants include alarm substances and death substances. The latter provide a particularly good example of the extent to which much insect behavior is stimulus-bound and rigidly stereotyped. Certain chemicals released from a decomposing ant act as pheromones that stimulate worker ants to pick up the carcass and carry it to a refuse pile outside the nest. When living ants are experimentally painted with these substances, workers pick them up and dump them on the refuse pile. The hapless victims of the experimenter promptly return to the nest, only to be thrown back on the refuse pile, again and again. The workers can surely see and feel that the object they are carrying is struggling in a most undead manner, but they disregard the evidence from all their other senses, respond only to the death pheromone, and continue to treat the painted ant as a carcass to be disposed of.

Releaser pheromones also occur in many other animal groups, notably mammals, most of which rely on smell much more than humans do. Male mammals can often tell by smell when the female is in heat because she secretes certain pheromones at that time. Besides playing a part in sexual recognition and reproduction, mammalian pheromones are often used to mark territories and home ranges, in the way that urine is used by dogs to mark familiar trees or fire hydrants. Many other mammals (e.g., hamsters, deer) use pheromones secreted by special scent glands for this purpose (Fig. 35.20).

Primer Pheromones Initiate Long-Term Physiological Changes

Primer pheromones produce relatively long-term alterations in the physiological condition of the recipient and thus change the effects that later stimuli will have on the recipient's behavior; they do not necessarily produce any immediate behavioral change.

Various experiments with lab mice have indicated that the odor of males may initiate and synchronize the estrous cycles of females, and that the odor of an unfamiliar male may block the pregnancy of a newly impregnated female mouse or block the estrous cycles of other females. That the removal of the olfactory bulbs from the brains of these mice restores the cycles to normal indicates that pheromones are involved.

Another type of primer pheromone is seen in social insects such as ants, bees, and termites. These pheromones, which are ingested rather than simply smelled, play an important role in caste determination. An example is queen substance, a pheromone secreted by a honeybee queen that prevents the workers from developing reproductive capabilities and also from rearing new queens. When the queen dies, or leaves with some of the workers to start a new colony, the concentration of pheromone circulating in the hive declines, and the remaining workers develop reproductive capabilities and also begin building large queen cells in which they will rear new queens.

35.21 Photomicrograph of male mosquito antennae Male mosquitoes have elaborate antennae that help them locate females. When a female mosquito of the same species flies near, her wing-beat frequency causes the antennae of males of the same species to resonate.

Sound Is Widely Used for Species Recognition and Mating

Being vocal animals, we are very familiar with the use of sound as a medium of communication. No other species has a sound language that even approaches the complexity and refinement of human spoken languages. But many other species can communicate an amazing amount of information through sound. Sound is a particularly useful form of communication in environments where vision is limited, as in turbid water, dense vegetation, or at night.

We have already mentioned that male *Aedes* mosquitoes are attracted by the buzzing sound produced by the female's wings during flight (or by sounds of a similar pitch emitted by devices such as tuning forks). But how does the male recognize and respond to the sound of the female's wings? The head of a male mosquito bears two antennae, each covered with long hairs. When sound waves of certain frequencies strike the antennae, these are caused to vibrate in unison (Fig. 35.21). The vibrations stimulate sensory cells at the base of each antenna. The male responds to such stimulation by homing in on the source of the sound, locating the female, and copulating with her. Furthermore, his built-in receptor system is species-specific; it is stimulated by sounds of the frequency characteristic for females of his own species, not by the frequencies characteristic of other mosquito species. Hence the sound produced by the female's wings functions both as mating call and species recognition signal.

Many other insects use sound in a similar way. For example, male crickets use calls produced by rasping together specialized parts of their wings. These calls function in species recognition, in attracting females and stimulating their reproductive behavior, and in warning away other males. So species-specific are the calls of crickets (Fig. 35.22) that in several cases we can distinguish different species best by listening to their calls; to us the species may be almost indistinguishable on an anatomical basis.

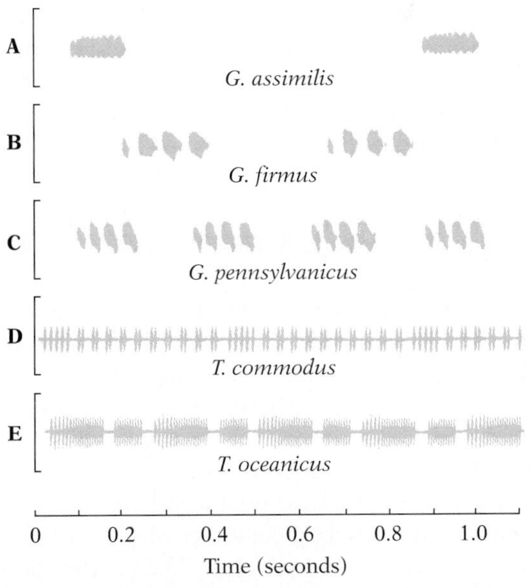

35.22 Cricket calling Male crickets produce a pulsed calling song that attracts females of their species. The pattern of trills for each species is clearly unique.

Bird Song Facilitates Species Recognition, Mating, and Territorial Defense

Of all the familiar animal sounds—the buzzing of mosquitoes, the calling of crickets and frogs, the barking, roaring, purring, and grunting of various mammals—perhaps none, with the exception of human speech, has received so much attention as the singing of birds. Bird song has a variety of functions. It acts as both a species recognition signal and as an individual recognition signal, since there are usually small individual differences in the song, which other individuals can learn to recognize. It also functions as a display that attracts females to the male and contributes to the synchronization of their reproductive behavior (increasing sexual motivation and decreasing attack and escape motivations). Finally, the song acts as a display important in defense of the territory. In its defensive function a bird's singing is certainly no indication of happiness or joy; if it were possible to properly apply such human-oriented concepts to birds, the singing would more accurately be taken as an indication of combativeness.

The role of singing in the establishment and defense of territories is an especially important one. A ***territory*** may be defined as an area defended by one member of a species against intrusion by other members of the same species (and occasionally against members of other species). A male bird chooses an unoccupied area and begins to sing vigorously within it, thus warning away other males. The boundaries between the territories of two males are regularly patrolled, and the two may sing loudly at each other across the border. Though during early spring there is often much shifting of boundaries as more and more males arrive and begin competing for territories, later in the season the boundaries usually become fairly well established and each male knows where they are. During the period when the boundaries are being established, the males that can sing most loudly and vigorously often successfully retain large territories or even expand their territories at the expense of other males that sing less impressively.

35.23 Highly motivated threat display of a wolf The animal makes himself look bigger (by standing high on his legs and raising his hair and ears) and stares directly at his opponent.

35.24 A toad directing a threat display at a snake Even this small animal, encountering a deadly predator, makes himself look as big as possible as he confronts his opponent.

Agonistic Behavior Reduces Physical Combat

Agonistic encounters between individuals of the same species may often be resolved by vocal and/or visual displays without any physical combat. ***Agonistic behavior*** embraces all aspects of behavior exhibited during hostile encounters between members of the same species; this includes threat, attack, appeasement, and fleeing. Two dogs or cats, for instance, may give ***threat displays,*** with hackles raised, teeth bared, ears laid back, body raised as high off the ground as possible, and movements stiff-legged and exaggerated (Fig. 35.23). The individual showing the higher attack motivation usually wins. In dogs a high attack motivation is often conveyed by directing the face straight at the antagonist and spreading and raising the body, making it look as large as possible (Figs. 35.24 and 35.25). The

35.25 Threat display of an Australian crayfish Note the similarities of this display with those of the wolf (Fig. 35.23). Again, the animal adopts a pose that makes it look as large as possible and faces the object of its threat, exposing its weapons.

35.26 A dog making a full appeasement display to another dog The loser in an agonistic encounter exposes his vulnerable underparts to the victor, who is thus inhibited from further attack.

loser of such an encounter responds with ***appeasement displays***—fur sleeked, tail tucked under, head down and often turned away from the antagonist, and legs bent (Fig. 35.26). Appeasement displays usually involve making the body appear as small as possible and turning the face away from the antagonist or exposing to the antagonist the appeaser's most vulnerable spot; such appeasement displays tend to inhibit further attack by the antagonist.

Analogous agonistic displays can be observed in many animals. It is rather rare for individuals of the same species to engage in combat serious enough to cause significant damage. The adaptive importance of a nonviolent settlement of differences is obvious. Agonistic encounters occur frequently, and if they often led to serious physical damage, even many of the winners might, in the long run, be losers. It is not surprising, therefore, that most animals have evolved other methods of resolving conflicts, methods usually involving displays by which the combatants convey to each other the intensity of their attack motivation. The individual showing the higher attack motivation is ordinarily the winner.

The interactions between wild stallions provide an instructive example of the adaptiveness of agonistic behavior. Daniel Rubenstein and Mace Hack of Princeton University studied the encounters of wild stallions living on an island off the coast of North Carolina. They found that during breeding season a stallion defends his group of mares from other stallions. The encounter between stallions involves a ritualized contest, and only rarely is there actual fighting (Fig. 35.27). In the contest, competing stallions approach each other, sniff each other's genitals, feces, and faces, and make squealing noises. The researchers found that the smell of dung identifies individuals, and the squeals signify the male's status. Both types of information are thought to convey information about fighting ability, and the probable winner. The stallion with the loudest and longest squeal is usually the best fighter and the dominant stallion, perhaps because the longer squeal reflects a greater lung capacity. This redundancy of signaling, using both olfactory and auditory cues, conveys information in an unambiguous manner.

Visual Displays Are Important in Reproductive Behavior

A ***display*** may be defined as a behavior that has evolved specifically as a signal. According to this definition, a song or a call is a display. Many animals have also evolved a variety of often complex actions that function as signals when seen by other individuals; such displays frequently include vocal elements. You may have seen male pigeons strutting, with tail spread and dragging on the ground, neck fluffed, and wings lowered, or courting songbirds in spring going through odd and seemingly senseless antics. If you watch a flock of ducks on a pond in

35.27 Agonistic encounters among stallions Wild stallions competing with one another for mares go through a ritualized contest in which the males approach one another and sniff each other's faces, genitals, and feces. During the encounter squeals are made, the loudest and longest squeals by the dominant males. The squealing and smelling convey information about the willingness and ability to fight. About half the time the contest ends quickly with one male running away. Very rarely is there overt fighting with stallions rearing on hind legs, biting, and kicking. Most of the contests are settled on the basis of the squeals, with the winner being the stallion with the loudest and longest squeal.

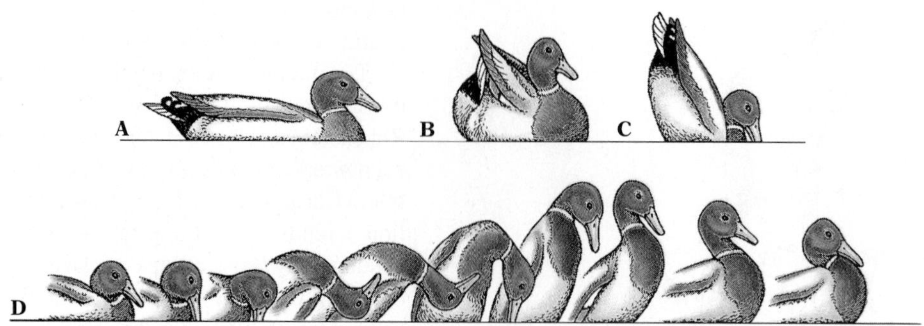

35.28 Courtship displays of a male mallard (A) Normal swimming posture. (B) The "Head-Up-Tail-Up" display. (C) Part of the "Down-Up" display. (D) The "Grunt-Whistle" display.

early spring, you may see a male give a loud whistle and raise both head and tail as high in the air as he can while also raising his wings (Fig. 35.28B), he may raise his stern in the air, dip his head in the water, and then abruptly raise it and whistle (Fig. 35.28C), or he may put his bill in the water and then quickly flick his head to the side, toss an arc of droplets into the air while arching his body upward, and follow this acrobatic feat with a whistle and a grunt (Fig. 35.28D).

Such carryings-on may seem senseless to a casual observer, but biologically they are far from senseless. They function in synchronizing the sexual physiology of the male and female and in making the female more receptive to the male. They also assure that the female will choose a male of her own species as her mate, since the males of other species in the same area give somewhat different displays.

COMMUNICATION IN HONEYBEES

Honeybee Dances Convey Information About Food

Scout honeybees have the amazing ability to inform the workers in the hive of the quality of a food source, as well as its direction and distance from the hive. This communication depends on displays that include auditory, visual, chemical, and tactile elements.

For many years, Karl von Frisch of the University of Munich, Germany, was interested in the ability of bees to distinguish between different colors and scents. In the course of his now classic experiments,[2] he would set up in the vicinity of a hive a table with sheets of paper on which he had smeared honey; he would then wait—sometimes for hours— for the bees to find the food. He noticed that when one bee finally discovered the feeding place, many others appeared at the table within a short time. It seemed likely that the first bee had somehow informed the others of the existence of the new feeding place.

In order to see what happened in the hive when a scout bee returns from a new food supply, von Frisch set up a special observation hive with glass sides. When a bee landed at the new feeding place and began to feed, von Frisch daubed a spot of paint on her thorax so that he could recognize her when she returned to the hive. He discovered that the returning bee first feeds several other bees and then performs what he called the ***round dance*** on the vertically oriented surface of the honeycomb. The dance consists of circling first to the left, then to the right, and repeating this pattern over and over with great vigor (Fig. 35.29A). Other bees in the vicinity of the dancer are excited by her movements, and they begin to follow, with their antennae held close to her. Suddenly, however, they would turn away one by one and leave the hive; a short time later they appeared at the feeding place. Apparently the round dance is a display that informs the other bees of the existence of the food supply. There was no evidence that the round dance indicated direction, and other similar experiments brought no evidence that it indicated distance. It seemed simply to say, "Fly out and seek in the neighborhood of the hive."

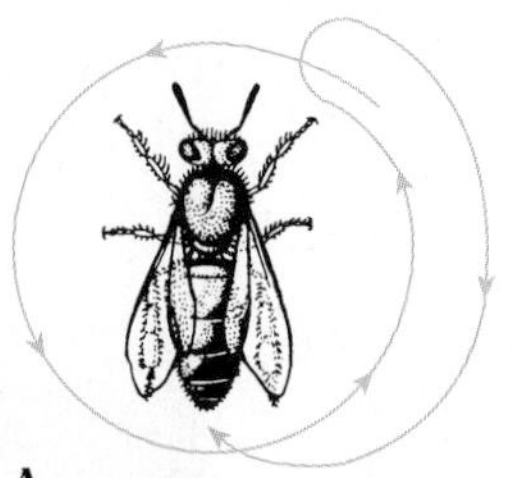

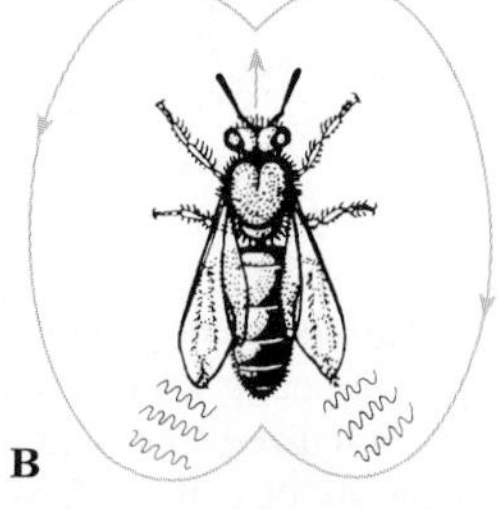

35.29 Round dance and waggle dance of scout honeybee (A) In the round dance the bee circles first one way and then the other, over and over again. The dance tells other bees that there is a source of nectar close by, within 100 meters of the hive. (B) In the waggle dance the scout runs forward in a straight line while waggling her abdomen, circles, runs forward again, circles in the other direction, and runs forward again. The orientation of the run indicates the direction from the hive of the nectar source, and the number of turns per unit time indicates the distance.

[2]In 1973 Karl von Frisch, Konrad Lorenz, and Niko Tinbergen, whose investigations have also been mentioned in this chapter, were jointly awarded the Nobel Prize, in recognition of their pioneering studies of animal behavior.

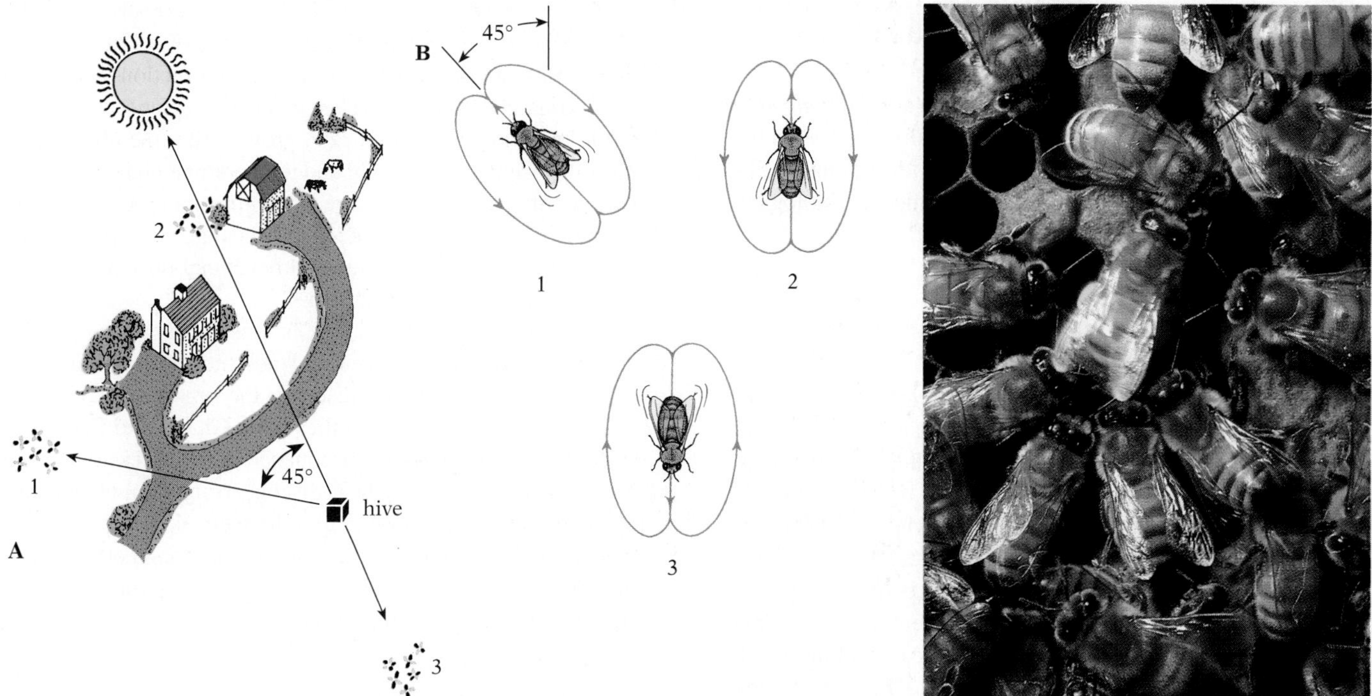

35.30 An example of bee language (A) Three different sources of nectar were located (1) 45 degrees to the left of the sun as seen from the hive; (2) straight toward the sun; and (3) straight away from the sun. (B) The dances performed on the vertical comb in the darkened hive by forager bees from these nectar sources were oriented 45 degrees to the left of vertical (1), straight up (2), and straight down (3). In short, the vertical direction on the comb symbolizes the sun. The dancing bee in the photograph, her legs bearing pollen, is closely surrounded by attenders.

Even though von Frisch had satisfied himself that the round dance indicates neither direction nor distance, he began to suspect that at times bees are able to communicate these kinds of information, presumably in some other way. In 1944 he performed the following experiment. He divided the foragers into two groups, marking the members of each with a different color of paint, and trained each group to visit a food source at a different location—one at 10 meters from the hive and the other at 300 meters. Von Frisch was then able to decode the dances by observing how the differences between the two groups' dance movements related to differences in the messages being conveyed.

Von Frisch observed that while the scouts from the dish 10 meters from the hive danced the familiar round dance, the scouts from the dish 300 meters away danced a different dance, one that he described as the ***waggle dance*** (Fig. 35.29B). In this dance the bee runs a short distance in a straight line on the vertical honeycomb while rapidly waggling her abdomen from side to side, then circles, runs forward again, circles in the other direction, and runs forward again. She repeats this dance many times. Distance can be determined from the duration of the waggle run or the number of waggles in each run. Each waggle specifies a particular increment of distance—about 40 meters in the case of von Frisch's honeybees.

Later, von Frisch found that bees can also communicate the direction of a food source using the waggle dance. The location of the food relative to the position of the sun as seen from the hive entrance is indicated by the direction of the straight portion of the waggle dance. In the dark hive a run straight up the vertical comb means that the food lies in the direction of the sun; a run straight down the vertical comb means that the food lies in the opposite direction from the sun; a run at an angle indicates that the food is to be found at that angle from the sun (e.g., a run 30 degrees to the right of straight up indicates that the feeding place is 30 degrees to the right of the sun). The honeybee dance is called a language because it refers to objects distant in both space and time (that is, the animal is not simply pointing and grunting) and because it is symbolic ("up," for instance, is a symbol in a dark hive for the sun's direction from the hive) (Fig. 35.30).

As far as we know, the honeybee dance language is unique among invertebrates in its ability to convey complex information. Though used to relay information about water, nectar, pollen, and new hive sites, it is basically a closed system under instinctive control: bees can perform or understand dances with no previous experience, but can use them only to specify the distance, the direction, and (in ways we have not discussed) the desirability of a location. In other highly social animals, however, communication may be more flexible, involving other types of neural mechanisms.

BIOLOGICAL CLOCKS

How is it that a honeybee can communicate accurate information about a food source even though the sun's position in the sky changes throughout the day? To adjust for the sun's position, it must have an accurate internal timer, a "biological clock."

All Living Organisms Have an Internal Timer

There are numerous examples of timekeeping by living organisms. The leaves and flowers of many plants show regular movements in a cycle of approximately 24 hours, even if kept under constant conditions; adult insects of many species emerge from the pupa at a particular time, whatever the age of the pupa or the conditions it has experienced; and many animals show activity rhythms that vary with a period of approximately 24 hours, even if the animals are kept under constant conditions and have no known external indication of the actual daily environmental cycle. Rhythms of this sort, with a period of approximately 24 hours, are called ***circadian rhythms*** (from the Latin *circa-*, "about," and *dies*, "day").

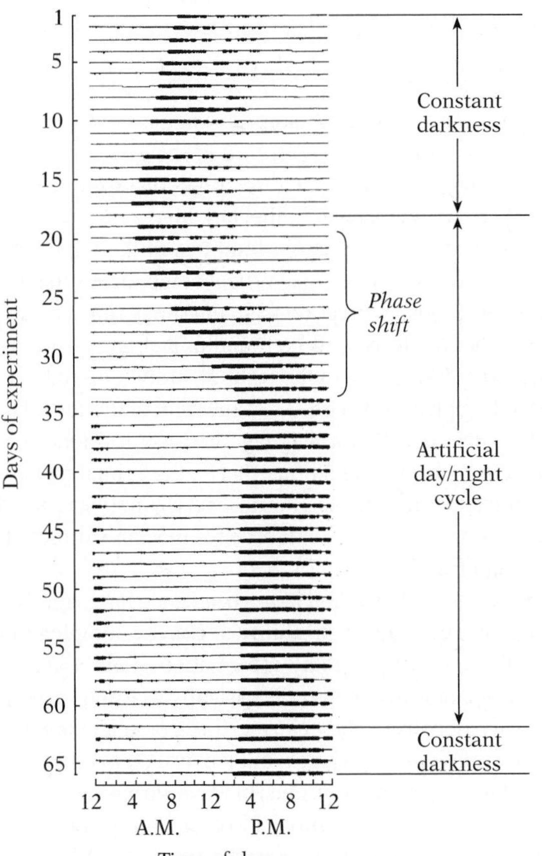

35.31 Record of a circadian-rhythm experiment In constant darkness (days 1–17) a flying squirrel (*Glaucomys*) continued to spend about 10 hours in foraging activity (dark bars) and then 14 hours resting (light bars). The time of foraging drifted slowly with respect to a 24-hour day. When an artificial day/night cycle was imposed (days 18–61), the activity rhythm gradually shifted into phase with it. When constant darkness was resumed (day 62), the period of foraging again began to drift. Circadian rhythms in mammals are controlled by a region of the hypothalamus. But under constant experimental conditions, the activity rhythms drift with respect to the 24-hour day. The rhythms of organisms can be reset by a species-specific hierarchy of cues, the most important of which is usually light.

Evidence indicates that clock phenomena are characteristic of all living things, whether individual cells or whole multicellular plants or animals. Aspects of cellular function that vary in approximately 24-hour cycles include enzyme activity, osmotic pressure, respiration rate, growth rate, membrane permeability, sensitivity to light and temperature, and reactions to various drugs. Physicians are becoming increasingly aware that the proper dosage of a drug may change with time of day; in some cases, what constitutes a beneficial dose at one time may actually be lethal at another.

Though the biological clock is innate, it is strongly influenced by environmental factors. Under normal conditions the clock is constantly being reset by the environmental cycle (Fig. 35.31). A person who flies from Chicago to Paris may suffer from "jet lag" as a consequence of the necessity to reset his internal clock. At first he is disconcertingly out of synchronism with the people around him; he feels like sleeping when they are wakeful, hungry when their minds are on other activities. After three or four days, however, his internal clock has shifted to Paris time, and his problems are over.

ORIENTATION AND NAVIGATION

Intriguing to scientist and nonscientist alike is the rather widespread ability of animals to find their way from place to place. Animals as diverse as butterflies, sea turtles, and hummingbirds migrate thousands of kilometers to places that they may have never before visited. Some types of orientation are now fairly well understood, while others remain unexplained and present a continuing challenge to biologists. We shall focus our attention on one type of orientation behavior: migration and homing in birds.

A subject of wonder for centuries has been the way many birds can travel thousands of kilometers from the place of their hatching in the high latitudes to the wintering grounds of their species in the lower latitudes, and back again the following spring. These migrations are not the result of random wandering. The members of each species usually follow a precise route characteristic of that species, often flying hundreds of kilometers each day or night; a few species make their entire migratory journey of several thousand kilometers nonstop. For example, the Pacific golden plover, a bird that cannot land on water, flies each fall from Alaska to its winter home in the Hawaiian Islands (Fig. 35.32). Even some small songbirds, such as warblers, depart in autumn from Nova Scotia or Cape Cod and fly nonstop over the Atlantic Ocean to South America.

The navigational abilities of birds are not restricted to the migratory seasons. Homing pigeons are renowned for their ability to return to their home lofts when released at distant points; they are frequently raced over distances from 150 to 1,000 kilometers, often averaging speeds of 80 or 90 kilometers (50 to 60 miles) an hour and sometimes making a 1,000-kilometer flight from an unfamiliar release point to their home lofts in a single day. Other birds can perform similar feats; a Manx shearwater captured on the west coast of England and flown by plane to Boston, where it was released, was back in its nest in England 12 days later, having flown 5,000 kilometers across open ocean.

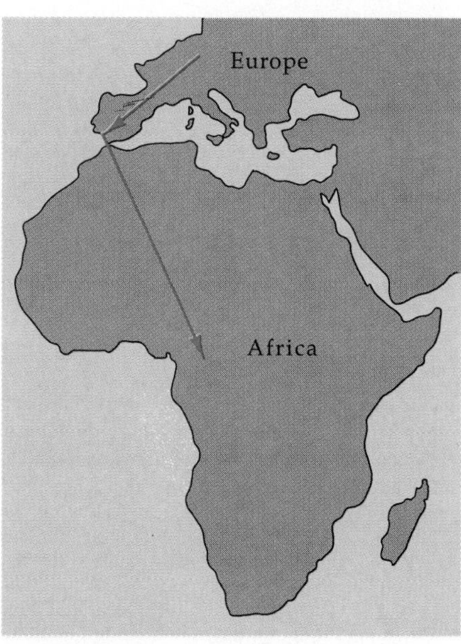

35.33 Warbler migration Some European garden warblers reach their winter grounds after a two-leg journey. They know at birth the two flight bearings they need, and how long to fly in each direction.

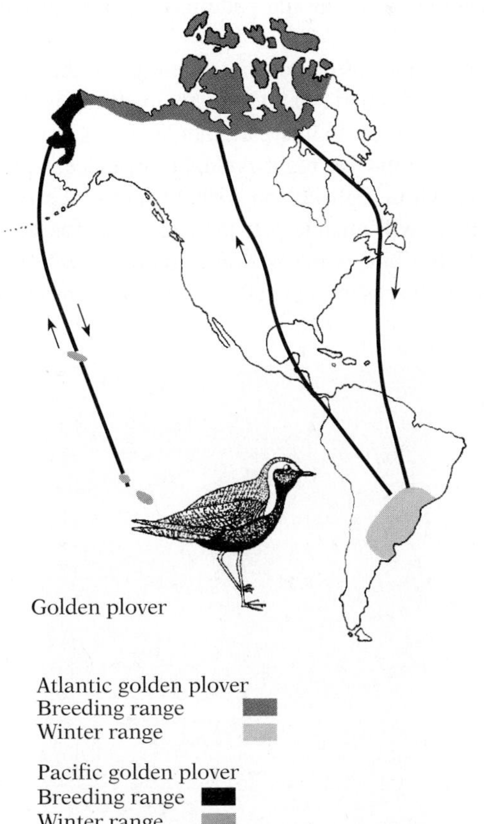

35.32 Distribution and migration of the golden plover The Pacific golden plover flies over open ocean from its breeding grounds in Alaska to its winter range in Hawaii, the Marquesas Islands, and the Low Archipelago. The Atlantic golden plover follows one route south in autumn to its winter range in South America and another route north in spring.

Some Birds Employ a Compass and Timer Strategy

It is clear from the golden plover and other birds that fly over open ocean that they are not relying on landmarks to navigate. Instead, two separate navigational strategies are evident. In one, the birds are preprogrammed to fly a certain course. In the fall European warblers from northern Germany will fly (or, in orientation cages, attempt to fly) southwest for 40 days, and then southeast for 20 to 30 days—following a course that, in the wild, carries them down through Spain, across Gibraltar, and into their winter ranges in Africa (Fig. 35.33). The animals seem to know at birth the direction they need to take, and how long to fly in each direction—in other words, they have a built-in compass and timer.

Many Birds Use Both a Compass and Map Sense

The compass-and-timer strategy is good enough when the target is a continent, but will hardly serve when the goal is small (such as a forest or specific nest site). Many migrating animals need to know precisely where they are even when in an area for the first time. This need is filled by a mysterious but very real ability known as ***map sense,*** which represents something quite different from a mental map of a familiar area. An animal with a map sense behaves as though always aware of longitude and latitude. The nature of this map sense is one of the most intriguing mysteries in modern biology.

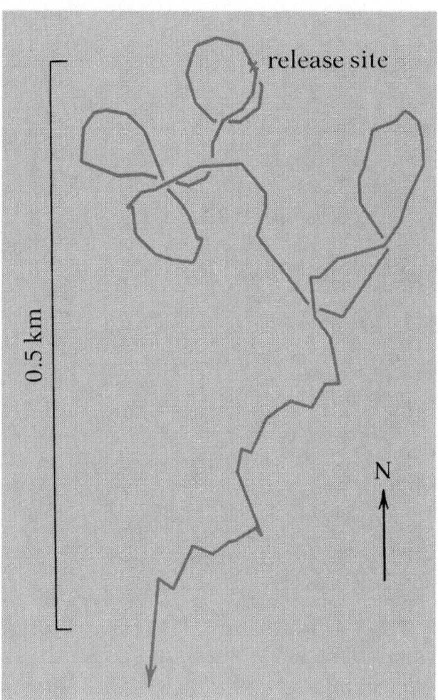

35.34 Pigeon homing Pigeons usually begin their journey home by circling the release site, but quickly set off along an irregular course for home. The actual routes are rarely straight and direct, and indicate that new map measurements must be taken from time to time to make midcourse corrections.

Homing pigeons have been widely studied for their navigational ability. A good "homer" can be taken from its loft and transported hundreds or even thousands of kilometers in total darkness, and when released it will circle briefly and then fly off in roughly the direction of home (Fig. 35.34). Homing pigeons appear to have both a compass sense and a map sense. To understand why they need both, imagine yourself kidnapped, taken 100 kilometers away in a windowless vehicle, and released. If you had a compass you would know which direction was north, but unless you knew in which direction you had been taken, that information would be useless. On the other hand, if you had a map sense, you would know that you were south of home, but without a compass you would be at a loss to put that information to use.

The Sun or Stars Act as Compasses for Many Birds

The workings of the compass sense in pigeons and migratory birds are now fairly well understood. Like many insects, pigeons and other daytime birds use the position of the sun as their standard cue. Of course, the sun's position depends on the time of day, and birds appear to have an internal time sense enabling them to allow for the westward movement of the sun from morning to night. Similarly, nocturnal migrants use a learned picture of the stars, and the pattern of star rotation, to set their course. The roles of the sun and stars in avian navigation are demonstrated by the behavior of caged birds under artificial skies. During migration season, the animals display an intense desire to escape in the direction that wild members of the same species are flying. A sudden shift of the artificial sun or pattern of stars results in a compensatory change in the direction in which the caged birds are struggling to go.

Experiments by Steven T. Emlen at Cornell University have shown that nocturnal migrants such as indigo buntings memorize the constellations while they are still nestlings, using the North Star, around which all other stars in the night sky appear to rotate, as their point of reference. As a result, they are able to infer north from even a small patch of sky (Fig. 35.35). Furthermore, Emlen has found that there is a short sensitive phase in which the young buntings must learn to read the star compass. Buntings that have not seen the starry sky during the few weeks before the start of their first migratory season in September never learn to use the star compass, no matter how often they see the stars thereafter. Such birds can be raised under an artificial sky, with an arbitrary pattern of stars rotating about a pole at any chosen compass point. When the time to migrate arrives they then attempt to set off in the appropriate direction relative to the star patterns they observed during the sensitive phase.

Homing pigeons demonstrate their use of the sun compass in an equally dramatic way. Correctly interpreting the sun's direction depends on an internal timer, which is sensitive to manipulations of the day/night cycle. For instance, a pigeon kept in a room whose lights go on six hours early, at midnight, and off at noon, will misinterpret the sun's position accordingly. When such a bird is released at true noon, its internal clock reads 6 P.M. It sees the sun in the south, but because it has been

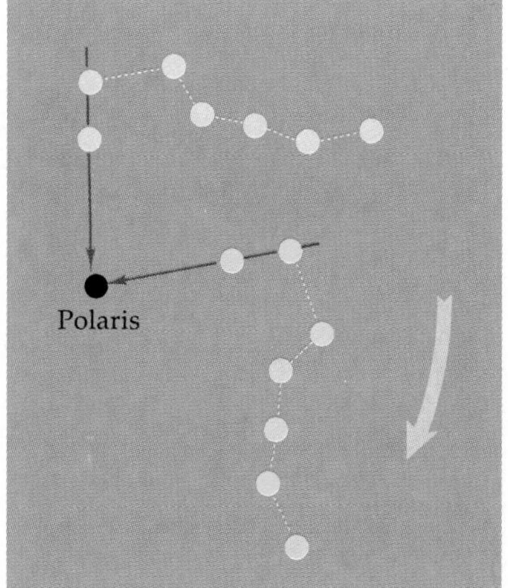

35.35 An example of how north can be located by star patterns Red arrows through the two end stars in the cup of the Big Dipper (Ursa Major) point toward Polaris, the North Star. Though the position of the constellation changes during the night, the same stars always determine an arrow pointing toward Polaris; hence directions can be determined without need of time compensation. Many different star patterns could be used for finding direction in this way.

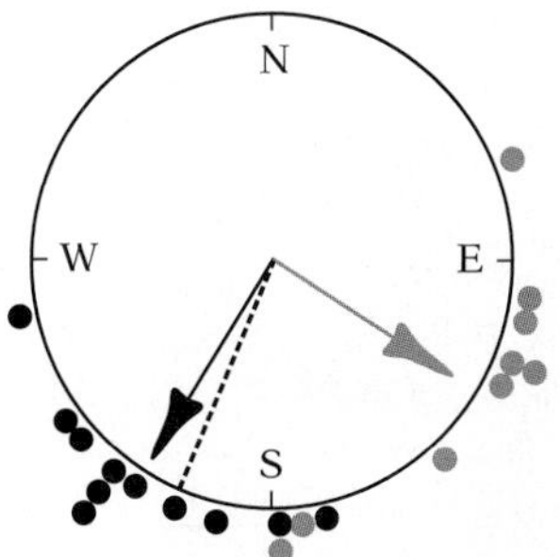

35.36 Effect of a six-hour-fast clock shift on the initial bearings chosen by homing pigeons on a sunny day Each dot indicates the bearing chosen by one bird; black dots represent control birds, brown dots experimental birds. The dashed line marks the proper homeward direction. The arrows show the mean bearing (average direction) for each group. The length of each arrow is proportional to the degree of clustering of the dots: if all the birds flew off in the same direction, the resulting arrow would touch the circle; if the departure directions were widely scattered, the arrow would be very short. In this experiment the mean bearing of the clock-shifted birds (brown arrow) is about 90 degrees to the left of the mean bearing of the control birds (black arrow). The experimental birds have been clock-shifted a quarter of a day, and they have made an error of a quarter of a circle in reading the sun compass.

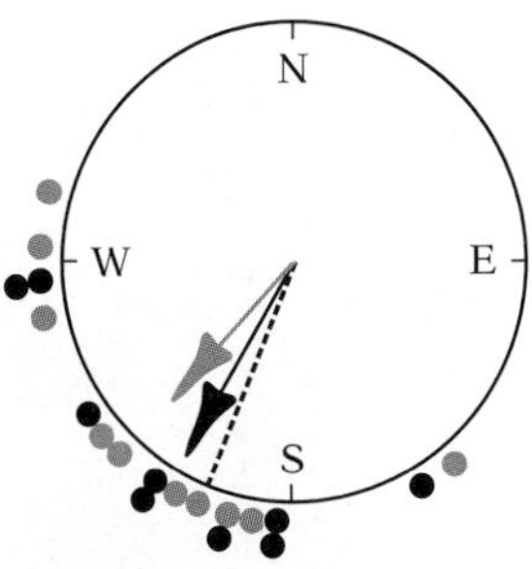

35.37 Effect of a six-hour-fast clock shift on the initial bearings chosen by homing pigeons on a totally overcast day When the sun is not visible, the experimental birds choose bearings (brown dots) not significantly different from those of control birds (black dots); the 90-degree deflection of their bearings seen on sunny days (compare Fig. 35.36) is not evident. In the absence of the sun compass, the birds appear to orient by some other system, which does not require time compensation.

clock-shifted, interprets the sun's position as indicating west. Therefore, if its home is to the south, it will fly 90 degrees to the *left* of the sun; attempting to fly south it heads east (Fig. 35.36).

But pigeons can also home under an overcast sky. If they use the sun as their compass, what guides them when it is invisible? William T. Keeton of Cornell University attacked this question by releasing both normal and clock-shifted birds on cloudy days (Fig. 35.37). The results are clear and dramatic: pigeons are able to home on overcast days, using cues that are not time-dependent, for the departure bearings of clock-shifted birds are not rotated. Obviously pigeons have a backup system, and Keeton guessed that the second compass might be magnetic.

The clearest demonstration that pigeons use a magnetic compass was performed by Charles Walcott, also of Cornell University. He fitted birds with tiny, head-mounted coils of wire. By passing a current from a battery through the coils, he was able to reverse the magnetic field. On cloudy days these birds flew away from home, while pigeons whose batteries were not connected to the coils homed normally (Fig. 35.38). On sunny days there was no effect, because the pigeons used the sun's position as their primary navigational cue.

Though research on the compass senses of animals has progressed steadily, understanding of the map sense is still limited. What other environmental cues might be involved? Recent evidence suggests that olfactory cues may be important, and several investigators suggested that meteorological

A

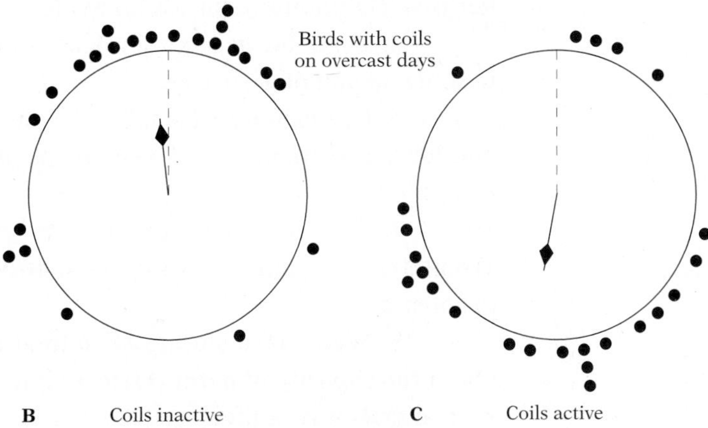

35.38 Effect of magnetic coils on pigeon homing On overcast days, birds wearing active coils (A) that reversed the direction of the earth's field in their heads flew away from home (C), while those with inactive coils behaved normally (B). On sunny days there was no effect.

35.39 Hatchling sea turtles leaving their nests in the sand

phenomena (atmospheric pressure patterns, wind direction, wind turbulence, etc.) might provide directional information. Melvin Kreithen at the University of Pittsburgh and Keeton have found that pigeons have a wide range of sensory capabilities (barometric-pressure detection, polarized-light detection, ultraviolet-light detection, infrasound detection) that might help them find their way. But despite all the progress in this field of research during the last three decades, a full solution to the mystery continues to elude scientists.

A wide variety of other animals have demonstrated the ability to orient. These range from Pacific sea slugs and spiny lobsters to salamanders and sea turtles. The sea turtle story provides a good illustration of how animals can rely on several cues to hold courses. Hatchling sea turtles dig out of untended nests (Fig. 35.39), locate the ocean within minutes, and begin an offshore migration that takes them away from predator-rich nearshore waters. These offshore migrations can ultimately lead turtles hundreds or thousands of kilometers from their nesting beaches. Once offshore they reside in oceanic current systems that serve as nursery areas. Work by Michael Salmon, Jeanette Wyneken, and Kenneth Lohmann has shown that initially the hatchlings rely on visual cues to locate the ocean, but once in the water, they abandon visual guide posts and use waves as the cues that lead them offshore. After swimming offshore for several hours sea turtle hatchlings abandon wave cues and use a magnetic compass to maintain their offshore orientation. It is likely, but not yet certain, that juvenile sea turtles rely primarily on their magnetic sense to complete a migration back to coast waters and, eventually, to the site where they were hatched.

MAKING CONNECTIONS

1. What evidence suggests that chemical communication evolved earlier than the auditory and visual modes?

2. Describe the roles of releaser and primer hormones in reproduction in some mammalian species.

3. Cite examples from the section on communication of how chemical, auditory, and visual signals serve as intrinsic isolating mechanisms (see Chapter 32) in various animal species.

4. Why do people experience "jet lag" after a long airline flight? Would traveling in a northward or southward direction be likely to cause this problem?

5. What have studies of animal orientation behavior demonstrated about the capacity of natural selection to adapt the sensory capabilities of an animal to its way of life?

SOCIAL BEHAVIOR (SOCIOBIOLOGY)

In some animal species the individuals pass much of their lives without any cooperative activity, their intraspecific interactions (aside from mating) being largely restricted to antagonistic ones. But in many species some degree of intraspecific cooperation is evident. The cooperative interaction may be relatively simple and of limited duration, as in the winter flocks formed by chickadees (which are not found in flocks during the breeding season); such flocking probably aids in locating patchily distributed food supplies and in spotting and eluding predators (Fig. 35.40). An example of transient cooperation with a different function is the huddling together of coveys of quail in winter, which enables the birds to conserve body heat and thus withstand low temperatures that would kill isolated birds. At the other end of the cooperativity scale are such highly evolved and long-lasting societies as those of honeybees and human beings, in which almost every aspect of each individual's life depends on the activities of others. The study of the biological basis of all social behavior is the discipline called ***sociobiology.***

35.40 Flocking defense against a falcon by flying starlings Starlings usually fly in a loose formation, and continue to do so even in the presence of a falcon if they are above the falcon. But if they are below, they form a tight flock. The falcon cannot dive at a bird in such a flock without risking serious damage to itself by crashing into other birds. The falcon will swoop only if an individual starling becomes separated from the flock.

Spatial Factors Influence Sociality

Social behavior is strongly influenced by the way the individuals of a species are organized in space, which is in turn strongly dependent on the distribution of the food supply and on the animal's method of exploiting it. We shall here mention three aspects of spatial organization: individual distance, territory, and home range.

In many species each individual may be said to carry around itself a small volume of space that it tends to treat as uniquely its own. The animal shows signs of agitation, or even overt aggression, when another individual comes too close and breaches this ***individual distance.*** One consequence of the effort of animals to maintain their individual distances from all others is the remarkably regular spacing often seen where many individuals are gathered together, as in a flock of birds lined up on a telephone wire (Fig. 35.41). Maintenance of relatively inviolate individual distance is often seen in animals that are only moderately social; it tends to be compromised (though still present) in highly social animals such as baboons, chimpanzees, and many other primates, where mutual grooming and other forms of bodily contact are common; and it is apparently nonexistent in social insects, where intimate bodily contact within the nest is continual.

A much larger unit of space is the ***territory,*** the area actively defended by an animal or group of animals against other

35.41 Birds on telephone wires Each bird attempts to maintain a small volume of space—the individual distance—around itself with the result that the birds space themselves out regularly on the wire.

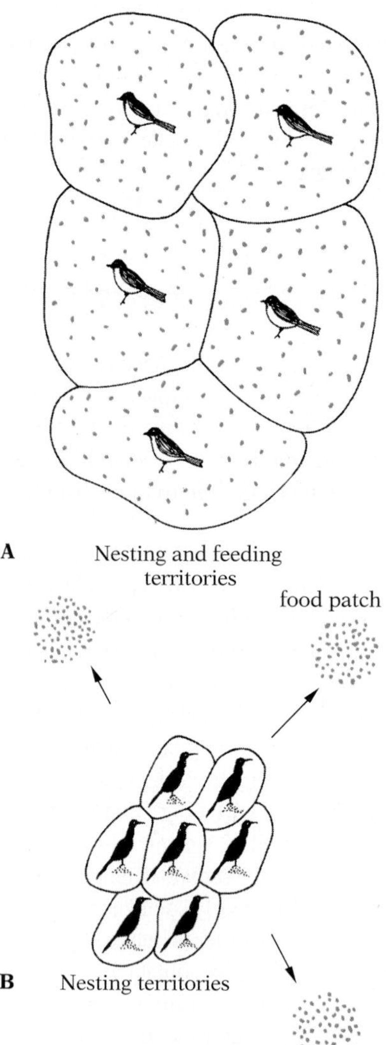

35.42 **Effect of food distribution on blackbird social systems** (A) When food is distributed more or less evenly throughout the area, it may be energetically efficient for each individual to occupy and defend its own territory for both feeding and breeding. This pattern is typical of redwing blackbirds, which occupy territories in marshes and feed on emerging insects. (B) If food occurs in unpredictable patches, often far from good nesting sites, it may be more efficient for the individuals to occupy only small nesting territories in a colony and to forage as a group. This type of organization is seen in tricolor blackbirds, which nest together for mutual protection and feed as a flock. They forage on ripening grass and grain seeds.

members of the same species; we have already discussed the role of displays in territorial defense. Each territory, or at least each hotly contested one, has a local abundance of some critical resource—food, mates, nesting sites, or whatever. Territories may be established for mating, nesting, or feeding (Fig. 35.42). Whatever the reason, territoriality seems to function by spacing individuals in such a manner that the most severe aspects of competition and individual antagonism are minimized while social stability is improved.

A still larger unit of space is the ***home range,*** the total area in which an animal (or group of animals) travels during the course of its normal activities. In some cases the home range of an animal may be identical to its territory, but often it embraces additional areas that are not defended and may be shared with other territory holders (Fig. 35.43). Many large mammals, especially herbivorous ones that must forage widely, have no territory and exhibit only a home range.

Both individual distance and territoriality, when they occur, make it difficult for one individual to approach another. But obviously individuals must approach one another if mating is to occur; a major function of courtship displays is thus the diminution of agonistic motivations aroused when the partners approach each other.

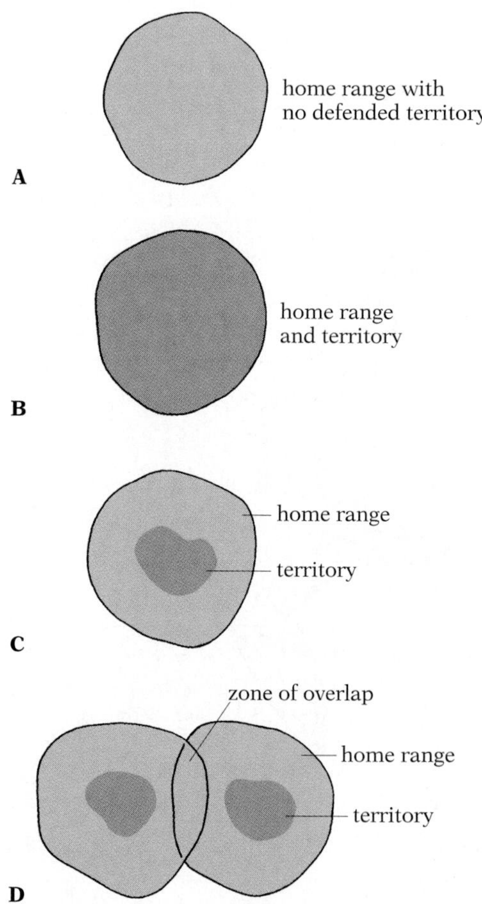

35.43 **Relationships between home range and territory** (A) Some animals have a home range but no defended territory. (B) In some cases an animal's home range may be the same as its territory. (C) More often the home range is larger than the territory. (D) Home ranges often overlap, but territories seldom do.

35.44 A ground squirrel giving an alarm call Alarm calls are sounded at the approach of airborne or terrestrial predators; there is a different call for each. The calls, which involve risk in that they draw attention to the individual giving the alarm, are produced primarily by females with close relatives living nearby.

ALTRUISTIC BEHAVIOR

Some social animals apparently display ***altruistic behavior,*** where one individual sacrifices itself for another, or more precisely, an individual lowers its own fitness—its personal reproductive potential—in order to raise that of another's. It is essential to keep in mind, however, that each member in a group is unconsciously acting in its own self-interest; helping another is selected for not because it is "good for the species," but because it pays the individual to cooperate. In each case, whether it is birds caring for their young, bees dying defending their hive, or primates sharing food, the individuals involved are turning a genetic profit, or at least minimizing a potential loss.[3] This is especially obvious in the context of parental care. The young are the parents' stake in the next generation, the sole index of evolutionary success. Investment in the young is an investment in immortality, the ultimate form of genetic selfishness.

The classic example of altruism is the caste system of social insects. The tens of thousands of female honeybees in a hive do not reproduce; instead they devote their energies to rearing the offspring of their mother and queen. But how would such behavior evolve? It would seem to be maladaptive; i.e., rather than enhance the altruist's chances of perpetuating its genes, it would seem to reduce those chances. An animal that gives an alarm call to warn others calls attention to itself, thereby putting itself in danger (Fig. 35.44). It would seem that individuals whose genetic makeup favors altruistic behavior should have less reproductive success than individuals with genes for more selfish behavior. Through the process of natural selection the genes for altruism, therefore, should eventually disappear from the population. Explaining the evolution of altruistic behavior has been one of the major challenges of sociobiology.

The altruistic behavior of a parent toward its offspring (e.g., in defending them against predators) is a special case. When a parent endangers its own life and its chances of producing offspring in the future by engaging in altruistic behavior, it may not, in fact, be reducing its personal reproductive success at all. For this animal, the best way to ensure the future continuance of its genes may be to risk defending those young rather than to gamble on being able to replace them in the future (Fig. 35.45). Rather than being maladaptive, then, such behavior may actually increase the altruist's individual fitness.

The matter becomes more complicated when the beneficiaries of the altruistic actions are not the altruist's offspring. Yet many examples of such actions among social animals have been documented. Among monkeys and apes, childless adults often care for the infants of others for short periods; they may hold the young, groom them, or play with them. African wild dogs sometimes bring fresh meat to the den and give it to other adults that remained with the cubs. How can the evolution of this kind of altruistic behavior be explained?

35.45 Wildebeest defending young The young wildebeest has been attacked by a hyena and the mother is charging in to attack the hyena and protect its young. In acting altruistically the mother is protecting her genetic investment in the next generation.

[3] Such terms as "self-interest," "investment," "profit," "strategy," etc., often used in evolutionary contexts, do not imply any conscious weighing of alternatives by the animals. The animals are not concerned with ways of maximizing their genetic contribution to future generations. The terms merely reflect the probabilistic aspects of the selection pressures leading to the evolution of certain traits.

Kin Selection Operates in Many Cases of Altruism

In recent years, attempts to explain the evolution of many kinds of altruistic behavior have focused on the concept of genetic relatedness. Such explanations are made in terms of ***individual fitness,*** which is an organism's probable genetic contribution to succeeding generations. Genes that cause one animal to be altruistic and forgo reproduction will still survive in the population if the resulting altruistic behavior sufficiently enhances the fitness of other individuals that carry the same genes. In sexually reproducing populations, for instance, an individual shares genes with its offspring and with other members of its kinship group—its siblings, nieces and nephews, cousins, and so forth. The individual therefore has some degree of evolutionary investment in all its genetic relatives. It follows that altruistic behavior that benefits the altruist's kin improves the chances that some of the altruist's genes will be perpetuated, since close relatives share some of the same genes. The closer the relationship, the higher the percentage of shared genes. For instance, an individual shares 50 percent of its genes with each of its own offspring, an average of 25 percent of its genes with its nieces and nephews, and an average of 12.5 percent of its genes with its cousins. Hence an altruistic act benefiting a relative enhances the survival of the altruist's genes, and the genes for altruism toward relatives will be selected for (as long as the relative is close and the cost to the altruist is not too high). It is not surprising, therefore, that the majority of documented instances of altruism occur within social groupings where all the members are closely related. That an altruist's fitness can be improved by the preservation of common genes in related individuals is called ***kin selection.***

Reciprocity and Selfishness Explain Some Types of Altruism

After kin selection, reciprocity is probably the most common motive force behind altruism. Behavior in which an animal confers a favor on another animal, not necessarily a relative, in the expectation of eventual repayment, is called ***reciprocal altruism.*** For example, two social animals may cooperate to groom places on each other that neither can reach on itself (Fig. 35.46). Such a mechanism, of course, could operate only in societies where individuals recognize each other, which seems in fact to be the case.

In still other instances, what at first appears to be completely altruistic behavior proves, on closer analysis, to serve ***selfish*** ends. Breeding pairs of the Florida scrub jay are regularly assisted by helpers, which are other jays that are not mated and have no territories of their own. Although caring for the young of another mated pair may seem like altruism, these helpers may actually be helping themselves. Some helpers are young offspring of the pair, in which case kin selection may be operating. But others are not kin, and they, by being in a desirable territory, are in a position to take over the territory if anything happens to the breeders. Through this helping behavior they are probably improving their own personal fitness.

35.46 Mutual grooming in penguins Each animal grooms an area the other cannot reach.

MATING STRATEGIES

Another set of behavioral alternatives subject to strong evolutionary selection pressure involves mate choice. The reason is clear: in diploid animals (those with two sets of chromosomes), half the genes of any individual's offspring come from another individual. Any behavior, then, that serves to secure genes of greater quality will be favored by natural selection. Mating strategies differ from species to species and range from individuals in some species not choosing mates (female mosquitoes, for example, copulate with practically any male when the time is right) to other species where choosing involves elaborate rituals.

Parental Investment Influences Mating Strategies

Different mating systems are found among the various groups of vertebrates. ***Monogamy,*** in which one male mates with one female, is common in birds, but relatively rare in mammals. Of the two kinds of ***polygamy,*** in which one male mates with many females (***polygyny***) or one female mates with many

35.47 Sage grouse on their mating grounds In this species many males (white breasts) display simultaneously on a compact mating area called a lek; each male, however, actively defends his own spot within that mating area. The females (smaller, browner birds) congregate in the lek and, after observing the males' displays, choose the ones from whom they will solicit copulation.

males (***polyandry***), polygyny is by far the more common. Many species form no pair bonds at all, and their mating is often called ***promiscuous,*** though the term may be misleading if one takes it to mean random and disorderly mating; the choice of mates is usually far from a random process (Fig. 35.47).

The evolution of different mating systems in different species is often dependent on ***parental investment***—the cost to the parent, in terms of its ability to produce more offspring. Generally, a high parental investment favors any behavior that enhances a particular offspring's survival and future reproduction. A fish that releases thousands of eggs into the surrounding water and then swims off has much less parental investment in any one of the eggs than a bird that constructs an elaborate nest, lays only a few eggs in it, and then remains to incubate the eggs and tend the young when they hatch. By studying the parental investment of different vertebrate species, together with the way of life they pursue (the places where they are found, the food they use, the predators that feed on them, etc.), we can better understand the patterns seen among the different types of vertebrate mating systems.

Monogamy Is Common in Birds

Why is monogamy more common in birds than in mammals? The parenting of young birds requires tremendous energy, from incubating and protecting the eggs to meeting the nestlings' prodigious appetite. Without the cooperation of both male and female parents, offspring would be unlikely to survive. It is therefore to each parent's evolutionary advantage to form a monogamous pair bond and make a large investment in the young. In fact, 93 percent of bird species form monogamous pair bonds and divide the labor of rearing the young more or less equally between the sexes.

Polygyny Is Typical of Most Mammals

In mammals, by contrast, the young are first nourished on milk, so the male's role is reduced, especially among large herbivorous species where there is often little or no carrying of food to the young. If the males play any role at all after fertilizing the eggs, it is usually to guard the female and the young or to defend a suitable territory. But a dominant male may be able to accomplish this task for several females at once; hence the chance of his leaving descendants is enhanced by polygyny. In only 10 percent of the mammalian species do males form long-term bonds with females and help with rearing the young.

For mammals the male and female have very unequal parental investments in the young. The female must carry the fetus from conception until birth, and must suckle the young for an extended period following birth. She is therefore more limited than the male in how many young she can produce, and it may be to her advantage evolutionarily to remain with one male who can provide good protection and a territory with a rich food supply for her young. Her chances of leaving offspring hinge to a large extent on her choice of a mate; effective reproductive strategy calls for her being very selective.

This discussion of vertebrate mating systems underscores an important point—in the study of mating behavior, as in all the other sorts of behavior we have examined, an evolutionary analysis reveals order underlying bewildering diversity.

SOCIAL ANIMALS

We will now compare some of the strategies that different species employ to shape their respective social systems. The unit of social organization in the social insects is the family, which typically consists of a single reproductive female—the queen—and her daughters (also sometimes her sons). In bees and wasps the daughters serve as workers; in ants some also serve as soldiers. The workers and soldiers are nonreproductive individuals; the society so influences their development that they are physically unlike the queen and constitute separate sterile castes.

Socially Complex Honeybee Societies Display Intricate Communication

Containing as many as 80,000 individuals, a colony of honeybees may persist for many years. The hive has a single queen, who may live for seven years or more and whose only function is to lay eggs. During the course of her life she lays hundreds of thousands of fertilized eggs, most of which hatch into larvae that develop into the worker bees that carry out the tasks of the colony.

The workers, which represent the labor force in an insect caste system, pass through a series of age-dependent tasks from cleaning cells, to nursing larvae and the queen, to building comb, to guarding the hive from predators, and finally, at about three weeks of age, foraging for nectar and pollen. The individual bee has no choice of roles in this system; her role is determined by a combination of caste, stage of development, and the conditions of the hive.

Extensive and complex communication is essential to the maintenance of the large and intricate honeybee social organization. The dances by which scout bees inform the foragers of the location of food sources are one of many possible communications by dances. For example, dances are also important in the establishment of new colonies. When a colony has become large, the queen may leave, taking with her many of the workers in a dense swarm. The swarm usually lands on the branch of a nearby tree, and remains there for some days while scout bees fly out in search of potential new nesting sites (Fig. 35.48). Maintenance of the swarm depends on secretion of a swarming pheromone by the queen.

35.48 A honeybee swarm The bees form a huge oval mass on a branch. Scout bees fly out from the swarm in search of possible new nesting sites.

Chemicals Coordinate Hive Activities

Chemical communication plays a fundamental role in the coordination of activities within the hive. Among the many pheromones used are swarming substance, queen substance (a substance emitted by the queen to prevent the hive from rearing rival queens), trail substance, alarm substances, various kinds of attractants, and hive-identification substance. In addition, constant regurgitative feeding among the workers provides all members of the colony with continuous information about the nutritional condition of the whole colony.

The honeybee colony is a smoothly operating corporation of individuals that fully exploits the many advantages of social living. Sociality makes possible the maintenance of the nest at a relatively precise temperature and humidity, and in honeybees, division of labor creates several subpopulations of efficient specialists. The protective cavity and the large colony size permit effective group defense against even the largest of predators.

Vertebrate Societies Show More Individualistic Behavior

Like a honeybee society where the queen plus her offspring constitute a special kind of family, most vertebrate societies, too, are based either on families or on extended kinship groups. Their organization, however, is very different from that of insect societies. Typically, a vertebrate social group consists of a leader (usually a male), together with his mates and offspring (and perhaps their mates and offspring) (Fig. 35.49). Such a grouping benefits both from reduced aggression and from the

35.49 Hooker's sea lion bull and his harem The male establishes a clearly defined breeding territory and drives away all individuals other than mates or potential mates. Females select the male and his territory and mating occurs. The male vigorously guards his mates and their young.

35.50 A dominance ritual in mountain sheep Male mountain sheep work out a dominance hierarchy through a ritualized duel in which a run on hind legs culminates in a loud collision. The clash is followed by a head display. In many group-living species the results of such encounters determine the way individuals will interact.

tendency for cooperativity and altruism to evolve among individuals with shared genes.

Vertebrate societies differ from insect societies in several fundamental ways (Table 35.2). Division of labor is not based on biologically determined castes. Most adult members of the society take part in reproduction (or at least try to do so); there is no setting aside of one or two individuals to serve as reproductive machines for the entire group. The organization depends on the ability of individuals to recognize one another; no such requirement appears to exist in insect societies. And the behavior of social vertebrates, though it may contain many instinctive elements, is nonetheless far less rigid and less stereotyped than the behavior of social insects.

TABLE 35.2 *Comparison of insect and vertebrate societies*

Insect societies	*Vertebrate societies*
Division of labor based on a biologically determined caste system	Division of labor not based on a biologically determined caste system
Single reproductive individual (queen)	Most adults capable of reproduction
Individual nonreproductive members short-lived	Individual members long-lived
No personal recognition	Personal recognition important for social organization
Rigid, stereotyped behavior	Behavior more flexible, less stereotyped

Social Hierarchies Confer Order

Many vertebrate social systems revolve around contests that establish the likely winner of a fight with a minimum of risk. Mountain sheep, for example, engage in highly coordinated and stylized duels in which two males crash into each other's well-armored heads and back-curved horns (Fig. 35.50). The physics of the collision make the stronger male the likely winner, while the specially shaped horns act as bumpers, rather than weapons, so this test of strength is relatively safe for both opponents. The specialized weaponry developed for protection from, or predation on, other species is rarely used against members of the same species in any effective way: poisonous snakes wrestle without striking; fish lock jaws but do not bite; antelope push and fence with their horns but will not stab. Considering the risk of unconstrained fighting even to the probable winner, the selective advantage of this innate restraint is clear: what would be the use of winning an all-out contest if even the winner was left exhausted, injured, or easy prey to watchful predators? However, when two contestants are so evenly matched that there is no clear victor, a ritual contest can sometimes turn into a brutal fight (Fig. 35.51).

The genetic orchestration of aggression through behavioral and hormonally mediated strategies has enormous advantages. In the working out of territories, for example, ritualized contests usually reduce the risk of injury. In more social species, they actually lessen the frequency of serious fighting and stabilize the group through the formation of a stable ***dominance hierarchy,*** or pecking order, in which every member knows which individuals it can defeat, and which can defeat it (Fig.

35.51 East African lions in combat

35.52). The formation of such a hierarchy causes fighting practically to disappear—at least until some outsider tries to break in. Every individual keeps its place, since it is aware of the probable outcome of a challenge to animals higher on the scale.

THE EVOLUTION OF BEHAVIOR

Throughout this chapter we have assumed that behavior is a biological attribute, one to be investigated like anatomy or physiology, and subject to the same kinds of evolutionary processes. The corollary assumptions—that behavior is adaptive and that natural selection brings about an increase in well-adapted and a decrease in poorly adapted behavior patterns in the population—are basic to the elucidation of many types of behavior.

If behavior patterns evolve and can be studied in terms of the selection pressures that produce them, it follows that we should be able to make reasonable conjectures concerning the ancestral behavior patterns from which newer behavior patterns have arisen, just as we can make inferences about the ancestral structures from which our hands or other structures have evolved. And we should be able to study the genetics of behavior just as we study the genetics of physical characteristics. It is to this approach to the study of behavior that we now turn.

Comparative Studies Help Explain the Derivation of Behavior Patterns

By comparing the behavior patterns of a number of related species, we can study the evolutionary derivation of behavior. These patterns often represent different stages of development of the same basic behavior; consequently, they may reveal the ancestral pattern. Let us look at several examples.

There are certain species of flies (family Empididae) in which the males always present the females with a silken balloon before mating (Fig. 35.53A,B). This is a curious bit of behavior, and we could hardly guess what selection pressures brought it about, or from what ancestral beginnings, were it not for the fact that there are other species of empidid flies that exhibit this courtship pattern in various stages of development. In

35.52 Bonobos society The bonobos are a species of chimpanzee that lives deep in the forests of Zaire. This is an unusually peaceful society with a dominance hierarchy. Shown here are young and female adults gathered around a dominant female. Cooperation and sharing is the hallmark of this society and deadly aggression and rape are unheard of.

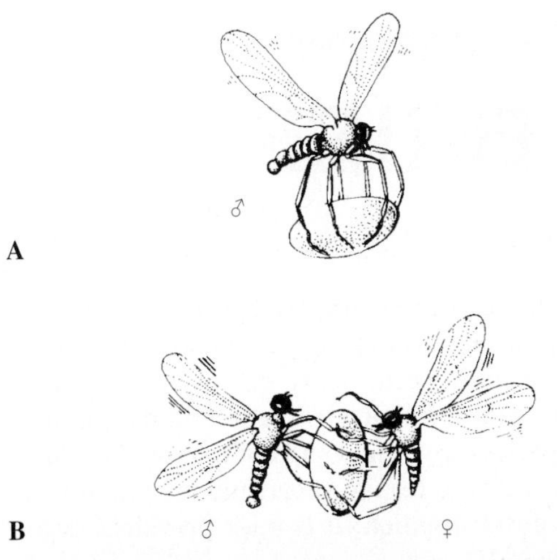

35.53 Presentation of a "gift" as part of courtship (A) Male balloon fly carries a silken balloon and (B) presents it to a female. (C) Male wolf spider approaches a female with a similar "gift."

many species of empidid flies where the male does not give anything to the female when courting her, she sometimes captures and devours him. In other slightly more advanced species, the male captures prey, and presents this to the female, and then mates with her while she is occupied with eating the prey. The selection pressure for this behavior seems easy to explain: males that divert the females' attention by giving them prey succeed in mating and escaping more frequently than those that do not. The males of still more advanced species of empidid flies wrap the prey in a ball of silk before presenting it to the female; the process of unwrapping it presumably gives the male more time to copulate. In still more advanced species, the male encloses only fragments of prey in the balloon, and the female does not actually eat the prey. In other words, at this stage in the evolution of the courtship behavior, the balloon has replaced the prey as the important element, becoming part of a display that functions in making the female more receptive to the male. We can understand, therefore, how in still more advanced species an empty balloon or even some other object can suffice. By examining a whole group of species in this way, we not only clarify the evolution of this particular behavior but also gain new insight into the evolution of animal behavior in general.

MAKING CONNECTIONS

1. Why would you feel irritated if, while you were sitting on a nearly empty bus, a stranger chose to sit beside you instead of in a totally vacant seat? Has your *territory* been invaded, or your *individual distance?*

2. A worker honeybee that stings an intruder in defending its hive will soon die as a result of the stinger being torn from its abdomen. Why is it a misunderstanding of evolution to say that such self-sacrificing acts are done "for the good of the species?"

3. The great evolutionary biologist J. B. S. Haldane once said that he would willingly lay down his life for two brothers or eight first cousins. What was his point?

4. Why do biologists feel justified in studying the evolution of behavior, just as they investigate the evolution of structural and physiological characteristics of organisms? Since behavior leaves no fossil record, how can its evolution be studied?

CONCEPTS IN BRIEF

1. The behavior of an animal—what it does and how it does it—is the product of the functions and interactions of its various control and effector mechanisms. The analysis of behavior poses a problem because it is difficult to avoid anthropomorphic interpretations.

2. Behavior that depends largely on inherited mechanisms is *innate* behavior whereas behavior that is acquired as a result of experience is *learned. Instinct* is the inheritable, genetically specified neural circuitry that organizes and guides behavior.

3. *Taxes, kineses,* and *reflexes* involve simple responses to stimuli. In a taxis, the response is oriented relative to the stimulus but a kinesis is not oriented. More complicated behavior patterns predominate in higher animals.

4. A signal that elicits a specific behavior response is a *sign stimulus.* The animal must possess neural mechanisms that are selectively sensitive to sign stimuli. Sign stimuli are known as *releasers* because they "release" (trigger) specific innate behavior patterns. Releasers can be chemical, auditory, or visual. Often it is possible to design *supernormal stimuli* even more effective in triggering a behavior than the natural sign stimulus.

5. *Fixed-action patterns* (or *motor programs*) are innate, stereotyped responses to releasers. They tend, once begun, to run to completion regardless of the situation.

6. Inheritance and learning are fundamental in determining the behavior of higher animals. In general, the inherited limits within which behavior patterns can be modified by learning are much narrower in the invertebrates than in the vertebrates, and narrower in the lower vertebrates than in the mammals.

7. Studying learning is difficult because one must determine whether the behavior results from learning or from maturation; whether an animal can learn one thing in one context but not another; and whether a sensitive phase or a latent period is involved.

8. *Habituation* is a simple type of learning in which the organism shows a gradual decline in response to repeated "insignificant" stimuli that bring no *reinforcement. Classical conditioning* is the association of a response with a stimulus with which it was not previously associated. In *operant conditioning* (trial-and-error learning) the animal learns to associate a certain activity with a reward or punishment.

9. In *imprinting* an animal learns to make a strong, long-lasting association with another organism (or object); imprinting occurs only during a short sensitive phase. Imprinting is important in parental recognition and in proper species recognition and interaction. *Insight* is the organism's ability to respond correctly the first time it encounters a novel situation; it is most prevalent in higher primates.

10. Animals may be motivated to do one thing now and another later. Current priorities can help the animal select among many behavioral possibilities.

11. Chemical communication is the most primitive form of communication. Many animals secrete chemical substances (*pheromones*) that influence the behavior or physiology of other members of the same species.

12. Many animals can communicate information by sound. Sound may function as a species recognition signal, as an individual recognition signal, as a display that attracts females to the male and helps synchronize their reproductive behavior, and as a display to defend territory.

13. Visual displays are frequently important in reproductive behavior. They function both in bringing together and synchronizing the two sexes in the mating act, and in avoiding mating errors. Agonistic encounters between individuals of the same species may often be resolved by displays without physical combat. Through displays the combatants convey their attack motivation.

14. Scout honeybees use language to communicate to workers in the hive, largely through dances, the quality of a food source as well as its distance and direction from the hive.

15. All living things appear to have an internal timer—a "biological clock," and many organisms show *circadian rhythms.* Though the basic period of the clock is innate, it is constantly being reset by the environmental cycle.

16. Migrating birds and homing pigeons have navigational abilities. Some birds are preprogrammed to fly a certain course for a certain time. Others can tell compass directions by observing the sun or stars. Pigeons can use magnetic cues in addition to sun cues for orientation.

17. Many animal species show some degree of intraspecific cooperation. The cooperation varies from relatively simple and transient forms to highly evolved and long-lasting societies in which almost every aspect of each

individual's life depends on the activities of others. Effective communication between individuals is a prerequisite for such cooperation.

18. Many social animals exhibit *altruistic behavior,* behavior that may reduce the personal reproductive success of the individual exhibiting the behavior while increasing the reproductive success of others. Behavior that benefits the altruist's kin may improve the chances that some of the altruist's genes will be perpetuated, since close relatives share some of the same genes (*kin selection*). In some instances *reciprocity* may be an explanation; in other cases, an act that seems altruistic may actually serve *selfish* ends.

19. One of the principal determining factors in the evolution of mating systems is *parental investment*—the cost of the parent of behavior enhancing the likelihood that the offspring will survive and reproduce. Monogamy is common among birds where there is a large parental investment, whereas polygyny is common among mammals where the investment for males is less than that of females.

20. In the social insects the unit of organization is the family, which typically consists of a single reproductive female—the queen—and her daughters (sometimes her sons). A biologically determined caste system is involved.

21. A vertebrate social group typically consists of a male leader together with his mates and offspring. There is no biologically determined caste system, and all adults are potential reproducers. An important aspect of vertebrate social organization is the *dominance hierarchy,* a series of dominance-subordination relationships that tend to give order and stability to the group.

22. Behavior is adaptive; natural selection brings about an increase in well-adapted and a decrease in poorly adapted behavior patterns in the population. Behavior patterns evolve and can be studied in terms of the selection pressures that produce them. By comparison of the behavior patterns of a whole group of related species, the evolutionary derivation of the patterns can be clarified.

STUDY QUESTIONS

1. How are *instinct* and *innate behavior* related? How does innate behavior differ from *learned behavior?* Can you think of any behavior that can be classified as strictly innate or strictly learned? (p. 804)
2. A territorial male European robin will attack a ball of red feathers mounted on a wire, treating it as an intruder despite its lack of resemblance to a bird. What is the adaptive advantage to the bird in responding to such a simple *sign stimulus* (*releaser*)? Would the bird necessarily respond to the stimulus if it were not defending its territory? Are sign stimuli always visual? Give examples. (pp. 805–807)
3. Describe the egg-retrieval behavior of geese. What aspects of this behavior mark it as a *fixed-action pattern?* How are releasers and fixed-action patterns related? (pp. 807–808)
4. What do Thorpe's experiments on the development of song in the European chaffinch reveal about the interaction of inheritance and learning in animal behavior, and about the difficulties of studying learning? (pp. 809–810)
5. Describe and give examples of these types of learning: *habituation, classical conditioning, trial-and-error learning, imprinting,* and *insight learning.* (pp. 810–812)
6. Bloodhounds can recognize individual human beings by their scent. Should the chemicals that these dogs detect be considered human *pheromones?* Which of these is a *releaser* and which a *primer:* (a) the queen substance that suppresses the sexual development of worker bees; (b) the chemical released by a queen bee to attract drones on her mating flight? (pp. 815–817)
7. The theme of Percy Bysshe Shelley's poem "To a Skylark" is that the song of the lark expresses a joy so intense that it is unattainable by human beings. How would an ethologist criticize Shelley's interpretation of bird song? (p. 803)
8. What is a *display?* How are *agonistic* displays adaptive? If you encounter an unfamiliar and possibly dangerous dog, how can you judge whether it is highly motivated to attack? (pp. 818–819)
9. Suppose a scout bee returns to the hive from a food source at 6 A.M. and performs a waggle dance on a vertical comb. If the waggle run of the dance is oriented straight down, in which direction is the food source? How is distance to the food source communicated by the dance? (pp. 820–821)

10. **Why is it necessary for a homing pigeon to have both "map" and "compass" information in order to return to its home from an unfamiliar location? What kinds of cues do pigeons and other birds use for compass information? What is known about the map sense of birds? (pp. 823–824)**

11. **Suppose one member of a flock of blackbirds detects a hawk and gives an alarm call that alerts the other flock members to the danger. How is this behavior *altruistic?* What explanations have been offered for the evolution of such behavior? (p. 829)**

12. **How does the concept of *parental investment* help to explain why birds and mammals typically have different mating systems? (pp. 830–831)**

13. **Compare the organization of a vertebrate society (such as a wolf pack) with that of an insect society (such as a honeybee colony). (pp. 831–833)**

P A R T   F I V E

The Genesis and Diversity of Organisms

CHAPTER 36

THE ORIGIN AND EARLY EVOLUTION OF LIFE

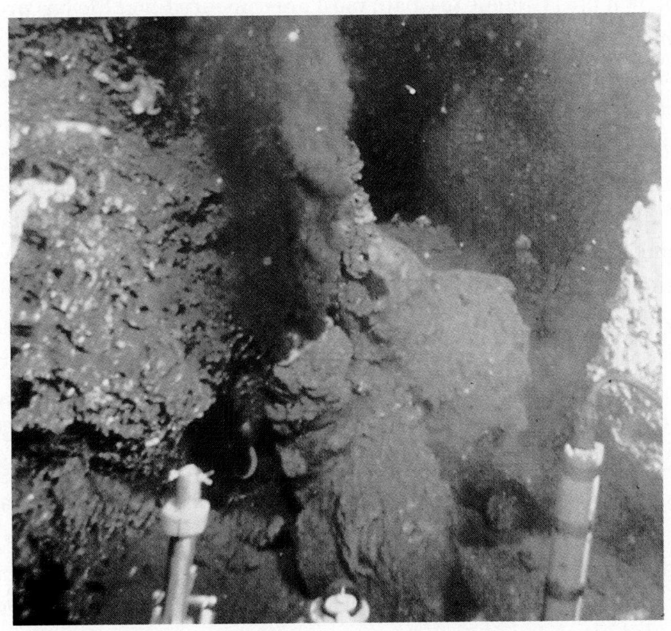

Few mysteries have exercised the human imagination as much as the origin of life. Religion, mythology, and philosophy have proposed a great variety of explanations, but as different as these explanations are, most share the assumption that life's origins must be attributed to an agency outside nature, a creator. In the same way, the diversity of species was conceived as resulting from separate, deliberate acts of creation. Not until the latter part of the 19th century was the theory of evolution able to account for the origin of species without invoking a supernatural agency. Can 20th-century science do the same for the origin of life itself?

THE ORIGIN OF LIFE

We discussed the principle of biogenesis—that life can arise only from life—in Chapter 5, and cited Pasteur's classic experiments in support of this principle. Pasteur's work effectively put to rest, as far as most biologists were concerned, the long-held idea of spontaneous generation. No longer could men of science seriously entertain the notion that the maggots in decaying meat arise *de novo* from the meat, for instance, or that microorganisms appear spontaneously in spoiling broth. It may seem strange, then, that in the last quarter of the 20th century spontaneous generation should be a topic of major interest in biology. But there is a radical difference between the modern ideas of spontaneous generation and those of Pasteur's day. The modern theorists do not suggest that life can arise spontaneously under present conditions on earth; indeed, most are convinced that this cannot happen. What they do suggest is that life could and did arise spontaneously from nonliving matter under the conditions prevailing on the early earth, and that it is from such beginnings that all present earthly life has descended.

This chapter will outline a theory of the origin of life now widely held by scientists. The basis for this theory was first put forth by the Russian biochemist A. I. Oparin in 1936. Be aware, however, that though the broad outlines of the theory have wide support, many of the details are disputed. Further, we have no direct evidence concerning the origin of life. We cannot be sure how life did arise; we can only gather indirect evidence to show how it could have arisen. (See Exploring Further: The Possibility of Life on Other Planets, p. 840).

Chapter opening photo: Hot spring located 2.5 km below the surface of the Pacific Ocean. Similar springs on the primordial earth may have provided nutrients for primitive life forms.

EXPLORING FURTHER

THE POSSIBILITY OF LIFE ON OTHER PLANETS

If life could arise spontaneously from nonliving matter on the primordial earth, might it also have arisen elsewhere in the universe? Few scientists concerned with the origin of organisms think that life is unique to earth; most are convinced that life has probably arisen many times in many places. They point out that no unduplicable event was necessary to the origin of life on earth. On the contrary, all the events now hypothesized and all the known characteristics of life seem to fall well within the general laws of the universe; they are natural phenomena susceptible to duplication. Indeed, some have argued recently that biochemical evolution (i.e., life) is an inevitable part of the overall evolution of matter in the universe. Given the immense size of the universe, they argue, it would actually be unreasonable to think that life is restricted to one small planet in one minor solar system.

One interesting series of calculations on the probability of life elsewhere in the universe was made by Harlow Shapley of Harvard University. Shapley said that at least 10^{20} stars are visible to us with present-day telescopes (to say nothing about the vast numbers beyond the reach of our telescopes). Many of these stars probably lack planets, but Shapley thought it reasonable to assume that at least one star in every thousand has a planetary system, which gives a total of 10^{17} stars with planets. Now, if it is assumed that life wherever it occurs is at least roughly similar to earthly life in its basic chemistry, it must be concluded that only planets with moderate temperatures could support life. The planetary systems of many stars may not include such a planet. Shapley suggested that, by a modest estimate, at least one in a thousand of the stars with planets has an appropriate planet; this gives a total of 10^{14} stars with at least one planet of the right temperature (Fig. A). Now, to support life a planet must not only have moderate temperatures but also be within a certain size range, to hold a suitable atmosphere. If one out of every thousand of the planets of the right temperature is also of the right size, this gives a total of 10^{11}. But even if a planet has an appropriate temperature and atmosphere, life still might not have arisen, for any number of reasons. Again Shapley used an estimate of one in a thousand, which gives 10^8 (100 million) planets on which life may well have arisen. Today biologists working in this field consider Shapley's estimate too conservative; they suggest that the figure should be 10^{16} or more.

If one admits the possibility that life has arisen on huge numbers of other planets, the question immediately arises whether civilizations as advanced as, or more advanced than, our own might have developed elsewhere. Here there is little on which to base an estimate. Nevertheless, many scientists believe that there are numerous civilizations in the universe more advanced than our own. Shapley's estimate was that there are at least 100,000 such civilizations; recent estimates have ranged as high as 5×10^{14} (500 trillion) civilizations.

So confident are some scientists that these other civilizations exist that they have used huge radio receivers to listen for signals from outer space, and they have included in some

recent space-probe capsules information designed to be interpretable by beings that, though unfamiliar with earthly languages, can decode messages written in mathematical terms (Fig. B). They think it possible that someday human beings may be able to make contact with intelligent organisms living on planets in other parts of the universe.

Though the universe may contain vast numbers of other intelligent creatures with technologically advanced civilizations, human beings are most unlikely to have evolved on any planet but earth. Totally separate evolutionary developments can never duplicate each other. Indeed, if human beings should become extinct on earth, and should intelligent civilization-building life arise a second time, that life would surely not be human. Loren Eiseley of the University of Pennsylvania has put the matter eloquently: "There may be wisdom; there may be power; somewhere across space great instruments, handled by strong, manipulative organs, may stare vainly at our floating cloud wrack, their owners yearning as we yearn. Nevertheless, in the nature of life and in the principles of evolution we have had our answer. Of men, elsewhere and beyond, there will be none forever."

A The rosette nebula Astronomers estimate that there are approximately 10^8 (100 million) planets on which life might have arisen.

B The Arecibo telescope, Puerto Rico.

As the Earth Formed, Materials Became Stratified

We are not certain how the solar system formed; we have only hypotheses. But as astronomers probe ever deeper into the secrets of the universe and gather more evidence, these hypotheses become increasingly convincing. The one most widely held today is that the universe is 15 to 20 billion years old, and that the sun and its planets formed about 4.6 billion years ago from a cloud of cosmic dust and gas. Most of this material condensed into a single compact mass, the sun. Within the remainder of the dust and gas cloud, lesser centers of condensation began to form. These became the planets, of which the earth is one.

As the earth condensed, a stratification of its components took place, heavier materials, such as iron and nickel, moving toward the center and lighter substances becoming more concentrated nearer the surface. Among these lighter materials must have been hydrogen and helium, which formed the first atmosphere. But unlike larger planets such as Jupiter and Saturn, the earth was too small, and its gravitational field too weak, to retain this first atmosphere and eventually all the gases escaped into space.

As time passed, the components became further stratified into three distinct regions, with the dense iron and nickel accumulating in the center to form the core, the less dense silicates of iron and magnesium forming a partly molten mantle surrounding the core, and the lighter substances remaining near the surface (Fig. 36.1). As the surface of the earth cooled, the surface materials, composed primarily of the lighter silicates, solidified to form a crust. This crust is quite thin compared to the diameter of the earth—about the thickness of an eggshell compared to the diameter of an egg. The crust solidified into massive plates resting on the molten mantle. These plates are moved about by the upwelling of new crust in some places and sinking of old crust in others, as we learned in Chapter 34 (see Fig. 34.48, p. 794). The intense heat in the interior of the earth also tended to drive out various gases, which escaped primarily by volcanic action. These gases formed a second atmosphere for the earth.

The Early Atmosphere Contained No Free Oxygen

We must know something about the probable early composition of this second atmosphere to understand the conditions under which life arose. Our present atmosphere contains about 78 percent molecular nitrogen (N_2), 21 percent molecular oxygen (O_2), 1 percent argon, and 0.033 percent carbon dioxide (CO_2), as well as traces of rarer gases such as helium and neon. But available evidence indicates that when the atmosphere first formed it contained virtually no free oxygen and was therefore not an oxidizing atmosphere, as the present one is.

The most widely accepted model of the earth's early atmosphere assumes that it was made up primarily of the gases known to be produced by present-day volcanoes. These gases are H_2O, CO, CO_2, H_2S, N_2, and H_2; in such a mixture, hydrogen cyanide (HCN) is easily formed and would probably also have been present. Much of the water vapor present in the primitive atmosphere may have come from comets colliding with the earth, since comets are largely water. Methane is also believed to have been present, since there are literally oceans

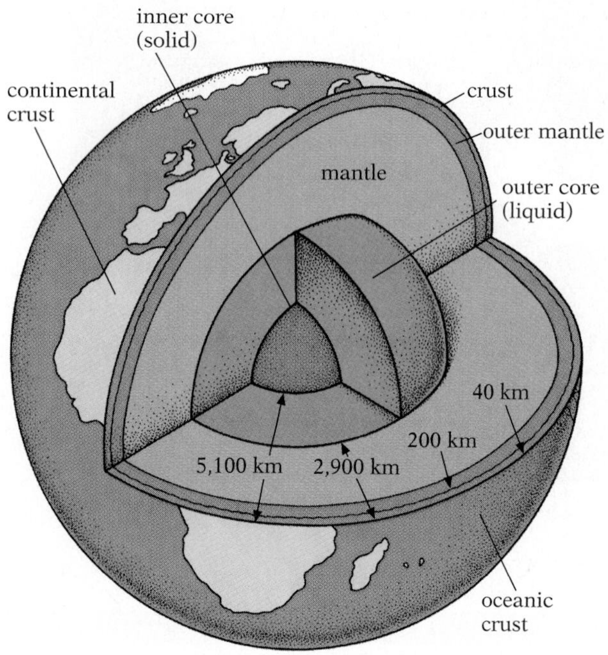

36.1 Interior of the earth The earth consists of three zones that vary in chemical composition: the core, mantle, and crust. The core consists mostly of iron, with some nickel. The inner core is solid whereas the outer core is liquid (i.e., molten). The mantle is the largest zone; it is composed primarily of iron and magnesium silicates. The upper layer of the mantle is solid and rigid while the inner layer is partly molten and plastic. The crust forms a thin skin over the earth's surface. It is differentiated into the oceanic crust, under the sea floor, and the thicker continental crust. The crust varies in composition. The thickness of the crust is exaggerated in this diagram.

of methane gas out in space and surrounding some stars. Note that this model envisions an early atmosphere in which there is no free oxygen but an abundance of hydrogen.

Centuries of Rains Formed the Oceans

Initially, most of the earth's water was probably present as vapor in the atmosphere, a condition leading to torrential rains as the earth cooled and the water vapor condensed. The centuries-long rains would have filled the low places on the crust with water and given rise to the first oceans. As rivers rushed down the slopes of early continents, they must have dissolved away and carried with them salts and minerals of various sorts, which slowly accumulated in the seas. Atmospheric gases probably also dissolved in the waters of the newly formed oceans.

Small Organic Molecules Formed Abiotically

If the early earth had an oxygen-poor atmosphere, as many astronomers and geochemists believe, and the primitive seas contained a mixture of salts, CO_2, H_2S, HCN, CH_4, NH_3, and N_2, how were the more complex organic molecules formed? The mixture thought to have been formed in the primitive seas is stable; there is no tendency for these materials to react spontaneously with each other to form other compounds. Yet for life to have arisen it would seem that at the very least the critical building-block materials, particularly amino acids and the purine and pyrimidine bases, would have been necessary. How might these compounds have been formed on the primitive earth?

There are two basic hypotheses to account for the accumulation of complex organic compounds on the early earth. The first suggests that the complex organic molecules came from asteroids and meteors striking the earth. Many of these extraterrestrial objects are rich in complex organic molecules created during the formation of the solar system billions of years ago (Fig. 36.2). Because the early atmosphere was denser than today's, incoming objects would have been slowed before striking the earth's surface. Some astronomers estimate that from 10^6 to 10^7 kilograms of complex organic molecules could have survived impact annually. It is possible, therefore, that the early earth had a vast supply of the complex molecules necessary for the evolution of life without any need to synthesize them out of simpler substances.

The second, more conventional, view of the development of organic molecules assumes no extraterrestrial input. According to this hypothesis, complex organic molecules were generated from the small inorganic compounds already present. The small molecules reacted with one another to form large, more complex organic molecules. To do so, some external source of energy must have been acting on the mixture. One possible source might have been solar radiation, including visible light, ultraviolet (UV) light, and X rays; of these, ultraviolet light would probably have been the most important. Remember that the primitive atmosphere contained virtually no oxygen and therefore no ozone layer; it is the ozone layer that screens out much of the ultraviolet radiation from the sun. The ultraviolet radiation would have been much more intense on the early earth than it is today. A second important possibility is energy from electrical discharges, such as lightning, while a third is heat from the earth's core and the sun.

How can it be demonstrated that ultraviolet radiation or electrical discharges or heat or a combination of these are capable of causing reactions that produce complex organic compounds? An answer was provided in 1953 by Stanley L. Miller,

A

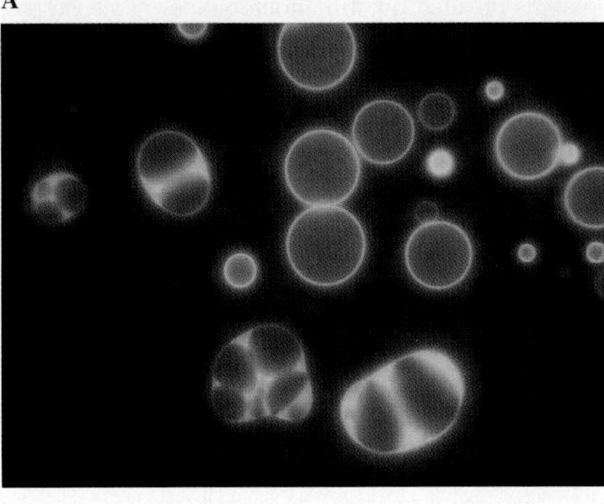

B

36.2 A carbonaceous chondrite meteorite (A) This golf ball–sized fragment is part of a meteorite that fell near Murchison, Australia, in 1969. Tiny particles of organic compounds, accounting for 1 to 2 percent of the fragment's weight, are scattered throughout the stone. (B) When the organic material is extracted, some of the molecules self-assemble into vesicles. The yellow-green color is produced by the fluorescence of polycyclic aromatic hydrocarbons, a class of extremely complex organic molecules.

then a graduate student at the University of Chicago. Miller set up an airtight apparatus in which a mixture of ammonia, methane, water, and hydrogen gases was circulated past electrical discharges from tungsten electrodes (Fig. 36.3). He kept the gases circulating continuously in this way for one week, and then analyzed the contents of his apparatus. He found that an amazing number and variety of organic compounds had been synthesized. Among these were some of the biologically most important amino acids and also such substances as urea, hydrogen cyanide, acetic acid, and lactic acid. To rule out the possibility that microorganisms had contaminated his gas mixture and synthesized the compounds, Miller in another experiment circulated the gases in the same way, but without any electrical discharges; no significant yield of complex organic compounds resulted. In still another experiment, he prepared the apparatus with the gas mixture inside and then sterilized it at 130°C for 18 hours before starting the sparking. The yields of complex organic compounds were the same as in his first experiments. Clearly, the synthesis was not brought about by microorganisms, but was ***abiotic***—a synthesis in the absence of any living organisms, under conditions presumably similar to those on the primitive earth. This experiment by Miller, which gave the first conclusive evidence that some of the steps hypothesized by Oparin could really occur, marked a turning point in the scientific approach to the problem of how life began.

In the years since 1953 many investigators have used mixtures of gases characteristic of volcanic emissions, plus hydrogen cyanide—on the widely accepted assumption that the early atmosphere contained these or similar gases—and have also achieved positive results. Those who have turned to other energy sources, such as ultraviolet light or heat or both, have also obtained large yields; these results are important, because on the early earth ultraviolet light was probably more available as a source of energy than lightning discharges (Fig. 36.4). Significantly, the amino acids most easily synthesized in all these experiments performed under abiotic conditions are the very ones that are most abundant in proteins today, and the most important nitrogenous base (adenine) is also the one most readily produced abiotically (Fig. 36.5).

The wide variety of conditions under which abiotic synthesis of the organic compounds essential to life has been achieved makes it possible to conclude that, even if conditions on the primitive earth were only roughly similar to the ones postulated in the model outlined above, these compounds actually appeared and became dissolved in the waters of the seas. In other words, synthesis of such organic compounds on the early earth was not only possible but highly probable.

It may be argued, however, that even if organic compounds were synthesized abiotically on the primordial earth, they would have been destroyed too fast to accumulate in quantities sufficient for the later origin of living things. After all, most of these organic compounds are known to be highly perishable. But why are they perishable? One reason is that they tend to

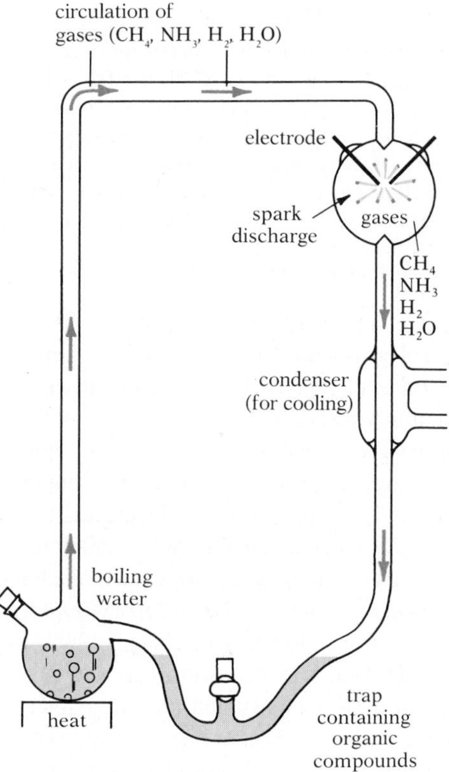

36.3 A simplified drawing of Miller's apparatus for synthesizing organic molecules under abiotic conditions The closed system contained CH_4, NH_3, H_2, and H_2O, gases that would have been abundant if the early atmosphere were reducing. Water in the lower flask was boiled to circulate gases past the electrical sparks in the upper flask. As the products of the reaction passed through the condenser, they were cooled and condensed into liquid that accumulated in the trap, forming compounds like urea, hydrogen cyanide, acetic acid, and lactic acid. Similar results have been obtained with volcanic gases.

36.4 Volcanoes and lightning One of the effective combinations for producing complex organic molecules abiotically is volcanic gas and electrical discharges. Here nature performs a similar experiment during the eruption of Surtsey, off the coast of Iceland.

$HC{\equiv}N$
$HC{\equiv}N$
$HC{\equiv}N$
$HC{\equiv}N$
$HC{\equiv}N$ $\longrightarrow$ [adenine structure: NH_2, C, N, C, N, CH, NH, HC, N, C]

Hydrogen cyanide — Adenine

36.5 Abiotic production of a nitrogenous base Hydrogen cyanide (HCN) can be easily formed from gases believed to have been present in the early atmosphere. Even in the absence of a catalyst, hydrogen cyanide can react with itself at a low rate to produce adenine, an important component of ATP, NAD, RNA, and DNA.

react slowly with molecular oxygen and become oxidized. Another is that they are broken down by organisms of decay, primarily microorganisms. But the primitive atmosphere contained virtually no free oxygen, and there were no organisms of any kind. Therefore, neither oxidation nor decay could have destroyed the organic molecules, and they could have accumulated in the seas over hundreds of millions of years. No such accumulation would be possible today.

Macromolecules Formed From Building Block Molecules Abiotically

Let us suppose, then, that a variety of hydrocarbons, fatty acids, amino acids, purine and pyrimidine bases, simple sugars, and other relatively small organic building block compounds slowly accumulated in the ancient seas. That is still not a sufficient basis for the beginning of life. Macromolecules are needed, particularly proteins (polypeptides) and nucleic acids. Proteins, for instance, are composed of many amino acids joined together by condensation reactions catalyzed by enzymes, but amino acids are stable molecules that do not easily join together outside of the living cell. How could these polymers have formed from the building-block substances present in the "soup" of the ancient oceans? This is not easy to answer, and several hypotheses are currently supported by different investigators.

Some think that the concentration of organic material in the seas was high enough for chance bondings between simpler molecules to give rise, over a period of hundreds of millions of years, to considerable quantities of macromolecules. Other investigators, however, have been unwilling to agree that organic material in the early oceans was sufficiently abundant for the occurrence of chance polymerizations. They have suggested concentration mechanisms that would have speeded up chemical reactions. Building-block compounds, for instance, could have become adsorbed on hydrophilic surfaces of clay particles. Alternatively, small amounts of dilute solutions of building-block compounds could have washed up and formed puddles on the beaches of lagoons and ponds. Solar radiation might have evaporated much of the water from the clay surfaces or puddles, concentrating the organic chemicals, and providing energy for polymerization reactions. The resulting polymers might then have been washed back into the seas or ponds. Such a process could slowly have built up a supply of macromolecules in the water. The hypothesis seems a reasonable one, for Sidney W. Fox of the University of Miami has shown that if a nearly dry mixture of amino acids is heated,

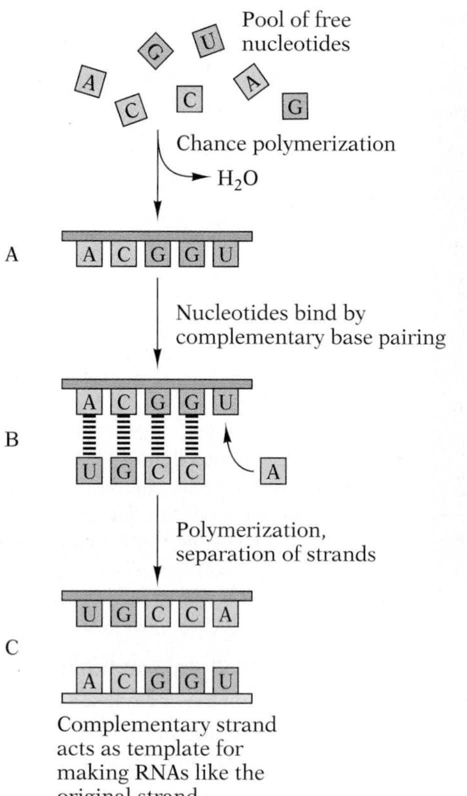

36.6 Abiotic replication of RNA Short chains of RNA could have formed abiotically in the primitive soup (A). If RNA is added to a mixture of activated nucleotides, the nucleotides will join weakly to the RNA by complementary base pairing (B). Polymerization of the complementary chain could have taken place under the conditions that existed on the early earth. The complementary chain could then act as a template for making RNA molecules identical to the original (C). Each molecule can act as a template for making many copies.

polypeptide molecules are indeed synthesized (particularly if phosphates are present). Likewise, short chains of nucleic acids such as RNA can be formed from various mixtures of nucleotides. Short RNA sequences may have formed abiotically in the primitive soup; these might have been able to reproduce themselves exactly if a sufficient pool of nucleotides was present, and if there were appropriate catalysts (even weak ones). A model for RNA replication is shown in Figure 36.6.

Complex Droplets Formed in the "Soup"

We have now reached a point in our model for the origin of life when the "soup" of the ancient seas, or at least that in some estuaries and lagoons, contained a mixture of salts and organic molecules, including polymers such as proteins and perhaps nucleic acids. How could the orderliness that characterizes living things have emerged from this mixture?

Oparin pointed out that, under appropriate conditions of temperature, ionic composition, and pH, collections of macromolecules tend to give rise to complex units called ***coacervate*** droplets. Each such droplet is a cluster of macromolecules surrounded by a shell of water molecules. There is thus a definite boundary between the coacervate droplet and the liquid in which it floats. In a sense, the shell of oriented water molecules forms a membrane around the droplet. Droplets can be formed with enzymes within them; in such cases the enzymes catalyze reactions within the droplet.

Coacervate droplets have a marked tendency to adsorb and incorporate various substances from the surrounding solution; sometimes this tendency, which is selective, is so pronounced that the droplets may almost completely remove some materials from the medium. In this way the droplets may grow at the expense of the surrounding liquid. And coacervate droplets have a strong tendency toward formation of definite internal structure, that is, the molecules within the droplet tend to become arranged in an orderly manner instead of being randomly scattered. Thus, though coacervate droplets are not alive in the usual sense of the word, they do exhibit many properties ordinarily associated with living organisms, and may represent early "prebiological" systems (i.e., prebionts).

Fox, like Oparin, has envisioned that such prebiological systems led to development of the first cells as microscopic multimolecular droplets, but he suggests ***proteinoid microspheres*** rather than coacervate droplets. Fox's microspheres are droplets that form spontaneously when hot aqueous solutions of polypeptides are cooled. The microspheres are more stable than coacervate droplets and exhibit many properties characteristic of cells, including swelling in a hypotonic medium and shrinking in a hypertonic one, formation of a double-layered outer boundary, growth in size and increase in complexity, budding in a manner superficially similar to that seen in yeasts, electrical differences across the boundary, and a tendency to aggregate in clusters of various types resembling those seen in many bacteria (Fig. 36.7).

Vast numbers of different prebiological systems of this kind may have arisen in the seas of the early earth. Although most would probably have been too unstable to last long, some of the early droplets may have contained particularly favorable combinations of materials, especially complexes with catalytic activity, and may thus have developed unusually harmonious interactions between the molecules occurring within them. As such droplets increased in size, they would have been more susceptible to fragmentation, which would have produced new smaller droplets with composition and properties essentially similar to those of the original droplet. These, in turn, would have grown and fragmented again. This primitive reproduction would not initially have been under the control of nucleic acids, even though these compounds could have been synthesized under abiotic conditions and may well have been incorporated into some of the droplets. The next essential step was the evolution of a genetic control system.

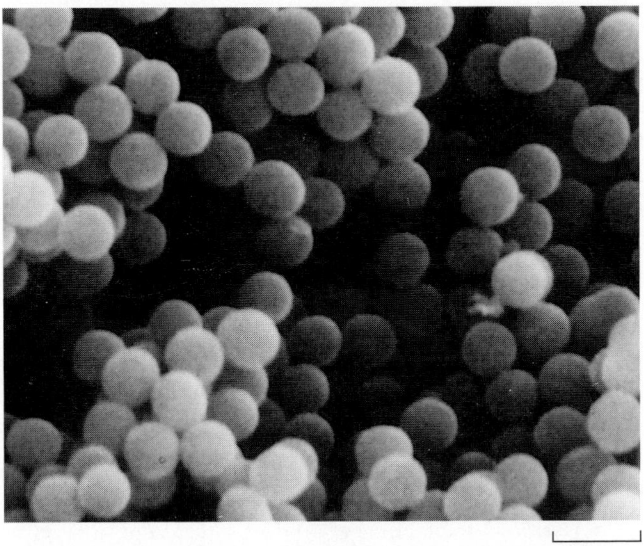

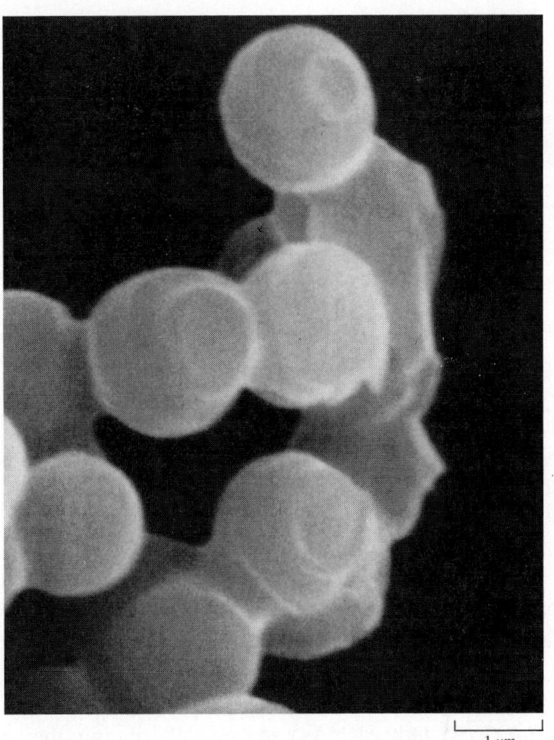

36.7 Scanning electron micrographs of proteinoid microspheres Left: The spheres are remarkably uniform in size, though a few are as much as four times larger than the others. Right: At higher magnification, connecting bridges between the spheres can be seen. The scars on the surface of some of the spheres indicate locations where bridges have been broken.

The First Genes Were Probably Composed of RNA

Although all living cells use DNA for storage of genetic information, the first cells probably used RNA. RNA is single-stranded, and consists of four different nucleotides joined together in varying sequences to form chains. Such chains could have been self-replicating, provided there was a sufficient supply of nucleotides, energy, and catalysts (Fig. 36.6). Today, however, RNA synthesis in cells is catalyzed by a whole array of enzymes working together. Since such enzymes would not have been present in the primordial soup, how could

MAKING CONNECTIONS

1. Why is the concept of biogenesis discussed in Chapter 5 consistent with the belief that life could and did arise spontaneously from nonliving matter on the early earth?

2. How is the internal structure of the earth related to the process of continental drift, described in Chapter 34?

3. What sources of energy would have been available on the primitive earth to drive the chemical reactions that created organic molecules such as amino acids and nitrogenous bases? Why is the absence of an ozone layer in the primitive atmosphere significant?

4. What type of chemical reaction joins together small organic molecules such as amino acids to form polymers?

5. As you learned in Chapter 11, RNA is made in all cellular organisms from a DNA template. How might it have been synthesized abiotically in the "primitive soup"?

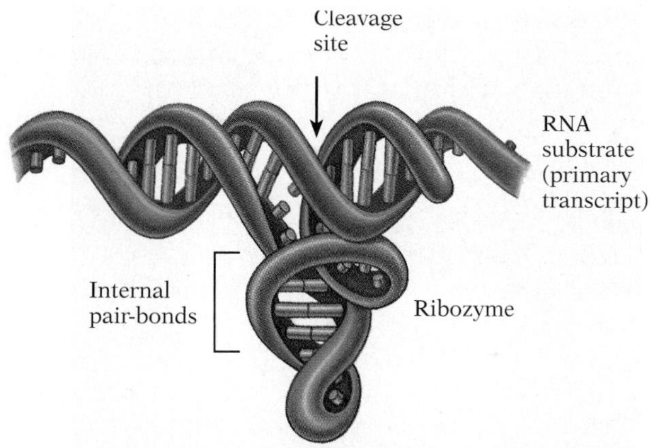

36.8 A ribozyme The 39-nucleotide ribozymes (blue) and mRNA (red) that aid in intron editing bind to the target site on the primary transcript and then cleave a particular bond.

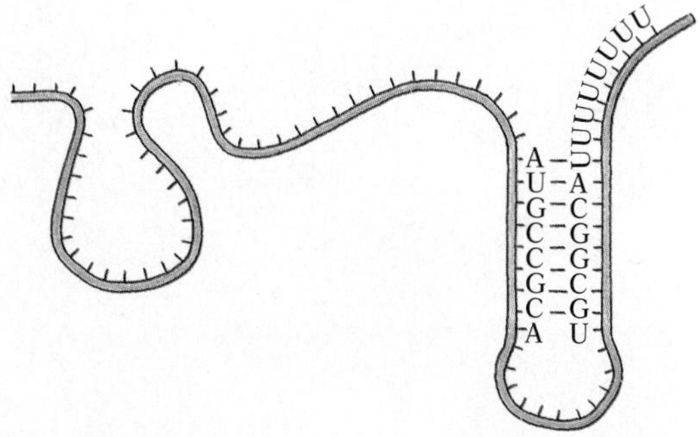

36.9 The folded structure of an RNA molecule Complementary base pairing between parts of an RNA molecule gives each molecule a unique, folded shape.

replication occur? Recently it has been found that some present-day RNAs have catalytic activity. Called ***ribozymes,*** these RNA molecules catalyze specific reactions involving other RNA molecules, such as the cutting of an RNA chain at certain sites (Fig. 36.8). Other RNAs carry on self-splicing: they catalyze the removal of specific nucleotide sequences and then join together the resulting pieces. It is likely that the first organic catalysts were RNA molecules. At an early stage of evolution, then, RNA may have served as both genes and catalysts, only later giving up those roles to DNA and enzymes.

The next step—and the most difficult one for theorists—was the evolution of translation, allowing living systems to use the more versatile proteins rather than RNA as enzymes and structural elements. Just how a correlation between nucleotide sequences in nucleic acids and amino acid sequences in proteins could have arisen remains a mystery. Some ancient RNA systems may have had the capability to catalyze the synthesis of primitive proteins. The key may lie in RNA's structure; because it is single-stranded, different parts of the chain can join by complementary base pairing to produce a molecule that has a precise three-dimensional shape (Fig. 36.9). Perhaps the R groups of different amino acids formed weak attachments with specific sites on particular RNA molecules; once in position the amino acids could more easily be joined together to form a polypeptide. Whatever the mechanism, RNA somehow became involved very early in evolution in directing protein synthesis, for in all living cells it plays a key role in this process: it is messenger RNA that provides the instructions for sequence of amino acids; it is transfer RNAs that bring the amino acids to the messenger RNA; and it is on the ribosomes, themselves composed largely of RNA, that protein synthesis takes place.

The genetic code must have evolved very early, because as far as is known, the genetic code is almost universal. With minor exceptions, the codons have the same amino acid translations in all organisms, from microorganisms to human beings. That so little variation should have arisen in the whole course of cellular existence argues against mere happenstance dictating which codon would go with which amino acid. A careful look at the dictionary of the genetic code (see Table 11.1, p. 216) shows that all codons in which U is the second letter code for hydrophobic amino acids, and that all four electrically charged amino acids (see Fig. 3.19, p. 65) have codons in which the middle letter is a purine (A or G). In short, an ordered chemical system seems to underlie the pairing of codons and amino acids. Though the chemical explanation for that system is unknown, the presumption is that, as it gradually came into being, it made possible both more accurate duplication during the reproductive process and more precise control over the chemical reactions taking place within the droplets. Finally, probably by reverse transcription, DNA came to code for RNA and DNA became the final repository for the genetic information (Fig. 36.10). (DNA, because it is double-stranded, is more stable than RNA, making replication errors less likely.)

The First Primitive Cells Arose About 3.8 Billion Years Ago

Once an accurate genetic control system evolved so that prebionts could faithfully replicate themselves, those with particularly favorable characteristics could have slowly developed into the first primitive cells. Notice that there was almost certainly no abrupt transition from "nonliving" prebionts to "living" cells. The attributes we normally associate with life were acquired gradually. At this stage, the boundary between living and nonliving is an arbitrary one. The first primitive cells may have originated as long ago as four billion years, although the earliest fossils are only about 3.5 billion years old. Life appears to have originated very early in the history of the earth.

According to the view now most widely accepted, the first organisms were single-celled and procaryotic; they did not have a membrane-surrounded nucleus and lacked other mem-

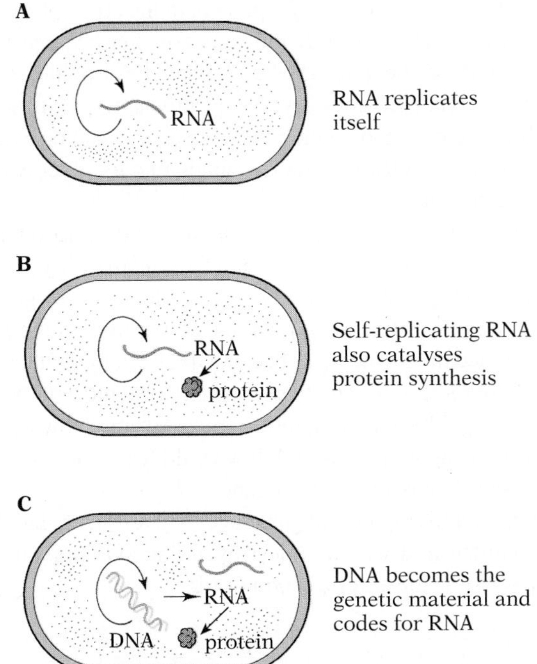

36.10 Evolution of genetic systems (A) The first cells probably used RNA for storage of genetic information. The RNA could have been self-replicating, if the raw materials and an energy source were present. (B) Some self-replicating RNA systems may have developed the ability to catalyze the synthesis of simple proteins, and a primitive form of translation evolved. (C) Finally, DNA, which is more stable than RNA, became the final repository for genetic information and came to code for RNA. Proteins became the major catalysts of the cell, and RNA functions as the middleman between the two.

branous structures present in the cells of all other types of organisms. The cells must have been anaerobic and probably heterotrophic, since they could utilize the carbohydrates, amino acids, and other organic compounds free in the environment in which they lived. In other words, they depended on previous abiotic synthesis of organic compounds. But, as the organisms became more abundant and more efficient at removing nutrients from the medium, they began to deplete the supply of nutrients. The rate of spontaneous formation of organic matter from inorganic raw materials was most likely never very high, and it must have taken many millions of years for a moderate supply of nutrients to accumulate. With the supply being depleted at an ever increasing rate, and the drain on available resources becoming more and more severe, the competition between organisms for available nutrients must have increased, too. Forms inefficient at obtaining nutrients perished; those more efficient survived in greater numbers. Natural selection would have favored any new mutation that enhanced the ability of its possessor to obtain or process food.

At first, the primitive organisms probably carried out relatively few complex biochemical transformations. They could obtain most of the materials they needed ready-made. But it would have been these very materials—materials that could be used directly with little alteration—that would have dwindled most rapidly. There would therefore have been strong selection for any organisms that could use alternative nutrients. Suppose, for example, that Compound *A*, which was necessary for the life of cells, was initially available in the medium, but its supply was being rapidly exhausted. If some cells possessed a gene that coded for an enzyme *a* that catalyzed synthesis of *A* from another compound, *B*, in greater supply in the medium, then those cells would have had an adaptive advantage over cells that lacked the gene. They could survive even when *A* was no longer available in the medium by carrying out the reaction

$$B \xrightarrow{a} A$$

But then there would have been increasing demand for free *B*, and the rate of its utilization would soon have exceeded the rate of its abiotic synthesis. Thus the supply of *B* would have dwindled, and there would have been strong selection for any cells possessing another gene that coded for an enzyme *b*, catalyzing synthesis of *B* from *C*. These cells would not have been dependent on a free supply of either *A* or *B*, because they could make both *A* and *B* for themselves as long as they could obtain enough *C*:

$$C \xrightarrow{b} B \xrightarrow{a} A$$

This process of evolution of synthetic ability might have continued until eventually most cells made all the *A* they required by carrying out a long chain of chemical reactions:

$$E \xrightarrow{d} D \xrightarrow{c} C \xrightarrow{b} B \xrightarrow{a} A$$

In this way the primitive cells would slowly have evolved more elaborate biochemical capabilities. The earliest form of metabolism was almost surely glycolysis and fermentation, processes universal in living organisms today (of the three main energy-yielding systems—glycolysis, aerobic respiration, and photosynthesis—only glycolysis and fermentation are universal, occurring in all living cells).

Some Procaryotic Cells Evolved Autotrophic Pathways

Even though the primitive heterotrophs probably evolved more and more elaborate biochemical pathways that enabled them to use a greater variety of the organic compounds free in their environment, and even though some of them probably evolved diverse methods of feeding on living and dead organic material, life would eventually have ceased if all nutrition had remained heterotrophic. The reason is not only that nutrients must have been used up much faster than they were being synthesized but also that the organisms themselves must have been altering the environment in ways that decreased the rate of abiotic synthesis of organic compounds. For example, their metabolism, which would have involved fermentation in the absence of molecular oxygen (see p. 165), would have released carbon dioxide into the atmosphere. Abiotic synthesis of complex organic compounds from carbon dioxide is much less likely than from methane or hydrogen cyanide as found in the early atmosphere.

That life did not become extinct as the supply of free organic compounds dwindled is attributable to the evolution of organisms capable of synthesizing their own food from inorganic nutrients. The first such autotrophic pathways were almost certainly chemosynthetic, utilizing the energy in inorganic molecules such as molecular hydrogen (H_2), ammonia, nitrite, or sulfur. The energy released in chemical reactions involving these compounds was used to synthesize many of the organic compounds no longer available from the "soup," as well as many new compounds. Chemosynthetic autotrophs are still found today, particularly in bogs and at volcanic vents on the ocean floor.

Sometime around 3.6 billion years ago, an enormously important biochemical event occurred: certain organisms acquired the ability to capture the sun's energy directly and use it to synthesize ATP. The first photosynthetic pathway was probably similar to that of cyclic photophosphorylation, during which light energy is used indirectly to synthesize ATP. The first photosynthetic autotrophs did not use chlorophyll as their light-absorbing pigments, nor did they split water and produce oxygen. Instead they evolved a number of different light-absorbing pigments and carried on a process similar to cyclic photophosphorylation. The present-day anaerobic photosynthetic bacteria may be the direct descendants of those organisms.

Later, the much more complex pathways of noncyclic photophosphorylation and carbon dioxide fixation (see pp. 153–155) evolved, probably appearing first in ancestors of cyanobacteria as early as 3 to 3.5 billion years ago. These organisms, like true plants but unlike most photosynthetic bacteria, use water as the electron source in noncyclic photophos-phorylation and therefore release molecular oxygen as a by-product. From this time onward, the continuation of life on earth depended on the activity of the photosynthetic autotrophs.

The evolution of noncyclic photophosphorylation using water as the electron donor probably administered the *coup de grace* to significant abiotic synthesis of complex organic compounds. An important by-product of such photosynthesis is molecular oxygen (O_2). At first the molecular oxygen released by photosynthesis did not accumulate in the water and atmosphere; instead it combined with iron dissolved in the oceans to form iron oxides, which precipitated and settled on the ocean floor. Oxygen began to accumulate at low levels in certain habitats by about 2.8 billion years ago and is thought to have reached its present level of 21 percent in the atmosphere about 540 million years ago. Once molecular oxygen became a major component of the atmosphere, both heterotrophic and autotrophic organisms could evolve the biochemical pathways of aerobic respiration, by which far more energy can be extracted from nutrient molecules than is obtainable by glycolysis and fermentation alone.

As molecular oxygen accumulated, the atmosphere became an oxidizing atmosphere, and the so-called ***oxygen revolution*** took place. Some of the molecular oxygen was converted into ozone (O_3), forming a layer high in the atmosphere. Once the ozone layer became thick enough (at least 500 million years ago), it effectively screened out most of the lethal high-energy ultraviolet radiation from the sun. Now land was safe for living organisms and they could leave the ultraviolet-absorbing oceans and move to land.

Note that living organisms, once they arose, changed their environment in a way that destroyed the conditions that had made possible the abiotic origin of life.

The Earliest Fossils Are Procaryotes

The oldest fossils known today are from two sites: one in Western Australia and the other in southern Africa; both are dated at about 3.5 billion years (Fig. 36.11). They appear to be bacteria—procaryotic cells, as would be expected of the oldest fossils. Many of these early procaryotes lived in complex mi-

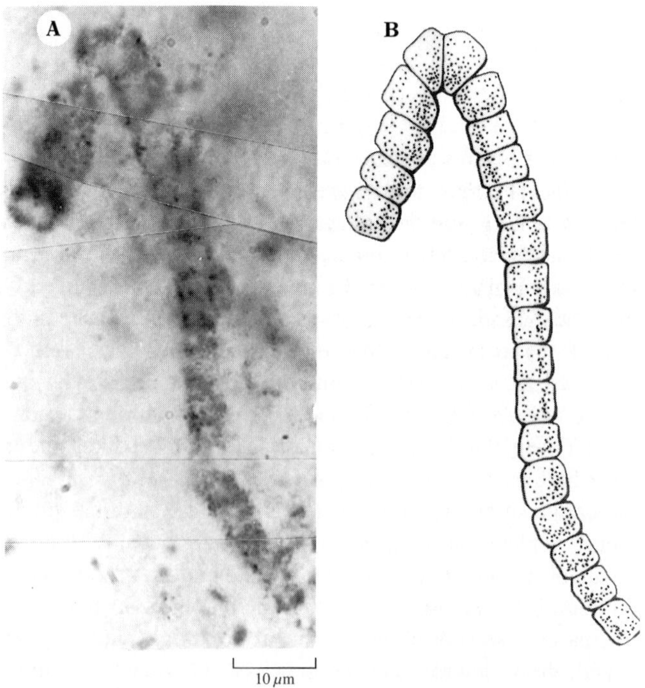

36.11 Filamentous procaryotic microfossil from Western Australia The bacteriumlike organism shown in this photograph (A) and corresponding drawing (B) lived about 3.6 billion years ago.

MAKING CONNECTIONS

1. Why is it thought that RNA molecules were both the first genetic material and the first organic catalysts?

2. What is the evidence that the genetic code evolved very early and that the pairing of specific codons and amino acids did not evolve by chance?

3. Why is DNA superior to RNA as a repository for genetic information?

4. Explain why natural selection led to the evolution of increasingly complex biochemical pathways in primitive organisms.

5. Why would fermentation carried out by primitive heterotrophs have led to an increased carbon dioxide concentration in the atmosphere? How would this increase have affected abiotic synthesis of organic compounds?

6. How did the evolution of noncyclic photophosphorylation revolutionize the ecology of life on earth?

7. In environments in which oxygen is available, what is the advantage of aerobic respiration compared to glycolysis and fermentation?

crobial communities, called bacterial mats, that covered the bottoms of the shallow seas and tidal pools. Scattered patches of these bacterial mats formed layered mounds that, when fossilized, are called ***stromatolites.*** The stromatolite ecosystems probably appeared about 3.5 billion years ago, and resemble the bacterial mats found in certain harsh environments today (Fig. 36.12). These sedimentary mounds form because the bacterial mats, which contain diverse bacteria and cyanobacteria, secrete a sticky substance that traps materials sweeping over them. Eventually the sediments cover the bacterial layer, and the bacteria multiply upward through the sediments to the surface and establish a new layer. The process repeats itself year after year, producing a mound of banded sediment layers that may become infiltrated with enough calcium carbonate to make them hard. Populations of cyanobacteria and other bacteria different from the mat organisms probably existed in the open water surrounding the ancient bacterial mats.

For the next two to 2.5 billion years the only living organisms on earth were procaryotes—archaebacteria, bacteria, and cyanobacteria—which occupied a variety of diverse environments.

36.12 Stromatolites Stromatolites are layered deposits of bacterial mats and trapped sediments that were once abundant but are now formed only in certain suitable environments such as the hot and salty environment of Western Australia's Shark Bay. The mats of photosynthetic cyanobacteria secrete a sticky substance that traps materials sweeping over them. When the sediments cover the bacterial layer, the bacteria move to the surface and establish a new layer. After many years a mound of banded sediment layers is produced, such as these shown at the left.

Eucaryotic Cells Evolved About 1.8 Billion Years Ago

Although more and more fossils of procaryotic organisms from the first 2.5 billion years or so of life are being found, there is virtually no direct evidence concerning the evolution of the first eucaryotic cells. The oldest fossils of eucaryotic cells are at least 1.7 billion years old, but these are already rel-

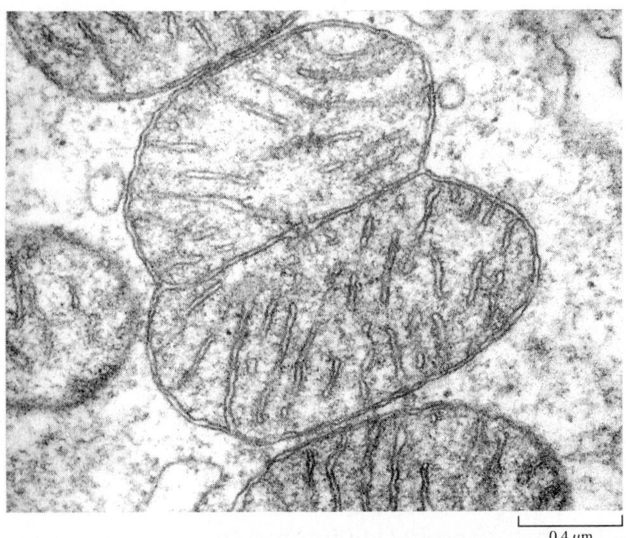

36.13 Electron micrograph of a dividing mitochondrion The partition between the two daughter mitochondria is nearly complete.

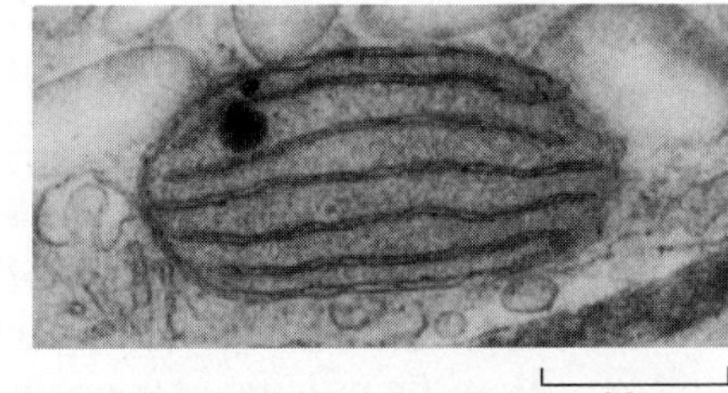

36.14 Chloroplast of a red alga compared with a whole cyanobacterium The red algal chloroplast (top) resembles an entire cyanobacterium (bottom); in both structures the photosynthetic pigments are on membranous lamellae that are free in the cytoplasm.

atively complex, and hence shed little light on the question of their derivation. Nonetheless, a growing understanding of the special properties of eucaryotic organelles such as chloroplasts and mitochondria has led to an intriguing hypothesis regarding the origin of the complicated kind of cell that is the structural unit of higher plants and animals.

As we saw in Chapters 6 and 11, chloroplasts and mitochondria are self-replicating bodies (Fig. 36.13). They are about the size of procaryotes (from one to 10 micrometers in diameter), lack nuclear membranes, have a single circular chromosome, and can carry out protein synthesis on their own ribosomes. In short, they have many features in common with free-living procaryotic organisms. Might these organelles be modern descendants of ancient procaryotic cells that were "swallowed" by host cells and became permanent residents, evolving in concert with their hosts ever since? Several investigators think the answer is yes. It was noted as long ago as the 19th century, when chloroplasts were first studied under the microscope, that they resembled certain free-living cyanobacteria, and the suggestion was made that they might have originated as such organisms (Fig. 36.14). A bacterial origin for mitochondria was proposed in the 1920s. More recent research findings have tended to support the suggestion that chloroplasts are derived from photosynthetic procaryotes and that mitochondria are derived from aerobic bacteria. Let's review and summarize here the evidence for the ***endosymbiont hypothesis,*** as put forward by Lynn Margulis of the University of Massachusetts.

1. The enzymes for synthesis of DNA, RNA, and protein in chloroplasts and mitochondria are qualitatively like those of procaryotic cells and correspondingly unlike those in the rest of the plant cell.
2. Unlike the nuclear DNA of eucaryotic cells, which is wound on nucleosomes, the DNA of chloroplasts and mitochondria is nearly naked, like that of procaryotic cells.
3. The single chromosome of chloroplasts and mitochondria is usually circular, like the bacterial chromosome.
4. The ribosomes of chloroplasts and mitochondria, like those of procaryotes, tend to be smaller than the cytoplasmic ribosomes of eucaryotes.
5. Gene expression in chloroplasts and mitochondria is similar to that of procaryotes; both use the same control sequences and the genes in both organelles lack the introns that are characteristic of eucaryotic genes.

In summary, a growing list of characteristics attests to a possible derivation of these two organelles from procaryotic ancestors. If chloroplasts and mitochondria did indeed arise as endosymbionts, the symbiosis has clearly been a mutualistic one. The resident procaryotes enabled the anaerobic host cell to become tolerant of oxygen, and to use the oxygen for aerobic respiration, and, if chloroplasts were present, to use light energy to synthesize their own organic compounds. In return, the host organism provided a stable, protective environment for the smaller organisms and eventually performed other functions, such as protein synthesis, for them. This mutualistic symbiosis evolved eventually into a permanent relationship, the eucaryotic cell.

The capacity of unicellular organisms to live symbiotically inside the cells of other organisms—as the ancestors of chloroplasts and mitochondria are presumed to have done—can be amply demonstrated by present-day examples. A variety of protozoans contain single-celled algae, and the cells lining the gastrovascular cavity of *Chlorohydra* and many species of sea anemones and coral also contain algal cells (see Fig 38.14, p. 888). Even some complex multicellular animals, especially several species of molluscs, regularly have intracellular algal symbionts.

The cyanobacteria have always seemed to be the likely ancestors of most chloroplasts, since they have chlorophyll *a* and use the same noncyclic photophosphorylation pathways used in chloroplasts. In 1976, however, a new division of procaryotes, the Prochlorophyta, was discovered. These bright green unicells have both chlorophyll *a* and *b* (cyanobacteria have only *a*) and carotenoids, and show no traces of the red or blue pigments found in cyanobacteria. In short, these organisms resemble eucaryotic chloroplasts much more closely than the cyanobacteria do, and have a pigment system like that of the green algae and land plants. Could the prochlorophytes have been the ancestors of chloroplasts? Until recently it was thought not, because the only known representatives of the group were two exceedingly rare forms. Recently though, oceanographers have identified a minute prochlorophyte that is an abundant component of the marine plankton (Fig. 36.15). Found in concentrations as large as 100,000 cells per milliliter, this organism is so tiny that it was not picked up by the usual microscopic techniques. These organisms have now been found in almost all of the earth's oceans. With this finding there is renewed interest in the prochlorophytes as the ancestor of the chloroplast. Studies comparing the DNA sequences of cyanobacteria, prochlorophytes, and chloroplasts are now under way in an attempt to solve the problem.

Mitochrondria were probably derived from aerobic bacteria; the inner mitochrondrial membrane may be viewed as the equivalent of the bacterial plasma membrane, since the two are biochemically similar. The outer mitochondrial membrane resembles the endoplasmic reticulum, and was probably derived from the host cell during the period of common evolution.

Though the organelles discussed here contain DNA, and divide, grow, and differentiate partly on their own, they are not fully autonomous. Many of their proteins are specified by nuclear genes. Presumably the symbionts gave up much of their genetic control to the host during the hundreds of millions of years since they last lived as free cells. This surrender of control to the nuclear genes is probably the reason why the minichromosomes of chloroplasts and mitochondria are less than one tenth the size of typical procaryotic chromosomes.

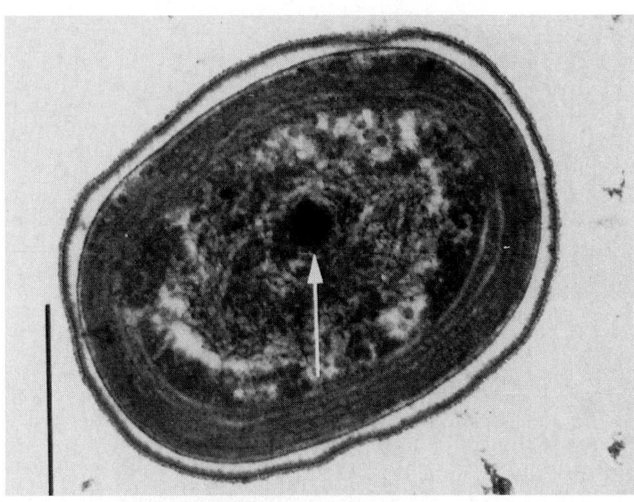

36.15 A prochlorophyte The single-celled prochlorophytes have the same pigments found in land plants (chlorophylls *a* and *b* and carotenoids) and may have been the ancestors of the chloroplasts in eucaryotic cells. The arrow points to an area that contains a key autotrophic enzyme.

In contrast to the wealth of data and speculation on chloroplasts and mitochondria, there is little evidence on the origin of the nuclear membrane and the associated endoplasmic reticulum (ER). Since membranes flow from the nuclear membrane to the ER to the Golgi apparatus (see Fig 6.11, p. 121), the nuclear membrane may have been the first to evolve, with the other parts arising from it later (Fig. 36.16). How the nuclear membrane itself arose is still a mystery. Most researchers believe that the nuclear membrane was an early acquisition, before the symbiotic union that gave rise to mitochondria. This is supported by the existence of a number of single-celled anaerobic eucaryotic organisms living today (Fig. 36.17). These are among the oldest known nucleated cells, their ancestors having diverged from the mainstream of eucaryotic evolution about 1.8 billion years ago. The endosymbiotic acquisition of mitochondria and the development of the ER and Golgi are likely to have come later, with chloroplasts being the most recent additions.

The evolution of the eucaryotic cell represents another landmark in the evolution of life, for the eucaryotic cell, with its potential for genetic diversity, is the cell that gave rise to all the eucaryotic lineages: protists, plants, animals, and fungi. With the origin of this cell, sexual reproduction became possi-

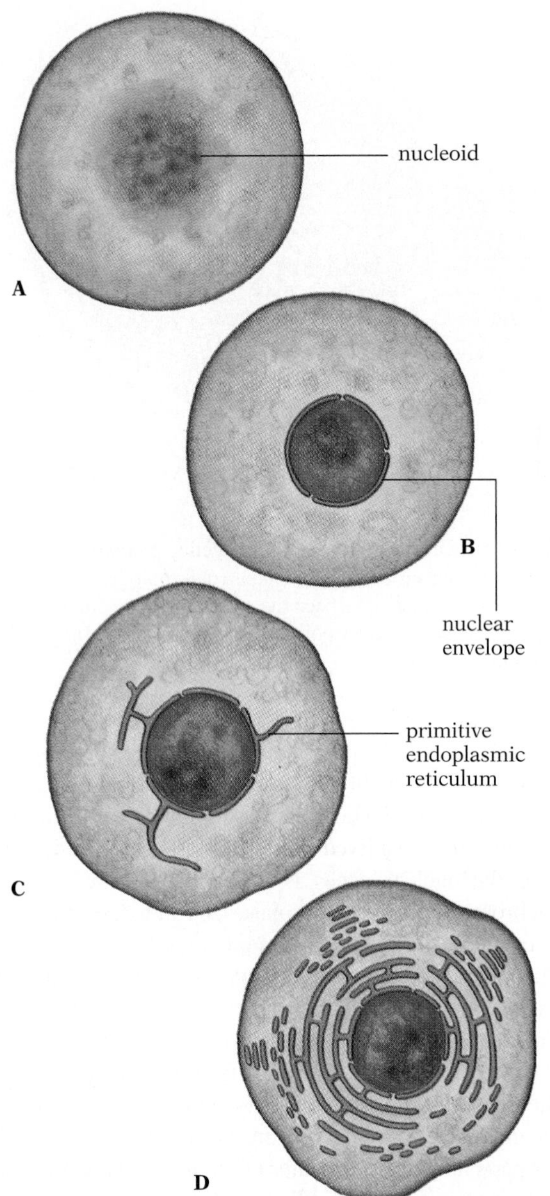

36.16 Model for the evolution of the nuclear envelope and the cytoplasmic membrane system (A) The ancestral cell is presumed to have been procaryotic—having a nucleoid with no boundary membrane. (B) Membranous vesicles may have formed as new structures surrounding the nuclear area, giving rise to the double nuclear envelope. (C) Evaginations from the outer membrane of the nuclear envelope may have given rise to the endoplasmic reticulum, which itself became elaborated (D) into an extensive cytoplasmic membrane system.

ble. While there are primitive forms of sexual recombination in some bacteria, most reproduction is asexual, by binary fission (see Fig 9.2, p. 180). In such organisms, mutation provides the only variation. With the eucaryotic cell, however, came sexual reproduction with the attendant processes of meiosis and fertilization, and genetic variability—the raw material for natural selection—increased.

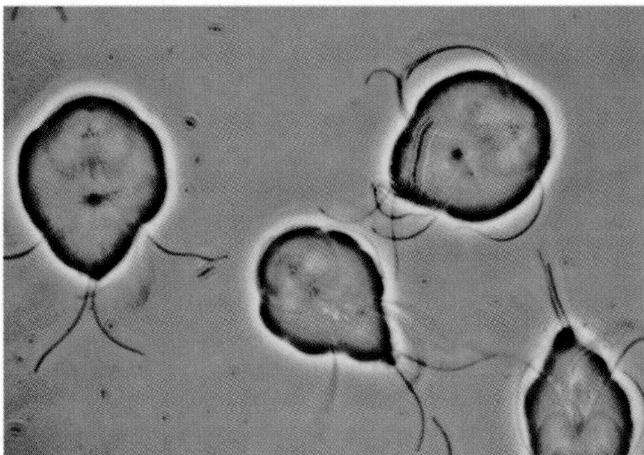

36.17 *Giardia lamblia,* a single-celled anaerobic eucaryote *Giardia* is a representative of a lineage that dates back to the oldest known nucleated cells. It has a nuclear membrane and is therefore eucaryotic, but lacks mitochondria and is anaerobic. *Giardia* can cause serious intestinal infections in humans.

A Great Diversity of Animals Appeared During the Cambrian Period

Although there are many fossils of procaryotic organisms during the Precambrian, the oldest geologic period from which fossils of multicellular forms of life are fairly abundant is the ***Cambrian,*** which began about 590 million years ago (Fig. 36.18). Near the beginning of this period complex animals "suddenly" increased in size and diversity in a veritable explosion of animal diversity—the "Big Bang" of animal evolution (see Fig. 32.25, p. 726). The diversity of design among the various animals that arose is staggering. Many of the Cambrian fossils are of relatively complex organisms—how complex is suggested by the fact that most of the invertebrate animal phyla existing today appeared at that time. The ancient body plans that evolved during the Cambrian persist, with modifications, to this day; few new body plans have appeared since then.

By the end of the Cambrian, the ozone layer was well developed and it was safe for organisms to live on land. Plants invaded the land about 450 million years ago, and invertebrate animals followed. Land was an "open niche" and a great burst of evolutionary diversification occurred. Some of the major events in the evolution of life, including mass extinctions, are summarized in the time line shown in Figure 36.19.

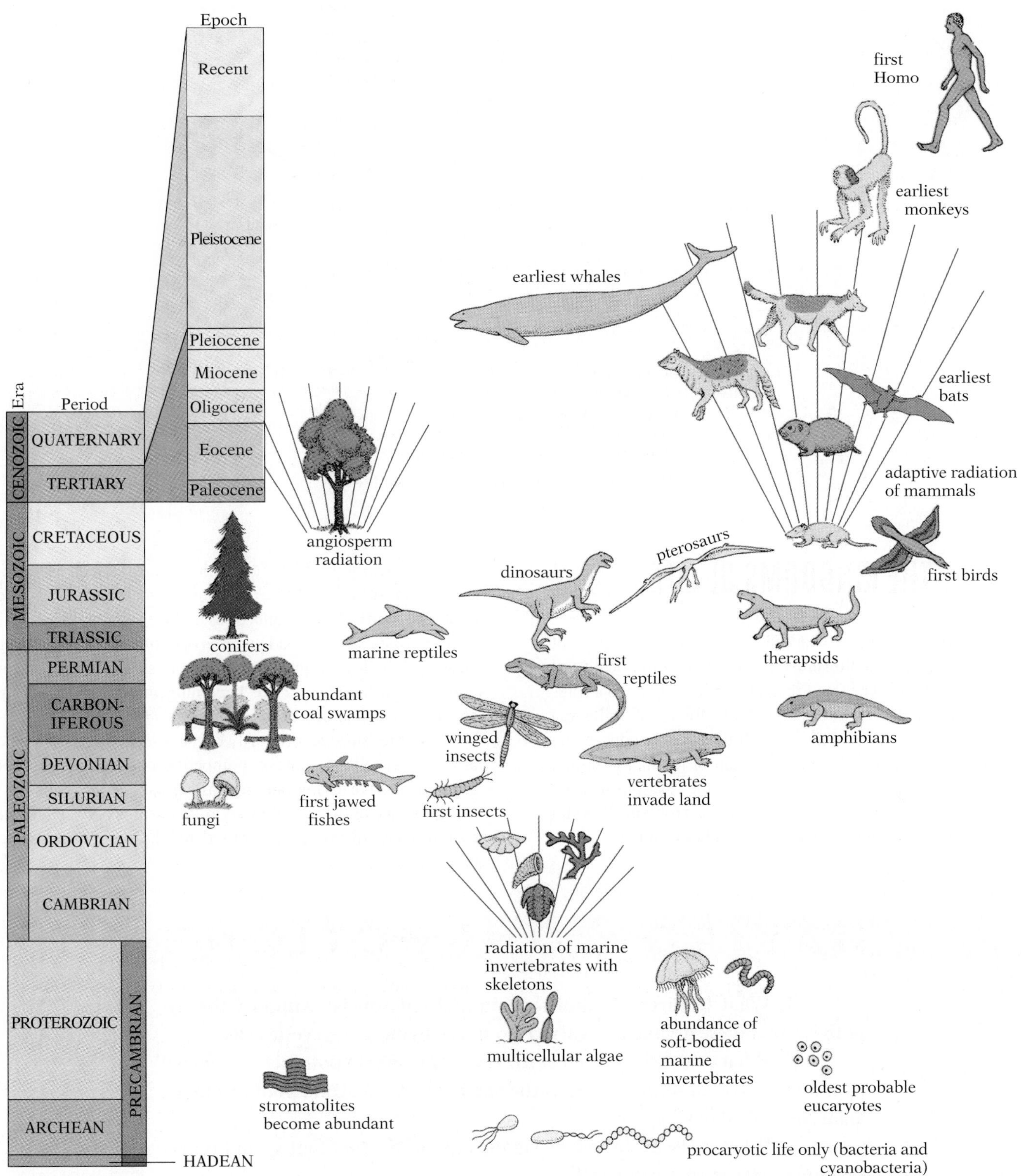

36.18 Timeline of the history of life The history of life dates back more than 3.5 billion years, when the first procaryotic cells evolved. Procaryotes were the only life forms until about 1.8 billion years ago when the first eucaryotes evolved. Most of the diversity of life as we know it arose during the last 500 million years. Diversity was reduced by periodic mass extinctions.

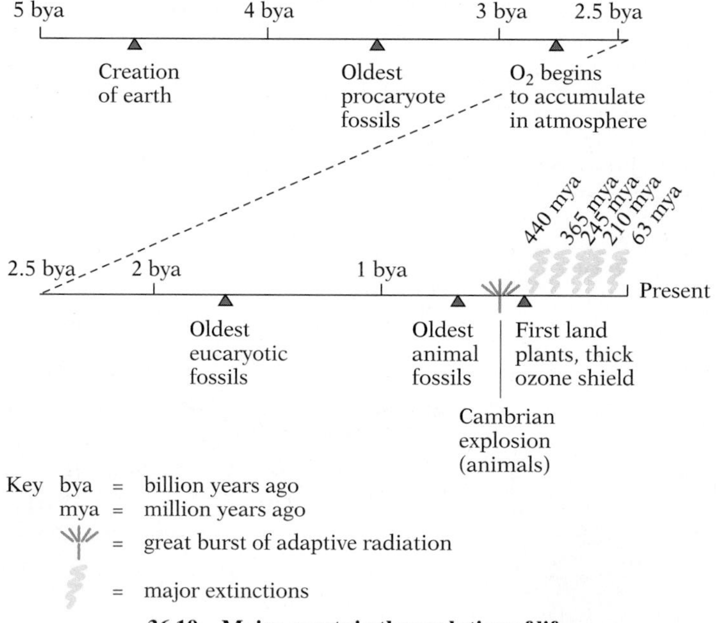

36.19 Major events in the evolution of life

THE KINGDOMS OF LIFE

Ideally, living organisms should be classified on the basis of their evolutionary relationships, i.e., on their commonality of ancestry. Because very few Precambrian fossils of eucaryotes have survived the destructive geologic forces of millennia (Fig. 36.20), however, we have very little evidence concerning the early radiation of life and the origin of the major animal phyla and plant divisions. For instance, the relationships between the major groups of algae remain a problem. It is not known whether fungi evolved from photosynthetic green algae, or directly from heterotrophic organisms such as bacteria, or from some other stock. It is uncertain how protozoa are related to multicellular plants or to multicellular animals.

Nevertheless, the broad patterns of evolution are becoming clear, and the logical groupings that any system of classification seeks to emphasize are becoming more apparent (Fig. 36.21). The genealogy of life resembles a tree; few of the earliest branches survive, primarily because most die in the shade of later branches—that is, competition from more recent groups can drive earlier ones to extinction.

Ignorance of the evolutionary relationships between major groups of organisms has never hindered the age-old attempt to

MAKING CONNECTIONS

1. What features of mitochondria and chloroplasts support the hypothesis that these organelles evolved from free-living procaryotic cells?

2. What evidence to support the endosymbiotic hypothesis is provided by mutualistic relationships such as the one between coral and photosynthetic algae, described in Chapter 34?

3. What is the evolutionary significance of the fact that sexual reproduction occurs in eucaryotic cells?

4. How did the development of the atmospheric ozone layer contribute to the diversification of life?

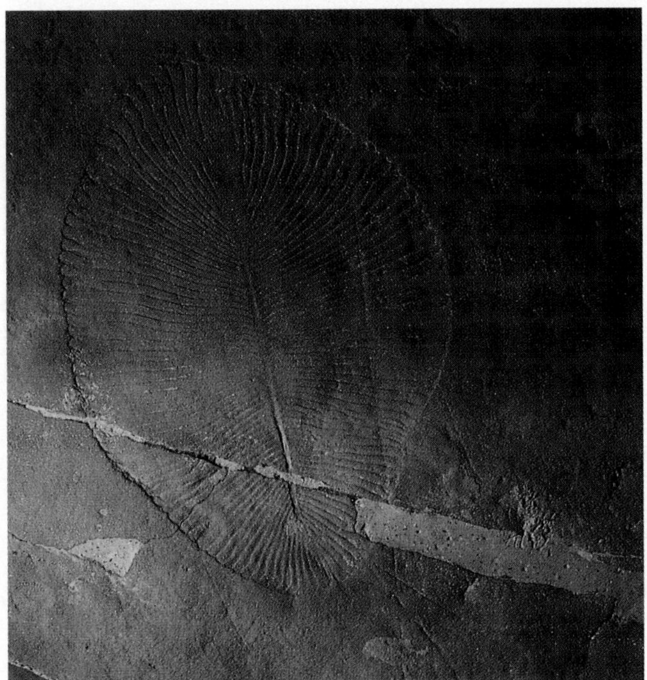

36.20 Fossil of a Precambrian animal (*Dickinsonia costata*) from South Australia This animal, which appears to have been a segmented worm of some sort, lived some 600 million years ago. Reproduced natural size.

assign all living things to one or another of a few large categories called kingdoms. Because the evolutionary relationships are often far from clear, other criteria, such as overall similarity and unique characters, are often used to create logical groupings. New molecular techniques such as electrophoresis, DNA hybridization, amino acid sequencing, and DNA and RNA sequencing are providing additional data for determining relationships at the molecular level; eventually these techniques may resolve the fundamental questions about evolutionary relationships as well. Depending on the criteria chosen, a number of different systems of classification are possible. The system used in this book recognizes six kingdoms: Eubacteria and Archaebacteria (formerly classified together in the kingdom Monera), Protista, Plantae, Fungi, and Animalia.

Members of the Kingdoms Eubacteria and Archaebacteria Are Procaryotic Cells

One characteristic used to classify organisms into kingdoms is the type of cell present. The various kinds of bacteria differ from all other forms of life in a very fundamental way: they are procaryotic cells, whereas the cells of all other organisms are eucaryotic. Molecular comparisons have provided data indicating that there are actually two major groups of bacteria, the archaebacteria and the eubacteria, each as distantly related to the other as they are to the eucaryotes. We therefore will recognize two kingdoms of procaryotes, the Archaebacteria and the Eubacteria. The two groups are thought to have arisen from a common ancestral cell and to have diverged very early in the evolution of life.

The Eucaryotic Groups Are Classified on the Basis of Level of Organization and Nutritional Mode

One major problem is the classification of eucaryotic unicellular organisms. The unicellular eucaryotes that zoologists have

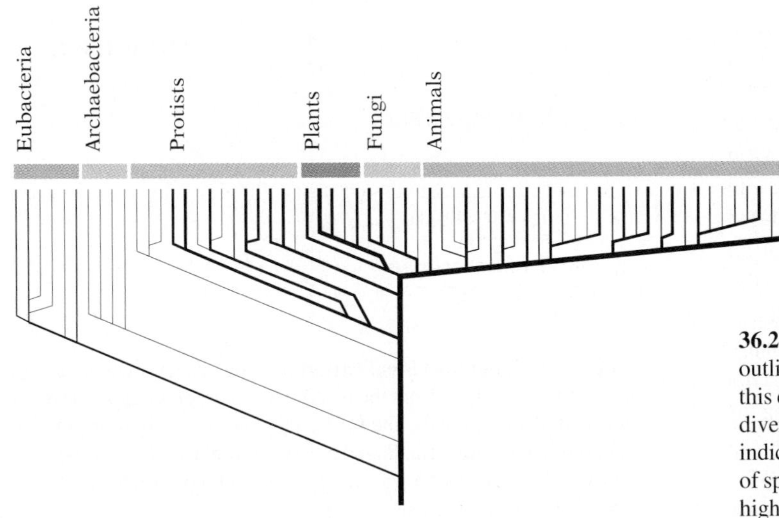

36.21 Schematic view of the evolution of major phyla The basic outlines of the evolutionary history of modern phyla are indicated in this diagram. Extinctions create a bias toward more recently evolved diversity. The kingdom groupings that will be used in this text are indicated at the top. Line thickness suggests the approximate number of species in each phyla. The details of the branching patterns are highly tentative.

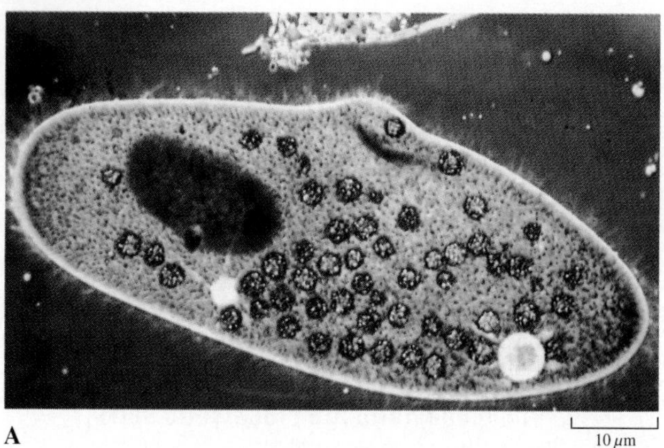

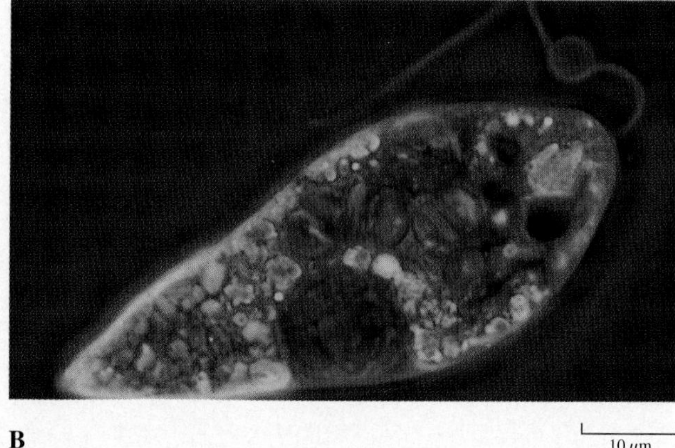

36.22 Protist diversity *Paramecium* has cilia and is heterotrophic (A), whereas *Euglena* has flagella and chloroplasts (B).

traditionally called protozoans have been particularly troublesome, especially the protozoans known as flagellates (Fig. 36.22). These creatures have long flagella that enable them to swim actively in a manner intuitively felt to be "animallike." Yet some of them possess chlorophyll and carry out photosynthesis, a characteristic ordinarily considered decidedly "plantlike." How can organisms such as these be classified? Should they be put in the plant or animal kingdom? Whatever the criteria chosen, it is impossible to make a clean separation between plants and animals at the unicellular level. Unicellular organisms (and some multicellular ones) are at an evolutionary stage where it is essentially meaningless to talk about "fungi," "plants," and "animals"—artificial human categories not dictated by the rules of nature. At the lowest evolutionary levels, about the only distinction between plants and animals that stands up is this: plants are living things studied by people who say they are studying plants (botanists), and animals are living things studied by people who say they are studying animals (zoologists).

Facetious as this distinction sounds, it is basically accurate. The unicellular green flagellates may reasonably be classified as both plants and animals because they are studied by both botanists and zoologists, the first calling them algae and the second calling them protozoa. An alternative that has won much support is to assign the eucaryotic unicellular and simple multicellular organisms to a separate kingdom, designated ***Protista.*** This has the advantage of allowing a clear separation between fungi, plant, and the animal kingdoms, because all the hard to categorize forms are lumped together in the Protista.

The protistan kingdom includes a variety of groups, many of which are probably not closely related evolutionarily. This is a grouping of convenience, placing together all the organisms at a similar level of organization, i.e., those eucaryotic organisms that are unicellular during most or all of their lives. Also included in the Protista as we define it are those plantlike multicellular organisms whose bodies show relatively little distinction between tissues.

The three multicellular eucaryotic lineages are classified into kingdoms on the basis of the exploitation of three different modes of nutrition—photosynthetic autotrophism by plants (kingdom Plantae), absorptive heterotrophism by fungi (kingdom Fungi), and ingestive heterotrophism by animals (kingdom Animalia).

The six-kingdom arrangement we will use in discussing the diversity of life is shown diagrammatically in Figure 36.23.

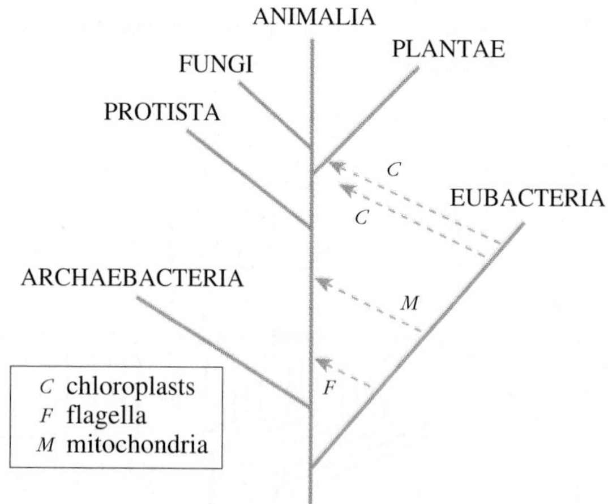

36.23 Six-kingdom classification scheme The protists are seen as a group that diverged from the rest of the eucaryotes *before* the divergence of plants, animals, and fungi. The tree also assigns origins for eucaryotic mitochondria, flagella, and chloroplasts. There is no consensus about how the various groups of photosynthetic protists obtained their chloroplasts.

CONCEPTS IN BRIEF

1. Scientists today believe that life could and did arise spontaneously from nonliving matter under the conditions that prevailed on the early earth, and that all present earthly life has descended from these beginnings.

2. The solar system was probably formed between 4.5 and 5 billion years ago from a cloud of cosmic dust and gas. As the earth condensed, the heavier materials moved toward the center and lighter substances were concentrated at the surface.

3. The lighter gases of the earth's first atmosphere escaped into space. The intense heat in the interior of the earth drove out various gases through volcanic action; these formed a second atmosphere. The atmosphere of the early earth was quite different from today's oxidizing atmosphere; it contained no molecular oxygen.

4. As the earth's crust cooled, the water vapor in the atmosphere condensed into rain and began to form the oceans, in which gases from the atmosphere and salts and minerals from the land dissolved. Ultraviolet radiation, lightning, and heat could have provided the energy for reactions that produced organic building-block molecules from the combination of these substances. These organic compounds could have accumulated slowly in the seas over millions of years since they would not have been destroyed by oxidation or decay.

5. The building-block molecules accumulating in the hot soup could have joined together by chance binding or through some concentrating mechanisms to form complex organic molecules such as polypeptides, nucleic acids, carbohydrates, and lipids. Short RNA sequences may have formed abiotically; these might have been able to reproduce themselves exactly if nucleotides and appropriate catalysts were present.

6. Coacervate droplets and proteinoid microspheres are examples of prebiological systems that show many properties of living cells. Systems similar to these may have formed in the ancient seas.

7. The first genes were probably composed of RNA rather than DNA. Some RNAs have catalytic activity, cutting other RNA chains at certain sites; others are self-splicing. Primitively, RNA may have served as both genes and catalysts, only later giving up those roles to DNA and enzymes.

8. Eventually, the sequences of the nucleotides in the nucleic acids came to code for the sequences of amino acids in protein, and transcription and translation evolved. Later DNA, which is more stable than RNA, became the repository for genetic information.

9. The earliest organisms were anaerobic, procaryotic heterotrophs that obtained energy from the nutrients available in the early ocean. As the nutrients disappeared, competition between the organisms must have increased. Natural selection favored any new mutation that enhanced an organism's ability to obtain or process food.

10. As the free nutrients were used up, some organisms evolved the ability to use the energy stored in inorganic molecules, and chemosynthetic organisms evolved. Later, some organisms evolved the ability to use the sun's energy to make ATP. Cyclic photophosphorylation probably evolved first. Later, the more complex pathways of noncyclic photophosphorylation and carbon dioxide fixation evolved. In these processes energy from sunlight is used in synthesis of carbohydrate from carbon dioxide and water. From this time onward, the continuation of life on earth depended on the activity of the photosynthetic autotrophs.

11. An important by-product of noncyclic photophosphorylation is molecular oxygen. The O_2 released by photosynthesis converted the atmosphere into an oxidizing atmosphere in the *oxygen revolution.* The oxygen also gave rise to a layer of ozone in the upper atmosphere that shields the earth's surface from intense ultraviolet radiation. In other words, living organisms, once they arose, changed their environment in a way that destroyed the conditions that had made possible the origin of life.

12. Once molecular oxygen became a major component of the atmosphere, both heterotrophic and autotrophic organisms could evolve the biochemical pathways of aerobic respiration, by which far more energy can be extracted from nutrient molecules than is obtainable by glycolysis alone.

13. The oldest cellular fossils are of procaryotic cells and are about 3.5 billion years old. The ancestors of photosynthetic bacteria probably evolved as early as 3.6 billion years ago, perhaps even earlier. Eucaryotic cells probably

arose about 1.8 billion years ago. The first fossils of multicellular animals are about 600 million years old.

14. Chloroplasts and mitochondria may be modern descendants of ancient procaryotic cells that became permanent residents within other cells and have evolved in concert with their hosts ever since. The evolution of the eucaryotic cell represents an important landmark in the evolution of life; it is this cell that gave rise to all the eucaryotic lineages: protists, plants, animals, and fungi.

15. The classification system used in this text recognizes six kingdoms. The kingdoms *Archaebacteria* and *Eubacteria* include the single-celled procaryotic organisms—the archaebacteria and true bacteria. Those eucaryotic organisms that are at the unicellular, colonial, or simple multicellular level are placed in the kingdom *Protista.* Each of the three lineages of multicellular eucaryotes exploits a different mode of nutrition and is placed in a separate kingdom. Photosynthetic autotrophism is used by plants (kingdom *Plantae*), absorptive heterotrophism by fungi (kingdom *Fungi),* and ingestive heterotrophism by animals (kingdom *Animalia*).

STUDY QUESTIONS

1. When and how did the earth and its atmosphere form? Why was the earth's first atmosphere lost? Name the major constituents of the second atmosphere. Was any molecular oxygen present? (p. 842)
2. Describe two hypotheses to account for the accumulation of complex organic compounds on the early earth. Which hypothesis is supported by Miller's 1953 experiment? Why was this experiment a landmark in the study of the origin of life? (pp. 843–845)
3. How could small organic molecules have polymerized to form macromolecules such as proteins and nucleic acids? (pp. 845–846)
4. Compare *coacervate droplets* and *proteinoid microspheres* as models of prebiological systems. What properties characteristic of cells do they exhibit? What properties do they lack? (p. 846)
5. Describe how the genetic control system of cells probably evolved. (pp. 846–847)
6. What were the characteristics of the first cells? How did they obtain energy? (pp. 848–850)
7. In what order did these nutritional pathways evolve: *noncyclic photophosphorylation, cyclic photophosphorylation, aerobic respiration,* and *chemosynthesis?* Which caused the *oxygen revolution?* (p. 850)
8. Describe how chloroplasts and mitochondria arose according to the *endosymbiotic hypothesis.* What evidence supports this hypothesis? (pp. 851–853)
9. Make a time line showing the major events in the evolution of life, including mass extinctions. (p. 855)
10. Name and describe the basic characteristics of the six kingdoms recognized by the classification system used in this text. Which of these kingdoms probably includes organisms not closely related evolutionarily? (pp. 856–858)

CHAPTER 37

VIRUSES AND BACTERIA

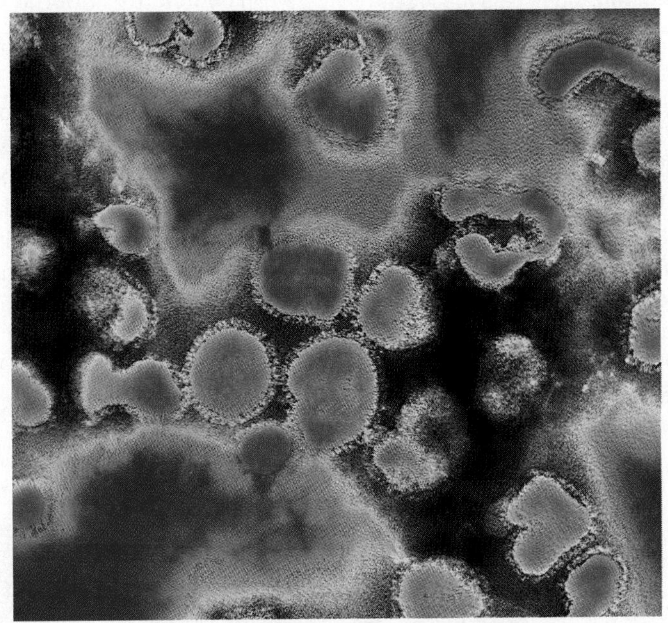

Viruses ordinarily are not formally classified as living organisms. Yet biologists and nonbiologists tend to think of them as living, using such expressions as "live-virus vaccine" and "killed-virus vaccine." The inconsistency is understandable: viruses lack all cellular organelles and cannot reproduce in the absence of a host cell, yet they possess genes that encode sufficient information for the production of new viruses with the same characteristics as the original viruses; and reproduction with gene-controlled heredity is one of the most basic attributes of life.

Whether viruses are classed as living, nonliving, or something in between, it is instructive to compare them with the simplest free-living organisms, the bacteria. Both these groups are the subject of this chapter.

THE VIRUSES

Viruses Are Filterable and Can Be Crystallized

Though viruses were not discovered until early in this century, evidence of their existence can be traced back thousands of years. The Chinese described a disease similar to smallpox in the 10th century B.C., some Egyptian mummies show signs of poliomyelitis (infantile paralysis), and yellow fever has been known for centuries in tropical Africa.

The discovery of viruses was dependent upon an improved understanding of their role in human diseases. By the latter part of the 19th century the idea had become firmly established that many diseases are caused by microorganisms. Pioneer bacteriologists, among them Louis Pasteur and Robert Koch, had isolated various disease-causing bacteria. But for some diseases, notably smallpox, biologists could find no causal microorganism. It has been known as early as 1796 that smallpox could be induced in a healthy person by something in the pus from a smallpox victim, and Edward Jenner had demonstrated that a person vaccinated with material from cowpox lesions developed an immunity to smallpox. Yet no bacterial agent could be found. It was later discovered that many other diseases of both plants and animals were also caused by unknown infectious agents. Further experiments showed that these infectious agents were so small that they could pass through ultrafine filters and could not been seen with even the best light microscopes; they came to be called filterable viruses, or simply viruses, and were still assumed to be very small bacteria.

Chapter opening photo: Colored transmission electron micrograph of Beijing influenza viruses with their spiked protein coats and cores of ribonucleic acid (orange).

However, there were hints that viruses might be something quite different from bacteria. First, all attempts to culture viruses on media customarily used for bacteria failed. Second, unlike bacteria, the virus material could be precipitated from an alcoholic suspension without losing its infectious power. Finally, in 1935 it was conclusively demonstrated that viruses and bacteria are two very different things. In that year W. M. Stanley of the Rockefeller Institute isolated and crystallized tobacco mosaic virus. If the crystals were injected into tobacco plants, they again became active, multiplied, and caused disease symptoms in the plants. That viruses could be crystallized showed that they were not cells but much simpler chemical entities.

Viruses Consist of a Nucleic Acid Core Surrounded by a Protein Coat

Viruses are very small entities, ranging in size from 0.02 to 0.30 micrometers. Each virus particle, or virion, consists of a core of a single type of nucleic acid surrounded by a protein coat, or ***capsid.*** The protein coat may be complicated, as in the bacteriophage T4 (Fig. 37.1A) with its tail and long, leglike fibers, or it may be a simple polyhedron or rod (actually a helix of protein molecules) (Figs. 37.1B, C, and 37.2). Many animal viruses, some plant viruses, and a very few bacteriophages have a membranous ***envelope*** surrounding the protein coat. Sometimes this envelope is derived from the plasma membrane of the host cell in which the virus was produced, and sometimes it is synthesized in the host cell's cytoplasm, but in either case it usually contains some proteins that are specific to that virus. In addition to proteins, some viruses—retroviruses in particular—possess a small number of enzymes.

The Genetic Material of Viruses Can Be Either DNA or RNA

The viral nucleic acid is usually a single molecule and may be composed of either DNA or RNA, but not both. Many have double-stranded DNA; some have single-stranded DNA. Other viruses differ from all cellular organisms in having RNA genes. In most instances the RNA is single-stranded, but in some the RNA is double-stranded. Most plant viruses are RNA viruses, as are many important human viruses, such as HIV and the polio virus.

Viruses are not cells. Even the most complex viruses, which carry an array of enzymes, lack cytoplasm and cellular

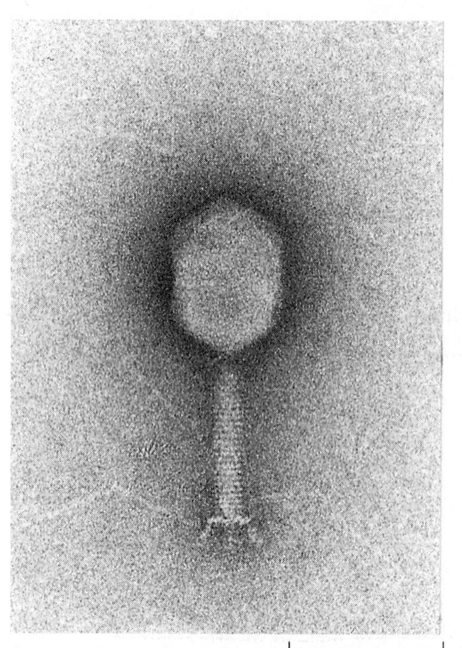

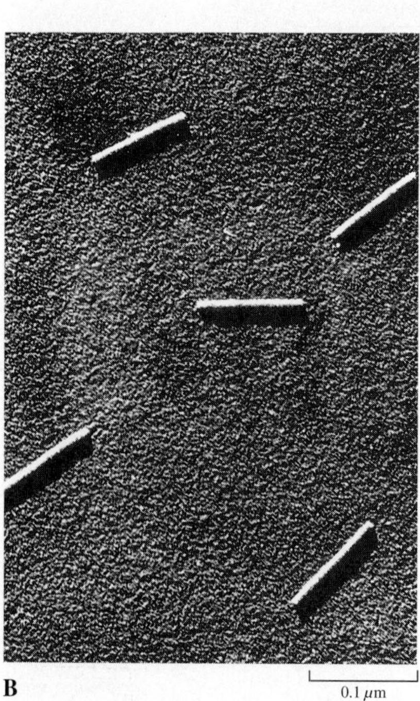

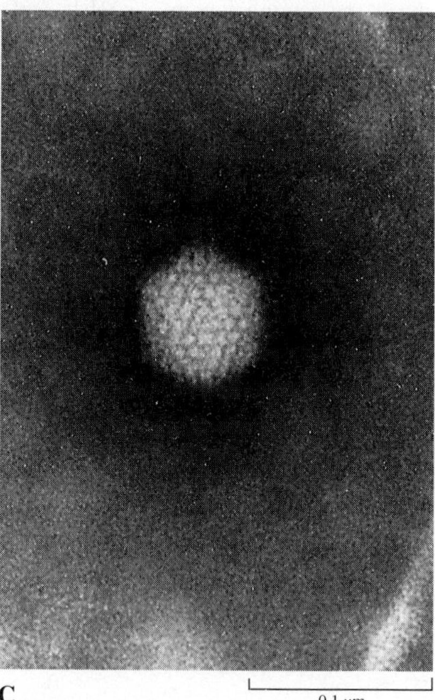

37.1 Electron micrographs of three types of viruses (A) T4 bacteriophage, a complex virus with a "head" containing its DNA, a "tail," and six leglike fibers which aid in attaching it to a host cell (normally *E. coli*). (B) Tobacco mosaic virus, a rod-shaped virus. (C) Adenovirus, a polyhedral virus that multiplies in the upper respiratory tract to produce coldlike symptoms. The adenovirus is actually an icosahedron, having 20 equal sides.

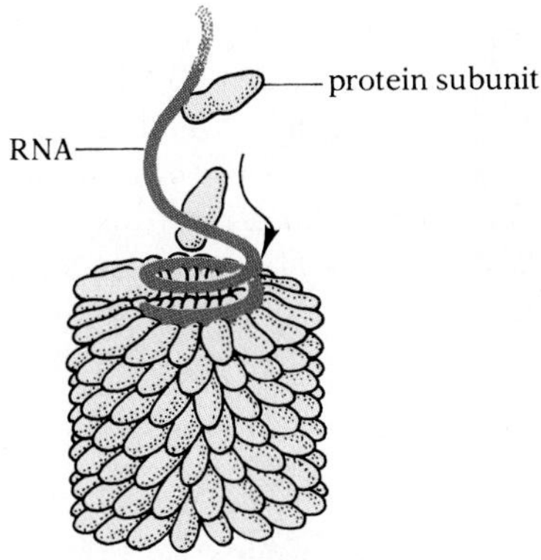

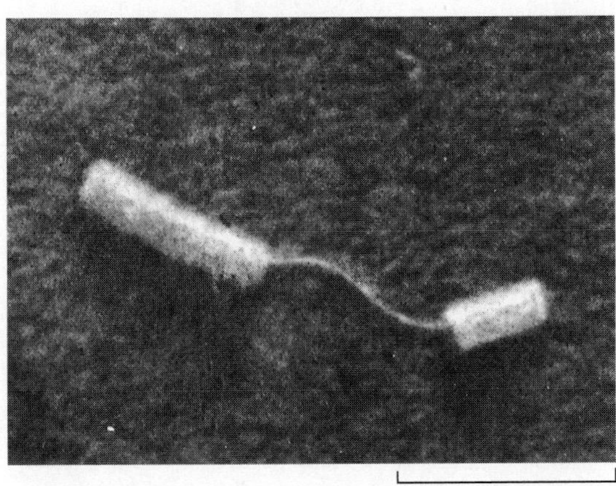

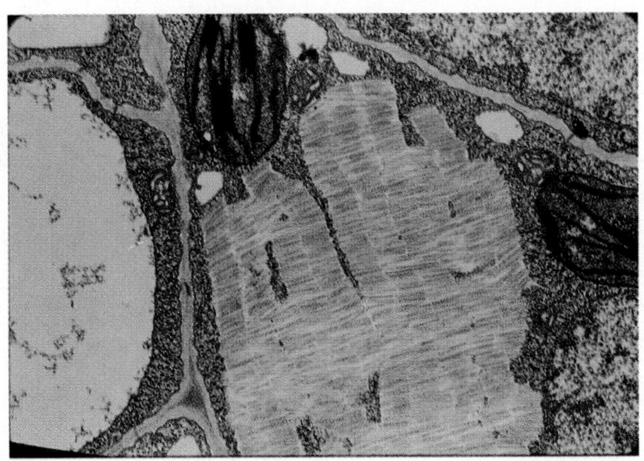

37.2 Tobacco mosaic virus Left top: Diagram of part of a tobacco mosaic virion, showing its structure. At the center of the cylindrical virion is a long RNA molecule. Protein subunits are packed around the coiled RNA; they are added in sequence as the virion is assembled. Left bottom: Scanning electron micrograph of the rod-shaped tobacco mosaic virus. Part of the protein coat has been stripped away from the center portion of the virion, exposing the nucleic acid core. Above: Transmission electron micrograph of mesophyll cell infected with tobacco mosaic virus. The dark oval structure at top left is a chloroplast. Hundreds of virions can be seen as tiny fine lines in the area below and to the right of the chloroplast. Each fine line represents one rod-shaped virion.

organelles, including ribosomes. They do not therefore have independent metabolism and cannot generate their own ATP, nor can they reproduce themselves. Since viruses are not cells, the question of their origin arises. Do they represent a primitive "nearly living" stage in the evolution of life? Or are they organisms that have reached the extreme of evolutionary specialization for parasitism, having lost all cellular components except the nucleus? Could viruses simply be fragments of genetic material derived from cellular organisms? No one really knows the answers to these questions, but we do know that viruses have been around for a long time, and that virtually every form of life is susceptible to viral attack.

It Is the Host Cell That Manufactures New Viruses

Viruses are ***obligate parasites*** of living cells—that is, they cannot reproduce outside of living cells. It is the host cell, not the virus, that manufactures new virus particles from the instructions the old virus provides. Consequently viruses cannot be cultured on standard culture media; they require living host cells. Pharmaceutical companies and research laboratories often grow them in fertilized chicken eggs or in cells in tissue cultures.

Reproduction of a bacteriophage (phage) virus is described in Chapter 10 (p. 205). Briefly, during reproduction, a free phage particle becomes attached by the tip of its tail to a specific site on the wall of a susceptible bacterial cell, and the phage nucleic acid is then injected into the host cell while the protein coat remains outside (Fig. 37.3). Once inside the bacterial cell, the phage genes take over the metabolic machinery of the cell, which is put to work making new viral nucleic acid and proteins for the coat. The new viral nucleic acid and proteins self-assemble into new bacteriophages, which are released by bursting, or lysis, of the bacterial cell. The whole series of events is known as the ***lytic cycle.***

Sometimes viral DNA does not immediately take control of the host cell's metabolic machinery and put it to work making new virus particles, but instead inserts itself into the host chromosome, in which case it is known as a ***provirus.*** While in the provirus form the viral DNA behaves as an additional part of the chromosome (Fig. 37.4); it is replicated with the rest of the cell and may have phenotypic effects on the host cell. The virus remains in its integrated state until the host cell is ex-

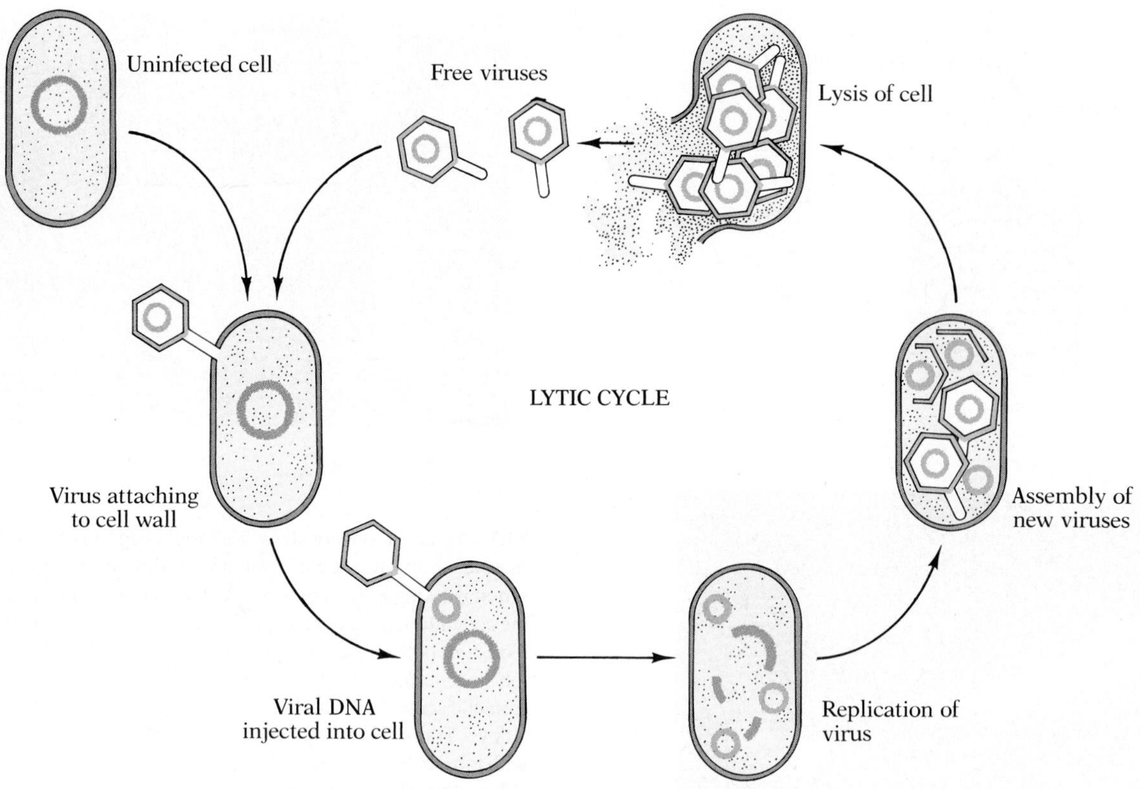

37.3 The lytic cycle of a bacteriophage

posed to some environmental insult, such as ultraviolet light or radiation, at which time the virus leaves the host chromosome and resumes normal replication.

The Entire Virus Particle of Plant and Animal Viruses Enters the Host Cell

Plant and animal viruses function somewhat differently from bacteriophages (viruses that only attack bacteria) in that the entire virus particle of a plant or animal virus, including the protein coat, enters the host cell. Most plant viruses depend on insect carriers called ***vectors*** to inject them through the thick plant cell walls into the host-cell cytoplasm.[1] Animal viruses, on the other hand, simply attach to special receptor sites on the cell membrane. Much of the host specificity of such viruses comes from their attraction to specific receptor sites. A given virus can infect only those cells that have a receptor site for that virus. For instance, the HIV virus, the virus causing AIDS (human acquired immune deficiency syndrome), can attack only those cells that have a specific receptor (CD4) complex (i.e., helper T cells and certain brain, skin, and intestinal cells). Once a virus is attached to its receptor site, it is taken into the cell by endocytosis. There the viral protein coat is broken down by cellular enzymes and the nucleic acid is released (Fig. 37.5).

RNA Viruses Require Special Enzymes for Replication

If the viral nucleic acid is DNA, it enters the nucleus of the host cell and acts as a template for the synthesis of both messenger RNA and new viral DNA, using the enzymes of the host cell (Fig. 37.5). If the viral nucleic acid is RNA, replication usually takes place in the cytoplasm. Special enzymes, ***RNA replicases,*** are necessary for this RNA-to-RNA synthesis. Some RNA viruses bring the RNA replicases with them into the cell; others use part of the viral RNA as messenger RNA on the host cell's ribosomes, providing instructions for the synthesis of RNA replicases by the cell (Fig. 37.6). In either case, then, the

[1] Attempts to control viral diseases in plants are often directed at controlling the insects that transmit the virus rather than at the virus itself.

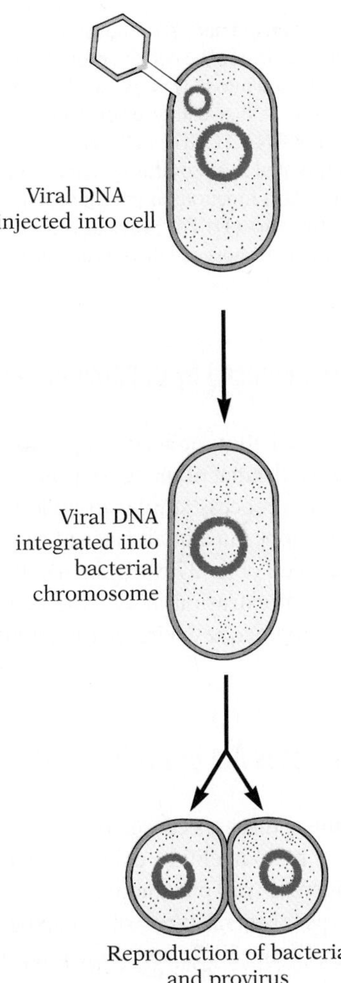

37.4 Integration of bacteriophage DNA as a provirus Some bacteriophages can insert their chromosome into the host chromosome, in which case it is known as a provirus. While in the provirus form the viral DNA behaves as an additional part of the chromosome. It is replicated with the rest of the cell and has phenotypic effects on the host cell.

secret of RNA→RNA replication by such viruses is RNA replicase, which is not present in uninfected host cells.

Some RNA viruses have another, more complicated mechanism of replication. A group of RNA viruses called ***retroviruses*** do not carry out RNA→RNA transcription. Instead, their RNA is transcribed into a DNA intermediate—a *reverse* of the transcription process. This reaction is catalyzed by a special enzyme called ***reverse transcriptase,*** which the viruses bring with them into the host cells. The newly formed DNA then becomes integrated as a provirus into the host's chromosomes, where it may remain for a long time, even being passed to the host's descendants, generation after generation (Fig. 37.7). Sometimes no new viruses are produced by the provirus, but in other instances the proviral DNA is transcribed into

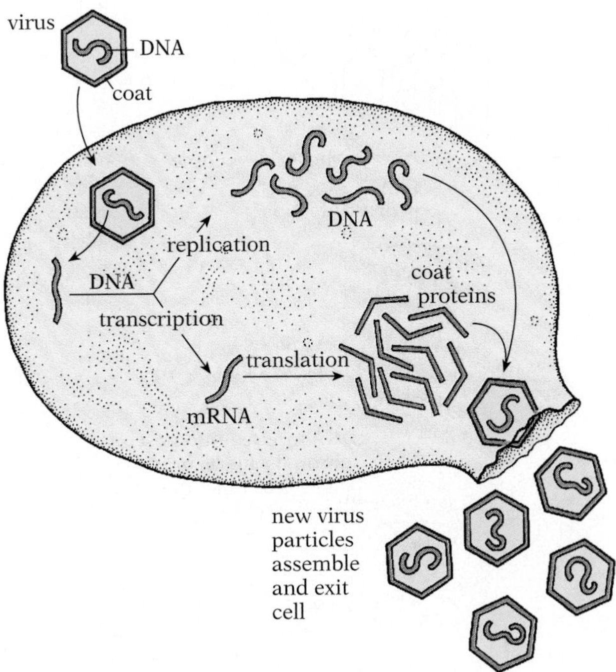

37.5 Infection of a cell by a DNA virus The DNA virus enters a host cell by a type of endocytosis. Once inside the cell the protein coat is stripped off, and viral DNA is released. The viral DNA is both replicated, to make new DNA copies for new viruses, and transcribed into mRNA. The mRNA is then translated to form proteins for the viral coat. The viral DNA and coat proteins self assemble to form new virus particles that leave the cell by lysis or extrusion.

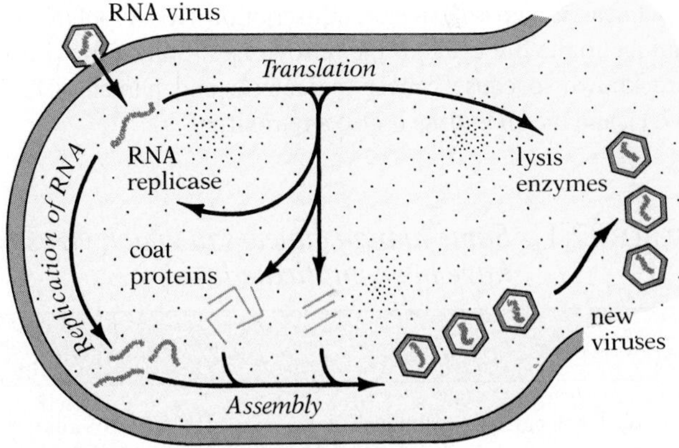

37.6 Replication of some RNA viruses In order to carry out RNA-to-RNA replication, RNA replicase is needed. Some viruses bring the enzyme with them into the cell, others, as shown above, use their viral RNA as messenger RNA, and the mRNA is translated on the ribosomes to produce RNA replicase and other viral proteins. Once present, RNA replicase catalyzes production of new copies of the viral RNA. New viruses then self-assemble and leave the cell by lysis or extrusion.

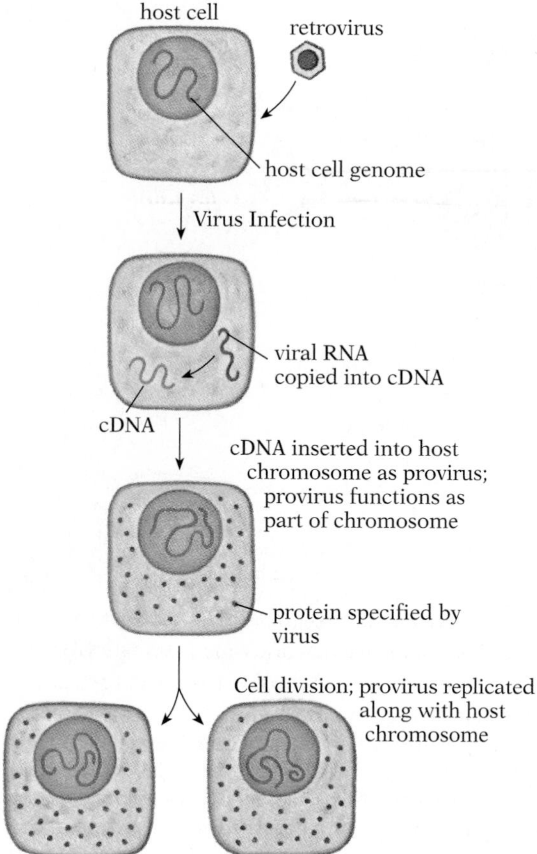

37.7 Infection of a cell by a retrovirus (A) A retrovirus enters a host cell by a type of endocytosis. (B) Inside the cell the protein coat is stripped off, and viral RNA (red) is released into the cytoplasm. The reverse transcriptase that is carried by the virus then catalyzes the formation of a cDNA copy (green) of the viral RNA. (C) The cDNA copy then moves to the nucleus where it is incorporated as a provirus into one of the host cell chromosomes (blue). The provirus functions as part of the chromosome and codes for viral proteins (brown). (D) The provirus is replicated along with the host chromosome when the host cell divides and is passed on to the daughter cells.

RNA and new viruses are produced. These viruses can be released from the cell by a kind of budding process, which, unlike lysis of host cells by bacteriophage, does not kill the cell. As far as is known, reverse transcription (RNA→cDNA) is unique to the life cycle of the retroviruses. Some retroviruses are known to cause cancer in animals and humans (Table 37.1), and the HIV virus is also a retrovirus.

TABLE 37.1 *Some human cancers in which viruses have been implicated*

Type of cancer	*Virus*	*Genetic Material*
Adult T-cell carcinoma	Human T-cell leukemia virus	RNA (retrovirus)
Burkitt's lymphoma	Epstein-Barr virus	DNA (herpesvirus)
Nasophyngeal carcinoma	Epstein-Barr virus	DNA (herpesvirus)
Liver cancer (hepatocellular)	Hepatitis B virus	DNA
Cervical and skin cancers	Papilloma virus	DNA

Viruses Are Released by Budding or Lysis

Most viruses are released after replication by lysis of the host cell, but in other cases, like that of the retroviruses, viruses are released by extrusion, a process similar to budding whereby the virus becomes enveloped in a small piece of cell membrane as it moves out of the cell (Fig. 37.8). Lysis results in the destruction of the cell, but extrusion allows the cell to remain alive and continue to produce new viruses for a long period of time.

Viral Diseases Are Difficult to Treat

Among the many human diseases caused by viruses are smallpox, yellow fever, poliomyelitis, chicken pox, mumps, measles, rabies, influenza, viral pneumonia, the common cold, oral and genital herpes, several types of encephalitis, infectious hepatitis, AIDS (Table 37.2), and some kinds of cancers (see Table 37.1).

TABLE 37.2 *Some important viruses infecting humans*

Virus	*Diseases caused*
Double-stranded DNA	
Adenovirus	Respiratory infections
Papovavirus	Common warts
Herpesvirus	Oral and genital cold sores
Poxvirus	Chicken pox, smallpox
Double-stranded RNA	
Reovirus	Infantile diarrhea
Single-stranded RNA	
Picornovirus	Polio
HIV and other retroviruses	AIDS
Orthomyxovirus	Influenza
Rhabdovirus	Rabies

Some of the viruses have a membranous envelope (yellow), others are naked.

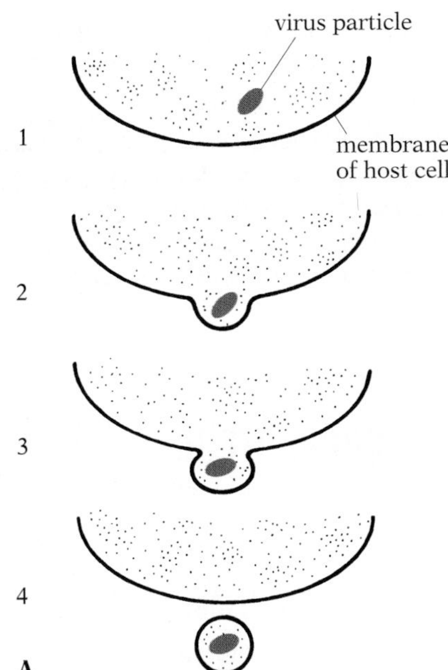

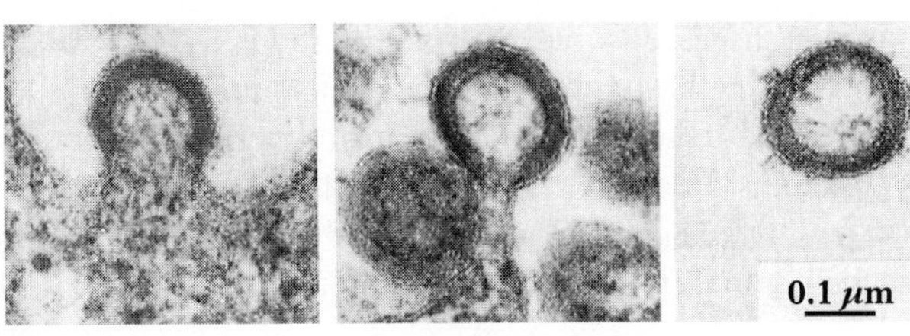

37.8 Extrusion of a virus from the host cell (A) In some cases a newly assembled virus particle becomes closely associated with the membrane of the host cell (1). The membrane then forms an evagination into which the particle moves (2, 3). Finally the evaginated membrane is pinched off as an envelope around the free virus particle (4). (B) Particles of Visna virus budding off cells of infected sheep.

Unfortunately, most viral infections do not respond to treatment with antibiotics. Most antibiotics act by inhibiting protein or nucleic acid synthesis, by disrupting metabolic pathways, by disrupting cell membranes, or by interfering with cell-wall synthesis. Since viruses are not cells and have neither metabolic pathways nor cell walls, such antibiotics are useless. Then too, viruses are replicating *inside* living cells, making it very difficult for an antibiotic to kill or inhibit viruses without harming the patient's cells at the same time. For this reason there are only a few antibiotics that are useful against viruses, and these tend to be highly toxic agents with serious side effects. But some of the most serious viral diseases, notably smallpox, measles, mumps, and polio, are preventable by means of vaccines (see p. 446). Indeed, smallpox has been eradicated as a result of intensive worldwide vaccination programs. Major research efforts today are directed at trying to develop a vaccine against the HIV virus.

The immune response acts to defend the body against foreign antigens such as viruses (Chapter 21). But much of the time viruses are found inside living cells where cytotoxic T cells and antibodies cannot reach them. Only when the virus is moving from one cell to the next, or if the virus alters the membrane of the host cell in some way, can the immune system respond to its presence. This is one of the reasons why recovery from a viral infection can be so prolonged. However, the human body does have a special weapon to use against viruses; infected cells produce specific proteins called ***interferons.*** Interferon is produced by host cells in response to invading

MAKING CONNECTIONS

1. What hypotheses for the origin of viruses have been suggested?

2. Plant cells have walls whereas animal cells do not. How does this difference affect how plant and animal viruses enter their host cells?

3. Why do animal viruses such as HIV (the AIDS virus) only attack a few specific types of cells in their host?

4. How does information transfer in RNA viruses and retroviruses differ from that in cellular organisms, as described in Chapter 11?

5. How has the reverse transcriptase of retroviruses been useful in the development of recombinant DNA technology? (See Chapter 13.)

6. Why are viruses sometimes "invisible" to the immune system of their host?

viruses. It cannot save those cells, but when interferon is released into the surrounding tissue fluid and encounters uninfected cells, it interacts with these cells and stimulates them to produce antiviral proteins. In other words, the interferon acts as a messenger from infected cells to uninfected cells, telling them to mobilize their defenses against viral infection.

THE PROCARYOTIC KINGDOMS

The procaryotes are the most ancient of organisms. Unlike viruses, procaryotes are cellular. They always contain both RNA and DNA; they have ribosomes; they possess integrated enzyme systems (viruses carry, at most, only a few individual enzymes); they can generate ATP and use it in the synthesis of other organic compounds; and they provide both the raw material and all the metabolic machinery for their own reproduction. Viruses lack these attributes.

All the organisms whose cells are procaryotic—that is, cells that lack a nuclear membrane, mitochondria, endoplasmic reticulum, Golgi apparatus, lysosomes, and other membranous organelles—are placed in one of the two procaryotic kingdoms: the Archaebacteria, or "ancient" bacteria, and Eubacteria, the much more common "true" bacteria (Fig. 37.9). Molecular sequencing of ribosomal RNAs from a variety of organisms suggests that the archaebacteria and the eubacteria are related as distantly to each other as they are to the eucaryotes; hence they are placed in different kingdoms. We now discuss some of the distinctive characteristics of these two kingdoms.

Kingdom Archaebacteria

Archaebacteria Are Unique and Live in Extreme Environments

Archaebacteria differ from eubacteria in a number of ways (Table 37.3): their cell walls have a markedly different chemical composition; the lipids in their cell membranes are

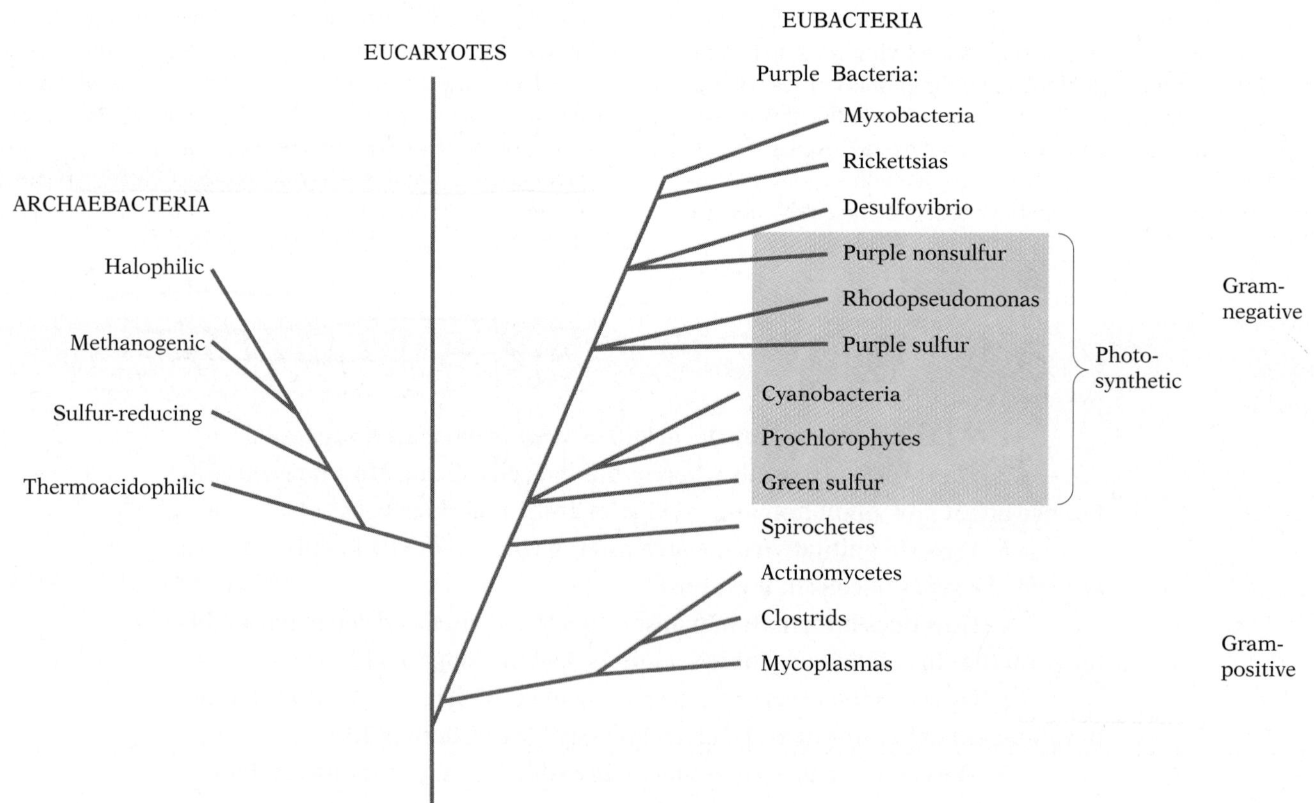

37.9 Probable phylogenetic relationships of modern bacteria The appropriate branching positions of mycoplasmas, rickettsias, and Gram-positive bacteria are especially uncertain.

TABLE 37.3 *A comparison of the archaebacteria, eubacteria, and eucaryotic cells*

Characteristic	*Archae-bacteria*	*Eubacteria*	*Eucaryotic cells*
Cell type	Procaryotic	Procaryotic	Eucaryotic
Cell size	Small, about 1 μm	Small, about 1 μm	Larger, >10 μm
Cell wall	No muramic acid	Muramic acid present	Usually cellulose if present
Membrane lipids	Contain branched-chain fatty acids	Contain straight-chain fatty acids	Contain straight-chain fatty acids
Ribosomes	Small	Small	Large
Chromosome(s)	Single, circular	Single, circular	Many, linear

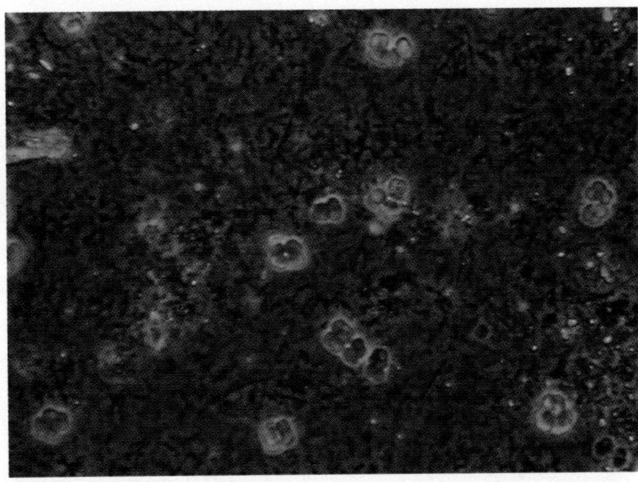

37.10 Methane-producing bacteria from a sewage digestor The larger organisms in this phase contrast photomicrograph probably belong to the genus *Methanosarcina,* which can convert acetic acid to methane. Also present are some smaller methanogen cells, which may use one of a number of different substrates to produce methane. Methane production occurs only under anaerobic conditions.

branched rather than straight; and most are anaerobic. Because they differ from the eubacteria in these very fundamental ways, the two procaryotic groups are separated into different kingdoms. Though the name means ancient bacteria, their kingdom appears to be somewhat younger than that of the eubacteria.

The archaebacteria live only in very harsh environments, thriving under anaerobic conditions, high salinity, or high temperature and acidity. There are three main groups of archaebacteria: the methanogens (methane-producers), the extreme halophiles ("salt-loving"), and the thermoacidophiles ("heat- and acid-loving").

The methanogens belong to an interesting group. They are all anaerobic chemosynthesizers and live in bogs or other habitats rich in decaying vegetation as well as in the intestines of humans and ruminant cattle. They produce methane ("marsh gas") as a by-product of their metabolism, which is thought to be one of the contributors to the "greenhouse effect." Because of their methane production, these archaebacteria have potential commercial importance; they are being used in sewage digesters to produce methane for use as fuel (Fig. 37.10).

The extreme halophilic archaebacteria (Fig. 37.11) require highly concentrated (over 20 percent) salt water like that found in the (otherwise) Dead Sea; they carry out a unique type of photosynthesis that does not use chlorophyll. The thermoacidophilic archaebacteria grow in hot sulfur springs and volcanic vents, at temperatures 65° to 80°C and pHs of 1 to 2 (Fig. 37.12). They require elemental sulfur in their metabolism and are often described as being sulfur-dependent. These are the archaebacteria that live in the hot sulfur springs of Yellowstone National Park.

That the archaebacteria live primarily in such extreme habitats suggests that they have a set of adaptations that protect them from their surroundings but put them at a disadvantage in competition with conventional organisms. And because the conditions under which many of them live are thought to be similar to the conditions under which life evolved on the early earth, the archaebacteria may be evolutionary relics of ancient forms of life that still survive in the same habitats.

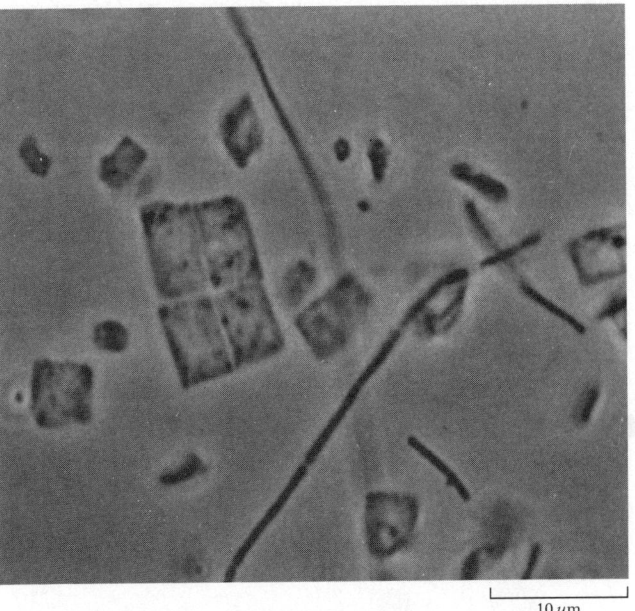

37.11 Halophilic archaebacteria from a brine pond This species grows as thin square units that resemble miniature postage stamps.

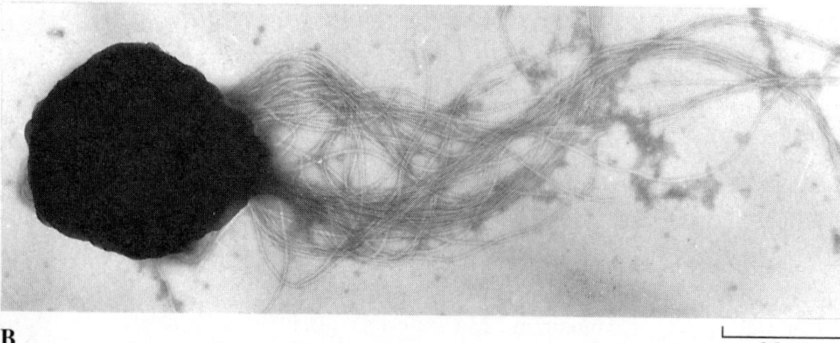

37.12 Scanning electron micrographs of thermoacidophilic archaebacteria (A) This dense mat is formed by chemosynthetic archaebacteria growing on a mussel shell near an underwater volcanic vent in the Pacific Ocean. (B) Close-up view of a single chemosynthetic archaebacterium, collected from the hot water of a submarine hydrothermal vent in the eastern Pacific.

Kingdom Eubacteria

Eubacteria have been evolving and diversifying for billions of years and so span a range of habitats and lifestyles greater than that of any eucaryotic kingdom. There is no consensus about the course of eubacterial evolution, and no single accepted scheme for classifying this group of organisms. One of many possible schemes is shown in Figure 36.9.

Eubacterial Cells Are Relatively Simple in Structure

Most eubacteria are very tiny (Fig. 37.13), far smaller than the individual cells in the body of the multicellular plant or animal (Table 37.3). Though some (e.g., the Rickettsiae and the Chlamydiae) are as small as some of the largest viruses, even the smallest eubacteria are fundamentally different from viruses. The cells of most eubacteria are classified according to their shape: spherical, the ***cocci;*** rod-shaped, the ***bacilli*** (Fig. 37.14); or helically coiled, the ***spirilla*** (Fig. 37.15). When cell

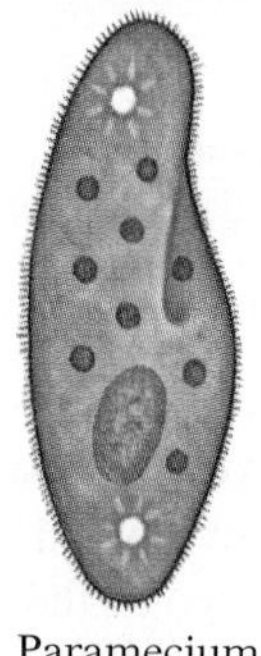

Paramecium
30 x 75 μm

Cyanobacteria 10 μm diameter

E. coli 1 x 2 m

Mycroplasma 0.3-0.8 μm diameter

Bacteriophage 0.07 x 0.2 μm

Lymphocyte 10 μm diameter

37.13 Sizes of viruses, bacteria, and eucaryotes compared

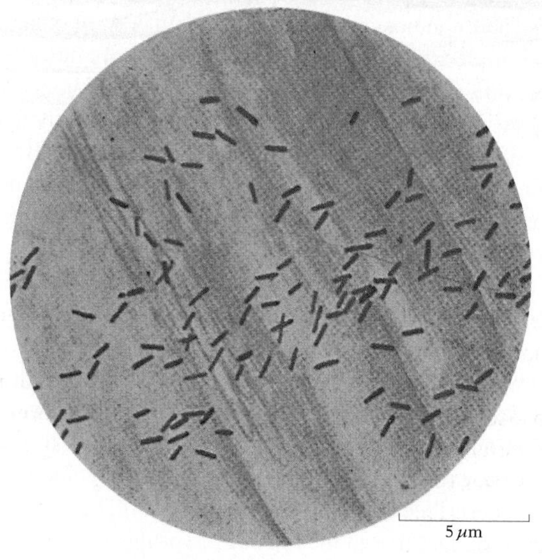

37.14 *Escherichia coli,* a rod-shaped purple eubacterium found in the intestines of vertebrates Top: Electron micrograph of a colony. Bottom: Stained cells as they appear under a high-power light microscope.

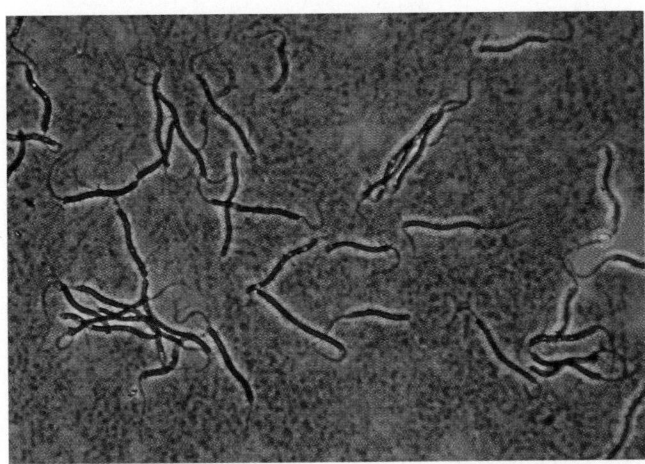

37.15 Spiral bacteria These bacteria, *Spirillium voluntans*, are rigid spirals. They live in fresh water and are one of the largest bacteria known. They move by means of flagella.

37.17 *Staphylococcus aureus*, a eubacterium common on human skin The round cells occur in grapelike clusters. They are the cause of pimples.

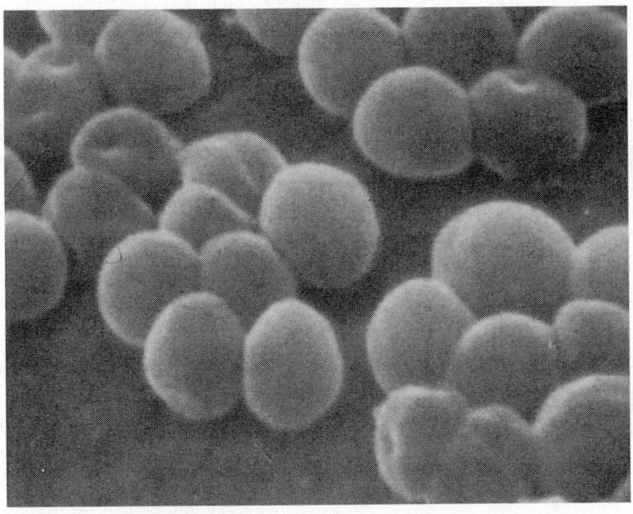

37.16 *Streptococcus salivarius*, a eubacterium common on the human mouth All streptococci are spherical (coccal) eubacteria; they are normally grouped in chainlike clusters.

division takes place, the daughter cells of some species remain attached and form characteristic groupings. Thus cells of the eubacterium that causes pneumonia are often found in pairs (diplococci). The cells of some spherical species form long chains (streptococci) (Fig. 37.16), while others form grapelike clusters (staphylococci) (Fig. 37.17). Each cell in a diplococcal, streptococcal, or staphylococcal grouping is an independent organism.

Being procaryotic, eubacteria lack a membrane-surrounded nucleus and most of the cytoplasmic organelles found in eucaryotic cells (see Table 6.1, p. 135). Within the cytoplasm are ribosomes and a nuclear region (nucleoid) containing a single circular chromosome (Fig. 37.18). Most eubacterial cells

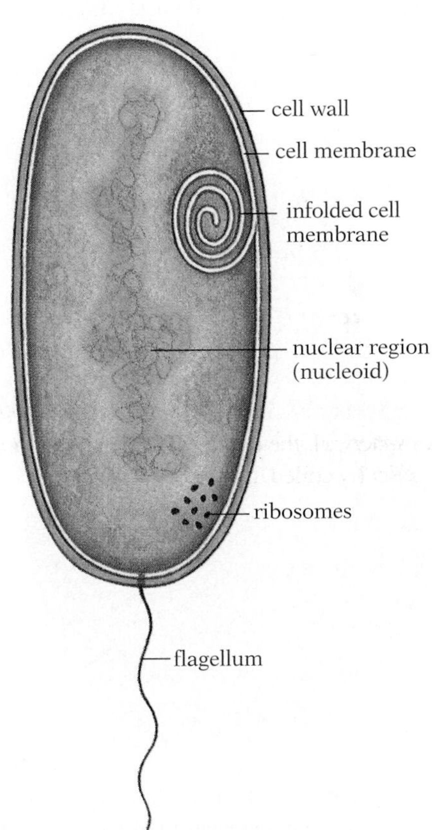

37.18 A typical eubacterial cell The cell consists of a cell wall, cell membrane, and granular cytoplasm containing ribosomes and a nuclear region. The nuclear region (nucleoid) contains the single circular chromosome made of DNA. Some eubacteria have one or more flagella, and many have a polysaccharide capsule (not shown) outside the cell wall.

are surrounded by a semirigid cell wall, inside of which is a cell membrane that may have infoldings that provide additional membranous surfaces for enzymatic activities.

Eubacterial Cell Walls Differ From Eucaryotic Walls

Eubacterial cell walls, like those of plants, protect the cell both from physical damage and from osmotic disruption. But the walls of eubacteria differ significantly in composition from those of plants. The walls of plant cells are made of cellulose and related compounds (or of chitin, in fungi), whereas those of eubacteria are made of murein, a huge polymer composed of polysaccharide chains linked together by short chains of amino acids. Muramic acid, one of the constituents of murein, never occurs in the walls of eucaryotic cells. This important difference in chemical composition between procaryotic and eucaryotic cells is the basis for the selective activity of some antibiotic drugs, such as penicillin. Penicillin is toxic to growing eubacteria because it inhibits formation of murein, and thus interferes with bacterial multiplication (Fig. 37.19).

Eubacterial cell walls show characteristic reactions to a variety of stains. One important stain is the Gram stain (Fig. 37.20). All eubacteria can be divided into Gram positive or Gram negative on the basis of whether or not a chemical in their cell walls binds Gram's solution: if it does, they are Gram positive; if not, they are Gram negative. Since there are few visible structural characters that can be used in identifying eubacteria, diagnostic staining is an important laboratory tool.

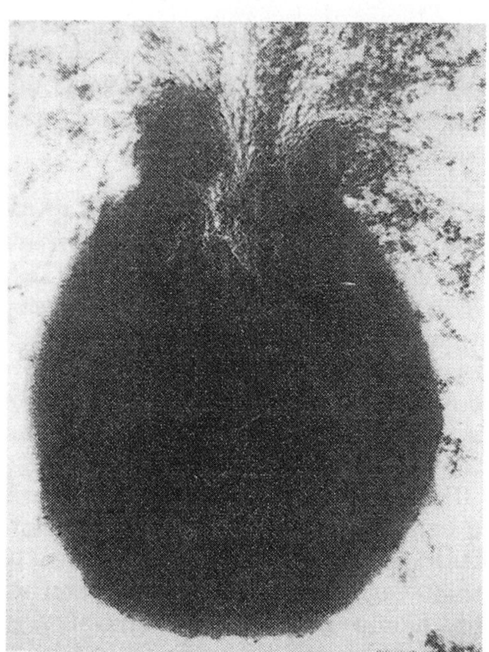

37.19 Exploding eubacterial cell Eubacteria generally live in a hypotonic environment and hence tend to take in water by osmosis, building up a high turgor pressure within the cell. The cell wall normally constrains the cytoplasm and keeps the cell from bursting. The cell shown here was exposed to penicillin, which interferes with cell-wall synthesis; as a result, the cell wall was weakened, and the high turgor pressure inside caused it to burst. ×66,600.

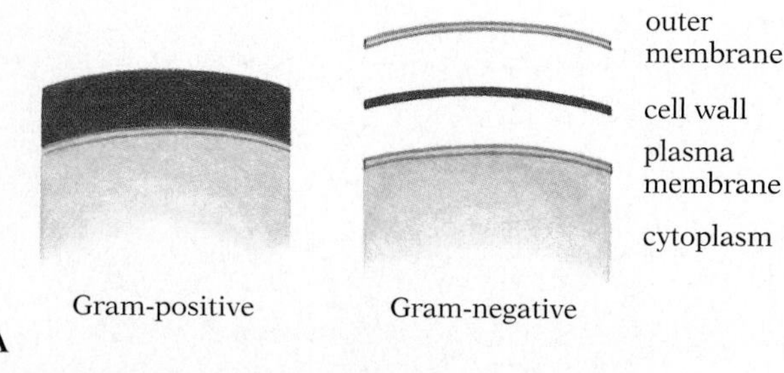

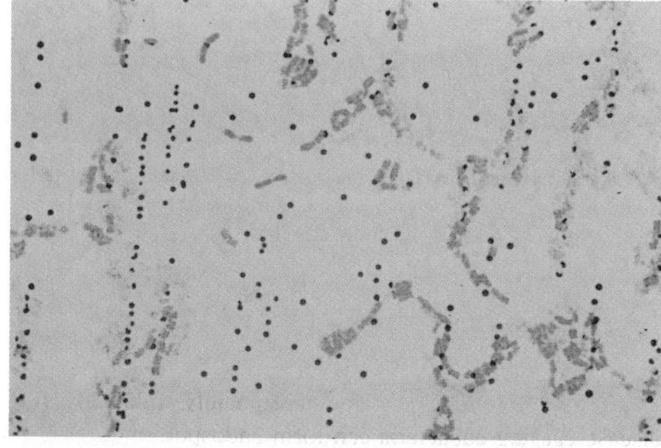

37.20 Cell walls of eubacteria (A) Gram-positive eubacteria have a thick layer of murein in their cell wall, which readily binds Gram's stain. Gram-negative eubacteria have an outer membrane of lipopolysaccharides and a thinner murein wall that does not retain Gram's stain. Many eubacteria secrete a hard or sticky capsule outside of these structures. (B) Photograph of Gram-negative bacilli (red) and Gram-positive cocci (purple). ×1,000.

Many eubacterial cells secrete polysaccharide materials that accumulate on the outer surface of the cell wall and form a ***capsule.*** The capsule makes the cell more resistant to the defenses of host organisms; hence encapsulated strains of a given eubacterial species are more likely to cause disease than unencapsulated strains.

Some Eubacteria Produce Resistant Spores

Some eubacteria (mostly rod-shaped ones) can form special resting cells called ***endospores,*** which can withstand conditions that would quickly kill the normal, active cell. Each small endospore develops inside a bacterial cell and contains DNA plus a limited amount of other essential materials from that cell

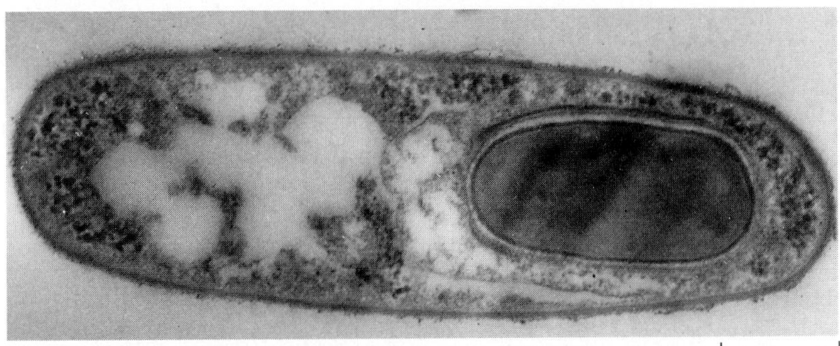

37.21 Electron micrograph of a sporulating bacillus The spore is the dark oval at the right side of the cell. The developing spore coat is clearly visible. The white areas at the left side of the cell are not vacuoles, but areas filled in with fatty material.

(Fig. 37.21). It is enclosed in an almost indestructible spore coat. Once the endospore has fully developed, the remainder of the cell in which it formed may disintegrate. Because of their very low water content and resistant coats, spores of many species can survive an hour or more of boiling or being in a hot oven. They can be frozen for decades or perhaps for centuries without harm. They can survive long periods of drying. And they can even withstand treatment with strong disinfectant solutions. When conditions again become favorable, the spores may germinate, giving rise to normal eubacterial cells that resume growing and dividing. Fortunately, relatively few disease-causing eubacteria can form endospores. The organisms causing tetanus, gas gangrene, botulism (a type of food poisoning), and anthrax are examples of pathogenic spore-formers.

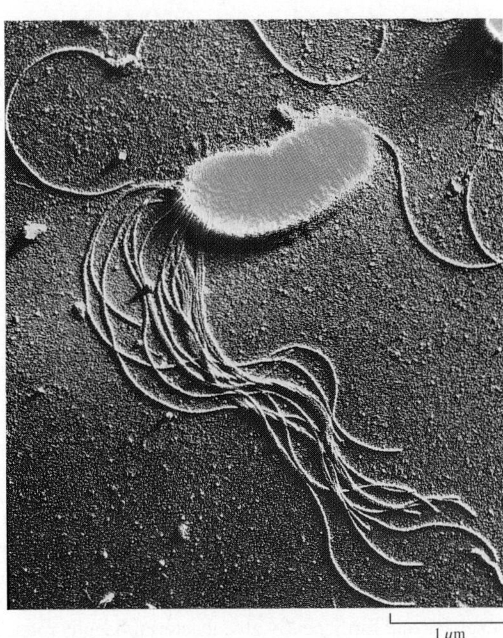

37.22 Scanning electron micrograph showing flagella of the soil bacterium *Pseudomonas fluorescens* Though bacterial flagella are organelles of locomotion, most differ from the flagella of eucaryotic cells in lacking the internal microtubules and in their mechanism of movement.

Many Eubacteria Move by Rotating Flagella

Many eubacteria can move about actively. In most cases the motion is produced by the rotation of flagella (Fig. 37.22). Eubacterial flagella are structurally very different from eucaryotic flagella; they do not contain the nine peripheral and two central microtubules found in all flagella of eucaryotic cells.

MAKING CONNECTIONS

1. What characteristics of the archaebacteria suggest that they may be evolutionary relics of ancient forms of life?

2. How do the methanogens contribute to the "greenhouse effect"?

3. Why is penicillin an effective antibiotic against many bacteria?

4. How does the anaerobic bacterium that causes tetanus (lockjaw) survive in aerobic conditions outside the body (for example, on the surface of a rusty nail)?

5. How are the flagella of eubacteria different from the eucaryotic cilia and flagella described in Chapter 6?

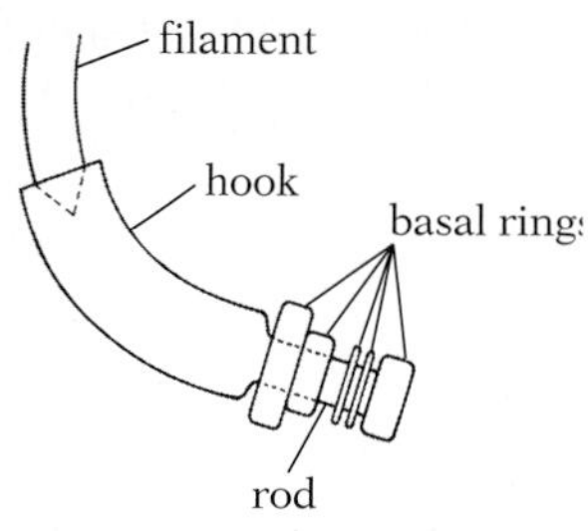

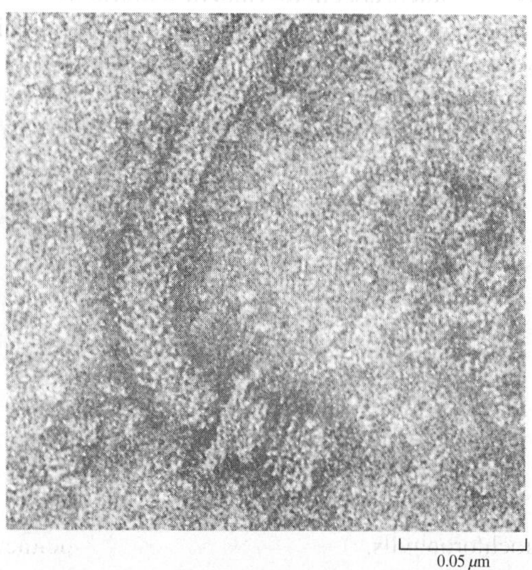

37.23 The basal portion of a typical eubacterial flagellum The electron micrograph shows the base of a flagellum from *Caulobacter crescentus.* The interpretive drawing elucidates the complex system of rings, rod, hook, and filament. Unlike the flagella of eucaryotes, which beat back and forth, nearly all eubacterial flagella rotate. Their complicated structure provides the necessary anchor, joints, and couplings for such rotational motion.

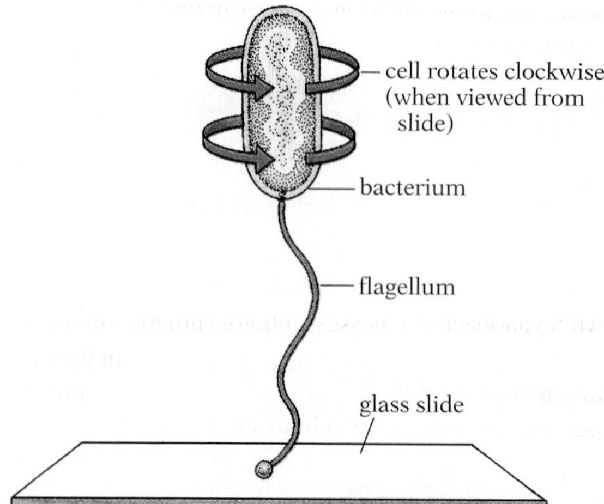

37.24 Rotary motion of a eubacterial flagellum Bacterial flagella use a rotary propellerlike motion to propel cells. If a bacterial flagellum is attached to a glass slide, the whole bacterium will rotate. The bacterium will rotate first in one direction, then in the other.

The motion of eubacterial flagella is not the whiplike beating characteristic of eucaryotes, but rather a rotary propellerlike motion made possible by an amazingly complex attachment (Fig. 37.23). In fact, if you tether a eubacterial flagellum to a glass slide, the whole bacterium will rotate (Fig. 37.24).

Eubacteria Have Enormous Reproductive Potential

As already noted, the eubacterial cell lacks a membrane-surrounded nucleus, but does have a nuclear region, called a ***nucleoid.*** Electron microscopy reveals genes in the nucleoid arranged in sequence along a single circular chromosome composed primarily of DNA. Most eubacteria reproduce by ***binary fission,*** a type of cell division in which two equal daughter cells are produced without mitosis (see Fig. 9.1, p. 180).

Eubacteria commonly have enormous reproductive potential. Many species may divide as often as once every 20 minutes under favorable conditions. If all the descendants of a cell of this type survived and divided every 20 minutes, the single initial cell would have about 500,000 descendants at the end of six hours, and by the end of 25 hours the total weight of its descendants would be nearly 2,000,000 kilograms. Though increases of this magnitude do not actually occur, the real increases are frequently huge, which helps explain the rapidity with which food sometimes spoils or a disease develops.

Although the reproductive process itself is asexual, genetic recombination does occur occasionally in some eubacteria. One particularly important mechanism of recombination is ***conjugation*** (Fig. 37.25), in which two eubacterial cells come to lie very close to one another and a cytoplasmic connection forms between them, through which part of the bacterial chromosome or a plasmid can be transferred.

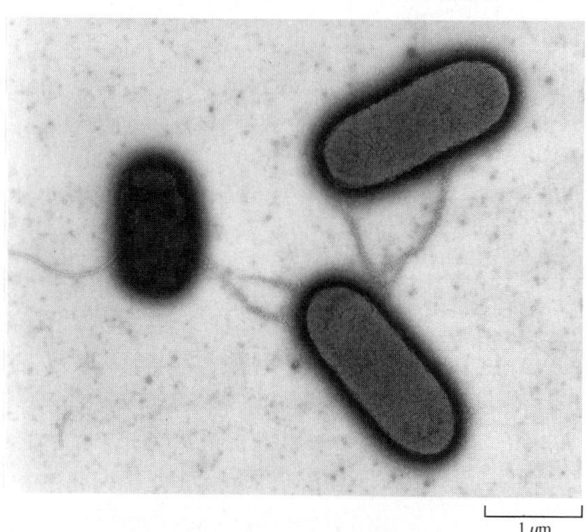

37.25 Conjugating eubacteria During conjugation, special appendages from one eubacterial cell attach to another cell. DNA can be passed from one cell to another through such cytoplasmic bridges.

Eubacteria Differ in Their Nutritional and Oxygen Requirements

Most eubacteria are absorptive heterotrophs, being either saprophytes or parasites. Like animals, the majority of these eubacteria are aerobic; they cannot live without molecular oxygen (O_2), which they use in the respiratory breakdown of carbohydrates and other food materials to carbon dioxide and water.

To some eubacteria that obtain all their energy by fermentation, however, oxygen is lethal. Such eubacteria are called ***obligate anaerobes***. *Clostridium botulinum*, the causal agent of the most dangerous kind of food poisoning, botulism, is an obligate anaerobe; it grows well in tightly closed food containers containing improperly processed food.

Other eubacteria, called ***facultative anaerobes,*** are more flexible and can live in either the presence or absence of O_2. Some of the facultative anaerobes are simply indifferent to O_2, obtaining all their energy from fermentation whether O_2 is present or not; others obtain their energy by fermentation when O_2 is not available, but carry out respiration (via the Krebs cycle) when O_2 is present, and may grow faster under these conditions. Respiration is a much more efficient energy-yielding process than fermentation.

Eubacteria differ considerably in their nutrient requirements. No matter how unfavorable an environment appears, some species of eubacteria are capable of living there. Their tremendous nutritional versatility allows them to fill a variety of niches. The differences in the amino acid and vitamin requirements of various eubacteria, and in their ways of utilizing nutrients, provide valuable diagnostic characters for workers attempting to identify unknown specimens. By growing samples of the organisms to be identified on a variety of nutrient media cultured at standard temperatures, it is often possible to match the colonies of unknown organisms to those of known species. Comparisons are based on the media the organisms will grow or not grow on, and the color, texture, and other characteristics of the colony they produce when they do grow (Fig. 37.26).

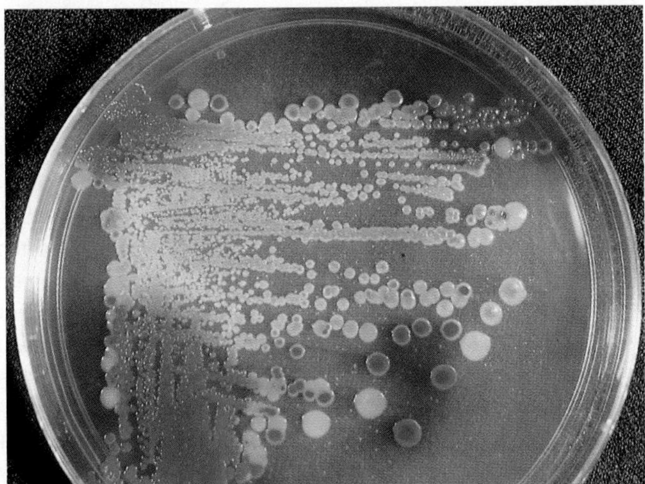

37.26 Colonies of two kinds of eubacteria growing on the same culture medium, with different effects on the medium Colonies of *E. coli* on Endo agar carry out fermentation, making the colonies red, whereas *Salmonella* bacteria do not carry out fermentation and the colonies are colorless.

Though most eubacteria are heterotrophs, some are chemosynthetizers and use *inorganic* chemicals as energy sources. Still others are photosynthetic autotrophs (see below). The astonishing diversity of energy-yielding metabolic pathways seen in eubacteria—greater by far than in all other forms of life combined—is interpreted by many investigators as reflecting the variety of evolutionary "experiments" with metabolism during the long period when procaryotes were the only living things on the earth.

The Photosynthetic Eubacteria Differ in Their Pigments

Many photosynthetic eubacteria (i.e., those belonging to the purple and green groups of bacteria) do not have chlorophylls *a* or *b*, the chief light-trapping pigments in higher plants. Instead these bacterias have other types of chlorophyll, called bacteriochlorophylls. And unlike higher plants, neither purple nor green bacteria use water as the electron donor in photosynthesis; hence neither produces molecular oxygen. Photosynthesis for the green and the purple bacteria is a strictly anaerobic process; it cannot even take place in the presence of O_2.

Eubacteria of the other two photosynthetic divisions, the cyanobacteria and prochlorophyta, possess chlorophyll *a*[2], use water as the electron donor in photosynthesis, and generate O_2 as a by-product of their photosynthesis. As we saw in Chapter 36, it was probably the ancestors of cyanobacteria that contributed most to the creation of an oxidizing atmosphere some 2.3 billion years ago. And, according to the endosymbiotic hypothesis, one of these types of photosynthetic eubacteria may have been the source of eucaryotic chloroplasts.

The cyanobacteria are the largest group of photosynthetic eubacteria. They are procaryotic unicellular or filamentous photosynthetic organisms (Fig. 37.27). The cells are typically larger than eubacterial cells, but smaller than most eucaryotic cells, though the ranges overlap (Fig. 37.13).

All cyanobacteria possess photosynthetic pigments (including carotenoids and chlorophyll *a*) located in the flattened membranous disks called thylakoids (Fig. 37.28), but the thylakoids are not contained within chloroplasts. Many of the cyanobacteria also contain blue pigment, and some contain red pigment. The presence of blue pigment and chlorophyll gives

[2]The Prochlorophyta also possess chlorphyll *b* and the accessory pigments found in higher plants.

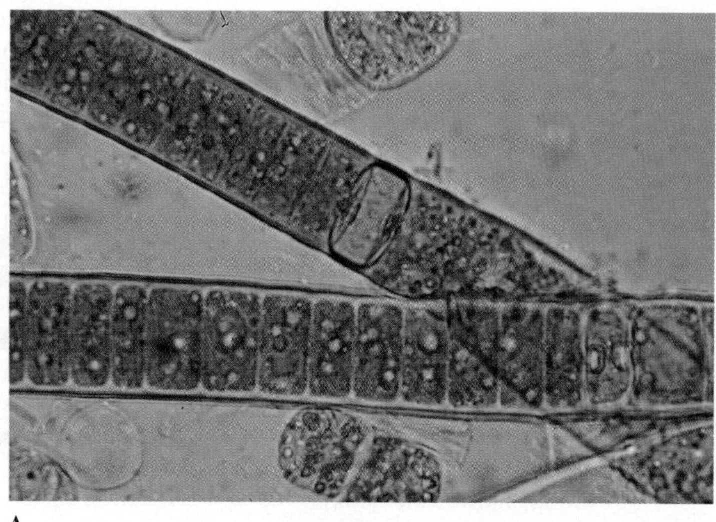

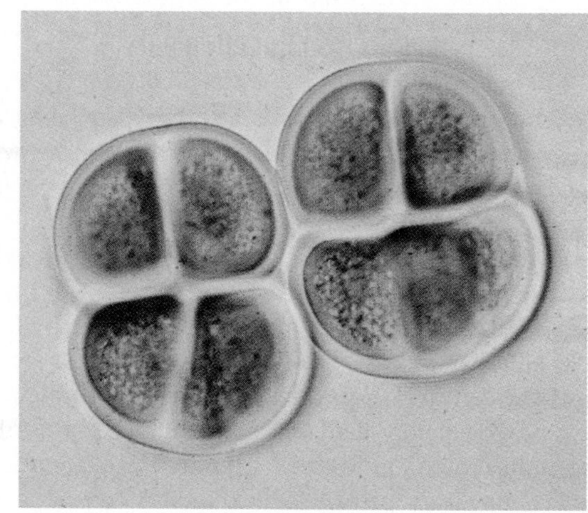

37.27 Some representative cyanobacteria (A) *Scytonema,* filamentous bacterium, has oval, nearly rectangular cells. The lighter colored cell in the diagonal filament is a heterocyst, a cell specialized for carrying out nitrogen fixation, converting molecular nitrogen into a biologically useful form. (B) Cyanobacteria (*Chroococcus*) from the Pine Barrens of New Jersey. Small groups of cells are enclosed within common gelatinous sheaths.

37.28 A cyanobacterium, *Plectonema boryanum,* undergoing cell division The photosynthetic membranes (thylakoids) lie near the periphery of the cell. Walls have begun to grow inward at a point midway along the cell's length.

some of them their characteristic blue-green color, but black, yellow, red, grass green, and other colors also occur. The periodic redness of the Red Sea is due to a species that contains a particularly large amount of red pigment.

One especially important property of many cyanobacteria is their ability to fix atmospheric nitrogen (N_2). The species most active in N_2 fixation are usually filamentous forms that produce a few highly specialized cells called heterocysts, where, under anaerobic conditions, N_2 fixation takes place (Figs. 37.27A and 37.29). The occurrence of cyanobacteria in some environments so bleak that no other organisms can inhabit them is probably due, in part, to the fact that the nitrogen-fixing species have the simplest nutritional requirements of any known organisms. In addition to N_2 and carbon dioxide, which are readily available from the atmosphere, they need only water, light, and certain minerals.

Many cyanobacteria live in fresh water, and a few are marine. Some species are very common on or in soil—among them species that live in the extremely nitrogen-poor soil often found in the tropics. Other species are often found growing on the sides of damp rocks and flowerpots and on the back of trees. Cyanobacteria tend to be particularly abundant on wet cliffs and ledges. Ponds or lakes containing a rich supply of organic matter, particularly nitrogenous compounds, often develop huge populations ("blooms," as they are called) of cyanobacteria, which may make the water so green that objects only a few centimeters below the surface are entirely invisible. Such blooms may give the water an objectionable "pigpen" odor, clog filters of water supplies, and even be toxic to livestock. Some species of cyanobacteria live mutualistically with fungi in the compound plants called lichens. A few species

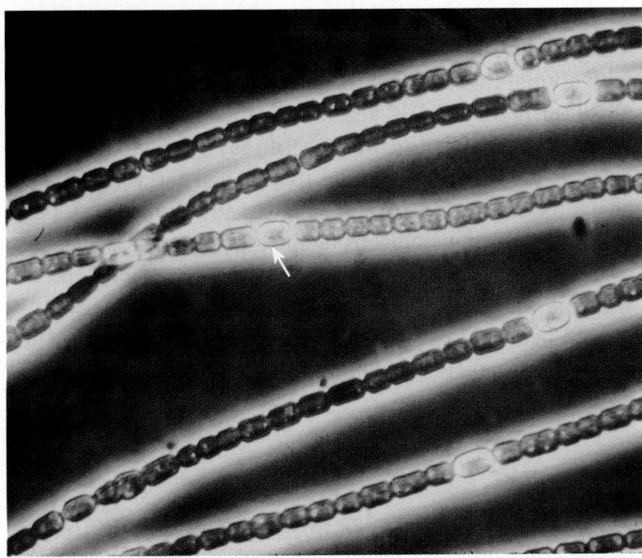

37.29 The effect of environmental nitrogen on heterocyst formation by a cyanobacterium, *Anabaena* When grown in a culture containing sodium nitrate, the filaments form no heterocysts. On a medium containing N_2 as the only nitrogen source, the filaments form numerous light-colored heterocysts (arrow).

(along with some archaebacteria) live in habitats that rank among the most inhospitable known—the hot springs that occur in various parts of the world.

Most Eubacteria Are Beneficial

Contrary to the popular impression, beneficial eubacteria outnumber harmful ones. In Chapter 34 we mentioned the nitrogen-fixing bacteria and the role of bacteria as decomposers—a role that not only prevents the accumulation of dead bodies and metabolic wastes but also converts materials such as the nitrogen of proteins into forms usable by other living things. Both plant and animal life are dependent on the activities of these bacteria, for without them the various nutrient cycles would cease. There are also bacteria in our intestines that synthesize vitamins absorbed by the body and that aid in the digestion of certain materials. Anyone who has been given doses of antibiotics massive enough to exterminate the intestinal flora can testify to the ensuing disturbances in normal intestinal activity.

Eubacteria are of great importance in many industrial processes. Manufacturers often find it easier and cheaper to use cultured microorganisms in certain difficult syntheses than to try to perform the syntheses themselves. Among the many substances manufactured commercially by means of eubacteria are acetic acid (vinegar), acetone, butanol, lactic acid, and several vitamins. Eubacteria are also used to prepare flax and hemp for making linen, other textiles, and rope. Commercial preparation of skins for making leather goods often involves use of eubacteria, as does the curing of tobacco.

Many branches of the food industry depend on eubacteria. You are probably aware of the central role of eubacteria in the making of dairy products, such as butter, yogurt, and the various kinds of cheeses; the characteristic flavor of Swiss cheese, for example, is due in large part to the propionic acid produced by eubacteria.

Farmers also depend on eubacterial action in the making of silage for use as cattle feed. And in the future eubacteria pathogenic for destructive insects may eventually be developed to replace some insecticides.

The role of eubacteria in the production of antibiotics that can help control other eubacteria is especially interesting.

MAKING CONNECTIONS

1. How does binary fission of bacteria differ from eucaryotic cell division? (See Chapter 9.)

2. What is the genetic significance of conjugation in eubacteria?

3. What does the great diversity of energy-yielding metabolic pathways found in modern eubacteria indicate about the early evolution of life on earth?

4. Why is no oxygen produced by photosynthesis as carried on by purple and green bacteria?

5. According to the endosymbiotic hypothesis, which groups of photosynthetic bacteria may have been ancestral to the chloroplasts of eucaryotes?

6. Why are cyanobacteria essential to the nitrogen cycle discussed in Chapter 34?

Most of the antibiotic drugs in use today (but not penicillin) are produced by eubacteria or, if synthesized artificially, were discovered in these organisms. Among these drugs are streptomycin, tetracycline, bacitracin, and neomycin.

Most recently, eubacteria—particularly *E. coli*—have become the workhorses of genetic engineering, turning out large quantities of enzymes or other products like insulin encoded by genes removed from other organisms and inserted into a eubacterial host (see Chapter 13).

Some Eubacteria Cause Disease

Perhaps the best-known eubacteria are the ones that cause disease in human beings, domestic animals, and cultivated plants. Most so-called germs are either bacteria or viruses (though a few are fungi, protozoans, or parasitic worms). As we have said, the idea that eubacteria can cause disease—often called the germ theory of disease—was first developed by Louis Pasteur in the late 19th century. Initially scorned, the idea soon gained the support of prominent scientists and physicians of the day.

We now know that eubacteria cause many human diseases, including bubonic plague, cholera, diphtheria, syphilis, gonorrhea (Fig. 37.30), leprosy, scarlet fever, tetanus, tuberculosis, typhoid fever, whooping cough, bacterial pneumonia, bacterial dysentery, meningitis, strep throat, boils, and abscesses. Equally long lists could be compiled of bacterial diseases of other animals and of plants.

Microorganisms cause disease symptoms in a variety of ways. In some cases their immense numbers simply place such a tremendous burden on the host's tissues that they interfere with normal function. In other cases the microorganisms actually destroy cells and tissues. In still other cases eubacteria produce poisons, called ***toxins***. These may be exotoxins, which are poisons released from the living eubacterial cell into the host's tissues, as in diphtheria and tetanus, or they may be endotoxins, which are poisons retained in the cells of the eubacteria that produce them and are only released into the host when the eubacteria die and disintegrate.

In Chapter 21 we discussed the role of the specific and nonspecific immune response in defending the body against foreign microbes.

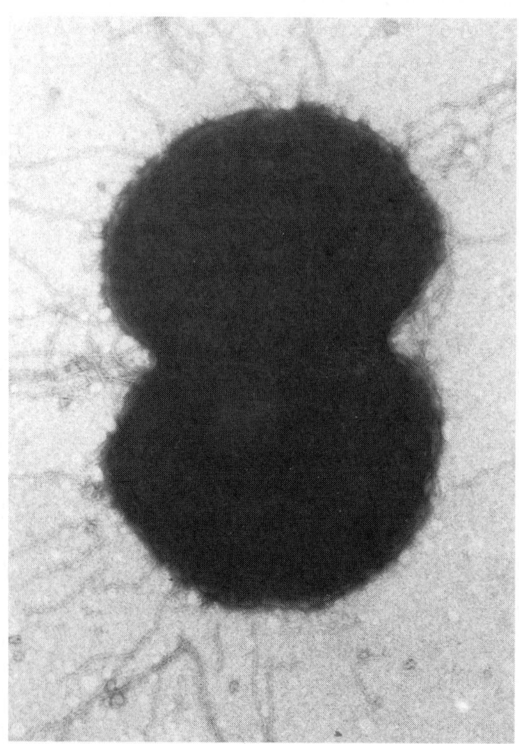

37.30 The gonorrhea eubacterium A diplococcal pair of the gonorrhea eubacterium, *Neisseria gonorrhoeae,* showing many cytoplasmic extensions called pili, which appear to play a role in the attachment of the bacterium to human mucosal cells. This eubacterium infects at least one million people in the United States annually. Unchecked infections can cause blindness in newborns and sterility in adults.

Eubacteria Can Evolve Resistance to Antibiotics

We know that antibiotics act by interfering with some physiological activity within the microorganism. Because eubacteria are procaryotic cells, many antibiotics specifically inhibit activities within the eubacteria while not harming the eucaryotic cells of the patient. As we saw in Chapter 31 (see p. 682), however, microorganisms can become resistant to antibiotics through natural selection: eubacteria that have a gene for an enzyme that inactivates a particular antibiotic have a selective advantage when exposed to that antibiotic, and survive in greater numbers. Since it is they that reproduce and perpetuate the population (the susceptible individuals having been killed), the next generation shows a marked resistance to that antibiotic. Drug resistance among eubacteria has become a serious problem. For example, in the 1950s gonorrhea was treated with a few hundred thousand units of penicillin, but today the dosage required is several million units, and there are now forms of gonorrhea-causing eubacteria that are completely resistant to penicillin.

Why has drug resistance become so widespread? One reason is that antibiotics have been used indiscriminately for many years. Such drugs have been prescribed for everything from the common cold to influenza—diseases caused by viruses for which antibiotics are useless. Another major use of antibiotics has been in livestock production. Over half the antibiotics produced in this country are used in animal feeds, particularly those for beef cattle, pigs, and poultry. Animals fed on antibiotic-supplemented feed grow faster than those on regular feed, making such supplements economically useful. Unfortunately such widespread use of antibiotics has led to the evolution of antibiotic-resistant strains of bacteria in animals, and these strains can be directly transmitted to humans. Making the problem even more serious, the genes specifying drug resistance in eubacteria may be on the bacterial chromosome or carried in plasmids. Those carried in plasmids are particularly important since plasmids can be transferred from one cell to another during conjugation; consequently drug resistance can be rapidly spread. It has, for example, been estimated that about half the Gram-negative eubacteria living as normal inhabitants of our intestines are resistant to one or more antibiotics. Thus, overuse of antibiotics coupled with the rapid transmission of genetic information for resistance constitutes a major potential health hazard for our future.

CONCEPTS IN BRIEF

1. A *virus* consists of a nucleic acid core covered with a protein coat. The nucleic acid can be DNA or RNA, double-stranded or single-stranded. Some viruses have a membranous envelope around the protein coat.

2. Viruses cannot reproduce themselves; they rely upon the host cell to manufacture new virus particles, using the genetic instructions provided by the old virus. Once new viral components have been synthesized by the host cell, they assemble into new viruses that are released by lysis of the host cell or by budding.

3. Sometimes viral DNA is integrated into the host chromosome as a *provirus,* which behaves as an additional part of the chromosome and is replicated with it. The virus remains there until the host cell is exposed to some stress, then the virus leaves the host chromosome and resumes normal replication.

4. DNA viruses use the enzymes of the host cell for their replication, but RNA viruses require special enzymes. *RNA replicase* catalyzes RNA-to-RNA replication. In retroviruses, *reverse transcriptase* transcribes RNA into DNA, which then becomes a provirus. Later it is transcribed into RNA for new viruses.

5. Viruses cause many diseases in various organisms, and have been associated with some kinds of cancer. Most viral infections do not respond to antibiotic treatment. A protein called *interferon* helps prevent the spread of the viruses within the body.

6. The procaryotes are placed in two kingdoms, the archaebacteria, or "ancient" bacteria, and the eubacteria, the "true" bacteria. In contrast to the viruses, bacteria are cellular; they have the metabolic machinery to generate ATP and to reproduce themselves; and they have ribosomes for protein synthesis.

7. The archaebacteria live in very harsh environments: in high temperatures, high salinity, and high acidity. There are three main groups of archaebacteria: the methanogens (methane-producers), the extreme halophiles ("salt-loving"), and the thermoacidophiles ("heat- and acid-loving").

8. Most bacterial cells are surrounded by a semirigid cell wall, inside of which is a cell membrane. Bacterial cell walls protect the cell both from physical damage and from osmotic disruption. Differences in cell wall chemistry of different eubacteria are associated with differences in their staining properties; these can be used for identification.

9. Some eubacteria can form special resting cells called *endospores,* which enable them to withstand adverse conditions. When conditions improve, the endospores produce new eubacterial cells.

10. Bacteria reproduce by *binary fission.* The single circular DNA chromosome found in the *nucleoid* area replicates before division. Many eubacteria obtain new genetic material by conjugation.

11. Most eubacteria are heterotrophic (either saprophytic or parasitic), but some are *chemosynthetic autotrophs* and others are *photosynthetic autotrophs.* The oxygen requirement varies; most are *aerobic* but some are *facultative anaerobes* and others are *obligate anaerobes.*

12. The green and purple bacteria have their own types of bacteriochlorophyll and do not use water or produce molecular oxygen. The cyanobacteria do possess chlorophyll *a,* the pigment found in higher plants, and generate molecular oxygen during photosynthesis. Many cyanobacteria can fix atmospheric nitrogen, allowing them to survive in minimal environments. The cyanobacteria or prochlorophyta may have been the ancestors of the chloroplasts of eucaryotes.

13. Eubacteria cause disease symptoms by interfering with normal function, destroying cells and tissues, or by producing poisons, called *toxins.* Antibiotics are useful against eubacteria but unfortunately many bacteria have evolved resistance to antibiotics.

STUDY QUESTIONS

1. **Describe the structure of a virus. How is it different from a cell? (p. 862)**
2. **How do viruses reproduce? Describe the *lytic cycle* of bacteriophage viruses. What is a *provirus*? Contrast the methods by which bacteriophage, plant, and animal viruses enter their host cells. How are viruses released from cells? (pp. 863–866)**
3. **Compare how DNA viruses, RNA viruses, and retroviruses replicate their genetic material. (pp. 864–865)**
4. **Why are antibodies generally ineffective against viruses? How does *interferon* help to protect cells against viral attack? (pp. 866–867)**
5. **Describe the three main groups of archaebacteria. What characteristics do all archaebacteria share with eubacteria? Why are they placed in a kingdom of their own? (pp. 868–869)**
6. **What are the three most common shapes of bacteria? Describe the internal organization and cell-wall structure of a eubacterial cell. How are differences in the composition of their cell walls useful in identifying eubacteria? (pp. 870–872)**
7. **Describe the various means by which eubacteria obtain energy and nutrients. Distinguish between *obligate anaerobes* and *facultative anaerobes*. (p. 875)**
8. **Discuss at least four ways in which eubacteria are ecologically or economically beneficial. (pp. 876–878)**
9. **How do eubacteria produce disease symptoms? How is it possible for antibiotics to kill bacteria while not harming the cells of the patient? Why has eubacterial resistance to antibiotics become a serious problem? (pp. 878–879)**

CHAPTER 38

THE PROTISTAN KINGDOM

For almost two billion years the only living organisms on earth were the single-celled procaryotes. About 1.8 billion years ago, the first eucaryotic organisms evolved. In this chapter we consider the descendants of these early eucaryotes. They are essentially single-celled or colonial[1] eucaryotic organisms, though some are multicellular organisms whose bodies show relatively little distinction between tissues (i.e., the plantlike multicellular algae).

Though some of the protistan groups tend to be animallike, others plantlike, and still others funguslike, they share many characteristics: whether they are unicellular, colonial, or multicellular, all are eucaryotic and lack tissue differentiation (i.e., they do not develop specialized tissues and organs). Although we shall treat the protistans in three sections here, according to the evolutionary thrusts—animal, plant, or fungal—these divisions are in many ways artificial. A formal outline of the classification of the protists is found in the Appendix. Table 38.1 relates the traditional groupings to the major formal classifications.

ANIMALLIKE PROTISTA: THE PROTOZOA

Protozoans Are Sometimes Considered Acellular Organisms

Protozoans are usually said to be unicellular, but a protozoan behaves as if it were a complete organism. Instead of organs, they have functionally equivalent organelles. In recognition of the complexity of protozoa, which often far exceeds that of individual cells in multicellular organisms, many biologists prefer to call them ***acellular*** organisms, i.e., organisms whose bodies do not exhibit the usual construction of cells (see Fig. 17.10, p. 354).

Protozoans occur in a great variety of habitats, including the sea, fresh water, soil, and the bodies of other organisms—in fact, wherever there is moisture. They are heterotrophic; they ingest small food particles or cells. Most are free-living, but some are commensalistic, mutualistic, or parasitic.

The protozoans usually digest food particles in food vacuoles. There are no special organelles for gas exchange, with the general cell membrane serving as the exchange surface. Many species, particularly those living in hypotonic media such as fresh water, possess contractile vacuoles, which function primarily in eliminating excess water. Small amounts of nitrogenous waste may also be expelled by the contractile vacuoles, but most of this is released as ammonia by diffusion across the general cell surface. Movement is by means of pseudopodia or by means of beating cilia or flagella.

[1] Colonial refers to organisms that live together in a group (colony).

Chapter opening photo: A foraminiferan. Like the amoebae to which they are related, the foraminiferans have pseudopodia (spine-like in this animal) for capturing prey. Foraminiferans construct a shell of secreted and/or gathered materials.

TABLE 38.1 *Major groups of organisms belonging to the kingdom Protista*

Animallike Protists: The Protozoa (primarily ingestive heterotrophs)
Phylum Kinetoplastida: the flagellates and sarcodinians (amoeboid organisms)
Phylum Ciliata: the ciliates

Funguslike Protists (absorptive heterotrophs)
Division Ampiplexa: the sporozoans
Division Mycetozoa: the slime molds
Division Oomycota: the water molds

Plantlike Protists (primarily photosynthetic autotrophs)
Unicellular algae
Division Euglenophyta: the euglenoids
Division Dinoflagellata: the dinoflagellates
Division Chrysophyta: the yellow-green and golden-brown algae, and diatoms
Multicellular algae
Division Phaeophyta: the brown algae
Division Rhodophyta: the red algae

Reproduction is sometimes sexual and sometimes asexual.

The protozoans are divided into two groups: Kinetoplastida and Ciliata. These are groupings of structurally similar, but probably not closely related, organisms. The relationships of the groups to one another are unclear.

The Kinetoplastids Possess Flagella or Pseudopodia

The Kinetoplastida is the largest phylum of protozoa. It is divided into two subphyla, the flagellates (Mastigophora), which possess flagella, and the amoeboid protozoans (Sarcodina), which possess pseudopodia. These two groups are thought to be closely related because some flagellates undergo amoeboid phases, and, conversely, some sarcodines have flagellated stages.

The Mastigophora appear to be the most primitive of all the protozoa, and it seems likely that some (and possibly all) of the other protozoan groups arose from them. Many biologists (though certainly not all) believe that the flagellate protozoans are also the ancestors of the multicellular plants and animals. There is good reason to think that flagellated unicells played a key role in the evolution of life on earth. Most groups of higher organisms have flagellated stages in their life cycle; only three major groups, the angiosperms, conifers, and fungi, have lost them entirely.

A few of the flagellates are free-living in salt or fresh water, but most live symbiotically in the bodies of higher plants or animals. Several species, for instance, are found in the gut of termites, where they participate mutualistically in the digestion of the cellulose consumed (Fig. 38.1).

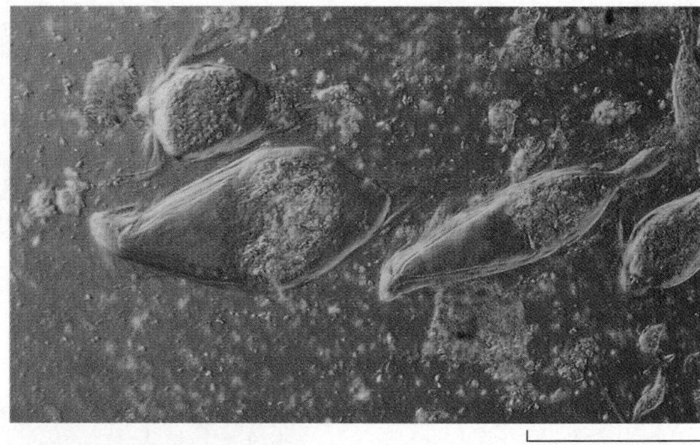

38.1 *Trichonympha,* a flagellate that inhabits the gut of termites The flagellate helps digest the cellulose in the termite's diet.

Pseudopods Function in Movement and Feeding

The most familiar of the amoeboid protozoa (subphylum Sarcodina) are the freshwater species of the genera *Amoeba* and *Pelomyxa* (see Fig. 17.8, p. 353), which have pseudopods. The pseudopods, which are large and have rounded or blunt ends, function both in movement and in feeding by phagocytosis.

Also included in the Sarcodina are several groups of protozoans that secrete hard shells around themselves. These shells, often quite elaborate and complex, can be used in species identification. The pseudopods of shelled sarcodines are usually thin and pointed (Fig. 38.2); a sticky secretion on the outside of the pseudopods traps prey that touch them. Two groups of shelled sarcodines, the foraminiferans and the radiolarians, have played major roles in the geologic history of the earth.

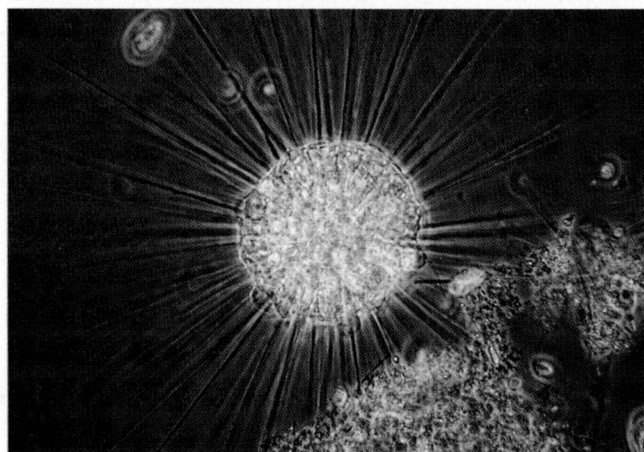

38.2 A sarcodine *(Actinosphaerium)* with long pointed pseudopods

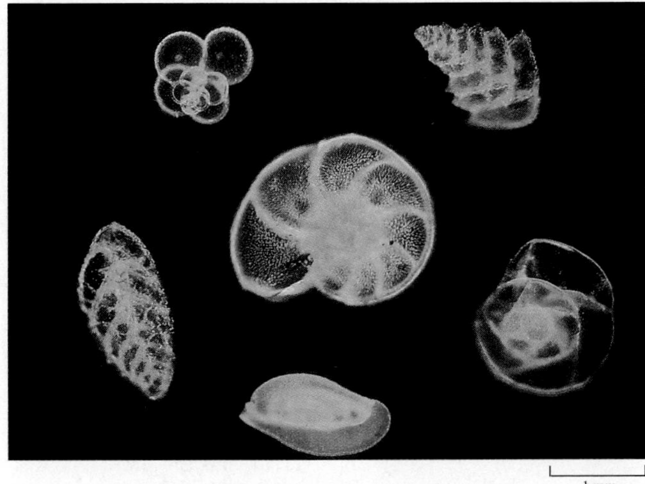

38.3 A group of calcareous foraminiferan shells

38.4 A group of siliceous radiolarian shells

Both groups are extremely abundant in the oceans, and when the individuals die, their shells become important components of the bottom mud. The shells of foraminiferans (Fig. 38.3), which contain calcium, are especially prevalent in the mud at depths of 2,500 to 4,500 meters, whereas the bottom ooze in deeper parts of the ocean is composed chiefly of the silica-containing shells of radiolarians (Fig. 38.4). Much of the limestone and chalk (such as the famous White Cliffs of Dover, England) were formed from deposits of foraminiferan shells, while radiolarian shells have contributed to the formation of silica-containing rocks.

The Ciliates Are the Most Complex Protozoans

The phylum Ciliata is the most homogeneous of the protozoan phyla, and all members probably share a common ancestor. Ciliates are unique among protozoans in having two types of nuclei: a large macronucleus, which controls the normal metabolism of the cell, and one or more small micronuclei, which are involved with reproduction and with giving rise to the macronucleus. This group includes the well-known *Paramecium*.

As their name implies, ciliates possess numerous cilia, which they use for movement (Figs. 38.5 and 38.6). They have

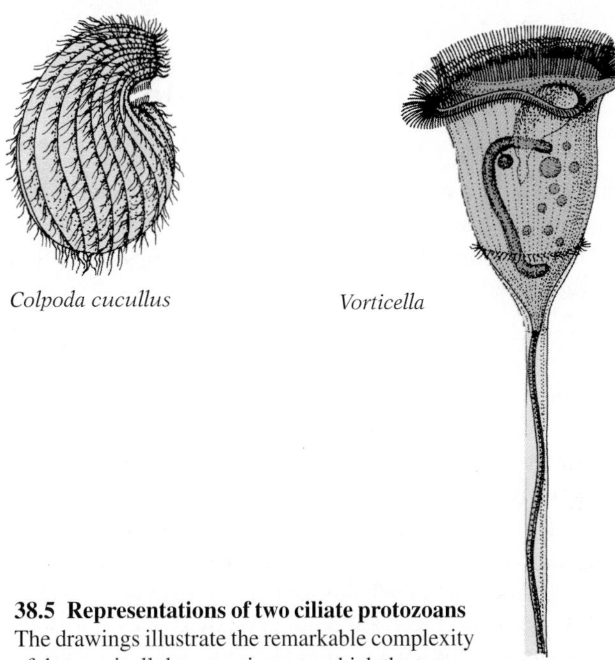

38.5 Representations of two ciliate protozoans
The drawings illustrate the remarkable complexity of these unicellular organisms, to which the term "acellular" applies especially well.

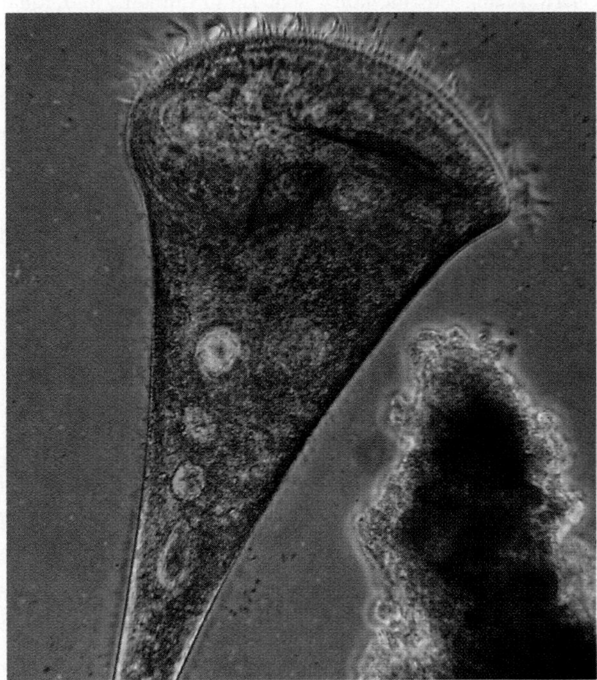

38.6 Photographs of *Stentor*, a representative freshwater ciliate
Stentor is long and funnel shaped. It attaches itself by its base to various objects, but can swim about. Cilia are used to sweep food into its oral groove.

the most elaborate organelles of any protozoan, and the term "acellular" applies best to them. As we saw with *Paramecium,* they may have a special oral groove and cytopharynx into which food particles are drawn in currents produced by beating cilia, and they have a fixed anal pore, through which indigestible wastes are expelled from food vacuoles (see Fig. 17.10, p. 354).

FUNGUSLIKE PROTISTA

There are several major groups of fungal-type organisms that are at the protistan level of complexity, but for our purposes we shall consider just three groups, the sporozoans, slime molds, and water molds.

All Sporozoans Are Internal Parasites

The sporozoans (phyla Apicomplexa and Microspora), though often classified with the protozoans, are not closely related as revealed by DNA and RNA sequence analysis. Unlike protozoans, they have no organelles for movement and are funguslike in many of their attributes. As their name implies, sporozoans usually have a sporelike, infective-cyst stage in their life cycles. All are internal parasites, and cause a variety of serious diseases including malaria in humans, and equally serious conditions in domestic and wild animals.

Malaria, which is caused by a sporozoan of a species of the genus *Plasmodium,* is transmitted from host to host by female *Anopheles* mosquitoes. The life cycle of *Plasmodium* is outlined in Figure 38.7. The host experiences attacks of chills and fever each time infected red cells burst and release the parasites that have multiplied within them; the symptoms apparently stem from toxins discharged into the blood by the rupturing red cells.

The World Health Organization lists malaria as one of the most serious infectious diseases. Efforts at control of this disease have largely been aimed at eradicating *Anopheles* mosquitoes, either directly by use of insecticides or indirectly by destroying their breeding places. The insecticide DDT was widely used for mosquito control; now that restrictions have been placed on DDT use, mosquito populations are increasing and malaria is once again becoming a very serious health problem.

Slime Molds Have an Animallike Amoeboid Stage

Slime molds are not true fungi; rather, they are curious organisms that are animallike at some stages in their life cycle and funguslike at others. They are generally found growing on damp soil, rotting logs, leaf mold, or other decaying organic matter in moist woods, where they look like glistening, viscous masses of slime, sometimes white, but often yellow or red.

Two very different groups of slime molds (division[2] Mycetozoa) are recognized, the ***cellular slime molds*** and the ***true slime molds,*** which are not closely related. Though funguslike in many ways, they are quite distant evolutionarily from the fungi and from each other. The cellular slime molds are an ancient group that probably branched off the main eucaryotic line that leads to plants, animals, and fungi. The true slime molds arose later, about 1.2 billion years ago, as a separate lineage. The life cycles of these two groups differ in many details, but both have an amoeboid stage where they move

[2] "Division" and "phylum" are equivalent terms. Whereas "phylum" is applied to animals and animallike organisms, "division" is applied to fungi, plants, and plantlike organisms.

MAKING CONNECTIONS

1. How would the high surface-to-volume ratio of the unicellular Protista make the maintenance of homeostasis easier for them than for larger organisms? How would it make it more difficult?

2. What evidence suggests that flagellated unicells may have been the ancestors of multicelluar plants and animals?

3. What is the link between malaria and the prevalence of the allele for sickle-cell anemia in some parts of Africa? (See Chapter 31.)

4. Which two characteristics of water molds distinguish them from the fungi discussed in Chapter 40?

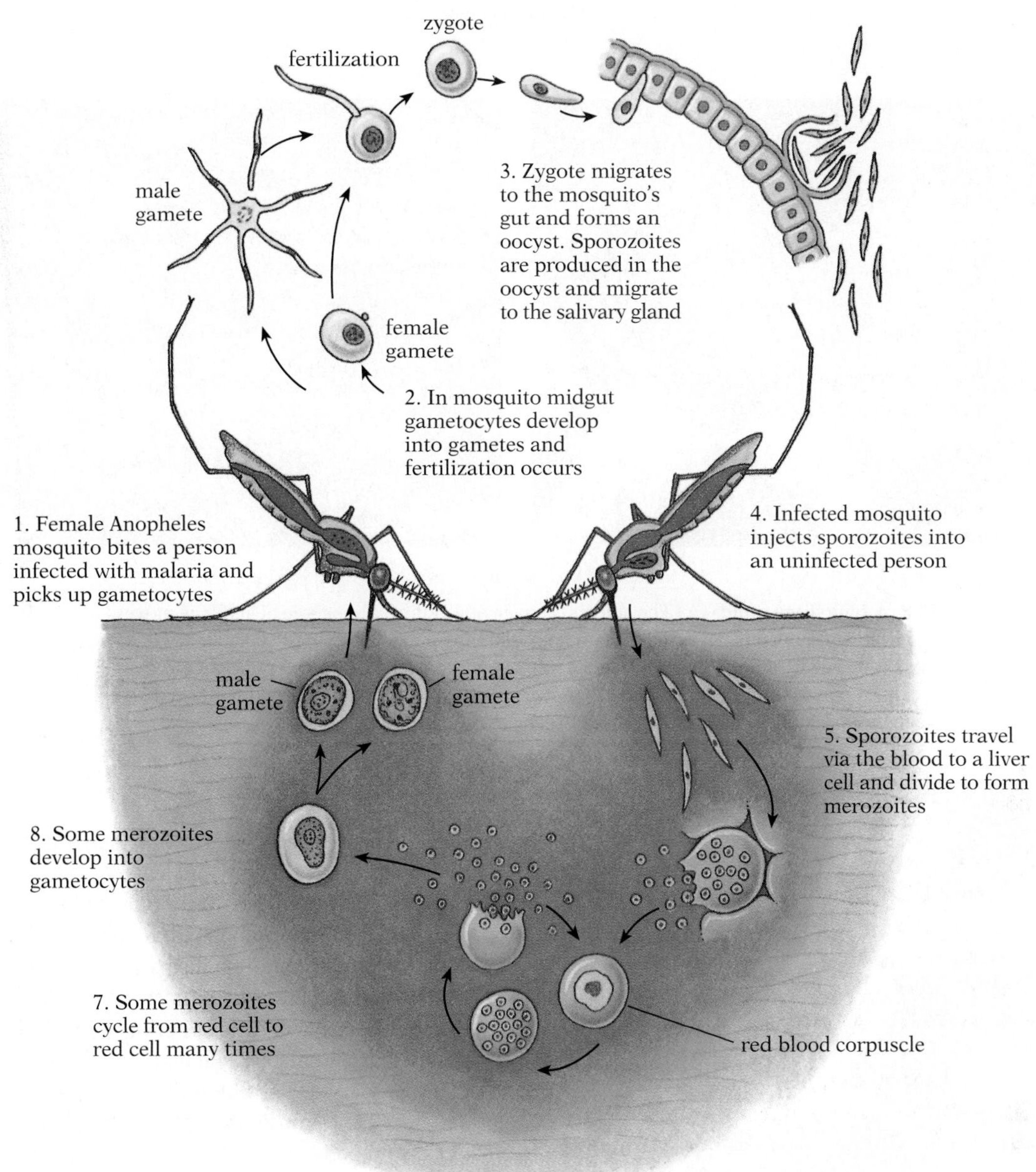
zygote
fertilization
male gamete
3. Zygote migrates to the mosquito's gut and forms an oocyst. Sporozoites are produced in the oocyst and migrate to the salivary gland
female gamete
2. In mosquito midgut gametocytes develop into gametes and fertilization occurs
1. Female Anopheles mosquito bites a person infected with malaria and picks up gametocytes
4. Infected mosquito injects sporozoites into an uninfected person
male gamete
female gamete
5. Sporozoites travel via the blood to a liver cell and divide to form merozoites
8. Some merozoites develop into gametocytes
7. Some merozoites cycle from red cell to red cell many times
red blood corpuscle
6. Merozoites infect red blood cells and divide asexually

38.7 The life cycle of *Plasmodium,* the malarial parasite

around like a large amoeboid protozoan and ingest particulate food (Fig. 38.8A). This is followed by a sedentary funguslike stage in which a fruiting body develops and spores are formed (Fig. 38.8B).

Water Molds Have Flagellated Reproductive Cells

As their common name, "water molds," suggests, most members of the division Oomycota are aquatic. Some, however, live in the soil, and some are parasites of higher plants. Among the latter group is the historic scourge *Phytophthora infestans,* which causes blight in potatoes and was responsible for such devastations as the Irish potato famine of 1845–47, which did much to change the course of Irish history. Other members of this group of fungi—notably the agents of downy mildew of sugar beets and grapes—can also cause severe economic damage. It was downy mildew that almost destroyed the wine grapes in France in the late 1870s.

The water molds are unicellular or filamentous (Fig. 38.9). They differ from the true fungi in that they produce motile, asexual reproductive cells called ***zoospores*** that have flagella and closely resemble the flagellated protozoa. The zoospores germinate into threadlike hyphae, some of which will form a reproductive organ called a sporangium in which new zoospores are produced, thereby completing the asexual cycle; complex sexual reproduction also occurs. Water molds also differ from fungi in lacking chitin in their cell walls—instead, some cellulose is present.

A

B

38.8 A true slime mold (A) The amoeboid stage of a yellow slime mold. This stage moves about slowly and feeds on particles of organic matter by phagocytosis. The behavior of this stage is decidedly animallike. (B) Under certain conditions, the amoeboid stage becomes sedentary and develops fruiting bodies (shown above), in which spores are formed. At this stage the appearance and behavior of the organism is funguslike.

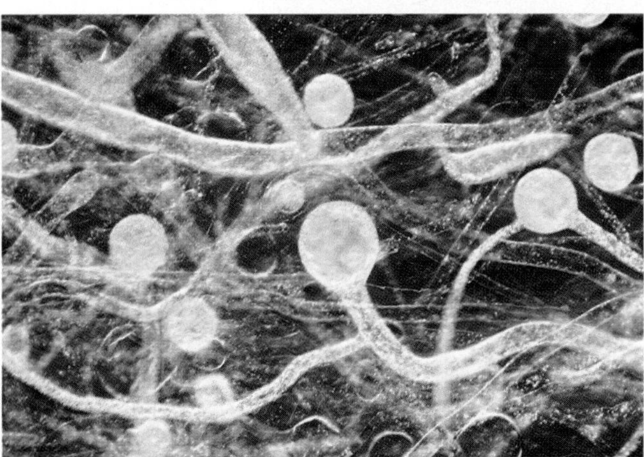

38.9 A water mold (Saprolegnia) Most water molds are aquatic and are found in most bodies of fresh water. Many are parasitic. The round structures shown above are oogonia, the reproductive structures in which fertilization will occur. Water molds also reproduce asexually through the production of flagellated zoospores.

PLANTLIKE PROTISTA: THE UNICELLULAR ALGAE

The algae are a diverse assemblage of organisms. They range in size from microscopic unicells to large seaweeds that may be hundreds of feet long. Some are single-celled or colonial, others are arranged end to end to form filaments, and still others are truly multicellular. All are eucaryotic, primarily photosynthetic, and lack the tissue differentiation seen in higher plants. There are no xylem and phloem tissues in the algae, and thus no true roots, stems, and leaves.

In our earlier discussions of the endosymbiotic hypothesis we suggested that mitochondria and chloroplasts are descendants of procaryotic organisms that took up permanent residence inside other procaryotic host cells. Molecular studies of the chloroplast and mitochondrial DNAs and RNAs have suggested that a number of cell lines may have acquired photosynthetic procaryotes as chloroplasts quite independently of one another. Thus the various groups of algae are probably not closely related (Fig. 38.10); they not only evolved from differ-

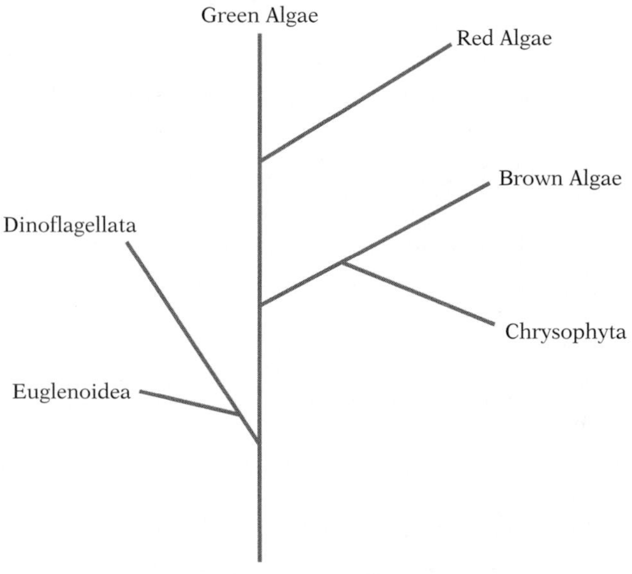

38.10 Evolutionary relationships among the algae

TABLE 38.2 *Some characteristics of the algae*

Group	*Structure*	*Habitat*	*Pigments*	*Cell wall*
Euglenoids	Unicellular	Fresh and seawater	Chlorophylls *a, b,* carotenoids	Absent
Golden-brown, yellow-green, diatoms	Unicellular	Fresh and seawater, soil	Chlorophylls *a, c,* carotenoids	Silica impregnated
Dinoflagellates	Unicellular	Fresh and seawater, soil	Chlorophylls *a, c,* carotenoids	Cellulose plates
Brown algae	Multicellular	Seawater	Chlorophylls *a, c,* carotenoids	Cellulose
Red algae	Multicellular	Seawater	Chlorophylls *a, d,* red, blue pigments	Cellulose
Green algae	Unicellular to multicellular	Fresh and seawater, soil	Chlorophylls *a, b,* carotenoids	Cellulose

ent eucaryotic cell lineages but also acquired different bacterial lineages as chloroplasts. Some of the important characteristics of the various algal lineages are summarized in Table 38.2.

Let us begin by looking at several groups of primarily unicellular, often flagellated, algae.

The Euglenoids Show Both Plantlike and Animallike Characteristics

The euglenoids (division Euglenophyta) are flagellated, unicellular organisms that are not closely related to the other plantlike protists or to the green plants. Most live in fresh water, but a few are found in soil, on damp surfaces, or even in the digestive tracts of certain animals. They are plantlike in that many species have chlorophyll and are photosynthetic; they are animallike in lacking a cell wall and being highly motile. Some species lack chloroplasts and are heterotrophic, like animals. Their pigments (chlorophylls *a* and *b* and carotenoids) are similar to those of the green algae and land plants and different from those of most other algae, but their chloroplasts have three membranes rather than the two of the green algae and land plants. The euglenophytes are closely related to the flagellates, and are very similar to them in every way except in possession of chloroplasts. They are thought to have evolved from flagellates that acquired photosynthetic procaryotes as chloroplasts.

A representative genus, *Euglena,* has an elongate, ovoid cell with a long flagellum emerging from an anterior indentation (Fig. 38.11). Because the cell lacks a wall, it is fairly flexible, and its shape may change somewhat as it swims about.

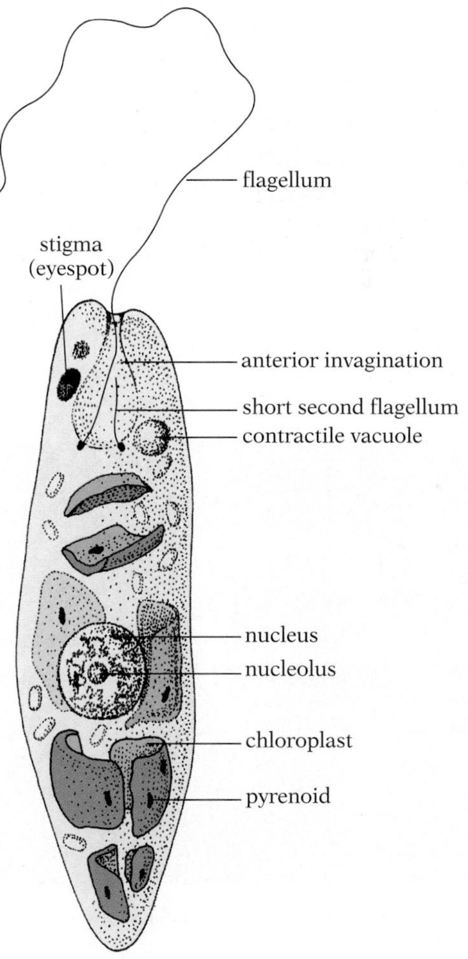

38.11 *Euglena* *Euglena* has two flagella, one long and one short. Note the unusually shaped chloroplasts and the contractile vacuole, which is used to expel excess water. The orange stigma functions in light detection and helps mediate the organism's positive response to light. Most *Euglena* have a special organelle, the pyrenoid, that functions in the production of a unique polysaccharide storage product.

Dinoflagellates Are Abundant in the Marine Plankton

The dinoflagellates (division Dinoflagellata) are small, usually unicellular organisms (Fig. 38.12) that are second only to the cyanobacteria and diatoms (discussed below) as primary producers of organic matter in the marine environment. Most dinoflagellates are photosynthetic but there are many colorless species, which are heterotrophic. Photosynthetic species usually have a yellowish-green to brown color due to the presence of large amounts of carotenoid and other pigments in addition to chlorophylls *a* and *c*. Most species have two flagella of unequal length, which lie in two grooves (Fig. 38.13). Their wall is very unusual, consisting of sculptured cellulose plates imbedded in the cell membrane.

Some photosynthetic dinoflagellates live symbiotically within the cells of a variety of marine invertebrate animals, including many corals, which get their distinctive colors from these algae. In many cases the animals cannot survive without them. Some types of coral, for instance, contain about two million algae cells per square centimeter. The dinoflagellate symbionts (called zooxanthellae) supply the coral with oxygen and food, while the coral provides a home and inorganic nutrients such as nitrogen and phosphorus. In fact, the coral takes as much as 60 percent of the organic material produced by the zooxanthellae. When placed under environmental stress (e.g., exposed to excessive heat or cold, physical damage, or pollution), the coral loses its zooxanthellae. Without them, the coral tissue is almost colorless, so its white calcium carbonate skeleton stands out (Fig. 38.14). This process, called "bleaching," occurred over a large part of the coral reefs in the tropics around the world during the 1980s. Unless the process is reversed, the coral will die, endangering the fragile coral reef ecosystems, which are the most biologically productive of the marine ecosystems.

Some species of dinoflagellates can produce light and cause much of the bioluminescence often seen in ocean water at night, and many produce toxins that are poisonous. Many of the poisonous varieties contain red pigments, and when they occur in great abundance they turn the water red. Their potent toxin (saxitomin) acts on nerves by blocking sodium channels. During one of these periodic "blooms," known commonly as a red tide, the toxin can kill millions of fish. Red tides are fairly common in the warm waters of the Gulf of Mexico off the coast of Florida.

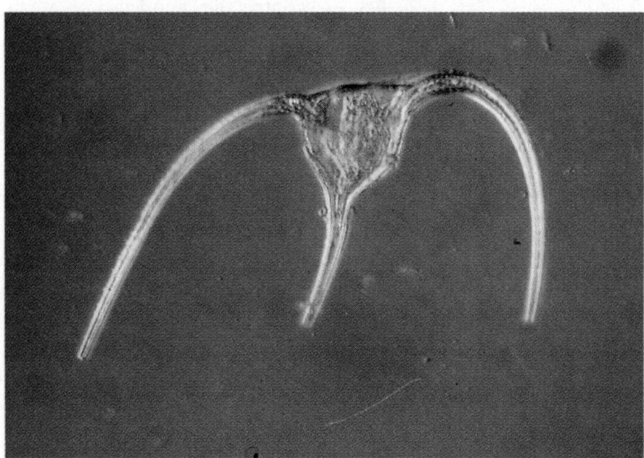

38.12 Photograph of *Ceratium*, a dinoflagellate

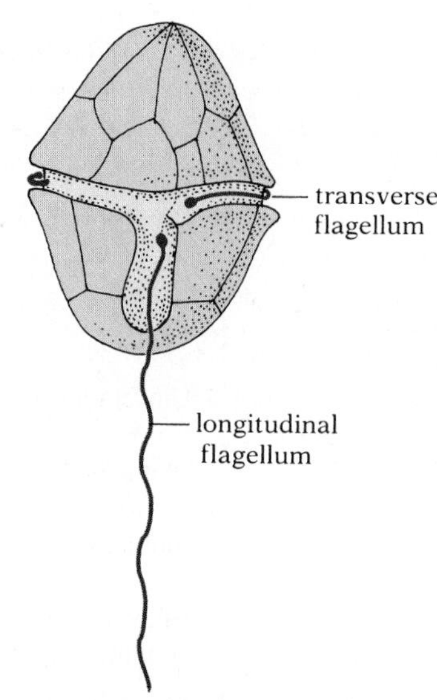

38.13 Drawing of a freshwater dinoflagellate *(Glenodinium cinctum)* The two flagella lie in distinctive grooves on the surface of the cell. The dinoflagellates appear to be armored due to the presence of overlapping cellulose plates.

38.14 Bleached elkhorn coral in the Florida Keys Coral gains its color from the endosymbiotic algae (zooxanthellae) that live within it. The white area in the center, above, represents the "bleached" region where the algae has left the coral. Such bleaching is occurring on a worldwide basis, but researchers are not yet certain why it is happening.

The Chrysophyta Are Distinguished by Their Golden Color

As the names of many algal divisions indicate, the earliest classifications of algae were based on their color, which depends on the sorts of pigments the cells contain. Later study of other important characters (e.g., energy-storage materials, cell wall chemistry) showed that algae of like pigmentation usually share such characters as well, and so the old color classification still stands. Most of the species of algae placed in the division Chrysophyta are some shade of yellow or brown due to the presence of large amounts of carotenoids. They also have chlorophylls *a* and *c* (but no *b*). Many are motile; two anteriorly attached flagella of unequal length are common (Fig. 38.15). Most of the yellow-green and golden-brown algae live in fresh water, but a few are marine.

One group of chrysophytes of special importance is the diatoms. They are abundant in both freshwater and marine habitats. Their cell walls, commonly called shells, consist of two parts that fit together like a box with its lid. The shells contain silica and are glasslike, often resembling jewels (Fig. 38.16). The classification of the diatoms is based almost entirely on the characters of the shells.

When diatoms die, their shells sink to the bottom, where they may accumulate in large numbers, forming deposits of a material called diatomaceous earth. This material is used in many commercial preparations, including toothpastes, detergents, polishes, paint removers, decolorizing and deodorizing oils, and fertilizers. It is also extensively used as a filtering agent and as a component in insulating and soundproofing products.

The diatoms are an abundant component of marine plankton and are important producers in freshwater and marine food webs. For example, it is not unusual for a gallon of seawater to contain as many as one or two million diatoms. ***Plankton*** consists, by definition, of small organisms floating or drifting near the surface. Planktonic organisms are generally divided into two groups—phytoplankton (photosynthetic plankton) and zooplankton (animallike plankton). The organisms constituting phytoplankton are the principal photosynthetic producers in marine communities.

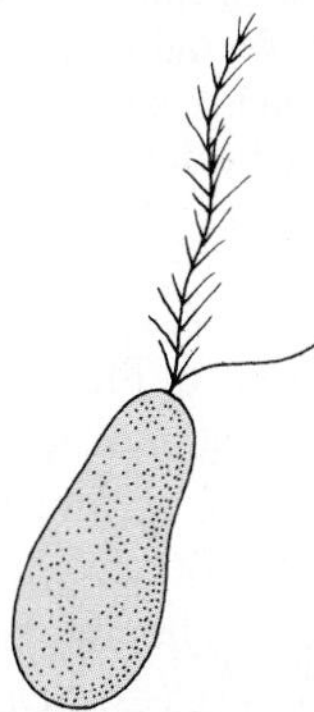

38.15 A unicellular chrysophyte with characteristically unequal flagella

38.16 Some representative diatoms

PLANTLIKE PROTISTA: THE MULTICELLULAR ALGAE

Included in the kingdom Protista are two divisions of algae (the brown and red algae) whose members are predominantly multicellular. These divisions are included among the protists because their bodies usually show little if any tissue differentiation—i.e., their cells are relatively unspecialized. Algae are aquatic, obtaining water, oxygen, nutrients, and even mechanical support directly from the surrounding medium. As a result, algae need little in the way of anatomical differentiation. Because there are no xylem and phloem tissues, there is no anatomical basis for distinguishing roots, stems, or leaves, and the entire plant is considered to consist of a single tissue, and is known as a ***thallus.***

The brown and red algae are thought to have evolved independently and to be not closely related, although they share many features in common. A third algal division, the green algae (Chlorophyta), also contains many multicellular forms. This division, however, shares a close evolutionary relationship with the land plants, and is therefore included in the plant kingdom.

Algal Life Cycles Vary Among the Different Groups

Most algae reproduce both sexually and asexually, but there is considerable variation in the nature and timing of the different stages. Let's review the different types of life cycles (see also pp. 198–199).

The algae so far discussed have a primitive type of sexual life cycle (Fig. 38.17A) where the stage in which the organism passes most of its life is haploid (n). During this haploid stage, the thallus, which is called the ***gametophyte*** ("gamete-producing plant"), produces haploid gametes by mitosis. Two gametes then fuse to produce a diploid zygote, which promptly divides by meiosis to generate haploid reproductive cells called ***spores.*** Each of the spores then gives rise to a haploid gametophyte stage, completing the cycle. In such a life cycle, the haploid phase of the cycle is dominant. The only diploid stage—the zygote—is very transitory. Dominance of the haploid stages is characteristic of most very primitive plantlike protists and fungi, and it seems clear that this was the ancestral condition.

Many of the multicellular algae and most plants show a sexual life cycle that is more complex, including both haploid multicellular stages and diploid multicellular stages (Fig. 38.17B). The cycle can be summarized as follows. Certain cells of the haploid multicellular thallus (the gametophyte) divide mitotically to produce gametes. Fusion of pairs of gametes produce diploid zygotes. Upon germination, the zygotes divide mitotically (not meiotically as in the primitive life cycle), producing a diploid multicellular thallus called the ***sporophyte*** ("spore-producing plant"). Eventually certain reproductive cells, the ***sporangia,*** of the sporophyte undergo meiosis, producing haploid ***spores,*** which divide mitotically to produce the haploid multicelluar gametophytes, beginning a new cycle. A life cycle of this type is said to exhibit ***alternation of generations*** in that a haploid multicelluar phase (gametophyte) alternates with a diploid multicelluar stage (sporophyte). *This type of life cycle is found in most multicellular algae and all land plants.*

Animals and a very few multicellular algae have a life cycle in which meiosis within the diploid organism produces gametes directly (Fig. 38.17C). In this cycle, certain cells of the diploid multicellular organism undergo meiosis, giving rise to haploid gametes that then fuse to form a diploid zygote, which develops into a new diploid organism.

Among the algae, all three life cycles are represented. Most of the unicellular algae show the more primitive life cycle, whereas most of the multicellular algae have a life cycle with alternation of generations. A very few algae (e.g., some brown algae) exhibit a life cycle in which meiosis produces gametes directly.

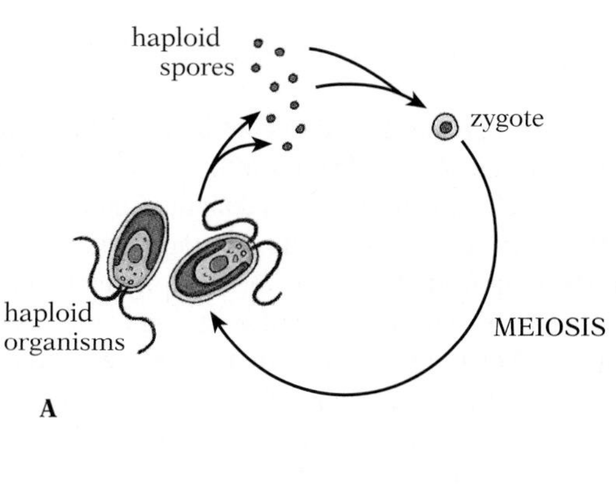

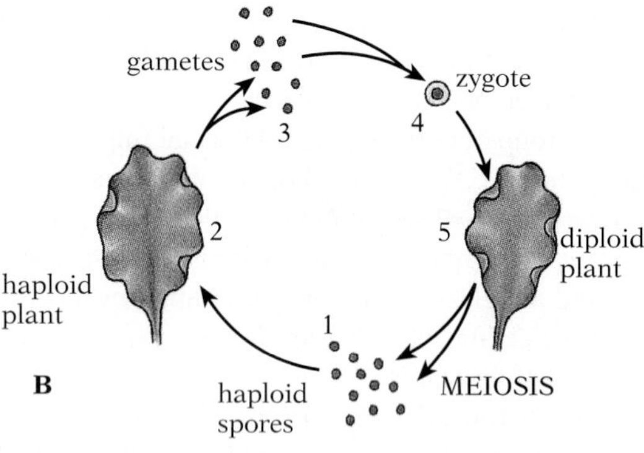

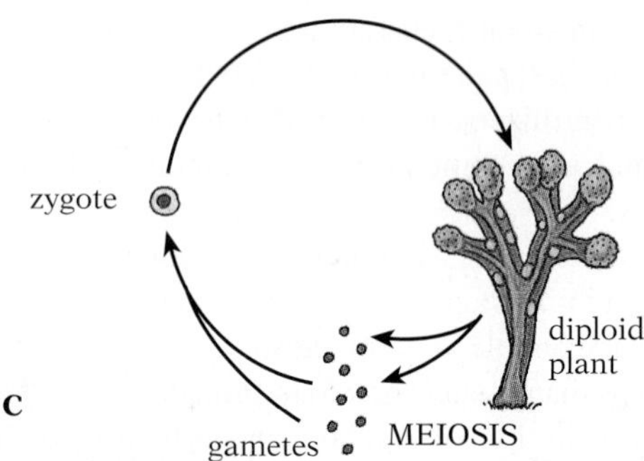

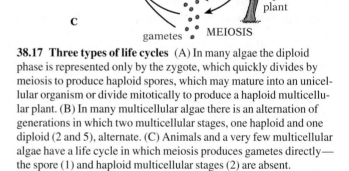

38.17 Three types of life cycles (A) In many algae the diploid phase is represented only by the zygote, which quickly divides by meiosis to produce haploid spores, which may mature into an unicellular organism or divide mitotically to produce a haploid multicellular plant. (B) In many multicellular algae there is an alternation of generations in which two multicellular stages, one haploid and one diploid (2 and 5), alternate. (C) Animals and a very few multicellular algae have a life cycle in which meiosis produces gametes directly—the spore (1) and haploid multicellular stages (2) are absent.

All Brown Algae Are Multicellular, Mostly Marine, and Large

The brown algae, division Phaeophyta, are mostly marine, and include the plants called seaweeds and kelps. They are most common along rocky coasts of the cooler parts of the oceans; they may be seen in great abundance covering the rocks exposed at low tide along the New England coast (Fig. 38.18). A few species, such as the famous *Sargassum,* are found far offshore, where they form dense floating mats that cover immense areas, like the so-called Sargasso Sea, which occupies some 6.5 million square kilometers of ocean between the West Indies and North Africa.

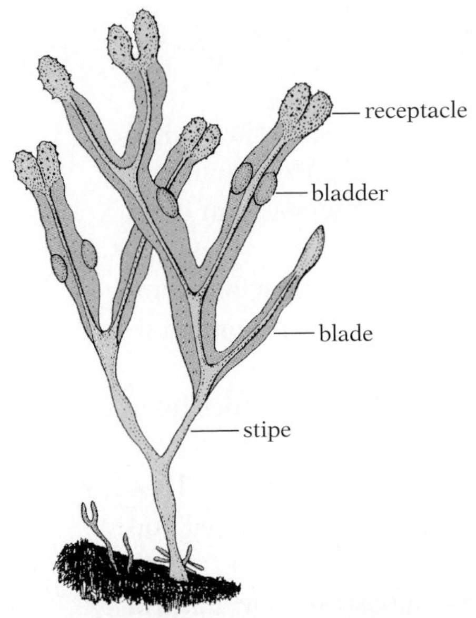

38.18 *Fucus* (often called rockweed), a brown alga common along northern coasts The plant body or thallus is large and flattened and consists of a rootlike holdfast, which anchors it, a stemlike stipe, and an expanded leaflike blade. Some species (including the one shown here) have gas-filled bladders that function as floats. The stipe of some kelps has a phloemlike conductive tissue. These tubes are analogous, not homologous, to the sieve tubes of plants. Swollen reproductive structures called receptacles develop at the tips of some thalluses. The surface of the receptacles is pocked by many tiny openings that lead into cavities where the sex organs are located. Bottom: the receptacles can be seen especially well here.

The brown algae have a set of pigments similar to those found in the golden algae (Chrysophyta), to which they are probably related. They have chlorophylls *a* and *c,* and large amounts of a pigment called ***fucoxanthin,*** which is the pigment that gives these algae their characteristic brown color. The principal storage product is not starch but another complex polysaccharide like that found in the golden algae.

All brown algae are multicellular, and most are large, some growing as long as 45 meters or more. The body is often a rather complex three-dimensional structure (Fig. 38.18). The individual cells have cell walls composed of cellulose and a gummy material called alginic acid. Brown algae are commercially harvested for alginic acid, which is used in cosmetics, in various drugs, for food, and as a stabilizer in most ice creams.

Let us look more closely at a group of typical brown algae, those commonly called kelps. The thallus is large (perhaps as long as 100 meters) and consists of a rootlike ***holdfast,*** which anchors it, a stemlike ***stipe,*** and an expanded leaflike ***blade*** (Fig. 38.18). Some species have gas-filled bladders (Fig. 38.18) that function as floats. Though the thallus generally lacks tissue differentiation, the stipe of some kelps has a phloemlike conductive tissue. These tubes are analogous, not homologous, to the sieve tubes of plants.

Reproduction in the brown algae may be sexual or asexual; the sexual life cycle is usually characterized by an alternation of generations (Fig. 38.17B), but some brown algae have a life cycle in which gametes are produced directly by the diploid plant (Fig. 38.17C).

Red Algae Live at Depths Where Other Algae Cannot Survive

The red algae, division Rhodophyta, are mostly marine seaweeds (Fig. 38.19), but a few live in fresh water or on land.

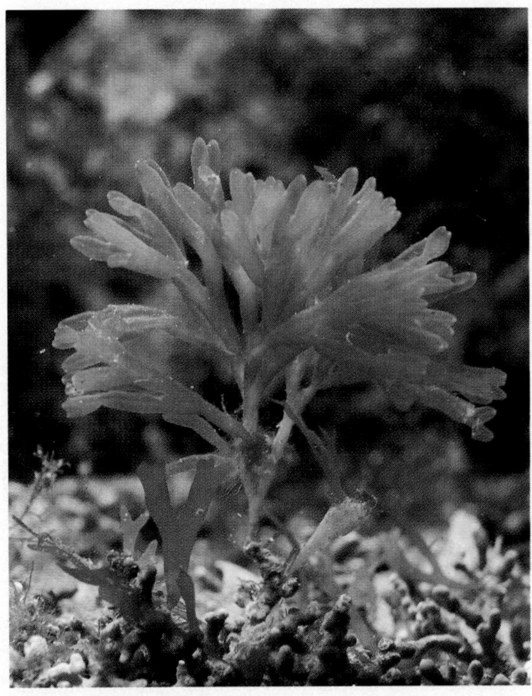

38.19 Example of red algae The seaweed *Scinaia furcellata.*

MAKING CONNECTIONS

1. What evidence suggests that the ancestors of the various algal groups acquired chloroplasts by endosymbiosis independently of one another?

2. Describe how dinoflagellates are linked to the phenomena of coral "bleaching" and "red tide."

3. Which two groups of unicellular marine algae are important components of the phytoplankton, the principal producers in marine communities?

4. How is the lack of tissue differentiation shown by the multicellular algae related to their environment?

5. Among protists, meiosis may produce either spores or gametes. How is the product of meiosis related to the relative importance of the diploid and haploid stages in the life cycle?

6. How do the pigments found in red algae demonstrate how natural selection has adapted photosynthetic organisms to their environment?

Most are multicellular and are attached to the bottom by a holdfast. The red algae are usually smaller than the brown algae and often occur at greater depths.

The red algae possess chlorophylls *a* and *d* (*d* is not found in another other group). They also contain blue and red pigments that are found only in the red algae and certain cyanobacteria, from which the red algae are thought to have evolved. The "red algae" are not always red; many are black, violet, brownish, yellow, or even green. The red and blue pigments play an important role in absorbing light for photosynthesis. These pigments can absorb the wavelengths of light that penetrate deep into the water and pass the energy to chlorophyll *a*. (The wavelengths absorbed best by chlorophyll *a* do not penetrate to the deeper waters of the oceans.) Thus the accessory pigments make it possible for red algae to live at depths where other algae, lacking these pigments, cannot survive.

Reproduction is complex in these organisms; most have an alternation of generations. One unusual feature is that there are no flagellated cells; even the sperm cells lack flagella and must be carried to the egg cells by water currents.

Red algae are an important source of commercial colloids—among others, agar used in culturing bacteria, suspending agents using in chocolate milk and puddings, stabilizers used in ice creams, some cheeses, and salad dressings, and moisture retainers used in icings, creams, and marshmallows.

Though most red algae are filamentous in nature, sometimes the filaments are so densely interwoven and intertwined that they superficially resemble the leaflike blade of the brown algae, though the two divisions are not closely related.

The third algal division that contains many multicellular forms, the green algae, arose independently of the other two and will be considered in the next chapter.

CONCEPTS IN BRIEF

1. The kingdom Protista includes primarily unicellular or colonial eucaryotic organisms and the plantlike multicellular organisms whose bodies show relatively little distinction between tissues.

2. Though the protistans can be separated into three evolutionary lines—the animallike, funguslike, and plantlike organisms—there are many similarities among them.

3. The animallike protists, or protozoans, are usually said to be unicellular, though they are far more complex than other individual cells. Most are heterotrophic. Reproduction is usually asexual but may be sexual. Digestion is generally intracellular.

4. The kinetoplastids include the flagellates, which move by flagella, and the amoeboid protozoans, which have pseudopods used for locomotion or feeding. The flagellates appear to be the most primitive of the protozoa and may have given rise to other protozoan groups, and possibly to the multicellular plants and animals.

5. The ciliates possess numerous cilia for movement and feeding. Their other organelles are greatly elaborated. The ciliates differ from other protozoans in having a macronucleus and one or more micronuclei.

6. The funguslike sporozoans usually have a sporelike stage in their complex life cycle. All are nonmotile and parasitic. Malaria is caused by a sporozoan.

7. The slime molds are funguslike protists whose life cycle includes an animallike amoeboid stage and a funguslike sedentary stage in which a fruiting body develops and produces spores.

8. The water molds are funguslike in their possession of hyphae and heterotrophic nutrition, but animallike in their production of flagellated zoospores.

9. The several groups of largely unicellular organisms classified as plantlike protists show a combination of plantlike and animallike characteristics; like plants, many have a cell wall and chlorophyll and are photosynthetic, and like animals, they are highly motile. The various algae divisions are probably not closely related; they evolved from different eucaryotic cell lineages and acquired different bacterial lineages as chloroplasts.

10. The euglenoids are highly motile unicellular flagellates that lack cell walls but may have chlorophylls *a* and *b,* like higher plants.

11. The golden algae and the diatoms possess chlorophylls *a* and *c* (no *b*). The walls of many contain silica or calcium. The diatoms are the most abundant component of the marine *phytoplankton.* The organisms of the phytoplankton are among the principal photosynthetic producers in marine communities. Another group of unicellular algae, the dinoflagellates, are also important marine producers.

12. Many algae show a primitive life cycle in which the haploid gametophyte stage of the cycle is dominant, and the only diploid stage—the zygote—is very transitory. Dominance of the haploid stages is characteristic of most primitive plantlike protists and fungi; this was the ancestral condition.

13. Many of the multicellular algae and most plants show a life cycle that is more complex in that it exhibits *alternation of generations:* a haploid multicellular stage (gametophyte) alternates with a diploid multicellular stage (sporophyte). Meiosis produces *spores* in these organisms.

14. Animals and a very few multicellular algae have a life cycle in which meiosis produces gametes directly—the haploid gametes then fuse to form a diploid zygote, which develops into a new diploid organism.

15. The brown algae are large multicellular seaweeds. They possess chlorophylls *a* and *c* and fucoxanthin, which gives them their brown color. Some show tissue differentiation. They are complex plants that have convergently evolved many similarities to the vascular plants.

16. The red algae are multicellular and mostly marine. They possess chlorophylls *a* and *d,* and blue and red pigments. The latter are important in absorbing the light that penetrates into deep water and transferring the energy to chlorophyll *a* for photosynthesis. There are no flagellated cells in their life cycle.

STUDY QUESTIONS

1. What are the three evolutionary lines of the Protista? What features do they share that justify grouping them in a single kingdom? (p. 881)
2. Why are protozoans sometimes considered *acellular* organisms? In what respects are they animallike? (p. 881)
3. Why are flagellated and amoeboid protozoans placed in the same phylum (Kinetoplastida)? What features distinguish the Ciliata from this phylum? (p. 882)
4. Describe the life cycle of *Plasmodium,* the parasite that causes malaria. What features typical of sporozoans does this life cycle exhibit? (p. 884)

5. Compare the life cycles of the cellular slime molds and true slime molds. Are these two slime molds closely related evolutionarily? (pp. 884–885)

6. You observe three specimens of unicellular algae with the following characteristics: (1) yellow-green, two flagella, plates of cellulose embedded in the cell membrane; (2) green, single flagellum, flexible in shape; (3) yellow-brown, glasslike cell wall consisting of two parts. Identify the division to which each belongs. (pp. 886–889)

7. Compare the life cycles of (a) a primitive alga; (2) an alga having alternation of generations; (c) a brown alga with an animallike life cycle. Which type is found in most multicellular algae and all land plants? (pp. 889–891)

8. Why are the multicellular brown and red algae placed in the Protista? Compare these groups with regard to size, photosynthetic pigments, and typical habitat. What is the link between their pigments and their habitats? (pp. 889–892)

CHAPTER 39

THE PLANT KINGDOM

The kingdom Plantae includes all the organisms we readily recognize as plants—those that are primarily adapted for life on land as well as the green algae from which the land plants are believed to have evolved. In this chapter we examine the evolutionary sequence of plants, which has culminated in full independence from an aquatic medium, even for reproduction.

The green algae are difficult to place in a classification system that separates Protista from Plantae. Like land plants, the green algae possess chlorophylls *a* and *b* plus carotenoids, their storage product is starch, and their cell walls usually contain cellulose. Sequence analyses of DNAs and RNAs from green algae and land plants indicate they are closely related. Because it makes no evolutionary sense to assign the green algae to one kingdom and their multicellular relatives to another, we include the green algae within the plant kingdom.

Unlike the multicellular algae, whose cells are relatively unspecialized, land plants show far more tissue differentiation. They face enormous problems on land: supporting their own weight, protecting their tissues from drying out, obtaining water and nutrients from the soil and transporting them to the leaves, and so on. Moving out of water onto land has had an enormous impact on the design and physiology of both plants and animals, and this is reflected in the evolution of differentiated tissues and organs. We will begin this chapter by looking at the green algae. Then, after seeing how terrestrial adaptations evolved, we will look at modern land plants, the nonvascular bryophytes (mosses and liverworts) and the vascular plants.

GREEN ALGAE

The Green Algae Show Divergent Evolutionary Tendencies

The division Chlorophyta, to which the green algae belong, is probably the only algal division that has not been a phylogenetic dead end; it is generally regarded as the group from which the land plants arose. The majority of green algae live in fresh water, but some live in moist places on land, and there are numerous marine species.

Many divergent evolutionary tendencies, all probably beginning with walled and flagellated unicellular organisms, can be traced in the Chlorophyta: (1) the evolution of colonies,

Chapter opening photo: Water lily floating on a pond. The leaves of this plant have large air spaces that enable them to float on the surface, but the roots are embedded in the pond bottom.

both motile and nonmotile; (2) the evolution of multicellular filaments, both branched and unbranched, and (3) the evolution of three-dimensional leaflike thalluses.

Chlamydomonas May Resemble the Ancestral Green Alga

The unicellular green algae of the genus *Chlamydomonas* probably resemble the ancestral organisms from which the other green algae arose. Its many species are common in ditches, pools, and other bodies of fresh water and in soils. The individual organism is an oval haploid cell with two anterior flagella and a single, large, cup-shaped chloroplast (Fig. 39.1).

Chlamydomonas usually divides asexually to produce 4 (as in Fig. 39.2), 8, 16, or more daughter cells, which develop walls and flagella and are released from the parent cell as free ***zoospores.*** Zoospores are motile asexual reproductive cells, i.e., motile reproductive cells that are not gametes. Each zoospore soon grows to full size, completing the asexual reproductive cycle.

Under certain conditions *Chlamydomonas* may reproduce sexually, in a primitive type of life cycle (Fig. 39.2). The mature *Chlamydomonas* divides mitotically to produce several gamete cells, which develop walls and flagella and are released from the parent cell. Two gametes eventually fuse, producing a single diploid cell, the zygote, which develops a thick protec-

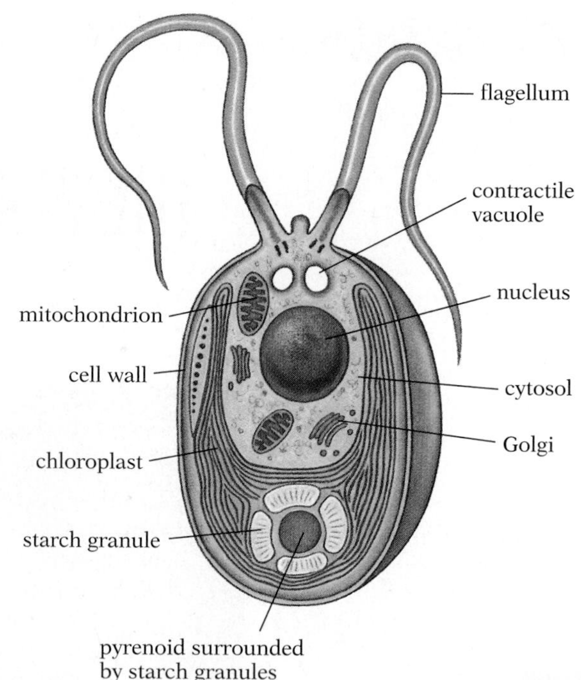

39.1 Mature cell of *Chlamydomonas* The individual organism is an oval haploid cell with two large anterior flagella and a single, large, cup-shaped chloroplast. The pyrenoid in the posterior portion of the chloroplast functions as the site of starch synthesis. Two small contractile vacuoles lie near the base of the flagella.

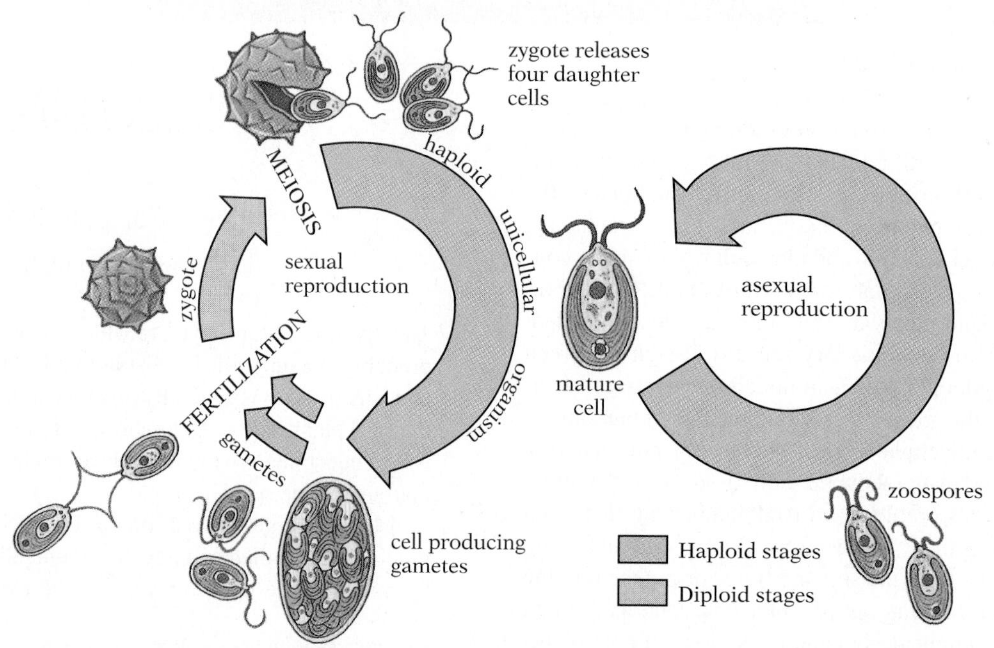

39.2 Life history of *Chlamydomonas* Diagram showing all stages of both the sexual and asexual cycles. Note that in the sexual cycle the zygote is the only diploid stage. This type of life cycle was probably characteristic of the first sexually reproducing unicellular organisms, and it may thus be the type from which all other types arose.

tive wall. When conditions are favorable, it germinates, dividing by meiosis to produce four (or eight) new flagellated haploid cells, which are released into the surrounding water. The new cells quickly mature, thus completing the sexual reproductive cycle.

Because sexual reproduction in most species of *Chlamydomonas* is at a very simple level, it provides insight into the way sexuality probably arose. The gametes are all similar in appearance; they cannot be separated into male gametes (sperm) and female gametes (eggs). Such a condition, where all gametes are alike, is called ***isogamy;*** it is probably the primitive (ancestral) condition in the algae. The isogametes of ***Chlamydomonas*** resemble the parent cell, except that they are smaller; they may be viewed as miniature ***Chlamydomonas*** cells that tend to fuse and act as gametes. This, too, is probably the ancestral condition; the specialization of gametes as distinctive (sperm or egg) cells—a characteristic of most higher plants and animals—is surely a later evolutionary development.

The Volvocine Series Shows How Multicellular Organization Could Have Evolved

An example of the evolutionary tendency to form elaborate colonial organizations can be traced in the Chlorophyta. The so-called volvocine or motile-colony series (Fig. 39.3) is a series of genera showing such a gradual progression from the unicellular condition of *Chlamydomonas* to an elaborate colonial organization. As you study the series in Figure 39.3 you will note a number of evolutionary trends: (1) the number of cells in the colonies increases; (2) there is increasing coordination of activity among the cells; (3) there is increasing division of labor, particularly between vegetative cells (which can carry out photosynthesis and other metabolic activities but cannot become reproductive cells) and reproductive cells; and (4) there is increasing interdependence among the vegetative cells, so that they cannot live apart from the colonies and the colonies cannot survive if disrupted.

Another important evolutionary trend in this series is the change from isogamy to oogamy. Sexual reproduction in *Gonium* is similar to that of *Chlamydomonas:* under certain conditions isogametes are produced and two fuse to form a zygote, which later undergoes meiosis to give rise to new daughter cells that will divide to form new colonies. In *Pandorina,* however, sexual reproduction is ***heterogamous;*** i.e., it involves two kinds of gametes, large "female" gametes and smaller "male" gametes. Both types of gametes have flagella and are free-swimming. In *Volvox,* sexual reproduction is also heterogamous, but the female gametes lack flagella, remain within the parental colony, and are fertilized there. This type of heterogamy, where only the male gamete is motile and the female gamete is a nonflagellated nonmotile egg cell, is called ***oogamy.*** Oogamy is characteristic of all land plants.

In tracing the series *Chlamydomonas–Gonium–Pandorina–Volvox,* we do not mean to imply that each genus evolved from the preceding one; the available evidence will not allow us to decide whether it did or not. But this series suggests how complex colonial forms may have evolved, and indicates how multicellularity may have arisen in plants.

Though multicellular animals certainly did not evolve from *Volvox* or any of the other genera discussed here, a similar evolutionary series, beginning with a unicellular organism, may have been the beginning of multicellularity in the animal kingdom.

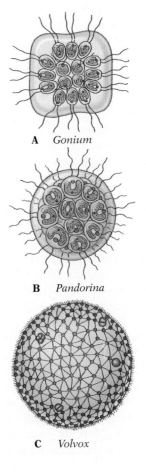

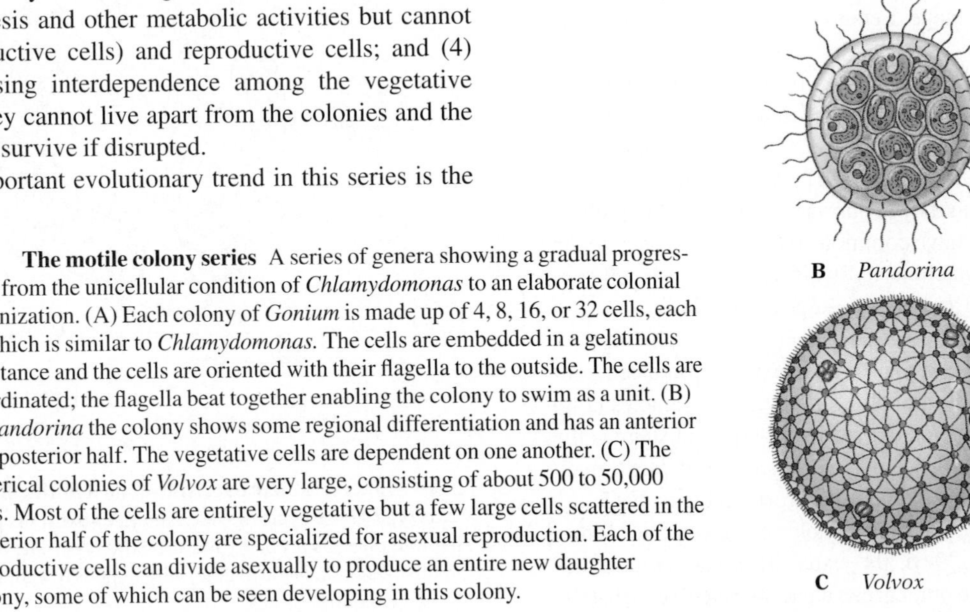

39.3 The motile colony series A series of genera showing a gradual progression from the unicellular condition of *Chlamydomonas* to an elaborate colonial organization. (A) Each colony of *Gonium* is made up of 4, 8, 16, or 32 cells, each of which is similar to *Chlamydomonas.* The cells are embedded in a gelatinous substance and the cells are oriented with their flagella to the outside. The cells are coordinated; the flagella beat together enabling the colony to swim as a unit. (B) In *Pandorina* the colony shows some regional differentiation and has an anterior and posterior half. The vegetative cells are dependent on one another. (C) The spherical colonies of *Volvox* are very large, consisting of about 500 to 50,000 cells. Most of the cells are entirely vegetative but a few large cells scattered in the posterior half of the colony are specialized for asexual reproduction. Each of the reproductive cells can divide asexually to produce an entire new daughter colony, some of which can be seen developing in this colony.

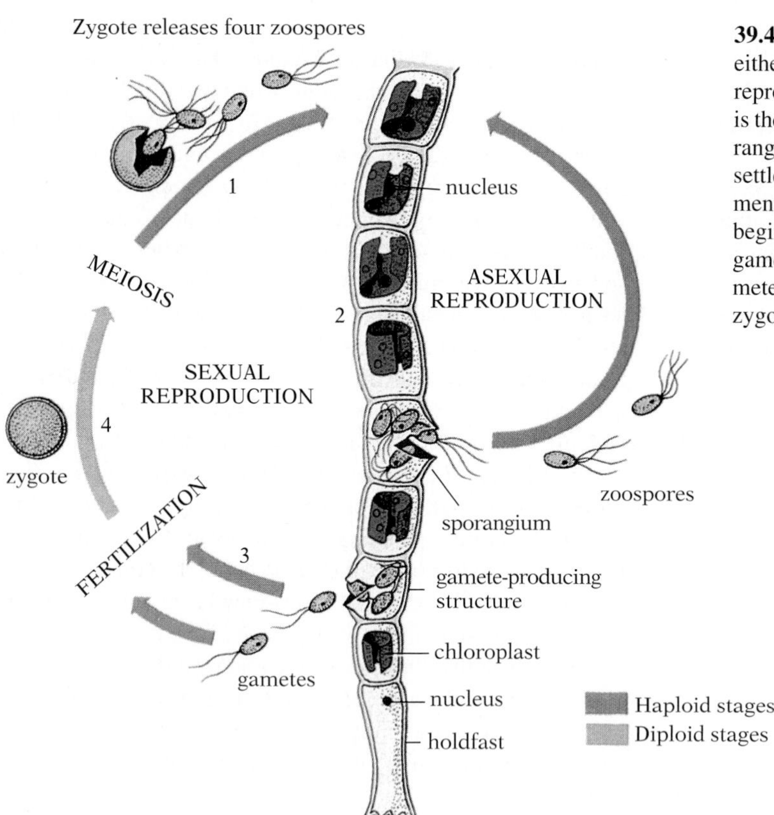

39.4 Life history of *Ulothrix* The haploid plant may reproduce either asexually or sexually (though a single filament would never reproduce in both ways at once as shown here). Asexual reproduction is the more common; certain cells of the filament develop into sporangia (spore-producing structure) and produce zoospores, which settle down and develop into new filaments. Under certain environmental conditions the filament may cease reproducing asexually and begin reproducing sexually; a cell becomes specialized as a gametangium (gamete-producing structure) and produces isogametes. Two such gametes may fuse in fertilization, producing a zygote, which divides meiotically and releases zoospores.

Many Green Algae Have a Multicellular Stage in Their Life Cycle

Other members of the Chlorophyta have a true multicellular stage in their life cycle. Some are filamentous, either branched or unbranched, others are tubular, and still others have a three-dimensional leaflike shape. Let us take *Ulothrix* as an example of a filamentous form.

Most species of *Ulothrix* live in fresh water though a few are marine (Fig. 39.4). The unbranched filament of each plant is a very small threadlike structure attached to the bottom by a holdfast. Except for the holdfast cell, all the cells of the filament are identical, and they are arranged end to end in a single series. Adjacent cells have common end walls—a basic step in the evolution of multicellularity in algae—and each cell contains a single nucleus and a single large chloroplast.

Ulothrix may reproduce by fragmentation (each fragment growing into a complete plant), by asexually produced zoospores, or by sexual reproduction involving isogametes (Fig. 39.4). As with *Chlamydomonas,* the only diploid stage is the zygote. This type of life cycle, characteristic of most green algae, is diagrammed in a generalized form in Figure 39.5.

Ulva, or sea lettuce, is an example of a green alga with a leaflike thallus (Fig. 39.6). Its sexual life cycle is more advanced than that of the other green algae. We have seen that the

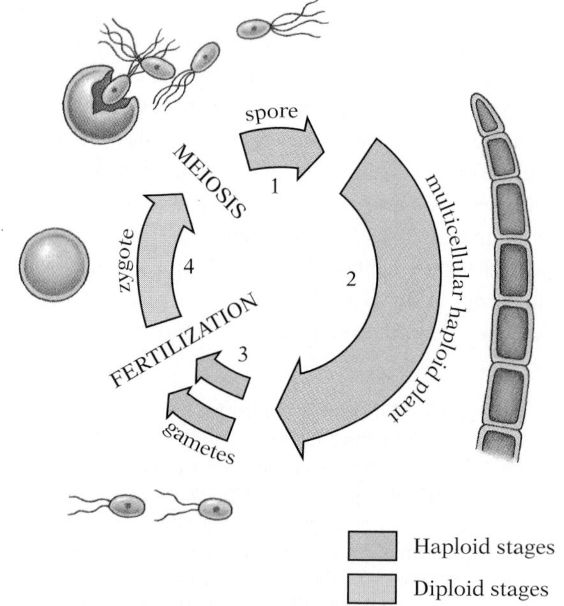

39.5 Life cycle characteristic of most multicellular green algae This type of life cycle in which multicellularity is present only in the haploid phase is probably the ancestral type. The only diploid stage in this life cycle is the zygote, which quickly undergoes meiosis to produce haploid spores.

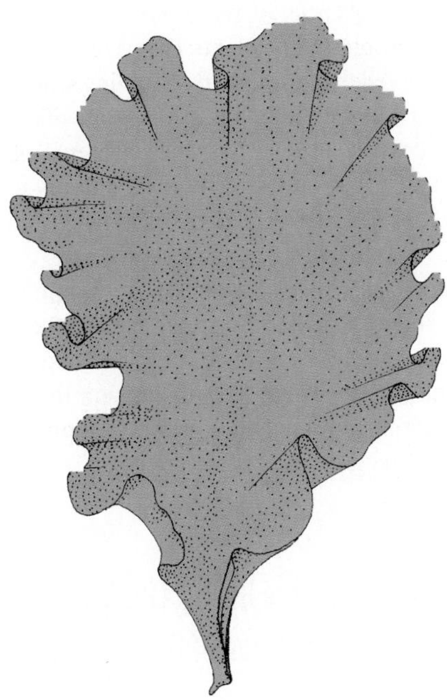

39.6 *Ulva*, a marine green alga with a leaflike thallus that is two cells thick

thalluses of most green algae are haploid, and that most have no sporophyte (multicellular diploid) stage. *Ulva*, however, has a life cycle that alternates between gametophyte and sporophyte stages (Fig. 39.7). Furthermore, the two stages are nearly equal in duration and appearance; in other words, the haploid portion of the life cycle is no longer dominant over the diploid. This type of life cycle, with a true alternation of generations, is the life cycle that is typical of land plants, as we shall see.

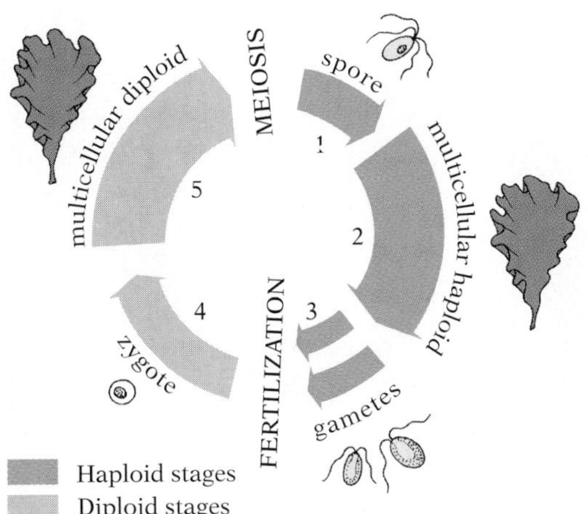

39.7 Life cycle of *Ulva* *Ulva* and its close relatives are unusual among the green algae in having a life cycle with alternation of generations. The gametophyte (haploid multicellular) stage alternates with the sporophyte (diploid multicellular) stage. In *Ulva*, both stages are equally prominent. Most of the green algae do not have a sporophyte stage and therefore no alternation of generations.

TERRESTRIAL PLANT ADAPTATIONS

Life on Land Presents Many Problems Requiring Special Adaptations

We now turn to the land plants, which have evolved numerous adaptations for life on land. The transition from an aquatic to a terrestrial way of life was apparently a difficult one, for algae existed for over a billion years before the first land plants appeared. Botanists are not sure when some lineage of green algae successfully made the transition to a land environment, but fossil evidence indicates that it may have been during the Silurian period, some 400 to 450 million years ago. The fossil record of the transition to land is difficult to interpret because these organisms lacked many of the structures likely to be preserved as fossils. The most likely ancestors of land plants are certain green algae similar in structure to the complex green alga *Chara* (Fig. 39.8). Two divergent lineages are thought to have evolved from this algal ancestor; one lineage leading to the nonvascular plants, the bryophytes, and the other to the vascular plants.

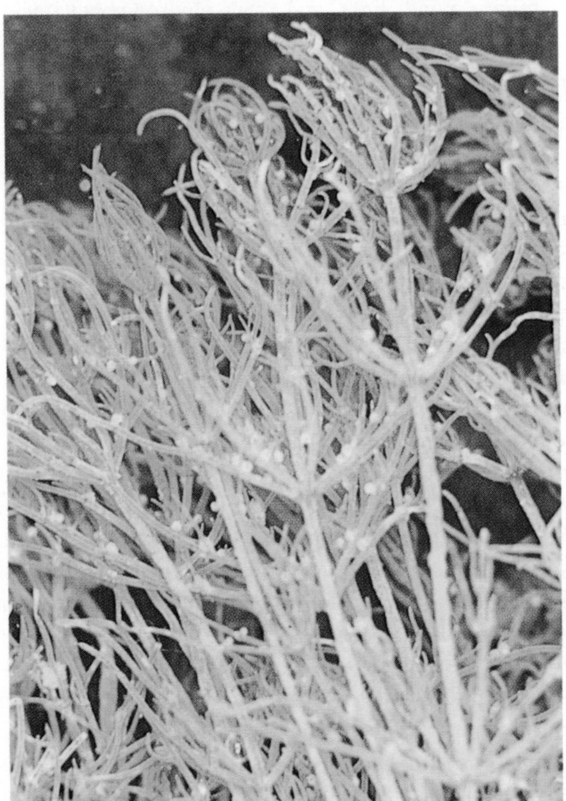

39.8 The green alga *Chara* Land plants are thought to have evolved from some extinct green alga that resembles *Chara*, a complex green alga that grows in shallow, freshwater lakes.

We have seen that life probably arose in water, and that the algae are still largely restricted to the aquatic environment. The few algae that live on land are not truly terrestrial, occurring only in very moist places and actually living in a film of moisture. The evolutionary move from an aquatic mode of existence to a terrestrial one was not simple, for the terrestrial environment is in many ways hostile to life. Among the many problems faced by land plants are the following:

1. Obtaining enough water when fluid no longer bathes the entire surface of the plant body.
2. Transporting water and dissolved substances from restricted areas of intake to other parts of the plant body, and transporting the products of photosynthesis to those parts of the plant that no longer carry out this process for themselves.
3. Maintaining an extensive moist surface for gas exchange when the surrounding medium is air instead of water.
4. Preventing excessive loss of water by evaporation.
5. Supporting a large plant body against the pull of gravity when the buoyancy of water is not available for support.
6. Carrying out reproduction when there is little water through which flagellated sperm can swim and when the zygote and early embryo are in severe danger of drying out.
7. Withstanding the extreme fluctuations in temperature, humidity, wind, light, and other environmental factors to which terrestrial organisms are often subjected.

Much of the evolution of plants can best be understood in terms of adaptations that help solve these problems.

Among the most important evolutionary innovations are those adaptations for reproduction on land. The reproductive structures of most algae are unicellular, and lack a protective wall or "jacket" of sterile (i.e., nonreproductive) cells. Such structures are fragile and subject to drying out. By contrast, the reproductive structures of land plants are always multicellular and have a jacket of sterile cells that helps protect the enclosed gametes. Male and female sex organs of this type are given special names: ***antheridia*** for the male structures where sperm are produced and ***archegonia*** for the female structures where eggs are produced (Fig. 39.9). All land plants are oogamous; the sperm fertilizes the egg within the archegonium. Furthermore, each zygote develops into an embryo while still inside the archegonium—an evolutionary advance over the algae, where the zygotes develop into embryos in the surrounding medium, outside the female reproductive structure. The plant embryo is protected within the archegonium and obtains some of its water and nutrients from the parent plant, and is thus a parasite on it. For this reason, land plants are often referred to as ***embryophytes.*** This type of embryonic development is undoubtedly an adaptation permitting the stages of development most susceptible to drying out to take place in a favorably moist environment, and is analogous to the internal development of mammalian embryos.

The relative dryness of the atmosphere and soil presents very serious problems for plants living on land. Many terrestrial plants solved this problem by evolving stomata for gas exchange and a waxy cuticle that waterproofs the epidermis and prevents excessive water loss. The problems of transport of water and food from one part of the plant body to another and support against gravity were solved by the evolution of vascular tissue.

Plants probably became terrestrial by first encroaching upon river and lake banks, swamps, and marshes. As the invasion of these habitats continued, the dead bodies of the colonizing plants produced a layer of litter on the surface of the bottom soil, contributing to the soil's development and making it suitable for the growth of other plants.

All Land Plants Have a Life Cycle With an Alternation of Generations

The life cycle of land plants always exhibits an alternation of generations in which a haploid multicellular stage, the gametophyte, alternates with a diploid multicellular stage, the sporophyte (Fig. 39.10). The gametes are produced within antheridia and archegonia, and fertilization and early development of the sporophyte embryo occurs within the archegonium.

We noted earlier that the larger and more complex algae,

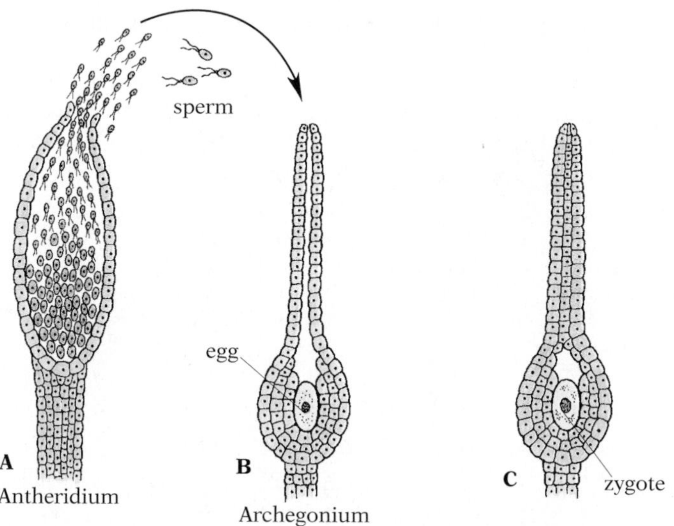

39.9 Reproductive structures of land plants (A) Male antheridium, the structure where sperm are produced. (B) Female archegonium, the structure where eggs are produced. Both the male and female reproductive structures are multicellular and have a jacket of sterile cells that helps protect the enclosed gametes. (C) The sperm fertilizes the egg within the archegonium and the zygote develops into an embryo while still inside the archegonium.

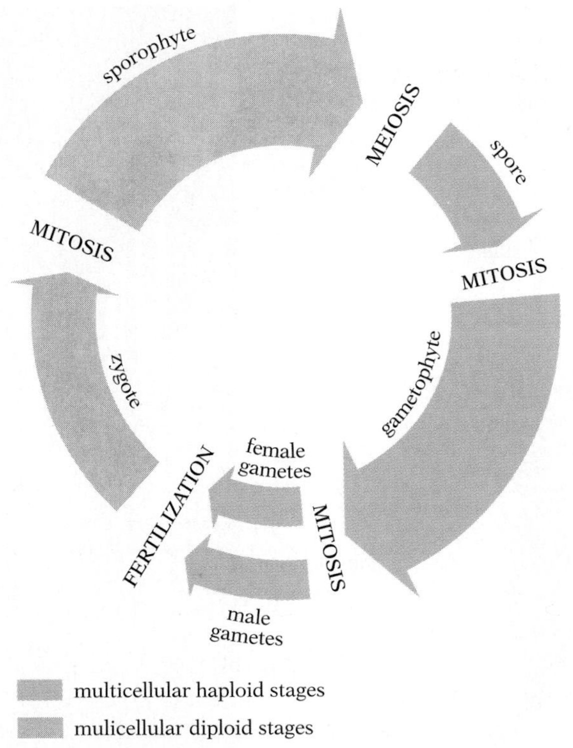

39.10 Generalized plant life cycle showing alternation of generations

such as the brown and red algae, demonstrate an evolutionary tendency toward reduction of the gametophyte stage and increasing dominance of the sporophyte stage. Indeed, the sporophyte stage is often more prominent than the gametophyte stage. The same evolutionary tendency is found in the land plants, where a transition from gametophyte to sporophyte dominance occurs. As we discuss the evolution of the various plant groups, note the gradual decrease in the size and importance of the gametophyte and the corresponding increase in the size and importance of the sporophyte.

Evolutionary innovations of the first land plants

- multicelluar reproductive structures with protective jacket cells
  - male antheridia
  - female archegonia
- oogamy
- egg fertilized within the archegonium
- early embryonic development inside archegonium
- life cycle with alternation of generations

MODERN LAND PLANTS

The Nonvascular Plants (Bryophytes)

Bryophytes Lack True Vascular Tissue

Traditionally, mosses, liverworts, and hornworts were classified together as "bryophytes." Systematists have come to understand, however, that the three groups represent different lineages, and have placed them in three separate divisions: liverworts (division Hepatophyta), hornworts (division Anthocerophyta), and mosses (division Bryophyta). The term "bryophyte" is used as a collective term for the three divisions.

For many years it was assumed that the vascular plants

MAKING CONNECTIONS

1. How is the role of zygote formed by sexual reproduction in *Chlamydomonas* similar to that of the endospores and cysts formed by eubacteria and protists? (See Chapters 37 and 38.)

2. What aspects of sexual reproduction in *Chlamydomonas* offer insight into the origin of sexuality?

3. How does the volvocine series aid our understanding of the evolution of multicellularity?

4. How is embryonic development in land plants analogous to the development of mammalian embryos, described in Chapter 26?

evolved from bryophyte ancestors, but the fossil evidence suggests that the bryophytes and vascular plants both evolved from a green algal ancestor and represent divergent evolutionary lines. Indeed, the first definite bryophyte fossils are only about 370 million years old whereas the earliest vascular plant fossils are about 430 million years old.

The bryophytes are relatively small plants that grow in moist places on land—on damp rocks and logs, on the forest floor, in swamps or marshes, or beside streams and pools. Some species can survive periods of drought, but only by becoming dormant and ceasing to grow. In short, the bryophytes live on land, but they have never become fully emancipated from their ancestral aquatic environment, and they have therefore never become a dominant group of plants. Their great dependence on a moist environment is linked to two characteristics: (1) they have flagellated sperm cells that must swim to the egg cells in the archegonia; and (2) they lack true vascular tissues—xylem and phloem—and therefore lack efficient long-distance internal transport of water and nutrients.[1] Bryophytes, then, do not have "true" roots, stems, and leaves, since, by definition, these structures must possess vascular tissue.

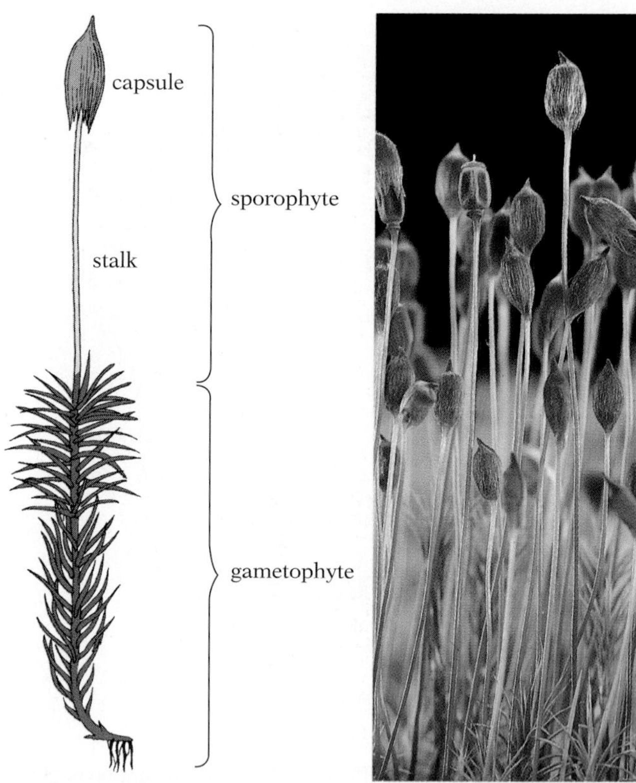

39.11 Gametophyte and sporophyte stages of a moss The haploid gametophyte (green) is the lower "leafy" plant; the "leaves," except at their midrib, are only one cell thick. The diploid sporophyte plant (reddish gold), which consists of a foot (not visible here), a stalk, and a capsule, is attached to the gametophyte and is to some degree parasitic on it. (The capsule of the sporophyte is here covered by a cap derived from the archegonium of the gametophyte; in time it will fall away, leaving the capsule fully exposed.)

Bryophytes Have a Dominant Gametophyte Stage

Although the evolutionary tendency in most land plants is toward reduction of the haploid gametophyte stage and increasing dominance of the diploid sporophyte stage, this tendency is not apparent in the bryophytes, where the haploid gametophyte stage is clearly dominant. The "leafy" green moss plant or liverwort is the gametophyte (Figs. 39.11 and 39.12). These plants bear antheridia and archegonia in which gametes are produced by mitosis. The flagellated sperm are released from the antheridia and swim through a film of moisture, such as rain or heavy dew, to archegonia, where they fertilize the egg cells, producing zygotes. Each zygote then develops into a diploid sporophyte plant.

In a moss the sporophyte (Figs. 39.11–39.13) is a relatively simple structure consisting of three parts: a ***foot*** embedded in the "leafy" green gametophyte, a ***stalk,*** and a ***capsule.*** Although the sporophyte carries out some photosynthesis, it also obtains nutrients parasitically from the gametophyte to which it is permanently attached. Meiosis occurs within the capsule, producing haploid spores, which are released. Each spore germinates into a filamentous plant, called a ***protonema*** (Fig. 39.14), that resembles a green alga. As the plant grows, it forms some branches (rhizoids) that enter the ground and function in anchoring the plant and absorbing water and nutrients;

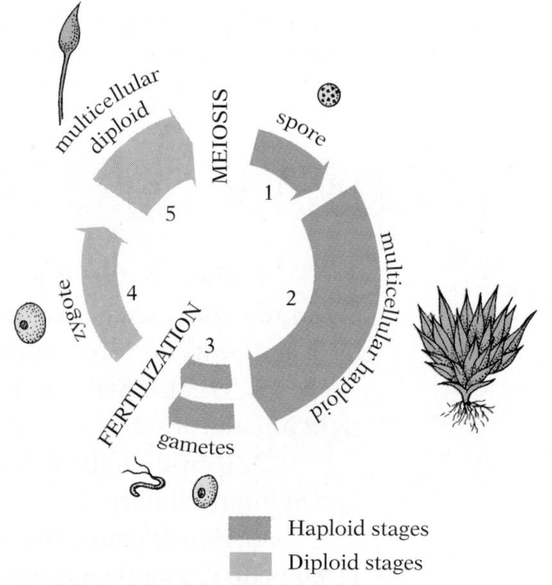

39.12 Life cycle of a bryophyte Both gametophyte (stage 2) and sporophyte (stage 5) are present. The former is dominant.

[1] Although some bryophytes do have specialized water and food-conducting cells, these are not considered to be xylem and phloem cells.

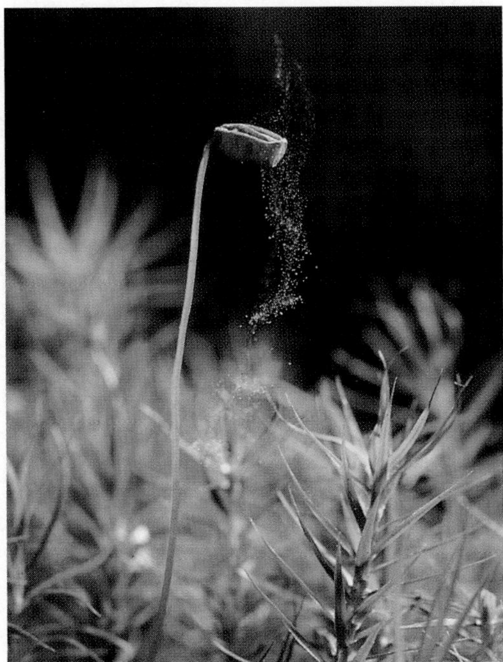

39.13 Spore release by a moss

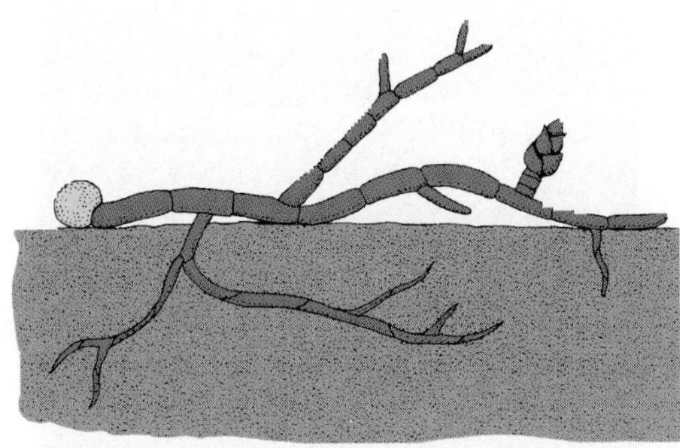

39.14 A young moss plant The spore (yellow) gives rise to a filamentous plant (called a protonema) that strikingly resembles a green alga. The protonema develops into the mature moss plant.

different branches form the upright shoots characteristic of the mature moss gametophyte.

The gametophyte plants of some liverworts are "leafy," but the ones most familiar to us grow as flat green structures lying on the ground (Fig. 39.15). They were given their name because of the liverlike shape of their "leaves," and because it was thought that these plants might therefore be used to treat diseases of the liver. Their life cycle is much like that of mosses, except that the sporophyte stage is even smaller (Fig. 39.15A). Liverworts can also reproduce asexually; special cells called ***gemmae*** are produced in cuplike structures on the surface of the gametophyte (Fig. 39.15B). Raindrops falling on the cups wash out the gemmae, which can then grow

A

B

39.15 The liverwort *Marchantia* (A) *Marchantia* female gametophyte plant with stalked receptacles bearing archegonia. Sexual reproduction in *Marchantia* occurs when sperm from the antheridia (scalloped structure at bottom) of one plant swim through a film of moisture to the female archegonium of another. The egg is fertilized within the archegonium and a small sporophyte develops. The sporophyte (yellow structures underneath the receptacle) of this liverwort consists of a foot, stalk, and capsule. Spores are produced within the capsule and are released when the capsule dries out. Each spore can grow into a new gametophyte plant. (B) The gametophyte plant of the liverwort *Marchantia* is a flat, green structure lying on the ground, anchored by rhizoids on the lower surface. The raised, cuplike structures on the upper surface are gemmae cups that function in asexual reproduction. Special cells called gemmae produced within the cups are washed out by rain; each can grow into a new gametophyte plant.

into new gametophyte plants. As in the mosses, the flagellated sperm must swim from the male antheridia to the female archegonia.

Evolutionary innovations of the bryophytes
- multicellular reproductive structures with protective jacket cells
- fertilization and embryonic development within the archegonium
- dominant gametophyte
- stomata (in mosses)

The Seedless Vascular Plants

Though most bryophytes live on land, in a sense they are not fully terrestrial because they are restricted to moist, protected environments. The vascular plants, by contrast, have evolved a host of adaptations to the terrestrial environment that have enabled them to invade all but the most inhospitable land habitats. Foremost among these adaptations is the presence of vascular tissue. Let us briefly trace the history of adaptation to life on land, beginning with the first vascular plants, the seedless plants.

The Evolution of Tracheids Was a Crucial Adaptation

The first recognized vascular plants belong to the genus *Cooksonia,* which apparently lived between 420 and 345 million years ago and is preserved in the fossil record. *Cooksonia* was a small, rather simple plant that lacked roots and leaves. Its naked stems showed a Y-shaped branching pattern—the stem divides and subdivides, each time into two equal branches (Fig. 39.16). Sporangia were located at the tips of the stems, indicating that the sporophyte generation was the dominant stage. *Cooksonia* also had a ***cuticle*** on aerial parts, produced spores that were similar in structure to those of a vascular plant, and (some) had tracheids in their stems. All of these are fundamental adaptations for life on land. Tracheids were a major evolutionary innovation for two reasons: (1) they permitted transport of water and nutrients from one part of the plant to another, and (2) their thick cell walls also provided support for the plant against the pull of gravity.

Several other types of early vascular plants have been described from fossils; all were small plants with simple branching patterns and no leaves. Some had a central strand of tracheids, forming a simple vascular system. One division of ancestral vascular plants, the Psilophyta, has some living members (Fig. 39.16B). Like their ancestors, the living psilophytes have a simple branching pattern and lack both roots and leaves. (They have a branching underground stem with rhizoids.)

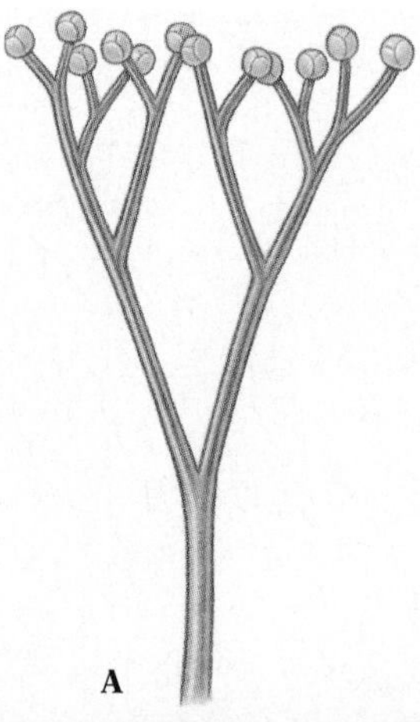
A

B

39.16 Ancestral vascular plants (A) A reconstruction of *Cooksonia,* the oldest known vascular plant. These were small, rather simple, plants that lacked roots and leaves. Note the Y-shaped branching pattern and the sporangia located at the tips of the stems. (B) Photograph of a *Psilotum* sporophyte. Note the branching pattern. The green scalelike structures resemble leaves but lack vascular tissue and are therefore not true leaves. The yellow structures are the sporangia.

Evolutionary innovations in the ancestral vascular plants
- true stems with Y-branching pattern
- primitive tracheids in stems
- waxy cuticle on aerial parts
- spores similar to those of modern seedless vascular plants
- dominant sporophyte
- stomata

Vascular Plants Diversified During the Devonian Period

The successful transition to land of a number of vascular plants led to an adaptive radiation from the late Silurian to the mid-Devonian periods (Fig. 39.17). Most of these plants were small (less than 50 centimeters in height) and restricted to moist habitats. But by the end of the Devonian, vascular plants had become quite diverse, with treelike communities and some plants as tall as 9 meters. These plants showed many adaptations for life on land, including increased amounts of vascular tissue (both xylem and phloem) and the development of true leaves and roots. Three important lineages of seedless plants stand out: the club mosses (Lycophyta), the horsetails (Sphenophyta), and the ferns (Pterophyta).

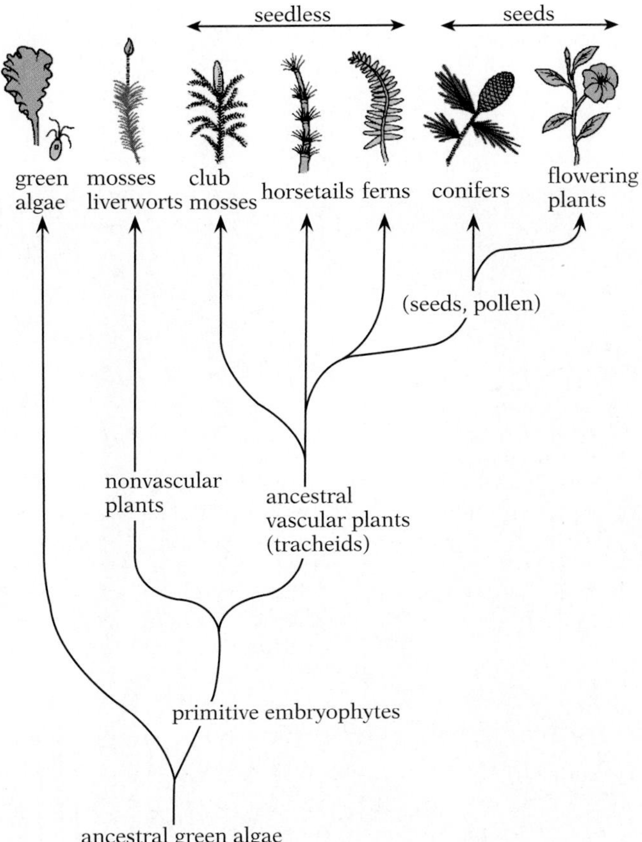

39.17 Evolutionary relationships among the plants

About 380 million years ago a group of vascular plants evolved that gave rise to the club mosses (division Lycophyta). This group has the richest evolutionary history of the vascular plants; the lineage begins in the mid-Devonian and continues to the present. The early members of this group still had the simple Y-shaped branching pattern, but possessed two major evolutionary innovations: true roots and leaves. These leaves apparently evolved as small scalelike outgrowths of the stem that later acquired vascular tissue and thus became "true" leaves (i.e., those with vascular tissue). Certain leaves became specialized for reproduction and bore sporangia on their surfaces. Such reproductive leaves are called ***sporophylls*** (spore-bearing leaves).

During the late Devonian the club mosses diversified, with some members becoming large and treelike and others retaining the small herbaceous form. By the Carboniferous period the treelike lycophytes were among the dominant plants on land (Fig. 39.18A). Some of them were very large trees that formed the earth's first forests. Toward the end of the Carboniferous period the climate became cooler and drier and the lycophyte trees abruptly died out; the smaller herbaceous forms persist to this day (Fig. 39.18B).

Another group to appear in the late Devonian comprised the horsetails (Sphenophyta). Like the lycophytes, the sphenophytes became a major component of the land flora during the Carboniferous period and then declined. Members of just one genus persist, the *Equisetum,* commonly called horsetails or scouring rushes (Fig. 39.19). Though most of these are small, some of the ancient sphenophytes were very large trees (Fig. 39.20). Many of these extinct plants had vascular cambium, and hence secondary growth. Much of the coal we use today was formed from the dead bodies of the ancient lycophyte and sphenophyte plants.

Evolutionary innovations in the club mosses and horsetails
- true roots
- needlelike leaves
- sporophylls; spore-bearing leaves

The third major lineage of seedless plants gave rise to the ferns and seed plants. Ferns (division Pterophyta), another im-

A

B

39.18 Two different lycophyte forms (A) Treelike lycophytes were the dominant plants in the late Devonian and Carboniferous periods. Many of these trees reached 30 to 35 meters in height. When these trees died out, their remains became compressed to form coal. (B) The club moss, *Lycopodium,* often called running pine or ground pine, is a modern lycophyte common in many parts of the United States. As in many of this group, the sporophylls are congregated on a short length of stem forming a club-shaped structure—hence the name "club moss." (Note, the club mosses are not related to the true mosses, which are bryophytes.)

39.19 *Equisetum* The stems of these plants are hollow and jointed, with whorls of leaves at each joint. Spores are produced in conelike structures, three of which can be seen here. Silicone compounds have been deposited within the epidermal cells, making the stems tough and abrasive. The pioneers often used them to scour pots and pans, hence the name "scouring rushes."

39.20 Carboniferous swamp forest Note the tree sphenophytes with their jointed stems and whorls of leaves. The trunk at right is a lycophte.

portant component of the coal-age forests, first appeared in the Devonian period and greatly increased in importance during the Carboniferous. Their decline was much less severe than that of the other seedless plants, and many modern species exist today.

Ferns have well-developed vascular systems and true roots, stems, and leaves. Unlike the small needlelike leaves of the club mosses and horsetails, the leaves of ferns are large and flat, with extensively branched vascular systems. These large leaves are thought to have evolved from branched and webbed stems, the spaces between the stems becoming filled in with tissue, which formed the flattened blade. Such large leaves provide a much greater surface area for photosynthesis. The leaves of ferns are often compound, being divided into numerous leaflets that may give the plant a lacy appearance. There are large tree ferns in the tropics, but most modern ferns have a horizontal stem on or in the soil, and the large leaves are the only parts normally seen.

The large leafy fern plant is the diploid sporophyte stage (Fig. 39.21). Spores are produced by meiosis in sporangia located in clusters on the underside of leaves in some species (Fig. 39.22), or on spikelike structures in others. The spores are literally flung out from the sporangia as they dry, and later develop into tiny, heart-shaped gametophytes that bear both antheridia and archegonia (Fig. 39.23). Water must be present for the flagellated sperm to swim from the antheridium to the archegonium, where the egg is fertilized and embryonic development of the sporophyte begins. Although most people are familiar with the sporophytes of ferns, few have ever seen a gametophyte, and even fewer would guess that it had anything to do with a fern. Small and obscure as it is, however, the fern gametophyte is an independent photosynthetic organism. Here, then, is a life cycle in which all five principal stages are

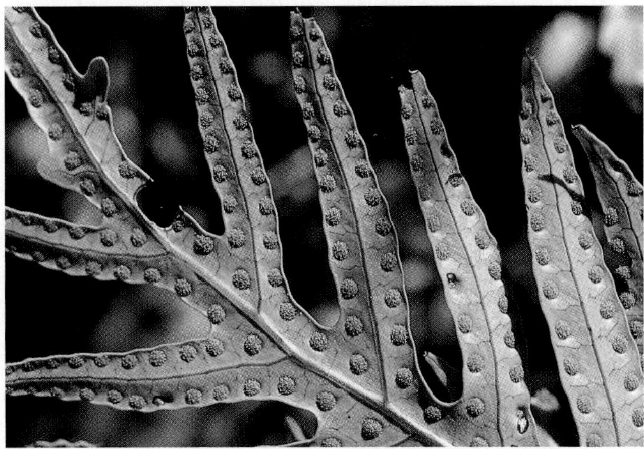

39.22 Undersurface of the sporophyll (spore-bearing leaf) of a fern Each of the round dots is a cluster of many tiny sporangia. Meiosis takes place within the sporangia, producing many tiny haploid spores. When the spores germinate they develop into tiny gametophyte plants that are independent photosynthetic organisms.

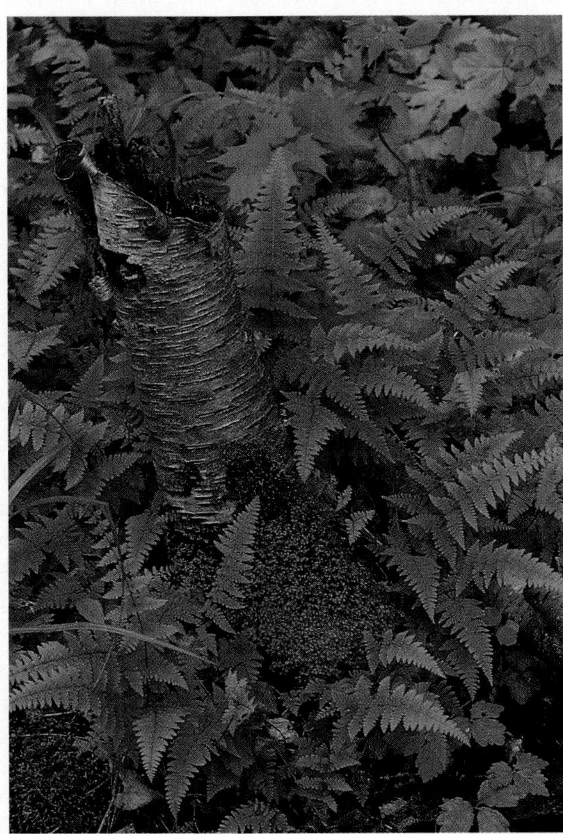

39.21 Marsh ferns in autumn The large leafy fern plant is the diploid sporophyte phase.

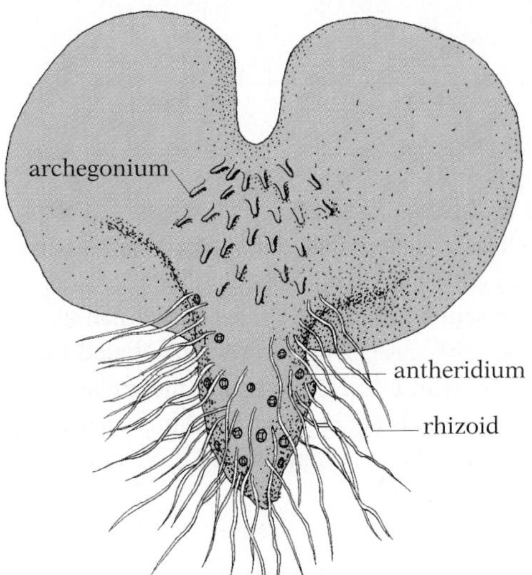

39.23 Fern gametophyte This much-magnified view shows the undersurface of the tiny heart-shaped organism. This nonvascular gametophyte plant is attached to the ground by rhizoids. Both antheridia and archegonia develop on the undersurface of the gametophyte. Water is required for the flagellated sperm produced within the antheridia to swim to the archegonia, where the egg is then fertilized.

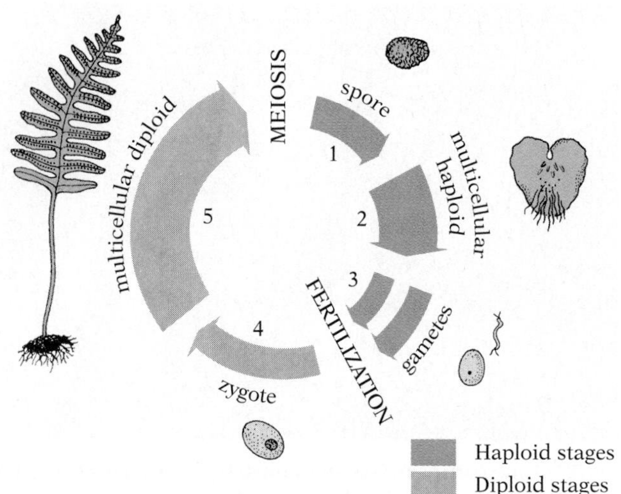

39.24 Life cycle of ferns Both gametophyte (stage 2) and sporophyte (stage 5) are present; the latter is much the more prominent. Compare this life cycle with that of bryophytes, shown in Figure 39.12.

present, but in which the haploid multicellular stage has been much reduced and the sporophyte stage is dominant (Fig. 39.24). Although we do not yet know much about the life cycles of the fossil species, it is thought that all of the seedless plants have a life cycle similar to that of ferns, with a small nonvascular gametophyte, flagellated sperm, and dominant sporophyte.

Evolutionary innovations in the ferns
- well-developed vascular system
- large leaves

In some respects, the seedless vascular plants are no better adapted for life on land than are the mosses. The presence of vascular tissue means that the sporophyte can live in drier places and grow bigger, but for a number of reasons—because their nonvascular free-living gametophytes can survive only in moist places, because their sperm are flagellated and must have moisture to swim to the egg cells in the archegonia, and because the embryo is not protected within a seed—these plants are most successful in habitats where there is at least a moderate amount of water.

The Seed Plants

The dominant plants during the Carboniferous period were the seedless vascular plants. As this period drew to a close, the climate on most of the continents changed, becoming cooler and drier. As a result, a new type of vascular plant, the seed-producing plant, had a selective advantage. They first appeared in the late Devonian and by the late Carboniferous they had replaced the club mosses and horsetails, becoming the dominant land plants. The seed plants have been by far the most successful in fully exploiting the terrestrial environment. Their great success derives largely from two important evolutionary innovations: pollen and seeds.

Pollen and Seeds: Significant Adaptations for Life on Land

Unlike most sporophytes of the seedless plants, which have a single type of sporangium, and produce one type of spore, the sporophytes of seed plants have two different kinds of sporangia. These, in turn, produce two different types of spores, large female ***megaspores*** and smaller male ***microspores.*** These spores are not released from the sporophyte plant, as in the seedless plants; instead, they develop into tiny female and male gametophytes while still within the tissues of the sporophyte. The female gametophyte lives its entire life within the sporophyte and is dependent upon it, but the male gametophyte plant matures to form a tiny ***pollen grain.*** The pollen grain has a resistant outer covering that can withstand dry conditions. The pollen is transported to another plant by some agent (e.g., wind, insect, bird) and the (nonflagellated) sperm cell is released directly into the female gametophyte, where it fertilizes the egg. *No longer must the sperm swim to the egg, so the need for external water is eliminated.* In addition, the young embryo, together with a rich supply of nutrients, is enclosed within a desiccation-resistant seed coat to form a ***seed,*** which can remain dormant for extended periods if environmental conditions are unfavorable. In short, the aspects of the reproductive process that are most vulnerable (i.e., flagellated sperm and fragile embryo) in the ancestral vascular plants have been eliminated in the seed plants.

Seeds evolved independently in several different lines of vascular plants, including the seed ferns, conifers, and flowering plants (Fig. 39.25). The oldest seeds are about 360 million years old, but it is the Mesozoic era in which the seed plants diversified and became dominant. The modern seed plants have traditionally been divided into two groups, the gymnosperms ("naked-seed" plants) and the angiosperms (flowering plants). In recent years, however, it has become increasingly clear that the relationships among the four divisions of gymnosperms are not particularly close and that these groups differ from one another as least as much as they differ from the angiosperms. We shall discuss only the largest division of gymnosperms, the conifers (division Coniferophyta), which includes such common species as pines, spruces, firs, cedars, hemlocks, yews, and larches. The leaves of most of these plants are small evergreen needles or scales (Fig. 39.26). This group first arose in the Carboniferous period and was very common during the

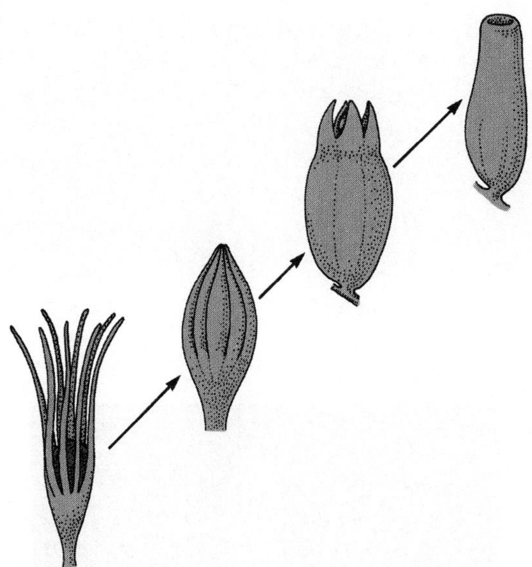

39.25 A model for the possible evolution of the seed Shown here are seeds of four species of extinct seed ferns (pteridosperms); note the progressive development of the outer covering. This sequence is thought to be similar to the one by which seeds of other plants evolved.

39.26 Branch of a larch tree in spring The leaves are needlelike. At this time of year the new female cones have a pinkish coloration.

Mesozoic era. The conifers remain an important part of the earth's flora.

Cones Are the Reproductive Structures of the Conifers

The term conifer means "cone-bearer." The cones are the reproductive structures, of which there are two kinds: small male cones and large female cones (Fig. 39.27). In both kinds of cones the sporangia are produced by the highly modified leaves (sporophylls) that form the scales of the cone.

The life cycle of a pine tree provides an example of the seed method of reproduction in the conifers. The large pine tree is in the diploid sporophyte stage, and bears both male and female cones. Each scale of a female cone has two sporangia on its upper surface (Fig. 39.28A). Each sporangium is encased in an ***integument*** with a small opening at one end (Fig. 39.28C). Meiosis takes place inside the sporangia, producing four large haploid ***megaspores,*** three of which soon disintegrate. The one remaining megaspore divides by mitosis to produce a mass of cells, which is the female gametophyte plant. When mature, the female gametophyte produces several tiny archegonia at the open end (Fig. 39.28C). Egg cells develop within the archegonia. Note that both the megaspore and female gametophyte remain embedded in the sporangium. The composite structure consisting of integument, sporangium, and female gametophyte is called an ***ovule.***

Meiosis also occurs within the sporangia of male cones, giving rise to tiny spores, or microspores, each of which develops into the male gametophyte plant—the pollen grain (Fig. 39.28B). A thick coat forms, which is highly resistant to loss of water, followed by winglike structures on each side, which aid its dispersal by wind. A single male cone may release millions of tiny yellow pollen grains, which may be carried many miles by the wind. Most of the millions of pollen grains released by a pine tree fail to reach a female cone. But of the few that sift down between the scales of a female cone, some land in a sticky secretion near the open end of the ovule. As this secretions dries, it pulls the pollen grain into the opening. The pollen grain then produces a tubular outgrowth, the pollen tube, and two sperm nuclei form. The pollen tube grows down through the tissues of the sporangium and into one of the archegonia of the female gametophyte. There it discharges its sperm cells, one of which fertilizes the egg cell. The resulting zygote then divides to produce a tiny embryo sporophyte. The embryo is still contained within the ovule. Finally, the entire ovule is shed from the cone as a seed, which consists of three main components: a seed coat derived from the integument, stored food derived from the tissue of the female gametophyte, and an embryo. The seeds lie uncovered ("naked") on the surface of the cone scales. They are released when the scales of the female cone dry out and separate. Thus, seeds rather than spores became the stage at which offspring were dispersed. The life cycle of the pine is summarized in Figure 39.29.

Evolutionary innovations in the conifers

- two kinds of sporangia → two types of spores (megaspores and microspores) → separate female and male gametophyte plants
- female gametophyte remains within the sporangium
- male gametophyte becomes the pollen grain
- no flagellated sperm cells so no need for external water
- embryo enclosed with food in a seed
- seeds borne "naked" on cone scale

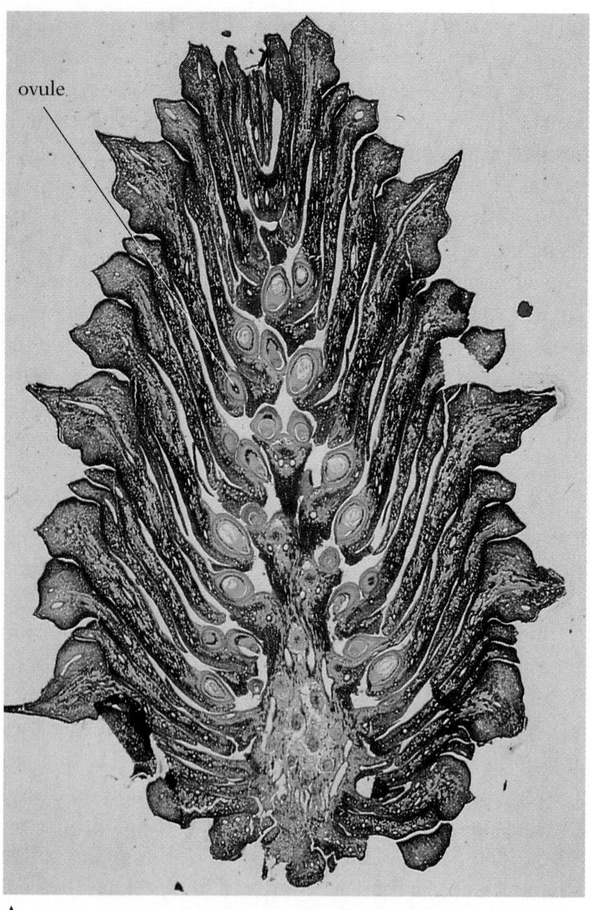

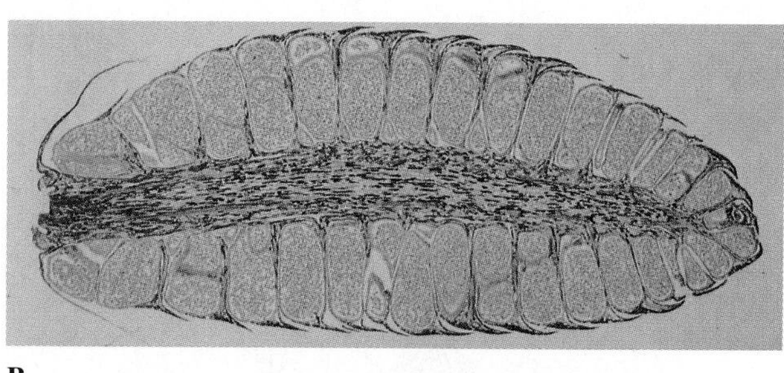

39.27 Photographs of sections of female and male pine cones (A) Female cone. Ovules can be seen on the surface of the sporophylls (cone scale) near their base. (B) Male cone. Each sporophyll (cone scale) bears a large sporangium that becomes a pollen sac. (C) Male and female cones can occur together. A large green female cone is seen at the bottom in this photo, with the male cones just above it.

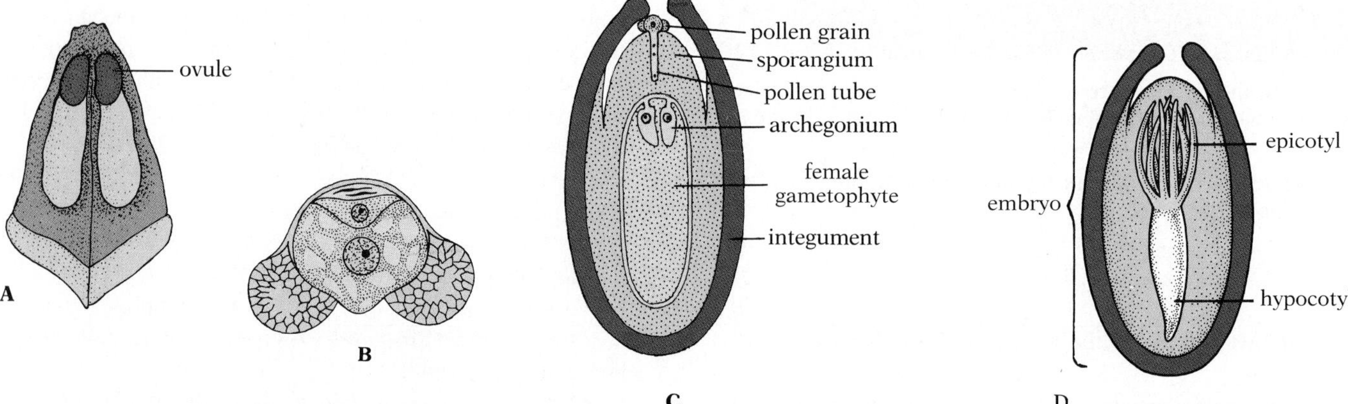

39.28 Ovules and pollen grains of pine (A) Scale from female cone. The two ovules, each containing a sporangium lie on the surface of the scale near its point of attachment. (B) Pollen grain, composed of four cells, the top two of which are small and degenerate. The winglike structures aid its dispersal by wind. (C) Section of an ovule with germinating pollen. (D) The seed with its developing embryo.

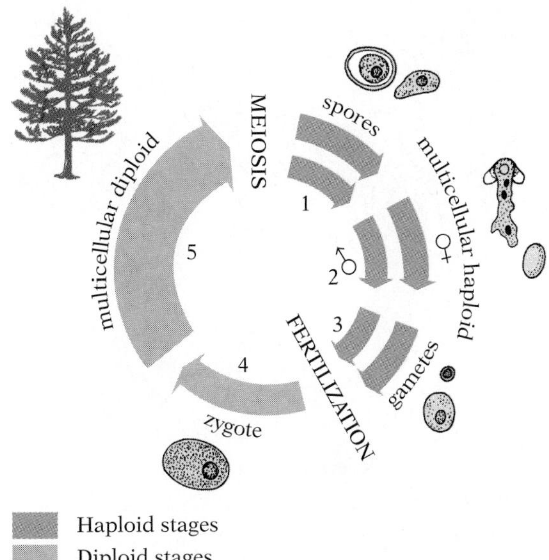

39.29 Life cycle of pine The familiar tree is the sporophyte (diploid multicellular) stage (5). Meiosis takes place within the male and female cones, producing spores that develop into tiny male and female gametophytes. The male gametophyte matures to form the pollen grain, which, when taken inside of the female sporangium, grows a pollen tube and releases sperm cells into the female archegonium. One of the sperm cells fertilizes the egg, which then divides to produce a tiny embryo sporophyte within a seed.

The Most Evolved Seed Plants Are the Flowering Plants

The flowering plants (division Anthophyta), commonly called angiosperms, first appeared about 120 million years ago. The group underwent great adaptive radiation, and these plants became the dominant land flora of the Cenozoic era, as they are today. Among the many traits contributing to their great success are the evolution of the flower for sexual reproduction, double-fertilization, and the enclosure of the ovule within a fruit.

We have seen that the reproductive structures of conifers are cones and that the seeds are borne "naked" on the surface of the cone scales. The reproductive structures of angiosperms, by contrast, are flowers. Flowers are an important adaptation for attracting pollinating insects and birds, as we learned in Chapter 31. In these plants the ovules are enclosed within modified leaves called ***carpels,*** which unite to form the ovary of the flower (Fig. 39.30). We have already discussed the processes of reproduction and development in the flowering plants in Chapter 24 (see pp. 492–505), but we shall review it briefly here.

Meiosis occurs within the ovules in the ovary, producing megaspores. One of the megaspores within each ovule divides mitotically to form an embryo sac, which is the small female gametophyte (Fig. 39.31A). The embryo sac is composed of only seven cells and is completely dependent on the sporophyte. One cell is much larger than the others and contains two haploid nuclei, the ***polar nuclei.*** One of the cells at the lower end will act as the egg cell. Meiosis also occurs within the sporangia of the anther of the flower, producing many haploid microspores. Each microspore becomes a two-nucleate male gametophyte—the pollen grain—which is later released from the anther.

When a pollen grain is deposited on the female part of the flower, a pollen tube forms (Fig. 39.31B) and grows downward, eventually entering the ovule (see Fig. 24.8, p. 496). Two sperm nuclei are released from the tip of the pollen tube directly into the embryo sac. ***Double fertilization*** then occurs: one haploid sperm fertilizes the egg cell, and the resulting zygote develops into an embryo sporophyte; the second sperm

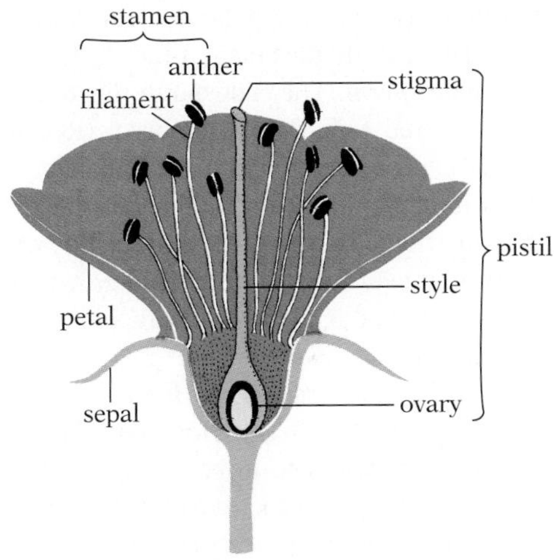

39.30 The major parts of a flower

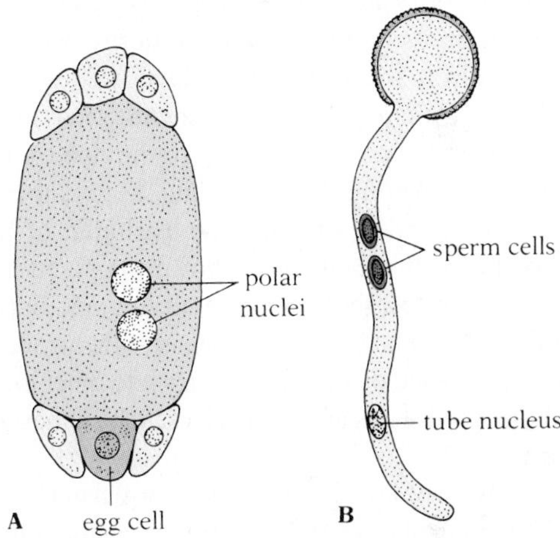

39.31 Gametophytes of an angiosperm (A) Female gametophyte (embryo sac), which is composed of seven cells. One cell is much larger than the others and contains the two polar nuclei. (B) Male gametophyte (pollen grain and tube).

combines with the two haploid polar nuclei to form a triploid ($3n$) nucleus. This cell divides many times to form a triploid tissue, called ***endosperm.*** The endosperm functions as a very rich source of stored food for the embryo. (Recall that the stored food for the conifer embryo is the female gametophyte tissue, which is haploid.) It is thought that the triploid endosperm has contributed to the success of the flowering plants since this rich source of food permits rapid growth of the embryo, which is important in plants with a rapid life cycle.

After fertilization, the ovule matures into a seed, which, as in pine, consists of seed coat, stored food (endosperm), and embryo (see Fig. 24.13, p. 498). However, the angiosperm seed differs from that of pine in being enclosed within the ovary of the flower. The ovary then develops into the ***fruit,*** usually enlarging greatly in the process (see Fig. 24.10, p. 496). Peaches, cherries, grapes, tomatoes, and blueberries are examples of fruits in which the fleshy portion of the fruit is the ovary wall. Sometimes other parts of the flower, such as the receptacle, are also incorporated into the fruit. Apples and pears, for examples, are fruits in which the main fleshy part, which we eat, is derived from the receptacle of the flower rather than the ovary. The ripe fruit may burst, expelling the seeds, as peas and beans do (where the pod is the fruit), or the ripe fruit with the seeds still inside may fall from the plant, as tomatoes, squash, apples, peaches, and acorns do. The evolution of fruit contributed to the great success of the flowering plants because it not only helps protect the seeds from drying out, but often facilitates their dispersal by various means—the wind, or an animal which, attracted by the fruit, carries it to other locations or eats both fruit and seeds and later releases the unharmed seeds in its feces.

The life cycle of flowering plants is shown in Figure 39.32.

The flowering plants (Anthophyta) are customarily divided into two classes, the Dicotyledones (dicots) and Monocotyledones (monocots). The dicots include oaks, maples, elms, willows, roses, beans, clover, tomatoes, asters, and dandelions; the monocots include many of our important cereals such as corn, wheat, rye, and oats, as well as grasses, daffodils, irises, lilies, tulips, and palms. The basic differences between the two groups are summarized in Figure 39.33.

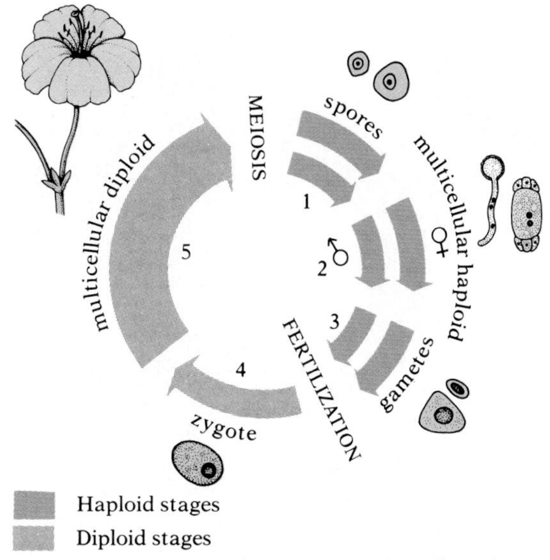

39.32 Life cycle of angiosperms The mature diploid plant produces gametes—pollen, which is dispersed, and eggs, which are retained in the ovary. Once fertilization has occurred, a seed is produced, from which the new plant develops.

Evolutionary innovations in the flowering plants
- flowers as the reproductive structure
- ovules contained in ovary
- double fertilization
- triploid endosperm
- seeds borne in fruits

MAKING CONNECTIONS

1. Why was the evolution of tracheids (whose structure is described in Chapter 19) a crucial step in the history of land plants?
2. Which groups of seedless vascular plants formed much of the coal we use today?
3. Compare the endosperm found in the seed of a flowering plant with the yolk of an animal's egg (described in Chapter 27) with respect to their structure and function.
4. How did the evolution of fruit contribute to the great success of the flowering plants?

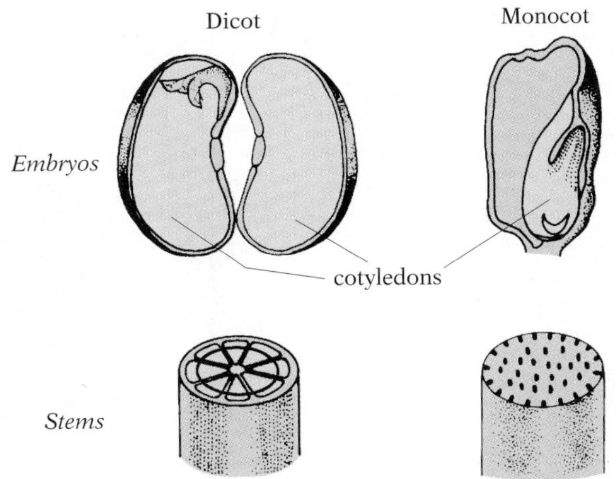

39.33 Differences between dicotyledons and monocotyledons The basic differences between the two groups are as follows. (1) As the names imply, the embryos of dicots have two cotyledons, whereas those of monocots have only one. (2) Dicots often have vascular cambium and secondary growth; monocots usually do not. (3) The vascular bundles in the stems of young dicots are arranged in a circle or fused to form a tubular vascular cylinder (shaded); monocots have more scattered vascular bundles. (4) Dicot leaves usually have a network of veins; monocot leaves usually have parallel veins. (5) Dicot leaves generally have petioles; monocots generally do not. Many monocots can be recognized by the way the leaf base clasps the stem (as in corn). (6) The flower parts of a dicot usually occur in fours or fives or multiples of these (e.g., four sepals, four petals, four stamens); those of a monocot usually occur in threes or multiples of three. The photo on the left is a nasturtium flower, a dicot, the photo on the right is a trillium flower, a monocot.

SUMMARY OF EVOLUTIONARY TRENDS

Let us summarize some evolutionary trends and innovations among the plant groups. One major trend, shown diagrammatically in Figure 39.34, is the progressive decrease in the size and importance of the gametophyte. Along with the reduction in size has come a gradual loss of independence. At the same time, the sporophyte has become increasingly larger and more independent.

Many of the evolutionary advances can be viewed as adaptations for life on land. The evolution of vascular tissue in the primitive vascular plants enabled plants to grow larger and survive in drier environments. However, these plants were not completely freed from their dependence on an aquatic envi-

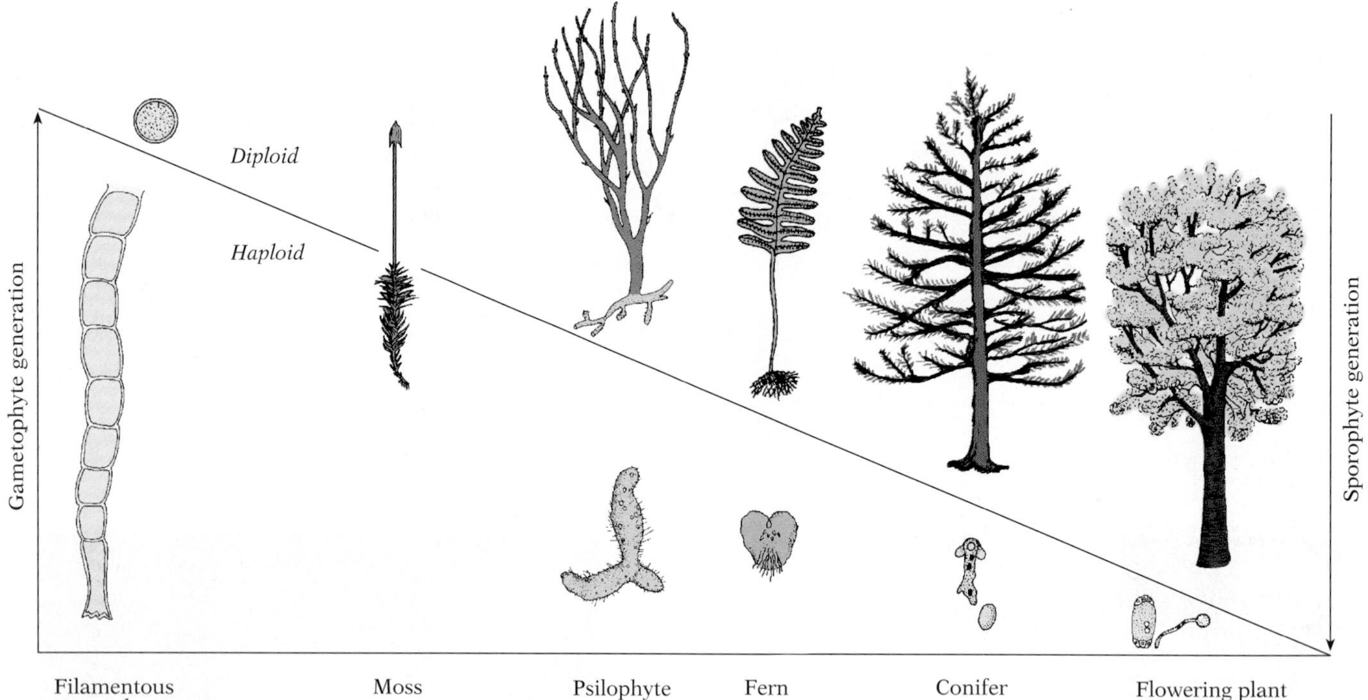

39.34 Transition from gametophyte to sporophyte dominance as shown by representative plant groups The gradual decrease in the size and importance of the gametophyte and the corresponding increase in the size and importance of the sporophyte is shown. In filamentous green algae and in mosses, the gametophyte stage is dominant and the sporophyte stage is much reduced. In the other plant groups, the sporophyte is dominant. In the conifers and flowering plants the gametophytes (pollen and embryo sacs) are tiny and are dependent on the sporophytes. The organisms are not drawn to scale.

ronment; water was still required for reproduction. The flowering plants and conifers solved this problem by the evolution of pollen and seeds. The development of the pollen grain eliminated the need for external water for fertilization since the sperm nuclei were deposited directly within the tissues of the female gametophyte. And the seed, with its protective coat and stored food, protected the embryo and enabled it to withstand unfavorable conditions. The development of the seed method of reproduction is probably responsible for the dominance of the seed plants today.

CONCEPTS IN BRIEF

1. The green algae (Chlorophyta) are probably the group from which the land plants arose. Like land plants, the green algae possess chlorophylls *a* and *b* and carotenoids, their storage product is starch, and most have cellulose in their walls.

2. Several evolutionary trends are seen in the green algae: a change from unicellular to colonial life, the evolution of multicellular filaments and three-dimensional leaflike thalluses, and a trend from isogamy to heterogamy to oogamy.

3. The life cycles of many green algae include a multicellular stage, often a filamentous thallus. The multicellular stage is generally haploid; the zygote is the only diploid stage. Sexual reproduction is usually *isogamous*. *Ulva* is a green alga with a leaflike thallus. It shows alternation of generations with a sporophyte stage alternating with a gametophyte stage.

4. Much of the evolution of the land plants can best be understood in terms of adaptation to solve the problems of terrestrial life. The sex organs (*antheridia* and *archegonia*) are surrounded by jacket cells to help prevent the enclosed gametes from drying out. Fertilization and early embryonic development take place within the protected environment of the archegonium.

5. Two lineages of land plants evolved from green algae: (1) the bryophytes, small plants with a dominant gametophyte that can grow only in moist places because they lack true vascular tissue and have flagellated sperm; and (2) the vascular plants, which possess xylem and other fundamental adaptations for terrestrial existence (stomata, cuticle).

6. All land plants have a life cycle with an alternation of generations. The haploid gametophyte stage is dominant in the bryophytes whereas the sporophyte is dominant in vascular plants.

7. The first vascular plants were simple plants that lacked true roots and leaves. Their naked stems showed a Y-shaped branching pattern. A cuticle and tracheids were present in the stems. The most primitive vascular plants living today belong to the division Psilophyta (the psilophytes); they also lack leaves and roots.

8. The club mosses (Lycophyta) and horsetails (Sphenophyta) are primitive vascular plants that have true roots, stems, and needlelike leaves. Although the living members of these divisions are small, many of the ancient forms were large trees; much of today's coal was formed from the bodies of these plants.

9. The ferns (Pterophyta) have well-developed vascular systems and true roots, stems, and leaves. The leaves of ferns are large and flat and are thought to have evolved from branched and webbed stems.

10. The psilophytes, club mosses, horsetails, and ferns are all seedless plants, and have a life cycle with a small nonvascular gametophyte, flagellated sperm, and dominant sporophyte.

11. The dominant and best-adapted land plants are the seed plants. These plants produce two types of spores (*microspores* and *megaspores*), which give rise to very small male and female gametophytes. The male gametophyte (*pollen grain*) is transported to the female gametophyte so water is not necessary for the transport of sperm. The *seed* consists of the embryo and stored food, surrounded by a seed coat.

12. The evergreens, or conifers (Coniferophyta), are a large and important division of seed plants. Their reproductive structures are *cones*. The seeds are borne exposed on the surface of the scale of the female cone.

13. The flowering plants (Anthophyta) are the dominant land plants. Their reproductive structures are *flowers*. *Double fertilization* occurs within the flower, after which the ovules mature into seeds. The seeds are enclosed in *fruits* that develop from the ovary of the flower.

STUDY QUESTIONS

1. What three evolutionary tendencies can be traced in the Chlorophyta (the green algae)? What evidence indicates that this group gave rise to the land plants? (pp. 895–896)
2. Compare *Ulothrix* and *Ulva* as examples of multicellular green algae. Which has a life cycle typical of land plants? (pp. 898–899)
3. What problems faced the ancestors of land plants in making the evolutionary move from the aquatic to the terrestrial habitat? Describe the adaptations by which land plants cope with these problems? (pp. 899–900)
4. How have the gametophyte and sporophyte stages of the life cycle changed in relative importance as the vascular plants have evolved? (pp. 900–901)
5. What adaptations to terrestrial life do the mosses and other bryophytes display? Why does their structure and method of reproduction confine them to moist habitats? (pp. 901–903)

6. **How were the structure and life cycle of the first vascular plants different from those of the bryophytes? Why were they better adapted to life on land? (p. 904)**
7. **Describe the major evolutionary innovations of the club mosses, horsetails, and ferns. In what respects are these seedless vascular plants no better adapted for terrestrial life than the mosses? (pp. 905–907)**
8. **Describe the life cycle of a conifer such as a pine tree. Which aspects of conifer reproduction make conifers better adapted to the land environment than earlier vascular plants? (pp. 909–910)**
9. **In what respects is the reproduction of flowering plants like that of conifers? In what ways is it different? (pp. 911–912)**

CHAPTER 40

THE FUNGAL KINGDOM

Fungi touch our lives in many ways. They, together with the bacteria, are the decomposers of the world, breaking down vast quantities of dead organic matter that would otherwise accumulate and make the earth uninhabitable. Through their activities, the minerals contained within the dead bodies of plants and animals are made available for recycling through the ecosystem. Their role as decomposers has its dark side, however, because they also cause spoilage of bread, fruit, vegetables, and other foodstuffs, and deterioration of leather goods, fabrics, paper, lumber, and other valuable products. Some fungi are parasitic on or in animals, including humans; many skin diseases, including ringworm and athlete's foot, are caused by fungi, and there are several serious fungal diseases of the lungs. Other fungi are parasitic on plants; they cause the majority of known plant diseases and hundreds of millions of dollars of damage to agricultural crops each year. On the other hand, many industrial processes depend upon fungi. Yeasts are used in fermenting beers and wines and causing bread dough to rise. Other fungi are important in the manufacture of many cheeses (e.g., "blue" cheese), organic acids, vitamins, and a number of antibiotics, such as penicillin. And, of course, mushrooms are cultivated for food.

Fungi Are Absorptive Heterotrophs

We learned in Chapter 17 that the fungi lack internal digestive systems and depend mainly on absorption as their mode of feeding. They are usually either saprophytic or parasitic. Digestion in fungi is always *extracellular;* digestive enzymes are secreted directly onto the food supply and the products of digestion are then absorbed.

Fungi are primarily filamentous and multicellular. The fungal body usually consists of microscopic, tubular threads that branch and spread over the surface of the food supply (see Fig. 17.7, p. 352). Each thread is known as a ***hypha;*** a mass of hyphae is called a ***mycelium.*** The hyphal threads have definite cell walls, which are often composed of chitin, a complex polysaccharide containing nitrogen. In most fungi the hyphae are divided by cross walls (septa) that form compartments or cells (Fig. 40.1). Sometimes, however, the partitions between the cells are incomplete or absent, so the cytoplasm is continuous. Further, the individual "cells" of fungi, unlike those of plants and animals, often have more than one nucleus.

Fungi Usually Reproduce Both Sexually and Asexually

Although most fungi reproduce both sexually and asexually, asexual reproduction is generally more important because of the enormous numbers of new individuals it produces. Whether reproduction is sexual or asexual, however, it is im-

Chapter opening photo: Underside of a mushroom showing gills. The above-ground portion of the mushroom is the structure in which sexual reproduction takes place. There is also an extensive growth of threadlike absorbing structures in the soil.

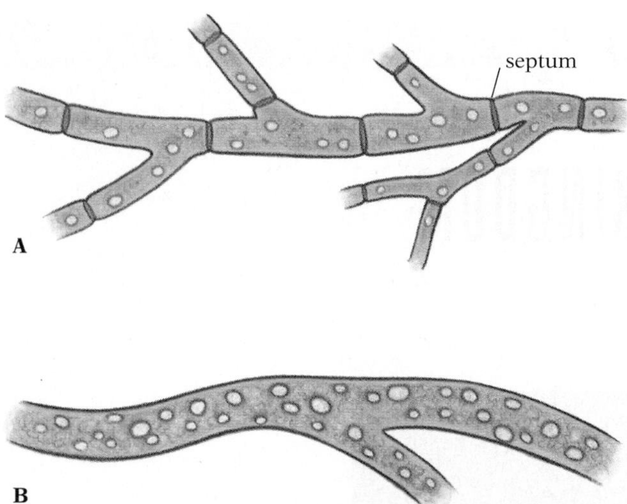

40.1 **Hyphae** The body of a fungus is composed of threadlike hyphae, which branch over the surface of the food supply. Some hypha have cross walls, or septa, that divide it into compartments or cells (A). Often, however, these partitions have pores so the cytoplasm is connected from one cell to the next. Other hyphae lack cross walls of any kind (B).

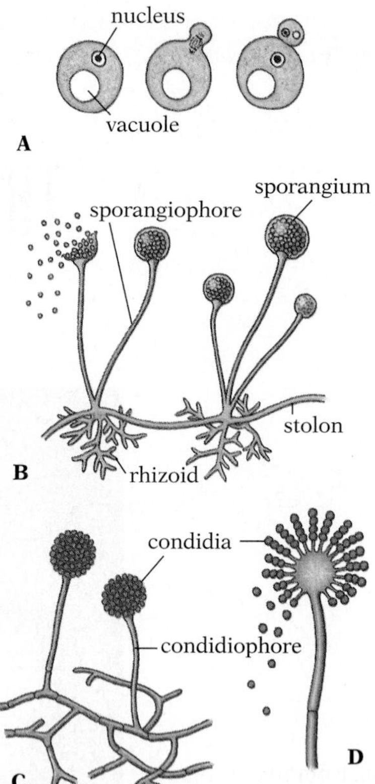

40.2 **Asexual reproduction in the fungi** (A) Budding in brewer's yeast. A small cell is pinched off a larger cell. (B) Sporangia in the black bread mold, *Rhizopus*. Sporangia are borne on the tips of special upright hyphae. Thousands of spores may be produced within a sporangium and each spore can germinate to form a new hypha. (C) and (D) Asexual reproductive structures of *Aspergillus*, a sac fungus. Conidia are produced at the tips or sides of special upright hyphae. (D) An enlarged section of an upright hypha showing that the structure bears numerous spores, called conidia, arranged in chains.

portant to remember that in fungi the haploid stage is the dominant stage—the nuclei within the fungal body are haploid.

There are three main ways that fungi reproduce asexually. First, ***fragmentation*** may occur—one part of the fungal body may break off and give rise to a new individual. Second, the fungus may reproduce by ***budding***—a small outgrowth (bud) of the parent cell becomes detached and grows into a new individual (Fig. 40.2A). Third, and most common, fungi reproduce by means of asexual ***spores,*** each of which can form a new individual. Spores are generally produced in a saclike ***sporangium*** (Fig. 40.2B). Sometimes, however, the spores are produced at the tips or sides of special hyphae rather than in a sporangium; in this case the spores are referred to as ***conidia*** (Fig. 40.2C, D). Conidia are produced in chains at the ends of special hyphae, where each can grow into a new fungus.

Sexual reproduction in the fungi is complex, and involves three basic stages: (1) haploid nuclei from two different cells are brought together into a single cell, forming a two-nucleate cell; (2) the two haploid nuclei eventually fuse to form a diploid zygote; and (3) the zygote then undergoes meiosis, restoring the haploid condition. The timing and duration of these three stages vary among the different groups, as does the reproductive structure. Indeed, the sexual reproductive structure (or lack thereof) is the basis for classifying the fungi into divisions. The four divisions we shall consider are listed in Table 40.1. Notice that the fungi of one group, the so-called imperfect fungi (division Deuteromycota), do not have a sexual ("perfect") stage; they reproduce only by asexual means. These are species for which the sexual stages have not yet been found, or do not exist; consequently they cannot be placed in one of the other divisions.

The Conjugation Fungi (Zygomycota) Reproduce Sexually by Conjugation

The Zygomycota are so named because during sexual reproduction they form a special type of sexual spore known as the ***zygospore.*** They are commonly referred to as "conjugation fungi" because it is the *conjugation* of cells from two special hyphae that produces the zygospore (Fig. 40.3).

The hyphae of the conjugation fungi lack cross walls, so the cytoplasm is continuous throughout, and each hypha contains many haploid nuclei. Cross walls appear only during the formation of reproductive structures. The hyphae grow rapidly into a well-developed mycelium that covers its food supply.

Like other fungi, the members of this group carry out both sexual and asexual reproduction, with asexual reproduction predominating. An example of a conjugation fungus is the common black bread mold, *Rhizopus,* which forms a whitish or grayish mycelium on bread. If the mycelium is examined mi-

TABLE 40.1 *Important characteristics of the fungi*

Division	*Hyphae*	*Sexual reproduction*	*Asexual reproduction*	*Economic importance*
Zygomycota: conjugation fungi	Usually lack cross walls (septa)	Conjugation between two hyphae to form zygospore	Sporangia	Black bread mold, cause of strawberry leak and soft rot of sweet potatoes in storage; some used to produce important industrial products
Ascomycota: sac fungi	Cross walls present, usually with perforations	Ascus formed	Usually by conidia	Yeasts, morels, truffles; the blue-green, brown, and red molds that cause food spoilage; powdery mildews, Dutch elm disease
Basidiomycota: club fungi	Cross walls present	Basidium formed	Budding, fragmentation of mycelium, or conidia	Mushrooms, toadstools, bracket fungi, puffballs; plant pathogens; rusts and smuts
Deuteromycota: imperfect fungi	Cross walls with perforations usually present	None	Conidia	Flavor cheeses, make soy sauce, tofu; produce antibiotics; causes ringworm, athlete's foot, thrush

croscopically, it can be seen to include three types of hyphae: horizontal hyphae that form a network on the surface of the bread; rootlike hyphae (called rhizoids) that penetrate into the bread and function both in anchoring the fungus and in absorbing nutrients; and upright hyphae that bear sporangia on their ends (Fig. 40.2B). Thousands of asexual spores produced in each sporangium may passively be carried long distances by wind, rain, or animals. If a spore lands in a suitable location,

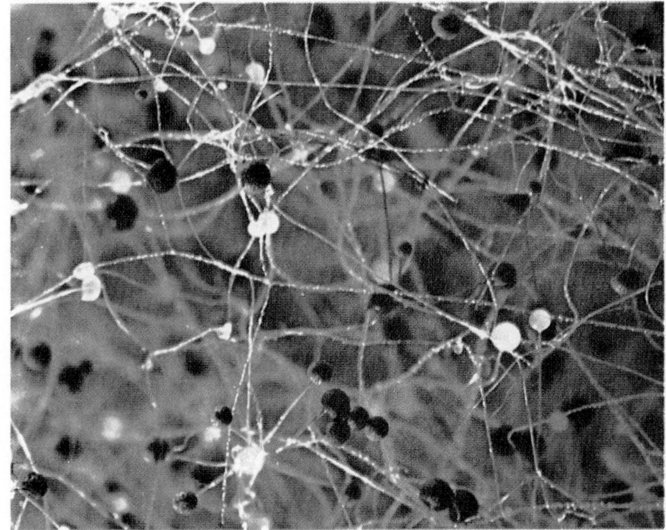

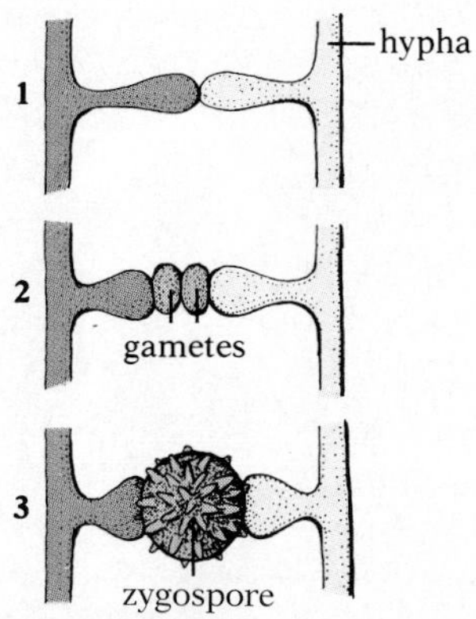

40.3 *Rhizopus (black bread mold)* (A) *Rhizopus* is made up of numerous threadlike filaments hung with what look like black and white balls. These are sporangia; the black ones are ripe and ready to release their spores. (B) Sexual reproduction by conjugation: (1) Short branches from two different hyphae meet. (2) The tips of the branch hyphae are cut off as gametes. (3) The gametes fuse in fertilization to form a zygote with a thick spiny wall called a zygospore.

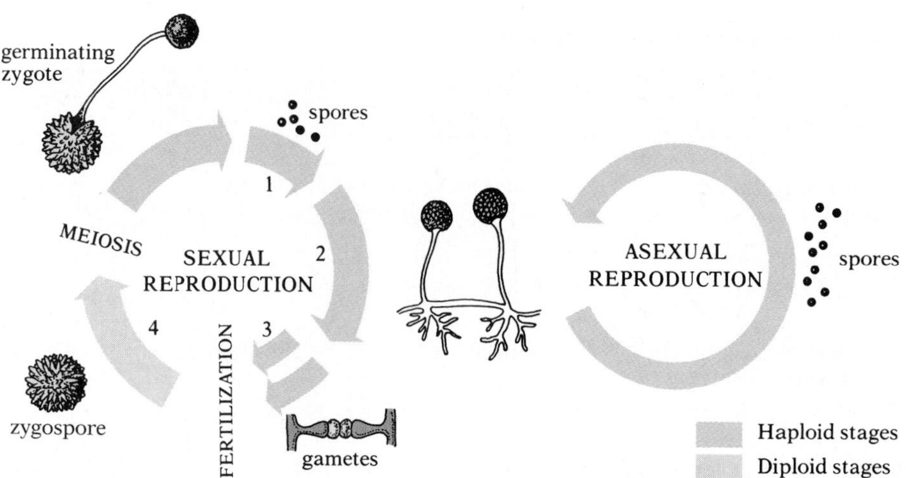

40.4 Life cycle of the conjugation fungus *Rhizopus* During sexual reproduction (left), short branches from two special horizontal hyphae (of different mating types) contact each other (3). Cross walls form just back of the tips to produce two gamete cells, each containing a single haploid nucleus. The two gametes and their nuclei then fuse to form a diploid cell called a zygospore (4), which develops a thick protective wall and enters a period of dormancy (usually one to three months). At germination, the nucleus of the zygospore undergoes meiosis, restoring the haploid condition, and a short hypha grows from the zygospore. This haploid hypha produces a sporangium at its tip, which releases asexual spores (1) that grow into new hyphae (2). During asexual reproduction (right), upright hyphae bear sporangia in which thousands of asexual spores are produced mitotically; each spore can germinate to produce a new mass of hyphae, completing the asexual cycle.

where conditions are warm and moist, it germinates and soon gives rise to a new mass of hyphae, thus completing the asexual cycle. Sexual reproduction occurs much less frequently, often when the conditions for growth become adverse. The life cycle is described in Figure 40.4.

Members of this group are widespread as saprophytes in soil and dung. Some, however, are parasites on plants or other fungi; others are important insect pathogens (Fig. 40.5).

The Sac Fungi (Ascomycota) Form an Ascus During Sexual Reproduction

The members of the Ascomycota are very diverse, varying all the way from unicellular yeasts through powdery mildews and cottony molds to morels, truffles, and cup fungi. The bodies of the more complex fungi are composed of many hyphae tightly

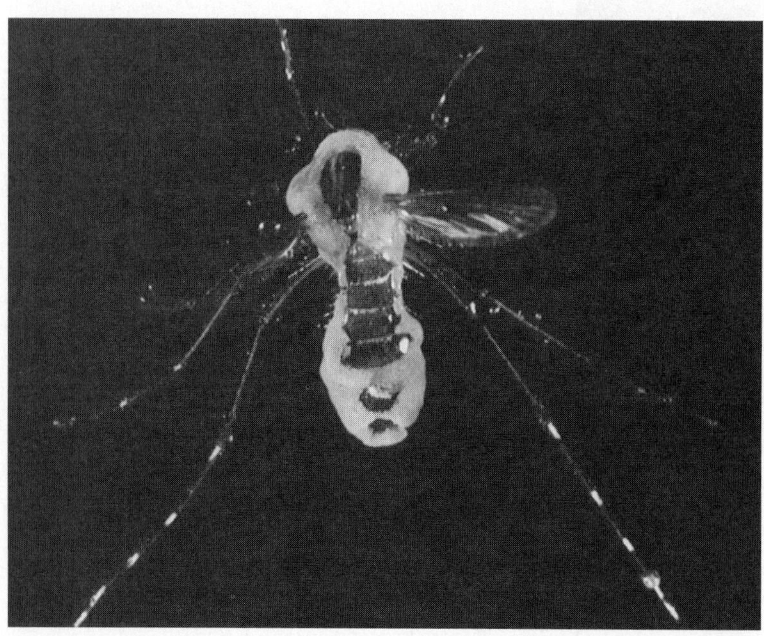

40.5 The yellow fever mosquito, *Aedes aegypti,* fatally parasitized by the fungus *Entomophthora culicis* The fungus (white) grows within the body cavity of the mosquito and emerges through membranous areas of the exoskeleton. Fungi of this sort play a role in the biological control of insects.

40.6 The cup fungus *Peziza* growing on a log in a rain forest in Central America The cup is actually a mycelium, composed of tightly packed hyphae. Cup fungi reproduce by ascospores, which are produced in special structures in the lining of the cup.

packed together (Fig. 40.6). The hyphae of this group, unlike those of the conjugation fungi, possess cross walls, which are usually incomplete, having large holes in their centers. The cytoplasm of adjacent cells is thus continuous and movement of materials and organelles along the hyphae is facilitated.

Though their vegetative structures differ, all sac fungi form a saclike reproductive structure called an ***ascus*** during their sexual cycle (Fig. 40.7). Meiosis occurs within the ascus, and

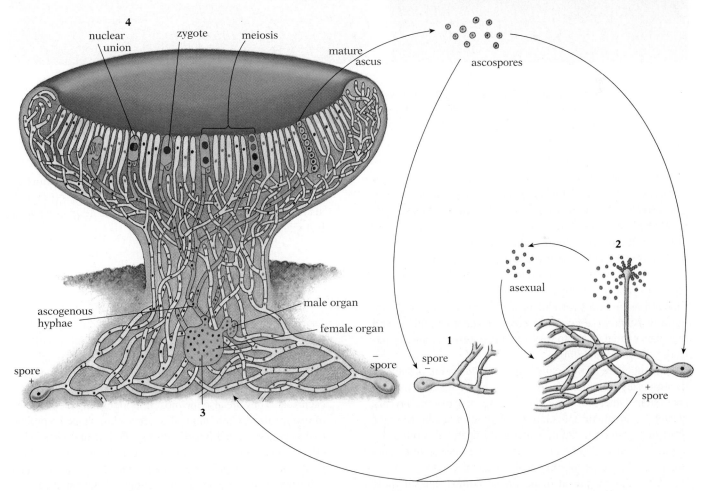

40.7 Life cycle of a sac fungus The life cycle begins with the germination of ascospores (1) to produce hyphae. The hyphae later form upright hyphae bearing conidia, which are released in enormous numbers and give rise to new individuals (2). The same mycelium that produces the conidia also produces the asci during sexual reproduction (3). The hyphae of one strain bears a female organ, another compatible strain produces a male organ, and haploid nuclei from the male organ are transferred into the female organ. The male and female nuclei remain separate. Hyphae with two nuclei in each cell emerge from this structure. Eventually the two nuclei will fuse in a special cell to form an ascus (4). Meiosis occurs immediately within the ascus, producing four haploid nuclei. Usually the nuclei divide once by mitosis, and produce eight haploid nuclei. A protective wall is laid down around each nucleus, forming eight ascospores, which remain lined up like peas in a pod. The spores are released when the mature ascus ruptures.

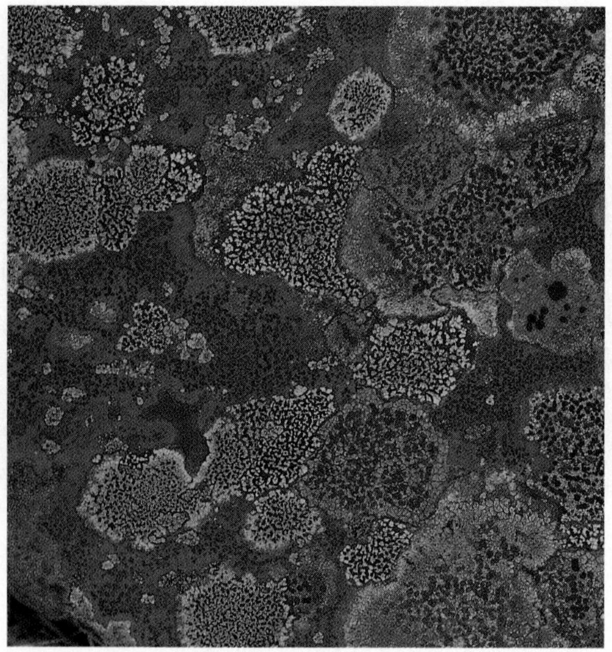

A B

40.8 Two types of lichens (A) Both *Lecidea atrata* (rust-red) and *Lecidea lithophila* (gray-black) grow as a crust, shown here on a boulder in the Cairngorms, Scotland. (B) British soldiers, an association of the fungus *Cladonia* and the alga *Trebouxia,* with their bright red tops, have a more upright growth.

haploid spores (usually eight, but sometimes four) called ***ascospores*** are produced. Each ascospore can germinate to form a new hypha. Sac fungi also reproduce asexually—most commonly by conidia formation (Fig. 40.2C) but also by budding or fragmentation, depending on the species.

It may seem strange that yeasts are sac fungi, since they are unicellular and usually reproduce by budding (Fig. 40.2A). However, under certain conditions a single yeast cell may function as an ascus, producing four spores. The spores are quite resistant to unfavorable environmental conditions and they may enable yeasts to survive temperature extremes or periods of prolonged drying.

Sac fungi are ecologically important. Many are prominent decomposers, functioning in the cycling of materials in the ecosystem. Others form the fungal component of some ***mycorrhizae***—mutualistic associations of fungi with the roots of most trees, which facilitate water and mineral absorption by the tree. Fossil evidence suggests that the evolution of mycorrhizae was key to permitting plants to colonize land.

The sac fungi are also the fungal components of most ***lichens,*** which are formed by mutualistic associations of a fungus and an alga (usually a green alga) or cyanobacterium (Fig. 40.8).

From a human standpoint, the sac fungi are among our worst fungal enemies. They are responsible for such important plant diseases as apple scab, ear rot of corn, powdery mildew, chestnut blight (which has almost eliminated the chestnut trees from North America), and Dutch elm disease (Fig. 40.9). They also cause important diseases in animals (e.g., ringworm, histoplasmosis).

The Club Fungi (Basidiomycota) Form a Basidium During Sexual Reproduction

Many of the largest and most conspicuous fungi—puffballs, mushrooms, toadstools, stinkhorns, and bracket fungi—are club fungi (Fig. 40.10), as are the rusts and smuts, important plant parasites that cause millions of dollars of damage to crops each year. They are also the most important fungal component of mycorrhizae. Club fungi all produce club-shaped reproductive structures called ***basidia*** during their sexual phase (Fig. 40.11). Nuclear fusion and meiosis occur within the basidium, producing four haploid spores called ***basidiospores.*** Each basidiospore can give rise to a new fungus. Club fungi can also reproduce asexually, by the usual processes of budding, fragmentation, and conidia formation. The hyphae of this group of fungi have cross walls.

A mushroom is a familiar example of a club fungus. Though the above-ground portion of a mushroom looks like a solid mass of tissue, and is differentiated into a stalk and a

40.9 Apple scab This fungal disease is caused by a sac fungus.

A

B

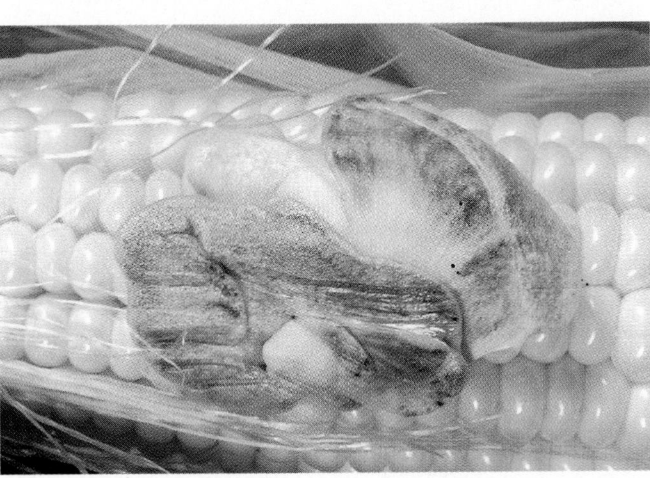

C

40.10 Two representative Basidiomycota (A) A tree fungus, *Cariolus versicolor.* (B) A puffball, *Lycopoerdon perlatum,* expelling spores after being struck by a drop of rain. (C) Corn smut (*Ustilago maydis*) i926

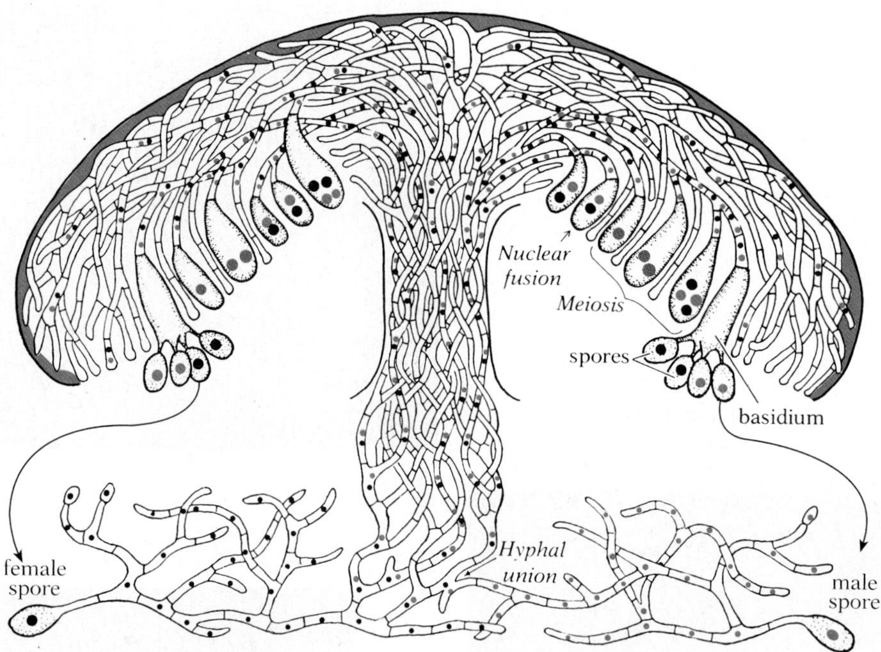

40.11 Life cycle of a mushroom, a club fungus Hyphae from two compatible mycelia (one of the plus, the other of the minus mating type) unite and give rise to hyphae with two nuclei in each cell. These two-nucleate hyphae form the fruiting body—the above ground part of the mushroom. The entire stalk and cap are composed of these hyphae tightly packed together. The two nuclei will fuse in certain terminal cells in the gills, forming a basidium. Its nucleus immediately undergoes meiosis, producing four haploid basidiospores. Each spore may give rise to a new mycelium.

prominent cap, it is nevertheless composed of hyphae, as are all fungi. The above-ground portion, or fruiting body, of many mushrooms is only a very small part of the total organism; there is an extensive mycelium in the soil. The mycelium tends to grow out in all directions from a central point, covering a circular area. The fruiting structures or mushrooms are often produced at the edges of the circle, producing what has been called a "fairy ring" of mushrooms. The basidia are produced within

MAKING CONNECTIONS

1. As recently as 25 years ago, the fungi were placed in the plant kingdom. What nutritional and functional characteristics would you cite to justify classifying them as a separate kingdom?

2. Why are fungi essential to the nutrient cycles described in Chapter 34?

3. Describe five ways in which your life is affected by fungi.

4. Why do fungi reproduce asexually in most circumstances?

5. How may the conditions in which *Rhizopus* reproduces sexually be related to the adaptive value of sexual reproduction, as explained in Chapter 9?

6. Which divisions of fungi participate in the mutualistic relationships known as mycorrhizae and lichens?

the mushroom; they are located in rows, or "gills" on the undersurface of the cap (Fig. 40.11).

The Imperfect Fungi (Deuteromycota) Do Not Reproduce Sexually

The imperfect fungi are not really a phylogenetic grouping but rather a taxonomic "wastebasket" into which are placed those species for which sexual reproduction (the "perfect stage") is absent or not yet found. These fungi have hyphae with cross walls and reproduce by means of conidia. Most of them are probably related to the sac fungi.

The organisms placed in this group have great economic and medical importance. For example, members of the genus *Penicillium* are used to produce the antibiotic penicillin, and other members of this genus are used to give Camembert and blue cheeses, like Roquefort, their characteristic flavor. Other imperfect species are involved in the production of soy sauce and tofu.

Many of the fungi pathogenic for humans are found in this group (Fig. 40.12); others cause various internal infections. For example, *Candida albicans* causes a disease called thrush in the mouths of infants and infections of the lungs and vagina in adults. Some of these fungi cause very serious internal infections in humans such as coccidioidomycosis, which can be fatal.

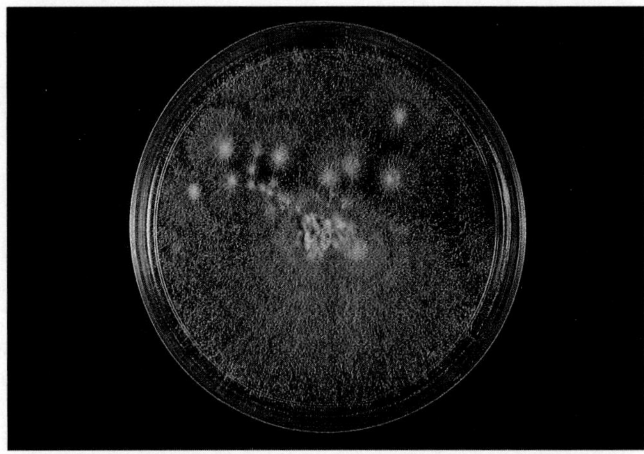

40.12 Culture of the fungus causing athlete's foot About 50 species of fungi cause human disease. This fungus grows on the surface of the skin, especially between the toes.

CONCEPTS IN BRIEF

1. Fungi are eucaryotic organisms that are absorptive heterotrophs and reproduce asexually by means of spores. They are parasitic or saprophytic. Fungi live on or in their food supply and most carry out extracellular digestion and then absorb the products.

2. The fungal body usually consists of threadlike *hyphae,* a mass of which is called a *mycelium.* Cell walls, which are often composed of chitin, are present and the hyphae are usually divided into cells by cross walls. The cross walls are often incomplete or absent, so the cytoplasm may be continuous. The individual cells often have more than one nucleus.

3. Most fungi carry out both sexual and asexual reproduction, but most reproduce more often by asexual means. The haploid stage is the dominant stage in fungi. Asexual reproduction may occur in the fungi by fragmentation, budding, conidia, and spore formation. Sexual reproduction in the fungi involves conjugation between two cells to form a two-nucleate cell, zygote formation, and then meiosis to restore the haploid condition.

4. The type of sexual reproductive structure is used to classify the fungi into different subdivisions. In the conjugation fungi (Zygomycota), conjugation leads to the formation of a diploid *zygospore,* which later undergoes meiosis. The sac fungi (Ascomycota) produce an *ascus* during sexual reproduction whereas the club fungi form a club-shaped *basidium.* The imperfect fungi (Deuteromycota) have no known form of sexual reproduction; they reproduce asexually by means of conidia.

5. Fungi are important as decomposers, but many cause important diseases in plants and animals. Some form mutualistic associations with the roots of plants to form mycorrhizae; others form mutualistic associations with certain algae to form lichens. Fungi are often used for industrial processes, to produce wine and beer, chemicals, and antibiotics.

STUDY QUESTIONS

1. **Describe the structure of a fungus. How do the cell walls of fungi differ from those of plants? How do fungi feed? (p. 917)**
2. **Compare the asexual and sexual methods of reproduction of fungi. Is the dominant stage in the life cycle diploid or haploid? (pp. 917–918)**
3. **Describe the life cycle of the black bread mold, *Rhizopus*. Why are Zygomycota called "conjugation fungi"? (pp. 918–920)**
4. **How would you identify a fungus as a member of the Ascomycota? How are members of this division ecologically and economically important? (pp. 920–922)**
5. **Picture the common white button mushroom. To which fungal division does it belong? What are its distinguishing characteristics? Explain how the aboveground portion of a mushroom is formed. What is the function of this portion? (pp. 922–925)**
6. **Why are the Deuteromycota called *imperfect fungi?* Why is this group medically and economically important? (p. 925)**

CHAPTER 41

THE ANIMAL KINGDOM: THE RADIATES AND THE PROTOSTOMES

As we have seen, the members of the animal kingdom are eucaryotic, multicellular organisms whose principal mode of feeding is ingestion. The diversity of animals is astounding; there are at least five times as many species of animals as there are species in the other five kingdoms combined. There are about 29 different phyla, all but one of which, the Chordata, comprise only invertebrates. In this chapter and the next we will examine the immense diversity of the animal kingdom and the evolutionary patterns within it. One focus of our discussion will be the adaptations that enabled animals to move from water to land. Bear in mind that the first animals to evolve ate plants, whose activity created the oxygen atmosphere upon which animals depend; carnivores, which preyed on the herbivores, evolved later.

Because the diversity of the animal kingdom is so great, this chapter and the next will concentrate on the largest and most important groups. Our aim will be to tie together the various phases of animal life discussed in previous chapters. The present discussion will deal with most of the invertebrate animal phyla; in the next chapter, we will examine the chordates and the phyla closely related to them.

A formal outline of the classification of the animal kingdom used in this book is given in the Appendix. This classification, though widely used, is not accepted by all biologists. Biologists vary in their interpretation of the evolution of animal groups. One of many possible interpretations of evolutionary relationships among the animal phyla, drawn in the traditional form of a phylogenetic tree, is shown in Figure 41.1. Note that there are two separate lineages evolving separately from the ancestral flagellates, one leading to the sponges and the other to the rest of the animal phyla. These other animal phyla are subdivided on the basis of their symmetry; two phyla are radially symmetrical and the remainder are primarily bilaterally symmetrical (Fig. 41.2).

Chapter opening photo: A large cnidarian jellyfish, Chrysaora. Some jellyfish, such as the one seen here, can be very large. The largest on record weighed almost a ton and was 3.7 meters in diameter with tentacles almost 37 meters long.

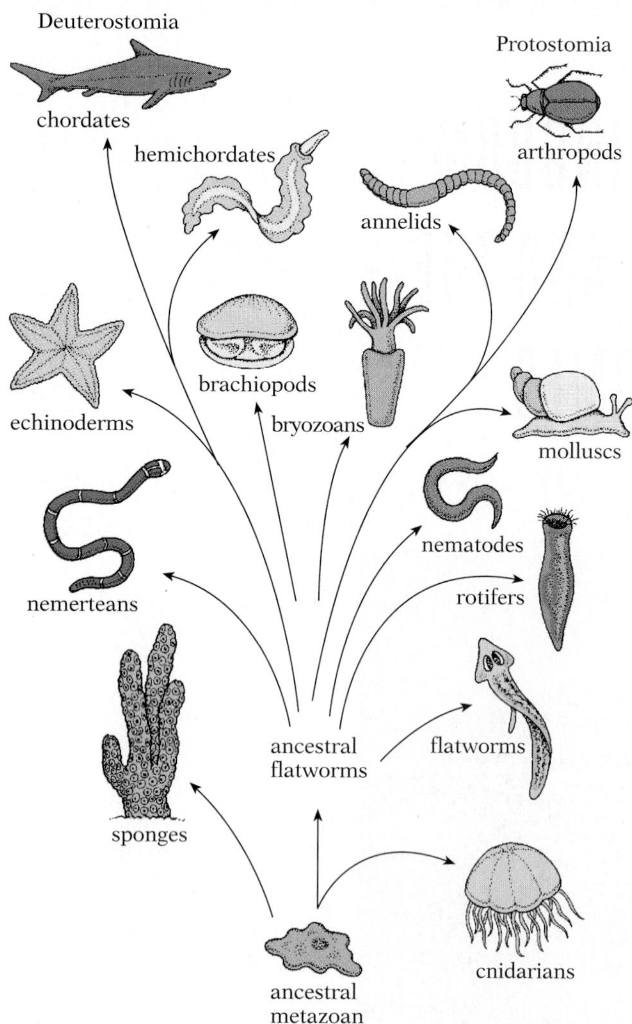

41.1 One interpretation of evolutionary relationships among the animal phyla

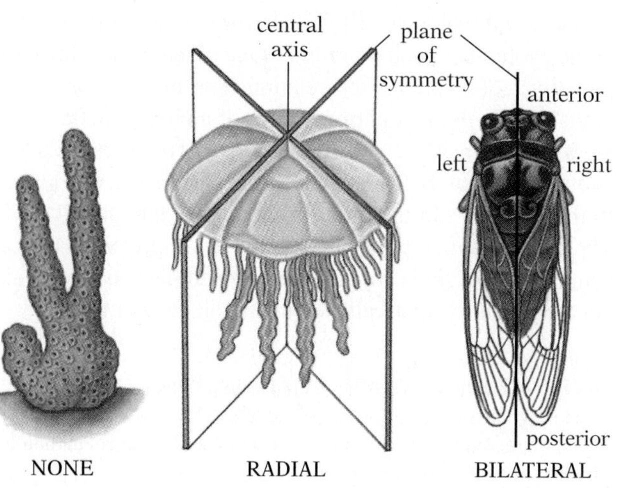

41.2 Types of body symmetry Some animals, such as many sponges (left), have asymmetrical bodies. Others, such as a cnidarian jellyfish (middle), have radially symmetrical bodies where the body parts are arranged regularly around a central axis, and any plane through the central axis divides the animals into mirror-image halves. Most animals, however, have a body plan with bilateral symmetry (right), with an elongated body and distinct dorsal (upper) and ventral (lower) surfaces, and distinct anterior (front) and posterior (rear) ends. Only one plane of symmetry (shown) can divide the animal into two mirror-image halves (i.e. right and left sides).

THE SPONGES (PHYLUM PORIFERA)

Sponges Are Filter Feeders at the Cellular Level of Organization

The organisms belonging to the phylum ***Porifera*** are aquatic, mostly marine animals (Fig. 41.3). Adult sponges are always sedentary and are usually attached to rocks, shells, or other submerged objects. They are multicellular and show some cellular differentiation; there are different types of cells specialized for different functions, but there is very little central coordination. For this reason they are said to be at the ***cellular level of organization.*** They have no mouth, no digestive system, no nervous system, and no circulatory system. In fact, they have no organs of any kind, and even their tissues are not well defined. Thus, they represent a very low level of organization.

The body of a sponge is rather like a perforated sac. Its wall is composed of three layers: an outer layer of flattened epidermal cells, a gelatinous middle layer with wandering amoeboid cells and needlelike spicules, and an inner layer of flagellated cells (Fig. 41.4). The cells of the inner layer are unusual in that the base of each flagellum is encircled by a delicate collar; such cells are called ***collar cells.***

The wall of a sponge is perforated by numerous pores. Water currents flow through the pores into the central cavity and out through a large opening (vent) at the top of the body; the flow is produced by the beating of the flagella of the collar cells. Sponges are ***filter feeders;*** microscopic food particles brought in by the water currents stick to the collar cells and are engulfed. The food may be digested in food vacuoles of the collar cells themselves or passed to the amoeboid cells for further digestion. The water currents also bring oxygen to the cells and carry away carbon dioxide and nitrogenous wastes (largely ammonia).

Sponges characteristically possess an internal skeleton secreted by the amoeboid cells. This skeleton is composed of crystalline ***spicules*** or proteinaceous fibers or both; the chemical composition and the shape of the spicules provide the basis

41.3 Colonial sponges Water, carrying microscopic food particles and oxygen, flows into the central body cavity of each of the sponges through the numerous pores in their body walls, and is discharged through the central openings (vents).

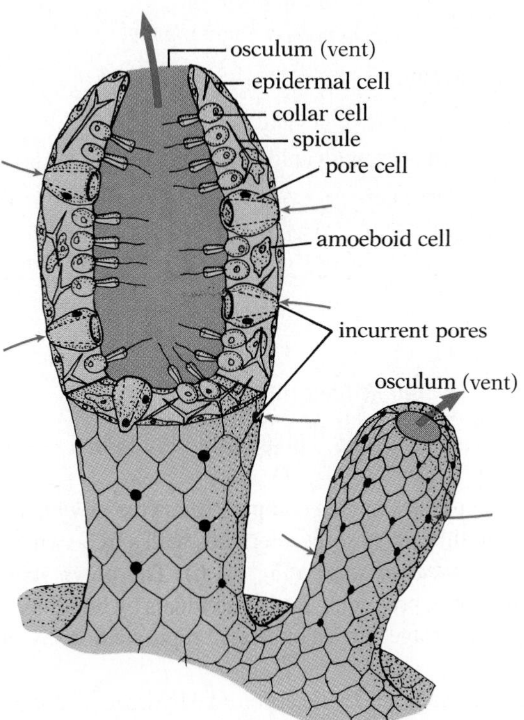

41.4 Detailed view of two colonial sponges One sponge is shown partially sectioned, the other in exterior view. The blue arrows indicate the path of the water through the sponges. These sponges have a simple tubular body. The majority of sponges have complexly folded walls in which water passes through a network of channels, but their basic body plan may be considered an elaboration of the simpler plan. Note that the large opening (vent) at the top is an excurrent opening, not a mouth as in all other multicellular animals.

for sponge classification. The proteinaceous skeletons of the bath sponges (*Spongia*) are cleaned and sold for many uses. A living bath sponge, which looks rather like a piece of raw liver, bears little resemblance to the familiar commercial object.

Sponges reproduce asexually by budding, or branching, and also carry out sexual reproduction. During certain seasons male and female gametes are produced; these are released and the sperm swim to the eggs. The fertilized egg divides to form a ciliated larva that swims about, eventually settling down and giving rise to a new sponge.

> *Key innovations of the Porifera*
> Multicellularity
> Cellular level of organization
> Filter feeding

The Sponges Arose Independently of Other Animals

Among the free-living flagellated protozoans are some collared organisms (Fig. 41.5) that closely resemble the collar cells of sponges—cells found in no other organisms. For this reason many biologists think that the Porifera evolved from collared flagellates. Other biologists suggest that sponges arose from a hollow, free-swimming colonial flagellate. In any

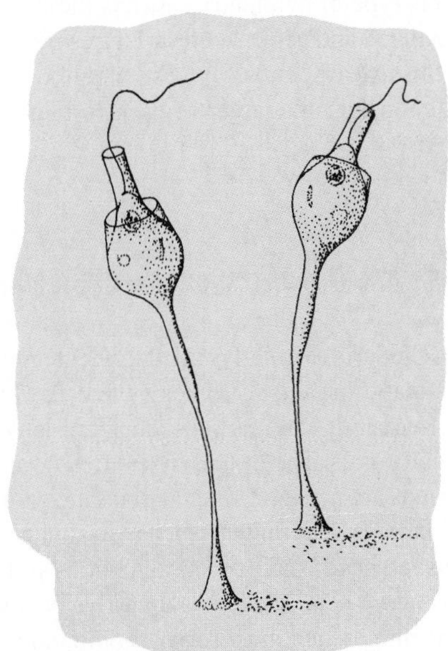

41.5 Two single-celled, collared flagellates These organisms strikingly resemble the collar cells of sponges, shown in Figure 41.4. The presence of a cytoplasmic collar in both the collar cells of sponges and the collared flagellates provides evidence suggesting that the sponges evolved from the collared flagellates.

case, because the sponge body is unique, it seems likely that sponges arose independently of the other multicellular animals. The phylum Porifera would then stand as an evolutionary development entirely separate from the rest of the animal kingdom, and it must be concluded that multicellular animals evolved at least twice from the protozoans.

Because the Porifera differ so greatly from all other multicellular animals and probably arose independently, they are often regarded as belonging to a separate subkingdom, the ***Parazoa.*** The other animal phyla belong to the subkingdom ***Eumetazoa*** (*eu-*, "true," *metazoa,* "many-celled animals"). The Eumetazoa are further divided into two groups on the basis of body symmetry; animals with radial symmetry are grouped together in the ***Radiata,*** while those with bilateral symmetry are collectively referred to as the ***Bilateria.*** We shall look first at the Radiata.

THE RADIATE PHYLA

Two animal phyla, Cnidaria and Ctenophora (comb jellies), are radially symmetrical and belong to the Radiata. Animals that display ***radial symmetry*** have similar body parts arranged regularly around a central axis, rather like spokes on a wheel. Such symmetry is often associated with a sedentary way of life. To an animal that spends much of its life attached to an immovable object or that moves about slowly, there are obvious advantages to being equally receptive to stimuli on all sides. However, this type of symmetry restricts the development of highly specialized and centralized structures because multiple repetitions of such structures would be required.

We shall consider the larger of the two radiate phyla, the Cnidaria.

Cnidaria Are at the Tissue Level of Organization

The phylum Cnidaria (formerly called Coelenterata) contains a variety of aquatic organisms, among them the hydras mentioned so frequently in earlier chapters, jellyfishes, sea anemones, and corals. The hydras live in fresh water, but most other cnidarians are marine. All cnidarians are carnivorous.

The Cnidaria have definite tissue layers, but no distinct internal organs, no head, and no central nervous system, though they possess nerve nets. There is a digestive cavity, but it has only one opening, which must serve as both mouth and anus; i.e., it is a ***gastrovascular cavity.***

Unique to the Cnidaria are specialized cells called ***cnidocytes,*** which contain stinging structures called ***nematocysts*** (see Fig. 17.12, p. 355). The nematocysts are used for capturing prey and for defense. Although the cnidocytes are found throughout the animal's outer surface, they are particularly concentrated on the tentacles.

The bodies of these animals were once thought to consist of only two layers of tissue—an outer epidermis (ectoderm) and an inner gastrodermis (endoderm)—but it is now known that a third layer (mesoderm), called ***mesolamella*** (or mesoglea), usually occurs between these two, just as it does in higher multicellular animals. However, this layer is not as well developed in the radiate phyla as in bilaterally symmetrical phyla.

The cnidarian body shows some cell specialization and division of labor. The outer epidermis contains a number of different types of cells that together act as a protective covering, whereas the cells of the inner gastrodermis are specialized for digestion and absorption. Although there is a division of labor among cells in cnidarians, it is never as complete as in most bilateral multicellular animals; and most functions performed by tissues derived from mesoderm in other animals are performed by ectodermal or endodermal cells in cnidarians.

> *Key innovations of the Cnidaria*
> Radial symmetry
> Tissue level of organization
> Three tissue layers
> Mouth
> Gastrovascular cavity
> Tentacles with cnidocytes and nematocysts
> Nerve net
> Polyp and medusa forms

Cnidarians Have Two Basic Body Forms, Polyps and Medusae

Many cnidarians have a complex life cycle in which a sedentary hydralike ***polyp stage*** alternates with a free-swimming jellyfishlike ***medusa stage*** (Fig. 41.6). The polyp stage is the asexual form, producing new individuals by budding, whereas the medusa stage carries on sexual reproduction, by means of gametes.

Let us examine *Obelia,* a colonial hydrozoan, as a cnidarian with a typical life cycle (Fig. 41.7). Much of the life of *Obelia* is passed as a sedentary branching colony of polyps. The colony arises from an individual hydralike polyp by asexual budding. The new polyps remain attached to each other by hollow, stemlike connections, and the gastrovascular cavities of all the polyps are thus interconnected via the cavity in the stems; partly digested food can pass from one polyp to another. A mature *Obelia* colony consists of two kinds of polyps: ***feed-***

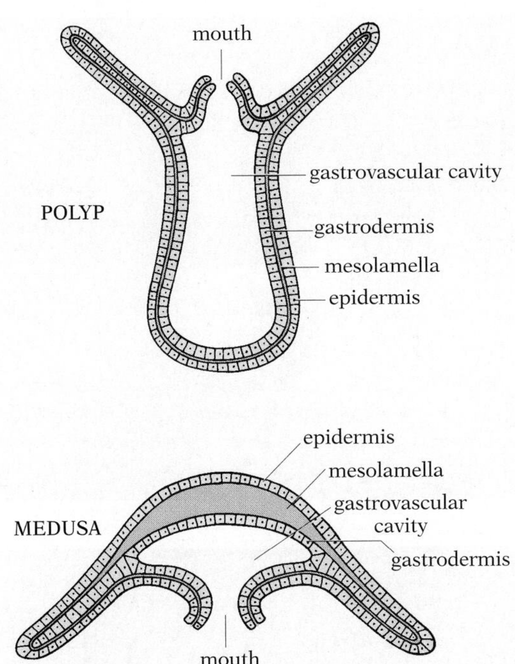

41.6 Diagram contrasting polyp and medusa The basic structure of these two forms is the same. A medusa is like a flattened polyp turned upside down.

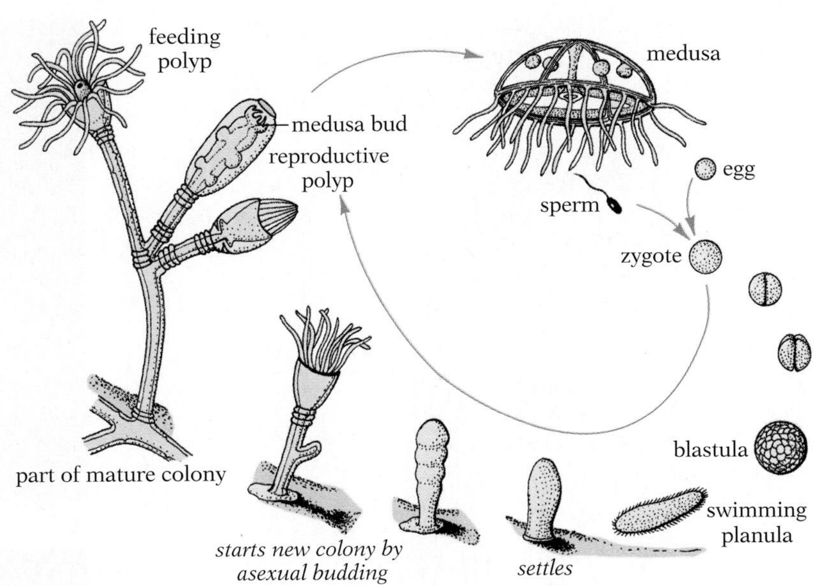

41.7 Life cycle of a colonial hydrozoan *(Obelia)* Since the medusas are of separate sexes, the eggs and sperm are produced by different individuals.

ing polyps, with tentacles and nematocysts; and ***reproductive polyps,*** without tentacles, which regularly bud off tiny transparent free-swimming medusae. A medusa is typically bell-shaped, with numerous tentacles hanging from the margin of the bell (Fig. 41.8).

Gonads beneath the bell produce either sperm or eggs, which are released into the surrounding water. Fertilization takes place and the resulting zygote develops into an elongate, ciliated larva called a ***planula.*** The planula eventually settles

41.8 A hydrozoan medusa, *Gonionemus* *Gonionemus* spends most of its life as a medusa—a weakly swimming bell-like creature with numerous tentacles.

to the bottom, attaches by one end to some object, and develops a mouth and tentacles at the other end, becoming a polyp that gives rise to a new colony. The life cycle is thus completed. Note that the alternation of polyp and medusa stages in a cnidarian like *Obelia* differs from the alternation of generations in plants in that both polyp and medusa are diploid; as in all multicellular animals, the only haploid stage in the life cycle is the gametes.

The phylum Cnidaria is divided into three classes: Hydrozoa, Scyphozoa, and Anthozoa. The classification is largely based on difference in emphasis in the life cycle. The Hydrozoa, like the *Obelia* already described, generally have well-developed polyp and medusa stages; the Scyphozoa, the true jellyfishes, have a large and well-developed medusa stage, but the polyp stage is reduced or absent (Fig. 41.9); and the Anthozoa, the sedentary polyp forms such as sea anemones and corals, have a prominent polyp stage, but the medusa stage is lacking (Fig. 41.10). The anthozoans are the most advanced members of the Cnidaria, and their body structure is much more complex than that of simple polyps like the hydras.

The hard corals (Fig. 41.11), anthozoans that secrete a hard, limy skeleton, have played a very important role in the geologic history of the earth, particularly in tropical oceans. As their skeletons have accumulated over the ages, they have formed many reefs, atolls, and islands, especially in the South Pacific. The Great Barrier Reef, a coral ridge many kilometers wide and about 2,000 kilometers long that lies off the eastern coast of Australia, is a particularly impressive example of the way these small animals can change the face of the earth and create a favorable environment for an exceptionally diverse collection of organisms. Moreover, most of the large oil de-

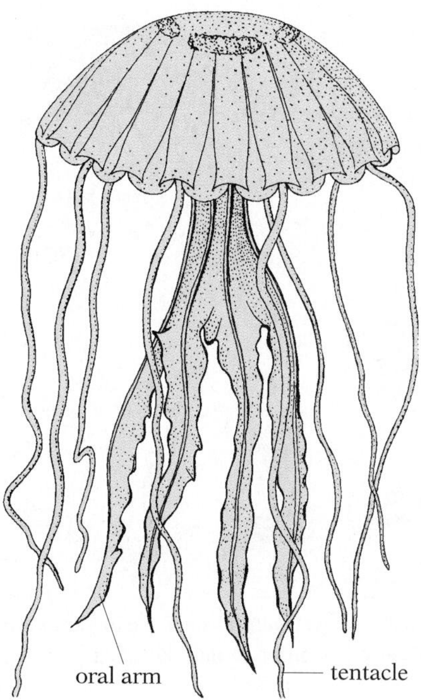

41.9 A true jellyfish (class Scyphozoa) Scyphozoan medusae resemble the hydrozoan medusae previously discussed except that they are larger and have long oral arms arising from the margin of the mouth. Many (as above) have long tentacles hanging from the margin of the bell. In the true jellyfishes the medusa stage is the dominant stage; the polyp is restricted to a small inconspicuous larva that produces the medusae asexually.

41.10 A sea anemone (class Anthozoa) The anthozoans, such as this sea anemone *Stephanauge,* exist only in the polyp form; there is no trace of a medusa stage. The basal portion of the anemone polyp attaches tightly to rocks and other submerged objects. The animal's mouth is surrounded by a ring of tentacles that are used to capture prey (invertebrates and fish). The nematocysts paralyze the prey and the tentacles then move the food into the mouth. Sea anemones are found in a variety of colors and are often called "the flowers of the sea."

A

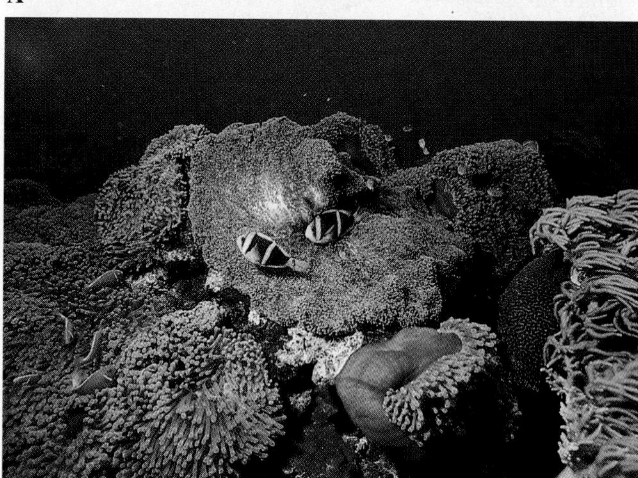

B

41.11 Coral reef The stony corals lay down calcium carbonate skeletons that constitute the basic structure of a reef and provide a multitude of microhabitats for the organisms it supports. (A) Schools of *Anthias* fish swim around this reef in the Red Sea, which includes a variety of stony corals. (B) Barred clownfish (center) in a reef in the West Pacific.

posits of the world are thought to be derived from the decay and burial of vast quantities of organic matter in enormous coral reefs.

THE ACOELOMATE BILATERIA

All other animal phyla contain animals that are primarily bilaterally symmetrical.[1] Bilateral symmetry is correlated with active movement; the bilateral form allows for a streamlined

[1]Bilateral symmetry is the property of having two similar sides. A bilaterally symmetrical animal has definite dorsal (upper) and ventral (lower) surfaces and definite anterior (head) and posterior (tail) ends.

body and consolidation of musculature for movement in one direction. The anterior end is usually the part of a bilateral animal that first meets the environment as the animal moves. Natural selection has therefore favored development of a particularly high concentration of sense organs in this region, a concentration which, in turn, led to the development of a coordinating center, the brain, at the anterior end. The combination of a streamlined shape and a brain to coordinate and direct movements enables an animal to respond quickly and effectively to environmental stimuli. Hence the bilateral body form has been very successful; beginning with the flatworms, all animals are bilaterally symmetrical (except the echinoderms, which develop radial symmetry as adults), and are collectively referred to as the ***Bilateria.***

The Bilateria are divided into three categories depending on the absence, or type of body cavity known as the ***coelom.*** Some animals are ***acoelomate***—that is, there is no space or body cavity between the body wall and internal organs; instead there is a solid network of mesodermal cells between the digestive tract and outer body wall (Fig. 41.12A). Other animals, however, have a fluid-filled body cavity that contains the digestive and reproductive organs. The cavity protects and cushions the internal organs, and allows the organs to grow and move within the cavity. The fluid-filled coelom can also act as a primitive circulatory system, transporting materials from one part of the body to another, or it can act as a hydrostatic skeleton, as we saw in earthworms (see p. 651). If this cavity is not completely lined by mesoderm (i.e., it is partly bounded by ectoderm and endoderm) it is called a ***pseudocoelom*** (Fig. 41.12B); if it develops with the mesoderm and is enclosed entirely by a mesodermal lining it is said to be a ***true coelom*** (Fig. 41.12C).

The two phyla Platyhelminthes (the flatworms) and Nemertea are the most primitive bilaterally symmetrical animals. Their bodies are composed of three well-developed tissue layers—ectoderm, mesoderm, and endoderm—but are solid and without a coelom (i.e., they are acoelomate).

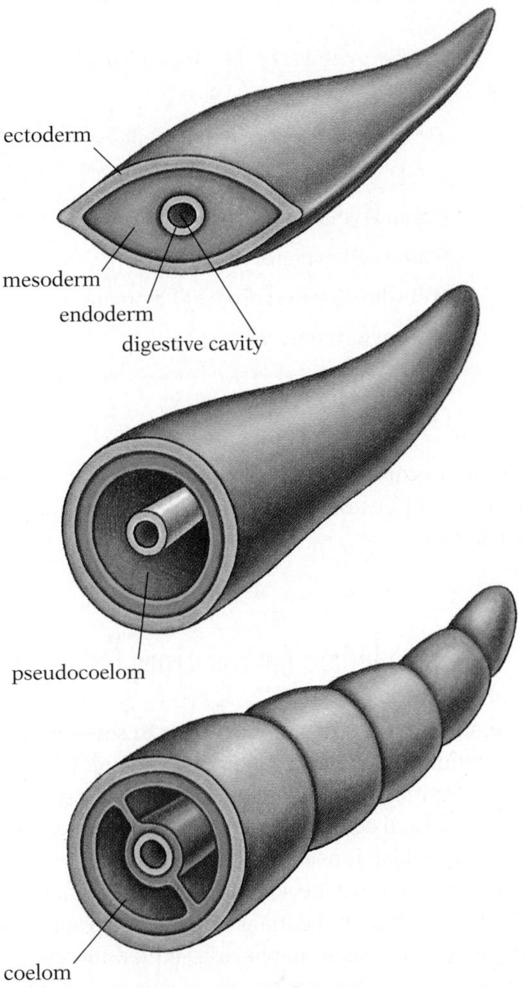

41.12 Diagrams of acoelomate, pseudocoelomate, and coelomate body types Top: Acoelomate body. There is no body cavity; the entire space between the ectoderm and endoderm is filled by a solid mass of mesoderm. Middle: Pseudocoelomate body. There is a functional body cavity, but it is not entirely bounded by mesoderm. Bottom: Coelomate body. The body cavity is completely bounded by mesoderm.

The Flatworms Are at the Organ Level of Organization

The flatworms[2] (phylum Platyhelminthes), as their name implies, are flattened, elongate animals. Their digestive cavity (not always present) resembles that of cnidarians, i.e., it is a gastrovascular cavity—a cavity with a single opening that must serve as both mouth and anus. However, there is a muscular pharynx leading into the cavity, and the cavity itself is often profusely branched (Fig. 17.13, p. 356). Respiratory and circulatory systems are absent since these organisms are still small enough and flat enough to survive by exchanging gases directly with the water. However, there is a flame-cell excretory system (see p. 465 and Fig. 22.12), and there are well-developed reproductive organs (usually both male and female in each individual). That both an excretory system and reproductive organs are present signifies that the flatworms have advanced beyond the tissue level of construction seen in the Cnidaria to an organ level of construction. The more extensive development of mesoderm, leading to greater division of labor, was probably a major factor in making this possible. Mesodermal muscles are well developed. Several longitudinal nerve cords run the length of the body and a tiny "brain" ganglion located in the head constitute a central nervous system (see p. 593 and Fig. 28.8).

[2]The term "worm" is applied to a great variety of unrelated animals. It is a descriptive, not a taxonomic, term that denotes possession of a slender elongate body, usually without legs or with very short ones.

Key innovations of the Platyhelminthes
Bilateral symmetry
Three well-developed tissue layers
Acoelomate
Organ level of construction
Flame cell excretory system
Well-developed reproductive organs
Central nervous system

The flatworms are divided into three classes: Turbellaria, Trematoda, and Cestoda. The trematodes and cestodes are entirely parasitic.

The Turbellarians Are Free-Living Flatworms

The members of this class, of which the freshwater planarians often mentioned in earlier chapters are examples, are free-living (nonparasitic) flatworms ranging from microscopic size to a length of several centimeters. The body is covered by an epidermal layer, which is usually ciliated. Though a few turbellarians live on land, most are aquatic (the majority marine) (Fig. 41.13). The smaller turbellarians move about by means of cilia but larger worms, such as planaria, utilize muscular movements. They are carnivores or scavengers; their tubular pharynx sucks the food into the gastrovascular cavity for digestion.

41.13 An aquatic turbellarian, *Prosthecereus vittatus* The free-living flatworms are often brightly colored. Movement is accomplished by the action of both muscles and cilia.

The turbellarians are believed to be the ancestral group from which the flukes and tapeworms evolved.

The Flukes Possess Adaptations for Internal Parasitism

The flukes (class Trematoda) are parasitic flatworms that have become specialized for living inside other animals. The outer epidermis has become modified to protect it from digestion by the host's enzymes—an important adaptation to a parasitic way of life. Adult flukes have lost external cilia but characteristically possess suckers, usually two or more, by which they attach themselves to their host (Fig. 41.14). The flukes feed on the tissues of the host, digesting the material in their gastrovas-

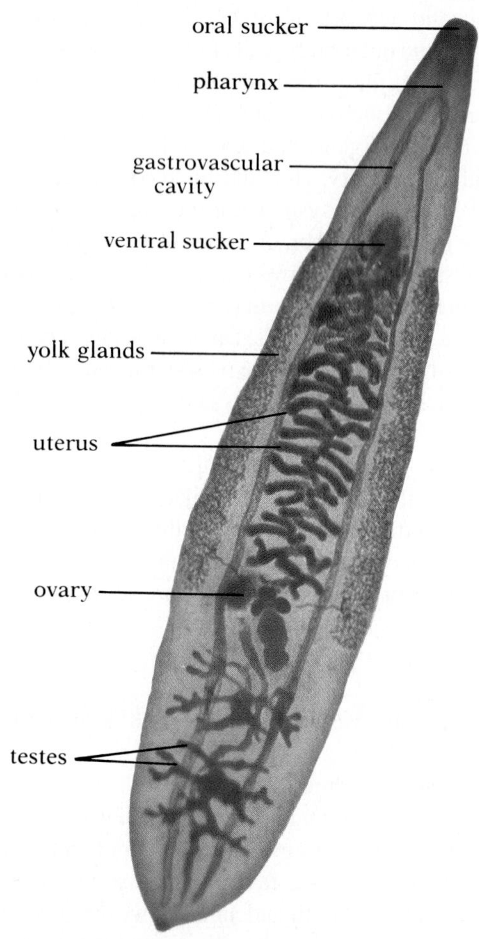

41.14 A Chinese liver fluke, *Opishthorchis sinensis* There are generally two suckers for attachment of the animal to its host: an oral sucker surrounding the mouth and a ventral sucker. The gastrovascular cavity consists of two branches and is much less subdivided than that of the turbellarians. Notice that most of the space within the animal is taken up by reproductive organs, and that both male and female organs are present.

cular cavity. A large proportion of their bodies is occupied by reproductive organs, both male and female.

Many flukes have very complicated life cycles involving two to four different kinds of hosts. The blood fluke, *Schistosoma japonicum,* common in China, Japan, Taiwan, and the Philippines, is a species with two hosts—snails (the intermediate host) and humans (the final host).

The adult blood fluke (Fig. 41.15) inhabits blood vessels near the intestine of a human being. When ready to lay its eggs, it pushes its way into one of the very small blood vessels in the wall of the intestine. There it deposits so many eggs that the vessel ruptures, discharging the eggs into the intestinal cavity, whence they are carried to the exterior in the feces. If there is a modern sewage system, that is the end of the story. In many Asiatic countries, however, human feces are regularly used as fertilizer. Thus the eggs get into water in rice fields, irrigation canals, or rivers, where they hatch into tiny ciliated larvae. A larva swims about until it finds a snail of a certain species; it dies if it cannot soon locate the correct species. When it finds such a snail, the larva bores into it and feeds on its tissues. It then reproduces asexually, and the new individuals thus produced leave the snail and swim about until they come into contact with the skin of a human being, such as a farmer wading in a rice paddy or a child swimming in a pond. They attach themselves to the skin and digest their way through it and into a blood vessel. Carried by the blood to the heart and lungs, they eventually reach the vessels of the intestine, where they settle down, mature, and lay eggs, thus initiating a new cycle.

Schistosomes in the body of a person can cause a serious disease called ***schistosomiasis,*** which is characterized initially by a cough, rash, and body pains, followed by severe dysentery and anemia. The disease so saps the strength of its victims that they become weak and emaciated and often die of other diseases to which their weakened condition makes them susceptible. There is as yet no cure for this disease. Schistosomiasis is one of the most widespread and debilitating human diseases in the world today, but because it is confined to the warmer regions of the earth, most inhabitants of North America and Europe are hardly aware of its existence.

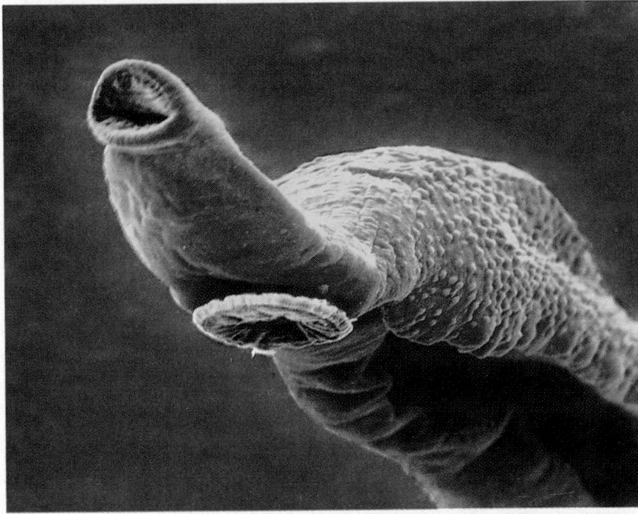

41.15 ***Schistosoma mansoni,* a blood fluke responsible for schistosomiasis** Note the oral and ventral suckers by which it attaches to its host.

The Tapeworms Are Intestinal Parasites of Vertebrates

Adult tapeworms (class Cestoda) are long, flat, ribbonlike animals (Fig. 41.16) that live as internal parasites of vertebrates, almost always in the intestine. However, the life cycle usually involves one or two intermediate hosts, which may be invertebrate or vertebrate, depending on the species. The life cycle of the beef tapeworm, in which the intermediate host is a cow and the final host is a human being, was outlined in Figure 31.37.

Tapeworms exhibit many special adaptations for their parasitic way of life, and are the most highly specialized of the flatworms. Like the flukes, they have a resistant epidermis. However, they have neither mouth nor digestive tract. Bathed by the food in their host's intestine, they absorb predigested nutrients across their general body surface, which has been modified by the presence of microvilli to facilitate absorption. Diffusion, probably augmented by active transport, suffices to provision all the cells, because none are far from the surface.

The head of a tapeworm is a small knoblike structure called a ***scolex,*** which usually bears suckers—and often also hooks—by which the worm attaches to the wall of the host's intestine (Figs. 41.16 and 41.17). Immediately behind the scolex is a neck, or growing region, which produces new body segments, the ***proglottids*** (Figs. 41.16 and 41.17). Tapeworms may grow very long; beef tapeworms occasionally grow to 23 meters, fish tapeworms to 18 meters, and pork tapeworms to eight meters.

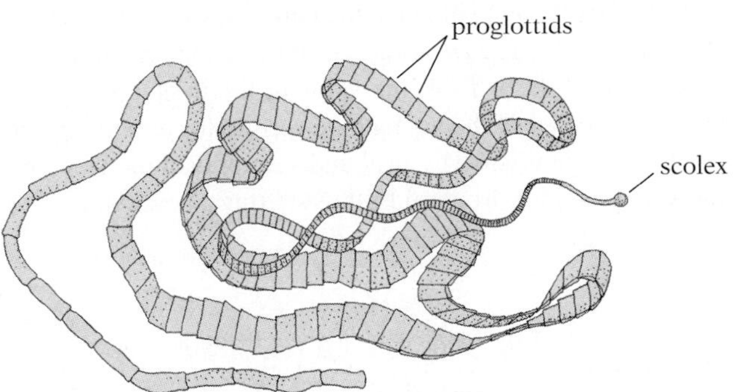

41.16 A tapeworm The body is composed of a small head (scolex) and neck, followed by a large number of segments called proglottids. As the proglottids ripen, they break off and pass with the host's feces to the outside. New proglottids are produced just behind the neck.

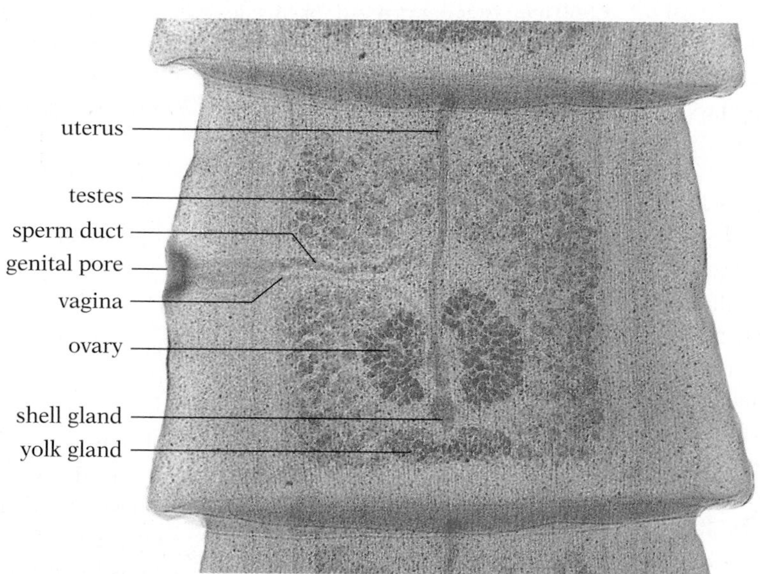

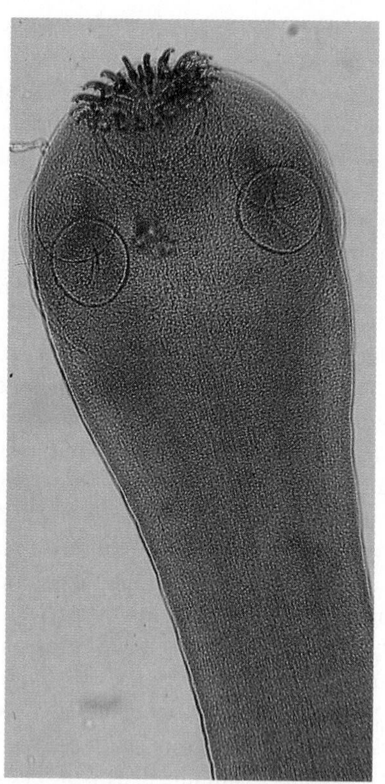

41.17 Scolex and proglottids of the dog tapeworm, *Taenia pisiformis* Left: A mature proglottid. Note that the only organs present have to do with reproduction. Right: Scolex, neck, and some of the proglottids. Note the hooks and suckers on the scolex.

Each proglottid is essentially a reproductive sac, containing both male and female organs. Sperm cells, usually from a more anterior proglottid of the same animal, fertilize the egg cells, which are stored in a uterus. Eventually the ripe proglottid detaches from the worm and passes out of the host's body with the feces. A single ripe proglottid may contain more than 100,000 eggs, and the annual output of one worm may be more than 600 million.

If an appropriate intermediate host eats food or vegetation contaminated with tapeworm eggs, its enzymes digest the shells of the eggs. The microscopic embryos thus released bore through the wall of the host's intestine, enter a blood vessel, and are carried by the blood to the muscles, where they encyst. If a person eats the raw or rare meat of this intermediate host (e.g., beef, pork, or fish), the walls of the cysts are digested away; the scolex of the young tapeworms attach to the intestinal wall and, nourished by an abundant supply of food, begin to grow and produce proglottids, thus starting a new cycle.

41.18 A ribbon worm Nemertean worms are mostly marine worms that burrow in sand or mud in shallow water or shelter in protected places. They are often called ribbon worms because they are long, slender worms with soft, flat bodies. They have a complete digestive tract that is a ciliated tube extending the length of the body, and a very long retractable proboscis that can be extended from the anterior end of the body for capturing prey or for use in self-defense.

The Proboscis Worms Have a Complete Digestive Tract

The members of the phylum Nemertea are long ribbonlike worms (Fig. 41.18) characterized by a very long retractable tube (proboscis) enclosed in a cavity at the anterior end of the body. The proboscis can be extended to capture prey or act in self-defense.

Nemerteans are of interest from an evolutionary standpoint because they resemble turbellarian flatworms (their probable ancestors) in their nervous systems and in many other ways. But they differ from turbellians and other animals considered thus far in two important characteristics. First, they have a ***complete digestive tract***—one that has two openings, a mouth and an anus. Such a system makes possible specialization of sequentially arranged chambers for different functions and thus permits an assembly-line processing of food, as we saw in Chapter 17. Second, they have a simple blood circulatory system consisting of three longitudinal blood vessels, which presumably facilitates transport of materials from one part of the body to another. There is no heart or definite circulatory plan; movements of the body wall and contractile action of muscles in the vessels serve to circulate the blood.

Key innovations of the Nemertea
Complete digestive system
Simple blood circulatory system

THE PROTOSTOMIA AND DEUTEROSTOMIA

We saw in Chapter 27 that a single opening, the blastopore, forms during gastrulation. In animals like cnidarians and flatworms, where the digestive tract is a gastrovascular cavity, the blastopore becomes the combined mouth and anus. In nemertean worms and the other more complex animals that have complete digestive systems, does the blastopore become the mouth, or does it become the anus? Embryologists have shown that in nemerteans the site of the embryonic blastopore becomes the mouth and that the anus is an entirely new opening. This is also the case in many other animals, including nematode worms, molluscs, annelids, and insects. But in a few phyla, among them two large and important ones—the Echinodermata and the Chordata—the situation is reversed: the embryonic blastopore becomes the anus, and the mouth is the new opening.

This fundamental difference in embryonic development suggests that a major split occurred in the animal kingdom soon after the origin of a bilateral ancestor. One evolutionary line led to all the phyla in which the blastopore becomes the mouth; these phyla are often called the ***Protostomia*** (from the Greek *proto-*, "first," and *stoma*, "mouth"). The other evolutionary line led to the phyla in which the blastopore becomes the anus and a new mouth is formed; these phyla are called the ***Deuterostomia*** (from the Greek *deutero-*,. "second").

As might be expected if the Protostomia and Deuterostomia diverged at a very early stage of their evolution, they differ in a number of fundamental characters besides the mode of formation of mouth and anus. A further essential difference between them is that the early cleavage stages are usually ***determinate*** in protostomes and ***indeterminate*** in deuterostomes. That is, the developmental fates of the first few cells of a protostome embryo are usually already at least partly determined, and if these cells are separated no one of them can form a complete individual, whereas the fates of the first few cells of a deuterostome embryo are not determined, and each cell, if separated, can develop into a normal individual (i.e., there can be identical twinning). Furthermore, the two groups exhibit strikingly different patterns of cleavage; the early cleavages in protostomes give rise to a spiral arrangement of cells, whereas the early cleavages in deuterostomes give rise to a so-called radial arrangement of cells (Fig. 41.19). The basic larval types are also different in the two groups, as we shall see in the next chapter.

MAKING CONNECTIONS

1. In what order did plants and animals evolve? Why?

2. How does the alternation of polyp and medusa stages in the life cycle of cnidarians differ from the alternation of generations in plants?

3. How have the hard corals been significant in the geologic history of the earth?

4. What is the link between bilateral symmetry and the evolutionary development of the brain?

5. How are tapeworms adapted to obtain nutrients from the intestinal tract of their host? Would you expect aerobic or anaerobic respiration to predominate in these parasites?

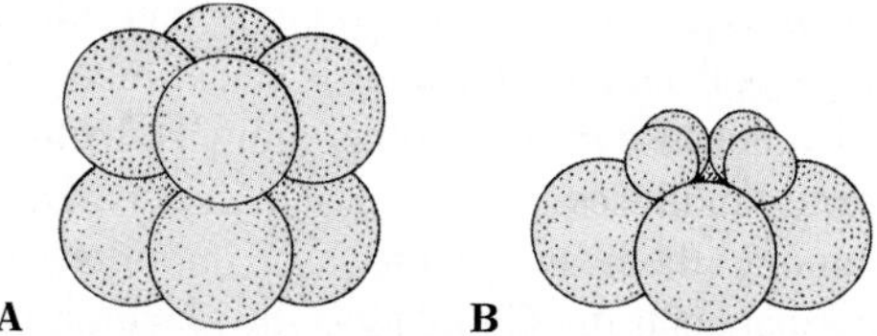

41.19 Radial and spiral cleavage patterns (A) Radial cleavage, characteristic of deuterostomes. The cells of the two layers are arranged directly above each other. (B) Spiral cleavage, characteristic of protostomes. The cells in the upper layer are located in the angles between the cells of the lower layer.

Another fundamental difference between the protostome and deuterostome phyla is seen in the manner of origin of the mesoderm in the embryo. Most of the mesoderm in protostomes and all of it in deuterostomes is derived from endoderm. However, in protostomes this mesoderm arises as a solid ingrowth of cells from a single initial cell located near the blastopore (Fig. 41.20, left), whereas in deuterostomes it usually arises by a saclike outfolding of the gut wall (Fig. 41.20, right).

Still another difference, correlated with the preceding one, has to do with the method of formation of the coelom, if one is present. All of the animals above the level of the nemerteans have some type of fluid-filled body cavity—either a pseudocoelom or a true coelom. The manner in which the coelom forms differs in the protostomes and deuterostomes (Fig. 41.20).

The differences, however, are not quite so clear-cut as we may have implied. Most of the contrasting characters are subject to exceptions, and some are far less distinct in the animals themselves than descriptions and diagrams tend to suggest. Indeed, some authors have seriously questioned whether the protostome-deuterostome dichotomy is real or imagined. Some of the important Protostome-Deuterostome differences are summarized in Table 41.1.

TABLE 41.1 *A comparison of the Protostomes and Deuterostomes*

Characteristic	*Protostomes*	*Deuterostomes*
Type of cleavage	Determinate, spiral	Indeterminate, radial
Symmetry	Bilateral	Primarily bilateral
Mesoderm formation	From single endodermal cell near blastopore	By outpouching of the gut wall
Type of coelom	Acoelomate, pseudocoelom, or true coelom	True coelom
Formation of coelom	Split in the solid mass of mesoderm	Cavities of the mesodermal pouches become coelom
Fate of blastopore	Becomes mouth	Becomes anus

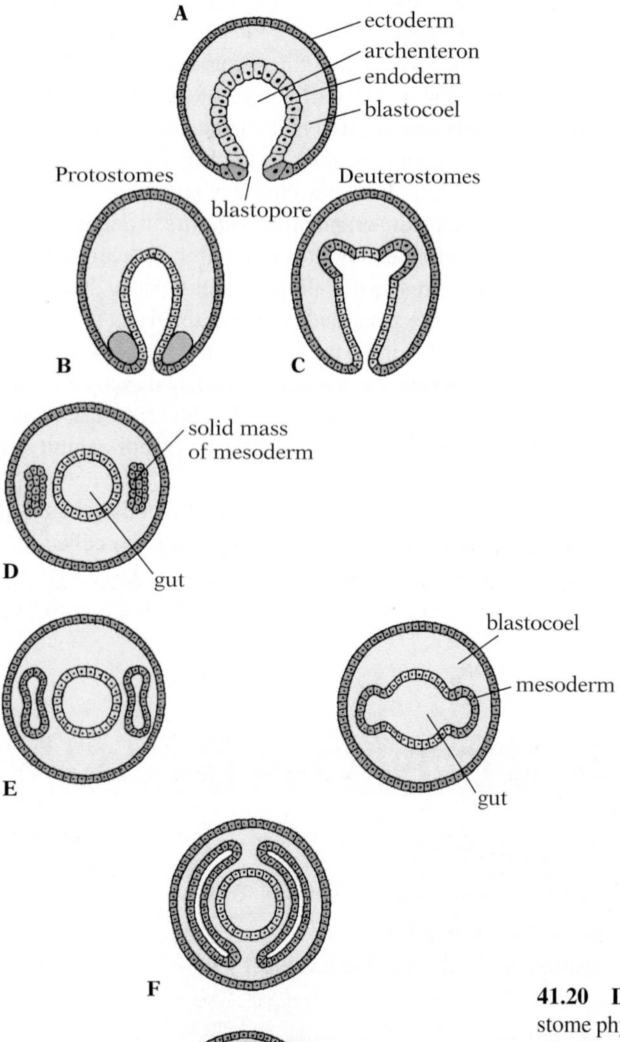

41.20 Different modes of origin of mesoderm and the coelom In most of the protostome phyla (left), the mesoderm arises from a single initial cell (not shown), which divides repeatedly to form a solid mass of mesodermal cells (D). In the deuterostome phyla (right), the mesoderm usually arises as a pair of pouches off the endodermal wall of the archenteron (C). The differences in the origin of the mesoderm correlate with the manner in which the coelom forms; in the protostomes it arises as a split in the solid mass of mesoderm (E, left); in the deuterostomes the coelom arises as the cavity in the mesodermal sacs (E, right). In both lines the body cavity is a true coelom, a cavity completely lined by mesoderm (G).

THE PSEUDOCOELOMATE PROTOSTOMIA

In several protostome phyla the body cavity functions in a manner similar to a true coelom but is not completely lined by mesoderm—it is a pseudocoelom. The two largest and most important of the six phyla of pseudocoelomate protostomes are the phyla Rotifera (the rotifers) and Nemata (the roundworms).

Rotifers (Phylum Rotifera) Are Characterized by a Ciliated Crown

The rotifers—or wheel animalcules, as they are commonly called—are microscopic, usually free-living aquatic animals with a crown of cilia at the anterior end (Fig. 41.21). Most of them live in fresh water, where they are an important source of food for other animals. The cilia are generally arranged in a circle, and when beating they often give the appearance of a rotating wheel; hence the name of the class. When feeding, rotifers attach themselves by a tapering posterior "foot," and the beating cilia draw a current of water into the mouth. In this manner, very small protozoans and algae are swept into a complicated muscular pharynx, where they are ground up by seven hard, jawlike structures. The rotifers are believed to have evolved from turbellarian flatworm ancestors.

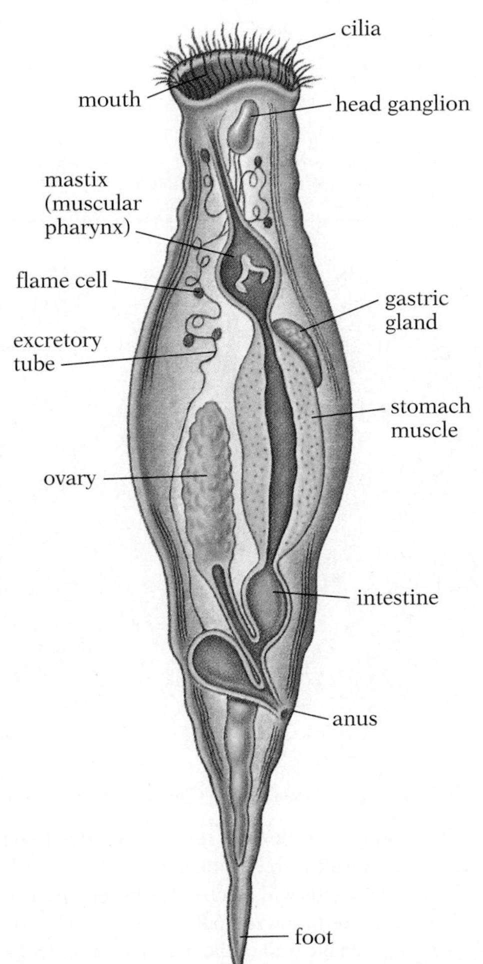

41.21 Section of a rotifer Rotifers are small, about the size of a *Paramecium.*

The freshwater rotifers are extremely abundant. Anyone examining a drop of water for Protozoa under a microscope is likely to see one or more of these interesting animals. In fact, most of them are no larger than protozoans, and it is often difficult, when encountering them for the first time, to realize that they are multicellular.

> *Key innovations of the Rotifera*
> Pseudocoelom
> Crown of cilia
> Jawlike structures

Nematode Worms Are Often Parasitic and Have High Internal Pressures

The organisms belonging to the Nemata (the nematodes) have round, elongate bodies that usually taper nearly to a point at each end (Fig. 41.22). They have no cilia or flagella; even the sperm lack flagella (they are amoeboid). The body is enclosed in a tough cuticle (Fig. 41.23). Just under the epidermal layer of the body wall are bundles of longitudinal muscles; there are no circular muscles. Between the intestine and the body wall is a fluid-filled body cavity (a pseudocoelom). A distinctive feature of the nematodes is their perfect cylindrical shape, which results from the extraordinarily high internal pressure of the fluids within the body cavity. While an earthworm may have an internal pressure of 1.5 to 21 mm Hg, the pressure in some nematodes may be as high as 225 mm Hg. The fluid-filled cavity acts as a hydrostatic skeleton against which the muscles can act. The high internal pressure, the stiff cuticle, and the lack of circular muscles severely limit the types of movements possible for these worms, and they usually thrash about with a characteristic whiplike motion.

Nematodes are extremely abundant, and occur in almost every type of habitat. Of the many that are free-living in soil or water, most are very tiny, often microscopic. A single spadeful of garden soil may contain a million or more. A single rotting apple may have 90,000. Many other nematodes are internal

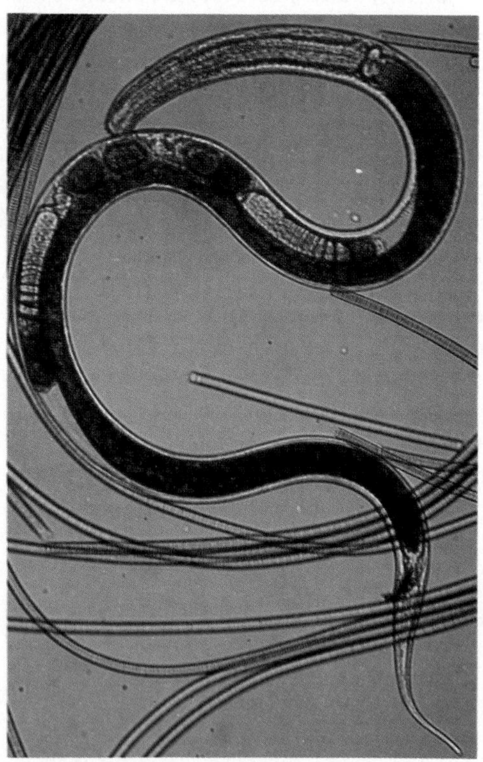

41.22 A living nematode worm viewed among cyanobacteria (*Oscillatoria*), its food

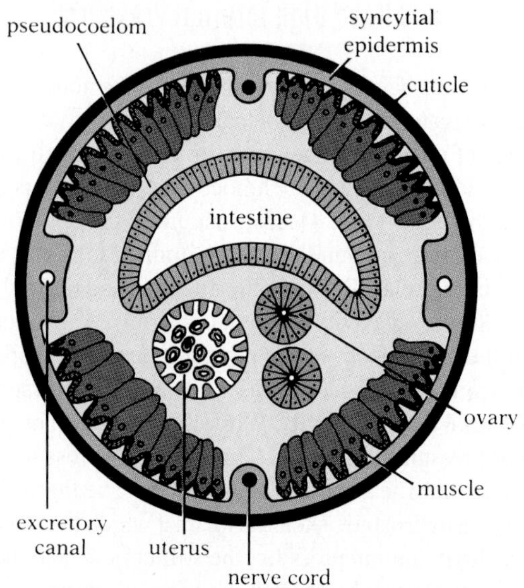

41.23 Diagrammatic cross section of a nematode worm

parasites of plants or animals; these also are often small, but some may attain a length of one meter.

Nematodes parasitic on cultivated plants cause an annual loss of millions of dollars. Nematode infections in humans are common; estimates indicated that more than 98 percent of the human population has had a nematode infection in their lifetime. Some cause serious diseases. *Trichinella spiralis,* for example, causes the disease ***trichinosis,*** often contracted by eating insufficiently cooked pork (Fig. 41.24). Other nematodes that parasitize humans are (1) *Ascaris,* a large worm (up to 30 centimeters long) that lives in the digestive tract, lays eggs that pass to the outside with the host's feces, and infects new hosts when vegetables grown in soil contaminated with feces are eaten without adequate washing; (2) hookworms, tiny worms, widespread in warm climates, that cause a severely debilitating disease in those who go barefoot on soil contaminated with infested feces; (3) pinworms, likewise tiny worms, common in schoolchildren, who usually get infected by putting unclean fingers with eggs on them into their mouths; and (4) filaria, microscopic worms that are spread by the bite of certain mosquitoes in tropical and subtropical areas. Filarial worms live in the lymphatic system, where they may accumulate in such numbers that they block the flow of lymph, causing accumulation of fluid and often enormous swelling (elephantiasis) of the infected part of the body (see Fig. 20.21, p. 425).

Key innovations of the Nemata
No cilia or flagella
Only longitudinal muscles

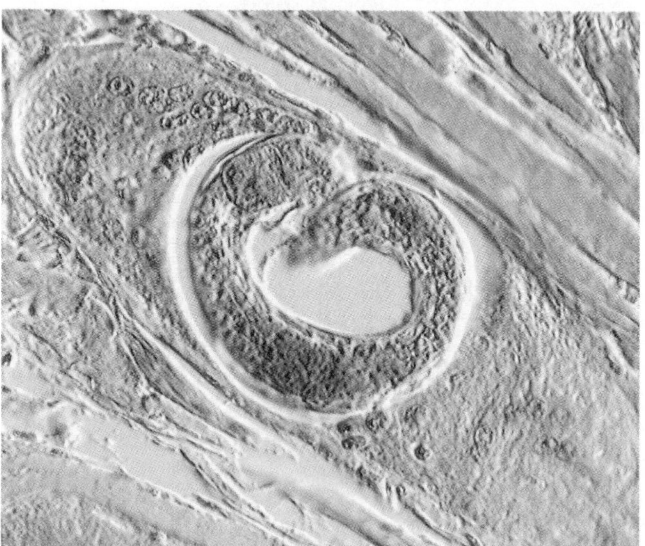

41.24 Micrograph of *Trichinella spiralis* encysted in pork muscle If a human eats insufficiently cooked pork that contains encysted juvenile worms, the worms will hatch out in the digestive tract, mature into the adult form, and reproduce sexually. The young worms burrow through the wall of the intestine and enter the bloodstream, eventually leaving the blood and burrowing into the muscles where they grow and encyst. The damage to the host occurs during the migration; as many as 500 million worms may be moving throughout the body and burrowing into the muscles at one time.

THE COELOMATE PROTOSTOMIA

All the protostome phyla except the ones discussed above possess *true coeloms.* All have a complete digestive tract, and most have well-developed circulatory, excretory, and nervous systems. We shall focus our attention on the three largest phyla of the coelomate Protostomia: the Mollusca, Annelida, and Arthropoda.

The Molluscs (Phylum Mollusca)

All Molluscs Share a Similar Body Plan and Have Undergone Extensive Adaptive Radiation

The phylum Mollusca (the name means "the soft ones") is the second largest in the animal kingdom; it contains nearly 100,000 living species and 35,000 fossil species. Among the best-known molluscs are snails and slugs, clams and oysters, and squids and octopuses.

The various groups of molluscs may differ considerably in outward appearance, but most have fundamentally similar body plans (Fig. 41.25). The soft body consists of four principal parts: (1) a ***head region,*** which contains the sense organs and brain; (2) the large ventral muscular ***foot,*** which can be extruded from the shell (if one is present) and functions in locomotion; (3) a ***visceral mass*** above the foot, which contains the digestive system, the excretory organs (nephridia), the heart, and other internal organs; and (4) the ***mantle,*** a heavy fold of tissue that covers the visceral mass and which in most species contains glands that secrete a shell. The mantle often overhangs the sides of the visceral mass, thus enclosing a ***mantle cavity,*** which usually contains gills. Most molluscs also have a toothed, tonguelike organ, the ***radula,*** which is used to rasp food off rocks. An open circulatory system is also characteristic; they usually have a well-developed heart but during part of each circuit the blood flows through large, open sinuses where it bathes the tissues directly.

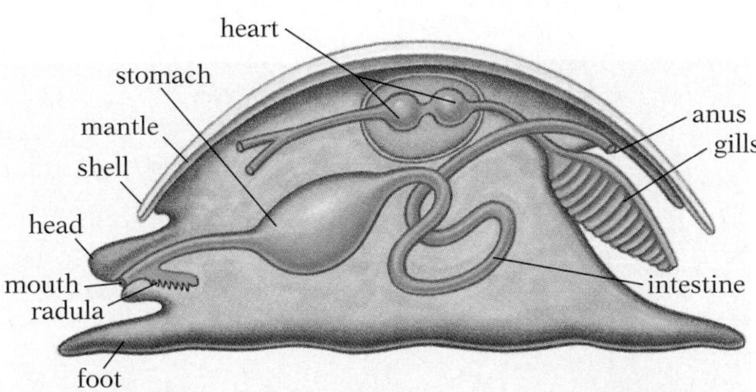

41.25 The basic molluscan body plan The hypothetical ancestral mollusc shown above illustrates the fundamental molluscan body plan. The mollusc creeps slowly about on a flattened muscular foot, using a radula to scrape food from the surface of rocks. The anterior portion of the foot contains the mouth and rudimentary sense organs and forms the head. The visceral mass, containing the internal organs, lay on top of the foot and is covered over by a mantle, which secretes a shell. Within the mantle cavity are gills for gas exchange.

During their evolutionary history this group has undergone extensive adaptive radiation, illustrating the general principle that successful phyla are highly diversified in structural adaptations and ways of life. The molluscs are divided into seven classes; we shall refer to five of them. The basic molluscan body plan has been variously modified in the different groups (Fig. 41.26).

> *Key innovations of the Mollusca*
> Ventral foot
> Dorsal visceral mass
> Mantle
> Shell (in most)

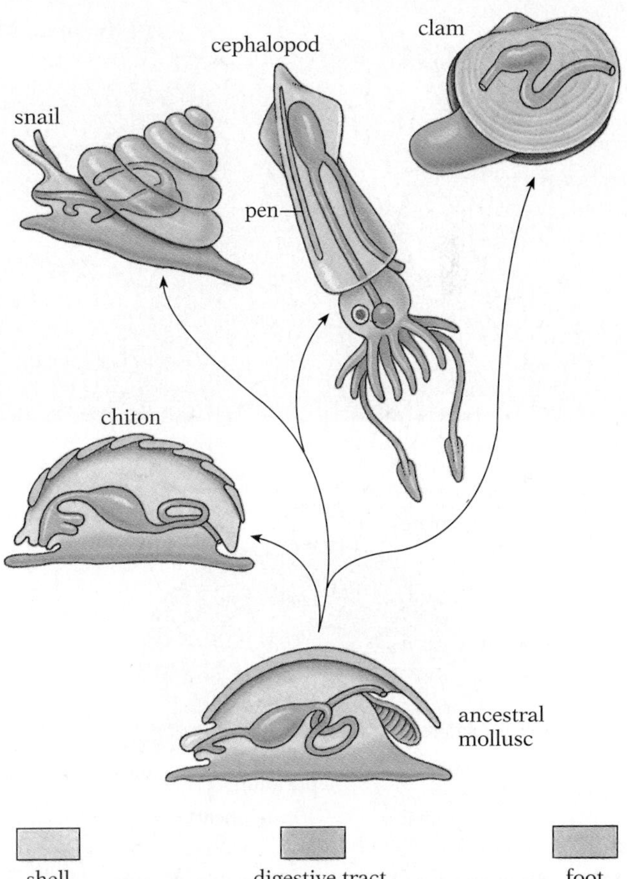

41.26 Modifications of the molluscan body plan Note changes in shell, foot, and digestive tract that adapt the animal to different lifestyles.

The Classes Polyplacophora and Monoplacophora May Resemble the Ancestral Molluscs

Because their hard shells preserve well, the molluscs are well represented in the fossil record and over 35,000 species have been identified, dating back into the Cambrian period. Although we do not know what the ancestral molluscs looked like, we can make some inferences about them from studying the primitive molluscs living today. The members of the classes ***Monoplacophora*** and ***Polyplacophora*** (the chitons) are generally regarded as the most primitive living members of the molluscs. As the names suggest, the monoplacophorans have a one-piece shell whereas the polyplacophorans have a shell consisting of many (actually eight) plates (Fig. 41.27). Members of both classes creep slowly about on their flattened muscular foot, using their radula to scrape food from the surface of rocks. All have gills for gas exchange and an anterior mouth and a posterior anus. The ancestral mollusc is thought to have shared these same characteristics (Figs. 41.25 and 41.27).

A

B

41.27 Photograph and longitudinal section of a chiton (A) Note the series of plates composing the chiton's shell. (B) Cross section of a chiton. Notice the close similarity in structure between the chiton and the hypothetical ancestral mollusc shown in Figure 41.25.

Members of the Monoplacophora have long been known as fossils; they were generally thought to be extinct until 1952, when ten living specimens (genus *Neopilina*) were dredged from a deep trench in the Pacific Ocean off the coast of Costa Rica. These specimens sparked a lively debate on the ancestry of the Mollusca, because they show some internal segmentation, a characteristic seen in no other members of the phylum. Since it was already known that the early cleavage pattern and larval type of molluscs show striking similarities to the corresponding developmental stages in the segmented worms (Annelida), the segmentation of *Neopilina* led many biologists to conclude that the ancestral molluscs were segmented animals, perhaps primitive annelids. Many other biologists, however, are convinced that the segmentation of *Neopilina* is a new development, not primitive, and that the original molluscan body was unsegmented. Whichever view is correct, it seems likely that the Mollusca and the Annelida arose from a common ancestor and are therefore closely related.

The ancestral molluscs are believed to have given rise to two principal evolutionary lines (Fig. 41.26), one leading to the gastropods and cephalopods, and the other to the bivalves and scaphopods (tusk shells).

Most Snails (Class Gastropoda) Have an Asymmetrical, Spiral Shell

The gastropods are the largest class of molluscs, and the most familiar. Most have a single spiral shell, a large foot, head, and a radula (Fig. 41.28). Two characteristics of the gastropods are particularly important: their bodies have undergone torsion and coiling has occurred. ***Torsion*** occurs during the development of the gastropod larva: the digestive tract bends downward and forward until the anus comes to lie close to the mouth, making the digestive tract U-shaped. Then the entire visceral mass rotates through an angle of 180 degrees, coming

41.28 A marine gastropod mollusc, *Trivia monaca* The animal moves along the substratum by means of the large muscular foot, here plainly seen. Notice also the head region, bearing prominent antennae, eyes, and a long siphon (top) through which water is brought to the gills.

41.29 Two nudibranch gastropods As the designation "nudibranch" suggests, the animals have no shells.

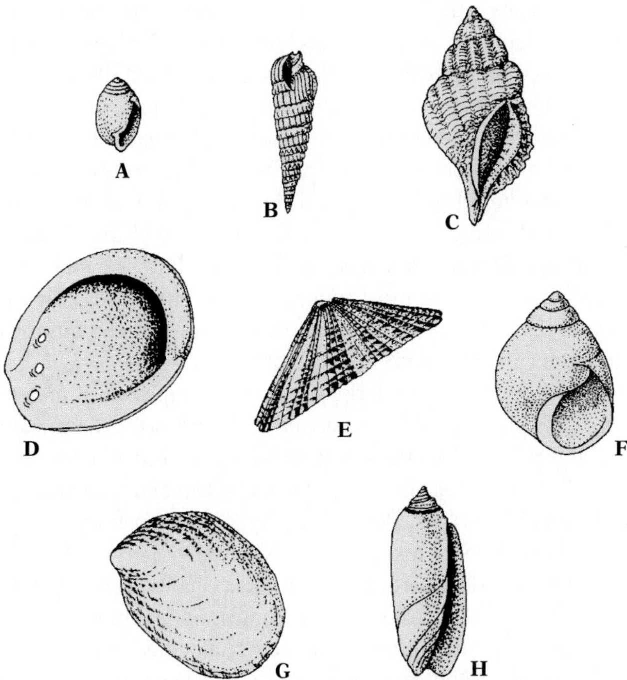

41.30 Shells of some representative gastropods (A) Salt-marsh snail (*Melampus*). (B) Auger shell (*Terebra*). (C) Oyster drill (*Urosalpinx*). (D) Abalone (*Haliotis*). (E) Californian keyhole limpet (*Diodora*). (F) Periwinkle (*Littorina*). (G) Boat shell (*Crepidula*). (H) Olive shell (*Oliva*).

to lie dorsal to the head in the anterior part of the body. The body then begins ***coiling;*** most of the visceral organs on one side (usually the left) atrophy, and growth proceeds asymmetrically, producing the characteristic spiral. The coiling is particularly noticeable in the shell, although in some cases the coiling is minimal. Some species— e.g., the nudibranchs and slugs—have lost the shell completely (Fig. 41.29).

Except for the peculiar twisting and spiraling of their bodies, gastropods are thought to be rather like the ancestral molluscs. They have a distinct head with well-developed sense organs, and most have a well-developed radula.

Gastropods occur in a great variety of habitats. The majority are marine, and their often large and decorative shells are among the most prized finds on a beach (Fig. 41.30), but there are also many freshwater species and some that live on land. The land snails are among the few groups of fully terrestrial invertebrates. In most of them, the gills are absent, but the mantle cavity has become very well supplied with blood vessels and functions as a lung.

> *Key innovations of the Gastropoda*
> Torsion and coiling of body
> Asymmetrical spiral shell

Cephalopods Are Adapted for Swimming and a Predatory Way of Life

The class Cephalopoda includes the squids and octopuses and their relatives. In these animals the foot is modified to form a number of tentacles (Fig. 41.31) and funnel for swimming.

41.31 A swimming octopus, *Octopus vulgaris* Note the large suckers on the tentacles, which are used in grasping prey and feeding. The eyes are well developed, with a cornea, iris, pupil, and lens.

The tentacles surround the well-developed head, hence the name cephalopod, which means "head-footed." Cephalopods swim by contracting the mantle, forcing water out through the funnel. The force of water leaving the funnel "jets" the animal in the opposite direction. Many of the cephalopods bear little outward resemblance to other molluscs because they are specialized for rapid swimming and a predatory way of life—for killing and eating large prey, such as fishes or crabs. Though fossil cephalopods often have large shells, these are much reduced or internal in most modern forms.[3] In squids, for instance, the only remnant of the shell is an internal, chitinous structure called the pen (Fig. 41.26).

Some species attain large size, often being several meters long. The giant squids (*Architeuthis*) of the North Atlantic are the largest living invertebrates; the biggest recorded individual was 17 meters long (including the tentacles) and weighed approximately two tons. Octopuses (Fig. 41.31) never grow anywhere near this size (except in Hollywood).

Cephalopods, particularly squids, have convergently evolved many similarities to vertebrates. For example, squids have internal cartilaginous supports analogous to the vertebrate skeleton, and they even have a cartilaginous braincase rather like a skull. They also have a closed circulatory system. Furthermore, they have an exceedingly well-developed nervous system with a large and complex brain. Perhaps the most striking of all the squids' similarities to vertebrates are their large camera-type eyes, which work the way ours do.

Key innovations of the Cephalopoda
Reduced or internal shell (in most)
Jet propulsion → rapid movement
Foot modified to form arms or tentacles
Well-developed nervous system
Camera-type eye
Closed circulatory system

The Bivalve Molluscs Are Filter Feeders

As the term bivalve indicates, the molluscs belonging to the class Bivalvia have a two-part shell, which is usually flattened for burrowing. The two parts, or valves, are usually similar in shape and size and are hinged on the animal's dorsal side (Fig. 41.32). The animal opens and shuts them by means of large muscles. Among the more common bivalves are clams, oysters, scallops, cockles, file shells, and mussels. Most lead rather sedentary lives as adults.

[3] *Nautilus,* a modern form with a well-developed shell, is an exception.

41.32 A scallop (*Chlamys opercularis*), a representative bivalve The shell is composed of two hinged valves. Note the numerous small eyes around the edges of the mantle.

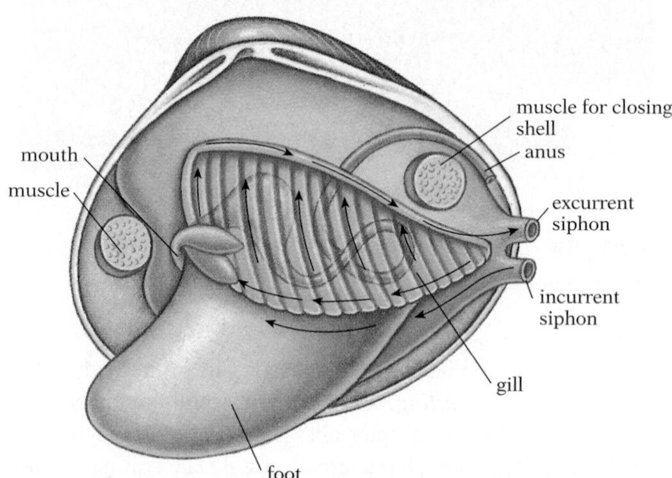

41.33 Filter feeding in a clam Water carrying minute food particles is drawn in through the incurrent siphon by ciliary action. Streams of mucus flowing over the gills trap the food particles and conduct them to the mouth. The water for gas exchange moves through minute openings into the interior of the gills, and out the excurrent siphon. Vertical arrows show the movement of water through the gills, lower horizontal arrows the movement of food particles over the gills and into the mouth.

The bivalves, which have no radula, are usually filter feeders, straining tiny food particles from the water flowing across their gills (Fig. 41.33). The gills thus function both in nutrient procurement and gas exchange.

Key innovations of the Bivalvia
Shell with two valves
Laterally compressed body
Filter feeding; no radula

The Annelids (Phylum Annelida)

The phylum Annelida (the term means "ringed") comprises the segmented worms, animals whose most characteristic structural feature is the ***segmentation*** of their bodies, both external and internal. Segmentation refers to the division of the body into a series of linearly arranged segments. All annelids show external segmentation, and many annelids have internal partitions between the segments as well. Many of the internal organs—those derived from mesoderm such as muscles, blood vessels, and excretory organs—are also segmented. In annelids segmentation probably evolved as an adaptation for burrowing and crawling—the muscles of each segment could be controlled independently, making possible localized movements (see Figure 30.3, p. 651). Other important annelidan characteristics are a closed circulatory system, a pair of excretory nephridia in each segment, a ventral nerve cord with ganglia in each segment, and a well-developed coelom that is used as a hydrostatic skeleton (see p. 466).

Most zoologists support the view that annelids and molluscs arose from common ancestral stock. It seems likely that the two groups diverged from a common ancestor, most likely a marine acoelomate worm, prior to the appearance of either molluscan or annelidan characteristics, especially segmentation.

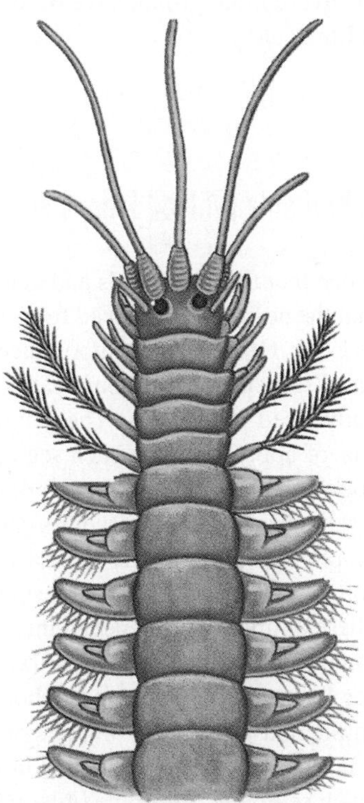

41.34 Head and anterior parapodia-bearing segments of a polychaete worm

The Members of the Class Polychaeta Have Parapodia With Many Setae

Polychaete worms are marine annelids with a well-defined head bearing eyes and antennae. Each of the numerous body segments usually bears a pair of lateral appendages called ***parapodia,*** which function in both locomotion and gas exchange (Fig. 41.34). There are numerous stiff setae (bristles) on the parapodia (the term Polychaeta means "many chaetae," i.e., many setae).

Some polychaetes swim or crawl about actively; others are more sedentary, usually living in tubes they construct in the mud or sand of the ocean bottom. These tubes may be simple, mucus-lined burrows, membranous structures, or elaborately constructed dwellings composed of sand grains cemented together. Many of the tube dwellers are beautiful animals, often colored bright red, pink, or green; some are iridescent. Among the most beautiful are the fanworms and peacock worms, which have a crown of colorful, highly branched fanlike or featherlike processes that they wave in the water at the entrance to their tubes (Fig. 41.35).

> *Key innovations of the Annelida*
> Segmentation, both internal and external
> Closed circulatory system

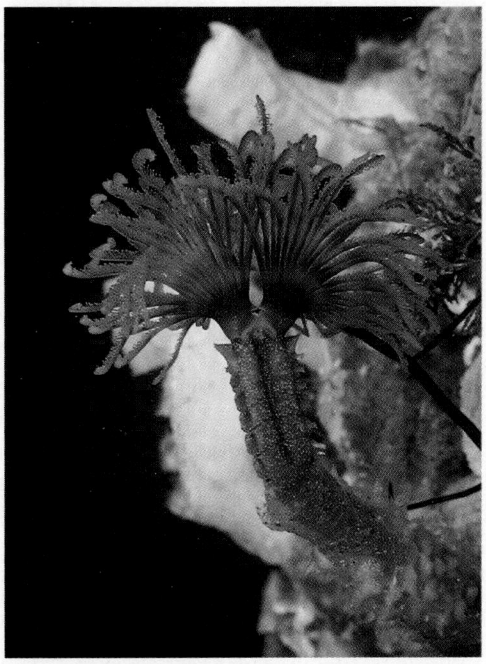

41.35 A fanworm, *Sabella crassicornis* The processes on this "feather-duster worm" help bring food and gases to the organism.

The Class Oligochaeta Contains the Earthworms and Many Freshwater Species

The oligochaete worms differ from polychaetes in that they lack a well-developed head and parapodia, and have fewer setae (the term Oligochaeta means "few setae"). These worms have a special structure, the ***clitellum,*** which produces a cocoon in which the embryonic development takes place. There is no ciliated larval stage in this group. We have described most of the important characteristics of earthworms in earlier chapters.[4]

The Leeches (Class Hirudinea) Are Highly Specialized Annelids

Probably the most familiar members of the Hirudinea are the bloodsuckers, which attack a variety of vertebrate and invertebrate hosts (Fig. 41.36). When a leech of this type attacks a host, it attaches itself by its posterior sucker, applies the anterior sucker very tightly to the skin, and either painlessly slits the skin with small bladelike jaws or dissolves an opening by means of enzymes. It then secretes a substance into the wound that prevents coagulation of the blood, and begins to suck the blood, usually consuming an enormous quantity at one feeding and then not feeding again for a fairly long time. Some leeches are predatory, capturing invertebrate prey such as worms, snails, and insect larvae and swallowing them whole.

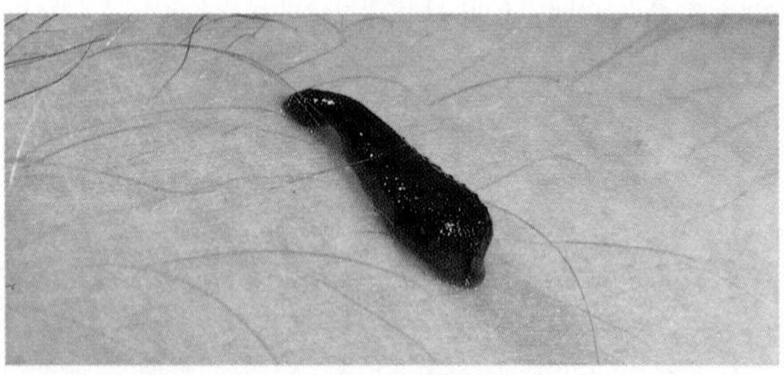
A

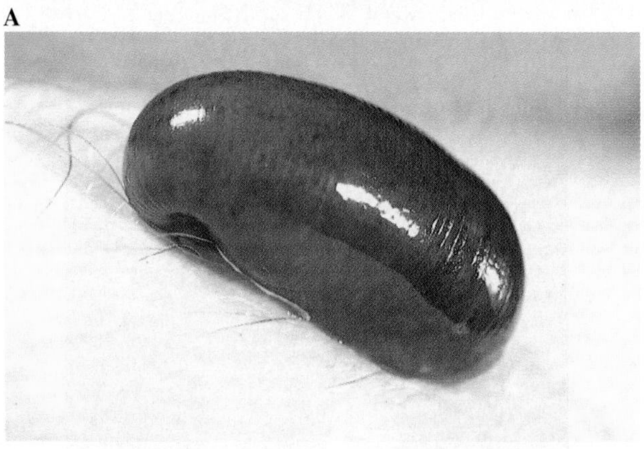
B

41.36 A terrestrial leech of Australia, sucking blood from a human arm Top: The leech has just begun feeding. Bottom: Somewhat later its body has become engorged with blood.

The hirudineans share many characteristics with the oligochaetes and are thought to have evolved from them. Leeches, however, lack setae, show almost no external segmentation, and the internal segmentation is reduced. Their first and last segments are modified to form suckers. The suckers may be used for locomotion, propelling the animal inchworm fashion, or for attaching to prey.

The Arthropods (Phylum Arthropoda)

The phylum Arthropoda is by far the largest of the phyla. Nearly a million species have been described, and there are doubtless hundreds of thousands more yet to be discovered. Probably more than 80 percent of all animal species on earth belong to this phylum. Certain classes of this phylum, more than any other invertebrate group, have become fully adapted to a terrestrial life cycle.

Arthropods Probably Evolved From a Polychaete Worm

There is evidence that the arthropods and annelids are closely related, and that the arthropods evolved from a segmented marine worm ancestor. Of particular importance is the fact that both phyla show segmentation, although it has been reduced in many of the advanced arthropods. The primitive arthropods also have a pair of appendages on each segment, just as each segment of a polychaete worm bears a pair of parapodia. They also share the same basic type of nervous system.

One small phylum of animals, the Onychophora, helps clarify the evolutionary relationship between the annelids and arthropods. These animals are of special interest because they have a combination of annelid and arthropod characters. Looking rather like caterpillars, onychophorans have a segmented, wormlike body with 14 to 43 pairs of short, unjointed legs (Fig. 41.37). They resemble annelids in having a pair of nephridia in each segment, a thin, flexible, permeable cuticle (though the cuticle is made of chitin, like that of arthropods), and unjointed appendages. In other characteristics, however,

[4] Recall that earthworms have a complete digestive tract (see Fig. 17.15, p. 357), a closed circulatory system (see Fig. 20.2, p. 409), and nephridia for the excretion of nitrogenous wastes (see Fig. 22.13, p. 466). Their nervous system shows a high degree of centralization (see Fig. 28.9, p. 594). Earthworms are hermaphroditic; that is, they have both male and female sex organs. During mating, two worms lie next to one another (see Fig. 26.2, p. 532) and cross-fertilization occurs.

41.37 An onychophoran Onychophorans have a mixture of annelid and arthropod characteristics. Note the number of short, unjointed legs and the prominent antennae.

they are more like arthropods. (They have claws, antennae, an open circulatory system, and appendages modified for feeding.) The onychophorans are regarded as an early evolutionary offshoot from the line leading to the arthropods from an ancient annelidlike ancestor (Fig. 41.38).

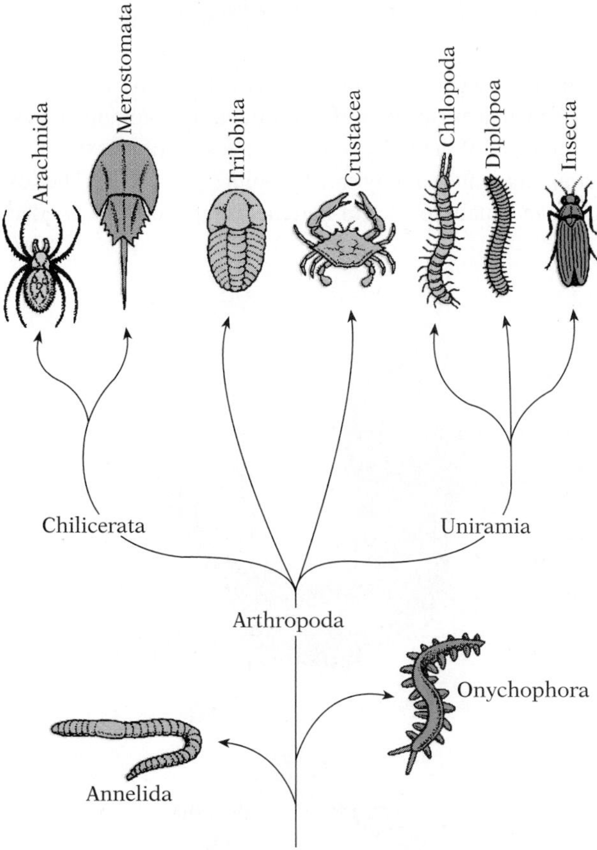

41.38 Diagram of possible relationships between the phylum Annelida, the phylum Onychophora, and the major subphyla and classes of the phylum Arthropoda Members of Trilobita are extinct, as are many unnamed groups.

Arthropods Have Many Characteristics That Adapt Them for Life on Land

Arthropods are characterized by a jointed, chitinous exoskeleton and jointed legs. The exoskeleton, which is secreted by the epidermis, functions both as a point of attachment for muscles and as a protective armor, but it imposes limitations on growth and must be periodically molted if the animal is to increase much in size (Fig. 41.39). The jointed exoskeleton was a key innovation in arthropod evolution. Not only did it provide support and protection for an animal living in an aquatic medium, but it also made possible the movement of the early arthropods onto land, a process that was well underway by 415 million years ago.

Together with their elaborate exoskeleton, arthropods have evolved a complex musculature quite unlike that of most other invertebrates. Most of the muscles are striated and are oriented in many different directions, making possible an extensive repertoire of movements.

The nervous system is very well developed. Similar in organization to the annelid nervous system, it consists of a dorsal brain and a ventral double nerve cord containing ganglia. Sense organs, including compound eyes, are many and varied in many arthropods.

As we have seen, arthropods have an open circulatory system (see Fig. 20.3, p. 409). There is usually an elongate dorsal blood vessel called the heart, which pumps the blood forward into arteries; from there the blood goes into open sinuses, where it bathes the tissues directly. Eventually the blood returns to the posterior portion of the heart.

41.39 A molting centipede Its new yellow-orange exoskeleton glistening, the animal is backing out of its old exoskeleton.

In most aquatic arthropods, excretion of nitrogenous wastes (primarily ammonia) is principally by way of the gills. However, the excretory organs in most groups of terrestrial arthropods are Malpighian tubules (see Fig. 22.14, p. 467), and the excretory product is uric acid. Because this tubule system is efficient at reabsorbing water, it is an excellent adaptation for conserving water and an important adaptation for life on land.

In almost all arthropods the sexes are separate. Fertilization is internal in all terrestrial and in most aquatic forms—an advantage that both conserves gametes and prevents their desiccation. Along with the exoskeleton and Malpighian tubules, internal fertilization helped make life on land possible. In most cases the females lay their fertilized eggs in the external environment and the eggs hatch into some sort of immature or larva stage that molts many times to form the adult.

Several Evolutionary Trends Are Apparent in the Arthropods

The arthropod body plan may be viewed as an elaboration and specialization of the segmented body of a presumed annelid ancestor. The evidence indicates that the first arthropods had long, wormlike bodies composed of many nearly identical segments, each bearing a pair of legs. All the legs were alike. Among the host of modifications of this ancestral body plan that have arisen in the various groups of arthropods during the millions of years of their evolution, four tendencies stand out: (1) reduction in the total number of segments; (2) grouping of segments into distinct body regions, such as a head and a trunk, or a head, a thorax, and an abdomen; (3) increasing cephalization, i.e., incorporation of more segments into the head and concentration of nervous control and sensory perception in or just behind the head; and (4) specialization and differentiation of the legs of some segments for a variety of functions other than locomotion, and complete loss of legs from many other segments. The ancestral arthropod legs consisted of two branches (rami), i.e., they were ***biramous*** in structure (Fig. 41.40). The outer branch typically had a gill for gas exchange whereas the inner branch, the walking leg, functioned in locomotion. Many arthropods show the ancestral condition (biramous legs), but others have become more specialized and have lost one of the two branches, and are thus ***uniramous*** (one-branched).

Key innovations of the Arthropoda
- Chitinous exoskeleton and jointed, biramous legs
- Complex musculature
- Reduction in number of segments
- Grouping of segments into body regions
- Loss or specialization of legs for different functions
- Well-developed nervous system with cephalization
- Open circulatory system
- Internal fertilization

The arthropods have been divided into four different subphyla: Trilobita, Chelicerata, Crustacea, and the Uniramia (Table 41.2). These groups are distinguished from each other primarily on the basis of their different appendages.

The Trilobites Are the Most Primitive Arthropods

Arthropods were very abundant in the Paleozoic seas, and fossils from that era are plentiful. Particularly common in rocks of the first half of the Paleozoic are the fossils of an extinct group belonging to the subphylum Trilobita (Fig. 41.41). The fossils show an usually oval and flattened shape and are divided by

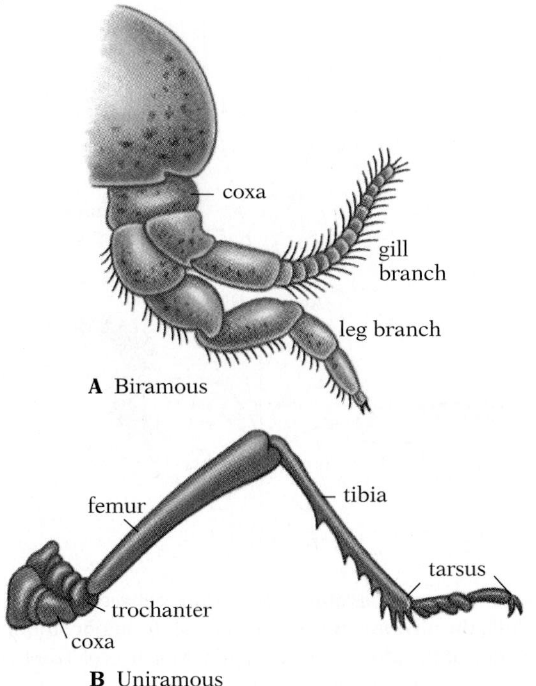

41.40 A comparison of the biramous and uniramous appendages (A) The ancestral arthropod appendage is biramous—divided into two branches. The outer branch is often modified as a gill and functions in gas exchange, while the inner branch, the walking leg, functions in locomotion. (B) In many arthropods, including insects, the legs have become specialized and one of the two branches is lost. These one-branched legs are called uniramous appendages.

TABLE 41.2 *A summary of some important characteristics of the arthropod subphyla*

Subphylum	*Habitat*	*Antennae*	*Mouthparts*	*Appendages*
Trilobita	Marine	One pair	Used bases of anterior legs	Pair of biramous legs per segment
Chelicerata	Land	None	Chelicerae and pedipalps	Four pairs of uniramous legs
Crustacea	Aquatic, mostly marine	Two pairs	Many mouth parts, mandibles, two pairs of maxillae	Many pairs of biramous legs
Uniramia	Land	One pair	Mandibles, maxillae, labium	Uniramous legs; many, or three pairs in insects

two longitudinal furrows that partition the animal's body into three longitudinal lobes, from which they get their name (*tri-*, "three," and *lobos,* "lobe"). The body consisted of a ***head,*** which bore a pair of slender antennae and, often, compound eyes; a ***thorax,*** which consisted of a variable number of separate segments; and an ***abdomen,*** which was composed of several fused segments. Trilobites, though they were surely different from the first arthropods, exhibiting specializations of their own (e.g., the longitudinal furrows and the fusion of the abdominal segments), nevertheless seem to approach the hypothetical arthropod ancestor more closely than any other known group. One ancestral character stands out—the lack of differentiation of the legs. The fossils show that every segment bore a pair of biramous legs, and that all these legs, including those of the four head segments, were nearly identical. There were no appendages specialized as mouthparts. This feature is true only of the subphylum Trilobita; in all three of the other arthropod subphyla—Chelicerata, Crustacea, and Uniramia—the appendages of the most anterior segments have been modified as mouthparts and no longer function in locomotion.

Chelicerates Have Mouthparts Called Chelicerae

The Chelicerata includes a great many animals familiar to you—spiders, ticks, scorpions, mites, and horseshoe crabs (Fig. 41.42). (The horseshoe crabs are not really crabs at all but living relicts of an ancient chelicerate class, most members of which have been extinct for millions of years.) Most chelicerates are terrestrial organisms, and have evolved many characteristics that allow them to survive on land—a tracheal

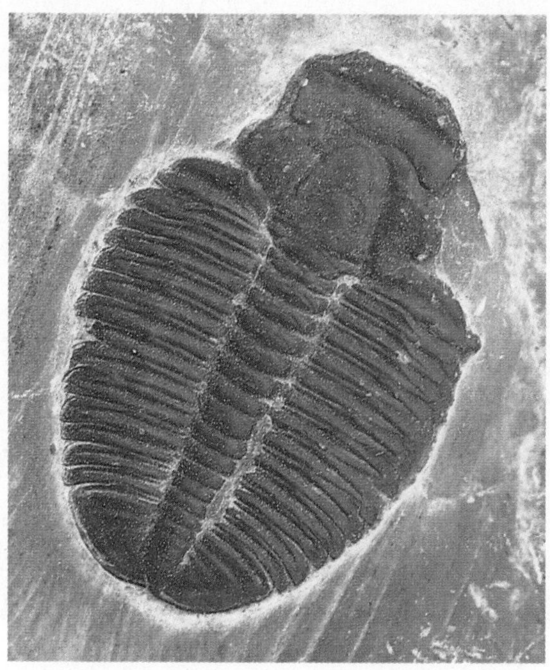

41.41 A fossil of a trilobite Note the two longitudinal furrows that partition the animal's body into a median lobe and two lateral lobes. It is this tripartite arrangement that suggested the name "Trilobita."

41.42 A group of spawning horseshoe crabs The cephalothorax is covered by the horseshoe-shaped carapace. Compound eyes are located on the outer side of the lateral ridges on the carapace, but there are no antennae. The chelicerae, pedipalps, and four pairs of walking legs are located on the ventral surface.

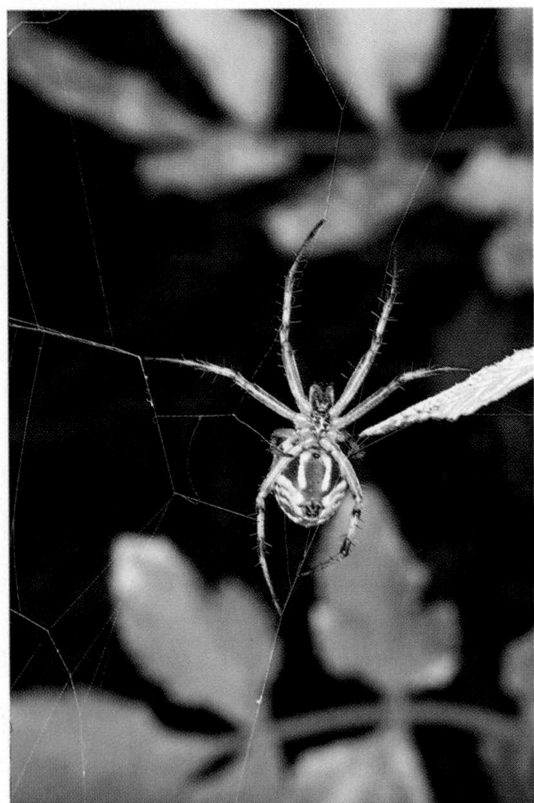

41.43 A spider, *Araneus marmoreus* The spiders, like most members of the class Arachnida, have two body regions: a cephalothorax and an abdomen, and six pairs of appendages: a pair of chelicerae, a pair of pedipalps, and four pairs of walking legs. There are often simple eyes on the cephalothorax but never any compound eyes or antennae. Note how the animal places its first pair of legs on strands of its web; it can detect vibrations when prey touch the web, and can often even distinguish, by the type of vibration, what sort of prey it is.

system, Malpighian tubules, internal fertilization, and a waterproof covering over their exoskeleton. The chelicerates derive their name from the modification of the first pair of legs in ancestral arthropods to form special mouthparts called ***chelicerae,*** which may be either pincerlike or fanglike. In addition to functioning as mouthparts, the chelicerae may also function in manipulating prey, or they may be modified as poison fangs, as in spiders (Fig. 41.43). Another unique characteristic of the chelicerates is the division of the body into two regions: a ***cephalothorax*** and an ***abdomen.*** The cephalothorax usually bears five pairs of uniramous appendages besides the chelicerae; in some groups these are all ***walking legs,*** while in others only the last four pairs are legs, the first pair being modified as feeding devices called ***pedipalps,*** which are often much longer than the chelicerae (Fig. 41.43). In most groups prey is seized and torn apart by the pedipalps. The legs of the abdominal segments have been either lost or modified into respiratory or sexual structures.

Key innovations of the Chelicerata

Chelicerae as mouthparts

Two body regions

Six pairs of legs (chelicerae, pedipalps, and walking legs)

No antennae

Crustaceans Are Mainly Aquatic Organisms With Modified Segmented Body Plans

Some representatives of this subphylum, such as crayfish, lobsters, shrimps, and crabs (Fig. 41.44), are well known to most people. But there are many other species of Crustacea that bear

A

B

41.44 A crab and a prawn, representatives of the larger, better-known crustaceans (A) The crab *Neolithodes grimaldii.* (B) A freshwater prawn.

A

B

41.45 Some representatives of the smaller crustaceans (A) A marine amphipod, *Gammarus.* (B) A freshwater crustacean, *Asellus aquaticus.*

little superficial resemblance to these familiar animals; among them are fairy shrimps, water fleas, brine shrimps, sand hoppers, barnacles, and sow bugs (Fig. 41.45). Most crustaceans are aquatic, mostly marine, though some (e.g., sow bugs) are terrestrial. The crustaceans are sometimes referred to as "the insects of the sea."

The members of this subphylum and the uniramians differ from chelicerates in having ***mandibles*** instead of chelicerae as their first pair of mouthparts (Fig. 41.46). Mandibles are modified from ancestral legs, and usually function in chewing. They are never clawlike or pincerlike, as chelicerae often are. There are also two additional pairs of mouthparts called ***maxillae,*** which are used in feeding and manipulating food.

Crustacea characteristically have two pairs of antennae, a pair of mandibles, and two pairs of maxillae. But their other appendages vary greatly from group to group, and whatever could be said about those of one group, such as crayfish and lobsters, would have little relevance to those of other groups. In fact, the Crustacea are an enormously diverse assemblage of animals that can hardly be characterized in any simple way. Some have a cephalothorax and an abdomen; others have a head and a trunk, or a head, a thorax, and an abdomen, or even a unified body. Most are active swimmers, but some, like barnacles, secrete a shell and are sedentary (Fig. 41.47). The majority are marine, but there are many freshwater species, and a few, such as sow bugs, are terrestrial and have a simple tracheal system. We could go on listing divergences, but the point has been made that this group shows amazing diversity. This is a subphylum in which the basic arrangement of a segmented body with numerous jointed appendages has been modified

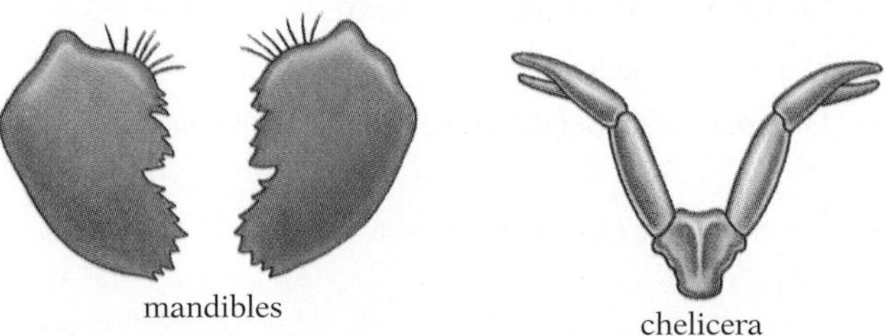

41.46 A comparison of chelicerae and mandibles Both chelicerae and mandibles are appendages modified to serve as mouthparts, but chelicerae are generally pincerlike or clawlike whereas the mandibles are thickened jawlike appendages that function in biting and chewing.

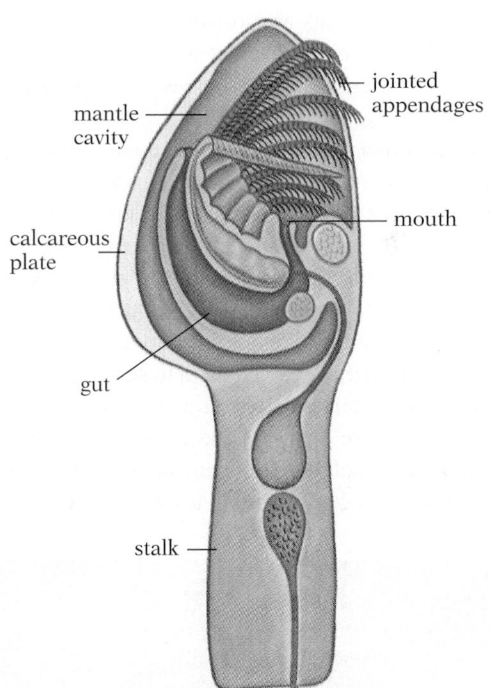

41.47 Interpretive drawing of a goose barnacle Barnacles were thought for many years to be molluscs because they have a mantle and external plates that superficially resemble shells. But the body is divided into a head and thorax, and there are mandibles and other mouthparts and jointed appendages on the thorax. The jointed appendages, called cirri, sweep through the water and trap small food particles. Most species lack the long stalk so prominent in the goose barnacles.

and exploited in countless ways as the members of the subphylum have diverged into different habitats and adopted different modes of life.

Key Innovations of the Crustacea

Two body regions

Mandibles and two pairs of maxillae

Two pairs of antennae

One pair of appendages per segment, primarily biramous

Compound eyes usually present

Gills

The Organisms Belonging to the Subphylum Uniramia Evolved on Land

The subphylum Uniramia includes the insects, centipedes, and millipedes (Fig. 41.48). This group is thought to have evolved independently of the chelicerates and crustaceans. As you might guess, uniramians are so-named because of their one-branched or uniramous appendages (Fig. 41.40B). Another distinctive characteristic is their possession of but a single pair of antennae.

Unlike the other subphyla, the uniramians evolved on land, and much of their evolution can be best understood in terms of adaptations for life on land. Three major problems associated with terrestrial life are (1) prevention of excessive water loss through evaporation, (2) support for the body when it is no longer buoyed up by water, and (3) reproduction in the absence of water. The uniramians, especially insects, have evolved a host of adaptations that enabled them to solve these problems. Their jointed exoskeleton, jointed appendages, and well-developed musculature supports the body against the pull of gravity and enables efficient movement on land. The problem of preventing water loss is met in a variety of ways. All uniramians have Malpighian tubules and use uric acid as their excretory product; consequently little water is lost in the process of excretion. Gases are exchanged internally in a tracheal system, where the humidity can be kept high without excessive water loss. And, in the insects, the exoskeleton is covered with a waxy, waterproof layer that helps prevent water loss across the body surface. Finally, all uniramians utilize some form of internal fertilization. In most species the females lay the fertilized eggs in the environment with some sort of protective outer covering, but in some species the female retains the fertilized eggs within her own body until hatching occurs. Notice that many of these characteristics (tracheal system, Malpighian tubules, waxy covering) are also found in the terrestrial chelicerates (spiders and relatives), but the shared features are thought to have originated independently.

41.48 A millipede The millipede body is divided into a head and a trunk. Each segment (except a few at the front and head) bears two pairs of legs.

Key innovations of the Uniramia

Unbranched (uniramous) appendages

Single pair of antennae

Tracheal system

Malpighian tubules

Mandibles

Uric acid as excretory product

Internal fertilization

The Insects (Class Insecta) Are Diverse and Highly Successful

The insects have undergone extensive adaptive radiation to form an enormous group of diverse animals that occupy almost every conceivable habitat on land and in fresh water. If numbers are the criterion by which to judge biological success, then the insects are the most successful group of animals that has ever lived; there are more species of insects than of all other animal groups combined. But there is one restriction on their dominant role: they do not occur in the sea (though a few species walk on the ocean surface or live in brackish water); the role played by insects on land is played in the sea by crustaceans.

There are a few insect fossils from the Devonian, but it was in the Carboniferous and Permian periods that insects took their place as one of the dominant groups of animals. By the end of the Paleozoic era, many of the modern orders had appeared, and the number of species was enormous. A second great period of evolutionary radiation began in the Cretaceous and continues to the present time; this second radition is correlated with the rise of flowering plants.

The insects are like the molluscs in that they evolved a

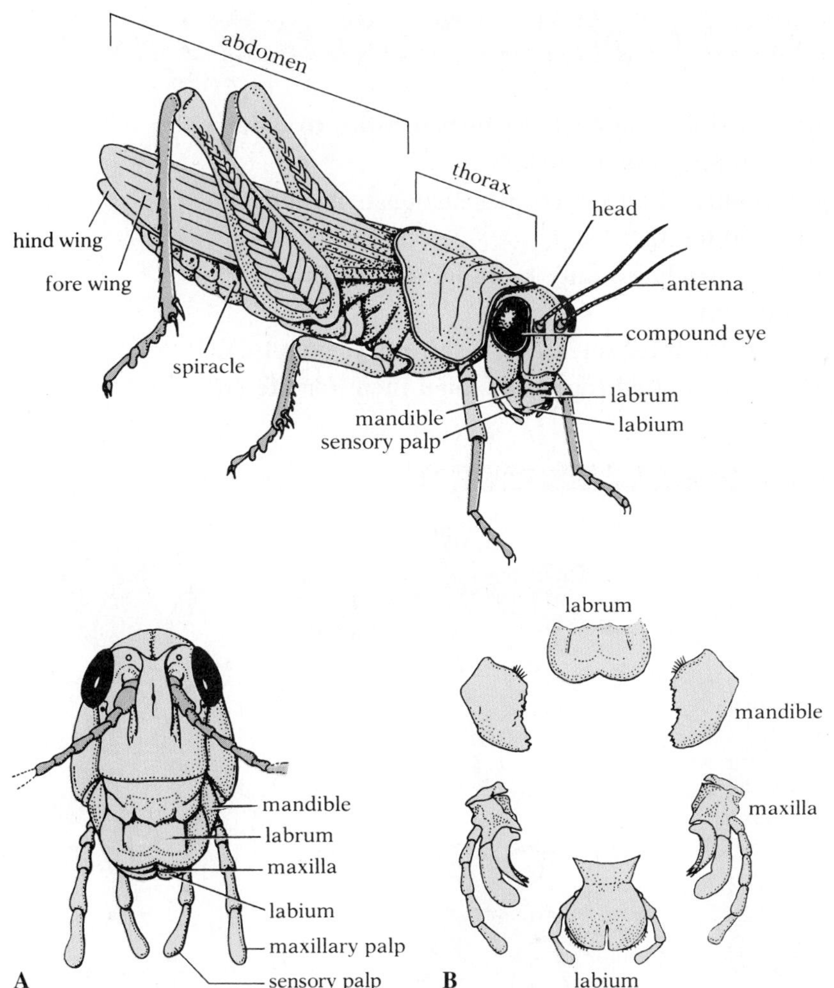

41.49 A grasshopper Spiracles are openings to the tracheal system.

41.50 The mouthparts of a grasshopper (A) Front view of head, with mouthparts in place. (B) The mouthparts removed from the head, but kept in their proper relative positions. The mandibles and probably the labrum (upper lip) are derived from the basal segments of ancestral legs; all the other segments of those legs have been lost. The maxillae and labium (lower lip) retain more of the segments of the legs from which they are derived; the basal segments are enlarged, but the distal segments form slender leglike structures called palps, which bear many sensory receptors.

basic body plan that could be easily modified to adapt to different environments. The insect body is divided into three regions: a head, a thorax, and an abdomen (Fig. 41.49). The head bears numerous sensory receptors, usually including compound eyes, one pair of sensory antennae, and three pairs of mouthparts derived from ancestral legs. The mouthparts include a pair of mandibles, a pair of maxillae, and a lower lip, or ***labium,*** formed by fusion of the two second maxillae (Fig. 41.50). The upper lip, or ***labrum,*** which has not traditionally been classified as a mouthpart, may also be derived from ancestral legs.

The thorax bears three pairs of uniramous walking legs and, in most groups, one or two pairs of wings. The insects are vigorous fliers and are the only invertebrates capable of flight (Fig. 41.51); they were the first animals that evolved the abil-

41.51 A desert locust in flight Like most flying insects except the flies, the locust—also called the grasshopper—has two pairs of wings. The front wings are leathery and, when the animal is at rest, provide a protective covering for the fragile pleated rear wings, which are the ones important for flight.

MAKING CONNECTIONS

1. How have biologists used developmental characteristics to determine evolutionary relationships among animal phyla?

2. How do cephalopod molluscs and vertebrates demonstrate convergent evolution, as described in Chapter 32?

3. What polysaccharide is produced only by fungi and arthropods? How is this polysaccharide significant?

4. What characteristics of the integumentary, respiratory, excretory, and reproductive systems of uniramian arthropods adapted them for life on land?

ity to fly. The evolution of flight gave the insects a distinct selective advantage; it made possible an entirely new way of life and led to an enormous adaptive radiation.

Key innovations of the Insecta
Three body regions
Three pairs of (uniramous) legs
One or two pairs of wings
Adaptations for flight
Compound eyes

We have already discussed the insects at considerable length in other chapters.[5] They are divided into approximately 26 orders, which are listed in the Appendix. Representatives of some of the more familiar orders are shown in Figure 41.52.

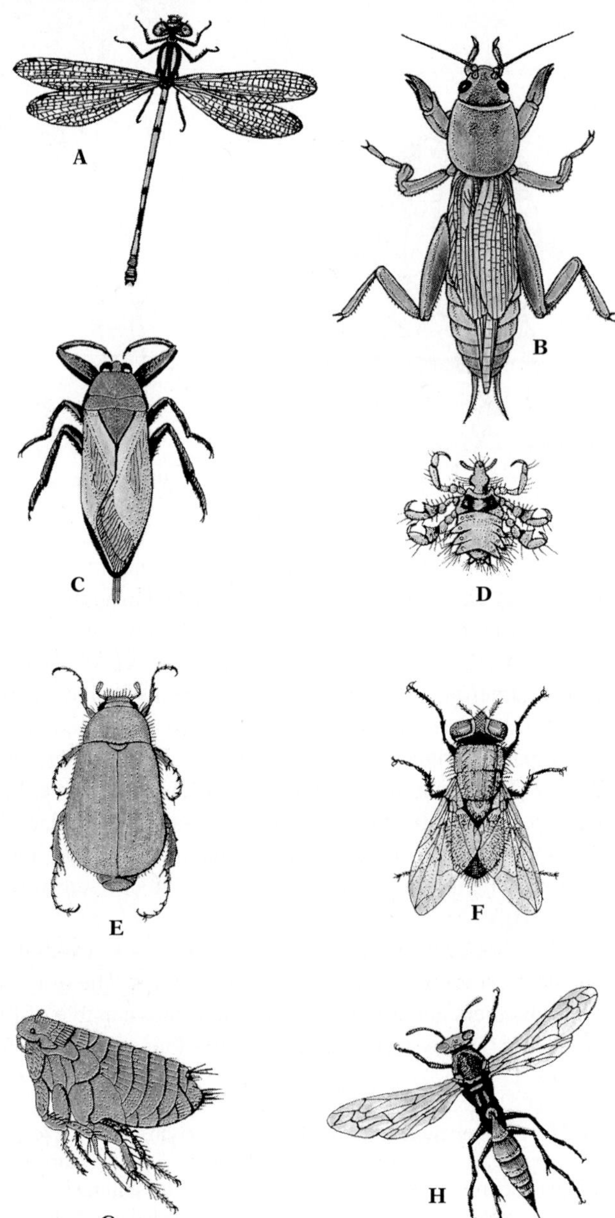

41.52 Some representatives of the major insect orders (A) Damselfly (order Odonata). (B) Mole-cricket (Orthoptera). (C) Bug (Hemiptera). (D) Louse (Anoplura). (E) Beetle (Coleoptera). (F) Fly (Diptera). (G) Flea (Siphonaptera). (H) Wasp (Hymenoptera). Not to scale.

[5] Recall that insects have an open circulation system that transports nutrients and wastes (see p. 409 and Fig. 20.3) and a tracheal system for gas exchange (see p. 379 and Fig. 18.11). Malpighian tubules are used for excretion of nitrogenous wastes (see p. 467 and Fig. 22.14); the wastes are eliminated as very dry material, an adaptation that conserves water. The nervous system is well developed; there is a high degree of centralization and a variety of sensitive sensory organs (see pp. 622–623 and Figs. 29.9 and 29.10; Fig. 29.11, p. 624). Like most land animals, insects have internal fertilization (see Fig. 26.4, p. 533) and lay eggs with a protective outer covering that minimizes water loss. The young go through metamorphosis, a series of developmental stages that convert the immature animal into an adult form (see p. 570 and Fig. 27.20).

CONCEPTS IN BRIEF

1. The sponges (*Porifera*) differ greatly from other multicellular animals and are believed to have evolved on a line of their own. They are multicellular animals at a primitive, cellular level of organization. The flagellated collar cells lining the interior are remarkably similar to protozoan collared flagellates.

2. The *Cnidaria* are radially symmetrical animals whose bodies have definite tissue layers but no distinct internal organs. They have a gastrovascular cavity. Their bodies consist of two well-developed tissue layers, the ectoderm and the endoderm, with a third layer, the mesolamella, between them. All have cnidocytes with nematocysts.

3. Many cnidarians have a life cycle in which a sedentary *polyp* stage alternates with a free-swimming jellyfishlike *medusa* stage, which produces the gametes. The zygote develops into a ciliated *planula* larva, which gives rise to the polyp.

4. The Platyhelminthes and Nemertea are regarded as the most primitive bilaterally symmetrical animals. Their bodies are composed of three well-developed tissue layers; there is no coelom.

5. The *Platyhelminthes,* or flatworms, are flattened, elongate animals with (usually) a gastrovascular cavity. The flatworms have advanced to the organ level of construction. The phylum is divided into three classes: Turbellaria, Trematoda, and Cestoda; the latter two are entirely parasitic.

6. Members of the phylum *Nemertea* have evolved a *complete digestive tract* and a simple circulatory system.

7. The fate of the embryonic blastopore differs in various animal groups. In the *Protostomia* the blastopore becomes the mouth, and a new anus develops; in the *Deuterostomia* the situation is reversed. The protostomes and deuterostomes also differ in the determinateness and pattern of the initial cleavages, the mode of origin of the mesoderm and coelom, and the type of larva.

8. A true coelom is a body cavity enclosed entirely by mesoderm. In several protostome phyla the body cavity is a *pseudocoelom;* it is partly bounded by ectoderm and endoderm. The *Nemata* are small, pseudocoelomate, wormlike animals without a definite head; they have a complete digestive tract.

9. Other protostome phyla (the coelomate Protostomia) possess true coeloms, which usually develop as a split in the mesoderm. All have a complete digestive tract, and most have well-developed circulatory, excretory, and nervous systems.

10. The *Mollusca* are soft-bodied animals with a basic body plan composed of a *head region,* ventral muscular *foot,* a *visceral mass,* and a *mantle,* which covers the visceral mass and usually contains glands that secrete a shell. The mantle often overhangs the visceral mass, thus enclosing a *mantle cavity,* which frequently contains gills. Molluscs have an open circulatory system.

11. The *Annelida* are the segmented worms. The polychaetes are marine annelids with a head bearing eyes and antennae. Each body segment bears a pair of *parapodia,* which function in movement and gas exchange. The Oligochaeta (earthworms) lack a well-developed head and parapodia. The Hirudinea (leeches) are the most specialized of the annelids and show little internal segmentation.

12. The *Arthropoda* are characterized by a jointed chitinous exoskeleton and jointed legs. They have elaborate musculature, a well-developed nervous system and sense organs, and an open circulatory system. Among evolutionary trends seen in the arthropods are reduction in the number of segments; grouping of segments into body regions; increasing cephalization; and specialization of some legs for functions other than movement, with loss of legs from other segments.

13. The major subphyla of the Arthropoda are the *Trilobita* (the extinct trilobites), *Chelicerata* (including arachnids, such as spiders), the *Crustacea* (including crabs, shrimps, barnacles, sow bugs), and the *Uniramia* (including centipedes, millipedes, and insects). These groups are distinguished from each other primarily on the basis of their different appendages.

14. The Uniramia evolved on land and show many adaptations for life on land, including a tracheal system, excretion by Malpighian tubules, waterproof exoskeleton, and internal fertilization.

15. The insects have undergone extensive adaptive radiation to form an enormous group of diverse animals that occupy almost every conceivable habitat on land and in fresh water. They are the first animals to have evolved the ability to fly, which opened a new way of life, and enabled them to occupy different niches.

STUDY QUESTIONS

1. Describe the structure and method of feeding of sponges. Why are they put in a separate subkingdom from all other animals? (pp. 928–929)
2. What are the distinguishing features of the Cnidaria? How is their symmetry adapted to their method of feeding? (pp. 932–933)
3. Make diagrams of *acoelomate, pseudocoelomate,* and *coelomate* body types. What are the functions of the coelom (or pseudocoelom)? (pp. 932–933)
4. What type of symmetry and what level of organization do the flatworms (Platyhelminthes) display? What are the identifying features of the three classes of this phylum? Describe the life cycle of the blood fluke, *Schistosoma mansoni.* (p. 933)
5. Describe two evolutionary innovations found in the proboscis worms (Nemertea). (pp. 936–937)
6. List five important differences between the Protostomia and the Deuterostomia. (pp. 937–938)
7. You are studying two wormlike animals, both of which have a body cavity. One has no segmentation and lacks both a circulatory system and circular muscles in its body wall; the other is segmented and has a closed blood circulatory system. To which phylum does each belong? (pp. 934–940 and 945–946)
8. Describe the basic molluscan body plan and the ways it has been modified in gastropods, cephalopods, and bivalves. (pp. 941–944)
9. What evidence indicates that the arthropods probably evolved from a polychaete worm? Why are the characteristics of the onychophorans relevant in this regard? (pp. 946–947)
10. What characteristics of arthropods differentiate them from all other invertebrates? Describe four tendencies that are evident in the evolution of arthropods. (pp. 947–948)
11. Place arthropods with the following characteristics in the proper subphylum: (1) three longitudinal lobes and undifferentiated legs on every segment; (2) two pairs of antennae and mandibles as mouthparts; (3) no antennae and fanglike mouthparts; (4) one pair of antennae, mandibles, and unbranched appendages. (pp. 948–953)
12. What features of insects distinguish them from all other arthropods? Which of these characteristics is probably most responsible for their enormous evolutionary success? (pp. 952–954)

CHAPTER 42

THE ANIMAL KINGDOM: THE DEUTEROSTOMES

We have already seen that among the bilaterally symmetrical animals (the Bilateria) two major lineages evolved, the Protostomia and Deuterostomia (see Fig. 41.1, p. 928), and we have studied the differences between the two lineages. The deuterostome phyla mark the transition from the invertebrate animals to the vertebrates (animals that possess a backbone). We shall consider three deuterostome phyla: the Echinodermata, whose most familiar representatives, the sea stars, sea urchins, and sea cucumbers, are clearly invertebrate, the Chordata, which contains many invertebrates and all of the vertebrates, and the Hemichordata, whose members fall somewhere between the vertebrates and invertebrates. These three phyla are thought to have evolved separately from a common deuterostome ancestor (Fig. 42.1).

It may seem strange that the Echinodermata form the major phylum generally considered most closely related to our own phylum, the Chordata. After all, sea stars, sea urchins, and sea cucumbers do not look like animals with which one would expect to claim kinship. But as we saw earlier when we discussed the differences between the Protostomia and Deuterostomia (see Table 41.1, p. 938), certain characteristics seem to link echinoderms, hemichordates, and chordates and set them apart from all the protostome phyla. These characteristics include formation of the anus from the blastopore, radial and indeterminate cleavage, origin of the mesoderm as pouches, and formation of the coelom as the cavities in the mesodermal pouches.

THE ECHINODERMS (PHYLUM ECHINODERMATA)

The echinoderms include the sea stars, brittle stars, sea urchins, sand dollars, sea cucumbers, and sea lilies. They are mostly bottom-dwelling animals and are common in all seas and at all depths from the intertidal zone to the ocean depths. One obvious feature of the echinoderm adults is their radial symmetry (actually ***five-part*** or pentaradial symmetry). Most

Chapter opening photo: A dolphin in the Sea of Cortez off Mexico. Although dolphins superficially resemble fish, they are mammals and their fishlike appearance is an example of convergent evolution.

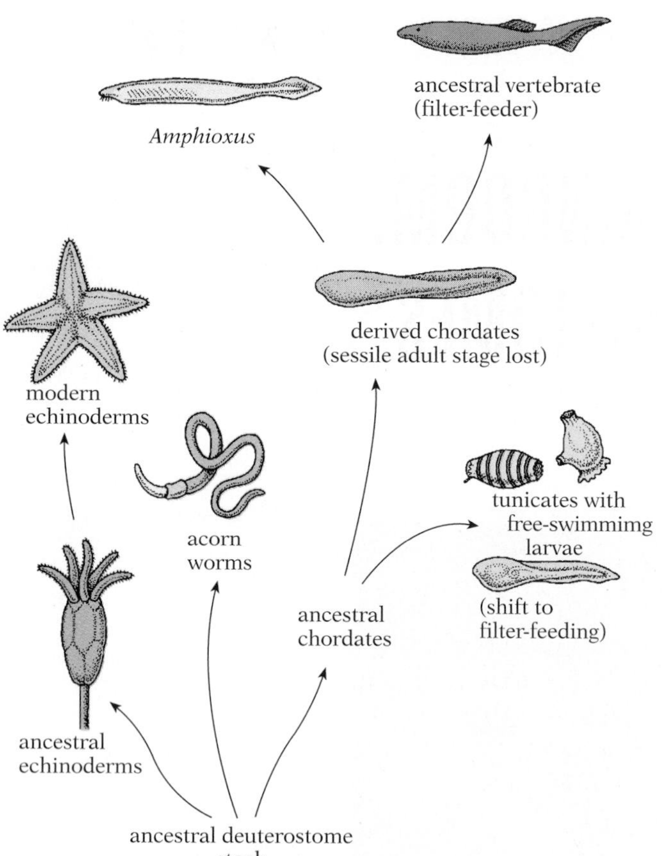

42.1 Deuterostome evolution There are at least three main lineages branching off from the hypothetical ancestral deuterostome: the echinoderms, the hemichordates, and the chordates. The hypothetical ancestor is thought to have been a filter feeder with a ring of ciliated tentacles surrounding the mouth.

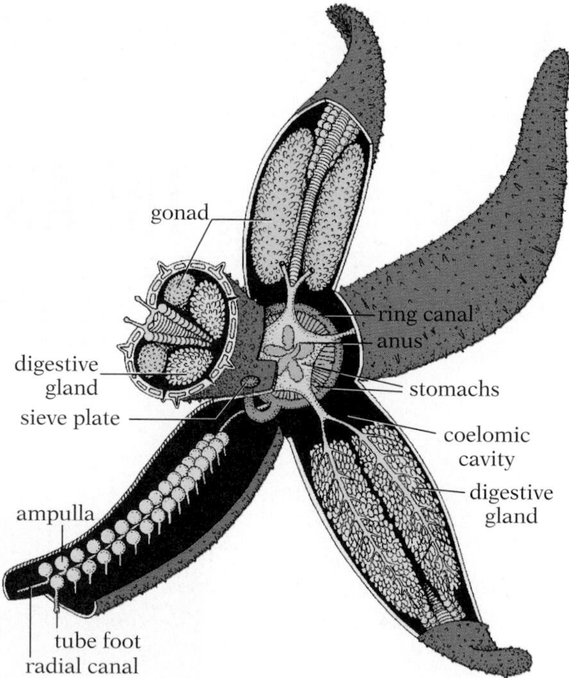

42.2 Dissection of a sea star (dorsal view) A characteristic unique to the echinoderms is the water-vascular system. Water enters through a sieve plate and enters a series of canals running along the base of each arm to the tube feet. The mouth is located on the undersurface. When the animal feeds, it everts part of its stomach and partially digested food is then taken into the dorsal portion of the stomach for further digestion. Each of the five arms has two rows of tube feet (blue) and a pair of gonads and digestive glands.

of their body parts are present in patterns of five and are arranged regularly around a central axis, rather like spokes on a wheel. Although the adults are radially symmetrical, the larval stage is bilaterally symmetrical, and it is generally thought that echinoderms evolved from ancient bilateral ancestors.

42.3 A sea urchin Sea urchins and their relatives have numerous movable spines on their hard, boxlike skeletons. In many species the spines contain a toxic chemical—if an unwary swimmer steps on an urchin the brittle spines penetrate the skin and break off, remaining embedded in the skin. The toxic substance acts as an irritant, and the wound can be quite painful.

Echinoderms Have a Unique Water-Vascular System

Almost all members of this phylum possess an internal skeleton composed of numerous calcium-containing plates embedded in the body wall. These plates may be separate, as in sea stars, or they may be fused to form a rigid, boxlike structure, as in sea urchins. The skeleton often bears many bumps or spines—they are particularly noticeable in sea urchins—that project from the surface of the animal (see Figs. 42.2 and 42.3). It is this characteristic that gives the animals the name "Echinodermata" (from the Greek *echino-*, "spiny," and *derma*, "skin").

A unique characteristic of echinoderms is their ***water-vascular system.*** This is a system of tubes (usually called canals) filled with fluid. Water enters the system through a sievelike plate on the surface of the animal (Fig. 42.2). A tube from this plate leads to a series of canals that run throughout the body. Many short side branches from the radial canals lead to hollow ***tube feet*** that project to the exterior. Each tube foot is a thin-walled, hollow cylinder, with a suction cup on its end. The animal can move about slowly by extending the tube feet and attaching the suction cups to a solid surface. By contracting the muscles in the walls of the tube feet, the feet are shortened and the animal is pulled forward. The tube feet also enable the animal to hold tightly to a rock or other object by applying suction; the sea star, which feeds on clams and oysters, uses this suction to pull open the molluscs' bivalve shells.

All Adult Echinoderms Have Five-Part Symmetry

The body of a sea star (starfish) illustrates many of the features of members of this phylum. It consists of a central ***disk*** and usually five rays, or ***arms,***[1] each with a groove bearing rows of tube feet running along the middle of its lower surface. The outer surface of the animal is studded with many short spines and numerous tiny skin gills, which are thin fingerlike projections of the body wall that protrude to the outside between the plates of the endoskeleton (Fig. 42.4).

The mouth is located in the center of the lower surface of the disk and the anus in the center of the upper surface. When the sea star feeds, it pushes part of its stomach out through the mouth, turning it inside out and placing it over food material, such as the soft body of a clam or oyster. The stomach secretes digestive enzymes onto the food, and digestion begins. The partly digested food is then taken into the rest of the digestive tract, where digestion is completed and the products are absorbed.

Though other members of the phyla, such as brittle stars (Fig. 42.5), basket stars (Fig. 42.5), sand dollars, sea urchins (Fig. 42.3), sea cucumbers (Fig. 42.6), and sea lilies, often

[1]Though most species of sea stars have five arms, some have more.

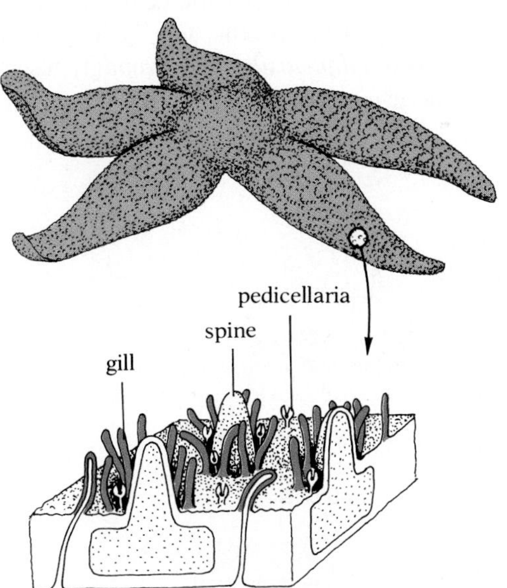

42.4 Sea star body wall The internal skeleton of the sea star is composed of plates of calcium carbonate, many of which have spines on them. The spines and pincerlike pedicellariae protect the tiny skin gills.

42.5 A brittle star (*Ophiopholis aculeata*) and a basket star (*Gorgonocephalus eucnemis*) Top: The body disk of the brittle star is relatively small, and the arms are long and slender. Bottom: The arms of the basket star branch repeatedly to produce a mass of coils resembling tentacles.

42.6 A sea cucumber (*Pseudocolochirus*) in the Coral Sea Notice the rows of tube feet. There are five of these rows, which correspond to the five arms of a sea star.

show little superficial resemblance to sea stars, their structure is fundamentally similar. For example, sea urchins and sea cucumbers lack the five arms of most sea stars, but they do have five bands of tube feet and thus show the same basic five-part symmetry.

> *Key innovations of the echinoderms*
> Water vascular system with tube feet
> Five-part symmetry
> Calcium-containing skeletal plates in skin

THE HEMICHORDATES (PHYLUM HEMICHORDATA)

The organisms belonging to the phylum Hemichordata lie somewhere between the vertebrates and invertebrates. The most common hemichordates are the acorn worms (Fig. 42.7), wormlike marine animals often found living in U-shaped burrows in sand or mud along coastlines. They are fairly large, ranging from nine to 43 centimeters in length, and their bodies consist of an anterior conical proboscis (thought by some to resemble an acorn—hence their name), a collar, and a long trunk (Fig. 42.8). A particularly important feature in the hemichordates is a series of openings, called ***pharyngeal slits,*** in the wall of the pharynx. Pharyngeal slits are found in all chordates at some stage in the life cycle. Many hemichordates use their pharyngeal slits for filter feeding; water carrying particulate food is drawn into the mouth and forced back into the pharynx and out through these slits. The food particles stick to the mucous coating the inside of the pharynx and are moved into the digestive system by cilia; the pharynx thus acts as a strainer or filter, separating food particles from water. Pharyngeal slits probably evolved primarily as a food-gathering device, but later assumed a gas-exchange function as well.

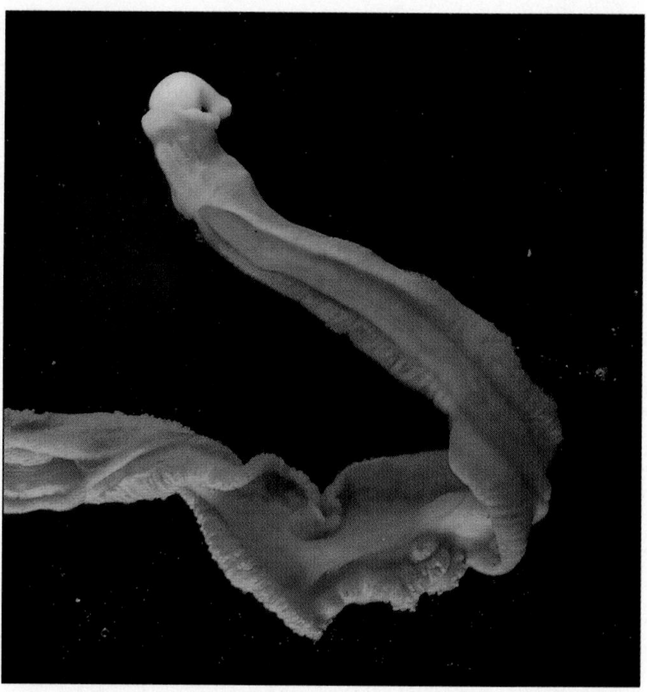

42.7 An acorn worm, *Balanoglossus* The short rounded proboscis is partially enveloped by the collar behind it. Only a part of the long trunk is shown.

Another important characteristic of hemichordates is the occurrence during development of a ciliated larval stage, sometimes called a ***dipleurula,*** that strikingly resembles the larvae of some echinoderms (Fig. 42.9).

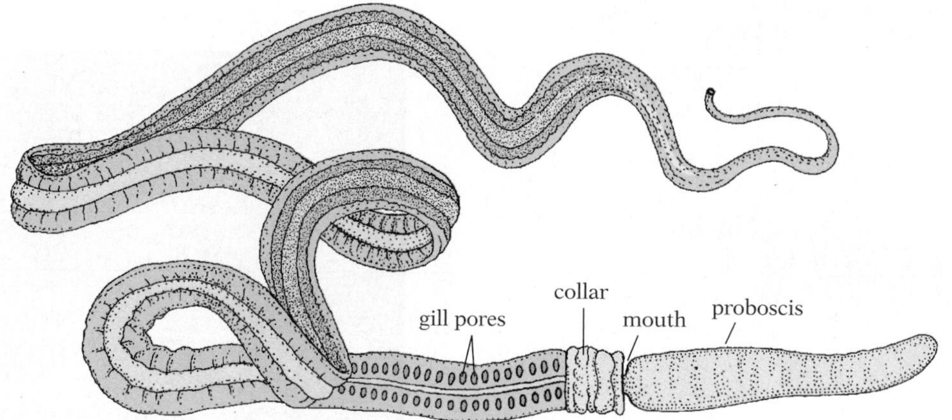

42.8 An adult acorn worm This particular genus (*Saccoglossus*) has a more elongated proboscis than *Balanoglossus,* shown in Figure 42.7.

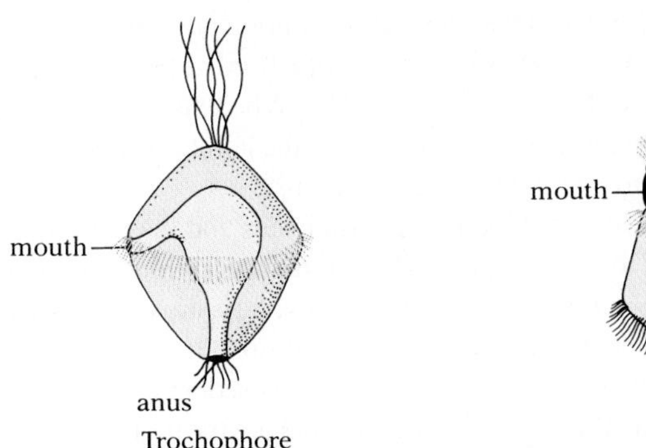

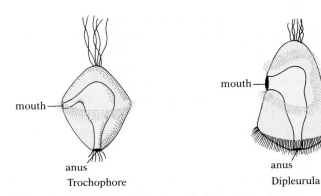

42.9 Trochophore and dipleurula larval types The band of cilia of the trochophore is located anterior to the mouth, whereas the corresponding band of the dipleurula encircles the mouth. The dipleurula-type larva is found only in the echinoderms and hemichordates, whereas the trochophore larva is characteristic of the protostomes and is found in some flatworms, the molluscs, and annelids.

> *Key innovations of the hemichordates*
> Bilateral wormlike body
> Pharyngeal slits
> Dipleurula larva

The Echinoderms, Hemichordates, and Chordates Share Certain Characteristics

The hemichordates have apparent affinities to both the echinoderms and the chordates, a relationship between those two large and important phyla. The ciliated larvae of hemichordates are so much like those of some echinoderms that they were mistaken for echinoderms when first discovered. This dipleurula type of larva is found only in the echinoderms and hemichordates. It has a band of cilia *encircling* the mouth, whereas the trochophore type of larva found in many protostomes (including some flatworms, molluscs, and annelids) has a band of cilia encircling the body *anterior* to the mouth (Fig. 42.9). The similar larvae of hemichordates and echinoderms, as well as the similarities in their early embryology, indicate that these two groups probably stem from a common ancestor.

The most obvious resemblance of hemichordates to chordates is their possession of pharyngeal slits which are found in all chordates but nowhere else in the animal kingdom. While the hemichordates are closer to the echinoderms than to the chordates, recognition of their ties with both Chordata and Echinodermata has helped clarify the evolutionary relationship between these two major groups. Note that there is no suggestion here that chordates evolved from echinoderms, but simply that the two groups diverged from a common ancestor at some remote time.

THE INVERTEBRATE CHORDATES (PHYLUM CHORDATA)

Throughout this book we have used the terms "vertebrate" and "invertebrate," and have assigned all the animals discussed to one or the other of the categories they designate. But this division of the animal kingdom is in many respects an odd one, because neither category coincides with any phylum or group of phyla. Indeed, the term vertebrate designates only a part of one phylum; the rest of that phylum and all other phyla fall under the heading invertebrate. The phylum that contains both invertebrate and vertebrate members is Chordata.

The phylum Chordata is customarily divided into three subphyla: Tunicata (tunicates), Cephalochordata (lancelets), and Vertebrata (vertebrates). These share three important characteristics: (1) all have pharyngeal slits (or pouches) at some stage in their development; (2) all have a single dorsal nerve cord, which is hollow; and (3) all have, at least during embryonic development, a structure called a ***notocord*** (hence the name "Chordata"), a flexible, supportive rod running longitudinally through the dorsal portion of the animal just ventral to the nerve cord.

The Tunicata and the Cephalochordata are both invertebrate, i.e., they have no backbone.

> *Key innovations of the chordates*
> Pharyngeal slits
> Dorsal hollow nerve cord
> Notochord

The Tunicates (Subphylum Tunicata)

Adult Tunicates Use Their Pharyngeal Slits for Filter Feeding

In the best-known class of tunicates, sometimes called sea squirts, the adults are sedentary marine animals that are embedded in a tough sac, or tunic (Fig. 42.10). They little resemble other chordates except in having pharyngeal slits. Water taken in through the mouth is filtered through the pharyngeal slits, which function in both gas exchange and feeding, acting

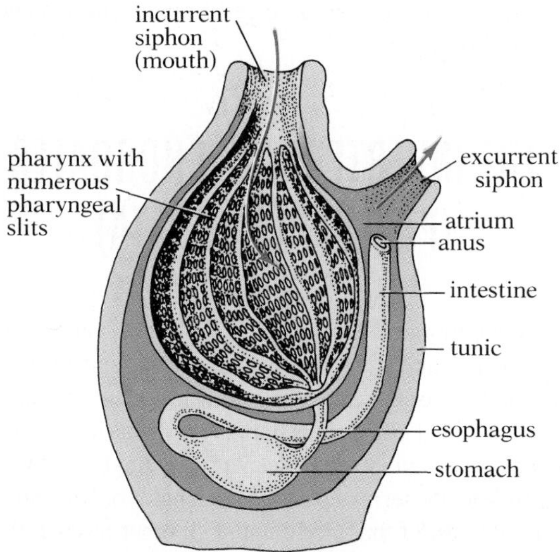

42.10 The adult tunicate The arrows show the path of water, which is drawn into the pharynx through the incurrent siphon, passes through the pharyngeal slits into the atrium, and then exits through the excurrent siphon. Oxygen is absorbed from the water across the walls of the pharyngeal slits, which thus serve as gills. Food particles drawn into the pharynx with the water do not pass through the slits, but instead are carried through the pharynx into the esophagus.

as a strainer to remove small food particles from the water flowing through them. Oxygen is absorbed from the water as it passes through.

The Tunicate Larva Resembles Other Chordates

The larval tunicates, which are motile, show much more resemblance to the other chordates than does the adult form. The larvae look rather like tadpoles and it is the larvae that possess the three chordate characteristics: pharyngeal slits, a dorsal nerve cord, which is hollow, and a notochord beneath the nerve cord in the tail region (Fig. 42.11). When the larvae settle down and undergo metamorphosis to the adult form, the notochord and most of the nerve cord are lost.

Some biologists hold that the tunicates and the vertebrates descended from a common ancestor that resembled a modern tunicate larva. If this is so, then the sedentary structure of modern adult tunicates is a later specialization. An alternative hypothesis is that the tunicate ancestor was sedentary, like adult tunicates, and that vertebrates evolved from the motile larva; in other words, in the line leading to the vertebrates, the larval stage increased in importance and duration, until finally it could reproduce without undergoing metamorphosis, and the ancestral sedentary stage dropped out of the life cycle entirely.

Key innovations of the tunicates

Sessile, filter-feeding adult

Larvae possessing chordate characteristics

The Lancelets (Subphylum Cephalochordata)

The Lancelets Are the First Deuterostomes to Swim in a Fishlike Manner

There are about 30 species of these small marine animals in the subphylum Cephalochordata. Lancelets spend most of their

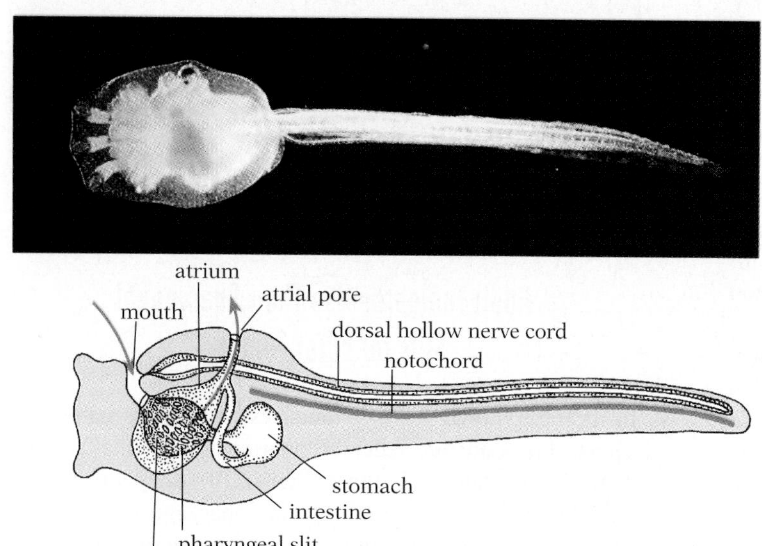

42.11 The larval tunicate Top: Photograph of a free-swimming larval tunicate. Bottom: Diagram of a larval tunicate. The arrows indicate the path of inflowing and outflowing water.

time buried tail down in sand in shallow water, with only their anterior end exposed. They too are filter feeders, taking in water through the mouth and straining it in the pharynx. Oxygen is removed from the water as it passes through the pharyngeal slits. A feature not seen in tunicates but characteristic of both cephalochordates and vertebrates is segmentation of the muscles, which are in Z-shaped, segmentally arranged bundles. Muscle segmentation and the presence of the notochord permit more efficient locomotion; lancelets are the first deuterostomes to swim in a fishlike manner. Both the dorsal hollow nerve cord and the notochord are well developed and retained for life.

The genus of lancelets most commonly studied is *Branchiostoma,* usually called amphioxus. The body of a typical specimen is about 5 centimeters long, translucent, and shaped rather like a fish (Fig. 42.12). Although amphioxus shows many similarities to vertebrates, it is quite unlike vertebrates in other features. The vertebrates and cephalochordates may have shared a common ancestor in the distant past, only to follow different evolutionary paths.

> *Key innovations of the cephalochordates*
> Fishlike body form
> Fishlike locomotion
> Segmented musculature

THE VERTEBRATE CHORDATES (SUBPHYLUM VERTEBRATA)

As the name Vertebrata implies, the animals in this subphylum of the Chordata are characterized by an internal skeleton (endoskeleton) that includes a backbone composed of a series of ***vertebrae*** as well as bones around the brain (the cranium). The vertebrae develop around the notochord, which in most vertebrates is present in the embryo only. The sequential arrangement of the vertebrae and the organization of the muscles are the principal tokens of segmentation. The dorsal hollow nerve cord runs through an opening in the dorsal portion of the vertebrae. The vertebrates have gills for gas exchange in their pharyngeal slits; the pharyngeal slits are therefore referred to as ***gill slits*** in vertebrates.

We discussed the anatomy, physiology, behavior, and development of vertebrates at length in other parts of this book. Here we shall be concerned primarily with the evolutionary history of the group.

> *Key innovations of the vertebrates*
> Vertebrae
> Bones surrounding brain (cranium)
> Gills in pharyngeal slits
> Closed circulatory system
> Three-part brain
> Well-developed sense organs

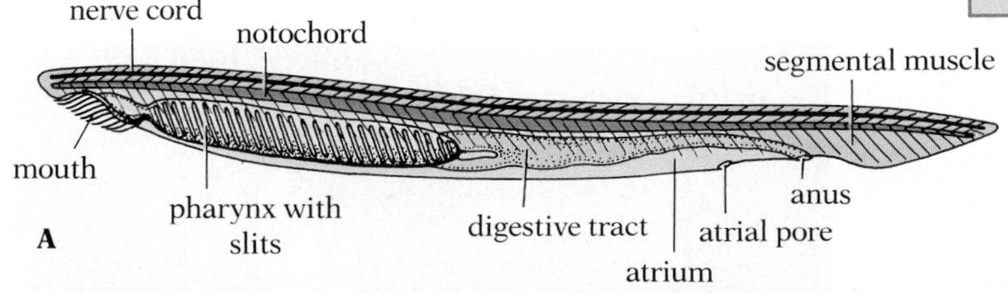

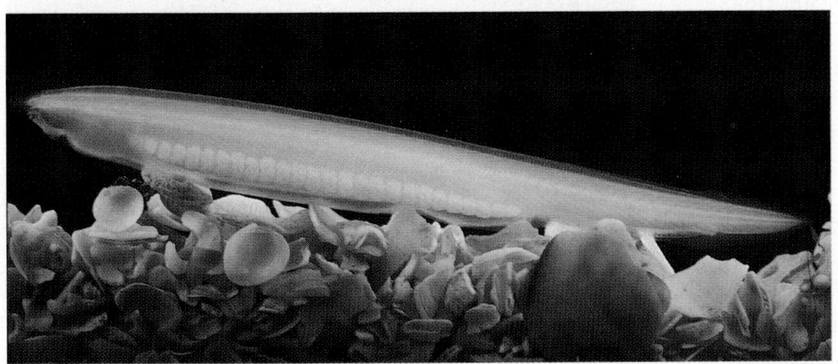

42.12 An adult lancelet (amphioxus) (A) Longitudinal view. Amphioxus is a filter feeder. The water enters the mouth and is propelled by cilia through the pharyngeal gill slits into a large chamber, the atrium, and then to the exterior through an atrial pore. Oxygen is removed from the water as it passes through the pharyngeal slits. Food particles do not pass through the pharyngeal slits, but move posteriorly into the digestive tract. (B) Photograph of a living amphioxus. The white notochord is visible along the dorsal surface. The white, serially arranged, blocklike structures near the ventral surface are the segmented gonads.

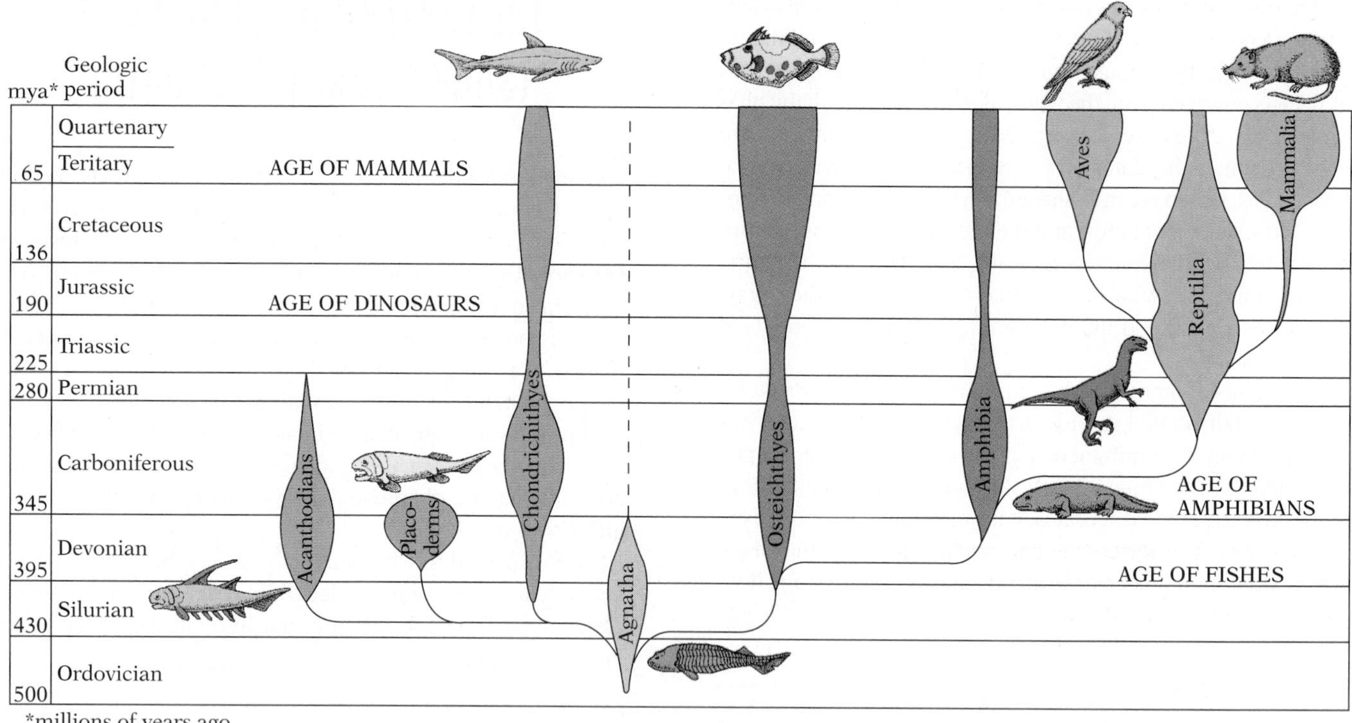

42.13 Evolution of the vertebrate classes Approximate dates of period boundaries are shown at left. The width of each band represents the relative abundance of fossils in each group.

The Fishes

The First Vertebrates Were Jawless Fishes

The oldest vertebrate fossils are from the Ordovician period, which began some 500 million years ago. Those first vertebrate fossils are of bizarre, fishlike animals covered by thick plates of bony material. Though they had an internal skeleton, they lacked an important character found in all later vertebrates—jaws. Furthermore, most of them had no paired fins. These ancient armored fishes constitute the class Agnatha (the term means "jawless"). Most were probably filter feeders, straining food material from mud and water flowing through their gill systems.

The Agnatha continued as an important group, sharing the seas with the already abundant sponges, coelenterates, molluscs (particularly gastropods and cephalopods), trilobites, and echinoderms. But about 400 million years ago the Agnatha began to decline, and they disappear from the fossil record by the end of the Devonian (Fig. 42.13).

A few peculiar species living today, the lampreys (Fig. 42.14) and the hagfishes, are generally classified as Agnatha, though they are quite unlike the ancient armored species. They have a soft body, having lost the external armor or scales; they have a cartilaginous skeleton, having lost all trace of bone; and their jawless mouth is modified as a round sucker that is lined with many horny teeth and accommodates a rasping tongue.

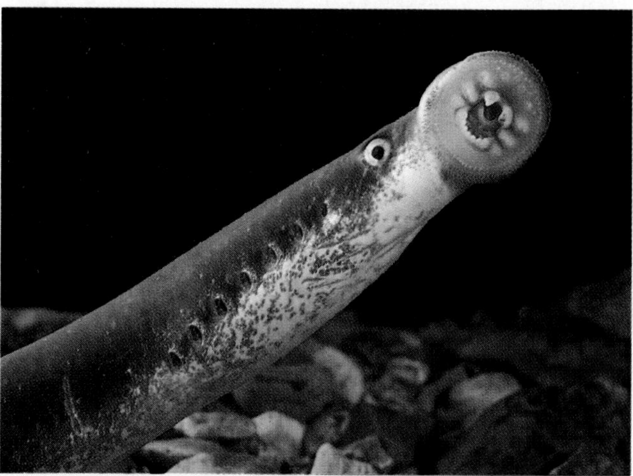

42.14 The head and pharyngeal region of a lamprey The animal has a large oral sucker instead of jaws, and seven prominent external gill openings. Adult lampreys are parasites on other large fish; they attach by their mouth to the prey and use the horny teeth in their mouth and on their tongue to rasp away the flesh. They have caused serious damage to other fish in the Great Lakes.

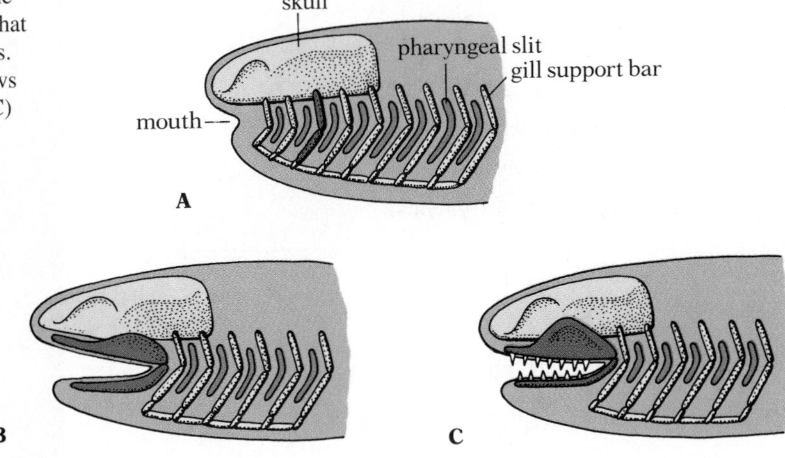

42.15 Evolution of the hinged jaws of vertebrates (A) The earliest vertebrates had no jaws. The structures (dark brown) that in their descendants would become jaws were gill support bars. (B) A pair of gill support bars has been modified into weak jaws (the two most anterior support bars, shown in A, were lost). (C) The jaws have become larger and stronger.

They feed by attaching themselves by their sucker to other fishes, rasping a hole in the skin of the prey, and sucking blood and other body fluids. The lampreys have a larval filter-feeding stage that strikingly resembles amphioxus.

> *Key innovations of the modern Agnatha (jawless fishes)*
> Loss of bone in skeleton
> Mouth and tongue with teeth
> Parasitic way of life

Some Descendants of the Ancient Fishes Evolved Jaws and Paired Fins

The decline of the ancient Agnatha coincided with the rise of four other classes of fishes (Fig. 42.13), two of which are now extinct (the acanthodians and the placoderms) and two of which persist to this day (the cartilaginous fishes and the bony fishes). These fishes mark a notable advance in vertebrate evolution in their possession of ***hinged jaws*** and ***paired fins.*** *The acquisition of hinged jaws was one of the most important events in the history of vertebrates, because it made possible a revolution in the method of feeding and hence in the entire mode of life of early fishes.* They became more active and wide-ranging animals and it is from the ancestral jawed fishes that the other fishes evolved. Many became ferocious predators. Even those that remained mud feeders were evidently adaptively superior to the ecologically similar agnaths, which they gradually replaced. Paired fins in the pelvic and pectoral (shoulder) regions were also an important innovation, since these fins have bony supports, and are the early forerunner of the anterior and posterior appendages.

Anatomical and embryological studies have convinced biologists that the hinged jaws of the early fishes developed from a set of gill support bars (Fig. 42.15). Notice that hinged jaws arose independently in two important animals groups, the arthropods and the vertebrates, but that, though they are functionally analogous structures, they arose in entirely different ways—in the one case from ancestral legs and in the other from gill support bars.

The acanthodians and placoderms were both armored fishes, often of rather bizarre body forms (Fig. 42.16). They have been extinct at least 230 million years, but the cartilaginous and bony fishes, which arose about the same time, have been very successful and persist to this day. Both of these groups lost most of the armored plates, but both have remnants of the ancient armor in the ***bony scales*** covering their bodies, and in their teeth.

42.16 Reconstruction of an extinct placoderm Notice the armored head, hinged jaws, and paired pelvic and pectoral fins.

Key innovations of the first jawed vertebrates
Hinged jaws from gill arches
Internal bony skeleton
Bony plates on body surface
Paired pelvic and pectoral fins

The Members of the Class Chondrichthyes Have Cartilaginous Skeletons

The modern Chondrichthyes (sharks, skates, rays, and their relatives) (Fig. 42.17) are distinguished by their cartilaginous skeletons; bone is unknown in the group. Though a cartilaginous skeleton might at first be taken as an ancestral trait, it is not thought to be one in Chondrichthyes. Their ancestors among the early jawed fishes probably had bony skeletons, and the loss of the bone must be regarded as an evolutionary specialization. Most members of this group have an unusual feature: internal fertilization. Males have evolved pelvic fins modified as "claspers" for introducing sperm into the body of the female.

Although sharks and their relatives have often been considered to be more "primitive" than bony fish, this is not the case—they are simply a different lineage that has evolved in a different direction.

Key innovations of the Chrondrichthyes (cartilaginous fishes)
Cartilaginous skeleton
Bony armor reduced to tiny scales
Internal fertilization
Active predators

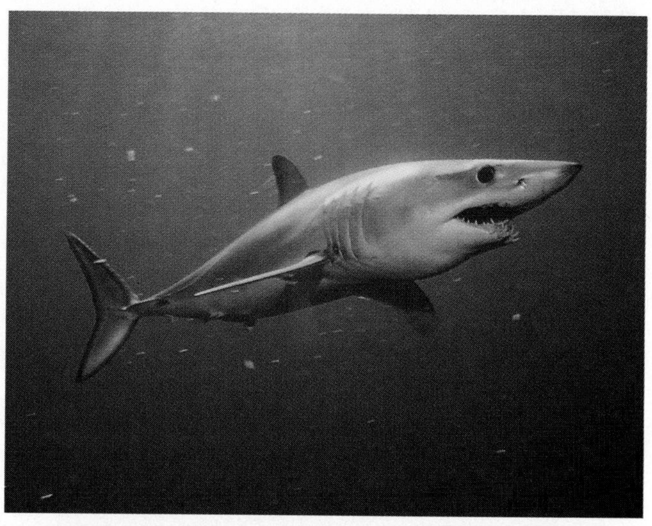

42.17 A mako shark (top) and a manta ray (bottom) The rays have flattened bodies adapted for life on the bottom. They swim by "flying" through the water, using the thin lateral parts of their bodies as wings. The manta ray has a further adaptation for its more mobile lifestyle: fleshy appendages on both sides of its mouth direct the microscopic food it eats into its mouth as it moves through the water.

The Bony Fishes (Class Osteichthyes) Are the Dominant Fishes Today

The other class that arose from the ancestral jawed fishes—the body fishes, or Osteichthyes—includes most of the fishes familiar to you. This is a large class, whose members are the dominant vertebrates in both fresh water and the oceans, as they have been since the Devonian—the so-called Age of Fishes.

Early in vertebrate evolution, the lining of the pharyngeal slits developed a highly folded surface, creating an enormous surface area for gas exchange (see Fig. 18.6, p. 377). In addition to functional gills, the earliest members of this class had prim-

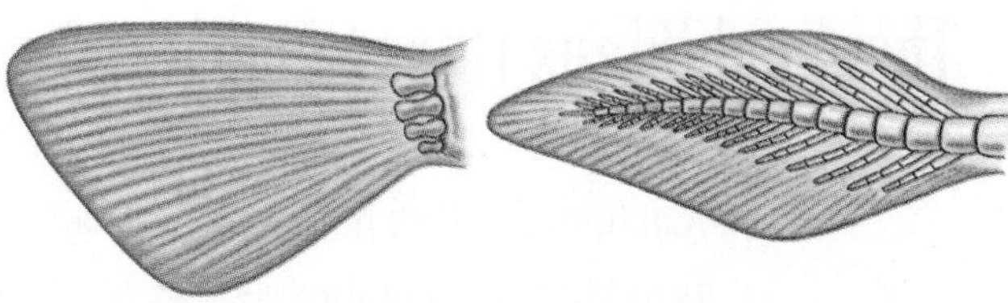

42.18 Two types of paired pectoral and pelivc fins The fossil and living fishes show two basic types of paired fin skeletons. (A) Fin of a sea bass, a ray-fin fish. Most bony fishes have fins in which the skeleton and muscle is found only in the base of the fin and the fins are supported by horny rays that extend in a fan-shaped arrangement. (B) Fin of the Australian lungfish, a fleshy-fin fish. This lineage has paired fins in which there is a main axis and side branches, rather featherlike in arrangement. There is more muscle in this type of fin, hence the name lobe-fin or fleshy fin.

itive lungs, which they probably used as supplementary gas-exchange devices when the water was stagnant and deficient in oxygen. There are still a few living relict species with lungs and gills.

Soon after the Osteichthyes arose, the class split into two divergent groups, one lineage with their paired fins that are "ray-fins" and the other with "fleshy fins" (Fig. 42.18). The ray-fin lineage underwent great evolutionary radiation, giving rise to nearly all the bony fishes alive today. The modern bony fishes no longer have lungs; instead the lungs have been modified to form a buoyancy device called a swim bladder. The fleshy-finned lineage radiated considerably but today is represented by only six relict species—five species of lungfishes (one in Australia, three in Africa, and one in South America) and one species of "lobe-fin," known only from deep waters off the southeast coast of Africa (Fig. 42.19). Despite its rarity, the lobe-fin has attracted special evolutionary interest because it has retained its lungs and belongs to the group thought to be ancestral to the land vertebrates.

42.19 The coelacanth, *Latimeria*, a modern lobe-fin fish There are two lateral pairs of lobe-fins: the most anterior pair are the pectoral fins and the fins in the middle, the pelvic fins. (The fin just ventral to the tail is the anal fin and does not have the lobe-fin structure.)

> *Key innovations of the Osteichthyes (bony fishes)*
> Bony armor reduced to interlocking bony scales
> Primitive lungs
> Gills in pharyngeal slits
> Two lineages: ray-finned and lobe-finned

The Amphibians Evolved From a Line of Lobe-Finned Fishes

Let us look more closely at the ancient lobe-finned fishes. This group has long been known from fossils, but until 1939 it was thought to have been extinct for some 75 million years. In that year a specimen was caught off the east coast of South Africa; since then, additional specimens of this living fossil, called the coelacanth (genus *Latimeria*), have been caught and studied (Fig. 42.19). The coelacanths are not the particular lobe-fins thought to be the ancestors of land vertebrates, but they resemble those ancestral forms in many ways.

In addition to lungs, the lobe-fins had another important preadaptation for life on land—the large, fleshy bases of their paired pectoral and pelvic fins. These fins have the basic bond plan that vertebrate legs do. At times, especially during droughts, lobe-fins living in fresh water probably used these leglike fins to pull themselves onto sandbars and mud flats (Fig. 42.20).

By the late Devonian period the land had already been colonized by plants, and some arthropods (crustaceans, arachnids, millipedes, and a few insects) had invaded land. Hence any animal that could survive on land would have had a whole new range of habitats and a good food supply open to it without competition and without predators. The aquatic environment, by contrast, was swarming with a variety of fishes including many active predators. Any lobe-finned fishes that had appendages slightly better suited for land locomotion than those of their fellows would have been able to exploit these habitats more fully; through selection pressure exerted over millions of years, the fins of these first vertebrates to walk (or rather crawl) on land would slowly have evolved into legs. Thus by the end of the Devonian, with a host of other adaptations for life on land evolving at the same time, one group of ancient lobe-finned fishes must have given rise to the first amphibians.

All of the animals we shall discuss from now on belong to the superclass ***Tetrapoda***, the four-legged vertebrates. There are four classes of Tetrapoda: the Amphibia (frogs, toads, and salamanders), Reptilia (snakes, lizards, turtles, and alligators), Aves (birds), and Mammalia (mammals).

42.20 The movement of vertebrates onto land Top: A model of a Devonian lobe-finned fish (*Eusthenopteron*), which probably pulled itself out of the water onto mud flats and sandbars. Bottom: A painting of an early amphibian (*Ichthyostega*). Its legs were better suited for locomotion on land than the lobe fins of *Eusthenopteron*, but it too probably spent most of its time in the water.

The Amphibians (Class Amphibia)

The Amphibians Were the First Terrestrial Vertebrates

Numerous fossils indicate that, as would be expected, the first amphibians were still quite fishlike. In fact, they probably spent most of their time in the water. But as they progressively exploited the ecological opportunities open to them on land, they slowly radiated and became a large and diverse group. So numerous were they during the Carboniferous that that period is often called the Age of Amphibians, just as the period before it, the Devonian, is called the Ages of Fishes. The amphibians were still abundant in the Permian, but the end of that period was a time of great change, both geological and biological. This so-called Permian extinction witnessed the mass extinction of many animal groups, including more than 95 percent of marine invertebrate species and most groups of amphibians. The only members of this once abundant and diverse class to survive were the immediate ancestors of the few small groups of modern Amphibia: the salamanders, some rate wormlike amphibians, and the frogs and toads (Fig. 42.21).

Although the amphibians live on land, they are not fully terrestrial, being restricted to moist environments and relying on environmental water for reproduction. Amphibians have thin, moist skin, which they use for gas exchange, and are in danger of drying out if conditions became very dry.[2] Furthermore, amphibians use external fertilization and lay fishlike eggs, i.e., eggs with no amnion or shell and which must therefore be deposited either in water or in very moist places on land

[2]All modern amphibians have thin, moist skin that functions as a respiratory organ (in addition to the gills and/or lungs), but this may not have been true of all ancient amphibians, which probably had scales.

MAKING CONNECTIONS

1. How does the development of echinoderms support the view that they evolved from a bilaterally symmetrical ancestor?

2. Describe two hypotheses regarding the evolutionary significance of the tunicate larva.

3. How have embryological and anatomical studies demonstrated that the hinged jaws of insects and vertebrates are analogous (as opposed to homologous) structures?

4. Describe two ways in which the lobe-fin fishes were preadapted for life on land.

42.21 Two modern amphibians Left: A bovine frog (*Rana palustris*) from New Jersey. Right: A banded salamander (*Salamandra salamandra*) from Europe.

to avoid drying up. The amphibian life cycle also has an aquatic larval stage. Amphibians are thus bound to the ancestral freshwater environment by the necessities of their mode of reproduction and their skin, which must be kept moist.

> *Key innovations of the amphibians*
> Terrestrial animals
> Better-developed lungs
> Positive pressure breathing
> Scaleless skin used for gas exchange
> Stronger skeleton with four limbs
> Three-chambered heart

The Reptiles (Class Reptilia)

The first reptiles had evolved from primitive amphibians by the late Carboniferous. The class expanded during the Permian, replacing its amphibian predecessors, and became a huge and dominant group during the Mesozoic era, which is often called the Age of Reptiles.

The Evolution of the Amniotic Egg Was a Crucial Innovation

One might wonder why the reptiles (Fig. 42.22) were so effectively able to displace the once-dominant amphibians. There were doubtless many reasons, but surely one of the most compelling was that the reptiles, unlike the amphibians, were terrestrial in the fullest sense of the word. Reptiles solved the problem of drying out by using uric acid as their excretory product and by evolving scales to cover their skin. These scales, composed of protein rather than bone as in fish, make the skin relatively impermeable and prevent water loss through evaporation. Reproduction on land was accomplished by developing mechanisms of internal fertilization, eliminating the aquatic larval stage, and laying special amniotic shelled eggs (Figs. 26.6 p. 534 and 42.23). The ***amniotic egg*** is often called the "land egg" because it provides a fluid-filled chamber in which the embryo may develop even when the egg itself is in a dry place. *Evolution of the amniotic egg was a feature as important in the conquest of land as the evolution of legs by the Amphibia.*

The reptiles had many other characteristics that made them better suited for terrestrial life than the amphibians. The legs of the ancient amphibians were small, weak, attached far up on the sides of the body, and oriented laterally; hence they were unable to support much weight, and the belly of the animal often dragged on the ground; walking was doubtless slow and labored, as it is in salamanders today. The legs of reptiles were usually larger and stronger and could therefore support more weight and effect more rapid locomotion; in many (though not all) species they were also attached lower on the sides and oriented more vertically, so that the animal's body cleared the ground. Whereas the lungs of amphibians were poorly developed and inefficiently ventilated, those of reptiles were fairly well developed, and greater rib musculature made their ventilation more efficient. And the amphibian heart was three-chambered (two atria and one ventricle), while that of reptiles is three- or four-chambered (the ventricle is often partially or completely divided by a partition); hence there was less chance

42.22 Representatives of the main group of modern reptiles (A) A tuatara (*Sphenodon*). (B) A lizard (*Chlamydosaurus*). (C) A snake (*Vipera*). (D) A turtle (*Pseudemys*). (E) A crocodile (*Crocodylus*).

42.23 Baby lizards hatching from their eggs The shells of reptilian eggs are usually leathery, not brittle like birds' eggs, as can be seen here from the way the shells have buckled and bent.

of mixing oxygenated and deoxygenated blood and therefore greater efficiency.

The ancestral reptiles belonged to an important group called the stem reptiles (Fig. 42.24). This group gave rise to several lineages, including one (therapsids) that ultimately led to the mammals (Fig. 42.25), one that led to the turtles, one that led to the reptiles that returned to the aquatic environment (Fig. 42.26), one leading to the modern snakes and lizards, and one (archosaurs) that in its turn gave rise to the flying reptiles (called pterosaurs), the great assemblage of reptiles called dinosaurs, the crocodilians, and the birds. The dinosaurs were extremely abundant and varied during the Jurassic and Cretaceous periods (Figs. 42.27 and 42.28).

> *Key innovations of the reptiles*
> Internal fertilization
> Amniotic egg
> Proteinaceous scales
> Uric acid as excretory product
> Stronger limbs
> Increased rib musculature for ventilating lungs

turtles
modern reptiles
lizards
snakes
crocodilians
birds
mammals
modern amphibians
pterosaurs
dinosaurs
therapsids
ichthyosaurs and pleiosaurs
thecodonts
stem reptiles (cotylosaurs)
ancestral amphibians

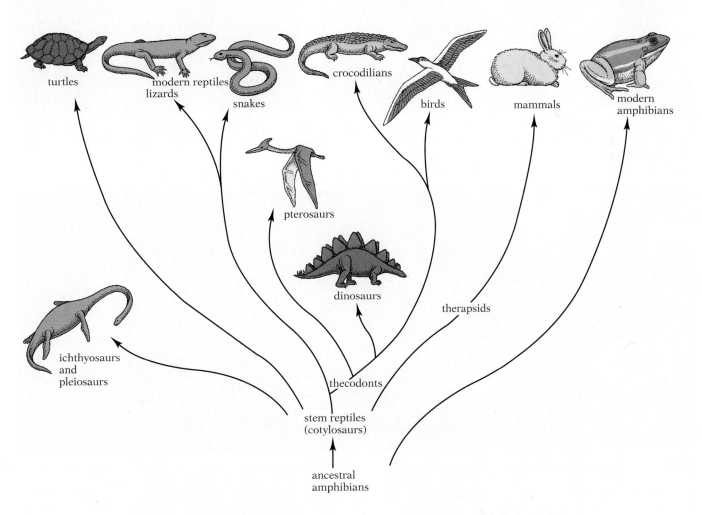

42.24 One interpretation of the evolution of the reptilian groups and their descendants

42.25 Restoration of a late Permian therapsid reptile from South Africa Fossil evidence indicates that the mammals evolved from the therapsids. Some investigators think the advanced therapsids had hair (not shown here).

42.26 Plesiosaurs The plesiosaur was representative of an ancient reptilian lineage that returned to an aquatic environment. All of the adaptations that enabled vertebrates to live on land had to be modified in order for them to efficiently move about in water and reproduce away from land.

42.27 Terrestrial dinosaur Great variation in body type among the fully terrestrial dinosaurs is evident in these two reconstructions. The scene at the left shows an encounter between two gracile species, a solitary *Oviraptor mongoliensis* and a trio of *Troodon mongoliensis.* The scene at the right shows a lumbering herd of single-horned, rhinoceroslike dinosaurs, classified as *Monoclonius apertus,* crossing a stream.

42.28 A gathering of dinosaurs The various types of dinosaurs depicted here are all representative of fossils found in the rocks of the Morrison formation in the western United States, dating from the late Jurassic period (approximately 145 million years ago). Included among the larger species are a herbivorous camptosaur, standing in the trees at the far left, a carnivorous allosaur, perched on the body of a fallen apatosaur at the center, and a herd of long-necked camarasaurs, coming on the scene from the right rear. Smaller species in the foreground include a primitive crocodilian (left), a few birdlike *Ornitholostes* (center), and a bizarrely outfitted stegosaur (right).

Many Reptilian Lines Died Out During a Mass Extinction

The decline of the dinosaurs was not as sudden as is often supposed; it took tens of millions of years. But it was a dramatic event in the history of life on earth nonetheless. Why previously successful animals should have died out on such a scale has never been satisfactorily explained. Extinction was not limited to reptiles; many invertebrates also disappeared. Yet many other groups living in the same sorts of habitats not only did not become extinct but did not even undergo significant change. In all, about 70 percent of the earth's species died out at the end of the Cretaceous period. Various hypotheses have been proposed to explain the Cretaceous extinction (see Chapter 32, pp. 722–724), but whether it resulted from an asteroid collision, comet impact, cycle of intense volcanic activity, or long-term climate change, its magnitude was devastating. It almost certainly created an evolutionary bottleneck of some sort, and likely initiated a series of widespread changes in the tempo and direction of evolution. (See Exploring Further: The Search for Dinosaur DNA, p. 974.)

A

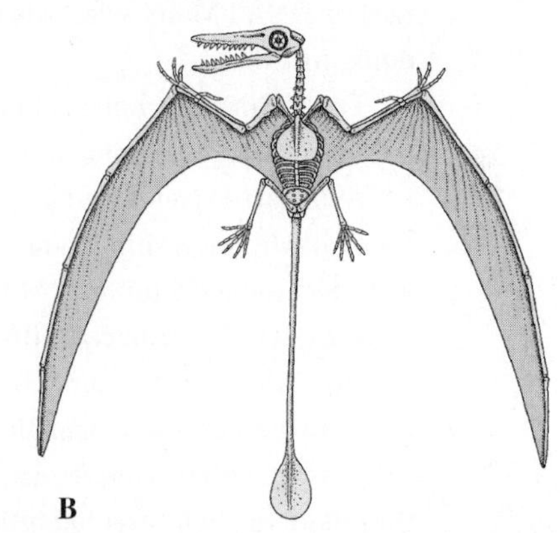

B

42.29 Pterosaur (A) The pterosaur was a giant, flying reptile. The coloration of the pterosaur is conjectural. (B) As this skeletal reconstruction shows, the pterosaur, like modern birds, had streamlined bodies and light, hollow bones adapted for flight. Also note the large membrane of skin stretched between the body and the enormously elongated arm and fourth finger.

The Birds (Class Aves)

The Forelimbs of Birds (Class Aves) Are Modified for Flight

At least two of the lineages of reptiles that descended from the archosaurs developed the power of flight. One of these lineages, the pterosaurs, included animals with wings modified for soaring; the wings consisted of a large membrane of skin stretched between the body and the enormously elongated arm and fourth finger; some species had wing-spreads as great as eight meters (Fig. 42.29). The pterosaurs were common for a time, but eventually became extinct. The other lineage developed wings of an entirely different sort, in which many long feathers, derived from the proteinaceous scales, were attached to the modified forelimbs. This line eventually became sufficiently different from the other reptiles to be designated by many taxonomists as a separate class, the Aves, otherwise known as birds. Many biologists believe, however, that dinosaurs never went extinct, but are still abundantly represented by modern birds.

The oldest known fossil bird (*Archaeopteryx*) still had many reptilian characters, notably teeth and a long, jointed tail (Fig. 42.30). Neither of these traits is present in modern birds,

42.30 Reconstruction of a skeleton of *Archaeopteryx*, the oldest known bird The skeleton is quite reptilian in character, having a long reptilian type of tail, teeth rather than a beak, and fingers with claws, but imprints of feathers preserved with the skeleton identify this specimen as a bird.

EXPLORING FURTHER

THE SEARCH FOR DINOSAUR DNA

Michael Crichton's novel *Jurassic Park* and the Steven Spielberg movie based on it are science fiction thrillers. In this story, dinosaur DNA is extracted from blood from the gut of a mosquito that had been preserved in amber, and the DNA is then used to clone new dinosaurs. This is science fiction at its best, but is it only fiction? Is it possible to extract dinosaur DNA? Many scientists think so, and a number of research groups are working on doing just that.

The technical problems in obtaining "dino DNA" are formidable. The dinosaurs died out 65 million years ago, and most researchers think the chances of any genetic material surviving are extremely unlikely. DNA is not a very stable molecule, and biochemical studies of its decomposition rates suggest that DNA could not persist for more than 50,000 years, let alone 65 million. Moreover, when bones are preserved as fossils, the bone's organic material is replaced with minerals, and the genetic material is destroyed. Still, some researchers think that some DNA may be preserved in bones if the bone is not completely mineralized and some organic material persists in protected areas, as in the center of a thick bone. Other workers are trying another approach—extracting DNA from blood in the gut of amber-preserved biting insects, as in Michael Crichton's novel (Fig. A).

Extracting undegraded dino DNA is only the beginning of the difficulties. To obtain the DNA, fossilized dinosaur bones are ground up and extracts are made and purified. The exquisitely sensitive polymerase chain reaction (PCR) is used to generate millions of copies. Unfortunately, PCR will also amplify any stray DNA that may be there, including that from a technician, bacteria, or fungus. The trick is to determine whether the extracted DNA is from a dinosaur or a contaminant. To do this the amplified DNA is sequenced, and the se-

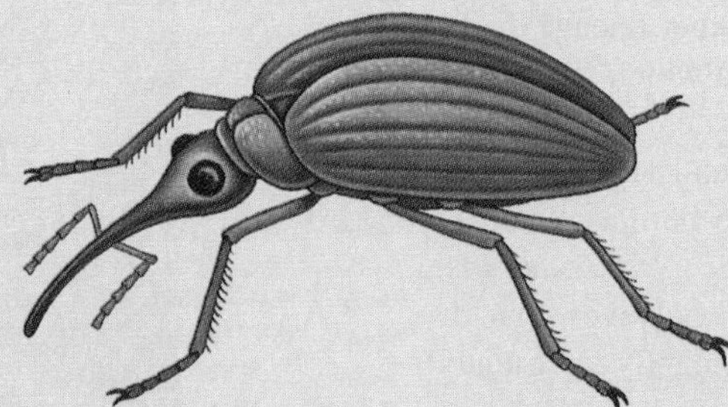

Fig. A Sketch of an extinct weevil from the age of dinosaurs Recently scientists extracted segments of DNA from this extinct weevil that is thought to have lived 125 million to 135 million years ago. This is the oldest DNA ever found, showing that DNA of this antiquity could be preserved, but the possibility of cloning dinosaur DNA still remains highly unlikely. The weevil is extremely well preserved because it apparently became trapped in tree resin that hardened into amber, not a likely scenario in the case of dinosaurs. Also, because weevils are herbivores, they did not feed on dinosaur blood.

quences are then compared to human, fungal, and bacterial DNA. If the sample matches one of these then it is a contaminant; if it doesn't, the sample *might* be dino DNA, but how can one know if it is? There are no living dinosaurs to compare it with, but there are close relatives—the birds and crocodiles. So, the sample DNA is compared to bird and crocodile DNA. If they are similar, then the sample *could* be dinosaur DNA. The experiment would have to be repeated and replicated a number of times to provide sufficient evidence. This in itself is a serious problem because DNA this old is in short supply, and the amount of DNA in a sample varies, so reproducing the results is an almost impossible task. The technical problems are formidable, but the work goes on.

In *Jurassic Park,* dinosaur DNA is used to clone dinosaurs, but no one is suggesting that this is possible. Too much remains to be learned about developmental biology to even dream about cloning a dinosaur.

which have a beak instead of teeth and only a tiny remnant of the ancestral tail bones (the tail of a modern bird consists only of feathers).

Along with wings, birds evolved a host of other adaptations for their very active way of life. Among the most important was the evolution of endothermy and homeothermy—the ability to maintain a high and constant metabolic rate, and hence great activity, despite fluctuations in environmental temperature. An anatomical feature that helped make possible the metabolic efficiency necessary for endothermy was an efficient gas-exchange system (see Fig. 18.19, p. 385) and the complete separation of the two ventricles of the heart; birds have completely four-chambered hearts. This means that the oxygenated blood is kept completely separate from deoxygenated blood, and that blood is pumped out under high blood pressure to all parts of the body. The insulation against heat loss provided by the body feathers plays an important role in temperature regulation; in modern birds all the scales except those of the legs and feet are modified as feathers, which serve as insulation. Among other adaptations for flight are a streamlined body and light, hollow bones. Birds also have very keen senses of vision, hearing, and equilibrium.

Key innovations of the birds
Forelimbs modified as wings
Feathers
Endothermy and homeothermy
Efficient gas-exchange system
Four-chambered heart
Skeleton modified for flight
Streamlined body

The Mammals (Class Mammalia)

Both birds and mammals are endothermic and evolved from reptiles that were probably at least partly endothermic, and both became highly successful groups of organisms. But the two groups did not arise from the same ancestral reptilian stock. The line leading to the mammals split off from the stem reptiles, while that leading to the birds diverged from the line leading to the crocodiles. Thus, many of the similarities that we see in birds and mammals, such as endothermy and a four-chambered heart, are thought to result from convergent evolution.

The mammals are believed to have evolved from the therapsid reptiles, some of which became very mammallike (Fig. 42.25); they may even have had hair. Precisely at what point therapsids ceased and mammals began is impossible to say. There was no sudden transformation of reptile into mammal, no dramatic event to mark the appearance of the first member of our class; when the characters that distinguish modern mammals from stem reptiles first appeared cannot be specified.

Mammals have a four-chambered heart and are both endothermic and homeothermic. They have a diaphragm, which increases breathing efficiency. There is increased separation (by the palate) of the gas-exchange and food passages. The body is covered with an insulating layer of hair. The limbs are oriented ventrally and lift the body high off the ground. The teeth are differentiated for a variety of functions. The brain, particularly the neopallium (see p. 639), is much larger than in reptiles, and behavior is more easily modifiable by experience. No eggs are laid (except in monotremes); embryonic development occurs in the uterus of the mother, and the young are born alive. After birth, the young are nourished on milk secreted by the mammary glands of the mother.

Key innovations of the mammals
Mammary glands
Placenta and uterus; young born alive
Endothermy and Homeothermy
Four-chambered heart
Hair/fur
Muscular diaphragm
Palate separating oral and nasal cavities
Differentiated teeth
Enlarged neopallium

42.31 Duck-billed platypus, an egg-laying mammal Though the platypus is well adapted for aquatic life, it lays its eggs on land.

The Monotremes Are a Mixture of Reptilian and Mammalian Traits

As indicated above, there is one small group of mammals—the monotremes—that is fundamentally different from all other members of the class. They lay eggs; yet they secrete milk. In many other ways they are a curious blend of reptilian traits, mammalian traits, and traits peculiar to themselves. It seems clear that they were a very early offshoot of the mammalian lineage and were not ancestral to the other mammals. Some biologists think they should be considered mammallike reptiles rather than reptilelike mammals. The only living monotremes are the spiny anteater and the duck-billed platypus (Fig. 42.31); both are found in Australia, and the anteater also in New Guinea.

There Are Two Main Lines of Mammalian Evolution

The main stem of mammalian evolution split into two parts very early, one leading to the ***marsupials*** and the other to the ***placentals.*** The characteristic difference between them is that marsupial embryos remain in the uterus for a relatively short time and then complete their development while attached to a nipple in an abdominal pouch of the mother (Fig. 42.32), whereas placental embryos complete their development in the uterus.

The oldest mammalian fossils identified as placentals are from the Jurassic. These fossils are of small, shrewlike creatures that are thought to have fed primarily on insects (see Fig. 32.20, p. 178). It seems likely that these ancient insectivores were the ancestors of all the other placental mammals (Fig.

MAKING CONNECTIONS

1. Both ferns and amphibians flourished during the latter part of the Paleozoic era, only to be largely eclipsed by seed plants and reptiles during the Mesozoic. What role did the reproductive methods of these four groups play in their reversal of fortunes?

2. How does the excretion of uric acid by reptiles help them to avoid desiccation? What is the link between this excretory product and the evolution of the amniotic egg? (See Chapter 22.)

3. Should the similarities of birds and mammals—such as endothermy and a four-chambered heart—be taken as evidence of common ancestry from a single ancestral reptilian stock? Explain your answer.

4. How is the adaptive radiation of mammals at the beginning of the Cenozoic era related to the demise of the dinosaurs?

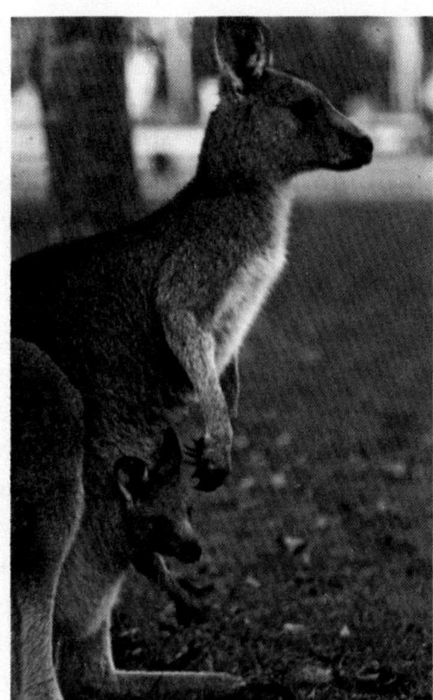

42.32 A kangaroo with young in pouch The young kangaroo (called a joey) seen here is hundreds of times larger than when it first entered its mother's pouch; it is no longer attached to a nipple, and it often comes out of the pouch for extended periods.

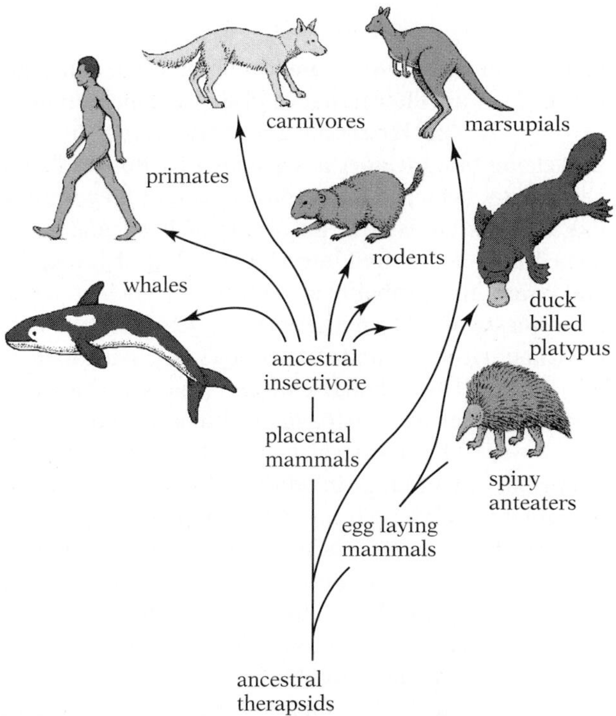

42.33 Evolution of the mammalian groups Notice the divergence between placentals and marsupials and that insectivores are ancestral to the primates.

42.33). The great radiation from the insectivore ancestors dates from the beginning of the Cenozoic era, as the mammals rapidly filled the many niches left open by the demise of the dinosaurs. The Cenozoic, which includes the present, is aptly termed the Age of Mammals.

The living placental mammals are classified in approximately 17 orders, several of which contain species familiar to almost everyone. The orders are listed in the Appendix (see p. A10). One of these orders, the Primata, is a direct descendant of the primitive insectivores, and is the order to which human beings belong. We will discuss the primates in the next chapter.

CONCEPTS IN BRIEF

1. The phyla Echinodermata, Hemichordata, and Chordata are in the Deuterostomia and share certain deuterostome characteristics, such as formation of the anus from the blastopore, radial and indeterminate cleavage, origin of the mesoderm as pouches, and formation of the coelom as the cavities in the mesodermal pouches.

2. The *Echinodermata* are exclusively marine, mostly bottom-dwelling animals. The adults are radially symmetrical; all show five-part symmetry. All members have an internal skeleton made of bony plates and spines, and a unique *water-vascular system.*

3. The *Hemichordata,* notably acorn worms, resemble the chordates in their possession of pharyngeal slits. However, their ciliated larva (a *dipleurula*) is almost identical to that of the echinoderms. The chordates and echinoderms are believed to have evolved from a common ancestor.

4. The phylum *Chordata* contains both invertebrate and vertebrate members. At some time in their life cycle, all chordates have pharyngeal slits, a dorsal nerve cord, which is hollow, and a *notochord.*

5. The adults of the subphylum *Tunicata,* the tunicates, are sessile filter feeders with pharyngeal slits. The larvae, however, are tadpolelike organisms with pharyngeal slits, a dorsal nerve cord, which is hollow, and a notochord.

6. Members of the subphylum *Cephalochordata,* the lancelets, are small marine fishlike organisms with

pharyngeal slits, a dorsal hollow nerve cord, and a notochord. They are filter feeders and show segmentation.

7. The *Vertebrata* are characterized by an endoskeleton that includes a segmented backbone. The first vertebrates belonged to the class *Agnatha;* they were fishlike animals encased in an armor of bony plates. They lacked jaws and paired lateral fins. A line of jawed fishes and paired fins probably arose from the ancient agnaths. Four lines evolved from the ancestral jawed ancestor; two, the Chondrichthyes and Osteichthyes, are living today.

8. The Chondrichthyes (sharks, skates, and rays) have evolved a cartilaginous endoskeleton.

9. The ancestral bony fishes belonging to the *Osteichthyes* had lungs in addition to gills. Early in their evolution the class split into two groups. One group (ray-fin) gave rise to most of the modern bony fishes alive today; the other group (fleshy-fin), now mostly extinct, includes the lungfishes and lobe-finned fishes. The lobe-fins had lungs and leglike fleshy paired fins. They appear to have been ancestral to the amphibians.

10. The first members of the class *Amphibia* were quite fishlike. Amphibians use their moist skin for gas exchange so must live in a moist environment. Also, they must return to water to reproduce, since they use external fertilization, lay fishlike eggs, and have aquatic larvae.

11. The *Reptilia* are the first truly terrestrial group. They use internal fertilization, lay amniotic shelled eggs, have no larval stage, and have evolved protein scales, producing a relatively impermeable skin. In addition, their legs are larger and stronger than those of amphibians, their lungs are better developed, and their heart is almost four-chambered.

12. The stem reptiles underwent a great deal of adaptive radiation, producing many different lineages of reptiles. One line gives rise to the dinosaurs, crocodiles, and birds. The *Aves,* or birds, are both endothermic and homeothermic; their heart is four-chambered, they have feathers for insulation, their bones are light, and their forelimbs are adapted as wings.

13. A line of reptiles called the therapsids gave rise to the mammals. In the *Mammalia,* embryonic development occurs within the female uterus; the young are born alive and are nourished by milk from the mammary glands. Mammals also are endothermic and homeothermic, and have a four-chambered heart. They have a diaphragm, hair for insulation, differentiated teeth, and an enlarged neopallium.

14. There are two major mammalian lines, the marsupials, in which the young complete their development while attached to a nipple in an abdominal pouch of the mother, and the placentals, which complete their development within the uterus.

STUDY QUESTIONS

1. What shared characteristics mark sea urchins (echinoderms), acorn worms (hemichordates), and humans (chordates) as members of the Deuterostomia? (p. 957)
2. How does an echinoderm such as a sea star use its water-vascular system for feeding and locomotion? What other unique features are shared by members of this phylum? (pp. 958–959)
3. What evidence indicates that hemichordates are related to both the echinoderms and the chordates (pp. 960–961)
4. Compare the structure and way of life of tunicates and lancelets. Why are they placed in the phylum Chordata along with the vertebrates? (pp. 961–963)
5. What were the key innovations of the first vertebrates, the jawless fish? How do modern agnathans differ from their early ancestors? (pp. 964–965)
6. Describe two evolutionary advances found in the four classes of fish that evolved from the ancient Agnatha? Which two of these classes survive today? How do they differ from one another? (pp. 964–967)
7. From what lineage of bony fish did amphibians evolve? How are amphibians adapted for terrestrial life? Why are they not fully terrestrial? (p. 967)
8. What characteristics of crocodiles make them more fully adapted to terrestrial life than frogs? Briefly describe how the stem reptiles gave rise to modern reptiles, as well as dinosaurs, birds, and mammals. (pp. 969–971)
9. Consider a swallow gracefully pursuing flying insects on a summer day. What features of the swallow's anatomy and physiology adapt it for flight? (pp. 973–975)
10. What traits do human beings share with other mammals? As placental mammals, how are we different from marsupials? (pp. 975–976)

CHAPTER 43

THE EVOLUTION OF PRIMATES

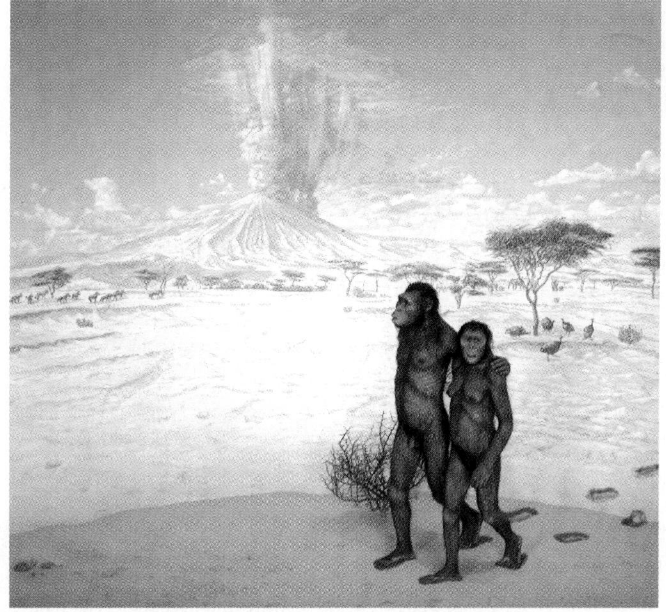

Humans are members of the mammalian order Primata (the primates). Naturally, we have a special interest in the evolutionary history of primates, especially concerning the origin of our own species. The fossil evidence suggests that ***Homo sapiens*** evolved from tree-dwelling ancestors that lived about 65 million years ago. The early primates continued to live in the trees until the Miocene, when the climate began to change and the subtropical forests gave way to more open woodland savannas and grasslands. Some of the primates descended from the trees and adapted to a more terrestrial way of life. It is from a stock of these ancient ground-dwelling primates that our species has descended.

Evolution of Primate Characteristics Is Related to Tree-Dwelling Ancestors

Fossil evidence indicates that the primates arose very early in the Cenozoic, from a stock of small, tree-dwelling insectivores (insect-eating mammals) not unlike the present-day tree shrews (Fig. 43.1). The group soon split into several evolutionary lines that have had independent histories ever since. Although the first primates were probably insect-eaters, their descendants began to utilize edible plant parts found in the tree canopy. With only a few exceptions, primates are herbivores and insectivores; they live in trees (out of the range of most predators) on omnivorous diets. Although the modern primates are a rather diverse lot, most of them share the following characteristics:

1. enlargement of the clavicle (collarbone), which is greatly reduced or lost in many other mammals;
2. development of a shoulder joint permitting relatively free movements in all directions, and an elbow joint permitting some rotational movement;
3. retention of five functional digits (fingers and toes) on each limb;
4. enhanced individual mobility of the digits, especially the thumb and big toe, which are usually opposable (that is, the thumb or big toe can be placed against the other digits of a hand or foot);
5. modification of the claws into flattened nails;
6. development of sensitive tactile pads on the digits;
7. shortening of the snout or muzzle;
8. development of three-dimensional (stereoscopic) vision;
9. enhancement of visual acuity and development of color vision;
10. enlargement of the brain, particularly the cerebral cortex;

Chapter opening photo: A painting of *Australopithecus afarensis,* "Lucy," and mate from the American Museum of Natural History.

43.1 A Malayan tree shrew The living tree shrews are members of the order Insectivora (the insect-eating mammals). They are intermediate in many of their traits between the Insectivora and the Primata, and are thought to resemble the early ancestors of the modern primates. They are not closely related to squirrels (members of the Rodentia), which they superficially resemble.

11. usually only two mammary glands and one young per pregnancy.

Notice that most of the traits are correlated with an arboreal (tree-dwelling) way of life.

In quadrupedal (four-legged) terrestrial mammals the legs have tended to evolve toward greater stability at the expense of freedom of movement. Think of the forelimbs of a dog or a horse: the clavicles are greatly reduced or lost; the two limbs are positioned close together under the animal, and their movement is restricted largely to one plane (i.e., they can move easily back and forth, but cannot be spread far to the side like human arms). By contrast, in an animal leaping about in the branches of a tree, the limbs function in grasping, clasping, and swinging. Mobility at the shoulder, elbow, and digit joints facilitates such activities, as does attachment of the limbs (braced by the clavicles) far apart at the sides of the body instead of underneath. Leaping from branch to branch also selected for hands capable of grasping, i.e., hands with mobile digits and opposable thumbs. Such hands are suited for grasping branches, but are also useful for manipulating food. An evolutionary trend toward ever-finer manual dexterity, culminating in the human hand, is shown in Figure 43.2.

The eyes of many quadrupedal terrestrial mammals (e.g., horses, cows, dogs) are located on the sides of the head. As a result, they have a very wide total visual field, but the fields of the two eyes overlap only slightly; hence the animals have little stereoscopic (three-dimensional) vision. But stereoscopic vision aids in localizing near objects, and an animal jumping from limb to limb obviously must be able to detect very accurately the position of the next limb. Hence the tree-dwelling way of life of the early primates doubtless led to selection for sharpened visual acuity and stereoscopic vision and, consequently, for eyes directed forward rather than laterally. This change, in turn, would have led to the distinctive flattened, forward-directed face of most higher primates. The evolution of the three-different types of cones permitted the development of fine-discrimination color vision, also a primate characteristic. Such vision would have helped early primates locate ripe fruits and other food items.

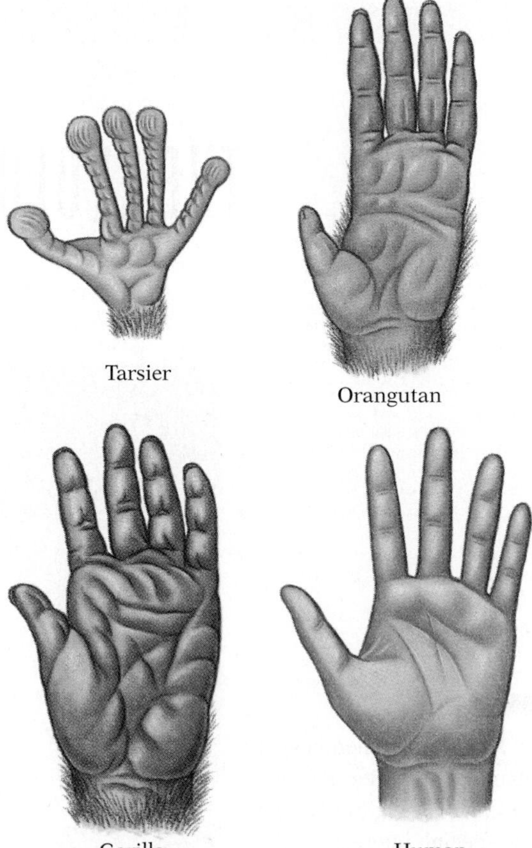

43.2 A comparison of primate hands Primate hands vary in their structure and manipulative ability. The hand of the tarsier has an opposable thumb and slender fingers with adhesive pads adapted for grasping branches. The hands of the orangutans are adapted for swinging from branch to branch; the fingers serve mainly as hooks for hanging on branches and the thumb is short so as not to get in the way. The hand of the gorilla has a longer, more opposable thumb than the orangutan, and the fingers are shorter. The human hand has the most manipulative ability; the long thumb is more opposable and the fingers more dexterous and capable of finer movements.

Now, hands capable of grasping the next limb and keen eyes with broadly overlapping fields of vision would not by themselves have met the requirements of life in the forest canopy. Essential, too, would have been neural and muscular mechanisms capable of very precise eye-hand coordination. This need was one of the factors that led to the early enlargement of the primate brain.

In short, *many of the traits most important to us as human beings first evolved because our distant ancestors lived in trees and hunted moving insects to supplement their diet of leaves and fruit.*

The Prosimians Were the First Primates

The living primates are usually classified into suborders: the prosimians—(Strepsirhini, or Prosimii)—and the tarsiers and anthropoids—(Haplorhini, or Anthropoidea). The prosimians ("premonkeys") were the first primates to evolve from the insectivore ancestors, and flourished about 50 million years ago. They are a miscellaneous group of more or less primitive primates, of which there are only a few isolated species living today—such as lemurs, lorises, and galagos. The lemurs were once widespread throughout Eurasia and probably Africa but are now found only on the island of Madagascar off the east coast of Africa. Most lemurs are fairly small tree-dwelling animals with a bushier coat than is usual among higher primates. They have fairly long, foxlike snouts and bushy tails, and hardly resemble the higher primates (Fig. 43.3). But they have opposable first digits, and the digits are usually provided with flattened nails. Their relatives, the lorises, pottos, and galagos, inhabit southeast Asia and tropical Africa.

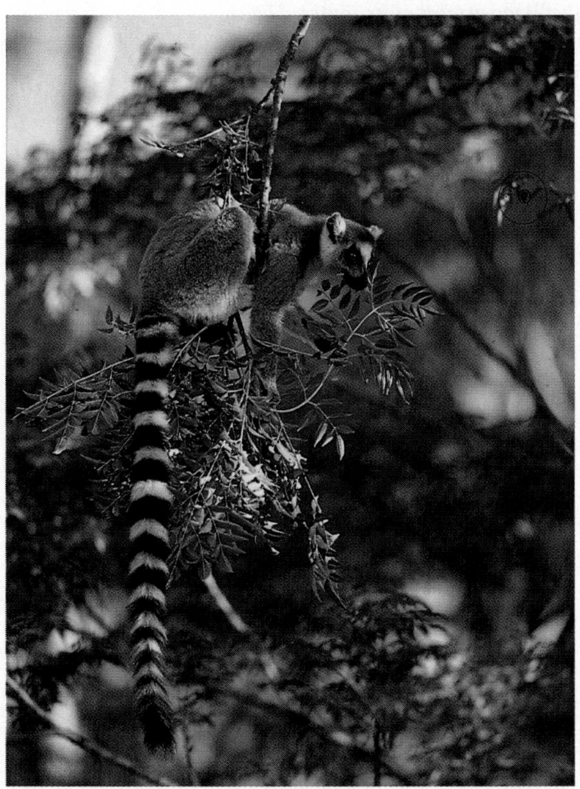

43.3 A ring-tailed lemur The animal has a long snout and a bushy tail—both uncharacteristic of the higher primates—but its hands and feet have opposable first digits.

43.4 A tarsier Tarsiers are small animals found in the Philippines and the East Indies. They are more specialized than the lemurs. They have a much shorter muzzle than a lemur, and thus a more distinct face. Note the enormous eyes, which are directed more completely forward than in lemurs. The hind limbs are long and specialized for leaping. The long tail is naked except at the end.

The Haplorhini Evolved From a Line of Prosimians

The second suborder, the Haplorhini, contains tarsiers, New World monkeys, Old World monkeys, and the hominoids (an informal grouping of apes and humans). The fossil record suggests that they evolved from a tarsierlike ancestor (Fig. 43.4). This ancestral line eventually diverged to form two lineages, one leading to the New World monkeys and the other to the Old World monkeys and hominoids. These hypothetical relationships are diagrammed in Figure 43.5.

There are many differences between New World and Old World monkeys, but we shall focus on just three here: (1) most New World monkeys have a prehensile tail that they use almost like another hand for grasping branches; the tail of Old World monkeys is short, or even absent, and is never prehensile; (2) the New World monkeys have flat noses and nostrils that are widely separated and are directed laterally; the nostrils of Old World monkeys are close together and are directed forward and down; (3) the New World monkeys tend to be rather small; the Old World monkeys are, on the average, considerably larger. Both lineages are believed to have evolved from a common ancestor about 35 million years ago.

Among the best-known New World monkeys are capuchins (the traditional organ-grinders' monkeys), howlers (Fig. 43.6), spider monkeys, and squirrel monkeys. Examples

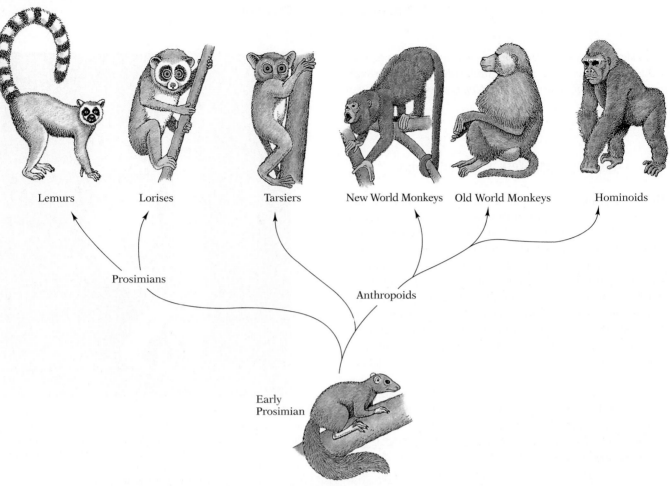

Lemurs
Lorises
Tarsiers
New World Monkeys
Old World Monkeys
Hominoids
Prosimians
Anthropoids
Early Prosimian

43.5 Hypothetical phylogenetic relationships of the living groups of primates The exact point at which the line leading to the tarsiers diverged is controversial. The term hominoid refers collectively to both apes and humans.

43.6 A New World and an Old World monkey Left: A howler (*Alouatta seniculus*), one of the New World monkeys, has a long, prehensile tail and separated, laterally directed nostrils. Right: The red patas (*Erythrocebus patas*), one of the Old World monkeys, lacks a prehensile tail and has nostrils that are close together and directed forward and down.

of Old World monkeys are macaques, mandrills, baboons, proboscis monkeys, mona monkeys, and the sacred hanuman monkeys of India. One of the macaques, commonly called the rhesus monkey, has been used extensively in physiological and psychological research; when physiologists or psychologists refer to "the monkey," this is usually the species they mean.

Apes and Humans Evolved From a Line That Led to the Old World Monkeys

There are two families of apes: the lesser apes, the group to which the gibbons belong, and the great apes. The gibbons (Fig. 43.7), of which several species are found in southeast Asia, represent a group that probably split from the others soon after the ancestral lineage arose. They are the smallest of the apes (about three feet tall when standing), and the most acrobatic. Their arms are exceedingly long, reaching the ground even when the animal is standing erect.

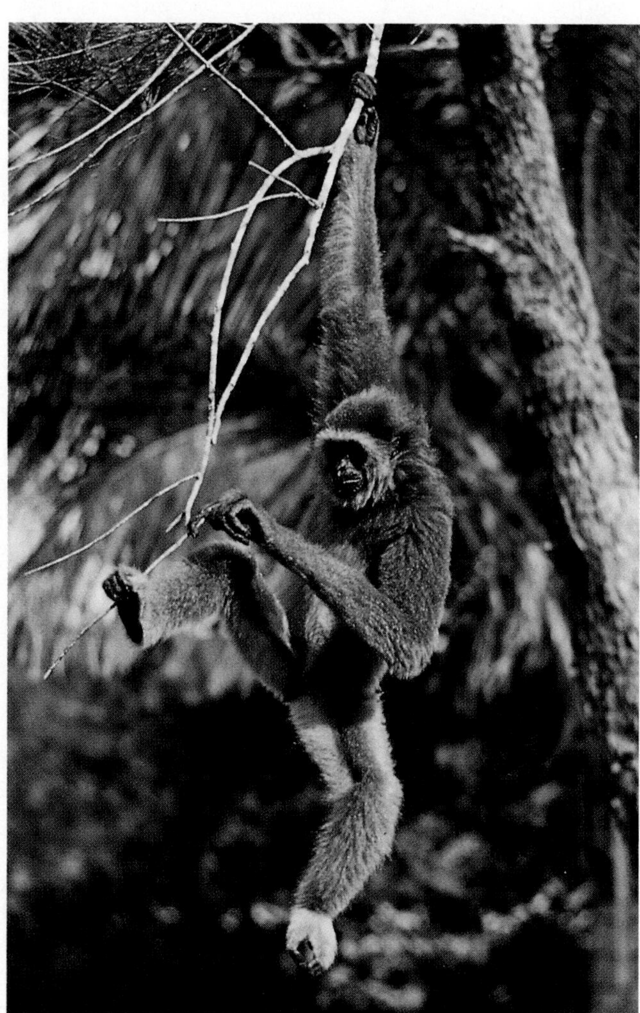

43.7 A gibbon (*Hylobates moloch*) from Sunda Island, Borneo Notice the extremely long arms.

43.8 An orangutan (*Pongo pygmaeus*) with young The one living species of orangutan is native to Sumatra and Borneo. Though the orangs are fairly large, and their movements slow and deliberate, they nevertheless spend most of their time in trees and only rarely descend to the ground.

The living great apes (family Pongidae) fall into three groups: orangutans (Fig. 43.8), gorillas, and chimpanzees. All are fairly large animals that have no tail, a relatively large skull and brain, and very long arms. All have a tendency, when on

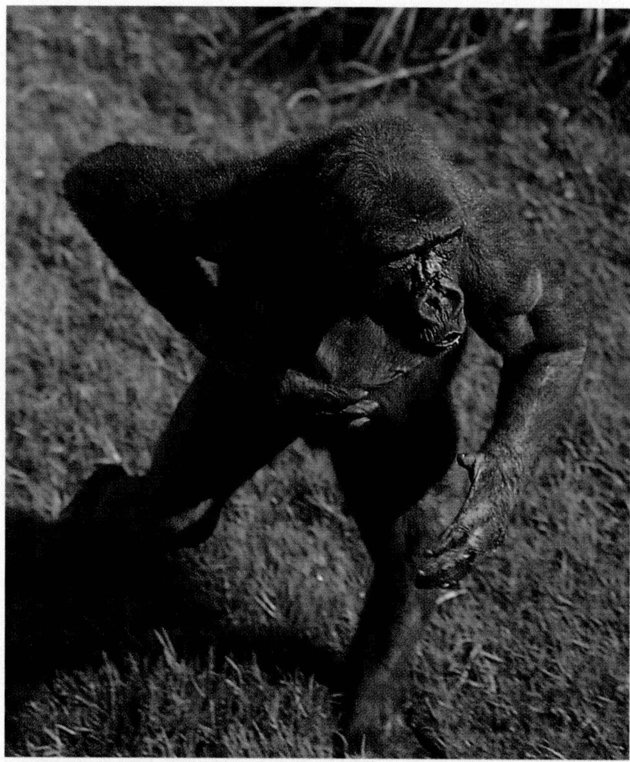

43.9 A young male lowland gorilla (*Gorilla gorilla*) thumping his chest

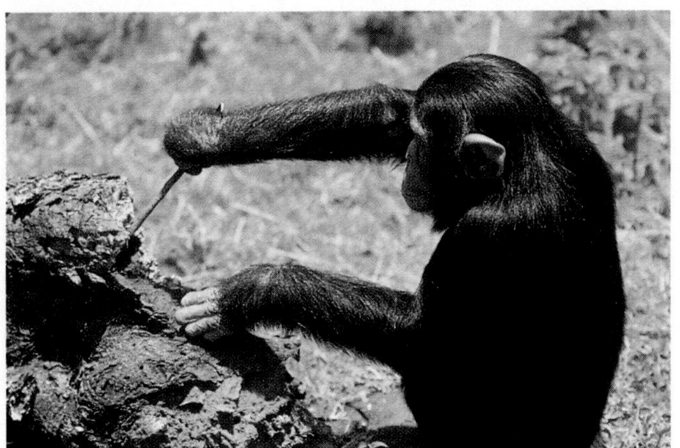

43.10 A chimpanzee using a twig as a tool to get termites out of their nest Chimps are quite intelligent and can learn to perform a variety of tasks. There appears to have been considerable recent success in teaching them to communicate with sign language, though some question remains as to whether they are actually using language or simply mimicking actions for rewards.

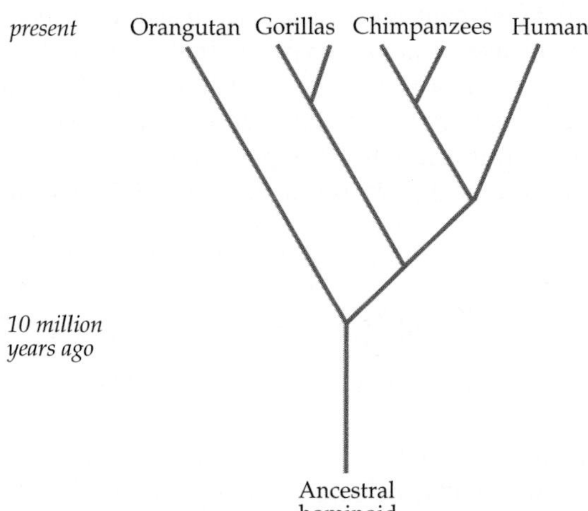

43.11 Hypothetical evolutionary relationships among the living apes and humans As the drawing indicates, there are two species each of gorillas and chimpanzees. Biochemical evidence suggests that humans are most closely related to the chimpanzees, the gorillas are less closely related, and the orangutans more distantly related.

the ground, to walk semierect, using the knuckles of the hand for support (sometimes referred to as knuckle-walking).

Gorillas are the largest of the great apes (Fig. 43.9); wild adult males may weigh as much as 450 pounds and stand six feet tall. Their arms, while proportionately much longer than human arms, are not as long as those of gibbons and orangs. Unlike gibbons and orangs, gorillas spend most of their time on the ground. Despite their fierce appearance, they are not usually aggressive.

Chimpanzees, which are native to tropical Africa, have been used extensively in psychological experiments. In general appearance they are the most human-looking of the living apes (see Fig. 43.10). Though they spend most of their time in trees, they descend to the ground more frequently than orangs, and sometimes even adopt a bipedal (two-legged) posture (their usual locomotion, however, is quadrupedal, with the knuckles of the hand used for support). They are quite intelligent and can learn to perform a variety of tasks, such as opening doors and manipulating household gadgets.

THE EVOLUTION OF HUMAN BEINGS

Humans Are Most Closely Related to Chimpanzees

The earliest humanlike species probably arose from the same ancestral stock that produced the gorillas and chimpanzees (Fig. 43.11). Both fossil evidence and biochemical data indicate that gorillas, chimpanzees, and humans are more closely related, in terms of recentness of common ancestry, than any one of them is to orangutans or gibbons. Indeed, comparisons of the amino acid sequences of their proteins and base sequences in their DNA have led some investigators to conclude that humans and chimps share about 99 percent of their genes. It has even been suggested that humans and chimps are so similar that a hybrid between the two species is possible. Indeed, the biochemical similarities among the gorillas, chimpanzees, and humans are such that many systematists believe all three should be placed in the same family, and perhaps even in the same genus. However, anatomical differences between humans and apes are so great that traditionally they have been placed in separate families; the human family is called the ***Hominidae*** (the hominids) and includes modern ***Homo sapiens*** and their fossil ancestors. The fossil evidence and the amino acid and DNA sequencing data suggest that the line leading to chimpanzees separated from the human line no more than four to six million years ago.

Many Anatomical Differences Separate Apes and Humans

A few of the many anatomical changes that occurred in the course of evolution from apelike ancestor to modern human are: (1) the jaw became shorter and more bow-shaped (rather than U-shaped as in apes), and the teeth became smaller; (2) the point of attachment of the skull to the vertebral column shifted from the rear of the braincase to a position under the braincase, which balances the skull more on top of the verte-

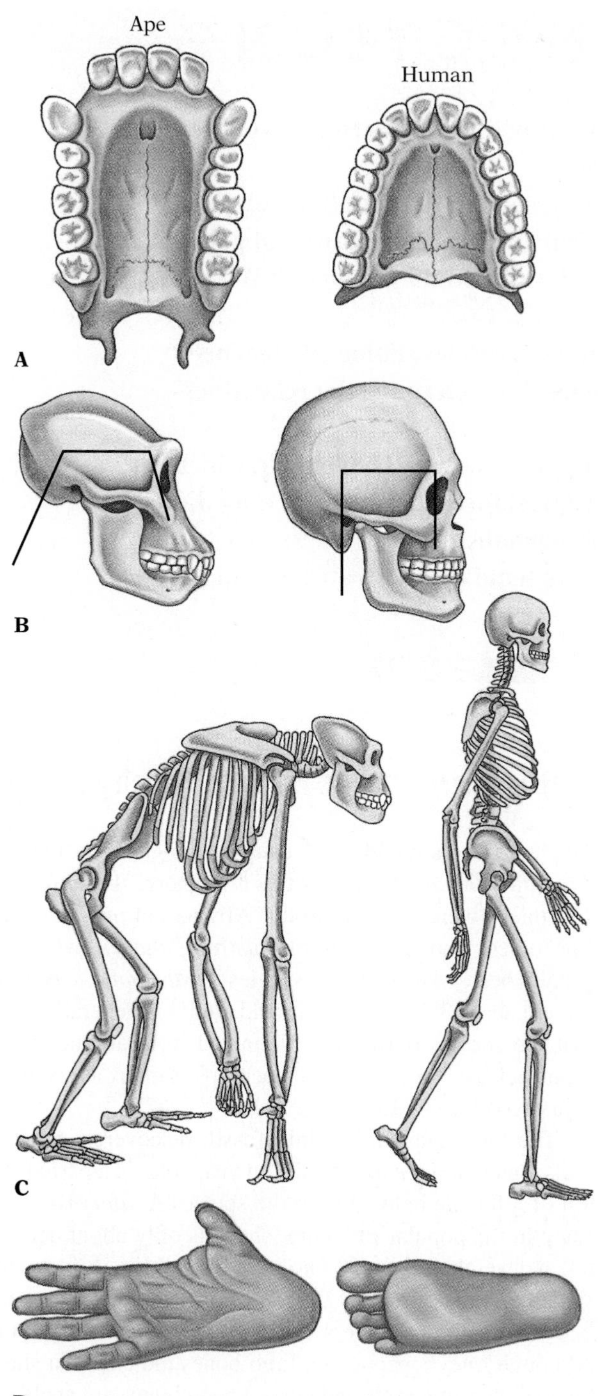

43.12 Anatomical changes between apes and humans
(A) Ape jaws are longer and more U-shaped compared to the shorter and bow-shaped human jaw. (B) A shift in the point of attachment from the rear of the braincase (apes) to a position under the braincase (humans) enables the skull to become balanced more on top of the vertebral column. Also note that the human face is flatter, with a prominent vertical forehead, and that the braincase is larger. (C) The stance of apes is quadrupedal whereas the stance of humans is upright, and bipedal. Note the differences in the shape and orientation of the pelvis, curvature of the vertebral column, and relative arm lengths. (D) The human foot is flat for bipedal walking and the big toe has moved back in line with other toes, ceasing to be opposable.

bral column; permitting a more upright, vertical posture; (3) the braincase became much larger, and, as it did, a prominent vertical forehead developed; (4) crests and ridges on the skull, such as the eyebrow ridges, were reduced as the muscles that once attached to them became smaller; (5) the nose became more prominent, with a distinct bridge and tip; (6) the arms (though probably never as long as in the modern apes) became shorter, and the legs became longer relative to the arms; (7) the posture became upright, and bipedal, which required a change in the curvature of the vertebral column (becoming more S-shaped), as well as modifications in the shape of the pelvis and positioning of the legs; (8) the feet became flattened for bipedal walking, and then an arch developed; (9) the big toes moved back into line with the other toes and ceased being opposable (Fig. 43.12).

Bipedal Walking Is a Hominid Characteristic

One of the most distinctive traits of human beings is their ***bipedal locomotion*** and upright posture. How might this trait, which is unique among mammals, have evolved? There is considerable controversy today concerning the selection pressures that led to the evolution of bipedal walking. For many years, the assumption that early hominids were primarily hunters led many theorists to believe that having both hands free for the manipulation of weapons might have been the most important selection pressure for bipedalism. However, the unusually thick tooth enamel found in early hominid fossils suggests that these animals were mostly herbivorous, probably feeding on roots, seeds, tubers, and nuts and other items picked up from the ground. Current opinion now favors the view that hunting arose much later, long after bipedalism evolved. Some theorists have suggested that bipedalism may actually have begun in the trees, with the animals balancing erect on branches, rather like a tightrope walker, much as the gibbons do today. It may then have arisen incrementally as a series of adaptations to ground feeding and the transporting of food to family members. How might this have happened?

The geological record shows that the climate underwent a change during the Miocene, and that the tropical forests gave way to more open woodland savannas and grasslands. When our ancestors moved from the forest to the savanna, their tree-dwelling forelimbs were almost certainly used for locomotion and other tasks, like carrying food. But the locomotion may well have been of the kind in which the knuckles rather than the palms are on the ground—the method used by gorillas and chimpanzees today. Now, knuckle-walking enables chimpanzees to carry objects in their hands as they walk—an important improvement in transport capabilities. And hands with opposable thumbs are preadapted not only for carrying, but also for obtaining and manipulating food. Foraging members of family groups could gather food and carry it back to other members, such as nursing mothers and young children. A

MAKING CONNECTIONS

1. Describe how primates' arboreal way of life influenced the evolution of their skeletal, visual, and nervous systems.

2. Suppose you are an art collector well acquainted with the evolutionary history of primates. If you saw a whimsical French 12th-century oil painting of a small monkey hanging from a tree by its tail, would you buy it? Why or why not?

3. How have the molecular techniques for determining phylogenetic relationships described in Chapter 32 been used to determine the relatedness of human beings and the great apes?

4. Because of climatic change during the Miocene, the biome in which our ancestors were evolving shifted from tropical forest to savanna. Why did this ecological change favor the evolution of bipedalism?

5. How does the structure of the human hand demonstrate the concept of preadaptation?

bipedal posture with hands free for carrying food, tools, infants, and other burdens is clearly advantageous. If such behavior became increasingly important to the early representatives of the hominid line, then there might have been selection for the evolution of upright posture. Another advantage of upright posture would have been the increased ease of maintaining surveillance—an advantage for an animal living on the ground in the open country.

The Human Fossil Record Is Remarkable in Its Extent

The first discovery of human fossil bones was reported in 1856 in Germany. In the years since, many bones of ancient humans have been found. Now there are literally thousands of specimens from over 600 individuals. At first, the tendency of anthropologists was to erect both a new genus and a new species for each new find, without regard to biological criteria. The result was a very long list of names that gave no indication of the probable relationships of the organisms they designated. More recently, however, the modern biological ideas concerning speciation and intraspecific variation have been increasingly applied to the study of fossil humans, and this, together with the discovery of many new fossils, particularly in Africa, has begun to improve our understanding of human evolution. But much is still unknown, and there is still considerable controversy over how the data should be interpreted; the brief sketch given here must be taken as only indicating some possible interpretations.

The Earliest Hominids Were the Australopithecines

The earliest hominids are usually assigned to the genus ***Australopithecus*** (from *australis,* "southern," because the first specimens were found in South Africa, and *pithecus* "ape"). The oldest known fossils of *Australopithecus,* which have usually been assigned to the species ***Australopithecus afarensis,*** are about 3.8 million years old, and it is generally thought that the species must have originated from an ancestral hominid at least four million (and perhaps as many as six million) years ago (Fig. 43.13).

The most famous hominid fossil, discovered in 1974 in rocks dated at three to 3.5 million years old, is a partial skeleton of a female belonging to the species *A. afarensis.* Called Lucy in the popular literature, she was only about one meter tall and weighed about 30 kilograms. Other fossils found in the same region suggest that the males of the species were much larger—up to 1.5 meters tall and 68 kilograms in weight. Although Lucy's pelvic and limb bones indicate that she was bipedal, her long arms and curved hand bones are apelike and suggest that she and her relatives may have slept in trees. Further evidence of bipedalism in these man-apes comes from the famous fossilized footprints found in ash deposits, dated between 3.6 and 3.8 million years old, at Laetoli in Tanzania. All the evidence indicates that bipedalism arose long before the acquisition of a large brain.

The fossil record collected over the last few years now suggest that the early *A. afarensis* populations diversified and split into at least three lineages that lived contemporaneously in South and East Africa for at least a million years. One lineage

43.13 A reconstruction of *Australopithecus africanus* standing Though this primitive human was fully bipedal, his stance was not as erect as that of our own species.

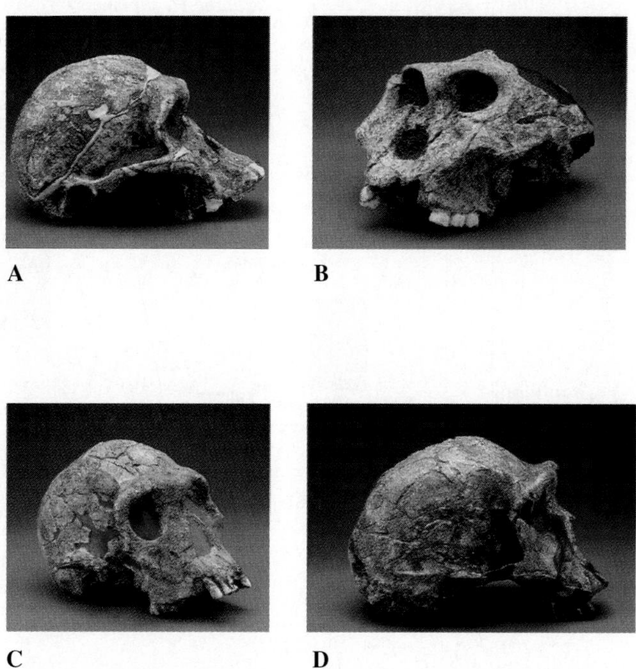

43.14 Skulls of prehistoric humans (A) *Australopithecus africanus,* skull of a child about five years old. (B) *A. robustus.* (C) *Homo habilis.* (D) *H. erectus,* often called Java man.

had smaller teeth and a more delicate ("gracile") form and is given the species name ***A. africanus*** (Fig. 43.14A). A second lineage had more robust forms, and was distinguished by large, gorillalike, bony crests on its skull and greater height (Fig. 43.14B). There were two species of robust australopithecines, the large ***A. robustus*** and the somewhat smaller ***A. boisei.*** The three species in these two lineages are believed to have lived in overlapping ranges.

The australopithecine species, whose fossil remains have been found in South and East Africa, were apparently fully bipedal, though their stance may not have been as upright as that of modern humans. They were apelike, with large jaws, almost no forehead or chin, and long arms. Their cranial capacity was only about 450 to 550 cubic centimeters, compared with 350 to 450 for normal chimps, and 1,200 to 1,600 (average about 1,360) for modern humans. The australopithecine species had large molar teeth, suggesting a diet of tough plant material, but they probably supplemented their diet with some animal food, particularly during the dry season when plant food was scarce. Some researchers suggest that they were scavengers, feeding on other animal's kills. The early hominids may also have captured small prey with their hands, much as baboons and chimpanzees do today.

The First *Homo* Species Was Still Very Apelike

The third lineage of hominids descended from the ancestral australopithecines diverged greatly and the fossil hominids from this lineage have been placed in our own genus ***Homo.*** These hominids, given the name ***Homo habilis,*** have been dated at two million to 1.8 million years old. The fossils were recovered from deposits that also contained bones of *A. africanus* and *A. robustus.* Since then, *A. boisei* remains have also been found with fossils of *H. habilis,* confirming that the australopithecines must have lived at the same time as *Homo. Australopithicus afarensis,* then, is now thought to be the ancestor of both the *Homo* and the *Australopithecus* lines (Fig. 43.15).

Homo habilis (Fig. 43.14C) had a larger cranial capacity than the australopithecines, usually from 500 to 700 cubic centimeters, but were still rather apelike in many of their skeletal

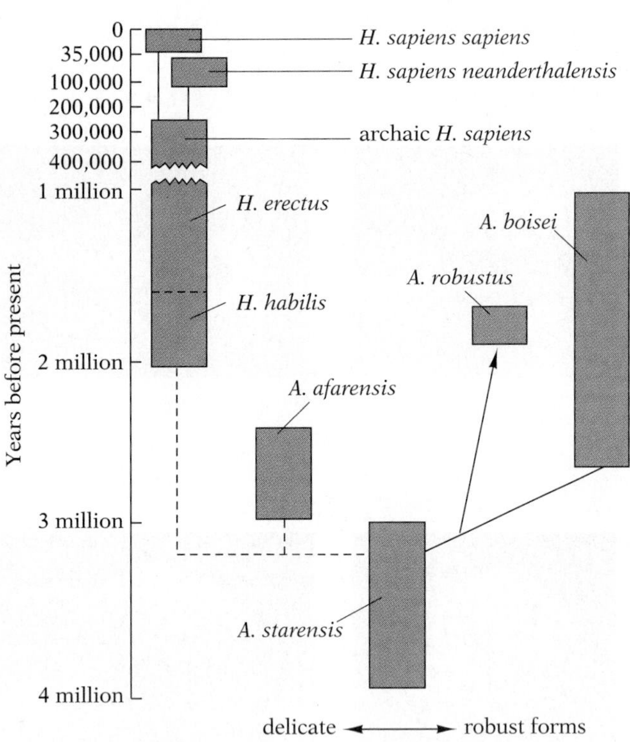

43.15 Hypothetical evolutionary relationships among the known hominids There is still no universal consensus upon which to base a phylogenetic tree of hominids. This figure shows one of several possible arrangements. The relationships among the early hominids is unclear, which is indicated by the dashed lines. The boxes represent the approximate length of time each species was thought to exist, based on fossil evidence. Notice that there are several periods in which there were two or more different species of hominids living at the same time.

features, such as possession of long arms and curved hand bones (Fig. 43.16). Like Lucy's species, there was a large difference in size between the males and females, with the females being rather small (around one meter) and the males almost twice as large. There is convincing evidence that *H. habilis* not only used stones as tools but also chipped and shaped them for various purposes. (The word *habilis* means "skillful" and refers to the ability to make tools.) No tools that could be clearly classified as hunting weapons have been found, however; instead there are tools that function as scrapers, hammerstones, and hand axes.

While the australopithecines became extinct, *H. habilis* survived and evolved rapidly, showing an increasing brain size and height. There was also a decrease in the size of the molar and premolar teeth, suggesting that the hominids were eating higher quality plant food or processing their food in some way, or perhaps turning more to meat to supply their dietary protein.

43.16 A comparison of reconstructed skeletons of six hominid species

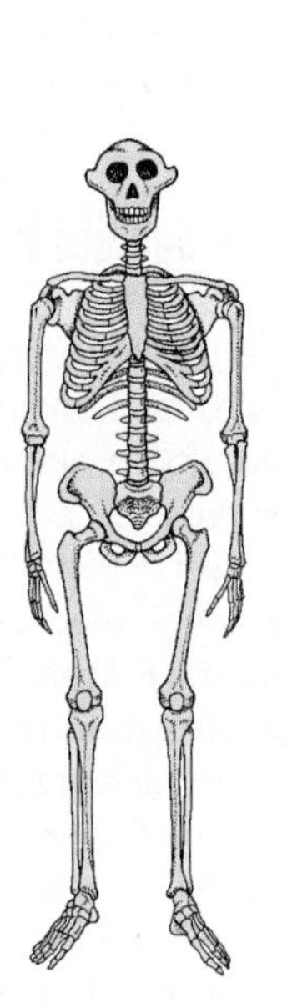

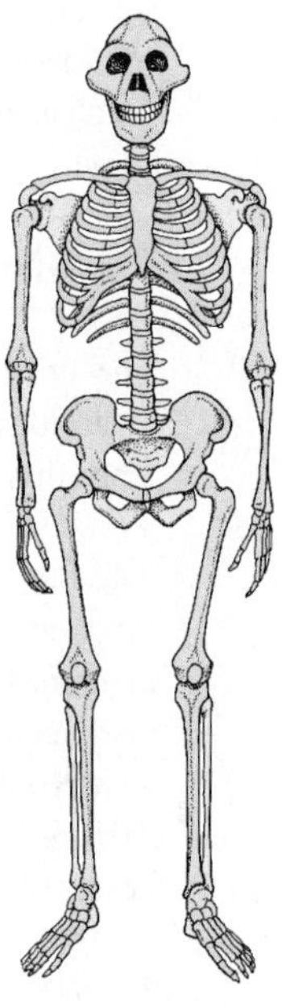

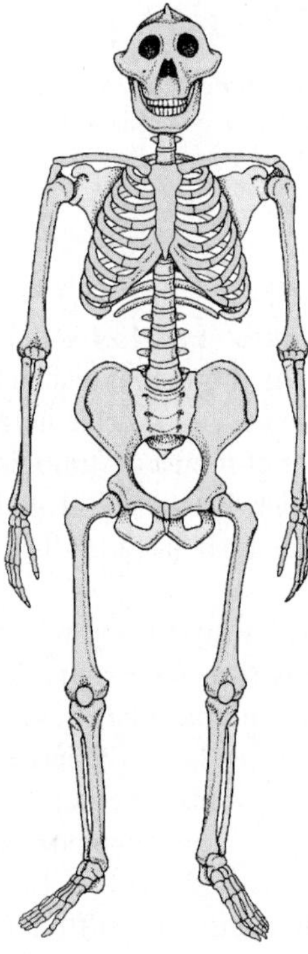

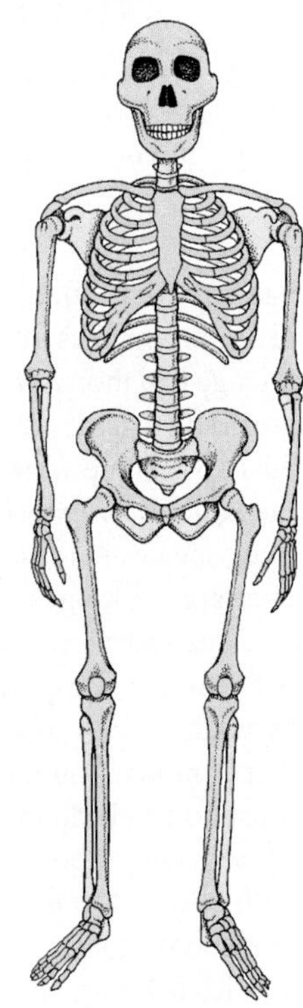

Homo erectus Evolved From *Homo habilis* About 1.6 Million Years Ago

A later stage in human evolution is represented by fossils that are classified as ***Homo erectus.*** This species, which almost certainly descended from *H. habilis,* first appeared in Africa about 1.6 million years ago. It was the first hominid to leave Africa; beginning about a million years ago it migrated across Europe and eventually reached Asia. Its cranial capacity was considerably larger than that of *H. habilis,* averaging about 900 cubic centimeters, which probably reflects a strong selection pressure for intelligence. Edward O. Wilson of Harvard University has estimated that the human brain grew at a rate of about one tablespoon every 100,000 years during the period from 2.25 million years until 250,000 years ago.

The limb and pelvic bones of *H. erectus* indicate that they walked with a more upright posture. However, the facial features remained primitive, with a projecting massive jaw, large teeth, almost no chin, a receding forehead, heavy, bony eyebrow ridges, and a broad, low-bridged nose (Fig. 43.14D). Grooves in the interior of the skulls indicate the presence of the speech areas of the brain; of course we have no way of knowing whether or how language was used. The extreme size difference between males and females seen in earlier hominids is diminished, suggesting that a new social structure may have evolved with reduced competition between males.

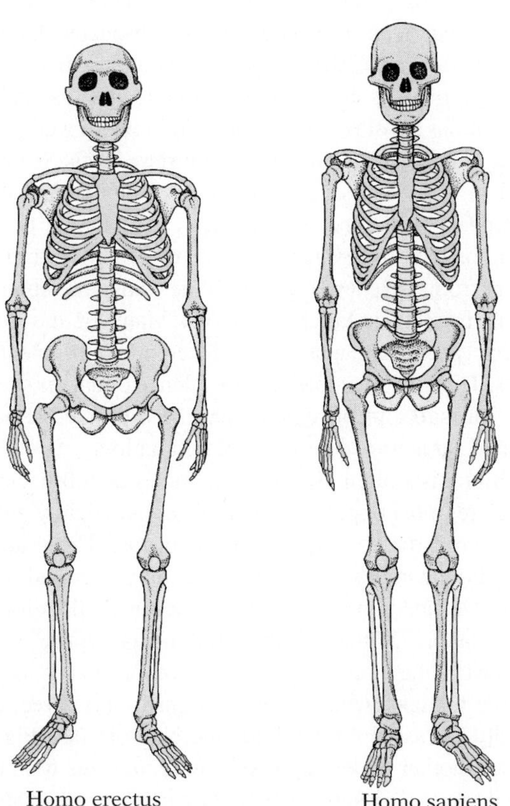

Not only did the members of this species make and use tools, but they also used fire, and meat probably formed an important part of their diet. ***H. erectus*** soon became the most dangerous animal on earth, hunting in groups with stone weapons to kill large animals like giraffe that had previously known little predation. The species apparently became extinct about 300,000 years ago.

The Earliest *Homo sapiens* Fossils Are About 300,000 Years Old

Modern humans are given the Latin name ***Homo sapiens*** ("wise man"). Early representatives of this species, known as ***archaic Homo sapiens,*** probably evolved from *H. erectus,* but it is uncertain where the early stages of this evolutionary transition occurred. The fossils of archaic *H. sapiens* share many primitive features with *H. erectus,* and it is difficult to separate the latest fossils of *H. erectus* from those of early *H. sapiens.* The oldest fossils of archaic humans are from England and continental Europe, but the species may well have migrated there from Africa or southern Asia.

During the period from about 130,000 to 32,000 years ago, a very distinctive form of *H. sapiens,* designated ***Homo sapiens neanderthalensis,*** Neanderthal man, evolved from archaic *H. sapiens.* The Neanderthals lived throughout most of Europe and also in parts of Asia. Because they were living at the time of the last ice age, they must have been cold-adapted. Neanderthals were about five to $5\frac{1}{2}$ feet tall, and had a receding forehead, prominent eyebrow ridges, and a receding chin, but their brain was actually bigger than ours—approximately 10 percent bigger. Their fossil record shows that they had very thick, heavy bones and were massively muscled, particularly in the neck and shoulders, and were almost certainly much stronger than modern humans—it has been estimated that the strongest individuals could probably lift weights of about 1,000 pounds! Their hands too were much more powerful than ours. Neantherthals made many kinds of stone tools, and they buried their dead (Fig. 43.17), which has been interpreted as a capacity on their part for abstract and religious thought. The presence of skeletons of old and severely impaired individuals suggests that Neanderthals cared for their sick and aged individuals.

Modern humans, ***Homo sapiens sapiens*** ("wise, wise man"), may have arisen from another isolated group of archaic *H. sapiens.* Despite a lively debate (see Exploring Further: Where and When Did Modern Human Beings Evolve, p. 991), we still do not where and when these modern humans evolved. Neanderthals abruptly disappeared around 32,000 years ago, after the modern form, the earliest representatives of which are often called Cro-Magnon man, arrived in their ranges. This modern form is the only kind of human found on the earth

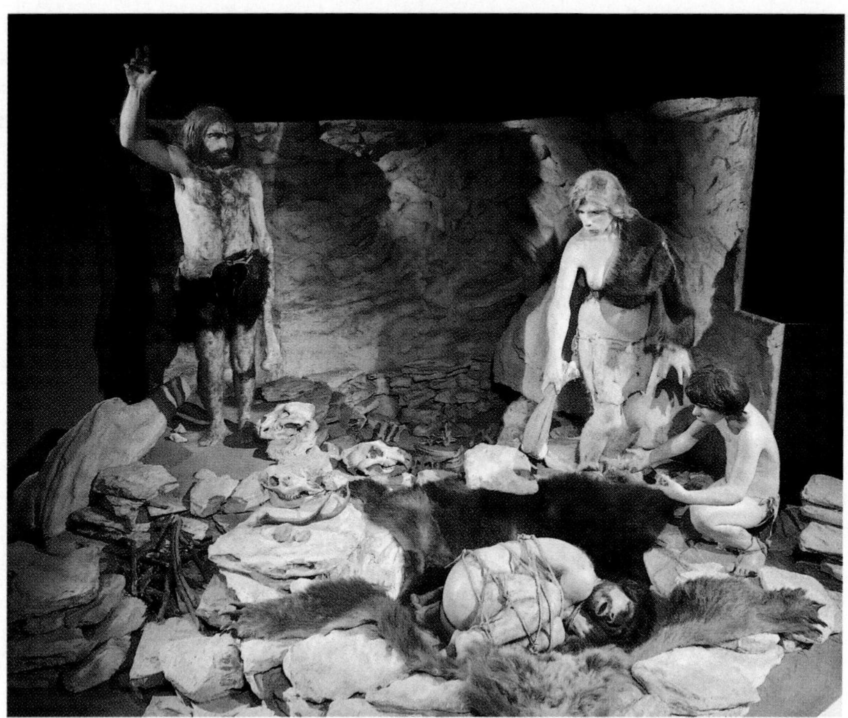

43.17 A recreation of a Neanderthal burial

today. Why the Neanderthals disappeared so rapidly is unclear. They may have been eliminated by combat or competition, or they may merely have been absorbed into *Homo sapiens sapiens* by interbreeding.

Our species evolved as hunter-gatherers, under strong selection pressure for high intelligence (or at least large brain volume), but the domestication of animals and then plants only 10,000 to 20,000 years ago made possible great changes in social structure. With domesticated herds and cultivated grain providing reliable sources of food, groups no longer had to be nomadic and small to follow game. The result was the beginnings of civilization as we see it now across most of the world, characterized by cities, division of labor, private wealth, and large-scale central authority.

Modern civilization has created a new environment for our species, to which we must adapt as best we can. It is logical to suppose that our societies represent a compromise between the behaviors that evolved under the pressure of selection for a hunter-gatherer niche, and the new behaviors required by the technology on which we now so completely depend. Conflicting demands from the evolutionary past and the cultural present may create many of the stresses we associate with modern civilization. But the fossil evidence of human evolution can tell us little about the more interesting questions of early human behavior. Though there is evidence that our ancestors must have been social, we have no idea what their social system may have been. And though their diet is well understood, we still do not know whether our evolution was shaped by a relative scarcity of animal protein, or carbohydrate, or salt—or indeed whether the critical determinants involved food at all.

The Human Population Shows Geographic Variation

As we saw in Chapter 32, widespread species often tend to become subdivided into geographical subspecies that differ genetically and have different ranges. Humans are no exception. *Homo sapiens* is an extremely variable species, and regional populations are often recognizably different (or were, prior to the great mobility of the last few centuries). Thus Scandinavians tend to have blue eyes and a fair complexion, while south Europeans tend to have brown eyes and a darker complexion. Eskimos look different from Mohawks, who in turn look different from Apaches. Pygmies of the Congo are obviously different from their taller neighbors. Many of the differences probably reflect adaptations to different environmental conditions. Thus, for example, the prevalence of darker skin in tropical and subtropical regions may be a protective adaptation against damaging ultraviolet solar radiation.

Biologists often use the term "races or subspecies" to describe regional populations that differ genetically but have no effective intrinsic isolating mechanisms. There are seldom sharp boundaries between them, and they intergrade over wide areas. Designation of races in most species is therefore an arbitrary matter, and there is no such thing as a "pure" race. Some authorities have chosen to recognize as many as 30 races among humans, while others recognize only three: the traditional Caucasoid, Mongoloid, and Negroid (see Fig. 32.6, p. 707). Another widely used system recognizes five: the traditional three, plus Native Americans and Australian aborigines.

No one of these systems has any more biological validity

EXPLORING FURTHER

WHERE AND WHEN DID MODERN HUMAN BEINGS EVOLVE?

About a million years ago *Homo erectus* migrated out of Africa and populated vast regions of Europe and Asia. The fossil record of this species shows it to have died out about 300,000 years ago. Because the oldest fossils of anatomically modern *Homo sapiens* are only about 100,000 years old, the fossil record is unclear regarding how, when, and where modern humans originated. The issue has generated two competing hypotheses among paleoanthropologists.

One hypothesis, proposed by the "Out-of-Africa" group, is that *H. erectus* gave rise to modern humans in Africa, and that modern humans continued evolving in Africa, eventually migrating out of Africa and spreading around the world, completely replacing all other human groups living at the time, including the Neanderthals (Fig. A). Support for this view comes from the study of DNA found in mitochondria. Mitochondrial DNA differs from nuclear DNA in several ways. Mitochondrial chromosomes are small, encoding only 37 genes, and are maternally inherited, which means they are passed from mother to child as a unit. (In contrast, half of the DNA in the nucleus comes from the father, and half from the mother.) This greatly simplifies the analysis procedure, because mutation provides the only source of variation in mitochondrial genes. Moreover, mitochondrial DNA has a higher mutation rate than nuclear DNA, so more changes can be observed. By sequencing and comparing mitochondrial DNA from people around the world, researchers can see how similar or different the sequences are and estimate how closely or distantly related they are.

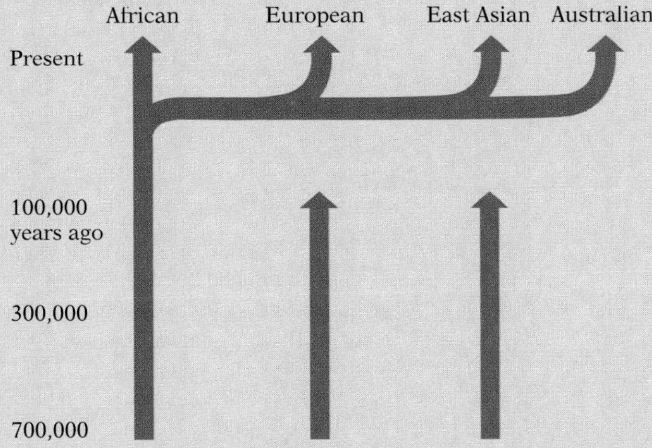

A Out-of-Africa replacement hypothesis.

Those individuals that are closely related would be expected to have very similar mitochondrial DNA sequences because there would have been little time to accumulate mutations, while those more distantly related should show more differences. Using such information, a tree of descent was formulated, tracing all the studied maternal lineages backward until they reached a common female ancestor (dubbed "Mitochondrial Eve"). The researchers were struck by the fact that mitochondrial DNA among Africans is much more diverse than that of any other group, suggesting that humans have lived in Africa longer than anywhere else. Eve, therefore, must have been an African. A measure of how long ago this woman lived was made using a "molecular clock." Assuming mutations occur at a constant rate, geneticists can estimate the number of mutations that could be expected to occur in a given time period. The clock was "calibrated" by comparing the genetic differences accumulated by humans with those of chimpanzees, which are known to have diverged from humans about five million years ago. Using the molecular clock researchers concluded that all living humans trace their ancestry back to a single African woman who lived about 200,000 years ago. Eve's descendants later left their African homeland and replaced the more primitive humans living in Europe and Asia.

The competing hypothesis is the "regional continuity" hypothesis, which maintains that humans left Africa at a much earlier time, at least a million years ago, and slowly evolved their modern form independently in various regions of the Old World (Fig. B). The regional populations did not evolve into separate species because there was always interbreeding and therefore gene flow between them. Proponents of this hypothesis point to the fossil record as the real evidence for human evolution, citing dated fossils at different sites that support the linked evolution of humans.

When and how the debate will be settled is still in doubt. The regional continuity theorists point out several problems with the Out-of-Africa hypothesis. One is the total replacement scenario: the idea that one group of humans could invade and then completely replace all other groups in every single climate and environment seems highly unlikely and unprecedented. They also point out that if the Out-of-Africa hypothesis is correct, the earliest modern humans in all areas should have African features, and that there should have been a rapid range expansion about 100,000 years ago, before the other populations were driven

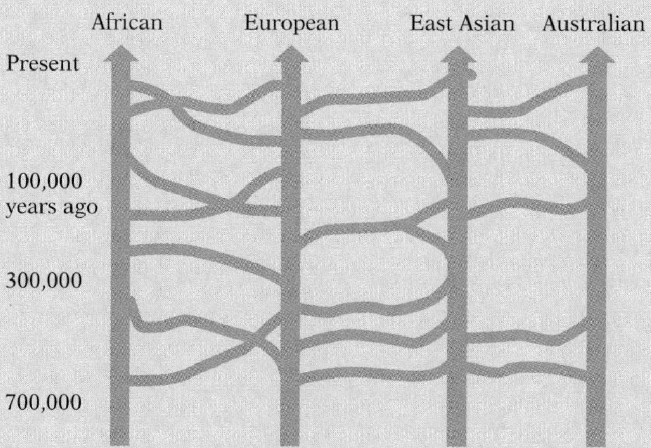

B Multiregional evolution hypothesis.

to extinction. This would produce discontinuities in the fossil record before and after replacement, but the fossil record does not appear to support any of this. Criticism also comes from molecular geneticists concerning the manner in which the computer program was used to generate the Eve family tree. Apparently the program can generate thousands of equally plausible family trees, and not all of them have an African origin. Still another problem lies in the accuracy of the mitochondrial molecular clock. Some researchers suggest that this clock is inaccurate; in some cases the rate of ticking seems too slow, in others too fast, and that the 200,000-year-ago origin of Eve is highly suspect. On the other side, the supporters of the Out-of-Africa hypothesis point out that the fossil record is spotty, and open to various interpretations and biases. Molecular data, they suggest, are objective, verifiable, and free from theoretical prejudices. Furthermore, genetic analyses of human nuclear DNA also show that Africans have much more variation in their DNA than other groups, supporting the African heritage of modern humans. Ultimately, molecular geneticists hope that analysis of DNA preserved in bone in ancient remains may help settle the question. Until then, the debate rages on.

than the others, since subspecies and races, as categories, are largely human inventions. What is biologically real is the geographic variation within the species ***Homo sapiens,*** a variation that will surely tend to break down as people move about more and more. The main barriers to interbreeding in many parts of the world are now cultural or social rather than geographic, and it seems very unlikely that such barriers will ever approach the effectiveness of the original geographic barriers. Hence, whatever races are recognized now, it seems probable that they will become less and less distinct as time goes on. This, too, is a phenomenon that has occurred countless times in other species.

Cultural and Biological Evolution Interact

One of the most interesting discoveries of modern anthropologists is that early hominids used tools long before their brains were much larger than those of apes. Hence the old idea that a

MAKING CONNECTIONS

1. How has the relationship between diet and dentition described in Chapter 17 been used to make inferences about the diet of fossil hominids?

2. How have size differences between male and female fossils been used to make inferences about the evolution of hominid social structure and behavior?

3. How did the ecological niche of *Homo erectus* differ from that of the earlier *H. habilis*?

4. How may the stresses that many people feel in modern civilization be related to our evolutionary past?

5. What is the biological basis for the attempt to divide human beings into races? Why has this effort led to so many different racial classification systems?

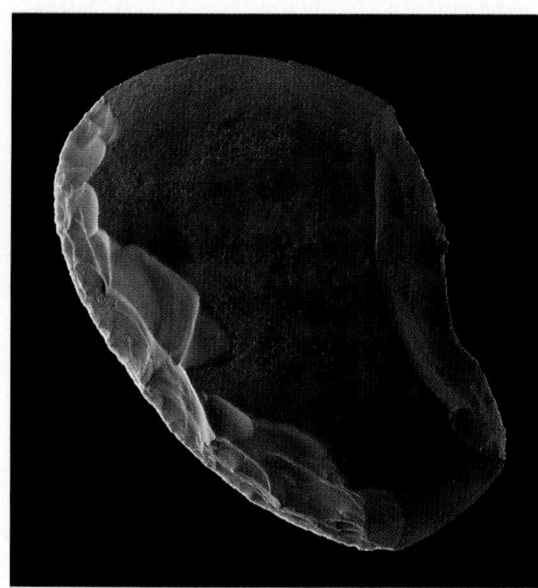

43.18 A primitive hide scraper This artifact represents a type of stone tool used by early humans to dress hides, for perhaps half a million years.

large brain and high intelligence were necessary prerequisites for the use of tools has been discredited. The early hominids' use of tools may, in fact, have been an important factor in leading to evolution of higher intelligence. Once the use and making of tools began (Fig. 43.18), individuals that excelled in these endeavors would surely have had an advantage over their less talented fellows. There would thus have been strong selection for neural mechanisms making possible improved fashioning and use of tools. So perhaps, instead of considering culture the crown of the fully evolved human intelligence, we should regard early cultural development and increasing intelligence as two faces of the same coin; in a sense, the highly developed brain of modern human beings may be as much a consequence as a cause of culture.

Cultural evolution can proceed at a far more rapid pace than biological evolution. Words as units of inheritance are much more effective than genes in spreading new developments and in giving dominance to new approaches originating with a few talented individuals. But the two types of evolution continue to be interwoven just as they were in the use of tools; they may well be even more so in the future.

Human beings, by their unrivaled ability to alter their environment, are influencing in profound ways the evolution of all species with which they come into contact. Thus, as we have seen in recent years, new strains of bacteria have evolved in response to the use of antibiotics; many species of insects have evolved new physiological and behavioral traits as a consequence of the intense selection resulting from the use of insecticides; the clearing of forests for agricultural purposes has led to drastic decline in the population densities of some species and to increase in others; long-established balances between prey species and their predators and parasites have been destroyed, often with far-reaching consequences for the entire ecosystem.

While some human actions on the environment have been deliberate, many have been unintentional. But whether deliberate or unintentional, they have precipitated a period of wholesale and rapid change unmatched since life began. Since disruption of the ecosystem will unavoidably grow as civilization expands, the great challenge to our species is to use our knowledge and technology to guide the change in ways that will benefit both our own species and the other organisms around us.

Humans Can Directly Affect Their Future Evolution

Not only can we humans influence the evolution of other species, but we now also have the ability to alter deliberately some aspects of the future evolution of our own species. Thus modern medicine, by saving people with gross genetic defects that would once have been fatal, permits perpetuation of genes that natural selection would formerly have eliminated. Does this mean that we should practice eugenics—deliberately restrict, by law or by social pressure, the perpetuation of some genetic traits and encourage the perpetuation of others? Some thinkers, concerned over what they see as the inevitable physical "decline" of the species, have urged just that. Others have pointed out that though there may indeed be an increase in traits that would once have been maladaptive, modern human beings live in an environment of their own design, in which those traits are no longer so harmful.

Our ability to control our own evolution may eventually no longer depend primarily on regulating reproduction. Now that the genetic code has been deciphered, the day will surely come when the DNA of genes can deliberately be altered in order to design, at least in part, new human beings. When that day comes, how do we decide what to design? What do we look for in human beings? We might all agree to rid our species of the genes for muscular dystrophy or sickle-cell anemia (at least in areas where there is no malaria), but once the techniques for achieving these apparently worthy ends are mastered, suggestions will surely be put forward for other alterations, to which we cannot all agree.

More immediately pressing, perhaps, is the problem of regulating the size of human populations now that we have interfered with the action of the many former natural regulating factors. Already some people are asking whether we should abandon the campaigns to eradicate malaria, to cure cancer and heart disease, or to slow the aging process. They point out that the current population problem is the result of major advances in the technology of death control, which societies have generally been as eager to accept as they have been reluctant to accept compensatory birth control. Since no species can continue indefinitely with its birth and death rates unbalanced, and since there is little evidence that our species will consent soon enough to the efficacious population-control measures that become increasingly necessary as the traditional killers are overcome, perhaps the only solution is less death control, heartless as that may seem.

These complex and unnerving questions—at once biological, economic, political, and moral—must be faced, and soon. Human beings have already gone too far toward modifying biological evolution to pull back. The answers we give to these questions in the next few generations may well have as profound an influence on the future of life as anything that has happened since the first cells materialized in the primordial seas.

CONCEPTS IN BRIEF

1. Fossil evidence indicates that the members of the mammalian order Primata arose from a tree-dwelling stock of small shrewlike insectivores. Important characteristics of the primates include grasping hands (with nails and opposable first digits), increased mobility of the shoulder joint (braced by the clavicle), rotational movement of the elbow, good stereoscopic vision, and expansion of the brain. These traits are associated with life in the trees.

2. The most primitive primates, the prosimians, gave rise to several lineages: the tarsiers, New World monkeys, and Old World monkeys. The apes evolved from the line leading to the Old World monkeys.

3. The living great apes are large animals with no tail, a large skull and brain, and long arms. All have a tendency to walk semierect. The earliest members of the family Hominidae (humans) probably arose from the same stock that produced the gorillas and chimpanzees.

4. In the course of human evolution from a man-ape ancestor, the jaw became shorter, the skull more balanced on top of the vertebral column, the braincase larger, the nose more prominent, the arms shorter, and the feet flatter and the big toe ceased being opposable. The ape and human lineages probably split between four and six million years ago.

5. The upright posture and bipedal locomotion of human beings may have evolved as adaptations for freeing the hands to carry objects or to gather or manipulate food, or may have made it easier to maintain surveillance.

6. The first truly humanlike fossils belong to the genus *Australopithecus*. Fossils indicate that these early hominids were bipedal but rather apelike in their skeletal features. There were apparently at least three species of Australopithecus living contemporaneously.

7. *Homo habilis* evolved from *Australopithecus* about two million years ago. They had larger brains and used stone tools. *H. habilis* was later replaced by *Homo erectus*. *Homo sapiens* probably evolved from this form about 300,000 years ago.

8. Two forms of *H. sapiens* arose: *H. sapiens neanderthalensis* in Europe and modern *H. sapiens sapiens*. Modern humans spread over Eurasia and replaced the Neanderthals about 30,000 years ago.

9. Cultural and biological evolution are interwoven, though cultural evolution can proceed more rapidly than biological evolution. Human beings profoundly influence the evolution of other species and now have the ability to influence and control their own evolution.

STUDY QUESTIONS

1. **What characteristics do humans share with other primates? How are many of these traits linked to the arboreal way of life of our primate ancestors? (pp. 979–980)**
2. **How do the New World and Old World monkeys differ? To which group are human beings more closely related? (pp. 981–984)**
3. **What characteristics do the great apes have in common? Which of them is most closely related to humans? (pp. 983–984)**
4. **Describe the anatomical changes that occurred in the course of evolution from an apelike ancestor to modern humans. (pp. 984–986)**

5. How are the upright posture and bipedal locomotion of human beings believed to have evolved? (pp. 985–986)

6. Describe the characteristics of the australopithecines. Where did they live? What was the earliest species? What lineages evolved from it? (pp. 986–987)

7. When did *Homo habilis* appear? How did it differ from the australopithecines? What species replaced *Homo habilis* and what were its characteristics? (pp. 987–989)

8. When did *Homo sapiens* evolve? Compare Neanderthal man and modern man. What hypotheses have been proposed to explain the disappearance of the Neanderthals? (pp. 989–990)

APPENDIX 1

THE CHEMICAL ELEMENTS AND THE PERIODIC TABLE

I*A*	II*A*	III*B*	IV*B*	V*B*	VI*B*	VII*B*	VIII			I*B*	II *B*	III*A*	IV*A*	V*A*	VI*A*	VII*A*	0
1 H 1.00794																	2 He 4.00260
3 Li 6.941	4 Be 9.01218											5 B 10.811	6 C 12.011	7 N 14.0067	8 O 15.9994	9 F 18.998403	10 Ne 20.1797
11 Na 22.98977	12 Mg 24.305											13 Al 26.98154	14 Si 28..0855	15 P 30.97376	16 S 32.066	17 Cl 35.4527	18 Ar 39.948
19 K 39.0983	20 Ca 40.078	21 Sc 44.9559	22 Ti 47.88	23 V 50.9415	24 Cr 51.996	25 Mn 54.9380	26 Fe 55.847	27 Co 58.9332	28 Ni 58.6934	29 Cu 63.546	30 Zn 65.39	31 Ga 69.723	32 Ge 72.61	33 As 74.9216	34 Se 78.96	35 Br 79.904	36 Kr 83.80
37 Rb 85.4678	38 Sr 87.62	39 Y 88.9059	40 Zr 91.224	41 Nb 92.9064	42 Mo 95.94	43 Tc (97.907)	44 Ru 101.07	45 Rh 102.9055	46 Pd 106.42	47 Ag 107.8682	48 Cd 112.411	49 In 114.82	50 Sn 118.71	51 Sb 121.757	52 Te 127.60	53 I 126.9045	54 Xe 131.29
55 Cs 132.9054	56 Ba 137.327	57–71 Rare Earths	72 Hf 178.49	73 Ta 180.9479	74 W 183.85	75 Re 186.207	76 Os 190.2	77 Ir 192.22	78 Pt 195.08	79 Au 196.9665	80 Hg 200.59	81 Tl 204.383	82 Pb 207.2	83 Bi 208.9804	84 Po (208.9804)	85 At (209.9871)	86 Rn (222.0176)
87 Fr (223.020)	88 Ra (226.0254)	89–103 Acti- nides	104 Rf (261.11)	105 Ha (262.114)	106 (263.118)	107 (262.12)	108 (265)	109 (266)									

Rare Earths (Lanthanides)	57 La 138.9055	58 Ce 140.15	59 Pr 140.9077	60 Nd 144.24	61 Pm (144.913)	62 Sm 150.36	63 Eu 151.965	64 Gd 157.25	65 Tb 158.9253	66 Dy 162.50	67 Ho 164.9303	68 Er 167.26	69 Tm 168.93421	70 Yb 173.04	71 Lu 174.967
Actinides	89 Ac (227.0278)	90 Th 232.0381	91 Pa 231.0359	92 U 238.0289	93 Np (237.0482)	94 Pu (244.064)	95 Am (243.061)	96 Cm (247.070)	97 Bk (247.070)	98 Cf (251.080)	99 Es (252.083)	100 Fm (257.095)	101 Md (258.10)	102 No (259.101)	103 Lr (262.11)

[a] In each box, the upper number is the atomic number. The lower number is the atomic mass, that is, the mass (in grams) of one mole or, alternatively, the mass (in atomic mass units) of one atom. Numbers in parentheses denote the atomic masses of the most stable or best-known isotope of the element; all other numbers represent the average masses of a mixture of several isotopes as found in naturally occurring samples of the element.

APPENDIX 2
BASIC UNITS OF MEASUREMENT

TABLE 1 *Standard prefixes of the metric system*

kilo- (k)	1.000	10^{3}
deci- (d)	0.1	10^{-1}
centi- (c)	0.01	10^{-2}
milli- (m)	0.001	10^{-3}
micro- (μ)	0.000001	10^{-6}
nano- (n)	0.000000001	10^{-9}

TABLE 2 *Common units of length, weight, and liquid capacity*

kilometer (km)	1,000 m	0.62137 mile
meter (m)		39.37 inches
centimeter (cm)	0.01 m	0.39 inch
millimeter (mm)	0.001 m	0.039 inch
micrometer* (μm)	10^{-6} m	
nanometer (nm)	10^{-9} m	
angstrom† (Å)	10^{-10} m	
kilogram (kg)	1,000 g	2.2 pounds
gram (g)		0.035 ounce
milligram (mg)	0.001 g	
microgram (μg)	10^{-6} g	
liter (l)	1,000 cm^3	1.057 quarts
milliliter (ml)	0.001 l	

*Formerly called micron.
†No longer used; nanometer used instead

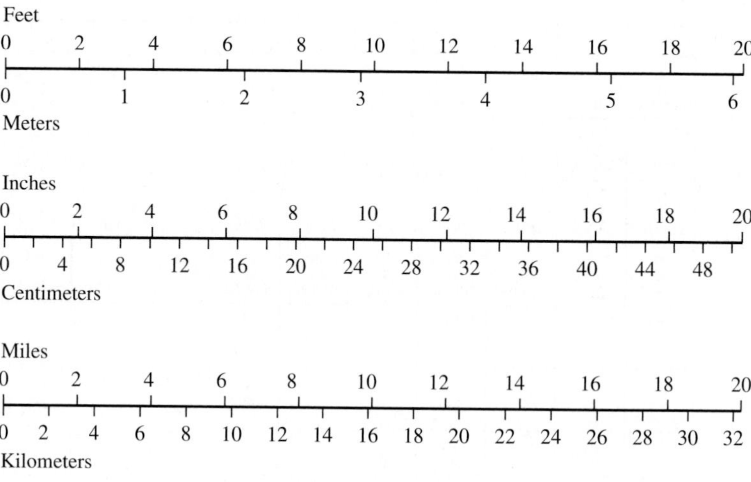

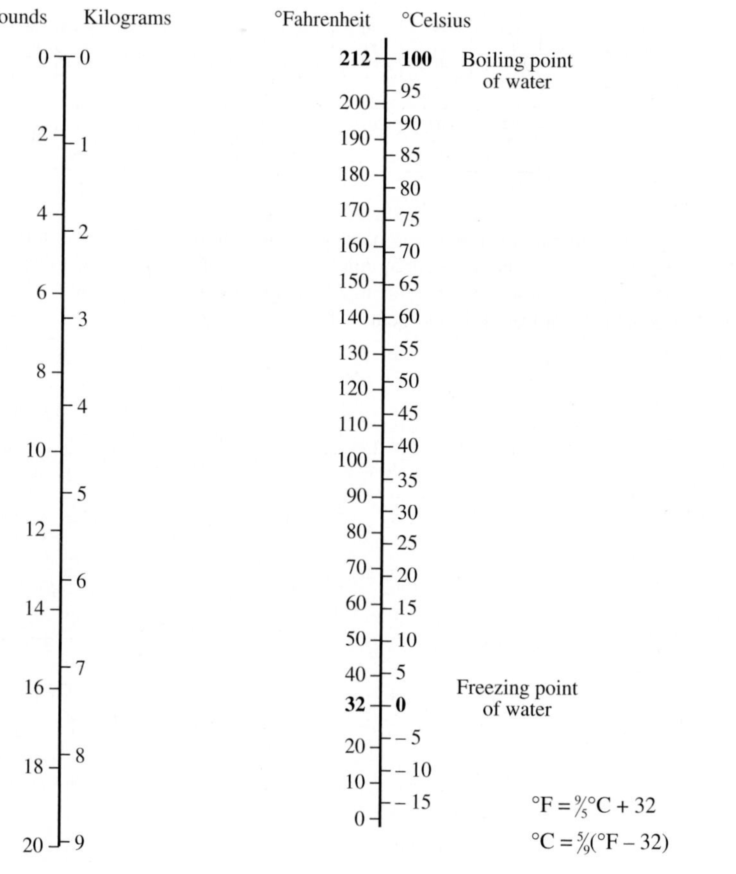

APPENDIX 3
GEOLOGIC TIME SCALE

*Era**	*Period**	*Epoch**		*Climates and major physical events*	*Life forms*
CENOZOIC	Quaternary	Recent	**0.01**	Alternating cold and mild climates. Four glaciations and retreats; uplift of Sierra Nevada	Age of humans. Large scale extinction of many large mammals and birds. Increase in number of herbs.
		Pleistocene	**2**		
	Tertiary	Pliocene	**5.1**	Cooler. Much uplift of mountains; glaciers in Northern Hemisphere. North and South America joined by uplift of Panama	First humans. Large carnivores, apes evolve.
		Miocene	**24.6**	Moderate climate. Extensive glaciation in Southern Hemisphere.	Spread of grasslands as forests shrink. Many grazing animals.
		Oligocene	**38**	Rise of Alps and Himalayas. South America separates from Antarctica.	Browsing mammals, monkey-like primates; many modern genera of plants evolve.
		Eocene	**54.9**	Mild to very tropical climate. Australia separates from Antarctica; India collides with Asia.	Extensive radiation of mammals and birds; initial formation of grasslands.
		Paleocene	**65**	Mild to cool climate. Wide, shallow continental seas largely disappear.	Early carnivores and primates.
MESOZOIC	Cretaceous		**144**	Rise of Rockies. Africa and South America separate.	Angiosperms arise. Second radiation of insects. Last of the dinosaurs.
	Jurassic		**213**	Mild climate. Most of Europe covered by seas. Pangaea breaks up.	Age of Reptiles. First birds. Dinosaurs abundant. Cycads, ferns dominant.
	Triassic		**248**	Continents mountainous, joined into a supercontinent. Large areas arid.	Age of Reptiles. Forests of conifers and ferns. First dinosaurs and first mammals.
PALEOZOIC	Permian		**286**	Extensive glaciation in Southern Hemisphere. Appalachians uplifted.	Origin of conifers and cycads. Reptiles diversify; amphibians decline.
	Carboniferous		**360**	Warm climate, lands low and swampy. Formation of coal deposits.	Age of amphibians. Origin of reptiles. First great insect radiation.
	Devonian		**408**	Sea over most of the land, with mountains in Europe.	Age of fishes. Expansion of early vascular plants; origin of first seed plants.
	Silurian		**438**	Mild climate. Continents generally flat.	Major extinction event. First jawed fishes. First vascular plants and arthropods invade land.
	Ordovician		**505**	Mild climate. Shallow seas cover flat continents.	Major extinction event. First vertebrates.
	Cambrian		**590**	Mild. Extensive seas cover present continents.	Explosive evolution of phyla and divisions. First chordates.
PRECAMBRIAN				Formation of earth's crust and continents.	Origin of procaryotes by 3.6 bya; eucaryotes, by 1.8 bya; multicellular animals by 700 mya[a]

*The numbers on the lines of the geologic time divisions indicate the age (in millions of years) at which it began.

[a]mya = millions of years ago; bya = billions of years ago.

APPENDIX 4
A CLASSIFICATION OF LIVING THINGS

The classification given here is one of many in current use and includes only those phyla and divisions mentioned in the text.

Botanists have traditionally used the term "division" for the major groups that zoologists have called phyla. In classifications recognizing only two or three kingdoms, this difference in terminology causes little difficulty, because usage can be consistent within each kingdom. But when a kingdom Protista is recognized, as it is here, consistency is achieved only at the expense of violating well-established usage. The Protista contain some plant-like and funguslike groups traditionally called divisions and some animal-like groups traditionally called phyla. These usages we have respected.

Some classes within a division or phylum are listed here, and for some classes—Insecta and Mammalia, for example—orders are given too. Except for a few extinct groups of particular evolutionary importance, like Placodermi, only groups with living representatives are included. A few of the better-known genera are mentioned as examples in each of the taxons.

Whenever possible, an estimate (a very rough one) of the number of living species is provided for higher taxons.

KINGDOM EUBACTERIA* (3,000)

DIVISION PURPLE BACTERIA. *Pseudomonas, Escherichia, Rhizobium, Rickettsia, Coxiella*

DIVISION PROCHLOROPHYTA. *Chloroxybacteria, Prochloron*

DIVISION CYANOBACTERIA. *Oscillatoria, Nostoc, Gloeocapsa*

DIVISION GREEN SULFUR BACTERIA. *Chlorobium*

DIVISION SPIROCHETES. *Treponema, Spirochaeta*

DIVISION GRAM-POSITIVE BACTERIA

Bacillus, Clostridium, Mycoplasma, Staphylococcus, Actinomyces, Streptomyces

KINGDOM ARCHAEBACTERIA (200)

DIVISION METHANOGENS. *Methanobacterium*

DIVISION HALOPHILES. *Halobacterium*

DIVISION THERMOACIDOPHILES. *Thermoplasma*

DIVISION SULFUR REDUCERS.

KINGDOM PROTISTA

ALGAL PROTISTS

DIVISION EUGLENOPHYTA Euglenoids (800). *Euglena*

DIVISION DINOFLAGALLATA. Dinoflagellates (1,000).

DIVISION CHRYSOPHYTA. Golden-brown and yellow green algae, diatoms (6, 650).

DIVISION PHAEOPHYTA. Brown algae (1,500). *Sargassum, Ectocarpus, Fucus, Laminaria*

DIVISION RHODOPHYTA. Red algae (4,000). *Chondrus, Nemalion*

FUNGAL PROTISTS

DIVISION OOMYCOTA. Water molds, white rusts, down mildews (400). *Saproglenia, Phytophthora, Albugo*

DIVISION MYCETOZOA. Cellular slime molds, true slime molds (430).

ANIMAL-LIKE PROTISTS

DIVISION SPOROZOA. Sporulating protozoa (3,600). *Plasmodium, Toxoplasma, Gregorina*

DIVISION KINETOPLASTIDA. (17,000).

Subphylum Mastigophora. Flagellates. *Trypanosoma*

Subphylum Sarcodina. Naked and shelled amoebae; formaminiferans. *Amoeba, Pelomyxa, Entamoeba*

DIVISION CILIATA. Cilliates (6,000). *Paramecium, Stentor, Vorticella*

KINGDOM PLANTAE

DIVISION CHLOROPHYTA. Green algae (7,300). *Chlamydomonas, Volovx, Ulothrix, Spirogyra, Oedogonium, Ulva, Chara, Nitella*

DIVISION HEPATOPHYTA. Liverworts (6,000). *Marchantia, Conocephalum, Riccia*

DIVISION ANTHOCEROPHYTA. The hornworts (100).

DIVISION BRYOPHYTA. Mosses (9,000). *Polytrichum, Sphagnum, Mnium*

DIVISION PSILOPHYTA. Psilopsids. *Psilotum*

DIVISION LYCOPHYTA. Club mosses (1,500). *Lycopodium, Selaginella*

DIVISION SPHENOPHYTA. Horsetails (25). *Equisetum*

DIVISION PTEROPHYTA. Ferns (1,000). *Polypodium*

DIVISION CONIFEROPHYTA. Conifers (550). *Pinus, Tsuga, Taxus, Sequoia*

DIVISION CYCADOPHYTA. Cycads (100).

DIVISION GINKGOPHYTA. Ginko. *Ginko*

DIVISION ANTHOPHYTA. Flowering plants (235,000).

Class, Monocotyledones. Monocots (65,000). *Lilium, Tulipa, Poa, Zea, Triticum*

Class Dicotyldeones. Dicots (170,000). *Magnolia, Quercus, Acer, Rosa, Chrysanthemum, Aster, Ligustrum, Ranunculus*

KINGDOM FUNGI

DIVISION ZYGOMYCOTA. Conjugation fungi (250). *Rhizopus, Mucor, Phycomyces*

DIVISION ASCOMYCOTA. Sac fungi (30,000). Yeasts, powdery mildews, fruit molds. *Aspergillus, Penicillium, Sordaria, Neurospora*

DIVISION BASIDIOMYCOTA. Club fungi (25,000). Toadstools, bracket fungi, mushrooms, puffballs, rusts, smuts. *Coprinus, Amanita, Agaricus, Phallus*

DIVISION DEUTEROMYCOTA. Imperfect fungi (17,000). *Penicillium, Candida, Aspergillus*

KINGDOM ANIMALIA*

SUBKINGDOM PARAZOA

PHYLUM PORIFERA. Sponges (5,000). *Scypha, Leucosolenia, Grantia, Spongilla*

SUBKINGDOM EUMETAZOA

SECTION RADIATA

PHYLUM CNIDARIA (or Coelenterata).

CLASS HYDROZOA. Hydrozoans (3,700). *Hydra, Obelia, Gonionemus, Physalia*

CLASS SCYPHOZOA. Jellyfish (200). *Aurelia, Pelagia*

CLASS ANTHOZOA. Sea anemones and corals (6,100). *Metridium, Astrangia*

PHYLUM CTENOPHORA. Comb jellies (90). *Pleurobrachia, Haeckelia*

SECTION PROTOSTOMIA

PHYLUM PLATYHELMINTHES. Flatworms (10,000).

CLASS TURBELLARIA. Free-living flatworms. Planaria, Dugesia

CLASS TREMATODA. Flukes. Fasciola, Schistosoma, Prosthogonimus

CLASS CESTODA. Tapeworms. Taenia, Dipylidium, Mesocestoides

PHYLUM NEMERTEA (or Rhynchocoela). Proboscis or ribbon worms (650).

PHYLUM ROTIFERA.* Rotifers (1,700). *Asplanchna, Hydatina, Rotaria*

PHYLUM NEMATA.* Roundworms or nematodes (12,000). *Ascaris, Trichinella, Necator, Enterobius, Ancylostoma, Heterodera*

PHYLUM BRYOZOA† (or Ectoprocta). Bryozoans, moss animals (4,000).

PHYLUM BRACHIOPODA.† LAMP SHELLS (300)

PHYLUM MOLLUSCA. Molluscs.

CLASS PHYLACOPHONA. Chitons (600).

CLASS MONOPLACOPHORA (8). Neopilina

CLASS GASTROPODA. Snails and their allies (univalve molluscs) (25,000). Helix, Busycon, Crepidula, Haliotis, Littorina, Doris, Limax

CLASS SCAPHOPODA. Tusk shells (350).

CLASS BIVALVA. Bivalve molluscs (7,500). Mytilus, Ostrea, Pecten, Mercenaria, Teredo, Tagelus, Unio, Anodonta

CLASS CEPHALOPODA. Squids, octopuses, etc. (600). *Loligo, Octopus, Nautilus*

PHYLUM ANNELIDA. Segmented worms.

CLASS POLYCHAETA Sandworms, tubeworms, etc. (8,000). *Nereis, Chaetopterus, Aphrodite, Arenicola*

CLASS OLIGOCHAETA. Earthworms and many freshwater annelids (3,100). *Tubifex, Enchytraeus, Lumbricus*

CLASS HIRUDINOIDEA. Leeches (500).

PHYLUM ONYCHOPHORA (65). *Peripatus. Peripatopsis*

PHYLUM ARTHROPODA (at least 2,000,000)

Subphylum Trilobita. No living representatives

Subphylum Chelicerata

CLASS MEROSTOMATA. Horseshoe crabs (4). *Limulus*

CLASS ARACHNIDA. Spiders, ticks, mites, scorpions, whip-scorpions, daddy longlegs, etc. (55,000; at least 500,000 undiscovered species of mites are thought to exist). Archaearanea, Latrodectus, Argiope, Centruroides, Chelifer, Mastigoproctus, Phalangium, Ixodes

Subphylum Uniramia (or Mandibulata)

CLASS CRUSTACEA (26,000). *Homarus, Cancer, Daphnia, Artemia, Cyclops, Balanus, Porcellio*

CLASS CHILOPODA. Centipedes (2,500)

CLASS DIPLOPODA. Millipedes (10,000; another 50,000 *species are thought to exist).*

CLASS INSECTA. Insects (900,000; another 1,000,000 *species are thought to exist)*

ORDER THYSANURA. Bristletails, silverfish, firebrats. *Machilis, Lepisma. Thermobia*

* The classification of Animalia has been adapted from *Synopsis and Classification of Living Organisms,* ed. S. P. Parker, McGraw-Hill, New York, 1982.

*Formerly considered a class of Phylum Aschelminthes.

† Bryozoa, Phoronida, and Brachiopoda are often referred to as the lophophorate phyla.

ORDER EPHEMERIDA Mayflies. *Hexagenia, Callibaetis, Ephemerella*

ORDER ODONATA. Dragonflies, damselflies. *Archilestes, Lestes, Aeshna, Gomphus*

ORDER ORTHOPTERA. Grasshoppers, crickets, etc. *Schistocerca, Romalea, Nemobius, Megaphasma*

ORDER PHASMATOPTERA. Walking sticks. *Phyllium*

ORDER BLATTARIA. Cockroaches. *Blatta, Periplaneta*

ORDER MANTODEA. Mantids. *Mantis*

ORDER GRYLLOBLATTARIA. *Grylloblatta*

ORDER ISOPTERA. Termites. *Reticulitermes, Kalotermes, Zootermopsis, Nasutitermes*

ORDER DERMAPTERA. EARWIGS, *Labia, Forficula, Prolabia*

ORDER EMBIIDINA (or Embiaria or Embioptera). *Oligotoma, Anisembia, Gynembia*

ORDER PLECOPTERA. Stoneflies. *Isoperla, Taeniopteryx, Capnia, Perla*

ORDER ZORAPTERA. *Zorotypus*

ORDER PSOCOPTERA. Book lice. *Ectopsocus, Liposcelis, Trogium*

ORDER MALLOPHAGA. Chewing lice. *Cuclogtogaster, Menacanthus, Menopon, Trichodectes*

ORDER ANOPLURA. Sucking lice. *Pediculus, Phthirius, Haematopinus*

ORDER THYSANOPTERA. Thrips. *Heliothrips, Frankliniella, Hercothrips*

ORDER HEMIPTERA. True bugs. *Belostoma, Lygaeus, Notonecta, Cimex, Lygus, Oncopeltus*

ORDER HOMPTERA. Cicadas, aphids, leafhoppers, scale insects, etc. *Magicicada, Circulifer, Psylla, Aphis, Saissetia*

ORDER NEUROPTERA. Dobsonflies, alderflies, lacewings, mantispids, snakeflies, etc. *Corydalus, Hemerobius, Chrysopa, Mantispa, Agulla*

ORDER COLEOPTERA. Beetles, weevils. *Copris, Phyllophaga, Harpalus, Scolytus, Melanotus, Cicindela, Dermestes, Photinus, Coccinella, Tenebrio, Anthonomus, Conotrachelus*

ORDER HYMENOPTERA. Wasps, bees, ants, sawflies. *Cimbex, Vespa, Glypta, Scolia, Bembix, Formica Bombus, Apis*

ORDER STREPSIPTERA. Endoparasites

ORDER MECOPTERA. Scorpionflies. *Panorpa, Boreus, Bittacus*

ORDER SIPHONAPTERA. Fleas. *Pluex, Nosopsyllus, Xenopsylla, Ctenocephalides*

ORDER DIPTERA. True flies, mosquitoes. *Aedes, Asilus, Sarcophaga, Anthomyia, Musca, Chironomus, Tabanus, Tipula, Drosophila*

SECTION DEUTEROSTOMIA

PHYLUM ECHINODERMATA

Subphylum Crinozoa Crinoids, sea lilies (630).

Subphylum Asterozoa Sea stars, brittle stars (2,600). *Asterias*

Subphylum Echinozoa Sea urchins, sand dollars, heart urchins (860) and Sea cucumbers (900).

PHYLUM HEMICHORDATA (90) Acorn worms. *Saccoglossus, Balanoglossus*

PHYLUM CHORDATA. Chordates

Subphylum Tunicata (or Urochordata). Tunicates (2,000)

Subphylum Cephalochordata. Lancelets, amphioxus (30). *Branchiostoma, Asymmetron*

Subphylum Vertebrata. Vertebrates

CLASS AGNATHA. Jawless fish (50). *Cephalaspis,* Pteraspis,* Petromyzon, Entosphenus, Myxine, Eptatretus*

CLASS ACANTHODII. No living representatives

CLASS PLACODERMI. No living representatives

CLASS CHONDRICHTHYES. Cartilaginous fish, including sharks and rays (800). *Squalus, Hyporion, Raja, Chimaera*

CLASS OSTEICHTHYES. Bony fish (18,000)

SUBCLASS SARCOPTERYGII lungfish.

SUBCLASS ACTINOPTERYGH. Ray-finned fish. *Amia, Cyprinus, Gadus, Perca, Salmo*

CLASS AMPHIBIA (3,100) Frogs, toads, and salamanders. *Rana, Hyla, Bufo Necturus, Triturus, Plethodon, Ambystoma*

CLASS REPTILIA (6,500)

ORDER TESTUDINES. Turtles. *Chelydra, Kinosternon, Clemmys, Terrapene*

ORDER RHYNCHOCEPHALIA. Tuatara, *Sphenodon*

ORDER CROCODYLIA. Crocodiles and alligators. *Crocodylus, Alligator*

ORDER LEPIDOSAURIA. Snakes and lizards, *Iguana, Anolis, Sceloporus, Phrynosoma, Natrix, Elaphe, Coluber, Thamnophis, Crotalus*

CLASS AVES. Birds (8,600). *Anas, Larus, Columba, Gallus, Turdus, Dendroica, Sturnus, Passer, Melospiza*

CLASS MAMMALIA. Mammals (4,100)

SUBCLASS PROTOTHERIA

ORDER MONOTREMATA. Egg-laying mammals. *Ornithorhynchus, Tachyglossus*

**Extinct.*

SUBCLASS THERIA. Marsupial and placental mammals

ORDER METATHERIA (or Marsupialia). Marsupials. *Didelphis, Sarcophilus, Notoryctes, Macropus*

ORDER INSECTIVORA. Insectivores (moles, shrews, etc.). *Scalopus, Sorex, Erinaceus*

ORDER DERMOPTERA. Flying lemurs. *Galeopithecus*

ORDER CHIROPTERA.* Bats, *Myotis, Eptesicus, Desmodus*

ORDER PRIMATA. Lemurs, monkeys, apes, humans, *Lemur, Tarsius, Cebus, Macacus, Cynocephalus, Pongo, Pan, Homo*

ORDER EDENTATA. Sloths, anteaters, armadillos, *Bradypus, Myrmecophagus, Dasypus*

ORDER PHOLIDOTA. Pangolin. *Manis*

ORDER LAGOMORPHA. Rabbits, hares, pikas, *Ochotona, Lepus, Sylvilagus, Oryctolagus*

ORDER RODENTIA. Rodents. *Sciurus, Marmota, Dipodomys, Microtus, Peromyscus, Rattus, Mus, Erethizon, Castor*

ORDER ODONTOCETA. Toothed whales, dolphins, porpoises. *Delphinus, Phocaena, Monodon*

ORDER MYSTICETA. Baleen whales. *Balaena*

ORDER CARNIVORA. Carnivores, Canis, Procyon, Ursus, Mustela, Mephitis, Felis, Hyaena, Eumetopias

ORDER TUBULIDENTATA. Aardvark. *Orycteropus*

ORDER PROBOSCIDEA. Elephants. *Elephas, Loxodonta*

ORDER HYRACOIEDEA. Hyraxes, conies. *Procavia*

ORDER SIRENIA. Manatees. *Trichechus, Halicore*

ORDER PERISSODACTYLA. Odd-toed ungulates. *Equus, Tapirella, Tapirus, Rhinoceros*

ORDER ARTIODACTYLA. Even-toed ungulates. *Pecari, Sus, Hippopotamus, Camelus, Cervus, Odocoileus, Giraffa, Bison, Ovis, Bos*

*Some authorities propose dividing the echo-locating bats and fruit-eating bats into separate orders.

APPENDIX 5
COMMON ANATOMICAL TERMS

The following terms are often used to denote orientation and direction in anatomical work; you may find it useful to learn these terms and refer to this list whenever necessary.

(1) right and left: the organism's right and left (not yours!)
(2) anterior: toward the head
(3) posterior: toward the tail
(4) caudal: toward the tail
(5) dorsal: toward the back; up
(6) ventral: toward the belly; down
(7) lateral: toward the side; (right or left)
(8) medial: toward the middle or center
(9) proximal: near the point of attachment or origin
(10) distal: far from the point of attachment or origin
(11) pectoral: referring to the shoulder region
(12) pelvic: referring to the hip region
(13) paired: having a mirror opposite; coming in twos
(14) longitudinal axis: the midline of the body; anterior-posterior axis
(15) cross section (transverse section): a cut at right angles to the longitudinal axis
(16) frontal section: a horizontal cut along the longitudinal axis
(17) sagittal section: a vertical cut through or parallel to the longitudinal axis

APPENDIX 6
GLOSSARY

The Glossary gives brief definitions of the most important terms that recur in the text, excluding taxonomic designations. For fuller definitions, consult the index, where italicized page numbers refer you to explanations of key terms in context.

Of the basic units of measurement, some are tabulated on p. A2 , others have their own alphabetical entries.

Interspersed alphabetically with the vocabulary are the main prefixes and combining forms used in biology. You will notice that, while they are generally of Greek or Latin origin, many of them have acquired a new meaning in biology (examples: ***blasto-***, ***-cyte***, ***caryo-***, ***-plasm***). Familiarity with these forms will make it easier for you to learn and remember the numerous terms in which they are incorporated.

a- Without, lacking.

ab- Away from, off.

abdomen [L belly] In mammals, the portion of the trunk posterior to the thorax, containing most of the viscera except heart and lungs. In other animals, the posterior portion of the body.

abscisic acid A plant hormone that promotes the transport of photosynthetic products to growing embryos in seeds, induces protein synthesis in seeds, and promotes dormancy. It also causes the stomata to close when water is in short supply.

abscission The dropping of leaves, seeds, or fruits.

abscission layer A thin-walled layer of cells that forms at the base of a petiole, permitting the leaf to fall off the stem.

absolute zero The temperature (-273° C) at which all thermal agitation ceases. The lowest possible temperature.

absorption spectrum A measure of the amount of energy at different wavelengths of light that is absorbed by a specific pigment. Each photosynthetic pigment has a characteristic absorption spectrum.

abyssal plains The deep ocean basins where no light penetrates.

acellular Not constructed on a cellular basis.

acetylcholine A transmitter chemical in the peripheral nervous system.

acetyl CoA An intermediate compound formed in the metabolism of fatty acids and glucose; entry compound for the Krebs cycle.

acid [L *acidus* sour] A substance that increases the concentration of hydrogen ions when dissolved in water; a hydrogen ion donor. It has a pH lower than 7.

acoelomate A body plan in which there is no cavity between the digestive tract and the body wall.

ACTH *See* adrenocorticotropic hormone.

action potential *See* potential.

activation energy The energy required to initiate a chemical reaction.

active site In an enzyme, the portion of the molecule that binds with a substrate molecule.

active transport Movement of a substance across a membrane by transport proteins requiring expenditure of energy by the cell.

ad- Next to, at, toward.

adaptation In evolution, any genetically controlled characteristic that increases an organism's fitness, usually by helping the organism to survive and reproduce in the environment it inhabits. In neurobiology, the process that results in a short-lasting decline in responsiveness of a sensory neuron after repeated firing; *cf.* habituation, sensitization.

adaptive radiation The rapid evolutionary splitting of species into many separate descendant species when organisms have new evolutionary opportunities or conditions change rapidly.

adenosine diphosphate (ADP) A doubly phosphorylated organic compound that can be further phosphorylated to form ATP.

adenosine monophosphate (AMP) A singly phosphorylated organic compound that can be further phosphorylated to form ADP.

adenosine triphosphate (ATP) A triply phosphorylated organic compound that functions as "energy currency" for organisms.

adipose [L *adeps* fat] Fatty.

ADP *See* adenosine diphosphate.

adrenal [L *renes* kidneys] An endocrine gland of vertebrates located near the kidney. It consists of two separated parts, the outer cortex and inner medulla.

adrenalin A hormone produced by the adrenal medulla that stimulates "fight-or-flight" reactions, also called epinephrine.

adrenocorticotropic hormone (ACTH) A hormone produced by the anterior pituitary that stimulates the adrenal cortex.

adsorb [L *sorbēre* to suck up] Hold on a surface.

advanced New, unlike the ancestral condition, derived.

aerobic [L *aer* air] With oxygen.

afferent Carrying substances or impulses from the periphery of the body or organ towards its center. Usually used with reference to neurons or blood vessels.

agnathans Jawless class of vertebrates.

agonistic behavior Embraces all aspects of behavior exhibited during hostile encounters between members of the same species. Combatants convey their attack or reproductive motivation through displays without physical combat; the individual showing the higher motivation is ordinarily the winner.

alcohol Any of a class of organic compounds in which one or more —OH groups are attached to a carbon backbone.

alkaline Having a pH of more than 7. *See* base.

all-, allo- [Gk *allos* other] Other, different.

allele Any of several alternative gene forms at a given chromosomal locus.

allopatric [L *patria* homeland] Having different ranges.

allopatric speciation A type of speciation that occurs when portions of an ancestral population are completely separated by a geographic barrier.

allosteric Of an enzyme: one that can exist in two or more conformations. *Allosteric control:* control of the activity of an allosteric enzyme by determination of the particular conformation it will assume.

altruistic behavior The willingness of an individual to sacrifice its fitness for the benefit of another. *Reciprocal altruism:* the performance of a favor by one individual in the expectation of a favor in return, as when two animals groom each other.

alveolus [L little hollow] A small cavity, especially one of the microscopic cavities that are the functional units of lungs.

amino acid An organic acid carrying an amino group ($—NH_2$); the building-block compound of proteins.

amniocentesis A technique in which a needle is inserted into the amniotic cavity of a pregnant woman to remove fetal cells for the purpose of determining whether certain genetic abnormalities are present.

amnion [Gk caul] An extraembryonic membrane that forms a fluid-filled sac containing the embryo in reptiles, birds, and mammals.

amniotic egg The land egg of reptiles, birds, and mammals in which the embryo develops within a fluid filled environment with a supply of yolk, and which is surrounded by protective extraembryonic membranes and (in reptiles and birds), a shell.

amoeboid [Gk *amoibē* change] Amoebalike in the tendency to change shape by protoplasmic flow.

AMP *See* adenosine monophosphate.

amylase [L *amylum* starch] A starch-digesting enzyme.

an- Without.

anabolism [Gk *ana-* upward; *metabolē* change] The biosynthetic building-up aspects of metabolism.

anaerobic [L *aer* air] Without oxygen.

analogous Of characters in different organisms: similar in function and often in superficial structure but of different evolutionary origins.

anemia A condition in which the blood has lower than normal amounts of hemoglobin or red blood corpuscles.

angio-, -angium [Gk *angeion* vessel] Container, receptacle.

anion A negatively charged ion.

annelids Segmented worms belonging to the phylum Annelida.

annuals Plants that complete their life cycle in a single year.

anterior Toward the front end.

antibiotic Substances produced by microorganisms (usually molds) that slow or prevent the growth of other microorganisms.

anticodon A sequence of three nucleotides in transfer RNA that is complementary to the three base codon on messenger RNA.

antheridium [Gk *anthos* flower] Male reproductive organ of a plant; produces sperm cells.

antibody A protein, produced by the B lymphocytes of the immune system, that binds to a particular antigen.

antigen A substance, usually a protein or polysaccharide, that activates an organism's immune system.

anus [L ring] Opening at the posterior end of the digestive tract, through which indigestible wastes are expelled.

aorta The main artery of the systemic circulation.

aphotic zone The lower portion of the ocean, usually below 200 meters, where light does not penetrate enough for photosynthesis to occur.

apical At, toward, or near the apex, or tip, of a structure such as a plant shoot.

apical meristem Areas of active cell division found at the tips of shoots and roots. Apical meristems produce primary tissues that enable plant to grow in length.

apo- Away from.

apoplast The network cell walls and intercellular spaces within a plant body; permits extensive extracellular movement of air and water within the plant.

aqueous solution A solution in which the solvent is water.

arch- [Gk *archein* to begin] Primitive, original.

archaebacteria One of the two lineages of procaryotic organisms; these bacteria are adapted to survive in certain harsh environments.

archegonium [Gk *archegonos* the first of a race] Female reproductive organ of an embryophyte; produces egg cells.

archenteron [Gk *enteron* intestine] The cavity in an early embryo that becomes the digestive cavity.

arteriole A small artery.

artery A blood vessel that carries blood away from the heart.

articulation A joint between bones. Articulating surfaces are those formed between bones and joints.

artifact A by-product of scientific manipulation rather than an inherent part of the thing observed.

ascus [Gk *askos* bag] The elongate spore sac of a fungus of the Ascomycota group.

asexual reproduction Without sex; a type of reproduction in which one parent gives rise to genetically identical offspring by division of a single cell, budding, or fission.

atherosclerosis An extremely widespread disease in which excessive cholesterol-rich fatty deposits, smooth muscle, and fibrous tissues are laid down within the inner lining of the blood-vessel walls. Eventually the walls become narrowed and hardened to form rigid tubes interfering with normal blood flow and causing blood clots to develop.

atmosphere (atm) (unit of pressure) The normal pressure of air at sea level: 101,325 newtons per square meter (approx. 14.7 pounds per square inch).

atom [Gk *atomos* indivisible] The smallest unit of an element, not divisible by ordinary chemical means.

atomic mass unit (amu) *See dalton.*

atomic weight The average mass of an atom of an element relative to ^{12}C, an isotope of carbon with six neutrons in the nucleus. The atomic mass of ^{12}C has arbitrarily been fixed as 12.

ATP *See* adenosine triphosphate.

atrium Chamber of the heart that receives blood from the veins.

auto- Self, same.

autonomic nervous system A portion of the vertebrate nervous system, comprising motor neurons that innervate internal organs and are not normally under direct voluntary control.

autosome [Gk *sōma* body] Any chromosome other than a sex chromosome.

autotrophic [Gk *trophē* food] Capable of manufacturing organic nutrients from inorganic raw materials.

auxin [Gk *auxein* to grow] Any of a class of plant hormones that promote cell elongation.

axon [Gk *axōn* axis] A fiber of a nerve cell that conducts impulses away from the cell body and can release transmitter substance.

bacillus A rod shaped bacterium.

bacteriophage [Gk *phagein* to eat] A virus that attacks bacteria, *abbrev.* phage.

Barr body A dark body in the nucleus of cells of female mammals which is the tightly condensed and inactivated X chromosome.

basal At, near, or toward the base (the point of attachment) of a structure such as a limb.

basal body A structure, identical to the centriole, found at the base of cilia and eucaryotic flagella; consists of nine triplet microtubules arranged in a circle.

base A substance that decreases the hydrogen ion concentration in solution; a hydrogen ion acceptor. It has a pH higher than 7.

basement membrane An extracellular layer of material that separates epithelial (surface) tissue from the underlying connective tissue.

basidium The spore-bearing structure of Basidiomycota (club fungi).

Batesian mimicry A type of mimicry in which a harmless or palatable species resembles one that is harmful or unpalatable.

bi- Two.

bilateral symmetry The property of having two similar sides, with definite upper and lower surfaces and anterior and posterior ends.

binary fission Reproduction by the division of a cell into two essentially equal parts by a nonmitotic process.

bio- [Gk *bios* life] Life, living.

biogenesis [Gk *genesis* source] Origin of living organisms from other living organisms.

biogeography The study of the distribution of organisms.

biological control The control of pests through biological rather than chemical means.

biological magnification Increasing concentration of relatively stable chemicals as they are passed up a food chain from initial consumers to top predators.

biomass The total weight of all the organisms, or of a designated group of organisms, in a given area.

biome Major world communities that are characterized by the predominant vegetation and climate (e.g. tropical rain forest, deciduous forest).

biosphere The region of the earth's surface, atmosphere, and ocean that is inhabited by living organisms; the global ecosystem.

biosynthesis The production of molecules by the living cell.

biotechnology The development of techniques for using biological processes for the production of substances useful in medicine or industry.

biotic Pertaining to life.

bipedal Walking upright, on two feet.

biramous appendages Arthropod appendages that consist of two branches (rami).

blasto- [Gk *blastos* bud] Embryo.

blastocoel [Gk *koilos* hollow] The cavity of a blastula.

blastopore [Gk *poros* passage] The opening from the cavity of the archenteron to the exterior in a gastrula.

blastula An early embryonic stage in animals, preceding the delimitation of the three principal tissue layers; frequently spherical and hollow.

B lymphocyte *See* lymphocyte.

bottleneck effect Genetic drift resulting from a sudden decrease in a population, often produced by extraordinary environmental conditions.

Bowman's capsule A cup-shaped structure at the beginning of each nephron in the kidney.

bronchiole In the lungs, the small tubes that subdivide off from the bronchi and branch repeatedly, terminating in the alveoli.

bryophytes A lineage of nonvascular terrestrial plants that includes the mosses, liverworts, and hornworts.

budding A form of asexual reproduction in which a small outgrowth (bud) of the parent's body becomes detached and develops into a new individual.

buffer A substance that binds H^+ ions when their concentration rises and releases them when their concentration falls, thereby minimizing fluctuations in the pH of a solution.

bulb A large, fleshy, underground plant organ that is a modified stem with short, fleshy leaves.

bundle sheath A layer or layers of cells that surround a vascular bundle.

C_3 plants Plants in which the Calvin cycle is the only pathway of CO_2 fixation. One of the products of photosynthesis is a three-carbon intermediary (C_3) of the Calvin cycle, from which the plants derive their name.

C_4 plants Also called **Kranz plants.** Plants in which one of the main early products of photosynthesis is a four-carbon compound (C_4). Kranz plants can carry out photosynthesis under conditions that are inhospitable to other plants.

caecum [L *caecus* blind] A blind diverticulum of the digestive tract.

calorie [L *calor* heat] The quantity of energy, in the form of heat, required to raise the temperature of one gram of pure water one degree from 14.5 to 15.5°C. The nutritionists' Calorie (capitalized) is 1,000 calories, or one kilocalorie.

Calvin-Benson cycle A series of enzymatically controlled photosynthetic reactions in which carbon dioxide is reduced to a three-carbon sugar (PGAL) and the starting compound, ribulose-bisphosphate (RuBP) is regenerated.

cambium [L *cambiare* to exchange] The principal lateral meristem of vascular plants; gives rise to most secondary tissue.

camera type eye The eye of vertebrates and cephalopods in which there is a single lens system focussing light on the light-sensitive retina.

cAMP *See* cyclic adenosine monophosphate.

capillarity [L *capillus* hair] The tendency of aqueous liquids to rise in narrow tubes with hydrophilic surfaces.

capillary [L *capillus*] A tiny blood vessel with walls one cell thick, across which exchange of materials between blood and the tissues takes place; receives blood from arteries and carries it to veins. Also, a similar vessel of the lymphatic system.

carbohydrate Any of a class of organic compounds composed of carbon, hydrogen, and oxygen in a ratio of about two hydrogens and one oxygen for each carbon; examples are sugar, starch, cellulose.

carbon fixation The process by which CO_2 is incorporated into organic compounds, primarily PGAL; energy usually comes from the ATP and $NADP_H$ generated by photophosphorylation, and the metabolic pathway utilizing this energy is usually the Calvin-Benson cycle.

carboxyl group The —COOH group characteristic of organic acids.

carcinogen A substance that causes cancer.

cardiac [Gk *kardia* heart] Pertaining to the heart.

carnivore [L *carnis* of flesh; *vorare* to devour] An organism that feeds on animals.

carpel A leaflike organ in angiosperms that bears the ovules.

carotenoid [L *carota* carrot] Any of a group of red, orange, and yellow accessory pigments of plants, found in plastids.

carrying capacity The maximum population that a given environment can support indefinitely.

cartilage A specialized type of dense fibrous connective tissue with a rubbery intercellular matrix.

caryo- [Gk *karyon* kernel] Nucleus.

Casparian strip A waterproof thickening in the radial and end walls of endodermal cells of plants.

cata- Down.

catabolism [Gk *katabolē* a throwing down] The degradational breaking-down aspects of metabolism, by which living things extract energy from food.

catalysis [Gk *katalyein* to dissolve] Acceleration of a chemical reaction by a substance that is not itself permanently changed by the reaction.

catalyst A substance that produces catalysis.

cation A positively charged ion.

caudal [L *cauda* tail] Pertaining to the tail.

cell cycle The cycle of cellular events from one mitosis through the next. Four stages are recognized, of which the last—distribution of genetic material to the two daughter nuclei—is mitosis proper.

cell-mediated immune system The portion of the vertebrate immune system that acts to destroy invading antigens through the production of specialized cells that attack the invaders; it primarily defends the body against infected cells.

cell wall A rigid outer layer lying just outside the plasma membrane of plants, fungi, and plant cells.

cell sap *See* sap.

cellulose [L *cellula* cell] A complex polysaccharide that is a major constituent of most plant cell walls.

centi- [L *centum* hundred] One hundredth.

central nervous system A portion of the nervous system that contains interneurons and exerts some control over the rest of the nervous system. In vertebrates, the brain and the spinal cord.

centri- [L *centrum* center] Center.

centrifugation [L *fugere* to flee] The spinning of a mixture at very high speeds to separate substances of different densities.

centriole One of a pair of cylindrical cytoplasmic organelles located just outside the nucleus of animal cells and the cells of some lower plants; associated with the spindle during mitosis and meiosis.

centromere [Gk *meros* part] A special region on a chromosome from which kinetochore microtubules radiate during mitosis or meiosis.

centrosome An area in the cell just outside the nucleus that contains a pair of centrioles; it acts as the microtubule-organizing center.

cephalization [Gk *kephalē* head] Localization of neural coordinating centers and sense organs at the anterior end of the body.

cerebellum [L small brain] A part of the hindbrain of vertebrates that controls muscular coordination.

cerebrum [L brain] Part of the forebrain of vertebrates, the chief coordination center of the nervous system.

channel *See* membrane channel.

character Any structure, functional attribute, behavioral trait, or other characteristic of an organism.

character displacement The rapid divergent evolution in sympatric species of characters that minimize competition and/or hybridization between them.

chelicerae The first pair of mouthparts in certain arthropods; are usually pincerlike or fanglike.

chemiosmotic gradient The combined electrostatic and osmotic concentration gradient generated by the electron-transport chains of mitochondria and chloroplasts; the energy in this gradient is used, for the most part, to synthesize ATP.

chemoreceptor A sensory receptor sensitive to the presence of certain chemicals that triggers an action potential in a sensory neuron.

chemosynthesis Autotrophic synthesis of organic materials, energy for which is derived from inorganic molecules.

chiasma (pl. chiasmata) The position on a pair of synapsed homologous chromosomes where crossing over is occurring during prophase of meiosis I.

chitin [Gk *chitōn* tunic] Polysaccharide that forms part of the hard exoskeleton of insects, crustaceans, and other invertebrates; also occurs in the cell walls of fungi.

chlorophyll [Gk *chlōros* greenish yellow; *phyllon* leaf] The green pigment of plants necessary for photosynthesis.

chloroplast A plastid containing chlorophyll.

chrom-, -chrome [Gk *chrōma* color] Colored; pigment.

chromatid A single chromosomal strand.

chromatin The mixture of DNA and protein (mostly histones in the form of nucleosome cores) that comprises eucaryotic nuclear chromosomes.

chromatography Process of separating substances by adsorption on media for which they have different affinities.

chromosome [Gk *sōma* body] A filamentous structure in the cell nucleus (or nucleoid), mitochondria, and chloroplasts, along which the genes are located.

cilium [L eyelid] A short hairlike locomotory organelle on the surface of a cell (*pl* cilia).

cisterna [L cistern] A cavity, sac, or other enclosed space serving as a reservoir.

classical conditioning *See* conditioning.

cleavage Division of a zygote or of the cells of an early embryo.

climax (ecological) A relatively stable stage reached in some ecological successions.

cline [Gk *klinein* to lean] Gradual variation, correlated with geography, in a character of a species.

cloaca [L sewer] Common chamber that receives materials from the digestive, excretory, and reproductive systems.

clone [Gk *klōn* twig] A group of cells or organisms derived asexually from a single ancestor and hence genetically identical.

co- With, together.

codon The unit of genetic coding, three nucleotides long, specifying an amino acid or an instruction to terminate translation.

coel-, -coel [Gk *koilos* hollow] Hollow, cavity; chamber.

coelom A body cavity surrounded by mesoderm in which the internal organs are located.

coenocytic [Gk *koinos* common] Having more than one nucleus in a single mass of cytoplasm.

coenzyme A nonproteinaceous organic molecule that plays an accessory role, but a necessary one, in the catalytic action of an enzyme.

coevolution Two or more organisms evolving, each in response to the other.

coleoptile [Gk *koleon* sheath, *ptilon* feather] A sheath around the young shoot of grasses.

collagen A fibrous protein; the most abundant protein in mammals.

collenchyma [Gk *kolla* glue] A supportive tissue in plants in which the cells usually have thickenings at the angles of the walls.

colloid [Gk *kolla*] A stable suspension of particles that, though larger than in a true solution, do not settle out.

colon The large intestine.

com- Together.

commensalism [L *mensa* table] A symbiosis in which one party is benefited and the other party receives neither benefit nor harm.

community In ecology, a unit composed of all the populations living in a given area.

competition In ecology, utilization by two or more individuals, or by two or more populations, of the same limited resource; an interaction in which both parties are harmed.

condensation reaction A reaction joining two compounds with resultant formation of water. Also called dehydration synthesis.

conditioning Associative learning. *Classical conditioning:* the association of a novel stimulus with an innately recognized stimulus. *Operant conditioning:* learning of a novel behavior as a result of reward or punishment; trial-and-error learning.

conformation (of a protein) [L *conformatio* symmetrical forming] The three-dimensional pattern according to which the polypeptide chains of a protein coil (secondary structure), fold (tertiary structure), and—if there is more than one chain—fit together (quarternary structure).

conjugation [L *jugare* to join, marry] Process of genetic recombination between two organisms (e.g., bacteria, algae) through a cytoplasmic bridge between them.

connective tissue A type of animal tissue whose cells are embedded in an extensive intercellular matrix; connects, supports, or surrounds other tissues and organs.

contractile vacuole An excretory and/or osmoregulatory vacuole in some cells, which, by contracting, ejects fluids from the cell.

convergent evolution The development of superficially similar structures in unrelated organisms, usually because of independent adaptations to similar environmental conditions.

cooperativity The phenomenon of enhanced reactivity of the remaining binding sites of a protein as a result of the binding of substrate at one site.

cork [L *cortex* bark] A waterproof tissue, derived from the cork cambium, that forms at the outer surfaces of the older stems and roots of woody plants; the outer bark or periderm.

cornea A transparent layer of tissue, continuous with the sclera, covering the anterior portion of the vertebrate eye.

corpus luteum [L yellow body] A yellowish structure in the ovary, formed from the follicle after ovulation, that secretes estrogen and progesterone (*pl.* corpora lutea).

cortex [L bark] In plants, tissue between the epidermis and the vascular cylinder of stems and roots. In animals, the outer barklike tissue of some organs, as *cerebral cortex, adrenal cortex,* etc.

cotyledon [Gk *kotyle* cup] A "seed leaf," a food-digesting and -storing part of a plant embryo.

countercurrent exchange A strategy in which two streams move past each other in opposite directions, facilitating the exchange of substances between them across a membrane. The gills in gas exchange and kidneys in the production of concentrated urine are two sites of countercurrent exchange.

covalent bond A chemical bond resulting from the sharing of a pair of electrons.

crossing over Exchange of parts between two homologous chromosomes.

cross section *See* section.

cryptic [Gk *kryptos* hidden] Concealing.

cuticle [L *cutis* skin] A waxy layer on the outer surface of leaves, insects, etc.

cyanobacteria Photosynthetic group of eubacteria possessing chlorophyll *a* and using water as an electron donor for photosynthesis, thereby producing molecular oxygen. Also called blue-green algae.

cyclic adenosine monophosphate (cyclic AMP or cAMP) Compound, synthesized in living cells from ATP, that functions as an intracellular mediator of hormonal action; also plays a part in neural transmission and some kinds of cellular control systems.

cyclic photophosophorylation The series of reactions in photosynthesis in which electrons from chlorophyll are energized by light energy and passed along an electron transport chain in the thylakoid membrane to return to the chlorophyll, and causing the production of ATP.

cyst [Gk *kystis* bladder, bag] (1) A saclike abnormal growth. (2) Capsule that certain organisms secrete around themselves and that protects them during resting stages.

-cyte, cysto- [Gk *kytos* container] Cell.

cytochrome Any of a group of iron-containing enzymes important in electron transport during respiration or photophosphorylation.

cytokinesis [Gk *kinēsis* motion] Division of the cytoplasm of a cell.

cytokinins A class of plant growth hormones that stimulate cell division.

cytoplasm All of a cell except the nucleus.

cytoskeleton The internal supporting skeleton of cells that consists of microfilaments, intermediate filaments, and microtubules.

cytosol The relatively fluid, less structured part of the cytoplasm of a cell, excluding organelles and membranous structures.

cytotoxic T cells T lymphocytes that kill infected and abnormal cells on contact. Also called killer T cells.

dalton A unit of mass equal to one twelfth the atomic weight of ^{12}C, or 1.66024 x 10^{-24} gram. Formerly called atomic mass unit (amu).

day-neutral plants Plants whose flowering is independent of the timing of the light and dark cycle.

deamination Removal of an amino group.

deciduous [L *decidere* to fall off] Shedding leaves each year.

decomposers Organisms that obtain their energy from breaking down dead organic matter.

dehydration reaction A condensation reaction.

deme [Gk *dēmos* population] A local interbreeding unit of population of any one species.

dendr-, dendro- [Gk *dendron* tree] Tree, branching.

dendrite A short unsheathed fiber of a nerve cell—often spiny, usually branched and tapering—that receives many synapses and carries excitation and inhibition toward the cell body.

denitrification A series of chemical reactions carried out by bacteria in the soil in which nitrogen-containing molecules are converted into nitrogen gas which is released to the atmosphere.

deoxyribonucleic acid (DNA) A nucleic acid found in most viruses, all bacteria, chloroplasts, mitochondria, and the nuclei of eucaryotic cells, characterized by the presence of a deoxyribose sugar in each nucleotide; the genetic material of all organisms except the RNA viruses.

-derm [Gk *derma* skin] Skin, covering; tissue layer.

determinate cleavage A type of cell division in embryonic development in which the cell division (cleavage) partitions critical cytoplasmic constituents unequally, and the developmental fate of the daughter cells is sealed so they may take very different developmental paths.

determination The progressive restriction of a cell's developmental potential such that it can develop only as a certain cell type.

deuterostomes A lineage of coelomate animals in which the embryonic blastopore forms the anus; includes the echinoderms and vertebrates.

di- Two.

diastole The period in the heart cycle when the ventricles are relaxing.

dicot A member of a subclass of the angiosperms, or flowering plants, characterized by the presence of two cotyledons in the embryo, a netlike system of veins in the leaves, and flower petals in fours or fives; *cf.* monocot. *Herbaceous dicot:* a perennial whose aboveground parts die annually. *Woody dicot:* a perennial whose aboveground parts—trunk and branches—remain alive and grow annually.

differentiation The process of developmental change from an immature to a mature form, especially in a cell.

diffusion The movement of dissolved or suspended particles from place to place as a result of their heat energy (thermal agitation).

digestion Hydrolysis of complex nutrient compounds into their building-block units.

dihybrid cross A genetic cross involving two different genes.

diploid [Gk *diploos* double] Having two of each type of chromosome.

directional selection Natural selection that selects for organisms at one end of the population distribution, thereby causing a population to evolve along a particular line.

disaccharide A double sugar, one composed of two simple sugars.

disruptive selection Natural selection in which there are opposing selective pressures that select the extreme variants in a distribution while those in the center of the distribution are selected against. The combined action of the opposing directional pressures is thus to produce two different variants within the population.

distal [L *distare* to stand apart] Situated away from some reference point (usually the main part of the body).

diverticulum [L *devertere* to turn aside] A blind sac branching off a cavity or canal.

DNA *See* **deoxyribonucleic acid.**

dominant (1) of an allele: exerting its full phenotypic effect despite the presence of another allele of the same gene, whose phenotypic expression it blocks or masks. *Dominant phenotype, dominant character:* one caused by a dominant allele. (2) Of an individual: occupying a high position in the social hierarchy.

dormancy [L *dormire* to sleep] The state of being inactive, quiescent. In plants, particularly seeds and buds, a period in which growth is arrested until environmental conditions become more favorable.

dorsal [L *dorsum* back] Pertaining to the back.

double fertilization A process unique to angiosperms in which one sperm fertilizes an egg, and a second sperm combines with two polar nuclei in the embryo sac to give rise to the endosperm.

drive *See* motivation.

duodenum [From a Latin phrase meaning 12 *(duodecin)* finger's-breadths long] The first portion of the small intestine of vertebrates, into which ducts from the pancreas and gallbladder empty.

ecology The study of the interactions between organisms and their environment, including both the physical and biotic environment.

ecosystem [Gk *oikos* habitation] The sum of physical features and organisms occurring in a given area.

ecto- Outside, external.

ectoderm The outermost tissue layer of an animal embryo. Also, tissue derived from the embryonic ectoderm.

ectothermic Organisms dependent on environmental temperature for body heat and incapable of precise self-regulation of body temperature. Also called cold-blooded or poikilothermic.

effector The part of an organism that produces a response, e.g., muscle, cilium, flagellum.

efferent Carrying substances or impulses away from the center of the body or organ toward the periphery. Usually used in reference to neurons or blood vessels.

egg An egg cell or female gamete. Also a structure in which embryonic development takes place, especially in birds and reptiles; consists of an egg cell, various membranes, and often a shell.

electrochemical gradient Combined electrostatic and osmotic-concentration gradient, such as the chemiosmotic gradient of mitochondria and chloroplasts.

electron A negatively charged primary subatomic particle.

electronegativity The formal measure of an atom's attraction for free electrons. Atoms with few electron vacancies in their outer shell tend to be more electronegative than those with more. In covalent bonds, the shared electrons are, on average, nearer the more electronegative atom; this asymmetry, in part, gives rise to the polarity of certain molecules.

electronic charge unit The charge of one electron, or 1.6021 $\times 10^{-19}$ coulomb.

electron-transport chain A series of enzymes found in the inner membrane of mitochondria and (with somewhat different components) in the thylakoid membrane of chloroplasts. The chain accepts high-energy electrons and uses their energy to create a chemiosmotic gradient across the membrane in which it is located.

electrostatic force The attraction (also called *electrostatic attraction*) between particles with opposite charges, as between a proton and an electron, or between H^+ and OH^- ; and the repulsion between particles with like charges, as between two H^+ ions.

electrostatic gradient The free-energy gradient created by a difference in charge between two points, generally the two sides of a membrane.

element Basic substances that cannot be decomposed into simpler substances by ordinary chemical reactions.

elimination (or defecation) The release of unabsorbed wastes from the digestive tract. *Cf.* excretion.

embryo A plant or animal in an early stage of development; generally still contained within the seed, egg, or uterus.

emulsion [L *emulsus* milked out] Suspension, usually as fine droplets, of one liquid in another.

-enchyma [Gk *parenchein* to pour in beside] Tissue.

end-, endo- Within, inside; requiring.

endergonic reaction [Gk *ergon* work] An energy-absorbing chemical reaction; endothermic.

endocrine [Gk *krinein* to separate] Pertaining to ductless glands that produce hormones.

endocytosis The process by which the cell membrane forms an invagination which becomes a vesicle, trapping extracellular material that is then transported within the cell; in general, the invagination is triggered by the binding of membrane receptors to specific substances used by the cell.

endoderm The innermost tissue layer of an animal embryo.

endodermis A plant tissue, especially prominent in roots, that surrounds the vascular cylinder; all endodermal cells have Casparian strips.

endoplasmic reticulum [L *reticulum* network] A system of membrane-bounded channels in the cytoplasm.

endoskeleton An internal skeleton.

endosperm [Gk *sperma* seed] A nutritive material in seeds.

endosymbiotic hypothesis Hypothesis that certain eucaryotic organelles—in particular mitochondria and chloroplasts—originated as free-living procaryotes that took up mutalistic residence in the ancestors of modern eucaryotes.

endotherm ("internally heated") Animals that retain the heat generated by their metabolism and hence maintain a constant high body temperature. Also called warm-blooded animals.

endothermic reaction In thermodynamics, energy-absorbing (endergonic) reactions.

entropy Measure of the disorder of a system.

enzyme [Gk *zymē* leaven] A compound, usually a protein, that acts as a catalyst.

epi- Upon, outer.

epicotyl A portion of the axis of a plant embryo above the point of attachment of the cotyledons; forms most of the shoot.

epidermis [Gk *derma* skin] The outermost portion of the skin or body wall of an animal.

epithelium An animal tissue that forms the covering or lining of all free body surfaces, both external and internal.

equilibrium constant The ratio of products of a reaction to the reactants after the reaction has been allowed to proceed until there is no further change in these concentrations.

erythrocyte [Gk *erythros* red] A red blood corpuscle, i.e., a blood corpuscle containing hemoglobin.

esophagus [Gk *phagein* to eat] An interior part of the digestive tract; in mammals it leads from the pharynx to the stomach.

essential fatty acid A fatty acid an organism needs but cannot synthesize, and so must obtain preformed (or in a precursor form) from its diet.

essential nutrient A nutrient that is required but must be obtained preformed in the diet because it cannot be synthesized by the organism.

estrogen [L *oestrus* frenzy] Any of a group of vertebrate female sex hormones.

estrous cycles [L *oestrus*] In female mammals, the higher primates excepted, a recurrent series of physiological and behavioral changes connected with reproduction.

estuary That portion of a river that is close enough to the sea to be influenced by marine tides.

ethology The study of the biology animal behavior.

ethylene A gaseous hormone in plants that promotes ripening and aging in plants.

eu- [Gk *eus* good] Most typical, true.

eucaryotic cell A cell containing a distinct membrane-bounded nucleus, characteristic of all organisms except bacteria.

euchromatin Chromatin in an expanded form that is active, i.e., being transcribed.

eutrophic A body of water that has an abundant supply of nutrients and high productivity.

evaginated [L *vagina* sheath] Folded or protruded outward.

eversible [L *evertere* to turn out] Capable of being turned inside out.

evolution [L *evolutio* unrolling] Change in the genetic makeup of a population with time.

ex-, exo- Out of, outside, producing.

excretion Release of metabolic wastes and excess water. *Cf.* elimination.

exergonic reaction [Gk *ergon* work] Energy-releasing chemical reaction; exothermic.

exocrine gland A gland, such as the liver or salivary glands, that discharge its secretion into tubes or ducts.

exocytosis The process by which an intracellular vesicle fuses with the cell membrane, expelling its contents into its surroundings.

exon A part of a primary transcript (and the corresponding part of a gene) that is ultimately either translated (in the case of mRNA) or utilized in a final product, such as tRNA.

exoskeleton An external skeleton.

extrinsic External to, not a basic part of; as in *extrinsic isolating mechanism.*

facilitated diffusion The diffusion of a substance across a membrane by specific permeases with no expenditure of energy by the cell.

fallopian tube The tube that conducts the egg from the ovary to the uterus. Also called oviduct.

fat A triglyceride; formed from a condensation reaction between three fatty acids and a glycerol molecule.

fatty acid A hydrocarbon chain, usually between 4 and 24 carbons in length, with a terminal carboxyl group. If the hydrocarbon chain has only single bonds, it is saturated, if there is at least one double bond, it is unsaturated.

fauna The animals of a given area or period.

feature detector A circuit in the nervous system that responds to a specific type of feature, such as a vertically moving spot or a particular auditory time delay.

feces [L *faeces* dregs] Indigestible wastes discharged from the digestive tract.

feedback The process by which a control mechanism is regulated through the very effects it brings about. *Positive feedback:* the process by which a small effect is amplified, as when a depolarization triggers an acton potential. *Negative feedback* (or feedback inhibition): the process by which a control mechanism is activated to restore conditions to their original state.

fermentation Anaerobic production of alcohol, lactic acid, or similar compounds from carbohydrates via the glycolytic pathway.

fertilization Fusion of nuclei of egg and sperm.

fetus [L *fetus* pregnant] An embryo in its later development, still in the egg or uterus. In humans, the embryo from eight weeks till birth.

filter feeding A method of feeding in which small food particles are sifted out of the surrounding water.

filtrate The liquid obtained from filtration, i.e., the liquid passing through the filter.

first law of thermodynamics The Law of Conservation of Energy: energy can be transformed from one form to another, but it cannot be created nor destroyed.

fitness The probable genetic contribution of an individual (or allele or genotype) to succeeding generations. *Inclusive fitness:* the sum of an individual's personal fitness plus the fitness of that individual's relatives devalued in proportion to their genetic distance from the individual.

fixation (1) Conversion of a substance into a biologically more usable form, as the conversion of CO_2 into carbohydrate by photosynthetic plants or the incorporation of N_2 into more complex molecules by nitrogen-fixing bacteria. (2) Process of treating living tissue for microscopic examination.

flatworms Worms belonging to the phylum Platyhelminthes.

flagellum [L whip] A long hairlike locomotory organelle on the surface of a cell.

flora The plants of a given area or period.

follicle [L *follis* bag] A jacket of cells around an egg cell in an ovary.

follicle-stimulating hormone (FSH) A gonadotropic hormone of the anterior pituitary that stimulates growth of follicles in the ovaries of females and function of the seminiferous tubules in males.

food chain Sequence of organisms, including producers, consumers, and decomposers, through which energy and materials may move in a community.

foot-candle Unit of illumination; the illumination of a surface produced by one standard candle at a distance of one foot; *cf.* lambert.

founder effect The difference between the gene pool of a population as a whole and that of a newly isolated population of the same species.

free energy Usable energy in a chemical system; energy available for producing change.

fruit A mature ovary or cluster of ovaries (sometimes with additional floral structures associated with the ovary).

fruiting body A spore-bearing structure (e.g., the aboveground portion of a mushroom).

FSH *See* follicle-stimulating hormone.

functional groups Specific grouping of atoms that are attached to the carbon-atom backbone and which have their own characteristic properties and help determine the molecule's solubility and reactivity.

fungi Multicellular absorptive heterotrophic organisms belonging to the kingdom Fungi.

gamete [Gk *gamete(s)* wife, husband] A sexual reproductive cell that must usually fuse with another such cell before development begins; an egg or sperm.

gametophyte [Gk *phyton* plant] A multicellular haploid plant that can produce gametes.

ganglion [Gk tumor] A structure containing a group of cell bodies of neurons (*pl.* ganglia).

gastr-, gastro- [Gk *gaster* belly] Stomach; ventral; resembling the stomach.

gastrovascular cavity An often branched digestive cavity, with only one opening to the outside, that conveys nutrients throughout the body; found only in animals with circulatory systems.

gastrula A two-layered, later three-layered, animal embryonic stage.

gastrulation The process by which a bastula develops into a gastrula, usually by an involution of cells.

gated channel A membrane channel that can open or close in response to a signal, generally a change in the electrostatic gradient or the binding of a hormone, transmitter, or other molecular signal.

gel Colloid in which the suspended particles form a relatively orderly arrangement; *cf.* sol.

gel electrophoresis Separation of molecules on the basis of size and electrical charge by exposing the molecules to an electrical field across a gel.

-gen; -geny [Gk *genos* birth, race] Producing; production, generation.

gene [Gk *genos*] The unit of inheritance; usually a portion of a DNA molecule that codes for some product such as a protein, tRNA, or rRNA.

gene amplification Any of the strategies that give rise to multiple copies of certain genes, thus facilitating the rapid synthesis of a product (such as rRNA for ribosomes) for which the demand is great.

gene flow The movement of genes from one part of a population to another, or from one population to another, via gametes.

gene pool Thes sum total of all the genes of all the individuals in a population.

gene regulation Any of the strategies by which the rate of expression of a gene can be regulated, as by controlling the rate of transcription.

generator potential *See* potential.

genetic drift Change in the gene pool as a result of chance and not as a result of selection, mutation, or migration.

genome The cell's total complement of DNA: in eucaryotes, the nuclear and organelle chromosomes; in procaryotes, the major chromosome, episomes, and plasmids. In viruses and viroids, the total complement of DNA or RNA.

genotype The particular combination of genes present in the cells of an individual.

germ cell A sexual reproductive cell; an egg or sperm.

gibberellin A plant hormone—one of its effects is stem elongation in some dwarf plants.

gill An evaginated area of the body wall of an animal, specialized for gas exchange.

gizzard A chamber of an animal's digestive tract specialized for grinding food.

glucose [Gk *glykys* sweet] A six-carbon sugar; plays a central role in cellular metabolism.

glycocalyx The layer of protein and carbohydrates just outside the plasma membrane of an animal cell; in general, the proteins are anchored in the membrane, and the carbohydrates are bound to the proteins.

glycogen [Gk *glykys*] A polysaccharide that serves as the principal storage form of carbohydrate in animals.

glycolysis [Gk *glykys*] Anaerobic catabolism of carbohydrates to pyruvic acid.

Golgi apparatus Membranous subcellular structure that plays a role in storage and modification particularly of secretory products.

gonadotropic Stimulatory to the gonads.

gonadotropin A hormone stimulatory to the gonads, a gonadotropic hormone.

gonads [Gk *gonos* seed] The testes or ovaries.

gradualism A hypothesis that large scale evolutionary change is due to the gradual accumulation of very small changes over long periods of time.

granum [L grain] A stacklike grouping of photosynthetic membranes in a chloroplast (*pl.* grana).

gravitropism The growth of plant organs in response to gravity.

greenhouse effect The warming of the earth's atmosphere, thought to be due to the increase of carbon dioxide and certain atmospheric pollutants that absorb and retain the reflected heat from the earth's surface.

gross primary productivity The total amount of energy that photosynthesis converts into organic matter.

growth factors Specialized proteins that stimulate the normal growth and development of certain types of cells.

guard cells Specialized epithelial cells of leaves and stems that occur in pairs and surround and regulate the size of the stomata.

habit [L *habitus* disposition] In biology, the characteristic form or mode of growth of an organism.

habitat [L it lives] The kind of place where a given organism normally lives.

habituation The process that results in a long-lasting decline in the receptiveness of interneurons (primarily) to the input from sensory neurons or other interneurons; *cf.* sensitization, adaptation.

hair cells Highly sensitive mechanoreceptor cells that are found in the lateral line system of fishes and the inner ear of many vertebrates. Hair cells have a hairlike tufts of very large microvilli projecting from one end of the cell.

haploid [Gk *haploos* single] Having only one of each type of chromosome.

Hardy-Weinberg equilibrium According to the Hardy-Weinberg law, both genotypic ratios and allelic frequencies will remain constant from generation to generation in sexually reproducing populations as long as certain conditions of stability are met.

Haversion system The structural unit of vertebrate compact bone; consists of a central Haversion canal containing blood vessels and nerves, surrounded by rings of bony matrix containing bone cells.

helper T cells This type of T lymphocytes plays a central role in mediating the immune response. When stimulated, they produce a variety of chemicals that activate macrophages and other T and B cells, amplifying the immune response. They also stimulate antigen-bound B cells to divide and produce antibodies.

hem-, hemat-, hemo- [Gk *haima* blood] Blood.

hematopoiesis [Gk *poiésis* making] The formation of blood.

hemoglobin A red iron-containing pigment in the blood that functions in oxygen transport.

hepatic [Gk *hēpar* liver] Pertaining to the liver.

herbaceous [L *herbaceus* grassy] Having a stem that remains soft and succulent; not woody.

herbaceous dicot *See* dicot.

herbivore [L *herba* grass; *vorare* to devour] An animal that eats plants.

Hertz A unit of frequency (as of sound waves) equal to one cycle per second.

hetero- [Gk *heteros* other] Other, different.

heterochromatin High condensed chromatin that is inactive; i.e., not transcribed.

heterogamy [Gk *gamos* marriage] The condition of producing gametes of two or more different types.

heterosporous Plants producing two types of spores, large megaspores and smaller microspores that develop into different gametophytes, female and male respectively.

heterotrophic [Gk *trophē* food] Incapable of manufacturing organic compounds from inorganic raw materials, therefore requiring organic nutrients from the environment.

heterozygous [Gk *zygōtos* yoked] Having two different alleles of a given gene.

Hg [L *hydrargyrum* mercury] The symbol for mercury. Pressure is often expressed in *mm Hg*—the pressure exerted by a column of mercury whose height is measured in millimeters (at 0 C, 1 mm Hg = 133.3 newtons per square meter).

hibernation A physiological sleeplike state in which the metabolic rate, heart rate, and body temperature are lowered, allowing the animal to survive the winter period.

hilum Region where blood vessels, nerves, ducts, enter an organ.

hist- [Gk *histos* web] Tissue.

histamine A chemical released from mast cells during allergic reactions that causes dilation and increased permeability of blood vessels.

histology The structure and arrangement of the tissues of organisms; the study of these.

histone One of a class of basic proteins serving as structural elements of eucaryotic chromosomes.

home range An area within which an animal tends to confine all or nearly all its activities for a long period of time.

homeo-, homo- [Gk *homoios* like] Like, similar.

homeostasis The tendency in an organism toward maintenance of physiological and psychological stability.

homeothermic [Gk *therme* heat] Maintaining a constant body temperature.

homeotic genes Genes in a variety of different organisms that control the development of the overall body plan. The products of these genes are generally control substances that bind to DNA and orchestrate the operation of many other genes.

homologous chromosomes Bearing genes for the same characters.

homology Characters in different organisms that are derived from the same ancestral structures; i.e. similarity by descent.

homosporous Plants in which a single type of spore is produced which develops into a gametophyte bearing both male and female reproductive structures.

homozygous [Gk *zygōtos* yoked] Having two copies of the same allele of a given gene.

hormone [Gk *horman* to set in motion] A control chemical secreted in one part of the body that affects other parts of the body.

humoral immune system The portion of the vertebrate immune system that acts to defend the body against invading antigens through the secretion of antibodies that help inactivate or destroy the antigen, i.e., antibody mediated immunity.

hybrid In evolutionary biology, a cross between two species. In genetics, a cross between two genetic types.

hybrid vigor The increased vigor shown by the offspring of a cross of genetically different parents.

hydr-, hydro- [Gk *hydor* water] Water; fluid; hydrogen.

hydration Formation of a sphere of water around an electrically charged particle.

hydrocarbon Any compound made of only carbon and hydrogen.

hydrogen bond A weak chemical bond formed when two polar molecules, at least one of which usually consists of a hydrogen bonded to a more electronegative atom (usually oxygen or nitrogen), are attracted electrostatically.

hydrolysis [Gk *lysis* loosing] Breaking apart of a molecule by addition of water.

hydrophilic Readily entering into solution by forming hydrogen bonds with water or other polar molecules.

hydrophobic Incapable of entering into solution by molecules that are neither ionic nor polar, and therefore cannot dissolve in water.

hydrostatic [Gk *statikos* causing to stand] Pertaining to the pressure and equilibium of fluids.

hydrostatic skeleton A supportive skeleton composed of fluid-filled body compartments.

hydroxide ion The OH^- ion.

hydroxyl group An —OH functional group which, when attached to a carbon atom, is an alcohol group.

hyper- Over, overmuch; more.

hypertonic Of a solution (or colloidal suspension): tending to gain water from some reference solution (or colloidal suspension) separated from it by a selectively permeable membrane—usually because it has a higher osmotic concentration than the reference solution.

hypertrophy [Gk *trophē* food] Abnormal enlargement, excessive growth.

hypha [Gk *hyphē* web] A fungal filament.

hypo- Under, lower; less.

hypocotyl The portion of the axis of a plant embryo below the point of attachment of the cotyledons; forms the base of the shoot and the root.

hypothalamus [Gk *thalamos* inner chamber] Part of the posterior portion of the vertebrate forebrain, containing important centers of the autonomic nervous system and centers of emotion.

hypotonic Of a solution (or colloidal suspension): tending to lose water to some reference solution (or colloidal suspension) separated from it by a selectively permeable membrane—usually because it has a lower osmotic concentration than the reference solution.

immune response The reaction of the body that involves the recognition and destruction of substances foreign to the body. It involves two interacting defense systems called nonspecific and specific immunity.

imprinting A kind of associative learning in which an animal rapidly learns during a particular critical period to recognize an object, individual, or location in the absence of overt reward; distinguished from most other associative learning in that it is retained indefinitely, being difficult or impossible to reverse.

independent assortment The Principle of Independent Assortment is frequently referred to as Mendel's second law. Genes found on different chromosomes, so-called unlinked genes, assort independently in meiosis unless they are recombined by crossing over.

indeterminate cleavage A type of cell division in embryonic development in which the cell division (cleavage) partitions critical cytoplasmic constituents equally, and the resulting daughter cells still have the capacity to develop into a complete embryo (i.e., they are totipotent).

inducer In embryology, a substance that stimulates differentiation of cells or development of a particular structure. In genetics, a substance that activates particular genes.

infection The condition in which the body is invaded by a disease-causing microorganism which then multiplies and produces injurious effects on the body.

inflammatory response The defense reaction of the body in response to physical injury, infection, or chemical agents. The classic signs are swelling, heat, redness, and pain.

ingestion The process of taking in food into a digestive cavity or, in cells, the process by which a cell takes in foreign particles.

inner cell mass A cluster of cells from the embryonic blastocyst in mammals that will give rise to the embryo.

inorganic compound A chemical compound not based on carbon.

in situ [L in place] In its natural or original position.

instinct Heritable, genetically specified neural circuitry that guides and directs behavior.

insulin [L *insula* island] A hormone produced by the β islet cells in the pancreas that helps regulate carbohydrate metabolism, especially conversion of glucose into glycogen.

integument [L *integere* to cover] A coat, skin, shell, rind, or other protective surface structure.

inter- Between (e.g., *interspecific,* between two or more different species).

interferon A chemical produced by virus-infected cells that helps other cells resist infection by viruses.

interleukins A group of chemicals produced by macrophages or helper T cells that activate other helper and cytotoxic T cells and nearby macrophages.

interneuron A neuron that receives input from and synapses on other neurons, as distinguished from a sensory neuron (which receives sensory information) and a motor neuron (which synapses on a muscle).

internode The portion of a plant stem between the sites where leaves are attached.

intra- Within (e.g., *intraspecific,* **within a single species).**

intrinsic Inherent in, a basic part of; as in *intrinsic isolating mechanism.*

intron A part of a primary transcript (and the corresponding part of a gene) that lies between exons, and is removed before the RNA becomes functional.

invaginated [L *vagina* sheath] Folded or protruded outward.

invertebrate [L *vertebra* joint] Lacking a backbone, hence an animal without bones.

in vitro [L in glass] Not in the living organism, in the laboratory.

in vivo [L in the living] In the living organism.

ion An electrically charged atom.

ionic bond A chemical bond formed by the electrostatic attraction between two oppositely charged ions.

iso- equal, uniform

isogamy [Gk *gamos* marriage] The condition of producing gametes of only one type, with no distinction existing between male and female.

isolating mechanism An obstacle to interbreeding, either extrinsic, such as a geographic barrier, or intrinsic, such as structural or behavioral incompatibility.

isomers Organic molecules that have the same molecular formula but different structural arrangements and therefore different chemical properties.

isotonic Of a solution (or colloidal suspension): tending neither to gain nor to lose water when separated from some reference solution (or colloidal suspension) by a selectively permeable membrane—usually because it has the same osmotic concentration as the reference solution.

isotope [Gk *topos* place] An atom differing from another atom of the same element in the number of neutrons in its nucleus.

karyotype An arrangement of all the human chromosomes in a cell on the basis of chromosomal size and shape.

keratin A type of fibrous protein found in hair, feathers, hooves, and horns.

kilo- A thousand.

kin-, kino- [Gk *kinema* motion] Motion, action.

kinase An enzyme that catalyzes the phosphorylation of a substrate by ATP.

kin selection Altruistic behavior towards closely related kin.

kinetic energy The energy of motion.

Krebs cycle A cyclical series of chemical reactions that is central in the aerobic metabolism of proteins, fats, and carbohydrates. Also called the citric acid cycle, the tricarboxylic acid cycle, or the TCA cycle.

***K*-selected species** Organisms whose life histories involve producing relatively few offspring with parental care, each of which has a good chance of surviving.

lacteal A lymph capillary in the center of each intestinal villus that absorbs fat.

lactic acid A three-carbon organic acid produced in animals and some microorganisms by fermentation.

lactose Milk sugar, a disaccharide composed of one glucose and one galactose molecule.

lamella [L thin plate] A thin platelike structure; a fairly straight intracellular membrane.

larva [L ghost, mask] Immature form of some animals that undergo radical transformation to attain the adult form.

law of independent assortment Mendel's second law which states that when two or more genes are involved in a cross, the alleles of one gene are inherited independently of the alleles for another gene.

law of segregation Mendel's first law which states during gamete formation, each of the two hereditary "factors" for each trait separate into different gametes.

learning The modification of behavior by experience.

lenticel [L *lenticella* small lentil] A porous region in the periderm of a woody stem through which gases can move.

leukocyte [Gk *leukos* white] A white blood cell; *cf.* lymphocyte, macrophage.

LH *See* **luteinizing hormone.**

ligament [L *ligare* to bind] A type of connective tissue linking two bones in a joint.

ligase An enzyme that catalyzes the bonding between adjacent nucleotides in DNA and RNA.

light reactions The light dependent reactions of photosynthesis that occur in the thylakoid membrane, resulting in the conversion of solar energy into chemical energy in the form of ATP and NADPH, and the production of molecular oxygen.

lignin [L *lignum* wood] An organic compound in wood that makes cellulose harder and more brittle.

linkage The presence of two or more genes on the same chromosome, which, in the absence of crossing over, causes the characters they control to be inherited together.

lip- [Gk *lipos* fat] Fat or fatlike.

lipase A fat-digesting enzyme.

lipid Any of a variety of compounds insoluble in water but soluble in ethers and alcohols; includes fats, oils, waxes, phospholipids, and steroids.

locus [L place] In genetics, a particular location on a chromosome, hence often used synonymously with gene (*pl.* loci).

long-day plant Plants that flower only when the dark period is shorter than a certain critical value.

lumen [L light, opening] The space or cavity within a tube or sac (*pl.* lumina).

lung An internal chamber specialized for gas exchange in an animal.

luteinizing hormone (LH) A gonadotropic hormone of the pituitary that stimulates conversion of a follicle into corpus luteum and secretion of progesterone by the corpus luteum; also stimulates secretion of sex hormone by the testes.

lymph [L *lympha* water] A fluid derived from tissue fluid and transported in special lymph vessels to the blood.

lymphocyte A white blood cell that responds to the presence of a foreign antigen. *B lymphocyte:* a cell that upon stimulation by an antigen secretes antibodies. *T lymphocyte:* a cell that attacks infected cells and modulates the activity of B lympohocytes.

-lysis, lyso- [Gk *lysis* loosing] Loosening, decomposition.

lysogenic Of bacteria: carrying bacteriophage capable of lysing, i.e., destroying, other bacterial cells.

lysosome A subcellular organelle that stores digestive enzymes.

lytic cycle Reproductive cycle of a virus that results in lysis and death of the host cell.

macro- Large.

macroevolution Large-scale evolutionary changes above the level of species; includes the processes of adaptive radiation, the development of evolutionary trends, the origin of innovative features, and the effect of extinctions.

macronutrients Nutrients required in relatively large amounts.

macrophage A phagocytic white blood cell that ingests material—particularly viruses, bacteria, and clumped toxins—bound by circulating antibodies.

malignant tumor A cancerous tumor.

Malpighian tubules Excretory organs in insects that consist of tubular outgrowths of the digestive tract.

mantle A layer of tissue in molluscs that covers the visceral mass and secretes the shell (if present).

marsupials A branch of mammals whose young complete their embryonic development in a pouch (marsupium).

mast cell Cells that are specialized for the secretion of histamine and other local chemical mediators as part of the immune response.

matrix [L *mater* mother] A mass in which something is embedded, e.g., the intercellular substance of a tissue.

matter Anything that has mass and takes up space.

mechanoreceptors Specialized receptor cells that are specialized for the detection of mechanical disturbances.

medulla [L marrow, innermost part] (1) The inner portion of an organ, e.g., *adrenal medulla,* (2) The *medulla oblongata,* a portion of the vertebrate hindbrain that connects with the spinal cord.

medusa [*after* Medusa, mythological monster with snaky locks] The free-swimming stage in the life cycle of a coelenterate.

mega- Large.

megaspore A spore that will germinate into a female gametophyte plant.

meiosis [Gk *meiōsis* diminution] A process of nuclear division in which the number of chromosomes is reduced by half.

membrane A structure, formed mainly by a double layer of phospholipids, which surrounds cells and organelles.

membrane channel A pore in a membrane through which certain molecules may pass.

membrane pump A permease that uses energy, usually from ATP, to move substances across the membrane against their osmotic-concentration or electrostatic gradients.

meristematic tissue [Gk *meristos* divisible] A plant tissue that functions primarily in production of new cells by mitosis.

meso- Middle.

mesoderm The middle tissue layer of an animal embryo.

mesophyll [Gk *phyllon* leaf] The parenchymatous middle tissue layers of a leaf.

messenger RNA RNA synthesized from a DNA template in the nucleus that carries the information to the ribosome where it specifies the primary structure of a protein.

meta- Posterior, later; change in.

metabolism [Gk *metabolē* change] The sum of the chemical reactions within a cell (or a whole organism), including the energy-releasing breakdown of molecules (catabolism) and the synthesis of complex molecules and new protoplasm (anabolism).

metamorphosis [Gk *morphē* form] Transformation of an immature animal into an adult. More generally, change in the form of an organ or structure.

metastasis The spread of cancer cells from one site to another.

micro- Small. Male. In units of measurement, one millionth.

microevolution A change in the allelic frequencies in a gene pool over time.

microfilament A long, thin structure, usually formed from the protein actin; when associated with myosin filaments, as in muscles, microfilaments are involved in movement.

micronutrients Nutrients required by organisms in very small amounts.

microorganism A microscopic organism, especially a bacterium, virus, or protozoan.

microspore A spore that will germinate into a male gametophyte.

microtubule A long, hollow structure formed from the protein tubulin; found in cilia, eucaryotic flagella, basal bodies/centrioles, and the cytoplasm.

microtubule-organizing center An area in the cell just outside the nucleus that contains a pair of centrioles and acts as the center for organizing the cytoskeleton or spindle.

middle lamella A layer of substance deposited between the walls of adjacent plant cells.

milli- One thousandth.

mineral In biology, any naturally occurring inorganic substance, excluding water.

mitochondrion [Gk *mitos* thread; *chondrion* small grain] Subcellular organelle in which aerobic respiration takes place.

mitosis [Gk *mitos*] Process of nuclear division in which complex movements of chromosomes along a spindle result in two new nuclei with the same number of chromosomes as the original nucleus.

mold Any of many fungi that produce a cottony or furry growth.

mole The amount of a substance that has a weight in grams numerically equal to the molecular weight of the substance. One mole of a substance contains 6.023×10^{23} molecules of that substance; hence one mole of a substance will always contain the same number of molecules as a mole of any other substance.

molecular weight The weight of a molecule calculated as the sum of the atomic weights of its constituent atoms.

molecule A chemical unit consisting of two or more atoms bonded together.

molting A process in arthropods during which the exoskeleton is periodically shed and a new one forms to allow for growth.

mono- One.

monocot A member of a subclass of angiosperms, or flowering plants, characterized by the presence of a single cotyledon in the embryo, parallel veins in the leaves, and flower petals in threes; *cf.* dicot.

monohybrid cross A genetic cross involving parents that differ only in one trait.

monosaccharide A simple sugar; the basic building blocks of the disaccharides and polysaccharides.

-morph, morpho- [Gk *morphē* form] Form, structure.

morphogenesis The establishment of shape and pattern in an organism.

morphology The form and structure of organisms or parts of organisms; the study of these.

motivation The internal state of an animal that is the immediate cause of its behavior; drive.

motor neuron A neuron, leading away from the central nervous system, that synapses on and controls an effector.

motor program A coordinated, relatively stereotyped series of muscle movements performed as a unit, either innate (as the movements of swallowing) or learned (as in speech); also, the neural circuitry underlying such behavior; *cf.* reflex.

mouthparts Structures or appendages near the mouth used in manipulating food.

mucosa Any membrane that secretes mucus (a slimy protective substance), e.g., the membrane lining the stomach and intestine.

Müllerian mimicry A type of mimicry in which two species, both of which are harmful or unpalatable, resemble one another.

muscle [Gk *musculus* small mouse, muscle] A contractile tissue of animals.

mutagen A chemical that causes a mutation in an organism's DNA.

mutation [L *mutatio* change] Any relatively stable heritable change in the genetic material.

mutualism A symbiosis in which both parties benefit.

myc-, myco- Pertaining to fungi.

mycelium [Gk *mykēs* fungus] A mass of hyphae forming the body of a fungus.

mycorrhizae A mutualistic association between the hyphae of a fungus and plant root.

myelin sheath A layer of lipid that surrounds and insulates the axons of neurons; it acts to speed up conduction.

myo- [Gk *mys* mouse, muscle] Muscle.

myoglobin A globular protein in muscle that stores oxygen.

myosin A protein filament with large globular head that bind to actin and cause myosin to slide past actin in the process of muscle contraction.

NAD$^+$ *See* nicotinamide adenine dinucleotide.

NADP$^+$ *See* nicotinamide adenine dinucleotide phosphate.

nano- [L *nanus* dwarf] One billionth.

natural selection Differential reproduction in nature, leading to an increase in the frequency of some genes or gene combinations and to a decrease in the frequency of others.

navigation The initiation and/or maintenance of movement toward a goal.

negative feedback *See* feedback.

nematocyst [Gk *nēma* thread; *kystis* bag] A specialized stinging cell in coelenterates; contains a hairlike structure that can be ejected.

neo- New.

neocortex Portion of the cerebral cortex in mammals, of relatively recent evolutionary origin; often greatly expanded in the higher primates and dominant over other parts of the brain.

nephr- [Gk *nephros* kidney] Kidney.

nephridium An excretory organ consisting of an open bulb and a tubule leading to the exterior; found in many invertebrates, such as segmented worms.

nephron The functional unit of a vertebrate kidney, consisting of Bowman's capsule, convoluted tubule, and loop of Henle.

nerve [L *nervus* sinew, nerve] A bundle of neuron fibers (axons).

nerve net A nervous system without any central control, as in coelenterates.

net primary productivity The gross primary productivity minus the energy plants used in their own metabolism and growth.

neuron [Gk nerve, sinew] A nerve cell.

neurotransmitter A chemical released from the terminal of a presynaptic axon that diffuses across the synapse and stimulates a postsynaptic neuron. Also called transmitter substance or transmitter chemical.

neutron An electrically neutral subatomic particle with approximately the same mass as a proton.

niche The functional role and position of an organism in the ecosystem; the way an organism makes its living, including, for an animal, not only what it eats, but when, where, and how it obtains food, where it lives, etc.

nicotinamide adenine dinucleotide (NAD$^+$) An organic compound that functions as an electron acceptor, e.g., in respiration.

nicotinamide adenine dinucleotide phosphate (NADP$^+$) An organic compound that functions as an electron acceptor, e.g., in biosynthesis.

nitrogen fixation Incorporation of nitrogen from the atmosphere into substances more generally usable by organisms.

node (of plant) [L *nodus* knot] Point on a stem where a leaf or bud is (or was) attached.

noncyclic photophosphorylation The series of reactions in photosynthesis in which electrons from water are energized by light energy and passed along an electron transport chain in the thylakoid membrane to NADP$^+$, resulting in the production of ATP, NADPH, and oxygen.

nondisjunction An abnormal meiotic division in which the members of a homologous pair of chromosomes fail to separate properly and both members end up in the same daughter cell.

nonhomologous Of chromosomes: two chromosomes that do not share the same genes and thus do not pair during meiosis.

nonpolar covalent bond A strong chemical bond in which the constituent atoms share electrons equally.

notochord [Gk *nōtos* back; *chordē* string] In the lower chordates and in the embryos of the higher vertebrates, a flexible supportive rod running longitudinally through the back just ventral to the nerve cord.

nucleic acid Any of several organic acids that are polymers of nucleotides and function in transmission of hereditary traits, in protein synthesis, and in control of cellular activities.

nuclear envelope The membrane in eucaryotic cells that surrounds the genetic material and separates it from the cytoplasm.

nucleoid A region, not bounded by a membrane, where the chromosome is located in a procaryotic cell.

nucleolus A dense body within the nucleus, usually attached to one of the chromosomes; consists of multiple copies of the genes for certain kinds of rRNA.

nucleosome A complex consisting of several histone proteins; which together form a "spool," and chromosomal DNA, which is wrapped around the spool.

nucleotide A chemical entity consisting of a five-carbon sugar with a phosphate group and a purine or pyrimidine attached; building-block unit of nucleic acids.

nucleus (of cell) [L kernel] A large membrane-bounded organelle containing the chromosomes.

nutrient [L *nutrire* to nourish] A food substance usable in metabolism as a source of energy or of building material.

nymph [Gk *nymphē* bride, nymph] Immature stage of insect that undergoes gradual metamorphosis.

obligate aerobes Organisms that require oxygen for metabolism and cannot survive without it.

obligate anaerobes Organisms that cannot survive in the presence of oxygen.

olfaction [L *olfacere* to smell] The sense of smell.

ommatidium A single light detecting unit of the arthropod compound eye.

omnivorous [L *omnis* all; *vorare* to devour] Eating a variety of foods, including both plants and animals.

oncogene A gene that causes one of the biochemical changes that lead to cancer.

ontogeny [Gk *ōn* being] The course of development of an individual organism.

oo- [Gk *ōion* egg] Egg.

oogamy A type of heterogamy in which the female gametes are large nonmotile egg cells.

oogenesis The production of ova (eggs) in the animal ovary.

oogonium Unjacketed female reproductive organ of a thallophyte plant.

open circulatory system A circulatory plan in which blood leaves the vessels and bathes the organs directly.

operant conditioning *See* **conditioning.**

operator A region on the DNA to which a control substance can bind, thereby altering the rate of transcription.

operon A grouping of genes consisting of the promoter, operator, and structural genes, that is controlled as a unit.

oral [L *oris* of the mouth] Relating to the mouth.

organ [Gk *organon* tool] A body part usually composed of several tissues grouped together into a structural and functional unit.

organelle A well-defined subcellular structure.

organic compound A chemical compound containing carbon.

organism An individual living thing.

organogenesis The period during embryonic development following neurulation when the organs are forming.

orientation The act of turning or moving in relation to some external feature, such as a source of light.

osmol Measure of osmotic concentration; the total number of moles of osmotically active particles per liter of solvent.

osmoregulation Regulation of the osmotic concentration of body fluids in such a manner as to keep them relatively constant despite changes in the external medium.

osmosis [Gk *ōsmos* thrust] Movement of a solvent (usually water in biology) through a selectively permeable membrane.

osmotic potential The free energy of water molecules in a solution or colloid under conditions of constant temperature and pressure; since this free energy decreases as the proportion of osmotically active particles rises, a measure of the tendency of the solution or colloid to lose water.

osmotic pressure The pressure that must be exerted on a solution or colloid to keep it in equilibrium with pure water when it is separated from the sater by a selectively permeable membrane; hence a measure of the tendency of the solution or colloid to take in water.

ov-, ovi- [L *ovum* egg] Egg.

ovary Female reproductive organ in which egg cells are produced.

oviduct The tube that conducts the egg from the ovary to the uterus. Also called the fallopian tube.

ovulation Release of an egg from the ovary.

ovule A plant structure, composed of an integument, sporangium, and megagametophyte, that develops into a seed after fertilization.

oxidation Energy-releasing process involving removal of electrons from a substance; in biological systems, generally by the removal of hydrogen (or sometimes the addition of oxygen).

oxidative phosphorylation The process by which ATP is formed in the mitochondrion using the chemiosmotic gradient generated by the electrons flowing along the electron transport chain.

paleontology The study of fossils.

pancreas In vertebrates, a large glandular organ located near the stomach that secretes digestive enzymes into the duodenum and also produces hormones.

papilla [L nipple] A small nipplelike protuberance.

para Alongside of.

parapodium [Gk *podion* little foot] One of the paired segmentally arranged lateral flaplike protuberances of polychaete worms.

parasitism [Gk *parasitos* eating with another] A symbiosis in which one party benefits at the expense of the other.

parasympathetic nervous system One of the two parts of the autonomic nervous system.

parathyroids Small endocrine glands of vertebrates located near the thyroid.

parenchyma A plant tissue composed of thin-walled, loosely packed, relatively unspecialized cells.

parthenogenesis [Gk *parthenos* virgin] Production of offspring without fertilization.

passive immunity The short term immunity that results from being given already synthesized antibodies

pathogen [Gk *pathos* suffering] A disease-causing organism.

pattern formation The spatial arrangement of those tissues to form particular structures and patterns during development.

pectin A complex polysaccharide that cross-links the cellulose fibrils in a plant cell wall and is a major constituent of the middle lamella.

pellicle [L *pellis* skin] A thin skin or membrane.

pepsin [Gk *pepsis* digestion] A protein-digesting enzyme of the stomach.

peptide bond A covalent bond between two amino acids resulting from a condensation reaction between the amino group of one acid and the acidic group of the other.

perennial A plant that lives for several years, as compared to annuals and biennials, which live for one and two years respectively.

peri- Surrounding.

pericycle A layer of cells inside the endodermis but outside the phloem of roots and stems.

periderm The corky outer bark of older stems and roots that replaces the epidermis when secondary growth occurs; includes the cork cells, cork cambium, and cork parenchyma.

peripheral nervous system The sensory and motor neurons that connect the central nervous system to the rest of the body.

peristalsis [Gk *stalsis* contraction] Alternating waves of contraction and relaxation passing along a tubular structure such as the digestive tract.

permeable [L *permeare* to go through] Of a membrane: permitting other substances to pass through.

permease [L *permeare*] A protein that allows molecules to move across a membrane; *cf.* gated channel, membrane channel, membrane pump.

peroxisomes Saclike organelles within eucaryotic cells that contain oxidative enzymes.

petiole [L *pediculus* small foot] The stalk of a leaf.

PGAL *See* phosphoglyceraldehyde.

pH Symbol for the logarithm of the reciprocal of the hydrogen ion concentration; hence a measure of acidity. A pH of 7 is neutral; lower values are acidic, higher values alkaline (basic).

phage *See* bacteriophage.

phagocytosis [Gk *phagein* to eat] The active engulfing of particles by a cell.

pharynx Part of the digestive tract between the oral cavity and the esophagus; in vertebrates, also part of the respiratory passage.

phenotype [Gk *phainein* to show] The physical manifestation of a genetic trait.

pheromone [Gk *pherein* to carry + hormone] A substance that, secreted by one organism, influences the behavior or physiology of other organisms of the same species when they sense its odor.

phloem [Gk *phloios* bark] A plant vascular tissue that transports organic materials; the inner bark.

-phore [Gk *pherin* to carry] Carrier.

phosphoglyceraldehyde (PGAL) A three-carbon phosphorylated carbohydrate, important in both photosynthesis and glycolysis.

phospholipid A compound composed of glycerol, fatty acids, a phosphate group, and often a nitrogenous group.

phosphorylation Addition of a phosphate group.

photic zone The upper region of the ocean where enough light penetrates for photosynthesis.

photo- [Gk *phōs* light] Light.

photon A discrete unit of radiant energy.

photoperiodism A response by an organism to the duration and timing of the light and dark conditions.

photophosphorylation The process by which energy from light is used to convert ADP into ATP.

photosynthesis Autotrophic synthesis of organic materials in which the source of energy is light; *cf.* photophosphorylation.

photosystem The light-absorbing unit of photosynthesis built into the thylakoid membrane; consists of antenna pigments and a reaction center molecule.

phototropism A bending of a plant shoot towards light.

-phyll [Gk *phyllon* leaf] Leaf.

phylogeny [Gk *phylē* tribe] Evolutionary history of an organism.

physiology [Gk *physis* nature] The life processes and functions of organisms; the study of these.

phyte, phyto [Gk *phyton* plant] Plant.

phytochrome A protein pigment of plants sensitive to red and farred light.

pineal gland A tiny endocrine gland in the forebrain of vertebrates that secretes the hormone melatonin, which is involved in coordinating functions of the body in response to the timing of the light/dark cycle.

pinocytosis [Gk *pinein* to drink] The active engulfing by cells of liquid or of very small particles.

pistil The female reproductive organ of a flower, composed of one or more megasporophylls.

pith A tissue (usually parenchyma) located in the center of a stem (rarely a root), internal to the xylem.

pituitary gland An endocrine gland located below the brain of vertebrates, it consists of two parts: a posterior pituitary that secretes the hormones vasopressin and oxytocin, and the anterior pituitary, which is known as the master gland because it secretes hormones that regulate the action of other endocrine glands.

placenta [Gk *plax* flat surface] An organ in mammals, made up of fetal and maternal components, that aids in exchange of materials between the fetus and the mother.

plankton Small organisms that float or drift passively with the current in a lake or ocean. There are two types of plankton: the photosynthetic phytoplankton, and the heterotrophic zooplankton.

plasm-, plasmo-, -plasm [Gk *plasma* something formed or molded] Formed material; plasma; cytoplasm.

plasma Blood minus the cells and platelets.

plasma membrane The outer membrane of a cell.

plasmid A small circular piece of DNA free in the cytoplasm of a bacterial or yeast cell and replicated independently of the cell's chromosome.

plasmodesma [Gk *desma* bond] A connection between adjacent plant cells through tiny openings in the cell walls (*pl.* plasmodesmata).

plasmolysis Shrinkage of a plant cell away from its wall when in a hypertonic medium.

plastid Relatively large organelle in plant cells that functions in photosynthesis and/or nutrient storage.

pleiotropic [Gk *pleion* more] Of a gene: having more than one phenotypic effect.

polar covalent bond A strong chemical bond in which the constituent atoms differ in their electronegativity and therefore share electrons unequally. The resulting molecule is partially positive at one end and partially negative at another.

polar molecule A molecule with oppositely charged sections; the charges, which are far weaker than the charges on ions, a rise from differences in electronegativity between the constituent atoms.

pollen grain [L *pollen* flour dust] A microgametophyte of a seed plant.

pollination The transfer of pollen from the stamen, the male part of the plant, to the pistil, the female part, by an animal or the wind.

poly- Many.

polygyny A type of polygamy in which one male mates with many females.

polymer [Gk *meros* part] A large molecule consisting of a chain of small molecules bonded together by condensation reactions or similar reactions.

polymerase An enzyme complex that catalyzes the polymerization of nucleotides; examples are DNA polymerase, which is involved in replication, and RNA polymerase, which is involved in transcription.

polymorphism [Gk *morphē* form] The simultaneous occurrence of several discontinuous phenotypes in a population.

polyp [Gk *polypous* many-footed] The sedentary stage in the life cycle of a Cnidarian.

polypeptide chain A chain of amino acids linked together by peptide bonds.

polyploid Having more than two complete sets of chromosomes.

polysaccharide Any carbohydrate that is a polymer of simple sugars.

population In ecology, group of individuals belonging to the same species living in a particular geographic area at a particular time.

portal system [L *porta* gate] A blood circuit in which two beds of capillaries are connected by a vein (e.g., *hepatic portal system).*

positive feedback *See* feedback.

posterior Toward the hind end.

potential Short for *potential difference:* the difference in electrical charge between two points. *Resting p.:* a relatively steady potential difference across a cell membrane, particularly of a nonfiring nerve cell or a relaxed muscle cell. *Action p.:* a sharp change in the potential difference across the membrane of a nerve or muscle cell that is propagated along the cell; in nerves, identified with the nerve impulse. *Generator p.:* a change in the potential difference across the membrane of a sensory cell that, if it reaches a threshold level, may trigger an action potential along the associated neural pathway.

preadapted A structure that evolved for one function but can at least minimally perform another function when placed in a new environment.

predation [L *praedatio* plundering] The feeding of free-living organisms on other organisms.

presumptive Describing the developmental fate of a tissue that is not yet differentiated. Presumptive neural tissue, for example, is destined to become part of the nervous system once it has differentiated.

primary consumers Plant eating animals; herbivores.

primary germ layers The three tissue layers (ectoderm, endoderm, and mesoderm) that form during gastrulation and that give rise to all the tissues of the body.

primary structure The sequence of amino acids and the location of the disulfide bonds in a polypeptide chain.

primary succession An ecological succession that occurs in a newly formed area where no other organisms existed previously.

primary tissues In plants, the tissues produced by the apical meristems, which result in a growth in the length of the stem or root.

primary transcript Newly synthesized RNA—generally mRNA—before the introns are moved.

primitive [L *primus* first] Old, like the ancestral condition.

primordium [L *primus; ordiri* to begin] Rudiment, earliest stage of development.

pro- Before.

proboscis [Gk *boskein* to feed] A long snout; an elephant's trunk. In invertebrates, an elongate, sometimes eversible process originating in or near the mouth that often serves in feeding.

procaryotic cell A type of cell that lacks a membrane-bounded nucleus; found only in bacteria.

progesterone [L *gestare* to carry out] One of the principal female sex hormones of vertebrates.

promoter The region of DNA to which the transcription complex binds.

prot-, proto- First, primary.

protease A protein-digesting enzyme.

protein Long polypeptide chain or chains that forms a precise 3-D structure.

protists A group of eucaryotic organisms, most unicellular, that are at a simple level of biological organization and do not fit the definition of plant, animal, or fungus.

proton A positively charged primary subatomic particle.

proto-oncogene A gene that can, after certain sorts of mutation or translocation, or after mutation or translocation in associated control regions, become an oncogene and cause one of the changes leading to cancer.

protoplasm Living substance, the material of cells.

protostomes A lineage of coelomate animals in which the embryonic blastopore forms the mouth.

provirus Viral nucleic acid integrated into the genetic material of a host cell.

proximal Near some reference point (often the main part of the body).

pseudo- False; temporary.

pseudocoelom A functional body cavity not entirely enclosed by mesoderm.

pseudogene An untranscribed region of the DNA that closely resembles a gene.

pseudopod, pseudopodium [L *podium* foot] A transitory cytoplasmic protrusion of an amoeba or an amoeboid cell.

pulmonary [L *pulmones* lungs] Relating to the lungs.

punctuated equilibrium An evolutionary theory that states that most evolution is marked by long periods of equilibrium, when the organisms change very little, punctuated by short periods of rapid change in small populations.

purine Any of several double-ringed nitrogenous bases important in nucleotides.

pyloric [Gk *pylōros* gatekeeper] Referring to the junction between the stomach and the intestine.

pyrimidine Any of several single-ringed nitrogenous bases important in nucleotides.

pyruvic acid A three-carbon compound produced by glycolysis.

race A subspecies.

radial symmetry A type of symmetry in which the body parts are arranged regularly around a central line (in animals, running through the oral-anal axis) rather than on the two sides of a plane.

radiation As an evolutionary phenomenon, divergence of members of a single lineage into different niches or adaptive zones.

radicle The part of the plant embryo that develops into the root system.

receptor In cell biology, a region, often the exposed part of a membrane protein, that binds a substance but does not catalyze a reaction in the chemical it binds; the membrane protein frequently has another region that, as a result of the binding, undergoes an allosteric change and so becomes catalytically active.

recessive Of an allele: not expressing its phenotype in the presence of another allele of the same gene, therefore expressing it only in homozygous individuals. *Recessive character, recessive phenotype:* one caused by a recessive allele.

reciprocal altruism *See* altruistic behavior.

recombinant DNA technology Techniques in which DNA from different organisms is combined and transferred into different cells where they can be expressed.

recombination In genetics, a novel arrangement of alleles resulting from sexual reproduction and from crossing over (or, in procaryotes and eucaryotic organelles, from conjugation). In gene evolution, a novel arrangement of exons resulting from a variety of processes that duplicate and transport segments of the chromosomes within the genome; these processes include transposition, unequal crossing over, and chromosomal breakage and fusion.

rectum [L *rectus* straight] The terminal portion of the intestine.

redox reaction [*from* *re*duction-*ox*idation] A reaction involving reduction and oxidation, which inevitably occur together; *cf.* reduction, oxidation.

reduction Energy-storing process involving addition of electrons to a substance; in biological systems, generally by the addition of hydrogen (or sometimes the removal of oxygen).

reflex [L *reflexus* bent back] An automatic act consisting, in its pure form, of a single simple response to a single stimulus, as when a tap on the knee elicits a knee jerk. Distinguished from a motor program, which involves a coordinated response of several muscles.

reflex arc A functional unit of the nervous system, involving the entire pathway from receptor cell to effector.

reinforcement (psychological) Reward for a particular behavior.

releaser *See* sign stimulus.

renal [L *renes* kidneys] Pertaining to the kidney.

repressor protein A protein produced by a regulatory protein that regulates the activity of an operon.

respiration [L *respiratio* breathing out] (1) The release of energy by oxidation of fuel molecules. (2) The taking of O_2 and release of CO_2; breathing.

resting potential *See* potential.

restriction enzyme An enzyme that breaks bonds within DNA only within a specific sequences of bases. Also called a restriction endonuclease.

reticulum [L little net] A network.

retina The tissue in the rear of the eye that contains the sensory cells of vision.

retinal A light absorbing pigment derived from vitamin A that is found in the rods of the vertebrate eye.

retrovirus An RNA virus that, by means of a special enzyme (reverse transcriptase), makes a DNA copy of its genome which is then incorporated into the host's genome.

rhizoid [Gk *rhiza* root] Rootlike structure.

ribonucleic acid (RNA) Nucleic acid characterized by the presence of a ribose sugar in each nucleotide. The primary classes of RNA are mRNA (messenger RNA, which carries the instructions specifying the order of amino acids in new proteins from the genes to the ribosomes where protein synthesis takes place), rRNA (ribosomal RNA, which is incorporated into ribosomes), and tRNA (transfer RNA, which carries amino acids to the ribosomes as part of protein synthesis).

ribosome A small cytoplasmic organelle that functions in protein synthesis.

RNA *See* ribonucleic acid.

RNA polymerase The enzyme complex that directs the synthesis of RNA from a DNA template.

root pressure The pressure developed in roots as a result of the osmosis of water into the stele.

***r*-selected species** Organisms whose life histories involve producing enormous numbers of offspring while conditions are favorable.

ruminants The hoofed mammals, such as sheep, cattle, goats, and deer, which are characterized by the possession of a four-chambered stomach.

salt Any of a class of generally ionic compounds that may be formed by reaction of an acid and a base, e.g., table salt, NaCI.

sap Water and dissolved materials moving in the xylem; less commonly, solutions moving in the phloem. *Cell sap:* the fluid content of a plant-cell vacuole

saprophyte [Gk *sapros* rotten] A heterotrophoic plant or bacterium that lives on dead organic material.

sarcomere [Gk *sarx* flesh; *meros* part] The region of a skeletal-muscle myofibril extending from one Z line to the next; the functional unit of skeletal-muscle contraction.

sarcoplasmic reticulum The form of endoplasmic reticulum found in muscle fibers.

sclerenchyma [Gk *sclēros* hard] A plant supportive tissue composed of cells with thick secondary walls.

second law of thermodynamics The principle that states that the total amount of free energy is declining in the universe. This occurs because each energy transfer generates heat that is then no longer available for doing work.

secondary structure The folding of a polypeptide chain into an alpha helix or a beta sheet, both of which are held together by hydrogen bonds between different parts of the chain.

secondary tissues Tissues produced by the lateral meristems, particularly the vascular cambium, that result in an increase in girth of the stem.

section *cross or transverse s.:* section at right angles to the longest axis. *Longitudinal s.: section parallel to the longest axis. Radial s.:* longitudinal section along a radius. *Sagittal s.:* vertical longitudinal section along the midline of a bilaterally symmetrical animal.

seed A plant reproductive entity consisting of an embryo and stored food enclosed in a protective coat.

segmentation The subdivision of an organism into more or less equivalent serially arranged units.

selection pressure In a population, the force for genetic change resulting from natural selection.

selectively permeable A membrane that is permeable to some substances but not to others.

senescence Aging.

sensitization The process by which an unexpected stimulus alerts an animal, reducing or eliminating any preexisting habituation; *cf.* adaptation, habituation.

sensory neuron A neuron, leading toward the central nervous system, that receives input from a receptor cell or is itself responsive to sensory stimulation.

sensory receptors Specialized structures, either cells or modified sensory neurons, that rare highly selective and maximally responsive to just one kind of stimulus. They transform the energy of the stimulus into the electrochemical energy of an action potential in the associated sensory neuron.

septum [L barrier] A partition or wall (*pl.* septa).

sessile [L *sessilis* of sitting, low] Of animals, sedentary. Of plants, without a stalk.

sex chromosomes A pair of chromosomes that is different between the sexes and determines the sex of the individual.

sexual dimorphism Morphological differences between the two sexes of a species, as in the size of tails of peacocks as compared to peahens.

sex-limited A gene that is only expressed in one sex.

sex-linked Of genes: located on the X chromosome.

sexual reproduction The form of reproduction that involves the fusion of two gametes in the process of fertilization.

sexual selection Selection for morphology or behavior directly related to attracting or winning mates. *Male-contest sexual selection:* selection for morphology or behavior that enables a male to win fights or contests for access to females, gaining a high position in a dominance hierarchy, for example, or possession of a territory. *Female-choice sexual selection:* selection for morphology or behavior that enables a male to attract females directly.

shoot A stem with its leaves, flowers, etc.

short day plant A plant that flowers only when the dark period is longer than a certain critical value.

sieve element A conductile cell of the phloem.

sign stimulus (or releaser) A simple cue that orients or triggers specific innate behavior.

sinus [L curve, hollow] (1) A channel for the passage of blood lacking the characteristics of a true blood vessel. (2) A hollow within bone or another tissue (e.g. the air-filled sinuses of some of the facial bones).

smooth muscle Spindle shaped uninucleate muscle fibers that are found in the internal organs and innervated by the autonomic nervous system; involuntary muscle.

sol Colloid in which the suspended particles are dispersed at random; *cf.* gel.

solute Substance dissolved in another (the solvent).

solution [L *solutio* loosening] A homogeneous molecular mixture of two or more substances.

solvent Medium in which one or more substances (the solute) are dissolved.

-soma, somat-, -some [Gk *soma* body] Body, entity.

somatic Pertaining to the body; to all cells except the germ cells; to the body wall. *Somatic nervous system:* a portion of the nervous system that is at least potentially under control of the will; *cf.* autonomic nervous system.

somatic nervous system The branch of the peripheral nervous system that controls the skeletal muscles; the voluntary system.

specialized Adapted to a special, usually rather narrow, function or way of life.

speciation The process of formation of new species.

species [L kind] (1) A group of organisms that more closely resemble each other, with respect to their physical appearance (morphology), physiology, behavior, and reproductive patterns, than they resemble any other organisms. (2) A genetically distinctive group of populations whose members are able to interbreed freely under natural conditions, and that are reproductively isolated from all other such groups.

sperm [Gk *sperma* seed] A male gamete.

spermatogenesis The production of sperm in the male reproductive structure.

sphincter [Gk *sphinktēr* band] A ring-shaped muscle that can close a tubular structure by contracting.

spindle A microtubular structure with which the chromosomes are associated in mitosis and meiosis.

spiracles Minute openings located over the surface of an insect's body that open into the tracheal system.

sporangium A plant structure that produces spores.

spore [Gk *spora* seed] In plants, a reproductive cell produced by meiosis that is capable of developing into a gametophyte without fusion with another cell. In fungi, an asexual reproductive cell, often a resting stage adapted to resist unfavorable environmental conditions.

sporophyll [Gk *phyllon* leaf] A modified leaf that bears spores.

sporophyte [Gk *phyton* plant] A diploid plant that produces spores.

stabilizing selection Natural selection that selects against the extreme variants in a distribution thereby favoring those in the middle.

stamen [L thread] A male sexual part of a flower; a microsporophyll of a flowering plant.

starch A glucose polymer, the principal polysaccharide storage product of vascular plants.

stele [Gk *stēlē* upright slab] The vascular cylinder in the center of a root or stem, bounded externally by the endodermis.

stereo- [Gk *stereos* solid] Solid; three-dimensional.

steroid Any of a number of complex, often biologically important compounds (e.g., some hormones and vitamins), composed of four interlocking rings of carbon atoms.

stigma The sticky tip of the pistil that receives the pollen grain.

stimulus Any environmental factor that is detected by a receptor.

stoma [Gk mouth] An opening, regulated by guard cells, in the epidermis of a leaf or other plant part (*pl.* stomata).

striated muscle Type of muscle fibers that make up skeletal muscles and are innervated by the somatic nervous system; voluntary muscle.

stroma [Gk *stroma* bed, mattress] The ground substance within such organelles as chloroplasts and mitochondria.

stromatolites Complex microbial communities of cyanobacteria found in bacterial mats that formed layered mounds which became impregnated by minerals and fossilized.

subspecies A genetically distinct geographic subunit of a species.

substrate (1) The base on which an organism lives, e.g., soil. (2) In chemical reactions, a substance acted upon, as by an enzyme.

substrate level phosphorylation The process by which a phosphate group is transferred from an energy rich substrate to ADP to form ATP.

succession In ecology, progressive change in the plant and animal life of an area.

sucrose A double sugar composed of a unit of glucose and a unit of fructose; table sugar.

suppressor T cells A type of T lymphocyte that suppresses the immune response.

surface tension The property of a liquid that makes it behave as if its surface is covered with an elastic film. In water, the property results from hydrogen bonds between the molecule which must be broken if the surface is to be broken.

suspension A heterogeneous mixture in which the particles of one substance are kept dispersed by agitation.

sym-, syn- Together.

symbiosis [Gk *bios* life] The living together of two organisms in an intimate relationship.

sympathetic nervous system One of the two parts of the autonomic nervous system.

sympatric [L *patria* homeland] Having the same range.

sympatric speciation The type of speciation that occurs when the daughter species shares the same range as the parent species, usually as a result of polyploidy.

symplast In a plant, the system constituted by the cytoplasm of cells interconnected by plasmodesmata.

synapse [Gk *haptein* to fasten] A juncture between two neurons.

synapsis The pairing of homologous chromosomes during meiosis.

synaptic terminal The swollen end of the axon that secretes transmitter chemicals. Also called the synaptic know or bouton.

synergistic [Gk *ergon* work] Acting together with another substance or organ to achieve or enhance a given effect.

system A grouping of organs that carries out body functions.

systematics The branch of biology that studies the classification of life. It includes taxonomy and phylogeny.

systemic circulation The part of the circulatory system supplying body parts other than the gas-exchange surfaces.

systole Contraction of the ventricles of the human heart.

-tactic Referring to a taxis.

taiga The biome dominated by the coniferous (evergreen) forests; characterized by an average low temperature and low to moderate precipitation.

taxis A simple continuously oriented movement in animals (e.g., phototaxis, geotaxis) (*pl.* taxes).

taxonomy [Gk *taxis* arrangement] The classification of organisms on the basis of their evolutionary relationships.

tendon [L *tendere* to stretch] A type of connective tissue attaching muscle to bone.

territory A particular area defended by an individual against instrusion by other individuals, particularly of the same species.

tertiary structure The complexely folded three dimensional structure of a protein formed by weak bonds between different R groups of the polypeptide chain.

testis Primary male sex organ in which sperm are produced (*pl.* testes).

testosterone Male sex hormone; androgen.

tetrapods Vertebrates with four legs (two pairs).

thalamus [Gk *thalamos* inner chamber] Part of the rear portion of the vertebrate forebrain, a center for integration of sensory impulses.

thallus [Gk *thallos* young shoot] A plant body exhibiting relatively little tissue differentiation and lacking true roots, stems, and leaves.

thermoregulation The ability to regulate the body's internal temperature; characteristic of birds and mammals but found in other groups as well.

thorax [Gk *thōrax* breastplate] In mammals, the part of the trunk anterior to the diaphragm, which partitions it from the abdomen. In insects, the body region between the head and the abdomen, bearing the walking legs and wings.

threshold The electrical potential a cell membrane of a neuron or muscle fiber must read for an action potential to be triggered.

thylakoid Membranous sacs within the chloroplast that are the site of the light-dependent reactions of photosynthesis.

thymus [Gk *thymos* warty excrescence] Glandular organ that plays an important role in the development of immunologic capabilities in vertebrates.

thyroid [Gk *thyreoeidēs* shield-shaped] An endocrine gland of vertebrates located in the neck region.

thyroxin A hormone, produced by the thyroid, that stimulates a speedup of metabolism.

tissue [L *texere* to weave] An aggregate of cells, usually similar in both structure and function, that are bound together by intercellular material.

T lymphocyte *See* lymphocyte.

totipotent Refers to embryonic cells that have full developmental potential and can form a new embryo.

toxin A proteinaceous substance produced by one organism that is poisonous to another.

trachea In vertebrates, the part of the respiratory system running from the pharynx into the thorax; the "windpipe." In land arthropods, an air duct running from an opening in the body wall to the tissues.

tracheae Tiny air ducts in insects that branch repeatedly to deliver oxygen to all parts of the insect body.

tracheid An elongate thick-walled tapering conductile cell of the xylem.

trans- Across; beyond.

transcription In genetics, the synthesis of RNA from a DNA template.

transduction [L *ducere* to lead] In genetics, the transfer of genetic material from one host cell to another by a virus. In neurobiology, the translation of a stimulus like light or sound into an electrical change in a receptor cell.

transformation The incorporation by bacteria of fragments of DNA released into the medium from dead cells.

translation In genetics, the synthesis of a polypeptide from an mRNA template.

translocation In botany, the movement of organic materials from one place to another within the plant body, primarily through the phloem. In genetics, the exchange of parts between nonhomologous chromosomes.

transpiration Release of water vapor from the aerial parts of a plant, primarily through the stomata.

transposition The movement of DNA from one position in the genome to another. *Transposon:* a mobile segment of DNA, usually encoding the enzymes necessary to effect its own movement.

triglyceride A fat; composed of three fatty acids joined to a glycerol molecule by condensation reactions.

-trophic [Gk *trophē* food] Nourishing; stimulatory.

tropic hormone A hormone produced by one endocrine gland that stimulates another endocrine gland.

tropism [Gk *tropos* turn] A turning response to a stimulus, primarily by differential growth patterns in plants.

tundra The northernmost biome of North America, Europe, and Asia where the subsoil is permanently frozen.

turgid [L *turgidus* swollen] Swollen with fluid.

turgor pressure [L *turgēre* to be swollen] The pressure exerted by the contents of a cell against the cell membrane or cell wall.

tympanic membrane [Gk *tympanon* drum] A membrane of the ear that picks up vibrations from the air and transmits them to other parts of the ear; the eardrum.

uniramous appendages Arthropod appendages that have only a single branch (ramus).

urea The nitrogenous waste product of mammals and some other vertebrates, formed in the liver by combination of ammonia and carbon dioxide.

ureter The duct carrying urine from the kidney to the bladder in higher vertebrates.

urethra The duct leading from the bladder to the exterior in higher vertebrates.

uric acid An insoluble nitrogenous waste product of most land arthropods, reptiles, and birds.

uterus In mammals, the chamber of the female reproductive tract in which the embryo undergoes much of its development; the womb.

vaccine [L *vacca* cow] Drug containing an antigen, administered to induce active immunity in the patient.

vacuole [L *vacuus* empty] A membrane-bounded vesicle or chamber in a cell.

vagina The tube leading from the uterus to the outside.

valence A measure of the bonding capacity of an atom, which is determined by the number of electrons in the outer shell.

valence shell The outermost energy level of an atom.

vascular cambium A lateral meristem located between the xylem and phloem; it produces secondary xylem and phloem and is responsible for a growth in the stem's girth.

vascular plants Plants possessing xylem and phloem, which includes all the land plants except the mosses and their relatives.

vascular tissue [L *vasculum* small vessel] Tissue concerned with internal transport, such as xylem and phloem in plants and blood and lymph in animals.

vaso- [L *vas* vessel] Blood vessel.

vector [L *vectus* carried] Transmitter of pathogens.

vegetative Of plant cells and organs: not specialized for reproduction. Of reproduction: asexual. Of bodily functions: involuntary.

vein [L *vena* blood vessel] A blood vessel that transports blood toward the heart.

vena cava [L hollow vein] One of the two large veins that return blood to the heart from the systemic circulation of vertebrates.

ventral [L *venter* belly] Pertaining to the belly or underparts.

vertebrates Animals possessing a backbone.

vessel element A highly specialized cell of the xylem, with thick secondary walls and extensively perforated end walls.

villus [L shaggy hair] A highly vascularized fingerlike process from the intestinal lining or from the surface of some other structure (e.g., a chorionic villus of the placenta) (*pl.* villi).

virus [L slime, poison] A submicroscopic noncellular, obligatorily parasitic entity, composed of a protein shell and a nucleic acid core, that exhibits some properties normally associated with living organisms, including the ability to mutate and to evolve.

viscera [L] The internal organs, especially those of the great central body cavity.

vitamin [L *vita* life] An organic compound, necessary in small quantities, that a given organism cannot synthesize for itself and must obtain prefabricated in the diet.

woody dicot *See* dicot.

X chromosome The female sex chromosome.

xylem [Gk *xylon* wood] A vascular tissue that transports water and dissolved minerals upward through the plant body.

Y chromosome The male sex chromosome.

yolk stored food material in an egg.

zoo- [Gk *zōion* animal] Animal, motile.

zoospore A ciliated or flagellated plant spore.

zygote [Gk *zygōtos* yoked] A fertilized egg cell.

zymogen [Gk *zymē* leaven] An inactive precursor of an enzyme.

APPENDIX 7
SUGGESTED READINGS

Chapter 1

COMROE, J. H., 1977. *Retrospectroscope.* Von Gehr Press, Menlo Park, Calif. *A fascinating study of how important scientific discoveries are made. The author concludes that great advances usually arise out of research directed at wholly unrelated problems.**

DARWIN, C., 1859. *The Origin of Species. Of the many reprints of this classic work, the edition by R. E. Leaky (Hill and Wang, New York, 1979) provides perhaps the best introduction and illustrations.**

GINGERICH, O., 1982. The Galileo affair, *Scientific American* 247 (2). *The complex interactions between eccelesiastical politics and Galileo's difficult personality are traced in illuminating detail.*

HERBERT, S., 1986. Darwin as a geologist, *Scientific American* 254 (5). *Readable account of Darwin's important contributions to nineteenth-century geology.*

KOESTLER, A., 1959. *The Sleepwalkers.* Macmillan, New York. *This informal and gossipy account of the Copernican revolution focuses on the personalities and motivations of Copernicus, Galileo (to whom Koestler is unsympathetic), and Kepler. It is a good supplement to Kuhn's book on Copernicus, listed next.**

KUHN, T. S., 1959. *The Copernican Revolution.* Random House, New York.*

KUHN, T. S., 1962. *The Structure of Scientific Revolutions.* University of Chicago Press, Chicago. *In this influential book Kuhn argues that science works in two ways—that of Normal Science (in which experiments are designed to investigate the dominant theory, or "paradigm") and that of Revolutionary Science (which arises when the dominant theory has accumulated so many anomalies that the field becomes unstable and a replacement is needed).**

Taylor, F. S., 1949. *A Short History of Science and Scientific Thought.* W. W. Norton, New York. *An excellent brief history with numerous excerpts from the major writings of important scientists.**

Chapter 2

DICKERSON, R. E., and I. GEIS, 1976. *Chemistry, Matter, and the Universe.* W. A. Benjamin, Menlo Park, Calif. *An excellent introduction to chemistry from a biological perspective.*

FRIEDEN, E., 1972. The chemical elements of life, *Scientific American,* 227 (1). *On procedures for determining whether an element is essential to life, with particular emphasis on four elements (fluorine, silicon, tin, and vanadium).*

Chapter 3

ATKINS, P. W., 1987. *Molecules.* Scientific American Library, New York. *A beautifully produced "molecular glossary" illustrating the chemical formula, three-dimensional structure, and biological action of many common or unusually interesting organic molecules.*

DOOLITTLE, R. F., 1985. Proteins, *Scientific American* 253 (4). *Reviews the properties of amino acids and the structure of proteins, and discusses the evolution of different modern proteins from common ancestral enzymes.*

KARPLUS, M., and MCCAMMON, J. A., 1986. The dynamics of proteins, *Scientific American* 254 (4).

STROUD, R. M., 1974. A family of protein-cutting proteins, *Scientific American* 231 (1). (Offprint 1301) *A good discussion of how enzymes like chymotrypsin work.*

STRYER, L., 1988. Biochemistry, 3rd ed. W. H. Freeman, San Francisco. *Beautifully produced, clearly written, but highly technical exposition of biochemistry.*

Chapter 4

ALBERTS, B., D. BRAY, J. LEWIS, M. RAFF, K. ROBERTS, and J. D. WATSON, 1989. *Molecular Biology of the Cell,* 2nd ed. Garland, New York. *A thorough, but highly readable book on cell biology.*

DRESSLER, D., and H. POTTER, 1991. *Discovering Enzymes.* W. H. Freeman, New York. *A well-written and illustrated history of the study of enzymes, with particular emphasis on how the digestive enzyme chymotrypsin works.*

KOSHLAND, D. E., 1973. Protein shape and biological control, *Scientific American* 229 (4). (Offprint 1280) *On the importance of protein conformation in determining enzymatic activity; how substances that cause changes in the shape of a protein can regulate its activity.*

STRYER, L., 1988 *Biochemistry,* 3rd ed. W. H. Freeman, San Francisco. *Beautifully produced, clearly written, but highly technical exposition of biochemistry.*

Chapter 5

BRETSCHER, M. S., 1985. The molecules of the cell membrane, *Scientific American* 253 (4). *Reviews the bilayer plasma membrane and membrane proteins, and the process of endocytosis.*

BRETSCHER, M. S., 1987. How animal cells move, *Scientific American* 257 (6). *The role of pinocytosis in the amoeboid movement of cells.*

BROWN, M. S., and J. L. GOLDSTEIN, 1984. How LDL receptors influence cholesterol and atherosclerosis, *Scientific American* 251 (5). (Offprint 1555)

DAUTRY-VARSAT, A., and H. F. LODISH, 1984. How receptors bring proteins and particles into cells, *Scientific American* 250 (5). (Offprint 1550) *The life cycle of coated pits.*

LIENHARD, G. E., J. W. SLOT, D. E. JAMES, and M. M. MUECKLER, 1992. How cells absorb glucose, *Scientific American* 266 (1). *On the structure and function of the transporter molecule and how insulin regulates it.*

NEHER, E., and B. SAKMANN, 1992. The patch clamp technique, *Scientific American* 266 (3). *Summary of what the procedure has revealed about ion channels and their functions.*

SHARON, N., and H. LIS, 1993. Carbohydrates in cell recognition,

* Available in paperback.

Scientific American 268 (1). *As primary markers, sugars will have practical applications to the prevention and treatment of disease.*

UNWIN, N., and R. HENDERSON, 1984. The structure of proteins in biological membranes, *Scientific American* 250 (2). (Offprint 1547)

Chapter 6

ALBERTS, B., D. BRAY, J. LEWIS, M. RAFF, K. ROBERTS, and J. D. WATSON, 1989. *Molecular Biology of the Cell,* 2nd ed. Garland, New York.

ALLEN, R. D., 1987. The microtubule as an intracellular engine, *Scientific American* 256 (2).

BRETSCHER, M. S., 1987. How animal cells move. *Scientific American* 257 (6). *On the role of endocytosis in movement.*

DEDUVE, C., 1983. Microbodies in the living cell, *Scientific American* 248 (5). (Offprint 1538) *On the class of specialized enzymatic organelles, such as peroxisomes, that are not produced by the Golgi.*

DEDUVE, C., 1986. *The Living Cell.* Scientific American Library, New York. *This two-volume set provides a well-illustrated, up-to-date tour of the cell.*

DUSTIN, P., 1980. Microtubules, *Scientific American* 243 (2). (Offprint 1477) *An excellent summary of the formation of microtubules and the diverse roles they play in the cell.*

GLOVER, D. M., C. GONZALEZ, and J. W. RAFF, 1993. The centrosome, *Scientific American* 268 (6). *On the discovery of details of its structure and function.*

ROTHMAN, J. E., 1985. The compartmental organization of the Golgi apparatus. *Scientific American,* 253 (3). *An incisive analysis of the fine structure of this important organelle.*

STOSSEL, T. P., 1990. How cells crawl. *American Scientist* 78, 407–423. *A detailed look at the mechanisms and control of amoeboid movement.*

WEBER, K., and M. OSBORN, 1985. The molecules of the cell matrix. *Scientific American* 253 (4). *Reviews the structure and the function of microfilaments and microtubules.*

Chapter 7

BASSHAM, J. A., 1962. The path of carbon in photosynthesis, *Scientific American* 206 (6). (Offprint 122) *An old but informative account of how the Calvin cycle was worked out.*

BAZZAZ, F. A., and E. D. FAJER, 1992. Plant life in a CO_2-rich world, *Scientific American* 266 (1). *On how increasing levels may alter the structure and function of ecosystems, while not necessarily benefitting plants.*

BJÖRKMAN, O., and J. BERRY, 1973. High-efficiency photosynthesis, *Scientific American* 229 (4). (Offprint 1281) *The photosynthetic pathway and leaf anatomy of a group of* C_4 *plants.*

GOVINDJEE and W. J. COLEMAN, 1990. How plants make oxygen, *Scientific American* 262(2). *On the operation of the water-splitting enzyme of noncyclic photosynthesis.*

YOUVAN, D. C., and B. L. MARRS, 1987. Molecular mechanisms of photosynthesis, *Scientific American* 256(6). *An account of the molecular events occurring during the first 200 microseconds following photon absorption.*

Chapter 8

CHILDRESS, J. J., H. FELBECK, and G. N. SOMERO, 1987. Symbiosis in the deep sea, *Scientific American* 256 (5). *How chemosynthetic bacteria manage to extract energy from the 250° C sulfurous water at deep sea vents.*

CLOUD, P., 1983. The biosphere, *Scientific American* 249 (3). *On the combined evolution of the earth, life, and the atmosphere, with particular emphasis on the role of oxygen concentration.*

HINKLE, P. C., and R. E. MCCARTY, 1978. How cells make ATP, *Scientific American* 238 (3). (Offprint 1383) *A difficult but rewarding explanation of how ATP is made.*

RACKER, E., 1968. The membrane of the mitochondrion, *Scientific American* 218 (2). (Offprint 1101)

STRYER, L., 1988. *Biochemistry,* 3rd ed. W. H. Freeman, New York. *Traces in great detail the biochemical pathways of respiration, and their regulation.*

Chapter 9

ALBERTS, B., et al., 1989. *Molecular Biology of the Cell* 2nd ed. Garland, New York. *Contains a brief but up-to-date discussion of cell division in molecular terms.*

GOULD, J. L., and C. G. GOULD, 1989. *Sexual Selection.* Scientific American Library, New York. *A wide-ranging, nontechnical account of the many theories that seek to account for the evolution of sex and gender.*

MCINTOSH, J. R., and K. L. MCDONALD, 1989. The mitotic spindle, *Scientific American* 261 (4).

MAZIA, D., 1974. The cell cycle, *Scientific American* 230 (1). (Offprint 1288) *The stages of interphase and mitosis proper; experiments conducted to ascertain the characteristics of these stages and the controls governing them.*

MURRAY, A. W., and M. W. KIRCHNER, 1991. What controls the cell cycle, *Scientific American* 264 (3). *On the biochemical control of cell division, with emphasis on cyclin and cdc.*

STAHL, F. W., 1987. Genetic recombination, *Scientific American* 256 (2).

Chapter 10

BAUER, W. R., F. H. C. CRICK, and J. H. WHITE, 1980. Supercoiled DNA, *Scientific American* 243 (1). (Offprint 1474) *On how the genetic material is compacted, so that DNA a thousand times the length of a bacterium can be packed into the space available, without tangling, and leaving most of the cell volume free for the cytoplasm.*

HOWARD-FLANDERS, P., 1981. Inducible repair of DNA, *Scientific American* 245 (5). (Offprint 1503) *On how cells recognize when the DNA has been damaged, how they switch on genes for repair enzymes, and how the enzymes work.*

RADMAN, M., and R. WAGNER, 1988. The high fidelity of DNA replication, *Scientific American* 259 (2). *On how the proofreading component of DNA polymerase works.*

STRYER, L., 1988. *Biochemistry,* 3rd ed. W. H. Freeman, San Francisco.

UPTON, A. C., 1982. The biological effects of low-level ionizing radiation, *Scientific American* 246 (2). (Offprint 1509) *A superb summary of how DNA is damaged by radiation.*

WATSON, J. D., 1980. *The Double Helix.* A Norton Critical Edition, ed. Gunther S. Stent, W. W. Norton, New York. *A fascinating account of the elucidation of the structure of DNA by one of the two discoverers. Watson spares neither himself nor others in giving a rare behind-the-scenes look at the dynamics of research in a very competitive field. This edition of the original 1968 book includes articles, relevant to the discovery, by other scientists.*

WATSON, J. D., N. H. HOPKINS, J. W. ROBERTS, J. A. STEITZ, and A. M. WEINER, 1987. *Molecular Biology of the Gene,* 4th ed. Benjamin Cummings, Menlo Park, Calif. *A particularly well-written text on molecular genetics. Makes even difficult topics easy to understand.*

YUAN, R., and D. L. HAMILTON, 1982. Restriction and modification of DNA by a complex protein, *American Scientist* 70, 61–69. *On how some endonucleases can cut DNA at a specific site if it is fully unmethylated, or finish methylating (and thereby protect) the same site if it is partially methylated.*

Chapter 11

CECH, T. R., 1986. RNA as an enzyme, *Scientific American* 255 (5). *On the enzymatic properties of ribosomal and, especially, snRNP RNA, and the possibility that RNA originally served the functions now taken over by DNA and protein.*

CHAMBON, P., 1981. Split genes, *Scientific American* 244 (5). (Offprint 1496) *On the organization of introns and exons.*

CRICK, F. H. C., 1962. The genetic code, *Scientific American* 207 (4). (Offprint 123) *Describes Crick's demonstration that the codon is three bases long.*

DARNELL, J. E., 1983. The processing of RNA, *Scientific American* 249 (4). (Offprint 1543) *An excellent summary of the topic.*

DARNELL, J. E., 1985. RNA, *Scientific American* 253 (4). *Reviews transcription, processing, translation, and transcriptional control.*

DARNELL, J. E., LODISH, and D. BALTIMORE, 1990. *Molecular Cell Biology.* W. H. Freeman, New York. *Excellent detailed study of transcription and translation.*

JUDSON, H., 1980. *The Eighth Day of Creation.* Simon & Schuster, New York. *Well-written history of molecular biology.*

LAKE, J. A., 1981. The ribosome, *Scientific American* 245 (2). (Offprint 1501) *On the three-dimensional structure of the ribosome and the details of translation.*

STEITZ, J. A., 1988. "Snurps," *Scientific American* 258 (6). *On the enzymes that remove the introns from eucaryotic primary transcripts.*

Chapter 12

BEARDSLEY, T., 1994. A war not won, *Scientific American* 270 (1). *War on cancer has not slowed deaths from the disease in the U.S.*

BISHOP, J. M., 1982. Oncogenes, *Scientific American* 246 (3). (Offprint 1513) *An illuminating look at the relationship between cancer genes carried by viruses and the similar, noncancerous genes in normal cells.*

CROCE, C. M., and G. KLEIN, 1985. Chromosome translocations and human cancer, *Scientific American* 252 (3). (Offprint 1558) *A clear discussion of the translocations involved in Burkitt's lymphoma.*

FELDMAN, M., and L. EISENBACH, 1988. What makes a tumor cell metastatic? *Scientific American* 259 (5). *About the oncogenes that permit tumor cells to stop adhering to other cells or structures, and so spread to other parts of the body.*

FELSENFELD, G., 1985. DNA, *Scientific American* 253 (4). *The role of DNA structure in the regulation of gene expression.*

GRUNSTEIN, M., 1992. Histones as regulators of genes, *Scientific American* 267 (4). *On their role in expression and suppression of genes.*

HUNTER, T., 1984. The proteins of oncogenes, *Scientific American* 251 (2). (Offprint 1553)

JAIN, R. K., 1994. Barriers to drug delivery in solid tumors, *Scientific American* 271 (1) *An interesting description of how cancerous tumors actively resist penetration by anti-cancer agents.*

JOHNSON, H. M., F. W. BAZER, B. E. SZENTE, and M. A. JARPE, 1994. How interferons fight disease, *Scientific American* 270 (5). *Interferons provide therapy for infectious diseases and cancers.*

LIOTTA, L. A., 1992. Cancer cell invasion and metastasis, *Scientific American* 266 (2). *On regulatory genes and proteins that control the spread of tumor cells and lead to new treatments.*

MCGINNIS, W., and M. KUZIORA, 1994. The molecular architects of body design, *Scientific American* 270 (2). *Nearly identical molecular mechanisms define body shapes in all animals.*

MCKNIGHT, S. L., 1991. Molecular zippers in gene regulation, *Scientific American* 264 (4). *On the operation of the leucine zipper.*

MOSES, P. B., and N. -H. CHUA, 1988. Light switches for plant genes, *Scientific American* 258 (4). *How light energy is used to activate the genes involved in photosynthesis.*

NICOLSON, G. L., 1979. Cancer metastasis, *Scientific American* 240 (3). (Offprint 1422)

NOMURA, M., 1984. The control of ribosome synthesis, *Scientific American* 250 (1). (Offprint 1546)

PTASHNE, M., 1989. How gene activators work, *Scientific American* 260 (1). *A very up-to-date summary of how promoters work in bacteria and yeast.*

PTASHNE, M., A. D. JOHNSON, and C. O. PABO, 1982. A genetic switch in a bacterial virus, *Scientific American* 247 (5). (Offprint 1526) *An excellent description of the details of the lytic/lysogenic switch of lambda virus.*

RHODES, D., and A. KLUG, 1993. Zinc fingers, *Scientific American* 268 (2). *On how they select and bind to DNA and the role they play in switching on genes.*

ROSS, J., 1989. The turnover of messenger RNA, *Scientific American* 260 (4). *On how the rate of degradation of different messengers is controlled.*

SAPIENZA, C., 1990. Parental imprinting of genes, *Scientific American* 263 (4). *On the inheritance of gene switches bound to the DNA of gametes.*

TIOLLAS, P., and M. A. BUENDIA (1991). Hepatitus B virus, *Scientific American* 263 (4). *On the life history of a cancer-promoting virus.*

WEINBERG, R. A., 1983. A molecular basis of cancer, *Scientific American* 249 (5). (Offprint 1544) *A good discussion of oncogenes and the discovery that a single-base change can transform a prepared cell into a cancer cell.*

WEINBERG, R. A., 1988. Finding the anti-oncogene, *Scientific American* 259 (3). *On the genes that restrain cell growth, focusing on retinoblastoma.*

WEINTRAUB, H., 1990. Antisense RNA and DNA, *Scientific American* 262 (1). *On the use of antisense RNA to regulate mRNA activity.*

Chapter 13

AHARONWITZ, Y., and G. COHEN, 1981. Microbial production of pharmaceuticals, *Scientific American* 245 (3). *On how recombinant DNA techniques are used to make microbes produce antibiotics, hormones, and other drugs. There is also an explanation of how antibiotics work to destroy bacteria, which suggests how plasmid genes may confer resistance.*

CHILTON, M-D., 1983. A vector for introducing new genes into plants. *Scientific American* 248 (6). (Offprint 1539) *On bacteria (as opposed to viruses) that transduce host cells.*

COHEN, S. N., and J. A. SHAPIRO, 1980. Transposable genetic elements, *Scientific American* 242 (2).

DEVORET, R., 1979. Bacterial tests for potential carcinogens, *Scientific American* 241 (2). (Offprint 1433) *The close relationship between mutations and cancer, and how to measure mutagenicity.*

FEDOROFF, N. V., 1984. Transposable genetic elements in maize, *Scientific American* 250 (6). *A modern interpretation of the transposition discovered by McClintock.*

GASSER, C. S., and R. T. FRALEY, 1992. Transgenic crops, *Scientific American* 266 (6). *Description of current methods to engineer plants genetically.*

GILBERT, W., and L. VILLA-KOMAROFF, 1980. Useful proteins from recombinant bacteria, *Scientific American* 242 (4). (Offprint 1466) *How recombinant methods can be used to create insulin-producing bacteria.*

GOSDON, G. N., 1985. Molecular approaches to malaria vaccines, *Scientific American* 252 (5). *An example of how haptens are used in designing vaccines.*

GRIVELL, L. A., 1983. Mitochondrial DNA, *Scientific American* 248 (3). (Offprint 1535) *On the procaryotelike organization of mitochondrial genes, and their unique modification of the genetic code.*

HOPWOOD, D. A., 1981. The genetic programming of industrial microorganisms, *Scientific American* 245 (3). *An excellent summary of how basic recombinant DNA techniques work.*

MCCONKEY, E. H., 1993. *Human genetics, the molecular revolution,* Jones and Bartlett, Boston. *An excellent, up to date and readable textbook focusing on new molecular techniques.*

NOVICK, R. P., 1980. Plasmids, *Scientific American* 243 (6). (Offprint 1486)

RENNIE, J., 1993. DNA's new twists, *Scientific American* 266 (3). *On the possibility of transfers of DNA from one species to another.*

RENNIE, J., 1994. Grading the gene tests, *Scientific American* 270 (6). *On concern for rapid growth of genetic testing and its consequences.*

VARMUS, H., 1987. Reverse transcription, *Scientific American* 257 (3). *A detailed description of the process that reverses the usual direction of information flow.*

VERMA, I. M., 1990. Gene therapy. *Scientific American* 263 (5). *On attempts to correct genetic defects.*

WEINBERG, R. A., 1985. The molecules of life, *Scientific American* 253 (4). *A good, very brief review of recombinant DNA techniques.*

Chapter 14

CROW, J. F., 1979. Genes that violate Mendel's rules, *Scientific American* 240 (2). *On alleles that manipulate the genome to enhance the probability of their own transmission.*

GOODENOUGH, U., 1978. *Genetics,* 2nd ed. Holt, Rinehart & Winston, New York, NY. *One of several very good introductory genetics texts currently available.*

HOLLIDAY, R., 1989. A different kind of genetic inheritance, *Scientific American* 260 (6). *On genetic "imprinting."*

HORGAN, J., 1993. Eugenics revisited, *Scientific American* 268 (6). *Linking phenomena, such as homosexuality, to specific genes.*

MANGE, E. J., and A. P. MANGE, 1994. *Human genetics,* Sinauer Associates, Sunderland, Mass. *A good, understandable book on human genetics.*

MCCONKEY, E. H., 1993. *Human genetics, the molecular revolution,* Jones and Bartlett, Boston. *An excellent, up to date and readable textbook focusing on new molecular techniques.*

WHITE, R., and J. M. LALOUEL, 1988. Chromosome mapping with DNA markers, *Scientific American* 258 (2). *On modern methods for mapping the human genome.*

Chapter 15

FAWCETT, D. W., 1986. *A textbook of histology.* 11th ed. Saunders, Philadelphia.

RAVEN, P. H., R. F. EVERT, and S. E. EICHHORN, 1992. *Biology of plants,* 5th ed. Worth, New York. *An excellent introductory botany textbook.*

WHEATER, P. R., H. G. BURKITT, and V. G. DANIELS. *Functional histology,* 2nd ed. Churchill Livingstone, New York. *A good, well illustrated histology book.*

Chapter 16

DENISON, W. C., 1973. Life in tall trees, *Scientific American* 228 (6). *On the interaction between lichens and the trees on which they live.*

EPSTEIN, E., 1973. Roots, *Scientific American* 228 (5). (Offprint 1271) *The mechanisms by which roots take up nutrients from the soil.*

GALSTON, A. W., P. J. DAVIES, and R. K. SLATER, 1980. *The Life of the Green Plants,* 3rd ed. Prentice-Hall, Englewood Cliffs, N. J. *Good short book on plant physiology.**

HESLOP-HARRISON, Y., 1978. Carnivorous plants, *Scientific American* 238 (2). (Offprint 1382)

RAVEN, P. H., R. F. EVERT, and S. E. EICHHORN, 1992. *Biology of plants,* 5th ed. Worth, New York. *An excellent introductory botany textbook.*

Chapter 17

DAVENPORT, H. W., 1972. Why the stomach does not digest itself, *Scientific American* 226 (1). (Offprint 1240)

DEGABRIELE, R., 1980. The physiology of the koala, *Scientific American* 243 (1). *On the nutritional problems of this postgastric fermentor, whose need to maximize nitrogen extraction while minimizing water loss can cause it to starve with a full stomach.*

FOX, S. I., 1990. *Human Physiology,* Wm. C. Brown, Dubuque, Iowa. *An introductory human physiology textbook with excellent illustrations and diagrams.*

GUYTON, A., 1991 *Textbook of Medical Physiology,* 3rd ed. W. B. Saunders, Philadelphia. *The "bible" of human physiology textbooks.*

HARPSTEAD, D. D., 1971. High-lysine corn, *Scientific American* 225 (2). (Offprint 1229) *About an attempt to breed corn with an amino acid profile more suitable for our species.*

JARNICK, J., C. H. NOLLER, and C. I. RHYKERD, 1976. The cycles of plant and animal nutrition, *Scientific American* 235 (3). *The movement of nutrients through plants, animals, and the nonliving environment, with a look at some of the agricultural practices that affect this cycling.*

KRETCHMER, N., 1972. Lactose and lactase, *Scientific American* 227 (4). (Offprint 1259) *On differences in human tolerance of milk sugar.*

ROSENTHAL, G. A., 1983. A seed-eating beetle's adaptations to a poisonous seed, *Scientific American* 249 (5). *Ploy and counterploy as plants attempt to protect themselves by sabotaging the digestion and metabolism of herbivores.*

SANDERSON, S. L., and R. WASSERSUG, 1990. Suspension-feeding vertebrates, *Scientific American* 262 (3). *On filter feeders.*

SCHMIDT-NIELSEN, K., 1970. *Animal Physiology,* 3rd ed. Prentice-Hall, Englewood Cliffs, N. J. *Well-written elementary text on animal physiology.*

YOUNG, V. R., and N. S. SCRIMSHAW, 1971. The physiology of starvation, *Scientific American* 225 (4). (Offprint 1232) *On the remarkable biochemical responses of a starving person's body, which permit relatively long-lasting mineral nutrition of the most important organs, particularly the brain.*

Chapter 18

AVERY, M. E., N. S. WANG, and H. W. TAEUSCH, 1973. The lung of the newborn infant, *Scientific American* 228 (4). *Fascinating account of the changes that take place at birth to prepare the infant lung for breathing.*

FEDER, M. E., and W. W. BURGGREN, 1985. Skin breathing in vertebrates, *Scientific American* 253 (5). *On how some vertebrates supplement or even replace lungs or gills in obtaining oxygen and eliminating carbon dioxide.*

FOX, S. I., 1990. *Human Physiology,* Wm. C. Brown, Dubuque, Iowa. *An introductory human physiology textbook with excellent illustrations and diagrams.*

GUYTON, A., 1991. *Textbook of Medical Physiology,* 3rd ed. W. B. Saunders, Philadelphia. *The "bible" of human physiology textbooks.*

RAHN, H., A. AR, and C. V. PAGANELLI, 1979. How bird eggs breathe,

Scientific American 240 (2). *On the unique gas-exchange problems faced by developing embryos in eggs.*

SCHMIDT-NIELSEN, K., 1971. How birds breathe, *Scientific American* 225 (6). (Offprint 1238) *On the remarkable phenomenon of uni directional flow in the respiratory system of birds.*

SCHMIDT-NIELSEN, K., 1972. *How Animals Work.* Cambridge University Press, New York. *A short, clearly written book that pays particular attention to gas exchange in animals.*

SCHMIDT-NIELSEN, K., 1981. Countercurrent systems in animals, *Scientific American* 244 (5). *Excellent analysis of how noses conserve water during breathing.*

Chapter 19

BIDDULPH, O., and S. BIDDULPH, 1959. The circulatory system of plants, *Scientific American* 200 (2). (Offprint 53) *Excellent discussion of the physiology of phloem by two investigators who carried out some of the early radioactive-tracer studies of mineral transport.*

RAVEN, P. H., R. F. EVERT, and S. E. EICHHORN, 1992. *Biology of plants,* 5th ed. Worth, New York. *An excellent introductory botany textbook.*

SALISBURY, R. B., and C. W. ROSS, 1992. *Plant physiology,* 4th ed. Wadsworth, Belmont, California.

ZIMMERMAN, M. H., 1963. How sap moves in trees, *Scientific American* 208 (3). (Offprint 154)

Chapter 20

ADOLPH, E. F., 1967. The heart's pacemaker, *Scientific American* 216 (3). (Offprint 1067) *How nodal tissue regulates the fundamental rhythm of the heart.*

FOX, S. I., 1990. *Human Physiology,* Wm. C. Brown, Dubuque, Iowa. *An introductory human physiology textbook with excellent illustrations and diagrams.*

GOLDE, D. W., 1991. The stem cell, *Scientific American* 265 (6). *On its central role in growth and maintenance of the human blood-producing and immune systems.*

GUYTON, A., 1991. *Textbook of Medical Physiology,* 3rd ed. W. B. Saunders, Philadelphia. *The "bible" of human physiology textbooks.*

KILGOUR, F. G., 1952. William Harvey, *Scientific American* 186 (6).

LAWN, R. M., 1992. Lipoprotein(a) in heart disease, *Scientific American* 266 (6). *How the particle's role in heart attacks may be a side effect of its possible involvement in the repair of torn blood vessels.*

LAWN, R. M., and G. A. VEHAR, 1986. The molecular genetics of hemophilia, *Scientific American* 254 (3). *Insights into how clotting works from studies of "bleeders."*

LILLYWHITE, H. B., 1988. Snakes, blood circulation, and gravity, *Scientific American* 259 (6). *Interesting comparison of circulatory adaptations of aquatic, terrestrial, and arboreal snakes.*

MAYERSON, H. S., 1963. The lymphatic system, *Scientific American* 208 (6). (Offprint 158)

PERUTZ, M. F., 1964. The hemoglobin molecule, *Scientific American* 211 (5). (Offprint 196) *The research that revealed the structure of hemoglobin, described by the major figure in that work.*

PERUTZ, M. F., 1978. Hemoglobin structure and respiratory transport, *Scientific American* 239 (6). (Offprint 1413)

ROBINSON, T. F., M. FACTOR, and E. H. SONNENBLICK, 1986. The heart as a suction pump, *Scientific American* 254 (6).

WOOD, J. E., 1968. The venous system, *Scientific American* 218 (1). (Offprint 1093) *How the constriction and dilation of veins help determine the distribution of blood in the human body.*

ZAPOL, W. M., 1987. Diving adaptations of the Weddell seal, *Scientific American* 256 (6). *On the many specializations in this mammal's circulatory system that permit lengthy dives to great depths.*

ZWEIFACH, B. W., 1959. The microcirculation of the blood, *Scientific American* 200 (1). (Offprint 64) *On capillaries, arterioles, and venules.*

Chapter 21

"Life, Death, and the immune system," special issue of *Scientific American* 269 (3) (September 1993). *Ten excellent articles on such topics as fighting cancer, tolerating grafts, AIDS, allergy, infection, and autoimmune diseases.*

ANDERSON, R. M., and R. M. MAY, 1992. Understanding the AIDS pandemic, *Scientific American* 266 (6). *On the epidemiology of the AIDS epidemic from an ecological and evolutionary point of view.*

BUISSERET, P. D., 1982. Allergy, *Scientific American* 247 (2). (Offprint 1522) *A clear explanation of how the immune system's overreaction to harmless antigens can create annoying and even dangerous allergies.*

COHEN, I. R., 1988. The self, the world, and autoimmunity, *Scientific American* 258 (4). *About the rare failures in the immune system's ability to "remember" and ignore self-antigens. It argues for a speculative hypothesis that assumes that the variable region of each kind of antibody is recognized, bound, and deactivated by another kind of antibody. Out of these complex interactions immunological responses are regulated, or autoimmune attacks generated.*

CUNNINGHAM, B. A., 1977. The structure and function of histocompatibility antigens, *Scientific American* 237 (4). (Offprint 1369) *On the role of the MHC antigens on the surface of normal cells, including transplant rejection on the one hand and defense against cancer and infection on the other.*

EDELSON, R. L., and J. M. FINK, 1985. The immunologic function of skin, *Scientific American* 252 (6). *On how cells in the skin interact with T cells.*

GREY, H. M., A. SETTE, and S. BUUS, 1989. How *T* cells see antigen, *Scientific American* 261 (5). *A clear discussion of how processed antigens are "presented" by MHC proteins, and recognized by T-cell antibodies.*

HASSELTINE, W. A., and F. WONG-STAAL, 1988. The molecular biology of the AIDS virus, *Scientific American* 259 (4). *An up-to-date description of how HIV-1 works, emphasizing gene regulation. This same issue has nine related articles on various aspects of the AIDS epidemic.*

HENLE, W., G. HENLE, and E. T. LENNETTE, 1979. The Epstein-Barr virus, *Scientific American* 241 (1). (Offprint 1431) *A wide-ranging discussion that pulls together immunology, research with monoclonal antibodies, viral biochemistry, and disease statistics to link Epstein-Barr virus with Burkitt's lymphoma.*

JOHNSON, H. M., J. K. RUSSELL, and C. H. PONTZER, 1992. Superantigens in human disease, *Scientific American* 266 (4). *On their immune-stimulating and immune-depressing properties.*

MARRACK, P., and J. KAPPLER, 1986. The *T* cell and its receptor, *Scientific American* 254 (2). *An excellent summary of how T cells interact with the other elements of the immune system.*

MAYER, M. M., 1973. The complement system, *Scientific American* 229 (5). (Offprint 1283) *On the way an intricate set of enzyme works with antibodies to make novel channels in the membrane of foreign cells, thereby destroying them.*

MILLS, J., and H. MASUR, 1990. AIDS-related infections, *Scientific American* 263 (2).

MILSTEIN, C., 1980. Monoclonal antibodies, *Scientific American* 243 (4). (Offprint 1479)

OLD, L. J., 1977. Cancer immunology, *Scientific American* 236 (5).

(Offprint 1358) *On the distinctive antigens on the surfaces of cancer cells and the problem of mobilizing the immune system to combat cancer.*

OLD, L. J., 1988. Tumor necrosis factor, *Scientific American* 258 (5). *On the many messenger chemicals immune-system cells use to regulate each other's activity.*

REDFIELD, R. R., and D. S. BURKE, 1988. HIV infection: the clinical picture, *Scientific American* 259 (4). *On the progression from infection to death, detailing how the immune system holds the disease at bay for so long.*

SNYDER, S. H., and D. S. BREDT, 1992. Biological roles of nitric oxide, *Scientific American* 266 (6). *On the many roles of this newly discovered transmitter, local chemical mediator, and toxic weapon.*

TONEGAWA, S., 1985. The molecules of the immune system, *Scientific American* 253 (4). *An excellent review of antibody structure, antigen binding, and B- and T-cell function. Does not discuss the interactions between T cells, B cells, macrophages, and the other elements of the immune system.*

YOUNG, J. D. -E., and Z. A. COHN, 1988. How killer cells kill, *Scientific American* 258 (1). *Compares the action of the complement system with the analogous strategy of killer cells in perforating the membranes of target cells.*

Chapter 22

FOX, S. I., 1990. *Human Physiology,* Wm. C. Brown, Dubuque, Iowa. *An introductory human physiology textbook with excellent illustrations and diagrams.*

GUYTON, A., 1991. *Textbook of Medical Physiology,* 3rd ed. W. B. Saunders, Philadelphia. *The "bible" of human physiology textbooks.*

SCHMIDT-NIELSEN, K., 1959. Salt glands, *Scientific American* 200 (1). *The salt-secreting glands of marine birds and turtles are described.*

SCHMIDT-NIELSEN, K., 1959. The physiology of the camel, *Scientific American* 201 (6). (Offprint 1096) *Fascinating discussion of the special adaptations that enable camels to survive and prosper in the desert.*

SCHMIDT-NIELSEN, K., 1983. *Animal Physiology: Adaptation and Environment,* 3rd ed. Cambridge University Press, New York.

SCHMIDT-NIELSEN, K., and B. SCHMIDT-NIELSEN, 1953. The desert rat, *Scientific American* 189 (1). (Offprint 1050) *The extraordinary osmoregulatory abilities that allow the desert rat to survive without ever drinking.*

SMITH, H. W., 1953. The kidney, *Scientific American* 188 (1). (Offprint 37)

SMITH, H. W., 1953. *From Fish to Philosopher.* Little, Brown, Boston. (Paperback edition by Doubleday Anchor Books, 1961.) *A little classic on vertebrate evolution in terms of osmoregulation and excretion, enlivened by liberal doses of personal philosophy.* *

VALTIN, H., 1983. *Renal function.* Boston, Little, Brown.

Chapter 23

ALBERSHEIM, P., 1975. The walls of growing plant cells, *Scientific American* 232 (4). (Offprint 1320) *Some insights into the special properties of cell walls, derived from study of the arrangements of the various polysaccharides they contain.*

EVANS, M. L., R. MOORE, and K.-H. HASENSTEIN, 1986. How roots respond to gravity, *Scientific American* 255 (6).

GALSTON, A. W., P. J. DAVIES, and R. L. SATTER, 1980. *The Life of the Green Plant.* Prentice Hall, Englewood Cliffs, N.J. *A brief elementary text.*

MOSES, P. B., AND N.-H. CHUA, 1988. Light switches for plant genes, *Scientific American* 258 (4).

RAVEN, P. H., R. F. EVERT, and S. E. EICHHORN, 1992. *Biology of plants,* 5th ed. Worth, New York. *An excellent introductory botany textbook.*

SALISBURY, R. B., and C. W. ROSS, 1992. *Plant physiology,* 4th ed. Wadsworth, Belmont, California.

VAN OVERBEEK, J., 1968. The control of plant growth, *Scientific American* 219 (1). (Offprint 1111)

Chapter 24

BOWLEY, J. D., and M. BLACK, 1985. *Seeds: Physiology of development and germination.* Plenum, New York.

GALSTON, A. W., P. J. DAVIES, and R. L. SATTER, 1980. *The life of the green plant,* 3rd ed. Prentice-Hall, Englewood Cliffs, N.J. *A readable plant-physiology text.*

GRANT, V., 1951. The fertilization of flowers, *Scientific American* 184 (6). (Offprint 12) *The special adaptations of flowers that help ensure their pollination.*

JENSEN, W. A., and F. B. SALISBURY, 1972. *Botany: An Ecological Approach.* Wadsworth, Belmont, Calif. *Very readable general text written from an evolutionary and ecological point of view.*

MOSES, P. B., and N-H. CHUA, 1988. "Light switches for plant genes," *Scientific American* 258 (4)

RAVEN, P. H., R. F. EVERT, and S. E. EICHHORN, 1992. *Biology of plants,* 5th ed. Worth, New York.

SALISBURY, F. B., and C. W. ROSS, 1992. *Plant physiology,* 4th ed. Wadsworth, Belmont, CA. *(See esp. Chapters 17–23.)*

Chapter 25

ATKINSON, M. A., and N. K. MACLAREN, 1990. What causes diabetes? *Scientific American* 263 (1).

BERRIDGE, M. J., 1985. The molecular basis of communication within the cell, *Scientific American* 253 (4). *Reviews the operation of the second messengers.*

CANTIN, M., and J. GENEST, 1986. The heart as an endocrine gland, *Scientific American* 254 (2). *On the role of ANF.*

CARMICHAEL, S. W., and H. WINKLER, 1985. The adrenal chromaffin cell, *Scientific American* 253 (2). *On the cellular and molecular bases of adrenalin secretion.*

FOX, S. I., 1990. *Human Physiology,* Wm. C. Brown, Dubuque, Iowa. *An introductory human physiology textbook with excellent illustrations and diagrams.*

GARDNER, L. I., 1972. Deprivation dwarfism, *Scientific American* 227 (1). (Offprint 1253) *Inadequate secretion of pituitary hormones, especially growth hormone, as a probable cause of stunted growth in some emotionally deprived children.*

GILLIE, R. B., 1971. Endemic goiter, *Scientific American* 224 (6). *The intriguing history of this disorder; its current distribution as primarily a disease of the poor.*

GUILLEMIN, R., and R. BURGUS, 1972. The hormones of the hypothalamus, *Scientific American* 227 (5). (Offprint 1260) *The discovery of the hypothalamic releasing hormones and their role in regulating the anterior pituitary.*

GUYTON, A., 1991. *Textbook of Medical Physiology,* 3rd ed. W. B. Saunders, Philadelphia. *The "bible" of human physiology textbooks.*

LIENHARD, G. E., J. W. SLOT, D. E. JAMES, and M. M. MUECKLER (1992). How cells absorb glucose, *Scientific American* 266 (1). *On the interactions between glucose channels and insulin.*

LINDER, M. E., and A. G. GILMAN, 1992. G proteins, *Scientific American* 267 (1). *On their role in cellular activities such as vision*

* Available in paperback.

and cognition and, when malfunctioning, their implication in diseases.

McEWEN, B. S., 1976. Interactions between hormones and nerve tissue, *Scientific American* 235 (1). (Offprint 1341) *The influence of steroid hormones on the infant's development of brain circuits that control later behavior.*

NOTKINS, A. L., 1979. The causes of diabetes, *Scientific American* 241 (5). (Offprint 1450)

O'MALLEY, B. W., and W. T. SCHRADER, 1976. The receptors of steroid hormones, *Scientific American* 234 (2). (Offprint 1334) *The mechanism by which steroid hormones are thought to act on their target cells.*

ORCI, L., J.-D. VASSALLI, and A. PERRELET, 1988. The insulin factory, *Scientific American* 259 (3). *Traces the synthesis, sorting, packaging, activation, and secretion of insulin in great detail.*

RASMUSSEN, H., 1989. The cycling of calcium as an intracellular messenger, *Scientific American* 261 (4).

ROSENTHAL, G. A., 1986. The chemical defenses of higher plants, *Scientific American* 254 (1). *Touches on instances in which plants use hormone mimics to sabotage the developmental program of the insects that feed on them.*

SCHNEIDERMAN, H. A., AND L. I. GILBERT, 1964. Control of growth and development in insects, *Science* 143, 325–33.

SNYDER, S. H., 1985. The molecular basis of communication between cells, *Scientific American* 253 (4). *An excellent review of hormones and local chemical mediators, and of the elaborate feedback system for controlling hormone levels.*

SNYDER, S. H., and D. S. BREDT (1992), Biological roles of nitric oxide, *Scientific American* 266 (6). *On the many roles of this newly discovered transmitter, local chemical mediator, and toxic weapon.*

UVNÄS-MOBERG, K., 1989. The gastrointestinal tract in growth and reproduction, *Scientific American* 261 (1). *How intestinal hormones prepare a pregnant woman for the extreme metabolic demands of pregnancy and lactation.*

WURTMAN, R. J., 1975. The effects of light on the human body, *Scientific American* 233 (1).

WURTMAN, R. J., and J. AXELROD, 1965. The pineal gland, *Scientific American* 213 (1). (Offprint 1015) *The long search for the function of the pineal.*

Chapter 26

BYNE, W., 1994. The biological evidence challenged, *Scientific American* 270 (5). *Physiologic evidence and genetic studies of causes of male homosexuality are debated.*

CREWS, D., 1994. Animal sexuality, *Scientific American* 270 (1). *A new framework for understanding the origin and function of sexuality.*

EPEL, D., 1977. The program of fertilization, *Scientific American* 237 (5). (Offprint 1372) *The numerous changes that occur in an egg cell as soon as a sperm cell reaches it.*

GORDON, R., and A. G. JACOBSON, 1978. The shaping of tissues in embryos, *Scientific American* 238 (6). (Offprint 1391)

HAYFLICK, L., 1980. The cell biology of human aging, *Scientific American* 242 (1). (Offprint 1457)

LEVAY, S., and D. H. HAMER, 1994. Evidence for a biological influence in male homosexuality, *Scientific American* 270 (5). *Studies indicate that genes and brain development play a significant role in determining sexual orientation.*

WASSARMAN, P. M., 1988. Fertilization in mammals, *Scientific American* 259 (6). *On how eggs manage to be fertilized by only one sperm.*

Chapter 27

BRYANT, P. J., S. V. BRYANT, and V. FRENCH, 1977. Biological regeneration and pattern formation, *Scientific American* 237 (1). (Offprint 1363) *On basic principles of the organization and growth of complex structures in animals.*

COOKE, J., 1988. The early embryo and the formation of body pattern, *Scientific American* 76, 35–41. *A wide-ranging review looking for common mechanisms.*

COWAN, W. M., 1979. The development of the brain, *Scientific American* 241 (3). (Offprint 1440) *At the peak of brain growth, hundreds of thousands of neurons are added each minute, and yet they are wired together correctly.*

DEROBERTS, E. M., G. OLIVER, and C. V. E. WRIGHT, 1990. Homeobox genes and the vertebrate body plan, *Scientific American* 263 (1). *Application of the principles from Drosophila development to vertebrates.*

EDELMAN, G. M., 1984. Cell-adhesion molecules: A molecular basis for animal form, *Scientific American* 250 (4). (Offprint 1549) *On the likely molecular basis of cell-to-cell adhesion and changes in adhesion during embryonic development.*

GARCIA-BELLIDO, A., P. A. LAWRENCE, and G. MORATA, 1979. Compartments in animal development, *Scientific American* 241 (1). (Offprint 1432) *An excellent discussion of imaginal discs and insect development.*

GEHRING, W. J., 1985. The molecular basis of development, *Scientific American* 253 (4). *Focuses exclusively on* Drosophila *development, with a nice discussion of homeotic mutations.*

GOODMAN, C. S., and M. J. BASTIANI, 1984. How embryonic nerve cells recognize one another. *Scientific American* 251 (6). (Offprint 1556) *An excellent description of how axons of invertebrates employ the stepping-stone strategy, following first gradients and then one preexisting axon after another to reach their targets.*

GURDON, J. B., 1968. Transplanted nuclei and cell differentiation, *Scientific American* 219 (6). (Offprint 1128)

LEVI-MONTALCINI, R., and P. CALISSANO, 1979. The nerve-growth factor, *Scientific American* 240 (6). (Offprint 1430) *About the best-understood molecule important in creating chemical gradients for axon growth and development.*

RUSTING, R. L., 1992. Why do we age? *Scientific American* 267 (6). *On the discovery of genes that contribute to deterioration and death.*

SELKOE, D. J., 1992. Aging brain, aging mind, *Scientific American* 267 (3).

Chapter 28

ALKON, D. L., 1989. Memory storage and neural systems, *Scientific American* 260 (1). *On the synaptic chemistry of learning.*

CHANGEUX, J.-P., 1993. Chemical signaling in the brain, *Scientific American* 269 (5). *On how studies of acetylcholine receptors in fish have generated insights into how neurons in the human brain communicate with one another.*

EVARTS, E. V., 1979. Brain mechanisms of movement, *Scientific American* 241 (3).(Offprint 1443) *How the brain and the muscles interact.*

GOTTLIEB, G. I., 1988. GABAergic neurons, *Scientific American* 258 (2). *On the workings of this major inhibitory system.*

IVERSEN, L. L., 1979. The chemistry of the brain, *Scientific American* 241 (3). (Offprint 1441) *A look at the workings of the roughly 30 different transmitter chemicals in the brain.*

JACOBS, B. L., 1987. How hallucinogenic drugs work, *American Scientist* 75 (4). *On the interaction between drugs and the serotonin receptor.*

KANDEL, E. R., 1979. Small systems of neurons, *Scientific American*

241 (3). (Offprint 1438) *A discussion of the circuits in* Aplysia *that are involved in gill-withdrawal behavior.*

KEYNES, R. D., 1958. The nerve impulse and the squid, *Scientific American* 199 (6). (Offprint 58) *The role of the giant axon of the squid in the development of the modern understanding of impulse conduction.*

KIMELBERG, H. K., and M. D. NORENBERG, 1989. Astrocytes, *Scientific American* 260 (4). *On an important kind of glia.*

LENT, C. M., and M. H. DICKINSON, 1988. The neurobiology of feeding in leeches, *Scientific American* 258 (6).

LESTER, H. A., 1977. The response to acetylcholine, *Scientific American* 236 (2). (Offprint 1352) *How the receptors for acetycholine are thought to work.*

LLINAS, R. R., 1982. Calcium in synaptic transmission. *Scientific American* 247 (4). (Offprint 1523)

NEHER, E., and B. SAKMANN 1992. The patch clamp technique, *Scientific American* 266 (3). *On the use of an isolated patch of membrane to study the operation of single ion channels.*

NICHOLLS, J. G., and D. VAN ESSEN, 1974. The nervous system of the leech, *Scientific American* 230 (1). (Offprint 1287) *Mapping the circuits in a relatively simple animal.*

SHEPHERD, G. M., 1978. Microcircuits in the nervous system, *Scientific American* 238 (2). (Offprint 1380) *A difficult but important article on dendritic interactions and other unconventional circuits.*

SNYDER, S. H., 1977. Opiate receptors and internal opiates, *Scientific American* 237 (3). (Offprint 1354) *How morphine and some morphinelike substances normally synthesized by certain nerve cells exert their effects on the brain.*

SNYDER, S. H., and D. S. BREDT (1992). Biological roles of nitric oxide, *Scientific American* 266 (6). *On the many roles of this newly discovered transmitter, local chemical mediator, and toxic weapon.*

STEVENS, C. F., 1979. The neuron, *Scientific American* 241 (3). (Offprint 1437) *An excellent description of how nerve cells work.*

Chapter 29

"Mind and Brain," special issue of *Scientific American* 267 (3) (September 1992). *Eleven articles by prominent investigators on such topics as the mind, visual imaging, learning, language, memory, sex differences in the brain, brain disorders, aging, and consciousness.*

ALBERTS, R., et al, 1989. *Molecular biology of the cell,* 2nd ed. Garland, New York.

FISCHBACH, G. D., 1992. Mind and brain, *Scientific American* 267 (3). *An overview of biological foundations of consciousness, memory, and other attributes of mind.*

FRENCH, J. D., 1957. The reticular formation, *Scientific American* 196 (5). (Offprint 66) *On the network that arouses and focuses our attention.*

GLICKSTEIN, M., 1988. The discovery of the visual cortex. *Scientific American* 259 (3). *Fascinating history of how war wounds led to the discovery of specialized brain areas.*

GOULD, J. L., 1982. *Ethology: The Mechanisms and Evolution of Behavior.* W. W. Norton, New York. *Discusses animal senses and their role in behavior.*

HINTON, G. E., D. C. PLAUT, and T. SHALLICE, 1993. Simulating brain damage, *Scientific American* 269 (4). *Injuries reproduced in computer models shed light on the way written language is processed in the brain.*

HUBEL, D., 1988. *Eye, Brain, and Vision.* Scientific American Books, New York. *Beautifully written and produced.*

KALIL, R. E., 1989. Synapse formation in the developing brain, *Scientific American* 261 (6). *On how correlated firing helps organize and tune brain circuits and maps.*

KALIN, N. H., 1993. The neurobiology of fear, *Scientific American* 268 (5). *Research on fear in monkeys may lead to new ways to treat anxiety in humans.*

KONISHI, M., 1993. Listening with two ears, *Scientific American* 268 (4). *Studies of barn owls reveal how brain combines acoustic signals from two sides of head into a single spatial perception.*

KORETZ, J. F., and G. H. HANDELMAN, 1988. How the human eye focuses, *Scientific American* 259 (1). *On the structure of the lens, how its shape is changed to allow focusing, and the process of aging.*

LEDOUX, J. E., 1994. Emotion, memory, and the brain, *Scientific American* 270 (6). *On tracing of neural routes underlying the formation of memories about emotional experiences.*

LEVINE, J. S., and E. F. MACNICHOL, 1982. Color vision in fishes, *Scientific American* 246 (2). (Offprint 1512) *How the choice and arrangement of cones within the retina suit each fish species' needs.*

MASLAND, R. H., 1986. The functional architecture of the retina, *Scientific American* 255 (6).

MELZACK, R., 1961. The perception of pain, *Scientific American* 204 (2). (Offprint 457) *A fascinating discussion of the subject.*

NAUTA, W. J. H., and M. FEIRTAG, 1979. The organization of the brain, *Scientific American* 241 (3). (Offprint 1439) *An excellent discussion of vertebrate—especially human—brain anatomy, with special emphasis on the major pathways of information flow within the brain.*

NEWMAN, E. A., and P. H. HARTLINE, 1982. The infrared "vision" of snakes, *Scientific American* 246 (3). *A look at the neural basis of the ability of pit vipers to see in the dark.*

PARKER, D. E., 1979. The vestibular apparatus, *Scientific American* 243 (5). (Offprint 1484) *How the otoliths and semicircular canals work.*

POGGIO, T., and C. KOCH, 1987. Synapses that compute motion, *Scientific American* 256 (5).

RAICHLE, M. E., 1994. Visualizing the mind, *Scientific American* 270 (4). *Cognitive science and brain imaging are used to explore the relation between the human mind and brain.*

RAPOPORT, J. L., 1989. The biology of obsessions and compulsions, *Scientific American* 260 (3). *On the chemical bases of psychological disorders.*

SELKOE, D. J., 1991. Amyloid protein and Alzheimer's disease, *Scientific American* 265 (5). *Understanding how plaques accumulate in the brain may lead to treatments.*

STRYER, L., 1987. The molecules of visual excitation, *Scientific American* 257 (1). *The molecular biology of rhodopsin and its second-messenger system.*

SUGA, N., 1990. Biosonar and neural computation in bats, *Scientific American* 262 (6). *On the sensory processing underlying the ability of bats to locate prey in the dark.*

TREISMAN, A., 1986. Features and objects in visual processing, *Scientific American* 255 (5). *On parallel processing in the visual system.*

Chapter 30

ALBERTS, R., et al, 1989. *Molecular biology of the cell,* 2nd ed. Garland, New York.

ALEXANDER, R., 1992. *The human machine,* Columbia University Press, New York.

COHEN, C., 1975. The protein switch of muscle contraction, *Scientific American* 233(5). (Offprint 1329) *A detailed account of the way calcium, troponin, and tropomyosin interact to control muscle contraction.*

HUXLEY, H. E., 1958. The contraction of muscle, *Scientific American* 199 (5). (Offprint 19) *A description by Huxley of his sliding-filament model, written shortly after he developed it.*

HUXLEY, H. E., 1965. The mechanism of muscular contraction, *Scientific American* 213 (6). (Offprint 1026)

MURRAY, J. M., and A. WEBER, 1974. The cooperative action of muscle proteins, *Scientific American* 230 (2). (Offprint 1290) *The structure and mode of action of the four major proteins that make up the microfilaments of muscle.*

RASMUSSEN, H., 1989. The cycling of calcium as an intracellular messenger, *Scientific American* 261 (4). *On the role of* Ca^{++} *in muscle contraction.*

SMITH, D. S., 1965. The flight muscles of insects, *Scientific American* 212 (6). (Offprint 1014) *The special arrangements of muscles enabling insect wings to beat hundreds of times per second.*

SMITH, K. K., and W. M. KIER, 1989. Trunks, tongues, and tentacles, *American Scientist* 77 (1). *On hydrostatic movement in vertebrates and cephalopods.*

Chapter 31

BISHOP, J. A., and L. M. COOK, 1975. Moths, melanism and clean air, *Scientific American* 232 (1). (Offprint 1314) *On the lessening of air pollution in Britain and the diminishing frequency of melanics in some moth populations.*

CLARKE, B., 1975. The causes of biological diversity, *Scientific American* 233 (2). (Offprint 1326)

COLES, C. J., 1984. Unisexual lizards, *Scientific American* 250 (1). *On the evolution of asexual species from sexual ones.*

DAWKINS, R., 1976. *The Selfish Gene.* Oxford University Press, New York. *A well-written exposition of the controversial idea that the gene, not the individual organism, is the unit of selection, the organism being merely the robot vehicle of its selfish genes.*

DOOLITTLE, R. F., and P. BORK, 1993. Evolutionarily mobile modules in proteins, *Scientific American* 269 (4). *On how modules are able to travel laterally across species lines.*

DEVRIES, P. J., 1992. Singing caterpillars, ants, and symbiosis, *Scientific American* 267 (4). *Adaptations of caterpillars allow them to attract ants to serve as their guardians.*

FUTUYMA, D. J., 1986. *Evolutionary Biology,* 2nd ed. Sinauer Associates, Sunderland, Mass.

GOULD, J. L., and C. G. GOULD, 1989. *Sexual Selection.* Scientific American Books, New York. *Well-illustrated treatment of the evolution of sex and mate choice.*

GRANT, V., 1951. The fertilization of flowers, *Scientific American* 184 (6). (Offprint 12) *The special adaptations of flowers that help ensure their pollination.*

KETTLEWELL, H. B. D., 1959. Darwin's missing evidence, *Scientific American* 200 (3). (Offprint 842) *The story of the industrial melanism of the peppered moth in England.*

LEWONTIN, R. C., 1978. Adaptation. *Scientific American* 239 (3). (Offprint 1408) *An excellent discussion of the process of adaptation, emphasizing that most features are compromises between different selection pressures, and that chance plays a role in evolution when more than one solution to a problem is possible.*

LI, W. H., and D. GRAUR, 1991. *Fundamentals of Molecular Evolution.* Sinauer, Sunderland, Mass. *An excellent account of gene evolution through duplication and exon recombination.*

MAYR, E., 1978. Evolution, *Scientific American* 239 (3). (Offprint 1400) *A nice history of evolutionary thought.*

RENNIE, J., 1992. Living together, *Scientific American* 266 (1). *On the coevolution of parasites and hosts and its significance to the coevolution of all organisms.*

SPENCER, C. H., 1987. Mimicry in plants, *Scientific American* 257 (3).

STROBEL, G. A., 1991. Biological control of weeds, *Scientific American* 265 (1). *Illustrates how weeds, like diseases, can reproduce unchecked in new habitats, whereas in their normal range they are in balance with locally evolved parasites (usually insects or fungi).*

Chapter 32

AYALA, F. J., 1978. The mechanisms of evolution, *Scientific American* 239 (3). (Offprint 1407) *On the large amount of hidden genetic variation in a species, and its consequences for speciation and taxonomy.*

CLARKE, B., 1975. The causes of biological diversity, *Scientific American* 233 (2). (Offprint 1326)

FUTUYMA, D. J., 1986. *Evolutionary Biology,* 2nd ed. Sinauer Associates, Sunderland, Mass.

GOULD, S. J., 1985. *The Flamingo's Smile: Reflections in Natural History.* W. W. Norton, New York. *A compellingly written collection of essays on evolutionary theory and other biological topics by the coauthor of the theory of punctuated equilibrium.*

GOULD, S. J., 1989. *Wonderful Life.* W. W. Norton, New York. *A punctuationist's view of the Burgess fauna.*

GRANT, P. R., 1991. Natural selection and Darwin's finches. *Scientific American* 265(4). *On the effects of a season of drought on the finch population in the Galapagos. A well-written and modern analysis of speciation.*

GRANT, V., 1985. *The Evolutionary Process.* Columbia University Press, New York. *Excellent treatment, with special emphasis on speciation.*

KIMURA, M., 1979. The neutral theory of molecular evolution, *Scientific American* 241 (5). *On the idea that most single-base mutations are neutral, and therefore provide an evolutionary "clock."*

LI, W. H., and D. GRAUR, 1991. *Fundamentals of Molecular Evolution.* Sinauer, Sunderland, Mass. *Excellent treatment of how molecular techniques can be applied to phylogeny.*

PAABO, S., 1993. Ancient DNA, *Scientific American* 269 (5). *On how the study of reconstituted fragments allows comparisons to be made between extant and ancient species.*

SCHOENER, T. W., 1982. The controversy over interspecific competition, *Scientific American* 70, 586–95.

SIBLEY, C. G., and J. E. AHLQUIST, 1986. Reconstructing bird phylogeny by comparing DNAs, *Scientific American* 254 (2). *On the DNA-hybridization technique.*

SIMPSON, G. G., 1983. *Fossils and the History of Life.* Scientific American Books, New York. *Clearly written and well-illustrated exposition of evolution.*

STANLEY, S. M., 1987. *Extinction.* Scientific American Books, New York. *Another excellent summary in this series.*

STEBBINS, G. L., and F. J. AYALA, 1985. The evolution of Darwinism, *Scientific American* 253 (1). *A modern summary of evolutionary theory, with particular attention to the claims for punctuated equilibrium.*

WILSON, A. C., 1985. The molecular basis of evolution, *Scientific American* 253 (4). *On methods for tracing evolution through similarities in nucleotide or amino acid sequences.*

Chapter 33

BEDDINGTON, J. R., and R. M. MAY, 1982. The harvesting of interacting species in a natural ecosystem, *Scientific American* 247 (5). (Offprint 1525)

BELL, R. H. V., 1971. A grazing ecosystem in the Serengeti, *Scientific American* 225 (1). (Offprint 1228) *The synchronization of the migrations of ungulates across the plains of Tanzania with the growth of certain grasses—a striking example of the precision with which organisms mesh within an ecosystem.*

CLUTTON-BROCK, T. H., 1985. Reproductive success in red deer, *Scientific American* 252 (2). *A classic example of a harem-based social system, with male contests, and sex-ratio manipulation.*

COOPER, C. F., 1961. The ecology of fire, *Scientific American* 204 (4). (Offprint 1099) *How some biotic communities depend on periodic burning for their continued existence, and why efforts to eliminate fires may be threatening some of our finest forests.*

DUELLMAN, W. E., 1992. Reproductive strategies of frogs, *Scientific American* 267 (1). *On strategies to colonize niches throughout the terrestrial environment.*

GOULD, J. L., and C. G. GOULD, 1989. *Sexual Selection.* New York, W. H. Freeman. *A good introduction to behavioral ecology.*

HORN, H. S., 1975. Forest succession, *Scientific American* 232 (5). (Offprint 1321)

LIGON, J. D., and S. H. LIGON, 1982. The cooperative breeding behavior of the green woodhoopoe, *Scientific American* 247 (1). *A good example of how altruism turns out to enhance fitness.*

MAY, R. M., 1978. The evolution of ecological systems, *Scientific American* 239 (3). (Offprint 1404)

MAY, R. M., 1983. Parasitic infections as regulators of animal populations, *American Scientist* 71, 36–45.

POWER, J. F., and R. F. FOLLETT, 1987. Monoculture, *Scientific American* 256 (3).

ROBEY, B., S. O. RUTSTEIN, and L. MORRIS, 1993. The fertility decline in developing countries, *Scientific American* 269 (6). *On effect of access to contraception as well as changes in cultural values and education.*

WENT, F. W., 1949. The plants of Krakatoa, *Scientific American* 181 (3). *The reinvasion of the island of Krakatoa by plants after all life on it had been destroyed by a volcanic eruption.*

WILKINSON, G. S., 1990. Food sharing in vampire bats, *Scientific American* 262 (2). *On a remarkable example of reciprocal altruism in bat roosts.*

Chapter 34

BERNER, R. A., and A. C. LASAGA, 1989. Modeling the geochemical carbon cycle, *Scientific American* 260 (3).

BORMANN, F. H., and G. E. LIKENS, 1970. The nutrient cycles of an ecosystem, *Scientific American* 223 (4). (Offprint 1202) *On the studies of the Hubbard Brook Forest discussed in the text.*

BRAY, F., 1994. Agriculture for developing nations. *Scientific American* 271 (1). *A fascinating article on the advantages of the sustainable polyculture practiced by the Asian rice farmers over the Western agricultural model.*

BRILL, W. J., 1977. Biological nitrogen fixation, *Scientific American* 236 (3). (Offprint 922) *How certain bacteria, the Cyanobacteria among them, act as the major suppliers of fixed nitrogen for the rest of the living world.*

BROWN, B. E., and J. C. OGDEN, 1993. Coral bleaching, *Scientific American* 268 (1). *On how unusually high seawater temperatures may signal global warming.*

CHARLSON, R. J., and T. M. L. WIGLEY, 1994. Sulfate aerosol and climatic change, *Scientific American* 270 (2). *Eliminating sulfur emissions could accelerate global warming by greenhouse gases.*

COFFIN, M. F., and O. ELDHOLM, 1993. Large igneous provinces, *Scientific American* 269 (4). *On how the formation of vast fields of lava may have triggered changes in the global environment.*

COX, P. A., and M. BALICK, 1994. The ethnobotanical approach to drug discovery, *Scientific American* 270 (6). On medicinal plants discovered by traditional societies as source of therapeutic drugs.

DIETZ, R. S., and J. C. HOLDEN, 1970. The breakup of Pangaea, *Scientific American* 223 (4). (Offprint 892)

GATES, D. M., 1971. The flow of energy in the biosphere. *Scientific American* 225 (3). (Offprint 664) *On the radiant energy the earth receives from the sun—and the relatively small percentage that is trapped by green plants and made available to biotic communities.*

GOSZ, J. R., R. T. HOLMES, G. E. LIKENS, and F. H. BORMANN, 1978. The flow of energy in a forest ecosystem. *Scientific American* 238 (3). (Offprint 1384)

GOULDING, M., 1993. Flooded forests of the Amazon, *Scientific American* 266 (3). *On special adaptations for surviving in this constantly changing environment.*

GRAEDEL, T. E., and P. J. CRUTZEN, 1989. The changing atmosphere, *Scientific American* 261 (3). *On the greenhouse effect.*

HALLAM, A., 1972. Continental drift and the fossil record, *Scientific American* 227 (5). (Offprint 903)

HOLLOWAY, M., 1994. Nurturing nature, *Scientific American* 270 (4). *An attempt to restore the environment of Florida's Everglades damaged by human activity.*

HOLLOWAY, M., 1993. Sustaining the Amazon, *Scientific American* 269 (1). *Scientists are seeking to reconcile the need for economic development with preservation of the ecology.*

KUSLER, J. A., W. J. MITSCH, and J. S. LARSON, 1994. Wetlands, *Scientific American* 270 (1). *On understanding their biodiversity and conservation efforts.*

MAY, R. M., 1992. How many species inhabit the earth? *Scientific American* 267 (4). *On how an accurate census is crucial to management of the environment.*

MOHNER, V. A., 1988. The challenge of acid rain, *Scientific American* 259 (2).

MYERS, N., 1984. *The Primary Source: Tropical Forests and Our Future.* W. W. Norton, New York.

POLLACK, H. N., and D. S. CHAPMAN, 1993. Underground records of changing climate, *Scientific American* 268 (6). *Ancient temperatures archived in continental crust may reveal climate of past eras.*

POST, W. M., 1990. The global carbon cycle, *American Scientist* 78, 310–326. *An extremely detailed and wide-ranging treatment.*

RICHARDS, P. W., 1973. The tropical rain forest, *Scientific American* 229 (6). (Offprint 1286) *On the human threat to its survival.*

SCHNEIDER, S. H., 1989. The changing climate, *Scientific American* 261 (3). *On global warming.*

TERBORGH, J., 1992. Why American songbirds are vanishing, *Scientific American* 266 (5).

WHITE, R. M., 1990. The great climate debate, *Scientific American* 263 (1). *On the controversy over the effects of increasing levels of carbon dioxide.*

WILSON, E. O., 1989. Threats to biodiversity, *Scientific American* 261 (3). *On human activities that are leading to large numbers of extinctions.*

Chapter 35

BENTLEY, D., and R. R. HOY, 1974. The neurobiology of cricket song. *Scientific American* 231 (2). (Offprint 1302) *An examination of the maturation of a neural circuit, and its genetic basis.*

CARTER, C. S., and L. L. GETZ, 1993. Monogamy and the prairie vole, *Scientific American* 268 (6). *On how hormones may be responsible for this monogamous behavior.*

DILGER, W. C., 1962. The behavior of lovebirds, *Scientific American* 206 (1). (Offprint 1049) *A comparative study of the courtship and nest-building behavior of several species of lovebirds and their hybrids. This classic study offers many insights into the evolution of behavior.*

EMLEN, S. T., 1975. The stellar-orientation system of a migratory bird, *Scientific American* 233 (2). (Offprint 1327) *Summarizes elegant planetarium experiments on the star-compass strategy of avian navigation.*

EWERT, J. P., 1974. The neural basis of visually guided behavior,

Scientific American 230 (3). (Offprint 1293) *How toads recognize and capture prey.*

FITZGERALD, G. J., 1993. The reproductive behavior of the stickleback, *Scientific American* 268 (4).

GOULD, J. L., 1982. *Ethology: The Mechanisms and Evolution of Behavior.* W. W. Norton, New York. *An introductory textbook on animal behavior.*

GOULD, J. L., and C. G. GOULD, 1988. *The Honey Bee.* Scientific American Books, New York. *Clear, well-illustrated descriptions of the sensory world, communication, navigation, and learning of this remarkable insect.*

GOULD, J. L., and P. MARLER, 1987. Learning by instinct. *Scientific American* 256 (1). *On innately guided learning.*

GWINNER, E., 1986. Internal rhythms in bird migration, *Scientific American* 254 (4). *On the role of annual rhythms and internal timers in guiding warbler migration.*

HAILMAN, J. P., 1969. How an instinct is "learned," *Scientific American* 221 (6). (Offprint 1165) *The classic reappraisal of how gull chicks know what to peck at and learn to recognize their parents.*

HASLER, A. D., A. T. SCHOLZ, and R. M. HORRALL, 1978. Olfactory imprinting and homing in salmon, *American Scientist* 66, 347–55.

HESS, E. H., 1972. "Imprinting" in a natural laboratory, *Scientific American* 227 (2). (Offprint 546)

KEETON, W. T., 1974. The mystery of pigeon homing, *Scientific American* 231 (6). (Offprint 1311)

KIRCHNER, W. H., and W. F. TOWNE, 1994. The sensory basis of the honeybee's dance language, *Scientific American* 270 (6). *Exploration of how sounds are involved in dance language.*

LOHMANN, K. J., 1992. How sea turtles navigate, *Scientific American* 266 (1). *How hatchling sea turtles use biological compass and props to navigate between feeding ground and nesting sites.*

SEYMOUR, R. S., 1991. The brush turkey, *Scientific American* 265 (6). *On its unique parenting strategy requiring adaptations of egg and hatchling.*

SHERMAN, P. W., J. U. M. JARVIS, and S. H. BRAUDE, 1992. Naked mole rats, *Scientific American* 267 (2). *Genetic and evolutionary roots of their social structure which resembles that of some insects.*

TUMLINSON, J. H., W. J. LEWIS, and L. E. M. VET, 1993. How parasitic wasps find their hosts, *Scientific American* 266 (3). *On how wasps learn to identify compounds released by the plant on which caterpillars feed.*

WILLIAMS, T. C., and J. M. WILLIAMS, 1978. Oceanic mass migration of land birds, *Scientific American* 239 (4). (Offprint 1411) *How small songbirds get to their wintering ranges in South America by flying out over the Atlantic.*

WILSON, E. O., 1963. Pheromones, *Scientific American* 208 (5) (Offprint 157)

Chapter 36

BOGORAD, L. 1975. Evolution of organelles and eukaryotic genomes, *Science* 188, 891–98. *An alternative to the endosymbiont hypothesis of the origin of eucaryotic cells.*

DICKERSON, R. E., 1978. Chemical evolution and the origin of life, *Scientific American* 233 (3). (Offprint 1401)

EIGEN, M., W. GARDINER, P. SCHUSTER, and R. WINKLER-OSWATITSCH, 1981. The origin of genetic information, *Scientific American* 244 (4). (Offprint 1495)

GLAESSNER, M. F., 1961. Pre-Cambrian animals, *Scientific American* 204 (3). (Offprint 837) *Some of the fascinating Precambrian invertebrate fossils found in Australia.*

KASTING, J. F., O. B. TOON, and J. B. POLLACK, 1988. How climate evolved on the terrestrial planets, *Scientific American* 258 (2).

KNOLL, A. H., 1991. End of the Proterozoic eon, *Scientific American* 265 (4). *On the change in fauna and flora 600 million years ago when multicellular life began to flourish.*

MARGULIS, L., 1971. Symbiosis and evolution, *Scientific American* 225 (2). (Offprint 1230) *Clear exposition of the endosymbiont hypothesis of the origin of eucaryotic cells by one of its principal proponents.*

MILLER, S. L., and L. E. ORGEL, 1974. *The Origins of Life on Earth.* Prentice-Hall, Englewood Cliffs, N.J. *Easy-to-read introduction to the subject. Miller's experiments of 1953 initiated the modern era of research on the beginnings of life.*

OPARIN, A. I., 1969. *Genesis and Evolutionary Development of Life.* Academic Press, New York. *Excellent summary by one of the founding fathers of this field of biology.*

SCHOPF, J. W., 1978. The evolution of the earliest cells, *Scientific American* 239 (3). (Offprint 1402) *The fossil evidence for the appearance of cellular life on earth; the special metabolic characteristics of bacteria that enabled them to prosper under conditions that shut out most higher forms of life.*

UZZEL, T., and C. SPOLSKY, 1974. Mitochondria and plastids as endosymbionts: A revival of special creation? *American Scientist* 62, 334–43. *A critique of the endosymbiont hypothesis, and a proposal for an alternative model of the origin of eucaryotic cells.*

VIDAL, G., 1984. The oldest eucaryotic cells, *Scientific American* 250 (2).

WHITTAKER, R. H., 1969. New concepts of kingdoms of organisms, *Science* 163, 150–60. *The five-kingdom system proposed and explained.*

WHITTAKER, R. H., and L. MARGULIS, 1978. Protist classification and the kingdoms of organisms, *Biosystems* 10, 3–18. *Alternative ways of applying the five-kingdom system.*

WOESE, C. R., 1981. Archaebacteria, *Scientific American* 244 (6). (Offprint 1516) *A clear exposition of the case for considering Archaebacteria a separate kingdom.*

Chapter 37

BERG, H. C., 1975. How bacteria swim, *Scientific American* 223 (2). *The structure and mode of action of bacterial flagella.*

BROCK, T. D., and M. T. MADIGAN, 1991. *Biology of microorganisms,* 6th ed. Prentice-Hall, Englewood Cliffs, N.J. *An excellent text with balanced coverage of all aspects of the field.*

CARMICHAEL, W. W., 1994. The toxins of cyanobacteria, *Scientific American* 270 (1). *On the hazards and benefits of this microorganism.*

COSTERTON, J. W., G. G. GEESLEY, and K-J. CHENG, 1978. How bacteria stick, *Scientific American* 238 (1). (Offprint 1379) *On the surface molecules of bacteria that enable these to adhere to host cells.*

DAVIS, B. D., R. DULBECCO, H. N. EISEN, and H. S. GINSBERG, 1990. *Microbiology,* 4th ed. Lippincott, Philadelphia. *An authoritative text emphasizing medical aspects of the field.*

DIENER, T. O., 1981. Viroids, *Scientific American* 244 (1). (Offprint 1488)

ECHLIN, P., 1966. The blue-green algae, *Scientific American* 214 (6). *On the cyanobacteria.*

EIGEN, M., 1993. Viral quasispecies, *Scientific American* 269 (1). *A statistical classification scheme offers insights into the evolution and behavior of viruses.*

EWALD, P. W., 1993. The evolution of virulence, *Scientific American* 268 (4). *On how human behavior influences whether pathogens evolve into benign or harmful forms.*

KOCH, A. L., 1990. Growth and form of the bacterial cell wall, *American Scientist* 78, 327–341.

PRUSINER, S. B., 1984. Prions, *Scientific American* 251 (4). (Offprint

1554) *On the peculiar structure of bacterial walls, and the way many antibiotics like penicillin block their synthesis.*

RIETSCHEL, E. T., and H. BRADE, 1992. Bacterial endotoxins, *Scientific American* 267 (2). *On efforts to block the bad effects and harness the disease-fighting capacity of these cell wall components.*

STANIER, R. Y., E. A. ADELBERG, and J. L. INGRAHAM, 1986. *The microbial world,* 5th ed. Prentice-Hall, Englewood Cliffs, N.J. *Excellent general microbiology text.*

STOECKENIUS, W., 1976. The purple membrane of salt-loving bacteria. *Scientific American* 234 (6). (Offprint 1340) *Rhodopsin as the light-trapping pigment of a newly discovered kind of photosynthesis carried out by certain Archaebacteria.*

TORTORA, G. J., B. R. FUNKE, and C. L. CASE, 1992. *Microbiology,* 4th ed. Benjamin/Cummings, Redwood City, CA. *A clearly written introductory text.*

TUOMANEN, E., 1993. Breaching the blood-brain barrier, *Scientific American* 268 (2). *On how bacteria penetrate to cause meningitis and use of this knowledge to treat other disorders.*

WALSBY, A. E., 1977. The gas vacuoles of blue-green algae, *Scientific American* 237 (2). (Offprint 1367) *How Cyanobacteria regulate their buoyancy.*

WOESE, C. R., 1981. Archaebacteria, *Scientific American* 244 (6). (Offprint 1516)

Chapter 38

BOLD, H. C., and M. J. WYNNE, 1978. *Introduction to the Algae.* Prentice-Hall, Englewood Cliffs, N.J. *Comprehensive, rather technical text that includes the protistan algae.*

BONNER, J. T., 1969. Hormones in social amoebae and mammals, *Scientific American* 220 (6). (Offprint 1145) *The role of cyclic AMP in the communication system of the cellular slime molds.*

BONNER, J. T., 1983. Chemical signals of social amoebae, *Scientific American* 248 (4). (Offprint 1537)

BROCK, T. D., and T. T. MADIGAN, 1991. *Biology of microorganisms,* 6th ed. Prentice-Hall, Englewood Cliffs, N.J. *An excellent text with balanced coverage of all aspects of the field.*

GRELL, K. G., 1973. *Protozoology.* Springer, Heidelberg. *Excellent comprehensive treatment of the Protozoa.*

RAGAN, M. A., and D. J. CHAPMAN, 1978. *A biochemical phylogeny of the protists.* Academic Press, New York.

Chapter 39

BANKS, H. P., 1970. *Evolution and Plants of the Past.* Wadsworth, Belmont, Calif. *Excellent short book on fossil plants, with special emphasis on the evolutionary relationships of the tracheophyte groups.*

BOLD, H. C., and M. J. WYNNE, 1978. *Introduction to the Algae.* Prentice-Hall, Englewood Cliffs, N.J. *Thorough, rather technical treatment of all the algal groups.*

COX, P. A., 1993. Water-pollinated plants, *Scientific American* 269 (4). *Adaptations enabled terrestrial plants to return to an aquatic environment.*

GRANT, V., 1951. The fertilization of flowers, *Scientific American* 184 (6). (Offprint 12) *The special adaptations of flowers that help ensure their pollination.*

JENSEN, W. A., and F. B. SALISBURY, 1972, *Botany: An Ecological Approach.* Wadsworth, Belmont, Calif. *Very readable general text written from an evolutionary and ecological point of view.*

NIKLAS, K. J., 1989. The cellular mechanics of plants, *American Scientist* 77, 344–349. *On how land plants support themselves against the force of gravity.*

RAVEN, P. H., R. F. EVERT, and S. E. EICHHORN, 1992. *Biology of plants,* 5th ed. Worth, New York. *An excellent introductory botany textbook.*

Chapter 40

ABMADJIAN, V., 1963. The fungi of lichens, *Scientific American* 208 (2)

COOKE, R. C., 1978. *Fungi, Man and his Environment.* Longman, London. *Short but fascinating treatment of the biology of fungi, with emphasis on the many ways these organisms affect humans.*

EMERSON, R., 1952. Molds and men, *Scientific American* 186 (1). (Offprint 115) *The diversity of the fungi, and their many effects on human lives.*

LITTEN, W., 1975. The most poisonous mushrooms, *Scientific American* 232 (3). *The members of the genus* Amanita *and the highly toxic compound they produce.*

RAVEN, P. H., R. F. EVERT, and S. E. EICHHORN, 1992. *Biology of plants,* 5th ed. Worth, New York. *An excellent introductory botany textbook.*

Chapter 41

BARNES, R. D., 1987. *Invertebrate Zoology,* 5th ed. Saunders, Philadelphia. *Rather technical, comprehensive treatment of all the invertebrate groups.*

BUCHSBAUM, R., 1987. *Animals Without Backbones,* 3rd ed. University of Chicago Press, Chicago. *One of the most readable and fascinating discussions of the invertebrates ever written. Not technical.*

BUCHSBAUM, R., and L. J. MILNE, 1960. *The Lower Animals: Living Invertebrates of the World.* Doubleday, Garden City, N.Y. *One of the "Living Animals of the World" books. Like the others in the series—on amphibians, birds, fish, insects, mammals, and reptiles—beautifully illustrated, well written, and nontechnical.*

DARWIN, C., 1842. *Structure and Distribution of Coral Reefs,* 1984 reissue. *University of Arizona Press, Tucson.*

EVANS, H. E., 1984. *Insect Biology.* Addison-Wesley, Reading, Mass. *Probably the best entomology textbook available.*

GOREAU, T. F., N. I. GOREAU, and T. J. GOREAU, 1979. Corals and coral reefs, *Scientific American* 241 (2). (Offprint 1434)

HADLY, N. F., 1986. The arthropod cuticle, *Scientific American* 255(1). *On the multipurpose exoskeleton, whose evolutionary development made arthropods the dominant form of life on the planet (as judged by species number).*

MCMENAMIN, M. A. S., 1987. The emergence of animals, *Scientific American* 256 (4). *On the first invertebrates.*

RICHARDSON, J. R., 1986. Brachiopods, *Scientific American* 255 (3).

Chapter 42

ALEXANDER, R. M., 1981. *The chordates,* 2nd ed. Cambridge University Press, New York. *Covers the invertebrate chordates as well as the vertebrates.*

ALVAREZ, W., and F. ASARO, 1990. What caused the mass extinctions? An extraterrestrial impact, *Scientific American* 263 (4).

BAKKER, R. T., 1975. Dinosaur renaissance, *Scientific American* 232 (4). (Offprint 916) *Reasons for believing that the dinosaurs were warm-blooded and that the birds descended from them.*

COLBERT, E. H., and M. MORALES, 1989, *Evolution of the vertebrates,* 4th ed. Wiley, New York.

COURTILLOTT, V. E., 1990. What caused the mass extinctions? A volcanic eruption, *Scientific American* 263 (4).

GLEN, W., 1990. What killed the dinosaurs? *American Scientist* 78, 354–370.

KENT, G. D., 1983. *Comparative anatomy of the vertebrates,* 5th ed. Mosby, St. Louis.

MARSHALL, L. G., 1994. The terror birds of South America, *Scientific American* 270 (2). *On the rise and fall of the dominant terrestrial carnivore of South America.*

POUGH, F. H., J. B. HEISER, and W. N. MCFARLAND, 1989. *Vertebrate life,* 3rd ed. Macmillan Co., New York.

ROMER, A. S., 1971. *The vertebrate story,* 4th ed. University of Chicago Press, Chicago.

ROMER, A. S., and T. S. PARSONS, 1986. *The vertebrate body,* 6th ed. Saunders, Philadelphia.

SIMPSON, G. G, 1983. *Fossils and the history of life.* W. H. Freeman, New York. *Excellent nontechnical treatment.*

VICKERS-RICH, P, and T. H. RICH, 1993. Australia's polar dinosaurs, *Scientific American* 269 (1). *Their adaptations may have enabled them to survive longer in the late Cretaceous period.*

Chapter 43

BLUMENSCHINE, R. J., and J. A. CAVALLO, 1992. Scavenging and human evolution, *Scientific American* 267 (3).

CAVALLI-SFORZA, L. L., 1991. Genes, people and languages, *Scientific American* 265 (5). *On the genetic, linguistic, and archaeological evidence favoring a recent African origin for modern humanity.*

CIOCHON, R., J. OLSEN, and J. JAMES, 1990. *Other origins.* Bantam Books, New York.

CONROY, G., 1990. *Primate Evolution.* W. W. Norton, New York.

COPPENS, Y., 1994. East side story: the origin of humankind, *Scientific American* 270 (5). *The Rift Valley in Africa reveals secret to divergence of hominids and emergence of human beings.*

EDEY, M. A., and D. C. JOHANSON, 1990. *Blueprints: Solving the mystery of evolution.* Penguin, New York.

HAY, R. L., and M. D. LEAKEY, 1982. The fossil footprints of Laetoli, *Scientific American,* February. (Offprint 1510) *Describes the site in Tanzania where thousands of animal tracks, including those of early hominids, are found.*

KLEIN, J., N. TAKAHATA, and F. J. AYALA, 1993. MHC polymorphism and human origins, *Scientific American* 269 (6). *On how the immune system is much older than the species it protects.*

LEAKEY, R. E., and R. LEWIN, 1992. *Origins revisited.* Doubleday, New York.

LEWIN, R., 1989. *Bones of contention: Controversies in the search for human origins.* Simon & Schuster, New York.

LEWONTIN, R. C., 1982. *Human diversity.* Freeman, New York. *Excellent.*

LOVEJOY, C. O., 1988. Evolution of human walking, *Scientific American* 259 (5). *Fascinating analysis of the "Lucy" fossil showing how anthropologists are able to conclude that our ancestors were bipedal 3 million years ago.*

MELLARS, P., (ed)., 1990. *The emergence of modern humans.* Cornell University Press, Ithaca, N.Y.

RENFREW, C., 1994. World linguistic diversity, *Scientific American* 270 (1). *Today's many tongues seem rooted in a few ancient ones.*

STRINGER, C. B., 1990. "The emergence of modern humans," *Scientific American* 263 (6). *A comparison of the DNA of various human groups is used to reconstruct and date the origin of our species.*

TATTERSALL, I., 1993. Madagascar's lemurs, *Scientific American* 268 (1). *With the closest resemblance to ancestors from which humans and apes branched, they are disappearing fast.*

TEMPLETON, A. R., 1992, and HEDGES, S. B., S. KUMAR, K. TAMURA, AND M. STONEKING, 1992. Human origins and analysis of mitochondrial DNA sequences, *Science,* 255, 737–39. *Two short "technical comments" that cast doubt on the mitochondrial DNA sequence data supporting the "mitochondrial Eve" hypothesis of human origins.*

THORNE, A. G., and M. H. WOLPOFF, 1992. The multiregional evolution of humans, *Scientific American* 266 (4). *Presents evidence against the "mitochondrial Eve" hypothesis.*

WEISS, M., and A. MANN, 1990. *Human biology and behavior,* 5th ed. Harper Collins, New York.

WILSON, A. C., and R. L. CANN, 1992. The recent African genesis of humans, *Scientific American* 266 (4). *A defense of the "mitochondrial Eve" hypothesis by two of its original proponents.*

APPENDIX 8
ANSWERS TO MAKING CONNECTIONS

Chapter 1

Page 13: 1. p. 1; 2. pp. 3–4; 3. p. 4; 4. pp. 5–6; 5. pp. 11–12. **Page 25:** 1. p. 13; 2. The diversity of life is the product of evolution directed by natural selection. Through this process, organisms have adapated to a vast array of habitats and ways of life (p. 12).

Chapter 2

Page 42: 1. p. 31; 2. p. 33; 3. p. 37; 4. p. 40. **Page 46:** 1. Weak bonds (ionic and hydrogen bonds, and hydrophobic interactions) are discussed on page 40. Such bonds are crucial in determining and stabilizing the structure of many important macromolecules (e.g., nucleic acids and proteins) and cellular structures (especially membranes). We will refer to them frequently in later chapters when discussing such topics as protein structure and function, cellular membranes, and molecular genetics; 2. As described on page 42, water is a superb solvent for a wide range of polar and charged substances. It is the medium in which all the chemical reactions essential for life take place. The importance of water for life will be a recurring theme appearing in one form or another in almost every chapter of this book, 3. The role of buffers in minimizing fluctuations in pH is described on page 45. We will see that they are crucially important to life in many ways, such as maintaining the conformation of proteins, as described in Chapter 3. They are also essential for maintaining the proper pH of digestive functions and of the blood and other fluids that surround our cells.

Chapter 3

Page 62: 1. p. 51; 2. pp. 53–54; 3. As explained on page 59, fats are excellent energy-storage molecules because they provide more than twice as much energy per unit of weight when broken down as carbohydrates. Thus, a person whose body weight included thirty pounds of fat would be more than thirty pounds heavier if carbohydrate instead of fat were used for energy storage; 4. pp. 60–61. **Page 74:** 1. As described on pages 64–65, proteins are usually extremely long polymers constructed from twenty possible building-block molecules—the amino acids. A typical carbohydrate or lipid is constructed from fewer possible subunits and is much smaller than a protein. The greater versatility in protein function as compared to the function of carbohydrates or lipids is a direct consequence of their greater structural diversity; 2. pp. 72–74; 3. Weak bonds are largely responsible for determining the conformation of proteins: secondary, tertiary, and (often) quaternary structure all depend on the existence of hydrogen bonds, ionic bonds, and hydrophobic interactions. In nucleic acids, hydrogen-bonding links the complementary nitrogenous bases.

Chapter 4

Page 81: 1. p. 78; 2. Many organisms, such as human beings, replenish lost energy by eating other organisms and thereby obtaining a supply of energy-rich organic molecules. Virtually all energy used in biological systems can be traced back to the sun, which supplies the light energy used by plants to produce energy-rich food molecules from low-energy raw materials; 3. pp. 79–80; 4. p. 81. **Page 86:** 1. p. 81; 2. pp. 82–83; 3. p. 83; 4. p. 84; 5. Feedback inhibition is explained on pages 85–86. This method of regulating the production of essential chemicals synthesized and used within a cell is an example of homeostasis (see Chapter 1) at the molecular level. We will see in later chapters that the same principle operates at higher levels of organization as well. For example, in the regulation of hormone production in the human body.

Chapter 5

Page 100: 1. As discussed on page 90, biogenesis is the theory that all life arises from pre-existing life. Most biologists believe, however, that in the primitive earth the spontaneous generation of life from nonliving matter occurred. This singular example of spontaneous generation of life is responsible for the existence of life on earth at the present time. If the basic unit of life is the cell, then the evolution of the first cell can reasonably be viewed as the most crucial event in the history of life; 2. The role of the surface-to-volume ratio in restricting cell size is discussed on pages 90–93. The geometrical principle that volume increases as a cubic function of an object's linear dimensions, whereas its surface area increases as a square function, has profound implications for the structure of organisms. In later chapters concerned with the structure and function of animals and plants, adaptations that increase the surface area for exchange of materials will be a recurring theme. 3. pp. 94–96; 4. pp. 96–98; 5. According to the fluid-mosaic model of the cell membrane described on pages 98–100, phospholipids arranged in a bilayer are the basic structural components of the membrane. Lipids and other hydrophobic substances are able to dissolve in the lipid bilayer and diffuse through it freely. Being nonpolar, lipid molecules do not interact electrically with the charged phosphate-containing "head" of each phospholipid. In contrast, ions are attracted or repelled by the phosphate groups and hence cannot easily pass through the interior of the bilayer. **Page 109:** 1. pp. 101–102; 2. p. 102; 3. As in the case of some enzyme inhibitors (see pages 85–86 in Chapter 4), the functioning of gated channels (described on page 101) depends on the ability of proteins to change their conformation. The shifts in conformation shown by such proteins are made possible by the fact that the weak bonds that stabilize a protein's three-dimensional structure are easily disrupted. As we will learn in later chapters, conformational changes in proteins and other molecules are an important means of communication between and within cells; 4. As mentioned on page 101, cells must often transport ions and building-block molecules across the cell membrane against a concentration gradient. This transport is necessary to supply the cell with energy and raw materials and to maintain the right balance of ions and other solutes inside the cell. The ability of the cell to maintain the proper concentrations of various substances, despite their spontaneous tendency to diffuse in or out until their concentrations equal those found in the surrounding medium, is an example of homeostasis (see page 26 in Chapter 1). That the cell must continually expend energy to maintain its internal order is a consequence of the Second Law of Thermodynamics, which states that (1) energy is constantly degraded to waste heat and therefore cannot be recycled and (2) systems left to themselves tend to become more and more disordered (see Central Concept No. 4, page 26 in Chapter 1); 5. The cells of plants normally exhibit turgor pressure (see page 109) because the water they absorb from the soil is hypotonic compared to the cell contents, and water therefore tends to enter the cell by osmosis. The outward-directed turgor pressure exerted against the cell wall keeps the cells rigid and helps to support the plant. If the plant lacks water, its cells will eventually exhibit less turgor pressure and may even plasmolyze as water loss causes the extracellular medium to become more osmotically concentrated. Without sufficient turgor pressure, the plant wilts. Watering the plant before permanent damage to its cells occurs will restore the normal osmotic relationship between the cells and their medium, and the plant will recover; 6. p. 110.

Chapter 6

Page 124: 1. p. 113; 2. p. 113; 3. As described on page 121, the digestive enzymes are synthesized in the rough ER, from which they pass first to the smooth ER and then (via transport vesicles) to the Golgi apparatus. Lysosomes containing the enzymes bud off the Golgi membrane. Proteins in the lysosome membrane enable it to recognize endocytotic vesicles containing food, which enters the cell either by pinocytosis (if liquid) or phagocytosis (if solid). After the lysosome and vesicle fuse, the digestive enzymes introduced by the lysosome hydrolyze the food macromolecules; 4. Both protozoans with contractile vacuoles and plant cells normally live in a hypotonic medium. Like a bilge pump, the protozoan's contractile vacuole continually expels the excess water entering the cell by osmosis and thus prevents it from bursting. As the vacuole of a plant cell accumulates water, it swells and transmits pressure through the cytoplasm to the cell wall. The resulting turgor pressure helps to support the plant. (See page 123 in this chapter). **Page 136:** 1. Cells must have a constant supply of energy available to do work to counteract the natural tendency of all orderly systems (including cells) to become disorganized, as predicted by the Second Law of Thermodynamics. In eucaryotic cells, mitochondria and chloroplasts are the principal suppliers of this energy; 2. p. 125; 3. (1) liver; (2) pancreas and intestine; (3) liver and kidney; (4) heart muscle; (5) muscle; (6) nerve cells; (7) trachea. (Note that these are not the only cells with the characteristics mentioned; for example, mitochondria are generally abundant in muscle cells, cilia are also found in cells lining the oviduct, and so on.); 4. According to the endosymbiotic hypothesis discussed on page 136, mitochondria and chloroplasts are the descendants of procaryotic cells captured by endocytosis by a host-cell precursor of eucaryotes. If so, the outer membrane of these organelles might represent the membrane of the vesicle in which they were enclosed, and the inner might represent the cell membrane of the engulfed procaryote.

Chapter 7

Page 142: 1. As discussed on page 139, the evolution of photosynthesis allowed organisms to tap into the energy of the sun and thus made them independent of the very limited amount of energy available from ready-made organic energy sources and from chemosynthesis. The later development of photosynthesis based on the use of water led to the accumulation of molecular oxygen in the atmosphere and set the stage for the evolution of organisms capable of aerobic respiration; 2. pp. 139–140; 3. p. 140; 4. p. 142. **Page 155:** 1. When either the atom in Fig. 24 or cholorphyll absorb a photon of light energy, an electron of the substance is raised from its ground state to a higher, unstable energy level. In the atom, this energy is emitted as the electron returns to its ground state. In chlorophyll within a photosynthetic unit, the energy given off as the excited electron returns to its ground state is transferred to another pigment molecule, one of whose electrons becomes similarly excited. Eventually, as described on page 145, the energy is transferred to a reaction center molecule where it is trapped and an excited electron is passed to an acceptor molecule. As shown in Fig. 7.8, photons of green light are poorly absorbed by photosynthetic pigments compared to photons at the red and violet ends of the visible spectrum. The photons that are not absorbed are reflected from the leaf and account for its green color; 2. Both the cell membrane and the thylakoid membrane consist of a phospholipid bilayer containing proteins. Both exhibit selective permeability and maintain concentration gradients of various molecules and ions across the membrane. In addition, both have channel proteins allowing movement of substances to which the membrane is otherwise impermeable. Each also has unique features reflecting their respective functions; thus the cell membrane has many glycoproteins and glycolipids needed for cell recognition, whereas the thylakoid membrane contains the photosystems, electron-transport proteins, and ATP synthetase complexes required for the light reactions of photosynthesis; 3. As a mobile molecule, PQ demonstrates the fluidity of membrane structure; 4. The sources of the H^+ ions that accumulate in the interior of the thylakoid are the splitting of water and the flow of electrons along the transport chain. Both of these events are driven by light energy absorbed by pigment molecules in the photosystems. **Page 158:** 1. p. 155; 2. pp. 156–158; 3. The ADP, Ⓟ, and $NADP^+$ are used in the light reactions to regenerate more ATP and NADPH. The availability of $NADP^+$ determines whether the cyclic or noncyclic pathway is used (though whether it is ever completely unavailable is a controversial point). (See Fig. 7.20 on page 157); 4. If the membrane of the thylakoids were freely permeable to H^+, no ADP would be synthesized, but all other aspects of the light reactions would be unaffected. PGAL is produced in the dark reactions, which are driven by NADPH and ATP. Without ATP supplied by the light reactions, PGAL production would cease; 5. As explained on pages 152–154, neither C_3 nor C_4 plants are likely to become more productive because of a higher atmospheric concentration of CO_2. A warmer climate should favor C_4 plants, which are more efficient photosynthesizers at high temperatures.

Chapter 8

Page 166: 1. p. 161; 2. p. 162; 3. p. 162; 4. The metabolic problem for anaerobic organisms is that the production of ATP via glycolysis results in the reduction of NAD^+ to NADH but provides no mechanism for regenerating NAD^+. Without NAD^+ to accept electrons, glycolysis would halt. They solve the problem by means of the same process used by aerobic organisms in the absence of oxygen: fermentation, in which NADH is oxidized to NAD^+ as pyruvic acid is reduced (pp. 165–166). **Page 173:** 1. As explained on page 167, the folds increase the total surface area of the inner membrane and thus provide more space for the molecules of the electron-transport chain and for the many protein channels and pumps that are essential to mitochondrial function. Because membranes are vital for exchange of materials (and, within cells, as "work surfaces"), structural adaptations for increasing surface area are a recurring theme in biology and will be mentioned often in later chapters; 2. Two-thirds of the CO_2 is produced in the Krebs cycle (4 out of 6 CO_2 molecules per glucose). The other third stems from the oxidation of pyruvic acid to acetyl CoA; 3. pp. 168–169; 4. If either membrane of the mitochondrion were freely permeable to H^+ ions, no concentation gradient could be build up between the inner and outer compartments of the mitochondrion, and no synthesis of ATP via chemiosmosis could occur. Thus glucose oxidation would produce only the four molecules of ATP that stem from substrate-level phosphorylation; 5. p. 173. **Page 173:** 1. The similarities and differences between chemiosmotic ATP synthesis in mitochondria and chloroplasts are discussed on page 173. A comparison of the inner membrane of the mitochondrion with the thylakoid membrane reveals some important differences related to the source of the energy that drives chemiosmosis in the two organelles. In the mitochondrion, the immediate source of energy is high-energy NADH and $FADH_2$ molecules, which, in turn, have derived *their* energy from the oxidation of food molecules. During their passage through the respiratory chain, the electrons taken from NADH and $FADH_2$ lose energy at every step until at the end of the chain they are accepted by oxygen to form water. Thus, from an energetic standpoint the movement of the electrons is steadily "downhill," with some of the released energy being used to pump H^+ ions across the inner membrane. In noncylic photophosphorylation occurring in the thylakoid of the chloroplast, by contrast, the electrons move from water to $NADP^+$, forming the high-energy molecule NADPH. The electrons gain the energy for this "uphill" movement and for the creation of the H^+ ion gradient across the thylakoid membrane from light striking the photosystems embedded in the membrane; 2. If more food is consumed than is required to supply the energy needs of the body, the excess intake of energy will be stored as energy-rich organic molecules. Recall from chapter 3 that fats are best suited for this function because they store more than twice as much energy as carbohydrates or proteins per unit of mass. As described on page 174, fats can be made from carbohydrates or amino acids, so the diet need not include fat in order for the body to synthesize it. This section of the text highlights the importance of the metabolic pathways linking the major classes or organic molecules. The homeostatic control of all of these catabolic and anabolic pathways is achieved primarily through hormones discussed in chapter 25; 3. p. 174; 4. Several factors discussed on pages 175–177 are relevant to this question. Since ectothermic animals depend on the environment for their heat, they would seem to be more vulnerable to extinction than endotherms, which might be better able to maintain a stable body temperature and activity level regardless of widely fluctuating environmental temperatures. But recall that many ectotherms, such as the desert lizards mentioned on page 176, are able to maintain a nearly constant internal temperature through behavioral adaptations across a wide range of external temperatures at a relatively low energetic cost. The lower demand for energy of ectotherms might also mean that if the fluctuating climate resulted in a reduction in the availability of a particular food, an endotherm dependent on that food source might be more likely to become extinct than a competing ectotherm. In summary, it depends!

Chapter 9

Page 188: 1. p. 179; 2. DNA replication occurs during the S phase of the cell cycle; therefore a cell in the G_2 phase (which follows the S phase) would have twice as much DNA as a cell in the G_1 phase (which precedes the S phase). A cell with an intermediate amount of DNA would be in the S phase; 3. p. 184; 4. p. 187. **Page 200:** 1. p. 189; 2. It is meiosis I (the reduction division) that the important unique features of meiosis occur: the pairing of homologous chromosomes (synapsis), crossing-over, the attachment of each synapsed pair of chromosomes to the same set of spindle microtubules, and the separation of the double-stranded chromosomes of each pair into each daughter cell. The result is that each cell produced by meiosis I receives the haploid number of chromosomes. The mechanics of meiosis II are similar to mitosis, except that the chromosome number is haploid; 3. pp. 196–198; 4. p. 199.

Chapter 10

Page 210: 1. If 23% of the nucleotides contain thymine, then another 23% must contain adenine, since these bases must pair with one another. The other 54% of the nucleotides must contain cytosine or guanine, which must also occur in equal proportions. Therefore 27% of the nucleotides would have cytosine as their base; 2. p. 207. See also Chapter 2; 3. p. 209. **Page 213:** 1. As discussed in Chapter 3, polymerization reactons such as the linking together of nucleotides always occur by condensation; 2. Temperature, substrate concentration, and pH could all affect the efficiency of the enzyme. An increase in the error rate in the incorporation of new nucleotides by DNA polymerase could occur if the conformation of the active site of the enzyme were affected by denaturation—for example by exposure to high temperature or to abnormally alkaline or acidic conditions. If the enzyme were completely denatured, it would be nonfunctional; 3. p. 211; 4. p. 211; 5. p. 213.

Chapter 11

Page 222: 1. p. 218; 2. pp. 218–222; 3. pp. 218–219; 4. p. 221; 5. As described in Chapter 6 on page 118, proteins synthesized on free ribosomes are released into the cytoplasm. Proteins that are synthesized on ribosomes of the rough ER are destined for export from the cell, for integration into a cellular membrane, or for inclusion in a cellular organelle such as a lysosome. In general, these proteins would pass from the ER to the Golgi apparatus. The Golgi apparatus buds off vesicles, including lysosomes that hold digestive enzymes and secretory vesicles that may contain "export" proteins. The latter fuse with the cell membrane and release their contents to the cell's surroundings. **Page 228:** 1. The answer is on page 227. Though such antibiotics do not harm the cells of the person taking them, they can nevertheless cause problems. Often they result in the decimation of the body's natural bacterial flora, which has such beneficial functions as vitamin K synthesis and protection from more pathogenic microorganisms. The result can be intestinal upset and increased risk of yeast infections, etc.; 2. The codons for phenylalanine are UUU and UUC. Substituting C for the second U gives serine. Substituting C for the first U (or A or G for the U or C in the third position) produces leucine. The mutations, of course, occur in DNA. Since AAA and AAG are the DNA triplets on the template strand that code for phenylalanine, the mutated DNA triplets would be: for serine, AGA or AGG; for leucine, GAA, GAG, AAT, or AAC. Figure 3.19, page 65, shows that the R groups of both phenylalanine and leucine are hydrophobic, whereas the R group of serine is hydrophilic. Replacing phenylalanine by serine would therefore be more likely to disrupt the tertiary structure of the protein and interfere with its function. **Page 230:** 1. The thinning of the ozone layer may result in decreased atmospheric absorpton of UV radiation, which can cause mutations (see page 228). The increased rate of mutation could lead to an increase in the rate of skin cancer in human beings and to other harmful environmental effects; 2. p. 228; 3. p. 288. An interesting and controversial aspect of the screening of additives and drugs concerns the fact that many of the chemicals that are natural components of foods such as vegetables have been found to be mutagenic on the basis of the Ames test. Such foods are, of course, legally sold, whereast mutagenic food additives are banned; 4. p. 229; 5. pp. 229–230.

Chapter 12

Page 236: 1. p. 233; 2. The many different cell types found in a complex multicellular organism such as a human all arise (with the exception of gametes or spores produced by meiosis) by mitotic cell division from a fertilized egg. The differing characteristics of these cells therefore cannot be due to differences in their genetic composition, but rather to differences in gene expression. An understanding of the control of gene expression is thus crucial to understanding the process of development. See p. 234; 3. p. 233; 4. As described on page 236, the activity of the tryptophan operon is controlled by the amount of tryptophan present in the cell. If there is a sufficient amount present, tryptophan, acting as a corepressor, combines with the repressor molecule and turns off the operon. If the level of tryptophan falls below a critical level, the operon is activated and enzymes are produced that result in more tryptophan being synthesized. Thus the level of tryptophan in the cell is stabilized within acceptable limits. *This ability to regulate the internal composition of the cell is the essence of homeostasis.* The control of gene activity exemplified by operons and essential to homeostasis occurs in both eucaryotic and procaryotic cells. **Page 243:** 1. p. 238; 2. pp. 239–240; 3. pp. 242–243. **Page 252:** 1. pp. 244–245; 2. As explained in Chapter 5 on page 110, the cell coat (glycocalyx) of an animal cell is composed of glycolipids and glycoproteins and provides the recognition sites that enable it to interact with other cells. One of the most important of these interactions is contact inhibition, by which crowded cells are induced to stop moving and dividing. The abnormal glycocalyx of cancer cells described on p. 246 disrupts intercellular communication and thus allows them to continue to divide in crowded conditions; 3. pp. 252–254; 4. p. 252.

Chapter 13

Page 264: 1. Both selective breeding and natural selection are based on the occurrence of genetically based variation in populations of organisms. In selective breeding, human beings choose the genetic characteristics of a plant or animal that are desirable from a human standpoint and allow only those individuals possessing those traits to breed. In natural selection, on the other hand, it is the environment of the organism that "selects" which characteristics are desirable (i.e., adaptive), as determined by the reproductive success of the individuals having those traits. In both processes, the rigor of the selective process acting on the population determines the rapidity of genetic change. Though there are many examples of rapid evolution via natural selection, selective breeding tends to be more rigorous and thus to produce faster change. (See page 257 and pages 11–13 in Chapter 1.); 2. pp. 258–259; 3. As in the case of other enzymes, the catalytic activity of a restriction enzyme depends on its ability to "recognize" a specific substrate—in this case, a particular palindromic sequence in DNA. If donor and vector DNA were cut with different restriction enzymes, they could not be joined by complementary "sticky ends" because they would have been cut at different palindromic sequences. (See page 260); 4. p. 260; 5. p. 264. **Page 274.** 1. p. 264; 2. pp. 268–271; 3. pp. 271–274; 4. pp. 274–275.

Chapter 14

Page 286: 1. As a cultivated plant, the pea was readily available and easily grown under controlled conditions. Its many genetically different strains enabled Mendel to find a number of traits that occurred in two contrasting, easily distinguishable forms. That the pea is normally self-pollinated but can be cross-pollinated by human action permitted Mendel to make crosses between particular plants with little possibility of error. Finally, because the pea produces many offspring, Mendel was able to perform statistical analysis of the results of his experimental crosses; 2. p. 279; 3. pp. 283–284. 4. pp. 285–286.

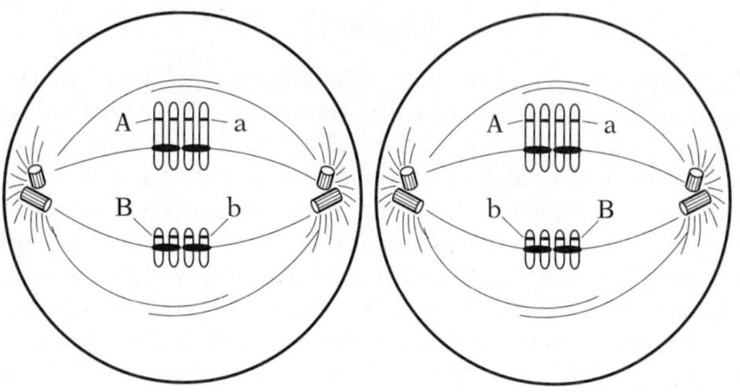

Page 293: 1. As described in Chapter 5, the cell coat (glycocalyx) of animal cells is composed of glycoproteins and glycolipids and is instrumental in the process of cell recognition. The alleles of the A-B-O system code for enzymes that determine the structure of the carbohydrate portion of one such glycoprotein or glycolipid. The immune system is able to recognize differences in this component of the glycocalyx of red blood cells by mechanisms described in Chapter 21; 2. pp. 286–288; 3. pp. 289–292; 4. p. 292; 5. pp. 292–293. **Page 301:** 1. pp. 293–294; 2. pp. 295–297; 3. Mutation is the ultimate source of genetic variation since it is the process by which new alleles are created. As described on page 299, crossing-over occurring during meiosis breaks the original linkages between alleles on a chromosome and thereby increases the number of genetic combinations that can be produced by sexual reproduction. Both processes contribute to the genetic variation in populations and hence increase the amount of "raw material" on which natural selection can act; 4. Since a diploid organism, having two copies of each chromosome, has a maximum of two different alleles for each gene, it follows that a tetraploid organism, with four copies of each chromosome, could have as many as four different alleles (see pages 301–303); 5. p. 303.

Chapter 15

Page 314: 1. p. 307; 2. p. 307; 3. p. 309; 4. pp. 312–313. **Page 320:** 1. pp. 315–317; 2. In Chapter 5 you learned that because the lipid portion of the cell membrane is relatively impermeable to ions, they primarily pass through the membrane via protein-lined channels. Since gap junctions are protein channels in cell membranes aligned so as to create a continuous passage between cells, it is logical that they would play a major role in the movement of ions between cells. (see page 315); 3. Epithelia that are primarily protective in function are likely to have several cell layers (i.e., they are stratified), whereas those through which absorption must occur generally have only one (i.e., they are simple). The greater risk of infection by the AIDS virus (HIV) via rectal intercourse as compared to vaginal intercourse is largely due to the difference in the epithelia of these structures. The rectal lining is a simple columnar epithelium that is more easily damaged by the friction of intercourse (and hence more vulnerable to penetration by HIV) than the tough stratified squamous epithelium that lines the vagina (see pages 315–317); 4. pp. 317–319; 5. p. 320.

Chapter 16

Page 327: 1. These ions would be nutrients for a plant because, as autotrophs, plants have the ability to use nitrates, sulfates, and phosphates obtained from the soil in the synthesis of proteins and nucleic acids. In contrast, animals such as ourselves depend on *organic* sources of nitrogen, sulfur, and phosphorus. Thus, these ions cannot serve as human nutrients (see pages 323–324); 2. p. 324; 3. p. 325; 4. pp. 325–326; 5. pp. 325–326; 6. p. 326. **Page 331:** 1. As described on pages 326–327, soil pH strongly influences the availability to plants of mineral ions. For example, if the soil contained too much lime (and hence was excessively alkaline), it would be difficult for plants to obtain nutrients such as iron, copper, manganese, and zinc. Many enzymes important in plant metabolism would be adversely affected by an insufficient supply of these particular micronutrients (see Table 16.1, page 326), so the plants would probably grow poorly. To counteract the effect of the lime, you could add to the soil an acidifying chemical (such as sulfur) or a fertilizer containing the nutrients known to be in short supply; 2. As explained on pages 328–329, legumes such as clover house nitrogen-fixing bacteria in nodules on their roots. By planting clover and plowing it under, the farmer adds the fixed nitrogen to the soil and thus improves its fertility; 3. p. 329; 4. pp. 330–331. **Page 340:** 1. pp. 337–338; 2. The adaptations by which roots and leaves provide enough surface area are described on pages 331–333 and page 338, respectively. Many of these same "strategies"—such as the evolution of highly folded, branched, filamentous, or spongy structures, all of which increase surface area enormously—will be discussed again in later chapters concerned with the structure and function of animals and fungi. In the case of a large woody plant, the problem of providing adequate surface area is eased by the fact that it mostly consists of dead tissue—the xylem (see chapter 19); 3. pp. 338–339; 4. pp. 339–340; 5. p. 341.

Chapter 17

Page 346: 1. As described in Chapter 1, each of the three kingdoms of multicellular eucaryotes is characterized by a different mode of nutrition. Plants are autotrophs; the fungi are absorptive heterotrophs; and animals are primarily ingestive heterotrophs (see page 343); 2. pp. 345–346; 3. p. 345; 4. pp. 346–347; 5. pp. 346–347. **Page 355:** 1. p. 347; 2. As discussed on page 348, our nutritional requirements suggest that the diet of early humans was omnivorous. The presence of meat in the diet led to the evolutionary loss of the ability to synthesize certain amino acids that were easily obtained from animal protein. Natural selection acts in accordance with current environmental conditions; it cannot predict future needs. In this instance, those of our human ancestors who spent valuable metabolic energy and raw materials synthesizing unneeded enzymes were evidently at a selective disadvantage in competition with those who had lost the ability to make such enzymes; 3. Thiamine, riboflavin, pantothenic acid, and nicotinamide function as coenzymes in respiration (see Table 17.2, page 349); 4. In both plants and animals, most mineral micronutrients activate enzymes (see page 348); 5. p. 352; 6. As mentioned on page 353 with particular reference to lysosomes, the processes by which cells engulf and digest food are essentially the same, whether the cell is a protozoan or is part of a multicellular animal. **Page 364:** 1. The extensively branched gastrovascular cavity of planaria provides more surface area for absorption of nutrients than its saclike equivalent in hydra. It is impossible to say, however, that one organism is "superior" to another because of the presence of a particular adaptation. Both hydra and planaria are evidently well adapted to their particular ways of life (see page 356); 2. p. 356; 3. pp. 358–359; 4. pp. 359–361; 5. p. 362; 6. As mentioned on page 361, the senses of smell and taste are very important in food selection. They help us to avoid potential food items that are toxic or spoiled and to choose those that have a high nutritional value; 7. Like root hairs, microvilli (described and pictured on page 363) are hairlike projections of single cells that add enormously to the total absorptive surface area. A comparison of Figure 16.8 with Figure 17.27 shows that root hairs are large enough to be seen by the naked eye, whereas microvilli are only detectable in electron micrographs. If the microvilli were as large as root hairs, they would be torn away by the movement of material passing through the small intestine. **Page 371:** 1. p. 364; 2. p. 364; 3. p. 365 and pp. 366–367; 4. The discussion of protein digestion on pages 367–368 points out that the various protein-digesting enzymes break peptide bonds at specific locations, or between particular pairs of amino acids. This specificity results from the fact that proteins are built from 20 kinds of amino acids, each with a different R group and hence a different shape. Thus it requires many enzymes to catalyze the hydrolysis of all the many peptide bonds that can occur in a protein; 5. pp. 368–369.

Chapter 18

Page 382: 1. p. 373; 2. pp. 378–379; 3. As explained on pages 378–379, the loss of water from the gas-exchange surface is the major problem for terrestrial animals, just as it is for terrestrial plants. The solution in both animals and plants is to have most gas exchange take place within almost completely enclosed chambers in which the humidity can be kept high to reduce the rate of water evaporation. In animals, these chambers are lungs or tracheae (in insects); in plant leaves, the same function is served by the air spaces of the spongy mesophyll. Lungs, tracheal systems, and the spongy mesophyll are also generally characterized by having a large internal surface area for gas exchange. One difference between plants and animals is that plants depend solely on diffusion to move gases to and from their respiratory surfaces, whereas terrestrial animals ventilate their lungs or (in the case of some insects) their tracheae; 4. Both the spiracles of insects described on pages 379–380 and the stomata of plants are openings to structures in which gas exchange occurs. Both openings can be shut, either by guard cells (stomata) or by valves (spiracles). Plants normally close their stomata to minimize the loss of water from the plant. Insects can shut their spiracles for protection—from insecticides or from drowning, for example. The control that plants and insects can exert over these openings is an aspect of homeostasis: the ability of an organism to regulate its internal environment; 5. pp. 380–381. **Page 386:** 1. The role of the ciliated cells in producing a "conveyor belt" of mucus that moves foreign particles away from the lungs is described on pages 382–383. The paralyzing effect of cigarette smoke on these cells allows mucus to stagnate and build up in the respiratory tract. This build-up of mucus can, at least potentially, interfere with gas exchange by hampering the free flow of air to and from the lungs. "Smoker's cough" is the cigarette smoker's method of clearing these passageways; 2. pp. 384–385; 3. p. 384; 4. As explained in Chapter 14, people with cystic fibrosis have a defective protein regulating chloride transport across their cell membranes. The result is an osmotic imbalance leading to excessive production of thick mucus; 5. As described on pages 385–386, amphibians such as frogs have less surface area in their lungs than reptiles or mammals of comparable size. A frog can get by with a relatively simple lung because (as explained on page 386) it also uses its skin and mouth cavity for the exchange of gases. As a reptile, a lizard is characterized by a dry scaly skin through which relatively little gas exchange can occur. Thus a lizard requires a greater surface area in its lungs. Because a mouse is an endotherm and has a far higher metabolic rate than either of the other animals, its lungs must have even more surface area to meet its high oxygen demand.

Chapter 19

Page 396: 1. p. 389; 2. p. 390; 3. The cells of the meristematic regions of the plant, described on pages 391–392, retain an embryonic character, in that they are capable of differentiating into many different cell types. *Cell differentiation invariably involves alterations in the pattern of gene expression in a cell,* by the mechanisms described in Chapter 12; 4. p. 394. **Page 401:** 1. p. 394; 2. pp. 399–400; 3. If water were nonpolar, hydrogen bonding would not occur between water molecules (see Fig. 2.14, page 42). Without hydrogen bonding, water would not exhibit cohesion (or adhesion—e.g., to the walls of tracheids and vessels), and the transport of water by the cohesion-tension mechanism described on pages 400–401 would not occur. Since this mechanism is the only one that can account for the movement of sap to the tops of trees, it follows that if water were nonpolar such transport would be impossible. **Page 404:** 1. The inner bark of the sapling is the phloem (see page 401), in which translocation of most carbohydrate and organic nitrogen compounds occurs. This layer of bark would therefore be a good food source for rabbits and other gnawing herbivores such as beavers and porcupines. As explained on pages 402–403, removing a ring of inner bark from the trunk ("girdling") interrupts carbohydrate translocation and kills the tree; 2. p. 403; 3. The sun's energy drives the transport of sap in the xylem, whereas the plant must expend energy in the form of ATP in phloem translocation (to actively transport sugars into and out of sieve elements) (see pages 397 and 403); 4. The amazingly high pressures found in sieve elements are possible because of their strong cell walls. The high pressure inside has made phloem cells difficult to study, since they "blow out" (like an automobile tire) when their cell walls are damaged. This problem has been overcome by using phloem-feeding insects called aphids. They feed by inserting their needle-like mouthparts (the stylet) through the phloem cell wall without triggering either a "blow out" or a wound-response reaction. By cutting a sucking aphid away from its stylet, researchers can sample the phloem juices.

Chapter 20

Page 412: 1. p. 407; 2. p. 408; 3. p. 409. **Page 417:** 1. p. 412; 2. p. 414; 3. pp. 414–415; 4. p. 415; 5. As described on pages 415–417, blood in the circulatory system flows because of the decline in pressure from one point in the system to the next. Thus the highest pressure in both the pulmonary and systemic circuits is found in the heart ventricles and the lowest pressure is found in the veins nearest the heart. The pressure in the system is created by the contraction of the ventricles, and the decline in pressure is the result of friction between the blood and the walls of the vessels. According to the pressure-flow hypothesis, the force responsible for the movement of fluid in the phloem is turgor pressure created in "source" regions by the active transport of sugars and other solutes into phloem cells and the resulting osmotic influx of water. The gradient is created by the low turgor pressure found in "sink" areas as solutes are removed from the phloem and water follows by osmosis. Ventilation of our lungs is achieved by negative-pressure breathing. During inhalation, the pressure inside the chest cavity is lowered as its volume is expanded by contractions of the diaphragm and rib cage musculature. Thus a pressure gradient is created between the atmosphere and the lungs and air flows into the lungs. When the diaphragm and rib muscles relax, additional (i.e., above atmospheric) pressure is put on the air in the lungs and it is expelled. **Page 422:** 1. As discussed on pages 417–418, pressure put on the veins by muscular activity is important in pushing the blood upward toward the heart from the legs. Standing still for a long period can cause the blood to pool in the area of the ankles and feet. In pregnant women, the greatly enlarged uterus containing the fetus can put pressure on veins carrying blood from the lower extremities and cause the return of blood to be impaired. Varicose veins (in which the walls of the veins are distended) often occur for the same reason; 2. p. 418; 3. p. 421; 4. Kwashiorkor causes the concentration of proteins in the blood to decline as they are broken down to supply the energy needs of the body. The edema results from a decrease in the osmotic concentration of the blood (because of the reduced protein concentration) and, therefore, decreased reabsorption of fluid from the tissues into the blood stream (see page 423). **Page 430:** 1. p. 424; 2. p. 424; 3. p. 426; 4. pp. 427–428; 5. pp. 428–429; 6. p. 429; 7. Although hemoglobin is not an enzyme, its binding site for oxygen is analogous to the active site of an enzyme. Just as some of the enzyme inhibitors described in Chapter 4 permanently combine with the active site, so carbon monoxide combines irreversibly with the oxygen-binding site of hemoglobin and thus disrupts its function (see page 429).

Chapter 21

Page 439: 1. Like the large intestine (see Chapter 17), the skin and other parts of the body open to the exterior are inhabited by vast numbers of bacteria. As explained on page 434, by fully occupying habitats that might otherwise be open to colonization by pathogenic microorganisms, these bacteria protect the body from disease. The relationship of human beings to our normal microbial flora is an example of mutualism, discussed in Chapter 31; 2. p. 436; 3. pp. 438–439. **Page 446:** 1. pp. 440–442; 2. p. 442, 444–446; 3. p. 446; **Page 447:** 1. pp. 446–447; 2. p. 447; 3. Recall from Chapter 20 that the passage of fluid through capillary walls depends on the action of two opposing forces: osmotic pressure (which draws fluid into the capillary) and blood pressure (which forces fluid out). The effect of histamine is to make the walls of capillaries so leaky that plasma—*including plasma protein*—is forced out by blood pressure into the surrounding tissues. The presence of protein in the tissue fluid lessens the osmotic gradient that normally returns fluid to the blood, and swelling of the tissues (edema) occurs. In the skin, for example, the result is hives.

Chapter 22

Page 456: 1. p. 453; 2. Such cellular activities as active transport to maintain proper ionic concentrations inside and outside the cell are homeostatic in function and require energy. Cellular respiration, discussed in Chapter 8, supplies the energy needed for homeostatic functions (as well as growth). As mentioned on page 453, the circulatory system delivers to the cells the oxygen and nutrients required for cellular respiration and carries away the wastes. In this chapter, we will discuss how animals maintain homeostasis of extracellular fluids by regulating their concentrations of nutrients and ions and eliminating wastes; 3. p. 454; 4. pp. 454–456; 5. The red blood cell would pass through three capillary beds—the first in the liver, the second in the lungs, and the third in the digestive tract. As explained on page 455, this circulatory arrangement places the liver in the ideal position to monitor the concentration of nutrients absorbed into the blood from the digestive tract. It is thus crucial to the liver's homeostatic role; 6. pp. 455–456. **Page 463:** 1. pp. 456–457; 2. p. 458; 2. As explained on page 459, the oxidation of alcohol to acetaldehyde in the liver requires NAD^+ to accept the hydrogen and electrons produced by this reaction. But recall from Chapter 8 that NAD^+ is also required as a coenzyme in cellular respiration of all nutrients, including fats. The top priority given by the liver to the breakdown of alcohol (which is toxic) can lead to an insufficient supply of NAD^+ for fat metabolism, and thus to the deposition of the fat in liver cells; 4. p. 462; **Page 471:** 1. p. 464; 2. pp. 466–467; 3. pp. 467–469; 4. As explained on pages 470–471 vasopressin increases the permeability of the collecting tubules in the kidneys to water, thereby stimulating its reabsorption into the blood. In its absence, virtually all the water originally in the collecting tubules is passed out of the kidneys into the ureters, and a large volume of dilute urine is produced. Overconsumption of alcohol upsets this homeostatic osmoregulatory mechanism by suppressing the release of vasopressin by the pituitary. Dehydration ensues because of excessive water loss from the body through urination; 5. p. 471; 6. p. 472.

Chapter 23

Page 476: 1. p. 475; 2. p. 507; 3. p. 507; 4. According to the acid growth hypothesis discussed on page 477, acidic conditions in the cell wall brought about by auxin cause enzymes to be activated. Recall from Chapter 4 that the conformation of an enzyme is crucial to its activity and can be altered by changes in pH or temperature. Thus the acidic pH probably activates the enzymes in the cell wall by changing their shape; 5. As described on page 477, auxin brings about the insertion of new wall materials in plant cells by activating certain genes. This effect of auxin illustrates the principle that the processes of growth and cell differentiation in multicellular organisms require that particular genes be switched on and off at various stages in the organism's life cycle. The action of auxin also illustrates the concept of positive control gene activity discussed in Chapter 12. **Page 486:** 1. p. 482; 2. As described on page 485, gibberellins are similar to the steroid hormones discussed in Chapter 25 in that they can pass through the membranes of cells and turn genes on and off. Thus their action, like auxin's, illustrates the critical role of differential gene activity in the regulation of development. Hormones such as gibberellins and steroids that can pass through cell membranes are generally nonpolar, since cell membranes are much more permeable to such molecules (see Chapter 5); 3. The interactions of auxins and cytokinins described on page 485 illustrate an important general principle about the action of both plant and animal hormones. As you will learn in Chapter 25, chemical control in vertebrate animals, just as in plants, frequently involves the activity of hormones with antagonistic or cooperative effects; 4. Although plant hormones are primarily regulators of growth and development, the action of abscisic acid in triggering stomatal closing when there is a shortage of water is an exception. By helping to regulate the internal fluid environment of the plant, abscisic acid plays a major homeostatic role (see pages 485–486).

Chapter 24

Page 494: 1. p. 490; 2. Through its role in detecting light, phytochrome is involved in the timing of many hormonally controlled processes such as seed germination and the breaking of dormancy in spring (both mentioned on page 492). Such mechanisms to insure the proper timing of the plant's activities are obviously critical to its survival; 3. p. 492; 4. pp. 492–493, p. 179, 196. **Page 499:** 1. p. 494; 2. As described in Chapter 26, the yolk in a bird's egg is the major source of stored food for the embryo; thus its role is similar to that of the endosperm described on page 495; 3. Because plants are sedentary, they depend on pollinating agents to transport the male gametophyte (pollen) to the stigma of the flower. Thereafter, the process of fertilization in angiosperms depends on the growth of a pollen tube to deliver the sperm to the egg, as described on page 495. As discussed in detail in Chapter 26, terrestrial vertebrates surmount the problem of the lack of water for sperm transport by utilizing internal fertilization. The mobility of animals makes this method possible; 4. p. 497; 5. pp. 498–499; 6. p. 499, 501. **Page 504:** 1. p. 498; 2. p. 503; 3. The resumption of totipotency by fully differentiated plant cells described on page 504 is evidence that, at least in plants, the eucaryotic regulatory mechanisms described in Chapter 12 produce reversible (i.e., not necessarily permanent) patterns of gene activity; 4. A major practical benefit of the ability to produce clones is that fruits and vegetables with particularly desirable genetic qualities (such as excellent flavor, texture, or disease resistance) can be produced without resorting to sexual reproduction, which would likely result in the breaking up of the desired genetic combination. Older methods of producing clones (e.g., by planting cuttings taken from trees) have long been used in the asexual propagation of desirable varieties of some fruits, such as apples and oranges. Thus, for example, all Macintosh apple trees are a clone that can be traced back to a single original tree. The scientific benefit of clones can be illustrated by considering an experiment designed to study the effect of fertilizer on the growth of carrots. The confusing influence of genetic variability among the plants can be eliminated by using a clone of carrots divided into control (not fertilized) and experimental (fertilized) groups; then any difference in growth rate of the plants must be due to environmental factors, such as the amount of fertilizer they received (see page 504).

Chapter 25

Page 512: 1. p. 507; 2. p. 507; 3. p. 508, p. 98; 4. p. 101, pp. 508–51; 5. p. 511; 6. As explained in Chapter 4, the specificity of enzymes depends on the occurrence of a precise physical and chemical fit between the enzyme and the substrate(s) of the reaction catalyzed by the enzyme. Similarly, the ability of a cell to respond to either a hydrophilic or hydrophobic hormone hinges on the presence in the cell (either in the cell membrane or in the cytoplasm) of a receptor that can bind the hormone because of its complementarity of shape, much as the proper key fits in a lock. The principle of recognition and/or stimulation of a response by the binding of molecules of complementary shape is widespread in cellular biology. Recall that it is also the basis of the antibody-antigen interactions discussed in Chapter 21 (see page 436). **Page 521:** 1. p. 513; 2. As explained in Chapter 22, the ability of the kidney to produce a hypertonic urine has an upper limit. Therefore, the presence of a large amount of glucose as a solute in the urine necessitates the excretion of an abnormally large quantity of water, even if the urine is maximally concentrated. For that reason, a person with uncontrolled diabetes has to urinate frequently and must drink large quantities of water to replace what is lost in the urine or risk dehydration (see page 514); 3. pp. 513–514; 4. pp. 518–520; 5. Biology is a science of exceptions; therefore it is not surprising that there are exceptions to the generalization that unlike plant hormones, which regulate growth and development, animal hormones have a homeostatic function. As described on page 520, thyroid hormone plays a role in regulating many aspects of development, and in this respect is similar to most plant hormones, As another example, pituitary growth hormone also plays an important role in the control of growth and development. **Page 527:** 1. p. 524; 2. As described on page 525, in both portal systems substances are picked up by the blood in capillaries at the beginning of the system and are removed from the blood in the second capillary bed. Both systems also have an important role in homeostasis, but in different ways. In the case of the hepatic portal system (described in Chapter 22), glucose and other nutrients absorbed in the digestive tract are the substances that may be removed by the second capillary bed in the liver in order to keep their blood concentration relatively stable. In contrast, the hypothalamic-anterior pituitary portal system serves a homeostatic function by carrying chemical messengers (releasing hormones) that are critical components of the negative feedback loops governing the behavior of several important endocrine glands; 3. Goiter is usually (though not always) associated with underproduction of thyroid hormone (TH) as a result of a dietary deficiency of iodine. The abnormally low blood concentration of TH triggers a stimulatory response by the hypothalamus and anterior pituitary, as discussed on pages 524–525. The high level of TSH produced by the anterior pituitary stimulates the thyroid to enlarge in a futile effort to supply enough of its hormone and the result is a goiter; 4. p. 525; 5. The detection of light and measurement of the photoperiod by a plant involves the combination of phytochrome and the plant's biological clock, as explained in detail in Chapter 24. Like many vertebrates, many species of plants use the length of the photoperiod for the timing of reproduction (among other functions). In vertebrates, reproductive responses to photoperiod are mediated through melatonin produced by the pineal. Interestingly, in many nonmammalian vertebrates the pineal is eyelike and responds to light directly, both by generating nervous impulses and by secreting melatonin. Both comparative anatomy and the fossil record support the view that the pineal originally served as a "third eye" in ancestral vertebrates (see page 526); 6. p. 526; 7. pp. 527–528.

Chapter 26

Page 538: 1. p. 531; 2. p. 531; 3. p. 533; 4. p. 535; 5. As explained on page 537, castration (the removal of the testes) eliminates the most important source of the male hormone, testosterone, whereas vasectomy (the cutting of the vasa deferentia) does not alter hormone levels. The elimination of the sex urge (and other behavioral changes, such as lessened aggressiveness) in castrated animals is just one of the many possible illustrations of the influence of hormones on the functioning of the nervous system. By the same token, our discussion of the endocrine system has made it clear that the nervous system helps to regulate hormonal levels (e.g., via the hypothalamus); 6. As a hydrophilic molecule, a steroid hormone enters a target cell and combines with a receptor in the cytoplasm. The hormone-receptor complex then moves into the nucleus and causes a change in the pattern of gene activity in the cell. The relatively show by long-lasting responses characteristic of steroid hormones are well demonstrated by the action of sex hormones in stimulating sexual maturation (see pages 537–538). **Page 549:** 1. Page 526 in Chapter 25 describes the sequence of intersections by which day length influences the secretion of melatonin by the pineal gland, and melatonin in turn affects the production of FSH and LH by the anterior pituitary. In animals living outside the tropics, seasonal changes in day length are the most commonly used cue to trigger reproductive behavior; 2. p. 547; 3. p. 548; 4. Infection would immediately elicit the nonspecific immune response, in which macrophages and other phagocytic cells consume the invading microbes. Infection would also trigger the slower specific immune response. As mentioned on page 544, gonorrhea is a bacterial disease, so the humoral response (antibody production by the B-cell system) would be more important in fighting the infection than the cell-mediated response. Despite the body's defensive measures, gonorrhea frequently persists and spreads (especially in women), causing serious complications such as sterility. Prompt treatment with antibiotics is thus imperative; 5. As explained on page 548, a major disadvantage of the rhythm method is that it is unreliable because of the unpredictability of the time of ovulation. A second disadvantage, not unique to the rhythm method, is that it offers no protection against sexually transmitted diseases. Its advantages include its reversibility, lack of side-effects, and low cost. **Page 554:** 1. p. 548; 2. The separation of maternal and fetal blood in the placenta (described on page 552) is essential to prevent an immune response by the mother to the presence of foreign tissue, as discussed in Chapter 21. As you recall from Chapter 14, Rh disease in new-born infants is due to mixing of fetal and maternal blood that occurred during a previous pregnancy (most likely at the time of birth) and that triggered anti-Rh antibody production by the mother; 3. pp. 552–553; 4. As explained on page 554, suckling stimulates release of oxytocin from the posterior pituitary of the mother. The oxytocin stimulates both the release of milk from the mammary glands and contractions of the smooth muscle in the wall of the uterus. These contractions constrict uterine blood vessels and thus lessen the risk of excessive bleeding after delivery.

Chapter 27

Page 561: 1. pp. 557–558; 2. p. 588; 3. p. 559; 4. As mentioned on page 559, the cell divisions that flow fertilization occur very rapidly and are not accompanied by cytoplasmic growth. Thus the G_1 stage, in which protein synthesis occurs, would be shortened or eliminated; 5. The hollow shape of the blastula places all of its cells in direct (or nearly direct) contact with the external environment, thus facilitating the exchange of gases and wastes (see pages 559–560); 6. pp. 560–561. **Page 573:** 1. pp. 562–563; 2. As explained on pages 566–567, the features shared by the embryos of all vertebrates—such as the notochord and pharyngeal gill pouches—are evidence of their evolutionary relatedness. The great importance of embryological evidence in determining evolutionary relationships will be stressed in Chapters 32 and 41; 3. pp. 570–572; 4. p. 573; 5. 574. **Page 583:** 1. pp. 574–575; 2. See page 574. Temperature and pH could influence cell differentiation by affecting protein conformation, as discussed in Chapters 3 and 4. Because proteins play many roles (e.g., as enzymes and as genetic control molecules) that determine cellular structure and function, any environmentally induced changes in their conformation could possible affect cell differentiation. An example of the effect of temperature on gene expression (in this case by affecting enzyme activity) is the coat pattern of Himalayan rabbits, described in Chapter 14; 3. pp. 579–580; 4. As explained on page 581, the products of homeotic genes are proteins that bind to DNA. Recall from Chapter 12 that it is the binding of such control molecules to enhancer regions and promoter regions of DNA that activates gene transcription in eucaryotes; 5. There has been speculation (in many articles, books, and a film by Woody Allen) that the experiments described on pages 576–577 could be extended to humans. In one scenario, the nuclei of many cells of an adult person would be injected into enucleated eggs *in vitro.* The eggs would then be implanted in the uteri of a number of women, develop to term, and thus give rise to a clone of individuals genetically identical to the donor of the nuclei. In reality, it still is not possible *in any species of animal* to use nuclei from the cells of an adult to promote the complete and normal development to adulthood of enucleated eggs. In some mammals—sheep, for example—it is possible to use nuclei taken from blastula cells to produce a clone. Whether this procedure would be successful with human embryos is unknown.

Chapter 28

Page 595: 1. pp. 587–588; 2. p. 587; 3. p. 592; 4. pp. 592–593; 5. p. 594. **Page 605:** 1. pp. 597–598; 2. pp. 598–599; 3. The sodium-potassium pump is ATP-driven, as explained on page 599. The nerve impulse itself requires no expenditure of energy because Na^+ and K^+ ions are both moving down their concentration gradients during an action potential; 4. pp. 604–605. **Page 613:** 1. p. 606; 2. As described on page 607, curare blocks acetylcholine receptor sites on skeletal muscle, preventing stimulation by motor neurons. An animal poisoned by curare would die because it would be unable to breathe. (Which muscles would be affected?) In contrast to the relaxed paralysis produced by curare, an acetylcholinesterase inhibitor permits uncontrolled stimulation of muscles by acetylcholine, resulting in muscular twitches or even—if the dose is strong enough—spastic muscle paralysis; 3. p. 608; 4. Many internal organs are controlled by the antagonistic action of the sympathetic and parasympathetic nervous systems, as described on page 610. The same principle is used in the hormonal control of the blood concentrations of sugar and calcium, as well as in other physiological contexts, such as the regulation of the immune response by helper and suppressor T cells (see chapter 21); 5. p. 612.

Chapter 29

Page 623: 1. p. 617; 2. pp. 618–619; 3. pp. 619–620; 4. In both the detection of a taste (or odor) and the production of an EPSP, the binding of a chemical to a receptor protein mounted in the cell membrane causes chemically gated Na^+ channels in the membrane to open, leading to partial depolarization of the membrane. In the case of the EPSP, the stimulatory molecule is a transmitter released by a presynaptic neuron; whereas in the case of the receptor, the stimulus is a chemical in the environment. This example demonstrates the basic similarity of postsynaptic potentials and generator potentials (see pages 620–621); 5. pp. 621–622; 6. p. 622. **Page 636:** 1. p. 626; 2. pp. 628–629; 3. pp. 630–631; 4. p. 631; 5. pp. 634–636; 6. Like the evolution of camera-type eyes in both certain molluscs and vertebrates (see page 624), the development of echolocation in bats and dolphins described on page 637 is an example of convergence—the independent evolution of similar "solutions" to an environmental challenge. This phenomenon has occurred frequently in the course of evolution and is discussed in more detail in Chapter 32. **Page 645:** 1. pp. 637–638; 2. pp. 638–639; 3. p. 638; 4. p. 641; 5. p. 644; 6. p. 644; 7. pp. 644–646.

Chapter 30

Page 655: 1. pp. 649–650; 2. p. 650; 3. As explained on page 652, the weight of an animal is a function of its volume, which increases as the cube of its linear dimensions. By contrast, the load-bearing capacity of a column such as a leg bone is a function of its cross-sectional area, which increases as the square of its linear dimensions. Thus, as mentioned on page 653, supporting bones in large animals must be disproportionately thick in comparison with small animals; 4. pp. 653–654. **Page 660:** 1. The similar all-or-none responsivity of muscle fibers and nerve cells is mentioned on page 656. Recall from Chapter 28 that detection of stimulus intensity requires that the nervous system produce graded responses. One way the nervous system achieves a graded response to stimulation is by having more cells respond to a stimulus of higher intensity. The same method is found in muscle, as described on pages 658–659; 2. p. 660; 3. pp. 660–661. **Page 666:** 1. p. 660; 2. p. 660; 3. p. 660; 4. p. 661; 5. p. 664; 6. The role of calcium ions in initiating muscle contraction by triggering the formation of cross-bridges between thick and thin filaments is described on pages 664–665. Recall from Chapter 28 that an influx of Ca^{++} ions also stimulates the release of transmitter molecules from presynaptic cells at synapses and neuromuscular junctions. There is also evidence, not mentioned in Chapter 28, that a deficiency of Ca^{++} ions in the extracellular fluid lowers the threshold at which voltage-gated Na^{+} channels open in neuron and muscle fiber membranes. Thus an abnormally low tissue concentration of Ca^{++} would make neurons and muscles hyperexcitable.

Chapter 31

Page 673: 1. p. 669; 2. As explained on page 670, natural selection only acts on genetic variation expressed phenotypically. In the case of a rare recessive allele, the vast majority of individuals carry only one copy of it, which would not be expressed phenotypically and hence would be hidden from selection. In addition, new mutations from the "normal" to the lethal allele may occur. Thus an equilibrium frequency for the allele is reached that depends on the balance between the rate at which homozygotes are eliminated by natural selection and the rate at which mutation produces new lethal alleles. 3. In crossing-over, corresponding sections of chromatids of a pair of homologous chromosomes are exchanged. The result is an enormous increase in the genetic diversity of the gametes created by meiosis, which supplies more of evolution's raw material: variation; 4. pp. 670–672; 5. p. 672; 6. p. 673. **Page 682:** 1. p. 674; 2. p. 675; 3. As explained on pages 676–678 and 680, genetic shift is an important evolutionary factor in small populations and tends to result in the loss of genetic diversity, as random fluctuations in the frequencies of alleles result in some of them being lost, while others are "fixed" as the only allele present. Chapter 32 addresses the significance for evolution of a "bottleneck," in which the population size of a species is drastically reduced; 4. The effect of inbreeding is to increase the frequency of homozygotes in the population. In the case of self-fertilization, all of the descendants of a homozygous parent would also be homozygous. In addition, one-half of the offspring of a heterozygous parent would be expected to be homozygous and thereafter would produce only homozygous descendants. Even though its allelic frequencies would be unchanged, the population would no longer be at Hardy-Weinberg equilibrium because of the rapid decline in the frequency of heterozygotes and the corresponding increase in homozygotes. Less extreme forms of inbreeding have similar effects on genotype frequencies, albeit not as rapid. The increased risk of producing offspring homozygous for a rare harmful recessive allele as a result of inbreeding was explained in Chapter 14. In addition, as you will learn on page 686, there are many cases known in which the heterozygous condition is adaptively superior to either the homozygous dominant or the homozygous recessive genotype; 5. Indiscriminate use of antibiotics unnecessarily increases the risk that antibiotic-resistant strains of pathogenic bacteria will evolve. The evolution of penicillin-resistant *Staphylococcus,* described on pages 681–683, is an excellent example of how strong selection pressure can quickly produce resistant varieties of bacteria. This topic is discussed again in greater detail in Chapter 37. **Page 699:** 1. The breeding of new varieties of domestic plants (or animals) with superior productivity, disease resistance, etc., is a type of artificial directional selection. As explained on pages 683–684, the new phenotypes that appear during this type of selection result primarily from the formation of new combinations of old genes rather than from new mutations. Hence success in artificial selection depends on the existence of a large store of genetic variability in the population; 2. p. 686; 3. p. 688; 4. As defined on page 695, mutualism is a symbiotic relationship beneficial to both participants. Recall from Chapter 16 that fungi and plant roots form associations called mycorrhizae, which aid mineral uptake by the plants and also benefit the fungi. Another mutualistic relationship is that of legumes and the nitrogen-fixing bacteria housed in nodules on their roots; 5. Since most varieties of domesticated plants and animals depend on human support for their survival and reproduction, and since these organisms are clearly beneficial to us, the relationship can be called mutualistic (pp. 695–696). For instance, there would be many fewer tomatoes if humans did not plant them. Many animal rights advocates argue, however, that many relationships of human beings to our domestic animals are beneficial only to our species; 6. The relationship of rabbits and myxomatosis is a classic case of host-parasite coevolution, as described on pages 697–699. By 1958 the mortality rate from the virus had declined to 54 percent. Studies showed not only that the surviving rabbits had become resistant to the virus, but the virus itself had become less deadly. Since the virus is transmitted by mosquitoes that bite only living rabbits, there was selection against the most virulent viral strains.

Chapter 32

Page 714: 1. pp. 706–707; 2. Since fitness is defined as an individual's probable genetic contribution to succeeding generations, a sterile organism such as a mule has zero fitness (see page 711); 3. Recall from Chapter 9 that homologous chromosomes are paired during prophase and metaphase of meiosis I (see pages 190–193); 4. p. 712; 5. p. 714; 6. The joining of two haploid sets of chromosomes from the parental species would give the hybrid 16 (7 + 9) chromosomes. For each chromosome to have a homologue with which to pair during meiosis, this number would have to be doubled. Hence the allopolyploid hybrid would have 32 chromosomes (see pages 712–713). **Page 720:** 1. pp. 714–716; 2. As defined on page 718, the niche of a species is its functional role in an ecosystem. The adaptive radiation of Darwin's finches produced many species, each adapted to a different niche, as shown by their varied feeding habits. The niche is a fundamental concept in ecology and is discussed further in Chapter 33; 3. p. 719; 4. p. 719; 5. p. 721. **Page 731:** 1. p. 721, 725; 2. Frequently, the best evidence for concluding that two structures in different species are homologous is determining that they develop from the same primordial structure in the embryo. If the structures are very different in appearance or function in the fully developed organism, studying their divergent embryological development may yield phylogenetic insights. By this method it was discovered, for example, that two of the mammalian middle-ear bones (the hammer and the anvil) evolved from jaw bones of our reptilian ancestors. As for the panda's "thumb," both embryology and comparative anatomy indicate that it is an extra digit derived from a wrist bone; hence it is only analogous to the human thumb (see pages 727–728); 3. As described on pages 730–731, the similarities of amino acid sequences in proteins and nucleotide sequences in DNA or ribosomal RNA are measures of the evolutionary relatedness of two species. Studies of the DNA of the giant and lesser pandas have convinced systematists that they are not closely related. The giant panda is a bear with recently acquired specializations (such as its opposable "thumb") related to its diet of bamboo. The lesser panda is more closely allied to raccoons than to bears; 4. p. 731.

Chapter 33

Page 746: 1. pp. 737–738; 2. pp. 739–740; 3. p. 740; 4. China's birth rate (*b*) is 22 per 1000, or 0.022; its death rate (*d*) is 7 per 1000, or 0.007. Since *r* (growth rate) = *b-d,* China's population growth rate is 0.015, or 1.5 percent per year. At this rate of growth, its population will double in about 47 years (see pages 740–741); 5. p. 745; 6. As defined on pages 740–741, exponential growth occurs when a population increases at a constant rate (*r*) per unit of time. One property of such growth is that the doubling time (i.e., the time required for a 100 percent increase in the population) is also a constant. Human population growth between 1650 and 1975 was thus, in a sense, *faster* than exponential, since its doubling time was decreasing. There is evidence that in recent years the rate of growth has begun to decline. An extrapolation of this trend into the future is the basis for the estimate on page 744 that the population of the world will stabilize by 2030. It should be noted, however, that demographic projections are not always accurate. **Page 753:** 1. Whereas interspecific competition can lead to character displacement and specialization (i.e., a "narrowing" of the niche of a species) so as to minimize competition, the tendency of intraspecific competition is to "broaden" the niche of a species, again to minimize competition. In the case of Darwin's finches, for example, a species typically exhibits a greater range of bill size when it occurs alone on an island than when it is sympatric with a close competitor (see pages 749–750); 2. pp. 749–750; 3. pp. 749–750; 4. pp. 751–752; 5. Recall from Chapter 3 that cellulose is a major constituent of plant cell walls and that it is an extremely stable polysaccharide. In view of its abundance in ecosystems and its resistance to rapid and complete breakdown to carbon dioxide and water, it is to be expected that it would be the most important contributor to humus. The contributions of humus to soil development are described on page 753. **Page 758:** 1. Lichens were described in Chapter 31 as an example of a particularly close mutualistic relationship between an alga and a fungus. The fungus benefits from the photosynthetic activity of the alga, and the alga benefits from the ability of the fungus to absorb and retain moisture. These properties and the others mentioned on page 754 make lichens well adapted to their bare-rock habitat; 2. As nonvascular plants, mosses have no true roots or stems and are severely limited in the size they can attain. They are thus able to compete successfully with vascular plants in conditions in which there is little or no soil in which roots can grow. As soil thickness increases, vascular plants, particularly small annuals, have an advantage over moss. Later, as the soil layer becomes even deeper, most annual plants are likely to be outcompeted by perennials that can grow more extensive roots and stems containing secondary tissue (e.g. the secondary xylem that constitutes the wood of shrubs and trees) (see pages 755–756); 3. As discussed in Chapter 18, water is a poor source of oxygen compared to air. Plant roots ordinarily obtain their oxygen from the air found in pores in the soil not filled with water. Plants adapted to grow in water or in water-logged soil must either have roots that are unusually tolerant of oxygen-deprivation, or special transport mechanisms to supply oxygen to the roots, or both. The "knees" of bald cypress trees that grow in southern swamps are a good example of such an adaptation. They are actually extensions of roots that project out of the water and absorb oxygen from the air. Their porous structure allows oxygen to diffuse relatively freely to the submerged parts of the roots (see page 375); 4. p. 758; 5. pp. 758–760.

Chapter 34

Page 772: 1. p. 763; 2. p. 764; 3. Eating a purely vegetarian diet makes the energy stored in producers directly available to human beings, whereas eating meat requires that the energy pass through at least one extra trophic level before reaching the human consumer. Each additional level results in a substantial loss in available energy, for the reasons explained on page 764. For example, it has been estimated that a person who ate nothing but beef cattle raised on a diet of grain and soybeans would indirectly consume enough of those plant products to support 22 vegetarians. The domestic herbivores that supply us with meat differ greatly in the energetic efficiency with which they convert their food into edible flesh. As you might guess from checking meat prices at the supermarket, chickens are several times more efficient than beef cattle. No herbivore, however, can violate the Second Law of Thermodynamics and be 100 percent efficient. 4. As explained on page 766, productivity pyramids reflect the *rate* at which energy is bound into new organic matter, whereas biomass pyramids indicate the weight of living material *present at a particular time.* Thus, for example, in some aquatic ecosystems in which microscopic, short-lived algae are the producers and long-lived fish are the top predators, the biomass of the fish may exceed that of the algae, even though the productivity of the algal trophic level is perhaps a thousand times greater than that of the predatory fish; 5. pp. 768–773. **Page 785:** 1. pp. 772–773; 2. pp. 773–774; 3. As described on page 773, nitrogen is primarily transported in organic compounds such as amino acids. As you learned in Chapter 19, most upward movement of organic molecules is in phloem. There is now good evidence, however, that in some plants, especially trees, organic nitrogen move through the xylem; 4. pp. 774–775; 5. pp. 778–779; 6. pp. 779–781; 7. If the earth's axis were not tilted, there would be no seasons. Consequently, organisms would never have evolved the myriad structural, physiological, and behavioral adaptations that enable them to cope with the cyclic changes in temperature, precipitation, etc., associated with seasonality. Since seasonal variation in solar radiation increases from the equator to the poles, organisms living far from the equator would generally be affected more strongly than tropical organisms (though seasonal changes also occur in the tropics) (see pages 781–783); **Page 799:** 1. pp. 787–788; 2. p. 788; 3. p. 793; 4. According to the theory of continental drift described on page 794, India moved northward after Gondwanaland fragmented. As the Indian land mass approached and finally collided with Asia about 40 million years ago, parts of the sea floor containing marine fossils were crumpled and elevated as the Himalayas formed. More generally, the common occurrence of such fossils on mountains and in areas far from the sea is striking evidence of the immensely powerful geological forces that have molded the earth over its long history and profoundly influenced the evolution and present distribution of life forms on the planet; 5. pp. 796–797; 6. pp. 797–798.

Chapter 35

Page 814: 1. p. 805; 2. As explained in Chapter 31, many characteristics of organisms represent the evolutionary result of conflicting selection pressures. In this case, we must ask whether the reproductive success of gulls depends more on their possessing a bill that: (1) enables them to collect food for their young efficiently, or (2) is maximally effective as a releaser of the chick's food-begging behavior. It seems reasonable to assume that the first choice is correct. A similar line of reasoning could probably account for the existence of other supernormal stimuli (see pages 805–807); 3. As noted on pages 807–808, fixed-action patterns are essentially motor programs. The latter term, however, is more inclusive than the former, which refers only to *innate* responses. As discussed on pages 807–808, there are many examples of learned motor programs in human beings and other animals; 4. The small size of insects (which results from the constraints of their tracheal system and exoskeleton) limits the total number of neurons in their nervous system and thus certainly restricts their learning potential. In addition, recall from Chapter 28 that the nervous systems of arthropods such as insects show far less centralization and cephalization than those of vertebrates; thus even the largest insects, though bigger than some vertebrates, have a much smaller brain.

Their small size also means that insects tend to be short-lived and produce many offspring, which receive little or no parental care. (These are characteristics of *r*-selected species, as defined in Chapter 33.) Thus they have neither the capacity nor the opportunity for extensive learning (see page 809); 5. p. 811. **Page 826:** 1. pp. 815–816; 2. pp. 816–817; 3. pp. 816–820; 4. p. 822; 5. pp. 822–826. **Page 835:** 1. pp. 827–828; 2. pp. 829–830; 3. On average, Haldane's siblings share 50 percent of their genes with him, whereas his cousins each are 12.5 percent. Thus, if we assume that the reproductive potential of his siblings and cousins is the same as his own, Haldane's probable genetic contribution to succeeding generations (i.e., his fitness) is the same whether he reproduces, or two of his siblings, or eight of his first cousins. This example demonstrates the principle of kin selection, discussed on page 830. To what extent such sociobiological concepts as kin selection and parental investment can (or should) be involved to explain *human* ethical and social behavior such as self-sacrifice and gender roles is a matter of lively controversy; 4. pp. 834–835.

Chapter 36

Page 847: 1. p. 839; 2. p. 842; 3. pp. 842–844; 4. p. 845; 5. pp. 847–848. **Page 851:** 1. pp. 847–848; 2. p. 848; 3. p. 848; 4. p. 849; 5. Recall from Chapter 8 that ethyl alcohol fermentation produces CO_2 as a byproduct. The rate of abiotic synthesis probably declined because forming complex organic molecules from this gas is much less likely than from methane or hydrogen cyanide, which were the two other sources of carbon probably present in the early atmosphere (see page 850); 6. p. 850; 7. p. 850. **Page 856:** 1. p. 852; 2. p. 853; 3. pp. 853–843; 4. p. 854.

Chapter 37

Page 867: 1. pp. 862–863; 2. p. 864; 3. p. 864; 4. You learned in Chapter 11 that transcription in all cellular organisms produces RNA from a DNA template. As explained on pages 864–865, RNA viruses carry out RNA→RNA transcription, and retroviruses perform reverse transcription (RNA→DNA). Both of these processes require special enzymes that are coded for by the viral genome; 5. Recall from Chapter 13 that complementary DNA (cDNA) is sometimes used to clone a gene. cDNA is made from a mRNA template in a reaction catalyzed by reverse transcriptase. The role of this enzyme in retroviral replication is explained on pages 864–866; 6. pp. 867–868. **Page 873:** 1. p. 869; 2. p. 869; 3. p. 872; 4. pp. 872–873; 5. pp. 873–874. **Page 877:** 1. p. 874; 2. p. 874; 3. p. 875; 4. pp. 875–876; 5. pp. 876–877.

Chapter 38

Page 884: 1. As mentioned on page 881, protists such as the protozoans require no special mechanisms for gas exchange or for the disposal of nitrogenous wastes. Thus their small size and high surface-to-volume ratio make these homeostatic functions relatively simple. On the other hand, these same features make it comparatively difficult for these unicellular organisms to maintain constant internal conditions in the face of environmental fluctuations (for example, temperature fluctuations). Like some eubacteria discussed in Chapter 37 that form resistant endospores, many protists respond to unfavorable conditions by becoming inactive cysts; 2. p. 882; 3. The relationship between malaria and the allele for sickle-cell anemia demonstrates the importance of infectious disease as a selective pressure on human evolution. Recall from Chapters 14 and 31 that individuals who are heterozygous for the sickle-cell allele have better resistance to malaria than those who are homozygous for the normal allele. In parts of Africa and other tropical regions, malaria has been such an important cause of death that heterozygotes are better adapted than homozygotes for their allele, and the sickle-cell allele occurs with an unusually high frequency—an excellent illustration of the principle that the environment determines the adaptiveness of a characteristic (see page 884); 4. p. 885. **Page 892:** 1. pp. 886–887; 2. p. 888; 3. p. 889; 4. p. 889; 5. By definition, gametes are haploid cells whose function is to unite to form a diploid zygote. Thus gametes can only be the products of meiosis in animal-like life cycles in which they constitute the only haploid stage. In life cycles with a more prolonged haploid stage, the products of meiosis are spores, which in turn produce by mitosis either single-celled haploid organisms or a multicellular haploid structure called a gametophyte (so named because *it* produces the gametes—by mitosis) (see page 890); 6. pp. 891–892.

Chapter 39

Page 901: 1. As described on pages 896–897, the zygote of *Chlamydomonas* has a thick wall that enables it to survive in a dormant state, protected from unfavorable environmental conditions. In this respect, the zygote is similar to the endosperms and cysts formed by eubacteria and protists. The prevalence of dormancy in these organisms is evidence of the difficulty that small size imposes on the maintenance of homeostasis when the external environment becomes harsh; 2. p. 897; 3. p. 897; 4. pp. 899–900. **Page 912:** 1. p. 904; 2. p. 905; 3. Both endosperm and yolk consist of nutritive material used to support development. But whereas yolk is a component of the zygote itself, endosperm is a multicellular triploid tissue genetically distinct from the plant embryo. And whereas yolk supports embryonic development of an animal beginning with the first cell division of the zygote, endosperm primarily supplies energy to the embryo during seed germination (though in many cases the endosperm has been absorbed by the cotyledon(s) during seed maturation); 4. pp. 911–912.

Chapter 40

Page 924: 1. pp. 917–918; 2. p. 917; 3. p. 917; 4. pp. 917–918; 5. As explained in Chapter 9, the major advantage of sexual reproduction is that it produces variable offspring, some of which may be better adapted than their parents to the new and hostile conditions that frequently arise in an unstable environment. The association of sexual reproduction with the ability to survive in an adverse or changeable environment is demonstrated by organisms, like *Rhizopus,* that are capable of both sexual and asexual reproduction but use the former method primarily when conditions are unfavorable (see pages 918–920); 6. pp. 920–922.

Chapter 41

Page 937: 1. p. 927; 2. pp. 930–931; 3. pp. 931–932; 4. pp. 932–933; 5. Tapeworms exhibit two structural features that increase their absorptive surface area. First, their flattened shape, inherited from their free-living ancestors, provides a high surface-to-volume ratio compares to a more cylindrical form. Second, as mentioned on pages 935–936, tapeworms have also evolved microvilli, a feature of the digestive tract of humans and other mammals discussed in Chapter 17, and an excellent example of convergent evolution. Since there is little oxygen present in the lumen of the intestinal tract, it is not surprising that tapeworms primarily depend on anaerobic respiration (glycolosis and fermentation) to meet their energy needs (see Chapter 8). **Page 954:** 1. p. 937; 2. pp. 943–944; 3. Recall from Chapter 39 that the cell wall of fungi is composed of chitin, the same polysaccharide found in the exoskeleton of arthropods. Since there is no common ancestor of these groups that synthesizes chitin, this is an example of biochemical convergence (see page 947); 4. p. 948.

Chapter 42

Page 968: 1. pp. 957–958; 2. pp. 961–962; 3. p. 965; 4. p. 967. **Page 976:** 1. Largely because of their superior adaptations for reproduction on land, both seed plants and reptiles were able to occupy a greater variety of terrestrial habitats and ecological niches than the ferns (and other nonseed plants) and amphibians. As explained in Chapter 39, ferns depend on water for sperm transport, and the young sporophyte plants grow out of the archegonium of a small, fragile gametophyte that can only survive in moist conditions. Amphibians also require water for sperm transport. Their eggs are small and fishlike, and embryonic development must take place in water or in very moist places on land. In contrast, reptiles use internal fertilization and lay amniotic shelled eggs that protect the embryo from dehydration. Similar adaptations evolved in seed plants, in which motile sperm no longer occur, and the sporophyte embryo develops in the ovule (which becomes the desiccation-resistant seed) (see pages 969–971); 2. Recall from Chapter 22 that uric acid is almost insoluble in water. This property enables reptiles to excrete it as a semisolid paste from which nearly all the water has been excreted and reabsorbed into the blood. Since an embryo inside a water-permeable egg must produce an insoluble waste product to avoid poisoning itself, it is doubtful that the amniotic egg could have evolved if reptiles had been unable to excrete uric acid (see page 969); 3. pp. 973–975; 4. pp. 975–977.

Chapter 43

Page 986: 1. pp. 979–981; 2. Because you are a scholar of both biology and art, you would save your money. Since only New World monkeys have prehensile tails, and since a French artist could not possibly have seen a New World monkey before the discovery of America in 1492, you would know that a painting from the 1110s that included a New World monkey is undoubtedly a forgery (see pages 981–983); 3. p. 984; 4. p. 985; 5. pp. 985–986. **Page 993:** 1. pp. 986–987; 2. p. 989; 3. p. 989; 4. pp. 993–994; 5. pp. 990–993.

APPENDIX 9
CREDITS

i ©Will and Demi McIntyre. **iii** © Leonard Lee Rue III/Photo Researchers, Inc. **(inset)** © Will and Demi McIntyre.

Table of Contents (Brief) **vi (Part One)** © Flip Nicklin/Minden Pictures. **(Part Two)** Francis Leroy/Biocosmos/Science Photo Library-Photo Researchers, Inc. **(Part Three)** John Visser/Bruce Coleman, Inc. **vii** Ciabou Hany/Nature Production, Tokyo. **(Part Four)** Howard Hall/HHP. **(Part Five)** © Frans Lanting/Minden Pictures.

Contents **iv** © Jim Brandenburg/Minden Pictures **x** Nasa Science Photo Library **xii** © Dwight R. Kuhn **xiv** Runk/Schoenberger, Grant Heilman Photography, Inc. **xv** Photo Researchers, Inc. **xvii** © David B. Fleetham, Innerspace Visions, Miami, Florida **xviii** © Professors P. M. Motta and S. Correr/Science Photo Library **xxi** © Daniel J. Cox/DJC & Associates **xxiii** © Flip Nicklin/Minden Pictures **xxiv** Bruce Coleman, Inc. New York, NY **xxvi** © Daniel J. Cox/DJC & Associates **xxvii** © A. B. Dowsett/Science Photo Library **xviii** © Gary Retherford/Science Source **xxix** © Howard Hall **xxx** © Frans Lanting/Minden Pictures

Chapter 1 **CO** Hans Reinhard/Bruce Coleman, Inc. **1.1** Courtesy NASA. **1.4** AIP/Emilio Segre Visual Archives. **1.5** © Susan McCartney/Photo Researchers. **1.6** SCALA/Art Resource, New York. **1.7** The Master and Fellows of Trinity College, Cambridge. **1.8** The Royal Society, London. **1.9** Rijksmuseum, Amsterdam. **1.10** Musee D'Orsay, © RMN. **1.11** Courtesy Department of Library Services, American Museum of Natural History, New York. **1.12a** Tass/Sovfoto. **1.12b** © 1978 F. Gohier/Photo Researchers, Inc. **1.13** © Bibl. Museum Hist. Nat., Paris. **1.15** The New York Public Library, Astor, Tilden, and Lenox Foundations. **1.16** © Tim Davis/Photo Researchers, Inc. **1.17 (top)** © Lennart Nilsson, from L. Nilsson, *Behold Man,* English translation © 1974, Albert Bonniers Alba, Stockholm, and Little, Brown, and Co. (Canada) Ltd. **1.17 (bottom)** © Lennart Nilsson from L. Nilsson, *A Child is Born,* Dell Publishing, New York, 1986. **1.18** (clockwise from left) **a** © Stephen Krasemann/Photo Researchers, Inc. **1.18b-e** © Reynolds Stock Photos. **1.19** Courtesy National Portrait Gallery, London. **1.20** (clockwise from left) **a** H. W. Jannasch and C. O. Wirsen, *Biological Science,* vol. 29, 1979; copyright © 1979 by the American Institute of Biological Science. **1.20b** M. I. Walker/Science Source-Photo Researchers, Inc. **1.20c** © G. R. Roberts **1.20d** © M. P. Kahl, 1972/Photo Researchers, Inc. **1.20e** Gene Ahrens/Bruce Coleman, Inc. **1.20f** © Jeff Rotman. **1.20g** CNRI/SPL/Photo Researchers, Inc. **1.21 (top)** © H. Chaumeton/Nature. **1.21 (bottom)** Elliot Scientific Corp. **1.22a** © Fred Hossler/Visuals Unlimited. **1.22b** © T. E. Adams/Visuals Unlimited. **1.23 (left)** From W. J. Jones, J. A. Leigh, F. Mayer, C. R. Woese, and R. S. Wolfe, *Arch. Microbiol.,* vol. 136, 1983, copyright 1983 by Springer-Verlag. **1.23 (right)** Courtesy Woods Hole Oceanographic Institution. **1.24** Ed Degginger/Bruce Coleman. **1.25** © H. Chaumeton/Nature **1.26** © H. Chaumeton/Nature. **1.27** Courtesy E. V. Grave. **1.28** Courtesy E. V. Grave. **1.29** © H. Chaumeton/Nature. **1.30** Adrian Davis/Bruce Coleman, Inc. **1.31** © Heather Angel. **1.32a** © Heather Angel. **1.32b** © 1965 Winton Patnode/Photo Researchers, Inc. **1.33a** Robert Carr/Bruce Coleman, Inc. **1.33b** Jane Burton/Bruce Coleman, Inc. **134a** W. H. Amos/Bruce Coleman, Inc. **1.34b** Courtesy D. G. Allen. **1.35** R. N. Mariscal/Bruce Coleman, Inc. **1.36ab** Oxford Scientific Films. **1.37** CNRI/Science Photo Library-Photo Researchers, Inc. **1.38** T. E. Adams/Visuals Unlimited. **1.39a** Courtesy J. H. Carmichael, Jr. **1.39b** © Fred Bavendam/Peter Arnold, Inc. **1.40** Oxford Scientific Films/Bruce Coleman, Inc. **1.41** E. R. Degginger/Bruce Coleman, Inc. **1.42a** © G. R. Roberts. **1.42b** © Harmon's. **1.42c** © H. Chaumeton/Nature. **1.43** © Jeffrey L. Rotman/Peter Arnold, Inc. **1.44a** Howard Hall/HHP. **1.44b** Ferrero/Nature. **1.44c** © 1991 M. H. Sharp/Photo Researchers, Inc. **1.44d** C. O. Rentmeester/Life Magazine © 1970 Time Inc.

Part One Courtesy Dr. Maria Burgal, Instituto de Investigaciones Citologicas de la Fundacion Valenciana de Investigaciones Biomedicas, Valencia, Spain. **(inset)** © Will and Demi McIntyre.

Chapter 2 **CO** © 1985 Stephen P. Parker/Photo Researchers, Inc. **2.2** © SIU/Visuals Unlimited. **2.10b** © Dwight Kuhn. **2.15** © Ken Graham/Bruce Coleman, Inc. **2.16** Courtesy Carl and Ann Purcell. **2.21** Courtesy USDA Forest Service. **2.22** © Nuridsany et Perennou/Photo Researchers, Inc. **2.24b** John Shaw/Tom Stack & Associates. **2.25** © William Curtsinger/Photo Researchers, Inc. **2.26** Courtesy NASA.

Chapter 3 **CO** Tom McHugh/Photo Researchers, Inc. **Table 3.3b** Adapted by permission from *The Structure and Action of Proteins* by Richard E. Dickerson and Irving Geis, W. A. Benjamin, Inc., Menlo Park, Calif., Publisher; copyright © 1969 by Dickerson and Geis. **3.9b** © Dwight R. Kuhn, 1980. **3.12 and 3.13** From R. G. Kessel and R. H. Kardon, *Tissues and Organs: A Text-Atlas of Scanning Electron Microscopy,* W. H. Freeman, San Francisco, copyright © 1979. **3.24** © Leonard Lee Rue III/Bruce Coleman, Inc. **3.25a** Adapted by permission from *The Structure and Action of Proteins* by Richard E. Dickerson and Irving Geis, W. A. Benjamin, Inc., Menlo Park, Calif., Publisher; copyright © 1969 by Dickerson and Gies. **3.27** Adapted by permission from *The Structure and Action of Proteins* by Richard E. Dickerson and Irving Geis, W. A. Benjamin, Inc., Menlo Park, Calif., Publisher; copyright © 1969 by Dickerson and Geis. **3.28** Adapted by permission from *The Structure and Action of Proteins* by Richard E. Dickerson and Irving Geis, W. A. Benjamin, Inc., Menlo Park, Calif., Publisher; copyright © 1969 by Dickerson and Geis.

Chapter 4 **CO** © Frans Lanting/Minden Pictures. **4.2** R. Clark and M. Goff/Photo Researchers, Inc.

Chapter 5 **CO** © Biology Media/Science Source-Photo Researchers, Inc. **5.2** © Groskinsky. **5:Box e-g** Courtesy J. D. Pickett-Heaps, University of Melbourne. **5:Box h** Courtesy B. Michel/IBM Research Division, Zurich. From M. Amrein, *Science,* 240:515, 1988. Courtesy American Association for the Advancement of Science

(AAAS) . **5.7** © David M. Phillips/Visuals Unlimited. **5.8** © 1992 Tul A. De Roy/Bruce Coleman, Inc. **5.16b** Courtesy Dorothy F. Bainton, University of California, San Francisco. **5.17b** J. Ross, J. Olmstead, and J. Rosenbaum, *Tissue and Cell,* vol. 7, 1975. **5.18a-d** M. M. Perry and A. B. Gilbert, *J. Cell Sci.,* vol. 39, 1979, by copyright permission of the Rockefeller University Press. **5.21** Courtesy Eva Frei and R. D. Preston, University of Leeds. **5.22** © Biophoto Associates/Science Source-Photo Researchers, Inc. **5.23** W. Cheng, International Paper Company. **5.24b** Courtesy W. G. Whaley et al., *J. Biophys., Biochem., Cytol.* (now *J. Cell Biol.*), vol. 5, 1959, by copyright permission of the Rockefeller University Press. **5.27b** H. Latta, W. Johnson, and T. Stanley, *J. Ultrastruct. Res.,* vol. 51, 1975.

CHAPTER 6 **CO** Courtesy J. V. Small and J. Rinnerthaler, Austrian Academy of Sciences. **6.1** © Biophoto Associates/Science Source-Photo Researchers, Inc. **6.2** © London Scientific Films/Oxford Scientific Films. **6.3a** Courtesy Barbara Hamkalo and J. B. Rattner, University of California, Irvine. **6.3b** Courtesy Victoria Foe. **6.5** W. G. Whaley, H. H. Mollenbauer, and J. H. Leech, *Am. J. Bot.,* vol. 47, 1960. **6.6a** R. L. Roberts, R. G. Kessel, and H. N. Tung, *Freeze Fracture Images of Cells and Tissues,* Oxford University Press, 1991. **6.6b** Courtesy Nigel Unwin, Stanford University School of Medicine. **6.7** Courtesy K. R. Porter, University of Colorado. **6.8 (left)** Courtesy D. S. Friend, University of California, San Francisco. **6:Box g** Courtesy Daniel Branton, Harvard University. **6.10** D. W. Fawcett/Visuals Unlimited. **6.12** Courtesy D. S. Friend, University of California, San Francisco. **6.14** S. E. Frederick and E. H. Newcomb, *J. Cell Biol.,* vol. 43, 1969. **6.15** © M. I. Walker, Photo Researchers, Inc. **6.17** Courtesy K. R. Porter, University of Colorado. **6.18 (top)** W. P. Wergin, courtesy E. H. Newcomb, University of Wisconsin. **6.18 (left)** M. C. Ledbetter/Photo Researchers, Inc. **6.19** Courtesy M. C. Ledbetter, Brookhaven National Laboratory. **6.20b** Courtesy Dr. Maria Burgal, Instituto de Investigaciones Citologicas de la Fundacion Valenciana de Investigaciones Biomedicas, Valencia, Spain. **6.23** © Lennart Nilsson/Boehringer Ingelheim International. **6.24b** U. Aebi, University of Basel. **6.26 (left)** H. Kim, L. I. Binder, and J. L. Rosenbaum, *J. Cell Biol.,* vol. 80, 1979, by copyright permission of the Rockefeller University Press. **6.26 (right)** Courtesy D. W. Fawcett, Harvard Medical School. **6.27** M. McGill, D. P. Highfield, T. M. Monahan, and B. R. Brinkley, *J. Ultrastruct. Res.,* vol. 57, 1976. **6.29a** Courtesy R. W. Linck, Harvard Medical School, and D. T. Woodrum. **6.30a** Courtesy E. R. Dirksen, University of California, Los Angeles. **6.30b** C. J. Brokaw, *Science,* vol. 178, 1972; © 1972 by AAAS. **6.31** K. Roberts, John Innes Institute, Norwich, England; from B. Alberts et al., *Molecular Biology of the Cell,* Garland Press, New York, 1983. **6.34** (clockwise from upper left) **a** Courtesy N. B. Gilula, Baylor College of Medicine. **6.34b** Courtesy A. Ryter, Institut Pasteur, Paris. **6.34c** Courtesy N. B. Gilula, Baylor College of Medicine. **6.34d** © D. W. Fawcett/Visuals Unlimited. **6.34e** Courtesy K. R. Porter, University of Colorado. **6.34f** M. McGill, D. P. Highfield, T. M. Monahan, and B. R. Brinkley, *J. Ultrastruct. Res.,* vol. 57, 1976. **6.34g** Courtesy A. B. Novikoff, Albert Einstein College of Medicine. **6.34h** Courtesy N. B. Gilula, Baylor College of Medicine. **6.35** (clockwise from upper left) **a** Courtesy M. C. Ledbetter, Brookhaven National Laboratory. **6.35b** Courtesy W. G. Whaley et al., *J. Biophys., Biochem., Cytol.,* (now *J. Cell Biol.,* vol. 5, 1959, by copyright permission of the Rockefeller University Press. **6.35c-g** Courtesy M. C. Ledbetter, Brookhaven National Laboratory. **6.36** Courtesy A. Ryter, Institut Pasteur, Paris. **6.37** Courtesy J. M. Whatley, Oxford University, London. **6.38** Courtesy J. Griffith, School of Medicine, University of North Carolina, Chapel Hill. **6.39** Dr. Tony Brain/Science Photo Library-Photo Researchers, Inc.

CHAPTER 7 **CO** George Bernard/Science Photo Library-Photo Researchers, Inc. **7.1** Courtesy Karl O. Stetter, Universität Regensburg. **7.3** © Martin Colbeck/Oxford Scientific Films. **7.6 (bottom)** Courtesy J. H. Troughton, Dept. of Scientific and Industrial Research, Wellington, New Zealand. **7.7 (left)** W. P. Wergin, Courtesy E. H. Newcomb, University of Wisconsin. **7.10** © Runk/Schoenberger/Grant Heilman Photography, Inc. **7:Box b** Adapted from Bazzar, F. A., and Fajer, E. D., *Sci. Am.,* Jan. 1992; copyright © 1992 by Scientific American, Inc.; all rights reserved.

CHAPTER 8 **CO** © Frans Lanting/Minden Pictures. **8.4 (left)** Courtesy Andree Abecasis. **8.4 (right)** Courtesy Bob Burch. **8.5a** © K. R. Porter, D. W. Fawcett/Visuals Unlimited. **8.5c** H. Fernandez-Moran, Courtesy E. Valdivia, University of Wisconsin. From Fernandez-Moran et al., *J. Cell Biol.,* 22:63-100, 1964. Reproduced by copyright permission of the Rockefeller University Press. **8.15** Courtesy David G. Campbell. **8.16 (top)** Courtesy Fred Bruemmer. **8.16 (bottom)** Hans Reinhard/Bruce Coleman, Inc.

PART TWO Jane Burton/Bruce Coleman, Inc. **(inset)** © Will and Demi McIntyre.

CHAPTER 9 **CO** CNRI/Science Photo Library-Photo Researchers, Inc. **9.3** Courtesy R. G. E. Murray, University of Western Ontario. **9.4f** Micrograph by W. Engler, Courtesy of G. F. Bahr, *Fed. Proc., Fed. Am. Soc. Exp. Biol.,* vol. 34, 1975. **9.4g** M. P. Marsden and U. K. Laemmli; *Cell,* 17:849-58, 1979, used by permission of MIT Press, Cambridge, Mass. **9.5a** Courtesy M. W. Shaw, University of Michigan, Ann Arbor. **9.6** Courtesy Richard H. Gross. **9.7** Courtesy James E. Cleaver. **9.11a-f** Courtesy A. S. Bajer, University of Oregon. **9.13a** From H. W. Beams and R. G. Kessel, *Am. Sci.,* vol. 64, 1976; reprinted by permission of *American Scientist,* Journal of Sigma Xi, the Scientific Research Society. **9.13b** H. W. Beams and R. G. Kessel, *Am. Sci.,* vol. 64, 1976; reprinted by permission of *American Scientist,* Journal of the Sigma Xi, the Scientific Research Society. **9.15 (top left, and a-d)** Courtesy A .S. Bajer, University of Oregon. **9.16** From W. G. Whaley et al., *Am. J. Bot.,* vol. 47, 1960. **9.17** S.I.U./Bruce Coleman, Inc. **9.20a** From D. von Wettstein, *Proc. Nat'l. Acad. Sci. USA,* vol. 68, 1961. **9.22** Courtesy James Kezer, University of Oregon. **9.24a-h** Courtesy A. S. Bajer, University of Oregon. **9.26a** Courtesy Ed Reschke **9.27** Professors P. M. Motta and S. Makabe, Science Photo Library/Photo Researchers, Inc. **9.29** © Lennart Nilsson, from L. Nilsson, *Behold Man,* English translation © 1974, Albert Bonniers Alba, Stockholm, and Little, Brown, and Co. (Canada) Ltd. **9.30 (right)** © Oxford Scientific Films.

CHAPTER 10 **CO** © Will and Demi McIntyre/Photo Researchers, Inc. **10.2 (top)** Courtesy M. Wurtz, University of Basel. **10.3a** Courtesy Lee D. Simon, Waksman Institute, Rutgers University. **10.3b** Courtesy M. Wurtz, University of Basel. **10.3c** Biozentrum, University of Basel, Science Photo Library. **10.8b** From J. D. Watson, *The Double Helix,* Atheneum, New York, 1968. © J. D. Watson. Photo Courtesy of Cold Spring Harbor Laboratory Archives. **10.12b** © Michael Freeman/Phototake NYC. **10.15** Courtesy J. Wolfson and D. Dressler, *Proc. Nat. Acad. Sci.,* vol. 69, 1972. **10.17** © R. Bisson, Stockshots.

CHAPTER 11 **CO** From Arents and Moudrianakis, *Proc. Nat'l. Acad. Sci.* (USA), 90, 10489 (1993). **11.2** Department of Special Collections, Stanford University Libraries. **11.3 (top, bottom)** D. M. Prescott, *Progress in Nucleic Acid Research and Molecular Biology,* 3:35, Academic Press, 1964. **11.5b** Jack R. Griffith, University of North Carolina, Chapel Hill. **11.13 (right)** From O. L. Miller, Jr., B. A. Hamkalo, and C. A. Thomas, Jr., *Science,* vol. 169, 1970; copyright © 1970 by A.A.A.S. **11.14** Modified from B. Alberts et al., *Molecular Biology of the Cell,* Garland Press, New York, 1983. **11.15** From M. A. Rould, J. J. Perona, D. Stoll, and T. A. Steitz, *Science,* 246, 1989; Copyright © 1989 by A.A.A.S. **11.18** Courtesy Nik Kleinberg. **11.20** Courtesy Bruce N. Ames, University of California,

Berkeley. **11.21** Courtesy A. C. Arnberg, Biochemical Laboratory, State University, Groningen, the Netherlands.

CHAPTER 12 CO © Omikron/Science Source-Photo Researchers, Inc. **12.1ab** Cold Spring Harbor Lab Archives. **12.9** Photo courtesy J. G. Gall, Carnegie Institution, reproduced from M. B. Roth and J. G. Gall, *J. Cell Biol.,* 105:1047-1054 (1987). Copyright © 1987 by Rockefeller University Press. **12.10** Michael Ashburner, Cambridge University. **12.14a** From Porter, Fonte, and Weiss, *Cancer Res.,* 34:1385-1394. **12.14b** Gilla Kaplan, the Rockefeller University. **12.16 (top and bottom)** Lance Liota, NCI. **12.18** Curve for retinablastoma based on data from H. W. Hethcote and A. G. Knudson, *Proc. Nat'l. Acad. Sci. U.S.A.,* vol. 75, 1978; curves for prostate and skin cancer based on data from Japanese Cancer Association, *Cancer Mortality and Morbidity Statistics,* Japanese Scientific Press, Tokyo, 1981.

CHAPTER 13 CO © Dwight Kuhn, 1987. **13.1** G. R. Roberts. **13.2** ARS-USDA Information Staff. **13.4a,c** Jack R. Griffith, University of North Carolina, Chapel Hill. **13.4b** Courtesy S. N. Cohen, Stanford University. **13.10a** Adapted from Gasser, C. S., and Fraley, R. T., *Sci, Am.,* June 1992; copyright © 1992 by Scientific American, Inc.; all rights reserved. **13.10b** Courtesy Dr. John Sanford, Cornell University. **13.11** Courtesy Genentech, Inc. **13.12a** John D. Cunningham/Visuals Unlimited. **13.13** Courtesy Steven Henikoff, Fred Hutchinson Cancer Research Center, Seattle, Washington. **13:Box a** Cellmark Diagnostics, Germantown, MD. **13:Box b** © Reiger Communications, Inc. **13.14ab** Calgene, Inc. **13.15a** ARS-USDA Information Staff. **13.15b** Philippe Plailly/Eurelios/Science Photo Library/Photo Researchers, Inc.

CHAPTER 14 CO Courtesy Claire Acher, National Organization of Albinism and Hypopigmentation, Philadelphia, PA. **14.1** Courtesy V. Orel, Mandelianum of Moravian Museum. **14.7** Jane Burton/Bruce Coleman, Inc. **14.10** Library of Congress. **14.11** Modified from A. M. Winchester, *Genetics,* 5th ed., Houghton Mifflin, Boston, 1977. **14:Box a,b** © Bill Marchel. **14:Box c** G. R. Roberts. **14.14** Larry Lefever/Grant Heilman Photography. **14.16a,b** Courtesy Marion I. Barnhart, Wayne State University Medical School, Detroit, Michigan. **14.18 (top)** Dr. Jeremy Burgess/Science Photo Library/Photo Researchers, Inc. **14.19** Courtesy M. L. Barr, *Can. Cancer Conf.,* vol. 2, Academic Press, New York, 1957. **14.21** © H. Chaumeton/Nature. **14.22** K. R. Dronamraju, in E. J. Gardner, *Principles of Genetics,* © 1975 by John Wiley & Sons, Inc., reproduced with their permission. **14.26** Courtesy R. A. Boolootian, Science Software Systems, Inc. **14.27** Modified from *Biological Science: An Ecological Approach,* 4th ed., Houghton Mifflin, Boston, 1982. Used by permission.

PART THREE Adapted from Fox, S. I., *Human Physiology,* Third Edition, copyright © 1990 Wm. C. Brown Communications, Inc., Dubuque, Iowa. All rights reserved. Reprinted by permission. **(inset)** © Will and Demi McIntyre

CHAPTER 15 CO Alfred Pasieka, Bruce Coleman, Inc. **15.2** Carolina Biological Supply Company. **15.3** Ward's Natural Science Establishment, Inc. **15.4** Courtesy Thomas Eisner, Cornell University. **15.5** © Biophoto Associates/Science Source/Photo Researchers, Inc. **15.6** Ed Reschke/Peter Arnold, Inc. **15.7a** Courtesy Nels R. Lersten, Iowa State University. **15.8a** Courtesy Victor B. Eichler. **15.8b** Courtesy Thomas Eisner, Cornell University. **15.9** Modified from V. A. Greulach and J. E. Adams, *Plants: An Introduction to Modern Botany,* Wiley, New York, 1962. **15.11** Courtesy Ed Reschke. **15.13a-c** Courtesy Ed Reschke. **15.14a** D. Claugher, courtesy of the Trustees, the British Museum (Natural History). **15.14b** Courtesy Jerome Gross, Massachusetts General Hospital. **15.15ab** Courtesy Ed Reschke. **15.17a** © Lennart Nilsson, from L. Nilsson, *Behold Man,* English translation © 1974, Albert Bonniers Alba, Stockholm, and Little, Brown, and Co. (Canada) Ltd. **15.17b** Carolina Biological Supply Company/Phototake-NYC. **15.19 (right)** Courtesy Ed Reschke.

CHAPTER 16 CO Geoff Dore/Bruce Coleman, Ltd. **16.5b** © Breck P. Kent/Animals Animals/Earth Scenes. **16.6** Courtesy Thomas Eisner. **16.8** G. R. Roberts. **16.9** Biophoto Associates/Science Source/Photo Researchers, Inc. **16.10 (top and bottom)** Courtesy Ed Reschke. **16.14 (top and bottom)** Carolina Biological Supply Company. **16.16** Adapted from Weier, T. E., Stocking, C. R., Barbour, M. G., *An Introduction to Plant Biology,* Fourth Edition, John Wiley & Sons, Inc., New York, 1970. **16.17** E. R. Degginger. **16.18** © Sean Morris/Oxford Scientific Films. **16.19** E. R. Degginger. **16.20** Courtesy J. H. Troughton, Department of Scientific and Industrial Research, Wellington, New Zealand. **16.21** Courtesy J. H. Troughton, Department of Scientific and Industrial Research, Wellington, New Zealand. **16.23 (top)** Courtesy Thomas Eisner, Cornell University. **16.23 (bottom)** G. R. Roberts.

CHAPTER 17 CO Laura Riley/Bruce Coleman, Inc. **17.1a** A. B. Dowsett/Science Photo Library/Photo Researchers, Inc. **17.1b** Runk/Schoenberger/Grant Heilman Photography. **17.2 (top)** © Mark Hamblin/Oxford Scientific Films. **17.2 (bottom)** © Norbert Rosing/Oxford Scientific Films. **17.5** Courtesy R. Farquharson, UNICEF. **17:Box (left and right)** From *The Vitamin Manual,* Courtesy the Upjohn Company. **17.8** Courtesy K. G. Grell, University of Tubingen. **17.12a** G. Lower/Photo Researchers, Inc. **17.12b,c** Modified from W. D. Russell-Hunter, *A Biology of Lower Invertebrates,* Macmillan Publishing Co., New York, 1968. **17.13** Courtesy E. V. Grave. **17.14** © Dickson Despommier/Photo Researchers, Inc. **17.15 (right)** Oxford Scientific Films/Bruce Coleman, Inc. **17.17** P. Sereno. **17.18a** Howard Hall/HHP. **17.18b** Courtesy Thomas Eisner, Cornell University. **17.20** Adapted from an original painting by Frank H. Netter, M.D., from *The CIBA Collection of Medical Illustrations,* copyright © 1959 by CIBA Pharmaceutical Company, division of CIBA-GEIGY Corp. **17.26** From Warren Andrew, *Textbook of Comparative Histology,* Oxford, The University Press, 1959. **17.27a** Courtesy Juan Vial, University of Colorado. **17.27b** Courtesy Susumu Ito, Harvard Medical School. **17.29** Professor J. James/Science Photo Library/Photo Researchers, Inc. **17.30** Adapted from Guyton, A. C., *Textbook of Medical Physiology,* Eighth Edition, W. B. Saunders Company, Philadelphia, 1991.

CHAPTER 18 CO © 1991 James D. Watt/Innerspace Visions. **18.2** Howard Hall/HHP. **18.6** Modified from Ralph Buchsbaum, *Animals Without Backbones,* by permission of the University of Chicago Press, copyright © 1948 by the University of Chicago. **18.10** © F. Sauer/Nature. **18.11b** Roman Vishniac Archives at the International Center for Photography, New York. **18.12 (left)** D. Claugher, courtesy of the Trustees, the British Museum (Natural History). **18.12 (right)** From Warren Andrew, *Textbook of Comparative Histology,* Oxford, The University Press, 1959. **18.13a** © 1972 Tom McHugh/Photo Researchers, Inc. **18.13b** © Joe McDonald/Animals Animals/Earth Scenes. **18.13c** © Zig Leszczynski/Animals Animals/Earth Scenes. **18.13d** © S. J. Krasemann/Peter Arnold, Inc. **18.17a** From R. G. Kessel and R. H. Kardon, *Tissues and Organs: A Text-Atlas of Scanning Electron Microscopy,* © W. H. Freeman and Company, 1979. **18.16** Adapted from Vander, A. J., Sherman, J. H., and Luciano, D. S., *Human Physiology: The Mechanisms of Body Function,* McGraw-Hill, Inc., New York, 1980.

CHAPTER 19 CO Alfred Pasieka/Bruce Coleman, Inc. **19.4 (bottom)** Courtesy Thomas Eisner, Cornell University. **19.5** Courtesy Nels R. Lersten, Iowa State University. **19.6ab** Courtesy Ed Reschke. **19.8** Courtesy Ed Reschke. **19.9** B. Bracegirdle, from W. Krommenhoek, J. Sebus, and G. J. van Esch, *Biological Structures,* copyright © 1979 by L. C. G. Malmberg, B.V., the Netherlands. **19.10** Courtesy Thomas Eisner, Cornell University. **19.12** Modified from V. A. Greulach and J. E. Adams, *Plants: An Introduction to Modern Botany,* Wiley, New York, 1962. **19.14a** Courtesy J. H. Troughton, Department of Scientific and Industrial Research, Wellington, New Zealand. **19.14bc** Courtesy B. G. Butterfield, Canterbury University, and B. A. Meylan, Department of Scientific and Industrial Research, Wellington, New Zealand. **19.15a-c** Courtesy Thomas Eisner, Cornell University.

CHAPTER 20 CO CNRI/Science Photo Library-Photo Researchers, Inc. **20.6a** From R. G. Kessel and R. H. Kardon, *Tissues and Organs: A Text-Atlas of Scanning Electron Microscopy,* © W. H. Freeman and Company, San Francisco, 1979. **20.8** Modified from B. S. Guttman and J. W. Hopkins III, *Understanding Biology;* copyright © 1983 by Harcourt Brace Jovanovich, Inc.; used by permission of the publisher. **20.10** Adapted from Vander, A. J., Sherman, J. H., and Luciano, D. S., *Human Physiology: The Mechanisms of Body Function,* McGraw-Hill, Inc., New York, 1980. **20.11** © Albert Paglialunga/Phototake NYC. **20.14** Adapted from Fox, S. I., *Human Physiology,* Fourth Edition, copyright © 1993 by Wm. C. Brown Communications, Inc., Dubuque, Iowa. All rights reserved. **20.15** © Lennart Nilsson, from L. Nilsson, *Behold Man,* English translation © 1974, Albert Bonniers Alba, Stockholm, and Little, Brown, and Co. (Canada) Ltd. **20Box:a** Roman Vishniac Archives at the International Center of Photography, New York. **20Box:b (left, right)** Courtesy Ed Reschke. **20.16 (left)** Courtesy D. W. Fawcett, Harvard Medical School. **20.16 (right)** From R. G. Kessel and R. H. Kardon, *Tissues and Organs: A Text-Atlas of Scanning Electron Microscopy,* © W. H. Freeman and Company, San Francisco, 1979. **20.19** © Manfred Kage/Peter Arnold, Inc. **20.20b** Adapted from Van de Graff, Kent M., *Human Anatomy,* 2nd ed., Wm. C. Brown Communications, Inc., Dubuque, Iowa. All rights reserved. Reprinted by permission. **20.22** Courtesy Turtox/Cambosco, Macmillan Science Co., Inc. **20.24** Courtesy Eila Kairinen, Gillette Research Institute. **20.25** Courtesy J. R. Porter and Ginny Fonte, University of Colorado.

CHAPTER 21 CO David M. Phillips/Photo Researchers, Inc. **21.2** © Lennart Nilsson, from L. Nilsson, *Behold Man,* English translation © 1974, Albert Bonniers Alba, Stockholm, and Little, Brown, and Co. (Canada) Ltd. **21.4** Courtesy Peter Marks and Frederick Maxfield, from P. Marks and F. R. Maxfield, *J. Cell Biol.,* 110:43-52, (1992). Reproduced by copyright permission of the Rockefeller University Press. **21.5b** Computer graphic modelling and photography by A. J. Olson, The Scripps Research Institute, © 1992. **21:Box1 a** Courtesy Center for Disease Control, Atlanta. **21:Box2 b** Reproduced from D. Lawson, C. Fewtrell, B. Gomperts, and M. Raff, *Journal of Experimental Medicine,* 142:391-402, 1975. Used by copyright permission of the Rockefeller University Press.

CHAPTER 22 CO Professor P. Motta/Dept. of Anatomy/University "La Sapienza," Rome/Science Photo Library-Photo Researchers, Inc. **22.1** Institut de Pathologie Cellulaire/Hopital Kremlin-Bicêtre. **22.8** C. Lieber, *Sci. Am.,* vol. 234, 1976; copyright © 1976 by Scientific American, Inc.; all rights reserved. **22:Box a** Andrew Syred/Science Photo Library-Photo Researchers, Inc. **22:Box b** © 1993 George Steinmetz. **22.9** © Douglas Faulkner, Photo Researchers, Inc. **22.11a-c** Courtesy Thomas Eisner, Cornell University. **22.11** Modified from Ralph Buchsbaum, *Animals Without Backbones,* by permission of the University of Chicago Press, copyright © 1948 by the University of Chicago. **22.16 (left)** From R. G. Kessel and R. H. Kardon, *Tissues and Organs: A Text-Atlas of Scanning Electron Microscopy,* W. H. Freeman, San Francisco, copyright © 1979. **22.16 (right)** Modified from H. W. Smith, *The Kidney,* Oxford, The University Press, 1951. **22.17** M. G. Farquhar, University of California, San Diego. **22.19** © Hank Morgan/Science Source-Photo Researchers, Inc.

CHAPTER 23 CO Reprinted with permission from *Botany for Gardeners; An Introduction and Guide* by Brian Capon. © 1990 by Timber Press, Inc. All rights reserved. **23.1a** Courtesy Eva Frei and R. D. Preston, University of Leeds. **23.4** © Jerome Wexler/Photo Researchers, Inc. **23.5** Courtesy Donald Nevins, Iowa State University. **23.13** Courtesy C. R. Hawes, Oxford Polytechnic. **23.15** © Biophoto Associates/Science Source-Photo Researchers, Inc. **23.16 (left)** George D. Lepp/Bio-Tec Images. **23.16 (bottom)** Jack K. Clark, Bio-Tec Images. **23.17** Courtesy Robert Newman, University of Wisconsin. **23.18** Courtesy Sylvan Wittwer. **23.19** A.R.S.-USDA Information Staff. **23.20** Stephen Gladfelter, Stanford University, from research by P. W. Oeller, M.-W. Lu, L. P. Taylor, D. A. Pike, and A. Theologis at the Plant Gene Expression Center, University of California, Berkeley.

CHAPTER 24 CO Dr. Jeremy Burgess/Science Photo Library-Photo Researchers, Inc. **24.1** From *J. Ag. Res.,* 18:553-606 (1920). Courtesy of USDA Beltsville, MD. **24.6ab** Copr. Bruce Iverson. **24.8b** D. Claugher, by courtesy of the Trustees, the British Museum (Natural History). **24.9** Dr. Jeremy Burgess/Science Photo Library-Photo Researchers, Inc. **24.11** Jane Burton/Bruce Coleman, Inc. **24.12e** Redrawn from M. Schaffner, *The Ohio Naturalist,* 1906. **24.14b** G. R. Roberts. **24.17a** E. R. Degginger. **24.18** E. R. Degginger. **24.19** Carolina Biological Supply Company.

CHAPTER 25 CO J. Hoogesteger/Biofotos. **25.6** Painting by Frank H. Netter, M.D.; reprinted with permission from *The CIBA Collection of Medical Illustrations,* copyright © 1965 by the CIBA Pharmaceutical Company, division of CIBA-GEIGY Corporation; all rights reserved. **25.9** Carolina Biological Supply Company. **25.13** © John P. Kay/Peter Arnold, Inc. **25.15** From A. J. Carlson et al., *The Machinery of the Body,* University of Chicago Press, 1961.

CHAPTER 26 CO © Lennart Nilsson, from *The Incredible Machine,* National Geographic Society, Washington, D.C., 1986. **26.2** © Hans Pfletschinger/Peter Arnold, Inc. **26.3** © Michael Fogden/Bruce Coleman, Inc. **26.4** © Ken Highfill/Photo Researchers, Inc. **26.5** © K. H. Switak/Photo Researchers, Inc. **26.9a** Courtesy D. M. Phillips, Population Council, New York. **26.9b** David M. Phillips/The Population Council/Science Source/Photo Researchers, Inc. **26:Box** Reuters/Bettman. **26.16a** Courtesy Thomas Eisner, Cornell University. **26.16b** © C. Edelman/La Vilette/Photo Researchers, Inc.

CHAPTER 27 CO © Petit Format/Nestle/Science Source/Photo Researcher, Inc. **27.1b** David Phillips/Photo Researchers, Inc. **27.2 (right)** © Biophoto Associates/Photo Researchers, Inc. **27.4a-c** Courtesy R. G. Kessel and C.Y. Shih, *Scanning Electron Microscopy in Biology,* Springer-Verlag, New York, 1974. **27.7b** Arthur M. Siegelman. **27.10** Adapted from Postlethwait, J. H., and Hopson, J. L., *The Nature of Life,* McGraw-Hill, Inc., 1989. **27.14a-d** © Lennart Nilsson, from L. Nilsson, *Behold Man,* English translation © 1974, Albert Bonniers Alba, Stockholm, and Little, Brown, and Co. (Canada) Ltd. **27.15** © Phil Degginger/Bruce Coleman, Inc. **27.16** Modified from V. B. Wigglesworth, *The Principles of Insect Physiology,* Methuen, 1947. **27.20a-c** D. Overcash/Bruce Coleman, Inc. **27.20d, h** L. West. **27.20e** E. R. Degginger/Bruce Coleman, Inc. **27.20fg** Jeff Foote. **27.24b** D. A. Melton. **27.26a-c** From W. Krommenhoek, J. Sebus, and G. J. van Esch, *Biological Structures,* copyright © 1979 by L. C. G. Malmberg B.V., The Netherlands. **27.32** Carolina Biological Supply Company. **27.33ab** E. B. Lewis, California Institute of Technology, Pasadena.

CHAPTER 28 CO © Michael Delannoy/Visuals Unlimited. **28.3a** Courtesy H. F. Webster, as printed in W. Bloom and D. W. Fawcett, *A*

Textbook of Histology, 10th edition, W. B. Saunders Company, 1975. **28.3b** from painting by Naidi Wiebe. **28.7** Modified from T. H. Bullock and G. A. Horridge, *Structure and Function of the Nervous System of Invertebrates,* W. H. Freeman, San Francisco, 1965. **28.8** Modified from L. H. Hyman, *The Invertebrates,* vol. 2, McGraw-Hill Book Co., New York, copyright © 1951 and from Ralph Buchsbaum, *Animals Without Backbones,* University of Chicago Press, copyright © 1948 by the University of Chicago, used by permission of the publishers. **28.17** E. R. Lewis et al., *Science,* vol. 165, 1969; copyright © 1969 by A.A.A.S. **28.18b** R. L. Roberts, R. G. Kessel, and H. N. Tung, *Freeze Fracture Images of Cells and Tissues,* Oxford University Press, 1991. **28.22** Courtesy Ed Reschke.

CHAPTER 29 CO CNRI/Science Photo Library. **29.3** Courtesy Ed Reschke. **29.4** After Murray, 1973. **29.6** Jeff Lepore/Photo Researchers, Inc. **29.8 (top)** Dr. J. A. L. Cooke/Oxford Scientific Films. **29.8 (bottom)** Peter Parks/Oxford Scientific Films. After Murray, 1973. **29.9** Modified from R. E. Snodgrass, *Principles of Insect Morphology,* McGraw-Hill Book Co., copyright © 1935; used with permission of the McGraw-Hill Book Co. **29.10** J. A. L. Cooke/Oxford Scientific Films. **29.11 (top, bottom)** Courtesy J. L. Gould and C. G. Gould. **29.13** © Lennart Nilsson, from L. Nilsson, *Behold Man,* English translation © 1974, Albert Bonniers Alba, Stockholm, and Little, Brown, and Co. (Canada) Ltd. **29.14** Modified from D. Lack, *Darwin's Finches,* Cambridge, The University Press, 1947. **29.16 (bottom)** Omikron/Photo Researchers, Inc. **29.19 (top, bottom)** Courtesy Thomas Eisner, Cornell University. **29.21a** Courtesy Thomas Eisner, Cornell University. **29.21b** Courtesy E. S. Ross. **29.25a** Modified from D. Lack, *Darwin's Finches,* Cambridge, The University Press, 1947. **29.25b** Courtesy A. J. Hudspeth and R. A. Jacobs, from A. J. Hudspeth, *Nature,* 341:397-404 (1989). **29.26a** © Lennart Nilsson, from L. Nilsson, *Behold Man,* English translation © 1974, Albert Bonniers Alba, Stockholm, and Little, Brown, and Co. (Canada) Ltd. **29.29ab** Courtesy Widex, Inc. **29.30** Adapted from Fox, S. I., *Human Physiology,* Third Edition, copyright © 1990 Wm. C. Brown Communications, Inc., Dubuque, Iowa. All rights reserved. **29.31** © Merlin D. Tuttle. **29.32** Courtesy A. O. D. Willows, University of Washington. **29.33** Modified from A. S. Romerand, T. S. Parsons, *The Vertebrate Body,* 5th ed., copyright © 1977 by W. B. Saunders Company, reprinted by permission of CBS College Publishing. **29.38** Modified from W. Penfield and T. Rasmussen, *The Cerebral Cortex of Man,* Macmillan, New York, 1950. **29.39** Institute of Medicine, National Academy of Sciences Press. **29.40** Ralph M. Garruto, NIH-NINDS.

CHAPTER 30 CO Rod Borland/Bruce Coleman, Inc. **30.1 (top, bottom)** G. R. Roberts. **30.3** Photograph by D. Claugher, courtesy of the Trustees, the British Museum (Natural History). **30.5** © Biophoto Associates/Photo Researchers, Inc. **30.7a-c** © Ed Reschke. **30.14** Courtesy H. E. Huxley, Cambridge University. **30.17** Courtesy J. E. Heuser, Washington University Medical Center.

PART FOUR M. P. Kahl/Academy of Natural Sciences, Philadelphia. **(inset)** © Will and Demi McIntyre.

CHAPTER 31 CO Courtesy Edward S. Ross. **31.1** R. A. Lutz. **31.2a-f** Louise B. Van der Meid. **31.3ab** Courtesy J. L. Gould. **31.4** Adapted by permission from *The Structure and Action of Proteins* by Richard E. Dickerson and Irving Geis, W. A. Benjamin, Inc., Menlo Park, California, Publisher; copyright © 1969 by Dickerson and Geis. **31.5a** George Holton/Photo Researchers, Inc. **31.5b** Masud Quraishy/Bruce Coleman, Inc. **31.6 (top, bottom)** A. J. Ribbink, *S. African J. Zool.* vol. 18, 1985. **31.7** © G. Renson/Jacana/Photo Researchers, Inc. **31.14** © David G. Allen. **31.16** JohnWightman/Ardea London Ltd. **31.17** Stouffer Productions/Animals Animals/Earth Scenes. **31.18a** Courtesy E. S. Ross. **31.18b** Courtesy D. J. Howell, Purdue University. **31.20** Derek Washington/Bruce Coleman, Inc. **31:Box a** Dr. Rudiger Wehner. **31.21ab** Courtesy David Aneshansley and Thomas Eisner, Cornell University. **31.22** © Andrew Martinez/Photo Researchers, Inc. **31.23** Peter Parks/Oxford Scientific Films. **31.24a** Larry Lipsky/Bruce Coleman, Inc. **31.24b** Jane Burton/Bruce Coleman, Inc. **31.25a** David Overcash/Bruce Coleman, Inc. **31.25b** Jane Burton/Bruce Coleman, Inc. **31.26** Peter Ward/Bruce Coleman, Inc. **31.27** M. W. F. Tweedie/Bruce Coleman, Inc. **31.28 (top, bottom)** Breck P. Kent/Animals Animals/Earth Scenes. **31.29** E. R. Degginger. **31.30a** James L. Castner, University of Florida. **31.30b** Courtesy E. S. Ross. **31.31 (top)** D. Overcash/Bruce Coleman, Inc. **31.31 (bottom)** Kevin Byron/Bruce Coleman, Inc. **31.32ab** © J. H. Robinson/Photo Researchers, Inc. **31.33** C. H. Greenewalt/Academy of Natural Sciences, Philadelphia. **31.34** © Nancy Sefton/Photo Researchers, Inc. **31.35a** © Steinhardt Aquarium/Tom McHugh/Photo Researchers, Inc. **31.35b** © K. and K. Ammann. **31.36** R. Carr/Bruce Coleman, Inc. **31.38ab** Courtesy James H. Tumlinson, USDA.

CHAPTER 32 CO University of Chicago Hospitals. **32.1** Zig Leszczynski/Animals Animals/Earth Scenes. **32.2 (left)** H. Harrison/Photo Researchers, Inc. **32.2 (right)** Mike Hopiak, Cornell University Laboratory. **32.2 (bottom)** © R. Van Nostrand/Photo Researchers, Inc. **32.3a** John Cancalosi/Peter Arnold, Inc. **32.3b** Michael Giannechini/Photo Researchers, Inc. **32.4** Modified from J. Clausen et al., *Carnegie Inst. Washington Publ.,* no. 581, 1958. **32.5 (left)** © S.R. Maglione/Photo Researchers, Inc. **32.5 (right)** © 1990 Mark Boulton/Photo Researchers, Inc. **32.6** © 1994 Comstock Photography, Inc. **32.7** Adapted from Levinton, Jeffrey S., *Sci. Am.,* Nov. 1992. © 1992 by Scientific American, Inc. All rights reserved. **32.8 (top, bottom)** Pat and Tom Leeson/Photo Researchers, Inc. **32.11** John Shaw/Bruce Coleman, Inc. **32.12a** G. R. Roberts. **32.12b** L. West/Bruce Coleman, Inc. **32.12c** Jacana Scientific Control/Photo Researchers, Inc. **32.14ab** Courtesy T. K. Wood, University of Delaware. **32.15** Modified from D. Lack, *Darwin's Finches,* Cambridge, The University Press, 1947. **32.17** Drawing courtesy Sophie Webb. **32.19** Courtesy H. Douglas Pratt. **32.23** © Tim Davis/Photo Researchers, Inc. **32.25** Illustrations by Marianne Collins are reproduced from *Wonderful Life: The Burgess Shale and the Nature of History,* by Stephen Jay Gould, with permission of W. W. Norton & Co., Inc.; copyright © 1989 Stephen Jay Gould. **32.28a** Chicago Zoological Park/Tom McHugh/Photo Researchers, Inc. **32.28bc** Courtesy M. Morcombe. **32.28d** Jack Fields/Photo Researchers, Inc. **32.30** © 1993 Ed Bridges/AMNH.

CHAPTER 33 CO F. O. Jones, U.S. Geological Survey Photo Library, Denver, Colorado. **33.1** J. Scott Altenbach, University of New Mexico. **33.3** Courtesy NASA. **33.4ab** G. R. Roberts. **33.4c** M. P. Kahl/Academy of Natural Sciences, Philadelphia. **33.5** Richard Vaughan/Ardea London Ltd. **33.7** Modified from T. Carlson, *Biochem. Z.,* vol. 62, 1938. **33.8** Modified from J. Davidson, *Trans. R. Soc. South Aust.,* vol. 62, 1938. **33.9** © Andrew Cleave/Nature Photographers, Inc. **33.11** © 1992 Kim Taylor/Bruce Coleman, Inc. **33.12** Modified from E. P. Odum, *Fundamentals of Ecology,* W. B. Saunders, 1959, after Deevey. **33.13** Population Reference Bureau, Inc. Updated by Charles F. Westoff, Office of Population Research, Princeton University. **33.15** Stephen J. Krasemann/Photo Researchers, Inc. **33.17** Courtesy L. David Mech. **33.18** H. Reinhard/Bruce Coleman, Inc. **33.19a** Philip Boyer. **33.19b** R. Austing/Photo Researchers, Inc. **33.19c** Edgar T. Jones/Bruce Coleman, Inc. **33.20** © FAO Photo by G. Tortoli. **33.22a** © Jeff Foote/Bruce Coleman, Inc. **33.22b** Bill Ruth/Bruce Coleman, Inc. **33.23a-d** Courtesy E. J. Kormondy, *Smithsonian Magazine,* vol. 1, 1970. **33.24** S. Hurwitz/Bruce Coleman, Inc. **33.25** Based on data in E. P. Odum, *Fundamentals of Ecology,* W. B. Saunders, 1959. **33.26**

Redrawn from R. H. Whittaker, *Communities and Ecosystems,* 2nd. ed., Macmillan, New York, 1975; after B. Holt and G. M. Woodwell. **33.28** Gene Ahrens/Bruce Coleman, Inc.

CHAPTER 34 CO Courtesy NASA. **34.1** From Ricklefs, *The Economy of Nature,* copyright © 1933 by W. H. Freeman and Company. Used with permission. **34.6** Bruce Dale, © NGS. **34.7** Modified from E. P. Odum, *Fundamentals of Ecology,* E. B. Saunders, 1959. **34.12** From T. E. Graedel and P. J. Crutzen, *Sci. Am.,* September 1989; copyright © 1989 by Scientific American Inc.; all rights reserved. **34.14** From S. H. Schneider, *Sci. Am.,* September, 1989; copyright © 1989 by Scientific American Inc.; all rights reserved. **34.15** Courtesy Cold Spring Harbor Laboratory Archives. **34.17** Redrawn after K. P. Bowman, *Science,* vol. 239, 1988; copyright © 1988 by AAAS. **34.19** Courtesy U.S. Department of Agriculture. **34.21** Courtesy D. W. Schindler, *Science,* vol. 184, 1974; copyright © 1974 by AAAS. **34.22** © 1975 Aileen and W. Eugene Smith/Black Star. **34.24 (top, bottom)** Courtesy P. L. Ames. **34.25a** © David Newman/ Visuals Unlimited (VU). **34.25b** Larry Ditto/Bruce Coleman, Inc. **34.26** Adrian Davies/Bruce Coleman, Ltd. **34.27** Courtesy USDA Forest Service. **34.28** Redrawn from G. E. Likens et al., *Ecol. Monogr.,* vol. 40, 1970; copyright © 1970 by the Ecological Society of America. **34.29** Courtesy National Archives. **34.30** G. R. Roberts. **34.31** Oxford Scientific Films, Earth Scenes. **34.32** James Stanfield, © NGS. **34.34** C. J. Tucker, J. R. G. Towshend, and T. E. Goff, *Science,* vol. 227, 1985. **34.35 (left, right)** Courtesy E. S. Ross. **34.37** After Whittaker (1970). Reprinted with permission of Macmillan Publishing Co., Inc. from *Communities and Ecosystems* by Robert Whittaker. **34.38 (left)** Bob and Clara Calhoun/Bruce Coleman, Inc. **34.38 (right)** E. R. Degginger. **34.39 (left)** Jack Fields/Photo Researchers. **34.39 (right)** Wolfgang Bayer/Bruce Coleman, Inc. **34.40** © Zig Leszczynski/Animals Animals/Earth Scenes. **34.41 (left)** Ferrero/Nature. **34.41 (right)** N. DeVore III/Bruce Coleman, Inc. **34.42** E. R. Degginger. **34.43a** E. R. Degginger. **34.43b** Courtesy E. S. Ross. **34.45** Courtesy E. S. Ross. **34.47a** Howard Hall/HHP. **34.47b** Courtesy James D. Jordan. **34.49** R. E. Wallace/USGS Photo Library. **34.41** Redrawn from John Napier, *The Roots of Mankind,* copyright © 1970, Smithsonian Institution Press, Washington, D.C.; used by permission. **34.54** Ferrero/Nature.

CHAPTER 35 CO © Dr. Eckart Pott/Bruce Coleman, Inc. **35.1a** Courtesy Paul Trotschel. **35.1b** © Stephen Dalton/Photo Researchers, Inc. **35.2** Kim Taylor/Bruce Coleman, Inc. **35.3** Modified from N. Tinbergen, *The Study of Instinct,* Oxford, The University Press, 1951. **35.4** Modified from N. Tinbergen, *The Study of Instinct,* Oxford, The University Press, 1951. **35.5** © Frans Lanting/Minden Pictures. **35.6 (left, right)** Courtesy E. R. Willis, Illinois State University. **35.7 (top, bottom)** Courtesy Thomas Eisner, Cornell University **35.9** Adapted from K. Lorenz and N. Tinbergen, *Z. Tierpsychol.,* vol. 2, 1938. **35.10abde** From E. H. Hess, *Sci. Am.,* July 1956; copyright © 1956 by Scientific American, Inc.; all rights reserved. **35.10c** Wallace Kirkland, copyright © 1954 by Time, Inc. **35.12ab** G. R. Roberts. **35.13** From W. R. Miles, *J. Comp. Psychol.,* vol. 10, 1930; copyright © 1930 by the Williams and Wilkins Company, Baltimore, Maryland. **35.15a** © Stephen J. Krasemann/Photo Researchers, Inc. **35.15b** Nina Leen, *Life,* copyright © 1964 by Time, Inc. **35.16** Courtesy Klaus Immelmann, University of Bielefeld. **35.18** After Drees. **35.21** Courtesy Polaroid Corporation. **35.22** D. R. Bentley, Science, vol. 174, 1971; copyright © 1971 by AAAS. **35.25** Courtesy Michael Morcombe. **35.27** Courtesy D. Rubenstein, Princeton University. **35.30a** Redrawn from K. von Frisch, *Bees: Their Vision, Chemical Senses, and Language,* copyright © 1950 by Cornell University; used by permission of Cornell University Press. **35.30b** Courtesy Kenneth Lorenzen, University of California, Davis. **35.33** Adapted by permission from E. Gwinner and W. Wiltschko, *J. Comp. Physiol,* vol. 125, 1978; copyright © 1978 by Springer-Verlag, New York. **35.34** From B. Elsner, in *Animal Migration, Navigation, and Homing,* edited by K. Schmidt-Koenig and W. T. Keeton, Springer-Verlag, New York, 1978; from M. Michener and C. Walcott, *J. Exp. Biol.,* vol. 47, 1967. **35.36** Modified from N. Tinbergen and A. C. Perdeck, *Behaviour,* vol. 3, 1950. **35.37** Courtesy John Sparks, BBC (Natural History). **35.38** Russ Charif, courtesy Charles Walcott, Cornell University. **35.39** © Jeanette Wyneken. **35.41** A. Morris, Academy of Natural Sciences, Philadelphia; © VIREO. **35.42** Redrawn by permission from E. O. Wilson, *Sociobiology,* Harvard University Press, Cambridge, Massachusetts, 1975. **35.44** George D. Lepp, Bio-Tec Images. **35.45** H. Van Lawick/Nature Photographers, Ltd. **35.46** Roberto Bunge/ Ardea London Ltd. **35.47** Leonard Lee Rue III/Photo Researchers, Inc. **35.48** Courtesy E. S. Ross. **35.49** Francisco J. Erize/Bruce Coleman, Ltd. **35.50** Stouffer Productions/Animals Animals. **35.51** K. and K. Amman/Bruce Coleman, Inc. **35.52** © Frans Lanting, Minden Pictures. **35.53c** Hans Pfletschinger/Peter Arnold, Inc.

PART FIVE Jeffrey L. Rotman. **(inset)** © Will and Demi McIntyre.

CHAPTER 36 CO Dudley Foster/Woods Hole Oceanographic Institute. **36:Box a** Tony and Daphne Hallas. **36:Box b** The Arecibo Observatory is part of the National Astronomy and Ionosphere Center which is operated by Cornell University under a cooperative agreement with the National Science Foundation. **36.2ab** Courtesy D. W. Deamer, University of California, Davis. **36.4** Courtesy Sigurgeir Jonasson. **36.7 (left, right)** Courtesy Sidney W. Fox, University of Miami, and Steven Brooke Studios, Coral Gables, Florida. **36.11ab** Courtesy J. W. Schopf, University of California, Los Angeles. **36.12** Jan Taylor/Bruce Coleman, Inc. **36.13** W. J. Larsen, *J. Cell Biol.,* vol. 47, 1970; by copyright permission of the Rockefeller University Press. **36.14a** Courtesy R. E. Lee, University of Witwatersrand. **36.14b** M.M.J.P. Plant Res. Lab., East Lansing. **36.15** Photograph by John Waterbury, courtesy Sallie W. Chisholm, Massachusetts Institute of Technology. From Chisholm et al. A novel free-living prochlorophyte abundant in the oceanic euphotic zone. *Nature* 334: 6180, 1988. **36.17** © George J. Wilder/Visuals Unlimited. **36.19** G. R. Roberts. **36.22a** Courtesy E. V. Gravé. **36.22b** Courtesy E. B. Daniels, from K. W. Jeon (ed.) , *The Biology of Amoeba,* 1973, Academic Press, Orlando.

CHAPTER 37 CO NIBSC/Science Photo Library-Photo Researchers, Inc. **37.1a** Courtesy M. Wurtz, University of Basel. **37.1b** Courtesy R. C. Williams, University of California, Berkeley. **37.1c** Courtesy M. Gomersall, McGill University. **37.2 (right)** From K. Esau, *Viruses in plant hosts,* University of Wisconsin Press, 1968. Used by permission. **37.2 (bottom)** From K. Corbett, *Virology,* vol. 22, 1964; reprinted by permission of Academic Press, Inc., New York. **37.8b** M. A. Gonda et al., *Science,* vol. 227, 1985; copyright © 1985 by the American Association for the Advancement of Science. **37.10** Courtesy William Jewell, Cornell University. **37.11** Walther Stoeckenius, courtesy Carl Woese, University of Illinois. **37.12ab** From W. J. Jones, J. A. Leigh, F. Mayer, C. R. Woese, and R. S. Wolfe, *Arch. Microbiol.,* vol. 136, 1983; copyright © 1983 by Springer-Verlag. **37.14a** David Scharf/Peter Arnold, Inc. **37.14b** Center for Disease Control, Atlanta, Georgia. **37.15** © Michael Abbey/Photo Researchers, Inc. **37.16** Jeffrey L. Rotman. **37.17** Courtesy M. Gomersall, McGill University. **37.19** Courtesy V. Lorian, Bronx-Lebanon Hospital Center. **37.20b** Ward's Natural Science/Photo Researchers, Inc. **37.21** G. B. Chapman, *J. Bacteriol.,* vol. 71, 1956. **37.22** Dr. Tony Brain/Science Photo Library-Photo Researchers, Inc. **37.23** From R. C. Johnson, M. P. Walsh, B. Ely, and L. Shapiro, *J. Bacteriol.,* vol. 138, 1979. **37.24** Adapted from Vandemark, P. J., and Batzing, B. L., *The Microbes: An Introduction to Their Nature and Importance,* The Benjamin-Cummings Publishing Company, Redwood City, California, 1987. **37.25** Courtesy L. G. Caro,

FOLD 3

SPECIAL $19.95 OFFER

Biology Tutor Software for Macintosh or Windows

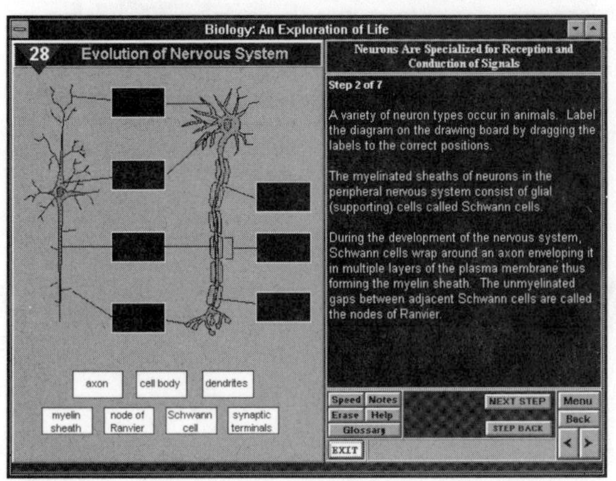

The powerful electronic review guide to Carol McFadden and William Keeton's *Biology: An Exploration of Life*.

Special features of The Biology Tutor include:

- Notecards explaining key concepts for each chapter of the text.
- Hundreds of animations that will help you visualize concepts in a way not possible by text alone.
- An interactive quiz feature with hundreds of questions that allow drill and practice before exams.
- A searchable glossary that allows easy access to definitions of key terms.

FOLD 2

Requirements:
The Windows version requires Windows and a 386 or higher processor with at least 2 megabytes of memory.

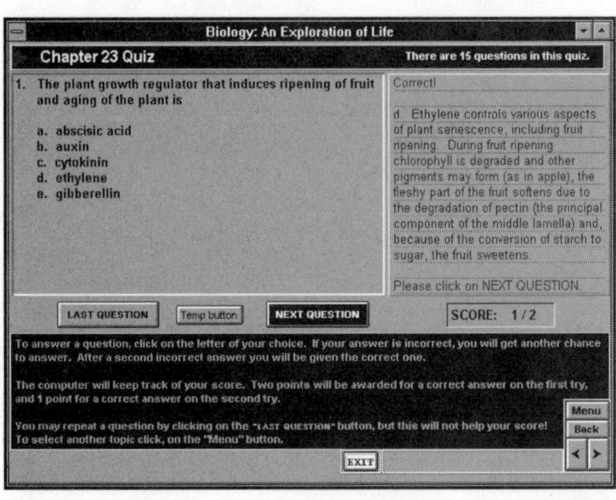

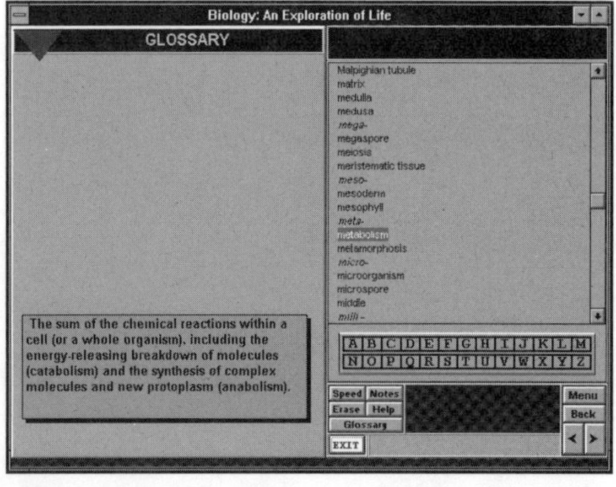

Tear along preforation, fold, and seal with tape. Do not glue or staple.

FOLD 1

YES, PLEASE SEND ME:

☐ The Biology Tutor for Macintosh [0-393-99949-1]—$19.95

☐ The Biology Tutor for Windows [0-393-99948-3]—$19.95

(residents of CA, NY, PA, IL must add their local sales tax)

☐ My check is enclosed. [Make checks payable to W. W. Norton & Company.]

Please charge to my: ☐ VISA ☐ MASTERCARD ☐ AMERICAN EXPRESS

Acct. # ______________________ Exp. Date __________ Telephone ______________

Name ______________________ Signature ______________________

Ship to:

Address ______________________

City ______________________

State/Prov. ____________ Zip/PC ____________

Billing Address (if different):

Address ______________________

City ______________________

State/Prov. ____________ Zip/PC ____________

Or call Customer Service at (800) 233-4830 to order by phone.

BUSINESS REPLY MAIL

FIRST CLASS PERMIT NO. 4008 NEW YORK, N.Y.

POSTAGE WILL BE PAID BY ADDRESSEE

W. W. NORTON & COMPANY, INC.
500 FIFTH AVENUE
NEW YORK, NEW YORK 10109-0145

University of Geneva, and R. Curtiss, University of Alabama. **37.26** Courtesy Elliot Scientific Corp. **37.27ab** T. E. Adams/Bruce Coleman, Inc. **37.28** Courtesy R. Malcolm Brown, Jr., University of Texas. **37.29** © S. Thompson/Visuals Unlimited. **37.30** Courtesy Zell A. McGee and E. N. Robinson, Jr., University of Utah.

CHAPTER 38 CO © Manfred Kage/Peter Arnold, Inc. **38.1** Eric V. Grave/Photo Researchers, Inc. **38.2** Eric V. Grave/Photo Researchers, Inc. **38.3** © M. Abbey/Photo Researchers, Inc. **38.4** Eric V. Grave/Photo Researchers, Inc. **38.6** Eric V. Grave/Photo Researchers, Inc. **38.8a** Patrick Grace/Science Source-Photo Researchers, Inc. **38.8b** L. West/Bruce Coleman, Inc. **38.9** © Cabisco/Visuals Unlimited. **38.12** Eric V. Grave/Photo Researchers, Inc. **38.14** © Larry Lipsky/Bruce Coleman, Inc. **38.16** E. R. Degginger/Bruce Coleman, Inc. **38.18** R. Carr/Bruce Coleman, Inc. **38.19** © H. Chaumeton/ Nature.

CHAPTER 39 CO Gene Ahrens/Bruce Coleman, Inc. **39.8** © John D. Cunningham/Visuals Unlimited. **39.11 (right)** Jane Burton/Bruce Coleman, Inc. **39.13** © Stephen Dalton/Photo Researchers, Inc. **39.14** Modified from H. J. Fuller and O. Tippo, *College Botany,* Holt, Rinehart, & Winston, Inc., New York, 1954. **39.15a** Adrian Davies/Bruce Coleman, Inc. **39.15b** G. R. Roberts. **39.16b** Courtesy E. S. Ross. **39.18b** E. R. Degginger. **39.19** Courtesy E. S. Ross. **39.20** Portion of group from the Carnegie Museum, Pittsburgh; used with permission. **39.21** Jane Shaw/Bruce Coleman, Inc. **39.22** G. R. Roberts. **39.23** Modified from H. J. Fuller and O. Tippo, *College Botany,* Holt, Rinehart, & Winston, Inc., New York, 1954. **39.25** Redrawn from H. N. Andrews, *Science,* vol. 142, 1963; copyright © 1963 by AAAS. **39.26** Roman Vishniac Archives at the International Center of Photography, New York. **39.27ab** Courtesy Thomas Eisner, Cornell University. **39.27c** Courtesy E. S. Ross. **39.28** Modified from H. J. Fuller and O. Tippo, *College Botany,* Holt, Rinehart, & Winston, Inc., New York, 1954. **39.33 (left)** Jane Burton/Bruce Coleman, Inc. **39.33 (right)** R. Carr/Bruce Coleman, Inc. **39.34** Modified from H. J. Fuller and O. Tippo, *College Botany,* Holt, Rhinehart, & Winston, Inc., New York, 1954.

CHAPTER 40 CO E. R. Degginger. **40.3a** W. H. Amos/Bruce Coleman, Inc. **40.5** Courtesy James P. Kramer, Cornell University. **40.6** M. P. L. Fogden/Bruce Coleman, Inc. **40.7** From L. W. Sharp, *Fundamentals of Cytology,* McGraw-Hill Book Co., New York, copyright © 1943; used by permission. **40.8a** Jane Burton/Bruce Coleman, Inc. **40.8b** L. West/Bruce Coleman, Inc. **40.9** G. R. Roberts. **40.10a** Masana Izawa/Nature Production, Tokyo. **40.10b** Satoshi Kuribyashi/Nature Production, Tokyo. **40.10c** Ed Degginger/Bruce Coleman, Inc. **40.11** From L. W. Sharp, *Fundamentals of Cytology,* McGraw-Hill Book Co., New York, copyright © 1943; used by permission. **40.12** Courtesy Edward J. Bottone, The Mount Sinai School of Medicine.

CHAPTER 41 CO Howard Hall/HHP. **41.3** Jeff Rotman. **41.8** Carolina Biological Supply Company. **41.10** R. N. Mariscal/Bruce Coleman, Inc. **41.11a** Jeffrey L. Rotman. **41.11b** © Robert Frerck. **41.13** © H. Chaumeton/Nature. **41.14** Ed Reschke. **41.15** CNRI/Science Photo Library-Photo Researchers, Inc. **41.17 (left)** Courtesy Ed Reschke. **41.17 (right)** © H. Chaumeton/Nature. **41.18** © H. Chaumeton/Nature. **41.21** Adapted from *Life* by W. K. Purves and G. H. Orians, Sinauer Associates, Sunderland, Massachusetts; copyright © 1983. **41.22** Courtesy T. E. Adams. **41.24** © H. Chaumeton/Nature. **41.27a** Peter Parks/Oxford Scientific Films. **41.28** © H. Chaumeton/Nature. **41.29a** © H. Chaumeton/Nature. **41.29b** Rod Borland/Bruce Coleman, Inc. **41.30** Based in part on drawings by Louise G. Kingsbury. **41.31** © H. Chaumeton/Nature. **41.32** © H. Chaumeton/Nature. **41.35** Ed Degginger/Bruce Coleman, Inc. **41.36ab** © John Cooke/Oxford Scientific Films. **41.37** Courtesy E. S. Ross. **41.39** Jane Burton/Bruce Coleman, Inc. **41.41** Courtesy Ed Reschke. **41.42** E. R. Degginger/Bruce Coleman, Inc. **41.43** © H. Chaumeton/Nature **41.44a** © H. Chaumeton/Nature. **41.44b** Hans Reinhard/Bruce Coleman, Inc. **41.45ab** © H. Chaumeton/ Nature. **41.46** Adapted from Barnes, R.D., *Invertebrate Zoology,* Fifth Edition, W. B. Saunders Company, Philadelphia, 1987. **41.48** A. Cosmos Blank/National Audubon Society/Photo Researchers, Inc. **41.50** Modified from T. I. Storer and R. L. Usinger, *General Zoology*, McGraw-Hill Book Co., New York, copyright © 1957; used by permission. **41.51** Courtesy Stephen Dalton/NHPA.

CHAPTER 42 CO Jack W. Dykinga/Bruce Coleman, Inc. **42.3** © Andrew J. Martinez/Photo Researchers, Inc. **42.5 (top)** © Robert Dunne/Photo Researchers, Inc. **42.5 (bottom)** Charlie Ott/Photo Researchers, Inc. **42.6** Alex Kerstitch. **42.7** © H. Chaumeton/Nature. **42.11 (top)** Courtesy Turtox/Cambosco, Macmillan Science Co., Inc. **42.12** © H. Chaumeton/Nature. **42.14** © Heather Angel. **42.15** Modified from A. S. Romer, *The Vertebrate Body,* W. B. Saunders, 1949. **42.16** Field Museum of Natural History, GEO 84533c, Chicago. **42.17 (top, bottom)** Howard Hall/HHP. **42.19** Courtesy the Royal Society, London. **42.20 (top, bottom)** Chip Clark, courtesy the National Museum of Natural History, Washington, D.C. **42.21 (left)** © Breck Kent/Animals Animals. **42.21 (right)** © Hans Reinhard/ Bruce Coleman, Inc. **42.22a** G. R. Roberts. **42.22b** Ferrero/Nature. **42.22cd** © Hans Reinhard/Bruce Coleman, Inc. **42.22e** © John Moss/Photo Researchers, Inc. **42.23** © Hans Pfletschinger/Peter Arnold, Inc. **42.25** From a painting by Naidi Wiebe, courtesy Field Museum of Natural History, GEO 81082, Chicago. **42.26** Alberta Culture and Multiculturalism Royal Tyrell Museum of Paleontology. **42.27 (left, right)** © Gregory S. Paul. **42.28** © Gregory S. Paul. **42.29a** © Gregory S. Paul. **42.30** Painting by Rudolf Freund, Carnegie Museum of Natural History. **42.31** Warren Garst/Tom Stack & Associates. **42.32** Courtesy Australian Information Service.

CHAPTER 43 CO D. Finnin & C. Chesek/American Museum of Natural History. **43.1** Courtesy American Museum of Natural History. **43.3** © Grospas/Nature. **43.4** M. P. L. Fogden/Bruce Coleman, Inc. **43.6 (left)** Jan Lindblad/Photo Researchers, Inc. **43.6 (right)** Peter Jackson/Bruce Coleman, Inc. **43.7** Jack Dermid/Bruce Coleman, Inc. **43.8** © Tom McHugh/Photo Researchers, Inc. **43.9** Brian Parker/Tom Stack & Associates. **43.10** Peter Davey/Bruce Coleman, Inc. **43.13** From R. E. Leakey and R. Lewins, *Origins,* Dutton, New York, 1977; reproduced by permission of Rainbird Publishing Group. **43.14a-d** Chip Clark, courtesy National Museum of Natural History, Washington, D.C. **43.17** Chip Clark, National Museum of Natural History, Washington, D.C. **43.18** David Brill, © 1985 National Geographic Society; photographed at the Institut de Quartenaire, Universite de Bordeaux, Talence, France.

INDEX